History of Biology

Year	Name	Country	Contribution
1927	Hermann J. Muller	United States	Proves that X rays cause mutations.
1929	Sir Alexander Fleming	Britain	Discovers the toxic effect of a mold product he called penicillin on certain bacteria.
1937	Konrad Z. Lorenz	Austria	Founds the study of ethology and shows the importance of imprinting as a form of early learning.
1937	Sir Hans A. Krebs	Britain	Discovers the reactions of a cycle that produces carbon dioxide during cellular respiration.
1940	George Beadle Edward Tatum	United States	Develop the one gene–one enzyme theory, based on red bread mold studies.
1944	O. T. Avery Maclyn McCarty Colin MacLeod	United States	Demonstrate that DNA alone from virulent bacteria can transform nonvirulent bacteria.
1945	Melvin Calvin Andrew A. Benson	United States	Discover the individual reactions of a cycle that reduces carbon dioxide during photosynthesis.
1950	Barbara McClintock	United States	Discovers transposons (jumping genes) while doing experiments with corn.
1952	Alfred D. Hershey Martha Chase	United States	Find that only DNA from viruses enters cells and directs the reproduction of new viruses.
1953	James Watson Francis Crick Rosalind Franklin	United States Britain	Establish that the molecular structure for DNA is a double helix.
1953	Harold Urey Stanley Miller	United States	Demonstrate that the first organic molecules may have arisen from the gases of the primitive atmosphere.
1954	Linus Pauling	United States	States that disease-causing abnormal hemoglobins are due to mutations.
1954	Jonas Salk	United States	Develops a vaccine that protects against polio.
1958	Matthew S. Meselson Franklin W. Stahl	United States	Demonstrate that DNA replication is semiconservative.
1961	Francois Jacob Jacques Monod	France	Discover that genetic expression is controlled by regulatory genes.
1964	Marshall W. Nirenberg Philip Leder	United States	Produce synthetic RNA, enabling them to break the DNA code.
1967	Christiaan Barnard	South Africa	Performs first human heart transplant operation.
1973	Stanley Cohen	United States	Uses recombinant DNA technique (genetic engineering) to place plant and animal genes in *Escherichia coli*.
1976	Georges Kohler Cesar Milstein	Britain	Fuse mouse leukemia cells with lymphocytes, developing clones, each of which produces only one type of "monoclonal" antibody.
1978	Peter Mitchell	Britain	Determines chemiosmotic mechanism by which ATP is produced in chloroplasts and mitochondria.
1982	William De Vries	United States	Performs first complete replacement of human heart with artificial heart on Dr. Barney B. Clark at University of Utah.
1989	Sidney Altman Thomas R. Check	United States	Independently discover that some RNA molecules can act as enzymes.
1990	R. Michael Blaese W. French Anderson Kenneth W. Culver	United States	Develop procedure to infuse genetically engineered blood cells for treatment of immune system disorder–first gene therapy used in a human.

Konrad Z. Lorenz

Barbara McClintock

Rosalind Franklin

Linus Pauling

Sylvia S. Mader

For My Children

Boston Burr Ridge, IL Dubuque, IA Madison, WI New York San Francisco St. Louis
Bangkok Bogotá Caracas Lisbon London Madrid
Mexico City Milan New Delhi Seoul Singapore Sydney Taipei Toronto

McGraw-Hill Higher Education
A Division of The **McGraw-Hill** *Companies*

INQUIRY INTO LIFE, NINTH EDITION

This book is printed on recycled, acid-free paper containing 10% postconsumer waste.

2 3 4 5 6 7 8 9 0 VNH/VNH 0 9 8 7 6 5 4 3 2 1 0

ISBN 0–697–36070–9

Vice president and editorial director: *Kevin T. Kane*
Publisher: *Michael D. Lange*
Sponsoring editor: *Patrick E. Reidy*
Senior developmental editor: *Suzanne M. Guinn*
Senior marketing manager: *Lisa L. Gottschalk*
Project manager: *Marilyn M. Sulzer*
Production supervisor: *Sandy Ludovissy*
Designer: *K. Wayne Harms*
Senior photo research coordinator: *Lori Hancock*
Senior supplement coordinator: *Audrey A. Reiter*
Compositor: *GTS Graphics, Inc.*
Typeface: *10/12 Palatino*
Printer: *Von Hoffmann Press, Inc.*

Cover photograph: © *Art Wolfe/Tony Stone Images*
Photo research: *Connie Mueller*

The credits section for this book begins on page C-1 and is considered an extension of the copyright page.

Library of Congress Cataloging-in-Publication Data

Mader, Sylvia S.
Inquiry into life / Sylvia S. Mader. — 9th ed.
p. cm.
Includes bibliographical references and index.
ISBN 0–697–36070–9
1. Biology. I. Title.
QH308.2.M24 2000
570—dc21 99–14024
CIP

www.mhhe.com

Brief Contents

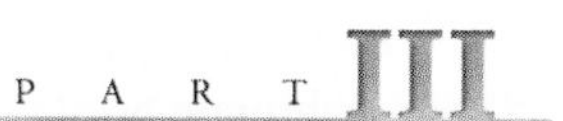

Contents

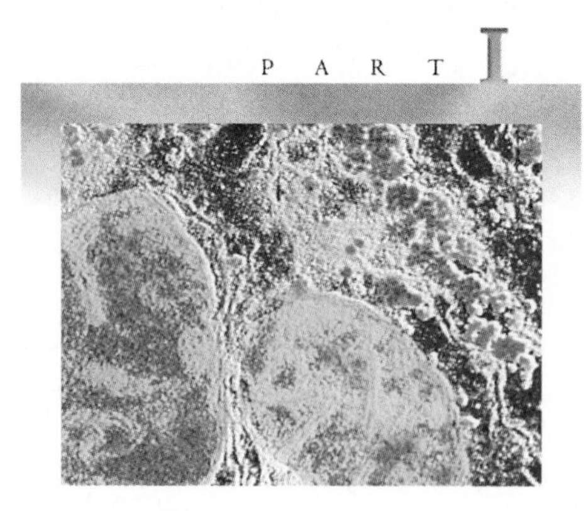

CHAPTER 9

Plant Organization and Growth 151

CHAPTER 10

Plant Physiology and Reproduction 169

PART III

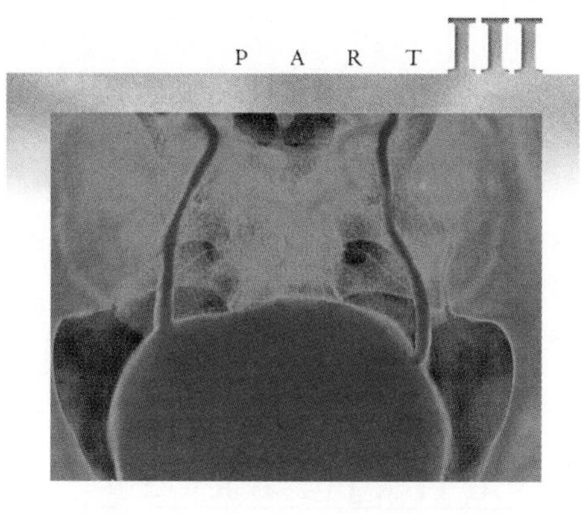

Maintenance of the Human Body 191

CHAPTER 11

Human Organization 193

CHAPTER 12

Digestive System and Nutrition 213

CHAPTER 13

Cardiovascular System 239

CHAPTER 14

Lymphatic System and Immunity 263

CHAPTER 15

Respiratory System 283

CHAPTER 16

Urinary System and Excretion 303

PART IV

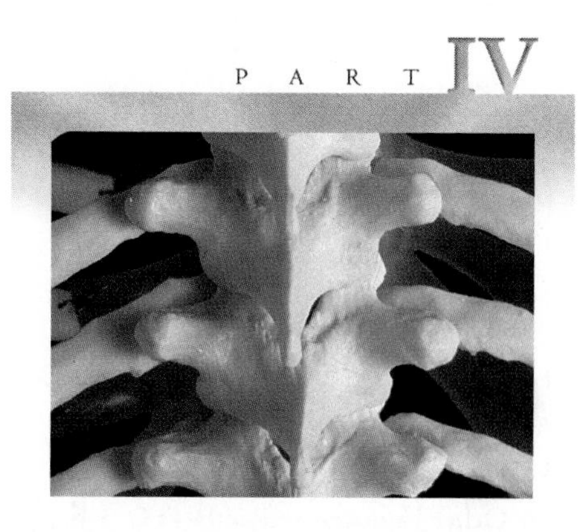

Integration and Control of the Human Body 319

CHAPTER 17

Nervous System 321

CHAPTER 18

Senses 347

Readings

Ecology Focus

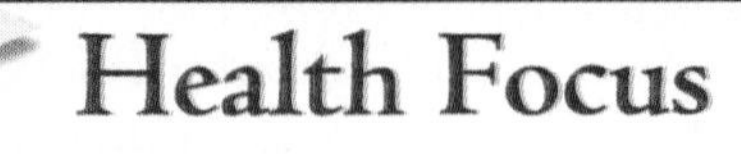

Health Focus

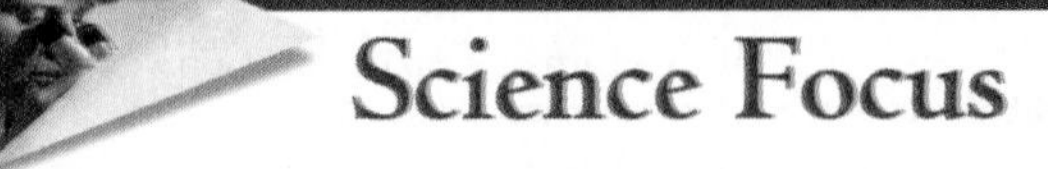

Science Focus

Visual Focus

Preface

Inquiry into Life is written for the introductory-level student who would like to develop a working knowledge of biology. While the text covers the whole field of general biology, it emphasizes the application of this knowledge to human concerns. Along with this approach concepts and principles are stressed, rather than detailed, high-level scientific data and terminology. Each chapter presents the topic clearly, simply, and distinctly so that students can achieve a thorough understanding of basic biology.

Major Themes

As with previous editions, the central theme of *Inquiry into Life* is understanding the workings of the human body and how humans fit into the world of living things. In keeping with this emphasis "Health Focus" readings review procedures and technology that can contribute to our well being and "Ecology Focus" readings show how the concepts of the chapter can be applied to ecological concerns. Concerned citizens need to realize that human beings are a part of a great, interrelated network called the biosphere.

Unlike any previous edition, *Inquiry into Life* now stresses the scientific process. This new emphasis is accomplished in several ways: The introductory chapter begins with an expanded explanation of the scientific process. Throughout the text, the part opening pages mention historical contributors to the concepts of that part. "Science Focus" readings describe more specifically the experiments and observations of modern biologists who have developed our present understanding of biological principles. For example, there is a "Science Focus" reading outlining the contributions of Susumu Tonegawa to our knowledge of immunity and another that describes the contributions of many neurologists to the expanding field of memory and learning. The end matter of each chapter contains Thinking Scientifically questions that ask students to reason as scientists do.

Bioethical Issues

This edition of *Inquiry into Life* also asks students to apply the concepts to the many and varied perplexing bioethical issues that face us every day. Each chapter ends with a description of a modern situation that calls for a value judgement on the part of the reader. Students are challenged to develop a point of view by answering a series of questions that pertain to the issue. The myriad of issues considered include genetic disease testing, human cloning, AIDS vaccine trials, animal rights, responsibility for one's health, and fetal research.

Vibrant, New Illustration Program

Almost every illustration in *Inquiry into Life* is new or has been revised to better interest students in the exciting world of biology. Students are visually motivated, and the new art program has many features they will find helpful. "Visual Focus" illustrations give a conceptual overview that relates structure to function. Color coordination includes assigning colors to the various classes of organic molecules and to the different human tissues and organs. Icons have been added to the cell, animal, and biosphere chapters to show how a portion relates to a general topic. An increased number of micrographs add realism to depicted structures.

Applications

Educational theory tells us that students are most interested in knowledge of immediate practical application. This text is consistent with and remains true to this approach. Each chapter begins with a short story that applies chapter material to real-life situations. The readings stress applications and so does the running text material. This edition features expanded treatment of such topics as eating disorders, allergies, pulmonary disorders, hepatitis infections, the human genome project, and gene therapy. Some topics such as the cloning of animals, xenotransplantation, and the role of apoptosis in immunity, development, and cancer are new.

New Chapters

A new part entitled Integration and Control of the Human Body contains rewritten chapters. The updated presentation of the nervous system and the senses includes new information on learning and memory. The endocrine system chapter better explains the homeostatic control of blood molecular and ionic levels. This part contains many new illustrations including several that depict the skeleton in greater detail.

The botanical and zoological chapters have been improved. New illustrations explain the cohesion-tension model of xylem transport and the pressure-flow theory of phloem transport. New evolutionary trees and icons are used in the animal diversity chapters. The evolution of humans has been updated.

The ecology chapters have been rewritten to have a modern approach to this important field. Coverage now logically flows from populations to communities to ecosystems. Adverse human influences, such as expanded human population and disruption of normal ecological cycles, are integrated into the general discussion where appropriate.

Pedagogical Features

As before, *Inquiry into Life* excels in pedagogical features. Each chapter begins with an integrated chapter outline that lists the chapter's concepts according to numbered sections of the chapter. This numbering system is continued in the chapter and summary so that instructors can assign just certain portions of the chapter, if they like. The text is paged so that major sections begin at the top of the page and illustrations are on the same or facing page as its reference. The questions at the end of the chapter are of both the essay and objective type. New to this edition, the objective questions are multiple choice with at least five choices for each one. Thinking Scientifically is a series of new questions that ask students to reason as a scientist does. All the boldfaced terms are listed and page referenced. A matching exercise tests student comprehension of the terms.

Technology

New to this edition, the free *Essential Study Partner CD-ROM,* accompanies the text. It contains high quality 3-D animations, interactive study activities, illustrated overviews of key topics in the text, and supplementary quizzing and exams that students will find extremely valuable. A CD-ROM icon has been placed throughout each chapter to remind students that this important learning tool can assist them in reviewing the concepts.

The *Dynamic Human 2.0,* which offers a pictorial review of each human system, has been revised to have greater student appeal. The Mader Home Page also contains interactive exercises to help students master the objectives of each chapter and provides further information on most topics discussed in the chapter.

Acknowledgments

The personnel at McGraw-Hill have always lent their talents to the success of *Inquiry into Life*. My publisher Michael Lange was always there to offer advice and my editor Patrick Reidy stepped in when needed to encourage us all. Suzanne Guinn, my developmental editor, served as a liaison between me and everyone else on the book team. Suzanne had many creative suggestions and was an inspiration to us all despite the long hours she labored.

Those in production also worked diligently toward the success of this edition. Marilyn Sulzer was the project manager, Jodi Banowetz, the art coordinator, and Lori Hancock was the photo research coordinator. And I especially want to thank Wayne Harms for the beautiful book he designed for all of us to enjoy. Everyone remained cheerful and helpful while going beyond the call of duty.

In my office Evelyn Jo Hebert has consistently provided support through several editions of the text, and Norma Costain's contributions have also made the success of *Inquiry* ninth edition possible. Kathleen Hagelston has been a wonderful resource for creative and expert input on illustrations through the editions of *Inquiry*.

New to This Edition

- The *Essential Study Partner CD-ROM* tutorial, which supports and enhances the concepts presented, is offered free with the text. A CD-ROM icon is used throughout the chapter to remind students to consult this useful learning tool.
- The scientific process is emphasized, and new to this edition are "Science Focus" readings and end-of-chapter "Thinking Scientifically" questions.
- A bioethical issue is discussed in a featured section at the end of each chapter. Challenging questions are provided that can be used as a basis for class discussion.
- A new illustration program adds vitality to the art and enhances the appeal of the text. Many new micrographs provide realism. "Visual Focus" illustrations give a pictorial overview of key topics. Color coding is used for both molecular structures and for human tissues and organs.
- A cell icon in the cell chapter, evolutionary tree icons in the animal chapters, and a biome map icon in the biosphere chapter help students relate the part to the whole.
- A new part entitled Integration and Control of the Human Body includes updated and rewritten chapters on the nervous system, the senses, the musculoskeletal system, and the endocrine system. The discussion is supported by many new illustrations.
- The ecology chapters have been rewritten to have a modern approach that still includes a discussion of adverse human activities on the biosphere.
- Relevancy of the text is increased with the inclusion or expanded treatment of topics like eating disorders, allergies, pulmonary disorders, hepatitis infections, xenotransplantation, human cloning, the human genome project, and gene therapy to treat cancer.

The Reviewers

Many instructors have contributed not only to this edition of *Inquiry into Life* but also to previous editions. I am extremely thankful to each one, for they have all worked diligently to remain true to our calling to provide a product that will be the most useful to our students.

In particular, it is appropriate to acknowledge the help of the following individuals for the ninth edition:

Lee Kats
Pepperdine University
Ken Kardong
Washington State University
Dr. John Harley
Eastern Kentucky University
Dr. Jane Aloi-Horlings
Saddleback College
George Spiegel
Mid Plaines Community College
Dr. Jean DeSaux
University of North Carolina at Chapel Hill
James Smith
California State University
Siu-Lam Lee
University of Massachusetts-Lowell
William Ambrose
Bates College
Dr. Kelly Williams
University of Dayton
Judith Schneidewent
Milwaukee Area Technical College
Janice Greene
Southwest Missouri State University
Lawrence Klotz
State University of New York College at Cortland
Joe Keen
Patrick Hewry Community College
John Capeheart
University of Houston-Downtown
Keith Overbaugh
Northwestern Michigan College
Virginia Naples
Northern Illinois University
Bruce Sundrud
Harrisburg Community College
Barbara Pleasants
Iowa State University
David Constantinos
Savannah State University
Nancy Prentiss
University of Maine at Farmington
Mike Lawson
Missouri Southern State College
Eileen Synnott
Bristol Community College
Phillip Eichman
University of Rio Grande
Linda Luck
Clarkson University
Neal McCord
Stephen F. Austin State University
David Hartsell
Phillips College
Catherine McCahill
Dean College
Fred Brock
Baptist Bible College of PA
Matthew McClure
Lamar University at Orange
John Erickson
Ivy Tech State College
Patricia Rugaber
Coastal Georgia Community College
Laszlo Hanzely
Northern Illinois University
Hugh Lefcort
Gonzaga University
Clementine de Angelis
Tarrant County Junior College
Jonathan Christie
Chemeketa Community College
David Cox
Mount Ida College
Gladys Crawford
University of Northern Texas
John Pleasants
Iowa State University
Arthur Jantz
Western Oklahoma State College
Solveig Krumins
Daytona Beach Community College West Campus
Andrea LeSchack
Florida Community College of Jacksonville
Dr. Forbes Davidson
Mesa State College
Charles Wert
Linn-Benton Community College
Gary Adams
Wabash Valley College
Gene Mesco
Savannah State Biology
Michael Westerhaus
Pratt Community College
Susan Smith
Massasoit Community College
Drew Howard
Becker College
J. Philip McLaren
Eastern Nazarene College
Wayne Vian
Central Community College-Grand Island
Lorena Blinn
Michigan State University
Cynthia Bottrell
Scott Community College
Anne Pacitti
Thomas Jefferson University
Dr. Marion Klaus
Northern Wyoming Community College District-Sheridan College
Sandra Horikami
Daytona Beach Community College West Campus
Beth Waters-Earhart
Columbia College
Anne Marie Helmenstine
Tusculum College
Elissa Ditto
Red Rocks Community College
Philip Nelson
Barstow College
Donald Hardy
Crown College
Maralyn Renner
College of the Redwoods
Randall Harris
William Carey College
Nan Perigo
Williamette University
Paul Boyer
University of Wisconsin-Parkside
Diane Doidge
Grand View College
Sandra Mitchell
Western Wyoming College
Joseph Wahome
Mississippi Valley State University
Dale Lambert
Tarrant County Junior College
Blair McMillian
Madison Area Technical College
Dalia Giedrimiene
Saint Joseph College
Gilbert Anderson
Central Carolina Technical College
Richard Hanke
Rose State College
Dr. Gibril Fadika
Saint Paul's College
Iona Baldridge
Lubbock Christian University
Deborah Hanson
Indiana University-Purdue University
Robert Turner
Western Oregon State College
Nancy Prentiss
University of Maine at Farmington
Venna Sallan
Owensboro Community College
Cherin Lee
University of Northern Iowa
Simon Cheung
Northeastern Illinois University
Mark Secord
Bee County College
Forbes Davidson
Mesa State College
Harry Kurtz
Sam Houston State University
Lori Rose
Sam Houston State University
Jordan Choper
Montgomery College
David Dallas
Northern Oklahoma A&M College
Kathleen Marr
Lakeland College
Dr. Mentor David
Barton County Community College
Amy Wernette
Hazard Community College
Dr. Steven Wheeler
Alvin Community College
Dr. Diane Dixon
Southeastern Oklahoma University
Howard Duncan
Norfolk State University

AIDS Booklet

This booklet describes how AIDS and related diseases are commonly spread so that readers can protect themselves and their friends against this debilitating and deadly disease.

Inquiry into Life Laboratory Manual

The ***Laboratory Manual,*** also written by Dr. Sylvia S. Mader, is a very resourceful accompaniment to ***Inquiry into Life.*** Most chapters in the text have an accompanying laboratory exercise in the manual (some chapters have more than one accompanying exercise). In this way, instructors are better able to emphasize particular portions of the curriculum. Every laboratory has been written to further help students appreciate the scientific method and to learn the fundamental concepts of biology and the specific content of each chapter.

Laboratory Resource Guide

More extensive information regarding preparation is found in this helpful guide. The guide includes suggested sources for materials and supplies, directions for making up solutions and otherwise setting up the laboratory, expected results for the exercises, and suggested answers to all questions in the laboratory manual.

250 Transparencies

A set of 250 full-color transparency acetates accompanies ***Inquiry into Life***. These acetates contain key illustrations from the text.

100 Micrograph Slides

This ancillary provides 35mm slides of many photomicrographs and all electron micrographs in the text.

Life Science Animations Videotape

Key biological processes are available on this videotape series. The animations bring visual movement to biological processes that are difficult to understand on the text page.

Visuals Testbank

A set of 50 transparency masters is available for use by instructors. These feature line art from the text with labels deleted for student quizzing or practice.

Instructor's Manual

The *Instructor's Manual,* prepared by Dr. Richard John Schrock with input from Dr. Sylvia S. Mader, includes Chapter Overview Introductions, Learning Objectives, Annotated Chapter Outlines, Ninth Edition Changes, Technology Correlations to McGraw-Hill Technology, Technology Resources, Lecture Enrichment Ideas, Topics and Projects, and Essay Questions.

Test Item File

The *Test Item File,* written by Dr. Richard John Schrock, includes more questions than any previous edition. Instructors can use the Classroom Testing Software to sort items in various ways, including by level and difficulty.

How to Study Science, Third Edition

This excellent workbook offers students helpful suggestions for meeting the considerable challenges of a college science course. It offers tips on how to take notes, how to get the most out of laboratories, and how to overcome science anxiety.

Schaum's Outlines: Biology

Updated to include the latest advances, ***Schaum's Outlines: Biology,*** features detailed illustrations of complex biologic systems and processes, ranging from the smallest elements of life to primates. Hundreds of problems with fully explained solutions cut down on study time and make important points easy to remember.

Classroom Testing Software (MicroTest)

This helpful testing software—in either Macintosh or Windows format—provides well-written and researched book-specific questions featured in the Test Item File. Items can be sorted in various ways, including by level and difficulty.

Basic Chemistry for Biology, Second Edition

Basic Chemistry for Biology is a self-paced supplement for students who need additional material to understand the basic concepts of chemistry. This text leads biology students through fundamental chemical concepts.

Critical Thinking Case Study Workbook

This ancillary provides 34 critical thinking case studies that are designed to immerse students in the "process of science" and challenge them to solve problems in the same way biologists do. An answer key accompanies this workbook.

Essential Study Partner CD-ROM

This interactive student study tool is packed with over 100 animations and more than 200 learning activities. From quizzes to interactive diagrams, your students will find that there has never been a more exciting way to study biology. A self-quizzing feature allows students to test their knowledge of a topic before moving on to a new module. Additional unit exams give students the opportunity to review an entire subject area. The quizzes and unit exams hyperlink students back to tutorial sections so they can easily review coverage for a more complete understanding. The text-specific *Essential Study Partner CD-ROM* supports and enhances the material presented in ***Inquiry into Life*** and has references in the text.

McGraw-Hill Course Solutions

Designed specifically to help you with your individual course needs, *Course Solutions* will assist you in integrating your syllabus with ***Inquiry into Life, ninth edition,*** and state-of-the-art new media tools.

At the heart of *Course Solutions* you'll find integrated multimedia, a full-scale Online Learning Center, and a Course Solutions Integration Guide. These unparalleled services are also available as a part of *Course Solutions:* McGraw-Hill Learning Architecture, McGraw-Hill Course Consultation Service, Visual Resource Library Image Licensing, McGraw-Hill Student Tutorial Service, McGraw-Hill Instructor Syllabus Service, PageOut Lite, PageOut: The Course Web Site Development Center, and other delivery options.

Inquiry Into Life Web Site

McGraw-Hill text-specific web sites allow students and instructors from all over the world to communicate. By visiting this site, students can access additional study aids—including quizzes and interactive activities—explore links to other relevant biology sites, and catch up on current information. Log on today!

www.mhhe.com/biosci/genbio/mader

Visual Resource Library CD-ROM

This helpful CD-ROM contains approximately 1,600 images and animations that can be easily imported into PowerPoint© to create multimedia presentations. Or, you may use the already prepared PowerPoint presentations.

Life Science Animations CD-ROM

Contains 125 animations of important biological concepts and processes for use in classroom presentations.

Microbes in Motion CD-ROM, Version 2.0

This interactive CD-ROM allows students to actively explore microbial structure and function. Great for self-study, preparation for class or exams, or for classroom presentations.

Explorations in Human Biology CD-ROM; and Explorations in Cell Biology and Genetics CD-ROM

These interactive CD-ROMs feature fascinating topics in biology. Both have 33 different modules that allow students to study a high-interest biological topic in an interactive format.

Life Science Living Lexicon CD-ROM

This CD-ROM contains a comprehensive collection of life science terms—including definitions of their roots, prefixes, and suffixes—as well as audio pronunciations and illustrations.

Virtual Biology Laboratory CD-ROM

Featuring ten simulations of the most common and important animal-based experiments ordinarily performed in introductory lab courses, the *Virtual Biology Laboratory CD-ROM* contains video, audio, and text to clarify complex physiological functions.

Biology Start-up Software

This software is a five-disk Macintosh tutorial that helps nonmajors master challenging biological concepts such as basic chemistry, photosynthesis, and cellular respiration.

The Dynamic Human CD-ROM, Version 2.0

This guide to anatomy and physiology interactively illustrates the complex relationships between anatomical structures and their functions in the human body. Realistic, three-dimensional visuals are the premier feature of this exciting learning tool.

Dynamic Human Videodisc

Enhance your classroom presentations with movement, sound, and motion of internal organs, cells, and systems. More than 80 premier 3-D animations covering all body systems from the outstanding *Dynamic Human CD-ROM* are included.

Before you begin, spend a little time looking over the next few pages. They provide a quick guide to the learning tools found throughout the text that have been designed to enhance your understanding of biology.

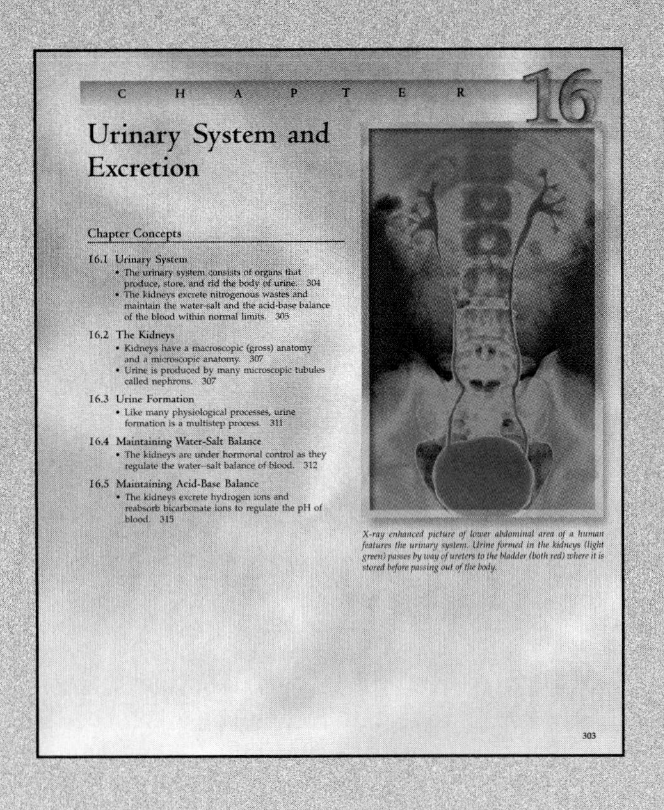

C H A P T E R 16

Urinary System and Excretion

Chapter Concepts

16.1 Urinary System
- The urinary system consists of organs that produce, store, and rid the body of urine. 304
- The kidneys excrete nitrogenous wastes and maintain the water-salt and the acid-base balance of the blood within normal limits. 305

16.2 The Kidneys
- Kidneys have a macroscopic (gross) anatomy and a microscopic anatomy. 307
- Urine is produced by many microscopic tubules called nephrons. 307

16.3 Urine Formation
- Like many physiological processes, urine formation is a multistep process. 311

16.4 Maintaining Water-Salt Balance
- The kidneys are under hormonal control as they regulate the water–salt balance of blood. 312

16.5 Maintaining Acid-Base Balance
- The kidneys excrete hydrogen ions and reabsorb bicarbonate ions to regulate the pH of blood. 315

X-ray enhanced picture of lower abdominal area of a human features the urinary system. Urine formed in the kidneys (light green) passes by way of ureters to the bladder (both red) where it is stored before passing out of the body.

303

Science Focus

Marine Enclosures for Whole-Ecosystem Studies

The MERL (Marine Ecosystem Research Laboratory) enclosures shown in Figure 36A provide marine researchers with a unique means to experiment with an entire marine ecosystem. The tanks measure 1.8 meters in diameter and 5 meters deep and are located outdoors, exposed to natural sunlight. To initiate a typical experiment, a benthic (bottom of the ocean) community is collected, usually from a silt-clay area in Narragansett Bay, off the coast of Rhode Island. A 37-cm-thick bed of sediment weighing roughly a ton is placed into each tank. Thirteen cubic meters of unfiltered seawater are transferred from the adjacent bay with nondisruptive displacement pumps. Mechanical mixing provides water movement in the enclosures to simulate wind and wave movement in the field. In the summer, cooling is provided, and in the winter, heat is provided to keep the enclosure temperatures similar to those in the bay.

When set up in this manner, unmanipulated enclosures maintain healthy ecosystems for many months and with properties that are similar to those actually found in the bay. Their large size and proximity to laboratory facilities allows repetitive sampling of all biological populations at relatively short time intervals. Because there are 14 enclosures, replication of experiments with controls is possible.

Experiments are done to determine the effect of a contaminant and to study the fate of chemicals within an entire coastal ecosystem. One set of MERL experiments addressed the problem of chronic additions of oil hydrocarbons to coastal waters. Water runoff from land, especially in urban areas, carries a continuous trickle of oil to coastal environments. The total amounts of petroleum hydrocarbons introduced into coastal

while the second 4-month experiment achieved about 0.1 ppm total hydrocarbons. Thereafter, recovery from the oil additions was studied for one year.

These levels of fuel hydrocarbons are below those that cause most tested marine species to die. One objective of the experiments was to develop an index system that could be used to indicate the health of an ecosystem. It was hypothesized that community diversity and primary productivity would be lower in the treated enclosures compared to the controls. The oil additions had a clear effect on the populations in the enclosures. Zooplankton and benthic macroorganisms were greatly reduced in abundance. Benthic populations remained depressed for at least a year after the treatments were stopped.

Even though the additions of oil clearly had a major impact on the communities in the enclosures, neither of the original hypotheses turned out to be correct. Although the population levels were quite different in treated and control tanks, measures of the diversity of benthic organisms in treated and control enclosures were indistinguishable. With oil treatments, there were increases in phytoplankton abundance and primary productivity instead of the expected decrease. In hindsight, the reason for the increase in phytoplankton abundance was clear. The oil was more toxic to the organisms that may graze the phytoplankton than to the phytoplankton itself. With the population of grazers reduced, the abundance and production of phytoplankton increased.

This is a good example of the interactions within an ecosystem, and the inherent difficulty of predicting how any component of an ecosystem will respond to stress.

Emphasis on the Scientific Process

This edition of ***Inquiry into Life*** places special emphasis on the scientific process. This added focus was accomplished in four ways:

- expanded explanation of the scientific process
- discussion of historical contributions
- "Scientifically Thinking" questions in each chapter
- "Science Focus" readings discuss biological research

16-4 Chapter 16 Urinary System and Excretion 307

16.2 The Kidneys

When a kidney is sliced lengthwise, it is possible to see the many branches of the renal artery and vein that reach inside the kidney (Fig. 16.3*a*). If the blood vessels are removed, it is easier to identify three regions of a kidney. The **renal cortex** is an outer granulated layer that dips down in between a radially striated, or lined, inner layer called the renal medulla. The **renal medulla** consists of cone-shaped tissue masses called renal pyramids. The **renal pelvis** is a central space, or cavity, that is continuous with the ureter (Fig. 16.3*b*).

Microscopically, the kidney is composed of over one million **nephrons,** sometimes called renal or kidney tubules (Fig. 16.3*c*). The nephrons produce urine and are positioned so that the urine flows into a collecting duct. Several nephrons enter the same collecting duct; the collecting ducts enter the renal pelvis.

Macroscopically, a kidney has three regions: renal cortex, renal medulla, and a renal pelvis that is continuous with the ureter. Microscopically, a kidney contains over one million nephrons.

The Essential Study Partner CD-ROM is an interactive student study tool referenced in the text and packed with over 100 animations and more than 200 learning activities. From quizzes to interactive diagrams, you will find that there has never been a more exciting way to study biology. A self-quizzing feature allows users to test their knowledge of a topic before moving on to a new module. Additional unit exams provide the opportunity to review an entire subject area. The quizzes and unit exams hyperlink back to tutorial sections so you can easily review material.

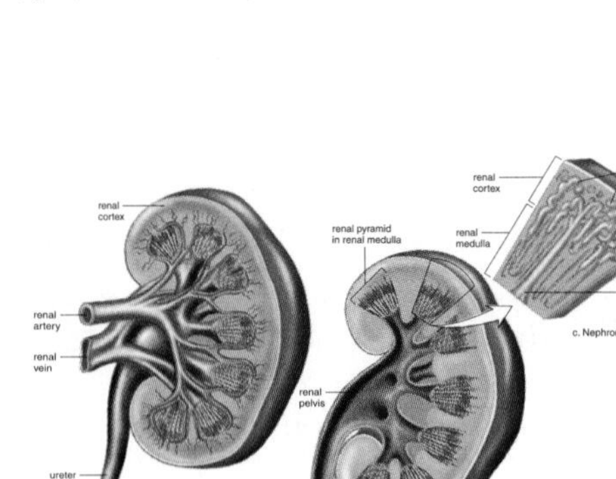

Figure 16.3 Gross anatomy of the kidney.
a. A longitudinal section of the kidney showing the blood supply. Note that the renal artery divides into smaller arteries, and these divide into arterioles. Venules join to form small veins, which join to form the renal vein. **b.** The same section without the blood supply. Now it is easier to distinguish the renal cortex, the renal medulla, and the renal pelvis, which connects with a ureter. The renal medulla consists of the renal pyramids. **c.** An enlargement showing the placement of nephrons.

New Illustration Program

Almost every illustration is new or revised. Icons have been added to the cell, animal, and biosphere chapters to help relate a topic to the big picture. And, new "Visual Focus" illustrations give a visual overview of key topics.

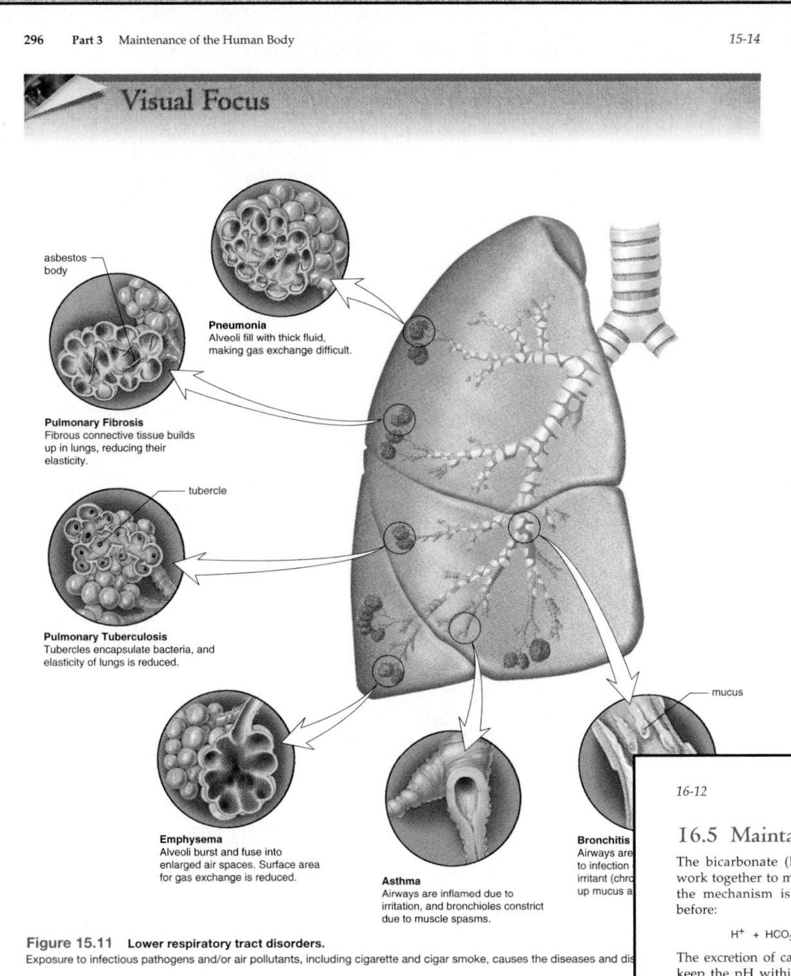

Figure 15.11 Lower respiratory tract disorders.
Exposure to infectious pathogens and/or air pollutants, including cigarette and cigar smoke, causes the diseases and dis

16-12 Chapter 16 Urinary System and Excretion 315

16.5 Maintaining Acid-Base Balance

The bicarbonate (HCO_3^-) buffer system and breathing work together to maintain the pH of the blood. Central to the mechanism is this reaction, which you have seen before:

$$H^+ + HCO_3^- \rightleftharpoons H_2CO_3 \rightleftharpoons H_2O + CO_2$$

The excretion of carbon dioxide (CO_2) by the lungs helps keep the pH within normal limits, because when carbon dioxide is exhaled this reaction is pushed to the right and hydrogen ions (H^+) are tied up in water. Indeed, when blood pH decreases, chemoreceptors in the carotid bodies (located in the carotid arteries) and in aortic bodies (located in the aorta) stimulate the respiratory center, and the rate and depth of breathing increases. On the other hand, when blood pH begins to rise, the respiratory center is depressed and the bicarbonate ion increases in the blood.

As powerful as this system is, only the kidneys can rid the body of a wide range of acidic and basic substances. The kidneys are slower acting than the buffer/breathing mechanism, but they have a more powerful effect on pH. For the sake of simplicity, we can think of the kidneys as reabsorbing bicarbonate ions and excreting hydrogen ions as needed to maintain the normal pH of the blood. If the blood is acidic, hydrogen ions are excreted and bicarbonate ions are reabsorbed. If the blood is basic, hydrogen ions are not excreted and bicarbonate ions are not reabsorbed. Since the urine is usually acidic, it shows that usually an excess of hydrogen ions are excreted. Ammonia (NH_3) provides a means for buffering these hydrogen ions in urine: ($NH_3 + H^+ \rightarrow NH_4^+$). Ammonia (whose presence is quite obvious in the diaper pail or kitty litter box) is produced in tubule cells by the deamination of amino acids. Phosphate provides another means of buffering hydrogen ions in urine.

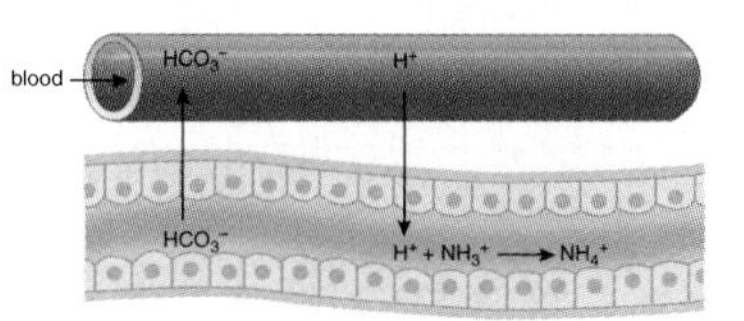

Figure 16.9 Acid-base balance.
In the kidneys, bicarbonate ions are reabsorbed and the hydrogen ions are excreted as needed to maintain the pH of the blood. Excess hydrogen ions are buffered, for example, by ammonia (NH_3), which is produced in tubule cells by the deamination of amino acids.

The acid-base balance of the blood is adjusted by the reabsorption of the bicarbonate ions (HCO_3^-) and the secretion of hydrogen ions (H^+) as appropriate.

Bioethical Issues

Each chapter of ***Inquiry into Life*** ends with a Bioethical Issue reading. No other general biology book covers as many issues or devotes as much space to this growing area of biology.

Bioethical Issue

As a society we are accustomed to thinking that as we grow older, diseases like urinary disorders will begin to occur. Almost everyone is aware that most males are subject to enlargement of the prostate as they age, and that cancer of the prostate is not uncommon among elderly men. However, like many illnesses associated with aging, medical science now knows how to treat or even cure prostate problems. Because of these successes, medical science has lengthened our life span. A child born in the United States in 1900 lived to, say, 47 years. If that same child were born today, it would probably live to at least 76. Even more exciting is the probability that scientists will improve the life span. People could live beyond 100 years and have the same vigor and vitality they had when they were young.

Most people are appreciative of living longer, especially if they can expect to be free of the illnesses and inconveniences associated with aging. But have we examined how we feel about longevity as a society? Whereas we are accustomed to considering that if the birthrate increases so does the size of a population, what about the death rate? If the birthrate stays constant and the death rate decreases, obviously population size also increases. Most experts agree that population growth depletes resources and increases environmental degradation. An older population can also put a strain on the economy if they are unable to meet their financial, including medical, needs without governmental assistance.

What is the ethical solution to this problem? Should we just allow the population to increase due to older people living longer? Should we decrease the birthrate? Should we reduce governmental assistance to older people so they realize that they must be able to take care of themselves? Should we call a halt to increasing the life span through advancements in medical science?

Questions

1. Do you feel that older people make a significant contribution to or a drain on society? Explain.
2. Would you be willing to have fewer children in order to hold the population in check if more people live longer? Why or why not?
3. Should the elderly expect governmental or family assistance as they age? Why or why not?

Summarizing the Concepts

The major headings are repeated in the summary because it is organized according to the major sections of the chapter. The summary helps students review the important concepts and topics discussed in each chapter.

Understanding the Terms

A matching exercise ensures that students understand the chapter's terms before proceeding to the next chapter. Answers are listed in Appendix A.

Testing Yourself

This section allows students to test their ability to fulfill the study objectives. Answers are listed in Appendix A.

316 Part 3 Maintenance of the Human Body 16-13

Summarizing the Concepts

16.1 Urinary System

The kidneys produce urine, which is conducted by the ureters to the bladder where it is stored before being released by way of the urethra.

The kidneys excrete nitrogenous wastes, including urea, uric acid, and creatinine, and they maintain the water-salt balance of the body and help keep the blood pH within normal limits.

16.2 The Kidneys

Macroscopically, the kidneys are divided into the renal cortex, renal medulla, and renal pelvis. Microscopically, they contain the nephrons.

Each nephron has its own blood supply; the afferent arteriole approaches the glomerular capsule and divides to become the glomerulus, a capillary tuft. The permeability of the glomerular capsule allows small molecules to enter the capsule from the glomerulus. The efferent arteriole leaves the capsule and immediately branches into the peritubular capillary network.

Each region of the nephron is anatomically suited to its task in urine formation. The spaces between podocytes of the glomerular capsule allow small molecules to enter the capsule from the glomerulus, a capillary knot. The cuboidal epithelial cells of the proximal convoluted tubule have many mitochondria and microvilli to carry out active transport (following passive transport) from the tubule to blood. In contrast, the cuboidal epithelial cells of the distal convoluted tubule have numerous mitochondria but lack microvilli. They carry out active transport from the blood to the tubule.

16.3 Urine Formation

Urine is composed primarily of nitrogenous waste products and salts in water. The steps in urine formation are glomerular filtration, tubular reabsorption, and tubular secretion, as explained in Figure 16.6.

16.4 Maintaining Water-Salt Balance

The kidneys regulate the water-salt balance of the body. Water is reabsorbed from all parts of the tubule. The ascending limb of the nephron loop establishes an osmotic gradient that draws water from the descending limb and also the collecting duct. The permeability of the collecting duct is under the control of the hormone ADH.

The reabsorption of salt increases blood volume and pressure because more water is also reabsorbed. Two other hormones, aldosterone and ANH, control the kidneys' reabsorption of sodium (Na^+).

16.5 Maintaining Acid-Base Balance

The kidneys keep blood pH within normal limits. They reabsorb HCO_3^- and excrete H^+ as needed to maintain the pH at about 7.4.

Studying the Concepts

1. State the path of urine and the function of each organ mentioned. 304
2. Explain how urination is controlled. 305
3. List and explain four functions of the urinary system. 305
4. Describe the macroscopic anatomy of a kidney. 307
5. Trace the path of blood about a nephron. 308
6. Name the parts of a nephron, and tell how the structure of the convoluted tubules suits their function. 309
7. State and describe the three steps of urine formation. 310–11
8. Where in particular is water and salt reabsorbed along the length of the nephron? Describe the contribution of the loop of the nephron. 312
9. Name and describe the action of antidiuretic hormone (ADH), the renin-aldosterone connection, and the atrial natriuretic hormone (ANH). 312–13
10. How do the kidneys maintain the pH of the blood within normal limits? 315

Testing Yourself

Choose the best answer for each question.

1. Which of these functions of the kidneys is mismatched?
   a. excretes metabolic wastes—rids the body of urea
   b. maintains the water-salt balance—helps regulate blood pressure
   c. maintains the acid-base balance—rids the body of uric acid
   d. secretes hormones—secretes erythropoietin
   e. All of these are properly matched.
2. Which of these is out of order first?
   a. glomerular capsule
   b. proximal convoluted tubule
   c. distal convoluted tubule
   d. loop of the nephron
   e. collecting duct
3. Which of these hormones is most likely to cause a rise in blood pressure?
   a. aldosterone
   b. antidiuretic hormone (ADH)
   c. renin
   d. atrial natriuretic hormone (ANH)
   e. Both a and c are correct.
4. If the blood is acidic,
   a. hydrogen ions are excreted and bicarbonate ions are reabsorbed.
   b. hydrogen ions are reabsorbed and bicarbonate ions are excreted.
   c. hydrogen ions and bicarbonate ions are reabsorbed.
   d. hydrogen ions and bicarbonate ions are excreted.
   e. urea, uric acid, and ammonia are excreted.
5. Excretion of a hypertonic urine in humans is associated best with the
   a. glomerular capsule.
   b. proximal convoluted tubule.
   c. loop of the nephron.
   d. distal convoluted tubule.
6. The presence of ADH (antidiuretic hormone) causes an individual to excrete
   a. sugars.
   b. less water.
   c. more water.
   d. Both a and c are correct.
7. In humans, water is
   a. found in the glomerular filtrate.
   b. reabsorbed from the nephron.
   c. in the urine.
   d. All of these are correct.

16-14 Chapter 16 Urinary System and Excretion 317

8. Filtration is associated with the
   a. glomerular capsule.
   b. distal convoluted tubule.
   c. collecting duct.
   d. All of these are correct.
9. Normally in humans, glucose
   a. is always in the filtrate and urine.
   b. is always in the filtrate, with little or none in urine.
   c. undergoes tubular secretion and is in urine.
   d. undergoes tubular secretion and is not in urine.
10. Label this diagram of a nephron.

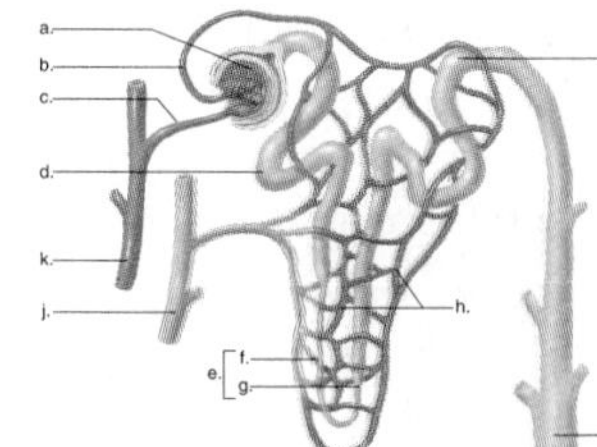

Thinking Scientifically

1. Considering filtration (page 311):
   a. Urine, not urea, is made by the kidneys. What is the difference between urine and urea?
   b. Blood pressure in the glomerulus favors filtration of molecules. The efferent arteriole (leaving the glomerulus) is narrower than the afferent arteriole (leading to the glomerulus). What effect does this have on blood pressure in the glomerulus?
   c. What force would oppose filtration of molecules from the glomerulus? (Hint, review the forces involved in capillary exchange.)
   d. If there is a loss of proteins from blood into the glomerular capsule, would the filtration rate increase or decrease? (Assume constant blood pressure.)
2. Considering urine formation (page 311):
   a. If 99% of water is reabsorbed, how can urine be 95% water.
   b. Carrier molecules work as fast as they can to return glucose to blood. Explain why excess glucose is not returned.
   c. When CO_2 is excreted by the lungs, does blood become more acidic or more basic? If the HCO_3^- is excreted by the kidneys, does blood become more acidic or more basic?
   d. The maintenance of normal blood pH is a very important function of the kidneys. What molecules in the cells are affected by pH changes.

Understanding the Terms

aldosterone 313
antidiuretic hormone (ADH) 312
atrial natriuretic hormone (ANH) 313
collecting duct 309
creatinine 305
distal convoluted tubule 309
diuretics 313
erythropoietin 305
excretion 305
glomerular capsule 309
glomerular filtrate 311
glomerular filtration 311
glomerulus 308
juxtaglomerular apparatus 313
kidney 304
loop of the nephron 309
nephron 307
peritubular capillary network 308
podocyte 309
proximal convoluted tubule 309
renal artery 304
renal cortex 307
renal medulla 307
renal pelvis 307
renal vein 304
renin 313
tubular reabsorption 311
tubular secretion 311
urea 305
ureter 304
urethra 304
uric acid 305
urinary bladder 304

Match the terms to these definitions:

a. ________ Portion of the nephron lying between the proximal convoluted tubule and the distal convoluted tubule that functions in water reabsorption.
b. ________ Movement of molecules from the contents of the nephron into blood at the proximal convoluted tubule.
c. ________ Hormone secreted by the adrenal cortex that regulates the sodium and potassium balance of the blood.
d. ________ Outer portion of the kidney that appears granular.
e. ________ Surrounds a nephron and functions in reabsorption during urine formation.

Using Technology

Your study of the urinary system and excretion is supported by these available technologies:

Essential Study Partner CD-ROM
Animals → Osmoregulation
Visit the Mader web site for related ESP activities.

Exploring the Internet
The Mader Home Page provides resources and tools as you study this chapter.
http://www.mhhe.com/biosci/genbio/mader

Dynamic Human 2.0 CD-ROM
Urinary System

Life Science Animations 3D Video
38 Kidney Function

Studying the Concepts

These questions, which are page-referenced, review important chapter material. Answers are available to instructors via the Online Learning Center.

Thinking Scientifically

In this section, students answer questions related to concepts they have learned to matters of scientific concern. Answers are listed in Appendix A.

Using Technology

References to McGraw-Hill technology point students to additional sources for more information.

C H A P T E R 1

The Study of Life

Chapter Concepts

Many biological experiments are performed in the laboratory where conditions can be more easily controlled.

About 200 years ago, an English country doctor, Edward Jenner, made medical history due to a keen observation. He noticed that milkmaids usually avoided contracting smallpox, a disfiguring and frequently fatal disease. Jenner reasoned that the women were somehow protected because they had previously had cowpox, a similar but much less serious illness. To test that hypothesis, Jenner injected a young boy with cowpox, and later exposed him to smallpox. When his volunteer remained healthy, the way was paved for the development of the modern version of a smallpox vaccine that has done away with the disease entirely.

Biology uses the same methodology today as was used by Jenner to come to conclusions that are accepted until proven false by future investigations. The information presented in this text is the result of work done by scientists over the ages. What is science and what methodology does science use?

1.1 The Scientific Process

Biology is the scientific study of life (Fig. 1.1). Religion, aesthetics, ethics, and, yes, science, are all ways that human beings have of finding order in the natural world. But science differs from other human ways of knowing and learning by its process.

The process of science is unique. **Science** is a human endeavor that considers only what is observable by the senses or by instruments that extend the ability of the senses. Microscopes extend the ability of sight way beyond what can be seen by the naked eye, for example. Observations can be made by any one of our five senses; we can observe with our noses that dinner is almost ready; observe with our fingertips that a surface is smooth and cold; and observe with our ears that a piano needs tuning. Various input allows scientists to formulate hypotheses.

Formulating a Hypothesis

Once a scientist becomes interested in an observable event, called a **phenomenon,** he or she will most likely study up on it. The goal will be to find any other past work in the area; the Internet, the library, other scientists will all be consulted to see what past ideas there may have been about this phenomenon. All of this input helps a scientist use imagination and creative thinking to formulate a hypothesis. A **hypothesis** is a tentative explanation of a scientific question based on observations and past knowledge.

Hypotheses, of course, vary in the level of their sophistication. A young person in elementary school might hypothesize that plants need light in order to grow; while a high school student might hypothesize that light allows plant leaves to produce sugar. In any case, all the past experiences of the individual no matter their source, will most likely influence the formulation of a hypothesis. Einstein said that aesthetic qualities like beauty and simplicity

Figure 1.1 Diversity of life.
Biology is the scientific study of life, which has diversified into the many forms we find on planet Earth. **a.** *Vorticella,* a protozoan; **b.** *Hibiscus,* a flowering plant; **c.** Sally lightfoot, a crab; **d.** Snow leopard, a mammal

helped him to formulate his hypotheses about the workings of the universe. Indeed, most every scientist looks favorably upon the simplest hypothesis for a phenomenon and tests that one first.

Science only considers hypotheses that can be tested. Moral and religious beliefs while very important to our lives differ between cultures and through time and they are not always testable by further observations and/or experimentation.

Testing a Hypothesis

Hypotheses are tested either in the laboratory setting or in a natural setting. The laboratory is a place that lends itself to carrying on **experiments,** which are artificial situations devised to test hypotheses. Experiments can also be carried out in a natural setting, called the field. David P. Barash, who was observing the mating behavior of mountain bluebirds, decided to perform an experiment to test the hypothesis that aggression of the male varies according to the reproductive cycle. He posted a male bluebird model near nests while the resident male was out foraging. The behavior of the male toward the model and toward his female mate were noted during the first ten minutes of the male's return. Aggression was most severe when the male model was presented before the first egg was laid, less severe when the model was presented after an egg was laid, and least severe after the eggs had hatched (Fig. 1.2).

Laboratory experiments are often preferred to experiments in the field because a laboratory setting eliminates extraneous variables. A **variable** is a factor that can cause an observable change during the progress of the experiment. The Science Focus on the next page describes a hypothetical experiment scientists conduct to determine whether substance S is a safe food additive. Some of the possible variables include the individual characteristics of the mice, availability of food and water, and environmental conditions such as the cleanliness of the cages and the temperature and humidity of the room, and so forth. The experimenters would try to keep all these variables constant except the presence or absence of substance

Experiments are considered more rigorous when they include a control group. Control in this context does not refer to the power of the experimenter—rather it refers to the validity of the experimental results. A **control group** experiences all the steps in the experiment except for the one that is being tested. Barash used a control group. He posted a male robin instead of a male bluebird near certain nests. (The male bluebirds did not respond at all to a robin model.) In the substance S experiment, the mice in the control group do not have substance S added to their food. When an experiment has a control group, we know that the results are due to the variable being tested and not due to some nonidentifiable chance event that occurred during the experiment.

a. Scientist makes observations, studies previous data, and formulates a hypothesis.

b. Scientist performs experiment and collects objective data.

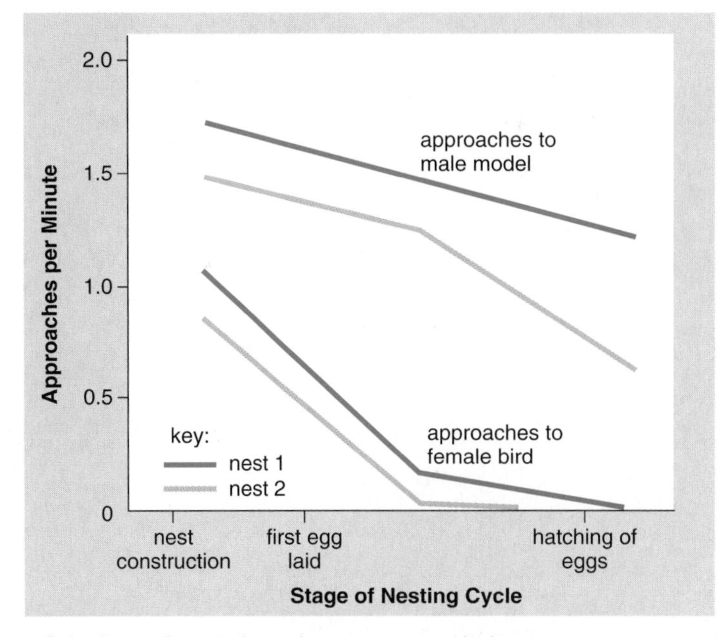

c. Scientist studies results and comes to a conclusion.

Figure 1.2 **Example of scientific method.**
Observation of male bluebird behavior allowed David Barash to formulate a testable hypothesis. Then, he collected the data displayed in a graph and came to a conclusion.

Science Focus

A Controlled Laboratory Experiment

Biologists want to determine if sweetener S is a safe food additive and design a controlled laboratory experiment. They place a certain number of randomly chosen inbred (genetically similar) mice into the two groups—say, 100 mice per group (Fig. 1A). If any of the mice are different from the others, it is hoped random selection will distribute them evenly among the groups. The researchers also make sure that all conditions, such as availability of water, cage setup, and temperature of the surroundings, are the same for both groups. The food for each group is exactly the same except for the amount of sweetener S.

In experiment 1, the researchers formulate a hypothesis that sweetener S is a safe food additive at any dietary intake. The control group receives no sweetener S and in the test group 50% of the diet is sweetener S. At the end of the experiment, both groups of mice are examined for bladder cancer. They find that one-third of the mice in the test group have bladder cancer, while none in the control group have bladder cancer. The results of this experiment falsify the hypothesis and the researchers conclude that sweetener S is not a safe food additive at 50% of dietary intake.

The researchers decide to refine their study, and in experiment 2 they hypothesize that sweetener S is safe if the diet contains a limited amount of sweetener S. They feed sweetener S to groups of mice at ever-greater concentrations until it reaches 50% of the diet.

Group 1: diet contains no sweetener S (the control)

Group 2: 5% of diet is sweetener S

Group 3: 10% of diet is sweetener S

↓

Group 11: 50% of diet is sweetener S

The researchers present their data in the form of a graph and statistically analyze their data to determine if the difference in the number of cases of bladder cancer between the various groups is significant and not due to simple chance. Finding that they are significant, the researchers conclude that they can now develop a recommendation concerning the intake of sweetener S in humans. They suggest that an intake of sweetener S up to 10% of the diet is relatively safe but thereafter an ever-greater incidence of bladder cancer should be expected (Fig. 1A).

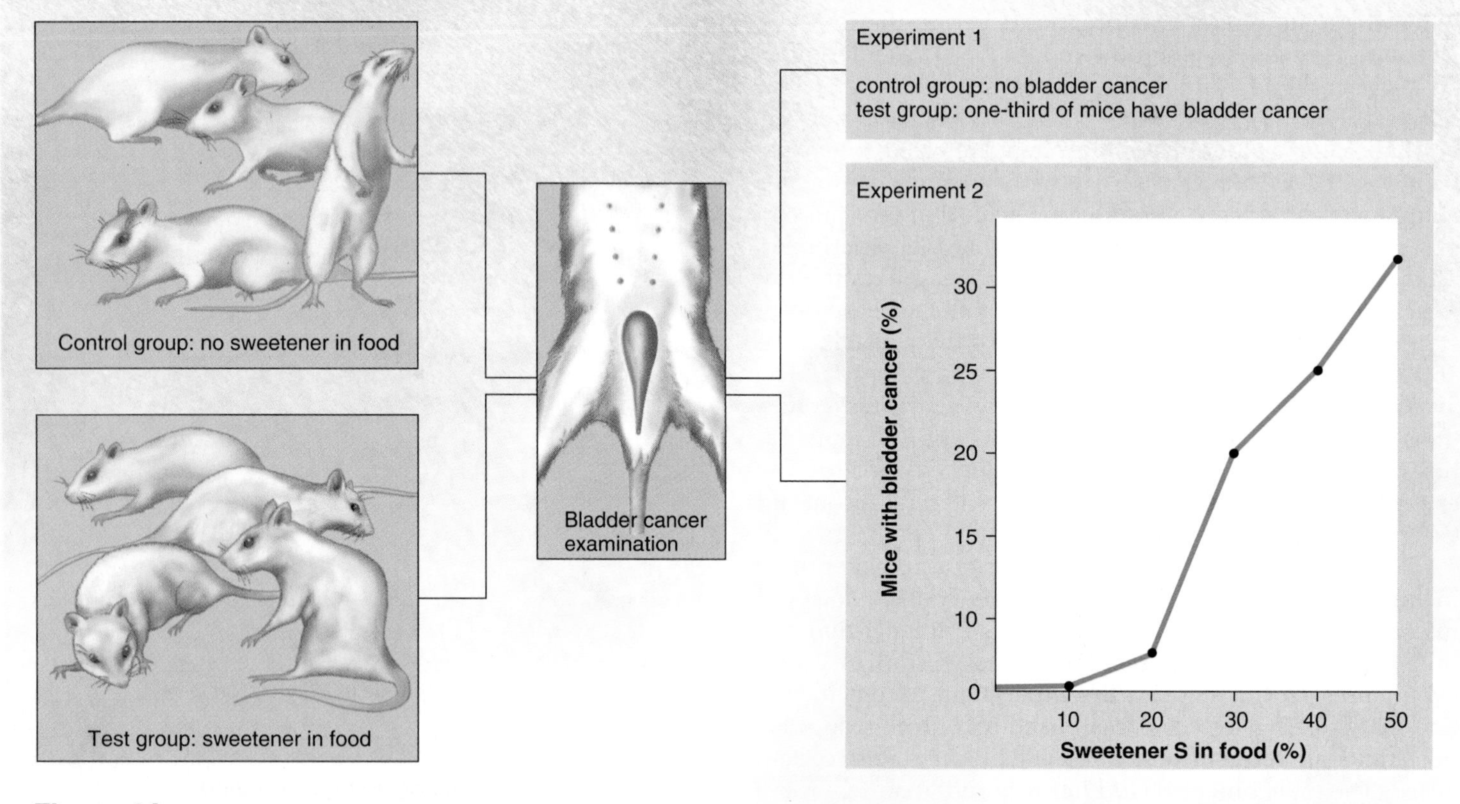

Figure 1A Design of a controlled experiment.
Genetically similar mice are randomly divided into a control group and (a) test group(s) that contain 100 mice each. All groups are exposed to same conditions, such as cage setup, temperature, and water supply. The control group is not subjected to sweetener S in the food. At the end of the experiment, all mice are examined for bladder cancer.

The Experimental Results

The results of an experiment are referred to as the **data.** Data often takes the form of a table or a graph that allows one to see the results in an organized manner. Barash's data are shown in Figure 1.2. Data should be objective rather than subjective, and mathematical data are conducive to objectivity. For example, Barash didn't report that it seemed to him that the male bluebirds were more aggressive before the eggs were laid (subjective); rather, he defined aggression as the "number of approaches per minute" (objective) and reported that number. When Mendel, the father of genetics, did a pea cross concerning height, he didn't report that he could tell at a glance there were more tall than short plants; he reported that the ratio was three tall to every short plant. In fact, Mendel repeated one particular experiment so many times that he counted 7,324 peas!

To ensure that the results are not due to chance or some unknown variable, experiments are often repeated not only by the original scientist but by others in the same area of expertise. Therefore, scientists must keep careful records of how they perform their experiments so that others can repeat them. If the same results are not obtained over and over again, the experiment is not considered valid.

The Conclusion

After studying the results, the experimenter comes to a **conclusion** whether the results support or **falsify** (show to be untrue) the hypothesis. The data allowed Barash to conclude that aggression in male bluebirds is related to their reproductive cycle. Therefore, his hypothesis was supported. If male bluebirds were always aggressive even toward male robin models, his hypothesis would have been proven false.

Science progresses and the hypothesis may have to be modified in the future. There is always the possibility that a more sophisticated experiment using perhaps more advanced technology might falsify the hypothesis. Therefore, a scientist never says that the data "prove" the hypothesis to be "true." Because of this feature, some think of science as what is left after alternative hypotheses have been rejected.

Scientists report their findings in scientific journals so that their methodology and results are available to the scientific community. Barash reported his experiment in the *American Naturalist.*[1] The reporting of experiments results in accumulated data that will help other scientists formulate hypotheses. Also, it results in a body of information that is made known to the general public through the publishing of books, such as this biology textbook.

[1]Barash, D. P. 1976. The male responds to apparent female adultery in the mountain bluebird, *Sialia currucoides:* An evolutionary interpretation. *American Naturalist* 110:1097–101.

Scientific Theories

The ultimate goal of science is to understand the natural world in terms of **scientific theories,** which are concepts that join together well-supported and related hypotheses. In a movie, a detective may claim to have a theory about the crime. Or you may say that you have a theory about the won-lost record of your favorite baseball team. But in science, the word *theory* is reserved for a conceptual scheme that is supported by a broad range of observations, experiments, and data.

Some of the basic theories of biology are:

Name of Theory	Explanation
Cell	All organisms are composed of cells.
Biogenesis	Life comes only from life.
Evolution	All living things have a common ancestor and are adapted to a particular way of life.
Gene	Organisms contain coded information that dictates their form, function, and behavior.

Evolution is the unifying concept of biology because it pertains to various aspects of living things. For example, the theory of evolution enables scientists to understand the history of life, the variety of living things, and the anatomy, physiology, and development of organisms—even their behavior.

Barash gave an evolutionary interpretation to his results. It was adaptive, he said, for male bluebirds to be less aggressive after the first egg is laid because by that time the male bird is "sure the offspring is his own" and maladaptive for the male bird to waste time and energy being too aggressive toward a rival and his mate after hatching because his offspring is already present. (When an organism is adapted to its environment, it is better able to survive and produce offspring.)

Fruitful theories are ones that help scientists generate new hypotheses, and the theory of evolution has been a very fruitful theory. In fact, it probably helped Barash develop the hypothesis he chose to test. Because the theory of evolution has been supported by so many observations and experiments for over 100 years, some biologists refer to the **principle** of evolution, suggesting that this is the appropriate term for theories that are generally accepted by an overwhelming number of scientists. The term **law** instead of principle is preferred by some. In another chapter concerning energy relationships, for example, we will examine the laws of thermodynamics.

The Scientific Method

The process of science is often described in terms of the **scientific method.** Some scientists object to outlining the steps of the scientific method as is done in Figure 1.3 because such a diagram suggests a rigid methodology. Actually, scientists approach their work in many different ways and even sometimes make discoveries by chance. The most famous case pertains to penicillin. When examining a petri dish in 1928, Alexander Fleming noticed an area around a mold that was free of bacteria. Upon investigating, Fleming found that the mold produced an antibacterial substance he called penicillin. Penicillin was later mass produced and is still a successfully used antibiotic in humans today. Some scientists do not test hypotheses—instead, they make observations and add to the known data. For example, a scientist might decide to find out what types of animals live on the ocean floor.

A discussion of the scientific process is not complete without an admission that scientists do have to accept certain assumptions. They have to believe, for example, that nature is real and understandable and knowable by observing it; that nature is orderly and uniform; that measurements yield knowledge of the thing measured; and that natural laws are not affected by time.

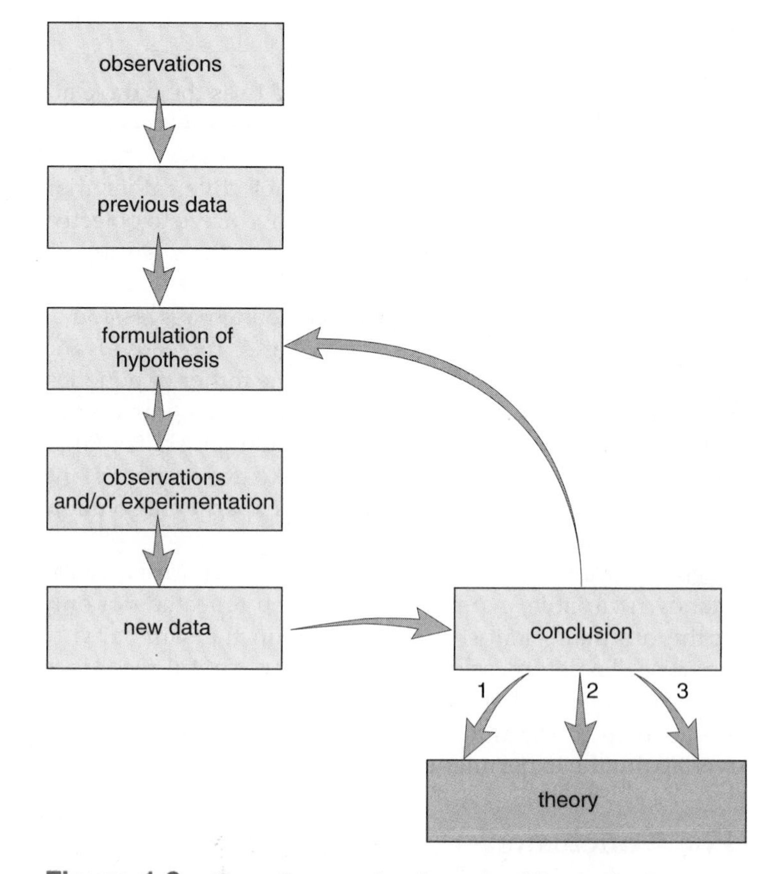

Figure 1.3 Flow diagram for the scientific method. On the basis of observations and previous data, a scientist formulates a hypothesis. The hypothesis is tested by further observations or a controlled experiment, and new data either support or falsify the hypothesis. The return arrow indicates that a scientist often chooses to retest the same hypothesis or to test a related hypothesis. Conclusions from many different but related experiments may lead to the development of a scientific theory. For example, studies in biology of development, anatomy, and fossil remains all support the theory of evolution.

Science and Social Responsibility

Science seeks only a natural cause for the origin and history of life. Doctrines of creation that have a mythical, philosophical, or theological basis are outside the realm of science because they cannot be tested by observation and/or experimentation. Creationism, which states that God created all species as they are today, cannot be considered science because explanations based on supernatural rather than natural causes involve faith rather than data.

There are many ways in which science has improved our lives. The discovery of antibiotics and vaccines has expanded the human life span. Cell biology research is helping us understand the causes of cancer. Genetic research has produced new strains of agricultural plants that have eased the burden of feeding our burgeoning world population. Still there are other instances in which science has resulted in technologies that have harmed the environment. **Technology** is a process, an instrument, or a structure that is developed or constructed using scientific principles. Biochemical knowledge was used to develop pesticides which have helped increase agricultural yields. Pesticides, as you may know, kill not only pests but also other types of organisms. The book *Silent Spring* was written to make the public aware of the harmful environmental effects of pesticide use.

Too often we blame science for these developments and think that scientists are duty bound to pursue only those avenues of research that are consistent with a certain system of values. But making value judgments is not a part of science. Ethical and moral decisions must be made by all people. The responsibility for how we use the fruits of science, including a given technology, must reside with people from all walks of life, not upon scientists alone. Scientists should provide the public with as much information as possible when such issues as the use of atomic energy, fetal research, and genetic engineering are being debated. Then they, along with other citizens, can help make decisions about the future role of these technologies in our society. All men and women have a responsibility to decide how to use scientific knowledge so that it benefits the human species and all living things.

1.2 The Characteristics of Life

Biological theories help us determine the characteristics of life. Despite its diversity (see Figure 1.1), certain properties characterize all living things. For example, all living things are organized.

Living Things Are Organized

Living things have levels of organization from the atoms that constitute all matter to an ecosystem in which they live (Fig. 1.4). Atoms join together to form molecules such as the DNA molecules that occur only within cells. The **cell** is the lowest level of biological organization to have the characteristics of life. A nerve cell is one of the types of cells in the body of a white-tailed deer. A **tissue** is a group of similar cells that perform a particular function. Nervous tissue has millions of nerve cells that transmit signals to all parts of the deer's body. Several tissues join together to form an **organ.** The main organs in the nervous system of a deer are the brain, the spinal cord, and the nerves. Organs work together to form an **organ system.** The brain sends messages to the spinal cord which, in turn, sends them to body parts by way of the spinal nerves. An **organism,** meaning an individual living thing, is a collection of organ systems. A deer contains a digestive system, a circulatory system, and several other systems in addition to a nervous system. Organisms usually live within a **population,** which is a group of interbreeding organisms in a particular locale. Several different populations interact within a **community.** For example, deer feed on many different types of living plants. An **ecosystem** includes a community and also the physical environment. Organisms not only interact with one another, they also interact with the physical environment, as when a deer takes a drink of water from a pond.

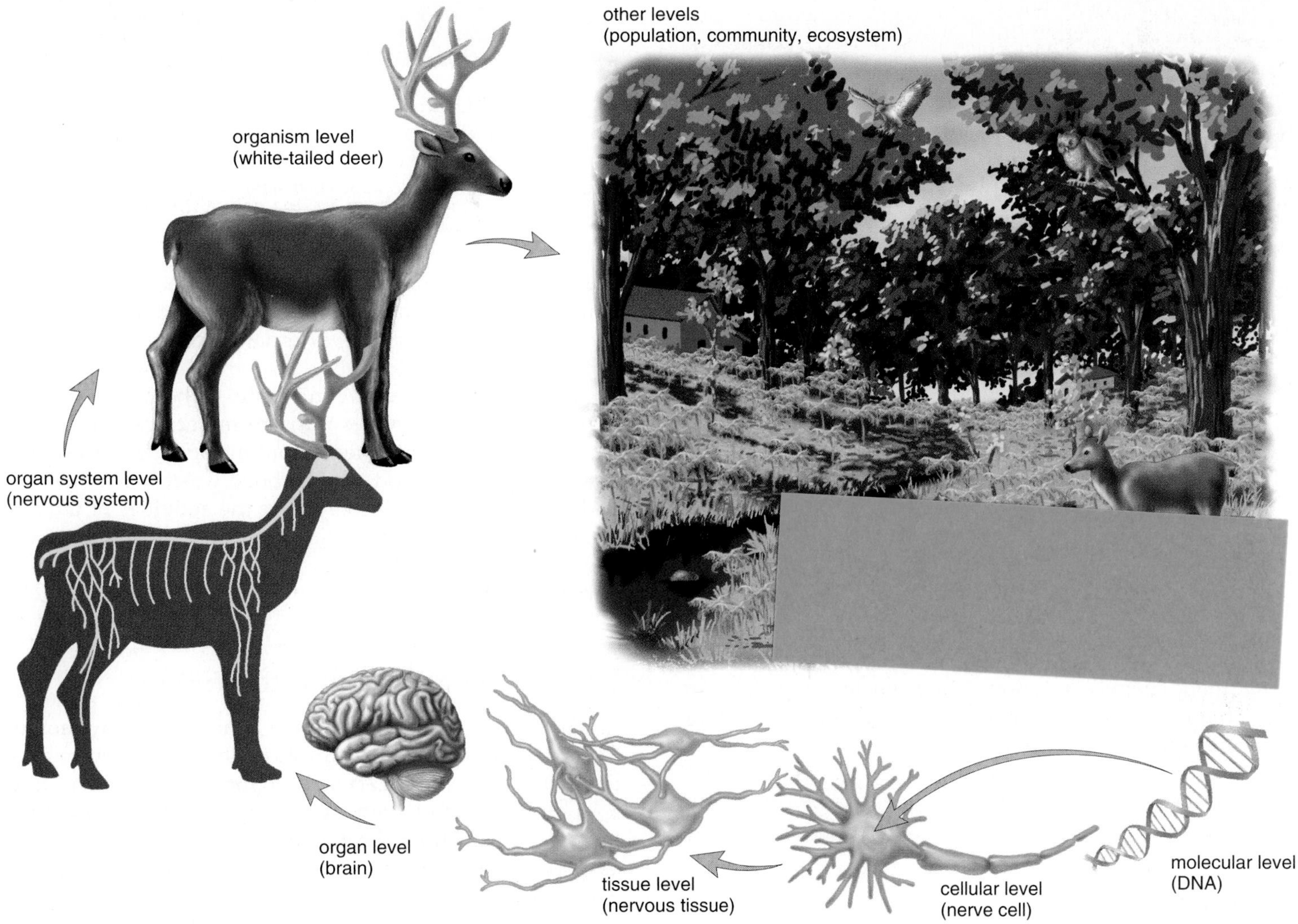

other levels
(population, community, ecosystem)
organism level
(white-tailed deer)
organ system level
(nervous system)
organ level
(brain)
tissue level
(nervous tissue)
cellular level
(nerve cell)
molecular level
(DNA)

Figure 1.4 Levels of biological organization.

Figure 1.5 Living things acquire materials and energy.
An osprey is a large bird that preys on fishes as its source of food.

Figure 1.6 Living things reproduce.
An osprey lays two to four eggs in a large nest located on the top of a tree, a rock pinnacle, or even a telephone pole. The fledglings are ready to leave the nest 40 to 50 days after hatching.

Living Things Acquire Materials and Energy

Living things cannot maintain their organization nor carry on life's other activities without an outside source of materials and energy. Photosynthesizers, such as trees, use carbon dioxide, water, and solar energy to make their own food. Human beings and other animals, like ospreys (Fig. 1.5), acquire materials and energy when they eat food.

Food provides nutrient molecules, which are used as building blocks or for energy. **Energy** is the capacity to do work, and it takes work to maintain the organization of the cell and of the organism. When nutrient molecules are used to make their parts and products, cells carry out a sequence of synthetic chemical reactions. Some nutrient molecules are broken down completely to provide the necessary energy to carry out synthetic reactions.

Most living things can convert energy into motion. Self-directed movement, as when we decide to rise from a chair, is even considered by some to be a characteristic of life.

Living Things are Homeostatic

Homeostasis means "staying the same." Actually, the internal environment stays *relatively* constant; for example, the human body temperature fluctuates slightly during the day. Also, the body's ability to maintain a normal internal temperature is somewhat dependent on the external temperature—we will die if the external temperature becomes overly hot or cold.

All human systems contribute to homeostasis. The digestive system provides nutrient molecules; the circulatory system transports them about the body; and the excretory system rids blood of metabolic wastes. The nervous and hormonal systems coordinate the activities of the other systems. One of the major purposes of this text is to show how all the systems of the human body help to maintain homeostasis.

Living Things Respond to Stimuli

Living things respond to external stimuli, often by moving toward or away from the stimulus. Movement in humans is dependent upon their nervous and muscular systems. Other living things use a variety of mechanisms in order to move. Leaves of plants track the passage of the sun during the day, and when a houseplant is placed near a window, its stem bends to face the sun.

The movement of an organism, whether self-directed or in response to a stimulus, constitutes a large part of an organism's **behavior.** Behavior largely is directed toward minimizing injury, acquiring food, and reproducing.

Figure 1.7 Living things grow and develop.
Stages in the development of an oak tree from an acorn to a seedling to an adult.

Living Things Reproduce

Life comes only from life. The presence of **genes** in the form of DNA molecules allows cells and organisms to **reproduce**—that is, make more of themselves (Fig. 1.6). DNA contains the hereditary information that directs the structure and **metabolism,** chemical reactions of a cell. Before reproduction occurs, genes are replicated and copies of genes are produced.

Unicellular organisms reproduce asexually simply by dividing. The new cells have the same genes and structure as the single parent. Multicellular organisms usually reproduce sexually. Each parent, male and female, contributes roughly one-half the total number of genes to the offspring, which then does not resemble either parent exactly.

Living Things Grow and Develop

Growth, recognized by an increase in size and often the number of cells, is a part of development. In humans, **development** includes all the changes that take place between conception and death. First, the fertilized egg develops into a newborn, and then a human goes through the stages of childhood, adolescence, adulthood, and aging. Development also includes the repair that takes place following an injury.

All organisms undergo development. Figure 1.7 illustrates that an oak tree progresses from an acorn to a seedling before it becomes an adult oak tree.

Living Things Are Adapted

Adaptations are modifications that make an organism suited to its way of life. Consider, for example, a bird like an osprey (see Fig. 1.5), which catches and eats fish. An osprey can fly in part because it has hollow bones to reduce its weight and flight muscles to depress and elevate the wings. When an osprey dives, its strong feet take the first shock of the water and then its long and sharp claws hold onto its slippery prey.

Adaptations come about through evolution. **Evolution** is the process by which characteristics of **species** (a group of similarly constructed organisms that successfully interbreed) change through time. When new variations arise that allow certain members of the species to capture more resources, these members tend to survive and to have more offspring than the other unchanged members. Therefore, each successive generation will include more members with the new variation. In the end, most members of a species have the same adaptations to their environment.

Evolution, which has been going on since the origin of life, explains both the unity and the diversity of life. All organisms share the same characteristics of life because their ancestry can be traced to the first cell or cells. Organisms are diverse because they are adapted to different ways of life.

Living Things Belong to a Population

Individual organisms belong to a **population,** all the members of a species that live within a particular **community.** The populations within a community interact among themselves and with the physical environment (soil, atmosphere, etc.), thereby forming an **ecosystem.** All ecosystems are included within the **biosphere,** a thin layer of life that encircles the earth.

Although an ecosystem like a tropical rain forest changes—some trees fall and some animals die—an ecosystem remains recognizable year after year. We say it is in dynamic balance. In many cases, even the extinction of species (and their replacement by new species through evolution) still allows the dynamic balance of the system to be maintained.

A major feature of the interactions between populations pertains to who eats whom. Plants produce food, and animals that eat plants are food for other animals. Such a sequence of organisms is called a food chain (Fig. 1.8). Both plants and animals interact with the physical environment, as when they exchange gases with the atmosphere.

Nutrients cycle within and between ecosystems. Plants take in inorganic nutrients, like carbon dioxide and water, and produce organic nutrients, such as carbohydrates, that are used by themselves and various levels of animal consumers. When these organisms die and decay, inorganic nutrients are made available to plants once more. The blue arrows in Figure 1.8 show how chemicals cycle through the various populations of an ecosystem.

In contrast, the yellow arrows in Figure 1.8 show how energy flows through an ecosystem: solar energy used by plants to produce organic food is eventually converted to heat when organisms, including plants, use organic food as an energy source. Therefore, a constant supply of solar energy is required for an ecosystem and for life to exist.

All living things share the same characteristics of life: they are organized; take materials and energy from the environment; are homeostatic; respond to stimuli; reproduce; grow and develop; are adapted to their way of life; and belong to populations.

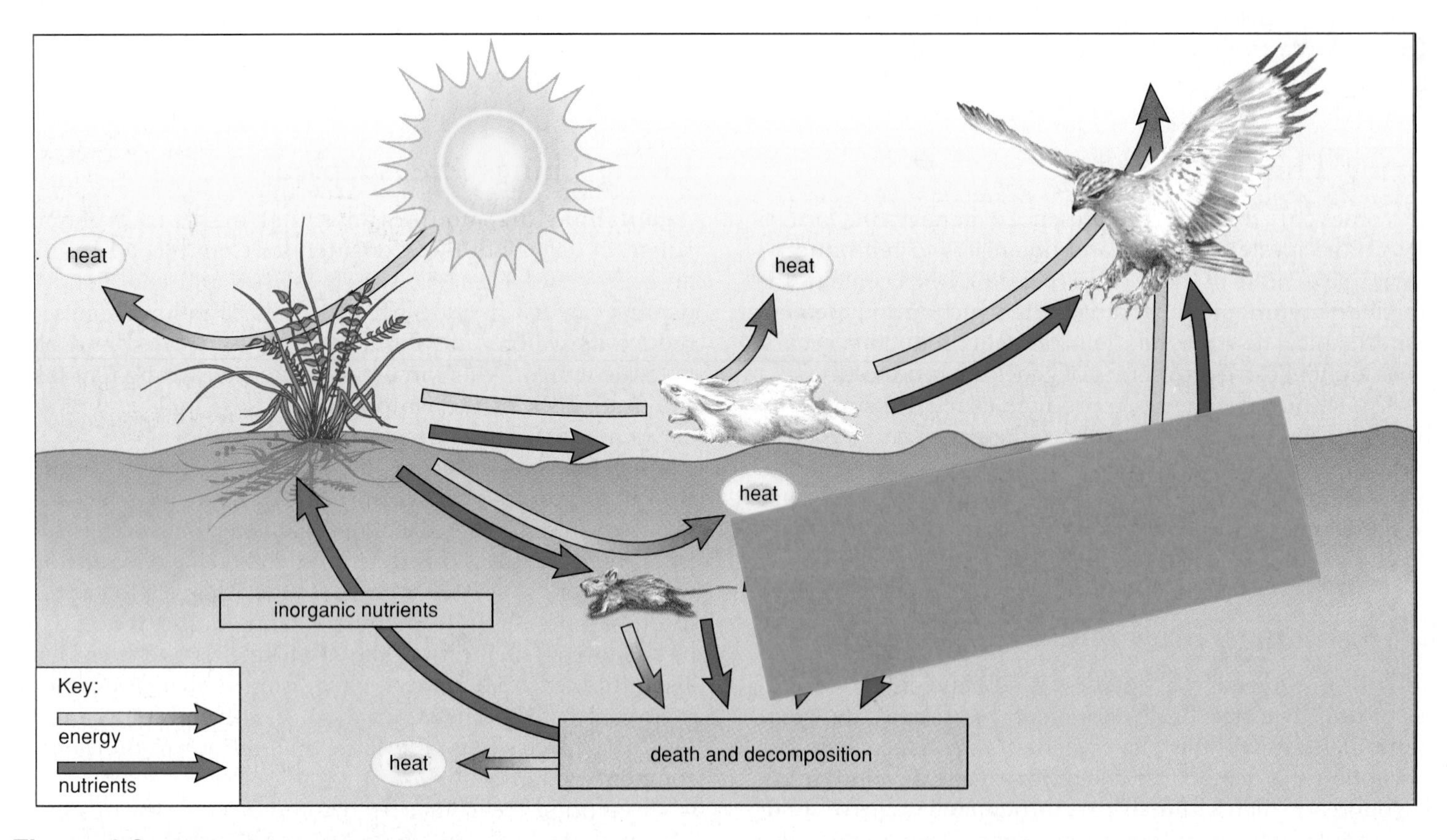

Figure 1.8 Ecosystem organization.
Within an ecosystem, nutrients cycle (see blue arrows); plants make and use their own organic food, and this becomes food for several levels of animal consumers, including humans. When these organisms die and decompose, the inorganic remains are used by plants as they produce organic food. Energy flows (see yellow arrows); solar energy used by plants to produce organic food is eventually converted to heat by all members of an ecosystem (therefore, a constant supply of solar energy is required for life to exist).

Tropical Rain Forest, a Terrestrial Ecosystem

Tropical rain forests are the most complex ecosystems in the world. They are found near the equator where there is plentiful sun and rainfall the entire year. Major rain forests are located in South America (Fig. 1.9), central and west Africa, and Southeast Asia. Rain forests have a multilayered canopy which consists of broad-leaved evergreen trees of different heights. Most animal populations live in the canopy where they interact with each other. Brightly colored birds, such as toucans and macaws, fly around eating fruits, buds, and pollen. Other birds, such as long-billed hummingbirds feed on nectar often taken from small plants that grow independently on the trees. Tree sloths and spider monkeys are mammals that live in the canopy and are preyed upon by jaguars. Other canopy animals include butterflies, tree frogs, and dart-poison frogs. Many canopy animals, such as bats, are active only at night. Snakes, spiders, and ants are animals that live on or near the ground and not in the canopy.

Figure 1.9 Tropical rain forest populations interact in the manner described in Fig. 1.8.

Ecology Focus

Tropical Rain Forests: Can We Live Without Them?

So far, nearly 2 million species of organisms have been discovered and named. Two-thirds of the plant species, 90% of the nonhuman primates, 40% of birds of prey, and 90% of the insects live in the tropics. Many more species of organisms (perhaps as many as 30 million) are estimated to live in the tropical rain forests but have not yet been discovered.

Tropical forests span the planet on both sides of the equator and cover 6–7% of the total land surface of the earth—an area roughly equivalent to our contiguous forty-eight states. Every year humans destroy an area of forest equivalent to the size of Oklahoma. At this rate, these forests and the species they contain will disappear completely in just a few more decades. Even if the forest areas now legally protected survive, 56–72% of all tropical forest species would still be lost.

The loss of tropical rain forests results from an interplay of social, economic, and political pressures. Many people already live in the forest, and as their numbers increase, more of the land is cleared for farming. People move to the forests because internationally financed projects build roads and open the forests up for exploitation. Small-scale farming accounts for about 60% of tropical deforestation, and this is followed by commercial logging, cattle ranching, and mining. International demand for timber promotes destructive logging of rain forests in Southeast Asia and South America. A market for low-grade beef encourages the conversion of tropical rain forests to pastures for cattle. The lure of gold draws miners to rain forests in Costa Rica and Brazil.

The destruction of tropical rain forests produces only short-term benefits but is expected to cause long-term problems. The forests soak up rainfall during the wet season and release it during the dry season. Without them, a regional yearly regime of flooding followed by drought is expected to destroy property and reduce agricultural harvests. Worldwide, there could be changes in climate that would affect the entire human race. On the other hand, the preservation of tropical rain forests offers benefits. For example, the rich diversity of plants and animals would continue to exist for scientific and pharmacological study. One-fourth of the medicines we currently use come from tropical rain forests. The rosy periwinkle from Madagascar has produced two potent drugs for use against Hodgkin disease, leukemia, and other blood cancers. It is hoped that many of the still-unknown plants will provide medicines for other human ills.

Studies show that if the forests were used as a sustainable source of nonwood products, such as nuts, fruits, and latex rubber, they would generate as much or more revenue while continuing to perform their various ecological functions. And biodiversity could still be preserved. Brazil is exploring the concept of "extractive reserves," in which plant and animal products are harvested but the forest itself is not cleared. Ecologists have also proposed "forest farming" systems, which mimic the natural forest as much as possible while providing abundant yields. But for such plans to work maximally, the human population size and the resource consumption per person must be stabilized.

Preserving tropical rain forests is a wise investment. Such action promotes the survival of most of the world's species—indeed, the human species, too.

The Human Population

The human population tends to modify existing ecosystems for its own purposes. As more and more ecosystems are converted to towns and cities, fewer of the natural cycles are able to function adequately to sustain an evergrowing human population. It is important to do all we can to preserve ecosystems, because only then can we be assured that we will continue to exist. The recognition that the workings of the ecosystems need to be preserved is one of the most important developments of our new ecological awareness.

Presently, there is great concern about preserving the world's tropical rain forests, as discussed in the reading on this page. The tropical rain forests perform many services for us. For example, they act like a giant sponge and absorb carbon dioxide, a pollutant that pours into the atmosphere from the burning of fossil fuels such as oil and coal. If the rain forests continue to be depleted as they are now, an increased amount of carbon dioxide in the atmosphere is expected to cause an increase in the average daily temperature. Problems with acid rain are also expected to increase, since carbon dioxide combines with water to form carbonic acid, a component of acid rain.

The present **biodiversity** (number and size of populations in a community) of our planet is being threatened. It has been estimated that the number of species in the biosphere may be as high as 80 million species, but thus far fewer than 2 million have been identified and named. Even so we may be presently losing from 24 to even 100 species a day due to human activities. The existence of the species featured in Figure 1.9 is threatened because tropical rain forests are being reduced in size. Most biologists are alarmed over the present rate of extinction and believe the rate may eventually rival the mass extinctions that have occurred during our planet's history.

The human population tends to modify existing ecosystems and to reduce biodiversity. Because all living things are dependent upon the normal functioning of the biosphere, ecosystems should be preserved.

Kingdoms of Life	Representative Organisms	Organization	Type of Nutrition	Representative Organisms
Monera	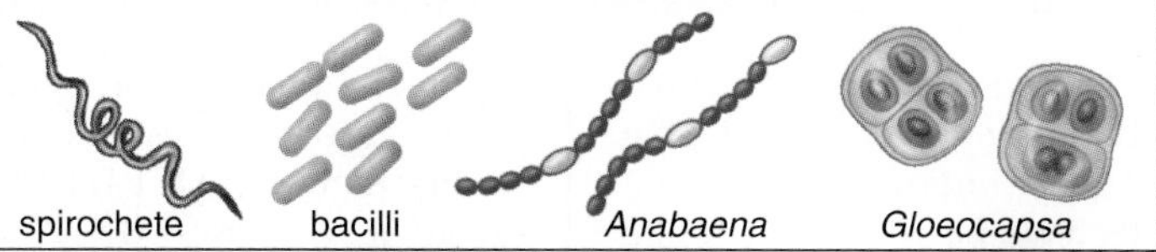 spirochete, bacilli, *Anabaena*, *Gloeocapsa*	Microscopic single cell (sometimes chains or mats)	Absorb food (some photo-synthesize)	Bacteria including cyanobacteria
Protista	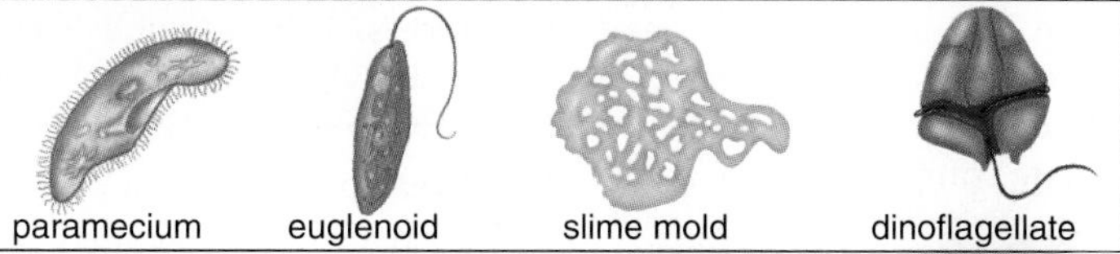 paramecium, euglenoid, slime mold, dinoflagellate	Complex single cell, some multicellular	Absorb, photosynthesize, or ingest food	Protozoans, algae, water molds, and slime mold
Fungi	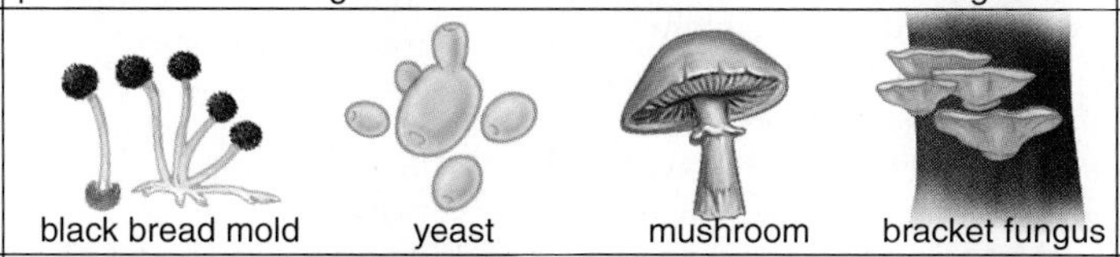 black bread mold, yeast, mushroom, bracket fungus	Some unicellular, most multicellular filamentous forms with specialized complex cells	Absorb food	Molds, yeast, and mushrooms
Plantae	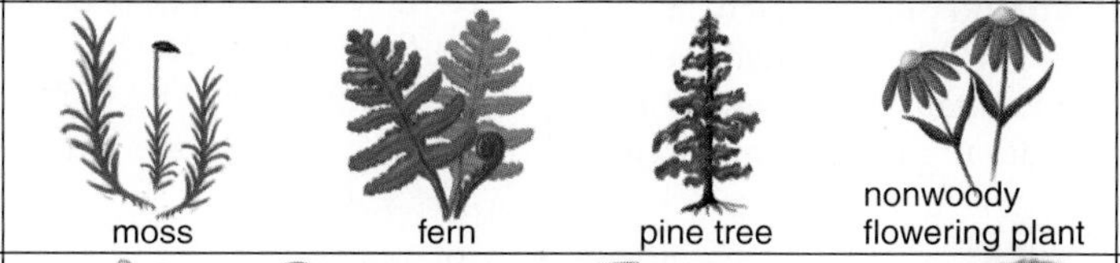 moss, fern, pine tree, nonwoody flowering plant	Multicellular form with specialized complex cells	Photosynthesize food	Mosses, ferns, nonwoody and woody flowering plants
Animalia	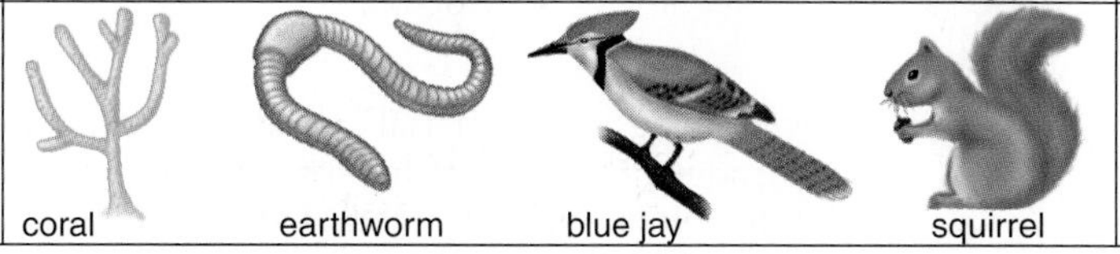 coral, earthworm, blue jay, squirrel	Multicellular form with specialized complex cells	Ingest food	Invertebrates, fishes, reptiles, amphibians, birds, and mammals

Figure 1.10 Classification of organisms.
In this text, organisms are classified into the five kingdoms illustrated in this table.

1.3 The Classification of Living Things

Taxonomy is that part of biology dedicated to naming, describing, and classifying organisms. Taxonomists use specific criteria to classify species, including human beings, into certain categories.

As we move from genus to kingdom, more and more different types of species are included in each successive category. Only human beings are in the genus *Homo,* but many different types of animals are in the animal kingdom. Notice that in the example given, species within the same genus share very similar characteristics, but those that are in the same kingdom have only general characteristics in common. In the same way, all species in the genus *Zea* look pretty much the same—that is, like corn plants—while species in the plant kingdom can be quite different, as is evident when we compare grasses to trees.

Categories	For Humans	Description
Kingdom	Animalia	Multicellular, moves, ingests food
Phylum	Chordata	Dorsal supporting rod and nerve cord
Class	Mammalia	Hair, mammary glands
Order	Primates	Adapted to climb trees
Family	Hominidae	Adapted to walk erect
Genus	*Homo*	Large brain, tool use
Species	*H. sapiens*	

Taxonomists give each species a scientific name in Latin. The scientific name is a binomial (*bi* means two; *nomen* means name). For example, the name for humans is *Homo sapiens,* and for corn, it is *Zea mays.* The first word is the genus, and the second word is a specific epithet for that species. (Note that both words are in italic but only the genus is capitalized.) Scientific names are universally used by biologists so as to avoid confusion. Common names tend to overlap and often are in the language of a particular country.

Taxonomy makes sense out of the bewildering variety of life on earth. Species are classified according to their presumed evolutionary relationship; those placed in the same genus are the most closely related, and those placed in separate kingdoms are the most distantly related. As more is known about evolutionary relationships between species, taxonomy changes. Presently many biologists recognize the five kingdoms listed in Figure 1.10. Others disagree not only about the number of kingdoms but also about which species should be placed in the various categories of classification.

Bioethical Issue

The Endangered Species Act requires the federal government to identify endangered and threatened species and to protect their habitats, even to the extent of purchasing their habitats. Developers feel that the act protects wildlife at the expense of jobs for US citizens. In an effort to allow development in sensitive areas, it is now possible to move forward after a Habitat Conservation Plan (HCP) is approved. An HCP permits, say, new-home construction or logging on a part of the land if wildlife habitat is conserved on another part. Conservation can also mean helping the government buy habitat some place else.

The nation's first HCP was approved in 1980. It permitted housing construction on San Bruno Mountain near San Francisco, if 97% of the habitat for the endangered mission blue butterfly was preserved. That sounds pretty good, but by this time hundreds of HCPs have been approved, and the conservation requirement may have slipped a bit. Tim Cullinan, a director of the National Audubon Society, recently found that logging companies in the Pacific Northwest are proposing the exchange of habitat on public land for the right to log privately owned old forests. In other words, nothing has been given up. By now, there are so many HCPs in the works they are being rubber-stamped by government officials with no public review at all.

Do you favor development over preservation of habitat or vice versa? Do you think the federal government should be in the business of trying to preserve endangered species? Do you think that public review of HCPs should be allowed, even if it slows down the approval process? Is it the public's responsibility to remain vigilant or is governmental review of HCPs sufficient?

Questions

1. What are the concerns of developers versus environmentalists with regard to natural areas?
2. Is it short-sighted to stress the importance of jobs over the rights of wildlife? Why or why not?
3. In what ethical ways can each side make their concerns known to the general public?

Summarizing the Concepts

1.1 The Scientific Process

When studying the world of living things, biologists and other scientists use the scientific process. Observations along with previous data are used to formulate a hypothesis. New observations and/or experiments are carried out in order to test the hypothesis. Scientists often do controlled experiments. The control sample does not go through the step being tested, and this acts as a safeguard against a wrong conclusion.

The new data may support a hypothesis or they may prove it false. Hypotheses cannot be proven true. Several conclusions in a particular area may allow scientists to arrive at a theory—generalizations such as the cell theory, gene theory, or the theory of evolution. Evolution is the unifying theory of biology.

Science is objective and uses conclusions based on data to arrive at theories about the natural world. Any explanation based on supernatural beliefs cannot be considered science because such beliefs are not tested in the usual scientific way.

Science does not answer ethical questions; we must do this for ourselves. Knowledge provided by science, such as the contents of this text, can assist us in making decisions that will be beneficial to human beings and to other living things.

1.2 The Characteristics of Life

Evolution accounts for both the diversity and the unity of life we see about us—all organisms share the same characteristics of life:

1. Living things are organized. The levels of biological organization extend from the cell to ecosystems: atoms and molecules → cells → tissues → organs → organ systems → organisms → populations → communities → ecosystems. In an ecosystem, populations interact with one another and the physical environment.
2. Living things take materials and energy from the environment; they need an outside source of nutrients.
3. Living things are homeostatic; internally they stay just about the same despite changes in the external environment.
4. Living things respond to stimuli; they react to internal and external events.
5. Living things reproduce; they produce offspring that resemble themselves.
6. Living things grow and develop; during their lives they change—most multicellular organisms undergo various stages from fertilization to death.
7. Living things are adapted; they have modifications that make them suited to a particular way of life.
8. All living things belong to a population, which can be defined as all the members of the same species that occur in a particular locale.

1.3 The Classification of Living Things

Living things are classified according to their evolutionary relationships into these ever more specific categories: kingdom, phylum, class, order, family, genus, species. Organisms in different kingdoms are only distantly related; organisms in the same genus are very closely related.

Studying the Concepts

1. What are the steps of the scientific process? Why can't this process prove a hypothesis true? 2
2. What is a controlled experiment? Why must a scientist test one variable at a time? have a control group? 3
3. What is the ultimate goal of science? Give an example that supports your answer. 5
4. Why isn't creationism considered a part of science by biologists? 6

5. Why doesn't science answer ethical and moral questions? 6
6. Name eight characteristics of life, and discuss each one. 7–10
7. Food provides which two necessities for living things? 8
8. Give an example of homeostasis. Tell how the digestive system contributes to homeostasis in humans. 8
9. Why is the phrase "lifelong developmental change" appropriate when speaking of humans? 9
10. How is an osprey adapted to its way of life? 9
11. How does evolution explain both the diversity and the unity of life? 9
12. What are the levels of biological organization beyond the organism? Define each level. 10
13. Name the categories of classification, from genus to kingdom. Which category contains more types of organisms having general characteristics in common? 13
14. Explain the scientific name of an organism. 13

Testing Yourself

Choose the best answer for each question.

1. Science always studies an event that
   a. has previously been published.
   b. lends itself to experimentation.
   c. is observable.
   d. fits in with an already existing theory.
   e. Both b and c are correct.
2. After formulating a hypothesis, a scientist
   a. proves the hypothesis true or false.
   b. tests the hypothesis.
   c. decides how to have a control.
   d. makes sure of his variables.
   e. formulates a theory.
3. A scientist cannot
   a. make value judgments like everyone else.
   b. prove a hypothesis true.
   c. contribute to a long-standing theory.
   d. make use of pre-existing mathematical data.
   e. be as objective as possible.

For questions 4–7, match the statements in the key with the sentences below.

Key:
a. Living things are organized.
b. Living things are homeostatic.
c. Living things respond to stimuli.
d. Living things reproduce.
e. Living things are adapted.

4. Genes made up of DNA are passed from parent to child.
5. Cells are made of molecules, tissues are made of cells, and organisms are made of tissues.
6. A herd of zebra will scatter when a lion approaches.
7. The long, sharp claws of an osprey can hold on to a fish.
8. An example of chemical cycling occurs when
   a. plants absorb solar energy and make their own food.
   b. energy flows through an ecosystem and becomes heat.
   c. osprey nest on the top of telephone poles.
   d. death and decay makes inorganic nutrients available to plants.
   e. we eat food and use the nutrients to grow/repair tissues.
9. Which of these is mismatched?
   a. Kingdom Monera—mosses, ferns, pine trees
   b. Kingdom Protista—protozoans, algae, water molds
   c. Kingdom Fungi—molds and mushrooms
   d. Kingdom Plantae—woody and nonwoody flowering plants
   e. Kingdom Animalia—fish, reptiles, birds, humans.
10. An investigator spills dye on a culture plate and then notices that the bacteria live despite exposure to sunlight. He hypothesizes that the dye protects bacteria against death by ultraviolet (UV) light. To test this hypothesis, he decides to expose two hundred culture plates to UV light. One hundred plates contain bacteria and dye; the other hundred plates contain only bacteria. Result: after exposure to UV light, the bacteria on both plates die. Fill in the right-hand portion of this diagram.

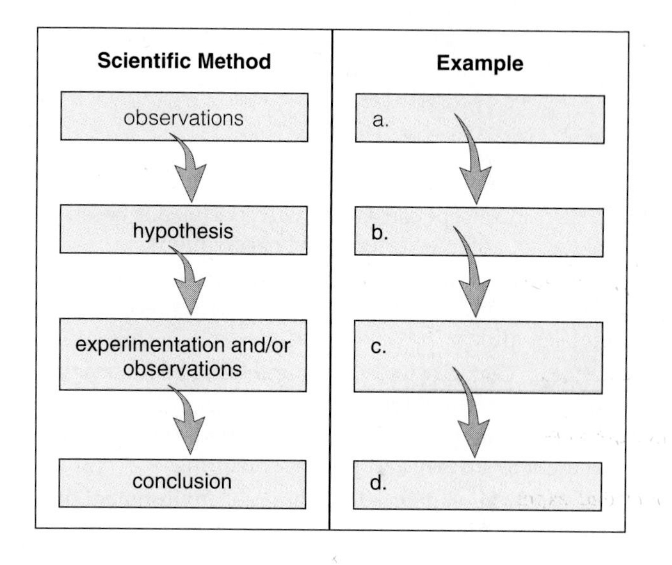

Thinking Scientifically

1. Considering controlled experiments (page 4):
   a. A variable is an element that changes. Why is sweetener S called the experimental variable in the experiment?
   b. With reference to Figure 1A, explain this definition: A control group goes through all the steps of an experiment except the one being tested.
   c. Why is bladder cancer the dependent variable in the described experiment?
   d. Does the experiment have elements that are constant and not variable? What are they?
   e. What is the value of including a control group in an experiment?
2. Considering the scientific process (pages 2–6):
   a. Scientific hypotheses must be falsifiable. Why is the hypothesis "Every human being has a guardian angel" not falsifiable?
   b. Why is the hypothesis "Biotin is required for good health" falsifiable?
   c. In what way are religious beliefs different from scientific beliefs?

Understanding the Terms

adaptation 9
behavior 8
biodiversity 12
biology 2
biosphere 10
cell 7
community 7
conclusion 5
control group 3
data 5
development 9
ecosystem 7
energy 8
evolution 9
experiment 3
falsify 5
gene 9
homeostasis 8
hypothesis 2
law 5
metabolism 9
organ 7
organism 7
organism system 7
phenomenon 2
population 7
principle 5
reproduce 9
science 2
scientific method 6
scientific theory 5
species 9
taxonomy 13
technology 6
tissue 7
variable 3

Match the terms to these definitions:

a. Theory Concept consistent with conclusions based on a large number of experiments and observations.
b. hypothesis Statement that is capable of explaining present observations and will be tested by further experimentation and observations.
c. energy Capacity to do work and bring about change; occurs in a variety of forms.
d. adaptation Suitability of an organism for its environment enabling it to survive and produce offspring.
e. homeostasis Maintenance of the internal environment of an organism within narrow limits.

Using Technology

Your study of biology is supported by these available technologies:

Essential Study Partner CD-ROM

Evolution & Diversity → Classification

Visit the Mader web site for related ESP activities.

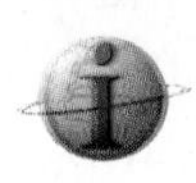

Exploring the Internet

The Mader Home Page provides resources and tools as you study this chapter.

http://www.mhhe.com/biosci/genbio/mader

Further Readings for Chapter 1

Balick, M. J., and Cox, P. A. 1996. *Plants, people, and culture: The science of ethnobotany*. New York: Scientific American Library. This well-illustrated book discusses the medicinal and cultural uses of plants, and the importance of rain forest conservation.

Barnard, C., et al. 1993. *Asking questions in biology.* Essex: Longman Scientific & Technical. First-year life science students are introduced to the skills of scientific observation.

Carey, S. S. 1997. *A beginner's guide to scientific method.* 2d ed. Belmont, Calif.: Wadsworth Publishing. The basics of the scientific method are explained.

Cox, G. W. 1997. *Conservation biology.* Dubuque, Iowa: Wm. C. Brown Publishers. This text examines the field of conservation, surveys basic principles of ecology and considers how biodiversity can be preserved.

Dobson, A. P. 1996. *Conservation and biodiversity.* New York: *Scientific American* Library. Discusses the value of biodiversity, and describes attempts to manage endangered species.

Drewes, F. 1997. *How to study science.* 2d ed. Dubuque, Iowa: Wm. C. Brown Publishers. Supplements any introductory science text; shows students how to study and take notes and how to interpret text figures.

Frenay, A. C. F., and Mahoney, R. M. 1997. *Understanding medical terminology*. 10th ed. Dubuque, Iowa: Wm. C. Brown Publishers. A structural approach to the study of medical terminology.

Johnson, G. B. 1996. *How scientists think.* Dubuque, Iowa: Wm. C. Brown Publishers. Presents the rationale behind 21 important experiments in genetics and molecular biology that became the foundation for today's research.

Kellert, S. R. 1996. *The value of life: Biological diversity and human society.* Washington D.C.: Island Press/Shearwater Books. The importance of biological diversity to the well-being of humanity is explored.

Marchuk, W. N. 1992. *A life science lexicon.* Dubuque, Iowa: Wm. C. Brown Publishers. Helps students master life sciences terminology.

Margulis, L., et al. 1998. *Five kingdoms: An illustrated guide to the phyla of life on earth.* New York: W. H. Freeman & Co. Introduces the kingdoms of organisms.

Minkoff, E. C., and Baker, P. J. 1996. *Biology today: An issues approach.* New York: The McGraw-Hill Companies, Inc. This introductory text emphasizes understanding of selected biological issues, and discusses each issue's social context.

Nemecek, S. August 1997. Frankly, my dear, I don't want a dam. *Scientific American* 277(2):20. The article discusses how dams affect biodiversity.

Primak, R. B. 1995. *A primer of conservation biology.* Sunderland, Mass.: Sinauer Associates. Addresses the loss of biological diversity throughout the world, and suggests remedies.

Schmidt, M. J. January 1996. Working elephants. *Scientific American* 274(1):82. In Asia, teams of elephants serve as an alternative to destructive logging equipment.

Serafini, A. 1993. *The epic history of biology.* New York: Plenum Press. This is a history of biology beginning with ancient Egyptian medicine.

PART I

Cell Biology

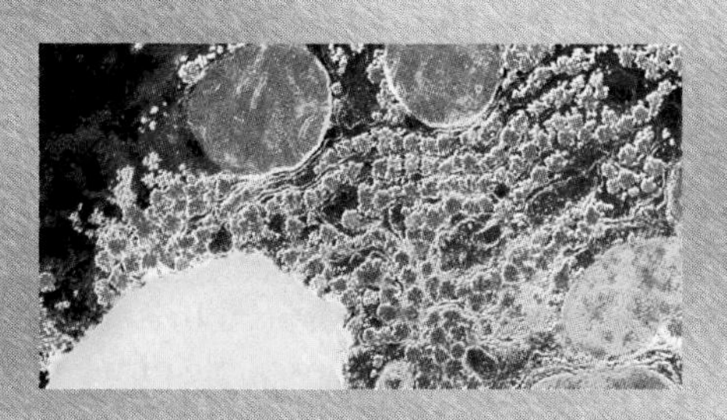

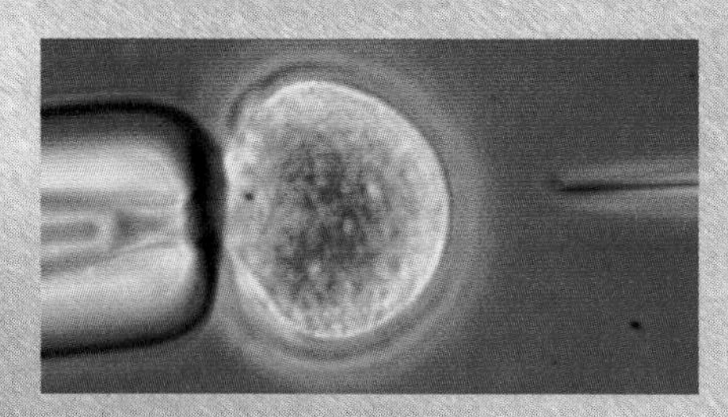

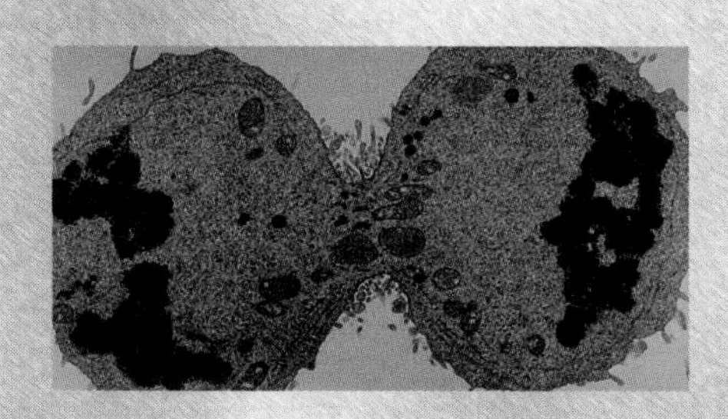

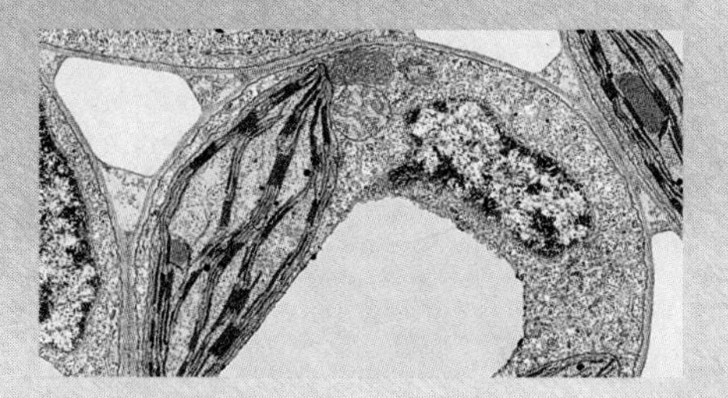

A hundred or so years ago, scientists believed that only nonliving things, like rocks and metals, consisted entirely of chemicals. They thought that living things, like roses and lions, had a vital force necessary for life. Scientific study, however, has repeatedly shown that both nonliving and living things have the same physical and chemical basis. Although cells, the fundamental unit of living things, contain molecules and carry out processes not found in inanimate objects, they must still be understood by studying basic chemical properties. The chapters in this part introduce you to the chemistry of molecules and the workings of the cell.

Careers in Cell Biology

Biochemist conducting an experiment

Biochemists study the chemical composition of living things at the cellular level. They decipher the chemical reactions and pathways involved in metabolism, reproduction, growth, and heredity. Much of the work in biotechnology is done by biochemists because this technology involves understanding the chemistry of life.

Physiologists study life functions of plants and animals, both in the whole organism and at the cellular and molecular level, under normal and abnormal conditions. Physiologists may specialize in functions such as growth, reproduction, photosynthesis, respiration, or movement, or in the physiology of a certain area or system of the body.

Microbiologist preparing a petri dish

Microbiologists study the general structure, metabolism, and genetics of microscopic organisms. They are concerned with host-microorganism relationships relevant to microbial diseases in humans and other animals. Microbiologists work with physicians to determine the mode of transmission, diagnosis, treatment, and prevention of human illnesses caused by microbes. They also study the beneficial roles of microorganisms in ecosystems and in the production of food and various industrial processes.

Laboratory technicians use the principles and theories of biochemistry to solve problems in research and development and to investigate, invent, and help improve products. Their jobs are more practically oriented than those of biochemists. Technicians set up, operate, and maintain laboratory instruments, monitor experiments, calculate and record results, and help develop conclusions. Some who work in production help produce and test products.

Scanning electron microscopist studying cell structure

Radiological technologists produce X-ray films of body parts for use in diagnosing medical problems. They know how to operate radiographic equipment and prevent unnecessary radiation exposure to themselves and the patient. Some radiographers operate computed tomography scanners to produce cross-sectional views of patients and may be called CT technologists. Others operate magnetic resonance imaging machines (giant magnets) and may be called MRI technologists.

C H A P T E R 2

The Molecules of Cells

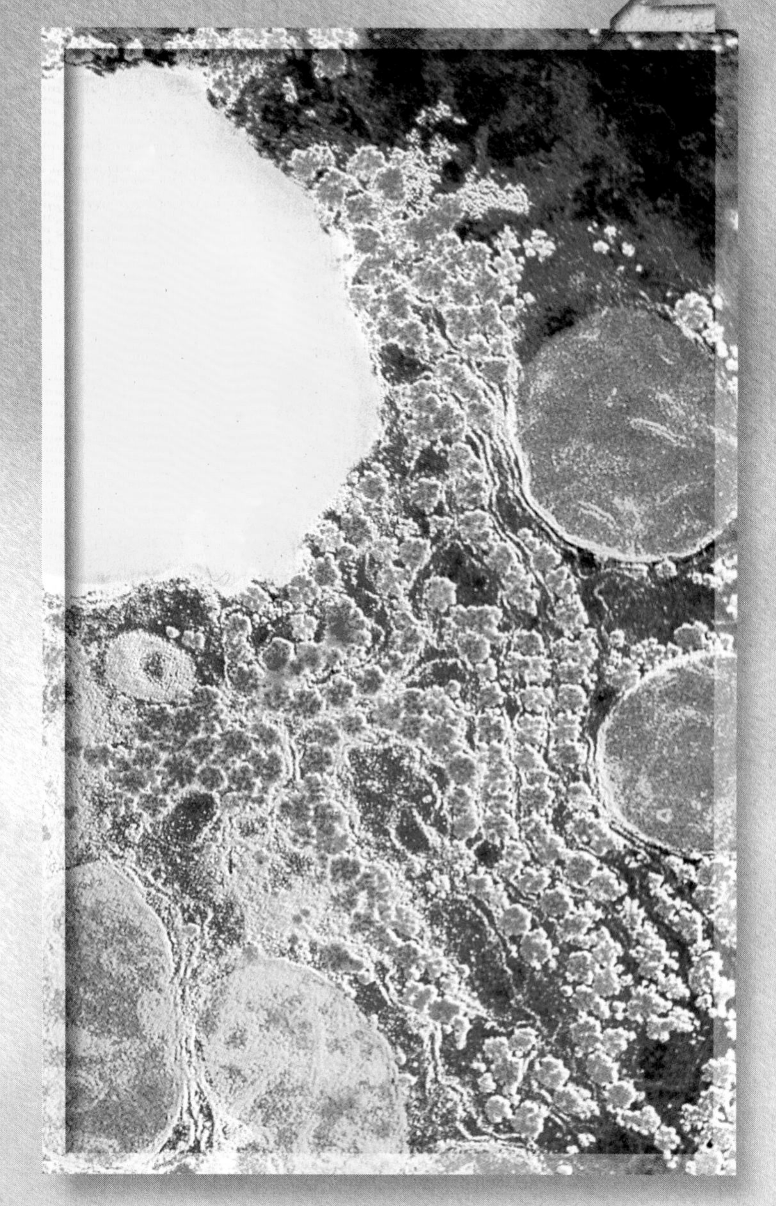

This electron micrograph of a liver cell has been colored to reveal its contents, which include many types of organic molecules. Energy is stored in glycogen granules (red) and in a fat vacuole (yellow).

Chapter Concepts

An oxygen molecule randomly bouncing in the air is suddenly sucked into the welcoming nostrils of a busily studying student. Down her throat and windpipe it goes, sweeping through increasingly narrow tubes until it comes up against a sticky, warm surface—the interior lining of her lungs. The oxygen molecule moves quietly into the blood and then no longer finds itself airborne but within a red blood cell filled with many other atoms and molecules. The current created by her pumping heart carries the blood from the lungs to the heart and then swiftly away through smaller and smaller blood vessels until the oxygen is released and enters the brain, where it is used to help generate energy for thinking.

Not just thinking, but all life processes rely on the atoms and molecules discussed in this chapter, which describes the basics of life's chemistry.

2.1 Elements and Atoms

Matter is anything that takes up space and has weight. All matter, living and nonliving, is composed of **elements.** Considering the variety of living and nonliving things in the word, it is quite remarkable that there are only 92 naturally occurring elements (see Appendix D). As indicated in Figure 2.1, only six elements—carbon, hydrogen, nitrogen, oxygen, phosphorus, and sulfur—make up most (about 98%) of the body weight of organisms. The acronym CHNOPS helps us remember these six elements. Instead of writing out the names of elements, scientists use symbols to identify them. The letter C stands for carbon and the letter N stands for nitrogen, for example. Some of the symbols used for elements are derived from Latin. For example, the symbol for sodium is Na (*natrium* in Latin means sodium).

Elements contain tiny particles called atoms. The same name is given to the element and its atoms. An **atom** is the smallest unit of matter to enter into chemical reactions. Even though an atom is extremely small, it contains even smaller subatomic particles called protons, neutrons, and electrons. Figure 2.2 shows a model of an atom. It has a central nucleus, where subatomic particles **protons** and **neutrons** are located, and shells, which are pathways about the nucleus where **electrons** orbit. The shells represent energy levels. The inner shell has the lowest energy level and can hold two electrons. The outer shell has a higher energy level and can hold eight electrons. An atom is most stable when the outer shell has eight electrons.

An atom has an atomic number; the **atomic number** is equal to the number of its protons. Notice in Table 2.1 that protons have a positive (+) electrical charge and electrons have a negative (−) charge. When an atom is electrically neutral, the number of protons equals the number of electrons. The carbon atom shown in Figure 2.3 has an atomic number of 6; therefore, it has six protons. Since it is electrically neutral, it also has six electrons. The inner shell has two electrons and the outer shell has four electrons.

In the periodic table of the elements (Fig. 2.1), atoms are horizontally arranged in order of increasing atomic number.

I	II	III	IV	V	VI	VII	VIII
1 **H** Hydrogen 1						**atomic number** **atomic symbol** **atomic weight**	2 **He** Helium 4
3 **Li** Lithium 7	4 **Be** Beryllium 9	5 **B** Boron 11	6 **C** Carbon 12	7 **N** Nitrogen 14	8 **O** Oxygen 16	9 **F** Fluorine 19	10 **Ne** Neon 20
11 **Na** Sodium 23	12 **Mg** Magnesium 24	13 **Al** Aluminum 27	14 **Si** Silicon 28	15 **P** Phosphorus 31	16 **S** Sulfur 32	17 **Cl** Chlorine 35	18 **Ar** Argon 40
19 **K** Potassium 39	20 **Ca** Calcium 40						

Figure 2.1 Periodic Table of the Elements (shortened).
Each element has an atomic number, an atomic symbol, and an atomic weight. The elements indicated by color make up most of the body weight of organisms. See Appendix D for a complete table.

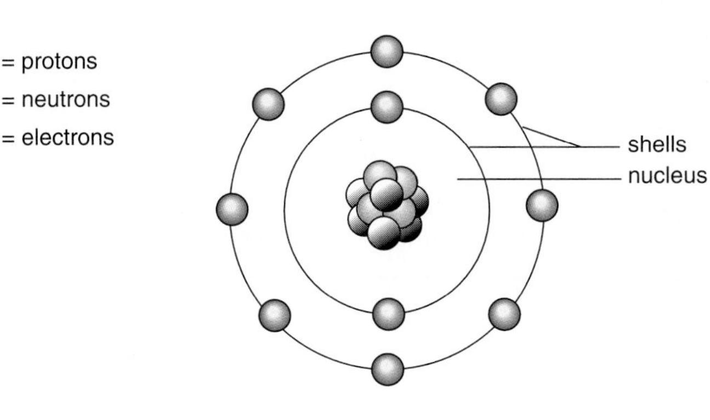

Figure 2.2 Model of an atom.
Protons and neutrons occur in the nucleus. Electrons occur at energy levels called shells.

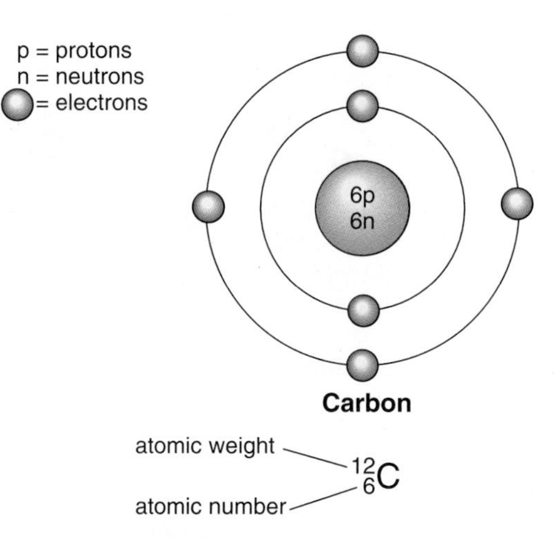

Figure 2.3 Carbon atom.
Carbon has an atomic number of 6; therefore, it has six protons and six electrons. Carbon has a weight of 12 atomic mass units. Therefore, it has six neutrons.

Table 2.1 Subatomic Particles

Name	Charge	Weight
Electron	One negative unit	Almost no mass
Proton	One positive unit	One atomic mass unit
Neutron	No charge	One atomic mass unit

They are vertically arranged according to the number of electrons in the outer shell. The numeral at the top of each column indicates how many electrons there are in the outer shell of the atoms in that column. An exception to this format is helium (He), which has only two electrons in the outer shell because it has only one shell. The number of electrons in the outer shell determines the chemical properties of an atom, including how readily it enters into chemical reactions.

Isotopes

The subatomic particles are so light that their weight is indicated by special units called atomic mass units. The **atomic weight** of each atom is noted in the periodic table beneath the atomic symbol. The atomic weight of an atom equals the number of protons plus the number of neutrons. Why should that be the case? Table 2.1 shows that electrons have almost no weight, but protons and neutrons each have a weight of one atomic mass unit. Since the atomic weight of carbon is twelve and it has six protons, it is easy to calculate that carbon has six neutrons.

The atomic weights given in the periodic table are the average weight for each kind of atom. This is because the atoms of one element may differ in the number of neutrons; therefore, the weight varies. Atoms that have the same atomic number and differ only in the number of neutrons are called **isotopes.** Isotopes of carbon can be written in the following manner, where the subscript stands for the atomic number and the superscript stands for the atomic weight:

$^{12}_{6}C$ $\quad$ $^{13}_{6}C$ $\quad$ $^{14}_{6}C$*

*radioactive

Carbon 12 has six neutrons, carbon 13 has seven neutrons, and carbon 14 has eight neutrons.

Isotopes have many uses. Because proportions of isotopes in various food sources are known, biologists can now determine the proportion of isotopes in mummified or fossilized human tissue to know what ancient peoples ate. Most isotopes are stable, but radioactive isotopes break down and emit radiation in the form of radioactive particles or radiant energy. It is the custom to use radioactive isotopes as tracers in biochemical experiments. And because carbon 14 breaks down at a known rate, the amount of this atom remaining is often used to determine the age of fossils.

All matter is composed of elements, each containing particles called atoms. Atoms have an atomic symbol, atomic number (number of protons), and atomic weight (number of protons and neutrons). The isotopes of some atoms are radioactive.

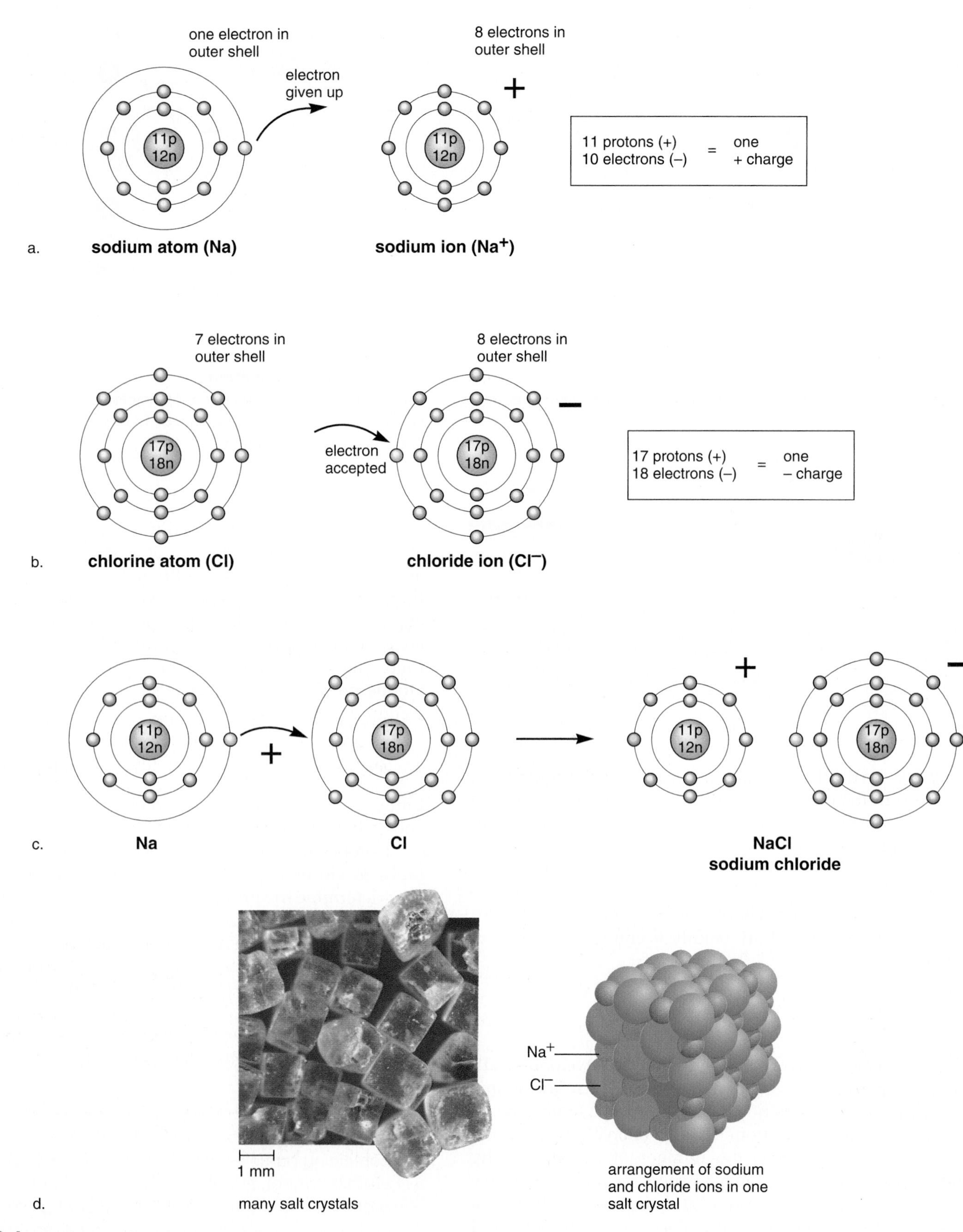

Figure 2.4 Ionic reaction.
a. When a sodium atom gives up an electron, it becomes a positive ion. **b.** When a chlorine atom gains an electron, it becomes a negative ion. **c.** When sodium reacts with chlorine, the compound sodium chloride (NaCl) results. In sodium chloride, an ionic bond exists between the ions. **d.** In a sodium chloride crystal, the ionic bonding between Na^+ and Cl^- causes the ions to form a three-dimensional lattice in which each sodium ion is surrounded by six chloride ions, and each chloride ion is surrounded by six sodium ions.

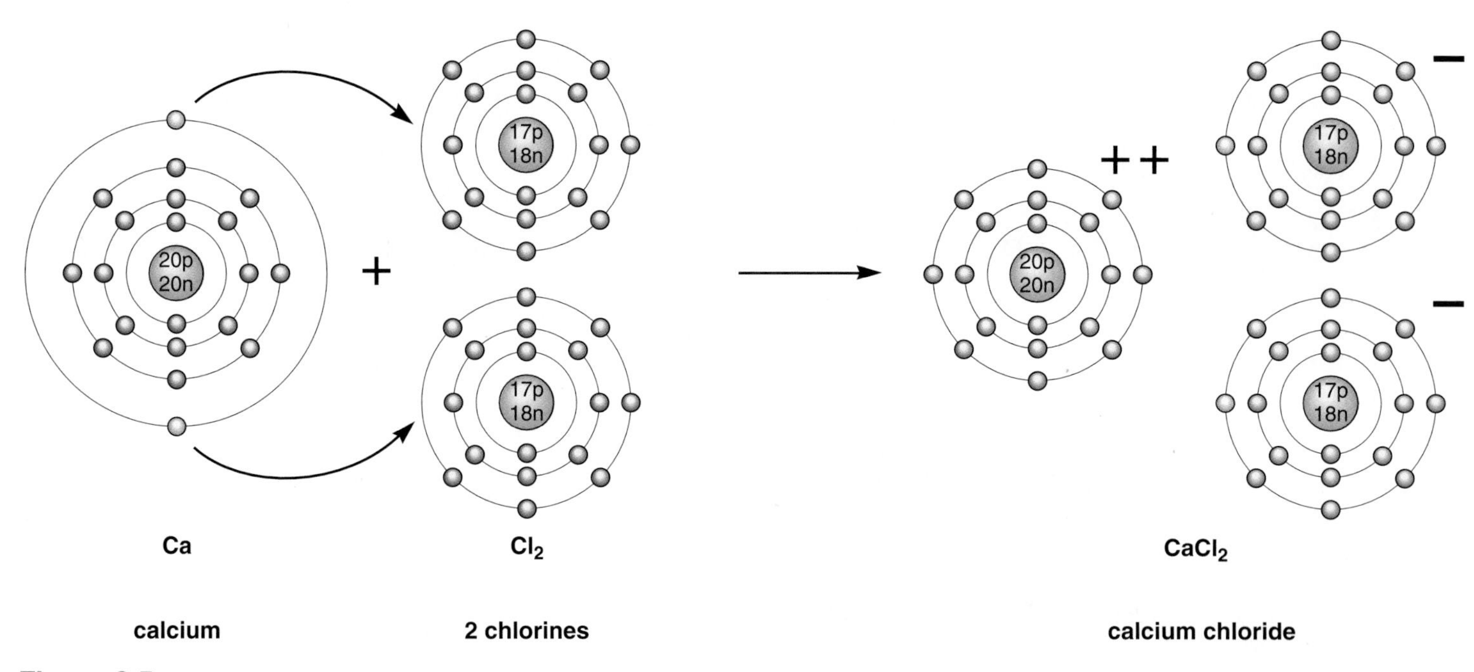

Figure 2.5 Ionic reaction.
The calcium atom gives up two electrons, one to each of two chlorine atoms. In the compound calcium chloride ($CaCl_2$), the calcium ion is attracted to two chloride ions.

2.2 Molecules and Compounds

Atoms often bond with each other to form a chemical unit called a **molecule.** A molecule can contain atoms of the same kind, as when an oxygen atom joins with another oxygen atom to form oxygen gas. Or the atoms can be different, as when an oxygen atom joins with two hydrogen atoms to form water. When the atoms are different, a compound results.

Two types of bonds join atoms: the ionic bond and the covalent bond.

Ionic Reactions

Recall that atoms (with more than one shell) are most stable when the outer shell contains eight electrons. During an ionic reaction, atoms give up or take on an electron(s) in order to achieve a stable outer shell.

Figure 2.4 depicts a reaction between a sodium (Na) and chlorine (Cl) atom in which chlorine takes an electron from sodium. **Ions** are particles that carry either a positive (+) or negative (−) charge. The sodium ion carries a positive charge because it now has one more proton than electrons, and the chloride ion carries a negative charge because it now has one fewer proton than electrons. The attraction between oppositely charged sodium ions and chloride ions forms an **ionic bond.** The resulting compound, sodium chloride, is table salt, which we use to enliven the taste of foods.

Figure 2.5 shows an ionic reaction between a calcium atom and two chlorine atoms. Notice that calcium with two electrons in the outer shell reacts with two chlorine atoms. Why? Because with seven electrons already, each chlorine requires only one more electron to have a stable outer shell. The resulting salt is called calcium chloride.

Significant ions in the human body are listed in Table 2.2. The balance of these ions in the body is important to our health. Too much sodium in the blood can cause high blood pressure; not enough calcium leads to rickets (a bowing of the legs) in children; too much or too little potassium results in heartbeat irregularities. Bicarbonate, hydrogen, and hydroxide ions are all involved in maintaining the acid–base balance of the body. If the blood is too acidic or too basic, the body's cells cannot function properly.

An ionic bond is the attraction between oppositely charged ions.

Table 2.2 Significant Ions in the Body

Name	Symbol	Special Significance
Sodium	Na^+	Found in body fluids; important in muscle contraction and nerve conduction.
Chloride	Cl^-	Found in body fluids.
Potassium	K^+	Found primarily inside cells; important in muscle contraction and nerve conduction.
Phosphate	PO_4^{3-}	Found in bones, teeth, and the high-energy molecule ATP.
Calcium	Ca^{2+}	Found in bones and teeth; important in muscle contraction.
Bicarbonate	HCO_3^-	Important in acid–base balance.
Hydrogen	H^+	Important in acid–base balance.
Hydroxide	OH^-	Important in acid–base balance.

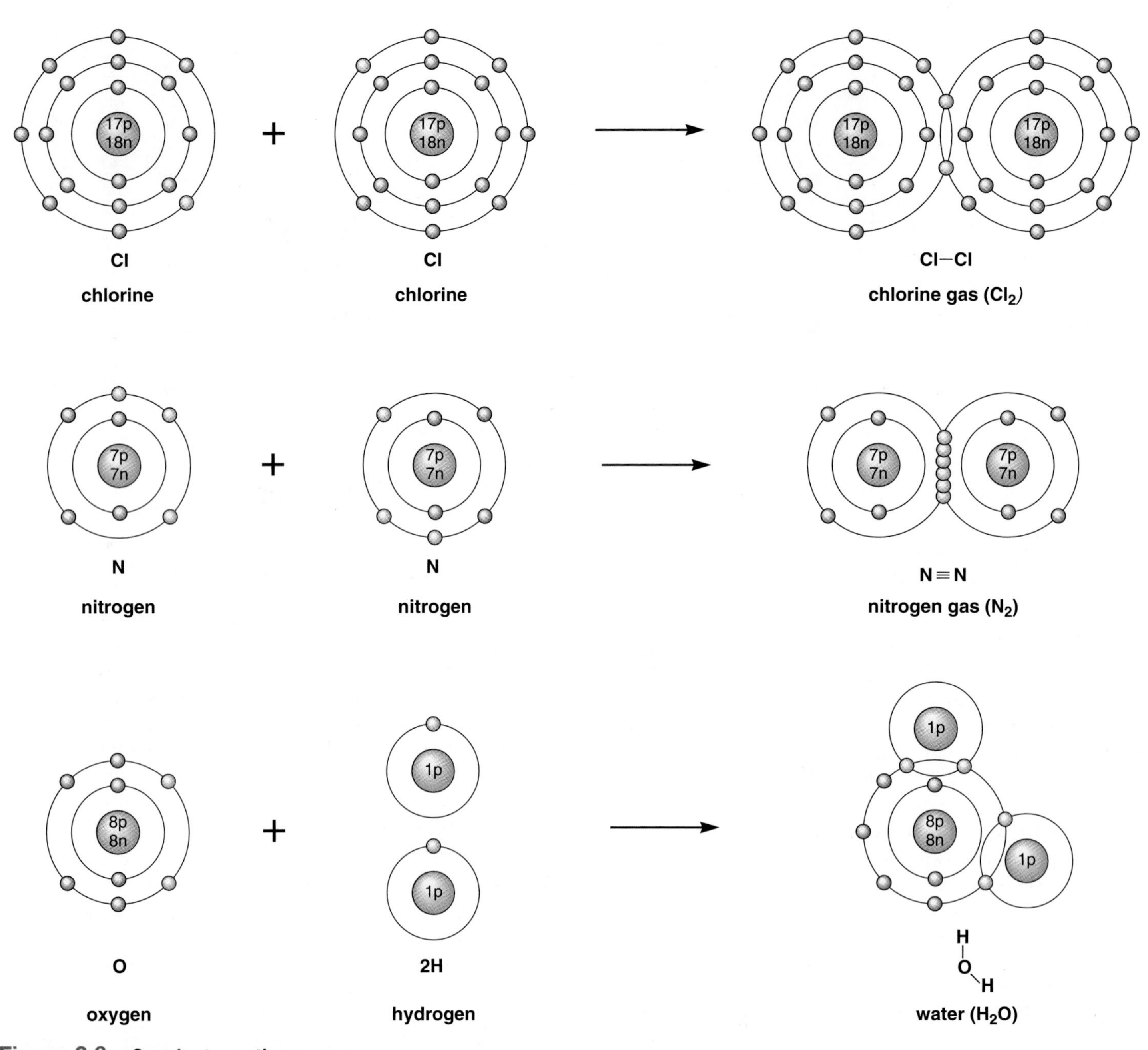

Figure 2.6 Covalent reactions.
After a covalent reaction, each atom will have filled its outer shell by sharing electrons. To show this, it is necessary to count the shared electrons as belonging to both bonded atoms. Oxygen and nitrogen are most stable with eight electrons in the outer shell; hydrogen is most stable with two electrons in the outer shell.

Covalent Reactions

After covalent reactions, the atoms share electrons in **covalent bonds** instead of losing or gaining them. Covalent bonds can be represented in a number of ways. The overlapping outermost shells in Figure 2.6 indicate that the atoms are sharing electrons. Just as two hands participate in a handshake, each atom contributes one electron to the pair that is shared. These electrons spend part of their time in the outer shell of each atom; therefore, they are counted as belonging to both bonded atoms.

Instead of drawing complex diagrams, electron-dot structures are sometimes used to depict covalent bonding between atoms. For example, in reference to Figure 2.7, each chlorine atom can be represented by its symbol, and the electrons in the outer shell can be designated by dots. The shared electrons are placed between the two sharing atoms, as shown here:

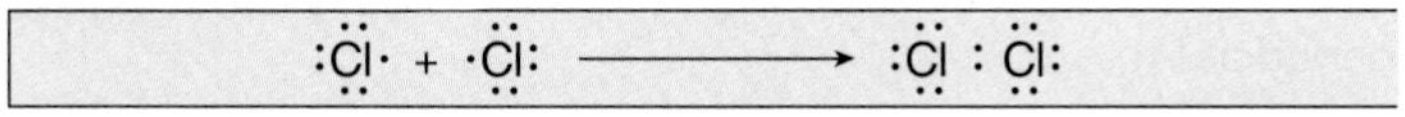

:C̈l· + ·C̈l: ⟶ :C̈l : C̈l:

As electron-dot structures are cumbersome, other representations are often used. Structural formulas use straight lines to show the covalent bonds between the atoms. Each line represents a pair of shared electrons. Molecular formulas indicate only the number of each type of atom making up a molecule.

Structural formula: Cl—Cl

Molecular formula: Cl_2

Additional examples of electron-dot, structural, and molecular formulas are shown in Figure 2.7.

Electron-Dot Formula	Structural Formula	Molecular Formula
:Ö::C::Ö: carbon dioxide	O=C=O carbon dioxide	CO_2 carbon dioxide
H :N̈:H H ammonia	H–N(–H)–H ammonia	NH_3 ammonia
H :Ö:H water	H–O–H water	H_2O water

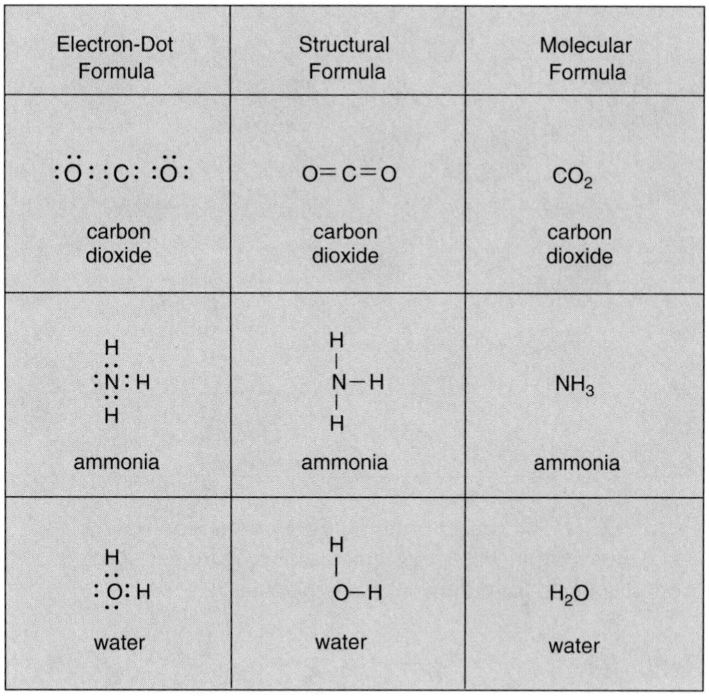

Figure 2.7 Electron-dot, structural, and molecular formulas.
In the electron-dot formula, only the electrons in the outer shell are designated. In the structural formula, the lines represent a pair of electrons being shared by two atoms. The molecular formula indicates only the number of each type of atom found within a molecule.

Double and Triple Bonds

Besides a single bond, in which atoms share only a pair of electrons, a double or a triple bond can form. In a double bond, atoms share two pairs of electrons, and in a triple bond, atoms share three pairs of electrons between them. For example, in Figure 2.6, each nitrogen atom (N) requires 3 electrons to achieve a total of 8 electrons in the outermost shell. Notice that 6 electrons are placed in the outer overlapping shells in the diagram and that three straight lines are in the structural formula for nitrogen gas (N_2).

A covalent bond arises when atoms share electrons. In double covalent bonds, atoms share two pairs of electrons, and in triple covalent bonds, atoms share three pairs of electrons.

Oxidation and Reduction Reactions

When oxygen (O) combines with a metal such as magnesium (Mg) or iron (Fe), oxygen receives electrons and forms ions that are negatively charged (O^{2-}); the metal loses electrons and forms ions that are positively charged (i.e., Mg^{2+}). When magnesium oxide (MgO) forms, it is obviously appropriate to say that the metal has been oxidized and that because of oxidation, the metal has lost electrons. On the other hand, oxygen has been reduced because it has gained negative charges (i.e., electrons).

Today, the terms oxidation and reduction are applied to many ionic reactions, whether or not oxygen is involved. Very simply, **oxidation** is the loss of electrons, and **reduction** is the gain of electrons. In the ionic reaction $Na + Cl \rightarrow NaCl$, sodium has been oxidized (loss of electron) and chlorine has been reduced (gain of electron). Because oxidation and reduction go hand-in-hand, the entire reaction is called a **redox reaction.**

The terms oxidation and reduction also apply to certain covalent reactions. In this case, however, oxidation is the loss of hydrogen atoms (H) and reduction is the gain of hydrogen atoms. A hydrogen atom contains one proton and one electron; therefore, when a molecule loses a hydrogen atom, it has lost an electron, and when a molecule gains a hydrogen atom, it has gained an electron. This form of oxidation-reduction is often seen during metabolic reactions within cells.

When oxidation occurs, an atom is oxidized (loses electrons). When reduction occurs, an atom is reduced (gains electrons). These two processes occur concurrently in oxidation-reduction reactions.

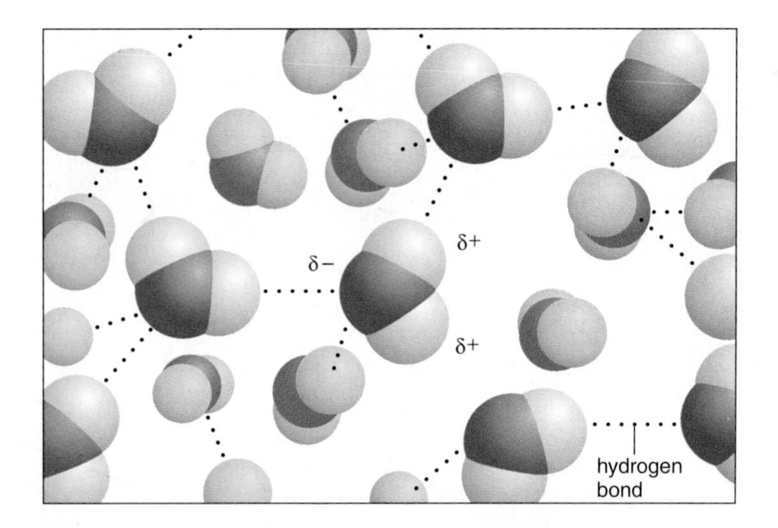

Figure 2.8 Hydrogen bonding between water molecules. The polarity of the water molecules allows hydrogen bonds (dotted lines) to form between the molecules.

2.3 Water and Living Things

Water is the most abundant molecule in living organisms, and it makes up about 60–70% of the total body weight of most organisms. We will see that the physical and chemical properties of water make life possible as we know it.

Water is a polar molecule; the oxygen end of the molecule has a slight negative charge, and the hydrogen end has a slight positive charge:

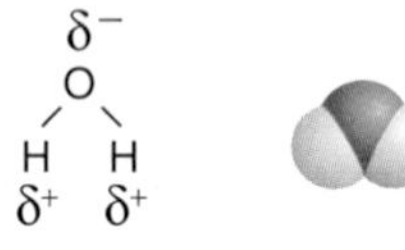

The diagram on the left shows the structural formula of water and the one on the right shows the space-filling model of water.

In polar molecules, covalently bonded atoms share electrons unevenly; that is, the electrons spend more time circling the nucleus of one atom than circling the other. In water, the electrons spend more time circling the larger oxygen (O) than the smaller hydrogen (H) atoms.

In water, the negative end and positive ends of the molecules attract one another. Each oxygen forms loose bonds to hydrogen atoms of two other water molecules (Fig. 2.8). These bonds are called hydrogen bonds. A **hydrogen bond** occurs whenever a covalently bonded hydrogen is positive and attracted to a negatively charged atom some distance away. A hydrogen bond is represented by a dotted line in Figure 2.8 because it is relatively weak and can be broken rather easily.

Properties of Water

Because of their polarity and hydrogen bonding, water molecules are cohesive and cling together. Polarity and hydrogen bonding causes water to have many characteristics beneficial to life.

1. Water is a liquid at room temperature. Therefore we are able to drink it, cook with it, and bathe in it.

Compounds with low molecular weights are usually gases at room temperature. For example, oxygen (O_2) with a molecular weight of 32 is a gas, but water with a molecular weight of 18 is a liquid. The hydrogen bonding between water molecules keeps water a liquid and not a gas at room temperature. Water does not boil and become a gas until 100°C, one of the reference points for the Celsius temperature scale. (See Appendix C.) Without hydrogen bonding between water molecules, our body fluids and indeed our bodies would be gaseous!

2. Water is the universal solvent for polar (charged) molecules and thereby facilitates chemical reactions both outside of and within our bodies.

When a salt such as sodium chloride (NaCl) is put into water, the negative ends of the water molecules are attracted to the sodium ions, and the positive ends of the water molecules are attracted to the chloride ions. This causes the sodium ions and the chloride ions to separate and to dissolve in water:

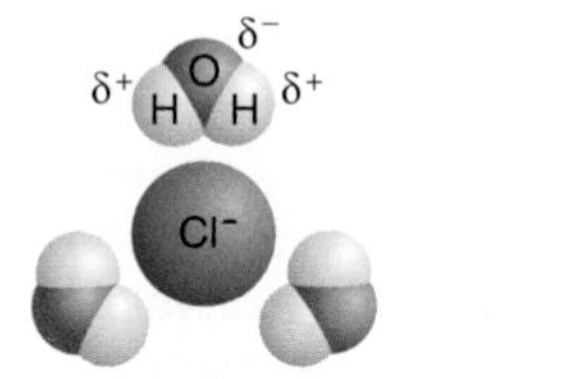

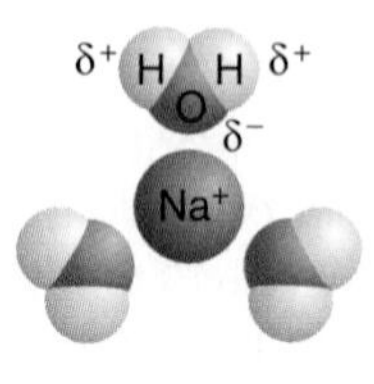

The salt NaCl dissolves in water

When ions and molecules disperse in water, they move about and collide, allowing reactions to occur. Therefore, water is a solvent that facilitates chemical reactions.

Ions and molecules that interact with water are said to be **hydrophilic.** Nonionized and nonpolar molecules that do not interact with water are said to be **hydrophobic.**

3. Water molecules are cohesive and therefore liquids will fill vessels.

Water molecules cling together because of hydrogen bonding, and yet, water flows freely. This property allows dissolved and suspended molecules to be evenly distributed throughout a system. Therefore, water is an excellent transport medium. Within our bodies, blood which fills our arteries and veins is 92% water. Blood transports oxygen and nutrients to the cells and removes wastes such as carbon dioxide.

a. b. c.

Figure 2.9 Characteristics of water.
a. Water boils at 100°C. If it boiled and was a gas at a lower temperature, life could not exist. **b.** It takes much body heat to vaporize sweat, which is mostly liquid water, and this helps keep bodies cool when the temperature rises. **c.** Ice is less dense than water, and it forms on top of water, making skate sailing possible.

4. The temperature of liquid water rises and falls slowly, preventing sudden or drastic changes.

The many hydrogen bonds that link water molecules cause water to absorb a great deal of heat before it boils (Fig. 2.9*a*). A **calorie** of heat energy raises the temperature of one gram of water 1°C. This is about twice the amount of heat required for other covalently bonded liquids. On the other hand, water holds heat, and its temperature falls slowly. Therefore, water protects us and other organisms from rapid temperature changes and helps us maintain our normal internal temperature. This property also allows great bodies of water, such as oceans, to maintain a relatively constant temperature. Water is a good temperature buffer.

5. Water has a high heat of vaporization, keeping the body from overheating.

It takes a large amount of heat to change water to steam (Fig. 2.9*a*). (Converting one gram of the hottest water to steam requires an input of 540 calories of heat energy.) This property of water helps moderate the earth's temperature so that life can continue to exist. Also, in a hot environment, animals sweat and the body cools as body heat is used to evaporate sweat, which is mostly liquid water (Fig. 2.9*b*).

6. Frozen water is less dense than liquid water so that ice floats on water.

As water cools, the molecules come closer together. They are densest at 4°C, but they are still moving about. At temperatures below 4°C, there is only vibrational movement, and hydrogen bonding becomes more rigid but also more open. This makes ice less dense. Bodies of water always freeze from the top down, making skate sailing possible (Fig. 2.9*c*). When a body of water freezes on the surface, the ice acts as an insulator to prevent the water below it from freezing. Aquatic organisms are protected, and they have a better chance of surviving the winter.

Because of its polarity and hydrogen bonding, water has many characteristics that benefit life.

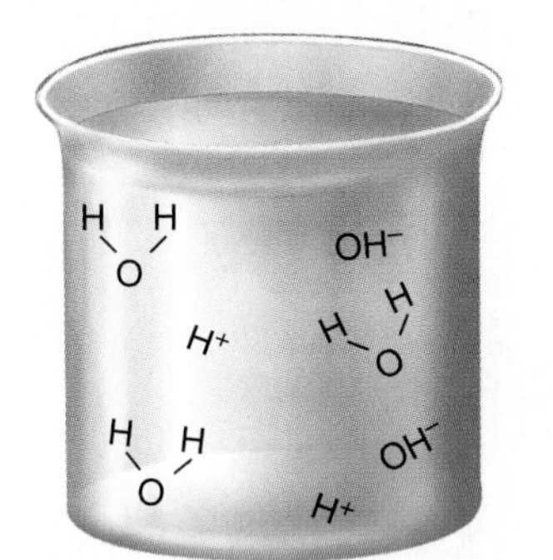

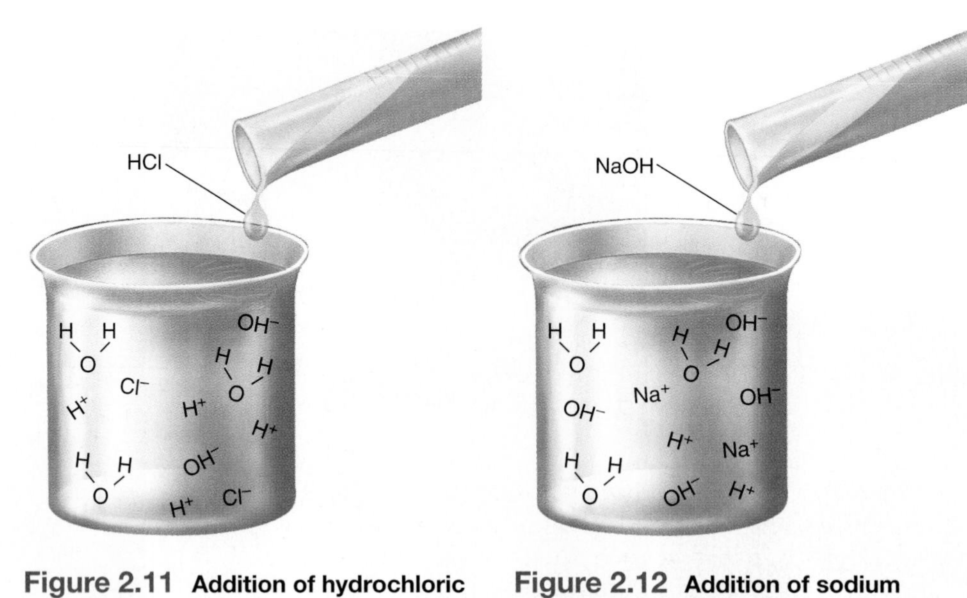

Figure 2.10 Dissociation of water molecules.
Dissociation produces an equal number of hydrogen ions (H^+) and hydroxide ions (OH^-). (These illustrations are not meant to be mathematically accurate.)

Figure 2.11 Addition of hydrochloric acid (HCl).
HCl releases hydrogen ions (H^+) as it dissociates. The addition of HCl to water results in a solution with more H^+ than OH^-.

Figure 2.12 Addition of sodium hydroxide (NaOH), a base.
NaOH releases OH^- as it dissociates. The addition of NaOH to water results in a solution with more OH^- than H^+.

Acidic and Basic Solutions

When water dissociates (breaks up), it releases an equal number of hydrogen ions (H^+) and hydroxide ions (OH^-).

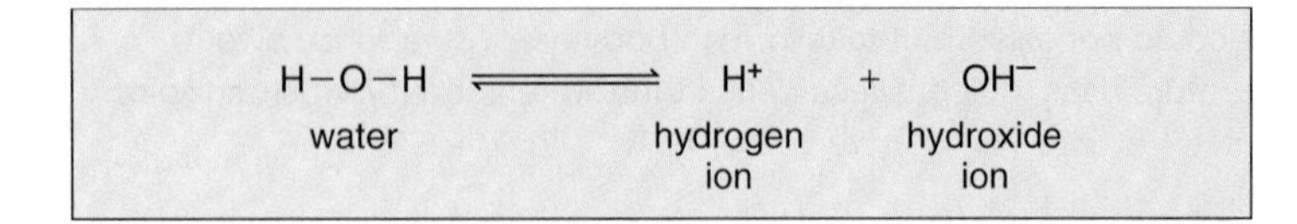

Only a few water molecules at a time are dissociated (Fig. 2.10). The actual number of ions is 10^{-7} moles/liter. A mole is a unit of scientific measurement for atoms, ions, and molecules.[1]

Acidic Solutions

Lemon juice, vinegar, tomato juice , and coffee are all familiar acidic solutions. What do they have in common? Acidic solutions have a sharp or sour taste, and therefore we sometimes associate them with indigestion. To a chemist, **acids** are molecules that dissociate in water, releasing hydrogen ions (H^+). For example, an important acid in the laboratory is hydrochloric acid (HCl), which dissociates in this manner:

$$HCl \rightarrow H^+ + Cl^-$$

Dissociation is almost complete; therefore, this is called a strong acid. When hydrochloric acid is added to a beaker of water, the number of hydrogen ions increases (Fig. 2.11).

Basic Solutions

Milk of magnesia and ammonia are common basic solutions that most people have heard of. Bases have a bitter taste and feel slippery when in water. To a chemist, **bases** are molecules that either take up hydrogen ions (H^+) or release hydroxide ions (OH^-). For example, an important inorganic base is sodium hydroxide (NaOH), which dissociates in this manner:

$$NaOH \rightarrow Na^+ + OH^-$$

Dissociation is almost complete; therefore sodium hydroxide is called a strong base. If sodium hydroxide is added to a beaker of water, the number of hydroxide ions increases (Fig. 2.12).

It is not recommended that you taste a strong acid or base, because they are quite destructive to cells. Any container of household cleanser like ammonia has a poison symbol and carries a strong warning not to ingest the product.

The Litmus Test

A simple laboratory test for acids and bases is called the litmus test. Litmus is a vegetable dye that changes color from blue to red in the presence of an acid and from red to blue in the presence of a base. The litmus test has become a common figure of speech, as when you hear a commentator say, "The litmus test for a Republican is . . ."

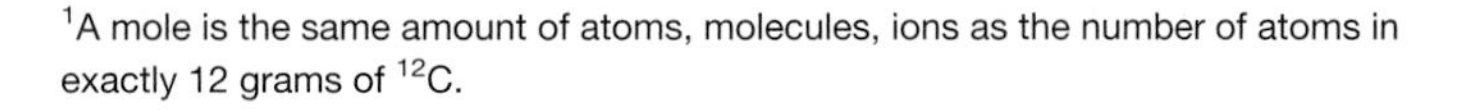

[1]A mole is the same amount of atoms, molecules, ions as the number of atoms in exactly 12 grams of ^{12}C.

The pH Scale

The **pH scale**[2] is used to indicate the acidity and basicity (alkalinity) of a solution. Since there are normally few hydrogen ions (H^+) in a solution, the pH scale was devised to eliminate the use of cumbersome numbers. For example, the possible hydrogen ion concentrations of a solution are on the left of this listing and the pH is on the right:

moles/liter

$1 \leftrightarrow 10^{-6}\,[H^+] = \text{pH } 6$ (an acid)
$1 \leftrightarrow 10^{-7}\,[H^+] = \text{pH } 7$ (neutral)
$1 \leftrightarrow 10^{-8}\,[H^+] = \text{pH } 8$ (a base)

Pure water (HOH) has an equal number of hydrogen ions (H^+) and hydroxide ions (OH^-); therefore, one of each is released when water dissociates. One mole of pure water contains only 10^{-7} moles/liter of hydrogen ions; therefore, a pH of exactly 7 is neutral pH. At a pH of 7 there is an equal number of hydrogen ions and hydroxide ions. Above pH 7 there are more hydroxide ions than hydrogen ions, and below pH 7 there are more hydrogen ions than hydroxide ions. Therefore any solution with a pH below 7 is an acidic solution and any solution with a pH above 7 is a basic solution. Also, as we move down the pH scale from 14 to 0, each unit has 10 times the $[H^+]$ of the previous unit (Fig. 2.13). As we move up the pH scale from 0 to 14, each unit has 10 times the $[OH^-]$ of the previous unit.

As discussed in the Ecology reading on page 30, there have been detrimental environmental consequences on nonliving and living things as rain and snow have become more acidic during modern times. In plants and animals, including ourselves, pH needs to be maintained within a narrow range or there are health consequences.

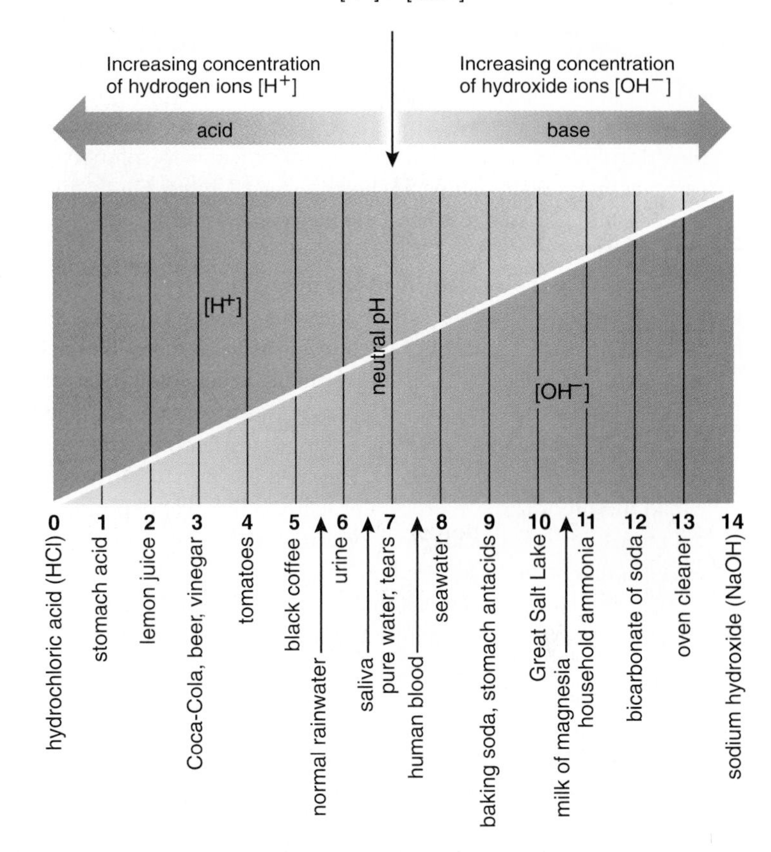

Figure 2.13 The pH scale.
The diagonal line indicates the proportionate concentration of hydrogen ions (H^+) to hydroxide ions (OH^-) at each pH value. Any pH value above 7 is basic, while any pH value below 7 is acidic.

Buffers

Buffers resist pH changes because they are chemicals or combinations of chemicals that can take up excess hydrogen ions (H^+) or hydroxide ions (OH^-). Many commercial products like Bufferin or shampoos or deodorants are buffered as an added incentive to have us buy them.

Our cells and body fluids are naturally buffered and for that reason the pH of our blood when we are healthy is always about 7.4. The main buffer in blood is a combination of carbonic acid and bicarbonate ions which are a dissociation product of carbonic acid. Carbonic acid (H_2CO_3) is a weak acid that minimally dissociates and then reforms in the following manner:

H_2CO_3	dissociates ⇌ reforms	H^+	+	HCO_3^-
carbonic acid		hydrogen ion		bicarbonate ion

When hydrogen ions (H^+) are added to blood, the following reaction occurs:

$$H^+ + HCO_3^- \rightarrow H_2CO_3$$

When hydroxide ions (OH^-) are added to blood, this reaction occurs:

$$OH^- + H_2CO_3 \rightarrow HCO_3^- + H_2O$$

These reactions prevent any significant change in blood pH.

Acids have a pH that is less than 7, and bases have a pH that is greater than 7. Buffers, which can combine with both hydrogen ions and hydroxide ions, resist pH changes.

[2]pH is defined as the negative logarithm of the molar concentration of the hydrogen ion $[H^+]$.

Ecology Focus

The Harm Done by Acid Deposition

Normally, rainwater has a pH of about 5.6 because the carbon dioxide in the air combines with water to give a weak solution of carbonic acid. Rain falling in northeastern United States and southeastern Canada now has a pH between 5.0 and 4.0. We have to remember that a pH of 4 is ten times more acidic than a pH of 5 to comprehend the increase in acidity this represents.

There is very strong evidence that this observed increase in rainwater acidity is a result of the burning of fossil fuels, like coal and oil, as well as gasoline derived from oil. When fossil fuels are burned, sulfur dioxide and nitrogen oxides are produced, and they combine with water vapor in the atmosphere to form the acids sulfuric acid and nitric acid. These acids return to earth contained in rain or snow, a process properly called wet deposition, but more often called acid rain. Dry particles of sulfate and nitrate salts descend from the atmosphere during dry deposition.

Unfortunately, regulations that require the use of tall smokestacks to reduce local air pollution only cause pollutants to be carried far from their place of origin. Acid deposition in southeastern Canada is due to the burning of fossil fuels in factories and power plants in the Midwest. Acid deposition adversely affects lakes, particularly in areas where the soil is thin and lacks limestone (calcium carbonate, $CaCO_3$), a buffer to acid deposition. It leaches aluminum from the soil, carries aluminum into the lakes, and converts mercury deposits in lake bottom sediments to soluble and toxic methyl mercury. Lakes not only become more acidic, but they also show accumulation of toxic substances. In Norway and Sweden, at least 16,000 lakes contain no fish, and an additional 52,000 lakes are threatened. In Canada, some 14,000 lakes are almost fishless, and an additional 150,000 are in peril because of excess acidity. In the United States, about 9,000 lakes (mostly in the Northeast and upper Midwest) are threatened, one-third of them seriously.

In forests, acid deposition weakens trees because it leaches away nutrients and releases aluminum. By 1988, most spruce, fir, and other conifers atop North Carolina's Mt. Mitchell were dead from being bathed in ozone and acid fog for years. The soil was so acidic, new seedlings could not survive. Nineteen countries in Europe have reported woodland damage, ranging from 5 to 15% of the forested area in Yugoslavia and Sweden to 50% or more in the Netherlands, Switzerland, and the former West Germany. More than one-fifth of Europe's forests are now damaged.

These aren't the only effects of acid deposition. Reduction of agricultural yields, damage to marble and limestone monuments and buildings, and even illnesses in humans have been reported. Acid deposition has been implicated in the increased incidence of lung cancer and possibly colon cancer in residents of the East Coast. Tom McMillan, Canadian Minister of the Environment, says that acid rain is "destroying our lakes, killing our fish, undermining our tourism, retarding our forests, harming our agriculture, devastating our heritage, and threatening our health."

There are, of course, things that can be done. We could

- whenever possible use alternative energy sources, such as solar, wind, hydropower, and geothermal energy.
- use low-sulfur coal or remove the sulfur impurities from coal before it is burned.
- require factories and power plants to use scrubbers, which remove sulfur emissions.
- require people to use mass transit rather than driving their own automobiles.
- reduce our energy needs through other means of energy conservation.

Figure 2A Effects of acid deposition.
Trees die and statues deteriorate due to the burning of fossil fuels. Gasoline is derived from oil, a fossil fuel. The combustion of fossil fuels results in atmospheric acids that return to the earth as acid rain.

2.4 Organic Molecules

Inorganic molecules constitute nonliving matter, but even so, inorganic molecules like salts (e.g., NaCl) and water play important roles in living things. The molecules of life are organic molecules. **Organic molecules** always contain carbon (C) and hydrogen (H). The chemistry of carbon accounts for the formation of the very large variety of organic molecules found in living things. A carbon atom has four electrons in the outer shell. In order to achieve eight electrons in the outer shell, a carbon atom shares electrons covalently with as many as four other atoms. Methane is a molecule in which a carbon atom shares electrons with four hydrogen atoms.

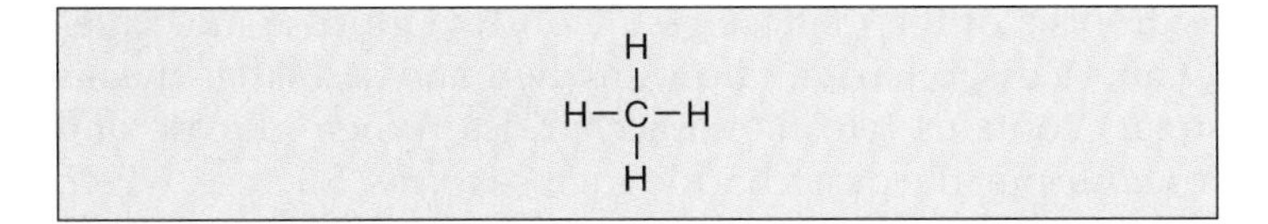

A carbon atom can share with another carbon atom, and in so doing, a long hydrocarbon chain can result:

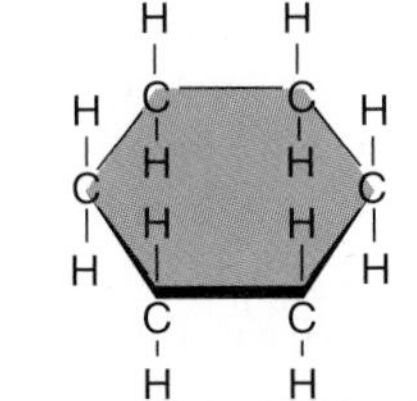

A hydrocarbon chain can also turn back on itself to form a ring compound:

So-called functional groups can be attached to carbon chains. A **functional group** is a particular cluster of atoms that always behaves in a certain way. One functional group of interest is the acidic (carboxyl) group —COOH because it can give up a hydrogen (H^+) and ionize to —COO^-.

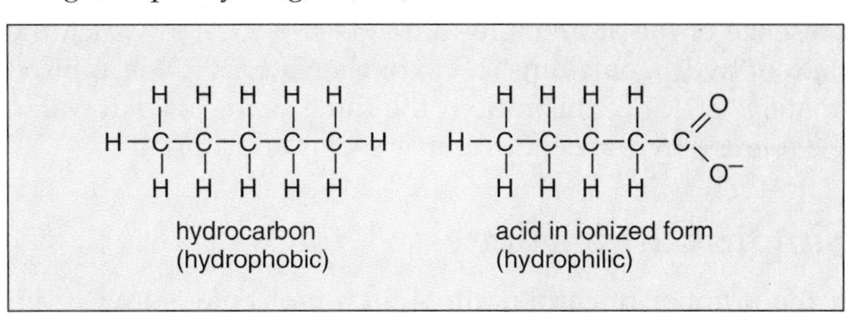

Whereas a hydrocarbon chain is *hydrophobic* (not attracted to water) because it is nonpolar, a hydrocarbon chain with an attached ionized group is *hydrophilic* (is attracted to water) because it is polar.

The molecules of life are divided into four classes: carbohydrates, lipids, proteins, and nucleic acids. Carbohydrates, lipids, and proteins are very familiar to you because certain foods are known to be rich in these molecules, as illustrated in Figures 2.14–2.16. The nucleic acid DNA makes up our genes which are hereditary units that control our cells and the structure of our bodies.

Many molecules of life are macromolecules. Just as atoms can join to form a molecule, so molecules can join to form a macromolecule. The smaller molecules are called monomers, and the macromolecule is called a polymer. A polymer is a chain of monomers.

Polymer	Monomer
polysaccharide	monosaccharide
protein	amino acid
nucleic acid	nucleotide

Figure 2.14 Foods rich in carbohydrates. Breads, pasta, rice, corn, and oats all contain complex carbohydrates.

Figure 2.15 Foods rich in lipids. Butter and oils contain fat, the most familiar of the lipids.

Figure 2.16 Foods rich in proteins. Meat, eggs, cheese, and beans have a high content of protein.

2.5 Carbohydrates

Carbohydrates first and foremost function for quick and short-term energy storage in all organisms, including humans. Carbohydrate molecules are characterized by the presence of the atomic grouping H—C—OH, in which the ratio of hydrogen atoms (H) to oxygen atoms (O) is approximately 2:1. Since this ratio is the same as the ratio in water, the name—hydrates of carbon—seems appropriate.

Simple Carbohydrates

If the number of carbon atoms in a molecule is low (from three to seven), then the carbohydrate is a simple sugar, or **monosaccharide.** The designation **pentose** means a 5-carbon sugar, and the designation **hexose** means a 6-carbon sugar. **Glucose,** a hexose, is blood sugar (Fig. 2.17); our bodies use glucose as an immediate source of energy. Other common hexoses are fructose, found in fruits, and galactose, a constituent of milk. These three hexoses (glucose, fructose, and galactose) all occur as ring structures with the molecular formula $C_6H_{12}O_6$, but the exact shape of the ring differs, as does the arrangement of the hydrogen (—H) and the hydroxyl groups (—OH) attached to the ring.

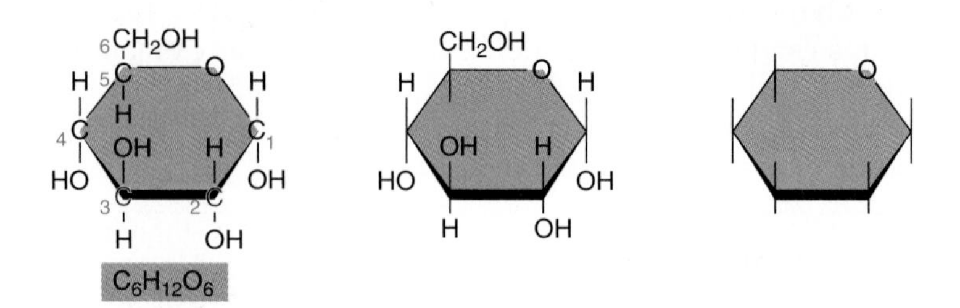

Figure 2.17 Three ways to represent the structure of glucose.
The *far left* structure shows the carbon atoms; $C_6H_{12}O_6$ is the molecular formula for glucose. The *far right* structure is the simplest way to represent glucose.

Glucose is a monomer for larger carbohydrates in the body. The body has a common way of joining monomers to build larger molecules. **Condensation synthesis** of a larger molecule is so called because synthesis means "making of" and condensation means that water has been removed as monomers are joined. Breakdown of the larger molecule is a **hydrolysis** reaction because water is used to split bonds between monomers. Polymers are synthesized and broken down in this manner:

$$\text{monomers} \rightleftharpoons \text{polymer} + H_2O \text{ molecules}$$

For convenience sake, Figure 2.18 shows how condensation synthesis results in a disaccharide called maltose and how hydrolysis of the maltose results in two glucose molecules again. A **disaccharide** (*di* means two and *saccharide* means sugar) contains two monosaccharides. When glucose and fructose join, the disaccharide sucrose forms. Sucrose, which is ordinarily derived from sugarcane and sugar beets, is commonly known as table sugar.

Starch and Glycogen

Starch and glycogen are ready storage forms of glucose in plants and animals, respectively. Starch and glycogen are **polysaccharides**; that is, they are polymers of glucose formed just as a necklace might be made using only one type of bead.

Some of the polymers in starch are long chains of up to 4,000 glucose units. Others are branched as is glycogen (Fig. 2.19). Starch has fewer side branches, or chains of glucose that branch off from the main chain, than does glycogen. **Starch** is the storage form of glucose inside plant cells. Flour, which we usually acquire by grinding wheat and use to bake bread and rolls, is high in starch. **Glycogen** is the storage form of glucose in humans. Figure 2.20 includes a micrograph of glycogen granules inside the liver.

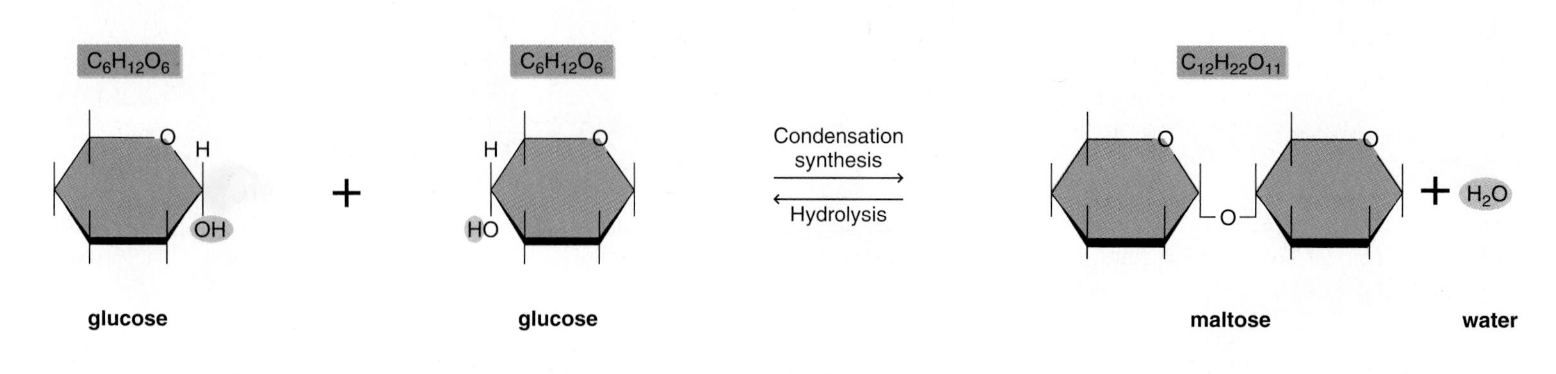

Figure 2.18 Condensation synthesis and hydrolysis of maltose, a disaccharide.
During condensation synthesis of maltose, a bond forms between the two glucose molecules and the components of water are removed. During hydrolysis, the components of water are added, and the bond is broken.

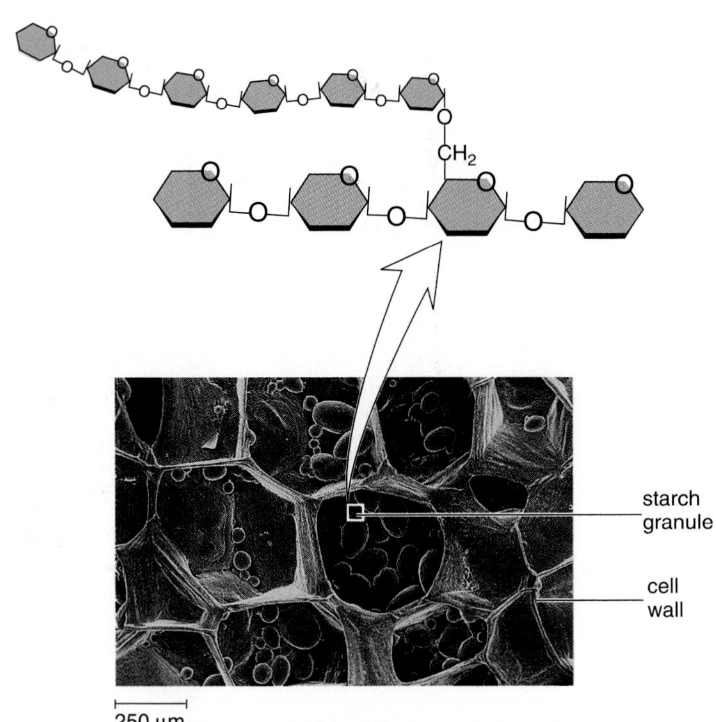

Figure 2.19 Starch structure and function.
Starch has straight chains of glucose molecules. Some chains are also branched as indicated. The electron micrograph shows starch granules in plant cells. Starch is the storage form of glucose in plants.

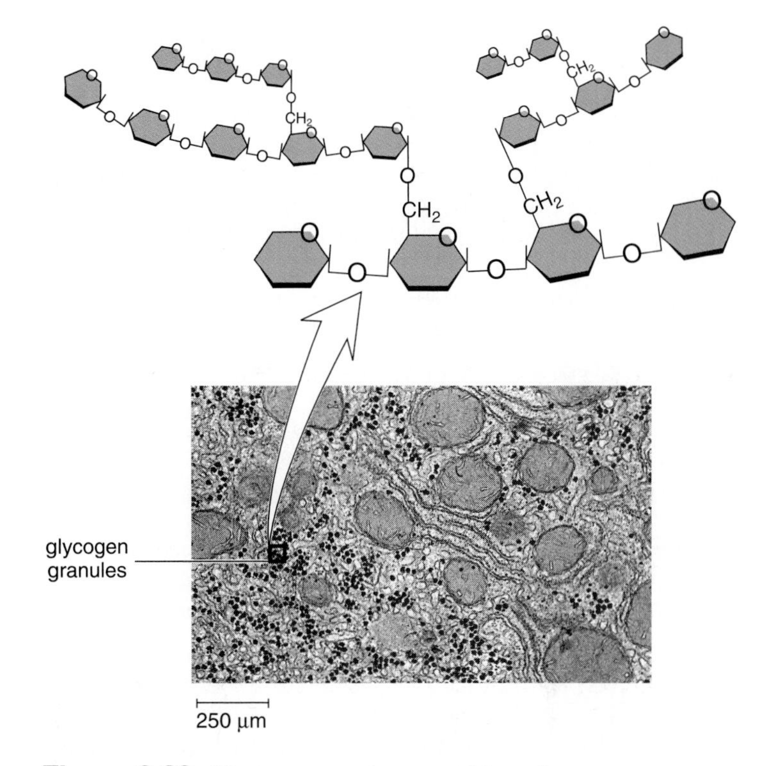

Figure 2.20 Glycogen structure and function.
Glycogen is a highly branched polymer of glucose molecules. The electron micrograph shows glycogen granules in liver cells. Glycogen is the storage form of glucose in animals.

After we eat starchy foods like bread, potatoes, and cake, glucose enters the bloodstream, and the liver stores glucose as glycogen. In between eating, the liver releases glucose so that the blood glucose concentration is always about 0.1%.

Cellulose

The polysaccharide **cellulose** is found in plant cell walls, and this accounts, in part, for the strong nature of these walls. In cellulose (Fig. 2.21), the glucose units are joined by a slightly different type of linkage than that in starch or glycogen. (Observe the alternating position of the oxygen atoms in the linked glucose units.) While this might seem to be a technicality, actually it is important because we are unable to digest foods containing this type of linkage; therefore, cellulose largely passes through our digestive tract as fiber, or roughage. Recently, it has been suggested that fiber in the diet is necessary to good health and may even help to prevent colon cancer.

Cells usually use the monosaccharide glucose as an energy source. The polysaccharides starch and glycogen are storage compounds in plant and animal cells, respectively, and the polysaccharide cellulose is found in plant cell walls.

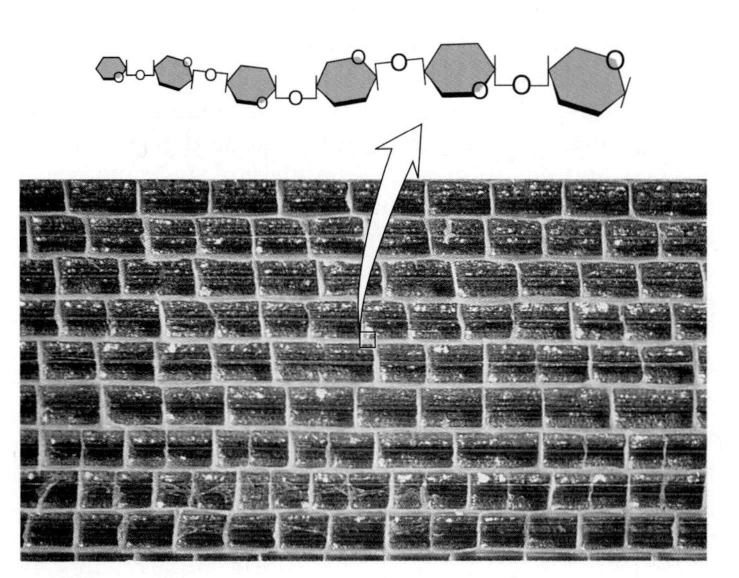

cattail leaf cell walls

Figure 2.21 Cellulose structure and function.
Cellulose contains a slightly different type of linkage between glucose molecules than that in starch or glycogen. Plant cell walls contain cellulose, and the rigidity of the cell walls permits nonwoody plants to stand upright as long as they receive an adequate supply of water.

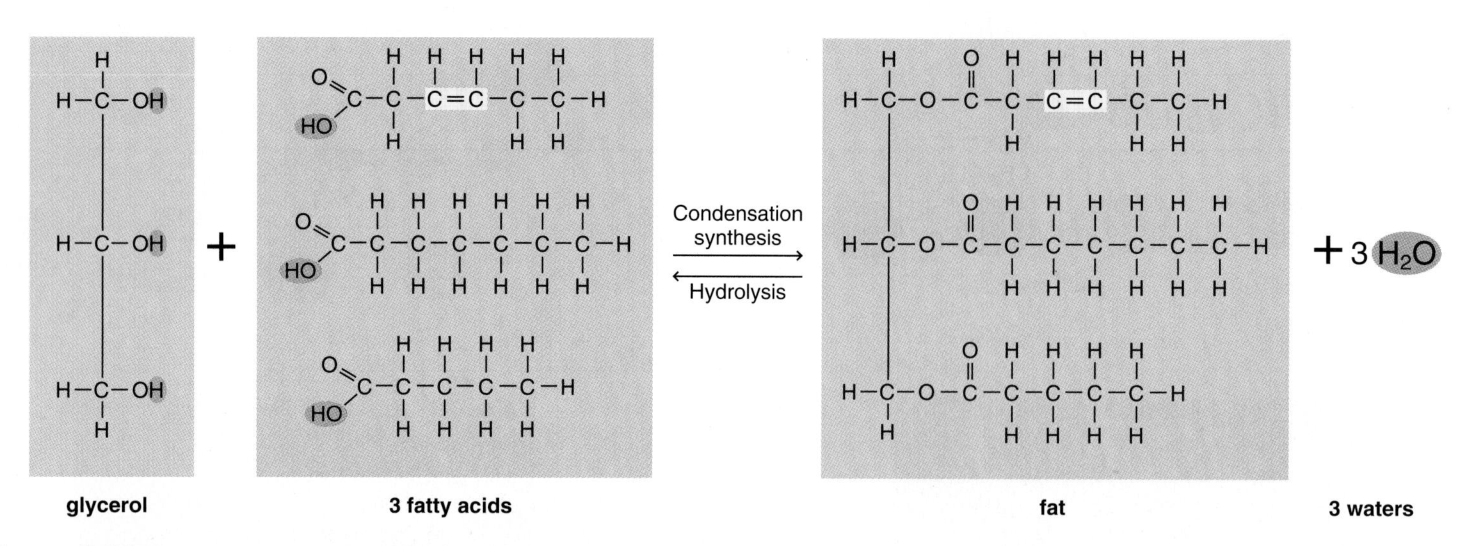

Figure 2.22 Condensation synthesis and hydrolysis of a fat molecule.
Fatty acids can be saturated (no double bonds between carbon atoms) or unsaturated (have double bonds, colored yellow, between carbon atoms). When a fat molecule forms, three fatty acids combine with glycerol, and three water molecules are produced.

2.6 Lipids

Lipids are diverse in structure and function, but they have a common characteristic: they do not dissolve in water.

Fats and Oils

The most familiar lipids are those found in fats and oils. **Fats,** which are usually of animal origin (e.g., lard and butter), are solid at room temperature. **Oils,** which are usually of plant origin (e.g., corn oil and soybean oil), are liquid at room temperature. Fat has several functions in the body: it is used for long-term energy storage, it insulates against heat loss, and it forms a protective cushion around major organs.

Fats and oils form when one glycerol molecule reacts with three fatty acid molecules. A fat is sometimes called a **triglyceride** because of its three-part structure, and the term neutral fat is sometimes used because the molecule is nonpolar (Fig. 2.22).

Saturated and Unsaturated Fatty Acids

A **fatty acid** is a hydrocarbon chain that ends with the acidic group —COOH (Fig. 2.22). Most of the fatty acids in cells contain 16 or 18 carbon atoms per molecule, although smaller ones with fewer carbons are also known.

Fatty acids are either saturated or unsaturated. **Saturated fatty acids** have no double bonds between carbon atoms. The carbon chain is saturated, so to speak, with all the hydrogens it can hold. Saturated fatty acids account for the solid nature at room temperature of butter and lard, which are derived from animal sources. **Unsaturated fatty acids** have double bonds between carbon atoms wherever the number of hydrogens is less than two per carbon atom. Unsaturated fatty acids account for the liquid nature of vegetable oils at room temperature. Hydrogenation of vegetable oils can convert them to margarine and products such as Crisco.

Soaps

Strictly speaking, soaps are not lipids, but they are considered here as a matter of convenience. A **soap** is a salt formed from a fatty acid and an inorganic base. For example,

NaOH	+	RCOOH	⟶	$RCOO^- Na^+$
sodium hydroxide		fatty acid		soap

Unlike fats, soaps have a polar end that is hydrophilic in addition to the nonpolar end that is hydrophobic (the hydrocarbon chain represented by *R*). Therefore, a soap does mix with water. When soaps are added to oils, the oils, too, mix with water because a soap positions itself about an oil droplet so that its nonpolar ends project into the fat droplet, while its polar ends project outward.

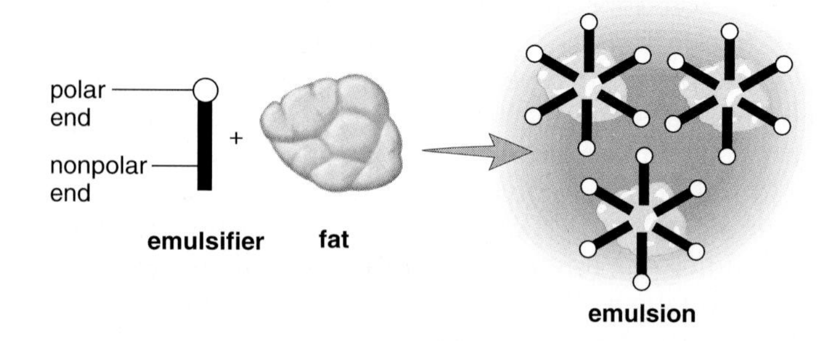

Now the droplet disperses in water, and it is said that **emulsification** has occurred. Emulsification occurs when dirty clothes are washed with soaps or detergents. Also, prior to the digestion of fatty foods, fats are emulsified by bile. A person who has had the gallbladder removed may have trouble digesting fatty foods because this organ stores bile for emulsifying fats prior to the digestive process.

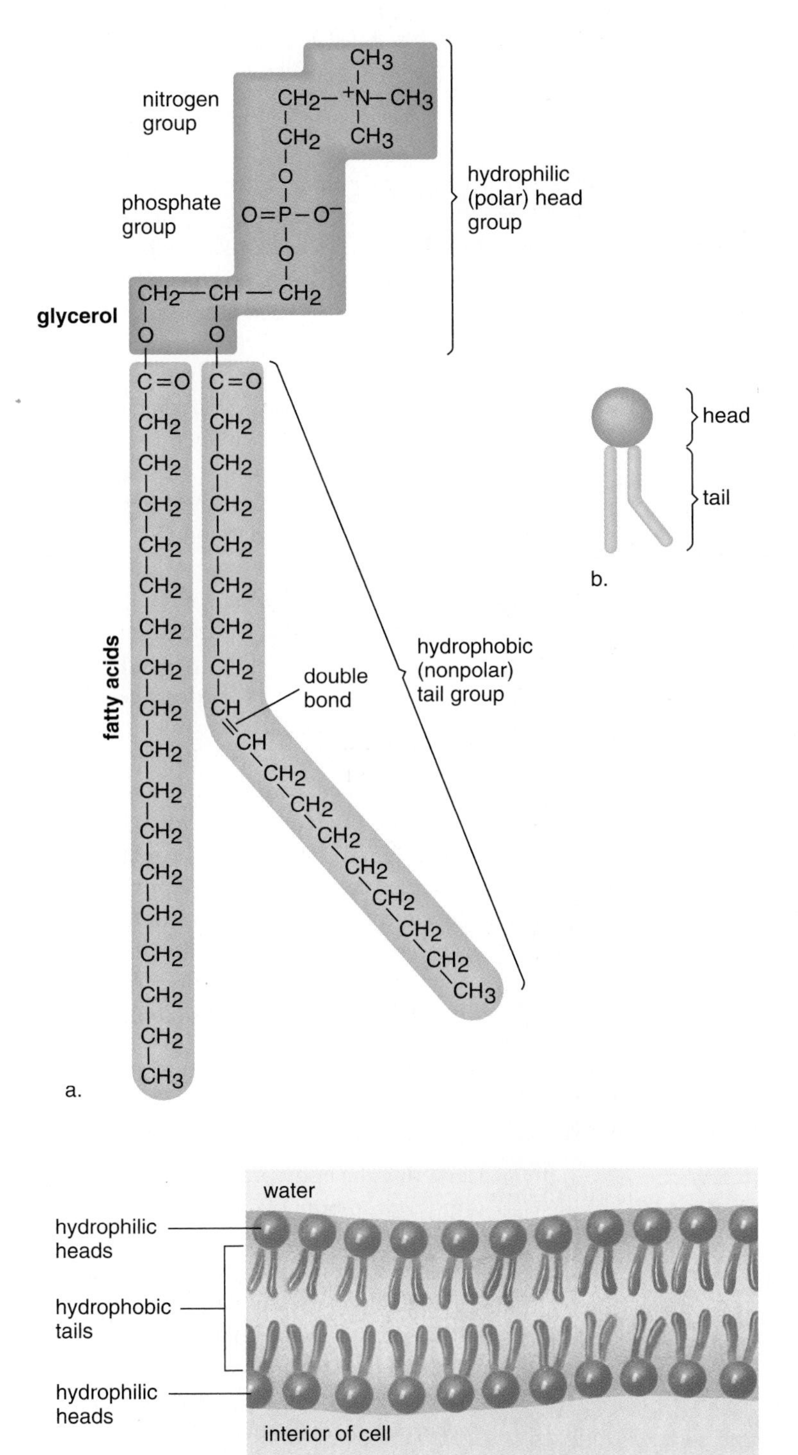

Figure 2.23 Phospholipid structure and shape.
a. Phospholipids are constructed like fats, except that they contain a phosphate group. This phospholipid also includes an organic group that contains nitrogen. **b.** The hydrophilic portion of the phospholipid molecule (head) is soluble in water, whereas the two hydrocarbon chains (tails) are not. **c.** This causes the molecule to arrange itself as shown when exposed to water.

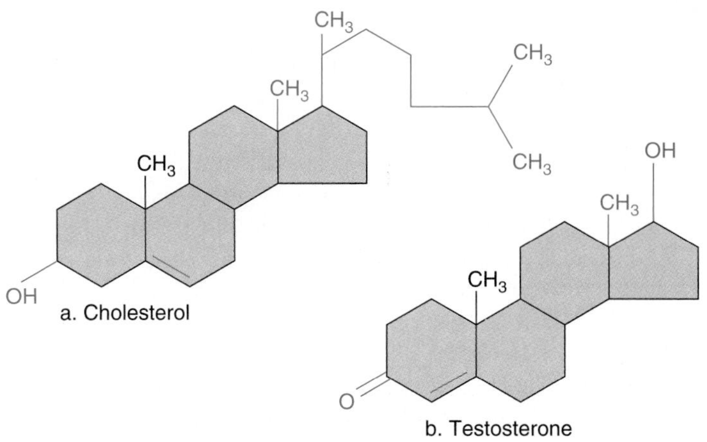

Figure 2.24 Steroid diversity.
a. Cholesterol, like all steroid molecules, has four adjacent rings, but the effects of steroids on the body largely depend on the attached groups indicated in red. **b.** Testosterone is the male sex hormone.

Phospholipids

Phospholipids, as their name implies, contain a phosphate group (Fig. 2.23). Essentially, they are constructed like fats, except that in place of the third fatty acid, there is a phosphate group or a grouping that contains both phosphate and nitrogen. These molecules are not electrically neutral as are fats because the phosphate and nitrogenous groups are ionized. It forms the so-called hydrophilic head of the molecule, while the rest of the molecule becomes the hydrophobic tails. The plasma membrane which surrounds cells is a phospholipid bilayer in which the heads face outward into a watery medium and the tails face each other because they are water repelling.

Steroids

Steroids are lipids having a structure that differs entirely from that of fats. Steroid molecules have a backbone of four fused carbon rings, but each one differs primarily by the arrangement of the atoms in the rings and the type of functional groups attached to them. Cholesterol is a component of an animal cell's plasma membrane and is the precursor of several other steroids, such as the sex hormones estrogen and testosterone (Fig. 2.24)

We know that a diet high in saturated fats and cholesterol can lead to circulatory disorders. This type of diet causes fatty material to accumulate inside the lining of blood vessels and blood flow is reduced. As discussed in the Science reading on page 36, nutrition labels are now required to list the calories from fat per serving and the percent daily value from saturated fat and cholesterol.

Lipids include fats and oils for long-term energy storage and steroids. Phospholipids, unlike other lipids, are soluble in water because they have a hydrophilic group.

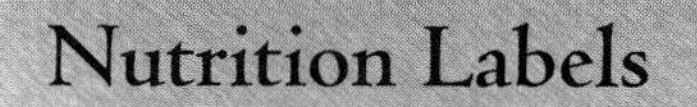

Nutrition Labels

As of May 1994, packaged foods have a nutrition label like the one depicted in Figure 2B. The nutrition information given on this label is based on the serving size (that is, 1¼ cup, 57 grams) of the cereal. A Calorie* is a measurement of energy. One serving of the cereal provides 220 Calories, of which 20 are from fat. It's also of interest that the cereal provides no cholesterol nor saturated fat. The suggestion that we study nutrition labels to determine how much cholesterol and fat (whether saturated or nonsaturated) they contain is based on innumerable statistical and clinical studies of three types:

(1) *Clinical trials that show that an elevated blood cholesterol level is a risk factor for coronary heart disease (CHD).* The Framingham Heart Study conducted in Framingham, Massachusetts, concludes that as the blood cholesterol level in over 5,000 men and women rises so does the risk of CHD. Elevated blood cholesterol appears to be one of the three major CHD risk factors along with smoking and high blood pressure. Other studies have shown the same. The Multiple Risk Factor Intervention Trial followed more than 360,000 men and found also that there was a direct relationship between an increasing blood cholesterol level and the risk of a heart attack.

(2) *Clinical trials indicate that lowering high blood cholesterol levels will reduce the risk of CHD.* For example, the Coronary Primary Prevention Trial found that a 9% reduction in total blood cholesterol levels produced a 19% reduction in CHD deaths and nonfatal heart attacks. The Cholesterol Lowering Atherosclerosis Study collected X-ray evidence that substantial cholesterol lowering produces slowed progression and regression of plaque in coronary arteries. Plaque is a buildup of soft fatty material including cholesterol beneath the inner linings of arteries. Plaque can accumulate to the point that blood can no longer reach the heart and a heart attack occurs.

(3) *Clinical studies show that there is a relationship between diet and blood cholesterol levels.* The National Research Council reviewed all sorts of scientific studies before concluding that the intake of saturated fatty acids raises blood cholesterol levels, while the substitution of unsaturated fats and carbohydrates in the diet lowers the blood cholesterol level. An Oslo Study, a Los Angeles Veterans Administration study, and a Finnish Mental Hospital Study showed that diet alone can produce 10–15% reductions in blood cholesterol level.

The desire of consumers to follow dietary recommendations led to the development of the type of nutrition label shown in 2B, which shows the amount of total fat, saturated fat, and cholesterol in a serving of food. The carbohydrate content in a food is of interest because carbohydrates aren't usually associated with health problems. In fact, carbohydrates should compose the largest proportion of the diet. Breads and cereals containing complex carbohydrates are preferable to candy and ice cream containing simple carbohydrates because they are likely to contain dietary fiber (nondigestible plant material). Insoluble fiber has a laxative effect and seems to reduce the risk of colon cancer; soluble fiber combines with the cholesterol in food and prevents the cholesterol from entering the body proper.

The amount of dietary sodium (as in table salt) is of concern because excessive sodium intake has been linked to high blood pressure in some people. It is recommended that the intake of sodium be no more than 2,400 mg per day. A serving of this cereal provides what percent of this maximum amount?

Vitamins are essential requirements needed in small amounts in the diet. Each vitamin has a recommended daily intake, and the food label tells what percent of the recommended amount is provided by a serving of this cereal.

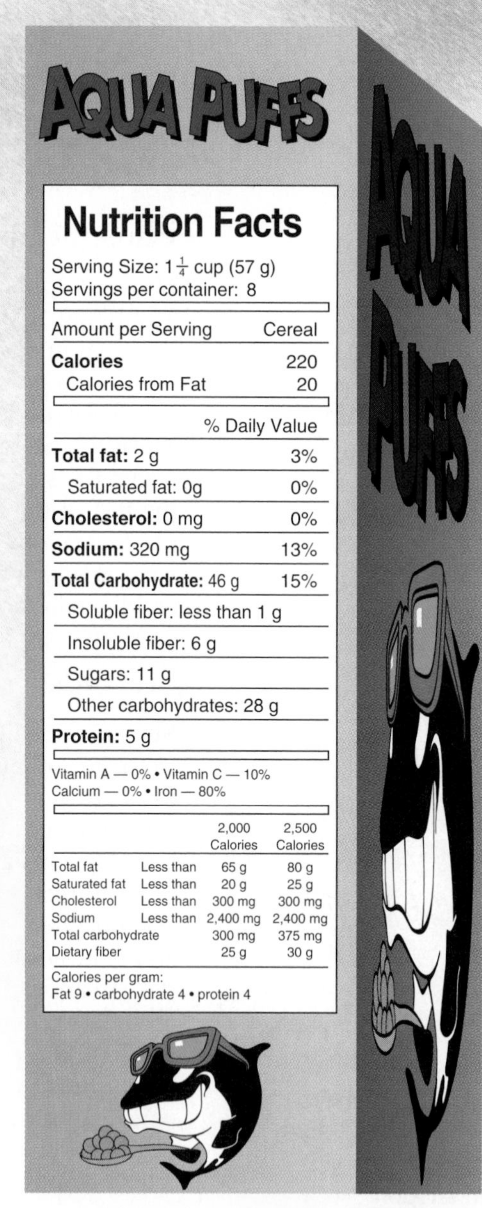

Nutrition Facts

Serving Size: 1¼ cup (57 g)
Servings per container: 8

Amount per Serving	Cereal
Calories	220
Calories from Fat	20
	% Daily Value
Total fat: 2 g	3%
Saturated fat: 0g	0%
Cholesterol: 0 mg	0%
Sodium: 320 mg	13%
Total Carbohydrate: 46 g	15%
Soluble fiber: less than 1 g	
Insoluble fiber: 6 g	
Sugars: 11 g	
Other carbohydrates: 28 g	
Protein: 5 g	

Vitamin A — 0% • Vitamin C — 10%
Calcium — 0% • Iron — 80%

		2,000 Calories	2,500 Calories
Total fat	Less than	65 g	80 g
Saturated fat	Less than	20 g	25 g
Cholesterol	Less than	300 mg	300 mg
Sodium	Less than	2,400 mg	2,400 mg
Total carbohydrate		300 mg	375 mg
Dietary fiber		25 g	30 g

Calories per gram:
Fat 9 • carbohydrate 4 • protein 4

Figure 2B Nutrition label on side panel of cereal box.

*A calorie is the amount of heat required to raise the temperature of 1 g of water 1°C. A Calorie (capital C), which is used to measure food energy, is equal to 1,000 calories.

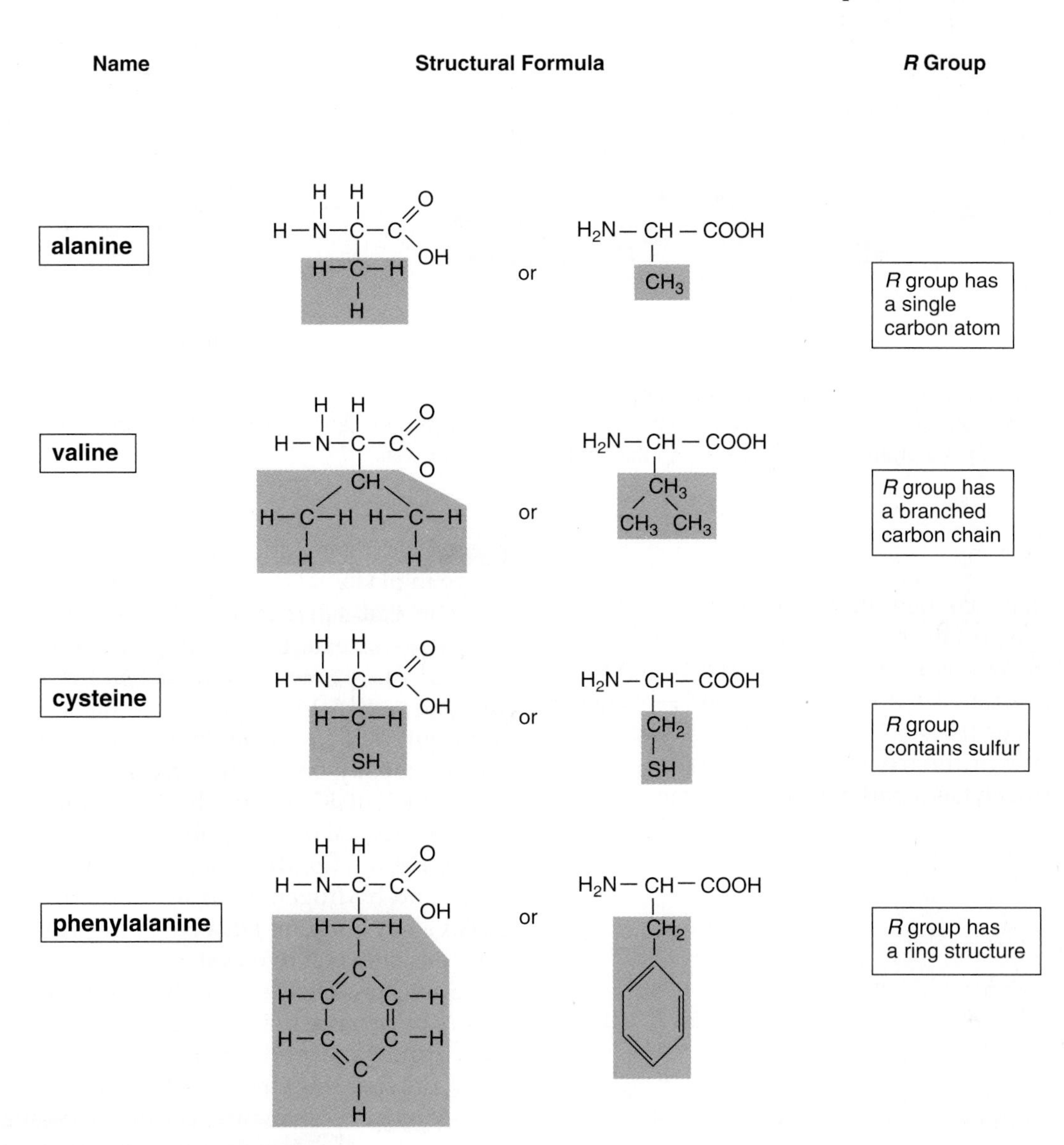

Figure 2.25 Representative amino acids.
Amino acids differ from one another by their *R* group; the simplest *R* group is a single hydrogen atom (H). The *R* groups that contain carbon vary as shown.

2.7 Proteins

Proteins sometimes have a structural function. For example, in humans, the protein keratin makes up hair and nails, whereas collagen is found in all types of connective tissue, including ligaments, cartilage, bones, and tendons. The muscles contain proteins, which account for their ability to contract.

Some proteins are enzymes, necessary contributors to the chemical workings of the cell and therefore of the body. **Enzymes** speed chemical reactions; they work so quickly that a reaction that normally takes several hours or days without an enzyme takes only a fraction of a second with an enzyme.

Proteins are macromolecules with amino acid monomers. An **amino acid** has a central carbon atom bonded to a hydrogen atom and three groups. The name of the molecule is appropriate because one of these groups is an amino group (—NH_2) and another is an acidic group (—COOH). The other group is called an *R* group because it is the *Remainder* of the molecule. Amino acids differ from one another by their *R* group; the *R* group varies from a single hydrogen (H) to a complicated ring (Fig. 2.25).

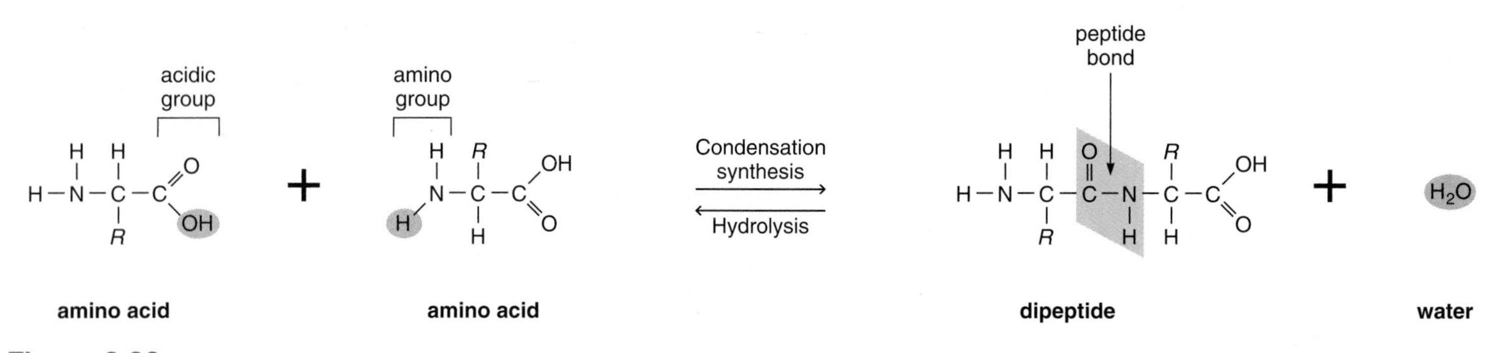

Figure 2.26 Condensation synthesis and hydrolysis of a dipeptide.
The two amino acids on the left-hand side of the equation differ by their *R* groups. As these amino acids join, a peptide bond forms, and a water molecule is produced. During hydrolysis, water is added, and the peptide bond is broken.

Peptides

Figure 2.26 shows that a condensation synthesis reaction between two amino acids results in a dipeptide and a molecule of water. A bond that joins two amino acids is called a **peptide bond.** The atoms associated with a peptide bond—oxygen (O), carbon (C), nitrogen (N), and hydrogen (H)—share electrons in such a way that the oxygen has a partial negative charge and the hydrogen has a partial positive charge.

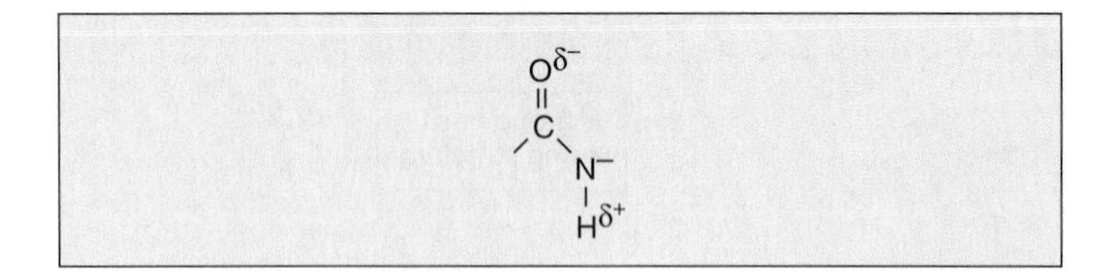

Therefore, the peptide bond is polar, and hydrogen bonding is possible between the C=O of one amino acid and the N—H of another amino acid in a polypeptide. A **polypeptide** is a single chain of amino acids.

Levels of Protein Organization

The structure of a protein has at least three levels of organization (Fig. 2.27*a–c*). The first level, called the *primary structure,* is the linear sequence of the amino acids joined by peptide bonds. Polypeptides can be quite different from one another. You will recall that the structure of a polysaccharide can be likened to a necklace that contains a single type "bead," namely, glucose. Polypeptides can make use of 20 different possible types of amino acids or "beads." Each particular polypeptide has its own sequence of amino acids. It can be said that each polypeptide differs by the sequence of its *R* groups and the number of amino acids in the sequence.

The *secondary structure* of a protein comes about when the polypeptide takes on a particular orientation in space. A coiling of the chain results in an alpha (α) helix, or a right-handed spiral, and a folding of the chain results in a pleated sheet. Hydrogen bonding between peptide bonds holds the shape in place.

The *tertiary structure* of a protein is its final three-dimensional shape. In muscles, the helical chains of myosin form a rod shape that ends in globular (globe-shaped) heads. In enzymes, the helix bends and twists in different ways. Invariably, the hydrophobic portions are packed mostly on the inside, and the hydrophilic portions are on the outside where they can make contact with water. The tertiary shape of a polypeptide is maintained by various types of bonding between the *R* groups; covalent, ionic, and hydrogen bonding all occur. One common form of covalent bonding between *R* groups is disulfide (S—S) linkages between two cysteine amino acids.

Some proteins have only one polypeptide, and some others have more than one polypeptide chain, each with its own primary, secondary, and tertiary structures. These separate polypeptides are arranged to give some proteins a fourth level of structure, termed the *quaternary structure* (Fig. 2.27*d*). Hemoglobin is a complex protein having a quaternary structure; most enzymes also have a quaternary structure.

The final shape of a protein is very important to its function. As we will discuss in chapter 6, for example, enzymes cannot function unless they have their usual shape. When proteins are exposed to extremes in heat and pH, they undergo an irreversible change in shape called **denaturation**. For example, we are all aware that the addition of acid to milk causes curdling and that heating causes egg white, which contains a protein called albumin, to coagulate. Denaturation occurs because the normal bonding between the *R* groups has been disturbed. Once a protein loses its normal shape, it is no longer able to perform its usual function.

Proteins, which contain covalently linked amino acids, are important in the structure and the function of cells. Some proteins are enzymes, which speed chemical reactions.

Visual Focus

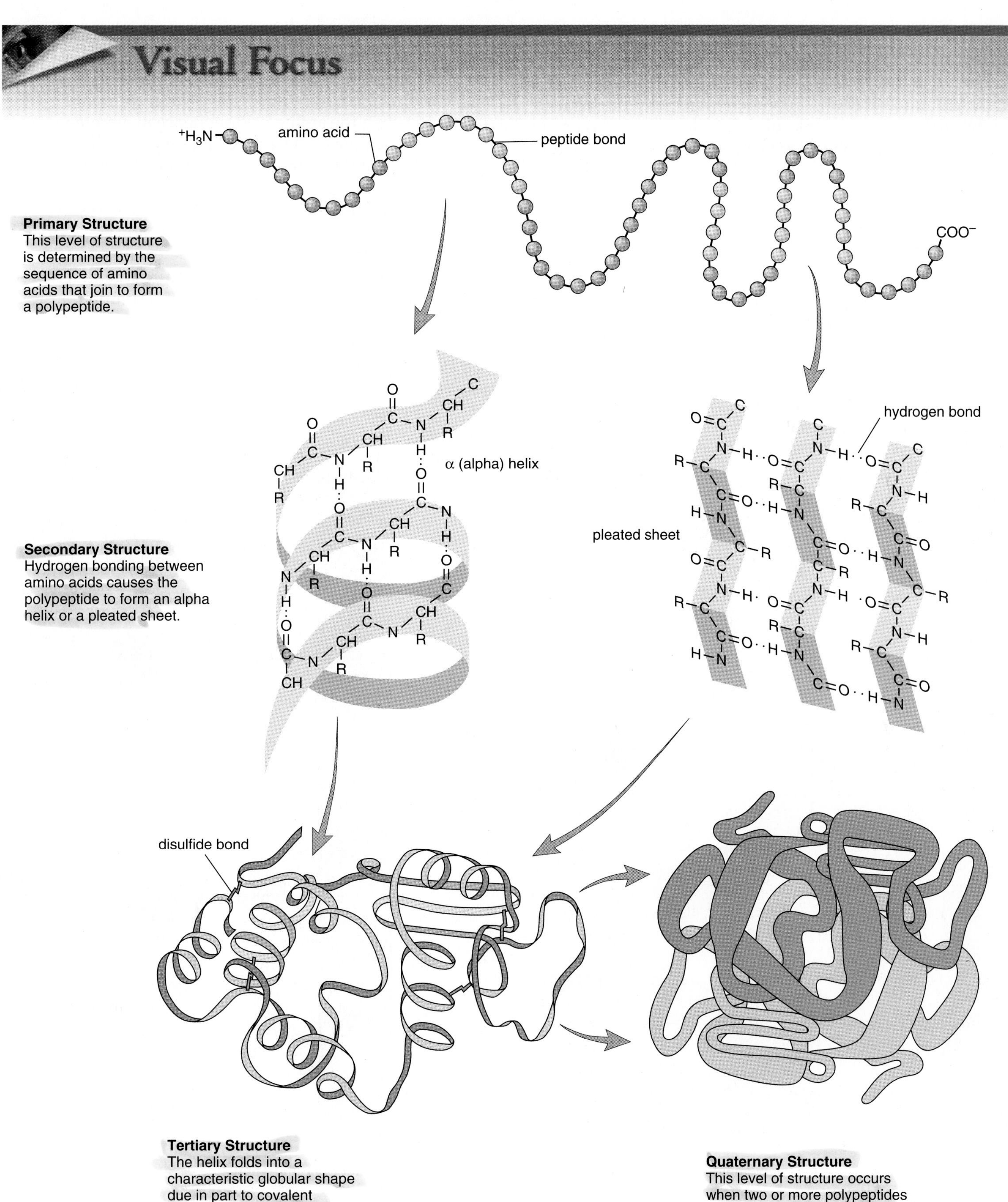

Figure 2.27 **Levels of protein organization.**

2.8 Nucleic Acids

Nucleic acids, such as **DNA (deoxyribonucleic acid)** and **RNA (ribonucleic acid),** are huge polymers of nucleotides. Every **nucleotide** is a molecular complex of three types of subunit molecules: phosphate (phosphoric acid), a pentose sugar, and a nitrogen-containing base:

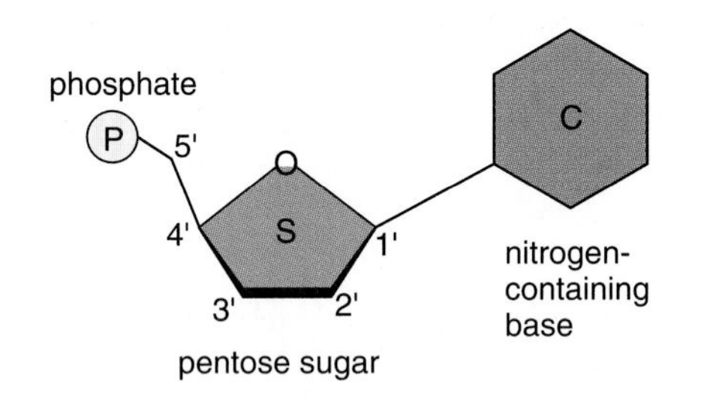

DNA makes up the genes and stores information regarding its own replication and the order in which amino acids are to be joined to form a protein. RNA is an intermediary in the process of protein synthesis, conveying information from DNA regarding the amino acid sequence in a protein.

The nucleotides in DNA contain the sugar deoxyribose, and in RNA they contain the sugar ribose; this difference accounts for their respective names (Table 2.3). As indicated in Figure 2.28, there are four different types of bases in DNA: A = **adenine,** T = **thymine,** G = **guanine,** and C = **cytosine.** The base can have two rings (adenine or guanine), or one ring (thymine or cytosine) These structures are called bases because their presence raises the pH of a solution. In RNA the base **uracil** replaces the base thymine.

Although the sequence can vary between molecules, any particular DNA or RNA has a definite sequence. The nucleotides form a linear molecule called a strand in

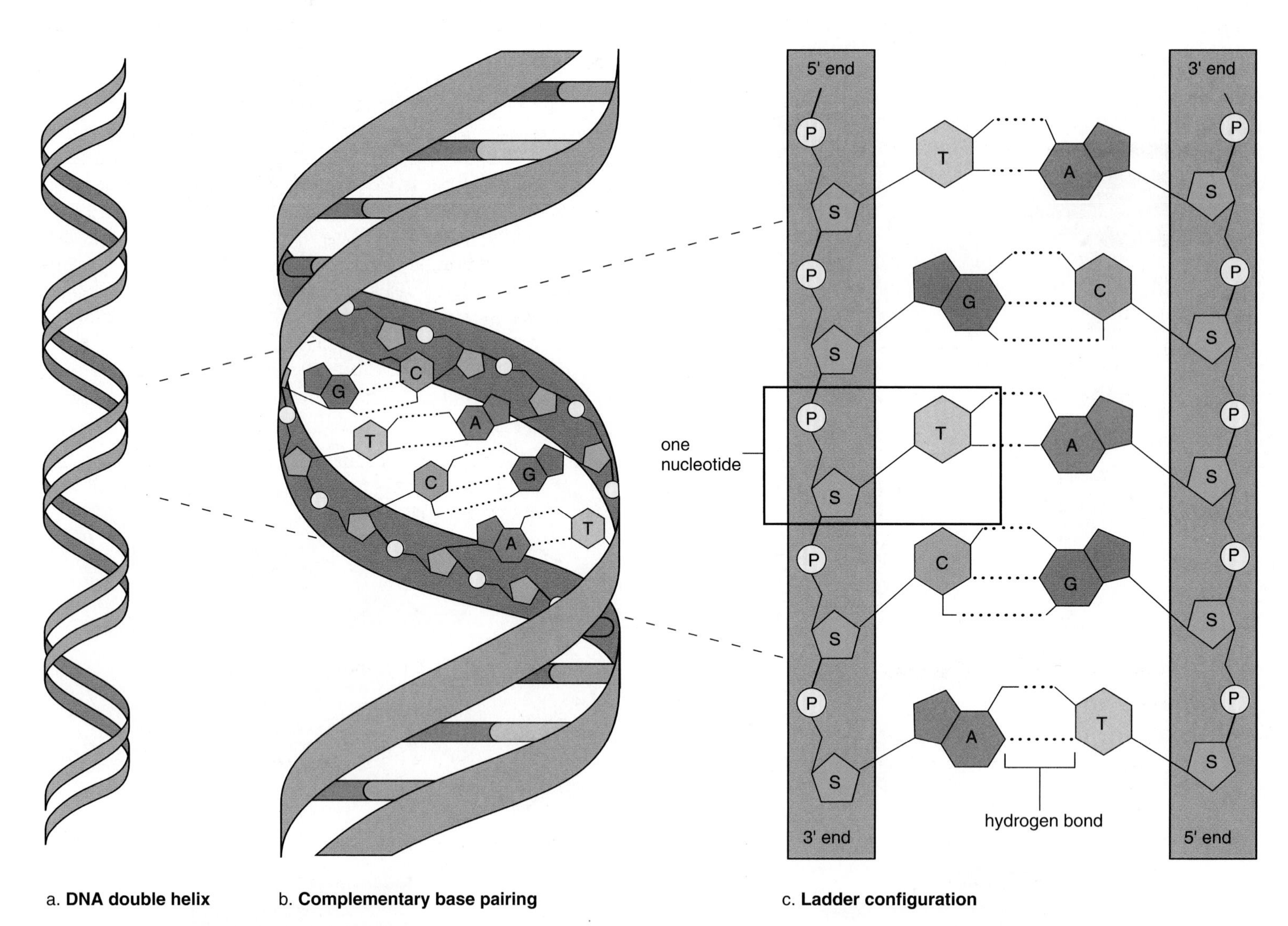

Figure 2.28 Overview of DNA structure.
a. Double helix. **b.** Complementary base pairing between strands. **c.** Ladder configuration. Notice that the uprights are composed of phosphate and sugar molecules and that the rungs are complementary paired bases.

which the backbone is made up of phosphate-sugar-phosphate-sugar, with the bases projecting to one side of the backbone. Since the nucleotides occur in a definite order, so do the bases.

RNA is usually single stranded, while DNA is usually double stranded, with the two strands twisted about each other in the form of a double helix. In DNA, the two strands are held together by hydrogen bonds between the bases. When unwound, DNA resembles a stepladder. The sides of the ladder are made entirely of phosphate and sugar molecules, and the rungs of the ladder are made only of complementary paired bases. Thymine (T) always pairs with adenine (A), and guanine (G) always pairs with cytosine (C) (Fig. 2.28). Complementary bases have shapes that fit together.

We shall see that complementary base pairing allows DNA to replicate in a way that assures the sequence of bases will remain the same. The sequence of the bases is the genetic information that specifies the sequence of amino acids in the proteins of the cell.

DNA has a structure like a twisted ladder: sugar and phosphate molecules make up the sides, and hydrogen-bonded bases make up the rungs of the ladder.

Table 2.3 DNA Structure Compared to RNA Structure

	DNA	RNA
Sugar	Deoxyribose	Ribose
Bases	Adenine, guanine, thymine, cytosine	Adenine, guanine, uracil, cytosine
Strands	Double stranded with base pairing	Single stranded
Helix	Yes	No

ATP (Adenosine Triphosphate)

In addition to being the monomers of nucleic acids, nucleotides alone have metabolic functions in cells. When adenosine (adenine plus ribose) is modified by the addition of three phosphate groups instead of one, it becomes **ATP (adenosine triphosphate),** an energy carrier in cells. A glucose molecule contains too much energy to be used as a direct energy source in cellular reactions. Instead, the energy of glucose is converted to that of ATP molecules. ATP contains an amount of energy that makes it useful to supply energy for chemical reactions in cells. As an analogy, consider that twenty dollar bills are more useful for everyday purchases than one hundred dollar bills.

ATP is sometimes called a high-energy molecule because the last two phosphate bonds are unstable and are easily broken. Usually in cells the terminal phosphate bond is hydrolyzed, leaving the molecule **ADP (adenosine diphosphate)** and a molecule of inorganic phosphate Ⓟ (Fig. 2.29). The terminal bond is sometimes called a high-energy bond, symbolized by a wavy line. But this terminology is misleading—the breakdown of ATP releases energy because the products of hydrolysis (ADP and Ⓟ) are more stable than ATP. It is the entire molecule that releases energy and not a particular bond.

The energy that is released by ATP breakdown is used by the cell for the synthesis of macromolecules like carbohydrates and proteins. In muscle cells, the energy is used for muscle contraction, and in nerve cells, it is used for the conduction of nerve impulses. ATP is called the energy currency of cells because when cells carry out an energy-requiring activity or build molecules, they often "spend" ATP as an energy source. After ATP breaks down, it is rebuilt by the addition of Ⓟ to ADP (Fig. 2.29).

ATP is a high-energy molecule. ATP breaks down to ADP + Ⓟ, releasing energy, which is used for all metabolic work done in a cell.

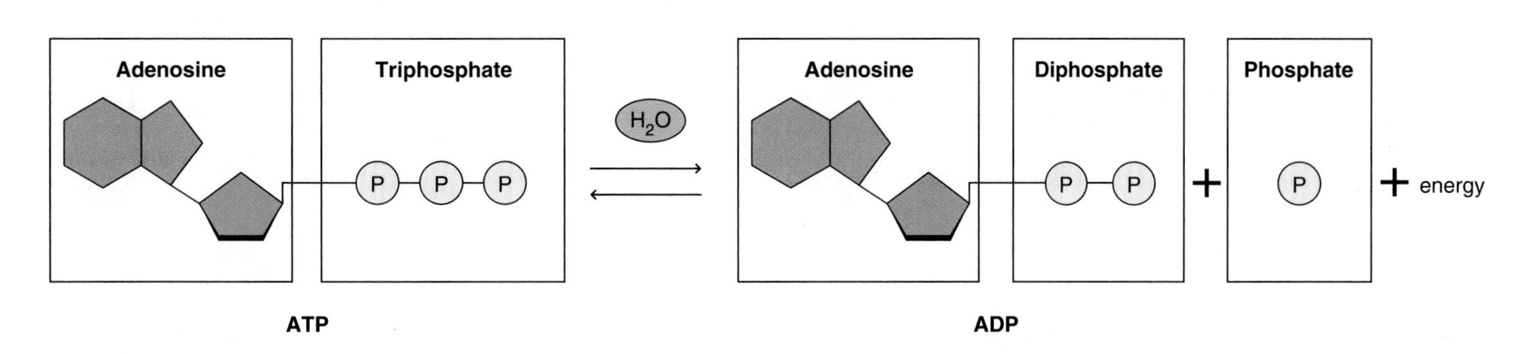

Figure 2.29 ATP reaction.
ATP, the universal energy currency of cells, is composed of adenosine and three phosphate groups. When cells require energy, ATP undergoes hydrolysis, producing ADP + Ⓟ, with the release of energy.

Bioethical Issue

Eric Stevenson and more than 110,000 veterans of the Gulf War are mysteriously ill. They complain of conditions like skin rashes, breathing difficulties, fatigue, diarrhea, muscle and joint pain, headaches and loss of memory. While the Pentagon doesn't recognize what is called the Gulf War Syndrome (GWS), it does admit that soldiers were exposed to at least four categories of chemicals:

- Petroleum products such as kerosene, diesel fuel, leaded gasoline and smoke from oil-well fires.
- Pesticides and insect repellents. These were applied to clothing and sprayed into the air.
- Drugs and vaccines. Pyridostigmine bromide was given to protect against nerve gas. Vaccines against anthrax and botulism were given in case of biological warfare.
- Biological and chemical weapons. Reluctantly, the Pentagon estimates that hundreds of thousands of solders may have been exposed to nerve gas released into the air when Iraqi ammunition depots were bombed.

While the military says that exposure can make you sick, they cling to the idea that toxic chemicals either kill you outright or you recover completely. Therefore, they suggest that the veterans are suffering from post-traumatic stress disorder. Lingering symptoms of stress are to be expected in a certain number of soldiers that come home from war.

Epidemiologist Robert Haley, however, has studied the effects of these chemicals on chickens, and found that their combination does produce Gulf War Syndrome symptoms. He says, "The Defense Department should agree that this is a physical injury, a brain injury, aided and abetted by acute stress."

For what reasons might the Pentagon be reluctant to accept the possibility that GWS is due to exposure of troops to chemicals? In so doing are they shirking their responsibility to Gulf War veterans?

Questions

1. Do you believe that veterans with GWS should be completely supported by the government? Why or why not?
2. The Pentagon has earmarked $42 million in grants for research of GWS. Do you approve of this action? Why or why not?
3. Should veterans be allowed to sue the government for their illness, much as tobacco users have sued tobacco companies? Why or why not?

Summarizing the Concepts

2.1 Elements and Atoms

All matter is composed of some 92 elements. Each element is made up of just one type atom. An atom has a weight, which is dependent on the number of protons and neutrons in the nucleus, and its chemical properties are dependent on the number of electrons in the outer shell.

2.2 Molecules and Compounds

Atoms react with one another by forming ionic bonds or covalent bonds. Ionic bonds are an attraction between charged ions. Atoms share electrons in covalent bonds, which can be single, double, or triple bonds. Oxidation is the loss of electrons (hydrogen atoms), and reduction is the gain of electrons (hydrogen atoms).

2.3 Water and Living Things

Water, acids, and bases are important inorganic molecules. The polarity of water accounts for it being the universal solvent; hydrogen bonding accounts for it boiling at 100°C and freezing at 0°C. Because it is slow to heat up and slow to freeze, it is liquid at the temperature of living things.

Pure water has a neutral pH; acids increase the hydrogen ion concentration $[H^+]$ but decrease the pH, and bases decrease the hydrogen ion concentration $[H^+]$ but increase the pH of water.

2.4 Organic Molecules

The chemistry of carbon accounts for the chemistry of organic compounds. Carbohydrates, lipids, proteins, and nucleic acids are macromolecules with specific functions in cells.

2.5 Carbohydrates

Glucose is the six-carbon sugar most utilized by cells for "quick" energy. Like the rest of the macromolecules to be studied, condensation synthesis joins two or more sugars, and a hydrolysis reaction splits the bond. Plants store glucose as starch, and animals store glucose as glycogen. Humans cannot digest cellulose, which forms plant cell walls.

2.6 Lipids

Lipids are varied in structure and function. Fats and oils, which function in long-term energy storage, contain glycerol and three fatty acids. Fatty acids can be saturated or unsaturated. Plasma membranes contain phospholipids that have a polarized end. Certain hormones are derived from cholesterol, a complex ring compound.

2.7 Proteins

The primary structure of a polypeptide is its own particular sequence of the possible 20 types of amino acids. The secondary structure is often an alpha (α) helix. The tertiary structure occurs when a polypeptide bends and twists into a three-dimensional shape. A protein can contain several polypeptides, and this accounts for a possible quaternary structure.

2.8 Nucleic Acids

Nucleic acids are polymers of nucleotides. Each nucleotide has three components: a sugar, a base, and phosphate (phosphoric acid). DNA, which contains the sugar deoxyribose, is the genetic material that stores information for its own replication and for the order in which amino acids are to be sequenced in proteins. DNA, with the help of RNA, specifies protein synthesis.

ATP, with its unstable phosphate bonds, is the energy currency of cells. Hydrolysis of ATP to ADP + Ⓟ releases energy that is used by the cell to do metabolic work.

Studying the Concepts

1. Name the subatomic particles of an atom; describe their charge, weight, and location in the atom. 20–21
2. Give an example of an ionic reaction, and explain it. 22–23
3. Diagram the atomic structure of calcium, and explain how it can react with two chlorine atoms. 23
4. Give an example of a covalent reaction, and explain it. 24–25
5. Relate the characteristics of water to its polarity and hydrogen bonding between water molecules. 26–27
6. On the pH scale, which numbers indicate a basic solution? An acidic solution? Why? 29
7. What are buffers, and why are they important to life? 29
8. Relate the variety of organic compounds to the bonding capabilities of carbon. 31
9. Name the four classes of organic molecules in cells, and relate them to macromolecules and also polymers. 31
10. Name some monosaccharides, disaccharides, and polysaccharides, and state some general functions for each. What is the most common monomer for polysaccharides? 32–33
11. How is a neutral fat synthesized? What is a saturated fatty acid? An unsaturated fatty acid? What is the function of fats? 34
12. Relate the structure of a phospholipid to that of a neutral fat. What is the function of a phospholipid? 35
13. What is the general structure and significance of cholesterol? 35
14. What are some functions of proteins? What is a peptide bond, a dipeptide, and a polypeptide? 37–38
15. Discuss the primary, secondary, and tertiary structures of globular proteins. 38
16. Discuss the structure and function of the nucleic acids, DNA and RNA. 40–41

Testing Yourself

Choose the best answer for each question.

1. The atomic number tells you the
   a. number of neutrons in the nucleus.
   b. number of protons in the nucleus.
   c. weight of the atom.
   d. number of protons in the outer shell.
   e. All of these are correct.
2. In the molecule

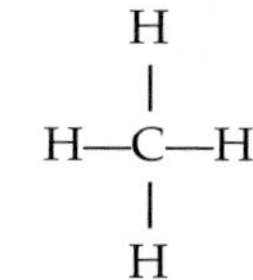

   a. all atoms have eight electrons in the outer shell.
   b. all atoms are sharing electrons.
   c. carbon could accept more hydrogen atoms.
   d. All of these are correct.
3. An atom that has two electrons in the outer shell would most likely
   a. share to acquire a completed outer shell.
   b. lose these two electrons and become a negatively charged ion.
   c. lose these two electrons and become a positively charged ion.
   d. gain two electrons and become a positively charged ion.
   e. gain two electrons and become a negatively charged ion.
4. Which of these properties of water is not due to hydrogen bonding between water molecules?
   a. stabilizes temperature inside and outside cell
   b. molecules are cohesive
   c. is a universal solvent
   d. ice floats on water
5. Which of these best describes the changes that occur when a solution goes from pH 5 to pH 8?
   a. The hydrogen ion concentration decreases as the solution goes from acidic to basic.
   b. The hydrogen ion concentration increases as the solution goes from basic to acidic.
   c. The hydrogen ion concentration decreases as the solution goes from basic to acidic.
   d. The hydrogen ion concentration increases as the solution goes from acidic to basic.
   e. The hydroxide ion concentration stays the same as the solution goes from acidic to basic.
6. Which of these molecules contains nitrogen (N)?
   a. glucose and fatty acids
   b. amino acids and ATP
   c. nucleotides and steroids
   d. cellulose and starch
   e. proteins and polysaccharides
7. Which of these makes cellulose nondigestible?
   a. It is a polymer of glucose subunits.
   b. It is a fibrous protein.
   c. The type of linkage between the glucose molecules.
   d. The peptide linkage between the amino acid molecules.
   e. The disulfide linkages between polymers.
8. A fatty acid is unsaturated if it
   a. contains hydrogen atoms.
   b. contains double bonds.
   c. contains an acidic group.
   d. bonds to glycogen.
   e. has a low melting point.
9. The difference between one amino acid and another is found in the
   a. amino group.
   b. carboxyl group.
   c. *R* group.
   d. peptide bond.
10. The shape of a polypeptide
    a. is maintained by bonding between parts of the polypeptide.
    b. is important to its function.
    c. is ultimately dependent upon the primary structure.
    d. involves hydrogen bonding.
    e. All of these are correct.
11. Nucleotides
    a. contain a sugar, a nitrogen-containing base, and a phosphate molecule.
    b. are the monomers for fats and polysaccharides.
    c. join together by covalent bonding between the bases.
    d. are found in DNA, RNA, and proteins.
    e. All of these are correct.

12. Label this diagram of condensation synthesis and hydrolysis.

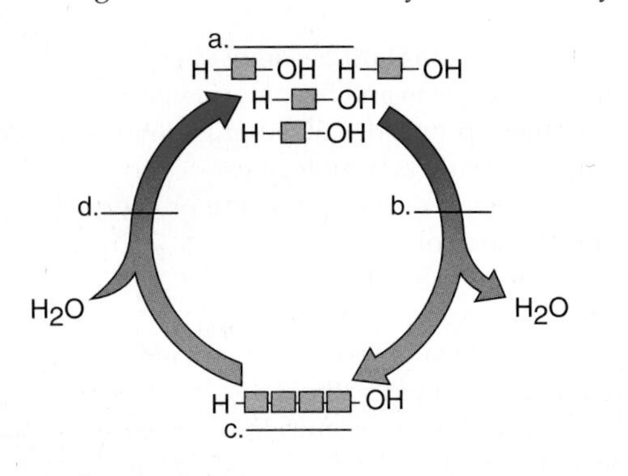

Thinking Scientifically

1. Usually we associate inorganic compounds with the nonliving world and organic compounds with the living world (page 31):
   a. What property of water makes it an inorganic compound? What property does water share with an organic compound?
   b. Show that the properties of water are essential to human life (pages 26–27).
2. Considering organic molecules:
   a. starch and glycogen are both composed of glucose. How do they differ (page 32)?
   b. oleic acid and linoleic acid are both unsaturated fatty acids. How do you predict they differ (page 34)?
   c. actin and myosin are both proteins found in muscles. How do you predict they differ? (page 38)

Understanding the Terms

acid 28
adenine 40
ADP (adenosine diphosphate) 41
amino acid 37
atom 20
atomic number 20
atomic weight 21
ATP (adenosine triphosphate) 41
base 28
buffer 29
calorie 27
carbohydrate 32
cellulose 33
condensation synthesis 32
covalent bond 24
cytosine 40
denaturation 38
disaccharide 32
DNA (deoxyribonucleic acid) 40
electron 20
element 20
emulsification 34
enzyme 37
fat 34
fatty acid 34
functional group 31
glucose 32
glycogen 32
guanine 40
hexose 32
hydrogen bond 26
hydrolysis 32
hydrophilic 26
hydrophobic 26
inorganic molecule 31
ion 23
ionic bond 23
isotope 21
lipid 34
matter 20
molecule 23
monosaccharide 32
neutron 20
nucleotide 40
oil 34
organic molecule 31
oxidation 25
pentose 32
peptide bond 38
pH scale 29
phospholipid 35
polypeptide 38
polysaccharide 32
protein 37
proton 20
redox reaction 25
reduction 25
RNA (ribonucleic acid) 40
saturated fatty acid 34
soap 34
starch 32
steroid 35
thymine 40
triglyceride 34
unsaturated fatty acid 34
uracil 40

Match the terms to these definitions:

a. ________ Polymer of many amino acids linked by peptide bonds.
b. ________ Organic catalyst that speeds up a reaction in cells due to its particular shape.
c. ________ Polysaccharide composed of glucose molecules; the chief constituent of a plant's cell wall.
d. ________ Subatomic particle that has weight of one atomic mass unit, carries no charge, and is found in the nucleus of an atom.
e. ________ Solution in which pH is less than 7; a substance that contributes or liberates hydrogen ions in a solution.

Using Technology

Your study of basic chemistry is supported by these available technologies:

Essential Study Partner CD-ROM
Cells → Chemistry
Visit the Mader web site for related ESP activities.

Exploring the Internet
The Mader Home Page provides resources and tools as you study this chapter.

http://www.mhhe.com/biosci/genbio/mader

Life Science Animations 3D Video
1 Atomic Structure and Covalent and Ionic Bonding

C H A P T E R 3

Cell Structure and Function

Artist's representation of a cell's interior illustrates its complexity. Vacuoles and the Golgi apparatus glide past a sectioned endoplasmic reticulum outside the nucleus.

Chapter Concepts

If you've ever tried to stuff a week's worth of clothing, toiletries, and other necessities into a piece of carry-on luggage, you would be amazed at what cells can pack into a space smaller than the period at the end of this sentence. Like the human body, which is composed of many trillions of cells working in harmony, cells have an internal skeleton that gives them shape and controls their movement. They harbor microscopic assembly lines that manufacture a wide array of proteins. Every cell even has its own power stations that produce energy. And nestled among such structures, often in a specialized area walled off from the rest of the cell, are the all-important chromosomes, which store the instruction manual for life. It took the development of powerful microscopes before biologists could visualize the innards of cells. Let us now take you on a guided tour.

3.1 The Cellular Level of Organization

Antonie van Leeuwenhoek of Holland, who lived in the seventeenth century, is now famous for making his own microscopes and observing all sorts of tiny things that no one had seen before. The Englishman Robert Hooke was the first to use the term **cell** for walled-off chambers he observed in cork, a material used to make stoppers today. A hundred years later—in the 1830s—the German microscopist Matthias Schleiden said that plants are composed of cells; his counterpart Theodor Schwann said that animals are also made up of cells. This was quite a feat, because aside from their own exhausting work, they had to be aware of the studies of many other microscopists. Rudolf Virchow, another German microscopist, later came to the conclusion that cells don't suddenly appear; rather, they come from preexisting cells. Today, the **cell theory,** which states that all organisms are made up of cells and that cells come only from preexisting cells, is a unifying concept of biology.

The cell marks the boundary between the nonliving and the living. The molecules that serve as food for a cell and the organic molecules that make up a cell are not alive, and yet the cell is alive. The answer to what life is will have to be found within the cell, because the smallest living organisms are unicellular, while larger organisms are multicellular—they are composed of many cells. The diversity of cells is exemplified by the many types in the human body. But despite variety of form and function, cells contain the same components. The basic elements that are common to all cells regardless of their specializations are the subject of this chapter.

All organisms are composed of self-reproducing units called cells. Microscopy revealed the presence of cells and even today is making known their detailed structure.

Three types of microscopes are most commonly used: the compound light microscope, transmission electron microscope, and scanning electron microscope (Fig. 3A). In a compound light microscope, light rays passing through a specimen are brought to a focus by a set of glass lenses, and the resulting image is then viewed by the human eye. In the transmission electron microscope, electrons passing through a specimen are brought to a focus by a set of magnetic lenses, and the resulting image is projected onto a fluorescent screen or photographic film.

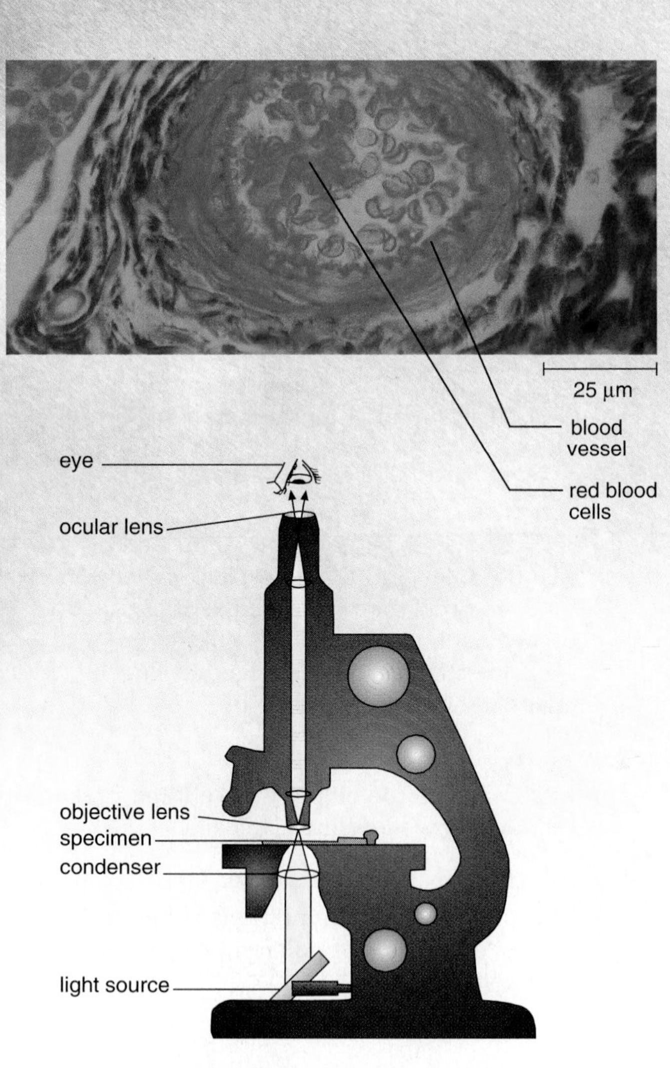

Figure 3A Blood vessels and red blood cells viewed with three different types of microscopes.

Science Focus

Microscopy of Today

The magnification produced by an electron microscope is much higher than that of a light microscope (50,000× compared to 1,000×). Also, the ability of the electron microscope to make out detail is much greater. The distance needed to distinguish two points as separate is much less for an electron microscope than a light microscope (10 nm compared to 200 nm[1]). The greater resolving power of the electron microscope is due to the fact that electrons travel at a much shorter wavelength than do light rays. However, because electrons only travel in a vacuum, the object is always dried out, whereas even living objects can be observed with a light microscope.

A scanning electron microscope provides a three-dimensional view of the surface of an object. A narrow beam of electrons is scanned over the surface of the specimen, which has been coated with a thin layer of metal. The metal gives off secondary electrons, which are collected to produce a television-type picture of the specimen's surface on a screen.

A picture obtained using a light microscope sometimes is called a photomicrograph, and a picture resulting from the use of an electron microscope is called a transmission electron micrograph (TEM) or a scanning electron micrograph (SEM), depending on the type of microscope used.

[1]See the metric system in Appendix C.

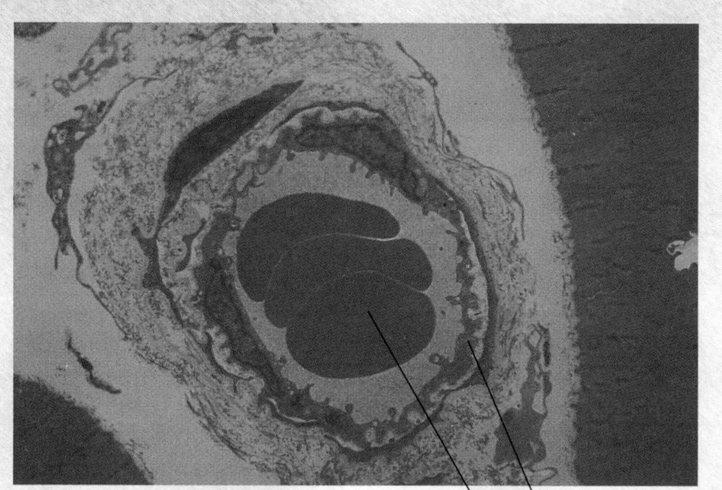

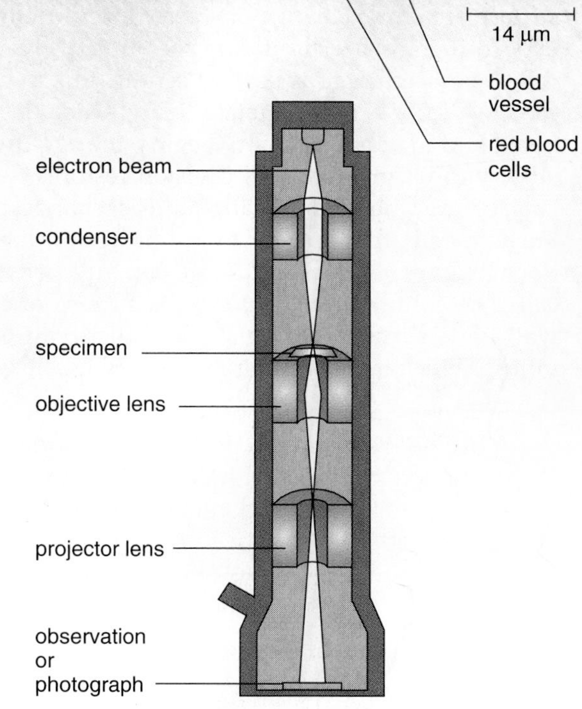

Transmission electron microscope

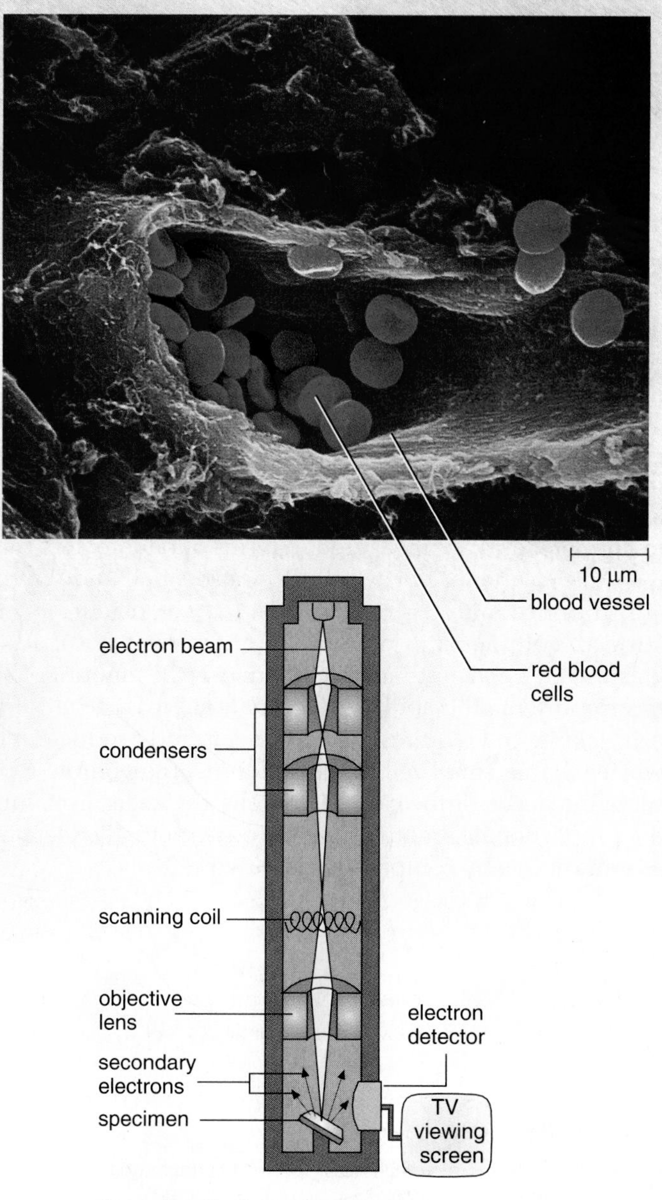

Scanning electron microscope

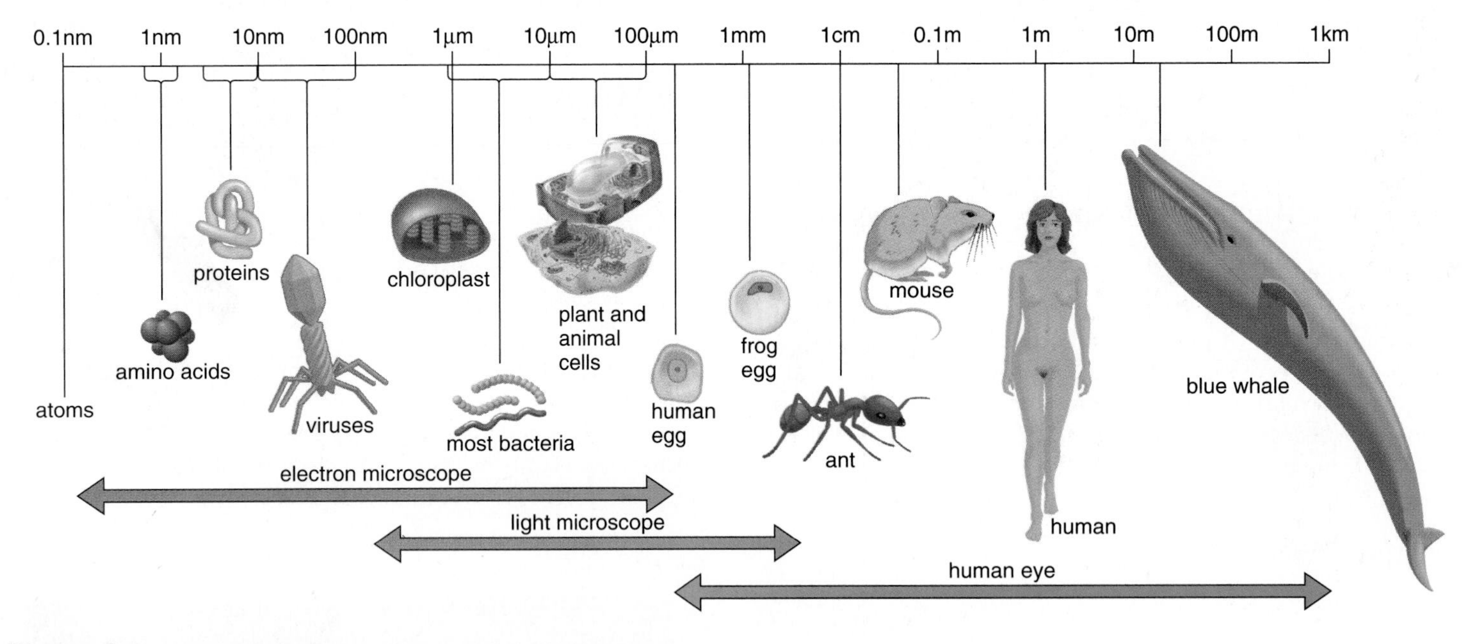

Figure 3.1 The sizes of living things and their components.
It takes a microscope to see most cells and lower levels of biological organization. Cells are visible with the light microscope, but not in much detail. It takes an electron microscope to see organelles in detail and to make out viruses and molecules. Notice that in this illustration each higher unit is 10 times greater than the lower unit. (In the metric system, 1 meter = 10^2 cm = 10^3 mm = 10^6 mm = 10^9 nm—see Appendix C.)

Cell Size

Figure 3.1 outlines the visual range of the eye, light microscope, and electron microscope. Cells are quite small. A frog's egg, at about one millimeter (mm) in diameter, is large enough to be seen by the human eye. Most cells are far smaller than one millimeter; some are even as small as one micrometer (µm)—one thousandth of a millimeter. Cell inclusions and macromolecules are even smaller than a micrometer and are measured in terms of nanometers (nm).

An explanation for why cells are so small and why we are multicellular is explained by considering the surface/volume ratio of cells. Nutrients enter a cell and wastes exit a cell at its surface; therefore, the amount of surface represents the ability to get material in and out of the cell. A large cell requires more nutrients and produces more wastes than a small cell. In other words, the volume represents the needs of the cell. Yet as cells get larger in volume, the proportionate amount of surface area actually decreases, as you can see by comparing these two cells:

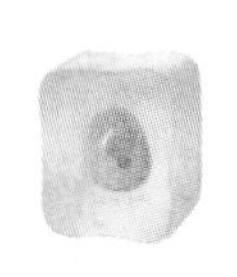

small cell—
more surface area
per volume

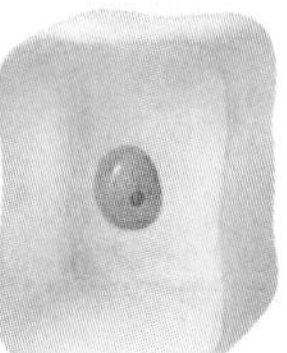

large cell—
less surface area
per volume

A small cube that is 1 mm tall has a surface area of 6 mm^2 and a volume of 1 mm^3. This is a ratio of surface area to volume of 6:1. But a cube that is 2 mm tall has a surface area of 24 mm^2 and a volume of 8 mm^3. This is a ratio of only 3:1. Therefore a small cell has more surface area per volume than does a large cell.[2]

Small cells, not large cells, are likely to have an adequate surface area for exchange of wastes for nutrients. We would expect, then, a size limitation for an actively metabolizing cell. A chicken's egg is several centimeters in diameter, but the egg is not actively metabolizing. Once the egg is incubated and metabolic activity begins, the egg divides repeatedly without growth. Cell division restores the amount of surface area needed for adequate exchange of materials. Further, cells that specialize in absorption have modifications that greatly increase the surface area per volume of the cell. The columnar cells along the surface of the intestinal wall have surface foldings called microvilli (sing., microvillus), which increase their surface area.

A cell needs a surface area that can adequately exchange materials with the environment. Surface-area-to-volume considerations require that cells stay small.

[2]surface area = 6 (1 × 1) = 6 mm^2
volume = 1 × 1 × 1 = 1 mm^2
surface area/volume = 6 : 1

surface area = 6 (2 × 2) = 24 mm^2
volume = 2 × 2 × 2 = 8 mm^2
surface area/volume = 3 : 1

3.2 Eukaryotic Cells

Eukaryotic cells have a nucleus, a large structure that controls the workings of the cell because it contains the genes.

Outer Boundaries of Animal and Plant Cells

All cells, including plant and animal cells, are surrounded by a **plasma membrane,** a phospholipid bilayer in which protein molecules are embedded.

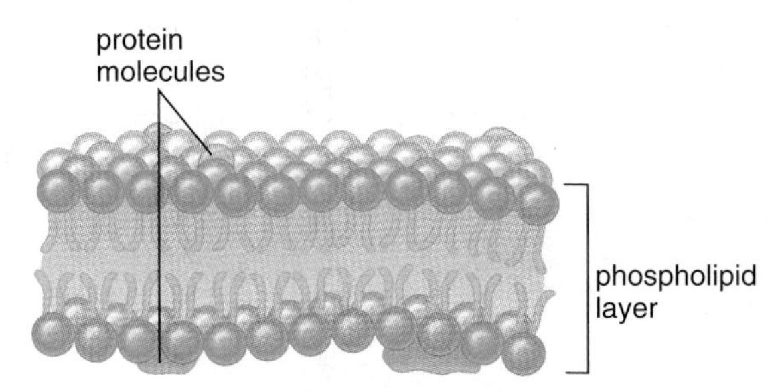

The plasma membrane is a living boundary that separates the contents of the cell from the surrounding environment. Inside the cell, the nucleus is surrounded by the **cytoplasm,** a semifluid medium that contains organelles. The plasma membrane regulates the entrance and exit of molecules into and out of the cytoplasm.

Plant cells (but not animal cells) have a permeable but protective **cell wall** in addition to a plasma membrane. Many plant cells have both a primary and secondary cell wall. A main constituent of a primary cell wall is cellulose molecules. Cellulose molecules form fibrils that lie at right angles to one another for added strength. A cell wall sometimes forms inside the primary cell wall. Secondary cell walls contain lignin, a substance that makes them even stronger than primary cell walls.

Organelles of Animal and Plant Cells

Animal and plant cells contain **organelles,** small bodies that have a specific structure and function. Originally the term organelle referred to only membranous structures, but we will use it to include any well-defined internal subcellular structure (Table 3.1). Still, membrane compartmentalizes the cell so that the various functions of the cell are kept separate from one another. Just as all the assembly lines of a factory are in operation at the same time, so all the organelles of a cell function simultaneously. Raw materials enter a factory and then are turned into various products by different departments. In the same way, chemicals are taken up by the cell and then processed by the organelles. The cell is a beehive of activity the entire twenty-four hours of every day.

Both animal (Fig. 3.2) and plant (Fig. 3.3) cells contain mitochondria, while only plant cells have chloroplasts. Only animal cells have centrioles. The color chosen to represent each structure in the plant and animal cell is used for that structure throughout the chapters of this part.

Table 3.1 Eukaryotic Structures in Animal Cells and Plant Cells

Name	Composition	Function
Cell wall*	Contains cellulose fibrils	Support and protection
Plasma membrane	Phospholipid bilayer with embedded proteins	Define cell boundary; regulation of molecule passage into and out of cells
Nucleus	Nuclear envelope surrounding nucleoplasm, chromatin, and nucleoli	Storage of genetic information; synthesis of DNA and RNA
Nucleolus	Concentrated area of chromatin, RNA, and proteins	Ribosomal subunit formation
Ribosome	Protein and RNA in two subunits	Protein synthesis
Endoplasmic reticulum (ER)	Membranous flattened channels and tubular canals	Synthesis and/or modification of proteins and other substances, and transport by vesicle formation
Rough ER	Studded with ribosomes	Protein synthesis
Smooth ER	Having no ribosomes	Various; lipid synthesis in some cells
Golgi apparatus	Stack of membranous saccules	Processing, packaging, and distribution of proteins and lipids
Vacuole and vesicle	Membranous sacs	Storage of substances
Lysosome	Membranous vesicle containing digestive enzymes	Intracellular digestion
Peroxisome	Membranous vesicle containing specific enzymes	Various metabolic tasks
Mitochondrion	Membranous cristae bounded by an outer membrane	Cellular respiration
Chloroplast*	Membranous grana bounded by two membranes	Photosynthesis
Cytoskeleton	Microtubules, intermediate filaments, actin filaments	Shape of cell and movement of its parts
Cilia and flagella	9 + 2 pattern of microtubules	Movement of cell
Centriole**	9 + 0 pattern of microtubules	Formation of basal bodies

*Plant cells only
**Animal cells only

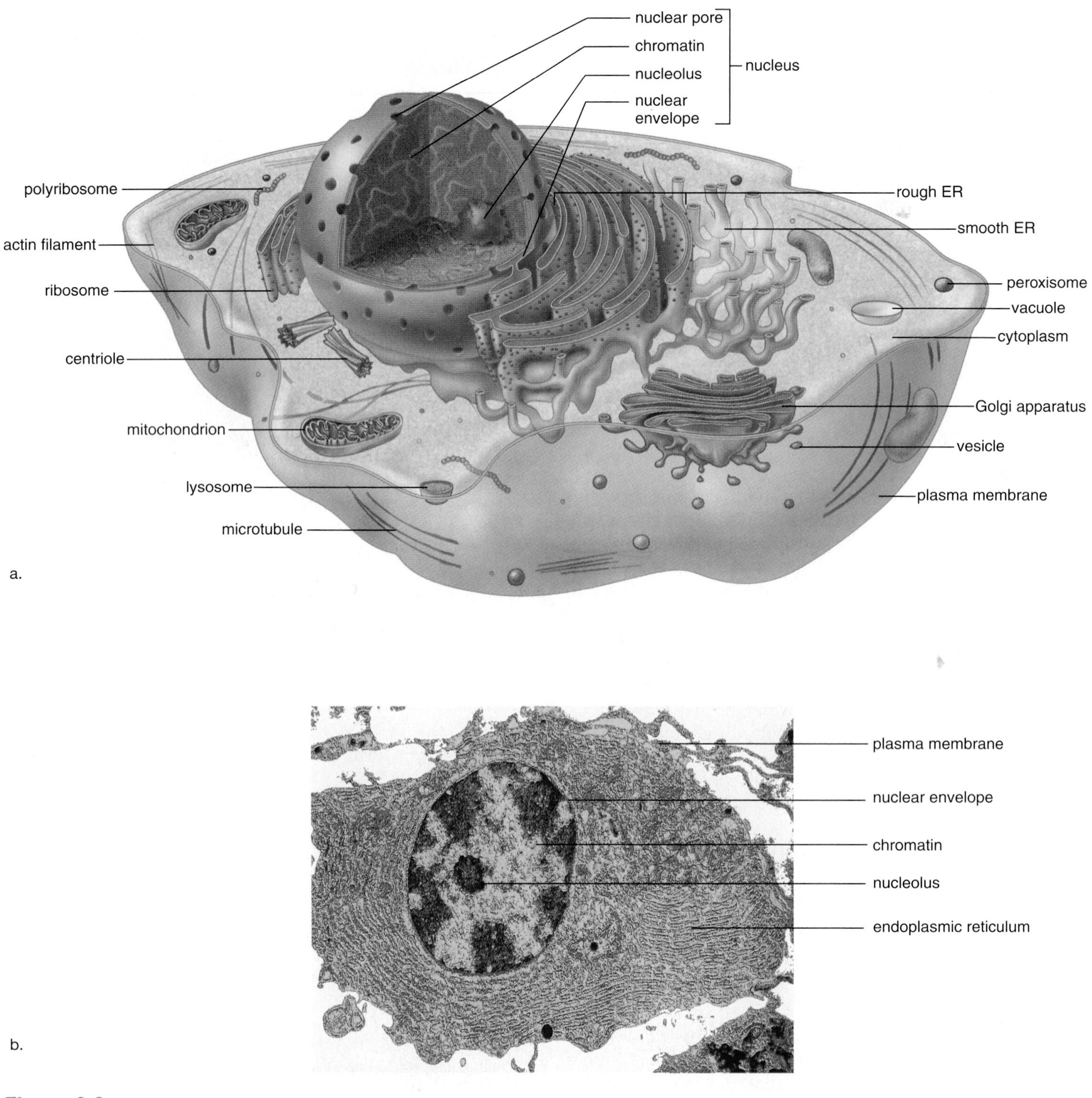

Figure 3.2 Animal cell anatomy.
a. Generalized drawing. **b.** Transmission electron micrograph. See Table 3.1 for a description of these structures, along with a listing of their functions.

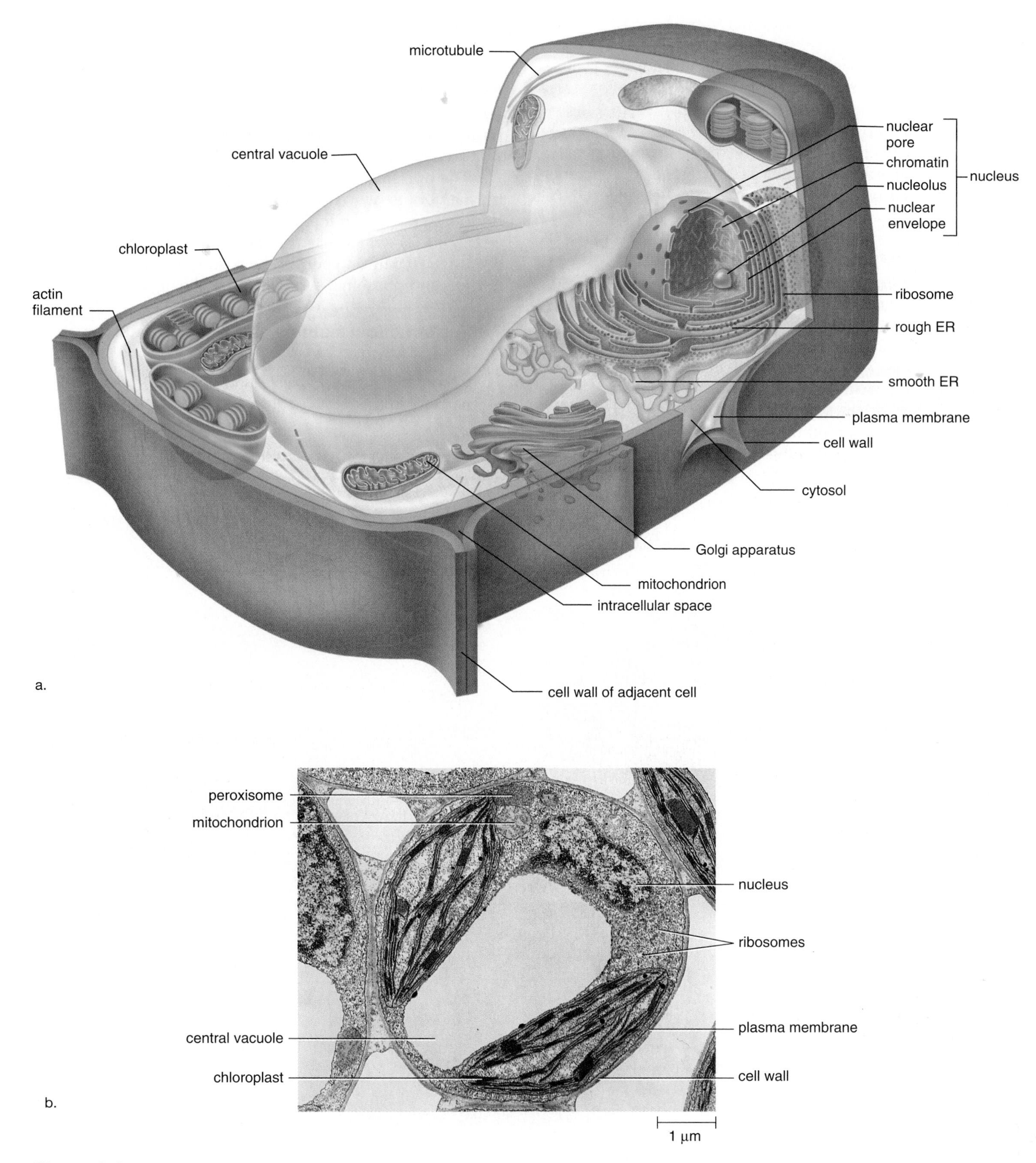

Figure 3.3 Plant cell anatomy.
a. Generalized drawing. **b.** Transmission electron micrograph of young leaf cell. See Table 3.1 for a description of these structures, along with a listing of their functions.

The Nucleus

The **nucleus,** which has a diameter of about 5 μm, is a prominent structure in the eukaryotic cell. The nucleus is of primary importance because it stores genetic information that determines the characteristics of the body's cells and their metabolic functioning. Every cell contains a complex copy of genetic information, but each cell type has certain genes, or segments of DNA, turned on, and others turned off. Activated DNA, with RNA acting as an intermediary, specifies the sequence of amino acids during protein synthesis. The proteins of a cell determine its structure and the functions it can perform.

When you look at the nucleus, even in an electron micrograph, you cannot see DNA molecules but you can see chromatin (Fig. 3.4). **Chromatin** looks grainy, but actually it is a threadlike material that undergoes coiling into rodlike structures called **chromosomes,** just before the cell divides. Chemical analysis shows that chromatin, and therefore chromosomes, contains DNA and much protein, and some RNA. Chromatin is immersed in a semifluid medium called the **nucleoplasm.** A difference in pH between the nucleoplasm and cytoplasm suggests that the nucleoplasm has a different composition.

Most likely, too, when you look at an electron micrograph of a nucleus, you will see one or more regions that look darker than the rest of the chromatin. These are nucleoli (sing., **nucleolus)** where another type of RNA, called ribosomal RNA (rRNA), is produced and where rRNA joins with proteins to form the subunits of ribosomes. (Ribosomes are small bodies in the cytoplasm that contain rRNA and proteins.)

The nucleus is separated from the cytoplasm by a double membrane known as the **nuclear envelope** which is continuous with the endoplasmic reticulum discussed on the next page. The nuclear envelope has **nuclear pores** of sufficient size (100 nm) to permit the passage of proteins into the nucleus and ribosomal subunits out of the nucleus.

The structural features of the nucleus include the following.

Chromatin:	DNA and proteins
Nucleolus:	chromatin and ribosomal subunits
Nuclear envelope:	double membrane with pores

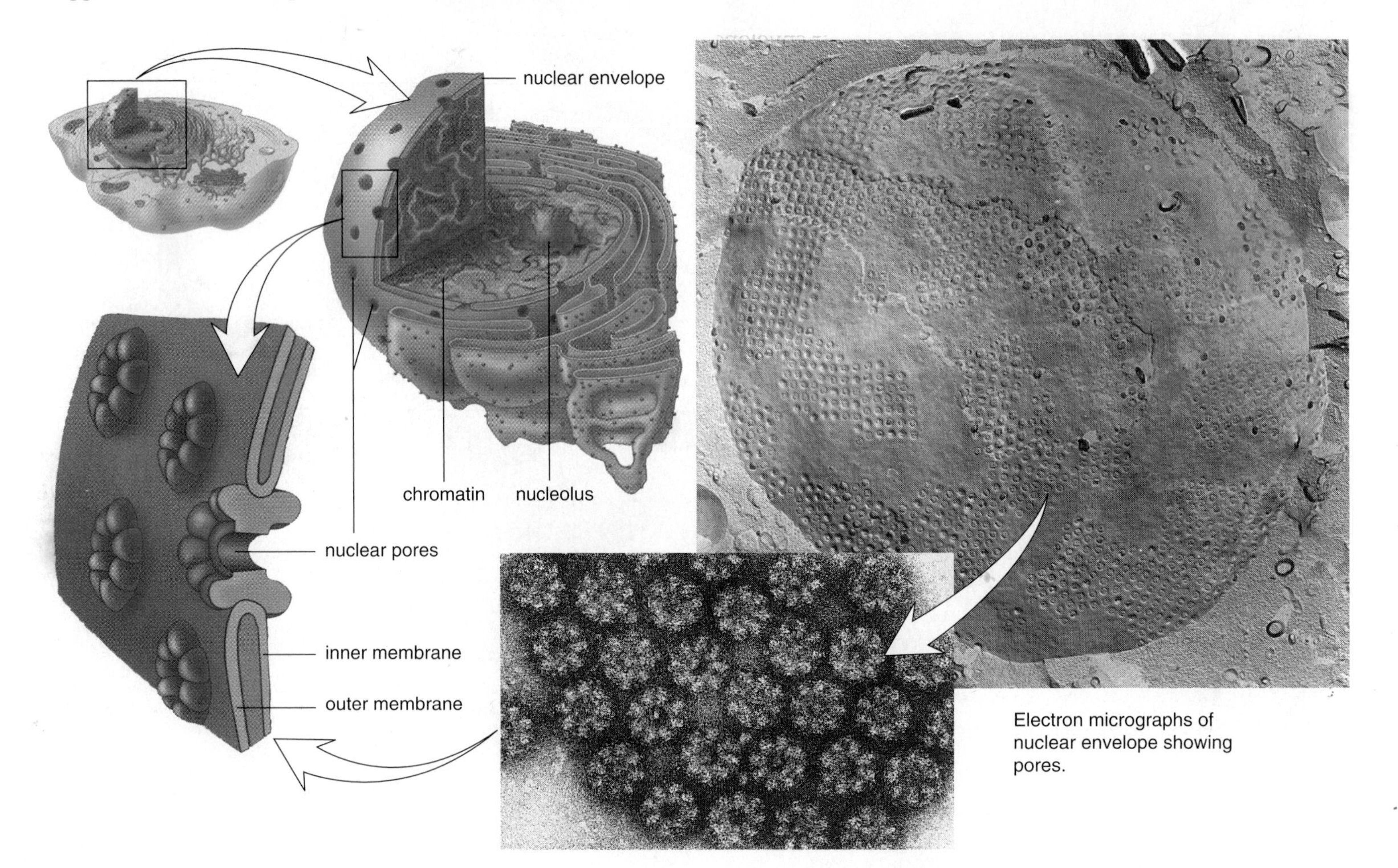

Figure 3.4 The nucleus and the nuclear envelope.
The nucleoplasm contains chromatin. Chromatin has a special region called the nucleolus, which is where rRNA is produced and ribosomal subunits are assembled. The nuclear envelope, consisting of two membranes separated by a narrow space, contains pores. The electron micrographs show that the pores cover the surface of the envelope.

Ribosomes

Ribosomes are composed of two subunits, one large and one small. Each subunit has its own mix of proteins and rRNA. Protein synthesis occurs on the ribosomes. Ribosomes occur free within the cytoplasm either singly or in groups called **polyribosomes.** Ribosomes are often attached to the endoplasmic reticulum, a membranous system of saccules and channels discussed in the next section. Proteins synthesized by cytoplasmic ribosomes are used in the cell, such as in the mitochondria and chloroplasts. Those produced by ribosomes attached to endoplasmic reticulum may eventually be secreted from the cell.

> Ribosomes are small organelles where protein synthesis occurs. Ribosomes occur in the cytoplasm, both singly and in groups (i.e., polyribosomes). Numerous ribosomes are attached to the endoplasmic reticulum.

The Endomembrane System

The endomembrane system consists of the nuclear envelope, the endoplasmic reticulum, the Golgi apparatus, and several **vesicles** (tiny membranous sacs). This system compartmentalizes the cell so that particular enzymatic reactions are restricted to specific regions. Membranes that make up the endomembrane system are connected by direct physical contact and/or by the transfer of vesicles from one part to the other.

The Endoplasmic Reticulum

The **endoplasmic reticulum** (ER), a complicated system of membranous channels and saccules (flattened vesicles), is physically continuous with the outer membrane of the nuclear envelope. Rough ER is studded with ribosomes on the side of the membrane that faces the cytoplasm (Fig. 3.5). Here proteins are synthesized and enter the ER interior where processing and modification begin. Smooth ER, which is continuous with rough ER, does not have attached ribosomes. Smooth ER synthesizes the phospholipids that occur in membranes and has various other functions depending on the particular cell. In the testes, it produces testosterone, and in the liver it helps detoxify drugs. Regardless of any specialized function, smooth ER also forms vesicles in which large molecules are transported to other parts of the cell. Often these vesicles are on their way to the plasma membrane or the Golgi apparatus.

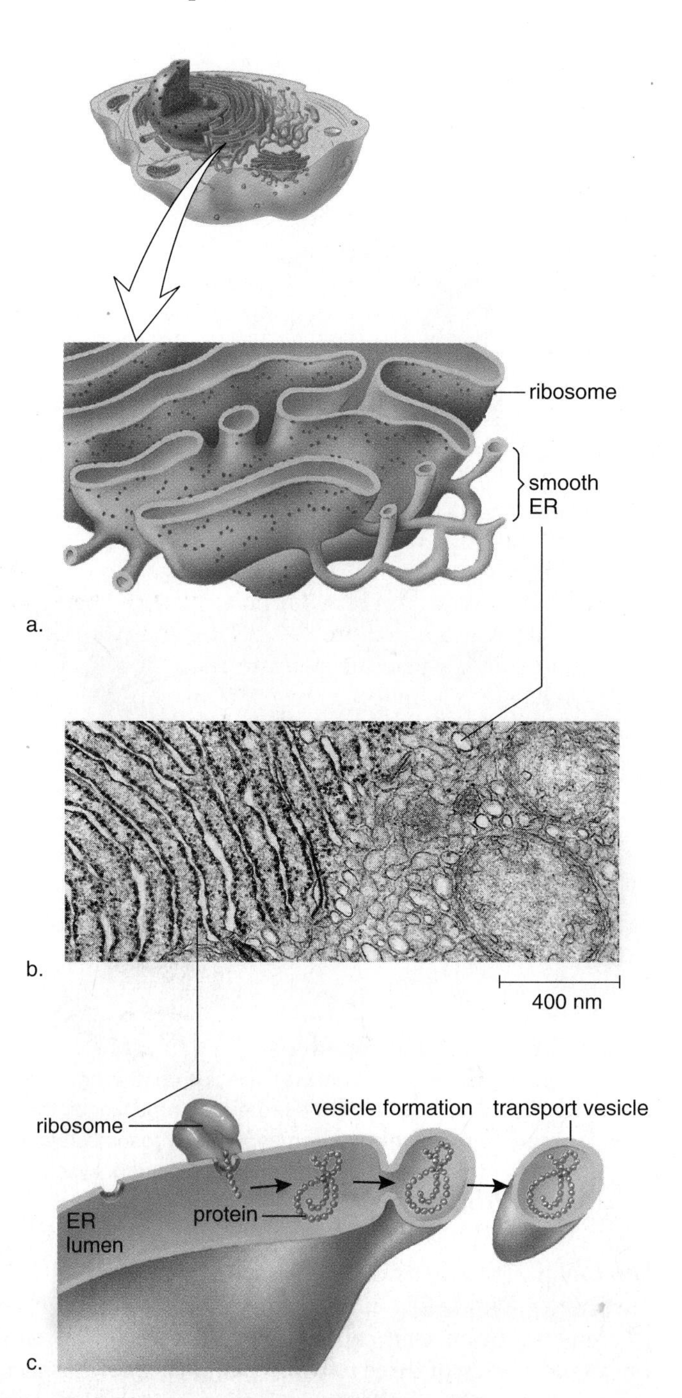

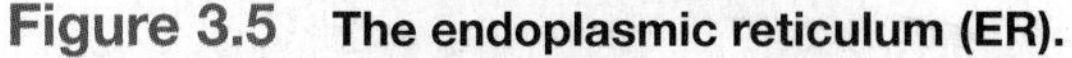

Figure 3.5 The endoplasmic reticulum (ER).
a. Rough ER has attached ribosomes but smooth ER does not. **b.** Rough ER appears to be flattened saccules, while smooth ER is a network of interconnected tubules. **c.** A protein made on a ribosome moves into the lumen of the system and eventually is packaged in a transport vesicle for distribution inside the cell.

> ER is involved in protein synthesis (rough ER) and various other processes such as lipid synthesis (smooth ER). Molecules that are produced or modified in the ER are eventually enclosed in vesicles that often transport them to the Golgi apparatus.

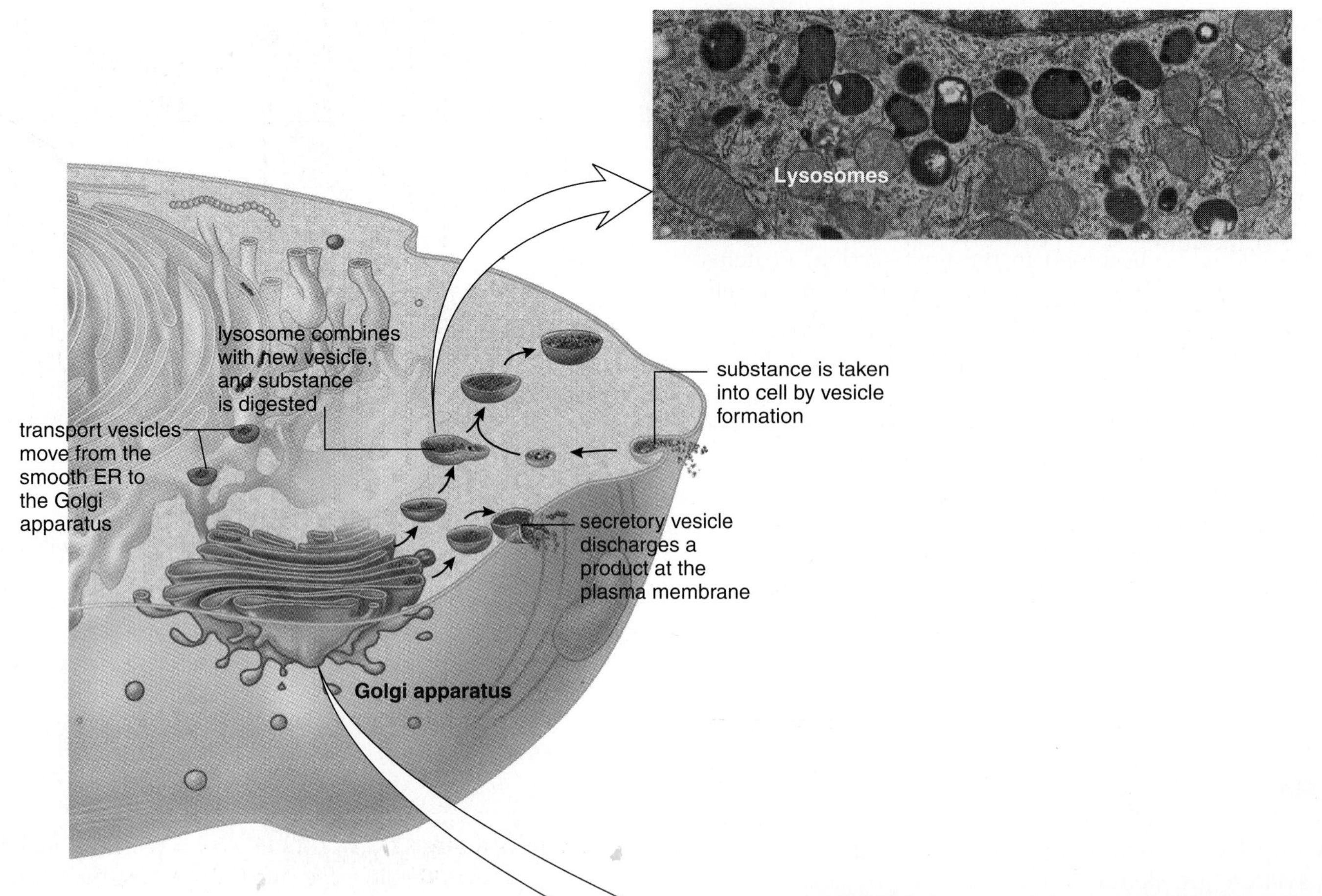

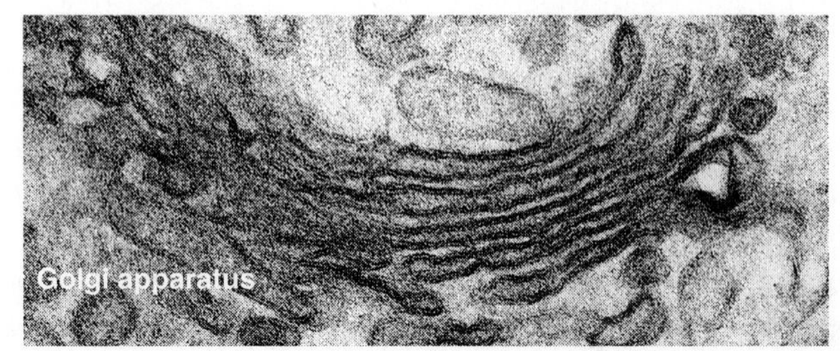

Figure 3.6 The Golgi apparatus.
The Golgi apparatus receives transport vesicles containing proteins from smooth ER. After modifying the proteins, it repackages them in either secretory vesicles or in lysosomes. When lysosomes combine with newly formed vesicles, their contents are digested. Lysosomes also break down cellular components.

The Golgi Apparatus

The **Golgi apparatus** is named for Camillo Golgi, who discovered its presence in cells in 1898. The Golgi apparatus consists of a stack of three to twenty slightly curved saccules whose appearance can be compared to a stack of pancakes (Fig. 3.6). In animal cells, one side of the stack (the inner face) is directed toward the ER, and the other side of the stack (the outer face) is directed toward the plasma membrane. Vesicles can frequently be seen at the edges of the saccules.

The Golgi apparatus receives protein and/or lipid-filled vesicles that bud from the ER. Some biologists believe that these fuse to form a saccule at the inner face and that this saccule remains as a part of the Golgi apparatus until the molecules are repackaged in new vesicles at the outer face. Others believe that the vesicles from the ER proceed directly to the outer face of the Golgi apparatus, where processing and packaging occurs within its saccules.

The Golgi apparatus contains enzymes that modify proteins and lipids. For example, it can add a chain of sugars to proteins, thereby making them glycoproteins and glycolipids, which are molecules found in the plasma membrane.

The vesicles that leave the Golgi apparatus move to different locations in the cell. Some vesicles proceed to the plasma membrane, where they discharge their contents. Because this is **secretion,** it is often said that the Golgi apparatus is involved in processing, packaging, and secretion. Other vesicles that leave the Golgi apparatus are lysosomes.

The Golgi apparatus processes, packages, and distributes molecules about or from the cell. It is also said to be involved in secretion.

Lysosomes

Lysosomes are membrane-bounded vesicles produced by the Golgi apparatus in animal cells and plant cells. Lysosomes contain hydrolytic digestive enzymes.

Sometimes macromolecules are brought into a cell by vesicle formation at the plasma membrane (Fig. 3.6). When a lysosome fuses with such a vesicle, its contents are digested by lysosomal enzymes into simpler subunits that then enter the cytoplasm. Some white blood cells defend the body by engulfing bacteria that are then enclosed within vesicles. When lysosomes fuse with these vesicles, the bacteria are digested. It should come as no surprise, then, that even parts of a cell are digested by its own lysosomes (called autodigestion). Normal cell rejuvenation most likely takes place in this manner, but programmed cell destruction occurs during development. For example, when a tadpole becomes a frog, lysosomes digest away the cells of the tail. The fingers of a human embryo are at first webbed, but they are freed from one another as a result of lysosomal action.

Occasionally, a child is born with a metabolic disorder involving a missing or inactive lysosomal enzyme. In these cases, the lysosomes fill to capacity with macromolecules that cannot be broken down. The cells become so full of these lysosomes that the child dies. Someday soon it may be possible to provide the missing enzyme for these children.

Lysosomes are produced by a Golgi apparatus, and their hydrolytic enzymes digest macromolecules from various sources.

Peroxisomes

Peroxisomes, similar to lysosomes, are membrane-bounded vesicles that enclose enzymes (Fig. 3.7). These enzymes were synthesized by free ribosomes and imported directly into a peroxisome. Peroxisomes contain enzymes for oxidizing certain organic substances with the formation of hydrogen peroxide (H_2O_2):

$$RH_2 + O_2 \rightarrow R + H_2O_2$$

Hydrogen peroxide, a toxic molecule, is immediately broken down to water and oxygen by another peroxisomal enzyme called catalase. Peroxisomes are abundant in cells that metabolize lipids and in liver cells that metabolize alcohol. They help detoxify alcohol.

Peroxisomes play additional roles in plants. In germinating seeds, they oxidize fatty acids into molecules that can be converted to sugars needed by the growing plant. In leaves, peroxisomes can carry out a reaction that is opposite to photosynthesis—the reaction uses up oxygen and releases carbon dioxide.

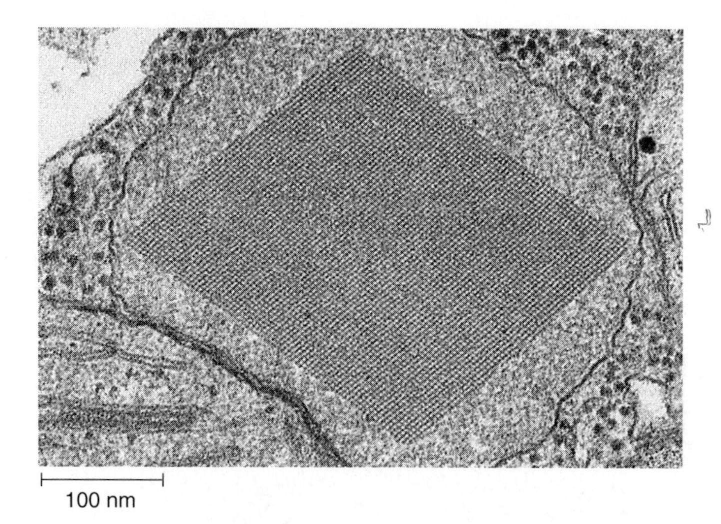

Figure 3.7 Peroxisome in a tobacco leaf.
Peroxisomes are vesicles that oxidize organic substances with a resulting build-up of hydrogen peroxide. The crystalline, squarelike core of a peroxisome contains the enzyme catalase, which breaks down hydrogen peroxide (H_2O_2) to water and oxygen.

Vacuoles

A **vacuole** is a large membranous sac. A vesicle is smaller than a vacuole. Animal cells have vacuoles, but they are much more prominent in plant cells. Typically, plant cells have a large central vacuole so filled with a watery fluid that it gives added support to the cell (see Fig. 3.3).

Vacuoles store substances. Plant vacuoles contain not only water, sugars, and salts but also pigments and toxic molecules. The pigments are responsible for many of the red, blue, or purple colors of flowers and some leaves. The toxic substances help protect a plant from herbivorous animals. The vacuoles present in unicellular protozoans are quite specialized, and they include contractile vacuoles for ridding the cell of excess water and digestive vacuoles for breaking down nutrients.

The organelles of the endomembrane system are as follows:

Endoplasmic reticulum (ER): synthesis and modification and transport of proteins and other substances
  Rough ER: protein synthesis
  Smooth ER: lipid synthesis in particular
Golgi apparatus: processing, packaging, and distribution of protein molecules
Lysosomes: intracellular digestion
Peroxisomes: various metabolic tasks
Vacuoles: storage areas

Energy-Related Organelles

Life is possible only because of a constant input of energy used for maintenance and growth. Chloroplasts and mitochondria are the two eukaryotic membranous organelles that specialize in converting energy to a form that can be used by the cell. **Chloroplasts** use solar energy to synthesize carbohydrates, and carbohydrate-derived products are broken down in mitochondria (sing., **mitochondrion**) to produce ATP molecules.

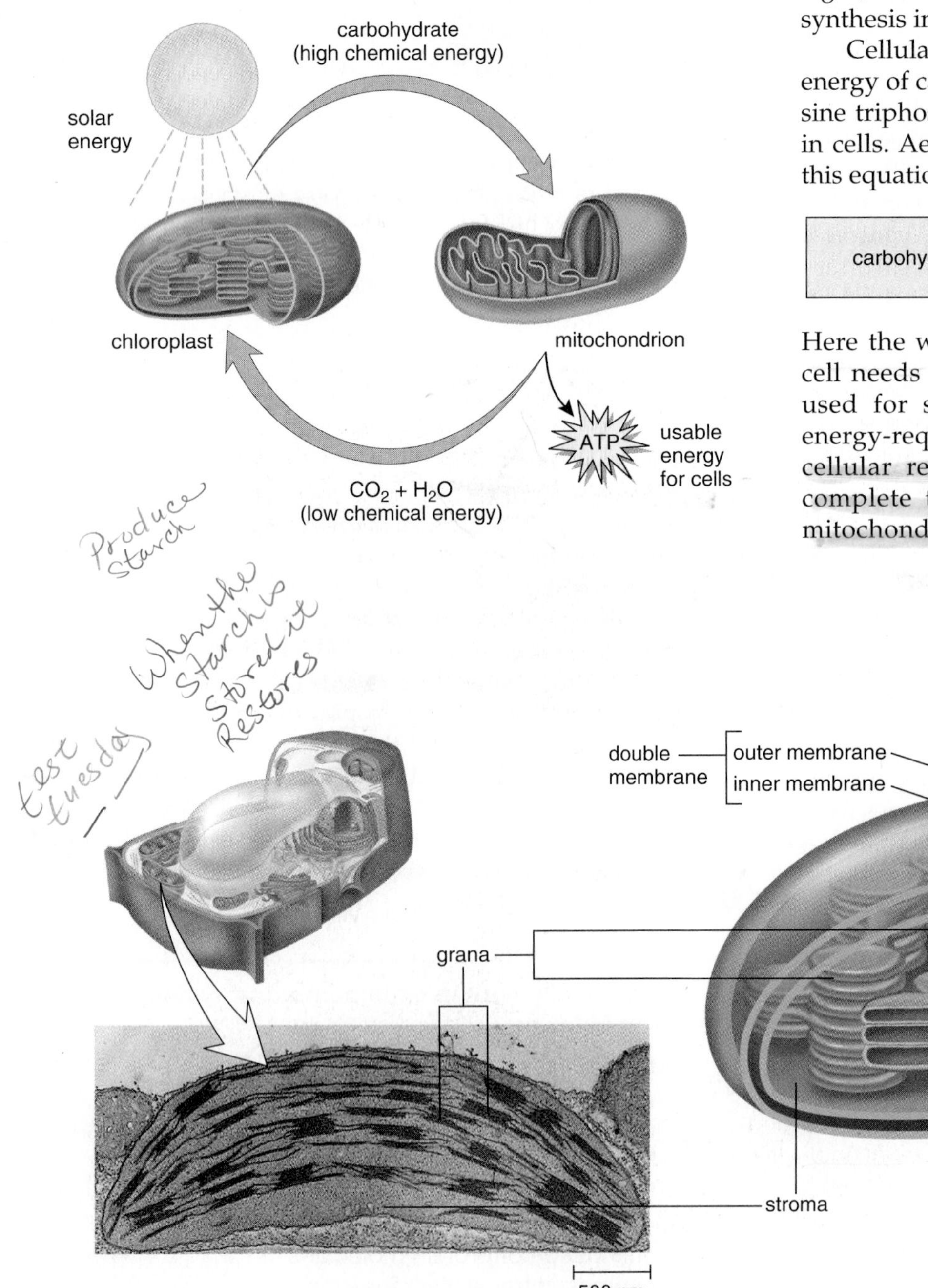

Photosynthesis, which occurs in chloroplasts, is the process by which solar energy is converted to chemical energy within carbohydrates. Photosynthesis can be represented by this equation:

light energy + carbon dioxide + water ⟶ carbohydrate + oxygen

Here the word *energy* stands for solar energy, the ultimate source of energy for cellular organization. Only plants, algae, and cyanobacteria are capable of carrying on photosynthesis in this manner.

Cellular respiration is the process by which the chemical energy of carbohydrates is converted to that of ATP (adenosine triphosphate), the common carrier of chemical energy in cells. Aerobic cellular respiration can be represented by this equation:

carbohydrate + oxygen ⟶ carbon dioxide + water + energy

Here the word *energy* stands for ATP molecules. When a cell needs energy, ATP supplies it. The energy of ATP is used for synthetic reactions, active transport, and all energy-requiring processes in cells. All organisms carry on cellular respiration, and all organisms except bacteria complete the process of aerobic cellular respiration in mitochondria.

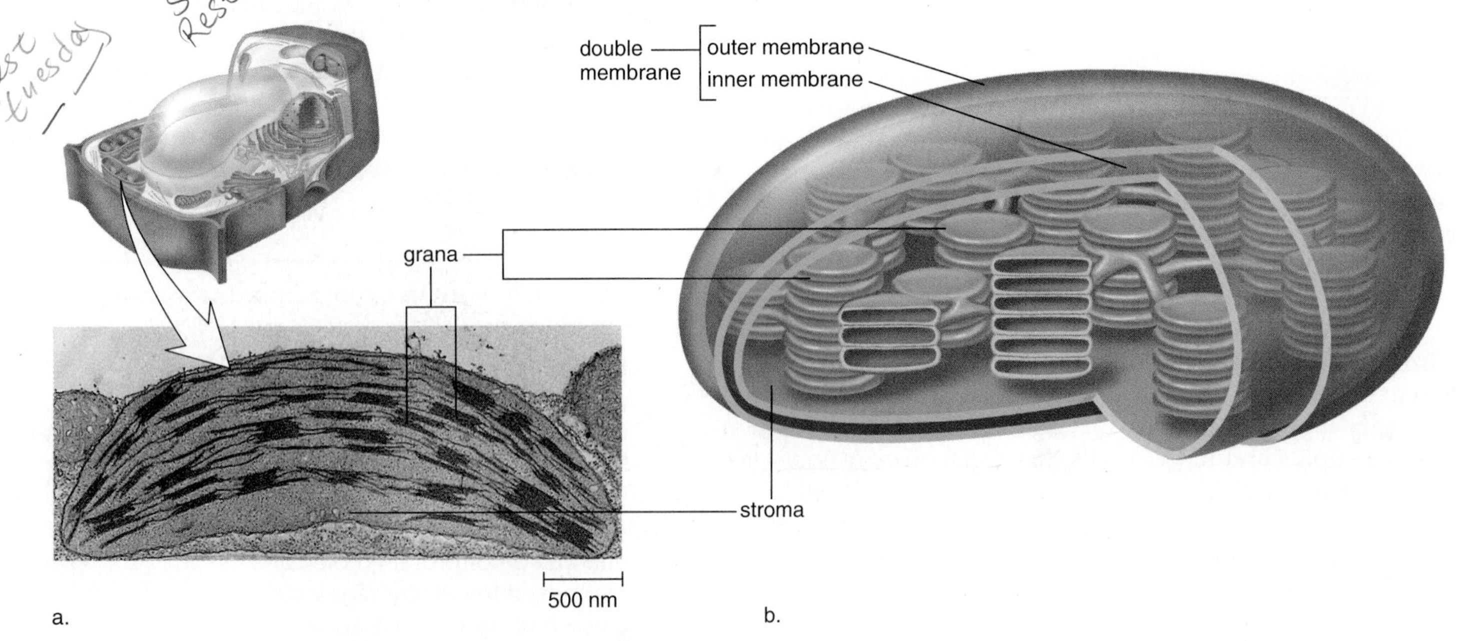

Figure 3.8 Chloroplast structure.
a. Electron micrograph. **b.** Generalized drawing in which the outer and inner membranes have been cut away to reveal the grana.

Chloroplasts

Plant cells contain chloroplasts, the organelles that allow them to produce their own organic food. Chloroplasts are about 4–6 μm in diameter and 1–5 μm in length; they belong to a group of organelles known as plastids. Among the plastids are also the *amyloplasts*, common in roots, which store starch; and the *chromoplasts*, common in leaves, which contain red and orange pigments. A chloroplast is green, of course, because it contains the green pigment chlorophyll.

A chloroplast is bounded by two membranes that enclose a fluid-filled space called the **stroma.** A membrane system within the stroma is organized into interconnected flattened sacs called **thylakoids.** In certain regions, the thylakoids are stacked up in structures called grana (sing., **granum**). There can be hundreds of grana within a single chloroplast (Fig. 3.8). Chlorophyll, which is located within the thylakoid membranes of grana, captures the solar energy that is needed to allow chloroplasts to produce carbohydrates. The stroma contains DNA, ribosomes, and enzymes that synthesize carbohydrates from carbon dioxide and water.

Mitochondria

All eukaryotic cells, including plant cells, contain mitochondria. This means that plant cells contain both chloroplasts and mitochondria. Most mitochondria are usually 0.5–1.0 μm in diameter and 2–5 μm in length.

Mitochondria, like chloroplasts, are bounded by a double membrane (Fig. 3.9). In mitochondria the inner fluid-filled space is called the **matrix.** The matrix contains DNA, ribosomes, and enzymes which break down carbohydrate products, releasing energy that is used for ATP production.

The inner membrane of a mitochondrion invaginates to form **cristae.** Cristae provide a much greater surface area to accommodate the protein complexes and other participants that produce ATP.

Mitochondria and chloroplasts are able to make some proteins, but others are imported from the cytoplasm.

Chloroplasts and mitochondria are membranous organelles whose structure lends itself to the processes that occur within them.

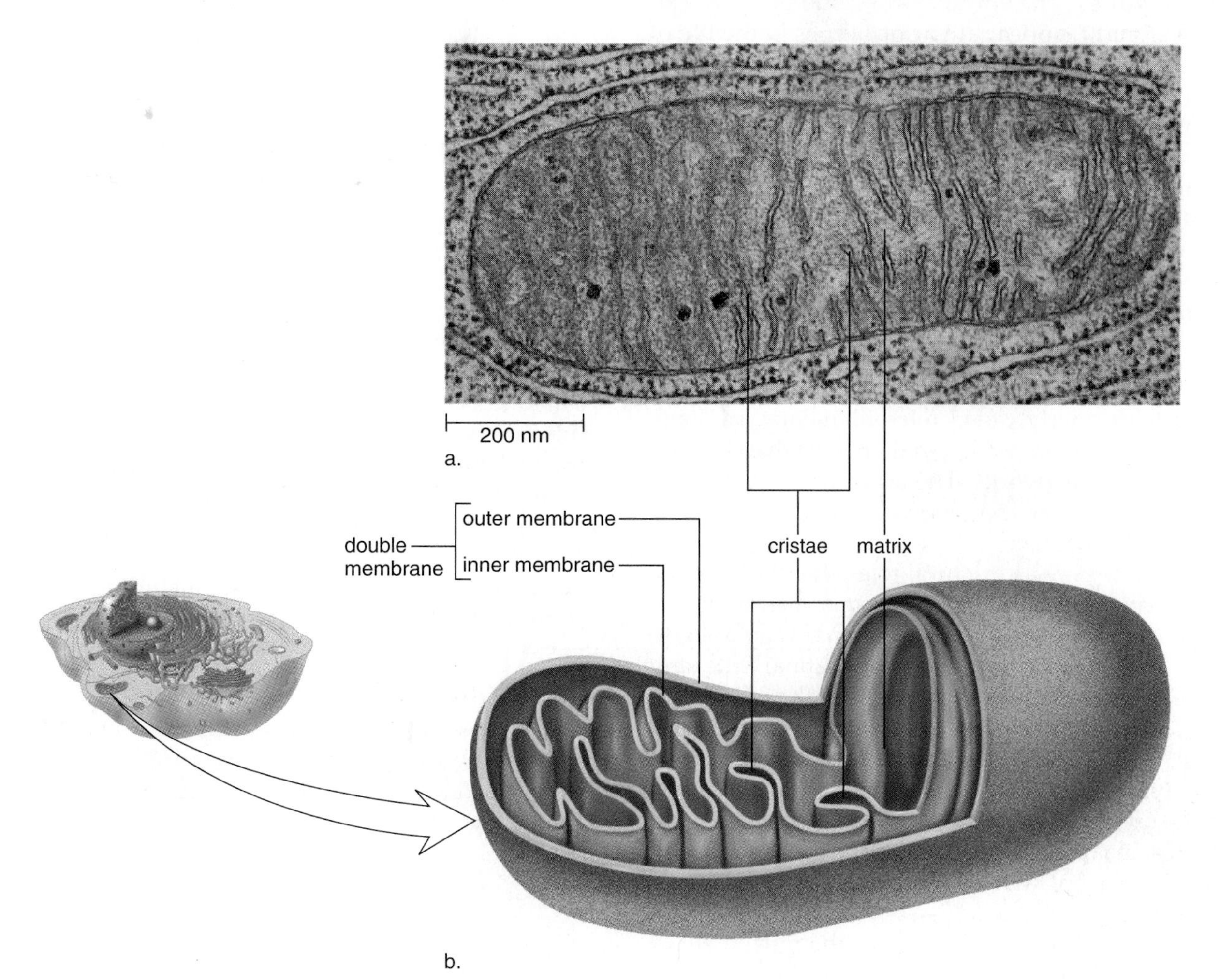

Figure 3.9 Mitochondrion structure.

a. Electron micrograph. **b.** Generalized drawing in which the outer membrane and portions of the inner membrane have been cut away to reveal the cristae.

The Cytoskeleton

The **cytoskeleton** is a network of interconnected filaments and tubules that extends from the nucleus to the plasma membrane in eukaryotic cells. Prior to the 1970s, it was believed that the cytoplasm was an unorganized mixture of organic molecules. Then, high-voltage electron microscopes, which can penetrate thicker specimens, showed that the cytoplasm was instead highly organized. The technique of immunofluorescence microscopy identified the makeup of specific protein fibers within the cytoskeletal network (Fig. 3.10).

The name *cytoskeleton* is convenient in that it allows us to compare the cytoskeleton to the bones and muscles of an animal. Bones and muscles give an animal structure and produce movement. Similarly, we will see that the elements of the cytoskeleton maintain cell shape and cause the cell and its organelles to move. The cytoskeleton is dynamic; elements undergo rapid assembly and disassembly by monomers continuously entering or leaving the polymer. These changes occur at rates that are measured in seconds and minutes. The entire cytoskeletal network can even disappear and reappear at various times in the life of a cell. Before a cell divides, for instance, the elements disassemble and then reassemble into a structure called a spindle that distributes chromosomes in an orderly manner. At the end of cell division, the spindle disassembles and the elements reassemble once again into their former array.

The cytoskeleton contains three types of elements: actin filaments, intermediate filaments, and microtubules, which are responsible for cell shape and movement.

Actin Filaments

Actin filaments (formerly called microfilaments) are long, extremely thin fibers (about 7 nm in diameter) that occur in bundles or meshlike networks. The actin filament contains two chains of globular actin monomers twisted about one another in a helical manner.

Actin filaments play a structural role when they form a dense complex web just under the plasma membrane, to which they are anchored by special proteins. They are also seen in the microvilli that project from intestinal cells, and their presence most likely accounts for the ability of microvilli to alternately shorten and extend into the intestine. In plant cells, they apparently form the tracks along which chloroplasts circulate or stream in a particular direction. Also, the presence of a network of actin filaments lying beneath the plasma membrane accounts for the formation of pseudopods, extensions that allow certain cells to move in an amoeboid fashion.

How are actin filaments involved in the movement of the cell and its organelles? They interact with **motor molecules,** which are proteins that move along either actin filaments or microtubules. These motor molecules accomplish this by attaching, detaching, and reattaching farther along the actin filament or microtubule. In the presence of ATP, the motor molecule myosin attaches, detaches, and reattaches to actin filaments. Myosin has both a head and a tail. In muscle cells, the tails of several muscle myosin molecules are joined to form a thick filament. In nonmuscle cells, cytoplasmic myosin tails are bound to membranes but the heads still interact with actin.

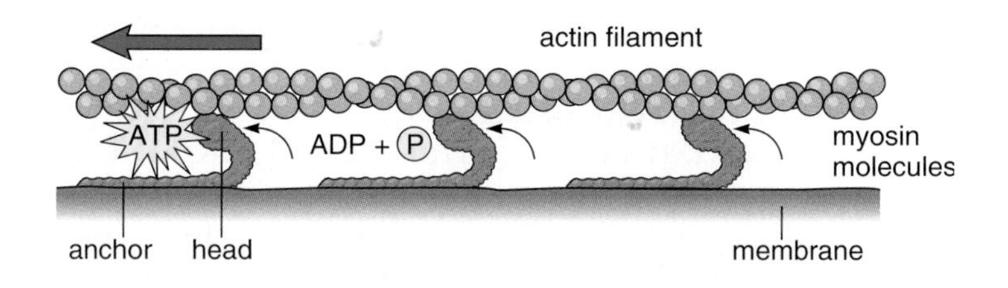

During animal cell division the two new cells form when actin, in conjunction with myosin, pinches off the cells from one another.

Microtubules

Microtubules are small hollow cylinders about 25 nm in diameter and from 0.2–25 μm in length.

Microtubules are made of a globular protein called tubulin. When microtubules assemble, tubulin molecules come together as dimers and the dimers arrange themselves in rows. Microtubules have 13 rows of tubulin dimers surrounding what appears in electron micrographs to be an empty central core.

In many cells the regulation of microtubule assembly is under the control of a microtubule organizing center, called the **centrosome,** which lies near the nucleus. Microtubules help maintain the shape of the cell and act as tracks along which organelles can move. Whereas the motor molecule myosin is associated with actin filaments, the motor molecules kinesin and dynein move along microtubules. One type of kinesin is responsible for moving vesicles along microtubules, including those that arise from the ER.

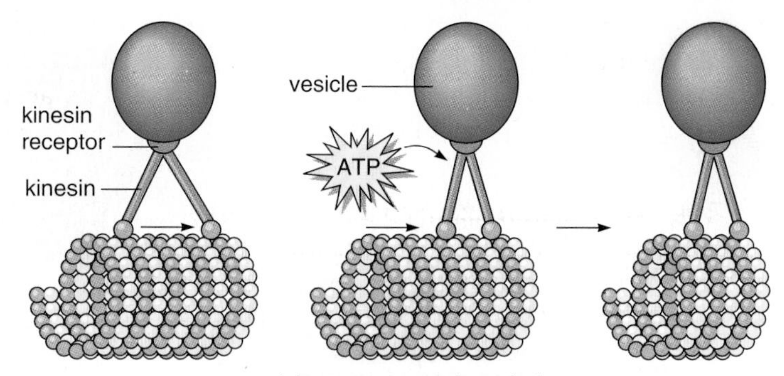

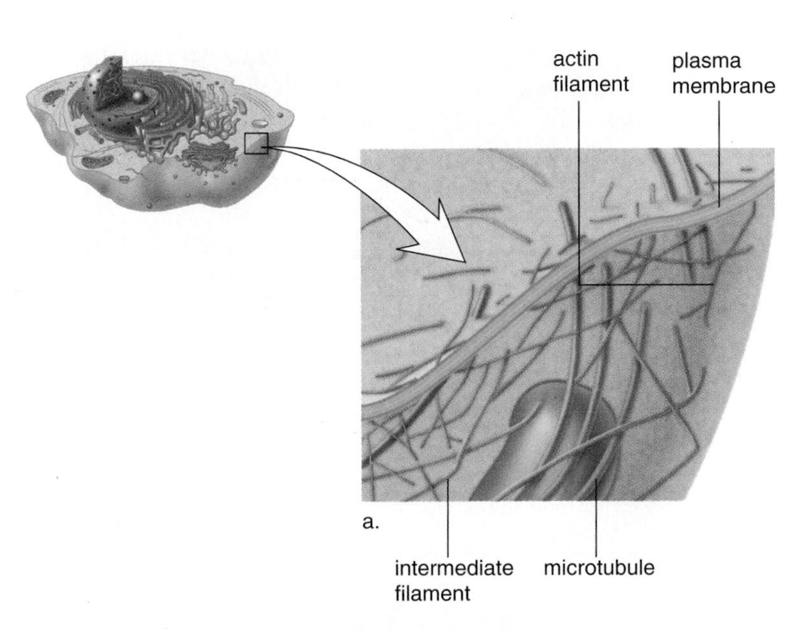

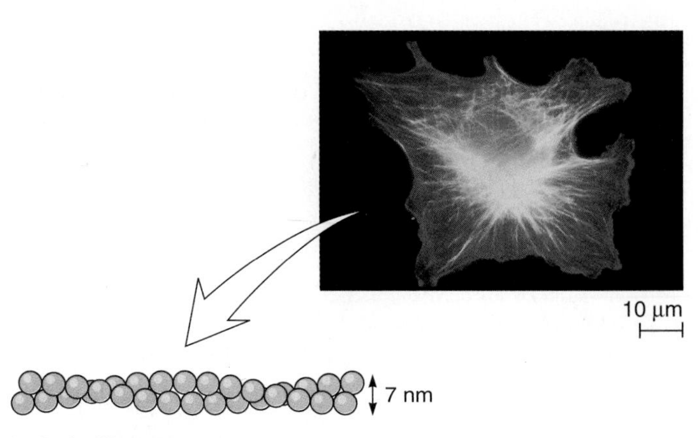

b. **Actin filament**

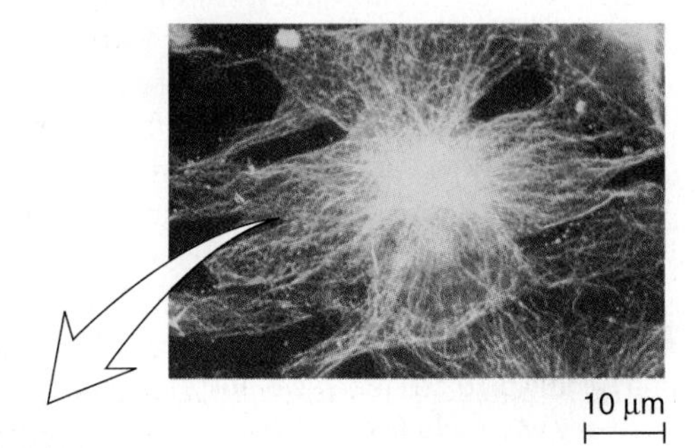

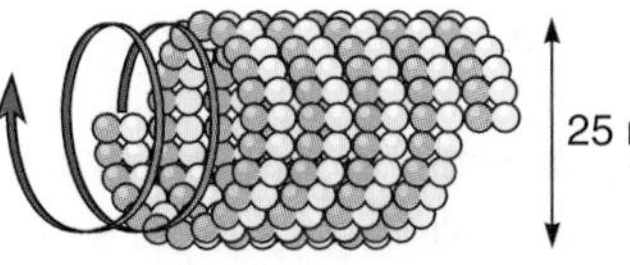

c. **Microtubule**

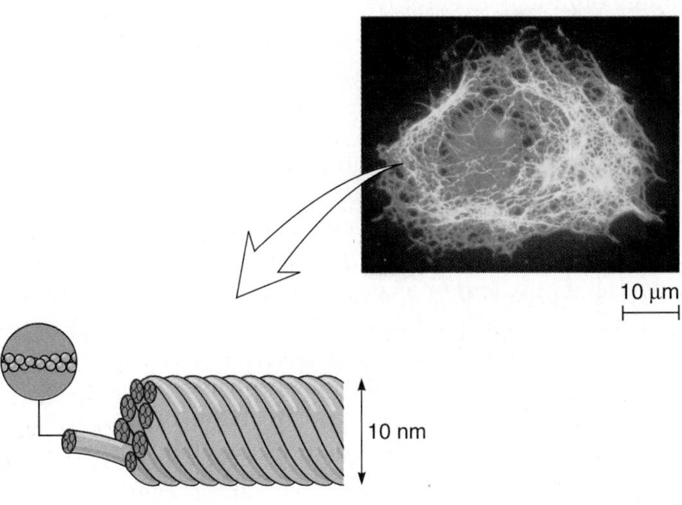

d. **Intermediate filament**

Figure 3.10 The cytoskeleton.
a. Diagram comparing the size relationship of actin filaments, intermediate filaments, and microtubules. **b.** Actin filaments as they appear in a cell and in diagram. **c.** Microtubules as they appear in the cell and in diagram. The filaments and tubules are visible following immunofluorescence, a technique that binds fluorescent antibodies to specific proteins in cells. **d.** Intermediate filaments as they appear in the cell and in diagram.

There are different types of kinesin proteins, each specialized to move one kind of vesicle or cellular organelle. A second type of cytoplasmic motor molecule is called cytoplasmic dynein because it is closely related to the molecule dynein found in flagella.

Intermediate Filaments

Intermediate filaments (8–11 nm in diameter) are intermediate in size between actin filaments and microtubules. They are a ropelike assembly of fibrous polypeptides that support the nuclear envelope and the plasma membrane. In the skin, intermediate filaments made of the protein keratin give great mechanical strength to skin cells. Recent work has shown intermediate filaments to be highly dynamic. They also are able to assemble and disassemble in the same manner as actin filaments and microtubules.

The cytoskeleton contains actin filaments, intermediate filaments, and microtubules. These maintain cell shape and allow organelles to move within the cytoplasm. Sometimes they are also involved in movement of the cell itself.

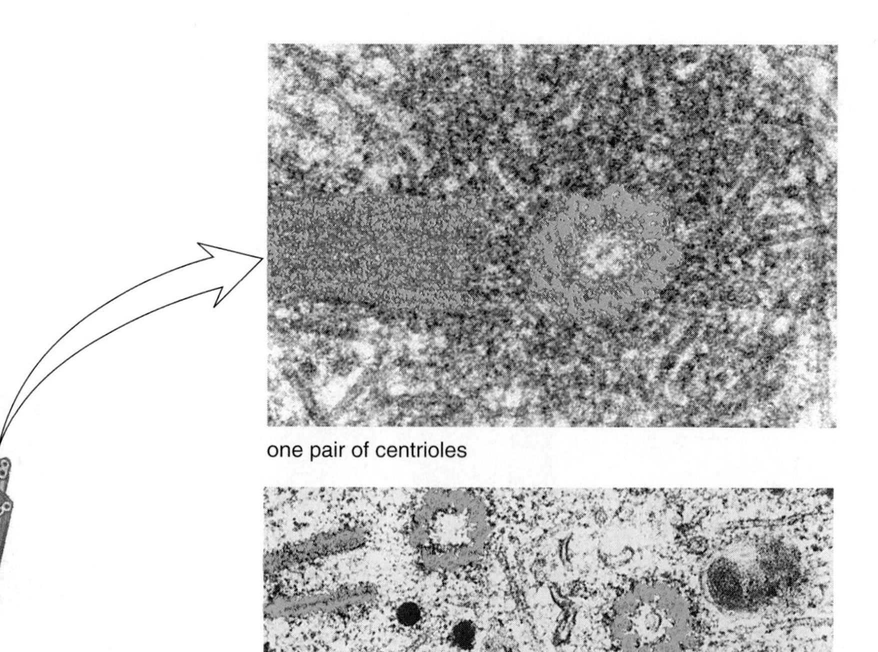

Figure 3.11 Centrioles.
In a nondividing cell there is a pair of centrioles in a centrosome outside the nucleus. Just before a cell divides, the centrosome divides so that there are two pairs of centrioles. During cell division, the centrosomes separate so that each new cell has one pair of centrioles.

Centrioles

Centrioles are short cylinders with a 9 + 0 pattern of microtubule triplets—that is, a ring having nine sets of triplets with none in the middle (Fig. 3.11). In animal cells and most protists, a centrosome contains two centrioles lying at right angles to each other. The centrosome is the major microtubule organizing center for the cell. Therefore, it is possible that centrioles are also involved in the process by which microtubules assemble and disassemble.

Before an animal cell divides, the centrioles replicate and the members of each pair are at right angles to one another (Fig. 3.11). Then each pair becomes part of a separate centrosome. During cell division the centrosomes move apart and may function to organize the mitotic spindle. In any case, each new cell has its own centrosome. Plant cells have the equivalent of a centrosome but it does not contain centrioles, suggesting that centrioles are not necessary to the assembly of cytoplasmic microtubules.

In cells with cilia and flagella, centrioles are believed to give rise to basal bodies that direct the organization of microtubules within these structures. In other words, a basal body may do for a cilium (or flagellum) what the centrosome does for the cell.

> Centrioles, which are short cylinders with a 9 + 0 pattern of microtubule triplets, may be involved in microtubule organization and in the formation of cilia and flagella.

Cilia and Flagella

Cilia and **flagella** are hairlike projections that can move either in an undulating fashion, like a whip, or stiffly, like an oar. Cells that have these organelles are capable of movement. For example, unicellular paramecia move by means of cilia, whereas sperm cells move by means of flagella. The cells that line our upper respiratory tract have cilia that sweep debris trapped within mucus back up into the throat, where it can be swallowed. This action helps keep the lungs clean.

In eukaryotic cells, cilia are much shorter than flagella, but they have a similar construction. Both are membrane-bounded cylinders enclosing a matrix area. In the matrix are nine microtubule doublets arranged in a circle around two central microtubules. Cilia and flagella move when the microtubule doublets slide past one another (Fig. 3.12).

As mentioned, each cilium and flagellum has a basal body lying in the cytoplasm at its base. Basal bodies have the same circular arrangement of microtubule triplets as centrioles and are believed to be derived from them. It is possible that basal bodies organize the microtubules within cilia and flagella, but this is not supported by the observation that cilia and flagella grow by the addition of tubulin dimers to their tips.

> Cilia and flagella, which have a 9 + 2 pattern of microtubules, are involved in the movement of cells.

Visual Focus

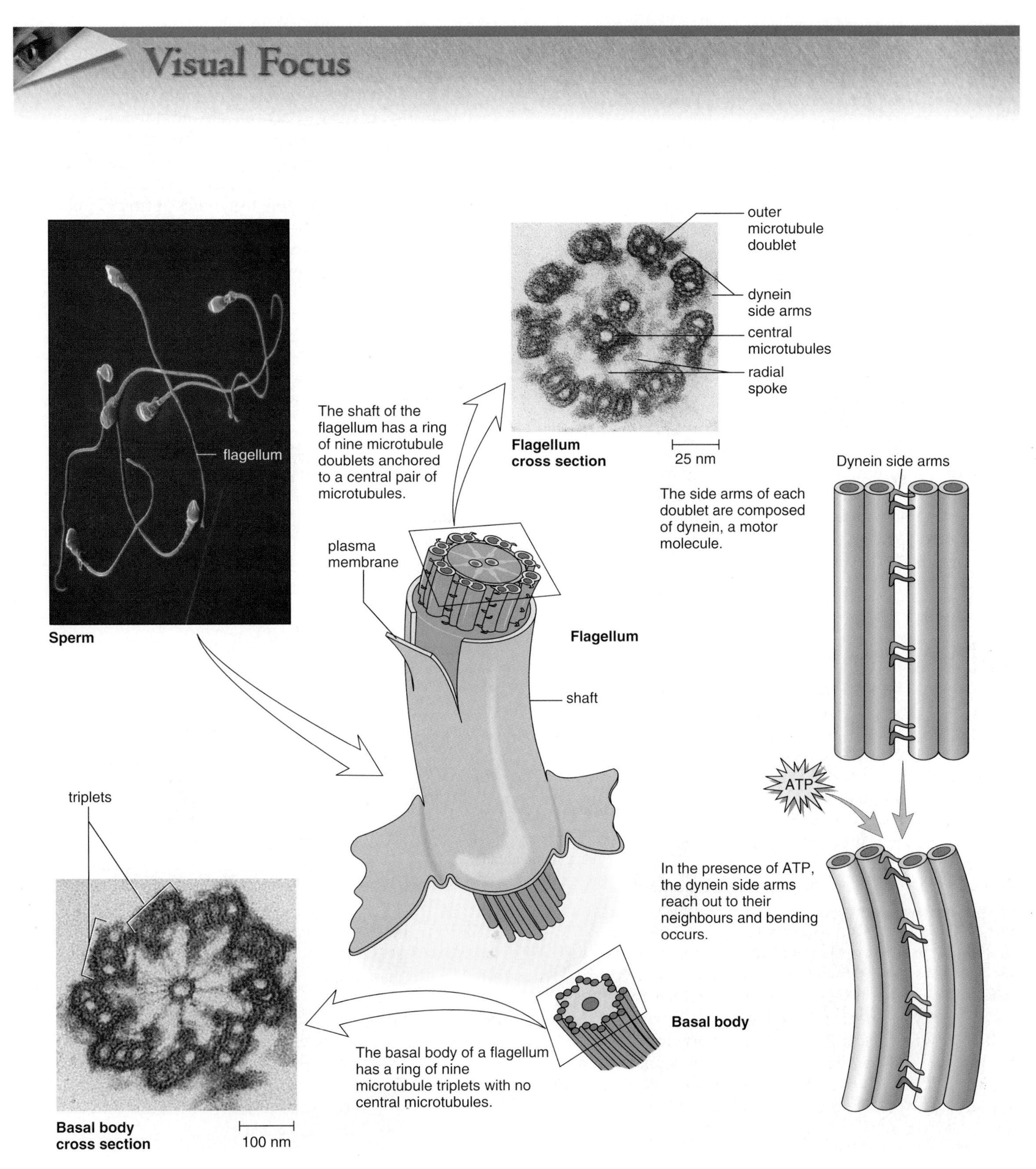

Figure 3.12 Structure of a flagellum or cilium.
A basal body derived from a centriole is at the base of a flagellum or cilium. The shaft of a flagellum (or cilium) contains microtubule doublets whose side arms are motor molecules that cause the flagellum (such as those of sperm) to move. Without the ability of sperm to move to the egg, human reproduction would not be possible.

3.3 Prokaryotic Cells

Bacteria are **prokaryotic cells** in the kingdom Monera. Most bacteria are 1–10 μm in size; therefore, they are just visible with the light microscope.

Figure 3.13 illustrates the main features of bacterial anatomy. The **cell wall** contains peptidoglycan, a complex molecule with chains of a unique amino disaccharide joined by peptide chains. In some bacteria, the cell wall is further surrounded by a **capsule** and/or gelatinous sheath called a **slime layer.** Motile bacteria usually have long, very thin appendages called flagella (sing., **flagellum**) that are composed of subunits of the protein called flagellin. The flagella, which rotate like propellers, rapidly move the bacterium in a fluid medium. Some bacteria also have *fimbriae,* which are short appendages that help them attach to an appropriate surface.

A membrane called the **plasma membrane** regulates the movement of molecules into and out of the cytoplasm, the interior of the cell. Cytoplasm in a prokaryotic cell consists of a semifluid medium, and thousands of granular inclusions called **ribosomes** that coordinate the synthesis of proteins. In prokaryotes, most genes are found within a single chromosome (loop of DNA, or deoxyribonucleic acid) located within the **nucleoid,** but they may also have small accessory rings of DNA called **plasmids.** In addition, the photosynthetic cyanobacteria have light-sensitive pigments, usually within the membranes of flattened disks called **thylakoids.**

Although bacteria seem fairly simple, they are actually metabolically diverse. Bacteria are adapted to living in almost any kind of environment and are diversified to the extent that almost any type of organic matter can be used as a nutrient for some particular bacterium. Given an energy source, most bacteria are able to synthesize any kind of molecule they may need. Therefore, the cytoplasm is the site of thousands of chemical reactions and bacteria are more metabolically competent than are human beings. Indeed, the metabolic capability of bacteria is exploited by humans who use them to produce a wide variety of chemicals and products for human use.

Bacteria are prokaryotic cells with these constant features.

Outer boundary:	cell wall
	plasma membrane
Cytoplasm:	ribosomes
	thylakoids (cyanobacteria)
	innumerable enzymes
Nucleoid:	chromosome (DNA only)

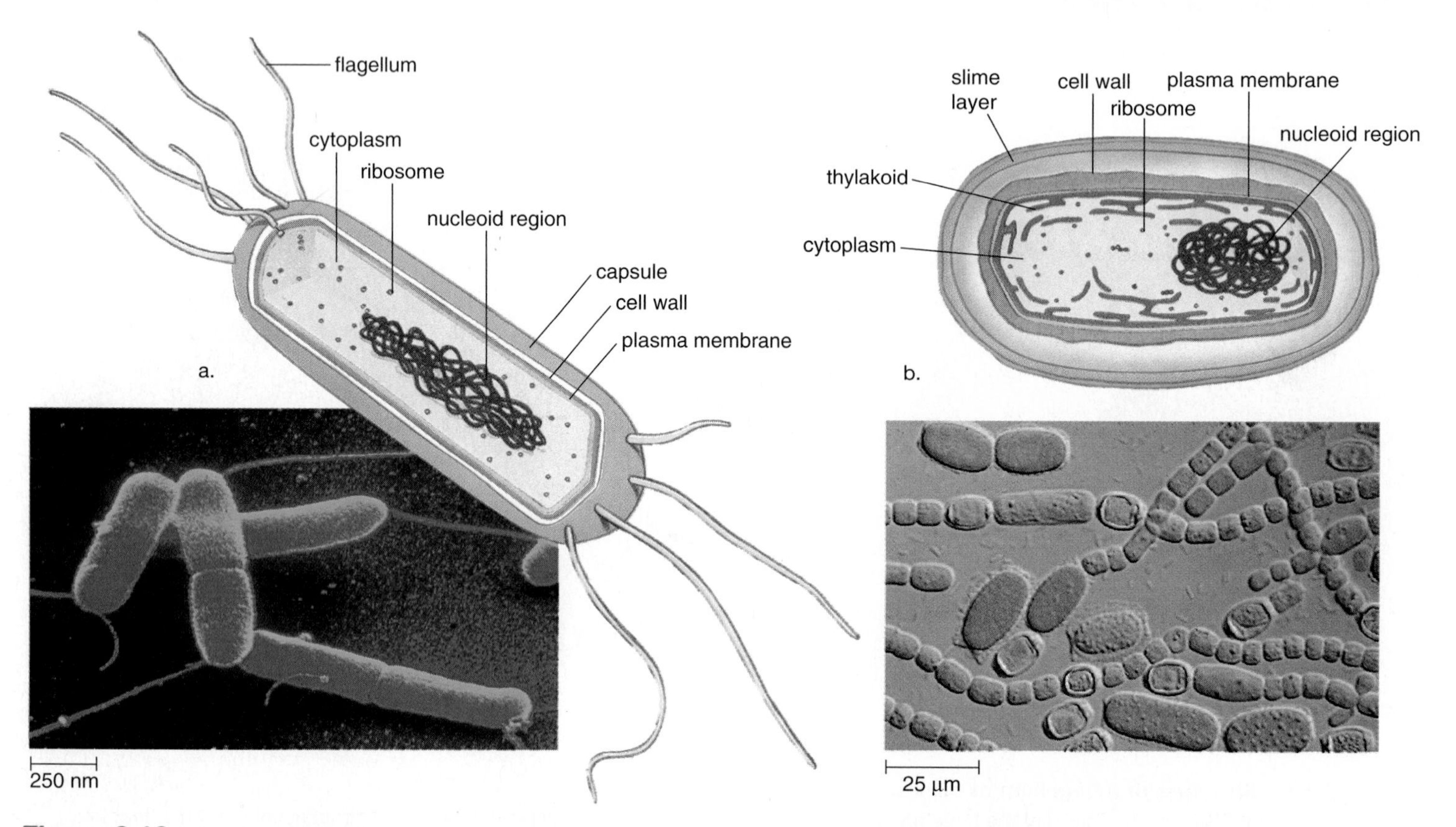

Figure 3.13 Prokaryotic cells.
a. Nonphotosynthetic bacterium. **b.** Cyanobacterium, a photosynthetic bacterium, formerly called a blue-green alga.

3.4 Evolution of the Eukaryotic Cell

Invagination of the plasma membrane might explain the origin of the nuclear envelope and organelles such as the endoplasmic reticulum and the Golgi apparatus. Some believe that the other organelles could also have arisen in this manner. But another hypothesis has been put forth. It has been observed that in the laboratory an amoeba infected with bacteria can become dependent upon them. Some investigators believe that mitochondria and chloroplasts are derived from prokaryotes that were taken up by a much larger cell (Fig. 3.14). Perhaps mitochondria were originally aerobic heterotrophic bacteria and chloroplasts were originally cyanobacteria. The host cell would have benefited from an ability to utilize oxygen or synthesize organic food when by chance the prokaryote was not destroyed. Therefore, after these prokaryotes entered by *endocytosis*, a *symbiotic* relationship was established. Some of the evidence for the **endosymbiotic hypothesis** is as follows:

1. Mitochondria and chloroplasts are similar to bacteria in size and in structure.
2. Both organelles are bounded by a double membrane—the outer membrane may be derived from the engulfing vesicle, and the inner one may be derived from the plasma membrane of the original prokaryote.
3. Mitochondria and chloroplasts contain a limited amount of genetic material and divide by splitting. Their DNA (deoxyribonucleic acid) is a circular loop like that of bacteria.
4. Although most of the proteins within mitochondria and chloroplasts are now produced by the eukaryotic host, they do have their own ribosomes and they do produce some proteins. Their ribosomes resemble those of bacteria.
5. The RNA (ribonucleic acid) base sequence of their ribosomes suggests a eubacterial origin for chloroplasts and mitochondria.

It's just possible also that the flagella of eukaryotes are derived from an elongated bacterium that became attached to a host cell (Fig. 3.14). However, it is important to remember that the flagella of eukaryotes are constructed differently. In any case, the acquisition of basal bodies, which could have become centrioles, may have led to the ability to form a spindle during cell division.

According to the endosymbiotic hypothesis, heterotrophic bacteria became mitochondria and cyanobacteria became chloroplasts after being taken up by precursors to modern-day eukaryotic cells.

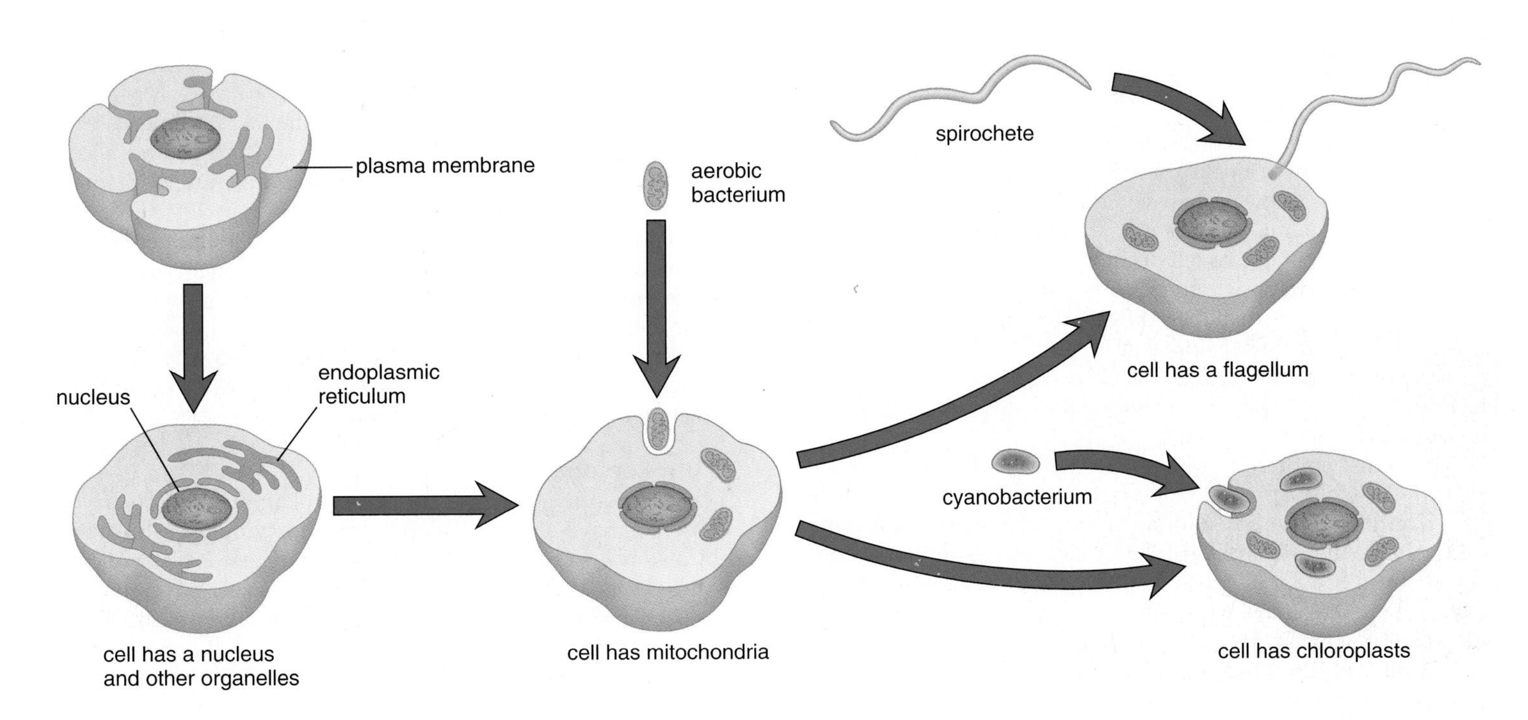

Figure 3.14 Evolution of the eukaryotic cell.
Invagination of the plasma membrane could account for the formation of the nucleus and certain other organelles. The endosymbiotic hypothesis suggests that mitochondria, chloroplasts, and flagella are derived from prokaryotes that were taken up by a much larger eukaryotic cell.

Bioethical Issue

As the cell theory tells us, cells can't be manufactured. Therefore, unlike medications such as certain hormones and vaccines, which are now biotechnology products, the only way to get human cells is from previous human cells. No wonder, then, there is a market in blood cells which are easily accessible. Viacord is in the business of storing umbilical blood for possible future use by expectant parents. It only costs an initial outlay of $2,500 and a yearly fee of $95.

The blood in a baby's umbilical cord is rich in stem cells, which give rise to all the other types of blood cells within an adult's red bone marrow. Should the parent's or the child's red bone marrow be destroyed by disease or cancer treatment, these stem cells will bring the red bone marrow back to life. Viacord says that stem-cell banking is a kind of insurance, but there is no doubt it capitalizes on our fear of future illness.

The banking of umbilical cord blood is a new idea. Suppose you are an elderly cancer patient, could you buy stem cells from someone else? The U.S. National Organ Transplant Act (NOTA) bans interstate commerce in the sale of transplant organs but excludes tissues such as blood or sperm, which are replenishable. The answer, then, is most likely yes, especially if the parents now find themselves in need of cash.

Where would you draw the line in the traffic of human parts? In 1984, John Moore sued his doctor for using, without his consent, tissue from his cancerous spleen to create a commercial cell line now valued at about $3 billion in profits. The California Supreme Court ruled against him because it might deter the work of research scientists. Was that right and proper?

Questions

1. Should there be any restrictions in the buying and selling of transplant organs, blood, or blood products? Why or why not?
2. Who should have the rights to tissues that are used to start up commercial cell lines, the person who donated the cells or the investigator who established the line, or both?
3. Who should get to use stored umbilical blood? Only family members, or also a paying customer? After all, some parents have several children.

Summarizing the Concepts

3.1 The Cellular Level of Organization

All organisms are composed of cells, the smallest units of living matter. Cells are capable of self-reproduction, and existing cells come only from preexisting cells. Cells are very small and are measured in micrometers. The plasma membrane regulates exchange of materials between the cell and the external environment. Cells must remain small in order to have an adequate amount of surface area per cell volume.

3.2 Eukaryotic Cells

The nucleus of eukaryotic cells, represented by animal and plant cells, is bounded by a nuclear envelope containing pores. These pores serve as passageways between the cytoplasm and the nucleoplasm. Within the nucleus, the chromatin undergoes coiling into chromosomes at the time of cell division. The nucleolus is a special region of the chromatin where rRNA is produced and where proteins from the cytoplasm gather to form ribosomal subunits. These subunits are joined in the cytoplasm.

Ribosomes are organelles that function in protein synthesis. They can be bound to ER or exist within the cytoplasm singly or in groups called polyribosomes.

The endomembrane system includes the ER (both rough and smooth), the Golgi apparatus, the lysosomes, and other types of vesicles and vacuoles. The endomembrane system serves to compartmentalize the cell and keep the various biochemical reactions separate from one another. Newly produced proteins enter the ER lumen, where they may be modified before proceeding to the interior of the smooth ER. The smooth ER has various metabolic functions depending on the cell type, but it also forms vesicles that carry proteins and lipids to different locations, particularly to the Golgi apparatus. The Golgi apparatus processes proteins and repackages them into lysosomes, which carry out intracellular digestion, or into vesicles that fuse with the plasma membrane. Following fusion, secretion occurs. The endomembrane system also includes peroxisomes that have special enzymatic functions. Peroxisomes contain enzymes that oxidize molecules by producing hydrogen peroxide that is subsequently broken down. The large single plant cell vacuole not only stores substances but lends support to the plant cell.

Cells require a constant input of energy to maintain their structure. Chloroplasts capture the energy of the sun and carry on photosynthesis, which produces carbohydrate. Carbohydrate-derived products are broken down in mitochondria as ATP is produced. This is an oxygen-requiring process called aerobic respiration.

The cytoskeleton contains actin filaments, intermediate filaments, and microtubules. These maintain cell shape and allow it and the organelles to move. Actin filaments, the thinnest filaments, interact with the motor molecule myosin in muscle cells to bring about contraction; in other cells, they pinch off daughter cells and have other dynamic functions. Intermediate filaments support the nuclear envelope and the plasma membrane and probably participate in cell-to-cell junctions. Microtubules radiate out from the centrosome and are present in centrioles, cilia, and flagella. They serve as tracks along which vesicles and other organelles move due to the action of specific motor molecules.

3.3 Prokaryotic Cells

There are two major groups of cells: prokaryotic and eukaryotic. Both types have a plasma membrane and cytoplasm. Eukaryotic cells also have a nucleus and various organelles. Prokaryotic cells have a nucleoid that is not bounded by a nuclear envelope. They also lack most of the other organelles that compartmentalize eukaryotic cells.

3.4 Evolution of the Eukaryotic Cell

The nuclear envelope most likely evolved through invagination of the plasma membrane, but mitochondria and chloroplasts may have arisen through endosymbiotic events.

Studying the Concepts

1. What are the two basic tenets of the cell theory? 46
2. Why is it advantageous for cells to be small? 48
3. Distinguish between the nucleolus, rRNA, and ribosomes. 52–53
4. Describe the structure and the function of the nuclear envelope and the nuclear pores. 52
5. Trace the path of a protein from rough ER to the plasma membrane. 53
6. Give the overall equations for photosynthesis and cellular respiration, contrast the two, and tell how they are related. 56
7. What are the three components of the cytoskeleton? What are their structures and functions? 58–59
8. What similar features do prokaryotic cells and eukaryotic cells have? What is their major difference? 62
9. Contrast the structure of chloroplasts and mitochondria. 63
10. Describe the endosymbiotic hypothesis and the evidence to support it. 63

Testing Yourself

Choose the best answer for each question.

1. The small size of cells is best correlated with
   a. the fact they are self-reproducing.
   b. their prokaryotic versus eukaryotic nature.
   c. an adequate surface area for exchange of materials.
   d. their vast versatility.
   e. All of these are correct.
2. Which of these is not found in the nucleus?
   a. functioning ribosomes
   b. chromatin that condenses to chromosomes
   c. nucleolus that produces rRNA
   d. nucleoplasm instead of cytoplasm
   e. DNA making up the genes.
3. Vesicles from the smooth ER most likely are on their way to the
   a. rough ER.
   b. lysosomes.
   c. Golgi apparatus.
   d. plant cell vacuole only.
   e. cell wall of adjoining cells.
4. Lysosomes function in
   a. protein synthesis.
   b. processing and packaging.
   c. intracellular digestion.
   d. lipid synthesis.
   e. All of these are correct.
5. Mitochondria
   a. are involved in cellular respiration.
   b. break down ATP to release energy for cells.
   c. contain grana and cristae.
   d. have a convoluted outer membrane.
   e. All of these are correct.
6. Which organelle releases oxygen?
   a. ribosome
   b. Golgi apparatus
   c. mitochondrion
   d. peroxisome
7. Which of these is not true?
   a. Actin filaments are found in muscle cells.
   b. Microtubules radiate out from the ER.
   c. Intermediate filaments sometimes contain keratin.
   d. Motor molecules use microtubules as tracks.
   e. Cilia and flagella are constructed similarly.
8. Cilia and flagella
   a. bend when microtubules try to slide past one another.
   b. contain myosin that pulls on actin filaments.
   c. are organized by basal bodies derived from centrioles.
   d. are of the same length.
   e. Both a and c are correct.
9. Which organelle would not have originated by endosymbiosis?
   a. mitochondria
   b. flagella
   c. nucleus
   d. chloroplasts
   e. All of these are correct.
10. Which structure would be found in a prokaryotic cell?
    a. cell wall, ribosomes, thylakoids, chromosome
    b. cell wall, plasma membrane, nucleus, flagellum
    c. nucleoid region, ribosomes, chloroplasts, capsule
    d. plasmid, ribosomes, enzymes, DNA, mitochondria
    e. chlorophyll, enzymes, Golgi apparatus, plasmids
11. Label these parts of the cell that are involved in protein synthesis and modification. Give a function for each structure.

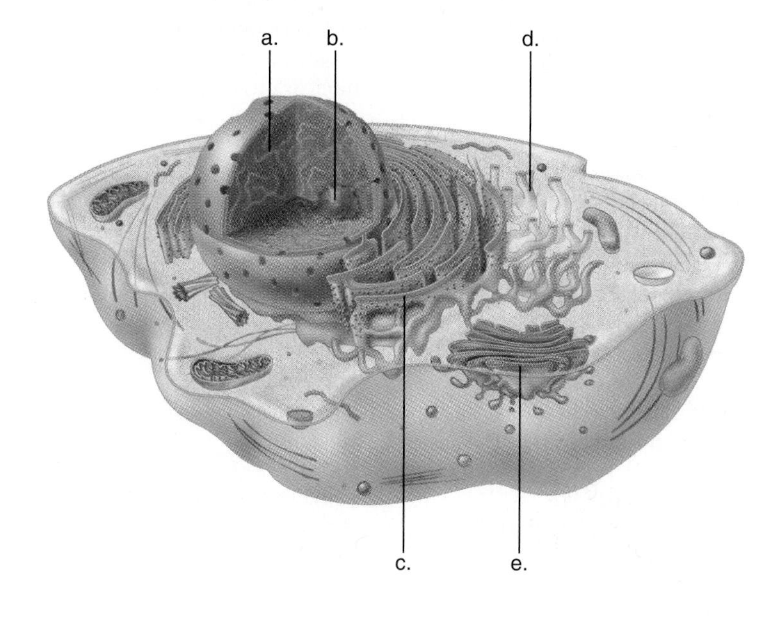

Thinking Scientifically

1. To study movement of molecules in the cell:
   a. It is possible to bathe a cell in radioactively tagged amino acids and then later to detect photographically the location of radiation, and therefore the amino acids in the cell. Why would you suggest using radioactive sulfur (i.e., the amino acids cysteine and methionine contain sulfur) rather than radioactive carbon?
   b. An investigator uses the procedure outlined in question a to support the belief that proteins move from the cytoplasm into the nucleus. If so, where will radiation first appear, and where will it subsequently appear in the cell? (page 52)
   c. An investigator uses this same procedure to support the belief that proteins move from the rough ER to secretory vesicles. Where will the radiation first appear, and where will it subsequently appear in the cell? (page 53)
2. A microtubular spindle apparatus appears in plant and animal cells at the time of cell division.
   a. Why does this suggest that centrioles are not necessary to the formation of the spindle apparatus?
   b. What evidence is there to suggest centrioles are necessary to microtubule organization in animal cells?
   c. In animal cells, each newly formed cell receives a pair of centrioles. Why might centrioles be necessary to animal cells but not to most plant cells? (page 60)

Understanding the Terms

bacteria 62
capsule 62
cell 46
cell theory 46
cell wall 49
centriole 60
centrosome 58
chloroplast 56
chromatin 52
chromosome 52
cilium (pl., cilia) 60
cristae 57
cytoplasm 49
cytoskeleton 58
endoplasmic reticulum 53
endosymbiotic hypothesis 63
eukaryotic cell 49
flagellum (pl., flagella) 60
Golgi apparatus 54
granum 57
lysosome 55
matrix 57
microtubule 58
mitochondrion 56
motor molecule 58
nuclear envelope 52
nuclear pore 52
nucleoid 62
nucleolus (pl., nucleoli) 52
nucleoplasm 52
nucleus 52
organelle 49
peroxisome 55
plasma membrane 49
plasmid 62
polyribosome 53
prokaryotic cell 62
ribosome 53
secretion 54
slime layer 62
stroma 57
thylakoid 57
vacuole 55
vesicle 53

Match the terms to these definitions:

a. ________ Area in prokaryotic cell where DNA is found.
b. ________ Dark-staining, spherical body in the nucleus that produces ribosomal subunits.
c. ________ Internal framework of the cell, consisting of microtubules, actin filaments, and intermediate filaments.
d. ________ Organelle consisting of saccules and vesicles that processes, packages, and distributes molecules about or from the cell.
e. ________ System of membranous saccules and channels in the cytoplasm, often with attached ribosomes.

Using Technology

Your study of cell structure and function is supported by these available technologies:

Essential Study Partner CD-ROM

Cells → Cell Structures

Visit the Mader web site for related ESP activities.

Exploring the Internet

The Mader Home Page provides resources and tools as you study this chapter.

http://www.mhhe.com/biosci/genbio/mader

Dynamic Human 2.0 CD-ROM

Human Body → Clinical Concepts Cell Components

Life Science Animations 3D Video

3 Cellular Secretion

C H A P T E R 4

Membrane Structure and Function

Chapter Concepts

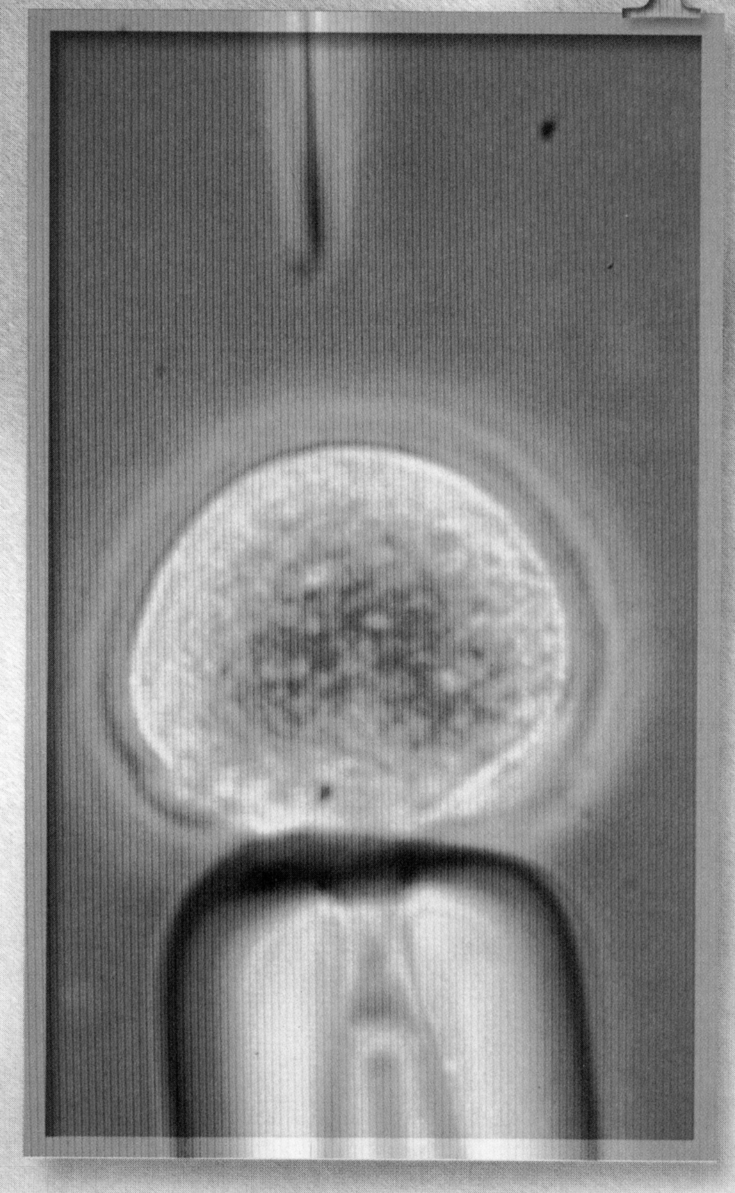

A single cell about to be pierced by a fine probe so that DNA can be removed by the suction tube on the bottom. An intact plasma membrane is necessary to the life of any cell and if it is ruptured the cell cannot continue to exist.

Banners flying on a castle wall mark off the community within from the surrounding countryside. Inside, residents go about their appointed tasks for the good of the community. Commands passed along from royalty to knights to workers are obeyed by all. The almost impenetrable wall prevents the enemy without from entering and disturbing the peace within. Only certain small creatures can pass through the open slitlike windows, and the drawbridge must be lowered for most needed supplies.

The plasma membrane, which carries markers identifying it as belonging to the individual, can be likened to the castle wall. Under the command of the nucleus, the organelles carry out their specific functions and contribute to the working of the cell as a whole. Very few molecules can freely cross the membrane, and most nutrients must be transported across by special carriers. The cell uses these nutrients as a source of building blocks and energy to maintain the cell. The operations of the cell will continue only as long as the plasma membrane selectively permits specific materials to enter and leave and prevents the passage of others.

4.1 Plasma Membrane Structure and Function

The plasma membrane is a phospholipid bilayer in which protein molecules are either partially or wholly embedded (Fig. 4.1). The phospholipid bilayer has a *fluid* consistency, comparable to that of light oil. The proteins are scattered throughout the membrane; therefore they form a *mosaic* pattern. This description of the plasma membrane is called the **fluid-mosaic model** of membrane structure.

Phospholipids spontaneously arrange themselves into a bilayer. The hydrophilic (water loving) polar heads of the phospholipid molecules face the outside and inside of the cell where water is found, and the hydrophobic (water fearing) nonpolar tails face each other (Fig. 4.1). In addition to phospholipids, there are two other types of lipids in the plasma membrane. **Glycolipids** have a structure similar to phospholipids except that the hydrophilic head is a variety of sugars joined to form a straight or

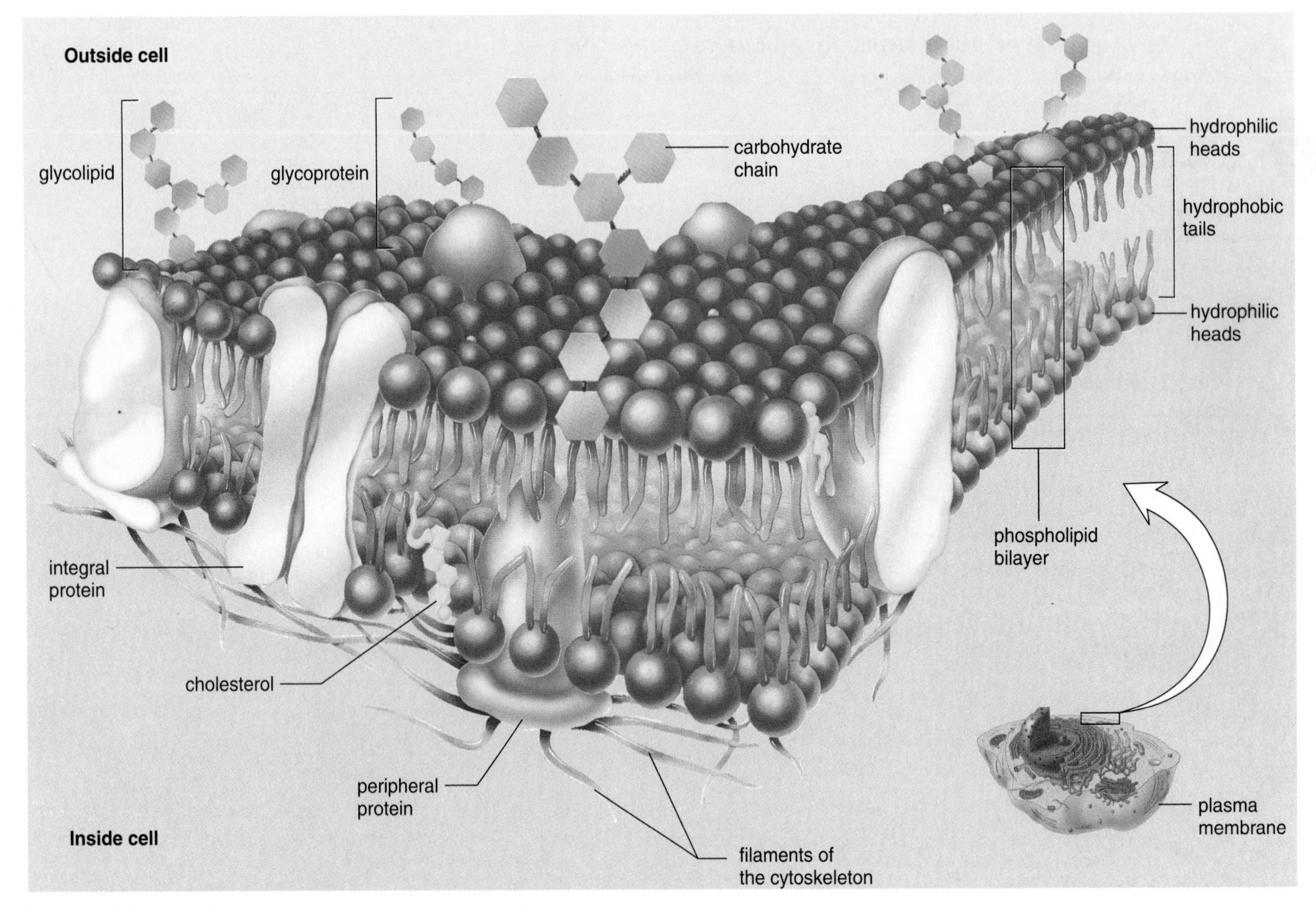

Figure 4.1 Fluid-mosaic model of plasma membrane structure.
The membrane is composed of a phospholipid bilayer in which proteins are embedded. The hydrophilic heads of phospholipids are a part of the outside surface and the inside surface of the membrane. The hydrophobic tails make up the interior of the membrane. Note the plasma membrane's asymmetry—carbohydrate chains are attached to the outside surface and cytoskeleton filaments are attached to the inside surface.

branching carbohydrate chain. **Cholesterol** is a lipid that is found in animal plasma membranes; related steroids are found in the plasma membrane of plants. Cholesterol reduces the permeability of the membrane to most biological molecules.

The proteins in a membrane may be peripheral proteins or integral proteins. **Peripheral proteins** occur either on the outside or the inside surface of the membrane. Some of these are anchored to the membrane by covalent bonding. Still others are held in place by noncovalent interactions that can be disrupted by gentle shaking or by change in the pH.

Integral proteins are found within the membrane and have hydrophobic regions embedded within the membrane and hydrophilic regions that project from both surfaces of the bilayer:

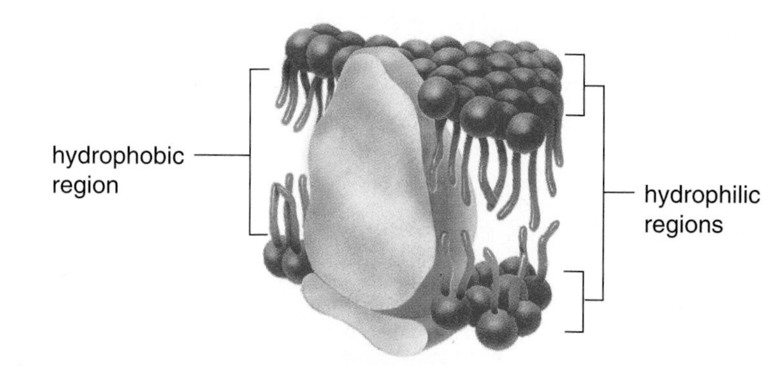

Many integral proteins are **glycoproteins,** which have an attached carbohydrate chain. As with glycolipids, the carbohydrate chain of sugars projects externally. Therefore it can be said that the plasma membrane is "sugar-coated."

The plasma membrane is asymmetrical: the two halves are not identical. The carbohydrate chains of the glycolipids and proteins occur only on the outside surface and the cytoskeletal filaments attach to proteins only on the inside surface.

The plasma membrane consists of a phospholipid bilayer. Peripheral proteins are found on the outside and inside surface of the membrane. Integral proteins span the lipid bilayer and often have attached carbohydrate chains.

The Fluidity of the Plasma Membrane

At body temperature, the phospholipid bilayer of the plasma membrane has the consistency of olive oil. The greater the concentration of unsaturated fatty acid residues, the more fluid is the bilayer. In each monolayer, the hydrocarbon tails wiggle, and the entire phospholipid molecule can move sideways at a rate averaging about 2 µm—the length of a prokaryotic cell—per second. (Phospholipid molecules rarely flip-flop from one layer to the other, because this would require the hydrophilic head to move through the hydrophobic center of the membrane.) The fluidity of a phospholipid bilayer means that cells are pliable. Imagine if they were not—the long nerve fibers in your neck would crack whenever you nodded your head!

Although some proteins are often held in place by cytoskeletal filaments, in general proteins are free to drift laterally in the fluid lipid bilayer. This has been demonstrated by fusing mouse and human cells, and watching the movement of tagged proteins (Fig. 4.2). Forty minutes after fusion, the proteins are completely mixed. The fluidity of the membrane is needed for the functioning of some proteins such as enzymes which become inactive when the membrane solidifies.

The fluidity of the membrane, which is dependent on its lipid components, is critical to the proper functioning of the membrane's proteins.

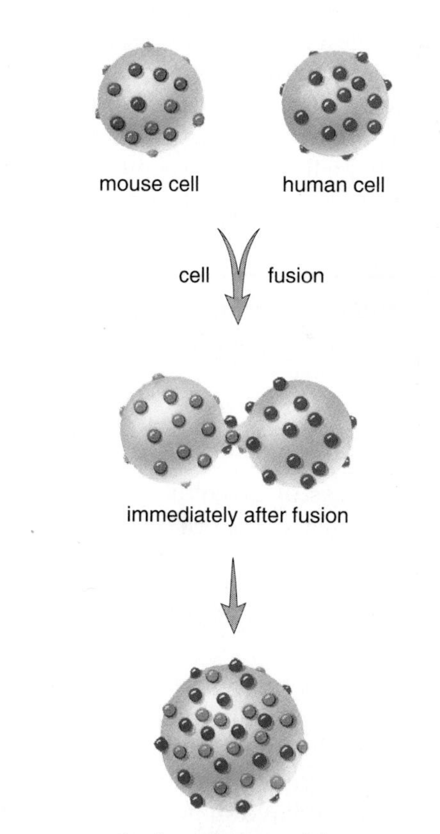

Figure 4.2 Experiment to demonstrate lateral drifting of plasma membrane proteins.
After human and mouse cells fuse, the plasma membrane proteins of the mouse (blue circles) and of the human cell (red circles) mix within a short time.

The Mosaic Quality of the Membrane

The plasma membranes of various cells and the membranes of various organelles each have their own unique collections of proteins. The proteins form different patterns according to the particular membrane and also within the same membrane at different times. When you consider that the plasma membrane of a red blood cell contains over 50 different types of proteins, you can see why the membrane is said to be a mosaic.

The integral proteins largely determine a membrane's specific functions. As we will discuss in more detail, certain plasma membrane proteins are involved in the passage of molecules through the membrane. Some of these are **channel proteins** through which a substance can simply move across the membrane; others are **carrier proteins** that combine with a substance and help it to move across the membrane. Still others are receptors; each type of **receptor protein** has a shape that allows a specific molecule to bind to it. The binding of a molecule, such as a hormone (or other signal molecule), can cause the protein to change its shape and bring about a cellular response. Some plasma membrane proteins are **enzymatic proteins** that carry out metabolic reactions directly. The peripheral proteins associated with the membrane often have a structural role in that they help stabilize and shape the plasma membrane.

Figure 4.3 depicts the various functions of membrane proteins.

The mosaic pattern of a membrane is dependent on the proteins, which vary in structure and function.

Cell–Cell Recognition

The carbohydrate chains of glycolipids and glycoproteins serve as the "fingerprints" of the cell. The possible diversity of the chain is enormous; it can vary by the number of sugars (15 is usual, but there can be several hundred), by whether the chain is branched, and by the sequence of the particular sugars.

Glycolipids and glycoproteins vary from species to species, from individual to individual of the same species, and even from cell to cell in the same individual. Therefore, they make cell–cell recognition possible. Researchers working with mouse embryos have shown that as development proceeds, the different type cells of the embryo develop their own carbohydrate chains and that these chains allow the tissues and cells of the embryo to sort themselves out.

As you probably know, transplanted tissues are often rejected by the body. This is because the immune system is able to recognize that the foreign tissue's cells do not have the same glycolipids and glycoproteins as the rest of the body's cells. We also now know that a person's particular blood type is due to the presence of particular glycoproteins in the membrane of red blood cells.

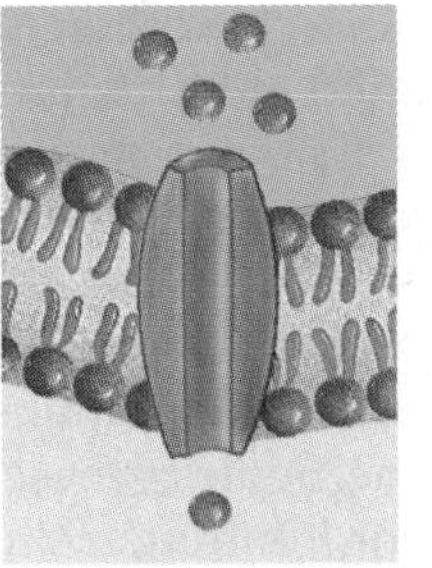

Channel Protein
Allows a particular molecule or ion to cross the plasma membrane freely. Cystic fibrosis, an inherited disorder, is caused by a faulty chloride (Cl^-) channel; a thick mucus collects in airways and in pancreatic and liver ducts.

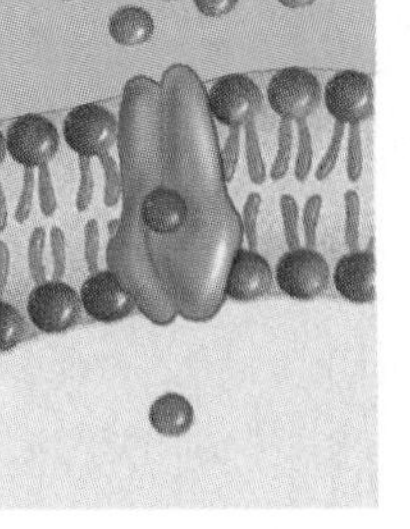

Carrier Protein
Selectively interacts with a specific molecule or ion so that it can cross the plasma membrane. The inability of some persons to use energy for sodium-potassium (Na^+–K^+) transport has been suggested as the cause of their obesity.

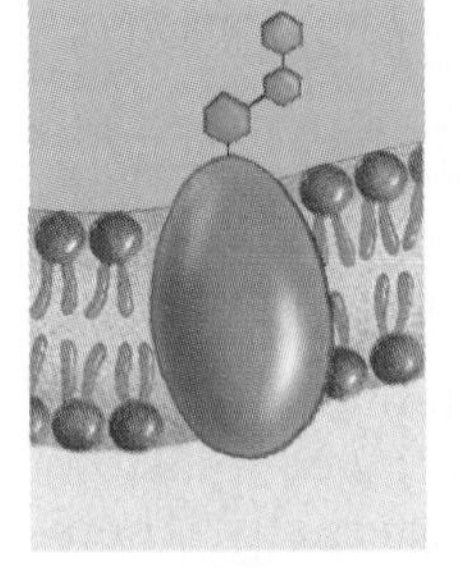

Cell Recognition Protein
The MHC (major histocompatibility complex) glycoproteins are different for each person, so organ transplants are difficult to achieve. Cells with foreign MHC glycoproteins are attacked by blood cells responsible for immunity.

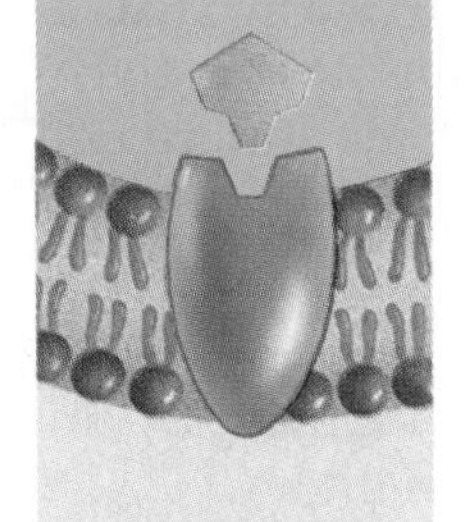

Receptor Protein
Is shaped in such a way that a specific molecule can bind to it. Pygmies are short, not because they do not produce enough growth hormone, but because their plasma membrane growth hormone receptors are faulty and cannot interact with growth hormone.

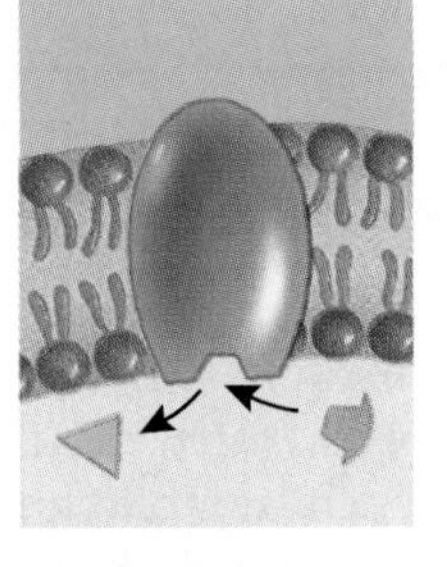

Enzymatic Protein
Catalyzes a specific reaction. The membrane protein, adenylate cyclase, is involved in ATP metabolism. Cholera bacteria release a toxin that interferes with the proper functioning of adenylate cyclase; sodium ions and water leave intestinal cells and the individual dies from severe diarrhea.

Figure 4.3 Membrane protein diversity.
These are some of the functions performed by proteins found in the plasma membrane.

Science Focus

The Growing Field of Tissue Engineering

Tissue culture, the growing of animal cells in laboratory glassware, has been done for quite some time, but researchers always used cancer cells that divide without coaxing. Now researchers have learned how to grow all sorts of human cells in tissue culture and have hopes that they can even make the need for organ transplantation obsolete. Organ transplantation encounters two hurdles that are hard to overcome: (1) there is an overwhelming need, but few human organs are available to be transplanted; and (2) immunosuppressive drugs must be administered even if the organs are carefully matched to the recipient, because the body tends to reject foreign organs. To address these problems, some researchers have turned to pigs as a source of organs for humans. Through genetic engineering, they have crippled the enzymes that produce plasma membrane carbohydrate chains on pig cells; therefore, the human body is unable to recognize a pig organ as being foreign. Pigs carry viruses such as the one that causes swine flu, but pig viruses are not expected to cause infections in humans. Therefore, it is predicted that pig-to-human transplants will someday be safely done.

Tissue engineering offers another possible solution to the transplant problem. Tissue engineering is an endeavor that produces manufactured bioproducts that can replace normal structures in the body. Integra is an artificial skin that consists of a porous matrix made of the protein collagen and a derivative of shark cartilage. This product, which is available in unlimited quantity, will not cause an immune reaction. Integra is used to cover extensive burns. Once the bottom layer of skin (the dermis) regrows, a graft of the patient's own outer layer of skin (the epidermis) replaces the artificial matrix.

Researchers have also had success growing human cartilage for knee operations. In one study, 23 patients who were experiencing pain because of a lack of cartilage received a batch of their own chondrocytes (cartilage cells) grown in the laboratory. All patients reported that they were doing much better following the procedure.

Other procedures have also been tried. It is possible to grow tissues to bolster weak ureters that take urine back to the kidneys instead of to the bladder where it belongs. And artificial tissue can be stitched into a bladder to increase its capacity. If research continues to be successful, nearly every human tissue is expected to undergo tissue engineering. Several groups are working on methods to reconstruct breast tissue after mastectomy so that one day women may have an alternative to silicone breast implants. Epithelial-lined plastic blood vessels are being developed because the walls of plastic blood vessels now used to replace weakened arteries sometimes cause the blood to clot.

Certain organs produce chemicals that are needed by other cells. Diabetes mellitus occurs when the pancreas is no longer producing insulin, a molecule that causes all cells to take up glucose and the liver to store glucose as glycogen. Tissue engineering can possibly come to the rescue. Insulin-producing pancreatic cells from a pig can be grown in the laboratory. The cells are encased in plastic capsules called microreactors, because reactors are typically large vats where chemicals are produced (Fig. 4A). These capsules are so small they can be placed into the abdomen where they will float freely and produce insulin as needed. The membrane of the capsule contains pores that are large enough to allow oxygen and nutrients to flow in and wastes and insulin to flow out by *diffusion*. But the membrane of a microreactor will prevent immune cells from coming into contact with the enclosed pancreatic cells. Unless immune cells actually come in contact with transplanted cells, they cannot recognize them as foreign and destroy them. Researchers are even busily growing implantable liver tissue. They use a spongy material that can be seeded with the patient's own liver cells.

Human embryonic cells are grown in tissue culture, and if differentiation can one day be achieved, it may be possible to supply Alzheimer patients with nerve cells, and cardiac patients with heart cells, and so forth.

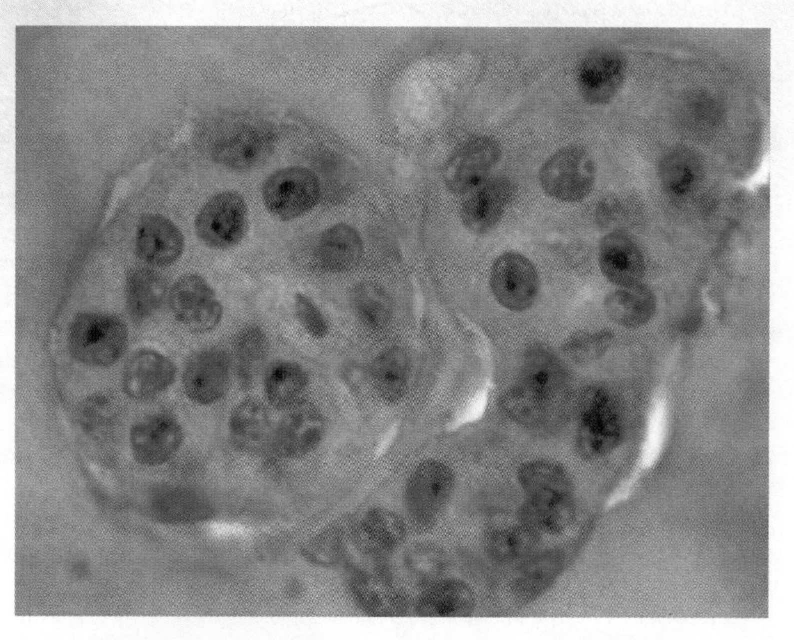

Figure 4A Microreactors.
Microreactors filled with insulin-producing pancreatic cells from pigs flourished for 10 weeks in a diabetic mouse without immune system-suppressing drugs.

Table 4.1 Passage of Molecules into and out of Cells

	Name	Direction	Requirement	Examples
Passive Transport Means	DIFFUSION	Toward lower concentration	Concentration gradient only	Lipid-soluble molecules, water, and gases
	FACILITATED TRANSPORT	Toward lower concentration	Carrier and concentration gradient	Some sugars and amino acids
Active Transport Means	ACTIVE TRANSPORT	Toward greater concentration	Carrier plus cellular energy	Other sugars, amino acids, and ions
	EXOCYTOSIS	Toward outside	Vesicle fuses with plasma membrane	Macromolecules
	ENDOCYTOSIS			
	Phagocytosis	Toward inside	Vacuole formation	Cells and subcellular material
	Pinocytosis (includes receptor-mediated endocytosis)	Toward inside	Vesicle formation	Macromolecules

4.2 The Permeability of the Plasma Membrane

The plasma membrane is **differentially** (selectively) **permeable.** Some substances can move across the membrane and some cannot (Fig. 4.4). Macromolecules cannot diffuse across the membrane because they are too large. Ions and charged molecules cannot cross the membrane because they are unable to enter the hydrophobic phase of the lipid bilayer.

Noncharged molecules such as alcohols and oxygen are lipid-soluble and therefore can cross the membrane with ease. They are able to slip between the hydrophilic heads of the phospholipids and pass through the hydrophobic tails of the membrane. Small polar molecules such as carbon dioxide and water also have no difficulty crossing through the membrane. These molecules follow their **concentration gradient** which is a gradual decrease in concentration over distance. To take an example, oxygen is more concentrated outside the cell than inside the cell because a cell uses oxygen during aerobic cellular respiration. Therefore oxygen follows its concentration gradient as it enters a cell. Carbon dioxide, on the other hand, which is produced when a cell carries on cellular respiration, is more concentrated inside the cell than outside the cell, and therefore it moves down its concentration gradient as it exits a cell.

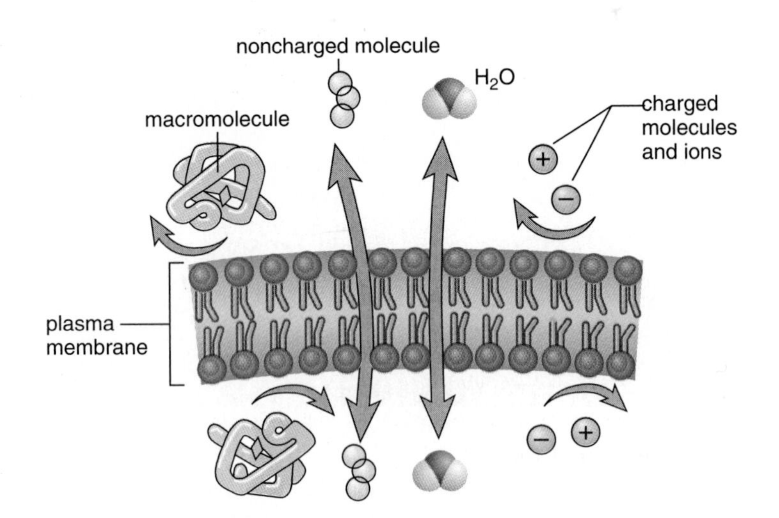

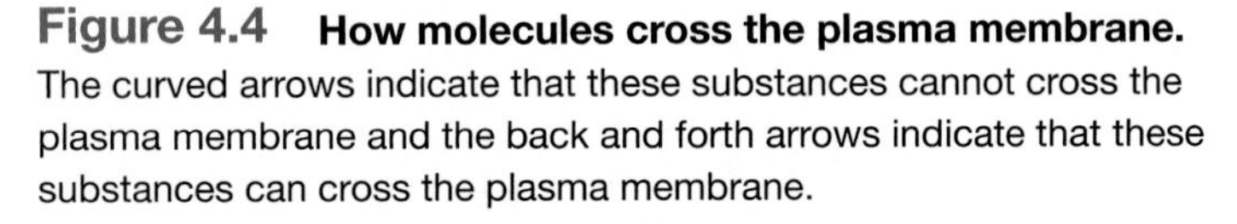

Figure 4.4 How molecules cross the plasma membrane. The curved arrows indicate that these substances cannot cross the plasma membrane and the back and forth arrows indicate that these substances can cross the plasma membrane.

Special means are sometimes used to get ions and charged molecules into and out of cells. Macromolecules can cross a membrane when they are taken in or out by vesicle formation (Table 4.1). Ions and molecules like amino acids and sugars are assisted across by one of two classes of transport proteins. Carrier proteins combine with an ion or molecule before transporting it across the membrane. Channel proteins form a channel that allows an ion or charged molecule to pass through. Our discussion in this chapter is largely restricted to carrier proteins. Carrier proteins are specific for the substances they transport across the plasma membrane.

Ways of crossing a plasma membrane are classified as passive or active (Table 4.1). *Passive ways,* which do not use chemical energy, involve diffusion or facilitated transport. These passive ways depend on the motion energy of ions and molecules. *Active ways,* which do require chemical energy, include active transport, endocytosis, and exocytosis.

The plasma membrane is differentially permeable. Certain substances can freely pass through the membrane and others must be transported across either by carrier proteins or by vacuole formation.

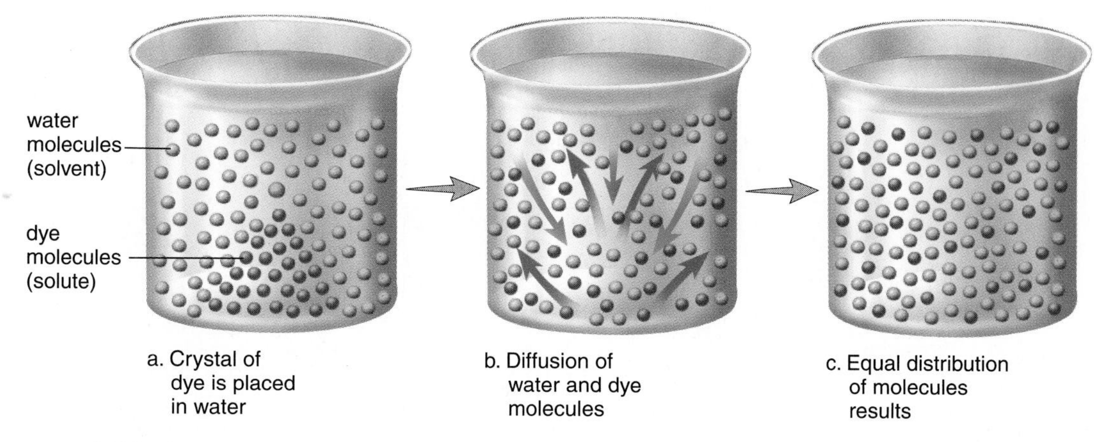

Figure 4.5 Process of diffusion.
Diffusion is spontaneous, and no chemical energy is required to bring it about. **a.** When dye crystals are placed in water, they are concentrated in one area. **b.** The dye dissolves in the water, and there is a net movement of dye molecules from higher to lower concentration. There is also a net movement of water molecules from a higher to a lower concentration. **c.** Eventually, the water and the dye molecules are equally distributed throughout the container.

4.3 Diffusion and Osmosis

Diffusion is the movement of molecules from a higher to a lower concentration—that is, down their concentration gradient—until equilibrium is achieved and they are distributed equally. Diffusion is a physical process that can be observed with any type of molecule. For example, when a crystal of dye is placed in water (Fig. 4.5), the dye and water molecules move in various directions, but their net movement, which is the sum of their motion, is toward the region of lower concentration. Therefore, the dye is eventually dissolved in the water, resulting in a colored solution. A solution contains both a solute, usually a solid, and a solvent, usually a liquid. In this case, the **solute** is the dye and the **solvent** is the water molecules. Once the solute and solvent are evenly distributed, they continue to move about, but there is no net movement of either one in any direction.

As discussed, the chemical and physical properties of the plasma membrane allow only a few types of molecules to enter and exit a cell simply by diffusion. Gases can diffuse through the lipid bilayer; this is the mechanism by which oxygen enters cells and carbon dioxide exits cells. Also, consider the movement of oxygen from the alveoli (air sacs) of the lungs to blood in the lung capillaries (Fig. 4.6). After inhalation (breathing in), the concentration of oxygen in the alveoli is higher than that in the blood; therefore, oxygen diffuses into the blood. The principle of diffusion can be employed in the treatment of certain human disorders, as is discussed in the Science Focus on page 71.

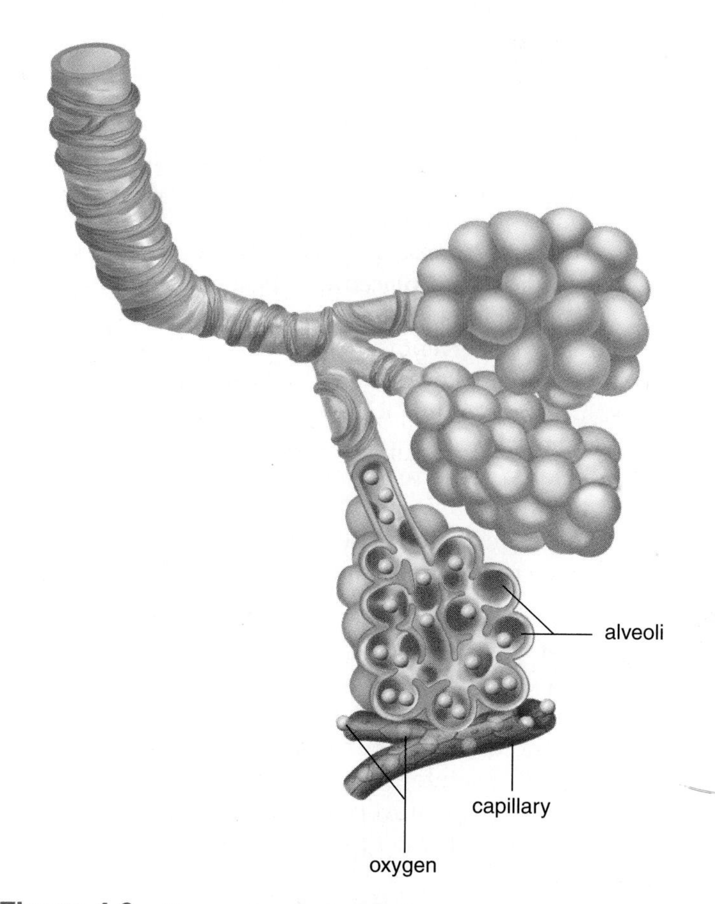

Figure 4.6 Gas exchange in lungs.
Oxygen (O_2) diffuses into the capillaries of the lungs because there is a higher concentration of oxygen in the alveoli (air sacs) than in the capillaries.

Molecules diffuse down their concentration gradients. A few types of small molecules can simply diffuse through the plasma membrane.

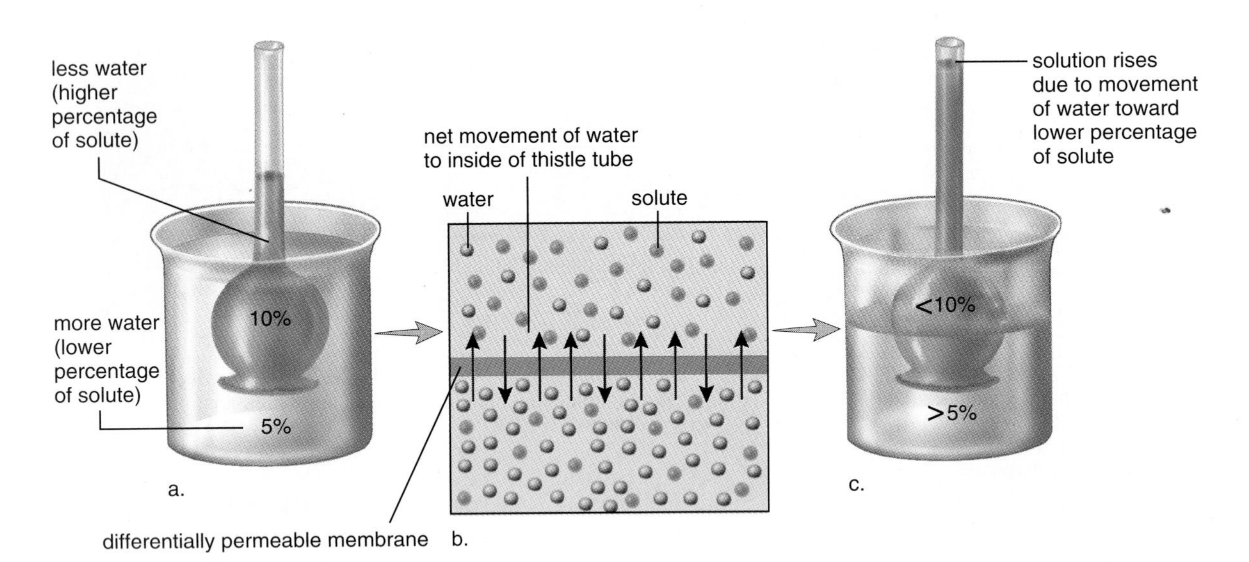

Figure 4.7 Osmosis demonstration.
(Far left) A thistle tube, covered at the broad end by a differentially permeable membrane, contains a 10% sugar solution. The beaker contains a 5% sugar solution. **(Middle)** The solute (green circles) is unable to pass through the membrane, but the water (blue circles) passes through in both directions. There is a net movement of water toward the inside of the thistle tube, where there is a lower percentage of water molecules. **(Far right)** Due to the incoming water molecules, the level of the solution rises in the thistle tube.

Osmosis

Osmosis is the diffusion of water into and out of cells. To illustrate osmosis, a thistle tube containing a 10% sugar solution[1] is covered at one end by a differentially permeable membrane and is then placed in a beaker containing a 5% sugar solution (Fig. 4.7). The beaker contains more water molecules (lower percentage of solute) per volume, and the thistle tube contains fewer water molecules (higher percentage of solute) per volume. Under these conditions, there is a net movement of water from the beaker to the inside of the thistle tube across the membrane. The solute is unable to pass through the membrane; therefore, the level of the solution within the thistle tube rises (Fig. 4.7*c*).

Notice the following in this illustration of osmosis:

1. A differentially permeable membrane separates two solutions. The membrane does not permit passage of the solute.
2. The beaker has more water (lower percentage of solute), and the thistle tube has less water (higher percentage of solute).
3. The membrane permits passage of water, and there is a net movement of water from the beaker to the inside of the thistle tube.
4. In the end, the concentration of solute in the thistle tube is less than 10%. Why? Because there is now less solute per volume. And the concentration of solute in the beaker is greater than 5%. Why? Because there is now more solute per volume.

Water enters the thistle tube due to the osmotic pressure of the solution within the thistle tube. **Osmotic pressure** is the pressure that develops in a system due to osmosis[2]. In other words, the greater the possible osmotic pressure the more likely water will diffuse in that direction. Due to osmotic pressure, water is absorbed from the human large intestine, is retained by the kidneys, and is taken up by capillaries from tissue fluid.

Tonicity

Tonicity refers to the strength of a solution in relationship to osmosis. In the laboratory, cells are normally placed in **isotonic solutions;** that is, the solute concentration is the same on both sides of the membrane, and therefore there is no net gain or loss of water (Fig. 4.8). The prefix *iso* means the same as, and the term tonicity refers to the strength of the solution. A 0.9% solution of the salt sodium chloride (NaCl) is known to be isotonic to red blood cells. Therefore, intravenous solutions medically administered usually have this tonicity.

Solutions that cause cells to swell, or even to burst, due to an intake of water are said to be **hypotonic solutions.** The prefix *hypo* means less than, and refers to a solution with a lower percentage of solute (more water) than the cell. If a cell is placed in a hypotonic solution, water enters the cell; the net movement of water is from the outside to the inside of the cell.

Any concentration of a salt solution lower than 0.9% is hypotonic to red blood cells. Animal cells placed in such a solution expand and sometimes burst due to the buildup of pressure. The term *lysis* is used to refer to disrupted cells; hemolysis, then, is disrupted red blood cells.

[1]Percent solutions are grams of solute per 100 ml of solvent. Therefore, a 10% solution is 10 g of sugar with water added to make up 100 ml of solution.

[2]Osmotic pressure is measured by placing a solution in an osmometer and then immersing the osmometer in pure water. The pressure that develops is the osmotic pressure of a solution.

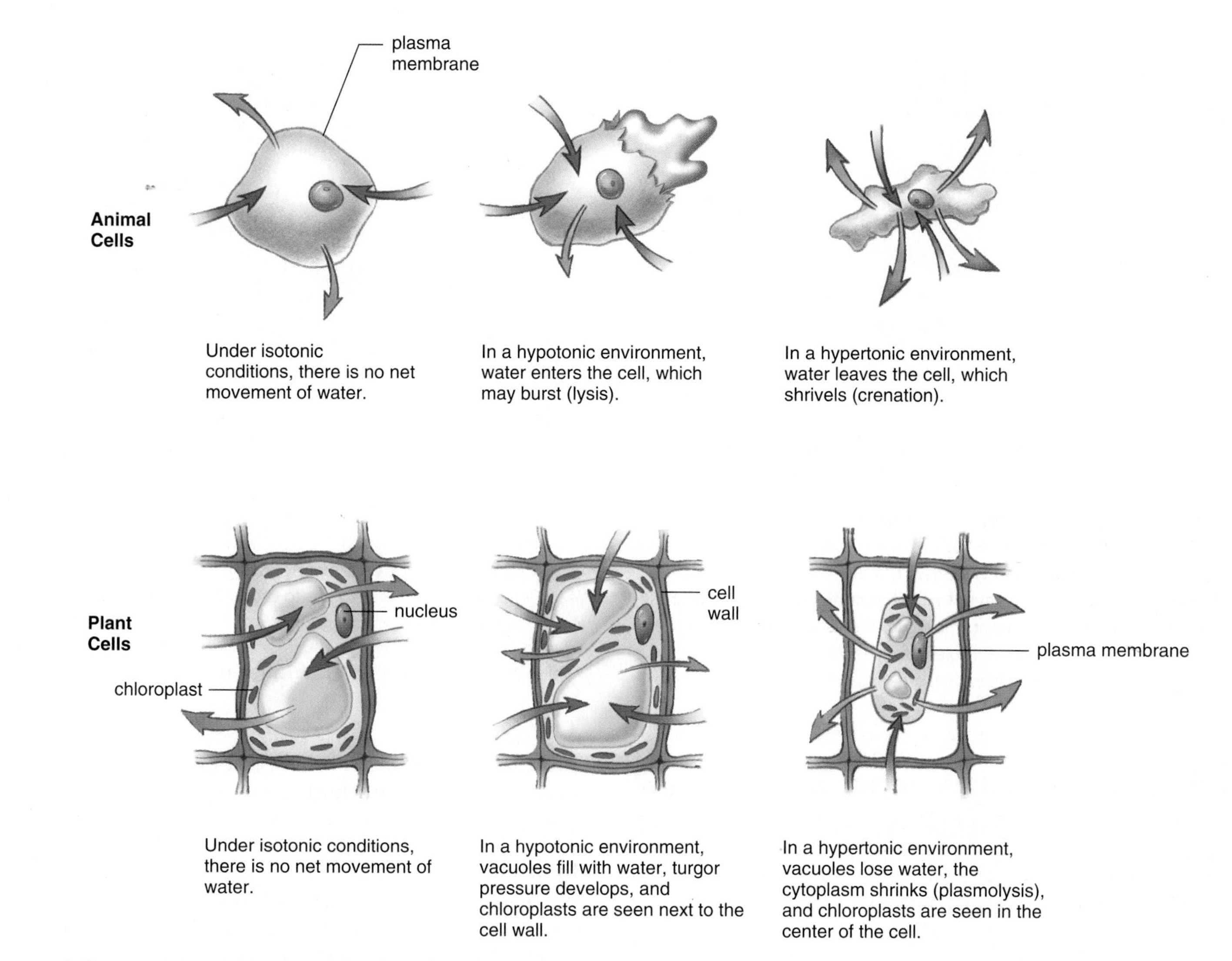

Figure 4.8 Osmosis in animal and plant cells.
The arrows indicate the net movement of water. In an isotonic solution, a cell neither gains nor loses water; in a hypotonic solution, a cell gains water; and in a hypertonic solution, a cell loses water.

The swelling of a plant cell in a hypotonic solution creates **turgor pressure.** When a plant cell is placed in a hypotonic solution, we observe expansion of the cytoplasm because the large central vacuole gains water and the plasma membrane pushes against the rigid cell wall. The plant cell does not burst because the cell wall does not give way. Turgor pressure in plant cells is extremely important to the maintenance of the plant's erect position. If you forget to water your plants they wilt due to decreased turgor pressure.

Solutions that cause cells to shrink or to shrivel due to a loss of water are said to be **hypertonic solutions.** The prefix *hyper* means more than, and refers to a solution with a higher percentage of solute (less water) than the cell. If a cell is placed in a hypertonic solution, water leaves the cell; the net movement of water is from the inside to the outside of the cell.

Any solution with a concentration higher than 0.9% sodium chloride is hypertonic to red blood cells. If animal cells are placed in this solution, they shrink. The term *crenation* refers to red blood cells in this condition. Meats are sometimes preserved by salting them. The bacteria are not killed by the salt but by the lack of water in the meat.

When a plant cell is placed in a hypertonic solution, the plasma membrane pulls away from the cell wall as the large central vacuole loses water. This is an example of **plasmolysis,** a shrinking of the cytoplasm due to osmosis. Dead plants you see along a salted roadside after the winter died because they were exposed to a hypertonic solution.

In an isotonic solution, a cell neither gains nor loses water. In a hypotonic solution, a cell gains water. In a hypertonic solution, a cell loses water and the cytoplasm shrinks.

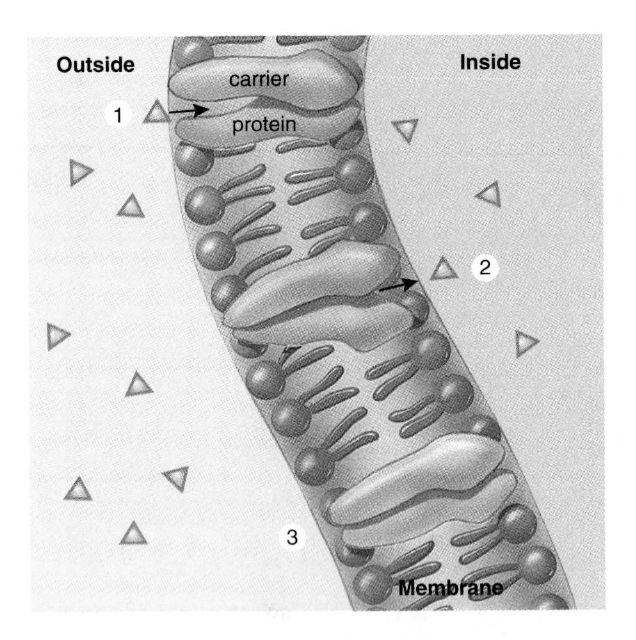

Figure 4.9 Facilitated transport.
A carrier protein speeds the rate at which a solute crosses a membrane from higher solute concentration to lower solute concentration. (1) Molecule enters carrier. (2) Molecule is transported across the membrane and exits on inside. (3) Carrier returns to its former state.

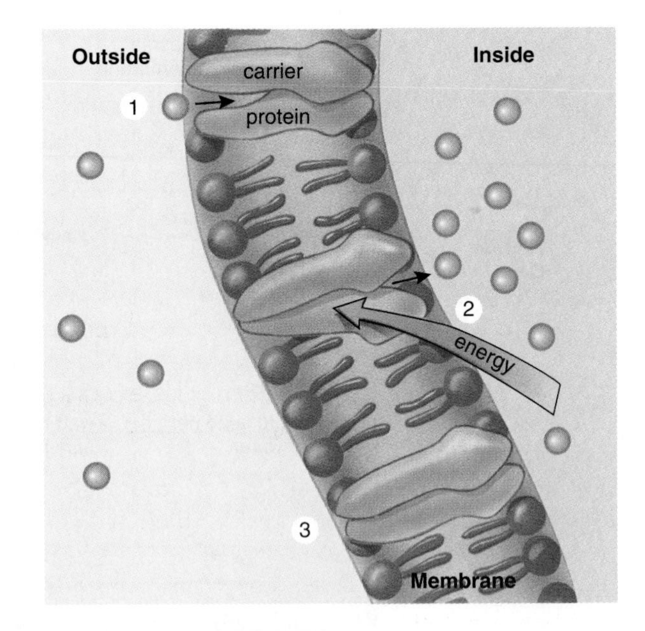

Figure 4.10 Active transport.
Active transport allows a solute to cross the membrane from lower solute concentration to higher solute concentration. (1) Molecule enters carrier. (2) Chemical energy of ATP is needed to transport the molecule which exits inside of cell. (3) Carrier returns to its former state.

4.4 Transport by Carrier Proteins

The plasma membrane impedes the passage of all but a few substances. Yet, biologically useful molecules do enter and exit the cell at a rapid rate because there are carrier proteins in the membrane. **Carrier proteins** are specific; each can combine with only a certain type of molecule, which is then transported through the membrane. It is not completely understood how carrier proteins function; but after a carrier combines with a molecule, the carrier is believed to undergo a change in shape that moves the molecule across the membrane. Carrier proteins are required for facilitated and active transport (see Table 4.1).

Some of the proteins in the plasma membrane are carriers; they transport biologically useful molecules into and out of the cell.

Facilitated Transport

Facilitated transport explains the passage of such molecules as glucose and amino acids across the plasma membrane, even though they are not lipid soluble. The passage of glucose and amino acids is facilitated by their reversible combination with carrier proteins, which in some manner transport them through the plasma membrane. These carrier proteins are specific. For example, various sugar molecules of identical size might be present inside or outside the cell, but glucose can cross the membrane hundreds of times faster than the other sugars. This is a good example of the differential permeability of the membrane.

The carrier for glucose has been isolated and a model has been developed to explain how it works (Fig. 4.9). It seems likely that the carrier has two conformations and that it switches back and forth between the two states. After glucose binds to the open end of a carrier, it closes behind the glucose molecule. As glucose moves along, the constricted end of the carrier opens in front of the molecule. After glucose is released into the cytoplasm of the cell, the carrier changes its conformation so that the binding site for glucose is again open. This process can occur as often as 100 times per second. Apparently, the cell has a pool of extra glucose carriers. When the hormone insulin binds to a plasma membrane receptor, more glucose carriers ordinarily appear in the plasma membrane. Some forms of diabetes are caused by insulin insensitivity; that is, the binding of insulin does not result in extra glucose carriers in the membrane.

The model shows that after a carrier has assisted the movement of a molecule to the other side of the membrane, it is free to assist the passage of other similar molecules. Neither diffusion, explained previously, nor facilitated transport requires an expenditure of chemical energy because the molecules are moving down their concentration gradient in the same direction they tend to move anyway.

Active Transport

During **active transport,** ions or molecules move through the plasma membrane, accumulating either inside or outside the cell. For example, iodine collects in the cells of the thyroid gland; nutrients are completely absorbed from the gut by the cells lining the digestive tract, and sodium ions (Na^+) can be almost completely withdrawn from urine by cells lining the kidney tubules. In these instances, substances have moved to the region of higher concentration, exactly opposite to the process of diffusion. It has been estimated that up to 40% of a cell's energy supply may be used for active transport of solute across its membrane.

Both carrier proteins and an expenditure of energy are needed to transport molecules against their concentration gradient (Fig. 4.10). In this case, energy (ATP molecules) is required for the carrier to combine with the substance to be transported. Therefore, it is not surprising that cells involved primarily in active transport, such as kidney cells, have a large number of mitochondria near the membrane through which active transport is occurring.

Proteins involved in active transport often are called *pumps,* because just as a water pump uses energy to move water against the force of gravity, proteins use energy to move a substance against its concentration gradient. One type of pump that is active in all cells, but is especially associated with nerve and muscle cells, moves sodium ions (Na^+) to the outside of the cell and potassium ions (K^+) to the inside of the cell. These two events are presumed to be linked, and the carrier protein is called a **sodium-potassium pump.** A change in carrier shape after the attachment, and again after the detachment, of a phosphate group allows the carrier to combine alternately with sodium ions and potassium ions (Fig. 4.11). The phosphate group is donated by ATP, which is broken down enzymatically by the carrier.

The passage of salt (NaCl) across a plasma membrane is of primary importance in cells. The chloride ion (Cl^-) usually crosses the plasma membrane because it is attracted by positively charged sodium ions (Na^+). First, sodium ions are pumped across a membrane, and then chloride ions simply diffuse through channels that allow their passage. As noted in Figure 4.3, the chloride ion channels malfunction in persons with cystic fibrosis, and this leads to the symptoms of this inherited (genetic) disorder.

During facilitated transport, substances follow their concentration gradient. During active transport, substances are moved against their concentration gradient.

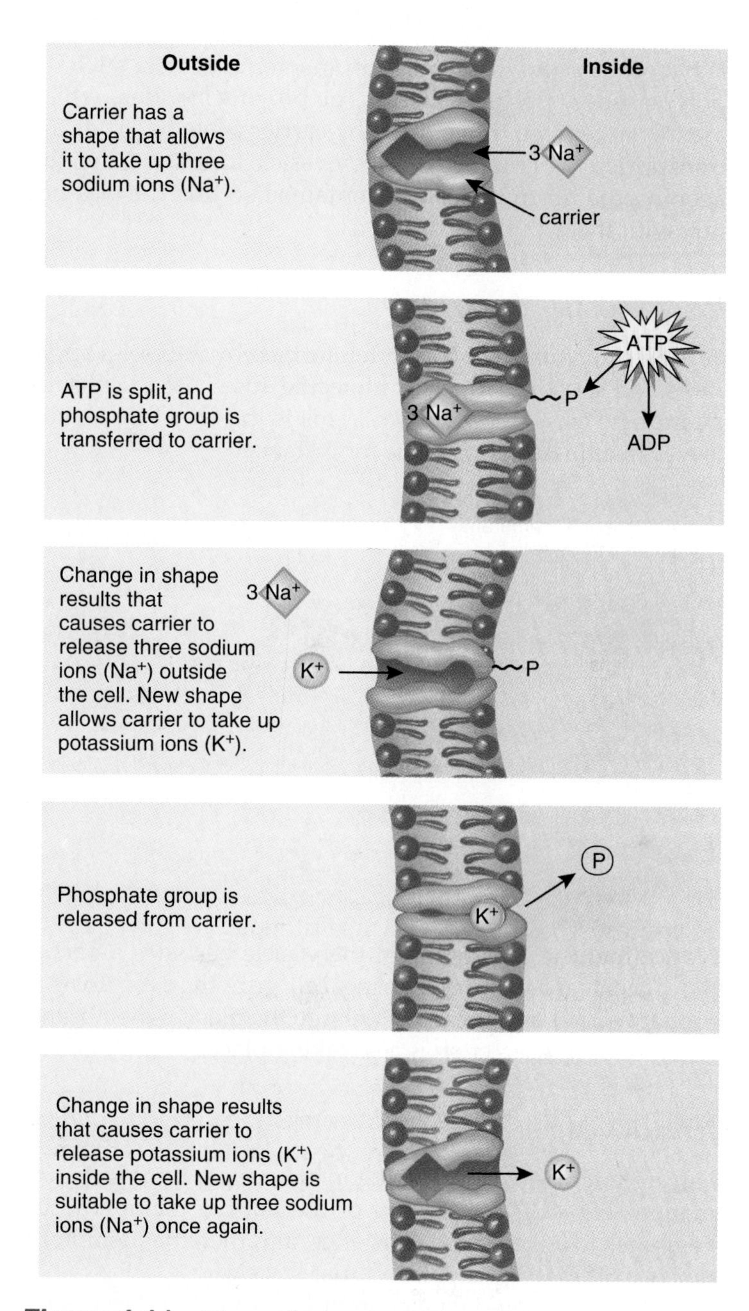

Figure 4.11 The sodium-potassium pump.
A carrier protein actively moves three sodium ions (Na^+) to the outside of the cell for every potassium ion (K^+) pumped to the inside of the cell. Note that chemical energy of ATP is required.

4.5 Exocytosis and Endocytosis

What about the transport of macromolecules such as polypeptides, polysaccharides, or polynucleotides, which are too large to be transported by carrier proteins? They are transported in or out of the cell by vesicle formation, thereby keeping the macromolecules contained so that they do not mix with those in the cytoplasm.

Exocytosis

During **exocytosis,** vesicles often formed by the Golgi apparatus and carrying a specific molecule, fuse with the plasma membrane as secretion occurs. This is the way that insulin leaves insulin-secreting cells, for instance.

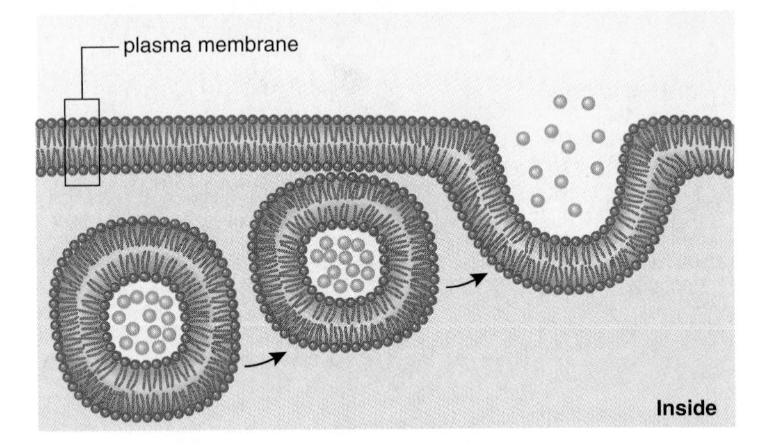

Notice that the membrane of the vesicle becomes a part of the plasma membrane. During cell growth, exocytosis is probably used as a means to enlarge the plasma membrane, whether or not secretion is also taking place.

Endocytosis

During **endocytosis,** cells take in substances by vesicle formation (Fig. 4.12). A portion of the plasma membrane invaginates to envelop the substance, and then the membrane pinches off to form an intracellular vesicle.

When the material taken in by endocytosis is large, such as a food particle or another cell, the process is called phagocytosis. **Phagocytosis** is common in unicellular organisms like amoebas and in amoeboid cells like macrophages, which are large cells that engulf bacteria and worn-out red blood cells in mammals. When the endocytic vesicle fuses with a lysosome, digestion occurs.

Pinocytosis occurs when vesicles form around a liquid or very small particles. Blood cells, cells that line the kidney tubules or intestinal wall, and plant root cells all use this method of ingesting substances. Whereas phagocytosis can be seen with the light microscope, the electron microscope must be used to observe pinocytic vesicles, which are no larger than 1–2 μm.

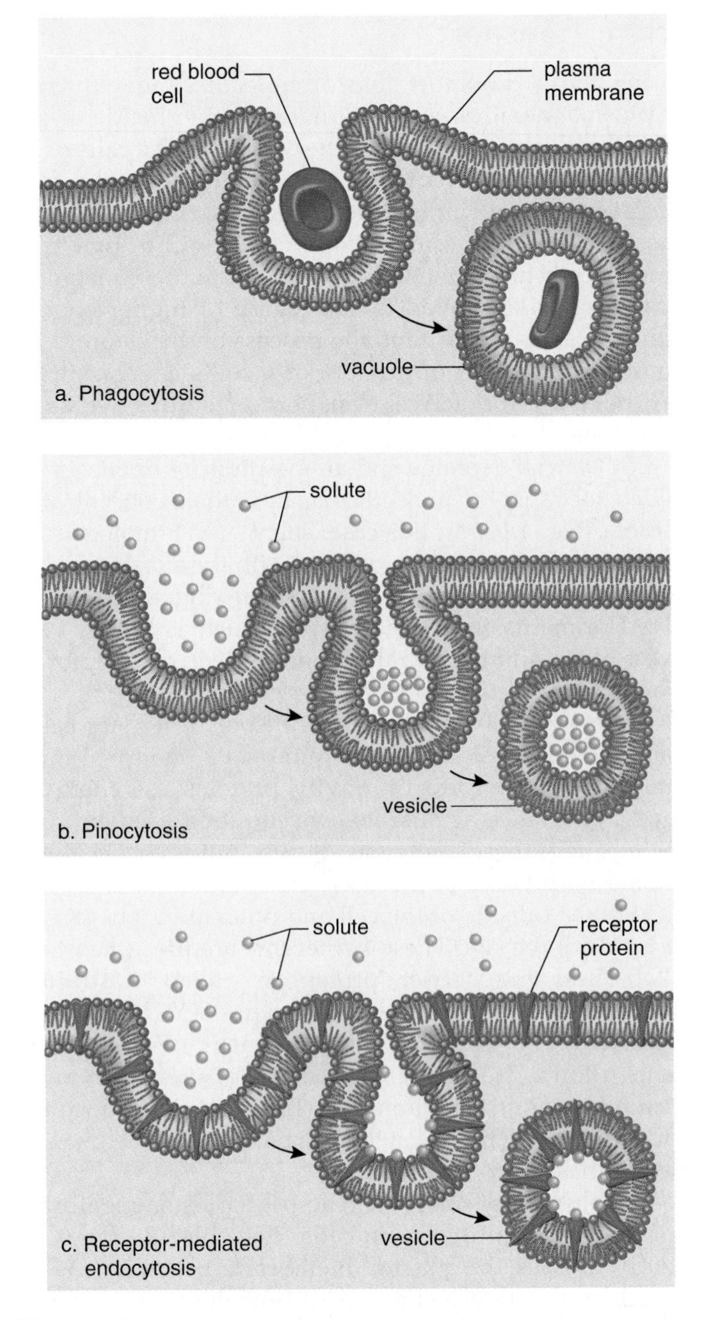

Figure 4.12 Three methods of endocytosis.
a. Phagocytosis occurs when the substance to be transported into the cell is large; certain specialized cells in the body can engulf worn-out red blood cells by phagocytosis. Digestion occurs when the resulting vacuole fuses with a lysosome. **b.** Pinocytosis occurs when a macromolecule such as a polypeptide is to be transported into the cell. The result is a small vacuole or vesicle. **c.** Receptor-mediated endocytosis is a form of pinocytosis. The substance to be taken in (a ligand) first binds to a specific receptor protein which migrates to a pit or is already in a pit. The vesicle that forms contains the ligand and its receptor. Sometimes the receptor is recycled, as shown in Figure 4.13.

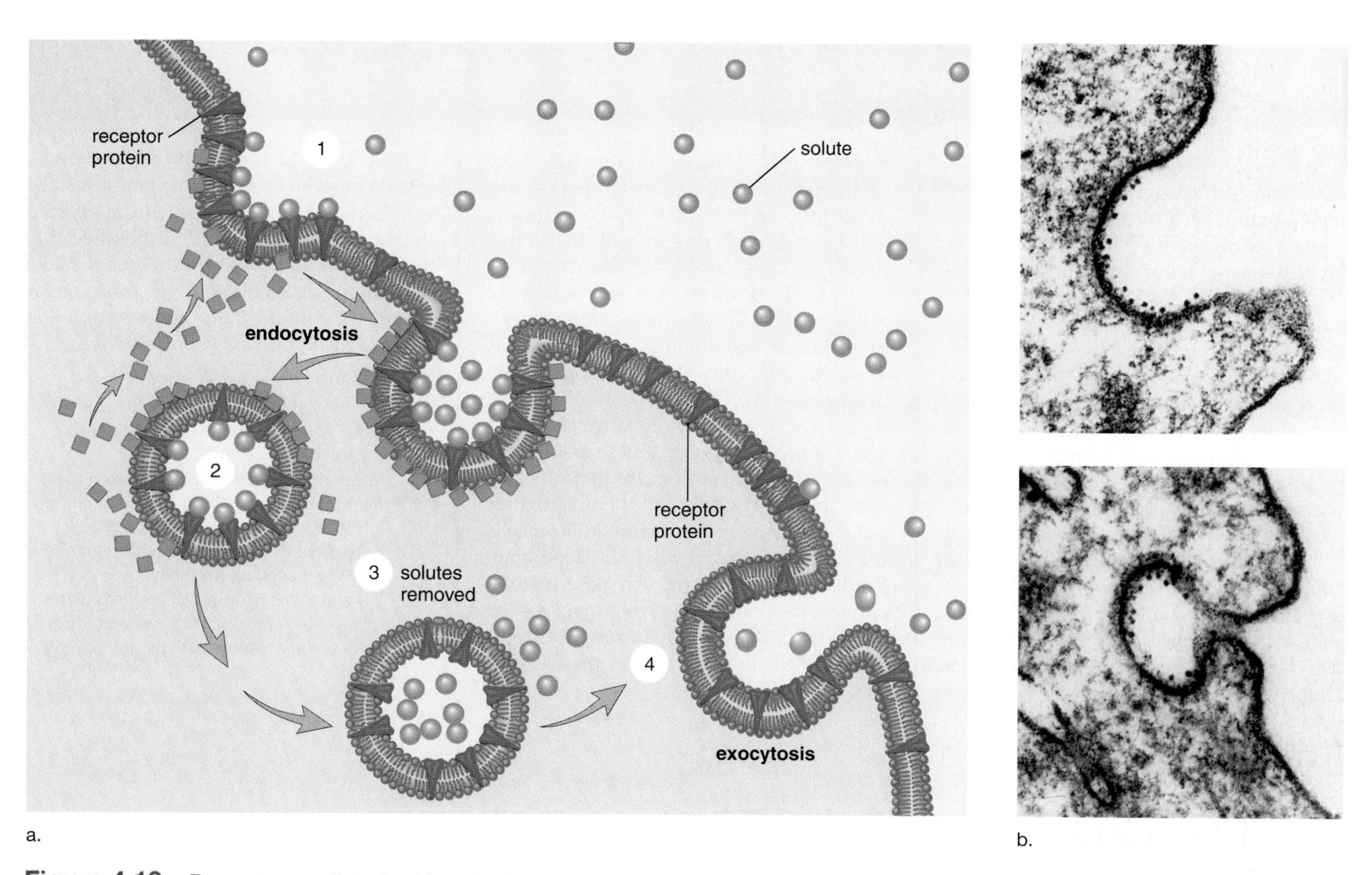

Figure 4.13 **Receptor-mediated endocytosis.**
a. (1) The receptors in the coated pits combine only with a solute. (2) The vesicle that forms is at first coated with a fibrous protein (blue squares), but soon the vesicle loses its coat. (3) Solutes leave the vesicle. (4) When exocytosis occurs, membrane and therefore receptors are returned to the plasma membrane. **b.** Electron micrographs of a coated pit in the process of forming a vesicle.

Receptor-mediated endocytosis is a form of pinocytosis that is quite specific because it involves the use of a receptor protein shaped in such a way that a specific molecule such as vitamins, peptide hormones, and lipoproteins can bind to it. The binding of a solute to the receptors causes the receptors to gather at one location. This location is called a coated pit because there is a layer of fibrous protein on the cytoplasmic side (see step 1, Fig. 4.13). Once the vesicle is formed, the fibrous coat is released and the vesicle appears uncoated (see step 2). The fate of the vesicle and its contents depends on the kind of solute it contains. Sometimes the solute simply enters the cytoplasm (step 3). A spent hormone, on the other hand, may be digested when the vesicle fuses with a lysosome. The membrane of the vesicle and, therefore, the receptors are returned to the plasma membrane (step 4), or the vesicle can go to other membranous locations.

Aside from simply allowing substances to enter cells selectively from an extracellular fluid, coated pits are also involved in the transfer and exchange of substances between cells. Such exchanges take place when the substances move from maternal blood into fetal blood at the placenta, for example.

The importance of receptor-mediated endocytosis is demonstrated by a genetic disorder called familial hypercholesterolemia. Cholesterol is transported in blood by a complex of lipids and proteins called low-density lipoprotein (LDL). These individuals have inherited a gene that causes them to have a reduced number and/or defective receptors for LDL in their plasma membranes. Instead of cholesterol entering cells, it accumulates in the walls of arterial blood vessels, leading to high blood pressure, occluded (blocked) arteries, and heart attacks.

Substances are secreted from a cell by exocytosis. Substances enter a cell by endocytosis. Receptor-mediated endocytosis allows cells to take up specific kinds of molecules and then sort them within the cell.

Bioethical Issue

Such celebrities as Mohammad Ali, a former heavyweight boxing champion, Janet Reno, the attorney general of the United States, and Michael J. Fox, a favorite movie actor, have Parkinson disease. By age 65, Parkinson disease (PD) affects roughly one of every 100 Americans. Due to the death of brain cells that produce a substance called dopamine, motor control is not as smooth as it should be. The three obvious signs of Parkinson are slowness of movement, tremor, and rigidity. As the disease worsens, patients become unable to carry out even the simplest activities.

You might think that the condition could be cured by simply giving a patient dopamine, but dopamine, like many other chemical substances, cannot cross the blood-brain barrier. The blood-brain barrier is simply due to the impermeability of the capillaries serving the brain. Nutrients, such as glucose, and essential amino acids can only pass through due to facilitated transport. Most drugs can't get through at all.

Luckily a precursor of dopamine called L-dopa can get through the blood-brain barrier, and when L-dopa is given as a medication, it will be changed into dopamine until there are few cells left to do the job. Along the way, physicians and patients are faced with a wide assortment of adjunct remedies. Some of these are surgical procedures. Michael J. Fox opted for pallidotomy, a procedure that kills off cells that go out of control when the dopamine-producing cells die off. An experimental surgical procedure, however, involves the transplantation of dopamine-producing fetal tissue into the brains of people with PD. People who have received such transplants report a lessening of symptoms.

Is it ethical to use tissue from aborted fetuses for transplants? Is it possible that women would have abortions just to make fetal tissue available to loved ones, or for payment? Should there be governmental safeguards to prevent such a possibility?

Questions

1. Is it ethical to use fetal tissue to prevent older people from having a debilitating disorder? Why or why not?
2. Suppose you had a choice between using fetal tissue (no payment required) and adult cells bioengineered to produce dopamine (payment required), which would you choose and why?
3. Do you favor banning all research using fetal tissue, or doing such research under certain circumstances? Explain.

Summarizing the Concepts

4.1 Plasma Membrane Structure and Function

There are two components of the plasma membrane: lipids and proteins. In the lipid bilayer, phospholipids are arranged with their hydrophilic heads at the surfaces and their hydrophobic tails in the interior. The lipid bilayer has the consistency of oil, and therefore proteins can move laterally in the membrane. Glycolipids and glycoproteins are involved in marking the cell as belonging to a particular individual and tissue.

The hydrophobic portion of an integral protein lies in the lipid bilayer of the plasma membrane, and the hydrophilic portion lies at the surface. Proteins act as receptors, carry on enzymatic reactions, join cells together, form channels, or act as carriers to move substances across the membrane.

4.2 The Permeability of the Plasma Membrane

Some substances like gases and water are free to cross a plasma membrane, and others, particularly ions, charged molecules, and macromolecules, have to be assisted across. Passive ways of crossing a plasma membrane (diffusion and facilitated transport) do not require an expenditure of chemical energy. Active ways of crossing a plasma membrane (active transport and vesicle formation) do require an expenditure of chemical energy.

4.3 Diffusion and Osmosis

Lipid-soluble compounds, water, and gases simply diffuse across the membrane from the area of higher concentration to the area of lower concentration.

The diffusion of water across a differentially permeable membrane is called osmosis. Water moves across the membrane into the area of lower water (higher solute) content. When cells are in an isotonic solution, they neither gain nor lose water; when they are in a hypotonic solution, they gain water; and when they are in a hypertonic solution, they lose water.

4.4 Transport by Carrier Proteins

Some molecules are transported across the membrane by carrier proteins that span the membrane.

During facilitated transport, a carrier protein assists the movement of a molecule down its concentration gradient. No energy is required.

During active transport, a carrier protein acts as a pump that causes a substance to move against its concentration gradient. The sodium-potassium pump carries Na^+ to the outside of the cell and K^+ to the inside of the cell. Energy in the form of ATP molecules is required for active transport to occur.

4.5 Exocytosis and Endocytosis

Larger substances can enter and exit a membrane by endocytosis and exocytosis. Exocytosis involves secretion. Endocytosis includes phagocytosis and pinocytosis which includes receptor-mediated endocytosis. Receptor-mediated endocytosis makes use of receptor molecules in the plasma membrane. Once specific solutes bind to their receptors, the coated pit becomes a coated vesicle. After losing the coat, the vesicle can join with the lysosome, or after freeing the solute, the receptor-containing vesicle can fuse with the plasma membrane.

Studying the Concepts

1. Describe the structure of the plasma membrane, including the phospholipid bilayer and the various types of proteins. 68–70
2. Why is a plasma membrane called differentially permeable? 72
3. What are the mechanisms by which substances enter and exit cells? Which are passive ways, and which are active ways? 72
4. Define diffusion, and give an example. 73
5. Define osmosis. Define isotonic, hypertonic, and hypotonic solutions, and give examples of how these concentrations affect red blood cells. 74–75
6. Draw a simplified diagram of a red blood cell before and after being placed in these solutions. What terms are used to refer to the condition of the red blood cell in a hypertonic solution and in a hypotonic solution? 75
7. Draw a simplified diagram of a plant cell before and after being placed in these solutions. Describe the cell contents under these conditions. 75
8. How does facilitated transport differ from simple diffusion across the plasma membrane? 76
9. How does active transport differ from facilitated transport? Give an example. 77
10. Diagram and define endocytosis and exocytosis. Describe and contrast three methods of endocytosis. 78–79

Testing Yourself

Choose the best answer for each question.

1. Label this diagram of the plasma membrane.

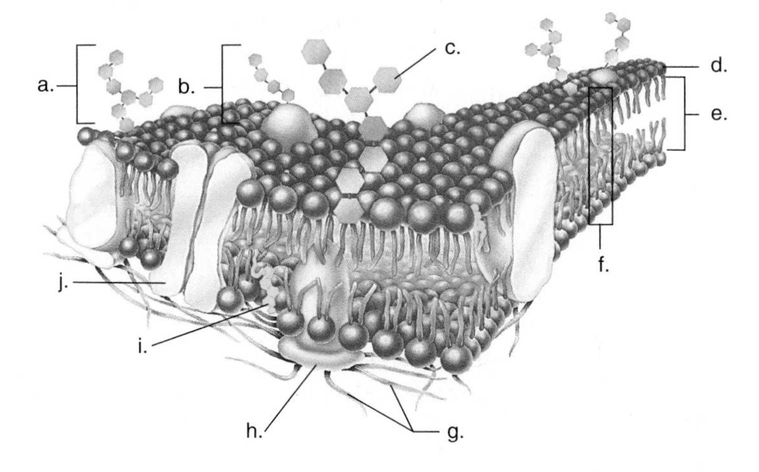

2. The fluid-mosaic model of membrane structure refers to
   a. the fluidity of proteins in the membrane and the pattern of phospholipids in the membrane.
   b. the fluidity of phospholipids and the pattern of proteins in the membrane.
   c. the fluidity of cholesterol and the pattern of sugar chains outside the membrane.
   d. the lack of fluidity of internal membranes compared to the plasma membrane, and the ability of the proteins to move laterally in the membrane.
   e. the fluidity of hydrophobic regions, proteins and mosaic pattern of hydrophilic regions.
3. A phospholipid molecule has a head and two tails. The tails are found
   a. at the surfaces of the membrane.
   b. in the interior of the membrane.
   c. spanning the membrane.
   d. where the environment is hydrophilic.
   e. Both a and b are correct.
4. During diffusion,
   a. all molecules move only from the area of higher to lower concentration.
   b. solvents move from the area of higher to lower concentration.
   c. there is a net movement of molecules from the area of higher to lower concentration.
   d. a cell must be present for any movement of molecules to occur.
   e. molecules move against their concentration gradient if they are small and charged.
5. When a cell is placed in a hypotonic solution,
   a. solute exits the cell to equalize the concentration on both sides of the membrane.
   b. water exits the cell toward the area of lower solute concentration.
   c. water enters the cell toward the area of higher solute concentration.
   d. solute exits and water enters the cell.
   e. Both c and d are correct.
6. When a cell is placed in a hypertonic solution,
   a. solute exits the cell to equalize the concentration on both sides of the membrane.
   b. water exits the cell toward the area of lower solute concentration.
   c. water exits the cell toward the area of higher solute concentration.
   d. solute exits and water enters the cell.
   e. Both a and c are correct.
7. Active transport
   a. requires a carrier protein.
   b. moves a molecule against its concentration gradient.
   c. requires a supply of energy.
   d. does not occur during facilitated transport.
   e. All of these are correct.
8. The sodium-potassium pump
   a. helps establish an electrochemical gradient across the membrane.
   b. concentrates sodium on the outside of the membrane.
   c. utilizes a carrier protein and energy.
   d. is present in the plasma membrane.
   e. All of these are correct.
9. A scientist observing a protozoan notices a vacuole discharging its contents at the plasma membrane. This is an example of
   a. phagocytosis and vacuole formation.
   b. endocytosis and active transport.
   c. exocytosis and secretion.
   d. active transport and vacuole release.
   e. Both c and d are correct.

10. Receptor-mediated endocytosis
    a. is no different from phagocytosis.
    b. brings specific substances into the cell.
    c. helps to concentrate proteins in vesicles.
    d. All of these are correct.
11. Write hypotonic solution or hypertonic solution beneath each cell. Justify your conclusions.

a. ______________________

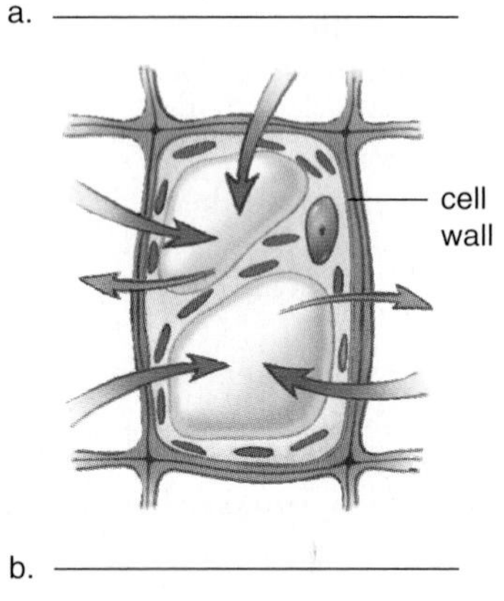

b. ______________________

Thinking Scientifically

1. Considering the movement of molecules across the plasma membrane:
   a. Contrast the manner in which alcohol and water enter a cell (page 72).
   b. Contrast the manner in which sodium ions (Na^+) and chloride ions (Cl^-) exit a cell (Fig. 4.3).
   c. Contrast the manner in which amino acids and proteins enter a cell (page 72).
   d. How might the proteins from question c be digested (chapter 3)?
2. Exocytotic vesicles add plasma membrane to the cell, and endocytotic vesicles remove plasma membrane.
   a. In a cell in which the amount of plasma membrane stays constant, how many exocytotic vesicles per endocytotic vesicles would you expect?
   b. Imagine a cell that is moving from left to right. If vesicle formation is facilitating movement, where would you expect exocytosis to be occurring? Where would you expect endocytosis to be occurring?
   c. Receptor-mediated endocytosis is a process by which a substance combines with a receptor before endocytosis brings the entire complex into the cell. Imagine a virus that enters a cell in this manner. If so, what additional step is needed for the virus to enter the cell proper?

Understanding the Terms

active transport 77
carrier protein 70
channel protein 70
cholesterol 68
concentration gradient 72
differentially permeable 72
diffusion 73
endocytosis 78
enzymatic protein 70
exocytosis 78
facilitated transport 76
fluid-mosaic model 68
glycolipid 68
glycoprotein 69
hypertonic solution 75
hypotonic solution 74
integral protein 69
isotonic solution 74
osmosis 74
osmotic pressure 74
peripheral protein 69
phagocytosis 78
pinocytosis 78
plasmolysis 75
receptor-mediated endocytosis 79
receptor protein 70
sodium-potassium pump 77
solute 73
solvent 73
tonicity 74
turgor pressure 75

Match the terms to these definitions:

a. __________ Movement of molecules from a region of higher concentration to a region of lower concentration.
b. __________ Internal pressure that adds to the strength of a cell and builds up when water moves by osmosis into a plant cell.
c. __________ Solution that contains the same concentration of solute and water as the cell.
d. __________ Passive transfer of a substance into and out of a cell along a concentration gradient by a process that requires a carrier.
e. __________ Process in which an intracellular vesicle fuses with the plasma membrane so that the vesicle's contents are released outside the cell.

Using Technology

Your study of membrane structure and function is supported by these available technologies:

Essential Study Partner CD-ROM
Cells → Cell Membrane
Visit the Mader web site for related ESP activities.

Exploring the Internet
The Mader Home Page provides resources and tools as you study this chapter.

http://www.mhhe.com/biosci/genbio/mader

Virtual Physiology Laboratory CD-ROM
Diffusion, Osmosis, & Tonicity
Enzyme Characteristics

Life Science Animations 3D Video
4 Diffusion
5 Osmosis

CHAPTER 5

Cell Division

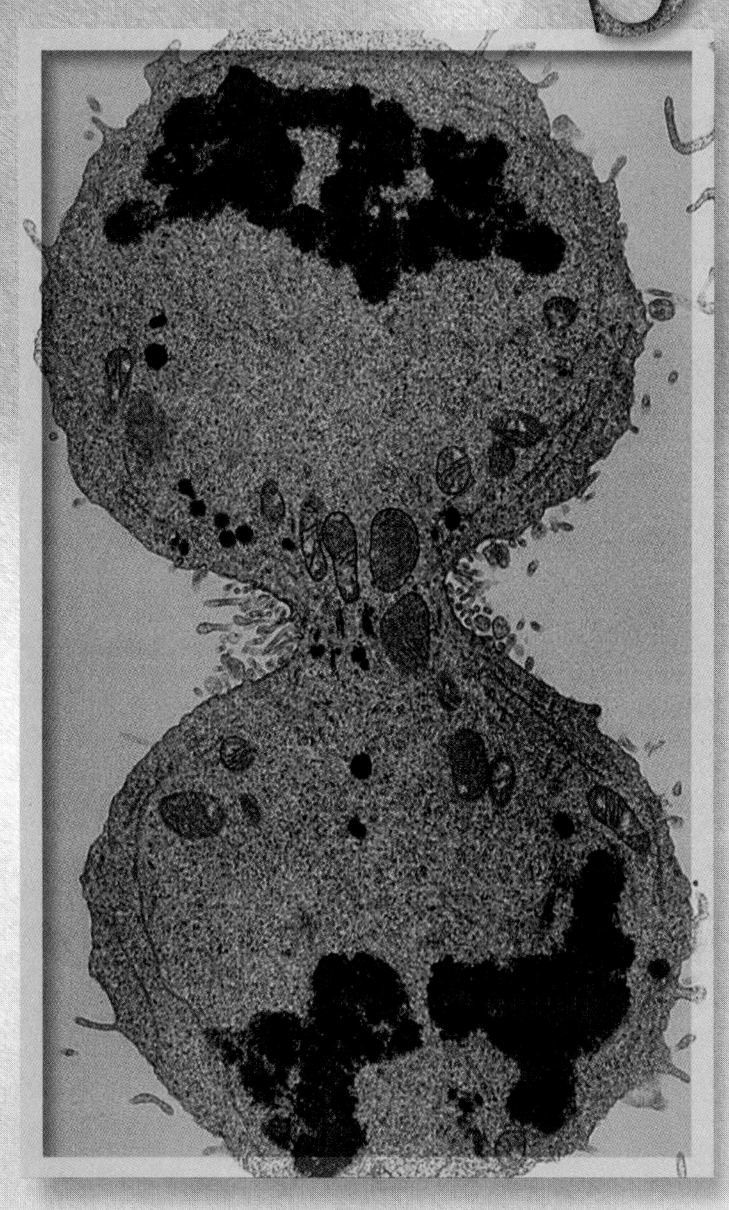

During mitotic cell division, the daughter cells receive a full complement of chromosomes. The process is very orderly; it has to be to ensure that the daughter cells receive one of each chromosome and not two of one kind and none of another. Cell division will produce new cells so that the organism can grow.

Chapter Concepts

Ten million times a day a beautiful dance finishes inside Laura's body. It's the intricate movement of chromosomes that occurs whenever one of her cells divides into two. The cell has already duplicated its chromosomes. Then, in a carefully choreographed set of maneuvers, the two parts of the chromosome separate so that the resulting two cells each have a complete chromosomal complement. Through this amazing reproductive process, skin cells can proliferate to repair a wound and the immune system can quickly amass an army of cells to defeat an infection. And through a refinement of the basic cell division process, sperm and egg cells arise, ready to join and create a new life. This chapter will provide a guided tour of cells as they divide.

5.1 Maintaining the Chromosome Number

When a eukaryotic cell is not undergoing division, the DNA (and associated proteins) within a nucleus is a tangled mass of thin threads called **chromatin.** At the time of cell division, chromatin coils, loops, and condenses to give highly compacted structures that are called **chromosomes.** The Science Focus on the next page describes the transition from chromatin to chromosomes in greater detail.

When the chromosomes are highly coiled and condensed at the time of cell division, it is possible to photograph and count them. Each species has a characteristic chromosome number; for instance, human cells contain 46 chromosomes, corn has 20 chromosomes, and a crayfish has 200! This is called the full or **diploid (2n)** number of chromosomes that occurs in all cells of the body. The diploid number includes two chromosomes of each kind. Half the diploid number is called the **haploid (n)** number of chromosomes, representing only one of each kind of chromosome. In the life cycle of many animals, only sperm and eggs have the haploid number of chromosomes.

Cell division in most eukaryotes involves nuclear division and **cytokinesis,** which is division of the cytoplasm. **Somatic,** or body, **cells** undergo **mitosis**—that is, nuclear division in which the chromosome number stays constant (Fig. 5.1). In diploid organisms such as ourselves, a diploid nucleus divides to produce daughter nuclei that are also diploid. Some organisms are haploid as adults. In that case, the haploid nucleus divides to produce daughter nuclei that are also haploid.

Mitosis is the type of nuclear division that occurs in growth and repair of the body. Humans begin life as a single cell but they eventually have one hundred trillion cells as adults due to mitosis. Even then, mitosis does not stop. As adults, however, certain tissues such as the epidermis of the skin, the lining of the digestive and respiratory tracts, and the lymphoid tissue that produces blood cells account for much of the cell division that occurs. Other tissues, namely

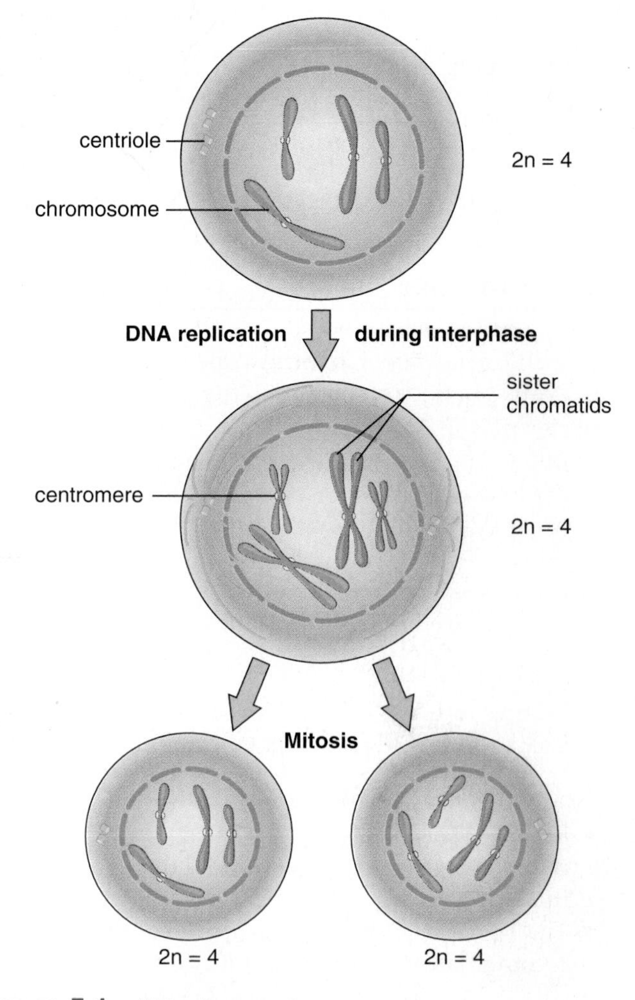

Figure 5.1 Mitosis overview.
Following DNA replication during interphase, each chromosome in the parental nucleus is duplicated and consists of two sister chromatids. During mitosis, the centromeres divide and the sister chromatids separate, becoming daughter chromosomes that move into the daughter nuclei. Therefore, daughter cells have the same number and kinds of chromosomes as the parental cell. (The blue chromosomes were inherited from the father, and the red chromosomes were inherited from the mother.)

skeletal muscle and nervous tissue, do not divide and therefore cannot be renewed. Regardless, cell division is controlled and does not usually occur helter-skelter, the notable exception occurs when cancer, a cell division disease, is present. Before mitosis takes place, DNA replicates; thereafter, each chromosome is duplicated and has two identical parts called sister chromatids. **Sister chromatids** are constricted and attached to each other at a region called the **centromere.** During nuclear division the centromere divides and the sister chromatids separate. Once separation occurs, they are called **daughter chromosomes.** These chromosomes, which consist of only one chromatid, are distributed equally to the daughter nuclei. In this way, each daughter cell gets a copy of each chromosome.

Science Focus

What's in a Chromosome?

When early investigators decided that the genes are contained in the chromosomes, they had no idea of chromosome composition. By the mid-1900s, it was known that chromosomes are made up of both DNA and protein. Only in recent years, however, have investigators been able to produce models suggesting how chromosomes are organized.

A eukaryotic chromosome is more than 50% protein. Many of these proteins are concerned with DNA and RNA synthesis, but a large proportion, termed histones, seem to play primarily a structural role. There are five primary types of histone molecules, designated H1, H2A, H2B, H3, and H4. Remarkably, the amino acid sequences of H3 and H4 vary little between organisms. For example, the H4 of peas is only two amino acids different from the H4 of cattle. This similarity suggests that there have been few mutations in the histone proteins during the course of evolution and that the histones therefore have very important functions.

A human cell contains at least 2 meters of DNA. Yet all of this DNA is packed into a nucleus that is about 5 μm in diameter. The histones are responsible for packaging the DNA so that it can fit into such a small space. First the DNA double helix is wound at intervals around a core of eight histone molecules (two copies each of H2A, H2B, H3, and H4), giving the appearance of a string of beads (Fig. 5A*a* and *b*). Each bead is called a nucleosome, and the nucleosomes are said to be joined by "linker" DNA. This string is coiled tightly into a fiber that has six nucleosomes per turn (Fig. 5A*c*). The H1 histone appears to mediate this coiling process. The fiber loops back and forth (Fig. 5A*d* and *e*) and can condense to produce a highly compacted form (Fig. 5A*f*) characteristic of metaphase chromosomes. No doubt, compact chromosomes are easier to move about than extended chromatin.

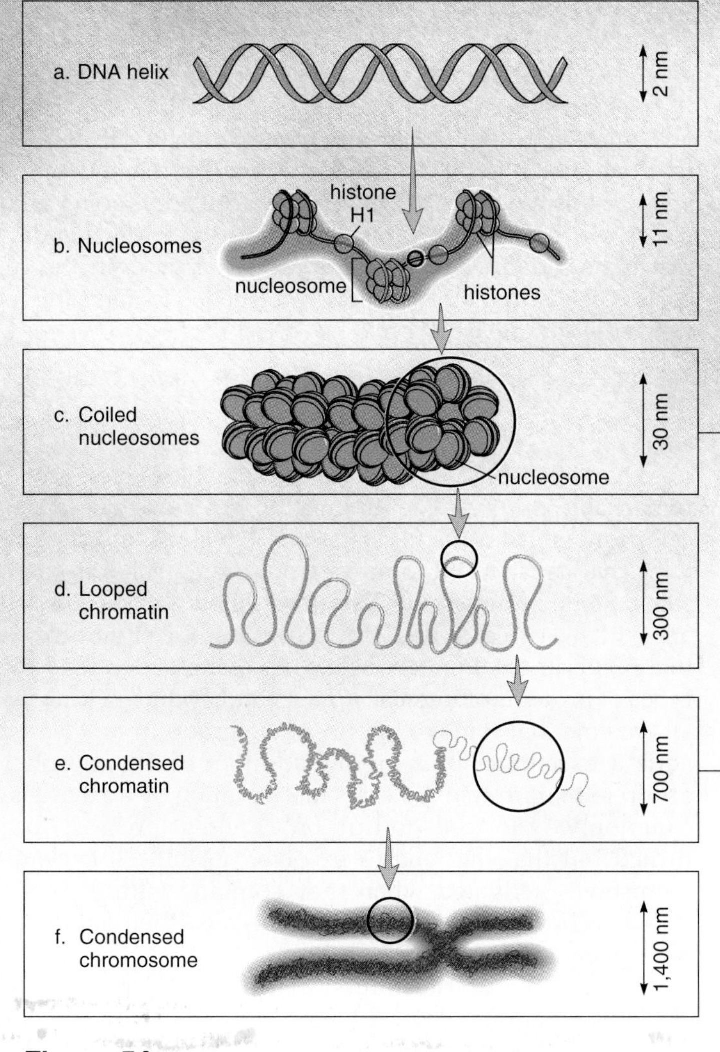

Figure 5A Levels of chromosome structure.
Each drawing has a scale giving a measurement of length for that drawing. Notice that each measurement represents an ever-increasing length; therefore, it would take a much higher magnification to see the structure in **(a)** than in **(f)**.

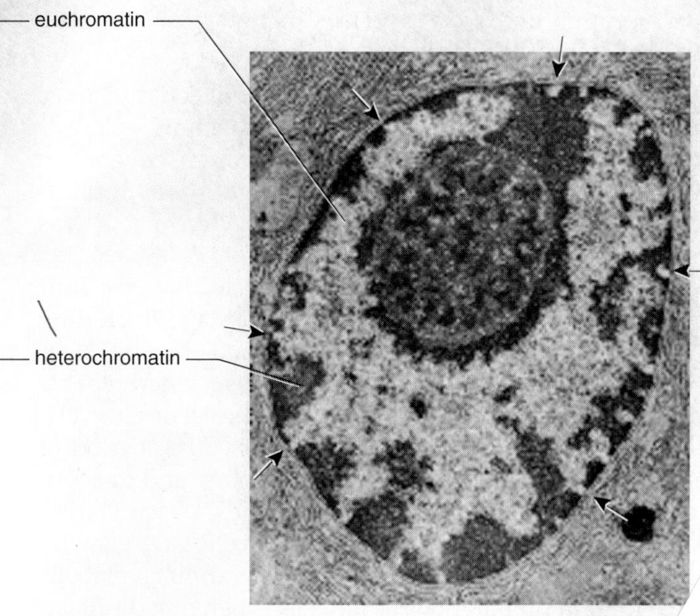

Figure 5B Eukaryotic nucleus.
The nucleus contains chromatin, DNA at two different levels coiling and condensation. Euchromatin is at the level of loc chromatin, and heterochromatin is at the level of conden chromatin in Figure 5A. Arrows indicate nuclear pores.

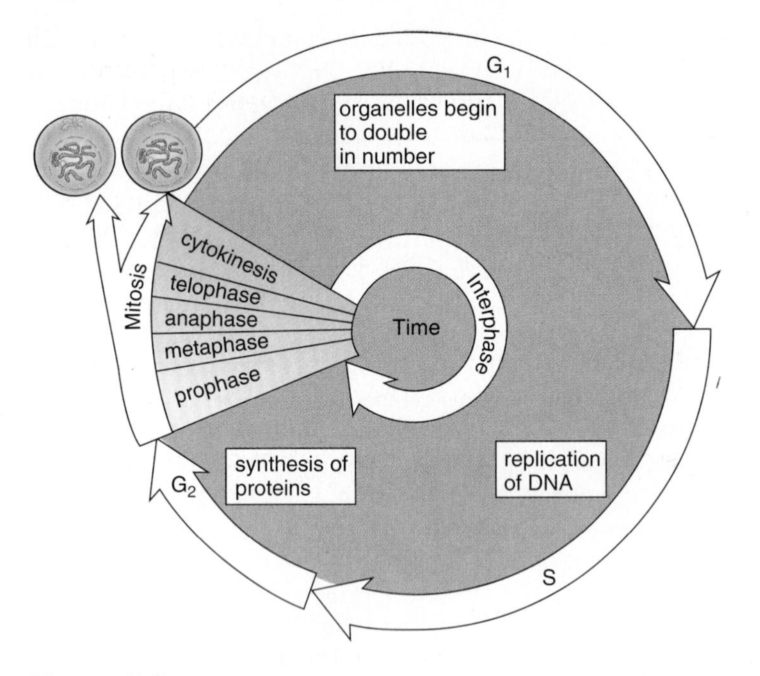

Figure 5.2 The cell cycle.
Cells go through a cycle that consists of four stages. The length of time that cells take to finish the cell cycle varies and some specialized cells like nerve cells and skeletal muscle cells no longer progress through the cycle. Mammalian cells usually take at least 16 hours to complete the cell cycle.

The Cell Cycle

Because there was little visible activity between divisions, early investigators dismissed this period of time as a resting state termed **interphase.** When it was discovered in the 1950s that DNA replication and chromosome duplication occur during interphase, the **cell cycle** concept was proposed.

Cells grow and divide during a cycle that has four stages (Fig. 5.2). The entire cell division stage, including both mitosis and cytokinesis, is termed the *M stage* (M = mitosis). The period of DNA synthesis when replication occurs is termed the *S stage* (S = synthesis) of the cycle.

There are two other stages of the cycle. The period of time prior to the S stage is termed the *G_1 stage,* and the period of time prior to the M stage is termed the *G_2 stage.* During the G_1 stage, the cell grows in size and the cellular organelles increase in number. During the G_2 stage, various metabolic events occur in preparation for mitosis. When first designated, G meant gap, but some biologists now prefer G = growth.

Cells differ in the length of time it takes them to complete the cell cycle. The difference seems to depend on how long they spend in G_1. There are even some human cells such as nerve cells and skeletal muscle cells that become permanently arrested in G_1, and these cells are said to have entered a G_0 stage. Once a mammalian cell enters the S stage, it usually only takes about 12 to 24 hours to finish the cell cycle.

The term interphase is now used to mean all the stages of the cell cycle (i.e., stages G_1, S, and G_2) except mitosis and cytokinesis. One thing to keep clearly in mind is that DNA replication occurs during the S stage—that is during interphase.

Cells undergo a cycle that includes the G_1, S, G_2, and M stages.

Proteins Regulate the Cell Cycle

If a cell arrested in the G_0 stage is placed in the cytoplasm of an S-stage cell, it will go on and finish the cell cycle. Similarly, there are cells that enter and never get beyond the G_2 stage. If this type of cell is fused with a cell undergoing mitosis, it will go ahead and undergo mitosis. From this data, researchers deduced that there are two places in the cell cycle where stimulatory proteins are needed to make the cell finish the complete cell cycle:

G_1 stage → S stage when DNA is synthesized

G_2 stage → M stage when mitosis occurs

Over the past few years, researchers have made remarkable progress identifying the proteins that cause a cell to move from the G_1 stage to the S stage and the proteins that cause a cell to move from the G_2 stage to the M stage. Some of these biologists worked with frog eggs, others with yeast cells, and still others used cell cultures as their experimental material. The researchers identified two types of proteins of interest: kinases and cyclins. A **kinase** is an enzyme that removes a phosphate group from ATP and adds it to another protein. The addition of the phosphate group to that protein activates it. Activation by a kinase is a common way for cells to turn on a cellular process, but it turned out that the kinases involved in the cell cycle are themselves activated when they combine with a protein called **cyclin.** Cyclins are so named because their quantity is not constant in the cell.

Figure 5.3 is an illustration that shows the process in a clockwise diagram. After S-kinase combines with S-cyclin, the kinase phosphorylates a protein that causes the cell to move from the G_1 stage to the S stage when DNA is synthesized (replicated). After that occurs, S-cyclin is destroyed, and S-kinase is no longer active.

Similarly, after M-kinase combines with M-cyclin, the kinase phosphorylates a protein that causes the cell to move from the G_2 stage to the M stage when mitosis occurs. Three things occur: (1) chromosomes condense, (2) the nuclear envelope disassembles, and (3) the spindle forms. (The spindle is the structure involved in chromosome movement during mitosis.) Now M-cyclin is destroyed.

Until recently, the mechanics of the cell cycle and the causes of cancer were thought to be distantly related. Now they appear to be intimately related. Growth factors are molecules that attach to plasma membrane receptors and thereby bring about cell growth. Ordinarily, a cyclin might combine with its kinase only when a growth factor is present. But a cyclin that has gone awry might combine with its kinase even when a growth factor is not present. The result would be a tumor. There are genes called tumor-suppressor genes that usually function to prevent cancer from occurring. It has been shown that the product of one major tumor-suppressor gene (known as *p53*) brings about the production of a protein that combines with kinases and prevents them from becoming activated.

Proteins regulate the cell cycle. A kinase combines with a cyclin at two critical checkpoints in the cell cycle: the beginning of the S stage when DNA is synthesized and the M stage with mitosis begins.

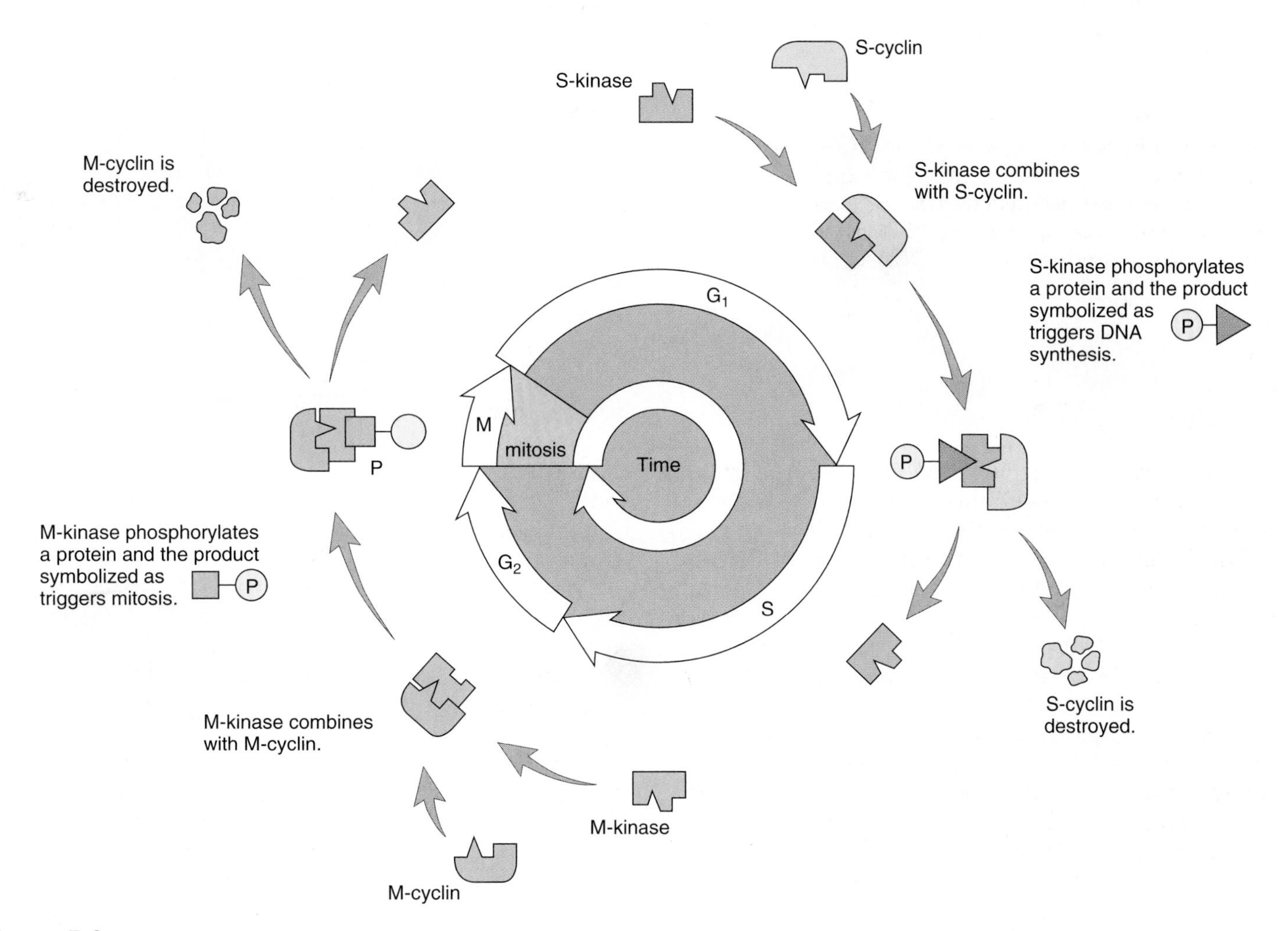

Figure 5.3 Control of the cell cycle.
At two critical checkpoints a kinase combines with a cyclin. Just before the S stage, S-kinase combines with S-cyclin and synthesis (replication) of DNA takes place. Just before the M stage, M-kinase combines with M-cyclin and mitosis takes place.

5.2 Mitosis in Detail

Mitosis is nuclear division that produces two daughter nuclei, each with the same number and kinds of chromosomes as the parental nucleus.

During mitosis, a **spindle** brings about an orderly distribution of chromosomes to the daughter cell nuclei. The spindle contains many fibers, each composed of a bundle of microtubules. Microtubules are hollow cylinders found in the cytoplasm which can assemble and disassemble. When microtubules assemble, tubulin protein dimers come together, and when they disassemble, the tubulin dimers separate.

The centrosome, which is the main microtubule-organizing center of the cell, divides before mitosis begins (Fig. 5.4). It's believed that centrosomes are responsible for organizing the spindle. Each centrosome contains a pair of barrel-shaped organelles called centrioles and an **aster** which is an array of short microtubules that radiate from the centrosome. The fact that plant cells lack centrioles suggests that centrioles are not required for spindle formation.

Mitosis in Animal Cells

Mitosis is a continuous process that is arbitrarily divided into four phases for convenience of description. These phases are prophase, metaphase, anaphase, and telophase (Fig. 5.5).

Prophase

It is apparent during early **prophase** that cell division is about to occur. The centrosomes begin moving away from each other toward opposite ends of the nucleus. Spindle fibers appear between the separating centrosomes as the nuclear envelope begins to fragment, and the nucleolus begins to disappear.

The chromosomes are now visible. Each is duplicated and composed of sister chromatids held together at a centromere.

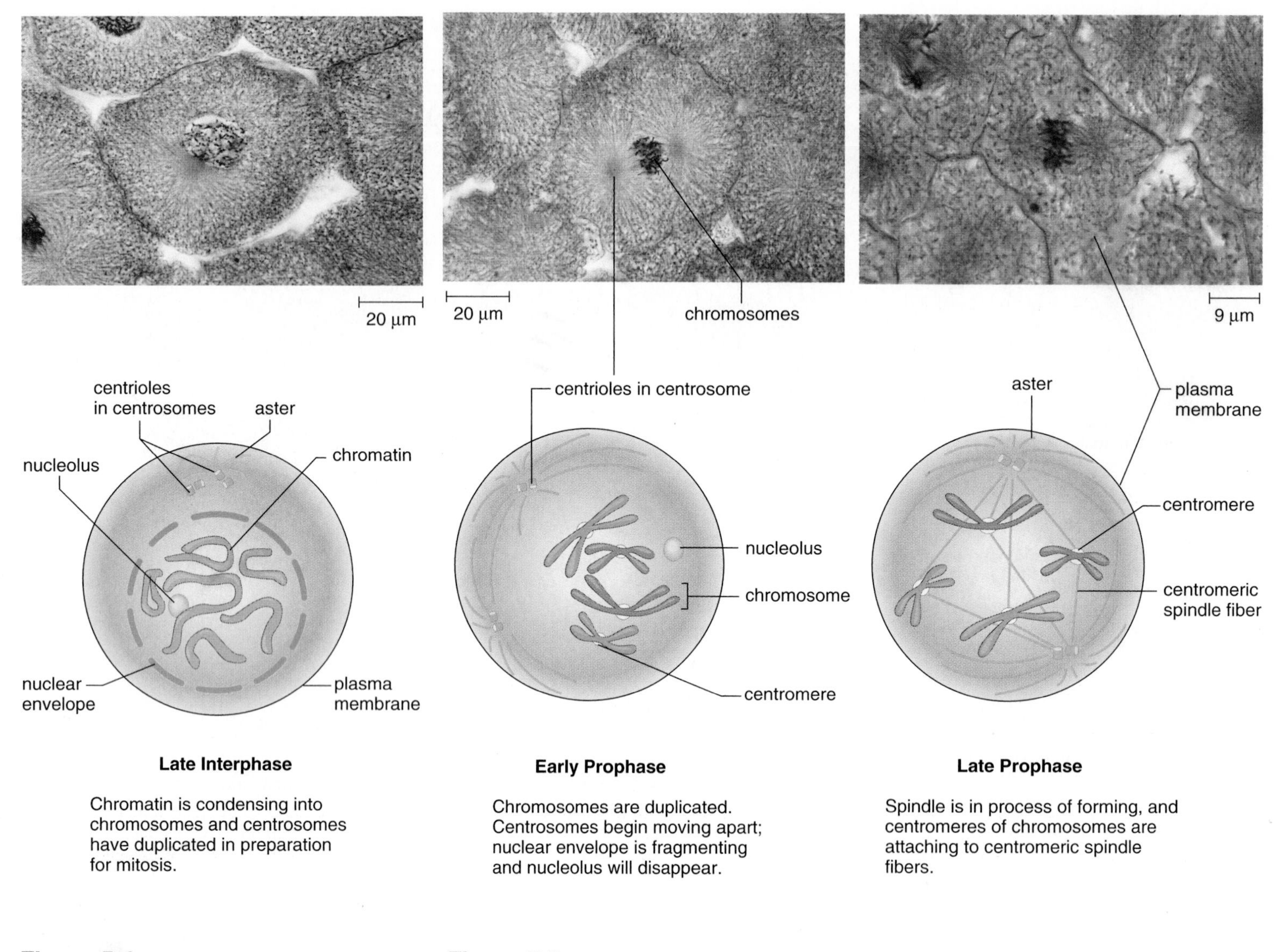

Figure 5.4 **Late interphase.**

Figure 5.5 **Phases of animal cell mitosis.**

Counting the number of centromeres in diagrammatic drawings tells you the number of chromosomes in a cell.

The spindle begins forming during late prophase and the chromosomes become attached to the spindle fibers. Their centromeres attach to fibers called centromeric (also called kinetochore) fibers. As yet, the chromosomes have no particular orientation as they move first one way and then the other.

Metaphase

By the time of **metaphase,** the fully formed spindle consists of poles, asters, and fibers. The chromosomes attached to centromeric spindle fibers are aligned at the metaphase plate (also called the equator) of the spindle. Polar spindle fibers reach beyond the metaphase plate and overlap. At the close of metaphase, the centromeres uniting the sister chromatids divide.

Anaphase

During **anaphase,** the centromeres divide. The sister chromatids separate, becoming two daughter chromosomes that move toward the opposite poles of the spindle. The daughter chromosomes have a centromere and single chromatid. What accounts for the movement of the daughter chromosomes? First, the centromeric spindle fibers disassemble at the region of the kinetochore, and this pulls the daughter chromosomes to the poles. Second, the polar spindle fibers lengthen as they slide past one another.

Telophase

During **telophase,** the spindle disappears and nuclear envelope components reassemble around the daughter chromosomes. Each daughter nucleus contains the same number and kinds of chromosomes as the original parental cell. Remnants of the polar spindle fibers are still visible between the two nuclei.

The chromosomes become more diffuse chromatin once again, and a nucleolus appears in each daughter nucleus. Cytokinesis is nearly complete, and soon there will be two individual daughter cells, each with a nucleus that contains the diploid number of chromosomes.

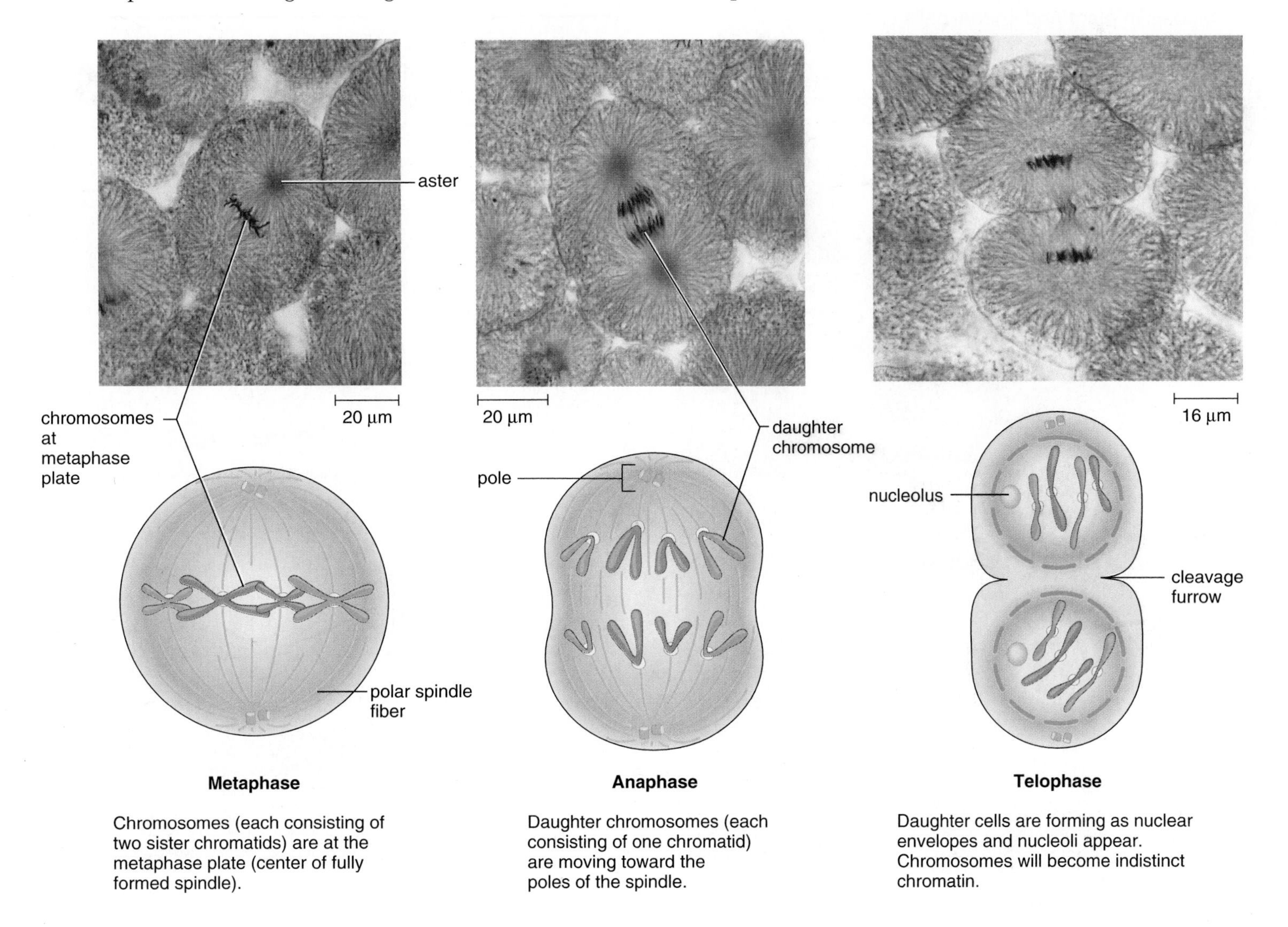

Metaphase

Chromosomes (each consisting of two sister chromatids) are at the metaphase plate (center of fully formed spindle).

Anaphase

Daughter chromosomes (each consisting of one chromatid) are moving toward the poles of the spindle.

Telophase

Daughter cells are forming as nuclear envelopes and nucleoli appear. Chromosomes will become indistinct chromatin.

How Plant Cells Divide

As with animal cells, mitosis in plant cells permits growth and repair. Certain plant tissue, called meristematic tissue, retains the ability to divide throughout the life of a plant. Meristematic tissue is found in root tip and shoot tip of stems. Lateral meristem accounts for the ability of trees to increase their girth each growing season.

Figure 5.6 illustrates mitosis in plant cells; exactly the same phases are seen in plant cells as in animal cells. During early prophase, the chromatin condenses into scattered previously duplicated chromosomes and the spindle forms; during late prophase, chromosomes attach to spindle fibers; during metaphase, the chromosomes are at the metaphase plate of the spindle; during anaphase, the sister chromatids separate becoming daughter chromosomes that move into the daughter nuclei; and during anaphase, cytokinesis begins. Although plant cells have a centrosome, and spindle, there are no centrioles nor asters during cell division.

Mitosis in plant and animal cells ensures the daughter cells have the same number and kinds of chromosomes as the parental cell.

Cytokinesis in Plant and Animal Cells

Cytokinesis, or cytoplasmic cleavage, usually accompanies mitosis. Division of the cytoplasm begins in anaphase, continues in telophase, but does not reach completion until just before the following interphase. By that time, the newly forming cells have received a share of the cytoplasmic organelles which duplicated during the previous interphase.

Cytokinesis in Plant Cells

Cytokinesis in plant cells occurs by a process different from that seen in animal cells (Fig. 5.7). The rigid cell wall that surrounds plant cells does not permit cytokinesis by furrowing. Instead, the Golgi apparatus produces membranous sacs called vesicles, which move along the microtubules to the midpoint between the two daughter nuclei. These vesicles fuse, forming a **cell plate.** Their membrane completes the plasma membrane for both cells. They also release molecules that signal the formation of plant cell walls, which are strengthened by the addition of cellulose fibrils.

A spindle forms during mitosis in plant cells, but there are no centrioles or asters. Cytokinesis in plant cells involves the formation of a cell plate.

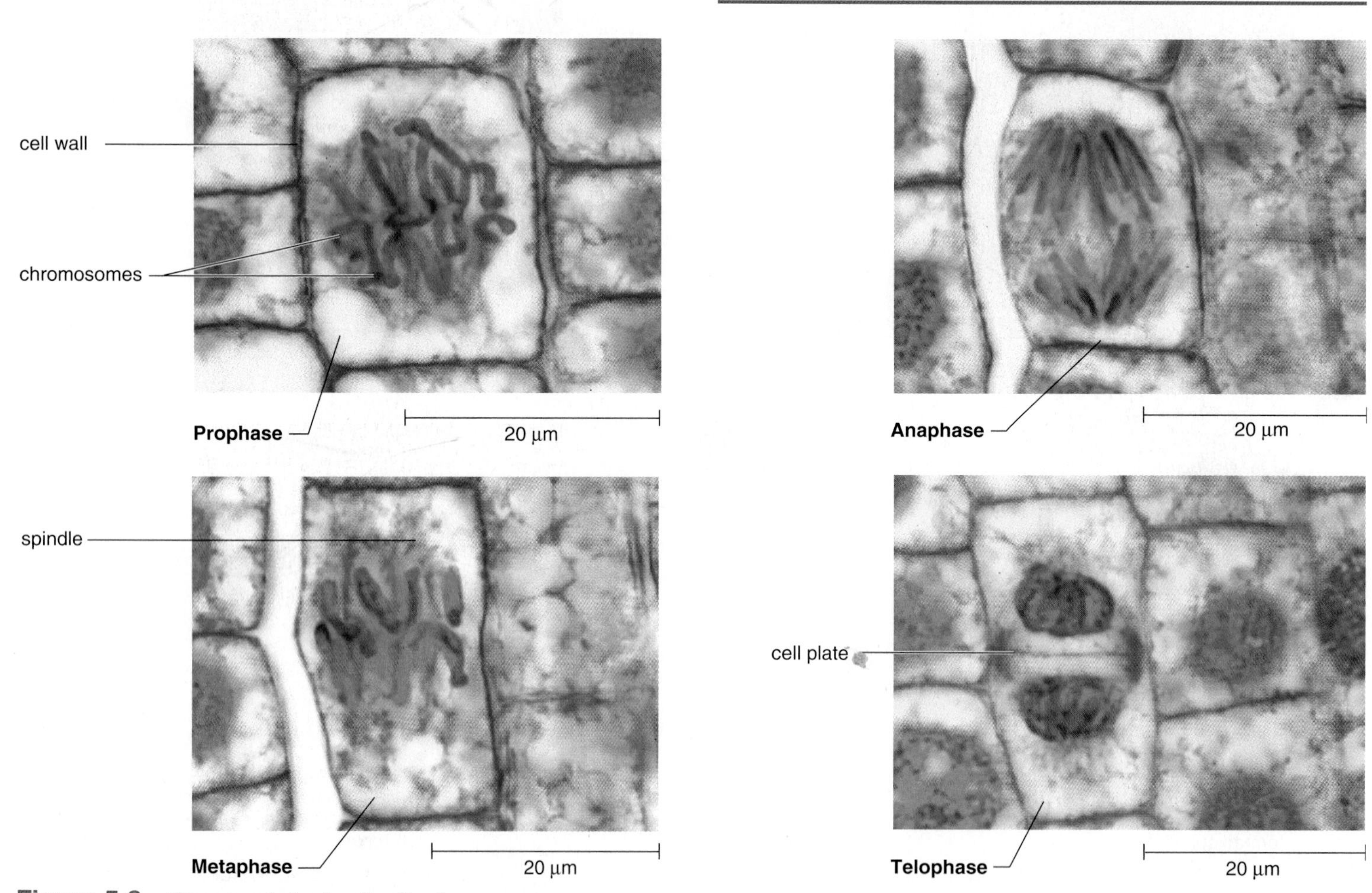

Figure 5.6 Phases of plant cell mitosis
Note the absence of centrioles and asters and the presence of the cell wall. In telophase, a cell plate develops between the two daughter cells. The cell plate marks the boundary of the new daughter cells, where new plasma membrane and a new cell wall is forming for each cell.

Cytokinesis in Animal Cells

In animal cells, a **cleavage furrow,** which is an indentation of the membrane between the two daughter nuclei, begins as anaphase draws to a close. The cleavage furrow deepens when a band of actin filaments, called the contractile ring, slowly forms a constriction between the two daughter cells. The action of the contractile ring can be likened to pulling a drawstring ever tighter about the middle of a balloon. As the drawstring is pulled tight, the balloon constricts in the middle.

A narrow bridge between the two cells can be seen during telophase, and then the contractile ring continues to separate the cytoplasm until there are two daughter cells (Fig. 5.8).

Cytokinesis in animal cells is accomplished by a furrowing.

Cell Division in Prokaryotes

Asexual reproduction requires a single parent, and the offspring are identical to the parent because they contain the same genes. The process of asexual reproduction in prokaryotes is termed binary fission because division (fission) produces two (binary) daughter cells that are identical to the original parental cell. Before division occurs, DNA replicates and the single chromosome is duplicated. Thus, there are two chromosomes that separate as the cell elongates. When the cell is approximately twice its original length, the plasma membrane grows inward and a new cell wall forms, dividing the cell into two approximately equal portions.

Asexual reproduction in prokaryotes is by binary fission. Following DNA replication, the two resulting chromosomes separate as the cell elongates.

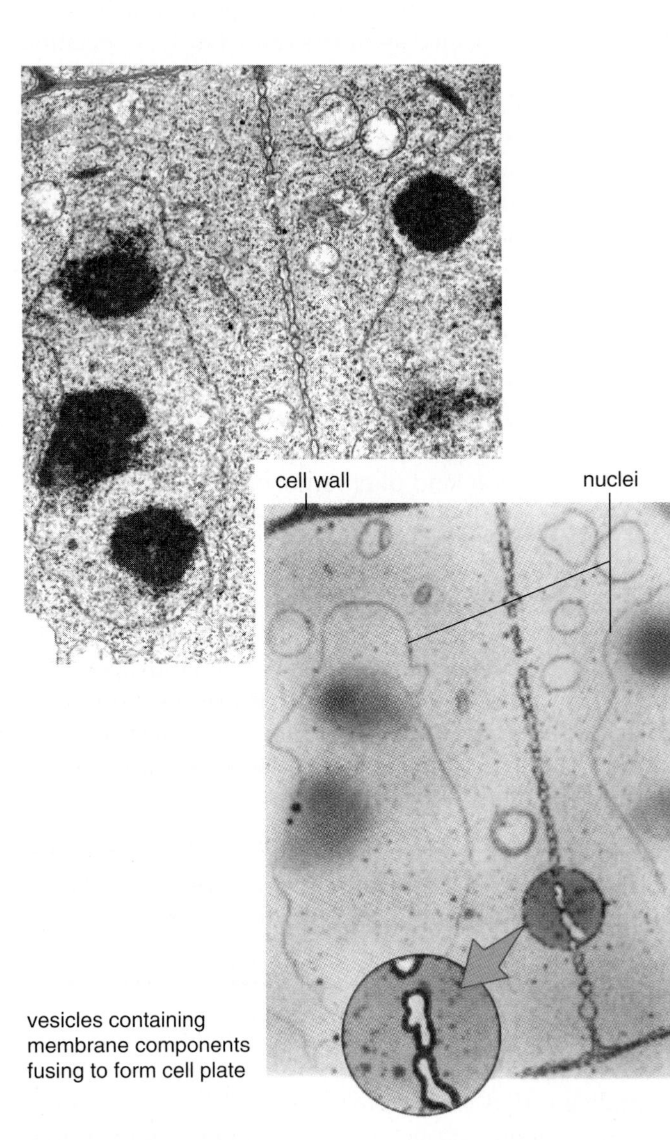

Figure 5.7 Cytokinesis in plant cells.
During cytokinesis in a plant cell, the cell plate forms midway between the two daughter nuclei and extends to the plasma membrane.

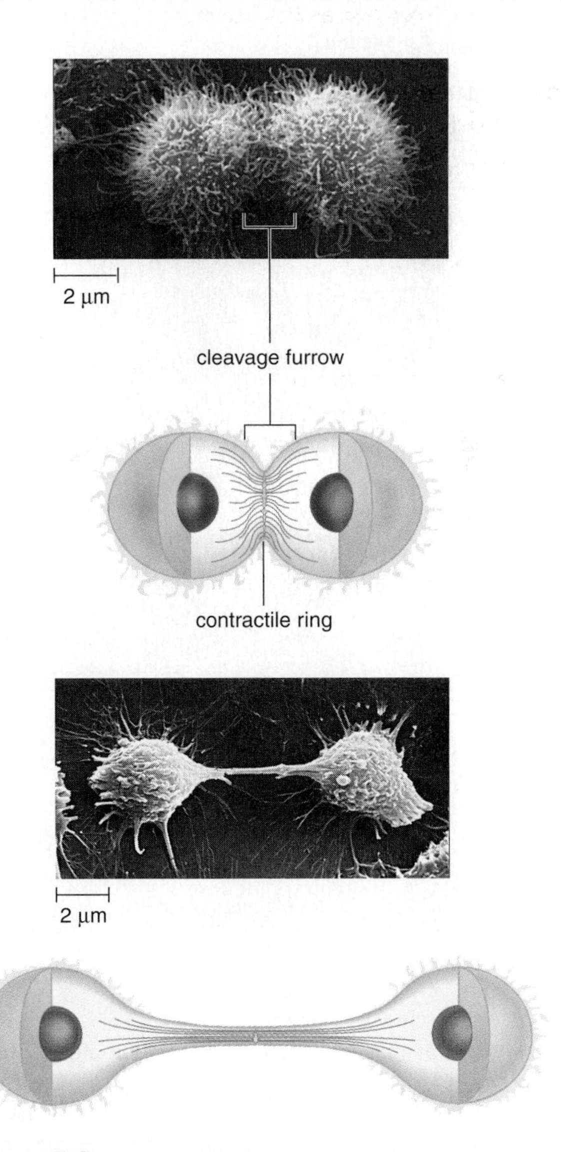

Figure 5.8 Cytokinesis in an animal cell.
A single cell becomes two cells by a furrowing process. A contractile ring composed of actin filaments gradually gets smaller, and the cleavage furrow pinches the cell into two cells.

5.3 Reducing the Chromosome Number

Meiosis occurs in any life cycle that involves sexual reproduction. **Meiosis** reduces the chromosome number in such a way that the daughter nuclei receive only one member of each homologous pair. The process of meiosis ensures that the next generation of individuals will have a combination of traits that are different from either parent.

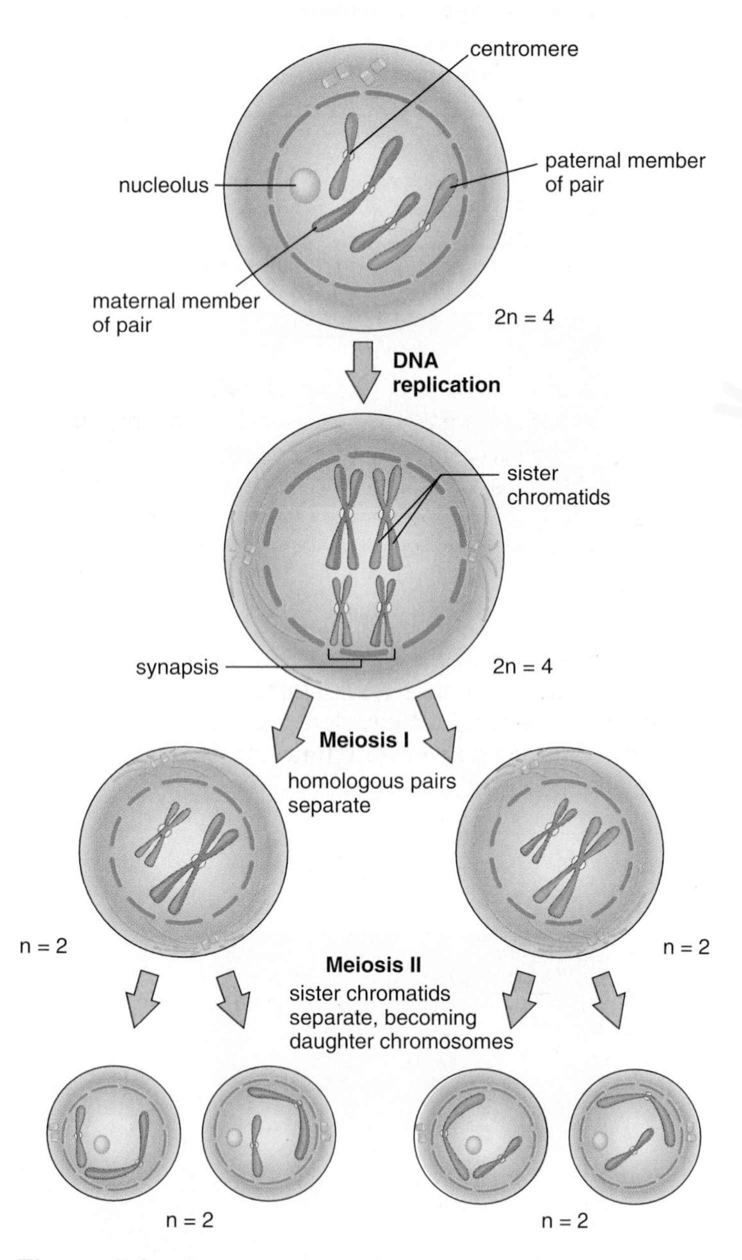

Figure 5.9 Overview of meiosis.
Following DNA replication, each chromosome is duplicated. During meiosis I, the homologous chromosomes pair during synapsis and then separate. During meiosis II, the centromeres divide and the sister chromatids separate, becoming daughter chromosomes that move into the daughter nuclei.

Overview of Meiosis

Meiosis requires two nuclear divisions and produces four haploid daughter cells, each having one of each kind of chromosome and therefore half the total number of chromosomes present in the diploid parental nucleus. The parental cell has the diploid number of chromosomes, while the daughter cells have the haploid number of chromosomes.

Recall that when a cell is 2n or diploid, the chromosomes occur in pairs. For example, the 46 chromosomes of humans occur in 23 pairs of chromosomes. The members of each pair are called **homologous chromosomes** or **homologues.**

Figure 5.9 presents an overview of meiosis, indicating the two cell divisions, meiosis I and meiosis II. Prior to meiosis I, DNA replication has occurred and the chromosomes are duplicated. Each chromosome consists of two chromatids held together at a centromere. During meiosis I the homologous chromosomes come together and line up side by side. This so-called **synapsis** results in an association of four chromatids that stay in close proximity during the first two phases of meiosis I.

Due to synapsis there are pairs of homologous chromosomes at the metaphase plate during meiosis I. Then, the members of these pairs separate and each daughter cell receives one member of each pair. Therefore, the daughter cells have the haploid number of chromosomes, as you can verify by counting the number of centromeres. Each chromosome, however, is still duplicated.

During meiosis I, homologous chromosomes pair up and then separate. Each daughter cell receives one copy of each kind of chromosome.

No replication of DNA is needed between meiosis I and meiosis II because the chromosomes are already duplicated: they already have two sister chromatids. During meiosis II, the centromeres divide and the sister chromatids separate, becoming daughter chromosomes that are distributed to daughter nuclei. In the end, each of four daughter cells has the haploid number of chromosomes and each chromosome consists of one chromatid.

Following meiosis II, there are four haploid daughter cells and each chromosome consists of one chromatid.

In some life cycles, such as that of humans (see Fig. 5.15), the daughter cells mature into **gametes** (sex cells—sperm and egg) that fuse during **fertilization.** Fertilization restores the diploid number of chromosomes in a cell that will develop into a new individual.

Genetic Recombination

Meiosis helps ensure that genetic recombination occurs through two key events: crossing-over and independent assortment of homologous chromosomes. In order to appreciate the significance of these events, it is necessary to realize that the members of a homologous pair can carry slightly different instructions for the same genetic trait. For example, one homologue may carry instructions for brown eyes while the corresponding homologue may carry instructions for blue eyes.

Crossing-over of Nonsister Chromatids

It's often said that we inherit half our chromosomes from our mother and half from our father, but this is not strictly correct because of crossing-over. During synapsis, the homologous chromosomes come together and line up side by side. Now an exchange of genetic material may occur between the nonsister chromatids of the homologous pair (Fig. 5.10). **Crossing-over** means that the genetic instructions from a mother and father are mixed and the chromatids held together by a centromere are no longer identical. When the chromatids separate during meiosis I, the daughter cells receive chromosomes with recombined genetic material.

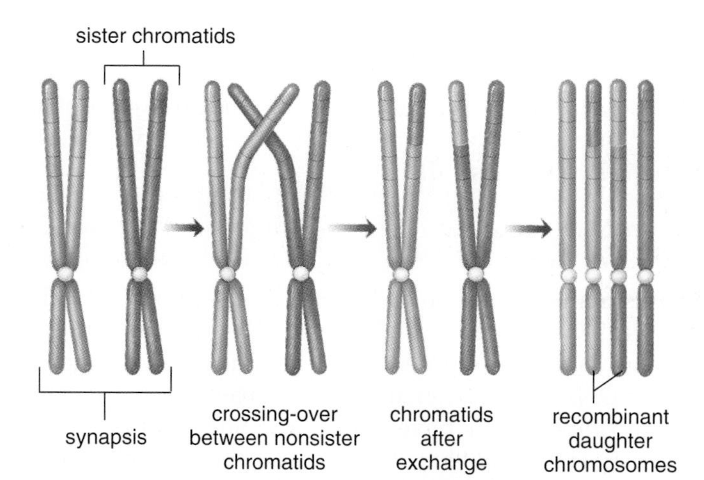

Figure 5.10 Synapsis and crossing-over.
During meiosis I, from left to right, duplicated homologous chromosomes undergo synapsis; nonsister chromatids break and then rejoin, so that two of the resulting daughter chromosomes have a different combination of genes.

Independent Assortment of Homologous Chromosomes

Independent assortment means that the homologous chromosomes separate independently or in a random manner. When homologues align at the metaphase plate, the maternal or paternal homologue may be orientated toward either pole. Figure 5.11 shows four possible orientations for a cell that contains only three pairs of chromosomes. Each orientation results in gametes that have a different combination of maternal and paternal chromosomes. Once all possible orientations are considered, the result will be 2^3 or eight combinations of maternal and paternal chromosomes in the resulting gametes from this cell. In humans, where there are 23 pairs of chromosomes, the possible chromosomal combinations in the gametes is a staggering 2^{23}, or 8,388,608. And this does not even consider the genetic variations that are introduced due to crossing-over.

During meiosis, crossing-over mixes the genetic information of maternal and paternal chromosomes and independent assortment leads to different combinations of these chromosomes in the gametes and offspring.

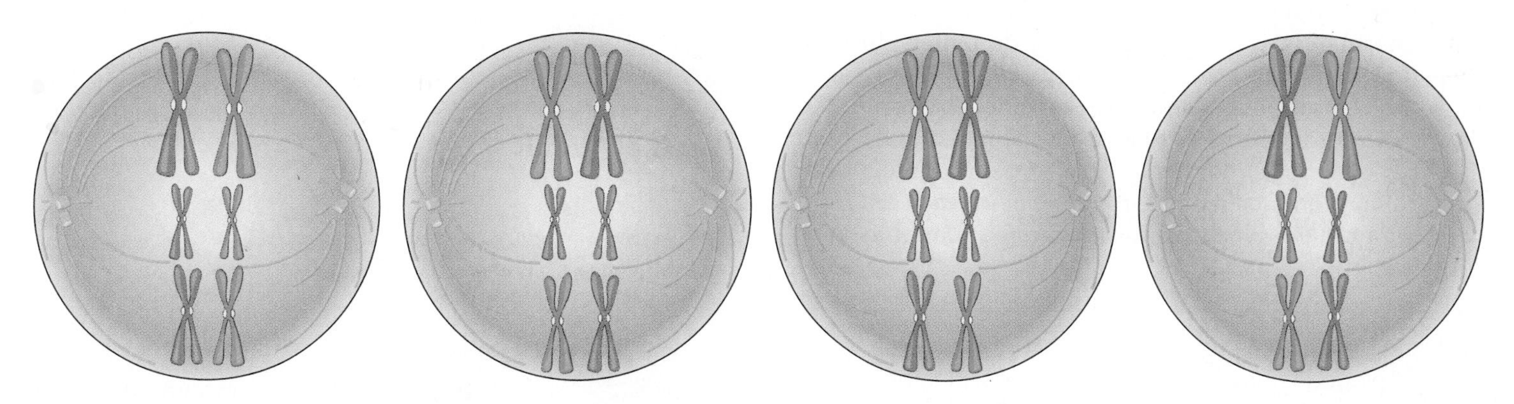

Figure 5.11 Independent assortment.
Four possible orientations of homologue pairs at the metaphase plate are shown. Each of these will result in daughter nuclei with a different combination of parental chromosomes. When a cell has three pairs of homologous chromosomes, there are 23 possible combinations of parental chromosomes in the daughter nuclei.

5.4 Meiosis in Detail

The same four phases seen in mitosis—prophase, metaphase, anaphase, and telophase—occur during both meiosis I and meiosis II.

The First Division

Phases of meiosis for an animal cell are diagrammed in Figure 5.12. During prophase I, the spindle appears while the nuclear envelope fragments and the nucleolus disappears. Due to DNA duplication during interphase, the homologous chromosomes each have two sister chromatids. During synapsis, crossing-over can occur. If so, the sister chromatids of a duplicated chromosome are no longer identical.

During metaphase I, homologous pairs are aligned at the metaphase plate. The maternal homologue may be orientated toward either pole, and the father homologue may be aligned toward either pole. This means that all possible combinations of chromosomes can occur in the daughter nuclei. During anaphase I, homologous chromosomes separate and move to opposite poles of the spindle. Each chromosome still consists of two chromatids.

In some species, there is a telophase I phase at the end of meiosis I. If so, the nuclear envelopes re-form and nucleoli appear. This phase may or may not be accompanied by cytokinesis, which is separation of the cytoplasm.

No replication of DNA occurs during a period of time between divisions called **interkinesis.**

The Second Division

Phases of meiosis II for an animal cell are diagrammed in Figure 5.13. At the beginning of prophase II, a spindle appears while the nuclear envelope dissembles and the nucleolus disappears. Each duplicated chromosome is attached to the spindle and lines up independently at the metaphase plate during metaphase II. At the close of metaphase II, the centromeres divide. During anaphase II, sister chromatids separate, becoming daughter chromosomes that move into the daughter nuclei. In telophase II, the spindle disappears as nuclear envelopes re-form. The plasma membrane furrows to give two complete cells, each of which has the haploid number of chromosomes. Each chromosome consists of one chromatid. Since each cell from meiosis I undergoes meiosis II, there are four daughter cells altogether.

During meiosis I, crossing-over occurs. Homologous chromosomes, each consisting of two sister chromatids, separate, and the daughter cells are haploid. Following meiosis II, there are four haploid daughter cells, and each chromosome has only one chromatid.

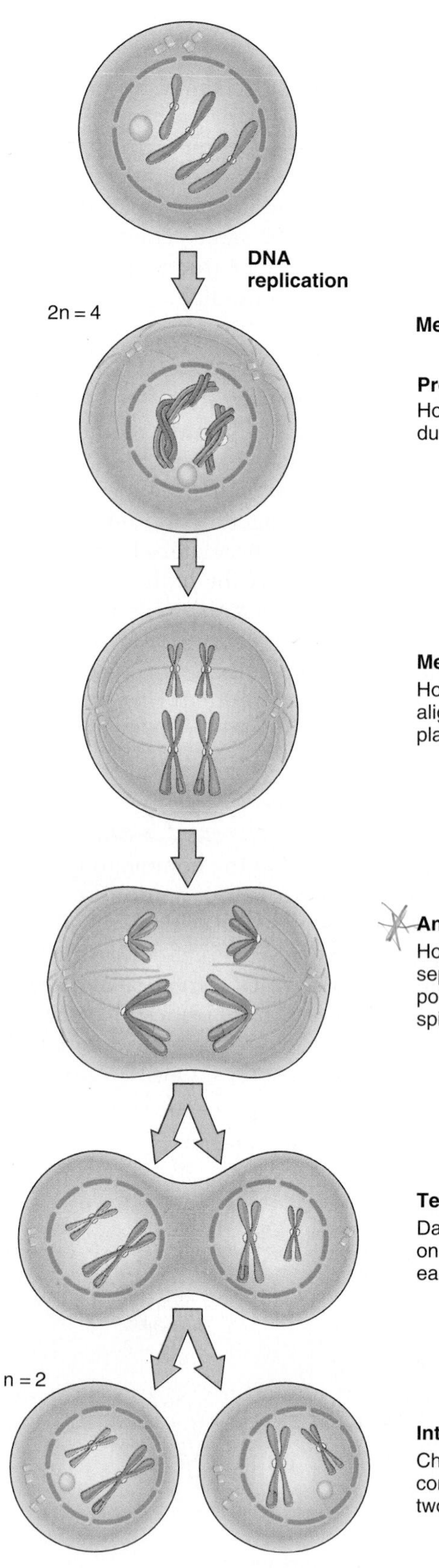

Figure 5.12 Meiosis I.
The exchange of color between nonsister chromatids represents crossing-over.

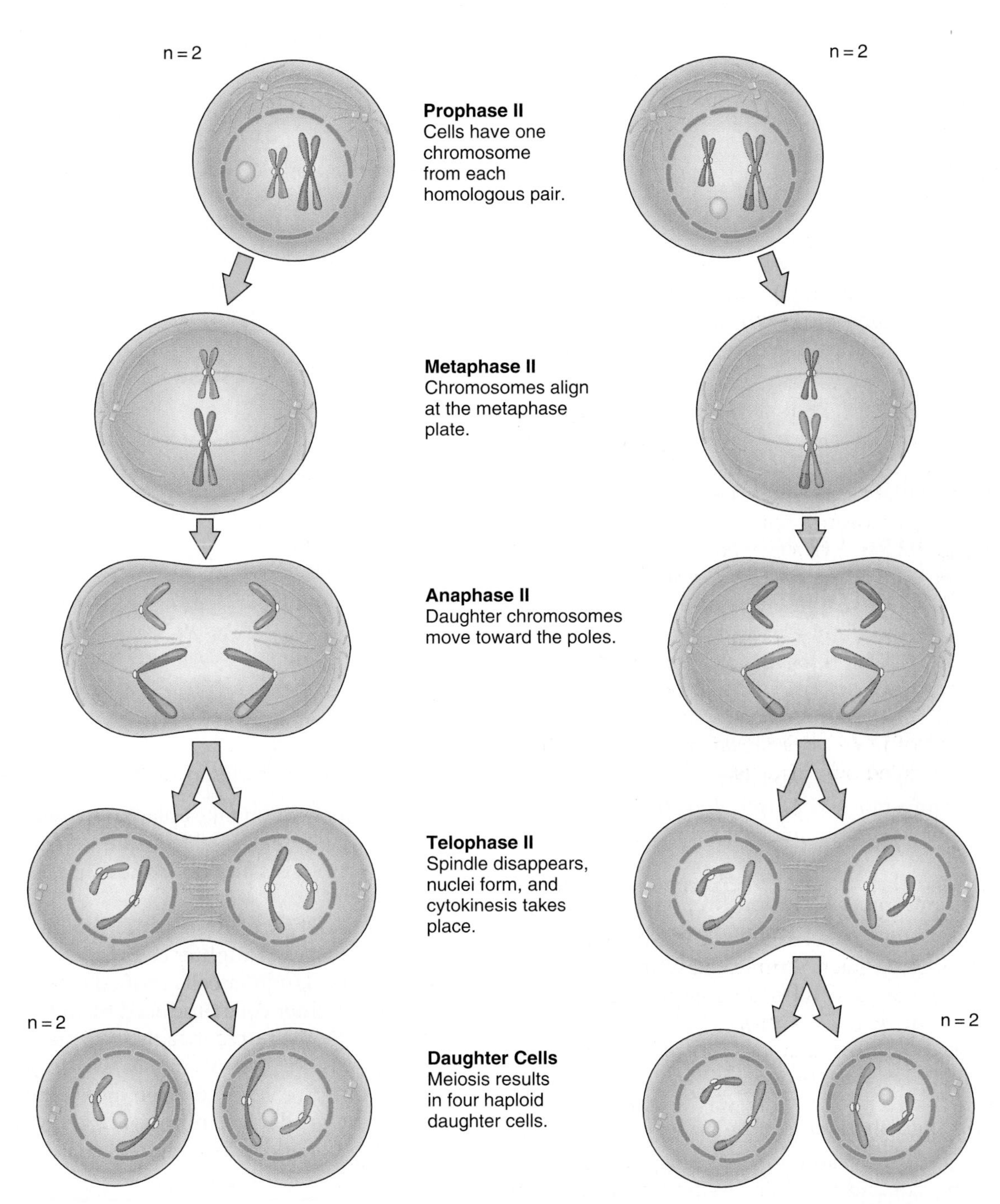

Figure 5.13 Meiosis II.
During meiosis II sister chromatids separate, becoming daughter chromosomes that are distributed to the daughter nuclei. Following meiosis II, there are four haploid daughter cells. Comparing the number of centromeres in the daughter cells with the number in the parental cell at the start of meiosis I verifies that the daughter cells are haploid.

5.5 Comparison of Meiosis with Mitosis

Figure 5.14 compares mitosis to meiosis. The differences between these cellular divisions can be categorized according to occurrence and process.

Occurrence

Meiosis occurs only at certain times in the life cycle of sexually reproducing organisms. In humans, meiosis occurs only in the reproductive organs and produces the gametes. Mitosis is more common because it occurs in all tissues during growth and repair.

Process

We will compare both meiosis I and meiosis II to mitosis.

Comparison of Meiosis I to Mitosis

The following are distinctive differences between the processes of meiosis I and mitosis.

Meiosis I	Mitosis
Prophase I Pairing of chromosomes	*Prophase* No pairing of chromosomes
Metaphase I Homologous chromosomes at metaphase plate	*Metaphase* Duplicated chromosomes at metaphase plate
Anaphase I Homologous chromosomes separate	*Anaphase* Sister chromatids separate, becoming daughter chromosomes that move to the poles
Telophase I Daughter cells are haploid	*Telophase* Daughter cells are diploid

These events distinguish meiosis I from mitosis.

1. Homologous chromosomes pair and undergo crossing-over during prophase I of meiosis but not during mitosis.
2. Paired homologous chromosomes align at the metaphase plate during metaphase I in meiosis. Individual (duplicated chromosomes) align at the metaphase plate during metaphase in mitosis.
3. Homologous chromosomes (with centromeres intact) separate and move to opposite poles during anaphase I in meiosis. Sister chromatids separate, becoming daughter chromosomes that move to opposite poles during anaphase in mitosis.

Comparison of Meiosis II to Mitosis

The events of meiosis II are just like those of mitosis except that in meiosis II, the nuclei contain the haploid number of chromosomes. The following listing compares meiosis II to mitosis:

Meiosis II	Mitosis
Prophase II No pairing of chromosomes	*Prophase* No pairing of chromosomes
Metaphase II Haploid number of duplicated chromosomes at metaphase plate	*Metaphase* Diploid number of duplicated chromosomes at metaphase plate
Anaphase II Sister chromatids separate, becoming daughter chromosomes that move to the poles	*Anaphase* Sister chromatids separate, becoming daughter chromosomes that move to the poles
Telophase II Four daughter cells	*Telophase* Two daughter cells

The following are differences between meiosis and mitosis:

1. DNA replication takes place only once during both meiosis and mitosis. There are two nuclear divisions during meiosis and only one nuclear division during mitosis.
2. Four daughter cells are produced by meiosis. Mitosis results in two daughter cells.
3. The four daughter cells formed by meiosis are haploid. The daughter cells produced by mitosis have the same chromosome number as the parental cell.
4. The daughter cells from meiosis are not genetically identical to each other or to the parental cell. The daughter cells from mitosis are genetically identical to each other and to the parental cell.

Meiosis is a specialized process that reduces the chromosome number and occurs only during the production of gametes. Mitosis is a process that occurs during growth and repair of all tissues.

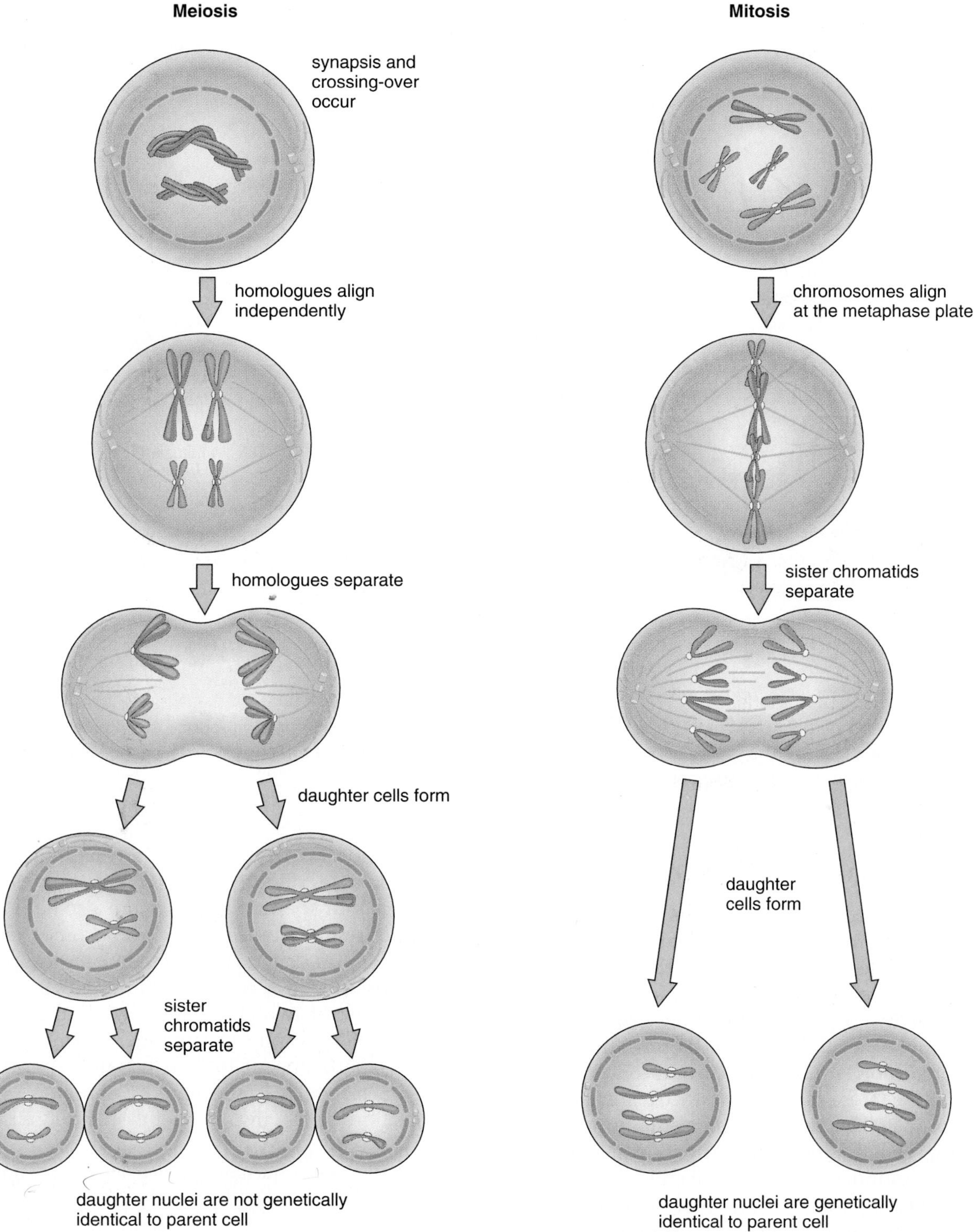

Figure 5.14 Meiosis compared to mitosis.
Why does meiosis produce haploid daughter cells while mitosis produces diploid daughter cells? Compare metaphase I of meiosis to metaphase of mitosis. Only in metaphase I are the homologous chromosomes paired at the metaphase plate. Members of the homologous chromosomes separate during anaphase I, and therefore the daughter cells are haploid. The blue chromosomes were inherited from one parent and the red chromosomes were inherited from the other parent. The exchange of color between nonsister chromatids represents crossing-over during meiosis I.

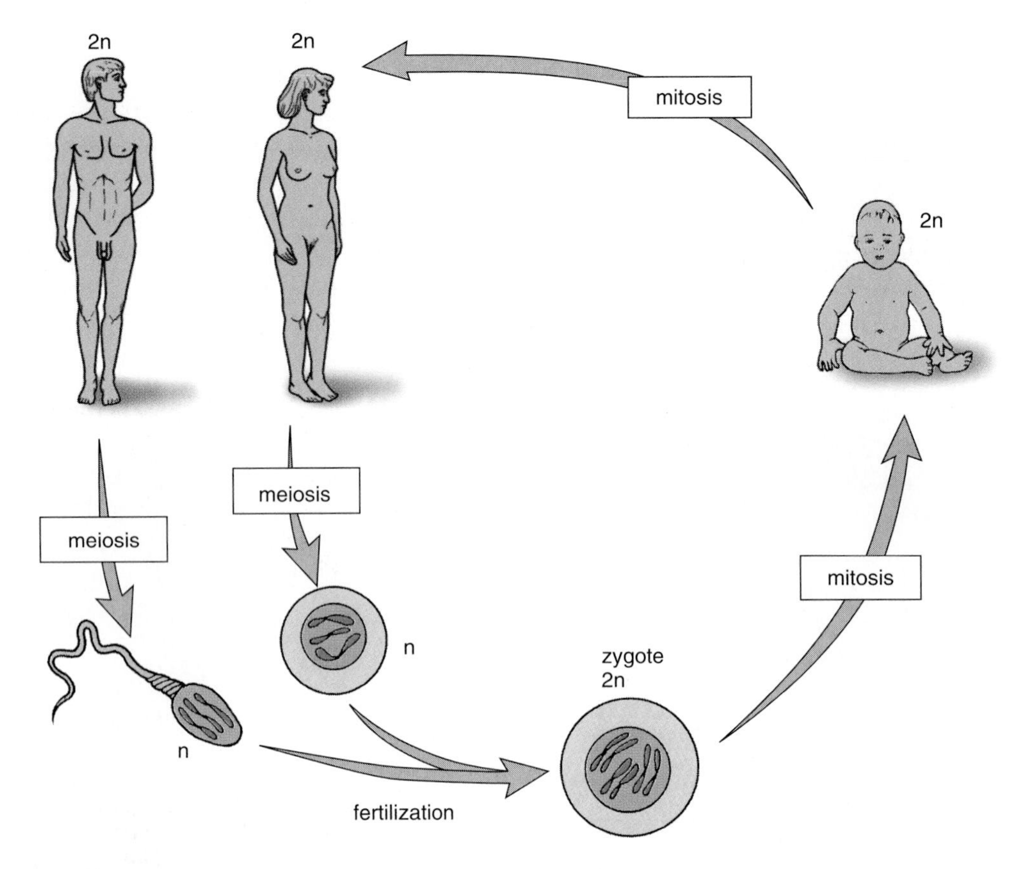

Figure 5.15 Life cycle of humans.
Meiosis in human males is a part of sperm production, and meiosis in human females is a part of egg production. When a haploid sperm fertilizes a haploid egg, the zygote is diploid. The zygote undergoes mitosis as it develops into a newborn child. Mitosis continues after birth until the individual reaches maturity; then the life cycle begins again.

5.6 The Human Life Cycle

The human life cycle requires both meiosis and mitosis (Fig. 5.15). In human males, meiosis is a part of spermatogenesis which occurs in the testes and produces sperm. In human females, meiosis is a part of oogenesis which occurs in the ovaries and produces eggs (Fig. 5.15). A haploid sperm and a haploid egg join at fertilization and the resulting **zygote** has the full or diploid number of chromosomes. During development of the fetus, which is the stage of development before birth, mitosis keeps the chromosome number constant in all the cells of the body. After birth, mitosis is involved in the continued growth of the child and repair of tissues at any time. As a result of mitosis, each somatic cell in the body has the same number of chromosomes.

Spermatogenesis and Oogenesis in Humans

Spermatogenesis is the production of sperm in males, and **oogenesis** is the production of eggs in females. In the testes of human males, primary spermatocytes, which are diploid (2n), divide during the first meiotic division to form two secondary spermatocytes, which are haploid (n). Secondary spermatocytes divide during the second meiotic division to produce four spermatids which are also haploid (n). What's the difference between the chromosomes in haploid secondary spermatocytes and those in haploid spermatids? The chromosomes in secondary spermatocytes are duplicated and consist of two chromatids, while those in spermatids consist of only one chromatid. Spermatids mature into sperm (spermatozoa). In human males, sperm have 23 chromosomes, which is the haploid number. The process of meiosis in males always results in four cells that become sperm.

In ovaries of human females, a primary oocyte, which is diploid (2n), divides during the first meiotic division into two cells, each of which is haploid but the chromosomes are duplicated. One of these cells, termed the **secondary oocyte,** receives almost all the cytoplasm. The other is a polar body. A **polar body** is a nonfunctioning cell that occurs during oogenesis. It contains little cytoplasm and will eventually disintegrate. The secondary oocyte begins the second meiotic division but stops at metaphase II. The secondary oocyte leaves the ovary and enters an oviduct where it may be

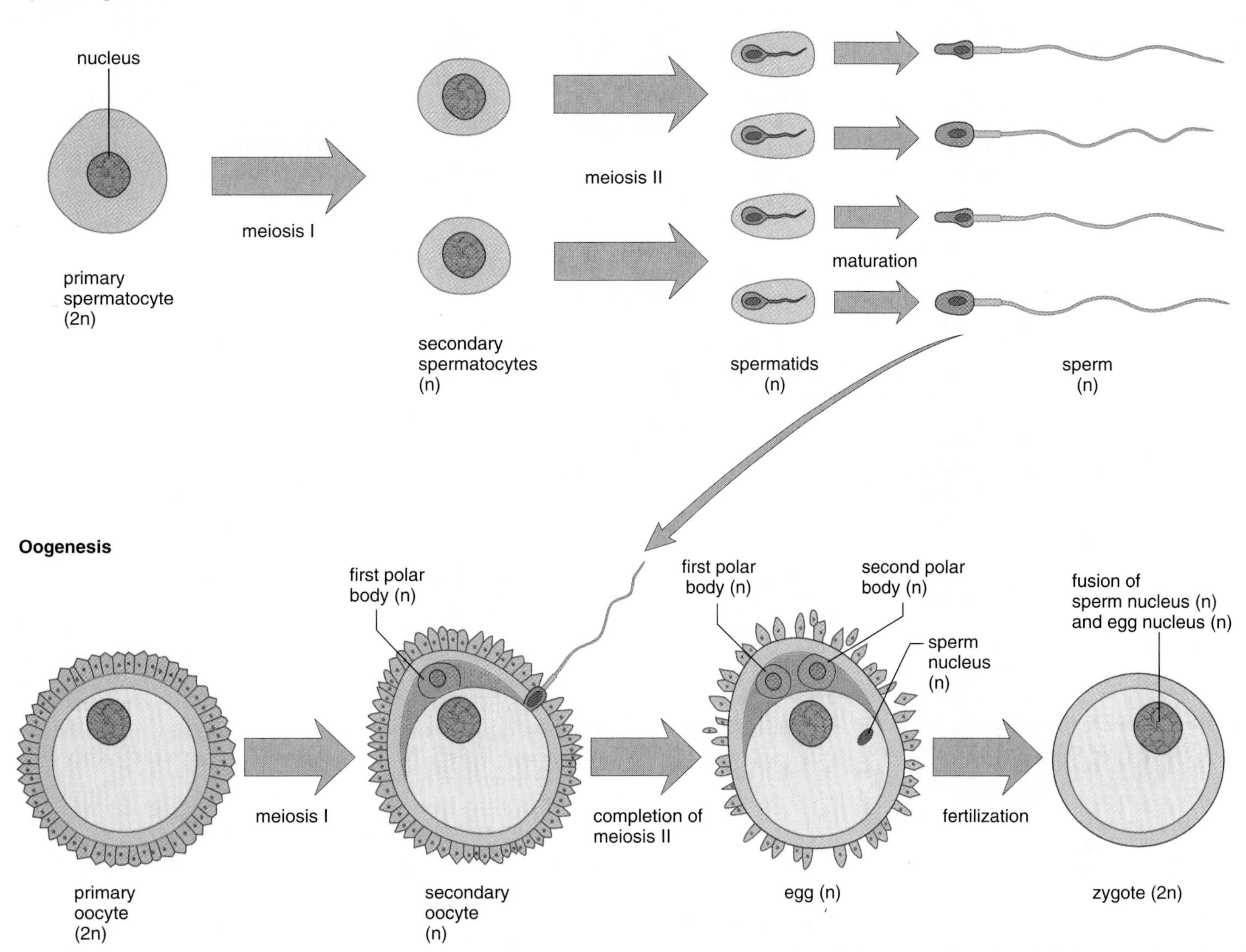

Figure 5.16 Spermatogenesis and oogenesis.
Spermatogenesis produces four viable sperm, whereas oogenesis produces one egg and two polar bodies. In humans, both sperm and egg have 23 chromosomes each; therefore, following fertilization, the zygote has 46 chromosomes.

approached by a sperm. If a sperm does enter the oocyte, the oocyte is activated to complete the second meiotic division. The mature egg has 23 chromosomes, each consisting of one chromatid. In human females, meiosis produces only one egg and two polar bodies. The polar bodies are a way to discard unnecessary chromosomes while retaining much of the cytoplasm in the egg. The cytoplasm serves as a source of nutrients for the developing embryo.

Genetic Recombination in Humans

Notice that fertilization is another means by which chromosomes are recombined in the next generation. Because each child receives both paternal and maternal chromosomes, no child is exactly like either parent. Altogether there are three ways in which meiosis ensures that a child has a different combination of genes than either parent.

1. Independent assortment of chromosomes means that all possible combinations of chromosomes occur in the gametes.
2. Crossing-over recombines genetic material so that sister chromatids are genetically dissimilar.
3. Upon fertilization, recombination of chromosomes occurs.

Sexual reproduction ensures that each generation has the same number of chromosomes, and that each individual has a different genetic makeup than either parent.

Bioethical Issue

Cloning is making exact multiple copies of DNA or a cell or an organism. The first two procedures have been around for some time. Through biotechnology, bacteria produce cloned copies of human DNA. When a single bacterium reproduces asexually on a petri dish, a colony results. Each member of the colony is a clone of the original cell. Now for the first time in our history, it is possible to produce a clone of a vertebrate. No sperm and egg are required. The DNA of an adult cell is placed in an egg that undergoes development to become an exact copy of the organism that donated the DNA. Some people fear that billionaires and celebrities will hasten to make multiple copies of themselves. Others feel that this is unlikely. Rather, they fear a different type of cloning.

Suppose it were possible to use the DNA of a burn victim to produce embryonic cells that are cajoled to become skin cells. These cells could be used to provide grafts of brand new skin. Would this be a proper use of cloning in humans?

Or suppose parents want to produce a child free of a genetic disease. Scientists produce a zygote through in vitro fertilization, and then they clone the zygote to produce any number of cells. Genetic engineering to correct the defect doesn't work on all the cells—only a few. They implant just those few in the uterus where development continues to term. Would this be a proper use of cloning in humans?

Well, what if science progressed to producing children with increased intelligence or athletic prowess in the same way? Would this be an acceptable use of cloning in humans?

Questions

1. Presently, research in the cloning of humans is banned. Should it be? Why or why not?
2. Under what circumstances might cloning in humans be acceptable? Explain.
3. Is cloning to produce improved breeds of farm animals acceptable? Why or why not?

Summarizing the Concepts

5.1 Maintaining the Chromosome Number

Each species has a characteristic number of chromosomes. The total number is the diploid number, and half this number is the haploid number. Among eukaryotes, cell division involves nuclear division and division of the cytoplasm (cytokinesis).

Replication of DNA precedes cell division. The duplicated chromosome is composed of two sister chromatids held together at a centromere. During mitosis the centromeres divide, and daughter chromosomes go into each new nucleus.

The cell cycle has four stages. During the G_1 stage the organelles increase in number; during the S stage, DNA replication occurs; during the G_2 stage, various proteins are synthesized; and during the M stage, mitosis occurs. It is now known that regulation of the cell cycle involves various proteins known as kinases and cyclins.

5.2 Mitosis in Detail

Mitosis has the following phases: prophase—in early prophase chromosomes have no particular arrangement, and in late prophase the chromosomes are attached to spindle fibers; metaphase, when the chromosomes are aligned at the metaphase plate; anaphase, when the chromatids separate, becoming daughter chromosomes that move toward the poles; and telophase, when new nuclear envelopes form around the daughter chromosomes and cytokinesis begins.

5.3 Reducing the Chromosome Number

Meiosis is found in any life cycle that involves sexual reproduction. During meiosis I, homologues separate, and this leads to daughter cells with half or the haploid number of homologous chromosomes. Crossing-over and independent assortment of chromosomes during meiosis I ensure genetic recombination in daughter cells. During meiosis II, chromatids separate, becoming daughter chromosomes that are distributed to daughter nuclei. In some life cycles, the daughter cells become gametes, and upon fertilization, the offspring have the diploid number of chromosomes, the same as their parents.

5.4 Meiosis in Detail

Meiosis utilizes two nuclear divisions. During meiosis I, homologous chromosomes undergo synapsis, and crossing-over between nonsister chromatids occurs. When the homologous chromosomes separate during meiosis I, each daughter nucleus receives one member from each pair of chromosomes. Therefore, the daughter cells are haploid. Distribution of daughter chromosomes derived from sister chromatids during meiosis II then leads to a total of four new cells, each with the haploid number of chromosomes.

5.5 Comparison of Meiosis with Mitosis

Figure 5.14 contrasts the phases of mitosis with the phases of meiosis.

5.6 The Human Life Cycle

The human life cycle involves both mitosis and meiosis. Mitosis ensures that each somatic cell will have the diploid number of chromosomes.

Meiosis is a part of spermatogenesis and oogenesis. Spermatogenesis in males produces four viable sperm, while oogenesis in females produces one egg and two polar bodies. Oogenesis does not go on to completion unless a sperm fertilizes the developing egg.

Among sexually reproducing organisms, such as humans, meiosis results in genetic recombination due to independent assortment of homologous chromosomes and crossing-over. Fertilization also contributes to genetic recombination.

Studying the Concepts

1. Describe the chromosome number using the terms diploid and haploid. 84
2. Explain how mitosis maintains the chromosome number in all the somatic cells of an individual. 84
3. Describe the cell cycle, including a description of interphase. 86
4. Describe the phases of animal mitosis, including in your description the terms centrosome, nucleolus, spindle, and cleavage furrow. 88–91

5. Name two differences between plant cell mitosis and animal cell mitosis. 91
6. Give an overview of meiosis and the manner in which it reduces the chromosome number. 92
7. How does meiosis ensure genetic recombination in the daughter cells? Explain in detail. 93
8. Describe the phases of meiosis I and meiosis II in detail. 94
9. Compare meiosis I and meiosis II to mitosis. 96
10. Explain why spermatogenesis results in four sperm and oogenesis produces one mature egg. 98–99
11. What are three events that ensure children have a different combination of genes than their parents? 99

Testing Yourself

Choose the best answer for each question.

1. The cell cycle ensures that
   a. the cell grows prior to cell division.
   b. DNA replicates prior to cell division.
   c. the chromatids separate, becoming the daughter chromosomes.
   d. the cytoplasm divides.
   e. All of these are correct.
2. In human beings, mitosis is necessary to
   a. growth and repair of tissues.
   b. formation of the gametes.
   c. maintaining the chromosome number in all body cells.
   d. the death of unnecessary cells.
   e. Both b and c are correct.

For questions 3–5 match the descriptions that follow to the terms in the key.

Key:
a. centriole
b. chromatid
c. chromosome
d. centromere

3. point of attachment for sister chromatids
4. found at a pole in the center of an aster
5. coiled and condensed chromatin
6. If a parent cell has fourteen chromosomes prior to mitosis, how many chromosomes will the daughter cells have?
   a. twenty-eight
   b. fourteen
   c. seven
   d. any number between seven and twenty-eight
7. In which phase of mitosis are chromosomes moving toward the poles?
   a. prophase
   b. metaphase
   c. anaphase
   d. telophase
   e. Both b and c are correct.
8. If a parent cell has twelve chromosomes, then the daughter cells following meiosis will have
   a. twelve chromosomes.
   b. twenty-four chromosomes.
   c. six chromosomes.
   d. Any one of these could be correct.
9. At the metaphase plate during metaphase I of meiosis, there are
   a. single chromosomes.
   b. unpaired duplicated chromosomes.
   c. homologous pairs.
   d. always twenty-three chromosomes.
10. Crossing-over occurs between
    a. sister chromatids of the same chromosomes.
    b. two different bivalents.
    c. nonsister chromatids of a homologous pair.
    d. two daughter nuclei.
11. Which of these is not a difference between spermatogenesis and oogenesis in humans?

Spermatogenesis	**Oogenesis**
occurs in males.	occurs in females.
produces four sperm per meiosis.	produces one egg per meiosis.
produces haploid eggs	produces diploid cells
always goes to completion	does not always go to completion

12. Label this diagram of a cell in early prophase of mitosis.

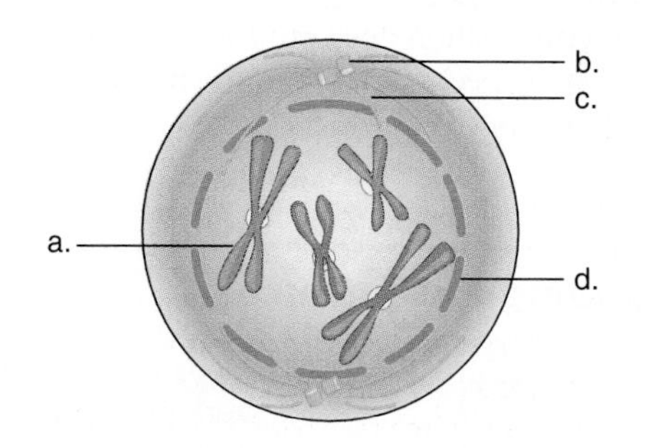

13. Which of these drawings represents metaphase I of meiosis? How do you know?

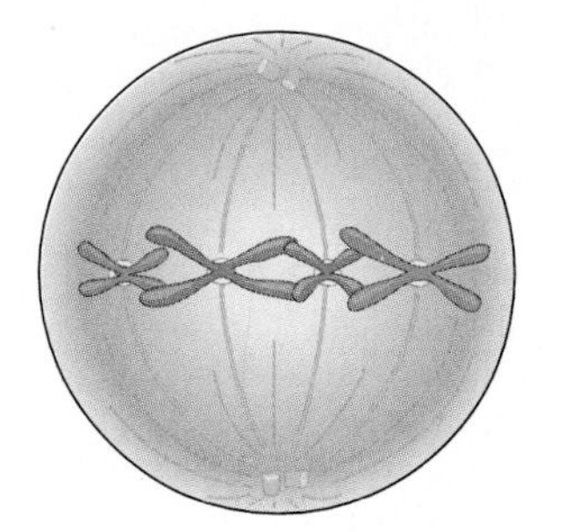

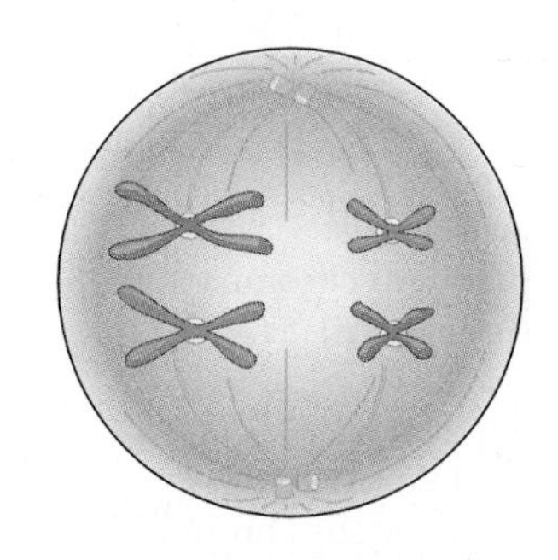

Thinking Scientifically

1. Concerning the genetic material:
   a. Which form—chromatin or chromosomes—would you expect to be metabolically active? Why? (page 84)
   b. The genes are part of the chromosomes. Ordinarily, a person inherits two copies of a gene—one from the mother and one from the father. Does it seem reasonable that the two copies of the gene might be different forms of the gene?
   c. The genes direct protein (enzyme) synthesis. Why do you suppose a person with three chromosomes of the same kind might suffer from various disorders?
   d. Not all genes are active in all cells. For example, why might a defective gene not adversely affect the maturation of a sperm or an egg?
2. Concerning cell division (page 96):
   a. The drug colchicine prevents cell division from finishing. What part of a dividing cell do you suppose it disrupts?
   b. Asexual reproduction ordinarily does not produce genetic variation. Why not?
   c. Sexual reproduction does produce genetic variation. Why?
   d. Would you expect sexual reproduction to aid the evolutionary process? Why?

Understanding the Terms

anaphase 89
aster 88
cell cycle 86
cell plate 90
centromere 84
chromatin 84
chromosome 84
cleavage furrow 91
crossing-over 93
cyclin 86
cytokinesis 84
daughter chromosomes 84
diploid (2n) 84
fertilization 92
gamete 92
haploid (n) 84
homologous chromosome 92
homologue 92
independent assortment 93
interkinesis 94
interphase 86
kinase 86
meiosis 92
metaphase 89
mitosis 84
oogenesis 98
polar body 98
prophase 88
secondary oocyte 98
sister chromatid 84
somatic cell 84
spermatogenesis 98
spindle 88
synapsis 92
telophase 89
zygote 98

Match the terms to these definitions:

a. ________ Production of sperm in males by the process of meiosis and maturation.
b. ________ In oogenesis, a nonfunctional product; two to three meiotic products are of this type.
c. ________ In oogenesis, the functional product of meiosis I; becomes the egg.
d. ________ Pairing of homologous chromosomes during meiosis I.
e. ________ Production of eggs in females by the process of meiosis and maturation.

Using Technology

Your study of cell division is supported by these available technologies:

Essential Study Partner CD-ROM

Cells → Cell Division

Visit the Mader web site for related ESP activities.

Exploring the Internet

The Mader Home Page provides resources and tools as you study this chapter.

http://www.mhhe.com/biosci/genbio/mader

Life Science Animations 3D Video

10 Mitosis
11 Meiosis
12 Crossing Over

C H A P T E R 6

Metabolism: Energy and Enzymes

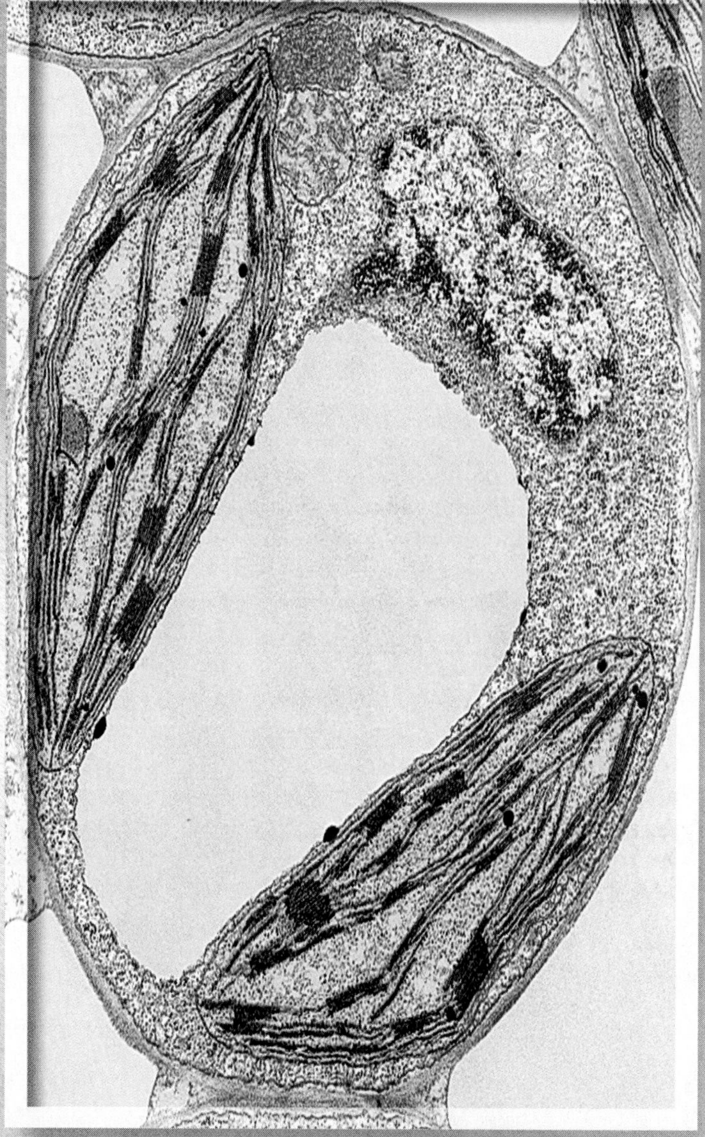

Plant cells carry on both photosynthesis in chloroplasts and aerobic cellular respiration in mitochondria. These metabolic pathways consist of a number of enzymatic reactions that involve energy transformations. Without enzymes and energy, cells could not continue to exist.

Chapter Concepts

Take a look around the room you're in. How many things are powered by batteries or plugged into electrical outlets? Just as electricity drives all those appliances, lights, etc., a versatile molecule called ATP provides cells with the energy to move, build proteins, perform chemical reactions, and carry out any other necessary duties. ATP doesn't work solo in the cell, however. Assistants known as enzymes help molecules interact with each other, speeding the cell's chemistry and making it more energy-efficient. Together, ATP and enzymes govern a cell's metabolism, as this chapter will explain.

6.1 Energy

Living things can't grow, reproduce, or exhibit any of the characteristics of life without a ready supply of energy. **Energy,** which is the capacity to do work, occurs in many forms: light energy comes from the sun; electrical energy powers kitchen appliances; and heat energy warms our houses. **Kinetic energy** is the energy of motion. All moving objects have kinetic energy. Thrown baseballs, falling water, and contracting muscles have kinetic energy. **Potential energy** is stored energy. Water behind a dam, or a rock at the top of a hill, or ATP, has potential energy that can be converted to kinetic energy. **Chemical energy** is in the interactions of atoms, one to the other, in a molecule. Molecules have varying amounts of potential energy. Glucose has much more energy than its breakdown products, carbon dioxide and water.

Figure 6.1 Energy for life.
All of the energy needed to move this athlete is provided by the food he has eaten. Once food has been processed in the digestive tract, nutrients are transported about the body, including to the muscles. The energy of nutrient molecules is converted to that of ATP molecules which power muscle contraction.

Two Laws of Thermodynamics

Early researchers who first studied energy and its relationships and exchanges formulated two **laws of thermodynamics.** *The first law, also called the "law of conservation of energy," says that energy cannot be created or destroyed but can only be changed from one form to another.* Think of the conversions that occur when coal is used to power a locomotive. First, the chemical energy of coal is converted to heat energy and then heat energy is converted to kinetic energy in a steam engine. Similarly, the potential energy of coal or gas is converted to electrical energy by power plants. Do energy transformations occur in the human body? As an example, consider that the chemical energy in the food we eat is changed to the chemical energy of ATP, and then this form of potential energy is converted to the mechanical energy of muscle contraction (Fig. 6.1).

The second law of thermodynamics says that energy cannot be changed from one form to another without a loss of usable energy. Only about 25% of the chemical energy of gasoline is converted to the motion of a car; the rest is lost as heat. Heat, of course, is a form of energy, but heat is the most random form of energy and quickly dissipates into the environment. When muscles convert the chemical energy within ATP to the mechanical energy of contraction, some of this energy becomes heat right away. With conversion upon conversion, eventually all usable forms of energy become heat that is lost to the environment. And because heat dissipates, it can never be converted back to a form of potential energy. The reading on the next page discusses how ecosystems also obey the second law of thermodynamics.

Entropy

Entropy is a measure of randomness or disorder. An organized, usable form of energy has a low entropy, whereas an unorganized, less stable form of energy such as heat has a high entropy. A neat room has a much lower entropy than a messy room. We know that a neat room always tends toward messiness. In the same way, energy conversions eventually result in heat, and therefore the entropy of the universe is always increasing.

How does an ordered system such as a neat room or an organism come about? You know very well that it takes an input of usable energy to keep your room neat. In the same way, it takes a constant input of usable energy from the food you eat to keep you organized. This input of energy goes through many energy conversions, and the output is finally heat, which increases the entropy of the universe.

The laws of thermodynamics explain why the entropy of the universe spontaneously increases and why organisms need a constant input of usable energy to maintain their organization.

Ecology Focus

Ecosystems and the Second Law of Thermodynamics

A cell converts the energy of one chemical molecule into another but so do ecosystems. In an ecosystem, the energy stored in the members of one population is used by another to maintain the organization of its members. Because of this, energy flows through an ecosystem (Fig. 6A). As transformations of energy occur, useful energy is lost to the environment in the form of heat, until finally useful energy is completely used up. Since energy cannot recycle, there is a need for an ultimate source of energy. This source, which continually supplies almost all living things with energy, is the sun. The entire universe is tending toward disorder, but in the meantime, solar energy is sustaining living things.

Human beings are also a population that feeds on other organisms. We feed directly on plants, such as corn, or on animals like poultry and cattle that have fed on corn. In the United States, however, much supplemental energy in addition to solar energy is used to produce food. Even before planting time, there is an input of fossil fuel energy for the processing of seeds, and the making of tools, fertilizers, and pesticides. Then, fossil fuel energy is used to transport these materials to the farm. At the farm, fuel is needed to plant the seeds, to apply fertilizers and pesticides, and to irrigate, harvest, and dry the crops. After harvesting, still more fuel is used to process the crops to make the neatly packaged products we buy in the supermarket. Most of the food we eat today has been processed in some way. Even farm families now buy at least some of their food from supermarkets in nearby towns.

Since 1940 the amount of supplemental fuel used in the American food system has greatly increased until now the amount of supplemental energy is at least three or four times that of the caloric content of the food produced! This is partially due to the trend toward producing more food on less land by using high-yielding hybrid wheat and corn plants. These plants require more care and about twice as much supplemental energy as the traditional varieties of wheat and corn. Cattle confined to feedlots and fed grain that has gone through the whole production process require about twenty times the amount of supplemental energy as do range-fed cattle. Our food system has been labeled energy-intensive because it requires such a large input of supplemental energy.

Our energy-intensive food system is a matter for concern because it increases the cost of food and the burning of fossil fuels adds pollutants to the atmosphere. What can be done? First of all, we could grow crops that do not require so much supplemental energy. And second, we could eat primarily vegetables and grains. It is estimated that only about 10% of the energy contained in one population is actually taken up by the next population. (About 90% is lost as heat.) This means that about ten times the number of people can be sustained on a diet of vegetables and grain rather than a diet of meat. And when we do eat meat we could depend more on range-fed cattle. Cattle kept close to farmland supply manure that can substitute, in part, for chemical fertilizer. Biological control, the use of natural enemies to control pests, would cut down on pesticide use. Solar and wind energy could be used instead of fossil fuel energy, particularly on the farm. For example, wind-driven irrigation pumps are feasible.

Finally, of course, consumers could help matters. We could overcome our prejudice against vegetables that have slight blemishes. We could consume less processed foods and buy cheaper cuts of beef, which have come from range-fed cattle. And we could avoid using electrically powered gadgets when preparing food at home.

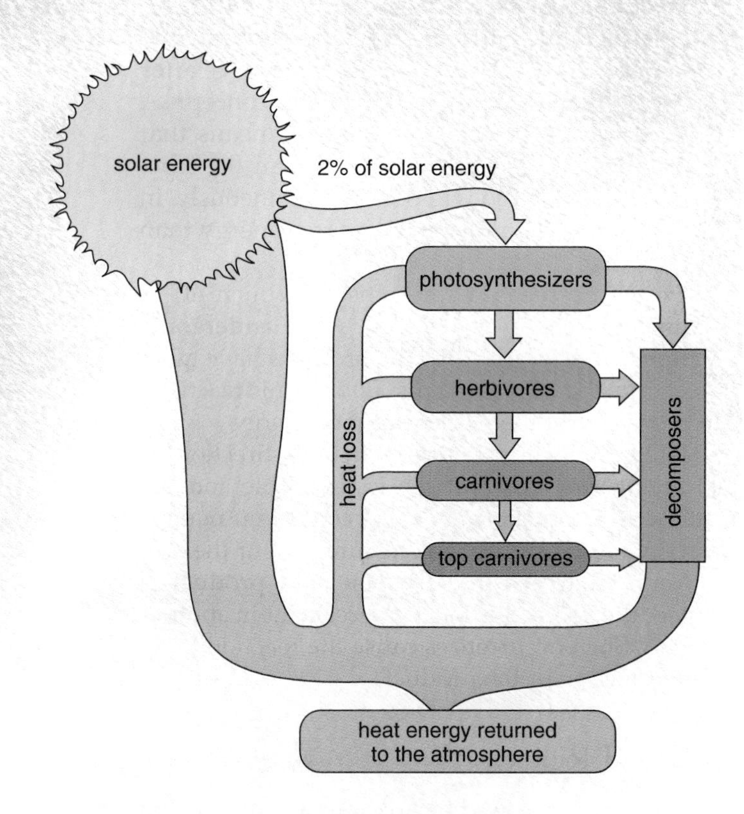

Figure 6A **Energy loss in an ecosystem.**
Ordinarily about 2% of the solar energy reaching the earth is taken up by photosynthesizers (plants and algae). This is the energy that allows them to make their own food. Herbivores obtain their food by eating plants, and carnivores obtain food by eating other animals. Whenever the energy content of food is used by organisms, it is eventually converted to heat. With death and decay by decomposers, all the energy temporarily stored in organisms returns as heat to the atmosphere. In order to support a very large population, human beings supplement solar energy with fossil fuel energy to grow crops. Usually, humans feed on crops directly or on animals (herbivores) that have been fed on crops.

6.2 Metabolic Reactions and Energy Transformations

Metabolism is the sum of all the reactions that occur in a cell. **Reactants** are substances that participate in a reaction, while **products** are substances that form as a result of a reaction. In the reaction A + B → C + D, A and B are the reactants while C and D are the products. How would you know that this reaction will occur spontaneously—that is, without an input of energy? Using the concept of entropy, it is possible to state that a reaction will occur spontaneously if it increases the entropy of the universe. But this is not very helpful in cell biology because we don't wish to consider the entire universe. We simply want to consider this reaction. In such instances, cell biologists use the concept of free energy. **Free energy** is the amount of energy available—that is, energy that is still "free" to do work after a chemical reaction has occurred. Free energy is denoted by the symbol G after Josiah Gibbs who first developed the concept. A negative ΔG (change in free energy) means that the products have less free energy than the reactants and the reaction will occur spontaneously. In our reaction, if C and D have less free energy than A and B, then the reaction will "go."

Exergonic reactions are ones in which ΔG is negative and energy is released, while **endergonic reactions** are ones in which the products have more free energy than the reactants. Endergonic reactions can only occur if there is an input of energy.

If the change in free energy in both directions is just about zero, the reaction is reversible and the reaction is at equilibrium. How could you make a reversible reaction "go" in one direction or the other? Very often in cells, as soon as a product is formed, the product is used as a reactant in another reaction. Such occurrences cause the reaction to go in the direction of the product.

Coupled Reactions

Can the energy released by an exergonic reaction be used to "drive" an endergonic reaction? In the body many reactions such as protein synthesis, nerve conduction, or muscle contraction are endergonic: they require an input of energy. On the other hand, the breakdown of ATP to ADP + Ⓟ is exergonic and energy is released (Fig. 6.2).

In **coupled reactions,** the energy released by an exergonic reaction is used to drive an endergonic reaction. ATP breakdown is often coupled to cellular reactions that require an input of energy. Coupling, which requires that the exergonic reaction and the endergonic reaction be closely tied, can be symbolized like this:

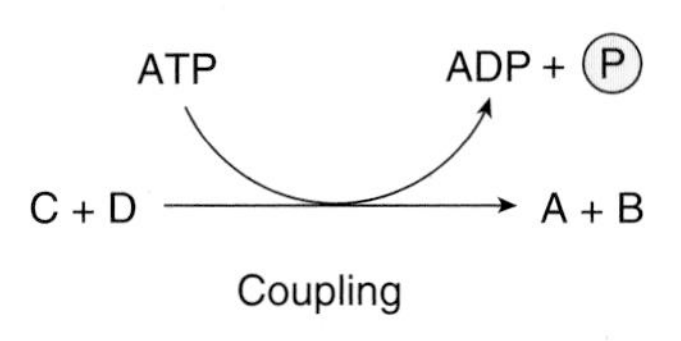

How is a cell assured of a supply of ATP? Recall that glucose breakdown during aerobic cellular respiration provides the energy for the buildup of ATP in mitochondria. Only

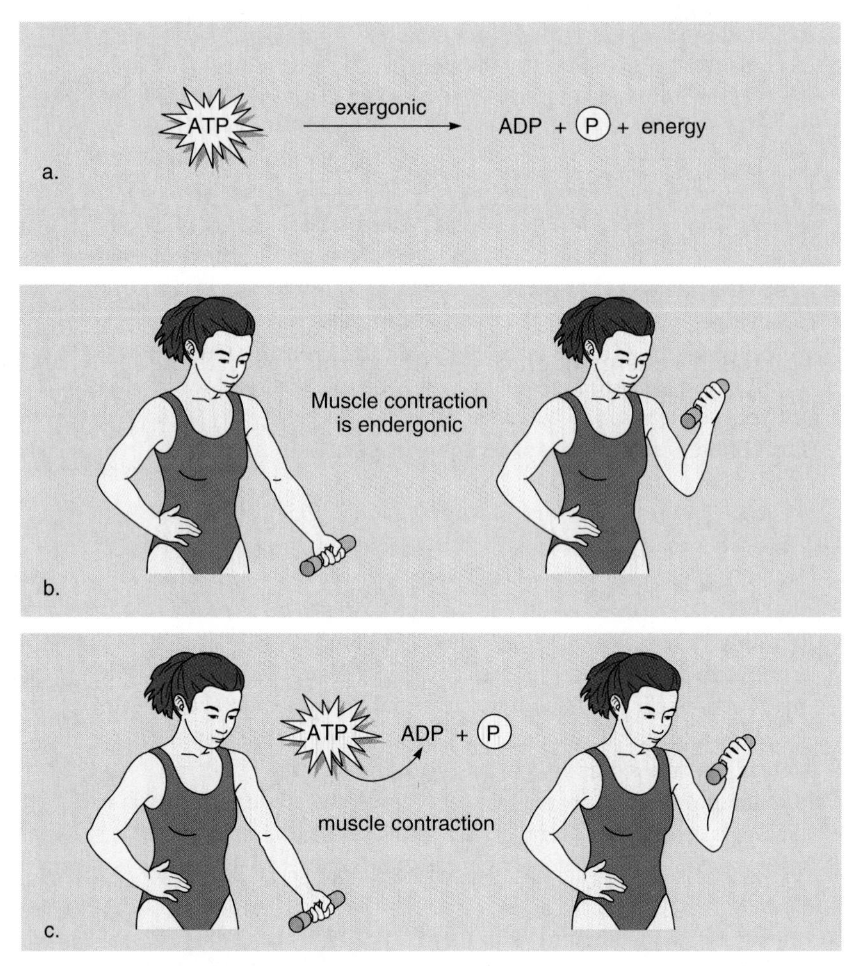

Figure 6.2 Coupled reactions.
a. The breakdown of ATP is exergonic. **b.** Muscle contraction is endergonic and therefore cannot occur without an input of energy. **c.** Muscle contraction is coupled to ATP breakdown, making the overall process exergonic. Now muscle contraction can occur.

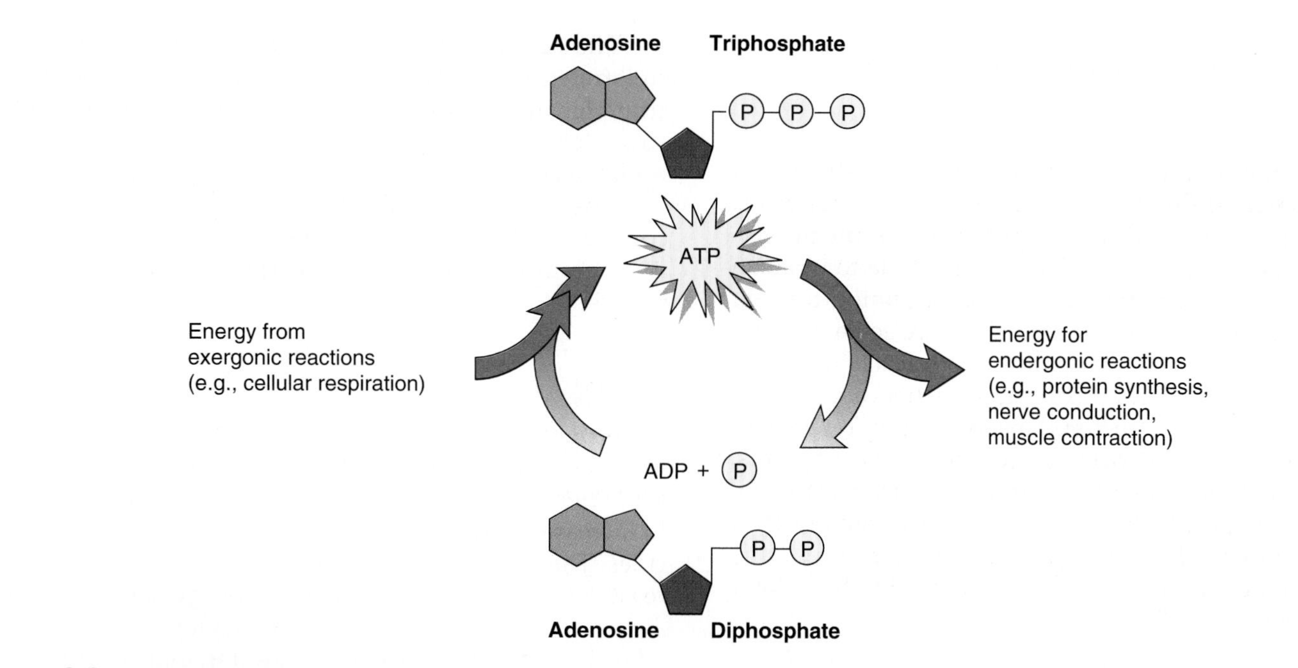

Figure 6.3 **The ATP cycle.**
In cells, the exergonic breakdown of glucose is coupled to the buildup of ATP, and then the exergonic breakdown of ATP is coupled to endergonic reactions in cells. When a phosphate group is removed by hydrolysis, ATP releases the appropriate amount of energy for most metabolic reactions. The high-energy content of ATP comes from the complex interaction of the atoms within the molecule.

39% of the free energy of glucose is transformed to ATP; the rest is lost as heat. When ATP breaks down to drive the reactions mentioned, some energy is lost as heat and the overall reaction becomes exergonic.

ATP: Energy for Cells

ATP (adenosine triphosphate) is the common energy currency of cells: when cells require energy, they "spend" ATP. You may think that this causes our bodies to produce a lot of ATP, and it does; however, the amount on hand at any one moment is minimal because ATP is constantly being generated from **ADP (adenosine diphosphate)** and Ⓟ (Fig. 6.3).

The use of ATP as a carrier of energy has some advantages: (1) It provides a common energy currency that can be used in many different types of reactions. (2) When ATP becomes ADP + Ⓟ, the amount of energy released is just about enough for the biological purposes mentioned in the following section, and so little energy is wasted. (3) ATP breakdown is coupled to endergonic reactions in such a way that it minimizes energy loss.

Function of ATP

Recall that at various times we have mentioned at least three uses for ATP.

Chemical work. Supplies the energy needed to synthesize macromolecules that make up the cell.

Transport work. Supplies the energy needed to pump substances across the plasma membrane.

Mechanical work. Supplies the energy needed to permit muscles to contract, cilia and flagella to beat, chromosomes to move, and so forth.

Structure of ATP

ATP is a nucleotide composed of the base adenine and the sugar ribose (together called adenosine) and three phosphate groups. ATP is called a "high-energy" compound because a phosphate group is easily removed. Under cellular conditions, the amount of energy released when ATP is hydrolyzed to ADP + Ⓟ is about 7.3 kcal per mole.[1]

ATP is a carrier of energy in cells. It is the common energy currency because it supplies energy for many different types of reactions.

---

[1]A mole is the number of molecules present in the molecular weight of a substance (in grams).

6.3 Metabolic Pathways and Enzymes

Reactions do not occur haphazardly in cells; they are usually a part of a **metabolic pathway,** a series of linked reactions. Metabolic pathways begin with a particular reactant and terminate with an end product. While it is possible to write an overall equation for a pathway as if the beginning reactant went to the end product in one step, there are actually many specific steps in between. In the pathway, one reaction leads to the next reaction, which leads to the next reaction, and so forth in an organized, highly structured manner. This arrangement makes it possible for one pathway to lead to several others, because various pathways have several molecules in common. Also, metabolic energy is captured and utilized more easily if it is released in small increments rather than all at once.

A metabolic pathway can be represented by the following diagram:

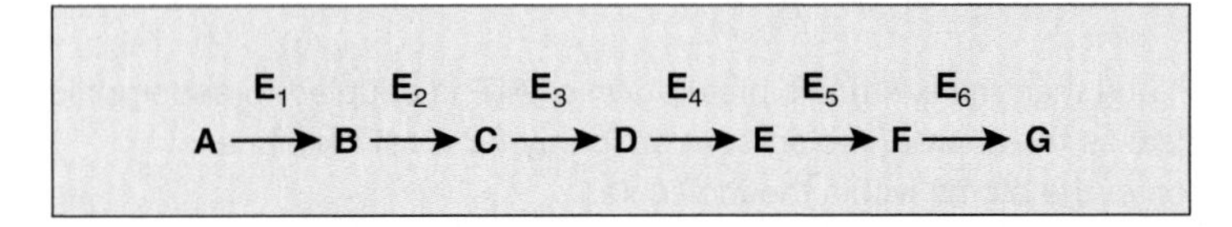

In this diagram, the letters A–F are reactants and letters B–G are products in the various reactions. The letters E_1–E_6 are enzymes.

An **enzyme** is a protein molecule[2] that functions as an organic catalyst to speed a chemical reaction. In a crowded ballroom, a mutual friend can cause particular people to interact. In the cell, an enzyme brings together particular molecules and causes them to react with one another.

The reactants in an enzymatic reaction are called the **substrates** for that enzyme. In the first reaction, A is the substrate for E_1 and B is the product. Now B becomes the substrate for E_2, and C is the product. This process continues until the final product G forms.

Any one of the molecules (A–G) in this linear pathway could also be a substrate for an enzyme in another pathway. A diagram showing all the possibilities would be highly branched.

Energy of Activation

Molecules frequently do not react with one another unless they are activated in some way. In the absence of an enzyme, activation is very often achieved by heating the reaction flask to increase the number of effective collisions between molecules. The energy that must be added to cause molecules to react with one another is called the **energy of activation** (E_a). Figure 6.4 compares E_a when an enzyme is not present to when an enzyme is present, illustrating that enzymes lower the amount of energy required for activation to occur.

In baseball, a home-run hitter must not only hit the ball to the fence, but over the fence. When enzymes lower the energy of activation, it is like removing the fence; then it is possible to get a home run by simply hitting the ball as far as the fence was.

Enzyme-Substrate Complexes

The following equation, which is pictorially shown in Figure 6.5, is often used to indicate that an enzyme forms a complex with its substrate:

[2]Catalytic RNA molecules are called ribozymes and are not enzymes.

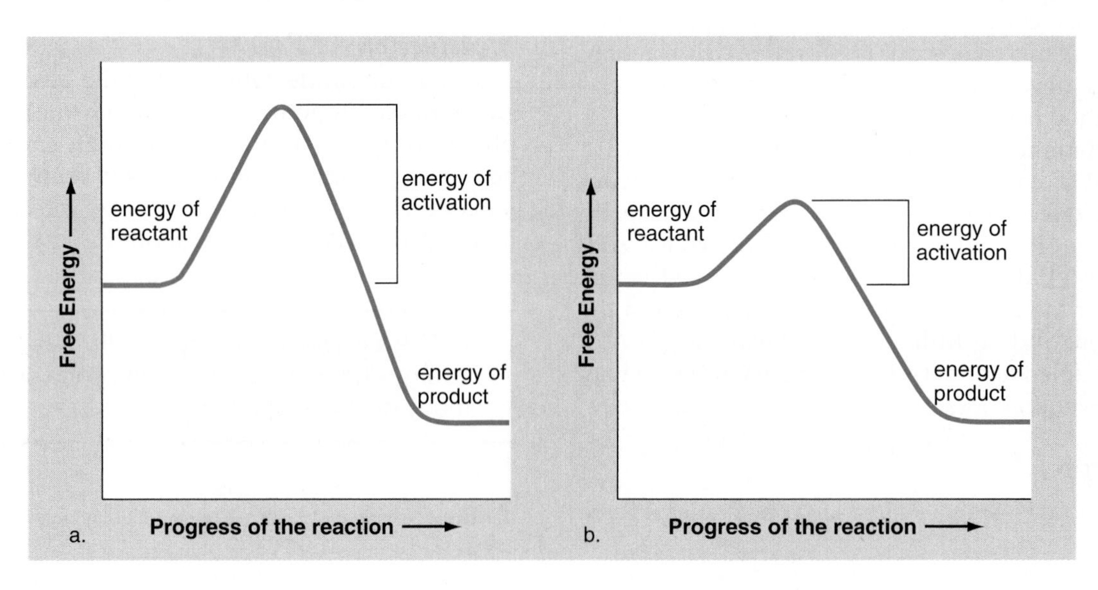

Figure 6.4 Energy of activation (E_a).
Enzymes speed the rate of chemical reactions because they lower the amount of energy required to activate the reactants. **a.** Energy of activation when an enzyme is not present. **b.** Energy of activation when an enzyme is present. Even spontaneous reactions like this one speed up when an enzyme is present.

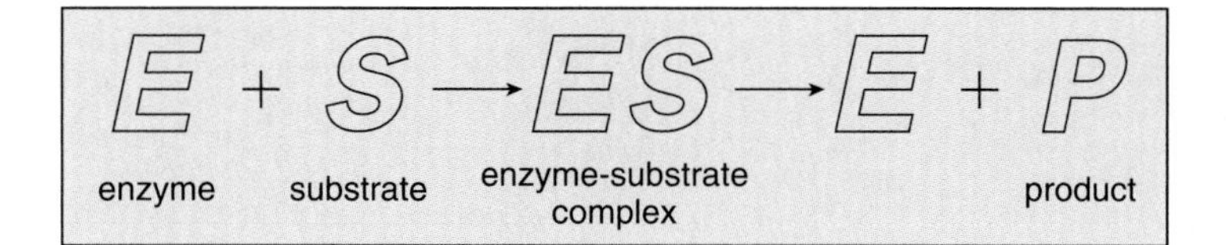

In most instances only one small part of the enzyme, called the **active site,** complexes with the substrate(s). It is here that the enzyme and substrate fit together, seemingly like a key fits a lock; however, it is now known that the active site undergoes a slight change in shape in order to accommodate the substrate(s). This is called the **induced-fit model** because the enzyme is induced to undergo a slight alteration to achieve optimum fit.

The change in shape of the active site facilitates the reaction that now occurs. After the reaction has been completed, the product(s) is released, and the active site returns to its original state, ready to bind to another substrate molecule. Only a small amount of enzyme is actually needed in a cell because enzymes are not used up by the reaction.

Some enzymes do more than simply complex with their substrate(s); they actually participate in the reaction. Trypsin digests protein by breaking peptide bonds. The active site of trypsin contains three amino acids with *R* groups that actually interact with members of the peptide bond—first to break the bond and then to introduce the components of water. This illustrates that the formation of the enzyme-substrate complex is very important in speeding up the reaction.

Sometimes it is possible for a particular reactant(s) to produce more than one type of product(s). The presence or absence of an enzyme determines which reaction takes place. If a substance can react to form more than one product, then the enzyme that is present and active determines which product is produced.

Every reaction in a cell requires its specific enzyme. Because enzymes only complex with their substrates, they are named for their substrates, as in the following examples:

Substrate	Enzyme
Lipid	Lipase
Urea	Urease
Maltose	Maltase
Ribonucleic acid	Ribonuclease
Lactose	Lactase

Most enzymes are protein molecules. Enzymes speed chemical reactions by lowering the energy of activation. They do this by forming an enzyme-substrate complex.

Factors Affecting Enzymatic Speed

Enzymatic reactions proceed quite rapidly. Consider, for example, the breakdown of hydrogen peroxide (H_2O_2) as catalyzed by the enzyme catalase: $2\ H_2O_2 \rightarrow 2\ H_2O + O_2$. The breakdown of hydrogen peroxide can occur 600,000 times a second when catalase is present. To achieve maximum product per unit time, there should be enough substrate to fill active sites most of the time. Temperature and optimal pH also increase the rate of an enzymatic reaction.

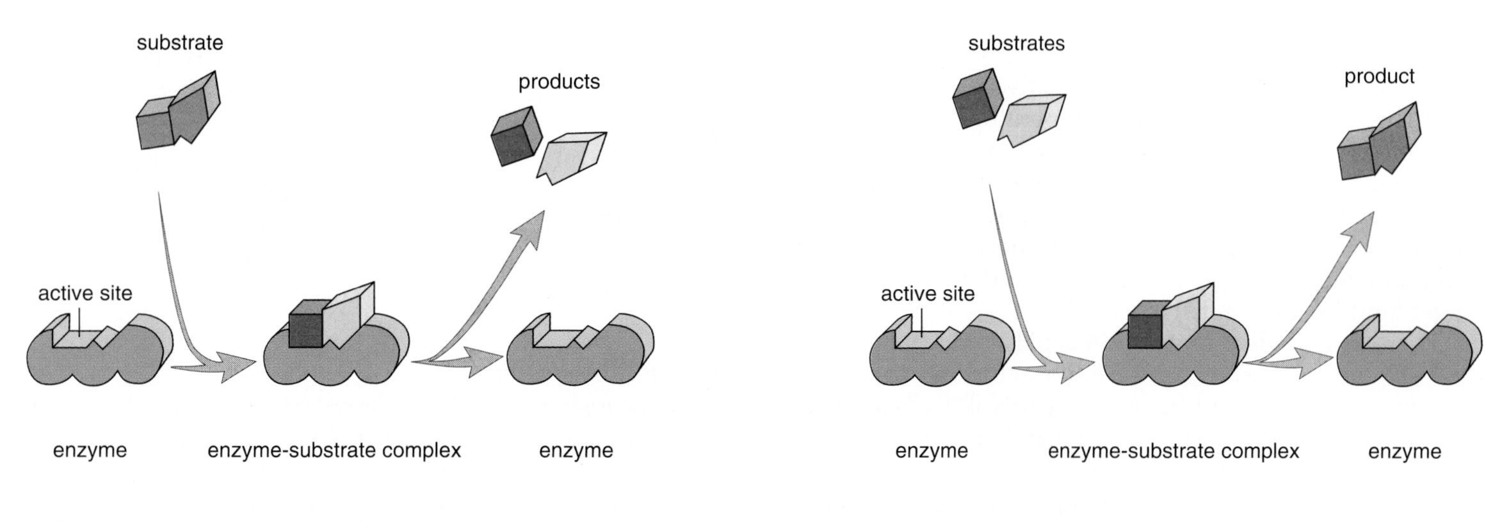

Figure 6.5 Enzymatic action.
An enzyme has an active site, which is where the substrates and enzyme fit together in such a way that the substrates are oriented to react. Following the reaction, the products are released and the enzyme is free to act again. **a.** Some enzymes carry out degradative reactions in which the substrate is broken down to smaller molecules. **b.** Other enzymes carry out synthetic reactions in which the substrates are joined to form a larger molecule.

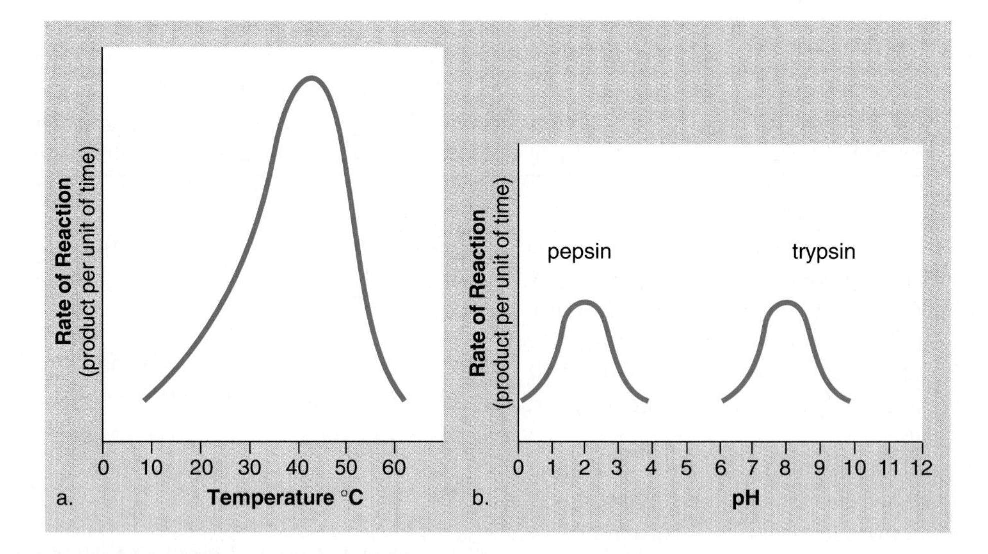

Figure 6.6 Rate of an enzymatic reaction as a function of temperature and pH.
a. At first, as with most chemical reactions, the rate of an enzymatic reaction doubles with every 10°C rise in temperature. In this graph, the rate of reaction is maximum at about 40°C; then it decreases until the reaction stops altogether, because the enzyme has become denatured. **b.** Pepsin, an enzyme found in the stomach, acts best at a pH of about 2, while trypsin, an enzyme found in the small intestine, performs optimally at a pH of about 8. The shape that enables these proteins to bind with their substrates is not properly maintained at other pHs.

Substrate Concentration

Generally, enzyme activity increases as substrate concentration increases because there are more collisions between substrate molecules and the enzyme. As more substrate molecules fill active sites, more product results per unit time. But when the enzyme's active sites are filled almost continuously with substrate, the enzyme's rate of activity cannot increase anymore. Maximum rate has been reached.

Temperature and pH

As the temperature rises, enzyme activity increases (Fig. 6.6*a*). This occurs because as the temperature rises there are more effective collisions between enzyme and substrate. However, if the temperature rises beyond a certain point, enzyme activity eventually levels out and then declines rapidly because the enzyme is **denatured.** An enzyme's shape changes during denaturation, and then it can no longer bind its substrate(s) efficiently.

Each enzyme also has an optimal pH at which the rate of the reaction is highest. Figure 6.6*b* shows the optimal pH for the enzymes pepsin and trypsin. At this pH value, these enzymes have their normal configurations. The globular structure of an enzyme is dependent on interactions, such as hydrogen bonding, between *R* groups. A change in pH can alter the ionization of these side chains and disrupt normal interactions, and under extreme conditions of pH, denaturation eventually occurs. Again, the enzyme has an altered shape and is then unable to combine efficiently with its substrate.

Enzyme Concentration

Since enzymes are specific, a cell regulates which enzymes are present and/or active at any one time. Otherwise enzymes may be present that are not needed, or one pathway may negate the work of another pathway.

Genes must be turned on to increase the concentration of an enzyme and must be turned off to decrease the concentration of an enzyme.

Another way to control enzyme activity is to activate or deactivate the enzyme. Phosphorylation is one way to activate an enzyme. Molecules received by membrane receptors often turn on kinases, which then activate enzymes by phosphorylating them:

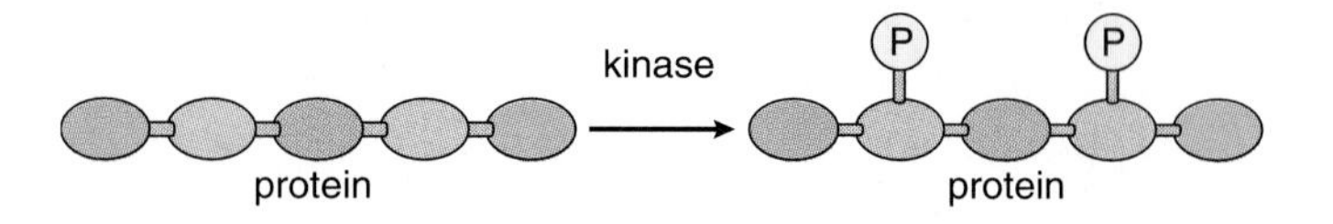

Enzyme Inhibition

Actually, **enzyme inhibition** is a common means by which cells regulate enzyme activity. In *competitive inhibition,* another molecule is so close in shape to the enzyme's substrate that it can compete with the true substrate for the enzyme's active site. This molecule inhibits the reaction because only the binding of the true substrate results in a product. In *noncompetitive inhibition,* a molecule binds to an enzyme, but not at the active site. The other binding site is called the allosteric site. In this instance, inhibition occurs when binding of a molecule causes a shift in the three-dimensional structure so that the substrate cannot bind to the active site.

The activity of almost every enzyme in a cell can be regulated by its product. When a product is in abundance, it binds competitively with its enzyme's active site; as the product is used up, inhibition is reduced and more product can be produced. In this way, the concentration of the product is always kept within a certain range. Most metabolic pathways are regulated by **feedback inhibition,** but the end product of the pathway binds at an allosteric site on the first enzyme of the pathway (Fig. 6.7). This binding shuts down the pathway, and no more product is produced.

In inhibition, a product binds to the active site or binds to an allosteric site on an enzyme.

Poisons are often enzyme inhibitors. Cyanide is an inhibitor for an essential enzyme (cytochrome *c* oxidase) in all cells, which accounts for its lethal effect on humans. Penicillin blocks the active site of an enzyme unique to bacteria. When penicillin is taken, bacteria die but humans are unaffected.

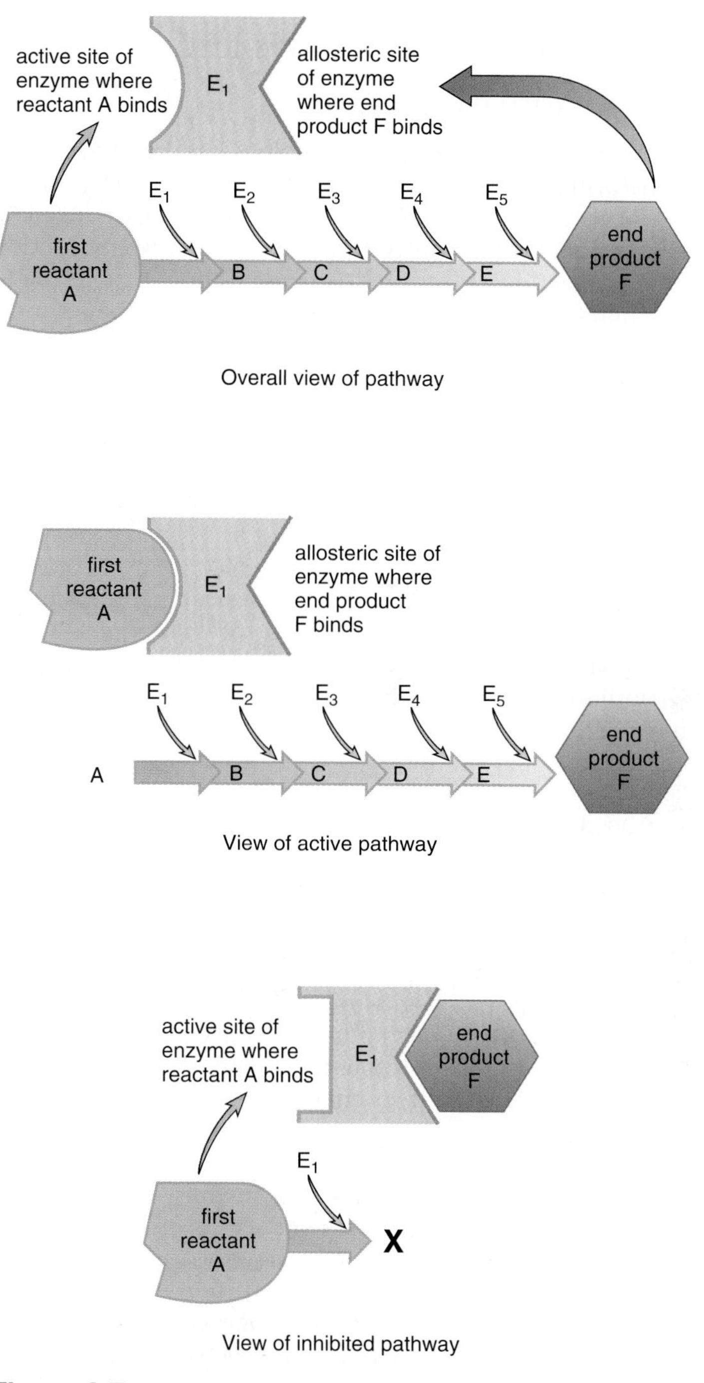

Figure 6.7 Feedback inhibition.
This hypothetical metabolic pathway is regulated by feedback inhibition. When reactant A binds to the active site of E_1, the pathway is active and the end product is produced. Once there is sufficient end product, some binds to the allosteric site of E_1. Now a change of shape prevents reactant A from binding to the active site of E_1 and the end product is no longer produced.

Enzyme Cofactors

Many enzymes require an inorganic ion or organic but non-protein molecule to function properly; these necessary ions or molecules are called **cofactors.** The inorganic ions are metals such as copper, zinc, or iron. The organic, nonprotein molecules are called **coenzymes.** These cofactors assist the enzyme and may even accept or contribute atoms to the reactions.

It is interesting that vitamins are often components of coenzymes. **Vitamins** are relatively small organic molecules that are required in trace amounts in our diet and in the diet of other animals for synthesis of coenzymes that affect health and physical fitness. The vitamin becomes a part of the coenzyme's molecular structure. These vitamins are necessary to formation of the coenzymes listed:

Vitamin	Coenzyme
Niacin	NAD^+
B_2 (riboflavin)	FAD
B_1 (thiamine)	Thiamine pyrophosphate
Pantothenic acid	Coenzyme A (CoA)
B_{12} (cobalamin)	B_{12} coenzymes

A deficiency of any one of these vitamins results in a lack of the coenzyme listed and therefore a lack of certain enzymatic actions. In humans, this eventually results in vitamin-deficiency symptoms: niacin deficiency results in a skin disease called pellagra, and riboflavin deficiency results in cracks at the corners of the mouth.

Enzymes speed a reaction by forming a complex with the substrate. Various factors affect enzymatic speed, including substrate concentration, temperature, pH, enzyme concentration, the presence of inhibitors or necessary cofactors.

6.4 Metabolic Pathways and Oxidation-Reduction

As we have noted, chemical reactions can involve energy transformations from one molecule to another, as when the potential energy stored in ATP molecules is used to synthesize macromolecules. In oxidation-reduction (redox) reactions, electrons also pass from one molecule to another. **Oxidation** is the loss of electrons and **reduction** is the gain of electrons. Oxidation and reduction always takes place at the same time because one molecule accepts the electrons given up by another. Oxidation-reduction reactions occur during photosynthesis and aerobic cellular respiration.

Photosynthesis

In living things, hydrogen ions often accompany electrons, and if so, oxidation is a loss of hydrogen atoms ($e^- + H^+$) and reduction is a gain of hydrogen atoms. For example, the overall reaction for photosynthesis can be written like this:

$$\underset{\text{Carbon dioxide}}{6CO_2} + \underset{\text{Water}}{6H_2O} + \text{Energy} \rightarrow \underset{\text{Glucose}}{C_6H_{12}O_6} + \underset{\text{Oxygen}}{6O_2}$$

This equation shows that during photosynthesis hydrogen atoms are transferred from water to carbon dioxide and glucose is formed. Therefore, water has been oxidized and carbon dioxide has been reduced. Since water is a low-energy molecule and glucose is a high-energy molecule, energy is needed to form glucose. This energy is supplied by solar energy. Chloroplasts are able to capture solar energy, and convert it to chemical energy of ATP molecules that are used along with hydrogen atoms to reduce glucose.

A coenzyme of oxidation-reduction called **NADP** (nicotinamide adenine dinucleotide phosphate) is active during photosynthesis. NADP carries a positive charge and, therefore, is written as $NADP^+$. During photosynthesis, $NADP^+$ accepts electrons and hydrogen ions derived from water and passes by way of a metabolic pathway to carbon dioxide.

Aerobic Cellular Respiration

The overall equation for aerobic cellular respiration is the opposite of the one we used to represent photosynthesis:

$$\underset{\text{Glucose}}{C_6H_{12}O_6} + \underset{\text{Oxygen}}{6O_2} \rightarrow \underset{\text{Carbon dioxide}}{6CO_2} + \underset{\text{Water}}{6H_2O} + \text{Energy}$$

In this reaction glucose has lost hydrogen atoms (been oxidized) and oxygen has gained hydrogen atoms (been reduced). When oxygen gains electrons it becomes water. Since glucose is a high-energy molecule and water is a low-energy molecule, energy has been released. You will remember that mitochondria in cells use the energy released from glucose breakdown to build ATP molecules.

In metabolic pathways, most oxidations such as those that occur during aerobic cellular respiration involve a coenzyme called **NAD** (nicotinamide adenine dinucleotide). NAD is a coenzyme of oxidation-reduction that accepts electrons from glucose products and then later passes them on to a metabolic pathway that reduces oxygen to water. NAD carries a positive charge and therefore is represented as NAD^+. During oxidation reactions, NAD^+ accepts two electrons but only one hydrogen ion. The reaction is:

$$NAD^+ + 2H \rightarrow NADH + H^+$$

The Cycling of Matter and the Flow of Energy

During photosynthesis, chloroplasts, present in plants, capture solar energy and use it to convert water and carbon dioxide into carbohydrates which serve as food for all living things. Oxygen is a by-product of photosynthesis (Fig. 6.8).

Mitochondria, present in both plants and animals, complete the breakdown of carbohydrates and use the released energy to build ATP molecules. Aerobic cellular respiration consumes oxygen and produces carbon dioxide and water, the very molecules taken up by chloroplasts.

This cycling of molecules between chloroplasts and mitochondria allows a flow of energy from the sun through all living things. This flow of energy maintains the levels of biological organization from molecules to ecosystems. In keeping with the laws of thermodynamics, energy is dissipated with each chemical transformation and eventually the solar energy captured by plants is lost in the form of heat. Therefore, most living things are dependent upon an input of solar energy.

Human beings are also involved in the cycling of molecules between plants and animals and in the flow of energy from the sun. We inhale oxygen and eat plants and their stored carbohydrates, or we eat other animals that have eaten plants. Oxygen and nutrient molecules enter our mitochondria which produce ATP and release carbon dioxide and water, the molecules used by plants to produce carbohydrates. Without a supply of energy-rich molecules produced by plants, we could not produce the ATP molecules needed to maintain our bodies.

Oxidation-reduction reactions are involved in the pathways of photosynthesis, which take place in chloroplasts, and of aerobic cellular respiration, which take place in mitochondria. These pathways permit a flow of energy from the sun through all living things.

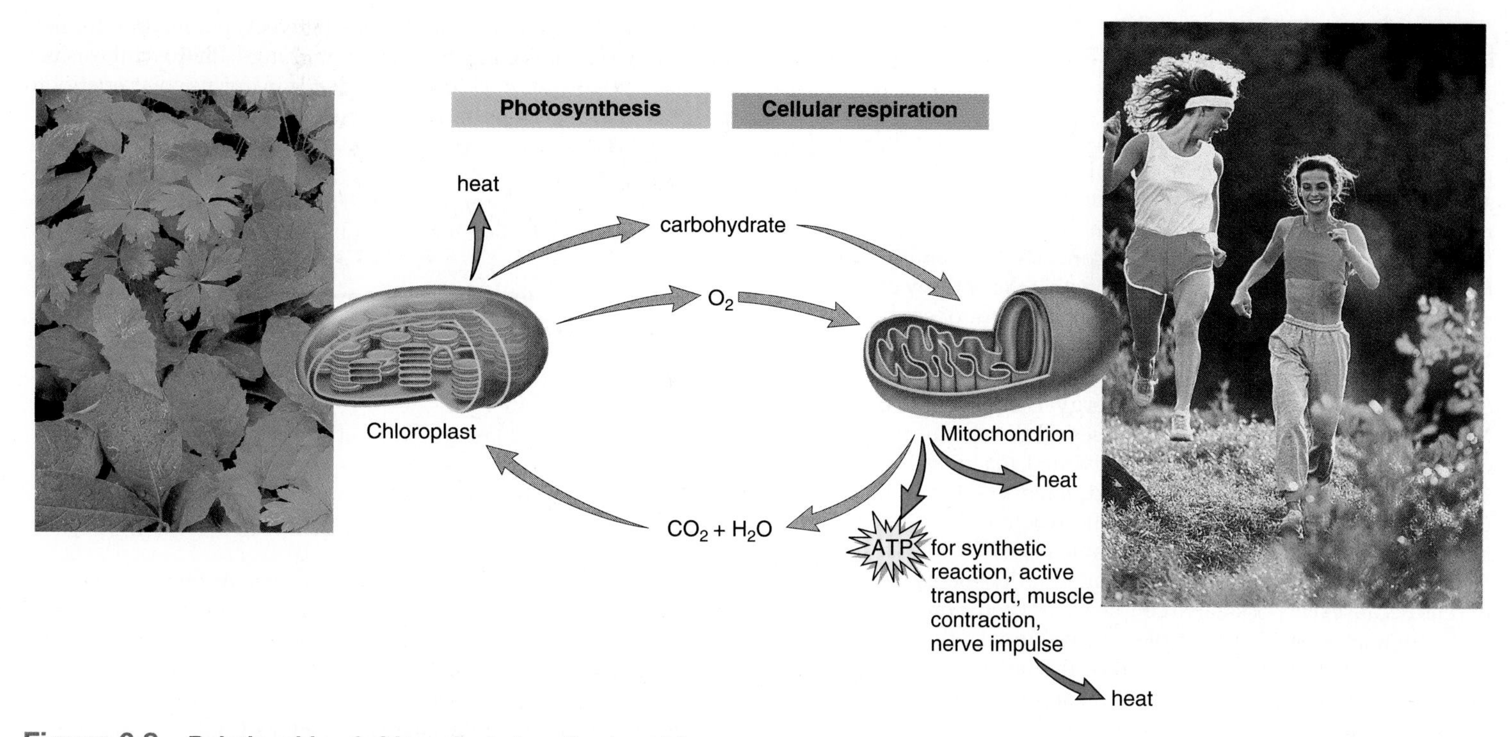

Figure 6.8 Relationship of chloroplasts to mitochondria.
Chloroplasts produce energy-rich carbohydrates. These carbohydrates are broken down in mitochondria, and the energy released is used for the buildup of ATP. There is a loss of usable energy due to the energy conversions of photosynthesis and aerobic respiration; and then, when ATP is used as an energy source, all usable energy is converted to heat.

Bioethical Issue

In the United States, solar energy to grow food is greatly supplemented by fossil fuel energy. Even before crops are sowed, there is an input of fossil fuel energy for the production of seeds, tools, fertilizers, pesticides, and their transportation to the farm. At the farm, fuel is needed to plant the seeds, to apply fertilizers and pesticides, to irrigate, and to harvest and dry crops. After harvesting, still more fuel is used to process crops and put it in those neatly packaged products we buy in the supermarket.

At this time, the supplemental energy to grow food is several hundred times its caloric content because we devote a limited amount of land to agriculture, and we use high-yielding plants that require more care anyway. It takes about twenty times the amount of energy to keep cattle in feedlots and feed them grain as it does to range-feed them. Because the combustion of fossil fuel energy contributes to environmental problems such as global warming and air pollution, it behooves us to take steps to cut down on supplemental energy to grow food. What can be done? First of all we could devote as much land as possible to farming and animal husbandry. Plant breeders could sacrifice some yield to develop plants that would require less supplemental energy. And we could range-feed cattle. If cattle are kept close to farmland, manure can substitute in part for chemical fertilizers. Biological control, the use of natural enemies to control pests, would cut down on pesticide use and possibly improve the health of farm families. Solar and wind energy could be used instead of fossil fuel energy; for example, wind-driven irrigation pumps are feasible.

Finally. consumers could help. We could overcome our prejudices against slight blemishes on our fruits and vegetables. We could cut down on our consumption of processed foods, eat less meat, and buy cheaper cuts. And we could avoid using electrically powered gadgets when preparing food at home.

Questions

1. Are you in favor of taking all possible steps to reduce the input of supplemental energy to grow food? Why or why not?
2. The way we grow food contributes to air, water, and land pollution. Should this become common knowledge? Why or why not?
3. Are you willing to make sacrifices to improve the quality of the environment? Why or why not?

Summarizing the Concepts

6.1 Energy
There are two energy laws that are basic to understanding energy-use patterns at all levels of biological organization. The first law states that energy cannot be created or destroyed, but can only be transferred or transformed. The second law states that one usable form of energy cannot be completely converted into another usable form. As a result of these laws, we know that the entropy of the universe is increasing and that only a constant input of energy maintains the organization of living things.

6.2 Metabolic Reactions and Energy Transformations
Metabolism is a term that encompasses all the chemical reactions occurring in a cell. Considering individual reactions, only those that result in a negative free energy difference—that is, the products have less usable energy than the reactants—occur spontaneously. Such reactions, called exergonic reactions, release energy. Endergonic reactions, which require an input of energy, occur only in cells because it is possible to couple an exergonic process with an endergonic process. For example, glucose breakdown is an exergonic metabolic pathway that drives the buildup of many ATP molecules. These ATP molecules then supply energy for cellular work. Thus, ATP goes through a cycle in which it is constantly being built up from, and then broken down, to ADP + Ⓟ.

6.3 Metabolic Pathways and Enzymes
A metabolic pathway is a series of reactions that proceed in an orderly, step-by-step manner. Each reaction requires a specific enzyme. Reaction rates increase when enzymes form a complex with their substrates. Generally, enzyme activity increases as substrate concentration increases; once all active sites are filled, maximum rate has been achieved.

Any environmental factor, such as temperature and pH, affects the shape of a protein and, therefore, also affects the ability of an enzyme to do its job. Cellular mechanisms regulate enzyme quantity and activity. The activity of most metabolic pathways is regulated by feedback inhibition. Many enzymes have cofactors or coenzymes that help them carry out a reaction.

6.4 Metabolic Pathways and Oxidation-Reduction
The overall equation for photosynthesis is the opposite of that for aerobic respiration. Both processes involve oxidation-reduction reactions. During photosynthesis, $NADP^+$ is a coenzyme that reduces carbon dioxide to glucose, and during aerobic respiration, NAD^+ is a coenzyme that oxidizes glucose products so that carbon dioxide is released. Redox reactions are a major way in which energy transformation occurs in cells.

There is a cycling of molecules between plants and animals and a flow of energy through all living things. Photosynthesis is a metabolic pathway in chloroplasts that transforms solar energy to the chemical energy within carbohydrates, and aerobic respiration is a metabolic pathway completed in mitochondria that transforms this energy into that of ATP molecules. Eventually the energy within ATP molecules becomes heat. The world of living things is dependent on a constant input of solar energy.

Studying the Concepts

1. State the first law of thermodynamics and give an example. 104
2. State the second law of thermodynamics and give an example. 104
3. Explain why the entropy of the universe is always increasing and why an organized system like an organism requires a constant input of useful energy. 104
4. What is the difference between exergonic reactions and endergonic reactions? Why can exergonic but not endergonic reactions occur spontaneously? 106
5. Define coupling and write an equation that shows an endergonic reaction being coupled to ATP breakdown. 106
6. Why is ATP called the energy currency of cells? What is the ATP cycle? 107
7. Diagram a metabolic pathway. Label the reactants, products, and enzymes. 108
8. Why is less energy needed for a reaction to occur when an enzyme is present? 108
9. Why are enzymes specific, and why can't each one speed up many different reactions? 109
10. Name and explain the manner in which at least three factors can influence the speed of an enzymatic reaction. How do cells regulate the activity of enzymes? 110–11
11. What are cofactors and coenzymes? 111
12. Describe how oxidation-reduction occurs in cells and discuss the overall equations for photosynthesis and aerobic cellular respiration in terms of oxidation-reduction. 112
13. How do chloroplasts and mitochondria permit a flow of energy through the world of living things? 112

Testing Yourself

Choose the best answer for each question.

1. When ATP becomes ADP + Ⓟ,
   a. some usable energy is lost to the environment.
   b. energy is created according to the first law of thermodynamics.
   c. an enzyme is required because the reaction does not occur spontaneously.
   d. the entropy of the universe is increased.
   e. All of these are correct.
2. If $A + B \rightarrow C + D +$ energy occurs in a cell,
   a. this reaction is exergonic.
   b. an enzyme could still speed the reaction.
   c. ATP is not needed to make the reaction go.
   d. A and B are reactants; C and D are products.
   e. All of these are correct.
3. Which of these does not utilize ATP?
   a. synthesis of molecules in cells
   b. active transport of molecules across the plasma membrane
   c. muscle contraction
   d. nerve conduction
   e. sweating to lose excess heat
4. Energy of activation
   a. is the amount of entropy in a system.
   b. is the amount of energy given off by a reaction.
   c. converts kinetic energy to potential energy.
   d. is the energy needed to start a reaction.
   e. is a way for cells to compete with one another.
5. The active site of an enzyme is
   a. similar to that of any other enzyme.
   b. the part of the enzyme where its substrate can fit.
   c. can be used over and over again.
   d. not affected by environmental factors like pH and temperature.
   e. Both b and c are correct.
6. If you wanted to increase the amount of product per unit time of an enzymatic reaction, do not increase
   a. the amount of substrate.
   b. the amount of enzyme.
   c. the temperature somewhat.
   d. the pH.
   e. All of these are correct.
7. An allosteric site on an enzyme is
   a. the same as the active site.
   b. nonprotein in nature.
   c. where ATP attaches and gives up its energy.
   d. often involved in feedback inhibition.
   e. All of these are correct.
8. Coenzymes
   a. have specific functions in reactions.
   b. have an active site just like enzymes do.
   c. can be a carrier for proteins.
   d. always have a phosphate group.
   e. are used in photosynthesis but not cellular respiration.
9. During photosynthesis, carbon dioxide
   a. is oxidized to oxygen.
   b. is reduced to glucose.
   c. gives up water to the environment.
   d. is a coenzyme of oxidation-reduction.
   e. All of these are correct.
10. The oxygen given off by photosynthesis
    a. is used by animal cells, but not plant cells, to carry on cellular respiration.
    b. is used by both plant and animal cells to carry on cellular respiration.
    c. is an example of the flow of energy through living things.
    d. is an example of the cycling of matter through living things.
    e. Both b and d are correct.
11. Use these terms to label this diagram: substrates, enzyme (used twice), active site, product, and enzyme-substrate complex. Explain the importance of an enzyme's shape to its activity.

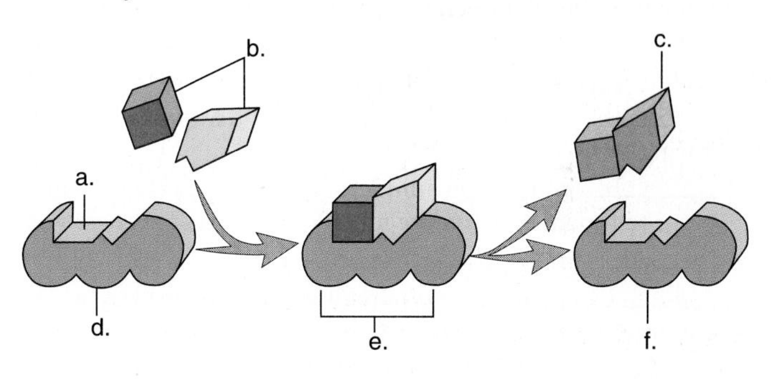

Thinking Scientifically

1. Pepsin is an enzyme that breaks down protein.
   a. A student has a test tube that contains pepsin, egg white, and water. What optimal conditions would you recommend to ensure digestion of the egg white? (page 110)
   b. If all the conditions are optimal, how could you increase the yield (i.e., amount of product—amino acids—per unit of time)? (page 110)
   c. The instructor adds an inhibitor to the test tube. How could the student tell if inhibition is reversible or irreversible? (page 110)
2. A lack of oxygen causes death. Explain why by referring to
   a. the overall equation for aerobic cellular respiration. (page 112)
   b. the necessity of ATP for muscle contraction.
   c. the needs of ordinary body cells.
   d. brain activity as a test of death.

Understanding the Terms

Match the terms to these definitions:

a. ________ All of the chemical reactions that occur in a cell during growth and repair.

b. ________ Nonprotein adjunct required by an enzyme in order to function; many are metal ions, others are coenzymes.

c. ________ Energy associated with motion.

d. ________ Essential requirement in the diet, needed in small amounts. They are often part of coenzymes.

e. ________ Loss of an enzyme's normal shape so that it no longer functions; caused by an extreme change in pH and temperature.

Using Technology

Your study of Metabolism: Energy and Enzymes is supported by these available technologies:

Essential Study Partner CD-ROM

Cells → Metabolism

Visit the Mader web site for related ESP activities.

Exploring the Internet

The Mader Home Page provides resources and tools as you study this chapter.

http://www.mhhe.com/biosci/genbio/mader

Virtual Physiology Laboratory CD-ROM

Enzyme Characterisitics

Life Science Animations 3D Video

7 Enzyme Action

CHAPTER 7

Cellular Respiration

The well-developed muscles of these swimmers give them strength. Muscles are powered by the energy of ATP, and mitochondria are the organelles that convert the energy of various organic compounds to the energy of ATP.

Chapter Concepts

The long-distance runner breathes deeply and rhythmically as his legs keep up a steady beat along the track. Internally, his body responds smoothly to match the pace of his motions. Oxygen flowing into his lungs enters the blood, passes through the heart and then ever smaller blood vessels, until it is delivered to the cells of his tissues. The intake of oxygen by the lungs and delivery by the blood has been enhanced by his previous training. His lung capacity is greater, his blood contains more red blood cells, and the blood supply to his muscles is greater than in those who do not train. His muscle cells are ready to receive oxygen; they contain an abundance of mitochondria, which start producing ATP as soon as oxygen has appeared. Mitochondria are fueled by glucose breakdown. Glucose is present because it has been stored in his muscle cells as glycogen—the runner loaded up on carbohydrates a couple of days before the race.

Respiration involves breathing and the transport of oxygen to cells, but this chapter concerns only the use of energy from the breakdown of organic molecules for the production of ATP.

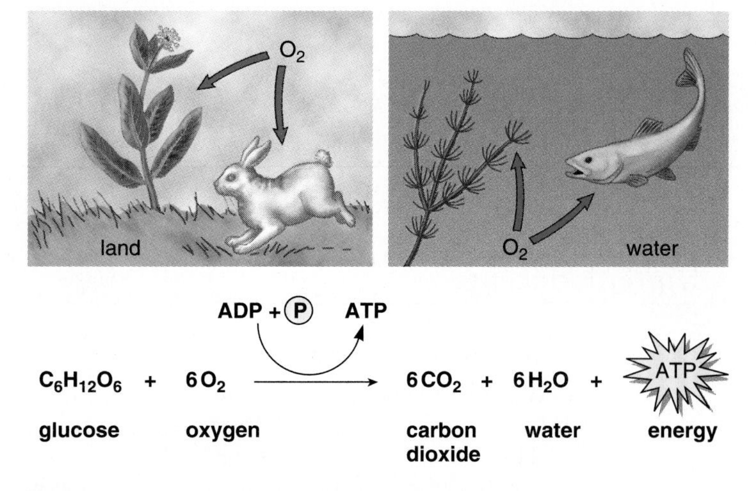

$$\underset{\text{glucose}}{C_6H_{12}O_6} + \underset{\text{oxygen}}{6\,O_2} \xrightarrow{\text{ADP} + \text{Ⓟ} \rightarrow \text{ATP}} \underset{\text{carbon dioxide}}{6\,CO_2} + \underset{\text{water}}{6\,H_2O} + \underset{\text{energy}}{\text{ATP}}$$

Figure 7.1 Aerobic cellular respiration.
Almost all organisms, whether they reside on land or in the water, carry on aerobic cellular respiration, which is most often glucose breakdown coupled to ATP buildup.

7.1 Aerobic Cellular Respiration

Aerobic cellular respiration includes all the various metabolic pathways by which carbohydrates and other molecules are broken down with the concomitant buildup of ATP. The expression refers to pathways that require oxygen, as indicated by the word **aerobic,** and results in a complete breakdown of molecules to carbon dioxide (CO_2) and water (H_2O). Figure 7.1 shows an overall equation for aerobic cellular respiration, when glucose is the first reactant.

Glucose is a high-energy molecule, and its breakdown products, CO_2 and H_2O, are low-energy molecules. Therefore, we would expect the process to be exergonic, that is, releases energy:

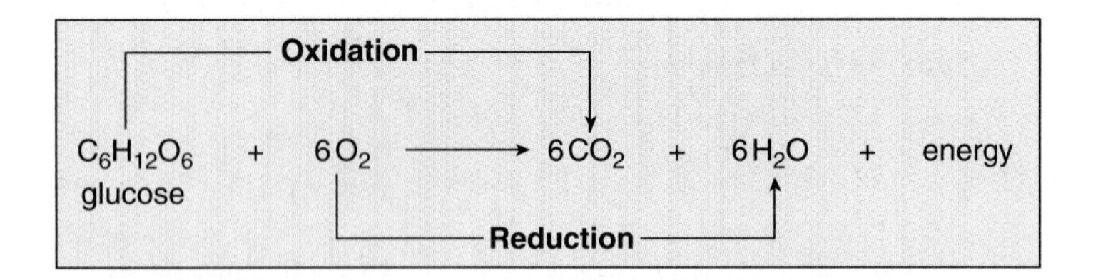

As breakdown occurs, electrons are removed from substrates and eventually are received by O_2, which then combines with H^+ and becomes H_2O. The equation shows changes in regard to hydrogen atom (H) distribution. A hydrogen atom consists of a hydrogen ion plus an electron ($H^+ + e^-$). When hydrogen atoms are removed from glucose, so are electrons. Since *oxidation* is the loss of electrons, and *reduction* is the gain of electrons, glucose breakdown is an oxidation-reduction reaction. Glucose is oxidized and O_2 is reduced.

The buildup of ATP is an endergonic reaction, that is, it requires energy. The overall equation for aerobic cellular respiration (Fig. 7.1) shows the coupling of glucose breakdown to ATP buildup. As glucose is broken down, ATP is built up, and this is the reason the ATP reaction is drawn using a curved arrow above the glucose reaction arrow. The breakdown of one glucose molecule results in a maximum of 36 or 38 ATP molecules. This represents about 40% of the potential energy within a glucose molecule; the rest of the energy is lost as heat. This conversion is more efficient than many others; for example, only about 25% of the energy within gasoline is converted to the motion of a car.

Phases of Aerobic Cellular Respiration

Aerobic cellular respiration does not occur all at once as indicated by the overall reaction. If it did, all of the energy within glucose would be given off at once and much of it would be lost as heat. Instead, glucose breakdown involves the four phases shown in Figure 7.2. Three of these are metabolic pathways and one is an individual reaction:

- **Glycolysis** is the breakdown of glucose to two molecules of **pyruvate.** Oxidation by removal of hydrogen atoms provides enough energy for the immediate buildup of two ATP. Glycolysis takes place outside the mitochondria and does not utilize

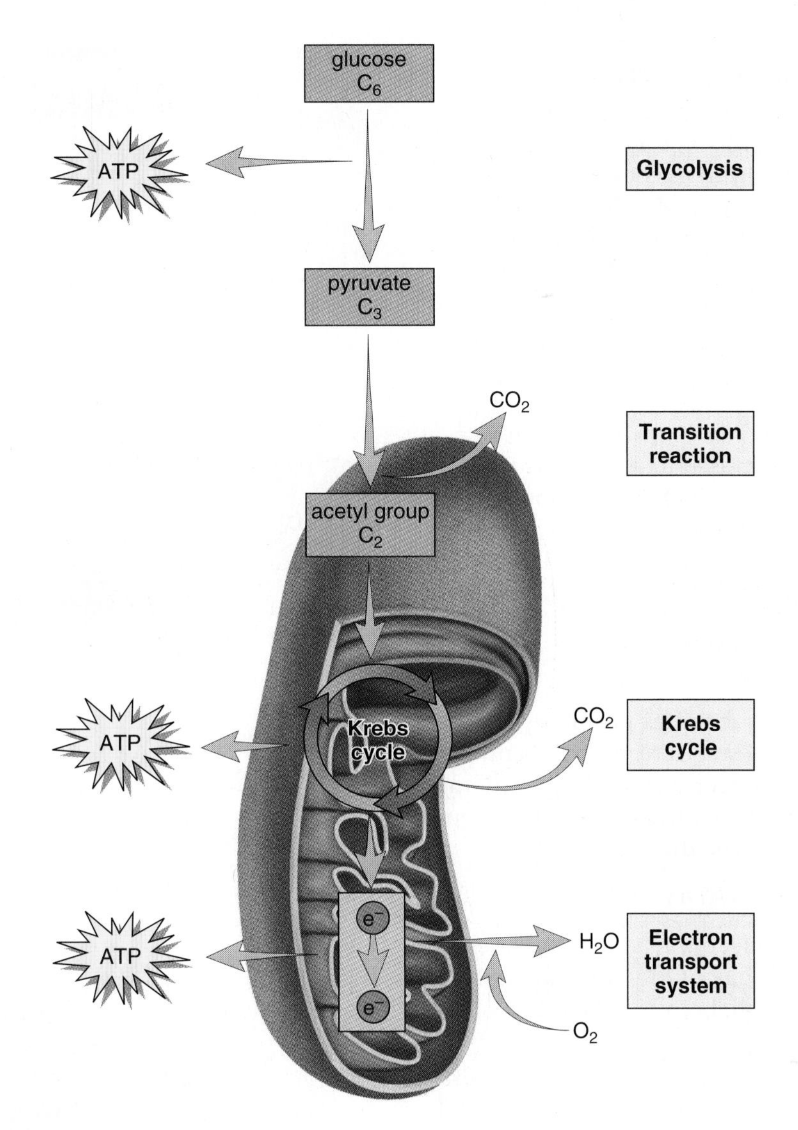

Figure 7.2 Steps of aerobic cellular respiration.
Glycolysis, the transition reaction, the Krebs cycle, and the electron transport system are the four phases of aerobic cellular respiration which results in 36 or 38 ATP per glucose molecule.

oxygen. The other stages of aerobic cellular respiration take place inside the mitochondria, where oxygen is utilized.

- During the **transition reaction,** pyruvate is oxidized to an acetyl group carried by CoA, and CO_2 is removed. Since glycolysis ends with two molecules of pyruvate, the transition reaction occurs twice per glucose molecule.
- The **Krebs cycle** is a cyclical series of oxidation reactions that give off CO_2 and produce one ATP. The Krebs cycle turns twice because two acetyl CoA molecules enter the cycle per glucose molecule. Altogether then, the Krebs cycle accounts for two immediate ATP molecules per glucose molecule.
- The **electron transport system** is a series of carriers that accept the electrons removed from glucose and pass them along from one carrier to the next until they are finally received by O_2. As the electrons pass from a higher energy to a lower energy state, energy is released and used for ATP buildup. The electrons from one glucose result in 32 or 34 ATP, depending on certain conditions.

Aerobic cellular respiration involves (1) glycolysis; (2) the transition reaction; (3) the Krebs cycle; and (4) the electron transport system. A total of 36 or 38 ATP molecules are produced per glucose molecule in aerobic cellular respiration.

7.2 Outside the Mitochondria: Glycolysis

Glycolysis, which takes place within the cytoplasm outside the mitochondria, is the breakdown of glucose to two pyruvate molecules. Since glycolysis is universally found in organisms, it most likely evolved before the Krebs cycle and the electron transport system. This may be why glycolysis occurs in the cytoplasm and does not require oxygen.

Energy Investment Steps

As glycolysis begins, two ATP are used to activate glucose, a C_6 (six-carbon) molecule that splits into two C_3 molecules, each of which carries a phosphate group. From this point on, each C_3 molecule undergoes the same series of reactions. Note that the left and right sides of Figure 7.3 are exactly the same.

Energy Harvesting Steps

During glycolysis, oxidation of substrates occurs by the removal of electrons. However, these electrons are accompanied by a hydrogen ion, so in effect two hydrogen atoms ($2e^- + 2H^+$) are removed. These are picked up by the coenzyme **NAD^+ (nicotinamide adenine dinucleotide)**:

$$NAD^+ + 2H \rightarrow NADH + H^+$$

Later, when NADH passes two electrons on to another electron carrier, it becomes NAD^+ again. Only a small amount of NAD^+ need be present in a cell, because like other coenzymes it is used over and over again.

Oxidation results in high-energy phosphate molecules that allow the formation of four ATP. This is called **substrate-level phosphorylation** because a substrate passes a high-energy phosphate to ADP, and ATP results. Subtracting the two ATP that were used to get started, there is a net gain of two ATP from glycolysis (Fig. 7.3).

When glycolysis is a part of aerobic cellular respiration, the end product pyruvate enters the mitochondria, where oxygen is utilized. However, glycolysis does not need to be a part of aerobic cellular respiration and can occur even when oxygen is not present in cells. As we will see on page 129, glycolysis is also a part of fermentation. Fermentation is an **anaerobic** process; it does not require oxygen.

Altogether the inputs and outputs of glycolysis are as follows:

Glycolysis

inputs	outputs
glucose	2 pyruvate
2 NAD^+	2 NADH
2 ATP	4 ATP (net 2 ATP)
4 ADP + 2 Ⓟ	

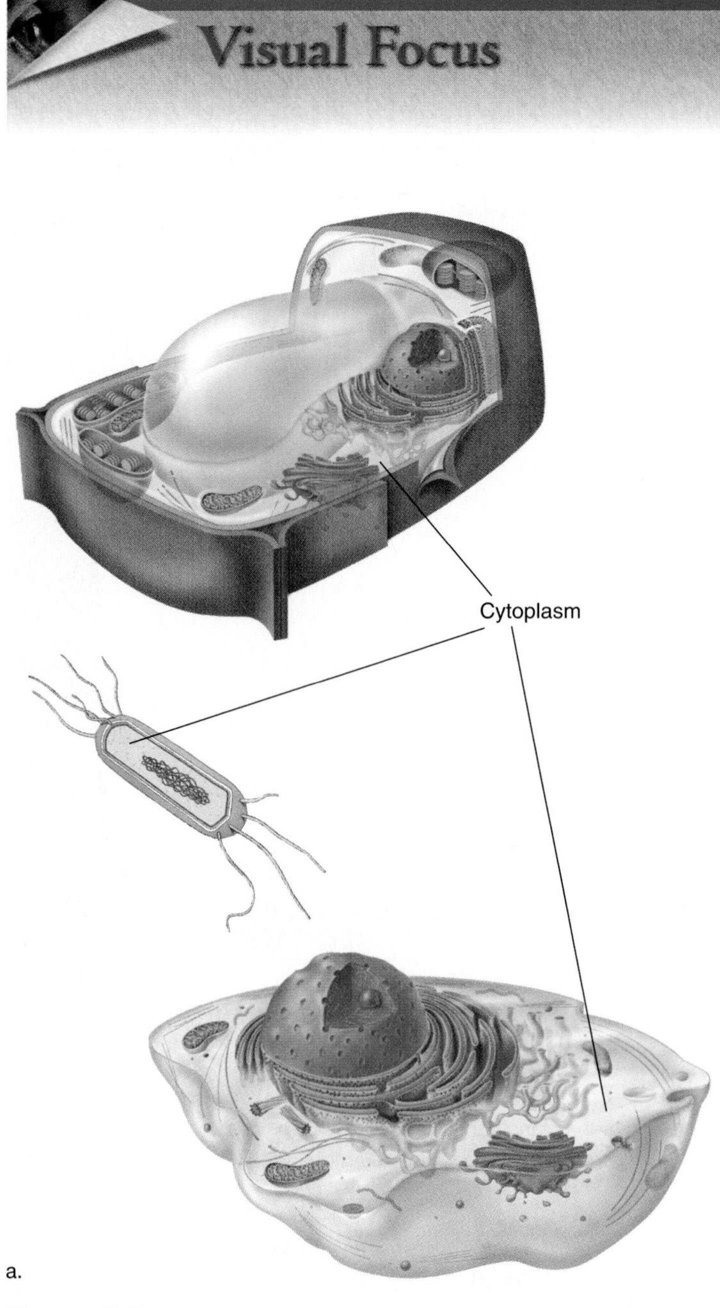

Figure 7.3 Glycolysis.
(a) Glycolysis takes place in cytoplasm. **(b)** This pathway begins with glucose and ends with two pyruvate molecules; there is a gain of two NADH, and a net gain of two ATP from glycolysis.

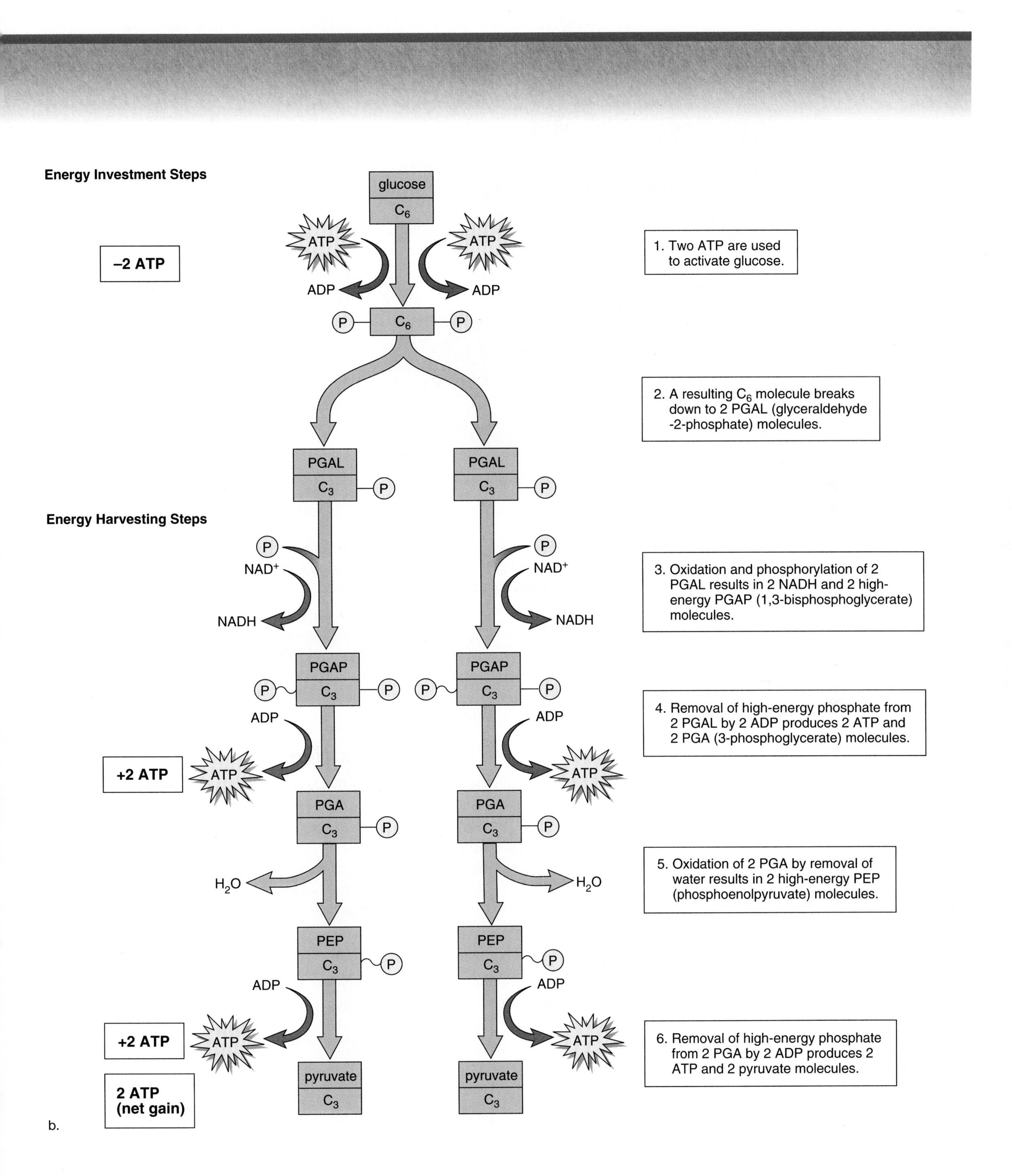

Energy Investment Steps
glucose
C_6
–2 ATP
ATP
ATP
ADP
ADP
P
C_6
P
1. Two ATP are used to activate glucose.
2. A resulting C_6 molecule breaks down to 2 PGAL (glyceraldehyde -2-phosphate) molecules.
PGAL
C_3
P
PGAL
C_3
P
Energy Harvesting Steps
P
NAD^+
NADH
P
NAD^+
NADH
3. Oxidation and phosphorylation of 2 PGAL results in 2 NADH and 2 high-energy PGAP (1,3-bisphosphoglycerate) molecules.
PGAP
P
C_3
P
PGAP
P
C_3
P
ADP
ADP
4. Removal of high-energy phosphate from 2 PGAL by 2 ADP produces 2 ATP and 2 PGA (3-phosphoglycerate) molecules.
+2 ATP
ATP
ATP
PGA
C_3
P
PGA
C_3
P
H_2O
H_2O
5. Oxidation of 2 PGA by removal of water results in 2 high-energy PEP (phosphoenolpyruvate) molecules.
PEP
C_3
P
PEP
C_3
P
ADP
ADP
+2 ATP
ATP
ATP
6. Removal of high-energy phosphate from 2 PGA by 2 ADP produces 2 ATP and 2 pyruvate molecules.
2 ATP (net gain)
pyruvate
C_3
pyruvate
C_3
b.

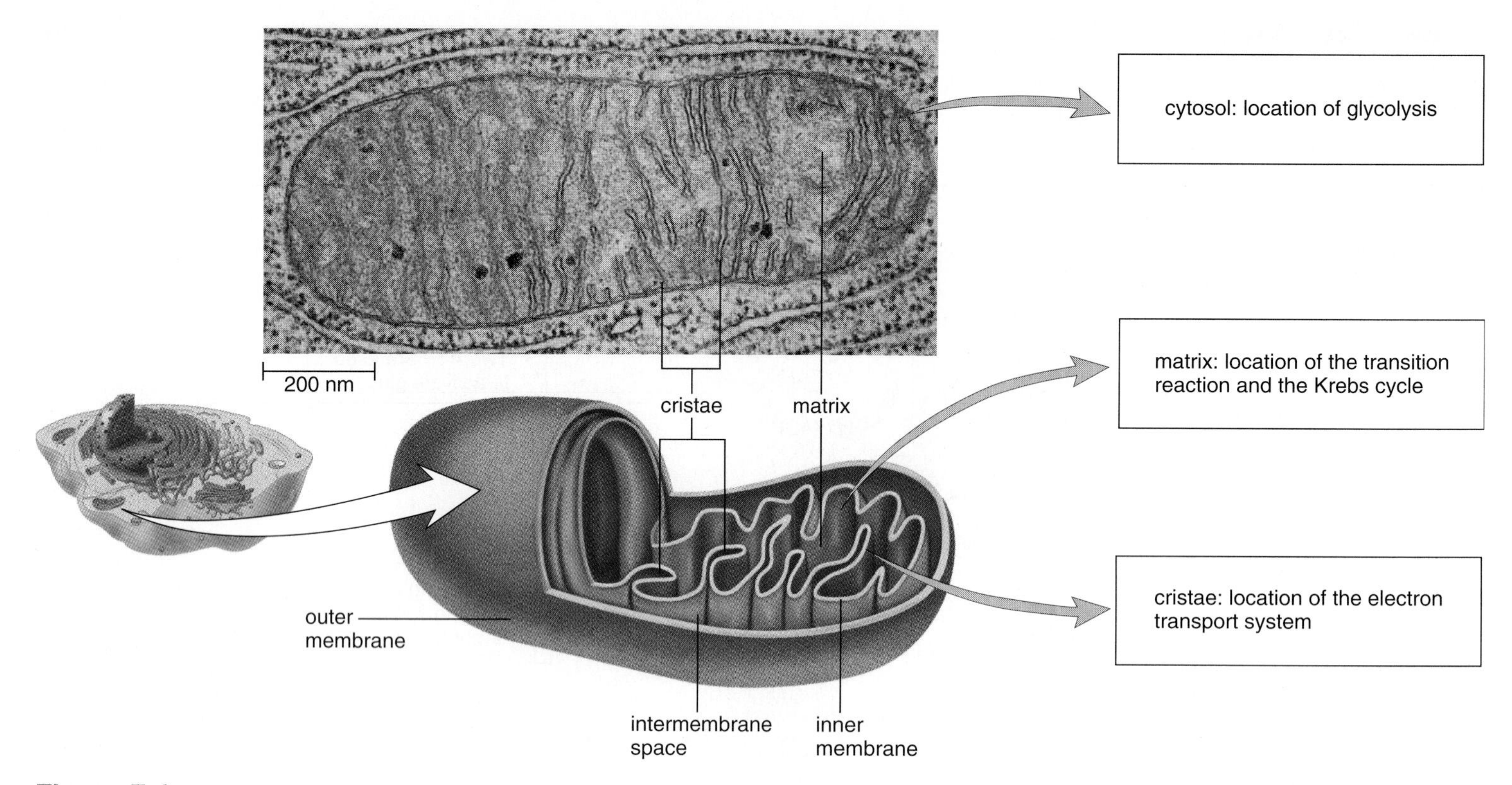

Figure 7.4 Mitochondrion structure and function.
A mitochondrion is bounded by a double membrane, with an intermembrane space. The inner membrane invaginates to form the shelflike cristae. The matrix is the fluid-filled interior of the mitochondrion. Note that glycolysis occurs outside the mitochondrion, whereas the Krebs cycle occurs in the matrix and the electron transport system is located on the cristae, both within the mitochondrion.

7.3 Inside the Mitochondria

Whereas glycolysis takes place in the cytoplasm, the transition reaction, the Krebs cycle, and the electron transport system all take place inside the cellular organelle called a **mitochondrion.** A mitochondrion has a double membrane, with an *intermembrane space* (between the outer and inner membrane). *Cristae* are folds of inner membrane that jut out into the *matrix,* the innermost compartment, which is filled with a gel-like fluid (Fig. 7.4). The transition reaction and the Krebs cycle enzymes are located in the matrix, and the electron transport system is located in the cristae. Most of the ATP produced during cellular respiration is produced in mitochondria; therefore, mitochondria are often called the powerhouses of the cell.

It is interesting to think about how our bodies provide the reactants for aerobic cellular respiration. The air we breathe contains oxygen, and the food we eat contains glucose. These enter the bloodstream, which carries them about the body, and they move into each and every cell. The end product of glycolysis, pyruvate, enters the mitochondria, where the transition reaction, the Krebs cycle, and electron transport system occur. Eventually, pyruvate is completely broken down to CO_2 and H_2O as ATP is produced. The CO_2 and ATP diffuse out of mitochondria into the cytoplasm. The ATP is utilized in the cell for energy-requiring processes. Carbon dioxide diffuses out of the cell and enters the bloodstream. The bloodstream takes the CO_2 to the lungs, where it is exchanged. The H_2O can remain in the mitochondria, or the cell, or enter the blood and be excreted by the kidneys as need be.

Transition Reaction

The **transition reaction** is so-called because it connects glycolysis to the Krebs cycle. In this reaction, pyruvate is converted to a two-carbon *acetyl group* attached to *coenzyme A,* or CoA, and CO_2 is given off. This is an oxidation reaction in which electrons are removed from pyruvate by an enzyme that uses NAD^+ as a coenzyme. NAD^+ goes to $NADH + H^+$ as **acetyl-CoA** forms. This reaction occurs twice per glucose molecule.

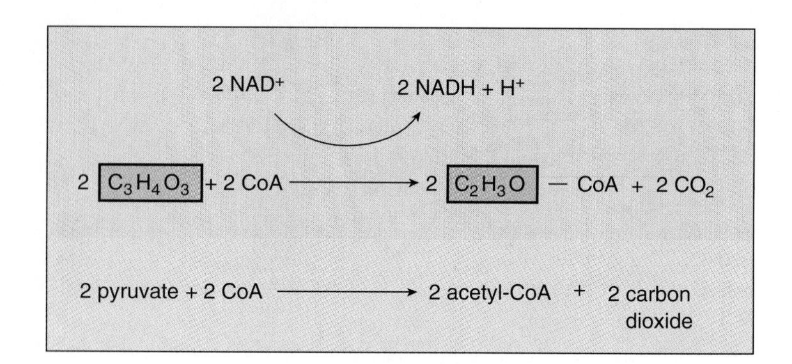

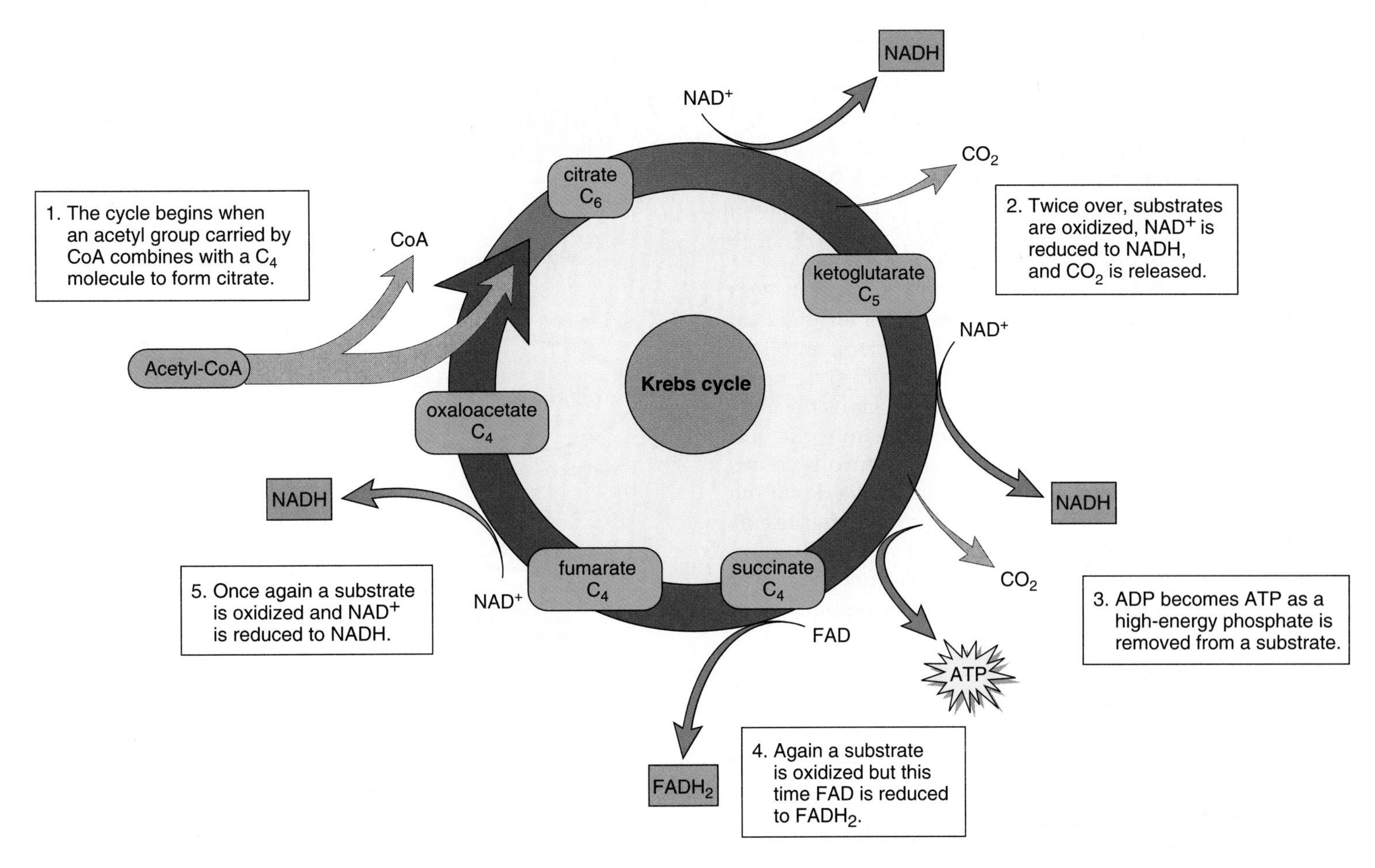

Figure 7.5 Krebs cycle.
The net result of this cycle of reactions is the oxidation of an acetyl group to two molecules of CO_2 along with a transfer of electrons to NAD^+ and FAD and a gain of one ATP. The Krebs cycle turns twice per glucose molecule.

Krebs Cycle

The **Krebs cycle** is a cyclical metabolic pathway located in the matrix of mitochondria. The Krebs cycle, named for Sir Hans Krebs, a British scientist, begins with the molecule citrate. For this reason, it is also known as the *citric acid cycle* (Fig. 7.5). During the Krebs cycle, the acetyl group that results from the transition reaction undergoes oxidation by the removal of hydrogen atoms as two molecules of CO_2 are given off.

At the start of the Krebs cycle, the C_2 acetyl group carried by CoA joins with a C_4 molecule, and citrate results. Oxidation occurs during the cycle not only by NAD^+ but also by FAD. **FAD** (flavin adenine dinucleotide) is another coenzyme of oxidation-reduction which is sometimes used instead of NAD^+. After FAD accepts two electrons, it becomes $FADH_2$. During the Krebs cycle, electrons are accepted by NAD^+ in three instances; FAD is used only once.

The two carbons of the acetyl group come off as CO_2 as a part of the oxidation process. Substrate-level phosphorylation is an important event of the Krebs cycle. When a high-energy phosphate molecule gives up its phosphate group, ATP eventually results.

The Krebs cycle turns twice for each original glucose molecule. Therefore, the inputs and outputs of the Krebs cycle per glucose molecule are as follows:

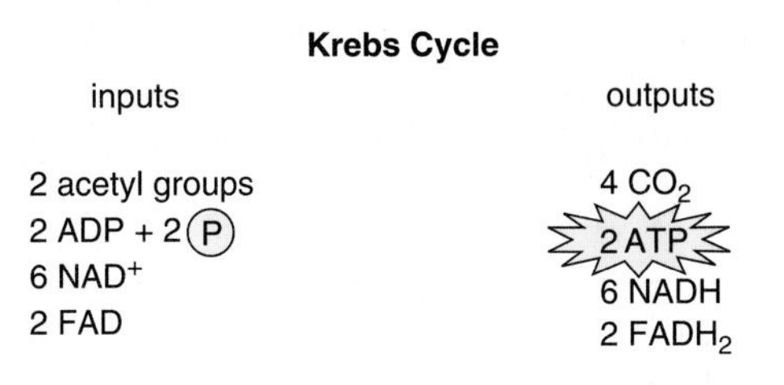

We have now accounted for six carbon dioxide molecules—two from the transition reaction and four from the Krebs cycle. Carbon dioxide is one of the end products of aerobic cellular respiration.

Electron Transport System

The **electron transport system** located in the cristae of the mitochondria is a series of carriers that pass electrons from one to the other. Some of the **electron carriers** of the system are cytochromes; therefore, the system is also termed a *cytochrome system.* **Cytochromes** are protein molecules with a heme group that contains an iron atom (Fe) capable of being reduced and oxidized in a reversible manner.

The electrons that enter the electron transport system are carried by NAD^+ and FAD. Figure 7.6 is arranged to show that high-energy electrons enter the system, and low-energy electrons leave the system. When NADH gives up its electrons, it becomes NAD^+; the next carrier gains the electrons and is reduced. This oxidation-reduction reaction starts the process, and each of the carriers in turn becomes reduced and then oxidized as the electrons move down the system. As the pair of electrons is passed from carrier to carrier, energy is released and used to form ATP molecules. Oxygen receives the energy-spent electrons from the last of the carriers. After receiving electrons, oxygen combines with hydrogen ions and water forms:

$$\frac{1}{2}O_2 + 2e^- + 2H^+ \longrightarrow H_2O$$

This manner of producing ATP is sometimes called **oxidative phosphorylation** because O_2 must be present to receive electrons or the electron transport system does not work.

When NADH delivers electrons to the first carrier of the electron transport system, enough energy is released by the time the electrons are received by O_2 to permit the production of three ATP molecules. When $FADH_2$ delivers electrons to the electron transport system, only two ATP are produced.

The cell needs only a limited supply of the coenzymes NAD^+ and FAD because they are constantly being recycled and reused. Therefore, once NADH has delivered electrons to the electron transport system, it is "free" to return and pick up more hydrogen atoms. In the same manner, the components of ATP are recycled in cells. Energy is required to join ADP + Ⓟ, and then when ATP is used to do cellular work, ADP and Ⓟ result once more. The recycling of coenzymes and ADP increases cellular efficiency since it does away with the necessity to synthesize NAD^+, FAD, and ADP anew.

As electrons pass down the electron transport system, energy is released and ATP is produced.

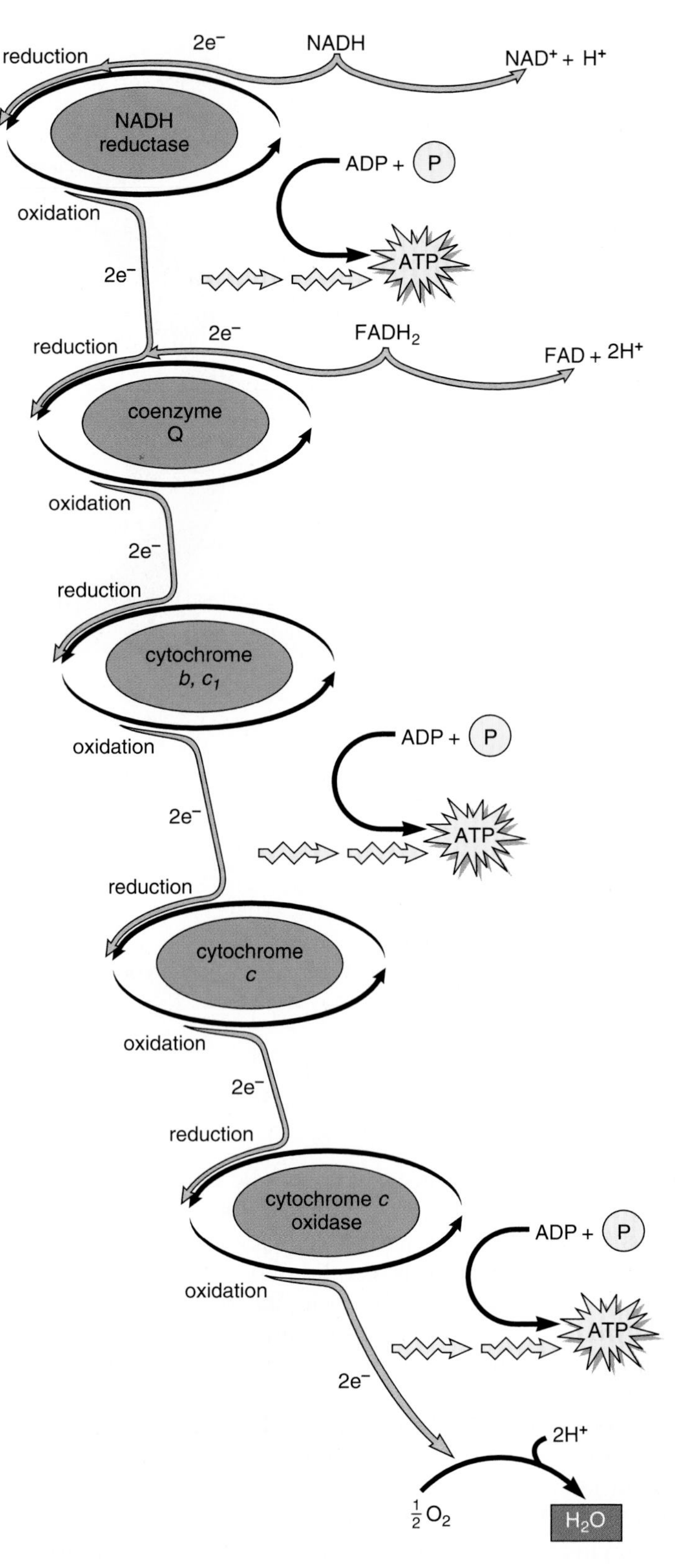

Figure 7.6 Overview of the electron transport system. NADH and $FADH_2$ bring electrons to the electron transport system. As the electrons move down the system, energy is released and used to form ATP. For every pair of electrons that enters by way of NADH, three ATP result. For every pair of electrons that enters by way of $FADH_2$, two ATP result. Oxygen, the final acceptor of the electrons, becomes a part of water.

Organization of Cristae

Figure 7.6 is a simplified overview of the electron transport system. Figure 7.7 shows how the electron transport system consists of three protein complexes and two mobile carriers. The mobile carriers transport electrons between the complexes, which also contain electron carriers.

Thus far we have been stressing that the carriers accept electrons, which they pass from one to the other. What happens to the hydrogen ions (H^+) carried by NADH and $FADH_2$? The carriers of the electron transport system use the energy released by electrons as they move down the electron transport system to pump hydrogen ions into the intermembrane space of a mitochondrion. The vertical arrows in Figure 7.8 show that the first complex, called the NADH dehydrogenase complex, the cytochrome b–c_1 complex, and the cytochrome c oxidase complex all pump H^+ into the intermembrane space. This establishes a very strong electrochemical gradient; there are many hydrogen ions in the intermembrane space and few in the matrix of a mitochondrion.

The cristae also contain an *ATP synthase complex* through which hydrogen ions flow down their gradient from the intermembrane space into the matrix. As hydrogen ions flow from high to low concentration, energy is released, allowing the enzyme ATP synthase to synthesize ATP from ADP + Ⓟ. Just how H^+ flow drives ATP synthesis is not known; perhaps the hydrogen ions participate in the reaction, or perhaps they cause a change in the shape of ATP synthase and this brings about ATP synthesis. Mitochondria produce ATP by **chemiosmosis,** so-called because ATP production is tied to an electrochemical gradient that is an H^+ gradient.

Once formed, ATP molecules diffuse out of the mitochondrial matrix by way of a channel protein.

Mitochondria utilize ATP synthesis by chemiosmosis. ATP production is dependent upon an electrochemical gradient established by the pumping of H^+ into the intermembrane space.

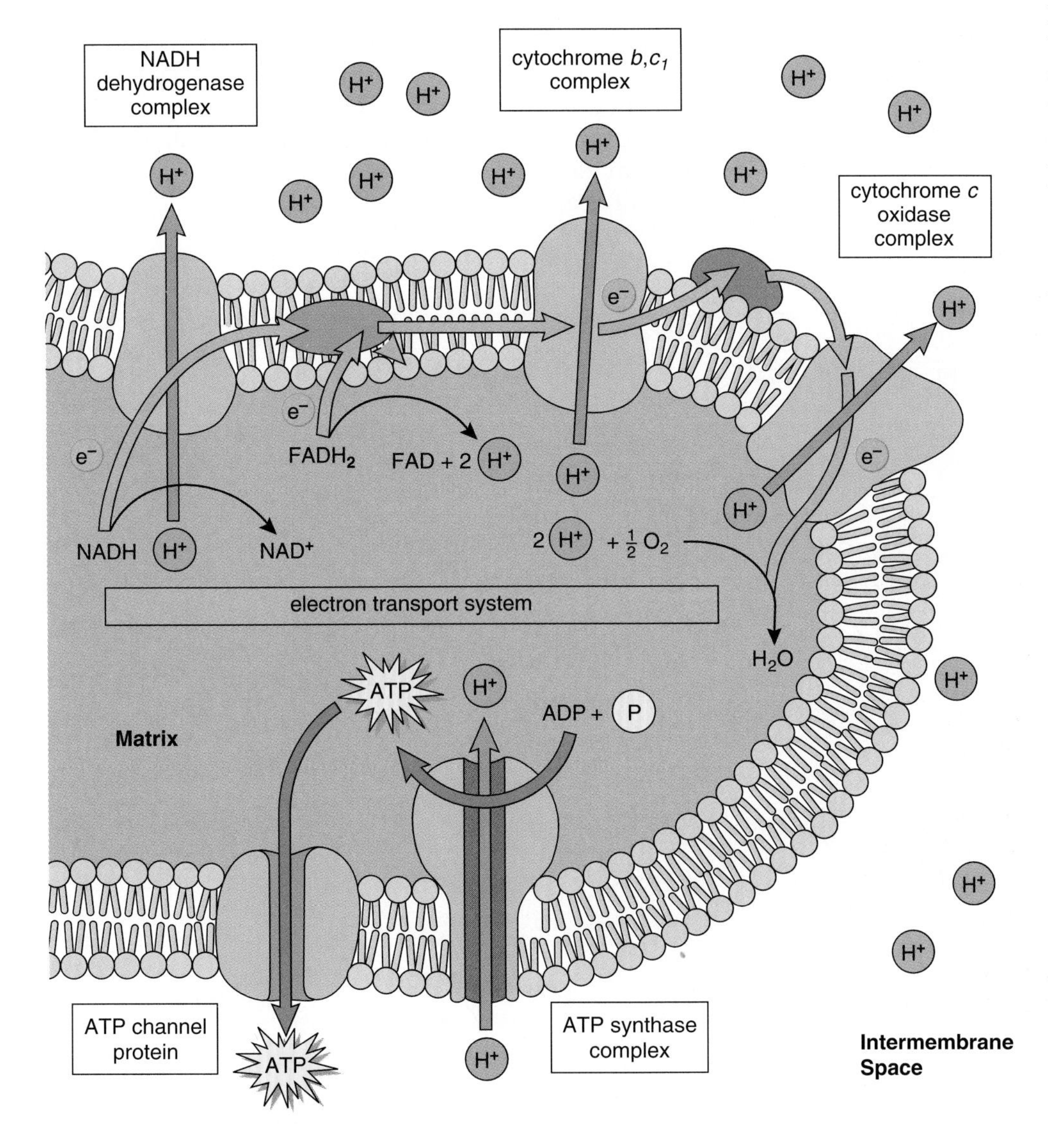

Figure 7.7 Organization of cristae.
The electron transport system is located in the cristae. As electrons (e^-) move from one complex to the other, hydrogen ions (H^+) are pumped from the matrix into the intermembrane space. As hydrogen ions flow down their concentration gradient from the intermembrane space into the matrix, ATP is synthesized by an ATP synthase. ATP leaves the matrix by way of a channel protein.

Energy Yield From Glucose Metabolism

Figure 7.8 calculates the ATP yield for the complete breakdown of glucose to CO_2 and H_2O. Notice that the diagram includes the number of ATP produced directly by glycolysis and the Krebs cycle and the number that is produced as a result of electrons passing down the electron transport system.

Per glucose molecule, there is a net gain of two ATP from glycolysis, which takes place in the cytoplasm. The Krebs cycle, which occurs in the matrix of mitochondria, accounts for two ATP per glucose molecule. This means that there is a total of four ATP formed outside the electron transport system.

Most ATP is produced by the electron transport system. Per glucose molecule, ten NADH and two $FADH_2$ take electrons to the electron transport system. For each NADH formed *inside* the mitochondria by the Krebs cycle, three ATP result, but for each $FADH_2$, there are only two ATP produced. Figure 7.6 explains the reason for this difference: $FADH_2$ delivers its electrons to the transport system after NADH, and therefore these electrons cannot account for as much ATP production.

What about the ATP yield of NADH generated *outside* the mitochondria by the glycolytic pathway? NADH cannot cross mitochondrial membranes, but there is a "shuttle" mechanism that allows its electrons to be delivered to the electron transport system inside the mitochondria. The shuttle consists of an organic molecule, which can cross the outer membrane, accept the electrons, and in most but not all cells, deliver them to a FAD molecule in the inner membrane. If FAD is used, only two ATP result because the electrons have not entered the start of the electron transport system.

Efficiency of Aerobic Respiration

It is interesting to calculate how much of the energy in a glucose molecule eventually becomes available to the cell. The difference in energy content between the reactants (glucose and O_2) and the products (CO_2 and H_2O) is 686 kcal. An ATP phosphate bond has an energy content of 7.3 kcal, and 36 of these are usually produced during glucose breakdown; 36 phosphates are equivalent to a total of 263 kcal. Therefore, 263/686, or 39%, of the available energy is usually transferred from glucose to ATP. The rest of the energy is lost in the form of heat.

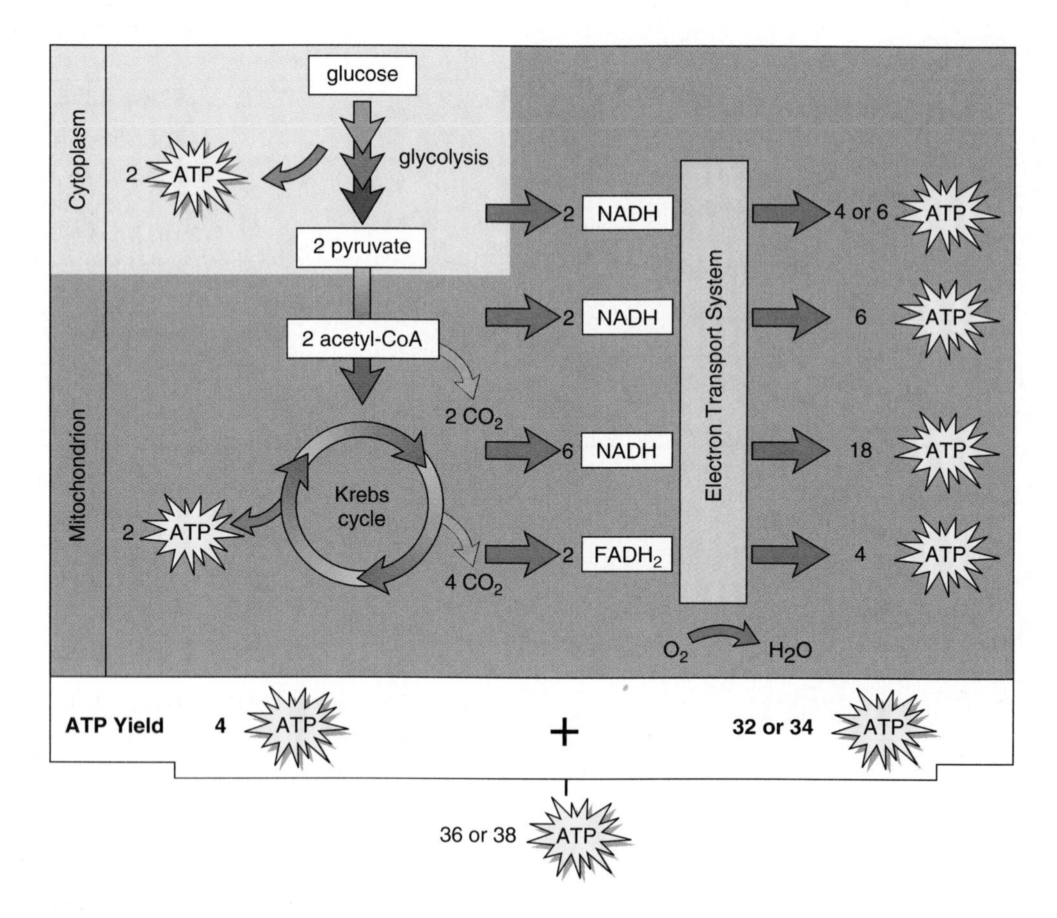

Figure 7.8 Accounting of energy yield per glucose molecule breakdown.
Substrate-level phosphorylation during glycolysis and the Krebs cycle accounts for four ATP. Oxidative phosphorylation accounts for 32 or 34 ATP, and the grand total of ATP is therefore 36 or 38 ATP. Cells differ as to the delivery of the electrons from NADH generated outside the mitochondria. If they are delivered by a shuttle mechanism to the start of the electron transport system, 6 ATP result; otherwise, 4 ATP result.

7.4 Metabolic Pool and Biosynthesis

Degradative reactions, which participate in **catabolism,** break down molecules and tend to be exergonic. Synthetic reactions, which participate in **anabolism,** tend to be endergonic. It is correct to say that catabolism drives anabolism because catabolism results in an ATP buildup that is used by anabolism.

Catabolism

We already know that glucose is broken down during aerobic cellular respiration. However, other molecules can also undergo catabolism. When a fat is used as an energy source, it breaks down to glycerol and three fatty acids. As Figure 7.9 indicates, glycerol is converted to PGAL, a metabolite in glycolysis. The fatty acids are converted to acetyl-CoA, which enters the Krebs cycle. An 18-carbon fatty acid results in nine acetyl-CoA molecules. Calculation shows that respiration of these can produce a total of 216 ATP molecules. For this reason, fats are an efficient form of stored energy—there are three long fatty acid chains per fat molecule.

The carbon skeleton of amino acids can also be broken down. The carbon skeleton is produced in the liver when an amino acid undergoes **deamination,** or the removal of the amino group. The amino group becomes ammonia (NH_3), which enters the urea cycle and becomes part of urea, the primary excretory product of humans. Just where the carbon skeleton begins degradation is dependent on the length of the *R* group, since this determines the number of carbons left after deamination.

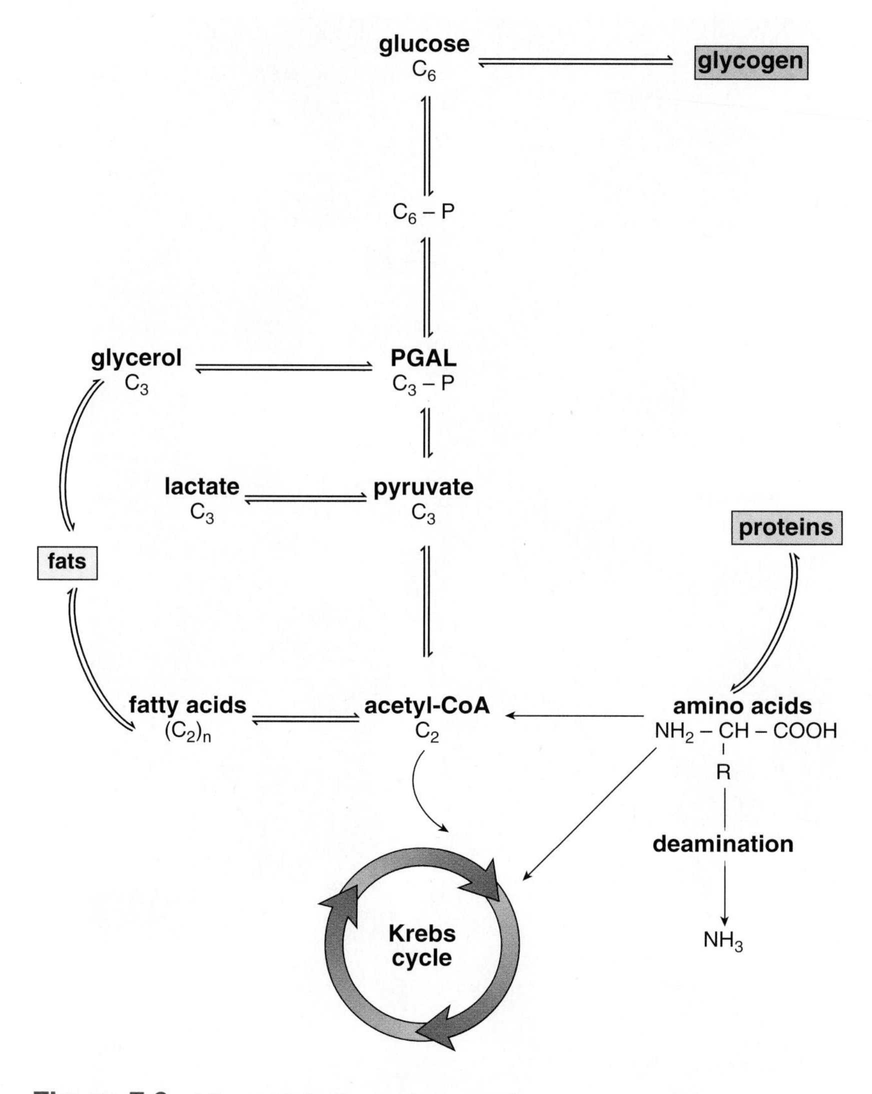

Figure 7.9 The metabolic pool concept.
Carbohydrates, fats, and proteins can be used as energy sources, and they enter degradative pathways at specific points. Catabolism produces molecules that can also be used for anabolism of other compounds.

Anabolism

We have already mentioned that the ATP produced during catabolism drives anabolism. But there is another way catabolism is related to anabolism. The substrates making up the pathways in Figure 7.9 can be used as starting materials for synthetic reactions. In other words, compounds that enter the pathways are oxidized to substrates that can be used for biosynthesis. This is the cell's **metabolic pool,** in which one type of molecule can be converted to another. In this way, carbohydrate intake can result in the formation of fat. PGAL can be converted to glycerol, and acetyl groups can be joined to form fatty acids. Fat synthesis follows. This explains why you gain weight from eating too much candy, ice cream, and cake.

Some substrates of the Krebs cycle can be converted to amino acids through transamination, the transfer of an amino group to an organic acid, forming a different amino acid. Plants are able to synthesize all of the amino acids they need. Animals, however, lack some of the enzymes necessary for synthesis of all amino acids. Adult humans, for example, can synthesize 11 of the common amino acids, but they cannot synthesize the other 9. The amino acids that cannot be synthesized must be supplied by the diet; they are called the essential amino acids. The nonessential amino acids can be synthesized. It is quite possible for animals to suffer from protein deficiency if their diets do not contain adequate quantities of all the essential amino acids.

All the reactions involved in aerobic cellular respiration are a part of a metabolic pool; the molecules from the pool can be used for catabolism or for anabolism.

Health Focus

Exercise: A Test of Homeostatic Control

Exercise is a dramatic test of the body's homeostatic control systems—there is a large increase in muscle oxygen (O_2) requirement, and a large amount of carbon dioxide (CO_2) is produced. These changes must be countered by increases in breathing and blood flow to increase O_2 delivery and remove the metabolically produced CO_2. Also, heavy exercise can produce a large amount of lactic acid due to the utilization of fermentation, an anaerobic process.

Both the accumulation of CO_2 and lactic acid can lead to an increase in intracellular and extracellular activity. Further, during heavy exercise, the working muscles produce large amounts of heat that must be removed to prevent overheating. In a strict sense, the body rarely maintains true homeostasis while performing intense exercise or during prolonged exercise in a hot or humid environment. However, a better maintenance of homeostasis is observed in those who have had endurance training.

The number of mitochondria increases in the muscles of persons who train; therefore, there is greater reliance on the Krebs cycle and the electron transport system to generate energy. Muscle cells with few mitochondria must have a high ADP concentration to stimulate the limited number of mitochondria to start consuming O_2. After an endurance training program, the large number of mitochondria start consuming O_2 as soon as the ADP concentration starts rising due to muscle contraction and subsequent breakdown of ATP. Therefore, a steady state of O_2 intake by mitochondria is achieved earlier in the athlete. This faster rise in O_2 uptake at the onset of work means that the O_2 deficit is less, and the formation of lactate due to fermentation is less. Further, any lactate that is produced is removed and processed more quickly.

Training also results in greater reliance on the Krebs cycle and increased fatty acid metabolism, because fatty acids are broken down to acetyl-CoA, which enters the Krebs cycle. This preserves plasma glucose concentration and also helps the body maintain homeostasis.

In athletes, there is:

- a smaller O_2 deficit due to a more rapid increase in O_2 uptake at the onset of work;
- an increase in fat metabolism that spares blood glucose;
- a reduction in lactate and H^+ formation;
- an increase in lactate removal.

7.5 Fermentation

Fermentation consists of glycolysis plus one other reaction—the reduction of pyruvate to either lactate or alcohol and CO_2 (Fig. 7.10). The pathway operates anaerobically because after NADH transfers its electrons to pyruvate, it is "free" to return and pick up more electrons during the earlier reactions of glycolysis.

Certain anaerobic bacteria such as lactic-acid bacteria, which help us manufacture cheese, consistently produce lactate in this manner. Other bacteria anaerobically produce chemicals of industrial importance: isopropanol, butyric acid, propionic acid, and acetic acid. Yeasts are good examples of organisms that generate alcohol and CO_2. Yeast is used to leaven bread; the CO_2 produced makes bread rise. Yeast is also used to ferment wine; in that case, it is the ethyl alcohol that is desired. Eventually yeasts are killed by the very alcohol they produce.

Animal, including human, cells are similar to lactic acid bacteria in that pyruvate, when produced faster than it can be oxidized through the Krebs cycle, is reduced to lactate.

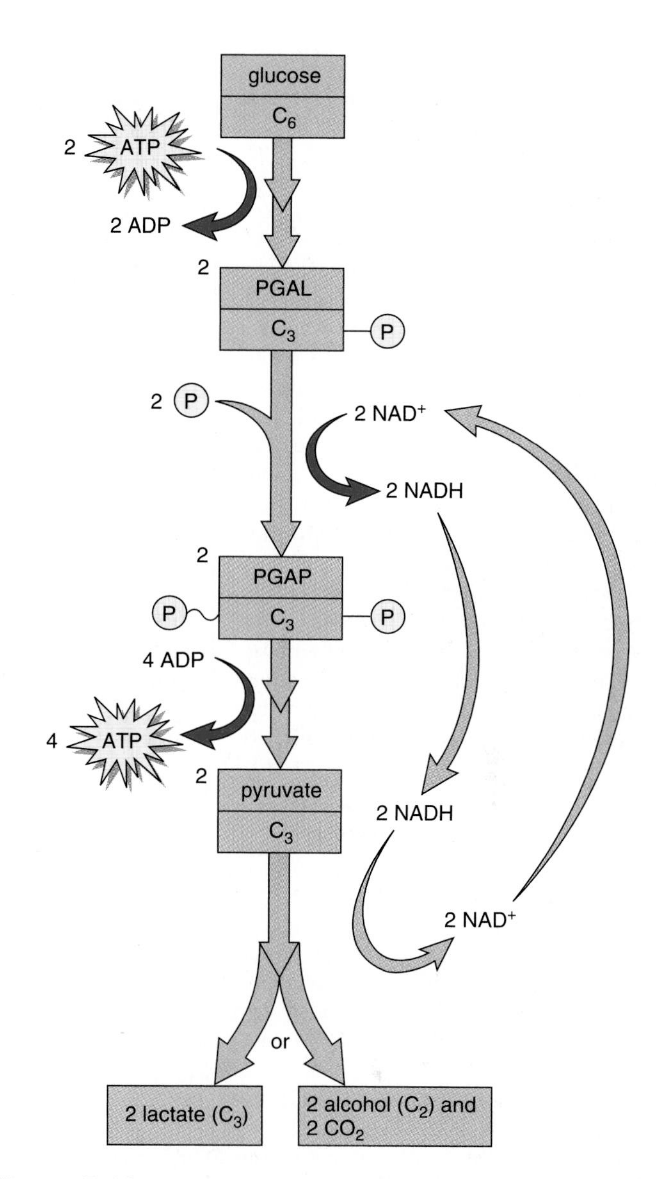

Figure 7.10 Fermentation.
Fermentation consists of glycolysis followed by a reduction of pyruvate. This "frees" NAD^+, and it returns to the glycolytic pathway to pick up more electrons.

Advantages and Disadvantages of Fermentation

Despite its low yield of only two ATP, fermentation is essential to humans because it can provide a rapid burst of ATP; muscle cells more than other cells are apt to carry on fermentation. When our muscles are working vigorously over a short period of time, as when we run, fermentation is a way to produce ATP even though oxygen is temporarily in a limited supply.

Lactate, however, is toxic to cells. At first, blood carries away all the lactate formed in muscles. Eventually, however, lactate begins to build up, changing the pH and causing the muscles to fatigue so that they no longer contract. When we stop running, our bodies are in **oxygen debt,** as signified by the fact that we continue to breathe very heavily for a time. Recovery is complete when the lactate is transported to the liver, where it is reconverted to pyruvate. Some of the pyruvate is respired completely, and the rest is converted back to glucose.

Efficiency of Fermentation

The two ATP produced per glucose molecule during fermentation is equivalent to 14.6 kcal. Complete glucose molecule breakdown to CO_2 and H_2O represents a possible energy yield of 686 kcal per molecule. Therefore, the efficiency for fermentation is only 14.6/686, or 2.1%. This is much less efficient than the complete breakdown of glucose. The inputs and outputs of fermentation are as shown in the illustration to the right.

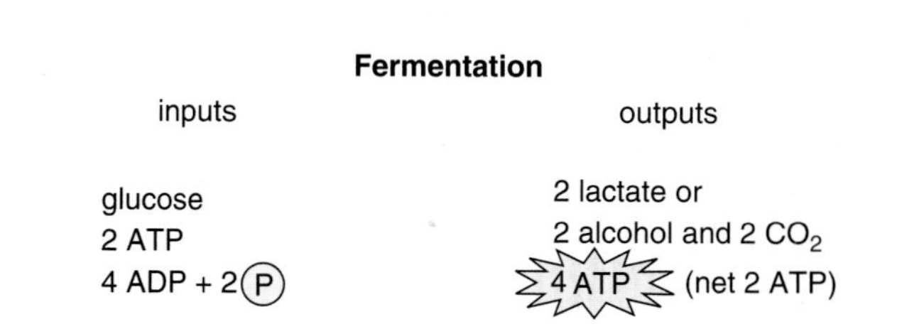

Bioethical Issue

Feeling tired and run down? Want to jump start your mitochondria? If you have iron deficiency anemia, you could take iron tablets to encourage your body to build more hemoglobin, the substance that carries oxygen in your blood. If you are diabetic, medications are available to make sure glucose is entering your cells. However, if you seem to have no specific ailment, you might be tempted to turn to what is now called alternative medicine. Alternative medicine includes such nonconventional therapies as herbal supplements, acupuncture, chiropractic therapy, homeopathy, osteopathy, and therapeutic touch (e.g., the laying on of hands).

Advocates of alternative medicine have made some headway in having alternative medicine practices accepted by most anyone. In 1992, congress established what is now called the National Center for Complementary and Alternative Medicine (NCCAM), whose budget has grown from \$2 million to its present \$50 million for 1999. Then in 1994, the Dietary Supplement Health and Education Act allowed the marketing of vitamins, minerals, and herbs without the requirement that they be approved by the Food and Drug Administration (FDA).

Many scientists feel that alternative medicine should be subjected to the same rigorous clinical testing as traditional medicine. They approve of a study that is now testing the efficacy of the herb St. John's wort, for example. In the study, 336 patients diagnosed with moderate depression will be divided into three groups. One group will receive the herb St. John's wort; another will be given sertraline, a traditional medicine, and a third group will receive a placebo. The results are expected to tell whether St. John's wort, a relatively inexpensive remedy, works just as well as a relatively expensive drug produced by pharmaceutical firms.

Questions

1. Do you believe that alternative medical practices should be subjected to clinical testing, or do you believe the public should simply rely on "word of mouth" recommendations?
2. What changes do you anticipate if alternative medical practices were to become a regular part of traditional medicine?
3. Would it be acceptable to you if it could be shown that alternative medical practices succeed simply for psychological reasons rather than for their physical benefits?

Summarizing the Concepts

7.1 Aerobic Cellular Respiration

During aerobic cellular respiration, glucose is oxidized to CO_2 and H_2O. This exergonic reaction drives ATP buildup, an endergonic reaction. Four phases are required: glycolysis, the transition reaction, the Krebs cycle, and the electron transport system. Oxidation occurs by the removal of hydrogen atoms ($e^- + H^+$) from substrate molecules.

7.2 Outside the Mitochondria: Glycolysis

Glycolysis, the breakdown of glucose to two pyruvate, is a series of enzymatic reactions that occur in the cytoplasm. Oxidation by NAD^+ releases enough energy immediately to give a net gain of two ATP by substrate-level phosphorylation. Two NADH are formed.

7.3 Inside the Mitochondria

Pyruvate from glycolysis enters a mitochondrion, where the transition reaction takes place. During this reaction, oxidation occurs as CO_2 is removed. NAD^+ is reduced, and CoA receives the C_2 acetyl group that remains. Since the reaction must take place twice per glucose, two NADH result.

The acetyl group enters the Krebs cycle, a cyclical series of reactions located in the mitochondrial matrix. Complete oxidation follows, as two CO_2, three NADH, and one $FADH_2$ are formed. The cycle also produces one ATP. The entire cycle must turn twice per glucose molecule.

The final stage of glucose breakdown involves the electron transport system located in the cristae of the mitochondria. The electrons received from NADH and $FADH_2$ are passed down a chain of carriers until they are finally received by O_2, which combines with H^+ to produce H_2O. As the electrons pass down the chain, ATP is produced. The term oxidative phosphorylation is sometimes used for ATP production by the electron transport system.

The carriers of the electron transport system are located in protein complexes on the cristae of the mitochondria. Each protein complex receives electrons and pumps H^+ into the intermembrane space, setting up an electrochemical gradient. When H^+ flows down this gradient through the ATP synthase complex, energy is released and used to form ATP molecules from ADP and Ⓟ. This is ATP synthesis by chemiosmosis.

To calculate the total number of ATP per glucose breakdown, consider that for each NADH formed inside the mitochondrion, three ATP are produced. In most cells, each NADH formed in the cytoplasm results in only two ATP. This is because the carrier that shuttles the hydrogen atoms across the mitochondrial outer membrane usually passes them to an FAD. Each molecule of $FADH_2$ results in the formation of only two ATP because the electrons enter the electron transport system at a lower energy level than NADH. Of the 36 or 38 ATP formed by aerobic cellular respiration, four occur outside the electron transport system: two are formed directly by glycolysis and two are formed directly by the Krebs cycle. The rest are produced by the electron transport system.

7.4 Metabolic Pool and Biosynthesis

Carbohydrate, protein, and fat can be broken down by entering the degradative pathways at different locations. These pathways also provide molecules needed for the synthesis of various important substances. Catabolism and anabolism, therefore, both utilize the same metabolic pool of reactants.

7.5 Fermentation

Fermentation involves glycolysis, followed by the reduction of pyruvate by NADH to either lactate or alcohol and CO_2. The reduction process "frees" NAD^+ so that it can accept more electrons during glycolysis.

Although fermentation results in only two ATP, it still serves a purpose: In humans, it provides a quick burst of ATP energy for short-term, strenuous muscular activity. The accumulation of lactate puts the individual in oxygen debt because oxygen is needed when lactate is completely metabolized to CO_2 and H_2O.

Studying the Concepts

1. What is the equation for aerobic cellular respiration? Explain how this is an oxidation-reduction reaction. Why is the reaction able to drive ATP buildup? 118
2. What are the four phases occurring during aerobic cellular respiration? 119
3. Define glycolysis; what are its inputs and outputs? 120
4. What portions of aerobic cellular respiration occur inside mitochondria? How does a human being acquire the needed substrates, and what happens to the products? 122
5. Give the substrates and products of the transition reaction. Where does it take place? 122
6. What happens to the acetyl group that enters the Krebs cycle? What are the other steps in this cycle? 123
7. What are NAD^+ and $FADH_2$, and what role do they play in aerobic cellular respiration? 120, 123
8. What is the electron transport system, and what are its functions? 124
9. Describe the organization of protein complexes within the cristae. 125
10. Calculate the energy yield of aerobic cellular respiration per glucose molecule. 126
11. Give examples to support the concept of the metabolic pool. 127
12. What is fermentation and how does it differ from glycolysis? Mention the benefit of pyruvate reduction during fermentation. What types of organisms carry out lactate fermentation, and what types carry out alcoholic fermentation? 129

Testing Yourself

Choose the best answer for each question. For questions 1–3, identify the pathway involved by matching them to the terms in the key.

Key:
a. glycolysis
b. Krebs cycle
c. electron transport system

1. carbon dioxide (CO_2) given off
2. PGAL
3. cytochrome carriers
4. The greatest contributor of electrons to the electron transport system is
   a. oxygen.
   b. glycolysis.
   c. the Krebs cycle.
   d. the transition reaction.
   e. fermentation.
5. Substrate-level phosphorylation takes place in
   a. glycolysis and the Krebs cycle.
   b. the electron transport system and the transition reaction.
   c. glycolysis and the electron transport system.
   d. the Krebs cycle and the transition reaction.
6. Fatty acids are broken down to
   a. pyruvate molecules, which take electrons to the electron transport system.
   b. acetyl groups, which enter the Krebs cycle.
   c. glycerol, which is found in fats.
   d. amino acids, which excrete ammonia.
   e. All of these are correct.
7. Which of these is not true of fermentation?
   a. net gain of only two ATP
   b. occurs in cytoplasm
   c. NADH donates electrons to electron transport system
   d. begins with glucose
   e. carried on by yeast

For questions 8–10, match the items below to one of the locations in the key.

Key:
a. matrix of the mitochondrion
b. cristae of the mitochondrion
c. the intermembrane space of mitochondrion
d. in the cytoplasm
e. None of these are correct.

8. electron transport system
9. glycolysis
10. accumulation of hydrogen ions (H^+)
11. Label this diagram of a mitochondrion.

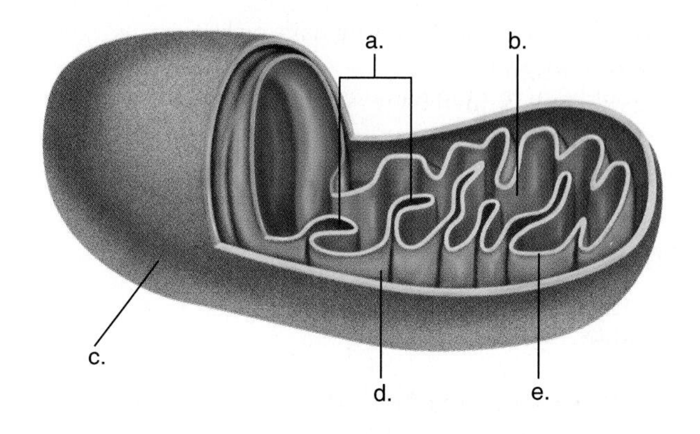

Understanding the Terms

acetyl-CoA 122
aerobic 118
aerobic cellular respiration 118
anabolism 127
anaerobic 120
catabolism 127
chemiosmosis 125
cytochrome 124
deamination 127
electron carrier 124
electron transport system 119
FAD (flavin adenine dinucleotide) 123
fermentation 129
glycolysis 119
Krebs cycle 119
metabolic pool 127
mitochondrion 122
NAD^+ (nicotinamide adenine dinucleotide) 120
oxidative phosphorylation 124
oxygen debt 129
pyruvate 119
substrate-level phosphorylation 120
transition reaction 119

Match the terms to these definitions:

a. ________ Cycle of reactions in mitochondria that begins with citric acid; it produces CO_2, ATP, NADH, and $FADH_2$; also called the citric acid cycle.

b. ________ Growing or metabolizing in the absence of oxygen.

c. ________ Anaerobic breakdown of glucose that results in a gain of two ATP and end products such as alcohol and lactate.

d. ________ End product of glycolysis; pyruvic acid.

e. ________ Passage of electrons along a series of membrane-bounded carrier molecules from a higher to lower energy level; the energy released is used for the synthesis of ATP.

Thinking Scientifically

1. Considering aspects of cellular respiration (page 118)
   a. The body breathes in oxygen (O_2). What exact function does oxygen perform in the body?
   b. The body breathes out carbon dioxide (CO_2). Exactly how does the body produce carbon dioxide?
   c. Just before a competition, athletes eat carbohydrates. Why does the intake of carbohydrates ensure them a supply of energy for the athletic event?
2. The drug dinitrophenol makes the inner mitochondrial membrane leaky to hydrogen ions (H^+),
   a. In which direction would you expect the H^+ to leak, from the intermembrane space to the matrix or the reverse? Why?
   b. What effect would this drug have on ATP production? Why?
   c. What effect would this drug have on stored glycogen and lipid supplies?
   d. Would you recommend this drug for weight loss? Why or why not?

Using Technology

Your study of cellular respiration is supported by these available technologies:

Essential Study Partner CD-ROM

Cells → Respiration

Visit the Mader web site for related ESP activities.

Exploring the Internet

The Mader Home Page provides resources and tools as you study this chapter.

http://www.mhhe.com/biosci/genbio/mader

Life Science Animations 3D Video

9 Electron Transport Chain

Further Readings for Part 1

Armbruster, P., and Hessberger, F. P. September 1998. Making new elements. *Scientific American* 279(3):72. The process of creating new artificial elements is examined.

Bayley, H. September 1997. Building doors into cells. *Scientific American* 277(3):62. Protein engineers are designing artificial pores for drug delivery.

Becker, W. M., and Deamer, D. W. 1996. *The world of the cell.* 3d ed. Redwood City, Calif.: Benjamin/Cummings Publishing. Presents an overview of cell biology.

Caret, R. L., et al. 1997. *Principles and applications of organic and biological chemistry.* 2d ed. Dubuque, Iowa: Wm. C. Brown Publishers. For undergraduates in the allied health fields, this text emphasizes material unique to health-related studies.

Chang, R. 1998. *Chemistry.* 6th ed. Dubuque, Iowa: McGraw-Hill. This general chemistry text provides a foundation in chemical concepts and principles, and presents topics clearly.

Chapman, C. 1999. *Basic chemistry for biology.* 2d ed. WCB/McGraw-Hill. The goal of this workbook is to provide a review of basic principles for biology students.

Ford, B. J. April 1998. The earliest views. *Scientific American* 278(4):50. Presents experiments of early microscopists.

Frank, J. September/October 1998. How the ribosome works. *American Scientist* 86(5):428. New imaging techniques using cryo-electron microscopy allows researchers to study a three-dimensional map of the ribosome.

Gerstein, M., and Levitt, M. November 1998. Simulating water and the molecules of life. *Scientific American* 279(5):100. Computer models show how water affects the structure and movement of proteins and other biological molecules.

Ingber, D. E. January 1998. The architecture of life. *Scientific American* 278(1):48. Simple mechanical rules may govern cell movements, tissue organization, and organ development.

Lang, F., and Waldegger, S. September/October1997. Regulating cell volume. *American Scientist* 85(5):456. Changes in cell volume may threaten organ or tissue function.

Lasic, D. D. May/June 1996. Liposomes. *Science & Medicine* 3(3):34. Liposomes can deliver genes or drugs for gene therapy.

Nemecek, S. October 1997. Gotta know when to fold 'em. *Scientific American* 277(4):28. Details about how proteins fold are discussed.

Nurse, P., et al. October 1998. Understanding the cell cycle. *Nature Medicine* 4(1):1103. Gives the relevance of cell-cycle research.

Ojcius, D. M., et al. January/February 1998. Pore-forming proteins. *Science & Medicine* 5(1):44. Peptide molecules that cause pore formation in membranes are similar in structure and function.

Scerri, E. R. November/December 1997. The periodic table and the electron. *American Scientist* 85(6):546. Electron configurations may only give an approximate explanation of the periodic table.

Scerri, E. R. September 1998. The evolution of the periodic system. *Scientific American* 279(3):78. Article discusses the history and evolution of the periodic table.

Schwartz, A. T., et al. 1997. *Chemistry in context: Applying chemistry to society.* 2d ed. Dubuque, Iowa: Wm. C. Brown Publishers. This introductory text is designed for students in the allied health fields.

Sperelakis, N., editor. 1998. *Cell physiology source book.* 2d ed. San Diego: Academic Press. For advanced biology students, this is a comprehensive and authoritative text covering topics in cell physiology written by experts in the field.

Stix, G. October 1997. Growing a new field. *Scientific American* 277(4):15. Tissue engineers try to grow organs in the laboratory.

Zubay, G. L. 1998. *Biochemistry.* 4th ed. Dubuque, Iowa: Wm. C. Brown Publishers. This text for chemistry majors relates biochemistry to cell biology, physiology, and genetics.

PART II

Plant Biology

By the end of the nineteenth century, scientists knew the overall reaction for photosynthesis. In the presence of sunlight, a plant uses carbon dioxide and water to produce carbohydrate—oxygen is given off as a by-product. The modern uses of cell fractionation, radioactive tracers, and electron microscopy have added to our knowledge of how algae and plants produce the food that feeds the biosphere. The human population is sustained by the crops we plant. As sources of fabrics, paper, lumber, fuel, and pharmaceuticals, plants make our modern society possible. And let us not forget the very many ways plants bring beauty into our lives.

Careers in Plant Physiology

Plant physiologist examining a plant.

Plant physiologists may work in laboratories and use greenhouse plants, electron microscopes, computers, electronic instruments, or a wide variety of other equipment to conduct their research. A good deal of research, however, is performed outside of laboratories. Physiologists may specialize in functions such as growth, reproduction, photosynthesis, respiration, or in the physiology of a certain structure or system of the plant.

Botanists study plants and their environments. Some study all aspects of plant life; others specialize in areas such as identification and classification of plants, the structure and function of plant parts, the biochemistry of plant processes, or the causes and cures of plant diseases.

Botanist working in a greenhouse.

Agricultural scientists study farm crops and animals and develop ways of improving their quantity and quality. They look for ways to increase crop yield and quality with less labor, control pests and weeds more safely and effectively, and conserve soil and water.

Another important area of agricultural science is plant science, which includes the disciplines of agronomy, crop science, and soil science. Agronomists and crop scientists not only help increase productivity, they also study ways to improve the nutritional value of crops and the quality of seed. Some crop scientists study the breeding, physiology, and management of crops and use genetic engineering to develop crops which are resistant to insects and drought.

Agricultural scientist inspecting crops.

Foresters manage, develop, and help protect forest resources. They advise on the type, number, and placement of trees to be planted. Foresters monitor the trees to ensure healthy growth and to determine the best time for harvesting. If foresters detect signs of disease or harmful insects, they decide on the best course of treatment to prevent contamination or infestation of healthy trees.

Horticulturists and florists grow and care for plants and flowers for retail and wholesale marketing. On horticultural specialty farms, operators oversee the production of ornamental plants, nursery products (such as flowers, bulbs, shrubbery, and sod), and fruits and vegetables grown in greenhouses.

C H A P T E R 8

Photosynthesis

Chapter Concepts

8.1 Radiant Energy

- Plants make use of solar energy in the visible light range when they carry on photosynthesis. 136

8.2 Structure and Function of Chloroplasts

- Photosynthesis takes place in chloroplasts, organelles that have two parts: membranous thylakoids are surrounded by a fluid-filled stroma. 138
- Photosynthesis has two sets of reactions: the light-dependent reactions and the light-independent reactions. 138–39

8.3 Solar Energy Capture

- Solar energy energizes electrons and permits a buildup of ATP. 140

8.4 Carbohydrate Synthesis

- Carbon dioxide reduction requires energized electrons and ATP. 143

8.5 Other Aspects of Photosynthesis

- Photosynthesis provides virtually all the food for the biosphere. 145
- Plants use either C_3, C_4, or CAM photosynthesis, named for the manner in which CO_2 is fixed. 146

8.6 Photosynthesis Versus Aerobic Cellular Respiration

- In photosynthesis, carbon dioxide is reduced to glucose and water is oxidized, releasing oxygen gas. In aerobic cellular respiration, carbohydrate is oxidized to carbon dioxide, and oxygen is reduced to water. 147

The sun provides the energy that allows photosynthesizers, like trees, to produce their own organic food.

Silently, the trees stand with limbs outstretched to a sun that radiates back heat and light energy. Unseen, oxygen gas floats away from leaves that are soaking up carbon dioxide. The picnickers enjoying a respite from the rush of the summer do not know that the trees are absorbing their carbon dioxide and returning to them the oxygen they require. The very food the picnickers eat is due to photosynthesis, a process that uses the energy of the sun to convert carbon dioxide and water into complex organic molecules. In the winter, the tree that has given sojourners shade in the summer may be cut down to serve as firewood to keep them warm. No wonder there are bumper stickers that read "Have You Thanked a Green Plant Today?"

8.1 Radiant Energy

The food produced through **photosynthesis** eventually becomes the food for the rest of the living world (Fig. 8.1). For example, humans and all other animals either eat plants directly or eat animals that have eaten plants. Another way to express this idea is to say that **autotrophs,** which have the ability to synthesize organic molecules from inorganic raw materials, not only feed themselves but also **heterotrophs,** which must take in preformed organic molecules. After food is eaten and digested by heterotrophs, they use the resulting small molecules either as building blocks for growth and repair or as a source of chemical energy. Plants also supply energy in another sense—their bodies become coal that is a fossil fuel resource today.

For the following reasons, then, it is correct to say that almost all life is ultimately dependent on solar energy (the energy of the sun).

1. Solar energy is used for photosynthesis.
2. Photosynthetic organisms produce food for the biosphere. This food is used not only for growth, but also as an energy source.
3. The bodies of plants become the fossil fuel coal, upon which we are still dependent today.
4. Solar energy can be captured to heat buildings and produce electricity.

Figure 8.1 Methods of acquiring organic food.
Autotrophs **(a)**, represented by green plants, algae, and a few types of bacteria (the latter two presented in circles) produce food for themselves and for **(b)** herbivores, which feed directly on plants or plant products, **(c)** carnivores, which feed on herbivores or other carnivores, and **(d)** omnivores, which feed on all of these.

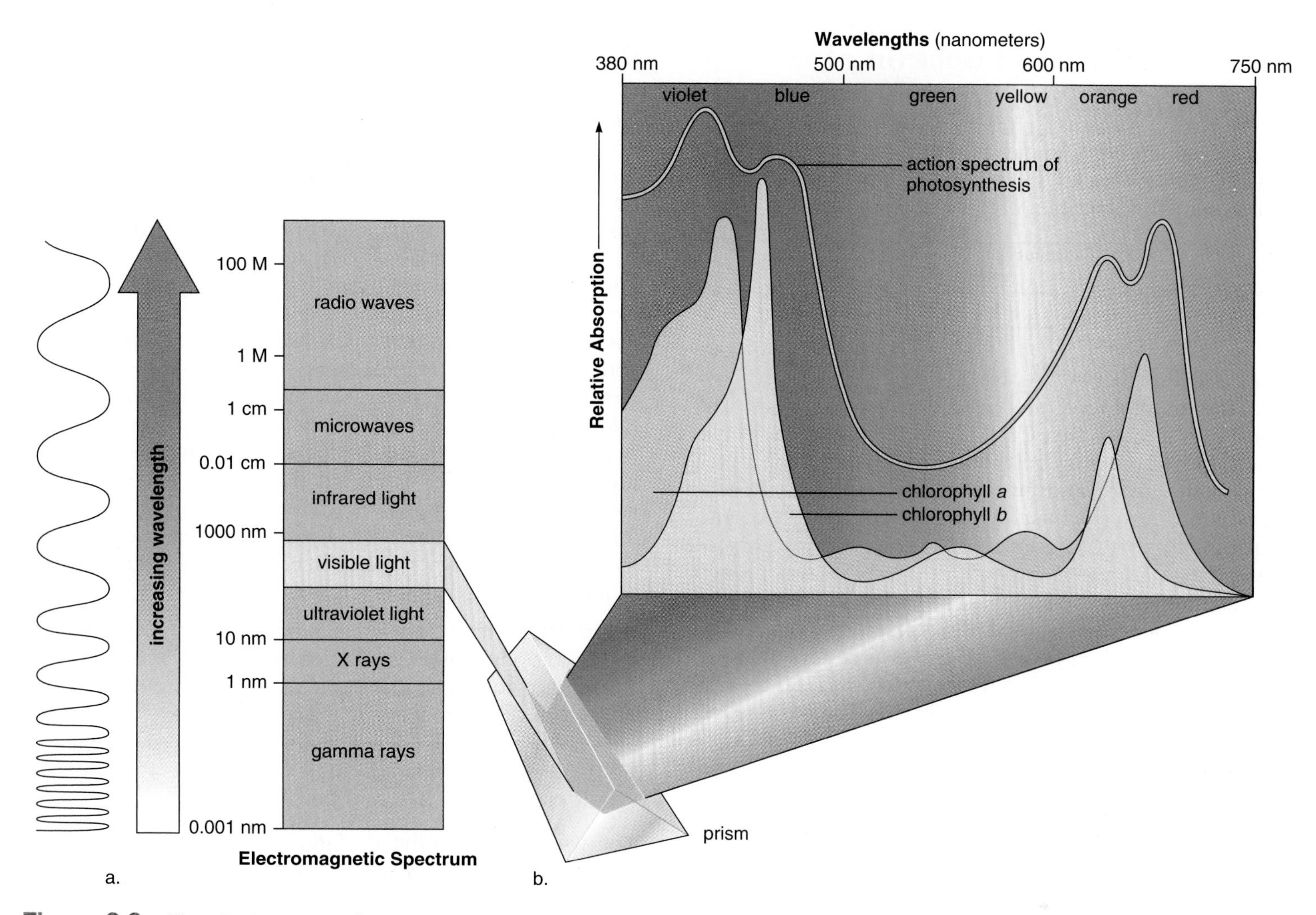

Figure 8.2 The electromagnetic spectrum and chlorophylls *a* and *b*.
a. The electromagnetic spectrum contains forms of energy that differ according to wavelength. Visible light is only a small portion of the electromagnetic spectrum. **b.** Chlorophylls *a* and *b* absorb certain wavelengths within visible light, which accounts for the absorption spectrum of (wavelengths necessary for) photosynthesis.

Sunlight

Radiant energy from the sun can be described in terms of its wavelength and its energy content. Figure 8.2*a* lists the different types of radiant energy, from the shortest wavelength, gamma rays, to the longest, radio waves. The shorter wavelengths contain more energy than the longer ones. White or *visible light* is only a small portion of this spectrum. Visible light itself contains various wavelengths of light, as can be proven by passing it through a prism; then we see all the different colors that make up visible light. (Actually, of course, it is our eyes that interpret these wavelengths as colors.) The colors in visible light range from violet (the shortest wavelength) to blue, green, yellow, orange, and red (the longest wavelength). The energy content is highest for violet light and lowest for red light.

Only about 42% of the solar radiation that hits the earth's atmosphere ever reaches the surface, and most of this radiation is within the visible-light range. Higher energy wavelengths are screened out by the ozone layer in the atmosphere, and lower energy wavelengths are screened out by water vapor and carbon dioxide (CO_2) before they reach the earth's surface. The conclusion is, then, that both the organic molecules within organisms and certain life processes, such as vision and photosynthesis, are chemically adapted to the radiation that is most prevalent in the environment.

The pigments found within photosynthesizing cells, the **chlorophylls** and **carotenoids,** are capable of absorbing various portions of visible light. The absorption spectrum for chlorophyll *a* and chlorophyll *b* is shown in Figure 8.2*b*. Both chlorophyll *a* and chlorophyll *b* absorb violet, blue, and red light better than the light of other colors. Because green light is only minimally absorbed, leaves appear green to us. Accessory pigments such as the carotenoids are yellow or orange and are able to absorb light in the violet-blue-green range. These pigments and others become noticeable in the fall when chlorophyll breaks down.

8.2 Structure and Function of Chloroplasts

Chloroplasts are the organelles found in a plant cell that carry on photosynthesis. The overall equation for photosynthesis is sometimes written in this manner:

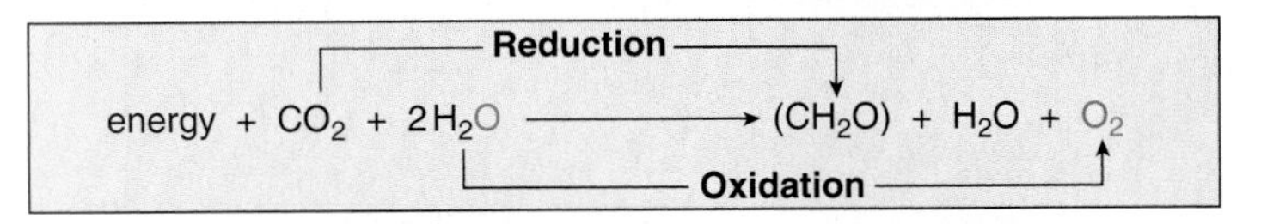

In this equation, (CH_2O) represents a generalized carbohydrate, the organic food produced by photosynthesis. The parentheses indicate that CH_2O is not a specific molecule. Water (H_2O) appears on both sides of the equation because H_2O is both utilized and produced during photosynthesis. This equation also has the advantage of keeping the chemical arithmetic correct because it is known that *the oxygen given off by photosynthesis comes from water.* This was proven experimentally by exposing plants first to CO_2 and then to H_2O that contained an isotope of oxygen called heavy oxygen (^{18}O). Only when heavy oxygen was a part of water (indicated by the color red in the equation) did this isotope appear in O_2 given off by the plant (Fig. 8.3). Therefore, O_2 released by chloroplasts comes from H_2O, and not from CO_2.

The overall equation of photosynthesis can also be written in this manner:

$$\text{energy} + 6CO_2 + 6H_2O \longrightarrow C_6H_{12}O_6 + 6O_2$$

This equation has the advantage of being the exact opposite of the equation for aerobic cellular respiration. During photosynthesis, water molecules are oxidized; they lose electrons (e^-) along with hydrogen ions (H^+). On the other hand, CO_2 is reduced; this molecule gains electrons given up by H_2O. But in the meantime, these electrons have been energized by the sun!

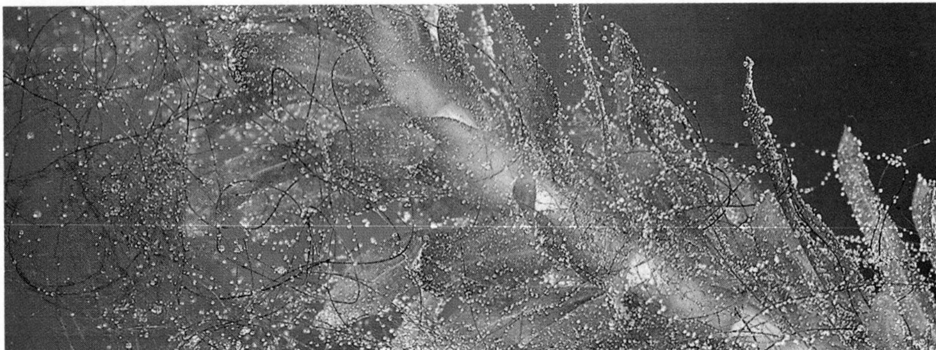

Figure 8.3 Photosynthesis releases oxygen.
Oxygen gas is making these bubbles you see about this sprig of *Elodea*, an aquatic plant. Release of O_2 can be used as a criterion that photosynthesis is occurring.

Structure of Chloroplasts

Chloroplasts primarily occur in leaves, whose structure is shown in Figure 8.4. The mesophyll cells in the center of a leaf receive water from vessels that extend from the leaves to the roots. Mesophyll cells are protected from drying out by epidermal tissue that is covered by a waxy cuticle. Pores called stomates allow passage of CO_2 and O_2 through this barrier. The many chloroplasts within mesophyll cells are bounded by double membranes. The inner membrane encloses a large central space called the **stroma.** The stroma contains an enzyme-rich solution in which CO_2 is reduced, converting it to an organic compound. A membranous system within the stroma forms flattened sacs called **thylakoids,** which in some places are stacked to form grana. All thylakoids are believed to be connected so that there is a single inner compartment within chloroplasts called the *thylakoid space.* Chlorophyll and other pigments are found within the membranes of the thylakoids. These pigments absorb the solar energy, which will energize the electrons prior to reduction of CO_2 in the stroma.

Chlorophyll within thylakoids absorbs solar energy so that energized electrons are sent to stroma, where CO_2 is reduced.

Function of Chloroplasts

The word photosynthesis suggests that the process is divided into two sets of reactions: *photo,* which means light, refers to the reactions that capture solar energy, and *synthesis,* which means building-up, refers to the reactions that produce carbohydrate. The first set of reactions is called the **light-dependent reactions** because the reactions cannot take place unless light is present. Electrons are energized when chlorophyll located within the thylakoid membranes absorbs solar energy. These high-energy electrons move from chlorophyll down an electron transport system, which produces ATP from ADP and Ⓟ. Energized electrons are also taken up by $NADP^+$ (nicotinamide adenine dinucleotide phosphate), an electron carrier similar to NAD^+, the coenzyme active during cellular respiration. After $NADP^+$ accepts electrons, it becomes NADPH.

The light-dependent reactions capture solar energy.

The second set of reactions which take place in the stroma is called the **light-independent reactions** because they do not require light. The light-dependent reactions consist of the **Calvin cycle,** a series of reactions that

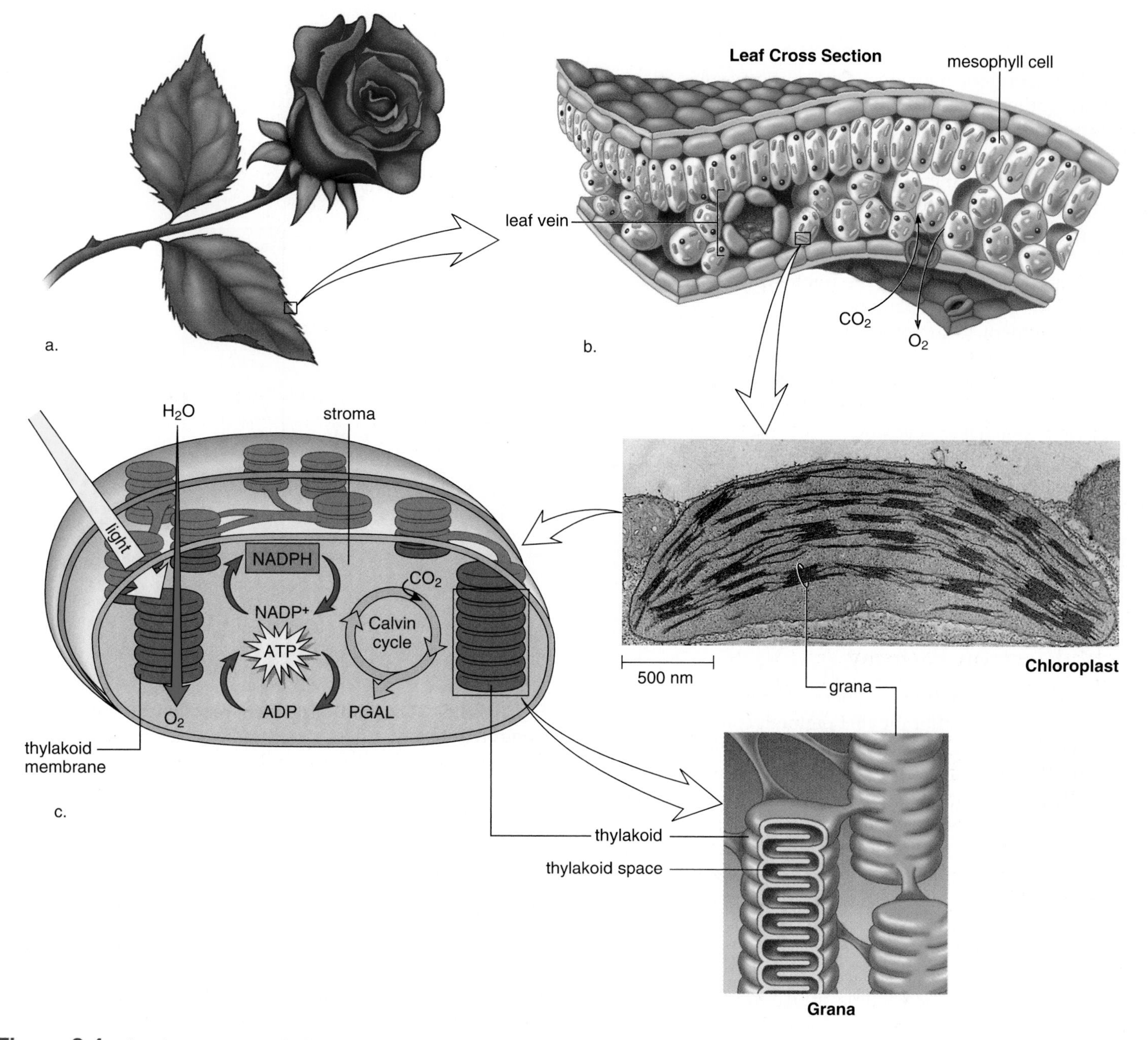

Figure 8.4 Leaf anatomy and photosynthesis.
a. Photosynthesis occurs in the leaves of plants. **b.** The mesophyll cells, which occur in the middle of a leaf, contain many chloroplasts, the organelles that carry on photosynthesis. Leaves carry on gas exchange, and they receive water from leaf veins. **c.** Within the chloroplasts, solar energy is absorbed by pigments within thylakoid membranes. During the light-dependent reactions, water is split, releasing oxygen; the energy carrier ATP and the electron carrier NADPH are formed. In the stroma of a chloroplast, during the light-independent reactions, the Calvin cycle uses ATP and NADPH to reduce CO_2 to a carbohydrate (PGAL). Then ADP and $NADP^+$ return to the thylakoid membrane to be recharged.

produce carbohydrate (CH_2O) before returning to the starting point once more. The cycle is named for Melvin Calvin, who, with colleagues, used the radioactive isotope ^{14}C to label CO_2 and thereby discovered the light-independent reactions. During the Calvin cycle, carbon dioxide is fixed (taken up) by a substrate and then reduced by the ATP and NADPH formed in the thylakoids. After ATP becomes ADP + Ⓟ and NADPH becomes $NADP^+$, they return and are energized once again during the light-dependent reactions.

The light-independent reactions synthesize carbohydrate.

8.3 Solar Energy Capture

The light-dependent reactions require the participation of two light-gathering units called photosystem I (PS I) and photosystem II (PS II). The photosystems are named for the order in which they were discovered and not for the order in which they occur in the thylakoid membrane. Each **photosystem** has a pigment complex composed of green pigments (chlorophyll *a* and chlorophyll *b*) and accessory pigments, especially the yellow and orange carotenoids. The closely packed pigment molecules in the photosystems serve as an "antenna" for gathering solar energy. Solar energy is passed from one pigment to the other until it is concentrated into one particular chlorophyll *a* molecule, the reaction-center chlorophyll. Electrons in the *reaction-center chlorophyll a* molecule become so excited that they escape and move to a nearby *electron-acceptor molecule.*

In a photosystem, the light-gathering antenna absorbs solar energy and funnels it to a reaction-center chlorophyll *a* molecule, which then sends energized electrons to an electron-acceptor molecule.

Cyclic Electron Pathway

The **cyclic electron pathway** generates only ATP (Fig. 8.5). This pathway begins after the PS I pigment complex absorbs solar energy. In this pathway, high-energy electrons (e^-) leave the PS I reaction-center chlorophyll *a* molecule but eventually return to it. Before they return, however, the electrons enter an **electron transport system,** a series of carriers that pass electrons from one to the other. Some of the carriers are cytochrome molecules; for this reason, the electron transport system is sometimes called a *cytochrome system.* As the electrons pass from one carrier to the next, energy that will be used to produce ATP molecules is released and stored.

Some photosynthetic bacteria utilize the cyclic electron pathway only; therefore, this pathway probably evolved early in the history of life. It is possible that in plants, the cyclic flow of electrons is utilized only when CO_2 is in such limited supply that carbohydrate is not being produced. At this time, there would be no need for additional NADPH, which is produced in addition to ATP by the noncyclic electron pathway.

It should also be noted that the second phase of photosynthesis, which occurs in the stroma, requires a larger number of ATP than NADPH. Perhaps the cyclic electron pathway routinely provides the extra ATP molecules required by reactions that occur throughout the cell.

The cyclic electron pathway, from PS I back to PS I, has only this one effect: production of ATP.

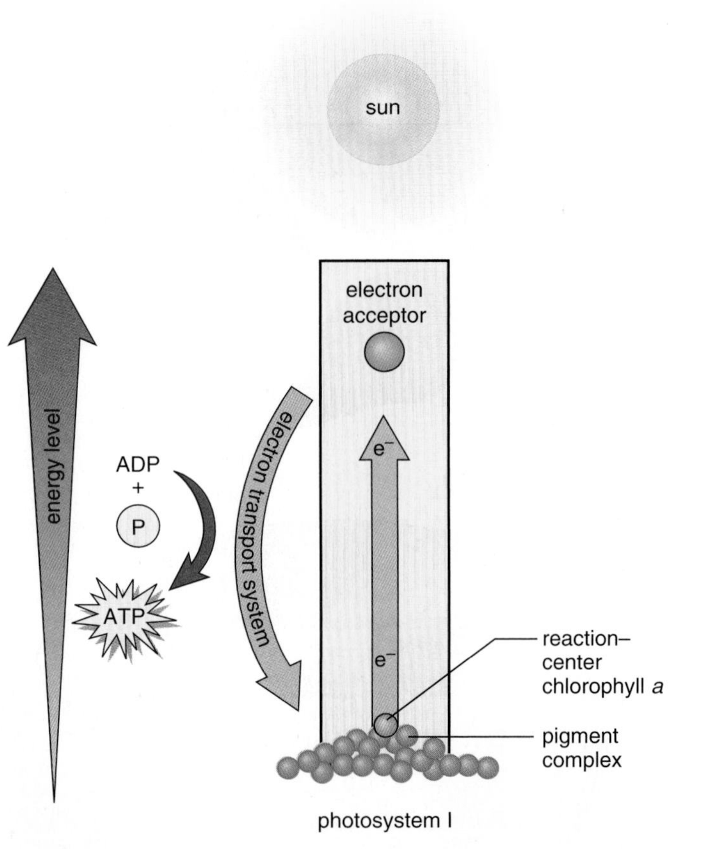

Figure 8.5 The light-dependent reactions: the cyclic electron pathway.
Energized electrons (e^-) leave the photosystem I (PS I) reaction-center chlorophyll *a* and are taken up by an electron acceptor, which passes them down an electron transport system before they return to PS I. Only ATP production results from this pathway.

Noncyclic Electron Pathway

The **noncyclic electron pathway** results in both ATP and NADPH molecules (Fig. 8.6). In this pathway, electrons move from H_2O through PS II to PS I and then on to $NADP^+$. This pathway begins when the PS II pigment complex absorbs solar energy and high-energy electrons (e^-) leave the reaction-center chlorophyll *a* molecule. PS II takes replacement electrons from water, which splits, releasing oxygen:

$$H_2O \longrightarrow 2\,H^+ + 2\,e^- + 1/2\,O_2$$

This oxygen evolves from the chloroplast and the plant as O_2 gas. The hydrogen ions (H^+) temporarily stay in the thylakoid space.

The high-energy electrons that leave PS II are captured by an electron acceptor, which sends them to an electron transport system. As the electrons pass from one carrier to the next, energy that will be used to produce ATP molecules

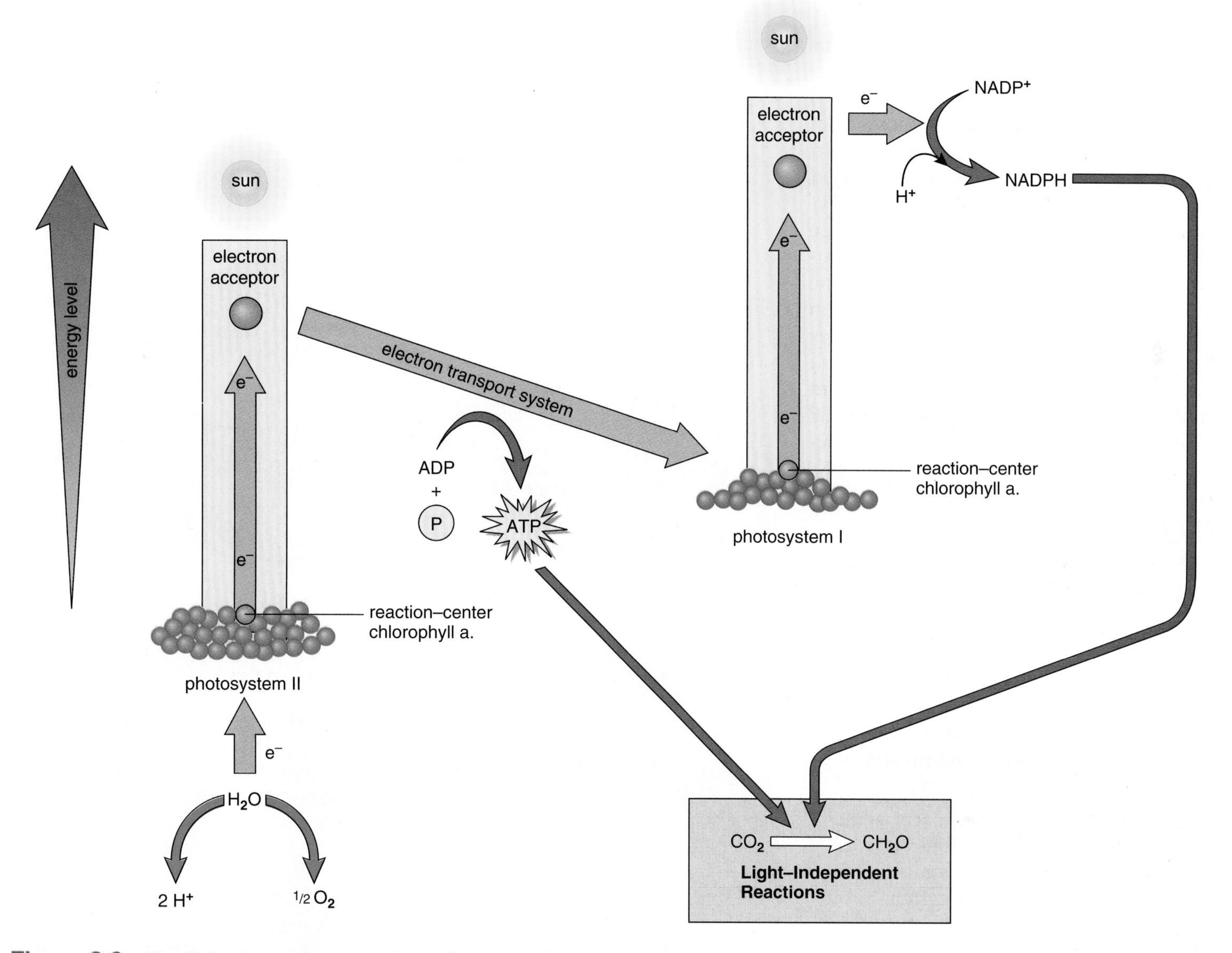

Figure 8.6 The light-dependent reactions: the noncyclic electron pathway.
Electrons, taken from water, move from photosystem II (PS II) to photosystem I (PS I) to $NADP^+$. The ATP at an electron transport system and NADPH produced by receiving the electrons will be used by the light-independent reactions to reduce carbon dioxide (CO_2) to a carbohydrate (CH_2O).

is released and stored. Low-energy electrons leaving the electron transport system enter PS I, replacing e^- in the reaction center of PS I.

The PS I pigment complex absorbs solar energy, and high-energy electrons leave the reaction-center chlorophyll *a* and are captured by an electron acceptor. This time, the electron acceptor passes the electrons on to $NADP^+$. $NADP^-$ now takes on an H^+ and becomes NADPH:

$$NADP^+ + 2e^- + H^+ \longrightarrow NAPDH$$

Results of noncyclic electron flow: Water is split, yielding H^+, e^-, and O_2. ATP is produced and $NADP^+$ becomes NADPH.

The NADPH and ATP produced by the noncyclic flow of electrons in the thylakoid membrane are used in the stroma during the light-independent reactions.

ATP Production

The thylakoid space acts as a reservoir for hydrogen ions (H^+). First, each time water is split, two H^+ remain in the thylakoid space. Second, as the electrons move from carrier to carrier in the electron transport system, they give up energy, which is used to pump H^+ from the stroma into the thylakoid space. Therefore, there is a large number of H^+ in the thylakoid space compared to the number in the stroma. The flow of H^+ from high to low concentration across the thylakoid membrane provides the energy that allows an *ATP synthase enzyme* to enzymatically produce ATP from ADP + Ⓟ. This method of producing ATP is called *chemiosmosis* because ATP production is tied to an electrochemical gradient.

The Thylakoid Membrane

Both biochemical and structural techniques have been used to determine that there are intact complexes (particles) in the thylakoid membrane (Fig. 8.7):

PS II consists of a protein complex and the light-gathering pigment complex shown to one side. PS II splits water and produces oxygen.

The cytochrome complex acts as the transporter of electrons between PS II and PS I. The pumping of H^+ occurs during electron transport.

PS I consists of a protein complex and a light-gathering pigment complex to one side. Notice that PS I is associated with the enzyme that reduces $NADP^+$ to NADPH.

ATP synthase complex has a H^+ channel and a protruding ATP synthase. As H^+ flows down its concentration gradient through this channel from the thylakoid space into the stroma, ATP is produced from ADP + Ⓟ.

The light-dependent reactions that occur in the thylakoid membrane produce ATP and NADPH and oxidize water so that oxygen is given off.

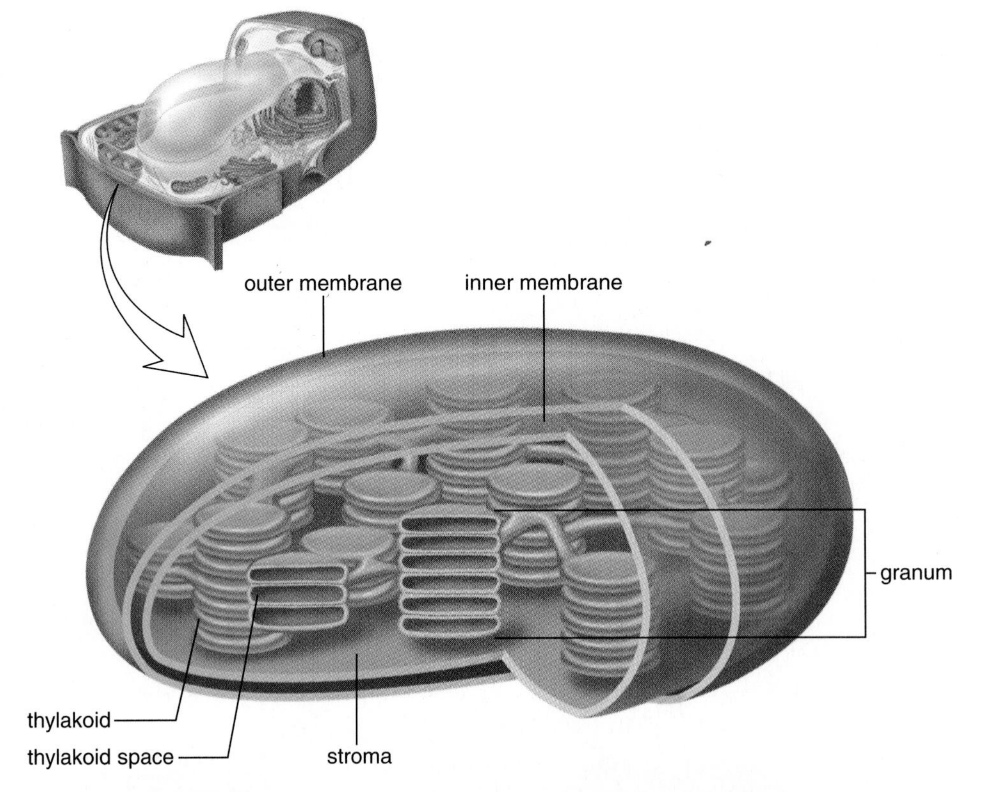

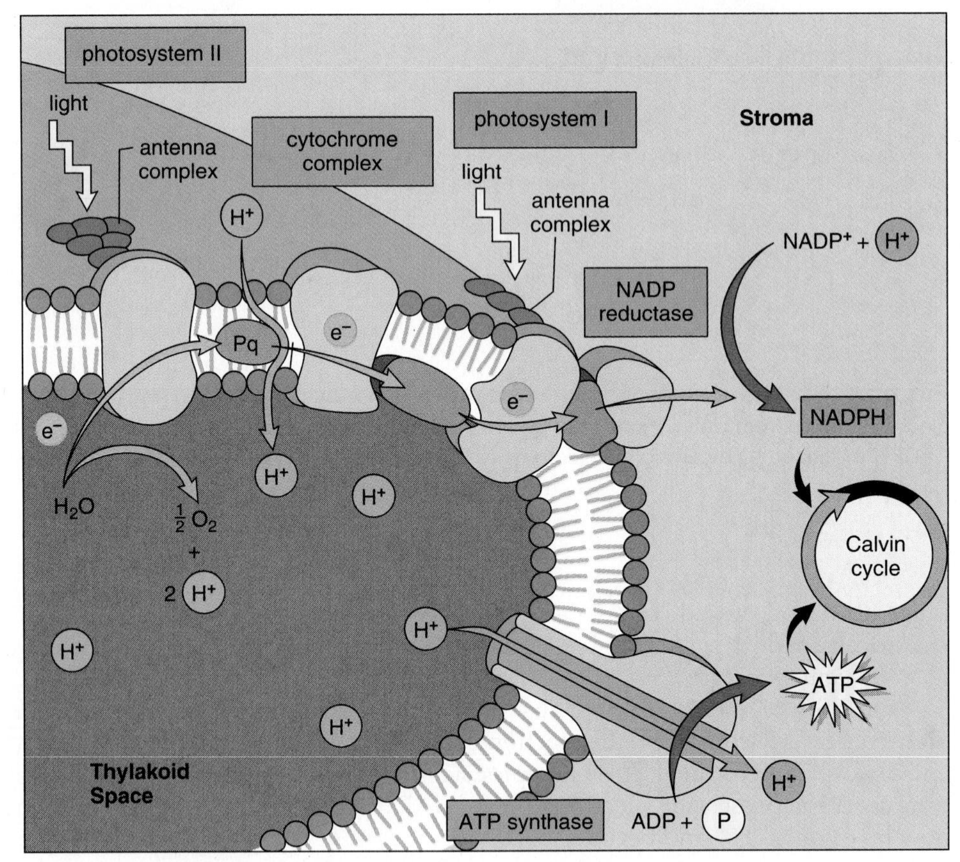

Figure 8.7 Organization of thylakoid.
Protein complexes within the thylakoid membrane pump hydrogen ions from the stroma into the thylakoid space. When hydrogen ions flow back out of the space into the stroma through the ATP synthase complex, ATP is produced from ADP + Ⓟ.

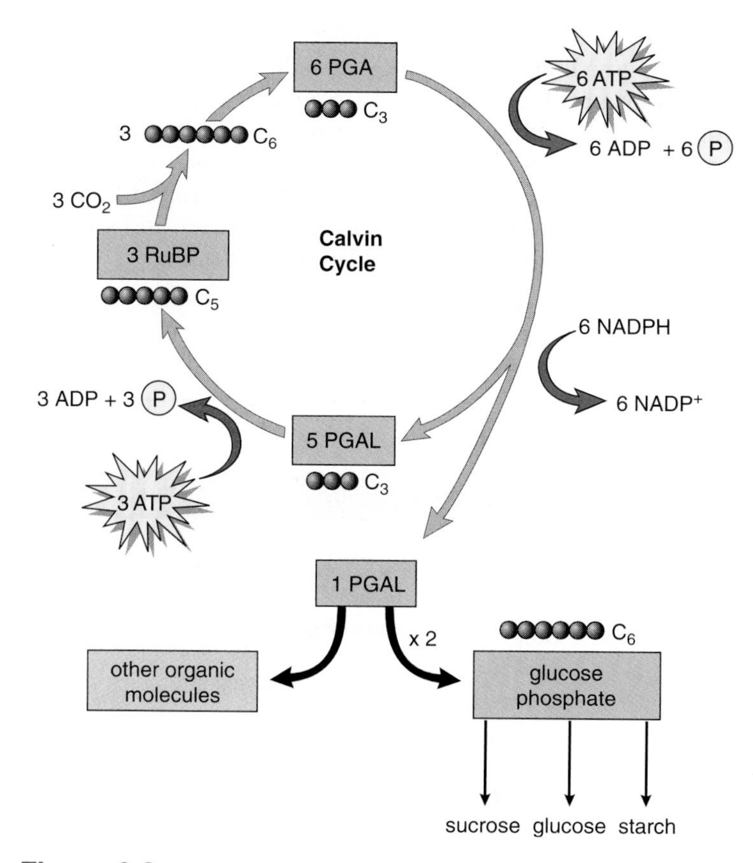

Figure 8.8 The light-independent reactions: the Calvin cycle (simplified).
RuBP accepts CO_2, forming a six-carbon molecule which immediately breaks down to two PGA. NADPH and ATP from the light-dependent reactions are used to reduce PGA to PGAL, the end product of the Calvin cycle, which can be converted to glucose. PGAL is also used to regenerate RuBP.

8.4 Carbohydrate Synthesis

The light-independent reactions are the second stage of photosynthesis. In this stage of photosynthesis, the NADPH and ATP produced during the light-dependent reactions are used to reduce carbon dioxide (CO_2): CO_2 becomes CH_2O, a carbohydrate molecule. Electrons and energy are needed for this reduction synthesis and these are supplied by NADPH and ATP.

The reduction of CO_2 occurs in the stroma of a chloroplast by means of a series of reactions known as the *Calvin cycle.* Although this cycle does not require light, it is most likely to occur during the day, when a plant is producing high levels of ATP and NADPH.

Overview of Calvin Cycle

Figure 8.8 is a simplified diagram of the reactions that occur as CO_2 is fixed and reduced. Carbon dioxide is taken up by a five-carbon sugar, **RuBP (ribulose bisphosphate).** The six-carbon molecule resulting from carbon dioxide fixation immediately breaks down to two PGA[1] (C_3) molecules.

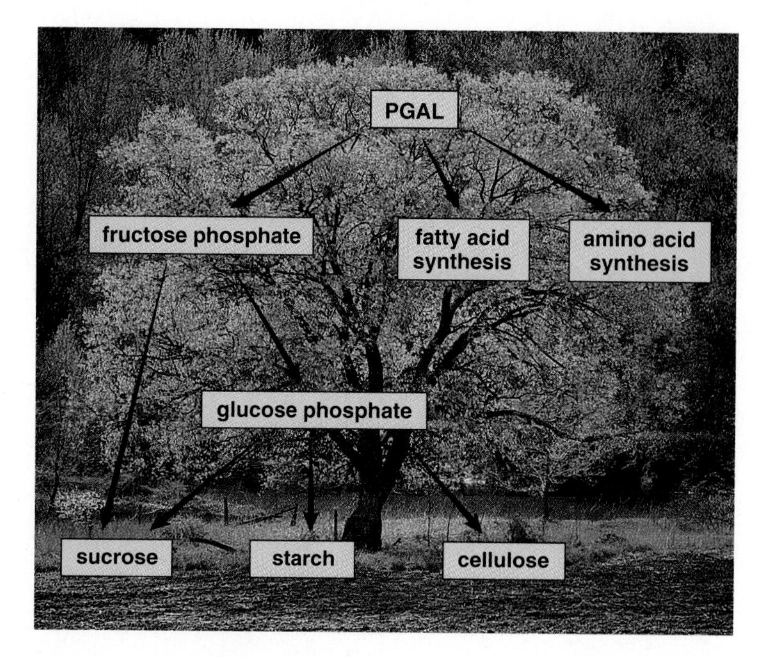

Figure 8.9 Fate of PGAL.
PGAL is the first reactant in a number of plant cell metabolic pathways. Two PGALs are needed to form glucose phosphate; glucose is often considered the end product of photosynthesis. Sucrose is the transport sugar in plants, starch is the storage form of glucose, and cellulose is a major constituent of plant cell walls.

Now each PGA is reduced to PGAL[2] using the NADPH and some of the ATP formed in the thylakoid membrane. These reductions represent the reduction of CO_2 and its conversion to a high-energy molecule. In other words, PGAL contains more hydrogen atoms than does PGA. Some of the PGAL is used within the chloroplast to re-form RuBP. This reaction also uses some of the ATP produced by light-dependent reactions.

PGAL Is a Reactant

PGAL (glyceraldehyde-3-phosphate), the product of the Calvin cycle, is converted to all sorts of organic molecules. Compared to animal cells, algae and plants have enormous biochemical capabilities. They use PGAL for the purposes described in Figure 8.9. As shown, *glucose phosphate* is among the organic molecules that result from PGAL metabolism. This is of interest to us because glucose is the molecule that plants and animals most often metabolize to produce the ATP molecules they require for their energy needs.

Glucose can be combined with fructose (and the phosphate removed) to form sucrose, the molecule that plants use to transport carbohydrates from one part of the body to the other. Starch is the storage form of glucose. Some starch is stored in chloroplasts, but most starch is stored in the plant roots. Cellulose is the molecule which is present in plant cell walls and becomes fiber in our diet because we are

[1]3-phosphoglycerate

[2]glyceraldehyde-3-phosphate

unable to digest it. A plant can utilize the hydrocarbon skeleton of PGAL to form fatty acids and glycerol, which are combined in plant oils. We are all familiar with corn oil, sunflower oil, or olive oil, which we use in cooking. Also, when nitrogen is added to the hydrocarbon skeleton derived from PGAL, amino acids are formed.

Stages of the Calvin Cycle

The reactions of the Calvin cycle are the light-independent reactions during which carbohydrate is synthesized. The Calvin cycle can be divided into: (1) CO_2 fixation; (2) CO_2 reduction; and (3) regeneration of RuBP.

Fixation of Carbon Dioxide

Carbon dioxide fixation, the attachment of CO_2 to an organic compound, is the first event in the Calvin cycle (Fig. 8.10). At that time, RuBP, a five-carbon molecule, combines with CO_2. The enzyme that speeds up this reaction is called RuBP carboxylase, a protein that makes up about 20–50% of the protein content in chloroplasts. The reason for its abundance may be that it is unusually slow (it processes only about three molecules of substrate per second, compared to about one thousand per second for a typical enzyme), and so there has to be a lot of it to keep the Calvin cycle going.

Carbon dioxide fixation occurs when CO_2 combines with RuBP.

Metabolites of the Calvin Cycle	
RuBP	ribulose bisphosphate
PGA	3-phosphoglycerate
PGAP	1,3-bisphosphoglycerate
PGAL	glyceraldehyde-3-phosphate

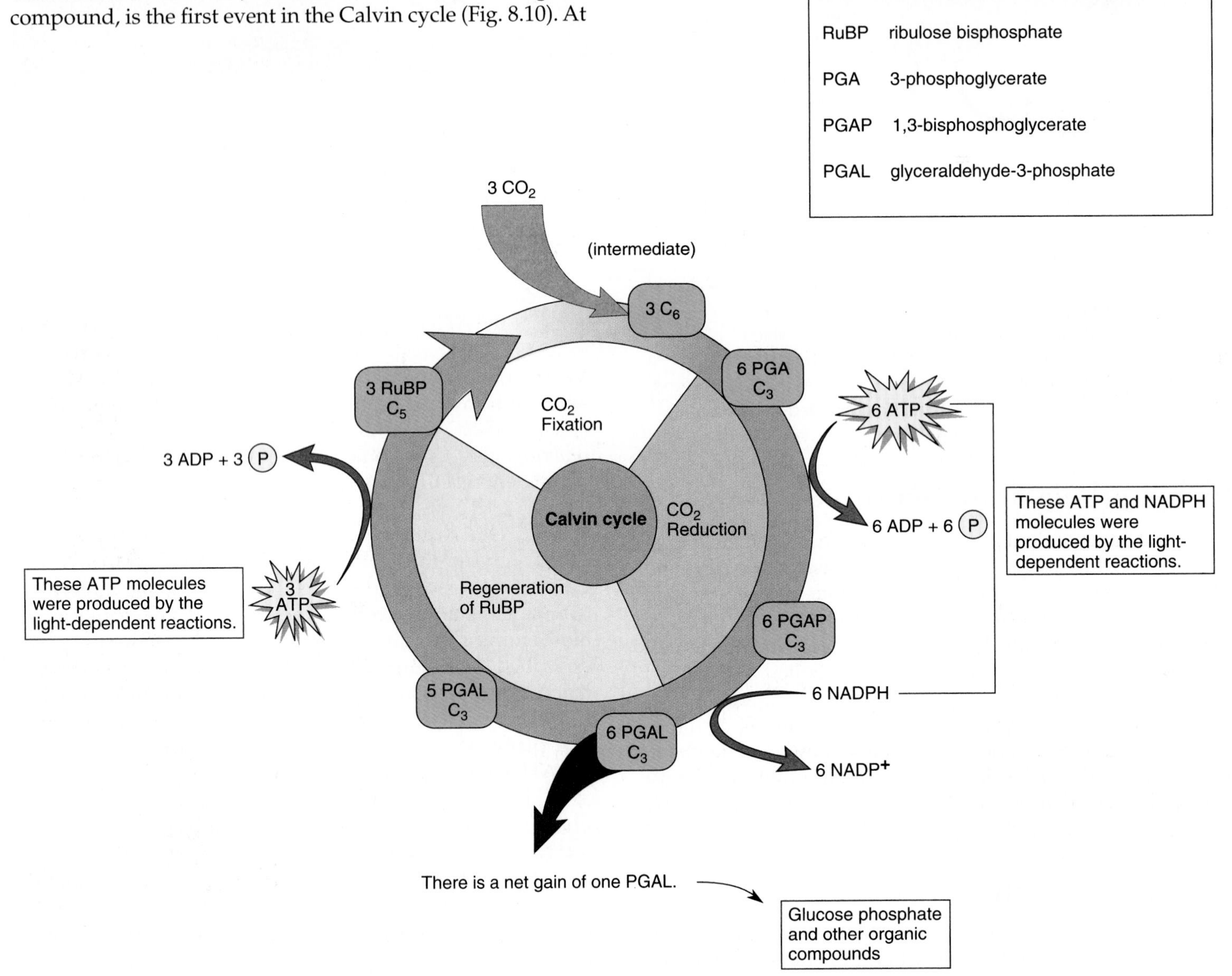

Figure 8.10 The light-independent reactions: the Calvin cycle (in detail).
The Calvin cycle is divided into three portions: CO_2 fixation, CO_2 reduction, and regeneration of RuBP. Because five PGAL are needed to re-form three RuBP, it takes three turns of the cycle to have a net gain of one PGAL, which can be used to form glucose.

Reduction of Carbon Dioxide

The six-carbon molecule resulting from CO_2 fixation immediately breaks down to form two PGA, which are C_3 molecules. Each of the two PGA molecules undergoes reduction to PGAL in two steps:

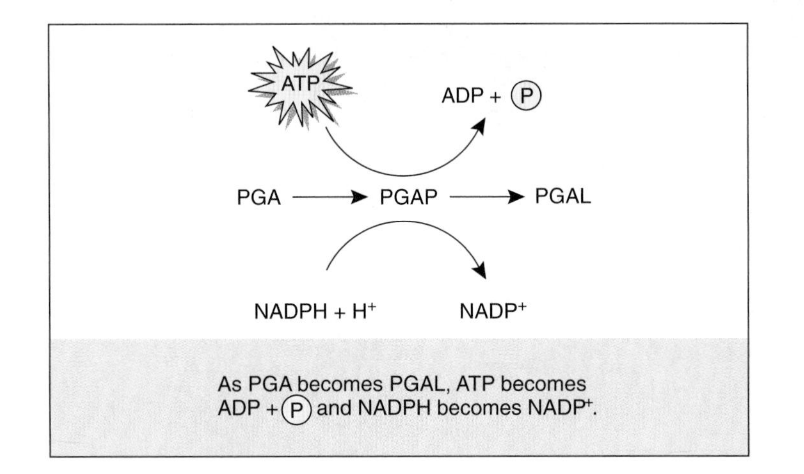

This is the actual reaction that uses NADPH and ATP from the light-dependent reactions, and it signifies the reduction of CO_2 to a carbohydrate (CH_2O). Hydrogen atoms and energy are needed for this reduction reaction, and these are supplied by NADPH and ATP, respectively. The reduction of CO_2 is a synthetic process because it requires the formation of new bonds.

Regeneration of RuBP

For every three turns of the Calvin cycle, five molecules of PGAL are used to re-form three molecules of RuBP so that the cycle can continue:

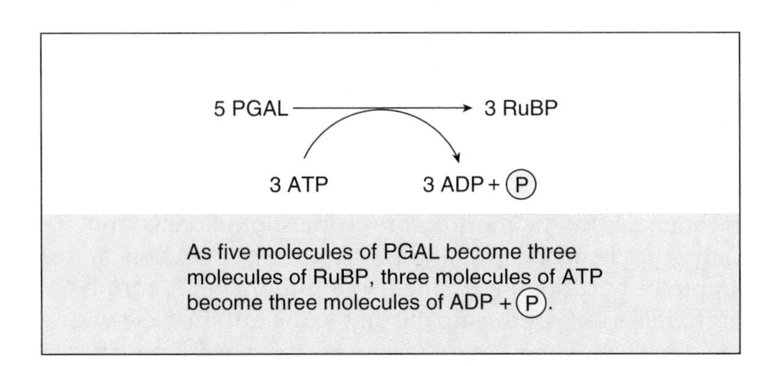

The net gain from the incorporation of 3 CO_2 in the Calvin cycle is one PGAL molecule. This reaction also utilizes some of the ATP produced by the light-dependent reactions.

Five out of six PGAL molecules of the Calvin cycle are used to regenerate three RuBP molecules and the cycle begins again.

8.5 Other Aspects of Photosynthesis

Figure 8.11 reviews the overall equation for photosynthesis. Animals and even plants themselves are dependent upon the oxygen produced during the light-dependent reactions. Most organisms are aerobic and need a constant supply of oxygen in order to exist. Oxygen produced by photosynthesis forms an ozone shield (O_3) in the upper atmosphere which filters out damaging ultraviolet (UV) rays of the sun. Living things wouldn't be able to live on land without the ozone shield, so there is much concern today that various pollutants have reduced the shield, particularly over the poles. You may have heard the expression *ozone holes* which refers to this reduction.

PGAL, which is produced during the light-independent reactions of photosynthesis, provides the hydrocarbon backbone and the energy necessary to allow plants to produce glucose ($C_6H_{12}O_6$) and all the organic molecules they require. These molecules are the ultimate food source for all living things. Consider that even when we eat meat, these animals have fed on plants. The light-independent reactions soak up carbon dioxide molecules that otherwise would contribute to global warming. Global warming threatens to drastically redistribute life on land due to its many effects.

Figure 8.11 Importance of photosynthesis.
Nearly all living things, including humans, are dependent on photosynthesis.

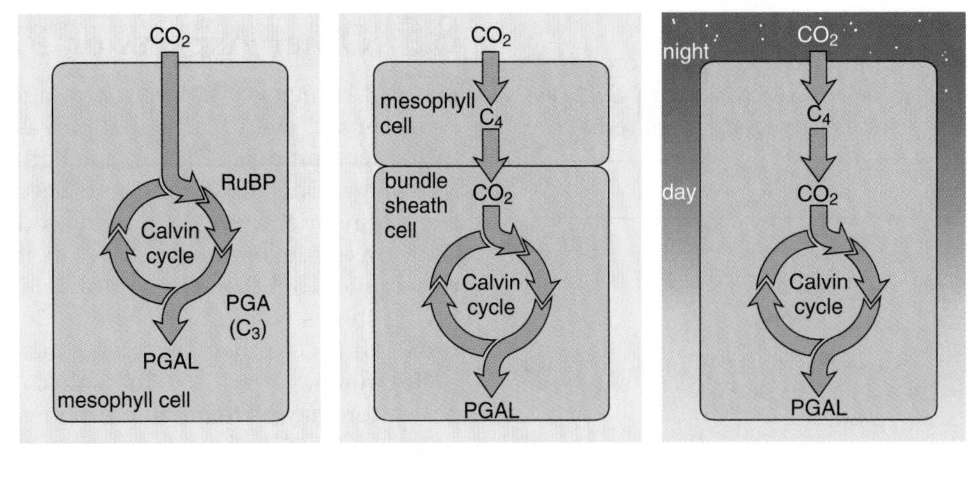

Figure 8.12 Carbon dioxide fixation.
Plants can be categorized according to the type of carbon dioxide fixation.

C_3 Versus C_4 Photosynthesis

The photosynthesis we have been discussing is called C_3 photosynthesis because a C_3 molecule is detectable immediately following CO_2 fixation. In **C_3 plants** like wheat, rice, and oats, the *mesophyll cells* contain well-formed chloroplasts and are arranged in parallel layers. In **C_4 plants,** the *bundle sheath cells* also contain chloroplasts. Further, the mesophyll cells are arranged concentrically around the bundle sheath cells:

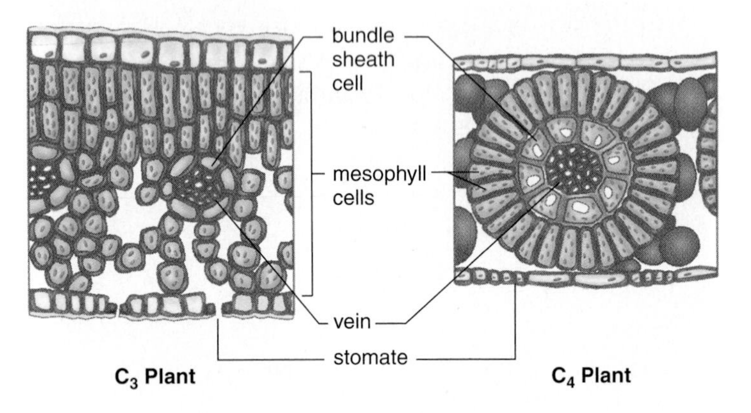

C_4 plants fix CO_2 by forming a C_4 molecule prior to the involvement of the Calvin cycle (Fig. 8.12). In C_4 plants like sugarcane and corn, CO_2 is taken up in mesophyll cells, and then a C_4 molecule, called oxaloacetate, is pumped into the bundle sheath cells, where it releases CO_2 to the Calvin cycle. It takes energy to pump molecules, and you would think that the C_4 pathway would be disadvantageous. Not so, however, if the weather is hot and dry. When stomates close in order to conserve water, the concentration of O_2, a by-product of photosynthesis, increases in leaf air spaces. In C_3 plants, this O_2 competes with CO_2 for the active site of RuBP carboxylase, the first enzyme of the Calvin cycle. In C_4 leaves, CO_2 is delivered to the Calvin cycle in the bundle sheath cells that are sheltered from leaf air spaces.

When the weather is moderate, C_3 plants like Kentucky bluegrass and creeping bent grass predominate in lawns in the cooler parts of the United States; but by midsummer, crabgrass, which is a C_4 plant, begins to take over.

CAM Photosynthesis

CAM plants also fix CO_2 by forming a C_4 molecule, but this occurs *at night* when stomates can open without much loss of water. CAM plants form a C_4 molecule, which is stored in large vacuoles in their mesophyll cells until the next day. CAM stands for crassulacean-acid metabolism; the Crassulaceae is a family of flowering succulent plants that live in warm, arid regions of the world. Now it is known that CAM photosynthesis is prevalent among most succulent plants that grow in desert environments, including the cactuses.

Whereas a C_4 plant represents partitioning in space—carbon dioxide fixation occurs in mesophyll cells and the Calvin cycle occurs in bundle sheath cells—CAM is an example of partitioning by the use of time. The C_4 formed at night releases CO_2 during the day to the Calvin cycle within the same cell. The primary reason for this partitioning has to do with the conservation of water. CAM plants open their stomates only at night to acquire CO_2. During the day, the stomates close to conserve water.

In C_4 plants, the Calvin cycle is located in bundle sheath cells where O_2 cannot accumulate when stomates close. CAM plants carry on carbon dioxide fixation at night and release CO_2 to the Calvin cycle during the day.

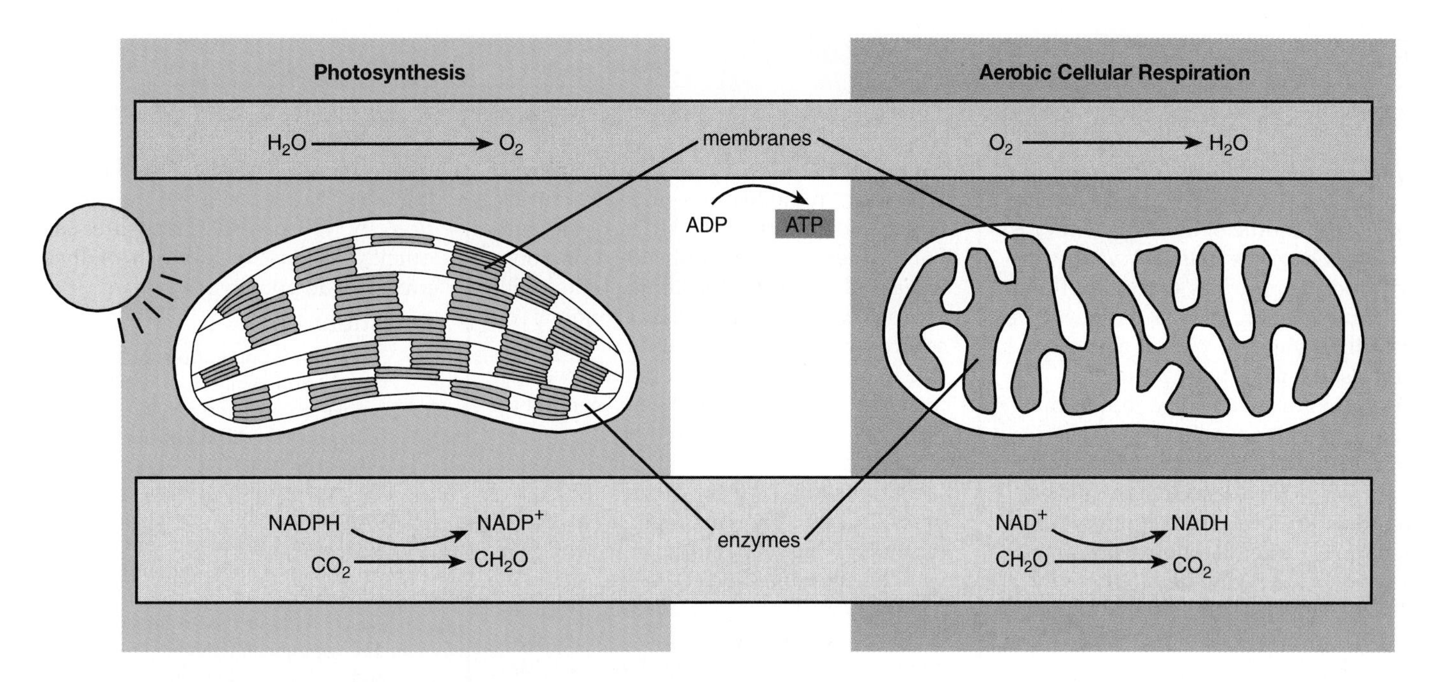

Figure 8.13 Photosynthesis versus aerobic cellular respiration.
Both processes have an electron transport system located within membranes where ATP is produced by chemiosmosis. In photosynthesis, water is oxidized and oxygen is released; in aerobic cellular respiration, oxygen is reduced to water. Both have enzyme-catalyzed reactions within the fluid interior. In photosynthesis, CO_2 is reduced to a carbohydrate; in aerobic cellular respiration, a carbohydrate is oxidized to CO_2.

8.6 Photosynthesis Versus Aerobic Cellular Respiration

Both plant and animal cells carry on aerobic cellular respiration, but only plant cells photosynthesize. The cellular organelle for aerobic respiration is the mitochondrion, while the cellular organelle for photosynthesis is the chloroplast.

The overall equation for photosynthesis is the opposite of that for aerobic cellular respiration. The reaction in the forward direction represents photosynthesis, and the word energy stands for solar energy. The reaction in the opposite direction represents aerobic cellular respiration, and the word energy then stands for ATP.

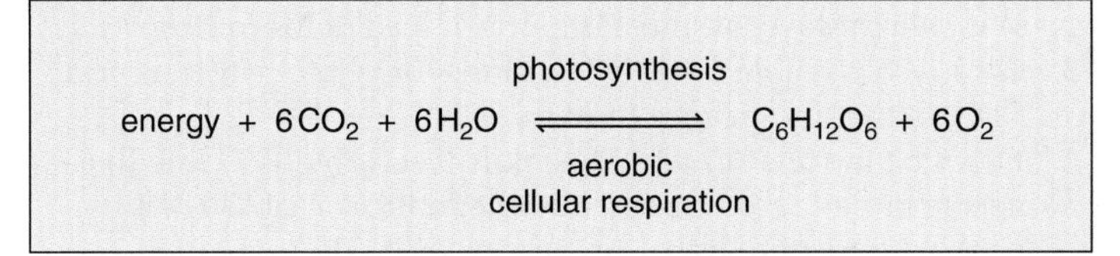

Obviously, photosynthesis is the building up of glucose, while aerobic cellular respiration is the breaking down of glucose.

Both photosynthesis and aerobic cellular respiration are metabolic pathways within cells and therefore consist of a series of reactions that the overall reaction does not indicate. Both pathways, which make use of an electron transport system located in membranes (Fig. 8.13), produce ATP by chemiosmosis. Both make use of an electron carrier—photosynthesis uses $NADP^+$, and aerobic cellular respiration uses NAD^+.

Both pathways utilize this reaction, but in opposite directions. For photosynthesis, read the reaction in the forward direction; and for aerobic cellular respiration, read the reaction in the opposite direction.

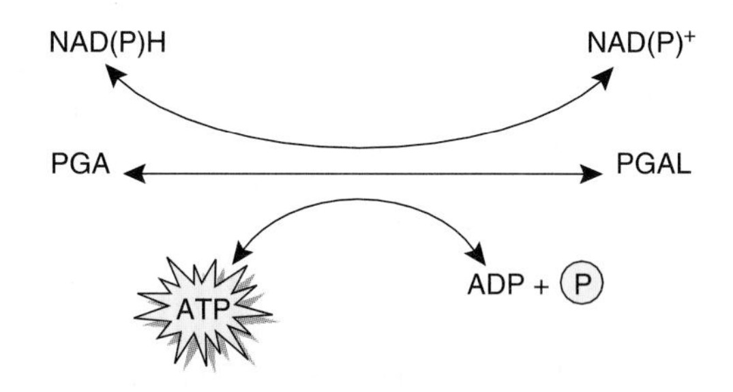

Both photosynthesis and aerobic cellular respiration occur in plant cells. While both of these occur during the daylight hours, only aerobic cellular respiration occurs at night. During daylight hours, the rate of photosynthesis exceeds the rate of aerobic cellular respiration, resulting in a net increase and storage of glucose. The stored glucose is used for cellular metabolism, which continues during the night. Also during daylight hours, the release of O_2 is greater than the release of CO_2, whereas at night the release of O_2 ceases and the release of CO_2 increases.

Bioethical Issue

Whether there will be enough food to feed the increase in population expected by the middle of the 21st century is unknown. Over the past 40 years, the world's food supply has expanded faster than the population, due to the development of high-yielding plants, and the increased use of irrigation, pesticides, and fertilizers. Unfortunately, modern farming techniques result in pollution of the air, water, and land. One of the most worrisome threats to food security is an increasing degradation of agricultural land. Soil erosion in particular is robbing the land of its top soil and reducing its productivity.

In 1986, it was estimated that humans already use nearly 40% of the Earth's terrestrial photosynthetic production and therefore we should reach maximum capacity in the middle of the 21st century when the population is projected to double its 1986 size. Most population growth will occur in the less developed countries which are countries in Africa, Asia, and Latin America that are only now becoming industrialized. Even the United States will face an increased drain on its economic resources and increased pollution problems due to population growth. In the developing countries the technical gains needed to prevent a disaster will be enormous.

Some people feel that technology will continue to make great strides for many years to come. They maintain science and technology hasn't begun to reach the limits of performance and therefore, they will solve the problems of increased population growth. Others feel that technology's successes are self defeating. The newly developed hybrid crops that led to enormous increases in yield per acre also cause pollution problems that degrade the environment. These scientists are in favor of calling a halt to an increasing human population by all possible measures and as quickly as possible. Which of these approaches do you favor?

Questions

1. The present population could very well double from 6 billion to 12 billion or higher in your lifetime. Do you approve of measures to limit population growth? Why or why not?
2. Do you believe we should continue to depend on technology to supply all the goods and services we need to maintain our present standard of living? Why or why not?
3. Do you believe the problems of the developing countries are their problems and nothing to do with us? Why or why not?

Summarizing the Concepts

8.1 Radiant Energy

During photosynthesis, solar energy is converted to chemical energy within carbohydrate. Photosynthesis uses solar energy in the visible-light range; specifically, chlorophylls *a* and *b* largely absorb violet, blue, and red wavelengths and reflect green wavelengths. This causes chlorophyll to appear green to us.

8.2 Structure and Function of Chloroplasts

The overall equation for photosynthesis indicates that H_2O is oxidized with the release of O_2, and CO_2 is reduced to a carbohydrate. A chloroplast contains two main portions: membranous grana made up of thylakoid sacs where chlorophyll absorbs solar energy, and a fluid-filled stroma where CO_2 is reduced.

Photosynthesis has two sets of reactions. The light-dependent reactions, which occur in the thylakoids, drive the light-independent reactions, which occur in the stroma.

8.3 Solar Energy Capture

In the cyclic electron pathway of the light-dependent reactions, electrons energized by the sun leave PS I and enter an electron transport system that produces ATP by chemiosmotic ATP synthesis and then the energy-spent electrons return to PS I. In the noncyclic electron pathway, electrons are energized in PS II before they enter the electron transport system and replace e^- lost in PS I. The electrons are then reenergized in PS I and then are passed to $NADP^+$, which becomes NADPH. Electrons from H_2O replace those lost in PS II.

8.4 Carbohydrate Synthesis

During the light-independent reactions, ATP and NADPH from the light-dependent reactions are used during the Calvin cycle to reduce PGA to PGAL. PGAL is metabolized to various molecules including carbohydrates (Fig. 8.9).

8.5 Other Aspects of Photosynthesis

Photosynthesis provides food and oxygen for the biosphere. Plants utilizing C_4 photosynthesis have a different construction than plants using C_3 photosynthesis. C_4 plants fix CO_2 in mesophyll cells and then deliver the CO_2 to the Calvin cycle in bundle sheath cells. Now O_2 cannot compete for the active site of RuBP carboxylase when stomates are closed due to hot and dry weather. CAM plants fix carbon dioxide at night when their stomates remain open.

8.6 Photosynthesis Versus Aerobic Cellular Respiration

Both photosynthesis and aerobic cellular respiration utilize an electron transport system and chemiosmotic ATP synthesis. However, aerobic cellular respiration oxidizes carbohydrate, and CO_2 is given off. Photosynthesis reduces CO_2 to a carbohydrate. The oxidation of H_2O releases O_2.

Studying the Concepts

1. Why are almost all living things dependent upon the process of photosynthesis and solar energy? 136
2. Which light rays are most important for photosynthesis? 137
3. Give two possible overall equations for photosynthesis and discuss the significance of each. 138
4. Describe the anatomy of the chloroplast, and associate the absorption of solar energy and the reduction of CO_2 with a particular portion of the organelle. 138–39
5. What are the two sets of reactions that occur during photosynthesis and how are the two pathways related? 140–41
6. What role do PS I and PS II play during the light-dependent reactions? 140–41
7. Trace the cyclic and noncyclic electron pathways. 140–41
8. Explain what is meant by chemiosmosis, and relate this process to the electron transport system present in the thylakoid membrane. 142

9. Describe the three stages of the Calvin cycle. Mention which stage utilizes the ATP and NADPH from the light-dependent reactions. 143–44
10. What is the difference between C_3, C_4, and CAM photosynthesis? 146
11. What are some major differences and similarities between photosynthesis and aerobic cellular respiration? 147

Testing Yourself

Choose the best answer for each question.

1. The absorption spectrum of chlorophyll *a* and chlorophyll *b*
   a. is not the same as that of carotenoids.
   b. approximates the action spectrum of photosynthesis.
   c. explains why chlorophyll is a green pigment.
   d. shows that some colors of light are absorbed more than others.
   e. All of these are correct.
2. The final acceptor of electrons during the noncyclic electron pathway is
   a. PS I.
   b. PS II.
   c. water.
   d. ATP.
   e. $NADP^+$.
3. A photosystem contains
   a. pigments, a reaction center, and an electron acceptor.
   b. ADP, Ⓟ, and hydrogen ions (H^+).
   c. protons, photons, and pigments.
   d. cytochromes only.
   e. Both b and c are correct.
4. Which of these should not be associated with the electron transport system?
   a. chloroplasts
   b. cytochromes
   c. movement of H^+ into the thylakoid space
   d. formation of ATP
   e. absorption of solar energy
5. C_4 photosynthesis
   a. is the same as C_3 photosynthesis because it takes place in chloroplasts.
   b. occurs in plants whose bundle sheath cells contain chloroplasts.
   c. takes place in plants like wheat, rice, and oats.
   d. is an advantage when the weather is hot and dry.
   e. Both b and d are correct.
6. The NADPH and ATP from the light-dependent reactions are used to
   a. split water.
   b. cause rubisco to fix CO_2.
   c. re-form the photosystems.
   d. cause electrons to move along their pathways.
   e. convert PGA to PGAL.
7. CAM photosynthesis
   a. stands for chloroplasts and mitochondria.
   b. is the same as C_4 photosynthesis.
   c. is an adaptation to cold environments in the Southern Hemisphere.
   d. is prevalent in desert plants that close their stomates during the day.
   e. occurs in plants that live in marshy areas.
8. Chemiosmosis depends on
   a. protein complexes in thylakoid membrane.
   b. a difference in H^+ concentration between the thylakoid space and the stroma.
   c. ATP breaking down to ADP + Ⓟ.
   d. the absorption spectrum of chlorophyll.
   e. Both a and b are correct.
9. Label this diagram of a chloroplast.

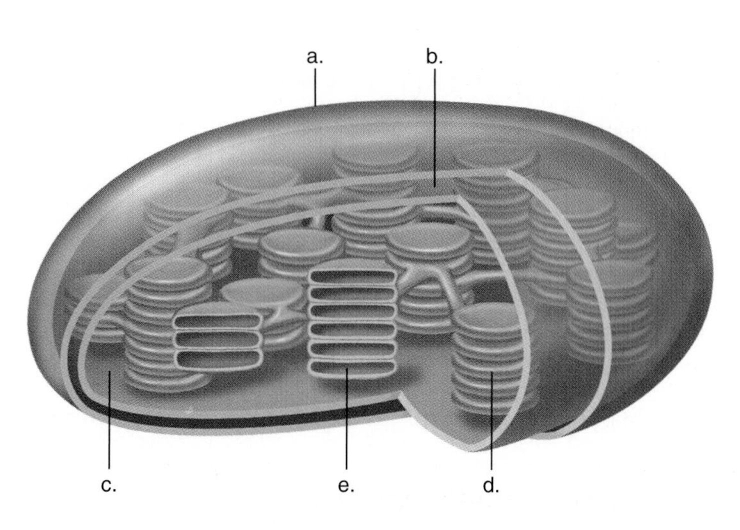

   f. The light-dependent reactions occur in which part of a chloroplast?
   g. The light-independent reactions occur in which part of a chloroplast?
10. Label this diagram using these labels: water, carbohydrate, carbon dioxide, oxygen, ATP, ADP + Ⓟ, NADPH, and $NADP^+$.

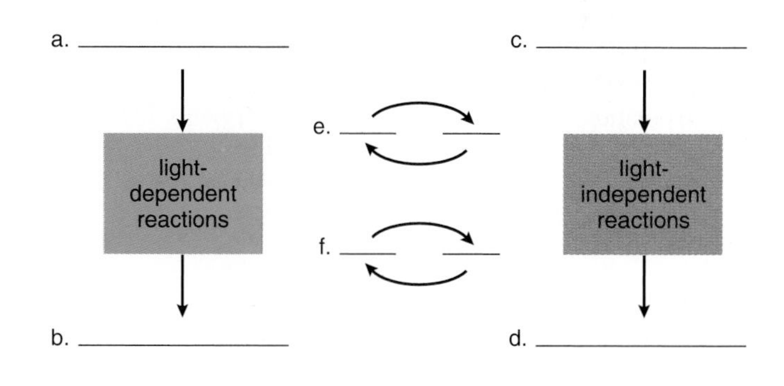

Thinking Scientifically

1. Considering the cycling of materials in the biosphere (page 147):
   a. A plant exchanges materials with its environment (surroundings). What substances does it take from the environment, and what substances does it give to the environment?
   b. What part of a chloroplast gives off O_2? What is the source of this oxygen? What part of a chloroplast takes up CO_2? What happens to the carbon dioxide.?
   c. How do plants and animals use the oxygen given off by a plant?
   d. In what ways does a plant use newly formed glucose?
2. Considering the flow of energy in the biosphere (page 147):
   a. How much energy do you suppose it takes to reduce carbon dioxide (CO_2) to glucose? From where does this energy ultimately come?
   b. Energy is needed to maintain cellular and organismal organization. The oxidation of glucose releases 686 kcal/mole. Specifically, what do cells do with the energy of glucose breakdown?
   c. Why would it be correct to say that photosynthesis drives cellular respiration?
   d. In general, what role is played by glucose in all organisms?

Understanding the Terms

autotroph 136
C_3 plant 146
C_4 plant 146
Calvin cycle 138
carbon dioxide fixation 144
carotenoid 137
chlorophyll 137
chloroplast 138
cyclic electron pathway 140
electron transport system 140
heterotroph 136
light-dependent reaction 138
light-independent reaction 138
noncyclic electron pathway 140
PGAL (glyceraldehyde-3-phosphate) 143
photosynthesis 136
photosystem 140
RuBP (ribulose bisphosphate) 143
stroma 138
thylakoid 138

Match the terms to these definitions:

a. ________ Energy-capturing portion of photosynthesis that takes place in thylakoid membranes of chloroplasts and cannot proceed without solar energy; it produces ATP and NADPH.
b. ________ Flattened sac with a granum whose membrane contains chlorophyll and where the light-dependent reactions of photosynthesis occur.
c. ________ Plant that directly uses the Calvin cycle; the first detected molecule during photosynthesis is PGA, a three-carbon molecule.
d. ________ Green pigment that absorbs solar energy and is important in photosynthesis.
e. ________ Photosynthetic unit where solar energy is absorbed and high-energy electrons are generated; contains a pigment complex and an electron acceptor.

Using Technology

Your study of photosynthesis is supported by these available technologies:

Essential Study Partner CD-ROM

Plants → Photosynthesis

Visit the Mader web site for related ESP activities.

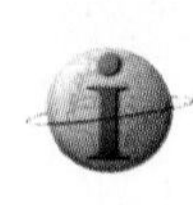

Exploring the Internet

The Mader Home Page provides resources and tools as you study this chapter.

http://www.mhhe.com/biosci/genbio/mader

Life Science Animations 3D Video

8 Photosynthesis

C H A P T E R 9

Plant Organization and Growth

Chapter Concepts

An onion plant with soil removed illustrates the three major organs of a plant: root, stem, and leaf.

The evergreen black bean tree (*Castanospermum australe*), whose glossy black leaves reach into the canopy of Australia's tropical rain forest, produces chestnutlike seeds. Used formerly as a source of timber to make prized furniture, its numbers are severely depleted. Its seeds, which have long been prized by aborigines as a food source, contain a toxic substance that kills insects. Who would have guessed that this substance would also dramatically affect the AIDS virus so that it cannot infect cells? An important research tool, the substance may become a medicine for various immunological ills. Such stories are now commonplace as we begin to learn more about various tropical rain forest plants.

Various types of plants, and not just those from tropical rain forest plants, supply us with food, medicines, spices, building materials, fabrics, and all sorts of chemicals. The study of plants, which we introduce in this chapter, is a matter of importance to us all.

9.1 The Flowering Plant

We will be studying **angiosperms,** the flowering plants, in this chapter. Nonflowering plants, such as mosses and ferns, are discussed in a later chapter.

The vegetative organs of a flowering plant—the root, the stem, and the leaf (Fig. 9.1)—allow a plant to live and grow and are not involved in reproduction.

Roots

The **root system** of a dicot flowering plant includes the main root and any and all of its lateral (side) branches. As a rule of thumb, the root system (the part of the plant below ground) is at least equivalent in size and extent to the **shoot system** (the part of the plant above ground). The root system of a plant anchors it in the soil and gives it support. It also provides an extensive surface area for the absorption of water and minerals from the soil. The products of photosynthesis are stored in the roots of some plants; for example, in carrots and sweet potatoes.

Stems

A **stem** is the main aboveground axis of a plant along with its lateral branches. The stem of a flowering plant supports leaves and, if upright as most are, supports leaves in such a way that each one is exposed to as much sunlight as possible. A **node** occurs where leaves are attached to the stem, and an **internode** is the region between two successive nodes. The presence of nodes and internodes identifies a stem, even if the stem happens to grow underground.

Aside from supporting the leaves, a stem has vascular tissue that transports water and minerals generally from the roots to the leaves and transports the products of photosynthesis generally in the opposite direction. Some stems also function in storage. The stem stores water in cactuses and in other plants. Tubers are horizontal stems that store nutrients.

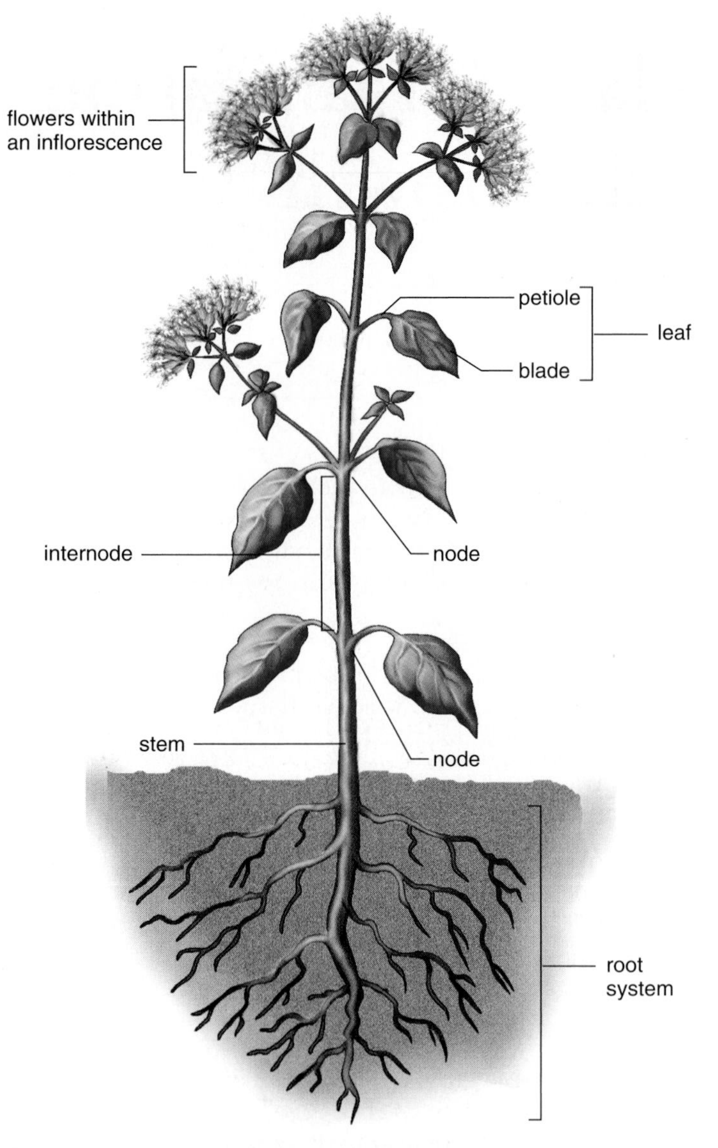

Figure 9.1 Organization of plant body.
Roots, stems, and leaves are vegetative organs. A flower is a reproductive structure.

Leaves

A **leaf** is that part of a plant that usually carries on photosynthesis, a process that requires water, carbon dioxide, and sunlight (Fig. 9.1). Leaves receive water from the root system by way of the stem. Broad, thin leaves have a maximum surface area for the absorption of carbon dioxide and the collection of solar energy. The wide portion of a leaf is called the *leaf blade;* the petiole is a structure that attaches the blade to the stem.

A flowering plant has three vegetative organs: roots anchor plants, and stems support leaves which carry on photosynthesis.

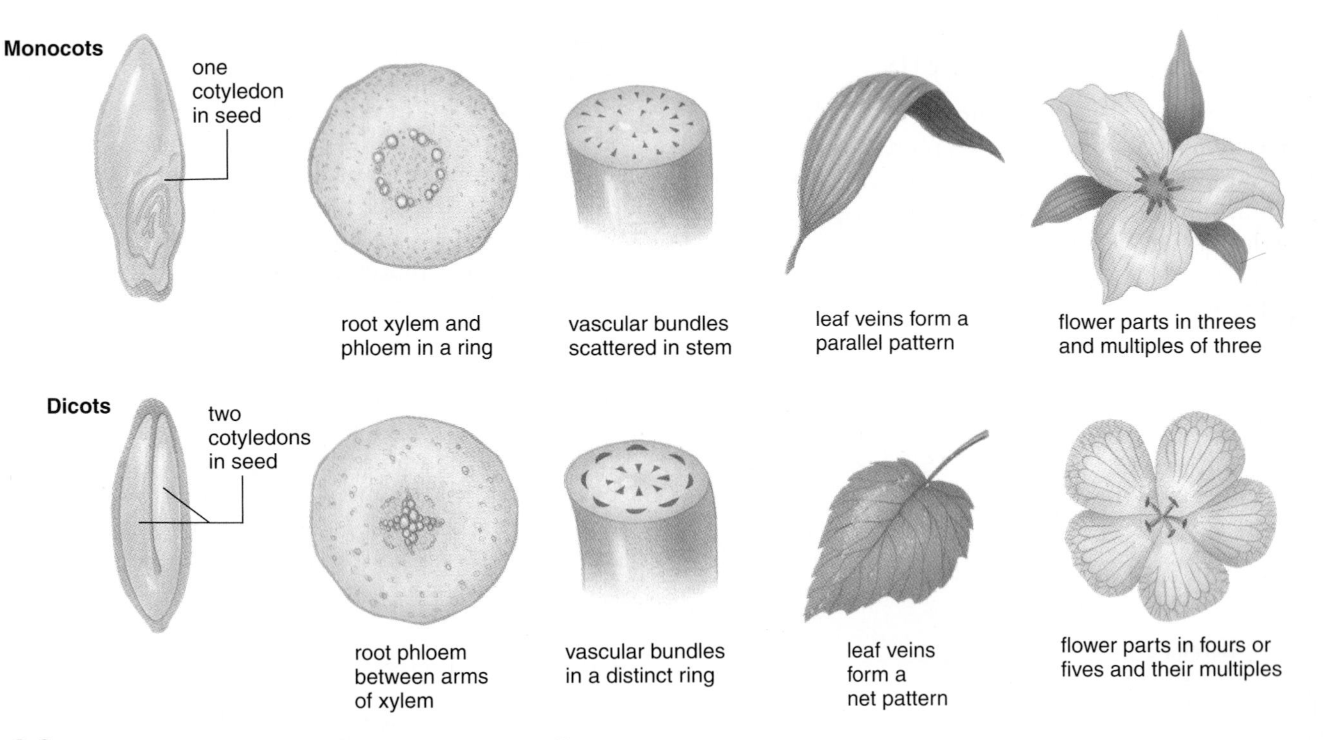

Figure 9.2 Flowering plants are either monocots or dicots.
Three features used to distinguish monocots from dicots are the number of cotyledons, the arrangement of vascular tissue in roots, stems, and leaves, and the number of flower petals.

Monocot Versus Dicot Plants

The structure of flowering plants differs according to whether they are monocots or dicots. Figure 9.2 gives you an overview of the topics we will be discussing in detail.

The term **monocot** and **dicot** refers to the number of cotyledons there are in the seed (Fig. 9.2). **Cotyledons** provide nutrient molecules for immature plants before the true leaves begin photosynthesizing. Plants, known as *monocotyledons* or monocots, have one cotyledon. From experience you know that a corn kernel has only one cotyledon because it cannot be taken apart. Other embryos have two cotyledons, and these plants are known as *dicotyledons,* or dicots. From experience you know that a peanut has two major parts. These parts are the cotyledons; the immature plant is found between the two halves of a peanut.

There is a different arrangement of vascular (transport) tissue in monocot and dicot roots. In the monocot root, *xylem,* the vascular tissue that transports water, occurs side by side in a ring; but in the dicot, it has a star shape. **Vascular bundles** contain vascular tissue surrounded by support tissue. In a monocot stem, the vascular bundles are scattered; in a dicot stem, they occur side by side in a ring. Figure 9.2 shows cross sections of stems; but keep in mind that vascular bundles extend lengthwise in the stem. Leaf **veins** are vascular bundles within a leaf. The veins in a monocot leaf are parallel, whereas in a dicot they have a net pattern.

Monocots and dicots also differ in their flower parts. Monocots have flower parts in threes and multiples of three, such as six or nine. Notice that the flower shown has three petals and three sepals (the whorl just beneath the petals). Dicots have flower parts in fours or fives and their multiples. Notice that the flower shown has five petals. Monocots and dicots have other structural differences, such as the number of apertures (thin areas in the wall) of pollen grains. Dicot pollen grains usually have three apertures and monocot pollen grains usually have one aperture.

Among angiosperms, the dicots are the larger group and include some of our most familiar flowering plants—from dandelions to oak trees. The monocots include grasses, lilies, orchids and palm trees, among others. Some of our most significant food sources are monocots, including rice, wheat, and corn.

Flowering plants are classified into monocots and dicots on the basis of structural differences.

9.2 Plant Tissues and Cells

A plant grows during its entire life because it has a tissue called **meristem** located in the stem and root tips. Meristem continually produces the three types of specialized tissue systems in the body of a plant: **dermal tissue, ground tissue,** and **vascular tissue** (Fig. 9.3).

Dermal Tissue

The entire body of a nonwoody (herbaceous) and young woody plant is protected and covered by a layer of **epidermis,** which contains closely packed epidermal cells (Fig. 9.4). The walls of epidermal cells that are exposed to air are covered with a waxy **cuticle** to minimize water loss. In roots, certain epidermal cells have long, slender projections called **root hairs** (see Figure 9.7). Root hairs increase the surface area of the root for absorption of water and minerals; they also help to anchor the plant firmly in place. In leaves, the lower epidermis in particular contains specialized cells, such as **guard cells,** which surround microscopic pores called **stomates** (see Figure 9.4). When stomates are open, gas exchange can occur. During the day, carbon dioxide diffuses in and oxygen diffuses out along with other gases.

In older woody plants, the epidermis of the stem is replaced by cork tissue. **Cork,** the outer covering of the bark of trees, is made up of dead cork cells that may be sloughed off (see Fig. 9.13). New cork cells are made by a meristem called cork cambium. As new cork cells mature, they become encrusted with *suberin,* a waterproof lipid material, and then the cells die. These nonliving cells protect the plant and make it resistant to attack by fungi, bacteria, animals, and environmental factors such as fire.

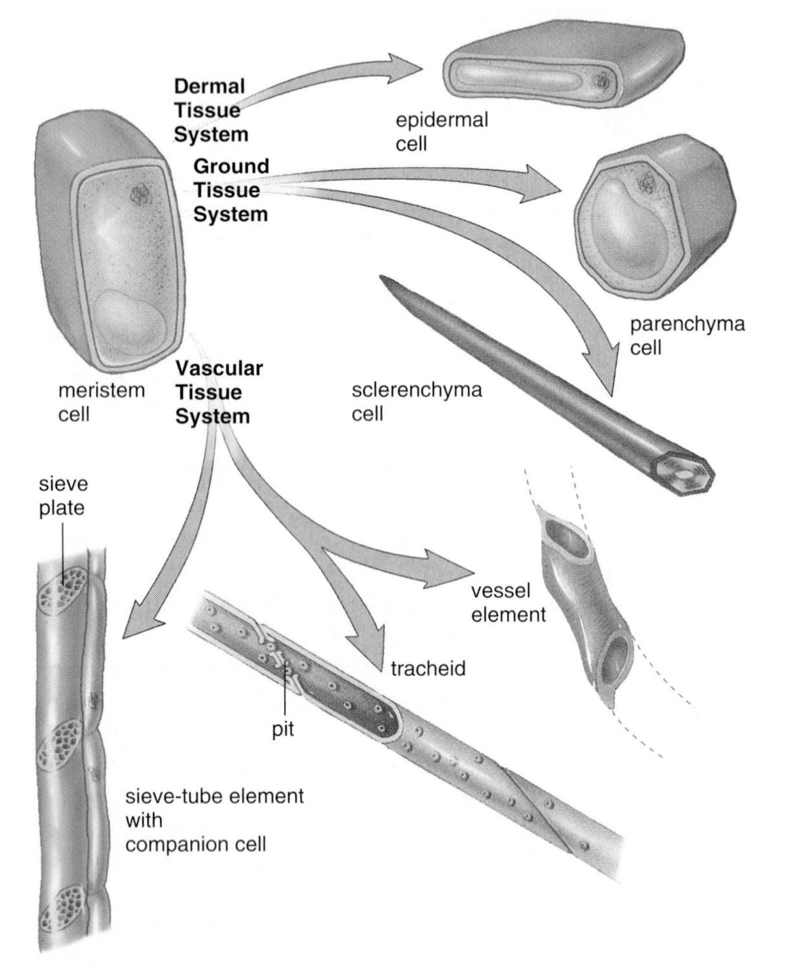

Figure 9.3 Some plant cell types.
Meristem produces new cells, and these cells differentiate into the specialized cells shown. The dermal tissue system consists of the epidermal cells of epidermis, the outermost tissue in all organs of the plant. The ground tissue system contains parenchyma and sclerenchyma cells found in cortex and pith, for example. The vascular tissue system contains the vessel elements and the tracheids found in xylem and the sieve-tube elements found in phloem.

Ground Tissue

The ground tissue system forms the bulk of a plant and contains parenchyma, collenchyma, and sclerenchyma cells. **Parenchyma cells** correspond best to the typical plant cell; they have a primary wall but no secondary wall (Fig. 9.5*a*). These are the least specialized of plant cell types and are found in all plant organs. They may contain chloroplasts and carry on photosynthesis, or they may contain colorless plastids that store the products of photosynthesis.

Collenchyma cells are like parenchyma cells except they have thicker primary walls. The thickness is uneven and usually involves the corners of the cell. They give flexible support to immature regions of a plant body. The familiar strands in celery stalks are composed mostly of collenchyma cells.

Sclerenchyma cells have thick secondary cell walls, usually impregnated with *lignin,* an organic substance that makes the walls tough and hard. Sclerenchyma cells are typically dead; their primary function is to support mature plant regions (Fig. 9.5*b*). Two types of sclerenchyma cells are sclereids and fibers. Sclereids are found in seed coats and nut shells; they give pears their gritty texture. Fibers can be commercially important: hemp fibers are used to make rope; flax fibers can be woven into linen.

Vascular Tissue

The vascular (transport) tissue system conducts water and nutrients in a plant. **Xylem** transports water and minerals from the roots up the stems to the leaves. Xylem contains two types of conducting cells, **tracheids** and **vessel elements** (Fig. 9.6*a*). Both of these types of conducting cells are hollow and nonliving at maturity. Water flows from tracheid to tracheid through **pits,** depressions where the secondary wall does not form. Water flows even more freely from one vessel element to the next because the elements have no end

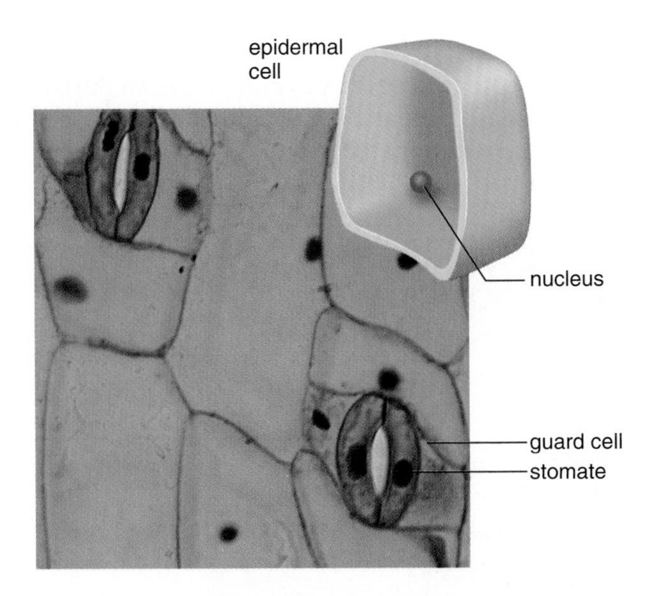

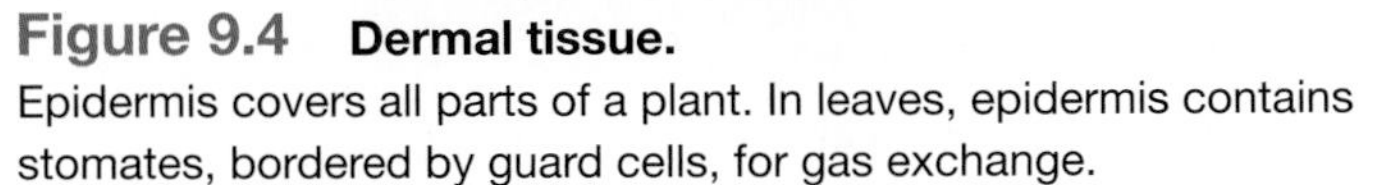

Figure 9.4 Dermal tissue.
Epidermis covers all parts of a plant. In leaves, epidermis contains stomates, bordered by guard cells, for gas exchange.

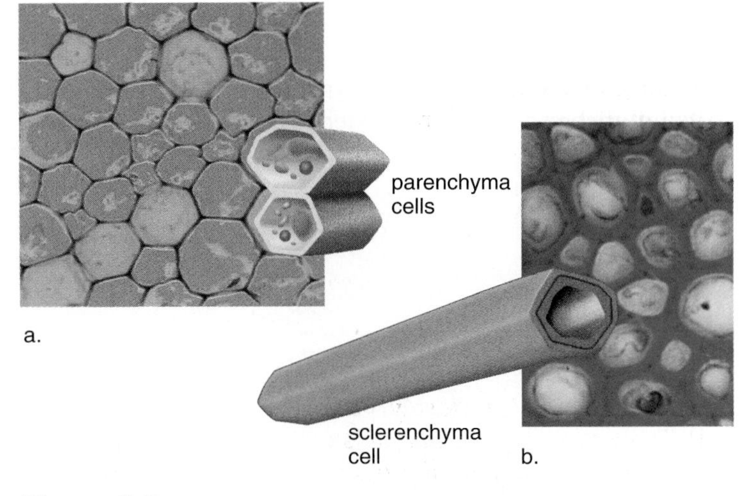

Figure 9.5 Ground tissue.
a. Parenchyma cells are the least specialized of the plant cells.
b. Sclerenchyma cells have very thick walls and are nonliving—their only function is to give strong support.

walls. **Phloem** transports organic nutrients, usually from the leaves to the rest of the plant. Phloem contains **sieve-tube elements,** each of which is associated with at least one **companion cell** (Fig. 9.6*b*). Sieve-tube elements, which have perforated end walls called **sieve plates,** contain cytoplasm but no nuclei. Strands of cytoplasm called plasmodesmata extend from one cell to the next through the sieve plates. Companion cells are much smaller than sieve-tube elements, but each has all the cellular components, including a nucleus. A companion cell most likely helps a sieve-tube element perform its function of transporting organic nutrients.

It is important to realize that vascular tissue (xylem and phloem) extends from the root to the leaves and vice versa. In the roots, the vascular tissue is usually located in the vascular cylinder; in the stem, it forms vascular bundles; and in the leaves, it is found in leaf veins.

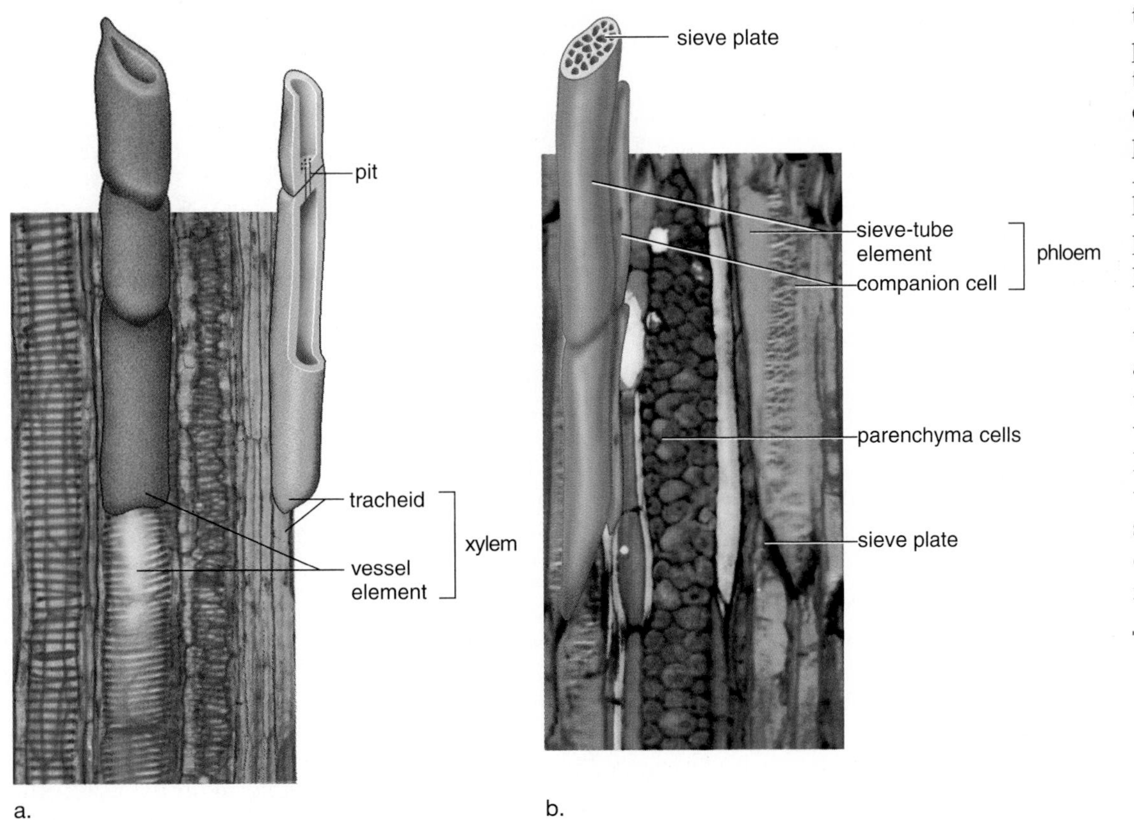

Figure 9.6 Vascular tissue.
a. Xylem contains vessel elements and tracheids, both of which transport water and minerals.
b. Phloem contains sieve-tube elements and companion cells. Sieve-tube elements transport organic nutrients, and companion cells assist sieve-tube elements.

The body of a plant is composed of tissues specialized to perform various functions. Meristem remains undifferentiated and is capable of continually dividing and producing new cells.

9.3 Organization of Roots

The shape of a root (Fig. 9.7) allows it to burrow through the soil and anchor a plant. The extensive branching of a root facilitates the absorption of water and minerals from the soil and anchors the plant. Some roots also store the products of photosynthesis received from the leaves.

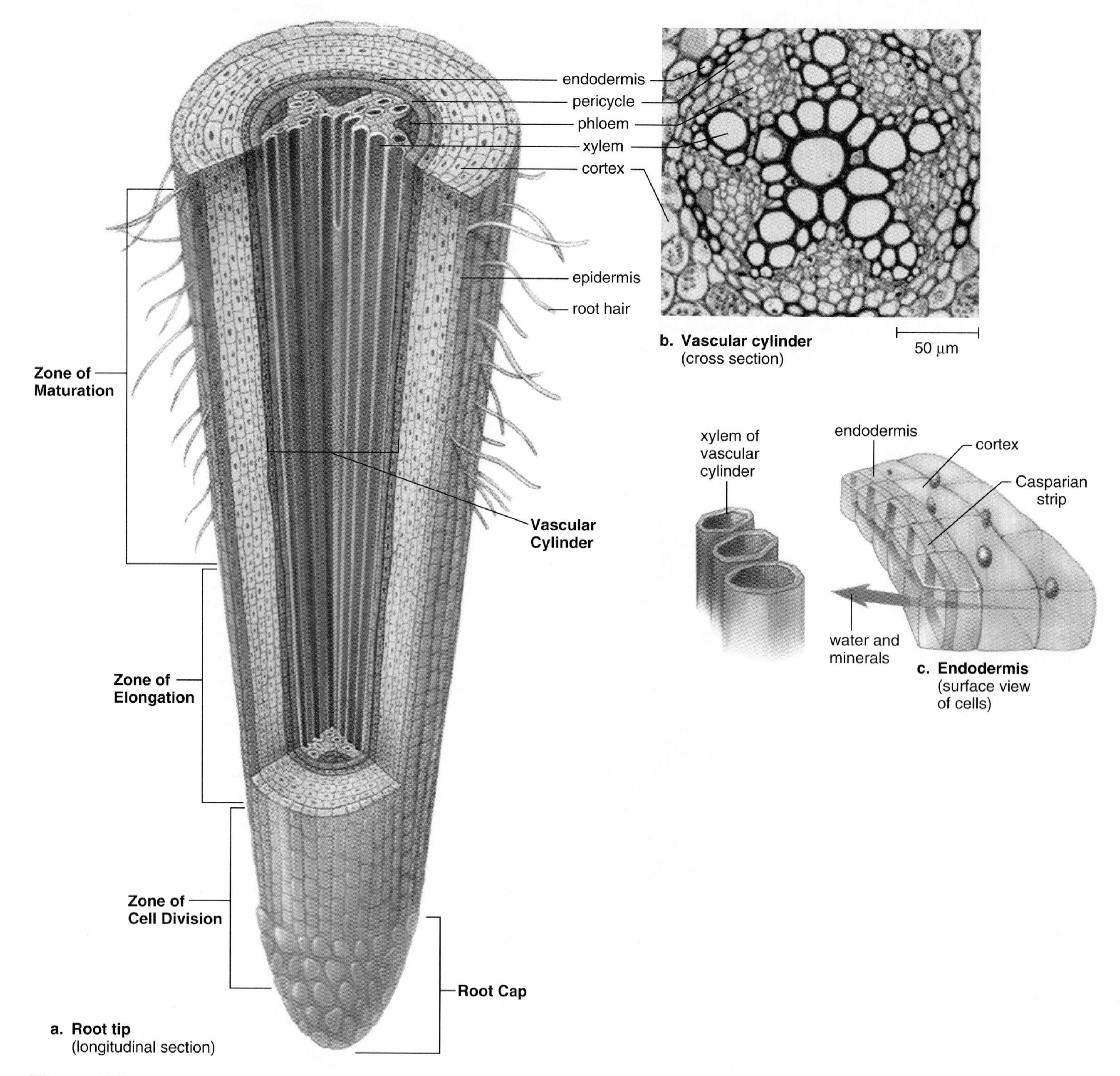

Figure 9.7 Dicot root tip.
a. The root tip is divided into three zones, best seen in a longitudinal section such as this. **b.** Cross section of root tip. The vascular cylinder contains the vascular tissue. Xylem is typically star-shaped, and phloem lies between the points of the star. **c.** Endodermis showing the Casparian strip, a layer of impermeable lignin and suberin in cell walls. Because of the Casparian strip, water and minerals must pass through the cytoplasm of endodermal cells. In this way, endodermal cells regulate the passage of minerals into the vascular cylinder.

Zones of a Dicot Root

Figure 9.7, a longitudinal section of a dicot root, reveals zones where cells are in various stages of differentiation as *primary growth* occurs. The root **apical meristem** is in the zone of cell division. Here cells are continuously added to the root cap below and the zone of elongation above. The **root cap** is a protective cover for the root tip. The cells in the root cap have to be replaced constantly, because they are ground off as the root pushes through rough soil particles. In the *zone of elongation,* the cells become longer as they become specialized. In the *zone of maturation,* the cells become mature and fully differentiated. This zone is recognizable even in a whole root because root hairs project from many epidermal cells.

Tissues of a Dicot Root

Figure 9.7 also shows a cross section of a root at the zone of maturation. The **vascular cylinder** consists of conductive tissues and one or more layers of cells called the **pericycle.** These specialized tissues are identifiable:

Epidermis The epidermis, which forms the outer layer of the root, consists of only a single layer of thin-walled and rectangular cells. In the zone of maturation, many epidermal cells have root hairs that project as far as 5–8 mm into the soil particles.

Cortex Moving inward, large, thin-walled parenchyma cells make up the **cortex,** which functions in food storage. Here, irregularly shaped cells are loosely packed and contain starch granules.

Endodermis The **endodermis** is a single layer of rectangular endodermal cells that form the innermost layer of the cortex. These cells fit snugly together and are bordered on four sides (but not the two sides that contact the cortex and the vascular cylinder) by a layer of impermeable lignin and suberin known as the **Casparian strip** (Fig. 9.7*c*). This strip does not permit water and mineral ions to pass between adjacent cell walls. Therefore, the only access to the vascular cylinder is through the endodermal cells themselves, as shown by the arrow in Figure 9.7*c*. It is said that the endodermis regulates the entrance of minerals and water into the vascular cylinder.

Vascular Tissue The pericycle, the first layers of cells within the vascular cylinder, has retained its capacity to divide and can start the development of branch (or secondary) roots (Fig. 9.8). The main portion of the vascular cylinder, though, contains vascular tissue. The xylem appears star-shaped in dicots because several arms of tissue radiate from a common center (Fig. 9.7*b*). The phloem is found in separate regions between the arms of the xylem.

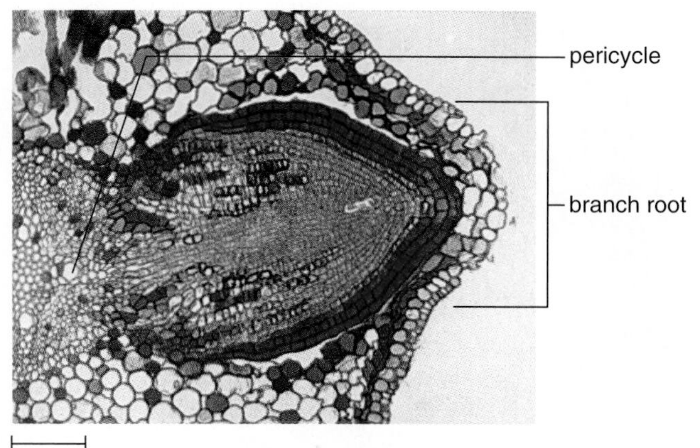

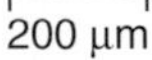

Figure 9.8 Organization of branch root.
Branch roots originate from the pericycle, a band of cells located inside the endodermis of the vascular cylinder.

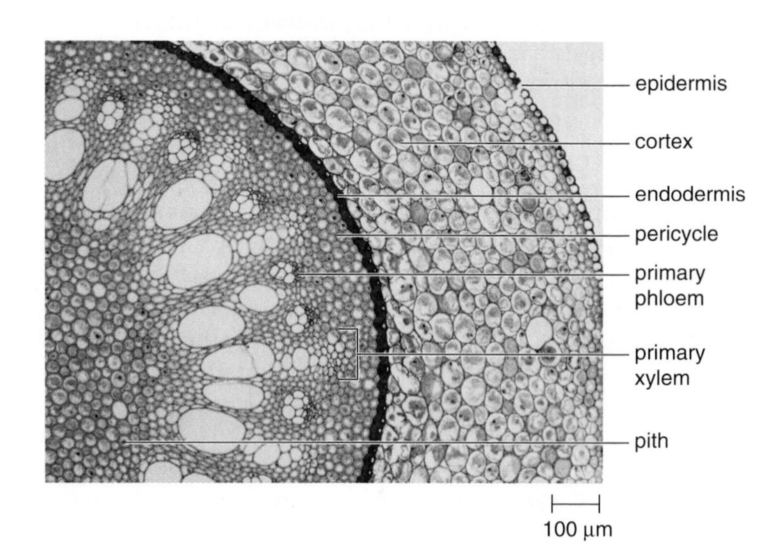

Figure 9.9 Cross section of monocot root.
In monocot roots, a vascular ring surrounds a central pith.

Vascular Ring of a Monocot Root

Monocot roots have the same growth zones as a dicot root, but the organization of their tissues is slightly different. In a monocot root, **pith,** a ground tissue, is surrounded by a vascular ring composed of alternating xylem and phloem tissues (Fig. 9.9). They also have pericycle, endodermis, cortex, and epidermis.

In the root of a dicot, the endodermis surrounds the vascular cylinder. In the root of a monocot, a vascular ring surrounds a central pith.

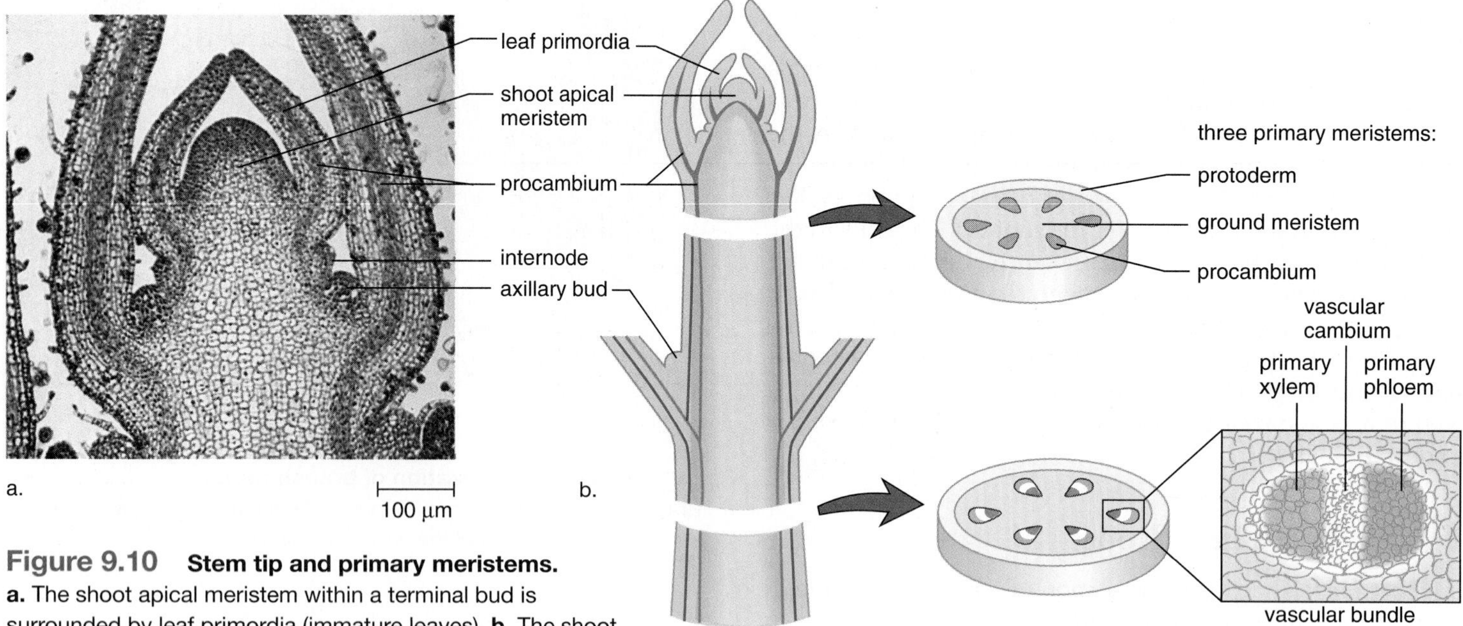

Figure 9.10 Stem tip and primary meristems. **a.** The shoot apical meristem within a terminal bud is surrounded by leaf primordia (immature leaves). **b.** The shoot apical meristem produces the primary meristems: protoderm gives rise to epidermis, ground meristem gives rise to pith and cortex, and procambium gives rise to xylem and phloem.

9.4 Organization of Stems

A stem supports leaves, flowers, and fruits; it conducts substances to and from the roots and the leaves; and it helps to store water and the products of photosynthesis. Growth of a stem can be compared to the growth of a root. During primary growth, the apical meristem at the stem tip called *shoot apical meristem* produces new cells that elongate and thereby increase the length of the stem. The shoot apical meristem, however, is protected within a terminal bud by **leaf primordia** (immature leaves), which envelop it (Fig. 9.10*a*). In the temperate zone, a terminal bud stops growing in the winter and is then protected by modified leaves called bud scales. In the spring, when growth resumes, these scales fall off and leave scars. The bud scale scars form a ring around the stem. You can tell the age of a stem by counting the bud scale scar rings.

Leaf primordia are produced by the apical meristem at regular intervals called nodes. The portion of a stem between two sequential nodes is called an internode. As a stem grows, the internodes increase in length. **Axillary buds,** which are usually dormant but may develop into branch shoots or flowers, are seen at the axes of the leaf primordia.

In addition to leaf primordia, three specialized types of primary meristem develop from shoot apical meristem (Fig. 9.10*b*). These primary meristems contribute to the length of a shoot. The *protoderm,* the outermost primary meristem, gives rise to epidermis. The *ground meristem* produces two tissues composed mostly of parenchyma cells—pith occurs in the center of the stem and cortex is located inside the epidermis. The *procambium,* seen as an obvious strand of tissue in Figure 9.10*a*, produces the first xylem cells, called **primary xylem,** and the first phloem cells, called **primary phloem.** Differentiation continues as certain cells become the first tracheids or vessel elements of the xylem within a vascular bundle. The first sieve-tube elements of the phloem do not have companion cells and are short-lived (some live only a day before being replaced). Mature phloem, consisting of sieve-tube elements and companion cells, then develops. **Vascular cambium,** a type of meristem tissue, occurs between the xylem and phloem of a vascular bundle.

Herbaceous Stems

Mature nonwoody stems, called **herbaceous stems,** exhibit only primary growth. The outermost tissue of herbaceous stems is the epidermis, which is covered by a waxy cuticle to prevent water loss. These stems have distinctive *vascular bundles,* where xylem and phloem are found. In each bundle, xylem is typically found toward the inside of the stem and phloem is found toward the outside.

In the herbaceous dicot stem, the vascular bundles are arranged in a distinct ring that separates the cortex from the central pith (Fig. 9.11). The cortex is sometimes green and carries on photosynthesis, and the pith may function as a storage site. In the monocot stem, the vascular bundles are scattered throughout the stem, and there is no well-defined cortex or pith (Fig. 9.12).

As a stem grows, the shoot apical meristem produces new leaves and primary meristems. The primary meristems produce the other tissues found in herbaceous stems.

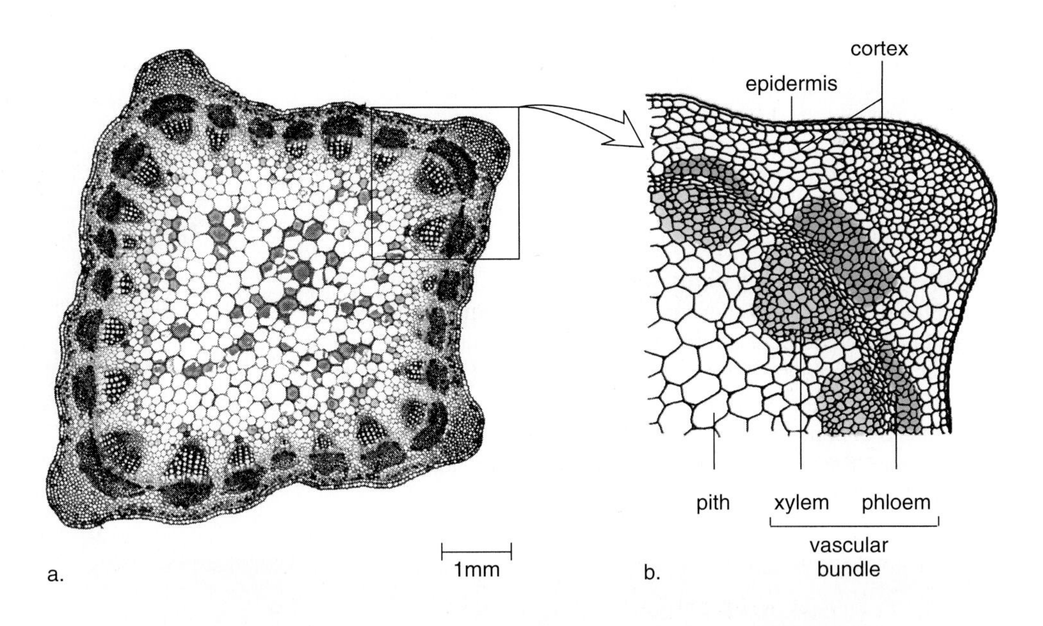

Figure 9.11 Herbaceous dicot stem anatomy.
a. A cross section of an alfalfa stem shows that the vascular bundles are in a ring. **b.** The drawing of a section of the stem identifies the tissues in the vascular bundle and the stem.

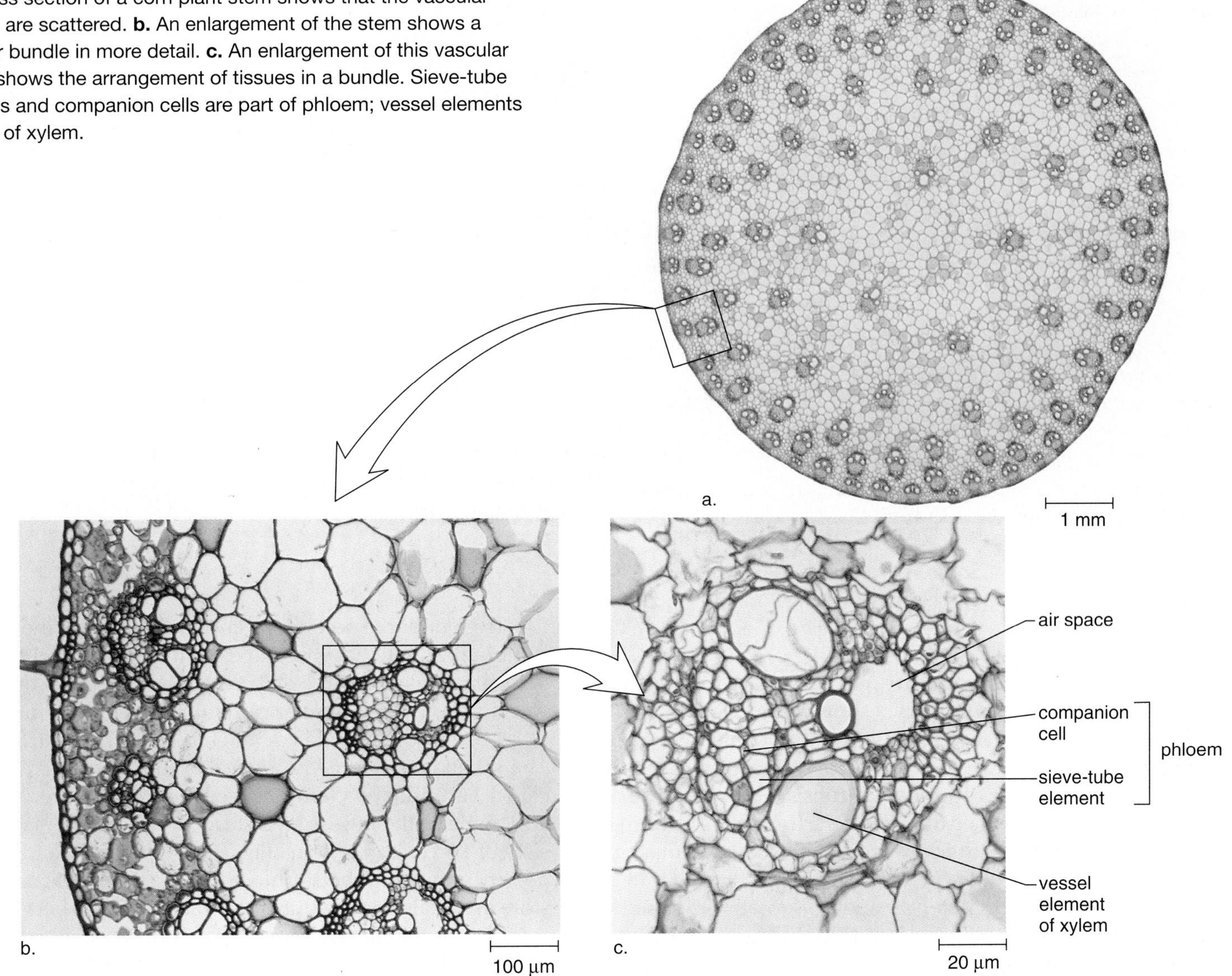

Figure 9.12 Monocot stem anatomy.
a. A cross section of a corn plant stem shows that the vascular bundles are scattered. **b.** An enlargement of the stem shows a vascular bundle in more detail. **c.** An enlargement of this vascular bundle shows the arrangement of tissues in a bundle. Sieve-tube elements and companion cells are part of phloem; vessel elements are part of xylem.

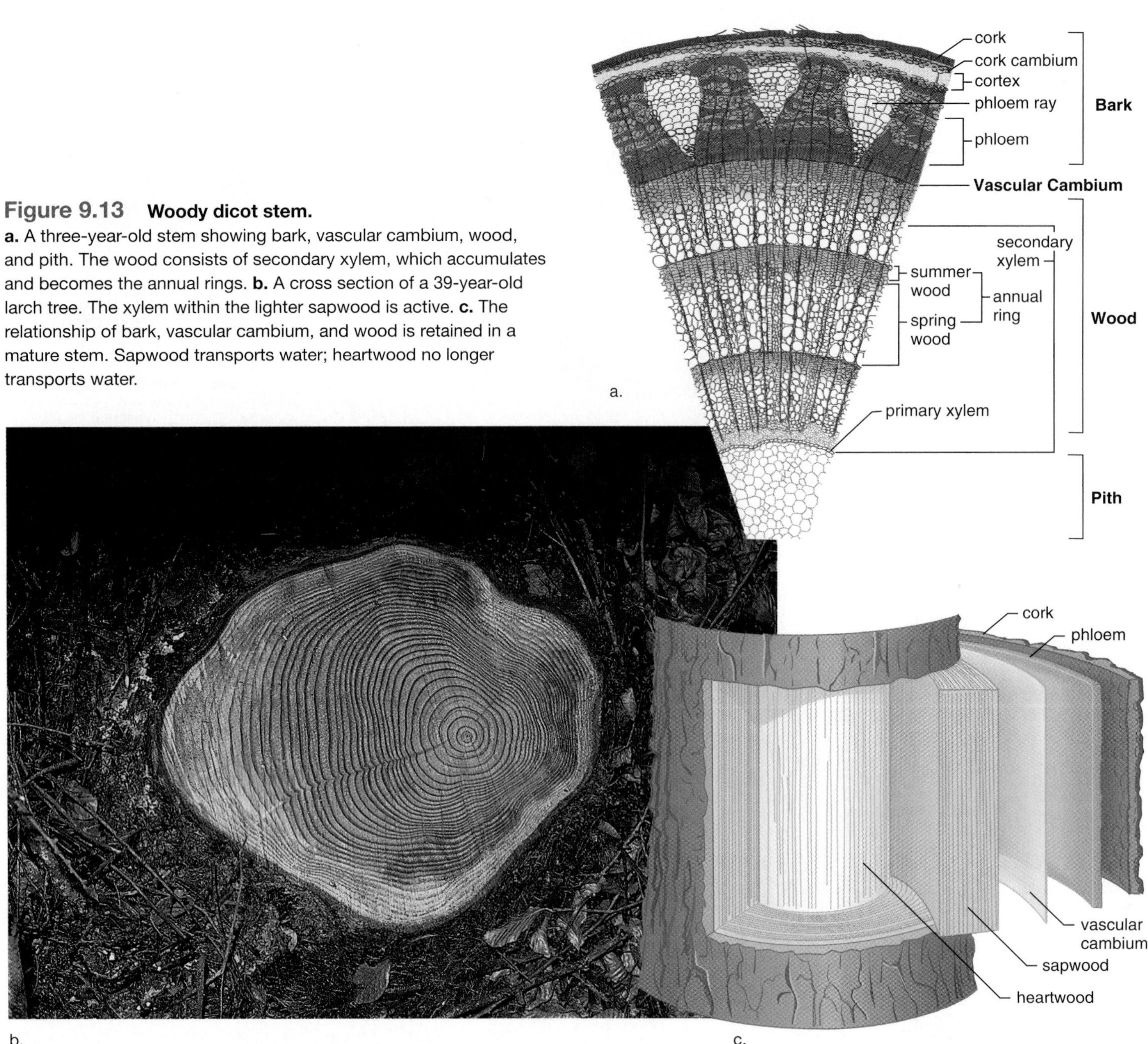

Figure 9.13 Woody dicot stem.
a. A three-year-old stem showing bark, vascular cambium, wood, and pith. The wood consists of secondary xylem, which accumulates and becomes the annual rings. **b.** A cross section of a 39-year-old larch tree. The xylem within the lighter sapwood is active. **c.** The relationship of bark, vascular cambium, and wood is retained in a mature stem. Sapwood transports water; heartwood no longer transports water.

Woody Stems

A woody plant has both primary and secondary tissues. Primary growth continues for only a short distance behind apical meristem; then secondary growth begins. **Primary growth** increases the length of a plant and **secondary growth** increases its girth. Trees and some shrubs are woody. Shrubs, and almost all trees other than conifers, are dicot flowering plants. As a result of secondary growth, a woody dicot stem has an entirely different type of organization than that of a herbaceous dicot stem. After secondary growth has continued for a time, it is no longer possible to make out individual vascular bundles. Instead, a woody stem has three distinct areas: the bark, the wood, and the pith (Fig. 9.13*a*).

Trees undergo secondary growth because of the vascular cambium. Recall that vascular cambium is a type of meristem between the xylem and phloem of each vascular bundle. In woody plants, the vascular cambium develops to form a ring of meristem that divides parallel to the surface of the plant. The secondary tissues produced by the vascular cambium, called **secondary xylem** and **secondary phloem,** therefore add to the girth of the stem instead of to its length. **Wood** consists of secondary xylem (Fig. 9.14). Another change is the presence of rays—both phloem rays and xylem rays—that store materials and conduct them radially for a short distance.

In trees that have a growing season, vascular cambium is dormant during the winter. In the spring, when moisture is plentiful, the xylem contains wide vessel elements with thin

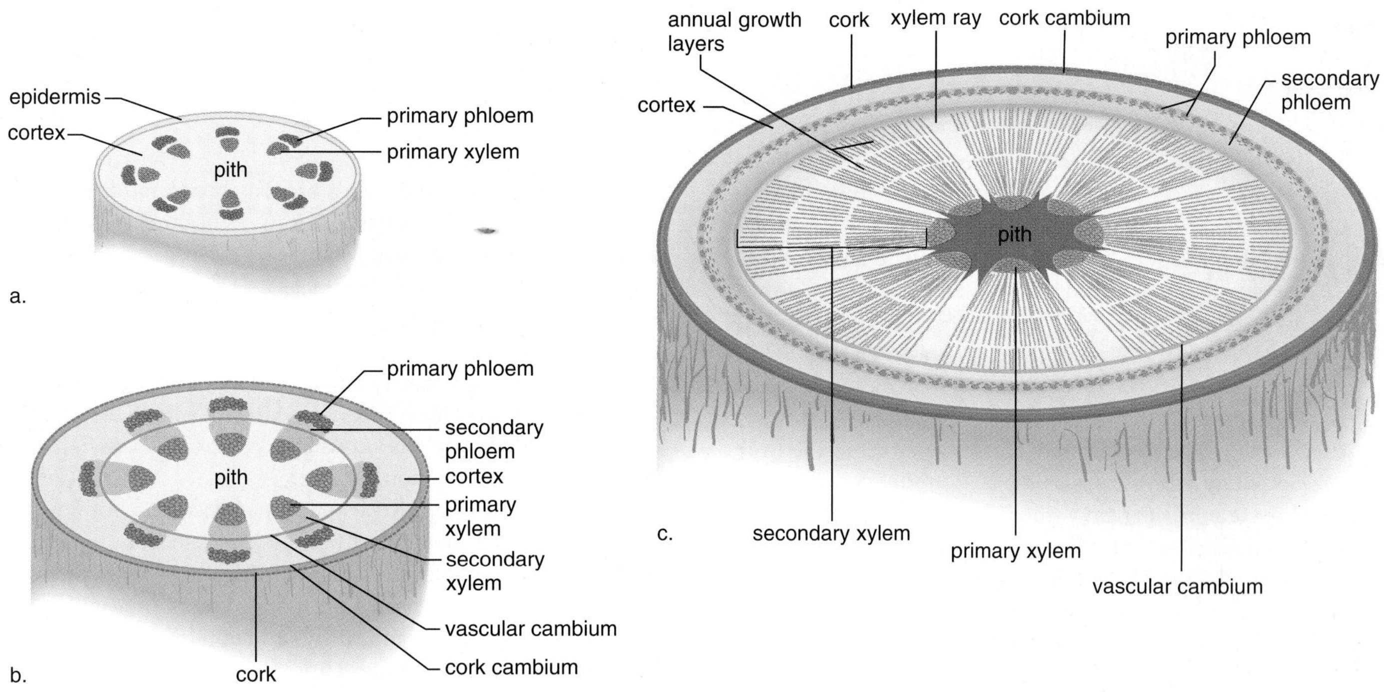

Figure 9.14 Secondary growth in a stem.
a. A dicot herbaceous stem before secondary growth begins. **b.** Secondary growth has begun. Cork replaces the epidermis. Vascular cambium produces secondary xylem and secondary phloem. **c.** A three-year-old stem. Cork cambium produces new cork. The primary phloem and cortex will eventually disappear and only the secondary phloem (within the bark), produced by vascular cambium, will be active that year. The secondary xylem, also produced by vascular cambium, builds up to become the annual rings.

walls. In this so-called spring wood or early wood, wide vessels transport sufficient water to the growing leaves. Later in the season, when moisture is scarce, the wood at this time, called summer wood or late wood, has a lower proportion of vessels. Instead, summer wood contains numerous fibers and thick-walled tracheids. At the end of the growing season, just before the cambium becomes dormant again, only heavy fibers with especially thick secondary walls may develop. When the trunk of a tree has spring wood followed by summer wood, the two together make up one year's growth, or an **annual ring** (Fig. 9.13*a*). You can tell the age of a tree by counting the annual rings. The outer annual rings, where most transport occurs, is called sapwood (Fig. 9.13*b*, *c*).

In older trees, the inner annual rings are called the heartwood. The cells become plugged with deposits, such as resins, gums, and other substances that inhibit the growth of bacteria and fungi. Heartwood may help support a tree, although some trees stand erect and live for many years after the heartwood has rotted away.

The **bark** of a tree contains cork, cork cambium, and phloem. Although secondary phloem is produced each year by vascular cambium, the soft cells are crushed, and therefore phloem does not build up for many seasons. The phloem tissue is soft, making it possible to remove the bark of a tree; however, this is very harmful because without phloem there is no conduction of organic nutrients.

Cork cambium is meristem located beneath the epidermis. When cork cambium begins to divide, it produces tissue that disrupts the epidermis and replaces it with cork cells. Cork cells, you will recall, are impregnated with suberin, a lipid material that makes them waterproof but also causes them to die. This is protective because now there is nothing nutritious for an animal to eat. But an impervious barrier means that gas exchange is impeded except at *lenticels*, which are pockets of loosely arranged cork cells not impregnated with suberin.

The first flowering plants to evolve may have been woody shrubs, and if so, herbaceous plants evolved later. Is it advantageous to be woody? If there is adequate rainfall, woody plants can grow taller and have more growth because they have adequate vascular tissue to support and service leaves. However, it takes energy to produce secondary growth and prepare the body for winter if the plant lives in the temperate zone. Also, there is a greater need for defense mechanisms, because a long-lasting plant that stays in one spot is likely to be attacked by herbivores and parasites. Then, too, trees don't usually reproduce until they have grown several seasons, by which time they may have succumbed to an accident or disease. Perhaps it is more advantageous for a plant to put most of its energy into simply reproducing rather than being woody.

Woody plants grow in girth due to the presence of vascular cambium and cork cambium. Their bodies have three main parts: bark (which contains cork, cork cambium, and phloem), wood (which contains xylem), and pith.

Ecology Focus

Paper: Can We Cut Back?

The word *paper* takes its origin from papyrus, the plant Egyptians used to make the first form of paper. The Egyptians manually made sheets from the treated stems of papyrus grass and then strung them together into scrolls. From that beginning some 5,500 years ago, the production of paper is now a worldwide industry of major importance. The process is fairly simple. Plant material is ground up mechanically and chemically treated to form a pulp that contains "fibers," which biologists know are the tracheids and vessel elements of a plant. The fibers automatically form a sheet when they are screened from the pulp. Today a revolving wire-screen belt is used to deliver a continuous wet sheet of paper to heavy rollers and heated cylinders that remove most of the remaining moisture and press the paper flat.

Each person in the United States consumes about 318 kilograms (699 pounds) of paper products per year, compared to only 2.3 kilograms (5 pounds) of paper per person in India. We are all aware that products in the United States are overpackaged and that the overabundance of our junk mail ends up in the trash almost immediately. Schools and businesses use far too much writing paper—even using two sides of a sheet instead of one side would be helpful!

Although eucalyptus plants from South America, bamboo from India, and even cotton are used to make paper, most paper is made from trees, with severe ecological consequences. Trees are sometimes clear-cut from natural forests, and if so, it will be many years before the forests will regrow. In the meantime, the community of organisms that depends on the forest ecosystem must relocate or die off. Natural ecosystems the world over have been replaced by giant tree plantations containing stands of uniform trees to serve as a source of wood. In Canada, there are temperate hardwood tree plantations of birch, beech, chestnut, poplar, and particularly aspen trees. Tropical hardwoods, which usually come from Southeast Asia and South America, are also sometimes used to make paper. In the United States, several species of pine trees have been genetically improved to have a higher wood density and to be harvestable five years earlier than ordinary pines. Southern Africa, Chile, New Zealand, and Australia also devote thousands of acres to growing pines for paper pulp production.

The making of paper uses energy and causes both air and water pollution. Caustic chemicals such as sodium hydroxide, sulfurous acid, and bleaches are used during the manufacturing process. These are released into the air and, along with paper mill wastes, also add significantly to the pollution of rivers and streams. Underground water supplies are poisoned by the ink left in the ground after paper biodegrades in landfills.

It is clear that we should all cut back on our use of paper so that less of it is made in the first place. In addition, we should recycle the paper that has already been made. When newspaper and office paper (including photocopies) are soaked in water, the fibers are released, and they can be used to make recycled paper and/or cardboard. Manufacturing recycled paper uses less energy and causes less pollution than making paper anew. Moreover, trees are conserved. It's estimated that recycling of Sunday newspapers alone would save an estimated 500,000 trees each week.

Figure 9A Paper production.
Machine No. 35 at Champion International's Courtland, Alabama, mill produces a 29-foot-wide roll of office paper every 60 minutes.

9.5 Organization of Leaves

Leaves are the organs of photosynthesis in vascular plants. As mentioned earlier, a leaf usually consists of a flattened **blade** and a **petiole** connecting the blade to the stem. The blade may be simple, or compound composed of several leaflets. Externally, it is possible to see the pattern of the *leaf veins,* which contain vascular tissue. Leaf veins have a net pattern in dicot leaves and a parallel pattern in monocot leaves (see Fig. 9.2).

Figure 9.15 shows a cross section of a typical dicot leaf of a temperate zone plant. At the top and bottom is a layer of epidermal tissue that often bears protective hairs and/or glands that produce irritating substances. These features may prevent the leaf from being eaten by insects. The epidermis characteristically has an outer, waxy **cuticle** that keeps the leaf from drying out. Unfortunately, the cuticle also prevents gas exchange because it is not gas permeable. However, the epidermis, particularly the lower epidermis, contains stomates that allow gases to move into and out of the leaf. Each stomate has two guard cells that regulate its opening and closing.

The body of a leaf is composed of **mesophyll** tissue, which has two distinct regions: **palisade mesophyll,** containing elongated cells, and **spongy mesophyll,** containing irregular cells bounded by air spaces. The parenchyma cells of these layers have many chloroplasts and carry on most of the photosynthesis for the plant. The loosely packed arrangement of the cells in the spongy layer increases the amount of surface area for gas exchange.

How does a leaf acquire the inputs for, and what happens to the outputs of, photosynthesis? A flattened blade exposes a wide surface area to the sun, and this facilitates the absorption of solar energy by chlorophyll pigments within chloroplasts. Water enters vascular tissue at the roots and is pulled upward to the leaves, where it exits at leaf veins. Thereafter, water enters leaf cells and chloroplasts.

The stomates within the lower epidermis allow carbon dioxide to enter the many air spaces of the spongy mesophyll. The palisade mesophyll carries on most of the photosynthesis. The tightly packed cells of the palisade mesophyll are specialized to absorb solar energy and fix carbon dioxide before it is reduced to a carbohydrate.

Carbohydrate produced by a leaf can be stored for a while, but then is transported in vascular tissue to all plant parts, where it is used as an energy source by mitochondria, which produce ATP molecules. All cells need a supply of ATP for metabolic purposes. Carbohydrate unneeded by plant cells is transported to the roots, where it is stored in plastids.

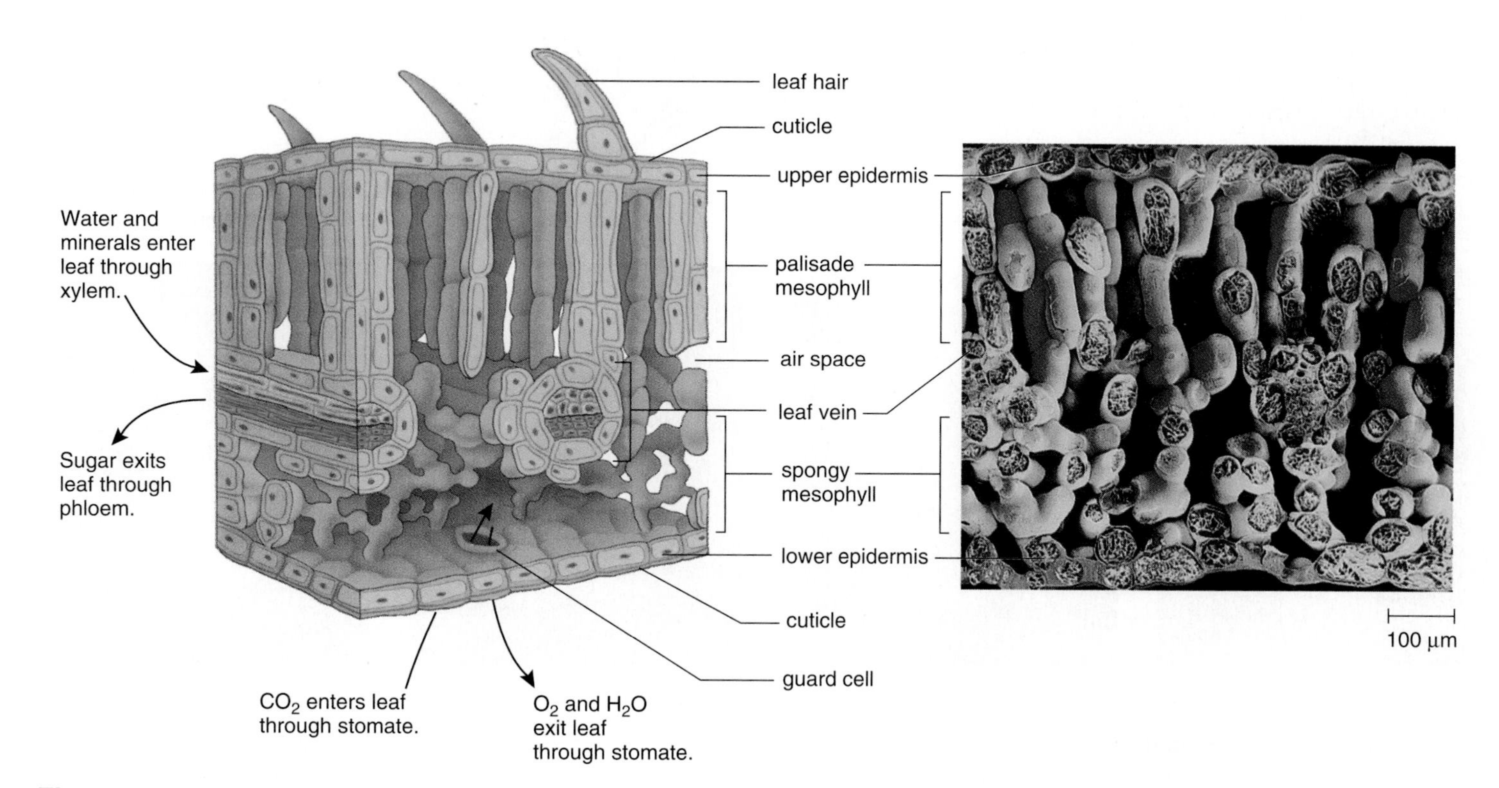

Figure 9.15 Leaf structure.
Photosynthesis takes place in the mesophyll tissue, which consists of palisade and spongy mesophyll. The veins contain xylem and phloem for the transport of water and solutes. The leaf is enclosed by epidermal cells covered with a waxy layer, the cuticle. The leaf hairs are also protective. A stomate is an opening in the lower epidermis that permits the exchange of gases.

9.6 Modified Roots, Stems, and Leaves

Roots are adapted for various functions. Some monocot plants have a **fibrous root system** consisting of a large number of slender roots. Most everyone has observed the fibrous root system of grasses and has noted how these types of roots can hold the soil. Other plants have a first or **primary root** that grows straight down and remains the dominant root of the plant. This so-called **taproot** is often fleshy and stores food. Carrots, beets, turnips, and radishes are taproots that we consume as vegetables.

Stem modifications are illustrated in Figure 9.16. Aboveground horizontal stems, called **stolons** or **runners,** produce new plants where nodes touch the ground. The strawberry plant is a common example of this type of stem. Vertical stems can also be modified. For example, cactuses have succulent stems modified for water storage. The tendrils of grape plants that twine around a support structure are modified stems.

Underground horizontal stems, **rhizomes,** may be long and thin, as in sod-forming grasses, or thick and fleshy, as in an iris. Rhizomes survive the winter and contribute to asexual reproduction because each node bears a bud. Some rhizomes have enlarged portions called tubers, which function in food storage. For example, Irish potatoes are *tubers.* The eyes of potatoes are buds that mark the nodes. *Corms* are bulbous underground stems that lie dormant during the winter, just as rhizomes do. They also produce new plants the next growing season. Gladiolus corms are referred to as bulbs by laypersons, but the botanist reserves the term *bulb* for a structure composed of modified fleshy leaves. An onion bulb is made up of leaves surrounding a short stem. Climbing leaves, such as those of peas and cucumbers, are modified into **tendrils** that can twist

Figure 9.16 Stem modifications.
a. A strawberry plant has aboveground, horizontal stems called stolons. Every other node produces a new shoot system. **b.** The underground horizontal stem of an iris is a fleshy rhizome. **c.** The underground stem of a potato plant has enlargements called tubers. We call the tubers potatoes. **d.** The corm of a gladiolus is a stem covered by papery leaves.

a. Cactus, *Opuntia*

b. Cucumber, *Cucumis*

c. Venus's-flytrap, *Dionaea*

Figure 9.17 Leaf modifications.
a. The spines of a cactus plant are modified leaves that protect the fleshy stem from animal consumption. **b.** The tendrils of a cucumber plant are modified leaves that twist around a physical support. **c.** The leaves of the Venus's-flytrap serve as a trap for prey such as an insect. When triggered by prey, the leaf snaps shut. Once shut, the leaf secretes digestive juices, which break down the soft parts of the prey's body

around nearby objects (Fig. 9.17*b*). The leaves of a few plants are specialized for catching prey such as insects. The leaves of a sundew have sticky epidermal hairs that trap insects and then secrete digestive enzymes. The Venus's-flytrap has hinged leaves that snap shut and interlock when an insect triggers sensitive hairs (Fig. 9.17*c*). The leaves of a pitcher plant resemble a pitcher and have downward-pointing hairs that lead insects into a pool of digestive enzymes. Insectivorous plants commonly grow in marshy regions, where the supply of soil nitrogen is severely limited. The digested insects provide the plants with a source of organic nitrogen.

Leaves are adapted to environmental conditions. Shade plants tend to have broad, wide leaves, and desert plants tend to have small leaves with sunken stomates. The leaves of a cactus are the spines attached to fleshy stem (Fig. 9.17*a*).

Bioethical Issue

Marbled murrelets are endangered seabirds that nest in huge redwood trees along the coast of the Pacific northwest. Pacific Lumber was given the right to fell many acres in exchange for a few that would be preserved. The government paid the owner of Pacific Lumber some $500 million so that 3,500 acres would be preserved. Activists thought the deal was unfair—they wanted 60,000 acres preserved. Any less and there would not be enough habitat to keep the birds from sinking into oblivion. Not to mention that the trees themselves cannot regrow to their present size for hundreds upon hundreds of years.

Activists feel done in by their own representatives in the government. After all, the owner of Pacific Lumber is a big contributor to the reelection campaigns of elected officials who approved the deal. Under the circumstances, what would you do to save the trees? Some activists climbed the trees and refused to come down even when the trees were being cut. In September of 1998, David Chain, 24, lost his life when a tree fell on him and crushed his skull.

What is the proper action for activist groups when they feel their government is letting them down? Should they defy the law, and if so, what is the proper response of governmental officials? Lawyers and legislators were supposed to be reconsidering the deal, but in the meantime Pacific Lumber did not stop cutting the trees. Who should be arrested—activists, Pacific Lumber CEOs, or both?

Questions

1. How much of our forests should be preserved—small token amounts or huge tracts that could make a difference?
2. What do you think of activists who take their lives into their own hands in order to preserve the environment? Are they any different from other types of terrorists?
3. How can the government balance the need for a healthy economy and the need to preserve what is left of our natural environment?

Table 9.1 Vegetative Organs and Major Tissues

	Root	Stem	Leaf
Function	Absorbs water and minerals Anchors plant Stores materials	Transports water and nutrients Supports leaves Helps store materials	Carries on photosynthesis
Tissue			
Epidermis*	Protects inner tissues Root hairs absorb water and minerals	Protects inner tissues	Protects inner tissues Stomates carry on gas exchange
Cortex†	Stores water and products of photosynthesis	Carries on photosynthesis, if green Some storage of products of photosynthesis	Not present
Endodermis†	Regulates passage of water and minerals in vascular tissue	Not present	Not present
Vascular‡	Transports water and nutrients	Transports water and nutrients	Transports water and nutrients
Pith†	Stores products of photosynthesis and water	Stores products of photosynthesis	Not present
Mesophyll†	Not present	Not present	Primary site of photosynthesis

Note: Plant tissues belong to one of three tissue systems:
*Dermal tissue system
†Ground tissue system
‡Vascular tissue system

Summarizing the Concepts

9.1 The Flowering Plant

A flowering plant has three vegetative organs. Roots anchor plants, absorb water and minerals, and store the products of photosynthesis. Stems support leaves, conduct materials to and from roots and leaves, and help store plant products. Leaves carry on photosynthesis.

Flowering plants are classified into dicots and monocots according to the number of cotyledons in the seed, the arrangement of vascular tissue in root, stems, and leaves, and the number of flower parts.

9.2 Plant Tissues and Cells

Dermal tissue consists of epidermis, which is modified in different organs of the plant. In the roots, epidermal cells bear root hairs; in the leaves, the epidermis contains guard cells. Cork replaces epidermis in woody plants.

Ground tissue contains parenchyma cells, which are thin walled and capable of photosynthesis when they contain chloroplasts. If they contain only colorless plastids, they serve as storage cells. Collenchyma cells have thicker walls for flexible support. Sclerenchyma cells are hollow, nonliving support cells with thick secondary walls.

Vascular tissue consists of xylem and phloem. Xylem contains vessel elements that have no end walls and tracheids that are tapered, with pitted end walls. Neither contains cytoplasm or nuclei. Xylem transports water and minerals. Phloem contains sieve-tube elements, each of which has a companion cell. Phloem transports organic nutrients.

9.3 Organization of Roots

A root tip shows three zones: the zone of cell division (apical meristem) protected by the root cap, the zone of elongation, and the zone of maturation. A cross section of a herbaceous dicot root reveals the epidermis (protects), the cortex (stores food), the endodermis (regulates the movement of water and minerals), and the vascular cylinder (transports) (Table 9.1). In the vascular cylinder of a dicot, the xylem appears star shaped, and the phloem is found in separate regions between the arms of the xylem. In contrast, a monocot root has a ring of vascular tissue with alternating bundles of xylem and phloem surrounding a pith.

9.4 Organization of Stems

Primary growth of a stem is due to the activity of the shoot apical meristem, which is protected within a terminal bud. A terminal bud contains leaf primordia at nodes. Nodes are separated by internodes. Lengthening of internodes results in stem growth. In cross section, the tissues of a herbaceous dicot are an outer epidermis, cortex, vascular bundles in a ring, and an inner pith. Monocot stems have scattered vascular bundles, and the cortex and pith are not well defined.

Secondary growth of a woody dicot stem is due to vascular cambium, which produces new xylem and phloem every year, and cork cambium, which produces new cork cells when needed. (Cork replaces epidermis in woody plants.) The cross section of a woody stem is divided into bark, wood, and pith. The bark contains cork and phloem. Wood contains annual rings of xylem. Pith is a ground tissue.

9.5 Organization of Leaves

A cross section of a leaf reveals the upper and lower epidermis, with stomates mostly in the lower epidermis. Vascular tissue is present within leaf veins. Leaves, which carry on photosynthesis, absorb solar energy, receive CO_2 by way of stomates, and receive H_2O by way of leaf veins. Carbohydrate is transported away from a leaf by vascular tissue, and O_2 exits at the stomates (Table 9.1).

9.6 Modified Roots, Stems, and Leaves

While fibrous roots have many fine branches, taproots have an enlarged central portion, sometimes modified for storage. Among various modifications, some stems are horizontal, with some occurring above the ground and others below the ground. Some leaves are modified according to the environment, as in cactuses, whose leaves are spines. Others have specific functions; the Venus's-flytrap even has leaves that trap and digest prey.

Studying the Concepts

1. Name and discuss the vegetative organs of a plant. 152
2. List four differences between monocots and dicots. 153
3. Give a function for epidermal tissue and tell how it is modified in roots, stems, and leaves. 154
4. Contrast the structure and function of parenchyma, collenchyma, and sclerenchyma cells. These cells occur in what type of plant tissue? 154
5. Contrast the structure and function of xylem and phloem. Xylem and phloem occur in what type of plant tissue? 155–56
6. Name the zones of a root tip and tell how each zone contributes to growth in length. 156–57
7. Describe the anatomy of a dicot root tip, both longitudinal and cross-sectional. How does the anatomy of a monocot root differ from this? 157
8. Describe the primary and secondary growth of a dicot stem. 160
9. Describe the anatomy of a monocot, a herbaceous dicot, and a woody stem. 159–60
10. Describe the anatomy of a typical dicot leaf and a typical monocot leaf in cross section. 163
11. Describe how a leaf acquires the reactants of photosynthesis and what it does with the products. 163
12. Describe ways in which roots, stems, and leaves are modified. 164–65

Testing Yourself

Choose the best answer for each question.

1. Which of these is an incorrect contrast between monocots and dicots?
   monocots—dicots
   a. one cotyledon—two cotyledons
   b. leaf veins parallel—net veined
   c. vascular bundles in a ring—vascular bundles scattered
   d. All of these are incorrect.
2. Which of these types of cells is most likely to divide?
   a. parenchyma
   b. meristem
   c. epidermis
   d. xylem
3. Which of these cells in a plant is apt to be nonliving?
   a. parenchyma
   b. collenchyma
   c. sclerenchyma
   d. epidermal
4. Root hairs are found in the zone of
   a. cell division.
   b. elongation.
   c. maturation.
   d. All of these are correct.
5. Cortex is found in
   a. roots, stems, and leaves.
   b. roots and stems.
   c. roots only.
   d. stems only.
6. Between the bark and the wood in a woody stem, there is a layer of meristem called
   a. cork cambium.
   b. vascular cambium.
   c. apical meristem.
   d. the zone of cell division.
7. Which part of a leaf carries on most of the photosynthesis of a plant?
   a. epidermis
   b. mesophyll
   c. epidermal layer
   d. guard cells
8. Annual rings are the number of
   a. internodes in a stem.
   b. rings of vascular bundles in a monocot stem.
   c. layers of secondary xylem in a stem.
   d. Both b and c are correct.
9. The Casparian strip is found
   a. between all epidermal cells.
   b. between xylem and phloem cells.
   c. on four sides of endodermal cells.
   d. within the secondary wall of parenchyma cells.
10. Which of these is a stem?
    a. taproot of carrots
    b. stolon of strawberry plants
    c. spine of cacti
    d. Both b and c are correct.
11. Label this root using the terms endodermis, phloem, xylem, cortex, and epidermis.

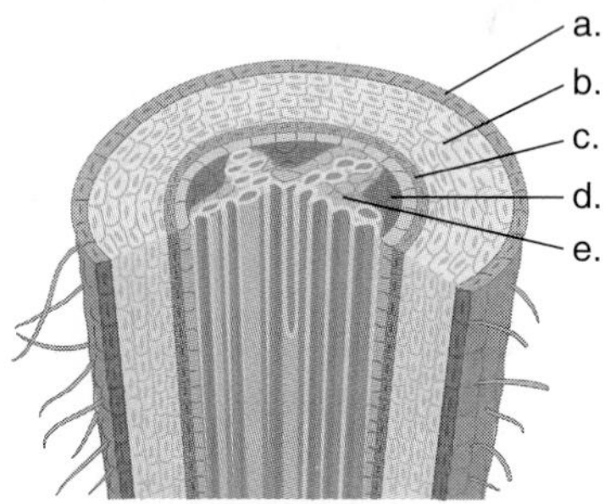
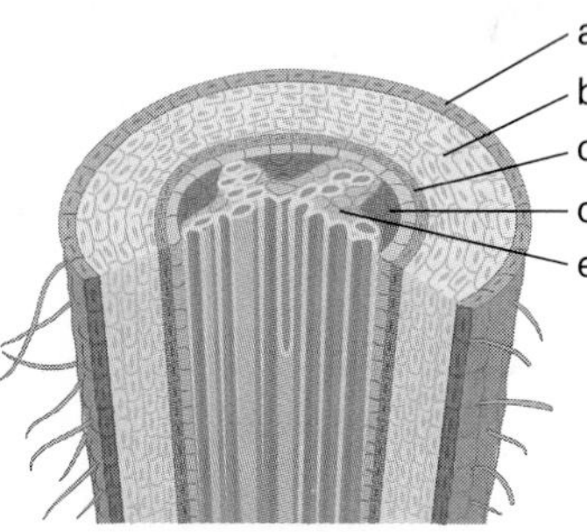

12. Label this woody stem using the terms annual ring, bark, cork, phloem, pith, vascular cambium, and xylem (wood):

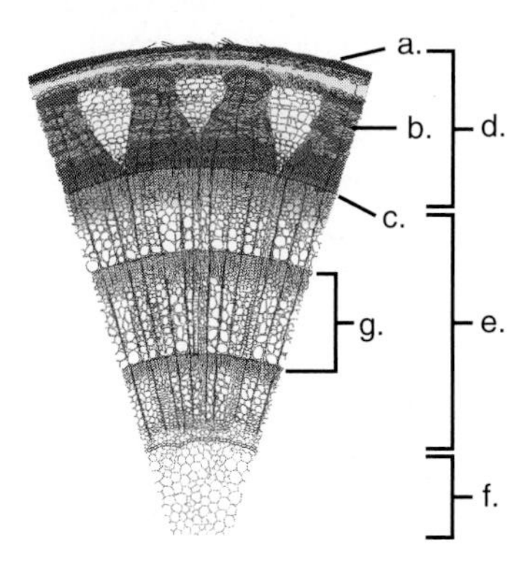

13. Label this leaf using the terms leaf vein, lower epidermis, palisade mesophyll, spongy mesophyll, upper epidermis.

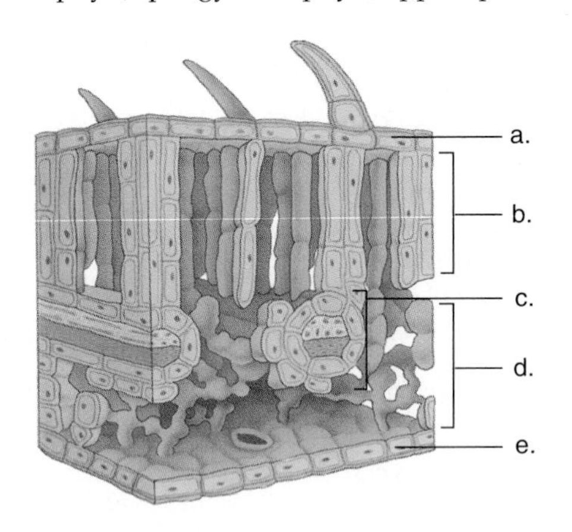

Thinking Scientifically

1. Considering the growth of plants:
   a. Why can plants grow an entire lifetime? (page 158)
   b. How does the growth of a woody plant differ from the growth of a human, with regard to height? With regard to new parts? (page 160)
   c. Of what benefit is the wood of a woody plant with regard to growth? (page 160)
   d. Which parts of a deciduous plant do not grow larger?
2. Considering the leaves of plants (page 163):
   a. What is the function of leaf epidermis, and how does the structure suit the function?
   b. What is the function of the spongy layer of mesophyll tissue, and how does the structure suit the function?
   c. What is the function of leaf veins, and how does the structure suit the function?
   d. What is the difference in structure between C_3 and C_4 leaves, and how does the structure suit the function?

Understanding the Terms

angiosperm 152
annual ring 161
apical meristem 157
axillary bud 158
bark 161
blade 163
Casparian strip 157
collenchyma cell 154
companion cell 155
cork 154
cork cambium 161
cortex 157
cotyledon 153
cuticle 154
dermal tissue 154
dicot 153
endodermis 157
epidermis 154
fibrous root system 164
ground tissue 154
guard cell 154
herbaceous stem 158
internode 152
leaf 152
leaf primordium 158
meristem 154
mesophyll 163
monocot 153
node 152
palisade mesophyll 163
parenchyma cell 154
pericycle 157
petiole 163
phloem 155
pit 154
pith 157
primary growth 160
primary phloem 158
primary root 164
primary xylem 158
rhizome 164
root cap 157
root hair 154
root system 152
runner 164
sclerenchyma cell 154
secondary growth 160
secondary phloem 160
secondary xylem 160
shoot system 152
sieve plate 155
sieve-tube element 155
spongy mesophyll 163
stem 152
stolon 164
stomate 154
taproot 164
tendril 164
tracheid 154
vascular bundle 153
vascular cambium 158
vascular cylinder 157
vascular tissue 154
vein 153
vessel element 154
wood 160
xylem 154

Match the terms to these definitions:

a. __________ Inner, thickest layer of a leaf consisting of palisade and spongy layers; the site of most photosynthesis.
b. __________ Outer covering of bark of trees; made up of dead cells that may be sloughed off.
c. __________ Extension of a root epidermal cell that increases the surface area for the absorption of water and minerals.
d. __________ Vascular tissue that conducts organic solutes in plants; contains sieve-tube elements and companion cells.
e. __________ In a plant leaf, this layer contains elongated cells with many chloroplasts.

Using Technology

Your study of plant organization and growth is supported by these available technologies:

Essential Study Partner CD-ROM

Plants → Plant Tissues, Plant Organs

Visit the Mader web site for related ESP activities.

Exploring the Internet

The Mader Home Page provides resources and tools as you study this chapter.

http://www.mhhe.com/biosci/genbio/mader

Life Science Animations 3D Video

32 Plant Transport

C H A P T E R 10

Plant Physiology and Reproduction

Flowering plants disperse their seeds in various ways. Milkweed relies on the wind to carry seeds to new locations after they are released from pods.

Chapter Concepts

A Japanese farmer raises a bowl of steaming rice to his lips as a Mexican mother serves freshly prepared tortillas and an American teenager reaches for another slice of white bread. Rice, corn, and wheat are the staples of diets throughout the world. Italians eat pasta, Irish and Germans consume potatoes, and Americans eat bread as a source of carbohydrates, the nutrient that supplies most of our energy needs. But plants also provide fats and minerals. Then, too, plants have cell walls of cellulose that provide much of the roughage in our diets. "Eat your vegetables" is not the nagging expression it seems—it is a recognition that plants provide us with all the essential nutrients that allow our cells to continue metabolizing. The dietary importance of plants gives ample reason for us to know something about their anatomy and physiology.

10.1 Water and Mineral Transport

Water and minerals enter a plant at the root, primarily through the root hairs. From there water, along with minerals, moves across the tissues of a root until it enters **xylem,** the vascular tissue that contains the hollow conducting cells called *tracheids* and *vessel elements* (Fig. 10.1). The vessel elements are larger than the tracheids, and they are stacked one on top of the other to form a pipeline that stretches from the roots to the leaves. It is an open pipeline because the vessel elements have no end walls separating one from the other. The tracheids, which are elongated with tapered ends, form a less obvious means of transport. Water can move across the end and side walls of tracheids because of pits, or depressions, where the secondary wall does not form.

Water entering root cells creates a positive pressure called root pressure that tends to push xylem sap upwards. It is estimated that root pressure, like atmospheric pressure, would be able to raise water to 10.4 meters. However, since some trees can be as tall as 120 meters, other factors must be involved.

Cohesion-Tension Theory of Xylem Transport

The cohesion-tension theory of xylem transport explains how water is transported to great heights against gravity (Fig. 10.2). Because water molecules are polar, they adhere to the walls of the vessel elements; and because of hydrogen bonding, water molecules are cohesive—they cling together. *Cohesion* of water molecules within the xylem pipeline is absolutely necessary for water transport in a plant. It causes water to fill the pipeline completely, from the roots to the leaves, and to resist any separation.

The other factor that causes water to rise in plants is **transpiration,** the loss of water by evaporation. Much of the water that is transported from the roots to the leaves evaporates and escapes from the leaf by way of the stomates, openings where gas exchange occurs. A single corn plant loses somewhere between 135 and 200 liters of water through transpiration during a growing season.

The water molecules that evaporate are replaced by other water molecules from the leaf veins. Therefore, transpiration exerts a driving force by creating a negative pressure—that is, a *tension*—which draws a column of water up the vessel elements from the roots to the leaves. As long as water molecules evaporate from the plant and the soil contains enough moisture, the water column is pulled continuously upward.

The tension created by transpiration is effective only because of the cohesive property of water. Therefore, this explanation for xylem transport is called the **cohesion-tension theory.**

Water is transported from the roots to the leaves in xylem. The water column remains intact as transpiration pulls water from the roots to the leaves.

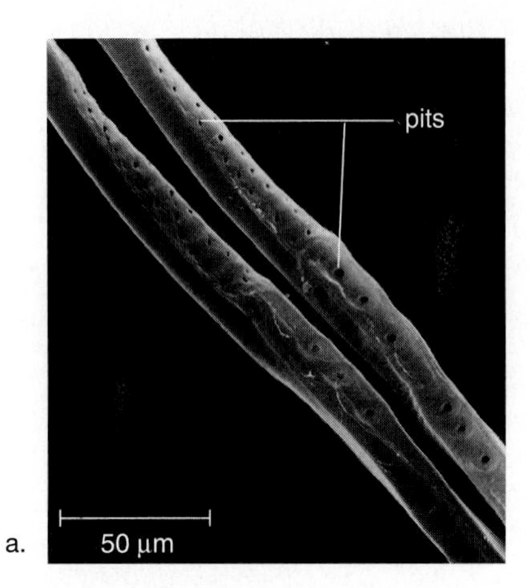

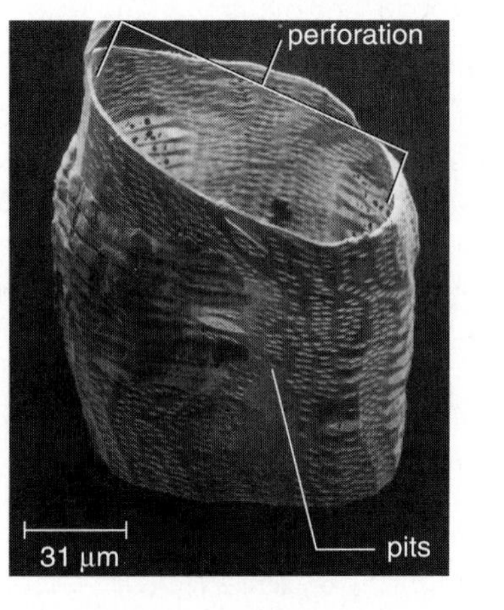

Figure 10.1 Conducting cells of xylem.
a. Tracheids are long, hollow cells with tapered ends. Water can move into and out of tracheids through pits only. **b.** Vessel elements are wider and shorter than tracheids. Water can move from vessel element to vessel element through perforations (a large hole at each end). Vessel elements can also exchange water with tracheids through pits.

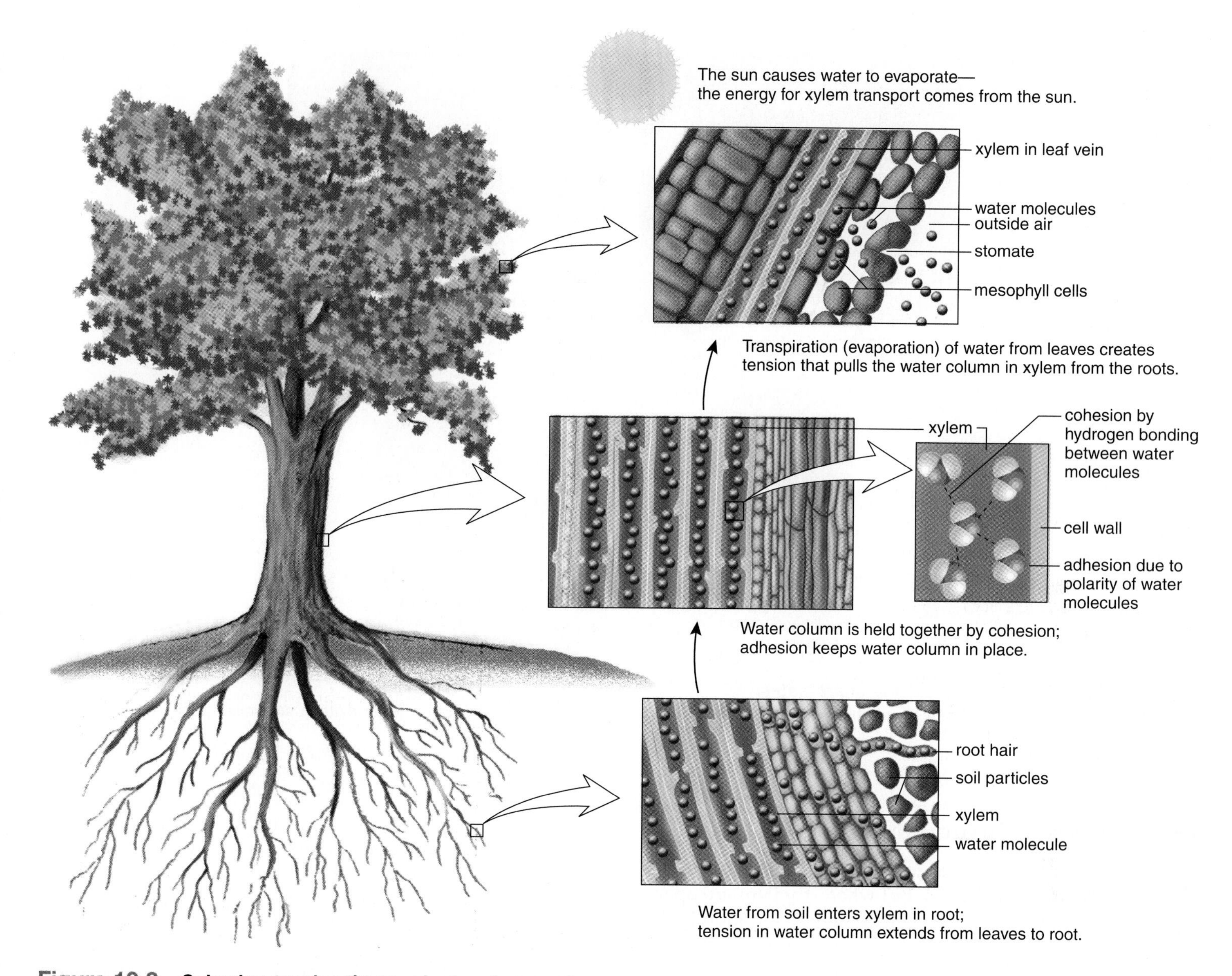

Figure 10.2 Cohesion-tension theory of xylem transport.
Tension (a negative pressure) created by evaporation (transpiration) at the leaves pulls water from the root along the length of the plant.

Mineral Transport

Plants need only inorganic nutrients in order to produce all the organic molecules that make up their bodies. Aside from carbon, hydrogen, and oxygen obtained from carbon dioxide, the **mineral** elements listed in Table 10.1 are required by plants. Minerals diffuse into root hairs, but eventually a plant uses active transport to increase its uptake of minerals which are carried along with water in xylem. A plant uses a great deal of ATP for active transport of minerals into root cells and xylem.

Human beings are fortunate that plants can concentrate minerals, for we often are dependent on them for our basic supply of such minerals as calcium to build bones and teeth and iron to help carry oxygen to our cells. Minerals like copper and zinc are cofactors for the functioning of enzymes.

Table 10.1 Inorganic Nutrients for Plants

Macronutrients	
$CaNO_3$	Ca, N (calcium, nitrogen)
$NH_4H_2PO_4$	N, P (nitrogen, phosphorus)
KNO_3	K, N (potassium, nitrogen)
$MgSO_4$	Mg, S (magnesium, sulfur)
Micronutrients	
Fe-EDTA	Fe (iron)
$ZnSO_4$	Zn (zinc)
KCl	Cl (chlorine)
$CuSO_4$	Cu (copper)
$MnSO_4$	Mn (manganese)

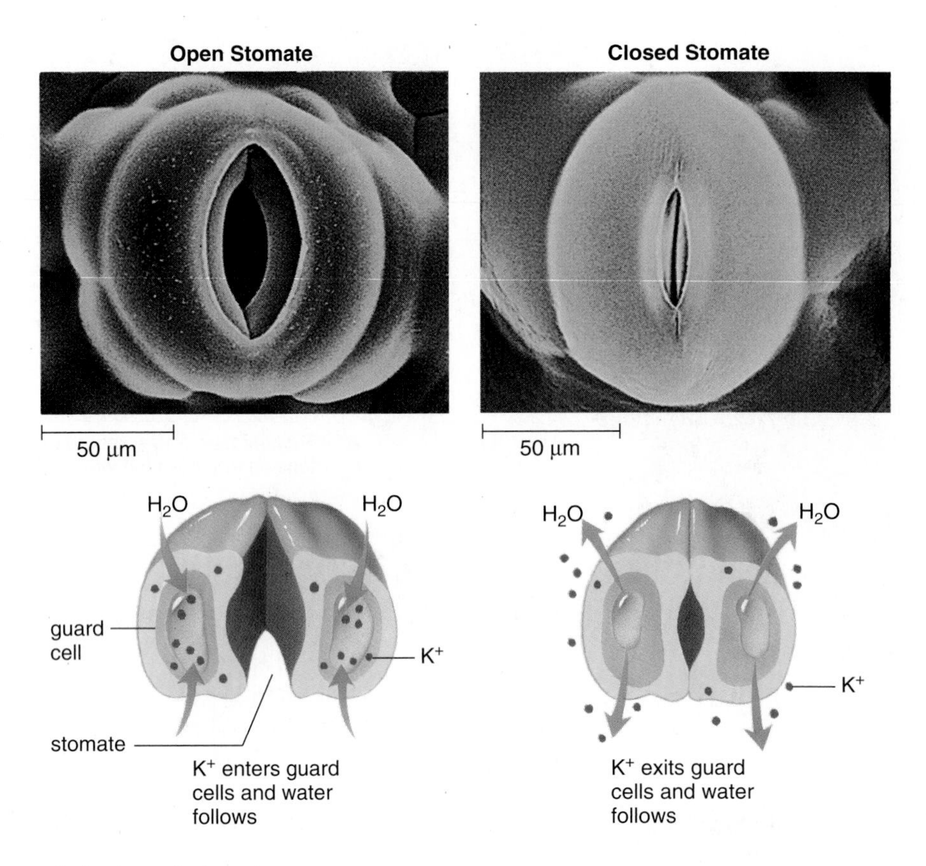

Figure 10.3 Opening and closing of stomates.
Stomates open when water enters guard cells and turgor pressure increases. Stomates close when water exits guard cells and there is a loss of turgor pressure.

Opening and Closing of Stomates

For a plant to photosynthesize, stomates must be open so that CO_2 will enter the leaves. But when stomates are open, water is also exiting the leaves because transpiration is occurring. When a plant is water stressed, stomates close to conserve water and then photosynthesis ceases.

The two **guard cells** on either side of a **stomate** regulate whether the stomate is open or closed. When water enters guard cells, their central vacuoles fill and the stomate opens. When water exits guard cells, the stomate closes. Notice in Figure 10.3 that the guard cells are attached to each other at their ends and that the inner walls are thicker than the outer walls. When water enters guard cells, cellulose microfibrils in the walls prevent radial expansion. Lengthwise expansion, which can occur, causes guard cells to buckle out from the region of their attachment and the stomate opens.

When a plant is photosynthesizing, an ATP-driven pump actively transports H^+ out of the cell. Now potassium ions (K^+) enter the guard cells and when water follows by osmosis, the stomate opens. When the pump is not working, K^+ moves into surrounding cells, the guard cells lose water, and the stomate closes.

What turns the H^+ pump on or off? It appears that the blue-light component of sunlight is a signal for stomates to open. There is evidence to suggest that a flavin pigment absorbs blue light, and then this pigment sets in motion the cytoplasmic response that leads to activation of the H^+ pump. Similarly, there could be a receptor in the plasma membrane of guard cells that brings about inactivation of the pump when carbon dioxide (CO_2) concentration rises, as might happen when photosynthesis ceases. Abscisic acid (ABA), which is produced by cells in wilting leaves, can also cause stomates to close. Although photosynthesis cannot occur, water is conserved.

If plants are kept in the dark, the stomates open and close on a 24-hour basis just as if they were responding to the presence of sunlight in the daytime and the absence of sunlight at night. This means that there must be some sort of internal *biological clock* that is keeping time. Circadian rhythms (a behavior that occurs every 24 hours) and biological clocks are areas of intense investigation at this time.

When stomates open, first K^+ and then water enters guard cells. Stomates open and close in response to environmental signals—the exact mechanism is being investigated.

Science Focus

Biodiversity and Nitrogen Uptake by Plants

G. David Tilman of the University of Minnesota was familiar with long-term experiments showing that amount of soil nitrogen usually limits biodiversity. The better competitors for soil nitrogen are expected to be prevalent to the point that other species do not have a chance to continue existing. Yet the grasslands he was studying often had more than 100 plant species coexisting in an area the size of a few hectares (10,000 square meters). This made him hypothesize that while competition for soil nitrogen might limit biodiversity, some other factor must be promoting biodiversity. First he studied the response of grassland plants to predation by insect and mammalian herbivores, light availability, and fire. None of these seemed to account for the biodiversity he observed.

Finally, he found that grassland plants differ in their ability to disperse to new sites. He discovered this by planting over 50 different plant species in several plots and recording how well they could germinate, grow, and reproduce there. The best competitors for nitrogen (bunch grasses) allocated 85% of their biomass to root growth and only 0.5% of their biomass to making seeds. Their massive root systems make them able to spread out where they are! Little bluestem is an example of a bunchgrass (Fig. 10A) The best dispersers allocated 40% of their biomass to root and 30% of their biomass to seed. Bent grass is an example of a poor competitor for soil nitrogen but a good disperser (Fig. 10B). The production of seeds allows these plants to occupy any new sites available.

Tilman developed a mathematical model which showed that such a tradeoff (root versus seed allocation) could explain the stable coexistence of a whole range of plant species that differ according to their ability to occupy present space or to disperse to new areas. Coexistence occurs because better competitors for soil nitrogen are poorer dispersers, and therefore they do not occupy all sites. Better dispersers are better at finding and occupying all possible sites. Because the grasses varied in their ability to take up nitrogen and disperse, a number of species could coexist.

Tilman studied another issue that relates to biodiversity. He and colleagues were annually sampling 207 permanent plots when there was a drought in 1987-88. The 1988 drought was the third worst in the past 150 years. They found that plots that contained only one to four species suffered a greater loss of overall biomass than plots that contained 16 to 26 species. This suggests to Tilman that high biodiversity does buffer an ecosystem against an environmental disturbance, and therefore it is wise to conserve the biodiversity of ecosystems in all areas—whether in Minnesota, New Jersey, Oregon, or the tropics.

Figure 10A
Little bluestem (*Schizachyrium scoparium*) is a North American bunchgrass that prefers sandy or rocky habitats. It is an excellent competitor for soil nitrogen because of its high allocation of roots.

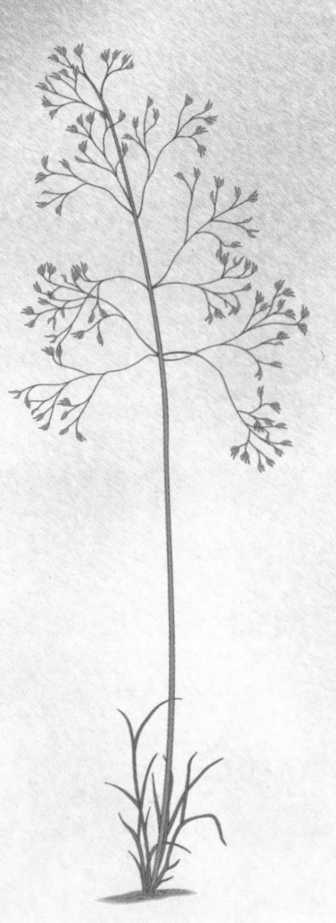

Figure 10B
Bent grass (*Agrostis scabra*) is a native North American prairie grass that is a poor competitor for soil nitrogen. It disperses rapidly into disturbed areas because of its high allocation of seed.

10.2 Organic Nutrient Transport

The leaves carry out photosynthesis and produce the sugar sucrose, which is transported in the vascular tissue called **phloem** to all parts of a plant. The movement of organic substances in phloem is termed translocation. Translocation makes sugars available to those parts of a plant that are actively metabolizing and growing. The conducting cells in phloem are sieve-tube elements, each of which typically has a *companion cell* (Fig. 10.4). Sieve-tube elements contain cytoplasm but have no nucleus. Their end walls have pores and resemble a sieve; therefore, the end walls are said to be sieve plates. The sieve-tube elements are aligned vertically, and strands of cytoplasm called plasmodesmata (sing., **plasmodesma**) extend from one cell to the other through the sieve plates. Therefore, there is a continuous pathway for organic nutrient transport throughout the plant.

The smaller companion cell, which does have a nucleus, is a more generalized cell than the sieve-tube element. It is speculated that the companion cell nucleus controls and maintains the lives of both itself and the sieve-tube element, and may help a sieve-tube element perform its translocating function.

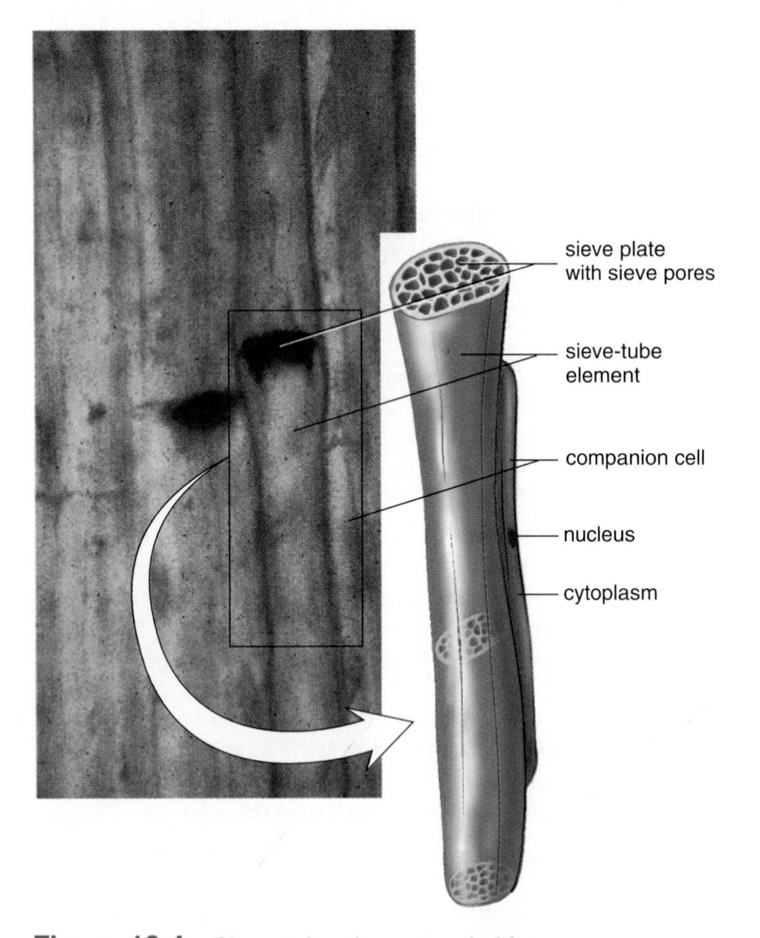

Figure 10.4 Sieve-tube elements of phloem.
Sieve-tube elements contain cytoplasm, and their flat end walls have a sieve plate with large sieve pores. Plasmodesmata connect the cells one with the other. Each sieve-tube element has a companion cell that has both cytoplasm and a nucleus.

Pressure-Flow Theory of Phloem Transport

Chemical analysis of phloem sap shows that it is composed chiefly of sugar and that the concentration of organic nutrients is 10–13% by volume. Samples for chemical analysis most often are obtained by using aphids, small insects that are phloem feeders. The aphid drives its stylet, a short mouthpart functioning like a hypodermic needle, between the epidermal cells and withdraws phloem sap from a sieve-tube element. The body of the aphid can be cut away carefully, leaving the stylet, which exudes phloem sap for collection and analysis.

During the growing season the leaves are a **source** of sugar—they are photosynthesizing and producing sugar. This sugar is actively transported into sieve-tube elements, and water follows passively by osmosis. Active transport is possible because sieve-tube elements have a living plasma membrane, and the necessary energy is provided by the companion cells. The buildup of water within the sieve-tube elements creates pressure, which starts a flow of phloem sap. The roots (and other growth areas) are a **sink** for sugar—the roots are removing sugar. It is actively transported out of the sieve-tube elements at this location. When water follows passively by osmosis, phloem sap flows from the leaves (source) to the roots (sink) (Fig. 10.5). This explanation for translocation in phloem is called the **pressure-flow theory.**

The pressure-flow model is supported by an experiment in which two bulbs are connected by a glass tube. The first bulb contains solute at a higher concentration than the second bulb. Each bulb is bounded by a differentially permeable membrane, and the entire apparatus is submerged in distilled water:

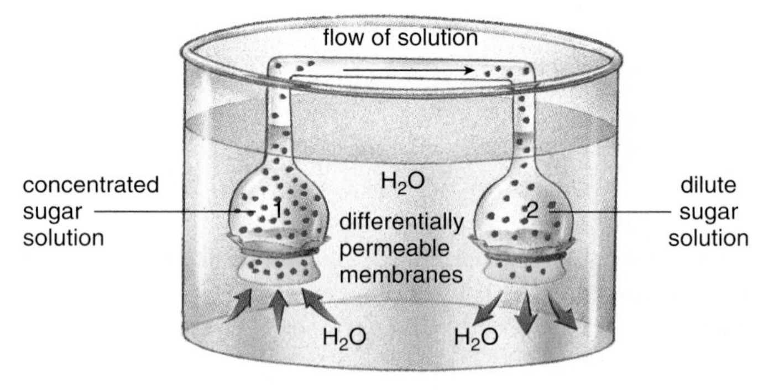

Distilled water flows into the first bulb because it has the higher solute concentration. In this way a pressure difference is created that causes water to flow from the first bulb to the second and even to exit from the second bulb. As the water flows, it carries solute with it from the first to the second bulb.

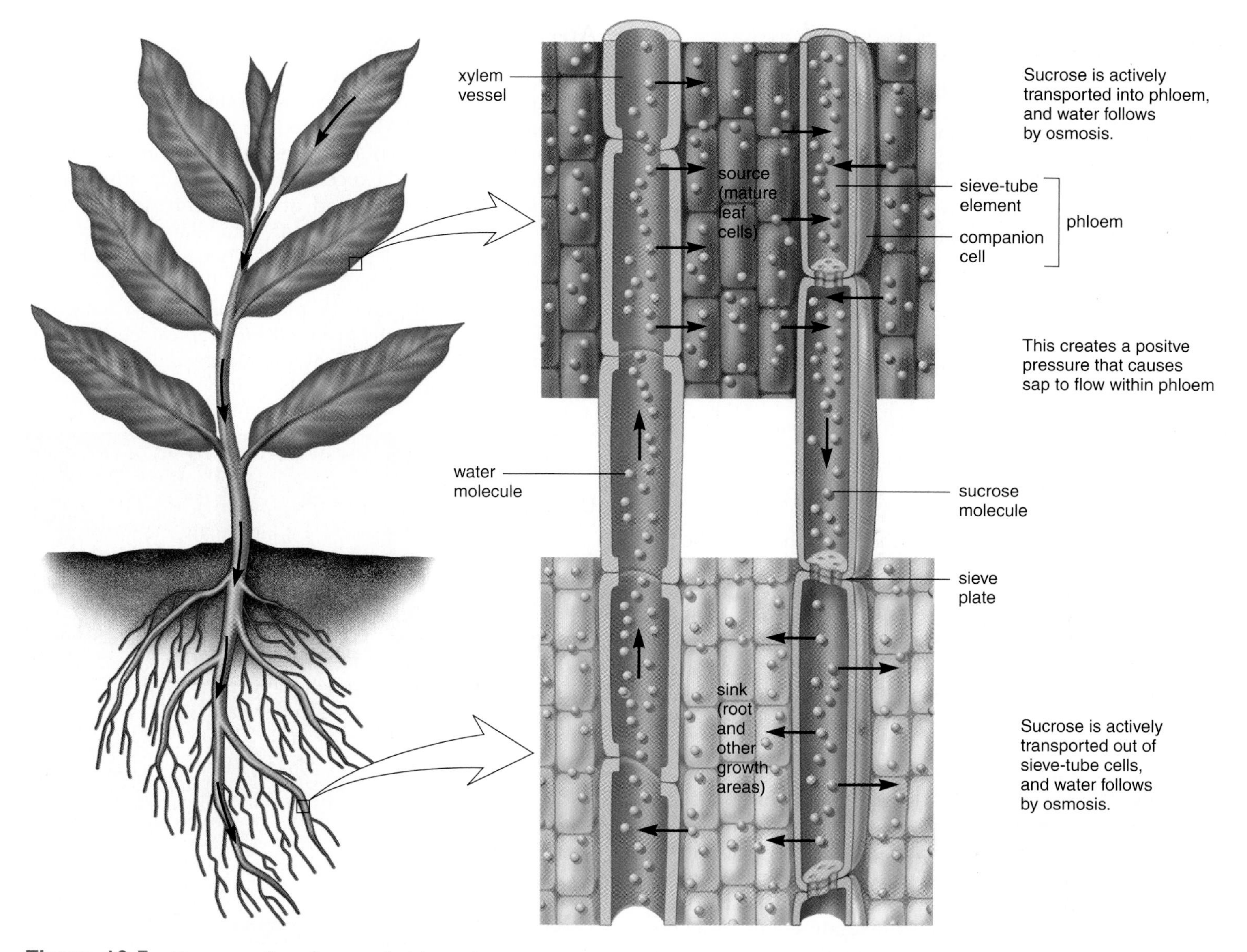

Figure 10.5 Pressure-flow theory of phloem transport.
Sugar and water enter sieve-tube elements at a source. This creates a positive pressure, which causes phloem contents to flow. Sieve-tube elements form a continuous pipeline from a source to a sink, where sugar and water exit sieve-tube elements.

The pressure-flow theory of phloem transport can account for any direction of flow in sieve tubes if we consider that the direction of flow is always from *source to sink.* In young seedlings, the cotyledons containing reserved food are a major source of sucrose, and roots are a sink. Therefore, the flow is from the cotyledons to the roots. In older plants, the most recently formed leaves can be a sink and they will receive sucrose from other leaves until they begin to maximally photosynthesize. When a plant is forming fruit, phloem flow is monopolized by the fruits, little goes to the rest of the plant, and vegetative growth is slow.

Because phloem sap flows from source to a sink, situations arise in which there is a bidirectional flow within phloem—not just at different times in the life cycle but even at the same time! Bidirectional flow is occurring because different sieve tubes can be conducting phloem sap in opposite directions.

Phloem transports organic nutrients in a plant. Typically, sugar and then water enter sieve-tube elements in the leaves. This creates pressure, which causes water to flow to the roots, carrying sugar with it.

10.3 Plant Responses to Environmental Stimuli

Plants respond to environmental stimuli (e.g., light, day length, gravity, and temperature) usually by changing their growth patterns. Plant growth toward or away from a directional stimulus is called a **tropism.** Three well-known tropisms, each named for the stimulus that causes the response, are

> Phototropism: a movement in response to a light stimulus
> Gravitropism: a movement in response to gravity
> Thigmotropism: a movement in response to touch

Growth toward a stimulus is called a positive tropism and growth away from a stimulus is called a negative tropism (Fig. 10.6). Tropisms are due to differential growth—one side of an organ elongates faster than the other, and the result is a curving toward or away from the stimulus.

Among the principal internal factors that regulate such responses are plant hormones. A **hormone** is a chemical messenger produced in very low concentrations that has physiological and/or developmental effects, usually in another part of the organism. Some plant hormones are made in meristem and transported to other tissues; others move directly from the tissue of origin to another tissue; and still others are used where they are made.

Table 10.2 lists the major types of plant hormones and their primary functions. Some of these are promoters of growth and some are inhibitors of growth. These hormones often interact to control physiological responses. Each naturally occurring hormone has a specific chemical structure. Other chemicals, some of which differ only slightly from the natural hormones, also affect the growth of plants. These and the naturally occurring hormones are sometimes grouped together and called plant growth regulators.

Auxin

Auxin is a plant hormone whose effects have been studied for a long time. The only known naturally occurring auxin is indoleacetic acid (IAA). It is produced in shoot apical meristem and is found in young leaves and in flowers and fruits. Therefore, you would expect that auxin would affect many aspects of plant growth and development. Apically produced auxin prevents the growth of axillary buds, a phenomenon called apical dominance. When a terminal bud is removed deliberately or accidentally, the nearest axillary buds begin to grow, and the plant branches. To achieve a fuller look, one generally prunes the top (apical meristem) of the plant. This removes apical dominance and causes more branching of the main body of the plant.

Figure 10.6 Positive phototropism.
The stem of a plant bends toward the light as it grows.

Table 10.2 Plant Hormones

Type	Primary Example	Notable Function
Growth Promoters		
Auxin	Indoleacetic acid (IAA)	Promotes cell elongation in stems; phototropism, gravitropism, apical dominance; formation of roots, development of fruit
Gibberellins	Gibberellic acid (GA)	Promote stem elongation; release some buds and seeds from dormancy
Cytokinins	Zeatin	Promote cell division and embryo development; prevent leaf senescence and promote bud activation
Growth Inhibitors		
Abscisic acid	Abscisic acid (ABA)	Resistance to stress conditions; causes stomatal closure; maintains dormancy
Ethylene	Ethylene	Promotes fruit ripening; promotes abscission and fruit drop; inhibits growth

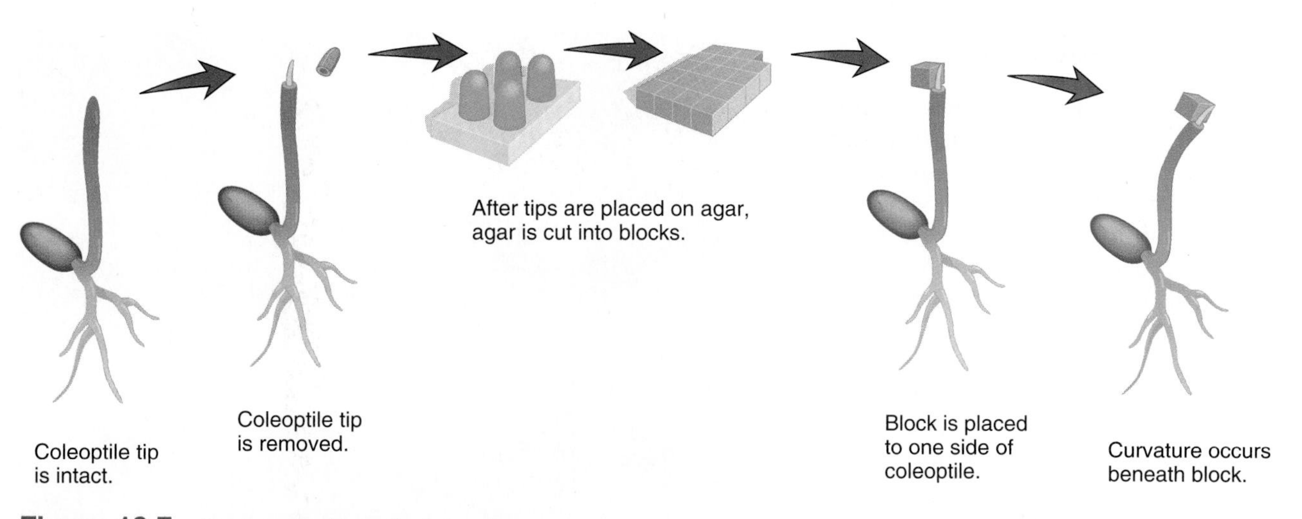

Figure 10.7 Demonstrating phototropism.
Oat seedlings are protected by a hollow sheath called a coleoptile. After a tip is removed and placed on agar, an agar block placed on one side of the coleoptile can cause it to curve.

The application of a weak solution of auxin to a woody cutting causes roots to develop. Auxin production by seeds also promotes the growth of fruit. As long as auxin is concentrated in leaves or fruits rather than in the stem, leaves and fruits do not fall off. Therefore, trees can be sprayed with auxin to keep mature fruit from falling to the ground.

Auxin is involved in gravitropism as well as phototropism. After gravity has been perceived, auxin moves to the lower surface of roots and stems. Thereafter, roots curve downward and stems curve upward.

Phototropism and Auxin

The role of auxin in the positive **phototropism** of stems has been studied for quite some time. The experimental material of choice has been oat seedlings with coleoptile intact. (A coleoptile is a protective sheath for the young leaves of the seedling.) For example, if the tips of the coleoptiles are cut off and placed on agar (a gelatinlike material), and then an agar block is placed to one side of a tipless coleoptile, the shoot will curve away from that side. The bending occurs even though the seedlings are not exposed to light (Fig. 10.7). Apparently, the agar blocks contain a chemical that was produced by the coleoptile tips.

Because blue light in particular causes phototropism to occur, it is believed that a yellow pigment related to the vitamin riboflavin acts as a photoreceptor for light. Following reception, auxin migrates from the bright side to the shady side of a stem. The cells on that side elongate faster than those on the bright side, causing the stem to curve toward the light. It is now known that when a plant is exposed to unidirectional light, auxin moves to the shady side, where it binds to receptors and activates an ATP-driven H^+ pump (Fig. 10.8). As hydrogen ions (H^+) are being pumped out of the cell, the cell wall becomes acidic, breaking hydrogen bonds within cellulose. Cellulose fibrils are weakened, and activated enzymes further degrade the cell wall. The electrochemical gradient established by the H^+ pump causes solutes to enter the cell, and water follows by osmosis. The turgid cell presses against the cell wall, stretching it so that elongation occurs.

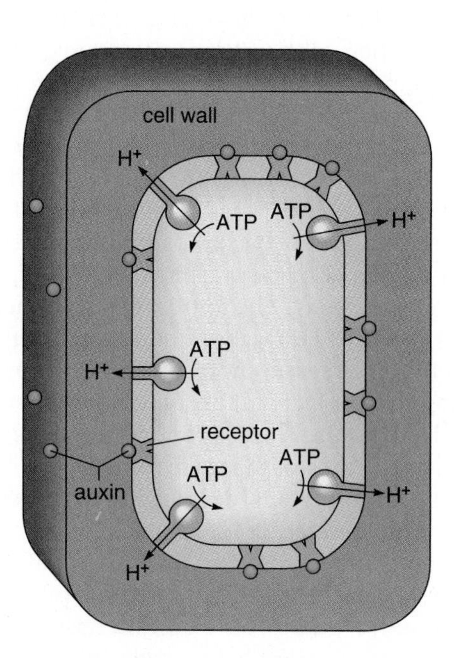

Figure 10.8 Auxin mode of action.
After auxin binds to a receptor, the combination stimulates an H^+ pump so that hydrogen ions (H^+) are transported out of the cytoplasm. The resulting acidity causes the cell wall to weaken, and the electrochemical gradient causes solutes to enter the cell. Water follows by osmosis and the cell elongates.

Other Plant Growth Regulators

Now that the formulas for many plant hormones are known, it is possible to synthesize them, as well as related chemicals, in the laboratory. Collectively, these substances are known as plant growth regulators. Many scientists hope plant growth regulators will bring about an increase in crop yield, just as fertilizers, irrigation, and pesticides have done in the past.

In high concentrations, some auxins are used widely in agriculture as herbicides to prevent the growth of broad-leaved plants. In addition to their use in weed control, the synthetic auxins known as 2,4D and 2,4,5T were used as defoliants during the Vietnam War. Even though 2,4D has been used for over 35 years, we do not yet know how it works. Apparently, it is structurally different enough from natural auxin that plants cannot store it in an inactive form and a plant's own enzymes cannot break it down. Its concentration rises until metabolism is disrupted, cellular order is lost, and the cells die.

Other plant hormones also have agricultural and commercial uses. When **gibberellins** are sprayed on grapefruit (Fig. 10.9), the peel stays youthful—green, firm, and unappealing to various types of fruit flies. Yet the treated fruit is fully ripe and sweet—perfect for juicing or eating. Gibberellins are used to stimulate seed germination and seedling growth of some grains, beans, and fruits. They also increase the size of some mature plants. Treatment of sugarcane with as little as 2 oz per acre increases the cane yield by more than 5 metric tons. The application of either auxins or gibberellins can cause an ovary and accessory flower parts to develop into fruit, even though pollination and fertilization have not taken place. In this way, it is sometimes possible to produce seedless fruits and vegetables, or bigger, more uniform bunches with larger fruit.

Because **cytokinins** retard the aging of leaves and other organs, they are sprayed on vegetables to keep them fresh during shipping and storage. Such treatment of holly, for example, allows it to be harvested many weeks prior to a holiday.

Plant growth regulators are used in tissue culture when seedlings are grown from a few cells in laboratory glassware. New varieties of food crops with particular characteristics—such as tolerance to herbicides and insects—are being developed by genetically engineering the original cells. One day it may be possible to endow most crops with the ability to utilize atmospheric nitrogen and/or make a wider range of proteins.

Several synthetic inhibitors are used to oppose the action of auxins, gibberellins, and cytokinins normally present in plants. Some of these can cause leaves and fruit to drop at a time convenient to the farmer. Removing leaves from cotton plants aids in their harvest, and thinning the fruit of young fruit trees results in larger fruit as the trees mature. Retarding the growth of other plants sometimes increases their hardiness. For example, an inhibitor has been used to reduce stem length in wheat plants, so that the plants do not fall over due to heavy winds and rain.

Figure 10.9 Grapefruit being sprayed with gibberellins. When sprayed with gibberellins, the peel stays green and firm so that fruit flies look for another place to lay their eggs. The tougher skin improves shipping and can be turned yellow by an application of ethylene.

The commercial uses of **ethylene** were greatly increased with the development of ethylene-releasing compounds. Ethylene gas is injected into airtight storage rooms to ripen bananas, honeydew melons, and tomatoes. It will also degreen oranges, lemons, and grapefruit when the rind would otherwise remain green because of a high chlorophyll level. When sprayed on certain fruit and nut crops, ethylene increases the chances that the fruit will detach when the trees are shaken at harvest time.

Today, fields and orchards are often sprayed with synthetic growth regulators.

Photoperiodism

A response based on the proportion of light to darkness in a 24-hour cycle is called **photoperiodism.** Photoperiodic responses in plants are particularly obvious in the temperate zone. In the spring, plants respond to increasing day length by initiating growth; in the fall, they respond to decreasing day length by ceasing growth processes. Day

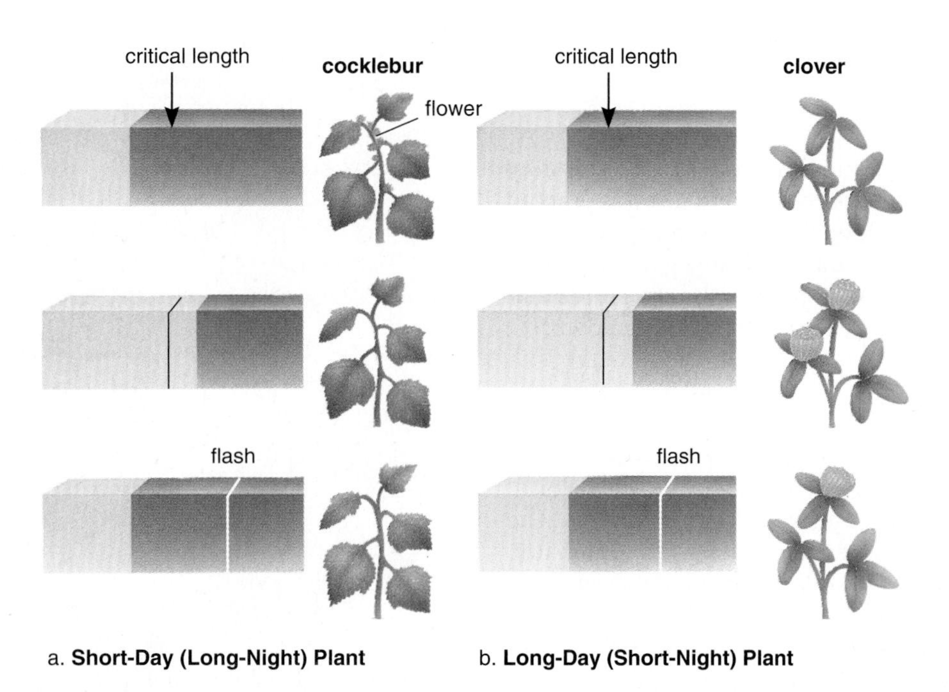

Figure 10.10 **Effect of day length on two types of plants.**
Some plants flower only during a particular season. The cocklebur flowers when days are short **(a),** and clover flowers when days are long **(b).** The length of the night is the determining factor, proven by interrupting a longer-than-critical-length night with a flash of light. Dependency on the photoperiod is exploited by commercial greenhouses, which use it to grow plants out of season.

length controls flowering in some plants; for example, violets and tulips flower in the spring, asters and goldenrods flower in the fall.

Long-day plants initiate flowering when the days get longer than a certain minimum value, or critical length. Short-day plants initiate flowering when the days get shorter than a critical length. Day-length-neutral plants are insensitive to the length of the day. The cocklebur is a short-day plant; if a long night is interrupted by a flash of light, it will not flower (Fig. 10.10). Clover, on the other hand, is a long-day plant; if a long night is interrupted by a flash of light, it will still flower. Interrupting the day with darkness has no effect. This shows that the length of continuous darkness, not the day length, actually controls flowering.

Florists use information about photoperiodism in order to provide us with flowers at a particular time of year or out of season. They only need to know how long it takes for a certain type of plant to flower, and then they start manipulating the photoperiod in order to have flowers ready on time. If chrysanthemums, which flower in the fall, are desired in the spring, they use blackout shades to artificially produce long nights. On the other hand, florists switch on incandescent lights so that irises, which are spring plants, are available in the winter. These same techniques make it possible to always have poinsettias at Christmas and lilies at Easter.

If flowering is dependent on night and day length, plants must have a photoreceptor to detect these periods. The photoreceptor for photoperiodism is **phytochrome,** a blue-green leaf pigment that alternately exists in these two forms:

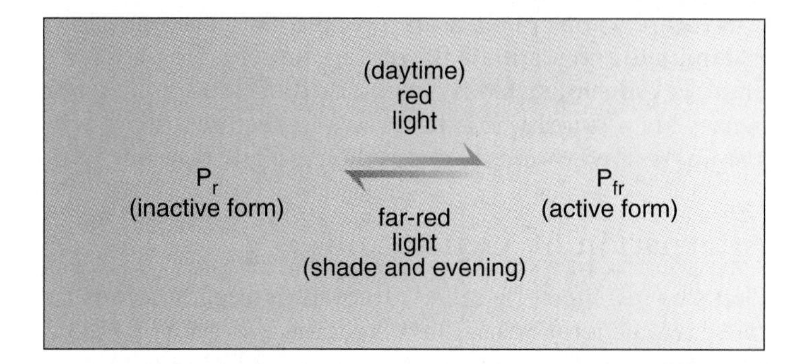

It could be that flowering is initiated in some plants according to the time of day phytochrome changes its structure. The pigment could be part of a biological clock system that controls flowering. Phytochrome is also believed to be involved in controlling a plant's growth pattern in a low-light versus a high-light environment.

Hormones often coordinate the response of plants to various stimuli. The proportion of stimulatory and inhibitory hormones often determines the particular response. Phytochrome is a pigment involved in response of plants to the photoperiod which affects flowering in some plants.

10.4 Sexual Reproduction in Flowering Plants

Sexual reproduction is the rule in flowering plants. This may come as a surprise to those who never thought of plants in terms of male and female. However, sexual reproduction is defined properly as reproduction requiring gametes, often an egg and a sperm. In a flowering plant, the structures that produce the egg and sperm are located within the flower.

Structure of Flowers

Figure 10.11 shows the parts of a typical **flower.** The **sepals,** most often green, form a whorl about the **petals,** the color of which accounts for the attractiveness of many flowers. The size, shape, and color of a flower attracts a specific pollinator. Flowers pollinated by the wind often have no petals at all.

In the center of the flower is a small vaselike structure, the **pistil,** which usually has three parts: the **stigma,** an enlarged sticky knob; the **style,** a slender stalk; and the **ovary,** an enlarged base. The ovary contains one or more **ovules,** which play a significant role in reproduction. Grouped about the pistil are the **stamens,** each of which has two parts: the **anther,** a saclike container, and the **filament,** a slender stalk.

Not all flowers have sepals, petals, stamens, and a pistil. Those that do are said to be complete and those that do not are said to be incomplete. Flowers with only stamens are called staminate flowers and those with only pistils are called pistillate flowers. If staminate flowers and pistillate flowers are on one plant, as in corn, the plant is monoecious. If staminate and pistillate flowers are on separate plants, the plant is dioecious. Holly trees are dioecious and, if red berries are a priority, it is necessary to acquire a plant with staminate flowers and another with pistillate flowers.

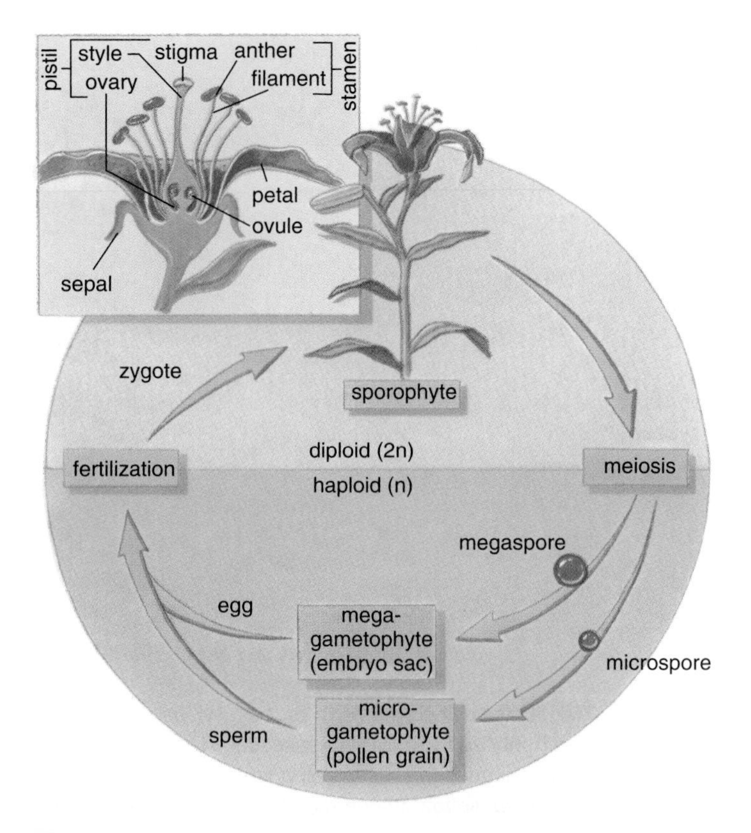

Figure 10.11 Alternation of generations in a flowering plant.
Flowering plants are heterosporous; they produce microspores and megaspores. A microspore becomes a pollen grain, which is a microgametophyte that produces sperm. A megaspore becomes an embryo sac, or megagametophyte, that produces an egg. When a sperm joins with an egg, the resulting zygote develops into an embryo, retained within a seed.

Alternation of Generations

Plants have a life cycle called **alternation of generations** because two generations are involved: the **sporophyte** and the **gametophyte.** The sporophyte is a diploid (2n) generation that produces haploid (n) **spores** by meiosis. A flower produces two types of spores: **microspores** and **megaspores.** Microspores are produced in the anthers of stamens, and megaspores are produced within ovules (Fig. 10.11). A microspore becomes a pollen grain, which upon maturity is a sperm-containing *microgametophyte,* also called the male gametophyte. A megaspore becomes an egg-containing embryo sac, which is a *megagametophyte,* also called the female gametophyte. Following fertilization, the zygote develops into an embryo located within a seed. When the seed germinates, the new sporophyte plant begins to grow.

Figure 10.12 shows these same steps in greater detail. Within an ovule, a megaspore (*mega* means large) parent cell undergoes meiosis to produce four haploid megaspores. Three of these megaspores disintegrate, leaving one functional megaspore, which divides mitotically. The result is the megagametophyte, or **embryo sac,** which typically consists of eight haploid nuclei embedded in a mass of cytoplasm. The cytoplasm differentiates into cells, one of which is an egg and another of which is the central cell with two nuclei (called the **polar nuclei**).

The anther has **pollen sacs,** which contain numerous microspore (*micro* means small) parent cells. Each parent cell undergoes meiosis to produce four haploid cells called microspores. The microspores usually separate, and each one becomes a **pollen grain.** At this point, the young pollen grain contains two cells, the **generative cells** and the **tube cell.**

Pollination occurs when pollen is windblown or carried by insects, birds, or bats to the stigma of the same type of plant. Only then does a pollen grain germinate and develop a long pollen tube. This pollen tube grows within the style until it reaches an ovule in the ovary. At this time, too, the generative cell divides mitotically, pro-

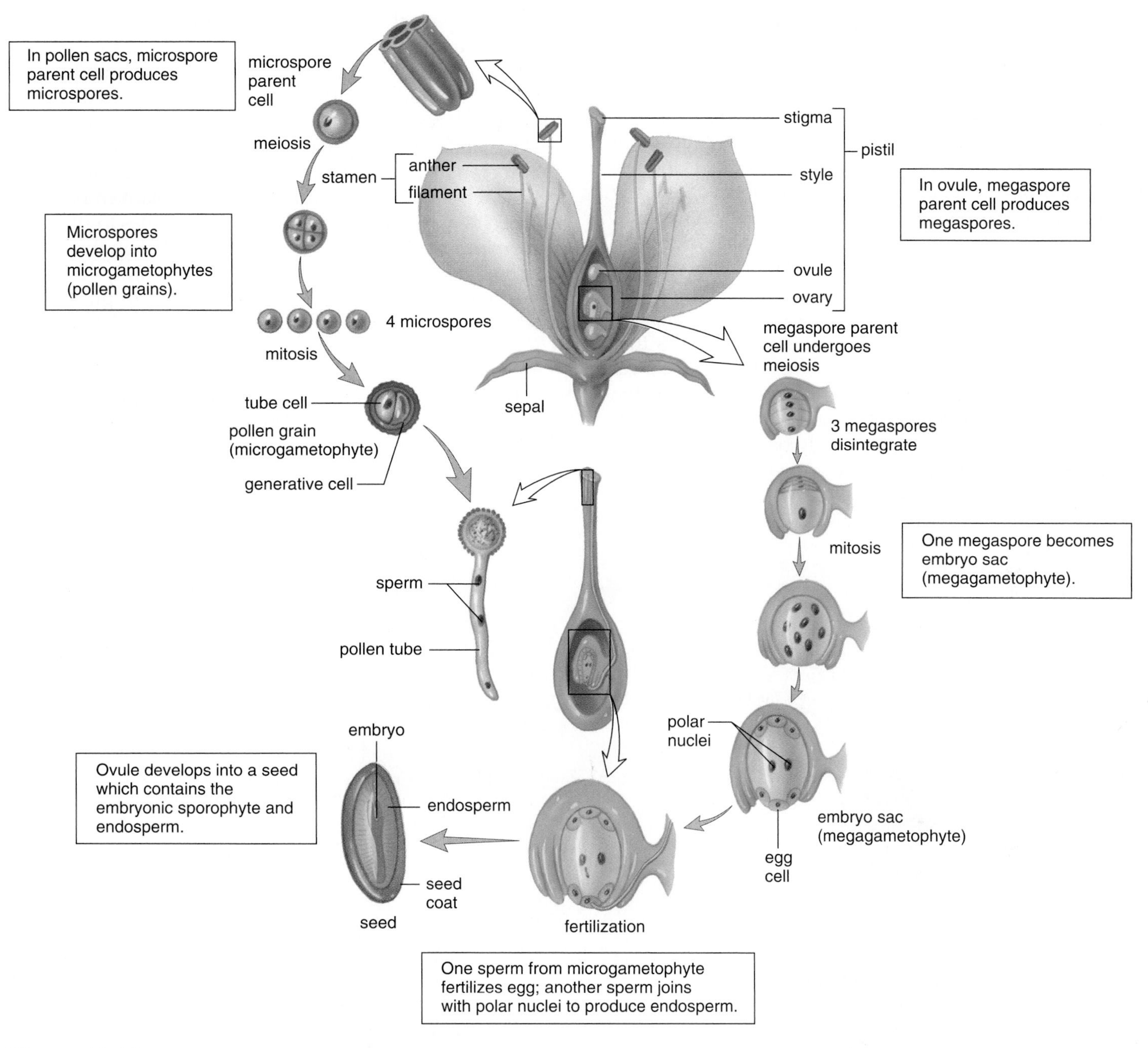

Figure 10.12 Life cycle of a flowering plant.
The flowering plant life cycle involves production of eggs and sperm by gametophyte generations and development of an embryo-containing seed.

ducing two sperm, which have no flagella. On reaching the ovule, the pollen tube discharges the sperm. These two fusions are known as **double fertilization.** One of the two sperm migrates to and fertilizes the **egg,** forming a **zygote,** and the other sperm migrates to and unites with the two polar nuclei, producing a 3n (triploid) endosperm nucleus.

The endosperm nucleus divides to form **endosperm,** food for the developing plant. The zygote develops into a multicellular **embryo** and the ovule wall hardens and becomes the **seed coat.** A **seed** is a structure formed by maturation of an ovule; it consists of a sporophyte embryo, stored food, and a seed coat. The ovary, and sometimes other floral parts, develops into a fruit.

Flowering plants produce an egg within each ovule of an ovary, and sperm within pollen grains. They have double fertilization—one fertilization produces the zygote; the other produces endosperm. The ovule now becomes a seed and the ovary becomes a fruit.

Development of the Plant Embryo

Stages in the development of a dicot embryo are shown in Figure 10.13. After double fertilization has taken place, the single-celled zygote lies beneath the endosperm nucleus. The endosperm nucleus divides to produce a mass of endosperm tissue surrounding the embryo. The zygote also divides, forming two parts: the upper part is the embryo, and the lower part is the suspensor, which anchors the embryo and transfers nutrients to it from the sporophyte plant. Soon the **cotyledons,** or seed leaves, can be seen. At this point, the dicot embryo is heart shaped. Later, when it becomes torpedo shaped, it is possible to distinguish the shoot apex and the root apex. These contain apical meristems, the tissues that bring about primary growth in a plant; the shoot apical meristem is responsible for aboveground growth, and the root apical meristem is responsible for underground growth.

Monocots, unlike dicots, have only one cotyledon. Another important difference between monocots and dicots is the manner in which nutrient molecules are stored in the seed. In a monocot, the cotyledon rarely stores food; rather, it absorbs food molecules from the endosperm and passes them to the embryo. During the development of a dicot embryo, the cotyledons usually store the nutrient molecules that the embryo uses. Therefore, in Figure 10.13 we can see that the endosperm seemingly disappears. Actually, it has been taken up by the two cotyledons. In a plant embryo, the **epicotyl** is above the cotyledon and contributes to shoot development; the **hypocotyl** is that portion below the cotyledon that contributes to stem development; and the **radicle** contributes to root development. The embryo plus stored food is now contained within a seed.

The plant embryo (which has gone through a set series of stages), plus its stored food, is contained within a seed.

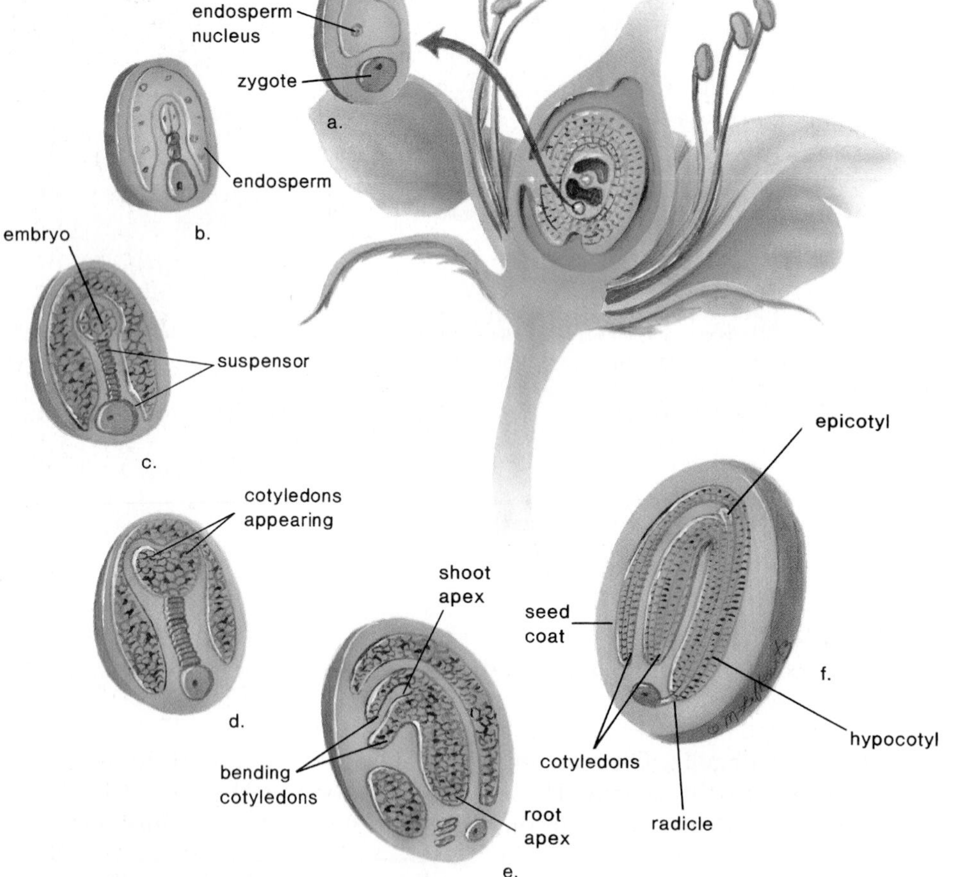

Figure 10.13 Development of a dicot embryo.
a. The unicellular zygote lies beneath the endosperm nucleus. **b, c.** The endosperm is a mass of tissue surrounding the embryo. The embryo is located above the suspensor. **d.** The embryo becomes heart shaped as the cotyledons begin to appear. **e.** There is progressively less endosperm as the embryo differentiates and enlarges. As the cotyledons bend, the embryo takes on a torpedo shape. **f.** The embryo consists of the epicotyl (represented here by the shoot apex), the hypocotyl, and the radicle which includes the root apex.

Figure 10.14 Fruit diversity.
a. The almond fruit is fleshy, with a single seed enclosed by a hard covering. **b.** The tomato is derived from a compound ovary. **c.** Each blackberry is from a flower with many ovaries.

Fruits and Seeds

In flowering plants, seeds are enclosed within a fruit, which develops from the ovary and at times, from other accessory parts. Although peas, beans, tomatoes, and cucumbers are commonly called vegetables, botanists categorize them as fruits. A **fruit** is a mature ovary that usually contains seeds.

As a fruit develops from an ovary, the ovary wall thickens to become the pericarp. In fleshy fruits, the pericarp is at least somewhat fleshy; peaches and plums are good examples of fleshy fruits. In almonds, the fleshy part of the pericarp is a husk removed before marketing. We crack the remaining portion of the pericarp to obtain the seed (Fig. 10.14*a*). An apple develops from a compound ovary, but much of the flesh comes from the receptacle, which grows around the ovary. It's more obvious that a tomato comes from a compound ovary, because in cross section, you can see several seed-filled cavities (Fig. 10.14*b*).

Dry fruits have dry pericarps. Legumes, such as peas and beans, produce fruits that split along two sides, or seams. Not all dry fruits split at maturity. The pericarp of a grain is tightly fused to the seed and cannot be separated from it. A corn kernel is a grain, as are the fruits of wheat, rice, and barley plants.

Some fruits develop from several individual ovaries. A blackberry is an aggregate fruit in which each portion is derived from a separate ovary of a single flower (Fig. 10.14*c*). The strawberry is also an aggregate fruit, but each ovary becomes a one-seeded fruit called an achene. The flesh of a strawberry is from the receptacle. In contrast, a pineapple comes from the fruit of many individual flowers attached to the same fleshy stalk. As the ovaries mature, they fuse to form a large, multiple fruit.

In flowering plants, the seed develops from the ovule and the fruit develops from the ovary.

Dispersal and Germination of Seeds

For plants to be widely distributed, their seeds have to be dispersed—that is, distributed preferably long distances from the parent plant. Following dispersal, the seeds germinate: they begin to grow so that a seedling appears.

Dispersal of Seeds

Plants have various means to ensure that dispersal takes place. The hooks and spines of clover, bur, and cocklebur attach to the fur of animals and the clothing of humans. Birds and mammals sometimes eat fruits, including the seeds, which are then defecated (passed out of the digestive tract with the feces) some distance from the parent plant. Squirrels and other animals gather seeds and fruits, which they bury some distance away.

The fruit of the coconut palm, which can be dispersed by ocean currents, may land many hundreds of kilometers away from the parent plant. Some plants have fruits with trapped air or seeds with inflated sacs that help them float in water. Many seeds are dispersed by wind. Woolly hairs, plumes, and wings are all adaptations for this type of dispersal. The seeds of an orchid are so small and light that they need no special adaptation to carry them far away. The somewhat heavier dandelion fruit uses a tiny "parachute" for dispersal. The winged fruit of a maple tree, which contains two seeds, has been known to travel up to 10 kilometers from its parent. A touch-me-not plant has seed pods that swell as they mature. When the pods finally burst, the ripe seeds are hurled out.

Animals, water, and wind help plants disperse their seeds.

Germination of Seeds

Some seeds do not germinate until they have been dormant for a period of time. For seeds, *dormancy* is the time during which no growth occurs, even though conditions may be favorable for growth. In the temperate zone, seeds often have to be exposed to a period of cold weather before dormancy is broken. In deserts, germination does not occur until there is adequate moisture. This requirement helps ensure that seeds do not germinate until the most favorable growing

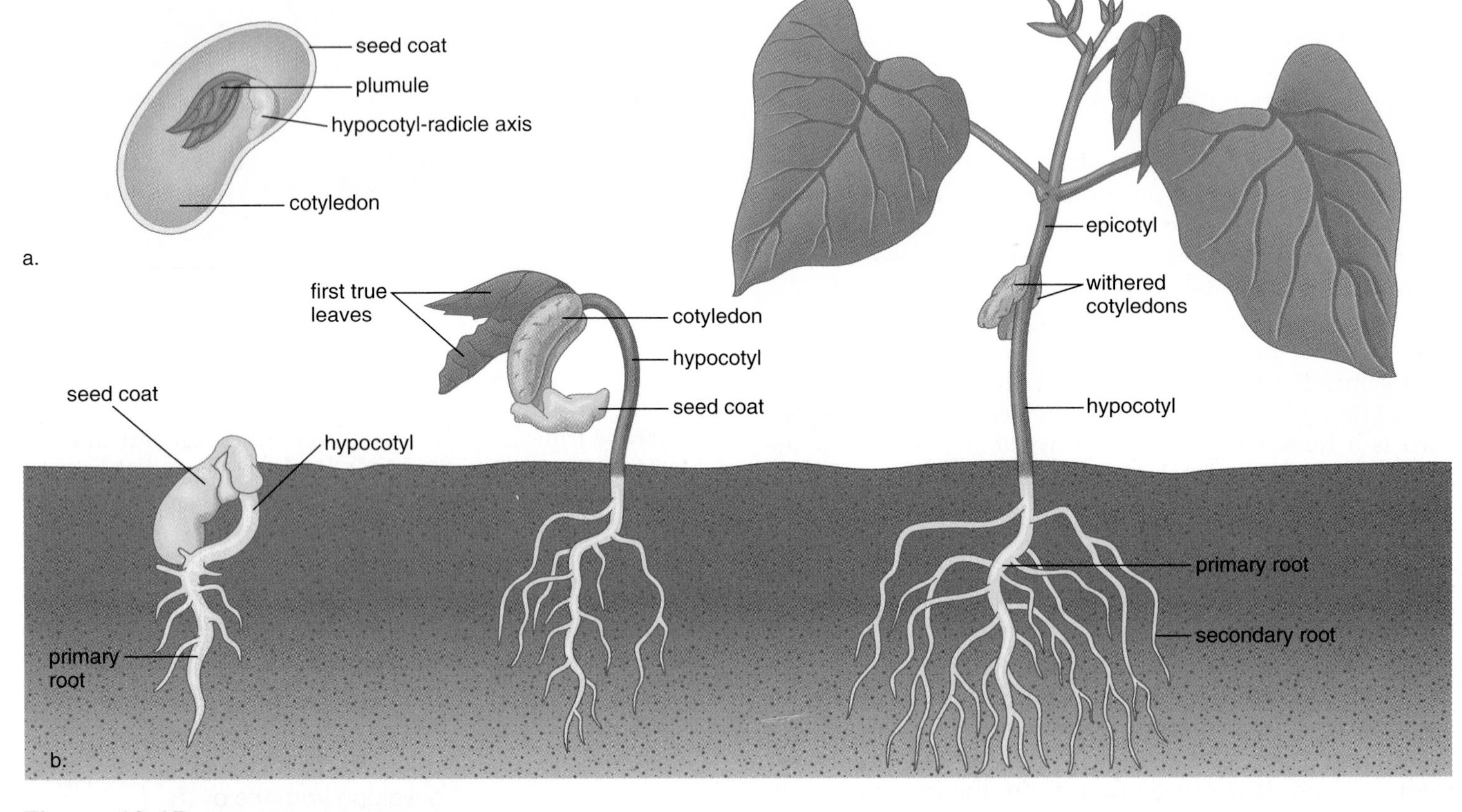

Figure 10.15 Common garden bean, a dicot.
a. Seed structure. **b.** Germination and development of the seedling. Notice that there are two cotyledons, and the leaves are net veined.

season has arrived. **Germination**—an event that takes place if there is sufficient water, warmth, and oxygen to sustain growth—requires regulation, and both inhibitors and stimulators are known to exist. It is known that fleshy fruits (e.g., apples, pears, oranges, and tomatoes) contain inhibitors so that germination does not occur until the seeds are removed and washed. In contrast, stimulators are present in the seeds of some temperate zone woody plants. Mechanical action may also be required. Water, bacterial action, and even fire can act on the seed coat, allowing it to become permeable to water. The uptake of water causes the seed coat to burst.

Dicots Versus Monocots

As mentioned, the embryo of a dicot has two seed leaves, called cotyledons. The cotyledons, which have absorbed the endosperm, supply nutrients to the embryo and seedling, and eventually shrivel and disappear. If the two cotyledons of a bean seed are parted, you can see a rudimentary plant (Fig. 10.15). The epicotyl bears young leaves and is called a **plumule.** As the dicot seedling emerges from the soil, the shoot is hook shaped to protect the delicate plumule. The hypocotyl becomes part of the stem and the radicle develops into the roots. When a seed germinates in darkness, it etiolates—the stem is elongated, the roots and leaves are small, and the plant lacks color and appears spindly. Phytochrome, a pigment that is sensitive to red and far-red light (page 179), regulates this response and induces normal growth once proper lighting is available.

A corn plant is a monocot that contains a single cotyledon. Actually, the endosperm is the food-storage tissue in monocots and the cotyledon does not have a storage role. Corn kernels are actually fruits, and therefore the outer covering is the pericarp (Fig. 10.16). The plumule and radicle are enclosed in protective sheaths called the coleoptile and the coleorhiza, respectively. The plumule and the radicle burst through these coverings when germination occurs.

Germination is a complex event regulated by many factors. The embryo breaks out of the seed coat and becomes a seedling with leaves, stem, and roots.

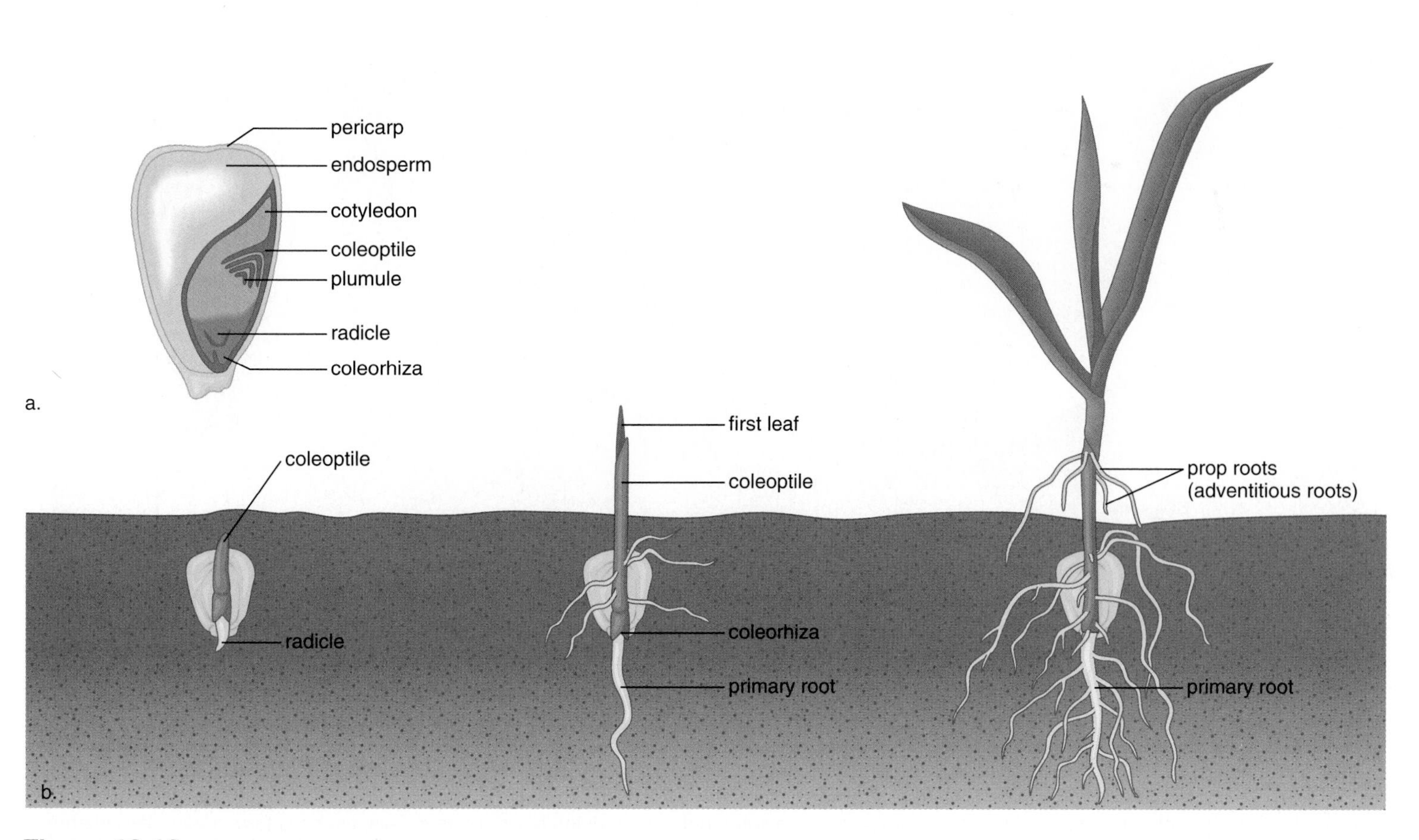

Figure 10.16 Corn, a monocot.
a. Grain structure. **b.** Germination and development of the seedling. Notice that there is one cotyledon and the leaves are parallel veined.

10.5 Asexual Reproduction in Flowering Plants

Because plants contain nondifferentiated meristem tissue, or cells which can revert to meristematic activity (totipotency), they routinely reproduce asexually by *vegetative propagation.* In asexual reproduction there is only one parent, instead of two as in sexual reproduction. Violets will grow from the nodes of rhizomes (underground horizontal stems), and complete strawberry plants will grow from the nodes of stolons (aboveground horizontal stems). White potatoes are actually portions of underground stems, and each eye is a bud that will produce a new potato plant if it is planted with a portion of the swollen tuber. Sweet potatoes are modified roots of sweet potato plants that can be propagated by planting sections of the root. You may have noticed that the roots of some fruit trees, such as cherry and apple trees, produce "suckers," small plants that can be used to grow new trees.

Propagation of Plants in Tissue Culture

Hybridization, the crossing of different varieties of plants or even species, is routinely done to produce plants with desirable traits. Hybridization, followed by vegetative propagation of the mature plants, will generate a large number of identical plants with these traits. But **tissue culture,** growth of a *tissue* in an artificial liquid *culture* medium, has led to *micropropagation,* a commercial method of producing thousands, even millions, of identical seedlings in a limited amount of space (Fig. 10.17). One favorite method to accomplish micropropagation is by *meristem culture.* If the correct proportions of auxin and cytokinin are added to a liquid medium, many new shoots will develop from a single callus, a group of nondifferentiated cells. When these are removed, more shoots form. Since the shoots are genetically identical, the adult plants that develop from them, called *clonal plants,* all have the same traits.

Figure 10.17 Micropropagation.
a. Sections of carrot root are cored, and thin slices are placed in a liquid nutrient medium. **b.** After a few days, the cells form a nondifferentiated callus. **c.** After several weeks, the callus begins sprouting cloned carrot plants. **d.** Eventually the carrot plants can be moved from the culture medium and potted.

The culturing of leaf, stem, or root tissues has led to a technique called *cell suspension culture.* Rapidly growing calluses are cut into small pieces and shaken in a liquid nutrient medium so that single cells or small clumps of cells break off and form a suspension. These cells will produce the same chemicals as the entire plant. For example, cell suspension cultures of *Cinchona ledgeriana* produce quinine, and those of *Digitalis lanata* produce digitoxin. Scientists envision that it will also be possible to maintain cell suspension cultures in bioreactors for the purpose of producing chemicals used in the production of drugs, cosmetics, and agricultural chemicals. If so, it will no longer be necessary to farm plants simply for the purpose of acquiring the chemicals they produce.

Genetic Engineering of Plants

Some plants will grow from protoplasts, naked cells that will take up foreign DNA in tissue culture medium. This form of genetic engineering has been most successful in dicots because monocots, such as corn and wheat, tend not to grow from protoplasts. But new methods have been developed to introduce DNA into plant cells that have a cell wall. For example, a device called a particle gun can bombard plant cells with DNA-coated microscopic metal particles. Many types of plant cells in tissue culture, including corn and wheat, have been genetically engineered using a particle gun. Later, adult plants are generated.

Various crops have been engineered to be resistant to viral infections, insect predation, and herbicides that are judged to be environmentally safe. If crops are resistant to a broad-spectrum herbicide and weeds are not, then the herbicide can be used to kill the weeds. The hope is that in the future it will be possible to produce plants that have a higher protein content and that require less water and fertilizer.

> The ability of plants to reproduce asexually has led to the generation of plants in tissue culture. This, in turn, has promoted the genetic engineering of plants.

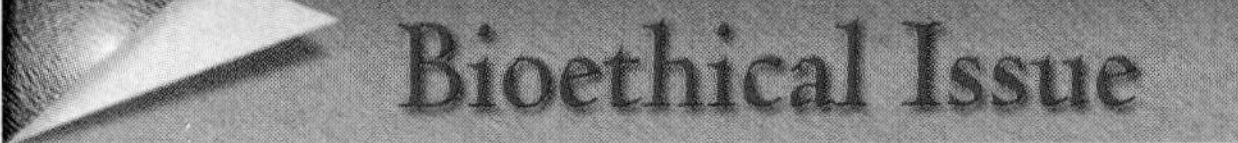

We are in the midst of a pollination crisis due to a decline in the population of honeybees and many other insects, birds, and small mammals that transfer pollen from stamen to stigma. Pollinator populations have been decimated by pollution, pesticide use, and destruction or fragmentation of natural areas. Belatedly, we have come to realize that various types of bees are responsible for pollinating such cash crops as blueberries, cranberries, and squash, and are partly responsible for pollinating apple, almond, and cherry trees.

Why are we so short-sighted when it comes to protecting the environment and living creatures like pollinators? Because pollinators are a resource held in common. The term "commons" originally meant a piece of land where all members of a village were allowed to graze their cattle. The farmer who thought only of himself and grazed more cattle than his neighbor was better off. The difficulty is, of course, that eventually the resource is depleted and everyone loses.

So, when a farmer or property owner uses pesticides he is only thinking of his field or his lawn, and not the good of the whole. The commons can only be protected if citizens have the foresight to enact rules and regulations by which all abide. DDT was outlawed in this country in part because it led to the decline of birds of prey. Similarly, we need legislation to protect pollinators from those factors that kill them off. As a society, we need legislation to protect pollinators because it helps protect the food supply for all of us.

Questions

1. Are you willing to stop using pesticides that kill pollinators if it means your lawn will suffer? Why or why not?
2. Are you willing to pressure your representatives for legislation to protect pollinators? Why or why not?
3. Are you willing to turn your lawn and garden into a haven for pollinators? Why or why not?

Summarizing the Concepts

10.1 Water and Mineral Transport

Water transport in plants occurs within xylem. The cohesion-tension model of xylem transport states that transpiration (evaporation of water at stomates) creates tension, which pulls water upward in xylem. This method works only because water molecules are cohesive.

Stomates open when guard cells take up potassium (K^+) ions and water follows by osmosis. Stomates open because the entrance of water causes the guard cells to buckle out.

10.2 Organic Nutrient Transport

Transport of organic nutrients in plants occurs within phloem. The pressure-flow theory of phloem transport states that sugar is actively transported into phloem at a source, and water follows by osmosis. The resulting increase in pressure creates a flow, which moves water and sucrose to a sink.

10.3 Plant Responses to Environmental Stimuli

Plant hormones control plant responses to environmental stimuli. Tropisms are growth responses toward or away from unidirectional stimuli. When a plant is exposed to light, auxin moves laterally from the bright to the shady side of a stem. Thereafter, the cells on the shady side elongate and the stem moves toward the light.

Plant hormones most likely control photoperiodism. Short-day plants flower when the days are shorter (nights are longer) than a critical length, and long-day plants flower when the days are longer (nights are shorter) than a critical length. Some plants are day-length neutral. Phytochrome, a plant pigment that responds to daylight, is believed to be a part of a biological clock system that in some unknown way brings about flowering.

10.4 Sexual Reproduction in Flowering Plants

Flowering plants have an alternation of generations life cycle, which includes separate microgametophytes and megagametophytes. The pollen grain, the microgametophyte, is produced within the stamens of a flower. The megagametophyte is produced within the ovule of a flower. Following pollination and fertilization, the ovule matures to become the seed and the ovary becomes the fruit. The enclosed seeds contain the embryo (hypocotyl, epicotyl, plumule, radicle) and stored food (endosperm and/or cotyledons). When a seed germinates, the root appears below and the shoot appears above.

10.5 Asexual Reproduction in Flowering Plants

Many flowering plants reproduce asexually, as when the nodes of stems (either aboveground or underground) give rise to entire plants, or when an isolated root produces new shoots. Micropropagation, the production of clonal plants utilizing tissue culture, is now a commercial venture. The particle-gun technique allows foreign genes to be introduced into plant cells, which then develop into adult plants with particular traits.

Studying the Concepts

1. Explain the cohesion-tension theory of water transport. 170
2. What mechanism controls the opening and closing of stomates by guard cells? 172
3. Explain the pressure-flow theory of phloem transport. 174
4. Name five plant hormones and state their functions. 176
5. How is auxin believed to bring about elongation of cells so that positive phototropism occurs. 176
6. With regard to photoperiodism, how was it shown that the length of darkness, not the day length, actually controls flowering? 179
7. What is phytochrome, and what are two possible functions of phytochrome in plants? 179
8. Describe how the megagametophyte (female gametophyte) forms in flowering plants. 180
9. Describe how the microgametophyte (male gametophyte) forms in flowering plants. 180
10. Contrast the monocot seed and its germination with the dicot seed and its germination. 185
11. What is one favorite method of micropropagating plants? 186

Testing Yourself

Choose the best answer for each question.

1. Stomates are usually open
   a. at night, when the plant requires a supply of oxygen.
   b. during the day, when the plant requires a supply of carbon dioxide.
   c. whenever there is excess water in the soil.
   d. All of these are correct.
2. What is the role of transpiration in water and mineral transport? Transpiration
   a. causes water molecules to be cohesive and cling together.
   b. occurs in the vessel elements and not the tracheids which do have end walls.
   c. is the force that causes sugar to be transported in xylem sap.
   d. creates the tension that draws a column of water up the vessel elements.
   e. Both a and d are correct.
3. The pressure-flow model of phloem transport states that
   a. phloem sap always flows from the leaves to the root.
   b. phloem sap always flows from the root to the leaves.
   c. water flow brings sucrose from a source to a sink.
   d. Both a and c are correct.
4. Root hairs do not play a role in
   a. oxygen uptake.
   b. mineral uptake.
   c. water uptake.
   d. carbon dioxide uptake.
5. After an agar block is placed on one side of an oat seedling, it bends only if
   a. unidirectional light is present.
   b. the agar block contains auxin.
   c. unidirectional light is present and the agar block contains auxin.
   d. the agar block is acidic.
   e. the agar block is acidic and unidirectional light is present and the agar block contains auxin.
6. Which of these is a correct statement?
   a. Both stems and roots show positive gravitropism.
   b. Both stems and roots show negative gravitropism.
   c. Only stems show positive gravitropism.
   d. Only roots show positive gravitropism.
7. Short-day plants
   a. are the same as long-day plants.
   b. are apt to flower in the fall.
   c. do not have a critical photoperiod.
   d. will not flower if a short-day is interrupted by bright light.
   e. All of these are correct.
8. A plant requiring a dark period of at least 14 hours will
   a. flower if a 14-hour night is interrupted by a flash of light.
   b. not flower if a 14-hour night is interrupted by a flash of light.
   c. not flower if the days are 14 hours long.
   d. Both b and c are correct.
9. How is the megaspore in the plant life cycle similar to the microspore?
   Both
   a. have the diploid number of chromosomes.
   b. become an embryo sac.
   c. become a gametophyte that produces a gamete.
   d. are necessary to seed production.
   e. Both c and d are correct.
10. Which of these is mismatched?
    a. polar nuclei—plumule
    b. egg and sperm—zygote
    c. ovule—seed
    d. ovary—fruit

11. Label the arrows and dots in this diagram. The icon represented by the dots plays what role in the opening of stomates?

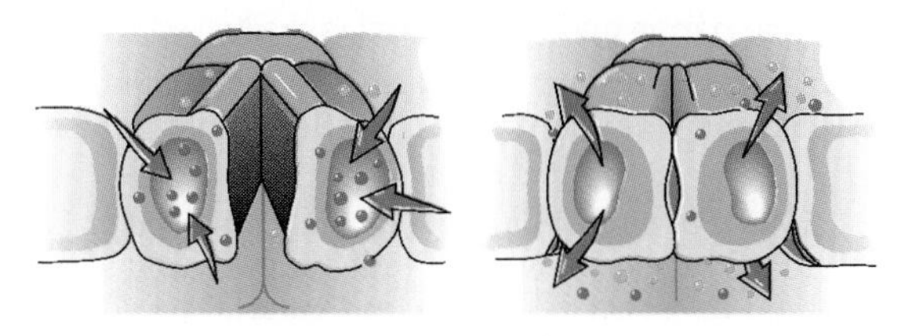

12. Label the following diagram of alternation of generations in flowering plants.

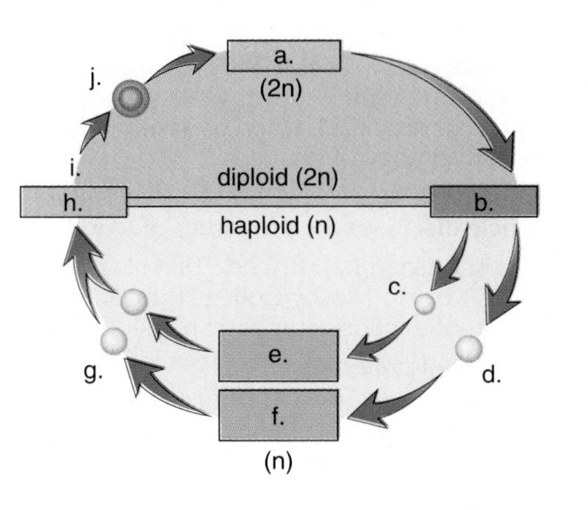

Thinking Scientifically

1. A twig with leaves is placed in the top of an open tube that contains water above mercury. 171
   a. Atmospheric pressure alone is sufficient to raise mercury only 76 cm (76 cm mercury = 10.5 m water). What is atmospheric pressure, and of what significance is this finding for a tree that is 120 m high?
   b. Why does the mercury rise higher than 76 cm when a twig with leaves is placed in the top of the tube?
   c. What does the experiment suggest about the ability of transpiration to raise water to the top of tall trees?
2. Scientists wish to produce a strain of corn that is resistant to a particular pesticide. 186–187
   a. Why would they choose to genetically engineer a diploid corn cell instead of an egg or sperm?
   b. How would they acquire diploid corn cells?
   c. How would they acquire adult plants from the genetically engineered corn cells?
   d. If they want to sell genetically altered corn seeds from these plants, how would they acquire seeds?

Understanding the Terms

alternation of generations 180
anther 180
auxin 176
cohesion-tension theory 170
cotyledon 182
cytokinin 178
double fertilization 181
egg 181
embryo 181
embryo sac 180
endosperm 181
epicotyl 182
ethylene 178
filament 180
flower 180
fruit 183
gametophyte 180
generative cell 180
germination 185
gibberellin 178
guard cell 172
hormone 176
hypocotyl 182
megaspore 180
microspore 180
mineral 171
ovary 180
ovule 180
petal 180
phloem 174
photoperiodism 178
phototropism 177
phytochrome 179
pistil 180
plasmodesma 174
plumule 185
polar nuclei 180
pollen grain 180
pollen sac 180
pollination 180
pressure-flow theory 174
radicle 182
seed 181
seed coat 181
sepal 180
sink 174
source 174
spore 180
sporophyte 180
stamen 180
stigma 180
stomate 172
style 180
tissue culture 186
transpiration 170
tropism 176
tube cell 180
xylem 170
zygote 181

Match the terms to these definitions:

a. __________ In plants, a growth response toward or away from a directional stimulus.

b. __________ In the life cycle of a plant, the haploid generation that produces gametes.

c. __________ Explanation for phloem transport; osmotic pressure following active transport of sugar into phloem brings about a flow of sap from a source to a sink.

d. __________ In seed plants, a small spore that develops into the sperm-producing microgametophyte; pollen grain.

e. __________ Plant pigment that induces a photoperiodic response in plants.

Using Technology

Your study of plant physiology and reproduction is supported by these available technologies:

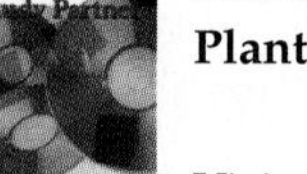

Essential Study Partner CD-ROM

Plants → Translocation, Plant Responses, Reproduction

Visit the Mader web site for related ESP activities.

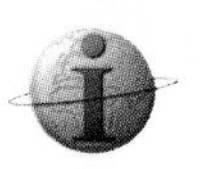

Exploring the Internet

The Mader Home Page provides resources and tools as you study this chapter.

http://www.mhhe.com/biosci/genbio/mader

Further Readings for Part 2

Arakawa, T., and Langridge W. R. H. May 1998. Plants are not just passive creatures! *Nature Medicine* 4(5):550. Plants are being used as bioreactors to produce foreign proteins for human immunity.

Balick, M. J., and Cox, P. A. 1996. *Plants, people, and culture: The science of ethnobotany.* New York: *Scientific American* Library. This interesting, well-illustrated book discusses the medicinal and cultural uses of plants, and the importance of rain forest conservation.

Busch, R. October/November 1998. The value of autumn leaves. *National Wildlife* 36(6):32. Article examines the physiology behind autumn leaf colors.

Canny, M. J. March/April 1998. Transporting water in plants. *American Scientist* 86(2):152. The cohesion-tension theory of water transport is discussed in detail.

Chrispeels, M., and Sadava, D. 1994. *Plants, genes and agriculture.* Boston: Jones and Bartlett Publishers. Teaches plant biology in an agricultural context.

Gibson, A. C. November 1998. Photosynthetic organs of desert plants. *Bioscience* 48(11):911. The specialized leaf and stem adaptations in desert plants may be to maximize photosynthetic rates and energy production, rather than to conserve water.

Glenn, E. P., et al. August 1998. Irrigating crops with seawater. *Scientific American* 279(2):76. Certain useful salt-tolerant plants can flourish on seawater irrigation.

Lee, D. W. January/February 1997. Iridescent blue plants. *American Scientist* 85(1):56. Some tropical plants produce a blue color, as do some insects, by using layered filters.

Levetin, E., and McMahon, K. 1996. *Plants and society.* Dubuque, Iowa: Wm. C. Brown Publishers. Basic botany and the impact of plants on society are topics covered in this introductory text.

Luoma, J. R. March 1997. The magic of paper. *National Geographic* 191(3):88. The paper-making process is discussed in this article.

Martinez del Rio, C. February 1996. Murder by mistletoe. *Natural History* 105(2):85. The quintral is a mistletoe that parasitizes cacti.

Mauseth, J. 1995. *Botany: An introduction to plant biology.* 2d ed. Philadelphia: Saunders College Publishing. Emphasizes evolution and diversity in botany and general principles of plant physiology and anatomy.

Milot, C. 17 July 1998. Plant biology in the genoma era. *Science* 281(5375):331. Genomic data can be used to make plants that produce more vitamins and minerals, and can help explain physiological processes, such as flowering.

Moore, R., and Clark, W. D., et al. 1998. *Botany.* 2d ed. Dubuque, Iowa: WCB/McGraw-Hill. This introductory botany text stresses the process of science while presenting the evolution, anatomy, and physiology of plants.

Niklas, K. J. February 1996. How to build a tree. *Natural History* 105(2):48. Article discusses the properties of wood.

Northington, D., and Goodin, J. R. 1996. *The botanical world.* 2d ed. St. Louis: Times-Mirror/Mosby College Publishing. This is an account of plant interactions and basic physiology.

Pitelka, L. F., et al. September/October 1997. Plant migration and climate change. *American Scientist* 85(5):464. A relationship between plant migration and climate change, as evidenced by the fossil record and computer models, is discussed.

Redington, C. 1994. *Plants in wetlands.* Dubuque, Iowa: Kendall/Hunt Publishing Company. Explains how specific plants interact within the wetlands ecosystem.

Schmiedeskamp, M. December 1997. Pollution-purging poplars. *Scientific American* 277(6):46. Hybrid poplar trees may be used to break down toxic organic compounds in the soil.

Seymour, R. S. March 1997. Plants that warm themselves. *Scientific American* 276(3):104. Some plants generate heat to keep blossoms at a constant temperature.

Stern, K. 1997. *Introductory plant biology.* 7th ed. Dubuque, Iowa: Wm. C. Brown Publishers. Presents basic botany in a clear, informative manner.

Stryer, L. 1995. *Biochemistry.* 5th ed. New York: W. H. Freeman and Company. Chapter 22 of this text presents an advanced but understandable treatment of photosynthesis.

Sze, P. 1998. *A biology of the algae.* 3d ed. Dubuque, Iowa: WCB/McGraw-Hill. A text that introduces algae morphology, evolution, and ecology to the botany major.

Walsh, R. May 1997. The chocolate bug. *Natural History* 106(4):54. The cacao tree's survival depends upon a tiny insect, which dictates where cacao will grow.

PART III

Maintenance of the Human Body

Claude Bernard, a French physiologist, realized in 1859 that while animals, such as humans, live in an external environment, the cells of the body are surrounded by an internal environment composed of tissue fluid, which is serviced by blood. Later, Walter Cannon, an American physiologist, introduced the term homeostasis to emphasize the dynamic equilibrium that keeps the composition of tissue fluid, and various vital signs like blood pressure and body temperature, within normal limits. The systems of the body discussed in this part either add substances to and/or remove substances from the blood. In this way, they contribute to homeostasis.

Careers in Human Anatomy & Physiology

Pharmacist filling prescriptions.

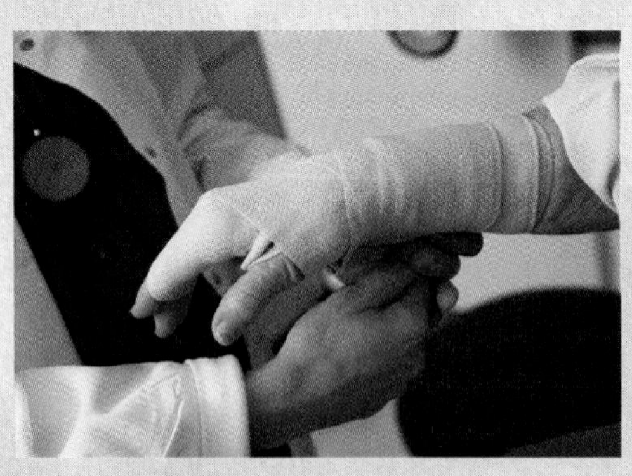
Physicians assistant preparing a cast.

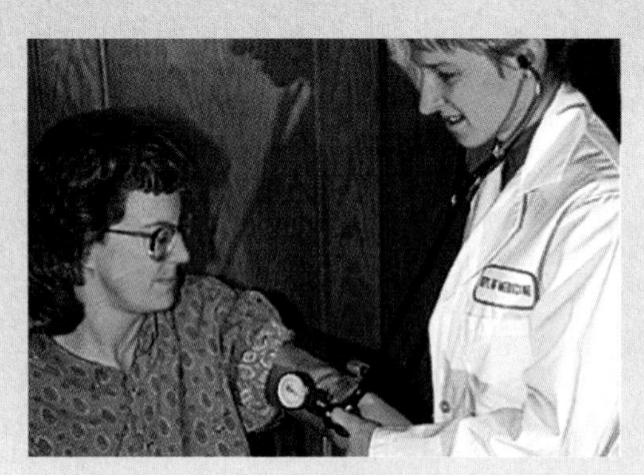
Registered nurse taking blood pressure.

Pharmacists measure, count, mix, and dispense drugs and medicines prescribed by physicians, physician assistants, and dentists, among others. Pharmacists must understand the use, composition, and effects of drugs, and how they are tested for purity and strength.

Respiratory therapists evaluate, treat, and care for patients with breathing disorders. They treat many types of patients from premature infants whose lungs are not fully developed to elderly people whose lungs are diseased. Therapists run ventilators and can check on them for mechanical problems.

Home health aides help elderly, disabled, or ill persons to live in their own homes instead of a health facility. They assist these patients with their daily routines, check their vital signs, oversee their exercise needs, and assist with medication routines.

Physician assistants (P.A.s) are formally trained to perform many of the routine but time-consuming tasks physicians usually do. They take medical histories, perform physical examinations, order laboratory tests and X rays, make preliminary diagnoses, and give inoculations. They also treat minor injuries by suturing, splinting, and casting.

Dental hygienists provide preventive dental care and teach patients how to practice good oral hygiene. They remove calculus, stain, and plaque from above and below the gumline; apply caries-preventive agents such as fluorides and pit and fissure sealants; and expose and develop dental X rays.

Registered nurses (R.N.s) care for the sick and injured and help people stay well. They observe, assess, and record symptoms, reactions, and progress; assist physicians during treatments and examinations; and administer medications. Licensed practical nurses (L.P.N.s) provide basic bedside care.

Dietitians and nutritionists plan nutrition programs and supervise the preparation and serving of meals in institutions such as hospitals and schools. Working in such places as public health clinics, home health agencies, and health maintenance organizations, dietitians and nutritionists determine individual needs, establish nutritional care plans, instructing both individuals and families.

C H A P T E R 11

Human Organization

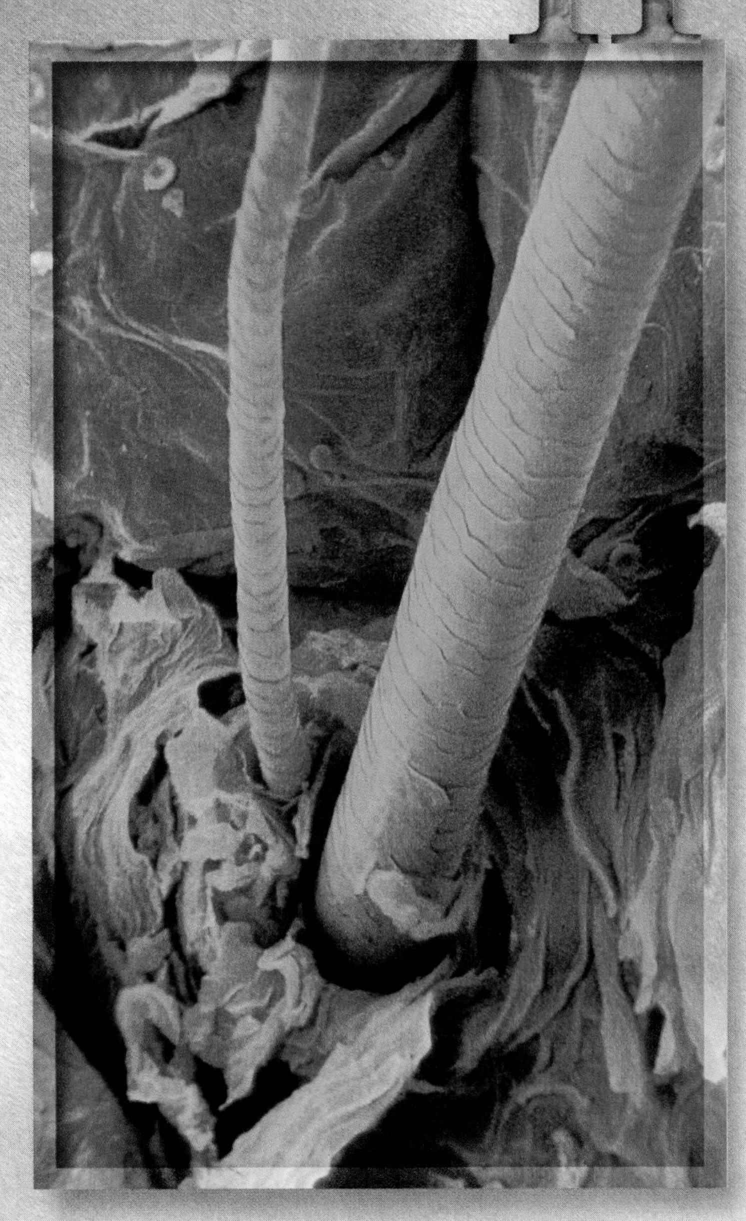

The cilia of epithelial cells lining the trachea sweep particles toward the throat. This action helps prevent impurities from reaching the lower air passages and helps prevent respiratory infections.

Chapter Concepts

Sally busily hammers nails into the roof of a new building. And except for the sweat dripping into her eyes, she hardly notices the heat of this midsummer morning even though it is about 100°F. She had skipped breakfast and then gulped down three jelly doughnuts and a cup of coffee the construction foreman had provided for the carpenters. At lunchtime, Sally heads for the local diner, where the temperature is a chilly 68°F. Remembering her diet, she orders a salad without dressing and a diet cola. Then she goes back to work.

The remarkable thing is that although Sally was exposed to environmental temperatures of both 68°F and 100°F, her body kept a relatively constant temperature of about 98.6°F. And even though she loaded her blood with sugar after she came to work and deprived it of sugar at lunch, her body maintained a relatively constant blood sugar level of about 100 mg/100 ml blood. And her blood remained at a pH of about 7.4 even though she consumed carbonic acid in the diet cola. As we will find out in this chapter, tissues and organs work together to maintain relatively constant levels of internal temperature and blood chemistry, regardless of whether the external temperature is hot or cold, or meals are high or low in sugar or acids. This chapter discusses the basic organization of the body, revealing the truly amazing way the body maintains normal conditions without us even thinking about it.

11.1 Types of Tissues

A **tissue** is composed of similarly specialized cells that perform a common function in the body. The tissues of the human body can be categorized into four major types: *epithelial tissue,* which covers body surfaces and lines body cavities; *connective tissue,* which binds and supports body parts; *muscular tissue,* which moves body parts; and *nervous tissue,* which receives stimuli and conducts impulses from one body part to another.

Cancers are classified according to the type of tissue from which they arise. **Carcinomas,** the most common type, are cancers of epithelial tissues; sarcomas are cancers arising in muscle or connective tissue (especially bone or cartilage); leukemias are cancers of the blood; and lymphomas are cancers of lymphoid tissue. The chance of a cancer developing in a particular tissue shows a positive correlation to the rate of cell division; new blood cells arise at a rate of 2,500,000 cells per second, and epithelial cells also reproduce at a high rate.

Epithelial Tissue

Epithelial tissue, also called epithelium, consists of tightly packed cells that form a continuous layer or sheet lining the entire body surface and most of the body's inner cavities. On the external surface, it protects the body from injury, drying out, and possible **pathogen** (virus and bacterium) invasion. On internal surfaces, epithelial tissue may be specialized for other functions in addition to protection. For example, epithelial tissue secretes mucus along the digestive tract and sweeps up impurities from the lungs by means of cilia (sing., **cilium**). It efficiently absorbs molecules from kidney tubules and from the intestine because of minute cellular extensions called **microvilli.**

There are three types of epithelial tissue (Fig. 11.1). **Squamous epithelium** is *composed of flattened cells* and is found lining the lungs and blood vessels. **Cuboidal epithelium** contains *cube-shaped cells* and is found lining the kidney tubules. **Columnar epithelium** has cells *resembling rectangular pillars* or *columns,* and nuclei are usually located near the bottom of each cell. This epithelium is found lining the digestive tract. Ciliated columnar epithelium is found lining the oviducts, where it propels the egg toward the uterus or womb.

An epithelium can be simple or stratified. Simple means the tissue has a single layer of cells, and stratified means that the tissue has layers of cells piled one on top of the other. The walls of the smallest blood vessels, called capillaries, are composed of a single layer of epithelial cells. The nose, mouth, esophagus, anal canal, and vagina are all lined by stratified squamous epithelium. As we shall see, the outer layer of skin is also stratified squamous epithelium, but the cells have been reinforced by keratin, a protein that provides strength.

Pseudostratified epithelium appears to be layered; however, true layers do not exist because each cell touches the base line. The lining of the windpipe, or trachea, is called *pseudostratified ciliated columnar epithelium.* A secreted covering of mucus traps foreign particles, and the upward motion of the cilia carries the mucus to the back of the throat, where it may either be swallowed or expectorated. Smoking can cause a change in mucous secretion and inhibit ciliary action, and the result is a chronic inflammatory condition called bronchitis.

A so-called **basement membrane** often joins an epithelium to underlying connective tissue. We now know that the basement membrane is glycoprotein, reinforced by fibers that are supplied by connective tissue.

An epithelium sometimes secretes a product, in which case it is described as glandular. A **gland** can be a single epithelial cell, as in the case of mucus-secreting goblet cells found within the columnar epithelium lining the digestive tract, or a gland can contain many cells. Glands that secrete their product into ducts are called *exocrine glands,* and those that secrete their product directly into the bloodstream are called *endocrine glands.* The pancreas is both an exocrine gland, because it secretes digestive juices into the small intestine via ducts, and an endocrine gland, because it secretes insulin into the bloodstream.

Epithelial tissue is named according to the shape of the cell. These tightly packed protective cells can occur in more than one layer, and the cells lining a cavity can be ciliated and/or glandular.

Visual Focus

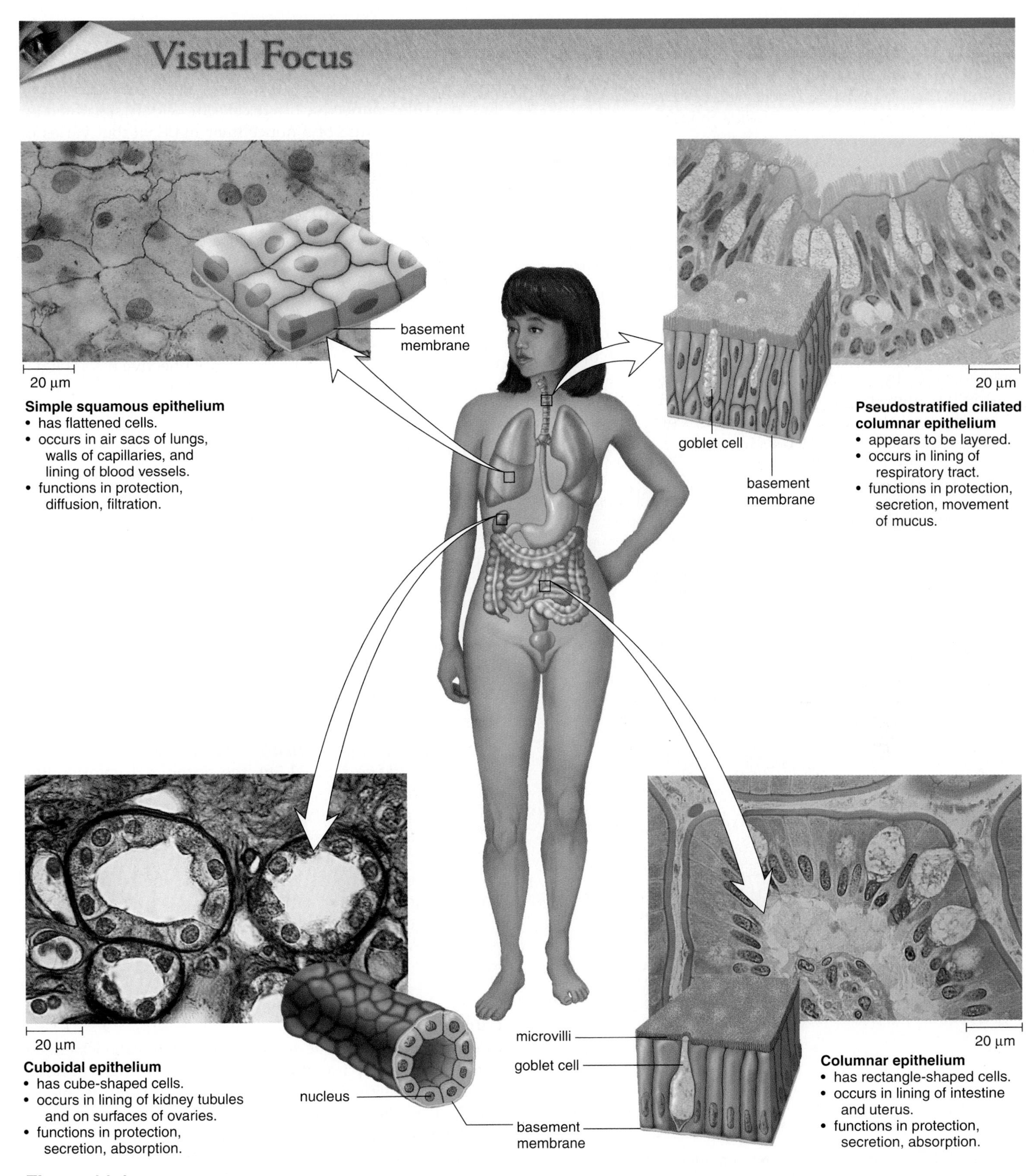

Figure 11.1 Epithelial tissue.
The three types of epithelial tissue—squamous, cuboidal, and columnar—are named for the shape of their cells. They all have a protective function, as well as the other functions noted.

Junctions Between Cells

The cells of a tissue can function in a coordinated manner when the plasma membranes of adjoining cells interact. The junctions that occur between cells help cells function as a tissue (Fig. 11.2). A **tight junction** forms an impermeable barrier because adjacent plasma membrane proteins actually join, producing a zipperlike fastening. In the intestine, the gastric juices stay out of the body, and in the kidneys, the urine stays within kidney tubules because epithelial cells are joined by tight junctions.

A **gap junction** forms when two adjacent plasma membrane channels join. This lends strength, but it also allows ions, sugars, and small molecules to pass between the two cells. Gap junctions in heart and smooth muscle ensure synchronized contraction. In an **adhesion junction** (desmosome), the adjacent plasma membranes do not touch but are held together by intercellular filaments firmly attached to buttonlike thickenings. In some organs—like the heart, stomach, and bladder, where tissues get stretched—adhesion junctions hold the cells together.

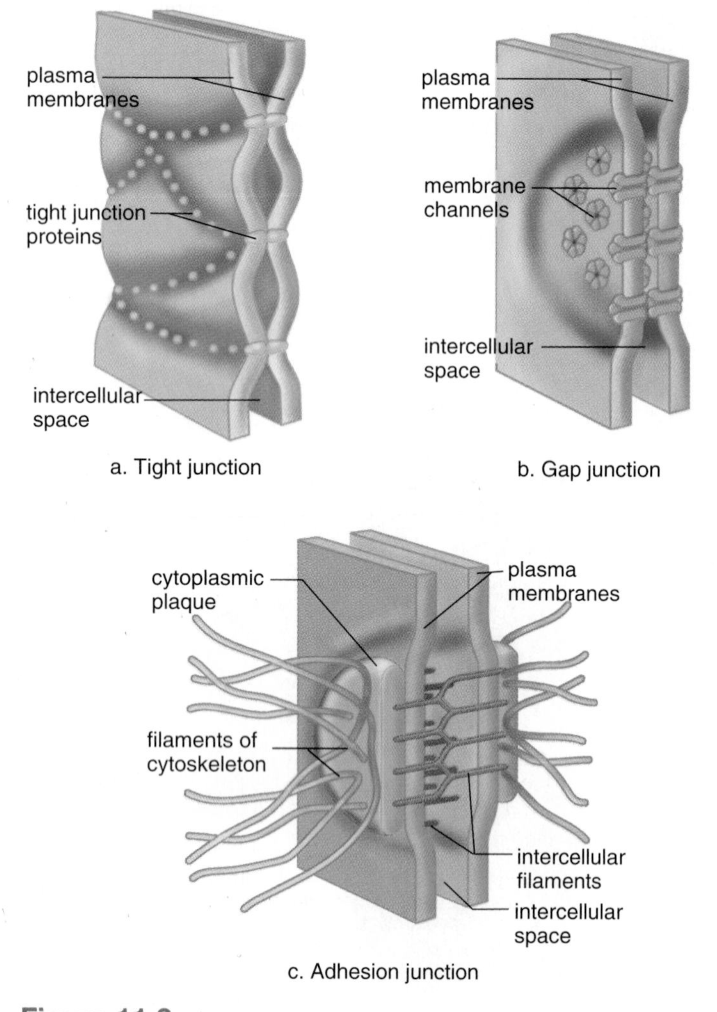

Figure 11.2 Junctions between epithelial cells.
Epithelial tissue cells are held tightly together by **(a)** tight junctions; **(b)** gap junctions that allow materials to pass from cell to cell; and **(c)** adhesion junctions that allow tissues to stretch.

Connective Tissue

Connective tissue binds organs together, provides support and protection, fills spaces, produces blood cells, and stores fat. As a rule, connective tissue cells are widely separated by a **matrix,** consisting of a noncellular material that varies in consistency from solid to semifluid to fluid. The matrix may have fibers of three possible types: **Collagen** (white) **fibers** contain collagen, a protein that gives them flexibility and strength. **Reticular fibers** are very thin collagen fibers that are highly branched and form delicate supporting networks. **Elastic** (yellow) **fibers** contain elastin, a protein that is not as strong as collagen but is more elastic.

Loose Fibrous and Dense Fibrous Tissues

Both loose fibrous and dense fibrous connective tissues have cells called **fibroblasts** that are located some distance from one another and are separated by a jellylike matrix containing white collagen fibers and yellow elastic fibers.

Loose fibrous connective tissue supports epithelium and also many internal organs (Fig. 11.3*a*). Its presence in lungs, arteries, and the urinary bladder allows these organs to expand. It forms a protective covering enclosing many internal organs, such as muscles, blood vessels, and nerves.

Dense fibrous connective tissue contains many collagen fibers that are packed together. This type of tissue has more specific functions than does loose connective tissue. For example, dense fibrous connective tissue is found in **tendons,** which connect muscles to bones, and in **ligaments,** which connect bones to other bones at joints.

Adipose Tissue and Reticular Connective Tissue

In **adipose tissue** (Fig. 11.3*b*), the fibroblasts enlarge and store fat. The body uses this stored fat for energy, insulation, and organ protection. Adipose tissue is found beneath the skin, around the kidneys, and on the surface of the heart. Reticular connective tissue, also called lymphoid tissue, is present in lymph nodes, the spleen, and the bone marrow. These organs are a part of the immune system because they store and/or produce white blood cells, particularly lymphocytes. All types of blood cells are produced in red bone marrow.

Cartilage

In **cartilage,** the cells lie in small chambers called lacunae (sing., **lacuna**), separated by a matrix that is solid yet flexible. Unfortunately, because this tissue lacks a direct blood supply, it heals very slowly. There are three types of cartilage, distinguished by the type of fiber in the matrix.

Hyaline cartilage (Fig. 11.3*c*), the most common type of cartilage, contains only very fine collagen fibers. The matrix has a white, translucent appearance. Hyaline cartilage is found in the nose and at the ends of the long bones and the ribs, and it forms rings in the walls of respiratory passages. The fetal skeleton also is made of this type of cartilage. Later, the cartilaginous fetal skeleton is replaced by bone.

Elastic cartilage has more elastic fibers than hyaline cartilage. For this reason, it is more flexible and is found, for example, in the framework of the outer ear.

Fibrocartilage has a matrix containing strong collagen fibers. Fibrocartilage is found in structures that withstand tension and pressure, such as the pads between the vertebrae in the backbone and the wedges found in the knee joint.

Bone

Bone is the most rigid connective tissue. It consists of an extremely hard matrix of inorganic salts, chiefly calcium salts, deposited around protein fibers, especially collagen fibers. The inorganic salts give bone rigidity, and the protein fibers provide elasticity and strength, much as steel rods do in reinforced concrete.

Compact bone makes up the shaft of a long bone (Fig. 11.3*d*). It consists of cylindrical structural units called osteons (Haversian systems). The central canal of each osteon is surrounded by rings of hard matrix. Bone cells, called *osteocytes,* are located in spaces called lacunae between the rings of matrix. Blood vessels in the central canal carry nutrients that allow bone to renew itself. The nutrients can reach all of the cells because *canaliculi* (minute canals) containing thin processes of the osteocytes connect the cells with one another and with the central canals.

The ends of a long bone contain spongy bone, which has an entirely different structure. **Spongy bone** contains numerous bony bars and plates, separated by irregular spaces. Although lighter than compact bone, spongy bone still is designed for strength. Just as braces are used for support in buildings, the solid portions of spongy bone follow lines of stress.

Connective tissues, which bind and support body parts, differ according to the type of matrix and the abundance of fibers in the matrix.

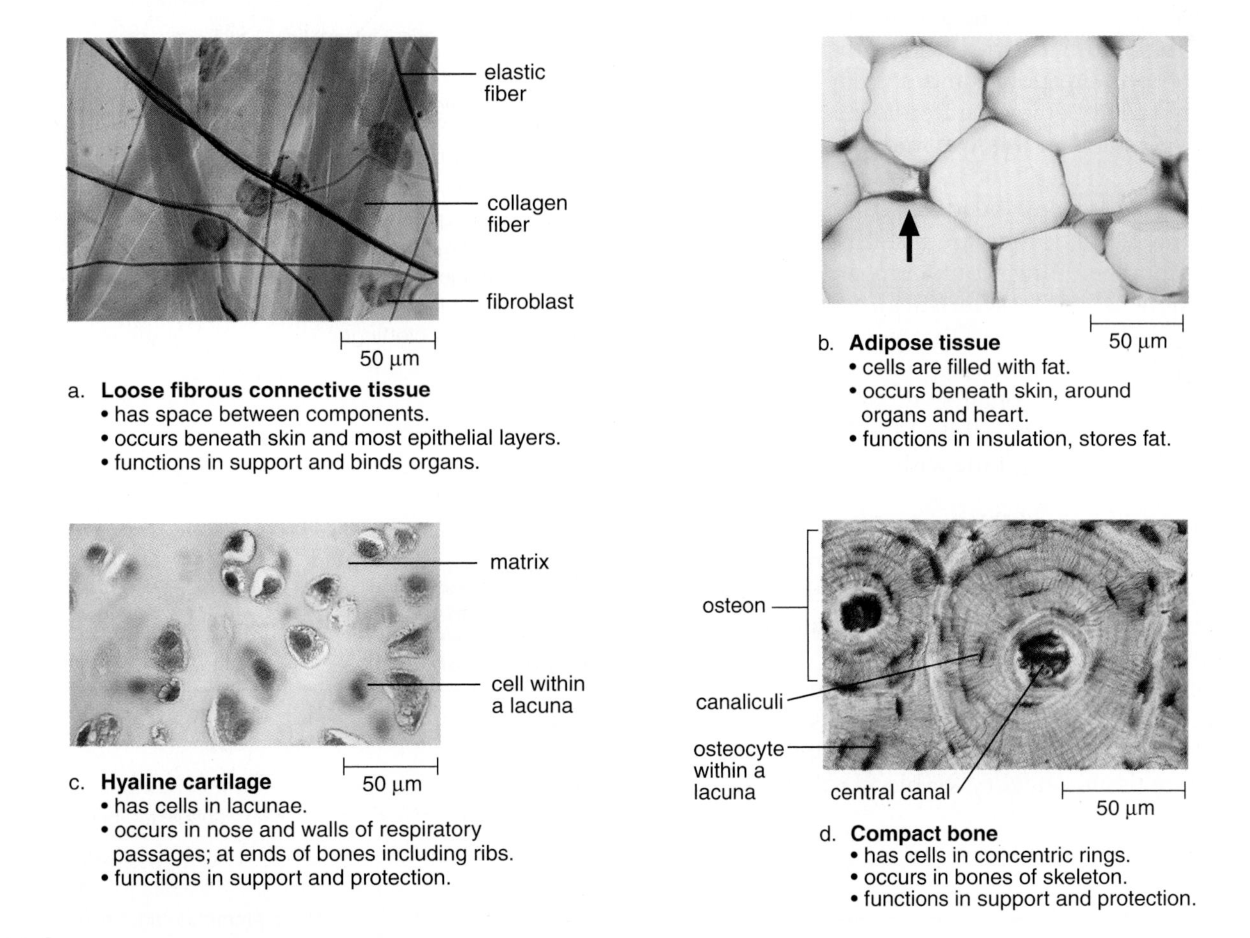

Figure 11.3 Connective tissue examples.
a. In loose connective tissue, cells called fibroblasts are separated by a jellylike matrix, which contains both collagen and elastic fibers.
b. Adipose tissue cells have nuclei (arrow) pushed to one side because the cells are filled with fat. **c.** In hyaline cartilage, the flexible matrix has a white, translucent appearance. **d.** In compact bone, the hard matrix contains calcium salts. Concentric rings of osteocytes in lacunae form an elongated cylinder called an osteon (Haversian system). An osteon has a central canal that contains blood vessels and nerve fibers.

Blood

The functions of blood include transporting molecules, regulating the tissues, and protecting the body. Blood transports nutrients and oxygen to cells and removes carbon dioxide and other wastes. It helps distribute heat and also plays a role in fluid, ion, and pH balance. Various components of blood, as discussed below, help protect us from disease, and its ability to clot prevents fluid loss.

If blood is transferred from a person's vein to a test tube and prevented from clotting, it separates into two layers (Fig. 11.4). The upper liquid layer, called **plasma,** represents about 55% of the volume of whole blood and contains a variety of inorganic and organic substances dissolved or suspended in water (Table 11.1). The lower layer consists of red blood cells (erythrocytes), white blood cells (leukocytes), and blood platelets (thrombocytes). Collectively, these are called the formed elements and represent about 45% of the volume of whole blood. Formed elements are manufactured in the red bone marrow of the skull, ribs, vertebrae, and ends of long bones.

The **red blood cells** are small, biconcave, disk-shaped cells without nuclei. The presence of the red pigment hemoglobin makes the cells red, and in turn, makes the blood red. Hemoglobin is composed of four units; each is composed of the protein globin and a complex iron-containing structure called heme. The iron forms a loose association with oxygen, and in this way red blood cells transport oxygen.

White blood cells may be distinguished from red blood cells by the fact that they are usually larger, have a nucleus, and without staining would appear to be translucent. White blood cells characteristically appear bluish because they have been stained that color. White blood cells, which fight infection, function primarily in two ways. Some white blood cells are phagocytic and engulf infectious pathogens, while other white blood cells produce antibodies, molecules that combine with foreign substances to inactivate them.

Platelets are not complete cells; rather, they are fragments of giant cells present only in bone marrow. When a blood vessel is damaged, platelets form a plug that seals the vessel and along with injured tissues release molecules that help the clotting process.

Blood is unlike other types of connective tissue in that the matrix (i.e., plasma) is not made by the cells. Some people do not classify blood as connective tissue; instead, they suggest a separate tissue category for blood called vascular tissue.

Blood is a connective tissue in which the matrix is plasma.

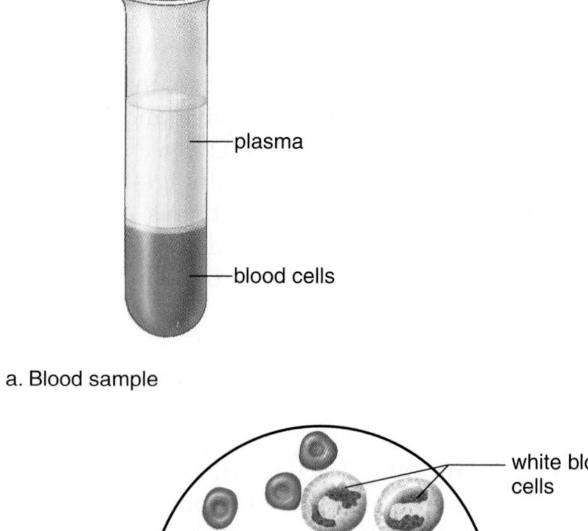

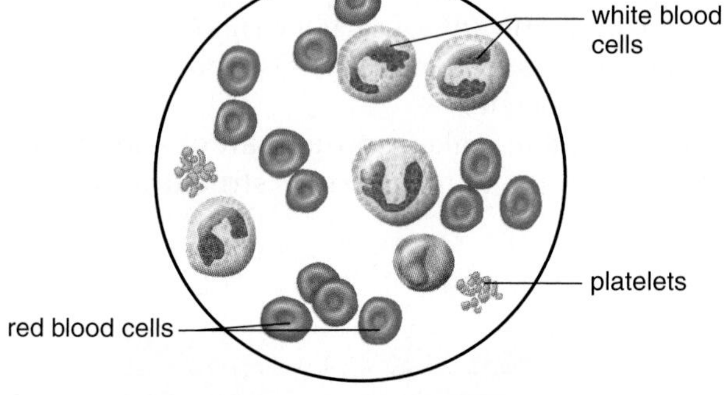

Figure 11.4 Blood, a fluid tissue.
a. In a test tube, a blood sample separates into its two components: blood cells and plasma. **b.** Microscopic examination of a blood smear shows that there are red blood cells, white blood cells, and platelets. Platelets are fragments of a cell. Red blood cells transport oxygen, white blood cells fight infections, and platelets are involved in initiating blood clotting.

Table 11.1 Blood Plasma

Water (92% of Total)	
Solutes (8% of Total)	
Inorganic ions (salts)	Na^+, Ca^{2+}, K^+, Mg^{2+}; Cl^-, HCO^{3-}, PO_4^{3-}, SO_4^{2-}
Gases	O_2, CO_2
Plasma proteins	Albumin, globulins, fibrinogen
Organic nutrients	Glucose, fats, phospholipids, amino acids, etc.
Nitrogenous waste products	Urea, ammonia, uric acid
Regulatory substances	Hormones, enzymes

Muscular Tissue

Muscular (contractile) tissue is composed of cells that are called muscle fibers. Muscle fibers contain actin filaments and myosin filaments, whose interaction accounts for movement. There are three types of vertebrate muscles: skeletal, smooth, and cardiac.

Skeletal muscle, also called voluntary muscle (Fig. 11.5*a*), is attached by tendons to the bones of the skeleton, and when it contracts, body parts move. Contraction of skeletal muscle is under voluntary control and occurs faster than in the other muscle types. Skeletal muscle fibers are cylindrical and quite long—sometimes they run the length of the muscle. They arise during development when several cells fuse, resulting in one fiber with multiple nuclei. The nuclei are located at the periphery of the cell, just inside the plasma membrane. The fibers have alternating light and dark bands that give them a **striated** appearance. These bands are due to the placement of actin filaments and myosin filaments in the cell.

Smooth (visceral) muscle is so named because the cells lack striations. The spindle-shaped cells form layers in which the thick middle portion of one cell is opposite the thin ends of adjacent cells. Consequently, the nuclei form an irregular pattern in the tissue (Fig. 11.5*b*). Smooth muscle is not under voluntary control and therefore is said to be involuntary. Smooth muscle, found in the walls of viscera (intestine, stomach, and other internal organs) and blood vessels, contracts more slowly than skeletal muscle but can remain contracted for a longer time. When the smooth muscle of the intestine contracts, food moves along its lumen (central cavity). When the smooth muscle of the blood vessels contracts, blood vessels constrict, helping to raise blood pressure.

Cardiac muscle (Fig. 11.5*c*) is found only in the walls of the heart. Its contraction pumps blood and accounts for the heartbeat. Cardiac muscle combines features of both smooth muscle and skeletal muscle. It has striations like skeletal muscle, but the contraction of the heart is involuntary for the most part. Cardiac muscle cells also differ from skeletal muscle cells in that they have a single, centrally placed nucleus. The cells are branched and seemingly fused one with the other, and the heart appears to be composed of one large interconnecting mass of muscle cells. Actually, cardiac muscle cells are separate and individual, but they are bound end to end at **intercalated disks,** areas where folded plasma membranes between two cells contain desmosomes and gap junctions.

All muscular tissue contains actin filaments and myosin filaments; these form a striated pattern in skeletal and cardiac muscle, but not in smooth muscle.

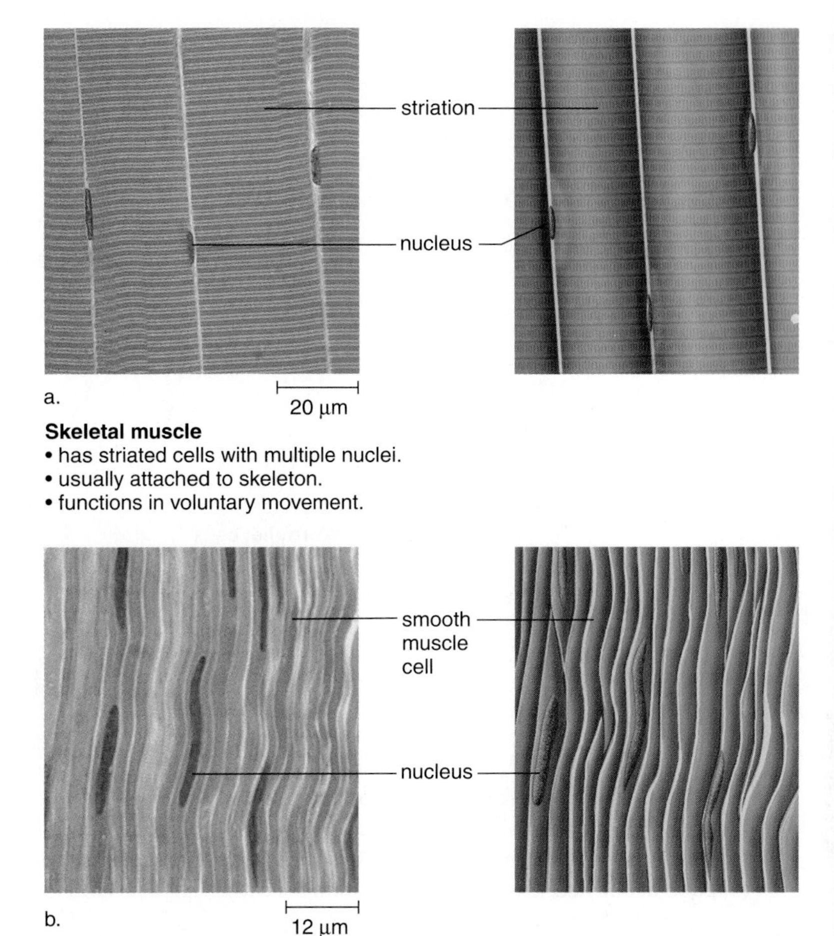

Skeletal muscle
- has striated cells with multiple nuclei.
- usually attached to skeleton.
- functions in voluntary movement.

Smooth muscle
- has spindle-shaped cells, each with a single nucleus.
- occurs in walls of hollow internal organs.
- functions in movement of substances in lumens of body.
- no cross striations, involuntary.

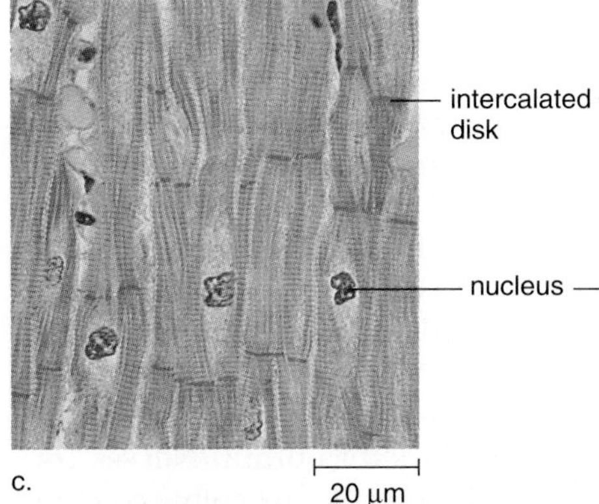

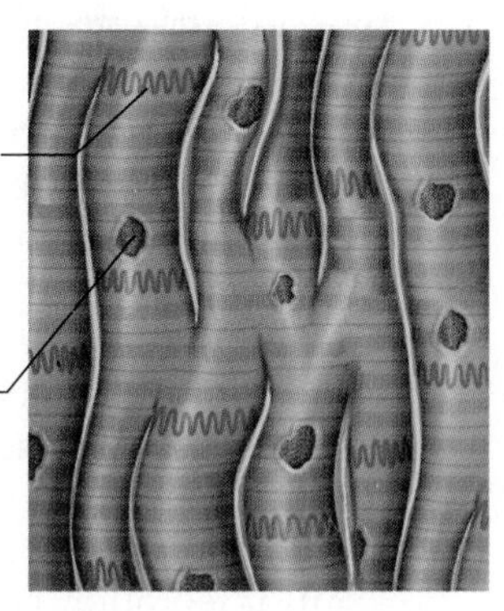

Cardiac muscle
- has branching striated cells, each with a single nucleus.
- occurs in the wall of the heart.
- functions in the pumping of blood.
- involuntary.

Figure 11.5 Muscular tissue.
a. Skeletal muscle is voluntary and striated. **b.** Smooth muscle is involuntary and nonstriated. **c.** Cardiac muscle is involuntary and striated. Cardiac muscle cells branch and fit together at intercalated disks.

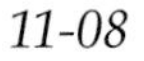

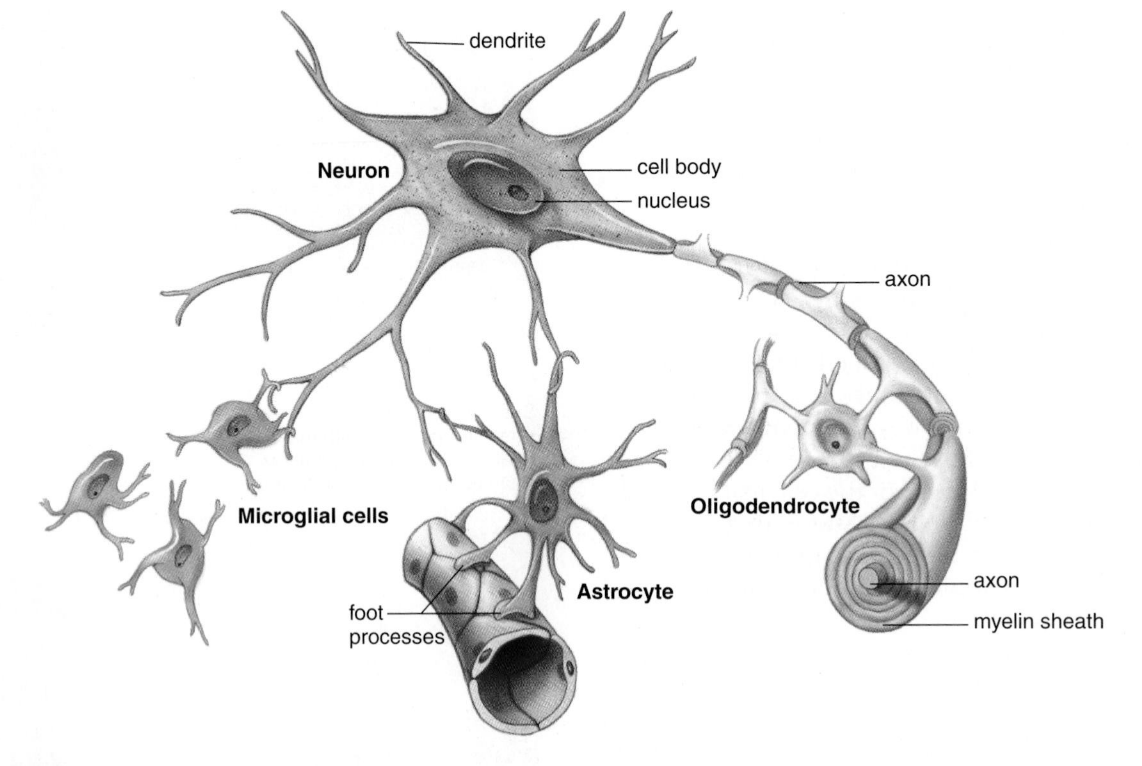

Figure 11.6 Neuron and neuroglial cells.
Neurons conduct nerve impulses. Neuroglial cells, which support and service neurons, have various functions: microglial cells are phagocytes that clean up debris. Astrocytes lie between neurons and a capillary; therefore, substances entering neurons from the blood must first pass through astrocytes. Oligodendrocytes form the myelin sheaths around fibers in the brain and spinal cord.

Nervous Tissue

Nervous tissue, which contains nerve cells called neurons, is present in the brain and spinal cord. A **neuron** is a specialized cell that has three parts: dendrites, cell body, and an axon (Fig. 11.6). A dendrite is a process that conducts signals toward the cell body. The cell body contains the major concentration of the cytoplasm and the nucleus of the neuron. An axon is a process that typically conducts nerve impulses away from the cell body. Axons can be quite long, and outside the brain and the spinal cord, long fibers, bound by connective tissue, form **nerves.**

The nervous system has just three functions: sensory input, integration of data, and motor output. Nerves conduct impulses from sensory receptors to the spinal cord and the brain where integration occurs. The phenomenon called sensation occurs only in the brain, however. Nerves also conduct nerve impulses away from the spinal cord and brain to the muscles and glands, causing them to contract and secrete, respectively. In this way, a coordinated response to the stimulus is achieved.

In addition to neurons, nervous tissue contains neuroglial cells.

Neuroglial Cells

There are several different types of neuroglial cells in the brain (Fig. 11.6), and much research is currently being conducted to determine how much "glial" cells contribute to the functioning of the brain. **Neuroglial cells** outnumber neurons nine to one and take up more than half the volume of the brain, but until recently, they were thought to merely support and nourish neurons. Three types of neuroglial cells are oligodendrocytes, microglial cells, and astrocytes. Oligodendrocytes form myelin; and microglial cells, in addition to supporting neurons, phagocytize bacterial and cellular debris. Astrocytes provide nutrients to neurons and produce a hormone known as glial-derived growth factor, which someday might be used as a cure for Parkinson's disease and other diseases caused by neuron degeneration. Neuroglial cells don't have a long process, but even so, researchers are now beginning to gather evidence that they do communicate among themselves and with neurons!

Nerve cells, called neurons, have fibers (processes) called axons and dendrites. Axons are found in nerves. Neuroglial cells support and service neurons.

11.2 Body Cavities and Body Membranes

The internal organs are located within specific body cavities (Fig. 11.7). During human development, there is a large ventral cavity called a **coelom,** which becomes divided into the thoracic (chest) and abdominal cavities. Membranes divide the thoracic cavity into the pleural cavities, containing the right and left lungs, and the pericardial cavity, containing the heart. The thoracic cavity is separated from the abdominal cavity by a horizontal muscle called the diaphragm. The stomach, liver, spleen, gallbladder, and most of the small and large intestines are in the upper portion of the abdominal cavity. The lower portion contains the rectum, the urinary bladder, the internal reproductive organs, and the rest of the large intestine. Males have an external extension of the abdominal wall, called the scrotum, containing the testes.

The dorsal cavity also has two parts: the cranial cavity within the skull contains the brain; and the vertebral column, formed by the vertebrae, contains the spinal cord.

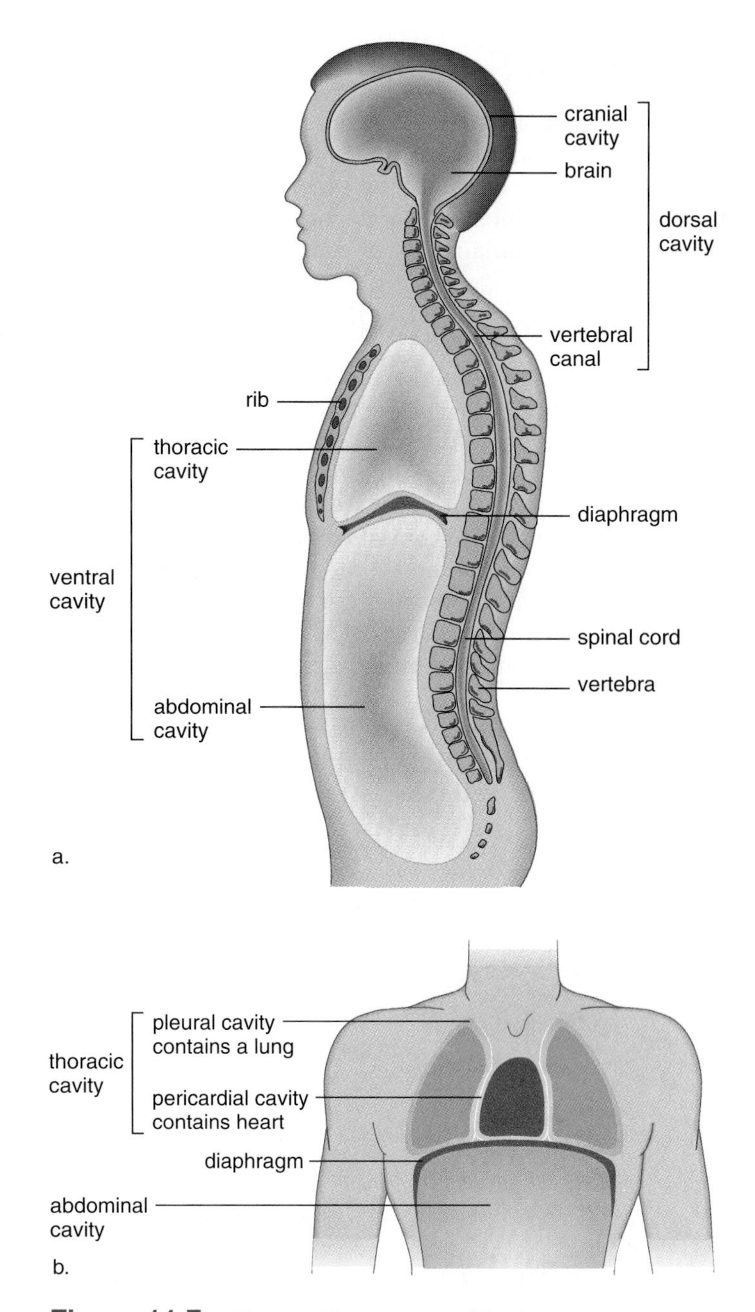

Figure 11.7 Mammalian body cavities.
a. Side view. There is a dorsal cavity, which contains the cranial cavity and the vertebral canal. The brain is in the cranial cavity, and the spinal cord is in the vertebral canal. There is a well-developed ventral cavity, which is divided by the diaphragm into the thoracic cavity and the abdominal cavity. The heart and lungs are in the thoracic cavity, and most other internal organs are in the abdominal cavity. **b.** Frontal view of the thoracic cavity.

Body Membranes

In this context, we are using the term *membrane* to refer to a thin lining or covering composed of an epithelium overlying a loose connective tissue layer. Body membranes line cavities and internal spaces of organs and tubes that open to the outside.

Mucous membranes line the tubes of the digestive, respiratory, urinary, and reproductive systems. The epithelium of this membrane contains goblet cells that secrete mucus. This mucus ordinarily protects the body from invasion by bacteria and viruses; hence, more mucus is secreted and expelled when a person has a cold and has to blow her/his nose. In addition, mucus usually protects the walls of the stomach and small intestine from digestive juices, but this protection breaks down when a person develops an ulcer.

Serous membranes line the thoracic and abdominal cavities and the organs that they contain. They secrete a watery fluid that keeps the membranes lubricated. Serous membranes support the internal organs and compartmentalize the large thoracic and abdominal cavities. This helps to hinder the spread of any infection.

The **pleural membranes** are serous membranes that line the pleural cavity and lungs. **Pleurisy** is a well-known infection of these membranes. The peritoneum lines the abdominal cavity and its organs. In between the organs, there is a double layer of peritoneum called mesentery. **Peritonitis,** a life-threatening infection of the peritoneum, is likely if an inflamed appendix bursts before it is removed.

Synovial membranes line freely movable joint cavities. They secrete synovial fluid into the joint cavity; this fluid lubricates the ends of the bones so that they can move freely. In rheumatoid arthritis, the synovial membrane becomes inflamed and grows thicker, restricting movement.

The **meninges** are membranes found within the dorsal cavity. They are composed only of connective tissue and serve as a protective covering for the brain and spinal cord. Meningitis is a life-threatening infection of the meninges.

11.3 Organ Systems

The body contains a number of organ systems (Fig. 11.8). The skin, which is sometimes called the **integumentary system,** is discussed in this chapter. The other organ systems contribute to either maintenance of the human body, integration and control of the human body, or continuance of the species.

Maintenance of the Body

The internal environment of the body consists of the blood within the blood vessels and the tissue fluid that surrounds the cells. Five systems add substances to and remove substances from the blood: the digestive, cardiovascular, lymphatic, respiratory, and urinary systems.

The **digestive system** consists of the mouth, esophagus, stomach, small intestine, and large intestine (colon) along with the associated organs: teeth, tongue, salivary glands, liver, gallbladder, and pancreas. This system receives food and digests it into nutrient molecules, which can enter the cells of the body.

The **cardiovascular system** consists of the heart and blood vessels that carry blood through the body. Blood transports nutrients and oxygen to the cells, and removes their waste molecules that are to be excreted from the body. Blood also contains cells produced by the lymphatic system.

The **lymphatic system** consists of lymphatic vessels, lymph, nodes, and other lymphoid organs. This system protects the body from disease by purifying lymph and storing lymphocytes, the white blood cells that produce antibodies. Lymphatic vessels absorb fat from the digestive system and collect excess tissue fluid, which is returned to the cardiovascular system.

The **respiratory system** consists of the lungs and the tubes that take air to and from the lungs. The respiratory system brings oxygen into the lungs and takes carbon dioxide out of the lungs.

The **urinary system** contains the kidneys and the urinary bladder. This system rids the body of nitrogenous wastes and helps regulate the fluid level and chemical content of the blood.

The digestive system, cardiovascular system, lymphatic system, respiratory system, and the urinary system all perform specific processing and transporting functions to maintain the normal conditions of the body.

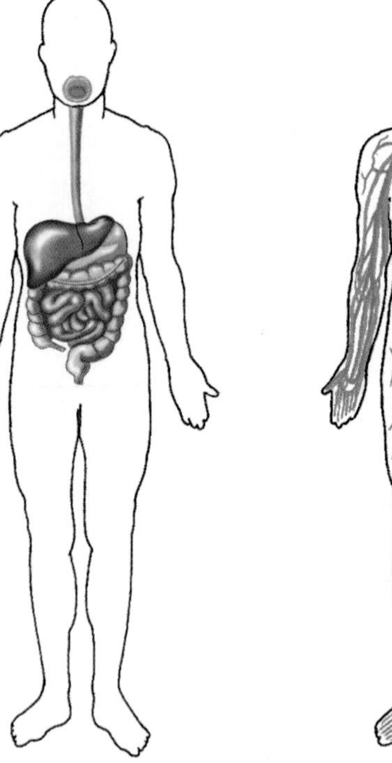

Digestive system
Breakdown and absorption of food materials

Cardiovascular system
Transport of nutrients to body cells, and transport of wastes from cells

Lymphatic system
Immunity; absorption of fats; drainage of tissue fluid

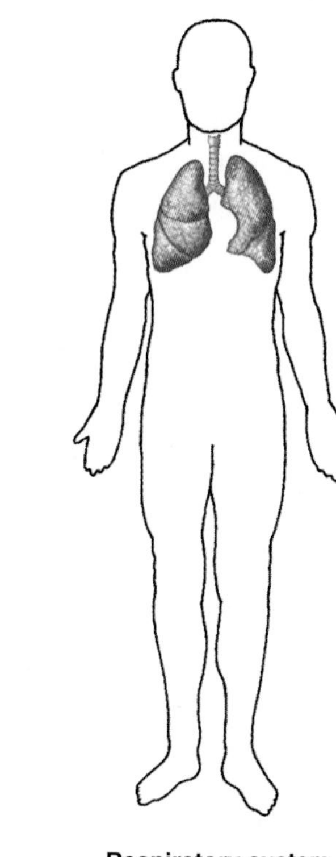

Respiratory system
Gaseous exchange between external environment and blood

Figure 11.8 **Organ systems of the body.**

Integration and Control of the Body

The **nervous system** consists of the brain, spinal cord, and associated nerves. The nerves conduct nerve impulses from sensory receptors to the brain and spinal cord. They also conduct nerve impulses from the brain and spinal cord to the muscles and glands, allowing us to respond to both external and internal stimuli.

The **musculoskeletal system,** consisting of the bones and muscles of the body, protects other body parts. For example, the skull forms a protective encasement for the brain, as does the rib cage for the heart and lungs. The skeleton, as a whole, serves as a place of attachment for the skeletal muscles. Contraction of muscles accounts for movement of the body and also body parts.

The **endocrine system** consists of the hormonal glands that secrete chemicals that serve as messengers between body parts. **Homeostasis** is a dynamic equilibrium of the internal environment. Both the nervous and endocrine systems help maintain homeostasis by coordinating and regulating the functions of the body's other systems. The endocrine system also helps maintain the proper functioning of male and female reproductive organs.

The nervous and endocrine systems coordinate and regulate the activities of the body's other systems, including the musculoskeletal system.

Continuance of the Species

The **reproductive system** involves different organs in the male and female. The male reproductive system consists of the testes, other glands, and various ducts that conduct semen to and through the penis. The female reproductive system consists of the ovaries, oviducts, uterus, vagina, and external genitals.

The reproductive system in males and in females carries out those functions that give humans the ability to reproduce.

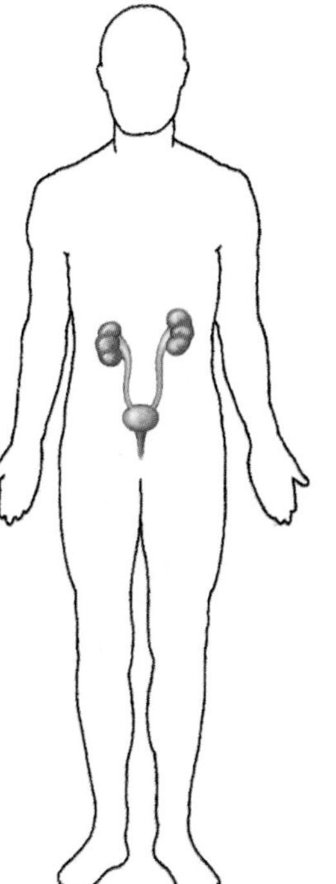

Excretory system
Filtration of blood; maintenance of volume and chemical composition of the blood

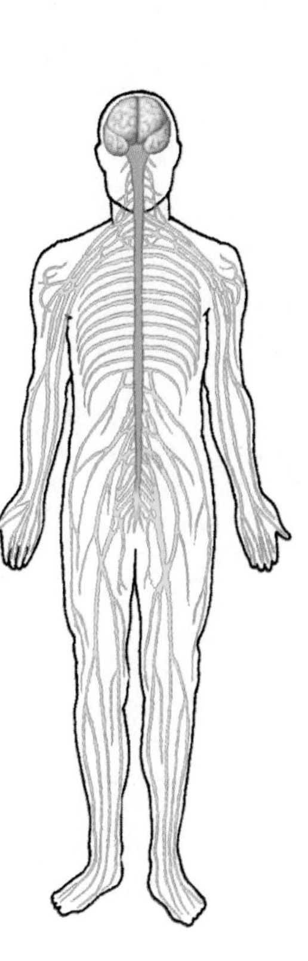

Nervous system
Regulation of all body activities; learning and memory

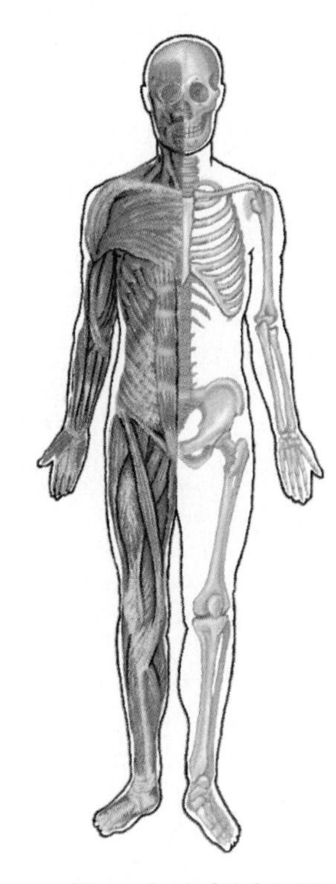

Musculoskeletal system
Internal support and protection; body movement

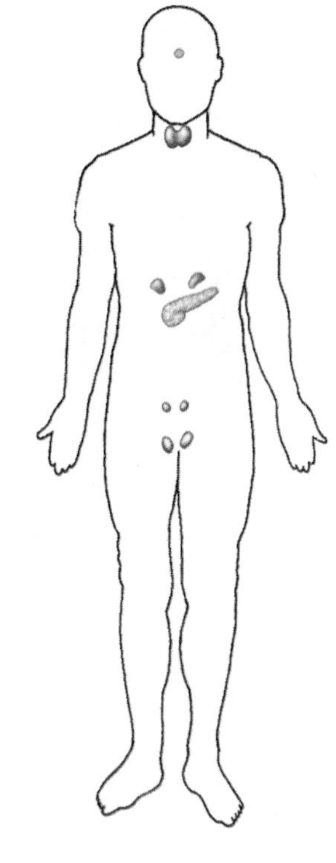

Endocrine system
Regulation of body activities; maintenance of reproductive system

Figure 11.8 **Organ systems of the body.**

11.4 Skin as an Organ System

The outer covering of the body, called **skin,** can be used as an example of an organ system (it is sometimes called the integumentary system) because it contains accessory structures such as nails, hair, and glands (Fig. 11.9). Skin covers the body, protecting underlying tissues from physical trauma, pathogen invasion, and water loss. Skin helps to regulate body temperature, and because it contains sensory receptors, skin also helps us to be aware of our surroundings and to communicate with others by touch. The skin even synthesizes certain chemicals such as vitamin D that affect the rest of the body.

Regions of the Skin

The skin has two regions: the epidermis and the dermis. A subcutaneous layer is found between the skin and any underlying structures, such as muscle or bone.

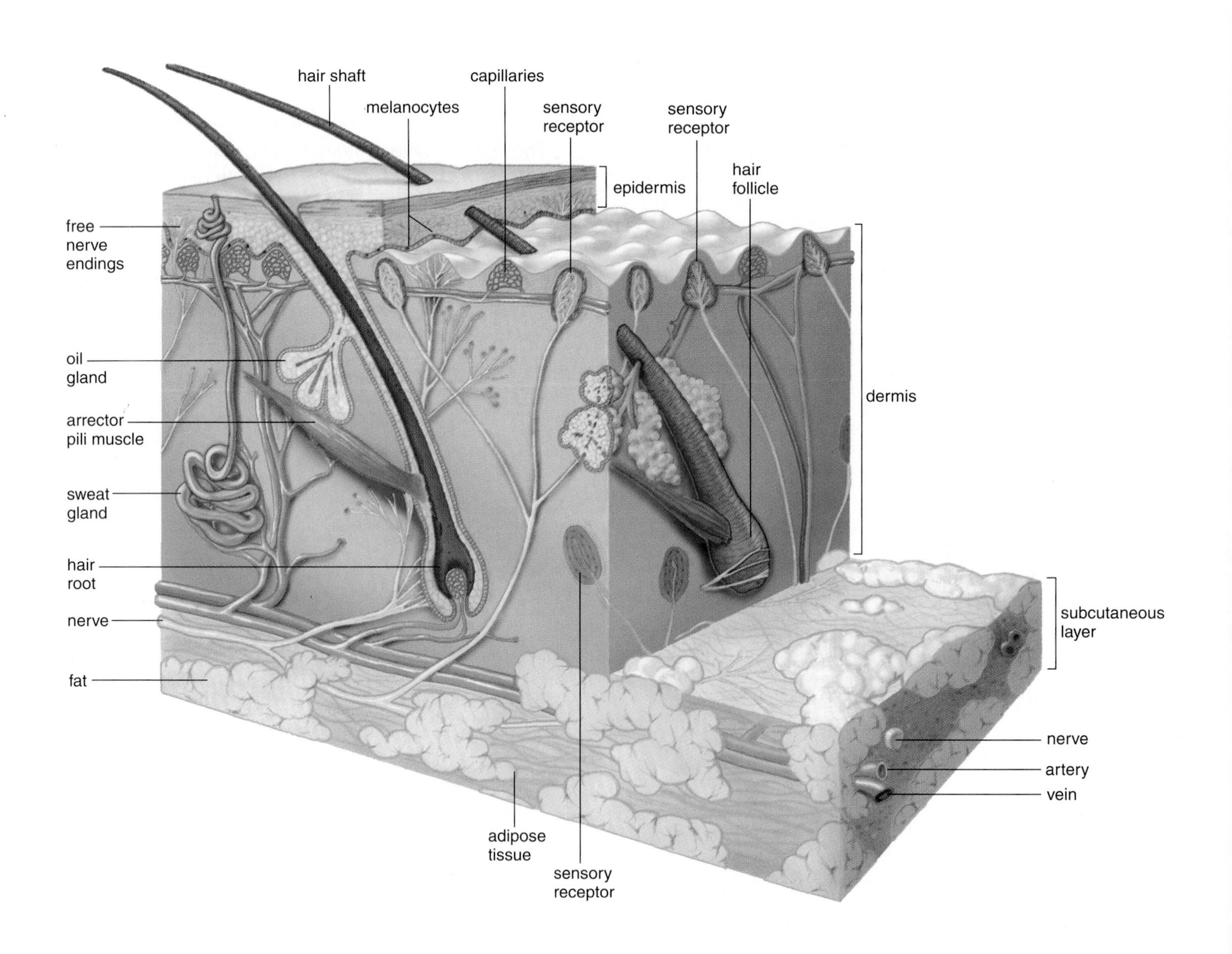

Figure 11.9 Human skin anatomy.
Skin consists of two regions, the epidermis and dermis. A subcutaneous layer lies below the dermis.

The **epidermis** of skin is made up of stratified squamous epithelium. New cells derived from basal cells become flattened and hardened as they push to the surface. Hardening occurs because the cells produce keratin, a waterproof protein. Dandruff occurs when the rate of keratinization is two or three times the normal rate. A thick layer of dead keratinized cells, arranged in spiral and concentric patterns, form fingerprints and footprints. Specialized cells in the epidermis called **melanocytes** produce melanin, the pigment responsible for skin color.

The **dermis** is a region of fibrous connective tissue beneath the epidermis. The dermis contains collagenous and elastic fibers. The collagenous fibers are flexible but offer great resistance to overstretching; they prevent the skin from being torn. The elastic fibers maintain normal skin tension but also stretch to allow movement of underlying muscles and joints. (The number of collagen and elastic fibers decreases with exposure to the sun, and the skin becomes less supple and is prone to wrinkling.) The dermis also contains blood vessels that nourish the skin. When blood rushes into these vessels, a person blushes, and when blood is minimal in them, a person turns "blue."

Sensory receptors are specialized nerve endings in the dermis that respond to external stimuli. There are receptors for touch, pressure, pain, and temperature. The fingertips contain the most touch receptors, and these add to our ability to use our fingers for delicate tasks.

The **subcutaneous layer,** which lies below the dermis, is composed of loose connective tissue and adipose tissue, which stores fat. Fat is a stored source of energy for the body. Adipose tissue helps to thermally insulate the body from either gaining heat from the outside or losing heat from the inside. A well-developed subcutaneous layer gives the body a rounded appearance and provides protective padding against external assaults. Excessive development of the subcutaneous layer accompanies obesity.

Skin has two regions: the epidermis and the dermis. A subcutaneous layer lies beneath the dermis.

Accessory Structures of the Skin

Nails, hair, and glands are structures of epidermal origin even though some parts of hair and glands are largely found in the dermis.

Nails grow from special epithelial cells at the base of the nail in the portion called the nail root. These cells become keratinized as they grow out over the nail bed. The visible portion of the nail is called the nail body. The cuticle is a fold of skin that hides the nail root. The whitish color of the half-moon-shaped base, or lunula, results from the thick layer of cells in this area (Fig. 11.10).

Hair follicles begin in the dermis and continue through the epidermis where the hair shaft extends beyond the skin. Epidermal cells form the root of hair, and their division causes a hair to grow. The cells become keratinized and dead as they are pushed farther from the root. Each hair follicle has one or more **oil** (sebaceous) **glands,** which secrete sebum, an oily substance that lubricates the hair within the follicle and the skin itself. If the sebaceous glands fail to discharge, the secretions collect and form "whiteheads" or "blackheads." The color of blackheads is due to oxidized sebum. Contraction of the arrector pili muscles attached to hair follicles causes the hairs to "stand on end" and causes goose bumps to develop.

Sweat (sudoriferous) **glands** are quite numerous and are present in all regions of skin. A sweat gland begins as a coiled tubule within the dermis, but then it straightens out near its opening. Some sweat glands open into hair follicles, but most open onto the surface of the skin. Acne is an inflammation of the sebaceous glands that most often occurs during adolescence. Hormonal changes during this time cause the sebaceous glands to become more active.

Regulation of Body Temperature

If the body temperature starts to rise, the blood vessels dilate so that more blood is brought to the surface of the skin and the sweat glands become active. Sweat absorbs body heat as it evaporates. If the outer temperature is cool, the blood vessels constrict so that less blood is brought to the surface of the skin. Whenever the body's temperature falls below normal, the muscles start to contract, causing shivering, which produces heat.

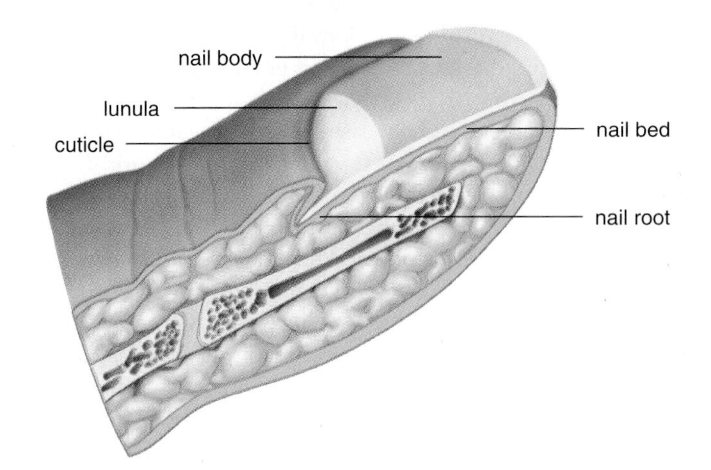

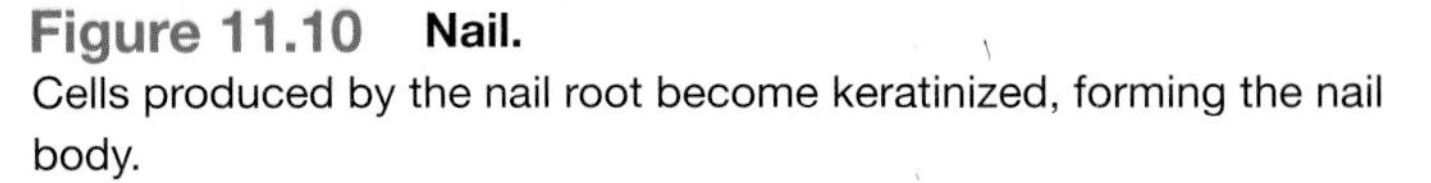
Figure 11.10 Nail.
Cells produced by the nail root become keratinized, forming the nail body.

Ecology Focus

Stratospheric Ozone Depletion Threatens the Biosphere

The earth's atmosphere is divided into layers. The troposphere envelops us as we go about our day-to-day lives. When ozone (O_3) is present in the troposphere (called ground-level ozone), it is considered a pollutant because it adversely affects a plant's ability to grow and our ability to breathe oxygen (O_2). In the stratosphere, some 50 kilometers above the earth, ozone forms a shield that absorbs much of the ultraviolet (UV) rays of the sun so that fewer rays strike the earth.

UV radiation causes mutations that can lead to skin cancer and can make the lens of the eyes develop cataracts. It also is believed to adversely affect the immune system and our ability to resist infectious diseases. Crop and tree growth is impaired, and UV radiation also kills off small plants (phytoplankton) and tiny shrimplike animals (krill) that sustain oceanic life. Without an adequate ozone shield, our health and food sources are threatened.

Depletion of the ozone layer within the stratosphere in recent years is, therefore, of serious concern. It became apparent in the 1980s that some worldwide depletion of ozone had occurred and that there was a severe depletion of some 40–50% above the Antarctic every spring. A vortex of cold wind (a whirlpool in the atmosphere) circles the pole during the winter months, creating ice crystals where chemical reactions occur that break down ozone. Severe depletions of the ozone layer are commonly called "ozone holes." Detection devices now tell us that the ozone hole above the Antarctic is about the size of the United States and growing. Of even greater concern, an ozone hole has now appeared above the Arctic as well, and ozone holes could also occur within northern and southern latitudes, where many people live. Whether or not these holes develop depends on prevailing winds, weather conditions, and the type of particles in the atmosphere. A United Nations Environment Program report predicts a 26% rise in cataracts and nonmelanoma skin cancers for every 10% drop in the ozone level. A 26% increase translates into 1.75 million additional cases of cataracts and 300,000 more skin cancers (see reading next page) every year, worldwide.

The cause of ozone depletion can be traced to the release of chlorine atoms (Cl) into the stratosphere (Fig. 11A). Chlorine atoms combine with ozone and strip away the oxygen atoms, one by one. One atom of chlorine can destroy up to 100,000 molecules of ozone before settling to the earth's surface as chloride years later. These chlorine atoms come from the breakdown of chlorofluorocarbons (CFCs), chemicals much in use by humans. The best known CFC is Freon, a heat transfer agent found in refrigerators and air conditioners. CFCs are also used as cleaning agents and foaming agents during the production of styrofoam found in coffee cups, egg cartons, insulation, and paddings. Formerly, CFCs were used as propellants in spray cans, but this application is now banned in the United States and several European countries.

Most countries of the world have agreed to stop using CFCs by the year 2000. The United States halted production in 1995. Computer projections suggest that an 85% reduction in CFC emissions is needed to stabilize CFC levels in the atmosphere. Otherwise they keep on increasing. Scientists are now searching for CFC substitutes that will not release chlorine atoms (nor bromine atoms) to harm the ozone shield.

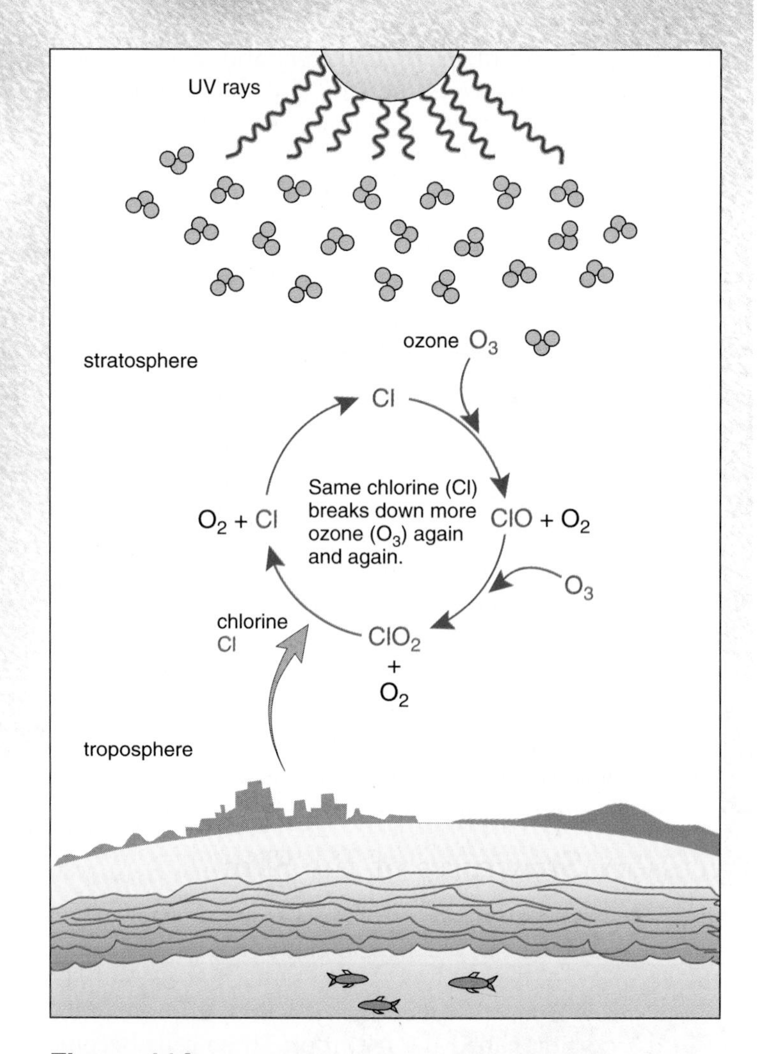

Figure 11A Ozone depletion.
CFCs release chlorine atoms that lead to the breakdown of ozone (O3) and the buildup of oxygen (O2) in the stratosphere. Oxygen does not absorb UV radiation and does not protect the earth.

Health Focus

Skin Cancer on the Rise

In the nineteenth century and earlier, it was fashionable for Caucasian women (those who did not labor outdoors) to keep their skin fair by carrying a parasol when they went out. But early in this century, some fair-skinned people began to prefer the golden-brown look, and they took up sunbathing as a way to achieve a tan. A few hours after exposure to the sun, pain and redness due to dilation of blood vessels occur. Tanning occurs when melanin granules increase in keratinized cells at the surface of the skin as a way to prevent any further damage by ultraviolet (UV) rays. The sun gives off two types of UV rays: UV-A rays and UV-B rays. UV-A rays penetrate the skin deeply, affect connective tissue, and cause the skin to sag and wrinkle. UV-A rays are also believed to increase the effects of the UV-B rays, which are the cancer-causing rays. UV-B rays are more prevalent at midday.

Skin cancer is categorized as either nonmelanoma or melanoma. Nonmelanoma cancers are of two types. Basal cell carcinoma, the most common type, begins when UV radiation causes epidermal basal cells to form a tumor, while at the same time suppressing the immune system's ability to detect the tumor. The signs of a tumor are varied. They include an open sore that will not heal, a recurring reddish patch, a smooth, circular growth with a raised edge, a shiny bump, or a pale mark (Fig. 11B). In about 95% of patients the tumor can be excised surgically, but recurrence is common.

Squamous cell carcinoma begins in the epidermis proper. Squamous cell carcinoma is five times less common than basal cell carcinoma, but if the tumor is not excised promptly it is more likely to spread to nearby organs. The death rate from squamous cell carcinoma is about 1% of cases. The signs of squamous cell carcinoma are the same as for basal cell carcinoma, except that the former may also show itself as a wart that bleeds and scabs.

Melanoma that starts in pigmented cells often has the appearance of an unusual mole. Unlike a mole that is circular and confined, melanoma moles look like spilled ink spots. A variety of shades can be seen in the same mole, and they can itch, hurt, or feel numb. The skin around the mole turns gray, white, or red. Melanoma is most apt to appear in persons who have fair skin, particularly if they have suffered occasional severe sun burns as children. The chance of melanoma increases with the number of moles a person has. Most moles appear before the age of 14, and their appearance is linked to sun exposure. Melanoma rates have risen since the turn of the century, but the incidence has doubled in the last decade. Most often, malignant moles are removed surgically; if the cancer has spread, chemotherapy and various other treatments are also available.

Since the incidence of skin cancer is related to UV exposure, scientists have developed a UV index to determine how powerful the solar rays are in different U.S. cities. In general, the more southern the city, the higher the UV index, and the greater the risk of skin cancer. Regardless of where you live, for every 10% decrease in the ozone layer, the risk of skin cancer rises 13–20%. To prevent the occurrence of skin cancer, observe the following:

- Use a broad-spectrum sunscreen, which protects you from both UV-A and UV-B radiation, with an SPF (sun protection factor) of at least 15. (This means, for example, that if you usually burn after a 20-minute exposure, it will take 15 times that long before you will burn.)
- Stay out of the sun altogether between the hours of 10 A.M. and 3 P.M. This will reduce your annual exposure by as much as 60%. Wear protective clothing. Choose fabrics with a tight weave and wear a wide-brimmed hat.
- Wear sunglasses that have been treated to absorb both UV-A and UV-B radiation. Otherwise, sunglasses can expose your eyes to more damage than usual because pupils dilate in the shade.
- Avoid tanning machines. Although most tanning devices use high levels of only UV-A, UV-A rays cause the deep layers of the skin to become more vulnerable to UV-B radiation when you are later exposed to the sun.

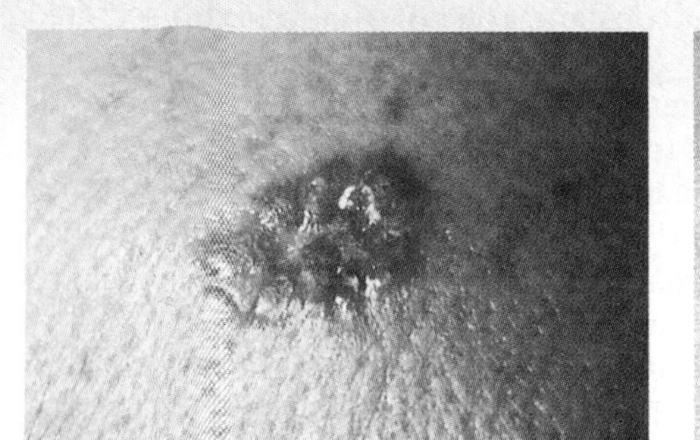

a. Basal cell carcinoma

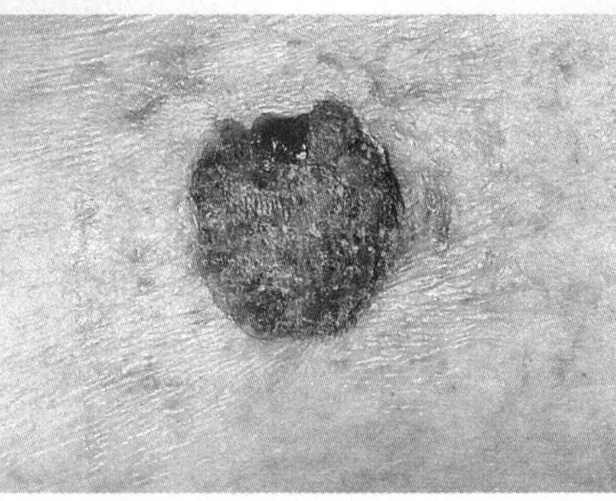

b. Squamous cell carcinoma

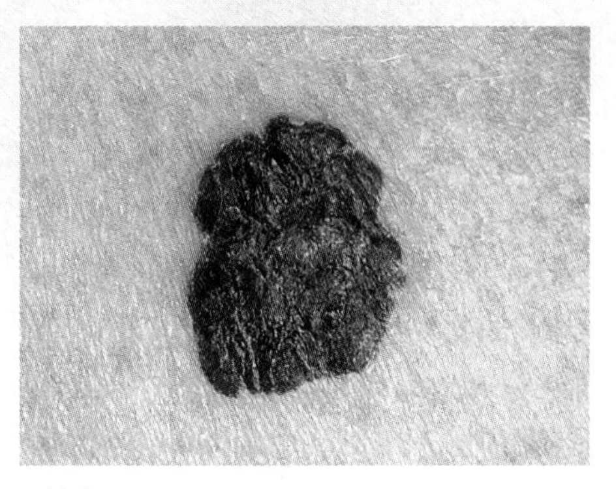

c. Melanoma

Figure 11B Skin cancer.
a. Basal cell carcinoma occurs when basal cells proliferate abnormally. **b.** Squamous cell carcinoma arises in epithelial cells derived from basal cells. **c.** Malignant melanoma is due to a proliferation of pigmented cells. About one-third develop from pigmented moles.

11.5 Homeostasis

Homeostasis means that the internal environment remains within normal limits or values, regardless of the conditions in the external environment. In humans, for example:

1. The blood glucose concentration remains at about 100 mg/100 ml.
2. The pH of blood is always near 7.4.
3. Blood pressure in the brachial artery averages near 120/80 mm Hg.
4. Body temperature averages around 37°C (98.6°F).

Because body conditions do fluctuate somewhat, homeostasis is often called a dynamic equilibrium of normal values. The ability of the body to keep the internal environment within a certain range allows humans to live in a variety of habitats, such as the Arctic regions, the deserts, or the tropics.

This internal environment consists of tissue fluid, which bathes all the cells of the body. Tissue fluid is refreshed when molecules such as oxygen and nutrients exit blood and wastes enter blood (Fig. 11.11). Tissue fluid remains constant only as long as blood composition remains constant. Although we are accustomed to using the word *environment* to mean the external environment of the body, it is important to realize that it is the internal environment of tissues that is ultimately responsible for our health and well-being.

The internal environment of the body consists of tissue fluid, which bathes the cells.

Most systems of the body contribute toward maintaining a relatively constant internal environment. The cardiovascular system conducts blood to and away from capillaries, the smallest of the blood vessels, whose thin walls permit exchanges to occur. Blood pressure aids the movement of water out of capillaries, and osmotic pressure aids the movement of water into capillaries. Blood pressure is created by the pumping of the heart, while osmotic pressure is maintained by the protein content of plasma. The formed elements also contribute to homeostasis. Red blood cells transport oxygen and participate in the transport of carbon dioxide. White blood cells fight infection, and platelets participate in the clotting process. The lymphatic system is accessory to the circulatory system. Lymphatic capillaries collect excess tissue fluid, and this is returned via lymphatic veins to the circulatory veins.

The digestive system takes in and digests food, providing nutrient molecules that enter blood and replace the nutrients that are constantly being used by the body cells. The respiratory system adds oxygen to and removes carbon dioxide from the blood. The chief regulators of blood composition are the liver and the kidneys. They monitor the chemical composition of plasma (see Table 11.1) and alter it as required. Immediately after glucose enters the blood, it can be removed by the liver for storage as glycogen. Later, the glycogen can be broken down to replace the glucose used by the body cells; in this way, the glucose composition of blood remains constant. The hormone insulin, secreted by the pancreas, regulates glycogen storage. The liver also removes toxic chemicals, such as ingested alcohol and other drugs. The liver makes urea, a nitrogenous end product of protein metabolism. Urea and other metabolic waste molecules are excreted by the kidneys. Urine formation by the kidneys is extremely critical to the body, not only because it rids the body of unwanted substances, but also because it offers an opportunity to carefully regulate blood volume, salt balance, and the pH of the blood.

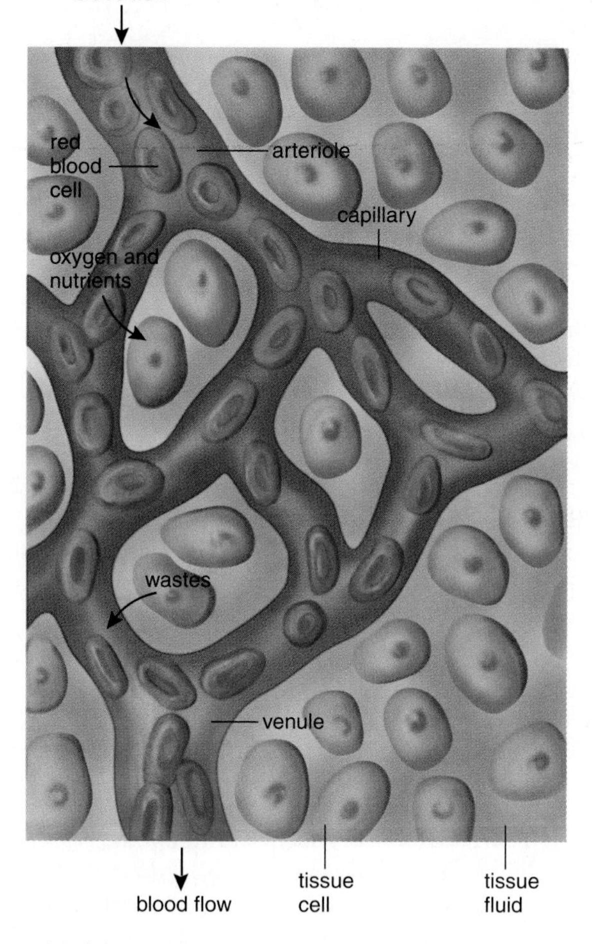

Figure 11.11 Tissue fluid composition.
Cells are surrounded by tissue fluid, which is continually refreshed because oxygen and nutrient molecules constantly exit, and waste molecules continually enter the bloodstream as shown.

Most systems of the body contribute to homeostasis, that is, maintaining the dynamic equilibrium of the internal environment.

Coordination of Organ Systems

The nervous system and endocrine system are ultimately in control of homeostasis. The endocrine system is slower acting than the nervous system, which rapidly brings about a particular response.

Previously, we mentioned that the liver is involved in homeostasis because it stores glucose as glycogen. But actually there is a hormone produced by an endocrine gland that regulates storage of glucose by the liver. When the glucose content of the blood rises after eating, the pancreas secretes insulin, a hormone that causes the liver to store glucose as glycogen. Now the glucose level falls, and the pancreas no longer secretes insulin. This is called control by **negative feedback** because the response (low blood glucose) negates the original stimulus (high blood glucose). In some instances, an endocrine gland is sensitive to the blood level of a hormone whose concentration it regulates. For example, the pituitary gland produces a hormone that stimulates the thyroid gland to secrete its hormone. When the blood level of this hormone rises to a certain level, the pituitary gland no longer stimulates the thyroid gland.

A negative feedback system can regulate itself because it has a sensing device which detects changes in environmental conditions. For example, consider the feedback mechanism that functions to maintain the room temperature of a house. In this feedback system, the thermostat is a device that is sensitive to room temperature. The furnace produces heat, and when the temperature of a room reaches a certain point, the thermostat signals a switching device that turns the furnace off. On the other hand, when the temperature falls below that indicated on the thermostat, it signals the switching device, which turns the furnace on again.

Figure 11.12*a* shows that in the body there are sensory receptors that fulfill the role of sensing devices. When a receptor is stimulated, it signals a regulatory center that then turns on an effector. The effector brings about a response that negates the original conditions that stimulated the receptor. In the absence of suitable stimulation, the receptor no longer signals the regulatory center.

Figure 11.12*b* gives an example involving the nervous system. When blood pressure rises, receptors signal a regulatory center, which then sends out nerve impulses to the arterial walls, causing them to relax, and the blood pressure now falls. Therefore, the sensory receptors are no longer stimulated, and the system shuts down. Notice that negative feedback control results in a fluctuation above and below an average. Thus, there is a dynamic equilibrium of the internal environment.

Positive feedback also occurs on occasion. In these instances, certain events increase the likelihood of a particular response. For example, once the childbirth process begins, each succeeding event makes it more likely that the process will continue until completion.

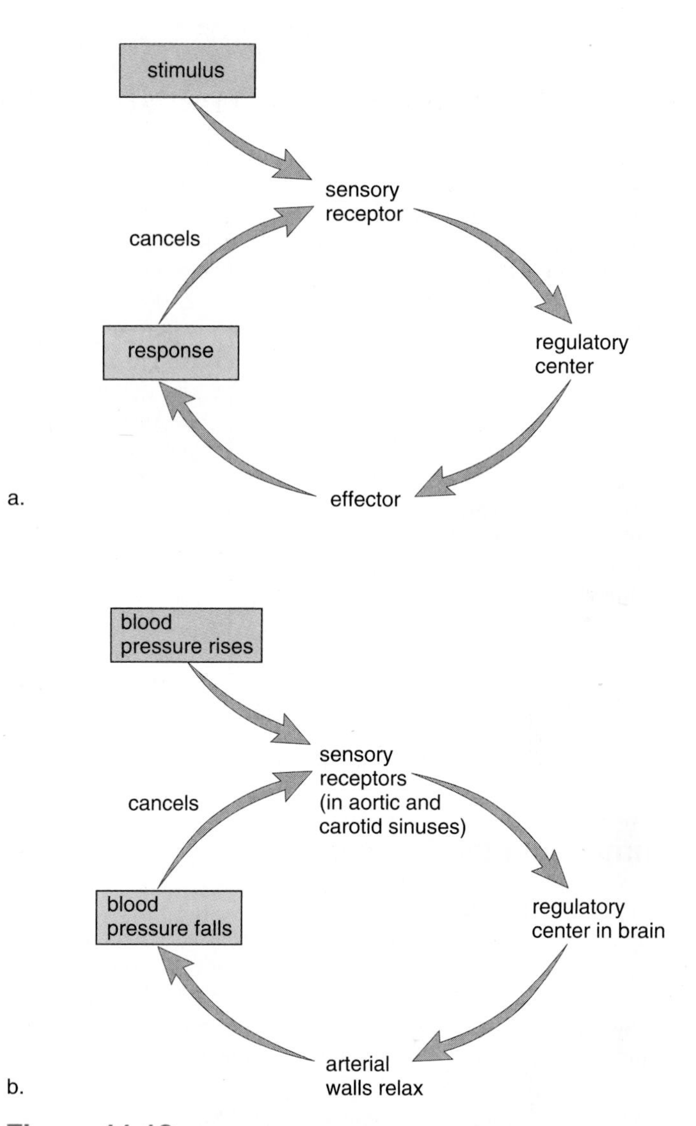

Figure 11.12 Negative feedback control.
a. A stimulus causes a receptor to signal a regulatory center in the brain. The regulatory center signals effectors to respond, and the response cancels the stimulus. **b.** For example, when blood pressure rises, special sensory receptors in blood vessels signal a particular center in the brain. The brain signals the arteries to relax, and blood pressure falls.

Homeostasis of internal conditions is a self-regulatory mechanism that usually results in slight fluctuations above and below an average.

Bioethical Issue

Transplantation of the kidney, heart, liver, pancreas, lung, and other organs is now possible due to two major breakthroughs. First, solutions have been developed that preserve donor organs for several hours. This made it possible for one young boy to undergo surgery for 16 hours, during which time he received five different organs. Second, rejection of transplanted organs is now prevented by immunosuppressive drugs; therefore, organs can be donated by unrelated individuals, living or dead. . After death, it is possible to give the "gift of life" to someone else—over 25 organs and tissues from one cadaver can be used for transplants. Survival rate after a transplant operation is good. So many heart recipients are now alive and healthy they have formed basketball and softball teams, demonstrating the normalcy of their lives after surgery.

One problem persists however, and that is the limited availability of organs for transplantation. At any one time, at least 27,000 Americans are waiting for a donated organ. Keen competition for organs can lead to various bioethical inequities. When the governor of Pennsylvania received a heart and lungs within a relatively short period of time, it appeared that his social status may have played a role. When Mickey Mantle received a liver transplant, people asked if it was right to give an organ to an older man who had a diseased liver due to the consumption of alcohol. If a father gives a kidney to a child, he has to undergo a major surgical operation that leaves him vulnerable to possible serious consequences in the future. If organs are taken from those who have just died, who guarantees that the individual is indeed dead? And is it right to genetically alter animals to serve as a source of organs for humans? Such organs will most likely be for sale, and does this make the wealthy more likely to receive a transplant than those who cannot pay?

Questions

1. Is it ethical to ask a parent to donate an organ to his or child? Why or why not?
2. Is it ethical to put a famous person at the top of the list for an organ transplant? Why or why not?
3. Is it ethical to remove organs from a newborn who is brain dead but whose organs are still functioning? Why or why not?
4. When xenotransplants (transplants for humans from other animals) are available, should they be for sale? Why or why not?

Summarizing the Concepts

11.1 Types of Tissues

Human tissues are categorized into four groups. Epithelial tissue covers the body and lines its cavities. The different types of epithelial tissue (squamous, cuboidal, and columnar) can be stratified and have cilia or microvilli. Also, columnar cells can be pseudostratified. Epithelial cells sometimes form glands that secrete either into ducts or into blood.

Connective tissues, in which cells are separated by a matrix, often bind body parts together. Loose fibrous connective tissue has both collagen and elastic fibers. Dense fibrous connective tissue, like that of tendons and ligaments, contains closely packed collagen fibers. In adipose tissue, the cells enlarge and store fat. Both cartilage and bone have cells within lacunae, but the matrix for cartilage is more flexible than that for bone, which contains calcium salts. In bone, the lacunae lie in concentric circles within an osteon (or Haversian system) about a central canal. Blood is a connective tissue in which the matrix is a liquid called plasma.

Muscular tissue is of three types. Both skeletal and cardiac muscle are striated; both cardiac and smooth muscle are involuntary. Skeletal muscle is found in muscles attached to bones, and smooth muscle is found in internal organs. Cardiac muscle makes up the heart.

Nervous tissue has one main type of conducting cell, the neuron, and several types of neuroglial cells. Each neuron has dendrites, a cell body, and an axon. The brain and spinal cord contain complete neurons, while the nerves contain only neuron fibers. Axons are specialized to conduct nerve impulses.

11.2 Body Cavities and Body Membranes

The internal organs occur within cavities; the thoracic cavity contains the heart and lungs; the abdominal cavity contains organs of the digestive, urinary, and reproductive systems, among others. Membranes line body cavities and internal spaces of organs. As an example, mucous membrane lines the tubes of the digestive system; serous membrane lines the thoracic and abdominal cavities and covers the organs they contain.

11.3 Organ Systems

The skin is sometimes called the integumentary system. The digestive, cardiovascular, lymphatic, respiratory, and urinary systems perform processing and transporting functions that maintain the normal conditions of the body. The nervous system receives sensory input from sensory receptors and directs the musculoskeletal system and glands to respond to outside stimuli. The musculoskeletal system supports the body and permits movement. The endocrine system produces hormones, some of which influence the functioning of the reproductive system, which allows humans to make more of their own kind.

11.4 Skin as an Organ System

The skin can be used as an example of an organ system because it contains accessory structures such as nails, hair, and glands. Skin protects underlying tissues from physical trauma, pathogen invasion, and water loss. Skin helps regulate body temperature, and because it contains sensory receptors, skin also helps us to be aware of our surroundings.

Skin is a two-layered organ that waterproofs and protects the body. The epidermis contains basal cells that produce new epithelial cells that become keratinized as they move toward the surface. The dermis, a largely fibrous connective tissue, contains epidermally derived glands and hair follicles, nerve endings, and blood vessels. Sensory receptors for touch, pressure, temperature, and pain are present. Sweat glands and blood vessels help control body temperature. A subcutaneous layer, which is made up of loose connective tissue containing adipose cells, lies beneath the skin.

11.5 Homeostasis

Homeostasis is the dynamic equilibrium of the internal environment. All organ systems contribute to the constancy of tissue fluid and blood. Special contributions are made by the liver, which keeps blood glucose constant, and the kidneys, which regulate the pH. The nervous and hormonal systems regulate the other body systems. Both of these are controlled by a negative feedback mechanism, which results in fluctuation above and below the desired levels. Body temperature is regulated by a center in the hypothalamus.

Studying the Concepts

1. Name the four major types of tissues. 194
2. Name the different kinds of epithelial tissue, and give a location and function for each. 194
3. What are the functions of connective tissue? Name the different kinds, and give a location for each. 196–98
4. What are the functions of muscular tissue? Name the different kinds, and give a location for each. 199
5. Nervous tissue contains what type of cell? Which organs in the body are made up of nervous tissue? 200
6. In what cavities are the major organs located? 201
7. Distinguish between plasma membrane and body membrane. 201
8. Describe the structure of skin, and state at least two functions of this organ. 204–05
9. What is homeostasis, and how is it achieved in the human body? 208–09
10. Give an example of a negative feedback system. 209

Testing Yourself

Choose the best answer for each question.

1. Which of these is mismatched?
   a. epithelial tissue—protection and absorption
   b. muscular tissue—contraction and conduction
   c. connective tissue—binding and support
   d. nervous tissue—conduction and message sending
   e. nervous tissue—neuroglial cells
2. Which of these is not epithelial tissue?
   a. simple cuboidal and stratified columnar
   b. bone and cartilage
   c. stratified squamous and simple squamous
   d. pseudostratified
   e. All of these are epithelial tissue.
3. Which tissue is more apt to line a lumen?
   a. epithelial tissue
   b. connective tissue
   c. nervous tissue
   d. muscular tissue
   e. epidermis tissue
4. Tendons and ligaments are
   a. connective tissue.
   b. associated with the bones.
   c. found in vertebrates.
   d. subject to injury.
   e. All of these are correct.
5. Which tissue has cells in lacunae?
   a. epithelial tissue
   b. fibrous connective
   c. cartilage
   d. bone
   e. Both c and d are correct.
6. Cardiac muscle is
   a. striated.
   b. involuntary.
   c. smooth.
   d. voluntary.
   e. Both a and b are correct.
7. Which of these components of blood fights infection?
   a. red blood cells
   b. white blood cells
   c. platelets
   d. plasminogen
   e. All of these are correct.
8. Which of these body systems contribute to homeostasis?
   a. digestive and excretory systems
   b. respiratory and nervous systems
   c. nervous and endocrine systems
   d. All of these are correct.
   e. Body systems are not involved in homeostasis.
9. With negative feedback,
   a. the output cancels the input.
   b. there is a fluctuation above and below the average.
   c. there is self-regulation.
   d. sensory receptors communicate with a regulatory center.
   e. All of these are correct.
10. Which of these correctly describes a layer of the skin?
    a. The epidermis is simple squamous epithelium in which hair follicles develop and blood vessels expand when we are hot.
    b. The subcutaneous layer lies between the epidermis and the dermis. It contains adipose tissue, which keeps us warm.
    c. The dermis is a region of connective tissue which contains sensory receptors, nerve endings, and blood vessels.
    d. The skin has a special layer, still unnamed, in which there are all the accessory structures like nails, hair, and various glands.
    e. All of these are correct.
11. Give the name, the location, and the function for each of these tissues.
    a. Type of epithelial tissue
    b. Type of muscular tissue
    c. Type of connective tissue

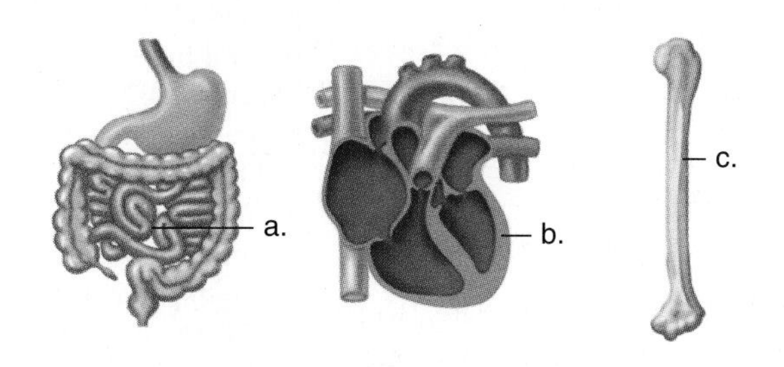

Thinking Scientifically

1. Considering the differentiation of tissue:
   a. How is the structure of an epithelial cell suited to its function?
   b. Tight junctions between epithelial cells are equivalent to which feature of endodermal cells in plants (page 157)? How are they equivalent?
   c. How is the structure of a skeletal muscle cell suited to muscle contraction? If, upon contraction, muscular tissue always shortens from right to left, what would happen to an object attached at the right?
   d. How is the structure of a nerve cell suited to its function?
2. Your task is to show that humans have a greater mental capacity than other animals. Mention these features as possible supportive data.
   a. number of neurons
   b. size of the overall brain
   c. organization of the brain
   d. organization of the nervous system

Understanding the Terms

adhesion junction 196
adipose tissue 196
basement membrane 194
blood 198
bone 197
carcinoma 194
cardiac muscle 199
cardiovascular system 202
cartilage 196
cilium 194
coelom 201
collagen fiber 196
columnar epithelium 194
compact bone 197
connective tissue 196
cuboidal epithelium 194
dense fibrous connective tissue 196
dermis 205
digestive system 202
elastic cartilage 197
elastic fiber 196
endocrine system 203
epidermis 205
epithelial tissue 194
fibroblast 196
fibrocartilage 197
gap junction 196
gland 194
hair follicle 205
homeostasis 203
hyaline cartilage 196
integumentary system 202
intercalated disk 199
lacuna 196
ligament 196
loose fibrous connective tissue 196
lymphatic system 202
matrix 196
melanocyte 205
meninges 201
microvillus 194
mucous membrane 201
muscular (contractile) tissue 199
musculoskeletal system 203
negative feedback 209
nerve 200
nervous system 203
nervous tissue 200
neuroglial cell 200
neuron 200
oil gland 205
pathogen 194
peritonitis 201
plasma 198
platelet 198
pleural membrane 201
pleurisy 201
positive feedback 209
red blood cell 198
reproductive system 203
respiratory system 202
reticular fiber 196
serous membrane 201
skeletal muscle 199
skin 204
smooth (visceral) muscle 199
spongy bone 197
squamous epithelium 194
striated 199
subcutaneous layer 205
sweat gland 205
synovial membrane 201
tendon 196
tight junction 196
tissue 194
urinary system 202
white blood cell 198

Match the terms to these definitions:

a. __________ Fibrous connective tissue that joins bone to bone at a joint.
b. __________ Outer region of the skin composed of stratified squamous epithelium.
c. __________ Having bands such as in cardiac and skeletal muscle.
d. __________ Self-regulatory mechanism that is activated by an imbalance and results in a fluctuation above and below a mean.
e. __________ Porous bone found at the ends of long bones where blood cells are formed.

Using Technology

Your study of human organization is supported by these available technologies:

Essential Study Partner CD-ROM

Animals → Body Organization

Visit the Mader web site for related ESP activities.

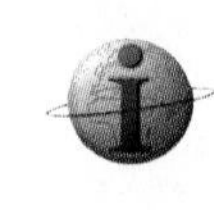

Exploring the Internet

The Mader Home Page provides resources and tools as you study this chapter.

http://www.mhhe.com/biosci/genbio/mader

Dynamic Human 2.0 CD-ROM

Human Body

CHAPTER 12

Digestive System and Nutrition

Chapter Concepts

12.1 The Digestive System

- The human digestive system is an extended tube with specialized parts between two openings, the mouth and the anus. 214
- Food is ingested and then digested to small molecules that are absorbed. Indigestible remains are eliminated. 214

12.2 Three Accessory Organs

- The pancreas, the liver, and the gallbladder are accessory organs of digestion because their activities assist the digestive process. 222

12.3 Digestive Enzymes

- The products of digestion are small molecules, such as amino acids and glucose, that can cross plasma membranes. 224
- The digestive enzymes are specific and have an optimum temperature and pH at which they function. 225

12.4 Nutrition

- Proper nutrition supplies the body with energy and nutrients, including the essential amino acids and fatty acids, and all vitamins and minerals. 226

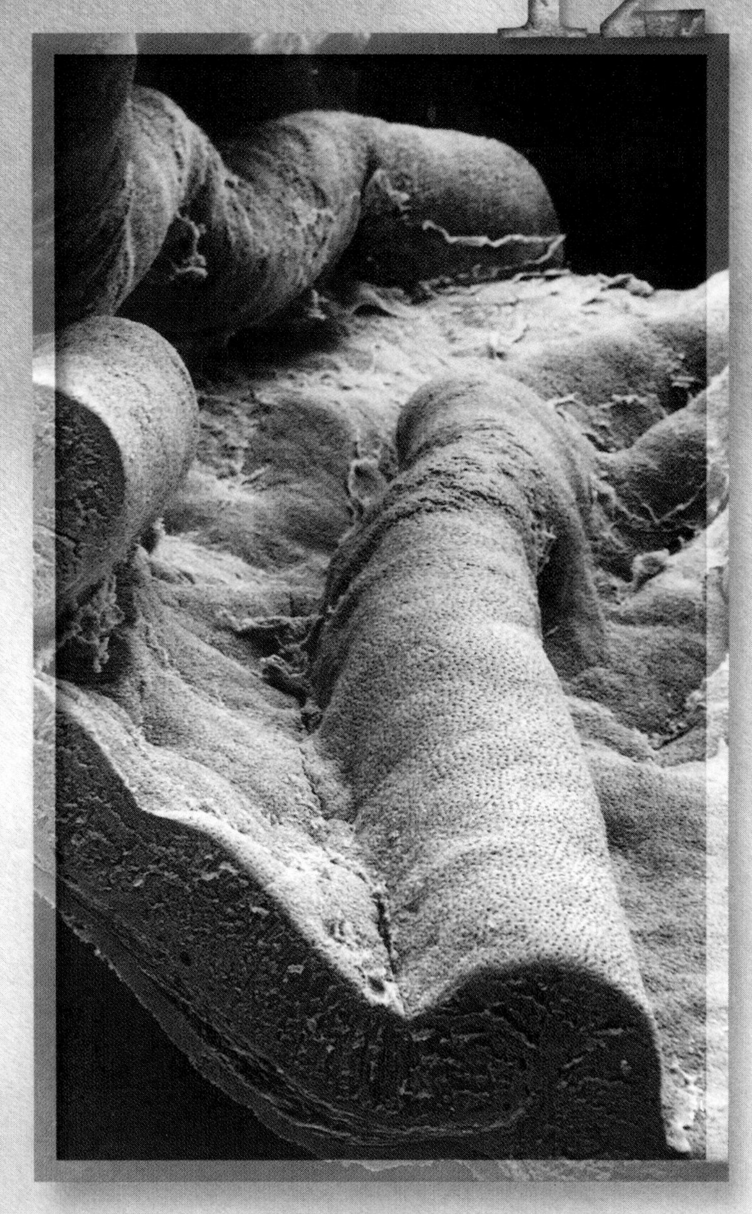

Scanning electron micrograph of the stomach shows how the wall is folded for easy expansion. There are many openings to gastric pits, where hydrochloric acid and an enzyme for protein digestion are produced.

Enjoying the summer night at an outdoor cafe, Evora washes down her last piece of garlic bread with a sip of wine. Even before Evora swallowed the bread, her mouth's saliva began to break it apart. The wine's alcohol is absorbed in the stomach, where the process of transforming Evora's meal into a nutrient-laden liquid begins. In the small intestine, wormlike projections from the intestinal wall absorb amino acids, sugars, and other needed molecules into Evora's bloodstream. Even the large intestine contributes by taking in needed water and salts. Her body now refueled, Evora heads off for a night of dancing. In this chapter, you will learn how the body digests food, and the importance of proper nutrition. Science is beginning to find the cellular basis for believing that fruits and vegetables, and yes, especially broccoli, can ensure a brighter and healthier future. The contents of this chapter should be of interest to everyone.

12.1 The Digestive System

Digestion takes place within a tube called the digestive tract, which begins with the mouth and ends with the anus (Fig. 12.1). The functions of the digestive system are to ingest food, digest it to nutrients that can cross plasma membranes, absorb nutrients, and eliminate indigestible remains.

The Mouth

The mouth, which receives food, is bounded externally by the lips and cheeks. The lips extend from the base of the nose to the start of the chin. The red portion of the lips is poorly keratinized and this allows blood to show through.

Most people enjoy eating food largely because they like its texture and taste. Sensory receptors called taste buds occur primarily on the tongue, and when these are activated by the presence of food, nerve impulses travel by way of cranial nerves to the brain. The tongue is composed of skeletal muscle whose contraction changes the shape of the tongue. Muscles exterior to the tongue cause it to move about. A fold of mucous membrane on the underside of the tongue attaches it to the floor of the oral cavity.

The roof of the mouth separates the nasal cavities from the oral cavity. The roof has two parts: an anterior (toward the front) **hard palate** and a posterior (toward the back) **soft palate** (Fig. 12.2*a*). The hard palate contains several bones, but the soft palate is composed entirely of muscle. The soft palate ends in a finger-shaped projection called the *uvula.* The tonsils are in the back of the mouth, on either side of the tongue and in the nasopharynx (called adenoids). The tonsils help protect the body against infections. If the tonsils become inflamed, the person has **tonsillitis.** The infection can spread to the middle ears. If tonsillitis recurs repeatedly, the tonsils may be surgically removed (called a tonsillectomy).

Three pairs of **salivary glands** send juices (saliva) by way of ducts to the mouth. One pair of salivary glands lies at the sides of the face immediately below and in front of the ears. These glands swell when a person has the mumps, a viral

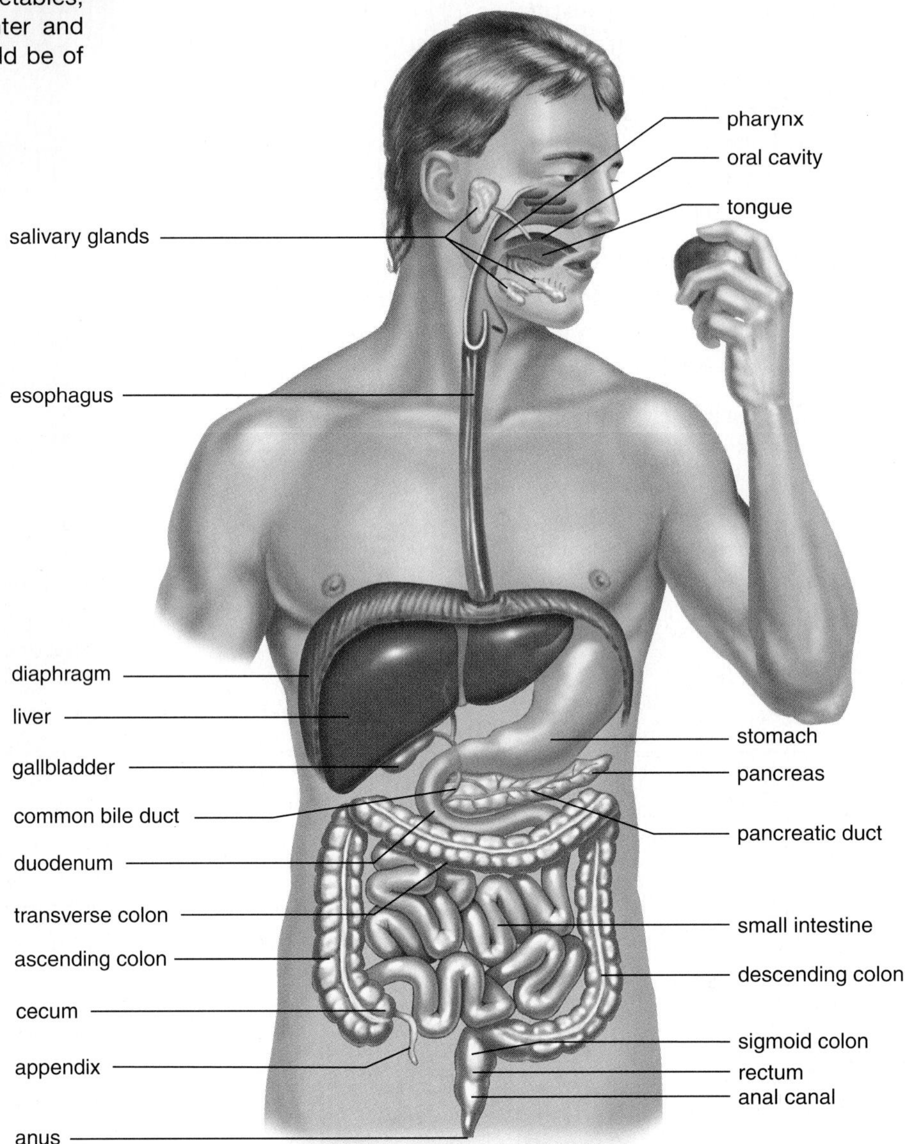

Figure 12.1 Digestive system.
Trace the path of food from the mouth to the anus. The large intestine consists of the cecum; ascending, transverse, descending, and sigmoid colons; plus the rectum and anal canal. Note also the location of the accessory organs of digestion: the pancreas, the liver, and the gallbladder.

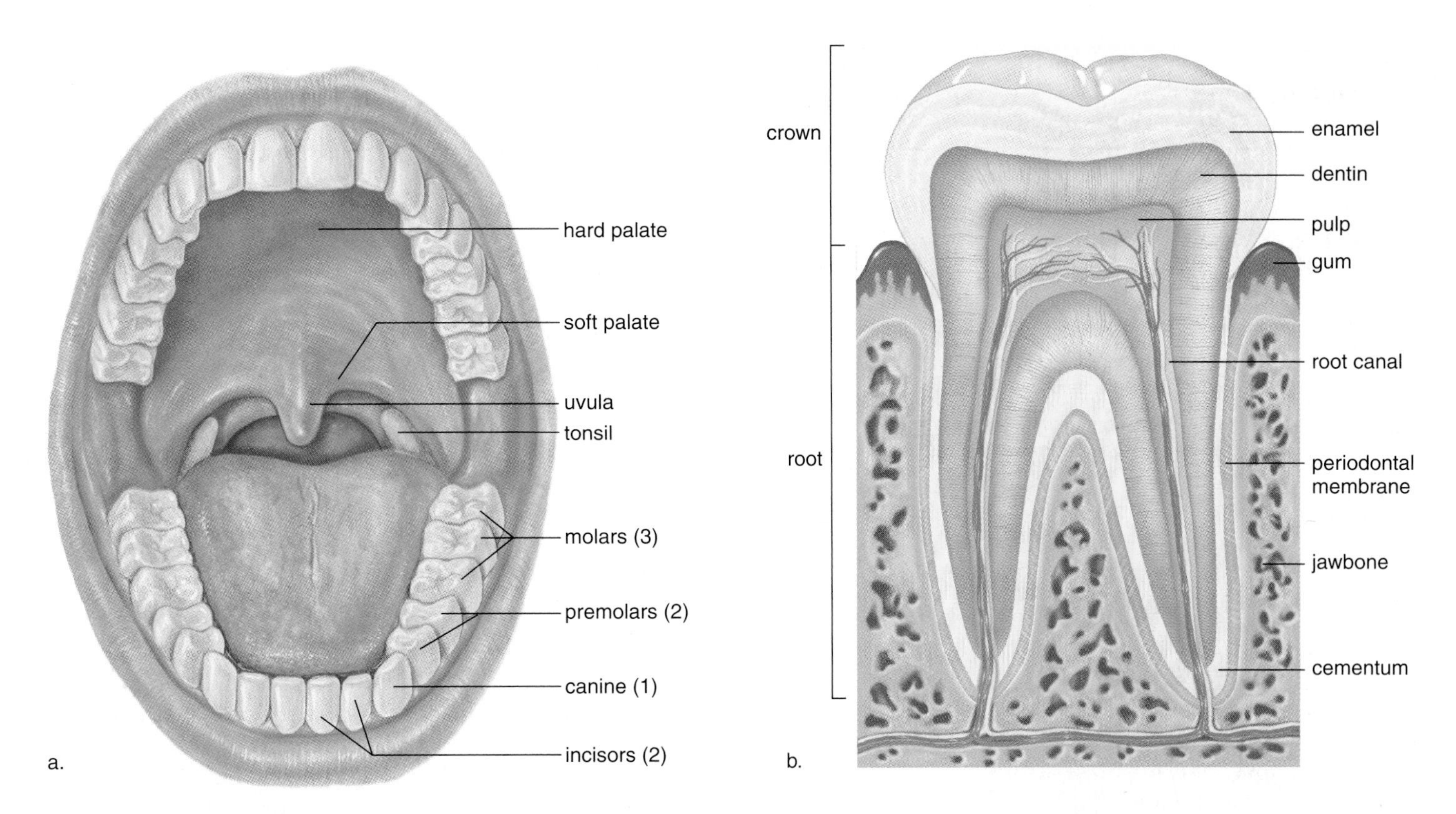

Figure 12.2 Adult mouth and teeth.
a. The chisel-shaped incisors bite; the pointed canines tear; the fairly flat premolars grind; and the flattened molars crush food. The last molar, called a wisdom tooth, may fail to erupt, or if it does, it is sometimes crooked and useless. Often dentists recommend the extraction of the wisdom teeth. **b.** Longitudinal section of a tooth. The crown is the portion that projects above the gum line and is sometimes replaced by a dentist. When a root canal is done, the nerves are removed. When the periodontal membrane is inflamed, the teeth can loosen.

infection most often seen in children. Salivary glands have ducts that open on the inner surface of the cheek at the location of the second upper molar. Another pair of salivary glands lies beneath the tongue, and still another pair lies beneath the floor of the oral cavity. The ducts from these salivary glands open under the tongue. You can locate the openings if you use your tongue to feel for small flaps on the inside of your cheek and under your tongue. Saliva contains an enzyme called salivary amylase that begins the process of digesting starch.

The Teeth

With our teeth we chew food into pieces convenient for swallowing. During the first two years of life, the smaller 20 deciduous, or baby, teeth appear. These are eventually replaced by 32 adult teeth (Fig. 12.2*a*). The third pair of molars, called the wisdom teeth, sometimes fail to erupt. If they push on the other teeth and/or cause pain, they can be removed by a dentist or oral surgeon.

Each tooth has two main divisions, a crown and a root (Fig. 12.2*b*). The crown has a layer of enamel, an extremely hard outer covering of calcium compounds; dentin, a thick layer of bonelike material; and an inner pulp, which contains the nerves and the blood vessels. Dentin and pulp are also found in the root.

Tooth decay, called **dental caries,** or cavities, occurs when bacteria within the mouth metabolize sugar and give off acids, which erode teeth. Two measures can prevent tooth decay: eating a limited amount of sweets and daily brushing and flossing of teeth. Fluoride treatments, particularly in children, can make the enamel stronger and more resistant to decay. Gum disease is more apt to occur with aging. Inflammation of the gums (*gingivitis*) can spread to the periodontal membrane, which lines the tooth socket. A person then has **periodontitis,** characterized by a loss of bone and loosening of the teeth so that extensive dental work may be required. Stimulation of the gums in a manner advised by your dentist is helpful in controlling this condition.

The tongue, which is composed of striated muscle and an outer layer of mucous membrane, mixes the chewed food with saliva. It then forms this mixture into a mass called a bolus in preparation for swallowing.

The salivary glands send saliva into the mouth, where the teeth chew the food and the tongue forms it into a bolus for swallowing.

Table 12.1 Path of Food

Organ	Function of Organ	Special Feature(s)	Function of Special Feature(s)
Oral cavity	Receives food; starts digestion of starch	Teeth Tongue	Chewing of food Formation of bolus
Esophagus	Passageway		
Stomach	Storage of food; acidity kills bacteria; starts digestion of protein	Gastric glands	Release gastric juices
Small intestine	Digestion of all foods; absorption of nutrients	Intestinal glands Villi	Release fluids Absorb nutrients
Large intestine	Absorption of water; storage of indigestible remains		

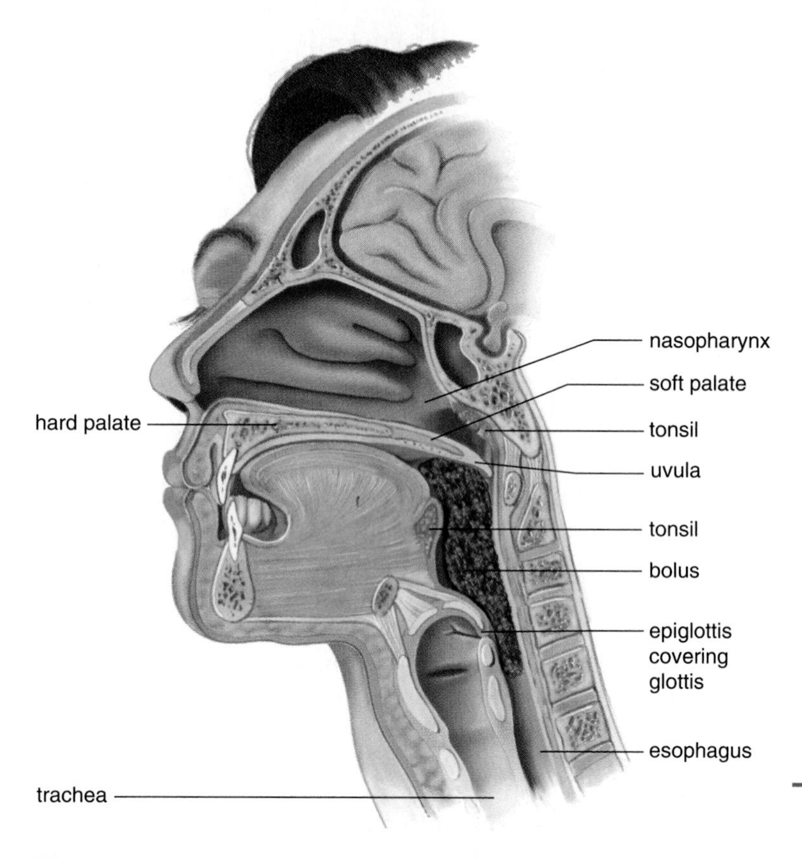

Figure 12.3 Swallowing.
When food is swallowed, the soft palate closes off the nasopharynx and the epiglottis covers the glottis, forcing the bolus to pass down the esophagus. Therefore, you do not breathe when swallowing.

The Pharynx

The **pharynx** is a region that receives food from the mouth and air from the nasal cavities (Table 12.1). The food passage and air passage cross in the pharynx because the trachea (windpipe) is ventral to (in front of) the esophagus, a long muscular tube that takes food to the stomach.

Swallowing, a process that occurs in the pharynx (Fig. 12.3), is a **reflex action** performed automatically, without conscious thought. During swallowing, food normally enters the esophagus because the air passages are blocked. Unfortunately, we have all had the unpleasant experience of having food "go the wrong way." The wrong way may be either into the nasal cavities or into the trachea. If it is the latter, coughing will most likely force the food up out of the trachea and into the pharynx again. Usually during swallowing, the soft palate moves back to close off the **nasopharynx,** and the trachea moves up under the **epiglottis** to cover the glottis. The **glottis** is the opening to the larynx (voice box). The up and down movement of the Adam's apple, the front part of the larynx, is easy to observe when a person swallows. We do not breathe when we swallow.

The air passage and the food passage cross in the pharynx. When you swallow, the air passage usually is blocked off, and food must enter the esophagus.

The Esophagus

The **esophagus** is a muscular tube that passes from the pharynx through the thoracic cavity and diaphragm into the abdominal cavity where it joins the stomach. The esophagus is ordinarily collapsed, but it opens and receives the bolus when swallowing occurs. A rhythmic contraction called **peristalsis** pushes the food along the digestive tract. Occasionally, peristalsis begins even though there is no food in the esophagus. This produces the sensation of a lump in the throat.

The esophagus plays no role in the chemical digestion of food. Its sole purpose is to conduct the food bolus from the mouth to the stomach. **Sphincters** are muscles that encircle tubes and act as valves; tubes close when sphincters contract, and they open when sphincters relax. The entrance of the esophagus to the stomach is marked by a constriction, often called a sphincter, although the muscle is not as developed as in a true sphincter. Relaxation of the sphincter allows the bolus to pass into the stomach, while contraction prevents the acidic contents of the stomach from backing up into the esophagus. **Heartburn,** which feels like a burning pain rising up into the throat, occurs when some of the stomach contents escape into the esophagus. When vomiting occurs, a contraction of the abdominal muscles and diaphragm propels the contents of the stomach upward through the esophagus.

The esophagus conducts the bolus of food from the pharynx to the stomach. Peristalsis begins in the esophagus and occurs along the entire length of the digestive tract.

The Wall of the Digestive Tract

The wall of the esophagus in the abdominal cavity is comparable to that of the digestive tract, which has these layers (Fig. 12.4):

Mucosa (mucous membrane layer) A layer of epithelium supported by connective tissue and smooth muscle lines the **lumen** (central cavity) and contains glandular epithelial cells that secrete digestive enzymes and goblet cells that secrete mucus.

Submucosa (submucosal layer) A broad band of loose connective tissue that contains blood vessels. Lymph nodules, called Peyer's patches, are in the submucosa. Like the tonsils, they help protect us from disease.

Muscularis (smooth muscle layer) Two layers of smooth muscle make up this section. The inner, circular layer encircles the gut; the outer, longitudinal layer lies in the same direction as the gut.

Serosa (serous membrane layer) Most of the digestive tract has a serosa, a very thin, outermost layer of squamous epithelium supported by connective tissue. The serosa secretes a serous fluid that keeps the outer surface of the intestines moist so that the organs of the abdominal cavity slide against one another. The esophagus has an outer layer composed only of loose connective tissue called the adventitia.

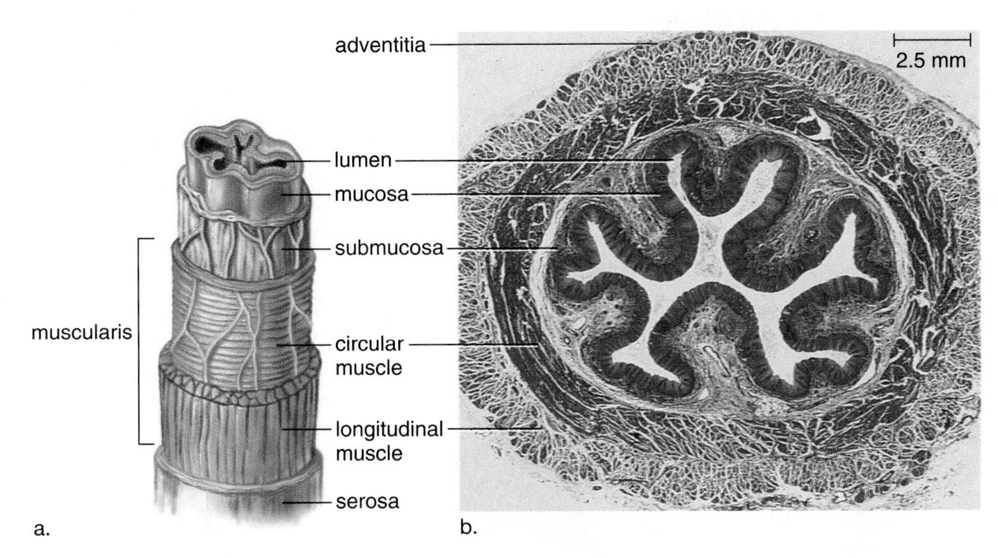

Figure 12.4 Wall of the digestive tract.
a. Several different types of tissues are found in the wall of the digestive tract. Note the placement of circular muscle inside longitudinal muscle.
b. Micrograph of the wall of the esophagus.

The Stomach

The **stomach** (Fig. 12.5) is a thick-walled, J-shaped organ that lies on the left side of the body beneath the diaphragm. The stomach is continuous with the esophagus above and the duodenum of the small intestine below. The stomach stores food and aids in digestion. The wall of the stomach has deep folds, which disappear as the stomach fills to an approximate capacity of one liter. Its muscular wall churns, mixing the food with gastric juice. The term *gastric* always refers to the stomach.

The columnar epithelial lining of the stomach has millions of gastric pits, which lead into **gastric glands.** The gastric glands produce gastric juice. Gastric juice contains an enzyme called **pepsin,** which digests protein, plus hydrochloric acid (HCl) and mucus. HCl causes the stomach to have a high acidity with a pH of about 2, and this is beneficial because it kills most bacteria present in food. Although HCl does not digest food, it does break down the connective tissue of meat and activates pepsin. The wall of the stomach is protected by a thick layer of mucus secreted by goblet cells in its lining. If, by chance, HCl penetrates this mucus, the wall can begin to break down, and an ulcer results. An **ulcer** is an open sore in the wall caused by the gradual disintegration of tissue. It now appears that most ulcers are due to a bacterial (*Helicobacter pylori*) infection that impairs the ability of epithelial cells to produce protective mucus.

Alcohol is absorbed in the stomach, but there is no absorption of food substances. Normally, the stomach empties in about 2–6 hours. When food leaves the stomach, it is a thick, soupy liquid called **chyme.** Chyme leaves the stomach and enters the small intestine in squirts by way of a sphincter that repeatedly opens and closes.

The stomach can expand to accommodate large amounts of food. When food is present, the stomach churns, mixing food with acidic gastric juice.

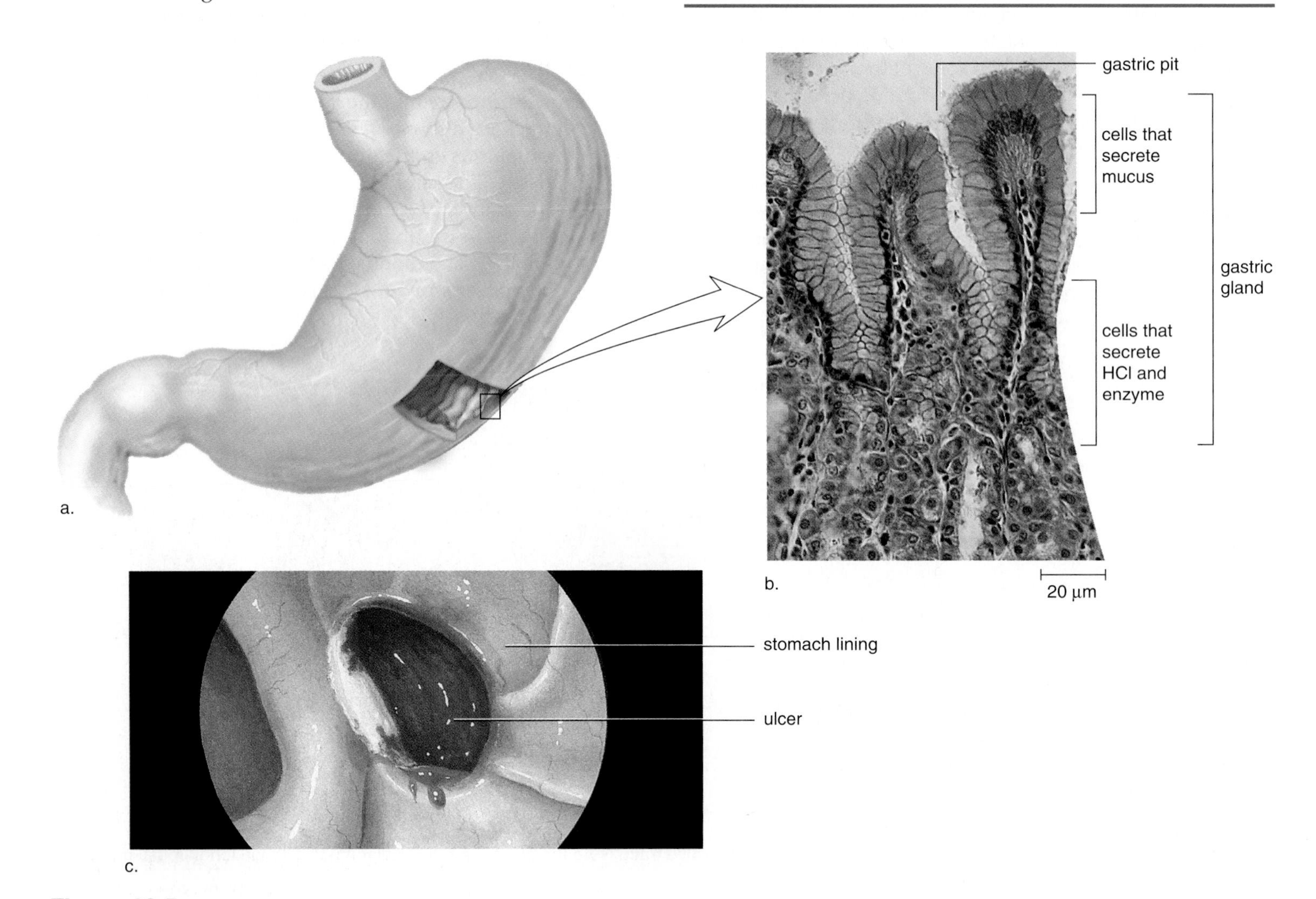

Figure 12.5 Anatomy and histology of the stomach.
a. The stomach has a thick wall with folds that allow it to expand and fill with food. **b.** The mucosa contains gastric glands, which secrete mucus and a gastric juice active in protein digestion. **c.** View of a bleeding ulcer by using an endoscope (a tubular instrument bearing a tiny lens and a light source) that can be inserted into the abdominal cavity.

The Small Intestine

The **small intestine** is named for its small diameter (compared to that of the large intestine); but perhaps it should be called the long intestine. In life, the small intestine averages about 3 meters (9 feet) in length, compared to the large intestine which is about 1.5 meters (4 ½ ft) in length. (After death, the small intestine becomes as long as 6 meters due to relaxation of muscles.)

The first 25 cm of the small intestine is called the **duodenum.** Ducts from the liver and pancreas join to form usually one duct that enters the duodenum (see Fig. 12.1). The small intestine receives bile from the liver and pancreatic juice from the pancreas via this duct. **Bile** emulsifies fat—emulsification causes fat droplets to disperse in water. The intestine has a slightly basic pH because pancreatic juice contains sodium bicarbonate ($NaHCO_3$), which neutralizes chyme. The enzymes in pancreatic juice and enzymes produced by the intestinal wall complete the process of digestion.

It's been suggested that the surface area of the small intestine is approximately that of a tennis court. What factors contribute to increasing its surface area? The wall of the small intestine contains fingerlike projections called **villi,** which give the intestinal wall a soft, velvety appearance (Fig. 12.6). Each villus has an outer layer of columnar epithelium and contains blood vessels and a small lymphatic vessel called a **lacteal.** The lymphatic system is an adjunct to the cardiovascular system—its vessels carry a fluid called lymph to the cardiovascular veins.

Each villus has thousands of microscopic extensions called microvilli. Collectively in electron micrographs, microvilli give the villi a fuzzy border known as a "brush border." Since the microvilli bear the intestinal enzymes, these enzymes are called brush-border enzymes. The microvilli greatly increase the surface area of the villus for the absorption of nutrients. Sugars and amino acids pass through the mucosa and enter a blood vessel. The components of fats (glycerol and fatty acids) rejoin in smooth endoplasmic reticulum and are combined with proteins in the Golgi apparatus before they enter a lacteal.

The small intestine is specialized to absorb the products of digestion. It is quite long (3 meters) and has fingerlike projections called villi, where nutrient molecules are absorbed into the cardiovascular (glucose and amino acids) and lymphatic (fats) systems.

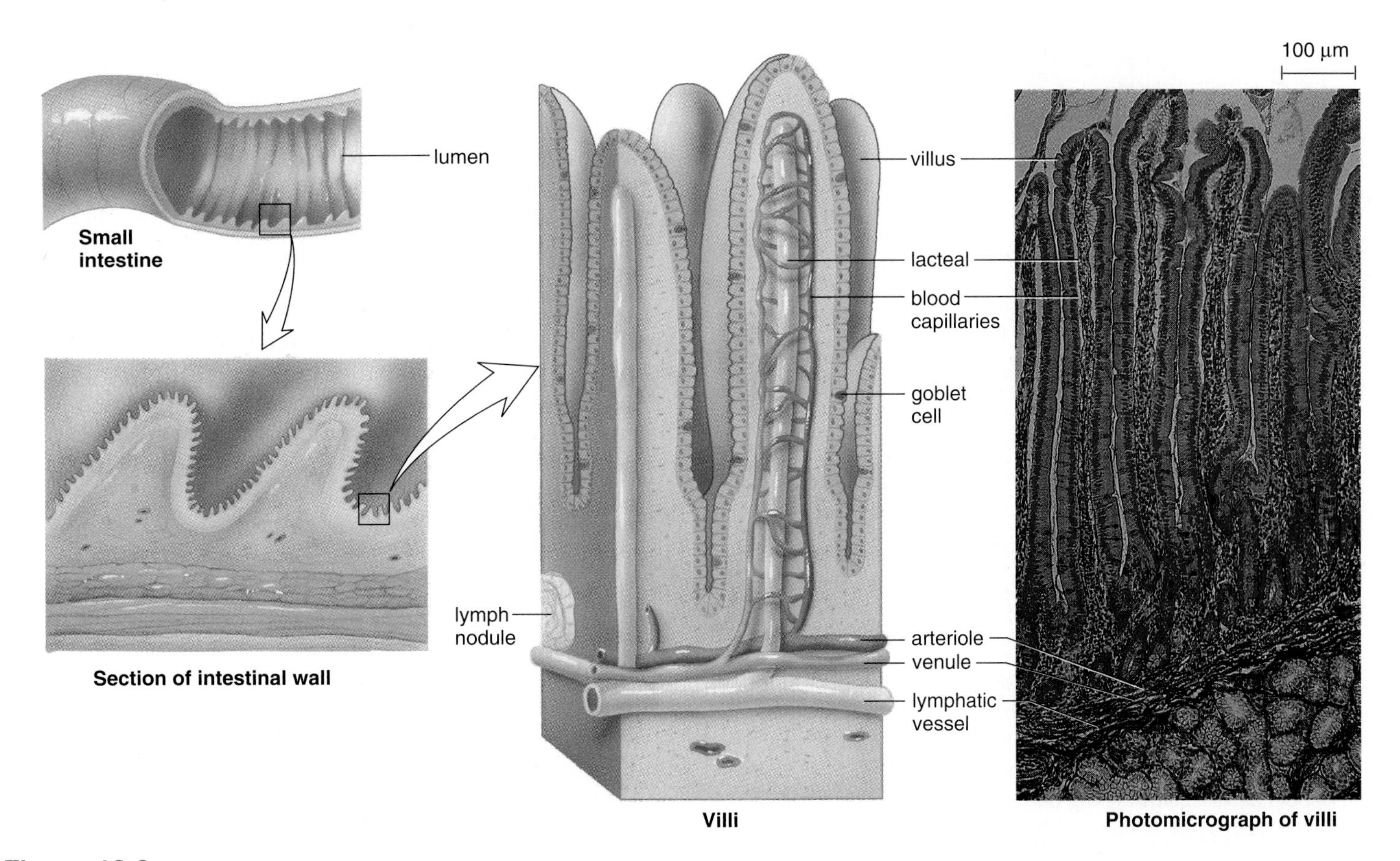

Figure 12.6 Anatomy of small intestine.
The wall of the small intestine has folds that bear fingerlike projections called villi. The products of digestion are absorbed by villi, which contain blood vessels and a lacteal. Each villus has many microscopic extensions called microvilli.

Regulation of Digestive Secretions

The nervous system promotes the secretion of digestive juices, but so do hormones (Fig. 12.7). A **hormone** is a substance produced by one set of cells that affects a different set of cells, the so-called target cells. Hormones are usually transported by the bloodstream. When a person has eaten a meal particularly rich in protein, the stomach produces the hormone gastrin. Gastrin enters the bloodstream, and soon the stomach is churning, and the secretory activity of gastric glands is increasing. A hormone produced by the duodenal wall, GIP (gastric inhibitory peptide), works opposite from gastrin: it inhibits gastric gland secretion.

Cells of the duodenal wall produce two other hormones that are of particular interest—secretin and CCK (cholecystokinin). Acid, especially hydrochloric acid (HCl) present in chyme, stimulates the release of secretin, while partially digested protein and fat stimulate the release of CCK. Soon after these hormones enter the bloodstream, the pancreas increases its output of pancreatic juice, which helps digest food, and the liver increases its output of bile. The gallbladder contracts to release bile.

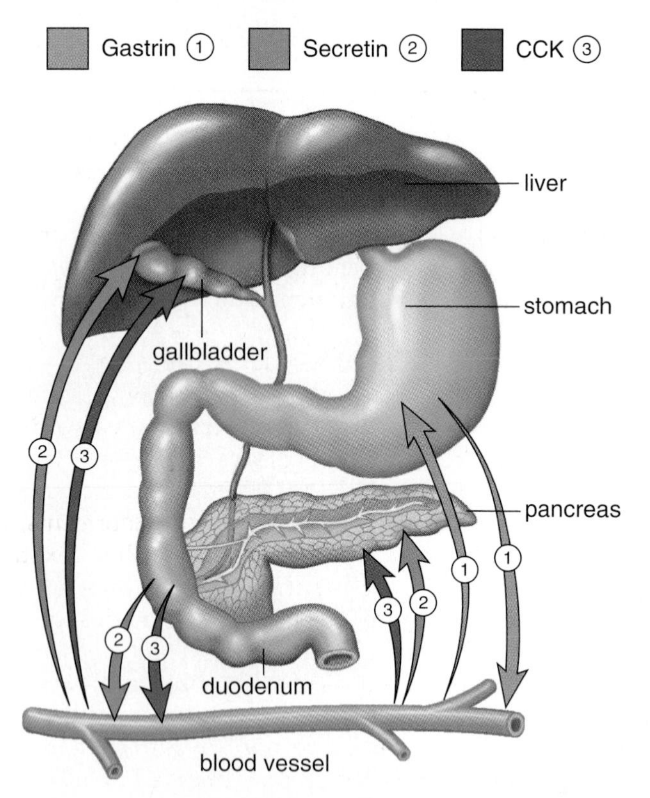

Figure 12.7 Hormonal control of digestive gland secretions. Gastrin ①, produced by the lower part of the stomach, enters the bloodstream and thereafter stimulates the upper part of the stomach to produce more digestive juice. Secretin ② and CCK ③, produced by the duodenal wall, stimulate the pancreas to secrete its digestive juice and the gallbladder to release bile.

The Large Intestine

The **large intestine,** which includes the cecum, the colon, the rectum, and the anal canal, is larger in diameter than the small intestine (6.5 cm compared to 2.5 cm), but it is shorter in length (1.5 meters compared to 3 meters) (see Fig. 12.1). The large intestine absorbs water, salts, and some vitamins. It also stores indigestible material until it is eliminated at the anus.

The **cecum,** which lies below the junction with the small intestine, is the blind end of the large intestine. The cecum has a small projection called the vermiform **appendix** (*vermiform* means wormlike) (Fig. 12.8). In humans, the appendix also may play a role in fighting infections. This organ is subject to inflammation, a condition called appendicitis. If inflamed, the appendix should be removed before the fluid content rises to the point that the appendix bursts, a situation that may cause **peritonitis,** a generalized infection of the lining of the abdominal cavity. Peritonitis can lead to death.

The **colon** includes the *ascending colon,* which goes up the right side of the body to the level of the liver; the *transverse colon,* which crosses the abdominal cavity just below the liver and the stomach; the *descending colon,* which passes down the left side of the body; and the sigmoid colon, which enters the *rectum,* the last 20 cm of the large intestine. The rectum opens at the **anus,** where **defecation,** the expulsion of *feces,* occurs. When feces are forced into the rectum by peristalsis, a defecation reflex occurs. The stretching of the rectal wall initiates nerve impulses to the spinal cord, and shortly thereafter contraction of the rectal muscles and relaxation of anal sphincters occur (Fig. 12.9). Feces are three-quarters water and one-quarter solids. Bacteria, fiber (indigestible remains), and other indigestible materials are in the solid portion. The brown color of feces is due to bilirubin (see page 222), and the odor is due to breakdown products as bacteria work on the nondigested remains. This bacterial action also produces gases.

For many years, it was believed that facultative bacteria (bacteria that can live with or without oxygen), such as *Escherichia coli,* were the major inhabitants of the colon, but new culture methods show that over 99% of the colon bacteria are obligate anaerobes (bacteria that die in the presence of oxygen). Not only do the bacteria break down indigestible material, they also produce some vitamins and other molecules that can be absorbed and used by us. In this way, they perform a service for us.

Water is considered unsafe for swimming when the coliform (nonpathogenic intestinal) bacterial count reaches a certain number. A high count is an indication that a significant amount of feces has entered the water. The more feces present, the greater the possibility that disease-causing bacteria are also present.

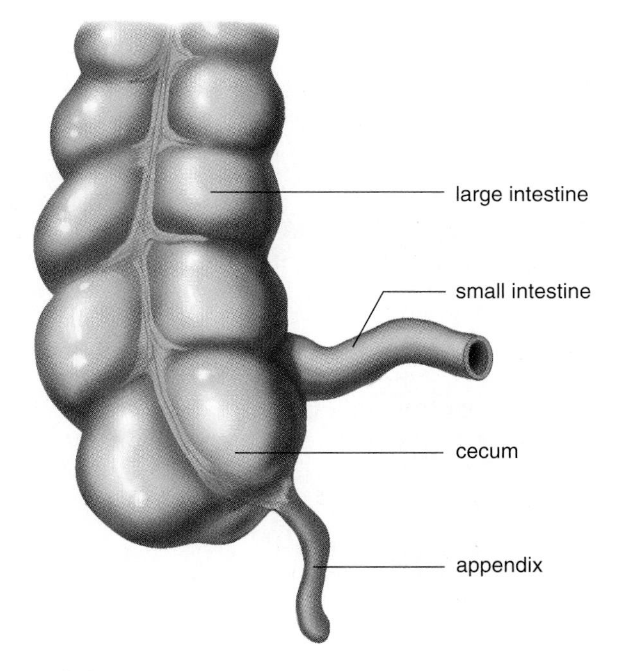

Figure 12.8 Junction of the small intestine and the large intestine.
The cecum is the blind end of the ascending colon. The appendix is attached to the cecum.

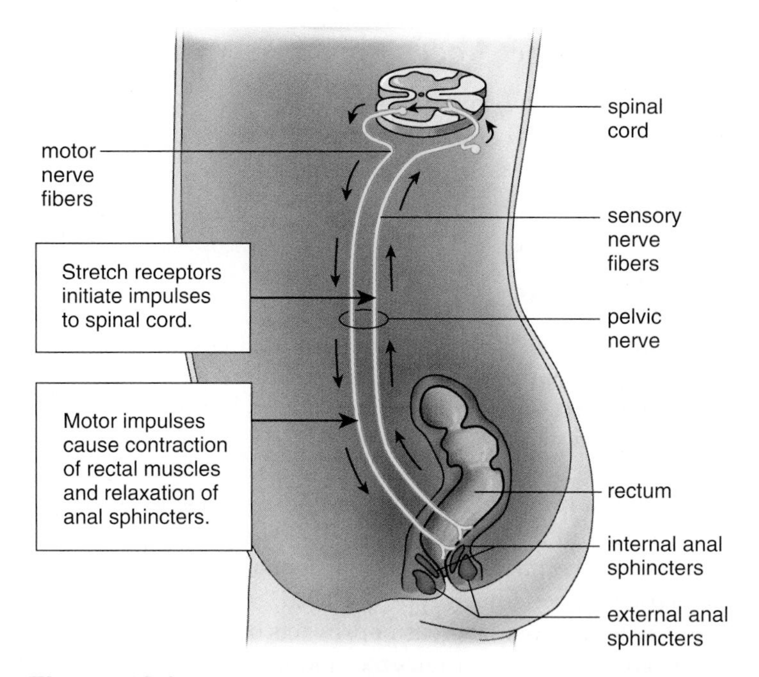

Figure 12.9 Defecation reflex.
The accumulation of feces in the rectum causes it to stretch, which initiates a reflex action resulting in rectal contraction and expulsion of the fecal material.

Polyps

The colon is subject to the development of **polyps,** small growths arising from the epithelial lining. Polyps, whether benign or cancerous, can be removed surgically. If colon cancer is detected while still confined to a polyp, the expected outcome is a complete cure. Some investigators believe that dietary fat increases the likelihood of colon cancer because dietary fat causes an increase in bile secretion. It could be that intestinal bacteria convert bile salts to substances that promote the development of cancer. On the other hand, fiber in the diet seems to inhibit the development of colon cancer. Dietary fiber absorbs water and adds bulk, thereby diluting the concentration of bile salts and facilitating the movement of substances through the intestine. Regular elimination reduces the time that the colon wall is exposed to any cancer-promoting agents in feces.

Diarrhea and Constipation

Two common everyday complaints associated with the large intestine are **diarrhea** and **constipation.** The major causes of diarrhea are infection of the lower tract and nervous stimulation. In the case of infection, such as food poisoning caused by eating contaminated food, the intestinal wall becomes irritated, and peristalsis increases. Water is not absorbed, and the diarrhea that results rids the body of the infectious organisms. In nervous diarrhea, the nervous system stimulates the intestinal wall, and diarrhea results. Prolonged diarrhea can lead to dehydration because of water loss and to disturbances in the heart's contraction due to an imbalance of salts in the blood.

When a person is constipated, the feces are dry and hard. One reason for this condition is that socialized persons have learned to inhibit defecation to the point that the desire to defecate is ignored. Two components of the diet that can help prevent constipation are water and fiber. Water intake prevents drying out of the feces, and fiber provides the bulk needed for elimination. The frequent use of laxatives is discouraged. If, however, it is necessary to take a laxative, a bulk laxative is the most natural because, like fiber, it produces a soft mass of cellulose in the colon. Lubricants, like mineral oil, make the colon slippery, and saline laxatives, like milk of magnesia, act osmotically—they prevent water from being absorbed and, depending on the dosage, may even cause water to enter the colon. Some laxatives are irritants; they increase peristalsis to the degree that the contents of the colon are expelled.

Chronic constipation is associated with the development of hemorrhoids, enlarged and inflamed blood vessels at the anus.

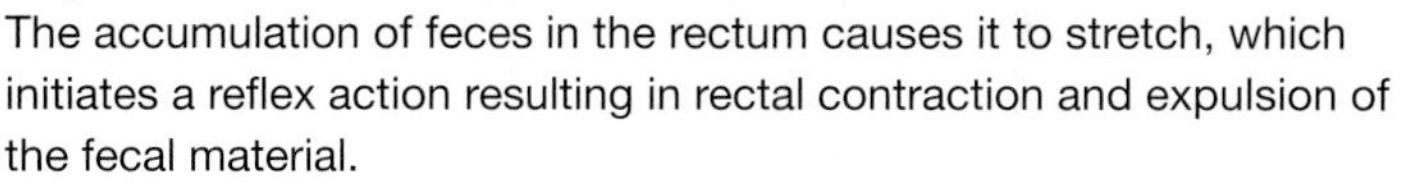
The large intestine does not produce digestive enzymes; it does absorb water, salts, and some vitamins.

12.2 Three Accessory Organs

The pancreas, liver, and gallbladder are accessory digestive organs. Figure 12.1 shows how the pancreatic duct from the pancreas and the common bile duct from the liver and gallbladder join before entering the duodenum.

The Pancreas

The **pancreas** lies deep in the abdominal cavity, resting on the posterior abdominal wall. It is an elongated and somewhat flattened organ that has both an endocrine and an exocrine function. As an endocrine gland it secretes insulin and glucagon, hormones that help keep the blood glucose level within normal limits. We are now interested in its exocrine function. Most pancreatic cells produce pancreatic juice, which contains sodium bicarbonate ($NaHCO_3$) and digestive enzymes for all types of food. Sodium bicarbonate neutralizes chyme; whereas pepsin acts best in an acid pH of the stomach, pancreatic enzymes require a slightly basic pH. **Pancreatic amylase** digests starch, **trypsin** digests protein, and **lipase** digests fat. In cystic fibrosis, a thick mucus blocks the pancreatic duct, and the patient must take supplemental pancreatic enzymes by mouth for proper digestion to occur.

The Liver

The **liver,** which is the largest organ in the body, lies mainly in the upper right section of the abdominal cavity, under the diaphragm (see Fig. 12.1). The liver has two main lobes, the right lobe and the smaller left lobe, which crosses the midline and lies above the stomach. The liver contains approximately 100,000 lobules that serve as the structural functional units of the liver (Fig. 12.10). Triads consisting of these three structures are located between the lobules: (1) a branch of the hepatic artery that brings oxygenated blood to the liver; (2) a branch of the hepatic portal vein that transports nutrients from the intestines; and (3) a bile duct that takes bile away from the liver. The central veins of lobules enter the hepatic vein. Note in Figure 12.11 that the liver lies between the hepatic portal vein (number 2 in the figure) and the hepatic vein (number 4 in the figure), which enters the vena cava.

In some ways, the liver acts as the gatekeeper to the blood. As the blood from the intestines passes through the liver, it removes poisonous substances and works to keep the contents of the blood constant. It also removes and stores iron and the fat-soluble vitamins A, D, E, and K. The liver makes the plasma proteins from amino acids, and lipids from fatty acids. It also produces cholesterol and helps regulate the quantity of this substance in the blood.

The liver maintains the blood glucose level at about 100 mg/100 ml (0.1%), even though a person eats intermittently. Any excess glucose that is present in the hepatic portal vein is removed and stored by the liver as glycogen. Between eating, glycogen is broken down to glucose, which enters the hepatic vein, and in this way, the blood glucose level remains constant.

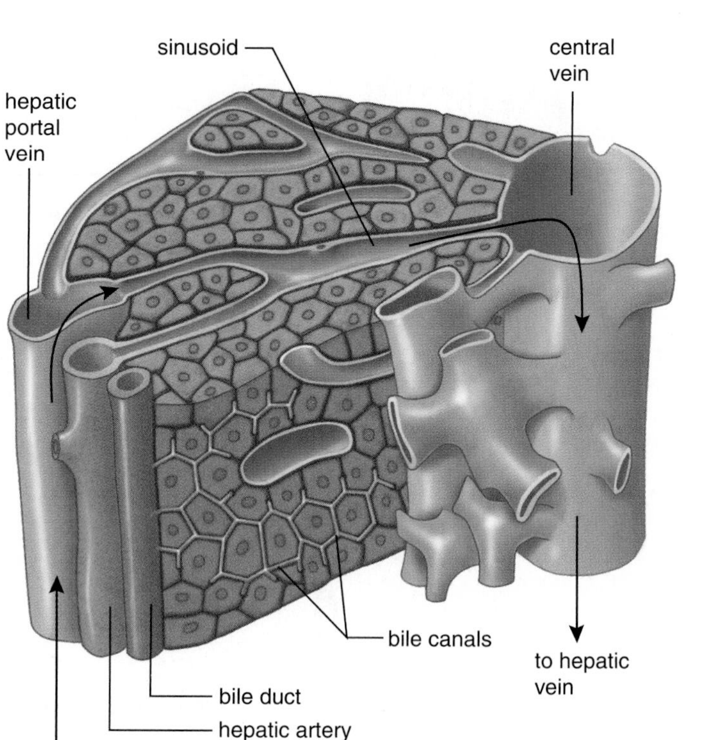

Figure 12.10 Hepatic lobules.
The liver contains over 100,000 lobules. Each lobule contains many cells that perform the various functions of the liver. They remove from and/or add materials to blood and deposit bile in bile ducts.

If the supply of glycogen is depleted, the liver will convert glycerol (from fats) and amino acids to glucose molecules. The conversion of amino acids to glucose necessitates deamination, the removal of amino acids. By a complex metabolic pathway, the liver then combines ammonia with carbon dioxide to form urea:

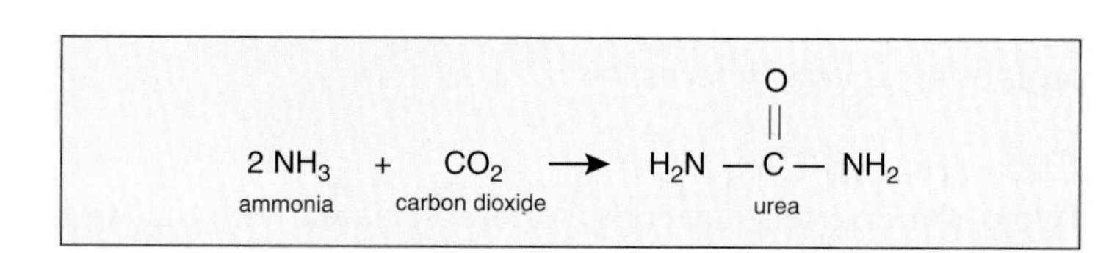

Urea is the usual nitrogenous waste product from amino acid breakdown in humans. After its formation in the liver, urea is excreted by the kidneys.

The liver produces bile, which is stored in the gallbladder. Bile has a yellowish green color because it contains the bile pigment bilirubin, derived from the breakdown of hemoglobin, the red pigment of red blood cells. Bile also contains bile salts, which are derived from cholesterol and emulsify fat in the small intestine. When fat is emulsified, it breaks

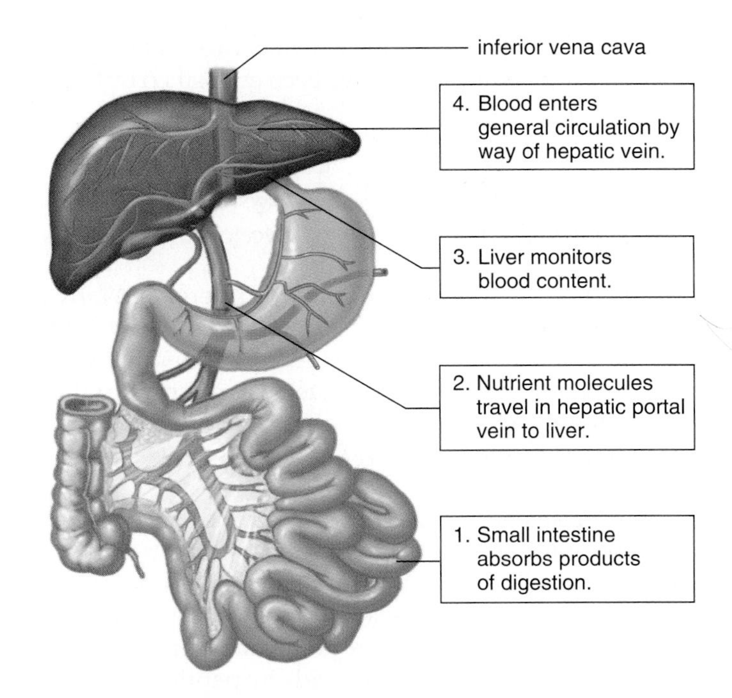

Figure 12.11 Hepatic portal system.
The hepatic portal vein takes the products of digestion from the digestive system to the liver, where they are processed before entering the cardiovascular system proper.

up into droplets, providing a much larger surface area, which can be acted upon by a digestive enzyme from the pancreas.

Altogether, the following are significant functions of the liver:

1. Detoxifies blood by removing and metabolizing poisonous substances.
2. Stores iron (Fe^{2+}) and the fat-soluble vitamins A, D, E, and K.
3. Makes plasma proteins, such as albumins and fibrinogen, from amino acids.
4. Stores glucose as glycogen after eating, and breaks down glycogen to glucose to maintain the glucose concentration of blood between eating periods.
5. Produces urea from the breakdown of amino acids.
6. Removes bilirubin, a breakdown product of hemoglobin from the blood, and excretes it in bile, a liver product.
7. Produces lipids from fatty acids; produces and helps regulate blood cholesterol level, converting some to bile salts.

Liver Disorders

Jaundice, hepatitis, and cirrhosis are three serious diseases that affect the entire liver and hinder its ability to repair itself. Therefore, they are life-threatening diseases. When a person has **jaundice,** there is a yellowish tint to the whites of the eyes and also to the skin of light-pigmented persons. Bilirubin is deposited in the skin due to an abnormally large amount in the blood. In *hemolytic jaundice,* red blood cells have been broken down in abnormally large amounts; in *obstructive jaundice,* bile ducts are blocked or liver cells are damaged.

Jaundice can also result from **hepatitis,** inflammation of the liver. Viral hepatitis occurs in several forms. Hepatitis A is usually acquired from sewage-contaminated drinking water. Hepatitis B, which is usually spread by sexual contact, can also be spread by blood transfusions or contaminated needles. The hepatitis B virus is more contagious than the AIDS virus, which is spread in the same way. Thankfully, however, there is now a vaccine available for hepatitis B. Hepatitis C, which is usually acquired by contact with infected blood and for which there is no vaccine, can lead to chronic hepatitis, liver cancer, and death.

Cirrhosis is another chronic disease of the liver. First the organ becomes fatty, and liver tissue is then replaced by inactive fibrous scar tissue. Cirrhosis of the liver is often seen in alcoholics due to malnutrition and to the excessive amounts of alcohol (a toxin) the liver is forced to break down.

The liver has amazing generative powers and can recover if the rate of regeneration exceeds the rate of damage. During liver failure, however, there may not be enough time to let the liver heal itself. Liver transplantation is usually the preferred treatment for liver failure, but artificial livers have been developed and tried in a few cases. One type is a cartridge that contains liver cells. The patient's blood passes through cellulose acetate tubing of the cartridge and is serviced in the same manner as with a normal liver. In the meantime, the patient's liver has a chance to recover.

The Gallbladder

The **gallbladder** is a pear-shaped, muscular sac attached to the surface of the liver (see Fig. 12.1). About 1,000 ml of bile are produced by the liver each day, and any excess is stored in the gallbladder. Water is reabsorbed by the gallbladder so that bile becomes a thick, mucus-like material. When needed, bile leaves the gallbladder and proceeds to the duodenum via the common bile duct.

The cholesterol content of bile can come out of solution and form crystals. If the crystals grow in size, they form gallstones. The passage of the stones from the gallbladder may block the common bile duct and cause obstructive jaundice. Then the gallbladder must be removed.

The pancreas produces pancreatic juice, which contains enzymes for the digestion of food. Among its many functions, the liver produces bile, which is stored in the gallbladder.

12.3 Digestive Enzymes

The digestive enzymes are **hydrolytic enzymes,** which break down substances by the introduction of water at specific bonds. Digestive enzymes, like other enzymes, are proteins with a particular shape that fits their substrate. They also have an optimum pH, which maintains their shape, thereby enabling them to speed up their specific reaction.

The various digestive enzymes present in the digestive juices, mentioned previously, help break down carbohydrates, proteins, nucleic acids, and fats, the major components of food. Starch is a carbohydrate, and its digestion begins in the mouth. Saliva from the salivary glands has a neutral pH and contains **salivary amylase,** the first enzyme to act on starch:

$$\text{starch} + H_2O \xrightarrow{\text{salivary amylase}} \text{maltose}$$

In this equation, salivary amylase is written above the arrow to indicate that it is neither a reactant nor a product in the reaction. It merely speeds the reaction in which its substrate, starch, is digested to many molecules of maltose, a disaccharide. Maltose molecules cannot be absorbed by the intestine; additional digestive action in the small intestine converts maltose to glucose, which can be absorbed.

Protein digestion begins in the stomach. Gastric juice secreted by gastric glands has a very low pH—about 2—because it contains hydrochloric acid (HCl). Pepsinogen, a precursor that is converted to the enzyme **pepsin** when exposed to HCl, is also present in gastric juice. Pepsin acts on protein to produce peptides:

$$\text{protein} + H_2O \xrightarrow{\text{pepsin}} \text{peptides}$$

Peptides vary in length, but they always consist of a number of linked amino acids. Peptides are usually too large to be absorbed by the intestinal lining, but later they are broken down to amino acids in the small intestine.

Starch, proteins, nucleic acids, and fats are all enzymatically broken down in the small intestine. Pancreatic juice, which enters the duodenum, has a basic pH because it contains sodium bicarbonate ($NaHCO_3$). Sodium bicarbonate neutralizes chyme, producing the slightly basic pH that is optimum for pancreatic enzymes. One pancreatic enzyme, pancreatic amylase, digests starch:

$$\text{starch} + H_2O \xrightarrow{\text{pancreatic amylase}} \text{maltose}$$

Another pancreatic enzyme, trypsin, digests protein:

$$\text{protein} + H_2O \xrightarrow{\text{trypsin}} \text{peptides}$$

Trypsin is secreted as trypsinogen, which is converted to trypsin in the duodenum.

Lipase, a third pancreatic enzyme, digests fat molecules in the fat droplets after they have been emulsified by bile salts:

$$\text{fat} \xrightarrow{\text{bile salts}} \text{fat droplets}$$

$$\text{fat droplets} + H_2O \xrightarrow{\text{lipase}} \text{glycerol} + \text{fatty acids}$$

The end products of lipase digestion, glycerol and fatty acid molecules, are small enough to cross the cells of the intestinal villi, where absorption takes place. As mentioned previously, glycerol and fatty acids enter the cells of the villi, and within these cells, they are rejoined and packaged as lipoprotein droplets before entering the lacteals (see Fig. 12.6).

Peptidases and **maltase,** two enzymes secreted by the small intestine, complete the digestion of protein to amino acids and starch to glucose, respectively. Amino acids and glucose are small molecules that cross into the cells of the villi. Peptides, which result from the first step in protein digestion, are digested to amino acids by peptidases:

$$\text{peptides} + H_2O \xrightarrow{\text{peptidases}} \text{amino acids}$$

Maltose, a disaccharide that results from the first step in starch digestion, is digested to glucose by maltase:

$$\text{maltose} + H_2O \xrightarrow{\text{maltase}} \text{glucose} + \text{glucose}$$

Other disaccharides, each of which has its own enzyme, are digested in the small intestine. The absence of any one of these enzymes can cause illness. For example, many people, including as many as 75% of African Americans, cannot digest lactose, the sugar found in milk, because they do not produce lactase, the enzyme that converts lactose to its components, glucose and galactose. Drinking untreated milk often gives these individuals the symptoms of **lactose intolerance** (diarrhea, gas, cramps), caused by a large quantity of nondigested lactose in the intestine. In most areas, it is possible to purchase milk made lactose-free by the addition of synthetic lactase or *Lactobacillus acidophilus* bacteria, which break down lactose.

Table 12.2 lists some of the major digestive enzymes produced by the digestive tract, salivary glands, or the pancreas. Each type of food is broken down by specific enzymes.

Digestive enzymes present in digestive juices help break down food to the nutrient molecules: glucose, amino acids, fatty acids, and glycerol. The first two are absorbed into the blood capillaries of the villi, and the last two re-form within epithelial cells before entering the lacteals as lipoprotein droplets.

Table 12.2 Major Digestive Enzymes

Food	Digestion	Enzyme	Optimum pH	Produced by	Site of Action
Starch	Starch + H_2O → maltose	Salivary amylase	Neutral	Salivary glands	Mouth
		Pancreatic amylase	Basic	Pancreas	Small intestine
	Maltose + H_2O → glucose	Maltase	Basic	Small intestine	Small intestine
Protein	Protein + H_2O → peptides	Pepsin	Acidic	Gastric glands	Stomach
		Trypsin	Basic	Pancreas	Small intestine
	Peptides + H_2O → amino acids	Peptidases	Basic	Small intestine	Small intestine
Nucleic Acid	RNA and DNA + H_2O → nucleotides	Nuclease	Basic	Pancreas	Small intestine
	Nucleotides → bases, sugars, phosphate ions	Nucleosidases	Basic	Small intestine	Small intestine
Fat	Fat droplets + H_2O → glycerol + fatty acids	Lipase	Basic	Pancreas	Small intestine

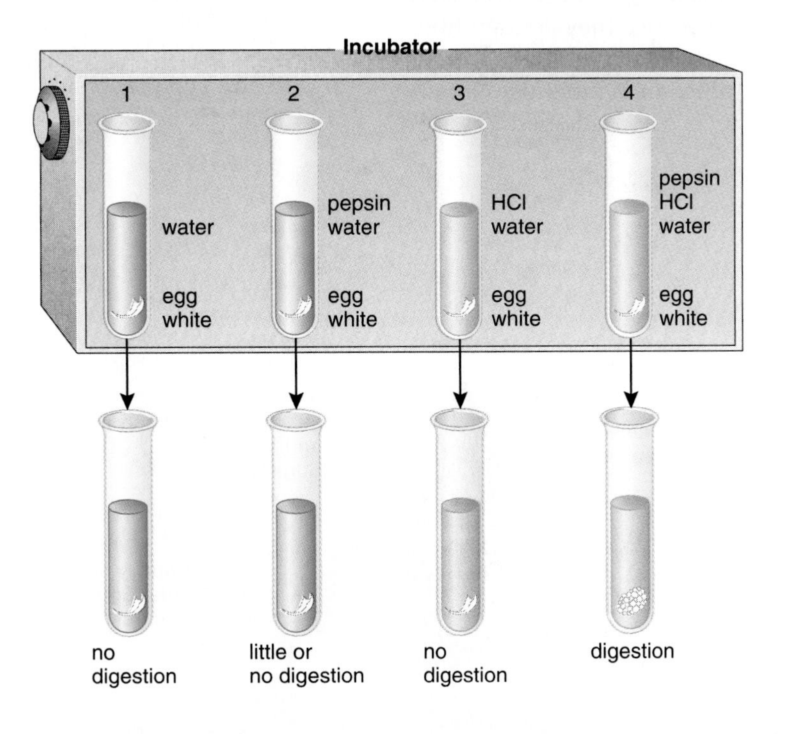

Figure 12.12 Digestion experiment.
This experiment is based on the optimum conditions for digestion by pepsin in the stomach. Knowing that the correct enzyme, optimum pH, optimum temperature, and the correct substrate must be present for digestion to occur, explain the results of this experiment.

Conditions for Digestion

Laboratory experiments can define the necessary conditions for digestion. For example, the four test tubes described in Figure 12.12 can be prepared and observed for the digestion of egg white, a protein digested in the stomach by the enzyme pepsin.

After all tubes are placed in an incubator at body temperature for at least one hour, the results depicted are observed. Tube 1 is a control tube; no digestion has occurred in this tube because the enzyme and HCl are missing. (If a control gives a positive result, then the experiment is invalidated.) Tube 2 shows limited or no digestion because HCl is missing, and therefore the pH is too high for pepsin to be effective. Tube 3 shows no digestion because although HCl is present, the enzyme is missing. Tube 4 shows the best digestive action because the enzyme is present and the presence of HCl has resulted in an optimum pH. This experiment supports the hypothesis that for digestion to occur, the substrate and enzyme must be present and the environmental conditions must be optimum. The optimal environmental conditions include a warm temperature and the correct pH.

12.4 Nutrition

The body requires three major classes of *macronutrients* in the diet: carbohydrate, protein, and fat. These supply the energy and the building blocks that are needed to synthesize cellular contents. *Micronutrients*—especially vitamins and minerals—are also required because they are necessary for optimum cellular metabolism.

Several modern nutritional studies suggest that certain nutrients can protect against heart disease, cancer, and other serious illnesses. These studies include an analysis of the eating habits of people in the United States and from around the world, especially those with and without heart disease and cancer. The result has been the dietary recommendations illustrated by a food pyramid (Fig. 12.13).

The bulk of the diet should consist of bread, cereal, rice, and pasta as energy sources. Whole grains are preferred over those that have been milled because they contain fiber and vitamins and minerals. Vegetables and fruits are another rich source of fiber, vitamins, and minerals. Notice, then, that a largely vegetarian diet is recommended. A vegetarian diet is not only healthy, it also makes sense in another way. As a rule of thumb, it is generally stated that only about 10 percent of the energy available in a food source is incorporated into the tissues of a consumer. This being the case, it can be reasoned that 100 lbs of grain could directly result in 10 human lbs, but if fed to cattle it would result in only one human lb. Therefore, a larger human population can be sustained on grain than on grain-consuming animals.

Animal products, especially meat, need only be minimally included in the diet; fats and sweets should be used sparingly. Dairy products and meats tend to be high in saturated fats, and an intake of saturated fats increases the risk of cardiovascular disease. Low-fat dairy products are available, but there is no way to take much of the fat out of beef, which has a relatively high fat content. Ironically, the affluence of people in the United States contributes to a poor diet and, therefore, possible illness. Only comparatively rich people can afford fatty meats from grain-fed cattle and carbohydrates that have been highly processed to remove fiber and to add sugar and salt.

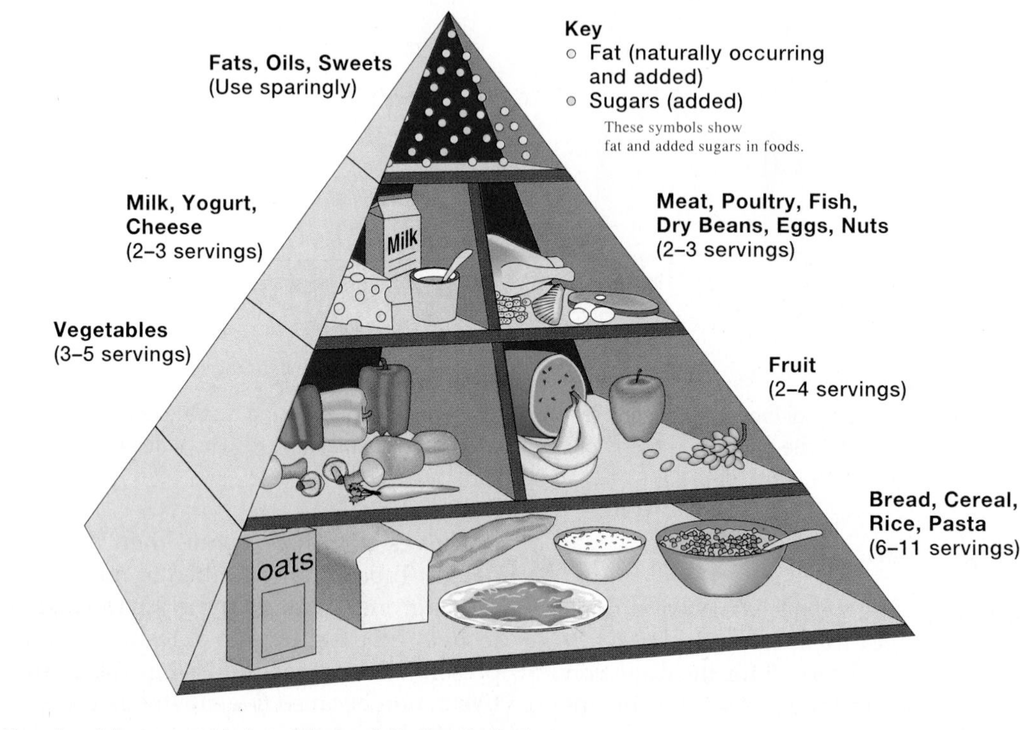

Figure 12.13 Food guide pyramid: A guide to daily food choices.
The U.S. Department of Agriculture uses a pyramid to show the ideal diet because it emphasizes the importance of including grains, fruits, and vegetables in the diet. Meats and dairy products are needed in limited amounts; fats, oils, and sweets should be used sparingly.
Source: Data from the U.S. Department of Agriculture.

Carbohydrates

The quickest, most readily available source of energy for the body is glucose. Carbohydrates are digested to simple sugars, which are or can be converted to glucose. Glucose is stored by the liver in the form of glycogen. Between eating periods, the blood glucose level is maintained at about 100 mg/100 ml of blood by the breakdown of glycogen or by the conversion of glycerol (from fats) or amino acids to glucose. If necessary, amino acids are taken from the muscles—even from the heart muscle. While body cells can utilize fatty acids as an energy source, brain cells require glucose. For this reason alone, it is necessary to include carbohydrates in the diet. According to Fig. 12.13, carbohydrates should make up the bulk of the diet. Further, these carbohydrates should be complex and not simple carbohydrates. Complex sources of carbohydrates include preferably whole-grain pasta, rice, bread, and cereal (Fig 12.14). Potatoes and corn, although considered vegetables, are also sources of carbohydrates.

Simple carbohydrates (e.g., sugars) are labeled "empty calories" by some dieticians because they contribute to energy needs and weight gain without supplying any other nutritional requirements. Table 12.3 gives suggestions on how to reduce dietary sugars (simple carbohydrates). In contrast to simple sugars, complex carbohydrates are likely to be accompanied by a wide range of other nutrients and by **fiber,** which is indigestible plant material.

The intake of fiber is recommended because it decreases the risk of colon cancer, a major type of cancer, and cardiovascular disease, the number one killer in the United States. Insoluble fiber, such as that found in wheat bran, has a laxative effect and may guard against colon cancer, because any cancer-causing substances are in contact with the intestinal wall for a limited amount of time. Soluble fiber, such as that found in oat bran, combines with bile acids and cholesterol in the intestine and prevents them from being absorbed. The liver now removes cholesterol from the blood and changes it to bile acids, replacing those that were lost. While the diet should have an adequate amount of fiber, a high-fiber diet can be detrimental. Some evidence suggests that the absorption of iron, zinc, and calcium is impaired by a diet too high in fiber.

Complex carbohydrates, which contain fiber, should form the bulk of the diet.

Table 12.3 Reducing Dietary Sugar

To reduce dietary sugar:

1. Eat fewer sweets, such as candy, soft drinks, ice cream, and pastry.
2. Eat fresh fruits or fruits canned without heavy syrup.
3. Use less sugar—white, brown, or raw—and less honey and syrups.
4. Avoid sweetened breakfast cereals.
5. Eat less jelly, jam, preserves
6. Drink pure fruit juices, not imitations.
7. When cooking, use spices like cinnamon instead of sugar to flavor foods.
8. Do not put sugar in tea or coffee.

Figure 12.14
Complex carbohydrates.
To meet our energy needs, dieticians recommend consuming foods rich in complex carbohydrates, like those shown here, rather than foods consisting of simple carbohydrates, like candy and ice cream. Simple carbohydrates provide monosaccharides but few other types of nutrients.

Proteins

Foods rich in protein include red meat, fish, poultry, dairy products, legumes (i.e., peas and beans), nuts, and cereals. Following digestion of protein, amino acids enter the bloodstream and are transported to the tissues. Ordinarily, amino acids are not used as an energy source. Most are incorporated into structural proteins found in muscles, skin, hair, and nails. Others are used to synthesize such proteins as hemoglobin, plasma proteins, enzymes, and hormones.

Adequate protein formation requires 20 different types of amino acids. Of these, eight are required from the diet in adults (nine in children) because the body is unable to produce them. These are termed the **essential amino acids.** The body produces the other 11 amino acids by simply transforming one type into another type. Some protein sources, such as meat, are *complete;* they provide all 20 types of amino acids. Vegetables and grains supply us with amino acids, but each vegetable or grain alone is an *incomplete* protein source because of a deficiency in at least one of the essential amino acids. Absence of one essential amino acid prevents utilization of the other 19 amino acids. Soybeans and tofu, made from soybeans, are rich in amino acids, but it is wise to combine foods to acquire all the essential amino acids. For example, the combinations of cereal with milk, or beans, a legume, with rice, a grain, will provide all the essential amino acids (Table 12.4).

Amino acids are not stored in the body, and a daily supply is needed. However, it does not take very much protein to meet the daily requirement. Two servings of meat a day (equal in total quantity to a deck of cards) is usually enough. Some meats (e.g., hamburger) are high in protein but also high in fat. Everything considered, it is probably a good idea to depend on protein from plant origins (e.g., whole-grain cereals, dark breads, and legumes) to a greater extent than is often the custom in the United States. This can be illustrated by the health statistics of native Hawaiians who no longer eat as their ancestors did (Fig. 12.15). The modern diet depends on animal rather than plant protein and is 42% fat. A statistical study showed that the island's native peoples now have a higher than average death rate from cardiovascular disease and cancer. Diabetes is also common in persons who follow the modern diet. But the health of those who have switched back to the ancient diet has improved immensely!

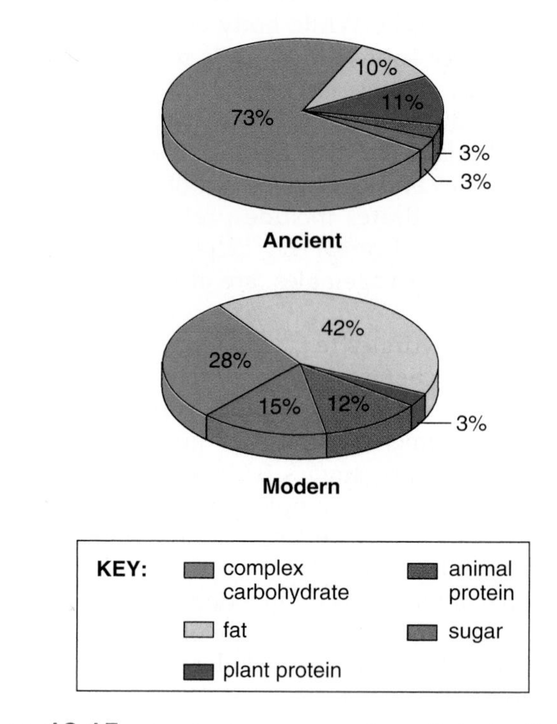

Figure 12.15 Ancient versus modern diet of native Hawaiians.
Among those native Hawaiians who have switched back to the native diet, the incidence of cardiovascular disease, cancer, and diabetes has dropped.

Table 12.4 Complementary Protein Combinations

For any particular meal, choose foods from two or more columns.

Legumes	*Grains*	*Vegetables*	*Seeds and Nuts*
Dried beans	Barley	Broccoli	Nut butters
Soy products	Cornmeal	Brussel sprouts	Sesame seeds
Dried lentils	Pasta	Leafy greens	Cashews
Peanuts	Oats		Sunflower seeds
Dried peas	Rice		Walnuts
	Whole-grain breads		

Nutritionists do not recommend the use of protein and/or amino acid supplements. Protein supplements that people take to build muscle or as part of a diet are not digested as well as protein-rich foods and they cost more than food. Amino acid supplements can be dangerous to your health. An excess of any particular amino acid can lead to a deficiency of absorption of other amino acids present in lesser amounts. Contrary to popular reports, the taking of lysine does not relieve or cure herpes sores. It is also unwise to take tryptophan to relieve pain or cure insomnia. Some who have taken supplements of tryptophan have come down with a blood disorder (eosinophilia-myalgia syndrome) characterized by severe muscle and joint pain and swelling of the limbs.

Lipids

Fat and cholesterol are both lipids. Fat is present not only in butter, margarine, and oils, but also in many foods high in animal protein. The body can alter ingested fat to suit the body's needs, except it is unable to produce linoleic acid, which is a polyunsaturated fatty acid. Saturated fatty acids have no double bonds; polyunsaturated fatty acids have many double bonds.

The current guidelines suggest that fat should account for no more than 30% of our daily calories. The chief reason is that an intake of fat not only causes weight gain, it also increases the risk of cancer and cardiovascular disease. Dietary fat apparently increases the risk of colon, hepatic, and pancreatic cancers. Although recent studies suggest no link between dietary fat and breast cancer, other researchers still believe that the matter deserves further investigation.

Cardiovascular disease is often due to arteries blocked by fatty deposits, called **plaque,** that contain saturated fats and cholesterol. Cholesterol is carried in the blood by two types of lipoproteins: low-density lipoprotein (LDL) and high-density lipoprotein (HDL). LDL is thought of as being "bad" because it carries cholesterol from the liver to the cells, while HDL is thought of as being "good" because it carries cholesterol to the liver, which takes it up and converts it to bile salts. Saturated fats, whether in butter or margarine, can raise LDL cholesterol levels, while monounsaturated (one double bond) fats and polyunsaturated (many double bonds) fats lower LDL cholesterol levels. Olive oil and canola oil contain mostly monounsaturated fats; corn oil and safflower oil contain mostly polyunsaturated fats. These oils have a liquid consistency and come from plants. Saturated fats, which are solids at room temperature, usually have an animal origin; two well-known exceptions are palm oil and coconut oil, which contain mostly saturated fats and come from the plants mentioned.

Nutritionists stress that it is more important for the diet to be low in fat rather than be overly concerned about which type fat is in the diet. Table 12.5 gives suggestions on how to reduce dietary fat.

Table 12.5 Reducing Lipids

To reduce dietary fat:

1. Choose poultry, fish, or dry beans and peas as a protein source.
2. Remove skin from poultry before cooking, and place on a rack so that fat drains off.
3. Broil, boil, or bake rather than fry.
4. Limit your intake of butter, cream, hydrogenated oils, shortenings, and tropical oils (coconut and palm oils).*
5. Use herbs and spices to season vegetables instead of butter, margarine, or sauces. Use lemon juice instead of salad dressing.
6. Drink skim milk instead of whole milk, and use skim milk in cooking and baking.
7. Eat nonfat or low-fat foods.

To reduce dietary cholesterol:

1. Avoid cheese, egg yolks, liver, and certain shellfish (shrimp and lobster). Preferably, eat white fish and poultry.
2. Substitute egg whites for egg yolks in both cooking and eating.
3. Include soluble fiber in the diet. Oat bran, oatmeal, beans, corn, and fruits such as apples, citrus fruits, and cranberries are high in soluble fiber.

*Although coconut and palm oils are from plant sources, they are mostly saturated fats.

Fake Fat

Olestra is a substance made to look, taste, and act like real fat but the digestive system is unable to digest it. Therefore it is commonly known as "fake fat." Unfortunately, the fat-soluble vitamins A, D, E, and K tend to be taken up by olestra and thereafter they are not absorbed by the body. Similarly, people using olestra have reduced amounts of carotenoids in the blood. In one study, just a handful of olestra-soaked potato chips caused a 20% decline in blood beta-carotene levels. Manufacturers fortify olestra-containing foods with the vitamins mentioned but not carotenoids.

More apparent, some people who consume olestra have developed anal leakage or underwear staining. Others experience diarrhea, intestinal cramping, and gas.

Dietary protein supplies the essential amino acids; proteins from plant origins generally have less accompanying fat. A diet no more than 30% fat is recommended because fat intake, particularly saturated fats, are known to be associated with various health problems.

Vitamins

Vitamins are organic compounds (other than carbohydrate, fat, and protein) that the body is unable to produce but uses for metabolic purposes. Many vitamins are portions of coenzymes, which are enzyme helpers. For example, niacin is part of the coenzyme NAD^+, and riboflavin is part of another dehydrogenase, FAD. Coenzymes are needed in only small amounts because each can be used over and over again. Not all vitamins are coenzymes; vitamin A, for example, is a precursor for the visual pigment that prevents night blindness.

If vitamins are lacking in the diet, various symptoms develop (Fig. 12.16). Altogether there are 13 vitamins, which are divided into those that are fat soluble (Table 12.6) and those that are water soluble (Table 12.7).

Antioxidants

Over the past 20 years, numerous statistical studies have been done to determine whether a diet rich in fruits and vegetables is protective against cancer. Cellular metabolism generates free radicals, unstable molecules that carry an extra electron. The most common free radicals in cells are the superoxide (O_2^-) and hydroxide (OH^-). In order to stabilize themselves, free radicals donate an electron to DNA, or proteins, including enzymes, or lipids, which are found in plasma membranes. Such donations most likely damage these cellular molecules and thereby may lead to disorders, perhaps even cancer.

Vitamins C, E, and A are believed to defend the body against free radicals, and therefore they are termed antioxidants. These vitamins are especially abundant in fruits and vegetables. The dietary guidelines shown in Figure 12.13 suggest that we eat a minimum of five servings of fruits and vegetables a day. To achieve this goal, include salad greens, raw or cooked vegetables, dried fruit, and fruit juice, in addition to traditional apples and oranges and such.

Dietary supplements may provide a potential safeguard against cancer and cardiovascular disease, but nutritionists do not think it is appropriate to take supplements instead of improving intake of fruits and vegetables. There are many beneficial compounds in fruits that cannot be obtained from a vitamin pill. These compounds enhance each other's absorption or action and also perform independent biological functions.

Vitamin D

Skin cells contain a precursor cholesterol molecule that first must be exposed to ultraviolet light and then when it enters the bloodstream must be acted on by the kidneys and liver before it finally becomes active vitamin D (calcitriol). Vitamin D promotes the absorption of calcium. Calcitriol promotes the absorption of calcium by the intestines. The lack of vitamin D leads to rickets in children (see Fig. 12.16*a*). Rickets, characterized by a bowing of the legs, is caused by defective mineralization of the skeleton. Most milk today is fortified with vitamin D, which helps prevent the occurrence of rickets.

Vitamins are essential to cellular metabolism; many are protective against identifiable illnesses and conditions.

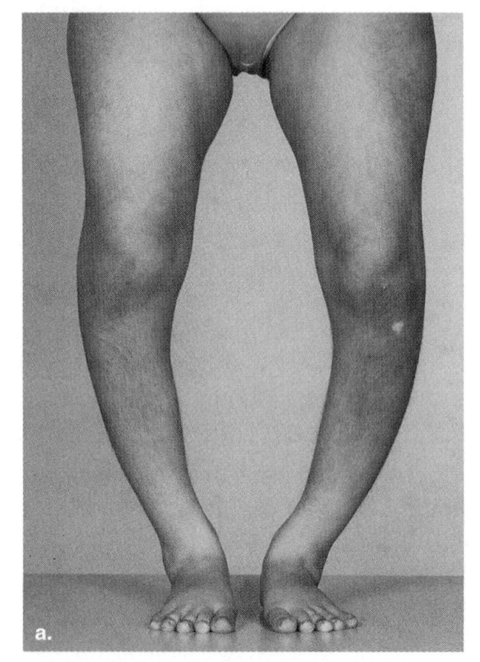

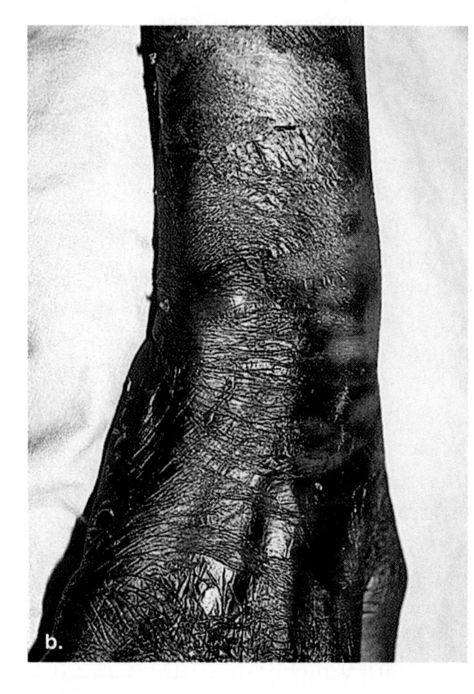

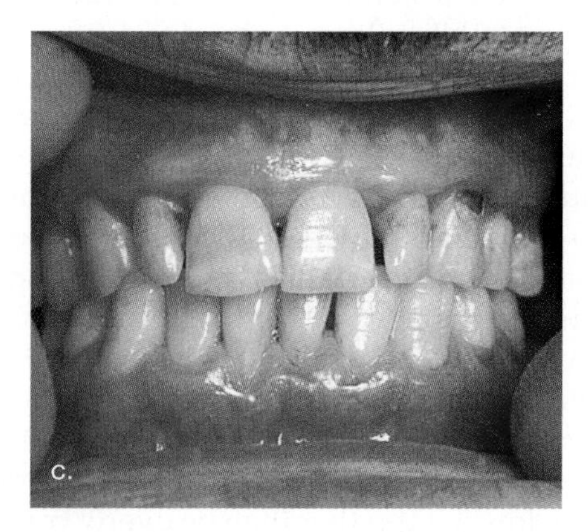

Figure 12.16 Illnesses due to vitamin deficiency.
a. Bowing of bones (rickets) due to vitamin D deficiency. **b.** Dermatitis (pellagra) of areas exposed to light due to niacin (vitamin B_3) deficiency. **c.** Bleeding of gums (scurvy) due to vitamin C deficiency.

Table 12.6 Fat-Soluble Vitamins

Vitamin	Functions	Food Sources	Conditions with	
			Too Little	Too Much
Vitamin A	Antioxidant synthesized from beta-carotene; needed for healthy eyes, skin, hair, and mucous membranes, and for proper bone growth	Deep yellow/orange vegetables, fruits, cheese, whole milk, butter, and eggs	Night blindness, impaired growth of bones and teeth	Headache, dizziness, nausea, hair loss
Vitamin D	A group of steroids needed for development and maintenance of bones and teeth	Milk fortified with vitamin D, fish liver oil; also made in the skin when exposed to sunlight	Rickets, bone decalcification and weakening	Calcification of soft tissues, diarrhea, and possible renal damage
Vitamin E	Antioxidant that prevents oxidation of vitamin A and polyunsaturated fatty acids	Leafy green vegetables, fruits, vegetable oils, nuts, whole-grain breads and cereals	Unknown	Diarrhea, nausea, headaches, fatigue, muscle weakness
Vitamin K	Needed for synthesis of substances active in clotting of blood	Leafy green vegetables, cabbage and cauliflower	Easy bruising and bleeding	Can interfere with anticoagulant medication

Table 12.7 Water-Soluble Vitamins

Vitamin	Functions	Food Sources	Conditions with	
			Too Little	Too Much
Vitamin C	Antioxidant; needed for forming collagen; helps maintain capillaries, bones, and teeth	Citrus fruits, leafy green vegetables, tomatoes, potatoes, cabbage	Scurvy, wounds heal slowly, infections	Gout, kidney stones, diarrhea, decreased copper
Thiamine (Vitamin B_1)	Part of coenzyme needed for cellular respiration; also promotes activity of the nervous system	Whole-grain cereals, dried beans and peas, sunflower seeds, and nuts	Beriberi, muscular weakness, heart enlarges	Interferes with absorption of other vitamins
Riboflavin (Vitamin B_2)	Part of coenzymes, such as FAD; aids cellular respiration, including oxidation of protein and fat	Nuts, dairy products, whole-grain cereals, poultry, and leafy green vegetables	Dermatitis, blurred vision, growth failure	Unknown
Niacin (Nicotinic acid)	Part of coenzymes NAD^+ and $NADP^+$; needed for cellular respiration, including oxidation of protein and fat	Peanuts, poultry, whole-grain cereals, leafy green vegetables, and beans	Pellagra, diarrhea, and mental disorders	High blood sugar and uric acid, vasodilation, etc.
Folacin (Folic acid)	Coenzyme needed for production of hemoglobin and the formation of DNA	Dark leafy green vegetables, nuts, beans, whole-grain cereals	Megaloblastic anemia, spina bifida	May mask B_{12} deficiency
Vitamin B_6	Coenzyme needed for the synthesis of hormones and hemoglobin; CNS control	Whole-grain cereals, bananas, beans, poultry, nuts, leafy green vegetables	Rarely, convulsions, vomiting, seborrhea, muscular weakness	Insomnia
Pantothenic acid	Part of coenzyme A needed for oxidation of carbohydrates and fats; aids in the formation of hormones and certain neurotransmitters	Nuts, beans, dark green vegetables, poultry, fruits, and milk	Rarely, loss of appetite, mental depression, numbness	Unknown
Vitamin B_{12}	Complex, cobalt-containing compound; part of the coenzyme needed for synthesis of nucleic acids and myelin	Dairy products, fish, poultry, eggs, fortified cereals	Pernicious anemia	Unknown
Biotin	Coenzyme needed for metabolism of amino acids and fatty acids	Generally in foods, especially eggs	Skin rash, nausea, fatigue	Unknown

Minerals

In addition to vitamins, various **minerals** are required by the body. Minerals are divided into macrominerals and microminerals. The body contains more than 5 grams of each macromineral and less than 5 grams of each micromineral (Fig. 12.17). The macrominerals are constituents of cells and body fluids and are structural components of tissues. For example, calcium (present as Ca^{2+}) is needed for the construction of bones and teeth and for nerve conduction and muscle contraction. Phosphorus (present as PO_4^{3-}) is stored in the bones and teeth and is a part of phospholipids, ATP, and the nucleic acids. Potassium (K^+) is the major positive ion inside cells and is important in nerve conduction and muscle contraction, as is sodium (Na^+). Sodium also plays a major role in regulating the body's water balance, as does chloride (Cl^-). Magnesium (Mg^{2+}) is critical to the functioning of hundreds of enzymes.

The microminerals are parts of larger molecules. For example, iron is present in hemoglobin, and iodine is a part of thyroxin, a hormone produced by the thyroid gland. Zinc, copper, and selenium are present in enzymes that catalyze a variety of reactions. Proteins, called zinc-finger proteins because of their characteristic shapes, bind to DNA when a particular gene is to be activated. As research continues, more and more elements are added to the list of microminerals considered to be essential. During the past three decades, for example, very small amounts of selenium, molybdenum, chromium, nickel, vanadium, silicon, and even arsenic have been found to be essential to good health. Table 12.8 lists the functions of various minerals and gives their functions, food sources, and signs of deficiency and toxicity.

Occasionally, individuals do not receive enough iron (especially women), calcium, magnesium, or zinc in their diet. Adult females need more iron in the diet than males (18 mg compared to 10 mg) because they lose hemoglobin each month during menstruation. Stress can bring on a magnesium deficiency, and due to its high-fiber content, a vegetarian diet may make zinc less available to the body. However, a varied and complete diet usually supplies enough of each type of mineral.

Calcium

Many people take calcium supplements (Fig. 12.17) to counteract osteoporosis, a degenerative bone disease that afflicts an estimated one-fourth of older men and one-half of older women in the United States. Osteoporosis develops because bone-eating cells called osteoclasts are more active than bone-forming cells called osteoblasts. Therefore, the bones are porous, and they break easily because they lack sufficient calcium. Due to recent studies that show consuming more calcium does slow bone loss in elderly people, the guidelines have been revised. A calcium intake of 1,000 mg a day is recommended for men and for women who are premenopausal or who use estrogen replacement therapy, and 1,300 mg a day is recommended for postmenopausal women who do not use estrogen replacement therapy. To achieve this amount, supplemental calcium is most likely necessary.

Estrogen replacement therapy and exercise in addition to calcium supplements are effective means to prevent osteoporosis. Medications are also available that slow bone loss while increasing skeletal mass. Etidronate disodium needs to be taken cyclically or else weak, abnormal bone instead of normal bone develops. Fosamax is a new drug that can be taken continually but may accumulate in the skeleton. Therefore, studies are needed to determine if the drug can safely be given for a long time.

Presently, calcium supplements, estrogen therapy for women, and exercise are thought to be the best ways to prevent osteoporosis.

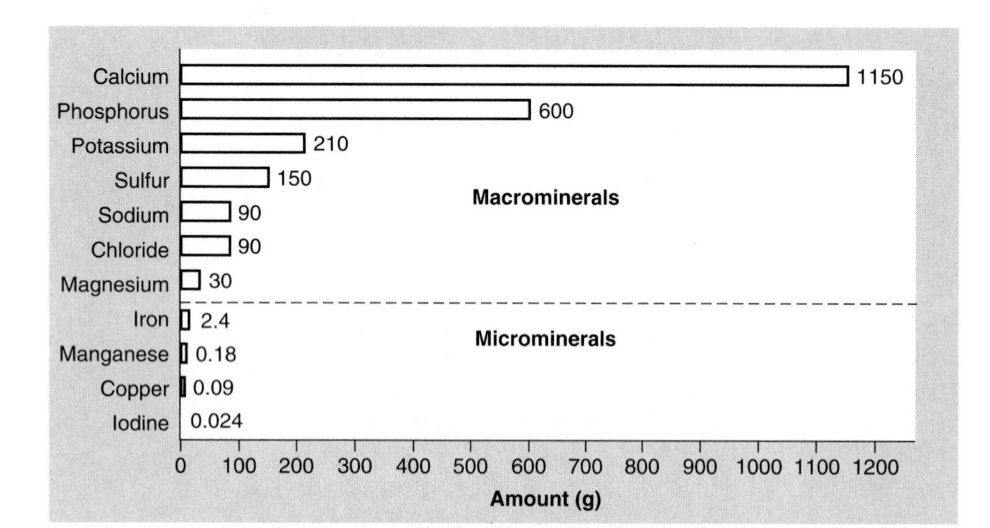

Figure 12.17 Minerals in the body.
This chart shows the usual amount of certain minerals in a 60-kilogram (135 lb) person. The macrominerals are present in amounts larger than 5 grams (about a teaspoon) and the microminerals are present in lesser amounts. The function of these minerals is given in Table 12.8.

Table 12.8 Minerals

Mineral	Functions	Food Sources	Conditions with: Too Little	Conditions with: Too Much
Macrominerals (more than 100 mg/day needed)				
Calcium (Ca^{2+})	Strong bones and teeth, nerve conduction, muscle contraction	Dairy products, leafy green vegetables	Stunted growth in children, low bone density in adults	Kidney stones, interferes with iron and zinc absorption
Phosphorus (PO_4^{3-})	Bone and soft tissue growth, part of phospholipids, ATP, and nucleic acids	Meat, dairy products, sunflower seeds, food additives	Weakness, confusion, pain in bones and joints	Low blood and bone calcium levels
Potassium (K^+)	Nerve conduction, muscle contraction	Many fruits and vegetables, bran	Paralysis, irregular heart beat, eventual death	Vomiting, heart attack, death
Sodium (Na^+)	Nerve conduction, pH and water balance	Table salt	Lethargy, muscle cramps, loss of appetite	Edema, high blood pressure
Chloride (Cl^-)	Water balance	Table salt	Not likely	Vomiting, dehydration
Magnesium (Mg^{2+})	Part of various enzymes for nerve and muscle contraction, protein synthesis	Whole grains, leafy green vegetables	Muscle spasm, irregular heartbeat, convulsions, confusion, personality changes	Diarrhea
Microminerals (less than 20 mg/day needed)				
Zinc (Zn^{2+})	Protein synthesis, wound healing, fetal development and growth, immune function	Meats, legumes, whole grains	Delayed wound healing, night blindness, diarrhea, mental lethargy	Anemia, diarrhea, vomiting, renal failure, abnormal cholesterol levels
Iron (Fe^{2+})	Hemoglobin synthesis	Whole grains, meats, prune juice	Anemia, physical, and mental sluggishness	Iron toxicity disease, organ failure, eventual death
Copper (Cu^{2+})	Hemoglobin synthesis	Meat, nuts, legumes	Anemia, stunted growth in children	Damage to internal organs if not excreted
Iodine (I^-)	Thyroid hormone synthesis	Iodized table salt, seafood	Thyroid deficiency	Depressed thyroid, function, anxiety
Selenium (SeO_4^{2-})	Part of antioxidant enzyme	Seafood, meats, eggs	Possible cancer development	Hair and fingernail loss, discolored skin

Sodium

The recommended amount of sodium intake per day is 500 mg, although the average American takes in 4,000–4,700 mg every day. In recent years, this imbalance has caused concern because high sodium intake has been linked to hypertension (high blood pressure) in some people. About one-third of the sodium we consume occurs naturally in foods; another one-third is added during commercial processing; and we add the last one-third either during home cooking or at the table in the form of table salt.

Clearly, it is possible for us to cut down on the amount of sodium in the diet. Table 12.9 gives recommendations for doing so.

Excess sodium in the diet can lead to hypertension; therefore, excess sodium intake should be avoided.

Table 12.9 Reducing Dietary Sodium

To reduce dietary sodium:

1. Use spices instead of salt to flavor foods.
2. Add little or no salt to foods at the table, and add only small amounts of salt when you cook.
3. Eat unsalted crackers, pretzels, potato chips, nuts, and popcorn.
4. Avoid hot dogs, ham, bacon, luncheon meats, smoked salmon, sardines, and anchovies.
5. Avoid processed cheese and canned or dehydrated soups.
6. Avoid brine-soaked foods, such as pickles or olives.
7. Read labels to avoid high salt products.

Figure 12.18 Recognizing obesity.

Eating Disorders

Authorities recognize three primary eating disorders: obesity, bulimia nervosa, and anorexia nervosa. Although they exist in a continuum as far as body weight is concerned, they all represent an inability to maintain normal body weight because of eating habits.

Obesity

As indicated in Figure 12.18, **obesity** is most often defined as a body weight 20% or more above the ideal weight for a person's height. By this standard, 28% of women and 10% of men in the United States are obese. Moderate obesity is 41–100% above ideal weight, and severe obesity is 100% or more above ideal weight.

Obesity is most likely caused by a combination of factors, including genetic, hormonal, metabolic, and social factors. It's known that obese individuals have more fat cells than normal, and when they lose weight the fat cells simply get smaller; they don't disappear. The social factors that cause obesity include the eating habits of other family members. Consistently eating fatty foods, for example, will make you gain weight. Sedentary activities, like watching television instead of exercising, also determine how much body fat you have. The risk of heart disease is higher in obese individuals, and this alone tells us that excess body fat is not consistent with optimal health.

The treatment depends on the degree of obesity. Surgery to remove body fat may be required for those who are moderately or greatly overweight. For most, a knowledge of good eating habits along with behavior modification may suffice, particularly if a balanced diet is accompanied by a sensible exercise program. A lifelong commitment to a properly planned program is the best way to prevent a cycle of weight gain followed by weight loss. Such a cycle is not conducive to good health. The Health reading on page 236 discusses the proper way to lose weight.

Bulimia Nervosa

Bulimia nervosa can coexist with either obesity or anorexia nervosa, which is discussed next. People with this condition have the habit of eating to excess (called binge eating) and then purging themselves by some artificial means, such as self-induced vomiting or use of a laxative. Bulimic individuals are overconcerned about their body shape and weight, and therefore they may be on a very restrictive diet. A restrictive diet may bring on the desire to binge, and typically the person chooses to consume sweets, like cakes, cookies, and ice cream (Fig. 12.19). The amount of food consumed is far beyond the normal number of calories for one meal, and the person keeps on eating until every bit is gone. Then, a feeling of guilt most likely brings on the next phase, which is a purging of all the calories that have been taken in.

Bulimia can be dangerous to your health. Blood composition is altered, leading to an abnormal heart rhythm, and damage to the kidneys can even result in death. At the very least, vomiting may result in inflammation of the pharynx and esophagus, and stomach acids can cause teeth to erode. The esophagus and stomach may even rupture and tear due to strong contractions during vomiting.

The most important aspect of treatment is to get the patient on a sensible and consistent diet. Again, behavioral modification is helpful and so perhaps is psychotherapy to help the patient understand the emotional causes of the behavior. Medications, including antidepressant medications, have sometimes been helpful to reduce the bulimic cycle and restore normal appetite.

Figure 12.19 **Recognizing bulimia nervosa.**

Anorexia Nervosa

In **anorexia nervosa,** a morbid fear of gaining weight causes the person to be on a very restrictive diet. Athletes such as distance runners, wrestlers, and dancers are at risk of anorexia nervosa because they believe that being thin gives them a competitive edge. In addition to eating only low-calorie foods, the person may induce vomiting and use laxatives to bring about further loss of weight. No matter how thin they have become, people with anorexia nervosa think they are overweight (Fig. 12.20). Such a distorted self-image may prevent recognition of the need for medical help.

Actually, the person is starving and has all the symptoms of starvation, such as low blood pressure, irregular heartbeat, constipation, and constant chilliness. Bone density decreases and stress fractures occur. The body begins to shut down; menstruation ceases in females; the internal organs including the brain don't function well, and the skin dries up. Impairment of the pancreas and digestive tract means that any food consumed does not provide nourishment. Death may be imminent. If so, the only recourse may be hospitalization and force-feeding. Eventually, it is necessary to use behavior therapy and psychotherapy to enlist the cooperation of the person to eat properly. Family therapy may be necessary, because anorexia nervosa in children and teens is believed to be a way for them to gain some control over their lives.

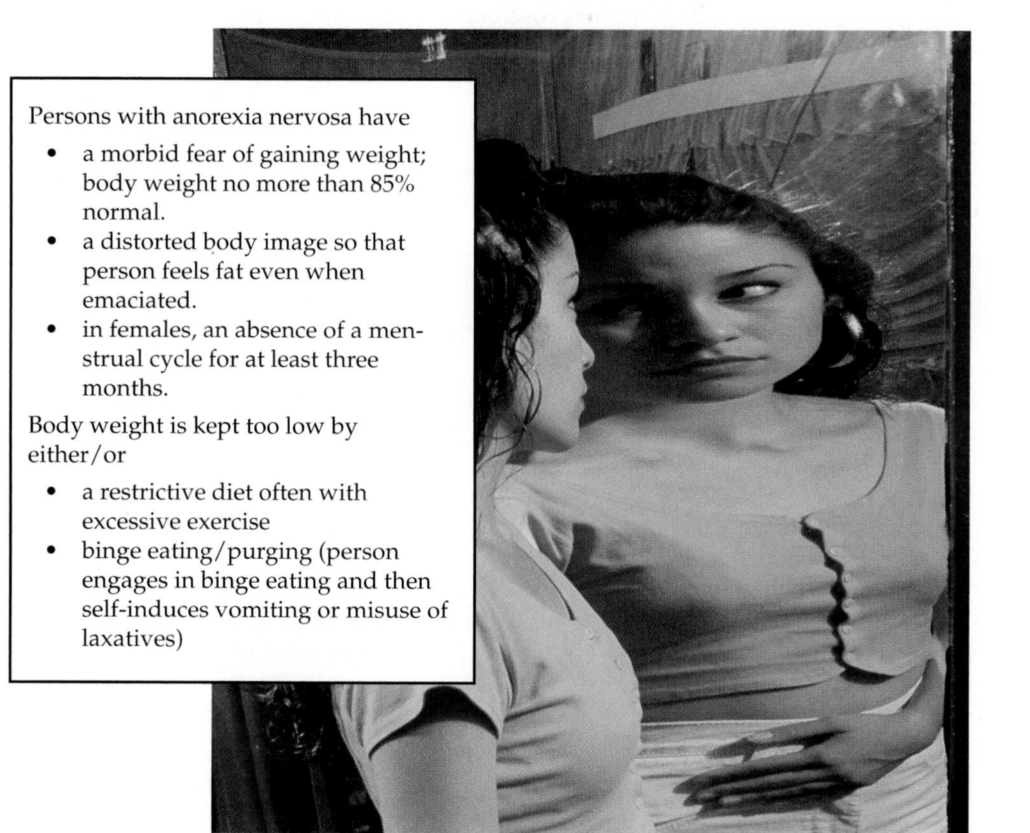

Figure 12.20 **Recognizing anorexia nervosa.**

In anorexia nervosa, the individual has a distorted body image and always feels fat. Medical help is often a necessity.

Health Focus

Weight Loss the Healthy Way

Those who need to lose weight should reduce their caloric intake and/or increase their level of exercise. However, a diet needs to be judged according to the principles of adequacy of nutrients; balance in regard to carbohydrates, proteins, and fats; caloric content; and variety of food sources. Unfortunately, many of the diets and gimmicks people use to lose weight are bad for their health. Unhealthy approaches include the following:

Pills The most familiar pills, and the only ones approved by the FDA, are those that claim to suppress the appetite. They may work at first, but the appetite soon returns to normal and weight lost is regained. Then the user has the problem of trying to get off the drug without gaining more weight. Other types of pills are under investigation and sometimes can be obtained illegally. But, as yet, there is no known drug that is both safe and effective for weight loss.

Low-Carbohydrate Diets The dramatic weight loss that occurs with a low-carbohydrate diet is not due to a loss of fat; it is due to a loss of muscle mass and water. Glycogen and important minerals are also lost. When a normal diet is resumed, so is the normal weight.

Liquid Diets Despite the fact that liquid diets provide proteins and vitamins, the number of calories is so restricted that the body cannot burn fat quickly enough to compensate, and muscle is still broken down to provide energy. A few people on this regime have died, probably because even the heart muscle was not spared by the body.

Single-Category Diets These diets rely on the intake of only one kind of food, either a fruit or vegetable or rice alone. However, no single type of food provides the balance of nutrients needed to maintain health. Some dieters on strange diets suffer the consequences—in one instance, an individual lost hair and fingernails.

Questions to Ask About a Weight-Loss Diet

1. Does the diet have a reasonable number of calories (Cal*), that is, usually no fewer than 1,000–1,200 Cal?
2. Does the diet provide the right amount of protein? For a 154 lb man, 56 grams is recommended. More than twice this amount is too much. For reference, 1c milk and 1 oz meat each has 8 grams protein.
3. Does the diet provide too much fat? No more than 20–30% of total Cal is recommended. For reference , a pat of butter has 45 Cal.
4. Does the diet provide enough carbohydrates? 100 grams = 400 Cal is the very least recommended per day. For reference, a slice of bread contains 14 grams of carbohydrate.
5. Does the diet provide a balanced assortment of foods?
6. Does the diet make use of ordinary foods that are available locally?

*Cal = 1,000 calories

Bioethical Issue

A fat-free fat, called olestra, is now being used to produce foods that are free of calories from fat. A slice of pie ordinarily contains 405 calories, but when the crust is made with olestra, it has only 252 calories. Just three chocolate chip cookies ordinarily adds 63 calories to your daily calorie count, but with olestra it's only 38 calories. And one brownie suddenly goes down from 85 to 49 calories. Olestra molecules made of six or eight fatty acids attached to a sugar molecule are much bigger than a triglyceride, and lipase can't break the molecule down. Olestra goes through the intestines without being absorbed. Proctor and Gamble, who developed olestra, expects to make one billion dollars within ten years on its product.

It is generally recognized that olestra can cause intestinal cramping, flatulence (gas), and diarrhea in a small segment of the population, and can prevent the absorption of some carotenoids, a nutrient that has health benefits in all of us. Even so, the Federal Drug Administration (FDA) approved it for use on the basis that olestra was reasonably harmless. Henry Blackburn, professor of public health at the University of Minnesota, thinks this standard is too low. Instead, the standard for approval ought to be, "Does this product contribute to the nutritional health of the nation?"

David Kessler, FDA commissioner, says, "Ask the American people if government should be off the backs of business, and you will get a resounding Yes!" However, what is the role of government in protecting public health? Do Americans want more protection than they already have from food additives? Or would they rather be free to make their own choices?

Questions

1. Should the government only approve food additives that will contribute to our health? Why or why not?
2. Are people responsible, themselves, for what they eat? Why or why not?
3. Is there too much emphasis in our culture on staying slim, despite what it may do to our health? Why or why not?

Summarizing the Concepts

12.1 The Digestive System

The salivary glands send saliva into the mouth, where the teeth chew the food and the tongue forms a bolus for swallowing. The air passage and food passage cross in the pharynx. When one swallows, the air passage is usually blocked off and food must enter the esophagus where peristalsis begins. The stomach expands and stores food. While food is in the stomach, it churns, mixing food with the acidic gastric juices. The walls of the small intestine have fingerlike projections called villi where nutrient molecules are absorbed into the cardiovascular and lymphatic systems. The large intestine consists of the cecum, colons (ascending, transverse, descending, and sigmoid), and the rectum, which ends at the anus. The large intestine does not produce digestive enzymes; it does absorb water, salts, and some vitamins.

12.2 Three Accessory Organs

The three accessory organs of digestion—the pancreas, liver, and gallbladder—send secretions to the duodenum via ducts. The pancreas produces pancreatic juice, which contains digestive enzymes for carbohydrate, protein, and fat.

The liver produces bile, which is stored in the gallbladder. The liver receives blood from the small intestine by way of the hepatic portal vein. It has numerous important functions, and any malfunction of the liver is a matter of considerable concern.

12.3 Digestive Enzymes

Digestive enzymes are present in digestive juices and break down food into the nutrient molecules glucose, amino acids, fatty acids, and glycerol (see Table 12.2). Glucose and amino acids are absorbed into the blood capillaries of the villi. Fatty acids and glycerol rejoin to produce fat, which enters the lacteals. Digestive enzymes have the usual enzymatic properties. They are specific to their substrate and speed up specific reactions at body temperature and optimum pH.

12.4 Nutrition

The nutrients released by the digestive process should provide us with an adequate amount of energy, essential amino acids and fatty acids, and all necessary vitamins and minerals.

The bulk of the diet should be carbohydrates (like bread, pasta, and rice) and fruits and vegetables. These are low in saturated fatty acids and cholesterol molecules, whose intake is linked to cardiovascular disease. The vitamins A, E, and C are antioxidants that protect cell contents from damage due to free radicals.

Studying the Concepts

1. List the organs of the digestive tract, and state the contribution of each to the digestive process. 214–20
2. Discuss the absorption of the products of digestion into the lymphatic and cardiovascular systems. 219
3. Name and state the functions of the hormones that assist the nervous system in regulating digestive secretions. 220
4. Name the accessory organs, and describe the part they play in the digestion of food. 222–23
5. Choose and discuss any three functions of the liver. 222–23
6. Name and discuss three serious illnesses of the liver. 223
7. Discuss the digestion of starch, protein, and fat, listing all the steps that occur to bring about digestion of each of these. 224–25
8. What is the chief contribution of each of these constituents of the diet: a. carbohydrates; b. proteins; c. fats; d. fruits and vegetables? 228–32
9. Why should the amount of saturated fat be curtailed in the diet? 230–32
10. Name and discuss three eating disorders. 234–35

Testing Yourself

Choose the best answer for each question.

1. If you were tracing the path of food, which is out of order first?
   a. mouth
   b. pharynx
   c. esophagus
   d. small intestine
   e. stomach
   f. large intestine
2. Which association is incorrect?
   a. mouth—starch digestion
   b. esophagus—protein digestion
   c. small intestine—starch, lipid, protein digestion
   d. stomach—food storage
   e. liver—production of bile
3. Which of these could be absorbed directly without need of digestion?
   a. glucose
   b. fat
   c. polysaccharides
   d. protein
   e. nucleic acid
4. Which association is incorrect?
   a. protein—trypsin
   b. fat—bile
   c. fat—lipase
   d. maltose—pepsin
   e. starch—amylase
5. Most of the products of digestion are absorbed across the
   a. squamous epithelium of the esophagus.
   b. striated walls of the trachea.
   c. convoluted walls of the stomach.
   d. fingerlike villi of the small intestine.
   e. smooth wall of the large intestine.
6. Bile
   a. is an important enzyme for the digestion of fats.
   b. cannot be stored.
   c. is made by the gallbladder.
   d. emulsifies fat.
   e. All of these are correct.
7. Which of these is not a function of the liver in adults?
   a. produce bile
   b. detoxify alcohol
   c. store glucose
   d. produce urea
   e. make red blood cells
8. The large intestine
   a. digests all types of food.
   b. is the longest part of the intestinal tract.
   c. absorbs water.
   d. is connected to the stomach.
   e. is subject to hepatitis.
9. Bulimia nervosa and anorexia nervosa have many elements in common, except
   a. restrictive diet often with excessive exercise.
   b. binge eating followed by purging.
   c. obsession about body shape and weight.
   d. distorted body image so person feels fat even when emaciated.
   e. health can be damaged by this complex.

10. Label each organ indicated in the diagram.
    For the arrows, use either glucose, amino acids, lipids, or water.

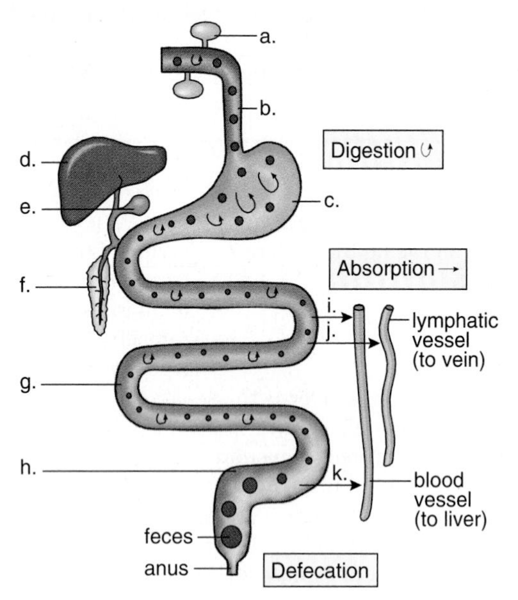

Thinking Scientifically

1. Considering homeostasis:
   a. What type of tissue found in the mucous membrane of the stomach and the small intestine produces digestive enzymes? (page 217)
   b. All systems of the body contribute to homeostasis. How does the digestive system contribute?
   c. The liver "fine tunes" the contribution to homeostasis by the digestive system. How?
2. Considering the experiment described in Figure 12.12 (page 225):
   a. Does the experiment show that the enzyme pepsin prefers an acidic rather than a basic pH? Describe an experiment that would show this.
   b. Does the experiment described in Figure 12.12 show that pepsin will digest only protein and not starch, for example? Describe an experiment that would show this.
   c. Does the experiment described in Figure 12.12 show that better digestion occurs at a warm temperature rather than a cold temperature? Describe an experiment that would show this.

Understanding the Terms

anorexia nervosa 235
anus 220
appendix 220
bile 219
bulimia nervosa 234
cecum 220
chyme 218
cirrhosis 223
colon 220
constipation 221
defecation 220
dental caries 215
diarrhea 221
duodenum 219
epiglottis 216
esophagus 217
essential amino acids 228
fiber 227
gallbladder 223
gastric gland 218
glottis 216
hard palate 214
heartburn 217
hepatitis 223
hormone 220
hydrolytic enzyme 224
jaundice 223
lacteal 219
lactose intolerance 224
large intestine 220
lipase 222
liver 222
lumen 217
maltase 224
mineral 232
nasopharynx 216
obesity 234
pancreas 222
pancreatic amylase 222
pepsin 218
peptidase 224
periodontitis 215
peristalsis 217
peritonitis 220
pharynx 216
plaque 229
polyp 221
reflex action 216
salivary amylase 224
salivary gland 214
small intestine 219
soft palate 214
sphincter 217
stomach 218
tonsillitis 214
trypsin 222
ulcer 218
villus 219
vitamin 230

Match the terms to these definitions:

a. ________ First portion of the small intestine into which secretions from the liver and pancreas enter.
b. ________ Muscle that surrounds a tube and closes or opens the tube by contracting and relaxing.
c. ________ Discharge of feces from the rectum through the anus.
d. ________ Fat-digesting enzyme secreted by the pancreas.
e. ________ Saclike organ associated with the liver that stores and concentrates bile.

Using Technology

Your study of the digestive system and nutrition is supported by these available technologies:

Essential Study Partner CD-ROM

Animals → Digestion

Visit the Mader web site for related ESP activities.

http://www.mhhe.com/biosci/genbio/mader/
(Click on *Inquiry into Life.*)

Exploring the Internet

The Mader Home Page provides resources and tools as you study this chapter.

http://www.mhhe.com/biosci/genbio/mader

Dynamic Human 2.0 CD-ROM

Human Body → Digestive System

Virtual Physiology Laboratory CD-ROM

Digestion of Fat
Enzyme Characteristics

C H A P T E R 13

Cardiovascular System

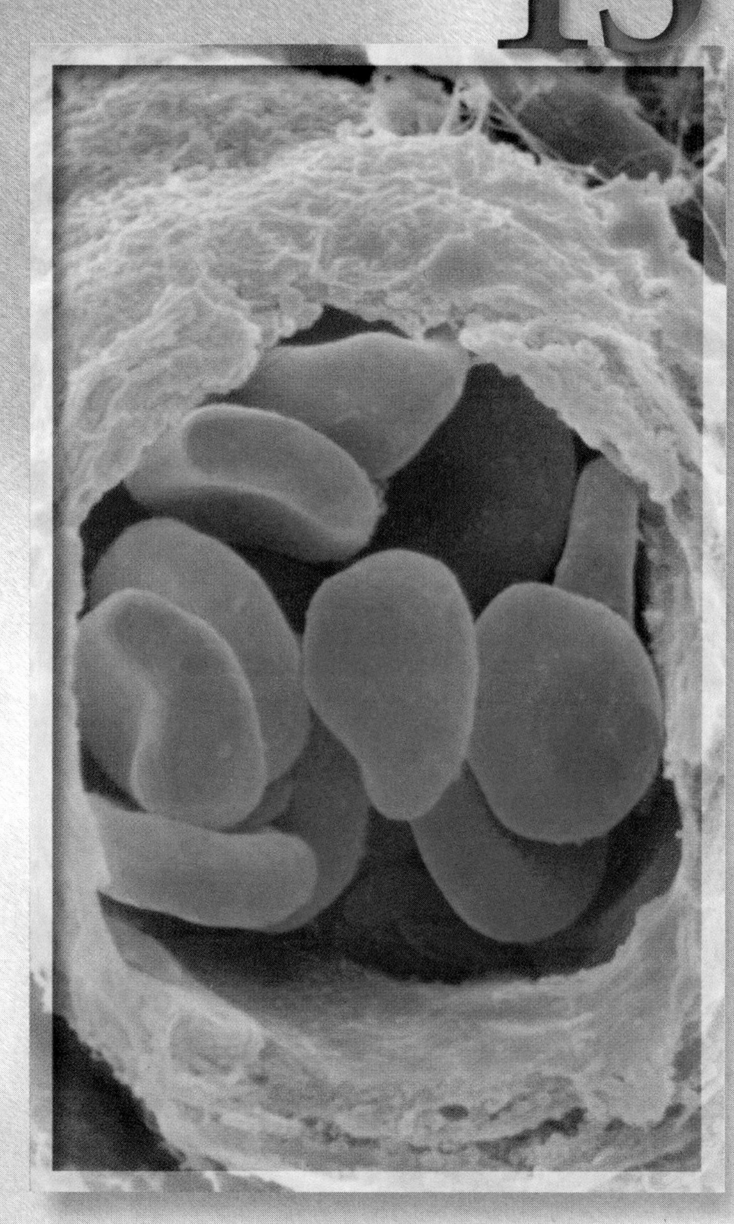

Red blood cells packed with the red respiratory pigment hemoglobin fill a blood vessel. Hemoglobin combines with oxygen and in this way red blood cells transport oxygen to the tissues.

Chapter Concepts

Jane stretched out her arm and watched as the nurse slipped a needle into a vein at the crook of her arm. She was thinking about her friend who needed the blood that was now coursing through a plastic tube into a bag. Jane had the same blood type as her friend who had been in an automobile accident. Blood, a vital fluid, carries oxygen from the lungs and nutrients from the intestines to the cells. Kept in motion by the pumping of the heart, it helps fight infection, helps regulate body temperature, coordinates body tissues, and carries wastes to the kidneys. A severe loss of blood must be replaced by transfusion if life is to continue.

This chapter discusses the cardiovascular system, which includes blood but also the heart and blood vessels. Humans have a closed system in that the blood never runs free and is conducted to and from the tissues by blood vessels. Only the capillaries have walls thin enough to allow exchange of molecules with the tissues. The heart is the organ that keeps the blood moving through the vessels to the capillaries. If the heart fails to pump the blood for even a few minutes, the individual's life is in danger. The body has various mechanisms for ensuring that blood remains in the vessels and under a pressure that will maintain the transport function of blood.

13.1 The Blood Vessels

The cardiovascular system has three types of blood vessels: the **arteries** (and arterioles), which carry blood away from the heart to the capillaries; the **capillaries,** which permit exchange of material with the tissues; and the **veins** (and venules), which return blood from the capillaries to the heart.

The Arteries

The *arterial wall has three layers* (Fig. 13.1*a*). The inner layer is a simple squamous epithelium called endothelium with a connective tissue basement membrane that contains elastic fibers. The middle layer is the thickest layer and consists of smooth muscle that can contract to regulate blood flow and blood pressure. The outer layer is fibrous connective tissue near the middle layer, but it becomes loose connective tissue at its periphery. Some arteries are so large that they require their own blood vessels.

Arterioles are small arteries just visible to the naked eye. The middle layer of arterioles has some elastic tissue

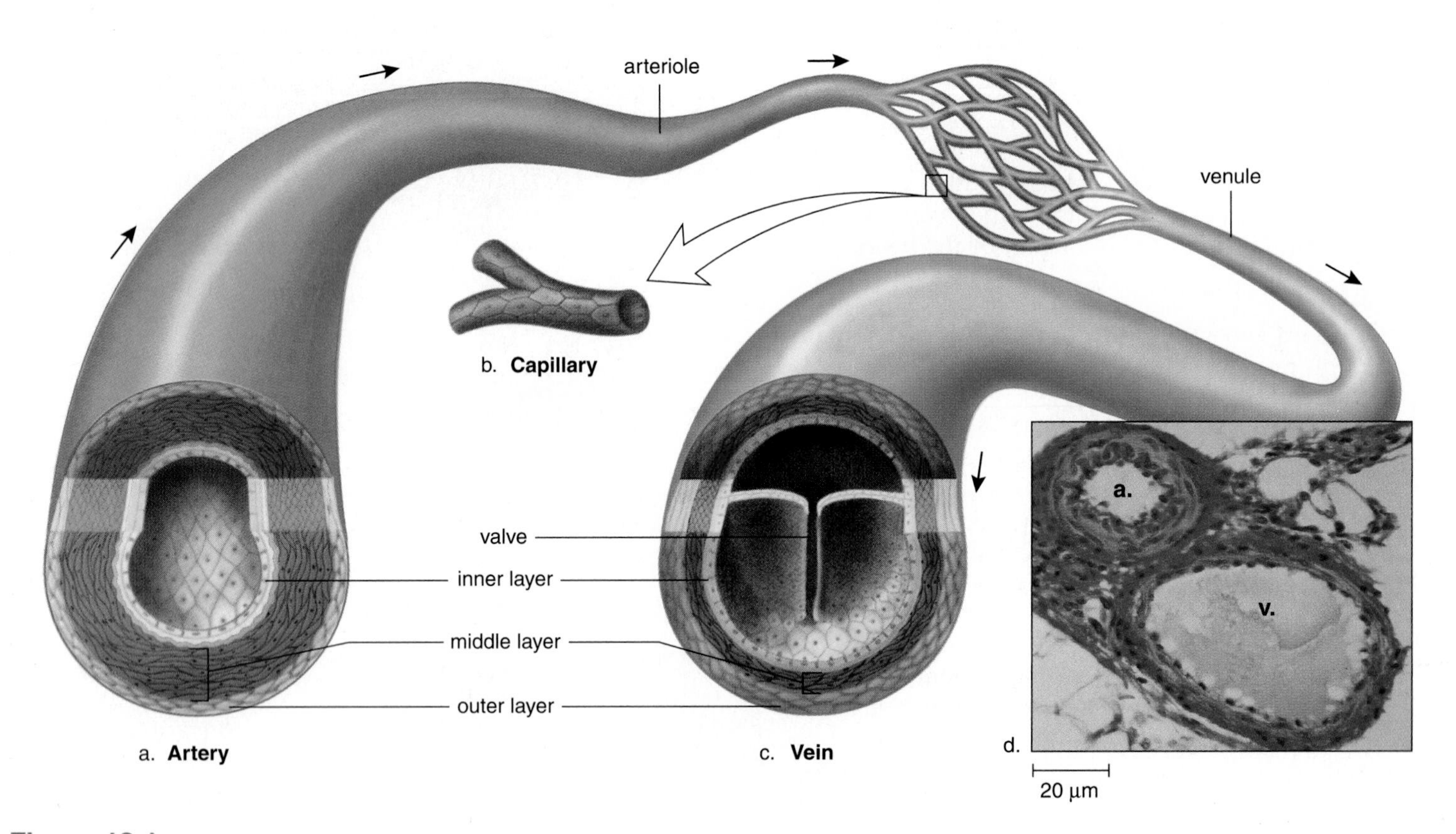

Figure 13.1 Blood vessels.
The walls of arteries and veins have three layers. The inner layer is composed largely of endothelium with a basement membrane that has elastic fibers; the middle layer is smooth muscle tissue; the inner layer is connective tissue (largely collagen fibers). **a.** Arteries have a thicker wall than veins because they have a larger middle layer than veins. **b.** Capillary walls are one-cell-thick endothelium. **c.** Veins are larger in diameter than arteries, so that collectively veins have a larger holding capacity than arteries. **d.** Scanning electron micrograph of an artery and vein.

but is composed mostly of smooth muscle whose fibers encircle the arteriole. When these muscle fibers are contracted, the vessel has a smaller diameter (is constricted); and when these muscle fibers relax, the vessel has a larger diameter (is dilated). Whether arterioles are constricted or dilated affects blood pressure. The greater the number of vessels dilated, the lower the blood pressure.

The Capillaries

Arterioles branch into capillaries (Fig. 13.1*b*). Each capillary is an extremely narrow, microscopic tube with one-cell-thick walls composed only of endothelium with a basement membrane. *Capillary beds* (networks of many capillaries) are present in all regions of the body; consequently, a cut to any body tissue draws blood. Capillaries are a very important part of the human cardiovascular system because an exchange of substances takes place across their thin walls. Oxygen and nutrients, such as glucose, diffuse out of a capillary into the tissue fluid that surrounds cells. Wastes, such as carbon dioxide, diffuse into the capillary.

Since the capillaries serve the cells, the heart and the other vessels of the cardiovascular system can be thought of as the means by which blood is conducted to and from the capillaries. Only certain capillaries are open at any given time. For example, after eating, the capillaries that serve the digestive system are open and those that serve the muscles are closed. When a capillary bed is closed, the precapillary sphincters contract, and the blood moves from arteriole to venule by way of an arteriovenous shunt (Fig. 13.2).

The Veins

Venules are small veins that drain blood from the capillaries and then join to form a vein. The walls of venules (and veins) have the same three layers as arteries, but there is less smooth muscle and connective tissue (Fig. 13.1*c*). Veins often have **valves,** which allow blood to flow only toward the heart when open and prevent the backward flow of blood when closed.

Since walls of veins are thinner, they can expand to a greater extent (Fig. 13.1*d*). At any one time about 70% of the blood is in the veins. In this way, the veins act as a blood reservoir.

Arteries and arterioles carry blood away from the heart toward the capillaries; capillaries join arterioles to venules; veins and venules return blood from the capillaries to the heart.

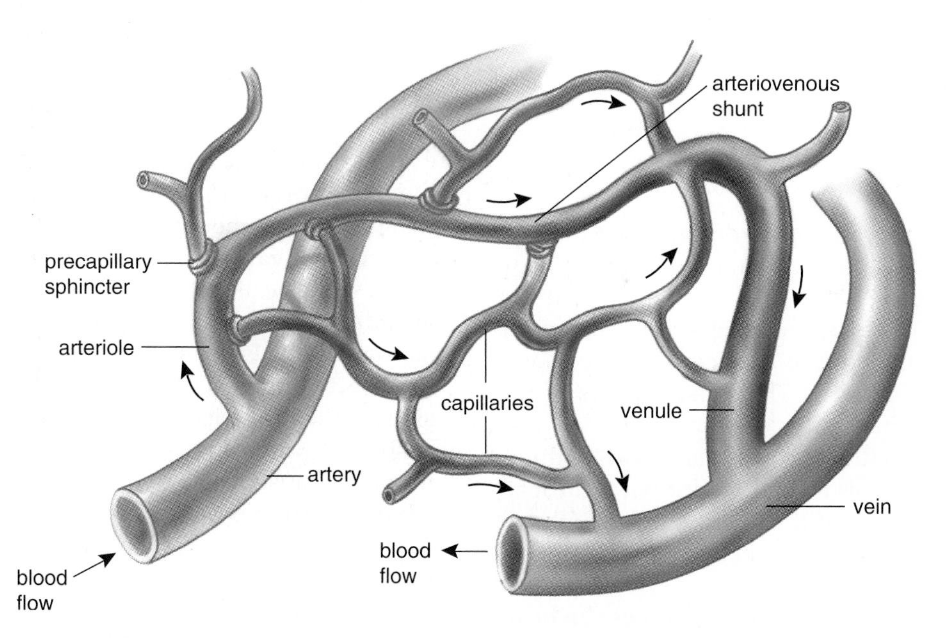

Figure 13.2 Anatomy of a capillary bed.
A capillary bed forms a maze of capillary vessels that lies between an arteriole and a venule. When sphincter muscles are relaxed, the capillary bed is open, and blood flows through the capillaries. When sphincter muscles are contracted, blood flows through a shunt that carries blood directly from an arteriole to a venule. As blood passes through a capillary in the tissues, it gives up its oxygen (O_2). Therefore, blood goes from carrying more oxygen in the arteriole (red color) to carrying less oxygen (blue color) in the vein.

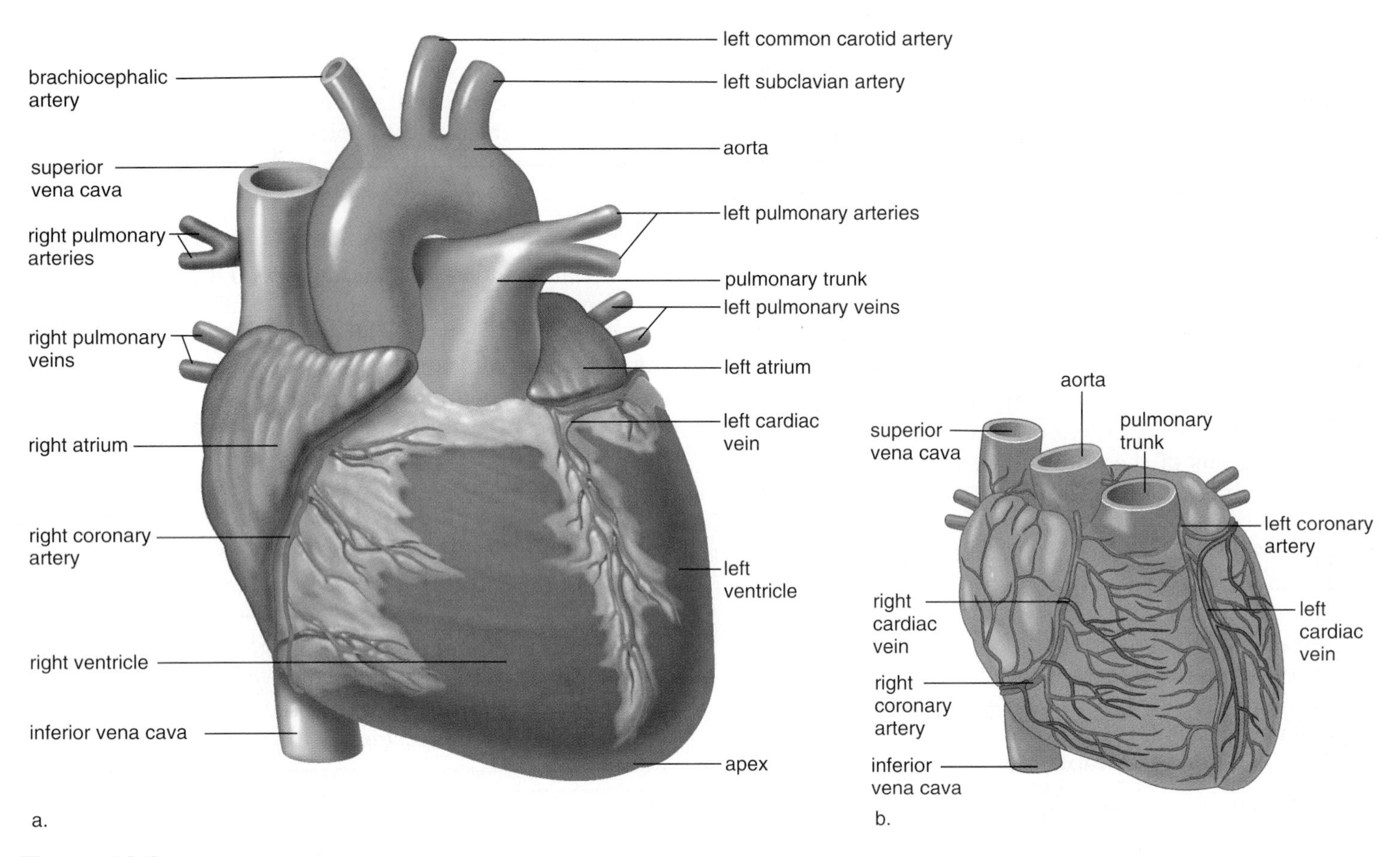

Figure 13.3 External heart anatomy.
a. The superior vena cava and the pulmonary arteries are attached to the right side of the heart. The aorta and left pulmonary veins are attached to the left side of the heart. The right ventricle forms most of the anterior surface of the heart, and the left ventricle forms most of the posterior.
b. The coronary arteries and cardiac veins pervade cardiac muscle. They bring oxygen and nutrients to cardiac cells, and return blood to the right atrium.

13.2 The Heart

The **heart** is a cone-shaped, muscular organ about the size of a fist (Fig. 13.3). It is located between the lungs directly behind the sternum (breastbone) and is tilted so that the apex (the pointed end) is oriented to the left. The major portion of the heart, called the **myocardium,** consists largely of cardiac muscle tissue. The muscle fibers of the myocardium are branched and tightly joined to one another. The heart lies within the **pericardium,** a thick, membranous sac that secretes a small quantity of lubricating liquid. The inner surface of the heart is lined with endocardium, which consists of connective tissue and endothelial tissue.

Internally, a wall called the **septum** separates the heart into a right side and a left side (Fig. 13.4) The heart has four chambers. The two upper, thin-walled atria (sing., **atrium**) have wrinkled protruding appendages called auricles. The two lower chambers are the thick-walled **ventricles,** which pump the blood.

The heart also has four valves, which direct the flow of blood and prevent its backward movement. The two valves that lie between the atria and the ventricles are called the **atrioventricular valves.** These valves are supported by strong fibrous strings called **chordae tendineae.** The chordae, which are attached to muscular projections of the ventricular walls, support the valves and prevent them from inverting when the heart contracts. The atrioventricular valve on the right side is called the tricuspid valve because it has three flaps, or cusps. The valve on the left side is called the bicuspid (or the mitral) because it has two flaps. The remaining two valves are the **semilunar valves,** whose flaps resemble half-moons, between the ventricles and their attached vessels. The pulmonary semilunar valve lies between the right ventricle and the pulmonary trunk. The aortic semilunar valve lies between the left ventricle and the aorta.

Humans have a four-chambered heart (two atria and two ventricles). A septum separates the right side from the left side.

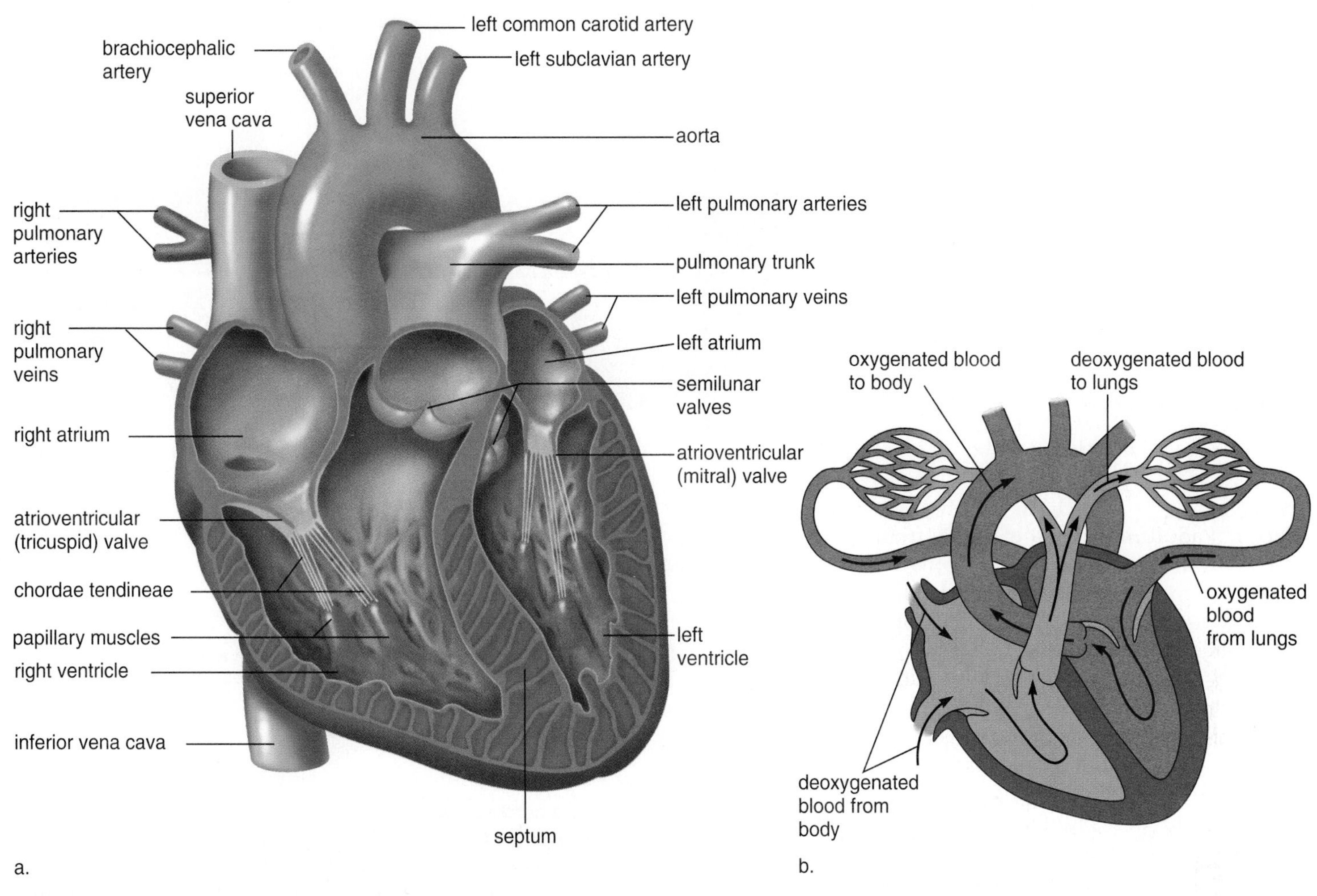

Figure 13.4 Internal view of the heart.
a. The heart has four valves. The atrioventricular valves allow blood to pass from the atria to the ventricles, and the semilunar valves allow blood to pass out of the heart. **b.** This diagrammatic representation of the heart allows you to trace the path of the blood.

Passage of Blood Through the Heart

We can trace the path of blood through the heart (Fig. 13.4) in the following manner:

The superior vena cava and the inferior **vena cava,** which carry blood that is relatively low in oxygen and relatively high in carbon dioxide, enter the right atrium.

The right atrium sends blood through an atrioventricular valve (the tricuspid valve) to the right ventricle.

The right ventricle sends blood through the pulmonary semilunar valve into the pulmonary trunk and the two **pulmonary arteries** to the lungs.

Four **pulmonary veins,** which carry blood that is relatively high in oxygen and relatively low in carbon dioxide, enter the left atrium.

The left atrium sends blood through an atrioventricular valve (the bicuspid or mitral valve) to the left ventricle.

The left ventricle sends blood through the aortic semilunar valve into the **aorta** to the body proper.

From this description, you can see that deoxygenated blood never mixes with oxygenated blood and that blood must go through the lungs in order to pass from the right side to the left side of the heart. In fact, the heart is a double pump because the right ventricle of the heart sends blood through the lungs, and the left ventricle sends blood throughout the body. Since the left ventricle has the harder job of pumping blood to the entire body, its walls are thicker than those of the right ventricle, which pumps blood a relatively short distance to the lungs.

The right side of the heart pumps blood to the lungs, and the left side of the heart pumps blood throughout the body.

The Heartbeat

Each heartbeat is called a **cardiac cycle** (Fig. 13.5). When the heart beats, first, the two atria contract at the same time; then the two ventricles contract at the same time. Then all chambers relax. The word **systole** refers to contraction of heart muscle, and the word **diastole** refers to relaxation of heart muscle. The heart contracts, or beats, about 70 times a minute, and each heartbeat lasts about 0.85 seconds.

Time	*Atria*	*Ventricles*
0.15 sec	Systole	Diastole
0.30 sec	Diastole	Systole
0.40 sec	Diastole	Diastole

A normal adult rate at rest can vary from 60 to 80 beats per minute.

When the heart beats, the familiar lub-dup sound occurs. The longer and lower-pitched lub is caused by vibrations occurring when the atrioventricular valves close due to ventricular contraction. The shorter and sharper dup is heard when the semilunar valves close due to back pressure of blood in the arteries. A heart murmur, or a slight slush sound after the lub, is often due to ineffective valves, which allow blood to pass back into the atria after the atrioventricular valves have closed. Rheumatic fever resulting from a bacterial infection is one cause of a faulty valve, particularly the bicuspid valve. Faulty valves can be surgically corrected.

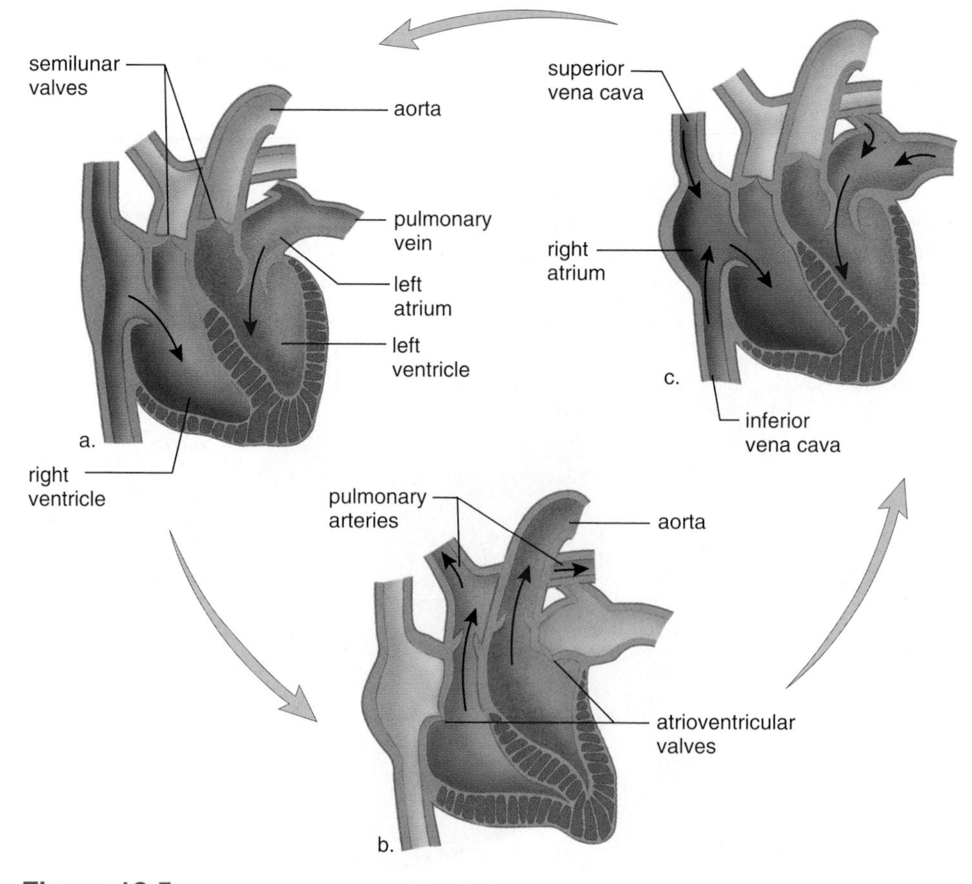

Figure 13.5 **Stages in the cardiac cycle.**
a. When the atria contract, the ventricles are relaxed and filling with blood. **b.** When the ventricles contract, the atrioventricular valves are closed, the semilunar valves are open, and the blood is pumped into the pulmonary trunk and aorta. **c.** When the heart is relaxed, both atria and ventricles are filling with blood.

The surge of blood entering the arteries causes their elastic walls to stretch, but then they almost immediately recoil. This alternating expansion and recoil of an arterial wall can be felt as a **pulse** in any artery that runs close to the body's surface. It is customary to feel the pulse by placing several fingers on a radial artery, which lies near the outer border of the palm side of the wrist. A carotid artery, on either side of the trachea in the neck, is another accessible location to feel the pulse. Normally, the pulse rate indicates the rate of the heartbeat because the arterial walls pulse whenever the left ventricle contracts.

Intrinsic Control of Heartbeat

The rhythmical contraction of the atria and ventricles is due to the instrinsic conduction system of the heart. Nodal tissue, which has both muscular and nervous characteristics, is a unique type of cardiac muscle located in two regions of the heart. The **SA (sinoatrial) node** is located in the upper wall of the right atrium; the other, the **AV (atrioventricular) node,** is located in the base of the right atrium very near the septum (Fig. 13.6*a*). The SA node initiates the heartbeat and automatically sends out an excitation impulse every 0.85 seconds; this causes the atria to contract. When impulses reach the AV node there is a slight delay that allows the atria to finish their contraction before the ventricles begin theirs. The signal for the ventricles to contract then travels through the two branches of the **atrioventricular bundle** before reaching the numerous and smaller **Purkinje fibers.** This ventricular portion of the conduction system consists of specialized cardiac muscle fibers that efficiently cause the ventricles to contract.

The SA node is called the **pacemaker** because it usually keeps the heartbeat regular. If the SA node fails to work properly, the heart still beats due to impulses generated by the AV node. But the beat is slower (40 to 60 beats per minute). To correct this condition, it is possible to implant an artificial pacemaker, which automatically gives an electric stimulus to the heart every 0.85 seconds.

The intrinsic conduction system of the heart consists of the SA node, the AV node, the atrioventricular bundle, and the Purkinje fibers.

Extrinsic Control of Heartbeat

The body has an extrinsic way to regulate the heartbeat. A cardiac control center in the medulla oblongata, a portion of the brain that controls internal organs, can alter the beat of the heart by way of the autonomic system, a division of the nervous system. This system has two divisions: the parasympathetic system, which promotes those functions we tend to associate with a restful state, and the sympathetic system, which brings about those responses we associate with increased activity and/or stress. The parasympathetic system decreases SA and AV nodal activity when we are inactive, and the sympathetic system increases SA and AV nodal activity when we are active or excited.

The hormones epinephrine and norepinephrine, which are released by the adrenal medulla, also stimulate the heart. During exercise, for example, the heart pumps faster and stronger due to sympathetic stimulation and due to the release of epinephrine and norepinephrine.

The body has an extrinsic way to regulate the heartbeat. The autonomic system and hormones can modify the heartbeat rate.

The Electrocardiogram

An **electrocardiogram (ECG)** is a recording of the electrical changes that occur in myocardium during a cardiac cycle. Body fluids contain ions that conduct electrical currents, and therefore the electrical changes in myocardium can be detected on the skin's surface. When an electrocardiogram is being taken, electrodes placed on the skin are connected by wires to an instrument that detects the myocardium's electrical changes. Thereafter a pen rises or falls on a moving strip of paper. Figure 13.6*b* depicts the pen's movements during a normal cardiac cycle.

When the SA node triggers an impulse, the atrial fibers produce an electrical change that is called the P wave. The P wave indicates that the atria are about to contract. After that, the QRS complex signals that the ventricles are about to contract. The electrical changes that occur as the ventricular muscle fibers recover produces the T wave.

Various types of abnormalities can be detected by an electrocardiogram. One of these, called ventricular fibrillation, is caused by uncoordinated contraction of the ventricles (Fig. 13.6*c*) Ventricular fibrillation is of special interest because it can be caused by an injury or drug overdose. It is the most common cause of sudden cardiac death in a seemingly healthy person. Once the ventricles are fibrillating, they have to be defibrillated by applying a strong electric current for a short period of time. Then the SA node may be able to reestablish a coordinated beat.

The electrocardiogram (ECG) is a recording of electrical changes occurring in the myocardium during a cardiac cycle.

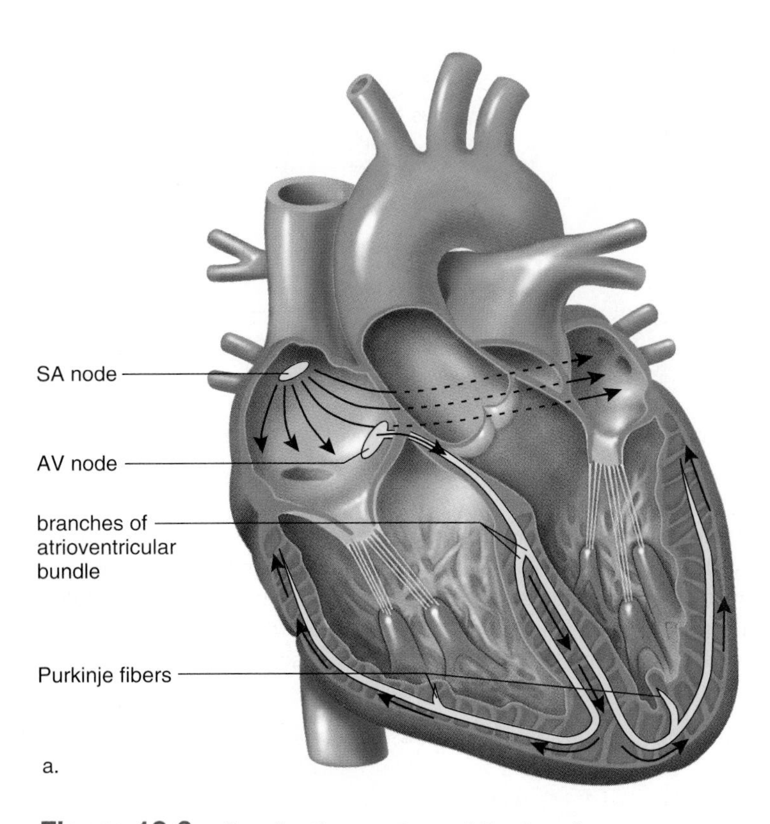

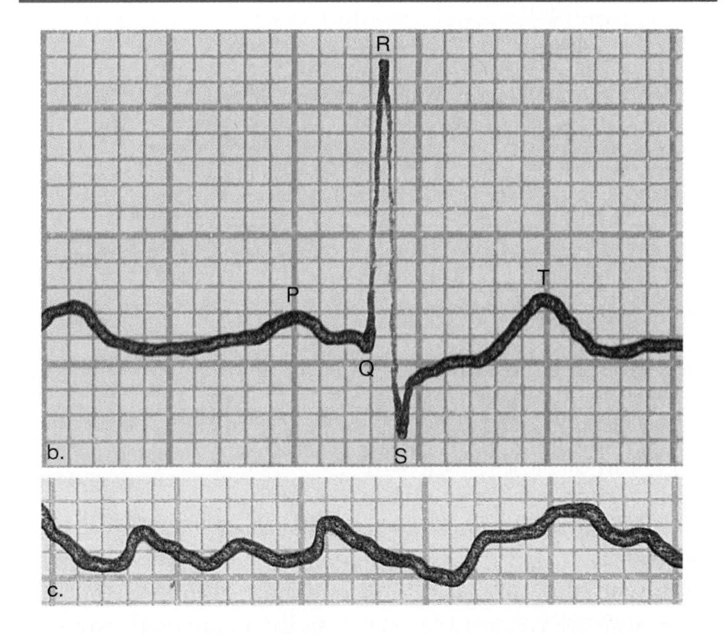

Figure 13.6 Conduction system of the heart.
a. The SA node sends out a stimulus, which causes the atria to contract. When this stimulus reaches the AV node, it signals the ventricles to contract. Impulses pass down the two branches of the atrioventricular bundle to the Purkinje fibers and thereafter the ventricles contract.
b. A normal ECG indicates that the heart is functioning properly. The P wave occurs just prior to atrial contraction; the QRS complex occurs just prior to ventricular contraction; and the T wave occurs when the ventricles are recovering from contraction. **c.** Ventricular fibrillation produces an irregular electrocardiogram due to irregular stimulation of the ventricles.

13.3 The Vascular Pathways

The cardiovascular system, which is represented in Figure 13.7, includes two circuits: the **pulmonary circuit,** which circulates blood through the lungs, and the **systemic circuit,** which serves the needs of body tissues.

The Pulmonary Circuit

The path of blood through the lungs can be traced as follows. Blood from all regions of the body first collects in the right atrium and then passes into the right ventricle, which pumps it into the pulmonary trunk. The pulmonary trunk divides into the right and left **pulmonary arteries,** which branch as they approach the lungs. The arterioles take blood to the pulmonary capillaries, where carbon dioxide is given off and oxygen is picked up. Blood then passes through the pulmonary venules, which lead to the four **pulmonary veins** that enter the left atrium. Since blood in the pulmonary arteries is relatively low in oxygen but blood in the pulmonary veins is relatively high in oxygen, it is not correct to say that all arteries carry oxygenated blood and all veins carry deoxygenated blood. It is just the reverse in the pulmonary circuit.

The pulmonary arteries take blood that is low in oxygen to the lungs, and the pulmonary veins return blood that is high in oxygen to the heart.

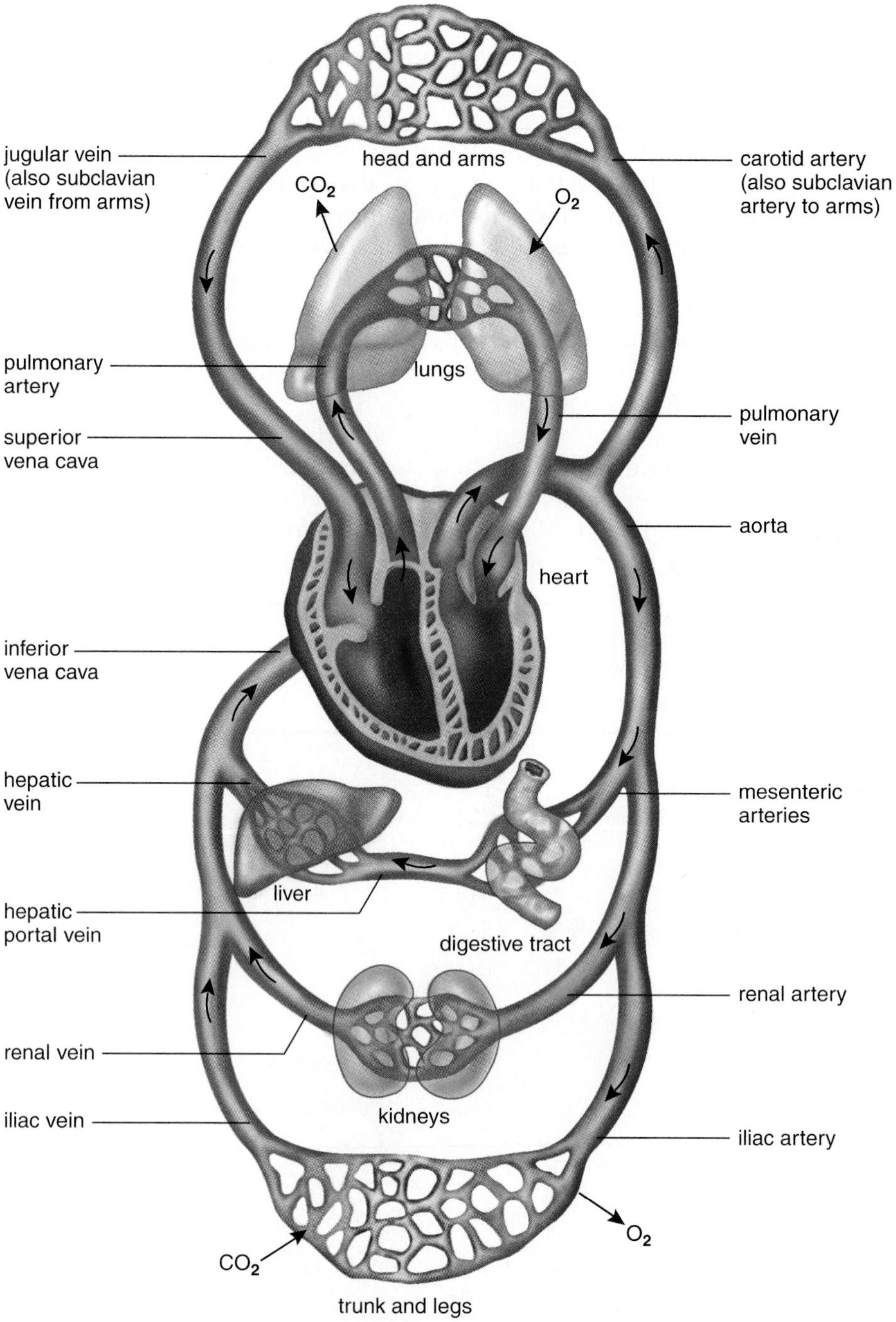

Figure 13.7 Cardiovascular system diagram.
The blue-colored vessels carry blood that is relatively low in oxygen, and the red-colored vessels carry blood that is relatively high in oxygen. The arrows indicate the flow of blood. Compare this diagram, useful for learning to trace the path of blood, to Figure 13.8 to realize that both arteries and veins go to all parts of the body. Also, there are capillaries in all parts of the body. No cell is located far from a capillary.

The Systemic Circuit

The systemic circuit includes all of the arteries and veins shown in Figure 13.8. The largest artery in the systemic circuit is the **aorta,** and the largest veins are the **superior** and **inferior venae cavae.** The superior vena cava collects blood from the head, the chest, and the arms, and the inferior vena cava collects blood from the lower body regions. Both enter the right atrium. The aorta and the venae cavae serve as the major pathways for blood in the systemic circuit.

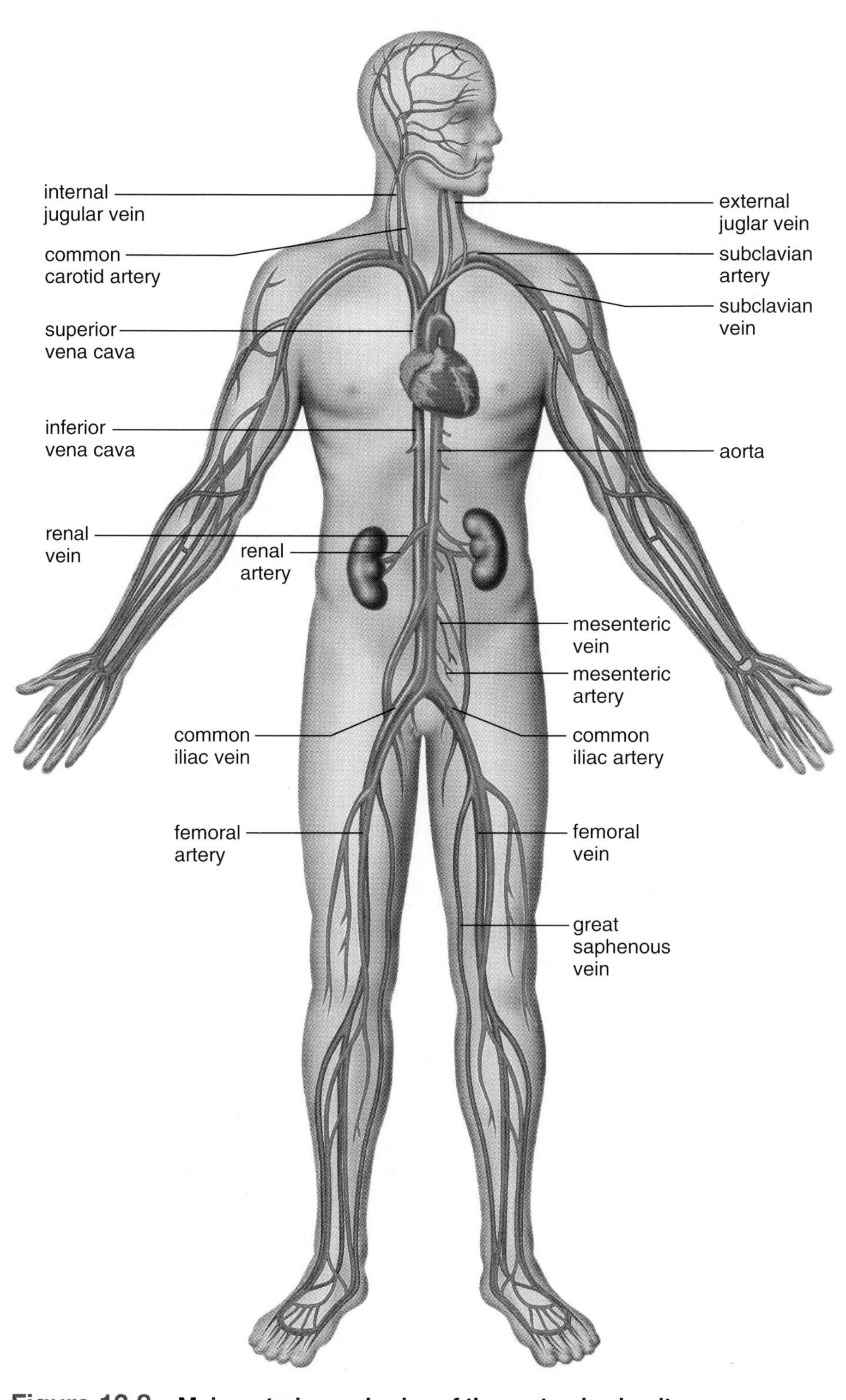

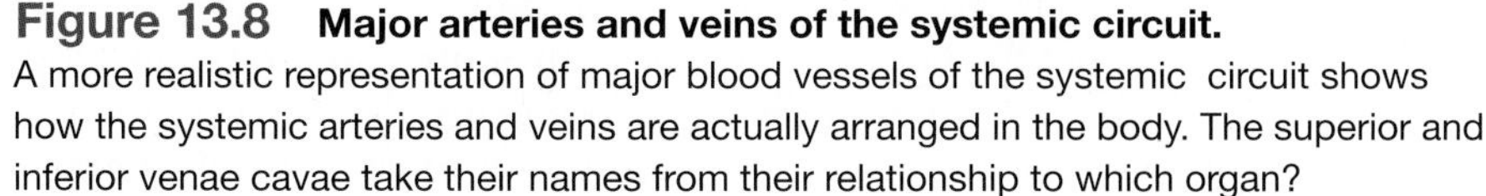

Figure 13.8 Major arteries and veins of the systemic circuit. A more realistic representation of major blood vessels of the systemic circuit shows how the systemic arteries and veins are actually arranged in the body. The superior and inferior venae cavae take their names from their relationship to which organ?

The path of systemic blood to any organ in the body begins in the left ventricle, which pumps blood into the aorta. Branches from the aorta go to the organs and major body regions. For example, this is the path of blood to and from the legs:

left ventricle—aorta—
common iliac artery—
legs—common iliac vein—
inferior vena cava—
right atrium

Notice when tracing blood, you need only mention the aorta, the proper branch of the aorta, the region, and the vein returning blood to the vena cava. In most instances, the artery and the vein that serve the same region are given the same name (Fig. 13.8). In the systemic circuit, arteries contain oxygenated blood and have a bright red color, but veins contain deoxygenated blood and appear a dark purplish color.

The **coronary arteries** (see Fig. 13.3) serve the heart muscle itself. (The heart is not nourished by the blood in its chambers.) The coronary arteries are the first branches off the aorta. They originate just above the aortic semilunar valve, and they lie on the exterior surface of the heart, where they divide into diverse arterioles. Because they have a very small diameter, the coronary arteries may become clogged, as discussed on page 258. The coronary capillary beds join to form venules. The venules converge to form the cardiac veins, which empty into the right atrium.

The body has a portal system called the **hepatic portal system,** which is associated with the liver. A portal system begins and ends in capillaries; in this instance, the first set of capillaries occurs at the villi of the small intestine and the second occurs in the liver. Blood passes from the capillaries of the intestinal villi into venules that join to form the **hepatic portal vein,** a vessel that connects the villi of the intestine with the liver, an organ that monitors the makeup of the blood. The **hepatic vein** leaves the liver and enters the inferior vena cava. While Figure 13.7 is helpful in tracing the path of blood, remember that all parts of the body receive both arteries and veins, as illustrated in Figure 13.8.

The systemic circuit takes blood from the left ventricle of the heart to the right atrium of the heart. It serves the body proper.

Blood Flow

Blood flow differs with regard to pressure and velocity in the different vessels of the cardiovascular system (Fig. 13.9).

Blood Flow in Arteries

Blood pressure created by the pumping of the heart accounts for the velocity of blood in the arteries. **Blood pressure,** which is simply the pressure of blood against the wall of a blood vessel, fluctuates in the arteries. **Systolic pressure,** which occurs when the ventricles contract, is higher than **diastolic pressure** when the ventricles relax. Normal resting blood pressure for a young adult is said to be 120 mm Hg (mercury) over 80 mm Hg, or simply 120/80. The higher number is the systolic pressure, and the lower number is the diastolic pressure. Actually, 120/80 is the expected blood pressure in the brachial artery of the arm.

Both systolic and diastolic blood pressure decrease with distance from the left ventricle because the total cross-sectional area of the blood vessels increase—there are more arterioles than arteries. The decrease in blood pressure causes the blood velocity to gradually decrease as it flows toward the capillaries.

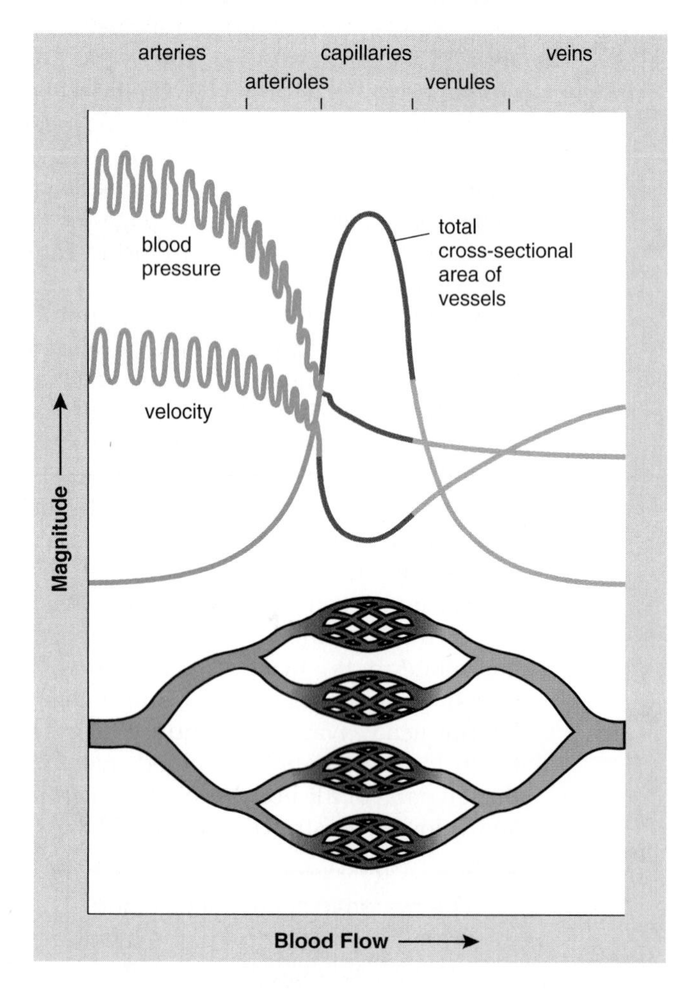

Figure 13.9 Cross-sectional area as it relates to blood pressure and blood velocity.
Blood pressure and blood velocity drop off in capillaries because capillaries have a greater cross-sectional area than arterioles.

Blood Flow in Capillaries

There are many more capillaries than arterioles, and blood moves slowly through the capillaries (Fig. 13.9). This is important because the slow progress allows time for the exchange of substances between blood in the capillaries and the surrounding tissues.

Blood Flow in Veins

Blood pressure is minimal in venules and veins (20–0 mm Hg). Instead of blood pressure, venous return is dependent upon three factors: skeletal muscle contraction, presence of valves in veins, and respiratory movements. When the skeletal muscles contract, they compress the weak walls of the veins. This causes blood to move past the next valve. Once past the valve, blood cannot flow backward (Fig. 13.1*c*). The importance of muscle contraction in moving blood in the venous vessels can be demonstrated by forcing a person to stand rigidly still for an hour or so. Frequently, fainting occurs because blood collects in the limbs, depriving the brain of needed blood flow and oxygen. In this case, fainting is beneficial because the resulting horizontal position aids in getting blood to the head.

When inspiration occurs, the thoracic pressure falls and abdominal pressure rises as the chest expands. This also aids the flow of venous blood back to the heart because blood flows in the direction of reduced pressure. Blood velocity increases slightly in the venous vessels due to a progressive reduction in the cross-sectional area as small venules join to form veins

Blood pressure accounts for the flow of blood in the arteries and the arterioles. Skeletal muscle contraction, valves in veins, and respiratory movements account for the flow of blood in the venules and the veins.

Dilated and Inflamed Veins

Varicose veins develop when the valves of veins become weak and ineffective due to the backward pressure of blood. Abnormal and irregular dilations are particularly apparent in the superficial (near the surface) veins of the lower legs. Crossing the legs or sitting in a chair so that its edge presses against the back of the knees can contribute to the development of varicose veins. Varicose veins also occur in the rectum, where they are called piles, or more properly, **hemorrhoids.**

Phlebitis, or inflammation of a vein, is a more serious condition, particularly when a deep vein is involved. Blood in an unbroken but inflamed vessel may clot, and the clot may be carried in the bloodstream until it lodges in a small vessel. If a blood clot blocks a pulmonary vessel, death can result.

13.4 Blood

If blood is transferred from a person's vein to a test tube and is prevented from clotting, it separates into two layers (Fig. 13.10). The lower layer consists of red blood cells (erythrocytes), white blood cells (leukocytes), and blood platelets (thrombocytes). Collectively, these are called the **formed elements.** Formed elements make up about 45% of the total volume of whole blood. The upper layer, called **plasma,** is the liquid portion of blood. Plasma, which accounts for about 55% of the total volume of whole blood, contains a variety of inorganic and organic substances dissolved or suspended in water.

FORMED ELEMENTS	Function and Description	Source
Red Blood Cells (erythrocytes) 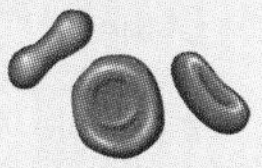 4 million–6 million per mm^3 blood	Transport O_2 and help transport CO_2 7–8 μm in diameter Bright-red to dark-purple biconcave disks without nuclei	Red bone marrow
White Blood Cells (leukocytes) 4,000–11,000 per mm^3 blood	Fight infection	Red bone marrow
Granular leukocytes		
• Basophil 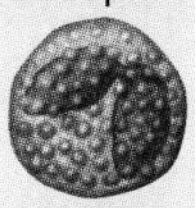 20–50 per mm^3 blood	10–12 μm in diameter Spherical cells with lobed nuclei; large, irregularly shaped, deep-blue granules in cytoplasm	
• Eosinophil 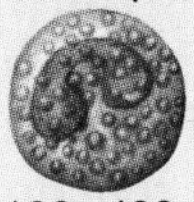 100–400 per mm^3 blood	10–14 μm in diameter Spherical cells with bilobed nuclei; coarse, deep-red, uniformly sized granules in cytoplasm	
• Neutrophil  3,000–7,000 per mm^3 blood	10–14 μm in diameter Spherical cells with multilobed nuclei; fine, pink granules in cytoplasm	
Agranular leukocytes		
• Lymphocyte  1,500–3,000 per mm^3 blood	5–17 μm in diameter (average 9–10 μm) Spherical cells with large round nuclei	
• Monocyte 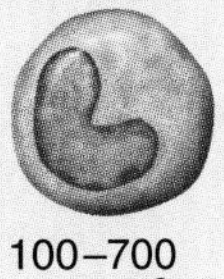 100–700 per mm^3 blood	10–24 μm in diameter Large spherical cells with kidney-shaped, round, or lobed nuclei	
• **Platelets** (thrombocytes)  150,000–300,000 per mm^3 blood	Aid clotting 2–4 μm in diameter Disk-shaped cell fragments with no nuclei; purple granules in cytoplasm	Red bone marrow

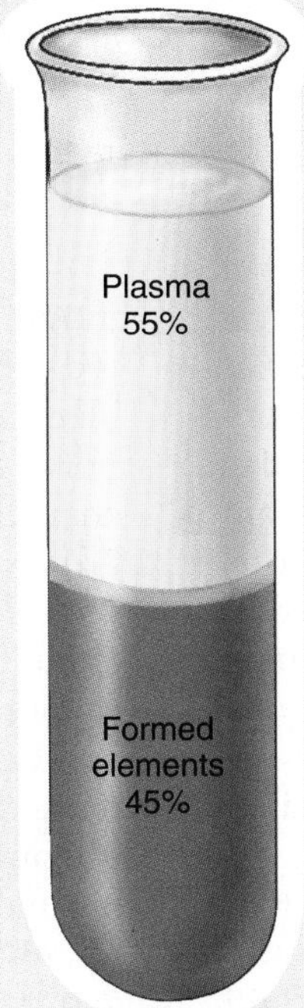

Plasma 55%

Formed elements 45%

PLASMA	Function	Source
Water (90–92% of plasma)	Maintains blood volume; transports molecules	Absorbed from intestine
Plasma proteins (7–8% of plasma)	Maintain blood osmotic pressure and pH	Liver
Albumin	Maintain blood volume and pressure	
Globulins	Transport; fight infection	
Fibrinogen	Clotting	
Salts (less than 1% of plasma	Maintain blood osmotic pressure and pH; aid metabolism	Absorbed from intestine
Gases		
Oxygen	Cellular respiration	Lungs
Carbon dioxide	End product of metabolism	Tissues
Nutrients	Food for cells	Absorbed from intestine
Fats Glucose Amino acids		
Nitrogenous waste	Excretion by kidneys	Liver
Urea Uric acid		
Other		
Hormones, vitamins, etc.	Aid metabolism	Varied

• with Wright's stain

Figure 13.10 Composition of blood.
When blood is transferred to a test tube and is prevented from clotting, it forms two layers. The transparent, yellow, top layer is plasma, the liquid portion of blood. The formed elements are in the bottom layer. The tables describe these components in detail.

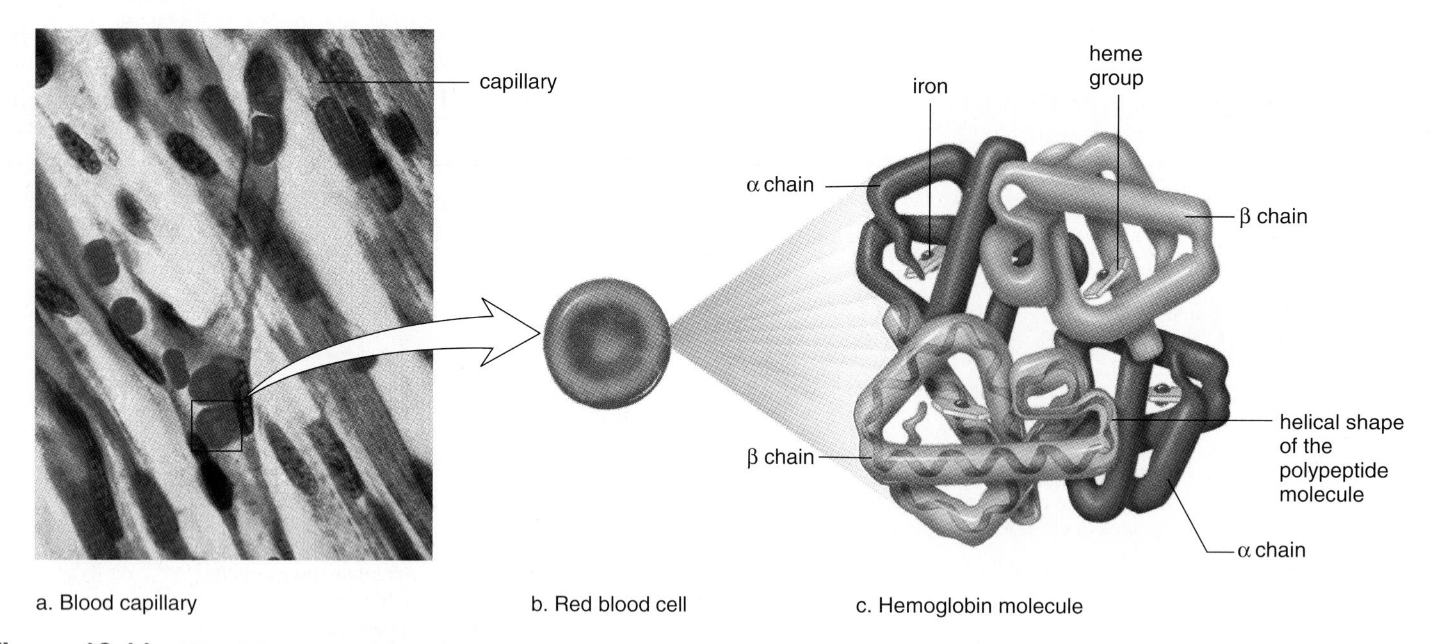

Figure 13.11 Physiology of red blood cells.
a. Red blood cells move single file through the capillaries. **b.** Each red blood cell is a biconcave disk containing many molecules of hemoglobin, the respiratory pigment. **c.** Hemoglobin contains four polypeptide chains (color coded blue), two of which are alpha (α) chains and two of which are beta (β) chains. The plane in the center of each chain represents an iron-containing heme group. Oxygen combines loosely with iron when hemoglobin is oxygenated. Oxygenated hemoglobin is bright red, and deoxygenated hemoglobin is a darker red color.

Plasma proteins, which make up 7–8% of plasma, assist in transporting large organic molecules in blood. For example, **albumin** transports bilirubin, a breakdown product of hemoglobin. The lipoproteins that transport cholesterol contain a type of protein called globulins. Plasma proteins also maintain blood volume because their size prevents them from readily passing through a capillary wall. Therefore, capillaries are always areas of lower water concentration compared to tissue fluid, and water automatically diffuses into capillaries. Certain plasma proteins have specific functions. As discussed later in the chapter, fibrinogen is necessary to blood clotting, and immunoglobulins are antibodies, which help fight infection.

We shall see that blood has numerous functions that help maintain homeostasis. Blood transports substances to and from the capillaries, where exchanges with tissue fluid take place. It helps guard the body against invasion by **pathogens** (microscopic infectious agents, such as bacteria and viruses), and it clots, preventing a potentially life-threatening loss of blood. It also regulates pH by the presence of buffers in blood.

The Red Blood Cells

Red blood cells (erythrocytes) are continuously manufactured in the red bone marrow of the skull, the ribs, the vertebrae, and the ends of the long bones. Normally, there are 4 to 6 million red blood cells per mm^3 of whole blood.

Red blood cells carry oxygen because they contain **hemoglobin,** the respiratory pigment. Since hemoglobin is a red pigment, the cells appear red, and their color also makes blood red. A hemoglobin molecule (Fig. 13.11) contains four polypeptide chains, making up the protein globin. Each chain is associated with heme, a complex iron-containing group. The iron portion of hemoglobin acquires oxygen in the lungs and gives it up in the tissues. Plasma carries only about 0.3 ml of oxygen per 100 ml of blood, but whole blood carries 20 ml of oxygen per 100 ml of blood. This shows that hemoglobin increases the oxygen-carrying capacity of blood more than 60 times. Unfortunately, as discussed in the reading on page 253, carbon monoxide combines with hemoglobin more readily than does oxygen, and it stays combined for several hours, making hemoglobin unavailable for oxygen transport.

The number of red blood cells increases whenever arterial blood carries a reduced amount of oxygen, as happens when an individual first takes up residence at a high altitude. Under these circumstances, the kidneys increase their production of a hormone called **erythropoietin,** which speeds the maturation of red blood cells. A **stem cell** is a cell that is ever capable of dividing and producing new cells that differentiate. The stem cell called an erythroblast produces red blood cells in red bone marrow (Fig. 13.12). Before they are released from the bone marrow into blood, red blood cells lose their nuclei and acquire hemoglobin.

Visual Focus

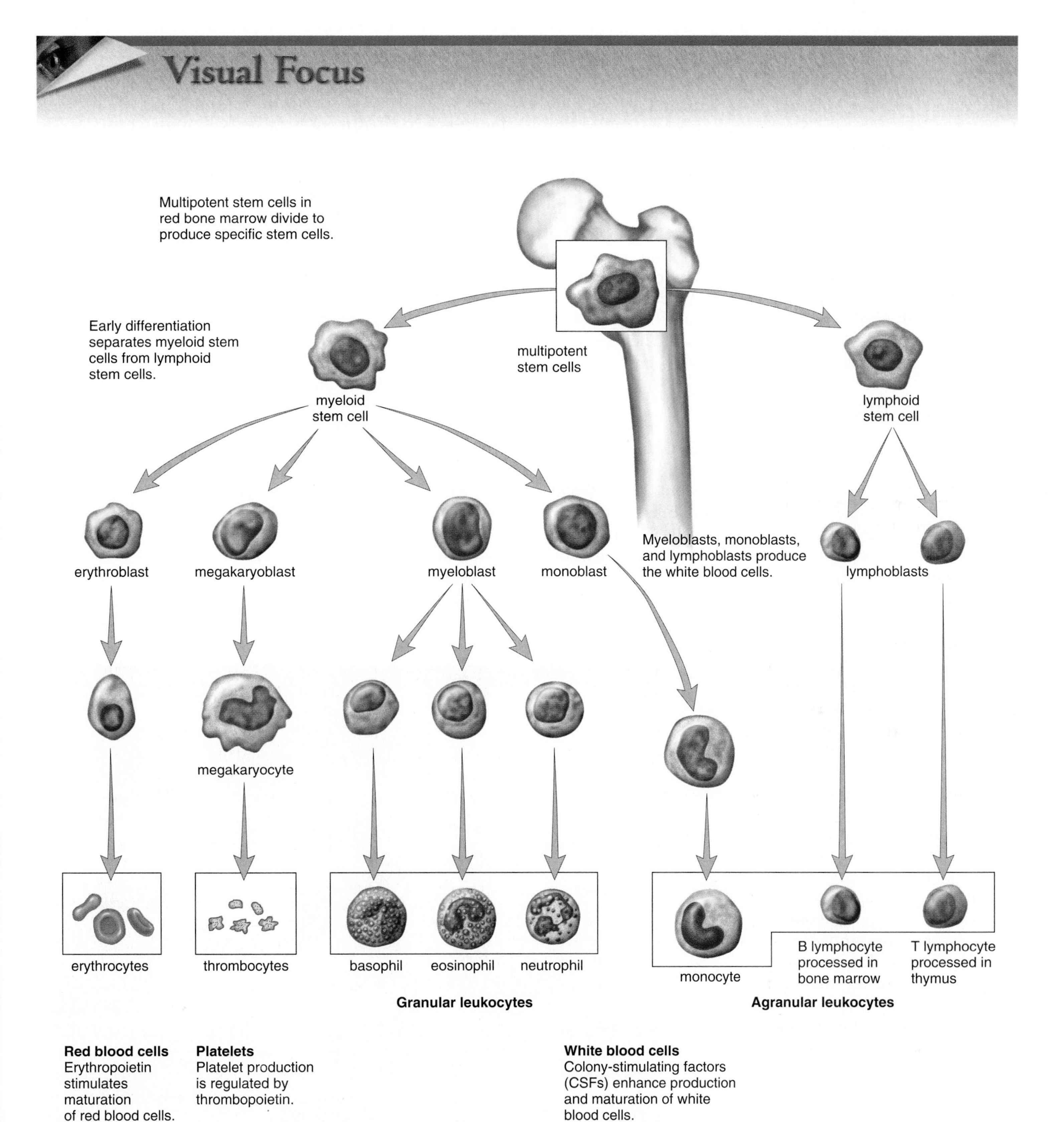

Figure 13.12 Blood cell formation in red bone marrow.
Multipotent stem cells give rise to specialized stem cells. The myeloid stem cell gives rise to still other cells, which become red blood cells, platelets, and all the white blood cells except lymphocytes. The lymphoid stem cell gives rise to lymphoblasts, which become lymphocytes.

Possibly because they lack nuclei, red blood cells live only about 120 days. They are destroyed chiefly in the *liver* and the *spleen,* where they are engulfed by large phagocytic cells. When red blood cells are broken down, the hemoglobin is released. The iron is recovered and returned to the bone marrow for reuse. The heme portion of hemoglobin undergoes chemical degradation and is excreted as bile pigments by the liver into the bile.

When there is an insufficient number of red blood cells or the red blood cells do not have enough hemoglobin, the individual suffers from **anemia** and has a tired, run-down feeling. In iron-deficiency anemia, the most common type, hemoglobin level is low, probably due to a diet that does not contain enough iron. Certain foods that are rich in iron, such as raisins and liver, can be added to the diet to help prevent iron-deficiency anemia.

The White Blood Cells

White blood cells (leukocytes) differ from red blood cells in that they are usually larger, have a nucleus, lack hemoglobin, and without staining appear translucent. White blood cells are not as numerous as red blood cells, with only 5,000–11,000 cells per mm^3. White blood cells fight infection and they play a role in the development of immunity, the ability to resist disease.

On the basis of structure, it is possible to divide white blood cells into **granular leukocytes** and **agranular leukocytes.** Granular leukocytes are filled with spheres that contain various enzymes and proteins, which help white blood cells do their job of defending the body against microbes. **Neutrophils** are granular leukocytes with a multilobed nucleus joined by nuclear threads. They are the most abundant of the white blood cells and are able to phagocytize and digest bacteria. The agranular leukocytes (monocytes and lymphocytes) typically have a spherical or kidney-shaped nucleus. After **monocytes,** the largest of the white blood cells, take up residence in the tissues, they differentiate into the even larger **macrophages** (Fig. 13.13). Macrophages phagocytize microbes and stimulate other white blood cells to defend the body. The **lymphocytes** are of two types, B lymphocytes and T lymphocytes, and each type has a specific role to play in immunity.

If the total number of white blood cells increases beyond normal, leukemia or an infection may be present. **Leukemia** is a form of cancer characterized by uncontrolled production of abnormal white blood cells. Sometimes an infection results in the increase or decrease of only one type of white blood cell, a condition that is detected with a differential white blood cell count. A blood sample is examined microscopically, and the number of each type of white blood cell is counted up to a total of 100 cells. A person with **infectious mononucleosis,** caused by the Epstein-Barr virus, has an excessive number of lymphocytes of the B type. A person with AIDS, caused by an HIV infection, has an abnormally low number of lymphocytes of the T type.

White blood cells are produced in the red bone marrow (see Fig. 13.12), from two types of stem cells. Macrophages and T lymphocytes are the most important source of **colony-stimulating factors (CSFs),** hormones that regulate the production of white blood cells. Many white blood cells live only a few days and are believed to die combating invading microbes. Others live months or even years.

White blood cells fight infection. They attack microbes that have invaded the body.

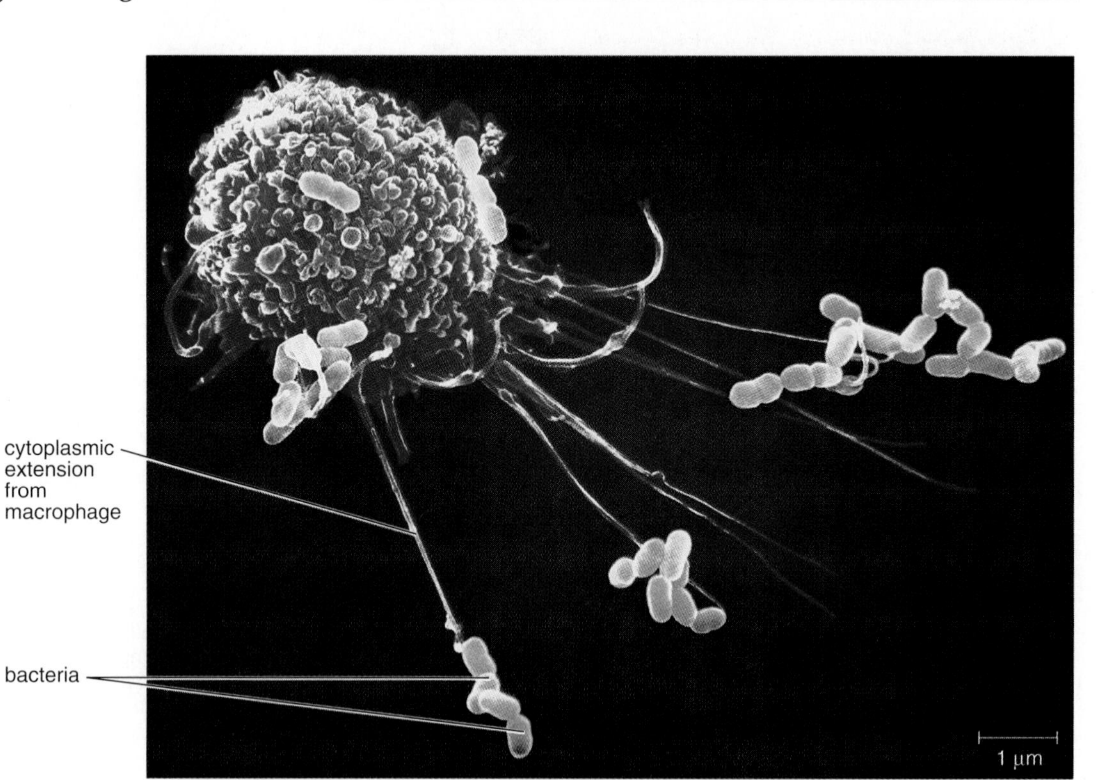

Figure 13.13 Macrophage (red) engulfing bacteria (green). Monocyte-derived macrophages are the body's scavengers. They engulf microbes and debris in the body's fluids and tissues, as illustrated in this colorized scanning electron micrograph.

Ecology Focus

Carbon Monoxide, A Deadly Poison

Carbon monoxide (CO) is an air pollutant that comes primarily from the incomplete combustion of natural gas and gasoline. Figure 13A shows that transportation contributes most of the carbon monoxide to our cities' air. But power plants, factories, waste incineration, and home heating also contribute to the carbon monoxide level. Cigarette smoke contains carbon monoxide and is delivered directly to the smoker's blood and also to nonsmokers nearby.

Because carbon monoxide is a colorless, odorless gas, people can be unaware that it is affecting their systems. But it binds to iron 200 times more tightly than oxygen. Hemoglobin contains iron and so does cytochrome oxidase, the carrier in the electron transport system that passes electrons on to oxygen. When these molecules bind to carbon monoxide preferentially, they cannot perform their usual functions. The end result is that delivery of oxygen to mitochondria is impaired and so is the functioning of mitochondria.

Flushed red skin, especially on facial cheeks, is a first sign of carbon monoxide poisoning, because hemoglobin bound to carbon monoxide is brighter red than oxygenated hemoglobin. Marked euphoria, then sleepiness, coma, and death follow. Removing a person from the carbon monoxide source is not sufficient treatment because carbon monoxide, unlike oxygen, remains tightly bound to iron for many hours. A transfusion of red blood cells will help to increase the carrying capacity of the blood, and pure oxygen given under pressure will displace some carbon monoxide. Despite good medical care, some people still die each year from CO poisoning.

The level of carbon monoxide in polluted air may not be sufficient to kill people, but it does interfere with the ability of the body to function properly. The elderly and those with cardiovascular disease are especially at risk. One study found that even levels once thought to be safe can cause angina patients to experience chest pains, because oxygen delivery to the heart is reduced.

We can all lessen air pollution and reduce the amount of carbon monoxide in the air by doing the following:

- Don't smoke (especially indoors).
- Walk or bicycle instead of driving.
- Use public transportation instead of driving.
- Heat your home with solar energy—not furnaces.
- Support the development of more efficient automobiles, factories, power plants, and home furnaces.
- Support the development of alternative fuels. When the gas hydrogen is burned, for example, the result is water, not carbon monoxide and carbon dioxide.

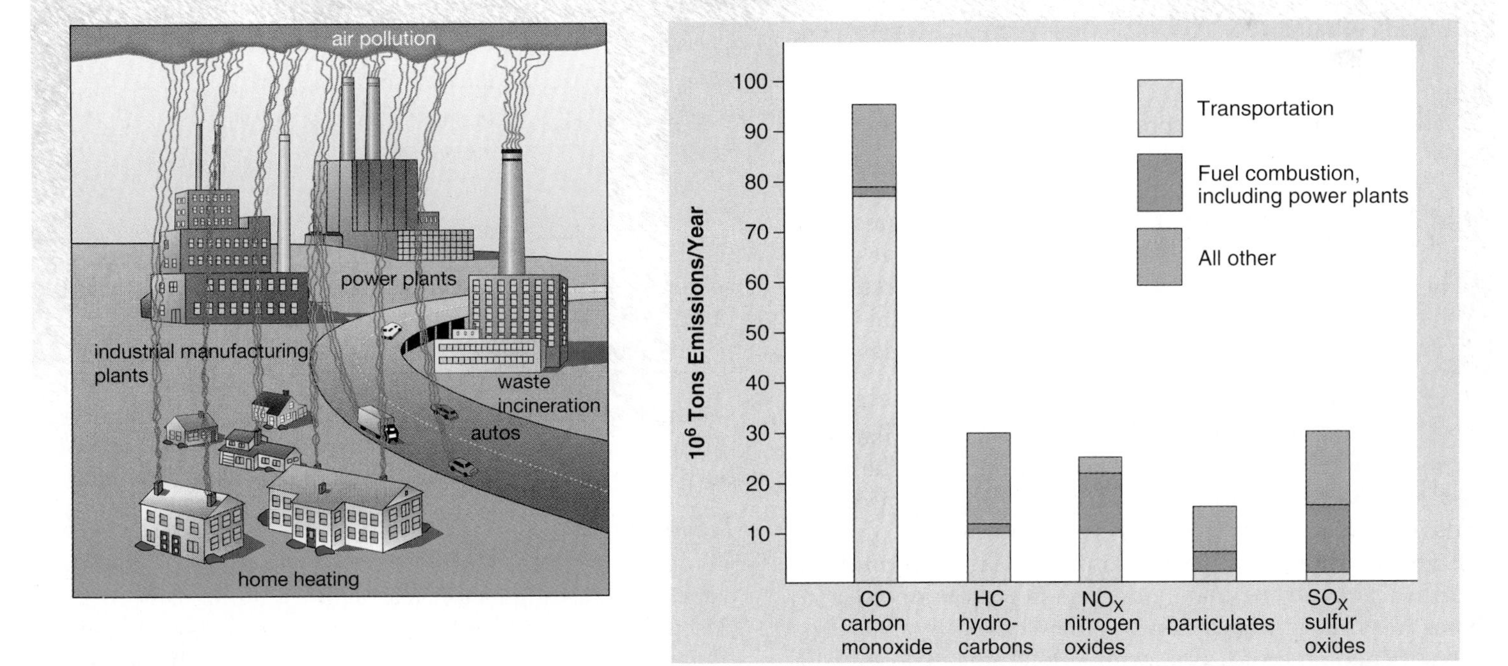

Figure 13A **Air pollution.**
Air pollutants enter the atmosphere from the sources noted.

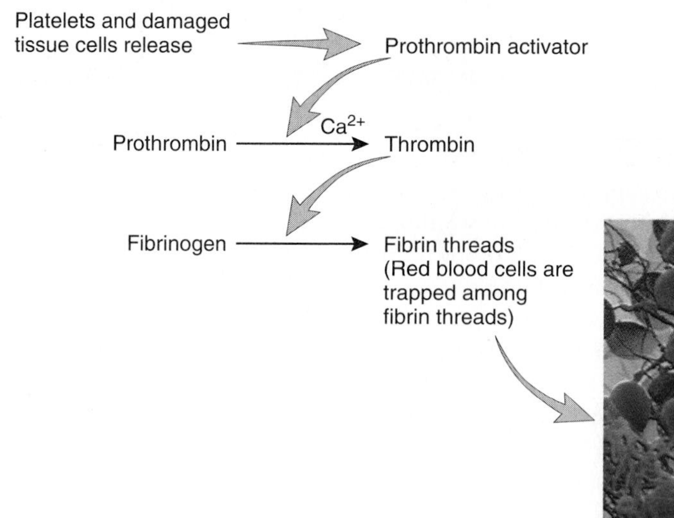

Figure 13.14 Blood clotting.
Platelets and damaged tissue cells release prothrombin activator, which acts on prothrombin in the presence of calcium ions (Ca^{2+}) to produce thrombin. Thrombin acts on fibrinogen in the presence of Ca^{2+} to form fibrin threads. The scanning electron micrograph of a blood clot shows red blood cells caught in the fibrin threads.

The Platelets and Blood Clotting

Platelets (thrombocytes) result from fragmentation of certain large cells, called **megakaryocytes,** in the red bone marrow. Platelets are produced at a rate of 200 billion a day, and the blood contains 150,000–300,000 per mm^3. These formed elements are involved in the process of blood **clotting,** or coagulation.

There are at least 12 clotting factors in the blood that participate in the formation of a blood clot. We will discuss the roles played by platelets, prothrombin, and fibrinogen. **Fibrinogen** and **prothrombin** are proteins manufactured and deposited in blood by the liver. Vitamin K, found in green vegetables and also formed by intestinal bacteria, is necessary for the production of prothrombin, and if by chance this vitamin is missing from the diet, hemorrhagic disorders develop.

Blood Clotting

When a blood vessel in the body is damaged, platelets clump at the site of the puncture and partially seal the leak. They and the injured tissues release a clotting factor called **prothrombin activator** that converts prothrombin to thrombin. This reaction requires calcium ions (Ca^{2+}). **Thrombin,** in turn, acts as an enzyme that severs two short amino acid chains from each fibrinogen molecule. These activated fragments then join end to end, forming long threads of **fibrin.** Fibrin threads wind around the platelet plug in the damaged area of the blood vessel and provide the framework for the clot. Red blood cells also are trapped within the fibrin threads; these cells make a clot appear red (Fig. 13.14). A fibrin clot is present only temporarily. As soon as blood vessel repair is initiated, an enzyme called plasmin destroys the fibrin network and restores the fluidity of plasma.

If blood is allowed to clot in a test tube, a yellowish fluid develops above the clotted material. This fluid is called **serum,** and it contains all the components of plasma except fibrinogen. Table 13.1 reviews the many different terms we have used to refer to various body fluids related to blood.

Table 13.1 Body Fluids

Name	*Composition*
Blood	Formed elements and plasma
Plasma	Liquid portion of blood
Serum	Plasma minus fibrinogen
Tissue fluid	Plasma minus most proteins
Lymph	Tissue fluid within lymphatic vessels

Hemophilia

Hemophilia is an inherited clotting disorder due to a deficiency in a clotting factor. The slightest bump can cause the affected person to bleed into the joints. Cartilage degeneration in the joints and resorption of underlying bone can follow. Bleeding into muscles can lead to nerve damage and muscular atrophy. The most frequent cause of death is bleeding into the brain with accompanying neurological damage.

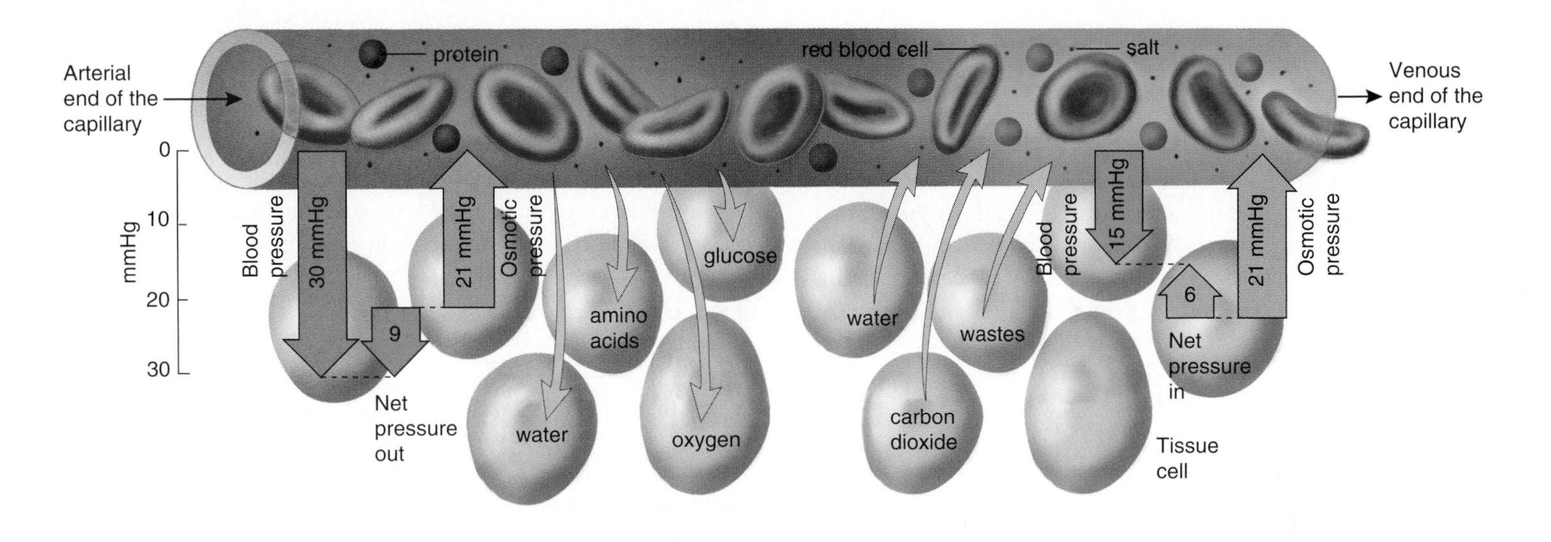

Figure 13.15 **Exchanges between blood and tissue fluid across a capillary wall.**

Capillary Exchange

Two forces primarily control movement of fluid through the capillary wall: osmotic pressure, which tends to cause water to move from tissue fluid to blood, and blood pressure, which tends to cause water to move in the opposite direction. At the arterial end of a capillary, blood pressure is higher than the osmotic pressure of blood (Fig. 13.15). Osmotic pressure is created by the presence of salts and the plasma proteins. Because blood pressure is higher than osmotic pressure at the arterial end of a capillary, water exits a capillary at this end.

Midway along the capillary, where blood pressure is lower, the two forces essentially cancel each other, and there is no net movement of water. Solutes now diffuse according to their concentration gradient—nutrients (glucose and oxygen) diffuse out of the capillary, and wastes (carbon dioxide) diffuse into the capillary. Red blood cells and almost all plasma proteins remain in the capillaries, but small substances leave. The substances that leave a capillary contribute to **tissue fluid,** the fluid between the body's cells. Since plasma proteins are too large to readily pass out of the capillary, tissue fluid tends to contain all components of plasma except lesser amounts of protein.

At the venule end of a capillary, where blood pressure has fallen even more, osmotic pressure is greater than blood pressure, and water tends to move into the capillary. Almost the same amount of fluid that left the capillary returns to it, although there is always some excess tissue fluid collected by the lymphatic capillaries (Fig. 13.16). Tissue fluid contained within lymphatic vessels is called **lymph.** Lymph is returned to the systemic venous blood when the major lymphatic vessels enter the subclavian veins in the shoulder region.

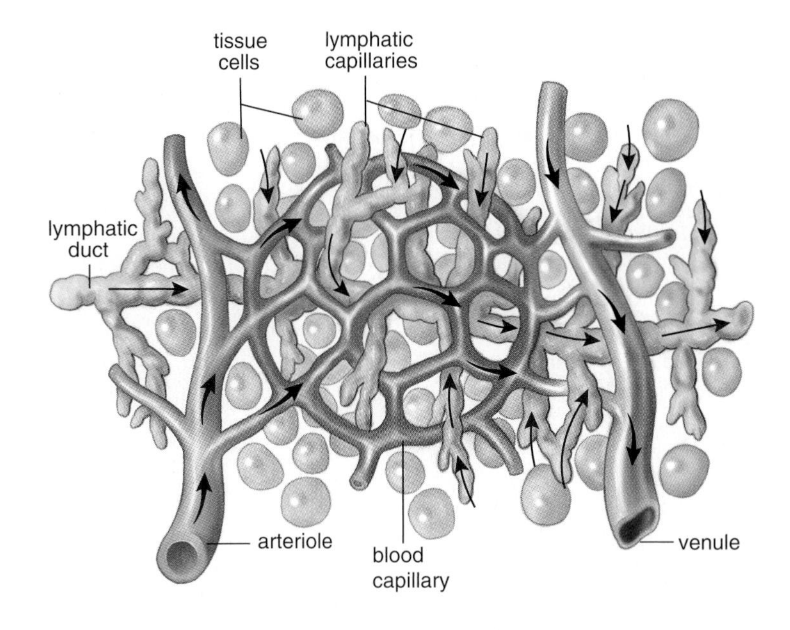

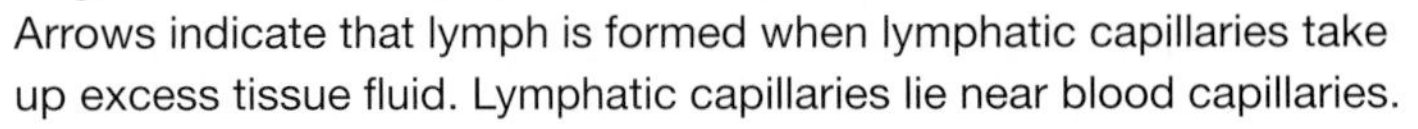

Figure 13.16 **Lymphatic vessels.**
Arrows indicate that lymph is formed when lymphatic capillaries take up excess tissue fluid. Lymphatic capillaries lie near blood capillaries.

Oxygen and nutrient substances exit a capillary near the arterial end; carbon dioxide and waste molecules enter a capillary near the venous end.

Health Focus

Prevention of Cardiovascular Disease

All of us can take steps to prevent the occurrence of cardiovascular disease, the most frequent cause of death in the United States. There are genetic factors that predispose an individual to cardiovascular disease, such as family history of heart attack under age 55, male gender, and ethnicity (African Americans are at greater risk). Those with one or more of these risk factors need not despair, however. It only means that they need to pay particular attention to these guidelines for a heart-healthy life-style.

The Don'ts

Smoking

Hypertension is well recognized as a major contributor to cardiovascular disease. When a person smokes, the drug nicotine, present in cigarette smoke, enters the bloodstream. Nicotine causes the arterioles to constrict and the blood pressure to rise. Restricted blood flow and cold hands are associated with smoking by most people. Now, the heart must pump harder to propel the blood through the lungs at a time when the oxygen-carrying capacity of the blood is reduced. (How smoking reduces the oxygen-carrying capacity is explained in the Ecology reading on page 253.) Smoking also damages the arterial wall and accelerates the formation of atherosclerosis and plaque (Fig. 13B).

Weight Gain

Hypertension also occurs more often in persons who are more than 20% above the recommended weight for their height. Because more tissue requires servicing, the heart must send extra blood out under greater pressure in those who are overweight. It may be very difficult to lose weight once it is gained, and therefore it is recommended that weight control be a lifelong endeavor. Even a slight decrease in weight can bring with it a reduction in hypertension. A 4.5-kilogram weight loss doubles the chance that blood pressure can be normalized without drugs.

The Do's

Healthy Diet

It was once thought that a low-salt diet was protective against cardiovascular disease, and it still may be in certain persons. Theoretically, hypertension occurs because the more salty the blood, the greater the osmotic pressure and the higher the water content. In recent years, the emphasis has switched to a diet low in saturated fats and cholesterol as protective against cardiovascular disease. Cholesterol is ferried in the blood by two types of plasma lipoproteins called LDL (low-density lipoprotein) and HDL (high-density lipoprotein). LDL (called "bad" lipoprotein) takes cholesterol from the liver to the tissues, and HDL (called "good" lipoprotein) transports cholesterol out of the tissues to the liver. When the LDL level in blood is abnormally high or the HDL level is abnormally low, cholesterol accumulates in the cells. When cholesterol-laden cells line the arteries, plaque develops, which interferes with circulation (Fig. 13.B).

It is recommended that everyone know his or her blood cholesterol level. Individuals with a high blood cholesterol level (240 mg/100 ml) should be further tested to determine their LDL cholesterol level. The LDL cholesterol level together with other risk factors such as age, family history, general health, and whether the patient smokes will determine who needs dietary therapy to lower their LDL. Drugs are to be reserved for high-risk patients.

Evidence is mounting to suggest a role for antioxidant vitamins (A, E, and C) in the prevention of cardiovascular disease.

13.5 Cardiovascular Disorders

Cardiovascular disease (CVD) is the leading cause of untimely death in the Western countries. Modern research efforts have resulted in improved diagnosis, treatment, and prevention. This section discusses the range of advances that have been made in these areas. The Health reading for this chapter emphasizes how to prevent CVD from developing in the first place.

Hypertension

It is estimated that about 20% of all Americans suffer from **hypertension,** which is high blood pressure. Hypertension is present when the systolic blood pressure is 140 or greater or the diastolic blood pressure is 90 or greater. While both systolic and diastolic pressures are considered important, it is the diastolic pressure that is emphasized when medical treatment is being considered.

Hypertension is sometimes called a silent killer because it may not be detected until a stroke or heart attack occurs. It has long been thought that a certain genetic makeup might account for the development of hypertension. Now researchers have discovered two genes that may be involved in some individuals. One gene codes for angiotensinogen, a plasma protein that is converted to a powerful vasoconstrictor in part by the product of the second gene. Persons with hypertension due to overactivity of these genes might one day be cured by gene therapy.

Antioxidants protect the body from free radicals that may damage HDL cholesterol through oxidation or damage the lining of an artery, leading to a blood clot that can block the vessel. Nutritionists believe that the consumption of at least five servings of fruit and vegetables a day may be protective against cardiovascular disease.

Exercise

Those who exercise are less apt to have cardiovascular disease. One study found that moderately active men who spent an average of 48 minutes a day on a leisure-time activity such as gardening, bowling, or dancing had one-third fewer heart attacks than peers who spent an average of only 16 minutes each day. Exercise helps to keep weight under control, may help minimize stress, and reduces hypertension. The heart beats faster when exercising, but exercise slowly increases its capacity. This means that the heart can beat slower when we are at rest and still do the same amount of work. One physician recommends that his cardiovascular patients walk for one hour, three times a week, and in addition, they are to practice meditation and yoga-like stretching and breathing exercises to reduce stress.

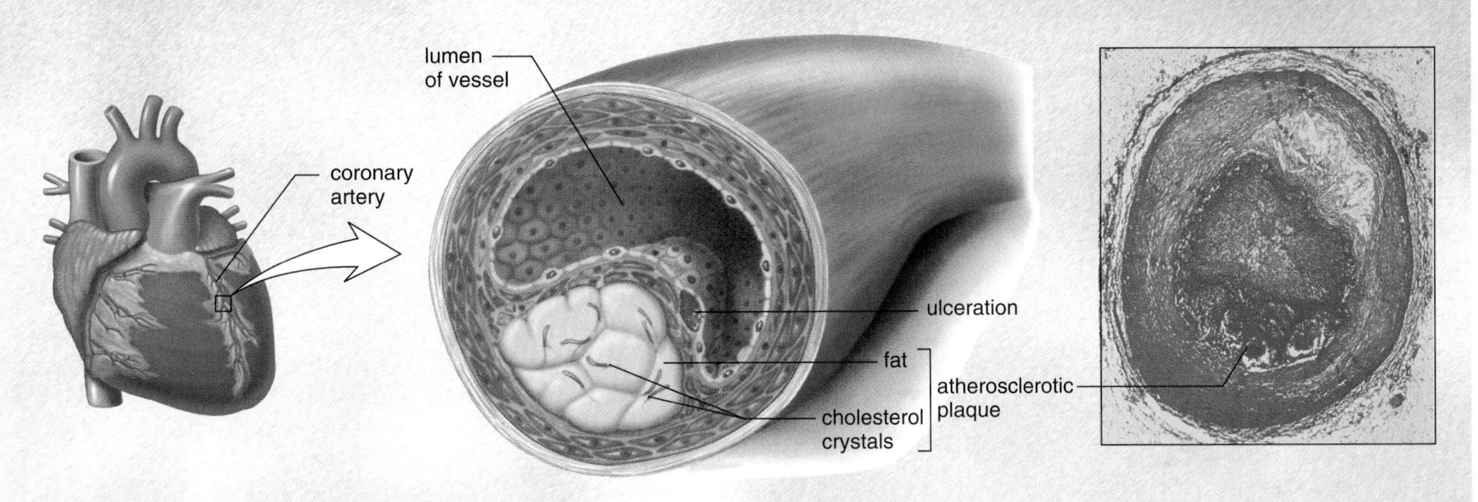

Figure 13B Coronary arteries and plaque.
Plaque (in yellow) is an irregular accumulation of cholesterol and other substances. When plaque is present in a coronary artery, a heart attack is more apt to occur because of restricted blood flow.

At present, however, the best safeguard against the development of hypertension is to have regular blood pressure checks and to adopt a life-style that lowers the risk of hypertension. The Health reading for this chapter discusses such health habits.

Atherosclerosis

Hypertension also is seen in individuals who have **atherosclerosis,** an accumulation of soft masses of fatty materials, particularly cholesterol, beneath the inner linings of arteries. Such deposits are called plaque. As it develops, plaque tends to protrude into the lumen of the vessel and interfere with the flow of blood. In certain families, atherosclerosis is due to an inherited condition such as familial hypercholesterolemia. The presence of the associated mutation can be detected, and this information is helpful if measures are taken to prevent the occurrence of the disease. In most instances, atherosclerosis begins in early adulthood and develops progressively through middle age, but symptoms may not appear until an individual is 50 or older. To prevent the onset and development of plaque, the American Heart Association and other organizations recommend a diet low in saturated fat and cholesterol and rich in fruits and vegetables.

Plaque can cause a clot to form on the irregular arterial wall. As long as the clot remains stationary, it is called a **thrombus,** but when and if it dislodges and moves along with the blood, it is called an **embolus.** If **thromboembolism** is not treated, complications can arise, as mentioned in the following section.

Stroke, Heart Attack, and Aneurysm

Stroke, heart attack, and aneurysm are associated with hypertension and atherosclerosis. A cerebrovascular accident (CVA), also called a **stroke,** often results when a small cranial arteriole bursts or is blocked by an embolus. A lack of oxygen causes a portion of the brain to die, and paralysis or death can result. A person sometimes is forewarned of a stroke by a feeling of numbness in the hands or the face, difficulty in speaking, or temporary blindness in one eye.

A myocardial infarction (MI), also called a **heart attack,** occurs when a portion of the heart muscle dies due to a lack of oxygen. If a coronary artery becomes partially blocked, the individual may then suffer from **angina pectoris,** characterized by a radiating pain in the left arm. Nitroglycerin or related drugs dilate blood vessels and help relieve the pain. When a coronary artery is completely blocked, perhaps because of thromboembolism, a heart attack occurs.

An **aneurysm** is a ballooning of a blood vessel, most often the abdominal artery or the arteries leading to the brain. Atherosclerosis and hypertension can weaken the wall of an artery to the point that an aneurysm develops. If a major vessel like the aorta should burst, death is likely. Since capillaries are the region of exchange with tissues, it is possible to replace a damaged or diseased portion of a vessel, such as an artery, with a plastic tube.

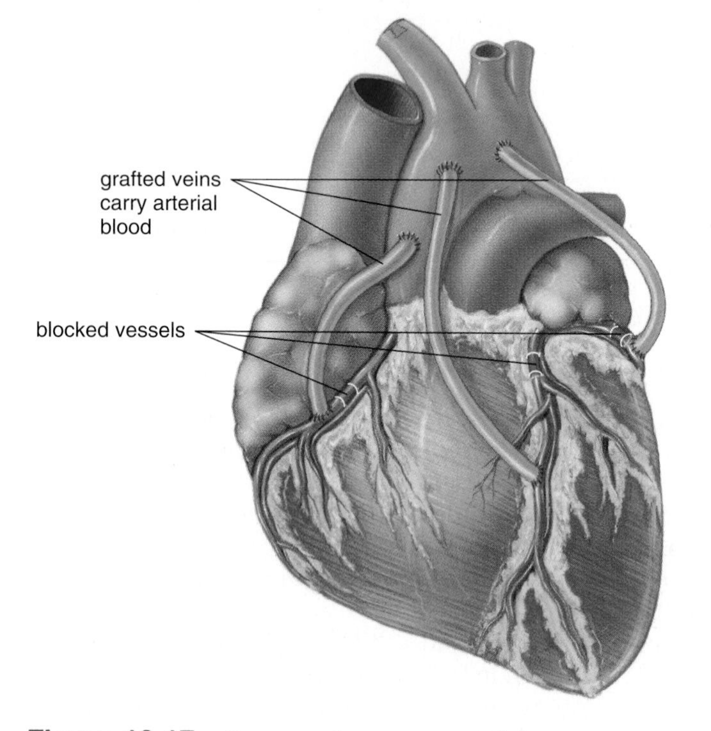

Figure 13.17 Coronary bypass operation.
During this operation, the surgeon grafts segments of another vessel, usually a small vein from the leg, between the aorta and the coronary vessels, bypassing areas of blockage. Patients who require surgery often receive two or five bypasses in a single operation.

Dissolving Blood Clots

Medical treatment for thromboembolism includes the use of t-PA, a biotechnology drug. This drug converts plasminogen, a molecule found in blood, into plasmin, an enzyme that dissolves blood clots. In fact, t-PA, which stands for tissue plasminogen activator, is the body's own way of converting plasminogen to plasmin. t-PA is also being used for thrombolytic stroke patients but with limited success because some patients experience life-threatening bleeding in the brain. A better treatment might be new biotechnology drugs that act on the plasma membrane to prevent brain cells from releasing and/or receiving toxic chemicals caused by the stroke.

If a person has symptoms of angina or a stroke, then aspirin may be prescribed. Aspirin reduces the stickiness of platelets and therefore lowers the probability that a clot will form. There is evidence that aspirin protects against first heart attacks, but there is no clear support for taking aspirin every day to prevent strokes in symptom-free people. Physicians warn that long-term use of aspirin might have harmful effects, including bleeding in the brain.

Clearing Clogged Arteries

Surgical procedures are available to treat clogged arteries. Each year thousands of persons have **coronary bypass surgery.** During this operation, surgeons take a segment of another blood vessel from the patient's body and stitch one end to the aorta and the other end to a coronary artery past the point of obstruction (Fig. 13.17). Another type of treatment has been tried. A laser is used to produce small pores

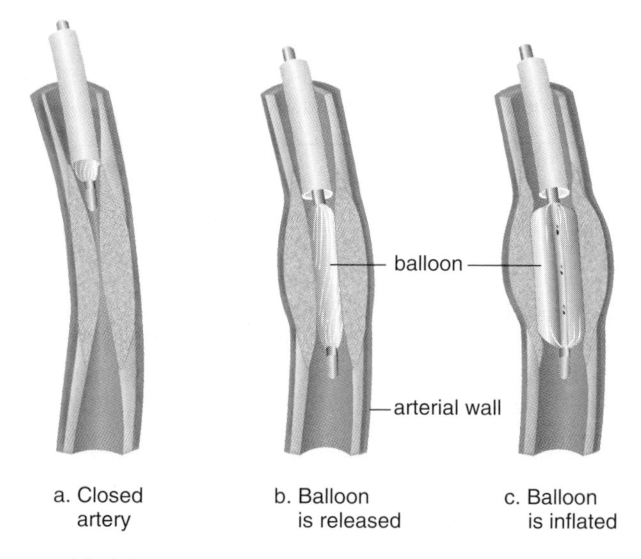

Figure 13.18 Angioplasty.
During this procedure **(a)**, a plastic tube is inserted into the coronary artery until it reaches the clogged area. **b.** A metal tip with balloon attached is pushed out the end of the plastic tube into the clogged area. **c.** When the balloon is inflated, the vessel opens. Sometimes metal coils or slotted tubes, called stents, are inserted to keep the vessel open.

in the ventricular musculature so that the blood from the ventricles can nourish the heart muscle.

In **angioplasty,** a cardiologist threads a plastic tube into an artery of an arm or a leg and guides it through a major blood vessel toward the heart. When the tube reaches the region of plaque in a coronary artery (Fig. 13.18), a balloon attached to the end of the tube is inflated, forcing the vessel open. However, the artery may not remain open because the trauma causes smooth muscle cells in the wall of the artery to proliferate and close it. Two lines of attack are being explored. Small metal devices—either metal coils or slotted tubes called stents—are expanded inside the artery to keep the artery open. When the stents are coated with heparin to prevent blood clotting and chemicals to prevent arterial closing, results have been promising.

Gene therapy for clogged arteries is another possibility. Someday, angioplasty might deliver a gene that prevents arteries from closing once they have been opened, or angioplasty could deliver a gene that codes for VEGF (vascular endothelial growth factor), which encourages new blood vessels to sprout out of an artery. If collateral blood vessels do form, they would transport blood past clogged arteries, making bypass surgery unnecessary.

Heart Transplants and Other Treatments

Persons with weakened hearts eventually may suffer from **congestive heart failure,** meaning the heart no longer is able to pump blood adequately, and blood backs up in the heart and lungs. Sometimes it is possible to repair a weak heart. For example, a back muscle can be wrapped around a heart to strengthen it. The muscle's nerve is stimulated with a kind of pacemaker that gives a burst of stimulation every 0.85 seconds. One day it may be possible to use cardiac cell transplants, because researchers who have injected live cardiac muscle cells into animal hearts find that they will contribute to the pumping of the heart.

On December 2, 1982, Barney Clark became the first person to receive an artificial heart that was driven by bursts of air received from a large external machine. Some clinically used artificial hearts today are driven by small battery-powered systems that can be carried about by a shoulder strap. The National Institutes of Health support the development of a hot-air engine that is fully implantable and is driven by an atomic heat source. The artificial heart is presently used only while the patient is waiting for a heart transplant. The difficulties with a heart transplant are, first, availability, and second, the tendency of the body to reject foreign organs. Recently researchers through genetic engineering altered the immune system of a strain of pigs with the hope that they will soon be a source of donor hearts.

Stroke, heart attack, and an aneurysm are associated with both hypertension and atherosclerosis. Various treatments are available, including both medical and surgical procedures. In the end, a heart transplant may be the only recourse for an ailing heart.

Bioethical Issue

According to a 1993 study, about one million deaths a year in the United States could be prevented if people adopted the healthy life style described in the reading on page 256. Tobacco, lack of exercise, and a high-fat diet probably cost the nation about $200 billion per year in health-care costs. To what lengths should we go to prevent these deaths and reduce health-care costs?

E. A. Miller, a meat-packing entity of ConAgra in Hyrum, Utah, charges extra for medical coverage of employees who smoke. Eric Falk, Miller's director of human resources says, "We want to teach employees to be responsible for their behavior." Anthem Blue Cross-Blue Shield of Cincinnati, Ohio, takes a more positive approach. They give insurance plan participants $240 a year in extra benefits, like additional vacation days, if they get good scores in five out of seven health-related categories. The University of Alabama, Birmingham, School of Nursing has a health-and-wellness program that counsels employees about how to get into shape in order to keep their insurance coverage. Audrey Brantley is in the program and says she has mixed feelings. She says, "It seems like they are trying to control us, but then, on the other hand, I know of folks who found out they had high blood pressure or were borderline diabetics and didn't know it."

Does it really work is another question. Turner Broadcasting System in Atlanta has a policy that affects all employees hired after 1986. They will be fired if caught smoking—whether at work or at home—but some admit they still manage to sneak a smoke.

Questions

1. Do you think employers who pay for their employees' health insurance have the right to demand, or encourage, or support a healthy life style?
2. Do you think all participants in a health insurance program should qualify for the same benefits, regardless of their life style?
3. What steps are ethical to encourage people to adopt a healthy life style?

Summarizing the Concepts

13.1 The Blood Vessels

Blood vessels include arteries (and arterioles) that take blood away from the heart; capillaries, where exchange of substances with the tissues occurs; and veins (and venules) that take blood to the heart.

13.2 The Heart

The movement of blood in the cardiovascular system is dependent on the beat of the heart. During the cardiac cycle, the SA node (pacemaker) initiates the beat and causes the atria to contract.

The AV node conveys the stimulus and initiates contraction of the ventricles. The heart sounds, lub-dup, are due to the closing of the atrioventricular valves, followed by the closing of the semilunar valves.

13.3 The Vascular Pathways

The cardiovascular system is divided into the pulmonary circuit and the systemic circuit. In the pulmonary circuit, two pulmonary arteries take blood from the right ventricle to the lungs, and four pulmonary veins return it to the left atrium. To trace the path of blood in the systemic circuit, start with the aorta from the left ventricle. Follow its path until it branches to an artery going to a specific organ. It can be assumed that the artery divides into arterioles and capillaries, and that the capillaries lead to venules. The vein that takes blood to the vena cava most likely has the same name as the artery that delivered blood to the organ. In the adult systemic circuit, unlike the pulmonary circuit, the arteries carry blood that is relatively high in oxygen and relatively low in carbon dioxide, and the veins carry blood that is relatively low in oxygen and relatively high in carbon dioxide.

Blood pressure accounts for the flow of blood in the arteries, but because blood pressure drops off after the capillaries, it cannot cause blood flow in the veins. Skeletal muscle contraction, presence of valves, and respiratory movements account for blood flow in veins. The reduced velocity of blood flow in capillaries facilitates exchange of nutrients and wastes.

13.4 Blood

Blood has two main parts: plasma and cells. Plasma contains mostly water (90–92%) and proteins (7–8%), but it also contains nutrients and wastes.

The red blood cells contain hemoglobin and function in oxygen transport. Defense against disease depends on the various types of white blood cells. Granular neutrophils and monocytes are phagocytic. Agranular lymphocytes are involved in the development of immunity to disease.

The platelets and two plasma proteins, prothrombin and fibrinogen, function in blood clotting, an enzymatic process that results in fibrin threads.

When blood reaches a capillary, water moves out at the arterial end, due to blood pressure. At the venule end, water moves in, due to osmotic pressure. In between, nutrients diffuse out and wastes diffuse in.

13.5 Cardiovascular Disorders

Hypertension and atherosclerosis are two cardiovascular disorders that lead to stroke, heart attack, and aneurysm. Medical and surgical procedures are available to control cardiovascular disease, but the best policy is prevention by following a heart-healthy diet, getting regular exercise, maintaining a proper weight, and not smoking cigarettes.

Studying the Concepts

1. What types of blood vessels are there? Discuss their structure and function. 240–41
2. Trace the path of blood in the heart, mentioning the vessels attached to, and the valves within, the heart. 243
3. Describe the cardiac cycle (using the terms systole and diastole), and explain the heart sounds. 244
4. Describe the cardiac conduction system and an ECG. Tell how an ECG is related to the cardiac cycle. 244–45
5. Trace the path of blood in the pulmonary circuit. Trace the path of blood to and from the kidneys in the systemic circuit. 246–47
6. In what type of vessel is blood pressure highest? Lowest? Why is the slow movement of blood in capillaries beneficial? What factors assist venous return of blood? 248
7. State the major components of blood, and give a function for each. 249–52
8. Name the steps that take place when blood clots. Which substances are present in the blood at all times, and which appear during the clotting process? 254
9. What forces operate to facilitate exchange of substances across the capillary wall? 255
10. What is atherosclerosis? Name two illnesses associated with hypertension and thromboembolism. 256–57
11. Discuss the medical and surgical treatment of cardiovascular disease. 258–59

Testing Yourself

Choose the best answer for each question.

1. Systemic arteries carry blood
   a. to the capillaries and away from the heart.
   b. away from the capillaries and toward the heart.
   c. lower in oxygen than carbon dioxide.
   d. higher in oxygen than carbon dioxide.
   e. Both a and d are correct.
2. Both the right side and the left side of the heart
   a. have semilunar valves between its chamber.
   b. consist of an atrium and ventricle.
   c. pump blood to the lungs and the body.
   d. communicate with each other.
   e. All of these are correct.
3. Systole refers to the contraction of the
   a. major arteries.
   b. SA node.
   c. atria and ventricles.
   d. major veins.
   e. All of these are correct.
4. During a heartbeat,
   a. the SA node initiates an impulse that passes to the AV node.
   b. first the atria contract and then the ventricles contract.
   c. the heart pumps the blood out into the attached arteries.
   d. all chambers rest for a while.
   e. All of these are correct.

5. Which of these does not correctly contrast the pulmonary circuit and the systemic circuit?
   Pulmonary circuit—Systemic circuit
   a. veins carry blood low in oxygen—veins carry blood high in oxygen.
   b. carries blood to and from the lungs—carries blood to and from the body.
   c. has a limited number of blood vessels—has a large number of blood vessels.
   d. goes between the right ventricle and the left atrium—goes between the left ventricle and the right atrium.
   e. Both b and d are correct.
6. The best explanation for the slow movement of blood capillaries is
   a. skeletal muscles press on veins, not capillaries.
   b. capillaries have much thinner walls than arteries.
   c. there are many more capillaries than arterioles.
   d. venules are not prepared to receive so much blood from the capillaries.
   e. All of these are correct.
7. Which of these associations is incorrect?
   a. white blood cells—infection fighting
   b. red blood cells—blood clotting
   c. plasma—water, nutrients, and wastes
   d. red blood cells—hemoglobin
   e. platelets—blood clotting
8. Water enters capillaries on the venule side as a result of
   a. active transport from tissue fluid.
   b. an osmotic pressure gradient.
   c. higher blood pressure on this side.
   d. higher blood pressure on the arterial side.
   e. higher red blood cell concentration on this side.
9. The last step in blood clotting
   a. is the only step that requires calcium ions.
   b. occurs outside the bloodstream.
   c. is the same as the first step.
   d. converts prothrombin to thrombin.
   e. converts fibrinogen to fibrin.
10. A thromboembolism
    a. can cause a heart attack or a stroke.
    b. causes hypertension in most individuals.
    c. is signaled by angina pectoris.
    d. is a blood clot usually caused by plaque within an artery.
    e. Both a and d are correct.
11. Label arrows as either blood pressure or osmotic pressure.

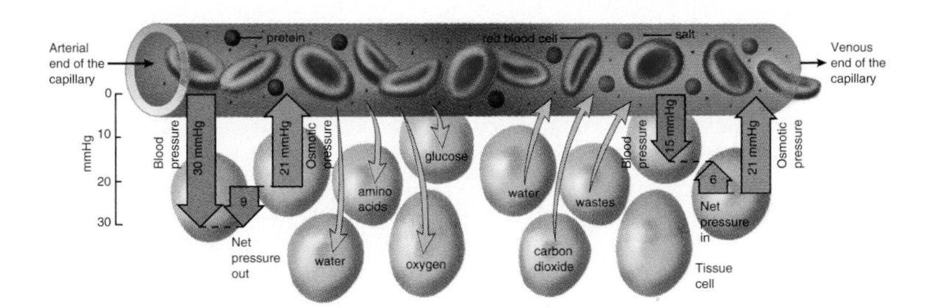

12. Label this diagram of the heart.

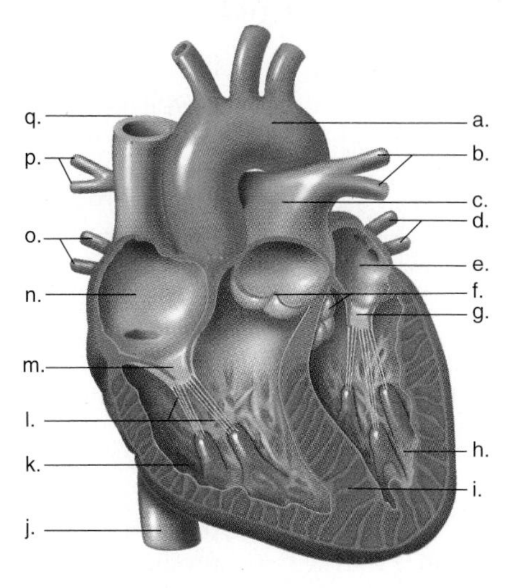

Think Scientifically

1. Assume that emboli are most apt to lodge in a capillary:
   a. If an embolus has formed in the iliac vein, in which organ would you expect it to lodge? (page 246)
   b. If an embolus has formed in the carotid artery, in which organ would you expect it to lodge?
   c. If an embolus has formed in the hepatic portal vein, in which organ would you expect it to lodge?
   d. Explain the term coronary thrombosis by telling why you would expect a thrombus and not an embolus formed in the veins to cause a heart attack.
2. Smoking cigarettes increases blood pressure. Considering this, what do you predict that smoking does to the (page 248)
   a. heartbeat rate?
   b. bore of the arteries?
   c. capillary beds in the fingers and the toes?
   d. resistance of blood flow through the lungs?

Understanding the Terms

agranular leukocyte 252
albumin 250
anemia 252
aneurysm 258
angina pectoris 258
angioplasty 259
aorta 246
arteriole 240
artery 240
atherosclerosis 257
atrioventricular bundle 244
atrioventricular valve 242
atrium 242
AV (atrioventricular) node 244
blood pressure 248
capillary 240
cardiac cycle 244
chordae tendineae 242
clotting 254
colony-stimulating factor (CSF) 252
congestive heart failure 259
coronary artery 247
coronary bypass surgery 258
diastole 244
diastolic pressure 248
electrocardiogram (ECG or EKG) 245
embolus 257
erythropoietin 250
fibrin 254
fibrinogen 254
formed element 249
granular leukocyte 252
heart 242
heart attack 258
hemoglobin 250
hemorrhoids 248
hepatic portal system 247
hepatic portal vein 247
hepatic vein 247
hypertension 256
infectious mononucleosis 252
inferior vena cava 246
leukemia 252
lymph 255
lymphocyte 252
macrophage 252
megakaryocyte 254
monocyte 252
myocardium 242
neutrophil 252
pacemaker 244
pathogen 250
pericardium 242
phlebitis 248
plasma 249
platelet 254
prothrombin 254
prothrombin activator 254
pulmonary artery 243
pulmonary circuit 246
pulmonary embolism 248
pulmonary vein 243
pulse 244
Purkinje fibers 244
red blood cell (erythrocyte) 250
SA (sinoatrial) node 244
semilunar valve 242
septum 242
serum 254
stem cell 250
stroke 258
superior vena cava 246
systemic circuit 246
systole 244
systolic pressure 248
thrombin 254
thromboembolism 257
thrombus 257
tissue fluid 255
valve 241
varicose veins 248
vein 240
vena cava 243
ventricle 242
venule 241
white blood cell (leukocyte) 252

Match the terms to these definitions:

a. __________ Relaxation of a heart chamber.

b. __________ Large systemic vein that returns blood to the right atrium of the heart.

c. __________ Plasma protein that is converted into fibrin threads during blood clotting.

d. __________ Iron-containing protein in red blood cells that combines with and transports oxygen.

e. __________ That part of the cardiovascular system that serves body parts and does not include the gas-exchanging surfaces in the lungs.

Using Technology

Your study of the digestive system and nutrition is supported by these available technologies:

Essential Study Partner CD-ROM

Animals → Circulatory System

Visit the Mader web site for related ESP activities.

Exploring the Internet

The Mader Home Page provides resources and tools as you study this chapter.

http://www.mhhe.com/biosci/genbio/mader

Dynamic Human 2.0 CD-ROM

Cardiovascular System

Virtual Physiology Laboratory CD-ROM

Effects of Drugs on Frog Heart
Electrocardiogram
Frog Muscle

HealthQuest CD-ROM

5 Cardiovascular Health
2 Fitness

C H A P T E R 14

Lymphatic System and Immunity

Cancer cells (blue) divide as they course through a lymphatic vessel, which also contains normal lymphocytes (red).

Chapter Concepts

Karlin casually rubs her nose, unaware she is infecting herself with a cold virus picked up by shaking hands with a coworker. The viral particles slip past the protective mucous barrier of her nasal cavities, enter cells, and begin to make copies of themselves. Just before the infected cells succumb, they secrete chemicals that alert her immune system to the invaders. As newly made viruses burst forth killing the cells, antibodies latch onto them and mark them for destruction. This is the job of amoebae-like immune cells that rush to the infection site and devour such complexes. Some other immune cells hereafter kill cells infected with the virus, and in this way prevent the production of more viruses. After a week of sniffles, Karlin's immune system wins its battle with the virus. This chapter explains how the immune system, working with the lymphatic system, fights off bacteria, viruses, even cancer, and how vaccines exploit these two systems to provide long-lasting protection from many diseases.

14.1 Lymphatic System

The **lymphatic system** consists of lymphatic vessels and the lymphoid organs. This system, which is closely associated with the cardiovascular system, has three main functions: (1) lymphatic capillaries take up excess tissue fluid and return it to the bloodstream; (2) lymphatic capillaries absorb fats at the intestinal villi and transport them to the bloodstream; and (3) the lymphatic system helps to defend the body against disease.

Lymphatic Vessels

Lymphatic vessels are quite extensive; most regions of the body are richly supplied with lymphatic capillaries (Fig.14.1). The construction of the larger lymphatic vessels is similar to that of cardiovascular veins, including the presence of valves. Also, the movement of lymph within these vessels is dependent upon skeletal muscle contraction. When the muscles contract, the lymph is squeezed past a valve that closes, preventing the lymph from flowing backwards.

The lymphatic system is a one-way system that begins with lymphatic capillaries. These capillaries take up fluid that has diffused from and has not been reabsorbed by the blood capillaries. **Edema** is localized swelling caused by the accumulation of tissue fluid. This can happen if too much tissue fluid is made and/or not enough of it is drained away. Once tissue fluid enters the lymphatic vessels, it is called **lymph.** The lymphatic capillaries join to form lymphatic vessels that merge before entering one of two ducts: the thoracic duct or the right lymphatic duct. The *thoracic duct* is much larger than the right lymphatic duct. It serves the lower extremities, the abdomen, the left arm, and the left side of both the head and the neck. The *right lymphatic duct* serves the right arm, the right side of both the head and the neck, and the right thoracic area. The lymphatic ducts enter the subclavian veins, which are cardiovascular veins in the thoracic region.

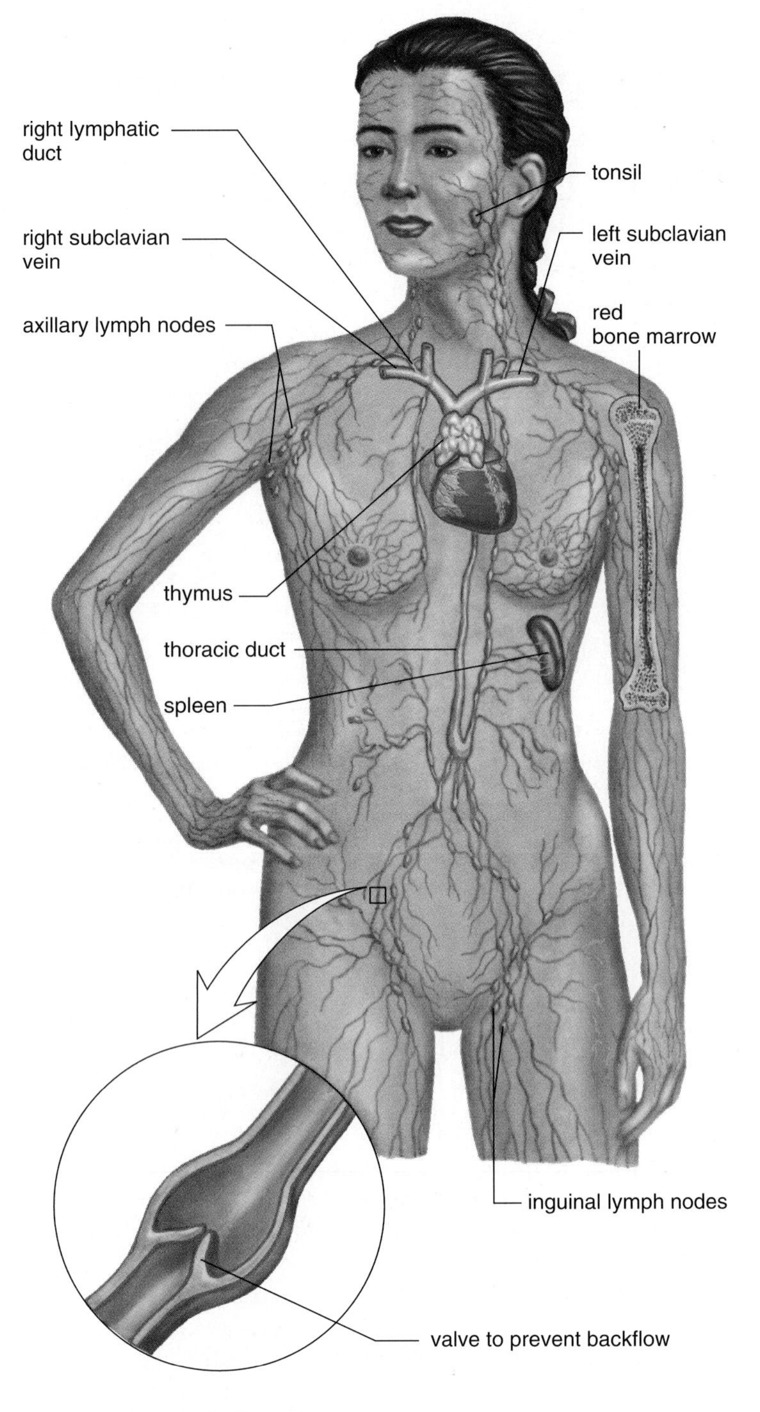

Figure 14.1 Lymphatic system.
The lymphatic vessels drain excess fluid from the tissues and return it to the cardiovascular system. The enlargement shows that lymphatic vessels have valves to prevent backward flow.

Lymph flows one way from a capillary to ever-larger lymphatic vessels and finally to a lymphatic duct, which enters a subclavian vein.

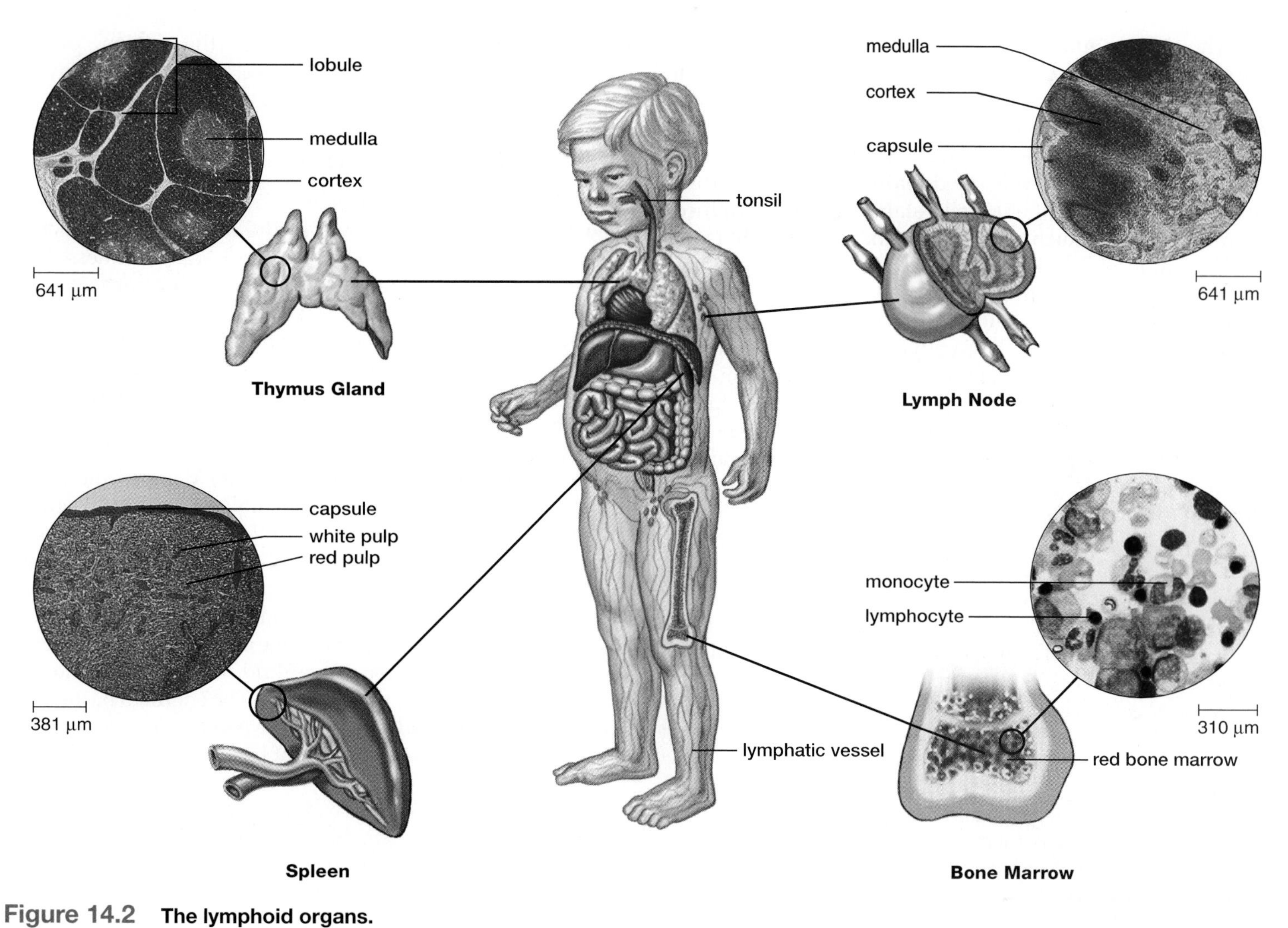

Figure 14.2 The lymphoid organs.
The lymphoid organs include the lymph nodes, the spleen, the thymus gland, and the red bone marrow, which all contain lymphocytes.

Lymphoid Organs

The lymphoid organs of special interest are the lymph nodes, the tonsils, the spleen, the thymus gland, and the bone marrow (Fig. 14.2).

Lymph nodes, which are small (about 1–25 mm) ovoid or round structures, are found at certain points along lymphatic vessels. A capsule surrounds two distinct regions known as the cortex and medulla which contain many lymphocytes. Macrophages, which occur along lymph capillaries called lymph sinuses, purify lymph of infectious organisms and any debris. Antigens which leak into the cortex and medulla activate the lymphocytes to mount an immune response to them. Lymph nodes are named for their location. Inguinal nodes are in the groin and axillary nodes are in the armpits. Physicians often feel for the presence of swollen, tender lymph nodes in the neck as evidence that the body is fighting an infection. This is a noninvasive preliminary way to help make such a diagnosis.

The **tonsils** are partially encapsulated lymphatic tissue located in a ring about the pharynx. The well-known pharyngeal tonsils are also called *adenoids,* while the larger palatine tonsils located on either side of the posterior oral cavity are most apt to be infected. The tonsils perform the same functions as lymph nodes inside the body, but because of their location they are the first to encounter pathogens and antigens that enter the body by way of the nose and mouth.

The **spleen** is located in the upper left region of the abdominal cavity just beneath the diaphragm. It is much larger than a lymph node, about the size of a fist. Whereas the lymph nodes cleanse lymph, the spleen cleanses blood. A capsule surrounds tissue known as white pulp and red pulp. White pulp contains lymphocytes and performs the immune functions of the spleen. The red pulp contains red blood cells and plentiful macrophages. The red pulp helps to purify blood that passes through the spleen by removing bacteria and worn-out or damaged red blood cells.

The spleen's outer capsule is relatively thin, and an infection and/or a blow can cause the spleen to burst. Although its functions are replaced by other organs, a person without a spleen is often slightly more susceptible to infections and may have to receive antibiotic therapy indefinitely.

The **thymus gland** is located along the trachea behind the sternum in the upper thoracic cavity. This gland varies in size, but it is larger in children than in adults and may disappear completely in old age. The thymus is divided into lobules by connective tissue. The T lymphocytes mature in these lobules. The interior (medulla) of the lobule, which consists mostly of epithelial cells, stains lighter. It produces thymic hormones, such as thymosin, that are thought to aid in maturation of T lymphocytes. Thymosin may also have other functions in immunity.

Red bone marrow is the site of origination for all types of blood cells, including the five types of white blood cells pictured in Figure 13.10. The marrow contains stem cells that are ever capable of dividing and producing cells that go on to differentiate into the various types of blood cells (see Figure 13.12). In a child, most bones have red bone marrow, but in an adult it is present only in the bones of the skull, the sternum (breastbone), the ribs, the clavicle, the pelvic bones, and the vertebral column. The red bone marrow consists of a network of connective tissue fibers, called reticular fibers, which are produced by cells called reticular cells. These and the stem cells and their progeny are packed around thin-walled sinuses filled with venous blood. Differentiated blood cells enter the bloodstream at these sinuses.

The lymphoid organs have specific functions that assist immunity. Lymph is cleansed in lymph nodes; blood is cleansed in the spleen; T lymphocytes mature in the thymus; and white blood cells are made in the bone marrow.

14.2 Nonspecific Defenses

Immunity is the ability of the body to defend itself against infectious agents, foreign cells, and even abnormal body cells, such as cancer cells. Thereby, the internal environment has a better chance of remaining stable. Immunity includes nonspecific and specific defenses. The four types of nonspecific defenses—barriers to entry, the inflammatory reaction, natural killer cells, and protective proteins—are effective against many types of infectious agents.

Barring Entry

Skin and the mucous membranes lining the respiratory, digestive, and urinary tracts serve as mechanical barriers to entry by pathogens. Oil gland secretions contain chemicals that weaken or kill certain bacteria on skin. The upper respiratory tract is lined by ciliated cells that sweep mucus and trapped particles up into the throat, where they can be swallowed, or expectorated (coughed out). The stomach has an acidic pH, which inhibits the growth of or kills many types of bacteria. The various bacteria that normally reside in the intestine and other areas, such as the vagina, prevent pathogens from taking up residence. A **pathogen** is any disease causing agent such as viruses and some bacteria.

Inflammatory Reaction

Whenever the skin is broken due to a minor injury, a series of events occurs that is known as the **inflammatory reaction.** The inflamed area has four outward signs: redness, heat, swelling, and pain. Figure 14.3 illustrates the participants in the inflammatory reaction. **Mast cells,** which occur in tissues, resemble **basophils,** one of the white cells found in the blood.

When an injury occurs, damaged tissue cells and mast cells release chemical mediators, such as **histamine** and **kinins,** which cause the capillaries to dilate and become more permeable. The enlarged capillaries cause the skin to redden, and the increased permeability allows proteins and fluids to escape and swelling results. A rise in temperature increases phagocytosis by white blood cells. The swollen area as well as kinins stimulate free nerve endings, causing the sensation of pain.

Neutrophils and monocytes migrate to the site of injury. They are amoeboid and can change shape to squeeze through capillary walls to enter tissue fluid. Neutrophils, and also mast cells, can phagocytize bacteria. The engulfed bacteria are destroyed by hydrolytic enzymes when the endocytic vesicle combines with a lysosome, one of the cellular organelles.

Monocytes differentiate into **macrophages,** large phagocytic cells that are able to devour a hundred bacteria or viruses and still survive. Some tissues, particularly connective tissue, have resident macrophages, which routinely act as scavengers, devouring old blood cells, bits of dead tissue, and other debris. Macrophages can also bring about an explosive increase in the number of leukocytes by liberating colony-stimulating hormones, which pass by way of blood to the red bone marrow, where they stimulate the production and the release of white blood cells, primarily neutrophils.

When a blood vessel ruptures, the blood clots to seal the break. The chemical mediators, mentioned earlier, and antigens move through the tissue fluid and lymph to the lymph nodes. Now lymphocytes can also be activated to react to the threat of an infection. As the infection is being overcome, some neutrophils may die. These—along with dead tissue, cells, bacteria, and living white blood cells—form *pus,* a whitish material. Pus indicates that the body is trying to overcome the infection.

Sometimes inflammation persists and the result is chronic inflammation that is often treated by the administration of anti-inflammatory agents such as aspirin, ibuprofen, or cortisone. They act against the chemical mediators released by the white blood cells in the area.

The inflammatory reaction is a "call to arms"—it marshals phagocytic white blood cells to the site of bacterial invasion and stimulates the immune system to react against a possible infection.

Visual Focus

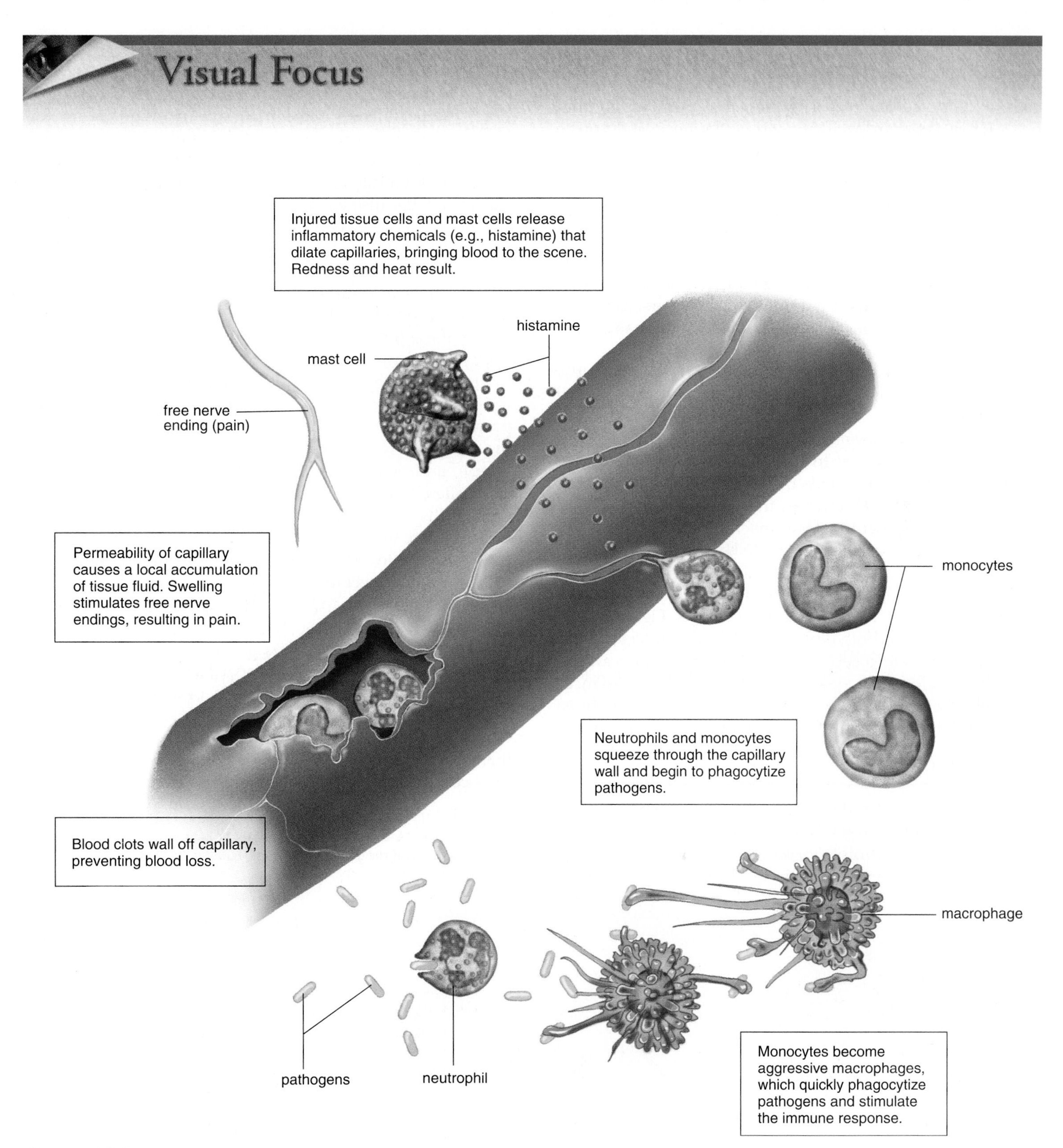

Figure 14.3 Inflammatory reaction.

Natural Killer Cells

Natural killer (NK) cells kill virus-infected cells and tumor cells by cell-to-cell contact. They are large granular lymphocytes. They have no specificity and no memory. Their number is not increased by immunization.

Protective Proteins

The **complement system,** often simply called complement, is a number of plasma proteins designated by the letter C and a subscript. A limited amount of activated complement protein is needed because a domino effect occurs: each activated protein in a series is capable of activating many other proteins.

Complement is activated when pathogens enter the body. It "complements" certain immune responses, which accounts for its name. For example, it is involved in and amplifies the inflammatory response because complement proteins attract phagocytes to the scene. Some complement proteins bind to the surface of pathogens already coated with antibodies, which ensures that the pathogens will be phagocytized by a neutrophil or macrophage.

Certain other complement proteins join to form a *membrane attack complex* that produces holes in bacterial cell walls and plasma membranes of bacteria. Fluids and salts then enter the bacterial cell to the point that it bursts (Fig. 14.4).

Interferon is a protein produced by virus-infected cells. Interferon binds to receptors of noninfected cells, causing them to prepare for possible attack by producing substances that interfere with viral replication. Interferon is specific to the species; therefore, only human interferon can be used in humans.

Immunity includes these nonspecific defenses: barriers to entry, the inflammatory reaction, natural killer cells, and protective proteins.

14.3 Specific Defenses

When nonspecific defenses have failed to prevent an infection, specific defenses come into play. An **antigen** is any foreign substance (often a protein or polysaccharide) that stimulates the **immune system** to react to it. Pathogens have antigens, but antigens can also be part of a foreign cell or a cancer cell. Because we do not ordinarily become immune to our own cells, it is said that the immune system is able to distinguish self from nonself.

Immunity usually lasts for some time. For example, once we recover from the measles, we usually do not get the illness a second time. Immunity is primarily the result of the action of the **B lymphocytes** and the **T lymphocytes.** B lymphocytes[1] mature in the *b*one marrow, and T lymphocytes mature in the *t*hymus gland. B lymphocytes, also called B cells, give rise to plasma cells, which produce **antibodies,** proteins that are capable of combining with and neutralizing antigens. These antibodies are secreted into the blood, lymph, and other body fluids. In contrast, T lymphocytes, also called T cells, do not produce antibodies. Instead, certain T cells directly attack cells that bear antigens. Other T cells regulate the immune response.

Lymphocytes are capable of recognizing an antigen because they have receptor molecules on their surface. The shape of the receptors on any particular lymphocyte is complementary to a specific antigen. It is often said that the receptor and the antigen fit together like a *lock and a key.* It is estimated that during our lifetime, we encounter a million different antigens, so we need a great diversity of lymphocytes to protect us against antigens. It is remarkable that diversification occurs to such an extent during the maturation process that there is a lymphocyte type for any possible antigen. Just how this occurs is discussed in the reading on page 271.

[1]Historically, the B stands for *bursa of Fabricius,* an organ in the chicken where these cells were first identified. As it turns out, however, the B can conveniently be thought of as referring to bone marrow.

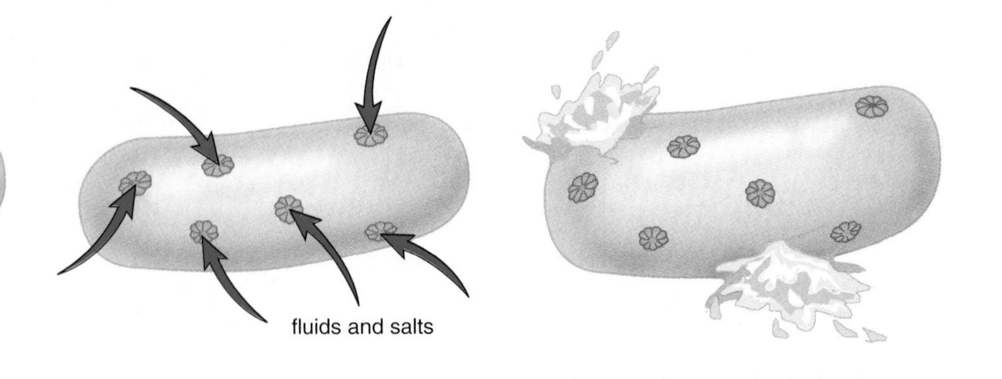

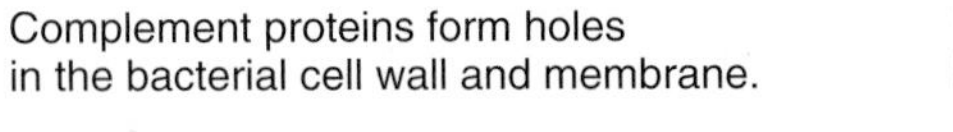
Complement proteins form holes in the bacterial cell wall and membrane.

Holes allow fluids and salts to enter the bacterium.

Bacterium expands until it bursts.

Figure 14.4 Action of the complement system against a bacterium.
When complement proteins in the plasma are activated by an immune reaction, they form holes in bacterial cell walls and plasma membranes, allowing fluids and salts to enter until the cell eventually bursts.

B Cells and Antibody-Mediated Immunity

Each type of B cell carries its specific antibody, as a membrane-bound receptor, on its surface. When a B cell in a lymph node or the spleen encounters a bacterial cell or a toxin bearing an appropriate antigen, it becomes activated to divide many times. Most of the resulting cells are plasma cells, which secrete antibodies against this antigen. A **plasma cell** is a mature B cell that mass-produces antibodies in lymph nodes and in the spleen.

The **clonal selection theory** states that the antigen selects which lymphocyte will undergo clonal expansion and produce more lymphocytes bearing the same type of receptor (Fig.14.5). Notice that a B cell does not divide until its antigen is present and binds to its receptors. B cells are also stimulated to divide and become plasma cells by helper T cell secretions, as is discussed in the next section. Some members of the clone become *memory cells* which are the means by which long-term immunity is possible. If the same antigen enters the system again, memory cells quickly divide and give rise to more lymphocytes capable of quickly producing antibodies.

Once the threat of an infection has passed, the development of new plasma cells ceases and those present undergo apoptosis. **Apoptosis** is a process of programmed cell death (PCD) involving a cascade of specific cellular events leading to the death and destruction of the cell. The methodology of PCD is still being worked out, but we know it is an essential physiological mechanism regulating the cell population within an organ system. PCD normally plays a central role in maintaining tissue homeostasis.

Defense by B cells is called **antibody-mediated immunity** because the various types of B cells produce antibodies. It is also called *humoral immunity* because these antibodies are present in blood and lymph. A *humor* is any fluid normally occurring in the body.

Characteristics of B cells:

- Antibody-mediated immunity
- Produced and mature in bone marrow
- Reside in spleen and lymph nodes, circulate in blood and lymph
- Directly recognize antigen and then undergo clonal selection
- Clonal expansion produces antibody-secreting plasma cells as well as memory B cells.

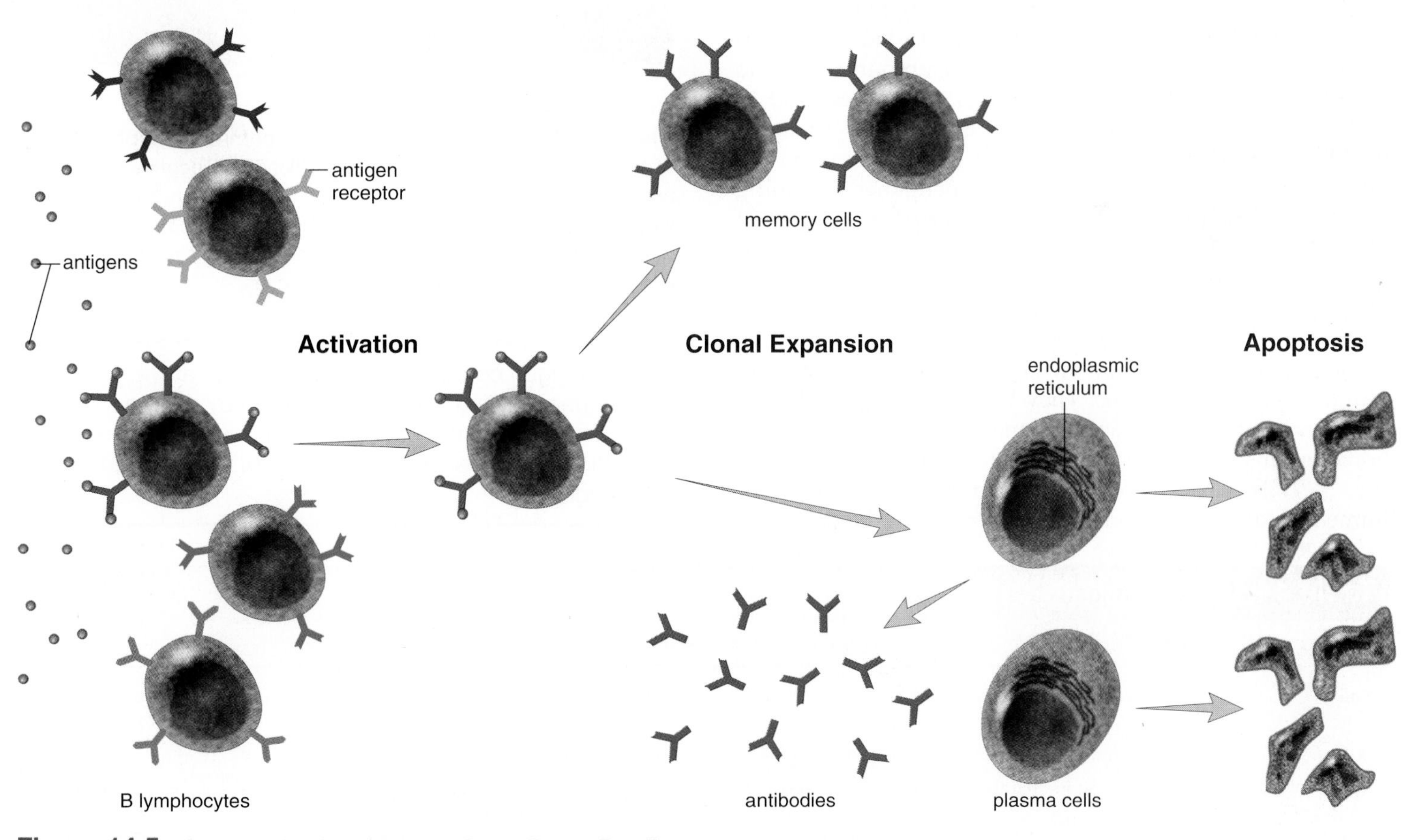

Figure 14.5 Clonal selection theory as it applies to B cells.
An antigen activates only the B cell whose receptors can combine with the antigen. This B cell then undergoes clonal expansion. During the process many plasma cells, which produce specific antibodies against this antigen, are produced. After the infection passes, they undergo apoptosis. Memory cells, which retain the ability to recognize this antigen, are retained in the body.

Structure of IgG

The most common type of antibody (IgG) is a Y-shaped protein molecule with two arms. Each arm has a "heavy" (long) polypeptide chain and a "light" (short) polypeptide chain. These chains have *constant regions,* where the sequence of amino acids is set, and *variable regions,* where the sequence of amino acids varies between antibodies (Fig. 14.6). The constant regions are not identical among all the antibodies. Instead, they are almost the same within different classes of antibodies. The variable regions form an antigen-binding site, and their shape is specific to a particular antigen, as discussed in the reading on the next page. The antigen combines with the antibody at the antigen-binding site in a lock-and-key manner.

The antigen-antibody reaction can take several forms, but quite often the reaction produces complexes of antigens combined with antibodies. Such antigen-antibody complexes, sometimes called immune complexes, mark the antigens for destruction. For example, an antigen-antibody complex may be engulfed by neutrophils or macrophages, or it may activate complement. Complement makes pathogens more susceptible to phagocytosis, as discussed previously.

Other Types of Antibodies

There are five different classes of circulating antibody proteins or **immunoglobulins (Igs)** (Table 14.1). IgG antibodies are the major type in blood, and lesser amounts are also found in lymph and tissue fluid. IgG antibodies bind to pathogens and their toxins. A *toxin* is a specific chemical (produced by bacteria, for example) that is poisonous to other living things. IgM antibodies are pentamers, meaning that they contain five of the Y-shaped structures shown in Figure 14.6*a*. These antibodies appear in blood soon after an infection begins and disappear before it is over. They are good activators of the complement system. IgA antibodies are monomers, dimers, or larger molecules containing two Y-shaped structures. They are the main type of antibody found in bodily secretions. They bind to pathogens before they reach the bloodstream. The main function of IgD antibodies seems to be to serve as receptors for antigens on mature B cells. IgE antibodies, which are responsible for immediate allergic responses, are discussed on page 277.

antigen-binding sites
V
V
C
V
V
C
C
C
light chain
heavy chain
a.

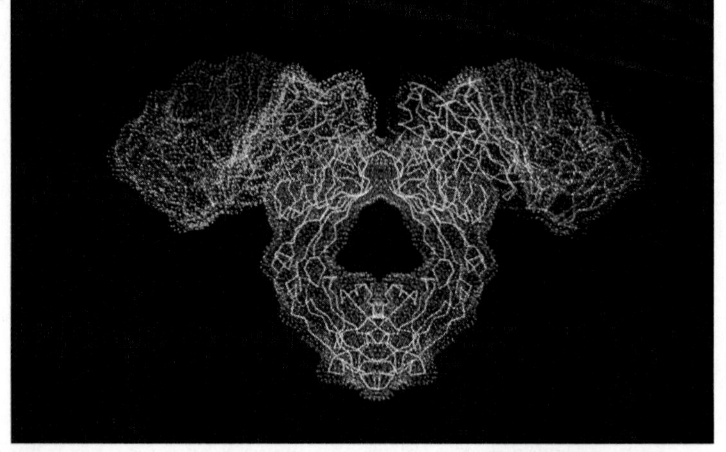

b.

Figure 14.6 Structure of the most common antibody (IgG). **a.** An IgG antibody contains two heavy (long) polypeptide chains and two light (short) chains arranged so there are two variable regions, where a particular antigen is capable of binding with the antibody. **b.** Computer model of an antibody molecule. In this model, the antigen combines with the two side branches.

An antigen combines with an antibody at the antigen-binding site in a lock-and-key manner. The reaction can produce antigen-antibody complexes, which contain several molecules of antibody and antigen.

Table 14.1 Antibodies

Classes	Presence	Function
IgG	Main antibody type in circulation	Binds to pathogens, activates complement, and enhances phagocytosis
IgM	Antibody type found in circulation; largest antibody	Activates complement; clumps cells
IgA	Main antibody type in secretions such as saliva and milk	Prevents pathogens from attaching to epithelial cells in digestive and respiratory tract
IgD	Antibody type found in circulation in extremely low quantity	Presence signifies maturity of B cell
IgE	Antibody type found as membrane-bound receptor on basophils in blood and on mast cells in tissues	Responsible for immediate allergic response and protection against certain parasitic worms

Science Focus

Susumu Tonegawa and Antibody Diversity

In 1987, Susumu Tonegawa became the first Japanese scientist to win the Nobel Prize in Physiology or Medicine. He had dedicated himself to finding the solution to an engrossing puzzle. Immunologists and geneticists knew that each B cell makes an antibody especially equipped to recognize the specific shape of a particular antigen. But they did not know how the human genome contained enough genetic information to permit the production of up to a million different antibody types needed to combat all of the pathogens we are likely to encounter during the course of our lives.

An antibody is composed of two light and two heavy polypeptide chains, which are divided into constant and variable regions. The constant region determines the antibody class and the variable region determines the specificity of the antibody, because this is where an antigen binds to a specific antibody (see Fig. 14.6). Each B cell must have a genetic way to code for the variable regions of both the light and heavy chains.

Tonegawa's colleagues say that he is a creative genius who intuitively knows how to design experiments to answer specific questions. In this instance, he examined the DNA sequences of lymphoblasts and compared them to mature B cells. He found that the DNA segments coding for the variable and constant regions were scattered throughout the genome in B lymphocyte stem cells and that only certain of these segments appeared in each mature antibody-secreting B cell where they randomly came together and coded for a specific variable region. Later, the variable and constant regions are joined to give a specific antibody (Fig. 14A*b*). As an analogy, consider that each person entering a supermarket chooses various items for purchase, and that the possible combination of items in any particular grocery bag is astronomical. Tonegawa also found that mutations occur as the variable segments are undergoing rearrangements. Such mutations are another source of antibody diversity.

Invariably some B cells with receptors that could bind to the body's own cell surface molecules arise. It is believed that these cells undergo apoptosis, or programmed cell death.

Tonegawa received his B.S. in chemistry in 1963 at Kyoto University and earned his Ph.D. in biology from the University of California at San Diego (UCSD) in 1969. After that he worked as a research fellow at UCSD and the Salk Institute. In 1971, he moved to the Basel Institute for Immunology and began the experiments that eventually led to his Nobel Prize-winning discovery. Tonegawa also contributed to the effort to decipher the receptors of T cells. This was an even more challenging area of research than the diversity of antibodies produced by B cells. Since 1981, he has been a full professor at the Massachusetts Institute of Technology (MIT), where he has a reputation for being an "aggressive, determined researcher" who often works late into the night.

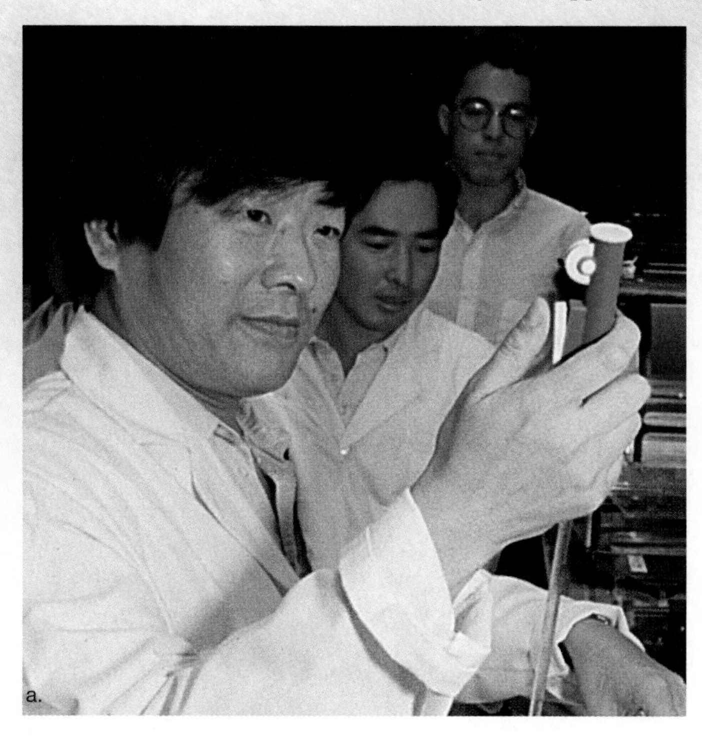

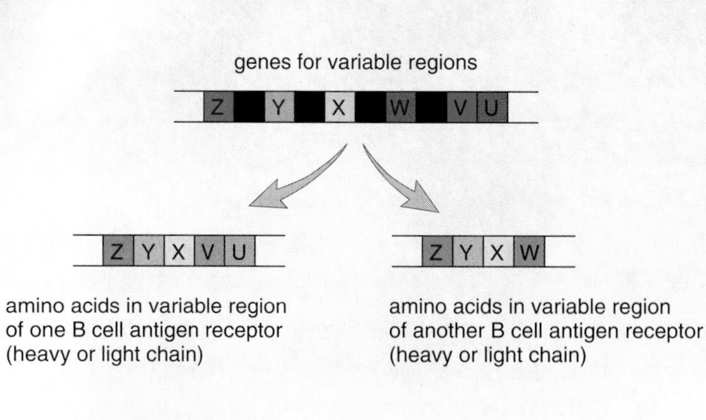

Figure 14A Antibody diversity.
a. Susumu Tonegawa received a Nobel Prize for his findings regarding antibody diversity. **b.** Different genes for the variable regions of heavy and light chains are brought together during the production of B lymphocytes so that their antigen receptors can combine with only a particular antigen.

T Cells And Cell-Mediated Immunity

The two main types of T cells are cytotoxic T cells and helper T cells. **Cytotoxic T cells** are the type of T cell responsible for **cell-mediated immunity,** so-called because T cells bring about the destruction of antigen-bearing cells, such as virus-infected or cancer cells. Cytotoxic T cells have storage vacuoles containing perforin molecules. **Perforin** molecules perforate a plasma membrane, forming a pore that allows water and salts to enter. The cell then swells and eventually bursts (Fig. 14.7).

Helper T cells regulate immunity by enhancing the response of other immune cells. When exposed to an antigen, they enlarge and secrete cytokines, stimulatory molecules that cause helper T cells to divide and other immune cells to perform their functions. For example, cytokines stimulate macrophages to phagocytize and stimulate B cells to become antibody-producing plasma cells. Because HIV, which causes AIDS, infects helper T cells and certain other cells of the immune system, it inactivates the immune response.

As we shall see, there are also memory T cells that remain in the body and can jump-start an immune reaction to an antigen previously present in the body.

Activation of T Cells

When T cells leave the thymus they have unique receptors just as B cells do. Unlike B cells, however, cytotoxic T cells and helper T cells are unable to recognize an antigen present in lymph, blood, or the tissues without help. The antigen must be presented to them by an **antigen-presenting cell (APC).** When an APC, usually a macrophage, engulfs a pathogen, the pathogen is broken down to fragments within an endocytic vesicle. These fragments are antigenic; that is, they have the properties of an antigen. The fragments are linked to a major histocompatibility complex (MHC) protein in the plasma membrane and then they can be presented to a T cell.

Human MHC proteins are called **HLA (human leukocyte associated) antigens.** Because they mark the cell as belonging to a particular individual, HLA proteins are self

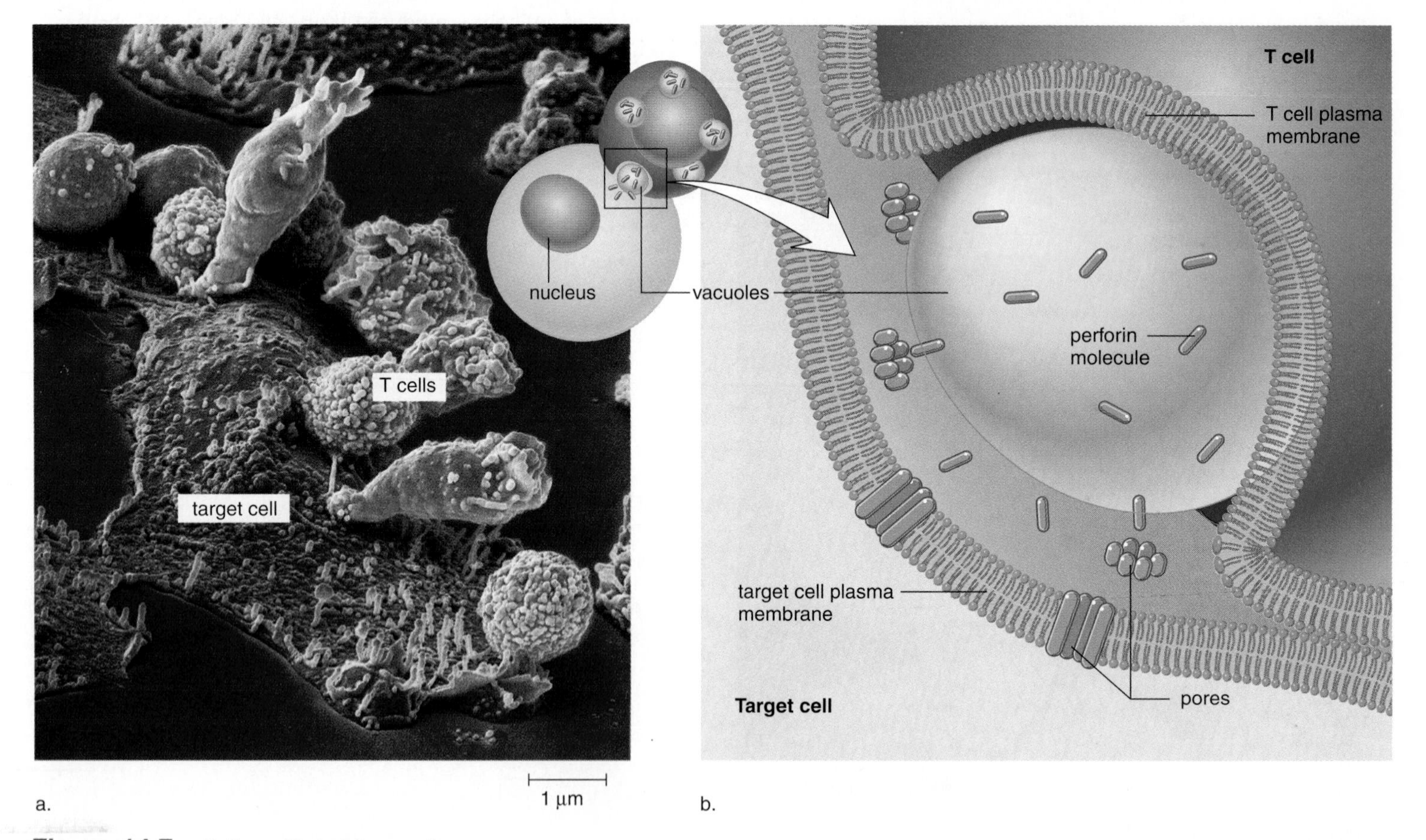

Figure 14.7 Cell-mediated immunity.
a. The scanning electron microscope shows cytotoxic T cells attacking and destroying a cancer cell. **b.** During the killing process, the vacuoles in a cytotoxic T cell release perforin molecules. These molecules combine to form pores in the target cell plasma membrane. Thereafter, fluid and salts enter so that the target cell eventually bursts.

antigens. The importance of self antigens in plasma membranes was first recognized when it was discovered that they contribute to the specificity of tissues and make it difficult to transplant tissue from one human (or animal) to another. In other words, when the donor and the recipient are histo (tissue)-compatible (the same or nearly so), a transplant is more likely to be successful.

Figure 14.8 shows a macrophage presenting an antigen to a T cell. Once a helper T cell recognizes an antigen and is stimulated to do so, it undergoes clonal expansion and produces cytokines that stimulate immune cells to remain active. Once a cytotoxic T cell is activated in this manner, it undergoes clonal expansion and destroys any cell infected with the same virus, if the cell bears the correct HLA. As the infection disappears, the immune reaction wanes and fewer cytokines are produced. Now, the activated T cells become susceptible to apoptosis. As mentioned previously, apoptosis is a programmed cell death that contributes to homeostasis by regulating the number of cells that are present in an organ, or in this case the immune system. A few of the clonally expanded T cells do not undergo apoptosis. The survivors are memory cells—T cells that can rapidly respond should the same antigen be present at a later time.

Apoptosis also occurs in the thymus as T cells are maturing. A T cell that bears a receptor with the potential to recognize a self antigen undergoes suicide. When apoptosis does not occur as it should, T-cell cancers (i.e., lymphomas and leukemias) can result.

Characteristics of T cells:

- Cell-mediated immunity
- Produced in bone marrow, mature in thymus
- Antigen must be presented in groove of an HLA molecule.
- Cytotoxic T cells destroy antigen-bearing cells.
- Helper T cells secrete cytokines that control the immune response.

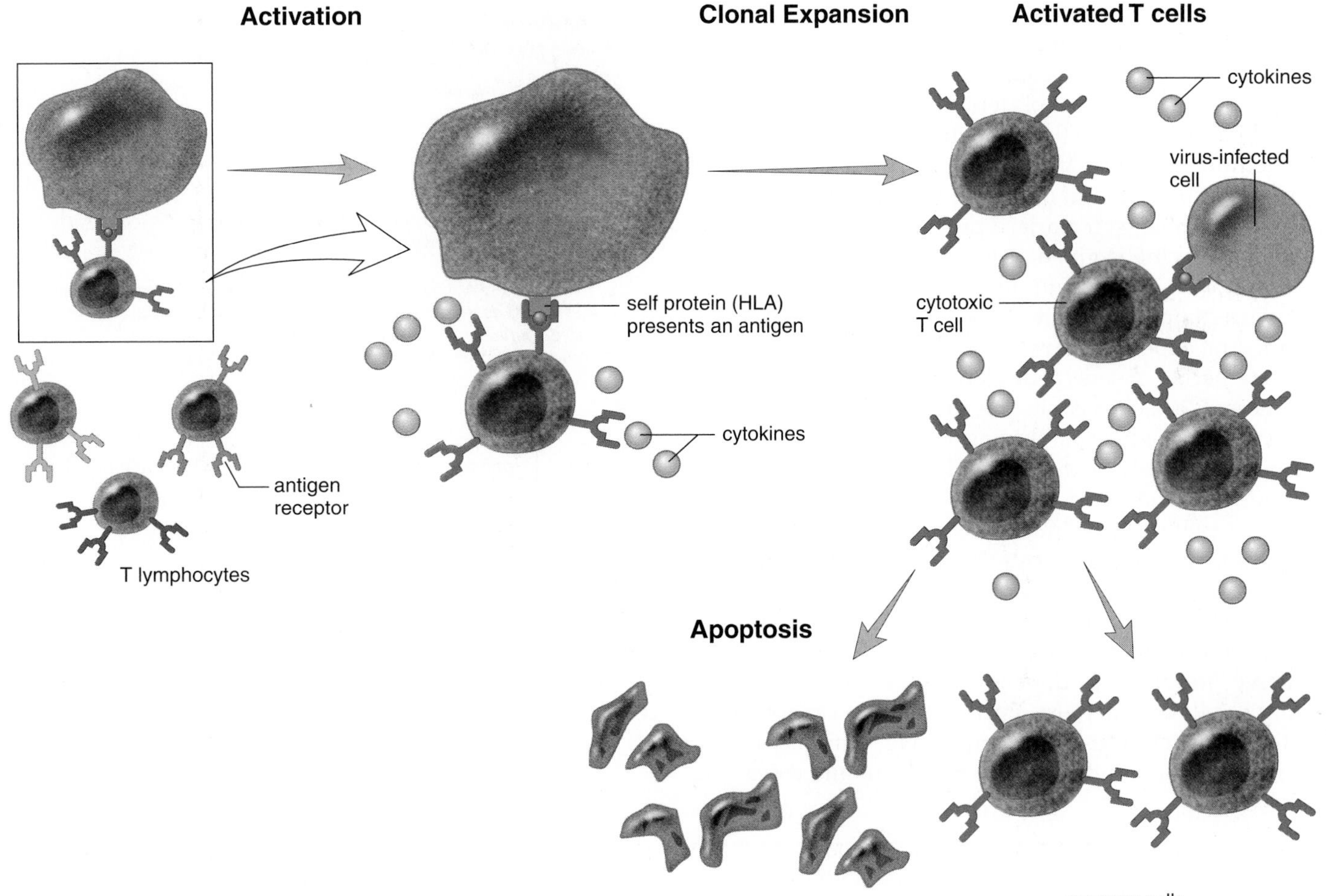

Figure 14.8 Clonal selection theory as it applies to T cells.
Each type of T cell bears a specific antigen receptor. When this receptor binds to an antigen in the groove of an HLA molecule and is stimulated to do so, it undergoes clonal expansion. After the immune response has been successful, the majority of T cells undergo apoptosis while a small number may become memory cells. Memory cells provide protection should the same antigen enter the body again at a future time.

14.4 Induced Immunity

Immunity occurs naturally through infection or is brought about artificially by medical intervention. There are two types of induced immunity: active and passive. In active immunity, the individual alone produces antibodies against an antigen; in passive immunity, the individual is given prepared antibodies.

Active Immunity

Active immunity sometimes develops naturally after a person is infected with a pathogen. However, active immunity is often induced when a person is well so that possible future infection will not take place. To prevent infections, people can be artificially immunized against them. The United States is committed to the goal of immunizing all children against the common types of childhood diseases listed in the immunization schedule given in Figure 14.9*a*.

Immunization involves the use of **vaccines,** substances that contain an antigen to which the immune system responds. Traditionally, vaccines are the pathogens themselves, or their products, that have been treated so they are no longer virulent (able to cause disease). Today, it is possible to genetically engineer bacteria to mass-produce a protein from pathogens, and this protein can be used as vaccine. This method now has been used to produce a vaccine against hepatitis B, a viral disease, and is being used to prepare a vaccine against malaria, a protozoan disease.

After a vaccine is given, it is possible to follow an immune response by determining the amount of antibody present in a sample of serum—this is called the *antibody titer.* After the first exposure to a vaccine, a primary response occurs. For a period of several days, no antibodies are present; then, there is a slow rise in the titer, followed by first a plateau and then a gradual decline as the antibodies bind to the antigen or simply break down (Fig. 14.9*b*). After a second exposure, a secondary response is expected. The titer rises rapidly to a plateau level much greater than before. The second exposure is called a "booster" because it boosts the antibody titer to a high level. The high antibody titer now is expected to help prevent disease symptoms even if the individual is exposed to the disease-causing antigen.

Active immunity is dependent upon the presence of memory B cells and memory T cells which are capable of responding to lower doses of antigen. Active immunity is usually long-lived, although a booster may be required every so many years.

Active (long-lived) immunity can be induced by the use of vaccines. Active immunity is dependent upon the presence of memory B cells and memory T cells in the body.

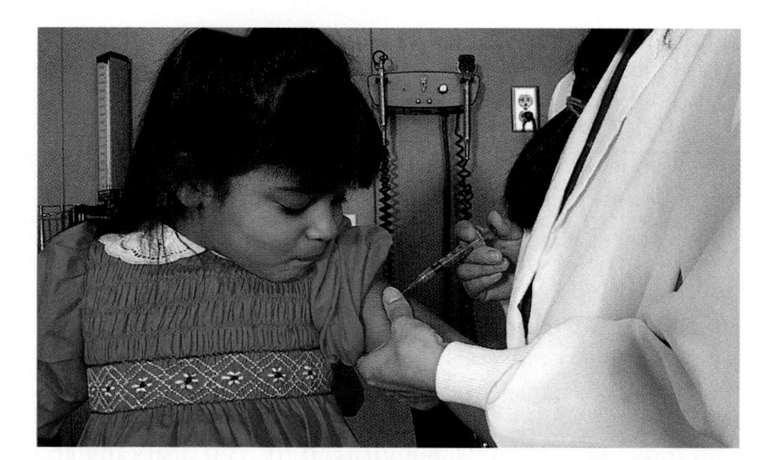

Vaccine	Age (Months)	Age (Years)
HepB* (hepatitis B)	Birth, 2, 4, 6, 12–15	11–12
DTP† (diphtheria, tetanus, whooping cough)	2, 4, 6, 15–18	4–6
Td‡ (adult tetanus)		11–12, 14–16
OPV§ (oral polio vaccine)	2, 4, 6, 12–15	4–6
Hib§ (Haemophilus influenza, type b)	2, 4, 6, 12–15	
MMR¶ (measles, mumps, rubella)	12–15 and 1 month later	4–6, 11–12

** Three doses will be required for kindergarten entry.*
† Five doses recommended for school entry.
‡ First Td needed 10 years after last DTP.
§ Doses 3 and 4 should be given according to manufacturer's guidelines.
¶ A second dose given at least 1 month after first dose required for kindergarten entry
Source: Iowa Department of Public Health, July 1998.

a.

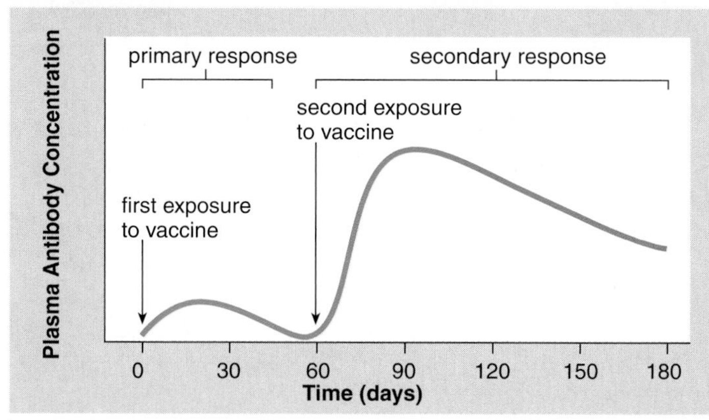

b.

Figure 14.9 Active immunity due to immunizations.
a. Suggested immunization schedule for infants and young children. **b.** During immunization, the primary response, after the first exposure to a vaccine, is minimal, but the secondary response, which may occur after the second exposure, shows a dramatic rise in the amount of antibody present in serum.

Passive Immunity

Passive immunity occurs when an individual is given prepared antibodies (immunoglobulins) to combat a disease. Since these antibodies are not produced by the individual's B cells, passive immunity is short-lived. For example, newborn infants are passively immune to some diseases because antibodies have crossed the placenta from the mother's blood. These antibodies soon disappear, however, so that within a few months, infants become more susceptible to infections. Breast-feeding prolongs the natural passive immunity an infant receives from the mother because antibodies are present in the mother's milk (Fig. 14.10).

Even though passive immunity does not last, it sometimes is used to prevent illness in a patient who has been unexpectedly exposed to an infectious disease. Usually, the patient receives a gamma globulin injection (serum that contains antibodies), perhaps taken from individuals who have recovered from the illness. In the past, horses were immunized, and serum was taken from them to provide the needed antibodies against such diseases as diphtheria, botulism, and tetanus. In the past, a patient who received these antibodies became ill about 50% of the time because the serum contained proteins that the individual's immune system recognized as foreign. This was called serum sickness. But problems can still occur with products produced in other ways. An immunoglobulin intravenous product called Gammagard was withdrawn from the market because of possible implication in the transmission of hepatitis.

Passive immunity provides immediate protection when an individual is in immediate danger of succumbing to an infectious disease. Passive immunity is short-lived because there are no memory cells.

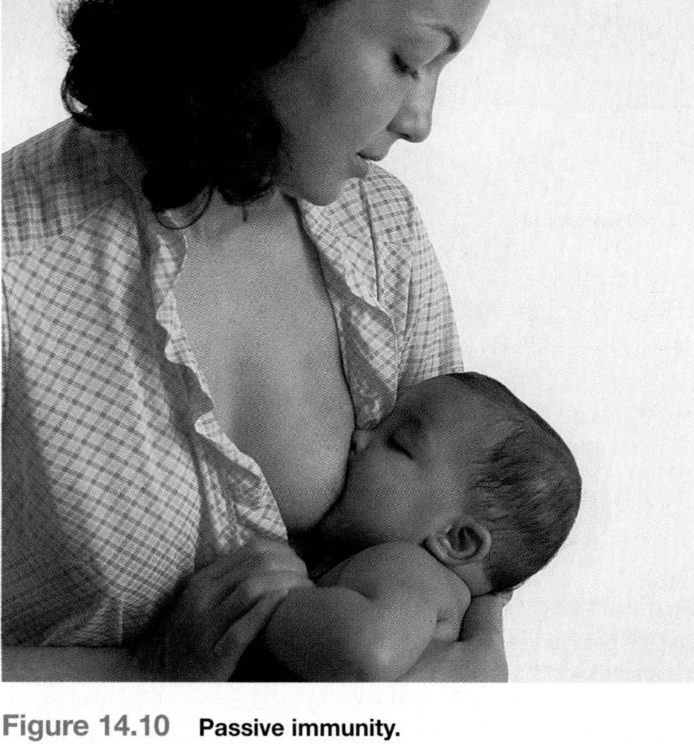

Figure 14.10 Passive immunity.
Breast-feeding is believed to prolong the passive immunity an infant receives from the mother because antibodies are present in the mother's milk.

Cytokines and Immunity

Cytokines are messenger molecules produced by lymphocytes, monocytes, or other cells. Because cytokines regulate white blood cell formation and/or function, they are being investigated as possible adjunct therapy for cancer and AIDS. Both interferon and **interleukins,** which are cytokines produced by various white blood cells, have been used as immunotherapeutic drugs, particularly to enhance the ability of the individual's own T cells (and possibly B cells) to fight cancer.

Interferon, discussed previously on page 268, is a substance produced by leukocytes, fibroblasts, and probably most cells in response to a viral infection. Interferon still is being investigated as a possible cancer drug, but so far it has proven to be effective only in certain patients, and the exact reasons for this as yet cannot be discerned.

When and if cancer cells carry an altered protein on their cell surface, they should be attacked and destroyed by cytotoxic T cells. Whenever cancer does develop, it is possible that the cytotoxic T cells have not been activated. In that case, cytokines might awaken the immune system and lead to the destruction of the cancer. In one technique being investigated, researchers first withdraw T cells from the patient and activate the cells by culturing them in the presence of an interleukin. The cells then are reinjected into the patient, who is given doses of interleukin to maintain the killer activity of the T cells.

Those who are actively engaged in interleukin research believe that interleukins soon will be used as adjuncts for vaccines, for the treatment of chronic infectious diseases, and perhaps for the treatment of cancer. Interleukin antagonists also may prove helpful in preventing skin and organ rejection, autoimmune diseases, and allergies.

The interleukins and other cytokines show some promise of potentiating the individual's own immune system.

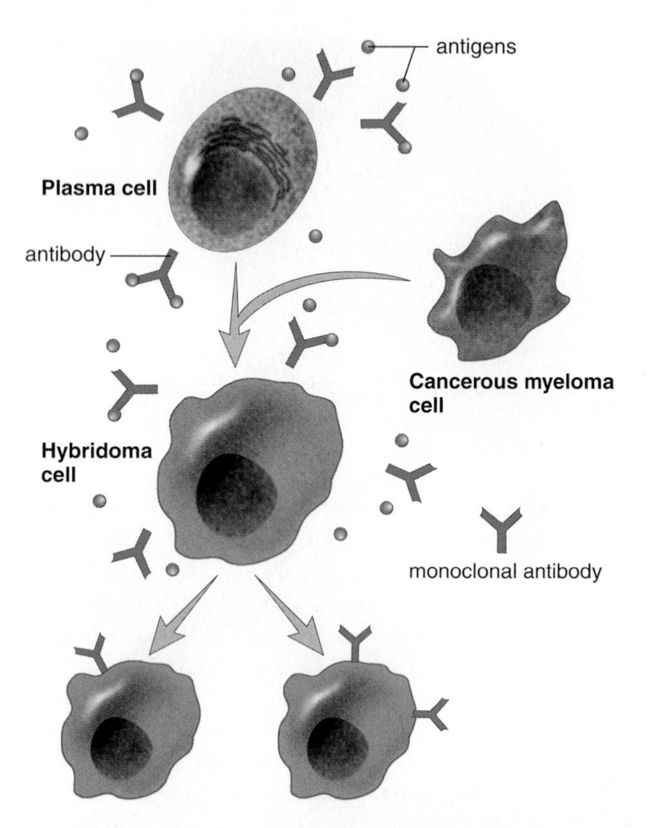

Figure 14.11 Production of monoclonal antibodies. Plasma cells (derived from immunized mice) are fused with myeloma (cancerous) cells, producing hybridoma cells that are "immortal." Hybridoma cells divide and continue to produce the same type of antibody, called monoclonal antibodies.

Monoclonal Antibodies

Every plasma cell derived from the same B cell secretes antibodies against a specific antigen. These are **monoclonal antibodies** because all of them are the same type and because they are produced by plasma cells derived from the same B cell. One method of producing monoclonal antibodies in vitro (outside the body in glassware) is depicted in Figure 14.11. B lymphocytes are removed from an animal (today, usually mice are used) and are exposed to a particular antigen. The activated B lymphocytes are fused with myeloma cells (malignant plasma cells that live and divide indefinitely). The fused cells are called hybridomas; *hybrid* because they result from the fusion of two different cells, and *oma* because one of the cells is a cancer cell.

At present, monoclonal antibodies are being used for quick and certain diagnosis of various conditions. For example, a particular hormone is present in the urine of a pregnant woman. A monoclonal antibody can be used to detect this hormone; if it is present, the woman knows she is pregnant. Monoclonal antibodies also are used to identify infections. And because they can distinguish between cancer and normal tissue cells, they are used to carry radioactive isotopes or toxic drugs to tumors so that they can be selectively destroyed.

14.5 Immunity Side Effects

The immune system usually protects us from disease because it can distinguish self from nonself. Sometimes, however, it responds in a manner that does harm to the body, as when individuals develop allergies, receive the wrong blood type, suffer tissue rejection, or have an autoimmune response.

Allergies

Allergies are hypersensitivities to substances such as pollen or animal hair that ordinarily would do no harm to the body. The response to these antigens, called **allergens,** usually includes some degree of tissue damage. There are four types of allergic responses, but we will consider only two of these: immediate allergic responses and delayed allergic responses.

Immediate Allergic Response

An **immediate allergic response** occurs within seconds of contact with the antigen. As discussed in the reading on page 277, coldlike symptoms are common. Anaphylactic shock is a severe reaction characterized by a sudden and life-threatening drop in blood pressure.

Immediate allergic responses are caused by antibodies known as IgE (see Table 14.1). IgE antibodies are attached to the plasma membrane of mast cells in the tissues and basophils in the blood. When an allergen attaches to the IgE antibodies on these cells, they release histamine and other substances that bring about the coldlike symptoms or, rarely, anaphylactic shock.

Allergy shots sometimes prevent the onset of an allergic response. It's been suggested that injections of the allergen may cause the body to build up high quantities of IgG antibodies, and these combine with allergens received from the environment before they have a chance to reach the IgE antibodies located in the membrane of mast cells and basophils.

Delayed Allergic Response

Delayed allergic responses are initiated by sensitized T cells at the site of allergen in the body. A sensitized T cell is one that is ready to respond to the antigen because it has been present in the body before. T cells initiate the response by recruiting the help of macrophages, which are able to phagocytize offending viral particles or infectious cells. The overall response is regulated by the cytokines secreted by both the T cells and macrophages.

A classic example of a delayed allergic response is the tuberculin skin test. When the result of the test is positive, there is a reddening and hardening of tissue where the antigen was injected. This shows that there was prior exposure to tubercle bacilli which cause TB. Contact dermatitis, such as occurs when one is allergic to poison ivy, jewelry, cosmetics, and so forth, is also an example of a delayed allergic response.

Health Focus

Immediate Allergic Responses

The runny nose and watery eyes of hay fever are often caused by an allergic reaction to the pollen of trees, grasses, and ragweed. Worse, the airways leading to the lungs constrict if one has asthma, resulting in difficult breathing characterized by wheezing. Windblown pollen, particularly in the spring and fall, brings on the symptoms of hay fever. Most people can inhale pollen with no ill effects. But others have developed a hypersensitivity, meaning that their immune system responds in a deleterious manner. The problem stems from a type of antibody called immunoglobulin E (IgE) that causes the release of histamine from mast cells and basophils whenever they are exposed to an allergen. Histamine is a chemical that causes mucosal membranes of the nose and eyes to release fluid as a defense against pathogen invasion. But in the case of allergy, copious fluid is released although no real danger is present.

Most food allergies are also due to the presence of IgE antibodies that bind usually to a protein in the food. The symptoms, such as nausea, vomiting, and diarrhea, are due to the mode of entry of the allergen. Skin symptoms may also occur, however. Adults are often allergic to shellfish, nuts, eggs, cows' milk, fish, and soybeans. Peanut allergy is a common food allergy in the United States possibly because peanut better is a staple in the United States. People seem to outgrow allergies to cows' milk and eggs more often than allergies to peanuts and soybeans.

Celiac disease occurs in people who are allergic to wheat, rye, barley, and sometimes oats—in short, any grain that contains gluten proteins. It is thought that the gluten proteins elicit a delayed cell-mediated immune response by T cells with the resultant production of cytokines. The symptoms of celiac disease can include diarrhea, bloating, weight loss, anemia, bone pain, chronic fatigue, and weakness.

People can reduce the chances of a reaction to airborne and food allergens by avoiding the offending substances.The reaction to peanuts can be so severe that airlines are now required to have a peanut-free zone for those allergic. The people in Figure 14B are trying to avoid windblown allergens. The taking of antihistamines can also be helpful. If these procedures are inadequate, patients can be tested to measure their susceptibility to any number of possible allergens. A small quantity of a suspected allergen is inserted just beneath the skin, and the strength of the subsequent reaction is noted. A wheal-and-flare response at the skin prick site demonstrates that IgE antibodies attached to mast cells have reacted to an allergen. In an immunotherapy called hyposensitization, ever-increasing doses of the allergen are periodically injected subcutaneously with the hope that the body will build up a supply of IgG. IgG, in contrast to IgE, does not cause the release of histamine after it combines with the allergen. If IgG combines first upon exposure to the allergen, the allergic response does not occur. Patients know they are cured when the allergic symptoms no longer occur. Therapy may have to continue for as long as two to three years.

Allergic-type reactions can occur without involving the immune system. Wasp and bee stings contain substances that cause swellings, even in those whose immune system is not sensitized to substances in the sting. Also, jellyfish tentacles and foods such as fish that is not fresh and strawberries contain histamine or closely related substances that can cause a reaction. Immunotherapy is also not possible in those who are allergic to penicillin and bee stings. High sensitivity has built up upon the first exposure, and when reexposed, anaphylactic shock can occur. Among its many effects, histamine causes increased permeability of the smallest blood vessels, called capillaries. In these individuals, there is a drastic decrease in blood pressure that can be fatal within a few minutes. People who know they are allergic to bee stings can obtain a syringe of epinephrine to carry with them. This medication can delay the onset of anaphylactic shock until medical help is available.

Figure 14B Protection against allergies.
The allergic reaction known as hay fever, and asthma attacks, can have many triggers, one of which is the pollen of a variety of plants. A dramatic solution to the problem has been found by these people.

Table 14.2 The ABO System

Blood Type	Antigen on Red Blood Cells	Antibody in Plasma	% U.S. African American	% U.S. Caucasian	% U.S. Asian	% North American Indians	% Americans of Chinese Descent
A	A	Anti-B	27	41	28	8	25
B	B	Anti-A	20	9	27	1	35
AB	A,B	None	4	3	5	0	10
O	None	Anti-A and anti-B	49	47	40	91	30

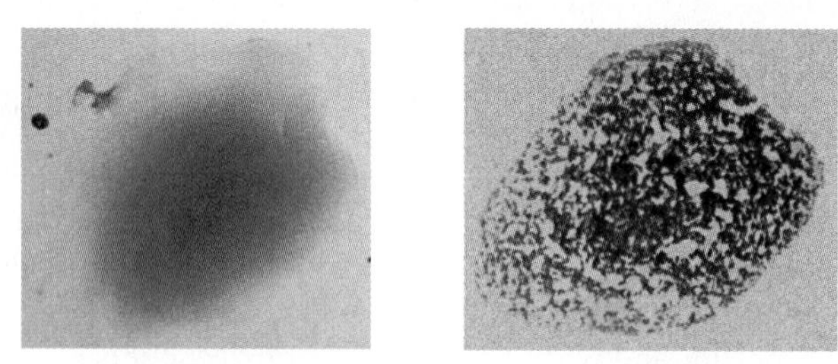

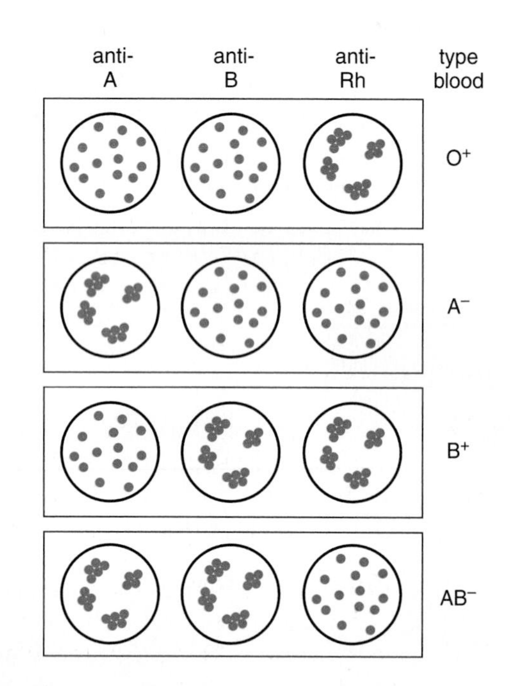

Figure 14.12 Blood typing.
The standard test to determine ABO and Rh blood type consists of putting a drop of anti-A antibodies, anti-B antibodies, and anti-Rh antibodies on a slide. To each of these, a drop of the person's blood is added. **a.** If agglutination occurs, as seen in the photo on the right, the person has this antigen on red blood cells. **b.** Several possible results.

Blood Types

When blood transfusions were first attempted, illness and even death sometimes resulted. Eventually, it was discovered that only certain types of blood are compatible because red blood cell membranes carry proteins or sugar residues that are antigens to blood recipients. The ABO system of typing blood is based on this principle.

ABO System

Blood typing in the ABO system is based on two antigens known as antigen A and antigen B. There are four blood types: O, A, B, and AB. Type O has neither the A antigen nor the B antigen on red blood cells; the other types of blood have antigen A, B, or both A and B, respectively (Table 14.2).

Within plasma, there are naturally occurring antibodies to the antigens not present on the person's red blood cells. This is reasonable, because if the same antigen and antibody are present in blood, **agglutination,** or clumping of red blood cells, occurs. Agglutination causes blood to stop circulating and red blood cells to burst.

Figure 14.12 shows a way to use the antibodies derived from plasma to determine the blood type. If agglutination occurs after a sample of blood is mixed with a particular antibody, the person has that type of blood.

Rh System

Another important antigen in matching blood types is the Rh factor. Persons with the Rh factor on their red blood cells are Rh positive (Rh^+); those without it are Rh negative (Rh^-). Rh-negative individuals normally do not have antibodies to the Rh factor, but they may make them when exposed to the Rh factor during pregnancy or blood transfusion.

If a mother is Rh negative and a father is Rh positive, a child may be Rh positive (Fig. 14.13). The Rh-positive red blood cells of the child may begin leaking across the placenta into the mother's circulatory system, as placental tissues normally break down before and at birth. This sometimes causes the mother to produce anti-Rh antibodies. In this or a subsequent pregnancy with another Rh-positive child, anti-Rh antibodies may cross the placenta and destroy the child's red blood cells. This condition is called hemolytic disease of the newborn (HDN).

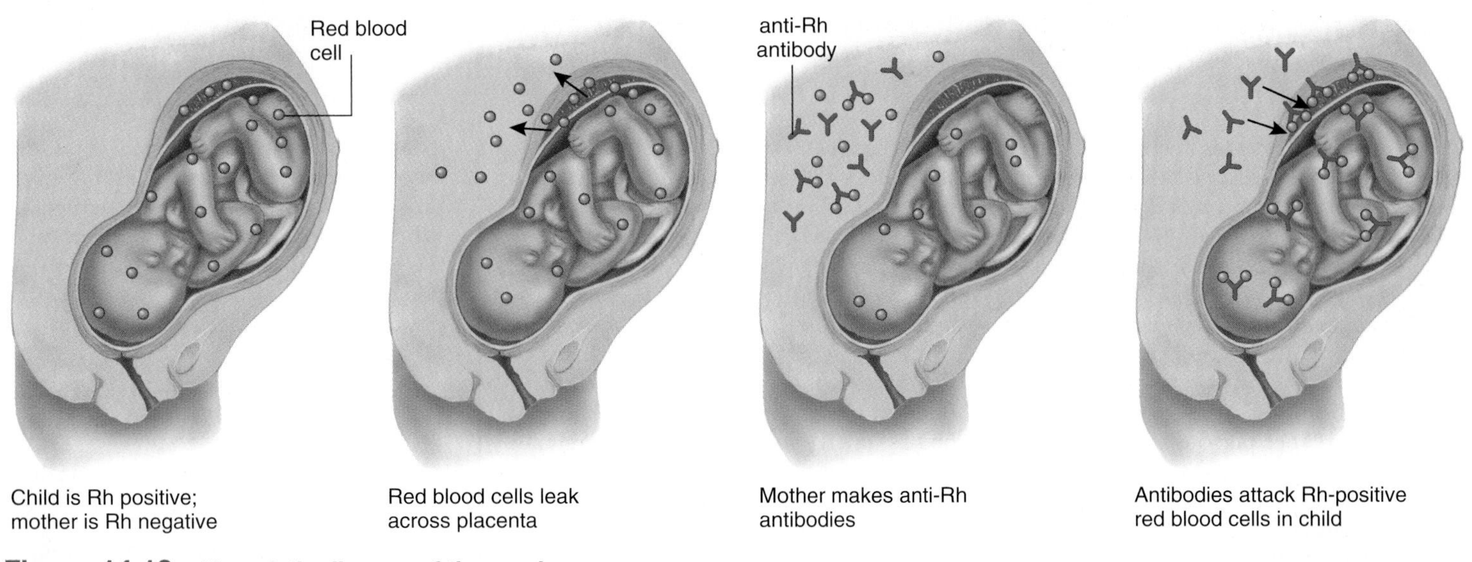

Figure 14.13 Hemolytic disease of the newborn.
Due to a pregnancy in which the child is Rh positive, an Rh-negative mother can begin to produce antibodies against Rh-positive red blood cells. In another pregnancy, these antibodies can cross the placenta and cause hemolysis (bursting of red blood cells) in an Rh-positive child's red blood cells.

The Rh problem has been solved by giving Rh-negative women an Rh-immunoglobulin injection (often a Rho-Gam injection) either midway through the first pregnancy or no later than 72 hours after giving birth to any Rh-positive child. This injection contains anti-Rh antibodies, which attack any of the child's red blood cells in the mother's blood before these cells can stimulate her immune system to produce her own antibodies. This injection is not beneficial if the woman has already begun to produce antibodies; therefore, the timing of the injection is most important.

Blood is often typed according to the ABO system combined with the Rh system. The possibility of hemolytic disease of the newborn exists when the mother is Rh negative and the father is Rh positive.

Autoimmune Diseases

When T cells or antibodies mistakenly attack the body's own cells as if they bore foreign antigens, the resulting condition is known as an **autoimmune disease.** Exactly what causes autoimmune diseases is not known. However, sometimes they occur after an individual has recovered from an infection.

In the autoimmune disease myasthenia gravis, neuromuscular junctions do not work properly and muscular weakness results. In multiple sclerosis, the myelin sheath of nerve fibers breaks down, and this causes various neuromuscular disorders. A person with systemic lupus erythematosus has various symptoms prior to death due to kidney damage. In rheumatoid arthritis, the joints are affected. Researchers suggest that heart damage following rheumatic fever and type I diabetes are also autoimmune illnesses. As yet there are no cures for autoimmune diseases, but they can be controlled with drugs.

Autoimmune diseases occur when antibodies and cytotoxic T cells recognize and destroy the body's own cells.

Tissue Rejection

Certain organs, such as skin, the heart, and the kidneys, could be transplanted easily from one person to another if the body did not attempt to *reject* them. Rejection occurs because cytotoxic T cells bring about destruction of foreign tissue in the body.

Organ rejection can be controlled by careful selection of the organ to be transplanted and the administration of immunosuppressive drugs. It is best if the transplanted organ has the same type of HLA antigens as those of the recipient, because cytotoxic T cells recognize foreign HLA antigens. The immunosuppressive drug cyclosporine has been used for many years. A new drug, tacrolimus (formerly known as FK-506), shows some promise, especially in liver transplant patients. However both drugs, which act by inhibiting the response of T cells to cytokines, are known to adversely affect the kidneys.

When an organ is rejected, the immune system has recognized and destroyed cells that bear HLA antigens different from those of the individual.

Bioethical Issue

The United Nations estimates that 16,000 people become newly infected with the human immunodeficiency virus (HIV) each day, or 5.8 million per year. Ninety percent of these infections occur in the less-developed countries[1] where infected persons do not have access to antiviral therapy. In Uganda, for example, there is only one physician per 100,000 people, and only $6.00 is spent annually on health care, per person. In contrast, in the United States $12,000–$15,000 is sometimes spent on treating an HIV infected person per year.

The only methodology to prevent the spread of HIV in a developing country is counseling against behaviors that increase the risk of infection. Clearly an effective vaccine would be most beneficial to these countries. Several HIV vaccines are in various stages of development, and all need to be clinically tested in order to see if they are effective. It seems reasonable to carry out such trials in developing countries, but there are many ethical questions.

A possible way to carry out the trial is this: vaccinate the uninfected sexual partners of HIV-infected individuals. After all, if the uninfected partner remains free of the disease, then the vaccine is effective. But is it ethical to allow a partner identified as having an HIV infection to remain untreated for the sake of the trial?

And should there be a placebo group—a group that does not get the vaccine? After all, if a greater number of persons in the placebo group become infected than those in the vaccine group, then the vaccine is effective. But if members of the placebo group become infected, shouldn't they be given effective treatment? For that matter, even participants in the vaccine group might become infected. Shouldn't any participant of the trial be given proper treatment if they become infected? Who would pay for such treatment when the trial could involve thousands of persons?

Questions

1. Should HIV vaccine trials be done in developing countries, which stand to gain the most from an effective vaccine? Why or why not?
2. Should the trial be carried out using the same standards as in developed countries? Why or why not?
3. Who should pay for the trial—the drug company, the participants, or the country of the participants?

[1]Country that has only low to moderate industrialization; usually located in the southern hemisphere.

Summary

14.1 Lymphatic System

The lymphatic system consists of lymphatic vessels and lymphoid organs. The lymphatic vessels collect fat molecules at intestinal villi and excess tissue fluid at blood capillaries, and carry these to the bloodstream.

Lymphocytes are produced and accumulate in the lymphoid organs (red bone marrow, lymph nodes, spleen, and thymus gland). Lymph is cleansed of pathogens and/or their toxins in lymph nodes, and blood is cleansed of pathogens and/or their toxins in the spleen. T lymphocytes mature in the thymus, while B lymphocytes mature in the red bone marrow where all blood cells are produced. White blood cells are necessary for nonspecific and specific defenses.

14.2 Nonspecific Defenses

Immunity involves nonspecific and specific defenses. Nonspecific defenses include barriers to entry, the inflammatory reaction, natural killer cells, and protective proteins.

14.3 Specific Defenses

Specific defenses require lymphocytes, which are produced in the bone marrow. B cells mature in the bone marrow. They undergo clonal selection with production of plasma cells and memory B cells after their specific plasma membrane receptors directly combine with a particular antigen. Plasma cells secrete antibodies and eventually undergo apoptosis. B cells are responsible for antibody-mediated immunity. IgG antibody is a Y-shaped molecule that has two binding sites for a specific antigen. Memory B cells remain in the body and produce antibodies if the same antigen enters the body at a later date.

T cells, which are responsible for cell-mediated immunity, mature in the thymus. The two main types of T cells are cytotoxic T cells and helper T cells. Cytotoxic T cells kill infected cells that bear a foreign antigen on contact; helper T cells stimulate other immune cells and produce cytokines. Like B cells, each T cell bears a specific receptor. However, for a T cell to recognize an antigen, the antigen must be presented by an antigen-presenting cell (APC), usually a macrophage, along with an HLA (human lymphocyte-associated) antigen. Thereafter the activated T cell undergoes clonal expansion until the infection has been stemmed. Then most of the activated T cells undergo apoptosis. A few cells remain, however, as memory T cells.

14.4 Induced Immunity

Immunity can be induced in various ways. Vaccines are available to induce long-lived active immunity, and antibodies sometimes are available to provide an individual with short-lived passive immunity.

Cytokines, including interferon, are used in an attempt to promote the body's ability to recover from cancer and to treat AIDS.

14.5 Immunity Side Effects

Allergic responses occur when the immune system reacts vigorously to substances not normally recognized as foreign. Immediate allergic responses, usually consisting of coldlike symptoms, are due to the activity of antibodies. Delayed allergic responses, such as contact dermatitis, are due to the activity of T cells.

Studying the Concepts

1. What is the lymphatic system, and what are its three functions? 264
2. Describe the structure and the function of lymph nodes, the spleen, the thymus, and red bone marrow. 265–66
3. What are the body's nonspecific defense mechanisms? 266–68
4. Describe the inflammatory reaction, and give a role for each type of cell and molecule that participates in the reaction. 266
5. What is the clonal selection theory? B cells are responsible for which type of immunity? 269
6. Describe the structure of an antibody, and define the terms variable regions and constant regions. 270
7. Name the two main types of T cells, and state their functions. 272
8. Explain the process by which a T cell is able to recognize an antigen. 272–73
9. How is active immunity achieved? How is passive immunity achieved? 274–75
10. What are cytokines, and how are they used in immunotherapy? 275
11. How are monoclonal antibodies produced, and what are their applications? 276
12. Discuss allergies, tissue rejection, and autoimmune diseases as they relate to the immune system. 276–77

Testing Yourself

Choose the best answer for each question.

1. Use these terms to label this IgG molecule: antigen-binding sites, light chain, heavy chain. d. What does V stand for in the diagram? e. What does C stand for in the diagram? f. What shape antigen would bind to this particular antigen-binding site?

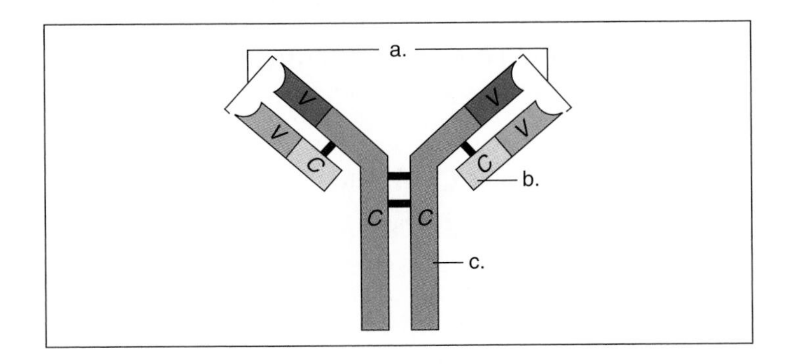

2. Complement
   a. is a general defense mechanism.
   b. is involved in the inflammatory reaction.
   c. is a series of proteins present in the plasma.
   d. plays a role in destroying bacteria.
   e. All of these are correct.
3. Which of these pertain(s) to T cells?
   a. have specific receptors
   b. are more than one type
   c. are responsible for cell-mediated immunity
   d. stimulate antibody production by B cells
   e. All of these are correct.
4. Which one of these does not pertain to B cells?
   a. have passed through the thymus
   b. have specific receptors
   c. are responsible for antibody-mediated immunity
   d. synthesize and liberate antibodies
5. The clonal selection theory says that
   a. an antigen selects certain B cells and suppresses them.
   b. an antigen stimulates the multiplication of B cells that produce antibodies against it.
   c. T cells select those B cells that should produce antibodies, regardless of antigens present.
   d. T cells suppress all B cells except the ones that should multiply and divide.
   e. Both b and c are correct.
6. Plasma cells are
   a. the same as memory cells.
   b. formed from blood plasma.
   c. B cells that are actively secreting antibody.
   d. inactive T cells carried in the plasma.
   e. a type of red blood cell.
7. For a T cell to recognize an antigen, it usually interacts with
   a. complement.
   b. a macrophage.
   c. a B cell.
   d. a thymus cell.
   e. All of these are correct.
8. Antibodies combine with antigens
   a. at variable regions.
   b. at constant regions.
   c. only if macrophages are present.
   d. only if T cells are present.
   e. Both a and c are correct.
9. Which one of these is mismatched?
   a. helper T cells—help complement react
   b. cytotoxic T cells—active in tissue rejection
   c. macrophage—activate T cells
   d. memory T cells—long-living line of T cells
   e. T cells—mature in thymus
10. Vaccines are
    a. the same as monoclonal antibodies.
    b. treated bacteria or viruses, or one of their proteins.
    c. short-lived.
    d. MHC proteins.
    d. All of these are correct.
11. The theory behind the use of cytokines in cancer therapy is that
    a. if cancer develops, the immune system has been ineffective.
    b. cytokines stimulate the immune system.
    c. cancer cells bear antigens that should be recognizable by cytotoxic T cells.
    d. cytokines can be isolated from the blood.
    e. All of these are correct.
12. During blood typing, agglutination indicates that the
    a. plasma contains certain antibodies.
    b. red blood cells carry certain antigens.
    c. plasma contains certain antigens.
    d. red blood cells carry certain antibodies.
    e. white blood cells fight infection.

Thinking Scientifically

1. Considering the action of B cells and T cells (page 268):
   a. A mouse is irradiated so that its bone marrow and thymus are destroyed. It then is resupplied only with bone marrow. The mouse is unable to form antibodies. Why?
   b. A mixture of B cells is exposed to a specific radiolabeled antigen in vitro (within laboratory glassware). Would you expect all B cells to bind with the antigen and to be radiolabeled?
   c. When B cells and T cells are incubated in vitro with a radiolabeled antigen, binding to certain B cells occurs but not to T cells. Why?
   d. Human beings communicate by sight, sound, and touch. How do immune cells communicate with one another?
2. In this text, antigen originally was defined as a foreign substance in the body.
   a. Expand this definition by telling what an antigen does in the body.
   b. It is possible to tag different types of monoclonal antibodies with different dyes so you can tell them apart. Knowing this, how would you produce and use monoclonal antibodies to distinguish helper T cells and cytotoxic T cells in a blood sample?
   c. How would you prove that a monoclonal antibody is specific to the herpes virus (HSV-2) that causes genital herpes, but not to the one (HSV-1) that causes cold sores?
   d. In a manufacturing process called affinity purification, a mixture that contains a desired substance is passed through a tube. In the tube, a large number of antibody molecules are fixed to a solid support. Why will this process result in purification of the product?

Understanding the Terms

agglutination 278
allergen 276
allergy 276
antibody 268
antibody-mediated immunity 269
antigen 268
antigen-presenting cell (APC) 272
apoptosis 269
autoimmune disease 279
B lymphocyte 268
basophil 266
cell-mediated immunity 272
clonal selection theory 269
complement system 268
cytokine 275
cytotoxic T cell 272
delayed allergic response 276
edema 264
helper T cell 272
histamine 266
HLA (human leukocyte associated) antigen 272
immediate allergic response 276
immune system 268
immunity 266
immunization 274
immunoglobulin (Ig) 270
inflammatory reaction 266
interferon 268
interleukin 275
kinin 266
lymph 264
lymphatic system 264
lymph node 265
macrophage 266
mast cell 266
monoclonal antibody 276
natural killer (NK) cell 268
pathogen 266
perforin 272
plasma cell 269
red bone marrow 266
spleen 265
T lymphocyte 268
thymus gland 266
tonsils 265
vaccine 274

Match the terms to these definitions:

a. __________ Antigens prepared in such a way that they can promote active immunity without causing disease.
b. __________ Fluid, derived from tissue fluid, that is carried in lymphatic vessels.
c. __________ Foreign substance, usually a protein or a polysaccharide, that stimulates the immune system to react, such as to produce antibodies.
d. __________ Process of programmed cell death involving a cascade of specific cellular events leading to the death and destruction of the cell.
e. __________ Lymphocyte that matures in the thymus and exists in three varieties, one of which kills antigen-bearing cells outright.

Using Technology

Your study of the lymphatic system and immunity is supported by these available technologies:

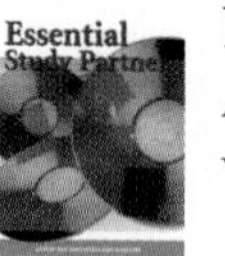

Essential Study Partner CD-ROM

Animals → Lymph and Immunity

Visit the Mader web site for related ESP activities.

Exploring the Internet

The Mader Home Page provides resources and tools as you study this chapter.

http://www.mhhe.com/biosci/genbio/mader

Dynamic Human 2.0 CD-ROM

Lymphatic System

HealthQuest CD-ROM

4 Communicable Diseases
6 Cancer

Life Science Animations 3D Video

33 Complement System
34 How T Lymphocytes Work
35 Clonal Selection

C H A P T E R 15

Respiratory System

Chapter Concepts

The human lung lying within the protective rib cage receives the pulmonary artery and its many branches.

Luckily the emergency medical technicians arrived just moments after Sammy was rescued from the pond by his mother. A technician immediately began CPR (cardiopulmonary resuscitation), alternately blowing into Sammy's mouth and then pressing on his chest until Sammy began to breathe on his own. All cells of the body require a constant supply of oxygen, and you have to keep breathing in order to bring this oxygen into the body. Any cessation of breathing is a cause for concern, and prolonged cessation usually results in death. The heart needs oxygen to pump the blood that will carry oxygen to all the cells of the body. The cells use oxygen in the process of replenishing their limited supply of ATP, without which they have no energy and cannot keep functioning. In this chapter, the structures and functions of the respiratory system are considered. Also, some of the conditions that decrease the functioning of the system will be discussed.

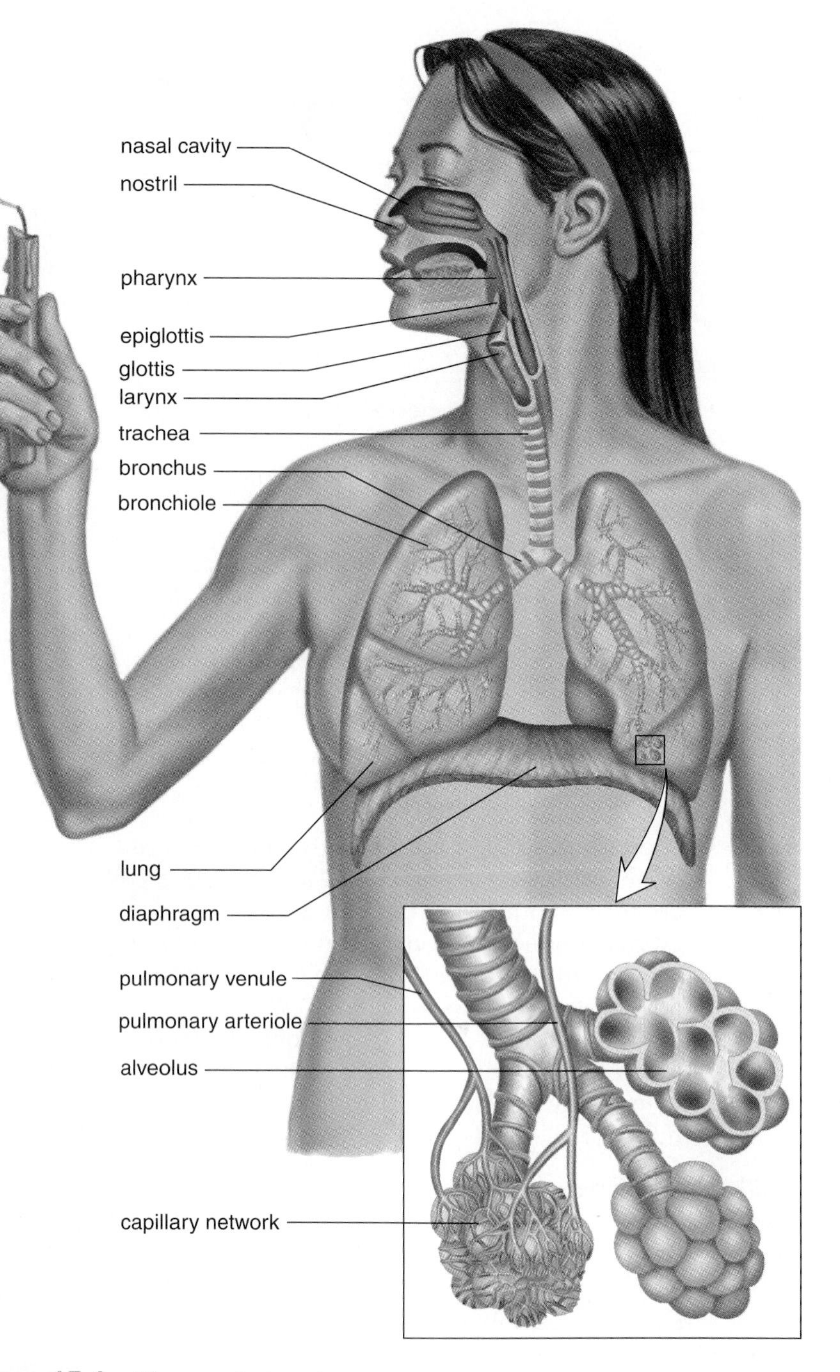

Figure 15.1 **The respiratory tract.**
The respiratory tract extends from the nose to the lungs, which are composed of air sacs called alveoli. Gas exchange occurs between air in the alveoli and blood within a capillary network that surrounds the alveoli.

15.1 Respiratory Tract

During **inspiration** or inhalation (breathing in) and **expiration** or exhalation (breathing out), air is conducted toward or away from the lungs by a series of cavities, tubes, and openings, illustrated in Figure 15.1.

As air moves in along the airways, it is filtered, warmed, and moistened. Filtering is accomplished by coarse hairs, cilia, and mucus in the region of the nostrils and by cilia alone in the rest of the nasal cavity and the airways of the lower respiratory tract. In the nose, the hairs and the cilia act as a screening device. In the trachea and other airways, the cilia beat upward, carrying mucus, dust, and occasional bits of food that "went down the wrong way" into the pharynx, where the accumulation can be swallowed or expectorated. The air is warmed by heat given off by the blood vessels lying close to the surface of the lining of the airways, and it is moistened by the wet surface of these passages.

Conversely, as air moves out during expiration, it cools and loses its moisture. As the air cools, it deposits its moisture on the lining of the windpipe and the nose, and the nose may even drip as a result of this condensation. The air still retains so much moisture, however, that upon expiration on a cold day, it condenses and forms a small cloud.

Air is filtered, warmed, and moistened as it moves from the nose toward the lungs.

Table 15.1 Path of Air

Structure	Description	Function
Nasal cavities	Hollow spaces in nose	Filter, warm, and moisten air
Pharynx	Chamber behind oral cavity and between nasal cavity and larynx	Connection to surrounding regions
Glottis	Opening into larynx	Passage of air into larynx
Larynx	Cartilaginous organ that contains vocal cords (voice box)	Sound production
Trachea	Flexible tube that connects larynx with bronchi (windpipe)	Passage of air to bronchi
Bronchi	Divisions of the trachea that enter lungs	Passage of air to lungs
Bronchioles	Branched tubes that lead from bronchi to the alveoli	Passage of air to each alveolus
Lungs	Soft, cone-shaped organs that occupy a large portion of the thoracic cavity	Gas exchange organs

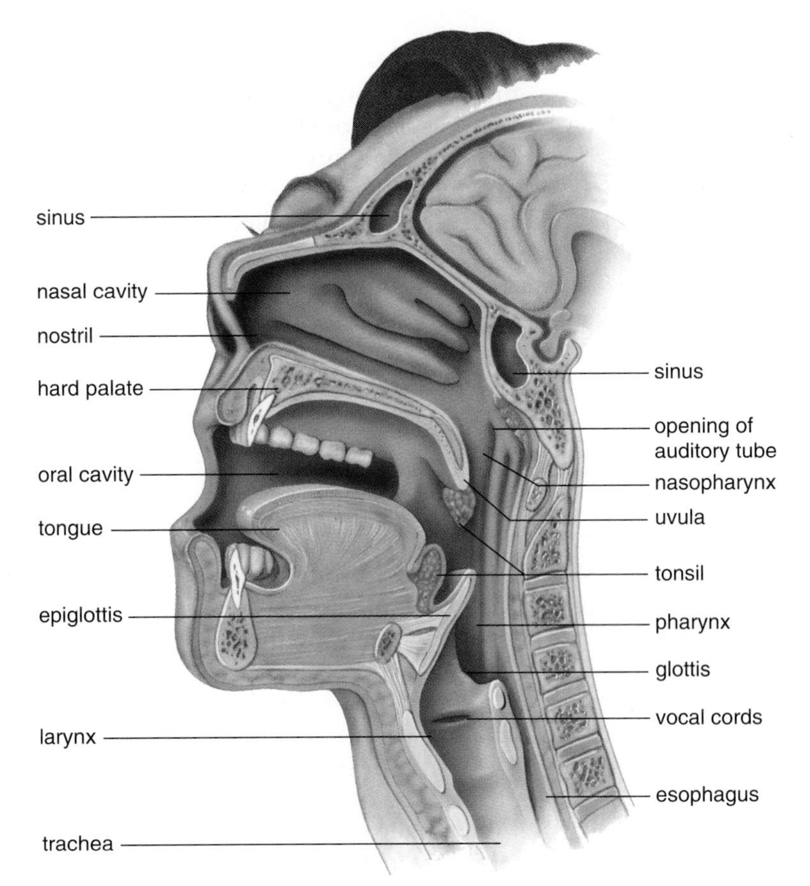

Figure 15.2 The upper respiratory tract.
The upper respiratory tract contains the nasal cavities, pharynx, and larynx.

The Nose

The nose contains two **nasal cavities** (Table 15.1), which are narrow canals separated from one another by a septum composed of bone and cartilage. Special ciliated cells in the narrow upper recesses of the nasal cavities (Fig. 15.2) act as odor receptors. Nerves lead from these cells to the brain, where the impulses generated by the odor receptors are interpreted as smell.

The tear (lacrimal) glands drain into the nasal cavities by way of tear ducts. For this reason, crying produces a runny nose. The nasal cavities also communicate with the cranial sinuses, air-filled mucosa-lined spaces in the skull. If inflammation due to a cold or an allergic reaction blocks the ducts leading from the sinuses, mucus may accumulate, causing a sinus headache.

The nasal cavities empty into the nasopharynx, the upper portion of the pharynx. The auditory tubes lead from the nasopharynx to the middle ears.

The nasal cavities, which receive air, open into the nasopharynx.

The Pharynx

The **pharynx** is a funnel-shaped passageway that connects the nasal and oral cavities to the larynx. Therefore, the pharynx, which is commonly referred to as the "throat," has three parts: the nasopharynx, where the nasal cavities open above the soft palate; the oropharynx, where the oral cavity opens; and the laryngopharynx, which opens into the larynx. In the pharynx, the air passage and the food passage cross because the larynx, which receives air, is ventral to the esophagus, which receives food. The larynx lies at the top of the trachea. The larynx and trachea are normally open, allowing the passage of air, but the esophagus is normally closed and opens only when swallowing occurs.

Air from either the nose or the mouth enters the pharynx, as does food. The passage of air continues in the larynx and then the trachea.

The Larynx

The **larynx** can be pictured as a triangular box whose apex, the Adam's apple, is located at the front of the neck. The Adam's apple is more prominent in men than women. At the top of the larynx is a variable-sized opening called the **glottis.** When food is swallowed, the larynx moves upward against the **epiglottis,** a flap of tissue that prevents food from passing into the larynx. You can detect this movement by placing your hand gently on your larynx and swallowing.

The larynx is called the voice box because the vocal cords are inside the larynx. The **vocal cords** are mucosal folds supported by elastic ligaments, which are stretched across the glottis (Fig. 15.3). When air passes through the glottis, the vocal cords vibrate, producing sound. At the time of puberty, the growth of the larynx and the vocal cords is much more rapid and accentuated in the male than in the female, causing the male to have a more prominent Adam's apple and a deeper voice. The voice "breaks" in the young male due to his inability to control the longer vocal cords. These changes cause the lower pitch of the voice in males.

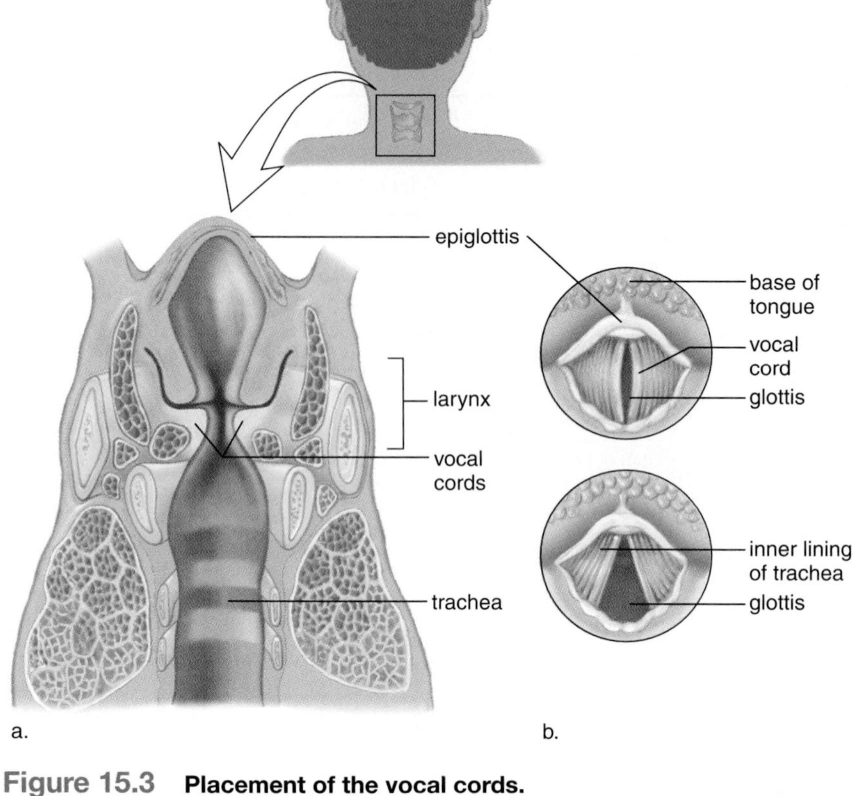

Figure 15.3 Placement of the vocal cords.
a. Frontal section of the larynx shows the location of the vocal cords inside the larynx. The vocal cords viewed from above are stretched across the glottis. When air passes through the glottis, the vocal cords vibrate, producing sound. **b.** The glottis is narrow when we produce a high-pitched sound (*top*) and widens as the pitch deepens (*bottom*).

The high or low pitch of the voice is regulated when speaking and singing by changing the tension on the vocal cords. The greater the tension, as when the glottis becomes more narrow, the higher the pitch. When the glottis is wider, the pitch is lower (Fig. 15.3*b*). The loudness, or intensity, of the voice depends upon the amplitude of the vibrations, that is, the degree to which vocal cords vibrate.

The Trachea

The **trachea,** commonly called the windpipe, is a tube connecting the larynx to the primary bronchi. The trachea lies ventral to the esophagus and is held open by C-shaped cartilaginous rings. The open part of the C-shaped rings faces the esophagus and this allows the esophagus to expand when swallowing. If the trachea is blocked because of illness or the accidental swallowing of a foreign object, it is possible to insert a tube by way of an incision made in the trachea. This tube acts as an artificial air intake and exhaust duct. The operation is called a **tracheostomy.**

The mucosa that lines the trachea has a layer of pseudostratified ciliated columnar epithelium. (Pseudostratified means that while the epithelium appears to be layered, actually each cell touches the basement membrane.) The cilia that project from the epithelium keep the lungs clean by sweeping mucus and debris toward the pharynx.

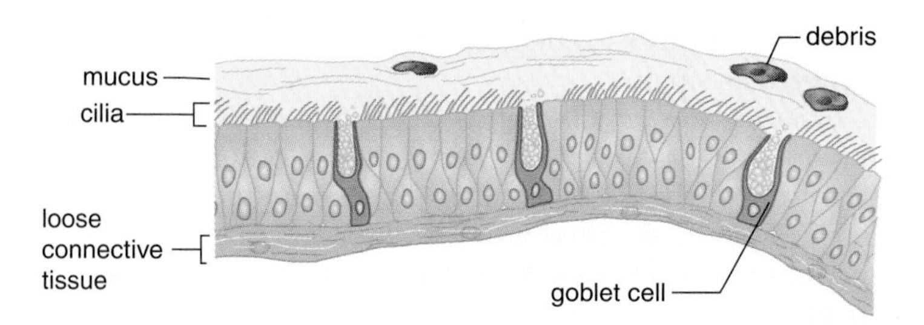

Smoking is known to destroy the cilia, and consequently the soot in cigarette smoke collects in the lungs. Smoking is discussed more fully at the end of this chapter.

The Bronchial Tree

The trachea divides into right and left primary bronchi (sing., **bronchus**), which lead into the right and left lungs (see Fig. 15.1). The bronchi branch into a great number of secondary bronchi that eventually lead to **bronchioles.** The bronchi resemble the trachea in structure, but as the bronchial tubes divide and subdivide, their walls become thinner, and the small rings of cartilage are no longer present. During an asthma attack, the smooth muscle of the bronchioles contracts, causing bronchiolar constriction and characteristic wheezing. Each bronchiole terminates in an elongated space enclosed by a multitude of air pockets, or sacs, called *alveoli* (sing., **alveolus**) (see Fig. 15.1). The alveoli make up the lungs.

The Lungs

The **lungs** are paired cone-shaped organs within the thoracic cavity. The right lung has three lobes, and the left lung has two lobes, allowing room for the heart, which is on the left side of the body. A lobe is further divided into lobules, and each lobule has a bronchiole serving many alveoli. The lungs lie on either side of the heart in the thoracic cavity. The base of each lung is broad and concave so that it fits the convex surface of the diaphragm. The other surfaces of the lungs follow the contours of the ribs and the diaphragm in the thoracic cavity.

The Alveoli

Each alveolar sac is made up of simple squamous epithelium surrounded by blood capillaries. Gas exchange occurs between air in the alveoli and blood in the capillaries (Fig. 15.4). Oxygen diffuses across the alveolar wall and enters the bloodstream, while carbon dioxide diffuses from the blood across the alveolar wall to enter the alveoli.

The alveoli of human lungs are lined with a surfactant, a film of lipoprotein that lowers the surface tension and prevents them from closing. The lungs collapse in some newborn babies, especially premature infants, who lack this film. The condition, called **infant respiratory distress syndrome,** is now treatable by surfactant replacement therapy.

There are approximately 300 million alveoli, with a total cross-sectional area of 50–70 m^2. This is the surface area of a typical classroom and at least 40 times the surface area of the skin. Because of their many air spaces, the lungs are very light; normally, a piece of lung tissue dropped in a glass of water floats.

The trachea divides into the primary bronchi, which divide repeatedly to give rise to the bronchioles. The bronchioles have many branches and terminate at the alveoli, which make up the lungs.

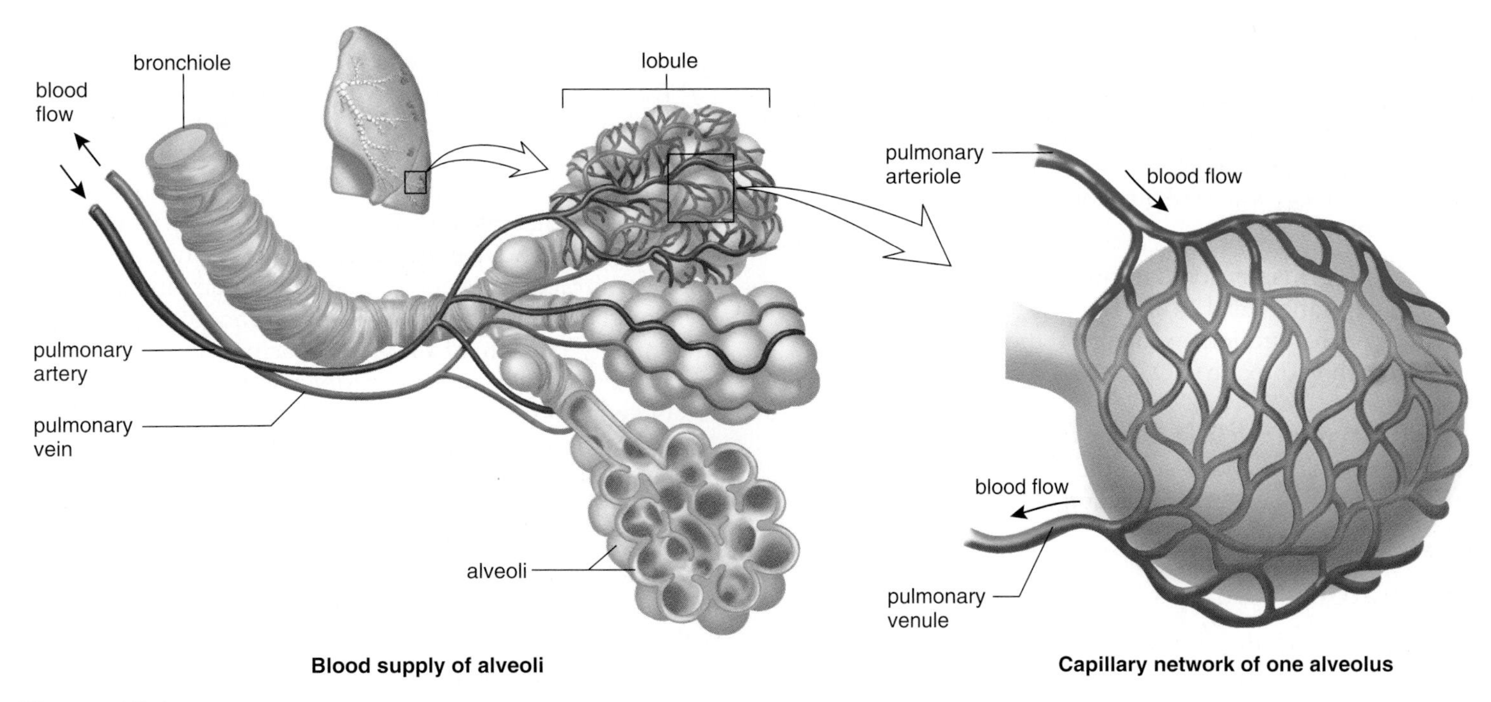

Figure 15.4 Gas exchange in the lungs.
The lungs consist of alveoli, surrounded by an extensive capillary network. Notice that the pulmonary arteriole carries blood low in oxygen (colored blue) and the pulmonary venule carries blood high in oxygen (colored red).

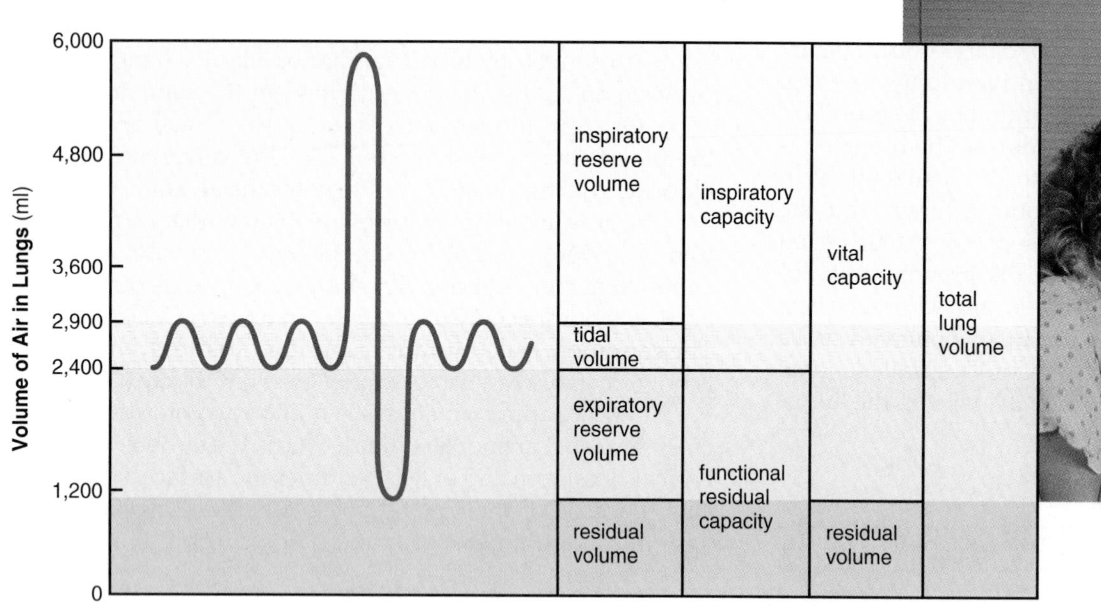

Figure 15.5 Vital capacity.
A spirometer measures the maximum amount of air that can be inhaled and exhaled when breathing by way of a tube connected to the instrument. During inspiration, a pen moves up, and during expiration, a pen moves down. The resulting pattern, such as the one shown here, is called a spirograph.

15.2 Mechanism of Breathing

The term respiration refers to the complete process of supplying oxygen to body cells for aerobic cellular respiration and the reverse process of ridding the body of carbon dioxide given off by cells. Respiration includes the following components:

1. Breathing: inspiration (entrance of air into the lungs) and expiration (exit of air from the lungs).
2. External respiration: exchange of the gases oxygen (O_2) and carbon dioxide (CO_2) between air and blood in the lungs.
3. Internal respiration: exchange of the gases O_2 and CO_2 between blood and tissue fluid.
4. Cellular respiration: production of ATP in cells.

Respiratory Volumes

When we breathe, the amount of air moved in and out with each breath is called the **tidal volume.** Normally, the tidal volume is about 500 ml, but we can increase the amount inhaled and exhaled by deep breathing. The maximum volume of air that can be moved in and out during a single breath is called the **vital capacity** (Fig. 15.5). First, we can increase inspiration by as much as 3,100 ml of air by forced inspiration. This is called the **inspiratory reserve volume.**

Even so, some of the inspired air never reaches the lungs; instead it fills the nose, trachea, bronchi, and bronchioles (see Fig. 15.1). These passages are not used for gas exchange, and therefore, they are said to contain dead space air. To ensure that inspired air reaches the lungs, it is better to breathe slowly and deeply. Similarly, we can increase expiration by contracting the abdominal and thoracic muscles. This is called the **expiratory reserve volume,** and it measures approximately 1,400 ml of air. Vital capacity is the sum of tidal, inspiratory reserve, and expiratory reserve volumes.

Note in Figure 15.5 that even after very deep breathing, some air (about 1,000 ml) remains in the lungs; this is called the **residual volume.** This air is no longer useful for gas exchange purposes. In some lung diseases, such as emphysema (p. 297), the residual volume builds up because the individual has difficulty emptying the lungs. This means that the vital capacity is reduced and the lungs tend to be filled with useless air.

The air used for gas exchange excludes both the air in the dead space of the respiratory tract and the residual volume in the lungs.

Ecology Focus

Photochemical Smog Can Kill

Most industrialized cities have photochemical smog at least occasionally. Photochemical smog arises when primary pollutants react with one another under the influence of sunlight to form a more deadly combination of chemicals. For example, the primary pollutants nitrogen oxides (NO_x) and hydrocarbons (HC) react with one another in the presence of sunlight to produce nitrogen dioxide (NO_2), ozone (O_3), and PAN (peroxyacetyl nitrate). Ozone and PAN are commonly referred to as oxidants. Breathing oxidants affects the respiratory and nervous systems, resulting in respiratory distress, headache, and exhaustion.

Cities with warm, sunny climates that are large and industrialized, such as Los Angeles, Denver, and Salt Lake City in the United States, Sydney in Australia, Mexico City in Mexico, and Buenos Aires in Argentina, are particularly susceptible to photochemical smog. If the city is surrounded by hills, a thermal inversion may aggravate the situation. Normally, warm air near the ground rises, so that pollutants are dispersed and carried away by air currents. But sometimes during a thermal inversion, smog gets trapped near the earth by a blanket of warm air (Fig. 15A). This may occur when a cold front brings in cold air, which settles beneath a warm layer. The trapped pollutants cannot disperse, and the results can be disastrous. In 1963, about 300 people died, and in 1966, about 168 people died in New York City when air pollutants accumulated over the city. Even worse were the events in London in 1957, when 700 to 800 people died, and in 1962, when 700 people died, due to the effects of air pollution.

Even though we have federal legislation to bring air pollution under control, more than half the people in the United States live in cities polluted by too much smog. We should place our emphasis on pollution prevention because, in the long run, prevention is usually easier and cheaper than pollution cleanup methods. Some prevention suggestions are as follows:

- Build more efficient automobiles or burn fuels that do not produce pollutants.
- Reduce the amount of waste to be incinerated by recycling materials.
- Reduce our energy use so that power plants need to provide less, and/or use renewable energy sources such as solar, wind, or water power.
- Require industries to meet clean air standards.

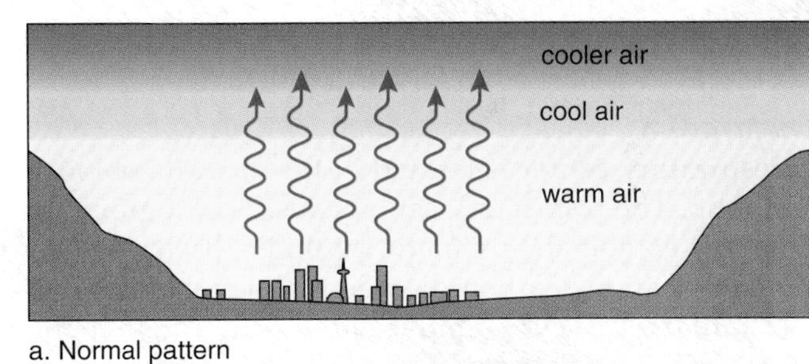

a. Normal pattern

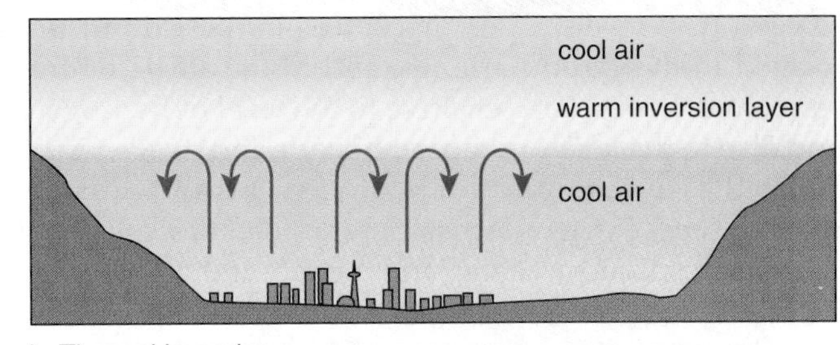

b. Thermal inversion

Figure 15A Thermal inversion.
a. Normally, pollutants escape into the atmosphere when warm air rises. **b.** During a thermal inversion, a layer of warm air (warm inversion layer) overlies and traps pollutants in cool air below. **c.** Los Angeles, a city of 8.5 million cars and thousands of factories, is particularly susceptible to thermal inversions, and this accounts for why this city is the "air pollution capital" of the United States.

Inspiration and Expiration

To understand **ventilation,** the manner in which air enters and exits the lungs, it is necessary to remember first that normally there is a continuous column of air from the pharynx to the alveoli of the lungs.

Secondly, the lungs lie within the sealed-off thoracic cavity. The **rib cage** forms the top and sides of the thoracic cavity. It contains the ribs, hinged to the vertebral column at the back and to the sternum (breastbone) at the front, and the intercostal muscles that lie between the ribs. The **diaphragm,** a dome-shaped horizontal sheet of muscle and connective tissue, forms the floor of the thoracic cavity.

The lungs are enclosed by two membranes called **pleural membranes.** An infection of the pleural membranes is called pleurisy. The parietal pleura adheres to the rib cage and the diaphragm, and the visceral pleura is fused to the lungs. The two pleural layers lie very close to one another, separated only by a small amount of fluid. Normally, the intrapleural pressure (pressure between the pleural membranes) is lower than atmospheric pressure by 4 mm Hg.

The importance of the reduced intrapleural pressure is demonstrated when, by design or accident, air enters the intrapleural space. The affected lobules collapse.

The pleural membranes enclose the lungs and line the thoracic cavity. Intrapleural pressure is lower than atmospheric pressure.

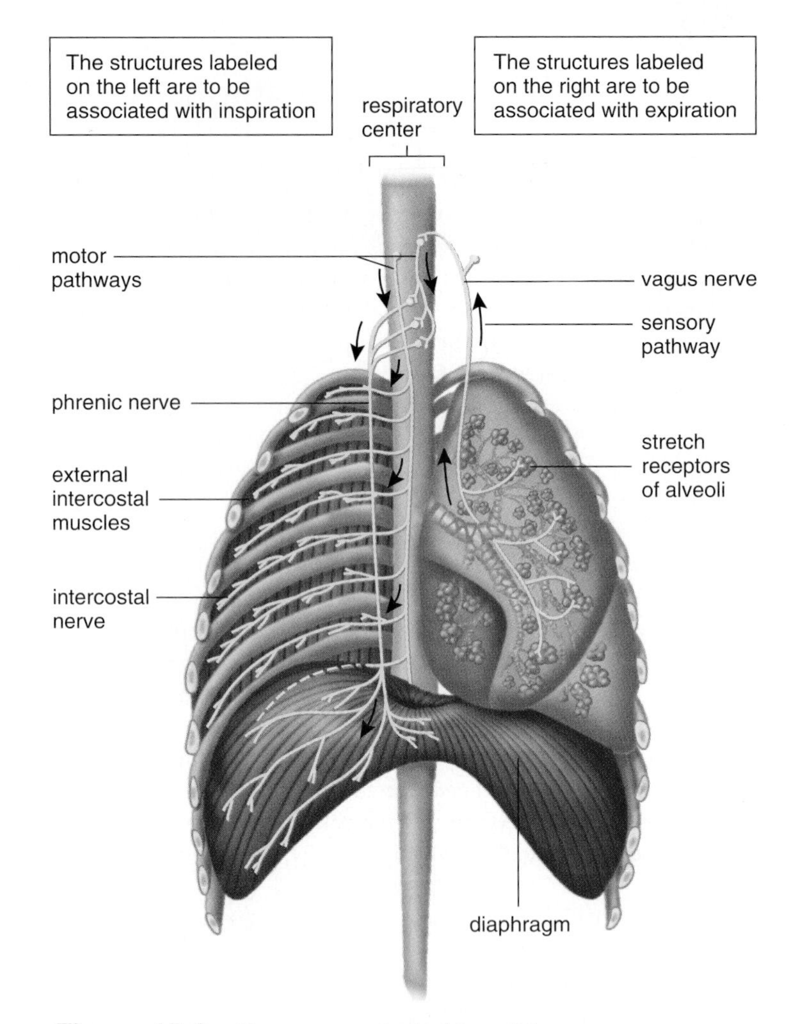

Figure 15.6 Nervous control of breathing.
During inspiration, the respiratory center stimulates the external intercostal (rib) muscles to contract via the intercostal nerves and the diaphragm to contract via the phrenic nerve. Should the tidal volume increase above 1.5 liters, stretch receptors send inhibitory nerve impulses to the respiratory center via the vagus nerve. In any case, expiration occurs due to a lack of stimulation from the respiratory center to the diaphragm and intercostal muscles.

Inspiration

A **respiratory center** is located in the medulla oblongata of the brain. The respiratory center consists of a group of neurons that exhibit an automatic rhythmic discharge that triggers inspiration. Carbon dioxide (CO_2) and hydrogen ions (H^+) are the primary stimuli that directly cause changes in the activity of this center. This center is not affected by low oxygen (O_2) levels. Chemoreceptors in the **carotid bodies,** located in the carotid arteries, and in the **aortic bodies,** located in the aorta, are sensitive to the level of hydrogen ions and also to the levels of carbon dioxide and oxygen in blood. When the concentrations of hydrogen ions and carbon dioxide rise (and oxygen decreases), these bodies communicate with the respiratory center, and the rate and depth of breathing increase.

The respiratory center sends out impulses by way of nerves to the diaphragm and the muscles of the rib cage (Fig. 15.6). In its relaxed state, the diaphragm is dome-shaped, but upon stimulation, it contracts and lowers. Also, the external intercostal muscles contract, causing the rib cage to move upward and outward. Now the thoracic cavity increases in size, and the lungs expand. As the lungs expand, air pressure within the enlarged alveoli lowers and air enters through the nose or the mouth.

Inspiration is the active phase of breathing (Fig. 15.7*a*). During this time, the diaphragm and the rib muscles contract, intrapleural pressure decreases, the lungs expand, and air comes rushing in. Note that air comes in because the lungs already have opened up; air does not force the lungs open. This is why it is sometimes said that *humans breathe by negative pressure.* The creation of a partial vacuum in the alveoli causes air to enter the lungs.

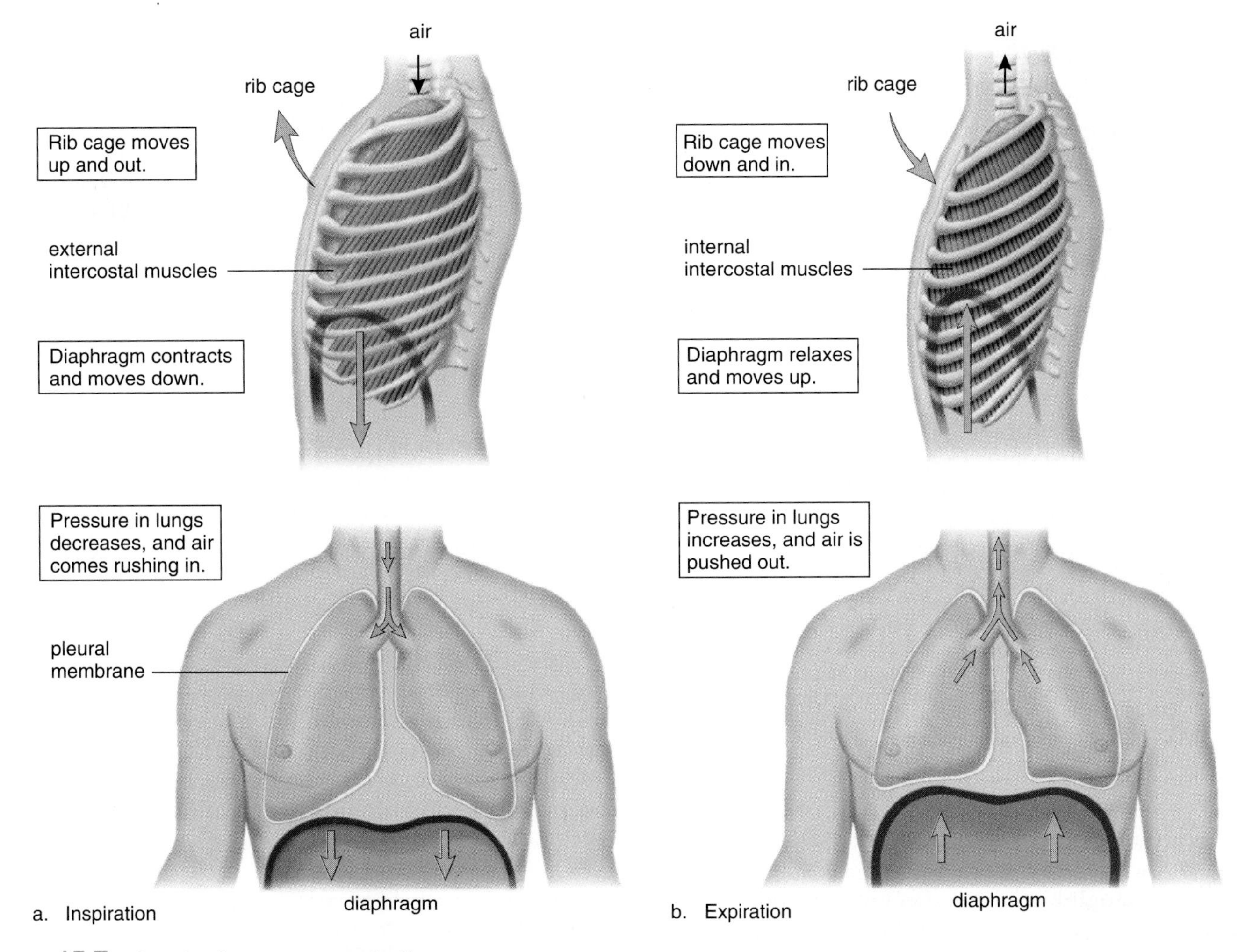

Figure 15.7 **Inspiration versus expiration.**
a. During inspiration, the thoracic cavity and lungs expand so that air is drawn in. **b.** During expiration, the thoracic cavity and lungs resume their original positions and pressures. Now, air is forced out.

Expiration

When the respiratory center stops sending nervous signals to the diaphragm and the rib cage, the diaphragm relaxes and it resumes its dome shape. The abdominal organs press up against the diaphragm, and the rib cage moves down and inward (Fig. 15.7*b*). Now, the elastic lungs recoil, and air is pushed out. The respiratory center acts rhythmically to bring about breathing at a normal rate and volume. If by chance we inhale more deeply, the lungs are expanded and the alveoli stretch. This stimulates stretch receptors in the alveolar walls, and they initiate inhibitory nerve impulses that travel from the inflated lungs to the respiratory center. This causes the respiratory center to stop sending out nerve impulses.

While inspiration is the active phase of breathing, expiration is usually passive—the diaphragm and external intercostal muscles are relaxed when expiration occurs. When breathing is deeper and/or more rapid, expiration can also be active. Contraction of internal intercostal muscles can force the rib cage to move downward and inward. Also, when the abdominal wall muscles are contracted, they push on the viscera, which push against the diaphragm, and the increased pressure in the thoracic cavity helps to expel air.

During inspiration, due to nervous stimulation, the diaphragm lowers and the rib cage lifts up and out. During expiration, due to a lack of nervous stimulation, the diaphragm rises and the rib cage lowers.

15.3 Gas Exchanges in the Body 

Figure 15.8 shows both external respiration and internal respiration. The principles of diffusion alone govern whether O_2 or CO_2 enters or leaves blood.

External Respiration

External respiration refers to the exchange of gases between air in the alveoli and blood in the pulmonary capillaries. Gases exert pressure, and the amount of pressure each gas exerts is its partial pressure, symbolized as P_{O_2} and P_{CO_2}. Blood flowing in the pulmonary capillaries has a higher P_{CO_2} than atmospheric air. *Therefore, CO_2 diffuses out of blood into the lungs.* Most of the CO_2 is being carried as bicarbonate ions (HCO_3^-). As the little remaining free CO_2 begins to diffuse out, the following reaction is driven to the right.

$$H^+ + HCO_3^- \longrightarrow H_2CO_3 \longrightarrow H_2O + CO_2\uparrow$$

hydrogen ion, bicarbonate ion, water, carbon dioxide

"Up" arrow indicates carbon dioxide is leaving the body.

The enzyme **carbonic anhydrase,** present in red blood cells, speeds up the reaction. As the reaction proceeds, the respiratory pigment **hemoglobin,** also present in red blood cells, gives up the hydrogen ions (H^+) it has been carrying; HHb becomes Hb. Hb is called deoxyhemoglobin.

The pressure pattern is the reverse for O_2. Blood flowing in the pulmonary capillaries is low in oxygen, and alveolar air contains a much higher partial pressure of oxygen. Therefore, *O_2 diffuses into blood and red blood cells in the lungs.* Hemoglobin takes up this oxygen and becomes **oxyhemoglobin.**

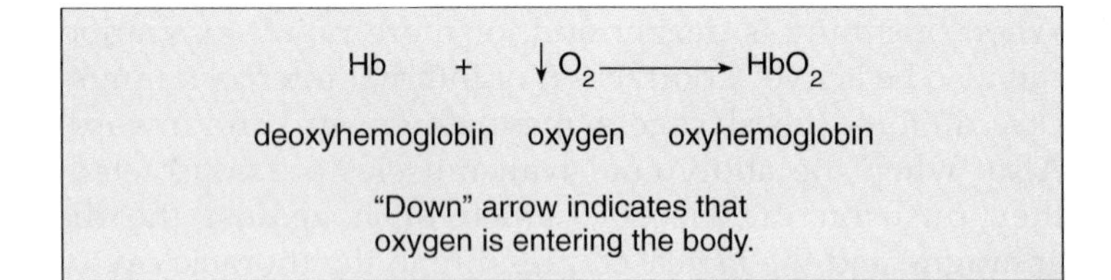

At high altitudes, people have breathing problems because the total air pressure is lower, making the P_{O_2} lower than normal; therefore, less O_2 diffuses into blood. In airplanes the cabin is pressurized to maintain sufficient partial pressures for normal breathing.

Internal Respiration

Internal respiration refers to the exchange of gases between blood in systemic capillaries and tissue fluid. Blood that enters the systemic capillaries is bright red in color because red blood cells contain oxyhemoglobin. Oxyhemoglobin gives up O_2, which diffuses out of blood into red blood cells and the tissues.

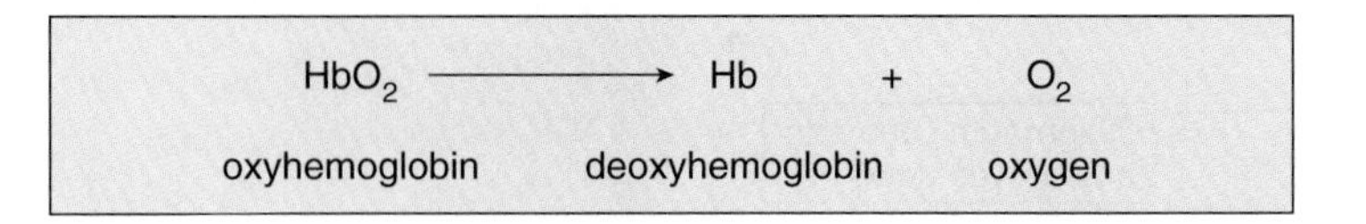

Oxygen diffuses out of blood into the tissues because the P_{O_2} of tissue fluid is lower than that of blood. The lower P_{O_2} is due to cells continuously using up oxygen in aerobic cellular respiration. *Carbon dioxide diffuses into blood from the tissues* because the P_{CO_2} of tissue fluid is higher than that of blood. Carbon dioxide, produced continuously by cells, collects in tissue fluid.

After CO_2 diffuses into blood, it enters the red blood cells, where a small amount is taken up by hemoglobin, forming **carbaminohemoglobin.** Most of the CO_2 combines with water, forming carbonic acid (H_2CO_3), which dissociates to hydrogen ions (H^+) and bicarbonate ions (HCO_3^-). The increased concentration of CO_2 in the blood causes the reaction to proceed to the right.

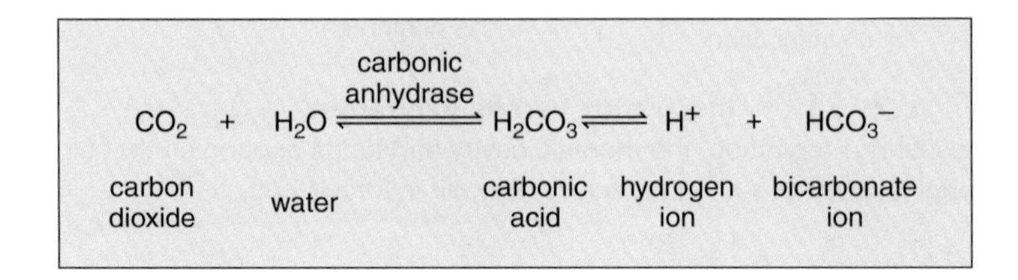

The enzyme carbonic anhydrase, present in red blood cells, speeds up the first portion of the overall reaction. Bicarbonate ions diffuse out of red blood cells and are carried in the plasma. Blood that leaves the capillaries is deep purple in color because red blood cells contain reduced hemoglobin. The globin portion of hemoglobin combines with excess hydrogen ions produced by the overall reaction, and Hb becomes HHb, called **reduced hemoglobin.** In this way, the pH of blood remains fairly constant.

External and internal respiration are the movement of gases between blood and the alveoli and between blood and the systemic capillaries, respectively. Both processes are dependent on the process of diffusion.

Visual Focus

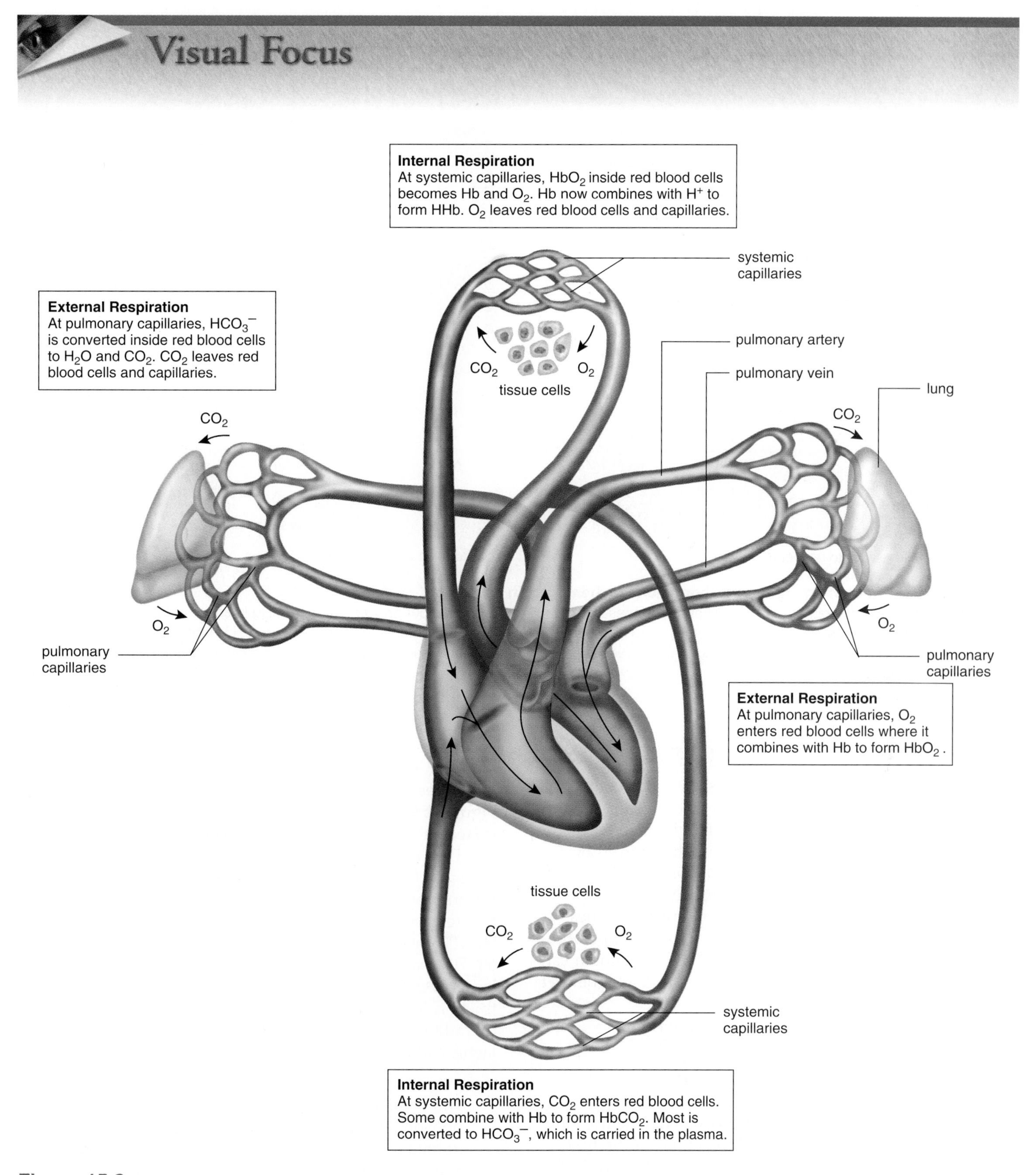

Figure 15.8 External and internal respiration.
During external respiration in the lungs, CO_2 leaves blood and O_2 enters blood. During internal respiration in the tissues, O_2 leaves blood and CO_2 enters blood.

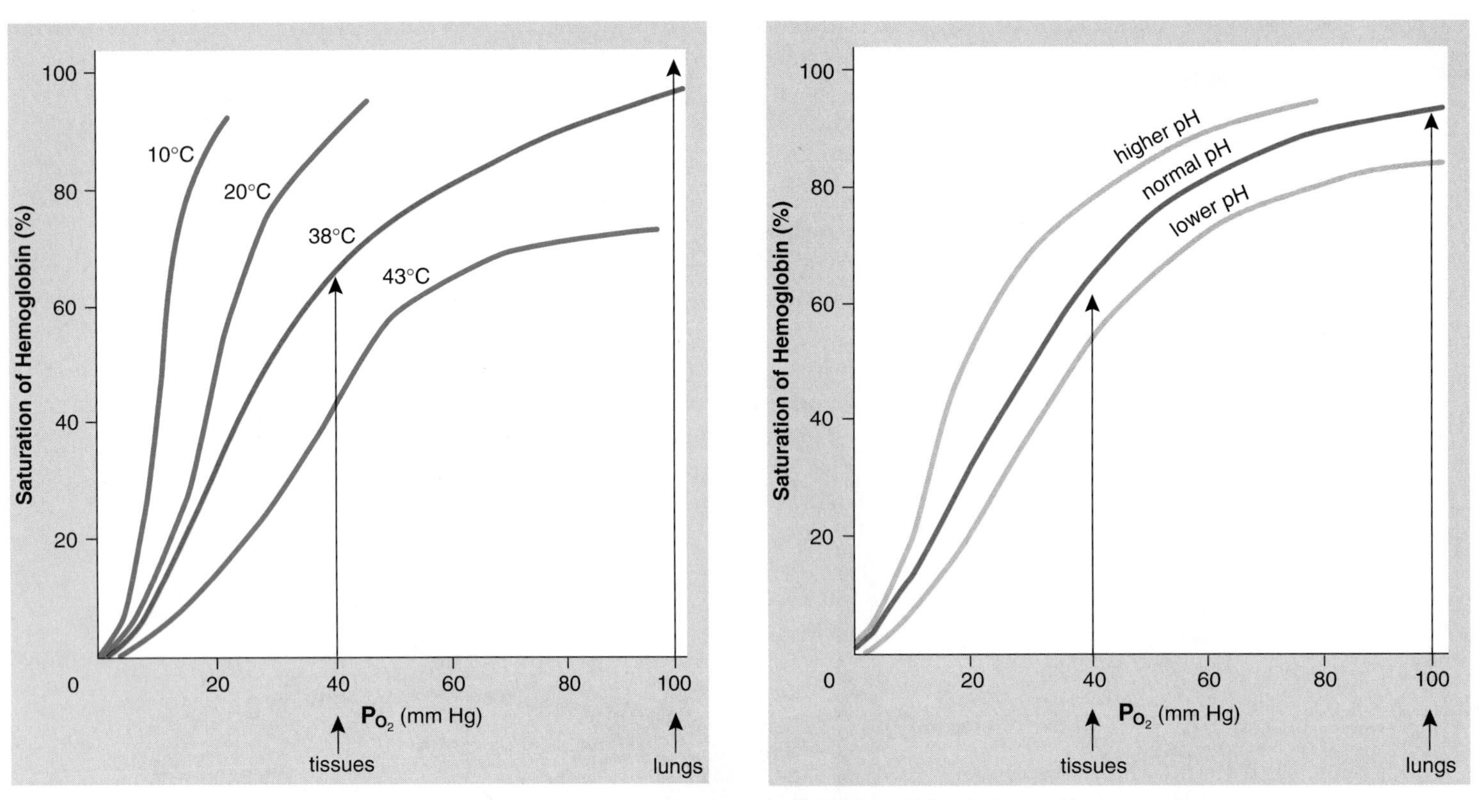

a. Saturation of Hb relative to temperature

b. Saturation of Hb relative to pH

Figure 15.9 Effect of environmental conditions on hemoglobin saturation.
The partial pressure of oxygen (P_{O_2}) in pulmonary capillaries is about 98–100 mm Hg, but only about 40 mm Hg in tissue capillaries. Hemoglobin is about 98% saturated in the lungs because of P_{O_2}, and also because **(a)** the temperature is cooler and **(b)** the pH is higher in lungs. On the other hand, hemoglobin is only about 60% saturated in the tissues because of the P_{O_2} and also because **(a)** the temperature is warmer and **(b)** the pH is lower in the tissues.

Binding Capacity of Hemoglobin

The binding capacity of hemoglobin is also affected by partial pressures. The P_{O_2} of air entering the alveoli is about 100 mm Hg, and at this pressure the hemoglobin in the blood becomes saturated with O_2. This means that iron in hemoglobin molecules has combined with O_2. On the other hand, the P_{O_2} in the tissues is about 40 mm Hg, causing hemoglobin molecules to release O_2, and O_2 to diffuse into the tissues.

In addition to the partial pressure of O_2, temperature and pH also affect the amount of oxygen hemoglobin can carry. The lungs have a lower temperature and a higher pH than the tissues:

	pH	Temperature
Lungs	7.40	37°C
Tissues	7.38	38°C

Both Figure 15.9*a* and *b* show that, as expected, hemoglobin is more saturated with O_2 in the lungs than in the tissues. This effect, which can be attributed to the difference in P_{O_2} between the lungs and tissues, is potentiated by the difference in temperature and pH between the lungs and tissues. Notice in Figure 15.9*a* that the saturation curve for hemoglobin is steeper at 10°C compared to 20°C, and so forth. Also, Figure 15.9*b* shows that the saturation curve for hemoglobin is steeper at higher pH than at lower pH.

This means that the environmental conditions in the lungs are favorable for the uptake of O_2 by hemoglobin, and the environmental conditions in the tissues are favorable for the release of O_2 by hemoglobin. Hemoglobin is about 98–100% saturated in the capillaries of the lungs and about 60–70% saturated in the tissues. During exercise, hemoglobin is even less saturated in the tissues because muscle contraction leads to higher body temperature (up to 103°F in marathoners!) and lowers the pH (due to the production of lactic acid).

The difference in P_{O_2}, temperature, and pH between the lungs and tissues causes hemoglobin to take up oxygen in the lungs and release oxygen in the tissues.

15.4 Respiration and Health

We have learned that the respiratory tract has a warm, wet mucosa, which is constantly exposed to environmental air. The quality of this air, as discussed in the Ecology reading on page 289, and whether it contains pathogens, can affect our health.

Upper Respiratory Tract Infections

The upper respiratory tract consists of the nose, the pharynx, and the larynx. Upper respiratory infections (URI) can spread from the nasal cavities to the sinuses, to the middle ears, and to the larynx (Fig. 15.10). Viral infections sometimes lead to secondary bacterial infections. What we call "strep throat" is a primary bacterial infection caused by *Streptococcus pyogenes* that can lead to a generalized upper respiratory infection and even a systemic (affecting the body as a whole) infection. While antibiotics have no effect on viral infections, they are successfully used for most bacterial infections, including strep throat.

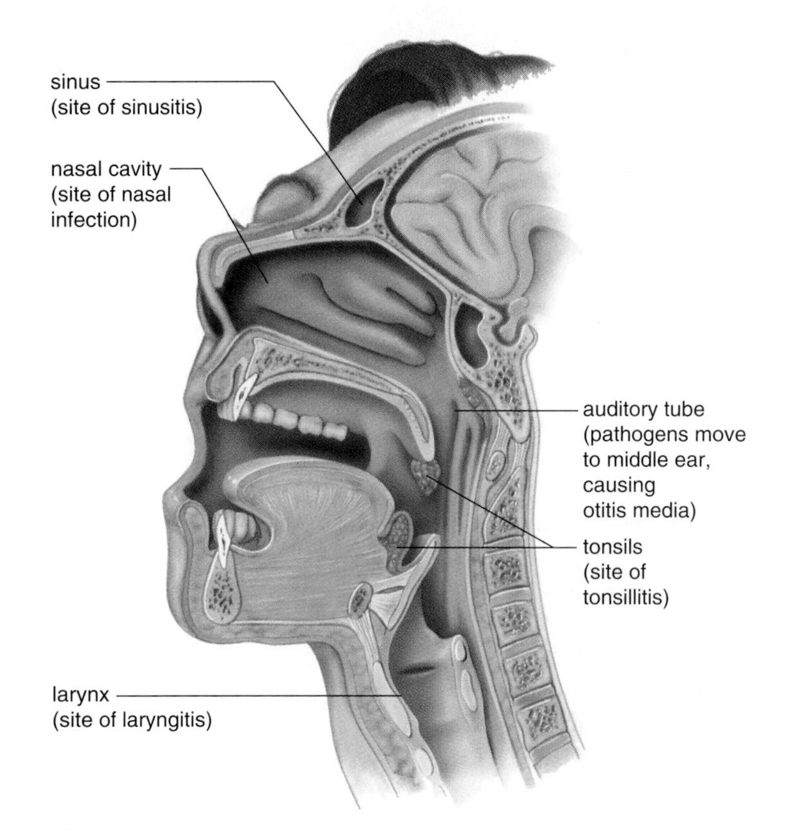

Figure 15.10 **Upper respiratory infections.**

Sinusitis

Sinusitis is an infection of the sinuses, cavities within the facial skeleton that drain into the nasal cavities. Only about 1–3% of upper respiratory infections are accompanied by sinusitis. Sinusitis develops when nasal congestion blocks the tiny openings leading to the sinuses. Symptoms include postnasal discharge as well as facial pain that worsens when the patient bends forward. Pain and tenderness usually occur over the lower forehead or over the cheeks. If the latter, toothache is also a complaint. Successful treatment depends on restoring proper drainage of the sinuses. Even a hot shower and sleeping upright can be helpful. Otherwise, spray decongestants are preferred over oral antihistamines, which thicken rather than liquefy the material trapped in the sinuses.

Otitis Media

Otitis media is a bacterial infection of the middle ear. The middle ear is not a part of the respiratory tract, but this infection is considered here because it is a complication often seen in children who have a nasal infection. Infection can spread by way of the **auditory tube** that leads from the nasopharynx to the middle ear. Pain is the primary symptom of a middle ear infection. A sense of fullness, hearing loss, vertigo (dizziness), and fever may also be present. Antibiotics almost always bring about a full recovery, and a recurrence is most likely due to a new infection. Drainage tubes (called tympanostomy tubes) are sometimes placed in the eardrum of children with multiple recurrences to help prevent the buildup of fluid in the middle ear and the possibility of hearing loss. Normally, the tubes slough out with time.

Tonsillitis

Tonsillitis occurs when tonsils become inflamed and enlarged. **Tonsils** are masses of lymphatic tissue that occur in the pharynx. The tonsils in the dorsal wall of the nasopharynx are often called adenoids. The tonsils remove many of the pathogens that enter the pharynx; therefore, they are a first line of defense against invasion of the body. If tonsillitis occurs frequently and enlargement makes breathing difficult, the tonsils can be removed surgically in a **tonsillectomy.** Fewer tonsillectomies are performed today than in the past because it is now known that tonsils serve an important function in defending the body against infection.

Laryngitis

Laryngitis is an infection of the larynx with an accompanying hoarseness leading to the inability to talk in an audible voice. Usually laryngitis disappears with treatment of the upper respiratory infection. Persistent hoarseness without the presence of an upper respiratory infection is one of the warning signs of cancer and therefore should be looked into by a physician.

Visual Focus

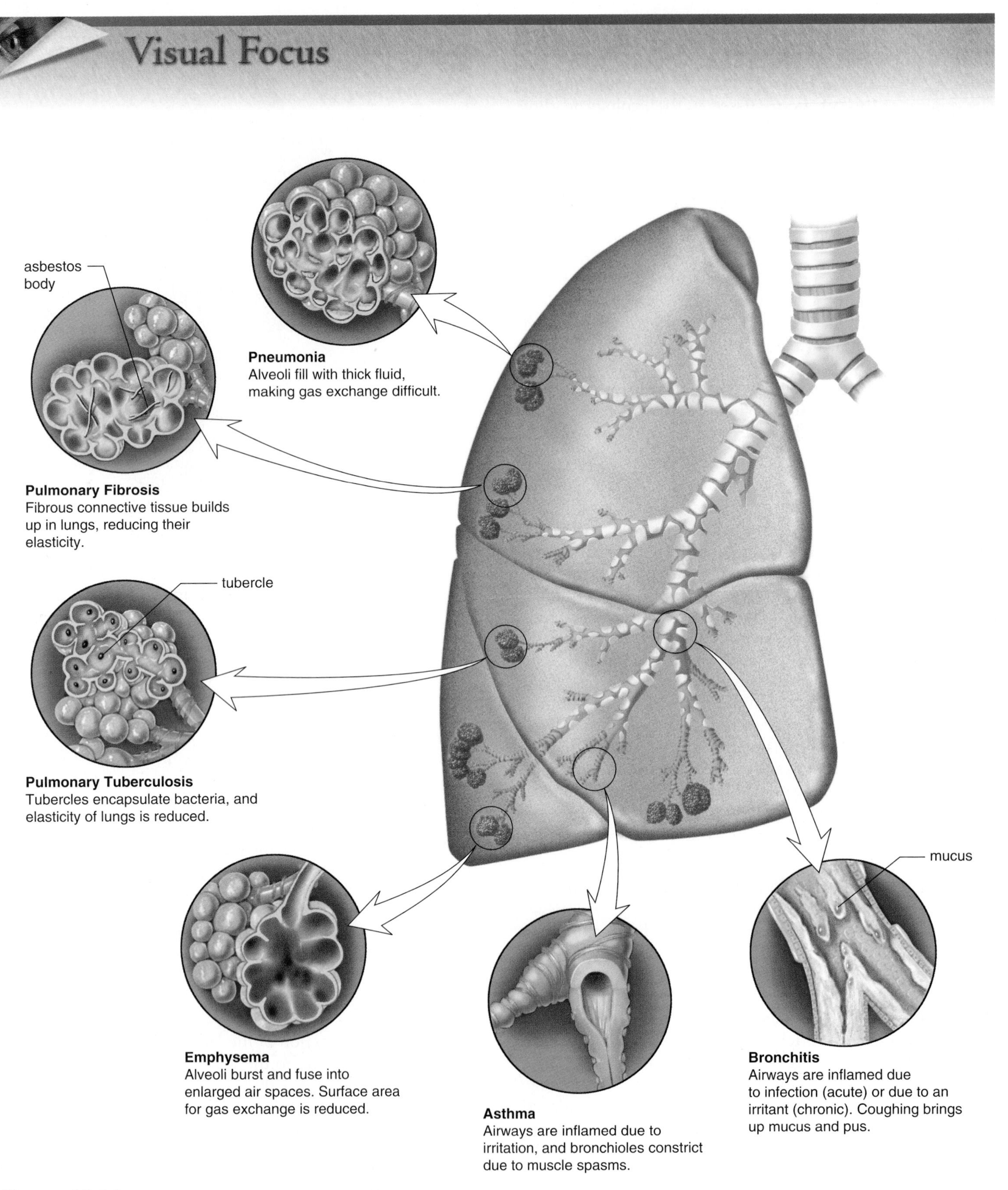

Figure 15.11 Lower respiratory tract disorders.
Exposure to infectious pathogens and/or air pollutants, including cigarette and cigar smoke, causes the diseases and disorders shown here.

Lower Respiratory Tract Disorders

Lower respiratory tract disorders, which are illustrated in Figure 15.11, include infections, restrictive pulmonary disorders, obstructive pulmonary disorders, and lung cancer.

Lower Respiratory Infections

Acute bronchitis, pneumonia, and tuberculosis are infections of the lower respiratory tract. **Acute bronchitis** is an infection of the primary and secondary bronchi. Usually it is preceded by a viral URI that has led to a secondary bacterial infection. Most likely, a nonproductive cough has become a deep cough that expectorates mucus and perhaps pus.

Pneumonia is a viral or bacterial infection of the lungs in which bronchi and alveoli fill with thick fluid. Most often it is preceded by influenza. Rather than being a generalized lung infection, pneumonia may be localized in specific lobules of the lungs. Obviously the more lobules involved, the more serious the infection. Pneumonia can be caused by a bacterium that is usually held in check, but that has gained the upper hand due to stress and/or reduced immunity. AIDS patients are subject to a particularly rare form of pneumonia caused by the protozoan *Pneumocystis carinii.* Pneumonia of this type is almost never seen in individuals with a healthy immune system. High fever and chills with headache and chest pain are symptoms of pneumonia.

Pulmonary tuberculosis is caused by the tubercle bacillus, a type of bacterium. It is possible to tell if a person has ever been exposed to tuberculosis with a skin test in which a highly diluted extract of the bacillus is injected into the skin of the patient. A person who has never been in contact with tubercle bacillus shows no reaction, but one who has developed immunity to the organism shows an area of inflammation that peaks in about 48 hours. When tubercle bacilli invade the lung tissue, the cells build a protective capsule about the foreigners, isolating them from the rest of the body. This tiny capsule is called a tubercle. If the resistance of the body is high, the imprisoned organisms die, but if the resistance is low, the organisms eventually can be liberated. If a chest X ray detects active tubercles, the individual is put on appropriate drug therapy to ensure the localization of the disease and the eventual destruction of any live bacterial organisms.

Tuberculosis was a major killer in the United States before the middle of this century, after which antibiotic therapy brought it largely under control. In recent years, however, the incidence of tuberculosis is on the rise, particularly among AIDS patients, the homeless, and the rural poor. Worse, the new strains are resistant to the usual antibiotic therapy. Therefore, some physicians would like to again quarantine patients in sanitariums.

Restrictive Pulmonary Disorders

In restrictive pulmonary disorders, vital capacity is reduced not because air does not move freely into and out of the lungs but because the lungs have lost their elasticity. Inhaling particles such as silica (sand), coal dust, asbestos, and, now it seems, fiberglass can lead to **pulmonary fibrosis,** a condition in which fibrous connective tissue builds up in the lungs. The lungs cannot inflate properly and are always tending toward deflation. Breathing asbestos is also associated with the development of cancer. Since asbestos has been used so widely as a fireproofing and insulating agent, unwarranted exposure has occurred. It is projected that two million deaths could be caused by asbestos exposure—mostly in the workplace—between 1990 and 2020.

Obstructive Pulmonary Disorders

In obstructive pulmonary disorders, air does not flow freely in the airways and the time it takes to inhale or exhale maximally is greatly increased. Vital capacity, however, is normal. Several disorders, including chronic bronchitis, emphysema, and asthma, are referred to as chronic obstructive pulmonary disorders (COPD) because they tend to recur and have flare-ups.

In **chronic bronchitis,** the airways are inflamed and filled with mucus. A cough that brings up mucus is common. The bronchi have undergone degenerative changes including the loss of cilia and their normal cleansing action. Under these conditions an infection is more likely to occur. Smoking cigarettes and cigars is the most frequent cause of chronic bronchitis. Exposure to other pollutants can also cause chronic bronchitis.

Emphysema is a chronic and incurable disorder in which the alveoli are distended and their walls damaged so that the surface area available for gas exchange is reduced. Emphysema is often preceded by chronic bronchitis. Air trapped in the lungs leads to alveolar damage and a noticeable ballooning of the chest. The elastic recoil of the lungs is reduced, so not only are the airways narrowed but the driving force behind expiration is also reduced. The victim is breathless and may have a cough. Because the surface area for gas exchange is reduced, oxygen reaching the heart and the brain is reduced. Even so, the heart works furiously to force more blood through the lungs, and an increased workload on the heart can result. Lack of oxygen to the brain can make the person feel depressed, sluggish, and irritable. Exercise, drug therapy, and supplemental oxygen, along with giving up smoking, may relieve the symptoms and possibly slow the progression of emphysema.

Asthma is a disease of the bronchi and bronchioles that is marked by wheezing, breathlessness, and sometimes cough and expectoration of mucus. The airways are unusually sensitive to specific irritants, which can include a wide range of allergens such as pollen, animal dander, dust, cigarette smoke, and industrial fumes. Even cold air, however, can be an irritant. When exposed to the irritant, the smooth muscle in the bronchioles undergoes spasms. It now appears that chemical mediators given off by immune cells in the bronchioles result in the spasms. Most asthma patients have some degree of bronchial inflammation that reduces the diameter of the airways and contributes to the seriousness of an attack. Asthma is not curable but is treatable. There are inhalers that control the inflammation and hopefully prevent an attack, but there are also inhalers that stop the muscle spasms should an attack occur.

Lung Cancer

Lung cancer used to be more prevalent in men than in women, but recently it has surpassed breast cancer as a cause of death in women. This can be linked to an increase in the number of women who smoke today. Autopsies on smokers have revealed the progressive steps by which the most common form of lung cancer develops. The first event appears to be thickening and callusing of the cells lining the airways. (Callusing occurs whenever cells are exposed to irritants.) Then there is a loss of cilia so that it is impossible to prevent dust and dirt from settling in the lungs. Following this, cells with atypical nuclei appear in the callused lining. A tumor (Fig. 15.12*b*) consisting of disordered cells with atypical nuclei is considered to be cancer in situ (at one location). A final step occurs when some of these cells break loose and penetrate other tissues, a process called metastasis. Now the cancer has spread. The original tumor may grow until a bronchus is blocked, cutting off the supply of air to that lung. The entire lung then collapses, the secretions trapped in the lung spaces become infected, and pneumonia or a lung abscess (localized area of pus) results. The only treatment that offers a possibility of cure is to remove a lobe or the lung completely before metastasis has had time to occur. This operation is called **pneumonectomy.**

Current research indicates that *involuntary smoking,* simply breathing in air filled with cigarette smoke, can also cause lung cancer and other illnesses associated with smoking. The Health reading on the next page lists the various illnesses that are apt to occur when a person smokes. If a person stops both voluntary and involuntary smoking, and if the body tissues are not already cancerous, they may return to normal over time.

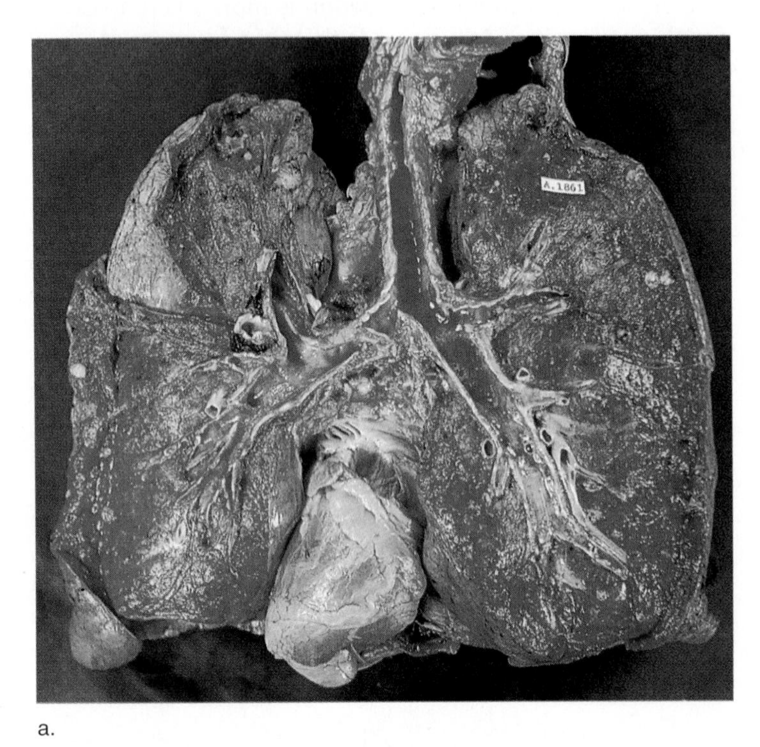

a.

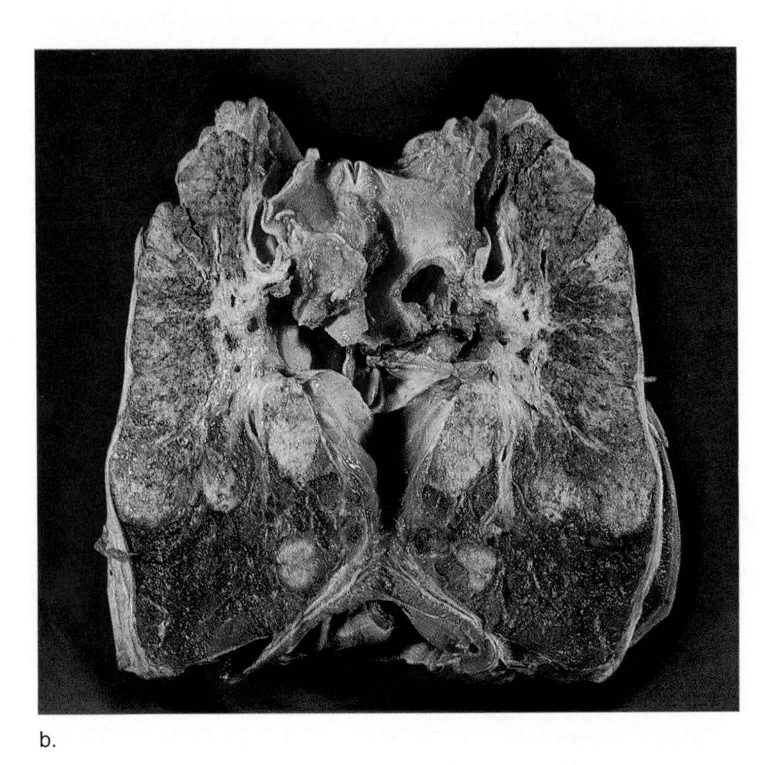

b.

Figure 15.12 **Normal lung versus cancerous lung. a.** Normal lung with heart in place. Note the healthy red color. **b.** Lungs of a heavy smoker. Notice how black the lungs are except where cancerous tumors have formed.

Health Focus

The Most Often Asked Questions About Tobacco and Health

Is there a safe way to smoke?

No. All forms of tobacco can cause damage and smoking even a small amount is dangerous. Tobacco is perhaps the only legal product whose advertised and intended use—that is, smoking it—will hurt the body. . . .

Does smoking cause cancer?

Yes, and not only lung cancer. . . . Besides causing lung cancer, smoking a pipe, cigarettes, or cigars is also a major cause of cancers of the mouth, larynx (voice box), and esophagus. In addition, smoking increases the risk of cancer of the bladder, kidney, pancreas, stomach, and the uterine cervix.

What are the chances of being cured of lung cancer?

Very low; the five-year survival rate is only 13%. . . . Fortunately, lung cancer is a largely preventable disease. That is, by not smoking it can probably be prevented. . . .

Does smoking cause other lung diseases?

Yes. . . . It leads to chronic bronchitis—a disease where the airways produce excess mucus, which forces the smoker to cough frequently. Smoking is also the major cause of emphysema—a disease that slowly destroys a person's ability to breathe. . . .

Why do smokers have "smoker's cough"?

. . . Normally, cilia (tiny hairlike formations that line the airways) beat outwards and "sweep" harmful material out of the lungs. Smoke, however, decreases this sweeping action, so some of the poisons in the smoke remain in the lungs. . . .

If you smoke but don't inhale, is there any danger?

Yes. Wherever smoke touches living cells, it does harm. So, even if smokers of pipes, cigarettes, and cigars don't inhale, they are at an increased risk for lip, mouth, and tongue cancer. . . .

Does smoking affect the heart?

Yes. Smoking increases the risk of heart disease, which is America's number one killer. . . . Smoking, high blood pressure, high cholesterol, and lack of exercise are all risk factors for heart disease. Smoking alone doubles the risk of heart disease. . . .

Is there any risk for pregnant women and their babies?

Pregnant women who smoke endanger the health and lives of their unborn babies. When a pregnant woman smokes, she really is smoking for two because the nicotine, carbon monoxide, and other dangerous chemicals in smoke enter the mother's bloodstream and then pass into the baby's body. . . .

Does smoking cause any special health problems for women?

Yes. . . . Women who smoke and use the birth control pill have an increased risk of stroke and blood clots in the legs as well. . . . In addition, women who smoke increase their chances of getting cancer of the uterine cervix.

What are some of the short-term effects of smoking cigarettes?

Almost immediately, smoking can make it hard to breathe. Within a short time, it can also worsen asthma and allergies. Nicotine reaches the brain only seven seconds after a smoker takes a puff where it produces a morphinelike effect.

Are there any other risks to the smoker?

Yes. There are many other risks. As we already mentioned briefly, smoking causes stroke, which is the third leading cause of death in America. Smoking causes lung cancer, but if a person smokes and is exposed to radon or asbestos, the risk increases even more. Smokers are also more likely to have and die from stomach ulcers than nonsmokers. . . .

What are the dangers of passive smoking?

. . . Passive smoking causes lung cancer in healthy nonsmokers. Children whose parents smoke are more likely to suffer from pneumonia or bronchitis in the first two years of life than children who come from smoke-free households. Passive smokers have a 30% greater risk for developing lung cancer than nonsmokers who live in a smoke-free house.

Are chewing tobacco and snuff safe alternatives to cigarette smoking?

No, they are not. Many people who use chewing tobacco or snuff believe it can't harm them because there is no smoke. Wrong. Smokeless tobacco contains nicotine, the same addicting drug found in cigarettes and cigars. Snuff dippers also take in an average of over ten times more cancer-causing substances than cigarette smokers. . . . While not inhaled through the lungs, the juice from smokeless tobacco is absorbed through the lining of the mouth. There it can cause sores and white patches, which often lead to cancer of the mouth. . . . .

Bioethical Issue

Since the introduction of the first antibiotics in the 1940s, there has been a dramatic decline in deaths due to respiratory illnesses like pneumonia and tuberculosis. Strep throat and ear infections have also been brought under control with antibiotics, which are chemicals that selectively kill bacteria and not host cells.

There are problems associated with antibiotic therapy, however. Aside from a possible allergic reaction, antibiotics not only kill off disease-causing bacteria, they also reduce the number of beneficial bacteria in the intestinal tract and other locations. These beneficial bacteria hold in check the growth of other microbes that now begin to flourish. Diarrhea can result, as can a vaginal yeast infection. The use of antibiotics can also prevent natural immunity from occurring, leading to the need for recurring antibiotic therapy. Especially alarming at this time is the occurrence of resistance. Resistance takes place when vulnerable bacteria are killed off by an antibiotic, and this allows resistant bacteria to become prevalent. The bacteria that cause ear, nose, and throat infections, and scarlet fever and pneumonia are becoming widely resistant because we have not been using antibiotics properly. Tuberculosis is on the rise, and the new strains are resistant to the usual combined antibiotic therapy. When a disease is caused by a resistant bacterium, it cannot be cured by the administration of any presently available antibiotic.

Although drug companies now recognize the problem and have begun to develop new antibiotics that hopefully will kill bacteria resistant to today's antibiotics, every citizen needs to be aware of our present crisis situation. Stuart Levy, a Tufts University School of Medicine microbiologist says that we should do what is ethical for society and ourselves. What is needed? Antibiotics kill bacteria, not viruses—therefore, we shouldn't take antibiotics unless we know for sure we have a bacterial infection. And we shouldn't take them prophylactically—that is, just in case we might need one. If antibiotics are taken in low dosages and intermittently, resistant strains are bound to take over. Animal and agricultural use should be pared down, and household disinfectants should no longer be spiked with antibacterial agents. Perhaps then, Levy says, vulnerable bacteria will begin to supplant the resistant ones in the population.

Questions

1. With regard to antibiotics, should each person think about the needs of society as well as themselves? Why or why not?
2. Should each person do what they can to help prevent the growing resistance of bacteria to disease? Why or why not?
3. Should you gracefully accept a physician's decision that an antibiotic will not help any illness you may have? Why or why not?

Summarizing the Concepts

15.1 Respiratory Tract

The respiratory tract consists of the nasal cavities, the nasopharynx, the pharynx, the larynx (which contains the vocal cords), the trachea, the bronchi, and the bronchioles. The bronchi, along with the pulmonary arteries and veins, enter the lungs, which consist of the alveoli, air sacs surrounded by a capillary network.

15.2 Mechanism of Breathing

Inspiration begins when the respiratory center in the medulla oblongata of the brain sends excitatory nerve impulses to the diaphragm and the muscles of the rib cage. As they contract, the diaphragm lowers and the rib cage moves upward and outward; the lungs expand, creating a partial vacuum, which causes air to rush in. The respiratory center now stops sending impulses to the diaphragm and muscles of the rib cage. As the diaphragm relaxes, it resumes its dome shape, and as the rib cage retracts, air is pushed out of the lungs during expiration.

15.3 Gas Exchanges in the Body

External respiration occurs when CO_2 leaves blood via the alveoli and O_2 enters blood from the alveoli. Oxygen is transported to the tissues in combination with hemoglobin as oxyhemoglobin (HbO_2). Internal respiration occurs when O_2 leaves blood and CO_2 enters blood at the tissues. Carbon dioxide is mainly carried to the lungs within the plasma as the bicarbonate ion (HCO_3^-). Hemoglobin combines with hydrogen ions from this reaction and becomes reduced (HHb).

15.4 Respiration and Health

A number of illnesses are associated with the respiratory tract. The disorders of the respiratory tract are divided into those that affect the upper respiratory tract and those that affect the lower respiratory tract. Infections of the nasal cavities, sinuses, throat, tonsils, and larynx are all well known. In addition, infections can spread from the nasopharynx to the ears.

The lower respiratory tract is also subject to infections such as acute bronchitis, pneumonia, and pulmonary tuberculosis. In restrictive pulmonary disorders, exemplified by pulmonary fibrosis, the lungs lose their elasticity. In obstructive pulmonary disorders, exemplified by chronic bronchitis, emphysema, and asthma, the bronchi (and bronchioles) do not effectively conduct air to and from the lungs. Smoking, which is associated with chronic bronchitis and emphysema, can eventually lead to lung cancer.

Studying the Concepts

1. List the parts of the respiratory tract. What are the special functions of the nasal cavity, the larynx, and the alveoli? 284–87
2. Name and explain the four parts of respiration. 288
3. What is the difference between tidal volume and vital capacity? Of the air we breathe, what part is not used for gas exchange? 288
4. What are the steps in inspiration and expiration? How is breathing controlled? 290–91
5. Discuss the events of external respiration, and include two pertinent equations in your discussion. 292

6. What two equations pertain to the exchange of gases during internal respiration? 292
7. State three factors that influence hemoglobin's O_2 binding capacity, and relate them to the environmental conditions in the lungs and tissues. 294
8. Name four respiratory tract infections other than cancer, and explain why breathing is difficult with these conditions. 295–98
9. What are emphysema and pulmonary fibrosis, and how do they affect a person's health? 297
10. List the steps by which lung cancer develops. 298

Testing Yourself

Choose the best answer for each question.

1. Which of these is anatomically incorrect?
   a. The nose has two nasal cavities.
   b. The pharynx connects the nasal and oral cavities to the larynx.
   c. The larynx contains the vocal cords.
   d. The trachea enters the lungs.
   e. The lungs contain many alveoli.
2. The maximum volume of air that can be moved in and out during a single breath is called the
   a. expiratory and inspiratory reserve volume.
   b. residual volume.
   c. tidal volume.
   d. vital capacity.
   e. functional residual capacity.
3. The enzyme carbonic anhydrase
   a. causes the blood to be more acidic in the tissues.
   b. speeds up the conversion of carbonic acid to carbon dioxide and water.
   c. actively transports carbon dioxide out of capillaries.
   d. is active only at high altitudes.
   e. All of these are correct.
4. Oxygen and carbon dioxide
   a. both exit and enter the blood in the lungs and tissues.
   b. are both carried by hemoglobin.
   c. are present only in arteries and not in veins.
   d. are both given off by mitochondria.
   e. Both a and c are correct.
5. Which of these statements is true?
   a. The Po_2, temperature, and pH are higher in the lungs.
   b. The Po_2, temperature, and pH are lower in the lungs.
   c. The Po_2 and temperature are higher and pH is lower in the lungs.
   d. The Po_2 and temperature are lower and the pH is higher in the lungs.
   e. The Po_2 and pH are higher but the temperature is lower in the lungs.
6. Carbon dioxide
   a. is carried as the bicarbonate ion in red blood cells.
   b. combines with hemoglobin in the tissues.
   c. is carried as the bicarbonate ion in the plasma.
   d. enters the capillaries in the lungs and exits capillaries in the tissues.
   e. Both b and c are correct.
7. Air enters the human lungs because
   a. atmospheric pressure is less than the pressure inside the lungs.
   b. atmospheric pressure is greater than the pressure inside the lungs.
   c. although the pressures are the same inside and outside, the partial pressure of oxygen is lower within the lungs.
   d. the residual air in the lungs causes the partial pressure of oxygen to be less than it is outside.
8. To trace the path of air in humans you would place the trachea
   a. directly after the nose.
   b. directly before the bronchi.
   c. before the pharynx.
   d. Both a and c are correct.
9. In humans, the respiratory center
   a. is stimulated by carbon dioxide.
   b. is located in the medulla oblongata.
   c. controls the rate of breathing.
   d. All of these are correct.
10. Which one of these is not an obstructive pulmonary disorder?
    a. pulmonary tuberculosis
    b. emphysema
    c. chronic bronchitis
    d asthma
    e. a disorder that keeps air from flowing freely into and out of the lungs
11. Label the diagram of the human respiratory tract.

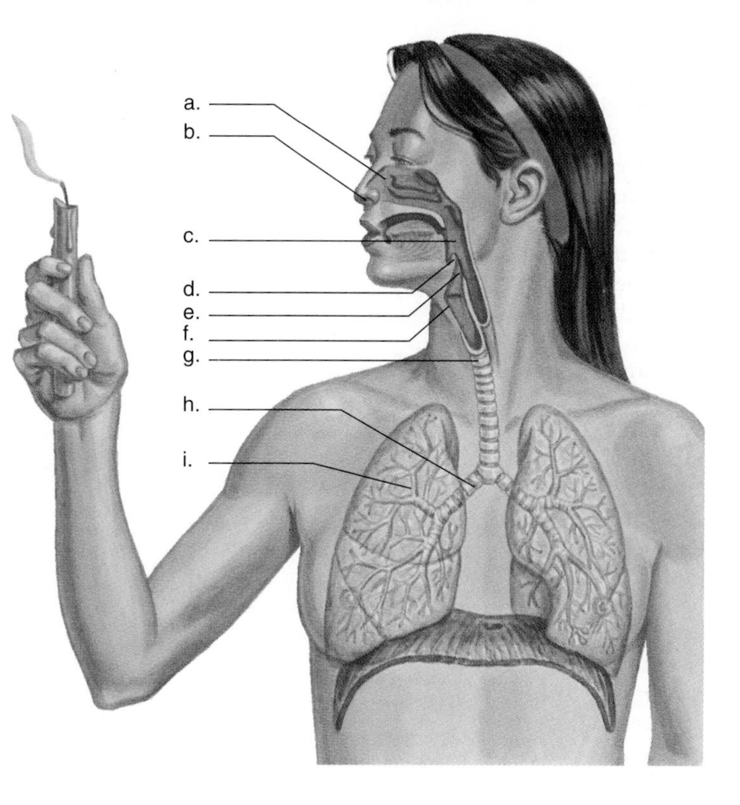

Thinking Scientifically

1. Consider these reactions:
   $$CO_2 + H_2O \rightleftharpoons H_2CO_3 \rightleftharpoons H^+ + HCO_3^-$$
   a. Why would you expect blood in the tissues to be more acidic than blood in the lungs? (page 294)
   b. Why would you expect blood in the lungs to be more basic than blood in the tissues?
   c. Carbonic acid (H_2CO_3) is a weak acid (dissociates only slightly). Why does this help to buffer blood if hydrogen ions (H^+) are added?
   d. Water also dissociates only slightly. What reaction would you expect if hydroxide ions (OH^-) are added to blood? How does this reaction help to buffer blood?
2. Considering the control of breathing (page 292):
   a. Why is it better to give a person a mixture of oxygen (O_2) and carbon dioxide (CO_2) gases to stimulate breathing?
   b. Why is it impossible for a person to commit suicide by holding his or her breath? (Hint—what gas builds up?)
   c. Why is it workable for the carotid and aortic bodies to be sensitive to the hydrogen ion concentration $[H^+]$ of the blood rather than O_2?
   d. Why would you predict that the respiratory center is sensitive to the *presence* of carbon dioxide rather than to the *absence* of oxygen in the blood?

Understanding the Terms

Match the terms to these definitions:

a. __________ Common passageway for both food intake and air movement, located between the mouth and the esophagus.
b. __________ Dome-shaped muscularized sheet separating the thoracic cavity from the abdominal cavity.
c. __________ Form in which most of the carbon dioxide is transported in the bloodstream.
d. __________ Stage during breathing when air is pushed out of the lungs.
e. __________ Terminal, microscopic, grapelike air sac found in lungs.

Using Technology

Your study of the digestive system and nutrition is supported by these available technologies:

Essential Study Partner CD-ROM
Animals → Respiration

Visit the Mader web site for related ESP activities.

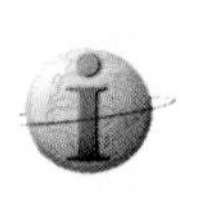

Exploring the Internet
The Mader Home Page provides resources and tools as you study this chapter.

http://www.mhhe.com/biosci/genbio/mader

Dynamic Human 2.0 CD-ROM
Respiratory System

Virtual Physiology Laboratory CD-ROM
Pulmonary Function
Respiration and Exercise

HealthQuest CD-ROM
2 Fitness
7 Tobacco

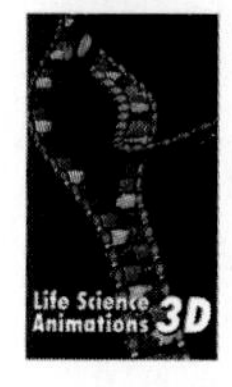

Life Science Animations 3D Video
37 Gas Exchange

C H A P T E R 16

Urinary System and Excretion

Chapter Concepts

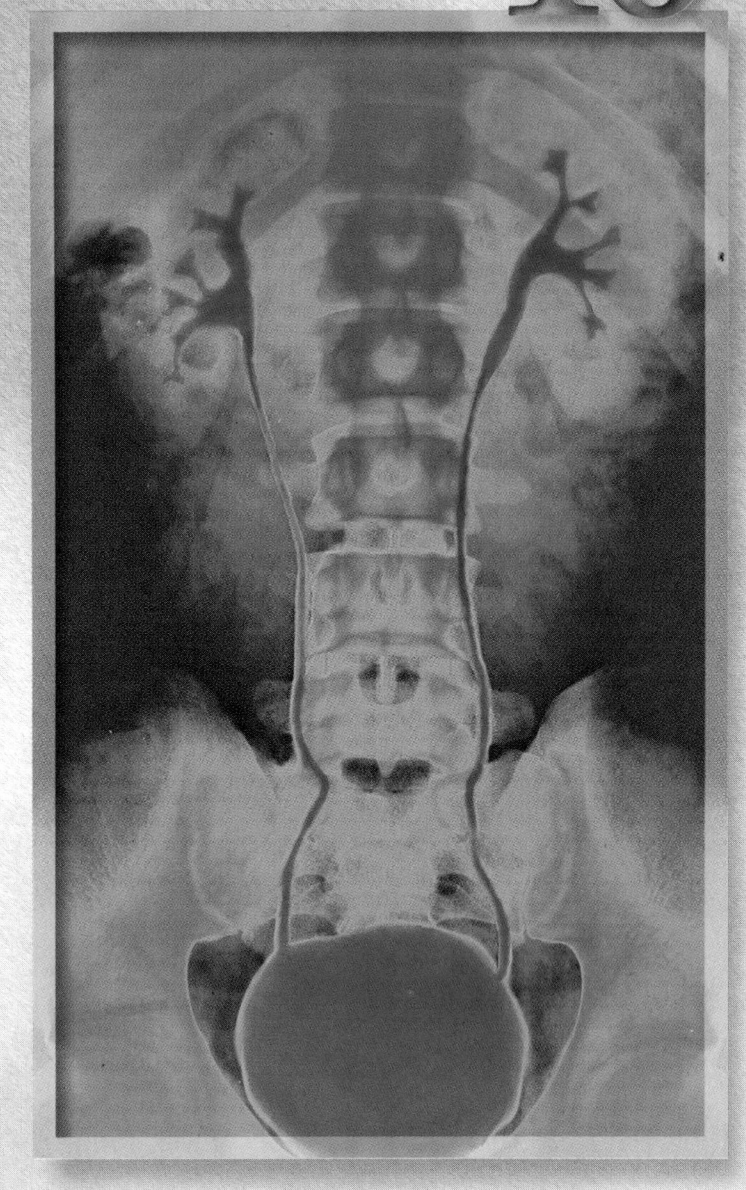

X-ray enhanced picture of lower abdominal area of a human features the urinary system. Urine formed in the kidneys (light green) passes by way of ureters to the bladder (both red) where it is stored before passing out of the body.

A helicopter lands on the roof of the hospital. Paramedics rush an insulated container from the aircraft to an operating room a few stories below. Brushing aside the ice in the container, a surgeon plucks out a fist-sized reddish mass, a kidney. Within hours, the organ, which has replaced the diseased kidneys inside a young girl's body, is busy producing urine. The transplanted organ, if not rejected, should save the girl from a difficult life of being periodically hooked up to dialysis machines. Rejection is unlikely because it has already been determined that the tissues of the donor are very compatible with those of the recipient.

A kidney is absolutely essential for a healthy life because it helps regulate the pH and the water-salt balance of blood, and it excretes nitrogenous wastes. By regulating the amount of salt and water in the blood, a kidney helps keep blood pressure within a normal range. By excreting nitrogenous wastes, it rids the body of toxic substances. One kidney alone is all we need, and therefore the donor of a kidney will suffer no ill consequences except the trauma of abdominal surgery. This chapter will detail exactly how a kidney performs its life-preserving functions.

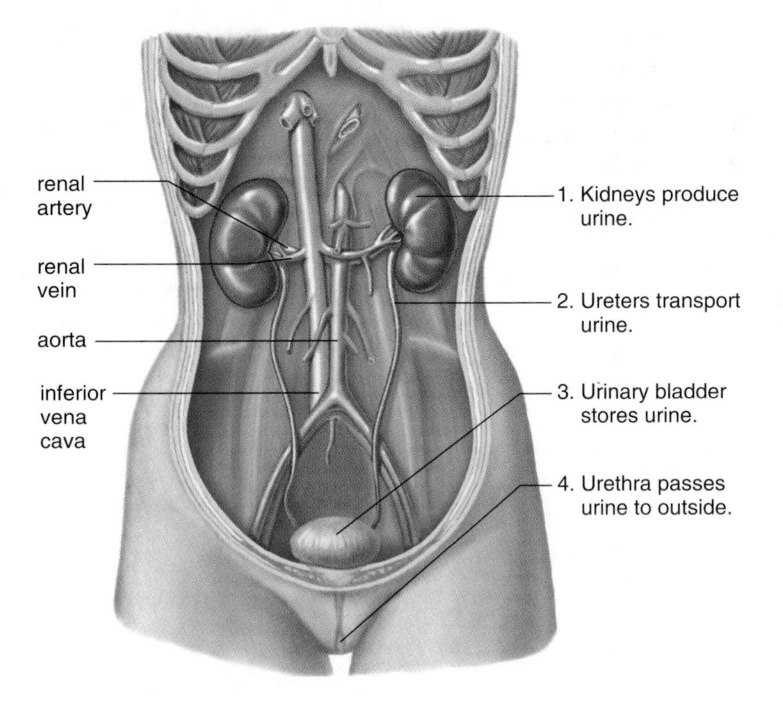

Figure 16.1 The urinary system.
Urine is found only within the kidneys, the ureters, the urinary bladder, and the urethra.

16.1 Urinary System

The urinary system includes the kidneys and associated structures, which are illustrated in Figure 16.1. The kidneys produce urine, which passes by way of the ureters to the bladder where it is stored. The urethra carries urine to outside the body.

The Urinary Organs

The kidneys are found on either side of the vertebral column, just below the diaphragm. They lie in depressions against the deep muscles of the back beneath the peritoneum, the lining of the abdominal cavity, where they also receive some protection from the lower rib cage. But the kidneys can be damaged by blows on the back—kidney punches are not allowed in boxing. Each kidney is usually held in place by connective tissue, called renal fascia. A sharp blow to the back can dislodge a kidney, which is then called a floating kidney.

The **kidneys** are bean-shaped, reddish brown organs, each about the size of a fist, which produce urine. They are covered by a tough capsule of fibrous connective tissue overlaid by adipose tissue. A depression (the hilum) on the concave side is where the **renal artery** enters and the **renal vein** and ureters exit.

The **ureters,** and indeed the entire urinary tract are lined by a mucosa. The ureters are tubes about 25 cm long that convey the urine from the kidneys toward the bladder by peristalsis. Urine enters the bladder by peristaltic contractions, in jets that occur at the rate of five per minute.

The **urinary bladder,** which can hold up to 600 ml of urine, is a hollow, muscular organ that gradually expands as urine enters. The wall of the bladder consists of connective tissue and smooth muscle. The bladder's outer serosa is continuous with the parietal peritoneum. (You get the urge to void when the bladder fills to about 250 milliliters, and you become uncomfortable at about 500 milliliters. When the bladder becomes overdistended, you may lose the urge to void.)

The **urethra** extends from the urinary bladder to an external opening called the external urethral orifice. The internal urethral sphincter occurs where the urethra leaves the bladder, and an external urethral sphincter is located where the urethra exits the pelvic cavity.

The urethra differs in length in the female and the male. In the female, the urethra is only about 4 cm long. The short length of the female urethra makes bacterial invasion easier, and explains why females are more prone to urinary tract infections than males. In the male, the urethra averages 20 cm when the penis is flaccid (limp, nonerect). As the urethra leaves the male urinary bladder, it is encircled by the prostate gland. In older men, enlargement of the prostate gland can restrict urination, a condition that usually can be corrected surgically.

There is no connection between the genital (reproductive) and urinary systems in females; there is a connection in males because the urethra also carries sperm during ejaculation. This double function does not alter the path of urine, and it is important to realize that urine is found only in those structures noted in Figure 16.2.

Urination and the Nervous System

When the urinary bladder fills with urine to about 250 ml, stretch receptors send sensory nerve impulses to the spinal cord. Subsequently, motor nerve impulses from the spinal cord cause the urinary bladder to contract and the sphincters to relax so that urination is possible. In older children and adults, the brain controls this reflex, delaying urination until a suitable time (Fig. 16.2).

> Only the urinary system, consisting of the kidneys, the urinary bladder, the ureters, and the urethra, holds urine.

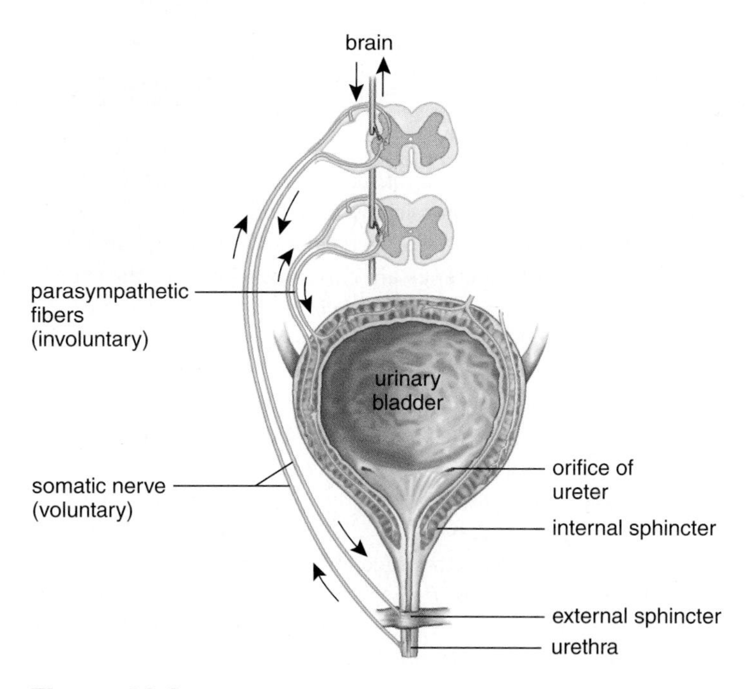

Figure 16.2 Urination.
As the bladder fills with urine, sensory impulses go to the spinal cord and then the brain. The brain can override the urge to urinate. When urination occurs, motor nerve impulses cause the bladder to contract and an internal sphincter to open. Nerve impulses also cause an external sphincter to open.

Functions of the Urinary System

The primary functions of the urinary system are carried out by the kidneys. The kidneys are organs of excretion. **Excretion** is the removal of metabolic wastes from the body. The kidneys also maintain the water-salt balance and the acid-base balance of the body. In addition, they have a hormonal function.

Metabolic Wastes

The kidneys are the primary organs for excretion of nitrogenous wastes including urea, creatinine, and uric acid.

Urea is the nitrogenous end product of amino acid metabolism. The breakdown of amino acids in the liver releases ammonia, which the liver combines with carbon dioxide to produce urea. Ammonia is very toxic to cells and urea is much less toxic. Urea is the primary nitrogenous end product of human beings.

Two other nitrogenous end products are excreted by the kidneys. **Creatinine** is the end product of creatine phosphate metabolism. Creatine phosphate is a high-energy phosphate reserve molecule in muscles. The breakdown of nucleotides produces **uric acid,** which is rather insoluble. If too much uric acid is present in blood, it precipitates out. Crystals of uric acid sometimes collect in the joints, producing a painful ailment called gout.

Water-Salt Balance

A principal function of the kidneys is to maintain the appropriate water-salt balance of the body. As we shall see, blood volume is intimately associated with the salt balance of the body. As you know, salts, such as NaCl, have the ability to cause osmosis, the diffusion of water—in this case into the blood. The more salts there are in the blood, the greater the blood volume and the greater the blood pressure. Therefore, the kidneys are also involved in regulating blood pressure.

The kidneys maintain the appropriate level of other ions such as potassium ions (K^+), bicarbonate ions (HCO_3^-), and calcium ions (Ca^{2+}) in the blood.

Acid-Base Balance

The kidneys also regulate the blood acid-base balance. In order for us to remain healthy, the blood pH should be just about 7.4. The kidneys monitor and control blood pH, mainly by excreting hydrogen ions (H^+) and reabsorbing bicarbonate ion (HCO_3^-) as needed. Urine usually has a pH of 6 or lower.

Hormonal Function

The kidneys assist the endocrine system. They secrete the hormone **erythropoietin,** which stimulates red blood cell production. They also modify a precursor molecule from the skin so that it becomes active vitamin D (calcitriol). Vitamin D promotes calcium (Ca^{2+}) reabsorption from the digestive tract.

The kidneys also secrete renin, a substance involved in the secretion of aldosterone from the adrenal cortex. Aldosterone causes the reabsorption of sodium ions (Na^+).

> The kidneys are major organs of homeostasis because they excrete nitrogenous wastes. They also regulate the water-salt balance and the acid-base balance of the blood.

Health Focus

Urinary Tract Infections Require Attention

Although males can get a urinary tract infection, the condition is 50 times more common in women. The explanation lies in a comparison of male and female anatomy (Fig. 16A). The female urethral and anal openings are closer together, and the shorter urethra makes it easier for bacteria from the bowels to enter and start an infection. Although it is possible to have no outward signs of an infection, usually urination is painful, and patients often describe a burning sensation. The urge to pass urine is frequent, but it may be difficult to start the stream. Chills with fever, nausea, and vomiting may be present.

Urinary tract infections can be confined to the urethra, in which case urethritis is present. If the bladder is involved, it is called cystitis. Should the infection reach the kidneys, the person has pyelonephritis. *Escherichia coli* (*E. coli*), a normal bacterial resident of the large intestine, is usually the cause of infection. Since the infection is caused by a bacterium, it is curable by antibiotic therapy. The problem is, however, that reinfection is possible as soon as antibiotic therapy is finished.

It makes sense to try to prevent infection in the first place. These tips might help.

Men and women should drink lots of water. Try to drink from 2–2.5 liters of liquid a day. Try to avoid caffeinated drinks, which may be irritating. Cranberry juice is recommended because it contains a substance that stops bacteria from sticking to the bladder wall once an infection has set in. If an attack occurs, testing and antibiotic therapy may be in order. Keep in mind that sexually transmitted diseases such as gonorrhea, chlamydia, or herpes can cause urinary tract infections. All personal behaviors should be examined carefully, and suitable adjustments should be made to avoid urinary tract infections.

Most women have a urinary tract infection for the first time shortly after they become sexually active. Honeymoon cystitis was coined because of the common association of urinary tract infections with sexual intercourse. Washing the genitals before having sex and being careful not to introduce bacteria from the anus into the urethra is recommended. Also, urinating immediately before and after sex will help to flush out any bacteria that are present. A diaphragm may press on the urethra and prevent adequate emptying of the bladder, and estrogen, such as in birth-control pills, can increase the risk of cystitis. A sex partner may have an asymptomatic (no symptoms) urinary infection that causes a woman to become infected repeatedly.

Women should wipe from the front to the back after using the toilet. Perfumed toilet paper and any other perfumed products that come in contact with the genitals may be irritating. Wearing loose clothing and cotton underwear discourages the growth of bacteria, while tight clothing, such as jeans and panty hose, provides an environment for the growth of bacteria.

Personal hygiene is especially important too at the time of menstruation. Hands should be washed before and after changing napkins and/or tampons. Superabsorbent tampons are not best if they are changed infrequently, as this may encourage the growth of bacteria. Also, sexual intercourse may cause menstrual flow to enter the urethra.

In males, the prostate is a gland that surrounds the urethra just below the bladder (Fig. 16A). The prostate contributes secretions to semen whenever semen enters the urethra prior to ejaculation. An infection of the prostate, called prostatitis, is often accompanied by a urinary tract infection. Fever is present and the prostate is tender and inflamed. The patient may have to be hospitalized and treated with a broad spectrum antibiotic. Prostatitis, which in a young person is often preceded by a sexually transmitted disease, can lead to a chronic condition. Chronic prostatitis may be asymptomatic or, as is more typical, there is irritation upon voiding and/or difficulty in voiding.

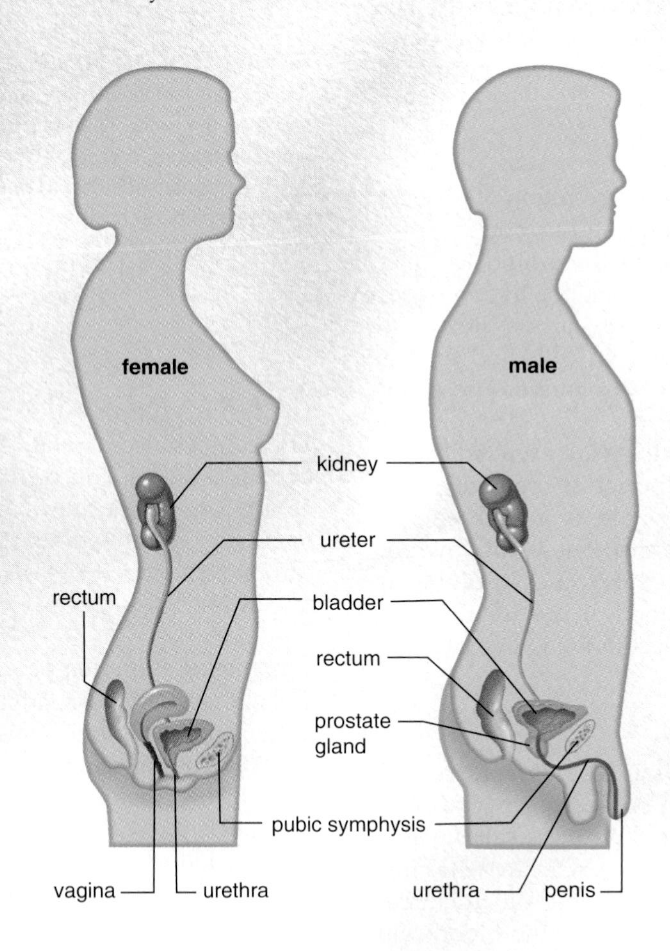

Figure 16A Female versus male urinary tract. Females have a short urinary tract compared to that of males. This means that it is easier for bacteria to invade the urethra and helps explain why females are 50 times more likely than males to get a urinary tract infection.

16.2 The Kidneys

When a kidney is sliced lengthwise, it is possible to see the many branches of the renal artery and vein that reach inside the kidney (Fig. 16.3*a*). If the blood vessels are removed, it is easier to identify three regions of a kidney. The **renal cortex** is an outer granulated layer that dips down in between a radially striated, or lined, inner layer called the renal medulla. The **renal medulla** consists of cone-shaped tissue masses called renal pyramids. The **renal pelvis** is a central space, or cavity, that is continuous with the ureter (Fig. 16.3*b*).

Microscopically, the kidney is composed of over one million **nephrons,** sometimes called renal or kidney tubules (Fig. 16.3*c*). The nephrons produce urine and are positioned so that the urine flows into a collecting duct. Several nephrons enter the same collecting duct; the collecting ducts enter the renal pelvis.

Macroscopically, a kidney has three regions: renal cortex, renal medulla, and a renal pelvis that is continuous with the ureter. Microscopically, a kidney contains over one million nephrons.

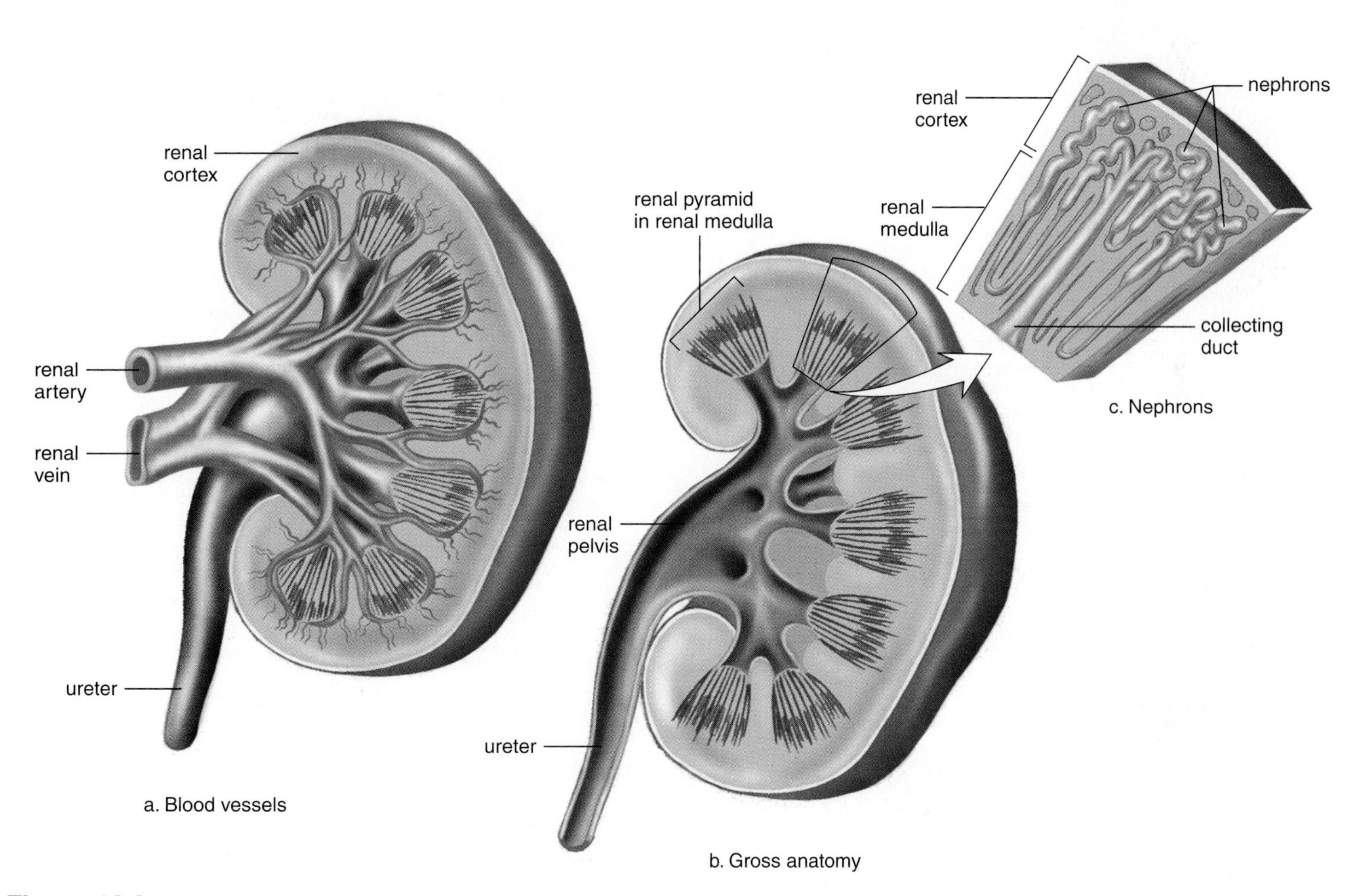

Figure 16.3 **Gross anatomy of the kidney.**
a. A longitudinal section of the kidney showing the blood supply. Note that the renal artery divides into smaller arteries, and these divide into arterioles. Venules join to form small veins, which join to form the renal vein. **b.** The same section without the blood supply. Now it is easier to distinguish the renal cortex, the renal medulla, and the renal pelvis, which connects with a ureter. The renal medulla consists of the renal pyramids. **c.** An enlargement showing the placement of nephrons.

Anatomy of a Nephron

Each nephron has its own blood supply, including two capillary regions (Fig 16.4). From the renal artery, an afferent arteriole leads to the **glomerulus,** a knot of capillaries inside the glomerular capsule. Blood leaving the glomerulus enters the efferent arteriole and then the **peritubular capillary network,** which surrounds the rest of the nephron. From there the blood goes into a venule that joins the renal vein.

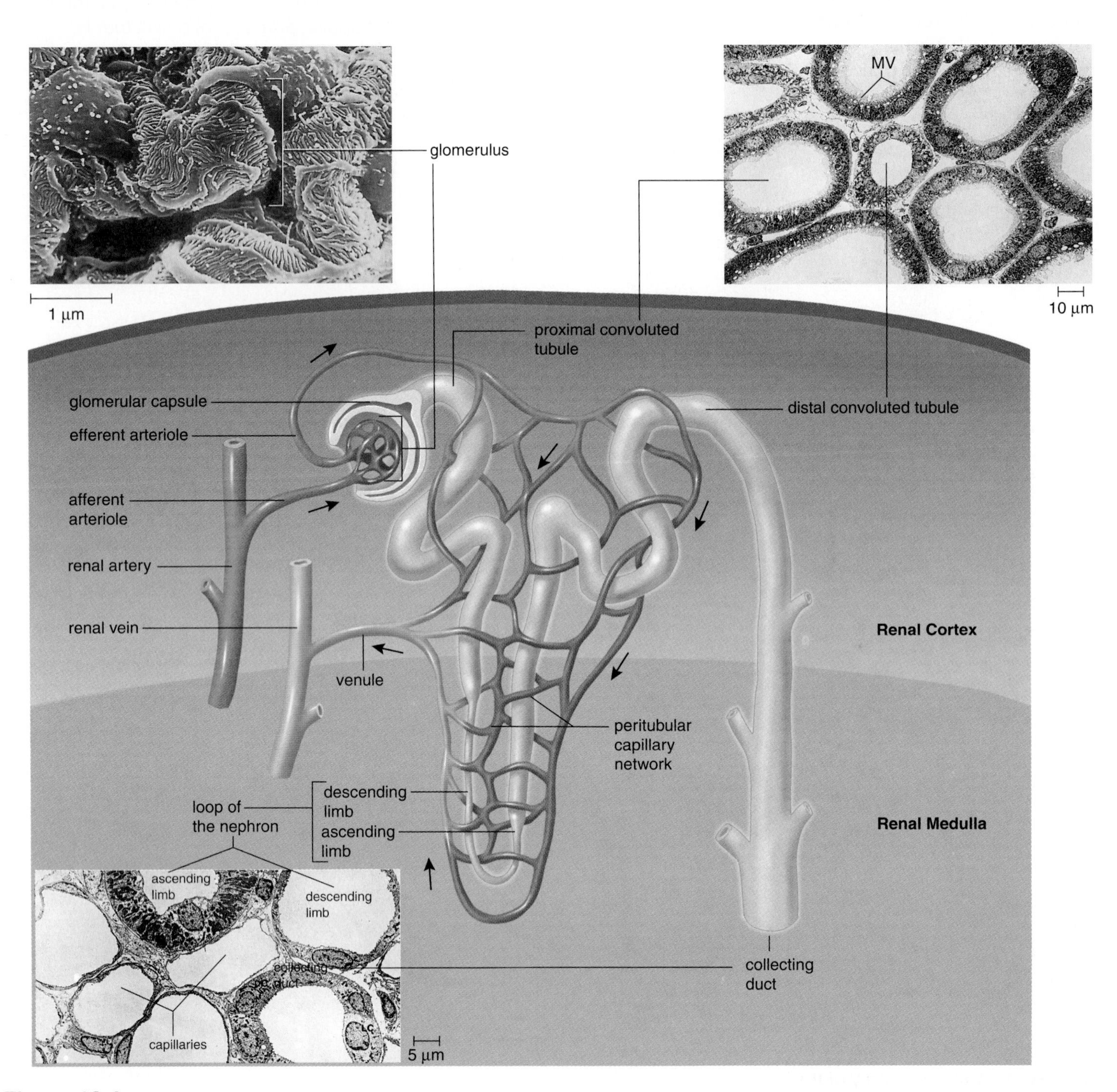

Figure 16.4 Nephron anatomy.
A nephron is made up of a glomerular capsule, the proximal convoluted tubule, the loop of the nephron, the distal convoluted tubule, and the collecting duct. The micrographs show these structures in cross section; MV = microvilli. You can trace the path of blood about the nephron by following the arrows.

Parts of a Nephron

Each nephron is made up of several parts (Fig. 16.4). The structure of each part suits its function.

First, the closed end of the nephron is pushed in on itself to form a cuplike structure called the **glomerular capsule** (Bowman's capsule). The outer layer of the glomerular capsule is composed of squamous epithelial cells; the inner layer is made up of **podocytes** that have long cytoplasmic processes. The podocytes cling to the capillary walls of the glomerulus and leave pores that allow easy passage of small molecules from the glomerulus to the inside of the glomerular capsule. This process, called *glomerular filtration,* produces a filtrate of blood.

Next, there is a **proximal** (meaning near the glomerular capsule) **convoluted tubule.** The cuboidal epithelial cells lining this part of the nephron have numerous microvilli, about 1 μm in length, that are tightly packed and form a brush border (Fig. 16.5). A brush border greatly increases the surface area for the *tubular reabsorption* of filtrate components. Each cell also has many mitochondria, which can supply energy for active transport of molecules from the lumen to the peritubular capillary network.

Simple squamous epithelium appears as the tube narrows and makes a U-turn called the **loop of the nephron** (loop of Henle). Each loop consists of a descending limb that allows water to leave and an ascending limb that extrudes salt (NaCl). Indeed, as we shall see, this activity facilitates the reabsorption of water by the nephron and collecting duct.

The cells of the **distal convoluted tubule** have numerous mitochondria, but they lack microvilli. This is consistent with the active role they play in moving molecules from the blood into the tubule, a process called *tubular secretion.* The distal convoluted tubules of several nephrons enter one collecting duct. A kidney contains many collecting ducts, which carry urine to the renal pelvis.

As shown in Figure 16.4, the glomerular capsule and the convoluted tubules always lie within the renal cortex. The loop of the nephron dips down into the renal medulla; a few nephrons have a very long loop of the nephron, which penetrates deep into the renal medulla. **Collecting ducts** are also located in the renal medulla, and they give the renal pyramids their lined appearance.

Each part of a nephron is anatomically suited to its specific function in urine formation.

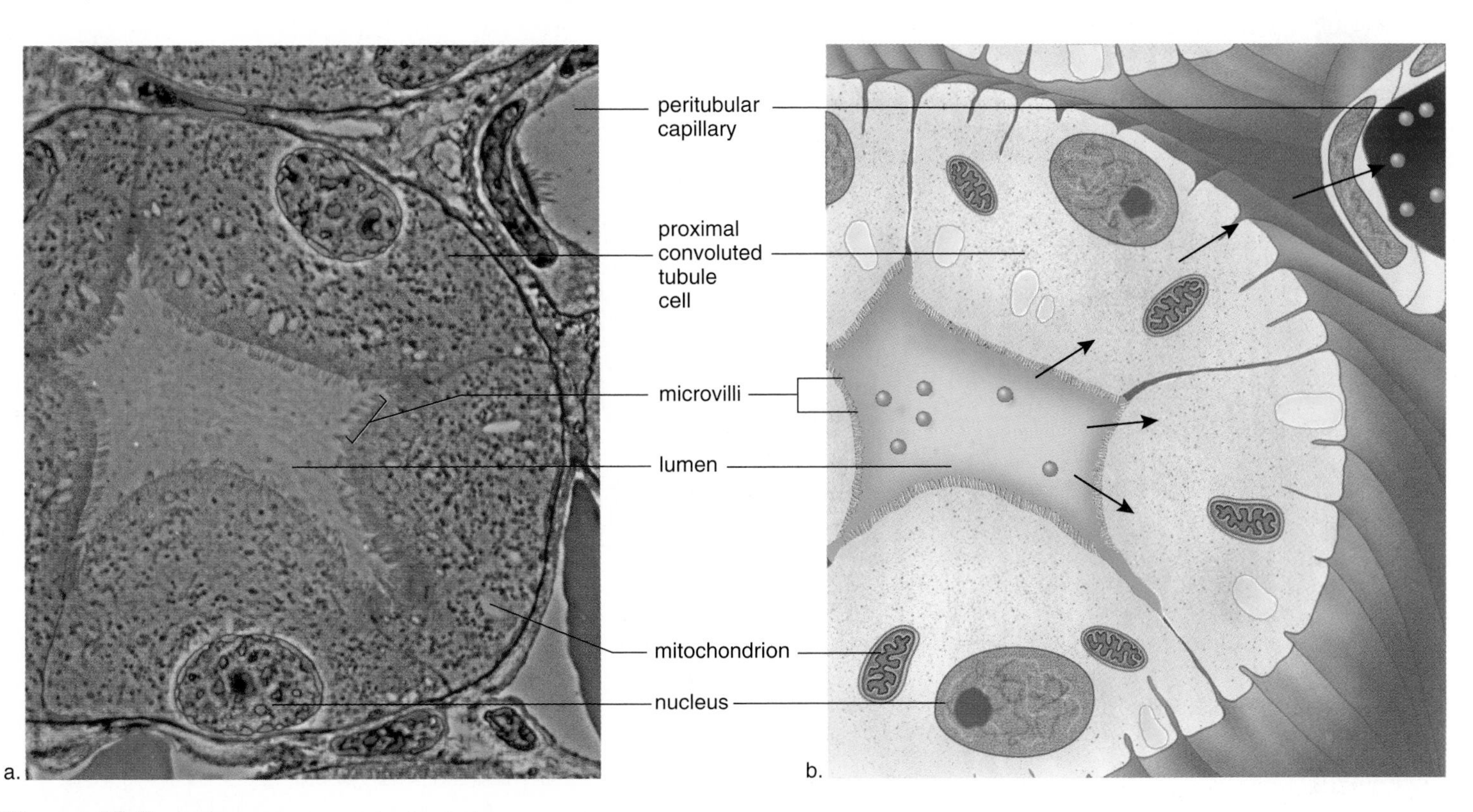

Figure 16.5 Proximal convoluted tubule.
a. This photomicrograph shows that the cells lining the proximal convoluted tubule have a brushlike border composed of microvilli, which greatly increases the surface area exposed to the lumen. The peritubular capillary network surrounds the cells. **b.** Diagrammatic representation of **(a)** shows that each cell has many mitochondria, which supply the energy needed for active transport, the process that moves molecules (green) from the lumen of the tubule to the capillary, as indicated by the arrows.

Visual Focus

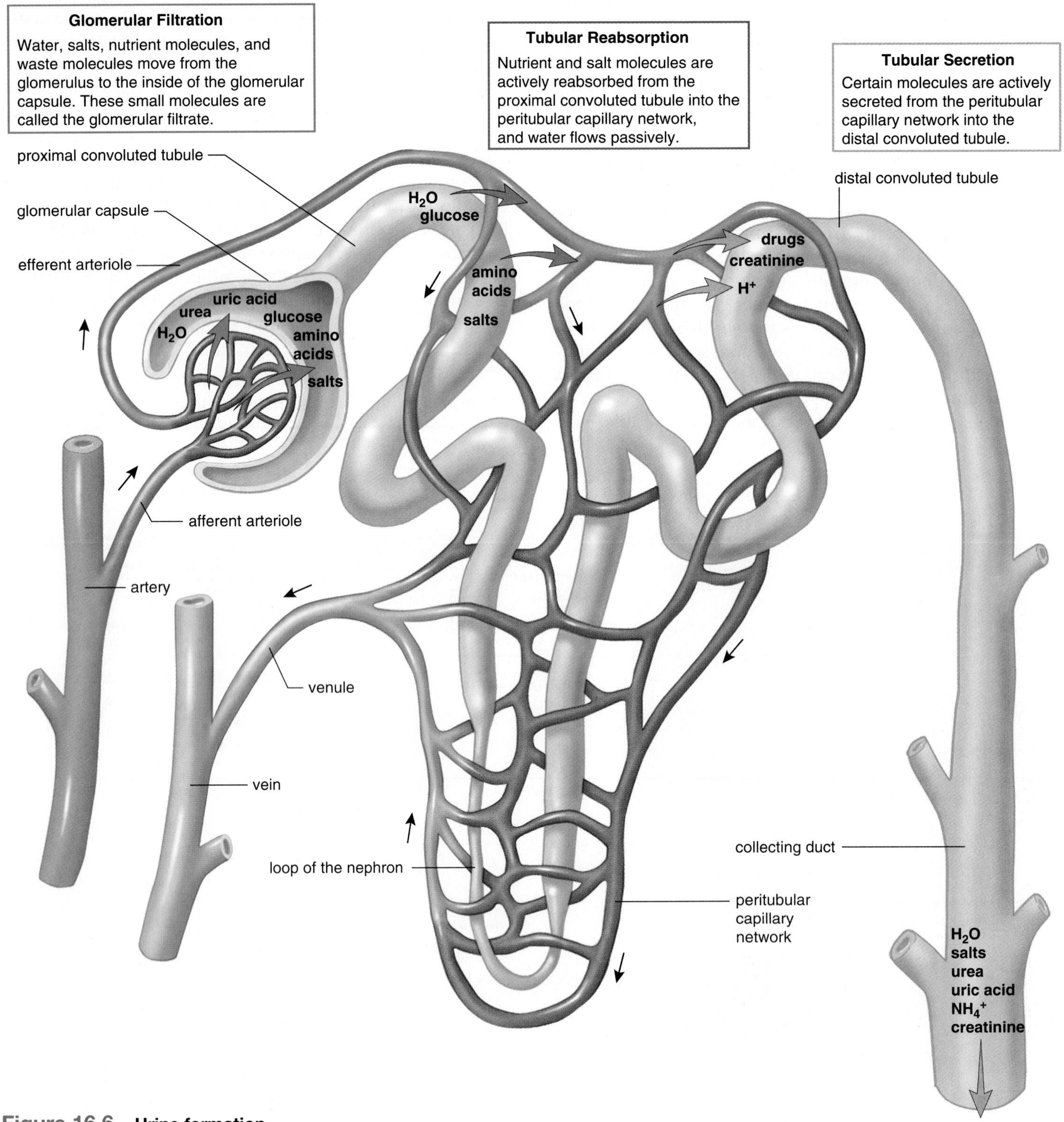

Glomerular Filtration

Water, salts, nutrient molecules, and waste molecules move from the glomerulus to the inside of the glomerular capsule. These small molecules are called the glomerular filtrate.

Tubular Reabsorption

Nutrient and salt molecules are actively reabsorbed from the proximal convoluted tubule into the peritubular capillary network, and water flows passively.

Tubular Secretion

Certain molecules are actively secreted from the peritubular capillary network into the distal convoluted tubule.

proximal convoluted tubule
glomerular capsule
efferent arteriole
uric acid
urea
H_2O
glucose
amino acids
salts
afferent arteriole
artery
H_2O
glucose
amino acids
salts
drugs
creatinine
H^+
distal convoluted tubule
venule
vein
loop of the nephron
collecting duct
peritubular capillary network
H_2O
salts
urea
uric acid
NH_4^+
creatinine

Figure 16.6 Urine formation.
The three steps in urine formation are numbered. Reabsorption of water is not an individual step because it occurs along the length of the nephron and also at the loop of the nephron and collecting duct. Excretion is not a step because it is the end result.

16.3 Urine Formation

Figure 16.6 gives an overview of urine formation, which is divided into these steps: glomerular filtration, tubular reabsorption, and tubular secretion.

Glomerular Filtration

Glomerular filtration occurs when whole blood enters the afferent arteriole and the glomerulus. Due to glomerular blood pressure, which is usually about 60 mm Hg, water and small molecules move from the glomerulus to the inside of the glomerular capsule. This is a filtration process because large molecules and formed elements are unable to pass through the capillary wall. In effect, then, blood in the glomerulus has two portions: the filterable components and the nonfilterable components.

Filterable Blood Components	Nonfilterable Blood Components
Water	Formed elements (blood cells and platelets)
Nitrogenous wastes	Proteins
Nutrients	
Salts (ions)	

The **glomerular filtrate** contains small dissolved molecules in approximately the same concentration as plasma. Small molecules that escape being filtered and the nonfilterable components leave the glomerulus by way of the efferent arteriole.

One hundred eighty liters of water are filtered per day along with a considerable amount of small molecules, such as glucose and amino acids. If the composition of urine were the same as that of the glomerular filtrate, the body would continually lose water, salts, and nutrients. Death from dehydration, starvation, and low blood pressure would quickly follow. Therefore, we can conclude that the composition of the filtrate must be altered as this fluid passes through the remainder of the tubule.

Tubular Reabsorption

Tubular reabsorption occurs as molecules and ions are both passively and actively reabsorbed from the nephron into the blood of the peritubular capillary network. The osmolarity of the blood is maintained by the presence of plasma proteins and also by salt. When sodium ions (Na^+) are actively reabsorbed, chloride ions (Cl^-) follow passively. The reabsorption of salt (NaCl) increases the osmolarity of the blood compared to the filtrate, and therefore water moves passively from the tubule into the blood. About 67% of Na^+ is reabsorbed at the proximal convoluted tubule.

Nutrients such as glucose and amino acids also return to the blood at the proximal convoluted tubule. This is a selective process because only molecules recognized by carrier molecules are actively reabsorbed. Glucose is an example of a molecule that ordinarily is completely reabsorbed because there is a plentiful supply of carrier molecules for it. However, every substance has a maximum rate of transport, and after all its carriers are in use, any excess in the filtrate will appear in the urine. For example, as reabsorbed levels of glucose approach 180–200 mg/100 ml plasma, the rest will appear in the urine. In diabetes mellitus, excess glucose occurs in the blood, and then in the filtrate, and then in the urine, because the liver and muscles fail to store glucose as glycogen and the kidneys cannot reabsorb all of it. The presence of glucose in the filtrate increases its osmolarity compared to blood, and therefore less water is reabsorbed into the peritubular capillary network. The frequent urination and increased thirst experienced by patients with uncontrolled diabetes mellitus is due to the fact that water is remaining in the filtrate and is not being reabsorbed.

We have seen that the filtrate that enters the proximal convoluted tubule is divided into two portions: components that are reabsorbed from the tubule into blood, and components that are nonreabsorbed and continue to pass through the nephron to be further processed into urine.

Reabsorbed Filtrate Components	Nonreabsorbed Filtrate Components
Most water	Some water
Nutrients	Much nitrogenous waste
Required salts (ions)	Excess salts (ions)

The substances that are not reabsorbed become the tubular fluid, which enters the loop of the nephron.

Tubular Secretion

Tubular secretion is a second way by which substances are removed from blood and added to the tubular fluid. Hydrogen ions, creatinine, and drugs such as penicillin are some of the substances that are moved by active transport from blood into the distal convoluted tubule. In the end, urine contains substances that underwent glomerular filtration but were not reabsorbed, and substances that underwent tubular secretion.

Urine formation requires glomerular filtration (small molecules enter tubule), tubular reabsorption (many molecules are reabsorbed), and tubular secretion (substances are actively added to tubule).

16.4 Maintaining Water-Salt Balance

The kidneys regulate the water-salt balance of the blood. In this way, they also maintain the blood volume and blood pressure. Most of the water and salt (NaCl) present in the filtrate is reabsorbed across the wall of the proximal convoluted tubule. Reabsorption also occurs along the remainder of the nephron.

Reabsorption of Water

The excretion of a hypertonic urine (one that is more concentrated than blood) is dependent upon the reabsorption of water from the loop of the nephron (loop of Henle) and the collecting duct.

A long loop of the nephron, which typically penetrates deep into the renal medulla, is made up of a *descending* (going down) *limb* and an *ascending* (going up) *limb.* Salt (NaCl) passively diffuses out of the lower portion of the ascending limb, but the upper, thick portion of the limb actively extrudes salt out into the tissue of the outer renal medulla (Fig. 16.7). Less and less salt is available for transport as fluid moves up the thick portion of the ascending limb. Because of these circumstances, the loop of the nephron establishes an *osmotic gradient* within the tissues of the renal medulla: the concentration of salt is greater in the direction of the inner medulla. (Note that water cannot leave the ascending limb because the limb is impermeable to water.)

Also, if you examine Figure 16.7 carefully, you can see that the innermost portion of the inner medulla has the highest concentration of solutes. This cannot be due to salt because active transport of salt does not start until the thick portion of the ascending limb. Urea is believed to leak from the lower portion of the collecting duct, and it is this molecule that contributes to the high solute concentration of the inner medulla.

Because of the osmotic gradient within the renal medulla, water leaves the descending limb of the loop of the nephron along its length. This is a countercurrent mechanism: as water diffuses out of the descending limb, the remaining solution within the limb encounters an even greater osmotic concentration of solute; therefore, water will continue to leave the descending limb from the top to the bottom.

Fluid entering a collecting duct comes from the distal convoluted tubule. This fluid is now isotonic to the cells of the cortex. This means that to this point, the net effect of reabsorption of water and salt is the production of a fluid that has the same tonicity as blood. However, the filtrate within the collecting duct also encounters the same osmotic gradient mentioned earlier (Fig. 16.7). Therefore, water diffuses out of the collecting duct into the renal medulla, and the urine within the collecting duct becomes hypertonic to blood plasma.

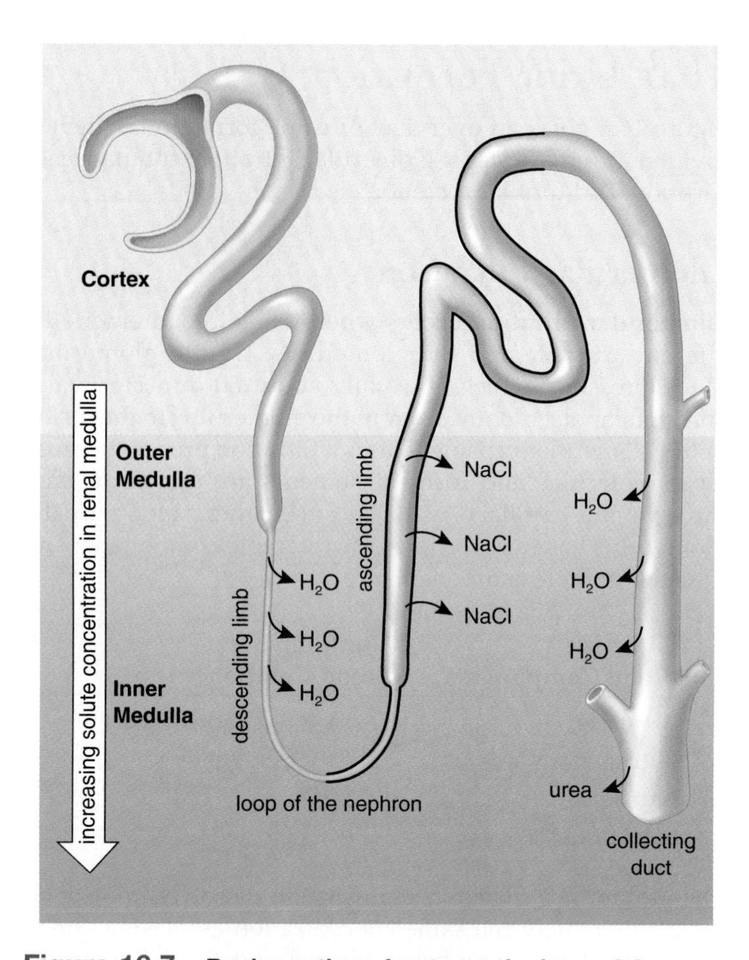

Figure 16.7 Reabsorption of water at the loop of the nephron and the collecting duct.
Salt (NaCl) diffuses and is actively transported out of the ascending limb of the loop of the nephron into the renal medulla; also, urea is believed to leak from the collecting duct and to enter the tissues of the renal medulla. This creates a hypertonic environment, which draws water out of the descending limb and the collecting duct. This water is returned to the cardiovascular system. (The thick line means the ascending limb is impermeable to water.)

Antidiuretic hormone (ADH) released by the posterior lobe of the pituitary plays a role in water reabsorption at the collecting duct. In order to understand the action of this hormone, consider its name. Diuresis means increased amount of urine, and antidiuresis means decreased amount of urine. When ADH is present, more water is reabsorbed (blood volume and pressure rise), and a decreased amount of urine results. In practical terms, if an individual does not drink much water on a certain day, the posterior lobe of the pituitary releases ADH, causing more water to be reabsorbed and less urine to form. On the other hand, if an individual drinks a large amount of water and does not perspire much, ADH is not released. Now more water is excreted, and more urine forms.

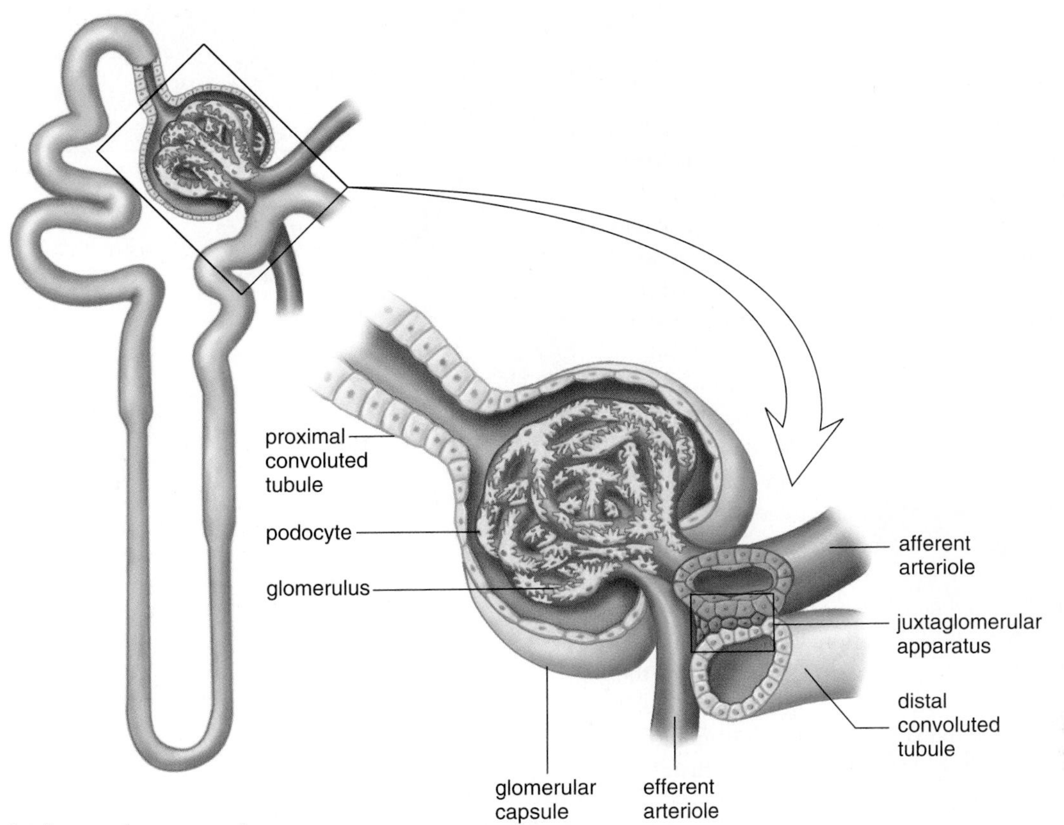

Figure 16.8 Juxtaglomerular apparatus.
This drawing shows that the afferent arteriole and the distal convoluted tubule usually lie next to each other. The juxtaglomerular apparatus occurs where they touch.

Reabsorption of Salt

Usually, more than 99% of sodium (Na^+) filtered at the glomerulus is returned to the blood. Most sodium (67%) is reabsorbed at the proximal tubule, and a sizable amount (25%) is extruded by the ascending limb of the loop of the nephron. The rest is reabsorbed from the distal convoluted tubule and collecting duct.

Hormones regulate the reabsorption of sodium at the distal convoluted tubule. **Aldosterone** is a hormone secreted by the adrenal cortex, the outer portion of the adrenal glands, which lie atop the kidneys. Aldosterone promotes the excretion of potassium ions (K^+) and the reabsorption of sodium ions (Na^+). The release of aldosterone is set in motion by the kidneys themselves. The **juxtaglomerular apparatus** is a region of contact between the afferent arteriole and the distal convoluted tubule (Fig. 16.8). When blood volume, and therefore blood pressure, is not sufficient to promote glomerular filtration, the juxtaglomerular apparatus secretes renin. **Renin** is an enzyme that changes angiotensinogen (a large plasma protein produced by the liver) into angiotensin I. Later, angiotensin I is converted to angiotensin II, a powerful vasoconstrictor that also stimulates the adrenal cortex to release aldosterone. The reabsorption of sodium ions is followed by the reabsorption of water. Therefore, blood volume and blood pressure increase.

Atrial natriuretic hormone (ANH) is a hormone secreted by the atria of the heart when cardiac cells are stretched due to increased blood volume. ANH inhibits the secretion of renin by the juxtaglomerular apparatus and the secretion of aldosterone by the adrenal cortex. Its effect, therefore, is to promote the excretion of Na^+, that is, natriuresis. When Na^+ is excreted, so is water, and therefore blood volume and blood pressure decrease.

These examples show that the kidneys regulate the salt balance in blood by controlling the excretion and the reabsorption of various ions. Sodium (Na^+) is an important ion in plasma that must be regulated, but the kidneys also excrete or reabsorb other ions, such as potassium ions (K^+), bicarbonate ions (HCO_3^-), and magnesium ions (Mg^{2+}), as needed.

Diuretics

Diuretics are agents that increase the flow of urine. Drinking alcohol causes diuresis because it inhibits the secretion of ADH. The dehydration that follows is believed to contribute to the symptoms of a hangover. Caffeine is a diuretic because it increases the glomerular filtration rate and decreases the tubular reabsorption of Na^+. Diuretic drugs developed to counteract high blood pressure in patients inhibit active transport of Na^+ at the loop of the nephron or at the distal convoluted tubule. A decrease in water reabsorption and a decrease in blood volume follow.

Science Focus

The Artificial Kidney

After a person suffers kidney damage, perhaps due to repeated infections, waste substances accumulate in the blood. This condition is called uremia because urea is one of the substances that accumulates. Although nitrogenous wastes can cause serious damage, some believe that it is the imbalance of ions in the blood that leads to loss of consciousness and to heart failure. Patients in renal failure most often seek a kidney transplant, but in the meantime they undergo hemodialysis utilizing an artificial kidney. Dialysis is the diffusion of dissolved molecules through a semipermeable membrane (an artificial membrane with pore sizes that allow only small molecules to pass through). These molecules, of course, move across a membrane from the area of greater concentration to one of lesser concentration. Substances more concentrated in blood diffuse into the dialysis solution, which is called the dialysate, and substances more concentrated in the dialysate diffuse into blood. Accordingly, the artificial kidney is utilized either to extract substances from blood, including waste products or toxic chemicals and drugs, or to add substances to blood—for example, bicarbonate ions (HCO_3^-) if blood is acidic.

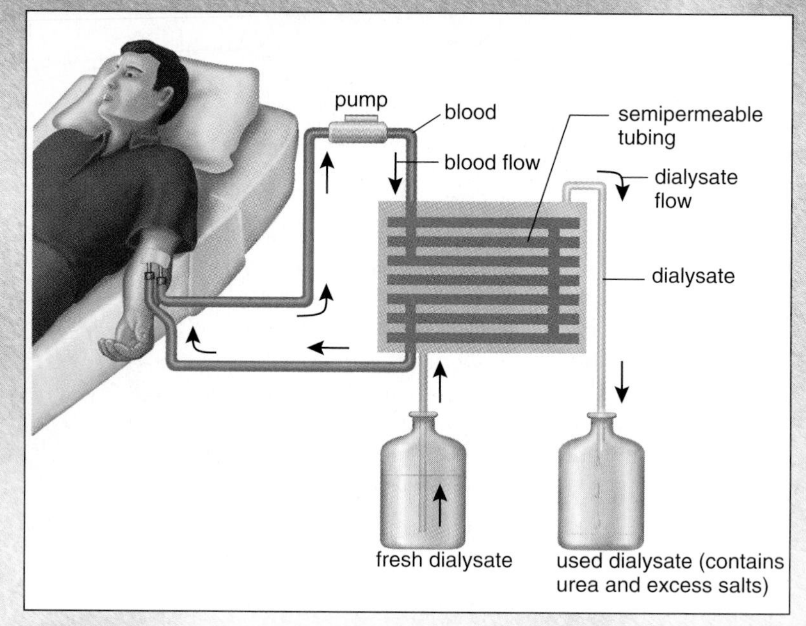

Figure 16B An artificial kidney machine.
As the patient's blood is pumped through dialysis tubing, it is exposed to a dialysate (dialysis solution). Wastes exit from blood into the solution because of a preestablished concentration gradient. In this way, blood is not only cleansed, but its water-salt and acid-base balance can also be adjusted.

The first clinically useful artificial kidney was devised by the Dutchman William J. Kolff in 1943. Based on the work of predecessors, Kolff hypothesized that he needed a machine with these specifications: (1) only a small volume of blood is ever out of the patient, (2) the blood should be exposed to a large surface area to lessen diffusion time, (3) the blood must be kept moving in tubes to and from the patient, and (4) the dialysate must be of a certain composition and kept fresh.

In Kolff's invention, 20 meters of cellophane tubing were wrapped around a drum. Rotation of the drum caused the blood to pass from one end of the tubing to the other. The lower half of the drum was immersed in a fluid. There was no pump and the tube taking blood from the patient to the machine was raised or lowered to let blood in or out of the patient. Kolff used an anticoagulant to prevent the patient's blood from clotting. The only patient treated with the first artificial kidney died when all possible entry points for his blood had been utilized.

Around 1960, Dr. Belding Scribner and Mr. Wayne Quinton in Seattle developed the arteriovenous shunt. Tubes permanently placed in an artery and vein of the arm were joined to a removable Teflon shunt. When the patient needed dialysis, the shunt was disconnected and the tubes were connected to the artificial kidney. Today, it is common to surgically join an artery and vein in the arm. The vein distends, and for each dialysis, one or two needles are inserted in the distended vein. The needles are connected to tubing leading to the artificial kidney, where fresh dialysate circulates past many meters of the tubing. A pump keeps the blood moving, and there is no need for the anticoagulant that Kolff had to use (Fig. 16B). Also, today's artificial kidney uses highly permeable dialysis tubing, making it possible to shorten dialysis time. Presently, dialysis usually occurs only three times per week for three hours or less. And there are portable machines that allow patients to dialyze at home or on trips!

Another group of Dutch investigators hypothesized that the peritoneum could serve as a dialyzing membrane. The peritoneal (abdominal) cavity has a large surface area and is richly vascularized. Therefore, an exchange does take place between a dialyzing fluid placed in the peritoneal cavity and the capillaries of the abdominal wall. A very careful mathematical analysis allowed the investigators to determine that only 1.2 liters of dialysate are needed and it should be kept in place for four hours. The old dialysate is then removed and a new batch is instilled via a permanently implanted tube. In the meantime the patient is free to go about his usual daily routine. This method of dialysis is appropriately called *continuous ambulatory peritoneal dialysis.*

Development of the artificial kidney and continuous ambulatory peritoneal dialysis exemplifies that science is intimately involved with the development of suitable instruments and is a community effort that builds on the work of those that have made contributions before.

16.5 Maintaining Acid-Base Balance

The bicarbonate (HCO_3^-) buffer system and breathing work together to maintain the pH of the blood. Central to the mechanism is this reaction, which you have seen before:

$$H^+ + HCO_3^- \rightleftharpoons H_2CO_3 \rightleftharpoons H_2O + CO_2$$

The excretion of carbon dioxide (CO_2) by the lungs helps keep the pH within normal limits, because when carbon dioxide is exhaled this reaction is pushed to the right and hydrogen ions (H^+) are tied up in water. Indeed, when blood pH decreases, chemoreceptors in the carotid bodies (located in the carotid arteries) and in aortic bodies (located in the aorta) stimulate the respiratory center, and the rate and depth of breathing increases. On the other hand, when blood pH begins to rise, the respiratory center is depressed and the bicarbonate ion increases in the blood.

As powerful as this system is, only the kidneys can rid the body of a wide range of acidic and basic substances. The kidneys are slower acting than the buffer/breathing mechanism, but they have a more powerful effect on pH. For the sake of simplicity, we can think of the kidneys as reabsorbing bicarbonate ions and excreting hydrogen ions as needed to maintain the normal pH of the blood. If the blood is acidic, hydrogen ions are excreted and bicarbonate ions are reabsorbed. If the blood is basic, hydrogen ions are not excreted and bicarbonate ions are not reabsorbed. Since the urine is usually acidic, it shows that usually an excess of hydrogen ions are excreted. Ammonia (NH_3) provides a means for buffering these hydrogen ions in urine: ($NH_3 + H^+ \rightarrow NH_4^+$). Ammonia (whose presence is quite obvious in the diaper pail or kitty litter box) is produced in tubule cells by the deamination of amino acids. Phosphate provides another means of buffering hydrogen ions in urine.

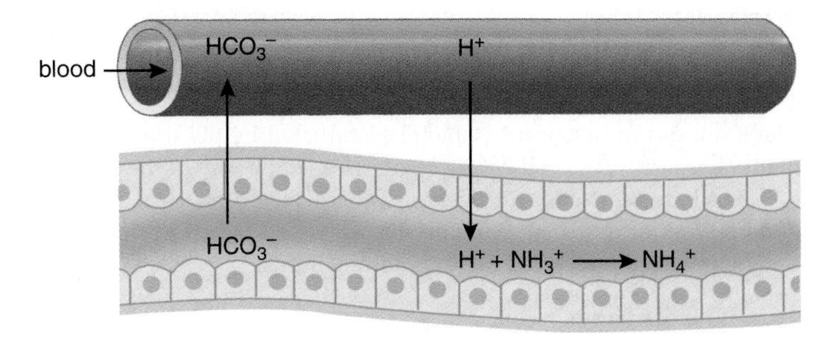

Figure 16.9 Acid-base balance.
In the kidneys, bicarbonate ions are reabsorbed and the hydrogen ions are excreted as needed to maintain the pH of the blood. Excess hydrogen ions are buffered, for example, by ammonia (NH_3), which is produced in tubule cells by the deamination of amino acids.

The acid-base balance of the blood is adjusted by the reabsorption of the bicarbonate ions (HCO_3^-) and the secretion of hydrogen ions (H^+) as appropriate.

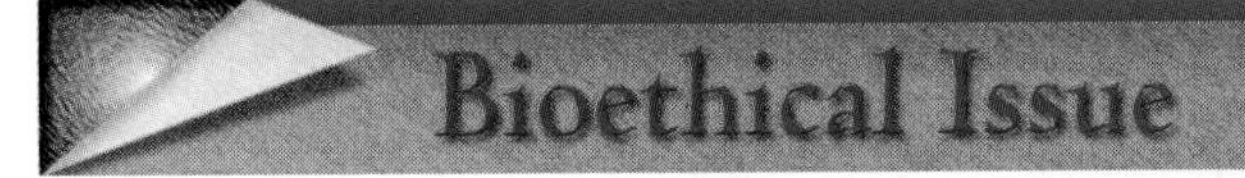

As a society we are accustomed to thinking that as we grow older, diseases like urinary disorders will begin to occur. Almost everyone is aware that most males are subject to enlargement of the prostate as they age, and that cancer of the prostate is not uncommon among elderly men. However, like many illnesses associated with aging, medical science now knows how to treat or even cure prostate problems. Because of these successes, medical science has lengthened our life span. A child born in the United States in 1900 lived to, say, 47 years. If that same child were born today, it would probably live to at least 76. Even more exciting is the probability that scientists will improve the life span. People could live beyond 100 years and have the same vigor and vitality they had when they were young.

Most people are appreciative of living longer, especially if they can expect to be free of the illnesses and inconveniences associated with aging. But have we examined how we feel about longevity as a society? Whereas we are accustomed to considering that if the birthrate increases so does the size of a population, what about the death rate? If the birthrate stays constant and the death rate decreases, obviously population size also increases. Most experts agree that population growth depletes resources and increases environmental degradation. An older population can also put a strain on the economy if they are unable to meet their financial, including medical, needs without governmental assistance.

What is the ethical solution to this problem? Should we just allow the population to increase due to older people living longer? Should we decrease the birthrate? Should we reduce governmental assistance to older people so they realize that they must be able to take care of themselves? Should we call a halt to increasing the life span through advancements in medical science?

Questions

1. Do you feel that older people make a significant contribution to or a drain on society? Explain.
2. Would you be willing to have fewer children in order to hold the population in check if more people live longer? Why or why not?
3. Should the elderly expect governmental or family assistance as they age? Why or why not?

Summarizing the Concepts

16.1 Urinary System

The kidneys produce urine, which is conducted by the ureters to the bladder where it is stored before being released by way of the urethra.

The kidneys excrete nitrogenous wastes, including urea, uric acid, and creatinine, and they maintain the water-salt balance of the body and help keep the blood pH within normal limits.

16.2 The Kidneys

Macroscopically, the kidneys are divided into the renal cortex, renal medulla, and renal pelvis. Microscopically, they contain the nephrons.

Each nephron has its own blood supply; the afferent arteriole approaches the glomerular capsule and divides to become the glomerulus, a capillary tuft. The permeability of the glomerular capsule allows small molecules to enter the capsule from the glomerulus. The efferent arteriole leaves the capsule and immediately branches into the peritubular capillary network.

Each region of the nephron is anatomically suited to its task in urine formation. The spaces between podocytes of the glomerular capsule allow small molecules to enter the capsule from the glomerulus, a capillary knot. The cuboidal epithelial cells of the proximal convoluted tubule have many mitochondria and microvilli to carry out active transport (following passive transport) from the tubule to blood. In contrast, the cuboidal epithelial cells of the distal convoluted tubule have numerous mitochondria but lack microvilli. They carry out active transport from the blood to the tubule.

16.3 Urine Formation

Urine is composed primarily of nitrogenous waste products and salts in water. The steps in urine formation are glomerular filtration, tubular reabsorption, and tubular secretion, as explained in Figure 16.6.

16.4 Maintaining Water-Salt Balance

The kidneys regulate the water-salt balance of the body. Water is reabsorbed from all parts of the tubule. The ascending limb of the nephron loop establishes an osmotic gradient that draws water from the descending limb and also the collecting duct. The permeability of the collecting duct is under the control of the hormone ADH.

The reabsorption of salt increases blood volume and pressure because more water is also reabsorbed. Two other hormones, aldosterone and ANH, control the kidneys' reabsorption of sodium (Na^+).

16.5 Maintaining Acid-Base Balance

The kidneys keep blood pH within normal limits. They reabsorb HCO_3^- and excrete H^+ as needed to maintain the pH at about 7.4.

Studying the Concepts

1. State the path of urine and the function of each organ mentioned. 304
2. Explain how urination is controlled. 305
3. List and explain four functions of the urinary system. 305
4. Describe the macroscopic anatomy of a kidney. 307
5. Trace the path of blood about a nephron. 308
6. Name the parts of a nephron, and tell how the structure of the convoluted tubules suits their function. 309
7. State and describe the three steps of urine formation. 310–11
8. Where in particular is water and salt reabsorbed along the length of the nephron? Describe the contribution of the loop of the nephron. 312
9. Name and describe the action of antidiuretic hormone (ADH), the renin-aldosterone connection, and the atrial natriuretic hormone (ANH). 312–13
10. How do the kidneys maintain the pH of the blood within normal limits? 315

Testing Yourself

Choose the best answer for each question.

1. Which of these functions of the kidneys is mismatched?
   a. excretes metabolic wastes—rids the body of urea
   b. maintains the water-salt balance—helps regulate blood pressure
   c. maintains the acid-base balance—rids the body of uric acid
   d. secretes hormones—secretes erythropoietin
   e. All of these are properly matched.
2. Which of these is out of order first?
   a. glomerular capsule
   b. proximal convoluted tubule
   c. distal convoluted tubule
   d. loop of the nephron
   e. collecting duct
3. Which of these hormones is most likely to cause a rise in blood pressure?
   a. aldosterone
   b. antidiuretic hormone (ADH)
   c. renin
   d. atrial natriuretic hormone (ANH)
   e. Both a and c are correct.
4. If the blood is acidic,
   a. hydrogen ions are excreted and bicarbonate ions are reabsorbed.
   b. hydrogen ions are reabsorbed and bicarbonate ions are excreted.
   c. hydrogen ions and bicarbonate ions are reabsorbed.
   d. hydrogen ions and bicarbonate ions are excreted.
   e. urea, uric acid, and ammonia are excreted.
5. Excretion of a hypertonic urine in humans is associated best with the
   a. glomerular capsule.
   b. proximal convoluted tubule.
   c. loop of the nephron.
   d. distal convoluted tubule.
6. The presence of ADH (antidiuretic hormone) causes an individual to excrete
   a. sugars.
   b. less water.
   c. more water.
   d. Both a and c are correct.
7. In humans, water is
   a. found in the glomerular filtrate.
   b. reabsorbed from the nephron.
   c. in the urine.
   d. All of these are correct.

8. Filtration is associated with the
   a. glomerular capsule.
   b. distal convoluted tubule.
   c. collecting duct.
   d. All of these are correct.
9. Normally in humans, glucose
   a. is always in the filtrate and urine.
   b. is always in the filtrate, with little or none in urine.
   c. undergoes tubular secretion and is in urine.
   d. undergoes tubular secretion and is not in urine.
10. Label this diagram of a nephron.

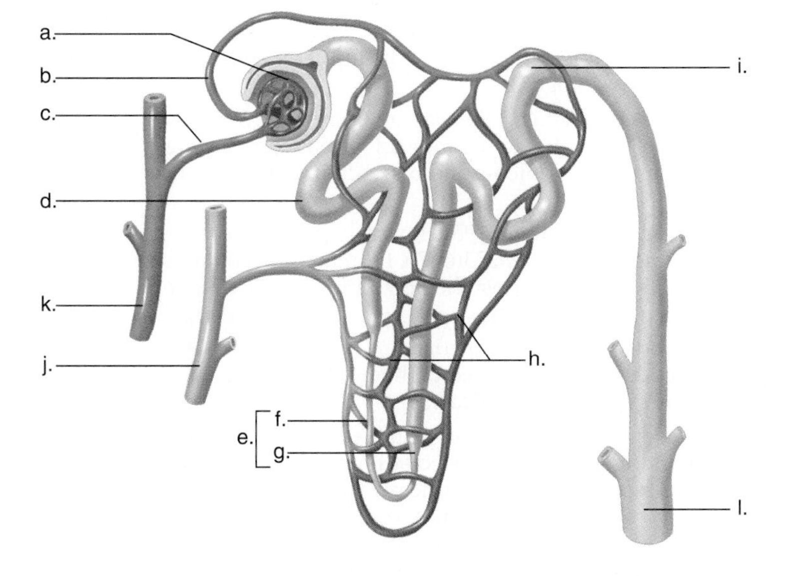

Thinking Scientifically

1. Considering filtration (page 311):
   a. Urine, not urea, is made by the kidneys. What is the difference between urine and urea?
   b. Blood pressure in the glomerulus favors filtration of molecules. The efferent arteriole (leaving the glomerulus) is narrower than the afferent arteriole (leading to the glomerulus). What effect does this have on blood pressure in the glomerulus?
   c. What force would oppose filtration of molecules from the glomerulus? (Hint, review the forces involved in capillary exchange.)
   d. If there is a loss of proteins from blood into the glomerular capsule, would the filtration rate increase or decrease? (Assume constant blood pressure.)
2. Considering urine formation (page 311):
   a. If 99% of water is reabsorbed, how can urine be 95% water.
   b. Carrier molecules work as fast as they can to return glucose to blood. Explain why excess glucose is not returned.
   c. When CO_2 is excreted by the lungs, does blood become more acidic or more basic? If the HCO_3^- is excreted by the kidneys, does blood become more acidic or more basic?
   d. The maintenance of normal blood pH is a very important function of the kidneys. What molecules in the cells are affected by pH changes.

Understanding the Terms

Match the terms to these definitions:

a. ________ Portion of the nephron lying between the proximal convoluted tubule and the distal convoluted tubule that functions in water reabsorption.

b. ________ Movement of molecules from the contents of the nephron into blood at the proximal convoluted tubule.

c. ________ Hormone secreted by the adrenal cortex that regulates the sodium and potassium balance of the blood.

d. ________ Outer portion of the kidney that appears granular.

e. ________ Surrounds a nephron and functions in reabsorption during urine formation.

Using Technology

Your study of the urinary system and excretion is supported by these available technologies:

Essential Study Partner CD-ROM

Animals → Osmoregulation

Visit the Mader web site for related ESP activities.

Exploring the Internet

The Mader Home Page provides resources and tools as you study this chapter.

http://www.mhhe.com/biosci/genbio/mader

Dynamic Human 2.0 CD-ROM

Urinary System

Life Science Animations 3D Video

38 Kidney Function

Further Readings for Part 3

Abraham, S. N. September/October 1997. Discovering the benign traits of the mast cell. *Science & Medicine* 4(5):46. Recent investigations suggest mast cells have roles in immune surveillance and control of the immune response.

Arakawa, T. and Langridge W. R. H. May 1998. Plants are not just passive creatures! *Nature Medicine* 4(5):550. Plants are being used as bioreactors to produce foreign proteins for human immunity.

Beck, G., and Habicht, G. S. November 1996. Immunity and the invertebrates. *Scientific American* 275(5):60. Nearly all aspects of the human immune system appear to have a cellular or chemical parallel among the invertebrates.

Becker, R. C. July/August 1996. Antiplatelet therapy. *Science & Medicine* 3(4):12. This article discusses the control of platelet aggregation.

Benjamini, E., and Leskowitz, S. 1996. *Immunology:* A short course. 3d ed. New York: John Wiley & Sons. Presents the essential principles of immunology.

Bikle, D. D. March/April 1995. A bright future for the sunshine hormone. *Science & Medicine* 2(2):58. Vitamin D receptors exist in many types of cells, suggesting therapeutic uses for the hormonally active metabolite of vitamin D.

Blaser, M. J. February 1996. The bacteria behind ulcers. *Scientific American* 274(2):104. Acid-loving microbes are linked to stomach ulcers and stomach cancer.

Brown, J. L., and Pollitt, E. February 1996. Malnutrition, poverty and intellectual development. *Scientific American* 274(2):38. Article discusses the complex role of essential nutrients in a child's mental development.

Dickman, S. July 1997. Mysteries of the heart. *Discover* 18(7):117. Article discusses why coronary arteries may still become blocked after treatment for atherosclerosis.

Gibbs, W. W. August 1996. Gaining on fat. *Scientific American* 275(2):88. Some weight problems are genetic or physiological in origin. New treatments might help.

Glausiusz, J. September 1998. Infected hearts. *Discover* 19(9):30. Infectious bacteria may play a role in heart disease; antibiotics could prevent the need for heart surgery.

Glausiusz, J. October 1997. The good bugs on our tongues. *Discover* 18(10):32. Without the friendly bacteria that live on our tongues, we would be vulnerable to bacteria such as Salmonella.

Guyton, A. C., and Hall, J. E. 1996. *Textbook of medical physiology.* Philadelphia: W. B. Saunders Co. Presents physiological principles for those in the medical fields.

Haen, P. J. 1995. *Principles of hematology.* Dubuque, Iowa: Wm. C. Brown Publishers. An introductory text for students planning a career in the medical sciences.

Hanson, L. A. November/December 1997. Breast-feeding stimulates the infant immune system. *Science & Medicine* 4(6):12. Long-lasting protection against some infectious diseases has been reported in breast-fed infants.

Hotez, P. J., and Pritchard, D. I. June 1995. Hookworm infection. *Scientific American* 272(6):68. Discusses how the biology of parasites offers clues to possible vaccines and also for new treatments for heart disease and immune disorders.

Klatsky, A. L. March/April 1995. Cardiovascular effects of alcohol. *Science & Medicine* 2(2):28. The effects of moderate and heavy alcohol use on the cardiovascular system is discussed.

Kooyman, G. L., and Ponganis, P. J. November/December 1997. The challenges of diving to depth. *American Scientist* 85(6):530. Marine animals have ways to control their oxygen supply, and do not experience problems associated with pressure at depth.

Little, R. C., and Little, W. C. 1989. *Physiology of the heart and circulation.* 4th ed. Chicago: Year Book Medical Publishers, Inc. A good reference that gives an in-depth look at cardiovascular physiology.

Mader, S. S. 1997. *Understanding anatomy and physiology.* 3d ed. Dubuque, Iowa: Wm. C. Brown Publishers. A text that emphasizes the basics for beginning allied health students.

Mader, S. S. 1998. *Human biology.* 6th ed. Dubuque, Iowa: WCB/McGraw-Hill, Inc. A student-friendly text that covers the principles of biology with emphasis on human anatomy and physiology.

Nature Medicine Vaccine Supplement, May 1998, Vol. 4 No. 5. Entire issue is devoted to the topic of vaccines, including history, recent developments and research in malaria, cancer, and HIV vaccines.

Newman, J. December 1995. How breast milk protects newborns. *Scientific American* 273(6):76. Human milk contains special antibodies that boost the newborn's immune system.

Nucci, M. L., and Abuchowski, A. February 1998. The search for blood substitutes. *Scientific American* 278(2):72. Artificial blood substitutes based on hemoglobin are being developed from synthetic chemicals.

Roitt, I., et al. 1998. *Immunology.* 5th ed. Mosby International, Ltd.: London. For the advanced student, this text features a clear description of the scientific principles involved in immunology, combined with clinical examples.

Superko, H. R. September/October 1997. The atherogenic lipid profile. *Science & Medicine* 4(5):36. Gel electrophoresis is used to relate the risk of heart disease to cholesterol lowering therapy.

Sussman, N. L., and Kelly, J. H. May/June 1995. The artificial liver. *Science & Medicine* 2(3):68. An artificial liver assists temporarily while the natural liver regenerates, restoring normal function.

Valtin, H. 1995. Renal function. 3d ed. Boston: Little, Brown and Co. A good reference resource that discusses renal mechanisms for preserving fluid and solute balance.

Wardlaw, G., et al. 1994. *Contemporary nutrition.* 2d ed. St. Louis: Mosby-Year Book, Inc. This text gives a clear understanding of nutritional information found on product labels.

Weindruch, R. January 1996. Caloric restriction and aging. *Scientific American* 274(1):64. Consuming fewer calories may increase longevity.

West, J. B. 1994. *Respiratory physiology—The essentials.* 5th ed. Baltimore: Williams & Wilkins. A good reference resource that discusses all aspects of respiratory physiology including breathing, and external and internal respiration.

White, R. J. September 1998. Weightlessness and the human body. *Scientific American* 279(3):58. Space medicine is providing new ideas about treatment of anemia and osteoporosis.

Yock, P., et al. September/October 1995. Intravascular ultrasound. *Science & Medicine* 2(3):68. Ultrasound images of coronary arteries helps diagnose atherosclerosis.

PART IV

Integration and Control of the Human Body

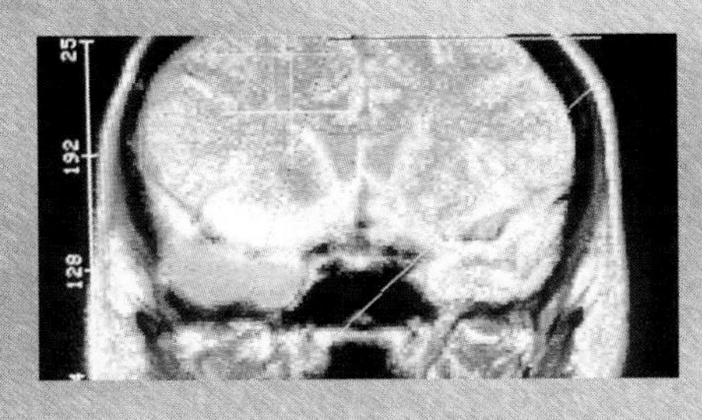

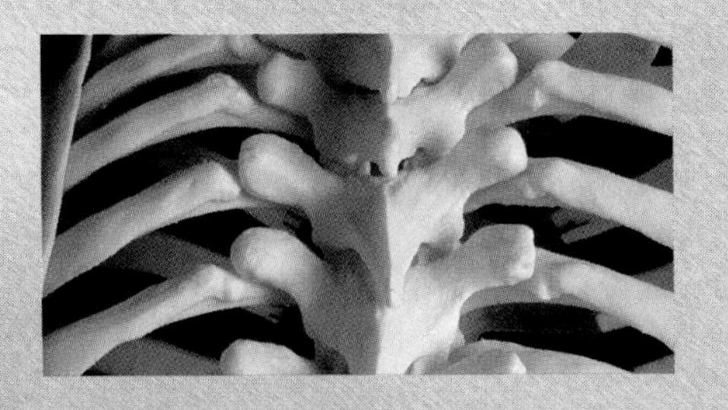

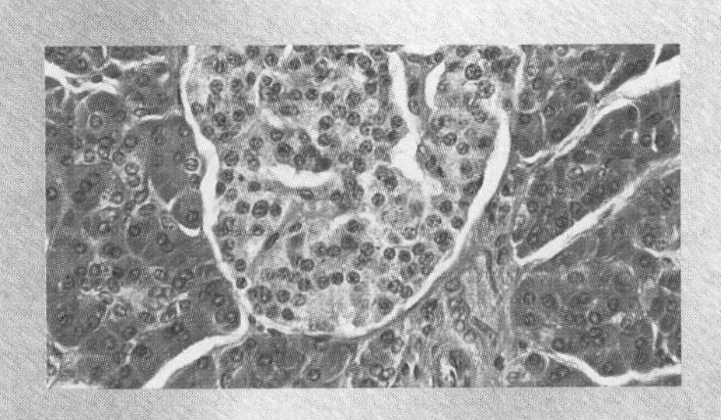

Luigi Galvani, an Italian, discovered in the eighteenth century that a nerve can be stimulated by an electric current, but it was not until the twentieth century that instrumentation was available to show that the nerve impulse is an electrochemical phenomenon—the movement of ions across the plasma membrane results in nerve impulses. While the nerve impulse is always the same, sensation and perception differ. This enigma, that is, puzzle, still cannot be fully explained, but neurophysiologists are making rapid strides in understanding how the brain turns nerve impulses into our knowledge of the world and ourselves.

Careers in Human Anatomy and Physiology

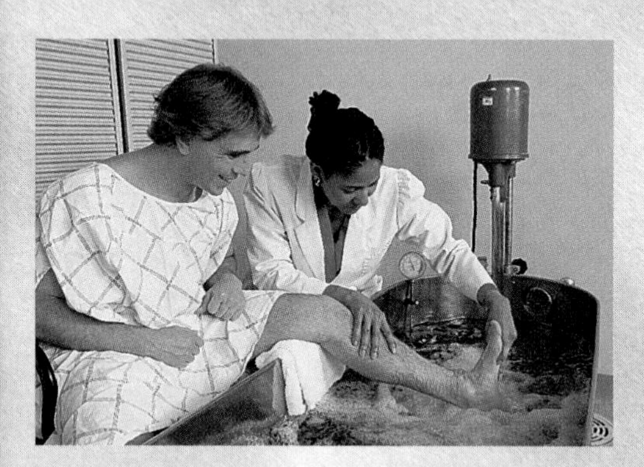

Physical therapist working on patient.

Psychologist listening to patient.

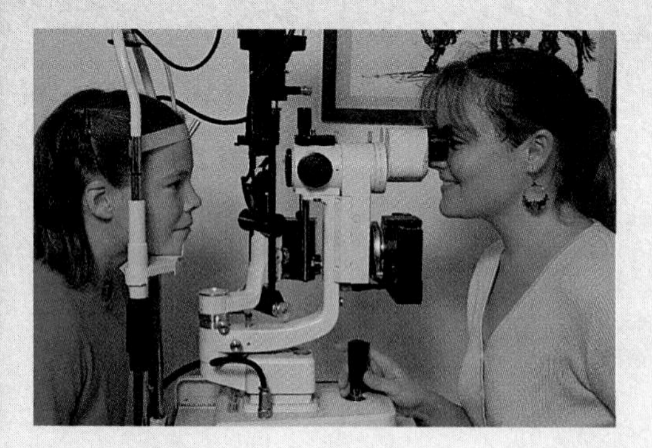

Optometrist testing eyesight of patient.

Chiropractors treat patients whose health problems are associated with the body's muscular, nervous, and skeletal systems, especially the spine. Chiropractors use natural, nondrug, nonsurgical health treatments and rely on the body's inherent recuperative abilities.

Physical therapists work to improve the mobility, relieve the pain, and prevent or limit the permanent physical disabilities of patients suffering from injuries or disease. Treatment often includes exercise for patients who have been immobilized and lack flexibility.

Psychologists meet with clients who are having difficulty coping with relationships and everyday problems. Psychologists allow patients to discuss their feelings in a friendly and secure atmosphere. The goal is to resolve inner conflicts, leading to a more productive and well-adjusted life-style.

Optometrists examine the eyes and related structures to determine the presence of vision problems and/or other abnormalities. They prescribe glasses or lenses when needed and routinely test for glaucoma and diseases of the retina. They may use visual training to preserve or restore vision to a maximum level of efficiency.

Biomedical laser technicians operate laser equipment in a proficient and safe manner and assist in the training of operating room personnel. Technicians keep abreast of the current laser research technologies and participate in basic research and data collection. They also order and are responsible for laser supplies and auxiliary equipment needed to perform the laser procedure.

Speech-language pathologists work with people who cannot speak clearly or who cannot understand language. They test patients to determine the nature and extent of impairment and to analyze speech irregularities. For people who cannot speak, they select an alternative communication system, such as sign language and teach patients how to use the system.

Neurophysiological technologists record brain waves using an EEG machine, and also perform related types of tests to diagnose brain disorders such as tumors, strokes, epilepsy, or the presence of Alzheimer disease. Some are specialists in speech disorders. Other technologists may choose to manage an EEG laboratory and/or become an instructor in EEG techniques.

CHAPTER 17

Nervous System

Chapter Concepts

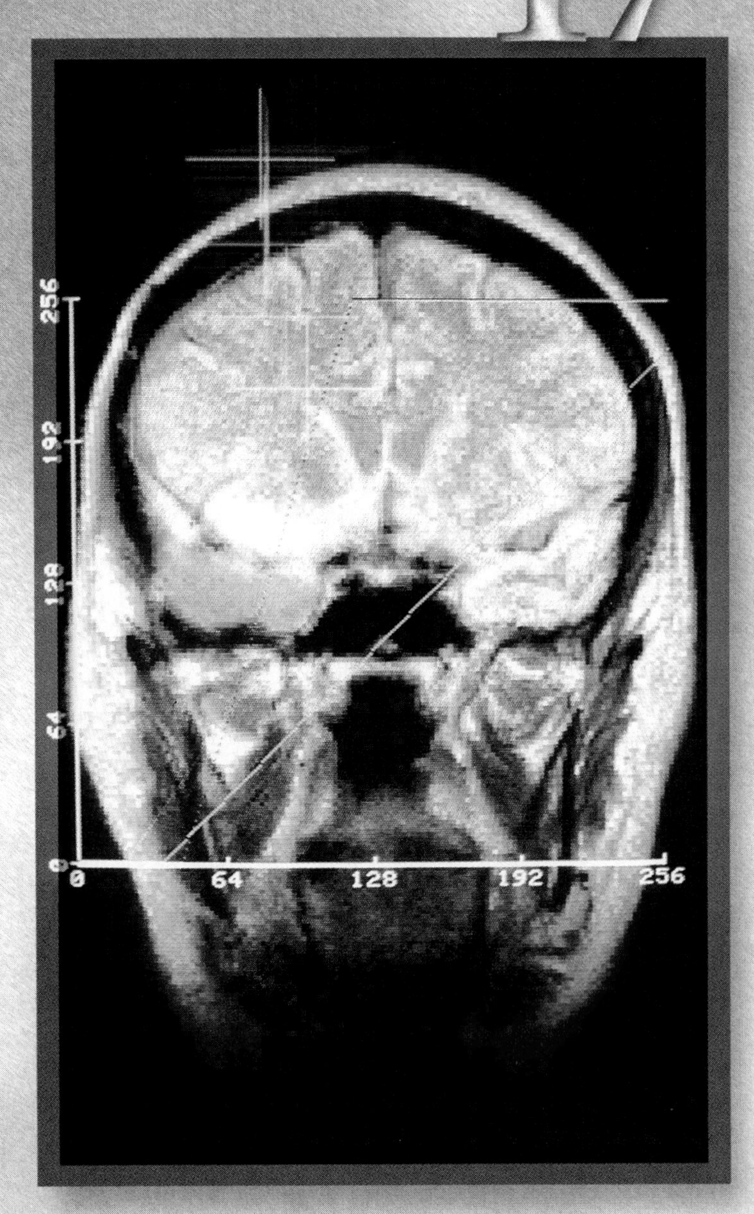

In magnetic resonance images, bone appears black while soft tissue, like the brain, are in color. To produce this image, a powerful magnet was used to align hydrogen ions in the body. When they come out of alignment, the hydrogen ions produce signals that are picked up by a computer.

It was a bright sunny day, without a cloud in the sky, when Malcolm's horse rose to clear the stone wall. As the rear hoofs caught on the wall, horse and rider went down with a tremendous thud. Other riders quickly dismounted and ran over to where Malcolm lay on the ground. He did not move, but his eyes were open and he was breathing; therefore, they knew he was alive. Xrays taken later showed a broken cervical vertebra and injury to the spinal cord.

The spinal cord is a ropelike bundle of long tracts that shuttle messages between the brain and the rest of the body. Together the brain and spinal cord compose the **central nervous system (CNS),** which interprets sensory input before coordinating a motor response. The **peripheral nervous system (PNS)** consists of nerves, which carry sensory information to the CNS and also motor commands from the CNS to the muscles and glands (Fig. 17.1).

When Malcolm hurt his spinal cord, the CNS lost its avenue of communication with the portion of his body located below the site of damage. He receives no sensation from most of his body nor can he command his arms and legs to move. But cranial nerves from his eyes and ears still allow him to see and hear and his brain still enables him to have emotions, remember, and reason. Also his internal organs still function, a sign that his injury was not as severe as it could have been. In this chapter we will examine the structure of the nervous system and how it carries out its numerous functions.

17.1 Neurons and How They Work

Nervous tissue contains two types of cells: neuroglial cells and neurons. **Neuroglial cells** support and service **neurons,** the cells that actually transmit nerve impulses (see Fig. 11.6).

Neuron Structure

Despite their varied appearance, all neurons have just three parts: dendrites, cell body, and an axon. It becomes apparent from studying the motor neuron in Figure 17.2 that **dendrites** are processes that send signals toward the cell body. The **cell body** is the part of a neuron that contains the nucleus and other organelles. An **axon** conducts nerve impulses along its entire length. Axons are sometimes referred to as long fibers.

There are three classes of neurons: sensory neurons, motor neurons, and interneurons, whose functions are best described in relation to the CNS (Fig. 17.2). A **sensory neuron** takes information from a sensory receptor to the CNS, and a **motor neuron** takes information away from the CNS to an effector (muscle fiber or gland). An **interneuron** conveys information between neurons in the CNS. An interneuron can receive input from sensory neurons and also from other interneurons in the CNS. Thereafter, they sum up all these signals before sending commands out to the muscles and glands by way of motor neurons.

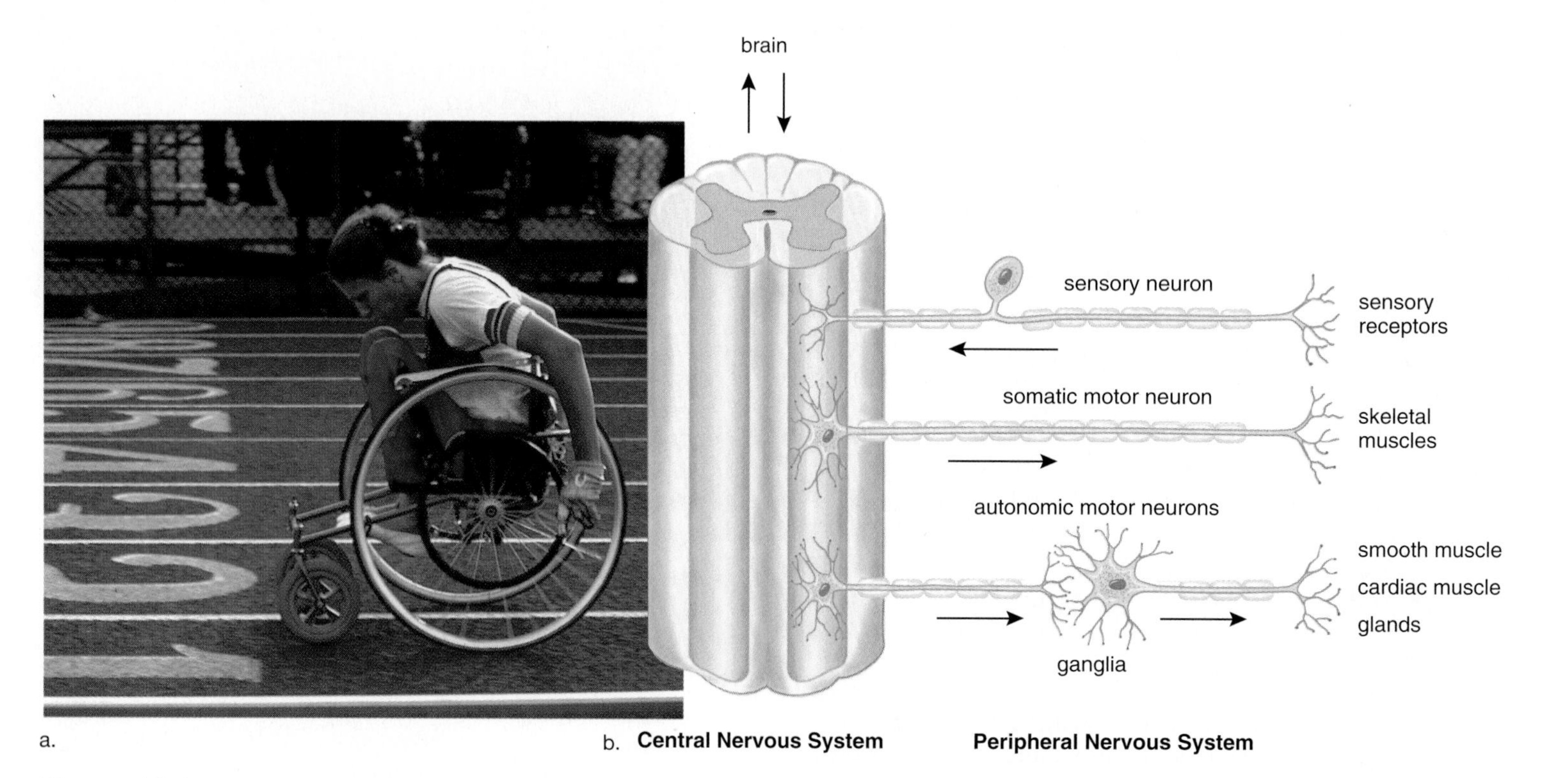

Figure 17.1 Organization of nervous system.
a. In paraplegics, messages no longer flow between the legs and the central nervous system (the spinal cord and brain). **b.** The sensory neurons of the peripheral nervous system take nerve impulses from sensory receptors to the central nervous system (CNS), and motor neurons (both somatic to skeletal muscles and autonomic to internal organs) take nerve impulses from the CNS to the organs listed.

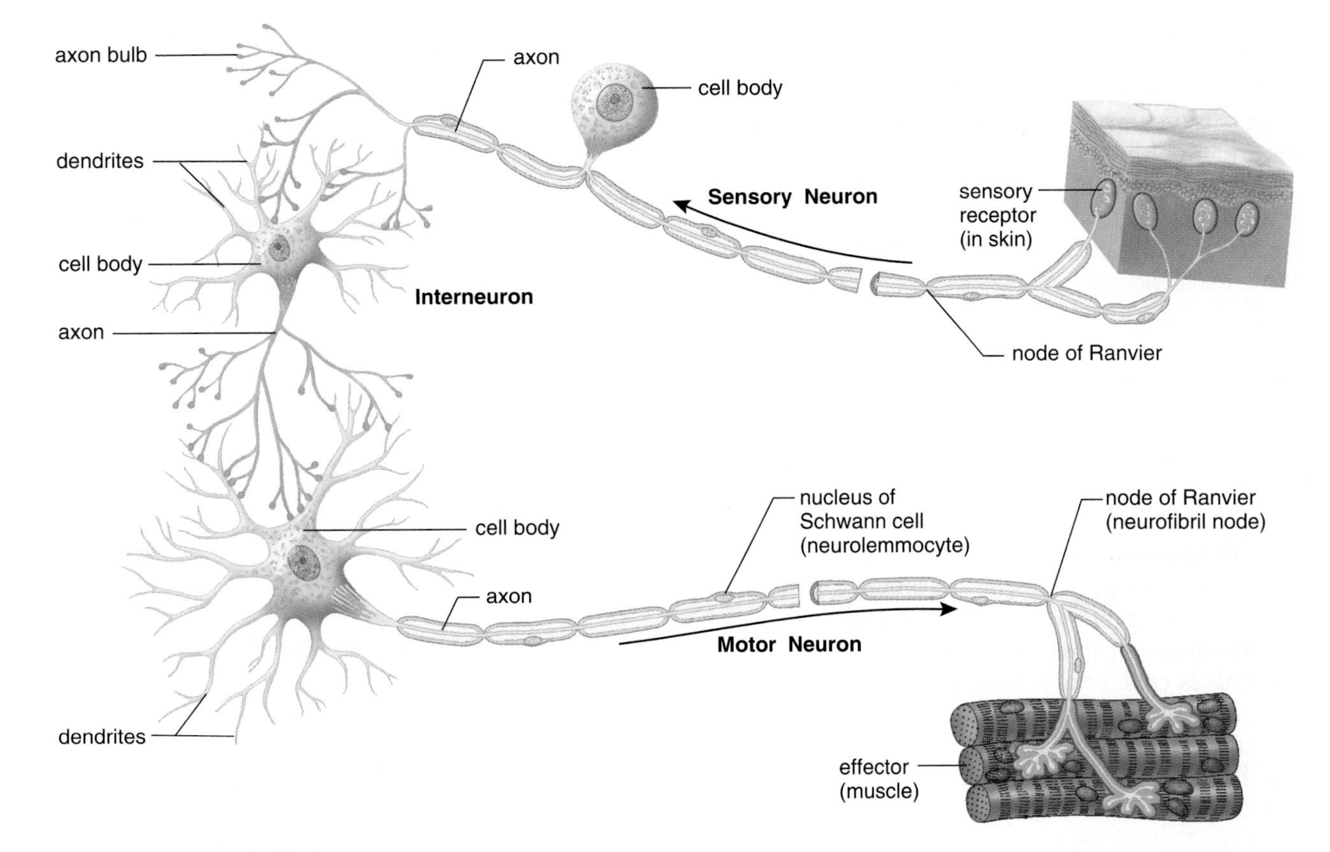

Figure 17.2 Types of neurons.
A sensory neuron, an interneuron, and a motor neuron are drawn here to show their arrangement in the body. (The breaks indicate that the fibers are much longer than shown.) How does this arrangement correlate with the function of each neuron?

Myelin Sheath

Long axons are covered by a protective **myelin sheath** formed by neuroglial cells called **Schwann cells** (neurolemmocytes). The myelin sheath develops when Schwann cells wrap themselves around an axon many times and in this way lay down several layers of plasma membrane. Schwann cell plasma membrane contains myelin, a lipid substance that is an excellent insulator. A myelin sheath, which is interrupted by gaps called **nodes of Ranvier,** gives nerve fibers their white, glistening appearance (Fig. 17.3).

Multiple sclerosis (MS) is a disease of the myelin sheath. Lesions develop and become hardened scars that interfere with normal conduction of nerve impulses, and the result is various neuromuscular symptoms. On the other hand, the myelin sheath plays an important role in nerve regeneration. If an axon is accidentally severed, the distal part of the axon degenerates but the myelin sheath remains and serves as a passageway for new fiber growth.

All neurons have three parts: dendrites, cell body, and axon. Sensory neurons take information to the CNS, and interneurons sum up sensory input before motor neurons take commands away from the CNS.

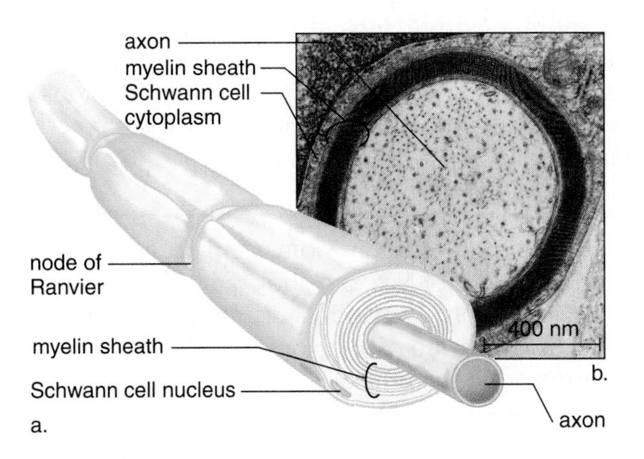

Figure 17.3 Myelin sheath.
a. A myelin sheath forms when Schwann cells wrap themselves around a nerve fiber. **b.** Electron micrograph of a cross section of an axon surrounded by a myelin sheath.

Nerve Impulse

A **nerve impulse** is the way a neuron transmits information. The nature of a nerve impulse has been studied by using excised axons and a voltmeter called an **oscilloscope.** Voltage, often measured in millivolts (mV), is a measure of the electrical potential difference between two points, which in this case are the inside and the outside of the axon. Voltage is displayed on the oscilloscope screen as a trace, or pattern over time.

Resting Potential

In the experimental setup shown in Figure 17.4, an oscilloscope is wired to two electrodes: one electrode is placed inside and the other electrode is placed outside an axon. The axon is essentially a membranous tube filled with axoplasm (cytoplasm of the axon). When the axon is not conducting an impulse, the oscilloscope records a potential difference across a membrane equal to about −65 mV. This reading indicates that the inside of the axon is negative compared to the outside. This is called the **resting potential** because the axon is not conducting an impulse.

The existence of this polarity (charge difference) can be correlated with a difference in ion distribution on either side of the axomembrane (plasma membrane of the axon). As Figure 17.4*a* shows, the concentration of sodium ions (Na^+) is greater outside the axon than inside, and the concentration of potassium ions (K^+) is greater inside the axon than outside. The unequal distribution of these ions is due to the action of the **sodium-potassium pump,** a membrane protein that actively transports Na^+ out of and K^+ into the axon. The work of the pump maintains the unequal distribution of Na^+ and K^+ across the membrane.

The pump is always working because the membrane is somewhat permeable to these ions, and they tend to diffuse toward their lesser concentration. Since the membrane is more permeable to K^+ than to Na^+, there are always more positive ions outside the membrane than inside. This accounts for the polarity recorded by the oscilloscope. Large, negatively charged organic ions in the axoplasm also contribute to the polarity across a resting axomembrane.

Because of the sodium-potassium pump there is a concentration of Na^+ outside an axon and K^+ inside an axon. An unequal distribution of ions causes the inside of an axon to be negative compared to the outside.

Action Potential

An **action potential** is a rapid change in polarity across an axomembrane as the nerve impulse occurs. If a stimulus causes the axomembrane to depolarize to a certain level, called **threshold,** an action potential occurs (Fig. 17.4). The strength of an action potential does not change, but an intense stimulus can cause an axon to fire (start an axon potential) at a greater frequency than a weak stimulus.

The action potential utilizes two types of gated channel proteins in the membrane. There is a gated channel protein that opens allowing Na^+ to pass through the membrane, and another that opens allowing K^+ to pass through the membrane. The sodium channel is faster to open than the potassium channel.

Sodium Gates Open When an action potential occurs, the gates of sodium channels open first and Na^+ flows into the axon. As Na^+ moves to inside the axon, the membrane potential changes from −65 mV to +40 mV. This is a *depolarization* because the charge inside the axon changes from negative to positive as Na^+ enters the axon.

Potassium Gates Open Second, the gates of potassium channels open and K^+ flows to outside the axon. As K^+ moves to outside the axon, the action potential changes from +40 mV back to −65 mV. This is a *repolarization* because the inside of the axon resumes a negative charge as K^+ exits the axon.

The nerve impulse consists of an electrochemical change that occurs across an axomembrane. During depolarization, Na^+ moves to inside the axon, and during repolarization K^+ moves to outside the axon.

Propagation of an Action Potential

When an action potential travels down an axon, each successive portion of the axon undergoes a depolarization and then a repolarization. Like a domino effect, each preceding portion causes an action potential in the next portion of an axon.

As soon as an action potential has moved on, the previous portion of an axon undergoes a **refractory period** during which the sodium gates are unable to open. This ensures that the action potential cannot move backwards and instead always moves down an axon toward its branches.

In myelinated axons, the gated ion channels that produce an action potential are concentrated at the nodes of Ranvier. Since ion exchange occurs only at the nodes, the action potential travels faster than in nonmyelinated axons. This is called saltatory conduction, meaning that the action potential "jumps" from node to node. Speeds of 200 meters per second (450 miles per hour) have been recorded.

An action potential travels along the length of an axon.

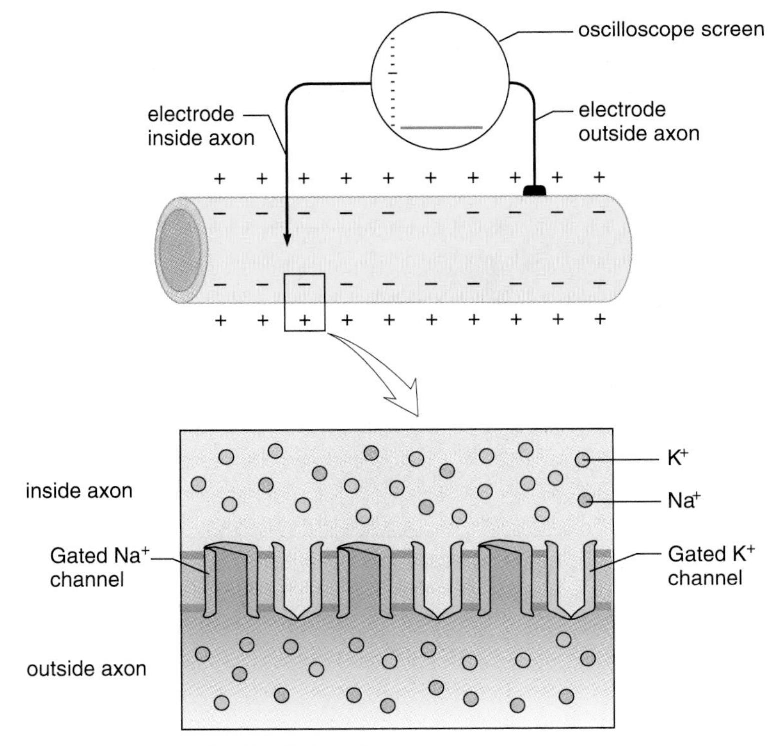

a. Resting Potential

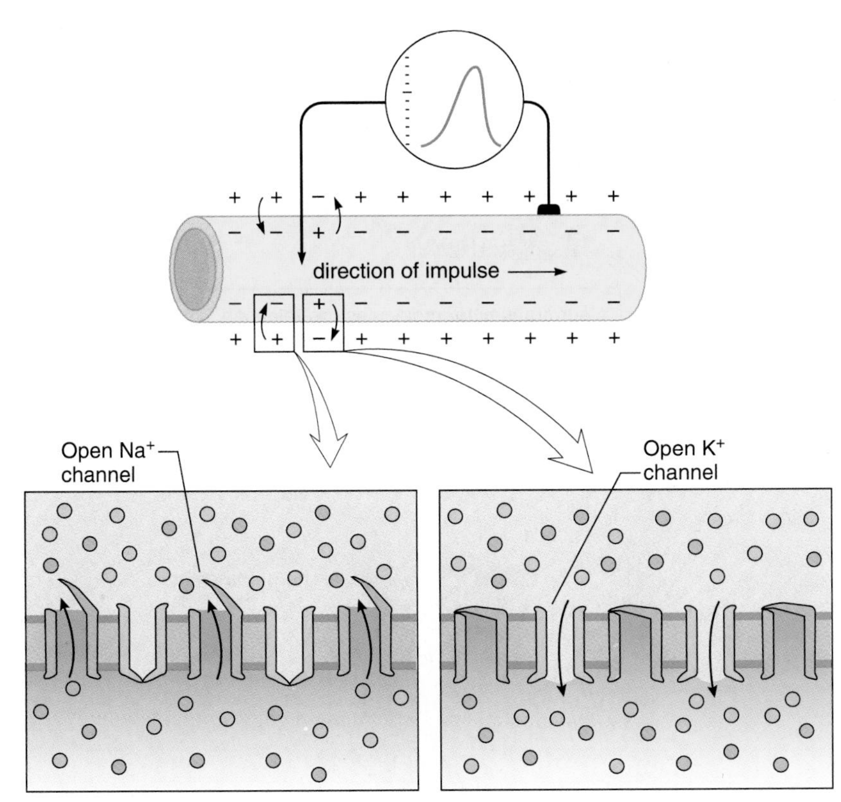

b. Action Potential

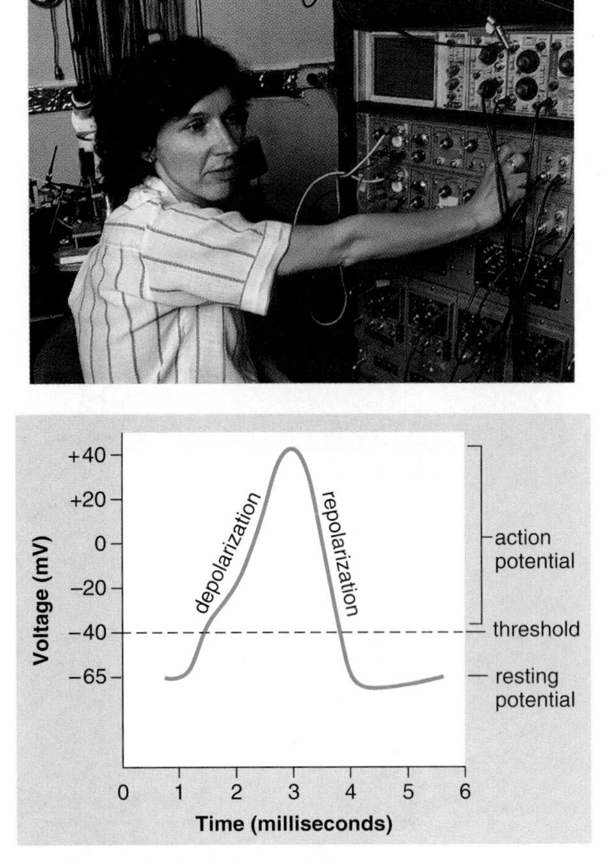

c. Enlargement of action potential

Figure 17.4 Resting and action potential.
a. Resting potential. An oscilloscope records a resting potential of −65 mV. There is a preponderance of Na^+ outside the axon and preponderance of K^+ inside the axon. The permeability of the membrane to K^+ compared to Na^+ causes the inside to be negative compared to the outside. **b.** Action potential. A depolarization occurs when Na^+ gates open and Na^+ moves to inside the axon, and a repolarization occurs when K^+ gates open and K^+ moves to outside the axon. **c.** Enlargement of the action potential, which is seen by an experimenter using an oscilloscope, an instrument that records voltage changes.

Visual Focus

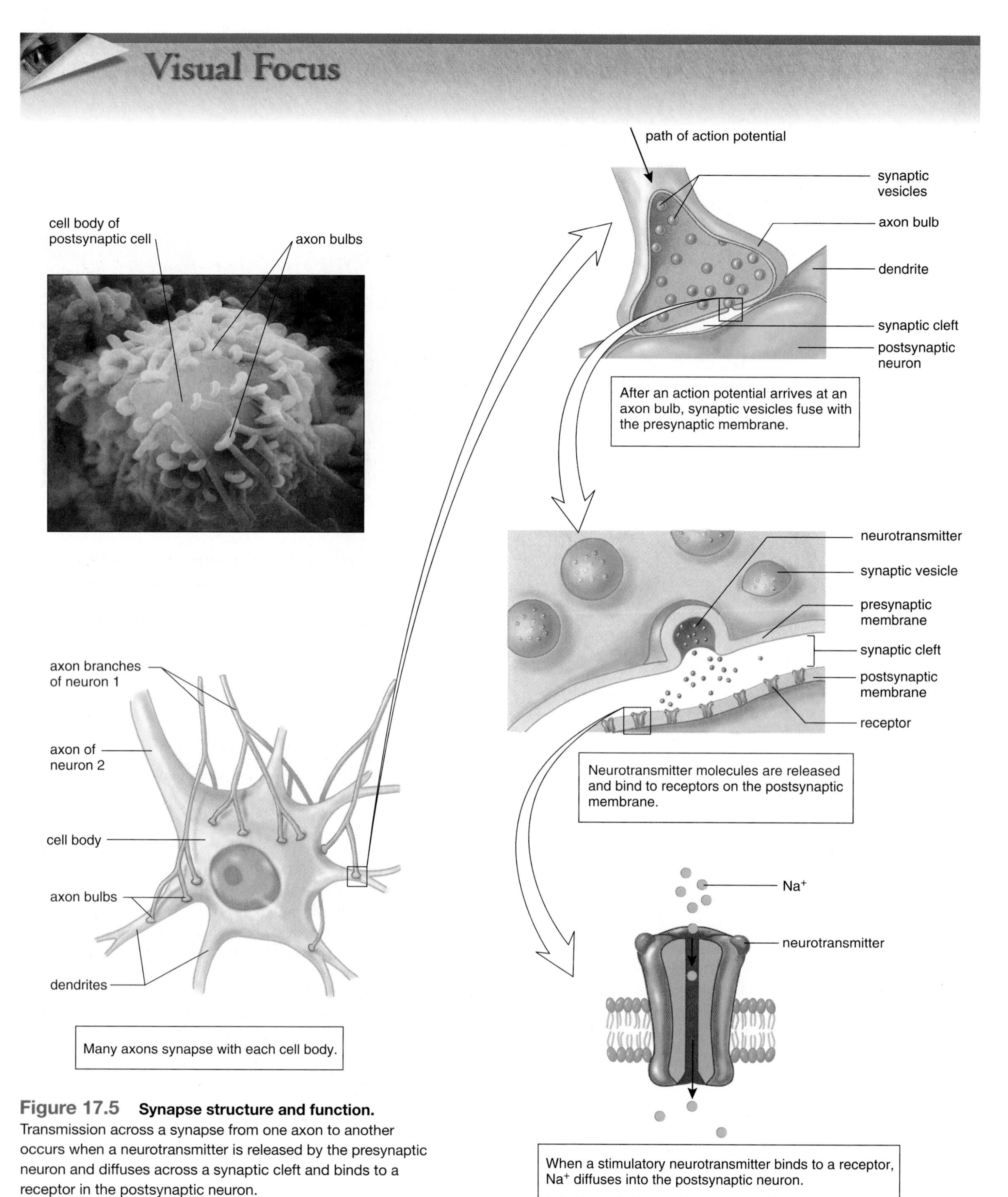

Figure 17.5 Synapse structure and function. Transmission across a synapse from one axon to another occurs when a neurotransmitter is released by the presynaptic neuron and diffuses across a synaptic cleft and binds to a receptor in the postsynaptic neuron.

Transmission Across a Synapse

Every axon branches into many fine endings, each tipped by a small swelling, called an **axon bulb** (Fig. 17.5). Each bulb lies very close to the dendrite (or the cell body) of another neuron. This region of close proximity is called a **synapse.** At a synapse, the membrane of the first neuron is called the *pre*synaptic membrane, and the membrane of the next neuron is called the *post*synaptic membrane. The small gap between is the **synaptic cleft.**

Transmission across a synapse is carried out by molecules called **neurotransmitters,** which are stored in synaptic vesicles (Fig. 17.5*b, c*). When nerve impulses traveling along an axon reach an axon bulb, gated channels for calcium ions (Ca^{2+}) open and calcium enters the bulb. This sudden rise in Ca^{2+} stimulates synaptic vesicles to merge with the presynaptic membrane, and neurotransmitter molecules are released into the synaptic cleft. They diffuse across the cleft to the postsynaptic membrane, where they bind with specific receptor proteins (Fig. 17.5*c*).

Depending on the type of neurotransmitter and/or the type of receptor, the response of the postsynaptic neuron can be toward excitation or toward inhibition. Excitatory neurotransmitters that utilize gated ion channels are fast acting. Other neurotransmitters affect the metabolism of the postsynaptic cell and therefore are slower acting.

Neurotransmitter Molecules

At least 25 different neurotransmitters have been identified, but two very well-known neurotransmitters are **acetylcholine (ACh)** and **norepinephrine (NE).**

Once a neurotransmitter has been released into a synaptic cleft and has initiated a response, it is removed from the cleft. In some synapses, the postsynaptic membrane contains enzymes that rapidly inactivate the neurotransmitter. For example, the enzyme **acetylcholinesterase (AChE)** breaks down acetylcholine. In other synapses, the presynaptic membrane rapidly reabsorbs the neurotransmitter, possibly for repackaging in synaptic vesicles or for molecular breakdown. The short existence of neurotransmitters at a synapse prevents continuous stimulation (or inhibition) of postsynaptic membranes.

It is of interest to note here that many drugs that affect the nervous system act by interfering with or potentiating the action of neurotransmitters. As described in Figure 17.18, drugs can enhance or block the release of a neurotransmitter, mimic the action of a neurotransmitter or block the receptor, or interfere with the removal of a neurotransmitter from a synaptic cleft.

Transmission across a synapse is dependent on the release of neurotransmitters, which diffuse across the synaptic cleft from one neuron to the next.

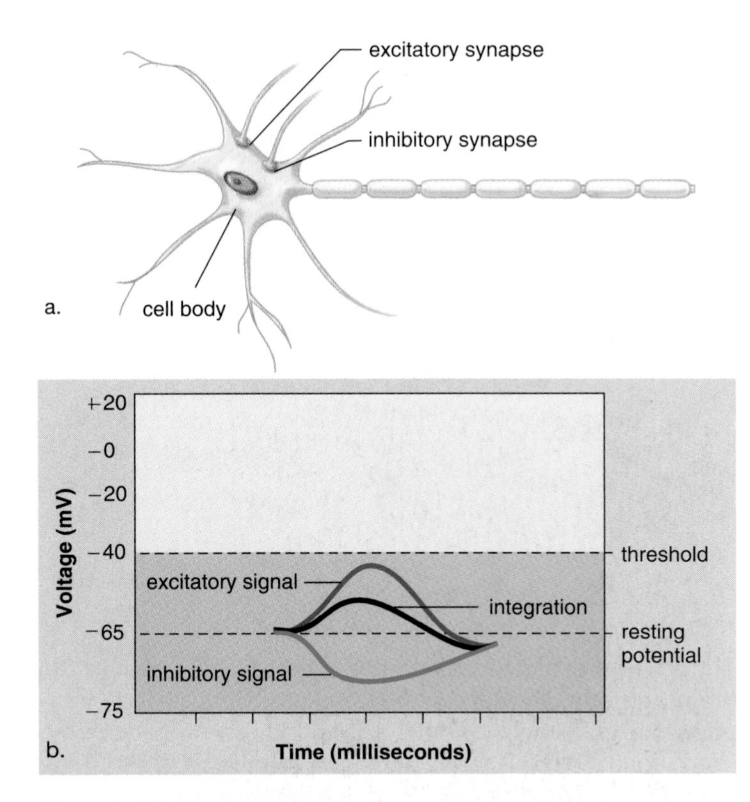

Figure 17.6 Integration.
a. Inhibitory signals and excitatory signals are summed up in the dendrite and cell body of the postsynaptic neuron. Only if the combined signals cause the membrane potential to rise above threshold does an action potential occur. **b.** In this example, threshold was not reached.

Synaptic Integration

A single neuron has many dendrites plus the cell body and both can have synapses with many other neurons. One thousand to ten thousand synapses per a single neuron is not uncommon. Therefore, a neuron is on the receiving end of many excitatory and inhibitory signals. An excitatory neurotransmitter produces a potential change called a signal that drives the neuron closer to an action potential, and an inhibitory neurotransmitter produces a signal that drives the neuron further from an action potential. Excitatory signals have a depolarizing effect, and inhibitory signals have a hyperpolarizing effect (Fig. 17.6).

Neurons integrate these incoming signals. **Integration** is the summing up of excitatory and inhibitory signals. If a neuron receives many excitatory signals (either from different synapses or at a rapid rate from one synapse), the chances are the axon will transmit a nerve impulse. On the other hand, if a neuron receives both inhibitory and excitatory signals, the summing up of these signals may prohibit the axon from firing.

Integration is the summing up of inhibitory and excitatory signals received by a postsynaptic neuron.

17.2 Peripheral Nervous System

The *peripheral nervous system (PNS)* lies outside the central nervous system and contains **nerves** which are bundles of axons. Axons that occur in nerves are also called nerve fibers.

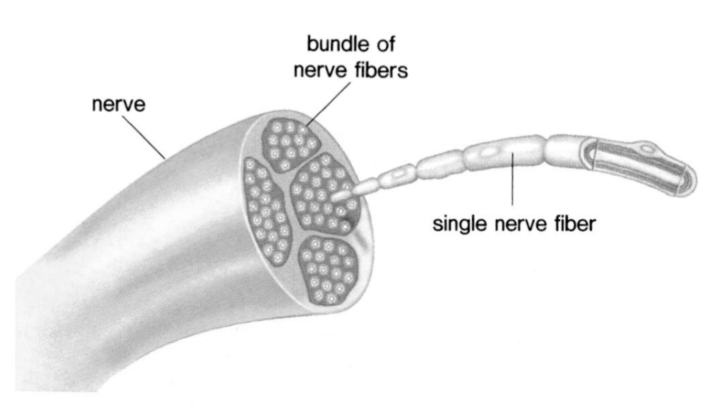

The cell bodies of neurons are found in the CNS—that is, the brain and spinal cord—or in ganglia. Ganglia (sing., **ganglion**) are collections of cell bodies within the PNS.

Humans have 12 pairs of **cranial nerves** attached to the brain (Fig. 17.7*a*). Some of these are sensory nerves; that is, they contain only sensory nerve fibers. Some are motor nerves that contain only motor fibers, and others are mixed nerves that contain both sensory and motor fibers. Cranial nerves are largely concerned with the head, neck, and facial regions of the body. However, the vagus nerve has branches not only to the pharynx and larynx, but also to most of the internal organs.

Humans have 31 pairs of **spinal nerves** (Fig. 17.7*b*). The paired spinal nerves emerge from the spinal cord by two short branches, or roots. The *dorsal root* contains the axons of sensory neurons, which are conducting impulses to the spinal cord from sensory receptors. The cell body of a sensory neuron is in the **dorsal-root ganglion.** The *ventral root* contains the axons of motor neurons, which are conducting impulses away from the cord to effectors. These two roots join to form a spinal nerve. All spinal nerves are mixed nerves that contain many sensory and motor fibers. Each spinal nerve serves the particular region of the body in which it is located.

In the PNS, cranial nerves take impulses to and/or from the brain, and spinal nerves take impulses to and from the spinal cord.

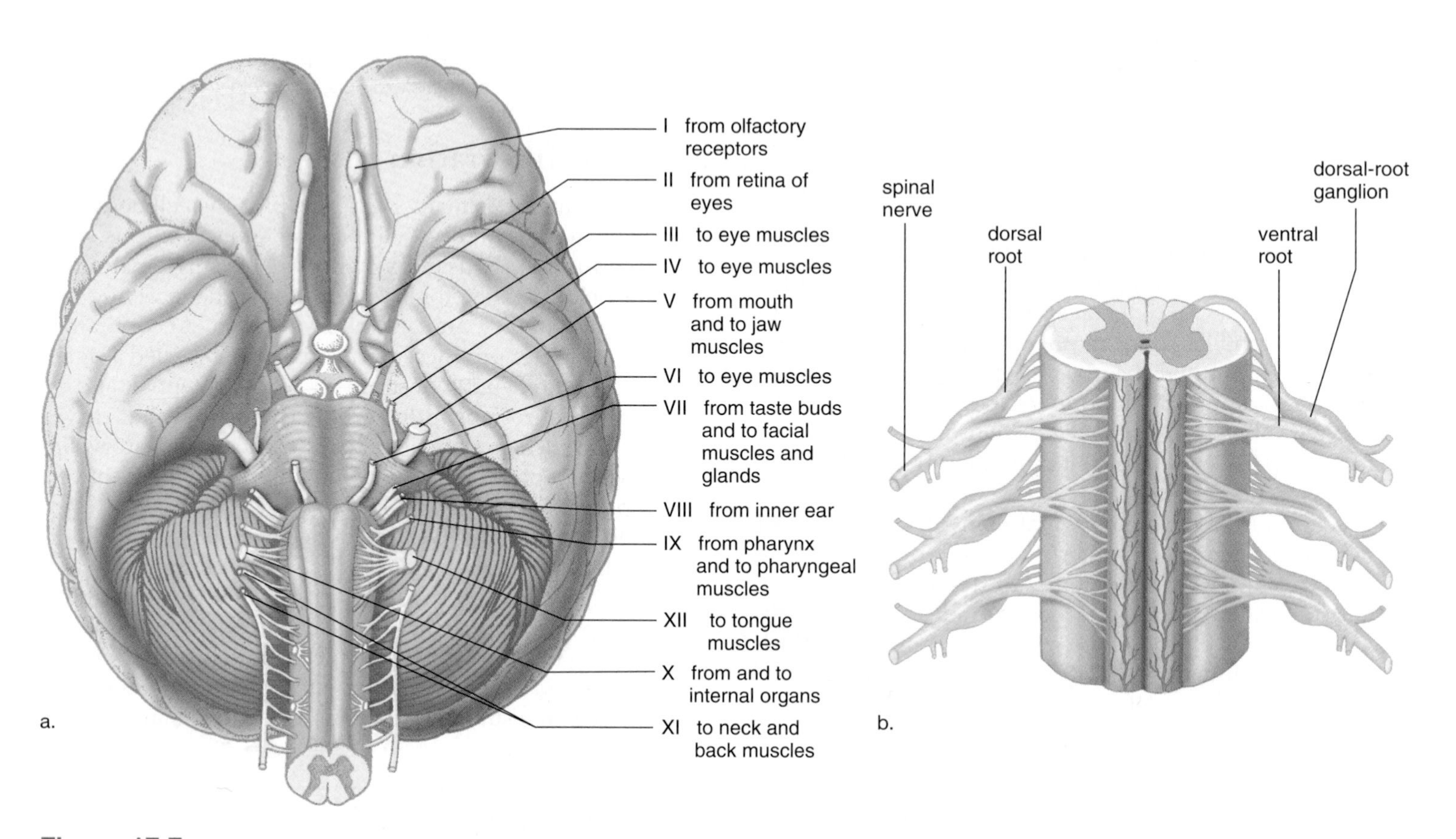

Figure 17.7 Cranial and spinal nerves.
a. Ventral surface of the brain showing the attachment of the 12 pairs of cranial nerves. **b.** Cross section of the spinal cord, showing a few spinal nerves. The human body has 31 pairs of spinal nerves and each spinal nerve has a dorsal root and a ventral root attached to the spinal cord.

Somatic System

The **somatic system** includes the nerves that take sensory information from external sensory receptors to the CNS and motor commands away from the CNS to skeletal muscles. Voluntary control of skeletal muscles always originates in the brain. Involuntary responses to stimuli, called **reflexes,** can involve either the brain or just the spinal cord. Flying objects cause eyes to blink and sharp tacks cause hands to jerk away even without us having to think about it.

The Reflex Arc

Figure 17.8 illustrates the path of a reflex that involves only the spinal cord. If your hand touches a sharp tack, sensory receptors in the skin generate nerve impulses that move along sensory axons toward the spinal cord. Sensory neurons which enter the cord dorsally pass signals on to many interneurons. Some of these interneurons synapse with motor neurons. The short dendrites and the cell bodies of motor neurons are in the spinal cord but their axons leave the cord ventrally. Nerve impulses travel along motor axons to an effector, which brings about a response to the stimulus. In this case a muscle contracts so that you withdraw your hand from the tack. Various other reactions are possible—you will most likely look at the tack, wince, and cry out in pain. This whole series of responses is explained by the fact some of the interneurons involved carry nerve impulses to the brain. The brain makes you aware of the stimulus and directs these other reactions to it.

In the somatic system, nerves take information from external sensory receptors to the CNS and motor commands to skeletal muscles. Involuntary reflexes allow us to respond rapidly to external stimuli.

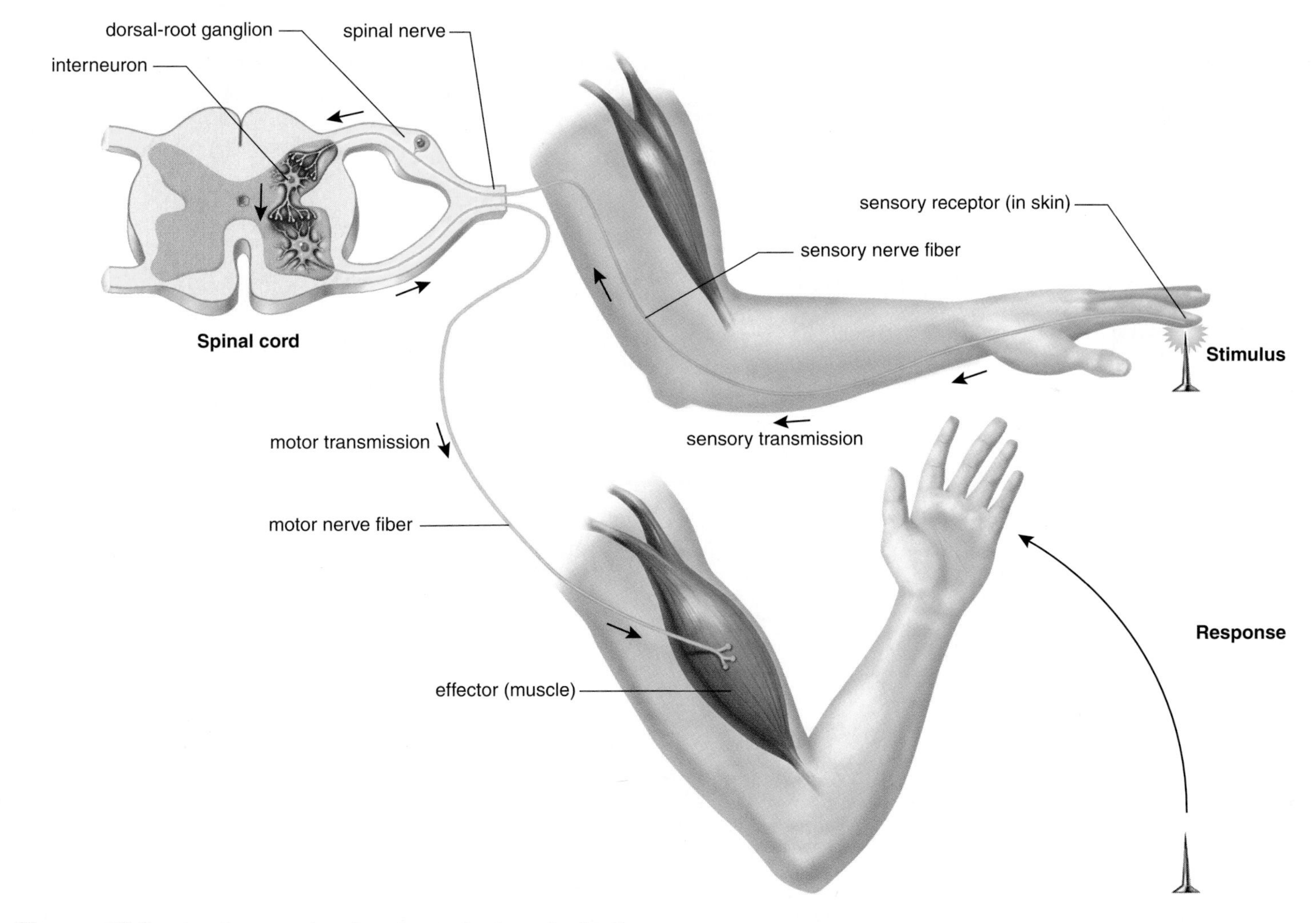

Figure 17.8 A reflex arc showing the path of a spinal reflex.
A stimulus (e.g., sharp tack) causes sensory receptors in the skin to generate nerve impulses that travel in sensory nerve fibers to the spinal cord. Interneurons integrate data from sensory neurons and then relay signals to motor neurons. Motor nerve fibers convey nerve impulses from the spinal cord to a skeletal muscle which contracts. Movement of the hand away from the tack is the response to the stimulus.

Visual Focus

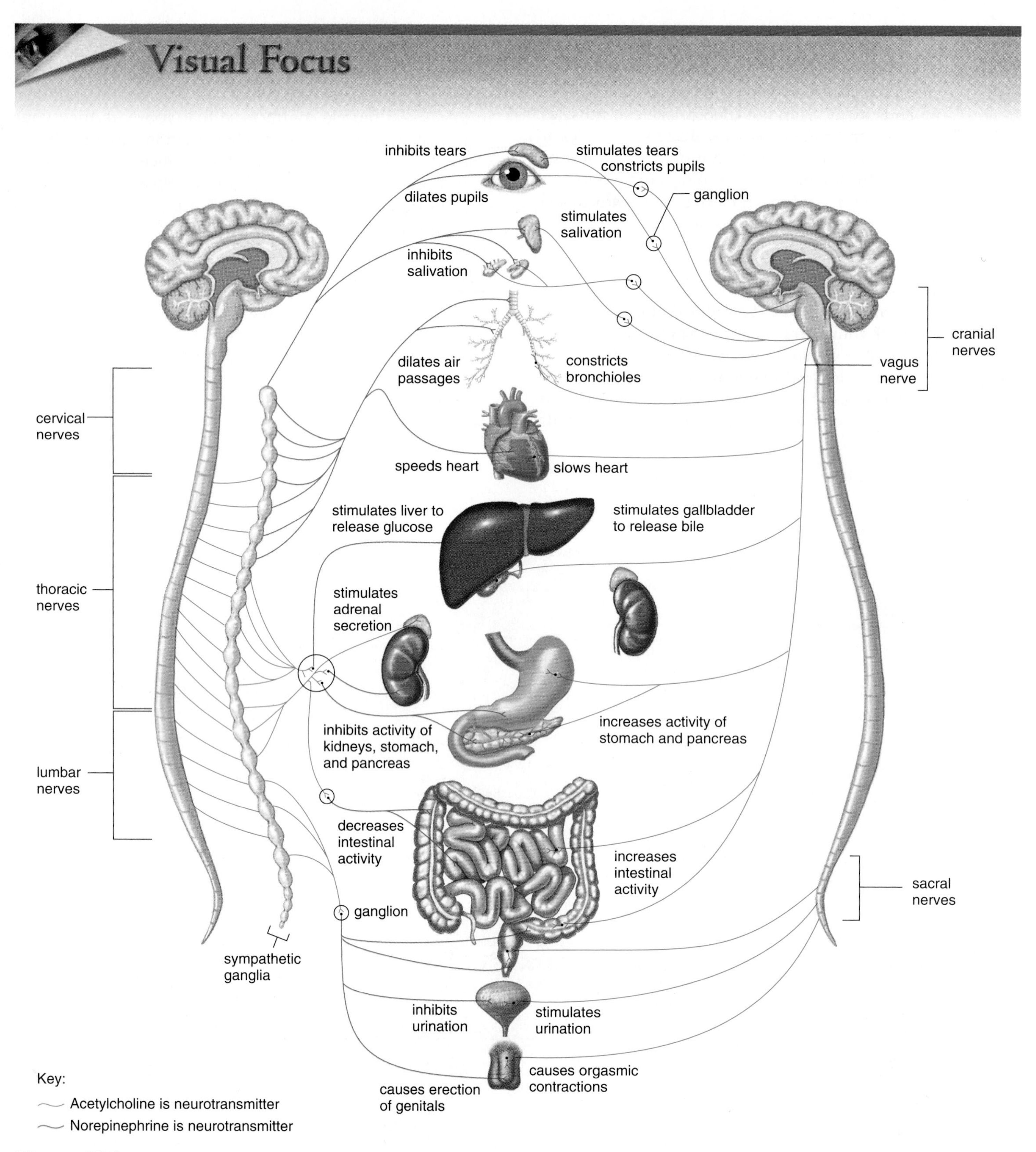

Figure 17.9 Autonomic system structure and function.
Sympathetic preganglionic fibers arise from the thoracic and lumbar portions of the spinal cord; parasympathetic preganglionic fibers arise from the brain and the sacral portion of the spinal cord. Each system innervates the same organs but has contrary effects as described.

Autonomic System

The **autonomic system** of the PNS regulates the activity of cardiac and smooth muscle and glands. The system is divided into the sympathetic and parasympathetic divisions (Fig. 17.9 and Table 17.1). Both of these divisions (1) function automatically and usually in an involuntary manner; (2) innervate all internal organs; and (3) utilize two neurons and one ganglion for each impulse. The first neuron has a cell body within the CNS and a *preganglionic fiber.* The second neuron has a cell body within the ganglion and a *postganglionic fiber.*

Reflex actions, such as those that regulate the blood pressure and breathing rate, are especially important to the maintenance of homeostasis. These reflexes begin when the sensory neurons in contact with internal organs send information to the CNS. They are completed by motor neurons within the autonomic system.

Sympathetic Division

Most preganglionic fibers of the **sympathetic division** arise from the middle, or *thoracic-lumbar,* portion of the spinal cord and almost immediately terminate in ganglia that lie near the cord. Therefore, in this division, the preganglionic fiber is short, but the postganglionic fiber that makes contact with an organ is long.

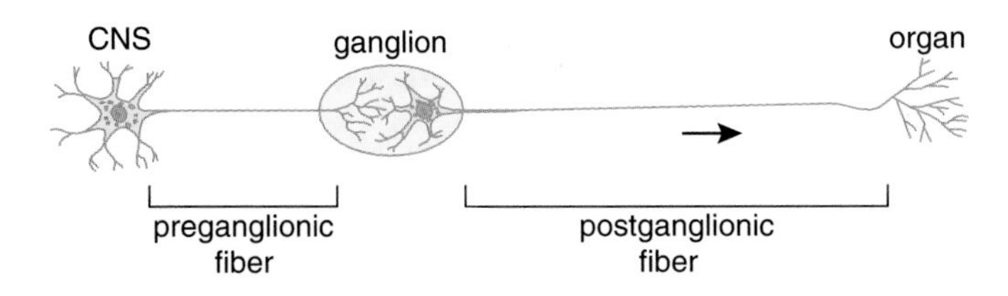

The sympathetic division is especially important during emergency situations and is associated with **"fight or flight."** If you need to fend off a foe or flee from danger, active muscles require a ready supply of glucose and oxygen. The sympathetic division accelerates the heartbeat and dilates the bronchi. On the other hand, the sympathetic division inhibits the digestive tract—digestion is not an immediate necessity if you are under attack. The neurotransmitter released by the postganglionic axon is primarily norepinephrine (NE). The structure of NE is like that of epinephrine (adrenaline), an adrenal medulla hormone that usually increases heart rate and contractility.

The sympathetic division brings about those responses we associate with "fight or flight."

Parasympathetic Division

The **parasympathetic division** includes a few cranial nerves (e.g., the vagus nerve) and also fibers that arise from the sacral (bottom) portion of the spinal cord. Therefore, this division often is referred to as the craniosacral portion of the autonomic system. In the parasympathetic division, the preganglionic fiber is long, and the postganglionic fiber is short because the ganglia lie near or within the organ.

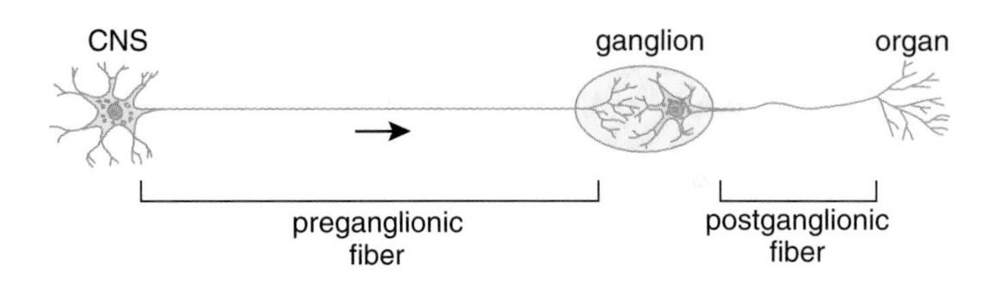

The parasympathetic division, sometimes called the "housekeeper division," promotes all the internal responses we associate with a relaxed state; for example, it causes the pupil of the eye to contract, promotes digestion of food, and retards the heartbeat. The neurotransmitter utilized by the parasympathetic division is acetylcholine (ACh).

The parasympathetic division brings about the responses we associate with a relaxed state.

Table 17.1 Comparison of Somatic Motor and Autonomic Motor Pathways

Items	Somatic Motor Pathway	Autonomic Motor Pathways	
		Sympathetic	Parasympathetic
Type of control	Voluntary/involuntary	Involuntary	Involuntary
Number of neurons per message	One	Two (preganglionic shorter than postganglionic)	Two (preganglionic longer than postganglionic)
Location of motor fiber	Most cranial nerves and all spinal nerves	Thoracolumbar spinal nerves	Cranial (e.g., vagus) and sacral spinal nerves
Neurotransmitter	Acetylcholine	Norepinephrine	Acetylcholine
Effectors	Skeletal muscles	Smooth and cardiac muscle, glands	Smooth and cardiac muscle, glands

17.3 Central Nervous System

The *central nervous system (CNS)* consists of the spinal cord and the brain where sensory information is received and motor control is initiated. Both the spinal cord and the brain are protected by bone; the brain is enclosed by the skull and the spinal cord is surrounded by vertebrae. Also, both the spinal cord and brain are wrapped in protective membranes known as **meninges** (sing., meninx). Meningitis is an infection of these coverings. The spaces between the meninges are filled with **cerebrospinal fluid,** which cushions and protects the CNS. A small amount of this fluid sometimes is withdrawn from around the cord for laboratory testing when a spinal tap (i.e., lumbar puncture) is performed.

Cerebrospinal fluid is also contained within the ventricles of the brain and in the central canal of the spinal cord. The brain's **ventricles** are interconnecting cavities that produce and serve as a reservoir for cerebrospinal fluid. Normally, any excess cerebrospinal fluid drains away into the circulatory system. However, blockages can occur. In an infant, the brain can enlarge due to cerebrospinal fluid accumulation, and this condition is called "water on the brain." If cerebrospinal fluid collects in an adult, the brain cannot enlarge and instead, the brain is pushed against the skull, possibly causing injury.

The CNS, which lies in the midline of the body and consists of the brain and the spinal cord, receives sensory information and initiates voluntary motor control.

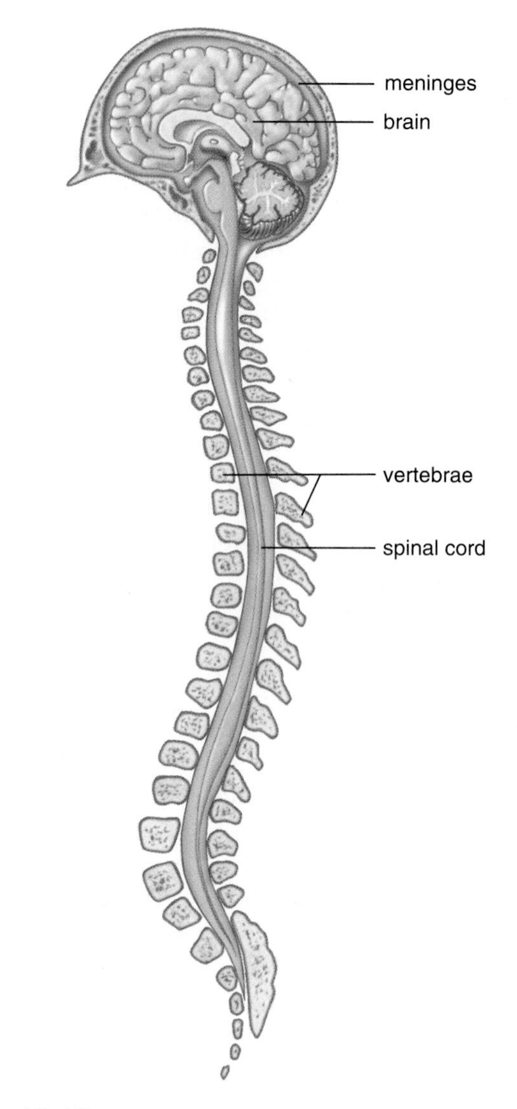

Figure 17.10 Central nervous system.
The central nervous system consists of the brain and spinal cord. The brain is protected by the skull and the spinal cord is protected by the vertebrae.

The Spinal Cord

The **spinal cord** extends from the base of the brain through a large opening in the skull called the foramen magnum and into the vertebral canal formed by the vertebrae (Fig. 17.10).

Structure of the Spinal Cord

Figure 17.11*a* shows how the individual vertebrae join to form the vertebral canal which protects the spinal cord. The spinal nerves which project from the cord pass through openings between the vertebrae of the vertebral canal.

A cross section of the spinal cord shows that the spinal cord has a central canal, gray matter, and white matter (Fig. 17.11*b, c*). The central canal contains cerebrospinal fluid, as do the meninges that protect the spinal cord. The **gray matter** is centrally located and shaped like the letter H. It is gray because it contains cell bodies and short, nonmyelinated fibers. Portions of sensory neurons and motor neurons are found here, as are interneurons that communicate with these two types of neurons. The dorsal root of a spinal nerve contains sensory nerve fibers entering the gray matter, and the ventral root of a spinal nerve contains motor nerve fibers exiting the gray matter. The dorsal and ventral roots join before the spinal nerve leaves the vertebral canal. Spinal nerves are a part of the PNS.

The **white matter** of the spinal cord occurs in areas around the gray matter. The white matter is white because it contains myelinated axons of interneurons that run together in bundles called **tracts.** Ascending tracts taking information to the brain are primarily located dorsally, and descending tracts taking information from the brain are primarily located ventrally. Because the tracts cross just after they enter and exit the brain, the left side of the brain controls the right side of the body, and the right side of the brain controls the left side of the body.

The spinal cord extends from the base of the brain into the vertebral canal formed by the vertebrae. A cross section shows that the spinal cord has a central canal, gray matter, and white matter.

Functions of the Spinal Cord

The spinal cord is the center for thousands of reflex arcs. Figure 17.8 indicates the path of a spinal reflex from sensory receptors to muscle effectors. Each interneuron in the spinal cord has synapses with many other neurons, and therefore they carry out integration of incoming information before sending signals to other interneurons and motor neurons (see Fig. 17.6).

The spinal cord provides a means of communication between the brain and the peripheral nerves that leave the cord. When someone touches your hand, sensory information passes from sensory receptors through sensory nerve fibers to the spinal cord and up ascending tracts to the brain. When we voluntarily move our limbs, motor impulses originating in the brain pass down descending tracts to the spinal cord and out to our muscles by way of motor nerve fibers. Therefore, if the spinal cord is severed, we suffer a loss of sensation and a loss of voluntary control—that is, we suffer a paralysis. If the spinal cord is completely cut across in the thoracic region, paralysis of the lower body and legs occurs. This condition is known as paraplegia. If the injury is in the neck region, the four limbs are usually affected. This condition is called quadriplegia.

The spinal cord is a center for reflex action. The spinal cord also serves as a means of communication between the brain and much of the body. Because tracts to and from the brain cross over, the left side of the brain controls the right side of the body and vice versa.

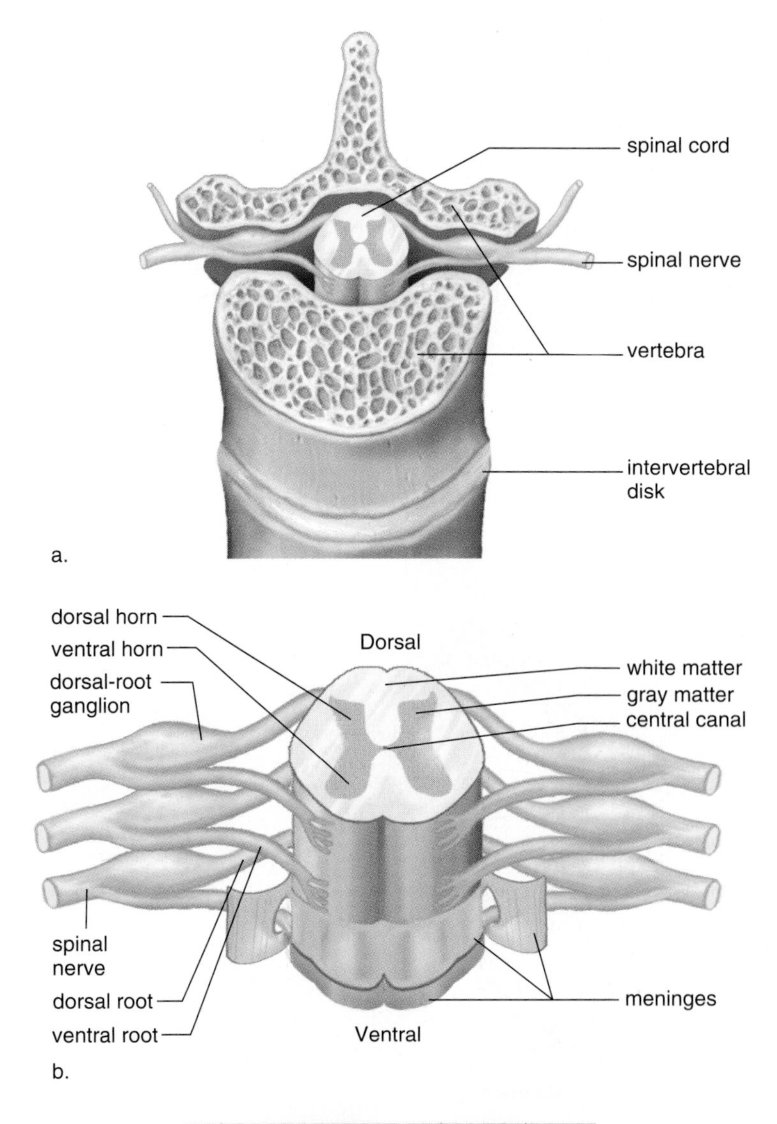

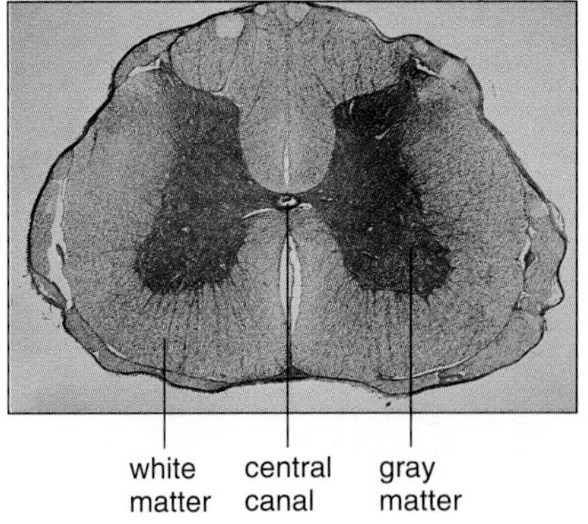

Figure 17.11 Spinal cord.
a. The spinal cord passes through the vertebral canal formed by the vertebrae. **b.** The spinal cord has a central canal filled with cerebrospinal fluid, H-shaped gray matter, and white matter. The white matter contains tracts that take nerve impulses to and from the brain. **c.** Photomicrograph.

The Brain

The human **brain** has been called the last great frontier of biology. The goal of modern neuroscience is to understand the structure and function of the brain's various parts so well that it will be possible to prevent or correct the more than 1,000 mental disorders that rob human beings of a normal life. This section gives only a glimpse of what is known about the brain and the modern avenues of research.

We will discuss the parts of the brain with reference to the brain stem, the diencephalon, and the cerebrum. The brain has four **ventricles** called, in turn, the fourth ventricle, the third ventricle, and the two lateral ventricles. It may be helpful to you to associate the brain stem with the fourth ventricle, the diencephalon with the third ventricle, and the cerebrum with the two lateral ventricles (Fig. 17.12*a*).

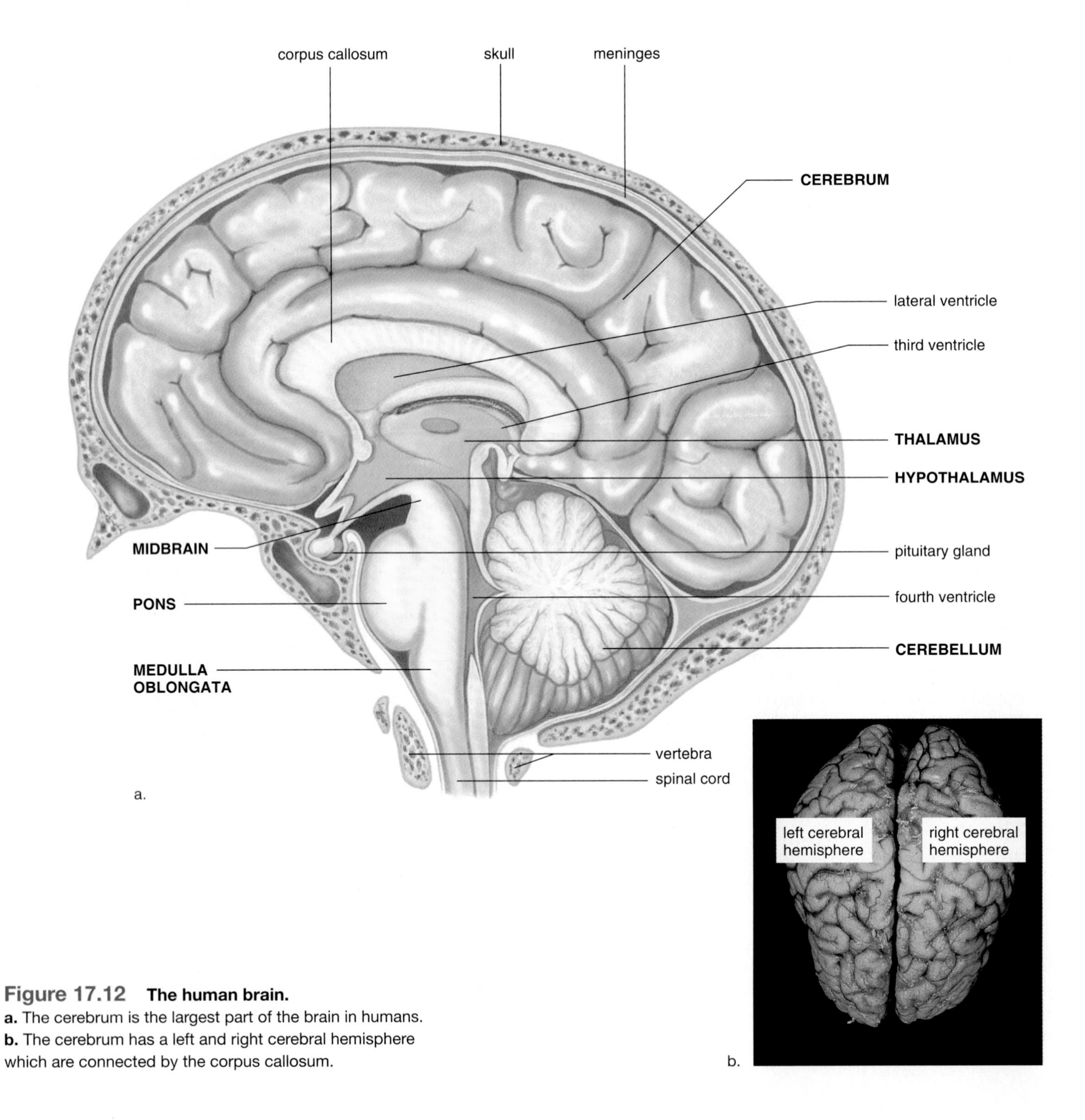

Figure 17.12 The human brain.
a. The cerebrum is the largest part of the brain in humans.
b. The cerebrum has a left and right cerebral hemisphere which are connected by the corpus callosum.

The Brain Stem

The **brain stem,** the initial portion of the brain, contains the medulla oblongata, the pons, and the midbrain. The **medulla oblongata** lies between the spinal cord and the pons. It contains a number of *vital centers* for regulating heartbeat, breathing, and vasoconstriction (blood pressure). It also contains the reflex centers for vomiting, coughing, sneezing, hiccuping, and swallowing. The medulla contains tracts that ascend or descend between the spinal cord and higher brain centers.

The Cerebellum

The **cerebellum** is separated from the brain stem by the fourth ventricle. The cerebellum has two portions that are joined by a narrow median portion. The surface of the cerebellum is gray matter, and the interior is largely white matter. The cerebellum integrates and passes on both sensory and motor information. It maintains normal muscle tone, posture, and balance and also ensures that all of the skeletal muscles work together to produce smooth and coordinated motions. The cerebellum is necessary for learning new motor skills like playing the piano or hitting a baseball.

The word **pons** means bridge in Latin, and true to its name, the pons contains bundles of axons traveling between the cerebellum and the rest of the CNS. In addition, the pons functions with the medulla to regulate breathing rate and has reflex centers concerned with head movements in response to visual and auditory stimuli.

The **midbrain** acts as a relay station for tracts passing between the cerebrum and the spinal cord or cerebellum. It also has reflex centers for visual, auditory, and tactile responses.

The Diencephalon

The hypothalamus and the thalamus are in the **diencephalon,** a region that encircles the third ventricle. The **hypothalamus** forms the floor of the third ventricle. The hypothalamus is the integrating center for the autonomic system. It also helps maintain homeostasis by regulating hunger, sleep, thirst, body temperature, and water balance. The hypothalamus controls the pituitary gland and thereby serves as a link between the nervous and endocrine systems.

The **thalamus** consists of two masses of gray matter located in the sides and roof of the third ventricle. The thalamus integrates sensory information and serves as a central relay station for sensory impulses traveling upward from other parts of the brain to the cerebrum. The thalamus is also involved in arousal and higher mental functions such as memory and emotion.

The pineal gland, which secretes the hormone melatonin, is located in the diencephalon. Presently there is much popular interest in the role of melatonin in our daily rhythms and whether it can help meliorate jet lag or insomnia. Scientists are also interested in the possibility that the hormone may regulate the onset of puberty.

The Cerebrum

The **cerebrum,** also called the telencephalon, is the foremost and largest portion of the brain in humans. Just as the human body has two halves, so does the cerebrum. These halves are called the left and right **cerebral hemispheres** (Fig. 17.12*b*). These two cerebral hemispheres are connected by a bridge of tracts within the corpus callosum.

The cerebrum is the highest center to receive sensory input and carry out integration before commanding voluntary motor responses. It is in communication with and coordinates the activities of the other parts of the brain. As we shall see, the cerebrum carries out higher thought processes required for learning and memory and for language and speech.

The Reticular Formation

The **reticular formation** is a complex network of nuclei and fibers that extend the length of the brain stem (Fig. 17.13). In this context, the term **nuclei** means masses of cell bodies in the CNS. The reticular formation receives sensory signals which it sends up to higher centers, and motor signals which it sends to the spinal cord.

One portion of the reticular formation called the reticular activating system (RAS) arouses the cerebrum via the thalamus and causes a person to be alert. It is believed to filter out unnecessary sensory stimuli and this may account for why you can study with the TV on. An inactive reticular formation results in sleep; a severe injury to the RAS can cause a person to be comatose.

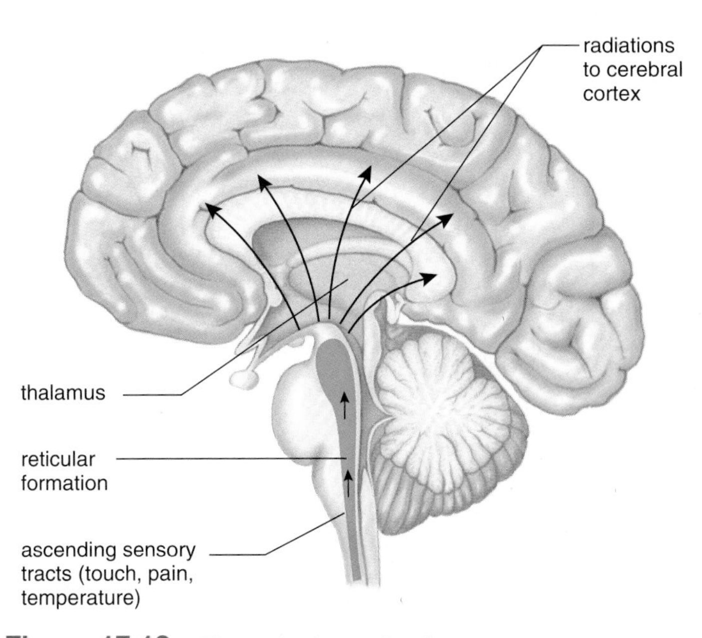

Figure 17.13 The reticular activating system.
The reticular formation receives and sends on motor and sensory information to various parts of the CNS. One portion, the reticular activating system (see arrows), arouses the cerebrum and in this way controls alertness versus sleep.

17.4 The Cerebral Hemispheres

A deep groove called the longitudinal fissure divides the left from the right cerebral hemisphere; shallow grooves called sulci (sing., sulcus) divide each hemisphere into lobes (Fig. 17.14). The frontal lobe is toward the front of a cerebral hemisphere and the parietal lobe is toward the back of a cerebral hemisphere. The occipital lobe is dorsal to (behind) the parietal lobe and the temporal lobe lies below the frontal and parietal lobes.

The Cerebral Cortex

The **cerebral cortex** is a thin but highly convoluted outer layer of gray matter that covers the cerebral hemispheres. The cerebral cortex contains over one billion cell bodies and is the region of the brain that accounts for sensation, voluntary movement, and all the thought processes we associate with consciousness.

The cerebral cortex contains motor areas and sensory areas and also association areas. The **primary motor area** is in the frontal lobe just ventral to (before) the central sulcus. Voluntary commands begin in the primary motor area and each part of the body is controlled by a certain section. Our versatile hand takes up an especially large area of the primary motor area. Ventral to the primary motor area is a *premotor area.* The premotor area organizes motor functions for skilled motor activities before the primary motor area sends signals to the cerebellum which integrates them. The unique ability of humans to speak is partially dependent upon *Broca's area,* a motor speech area located in the left frontal lobe. Signals originating here pass to the premotor area before reaching the primary motor area.

The **primary somatosensory area** is just dorsal to the central sulcus in the parietal lobe. Sensory information from the skin and skeletal muscles arrives here, where each part of the body is sequentially represented. A primary visual area in the occipital lobe receives information from our eyes, and a primary auditory area in the temporal lobe receives information from our ears. A primary taste area in the parietal lobe accounts for taste sensations.

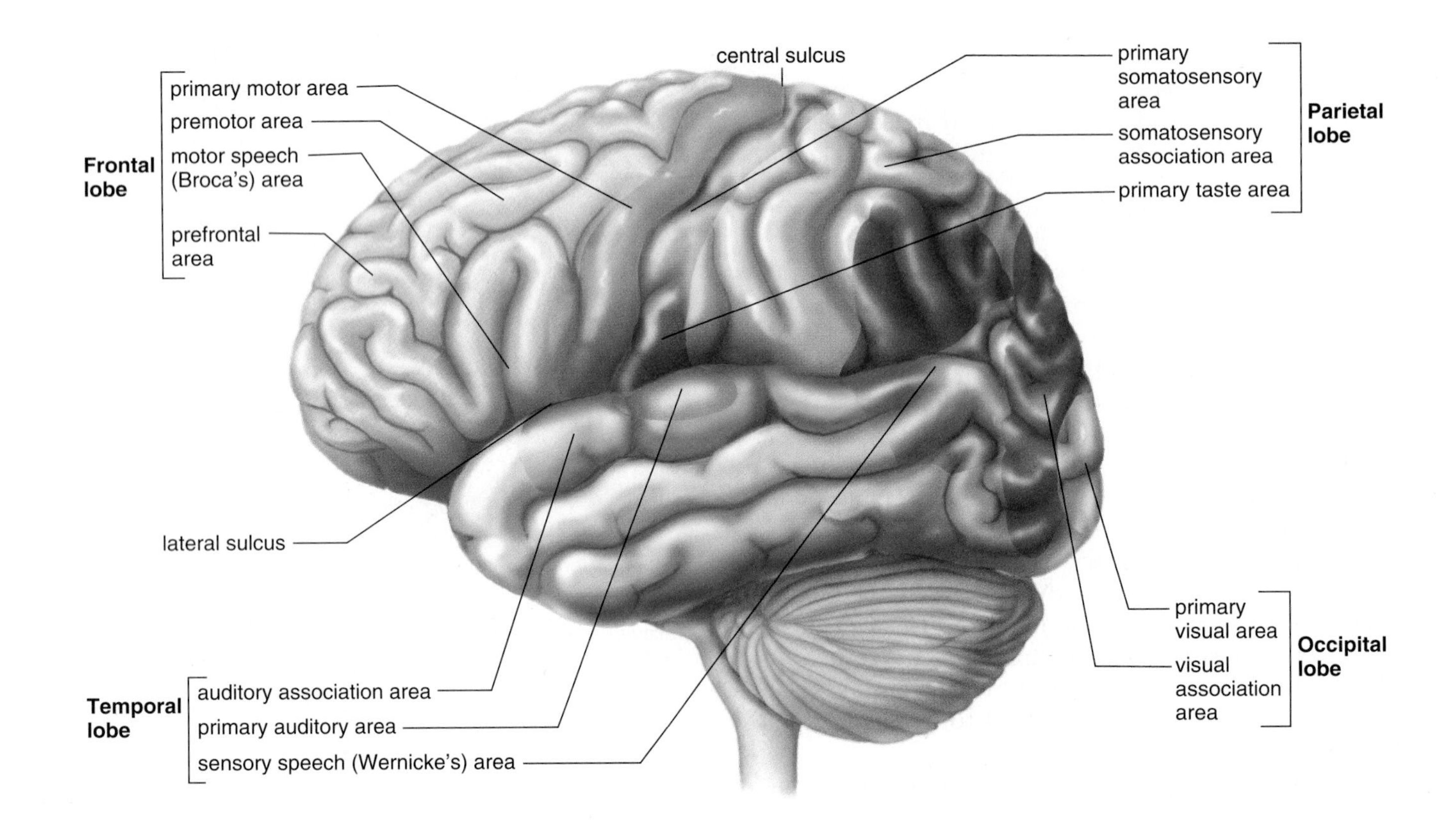

Figure 17.14 The cerebral cortex.
The cortex of the cerebrum is divided into four lobes: frontal, parietal, temporal, and occipital. The frontal lobe has motor areas and an association area called the prefrontal area. The other lobes have sensory areas and also association areas.

True to their name, **association areas** are places where cell bodies integrate information. The somatosensory association area is located just dorsal to the primary somatosensory area. This area processes and analyzes sensory information from the skin and muscles. The visual association area in the occipital lobe associates new visual information with previously received visual information. It might "decide," for example, if we have seen this face or tool or whatever before. The auditory association area in the temporal lobe performs the same functions with regard to sounds. The **prefrontal area,** an association area in the frontal lobe, receives information from the other association areas and uses this information to reason and plan our actions. Integration in this area accounts for our most cherished human abilities to think critically and to formulate appropriate behaviors.

The parietal, temporal, and occipital association areas meet near the dorsal end of the lateral sulcus. This region is called the general interpretation area because it receives information from all the sensory association areas and allows us to quickly integrate incoming signals and send them on to the prefrontal area so an immediate response is possible. Heroes are people that can quickly assess a situation and take actions that save others from dangerous situations.

Limbic System

The **limbic system** is a complex network of tracts and nuclei that incorporates medial portions of the cerebral lobes, subcortical nuclei, and the diencephalon (Fig. 17.15). The limbic system blends higher mental functions and primitive emotions into a united whole. It accounts for why activities like sexual behavior and eating seem pleasurable and also why, say, mental stress can cause high blood pressure.

Two significant structures within the limbic system are the hippocampus and the amygdala, which are essential for learning and memory. The hippocampus is well situated in the brain to make the prefrontal area aware of past experiences stored in association areas. The amygdala, in particular, can cause these experiences to have emotional overtones. The inclusion of the frontal lobe in the limbic system means that reason can keep us from acting out strong feelings.

The gray matter of the cerebrum consists of the cerebral cortex and also nuclei. The white matter consists of tracts. The limbic system is a unique combination of various portions of the brain which unify brain functions and sensations.

White Matter

Much of the rest of the cerebrum is composed of white matter. From your understanding of the spinal cord you know that white matter in the CNS consists of long myelinated fibers organized into tracts. Descending tracts from the primary motor area communicate with lower brain centers, and ascending tracts from lower brain centers send sensory information up to the primary somatosensory area. Tracts within the cerebrum take information between the different sensory, motor, and association areas pictured in Figure 17.14. The corpus callosum, you will recall, contains tracts that join the two cerebral hemispheres.

While the bulk of the cerebrum is composed of tracts, there are subcortical (below the cortex) nuclei deep within the white matter. The **basal nuclei** serve as relay stations for motor impulses from the primary motor area. The basal nuclei produce dopamine, an inhibitory neurotransmitter that helps control various skeletal muscle activities. Huntington disease and Parkinson disease, which are both characterized by uncontrollable movements, are believed to be due to malfunctioning of the basal nuclei.

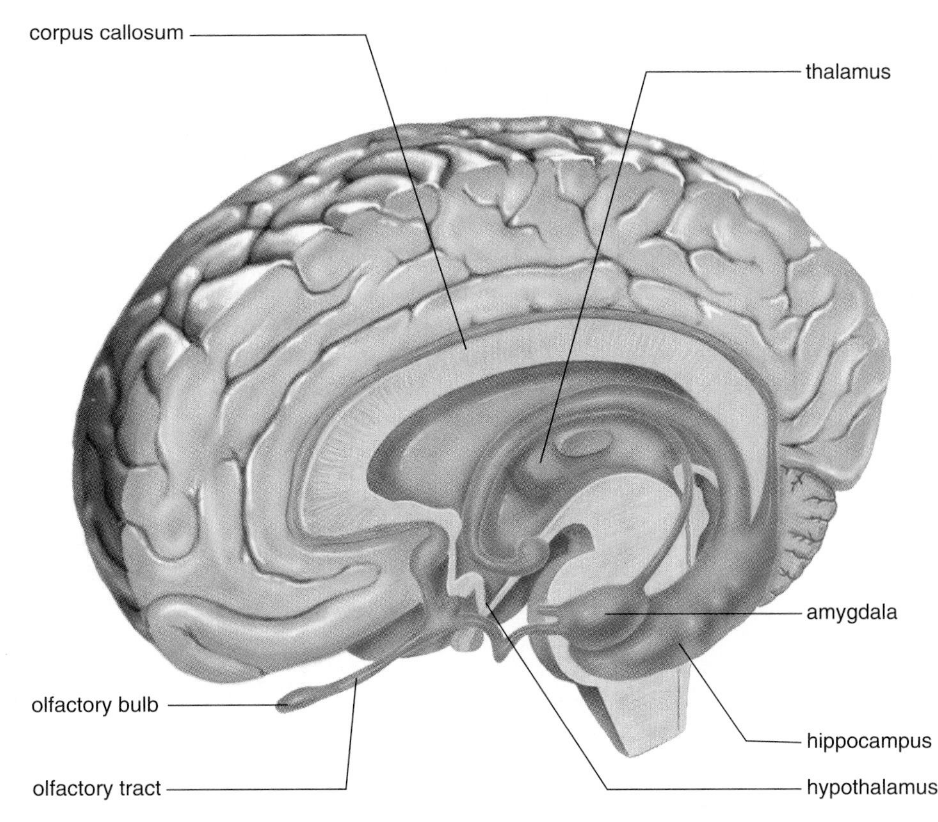

Figure 17.15 The limbic system.
Structures deep within the cerebral hemisphere and surrounding the diencephalon join higher mental functions like reasoning with more primitive feelings like fear and pleasure.

17.5 Higher Mental Functions

As in other areas of biological research, brain research has progressed due to technological breakthroughs. Neuroscientists now have a wide range of techniques at their disposal for studying the human brain. The reading on page 339 lays out several of the steps that have brought us to our current knowledge of higher mental functions, including our dependence on modern technologies that allow us to record the functioning of the brain.

Learning and Memory

Just as the corpus callosum gives evidence that the two cerebral hemispheres work together, so the limbic system indicates that cortical areas possibly work with lower centers to produce learning and memory (Fig. 17.16). **Memory** is the ability to hold a thought in mind or recall events from the past, ranging from a word we learned only yesterday to an early emotional experience that has shaped our lives. Learning takes place when we retain and utilize past memories.

Types of Memories

We have all had the experience of trying to remember a seven-digit telephone number for a very short period of time. If we say we are trying to keep it in the forefront of our brain, we are exactly correct because the prefrontal area is active during **short-term memory.** The prefrontal area is the association area that lies just dorsal to our forehead! There are some telephone numbers that you know by heart; in other words, they have gone into **long-term memory.** Think of a telephone number you know by heart and see if you can bring it to mind without also thinking about the place or person associated with that number. Most likely you cannot because typically long-term memory is a mixture of what is called **semantic memory** (numbers, words, etc.) and **episodic memory** (persons, events, etc.). The flow chart for long-term memory in Figure 17.16 has two sets of arrows—one for semantic memory and the other for episodic memory. Due to brain damage, some persons lose one type of memory ability but not the other. Without a working episodic memory they can carry on a conversation but have no recollection of recent events. If you are talking to them and leave the room, they don't remember you when you come back!

Skill memory is another type of memory that can exist independent of episodic memory. Skill memory is being able to perform motor activities like riding a bike or playing ice hockey. When a person first learns a skill, more areas of the cerebral cortex are involved than after the skill is perfected. In other words, you have to think about what you are doing when you learn a skill, but later the actions become automatic. This shows that the preprimary motor area can communicate with the primary motor area below the level of consciousness.

Memory Storage and Retrieval

The first step toward being able to cure memory disorders is to know what parts of the brain are functioning when we remember something. Investigators have been able to work it out pretty well. The **hippocampus,** a seahorse-shaped structure that lies deep in the temporal lobe, is in a unique position to serve as a bridge between the sensory association areas where memories are stored and the prefrontal area where memories are utilized. The prefrontal area communicates with the hippocampus when memories are stored and when these memories are brought to mind. Why are some memories so emotionally charged? The **amygdala** seems to be responsible for fear conditioning and associating danger with sensory stimuli received from both the diencephalon and the cortical sensory areas.

Our long-term memories are stored in bits and pieces throughout the sensory association areas of the cerebral cortex. Visions are stored in the vision association area, sounds are stored in the auditory association area, and so forth. The hippocampus gathers this information together for use by the prefrontal cortex when we remember Uncle Frank or our summer holiday. And the amygdala adds emotional overtones to memories.

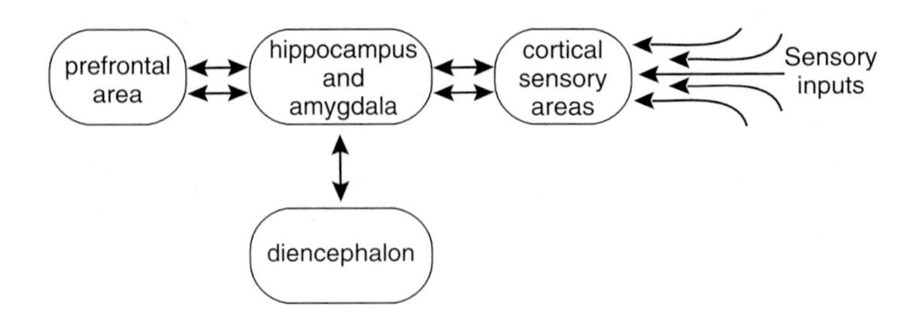

Figure 17.16 Memory circuits.
The hippocampus and amygdala are in communication with the cortical sensory areas where memories are stored, and with the prefrontal area where memories are utilized to plan future actions. Sensory information also reaches the hippocampus from the diencephalon.

Science Focus

How Memories Are Made

Neurobiologists want to understand higher mental functions like memory from the behavioral to the molecular level. They believe that knowing how the healthy brain goes about its business is essential if we are to ever cure disorders like Alzheimer disease or Parkinson disease, or even psychiatric disorders like schizophrenia.

It's been possible for quite some time to determine in general terms the activities of various parts of the brain. Accidents, illnesses, or birth defects that selectively destroy some particular region of the brain have allowed investigators to assign functions to brain parts. Let's consider, for example, the hippocampus, a small seahorse-shaped structure that lies in the medial temporal lobe. S. S. is a physicist who contracted a herpes simplex infection that destroyed his hippocampus. Although S. S. still has an IQ of 136, he forgets a recent experience in just a few minutes. Although he is able to remember childhood events, he is now unable to convert short-term memory into long-term memory. From the observation of S. S. and other patients like him, we can reason that the hippocampus must be involved in the storage of what is called episodic memories. Episodic memories pertain to episodes—that is, a series of occurrences.

Circumstantial data regarding the function of the hippocampus has been backed up by research in the laboratory. Monkeys can be taught to perform a delayed-response task, like choosing a particular type cup that always contains food. They will choose the correct cup even after a lapse of time between instruction and a subsequent testing period. In order to tell what parts of the brain are involved in a delayed-response task, investigators can use radioactive tracers. Radioactive glucose is a good choice because glucose is the primary source of molecular energy for neurons. After a monkey has performed a delayed-response task and the monkey is sacrificed, radioactivity is particularly detected in the prefrontal area and the hippocampus. From this we know that the prefrontal area is in communication with the hippocampus when an episodic memory is being stored.

It is possible that a monkey brain may be different from a human brain. But only the introduction of high-speed computers made it possible to view a functioning human brain. Data fed into the computers during positron emission tomography (PET) (Fig. 17A) and magnetic resonance imaging (MRI) produces brain scans—cross sections of the brain from all directions. Functional imaging occurs when the subjects undergoing PET or MRI are asked to perform some particular mental task, such as to study a few words, and then later to select these words from a long list. In PET, radioactively labeled water injected into the arm is taken up preferentially by metabolically active (rather than inactive) brain tissue. MRI does not require the use of radioactive material because the machine detects increases in oxygen levels. Blood vessels serving active neurons contain more oxygen than those serving inactive neurons. In a recent study utilizing functional MRI, the hippocampus was maximally active when human subjects were asked to remember a word cued by a picture, while the so-called parahippocampal region was maximally active when subjects were asked to remember pictures for later recall.

Once it became accepted that the hippocampus and surrounding regions are needed to store and retrieve memories, investigators began to study the activity of individual neurons in the region. What exactly are neurons doing when we store memories and bring them back? Neurobiologists have discovered that a long-lasting effect, called long-term potentiation (LTP), occurs when synapses are active within the hippocampus. Long-term potentiation is carried out by glutamate, a type of neurotransmitter. Investigators at the Massachusetts Institute of Technology first showed that it was possible to teach mice to find their way around a maze. Then, by a dramatic feat of genetic engineering, they were able to produce mice that lacked the receptor for glutamate only in hippocampal cells. These defective mice were unable to learn to run the maze. These studies are being hailed as the first time it has been possible to study memory at all levels, from behavior to a molecular change in one part of the brain.

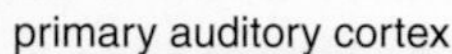

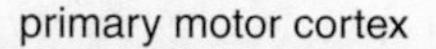

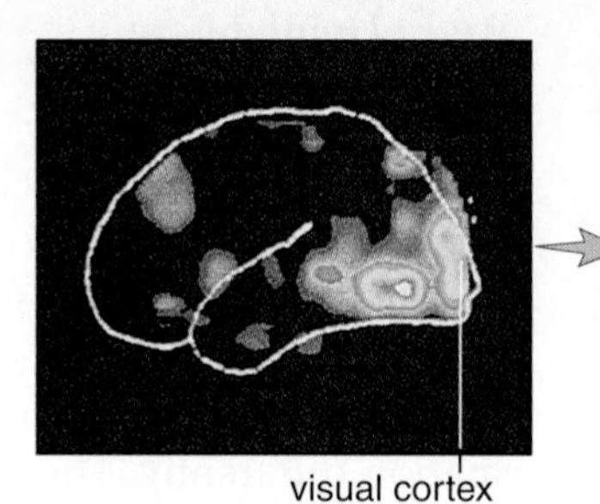

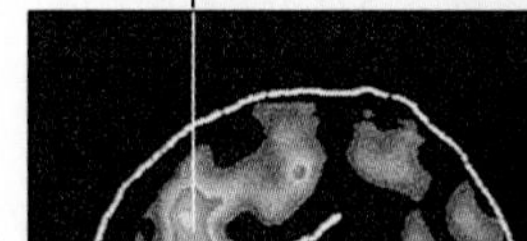

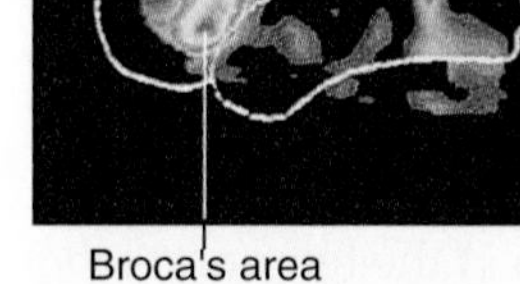

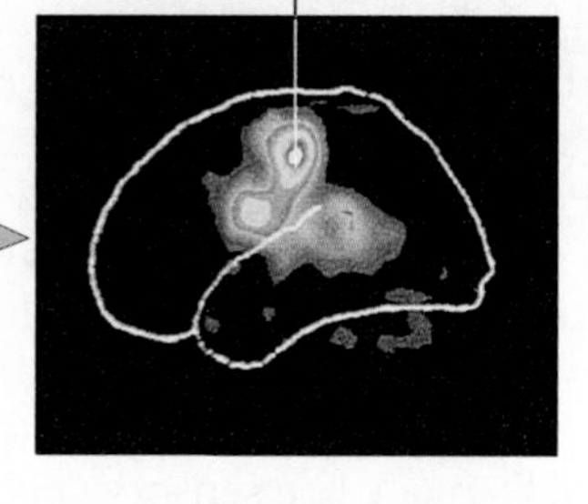

1. The word is seen in the visual cortex.

2. Information concerning the word is interpreted in Wernicke's area.

3. Information from Wernicke's area is transferred to Broca's area.

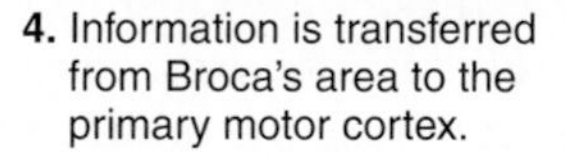

4. Information is transferred from Broca's area to the primary motor cortex.

Figure 17A Cortical areas for language.
These PET images show the cortical pathway for reading words and then speaking them. Red indicates the most active areas; blue indicates the least active areas.

Long-Term Potentiation

While it is helpful to know the memory functions of various portions of the brain, an important step toward curing mental disorders is understanding memory on the cellular level. **Long-term potentiation (LTP)** is an enhanced response at synapses within the hippocampus. LTP is most likely essential to memory storage but, unfortunately, it sometimes causes a postsynaptic neuron to become so excited, it undergoes apoptosis, a form of cell death. This phenomenon, called excitotoxicity, may develop due to a mutation. The longer we live the more likely that any particular mutation will occur.

Excitotoxicity is due to the action of the neurotransmitter glutamate which is active in the hippocampus. When glutamate binds with a specific type of receptor in the postsynaptic membrane, calcium (Ca^{2+}) may rush in too fast and the influx is lethal to the cell. A gradual extinction of brain cells in the hippocampus appears to be the underlying cause of Alzheimer disease (AD). Although it is not yet known how excitotoxicity is related to structural abnormalities of AD neurons, some researchers are trying to develop neuroprotective drugs that can possibly guard brain cells against damage due to glutamate. One type of neuroprotective drug blocks glutamate receptors and thereby prevents the entry of Ca^{2+} and excitotoxicity.

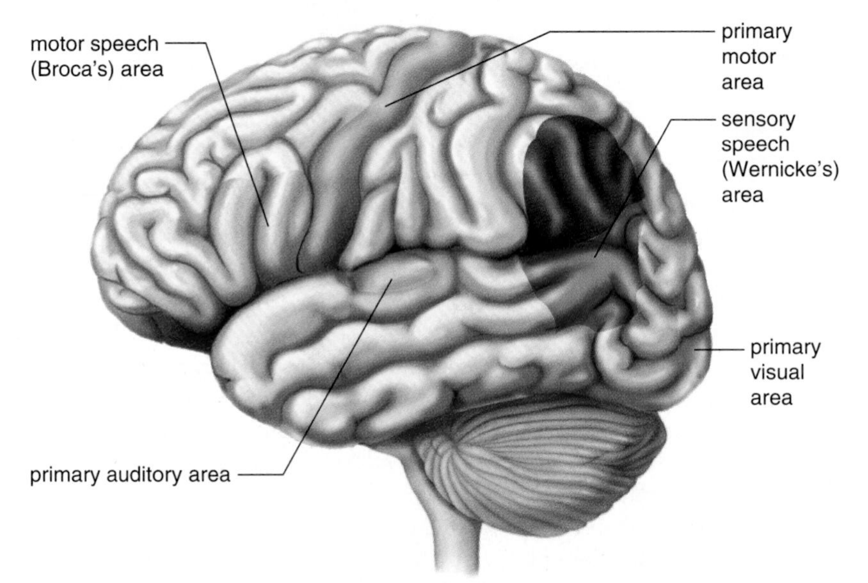

Figure 17.17 Language and speech.
Wernicke's area and Broca's area are thought to be involved in speech comprehension and use, as are the other labeled areas of the cerebral cortex.

Memory has been studied at various levels: behavioral, structural, and cellular. Knowledge of memory on the cellular level may offer treatments and cures for mental disorders.

Language and Speech

Language is obviously dependent upon semantic memory. Therefore, we would expect some of the same areas in the brain to be involved in both memory and language. Seeing and hearing words depends on sensory centers in the occipital and temporal lobes respectively. And generating and speaking words depends on motor centers in the frontal lobe. Functional imaging by a technique known as PET (positron emission tomography) supports these suppositions (see Fig. 17A).

From studies of patients with speech disorders, it has been known for some time that damage to a motor speech area called Broca's area results in an inability to speak. Broca's area is located just in front of the primary motor zone for speech musculature (lips, tongue, larynx, and so forth). Damage to a sensory speech area in the temporal lobe called Wernicke's area results in the inability to comprehend speech (Fig. 17.17). Any disruption of pathways between various regions of the brain could very well contribute to an inability to comprehend our environment and use speech correctly. Indeed, damage to lower centers, especially the thalamus, can also cause speech difficulties. Remember that the thalamus passes on sensory information, and if the cerebral cortex does not receive the necessary sensory input, the motor areas cannot formulate the proper output.

One interesting aside pertaining to language and speech is the recognition that the left brain and right brain have different functions. Only the left hemisphere and not the right contains a Broca's area and Wernicke's area! Indeed, the left hemisphere plays a role of great importance in language functions in general and not just in speech. In an attempt to cure epilepsy in the early 1940s, the corpus callosum was surgically severed in some patients. Later studies showed that these split-brain patients could name objects only if they were seen by the left hemisphere. If objects were viewed only by the right hemisphere, a split-brain patient could choose the proper object for a particular use but was unable to name it. Based on these and various types of studies, the popular idea developed that the left brain can be contrasted with the right brain along these lines:

Left Hemisphere	Right Hemisphere
Verbal	Nonverbal, visuo-spatial
Logical, analytical	Intuitive
Rational	Creative

Further, it became generally thought that one hemisphere was dominant in each person, accounting in part for personality traits. However, recent studies suggest that the hemispheres process the same information differently. The right hemisphere is more global, whereas the left hemisphere is more specific in its approach.

Language is dependent upon memory; special areas in the left hemisphere help account for our ability to comprehend and use speech.

17.6 Drug Abuse

A wide variety of drugs can be used to alter the mood and/or emotional state (see website). Drugs that affect the nervous system have two general effects: (1) they impact the limbic system, and (2) they either promote or decrease the action of a particular neurotransmitter (Fig. 17.18). Stimulants increase the likelihood of neuron excitation, and depressants decrease the likelihood of excitation. Increasingly, researchers believe that dopamine, a neurotransmitter in the brain, is primarily responsible for mood. Cocaine is known to potentiate the effects of dopamine by interfering with its uptake from synaptic clefts. Many new medications developed to counter drug addiction and mental illness affect the release, reception, or breakdown of dopamine.

Drug abuse is apparent when a person takes a drug at a dose level and under circumstances that increase the potential for a harmful effect. Drug abusers are apt to display either a psychological and/or a physical dependence on the drug. Dependence has developed when the person spends much time thinking about the drug or arranging to get it and often takes more of the drug than was intended. With physical dependence, formerly called an addiction to the drug, the person is tolerant to the drug—that is, must increase the amount of the drug to get the same effect and has withdrawal symptoms when he or she stops taking the drug.

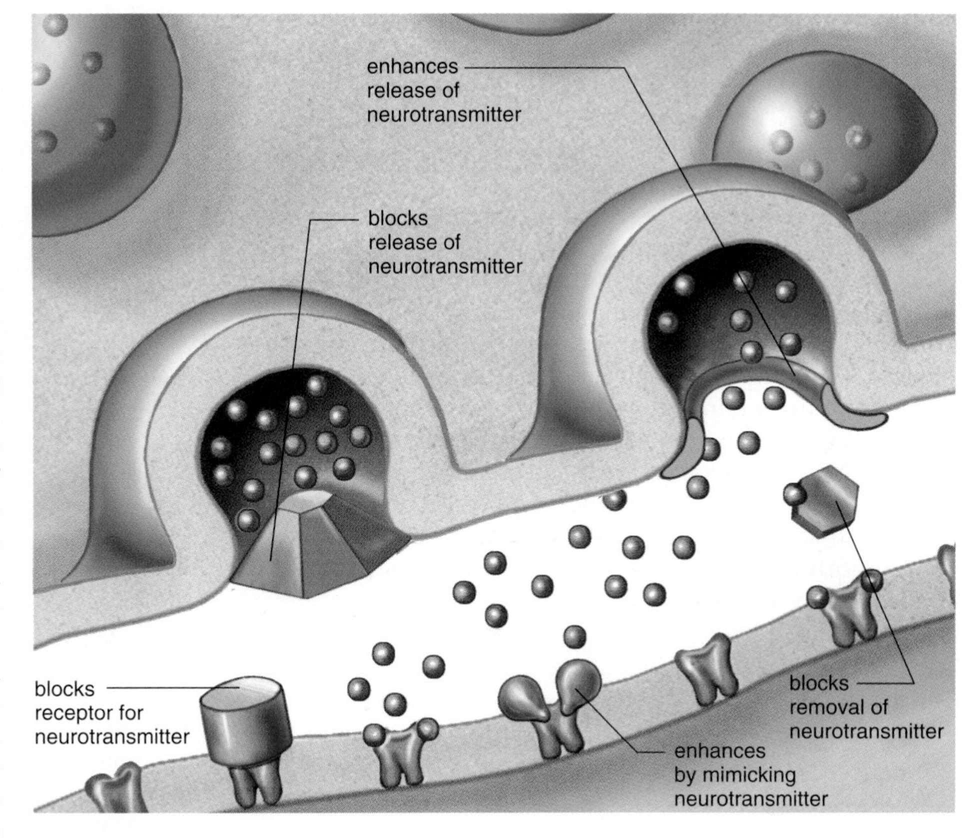

Figure 17.18 Drug actions at a synapse.
Each drug typically has one of these specific functions.

> Drugs that affect the nervous system can cause physical dependence and withdrawal symptoms.

Alcohol

It is possible that *alcohol* influences the action of GABA, an inhibiting transmitter, or glutamate, an excitatory neurotransmitter. Once imbibed, alcohol is primarily metabolized in the liver, where it disrupts the normal workings of glycolysis and the Krebs cycle. The cell begins to carry on fermentation, and lactic acid builds up. The pH of the blood decreases and becomes acidic.

Since the Krebs cycle is not working, fat cannot be broken down, and the liver turns fatty. Fat accumulation, the first stage of liver deterioration, begins after only a single night of heavy drinking. If heavy drinking continues, fibrous scar tissue appears during a second stage of deterioration. If heavy drinking stops, the liver can still recover and become normal once again. If not, the final and irrevocable stage, cirrhosis of the liver, occurs: liver cells die, harden, and turn orange (cirrhosis means orange).

The surgeon general recommends that pregnant women drink no alcohol at all. Alcohol crosses the placenta freely and causes *fetal alcohol syndrome,* which is characterized by mental retardation and various physical defects.

Good nutrition is difficult when a person is a heavy drinker. Alcohol contains little but calories; it does not supply amino acids, vitamins, or minerals as do most foods. Without adequate vitamins, red bone marrow cannot produce white blood cells. The immune system is depressed, and the chances of cancer increase. Without an adequate protein intake, muscles atrophy and weakness results. Fat deposits occur in the heart wall and hypertension develops. There is an increased risk of heart attack and stroke.

> Alcohol, which is the most abused drug in the United States, can lead to serious health consequences.

Nicotine

Nicotine, an alkaloid derived from tobacco, is a widely used neurological agent. When a person smokes a cigarette, nicotine is quickly distributed to the central and peripheral nervous systems. In the central nervous system, nicotine causes neurons to release dopamine, a neurotransmitter mentioned earlier. The excess of dopamine has a reinforcing effect that leads to dependence on the drug. In the peripheral nervous system, nicotine stimulates the same postsynaptic receptors as acetylcholine and leads to increased skeletal muscular activity. It also increases the heart rate and blood pressure, and digestive tract mobility.

Many cigarette smokers find it difficult to give up the habit because nicotine induces both physiological and psychological dependence. Withdrawal symptoms include headache, stomach pain, irritability, and insomnia. Tobacco contains not only nicotine but many other harmful substances. Cigarette and cigar smoking contributes to early death from cancer, including not only lung cancer but also cancer of the larynx, mouth, throat, pancreas, and urinary bladder. Now that women are as apt to smoke as men, lung cancer has surpassed breast cancer as a cause of death. Cigarette smoking in young women who are sexually active is most unfortunate because if they become pregnant, nicotine, like other psychoactive drugs, adversely affects a developing embryo and fetus.

Nicotine induces both physiological and psychological dependence. Most cases of lung cancer are due to smoking cigarettes and/or cigars.

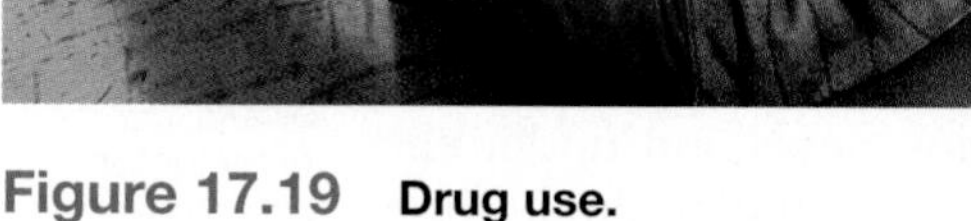

Figure 17.19 Drug use.
Users often smoke crack in a glass water pipe. The high produced consists of a "rush" lasting a few seconds, followed by a few minutes of euphoria. Continuous use makes the user extremely dependent on the drug.

Marijuana

The dried flowering tops, leaves, and stems of the Indian hemp plant *Cannabis sativa* contain and are covered by a resin that is rich in THC (tetrahydrocannabinol). The names *cannabis* and *marijuana* apply to either the plant or THC. Usually marijuana is smoked in a cigarette form called a joint.

The occasional user reports experiencing a mild euphoria along with alterations in vision and judgment, which result in distortions of space and time. Motor incoordination, including the inability to speak coherently, takes place. Heavy use can result in hallucinations, anxiety, depression, rapid flow of ideas, body image distortions, paranoid reactions, and similar psychotic symptoms. The terms *cannabis psychosis* and *cannabis delirium* refer to such reactions.

Recently, researchers have found that marijuana binds to a receptor for anandamide, a normal molecule in the body. Craving and difficulty in stopping usage can occur as a result of regular use. Some researchers believe that long-term marijuana use leads to brain impairment. *Fetal cannabis syndrome,* which resembles fetal alcohol syndrome, has been reported. Some psychologists believe that marijuana use among adolescents prevents them from dealing with the personal problems that often develop during that stage of life.

Although marijuana does not produce physical dependence, it does produce psychological dependence.

Cocaine

Cocaine is an alkaloid derived from the shrub *Erythroxylum cocoa.* Cocaine is sold in powder form and as *crack,* a more potent extract (Fig. 17.19). Cocaine prevents the synaptic uptake of dopamine and this causes the user to experience the sensation of a rush. The epinephrine-like effects of dopamine account for the state of arousal that lasts for some minutes after the rush experience.

With continued cocaine use, the body begins to make less dopamine to compensate for a seemingly excess supply. The user, therefore, now experiences *tolerance, withdrawal symptoms,* and an intense *craving* for cocaine. These are indications that the person is highly dependent upon the drug or, in other words, that cocaine is extremely addictive. Overdosing on cocaine can cause seizures and cardiac and respiratory arrest.

Individuals who snort the drug can suffer damage to the nasal tissues and even perforation of the septum between the nostrils. It is possible that long-term cocaine abuse causes brain damage; babies born to addicts suffer withdrawal symptoms and may suffer neurological and developmental problems. A cocaine binge can go on for days, after which the individual suffers a crash. During the binge period, the user is hyperactive and has little desire for food or sleep but has an increased sex drive. During the crash period, the user is fatigued, depressed, and irritable, has memory and concentration problems, and displays no interest in sex. Indeed, men are often impotent.

Heroin

Heroin is derived from morphine, an alkaloid of *opium*. After an intravenous injection, there is a feeling of euphoria along with relief of pain within 3 to 6 minutes. Side effects can include nausea, vomiting, dysphoria, and respiratory and circulatory depression leading to death.

Heroin binds to receptors meant for the **endorphins,** the special neurotransmitters that kill pain and produce a feeling of tranquility. With time, the body's production of endorphins decreases. *Tolerance* develops so that the user needs to take more of the drug just to prevent *withdrawal* symptoms. The euphoria originally experienced upon injection is no longer felt.

Heroin withdrawal symptoms include perspiration, dilation of pupils, tremors, restlessness, abdominal cramps, gooseflesh, defecation, vomiting, and increase in systolic pressure and respiratory rate. Those who are excessively dependent may experience convulsions, respiratory failure, and death. Infants born to women who are physically dependent also experience these withdrawal symptoms.

Cocaine and heroin produce a very strong physical dependence. An overdose of these drugs can cause death.

Methamphetamine (Ice)

Methamphetamine is related to amphetamine, a well-known stimulant. Both methamphetamine and amphetamine have been drugs of abuse for some time, but a new form of methamphetamine known as "ice" is now used as an alternative to cocaine. Ice is a pure, crystalline hydrochloride salt that has the appearance of sheetlike crystals. Unlike cocaine, ice can be illegally produced in this country in laboratories and does not need to be imported.

Ice, like crack, will vaporize in a pipe, so it can be smoked, avoiding the complications of intravenous injections. After rapid absorption into the bloodstream, the drug moves quickly to the brain. It has the same stimulatory effect as cocaine, and subjects report they cannot distinguish between the two drugs after intravenous administration. Methamphetamine effects, however, persist for hours instead of a few seconds. Therefore, it is the preferred drug of abuse by many.

Designer Drugs

Designer drugs are analogs; that is, they are slightly altered psychoactive drugs. One such drug is MPPP (1-methyl-4-phenylprionoxy-piperidine), an analog of the narcotic fentanyl. Even small doses of the drug are very toxic and can cause death.

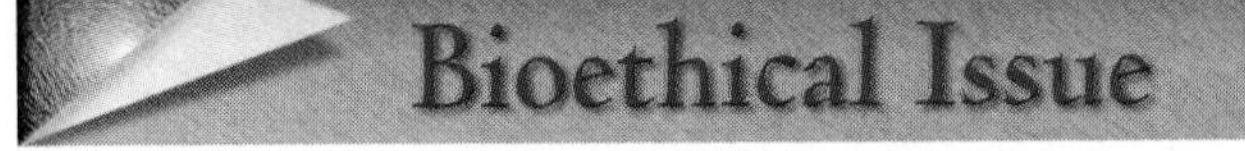

To control their weight, many people in the United States have turned to a new diet pill called Redux, which stimulates the production and availability of the neurotransmitter serotonin in the brain. This very same neurotransmitter is released when we eat a high carbohydrate-rich meal. Redux also prevents serotonin from being reabsorbed at presynaptic membranes. The result is spirits are lifted and appetite is squelched. It's not uncommon for patients to lose 20 or more pounds a week, simply because they have lost all interest in eating!

Unfortunately, Redux has side effects. Some, like fatigue, diarrhea, vivid dreams, and a dry mouth can be tolerated. Others, however, are low in incidence but very serious. Studies suggest that the drug increases the incidence of primary pulmonary hypertension from 1 to 2 per one million to as much as 46 per one million. Primary pulmonary hypertension destroys blood vessels in the lungs and heart, and can lead to death. Also, the drug causes significant, and possibly permanent, brain damage in lab animals. An abundance of serotonin makes the neurons that ordinarily produce the neurotransmitter swell, wither, and then die according to Dr. Mark Molliver, a Johns Hopkins neurologist.

Are you acting recklessly, and therefore unethically, if you take this drug? To decide, a risk-benefit analysis may be appropriate. Severe obesity puts people at risk for hypertension, heart attacks, diabetes, and some cancers. And these illnesses contribute to 300,000 deaths a year in the United States. The slim risk of serious side effects may be worth it for the obese, but the same conclusion does not hold for those who are merely overweight. Even so, most doctors will probably prescribe the drug for anyone who asks for it. Just three months after the introduction of Redux, doctors were writing 85,000 prescriptions a week!

Questions

1. Is there any difference in smoking to control weight gain and taking Redux to control weight gain? Explain your answer.
2. Is there any difference in drinking alcohol to feel better and taking Redux to feel better? Explain your answer.
3. Is it ever ethical to take drugs to control our everyday behavior? Why or why not?

Summarizing the Concepts

17.1 Neurons and How They Work

The nervous system contains neurons and neuroglial cells which service neurons. Sensory neurons take information from sensory receptors to the CNS; interneurons occur within the CNS, and motor neurons take information from the CNS to effectors (muscles or glands). A motor neuron can be used to demonstrate that neurons are composed of dendrites, a cell body, and an axon. Long axons are covered by a myelin sheath.

When an axon is not conducting a nerve impulse, the inside of an axon is negative (−65mV) compared to the outside. The sodium-potassium pump actively transports Na^+ out of an axon and K^+ to inside an axon. The resting potential is due to the leakage of K^+ to the outside of the neuron. When an axon is conducting a nerve impulse (action potential), Na^+ first moves into the axoplasm and then K^+ moves out of the axoplasm.

Transmission of the nerve impulse from one neuron to another takes place when a neurotransmitter molecule is released into a synaptic cleft. The binding of the neurotransmitter to receptors in the postsynaptic membrane causes either excitation or inhibition. Integration is the summing of excitatory and inhibitory signals. Neurotransmitter molecules are removed from the cleft by enzymatic breakdown or by reabsorption.

17.2 Peripheral Nervous System

The peripheral nervous system contains only nerves and ganglia. The brain is always involved in voluntary actions but reflexes are automatic, and some do not require involvement of the brain. In the somatic system, for example, a stimulus causes sensory receptors to generate nerve impulses which are conducted by sensory nerve fibers to interneurons in the spinal cord. Interneurons signal motor neurons which conduct nerve impulses to a skeletal muscle that contracts, giving the response to the stimulus.

The autonomic (involuntary) system controls smooth muscle of the internal organs and glands. The sympathetic division is associated with responses that occur during times of stress, and the parasympathetic system is associated with responses that occur during times of relaxation.

17.3 Central Nervous System

The CNS consists of the spinal cord and brain, which are both protected by bone. The CNS receives and integrates sensory input and formulates motor output. The gray matter of the spinal cord contains neuron cell bodies; the white matter consists of myelinated axons that occur in bundles called tracts. The spinal cord carries out reflex actions and sends sensory information to the brain and receives motor output from the brain. Because tracts cross over, the left side of the brain controls the right side of the body and vice versa.

In the brain, the medulla oblongata and pons have centers for vital functions, like breathing and the heartbeat. The cerebellum primarily coordinates muscle contractions. The hypothalamus controls homeostasis, and the thalamus specializes in sending on sensory input to the cerebrum. The cerebrum has two cerebral hemispheres connected by the corpus callosum. Sensation, reasoning, learning and memory, and also language and speech take place in the cerebrum.

17.4 The Cerebral Hemispheres

Each cerebral hemisphere contains a frontal, parietal, occipital, and temporal lobe. The cerebral cortex is a thin layer of gray matter covering the cerebrum. The primary motor area in the frontal lobe sends out motor commands to lower brain centers which pass them on to motor neurons. The primary somatosensory area in the parietal lobe receives sensory information from lower brain centers in communication with sensory neurons. A visual area occurs in the occipital lobe, an auditory area occurs in the temporal lobe, and so forth for the other senses. Association areas are located in all the lobes; the prefrontal area of the frontal lobe is especially necessary to higher mental functions.

The limbic system is a complex network of cortical and subcortical nuclei (masses of cell bodies) within the brain that is known to generate primitive emotions and function in higher mental functions.

17.5 Higher Mental Functions

Short-term memory and long-term memory are dependent upon the prefrontal area. The hippocampus acts as a conduit for sending information to long-term memory and retrieving it once again. The amygdala adds emotional overtones to memories. On the cellular level, long-term potentiation seems to be required for long-term memory. Unfortunately, long-term potentiation can go awry when neurons become overexcited and die. Neuroprotective drugs are being developed in the hope they will prevent disorders like Alzheimer disease. Language and speech are dependent upon Broca's area (a motor speech area) and Wernicke's area (a sensory speech area) that are in communication. Interestingly enough, these two areas are located only in the left hemisphere.

17.6 Drug Abuse

Although neurological drugs are quite varied, each type has been found to either promote or prevent the action of a particular neurotransmitter.

Studying the Concepts

1. With reference to a motor neuron, describe the structure and function of the three parts of a neuron. What are the three classes of neurons and what is their relationship to the CNS? 322
2. What is the sodium-potassium pump? What is resting potential, and how is it brought about? 324
3. Describe the two parts of an action potential and the changes that can be associated with each part. 324
4. What is a neurotransmitter, where is it stored, how does it function, and how is it destroyed? Name two well-known neurotransmitters. 327
5. The peripheral nervous system contains what three types of nerves? What is meant by a mixed nerve? 328
6. Trace the path of a somatic reflex. 329
7. What is the autonomic system, and what are its two major divisions? Give several similarities and differences between these divisions. 331
8. Describe the structure and function of the spinal cord. 332–33
9. Name the major parts of the brain, and give a function for each. What is the reticular formation? 335
10. Name the lobes of the cerebral hemispheres and describe the function of motor, sensory, and association areas. What is the limbic system? 336–37
11. Name several different types of memory and explain how long-term memory is thought to occur. What is long-term potentiation? 338–39

12. Language and speech require what portions of the brain? What is the left brain/right brain hypothesis? 340
13. Describe the physiological effects and mode of action of alcohol, marijuana, cocaine, and heroin. 341–43

Testing Yourself

Choose the best answer for each question.

1. Which of these are the first and last elements in a spinal reflex?
   a. axon and dendrite
   b. sensory receptor and muscle effector
   c. ventral horn and dorsal horn
   d. brain and skeletal muscle
   e. motor neuron and sensory neuron
2. A spinal nerve takes nerve impulses
   a. to the CNS.
   b. away from the CNS.
   c. both to and away from the CNS.
   d. from the CNS to the spinal cord.
3. Which of these correctly describes the distribution of ions on either side of an axon when it is not conducting a nerve impulse?
   a. more sodium ions (Na^+) outside and less potassium ions (K^+) inside
   b. more K^+ outside and less Na^+ inside
   c. charged protein outside; Na^+ and K^+ inside
   d. Na^+ and K^+ outside and water only inside
   e. chlorine ions (Cl^-) on outside and K^+ and Na^+ on inside
4. When the action potential begins, sodium gates open, allowing Na^+ to cross the membrane. Now the polarity changes to
   a. negative outside and positive inside.
   b. positive outside and negative inside.
   c. There is no difference in charge between outside and inside.
   d. neutral outside and positive inside.
   e. Any one of these could be correct.
5. Transmission of the nerve impulse across a synapse is accomplished by the
   a. movement of Na^+ and K^+.
   b. release of a neurotransmitter by a dendrite.
   c. release of a neurotransmitter by an axon.
   d. release of a neurotransmitter by a cell body.
   e. Any one of these is correct.
6. The autonomic system has two divisions called the
   a. CNS and PNS.
   b. somatic and skeletal systems.
   c. efferent and afferent systems.
   d. sympathetic and parasympathetic systems.
7. Synaptic vesicles are
   a. at the ends of dendrites and axons.
   b. at the ends of axons only.
   c. along the length of all long fibers.
   d. All of these are correct.
8. Which of these would be covered by a myelin sheath?
   a. short dendrites
   b. globular cell bodies
   c. long axons
   d. interneurons
   e. All of these are correct.
9. When you remove your hand from a hot stove, which system is least likely to be involved?
   a. somatic system
   b. parasympathetic system
   c. central nervous system
   d. sympathetic system
   e. peripheral nervous system
10. The spinal cord communicates with the brain via
    a. the gray matter of the cord and brain.
    b. sensory nerve fibers in a spinal nerve.
    c. the sympathetic system.
    d. tracts in the white matter.
    e. ventricles in the brain and the spinal cord.
11. Which two parts of the brain are least likely to work together?
    a. thalamus and the cerebrum
    b. cerebrum and the cerebellum
    c. hypothalamus and the medulla oblongata
    d. cerebellum and the medulla oblongata
    e. reticular formation and the thalamus
12. Which of these is not a proper contrast between the primary motor area and the primary somatosensory area?
    a. ventral to the central sulcus—dorsal to the central sulcus
    b. controls skeletal muscles—receives sensory information
    c. communicates directly with association areas in the parietal lobe—communicates directly with association areas in the frontal lobe
    d. has connections with the cerebellum—has connections with the thalamus
    e. All of these are contrasts between the two areas.
13. The limbic system
    a. involves portions of the cerebral lobes, subcortical nuclei, and the diencephalon.
    b. is responsible for our deepest emotions like pleasure, rage, and fear.
    c. is a system necessary to memory storage.
    d. is not responsible for reason and self control.
    e. All of these are correct.
14. Label this diagram.

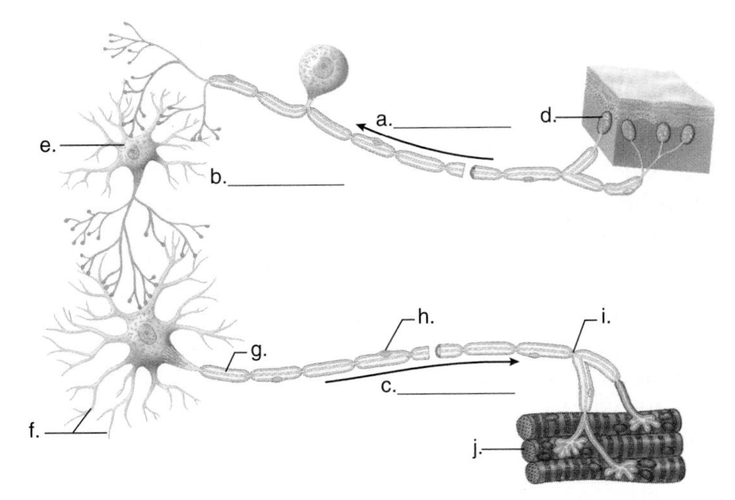

Thinking Scientifically

1. Considering the nerve impulse, electricity is the flow of electrons within a wire (page 324).
   a. How is the nerve impulse different from this?
   b. If a neurotransmitter substance is inhibitory, would you expect a higher or lower voltage reading compared to −65mV on the oscilloscope?
   c. In the laboratory, an axon segment can conduct a nerve impulse in either direction. Why do nerve impulses go only from axon to dendrite or cell body across a synapse in the body?
   d. Nerves cause muscles to contract. Assuming one nerve fiber per one muscle fiber of a muscle, explain how a nerve can bring about degrees of muscle contraction.
2. Considering a reflex arc (page 329):
   a. If you applied acid to the left leg of a frog, why might both legs respond?
   b. If you severed just the dorsal root ganglion of the left sciatic nerve that serves the leg, would either leg be able to respond?
   c. If you severed just the ventral root of the left sciatic nerve, would either leg be able to respond?
   d. If you destroyed just the spinal cord, would either leg be able to respond?

Understanding the Terms

acetylcholine (ACh) 327
acetylcholinesterase (AChE) 327
action potential 324
amygdala 338
association area 337
autonomic system 331
axon 322
axon bulb 327
basal nuclei 337
brain 334
brain stem 335
cell body 322
central nervous system (CNS) 322
cerebellum 335
cerebral cortex 336
cerebral hemisphere 335
cerebrospinal fluid 332
cerebrum 335
cranial nerve 328
dendrite 322
diencephalon 335
dorsal-root ganglion 328
drug abuse 341
endorphin 343
episodic memory 338
fight or flight 331
ganglion 328
gray matter 332
hippocampus 338
hypothalamus 335
integration 327
interneuron 322
limbic system 337
long-term memory 338
long-term potentiation (LTP) 340
medulla oblongata 335
memory 338
meninges (sing., meninx) 332
midbrain 335
motor neuron 322
myelin sheath 323
nerve 328
nerve impulse 324
neuroglial cell 322
neuron 322
neurotransmitter 327
node of Ranvier 323
norepinephrine (NE) 327
nuclei 335
oscilloscope 324
parasympathetic division 331
peripheral nervous system (PNS) 322
pons 335
prefrontal area 337
primary motor area 336
primary somatosensory area 336
reflex 329
refractory period 324
resting potential 324
reticular formation 335
Schwann cell 323
semantic memory 338
sensory neuron 322
short-term memory 338
skill memory 338
sodium-potassium pump 324
somatic system 329
spinal cord 332
spinal nerve 328
sympathetic division 331
synapse 327
synaptic cleft 327
thalamus 335
threshold 324
tract 333
ventricle 332
white matter 333

Match the terms to these definitions:

a. ________ Action potential (an electrochemical change) traveling along an axon.
b. ________ One of the large paired structures making up the cerebrum of the brain.
c. ________ The summing of excitatory and inhibitory signals received at synapses by a neuron.
d. ________ Neuron that takes information from the central nervous system to effectors.
e. ________ Network of nuclei and fibers extending from the brain stem to the thalamus that passes on sensory stimuli and arouses the cerebral cortex.

Using Technology

Your study of the nervous system is supported by these available technologies:

Essential Study Partner CD-ROM

Animals → Nervous System

Visit the Mader web site for related ESP activities.

Exploring the Internet

The Mader Home Page provides resources and tools as you study this chapter.

http://www.mhhe.com/biosci/genbio/mader

Dynamic Human 2.0 CD-ROM

Nervous System

Virtual Physiology Laboratory CD-ROM

Action Potential
Synaptic Transmission

HealthQuest CD-ROM

1 Stress Management and Mental Health
8 Alcohol
9 Other Drugs

C H A P T E R 18

Senses

A bike rider has to use all his senses. A sense of balance keeps the bike aligned, and a sense of sight allows the biker to see any obstacles ahead. The other senses also play important roles.

Chapter Concepts

Sally, keeping her eyes closed, opened her mouth, and her lab partner placed a piece of apple—or was it carrot—on her tongue. Joe, who was looking through a tube with one eye, saw a hole in his free hand with the other eye. As Mary moved the paper closer to her eye, a letter she knew was there disappeared. This laboratory session was a lot more fun than many of the others. The old expression "Don't believe everything you hear and only half of what you see" had more meaning after performing some of the exercises on sensory perception.

Sense organs at the periphery of the body are the "windows of the brain" because they keep the brain aware of what is going on in the external world. When stimulated, sensory receptors generate nerve impulses that travel to the central nervous system (CNS). Nerve impulses arriving at the cerebral cortex of the brain result in sensation. Stimulation of specific parts of the cerebral cortex result in seeing, hearing, tasting, and all our other senses. As the laboratory exercises mentioned above demonstrate, the brain's interpretation of data received from sensory receptors can lead to a mistaken perception of environmental circumstances. The brain has checks and balances, however. The prefrontal area can reason that a mistake may have been made.

18.1 Sensory Receptors and Sensations

Sensory receptors are specialized to receive certain types of stimuli (sing., **stimulus**), or particular forms of energy. Eventually these stimuli result in a sensation within the cerebral cortex.

Types of Sensory Receptors

Sensory receptors in humans can be classified into just four types.

Chemoreceptors respond to chemical substances in the immediate vicinity. The senses of taste and smell are well known to have this type of sensory receptor, but there are also chemoreceptors in various other organs that are sensitive to internal conditions. Chemoreceptors in certain blood vessels monitor the hydrogen ion concentration $[H^+]$ in the blood, and if the pH lowers, the breathing rate increases. As more carbon dioxide is expired, the blood pH will rise.

Pain receptors are a type of chemoreceptor. They are naked dendrites that respond to chemicals released by damaged tissues. Pain receptors are protective because they alert us to possible danger. Without the pain of appendicitis, we may never seek the medical help that is needed to avoid a ruptured appendix.

Mechanoreceptors are stimulated by mechanical forces, which are most often pressure of some sort. The sense of touch is dependent on pressure receptors that are sensitive to either strong or slight pressures. Pressure receptors located in certain arteries detect changes in blood pressure, and stretch receptors in the lungs detect the degree of lung inflation. Proprioceptors, which respond to the stretching of muscle fibers, tendons, joints, and ligaments make us aware of the position of our limbs. Even hearing is dependent on mechanoreceptors. In this case, the receptors are sensitive to pressure waves in inner ear fluid. Pressure receptors that provide information regarding equilibrium are also located in the inner ear.

Thermoreceptors are stimulated by changes in temperature. Those that respond when temperatures rise are called warmth receptors, and those that respond when temperatures lower are called cold receptors. There are internal thermoreceptors in the hypothalamus and surface thermoreceptors in the skin.

Photoreceptors respond to light energy. Our eyes contain photoreceptors that are sensitive to light and thereby provide us with a sense of vision. Stimulation of the photoreceptors, known as rod cells, results in black and white vision, while stimulation of the photoreceptors, known as cone cells, results in color vision.

The sensory receptors of humans respond to four types of stimuli: chemical, mechanical, temperature, and light.

How Sensation Occurs

Sensory receptors respond to environmental stimuli by generating nerve impulses. **Sensation** occurs when nerve impulses arrive at the cerebral cortex of the brain. **Perception** occurs when the cerebral cortex interprets the meaning of sensations.

Most sensory receptors are free nerve endings or encapsulated nerve endings, but some are specialized cells closely associated with neurons. The plasma membrane of a sensory receptor contains receptor proteins that react to the stimulus. For example, the receptor proteins in the plasma membrane of chemoreceptors bind to certain molecules. When this happens, ion channels open and ions flow across the plasma membrane. If the stimulus is sufficient, nerve impulses begin and are carried by a sensory nerve fiber to the CNS (Fig. 18.1). The stronger the stimulus, the greater the frequency of nerve impulses.

Although sensory receptors simply initiate nerve impulses, we have different senses. The brain, as you know, is responsible for sensation and perception. Nerve impulses that begin in the optic nerve eventually reach the visual areas of the cerebral cortex and, thereafter, we see objects. Nerve impulses that begin in the auditory nerve eventually reach the auditory areas of the cerebral cortex and, thereafter, we hear sounds. If it were possible to switch these nerves, then stimulation of the eyes would result in hearing! On the other hand, when a blow to the eye stimulates photoreceptors, we "see stars" because nerve impulses from the eyes can only result in sight.

Sensory receptors carry out **integration,** the summing of signals. In this instance, the signals are in the form of environmental stimuli. One example of integration is **sensory adaptation,** a decrease in response to a stimulus. We have all had the experience of smelling an odor when we first enter a room and then later not being aware of it at all. Some authorities believe that when sensory adaptation occurs, sensory receptors have stopped sending impulses to the brain. Others believe that interneurons in the CNS have filtered out the ongoing stimuli (see Fig. 17.13).

Sensation occurs when nerve impulses reach the cerebral cortex of the brain. Perception, which also occurs in the cerebral cortex, is an interpretation of the meaning of sensations.

Types of Senses

Senses are divided into the somatic senses and the special senses.

Somatic Senses

Senses associated with the skin, muscles, joints, and internal organs are called **somatic senses.** Mechanoreceptors in the skin give us a sense of touch and throughout the body a sense of pressure. A sense of temperature is due to receptors located in the skin and the brain. A sense of pain is due to pain receptors located in the skin and also among internal organs. Proprioception, which helps maintain posture and balance, is due to proprioceptors like Golgi tendon organs in tendons and muscle spindles in skeletal muscles.

Special Senses

The sense organs for taste, smell, vision, balance, and hearing contain a number of receptors all specialized to give us a particular sense. Table 18.1 lists the **special senses** and their associated sense organs.

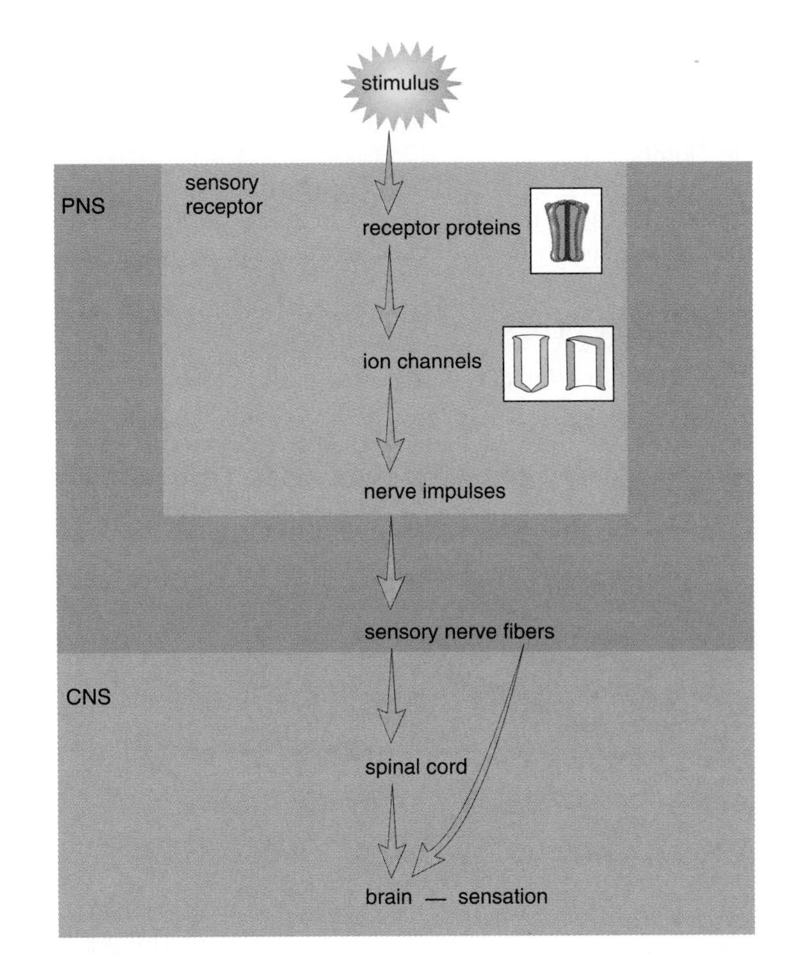

Figure 18.1 Sensation.
The stimulus is received by a receptor protein which causes ion channels to open and nerve impulses (action potentials) to be generated. Conduction of nerve impulses to the CNS by sensory nerve fibers results in sensation.

The sensory receptors, whether existing independently or in sense organs, account for our various senses.

Table 18.1 Special Sense Organs

Sense	Type of Sensory Receptor	Specific Sensory Receptors	Sense Organ
Taste	Chemoreceptor	Taste cells	Taste buds
Smell	Chemoreceptor	Olfactory cells	Olfactory epithelium
Vision	Photoreceptor	Rod cells and cone cells in retina	Eye
Hearing	Mechanoreceptor	Hair cells in spiral organ	Ear
Balance	Mechanoreceptor	Hair cells in utricle, saccule, and semicircular canals	Ear

18.2 Somatic Senses

Somatic senses are dependent upon receptors in the muscles, joints, and tendons and in the skin. The sensory receptors for somatic senses send nerve impulses to the spinal cord. From there, they travel up the spinal cord in tracts to the somatosensory areas of the cerebral cortex.

Proprioceptors

Proprioceptors detect the degree of muscle relaxation, the stretch of tendons, and the movement of ligaments. Muscle spindles act to increase and Golgi tendon organs act to decrease the degree of muscle contraction. The result is a muscle that has the proper length and tension, or muscle tone.

Figure 18.2 illustrates the activity of a muscle spindle. In a muscle spindle, sensory nerve endings are wrapped around thin muscle cells within a connective tissue sheath. When the muscle relaxes and undue stretching of the muscle spindle occurs, nerve impulses are generated. The rapidity of the nerve impulses generated by the muscle spindle is proportional to the stretching of a muscle. A reflex action occurs which results in contraction of muscle fibers adjoining the muscle spindle.

The information sent by muscle spindles to the CNS is used to maintain the body's balance and posture despite the force of gravity always acting upon the skeleton and muscles. Proprioception also helps us know the position of our limbs in space.

Proprioceptors are involved in reflex actions that maintain muscle tone and thereby the body's balance and posture.

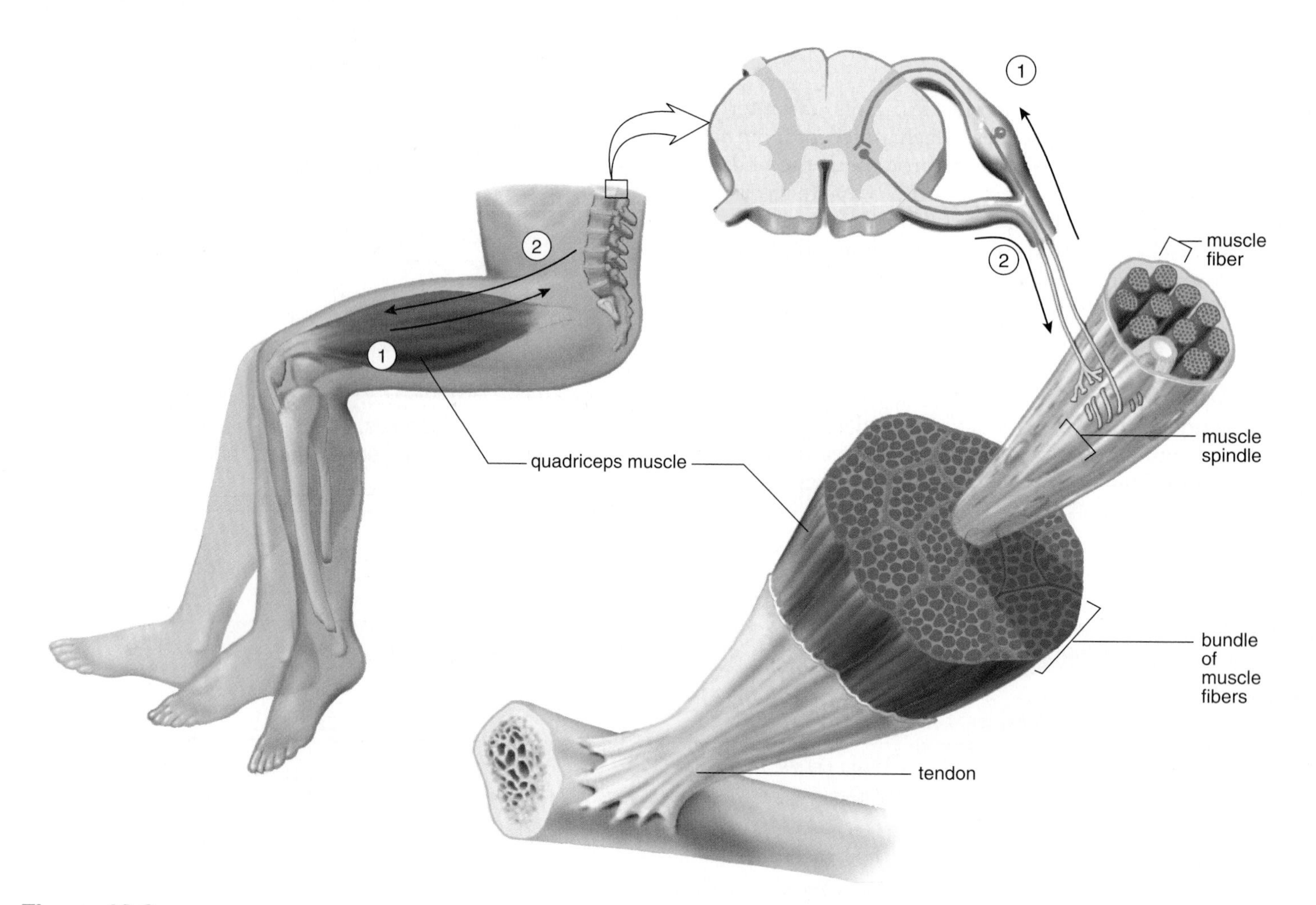

Figure 18.2 Muscle spindle.
① When a muscle relaxes, a muscle spindle is stimulated as it is stretched. ② A reflex action occurs, resulting in muscle fiber contraction so that muscle tone is maintained.

Cutaneous Receptors

The skin is composed of two layers: the epidermis and the dermis. In Figure 18.3, the artist has dramatically indicated these two layers by separating the epidermis from the dermis in one location. The epidermis is stratified squamous epithelium in which cells become keratinized as they rise to the surface where they are sloughed off. The dermis of skin is a thick connective tissue layer. The dermis contains receptors for touch, pressure, pain, and temperature. It is a mosaic of these tiny receptors, as you can determine by slowly passing a metal probe over your skin. At certain points, there is a feeling of pressure, and at others, there is a feeling of hot or cold (depending on the probe's temperature).

Two types of receptors are responsive to fine touch: the Meissner's corpuscles and the Merkel's disks. The Meissner's corpuscles are concentrated in the fingertips, the palms, the lips, the tongue, the nipples, the penis, and the clitoris. Their prevalence provides these regions with special sensitivity. Pacinian corpuscles, Ruffini's endings, and Krause end bulbs are three different types of pressure receptors. Pacinian corpuscles are onion-shaped sensory receptors that lie deep inside the dermis. Ruffini's endings and Krause end bulbs are encapsulated by sheaths of connective tissue and contain lacy networks of nerve fibers. Temperature and pain receptors are simply free nerve endings in the epidermis. Some free nerve endings are responsive to cold; others are responsive to warmth. Cold receptors are far more numerous than warmth receptors, but there are no known structural differences between the two.

Specialized receptors in the human skin respond to touch, pressure, pain, and temperature (warmth and cold).

Referred Pain

Like the skin, many internal organs have pain receptors. Sometimes, stimulation of internal pain receptors is felt as pain from the skin. This is called **referred pain.** Some internal organs have a referred pain relationship with areas located in the skin of the back, groin, and abdomen; pain from the heart is felt in the left shoulder and arm. This most likely happens when nerve impulses from the pain receptors of internal organs travel to the spinal cord and synapse with neurons also receiving impulses from the skin. The brain perceives this as pain in the skin.

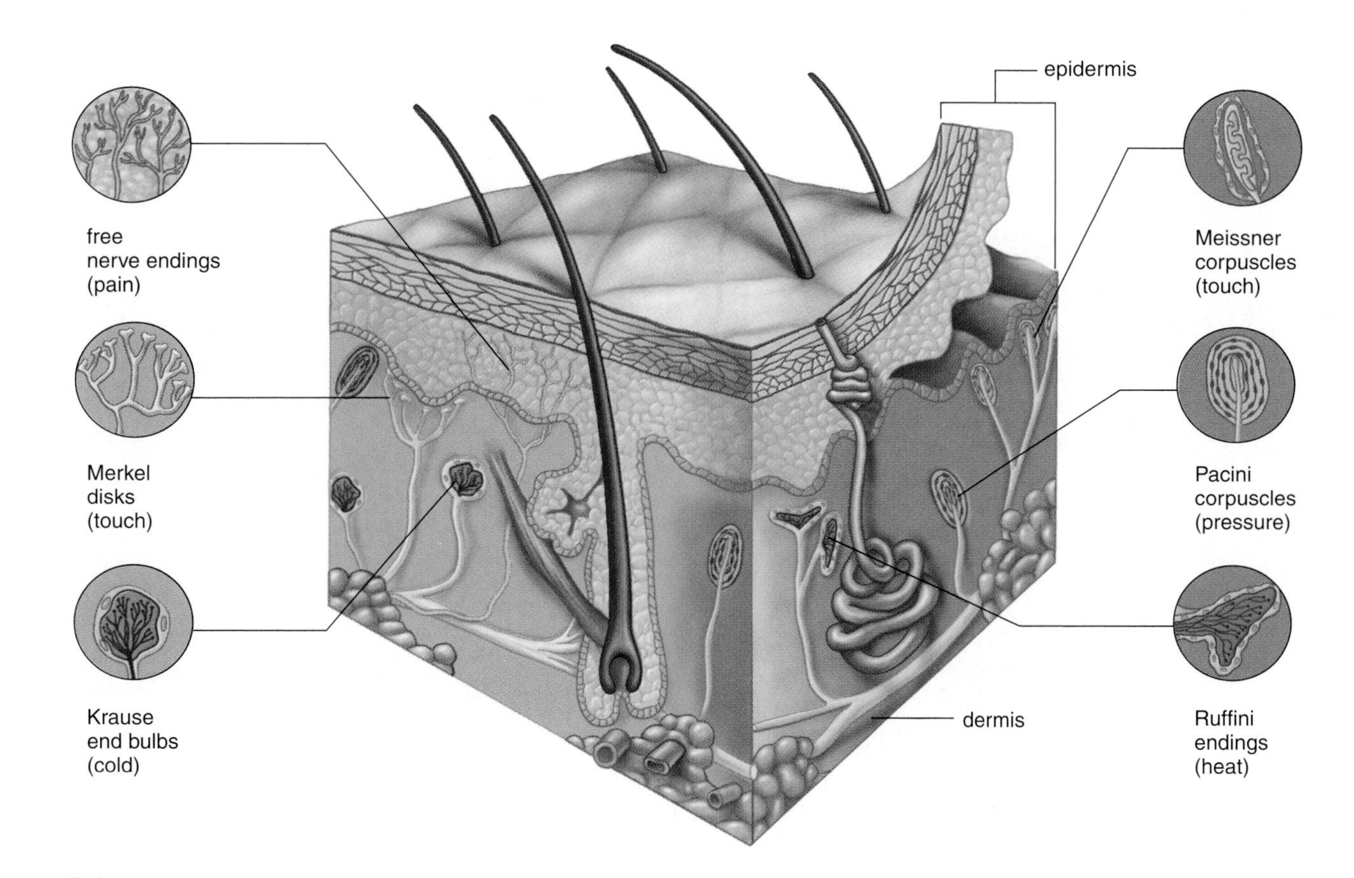

Figure 18.3 Receptors in human skin.
The classical view shown here is that each receptor has the main function indicated. However, investigations in this century indicate that matters are not so clear-cut. For example, microscopic examination of the skin of the ear shows only free nerve endings (pain receptors), and yet the skin of the ear is sensitive to all sensations. Therefore, it appears that the receptors of the skin are somewhat, but not completely, specialized.

18.3 Chemical Senses

There are chemoreceptors located in the carotid arteries and in the aorta that are primarily sensitive to the pH of the blood. These bodies communicate via sensory nerve fibers with the respiratory center located in the medulla oblongata. When the pH drops, they signal this center, and immediately thereafter the breathing rate increases. The expiration of CO_2 raises the pH of the blood.

Taste and smell are called the chemical senses because the receptors for these senses are sensitive to molecules in the food we eat and air we breathe.

Sense of Taste

Taste buds are sense organs located primarily on the tongue (Fig. 18.4). Many lie along the walls of the papillae, the small elevations on the tongue that are visible to the naked eye. Isolated ones are also present on the hard palate, the pharynx, and the epiglottis.

Taste buds are embedded in tongue epithelium and open at a taste pore. They have supporting cells and a number of elongated taste cells that end in microvilli. The microvilli bear receptor proteins for certain molecules. When these molecules bind to receptor proteins, nerve impulses are generated in associated sensory nerve fibers. These nerve impulses go to the brain including cortical areas which interpret them as tastes.

There are four primary types of tastes (sweet, sour, salty, bitter) and taste buds for each are concentrated on the tongue in particular regions (Fig. 18.4*a*). Sweet receptors are most plentiful near the tip of the tongue. Sour receptors occur primarily along the margins of the tongue. Salty receptors are most common on the tip and the upper front portion of the tongue. Bitter receptors are located toward the back of the tongue. Actually, the response of taste buds can result in a range of sweet, sour, salty, and bitter tastes. The brain appears to survey the overall pattern of incoming sensory impulses and to take a "weighted average" of their taste messages as the perceived taste.

Taste buds are sense organs that contain taste cells. The microvilli of taste cells have receptor proteins for molecules that cause the brain to distinguish between sweet, sour, salty, and bitter tastes.

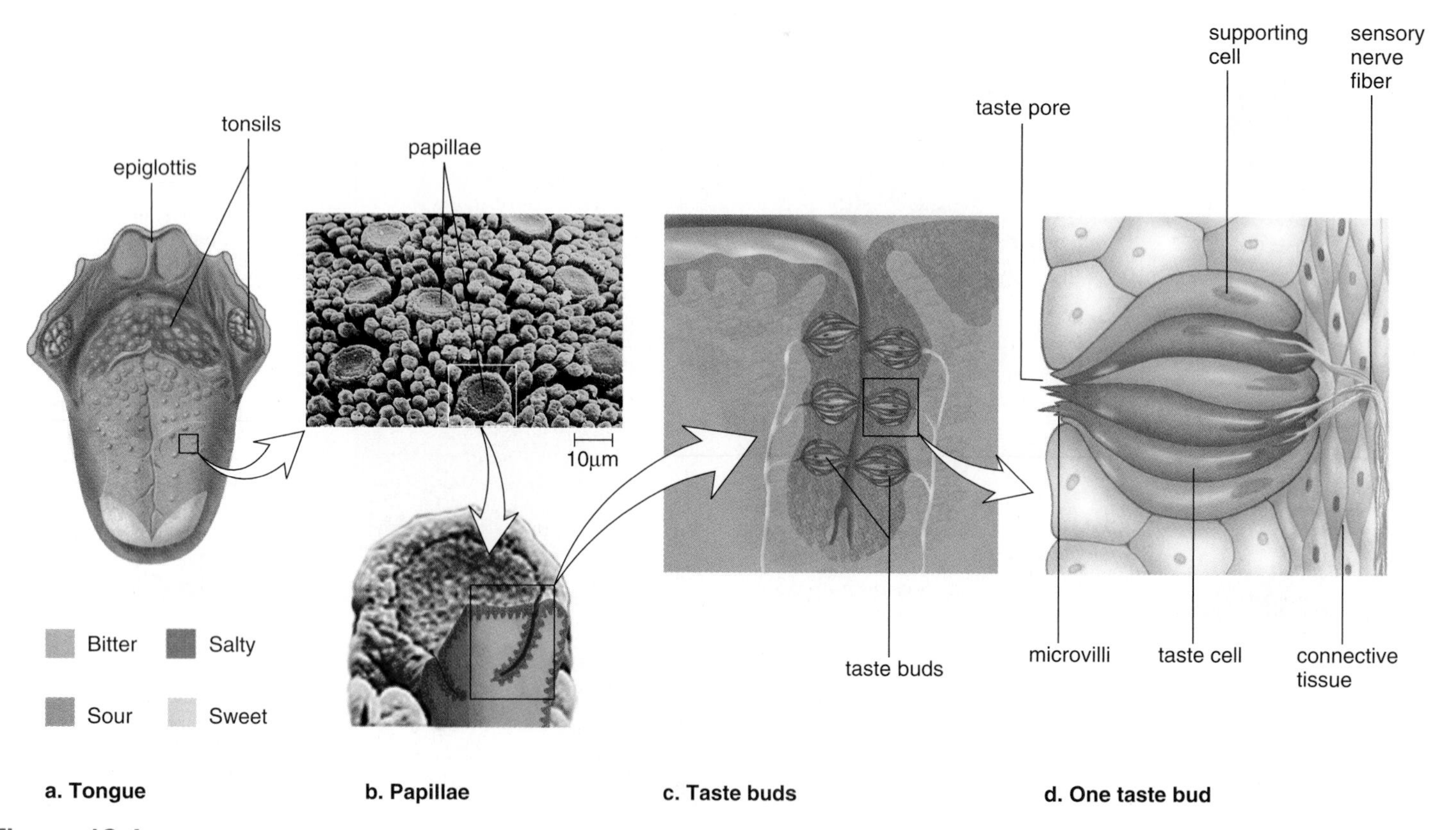

Figure 18.4 Taste buds.
a. Papillae on the tongue contain taste buds that are sensitive to sweet, sour, salty, and bitter tastes as indicated.
b. Enlargement of papillae. **c.** Taste buds occur along the walls of the papillae. **d.** Taste cells end in microvilli that bear receptor proteins for certain molecules. When molecules bind to the receptor proteins, nerve impulses are generated that go to the brain where the sensation of taste occurs.

Sense of Smell

Our sense of smell is dependent on **olfactory cells** located within olfactory epithelium high in the roof of the nasal cavity (Fig. 18.5). Olfactory cells are modified neurons. Each cell ends in a tuft of about five olfactory cilia which bear receptor proteins for odor molecules. Each olfactory cell has only one out of 1,000 different types of receptor proteins. Nerve fibers from like olfactory cells lead to the same interneuron in the olfactory bulb, an extension of the brain. An odor contains many odor molecules which activate a characteristic combination of receptor proteins. A rose might stimulate olfactory cell type A, B, C, while a daffodil might stimulate olfactory cell type B, C, E . An odor's signature in the olfactory bulb is determined by which corresponding interneurons are stimulated. When the interneurons communicate this information via the olfactory tract to the olfactory areas of the cerebral cortex, we know we have smelled a rose or a daffodil.

Have you ever noticed that an aroma will bring to mind a vivid memory of a person or place? A person's perfume may remind you of someone else, or the smell of boxwood may remind you of your grandfather's farm. Look again at Figure 17.15 and notice that the olfactory bulbs have direct connections with the limbic system and its centers for emotions and memory. One investigator showed that when subjects smelled an orange when viewing a painting, they not only remembered the painting, they had many deep feelings about it.

Actually, the sense of taste and the sense of smell work together to create a combined effect when interpreted by the cerebral cortex. For example, when you have a cold, you think that food has lost its taste, but most likely you have lost the ability to sense its smell. This method works in reverse also. When you smell something, some of the molecules move from the nose down into the mouth region and stimulate the taste buds there. Therefore, part of what we refer to as smell may in fact be taste.

The olfactory epithelium is the sense organ that contains olfactory cells. The cilia of olfactory cells have receptor proteins for odor molecules that cause the brain to distinguish odors.

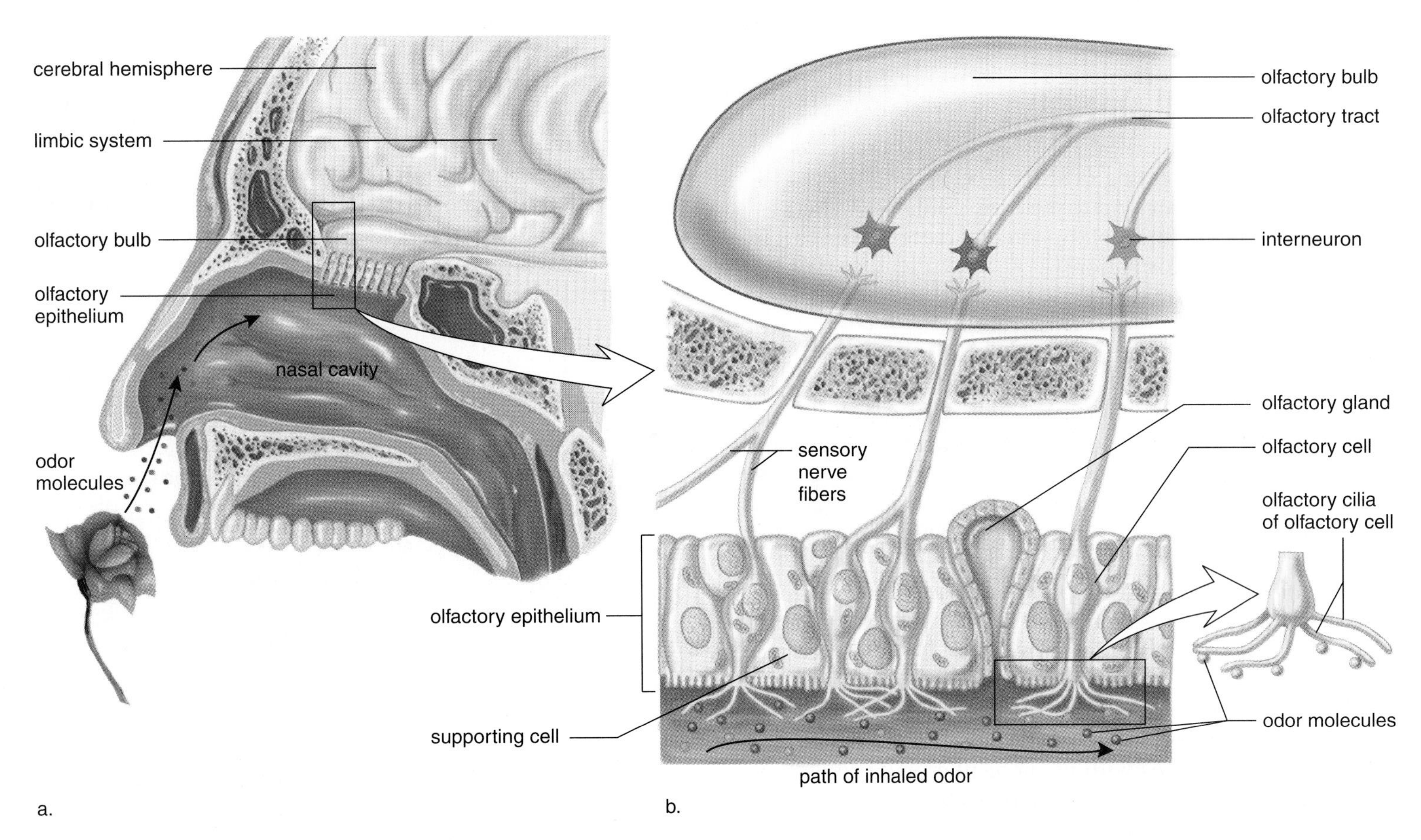

Figure 18.5 Olfactory cell location and anatomy.
a. The olfactory epithelium in humans is located high in the nasal cavity. **b.** Olfactory cells end in cilia that bear receptor proteins for specific odor molecules. The cilia of each olfactory cell can bind to only one type of odor molecule signified here by color. If a rose causes A, B, C type olfactory cells to be stimulated, then interneurons A, B, C in the olfactory bulb are activated. The primary olfactory area of the cerebral cortex interprets the pattern of interneurons stimulated as the scent of a rose.

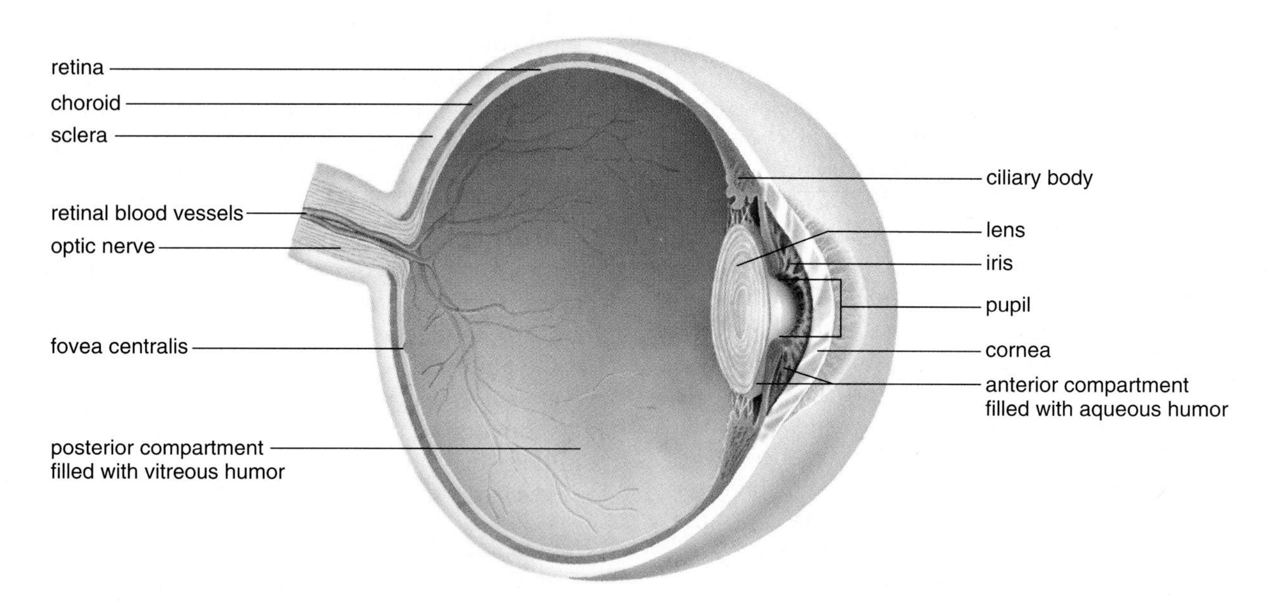

Figure 18.6 Anatomy of the human eye.
Notice that the sclera, the outer layer of the eye, becomes the cornea and that the choroid, the middle layer, is continuous with the ciliary body and the iris. The retina, the inner layer, contains the photoreceptors for vision; the fovea centralis is the region where vision is most acute.

18.4 Sense of Vision

Vision requires the work of the eyes and the brain. As we shall see, much processing of stimuli occurs in the eyes before nerve impulses are sent to the brain. Still, researchers estimate that at least a third of the cerebral cortex takes part in processing visual information.

Anatomy of the Eye

The eyeball, which is an elongated sphere about 2.5 cm in diameter, has three layers, or coats: the sclera, the choroid, and the retina (Fig. 18.6 and Table 18.2). The outer layer, the **sclera,** is white and fibrous except for the transparent **cornea** which is made of transparent collagen fibers. The cornea is the window of the eye.

The middle, thin, darkly pigmented layer, the **choroid,** is vascular and absorbs stray light rays that photoreceptors have not absorbed. Toward the front, the choroid becomes the donut-shaped **iris.** The iris regulates the size of the **pupil,** a hole in the center of the iris through which light enters the eyeball. The color of the iris (and therefore the color of your eyes) correlates with its pigmentation. Heavily pigmented eyes are brown, while lightly pigmented eyes are green or blue. Behind the iris, the choroid thickens and forms the circular ciliary body. The **ciliary body** contains the ciliary muscle, which controls the shape of the lens for near and far vision.

The **lens,** attached to the ciliary body by ligaments, divides the eye into two compartments; the one in front of the lens is the anterior compartment and the one behind the lens is the posterior compartment. The anterior compartment is filled with a clear, watery fluid called the **aqueous humor.** A small amount of aqueous humor is continually produced each day. Normally, it leaves the anterior compartment by way of tiny ducts. When a person has

Table 18.2 Function of Parts of the Eye

Part	Function
Sclera	Protects and supports eyeball
Cornea	Refracts light rays
Choroid	Absorbs stray light
Retina	Contains receptors for sight
Rods	Make black-and-white vision possible
Cones	Make color vision possible
Fovea centralis	Makes acute vision possible
Lens	Refracts and focuses light rays
Ciliary body	Holds lens in place, accommodation
Iris	Regulates light entrance
Pupil	Admits light
Humors	Transmit light rays and support eyeball
Optic nerve	Transmits impulse to brain

glaucoma, these drainage ducts are blocked, and aqueous humor builds up. If glaucoma is not treated, the resulting pressure compresses the arteries that serve the nerve fibers of the retina, where photoreceptors are located. The nerve fibers begin to die due to lack of nutrients, and the person becomes partially blind. Eventually, total blindness can result.

The third layer of the eye, the **retina,** is located in the posterior compartment which is filled with a clear gelatinous material called the **vitreous humor.** The retina contains photoreceptors called rod cells and cone cells. The rods are very sensitive to light but they do not see color; therefore, at night or in a darkened room we see only shades of gray. The cones, which require bright light, are sensitive to different wavelengths of light and, therefore, we have the ability to distinguish colors. The retina has a very special region called the **fovea centralis** where cone cells are densely packed. Light is normally focused on the fovea when we look directly at an object. This is helpful because vision is most acute in the fovea centralis. Sensory fibers from the retina form the **optic nerve** which takes nerve impulses to the brain.

> The eye has three layers: the outer sclera, the middle choroid, and the inner retina. Only the retina contains photoreceptors for light energy.

Focusing

When we look at an object, light rays pass through the pupil and are **focused** on the retina (Fig. 18.7*a*). The image produced is much smaller than the object because light rays are bent (refracted) when they are brought into focus. Focusing starts with the cornea and continues as the rays pass through the lens and the humors. Notice that the image on the retina is inverted (it is upside down) and reversed from left to right.

The lens provides additional focusing power as **visual accommodation** occurs for close vision. The shape of the lens is controlled by the ciliary muscle within the ciliary body. When we view a distant object, the ciliary muscle is relaxed, causing the suspensory ligaments attached to the ciliary body to be taut; therefore, the lens remains relatively flat (Fig. 18.7*b*). When we view a near object, the ciliary muscle contracts, releasing the tension on the suspensory ligaments, and the lens rounds up due to its natural elasticity (Fig. 18.7*c*). Because close work requires contraction of the ciliary muscle, it very often causes muscle fatigue known as eyestrain.

Usually after the age of 40, the lens loses some of its elasticity and is unable to accommodate. Bifocal lenses are then usually necessary for those who already have corrective lenses. Also with aging or possibly exposure to the sun (see the Health reading on page 360), the lens is subject to **cataracts.** The lens becomes opaque and therefore incapable of transmitting rays of light. Usually, today, the lens is replaced with an artificial lens during surgery. In the future, it may be possible to restore the original configuration of the proteins making up the lens.

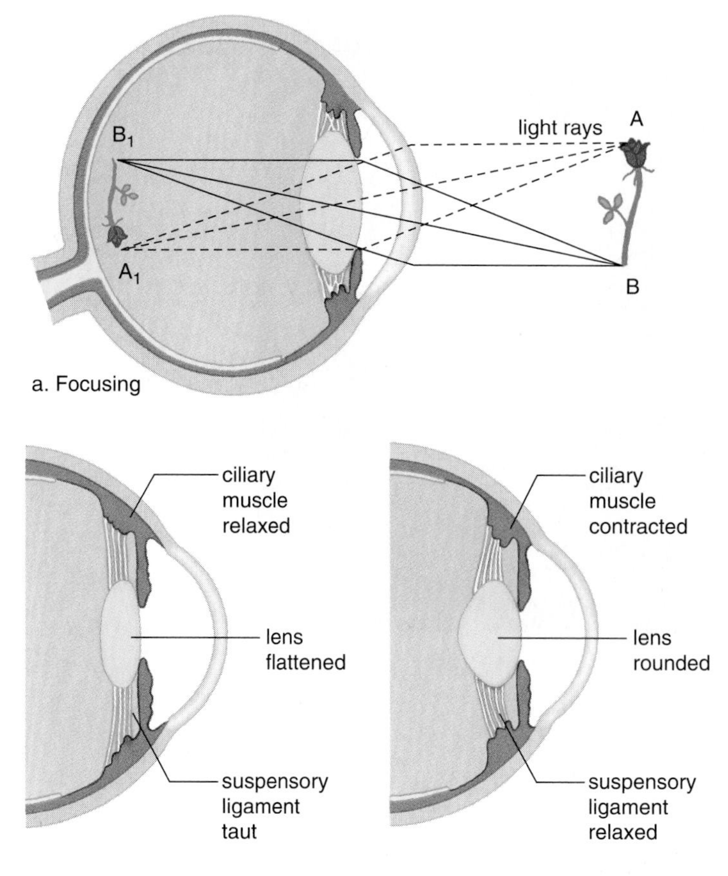

Figure 18.7 Focusing.
a. Light rays from each point on an object are bent by the cornea and the lens in such a way that an inverted and reversed image of the object forms on the retina. **b.** When focusing on a distant object, the lens is flat because the ciliary muscle is relaxed and the suspensory ligament is taut. **c.** When focusing on a near object, the lens accommodates; it becomes rounded because the ciliary muscle contracts, causing the suspensory ligament to relax.

> The lens, assisted by the cornea and the humors, focuses images on the retina.

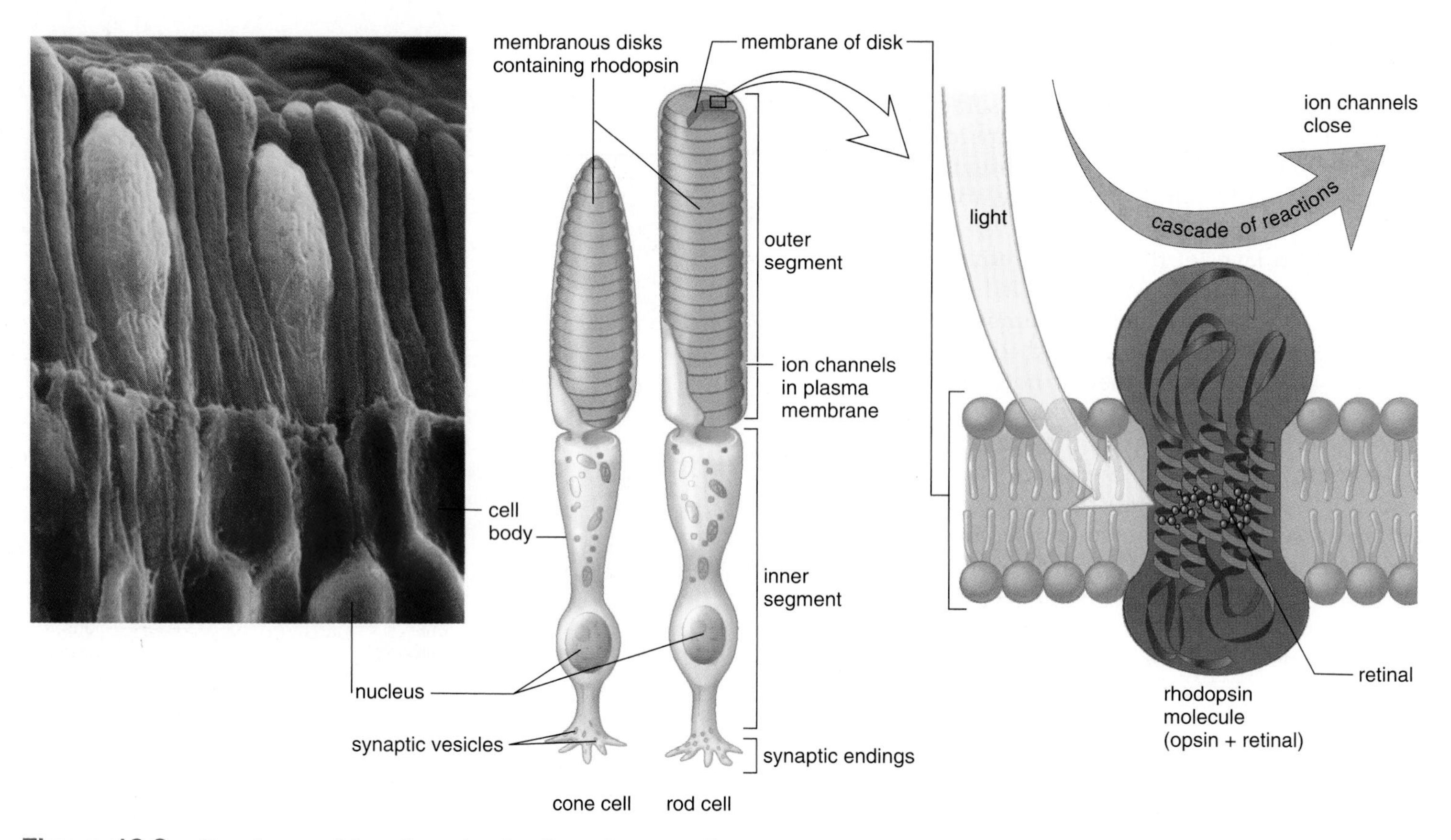

Figure 18.8 Structure and function of rod cells and cone cells.
The outer segment of rods and cones contains stacks of membranous disks, which contain visual pigments. In rods, the membrane of each disk contains rhodopsin, a complex molecule containing the protein opsin and the pigment retinal. When rhodopsin absorbs light energy, it splits, releasing opsin, which sets in motion a cascade of reactions that ends when ion channels in the plasma membrane close.

Photoreceptors

Vision begins once light has been focused on the photoreceptors in the retina. Figure 18.8 illustrates the structure of our photoreceptors called **rod cells** and **cone cells.** Both rods and cones have an outer segment joined to an inner segment by a stalk. The outer segment contains many membranous disks. And many pigment molecules are embedded in the membrane of these disks. Synaptic vesicles are located at the synaptic endings of the inner segment

The visual pigment in rods is a deep purple pigment called rhodopsin. **Rhodopsin** is a complex molecule made up of the protein opsin and a light absorbing molecule called retinal, which is a derivative of vitamin A. When a rod absorbs light, rhodopsin splits into opsin and retinal, leading to a cascade of reactions and the closure of ion channels in the rod cell's plasma membrane. This stops the release of inhibitory transmitter molecules from the rod's synaptic vesicles and starts the signals that result in nerve impulses going to the brain. Rods are very sensitive to light and therefore are suited to night vision. (Since carrots are rich in vitamin A, it is true that eating carrots can improve your night vision.) Rod cells are plentiful throughout the entire retina; therefore, they also provide us with peripheral vision and a perception of motion.

The cones, on the other hand, are located primarily in the fovea and are activated by bright light. They allow us to detect the fine detail and the color of an object. **Color vision** depends on three different kinds of cones, which contain pigments called the B (blue), G (green), and R (red) pigments. Each pigment is made up of retinal and opsin, but there is a slight difference in the opsin structure of each, which accounts for their individual absorption patterns. Various combinations of cones are believed to be stimulated by in-between shades of color.

The receptors for sight are the rods and the cones. The rods permit vision in dim light at night, and the cones permit vision in bright light needed for color vision.

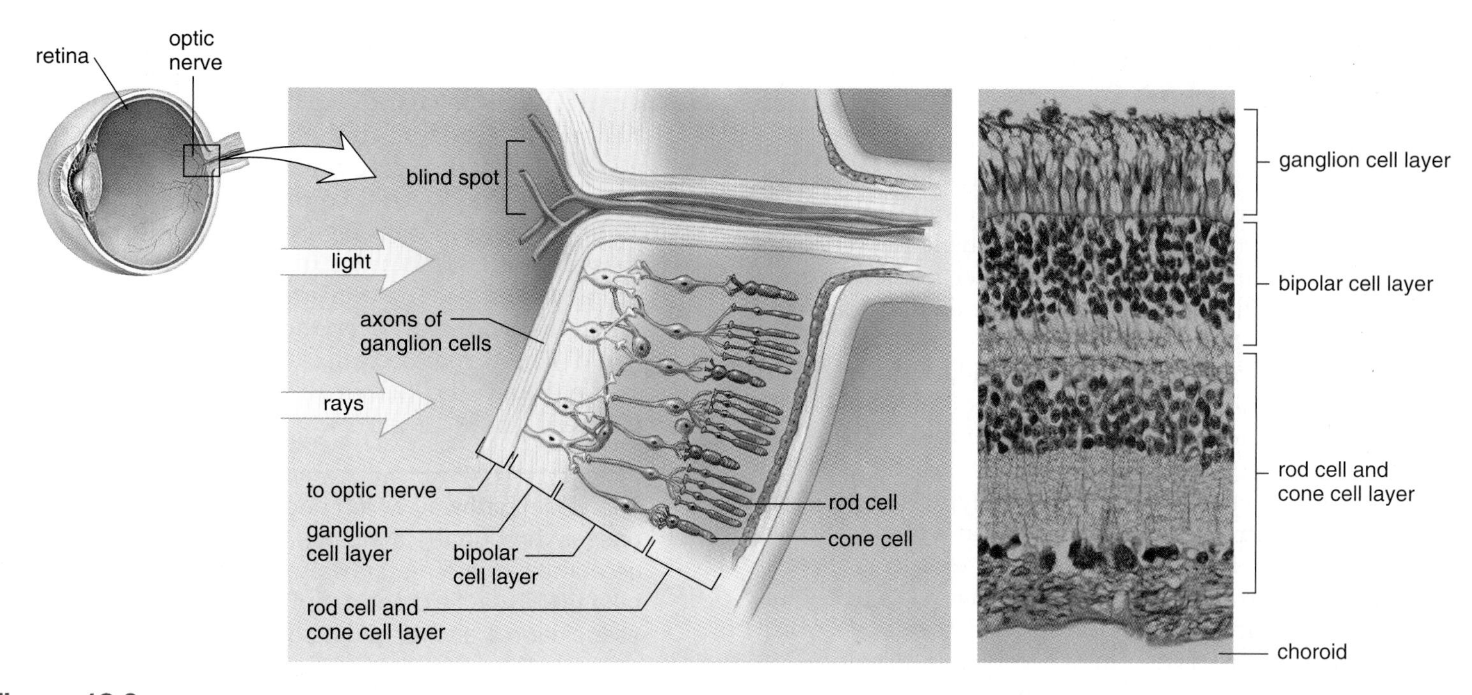

Figure 18.9 Structure and function of retina.
The retina is the inner layer of the eye. Rod cells and cone cells located at the back of the retina synapse with bipolar cells which synapse with ganglion cells. Integration of signals occurs at these synapses; therefore much processing occurs in bipolar and ganglion cells. Further, notice that many rod cells share one bipolar cell but cone cells do not. Each cone synapses with only one ganglion cell. Cone cells, therefore, distinguish more detail than do rod cells.

Integration of Visual Signals in the Retina

The retina has three layers of neurons (Fig. 18.9). The layer closest to the choroid contains the rods and cones; the middle layer contains bipolar cells; and the innermost layer contains ganglion cells, whose fibers become the optic nerve. Only the rod cells and the cone cells are sensitive to light, and therefore light must penetrate to the back of the retina before they are stimulated.

The rod cells and the cone cells synapse with the bipolar cells, which in turn synapse with ganglion cells that initiate nerve impulses. Notice in Figure 18.9 that there are many more rod cells and cone cells than ganglion cells. In fact, the retina has as many as 150 million rod cells and 6 billion cone cells but only one million ganglion cells. The sensitivity of cones in the fovea versus rods is mirrored by how directly they connect to ganglion cells. As many as 100 rods may synapse with the same ganglion cell. No wonder that stimulation of rods results in vision that is only blurred and indistinct. In contrast, each cone in the fovea synapses with only one ganglion cell. This accounts for why cones particularly in the fovea provide us with a sharper, more detailed image of an object.

As signals pass to bipolar cells and ganglion cells, integration occurs. Each ganglion cell receives signals from rod cells covering about one square millimeter of retina (about the size of a thumb tack hole). This region is called the ganglion cell's receptive field. Some time ago, scientists discovered that a ganglion cell is stimulated when light hits the center of its receptive field, and is inhibited when light hits the area of the receptive field surrounding the center. If all the rod cells in the receptive field are stimulated, the ganglion cell responds in a neutral way—it reacts only weakly or perhaps not at all. This supports the hypothesis that considerable processing occurs in the retina before nerve impulses are sent to the brain.

> Synaptic integration and processing begins in the retina before nerve impulses are sent to the brain.

Blind Spot

Figure 18.9 provides an opportunity to point out that there are no rods and cones where the optic nerve exits the retina. Therefore, no vision is possible in this area. You can prove this to yourself by putting a dot to the right of center on a piece of paper. Use the right hand to move the paper slowly toward the right eye while you look straight ahead. The dot will disappear at one point—this is your **"blind spot."**

Integration of Visual Signals in the Brain

Sensory fibers from ganglion cells assemble to form the optic nerves. At the ×-shaped optic chiasma, fibers from the right half of each retina converge and continue on together in the right optic tract, and fibers from the left half of each retina converge and continue on together in the left optic tract. This is the first way in which the brain divides up the visual field (Fig. 18.10).

Because the image is reversed, the left optic tract carries information about the right portion of the **visual field,** and the right optic tract carries information about the left portion of the visual field.

The optic tracts sweep around the hypothalamus, and most fibers synapse with interneurons in a nucleus (mass of neuron cell bodies) of the thalamus. Axons from the thalamic nucleus form an optic radiation that takes nerve impulses to the primary visual area of the cerebral cortex. Since each primary visual area receives information regarding only half the visual field, these areas must eventually share information to form a unified image. Also, the inverted and reversed image must be righted in the brain for us to correctly perceive the visual field.

The most surprising finding has been that the brain has a second way of taking the field apart. Each primary visual area of the cerebral cortex acts like a post office, parceling out information regarding color, form, motion, and possibly other attributes to different portions of the adjoining visual association area. Therefore, the brain has taken the field apart even though we see a unified visual field. The cerebral cortex is believed to rebuild the visual field and give us an understanding of it at the same time.

The visual pathway which begins in the retina passes through the thalamus before reaching the cerebral cortex. The pathway and the visual cortex take the visual field apart, but the visual association areas rebuild it so that we correctly perceive the entire field.

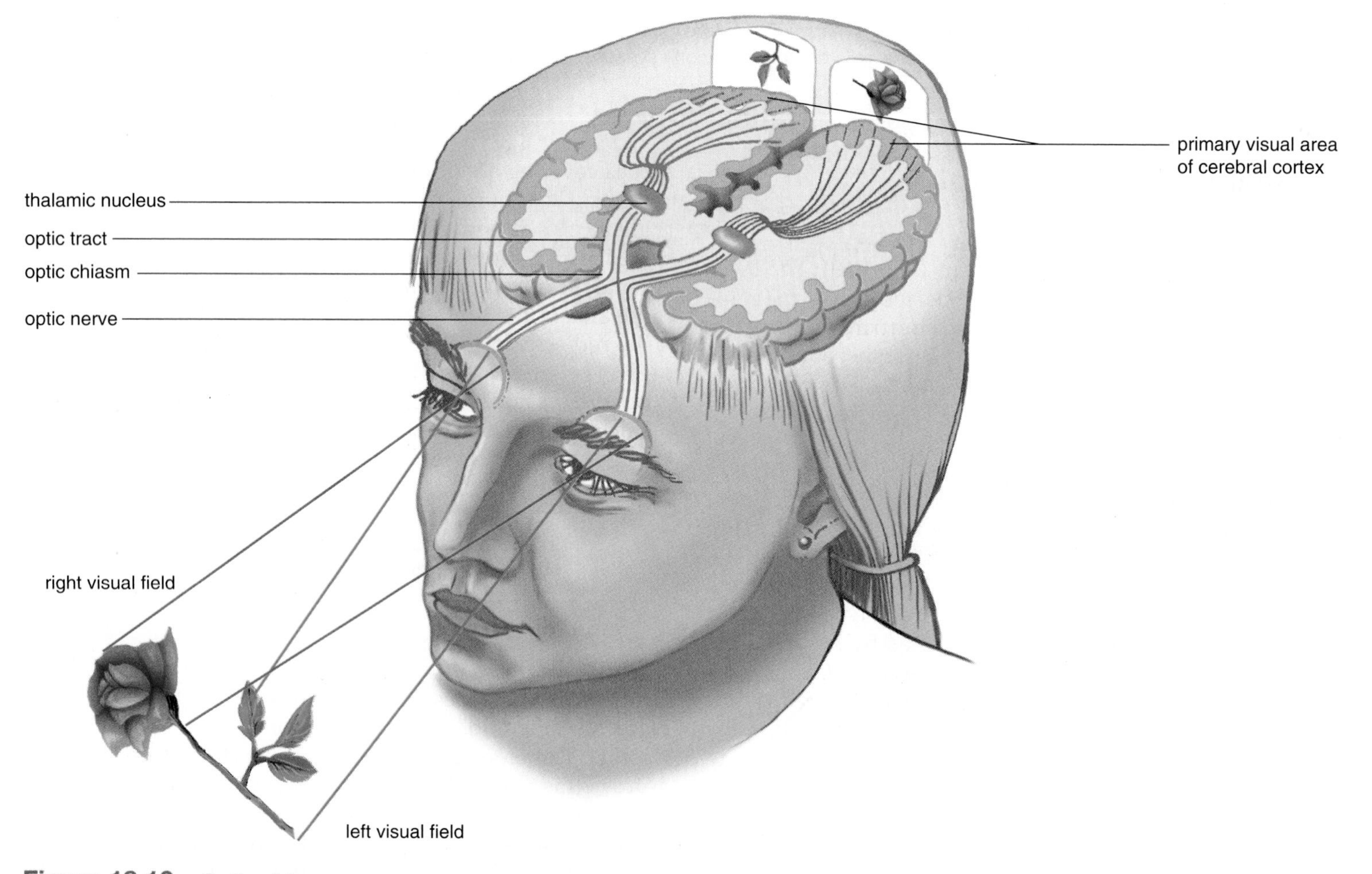

Figure 18.10 Optic chiasma.
Both eyes "see" the entire visual field. Because of the optic chiasma, data from the right half of each retina go to the right visual areas of the cerebral cortex, and data from the left half of the retina go to the left visual areas of the cerebral cortex. This data is combined to allow us to see the entire visual field. Note that the visual pathway to the brain includes the thalamus, which has the ability to filter sensory stimuli.

Abnormalities of the Eye

Color blindness and misshaped eyeballs are two common abnormalities of the eyes. More serious abnormalities are discussed in the Health reading on page 360.

Color Blindness

Complete color blindness is extremely rare. In most instances, a particular type of cone is lacking or deficient in number. The most common mutation is a lack of red or green cones. This abnormality affects 5–8% of the male population. If the eye lacks red cones, the green colors are accentuated, and vice versa.

Distance Vision

The majority of people can see what is designated as a size 20 letter 20 feet away, and so are said to have 20/20 vision. Persons who can see close objects but cannot see the letters from this distance are said to be nearsighted. Nearsighted people can see close objects better than they can see objects at a distance. These individuals have an elongated eyeball, and when they attempt to look at a distant object, the image is brought to focus in front of the retina (Fig. 18.11). They can see close objects because they can adjust the lens to allow the image to focus on the retina, but to see distant objects, these people must wear concave lenses, which diverge the light rays so that the image can be focused on the retina.

There is a treatment for nearsightedness called radial keratotomy, or RK. From four to eight cuts are made in the cornea so that they radiate out from the center like spokes in a wheel. When the cuts heal, the cornea is flattened. Lasers are now also used to flatten or change the shape of the cornea. Although some patients are satisfied with the result, others complain of glare and varying visual acuity.

Persons who can easily see the optometrist's chart but cannot see close objects well are farsighted; these individuals can see distant objects better than they can see close objects. They have a shortened eyeball, and when they try to see close objects, the image is focused behind the retina. When the object is distant, the lens can compensate for the short eyeball, but when the object is close, these persons must wear a convex lens to increase the bending of light rays so that the image can be focused on the retina.

When the cornea or lens is uneven, the image is fuzzy. The light rays cannot be evenly focused on the retina. This condition, called **astigmatism,** can be corrected by an unevenly ground lens to compensate for the uneven cornea.

The shape of the eyeball determines the need for corrective lenses; the inability of the lens to accommodate as we age also requires corrective lenses for close vision.

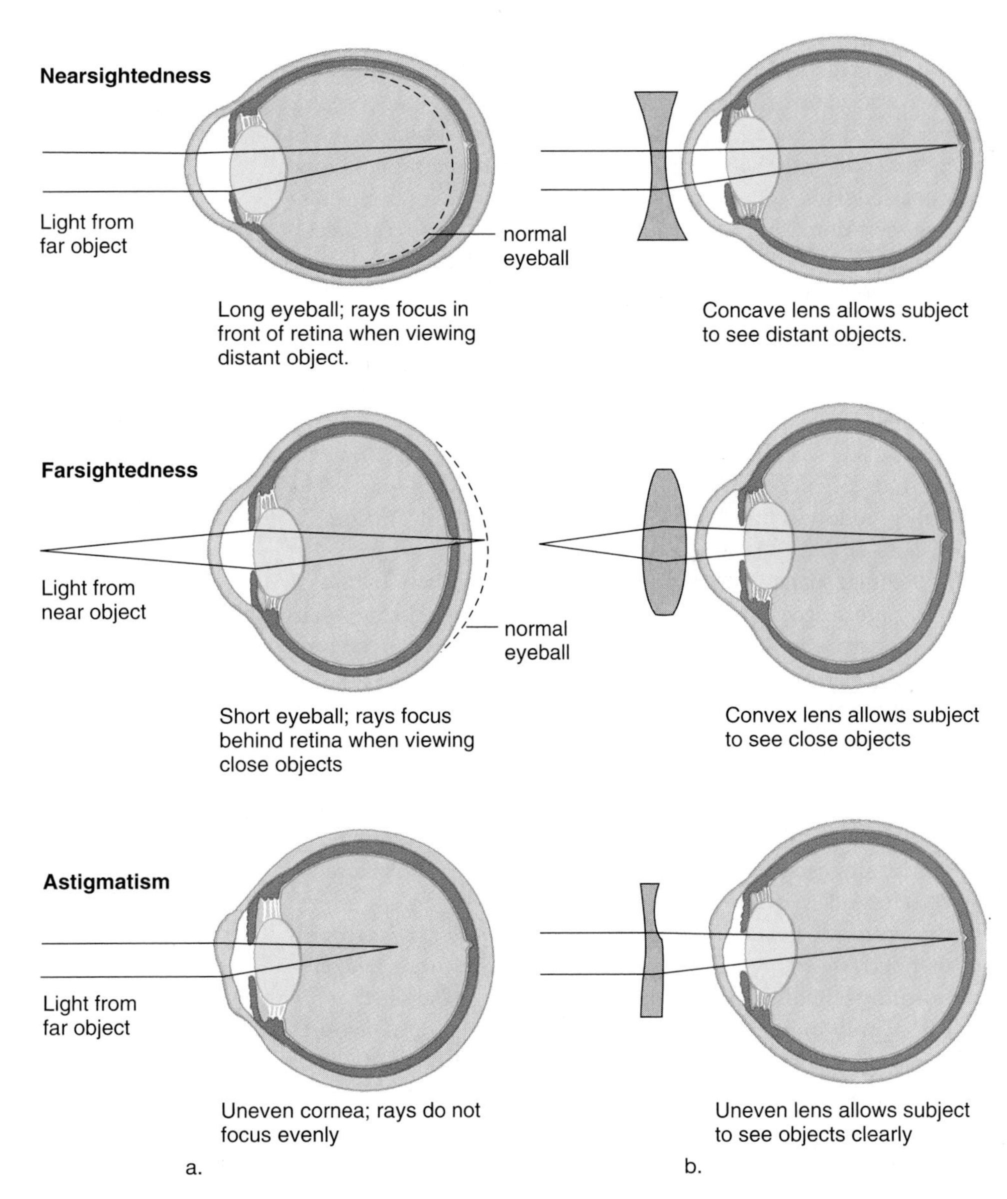

Figure 18.11 Common abnormalities of the eye, with possible corrective lenses.
a. The cornea and the lens function in bringing light rays (lines) to focus, but sometimes they are unable to compensate for the shape of the eyeball or for an irregular curvature of the cornea. **b.** In these instances, corrective lenses can allow the individual to see normally.

Health Focus

Protecting Vision and Hearing

Older age can be accompanied by a serious loss of vision and hearing. The time to start preventive measures for such problems, however, is when we are younger.

Preventing a Loss of Vision

The eye is subject to both injuries and disorders. Although flying objects sometimes penetrate the cornea and damage the iris, lens, or retina, careless use of contact lenses is the most common cause of injuries to the eye. Injuries cause only 4% of all cases of blindness; the most frequent causes are retinal disorders, glaucoma, and cataracts, in that order. Retinal disorders are varied. In diabetic retinopathy, which blinds many people between the ages of 20 and 74, capillaries to the retina burst and blood spills into the vitreous fluid. Careful regulation of blood glucose levels in these patients may be protective. In macular degeneration, the cones are destroyed because thickened choroid vessels no longer function as they should. Glaucoma occurs when the drainage system of the eyes fails, so that fluid builds up and destroys nerve fibers responsible for peripheral vision. Eye doctors always check for glaucoma, but it is advisable to be aware of the disorder in case it comes on quickly. Those who have experienced acute glaucoma report that the eyeball feels as heavy as a stone. In cataracts, cloudy spots on the lens of the eye eventually pervade the whole lens. The milky yellow-white lens scatters incoming light and blocks vision.

Regular visits to an eye-care specialist, especially by the elderly, are a necessity in order to catch conditions, such as glaucoma, early enough to allow effective treatment. It has not been proven, however, that consuming particular foods or vitamin supplements will reduce the risk of cataracts. Even so, it's possible that estrogen replacement therapy may be somewhat protective in postmenopausal women.

There are preventive measures that we can take to reduce the chance of defective vision as we age. Accumulating evidence suggests that both macular degeneration and cataracts, which tend to occur in the elderly, are caused by long-term exposure to the ultraviolet rays of the sun. It is recommended, therefore, that everyone, especially those who live in sunny climates or work outdoors, wear sunglasses that absorb ultraviolet light. Large lenses worn close to the eyes offer further protection. The Sunglass Association of America has devised the following system for categorizing sunglasses:

- Cosmetic lenses absorb at least 70% of UV-B and 20% UV-A and 60% of visible light. Such lenses are worn for comfort rather than protection.
- General purpose lenses absorb at least 95% of UV-B and 60% of UV-A and 60–92% of visible light. They are good for outdoor activities in temperate regions.
- Special purpose lenses block at least 99% of UV-B and 60% UV-A, and 20–97% of visible light. They are good for bright sun combined with sand, snow, or water.

Health care providers have found an increased incidence of cataracts in heavy cigarette smokers. In men, smoking 20 cigarettes or more a day, and in women, smoking more than 35 cigarettes a day doubles the risk of cataracts. It is possible that smoking reduces the delivery of blood and therefore nutrients to the lens.

Preventing a Loss of Hearing

Especially when we are young, the middle ear is subject to infections that can lead to hearing impairments if the infections are not treated promptly by a physician. The mobility of ossicles decreases with age, and in otosclerosis, new filamentous bone grows over the stirrup, impeding its movement. Surgical treatment is the only remedy for this type of conduction deafness. However, age-associated nerve deafness due to stereocilia damage from exposure to loud noises is preventable. Hospitals are now aware that even the ears of the newborn need to be protected from noise and are taking steps to make sure neonatal intensive care units and nurseries are as quiet as possible.

In today's society, exposure to the types of noises listed in Table 18A is a common occurrence. Noise is measured in decibels, and any noise above a level of 80 decibels could result in damage to the hair cells of the organ of Corti. Eventually, the stereocilia and then the hair cells disappear completely (Fig. 18A). If listening to city traffic for extended periods can damage hearing, it stands to reason that frequent attendance at rock concerts, constantly playing a stereo loudly, or using earphones at high volume is also damaging to hearing. The first hint of danger could be temporary hearing loss, a "full" feeling in the ears, muffled hearing, or tinnitus (e.g., ringing in the ears). If you have any of these symptoms, modify your listening habits immediately to prevent further damage. If exposure to noise is unavoidable, specially designed noise-reduction earmuffs are available, and it is also possible to purchase earplugs made from a compressible, spongelike material at the drugstore or sporting-goods store. These earplugs are not the same as those worn for swimming, and they should not be used interchangeably.

Aside from loud music, noisy indoor or outdoor equipment, such as a rug-cleaning machine or a chain saw, is also troublesome. Even motorcycles and recreational vehicles such as snowmobiles and motocross bikes can contribute to a gradual loss of hearing. Exposure to intense sounds of short duration, such as a burst of gunfire, can result in an immediate hearing loss. Hunters may have a significant hearing reduction in the ear opposite to the shoulder where the gun is carried. The butt of the rifle offers some protection to the ear nearest the gun when it is shot.

Finally, people need to be aware that some medicines are ototoxic. Anticancer drugs, most notably cisplatin, and certain antibiotics (e.g., streptomycin, kanamycin, gentamicin) make ears especially susceptible to a hearing loss. Anyone taking such medications needs to be especially careful to protect the ears from any loud noises.

Table 18A Noises That Affect Hearing

Type of Noise	Sound Level (decibels)	Effect
"Boom car," jet engine, shotgun, rock concert	over 125	Beyond threshold of pain; potential for hearing loss high
Discotheque, "boom box," thunderclap	over 120	Hearing loss likely
Chain saw, pneumatic drill, jackhammer, symphony orchestra, snowmobile, garbage truck, cement mixer	100–200	Regular exposure of more than one minute risks permanent hearing loss
Farm tractor, newspaper press, subway, motorcycle	90–100	Fifteen minutes of unprotected exposure potentially harmful
Lawn mower, food blender	85–90	Continuous daily exposure for more than eight hours can cause hearing damage
Diesel truck, average city traffic noise	80–85	Annoying; constant exposure may cause hearing damage

Source: National Institute on Deafness and Other Communication Disorders, January 1990, National Institute of Health.

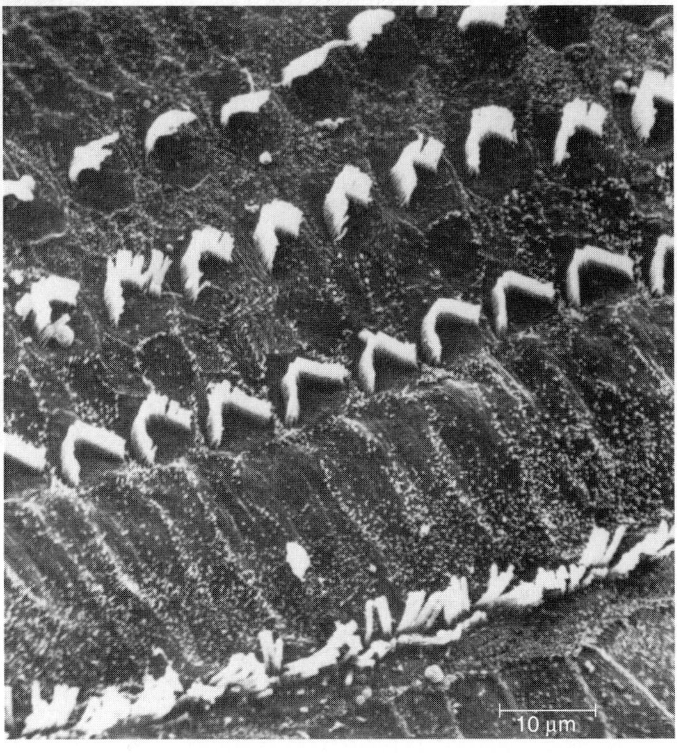

a.

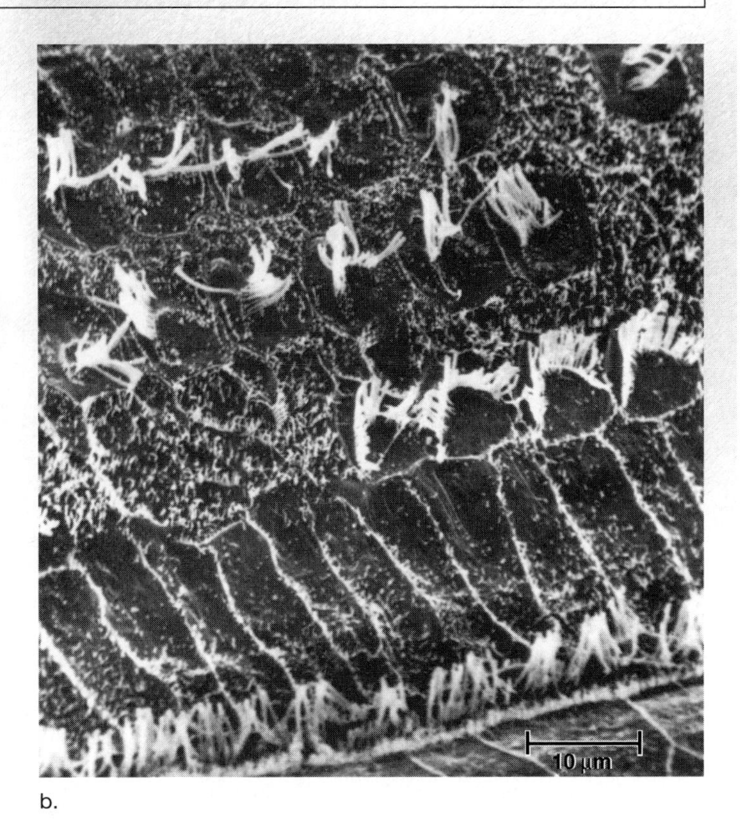

b.

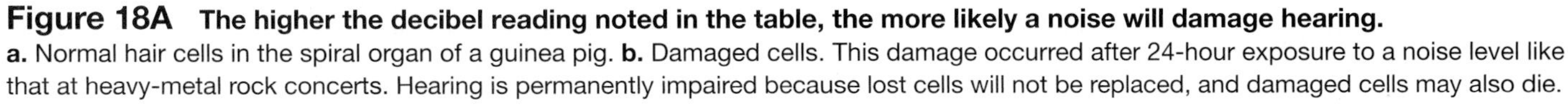

Figure 18A The higher the decibel reading noted in the table, the more likely a noise will damage hearing.
a. Normal hair cells in the spiral organ of a guinea pig. **b.** Damaged cells. This damage occurred after 24-hour exposure to a noise level like that at heavy-metal rock concerts. Hearing is permanently impaired because lost cells will not be replaced, and damaged cells may also die.

18.5 Sense of Hearing

The ear has two sensory functions: hearing and balance (equilibrium). The receptors for both of these are located in the inner ear, and each consists of **hair cells** with stereocilia that respond to mechanical stimulation.

Anatomy of the Ear

Figure 18.12 shows that the ear has three divisions: outer, middle, and inner. The **outer ear** consists of the pinna (external flap) and the auditory canal. The opening of the auditory canal is lined with fine hairs and sweat glands. Modified sweat glands are located in the upper wall of the canal; they secrete earwax, a substance that helps to guard the ear against the entrance of foreign materials, such as air pollutants.

The **middle ear** begins at the **tympanic membrane** (eardrum) and ends at a bony wall containing two small openings covered by membranes. These openings are called the **oval window** and the **round window.** Three small bones are found between the tympanic membrane and the oval window. Collectively called the **ossicles,** individually they are the **malleus** (hammer), the **incus** (anvil), and the **stapes** (stirrup) because their shapes resemble these objects. The malleus adheres to the tympanic membrane, and the stapes touches the oval window. An **auditory** (eustachian) **tube,** which extends from each middle ear to the nasopharynx, permits equalization of air pressure. Chewing gum, yawning, and swallowing in elevators and airplanes help to move air through the auditory tubes upon ascent and descent. As this occurs we often hear the ears "pop."

Whereas the outer ear and the middle ear contain air, the inner ear is filled with fluid. The **inner ear,** anatomically speaking, has three areas: the semicircular canals and the **vestibule** are concerned with equilibrium; the cochlea is concerned with hearing. The **cochlea** resembles the shell of a snail because it spirals.

Process of Hearing

The process of hearing begins when sound waves enter the auditory canal (Fig. 18.13). Just as ripples travel across the surface of a pond, sound waves travel by the successive vibrations of molecules. Ordinarily, sound waves do not carry much energy, but when a large number of waves strike the tympanic membrane, it moves back and forth (vibrates) ever so slightly. The malleus then takes the pressure from the inner surface of the tympanic membrane and passes it by means of the incus to the stapes in such a way that the pressure is multiplied about 20 times as it moves from the tympanic membrane to the stapes. The stapes strikes the membrane of the oval window, causing it to vibrate, and in

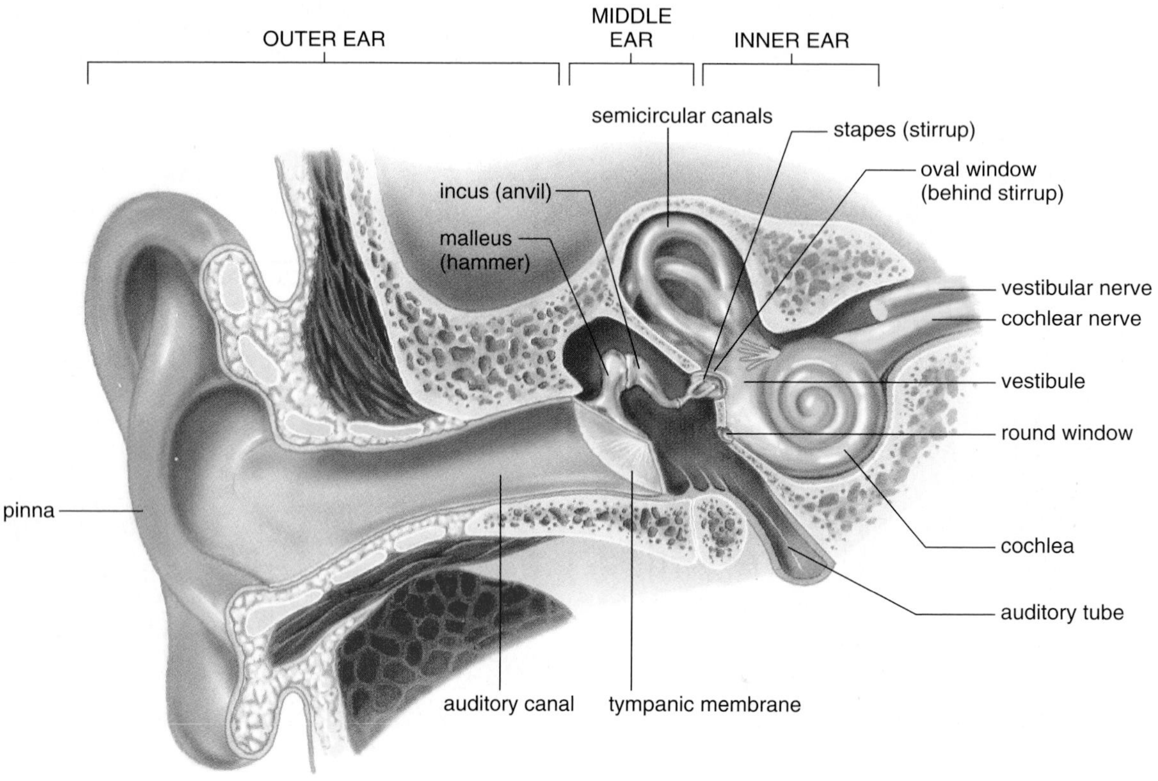

Figure 18.12 Anatomy of the human ear.
In the middle ear, the malleus (hammer), the incus (anvil), and the stapes (stirrup) amplify sound waves. The inner ear contains the sensory receptors for balance in the semicircular canals and the vestibule, and the sensory receptors for hearing in the cochlea.

this way, the pressure is passed to the fluid within the cochlea.

If the cochlea is unwound and examined in cross section (Fig. 18.13) you can see that it has three canals: the vestibular canal, the **cochlear canal,** and the tympanic canal. The vestibular canal connects with the tympanic canal, which leads to the round window membrane. Along the length of the basilar membrane, which forms the lower wall of the cochlear canal, are little hair cells whose stereocilia are embedded within a gelatinous material called the **tectorial membrane.** The hair cells of the cochlear canal, called the **spiral organ** (organ of Corti), synapse with nerve fibers of the **cochlear** (auditory) **nerve.**

When the stapes strikes the membrane of the oval window, pressure waves move from the vestibular canal to the tympanic canal and across the basilar membrane, and the round window bulges. As the basilar membrane moves up and down, the stereocilia of the hair cells embedded in the tectorial membrane bend. Then, nerve impulses begin in the cochlear nerve and travel to the brain stem. When they reach the auditory areas of the cerebral cortex they are interpreted as a sound.

Each part of the spiral organ is sensitive to different wave frequencies, or pitch. Near the tip, the spiral organ responds to low pitches, such as a tuba, and near the base, it responds to higher pitches, such as a bell or a whistle. The nerve fibers from each region along the length of the spiral organ lead to slightly different areas in the brain. The pitch sensation we experience depends upon which region of the basilar membrane vibrates and which area of the brain is stimulated.

Volume is a function of the amplitude of sound waves. Loud noises cause the fluid of the cochlea to vibrate to a greater degree, and this, in turn, causes the basilar membrane to move up and down to a greater extent. The resulting increased stimulation is interpreted by the brain as volume. It is believed that the tone of a sound is an interpretation of the brain based on the distribution of hair cells stimulated.

The sense receptors for sound are hair cells on the basilar membrane (the spiral organ). When the basilar membrane vibrates, the stereocilia of the hair cells bend, and nerve impulses are transmitted to the brain.

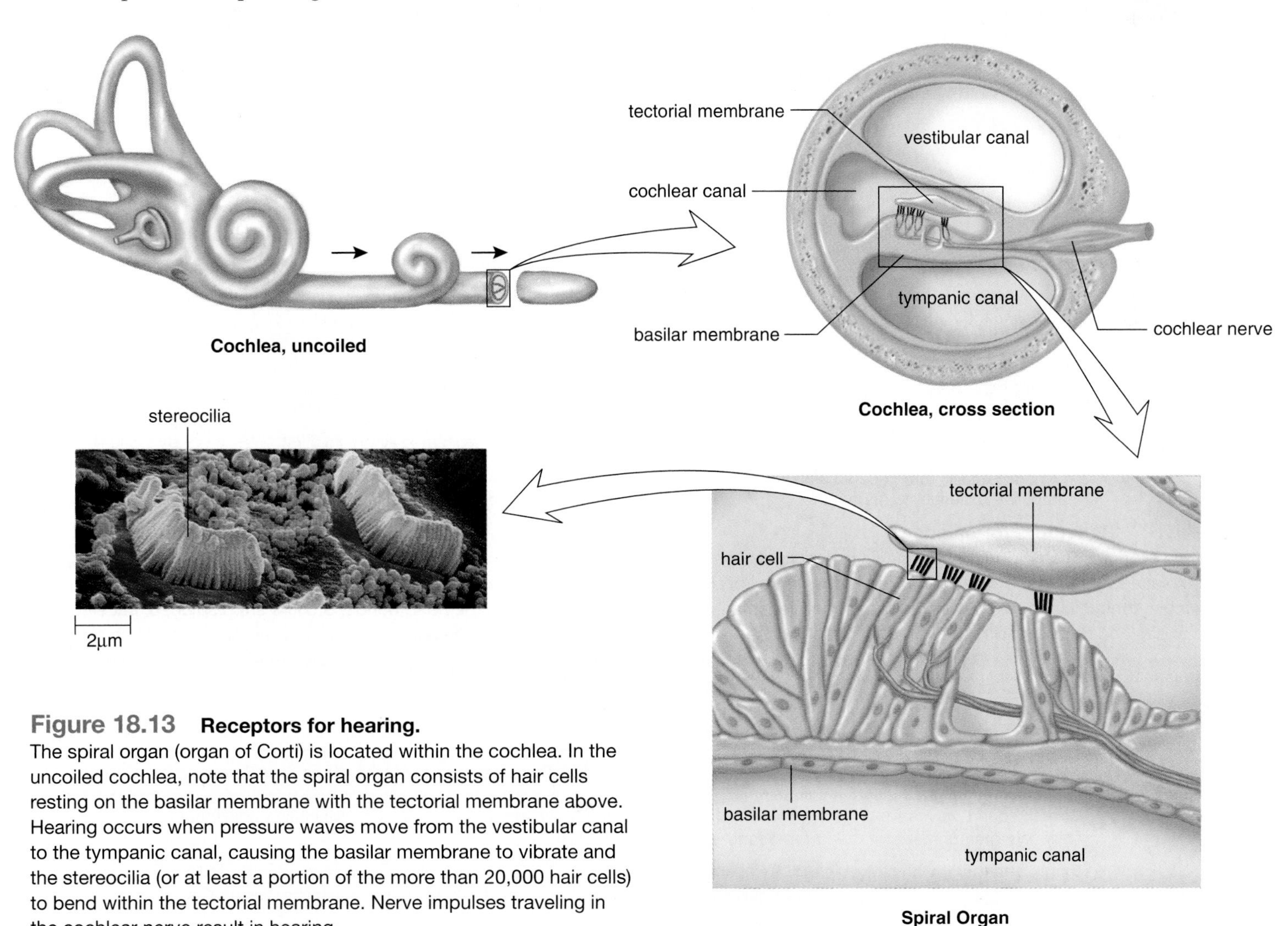

Figure 18.13 Receptors for hearing.
The spiral organ (organ of Corti) is located within the cochlea. In the uncoiled cochlea, note that the spiral organ consists of hair cells resting on the basilar membrane with the tectorial membrane above. Hearing occurs when pressure waves move from the vestibular canal to the tympanic canal, causing the basilar membrane to vibrate and the stereocilia (or at least a portion of the more than 20,000 hair cells) to bend within the tectorial membrane. Nerve impulses traveling in the cochlear nerve result in hearing.

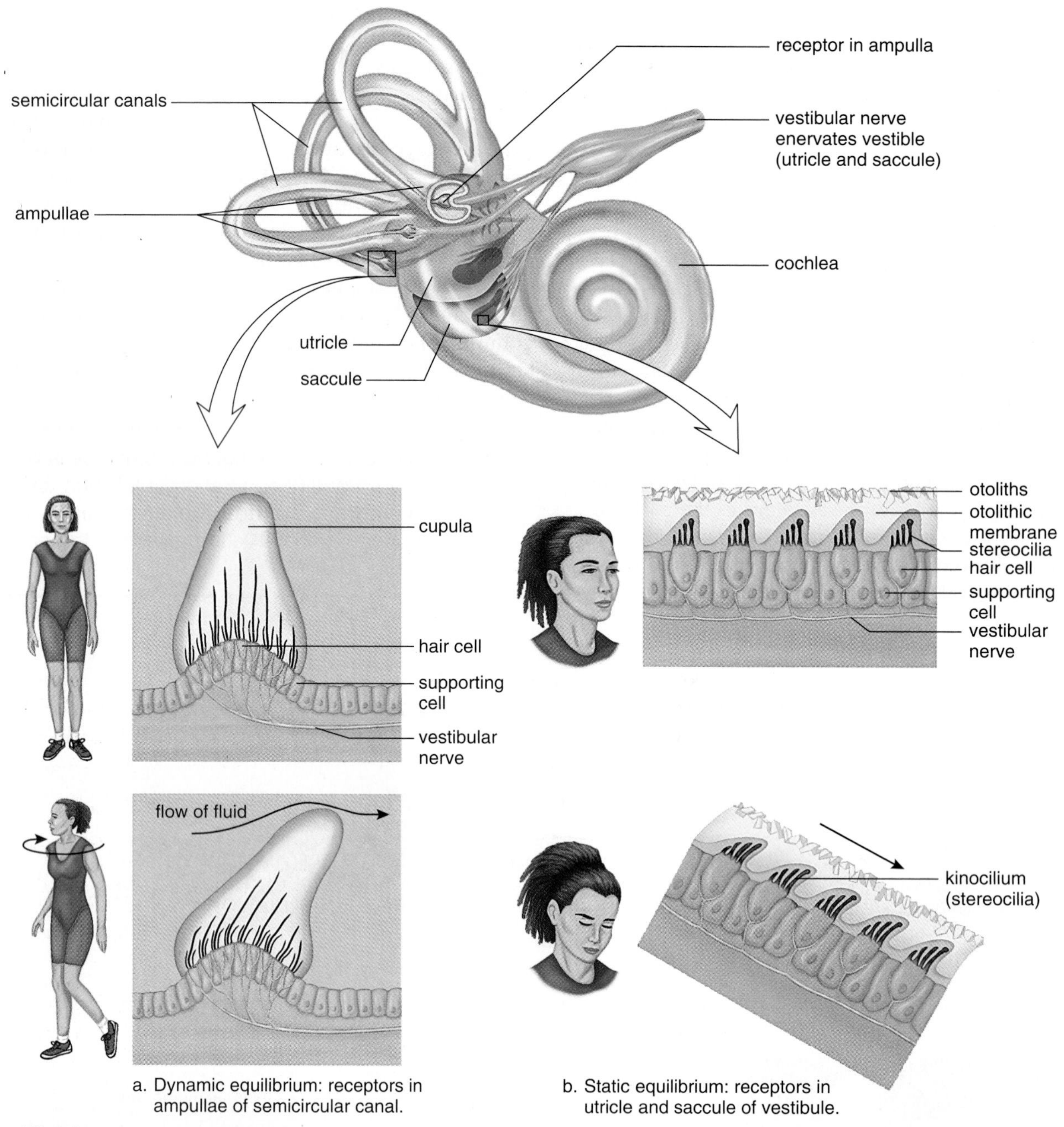

a. Dynamic equilibrium: receptors in ampullae of semicircular canal.

b. Static equilibrium: receptors in utricle and saccule of vestibule.

Figure 18.14 Receptors for balance.
a. Dynamic equilibrium. The ampullae of the semicircular canals contain hair cells with stereocilia embedded in a cupula. When the head rotates, the cupula is displaced, bending the stereocilia. Thereafter, nerve impulses travel in the vestibular nerve to the brain. **b.** Static equilibrium. The utricle and the saccule contain hair cells with stereocilia embedded in an otolithic membrane. When the head bends, otoliths are displaced, causing the membrane to sag and the stereocilia to bend. The rapidity of nerve impulses in the vestibular nerve tells the brain how much the head has moved.

Table 18.3 The Ear

	Outer Ear	Middle Ear	Inner Ear	
			Cochlea	***Sacs and Semicircular Canals***
Function	Directs sound waves to tympanic membranes	Picks up and amplifies sound waves	Hearing	Maintains equilibrium
Anatomy	Pinna; auditory canal	Tympanic membrane; ossicles	Spiral organ (organ of Corti)	Saccule and utricle; semicircular canals
Medium	Air	Air (auditory tube)	Fluid	Fluid

18.6 Sense of Balance

The sense of balance has been subdivided into two senses: **dynamic equilibrium,** involving angular and/or rotational movement of the head, and **static equilibrium,** involving movement of the head in one plane, either vertical or horizontal.

Dynamic Equilibrium

Dynamic equilibrium utilizes the **semicircular canals** which are arranged so that there is one in each dimension of space. The base of each of the three canals, called the **ampulla,** is slightly enlarged. Little hair cells whose stereocilia are embedded within a gelatinous material called a cupula are found within the ampullae. Because there are three semicircular canals, each ampulla responds to head rotation in a different plane of space. As fluid within a semicircular canal flows over and displaces a cupula, the stereocilia of the hair cells bend, and the pattern of impulses carried by the vestibular nerve to the brain changes. Continuous movement of fluid in the semicircular canals causes one form of motion sickness.

Vertigo is dizziness and a sensation of rotation. It is possible to simulate a feeling of vertigo by spinning rapidly and stopping suddenly. When the eyes are rapidly jerked back to a midline position, the person feels like the room is spinning. This shows that the eyes are also involved in our sense of balance.

Static Equilibrium

Static equilibrium depends on the **utricle** and **saccule,** two membranous sacs located in the vestibule. Both of these sacs contain little hair cells, whose stereocilia are embedded within a gelatinous material called an otolithic membrane. Calcium carbonate ($CaCO_3$) granules, or **otoliths,** rest on this membrane. The utricle is especially sensitive to horizontal movements and the bending of the head, while the saccule responds best to vertical (up-down) movements. When the body is still, the otoliths in the utricle and the saccule rest on the otolithic membrane above the hair cells (Fig. 18.14*b*). When the head bends or the body moves in the horizontal and vertical planes, the otoliths are displaced and the otolithic membrane sags, bending the stereocilia of the hair cells beneath. If the stereocilia move toward the kinocilium, the largest stereocilium, nerve impulses in the vestibular nerve increase. If the stereocilia move away from the kinocilium, nerve impulses in the vestibular nerve decrease. These data tell the brain the direction of the movement of the head.

Movement of a cupula within the semicircular canals contributes to the sense of dynamic equilibrium. Movement of the otolithic membrane within the utricle and the saccule accounts for static equilibrium.

Bioethical Issue

A cataract is a cloudiness of the lens that occurs in 50% of people between the ages of 65 and 74, and in 70% of those age 75 or older. The extent of visual impairment depends on the size, density of the cataract, and where it is located in the lens. A dense centrally placed cataract causes severe blurring of vision.

Are cataracts preventable? For most people the answer is yes, they are preventable. These factors have been identified as contributing to the chances of having a cataract:

- Smoking 20 or more cigarettes a day doubles the risk of cataracts in men—women have to smoke more than 30 cigarettes a day to increase their chances of a cataract.
- Exposure to the ultraviolet radiation in sunlight can more than double the risk of a cataract. In addition, this relationship is dose-dependent—the more sunlight, the higher the risk.
- Also, as many as one-third of cataracts may be caused by being overweight. Diet, rather than exercise, seems to reduce cataract formation, perhaps through lower blood sugar levels or improved antioxidant properties of the blood.

It's clear, then, that our own behavior contributes to the occurrence of cataracts for most of us. Should we be responsible in our actions and take all possible steps to prevent developing a cataract, such as not smoking, wearing sunglasses and a wide-brim hat, and watching our weight? Or should we simply rely on medical science to restore our eyesight? Most cataract operations today are performed on an outpatient basis with minimal postoperative discomfort and with a high expectation of restoration of sight. Perhaps it's better, though, to take all possible steps to prevent the occurrence of cataracts, just in case our experience is atypical?

Questions

1. To what extent do you feel each person is responsible for his or her own health? Explain your answer.
2. Should Medicare pay for cataract surgery in heavy smokers?
3. At what age should we make people aware that their behavior today can affect their health tomorrow?

Summarizing the Concepts

18.1 Sensory Receptors and Sensations

Each type of sensory receptor responds to a particular kind of stimulus. When stimulation occurs, sensory receptors initiate nerve impulses that are transmitted to the spinal cord and/or brain. Sensation occurs when nerve impulses reach the cerebral cortex. Perception is an interpretation of the meaning of sensations.

The senses are divided into the somatic (general) senses and the special senses (taste, smell, vision, hearing, balance).

18.2 Somatic Senses

Proprioception is illustrated by the action of muscle spindles which are stimulated when muscle fibers stretch. A reflex action, which is illustrated by the knee-reflex, causes the muscle fibers to contract. Proprioception helps maintain balance and posture.

The skin contains sensory receptors for touch, pressure, pain, and temperature (warmth and cold). The pain of internal organs is sometimes felt in the skin and is called referred pain.

18.3 Chemical Senses

Taste and smell are due to chemoreceptors that are stimulated by molecules in the environment. The taste buds contain taste cells that communicate with sensory nerve fibers, while the receptors for smell are neurons.

After molecules bind to plasma membrane receptor proteins on the microvilli of taste cells and the cilia of olfactory cells, nerve impulses eventually reach the cerebral cortex, which determines the taste and odor according to the pattern of receptors stimulated.

18.4 Sense of Vision

Vision is dependent on the eye, the optic nerves, and the visual areas of the cerebral cortex. The eye has three layers. The outer layer, the sclera, can be seen as the white of the eye; it also becomes the transparent bulge in the front of the eye called the cornea. The middle pigmented layer, called the choroid, absorbs stray light rays. The rod cells (sensory receptors for dim light, and the cone cells (sensory receptors for bright light and color) are located in the retina, the inner layer of the eyeball. The cornea, the humors, and especially the lens bring the light rays to focus on the retina. To see a close object, accommodation occurs as the lens rounds up.

When light strikes rhodopsin within the membranous disks of rod cells, rhodopsin splits into opsin and retinal. A cascade of reactions leads to the closing of ion channels in a rod cell's plasma membrane. Inhibitory transmitter molecules are no longer released and nerve impulses are carried in the optic nerve to the brain.

Integration occurs in the retina which is composed of three layers of cells: the rod and cone layer, the bipolar cell layer, and the ganglion cell layer. Integration also occurs in the brain, especially because the visual field is first taken apart by the optic chiasma and the primary visual area in the cerebral cortex parcels out signals for color, form, and motion to the visual association area.

18.5 Sense of Hearing

Hearing is a specialized sense dependent on the ear, the cochlear nerve, and the auditory areas of the cerebral cortex.

The ear is divided into three parts: outer, middle, and inner. The outer ear consists of the pinna and the auditory canal, which direct sound waves to the middle ear. The middle ear begins with the tympanic membrane and contains the ossicles (malleus, incus, stapes). The malleus is attached to the tympanic membrane, and the stapes is attached to the oval window, which is covered by membrane. The inner ear contains the cochlea and the semicircular canals, plus the utricle and saccule.

Hearing begins when the outer and middle portions of the ear convey and amplify the sound waves that strike the oval window. Its vibrations set up pressure waves within the cochlea, which contains the spiral organ, consisting of hair cells whose stereocilia are embedded within the tectorial membrane. When the stereocilia of the hair cells bend, nerve impulses begin in the cochlear nerve and are carried to the brain.

18.6 Sense of Balance

The ear also contains receptors for our sense of balance. Dynamic equilibrium is dependent on the stimulation of hair cells within the ampullae of the semicircular canals. Static equilibrium relies on the stimulation of hair cells within the utricle and the saccule.

Studying the Concepts

1. What are the four types of receptors in the human body? 348
2. Explain sensation from the reception of stimuli to the passage of nerve impulses to the brain. 348
3. What are the somatic senses? Explain how muscle spindles are involved in a reflex action. What are the cutaneous senses? 349–50
4. Describe the structure of a taste bud, and tell how a taste cell functions. 352
5. Describe the structure and function of the olfactory epithelium. How does the sense of smell come about? 353
6. Describe the anatomy of the eye, and explain focusing and accommodation. 354–55
7. Describe the structure and function of rod cells and cone cells. 356
8. Explain the process of integration in the retina and the brain. 357–58
9. Relate the need for corrective lenses to three possible eye shapes. 359
10. Describe the anatomy of the ear and how we hear. 362–63
11. Describe the role of the semicircular canals, the utricle, and the saccule in balance. 365

Testing Yourself

Choose the best answer for each question.

1. A receptor
   a. is the first portion of a reflex arc.
   b. initiates nerve impulses.
   c. can be internal or external.
   d. All of these are correct.

2. Which of these is not a proper contrast between olfactory receptors and equilibrium receptors?
   olfactory receptors—equilibrium receptors
   a. located in nasal cavities—located in the inner ear
   b. chemoreceptor—mechanoreceptor
   c. respond to molecules in air—respond to movements of the body
   d. communicate with brain via a tract—communicate with brain via vestibular nerve
   e. All of these contrasts are correct.
3. Which of these is not a proper contrast between proprioceptors and cutaneous receptors?
   proprioceptors—cutaneous receptors
   a. located in muscles and tendons—located in the skin
   b. mechanoreceptor—chemoreceptors
   c. respond to tension—respond to pain, hot, cold, touch, pressure
   d. type of somatic sense—a type of special sense
   e. Both b and d are correct.
4. Which of these gives the correct path for light rays entering the human eye?
   a. sclera, retina, choroid, lens, cornea
   b. fovea centralis, pupil, aqueous humor, lens
   c. cornea, pupil, lens, vitreous humor, retina
   d. cornea, fovea centralis, lens, choroid, rods
   e. optic nerve, sclera, choroid, retina, humors
5. Which gives an incorrect function for the structure?
   a. lens—focusing
   b. cones—color vision
   c. iris—regulation of amount of light
   d. choroid—location of cones
   e. sclera—protection
6. Which one of these wouldn't you mention if you were tracing the path of sound vibrations?
   a. auditory canal
   b. tympanic membrane
   c. ossicles
   d. semicircular canals
   e. cochlea
7. Which one of these correctly describes the location of the spiral organ?
   a. between the tympanic membrane and the oval window in the inner ear
   b. in the utricle and saccule within the vestibule
   c. between the tectorial membrane and the basilar membrane in the cochlear canal
   d. between the nasal cavities and the throat
   e. between the outer and inner ear within the semicircular canals
8. Which of these is mismatched?
   a. semicircular canals—inner ear
   b. utricle and saccule—outer ear
   c. auditory canal—outer ear
   d. cochlea—inner ear
   e. ossicles—middle ear
9. Retinal is
   a. a derivative of vitamin A.
   b. sensitive to light energy.
   c. a part of rhodopsin.
   d. found in both rods and cones.
   e. All of these are correct.
10. Both olfactory receptors and sound receptors have cilia, and they both
    a. are chemoreceptors.
    b. are a part of the brain.
    c. are mechanoreceptors.
    d. initiate nerve impulses.
    e. All of these are correct.
11. Label this diagram of an eye. State a function for each structure labeled.

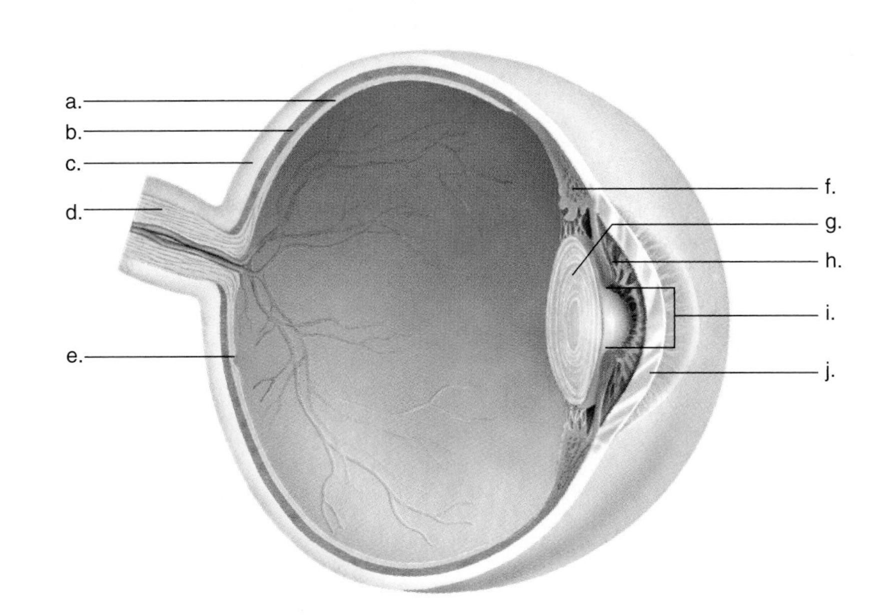

Thinking Scientifically

1. Considering photoreceptors, (page 355):
   a. Devise a categorization for the part of the eye (Table 18.2). Justify your system of categorization. When a person needs glasses, which of your categories is at fault?
   b. Phytochrome and chlorophyll (plant pigments) and retinal in our eyes are all pigments. What features do they have in common?
   c. The pineal gland, located on the edge of the third ventricle of the brain, is sensitive to the length of the day and the night. Hypothesize how the pineal gland detects when it is day or night.
2. Considering the human ear (page 362):
   a. Fishes do not have ears, but they do have a lateral line that contains mechanoreceptors that are sensitive to fluid pressure waves. What fossil evidence is needed to support the hypothesis that the lateral line of fishes evolved into the human ear?
   b. What part of the human ear (outer, middle, inner) would you expect to have evolved from the lateral line? What parts of the human ear would you expect to have evolved from other structures or to have been "added on," considering that humans are terrestrial? Why?

Understanding the Terms

ampulla 365
aqueous humor 354
astigmatism 359
auditory tube 362
blind spot 357
cataracts 355
chemoreceptor 348
choroid 354
ciliary body 354
cochlea 362
cochlear canal 363
cochlear nerve 363
color vision 356
cone cell 356
cornea 354
dynamic equilibrium 365
focus 355
fovea centralis 355
glaucoma 355
hair cell 362
incus 362
inner ear 362
integration 349
iris 354
lens 354
malleus 362
mechanoreceptor 348
middle ear 362
olfactory cell 353
optic nerve 355
ossicle 362
otolith 365
outer ear 362
oval window 362
pain receptor 348
perception 348
photoreceptor 348
proprioceptor 350
pupil 354
referred pain 351
retina 355
rhodopsin 356
rod cell 356
round window 362
saccule 365
sclera 354
semicircular canal 365
sensation 348
sensory adaptation 349
sensory receptor 348
somatic sense 349
special sense 349
spiral organ 363
stapes 362
static equilibrium 365
stimulus 348
taste bud 352
tectorial membrane 363
thermoreceptor 348
tympanic membrane 362
utricle 365
vertigo 365
vestibule 362
visual accommodation 355
visual field 358
vitreous humor 355

Match the terms to these definitions:

a. ________ Portion of the inner ear that resembles a snail's shell and contains the spiral organ, the sense organ for hearing.
b. ________ Sensory receptor that responds to chemical stimulation—for example, receptors for taste and smell.
c. ________ Visual pigment found in the rods whose activation by light energy leads to vision.
d. ________ An awareness of a stimulus.; occurs only in cerebral cortex.
e. ________ A mechanoreceptor that gives rise to nerve impulses when its stereocilia are bent or tilted.

Using Technology

Your study of the senses is supported by these available technologies:

Essential Study Partner CD-ROM
Animals → Sense Organs

Visit the Mader web site for related ESP activities.

Exploring the Internet
The Mader Home Page provides resources and tools as you study this chapter.

http://www.mhhe.com/biosci/genbio/mader

Dynamic Human 2.0 CD-ROM
Nervous System

C H A P T E R 19

Musculoskeletal System

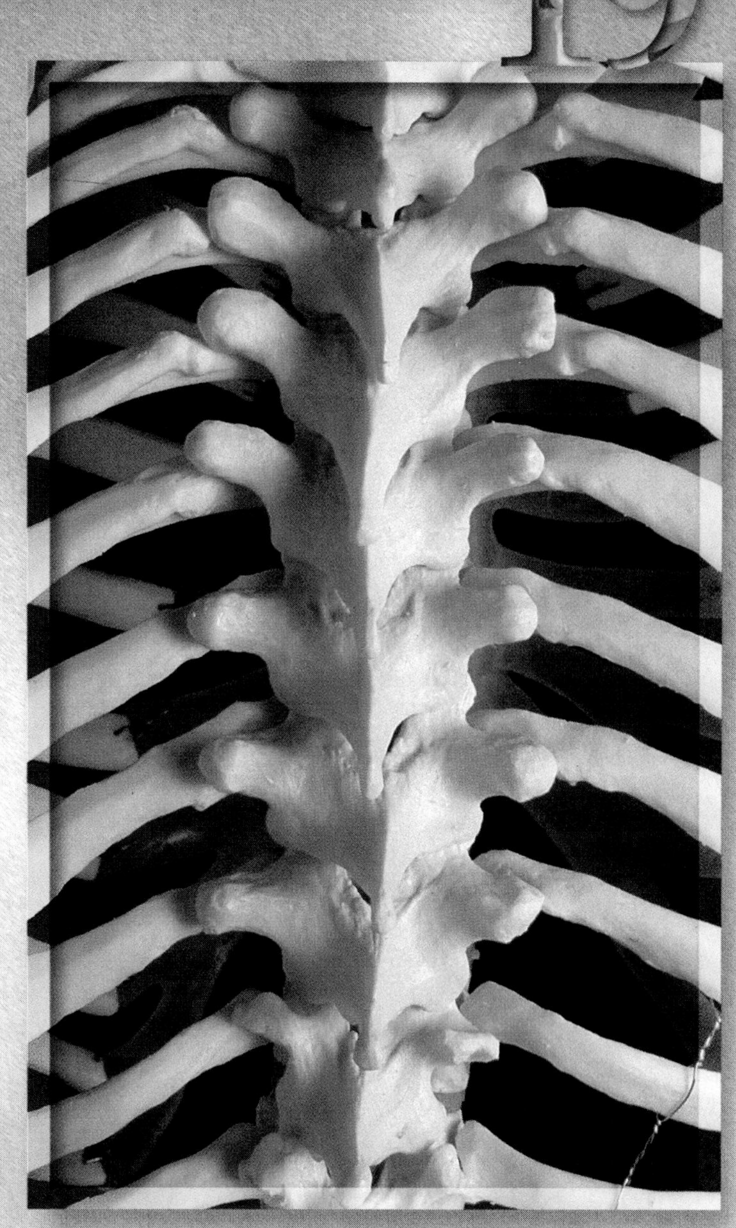

Humans have a strong but flexible skeleton. The vertebral column supports the head and trunk, yet it allows the body to bend forward, backward, and to the side.

Chapter Concepts

Two . . . Three . . . Four . . . the aerobics instructor yelled as the music blared. Mary, along with 14 others, dutifully stretched, bent, and squatted, despite her sore muscles from other recent workouts. She had decided to continue working out because she knew that exercise would benefit her heart and lungs and tone up her muscles.

The faster heartbeat and breathing rate Mary experiences when she works out ensures that oxygen will enter her body and be delivered to working muscle fibers. Glucose, absorbed by the digestive tract, is also needed by muscle fibers in order to produce ATP within mitochondria. When a muscle fiber contracts, ATP is broken down as one type of protein filament slides past another type. Mary's class is called aerobics because the pace is set to allow oxygen to be delivered to muscle fibers at a rate that prevents anaerobic ATP production.

Muscle contraction causes bones to move, and these two types of organs work together to cause locomotion, whether we are in an aerobics class or just going about our daily routine. This text, recognizing the close connection between the skeletal system and the muscular system, has joined the two systems under the common title of musculoskeletal system.

19.1 Anatomy and Physiology of Bones

The organs of the skeleton are largely composed of connective tissues. Connective tissue contains cells separated by a matrix that contains fibers.

Structure of Bone

Bones are strong because their matrix contains mineral salts, notably calcium phosphate. **Compact bone** is highly organized and composed of tubular units called **osteons.** In cross section of an osteon, bone cells called **osteocytes** lie in lacunae, which are tiny chambers arranged in concentric circles around a central canal (Fig. 19.1). The mineralized matrix, which also contains protein fibers, fills the spaces between the lacunae. Tiny canals called canaliculi run through the matrix, connecting the lacunae with each other and with the central canal. Central canals contain blood vessels, lymphatic vessels, and nerves. Canaliculi bring nutrients from the blood vessel in the central canal to the cells in the lacunae.

Compared to compact bone, **spongy bone** has an unorganized appearance (Fig. 19.1). Osteocytes are found in numerous thin plates separated by unequal spaces. Although the latter make spongy bone lighter than compact bone, spongy bone is still designed for strength. Just as braces are used for support in buildings, the plates of spongy bone follow lines of stress. The spaces of spongy bone are often filled with **red bone marrow,** a specialized tissue that produces all types of blood cells.

Tissues Associated With Bones

Cartilage and fibrous connective tissue are two types of tissues associated with bones.

Cartilage

Cartilage is not as strong as bone, but it is more flexible because the matrix is gel-like and contains many collagenous and elastic fibers. The cells lie within lacunae that are irregularly grouped. Cartilage has no blood vessels, and therefore injured cartilage is slow to heal. Hyaline cartilage is firm and somewhat flexible. The matrix appears uniform and glassy, but actually it contains a generous supply of collagenous fibers. Hyaline cartilage is found at the ends of long bones and in the nose, at the ends of the ribs, and in the larynx and trachea.

Fibrocartilage is stronger than hyaline cartilage because the matrix contains wide rows of thick collagenous fibers. Fibrocartilage is able to withstand both tension and pressure, and this type of cartilage is found where support is of prime importance—in the disks located between the vertebrae and also in the cartilage of the knee.

Elastic cartilage is more flexible than hyaline cartilage because the matrix contains mostly elastin fibers. This type of cartilage is found in the ear flaps and epiglottis.

Dense Fibrous Connective Tissue

Fibrous connective tissue contains rows of cells called fibroblasts separated by bundles of collagenous fibers. Fibrous connective tissue makes up the **ligaments** that connect bone to bone and the **tendons** that connect muscles to a bone at **joints,** which are also called articulations.

Structure of a Long Bone

The ends of a long bone are expanded regions composed largely of spongy bone. The ends of a long bone are coated with a thin layer of hyaline cartilage, which is called **articular cartilage** because it occurs at a joint.

The shaft, or main portion of the bone, has a large medullary cavity that is filled with fatty **yellow bone marrow** in adults. The walls of the medullary cavity are composed of compact bone.

Except for the articular cartilage on its ends, a long bone is completely covered by a layer of fibrous connective tissue called the **periosteum.** This covering contains blood vessels, lymphatic vessels, and nerves. Note in Figure 19.1 how a blood vessel penetrates the periosteum and the bone where it gives off branches within the central canals. The periosteum is continuous with ligaments and tendons.

Various types of connective tissue including bone, cartilage, and fibrous connective tissue make up the skeleton. A long bone contains all these tissues.

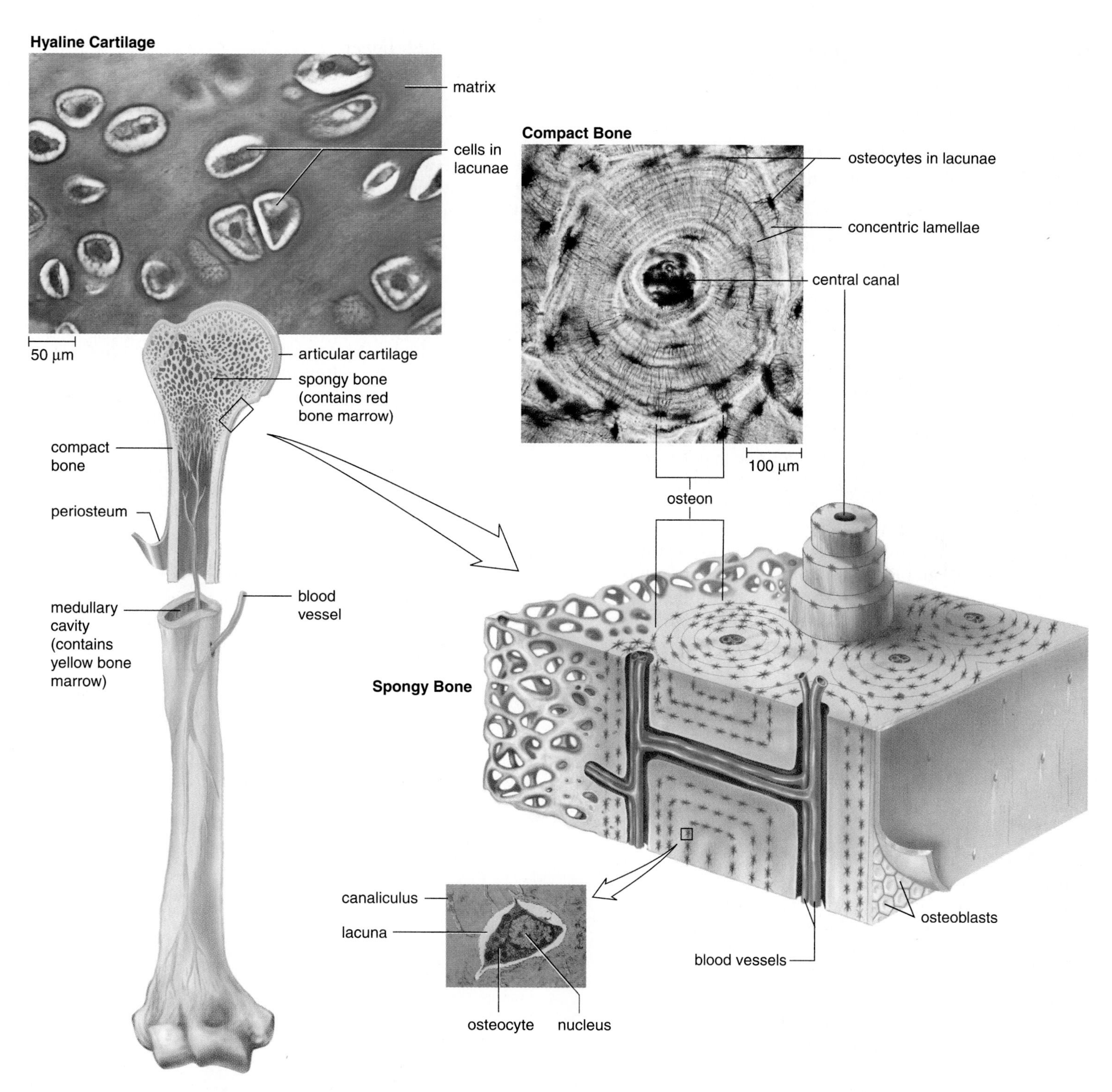

Figure 19.1 Anatomy of a long bone.
A long bone is encased by the periosteum (fibrous membrane) except where it is covered by hyaline (articular) cartilage (see micrograph). Spongy bone located at each end may contain red bone marrow. The central shaft contains yellow bone marrow and is bordered by compact bone, which is shown in the enlargement and micrograph.

Bone Growth and Repair

Bones are composed of living tissues as exemplified by their ability to grow and undergo remodeling.

Bone Development and Growth

The bones of the human skeleton, except those of the skull, first appear during embryonic development as hyaline cartilage models. The replacement of cartilaginous models by bone is called endochondral ossification (Fig. 19.2).

During endochondral ossification, the cartilage begins to break down in the center of a long bone, which is now covered by a periosteum. Osteoblasts invade the region and begin to lay down spongy bone in what is called a primary ossification center. Other osteoblasts lay down compact bone beneath the periosteum. As the compact bone thickens, the spongy bone is broken down by osteoclasts, and the cavity created becomes the medullary cavity.

The ends of developing bone continue to grow, but soon secondary ossification centers appear in these regions. Here spongy bone forms and does not break down. Also, a band of cartilage called a **growth plate** remains between the primary ossification center and each secondary center. The limbs keep increasing in length as long as the growth plates are still present. The rate of growth is controlled by hormones, such as growth hormones and the sex hormones. Eventually the growth plates become ossified, and the bone stops growing.

Remodeling of Bones

In the adult, bone is continually being broken down and built up again. **Osteoclasts,** which are derived from monocytes in red bone marrow, break down bone, remove worn cells, and deposit calcium in the blood. After a period of about three weeks, the osteoclasts disappear, and the bone is repaired by the work of osteoblasts. As they form new bone, osteoblasts take calcium from the blood. Eventually some of these cells get caught in the matrix they secrete and are converted to osteocytes, the cells found within the lacunae of osteons.

Because of continual remodeling, the thickness of bones can change. Physical use and hormone balance affect the thickness of bones. Strange as it may seem, adults apparently require more calcium in the diet (about 1,000 to 1,500 mg daily) than do children in order to promote the work of osteoblasts. Otherwise, osteoporosis, a condition in which weak and thin bones easily fracture, may develop. The likelihood of osteoporosis is greater in older women due to a reduction in estrogen levels after menopause. Although it is not known how estrogen acts on bone maintenance, it seems to play a role in calcium metabolism.

Bone is living tissue. It develops, grows, and remodels itself. In all these processes, osteoclasts break down bone, and osteoblasts build bone.

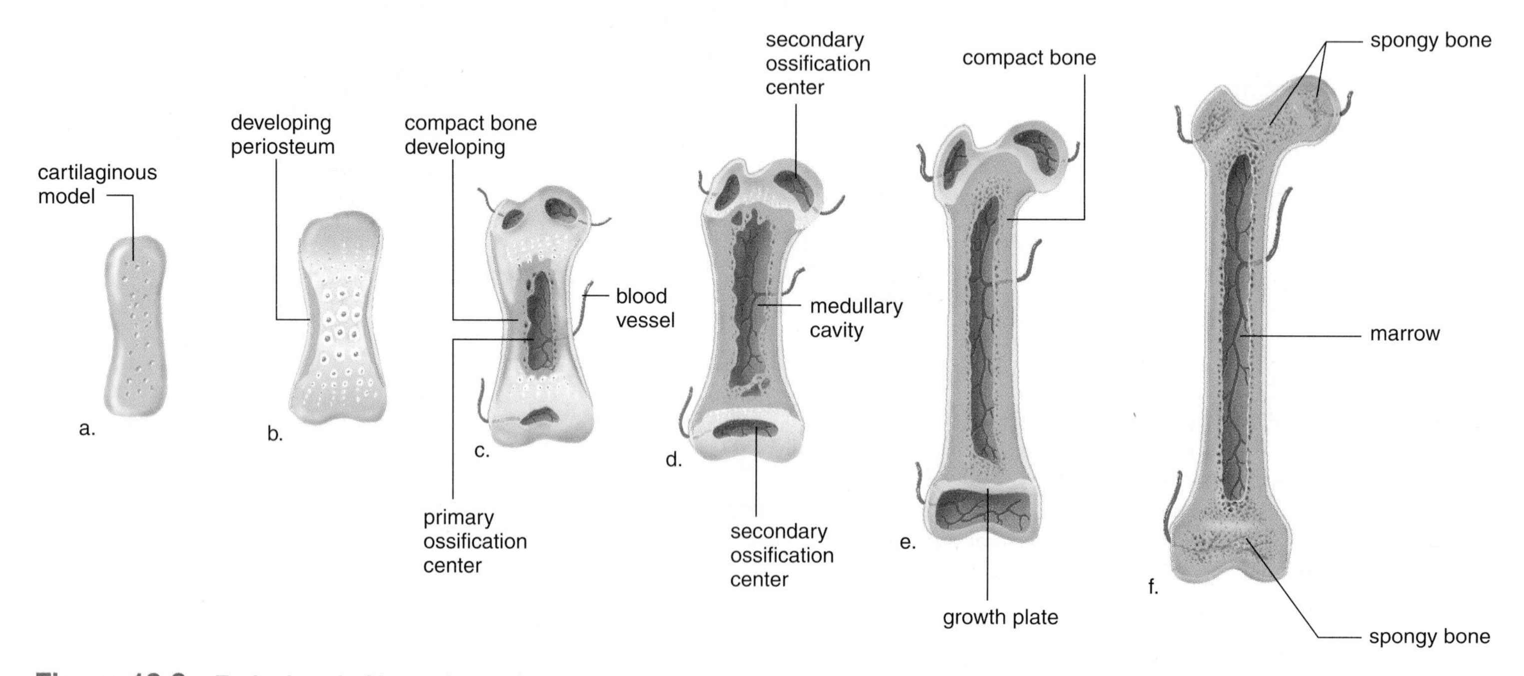

Figure 19.2 Endochondral bone formation.
a. A cartilaginous model develops during fetal development. **b.** A periosteum develops. **c.** A primary ossification center contains spongy bone surrounded by compact bone. **d.** Medullary cavity forms in the shaft, and secondary ossification centers develop at the ends of a long bone. **e.** Growth is still possible as long as cartilage remains at the growth plate. **f.** When the bone is fully formed, the growth plate disappears.

19.2 Bones of the Skeleton

Let's discuss the functions of the skeleton in relation to particular bones.

The skeleton supports the body. The bones of the legs (the femur in particular and also the tibia) support the entire body when we are standing, and the coxal bones of the pelvic girdle support the abdominal cavity.

The skeleton protects soft body parts. The bones of the skull protect the brain; and the rib cage, composed of the ribs, thoracic vertebrae, and sternum, protects the heart and lungs.

The skeleton produces blood cells. All bones in the fetus have spongy bone with red bone marrow that produces blood cells. In the adult, the flat bones of the skull, ribs, sternum, clavicles, and also the vertebrae and pelvis produce blood cells.

The skeleton stores minerals and fat. All bones have a matrix that contains calcium phosphate. When bones are remodeled, osteoclasts break down bone and return calcium ions and phosphorus ions to the bloodstream. Fat is stored in the yellow bone marrow.

The skeleton permits flexible body movement. While articulations (joints) occur between all the bones, we can associate body movement in particular with the bones of the legs (especially the femur and tibia) and the feet (tarsals, metatarsals, and phalanges) because we use them when walking.

Classification of the Bones

The bones are sometimes classified according to their shape. Long bones, exemplified by the humerus and femur, are longer than they are wide. Short bones, such as the carpals and tarsals, are cube shaped—their lengths and widths are about equal. Flat bones, like those of the skull, are platelike with broad surfaces. Round bones, exemplified by the patella, are circular in shape. Irregular bones, such as the vertebrae and facial bones, have varied shapes that permit connections with other bones.

Bones are also classified according to whether they occur in the axial skeleton or the appendicular skeleton. The axial skeleton is in the midline of the body, and the appendicular skeleton is the limbs along with their girdles (Fig. 19.3).

The bones of the skeleton are not smooth; they have articulating depressions and protuberances at various joints. And they have projections, often called processes, where the muscles attach. Also, there are openings for nerves and/or blood vessels.

The skeleton supports and protects internal organs; it produces blood cells and stores minerals and fat in addition to permitting flexible body movement. The skeleton is divided into the axial and appendicular skeleton.

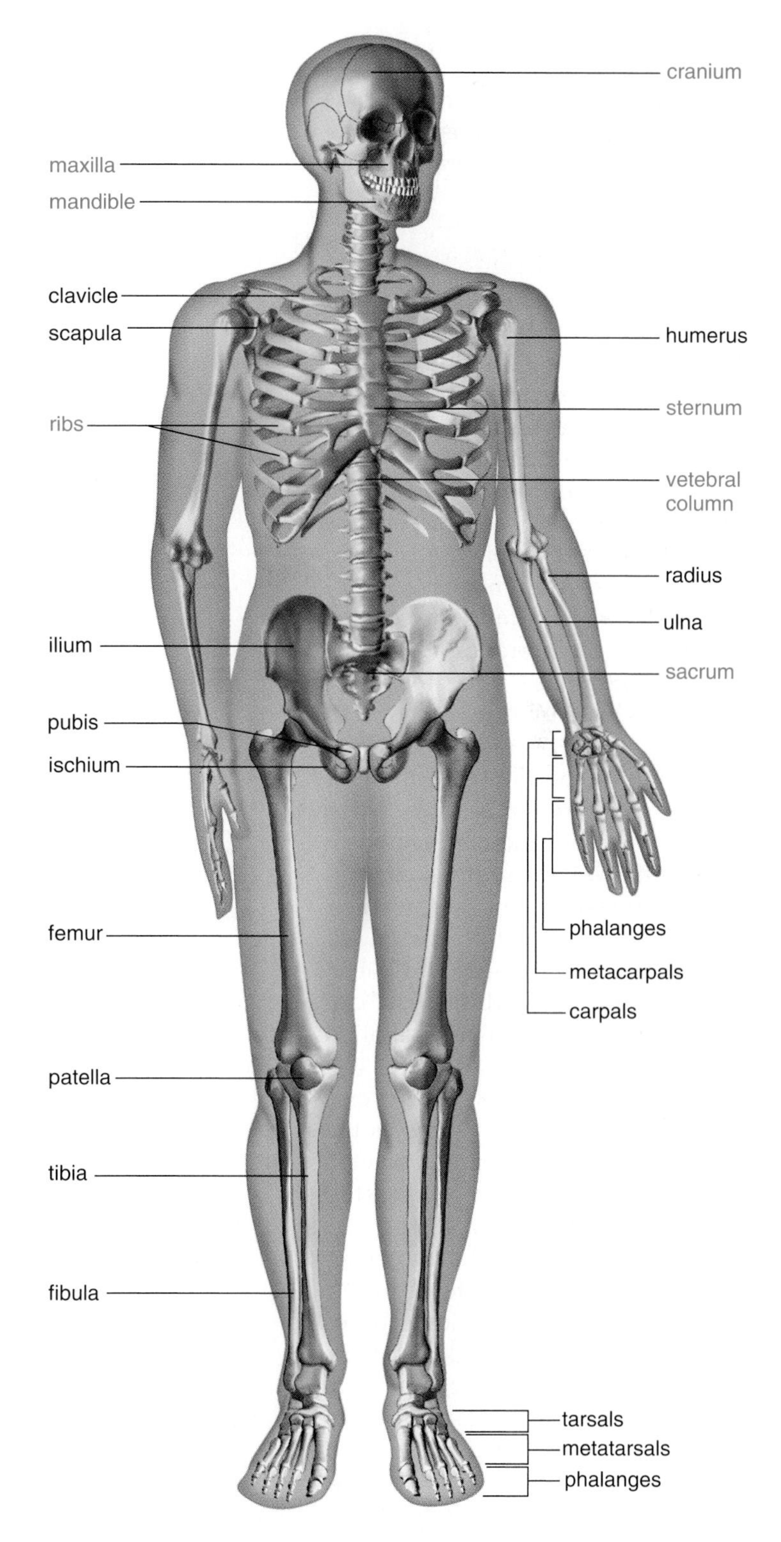

Figure 19.3 The skeleton.
The skeleton of a human adult contains bones that belong to the axial skeleton (red labels) and those that belong to the appendicular skeleton (black labels).

The Axial Skeleton

The **axial skeleton** lies in the midline of the body and consists of the skull, hyoid bone, vertebral column, and rib cage.

The Skull

The **skull** is formed by the brain case, called the cranium, and the facial bones. It can be noted, however, that the frontal bone is a part of the cranium even though it forms the forehead of the face.

The **cranium** protects the brain and is composed of eight flat bones fitted tightly together in adults. In newborns, certain bones are not completely formed and instead are joined by membranous regions called **fontanels.** The fontanels usually close by the age of 16 months.

Some of the bones of the cranium contain the **sinuses,** air spaces lined by mucous membrane, which reduce the weight of the skull and give a resonant sound to the voice. Two sinuses called the mastoid sinuses drain into the middle ear. **Mastoiditis,** a condition that can lead to deafness, is an inflammation of these sinuses.

The major bones of the cranium have the same names as the lobes of the brain: frontal, parietal, occipital, and temporal. On the top of the cranium (Fig. 19.4*a*), the **frontal bone** forms the forehead, the **parietal bones** extend to the sides, and the occipital bone curves to form the base of the skull. Here there is a large opening, the **foramen magnum** (Fig. 19.4*b*), through which the spinal cord passes and becomes the brain stem. Below the much larger parietal bones, each **temporal bone** has an opening (external auditory canal) that leads to the middle ear.

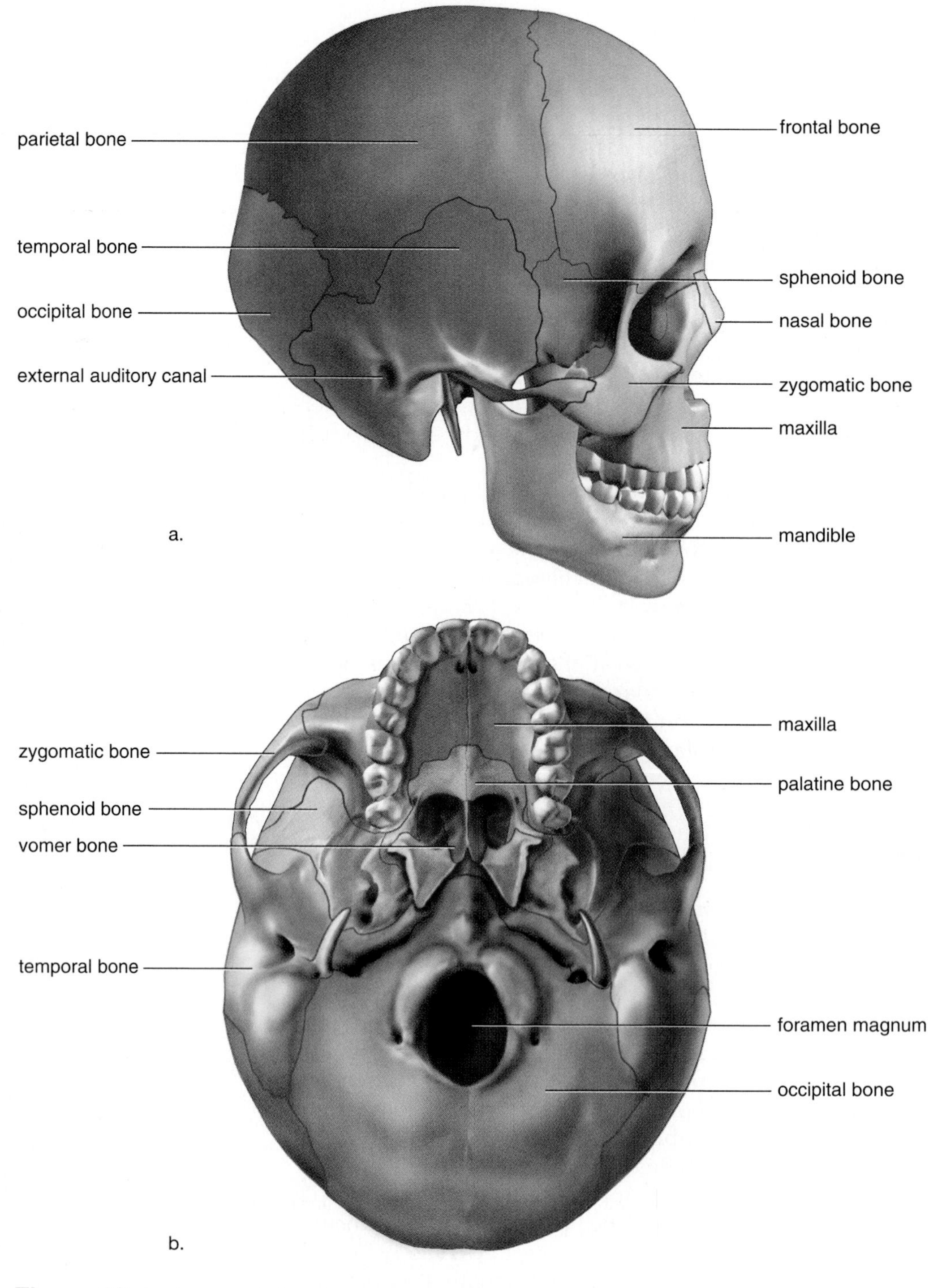

Figure 19.4 Bones of the skull.
a. Lateral view. **b.** Inferior view.

The **sphenoid bone,** which is shaped like a bat with wings outstretched, extends across the floor of the cranium from one side to the other. The sphenoid is considered to be the keystone bone of the cranium because all the other bones articulate with it. The sphenoid completes the sides of the skull and also contributes to forming of the eye sockets. Other bones, which lie in front of the sphenoid, complete the eye socket. Eye sockets are sometimes called orbits because the eyes can rotate.

The cranium contains eight bones: the frontal, two parietal, the occipital, two temporal, the sphenoid, and the ethmoid.

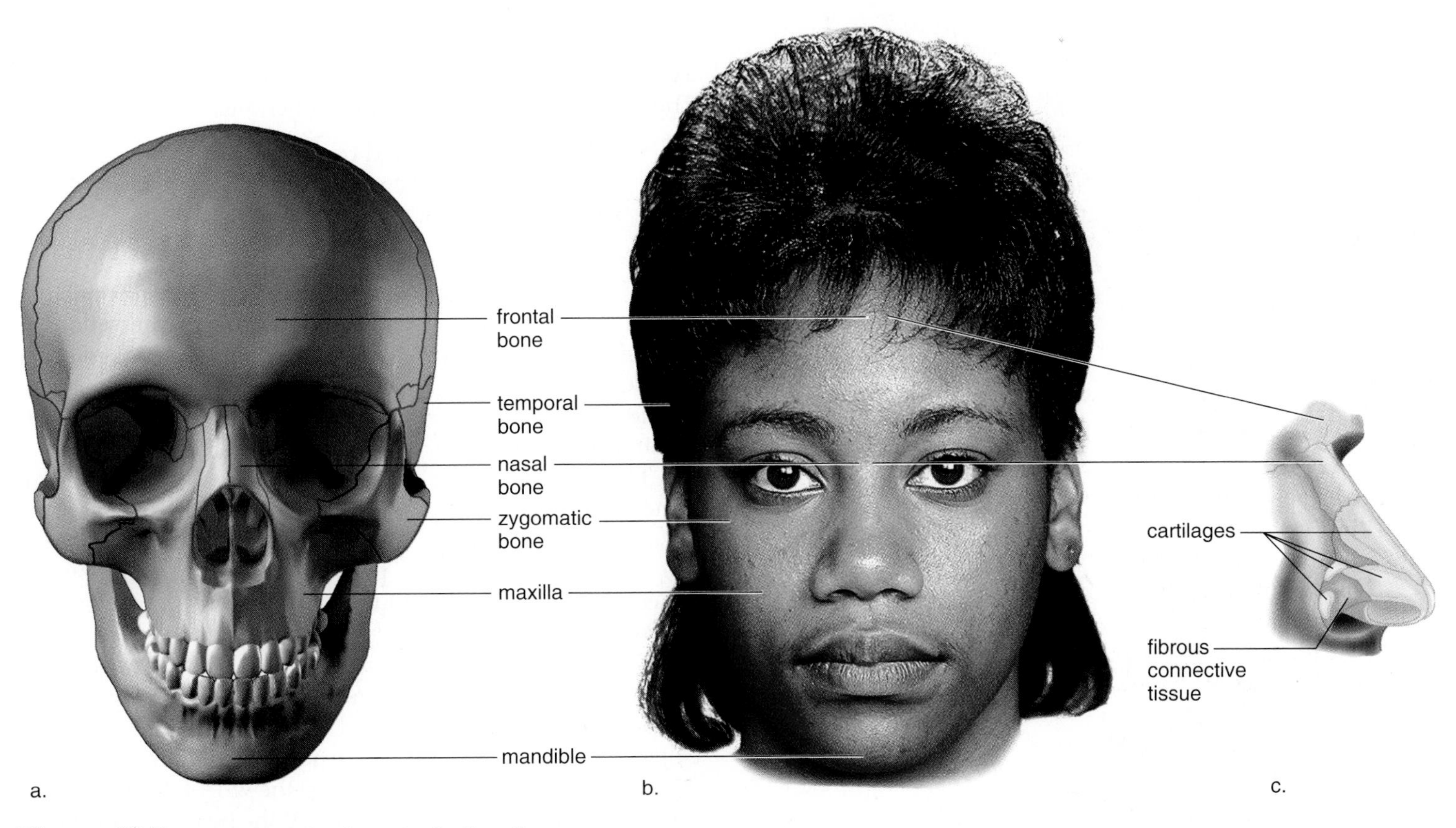

Figure 19.5 Bones of the face, including the nose.
a. The frontal bone forms the forehead and eyebrow ridges; the zygomatic bones form the cheekbones, and maxillae form the upper jaw. The maxillae are the most expansive facial bones, extending from the forehead to the lower jaw. The mandible has a projection we call the chin. **b.** The maxillae, frontal, and nasal bones help form the external nose. **c.** The rest of the nose is formed by cartilages and fibrous connective tissue.

The Facial Bones

The most prominent of the facial bones are the mandible, maxillae (maxillary bones), the zygomatic bones, and the nasal bones.

The **mandible,** or lower jaw, is the only movable portion of the skull (Fig. 19.5*a*), and its action permits us to chew our food. It also forms the "chin." Tooth sockets are located on the mandible and on the **maxillae,** the upper jaw that also forms the front portion of the hard palate. The back portion of the hard palate and the floor of the nasal cavity is formed by the palatine bones (see Fig. 19.4*b*).

The lips and cheeks have a core of skeletal muscle. The **zygomatic bones** are the cheekbone prominences, and the **nasal bones** form the bridge of the nose. Other bones make up the nasal septum which divides the nose cavity into two regions.

The temporal and frontal bones are cranial bones that contribute to the face. The temporal bones account for the flattened areas on each side of the forehead, which we call the temples. The frontal bone forms the forehead and has supraorbital ridges where the eyebrows are located. Glasses sit where the frontal bone joins the nasal bones.

While the ears are formed only by elastic cartilage and not by bone, the nose (Fig. 19.5*c*) is a mixture of bones and cartilages and fibrous connective tissue. The cartilages complete the tip of the nose, and fibrous connective tissue forms the flared sides of the nose.

Among the facial bones, the mandible is the lower jaw where the chin is located, the two maxillae form the upper jaw, the two zygomatic bones are the cheekbones, and the two nasal bones form the bridge of the nose.

The Hyoid Bone

Although the **hyoid bone** is not part of the skull, it will be mentioned here because it is a part of the axial skeleton. The larynx is the voice box at the top of the trachea in the neck region. The hyoid bone, which is located above the larynx, is the only bone in the body that does not articulate with another bone. It is attached to the temporal bones by muscles and ligaments. The hyoid bone anchors the tongue and serves as the site for the attachment of muscles associated with swallowing.

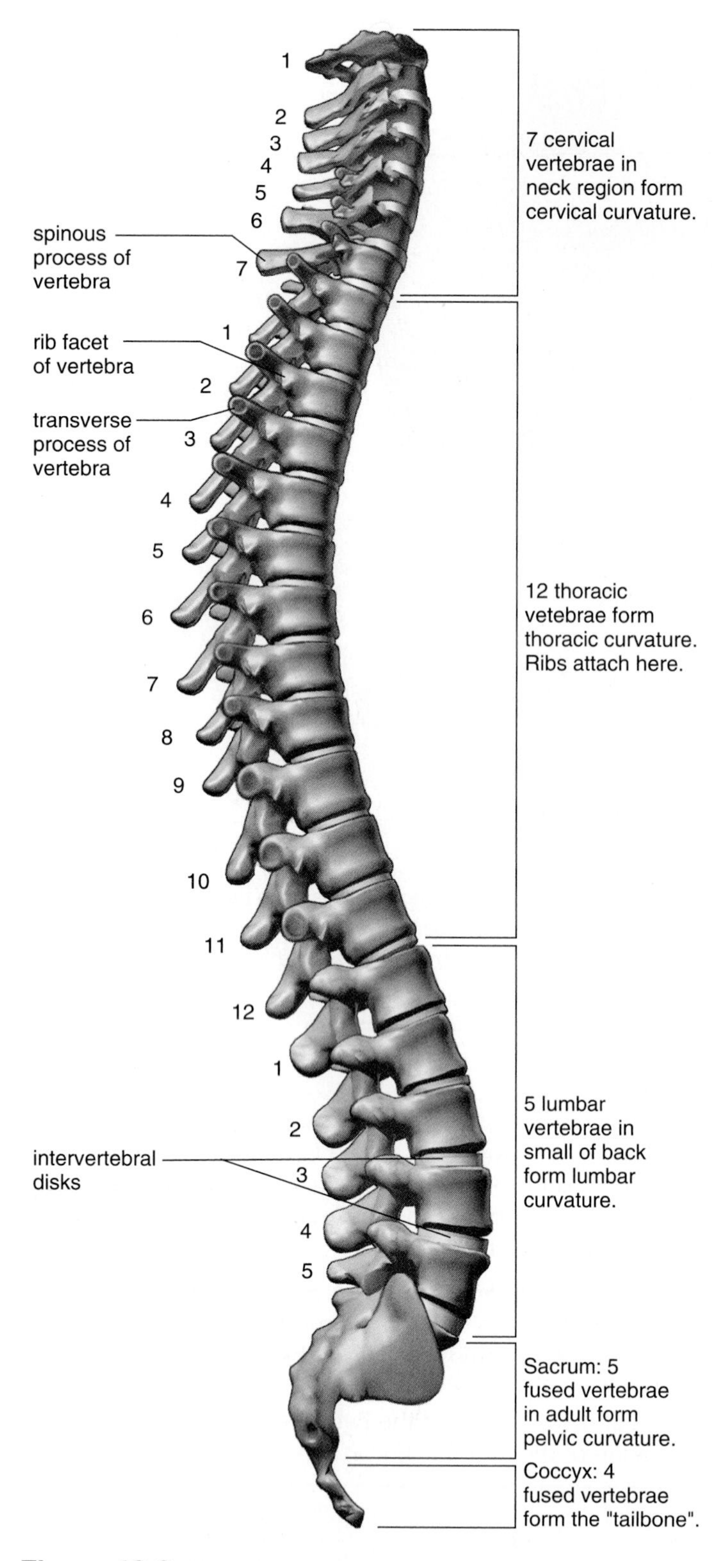

Figure 19.6 The vertebral column.
The vertebrae are named according to their location in the vertebral column, which is flexible due to the intervertebral disks. Note the presence of the coccyx, also called the tailbone.

The Vertebral Column

The **vertebral column** consists of 33 vertebrae (Fig. 19.6). The vertebral column has many functions, including:

- supports the head and trunk, allowing movement.
- protects the spinal cord and roots of spinal nerves.
- serves as a site for muscle attachment.

Normally, the vertebral column has four curvatures that provide more resiliency and strength in an upright posture than a straight column could. **Scoliosis** is an abnormal lateral (sideways) curvature of the spine. There are two other well-known abnormal curvatures: kyphosis is an abnormal curvature that often results in a hunchback, and lordosis is an abnormal curvature resulting in a swayback.

The vertebrae join together to form a canal through which the spinal cord passes. The spinous processes of the vertebrae can be felt as bony projections along the midline of the back. The spinous processes and the transverse processes serve as attachment sites for the muscles that move the vertebral column.

The various vertebrae are named according to their location in the vertebral column (Fig. 19.6). The cervical vertebrae are located in the neck. The first cervical vertebra, called the **atlas,** holds up the head. It is so-named because Atlas, of Greek mythology, held up the world. Movement of the atlas permits the "yes" motion of the head. It also allows the head to tilt from side to side. The second cervical vertebra is called the **axis** because it allows a degree of rotation as when we shake the head "no." Figure 19.7*a* shows the anatomy of the thoracic vertebrae which are found in the chest. The thoracic vertebrae have long, thin spinous processes, and articular facets (flat regions) for the attachment of the ribs. The lumbar vertebrae which are found in the small of the back have a large body and thick processes. The five sacral vertebrae are fused together in the sacrum, which is a part of the pelvis. The coccyx, or tailbone, is composed of four fused vertebrae.

Intervertebral disks composed of fibrocartilage that occur between the vertebrae act as a kind of padding. They prevent the vertebrae from grinding against one another and absorb shock caused by movements such as running, jumping, and even walking. The presence of the disks allows motion between vertebrae so that we can bend forward, backward, and from side to side. Unfortunately, these disks become weakened due to injuries and can even slip and rupture. Pain will result if a slipped disk presses against the spinal cord and/or spinal nerves. If so, surgical removal of the disk may relieve the pain.

The vertebral column consists of the vertebrae and serves as the backbone for the body. Disks between the vertebrae provide padding and account for flexibility of the column.

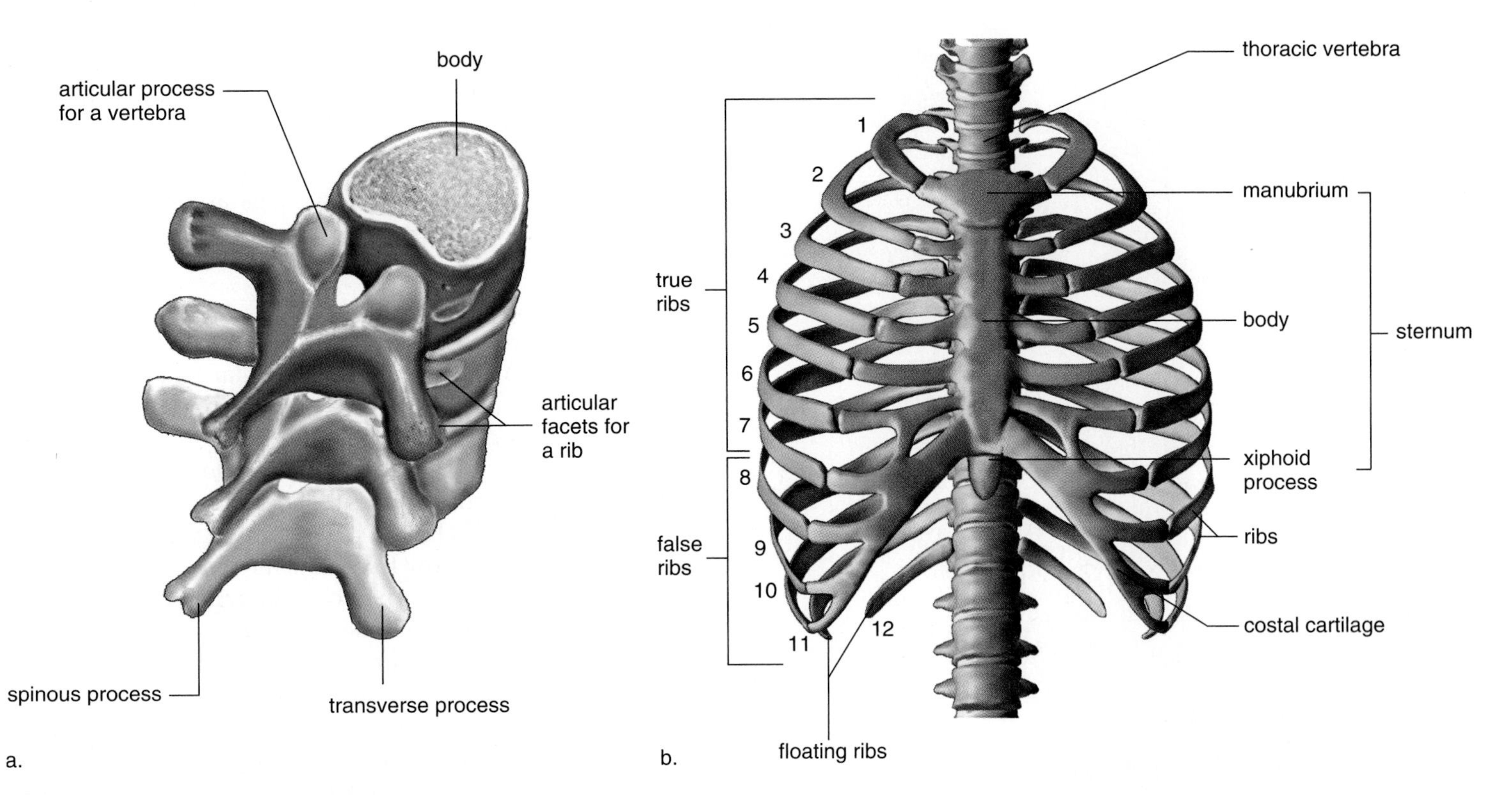

Figure 19.7 Thoracic vertebrae and the rib cage.
a. The thoracic vertebrae articulate with each other and with the ribs. A thoracic vertebra has two facets (flat regions) for articulation with a rib; one is on the body, and the other is on the transverse process. **b.** The rib cage consists of the thoracic vertebrae, the ribs, the costal cartilages, and the sternum.

The Rib Cage

The rib cage is composed of the thoracic vertebrae, the ribs and their associated cartilages, and the sternum (Fig. 19.7b).

The rib cage demonstrates how the skeleton is protective but also flexible. The rib cage protects the heart and lungs; yet it swings outward and upward upon inspiration and then downward and inward upon expiration.

The Ribs There are twelve pairs of **ribs.** All twelve pairs connect directly to the thoracic vertebrae in the back. A rib articulates with the body and transverse process of its corresponding thoracic vertebra. Each rib curves outward and then forward and down.

The upper seven pairs of ribs connect directly to the sternum by means of costal cartilages. These are called the "true ribs." The lower five pairs of ribs do not connect directly to the sternum and they are called the "false ribs." Three pairs of false ribs attach to the sternum by means of a common cartilage. The other two pairs are called "floating ribs" because they do not attach to the sternum at all.

The Sternum The **sternum,** or breastbone, is a flat bone that has the shape of a blade. The sternum, along with the ribs, helps protect the heart and lungs.

The sternum is composed of three bones that fuse during development. These bones are the manubrium, the body, and the xiphoid process. An elevation called the sternal angle occurs where the manubrium joins with the body of the sternum. This is an important anatomical landmark because it occurs at the level of the second rib and therefore allows the ribs to be counted. Counting the ribs is sometimes used to determine where the apex of the heart is located. The apex of the heart is usually between the fifth and sixth ribs.

The xiphoid process, the third part of the sternum, is composed of hyaline cartilage in the child but it becomes ossified in the adult. The diaphragm that divides the thoracic cavity from the abdominal cavity is attached to the xiphoid process.

The rib cage, consisting of the thoracic vertebrae, the ribs, and the sternum, protects the heart and lungs in the thoracic cavity.

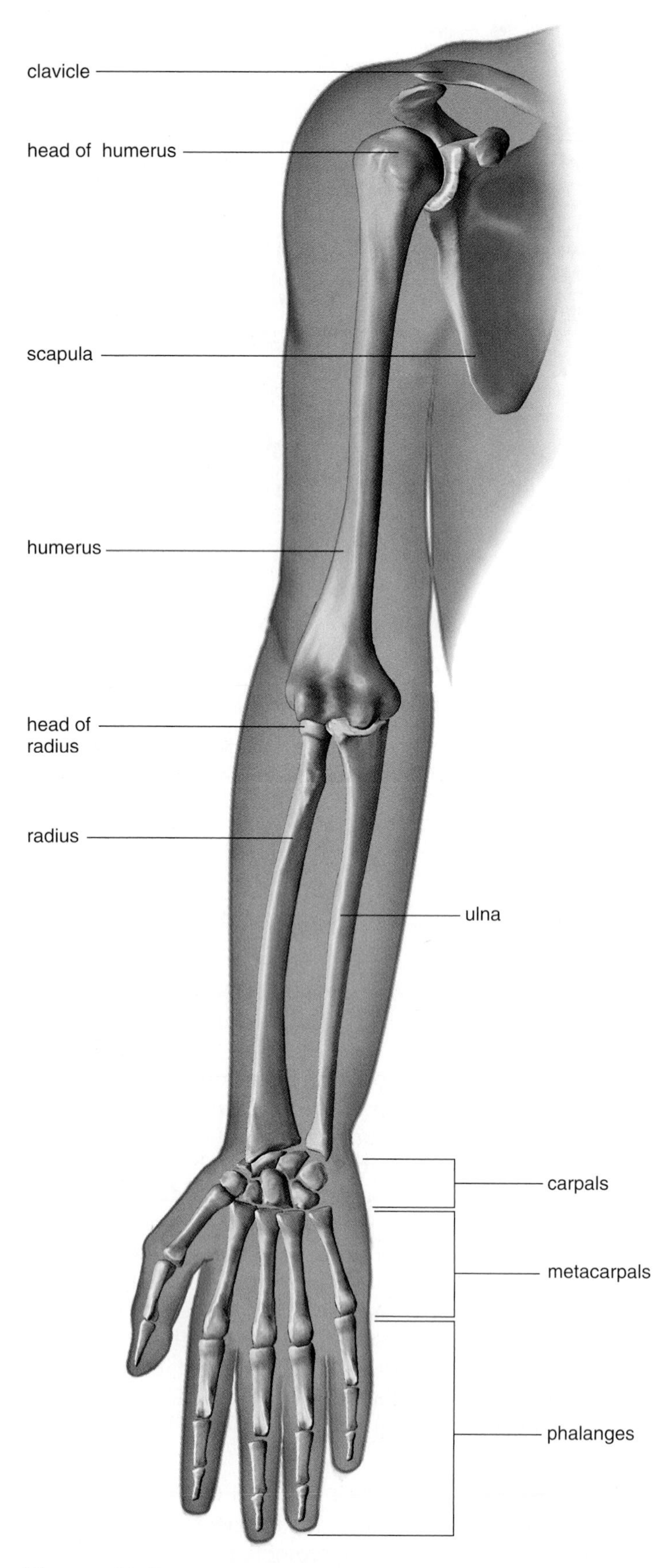

Figure 19.8 **Bones of the pectoral girdle and arm.**

The Appendicular Skeleton

The **appendicular skeleton** consists of the bones within the pectoral and pelvic girdles and their attached limbs. The pectoral (shoulder) girdle and arm are specialized for flexibility; the pelvic (hip) girdle and legs are specialized for strength.

The Pectoral Girdle and Arm

The **pectoral girdle** consists of four bones—two scapulae (shoulder blades) and two clavicles (collarbones) (Fig. 19.8). The **clavicles** extend across the top of the thorax; each articulates with a scapula and also with the sternum. The **scapulae,** which are quite visible in the back, are held in place only by the muscles of the arm and chest. The components of the pectoral girdle are loosely linked together by ligaments, and this allows the girdle to follow freely the movements of the arm.

A shallow cavity of the scapula articulates with the head of the **humerus,** the bone of the upper arm. The head of the humerus is much larger than this cavity. Although this means that the arm can move in almost any direction, there is little stability. Therefore, this is the joint that is most apt to dislocate. The shaft of the humerus has a protuberance where the deltoid, the prominent muscle of the chest, attaches. Upon death, enlargement of this protuberance can be used as evidence that the person did a lot of heavy lifting.

The far end of the humerus has two other protuberances which articulate respectively with the **ulna** and the **radius,** the bones of the lower arm. The "funny bone" of the elbow is a portion of the ulna.

When the arm is held so that the palm is turned upward, the radius and ulna are about parallel to one another. When the arm is turned so that the palm is next to the body, the radius crosses in front of the ulna, a feature that contributes to the easy twisting motion of the forearm.

The hand has many bones, and this increases its flexibility. The wrist has eight **carpal** bones, which look like small pebbles. From these, five **metacarpal** bones fan out to form a framework for the palm. The metacarpal bone that leads to the thumb is placed in such a way that the thumb can reach out and touch the other digits. (**Digits** is a term that refers to either fingers or toes.) Your knuckles are the enlarged ends of the metacarpals. Beyond the metacarpals are the **phalanges,** the bones of the fingers and the thumb. The phalanges of the hand are long, slender, and lightweight.

The pectoral girdle and arm are specialized for flexibility of movement.

The Pelvic Girdle and Leg

Figure 19.9 shows how the leg is attached to the pelvic girdle. The **pelvic** (hip) **girdle** consists of two heavy, large coxal bones (hipbones). The pelvic cavity is bordered by the bones of the pelvis: the sacrum, and the two coxal bones. The pelvis bears the weight of the body, protects the organs within the pelvic cavity, and serves as the place of attachment for the legs.

Each **coxal bone** has three parts: the ilium, the ischium, and the pubis, which are fused in the adult (Fig. 19.9). The hip socket occurs where these three bones meet. The *ilium* is the largest part of the coxal bones, and our hips occur where it flares out. We sit on the *ischium,* which has a posterior spine called the ischial spine that projects into the pelvic cavity. The *pubis* (referring to pubic hair) is the anterior part of a coxal bone. The two pubic bones are joined together by a fibrocartilage disk at the pubic symphysis.

The pelvic cavity differs in the male and female. In the female, the iliac bones are more flared; the pelvic cavity is more shallow, but the outlet is wider. These adaptations facilitate giving birth.

The **femur** (thighbone) is the longest and strongest bone in the body. Its head fits into the hip socket and its short neck better positions the legs for walking. The femur has large processes for attachment of the muscles of the legs and buttocks. The knee occurs where the femur articulates with the tibia of the lower leg. The **patella,** or kneecap, is held in place by tendons that attach it to the tibia.

The tibia and fibula are the bones of the lower leg. The tibia is the shin bone. At its far end, the **tibia** accounts for the inner bulge of the ankle. The **fibula** is the more slender bone in the lower leg. The fibula articulates with the tibia at its head and at its far end, where it accounts for the outer bulge of the ankle.

Each foot has an ankle, an instep, and five toes. The ankle contains seven **tarsal** bones, one of which (the talus) can move freely where it joins the tibia and fibula. Strange to say, the **calcaneus,** or heel bone, is also considered to be part of the ankle. The talus and calcaneus support the weight of the body.

The instep of the foot has five elongated **metatarsal** bones. The far end of the metatarsals form the ball of the foot. If the ligaments that bind the metatarsals together become weakened, flat feet are apt to result. The bones of the toes are called **phalanges,** just like those of the fingers, but in the foot, the phalanges are stout and extremely sturdy.

The pelvic girdle and leg are adapted to supporting the weight of the body. The femur is the longest and strongest bone in the body.

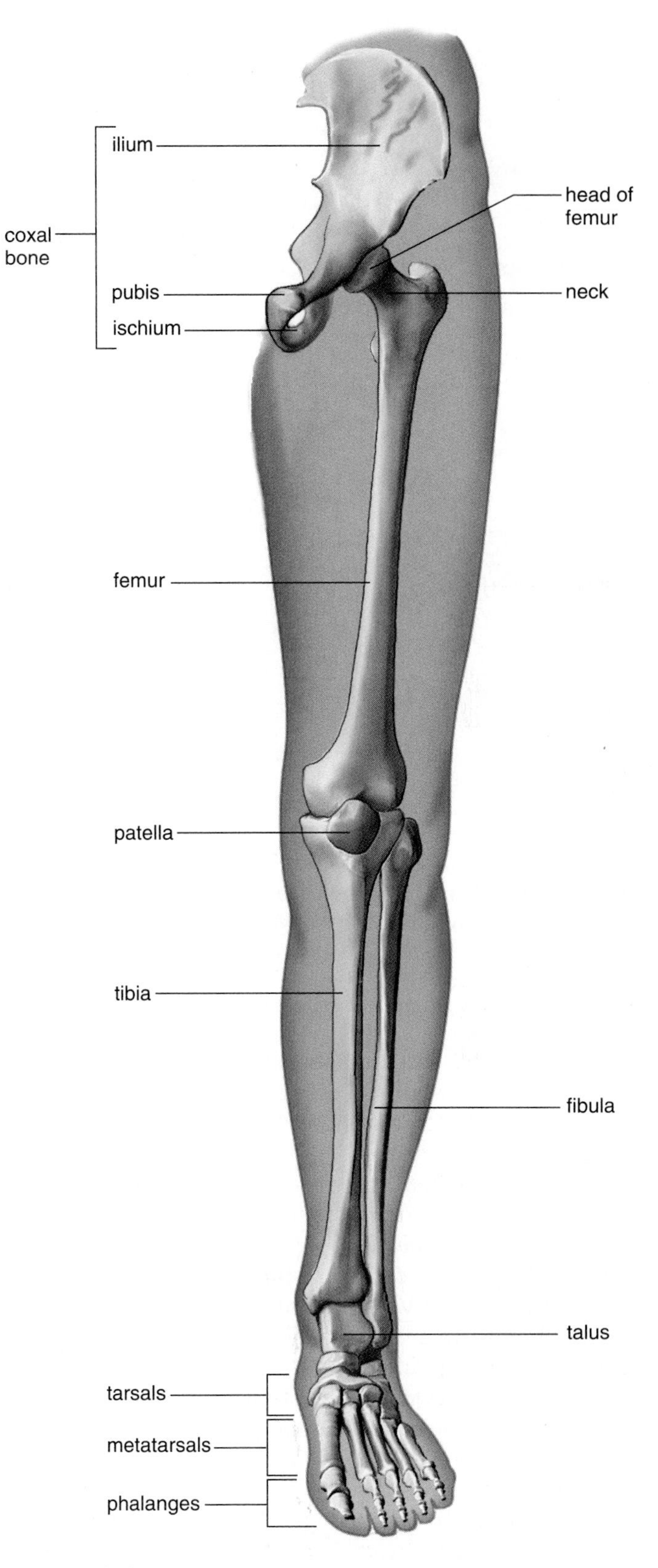

Figure 19.9 **Bones of the pelvic girdle and leg.**

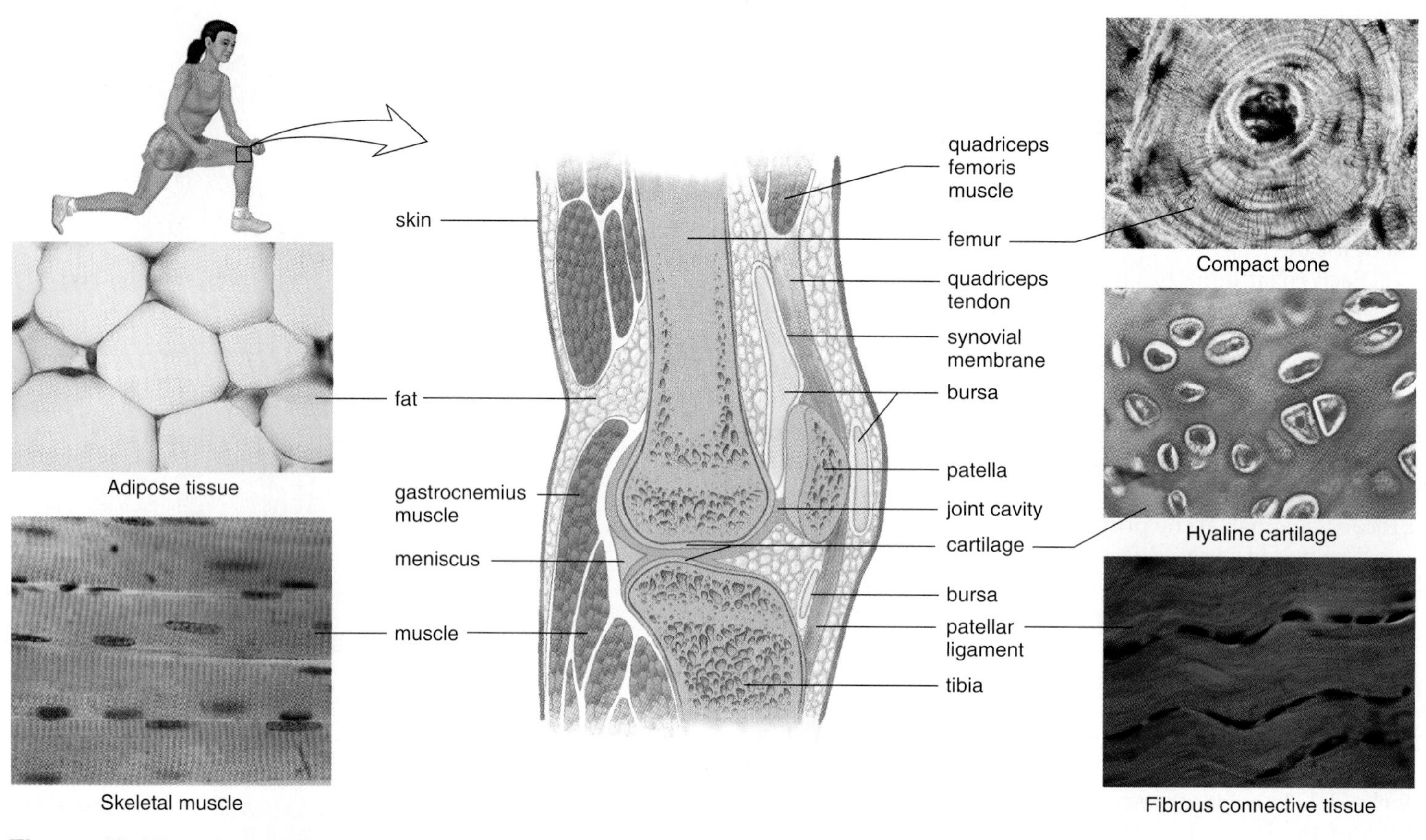

Figure 19.10 **Knee joint.**
The knee joint is a synovial joint. Notice the cavity between the bones, which is encased by ligaments and lined by synovial membrane. The patella (kneecap) serves to guide the quadriceps tendon over the joint when flexion or extension occurs.

Articulations

Bones are joined at the joints, which are classified as fibrous, cartilaginous, and synovial. Fibrous joints, such as the **sutures** between the cranial bones, are *immovable.* Cartilaginous joints are connected by hyaline cartilage, as in the costal cartilages that join the ribs to the sternum, or by fibrocartilage, as seen in the intervertebral disks. Cartilaginous joints are *slightly movable.*

In *freely movable* **synovial joints,** the two bones are separated by a cavity. Ligaments hold the two bones in place as they form a capsule. Tendons also help stabilize joints. The joint capsule is lined by a **synovial membrane,** which produces *synovial fluid,* a lubricant for the joint. The knee is an example of a synovial joint (Fig. 19.10). Aside from articular cartilage, the knee contains menisci (sing., **meniscus**), crescent-shaped pieces of cartilage between the bones. These give added stability and act as shock absorbers. Unfortunately, athletes often suffer injury of the menisci, known as torn cartilage. The knee joint also contains 13 fluid-filled sacs called bursae (sing., **bursa**), which ease friction between tendons and ligaments. Inflammation of the bursae is called **bursitis.** Tennis elbow is a form of bursitis.

There are different types of movable joints. The knee and elbow joints are **hinge joints** because, like a hinged door, they largely permit movement in one direction only. The joint between the radius and ulna is a pivot joint in which only rotation is possible. More movable are the **ball-and-socket joints;** for example, the ball of the femur fits into a socket on the hipbone. Ball-and-socket joints allow movement in all planes and even a rotational movement.

Synovial joints are subject to **arthritis.** In *rheumatoid arthritis,* the synovial membrane becomes inflamed and grows thicker. Degenerative changes take place that make the joint almost immovable and painful to use. Evidence indicates that these effects are brought on by an autoimmune reaction. In *osteoarthritis,* the cartilage at the ends of the bones disintegrates so that the two bones become rough and irregular. The pain of arthritis, however, is believed to come from the soft tissues associated with joints and not from bone.

Joints are regions of articulations between bones. Synovial joints are freely movable and allow particular types of movements. Unfortunately, synovial joints are subject to various disorders.

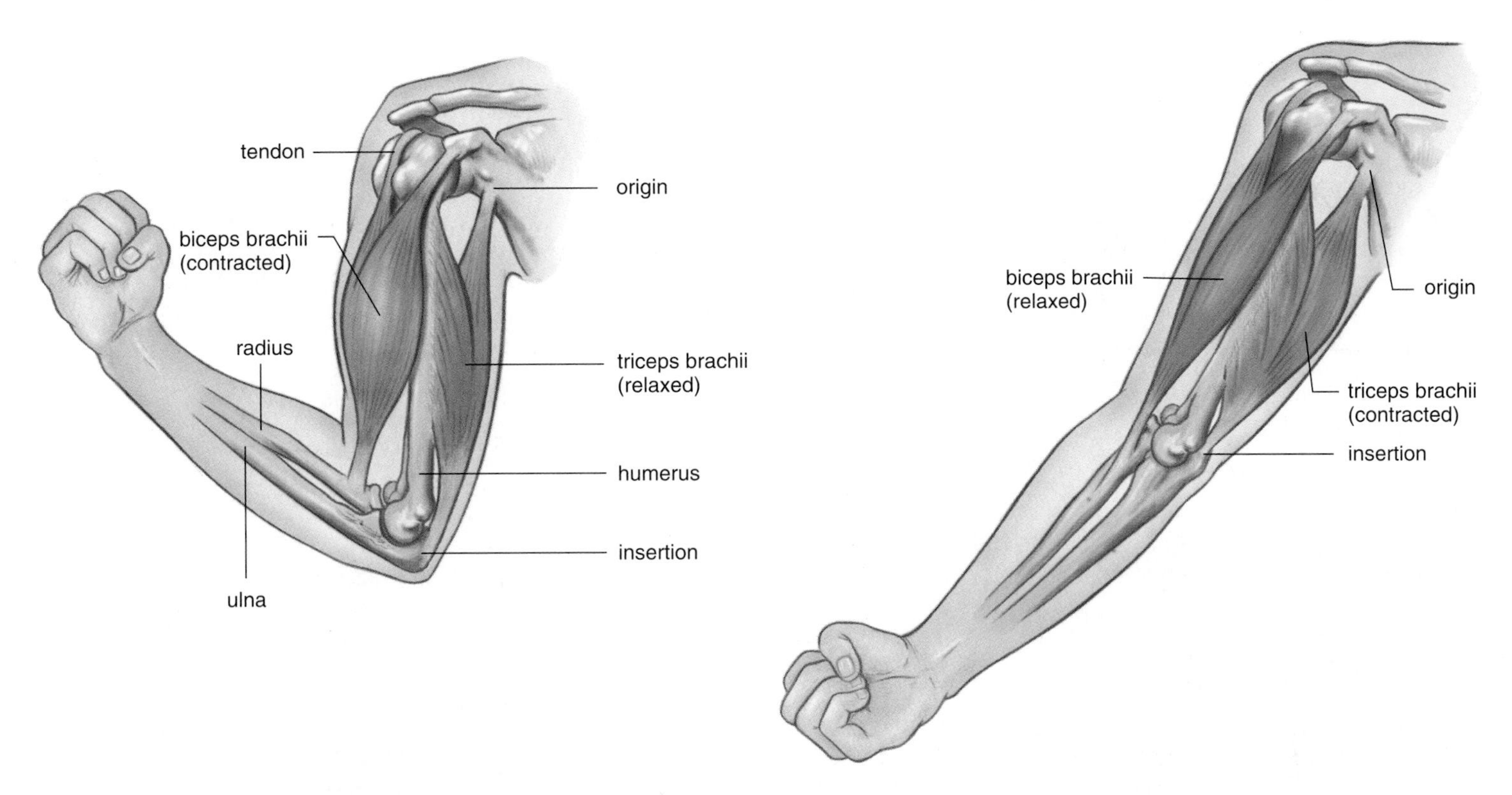

Figure 19.11 Attachment of the skeletal muscles
The origin of a muscle is on a bone that remains stationary, and the insertion of a muscle is on a bone that moves when the muscle contracts. The muscles in this drawing are antagonistic. When the biceps brachii contracts, the lower arm flexes, and when the triceps brachii contracts, the lower arm extends.

19.3 Skeletal Muscles

Muscles have various functions, which they are more apt to perform well if they are exercised regularly.

Skeletal muscles help support the body. Skeletal muscle contraction opposes the force of gravity and helps us to remain upright.

Skeletal muscles make bones move. Muscle contraction accounts not only for the movement of arms and legs but also for movements of the eyes, facial expressions, and breathing.

Skeletal muscles help maintain a constant body temperature. Skeletal muscle contraction causes ATP to breakdown, releasing heat that is distributed about the body.

Skeletal muscle contraction assists movement in cardiovascular and lymphatic vessels. The pressure of skeletal muscle contraction keeps blood moving in cardiovascular veins and lymph moving in lymphatic vessels.

Skeletal muscles help protect internal organs. Skeletal muscles cover the bones and underlying organs such as the kidneys and blood vessels.

Cardiac and smooth muscles have specific functions. The contraction of cardiac muscle causes the heart to pump blood. Smooth muscle contraction propels food in the digestive tract and urine in the ureters, for example.

Muscles Work in Pairs

This chapter is primarily concerned with skeletal muscles—those muscles that make up the bulk of the human body. Because tendons most often attach muscles to the far side of a joint, contraction of skeletal muscle causes the movement of bones at the joint. When muscles contract, one bone remains fairly stationary, and the other one moves. The **origin** of a muscle is on the stationary bone, and the **insertion** of a muscle is on the bone that moves.

Frequently, a body part is moved by a group of muscles working together. Even so, one muscle does most of the work and is called the *prime mover.* The assisting muscles are called the *synergists.* When muscles contract, they shorten; therefore, muscles can only pull; they cannot push.

Muscles have antagonists, and *antagonistic pairs* work opposite to one another to bring about movement in opposite directions. For example, the biceps brachii and the triceps brachii are antagonists; one flexes the forearm, and the other extends the forearm (Fig. 19.11).

When muscles cooperate to achieve movement, some act as prime movers, others are synergists, and still others are antagonists.

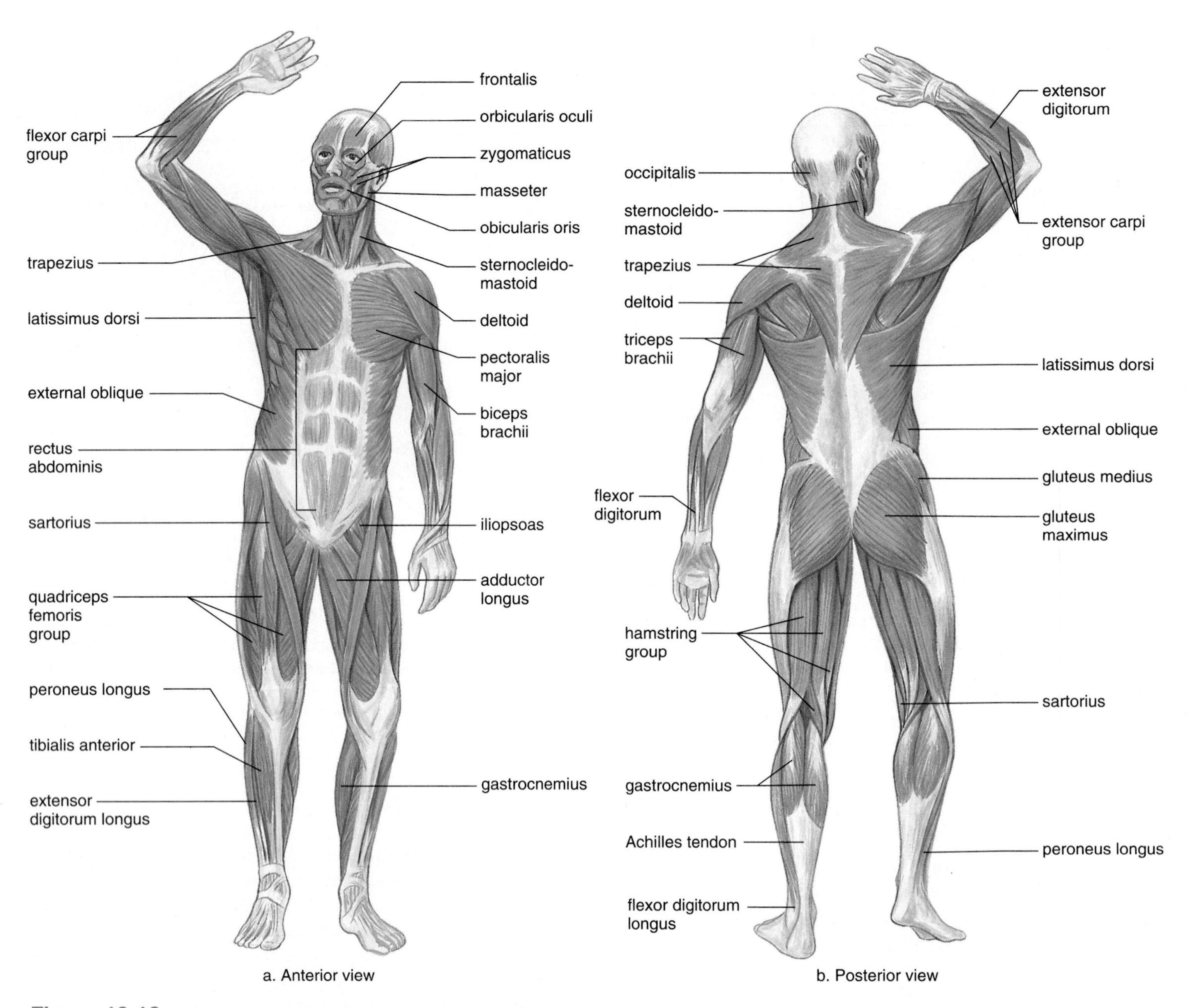

Figure 19.12 Human musculature.
Superficial skeletal muscles. In **(a)** anterior and **(b)** posterior view.

Skeletal Muscles of the Body

Skeletal muscles (Fig. 19.12 and Tables 19.1 and 19.2) are given names based on the following characteristics and examples:

1. Size. The gluteus maximus is the largest muscle that makes up the buttocks.
2. Shape. The deltoid is shaped like a delta, or triangle.
3. Location. The frontalis overlies the frontal bone.
4. Direction of fibers. The rectus abdominis is a longitudinal muscle of the abdomen (rectus means straight).
5. Number of attachments. The biceps brachii has two attachments, or origins.
6. Action. The extensor digitorum extends the fingers and digits. Extension increases the joint angle; flexion decreases the joint angle; abduction is the movement of a body part sideways away from the midline; adduction is the movement of a body part toward the midline.

Table 19.1 Muscles (anterior view)

Name	Function
Head and neck	
Frontalis	Wrinkles forehead and lifts eyebrows
Orbicularis oculi	Closes eye (winking)
Zygomaticus	Raises corner of mouth (smiling)
Masseter	Closes jaw
Orbicularis oris	Closes and protrudes lips (kissing)
Arms and trunk	
External oblique	Compresses abdomen; rotates trunk
Rectus abdominis	Flexes spine
Pectoralis major	Flexes and adducts shoulder and arm ventrally (pulls arm across chest)
Deltoid	Abducts and raises arm at shoulder joint
Biceps brachii	Flexes forearm and supinates hand
Leg	
Adductor longus	Adducts thigh
Iliopsoas	Flexes thigh
Sartorius	Rotates thigh (sitting cross-legged)
Quadriceps femoris group	Extends lower leg
Peroneus longus	Everts foot
Tibialis anterior	Dorsiflexes and inverts foot
Flexor and extensor digitorum longus	Flex and extend toes

Table 19.2 Muscles (posterior view)

Name	Function
Head and neck	
Occipitalis	Moves scalp backward
Sternocleidomastoid	Turns head to side; flexes neck and head
Trapezius	Extends head; raises and adducts shoulders dorsally (shrugging shoulders)
Arm and trunk	
Latissimus dorsi	Extends and adducts shoulder and arm dorsally (pulls arm across back)
Deltoid	Abducts and raises arm at shoulder joint
External oblique	Rotates trunk
Triceps brachii	Extends forearm
Flexor and extensor carpi group	Flex and extend hand
Flexor and extensor digitorum	Flex and extend fingers
Buttocks and legs	
Gluteus medius	Abducts thigh
Gluteus maximus	Extends thigh (forms buttocks)
Hamstring group	Flexes lower leg
Gastrocnemius	Extends foot (tiptoeing)

Visual Focus

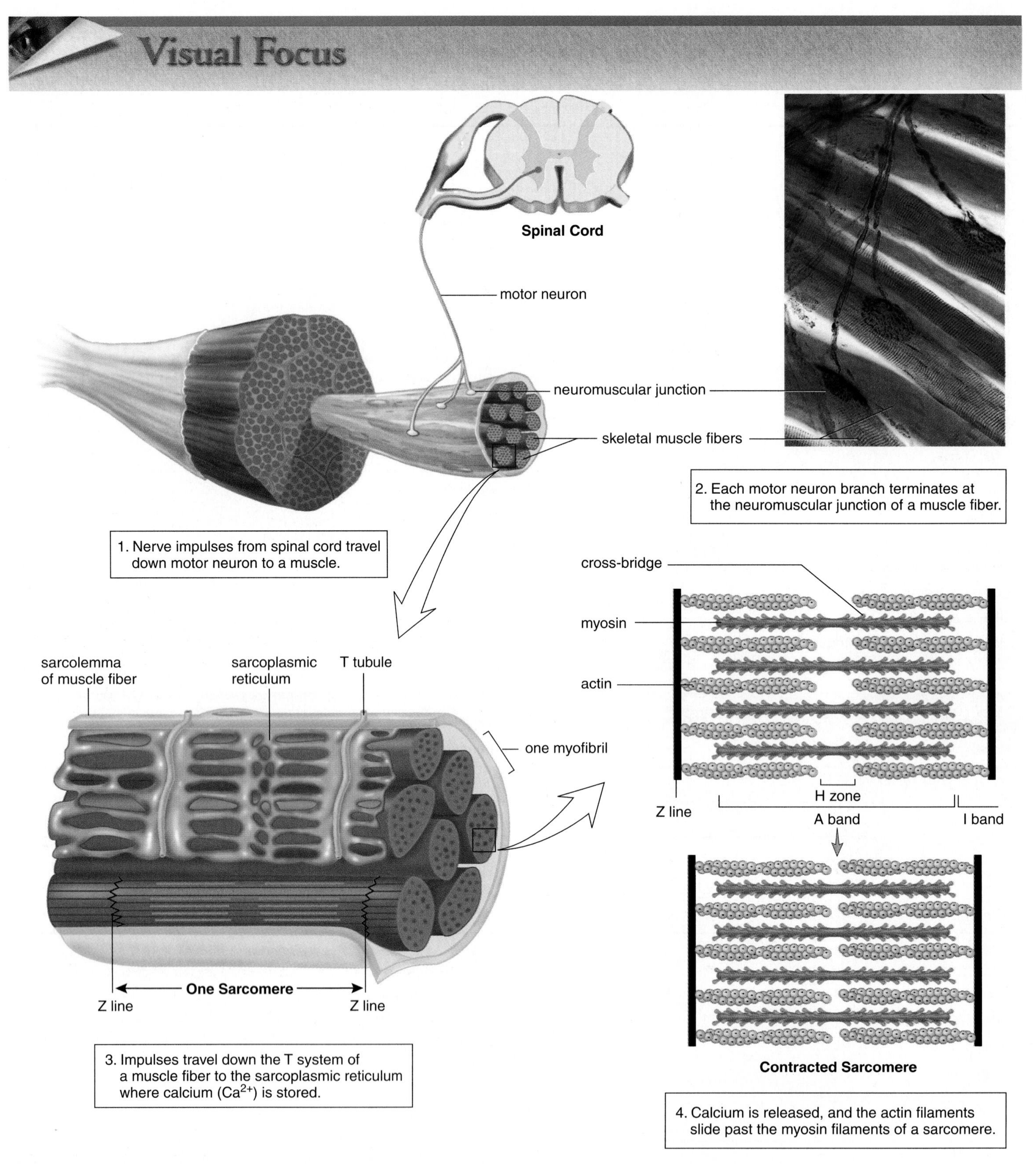

Figure 19.13 **Contraction of a muscle.**
A whole muscle contains bundles of skeletal muscle fibers. A motor neuron ends in neuromuscular junctions at these fibers. A fiber contains many myofibrils. Each myofibril is divided into sarcomeres, where actin and myosin filaments are precisely arranged. When a motor impulse reaches a neuromuscular junction, the impulse travels down the T system and calcium is released from the sarcoplasmic reticulum. Then the actin filaments slide past myosin filaments and the sarcomere contracts.

19.4 Mechanism of Muscle Fiber Contraction

A series of events leads up to muscle fiber contraction. After discussing these events, we will examine a neuromuscular junction and filament sliding in detail.

Overview of Muscular Contraction

Figure 19.13 shows the steps that lead to muscle contraction. Muscles are stimulated to contract by nerve impulses that begin in the brain or spinal cord. These nerve impulses travel down a motor neuron to a neuromuscular junction, a region where a motor neuron fiber meets a muscle fiber.

Each muscle fiber is a cell containing the usual cellular components, but special names have been assigned to some of these components. The plasma membrane is called the **sarcolemma,** the cytoplasm is the sarcoplasm, and the endoplasmic reticulum is the **sarcoplasmic reticulum.** A muscle fiber also has some unique anatomical characteristics. For one thing, the sarcolemma forms **T (transverse) tubules** that penetrate, or dip down, into the cell so that they come into contact—but do not fuse—with expanded portions of the sarcoplasmic reticulum. The expanded portions of the sarcoplasmic reticulum contain calcium ions (Ca^{2+}), which are essential for muscle contraction. The sarcoplasmic reticulum encases hundreds and sometimes even thousands of myofibrils, which are the contractile portions of the fibers.

Myofibrils and Sarcomeres

Myofibrils are cylindrical in shape and run the length of the muscle fiber. The light microscope shows that muscle fibers have light and dark bands called striations (Fig. 19.14). The electron microscope shows that the striations of myofibrils are formed by the placement of myofilaments within contractile units called **sarcomeres.** A sarcomere extends between two dark lines called the Z lines. A sarcomere contains two types of protein myofilaments. The thick filaments are made up of a protein called **myosin,** and the thin filaments are made up of a protein called **actin.** Other proteins in addition to actin are also present. The I band is light colored because it contains only actin filaments attached to a Z line. The dark regions of the A band contain overlapping actin and myosin filaments, and its H zone has only myosin filaments.

Sliding Filaments

Impulses generated at a neuromuscular junction travel down a T tubule, and calcium is released from the sarcoplasmic reticulum into the muscle fiber. Now the muscle fiber contracts as the sarcomeres within the myofibrils shorten. When a sarcomere shortens, the actin (thin) filaments slide past the myosin (thick) filaments and approach one another. This causes the I band to shorten and the H zone to almost or completely disappear. The movement of actin filaments in relation to myosin filaments is called the **sliding filament theory** of muscle contraction. During the sliding process, the sarcomere shortens even though the filaments themselves remain the same length.

The participants in muscle contraction have the functions listed in Table 19.3. ATP supplies the energy for muscle contraction. Although the actin filaments slide past the myosin filaments, it is the myosin filaments that do the work. Myosin filaments break down ATP and have cross-bridges that pull the actin filaments toward the center of the sarcomere.

Table 19.3 Muscle Contraction

Name	Function
Actin filaments	Slide past myosin, causing contraction
Ca^{2+}	Needed for myosin to bind the actin
Myosin filaments	Pull actin filaments by means of cross-bridges; are enzymatic and split ATP
ATP	Supplies energy for muscle contraction

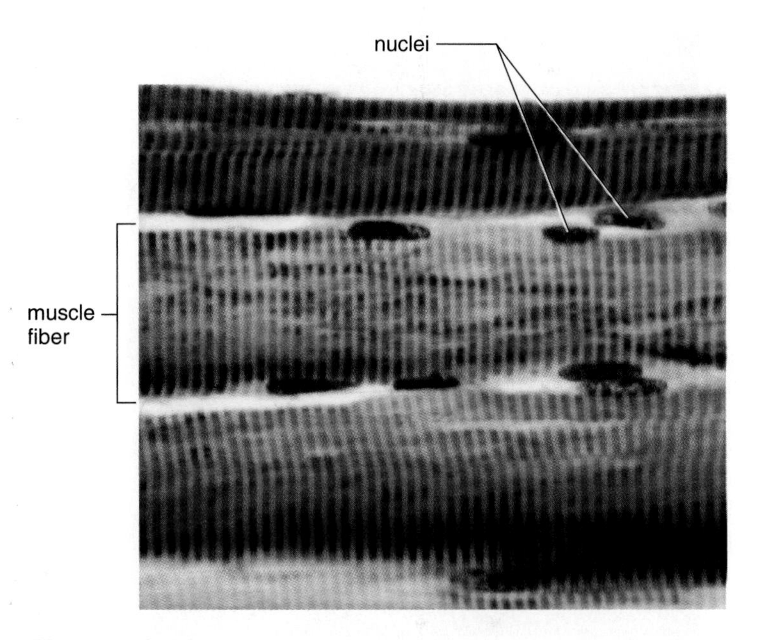

Figure 19.14 Light micrograph of skeletal muscle. The striations of skeletal muscle tissue are produced by alternating dark A bands and light I bands. See the sarcomere in Figure 19.13.

Muscle fibers are innervated, and when they are stimulated to contract, myofilaments slide past one another causing sarcomeres to shorten.

Muscle Innervation

Muscles are stimulated to contract by motor nerve fibers. Each nerve fiber has several branches ending in an axon bulb that lies in close proximity to the sarcolemma of a muscle fiber. A small gap, called a synaptic cleft, separates the axon bulb from the sarcolemma. This entire region is called a **neuromuscular junction** (Fig. 19.15).

Axon bulbs contain synaptic vesicles that are filled with neurotransmitter molecules called acetylcholine (ACh). When nerve impulses traveling down a motor neuron arrive at an axon bulb, the synaptic vesicles release ACh into the synaptic cleft. ACh quickly diffuses across the cleft and binds to receptor proteins in the sarcolemma. Now the sarcolemma generates impulses that spread over the sarcolemma and down T tubules to the sarcoplasmic reticulum. The release of calcium from the sarcoplasmic reticulum causes the myofibrils within a muscle fiber to contract as discussed previously.

At a neuromuscular junction, nerve impulses bring about the release of neurotransmitter molecules that signal a muscle fiber to contract.

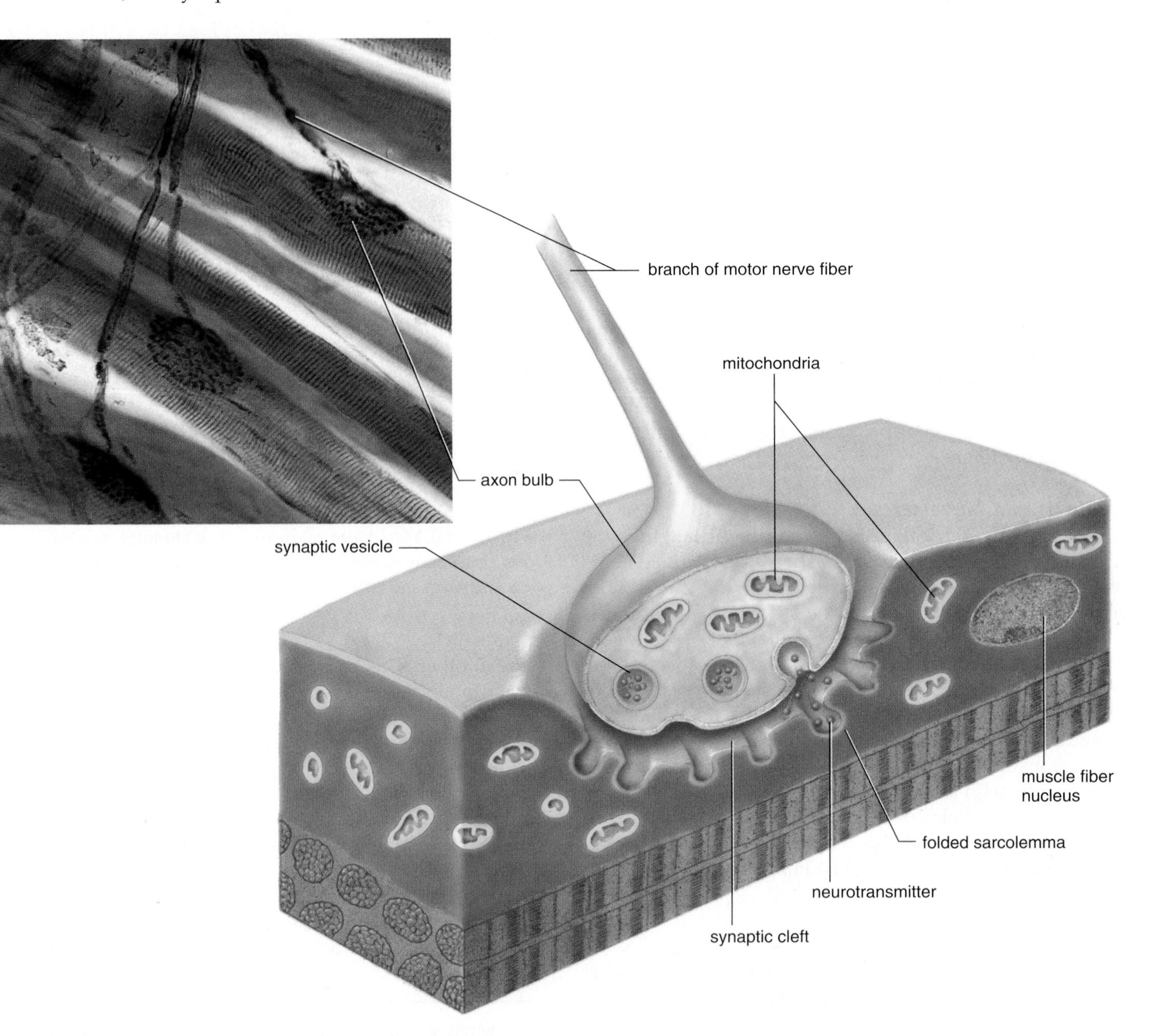

Figure 19.15 Neuromuscular junction.
The branch of a motor nerve fiber terminates in an axon bulb that meets but does not touch a muscle fiber. A synaptic cleft separates the axon bulb from the sarcolemma of the muscle fiber. Nerve impulses traveling down a motor neuron cause synaptic vesicles to discharge acetylcholine, which diffuses across the synaptic cleft. When this neurotransmitter is received by the sarcolemma of a muscle fiber, contraction follows.

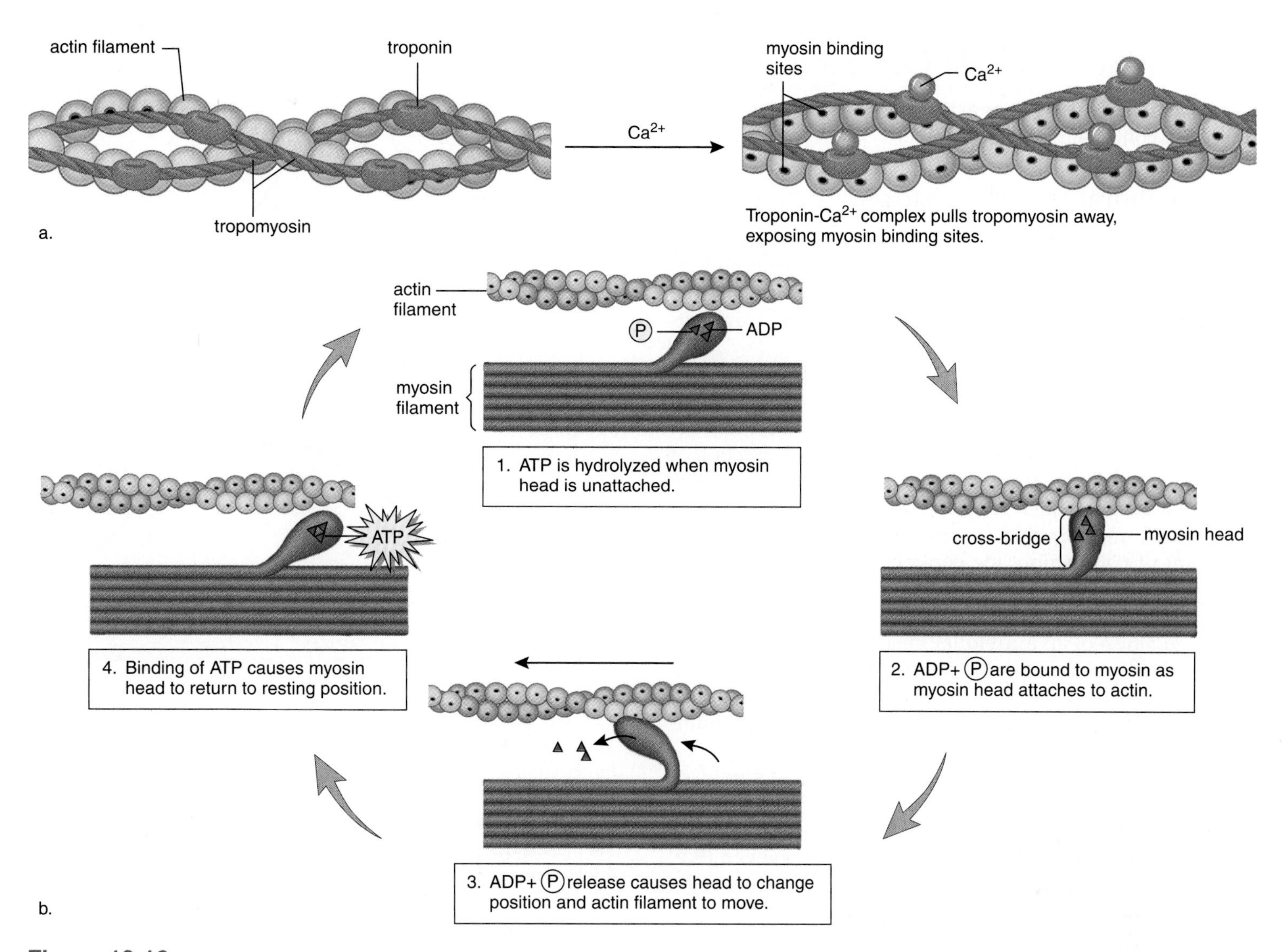

Figure 19.16 The role of calcium and myosin in muscle contraction.
a. Upon release, calcium binds to troponin, exposing myosin binding sites. **b.** After breaking down ATP, myosin heads bind to an actin filament, and later, a power stroke causes the actin filament to move.

Figure 19.16 shows the placement of two other proteins associated with a thin filament, which is composed of a double row of twisted actin molecules. Threads of **tropomyosin** wind about an actin filament, and **troponin** occurs at intervals along the threads. Calcium ions (Ca^{2+}) that have been released from the sarcoplasmic reticulum combine with troponin. After binding occurs, the tropomyosin threads shift their position, and myosin binding sites are exposed.

The thick filament is actually a bundle of myosin molecules, each having a double globular head with an ATP binding site. The heads function as ATPase enzymes, splitting ATP into ADP and Ⓟ. This reaction activates the head so that it will bind to actin. The ADP and Ⓟ remain on the myosin heads until the heads attach to actin, forming a cross-bridge. Now, ADP and Ⓟ are released, and this causes the cross-bridges to change their positions. This is the *power stroke* that pulls the thin filaments toward the middle of the sarcomere. When another ATP molecule binds to a myosin head, the cross-bridge is broken as the head detaches from actin. The cycle begins again; the actin filaments move nearer the center of the sarcomere each time the cycle is repeated.

Contraction continues until nerve impulses cease and calcium ions are returned to their storage sacs. The membranes of the sarcoplasmic reticulum contain active transport proteins that pump calcium ions back into the sarcoplasmic reticulum.

Myosin filament heads break down ATP and then attach to an actin filament, forming cross-bridges that pull the actin filament to the center of a sarcomere.

19.5 Whole Muscle Contraction

Observation of muscle contraction in the laboratory applies to muscle contraction in the body.

Basic Laboratory Observations

To study muscle contraction in the laboratory, the gastrocnemius (calf muscle) is usually removed from a frog and attached to a movable lever. The muscle is stimulated, and the mechanical force of contraction is recorded as a visual pattern called a **myogram.**

At first, the stimulus may be too weak to cause a contraction, but as soon as the strength of the stimulus reaches a threshold stimulus, the muscle contracts and then relaxes. This action—a single contraction that lasts only a fraction of a second—is called a **muscle twitch.** Figure 19.17*a* is a myogram of a twitch, which is customarily divided into the latent period, or the period of time between stimulation and initiation of contraction; the contraction period, when the muscle shortens; and the relaxation period, when the muscle returns to its former length.

Stimulation of an individual fiber within a muscle usually results in a maximal, *all-or-none contraction.* But the contraction of a whole muscle, as evidenced by the size of a muscle twitch, can vary in strength depending on the number of fibers contracting.

If a muscle is given a rapid series of threshold stimuli, it can respond to the next stimulus without relaxing completely. In this way, muscle contraction summates until maximal sustained contraction, called **tetanus,** is achieved (Fig. 19.17*b*). The myogram no longer shows individual twitches; rather, the twitches are fused and blended completely into a straight line. Tetanus continues until the muscle fatigues due to depletion of energy reserves. *Fatigue* is apparent when a muscle relaxes even though stimulation continues.

Muscle Tone in the Body

Tetanic contractions ordinarily occur in the body's muscles whenever skeletal muscles are actively used. But while some fibers are contracting, others are relaxing. Because of this, intact muscles rarely fatigue completely. Even when muscles appear to be at rest, they exhibit **tone** in which some of their fibers are contracting. Muscle tone is particularly important in maintaining posture. If all the fibers within the muscles of the neck, trunk, and legs were to suddenly relax, the body would collapse.

Maintenance of the right amount of tone requires the use of special receptors called muscle spindles. A muscle spindle consists of a bundle of modified muscle fibers, with sensory nerve fibers wrapped around a short, specialized region. A muscle spindle contracts along with muscle fibers, but thereafter it sends sensory nerve impulses to the central nervous system, which then regulates muscle contraction so that tone is maintained (see Fig. 18.2).

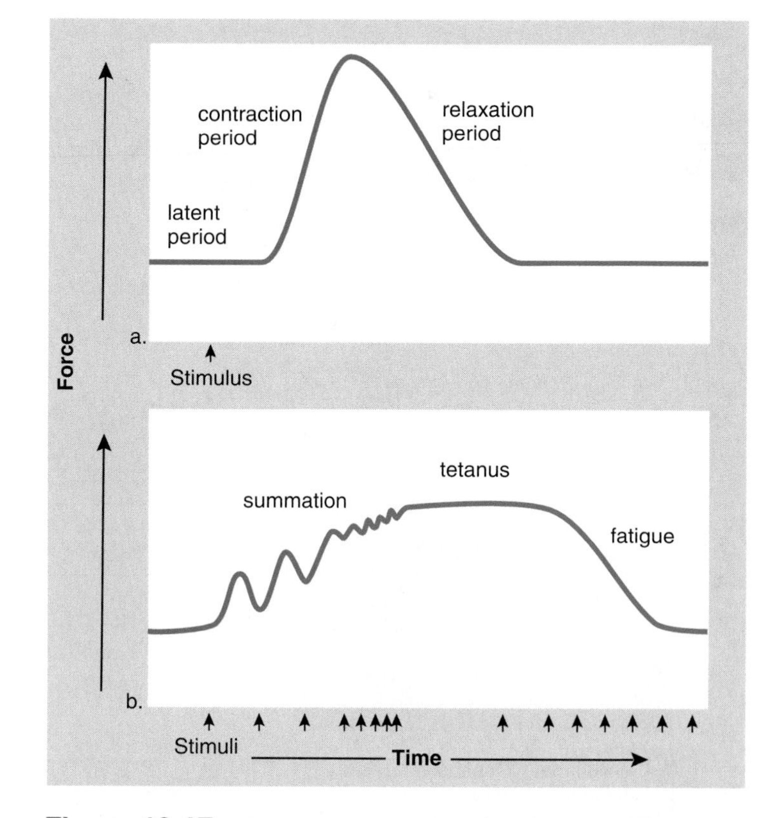

Figure 19.17 Physiology of skeletal muscle contraction. These are myograms, visual representations of the contraction of a muscle that has been dissected from a frog. **a.** Simple muscle twitch is composed of three periods: latent, contraction, and relaxation. **b.** Summation and tetanus. When stimulation frequency increases, the muscle does not relax completely between stimuli, and the contraction gradually increases in intensity. The muscle becomes maximally contracted until it fatigues.

Recruitment and the Strength of Contraction

Each motor neuron innervates several muscle fibers at a time. A motor neuron together with all of the muscle fibers that it innervates is called a **motor unit.** As the intensity of nervous stimulation increases, more and more motor units are activated. This phenomenon, known as recruitment, results in stronger and stronger contractions.

Another variable of importance is the number of fibers within a motor unit. In the ocular muscles that move the eyes, the innervation ratio is one neuron per 23 muscle fibers, while in the gastrocnemius muscle of the buttocks, the ratio is about one motor neuron per 1,000 muscle fibers. Moving the eyes requires finer control than moving the legs.

Observation of muscle contraction in the laboratory applies to muscle contraction in the body. A muscle at rest exhibits tone dependent on tetanic contractions; a contracting muscle exhibits degrees of contraction dependent on recruitment.

Energy for Muscle Contraction

ATP produced previous to strenuous exercise lasts a few seconds, and then muscles acquire new ATP in three different ways. There are two anaerobic ways by which muscles can acquire ATP immediately. The first method depends on **creatine phosphate,** a high-energy compound built up when a muscle is resting. Creatine phosphate cannot participate directly in muscle contraction. Instead, it is used to regenerate ATP by the following reaction:

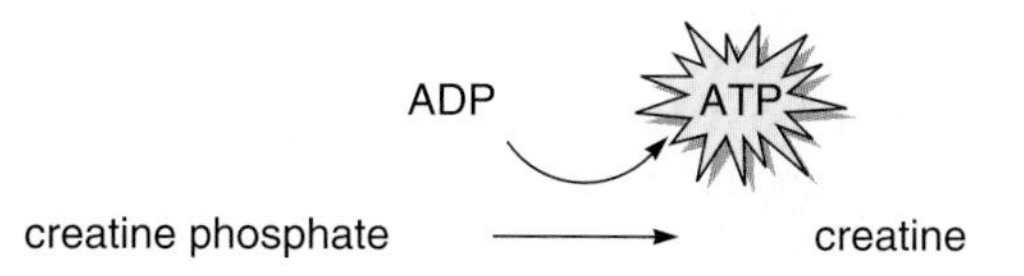

This reaction, which occurs in the midst of sliding filaments, is the speediest way to make ATP available to muscles. Creatine phosphate provides enough energy for only about eight seconds of intense activity, and then it is spent. Creatine phosphate is rebuilt when a muscle is resting by transferring a phosphate group from ATP to creatinine.

Fermentation also supplies ATP without consuming oxygen. During fermentation, glucose is broken down to lactate (lactic acid):

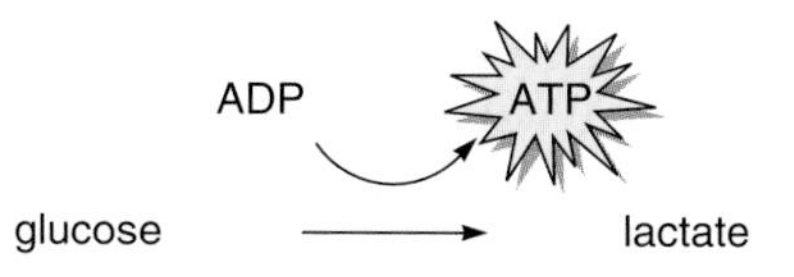

The accumulation of lactate in a muscle fiber makes the cytoplasm more acidic, and eventually enzymes will cease to function well. If fermentation continues longer than two or three minutes, cramping and fatigue will set in. Cramping seems to be due to lack of ATP needed to pump calcium ions back into the sarcoplasmic reticulum and to break the linkages between the actin and myosin filaments so that muscle fibers can relax.

Fortunately, aerobic respiration occurring in mitochondria usually provides most of a muscle's ATP. A muscle cell can make use of glycogen and fatty acids stored in the muscle as fuel, but still also needs a supply of oxygen in order to produce ATP:

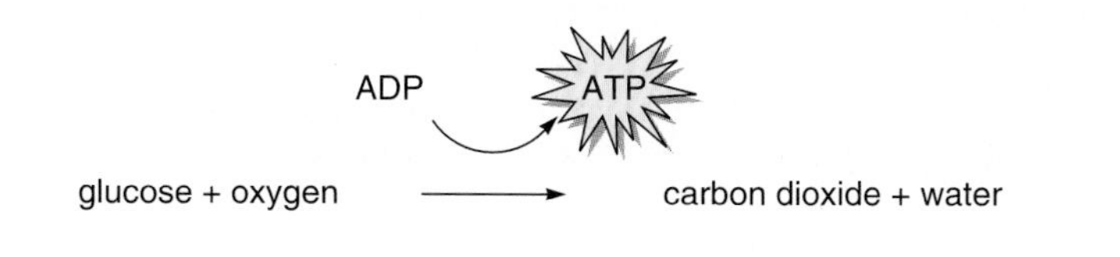

Myoglobin, an oxygen carrier similar to hemoglobin, is synthesized in muscle cells, and its presence accounts for the reddish brown color of skeletal muscle fibers. Myoglobin has a higher affinity of oxygen than does hemoglobin. Therefore, myoglobin can pull oxygen out of blood and make it available to muscle mitochondria which are carrying on aerobic respiration. Then, too, the ability of myoglobin to temporarily store oxygen reduces a muscle's immediate need for oxygen when aerobic respiration begins. The end products (carbon dioxide and water) can be rapidly disposed of, and the by-product heat keeps the entire body warm.

The three pathways for acquiring ATP work together during muscle contraction. But the anaerobic pathways are usually no longer needed as the body achieves an aerobic steady state. At this point, some lactate has accumulated but not enough to bring on exhaustion.

People who train rely even more heavily on aerobic respiration than people who do not train. In people who train, the number of muscle mitochondria increases, and so fermentation is not needed to produce ATP. Their mitochondria can start consuming oxygen as soon as ADP concentration starts rising during muscle contraction. Because mitochondria can break down fatty acid, instead of glucose, blood glucose is spared for the activity of the brain. (The brain, unlike other organs, can only utilize glucose to produce ATP.) Because less lactate is produced in people who train, the pH of the blood remains steady. And, there is less of an oxygen debt.

Oxygen Debt

When a muscle uses the anaerobic means of supplying energy needs, it incurs an **oxygen debt.** Oxygen debt is obvious when a person continues to breathe heavily after exercising. The ability to run up an oxygen debt is one of muscle tissue's greatest assets. Brain tissue cannot last nearly as long as muscles can without oxygen.

Repaying an oxygen debt requires replenishing creatine phosphate supplies and disposing of lactic acid. Lactic acid can be changed back to pyruvic acid and metabolized completely in mitochondria, or it can be sent to the liver to reconstruct glycogen. A marathon runner who has just crossed the finish line is not exhausted due to oxygen debt. Instead, the runner has used up all the muscles', and probably the liver's, glycogen supply. It takes about two days to replace glycogen stores on a high-carbohydrate diet.

Working muscles require a supply of ATP. Anaerobic creatine phosphate breakdown and glycolysis can quickly generate ATP. Aerobic respiration in mitochondria is best for sustained exercise.

Health Focus

Exercise, Exercise, Exercise

Exercise programs improve muscular strength, muscular endurance, and flexibility. Muscular strength is the force a muscle group (or muscle) can exert against a resistance in one maximal effort. Muscular endurance is judged by the ability of a muscle to contract repeatedly or sustain a contraction for an extended period. Flexibility is tested by observing the range of motion about a joint.

Exercise also improves cardiorespiratory endurance. The heart rate and capacity increase, and the air passages dilate so that the heart and lungs are able to support prolonged muscular activity. The blood level of high-density lipoprotein (HDL), the molecule that prevents the development of plaque in blood vessels, increases. Also, body composition, the proportion of protein to fat, changes favorably when you exercise.

Exercise also seems to help prevent certain kinds of cancer. Cancer prevention involves eating properly, not smoking, avoiding cancer-causing chemicals and radiation, undergoing appropriate medical screening tests, and knowing the early warning signs of cancer. However, studies show that people who exercise are less likely to develop colon, breast, cervical, uterine, and ovarian cancers.

Physical training with weights can improve bone density and strength and muscular strength and endurance in all adults regardless of age. Even men and women in their eighties and nineties make substantial gains in bone and muscle strength, which can help them lead more independent lives. Exercise helps prevent osteoporosis, a condition in which the bones are weak and tend to break. Exercise promotes the activity of osteoblasts in young people as well as older people. The stronger the bones when a person is young, the less chance of osteoporosis as a person ages. Exercise helps prevent weight gain not only because the level of activity increases but also because muscles metabolize faster than other tissues. As a person becomes more muscular, it is less likely that fat will accumulate.

Exercise relieves depression and enhances the mood. Some report exercise actually makes them feel more energetic, and after exercising, particularly in the late afternoon, they sleep better that night. Self-esteem rises not only because of improved appearance, but also due to other factors that are not well understood. It is known that vigorous exercise releases endorphins, hormone-like chemicals that are known to alleviate pain and provide a feeling of tranquility.

A sensible exercise program is one that provides all the benefits without the detriments of a too strenuous program. Overexertion can actually be harmful to the body and might result in sports injuries such as a bad back or bad knees. The programs suggested in Table 19A are tailored according to age and, if followed, are beneficial.

Dr. Arthur Leon at the University of Minnesota performed a study involving 12,000 men, and the results showed that only moderate exercise is needed to lower the risk of a heart attack by one-third. In another study conducted by the Institute for Aerobics Research in Dallas, Texas, which included 10,000 men and more than 3,000 women, even a little exercise was found to lower the risk of death from circulatory diseases and cancer. Increasing daily activity by walking to the corner store instead of driving and by taking the stairs instead of the elevator can improve your health.

Table 19A A Checklist for Staying Fit

Children, 7–12	Teenagers, 13–18	Adults, 19–55	Seniors, 55 and up
Vigorous activity 1–2 hours daily	Vigorous activity 1 hour 3–5 days a week, otherwise 1/2 hour daily moderate activity	Vigorous activity 1 hour 3 days a week, otherwise 1/2 hour daily moderate activity	Moderate exercise 1 hour daily 3 days a week, otherwise 1/2 hour daily moderate activity
Free play	Build muscle with calisthenics	Exercise to prevent lower back pain: aerobics, stretching, yoga	Plan a daily walk
Build motor skills through team sports, dance, swimming	Plan aerobic exercise to control buildup of fat cells	Take active vacations: hike, bicycle, cross-country ski	Daily stretching exercise
Encourage more exercise outside of physical education classes	Pursue tennis, swimming, horseback riding—sports that can be enjoyed for a lifetime	Find exercise partners: join a running club, bicycle club, outing group	Learn a new sport or activity: golf, fishing, ballroom dancing
Initiate family outings: bowling, boating, camping, hiking	Continue team sports, dancing, hiking, swimming		Try low-impact aerobics, boating, Before undertaking new exercises, consult your doctor

19.6 Exercise and Muscle Contraction

Aside from athletes who excel in a particular sport, there is a general interest today in staying fit by exercising.

Exercise and Size of Muscles

Muscles that are not used or that are used for only very weak contractions decrease in size, or atrophy. **Atrophy** can occur when a limb is placed in a cast or when the nerve serving a muscle is damaged. If nerve stimulation is not restored, muscle fibers gradually are replaced by fat and fibrous tissue. Unfortunately, atrophy can cause muscle fibers to shorten progressively, leaving body parts contracted in contorted positions.

Forceful muscular activity over a prolonged period causes muscle to increase in size as the number of myofibrils within muscle fibers increases. Increase in muscle size, called **hypertrophy,** occurs only if the muscle contracts to at least 75% of its maximum tension. Some athletes take anabolic steroids, either testosterone or related chemicals, to promote muscle growth. This practice has many undesirable side effects such as cardiovascular disease, liver and kidney dysfunction, impotency and sterility, and even an increase in rash behavior called "roid mania."

Exercise and Health

As discussed on the previous page, exercise is believed to have many benefits such as less risk of osteoporosis, cancer, and cardiovascular disease. Exercise also improves the spirits of many people, perhaps in part because exercise usually improves one's appearance.

Slow-Twitch and Fast-Twitch Muscle Fibers

We have seen that all muscle fibers metabolize both aerobically and anaerobically. Some fibers, however, utilize one method more than the other to provide myofibrils with ATP.

Slow-twitch fibers (Fig. 19.18) have a steadier tug and have more endurance despite motor units with a smaller number of fibers. These fibers are most helpful in sports like long-distance running, biking, jogging, and swimming. Because they produce most of their energy aerobically, they tire only when their fuel supply is gone. Slow-twitch fibers have many mitochondria and are dark in color because they contain myoglobin. They are also surrounded by dense capillary beds and draw more blood and oxygen than fast-twitch fibers. Slow-twitch fibers have a low maximum tension but are highly resistant to fatigue. Because slow-twitch fibers have a substantial reserve of glycogen and fat, their abundant mitochondria can maintain a steady, prolonged production of ATP when oxygen is available.

Fast-twitch fibers are those that tend to be anaerobic and seem to be designed for strength because their motor units contain many fibers. They provide explosions of energy and are most helpful in sports like sprinting, weight lifting, swinging a golf club, or putting a shot. Fast-twitch fibers are white in color because they have fewer mitochondria, little or no myoglobin, and fewer blood vessels than slow-twitch fibers do. Fast-twitch fibers can develop maximum tension more rapidly than slow-twitch fibers can, and their maximum tension is greater. However, their dependence on anaerobic energy leaves them vulnerable to an accumulation of lactic acid that causes them to fatigue quickly.

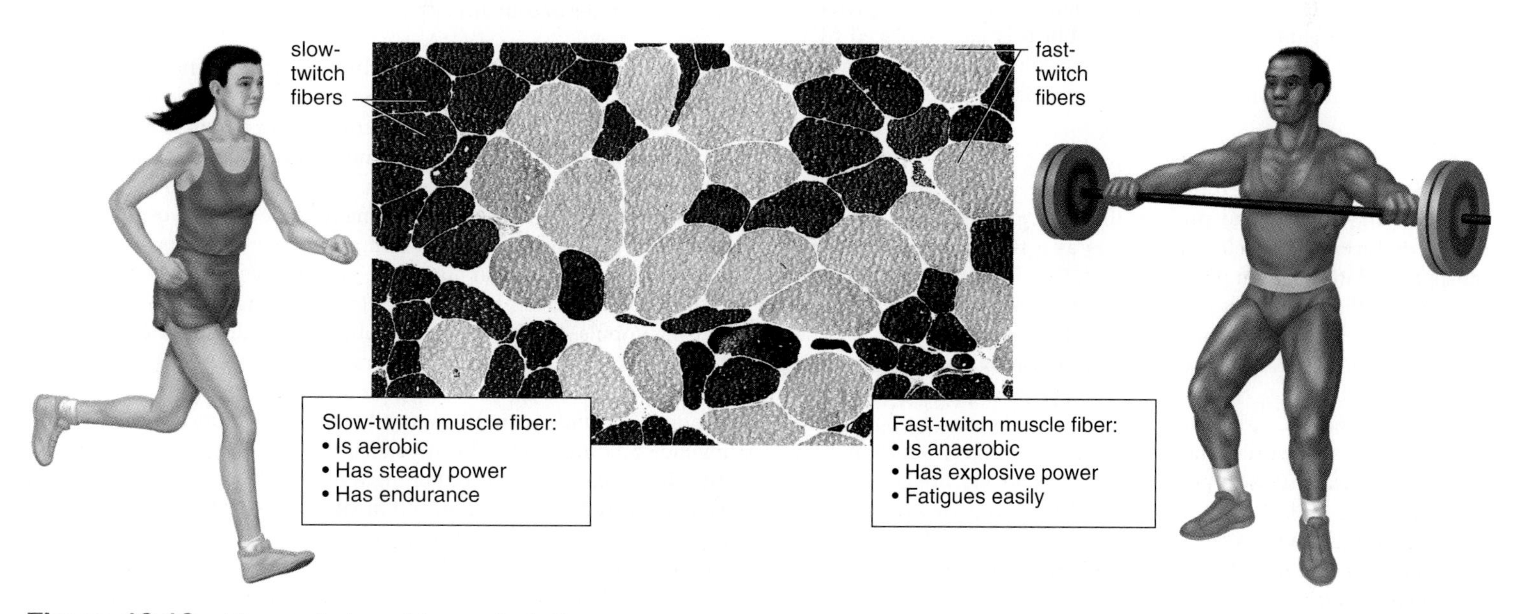

Figure 19.18 Slow-twitch and fast-twitch fibers.
If your muscles contain many slow-twitch fibers (dark color), you would probably do better at a sport like cross-country running. But if your muscles contain many fast-twitch fibers (light color), you would probably do better at a sport like weight lifting.

Bioethical Issue

On page 391, we learned that athletes are better at one sport or another, depending on whether their muscles contain fast- or slow-twitch fibers. A natural advantage of this sort does not bar an athlete from participating in and winning a medal in a particular sport at the Olympic games. Nor are athletes restricted to a certain amount of practice or required to eliminate certain foods from their diet.

Athletes are, however, prevented from participating in the Olympic games if they have taken certain performance enhancing drugs. There is no doubt that regular use of drugs like anabolic steroids leads to kidney disease, liver dysfunction, hypertension, and a myriad of other undesirable side effects. Even so, shouldn't the individual be allowed to take these drugs if they want to? Anabolic steroids are synthetic forms of the male sex hormone testosterone. Large doses along with strength training leads to much larger muscles than otherwise. Extra strength and endurance can give an athlete an advantage in certain sports such as racing, swimming, and weight lifting.

Should the Olympic committee outlaw the taking of anabolic steroids and, if so, on what basis? The basis can't be an unfair advantage, because some athletes naturally have an unfair advantage over other athletes. Should these drugs be outlawed on the basis of health reasons? Excessive practice alone and a purposeful decrease or increase in weight to better perform in a sport can also be injurious to one's health. In other words, how can you justify allowing some behaviors that enhance performance and not others?

Questions

1. Do you believe that the manner in which athletes wish to train and increase their performance should be regulated in any way? Why or why not?
2. Is it all right for athletes to endanger their health by excessive practice, gaining or losing pounds, or the taking of drugs? Why or why not?
3. Who should be in charge of regulating the behavior of athletes so that they do not do harm to themselves?

Summarizing the Concepts

19.1 Anatomy and Physiology of Bones

Bone and the associated tissues cartilage and fibrous connective tissue are found in the skeleton. Hyaline (articular) cartilage covers the ends, while periosteum (fibrous connective tissue) covers the rest of a long bone. Spongy bone, which may contain red bone marrow, is in the ends of long bones, and yellow bone marrow is in the medullary cavity of the shaft. The wall of the shaft is compact bone.

Bone is a living tissue that can grow and repair itself. The prenatal human skeleton is at first cartilaginous, but it is later replaced by a bony skeleton. During adult life, bone is constantly being broken down by osteoclasts and then rebuilt by osteoblasts that become osteocytes in lacunae.

19.2 Bones of the Skeleton

The skeleton supports and protects the body, permits flexible movement, produces blood cells, and serves as a storehouse for mineral salts, particularly calcium phosphate.

The axial skeleton lies in the midline of the body and consists of the skull, the hyoid bone, the vertebral column, and the rib cage. The skull contains the cranium, which protects the brain and the facial bones.

The appendicular skeleton consists of the bones of the pectoral girdle, arms, pelvic girdle, and legs. The pectoral girdle and arms are adapted for flexibility. The pelvic girdle and the legs are adapted for strength: the femur is the strongest bone in the body. However, the foot, like the hand, is flexible because it contains so many bones. Both fingers and toes are digits with bones called phalanges.

There are three types of joints: fibrous joints, like the sutures of the cranium, are immovable; cartilaginous joints, like those between the ribs and sternum and the pubic symphysis, are slightly movable; and the synovial joints, consisting of a membrane-lined (synovial membrane) capsule, are freely movable. There are different kinds of synovial joints based on the movements they permit.

19.3 Skeletal Muscles

Muscles have various functions: they help maintain posture, provide movement and heat, and protect underlying organs. Whole muscles work in antagonistic pairs; for example, the biceps flexes the lower arm and the triceps extends it.

Muscles are named for their size, shape, direction of fibers, location, number of attachments, and action.

19.4 Mechanism of Muscle Fiber Contraction

Nerve impulses travel down motor nerve fibers and meet muscle fibers at neuromuscular junctions. The sarcolemma of a muscle fiber forms T tubules that extend into the fiber and almost touch the sarcoplasmic reticulum, which stores calcium ions. When calcium ions (Ca^{2+}) are released into muscle fibers, actin filaments slide past myosin filaments within the sarcomeres of a myofibril.

At a neuromuscular junction, synaptic vesicles release acetylcholine (ACh), which binds to protein receptors on the sarcolemma, causing impulses to travel down T tubules and calcium to leave the sarcoplasmic reticulum. Myofibril contraction follows.

Calcium ions bind to troponin and cause tropomyosin threads winding around actin filaments to shift their position, revealing myosin binding sites. The myosin filament is composed of many myosin molecules, each containing a head with an ATP binding site. Myosin is an ATPase, and once it breaks down ATP, the myosin head is ready to attach to actin. The release of ADP + Ⓟ causes the head to change its position. This is the power stroke that causes the actin filament to slide toward the center of a sarcomere. When myosin catalyzes another ATP, the head detaches from actin, and the cycle begins again.

19.5 Whole Muscle Contraction

In the laboratory, muscle contraction is described in terms of a muscle twitch, summation, and tetanus. In the body, muscles exhibit tone, in which tetanic contraction involving a number of fibers is the rule.

A muscle fiber has three ways to acquire ATP after muscle contraction begins: (1) creatine phosphate, built up when a muscle is resting, donates phosphates to ADP, forming ATP; (2) fermentation with the concomitant accumulation of lactic acid quickly produces ATP; and (3) oxygen-dependent aerobic respiration occurs within mitochondria. Two of these ways (1 and 2) are anaerobic (they do not require oxygen). Fermentation can result in oxygen debt because oxygen is needed to complete the metabolism of lactate acid.

19.6 Exercise and Muscle Contraction

Certain sports like running and swimming can be associated with slow-twitch fibers, which rely on aerobic respiration to acquire ATP. They have a plentiful supply of mitochondria and myoglobin, which gives them a dark color. Other sports like weight lifting can be associated with fast-twitch fibers, which rely on an anaerobic means of acquiring ATP. They have few mitochondria and myoglobin, and their motor units contain more muscle fibers. Fast-twitching fibers are known for their explosive power, but they fatigue quickly.

Studying the Concepts

1. Describe the makeup of long bone. Describe endochondral ossification and remodeling of a bone. 370–72
2. List and discuss the functions of the skeletal system in humans. 373
3. Name and describe the bones of the axial and appendicular skeletons. 374–79
4. Which of the bones described in question 3 form the pectoral girdle and the pelvic girdle? 378–79
5. How are joints classified? Give examples of the different types of synovial joints and the movements they permit. 380
6. List and discuss the functions of the muscular system in humans. 381
7. Describe the steps resulting in muscle contraction by starting with the motor neuron and ending with the sliding of actin filaments. 385
8. Describe the structure and function of a neuromuscular junction. 386
9. Contrast a muscle twitch with summation and tetanus. 388
10. What is muscle tone, and how is it maintained? 388
11. What are the three ways a muscle fiber can acquire ATP for muscle contraction? Discuss a benefit of each method. 389
12. Contrast slow-twitch and fast-twitch fibers in as many ways as possible. 391

Testing Yourself

Choose the best answer for each question.

For questions 1–4, match the following items with the correct locations given in the key.

Key:

a. upper arm
b. lower arm
c. pectoral girdle
d. pelvic girdle
e. upper leg
f. lower leg

1. ulna
2. tibia
3. clavicle
4. femur
5. Spongy bone
   a. contains osteons.
   b. contains red bone marrow where blood cells are formed.
   c. lends no strength to bones.
   d. takes up most of a leg bone.
   e. All of these are correct.
6. Which of these is mismatched?
   a. immovable—cranial sutures
   b. slightly movable joint—vertebrae
   c. hinge joint—hip
   d. synovial joint—elbow
   e. immovable joint—sutures in cranium
7. In a muscle fiber
   a. the sarcolemma is connective tissue holding the myofibrils together.
   b. the T system consists of tubules.
   c. both filaments have cross-bridges.
   d. there is no endoplasmic reticulum.
   e. All of these are correct.
8. Nervous stimulation of muscles
   a. occurs at a neuromuscular junction.
   b. results in an action potential that travels down the T system.
   c. causes calcium to be released from expanded regions of the sarcoplasmic reticulum.
   d. All of these are correct.
9. When muscles contract
   a. sarcomeres increase in size.
   b. actin breaks down ATP.
   c. myosin slides past actin.
   d. the H zone disappears.
   e. calcium is taken up by the sarcoplasmic reticulum.
10. Which of these is a direct source of energy for muscle contraction?
    a. ATP
    b. creatine phosphate
    c. lactic acid
    d. glycogen
    e. Both a and b are correct.
11. When actin filaments move as a sarcomere contracts,
    a. myosin filaments are hydrolyzing ATP molecules.
    b. ADP + Ⓟ are released from myosin filament heads.
    c. ATP is binding to myosin heads.
    d. ADP + Ⓟ are attached to myosin heads.
    e. All of these in the sequence mentioned.

12. Label this diagram of a muscle fiber, using these terms: myofibril, mitochondrion, T tubule, sarcomere, sarcolemma, sarcoplasmic reticulum.

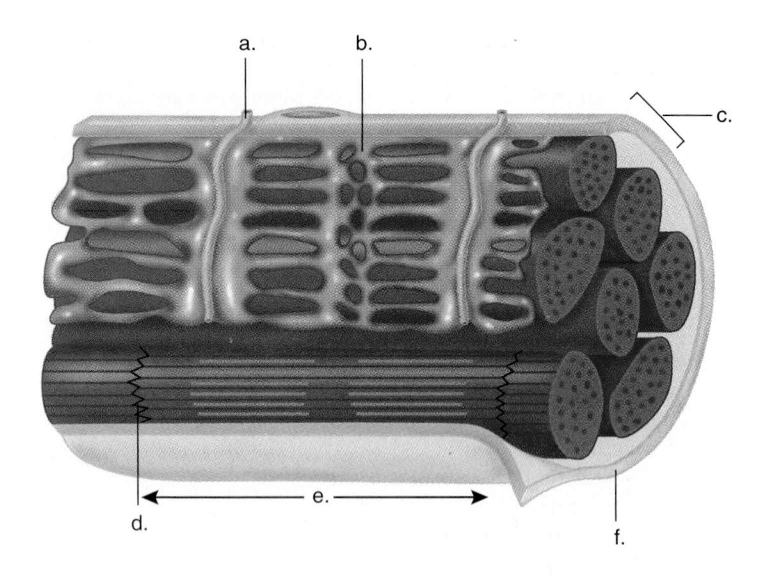

Thinking Scientifically

1. Considering skeletal bones (page 370):
   a. What evidence do you have that bone is living tissue?
   b. Why would you expect persons with stronger muscles to have stronger bones?
   c. Bones have all sorts of grooves and protuberances. What purpose might they have?
   d. The female pelvis is wider than the male pelvis. With what might this be associated?
2. A muscle fiber obeys the all-or-none law—that is, it contracts or it doesn't contract (page 385).
   a. Therefore, do all myofibrils in a muscle fiber contract at the same time?
   b. When a sarcomere contracts, does the Z line move? If so, in which direction?
   c. A respiratory pigment in muscle called myoglobin receives oxygen (O_2) from hemoglobin. Which of these two respiratory pigments has the higher affinity for oxygen?
   d. When we exercise, blood brings more oxygen to the muscles. What do the muscles specifically do with all this oxygen?

Understanding the Terms

actin 385
appendicular skeleton 378
arthritis 380
articular cartilage 370
atrophy 391
axial skeleton 374
ball-and-socket joint 380
bursa 380
bursitis 380
compact bone 370
creatine phosphate 389
fontanel 374
foramen magnum 374
growth plate 372
hinge joint 380
hypertrophy 391
insertion 381
intervertebral disk 376
joint 370
ligament 370
mastoiditis 374
menisci 380
motor unit 388
muscle twitch 388
myofibril 385
myoglobin 389
myogram 388
myosin 385
neuromuscular junction 386
origin 381
osteoclast 372
osteocyte 370
osteon 370
oxygen debt 389
pectoral girdle 378
pelvic girdle 379
periosteum 370
red bone marrow 370
sarcolemma 385
sarcomere 385
sarcoplasmic reticulum 385
scoliosis 376
sinus 374
sliding filament theory 385
spongy bone 370
suture 380
synovial joint 380
synovial membrane 380
T (transverse) tubule 385
tendon 370
tetanus 388
tone 388
tropomyosin 387
troponin 387
vertebral column 376
yellow bone marrow 370

Match the terms to these definitions:

a. __________ Sustained maximal muscle contraction.
b. __________ Fibrous connective tissue that joins bone to bone at a joint.
c. __________ Plasma membrane of a muscle fiber.
d. __________ Compound unique to muscles that contains a high-energy phosphate bond.
e. __________ Blood-cell forming tissue located in the spaces within spongy bone.

Using Technology

Your study of the musculoskeletal system is supported by these available technologies:

Essential Study Partner CD-ROM

Animals → Support / Locomotion

Visit the Mader web site for related ESP activities.

Exploring the Internet

The Mader Home Page provides resources and tools as you study this chapter.

http://www.mhhe.com/biosci/genbio/mader

Dynamic Human 2.0 CD-ROM

Skeletal System

Muscular System

Virtual Physiology Laboratory CD-ROM

Frog Muscle

C H A P T E R 20

Endocrine System

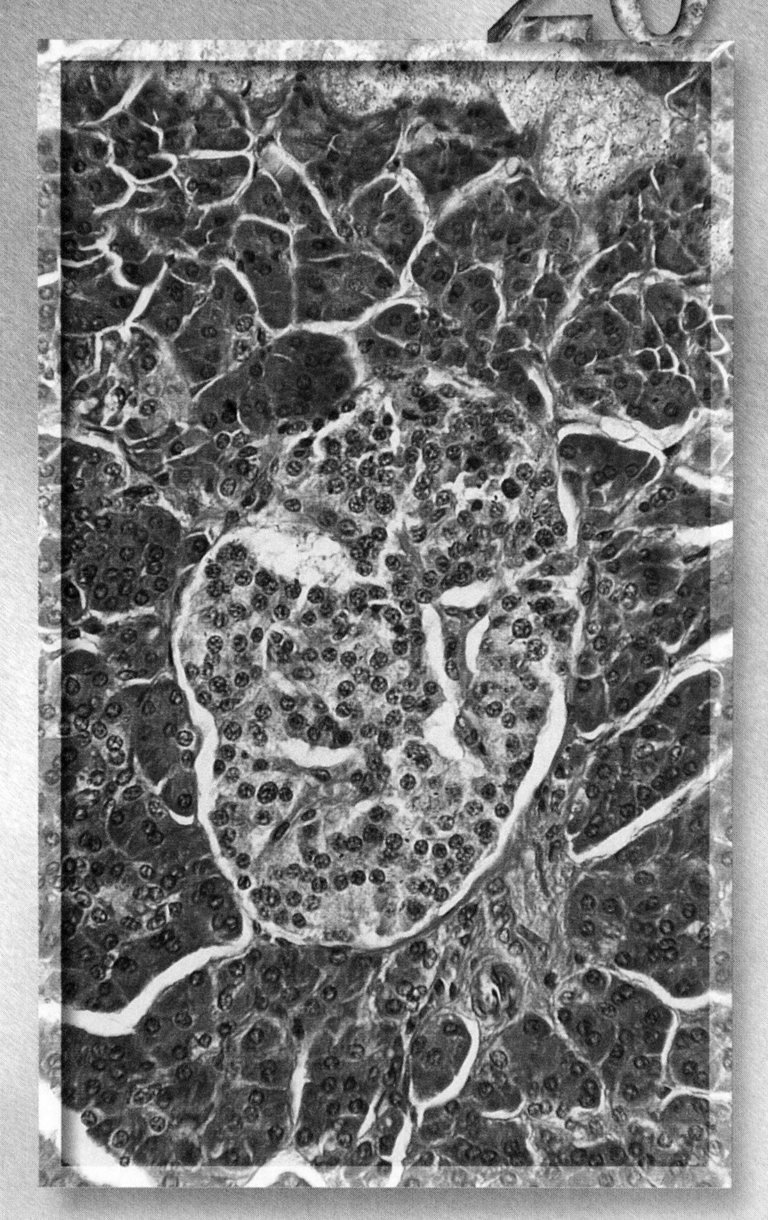

The pancreatic islets are specialized regions of the pancreas, an endocrine gland. The islets secrete the hormone insulin when the blood glucose level is high. Insulin increases the passage of glucose into cells, restoring the normal level.

Chapter Concepts

Harry stood before the mirror making sure his appearance was just right. Tonight was the big junior high dance, and he was taking Mary on his first date. He noticed that he was getting a slight growth of hair on his upper lip and chin, and he couldn't decide whether to shave or let it grow. But the facial hair didn't concern him nearly as much as his voice—the change of pitch, which he could not control, was quite embarrassing. Although some of the kids laughed when this happened, he hoped Mary would not be one of them. Harry even wondered why he cared what Mary thought; just last year he wouldn't have wanted to go to the dance, never mind go on a date with a girl.

Harry was experiencing normal signs of puberty brought on by an increase in sex hormones. Sex hormones are produced by the gonads which are a part of the endocrine system. Along with the nervous system, the endocrine system coordinates the various activities of body parts such as the changes that occur during puberty. The nervous system quickly transmits electrical signals along nerve fibers. The endocrine system acts more slowly because hormones have to be produced and then usually transported in the blood to target organs.

20.1 Environmental Signals

Even though the endocrine system and the nervous system have distinct differences, they both utilize chemical messengers—hormones versus neurotransmitter molecules. This realization supports a general interest in environmental signals that can be categorized in these three ways (Fig. 20.1).

Environmental signals that act at a distance between organisms. **Pheromones** are chemical messengers that pass between members of a species. As an example, ants lay down a pheromone trail to direct other ants to food, and female silkworm moths release a sex attractant that is received by male moth antennae even several kilometers away. Mammals also release pheromones, as when dogs use their urine to serve as a territorial marker. Some studies are being conducted to determine if humans have pheromones. In one investigation, it was observed that women who live in close quarters tend to have coinciding menstrual cycles. It is possible that this is due to a pheromone.

Environmental signals that act at a distance between body parts. This category includes **hormones** that are produced by endocrine glands or by neurosecretory cells of the hypothalamus. An overlap between the nervous and endocrine systems is also exemplified by the secretion of epinephrine and norepinephrine at sympathetic nerve endings and also by the adrenal medulla, an endocrine gland.

Environmental signals that act locally between adjacent cells. Neurotransmitters belong in this category, as do substances that are sometimes called local hormones. For example, when the skin is cut, histamine, released by mast cells, promotes the inflammatory response.

Today, hormones are categorized as one type of environmental signal, a term which includes molecules that work at a distance between individuals or body parts or locally between adjacent cells.

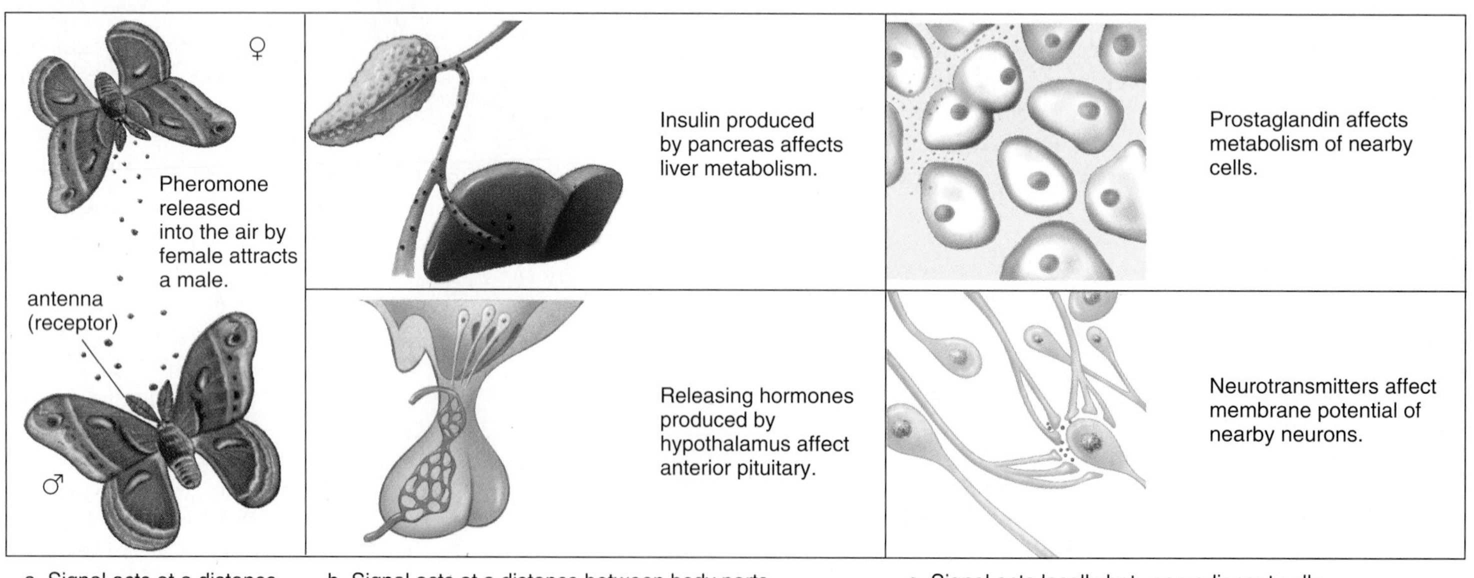

Figure 20.1 Environmental signals.
There are three categories of environmental signals (red dots). **a.** Pheromones are chemical messengers that act at a distance between individuals. **b.** Endocrine hormones and neurosecretions typically are carried in the bloodstream and act at a distance within the body of a single organism. **c.** Some chemical messengers have local effects only; they pass between cells that are adjacent to one another.

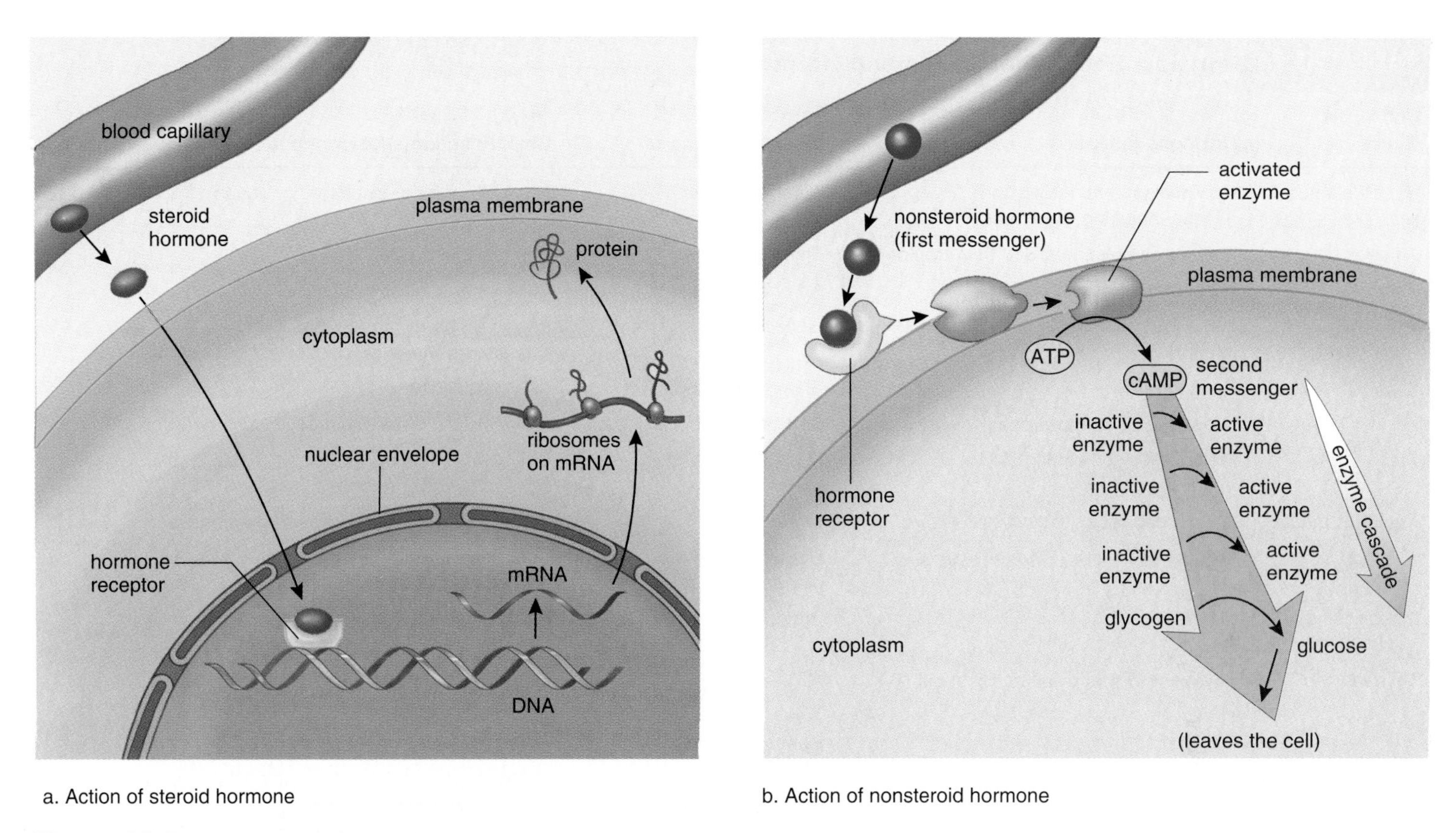

a. Action of steroid hormone

b. Action of nonsteroid hormone

Figure 20.2 Cellular activity of hormones.
a. After passing through the plasma membrane and nuclear envelope, a steroid hormone binds to a receptor protein inside the nucleus. The hormone-receptor complex then binds to DNA, and this leads to activation of certain genes and protein synthesis. **b.** Nonsteroid hormones, called first messengers, bind to a specific receptor protein in the plasma membrane. A protein relay ends when an enzyme converts ATP to cAMP, the second messenger, which activates an enzyme cascade.

The Action of Hormones

Hormones fall into two basic categories: (1) a **nonsteroid hormone** can be an amino acid, a peptide, or a protein composed of one or more polypeptides; (2) a **steroid hormone** is always the same complex of four carbon rings, but each type has different side chains. Because their effect is amplified through cellular mechanisms, either type of hormone can function at extremely low concentrations.

A hormone does not seek out a particular organ; rather the organ is awaiting the arrival of the hormone. Cells that can react to a hormone have hormone receptor proteins that combine with the hormone in a lock-and-key manner. Steroid hormones are lipids and therefore they cross cell membranes (Fig. 20.2*a*). Only after they are inside the nucleus do steroid hormones, such as estrogen and progesterone, bind to hormone receptor proteins. The hormone-receptor complex then binds to DNA, activating particular genes. Activation leads to production of a cellular enzyme in multiple quantities.

Most nonsteroid hormones cannot pass through the plasma membrane, and instead they bind to a receptor protein in the membrane (Fig. 20.2*b*). After epinephrine binds to a receptor protein, a relay system leads to the conversion of ATP to **cyclic AMP** (cyclic adenosine monophosphate). Cyclic AMP (cAMP) is made from ATP, but it contains only one phosphate group, which is attached to the adenosine portion of the molecule at two spots. Thus the molecule is cyclic. The nonsteroid hormone is called the *first messenger* and cAMP—or some other molecule—is called the *second messenger.* Calcium is also a common second messenger, and this helps explain why calcium regulation in the body is so important.

The second messenger sets in motion an *enzyme cascade.* In muscle cells epinephrine leads to the breakdown of glycogen to glucose (Fig. 20.2*b*). An enzyme cascade is so-called because each enzyme in turn activates another. Because enzymes work over and over, every step in an enzyme cascade leads to more reactions—the binding of a single nonsteroid hormone molecule can result even in a thousandfold response.

Hormones are chemical messengers that influence the metabolism of the cell either indirectly by regulating the production of a particular protein (steroid hormone) or directly by activating an enzyme cascade (nonsteroid hormone).

Table 20.1 Principal Endocrine Glands and Hormones

Endocrine Gland	Hormone Released	Target Tissues/Organs	Chief Function(s) of Hormone
Hypothalamus	Hypothalamic-releasing and release-inhibiting hormones	Anterior pituitary	Regulate anterior pituitary hormones
Anterior pituitary	Thyroid-stimulating (TSH, thyrotropic)	Thyroid	Stimulates thyroid
	Adrenocorticotropic (ACTH)	Adrenal cortex	Stimulates adrenal cortex
	Gonadotropic [follicle-stimulating (FSH), luteinizing (LH)]	Gonads	Egg and sperm production, and sex hormone production
	Prolactin (PRL)	Mammary glands	Milk production
	Growth (GH, somatotropic)	Soft tissues, bones	Cell division, protein synthesis, and bone growth
	Melanocyte-stimulating (MSH)	Melanocytes in skin	Unknown function in humans; regulates skin color in lower vertebrates
Posterior pituitary	Antidiuretic (ADH, vasopressin)	Kidneys	Stimulates water reabsorption by kidneys
	Oxytocin	Uterus, mammary glands	Stimulates uterine muscle contraction and release of milk by mammary glands
Pineal gland	Melatonin	Brain	Circadian and circannual rhythms; possibly involved in maturation of sex organs
Thyroid	Thyroxine (T_4) and triiodothyronine (T_3)	All tissues	Increases metabolic rate; regulates growth and development
	Calcitonin	Bones, kidneys, intestine	Lowers blood calcium level
Parathyroids	Parathyroid (PTH)	Bones, kidneys, intestine	Raises blood calcium level
Thymus	Thymosins	T lymphocytes	Production and maturation of T lymphocytes
Adrenal cortex	Glucocorticoids (cortisol)	All tissues	Raise blood glucose level; stimulate breakdown of protein
	Mineralocorticoids (aldosterone)	Kidneys	Reabsorb sodium and excrete potassium
	Sex hormones	Gonads, skin, muscles, bones	Stimulate sex characteristics
Adrenal medulla	Epinephrine and norepinephrine	Cardiac and other muscles	Emergency situations; raise blood glucose level
Pancreas	Insulin	Liver, muscles, adipose tissue	Lowers blood glucose level; promotes formation of glycogen
	Glucagon	Liver, muscles, adipose tissue	Raises blood glucose level
Gonads			
Testes	Androgens (testosterone)	Gonads, skin, muscles, bones	Stimulate secondary male sex characteristics
Ovaries	Estrogens and progesterone	Gonads, skin, muscles, bones	Stimulate female sex characteristics

Endocrine Glands

Endocrine glands can be contrasted with exocrine glands. The latter have ducts and secrete their products into these ducts for transport into body cavities. For example, the salivary glands send saliva into the mouth by way of the salivary ducts. **Endocrine glands** are ductless; they secrete their hormones directly into the bloodstream for distribution throughout the body.

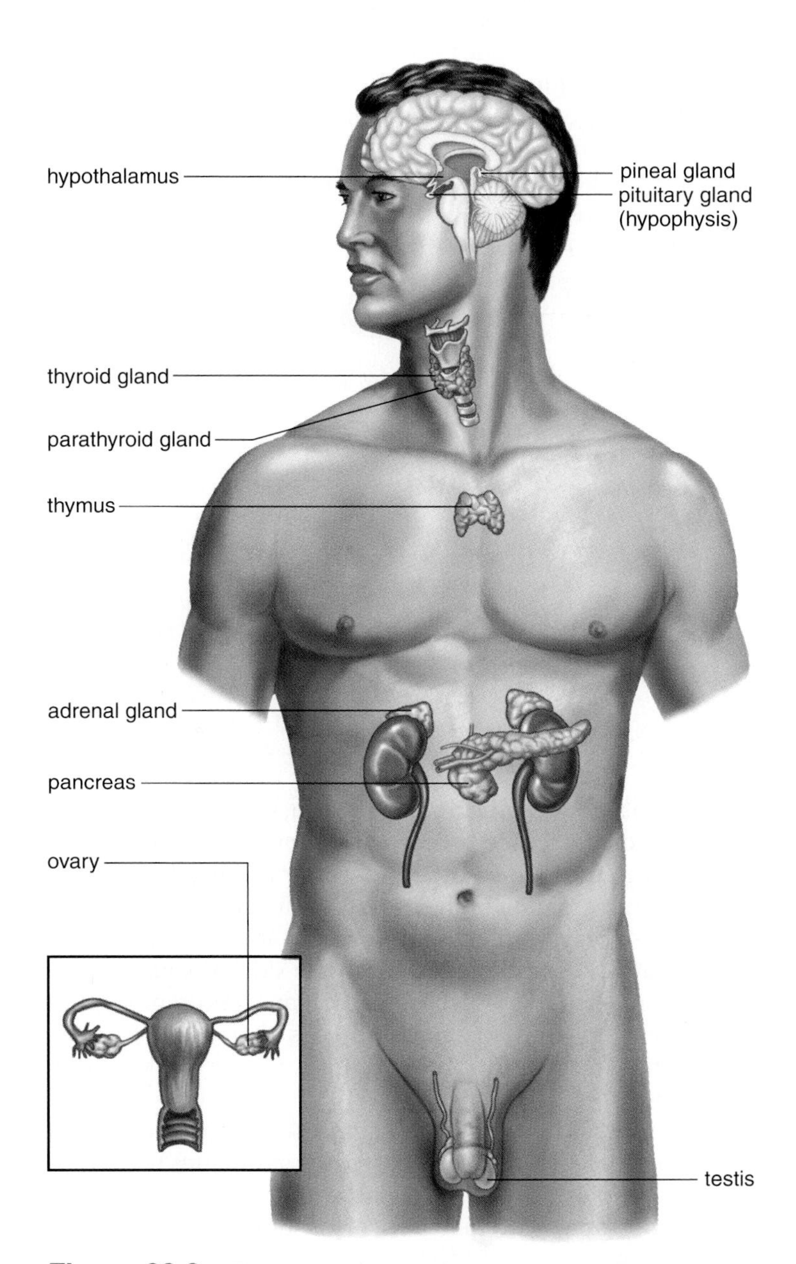

Figure 20.3 The endocrine system.
Anatomical location of major endocrine glands in the body.

Table 20.1 lists the hormones released by the principal endocrine glands, which are depicted in Figure 20.3. The hypothalamus, a part of the brain, is in close proximity to the pituitary. The hypothalamus controls the pituitary gland and this, too, exemplifies the close association between the nervous and endocrine systems. The pineal gland is also located in the brain. The thyroid and parathyroids are located in the neck, and the thymus lies just beneath the sternum, in the thoracic cavity. The adrenal glands and pancreas are located in the abdominal cavity. The gonads include the ovaries, located in the pelvic cavity, and the testes, located outside this cavity in the scrotum.

Like the nervous system, the endocrine system is especially involved in homeostasis, that is, the dynamic equilibrium of the internal environment. The internal environment is the blood and tissue fluid that surrounds the body's cells. Notice that several hormones directly affect the osmolarity of the blood. Others control the calcium and glucose levels. Several hormones are involved in the maturation and function of the reproductive organs. In fact, many people are most familiar with the effect of hormones on sexual functions.

There are two mechanisms that control the effect of endocrine glands. Quite often a negative feedback mechanism controls the secretion of hormones. An endocrine gland can be sensitive to either the condition it is regulating or to the blood level of the hormone it is producing. For example, when the blood glucose level rises, the pancreas produces insulin, which causes the cells to take up glucose and the liver to store glucose. The stimulus for the production of insulin has thereby been dampened, and therefore the pancreas stops producing insulin. On the other hand, when the blood level of thyroid hormones rises, the anterior pituitary stops producing thyroid-stimulating hormones. We will discuss these examples in more detail later.

The presence of contrary hormonal actions is a way the effect of a hormone is controlled. The action of insulin, for example, is offset by the production of glucagon by the pancreas. Notice there are other examples of contrary hormonal actions in Table 20.1. The thyroid lowers the blood calcium level, but the parathyroids raise the blood calcium level. We will also have the opportunity to point out other instances in which hormones work opposite to one another and thereby bring about the regulation of a substance in the blood.

The secretion of a hormone is often controlled by negative feedback, and the effect of a hormone is often opposed by a contrary hormone. The end result is homeostasis and the normal functioning of body parts.

20.2 Hypothalamus and Pituitary Gland

The **hypothalamus** regulates the internal environment through the autonomic system. For example, it helps control heart beat, body temperature, and water balance. The hypothalamus also controls the glandular secretions of the **pituitary gland.** The pituitary, a small gland about 1 cm in diameter, is connected to the hypothalamus by a stalklike structure. The pituitary has two portions: the anterior pituitary and the posterior pituitary (Fig. 20.4).

Posterior Pituitary

There are neurons in the hypothalamus called neurosecretory cells which produce the hormones **antidiuretic hormone (ADH)** and oxytocin. These hormones pass through axons into the **posterior pituitary** where they are stored in axon endings. ADH promotes the reabsorption of water from the collecting ducts attached to nephrons within the kidneys. There are neurons in the hypothalamus that act as a sensor because they are sensitive to the osmolarity of the blood. When these cells determine that the blood is too concentrated, ADH is released into the bloodstream from the axon endings in the posterior pituitary. Upon reaching the kidneys, ADH causes water to be reabsorbed. As the blood becomes dilute, ADH is no longer released. This is an example of control by negative feedback because the effect of the hormone (to dilute blood) acts to shut down the release of the hormone.

Inability to produce ADH causes diabetes insipidus (watery urine), in which a person produces copious amounts of urine with a resultant loss of ions from the blood. The condition can be corrected by the administration of ADH.

Oxytocin is the other hormone that is made in the hypothalamus and stored in the posterior pituitary. When labor begins during childbirth, pressure receptors in the uterine wall send nerve impulses to the hypothalamus, and thereafter oxytocin is released from the posterior pituitary. Oxytocin then causes the uterus to contract more forcefully. This is an example of control by positive feedback: uterine contractions (the condition) brings about a result that increases uterine contractions. Oxytocin also stimulates the release of milk from the mammary glands when a baby is nursing.

The posterior pituitary stores two hormones, ADH and oxytocin, both of which are produced by and released from neurosecretory cells in the hypothalamus.

Anterior Pituitary

A portal system, consisting of two capillary systems connected by a vein, lies between the hypothalamus and the anterior pituitary (Fig. 20.4). The hypothalamus controls the anterior pituitary by producing **hypothalamic-releasing hormones** and **hypothalamic-inhibiting hormones.** For example, there is a thyroid-releasing hormone (TRH) and a thyroid-inhibiting hormone (TIH). TRH stimulates the anterior pituitary to release thyroid-stimulating hormone, and TIH inhibits the pituitary from releasing thyroid-stimulating hormone.

Three of the six hormones produced by the **anterior pituitary** (hypophysis) have an effect on other glands: (1) **thyroid-stimulating hormone (TSH)** stimulates the thyroid to produce thyroid hormones; (2) **adrenocorticotropic hormone (ACTH)** stimulates the adrenal cortex to produce cortisol; and (3) **gonadotropic hormones (FSH** and **LH),** stimulate the gonads—the testes in males and the ovaries in females—to produce gametes and sex hormones. In each instance, the blood level of the last hormone in the sequence exerts negative feedback control over the secretion of the first two hormones:

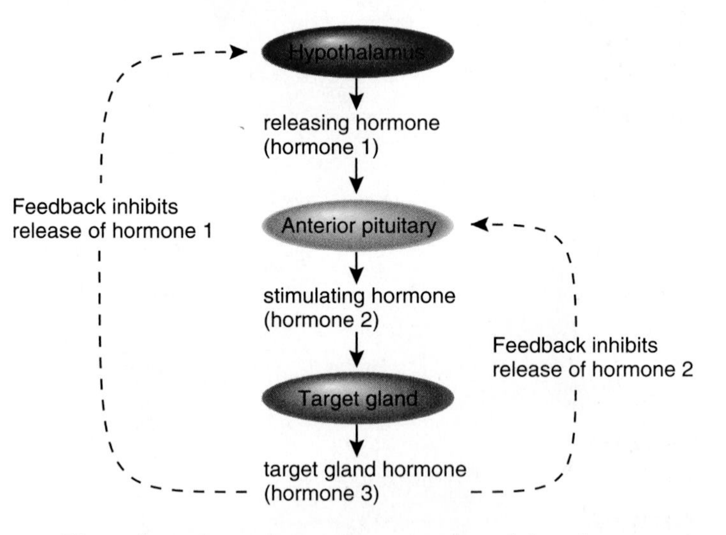

The other three hormones produced by the anterior pituitary do not affect other endocrine glands. **Prolactin (PRL)** is produced in quantity only after childbirth. It causes the mammary glands in the breasts to develop and produce milk. It also plays a role in carbohydrate and fat metabolism.

Melanocyte-stimulating hormone (MSH) causes skin-color changes in many fishes, amphibians, and reptiles that have melanophores, special skin cells that produce color variations. The concentration of this hormone in humans is very low.

Growth hormone (GH), or somatotropic hormone, promotes skeletal and muscular growth. It stimulates the rate at which amino acids enter cells and protein synthesis occurs. It also promotes fat metabolism as opposed to glucose metabolism.

The hypothalamus, the anterior pituitary, and other glands controlled by the anterior pituitary are all involved in a self-regulating negative feedback mechanism.

Visual Focus

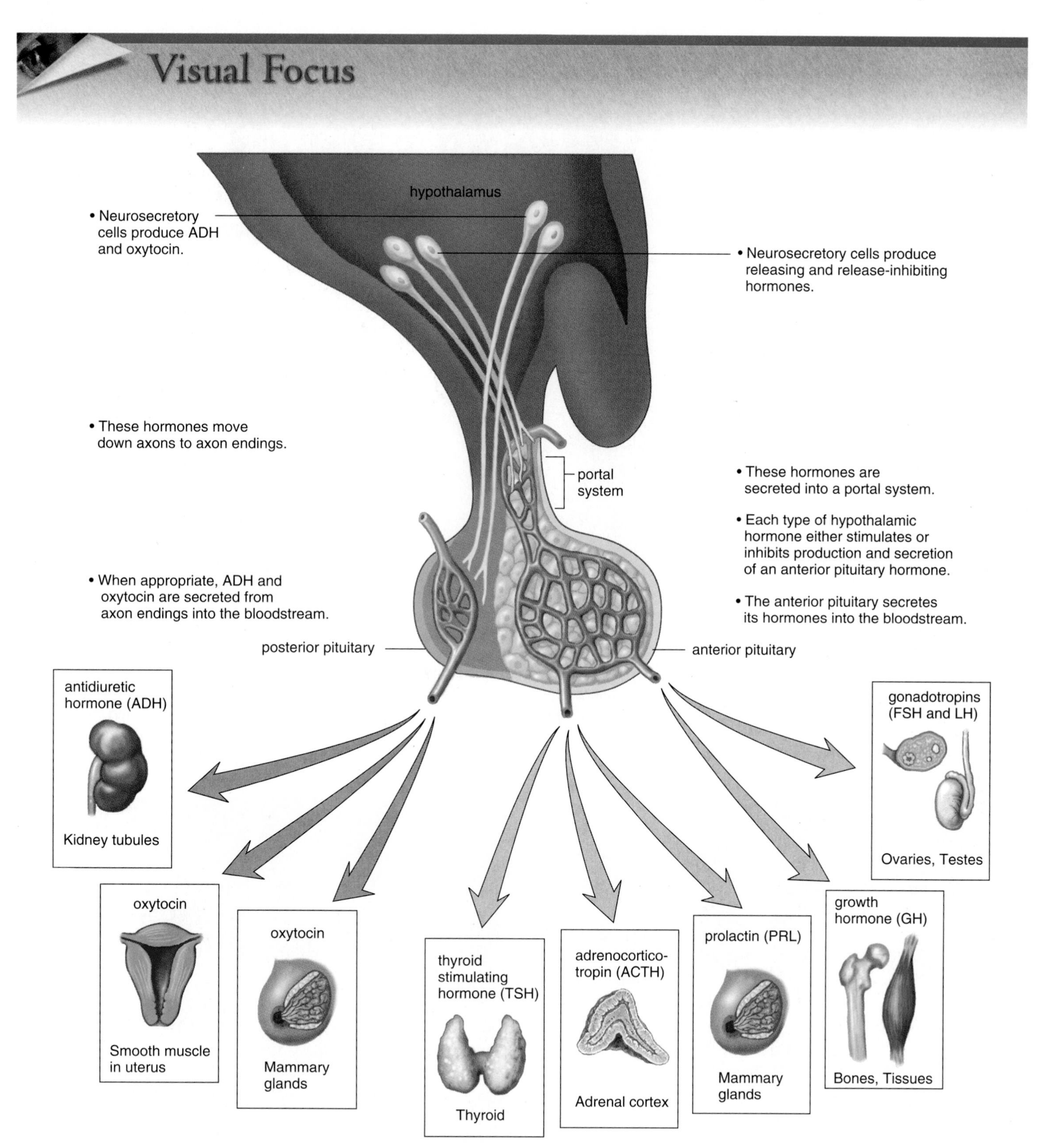

Figure 20.4 Hypothalamus and the pituitary.
The hypothalamus produces two hormones, ADH and oxytocin, which are stored and secreted by the posterior pituitary. The hypothalamus controls the secretions of the anterior pituitary, and the anterior pituitary controls the secretions of the thyroid, adrenal cortex, and gonads, which are also endocrine glands.

Effects of Growth Hormone

GH is produced in greatest quantities during childhood and adolescence, when most body growth is occurring. If too little GH is produced during childhood, the individual becomes a **pituitary dwarf,** characterized by perfect proportions but small stature. If too much GH is secreted, a person can become a giant (Fig. 20.5). Giants usually have poor health, primarily because GH has a secondary effect on the blood glucose level, promoting an illness called diabetes mellitus.

On occasion, there is overproduction of growth hormone in the adult, and a condition called **acromegaly** results. Long bone growth is no longer possible in adults. Only the feet, hands, and face (particularly the chin, nose, and eyebrow ridges) can respond, and these portions of the body become overly large (Fig. 20.6).

The amount of growth hormone during childhood and adulthood affects the height of an individual.

Figure 20.5 Effect of growth hormone.
The amount of growth hormone production during childhood affects the height of an individual. Excessive growth hormone produces very tall basketball players and even giants. Little growth hormone results in limited stature and even pituitary dwarfism.

Figure 20.6 Acromegaly.
Acromegaly is caused by overproduction of GH in the adult. It is characterized by an enlargement of the bones in the face, the fingers, and the toes of an adult.

20.3 Thyroid and Parathyroid Glands

The **thyroid gland** is a large gland located in the neck, where it is attached to the trachea just below the larynx (see Fig. 20.3). The parathyroid glands are imbedded in the posterior surface of the thyroid gland.

Thyroid Gland

The thyroid gland is composed of a large number of follicles, each a small spherical structure made of thyroid cells filled with **thyroxine** (T_4), which contains four iodine atoms and four **triiodothyronine** (T_3), which contains three iodine atoms.

Effects of Thyroid Hormones

To produce thyroxine and triiodothyronine, the thyroid gland actively acquires iodine. The concentration of iodine in the thyroid gland can increase to as much as 25 times that of blood. If iodine is lacking in the diet, the thyroid gland is unable to produce the thyroid hormones. In response to constant stimulation by the anterior pituitary, it enlarges, resulting in a **simple goiter** (Fig. 20.7). Some years ago it was discovered that the use of iodized salt allows the thyroid to produce the thyroid hormones, and therefore helps prevent simple goiter.

Thyroid hormones increase the metabolic rate. They do not have one target organ; instead, they stimulate all organs of the body to metabolize at a faster rate. More glucose is broken down and more energy is utilized.

If the thyroid fails to develop properly, a condition called **cretinism** results (Fig. 20.8). Individuals with this condition are short and stocky and have had extreme hypothyroidism since infancy or childhood. Thyroid hormone therapy can initiate growth, but unless treatment is begun within the first two months, mental retardation results. The occurrence of hypothyroidism in adults produces the condition known as **myxedema,** which is characterized by lethargy, weight gain, loss of hair, slower pulse rate, lowered body temperature, and thickness and puffiness of the skin. The administration of adequate doses of thyroid hormones restores normal function and appearance.

In the case of hyperthyroidism, or *Graves' disease,* the thyroid gland is enlarged and overactive, causing a goiter to form. The eyes protrude because of edema in eye socket tissues and swelling of muscles that move the eyes. This type of goiter is called **exophthalmic goiter.** The patient usually becomes hyperactive, nervous, irritable, and suffers from insomnia. Removal or destruction of a portion of the thyroid by means of radioactive iodine is sometimes effective in curing the condition. Hyperthyroidism can also be caused by a thyroid tumor, which is usually detected as a lump during physical examination. Again, the treatment is surgery in combination with administration of radioactive iodine. The prognosis for most patients is excellent.

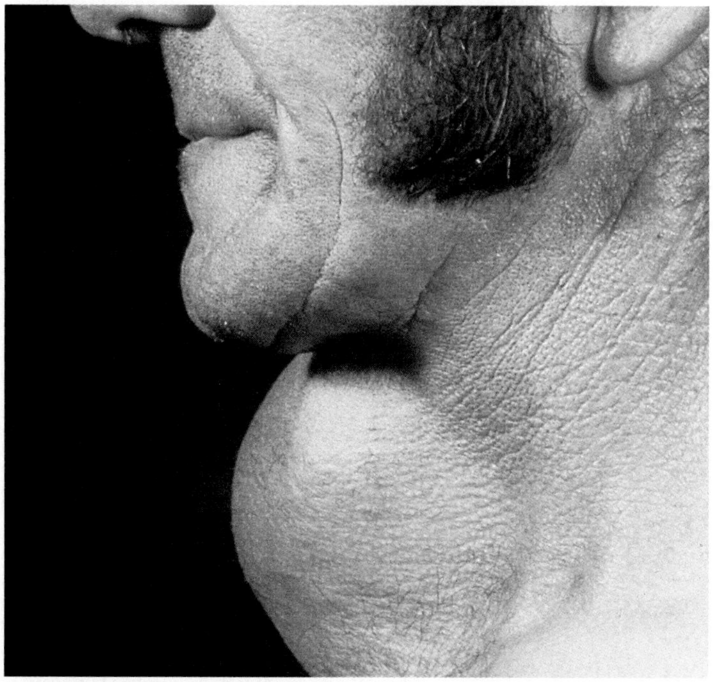

Figure 20.7 Simple goiter.
An enlarged thyroid gland often is caused by a lack of iodine in the diet. Without iodine, the thyroid is unable to produce thyroid hormones, and continued anterior pituitary stimulation causes the gland to enlarge.

Figure 20.8 Cretinism.
Individuals who have hypothyroidism since infancy or childhood do not grow and develop as others do. Unless medical treatment is begun, the body is short and stocky; mental retardation is also likely.

Calcitonin

Calcium (Ca^{2+}) plays a significant role in both neural conduction and muscle contraction. It is also necessary for blood clotting. The blood calcium level is regulated in part by **calcitonin,** a hormone secreted by the thyroid gland when the blood calcium level rises (Fig. 20.9). The primary effect of calcitonin is to bring about the deposit of calcium in the bones. It does this by temporarily reducing the activity and number of osteoclasts. When the blood calcium lowers to normal, the release of calcitonin by the thyroid is inhibited, but the low level stimulates the release of **parathyroid hormone (PTH)** by the parathyroid glands.

Parathyroid Glands

Many years ago, the four parathyroid glands were sometimes mistakenly removed during thyroid surgery because they are so small. Parathyroid hormone (PTH), the hormone produced by the **parathyroid glands,** causes the blood phosphate (HPO_4^{2-}) level to decrease and the blood calcium level to increase.

A low blood calcium level stimulates the release of PTH, a hormone that has a powerful effect on the body. PTH promotes the activity of osteoclasts and the release of calcium from the bones. PTH also promotes the reabsorption of calcium by the kidneys where it activates vitamin D. Vitamin D, in turn, stimulates the absorption of calcium from the intestine. These effects bring the blood calcium level back to the normal range so that the parathyroid glands no longer secrete PTH.

When an insufficient parathyroid hormone production leads to a dramatic drop in the blood calcium level, tetany results. In **tetany,** the body shakes from continuous muscle contraction. The effect is brought about by increased excitability of the nerves, which initiate nerve impulses spontaneously and without rest.

The contrary actions of calcitonin, from the thyroid gland, and parathyroid hormone, from the parathyroid glands, maintain the blood calcium level within normal limits.

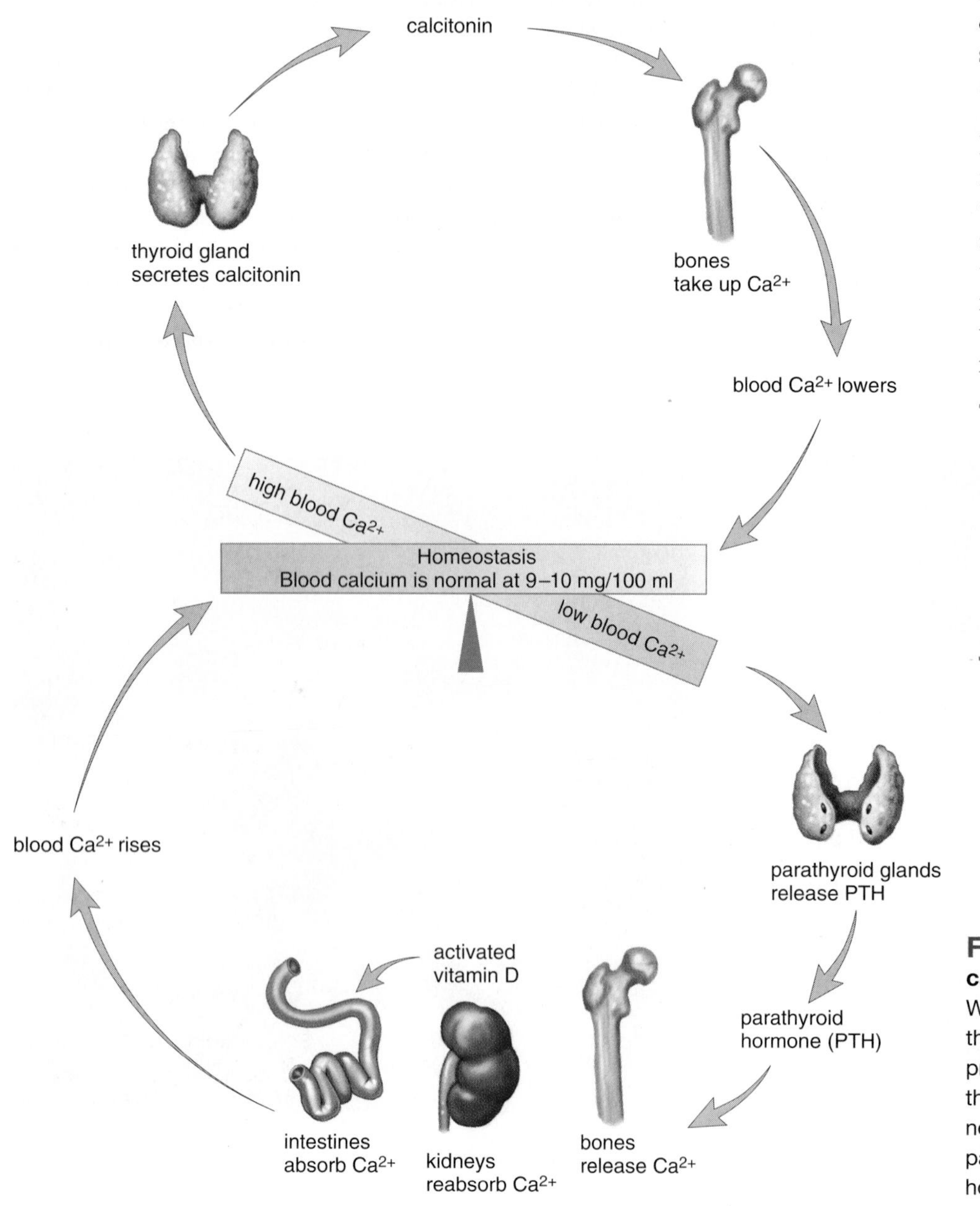

Figure 20.9 Regulation of blood calcium level.
When the blood calcium (Ca^{2+}) level is high, the thyroid gland secretes calcitonin. Calcitonin promotes the uptake of Ca^{2+} by the bones, and therefore the blood Ca^{2+} level returns to normal. When the blood Ca^{2+} level is low, the parathyroid glands release parathyroid hormone (PTH). PTH causes the bones to release Ca^{2+}, the kidneys to reabsorb Ca^{2+}, and the intestines to absorb Ca^{2+}; therefore, the blood Ca^{2+} level returns to normal.

20.4 Adrenal Glands

We have two **adrenal glands** that sit atop the kidneys (see Fig. 20.3). Each adrenal gland consists of an inner portion called the **adrenal medulla** and an outer portion called the **adrenal cortex.** These portions, like the anterior pituitary and the posterior pituitary, have no physiological connection with one another.

The hypothalamus exerts control over the activity of both portions of the adrenal glands. It can initiate nerve impulses that travel by way of the brain stem, spinal cord, and sympathetic nerve fibers to the adrenal medulla, which then secretes its hormones. The hypothalamus, by means of ACTH-releasing hormone, controls the anterior pituitary's secretion of ACTH, which, in turn, stimulates the adrenal cortex. Stress of all types, including both emotional and physical trauma, prompts the hypothalamus to stimulate the adrenal glands. The adrenal hormones increase during times of stress.

Epinephrine (adrenaline) and **norepinephrine** (noradrenaline) produced by the adrenal medulla rapidly bring about all the bodily changes that occur when an individual reacts to an emergency situation as listed in Figure 20.10. In contrast, the hormones produced by the adrenal cortex provide a sustained response to stress. The two major types of hormones produced by the adrenal cortex are the mineralocorticoids and the glucocorticoids. The **mineralocorticoids** regulate salt and water balance leading to increase in blood volume and blood pressure. The **glucocorticoids** regulate carbohydrate, protein, and fat metabolism, leading to an increase in blood glucose level. Cortisone, the medication that is often administered for inflammation of joints, is a glucocorticoid.

The adrenal cortex also secretes a small amount of male sex hormones and a small amount of female sex hormones in both sexes—that is, in the male, both male and female sex hormones are produced by the adrenal cortex, and in the female, both male and female sex hormones are also produced by the adrenal cortex.

The adrenal medulla is under nervous control, and the adrenal cortex is under the control of ACTH, a hormone of the anterior pituitary. Their hormonal secretions help us respond to stress.

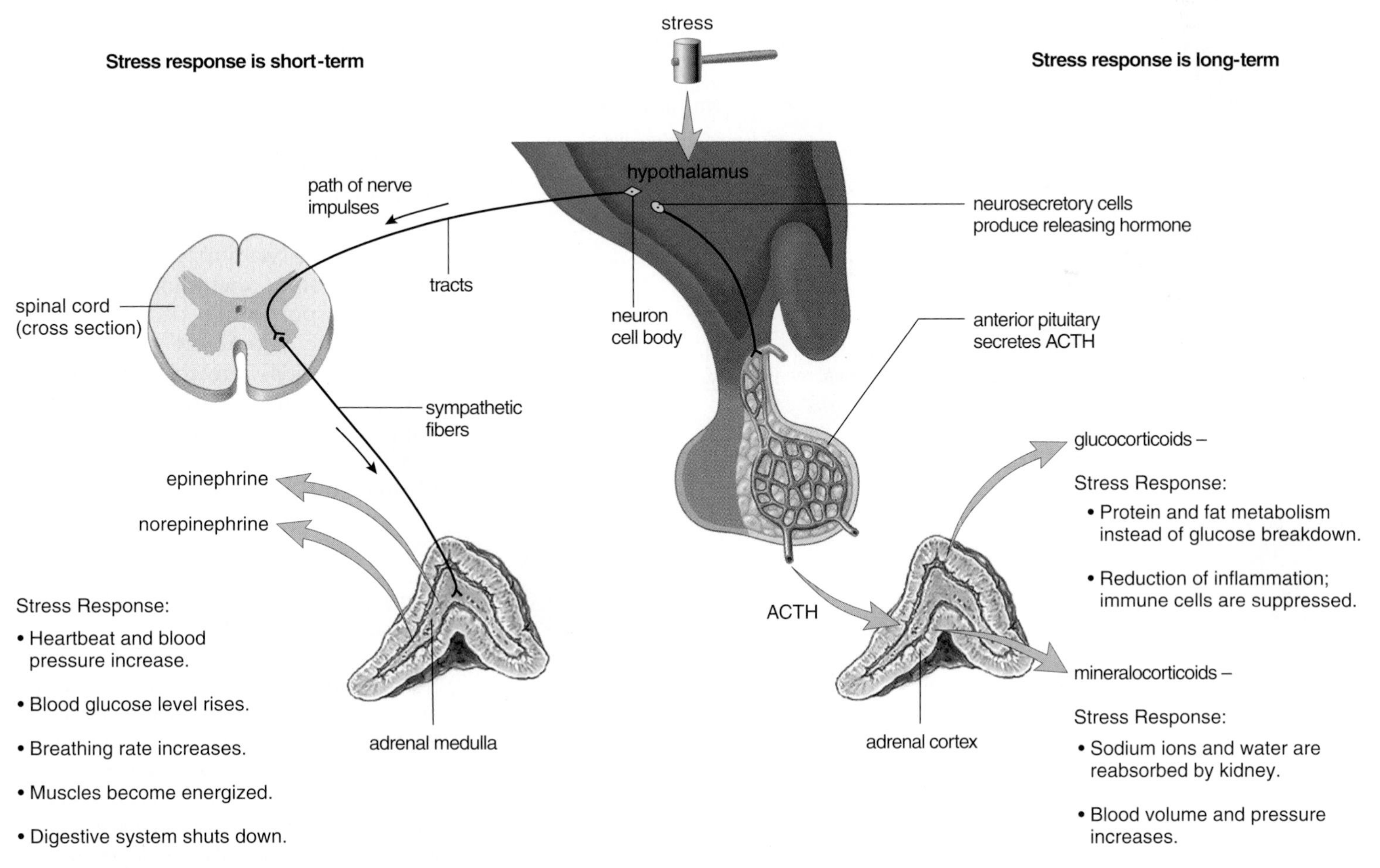

Figure 20.10 Adrenal glands.
Both the adrenal medulla and the adrenal cortex are under the control of the hypothalamus when they help us respond to stress. The adrenal medulla provides a rapid, but short-lived emergency response, while the adrenal cortex provides a sustained stress response.

Mineralocorticoids

Aldosterone is the most important of the mineralocorticoids. The primary target organ of aldosterone is the kidney, where it promotes renal absorption of sodium (Na^+) and renal excretion of potassium (K^+).

The secretion of mineralocorticoids is not under the control of the anterior pituitary. When the blood sodium level is low and therefore the blood pressure is low, the kidneys secrete **renin** (Fig. 20.11). Renin is an enzyme that converts the plasma protein angiotensinogen to angiotensin I, which is changed to angiotensin II by a converting enzyme found in the lungs. Angiotensin II stimulates the adrenal cortex to release aldosterone. The effect of this system, called the renin-angiotensin-aldosterone system, is to raise blood pressure in two ways. Angiotensin II constricts the arterioles, which increases blood pressure, and aldosterone causes the kidneys to reabsorb sodium. When the blood sodium level rises, water is reabsorbed in part because the hypothalamus secretes ADH (see page 400). Then blood pressure increases to normal.

There is a contrary hormone to aldosterone, as you might suspect. When the atria of the heart are stretched due to increased blood volume, cardiac cells release a hormone called **atrial natriuretic hormone (ANH),** which inhibits the secretion of aldosterone from the adrenal cortex. The effect of this hormone is, therefore, to cause the excretion of sodium, that is, *natriuresis.* When sodium is excreted, so is water, and therefore blood pressure lowers to normal.

Glucocorticoids

There are several glucocorticoids, of which **cortisol** is most important biologically. Cortisol promotes the hydrolysis of muscle protein to amino acids, which enter the bloodstream. This leads to a higher blood glucose level when the liver converts these amino acids to glucose. Cortisol also favors metabolism of fatty acids rather than carbohydrates. In opposition to insulin, therefore, cortisol raises the blood glucose level. Cortisol also counteracts the inflammatory response that leads to the pain and the swelling of joints in arthritis and bursitis. The administration of cortisol aids these conditions because it reduces inflammation.

Very high levels of glucocorticoids in the blood can suppress the body's defense system, including the inflammatory response that occurs at infection sites. Cortisone and other glucocorticoids can relieve swelling and pain from inflammation, but by suppressing pain and immunity they can also make a person highly susceptible to injury and infection.

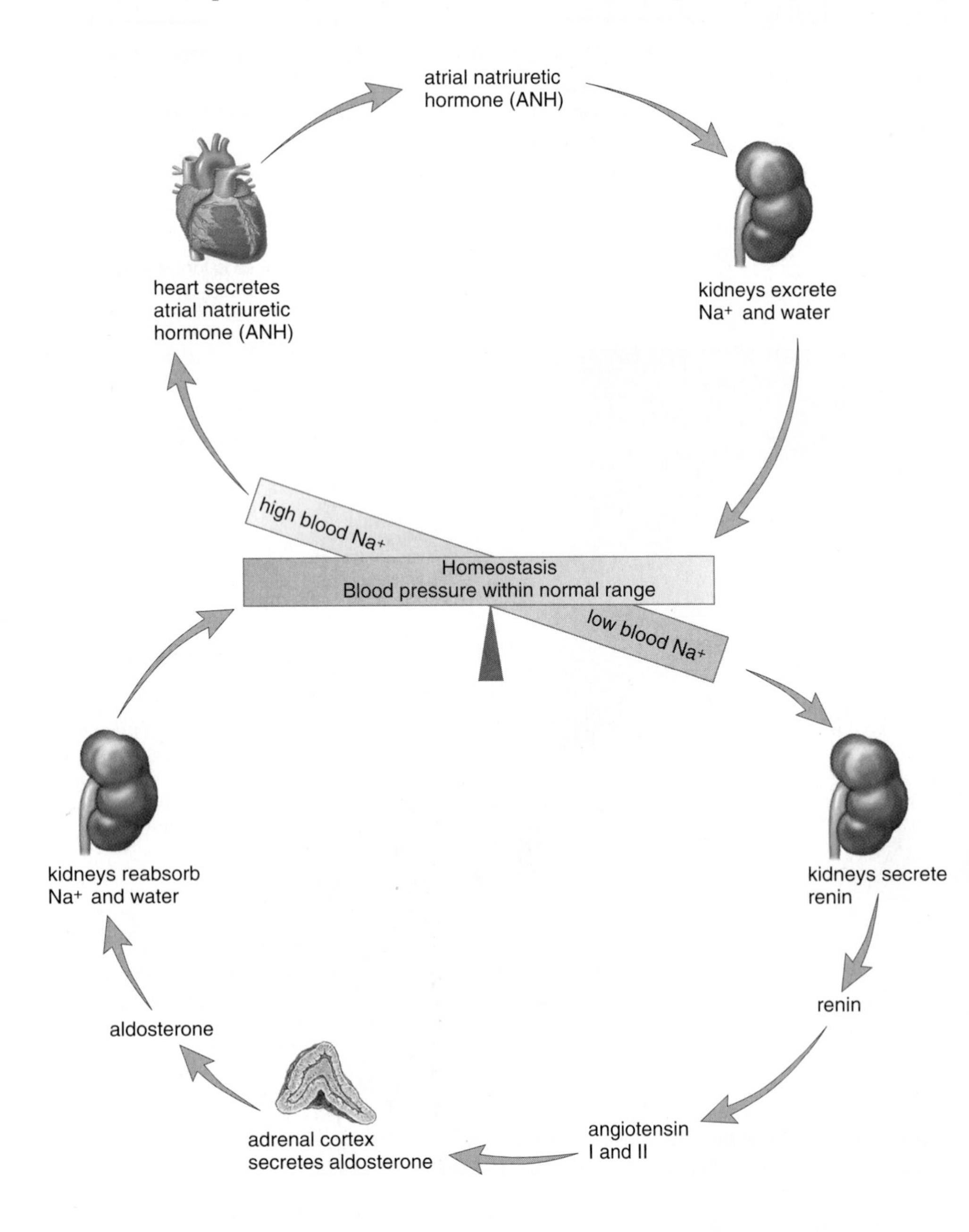

Figure 20.11 Regulation of blood pressure and volume.
When the blood sodium (Na^+) level is high, a high blood pressure causes the heart to secrete atrial natriuretic hormone (ANH). ANH causes the kidneys to secrete Na^+ and the blood volume and pressure return to normal. When the blood Na^+ is low, a low blood pressure causes the kidneys to secrete renin. Renin leads to the secretion of aldosterone from the adrenal cortex. Aldosterone causes the kidneys to reabsorb Na^+, and therefore the blood volume and pressure return to normal.

Malfunction of the Adrenal Cortex

When there is a low level of adrenal cortex hormones due to hyposecretion, a person develops **Addison disease.** The presence of ACTH, which is in excess but ineffective, causes a bronzing of the skin because ACTH, like MSH, can lead to a buildup of melanin (Fig. 20.12). The lack of cortisol results in an inability to replenish the blood glucose level when a stressful situation arises. Even a mild infection can lead to death. The lack of aldosterone results in a loss of sodium and water, and the development of low blood pressure and possibly severe dehydration. Left untreated, Addison disease can be fatal.

When there is a high level of adrenal cortex hormones due to hypersecretion, a person develops **Cushing syndrome** (Fig. 20.13). The excess of cortisol results in a tendency toward diabetes mellitus as muscle protein is metabolized and subcutaneous fat is deposited in the midsection. The trunk is obese while the arms and legs remain a normal size. An excess of aldosterone and reabsorption of sodium and water by the kidneys leads to a basic blood pH and hypertension. The face is moonshaped due to edema. Masculinization may occur in women because of excess adrenal male sex hormones.

The adrenal cortex hormones are essential to homeostasis. Addison disease is due to adrenal cortex hyposecretion, and Cushing syndrome is due to adrenal cortex hypersecretion.

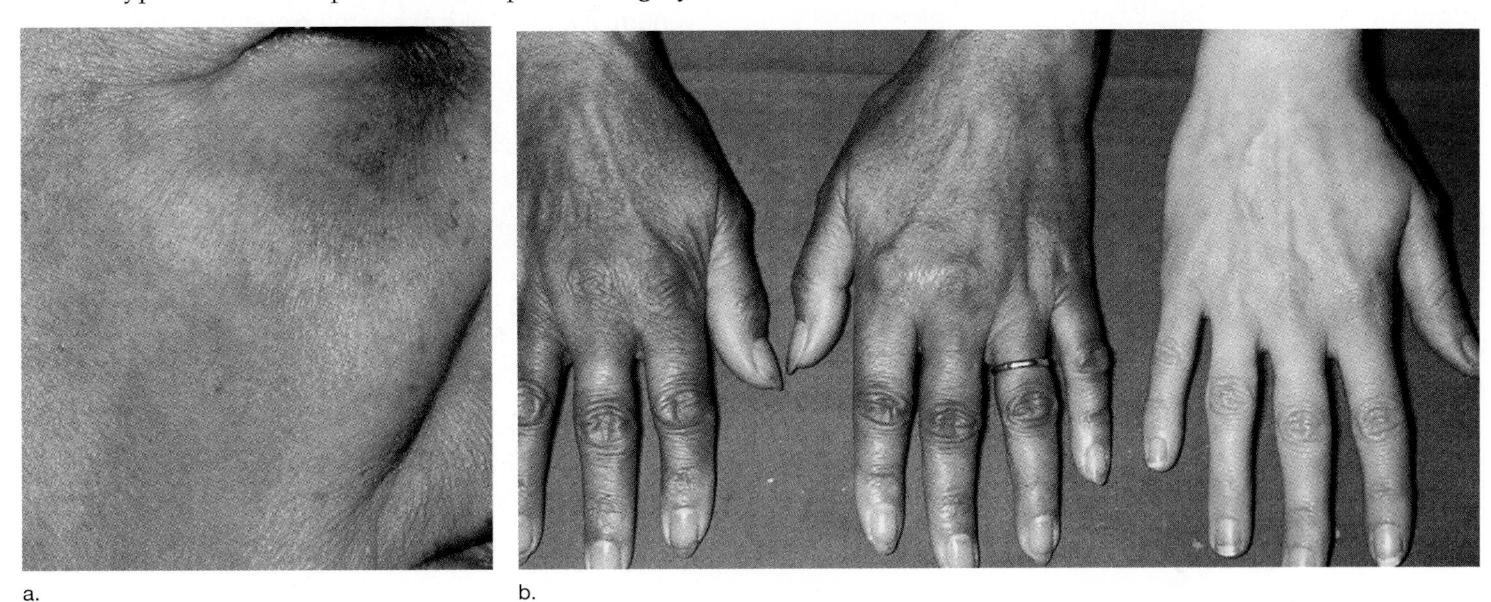

a. b.

Figure 20.12 Addison disease.
Addison disease is characterized by a peculiar bronzing of the skin, particularly noticeable in these light-skinned individuals. Note the color of **(a)** the face and **(b)** the hands compared to the hand of an individual without the disease.

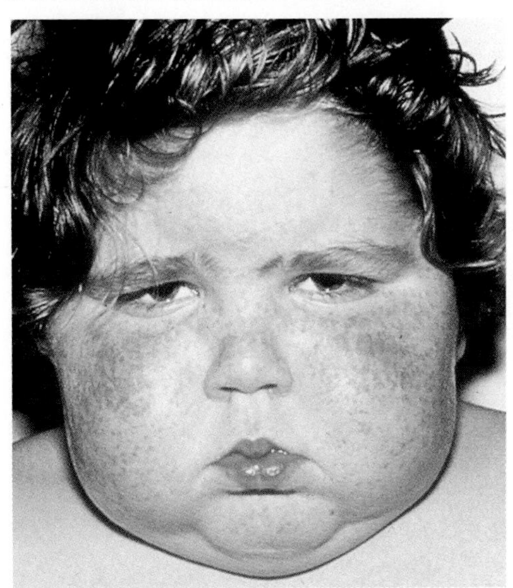

a.

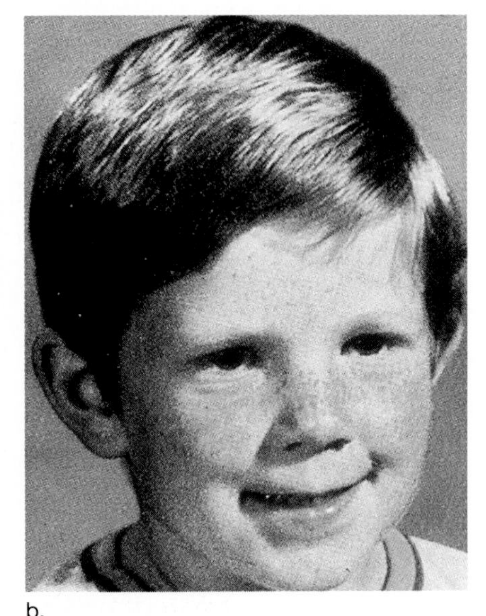

b.

Figure 20.13 Cushing syndrome.
Cushing syndrome results from hypersecretion due to an adrenal cortex tumor. **a.** First diagnosed with Cushing syndrome. **b.** Four months later, after therapy.

20.5 Pancreas

The **pancreas** is a long organ that lies transversely in the abdomen between the kidneys and near the duodenum of the small intestine. It is composed of two types of tissue. Exocrine tissue produces and secretes digestive juices that go by way of ducts to the small intestine. Endocrine tissue, called the **pancreatic islets** (islets of Langerhans), produces and secretes the hormones **insulin** and **glucagon** directly into the blood (Figure 20.14).

Insulin is secreted when there is a high blood glucose level, which usually occurs just after eating. Insulin stimulates the uptake of glucose by cells, especially liver cells, muscle cells, and adipose tissue cells. In liver and muscle cells, glucose is then stored as glycogen. In muscle cells the breakdown of glucose supplies energy for protein metabolism, and in fat cells the breakdown of glucose supplies glycerol for the formation of fat. In these various ways insulin lowers the blood glucose level.

Glucagon is secreted from the pancreas, usually in between eating, when there is a low blood glucose level. The major target tissues of glucagon are the liver and adipose tissue. Glucagon stimulates the liver to break down glycogen to glucose and to use fat and protein in preference to glucose as energy sources. Adipose tissue cells break down fat to glycerol and fatty acids. The liver takes these up and uses them as substrates for glucose formation. In these various ways glucose raises the blood glucose level.

> The two contrary hormones insulin and glucagon, both produced by the pancreas, maintain the normal level of glucose in the blood.

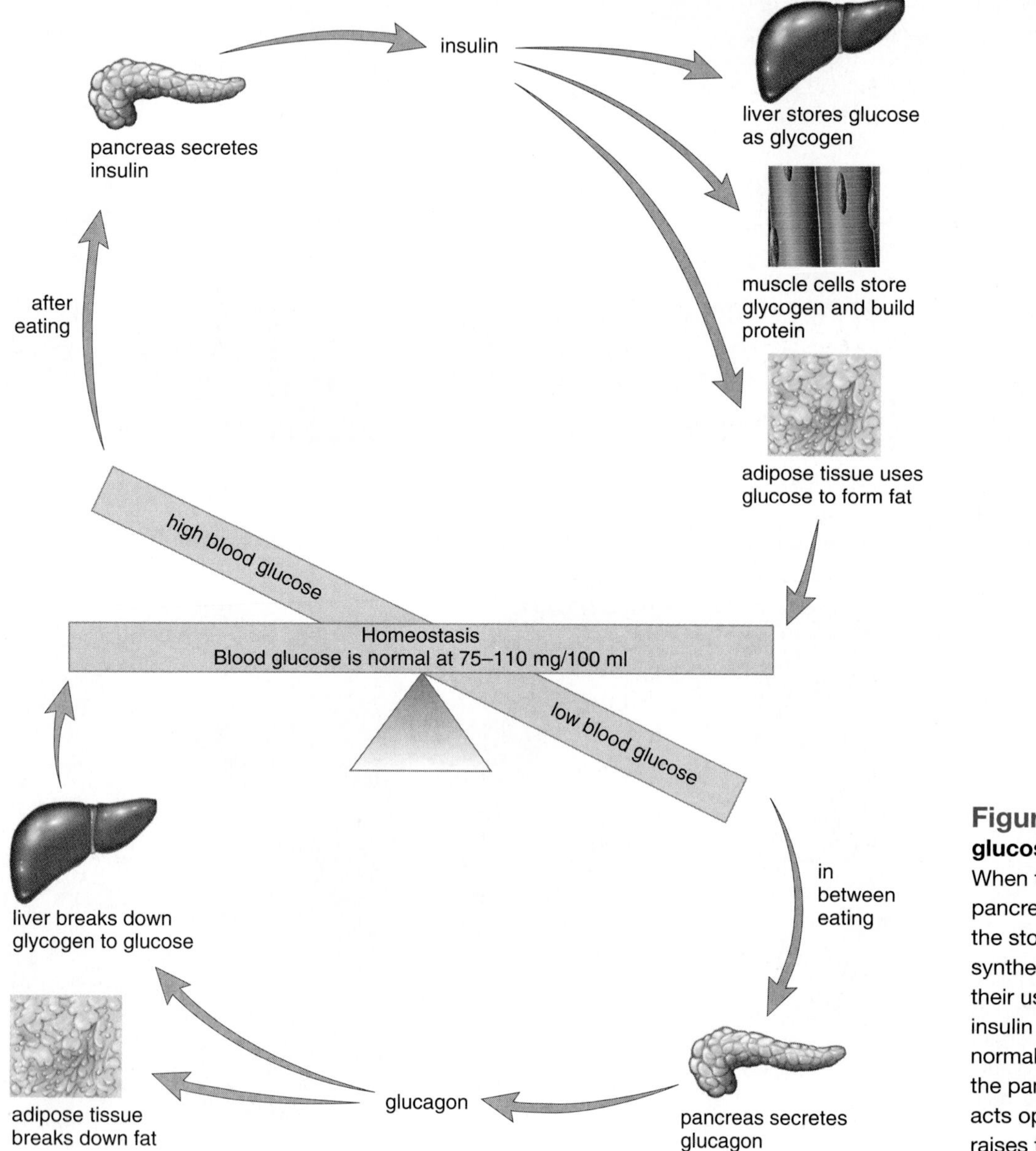

Figure 20.14 Regulation of blood glucose level.
When the blood glucose level is high, the pancreas secretes insulin. Insulin promotes the storage of glucose as glycogen and the synthesis of proteins and fats (as opposed to their use as energy sources). Therefore, insulin lowers the blood glucose level to normal. When the blood glucose level is low, the pancreas secretes glucagon. Glucagon acts opposite to insulin; therefore, glucagon raises the blood glucose level to normal.

Diabetes Mellitus

Diabetes mellitus is a fairly common hormonal disease in which liver, and indeed all body cells, do not take up and/or metabolize glucose. Therefore cellular famine exists in the midst of plenty. As the blood glucose level rises, glucose, along with water, is excreted in the urine. The loss of water in this way causes the diabetic to be extremely thirsty. Since glucose is not being metabolized, the body turns to the breakdown of protein and fat for energy. The metabolism of fat leads to the excessive presence of ketones in the blood and acidosis (acid blood) that can eventually cause coma and death.

There are two types of diabetes mellitus. In *type I (insulin-dependent) diabetes,* the pancreas is not producing insulin. The condition is believed to be brought on by exposure to an environmental agent, most likely a virus, whose presence causes cytotoxic T cells to destroy the pancreatic islets. As a result, the individual must have daily insulin injections. These injections control the diabetic symptoms but still can cause inconveniences, since either an overdose of insulin or the absence of regular eating can bring on the symptoms of hypoglycemia (low blood sugar) . These symptoms include perspiration, pale skin, shallow breathing, and anxiety. Because the brain requires a constant supply of glucose, unconsciousness can result. The cure is quite simple: immediate ingestion of a sugar cube or fruit juice can very quickly counteract hypoglycemia.

Of the 16 million people who now have diabetes in the United States, most have *type II (noninsulin-dependent) diabetes.* This type of diabetes mellitus usually occurs in people of any age who are obese and inactive. The pancreas produces insulin, but the liver and muscle cells do not respond to it in the usual manner. They may increasingly lack the receptor proteins necessary to detect the presence of insulin. If type II diabetes is untreated, the results can be as serious as type I diabetes. (Diabetics are prone to blindness, kidney disease, and circulatory disorders. Pregnancy carries an increased risk of diabetic coma, and the child of a diabetic is somewhat more likely to be stillborn or to die shortly after birth.) It is possible to prevent or at least control type II diabetes by adhering to a low-fat diet and exercising regularly. If this fails, oral drugs that stimulate the pancreas to secrete more insulin and enhance the metabolism of glucose in the liver and muscle cells are available.

Diabetes mellitus is caused by the lack of insulin or the insensitivity of cells to insulin. Insulin lowers blood glucose levels by causing the cells, particularly those of the liver and muscles, to take up glucose and convert it to glycogen.

Isolation of Insulin

In 1920, physician Frederick Banting decided to try to isolate insulin from the pancreas. Previous investigators had been unable to do this because the enzymes in the digestive juices made by the pancreas destroyed insulin during the isolation procedure. Banting hit upon the idea of tying off the pancreatic duct, which he knew from previous research would lead to the degeneration only of the cells that produce digestive juices and not of the pancreatic islets, where insulin is made. J. J. Macleod made a laboratory available to him at the University of Toronto and also assigned a graduate student, Charles Best, to assist him. Banting and Best had limited funds and spent that summer working, sleeping, and eating in the lab. By the end of the summer, they had obtained pancreatic extracts that did lower the blood glucose level in diabetic dogs. Banting and Best followed the required steps given in the chart to identify a chemical messenger. Macleod then brought in biochemists, who purified the extract. Insulin therapy for the first human patient began in 1922, and large-scale production of purified insulin from pigs and cattle followed. Banting and Macleod received a Nobel Prize for their work in 1923. The amino acid sequence of insulin was determined in 1953. Insulin is presently a biotechnology product.

Steps	Example
Identify the source of the chemical	Pancreatic islets are source
Identify the effect to be studied	Presence of pancreas in body lowers blood sugar
Isolate the chemical	Insulin isolated from pancreatic secretions
Show that the chemical alone has the effect	Insulin alone lowers blood sugar

Health Focus

Dangers of Anabolic Steroids

Anabolic steroids are synthetic forms of the male sex hormone testosterone. Trainers may have been the first to acquire anabolic steroids for weight lifters, bodybuilders, and other athletes such as professional football players. When taken in large doses (10 to 100 times the amount prescribed by doctors for illnesses) and accompanied by exercise, anabolic steroids promote larger muscles. Occasionally, steroid abuse makes the news because an Olympic winner tests positive for the drug and must relinquish a medal. Steroid use has been outlawed by the International Olympic Committee.

The U.S. Food and Drug Administration bans the importation of most steroids, but they are brought into the country illegally and sold through the mail or in gyms and health clubs. According to federal officials, 1 to 3 million Americans now take anabolic steroids. Their increased use by teenagers wishing to build bulk quickly is of special concern. Some attribute this to society's emphasis on physical appearance and the need of insecure youngsters to feel better about how they look.

Physicians, teachers, and parents are quite alarmed about anabolic steroid abuse. It's even predicted that two or three months of high-dosage use in a youngster can cause death two or three decades later. The many harmful effects of anabolic steroids on the body are listed in Figure 20A. In addition, these drugs increase aggression and make a person feel invincible. One abuser even had his friend videotape him as he drove his car at 40 miles per hour into a tree!

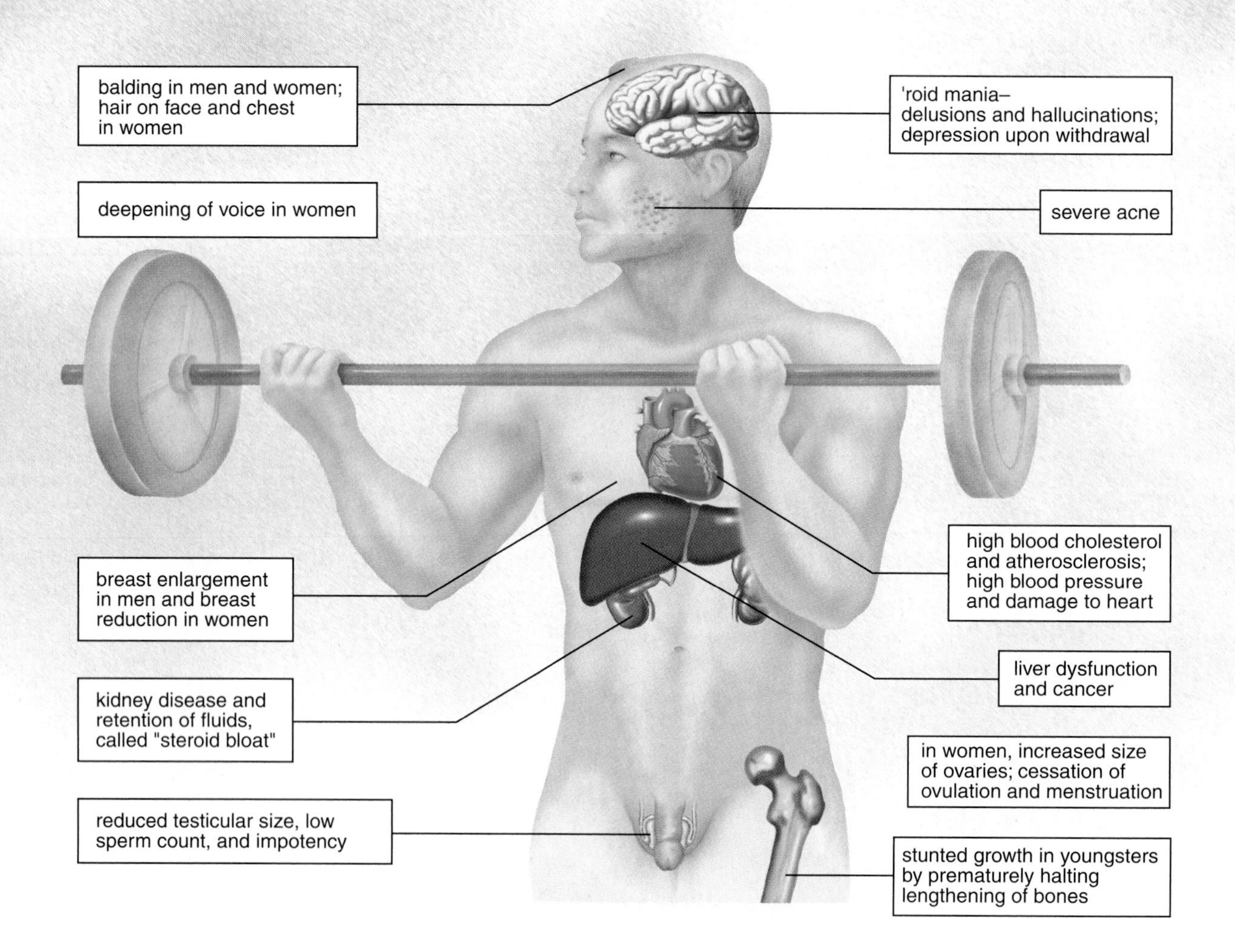

Figure 20A The effects of anabolic steroid use.

20.6 Other Endocrine Glands

The testes and ovaries are endocrine glands. There are also lesser known glands and some tissues that produce hormones.

Testes and Ovaries

The gonads are the testes in males and the ovaries in females. The **testes** are located in the scrotum, and the **ovaries** are located in the pelvic cavity. The testes produce **androgens** (e.g., **testosterone**), which are the male sex hormones, and the ovaries produce estrogens and progesterone, the female sex hormones. The hypothalamus and the pituitary gland control the hormonal secretions of these organs in the same manner that was previously described for the thyroid gland.

The male sex hormone, testosterone, has many functions. It is essential for the normal development and functioning of the sex organs in males. It is also necessary for the maturation of sperm.

Greatly increased testosterone secretion at the time of puberty stimulates the growth of the penis and the testes. Testosterone also brings about and maintains the secondary sex characteristics in males that develop at the time of puberty. Testosterone causes growth of a beard, axillary (underarm) hair, and pubic hair. It prompts the larynx and the vocal cords to enlarge, causing the voice to change. It is partially responsible for the muscular strength of males, and this is the reason some athletes take supplemental amounts of **anabolic steroids,** which are either testosterone or related chemicals. The contraindications of taking anabolic steroids are discussed in the Health reading on the previous page. Testosterone also stimulates oil and sweat glands in the skin; therefore, it is largely responsible for acne and body odor. Another side effect of testosterone is baldness. Genes for baldness probably are inherited by both sexes, but baldness is seen more often in males because of the presence of testosterone.

Testosterone is believed to be largely responsible for the sex drive. It may even contribute to the suggested aggressiveness of males.

The female sex hormones, **estrogens** and **progesterone,** have many effects on the body. In particular, estrogens secreted at the time of puberty stimulate the growth of the uterus and the vagina. Estrogen is necessary for egg maturation and is largely responsible for the secondary sex characteristics in females. It is responsible for female body hair and fat distribution. In general, females have a more rounded appearance than males because of a greater accumulation of fat beneath the skin. Also, the pelvic girdle is wider in females than in males, resulting in females having a larger pelvic cavity. Both estrogen and progesterone are required for breast development and regulation of the uterine cycle, which includes monthly menstruation (discharge of blood and mucosal tissues from the uterus).

Pineal Gland

The **pineal gland,** which is located in the brain (see Fig. 20.3), produces the hormone called **melatonin,** primarily at night. Melatonin is involved in daily cycles, or **circadian rhythms.** Normally we grow sleepy at night when melatonin levels increase and awaken once daylight returns and melatonin levels are low. Shift work is usually troublesome because it upsets this normal daily rhythm. Similarly, travel to another time zone, as when going to Europe from the United States, results in jet lag because the body is still producing melatonin according to the old schedule. Some people even have **seasonal affective disorder (SAD);** they become depressed and have an uncontrollable desire to sleep with the onset of winter. Receiving melatonin makes their symptoms worse, but exposure to a bright light improves them.

Based on animal research, it appears that melatonin also regulates sexual development. It is of interest that children whose pineal gland has been destroyed due to a brain tumor experience early puberty.

Thymus Gland

The **thymus** is a lobular gland that lies just beneath the sternum (see Fig. 20.3). This organ reaches its largest size and is most active during childhood. With aging, the organ gets smaller and becomes fatty. Lymphocytes that originate in the bone marrow and then pass through the thymus are transformed into T lymphocytes. The lobules of the thymus are lined by epithelial cells that secrete hormones called thymosins. These hormones aid in the differentiation of lymphocytes packed inside the lobules. Although the hormones secreted by the thymus ordinarily work in the thymus, there is hope that these hormones could be injected into AIDS or cancer patients where they would increase T lymphocyte function.

Nontraditional Sources

Some organs that are usually not considered endocrine glands do indeed secrete hormones. We have already mentioned that the heart produces atrial natriuretic hormone. And you will recall that the stomach and the small intestine produce peptide hormones that regulate digestive secretions. There are a number of other types of tissues that produce hormones.

Leptin

Leptin is a protein hormone produced by adipose tissue that acts on the hypothalamus where it signals satiety—that the individual has had enough to eat. Strange to say, the blood of obese individuals may be rich in leptin. The possibility exists that the leptin they produce is ineffective because of a genetic mutation or else their hypothalamic cells lack a suitable number of receptors for leptin.

Growth Factors

A number of different types of organs and cells produce peptide **growth factors,** which stimulate cell division and mitosis. They are like hormones in that they act on cell types with specific receptors to receive them. Some, like lymphokines, are released into the blood; others diffuse to nearby cells. Growth factors of particular interest are:

Granulocyte and macrophage colony-stimulating factor (GM-CSF) is secreted by many different tissues. GM-CSF causes a common stem cell to form either granulocyte or macrophage cells, depending on whether the concentration is low or high.

Platelet-derived growth factor is released from platelets and from many other cell types. It helps in wound healing and causes an increase in the number of fibroblasts, smooth muscle cells, and certain cells of the nervous system.

Epidermal growth factor and *nerve growth factor* stimulate the cells indicated by their names, as well as many others. These growth factors are also important in wound healing.

Tumor angiogenesis factor stimulates the formation of capillary networks and is released by tumor cells. One treatment for cancer is to prevent the activity of this growth factor.

Prostaglandins

Prostaglandins (PG) are chemical messengers that are produced and act locally. They are derived from fatty acids stored in plasma membranes as phospholipids. When a cell is stimulated by reception of a hormone or even by trauma, a series of synthetic reactions takes place in the plasma membrane, and PG is first released into the cytoplasm and then secreted from the cell. There are many different types of prostaglandins produced by many different tissues. In the uterus, prostaglandins cause muscles to contract; therefore, they are implicated in the pain and discomfort of menstruation in some women. Also, prostaglandins mediate the effects of pyrogens, chemicals that are believed to reset the temperature regulatory center in the brain. Aspirin reduces body temperature of a person and controls pain because of its effect on prostaglandins.

Certain prostaglandins reduce gastric secretion and have been used to treat ulcers; others lower blood pressure, and have been used to treat hypertension; and yet others inhibit platelet aggregation, and have been used to prevent thrombosis. Because the different prostaglandins can have contrary effects, it has been very difficult to standardize their use, and in most instances, prostaglandin therapy is still considered experimental.

Bioethical Issue

Hormone therapy saves lives. Diabetics and those with Addison disease would soon succumb if hormone therapy was not available for their conditions. Other hormone therapies are a matter of choice and raise ethical questions. If you were the parent of a child who was short for his or her age group, on what basis might you decide to have the child undergo hormone therapy?

Researchers have found that human growth hormone doesn't work. It causes puberty to arrive earlier than usual, and this does away with any gain in height that occurred before puberty. However, studies are being conducted with a synthetic growth hormone called somatotropin that seems to work better. Researchers in England conducted a survey of about 14,000 girls, and decided to treat seven girls who were short for their age. They compared their progress to a group of untreated girls. All girls started puberty at about the same age—13.5 years. The girls in the treated group received a daily injection of somatotropin between the ages of 8 and 14. They grew to 5′1″ by age 16, when most girls stop growing. That was 2.4 to 3.0 inches taller than the girls in the untreated group.

The researchers were interested in whether height affected psychological outlook. They found that girls who were 4′10″ appeared to be just as happy and well balanced as those who were taller. Is psychological outlook a good criterion to use when deciding whether to use hormone therapy to increase height? Is this criterion more appropriately applied to boys than to girls? When does short height become a disability in our society? Are tall people likely to have more successful careers than short people?

Questions

1. Do you approve of using hormone therapy to increase the growth of a child? Why or why not?
2. On what basis would you decide to use hormone therapy for your child in order to increase height? Should it matter whether the child is a girl or boy?
3. What assurances would you need before you enter your child in an experimental study?

Summarizing the Concepts

20.1 Environmental Signals

There are three categories of environmental signals: those that act at a distance between individuals (pheromones); those that act at a distance within the individual (traditional endocrine hormones and secretions of neurosecretory cells); and local messengers (such as prostaglandins, growth factors, and neurotransmitters). Since there is great overlap between these categories, perhaps the definition of a hormone should be expanded to include all of them.

Steroid hormones enter the nucleus and combine with a receptor hormone, and the complex attaches to and activates DNA. Transcription and translation lead to protein synthesis. Nonsteroid hormones are usually received by a hormone receptor located in the plasma membrane. Most often their reception leads to activation of an enzyme that changes ATP to cyclic AMP (cAMP). cAMP then activates an enzyme cascade. Hormones work in small quantities because their effect is amplified.

20.2 Hypothalamus and Pituitary Gland

Neurosecretory cells in the hypothalamus produce antidiuretic hormone (ADH) and oxytocin, which are stored in axon endings in the posterior pituitary until they are released.

The hypothalamus produces hypothalamic-releasing and hypothalamic-inhibiting hormones, which pass to the anterior pituitary by way of a portal system. The anterior pituitary produces at least six types of hormones, and some of these stimulate other hormonal glands to secrete hormones.

20.3 Thyroid and Parathyroid Glands

The thyroid gland requires iodine to produce thyroxine and triiodothyronine, which increase the metabolic rate. If iodine is available in limited quantities, a simple goiter develops; if the thyroid is overactive, an exophthalmic goiter develops. The thyroid gland also produces calcitonin, which helps lower the blood calcium level. The parathyroid glands secrete parathyroid hormone which raises the blood calcium and decreases the blood phosphate level.

20.4 Adrenal Glands

The adrenal glands respond to stress: immediately, the adrenal medulla secretes epinephrine and norepinephrine, which bring about responses we associate with emergency situations. On a long-term basis, the adrenal cortex produces the glucocorticoids (e.g., cortisol) and the mineralocorticoids (e.g., aldosterone). Cortisol stimulates hydrolysis of proteins to amino acids that are converted to glucose; in this way, it raises the blood glucose level. Aldosterone causes the kidneys to reabsorb sodium ions (Na^+) and excrete potassium ions (K^+). Addison disease develops when the adrenal cortex is underactive, and Cushing syndrome develops when the adrenal cortex is overactive.

20.5 Pancreas

The pancreatic islets secrete insulin, which lowers the blood glucose level, and glucagon, which has the opposite effect. The most common illness due to hormonal imbalance is diabetes mellitus, which is due to the failure of the pancreas to produce insulin or the cells to take it up.

20.6 Other Endocrine Glands

The gonads produce the sex hormones; the pineal gland produces melatonin, which may be involved in circadian rhythms and the development of the reproductive organs; and the thymus secretes thymosins, which stimulate T lymphocyte production and maturation.

Tissues also produce hormones. Adipose tissue produces leptin which acts on the hypothalamus, and various tissues produce growth factors. Prostaglandins are produced and act locally.

Studying the Concepts

1. Categorize chemical messengers into three groups based upon the distance between site of secretion and receptor site, and give examples of each group. 396
2. Give examples to show that there is an overlap between the mode of operation of the nervous system and that of the endocrine system. Explain why the traditional definition of a hormone may need to be expanded. 396
3. Explain how steroid hormones and nonsteroid hormones affect the metabolism of the cell. 397
4. Describe a mechanism by which the production of a hormone is regulated and another by which the effect of a hormone is controlled. 399
5. Explain the relationship of the hypothalamus to the posterior pituitary gland and to the anterior pituitary gland. List the hormones secreted by the posterior and anterior pituitary. 400
6. Give an example of the three-tier relationship among the hypothalamus, the anterior pituitary, and other endocrine glands. 400
7. Discuss the effect of growth hormone on the body as a result of there being too much or too little growth hormone when a young person is growing. What is the result if the anterior pituitary produces growth hormone in an adult? 402
8. What types of goiters are associated with a malfunctioning thyroid? Explain each type. 403
9. How do the thyroid and the parathyroid work together to control the blood calcium level? 404
10. How do the adrenal glands respond to stress? What hormones are secreted by the adrenal medulla, and what effects do these hormones have? 405
11. Name the most significant glucocorticoid and mineralocorticoid, and discuss their functions. Explain the symptoms of Addison disease and Cushing syndrome. 406–7
12. Draw a diagram to explain how insulin and glucagon maintain the blood glucose level. Use your diagram to explain three major symptoms of type I diabetes mellitus. 408–9
13. Name the other endocrine glands discussed in this chapter, and discuss the function of the hormones they secrete. 411
14. What are leptin, growth factors, and prostaglandins? How do these substances act? 411–12

Testing Yourself

Choose the best answer for each question.

1. One of the chief differences between pheromones and local hormones is
   a. the distance over which they act.
   b. that one is a chemical messenger and the other is not.
   c. that one is made by only invertebrates and the other is made by only vertebrates.
   d. All of these are correct.
2. Nonsteroid hormones
   a. are received by a receptor located in the plasma membrane.
   b. are received by a receptor located in the cytoplasm.
   c. bring about the transcription of DNA.
   d. Both b and c are correct.

Match the hormone in questions 3–7 to the correct gland in the key.

Key:
a. pancreas
b. anterior pituitary
c. posterior pituitary
d. thyroid
e. adrenal medulla
f. adrenal cortex

3. cortisol
4. growth hormone (GH)
5. oxytocin storage
6. insulin
7. epinephrine
8. The blood cortisol level controls the secretion of
   a. a releasing hormone from the hypothalamus.
   b. adrenocorticotropic hormone (ACTH) from the anterior pituitary.
   c. cortisol from the adrenal cortex.
   d. All of these are correct.
9. The anterior pituitary controls the secretion(s) of
   a. both the adrenal medulla and the adrenal cortex.
   b. both cortisol and aldosterone.
   c. thyroxine.
   d. All of these are correct.
10. Aldosterone
    a. is opposed by atrial natriuretic hormone (ANH).
    b. causes the kidneys to reabsorb sodium.
    c. causes the blood volume to increase.
    d. All of these are correct.
11. Diabetes mellitus is associated with
    a. too much insulin in the blood.
    b. too high a blood glucose level.
    c. blood that is too dilute.
    d. All of these are correct.
12. Which of these is not a pair of opposing hormones?
    a. insulin—glucagon
    b. calcitonin—parathyroid hormone
    c. cortisol—epinephrine
    d. aldosterone—atrial natriuretic hormone (ANH)
    e. thyroxine—growth hormone
13. Which hormone and condition is mismatched?
    a. growth hormone—acromegaly
    b. thyroxine—goiter
    c. parathyroid hormone—tetany
    d. cortisol—Addison disease
    e. insulin—diabetes
14. Fill in this diagram to explain the three-tiered relationship between the hypothalamus, the anterior pituitary, and the target gland.

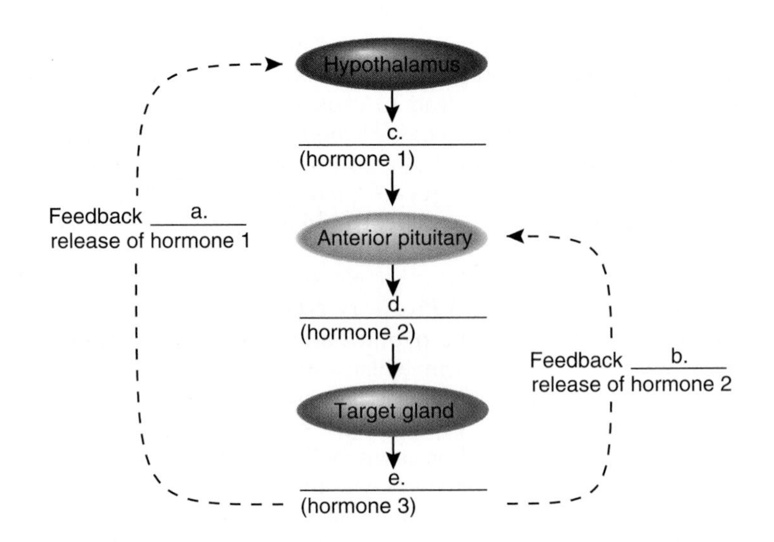

Thinking Scientifically

1. Environmental signals alter the behavior of target cells (page 396).
   a. How would you modify this definition to be consistent with Figure 20.1? Why?
   b. What makes a target cell sensitive to a particular environmental signal?
   c. Why is the term behavior a good one to use in the definition of an environmental signal? Give examples of change in behavior of the target cell, the organ, or the organism.
   d. When is the environmental signal insulin present? When is an environment signal like a neurotransmitter substance released?

2. Considering the function of the pancreas (page 408):
   a. If the pancreas is removed from an animal, what substance would you expect to find in the urine? Why?
   b. Would your findings support the contention that the pancreas is the source of insulin? What do they prove?
   c. How would you prove that insulin lowers blood sugar?
   d. Once you have shown that insulin lowers blood sugar, do your findings in question 2a support the belief that the pancreas is the source of insulin? Why?

Understanding the Terms

acromegaly 402
Addison disease 407
adrenal cortex 405
adrenal gland 405
adrenal medulla 405
adrenocorticotropic hormone (ACTH) 400
aldosterone 406
anabolic steroid 411
androgen 411
anterior pituitary 400
antidiuretic hormone (ADH) 400
atrial natriuretic hormone (ANH) 406
calcitonin 404
circadian rhythm 411
cortisol 406
cretinism 403
Cushing syndrome 407
cyclic AMP 397
diabetes mellitus 409
endocrine gland 399
epinephrine 405
estrogen 411
exophthalmic goiter 403
glucagon 408
glucocorticoid 405
gonadotropic hormone 400
growth factor 412
growth hormone (GH) 400
hormone 396
hypothalamic-inhibiting hormone 400
hypothalamic-releasing hormone 400
hypothalamus 400
insulin 408
leptin 411
melanocyte-stimulating hormone (MSH) 400
melatonin 411
mineralocorticoid 405
myxedema 403
nonsteroid hormone 397
norepinephrine (NE) 405
ovaries 411
oxytocin 400
pancreas 408
pancreatic islets (of Langerhans) 408
parathyroid gland 404
parathyroid hormone (PTH) 404
pheromone 396
pineal gland 411
pituitary dwarf 402
pituitary gland 400
posterior pituitary 400
progesterone 411
prolactin (PRL) 400
prostaglandin (PG) 412
renin 406
seasonal affective disorder (SAD) 411
simple goiter 403
steroid hormone 397
testes 411
testosterone 411
tetany 404
thymus 411
thyroid gland 403
thyroid-stimulating hormone (TSH) 400
thyroxine 403
triiodothyronine 403

Match the terms to these definitions:

a. ________ Organ that is in the neck and secretes several important hormones, including thyroxine and calcitonin.

b. ________ Condition characterized by high blood glucose level and the appearance of glucose in the urine.

c. ________ Hormone secreted by the anterior pituitary that stimulates activity in the adrenal cortex.

d. ________ Type of hormone that binds to a plasma membrane receptor and results in activation of an enzyme cascade.

e. ________ Hormone released by the posterior pituitary that causes contraction of uterus and milk letdown.

Using Technology

Your study of the endocrine system is supported by these available technologies:

Essential Study Partner CD-ROM

Animals → Endocrine System

Visit the Mader web site for related ESP activities.

Exploring the Internet

The Mader Home Page provides resources and tools as you study this chapter.

http://www.mhhe.com/biosci/genbio/mader

Dynamic Human 2.0 CD-ROM

Endocrine System → Clinical Concepts
→ Diabetes
→ Explorations
→ H-PT axis

HealthQuest CD-ROM

1 Stress Management and Mental Health → Gallery→ Stressors Affect Body

Life Science Animations 3D Video

41 Hormone Action

Further Readings for Part 4

Barkley, R. A. September 1998. Attention-deficit hyperactivity disorder. *Scientific American* 279(3):66. ADHD may result from neurological abnormalities with a genetic basis.

Beardsley, T. August 1997. The machinery of thought. *Scientific American* 277(2):78. Researchers have identified the area of the brain responsible for memory.

Debinski, W. May/June 1998. Anti-brain tumor cytotoxins. *Science & Medicine* 5(3):36. Delivery of bacterial toxins specifically to tumor cells is a new therapeutic strategy for brain tumor treatment.

Deyo, R. A. August 1998. Low-back pain. *Scientific American* 279(2):48. Treatment options for low-back pain which don't involve bed rest or surgery are improving.

Drollette, D. October 1997. The next hop: Can wallabies replace the lab rat? *Scientific American* 277(4):20. Wallaby embryos may be the ideal model in mammalian neurobiology.

Duke, R. C., et al. December 1996. Cell suicide in health and disease. *Scientific American* 275(6):80. Failures in the processes of cellular self-destruction may give rise to cancer, AIDS, Alzheimer disease, and some genetic diseases.

Gazzaniga, M. S. July 1998. The split brain revisited. *Scientific American* 279(1):50. Recent research on split brains has led to new insights into brain organization and consciousness.

Grillner, S. January 1996. Neural networks for vertebrate locomotion. *Scientific American* 274(1):64. Discoveries about how the brain coordinates muscle movement raise hopes for restoration of mobility for some accident victims.

Hadley, M. E. 1996. *Endocrinology.* 4th ed. Upper Saddle River, NJ: Prentice-Hall. This text discusses the role of chemical messengers in the control of neurological processes.

Halstead, L. S. April 1998. Post-polio syndrome. *Scientific American* 278(4):42. Recovered polio victims are experiencing fatigue, pain, and weakness, resulting from degeneration of motor neurons.

Harvard Health Letter. April 1998. A special report: Parkinson's disease. This overview presents the symptoms and diagnosis of Parkinson's disease, and discusses medications and surgical methods of treatment.

Jordan, V. C. October 1998. Designer estrogens. *Scientific American* 279(4):60. Selective estrogen receptor modulators (SERMs) may protect against breast and endometrial cancers, osteoporosis, and heart disease.

Julien, R. M. 1997. *A primer of drug action.* 8th ed. New York: W. H. Freeman and Company. A concise, nontechnical guide to the actions, uses, and side effects of psychoactive drugs.

Karch, S. B. 1996. *The pathology of drug abuse.* 2d ed. Boca Raton, Fl: CRC Press, Inc. This book, which can be used by students and medical professionals, contains the history, cultivation or manufacture, and effects of psychoactive drugs on the body.

Mader, S. S. 1997. *Understanding anatomy and physiology.* 3d ed. Dubuque, Iowa: Wm. C. Brown Publishers. A text that emphasizes the basics for beginning allied health students.

Marcus, D. M., and Camp, M. W. May/June 1998. Age-related macular degeneration. *Science & Medicine* 5(3):10. New therapies are needed for this common cause of vision loss in the elderly.

Mattson, M. P. March/April 1998. Experimental models of Alzheimer's disease. *Science & Medicine* 5(2):16. In Alzheimer's disease, mutations accelerate changes that occur during normal aging.

Maudgil, D. D., and Shorvon, S. D. September/October 1997. Locating the epileptogenic focus by MRI. *Science & Medicine* 4(5):26. Use of MRI should improve treatment of certain kinds of epilepsy.

McLachlan, J. A., and Arnold, S. F. September/October 1996. *American Scientist* 84(5):452. Environmental estrogens. Research suggests that ecoestrogens can mimic molecules involved in environmental signaling.

Nolte, J. 1993. *The human brain.* 3d ed. St. Louis: Mosby-Year Book, Inc. Beginners are guided through the basic aspects of brain structure and function.

Osorio, D. July/August 1997. The evolution of arthropod nervous systems. *American Scientist* 85(3):244. Nervous systems of insects and crustaceans have common features.

Rome, L. C. July/August 1997. Testing a muscle's design. *American Scientist* 85(4):356. Muscular systems adapt to specific functions, such as the specialized muscles found in frogs.

Sack, R. L. September/October 1998. Melatonin. *Science & Medicine* 5(5): 8. Certain mood and sleep disorders can be managed with melatonin treatments.

Schwartz, W. J. May/June 1996. Internal timekeeping. *Science & Medicine* 3(3):44. Article discusses circadian rhythm mechanisms, new insights into brain organization and consciousness.

Swerdlow, J. L. June 1995. The brain. *National Geographic* 187(6):2. New research leads to treatments for many age-old disorders.

Synder, S. H. 1996. *Drugs and the brain.* New York: Scientific American Library. How drugs affect brain function is clearly explained.

Thomas, E. D. September/October 1995. Hematopoietic stem cell transplantation. *Scientific American Science & Medicine* 2(5):38. Article discusses reconstituting marrow from cultured stem cells for bone marrow transplants.

White, R. J. September 1998. Weightlessness and the human body. *Scientific American* 279(3):58. Space medicine is providing new ideas about treatment of osteoporosis and anemia.

Wolkomir, R. August 1998. Oh, my aching back. *Smithsonian* 29(5):36. Researchers try to locate the source of back pain using a Virtual Corset to monitor the subject's activities.

Youdim, M. B., and Riederer, P. January 1997. Understanding Parkinson's disease. *Scientific American* 276(1):52. The tremors and immobility of Parkinson's disease can be traced to damage in a part of the brain that regulates movement.

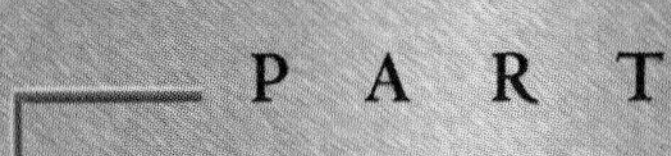

PART V

Continuance of the Species

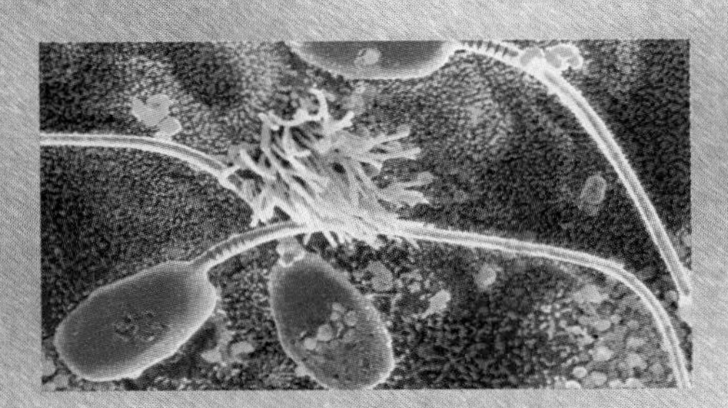

The reproductive systems of male and female produce the gametes, which unite to form a cell that develops into a human being.

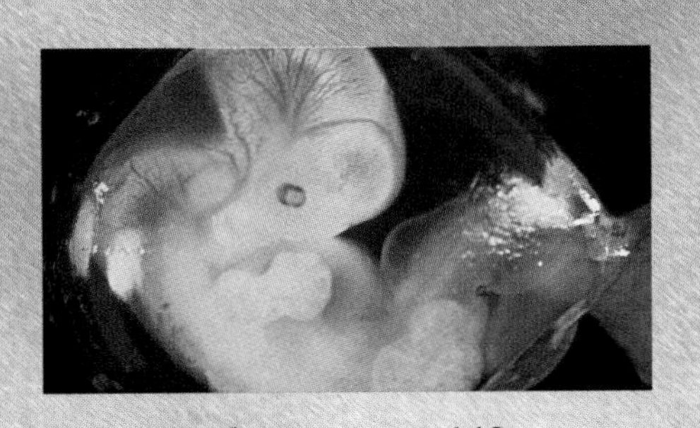

Development is a series of controlled events that produces a human being, which is like, and at the same time different from, its parents.

Genes are inherited from each parent and they bring about our characteristics, including any genetic disorders.

The genes are parts of chromosomes, and chromosomal inheritance has a marked effect on our individual characteristics.

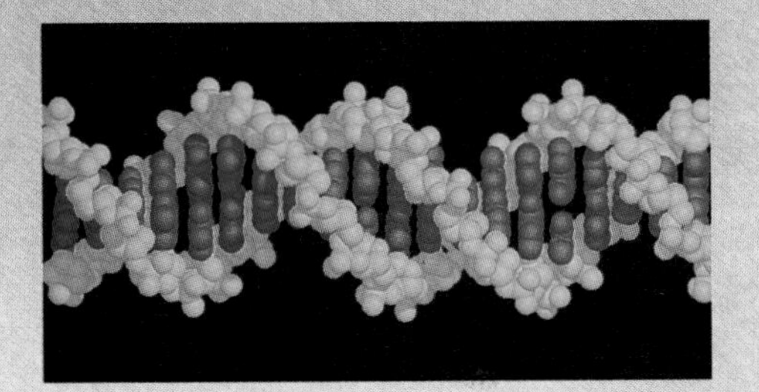

The genes are composed of DNA, the molecule that stores genetic information from generation to generation.

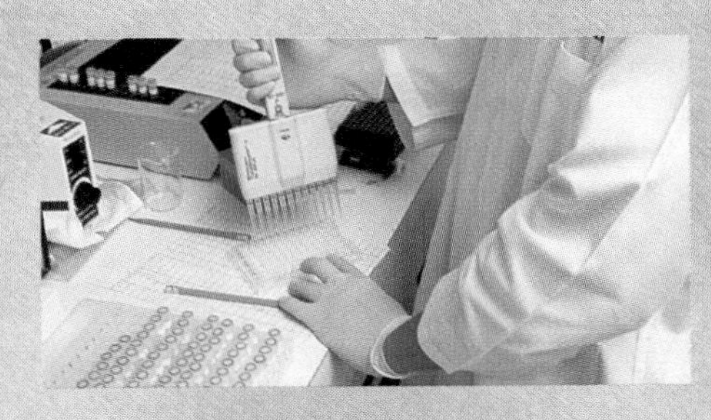

We have learned how to control DNA, and therefore the characteristics of cells and individuals.

In the eighteenth century, Gregor Mendel studied inheritance in peas and proposed that each parent donates hereditary factors to the offspring. The term gene was coined in the twentieth century when it was reasoned that the genes are parts of chromosomes. After X-raying bread mold, Beadle and Tatum gave us the one gene-one enzyme hypothesis. In other words, genes specify protein synthesis. Finally in 1944, scientists knew that genes are composed of DNA, and in 1953, Watson and Crick deduced the structure of DNA. Now, modern genetics has the ability to manipulate the genes of all organisms, including those of human beings.

Careers in Genetics

Biotechnologists doing DNA fingerprinting.

Forensic geneticists use DNA fingerprinting as a way to compare the DNA of persons with an unknown sample to see if there is a relationship. DNA from cells obtained at a crime scene is processed in a laboratory and made into a DNA print that can be compared with the suspect's DNA prints. DNA prints can also be evidence in paternity cases and in any circumstances in which information about ancestry is required.

Research geneticists study the DNA of organisms to understand how genes function and develop laboratory techniques that can be used by others. Some assist physicians in developing gene therapy techniques that will be used to treat genetic diseases. Many work on the ambitious Human Genome Project, designed to map the chromosomes and sequence every gene of the human species.

Geneticist weighing yield of corn strains.

Biotechnologists work in laboratories of biotechnology firms and in chemical plants where biotechnology products are produced.

Agricultural geneticists use genetic techniques to improve plant and animal food sources and to create plants and animals that will produce products desired by humans.

Pharmaceutical geneticists improve the drug-making capability of prokaryotic and eukaryotic cells, and use biotechnology techniques to design new and more powerful drugs. Some drugs that were previously available in limited quantity can now be mass-produced.

Genetic counselor and doctor talking to a couple.

Genetic counselors work with clients to determine if they have a genetic defect that can lead to a disorder in the future or can be passed on to their children. They also give advice about or arrange for genetic testing that detects such mutations. These tests can be administered to white blood cells from an adult, or fetal cells taken from the placenta or amniotic fluid.

C H A P T E R 21

Reproductive System

Chapter Concepts

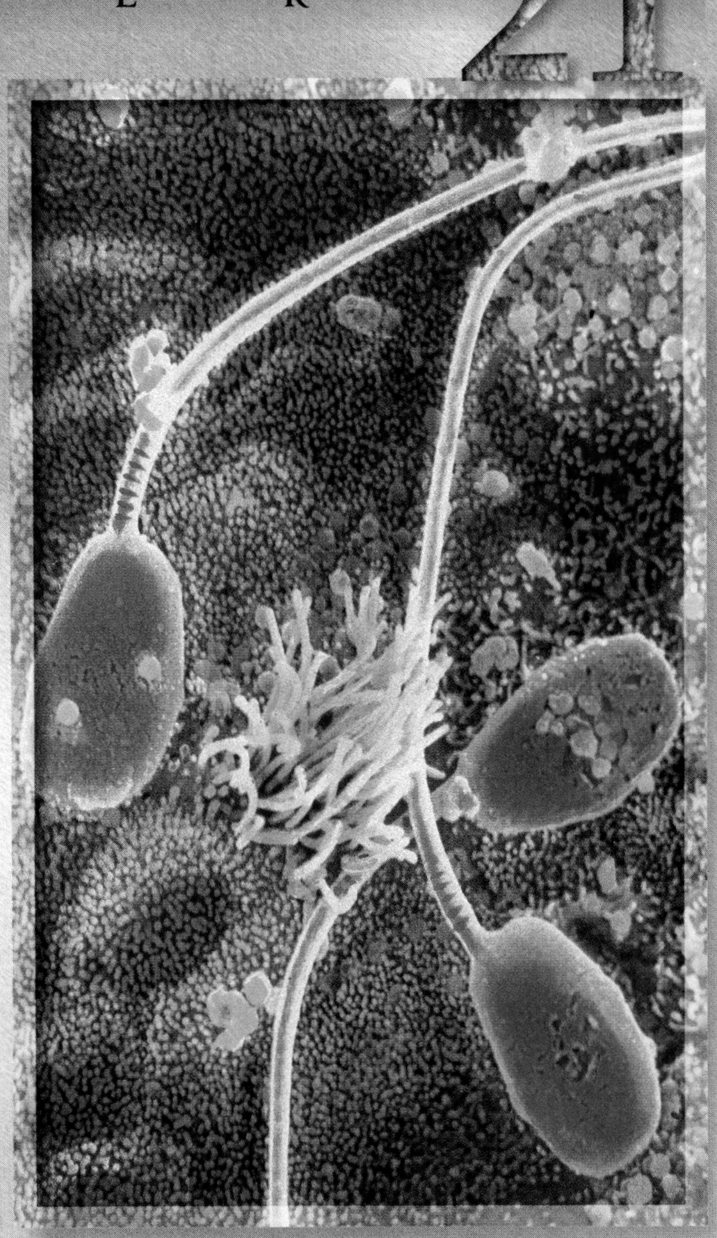

Human reproduction requires that the sperm make their way through the uterus to the oviduct where one of them will fertilize an egg.

Mark and Deidre had in vain tried to have a child for more than two years. They had even timed their lovemaking to follow Deidre's monthly cycle of producing an egg. They finally learned that while Mark's sperm were healthy, his testes were making very few of them. The couple turned to an infertility clinic for help. Using hormones, physicians stimulated Deidre's ovaries to mature a bunch of eggs at once. The doctors surgically removed those eggs and injected a single sperm of Mark's into each one. Those eggs that were successfully fertilized were then implanted into Deidre's uterus. Nine months later, she gave birth to triplets.

Mark and Deidre were reluctant to tell others how their children were conceived. They were afraid that some would think they should not interfere with the natural course of events. And how would the children be affected by being one of three instead of receiving the full attention of their parents? How do you feel about such questions? Is it beneficial or not that medical science has learned to control human reproduction?

This chapter will describe how the male and female reproductive systems normally bring about this miracle of new life. And it will consider alternative means of human reproduction as well.

21.1 Male Reproductive System

The male reproductive system includes the organs depicted in Figure 21.1 and listed in Table 21.1. The male **gonads** are paired testes (sing., **testis**), which are suspended within the *scrotal sacs* of the **scrotum.**

Sperm produced by the testes mature within the **epididymis** (pl., epididymides), which is a tightly coiled tubule lying just outside each testis. Maturation seems to be required for the sperm to swim to the egg. Each epididymis joins with a **vas deferens** (pl., vasa deferentia), which descends through a canal called the inguinal canal and enters the abdominal cavity where it curves around the bladder and empties into the **urethra.** Sperm are stored in both the epididymides and the vasa deferentia.

At the time of ejaculation, sperm leave the penis in a fluid called seminal fluid **(semen).** The pair of seminal vesicles, the prostate gland, and the bulbourethral glands (Cowper's glands) add secretions to seminal fluid. The **seminal vesicles** lie at the base of the bladder and each has a duct that joins with a vas deferens. The **prostate gland** is a single doughnut-shaped gland that surrounds the upper portion of the urethra just below the bladder. In older men,

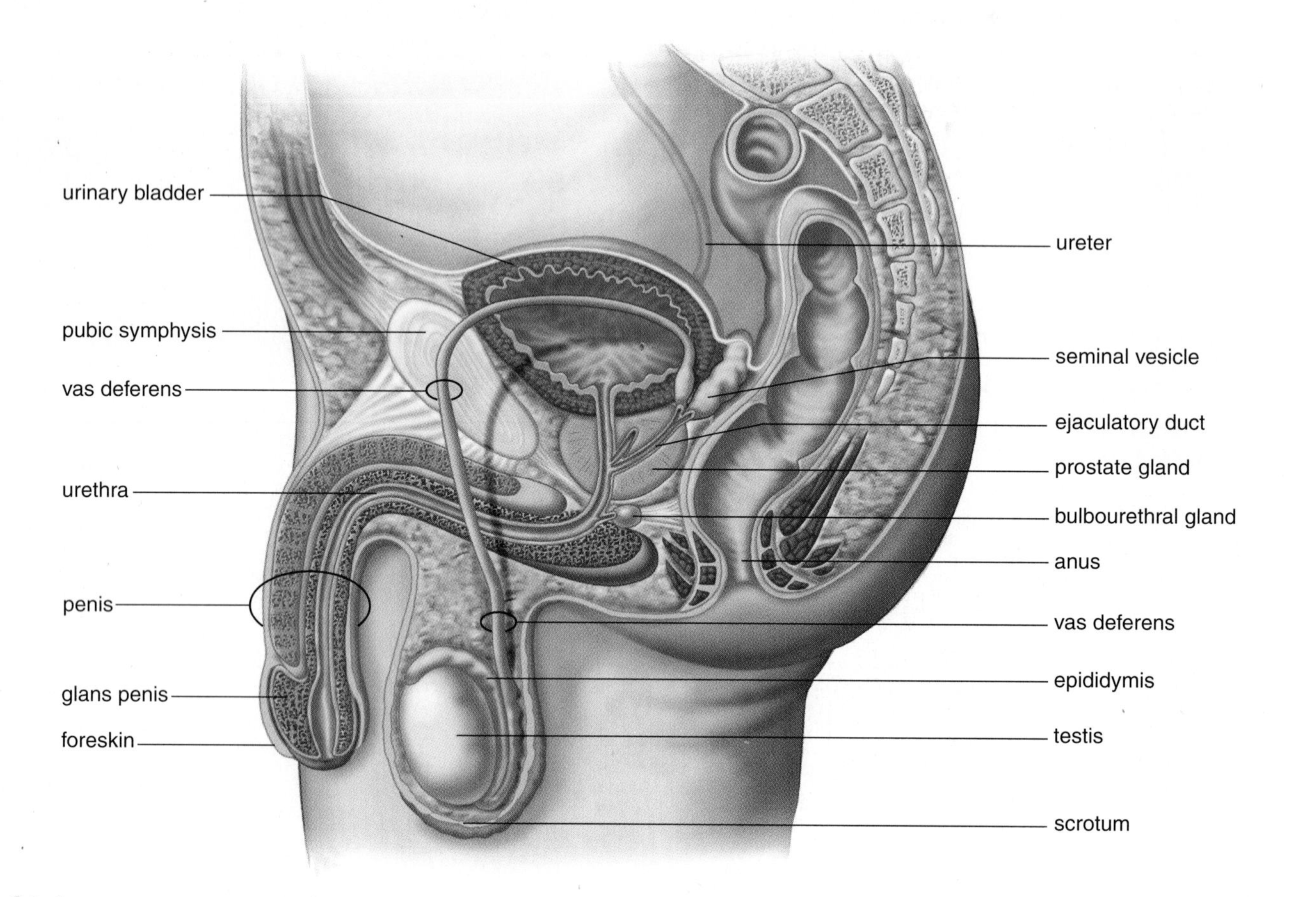

Figure 21.1 The male reproductive system.
The testes produce sperm. The seminal vesicles, the prostate gland, and the bulbourethral gland provide a fluid medium. Circumcision is the removal of the foreskin. Notice that the penis in this drawing is not circumcised because the foreskin is present.

the prostate can enlarge and squeeze off the urethra, making urination painful and difficult. The condition can be treated medically. **Bulbourethral glands** are pea-sized organs that lie posterior to the prostate on either side of the urethra.

Each component of seminal fluid seems to have a particular function. Sperm are more viable in a basic solution, and seminal fluid, which is milky in appearance, has a slightly basic pH (about 7.5). Swimming sperm require energy, and seminal fluid contains the sugar fructose, which presumably serves as an energy source. Seminal fluid also contains prostaglandins, chemicals that cause the uterus to contract. Some investigators believe that uterine contractions help propel the sperm toward the egg.

Orgasm in Males

The **penis** (Fig. 21.2) is the male organ of of sexual intercourse. The penis has a long shaft and an enlarged tip called the glans penis. The glans penis is normally covered by a layer of skin called the foreskin. Circumcision is the surgical removal of the foreskin, usually soon after birth.

Spongy, erectile tissue containing expandable blood spaces extends through the shaft of the penis. During sexual arousal, nerve impulses stimulate the release of cGMP (cyclic guanosine monophosphate), and the erectile tissue fills with blood. The veins that take blood away from the penis are compressed and the penis becomes erect. Impotency exists when the erectile tissue doesn't expand enough to compress the veins. The new drug Viagra inhibits an enzyme that breaks down cGMP, ensuring that a full erection will take place. Vision problems may occur because the same enzyme occurs in the retina.

As sexual stimulation intensifies, sperm enter the urethra from each vas deferens and the glands contribute secretions to seminal fluid (semen). Once seminal fluid is in the urethra, rhythmic muscle contractions cause it to be expelled from the penis in spurts. During ejaculation, a sphincter closes off the bladder so that no urine enters the urethra. (Notice that the urethra carries either urine or semen at different times.)

The contractions that expel seminal fluid from the penis are a part of male orgasm, the physiological and psychological sensations that occur at the climax of sexual stimulation. The psychological sensation of pleasure is centered in the brain, but the physiological reactions involve the genital (reproductive) organs and associated muscles, as well as the entire body. Marked muscular tension is followed by contraction and relaxation.

Following ejaculation and/or loss of sexual arousal, the penis returns to its normal flaccid state. After ejaculation, a male typically experiences a period of time, called the refractory period, during which stimulation does not bring about an erection. The length of the refractory period increases with age.

There may be in excess of 400 million sperm in the 3.5 ml of semen expelled during ejaculation. The sperm count can be much lower than this, however, and fertilization of the egg by a sperm still can take place.

Sperm are produced in the testes, mature in the epididymis, and pass from the vas deferens to the urethra. After glands add fluid to sperm, semen is ejaculated from the penis at the time of male orgasm.

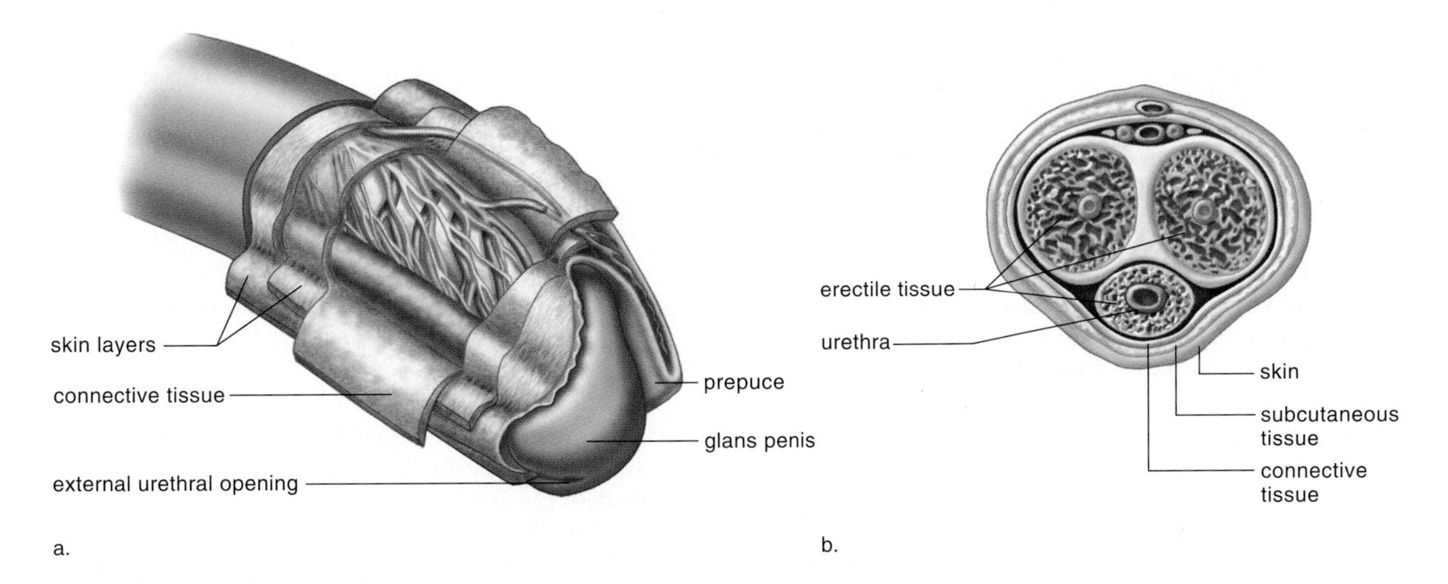

Figure 21.2 Penis anatomy.
a. Beneath the skin and the connective tissue lies the urethra, surrounded by erectile tissue. This tissue expands to form the glans penis, which in uncircumcised males is partially covered by the foreskin. **b.** Two other columns of erectile tissue in the penis are located dorsally.

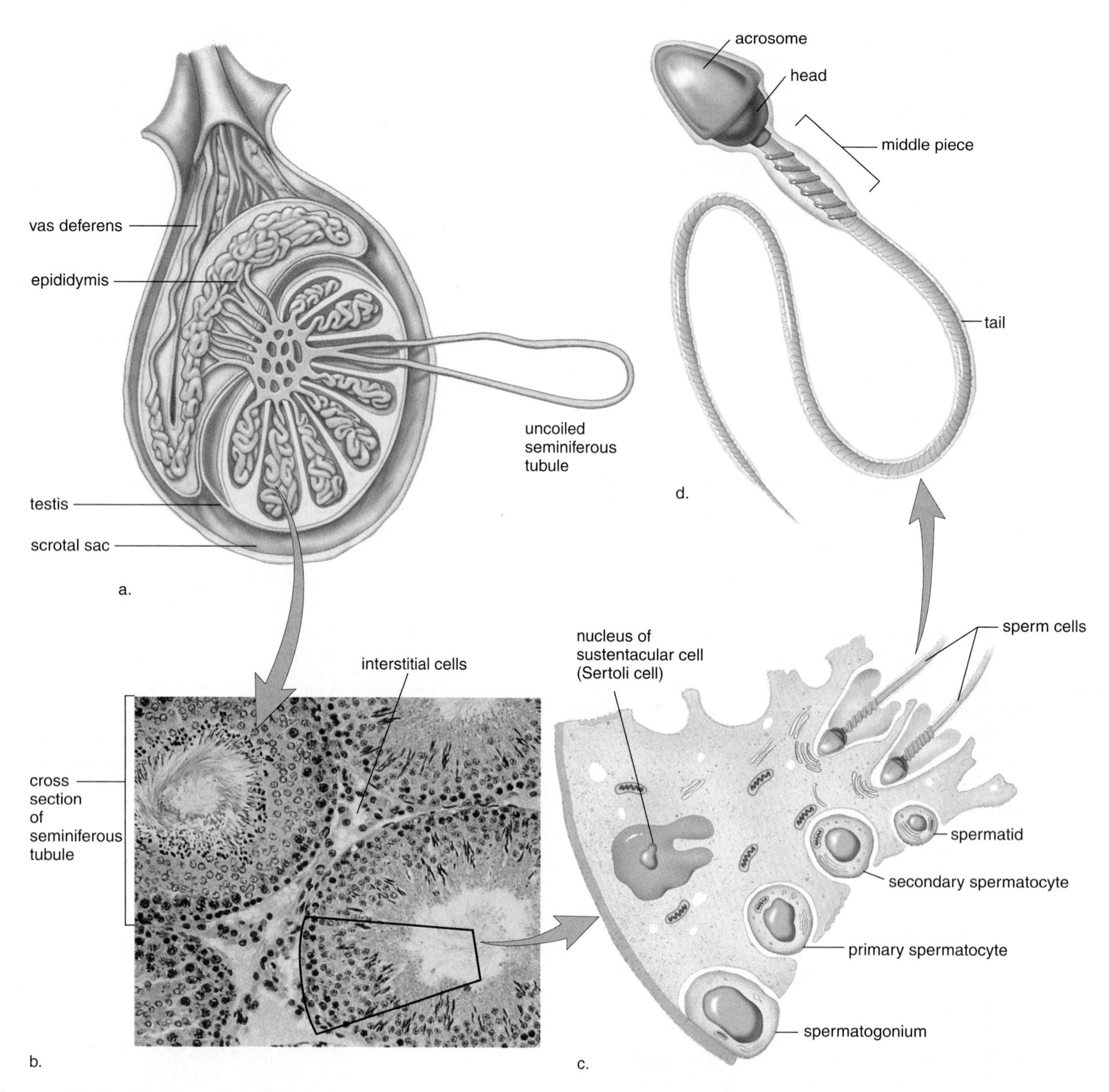

Figure 21.3 Testis and sperm.
a. The lobules of a testis contain seminiferous tubules. **b.** Light micrograph of cross section of seminiferous tubules where spermatogenesis occurs. **c.** Diagrammatic representation of spermatogenesis, which occurs in the wall of the tubules. **d.** A sperm has a head, a middle piece, and a tail. The nucleus is in the head, capped by the enzyme-containing acrosome.

The Male Gonads, the Testes

The testes lie outside the abdominal cavity of the male within the scrotum. The testes begin their development inside the abdominal cavity but descend into the scrotal sacs during the last two months of fetal development. If, by chance, the testes do not descend and the male is not treated or operated on to place the testes in the scrotum, sterility—the inability to produce offspring—usually follows. This is because the internal temperature of the body is too high to produce viable sperm. The scrotum helps regulate the temperature of the testes by holding them closer or farther away from the body.

A longitudinal section of a testis shows that it is composed of compartments called lobules, each of which contains one to three tightly coiled **seminiferous tubules** (Fig. 21.3*a*). Altogether, these tubules have a combined length of approximately 250 meters. A microscopic cross section of a

Table 21.1 Male Reproductive System

Organ	Function
Testes	Produce sperm and sex hormones
Epididymides	Maturation and some storage of sperm
Vasa deferentia	Conduct and store sperm
Seminal vesicles	Contribute fluid to semen
Prostate gland	Contributes fluid to semen
Urethra	Conducts sperm
Bulbourethral glands	Contribute fluid to semen
Penis	Organ of copulation

seminiferous tubule shows that it is packed with cells undergoing **spermatogenesis** (Fig. 21.3*b*), the production of sperm. Also present are **sustentacular (Sertoli) cells,** which support, nourish, and regulate the spermatogenic cells (Fig. 21.3*c*).

Mature **sperm,** or spermatozoa, have three distinct parts: a head, a middle piece, and a tail (Fig. 21.3*d*). There are mitochondria in the middle piece that provide the energy for the movement of the tail which has the structure of a flagellum. The head contains a nucleus covered by a cap called the **acrosome,** which stores enzymes needed to penetrate the egg. The ejaculated semen of a normal human male contains several hundred million sperm, assuring an adequate number for fertilization to take place. Only one sperm normally enters an egg.

Hormonal Regulation in Males

The hypothalamus has ultimate control of the testes' sexual function because it secretes a hormone called **gonadotropin-releasing hormone (GnRH)** that stimulates the anterior pituitary to secrete the gonadotropic hormones. There are two gonadotropic hormones—**follicle-stimulating hormone (FSH)** and **luteinizing hormone (LH)**—in both males and females. In males, FSH promotes the production of sperm in the seminiferous tubules, which also release the hormone inhibin.

LH in males is sometimes given the name *interstitial cell-stimulating hormone (ICSH)* because it controls the production of testosterone by the **interstitial cells,** which are found in the spaces between the seminiferous tubules. All these hormones are involved in a negative feedback relationship that maintains the fairly constant production of sperm and testosterone (Fig. 21.4).

Testosterone, the main sex hormone in males, is essential for the normal development and functioning of the organs listed in Table 21.1. Testosterone also brings about and maintains the male secondary sex characteristics that develop at the time of puberty. Males are generally taller than females and have broader shoulders and longer legs relative to trunk length. The deeper voice of males compared to females is due to a larger larynx with longer vocal cords. Since the so-called Adam's apple is a part of the larynx, it is usually more prominent in males than in females. Testosterone causes males to develop noticeable hair on the face, chest, and occasionally on other regions of the body such as the back. Testosterone also leads to the receding hairline and pattern baldness that occurs in males.

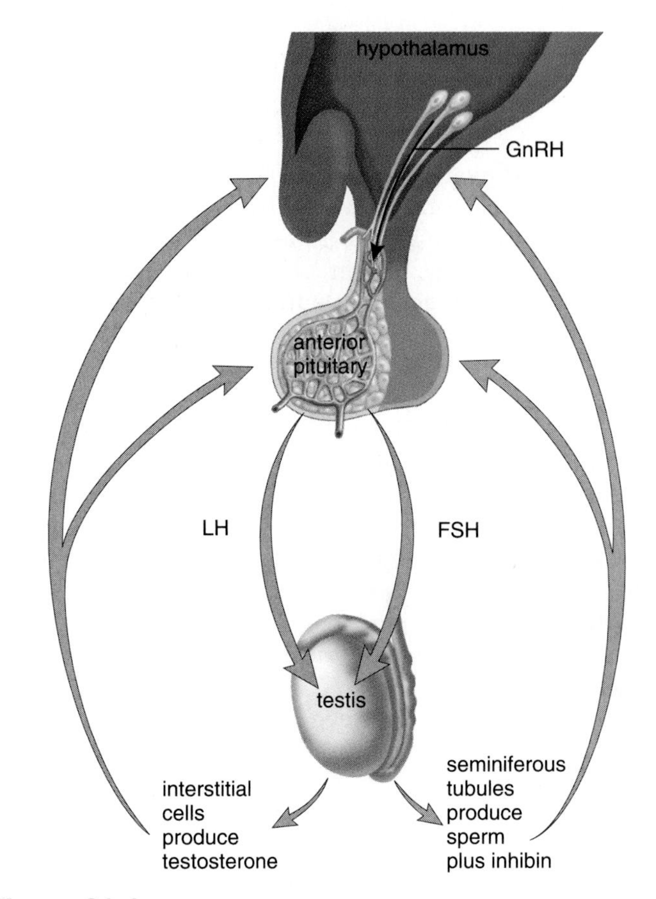

Figure 21.4 Hormonal control of testes.
GnRH (gonadotropin-releasing hormone) stimulates the anterior pituitary to secrete the gonadotropic hormones FSH and LH. FSH stimulates the testes to produce sperm, and LH stimulates the testes to produce testosterone. Testosterone and inhibin exert negative feedback control over the hypothalamus and the anterior pituitary, and this regulates the level of testosterone in blood.

Testosterone is responsible for the greater muscular development in males. Knowing this, males and females sometimes take anabolic steroids, which are either testosterone or related steroid hormones resembling testosterone. Health problems involving the kidneys, the circulatory system, and hormonal imbalances can arise from such use. The testes shrink in size, and feminization in regard to other male traits occurs.

The gonads in males are the testes, which produce sperm as well as testosterone, the most significant male sex hormone.

21.2 Female Reproductive System

The female reproductive system includes the organs depicted in Figure 21.5 and listed in Table 21.2. The female **gonads** are paired **ovaries** that lie in shallow depressions, one on each side of the upper pelvic cavity. **Oogenesis** is the production of an **egg,** the female gamete. The ovaries alternate in producing one egg (ovum) a month. **Ovulation** is the process by which an egg bursts from an ovary and usually enters an oviduct.

The Genital Tract

The **oviducts,** also called uterine or fallopian tubes, extend from the uterus to the ovaries; however, the oviducts are not attached to the ovaries. Instead, they have fingerlike projections called fimbriae (sing., **fimbria**) that sweep over the ovaries. When an egg bursts from an ovary during ovulation, it usually is swept into an oviduct by the combined action of the fimbriae and the beating of cilia that line the oviducts.

Once in the oviduct, the egg is propelled slowly by cilia movement and tubular muscle contraction toward the uterus. Fertilization and **zygote** formation occurs in an oviduct because the egg only lives approximately 6 to 24 hours. The developing embryo normally arrives at the uterus after several days and then embeds, or implants, itself in the uterine lining, which has been prepared to receive it.

The **uterus** is a thick-walled, muscular organ about the size and shape of an inverted pear. Normally, it lies above and is tipped over the urinary bladder. The oviducts join the uterus anteriorly, while posteriorly the

Table 21.2 Female Reproductive System

Organ	Function
Ovaries	Produce egg and sex hormones
Oviducts	Conduct egg; location of fertilization (fallopian tubes)
Uterus (womb)	Houses developing fetus
Cervix	Contains opening to uterus
Vagina	Receives penis during sexual intercourse; serves as birth canal, and as an exit for menstrual flow

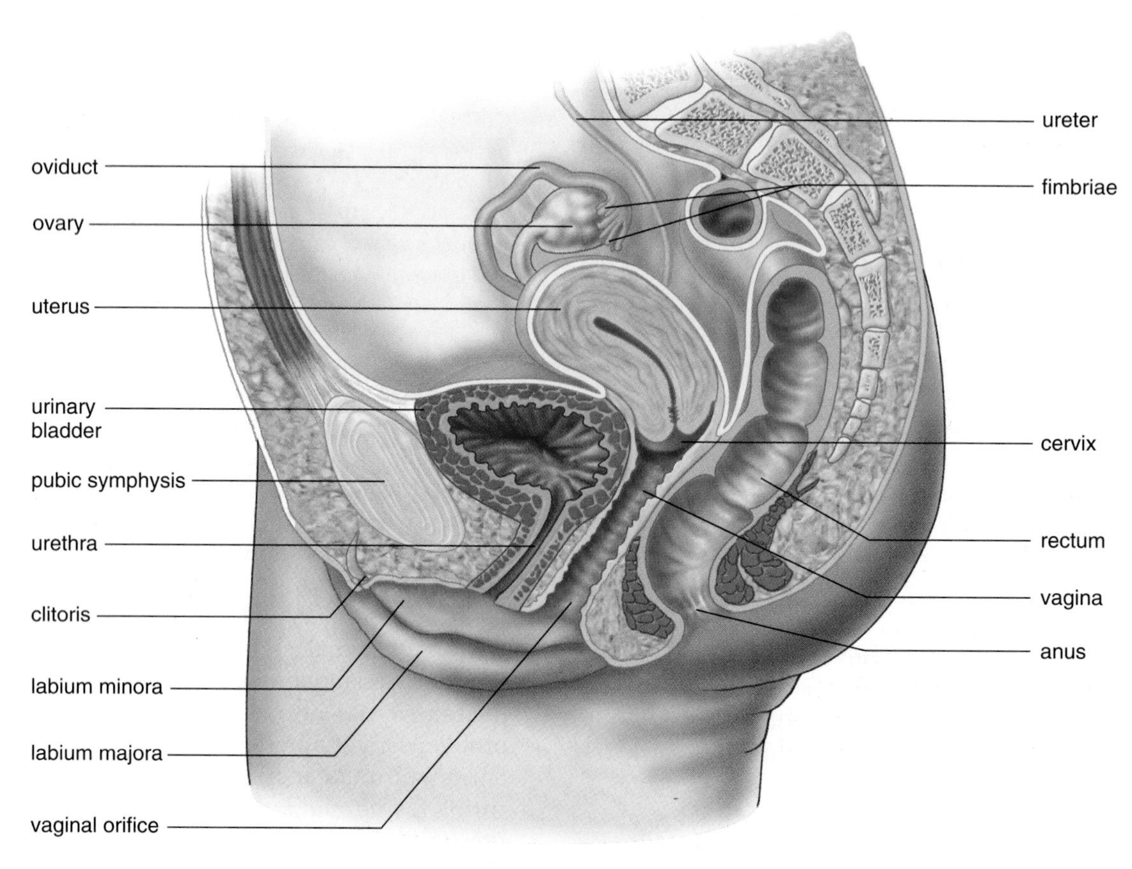

Figure 21.5 The female reproductive system.
The ovaries release one egg a month; fertilization occurs in the oviduct, and development occurs in the uterus. The vagina is the birth canal and the organ of sexual intercourse.

cervix enters the vagina nearly at a right angle. A small opening in the cervix leads to the vaginal canal. Development of the embryo normally takes place in the uterus. This organ, sometimes called the womb, is approximately 5 cm wide in its usual state but is capable of stretching to over 30 cm wide to accommodate the growing baby. The lining of the uterus, called the **endometrium,** participates in the formation of the placenta (p. 429) which supplies nutrients needed for embryonic and fetal development. The endometrium has two layers, a basal layer and an inner functional layer. In the nonpregnant female, the functional layer of the endometrium varies in thickness according to a monthly reproductive cycle, called the uterine cycle.

Cancer of the cervix is a common form of cancer in women. Early detection is possible by means of a **Pap test,** which requires the removal of a few cells from the region of the cervix for microscopic examination. If the cells are cancerous, a hysterectomy may be recommended. A hysterectomy is the removal of the uterus, including the cervix. Removal of the ovaries in addition to the uterus is termed an ovariohysterectomy. Because the vagina remains, the woman still can engage in sexual intercourse.

The **vagina** is a tube at a 45° angle with the small of the back. The mucosal lining of the vagina lies in folds and can extend. This is especially important when the vagina serves as the birth canal. It also facilitates sexual intercourse, when the vagina receives the penis, and acts as an exit for menstrual flow.

External Genitals

The external genital organs of the female are known collectively as the **vulva** (Fig. 21.6). The vulva includes two large, hair-covered folds of skin called the labia majora. They extend backward from the mons pubis, a fatty prominence underlying the pubic hair. The labia minora are two small folds lying just inside the labia majora. They extend forward from the vaginal opening to encircle and form a foreskin for the clitoris, an organ that is homologous to the penis. Although quite small, the clitoris has a shaft of erectile tissue and is capped by a pea-shaped glans. The glans clitoris also has sense receptors that allow it to function as a sexually sensitive organ.

The vestibule, a cleft between the labia minor, contains the openings of the urethra and the vagina. The vagina may be partially closed by a ring of tissue called the hymen. The hymen ordinarily is ruptured by initial sexual intercourse; however, it also can be disrupted by other types of physical activities. If the hymen persists after sexual intercourse, it can be surgically ruptured.

Notice that the urinary and reproductive systems in the female are entirely separate. For example, the urethra carries only urine, and the vagina serves only as the birth canal and the organ for sexual intercourse.

Orgasm in Females

Sexual response in the female may be more subtle than in the male, but there are certain corollaries. The clitoris is believed to be an especially sensitive organ for initiating sexual sensations. It is possible for the clitoris to become ever so slightly erect as its erectile tissues become engorged with blood, but vasocongestion is more obvious in the labia minora, which expand and deepen in color. Erectile tissue within the vaginal wall also expands with blood, and the added pressure in these blood vessels causes small droplets of fluid to squeeze through the vessel walls and to lubricate the vagina. Another possible source of lubrication is from mucus-secreting glands beneath the labia minora on either side of the vagina.

Release from muscular tension occurs in females, especially in the region of the vulva and vagina but also throughout the entire body. Increased uterine motility may assist the transport of sperm toward the oviducts. Since female orgasm is not signaled by ejaculation, there is a wide range in normalcy of sexual response.

Once each month, an egg produced by an ovary enters an oviduct. If fertilization occurs, the developing embryo is propelled by cilia to the uterus where it implants itself in the uterine lining. The vagina (which is also the birth canal) and the external genitals play an active role in the sexual response of females.

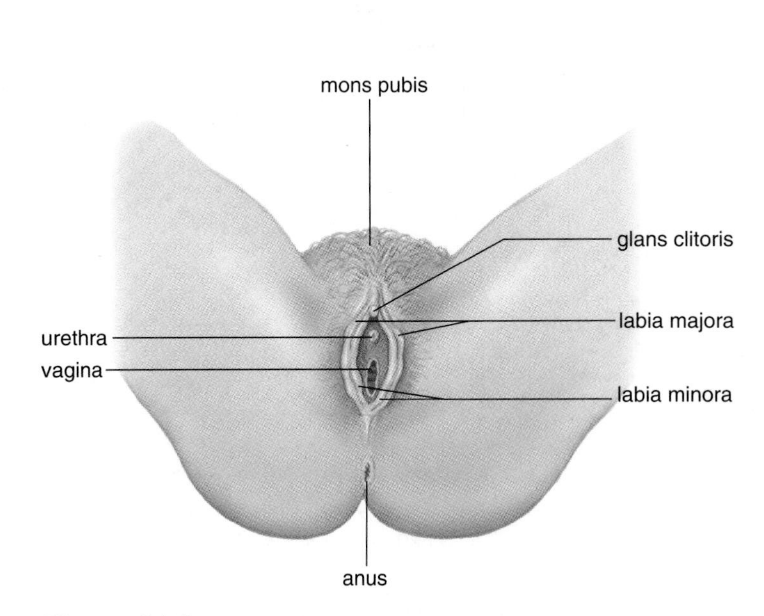

Figure 21.6 External genitals of the female.
At birth, the opening of the vagina is partially blocked by a membrane called the hymen. Physical activities and sexual intercourse disrupt the hymen.

Visual Focus

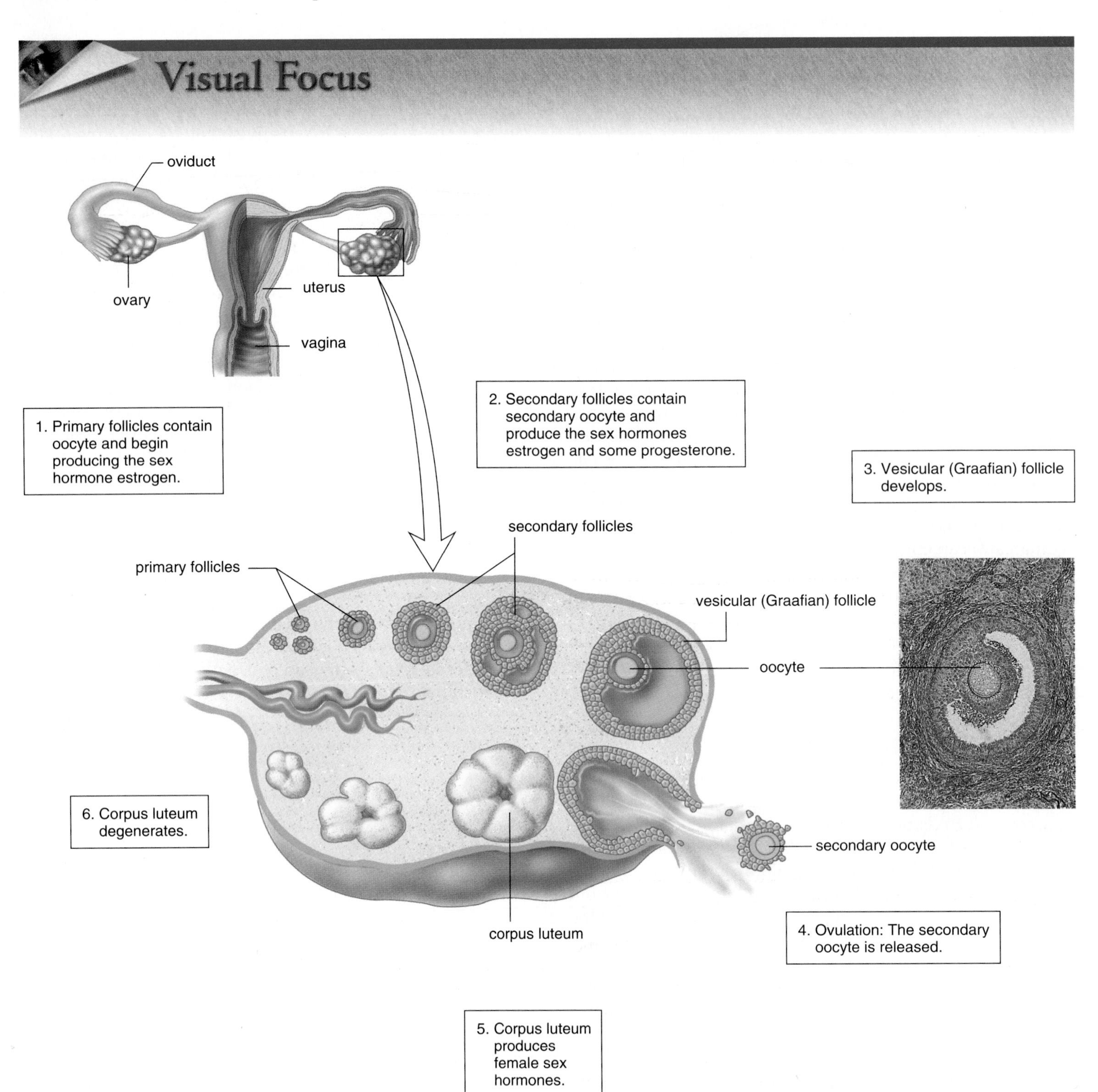

Figure 21.7 Anatomy of ovary and follicle.
As a follicle matures, the oocyte enlarges and is surrounded by layers of follicular cells and fluid. Eventually, ovulation occurs, the mature follicle ruptures, and the secondary oocyte is released. A single follicle actually goes through all stages in one place within the ovary.

21.3 Female Hormone Levels

Hormone levels cycle in the female on a monthly basis, and the ovarian cycle drives the uterine cycle as discussed in this section.

The Ovarian Cycle

A longitudinal section through an ovary shows that it is made up of an outer cortex and an inner medulla (Fig. 21.7). There are many **follicles** in the cortex and each one contains an immature egg, called an oocyte. A female is born with as many as 2 million follicles, but the number is reduced to 300,000–400,000 by the time of puberty. Only a small number of follicles (about 400) ever mature because a female usually produces only one egg per month during her reproductive years. Since oocytes are present at birth, they age as the woman ages. This may be one possible reason why older women are more likely to produce children with genetic defects.

As the follicle undergoes maturation, it develops from a primary follicle to a secondary follicle to a vesicular (Graafian) follicle. A secondary follicle contains a secondary oocyte pushed to one side of a fluid-filled cavity. In a *Graafian follicle,* the fluid-filled cavity increases to the point that the follicle wall balloons out on the surface of the ovary and bursts, releasing the secondary oocyte surrounded by a clear membrane and follicular cells. As mentioned, this is referred to as *ovulation.* Actually, the second meiotic division is not completed unless fertilization occurs. In the meantime, the follicle is developing into the **corpus luteum.** If pregnancy does not occur, the corpus luteum begins to degenerate after about ten days.

These events, called the **ovarian cycle,** are under the control of the gonadotropic hormones, *follicle-stimulating hormone (FSH)* and *luteinizing hormone (LH)* (Fig. 21.8). The gonadotropic hormones are not present in constant amounts but instead are secreted at different rates during the cycle. For simplicity's sake, it can be emphasized that during the first half, or *follicular phase,* of the cycle, FSH promotes the development of a follicle, which secretes estrogen. As the estrogen level in the blood rises, it exerts feedback control over the anterior pituitary secretion of FSH so that the follicular phase comes to an end.

Presumably, the high level of estrogen in the blood also causes the hypothalamus suddenly to secrete a large amount of GnRH. This leads to a surge of LH production by the anterior pituitary and to ovulation at about the 14th day of a 28-day cycle.

During the second half, or *luteal phase,* of the ovarian cycle, LH promotes the development of the corpus luteum, which secretes progesterone. Progesterone causes the uterine lining to build up. As the blood level of progesterone rises, it exerts feedback control over anterior pituitary secretion of LH so that the corpus luteum begins to degenerate. As the luteal phase comes to an end, menstruation occurs.

> One ovarian follicle per month produces a secondary oocyte. Following ovulation, the follicle develops into the corpus luteum.

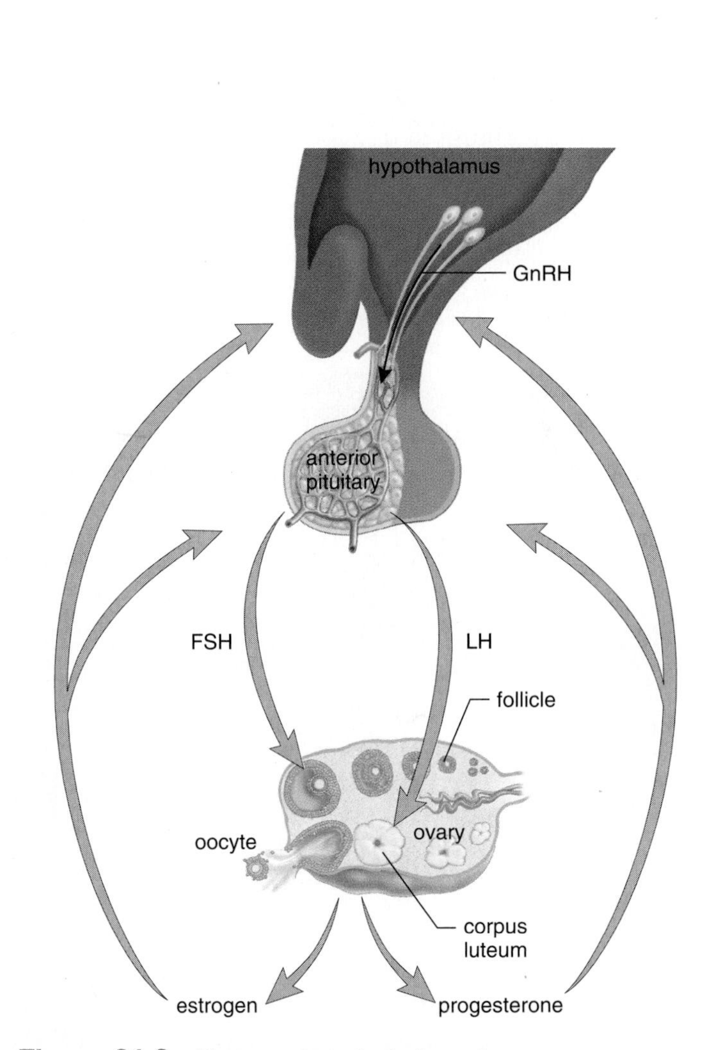

Figure 21.8 Hormonal control of ovaries.
The hypothalamus produces GnRH (gonadotropin-releasing hormone). GnRH stimulates the anterior pituitary to produce FSH (follicle-stimulating hormone) and LH (luteinizing hormone). FSH stimulates the follicle to produce estrogen, and LH stimulates the corpus luteum to produce progesterone. Estrogen and progesterone maintain the sex organs (e.g., uterus) and the secondary sex characteristics, and exert feedback control over the hypothalamus and the anterior pituitary.

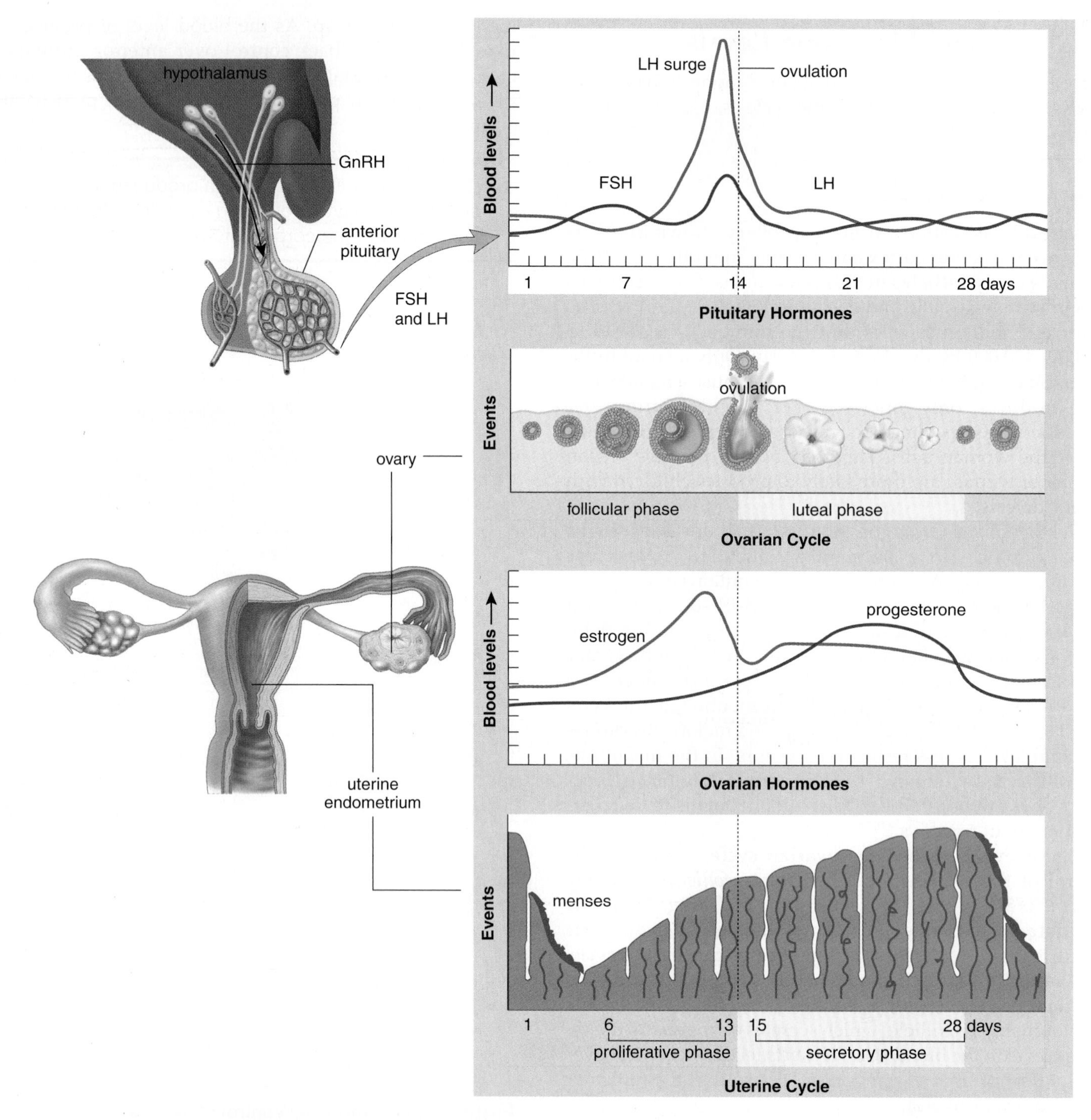

Figure 21.9 Female hormone levels.
During the follicular phase of the ovarian cycle, FSH released by the anterior pituitary promotes the maturation of a follicle in the ovary. The ovarian follicle produces increasing levels of estrogen, which causes the endometrium to thicken during the proliferative phase of the uterine cycle. After ovulation and during the luteal phase of the ovarian cycle, LH promotes the development of the corpus luteum. This structure produces increasing levels of progesterone, which causes the endometrial lining to become secretory. Menstruation begins when progesterone production declines to a low level.

Table 21.3 Ovarian and Uterine Cycles

Ovarian Cycle	Events	Uterine Cycle	Events
Follicular phase—Days 1–13	FSH	Menstruation—Days 1–5	Endometrium breaks down
	Follicle maturation	Proliferative phase—Days 6–13	Endometrium rebuilds
	Estrogen		
Ovulation—Day 14*	LH spike		
Luteal phase—Days 15–28	LH	Secretory phase—Days 15–28	Endometrium thickens and glands are secretory
	Corpus luteum		
	Progesterone		

*Assuming 28-day cycle.

The Uterine Cycle

The female sex hormones, **estrogen** and **progesterone,** have numerous functions. The effects of these hormones on the endometrium of the uterus cause the uterus to undergo a cyclical series of events known as the **uterine cycle** (Table 21.3 and Fig. 21.9). Twenty-eight-day cycles are divided as follows.

During *days 1–5,* a low level of female sex hormones in the body causes the endometrium to disintegrate and its blood vessels to rupture. On day one of the cycle, a flow of blood and tissues, known as the menses, passes out of the vagina during **menstruation,** also called the menstrual period.

During *days 6–13,* increased production of estrogen by a new ovarian follicle causes the endometrium to thicken and to become vascular and glandular. This is called the proliferative phase of the uterine cycle.

Ovulation usually occurs on the fourteenth day of the twenty-eight-day cycle.

During *days 15–28,* increased production of progesterone by the corpus luteum causes the endometrium to double in thickness and the uterine glands to mature, producing a thick mucoid secretion. This is called the secretory phase of the uterine cycle. The endometrium now is prepared to receive the developing embryo. If pregnancy does not occur, the corpus luteum degenerates and the low level of sex hormones in the female body causes the uterine lining to break down.

During the uterine cycle, the endometrium builds up and then is broken down when menstruation occurs.

Fertilization and Pregnancy

If fertilization does occur, an embryo begins development even as it travels down the oviduct to the uterus. The endometrium is now prepared to receive the developing embryo, which becomes embedded in the lining several days following fertilization (Fig. 21.10). The **placenta** originates from both maternal and fetal tissues. It is the region of exchange of molecules between fetal and maternal blood, although there is rarely any mixing of the two. At first, the placenta produces **human chorionic gonadotropin (HCG),** which maintains the corpus luteum until the placenta begins its own production of progesterone and estrogen. Progesterone and estrogen have two effects: they shut down the anterior pituitary so that no new follicles mature, and they maintain the lining of the uterus so that the corpus luteum is not needed. There is no menstruation during pregnancy.

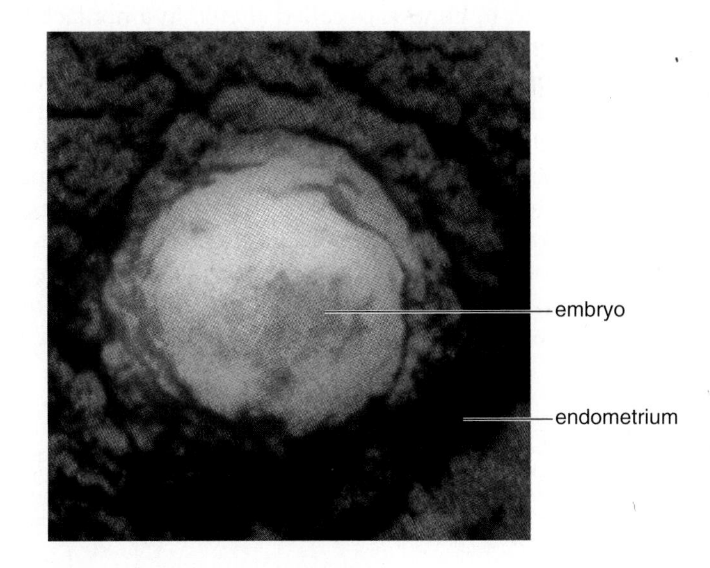

Figure 21.10 Implantation.
A scanning electron micrograph showing implantation on day 12 following fertilization.

Estrogen and Progesterone

Estrogen and progesterone affect not only the uterus but other parts of the body as well. Estrogen is largely responsible for the secondary sex characteristics in females, including body hair and fat distribution. In general, females have a more rounded appearance than males because of a greater accumulation of fat beneath the skin. Like males, females develop axillary and pubic hair during puberty. In females the upper border of pubic hair is horizontal, but in males it tapers toward the navel.

The pelvic girdle widens in females, so the pelvic cavity usually has a larger relative size compared to males. This means that females have wider hips than males and that the thighs converge at a greater angle toward the knees. Because the female pelvis tilts forward, females tend to have protruding buttocks, more lower back curve than men, an abdominal bulge, and a tendency to be somewhat knock-kneed.

Both estrogen and progesterone are also required for breast development.

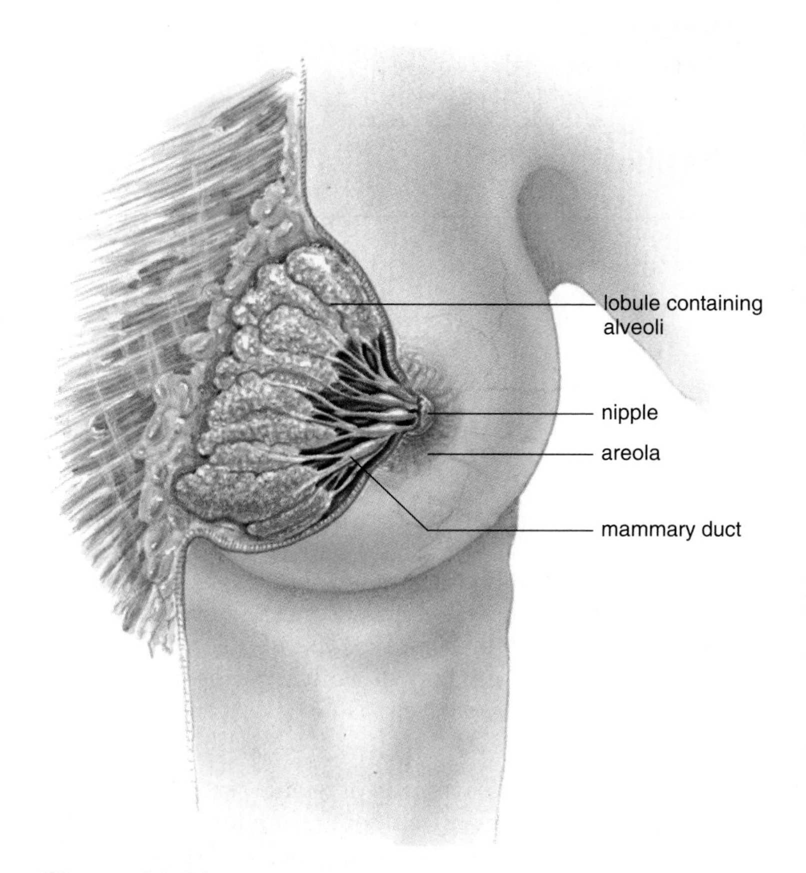

Figure 21.11 Anatomy of the breast.
The female breast contains lobules consisting of ducts and alveoli. The alveoli are lined by milk-producing cells in the lactating (milk-producing) breast.

Structure and Function of Breasts

A female breast contains one or two dozen lobules, each with its own mammary duct that opens at the nipple (Fig. 21.11). The nipple is surrounded by a pigmented area called the areola. Hair and sweat glands are absent from the nipples and areola, but glands are present that secrete a saliva-resistant lubricant to protect the nipples, particularly during nursing. Smooth muscle fibers in the region of the areola may cause the nipple to become erect in response to sexual stimulation or cold.

Each mammary duct divides into numerous other ducts, which end in blind sacs called *alveoli.* In a nonlactating breast, the ducts far outnumber the alveoli because alveoli are made up of cells that can produce milk. Milk is not produced during pregnancy. Prolactin is needed for lactation (milk production) to begin, and production of this hormone is suppressed by the feedback inhibition estrogen and progesterone have on the anterior pituitary during pregnancy. It takes a couple of days after delivery of a baby for milk production to begin. In the meantime, the breasts produce a watery, yellowish white fluid called **colostrum,** which is similar to milk but contains more protein and less fat.

Menopause

Menopause, the period in a woman's life during which the ovarian and uterine cycles cease, is likely to occur between ages 45 and 55. The ovaries are no longer responsive to the gonadotropic hormones produced by the anterior pituitary, and the ovaries no longer secrete estrogen or progesterone. At the onset of menopause, the uterine cycle becomes irregular, but as long as menstruation occurs, it is still possible for a woman to conceive. Therefore, a woman usually is not considered to have completed menopause until there has been no menstruation for a year.

The hormonal changes during menopause often produce physical symptoms, such as "hot flashes" (caused by circulatory irregularities), dizziness, headaches, insomnia, sleepiness, and depression. Again, there is great variation among women, and any of these symptoms may be absent altogether. Women sometimes report an increased sex drive following menopause. It has been suggested that this may be due to androgen production by the adrenal cortex.

21.4 Control of Reproduction

Several means are available to dampen or enhance our reproductive potential. **Contraceptives** are medications and devices that reduce the chance of pregnancy.

Birth-Control Methods

The most reliable method of birth control is abstinence, that is, the absence of sexual intercourse. This form of birth control has the added advantage of preventing transmission of a sexually transmitted disease. Other means of birth control used in this country are given in Table 21.4. The table gives the effectiveness for the various birth-control methods listed. For example, with the least effective method given in the table, we expect that within a year, 70 out of 100, or 70%, of sexually active women will not get pregnant, while 30 women will get pregnant.

Figure 21.12 features some of the most effective and commonly used means of birth control. *Oral contraception* **(birth-control pills)** usually involves taking a combination of estrogen and progesterone for 21 days of a 28-day cycle. The estrogen and progesterone in the birth-control pill effectively shut down the pituitary production of both FSH and LH so that no follicle begins to develop in the ovary; and since ovulation does not occur, pregnancy cannot take place. Since there are possible side effects, those taking birth-control pills should see a physician regularly.

An **intrauterine device (IUD)** is a small piece of molded plastic that is inserted into the uterus by a physician. IUDs are believed to alter the environment of the uterus and oviducts so that fertilization probably does not occur—but if fertilization should occur, implantation cannot take place. The type of IUD featured in Figure 21.12*b* has copper wire wrapped around the plastic.

The **diaphragm** is a soft rubber or latex cup with a flexible rim that lodges behind the pubic bone and fits over the cervix. Each woman must be properly fitted by a physician, and the diaphragm can be inserted into the vagina at most two hours before sexual relations. It must be used with spermicidal jelly or cream and should be left in place at least six hours after sexual relations. The cervical cap is a minidiaphragm.

Table 21.4 Common Birth-Control Methods

Name	Procedure	Methodology	Effectiveness	Risk
Abstinence	Refrain from sexual intercourse	No sperm in vagina	100%	None
Vasectomy	Vasa deferentia are cut and tied	No sperm in seminal fluid	Almost 100%	Irreversible sterility
Tubal ligation	Oviducts are cut and tied	No eggs in oviduct	Almost 100%	Irreversible sterility
Oral contraception	Hormone medication is taken daily	Anterior pituitary does not release FSH and LH	Almost 100%	Thromboembolism, especially in smokers
Depo-Provera injection	Four injections of progesterone-like steroid are given per year	Anterior pituitary does not release FSH and LH	About 99%	Breast cancer? Osteoporosis?
Contraceptive implants	Tubes of progestin (form of progesterone) are implanted under skin	Anterior pituitary does not release FSH and LH	More than 90%	Presently none known
Intrauterine device (IUD)	Plastic coil is inserted into uterus by physician	Prevents implantation	More than 90%	Infection (pelvic inflammatory disease, PID)
Diaphragm	Latex cup is inserted into vagina to cover cervix before intercourse	Blocks entrance of sperm to uterus	With jelly, about 90%	Presently none known
Cervical cap	Latex cap is held by suction over cervix	Delivers spermicide near cervix	Almost 85%	Cancer of cervix?
Male condom	Latex sheath is fitted over erect penis	Traps sperm and prevents STDs	About 85%	Presently none known
Female condom	Polyurethane tubing is fitted inside vagina	Blocks entrance of sperm to uterus and prevents STDs	About 85%	Presently none known
Coitus interruptus	Male withdraws penis before ejaculation	Prevents sperm from entering vagina	75%	Presently none known
Jellies, creams, foams	These spermicidal products are inserted before intercourse	Kills a large number of sperm	About 75%	Presently none known
Natural family planning	Day of ovulation is determined by record keeping; various methods of testing	Intercourse avoided on certain days of the month	About 70%	Presently none known
Douche	Vagina is cleansed after intercourse	Washes out sperm	Less than 70%	Presently none known

A male **condom** is a thin skin (lambskin) or latex sheath that fits over the erect penis. The ejaculate is trapped inside the sheath and, thus, does not enter the vagina. When used in conjunction with a spermicide, the protection is better than with the condom alone. The condom is generally recognized as giving protection against sexually transmitted diseases.

Contraceptive implants utilize a synthetic progesterone to prevent ovulation by disrupting the ovarian cycle. Six match-sized, time-release capsules are surgically inserted under the skin of a woman's upper arm. The effectiveness of this system may last five years. *Depo-Provera* injections utilize a synthetic progesterone that must be administered every three months.

There has been a revival of interest in *barrier methods* of birth control, including the male condom, because these methods offer some protection against sexually transmitted diseases. A female condom, now available, consists of a large polyurethane tube with a flexible ring that fits onto the cervix. The open end of the tube has a ring that covers the external genitals.

Investigators have long searched for a "*male pill.*" Analogues of gonadotropic-releasing hormone have been used to prevent the hypothalamus from stimulating the anterior pituitary. Inhibin has also been used to prevent the anterior pituitary from producing FSH. Testosterone and/or related chemicals have been used to inhibit spermatogenesis in males, but this hormone must be administered by injection.

Contraceptive vaccines are now being developed. For example, a vaccine developed to immunize women against HCG, the hormone so necessary to maintaining the **implantation** of the embryo, was successful in a limited clinical trial. Since HCG is not normally present in the body, no untoward autoimmunity reaction is expected, but the immunization does wear off with time. Others believe that it would also be possible to develop a safe antisperm vaccine that would be used in women.

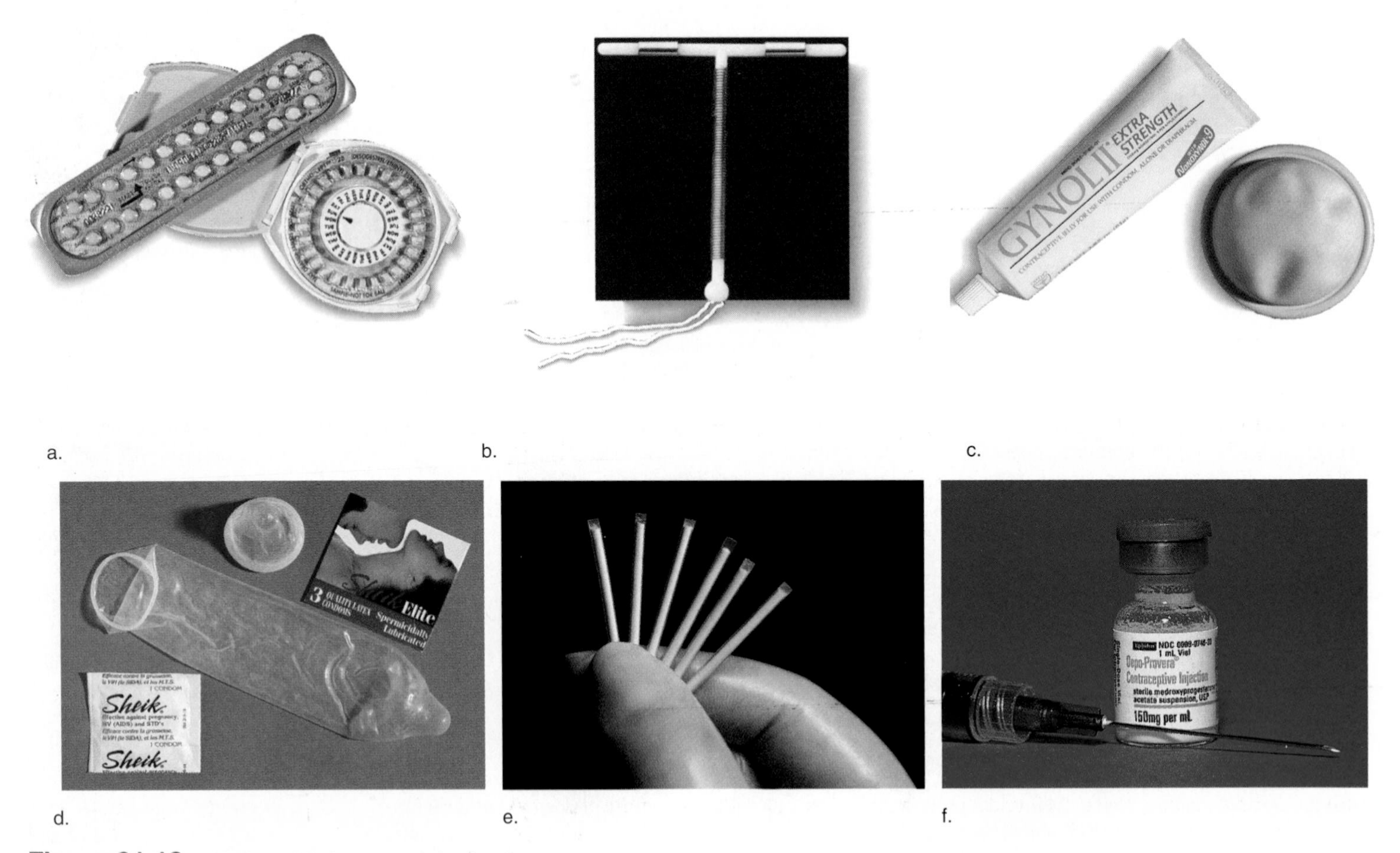

Figure 21.12 Various birth-control devices.
a. Oral contraception (birth-control pills). **b.** Intrauterine device. **c.** Spermicidal jelly and diaphragm. **d.** Male condom. **e.** Contraceptive implants. **f.** Depo-Provera injection.

Morning-after Pills

The expression "morning-after pill" refers to a medication that will prevent pregnancy after unprotected intercourse. The expression is a misnomer in that medication can begin one to several days after unprotected intercourse.

A kit, now called Preven, is made up of four synthetic progesterone pills; two are taken up to 72 hours after unprotected intercourse, and two more are taken 12 hours later. The medication upsets the normal uterine cycle, making it difficult for the embryo to implant itself in the endometrium. In a recent study, it was estimated that the medication was 85% effective in preventing unintended pregnancies.

Mifepristone, better known as RU-486, is a pill that is presently used to cause the loss of an implanted embryo by blocking the progesterone receptors of endometrial cells. Without functioning receptors for progesterone, the endometrium sloughs off, carrying the embryo with it. When taken in conjunction with a prostaglandin to induce uterine contractions, RU-486 is 95% effective. It is possible that some day this medication will also be a "morning-after pill," taken when menstruation is late without evidence that pregnancy has occurred.

There are numerous well-known birth-control methods and devices available to those who wish to prevent pregnancy. Their effectiveness varies. In addition, new methods are expected to be developed.

Figure 21.13 Mother and child.
Sometimes couples rely on medical intervention in order to experience parenthood.

Infertility

Sometimes couples do not need to prevent pregnancy; conception or fertilization does not occur despite frequent intercourse. The American Medical Association estimates that 15% of all couples in this country are unable to have any children and therefore are properly termed *sterile;* another 10% have fewer children than they wish and therefore are termed *infertile.* The latter assumes that the couple has been unsuccessfully trying to become pregnant for at least one year.

Causes of Infertility

The two major causes of infertility in females are blocked oviducts, possibly due to pelvic inflammatory disease (PID), discussed later in this chapter, and endometriosis. **Endometriosis,** the presence of uterine tissue outside the uterus, particularly in the oviducts and on the abdominal organs, can also contribute to infertility. Endometriosis occurs when the menstrual discharge flows up into the oviducts and out into the abdominal cavity. This backward flow allows living uterine cells to establish themselves in the abdominal cavity where they go through the usual uterine cycle, causing pain and structural abnormalities that make it more difficult for a woman to conceive.

Sometimes the causes of infertility can be corrected by medical intervention so that couples can have children (Fig. 21.13). If no obstruction is apparent and body weight is normal, it is possible to give females HCG, extracted from the urine of pregnant women, along with gonadotropins extracted from the urine of postmenopausal women. This treatment causes multiple ovulations and sometimes multiple pregnancies.

The most frequent cause of infertility in males is low sperm count and/or a large proportion of abnormal sperm. Disease, radiation, chemical mutagens, high testes temperature, and the use of psychoactive drugs can contribute to this condition.

When reproduction does not occur in the usual manner, many couples adopt a child. Others sometimes first try one of the alternative reproductive methods discussed in the following paragraphs. If all the alternative methods discussed are considered, it is possible for a baby to have five parents: (1) sperm donor, (2) egg donor, (3) surrogate mother, and (4) and (5) adoptive mother and father.

Ecology Focus

Ecoestrogens

Although biologists knew that males do have estrogen in their systems—for example, the adrenal cortex produces estrogens—they have been surprised to learn that estrogens are essential to male fertility. Apparently when deprived of estrogen, the epididymis and other portions of the genital tract do not function normally and the result is a very dilute ejaculate with an abnormally low sperm count.

This newfound knowledge may be pertinent to a noted rise in human infertility. For example, Danish endocrinologists combined the results of 61 separate studies of sperm count in men around the world and found that the average sperm count has fallen to about ½ of what it was in 1930. A New York City Fertility Research Foundation says that in 1960 only 8% of men who came in for consultation had a fertility problem. Now the number is up to 40%.

What could be causing a decline in male fertility that could conceivably affect most males within a century? In 1975 there was a spill of the pesticide Mirex and the men who helped clean it up developed a lowered sperm count. It's possible that exposure to many types of chemicals such as DDT, TCDD (the most toxic dioxin), PCBs (polychlorinated biphenyls), and BPA (bisphenol-A) a compound used to coat food cans) could have the same sort of effect once they enter the body. It appears that these chemicals bind to receptors that recognize natural estrogen in a male's body. Therefore, these chemicals are called ecoestrogens. Once ecoestrogens bind to receptors, they prevent natural estrogens from binding and fertility problems result.

There are thousands of studies showing that chemicals in heavily contaminated areas cause endocrine problems in wildlife. Only thirty to fifty panthers remain in the Florida Everglades. At one time, nearly all biologists of the U.S. Fish and Wildlife Service believed that the male panther's low sperm count was due to inbreeding. Now, however, they are considering the possibility that the problem stems from their diet. Because they feed on fish-eating raccoons, there is an abnormally high level of pesticides in their bodies.

Alternative Methods of Reproduction

Artificial Insemination by Donor (AID). During artificial insemination, sperm are placed in the vagina by a physician. Sometimes a woman is artificially inseminated by her husband's sperm. This is especially helpful if the husband has a low sperm count—the sperm can be collected over a period of time and concentrated so that the sperm count is sufficient to result in fertilization. Often, however, a woman is inseminated by sperm acquired from a donor who is a complete stranger to her. At times, a combination of husband and donor sperm is used.

A variation of AID is intrauterine insemination (*IUI*). IUI involves hormonal stimulation of the ovaries, followed by placement of the donor's sperm in the uterus rather than in the vagina.

In Vitro Fertilization (IVF). During IVF, conception occurs in laboratory glassware. The newer ultrasound machines can spot follicles that hold immature eggs; therefore, the latest method is to forgo the administration of fertility drugs and retrieve immature eggs by using a needle. The immature eggs are then brought to maturity in glassware before concentrated sperm from the male are added, After about 2 to 4 days, the embryos are inserted into the uterus of the woman, who is now in the secretory phase of her uterine cycle. If implantation is successful, development is normal and continues to term.

Gamete Intrafallopian Transfer (GIFT). Gamete intrafallopian transfer (GIFT) was devised as a means to overcome the low success rate (15–20%) of in vitro fertilization. The method is exactly the same as in vitro fertilization, except the eggs and the sperm are placed in the oviducts immediately after they have been brought together. This procedure is helpful to couples whose eggs and sperm never make it to the oviducts; sometimes the egg gets lost between the ovary and the oviducts, and sometimes the sperm never reach the oviducts. GIFT has an advantage in that it is a one-step procedure for the woman—the eggs are removed and are reintroduced all in the same time period. For this reason, it is less expensive—approximately $1,500 compared to $3,000 and up for in vitro fertilization.

Surrogate Mothers. In some instances, women are paid to have babies. These women are called surrogate mothers. Other individuals contribute sperm (or eggs) to the fertilization process in such cases.

Some couples are infertile due to various physical abnormalities. When corrective medical procedures fail, today it is possible to consider an alternative method of reproduction in order to be a parent.

21.5 Sexually Transmitted Diseases

There are many diseases that are transmitted by sexual contact and are therefore called sexually transmitted diseases (STDs). Our discussion centers on those that are most prevalent. AIDS, genital herpes, and genital warts are viral diseases; therefore, they are difficult to treat because viral infections do not respond to traditional antibiotics. Other types of drugs have been developed to treat these, however. And although gonorrhea and chlamydia are treatable with appropriate antibiotic therapy, they are not always promptly diagnosed. Unfortunately, the human body does not become immune to STDs, and as yet there are no vaccines available for any of them.

AIDS

Acquired immunodeficiency syndrome (AIDS) is caused by a group of related retroviruses known as HIV (human immunodeficiency viruses) that infect specific blood cells (Fig. 21.14). Infection spreads when infected cells in body secretions, such as semen, and in blood are passed to another individual. To date, as many as 40 million people worldwide may have contracted HIV and almost 12 million have died. A new infection is believed to occur every 15 seconds, the majority in heterosexuals. HIV infections are not distributed equally throughout the world. Most infected people live in Africa (66%) where the infection first began, but new infections are now occurring at the fastest rate in Southeast Asia and the Indian subcontinent.

HIV is transmitted by sexual contact with an infected person, including vaginal or rectal intercourse and oral/genital contact. Also, needle-sharing among intravenous drug users is high-risk behavior. Babies born to HIV-infected women may become infected before or during birth, or through breast-feeding after birth.

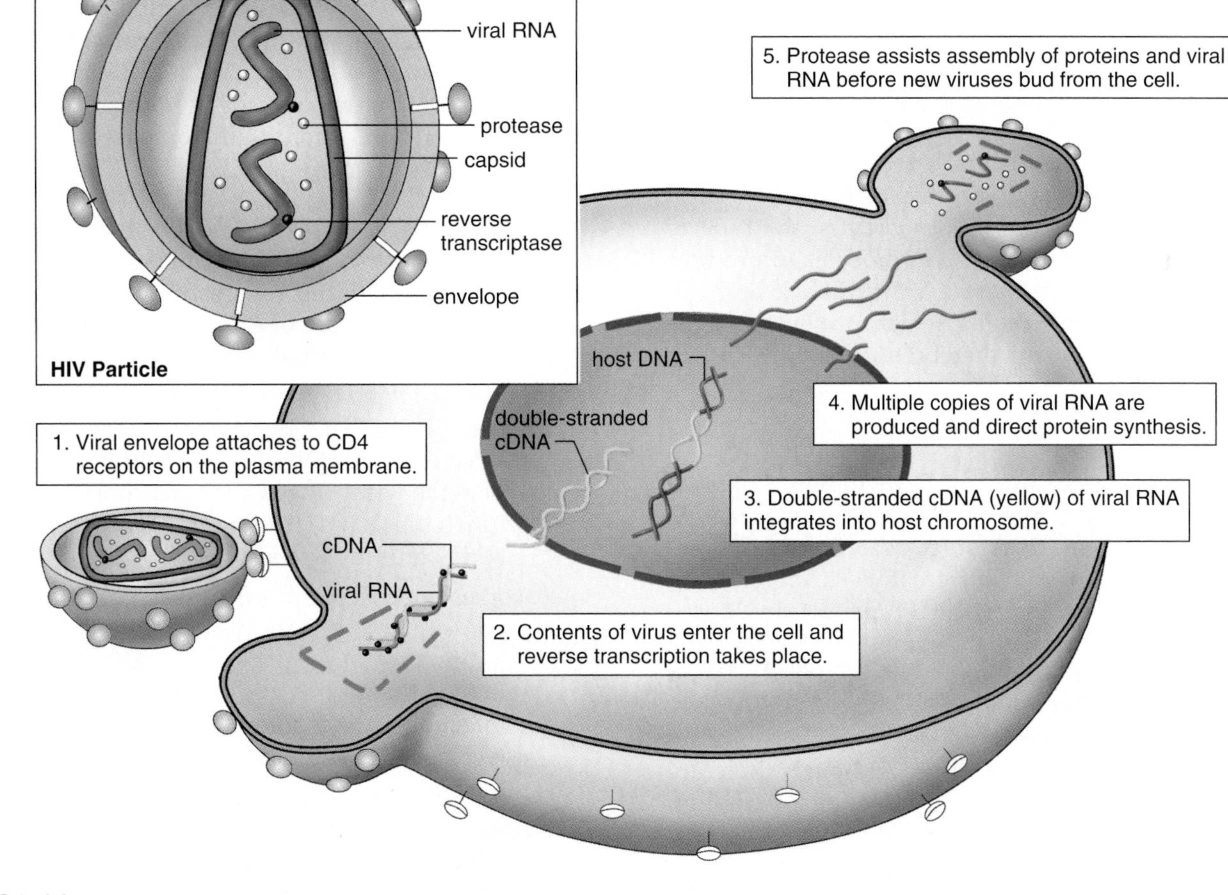

Figure 21.14 Structure and reproduction of HIV.
HIV is a retrovirus with an outer envelope containing proteins, some of which are spikes that attach to receptors in the host cell plasma membrane. The core of the virus contains two strands of RNA and the enzymes protease and reverse transcriptase. Reverse transcriptase makes cDNA (a DNA copy of viral RNA). During the reproductive cycle, double-stranded cDNA becomes integrated into host DNA and later is transcribed back to viral RNA when protein synthesis occurs.

HIV first spread through the homosexual community, and male-to-male sexual contact still accounts for the largest percentage of new AIDS cases in the United States. But the largest increases of HIV infections are occurring through heterosexual contact or by intravenous drug use. Now, women account for 20% of all newly diagnosed cases of AIDS. The rise in the incidence of AIDS among women of reproductive age is paralleled by a rise in the incidence of AIDS in children younger than 13.

Life Cycle of HIV

Much has been learned about the structure of HIV and its reproductive cycle (see Fig. 21.14). In order to enter a host cell, HIV attaches to a protein in the plasma membrane called a CD4 receptor. HIV is a *retrovirus,* meaning that its genetic material consists of RNA instead of DNA. Once inside the host cell, HIV uses a special enzyme called reverse transcriptase to make a DNA copy of its genetic material. Now the DNA copy integrates into a host chromosome, where it directs the production of more viral RNA. Each strand of viral RNA brings about synthesis of an outer protein coat called a capsid. The viral enzyme protease is necessary to the formation of capsids, which assemble with RNA strands prior to the viruses budding from the host cell.

The immune system, which consists of white blood cells and certain lymphoid organs, usually protects us from disease. Two ways in which the immune system does this is to produce *antibodies,* proteins that attack foreign proteins called antigens, and to attack infected cells outright. The primary host for HIV is *helper T lymphocytes,* the type of white blood cell that stimulates B lymphocytes to produce antibodies and *cytotoxic T lymphocytes* to attack and kill virus infected cells. Macrophages, which present antigens to helper T lymphocytes and thereby stimulate them, are also under attack.

Phases of AIDS

The Centers for Disease Control and Prevention recognize three stages of an HIV-1 infection called Category A, B, and C. During a Category A stage, the helper T lymphocyte count is 500 per mm^3 or greater. For a period of time after the initial infection with HIV, people don't usually have any symptoms at all. A few (1–2%) do have mononucleosis-like symptoms that may include fever, chills, aches, swollen lymph nodes, and an itchy rash. These symptoms disappear, however, and there are no other symptoms for quite some time. Although there are no symptoms, the person is highly infectious. Although there is a large number of viruses in the plasma, the HIV blood test is not yet positive because it tests for the presence of antibodies and not for the presence of HIV itself. This means that HIV can still be transmitted before the HIV blood test is positive.

Several months to several years after a nontreated infection, the individual will probably progress to category B in which the helper T lymphocyte count is 200 to 499 per mm^3. During this stage there will be swollen lymph nodes in the neck, armpits, or groin that persist for three months or more. Other symptoms that indicate category B are severe fatigue not related to exercise or drug use; unexplained persistent or recurrent fevers, often with night sweats; persistent cough not associated with smoking, a cold, or the flu; and persistent diarrhea.

When the individual develops non-life-threatening but recurrent infections, it is a signal that the disease is progressing. One possible infection is thrush, a fungal infection that is identified by the presence of white spots and ulcers on the tongue and inside the mouth. The fungus may also spread to the vagina, resulting in a chronic infection there. Another frequent infection is herpes simplex, with painful and persistent sores on the skin surrounding the anus, the genital area, and/or the mouth.

Previously, the majority of infected persons proceeded to category C, in which the helper T lymphocyte count is 200 per mm^3 and the lymph nodes degenerate. The patient, who is now suffering from "slim disease" (as AIDS is called in Africa)—characterized by severe weight loss and weakness due to persistent diarrhea and coughing—will most likely contract an opportunistic infection. An **opportunistic infection** is one that only has the *opportunity* to occur because the immune system is severely weakened. Persons with AIDS die from one or more opportunistic diseases and not from the HIV infection itself. Some of the opportunistic infections are the following:

Pneumocystis carinii pneumonia. The lungs become useless as they fill with fluid and debris due to an infection with this organism. There is not a single documented case of *P. carinii* pneumonia in a person with normal immunity.

Mycobacterium tuberculosis. This bacterial infection, usually of the lungs, is seen more often as an infection of lymph nodes and other organs in patients with AIDS. Of special concern, tuberculosis is spreading into the general populace and is multidrug resistant.

Toxoplasmic encephalitis is caused by a one-cell parasite that lives in cats and other animals as well as humans. Many persons harbor a latent infection in the brain or muscle, but in AIDS patients the infection leads to loss of brain cells, seizures, weakness, or decreased sensation on one side of the body.

Kaposi's sarcoma is an unusual cancer of blood vessels, which gives rise to reddish purple, coin-size spots and lesions on the skin.

Invasive cervical cancer. This cancer of the cervix spreads to nearby tissues. This condition has been added to the list because the incidence of AIDS has now increased in women.

Treatment for AIDS

There is no cure for AIDS, but a treatment called highly active antiretroviral therapy (HAART) is usually able to stop HIV replication to the extent the virus becomes undetectable in the blood. Therapy usually consists of two drugs that inhibit reverse transcriptase and one that inhibits protease, an enzyme needed for formation of a viral capsid. This multidrug therapy, when taken according to the manner prescribed, seems to usually prevent mutation of the virus to a resistant strain. Unfortunately, an HIV strain resistant to all known drugs has been reported—persons who become infected with this strain have no drug therapy available to them.

The sooner drug therapy begins after infection, the better the chances that the immune system will not be destroyed by HIV. And medication must be continued indefinitely. Investigators have found that when HAART is discontinued, the virus rebounds.

Many investigators are working on a vaccine for AIDS. Some are trying to develop a vaccine in the traditional way. Others are working on subunit vaccines that utilize just a single HIV protein as the vaccine. For example, the protein which forms the spike shown in Figure 21.14 can be produced by genetic engineering of bacteria. This protein is undergoing clinical trial as a vaccine in Thailand. Also, investigators have found that viral cDNA for this spike can act as a vaccine because it causes cells to produce and display the spike at the plasma membrane.

Transmission of AIDS

HIV is spread by passing virus-infected T lymphocytes found in body secretions or in blood from one person to another. The largest proportion of people with AIDS in the United States are homosexual men, but the proportions attributed to intravenous drug users and heterosexuals is rising. The fastest rate of increase is now seen in minority females. An infected woman can pass HIV to her unborn children by way of the placenta or to a newborn through breast milk. Transmission at birth can be prevented if the mother takes AZT, and delivers by planned cesarean section.

Hepatitis Infections

There are several types of hepatitis. Hepatitis A, caused by HAV (hepatitis A virus), is usually acquired from sewage-contaminated drinking water. However, hepatitis A can also be sexually transmitted through oral/anal contact. Hepatitis B, caused by HBV (hepatitis B virus) is usually transmitted by sexual contact. Hepatitis C, caused by HCV (hepatitis C virus), is called the post-transfusion form of hepatitis. Infection can lead to chronic hepatitis, liver cancer, and death. Still other types of hepatitis are now under investigation, and how many will be found is not yet known.

Hepatitis B

HBV is more likely to be spread by sexual contact than the HIV virus. Like HIV, it can also be spread by blood transfusions or contaminated needles. Because the mode of transmission is similar, it is common for an AIDS patient to also have an HBV infection. Also, like HIV, HBV can be passed from mother to child by way of the placenta.

Only about 50% of persons infected with HBV have flu-like symptoms, including fatigue, fever, headache, nausea, vomiting, muscle aches, and dull pain in the upper right of the abdomen. Jaundice, a yellowish cast to the skin, can also be present. Some persons have an acute infection that lasts only three to four weeks. Others have a chronic form of the disease that leads to liver failure and a need for a liver transplant.

To prevent an infection, the same directions given below to stop the spread of AIDS can be followed. However, inoculation with the HBV vaccine is the best protection. The vaccine, which is safe and does not have any major side effects, is now on the list of recommended immunizations for children.

Hepatitis B is an infection that can lead to liver failure. Because it is spread in the same way as AIDS, persons can be infected with both viruses at the same time.

To stop the spread of AIDS:

- Abstain from sexual intercourse, or develop a long-term monogamous (always the same partner) sexual relationship with a partner who is free of HIV and is not an intravenous drug user.
- Practice safe sex. If you do not know for certain that your partner has been free of HIV for the past five years, always use a latex condom during sexual intercourse. Be sure to follow the directions supplied by the manufacturer. Use of a spermicide containing nonoxynol-9 in addition to the condom can offer further protection because nonoxynol-9 immobilizes the virus and virus-infected lymphocytes.
- Avoid fellatio (kissing and insertion of the penis into a partner's mouth) and cunnilingus (kissing and insertion of the tongue into the vagina) because they may be a means of transmission. The mouth and gums often have cuts and sores that facilitate the entrance of infected T lymphocytes.
- Stop, if necessary, or do not start the habit of injecting drugs into veins. If you are a drug user and cannot stop your behavior, then always use a new sterile needle for injection or one that has been cleaned in bleach.
- Aside from intravenous drug use, do not use alcohol or any drugs in a way that may prevent you from being able to control your behavior.

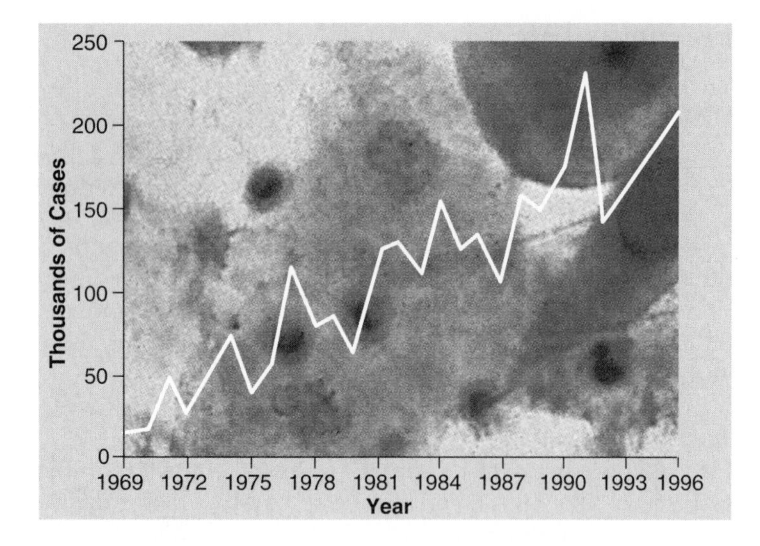

Figure 21.15 Genital herpes.
A graph depicting the incidence of new cases of genital herpes in the United States from 1969 to 1996 is superimposed on a photomicrograph of cells infected with the herpes virus.

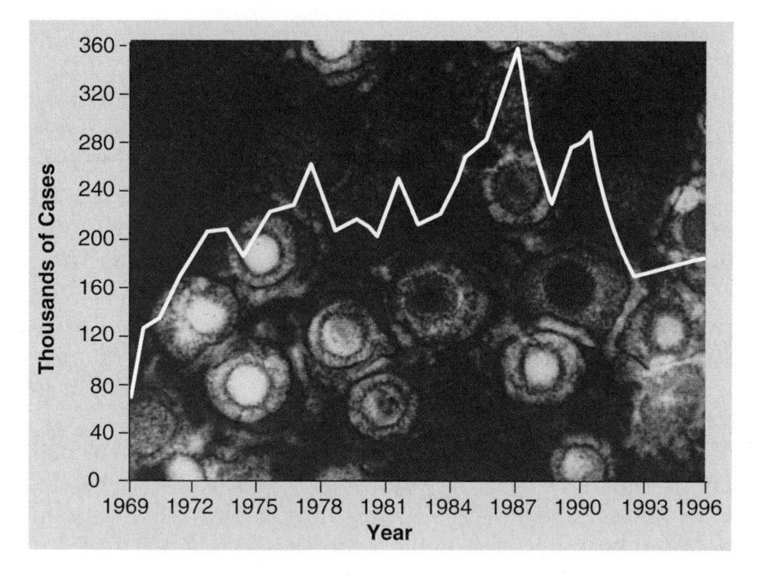

Figure 21.16 Genital warts.
A graph depicting the incidence of new cases of genital warts in the United States from 1969 to 1996 is superimposed on a photomicrograph of human papillomaviruses.

Genital Herpes

Genital herpes is caused by the herpes simplex virus of which there are two types: type 1, which usually causes cold sores and fever blisters, and type 2, which more often causes genital herpes. Cross-over infections do occur, however. That is, type 1 has been known to cause a genital infection, while type 2 has been known to cause cold sores and fever blisters.

Genital herpes is one of the more prevalent sexually transmitted diseases today (Fig. 21.15). At any one time, millions of persons could be having recurring symptoms. Immediately after infection, there are no symptoms, but the individual may experience a tingling or itching sensation before blisters appear at the infected site, usually within 2–20 days. Once the blisters rupture, they leave painful ulcers, which may take as long as three weeks or as little as five days to heal. These symptoms may be accompanied by fever, pain upon urination, and swollen lymph nodes.

After the ulcers heal, the disease is only dormant. Blisters can recur repeatedly at variable intervals. Sunlight, sexual intercourse, menstruation, and stress seem to cause the symptoms of genital herpes to recur. While the virus is dormant, it primarily resides in the ganglia of sensory nerves associated with the affected skin. Although type 2 was once thought to cause a form of cervical cancer, this is no longer believed to be the case.

Infection of the newborn can occur if the child comes in contact with a lesion in the birth canal. In 1–3 weeks, the infant is gravely ill and can become blind, have neurological disorders including brain damage, or die. Birth by cesarean section prevents these adverse developments.

Genital Warts

Genital warts are caused by the human papillomaviruses (HPVs), which are sexually transmitted. Over a million persons become infected each year with a form of HPV, but only a portion seek medical help (Fig. 21.16). Sometimes carriers do not have any sign of warts, although flat lesions may be present. When present, the warts are commonly seen on the penis and the foreskin of males and the vaginal orifice in females.

HPVs, rather than genital herpes, are now associated with cancer of the cervix as well as tumors of the vulva, the vagina, the anus, and the penis. Some researchers believe that the viruses are involved in 90–95% of all cases of cancer of the cervix. Teenagers who have or have had multiple sex partners seem to be particularly susceptible to HPV infections. More and more cases of cancer of the cervix are being seen in this age group.

Presently, there is no cure for an HPV infection, but warts can be effectively treated by surgery, freezing, application of an acid, or laser burning. A suitable medication to treat genital warts before cancer occurs is being sought. Efforts are also underway to develop a vaccine.

Gonorrhea

Gonorrhea is caused by the bacterium *Neisseria gonorrhoeae.* This bacterium is a diplococcus, meaning that two cells generally stay together.

The diagnosis of gonorrhea in the male is not difficult as long as he displays typical symptoms (as many as 20% of males may be asymptomatic). The patient complains of pain

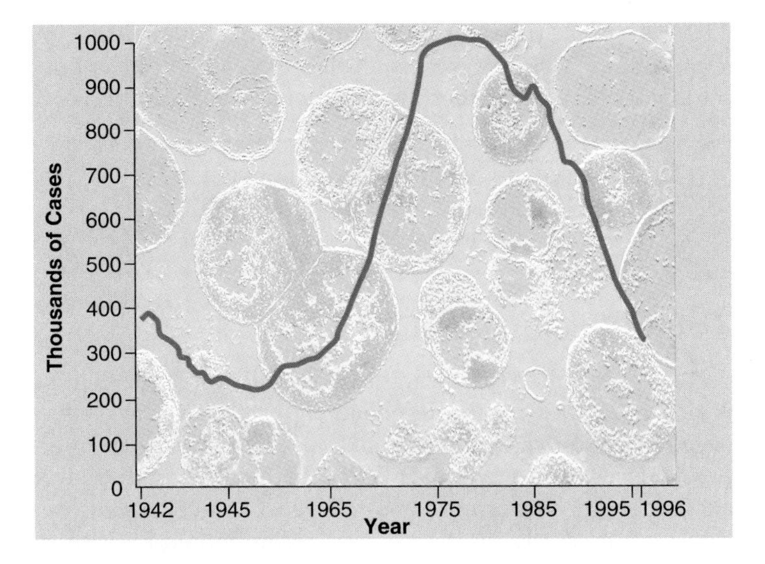

Figure 21.17 Gonorrhea.
A graph depicting the incidence of new cases of gonorrhea in the United States from 1942 to 1996 is superimposed on a photomicrograph of a urethral discharge from an infected male. Gonorrheal bacteria (*Neisseria gonorrhoeae*) occur in pairs; for this reason, they are called diplococci.

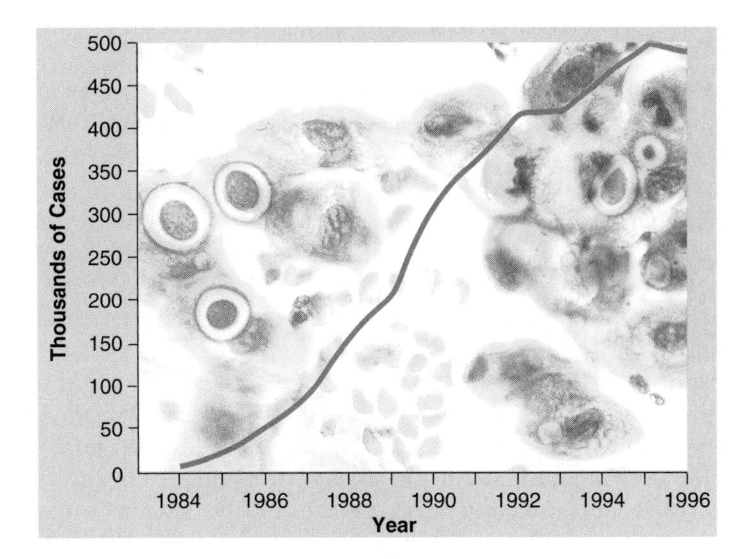

Figure 21.18 Chlamydial infection.
A graph depicting the incidence of new cases of chlamydia in the United States from 1984 to 1996 is superimposed on a photomicrograph of a cell containing different stages of the organism.

on urination and has a thick, greenish yellow urethral discharge 3–5 days after contact with an infected partner. In the female, the bacteria may first settle within the urethra or near the cervix, from which they may spread to the oviducts, causing **pelvic inflammatory disease (PID).** As the inflamed tubes heal, they may become partially or completely blocked by scar tissue. As a result, the female is sterile, or at best, subject to ectopic pregnancy. Similarly, there may be inflammation in untreated males followed by scarring of each vas deferens. Unfortunately, 60–80% of females are asymptomatic until they develop severe PID-induced pains in the abdominal region. The incidence of gonorrhea has been declining since an all-time high in 1978 (Fig. 21.17).

Homosexual males develop gonorrhea proctitis, or infection of the anus, with symptoms including pain in the anus and blood or pus in the feces. Oral sex can cause infection of the throat and the tonsils. Gonorrhea can also spread to other parts of the body, causing heart damage or arthritis. If, by chance, the person touches infected genitals and then his or her eyes, a severe eye infection can result.

Eye infection leading to blindness can occur as a baby passes through the birth canal. Because of this, all newborn infants receive eyedrops containing antibacterial agents such as silver nitrate, tetracycline, or penicillin as a protective measure.

Chlamydia

Chlamydia is named for the tiny bacterium that causes it, *Chlamydia trachomatis.* New chlamydial infections are more numerous than any other sexually transmitted disease (Fig. 21.18). About 8–21 days after infection, men experience a mild burning sensation on urination and a mucous discharge. Women may have a vaginal discharge, along with the symptoms of a urinary tract infection. If not properly treated, the infection can eventually cause PID, sterility, or ectopic pregnancy.

If a newborn is exposed to chlamydia during delivery, inflammation of the eyes or pneumonia can result. Some believe that chlamydial infections increase the possibility of premature births and stillbirths.

Detection and Treatment of Chlamydia

Expense sometimes prevents public clinics from testing for chlamydia. It's been suggested that these criteria could help physicians decide which women should be tested: no more than 24 years old; a new sex partner within the preceding two months; cervical discharge; bleeding during parts of the vaginal exam; and use of a nonbarrier method of contraception. Some doctors, however, are routinely prescribing additional antibiotics appropriate to treating chlamydia for anyone who has gonorrhea, because 40% of females and 20% of males with gonorrhea also have chlamydia.

As with AIDS, condoms protect against both gonorrheal and chlamydial infections. The concomitant use of a spermicide containing nonoxynol-9 enhances protection.

PID and sterility are common effects of a chlamydial infection in the female. This condition often accompanies a gonorrheal infection.

Syphilis

Syphilis is caused by a type of bacterium called *Treponema pallidum.* As with many other bacterial diseases, penicillin has been used as an effective antibiotic. Syphilis has three stages, which can be separated by latent stages during which the bacteria are resting before multiplying again. During the *primary stage,* a hard chancre (ulcerated sore with hard edges) indicates the site of infection. The chancre can go unnoticed, especially since it usually heals spontaneously, leaving little scarring. During the *secondary stage,* proof that bacteria have invaded and spread throughout the body is evident when the victim breaks out in a rash. Curiously, the rash does not itch and is seen even on the palms of the hands and the soles of the feet. There can be hair loss and infectious gray patches on the mucous membranes, including the mouth. These symptoms disappear of their own accord.

During the *tertiary stage,* which lasts until the patient dies, syphilis may affect the cardiovascular system: weakened arterial walls (aneurysms) are seen, particularly in the aorta. In other instances, the disease may affect the nervous system: an infected person may show psychological disturbances, for example. Gummas, large destructive ulcers, may develop on the skin or within the internal organs in another variety of the tertiary stage.

Congenital (present at birth) *syphilis* is caused by syphilitic bacteria crossing the placenta. The child is born blind and/or with numerous anatomical malformations.

The best preventive measure against these diseases is abstinence or monogamy with a partner who is free of them.

Two Other Infections

Females very often have vaginitis, or infection of the vagina, which is caused by the flagellated protozoan *Trichomonas vaginalis* or the yeast *Candida albicans.* The protozoan infection causes a frothy white or yellow foul-smelling discharge accompanied by itching, and the yeast infection causes a thick, white, curdy discharge, also accompanied by itching. **Trichomoniasis** is most often acquired through sexual intercourse, and the asymptomatic male is usually the reservoir of infection. *Candida albicans,* however, is a normal organism found in the vagina; its growth simply increases beyond normal under certain circumstances. For example, women taking birth-control pills are sometimes prone to yeast infections. Also, the indiscriminate use of antibiotics can alter the normal balance of organisms in the vagina so that a yeast infection flares up.

Bioethical Issue

The dizzying array of reproductive techniques has progressed from simple in vitro fertilization to the ability to freeze eggs or sperm or even embryos for future use. Older women who never had the opportunity to freeze their eggs can still have children if they use donated eggs—perhaps today harvested from a fetus.

Legal complications abound about which mother has first claim to the child—the surrogate mother, the woman who donated the egg, or the primary care giver—to which partner has first claim to frozen embryos following a divorce. Legal decisions about who has the right to use what techniques have rarely been discussed, much less decided upon. Some clinics will help anyone, male or female, no questions asked, as long as they have the ability to pay. And most clinics are heading toward doing any type of procedure, including guaranteeing the sex of the child, and making sure the child will be free from some particular genetic disorder. It would not be surprising if, in the future, zygotes could be engineered to have any particular trait desired by the parents.

Even today eugenic (good gene) goals are evidenced by the fact that reproductive clinics advertise for egg and sperm donors, primarily in elite college newspapers. The question becomes "Is it too late for us as a society to make ethical decisions about reproductive issues? Should we come to a consensus about what techniques should be allowed and who should be able to use them?" We all want to avoid, if possible, what happened to Jonathan Alan Austin. Jonathan, who was born to a surrogate mother, later died from injuries inflicted by his father. Perhaps if background checks were legally required, surrogate mothers would only make themselves available to individuals or couples who are known to have certain psychological characteristics.

Questions

1. As a society we have never been in favor of regulating reproduction. Should we regulate reproduction by an alternative method if we do not regulate unassisted reproduction? Why or why not?
2. Should the state be the guardian of frozen embryos and make sure they all get a chance to life? Why or why not?
3. Is it appropriate for physicians and parents to select which embryos will be implanted in the uterus? On the basis of sex? On the basis of genetic inheritance? Why or why not?

Summarizing the Concepts

21.1 Male Reproductive System

In males, spermatogenesis occurring in seminiferous tubules of the testes produces sperm that mature in the epididymides and may be stored in the vasa deferentia before entering the urethra, along with secretions produced by seminal vesicles, the prostate gland, and bulbourethral glands. Semen is ejaculated during male orgasm, when the penis is erect.

Hormonal regulation, involving secretions from the hypothalamus, the anterior pituitary, and the testes, maintains testosterone produced by the interstitial cells of the testes at a fairly constant level.

21.2 Female Reproductive System

In females, an egg produced by an ovary enters an oviduct, which leads to the uterus. The uterus opens into the vagina. The external genital area includes the vaginal opening, the clitoris, the labia minora, and the labia majora.

21.3 Female Hormone Levels

In the nonpregnant female, the ovarian and uterine cycles are under hormonal control of the hypothalamus, anterior pituitary, and the female sex hormones estrogen and progesterone.

If fertilization occurs, the corpus luteum is maintained because of HCG production. Progesterone production does not cease, and the embryo implants itself in the thick uterine lining. Estrogen and progesterone maintain the secondary sex characteristics of females, including less body hair than males, a wider pelvic girdle, a more rounded appearance, and development of breasts.

21.4 Control of Reproduction

Infertile couples are increasingly resorting to alternative methods of reproduction. Numerous birth-control methods and devices are available for those who wish to prevent pregnancy.

21.5 Sexually Transmitted Diseases

Sexually transmitted diseases include AIDS; herpes, which repeatedly flares up; genital warts, which lead to cancer of the cervix; gonorrhea and chlamydia, which cause pelvic inflammatory disease (PID); and syphilis, which has cardiovascular and neurological complications if untreated.

Studying the Concepts

1. Outline the path of sperm. What glands contribute fluids to semen? 420–21
2. Discuss the anatomy and physiology of the testes. Describe the structure of sperm. 422–23
3. Name the endocrine glands involved in maintaining the sex characteristics of males and the hormones produced by each. 423
4. Describe the organs of the female genital tract. Where do fertilization and implantation occur? Name two functions of the vagina. 424–25
5. Describe the external genitals in females. 425
6. Discuss the anatomy and the physiology of the ovaries. 426–27 Describe the ovarian cycle. 427–29
7. Describe the uterine cycle and relate it to the ovarian cycle. In what way is menstruation prevented if pregnancy occurs? 429
8. Name three functions of the female sex hormones. 430
9. Describe the anatomy and the physiology of the breast. 430
10. Discuss the various means of birth control and their relative effectiveness. 431–33
11. Describe the most common types of sexually transmitted diseases. 435–40

Testing Yourself

Choose the best answer for each question.

1. Label this diagram of the male reproductive system, and trace the path of sperm.

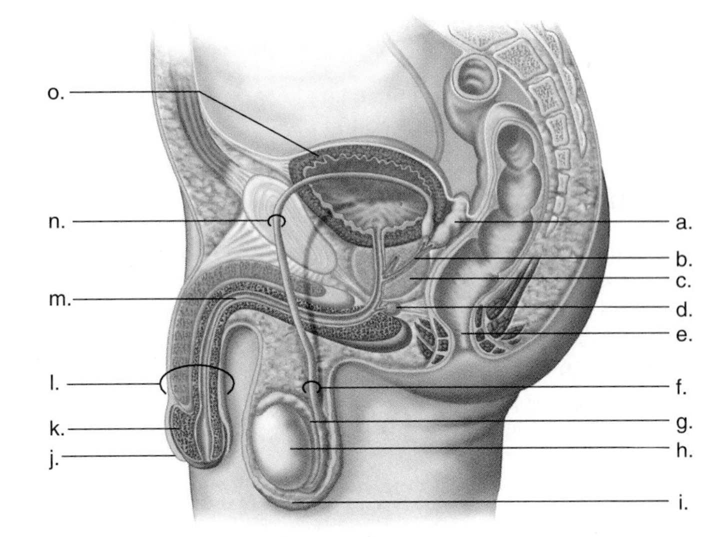

2. Which of these is mismatched?
   a. interstitial cells—testosterone
   b. seminiferous tubules—sperm production
   c. vasa deferentia—seminal fluid production
   d. urethra—conducts sperm
3. Follicle-stimulating hormone (FSH)
   a. is secreted by females but not males.
   b. stimulates the seminiferous tubules to produce sperm.
   c. secretion is controlled by gonadotropic-releasing hormone (GnRH).
   d. Both b and c are correct.
4. In tracing the path of sperm, you would mention vasa deferentia before
   a. testes.
   b. epididymides.
   c. urethra.
   d. uterus.
5. An oocyte is fertilized in the
   a. vagina.
   b. uterus.
   c. oviduct.
   d. ovary.

Match the method of protection in questions 6–8, with these means of birth control.

a. vasectomy
b. oral contraception
c. intrauterine device (IUD)
d. diaphragm
e. male condom
f. coitus interruptus

6. ________ Blocks entrance of sperm to uterus.
7. ________ Traps sperm and also prevents STDs.
8. ________ Prevents implantation of an embryo.

9. During pregnancy
   a. the ovarian and uterine cycles occur more quickly than before.
   b. GnRH is produced at a higher level than before.
   c. the ovarian and uterine cycles do not occur.
   d. the female secondary sexual characteristics are not maintained.

Match the description in questions 10–12 with these sexually transmitted diseases.

a. AIDS
b. hepatitis B
c. genital herpes
d. genital warts
e. gonorrhea
f. chlamydia
g. syphilis

10. _________ blisters, ulcers, pain on urination, swollen lymph nodes
11. _________ flu-like symptoms, jaundice, eventual liver failure possible
12. _________ males have a thick, greenish, yellow discharge; no symptoms in female can lead to PID.

Thinking Scientifically

1. Considering anabolic steroids (page 410):
   a. Using Figure 21.4, formulate a hypothesis to explain why anabolic steroids would shrink the size of the testes.
   b. How might you test your hypothesis, for example in laboratory mice?
   c. Would you predict that anabolic steroids raise or lower LDL blood levels (page 256), leading to increased risk of heart disease?
   d. How might you test your hypothesis, for example in laboratory mice?
2. Considering birth-control measures (page 431):
   a. Using Figure 21.8 as a guide, formulate a hypothesis to explain why a pill that contains estrogen and progesterone could be used as a birth-control pill.
   b. How might you test your hypothesis, for example in laboratory mice?
   c. In postmenopausal women, there are usually increased levels of FSH and LH, but because the ovaries are unable to respond, there are decreased levels of estrogen and progesterone. How would you expect these levels to change in postmenopausal women who took birth-control pills?
   d. How might you test your prediction, for example in postmenopausal women?

Understanding the Terms

acquired immunodeficiency syndrome (AIDS) 435
acrosome 423
birth-control pill 431
bulbourethral gland 421
cervix 425
chlamydia 439
colostrum 430
condom 432
contraceptive 431
corpus luteum 427
diaphragm 431
egg 424
endometriosis 433
endometrium 425
epididymis 420
estrogen 429
fimbria 424
follicle 427
follicle-stimulating hormone (FSH) 423
genital herpes 438
genital warts 438
gonad 420
gonadotropin-releasing hormone (GnRH) 423
gonorrhea 438
human chorionic gonadotropin (HCG) 429
implantation 432
interstitial cell 423
intrauterine device (IUD) 431
luteinizing hormone (LH) 423
menopause 430
menstruation 429
oogenesis 424
opportunistic infection 436
ovarian cycle 427
ovary 424
oviduct 424
ovulation 424
Pap test 425
pelvic inflammatory disease (PID) 439
penis 421
placenta 429
progesterone 429
prostate gland 420
scrotum 420
semen 420
seminal vesicle 420
seminiferous tubule 422
sperm 423
spermatogenesis 423
sustentacular (Sertoli) cell 423
syphilis 440
testis 420
testosterone 423
trichomoniasis 440
urethra 420
uterine cycle 429
uterus 424
vagina 425
vas deferens 420
vulva 425
zygote 424

Match the terms to these definitions:

a. _________ Organ that leads from the uterus to the vestibule and serves as the birth canal and organ of sexual intercourse in females.
b. _________ Lining of the uterus, which becomes thickened and vascular during the uterine cycle.
c. _________ Gland located around the male urethra below the urinary bladder; adds secretions to semen.
d. _________ Hormone secreted by the anterior pituitary gland that stimulates the development of an ovarian follicle in a female or the production of sperm in a male.
e. _________ External genitals of the female that surround the opening of the vagina.

Using Technology

Your study of the reproductive system is supported by these available technologies:

Essential Study Partner CD-ROM
Genetics → Reproduction
Visit the Mader web site for related ESP activities.

Exploring the Internet
The Mader Home Page provides resources and tools as you study this chapter.

http://www.mhhe.com/biosci/genbio/mader

Dynamic Human 2.0 CD-ROM
Reproductive System

C H A P T E R 22

Development

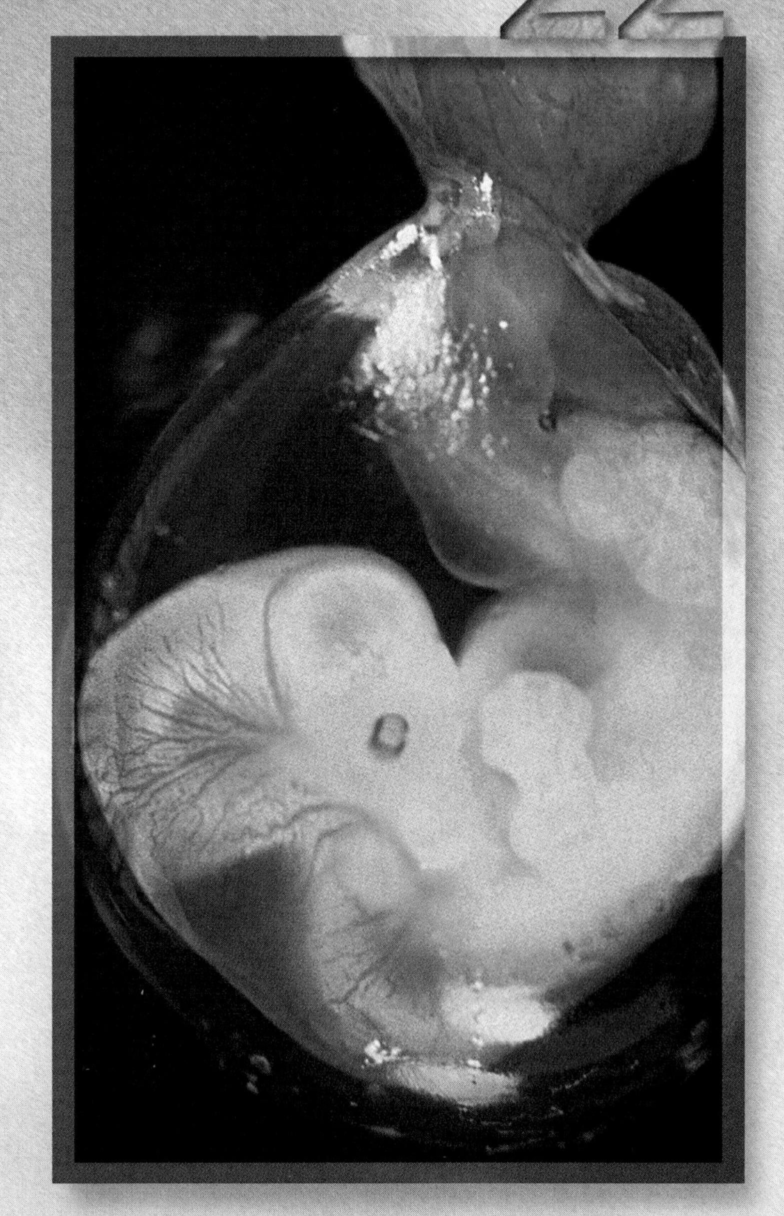

Development has a cellular basis. Union of the sperm and egg begins cell division, growth, and differentiation of cells that result in a human embryo such as the one shown here.

Chapter Concepts

The newborn baby nestled safely in its mother's arms, seeking the nipple in order to suckle. Just nine months before, a sperm had met and entered an egg and development had begun. At first there was only a ball of cells, but then tissues arose and organs formed. In only two months the embryo took on a human appearance and became a fetus still floating within a fluid medium inside its mother's womb. While there, the fetus was completely dependent upon a special structure called the placenta, through which nourishment and oxygen are supplied and wastes are removed. Upon birth, the fetus becomes an independently functioning human being.

The study of development concerns the events and the processes that occur as a single cell becomes a complex organism. These same processes are also seen as a newly born organism matures, as a lost part regenerates, as a wound heals, and even during aging. Therefore, today it is customary to stress that the study of development encompasses not only embryology (development of the embryo), but these other events as well.

22.1 Early Developmental Stages

Fertilization, which results in a zygote, requires that the sperm and egg interact. Figure 22.1 shows the manner in which an egg is fertilized by a sperm in sea stars. The sperm has three distinct parts: a head, a middle piece, and a tail. The tail is a flagellum, which allows the sperm to swim toward the egg, and the middle piece contains ATP-producing mitochondria. The head contains a haploid nucleus capped by a membrane-bounded acrosome containing enzymes that allow the sperm to penetrate the egg.

Several mechanisms have evolved to assure that fertilization takes place and in a species-specific manner (Fig. 22.1). A male releases so many sperm that the egg is literally covered by them. The sea star egg has a plasma membrane, a glycoprotein layer called the vitelline envelope, and a jelly coat. The acrosome enzymes digest away the jelly layer as the acrosome extrudes a filament that attaches to receptors located in the vitelline envelope. This is a lock-and-key reaction that is species-specific. Then, the egg plasma membrane and sperm plasma membrane fuse, allowing the sperm nucleus to enter. A zygote is present when the sperm nucleus fuses with the egg nucleus. Following fusion, the egg plasma membrane and the vitelline envelope undergo changes that prevent the entrance of any other sperm. The vitelline envelope is now called the fertilization envelope.

Most animals go through the same early developmental stages of zygote, morula, blastula, early gastrula, and late gastrula. As an example, we will consider the lancelet, a fish-like animal whose egg has little yolk. **Yolk** is dense nutrient material.

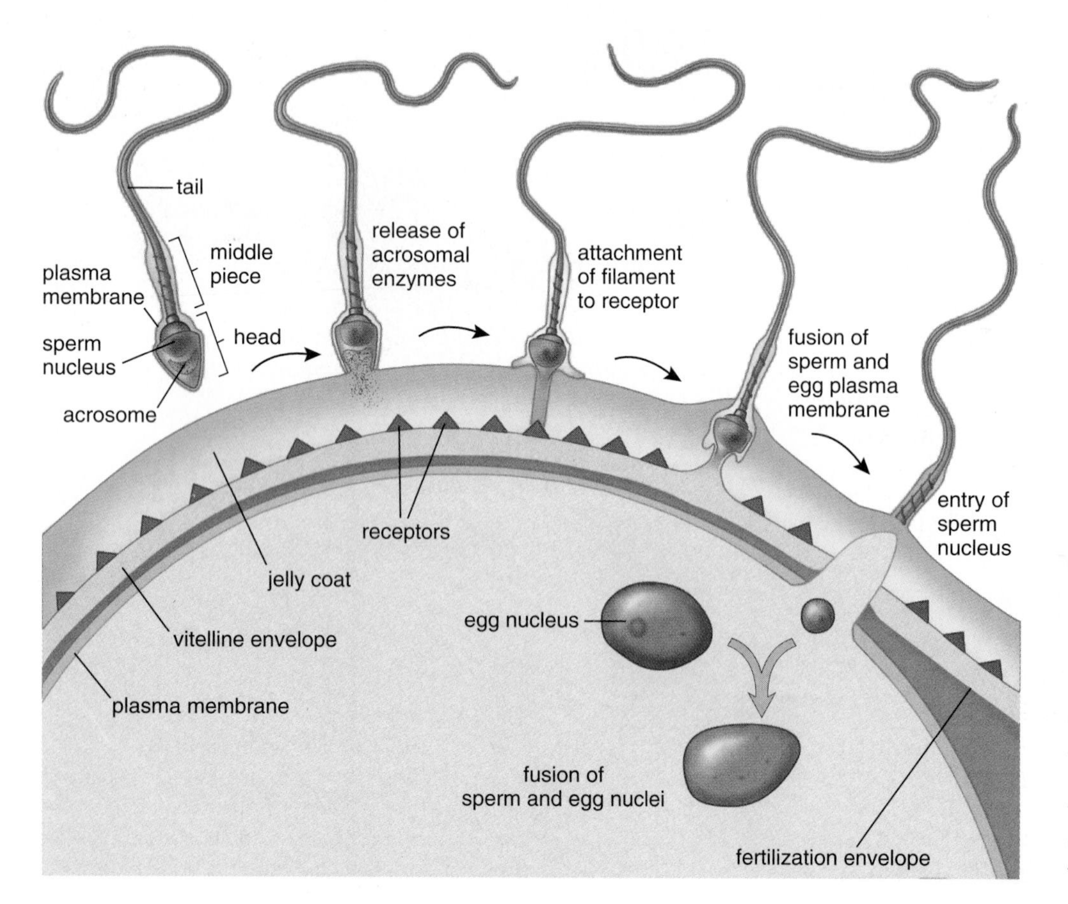

Figure 22.1 Fertilization of a sea star egg.
A head of a sperm has a membrane-bounded acrosome filled with enzymes. When released, these enzymes digest away the jelly coat around the egg, and the acrosome extrudes a filament that attaches to a receptor on the vitelline envelope. Now the sperm nucleus enters and fuses with the egg nucleus, and the resulting zygote begins to divide. The vitelline envelope becomes the fertilization envelope, which prohibits any more sperm from entering the egg.

Following fertilization, the zygote undergoes cleavage and becomes an **embryo. Cleavage** is cell division without growth (Fig. 22.2). DNA replication and mitotic cell division occur repeatedly, and the cells get smaller with each division. Because the lancelet has little yolk, the cell divisions are equal, and the cells are of uniform size in the resulting **morula.** Then a cavity called the **blastocoel** develops and a hollow ball of cells known as the **blastula** forms.

The **gastrula** stage is evident in a lancelet when certain cells begin to push, or invaginate, into the blastocoel, creating a double layer of cells. The outer layer is called the **ectoderm,** and the inner layer is called the **endoderm.** The space created by invagination will become the gut, but at this point it is termed either the primitive gut or the **archenteron.** The pore, or hole, created by invagination is the **blastopore,** and in a lancelet the blastopore eventually becomes the anus.

Gastrulation is not complete until three layers of cells are present. The third, or middle, layer of cells is called the **mesoderm.** In a lancelet, this layer begins as outpocketings from the primitive gut. These outpocketings grow in size until they meet and fuse. In effect then, two layers of mesoderm are formed, and the space between them is the coelom. The coelom is a body cavity that contains internal organs.

Ectoderm, mesoderm, and endoderm are called the embryonic **germ layers**. No matter how gastrulation takes place, the end result is the same: three germ layers are formed. It is possible to relate the development of future organs to these germ layers:

Embryonic Germ Layer	Vertebrate Adult Structures
Ectoderm (outer layer)	Epidermis of skin; epithelial lining of oral cavity and rectum; nervous system
Mesoderm (middle layer)	Skeleton; muscular system; dermis of skin; cardiovascular system; excretory system; reproductive system—including most epithelial linings; outer layers of respiratory and digestive systems
Endoderm (inner layer)	Epithelial lining of digestive tract and respiratory tract; associated glands of these systems; epithelial lining of urinary bladder

The three embryonic germ layers arise during gastrulation, when cells invaginate into the blastocoel. The development of organs can be related to the three germ layers: ectoderm, mesoderm, and endoderm.

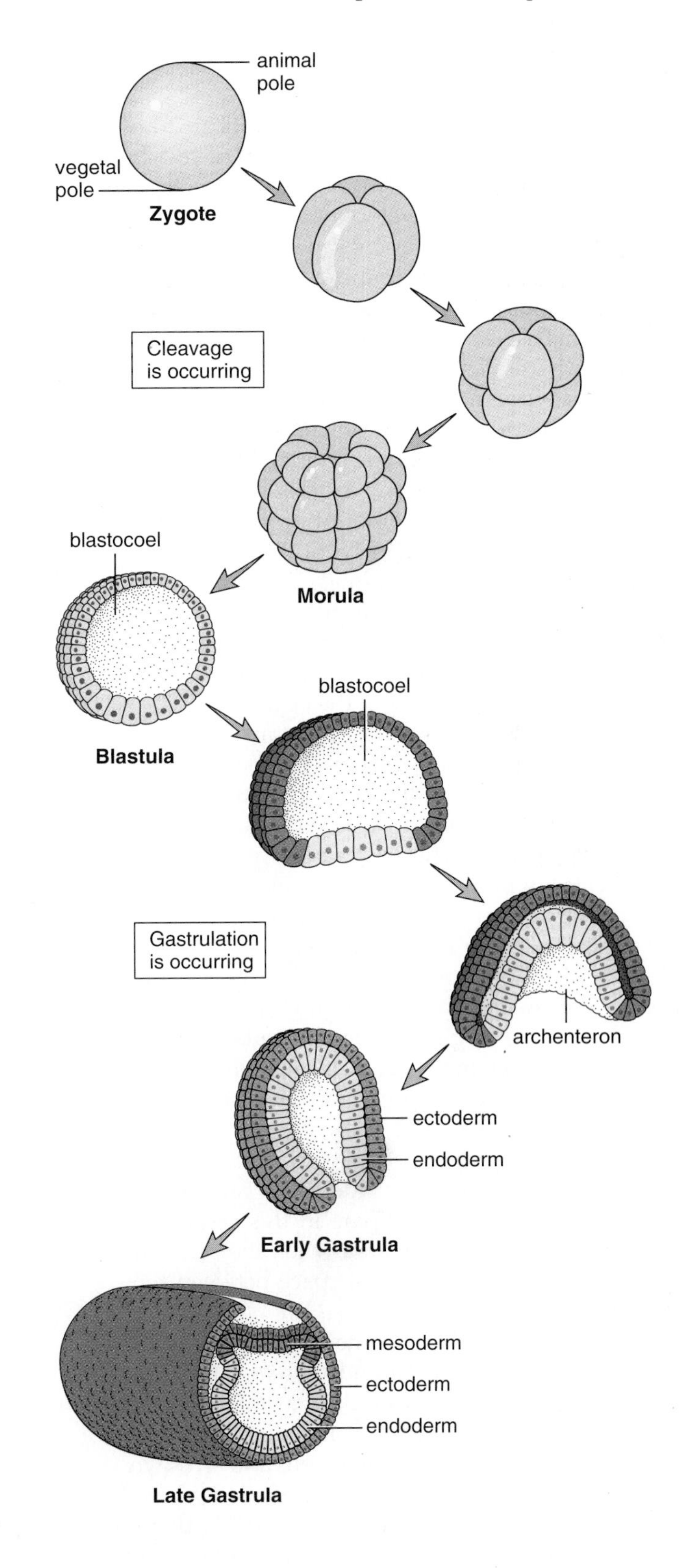

Figure 22.2 Lancelet early development.
A lancelet has little yolk as an embryo, and it can be used to exemplify the early stages of development in such animals. Cleavage produces a number of cells that form a cavity. Invagination during gastrulation produces the germ layers ectoderm and endoderm. Mesoderm arises from pouches that pinch off from the endoderm.

The Effect of Yolk

Table 22.1 indicates the amount of yolk in four types of embryos and relates the amount of yolk to the environment in which the animal develops. The lancelet and frog develop in water. They have less yolk than the chick because their development proceeds quickly to a swimming larval stage that can feed itself. The chick is representative of vertebrate animals that can develop on land because they lay a hard-shelled egg that contains plentiful yolk. Development continues in the shell until there is an offspring capable of land existence.

Early stages of human development resemble those of the chick embryo, yet this resemblance cannot be related to the amount of yolk because the human egg contains little yolk. The evolutionary history of these two animals can provide an answer for this similarity. Both birds and mammals are related to reptiles. This explains why all three groups develop similarly, despite a difference in the amount of yolk in the eggs.

Figure 22.3 compares the appearance of early developmental stages in the lancelet, the frog, and the chick. In the frog embryo, cells at the animal pole have little yolk while those at the vegetal pole contain more yolk. The presence of yolk causes cells to cleave more slowly; you can see that the cells of the animal pole are smaller than those of the vegetal pole. In the chick, cleavage is incomplete—only those cells lying on top of the yolk cleave. This means that although cleavage in the lancelet and the frog results in a morula, no such ball of cells is seen in the chick. Instead, during the morula stage the cells spread out on a portion of the yolk.

In the frog, the blastocoel is formed at the animal pole only. The heavily laden yolk cells of the vegetal pole do not participate in this step. In a chick, the blastocoel is created when the cells lift up from the yolk and leave a space between the cells and the yolk.

In the frog, the cells containing yolk do not participate in gastrulation and therefore they do not invaginate. Instead, a slitlike blastopore is formed when the animal pole cells begin to invaginate from above. Following this, other animal pole cells move down over the yolk, and the blastopore becomes rounded when these cells also invaginate from below. At this stage, there are some yolk cells temporarily left in the region of the pore; these are called the yolk plug. In the chick, there is so much yolk that endoderm formation does not occur by invagination. Instead, an upper layer of cells becomes ectoderm, and a lower layer becomes endoderm.

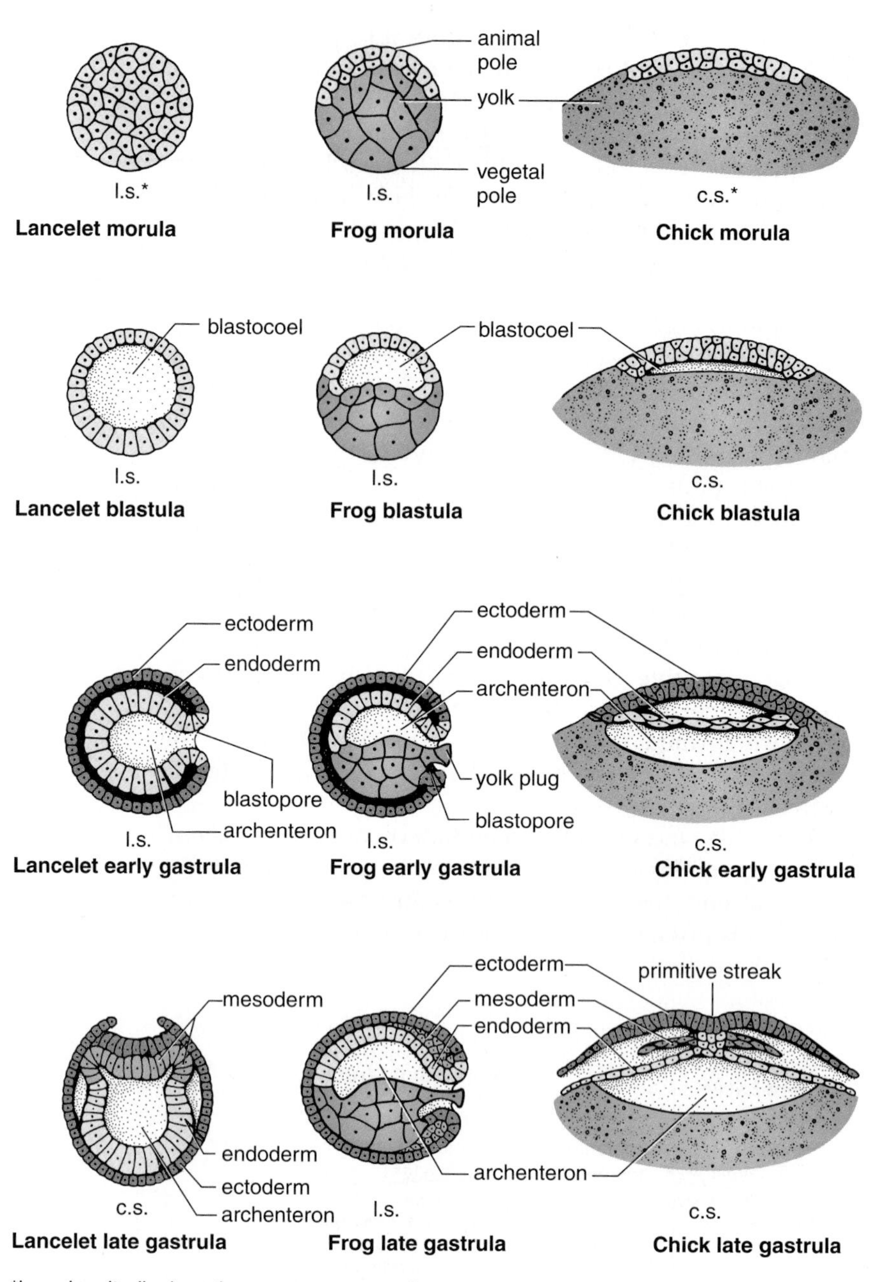

Figure 22.3 **Comparative animal development.**

Table 22.1 Amount of Yolk in Eggs Versus Location of Development

Animal	Yolk	Location of Development
Lancelet	Little	External in water
Frog	Some	External in water
Chick	Much	Within hard shell
Human	Little	Inside mother

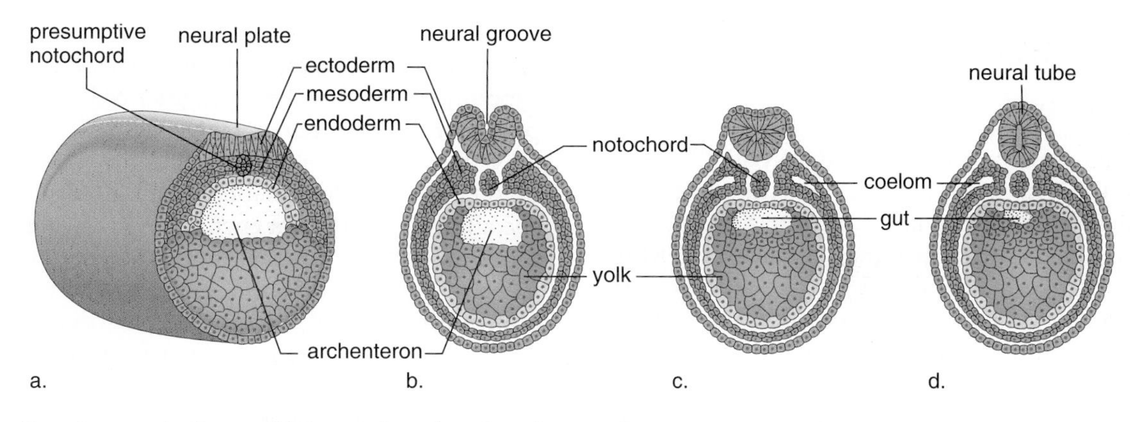

Figure 22.4 Development of neural tube and coelom in a frog embryo.
a. Ectoderm cells that lie above the future notochord (called presumptive notochord) thicken to form a neural plate. **b.** The neural groove and folds are noticeable as the neural tube begins to form. **c.** A splitting of the mesoderm produces a coelom, which is completely lined by mesoderm. **d.** A neural tube and a coelom have now developed.

In the frog, cells from the dorsal lip of the blastopore migrate between the ectoderm and endoderm, forming the mesoderm. Later, a splitting of the mesoderm creates the coelom. In the chick, the mesoderm layer arises by an invagination of cells along the edges of a longitudinal furrow in the midline of the embryo. Because of its appearance, this furrow is called the **primitive streak** (Fig. 22.3*d*). Later, the newly formed mesoderm will split to produce a coelomic cavity.

The amount of yolk typically affects the manner in which animals complete the first three stages of development.

Neurulation and the Nervous System

The mesoderm cells that lie along the main longitudinal axis of the animal coalesce to form a dorsal supporting rod called the **notochord.** The notochord persists in lancelets, but in frogs, chicks, and humans it is later replaced by the vertebral column. Therefore, they are called **vertebrates.**

The nervous system develops from midline ectoderm located just above the notochord. At first, a thickening of cells called the **neural plate** is seen along the dorsal surface of the embryo. Then, **neural folds** develop on either side of a neural groove, which becomes the **neural tube** when these folds fuse. Figure 22.4 shows cross sections of frog development to illustrate the formation of the neural tube. At this point, the embryo is called a **neurula.** Later, the anterior end of the neural tube develops into the brain.

Midline mesoderm cells that did not contribute to the formation of the notochord now become two longitudinal masses of tissue. These two masses become blocked off into the somites, which give rise to the body's muscles. The somites also produce the vertebral bones.

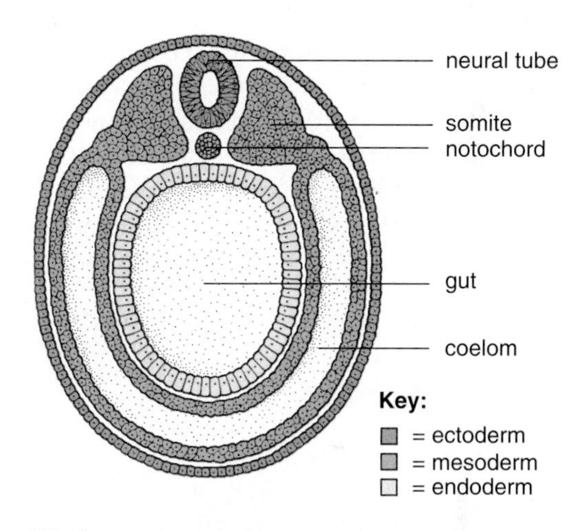

Figure 22.5 Chordate embryo, cross section.
At the neurula stage, each of the germ layers, indicated by color (see key), can be associated with the later development of particular parts. The somites give rise to the muscles of each segment and to the vertebrae, which replace the notochord.

The embryo in Figure 22.5 shows the location of various parts. This figure and the chart on page 445 will help you relate the formation of vertebrate structures and organs to the three embryonic layers of cells: the ectoderm, the mesoderm, and the endoderm.

During neurulation, the neural tube develops just above the notochord. At the neurula stage of development, a cross section of all vertebrate embryos is similar in appearance.

22.2 Differentiation and Morphogenesis

Development requires growth, cellular differentiation, and morphogenesis. **Cellular differentiation** occurs when cells become specialized in structure and function; a muscle cell looks different and acts differently than a nerve cell. **Morphogenesis** is a change in shape and form of a body part. There is a great deal of difference between your arm and leg, even though they contain the same types of tissues.

Differentiation

The process of differentiation most likely starts long before we can recognize different types of cells. Ectodermal, endodermal, and mesodermal cells in the gastrula look quite similar, but yet they must be different because they develop into different organs. What causes differentiation to occur, and when does it begin?

We know that differentiation cannot be due to a parceling out of genes into embryonic cells, because each cell in the body contains a full complement of chromosomes and, therefore, genes. However, we can note that the cytoplasm of a frog's egg is not uniform. It is polar and has both an anterior/posterior axis and a dorsal/ventral axis, which can be correlated with the **gray crescent,** a gray area that appears after the sperm fertilizes the egg (Fig. 22.6*a*). Normally, the first cleavage gives each daughter cell half of the gray crescent. In this case, each experimentally separated daughter cell develops into a complete embryo (Fig. 22.6*b*). However, if the egg divides so that only one daughter cell receives the gray crescent, only that cell becomes a complete embryo (Fig. 22.6*c*). We can therefore speculate that particular chemical signals within the gray crescent turn on the genes that control development in the frog.

It is hypothesized that genes are turned on/off due to **ooplasmic segregation,** which is the distribution of maternal cytoplasmic contents to the various cells of the morula:

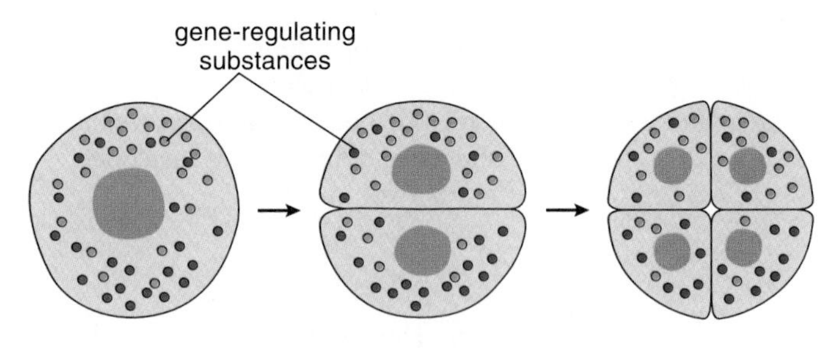

Cytoplasmic substances, parceled out during cleavage, initially influence which genes are activated and how cells differentiate.

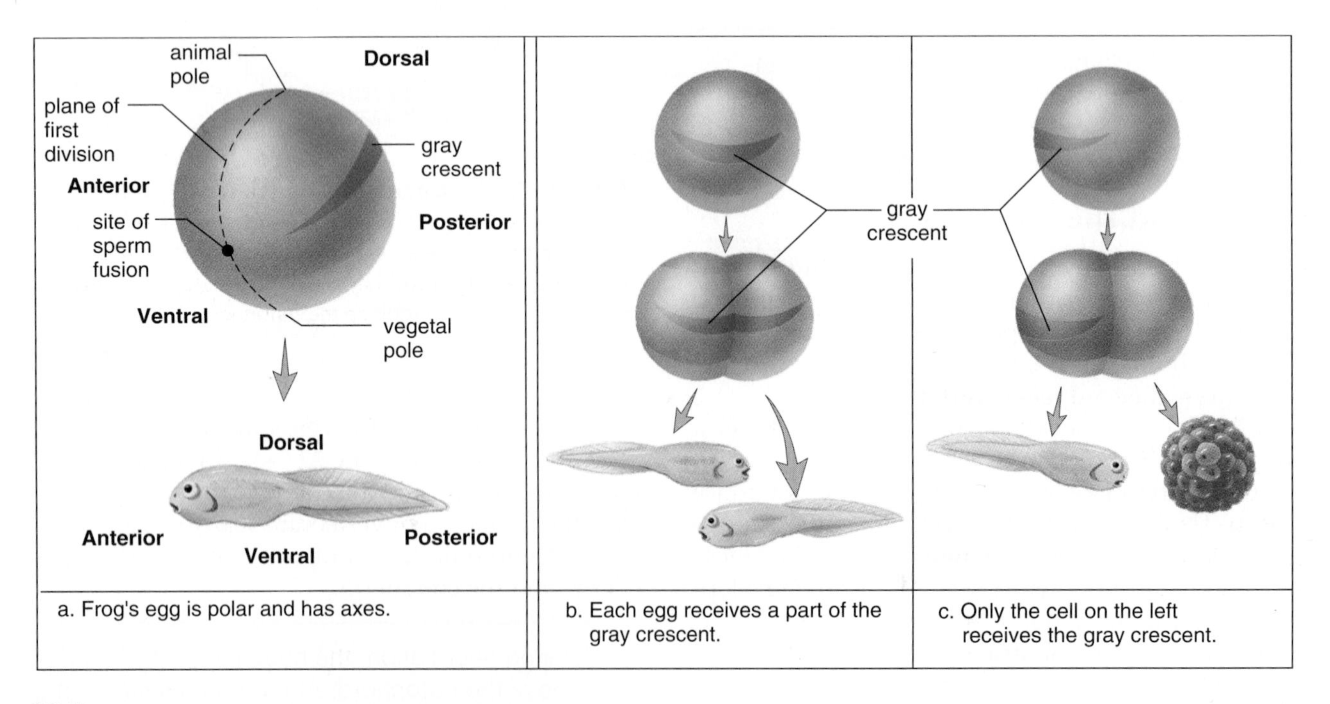

Figure 22.6 Cytoplasmic influence on development.
a. A frog's egg has anterior/posterior and dorsal/ventral axes that correlate with the position of the gray crescent. **b.** The first cleavage normally divides the gray crescent in half, and each daughter cell is capable of developing into a complete tadpole. **c.** But if only one daughter cell receives the gray crescent, then only that cell can become a complete embryo. This shows that chemical messengers are not uniformly distributed in the cytoplasm of frogs' eggs.

Morphogenesis

As development proceeds, a cell's differentiation is influenced not only by its cytoplasmic content, but also by signals given off by neighboring cells. Migration of cells occurs during gastrulation, and there is evidence that one set of cells can influence the migratory path taken by another set of cells. Some cells produce an extracellular matrix that contains fibrils. In the laboratory, it can be shown that the orientation of these fibrils influences migratory cells: the cytoskeletons of the migrating cells are oriented in the same direction as the fibrils. Although this may not be an exact mechanism at work during gastrulation, it suggests that formation of the germ layers is probably influenced by environmental factors.

More specific information is known about neurulation, the process by which the nervous system develops. Neurulation involves **induction,** the process by which one tissue influences the development of another tissue. Experiments have shown that the presumptive (potential) notochord induces formation of neural plates, the first sign of the nervous system (Fig. 22.7). If the presumptive nervous system, located just above the notochord, is cut out and transplanted to the belly region of the embryo, it will not form a neural plate. On the other hand, if presumptive notochord tissue is cut out and transplanted beneath what would be ectoderm, this ectoderm differentiates into a neural plate.

A well-known series of inductions accounts for the development of the vertebrate eye. An optic vesicle, which is a lateral outgrowth of the embryonic brain, induces the overlying ectoderm to thicken and become the lens of an eye. The lens induces an optic cup, where the retina develops. The optic cup in turn induces formation of other eye parts. Today, investigators believe that similar to this example, the process of induction goes on continuously—neighboring cells are always influencing one another. Either direct contact or the production of a chemical acts as a signal that activates certain genes in neighboring cells and brings about development of particular structures. The following diagram suggests that signals activate genes in cells, causing these cells to send out signals that activate genes in still other cells, and so forth.

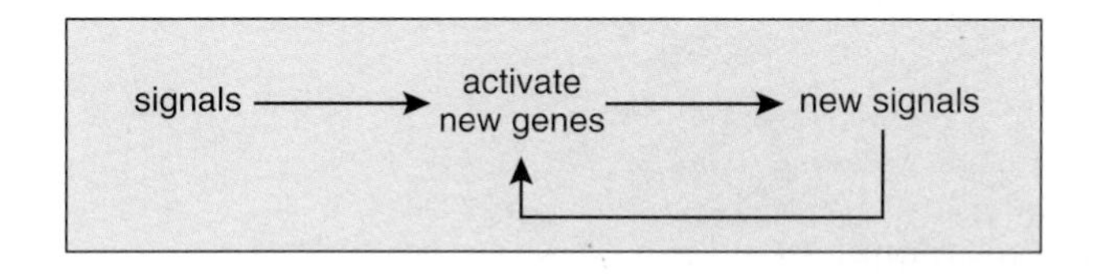

Morphogenesis is dependent upon induction. Signals (either contact or chemical) from cells cause differentiated cells located nearby to develop into particular structures.

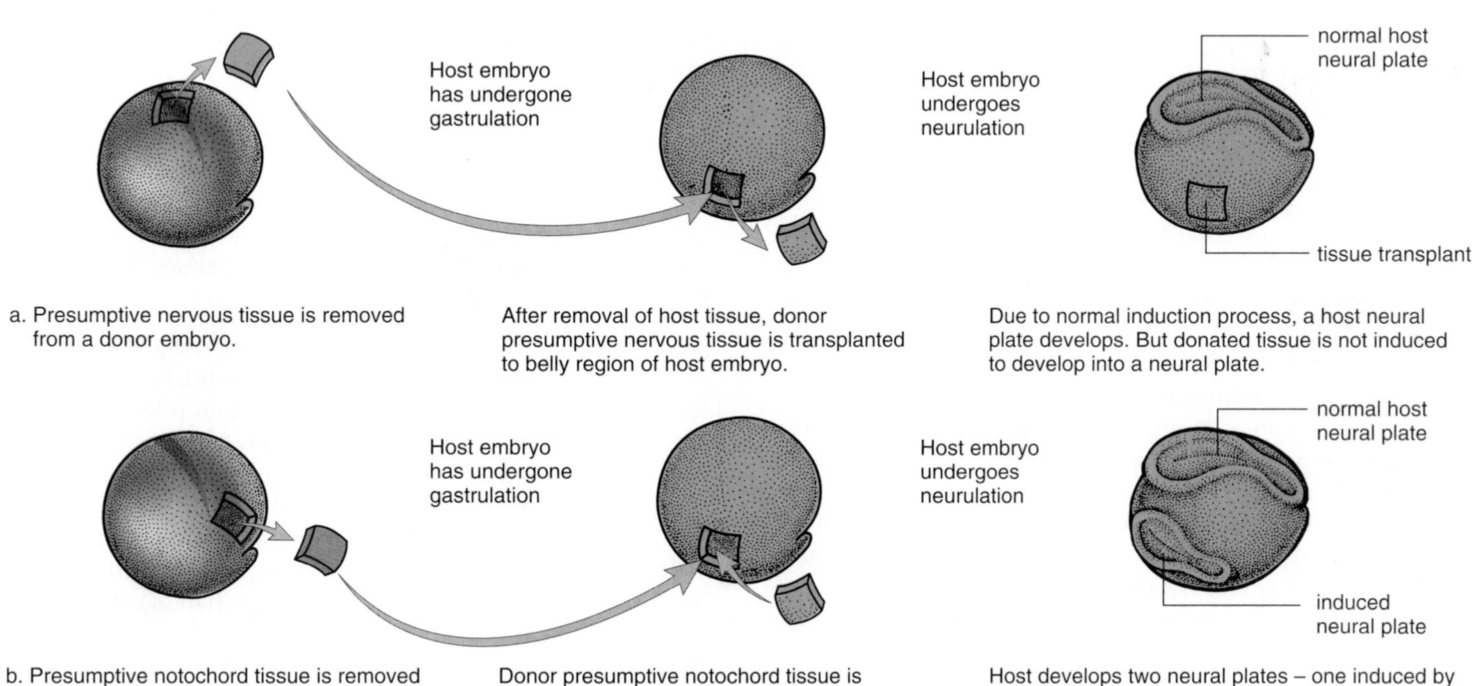

Figure 22.7 Control of nervous system development.
a. In this experiment, the presumptive nervous system does not develop into the neural plate if moved from its normal location. **b.** In this experiment, the presumptive notochord can cause even belly ectoderm to develop into the neural plate. This shows that the notochord induces ectoderm to become a neural plate, most likely by sending out chemical signals.

Pattern Formation

Investigators studying morphogenesis in *Drosophila* (fruit fly) have discovered that there are some genes that determine the animal's anterior/posterior and dorsal/ventral axes, others that determine the number and polarity of its segments, and still others, called *homeotic genes,* that determine how these segments develop. Homeotic genes have now been found in many other organisms and, surprisingly, they all contain the same particular sequence of nucleotides, called a **homeobox.**

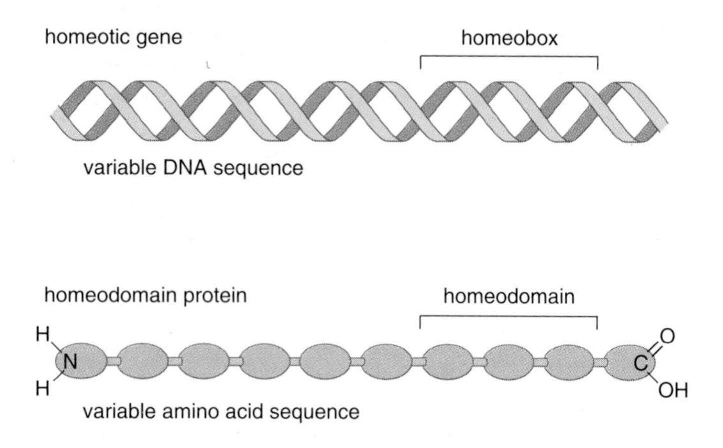

Figure 22.8 Homeotic mutations in *Drosophila*. Homeotic genes control pattern formation, an aspect of morphogenesis. If homeotic genes are activated at inappropriate times, abnormalities such as **(a)** a fly with four wings and **(b)** a fly with legs on its head can occur.

Since homeoboxes have been found in almost all eukaryotic organisms, it is believed that homeoboxes are derived from an original nucleotide sequence that has been largely conserved (maintained from generation to generation) because of its importance in the regulation of animal development.

In *Drosophila,* a homeotic mutation causes body parts to be misplaced—a homeotic mutant fly can have two pairs of wings or extra legs where antennae should be (Fig. 22.8). Similarly, in the frog *Xenopus,* if the expression levels of the homeotic genes are altered, headless and tailless embryos are produced.

Homeotic genes are clearly involved in **pattern formation**—that is, the shaping of an embryo so that the adult has a normal appearance. Homeotic genes are arranged in a definite order on a chromosome; those first in line determine the development of anterior segments of the animal, while those later in the sequence determine the development of posterior segments of the embryo. A homeotic gene codes for a homeodomain protein. Each of these proteins has a homeodomain, a sequence of 60 amino acids that is found in all other homeodomain proteins. Homeodomain proteins stay in the nucleus and regulate transcription of other genes during development. Researchers envision that a homeodomain protein produced by one homeotic gene binds to and turns on the next homeotic gene, and this orderly process determines the overall pattern of the embryo. It appears that homeotic genes also establish homeodomain protein gradients that affect the pattern development of specific parts, such as the limbs. One well-known gradient that determines wing formation in the chick involves retinoic acid, a chemical related to retinal, which is present in rods and cones of the eye.

Many laboratories are engaged in discovering much more about homeotic genes and homeodomain proteins. We need to know, for example, how homeotic genes are turned on and how protein gradients are maintained in tissues. The roles of the cytoskeleton and extracellular matrix are being explored.

Apoptosis. Investigators studying morphogenesis in the nematode *Caenorhabifitis elegans* have found that apoptosis (programmed cell death) plays an important role in development. At specific times in development, certain cells die because suicide genes coding for toxic proteins have been turned on. Shrinkage of the cytoplasm and nucleus turns cells into apoptotic bodies that are phagocytized by neighboring cells. If too few cells undergo apoptosis, abnormalities resulting from too many cells result. In humans, hands and feet first have the appearance of paddles, and only after apoptosis do fingers and toes result. Apoptosis is also necessary to normal development of the nervous system and the immune sytem.

Morphogenesis is dependent upon signals (either contact or chemical) from neighboring cells. These signals are believed to activate particular genes.

22.3 Human Embryonic and Fetal Development

In humans, the length of the time from conception (fertilization followed by **implantation**) to birth (parturition) is approximately nine months. It is customary to calculate the time of birth by adding 280 days to the start of the last menstruation, because this date is usually known, whereas the day of fertilization is usually unknown. Because the time of birth is influenced by so many variables, only about 5% of babies actually arrive on the forecasted date.

Human development is often divided into embryonic development (months 1 and 2) and fetal development (months 3–9). The **embryonic period** consists of early formation of the major organs, and fetal development is the refinement of these structures.

Before we consider human development chronologically, we must understand the placement of **extraembryonic membranes.** Extraembryonic membranes are best understood by considering their function in reptiles and birds. In reptiles, these membranes made development on land first possible. If an embryo develops in the water, the water supplies oxygen for the embryo and takes away waste products. The surrounding water prevents desiccation, or drying out, and provides a protective cushion. For an embryo that develops on land, all these functions are performed by the extraembryonic membranes.

In the chick, the extraembryonic membranes develop from extensions of the germ layers, which spread out over the yolk. Figure 22.9 shows the chick surrounded by the membranes. The **chorion** lies next to the shell and carries on gas exchange. The **amnion** contains the protective amniotic fluid, which bathes the developing embryo. The **allantois** collects nitrogenous wastes, and the **yolk sac** surrounds the remaining yolk, which provides nourishment.

Humans (and other mammals) also have these extraembryonic membranes. The chorion develops into the fetal half of the placenta; the yolk sac, which lacks yolk, is the first site of blood cell formation; the allantoic blood vessels become the umbilical blood vessels; and the amnion contains fluid to cushion and protect the embryo, which develops into a fetus. Therefore, the function of the membranes in humans has been modified to suit internal development, but their very presence indicates our relationship to birds and to reptiles. It is interesting to note that all animals develop in water, either in bodies of water or within amniotic fluid.

> The presence of extraembryonic membranes in reptiles made development on land possible. Humans also have these membranes, but their function has been modified for internal development.

Embryonic Development

Embryonic development includes the first two months of development.

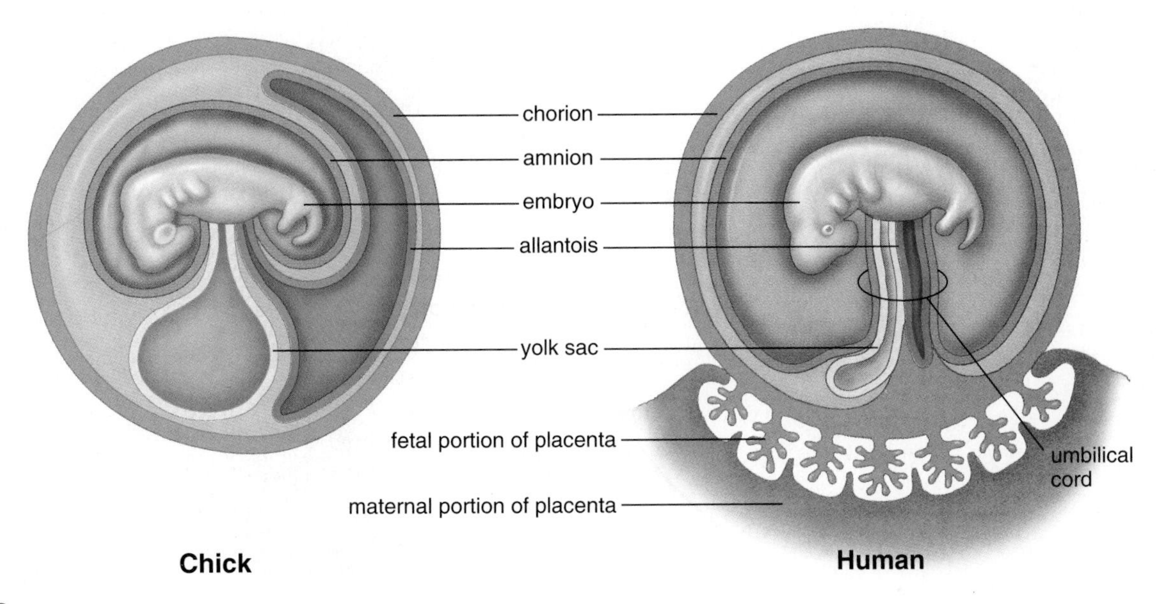

Figure 22.9 Extraembryonic membranes.
The membranes, which are not part of the embryo, are found during the development of chicks and humans, where each has a specific function.

The First Week

Fertilization occurs in the upper third of an oviduct (Fig. 22.10), and cleavage begins even as the embryo passes down this tube to the uterus. By the time the embryo reaches the uterus on the third day, it is a morula. The morula is not much larger than the zygote because, even though multiple cell divisions have occurred, there has been no growth of these newly formed cells. By about the fifth day, the morula is transformed into the blastocyst. The **blastocyst** has a fluid-filled cavity, a single layer of outer cells called the **trophoblast** and an inner cell mass. Later, the trophoblast, reinforced by a layer of mesoderm, gives rise to the *chorion*, one of the extraembryonic membranes (see Fig. 22.9). The *inner cell mass* eventually becomes the embryo, which develops into a fetus.

The Second Week

At the end of the first week, the embryo begins the process of *implanting* in the wall of the uterus. The trophoblast secretes enzymes to digest away some of the tissue and blood vessels of the uterine wall (Fig. 22.10). The embryo is now about the size of the period at the end of this sentence. The trophoblast begins to secrete **human chorionic gonadotropin (HCG),** the hormone that is the basis for the pregnancy test and that serves to maintain the corpus luteum past the time it normally disintegrates. Because of this, the endometrium is maintained and menstruation does not occur.

As the week progresses, the inner cell mass detaches itself from the trophoblast, and two more extraembryonic membranes form (Fig. 22.11*a*). The *yolk sac,* which forms below the embryonic disk, has no nutritive function as in

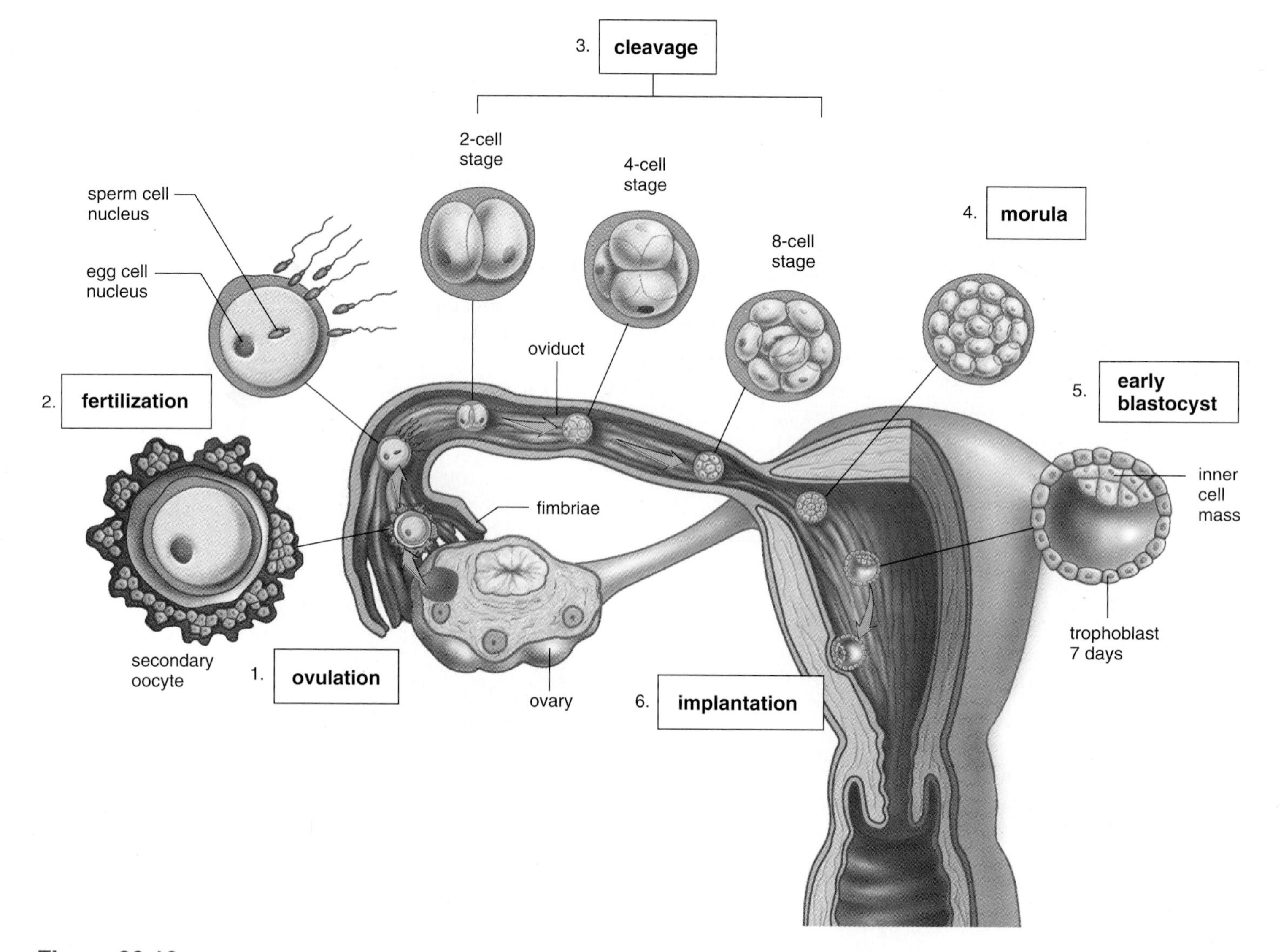

Figure 22.10 Human development before implantation.
Structures and events proceed counterclockwise. At ovulation (1), the secondary oocyte leaves the ovary. A single sperm penetrates the zona pellucida, and fertilization (2) occurs in the oviduct. As the zygote moves along the oviduct, it undergoes cleavage (3) to produce a morula (4). The blastocyst forms (5) and implants itself in the uterine lining (6).

chicks, but it is the first site of blood cell formation. However, the *amnion* and its cavity are where the embryo (later called the fetus) develops. In humans, amniotic fluid acts as an insulator against cold and heat and also absorbs shock, such as that caused by the mother exercising.

Gastrulation occurs during the second week. The inner cell mass now has flattened into the **embryonic disk,** composed of two layers of cells: ectoderm above and endoderm below. Once the embryonic disk elongates to form the *primitive streak,* similar to that found in birds, the third germ layer, mesoderm, forms by invagination of cells along the streak. The trophoblast is reinforced by mesoderm and becomes the chorion.

It is possible to relate the development of future organs to these germ layers (see page 445).

The Third Week

Two important organ systems make their appearance during the third week. The nervous system is the first organ system to be visually evident. At first, a thickening appears

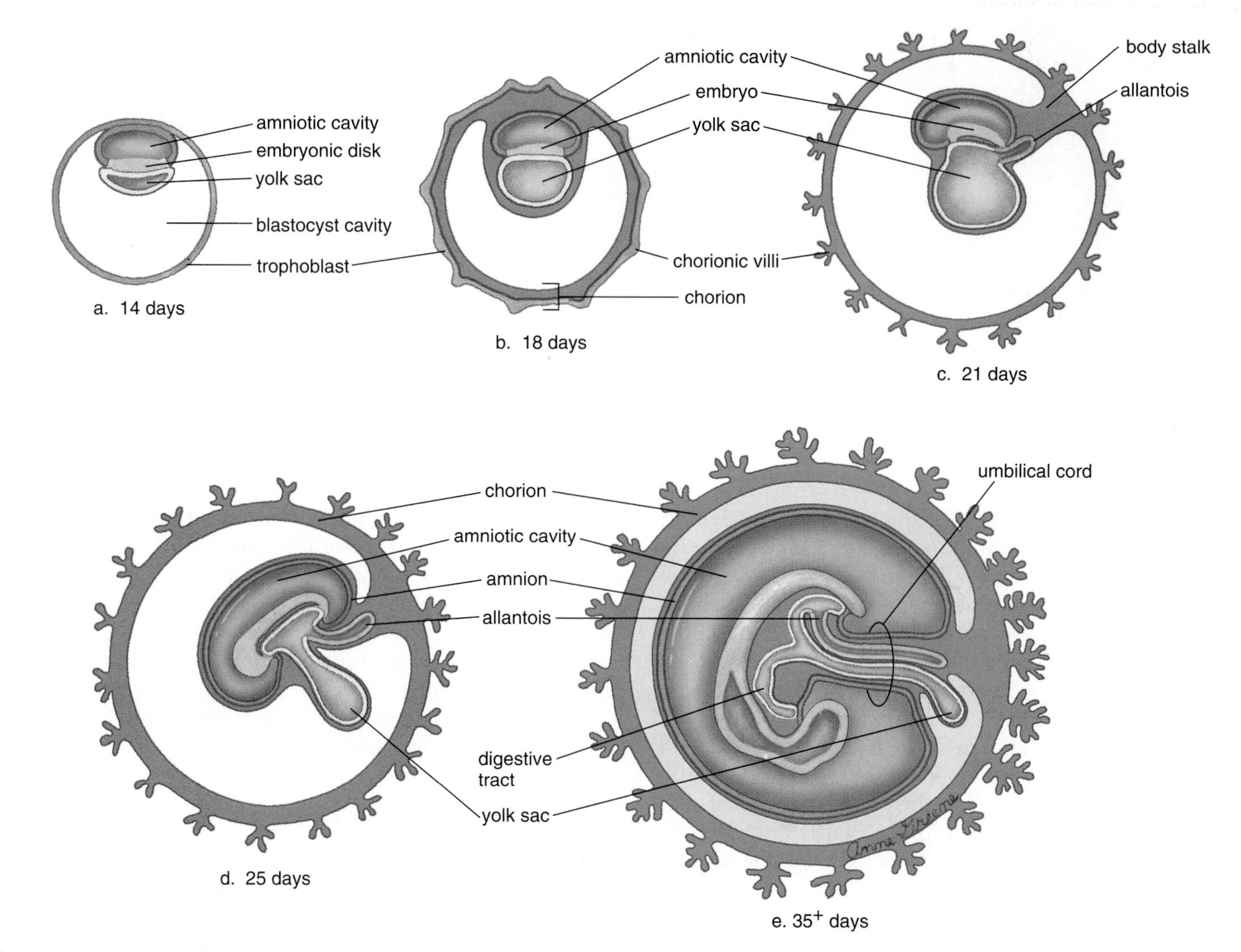

Figure 22.11 Human embryonic development.
a. At first, no organs are present in the embryo, only tissues. The amniotic cavity is above the embryo, and the yolk sac is below. **b.** The chorion is developing villi, so important to exchange between mother and child. **c.** The allantois and yolk sac are two more extraembryonic membranes. **d.** These extraembryonic membranes are positioned inside the body stalk as it becomes the umbilical cord. **e.** At 28 days, the embryo has a head region and a tail region. The umbilical cord takes blood vessels between the embryo and the chorion (placenta).

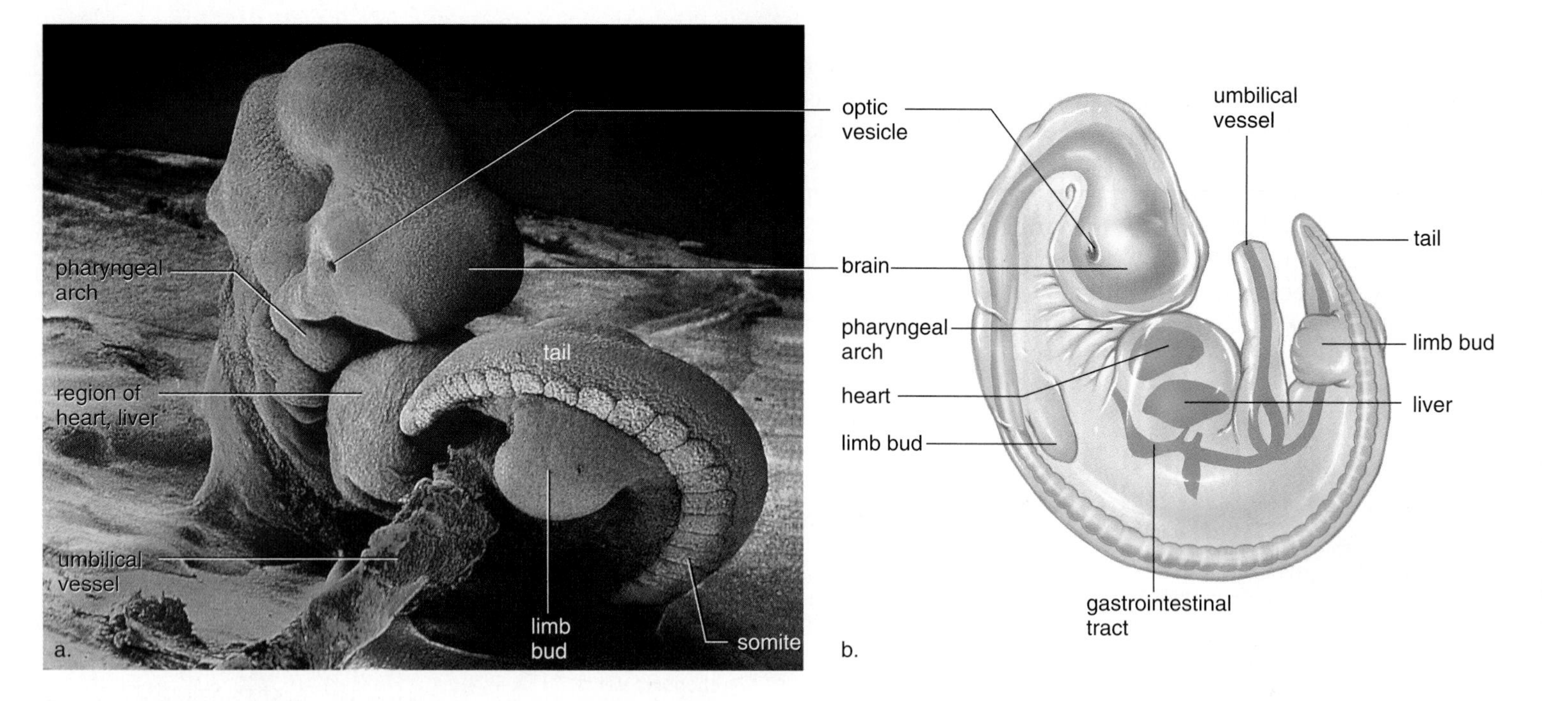

Figure 22.12 Human embryo at beginning of fifth week.
a. Scanning electron micrograph. **b.** The embryo is curled so that the head touches the heart, the two organs whose development is further along than the rest of the body. The organs of the gastrointestinal tract are forming, and the arms and the legs develop from the bulges that are called limb buds. The tail is an evolutionary remnant; its bones regress and become those of the coccyx (tailbone). The pharyngeal arches become functioning gills only in fishes and amphibian larvae; in humans, the first pair of pharyngeal pouches becomes the auditory tubes. The second pair becomes the tonsils, while the third and fourth become the thymus gland and the parathyroids.

along the entire dorsal length of the embryo, and then invagination occurs as neural folds appear. When the neural folds meet at the midline, the neural tube, which later develops into the brain and the nerve cord, is formed (see Fig. 22.4). After the notochord is replaced by the vertebral column, the nerve cord is called the spinal cord.

Development of the heart begins in the third week and continues into the fourth week. At first, there are right and left heart tubes; when these fuse, the heart begins pumping blood, even though the chambers of the heart are not fully formed. The veins enter posteriorly and the arteries exit anteriorly from this largely tubular heart, but later the heart twists so that all major blood vessels are located anteriorly.

The Fourth and Fifth Weeks

At four weeks, the embryo is barely larger than the height of this print. A bridge of mesoderm called the body stalk connects the caudal (tail) end of the embryo with the chorion, which has projections called chorionic villi (Fig. 22.11*b*). The fourth extraembryonic membrane, the *allantois,* is contained within this stalk, and its blood vessels become the umbilical blood vessels. The head and the tail then lift up, and the body stalk moves anteriorly by constriction (Fig. 22.11*d*). Once this process is complete, the **umbilical cord,** which connects the developing embryo to the placenta, is fully formed (Fig. 22.11*e*).

Little flippers called limb buds appear (Fig. 22.12); later, the arms and the legs develop from the limb buds, and even the hands and the feet become apparent. At the same time—during the fifth week—the head enlarges and the sense organs become more prominent. It is possible to make out the developing eyes, ears, and even nose.

The Sixth Through Eighth Weeks

There is a remarkable change in external appearance during the sixth through eighth weeks of development—a form that is difficult to recognize as a human becomes easily recognized as human. Concurrent with brain development, the head achieves its normal relationship with the body as a neck region develops. The nervous system is developed well enough to permit reflex actions, such as a startle response to touch. At the end of this period, the embryo is about 38 mm (1.5 inches) long and weighs no more than an aspirin tablet, even though all organ systems are established.

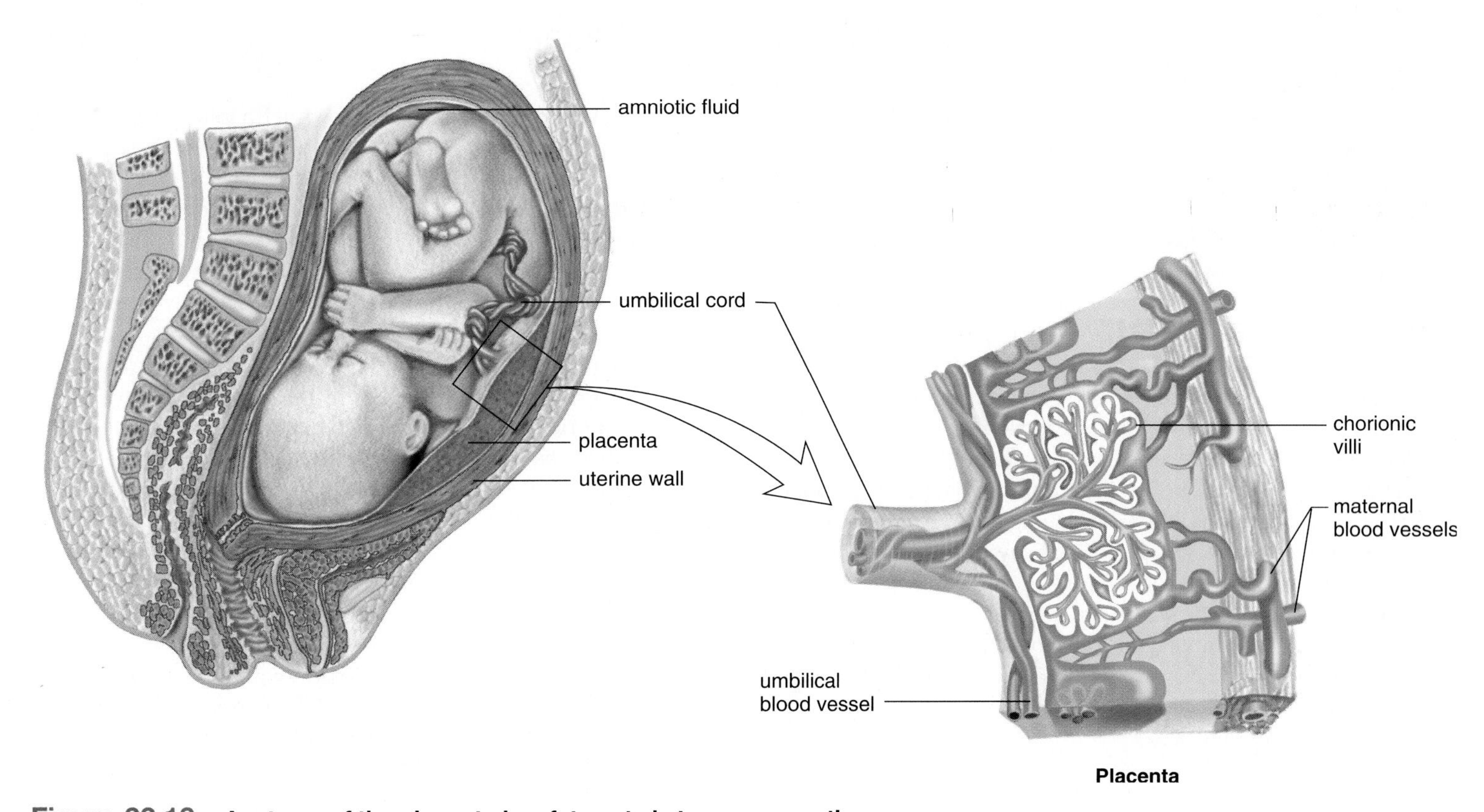

Figure 22.13 Anatomy of the placenta in a fetus at six to seven months.
The placenta is composed of both fetal and maternal tissues. Chorionic villi penetrate the uterine lining and are surrounded by maternal blood. Exchange of molecules between fetal and maternal blood takes place across the walls of the chorionic villi.

The Structure and Function of the Placenta

The **placenta** begins formation once the embryo is fully implanted. Treelike extensions of the chorion called **chorionic villi** project into the maternal tissues. Later, these disappear in all areas except where the placenta develops. By the tenth week, the placenta (Fig. 22.13) is fully formed and has already begun to produce progesterone and estrogen. These hormones have two effects: due to their negative feedback control of the hypothalamus and the anterior pituitary, they prevent any new follicles from maturing, and they maintain the lining of the uterus—now the corpus luteum is not needed. There is no menstruation during pregnancy.

The placenta has a fetal side contributed by the chorion and a maternal side consisting of uterine tissues. Notice in Figure 22.13 how the chorionic villi are surrounded by maternal blood sinuses; yet maternal and fetal blood never mix, since exchange always takes place across plasma membranes. Carbon dioxide and other wastes move from the fetal side to the maternal side, and nutrients and oxygen move from the maternal side to the fetal side of the placenta. The umbilical cord stretches between the placenta and the fetus. Although it may seem that the umbilical cord travels from the placenta to the intestine, actually the umbilical cord is simply taking fetal blood to and from the placenta. The umbilical cord is the lifeline of the fetus because it contains the umbilical arteries and vein, which transport waste molecules (carbon dioxide and urea) to the placenta for disposal and take oxygen and nutrient molecules from the placenta to the rest of the fetal circulatory system.

Harmful chemicals can also cross the placenta. This is of particular concern during the embryonic period, when various structures are first forming. Each organ or part seems to have a sensitive period during which a substance can alter its normal development. For example, if a woman takes the drug thalidomide, a tranquilizer, between days 27 and 40 of her pregnancy, the infant is likely to be born with deformed limbs. After day 40, however, the infant is born with normal limbs.

Health Focus

Preventing Birth Defects

It is believed that at least 1 in 16 newborns has a birth defect, either minor or serious, and the actual percentage may be even higher. Most likely, only 20% of all birth defects are due to heredity. Those that are hereditary can sometimes be detected before birth. Amniocentesis allows the fetus to be tested for abnormalities of development; chorionic villi sampling allows the embryo to be tested; and a new method has been developed for screening eggs to be used for in vitro fertilization (Fig. 22A).

It is recommended that all females take everyday precautions to protect any future and/or presently developing embryos and fetuses from defects. Proper nutrition is a must (deficiency in folic acid causes neural tube defects). X-ray diagnostic therapy should be avoided during pregnancy because X rays cause mutations in the developing embryo or fetus. Children born to women who received X-ray treatment are apt to have birth defects and/or to develop leukemia later. Toxic chemicals such as pesticides and many organic industrial chemicals, which are also mutagenic, can cross the placenta. Cigarette smoke not only contains carbon monoxide but also other fetotoxic chemicals. Babies born to smokers are often underweight and subject to convulsions.

Pregnant Rh^- women should receive an Rh immunoglobulin injection to prevent the production of Rh antibodies. These antibodies can cause nervous system and heart defects.

Sometimes, birth defects are caused by microbes. Females can be immunized before the childbearing years for rubella (German measles), which in particular causes birth defects such as deafness. Unfortunately, immunization for sexually transmitted diseases is not possible. The AIDS virus can cross the placenta, and over 1,500 babies who contracted AIDS while in their mother's womb are now mentally retarded. When a mother has herpes, gonorrhea, or chlamydia, newborns can become infected as they pass through the birth canal. Blindness and other physical and mental defects may develop. Birth by cesarean section could prevent these occurrences.

Pregnant women should not take any type of drug without a doctor's permission. Certainly illegal drugs, such as marijuana, cocaine, and heroin, should be completely avoided. "Cocaine babies" now make up 60% of drug-affected babies. Severe fluctuations in blood pressure which are produced by the use of cocaine temporarily deprive the developing brain of oxygen. Cocaine babies have visual problems, lack coordination, and are

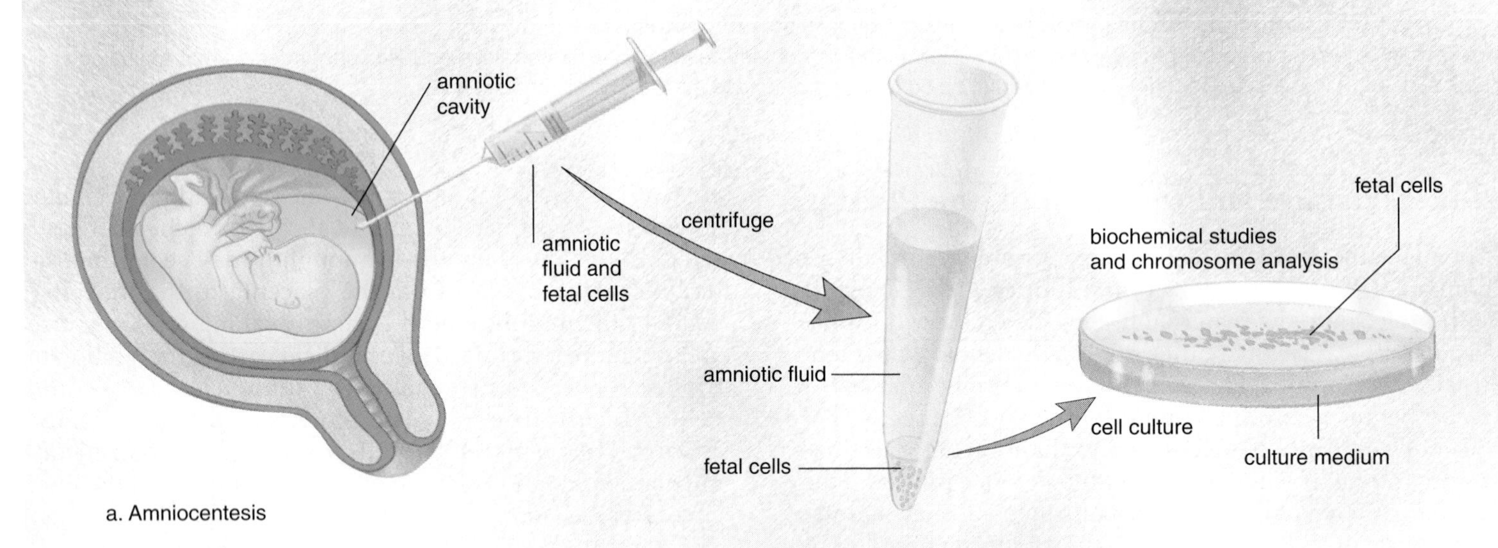

Figure 22A Three methods for genetic defect testing before birth.
a. Amniocentesis is usually performed from the 15th to the 17th week of pregnancy. A long needle is passed through the abdominal wall to withdraw a small amount of amniotic fluid, along with fetal cells. Since there are only a few cells in the amniotic fluid, testing may be delayed as long as four weeks until cell culture produces enough cells for testing purposes. About 40 tests are available for different defects. **b.** Chorionic villi sampling is usually performed from the 8th to the 12th week of pregnancy. The doctor inserts a long, thin tube through the vagina into the uterus. With the help of ultrasound, which gives a picture of the uterine contents, the tube is placed between the lining of the uterus and the chorion. Then a sampling of the chorionic villi cells is obtained by suction. Chromosome analysis and biochemical tests for genetic defects can be done immediately on these cells. **c.** Screening eggs for genetic defects is a new technique. Preovulatory eggs are removed by aspiration after a laparoscope (optical telescope) is inserted into the abdominal cavity through a small incision in the region of the navel. The first polar body is tested. If the woman is heterozygous (*Aa*) and the defective gene (*a*) is found in the polar body, then the egg must have received the normal gene (*A*). Normal eggs then undergo in vitro fertilization and are placed in the prepared uterus.

mentally retarded. The drugs aspirin, caffeine (present in coffee, tea, and cola), and alcohol should be severely limited. It is not unusual for babies of drug addicts and alcoholics to display withdrawal symptoms and to have various abnormalities. Babies born to women who have about 45 drinks a month and as many as 5 drinks on one occasion are apt to have fetal alcohol syndrome (FAS). These babies have decreased weight, height, and head size, with malformation of the head and face. Mental retardation is common in FAS infants.

Medications can also cause problems. When the synthetic hormone DES was given to pregnant women to prevent miscarriage, their daughters showed various abnormalities of the reproductive organs and an increased tendency toward cervical cancer. Other sex hormones, including birth-control pills, can possibly cause abnormal fetal development, including abnormalities of the sex organs. The tranquilizer thalidomide is well known for having caused deformities of the arms and legs in children born to women who took the drug. Therefore, a woman has to be very careful about taking medications while pregnant.

Now that physicians and laypeople are aware of the various ways in which birth defects can be prevented, it is hoped that the incidence of birth defects will decrease in the future.

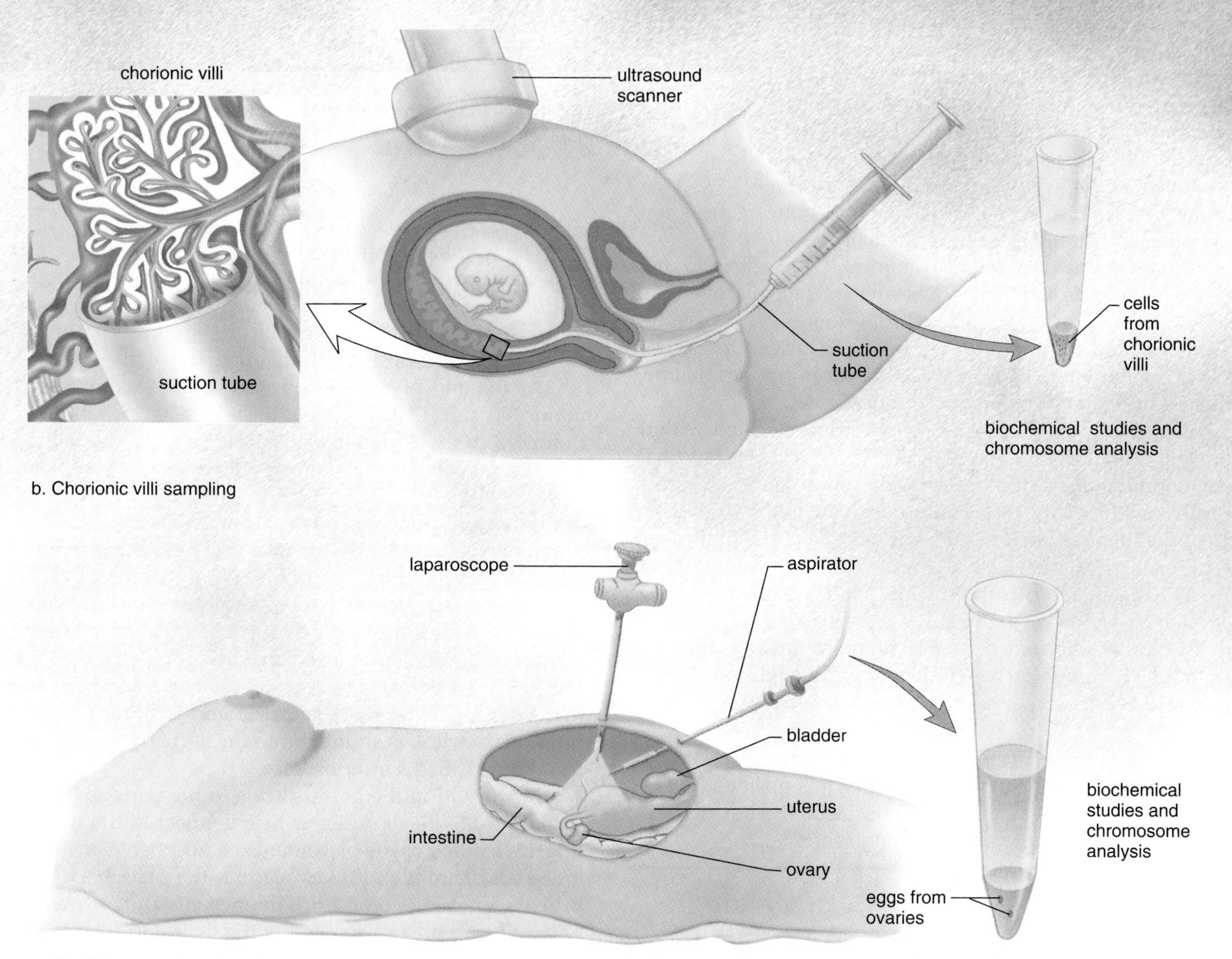

b. Chorionic villi sampling

c. Obtaining eggs for screening

Fetal Development and Birth

Fetal development includes the third through ninth months of development. At this time, the fetus looks human (Fig. 22.14).

The Third and Fourth Months

At the beginning of the third month, the fetal head is still very large, the nose is flat, the eyes are far apart, and the ears are distinctively present. Head growth now begins to slow down as the rest of the body increases in length. Epidermal refinements, such as eyelashes, eyebrows, hair on head, fingernails, and nipples, appear.

Cartilage begins to be replaced by *bone* as ossification centers appear in most of the bones. Cartilage remains at the ends of the long bones, and ossification is not complete until age 18 or 20 years. The skull has six large membranous areas called **fontanels,** which permit a certain amount of flexibility as the head passes through the birth canal and allow rapid growth of the brain during infancy. Progressive fusion of the skull bones causes the fontanels to usually close by 2 years of age.

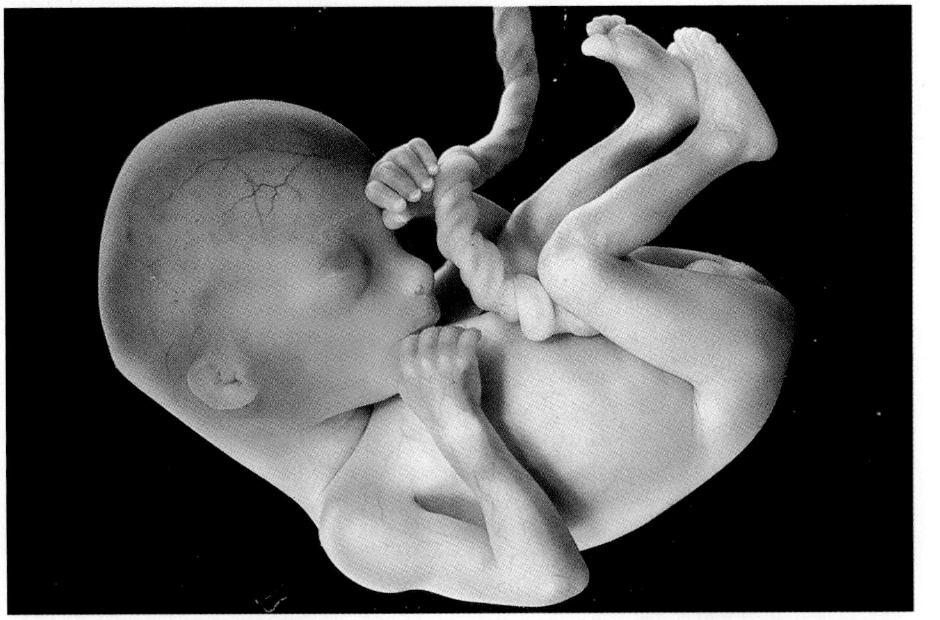

Figure 22.14 The three- to four-month-old fetus looks human. Face, hands, and fingers are well defined.

Sometime during the third month, it is possible to distinguish males from females. Researchers have discovered a series of genes on the X and Y chromosomes that cause the differentiation of gonads into testes and ovaries. Once these have differentiated, they produce the sex hormones that influence the differentiation of the genital tract.

At this time, either testes or ovaries are located within the abdominal cavity, but later, in the last trimester of fetal development, the testes descend into the scrotal sacs (scrotum). Sometimes the testes fail to descend, and in that case, an operation may be done later to place them in their proper location.

During the fourth month, the fetal heartbeat is loud enough to be heard when a physician applies a stethoscope to the mother's abdomen. By the end of this month, the fetus is about 152 mm (6 inches) in length and weighs about 171 grams (6 oz).

During the third and fourth months, it is obvious that the skeleton is becoming ossified. The sex of the individual can now be distinguished.

The Fifth Through Seventh Months

During the fifth through seventh months, the mother begins to feel movement. At first, there is only a fluttering sensation, but as the fetal legs grow and develop, kicks and jabs are felt. The fetus, though, is in the fetal position, with the head bent down and in contact with the flexed knees.

The wrinkled, translucent, pink-colored skin is covered by a fine down called **lanugo.** This in turn is coated with a white, greasy, cheeselike substance called **vernix caseosa,** which probably protects the delicate skin from the amniotic fluid. The eyelids are now fully open, however.

At the end of this period, the length has increased to about 300 mm (12 inches), and the weight is now about 1,380 grams (3 lb). It is possible that if born now, the baby will survive.

Fetal Circulation

The fetus has circulatory features that are not present in the adult circulation (Fig. 22.15). All of these features can be related to the fact that the fetus does not use its lungs for gas exchange. For example, much of the blood entering the right atrium is shunted into the left atrium through the **oval opening** (foramen ovale) between the two atria. Also, any blood that does enter the right ventricle and is pumped into the pulmonary trunk is shunted into the aorta by way of the **arterial duct** (ductus arteriosus).

Blood within the aorta travels to the various branches, including the iliac arteries, which connect to the **umbilical arteries** leading to the placenta. Exchange between maternal blood and fetal blood takes place at the placenta. The **umbilical vein** carries blood rich in nutrients and oxygen to the fetus. The umbilical vein enters the liver and then joins the **venous duct,** which merges with the inferior vena cava, a vessel that returns blood to the heart. It is interesting to note that the umbilical arteries and vein run alongside one another in the umbilical cord, which is cut at birth, leaving only the umbilicus (navel).

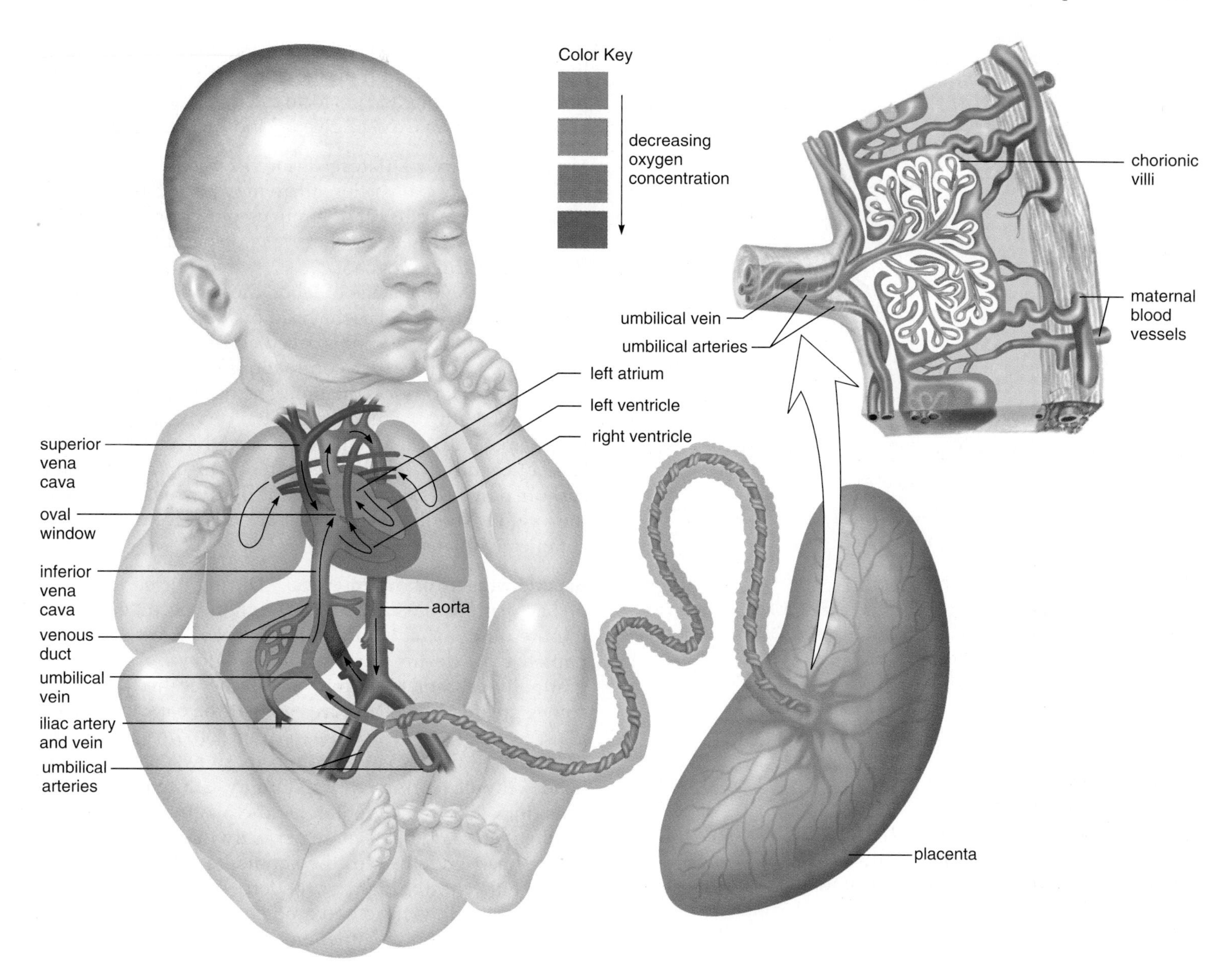

Figure 22.15 Fetal circulation and the placenta.
The lungs are not functional in the fetus, and the blood passes directly from the right atrium to the left atrium or from the right ventricle to the aorta. The umbilical arteries take fetal blood to the placenta where exchange of molecules between fetal and maternal blood takes place across the walls of the chorionic villi. Oxygen and nutrient molecules diffuse into the fetal blood, and carbon dioxide and urea diffuse from fetal blood. The umbilical vein returns blood from the placenta to the fetus.

The most common of all cardiac defects in the newborn is the persistence of the oval opening. With the tying of the cord and the expansion of the lungs, blood enters the lungs in quantity. Return of this blood to the left side of the heart usually causes a flap to cover the opening. Incomplete closure occurs in nearly one out of four individuals, but even so, passage of the blood from the right atrium to the left atrium rarely occurs because either the opening is small or it closes when the atria contract. In a small number of cases, the passage of impure blood from the right side to the left side of the heart is sufficient to cause a "blue baby." Such a condition now can be corrected by open heart surgery.

The arterial duct closes because endothelial cells divide and block off the duct. Remains of the arterial duct and parts of the umbilical arteries and vein later are transformed into connective tissue.

Fetal circulation shunts blood away from the lungs, toward and away from the placenta within the umbilical blood vessels located within the umbilical cord. Exchange of substances between fetal blood and maternal blood takes place at the placenta, which forms from the chorion, an extraembryonic membrane, and uterine tissue.

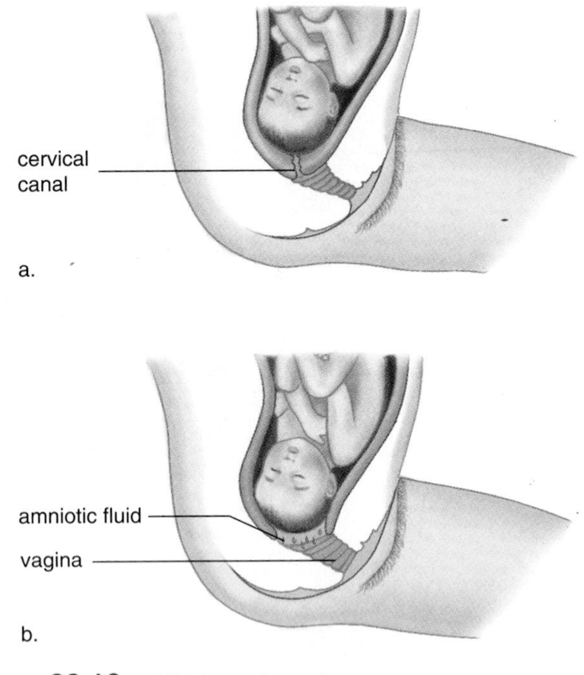

Figure 22.16 Dilation of cervix.
During the first stage of birth, the cervical canal slowly disappears as the head pushes on the lower part of the uterus. **a.** Cervix nondilated. **b.** Cervix dilated and "water breaks."

Birth Has Three Stages

At the end of nine months, the fetus is about 525 mm (20 ½ inches) long and weighs about 3,380 grams (7 ½ lb). Weight gain is largely due to the accumulation of fat beneath the skin.

As the time of birth approaches, the fetus rotates so that the head is pointed toward the cervix. If the fetus does not turn, then the likelihood of a breech birth (rump first) may call for a cesarean section. It is very difficult for the cervix to expand enough to accommodate this form of birth, and asphyxiation of the baby is more likely.

The uterus characteristically contracts throughout pregnancy. During the early months of pregnancy, light contractions that last about 20–30 seconds and occur every 15–20 minutes are often indiscernible. Near the end of pregnancy, contractions become stronger and more frequent, so that the woman may falsely think that she is in labor. The onset of true labor is marked by uterine contractions that occur regularly every 15–20 minutes and last for 40 seconds or more. **Parturition,** which includes labor and expulsion of the fetus, usually is considered to have three stages. During the first stage of parturition, the cervix dilates; during the second, the infant is born; and during the third, the afterbirth is expelled.

The events that cause parturition still are not known entirely, but there is now evidence suggesting the involvement of prostaglandins. It may be, too, that the prostaglandins cause the release of oxytocin from the mother's posterior pituitary. Both prostaglandins and oxytocin cause the uterus to contract, and either hormone can be given to induce parturition.

First Stage: Cervix Dilates. Prior to, or at the same time as, the first stage of parturition, there may be a "bloody show" caused by the expulsion of a mucus plug from the cervical canal. This plug prevents bacteria and sperm from entering the uterus during pregnancy.

During the first stage of labor, the cervical canal slowly disappears (Fig. 22.16) as the head pushes on the lower part of the uterus. This process is called effacement, or taking up the cervix. With further contractions, the head acts as a wedge to assist cervical dilation. The head usually has a diameter of about 10 cm; therefore, the cervix has to dilate to this diameter in order to allow the head to pass through. If it has not occurred already, the amniotic membrane is apt to rupture now, releasing the amniotic fluid, which escapes out the vagina. The first stage of labor ends once the cervix is completely dilated.

Second Stage: Infant Emerges. During the second stage of parturition, the uterine contractions occur every 1–2 minutes and last about one minute each. They are accompanied by a desire to push, or bear down. As the head gradually descends into the vagina, the desire to push becomes greater. The head turns so that the back of the head is uppermost when it appears (Fig. 22.17*a–c*). Since the vagina may not expand enough to allow passage of the head without tearing, an **episiotomy** is often performed. This incision, which enlarges the vaginal opening, is stitched later and heals more perfectly than a tear. As soon as the head is delivered, the shoulders rotate so that the face is either to the right or the left. The physician at this time may hold the head and guide it downward, while one shoulder and then the other emerges. The rest of the body follows easily.

Once the infant is breathing normally, the umbilical cord is cut and tied, severing the child from the placenta. The stump of the cord shrivels and leaves a scar, which is the navel.

Third Stage: Expelling Afterbirth. The placenta, or afterbirth, is delivered during the third stage of labor (Fig. 22.17*d*). About 15 minutes after delivery of the infant, uterine muscle contractions shrink the uterus and dislodge the placenta. This also helps to "close off" the mother's blood vessels that service the placenta, to prevent severe hemorrhaging. The placenta is then expelled into the vagina. As soon as the placenta and its membranes are delivered, the third stage of labor is complete.

During the first stage of birth, the cervix dilates; during the second, the infant is delivered; and during the third, the afterbirth is expelled.

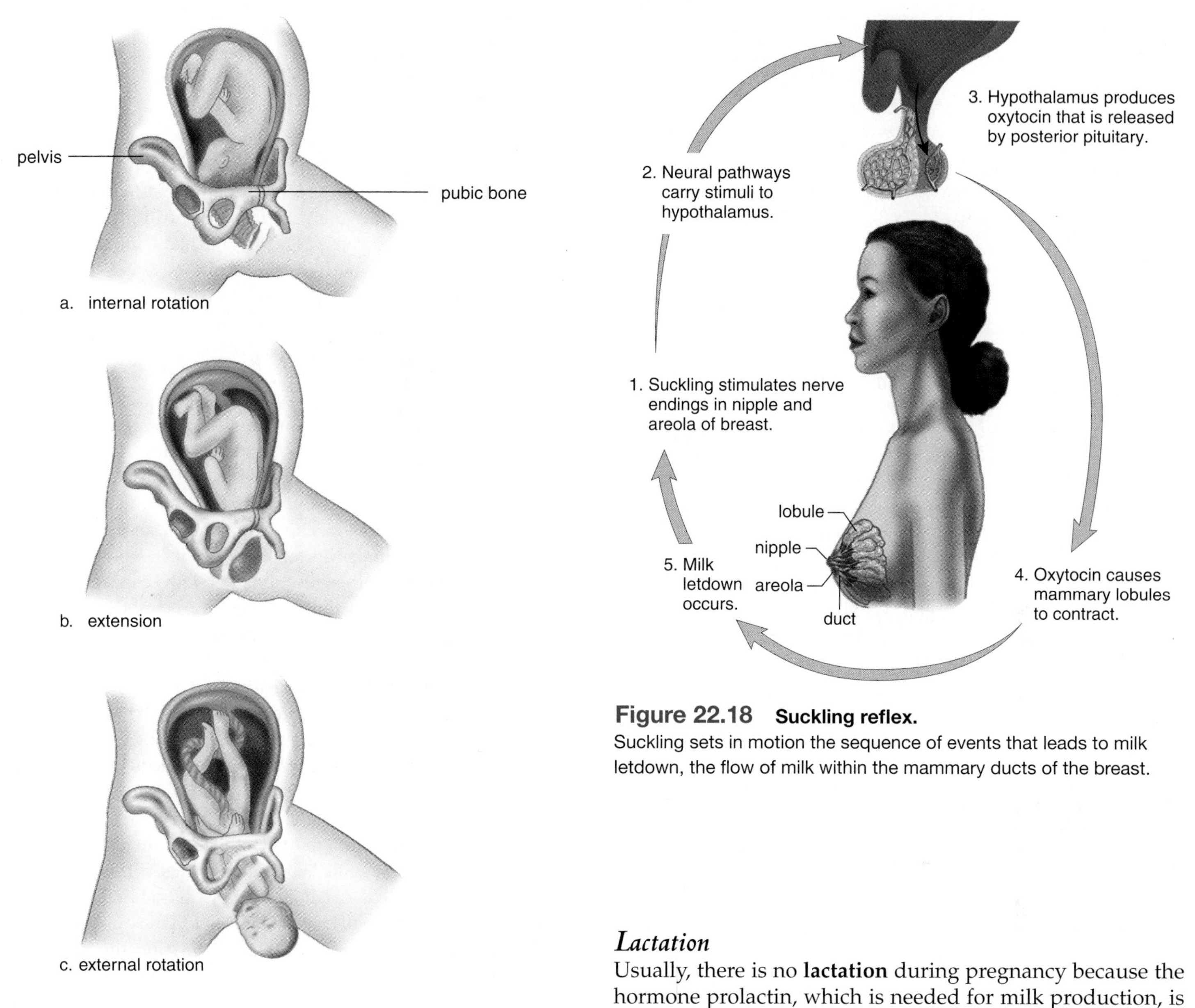

Figure 22.18 Suckling reflex.
Suckling sets in motion the sequence of events that leads to milk letdown, the flow of milk within the mammary ducts of the breast.

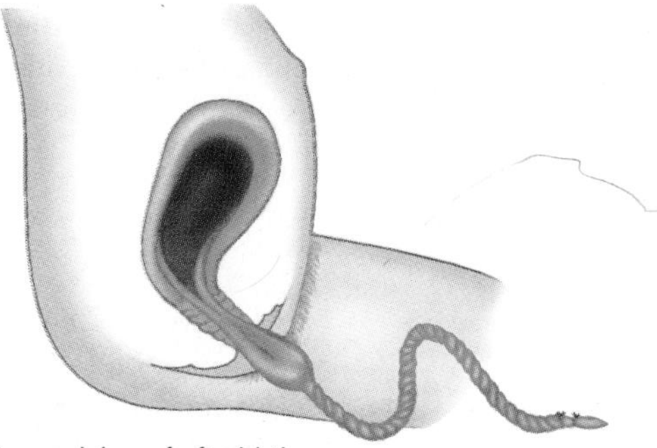

Figure 22.17 Delivery of infant and afterbirth.
a.–c. During the second stage of birth, the back of the head appears first, and then the shoulders rotate so that the body turns. **d.** During the third stage of birth, the placenta and amniotic sac are expelled from the vagina. These drawings show the woman sitting because a prone position for childbirth is not recommended by many.

Lactation

Usually, there is no **lactation** during pregnancy because the hormone prolactin, which is needed for milk production, is not being produced. The negative feedback exerted by high levels of estrogen and progesterone during pregnancy suppress anterior pituitary production not only of these hormones but also prolactin as well. Once birth occurs, however, the pituitary begins secreting prolactin. It takes a couple of days for milk production to begin; in the meantime the breasts produce **colostrum,** which is a thin yellowish fluid rich in protein and antibodies.

The **suckling reflex** is required for **milk letdown** to occur (Fig. 22.18). As discussed previously, the female breast contains lobules consisting of ducts and alveoli. The alveoli are lined by milk-producing cells in the lactating breast. When a breast is suckled, the nerve endings in the nipple and areola, which is a circular area of pigmented skin about the nipple, are stimulated. Nerve impulses travel along neural pathways from the breast to the hypothalamus, which then directs the posterior pituitary to release the hormone oxytocin. This hormone causes mammary lobules

Health Focus

Deciding Between Feeding by Breast or by Bottle

Breast-feeding has come back in vogue in a big way. Not 20 years ago, only one mother in five breast-fed her baby. Today, an estimated two out of every three newborns are being given breast milk. The shift has many advantages, not just for the infant but for the mother as well. One is that when the baby is ready to eat, the milk is ready for the drinking. There are no bottles to sterilize or formulas to measure. Breast milk is also less expensive than formula, even after you account for the cost of the extra food a mother needs to ensure that she will produce enough milk and that it will contain enough calories. In addition, hormones released during breast-feeding cause contractions in the uterus that assist it in shrinking closer to its prepregnancy size. And the calories a mother's body spends producing breast milk may help her to lose the 7 to 8 pounds—or more—of fat she puts on during pregnancy but does not "deliver" along with the baby, placenta, and amniotic and other fluids.

As for the baby, "human milk is unquestionably the best source of nutrition . . . during the first months of life," says the American Academy of Pediatrics. That's especially true if conditions in the home are unsanitary (breast milk, unlike formula, does not need to be kept "clean," because it goes directly from mother to child), or money is too scarce to ensure that formula will always be affordable, or the educational level of the parents is too low (or the emotional level of the household too "chaotic") for the family to read, fully understand, and consistently apply the rules of proper formula preparation and storage. But even fullterm infants born to reasonably well-educated, financially secure parents who live in clean environments may benefit.

It isn't that today's formulas are not reliable. Healthy babies can grow well on breast or bottle. But breast milk is an amazingly sophisticated substance (it contains more zinc in the first few weeks than later on to meet the newborn's higher zinc needs). And although scientists have been able to imitate it safely enough in store-bought formula, "there may still be some subtle, not-yet-discovered ways in which breast milk is preferable to bottle milk," says Ronald Kleinman, M.D., head of the Committee on Nutrition at the American Academy of Pediatrics. "There remain differences between the two," he adds, "whose significance isn't understood."

These differences may be behind such findings as one reported in the *British Medical Journal* that breast-fed infants appear better protected against wheezing during the first few months of life. The journal has also published a report suggesting that breast-fed newborns are less prone to develop stomach and intestinal illness during the first 13 weeks of life, and thus, suffer less than bottle-fed babies from vomiting and diarrhea.

Can the Bottle Ever Be A Better Choice?

Despite the many advantages of breast-feeding, it may not always be the appropriate way for a mother to nourish her child. Some women simply do not want to breast-feed because they will return to work soon after giving birth and would find "pumping" breast milk to be given to the infant while they are away from home too tiring. Others may not wish to breast-feed because they are uncomfortable with the sexuality of the process or with some other emotionally related aspect. All these reasons are considered valid. In other words, mothers who do not want to breast-feed should not be pressured or made to feel guilty about it. As family therapist and dietitian Ellyn Satter says in her book *Child of Mine: Feeding with Love and Good Sense* (Bull Publishing: Palo Alto, California), "You will have plenty of opportunities to feel guilty as a parent without feeling guilty about *that,* too." Besides, if a mother breast-feeds but hates doing it, her baby is going to sense that, and it will do greater harm than the breast milk will do good.

Indeed, more important than whether a mother breast-feeds, especially in countries like the United States, where the standard of living for the average family is such that breast-feeding is not a life-or-death matter as it is in some developing nations, is that she develops a relaxed, loving relationship with her child. "Even the sophisticated components of breast milk can't make up for that," Ms. Satter says. The closeness, warmth, and stimulation provided by an infant's caretakers are as important to his normal growth and development as the source of his food.

Source: *Tufts University Diet and Nutrition Letter* (ISSN 0747-4105) is published monthly by Tufts University Diet and Nutrition Letter, 53 Park Place, New York, NY 10007. This article extracted from a Special Report in the December 1990 issue. Reprinted by permission.

within a breast to contract and milk to flow (let down) into mammary ducts where it may be drawn out at the nipple by a suckling infant. The suckling reflex is required for lactation to continue; if suckling does not occur, milk production ceases. Also, the more suckling, the more oxytocin released, and the more milk there is for the child.

In addition to initiating milk letdown, oxytocin also causes the uterus to contract and this helps return the uterus to its nonpregnant state more quickly. Some women choose to breast-feed and some do not. The reading on this page discusses the benefits of both feeding methods.

Under the influence of the hormone prolactin, lactation begins after a child is born. The hormone oxytocin is involved in the suckling reflex which brings about milk letdown.

22.4 Human Development after Birth

Development does not cease once birth has occurred but continues throughout the stages of life: infancy, childhood, adolescence, and adulthood. **Aging** encompasses these progressive changes that contribute to an increased risk of infirmity, disease, and death (Fig. 22.19).

Today, there is great interest in **gerontology,** the study of aging, because there are now more older individuals in our society than ever before, and the number is expected to rise dramatically. In the next half-century, those over age 75 will rise from the present 13 million to 34–45 million, and those over age 80 will rise from 3 million to 6 million individuals. The human life span is judged to be a maximum of 110–115 years. The present goal of gerontology is not necessarily to increase the life span, but to increase the health span, the number of years that an individual enjoys the full functions of all body parts and processes.

Figure 22.19 Aging.
Aging is a slow process during which the body undergoes changes that eventually bring about death, even if no marked disease or disorder is present.

Theories of Aging

There are many theories about what causes aging. Three of these are considered here.

Genetic in Origin

Several lines of evidence indicate that aging has a genetic basis. (1) The number of times a cell divides is species-specific. The maximum number of times human cells divide is around 50. Perhaps as we grow older, more and more cells are unable to divide, and instead, they undergo degenerative changes and die. (2) Some cell lines may become nonfunctional long before the maximum number of divisions has occurred. Whenever DNA replicates, mutations can occur, and this can lead to the production of nonfunctional proteins. Eventually, the number of inadequately functioning cells can build up, which contributes to the aging process. (3) The children of long-lived parents tend to live longer than those of short-lived parents. Recent work suggests that when an animal produces fewer free radicals, it lives longer. Free radicals are unstable molecules that carry an extra electron. In order to stabilize themselves, free radicals donate an electron to another molecule like DNA or proteins (e.g., enzymes) or lipids found in plasma membranes. Eventually these molecules are unable to function and the cell is destroyed. There are genes that code for antioxidant enzymes that detoxify free radicals. This research suggests that animals with particular forms of these genes—and therefore more efficient antioxidant enzymes—live longer.

Whole-Body Process

A decline in the hormonal system can affect many different organs of the body. For example, type II diabetes is common in older individuals. The pancreas makes insulin, but the cells lack the receptors that enable them to respond. Menopause in women occurs for a similar reason. There is plenty of FSH in the bloodstream, but the ovaries do not respond. Perhaps aging results from the loss of hormonal activities and a decline in the functions they control.

The immune system, too, no longer performs as it once did, and this can affect the body as a whole. The thymus gland gradually decreases in size, and eventually most of it is replaced by fat and connective tissue. The incidence of cancer increases among the elderly, which may signify that the immune system is no longer functioning as it should. This idea is substantiated, too, by the increased incidence of autoimmune diseases in older individuals.

It is possible, though, that aging is not due to the failure of a particular system that can affect the body as a whole, but to a specific type of tissue change that affects all organs and even the genes. It has been noticed for some time that proteins—such as collagen, which makes up the white fibers and is present in many support tissues—become increasingly cross-linked as people age. Undoubtedly, this cross-linking contributes to the stiffening and the loss of elasticity characteristic of aging tendons and ligaments. It may also account for the inability of such organs as the blood vessels, the heart, and the lungs to function as they once did. Some researchers have now found that glucose has the tendency to attach to any type of protein, which is the first step in a cross-linking process. They are presently experimenting with drugs that can prevent cross-linking.

Extrinsic Factors

The current data about the effects of aging are often based on comparisons of the characteristics of the elderly to younger age groups; but perhaps today's elderly were not as aware when they were younger of the importance of, for example, diet and exercise to general health. It is possible, then, that much of what we attribute to aging is instead due to years of poor health habits.

Consider, for example, osteoporosis. This condition is associated with a progressive decline in bone density in both males and females so that fractures are more likely to occur after only minimal trauma. Osteoporosis is common in the elderly—by age 65, one-third of women will have vertebral fractures, and by age 81, one-third of women and one-sixth of men will have suffered a hip fracture. While there is no denying that a decline in bone mass occurs as a result of aging, certain extrinsic factors are also important. The occurrence of osteoporosis itself is associated with cigarette smoking, heavy alcohol intake, and perhaps inadequate calcium intake. Not only is it possible to eliminate these negative factors by personal choice, it is also possible to add a positive factor. A moderate exercise program has been found to slow down the progressive loss of bone mass.

Even more important, a proper diet that includes at least five servings of fruits and vegetables a day and a sensible exercise program will most likely help eliminate cardiovascular disease, the leading cause of death today. Experts no longer believe that the cardiovascular system necessarily suffers a large decrease in functioning ability with age. Persons 65 years of age and older can have well-functioning hearts and open coronary arteries if their health habits are good and they continue to exercise regularly.

Effect of Age on Body Systems

Data about how aging affects body systems should be accepted with reservations.

Skin

As aging occurs, skin becomes thinner and less elastic because the number of elastic fibers decreases and the collagen fibers undergo cross-linking, as discussed previously. Also, there is less adipose tissue in the subcutaneous layer; therefore, older people are more likely to feel cold. The loss of thickness accounts for the sagging and wrinkling of the skin.

Homeostatic adjustment to heat is also limited because there are fewer sweat glands for sweating to occur. There are fewer hair follicles, so the hair on the scalp and the extremities thins out. The number of oil (sebaceous) glands is reduced, and the skin tends to crack. Older people also experience a decrease in the number of melanocytes, making hair gray and skin pale. In contrast, some of the remaining pigment cells are larger, and pigmented blotches appear in skin.

Processing and Transporting

Cardiovascular disorders are the leading cause of death among the elderly. The heart shrinks because there is a reduction in cardiac muscle cell size. This leads to loss of cardiac muscle strength and reduced cardiac output. Still, it is observed that the heart, in the absence of disease, is able to meet the demands of increased activity. It can increase its rate to double or triple the amount of blood pumped each minute even though the maximum possible output declines.

Because the middle layer of arteries contains elastic fibers, which most likely are subject to cross-linking, the arteries become more rigid with time, and their size is further reduced by plaque, a buildup of fatty material. Therefore, blood pressure readings gradually rise. Such changes are common in individuals living in Western industrialized countries but not in agricultural societies. A low cholesterol and saturated fatty acid diet has been suggested as a way to control degenerative changes in the cardiovascular system.

There is reduced blood flow to the liver, and this organ does not metabolize drugs as efficiently as before. This means that as a person gets older, less medication is needed to maintain the same level in the bloodstream.

Cardiovascular problems often are accompanied by respiratory disorders, and vice versa. Growing inelasticity of lung tissue means that ventilation is reduced. Because we rarely use the entire vital capacity, these effects are not noticed unless there is increased demand for oxygen.

There is also reduced blood supply to the kidneys. The kidneys become smaller and less efficient at filtering wastes. Salt and water balance are difficult to maintain, and the elderly dehydrate faster than young people. Difficulties involving urination include incontinence (lack of bladder control) and the inability to urinate. In men, the prostate gland may enlarge and reduce the diameter of the urethra, making urination so difficult that surgery is often needed.

The loss of teeth, which is frequently seen in elderly people, is more apt to be the result of long-term neglect than aging. The digestive tract loses tone, and secretion of saliva and gastric juice is reduced, but there is no indication of reduced absorption. Therefore, an adequate diet, rather than vitamin and mineral supplements, is recommended. There are common complaints of constipation, increased amount of gas, and heartburn, and gastritis, ulcers, and cancer can also occur.

Integration and Coordination

It is often mentioned that while most tissues of the body regularly replace their cells, some at a faster rate than others, the brain and the muscles do not. No new nerve or skeletal muscle cells are formed in the adult. However, contrary to previous opinion, recent studies show that few neural cells of the cerebral cortex are lost during the normal aging process. This means that cognitive skills remain unchanged even though there is characteristically a loss in short-term memory. Although the elderly learn more slowly than the young, they can acquire and remember new material as well. It is noted that when more time is given for the subject to respond, age differences in learning decrease.

Neurons are extremely sensitive to oxygen deficiency, and if neuron death does occur, it may be due not to aging itself but to reduced blood flow in narrowed blood vessels. Specific disorders, such as depression, Parkinson disease, and Alzheimer disease, are sometimes seen, but they are not common. Reaction time, however, does slow, and more stimulation is needed for hearing, taste, and smell receptors to function as before. After age 50, there is a gradual reduction in the ability to hear tones at higher frequencies, and this can make it difficult to identify individual voices and to understand conversation in a group. The lens of the eye does not accommodate as well and also may develop a cataract. Glaucoma is more likely to develop because of a reduction in the size of the anterior cavity of the eye.

Loss of skeletal muscle mass is not uncommon, but it can be controlled by a regular exercise program. There is a reduced capacity to do heavy labor, but routine physical work should be no problem. A decrease in the strength of the respiratory muscles and inflexibility of the rib cage contribute to the inability of the lungs to expand as before, and reduced muscularity of the urinary bladder contributes to difficulties with urination.

As noted before, aging is accompanied by a decline in bone density. Osteoporosis, characterized by a loss of calcium and mineral from bone, is not uncommon, but there is evidence that proper health habits can prevent its occurrence. Arthritis, which causes pain upon movement of the joint, is also seen.

Weight gain occurs because the basal metabolism decreases and inactivity increases. Muscle mass is replaced by stored fat and retained water.

The Reproductive System

Females undergo menopause, and thereafter the level of female sex hormones in blood falls markedly. The uterus and the cervix are reduced in size, and there is a thinning of the walls of the oviducts and the vagina. The external genitals become less pronounced. In males, the level of androgens falls gradually over the age span of 50–90, but sperm production continues until death.

It is of interest that as a group, females live longer than males. Although their health habits may be better, it is also possible that the female sex hormone estrogen offers women some protection against cardiovascular disorders when they are younger. Males suffer a marked increase in heart disease in their forties, but an increase is not noted in females until after menopause, after which women lead men in the incidence of stroke. Men are still more likely than women to have a heart attack, however.

Figure 22.20 **Increasing the health span.**
The aim of gerontology is to allow the elderly to enjoy living. This requires studying the debilities that can occur with aging and then making recommendations as to how best to forestall or prevent their occurrence.

Conclusion

We have listed many adverse effects due to aging, but it is important to emphasize that while such effects are seen, they are not a necessary occurrence (Fig. 22.20). We must discover any extrinsic factors that precipitate these adverse effects and guard against them. Just as it is wise to make the proper preparations to remain financially independent when older, it is also wise to realize that biologically successful old age begins with the health habits developed when we are younger.

Bioethical Issue

The fetus is subject to harm by maternal use of medicines and drugs of abuse, including nicotine and alcohol. Also, various sexually transmitted diseases, notably an HIV infection, can be passed on to the fetus by way of the placenta. Women need to be aware of the need to protect their unborn child from harm. Indeed, their behavior should be protective if they are sexually active, even if they are using a recognized form of birth control. Harm can occur before a woman realizes she is pregnant!

Because we are now aware of the need for maternal responsibility *before* a child is born, there has been a growing acceptance of prosecuting women when a newborn has a condition such as fetal alcohol syndrome, a condition that can only be caused by the drinking habits of the mother. Employers have also become aware that they might be subject to prosecution. To protect themselves, Johnson Controls, a U.S. battery manufacturer, developed a fetal protection policy. No woman who could bear a child was offered a job that might expose her to toxins that could negatively affect the development of her baby. To get such a job, a woman had to show that she had been sterilized or was otherwise incapable of having children. In 1991, the U.S. Supreme Court declared this policy unconstitutional on the basis of sexual discrimination. The decision was hailed as a victory for women, but was it? The decision was written in such a way that women alone, and not an employer, are responsible for any harm done to the fetus by workplace toxins.

Some have noted that prosecuting women for causing prenatal harm can itself have a detrimental effect. The women may tend to avoid prenatal treatment, thereby increasing the risk to their children. Or they may opt for an abortion in order to avoid the possibility of prosecution. The women feel they are in a no-win situation. If they have a child that has been harmed due to their behavior, they feel they are bad mothers or if they abort, they feel they are also bad mothers.

Questions

1. Do you believe women should be prosecuted if their child is born with a preventable condition? Why or why not?
2. Is the woman or physician responsible when a women of child-bearing age takes a prescribed medication that does harm to the unborn? Is the employer or the woman responsible when a workplace toxin does harm to an unborn?
3. Should sexually active women who can bear a child be expected to avoid substances or situations that could possibly harm an unborn even if they are using birth control? Why or why not?

Summarizing the Concepts

22.1 Early Developmental Stages

During fertilization, the acrosome of a sperm releases enzymes that digest a hole in the jelly coat around the egg and then extrudes a filament that attaches to a receptor on the vitelline membrane. The sperm nucleus enters the egg and fuses with the egg nucleus.

During the early developmental stages, cleavage leads to a morula, which becomes the blastula when an internal cavity (the blastocoel) appears. Then, at the gastrula stage, invagination of cells into the blastocoel results in formation of the germ layers: ectoderm, mesoderm, and endoderm. During neurulation, the nervous system develops from midline ectoderm, just above the notochord. At this point, it is possible to draw a typical cross section of a vertebrate embryo (see Fig. 22.5).

22.2 Differentiation and Morphogenesis

Differentiation begins with cleavage, when the egg's cytoplasm is partitioned among the numerous cells. In a frog embryo, only a daughter cell that receives a portion of the gray crescent is able to develop into a complete embryo. Morphogenesis involves the process of induction, as when the notochord induces the formation of the neural tube in frog embryos. Today we envision that cells are constantly giving off signals that influence the genetic activity of neighboring cells.

22.3 Human Embryonic and Fetal Development

Human development can be divided into embryonic development (months 1 and 2) and fetal development (months 3–9). The extraembryonic membranes, including the chorion, amnion, yolk sac, and allantois, appear early in human development. In humans, fertilization occurs in an oviduct, and cleavage occurs as the embryo moves toward the uterus. The morula becomes the blastocyst before implanting in the uterine lining.

Organ development begins with neural tube and heart formation. There follows a steady progression of organ formation during embryonic development. During fetal development, refinement of features occurs, and the fetus adds weight. Birth occurs about 280 days after the start of the mother's last menstruation.

22.4 Human Development after Birth

Development after birth consists of infancy, childhood, adolescence, and adulthood. Young adults are at their prime, and then the aging process begins. Aging may be due to cellular repair changes, which are genetic in origin. Other factors that may affect aging are changes in body processes and certain extrinsic factors.

Studying the Concepts

1. Name the germ layers, and state organs derived from each. 445
2. Compare the process of cleavage and the formation of the blastula and gastrula in lancelets, frogs, and chicks. 446–47
3. Draw a cross section of a typical vertebrate embryo at the neurula stage, and label your drawing. 447
4. Give reasons for suggesting that differentiation begins with the embryonic stage of cleavage. 448
5. Describe an experiment that helped investigators conclude that the notochord induces formation of the neural plate. What is the significance of induction in general? 449
6. What are homeotic genes, and what do they do? 450
7. List the human extraembryonic membranes, give a function for each, and compare their functions to those in the chick. 451
8. Tell where fertilization, cleavage, the morula stage, and the blastocyst stage occur in humans. What does the blastocyst do when it arrives in the uterus? 452–53
9. Describe the structure and the function of the placenta in humans. 455
10. Describe fetal development in humans. 458
11. Trace the path of blood in the fetus from the right atrium to the aorta, using two different routes. 458–59
12. Describe the three stages of birth. What is the suckling reflex? 460–62
13. Discuss three theories of aging. What are the major changes in the body systems that have been observed as adults age? 463–64

Testing Yourself

Choose the best answer for each question.

1. Which of these stages is the first one out of sequence?
   a. cleavage
   b. blastula
   c. morula
   d. gastrula
   e. neurula
2. Which of these stages is mismatched?
   a. cleavage—cell division
   b. blastula—gut formation
   c. morula—ball of cells
   d. gastrula—three germ layers
   e. neurula—nervous system
3. Which of these germ layers is best associated with development of the heart?
   a. ectoderm
   b. mesoderm
   c. endoderm
   d. epidermis
   e. All of these are correct.
4. Differentiation begins at what stage?
   a. cleavage
   b. blastula
   c. morula
   d. gastrula
   e. neurula
5. Morphogenesis is best associated with
   a. overall growth.
   b. induction of one tissue by another.
   c. genetic mutations.
   d. apoptosis.
   e. Both b and d are correct.
6. In humans, the placenta develops from the chorion. This indicates that human development
   a. resembles that of the chick.
   b. is associated with extraembryonic membranes.
   c. cannot be compared to lower animals.
   d. is like that of nonplacental animals.
   e. begins only upon implantation.
7. In humans, the fetus
   a. is surrounded by four extraembryonic membranes.
   b. has developed organs and is recognizably human.
   c. is dependent upon the placenta for excretion of wastes and acquisition of nutrients.
   d. is no larger than the morula.
   e. Both b and c are correct.
8. Developmental changes
   a. require growth, differentiation, and morphogenesis.
   b. stop occurring when one is grown.
   c. are dependent upon a parceling out of genes into daughter cells.
   d. have no set sequence.
   e. Both a and c are correct.
9. Which of these is mismatched?
   a. brain—ectoderm
   b. gut—endoderm
   c. bone—mesoderm
   d. blood—mesoderm
   e. lens—endoderm
10. Fetal circulation
    a. bypasses the lungs by the presence of the oval opening between the atria.
    b. bypasses the placenta because the umbilical vein is connected to the umbilical artery.
    c. bypasses the heart by the presence of the venous duct.
    d. circulates maternal blood from the placenta to the heart as a way to provide the fetus with oxygen.
    e. All of these are correct.
11. Label this diagram illustrating the placement of the extraembryonic membranes, and give a function for each membrane in humans:

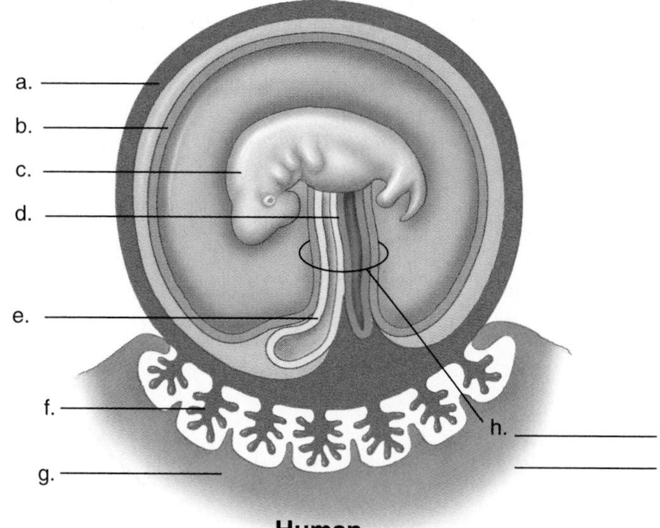

Thinking Scientifically

1. With the help of these questions, develop a scenario to explain why a particular type of cell gives off particular signals (page 448):
   a. What might be the effect on genes when a cell inherits a certain cytoplasmic composition?
   b. What do activated genes do?
   c. How might some of these proteins act?
2. With the help of these questions, develop a scenario to explain why embryonic development is so orderly.
   a. Tissue A has just become differentiated. What does it give off to affect tissue B?
   b. Having received a certain signal, what does tissue B do to affect tissue C?
   c. Having received a certain signal, what does tissue C do?

Understanding the Terms

aging 463
allantois 451
amnion 451
archenteron 445
arterial duct 458
blastocoel 445
blastocyst 452
blastopore 445
blastula 445
cellular differentiation 448
chorion 451
chorionic villi 455
cleavage 445
colostrum 461
ectoderm 445
embryo 445
embryonic disk 453
embryonic period 451
endoderm 445
episiotomy 460
extraembryonic membrane 451
fertilization 444
fontanel 458
gastrula 445
gastrulation 445
germ layer 445
gerontology 463
gray crescent 448
homeobox 450
human chorionic gonadotropin (HCG) 452
implantation 451
induction 449
lactation 461
lanugo 458
mesoderm 445
milk letdown 461
morphogenesis 448
morula 445
neural fold 447
neural plate 447
neural tube 447
neurula 447
notochord 447
ooplasmic segregation 448
oval opening 458
parturition 460
pattern formation 450
placenta 455
primitive streak 447
suckling reflex 461
trophoblast 452
umbilical artery 458
umbilical cord 454
umbilical vein 458
venous duct 458
vernix caseosa 458
vertebrate 447
yolk 444
yolk sac 451

Match the terms to these definitions:

a. __________ Process by which a chemical or a tissue influences the development of another tissue.

b. __________ Primary tissue layer of a vertebrate embryo; namely ectoderm, mesoderm, or endoderm.

c. __________ Movement of early embryonic cells to establish body outline and form.

d. __________ Structure formed during the development of placental mammals from the chorion and the uterine wall that allows the embryo, and then the fetus, to acquire nutrients and rid itself of wastes.

e. __________ Cell division without cytoplasmic addition or enlargement; occurs during first stage of development.

Using Technology

Your study of development is supported by these available technologies:

Essential Study Partner CD-ROM

Genetics → Development

Visit the Mader web site for related ESP activities.

Exploring the Internet

The Mader Home Page provides resources and tools as you study this chapter.

http://www.mhhe.com/biosci/genbio/mader

HealthQuest CD-ROM

5 Cardiovascular Health → Gallery → Leading Causes of Death

CHAPTER 23

Patterns of Gene Inheritance

The genes we inherit from our parents determine our characteristics, including our physical traits and even our behavior.

Chapter Concepts

The newborn wiggled and squirmed about, but they managed to collect a blood sample to test for PKU, an inborn error of metabolism. Mary and Jim, the infant's parents, were normal, but Mary was still anxious because her aunt had been mentally retarded due to PKU, the inability to metabolize an amino acid called phenylalanine. When abnormal breakdown products accumulate, the brain cannot go on to develop normally after birth. Later that day, the pediatrician reported that although the infant had tested positive for PKU, it was only necessary to keep him on a special diet for several years. The diet consists of a special formula and cereal during infancy, and foods very low in protein during childhood.

The study of inheritance most likely began as soon as people noticed the resemblance between parents and offspring. But Gregor Mendel was the first to systematically study inheritance (Fig. 23.1).

Figure 23.1 Mendel working in his garden.
Mendel grew and tended the pea plants he used for his experiments. For each experiment, he observed as many offspring as possible. For a cross that required him to count the number of round seeds to wrinkled seeds, he observed and counted a total of 7,324 peas!

23.1 Mendel's Laws

Gregor Mendel was an Austrian monk who in 1860 developed certain laws of heredity after doing crosses between garden pea plants. His studies form the basis for a particulate model of heredity, which assumes that genes are sections of chromosomes. In Figure 23.2, the letters on the homologous chromosomes stand for genes that control a trait, such as color of hair, type of fingers, or length of nose. The genes are in definite sequence and remain in their spots, or loci, on the chromosomes. Alternative forms of a gene having the same position on a pair of homologous chromosomes and affecting the same trait are called **alleles.** In Figure 23.2, *G* is an allele of *g,* and vice versa; also, *R* is an allele of *r,* and vice versa. *G* could never be an allele for *R* because *G* and *R* are at different loci.

Mendel's work is described in the reading on the next page. He said that pea plants have two factors for every trait, such as stem length. He observed that one of the factors controlling the same trait can be dominant over the other, which is recessive. For example, he found that a pea plant can be tall even if one factor was for shortness. In Mendel's experiments, a tall pea plant was sometimes the parent of a short plant. Therefore, he reasoned that while the individual plant has two factors for each trait, the gametes (i.e., sperm and egg) contain only one factor for each trait. This is now known as Mendel's law of **segregation.**

The law of segregation states the following:

- Each individual has two factors for each trait;
- The factors segregate (separate) during the formation of the gametes;
- Each gamete contains only one factor from each pair of factors;
- Fertilization gives each new individual two factors for each trait.

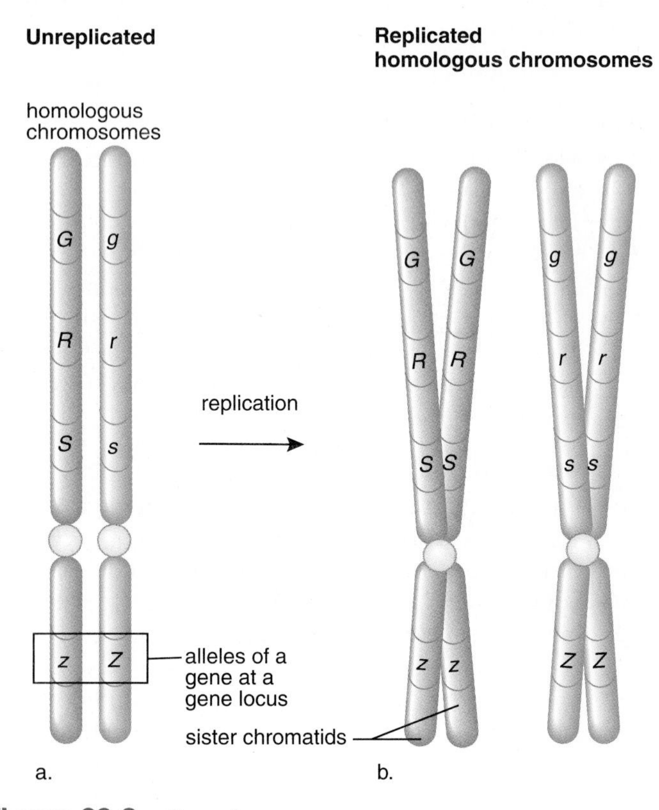

Figure 23.2 Gene locus.
a. Each allelic pair, such as *Gg* or *Zz,* is located on homologous chromosomes at a particular gene locus. **b.** Following replication, each sister chromatid carries the same alleles in the same order.

Science Focus

Gregor Mendel

Mendel's use of pea plants as his experimental material was a good choice, because pea plants are easy to cultivate, have a short generation time, and can be self-pollinated or cross-pollinated at will. Mendel selected certain traits for study and, before beginning his experiments, made sure his parental (P generation) plants bred true—he observed that when these plants self-pollinated, the offspring were like one another and like the parent plant. For example, a parent with yellow seeds always had offspring with yellow seeds; a plant with green seeds always had offspring with green seeds. Then, Mendel cross-pollinated the plants by dusting the pollen of plants with yellow seeds on the stigma of plants with green seeds whose own anthers had been removed, and vice versa (Fig. 23A). Either way, the offspring (called F_1, or first filial generation) resembled the parents with yellow seeds. Mendel then allowed the F_1 plants to self-pollinate. Once he had obtained an F_2 generation, he observed the color of the peas produced. He counted over 8,000 plants and found an approximate 3:1 ratio (about three plants with yellow seeds for every plant with green seeds) in the F_2 generation. Mendel realized that these results were explainable, assuming (a) there are two factors for every trait; (b) one of the factors can be dominant over the other, which is recessive; and (c) the factors separate when the gametes are formed.

He believed that the F_2 plants with yellow seeds carried a dominant factor because his results could be related to the binomial expression $a^2 + 2ab + b^2$. He said if $a = Y$ and $b = y$, then the four F_2 plants were $YY + 2Yy + yy$. And three plants with yellow seeds are expected for every plant with green seeds.

As a test to determine if the F_1 generation was indeed *Yy*, Mendel backcrossed it with the recessive parent, *yy*. His results of 1:1 indicated that he had reasoned correctly. Today, when a one-trait testcross is done, a suspected heterozygote is crossed with the recessive phenotype because this cross gives the best chance of producing a recessive phenotype.

Mendel performed a second series of experiments in which he crossed true-breeding plants that differed in two traits. For example, he crossed plants with yellow, round peas with plants with green, wrinkled peas. The F_1 generation always had both dominant characteristics; therefore, he allowed the F_1 plants to self-pollinate. Among the F_2 generation, he achieved an almost perfect ratio of 9:3:3:1. For example, for every plant that had green, wrinkled peas, he had approximately nine that had yellow, round peas, and so forth. Mendel saw that these results were explainable if pairs of factors separate independently from one another when the gametes form, allowing all possible combinations of factors to occur in the gametes. This would mean that the probability of achieving any two factors together in the F_2 offspring is the product of their chance of occurring separately. Therefore, since the chance of yellow peas was $3/4$ (in a one-trait cross) and the chance of round peas was $3/4$ (in a one-trait cross), the chance of their occurring together was $9/16$, and so forth.

Mendel achieved his success in genetics by studying large numbers of offspring, keeping careful records, and treating his data quantitatively. He showed that the application of mathematics to biology is extremely helpful in producing testable hypotheses.

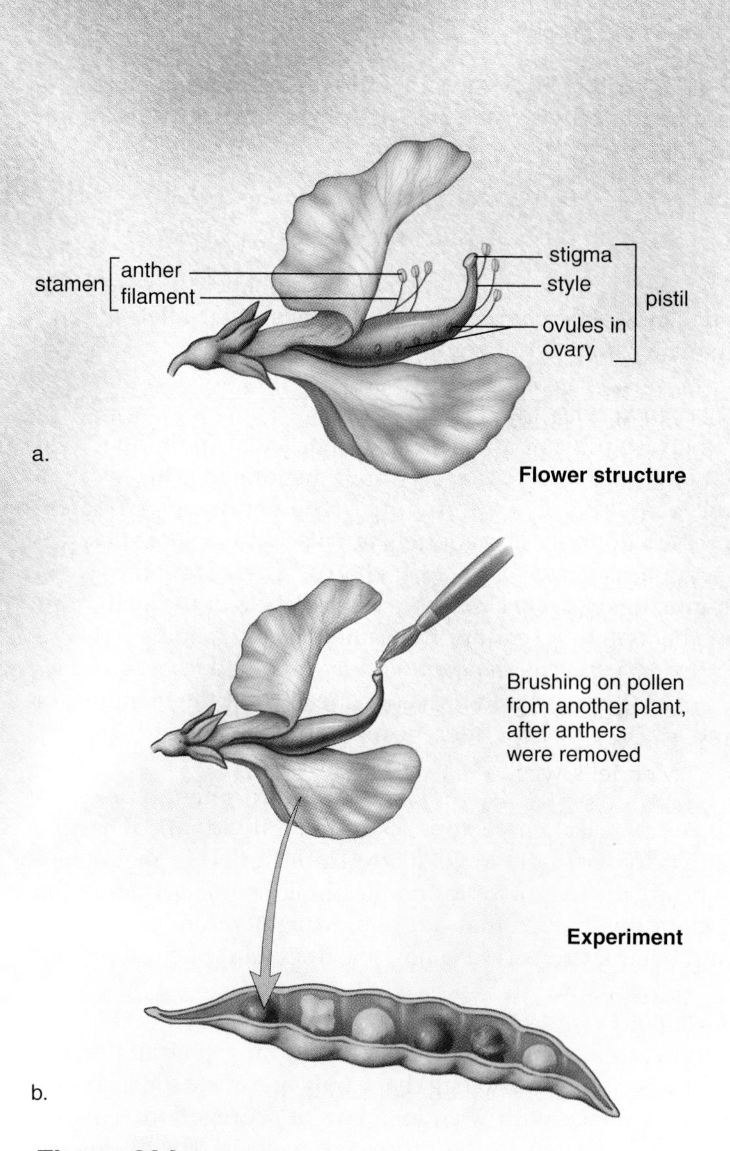

Figure 23A Garden pea anatomy and traits.
a. In the garden pea, pollen grains produced in the anther contain sperm, and ovules in the ovary contain eggs. **b.** When Mendel did crosses he brushed pollen from one plant on the stigma of another plant. After sperm fertilized eggs, the ovules developed into seeds (peas). The open pod shows all the results of a cross between plants with round, yellow seeds and plants with wrinkled, green seeds. Each plant produces just one of these seed types.

The Inheritance of a Single Trait

Mendel suggested the use of letters to indicate factors, now called alleles. A capital letter indicates a **dominant allele** and a lowercase letter indicates a **recessive allele.** The word *dominant* is not meant to imply that the dominant allele is better or stronger than the recessive allele. Dominant means that a certain trait will result if the organism has only one dominant allele. For the recessive trait to result, both alleles must be recessive. Mendel's procedures and laws are applicable not only to peas but to all diploid individuals. Therefore, we now take as our example not peas, but human beings.

Figure 23.3 illustrates the difference between a widow's peak and a straight hairline. In doing a problem concerning hairline, this key is suggested:

W = Widow's peak (dominant allele)
w = Straight hairline (recessive allele)

The key tells us which letter of the alphabet to use for the gene in a particular problem. It also tells which allele is dominant, a capital letter signifying dominance.

The Genotype and Phenotype

When we indicate the genes of a particular individual, two letters must be used for each trait mentioned. This is called the **genotype** of the individual. The genotype can be expressed not only using letters but also with a short descriptive phrase, as Table 23.1 shows. Therefore, the word **homozygous** means that the two members of the allelic pair in the zygote (*zygo*) are the same (*homo*); genotype *WW* is called *homozygous dominant* and *ww* is called *homozygous recessive.* The word **heterozygous** means that the members of the allelic pair are different (*hetero*); only *Ww* is heterozygous.

As Table 23.1 also indicates, the word **phenotype** refers to the physical characteristics of the individual—what the individual actually looks like. Also included in the phenotype are the microscopic and metabolic characteristics of the individual. Notice that both homozygous dominant (*WW*) and heterozygous (*Ww*) show the dominant phenotype.

Gamete Formation

Whereas the genotype has two alleles for each trait, the gametes (i.e., sperm and egg) have only one allele for each trait in accordance with Mendel's law of segregation. This, of course, is related to the process of meiosis. The alleles are present on homologous chromosomes, and these chromosomes separate during meiosis. Therefore, the members of each allelic pair separate during meiosis, and there is only one allele for each trait in each gamete. When doing genetic problems, keep in mind that no two letters in a gamete should be the same. For this reason *Ww* represents a possible genotype, and the gametes for this individual could contain either a *W* or a *w*. For easy recognition, we will circle the gametes.

a.

b.

Figure 23.3 Widow's peak.
In humans, widow's peak **(a)** is dominant over straight hairline **(b)**.

Practice Problems I*

1. For each of the following genotypes, give all possible gametes.
   a. *WW*
   b. *WWSs*
   c. *Tt*
   d. *Ttgg*
   e. *AaBb*
2. For each of the following, state whether a genotype or a gamete is represented.
   a. *D*
   b. *Ll*
   c. *Pw*
   d. *LlGg*

*Answers to Practice Problems appear in Appendix A.

Table 23.1 Genotype Related to Phenotype

Genotype	Genotype	Phenotype
WW	Homozygous dominant	Widow's peak
Ww	Heterozygous	Widow's peak
ww	Homozygous recessive	Straight hairline

One-Trait Crosses

It is now possible for us to consider a particular cross. If a homozygous woman with a widow's peak (Fig. 23.3*a*) reproduces with a man with a straight hairline (Fig. 23.3*b*), what kind of hairline will their children have?

In solving the problem, we use the key already established (p. 472) to indicate the genotype of each parent; we determine what the possible gametes are from each parent; we combine all possible gametes; and finally, we determine the genotypes and the phenotypes of all the offspring. In the format that follows, P stands for the *parental generation,* and the letters in the P row are the genotypes of the parents. The second row shows that each parent has only one type of gamete in regard to hairline, and therefore all the children (F = *filial generation*) have similar genotypes and phenotypes. The children are heterozygous (*Ww*) and show the dominant characteristic, a widow's peak.

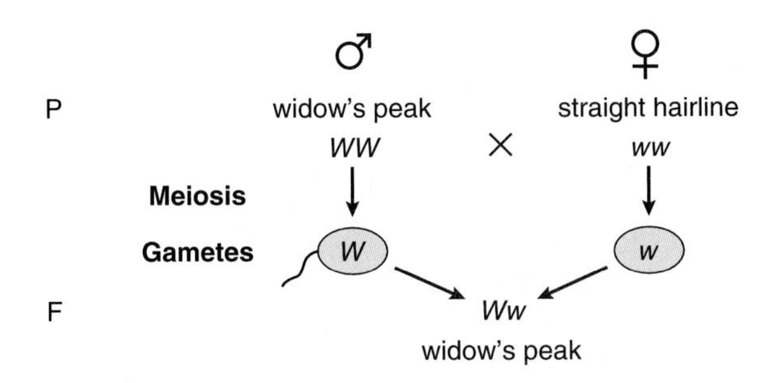

The children are **monohybrids;** that is, they are heterozygous for only one pair of alleles. If they reproduce with someone else of the same genotype, what type of hairline will their children have? In this problem (*Ww* × *Ww*), each parent has two possible types of gametes (*W* or *w*), and we must ensure that all types of sperm have equal chance to fertilize all possible types of eggs. One way to do this is to use a **Punnett square** (Fig. 23.4), in which all possible types of sperm are lined up vertically and all possible types of eggs are lined up horizontally (or vice versa), and every possible combination of gametes occurs within the squares.

After we determine the genotypes and the phenotypes of the offspring, we see that three are expected to have a widow's peak and one is expected to have a straight hairline. This 3:1 ratio is always expected for a monohybrid cross. The exact ratio is more likely to be observed if a large number of matings take place and if a large number of offspring result. Only then do all possible sperm have an equal chance to fertilize all possible eggs. Naturally, we do not routinely observe hundreds of offspring from a single type of cross in humans. The best interpretation of Figure 23.4 in humans is to say that each child has three chances out of four to have a widow's peak, or one chance out of four to have a straight hairline. It is important to realize that *chance has no memory;* for example, if two heterozygous parents already have three children with a widow's peak and are expecting a fourth child, this child still has a 75% chance of a widow's peak and a 25% chance of a straight hairline.

When solving a genetics problem, it is assumed that all possible types of sperm fertilize all possible types of eggs. The results can be expressed as a probable phenotypic ratio; it is also possible to state the chances of an offspring showing a particular phenotype.

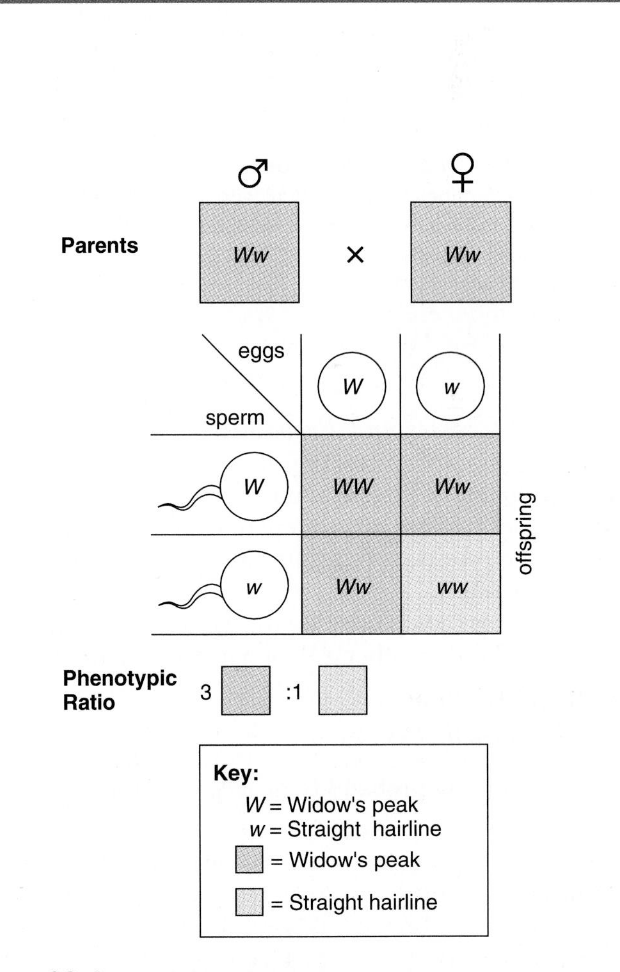

Figure 23.4 Monohybrid cross.
Use of a Punnett square to determine the results of a cross. When the parents are heterozygous, each child has a 75% chance of having the dominant phenotype and a 25% chance of having the recessive phenotype.

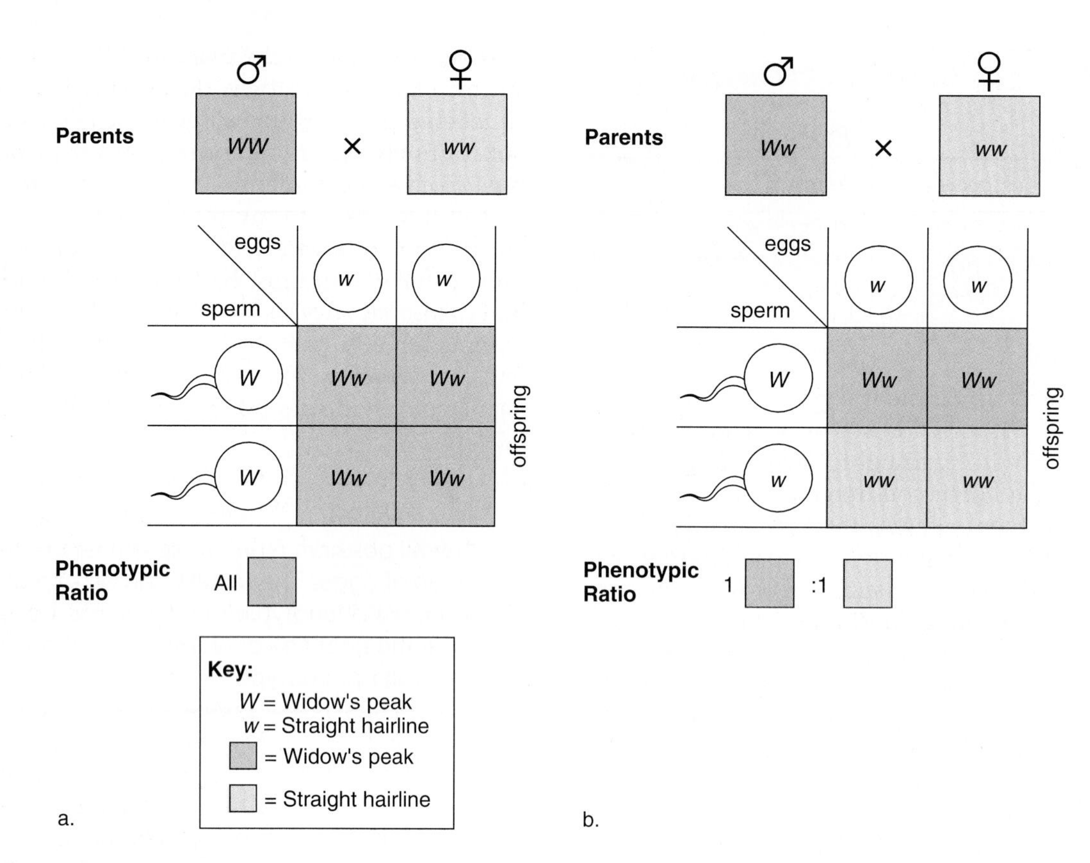

Figure 23.5 One-trait testcross.
A testcross determines if an individual with the dominant phenotype is homozygous or heterozygous. **a.** Because all offspring show the dominant characteristic, the individual is most likely homozygous as shown. **b.** Because the offspring show a 1:1 phenotypic ratio, the individual is heterozygous as shown.

One-Trait Crosses and Probability

Another method of calculating the expected ratios uses the laws of probability. First, we must know that the probability (or chance) of two or more independent events occurring together is the product (multiplication) of their chance of occurring separately.

In the cross just considered (*Ww* × *Ww*), what is the chance of obtaining either a *W* or a *w* from a parent?

The chance of *W* = ½
The chance of *w* = ½

Therefore, the probability of receiving these genotypes is as follows:

1. The chance of *WW* = ½ × ½ = ¼
2. The chance of *Ww* = ½ × ½ = ¼
3. The chance of *wW* = ½ × ½ = ¼
4. The chance of *ww* = ½ × ½ = ¼

Now we have to realize that the chance of an event that can occur in two or more independent ways is the sum (addition) of the individual chances. Therefore, the chance of offspring with a widow's peak (add chances of *WW*, *Ww*, or *wW* from the preceding) is ¾, or 75%. The chance of offspring with straight hairline (only *ww* from the preceding) is ¼, or 25%.

The One-Trait Testcross

A **testcross** occurs when an individual with the dominant phenotype is crossed with an individual having the recessive phenotype. It is not possible to tell by inspection if an individual expressing the dominant allele is homozygous dominant or heterozygous. However, the results of a testcross will most likely indicate which genotype it is.

Consider, for example, Figure 23.5. It shows two possible results when a man with a widow's peak reproduces with a woman who has a straight hairline. If the man is homozygous dominant, all his children will have a widow's peak. If the man is heterozygous, each child has a 50% chance of a straight hairline. The birth of just one child with a straight hairline indicates that the man is heterozygous.

A testcross utilizes the homozygous recessive rather than the heterozygote. The cross *Aa* × *aa* has a better chance of producing the recessive phenotype than the cross *Aa* × *Aa* (50% chance compared to 25% chance).

Practice Problems 2*

1. Both a man and a woman are heterozygous for freckles. Freckles are dominant over no freckles. What are the chances that their child will have freckles?
2. Both you and your sister or brother have attached earlobes, yet your parents have unattached earlobes. Unattached earlobes are dominant over attached earlobes. What are the genotypes of your parents?
3. A father has dimples, the mother does not have dimples, and all the children have dimples. Dimples (D) are dominant over no dimples (d). Give the probable genotypes of all persons concerned.

*Answers to Practice Problems appear in Appendix A.

The Inheritance of Many Traits

Although it is possible to consider the inheritance of just one trait, each individual actually passes on to his or her offspring an allele for each of many traits. In order to arrive at a general understanding of multitrait inheritance, the inheritance of two traits is considered. These genes are on different homologous chromosomes; therefore, the alleles are not linked. All the alleles on the same chromosome are said to form a linkage group.

Independent Assortment

When Mendel performed two-trait crosses, he noticed that his results were attainable only if sperm with every possible combination of factors fertilized eggs with every possible combination of factors. This caused him to formulate his second law, the law of **independent assortment.**

The law of independent assortment states the following:

- Each pair of factors segregates (assorts) independently of the other pairs;
- All possible combinations of factors can occur in the gametes.

Figure 23.6 illustrates that the law of segregation and the law of independent assortment hold because of the manner in which meiosis occurs. The law of segregation is dependent on the separation of members of homologous pairs. The law of independent assortment is dependent on the random arrangement of homologous pairs at the metaphase plate of the spindle. Because it matters not which member of a homologous pair faces which spindle pole, each pair of alleles segregates independently of all others during gamete formation.

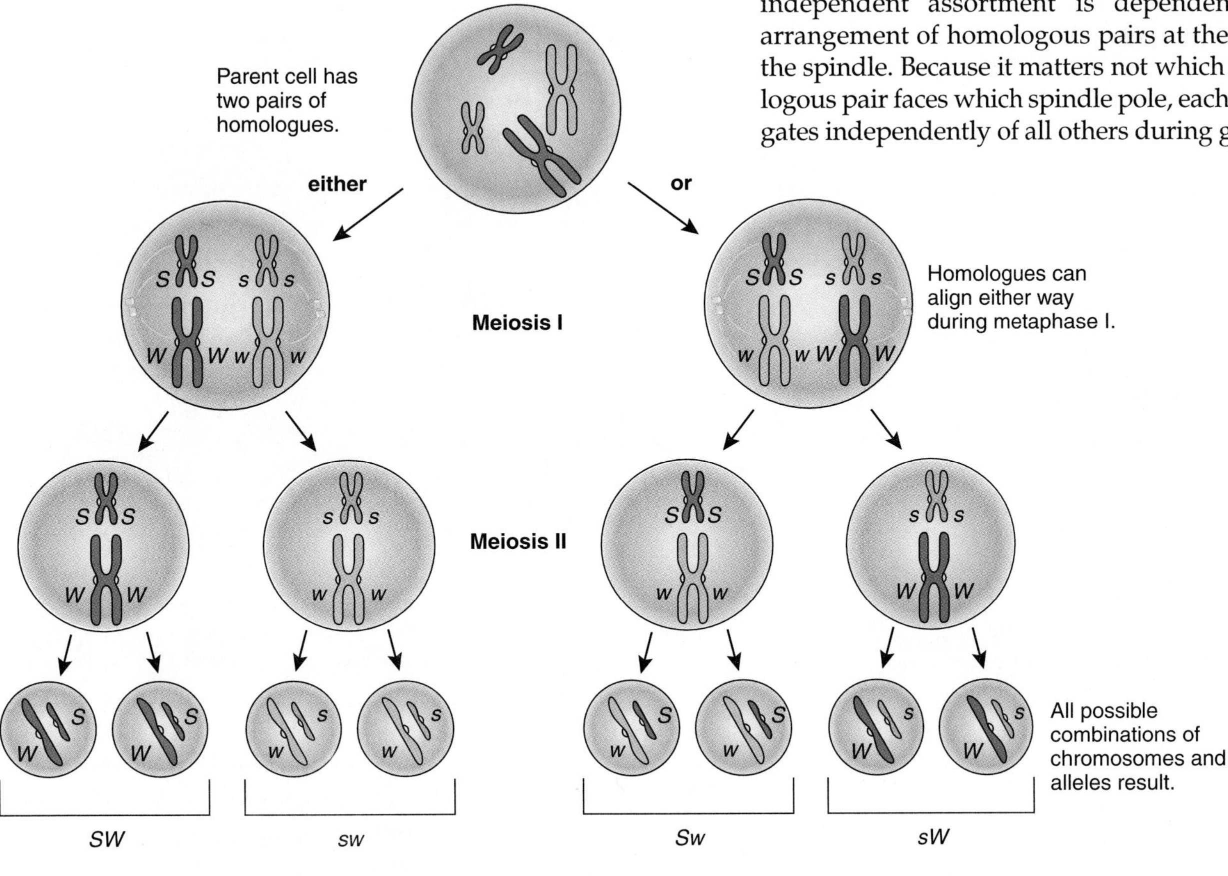

Figure 23.6 Segregation and independent assortment.
Segregation occurs because the homologous chromosomes separate during meiosis I. Also, independent assortment occurs. The homologous chromosomes line up randomly at the metaphase plate; therefore, the homologous chromosomes, and the alleles they carry, segregate independently during gamete formation. All possible combinations of chromosomes and alleles occur in the gametes.

Two-Trait Crosses

When doing a two-trait cross, the genotypes of the parents require four letters because there is an allelic pair for each trait. Also, the gametes of the parents contain one letter of each kind in every possible combination, in accordance with Mendel's law of independent assortment. Finally, in order to produce the probable ratio of phenotypes among the offspring, all possible matings are presumed to occur.

To give an example (Fig. 23.7), let us cross a person homozygous for widow's peak and short fingers (*WWSS*) with a person who has a straight hairline and long fingers (*wwss*). Because each parent has only one type of gamete, the F_1 offspring will all have the genotype *WwSs* and the same phenotype (widow's peak with short fingers). This genotype is called a **dihybrid** because the individual is heterozygous in two regards: hairline and fingers.

When a dihybrid reproduces with a dihybrid, each F_1 parent has four possible types of gametes because the alleles segregate independently during gamete formation.

An inspection of Figure 23.7 shows that the expected phenotypic ratio is:

9 widow's peak and short fingers:

3 widow's peak and long fingers:

3 straight hairline and short fingers:

1 straight hairline and long fingers.

This 9:3:3:1 phenotypic ratio is always expected for a dihybrid cross when simple dominance is present. We can use this expected ratio to predict the chances of each child receiving a certain phenotype. For example, the chance of getting the two dominant phenotypes together is 9 out of 16, and the chance of getting the two recessive phenotypes together is 1 out of 16.

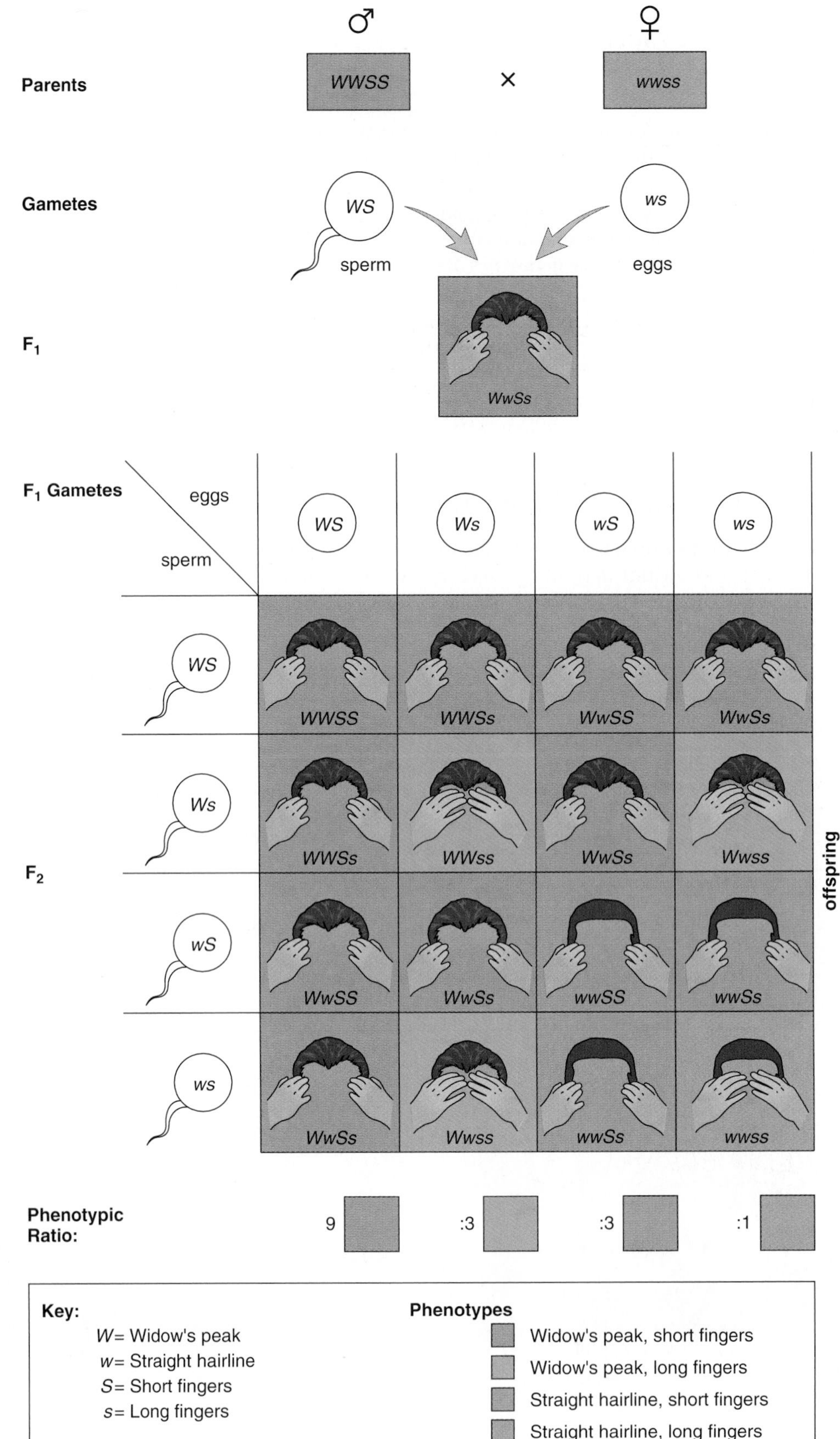

Figure 23.7 Dihybrid cross.
Since each F_1 parent can form four possible types of gametes, four different phenotypes occur among the offspring in the proportions shown.

Two-Trait Crosses and Probability

Instead of using a Punnett square to arrive at the chances of the different types of phenotypes in the cross under discussion, it is possible to use the laws of probability we discussed before. For example, we already know the results for two separate monohybrid crosses are as follows:

1. Probability of widow's peak = $\frac{3}{4}$
   Probability of short fingers = $\frac{3}{4}$
2. Probability of straight hairline = $\frac{1}{4}$
   Probability of long fingers = $\frac{1}{4}$

The probabilities for the dihybrid cross are, therefore, as follows:

Probability of widow's peak and short fingers =
$\frac{3}{4} \times \frac{3}{4} = \frac{9}{16}$

Probability of widow's peak and long fingers =
$\frac{3}{4} \times \frac{1}{4} = \frac{3}{16}$

Probability of straight hairline and short fingers =
$\frac{1}{4} \times \frac{3}{4} = \frac{3}{16}$

Probability of straight hairline and long fingers =
$\frac{1}{4} \times \frac{1}{4} = \frac{1}{16}$

The phenotypic ratio is 9:3:3:1.

The Two-Trait Testcross

A two-trait testcross occurs when an individual with the dominant phenotype for two traits is crossed with a homozygous recessive for both traits. It is impossible to tell by inspection whether an individual expressing the dominant allele for two traits is homozygous dominant or heterozygous in regard to these traits. A cross with the homozygous recessive for both traits gives the best possible chance of producing an offspring with the recessive phenotype for both traits.

For example, if a man is homozygous dominant for widow's peak and short fingers, then all his children will have the dominant phenotypes, even if his partner is homozygous recessive for both traits. However, if a man is heterozygous for both traits, then each child has a 25% chance of showing one or both recessive traits. A Punnett square (Fig. 23.8) shows that the expected ratio is 1 widow's peak with short fingers: 1 widow's peak with long fingers: 1 straight hairline with short fingers: 1 straight hairline with long fingers, or 1:1:1:1.

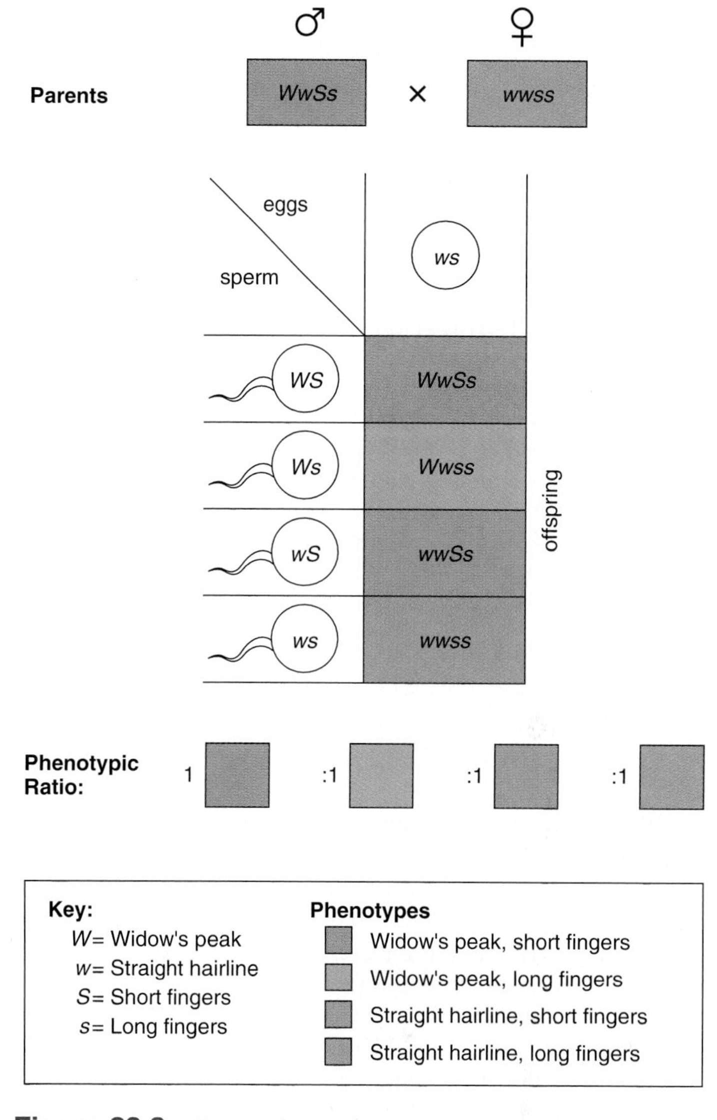

Figure 23.8 Two-trait testcross.
A testcross determines if the individual with a dominant phenotype is homozygous or heterozygous. If the individual is heterozygous as shown, there is a 25% chance for each possible phenotype.

Practice Problems 3*

1. What is the genotype of the offspring if a man homozygous recessive for type of earlobes and homozygous dominant for type of hairline reproduces with a woman who is homozygous dominant for earlobes and homozygous recessive for hairline?
2. If the offspring of this cross reproduces with someone of the same genotype, then what are the chances that this couple will have a child with a straight hairline and attached earlobes?
3. A person who has dimples and freckles reproduces with someone who does not. This couple has a child who does not have dimples or freckles. What is the genotype of all persons concerned?

*Answers to Practice Problems appear in Appendix A.

23.2 Genetic Disorders

It is now apparent that many human disorders are genetic in origin. Genetic disorders are medical conditions caused by alleles inherited from the parents. Some of these conditions are controlled by autosomal dominant or recessive alleles. An autosome is any chromosome other than a sex (X or Y) chromosome.

Patterns of Inheritance

When a genetic disorder is autosomal dominant, an individual with the alleles *AA* or *Aa* will have the disorder. When a genetic disorder is recessive, only individuals with the alleles *aa* will have the disorder. Genetic counselors often construct pedigree charts to determine whether a condition is dominant or recessive. A pedigree chart shows the pattern of

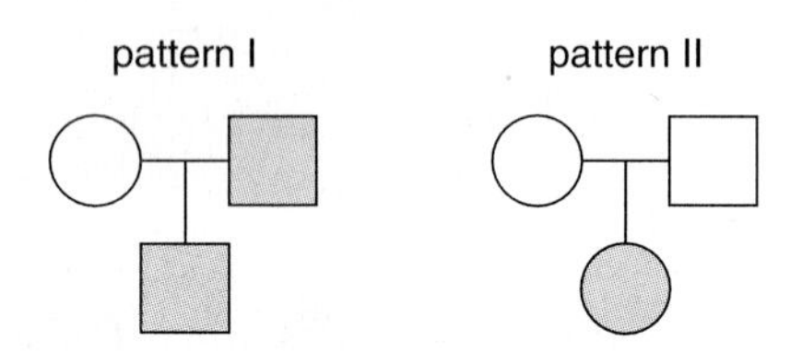

inheritance for a particular condition. Consider these two possible patterns of inheritance:

In both patterns, males are designated by squares and females by circles. Shaded circles and squares are affected individuals. A line between a square and a circle represents a union. A vertical line going downward leads, in these patterns, to a single child. (If there are more children, they are placed off a horizontal line.) Which pattern of inheritance do you suppose represents an autosomal dominant characteristic, and which represents an autosomal recessive characteristic?

In pattern I, the child is affected, as is one of the parents. When a disorder is dominant, an affected child usually has at least one affected parent. Of the two patterns, this one shows a dominant pattern of inheritance. Figure 23.9 illustrates other ways to recognize an autosomal dominant pattern of inheritance.

In pattern II, the child is affected but neither parent is; this can happen if the condition is recessive and the parents are *Aa*. Notice that the parents are **carriers** because they appear to be normal but are capable of having a child with a genetic disorder. See Figure 23.10 for other ways to recognize an autosomal recessive pattern of inheritance.

Dominant and recessive alleles have different patterns of inheritance.

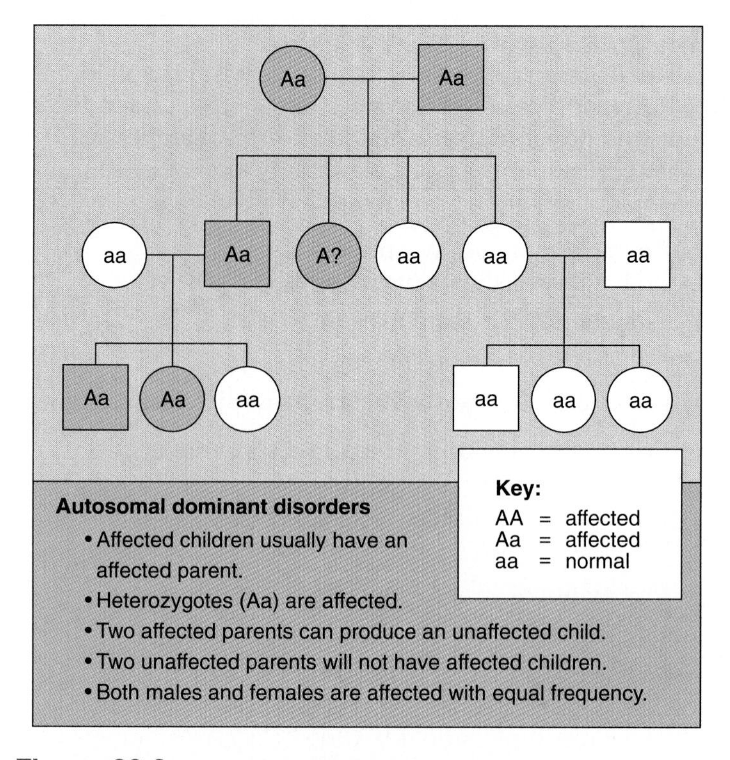

Figure 23.9 Autosomal dominant pedigree chart.
The list gives ways to recognize if an autosomal disorder is dominant.

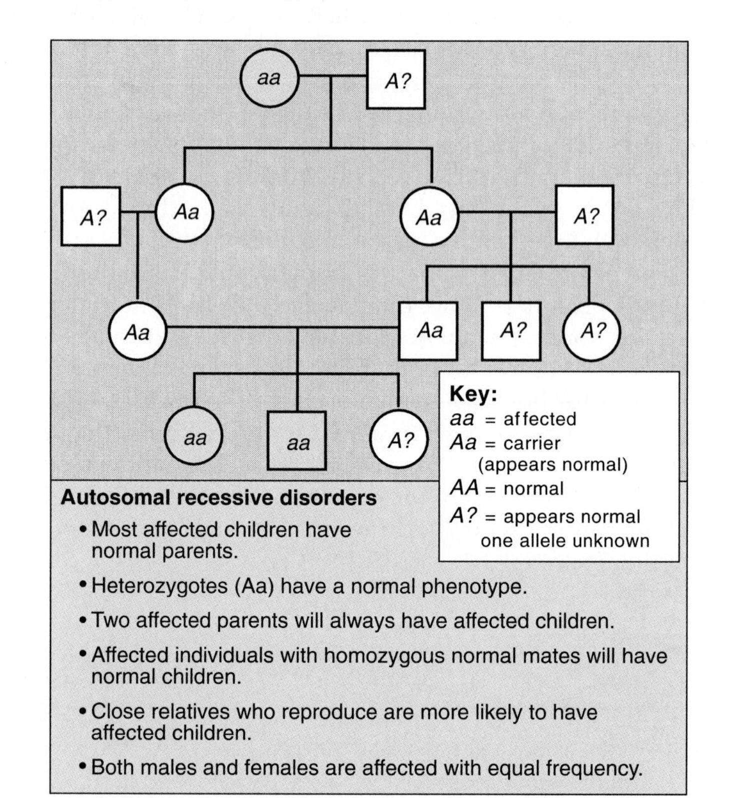

Figure 23.10 Autosomal recessive pedigree chart.
The list gives ways to recognize if an autosomal disorder is recessive.

Autosomal Dominant Disorders

Of the many autosomal (non-sex linked) dominant disorders, we will discuss only two.

Neurofibromatosis

Neurofibromatosis, sometimes called von Recklinghausen disease, is one of the most common genetic disorders. It affects roughly one in 3,000 people, including an estimated 100,000 in the United States. It is seen equally in every racial and ethnic group throughout the world.

At birth or later, the affected individual may have six or more large, tan spots on the skin. Such spots may increase in size and number and may get darker. Small, benign tumors (lumps) called neurofibromas may occur under the skin or in various organs. Neurofibromas are made up of nerve cells and other cell types.

This genetic disorder shows *variable expressivity.* In most cases, symptoms are mild and patients live a normal life. In some cases, however, the effects are severe. Skeletal deformities, including a large head, are seen, and eye and ear tumors can lead to blindness and hearing loss. Many children with neurofibromatosis have learning disabilities and are hyperactive.

In 1990, researchers isolated the gene for neurofibromatosis, which was known to be on chromosome 17. By analyzing the DNA (deoxyribonucleic acid), they determined that the gene was huge and actually included three smaller genes. This was only the second time that nested genes have been found in humans. The gene for neurofibromatosis is a tumor-suppressor gene active in controlling cell division. When it mutates, benign tumors develop.

Huntington Disease

One in 20,000 persons in the United States has *Huntington disease,* a neurological disorder that leads to progressive degeneration of brain cells, which in turn causes severe muscle spasms and personality disorders (Fig. 23.11). Most people appear normal until they are of middle age and have already had children who might also be stricken. Occasionally, the first signs of the disease are seen in these children when they are teenagers or even younger. There is no effective treatment, and death comes ten to fifteen years after the onset of symptoms.

Several years ago, researchers found that the gene for *Huntington disease* was located on chromosome 4. A test was developed for the presence of the gene, but few want to know if they have inherited the gene because as yet there is no treatment for Huntington disease. After the gene was isolated in 1993, an analysis revealed that it contains many repeats of the base triplet CAG (cytosine, adenine, and guanine). Normal persons have 11 to 34 copies of the triplet, but affected persons tend to have 42 to more than 120 copies. The more repeats present, the earlier the onset of Huntington disease and the more severe the symptoms. It also appears that persons most at risk have inherited the disorder from their fathers. The latter observation is consistent with a new hypothesis called **genomic imprinting**. The genes are imprinted differently during formation of sperm and egg, and therefore the sex of the parent passing on the disorder becomes important.

It is now known that there are a number of other genetic diseases whose severity and time of onset vary according to the number of triplet repeats present within the gene. Moreover, the genes for these disorders are also subject to genomic imprinting.

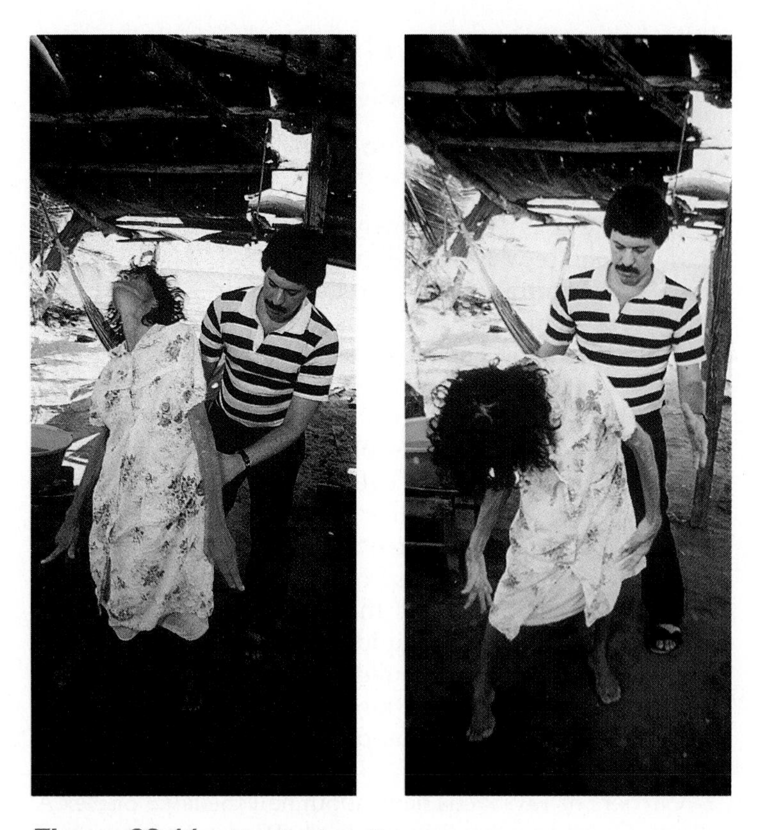

Figure 23.11 Huntington disease.
Persons with this condition gradually lose psychomotor control of the body. At first there are only minor disturbances, but the symptoms become worse over time.

Practice Problems 4*

1. A woman who is heterozygous for an autosomal dominant disorder reproduces with a normal man. What are the chances the child will develop the disorder?
2. A child has neurofibromatosis. The mother appears normal. What is the genotype of the father?

*Answers to Practice Problems appear in Appendix A.

Autosomal Recessive Disorders

Of the many autosomal recessive disorders, we will discuss only three.

Tay-Sachs Disease

Tay-Sachs disease is a well-known genetic disease that usually occurs among Jewish people in the United States, most of whom are of central and eastern European descent. At first, it is not apparent that a baby has Tay-Sachs disease. However, development begins to slow down between four months and eight months of age, and neurological impairment and psychomotor difficulties then become apparent. The child gradually becomes blind and helpless, develops uncontrollable seizures, and eventually becomes paralyzed. There is no treatment or cure for Tay-Sachs disease, and most affected individuals die by the age of three or four.

Tay-Sachs disease results from a lack of the enzyme hexosaminidase A (Hex A) and the subsequent storage of its substrate, a glycosphingolipid, in lysosomes. Although more and more lysosomes build up in many body cells, the primary sites of storage are the cells of the brain, which accounts for the onset, and the progressive deterioration, of psychomotor functions.

Carriers of Tay-Sachs have about half the level of Hex A activity found in normal individuals. Prenatal diagnosis of the disease also is possible following either amniocentesis or chorionic villi sampling.

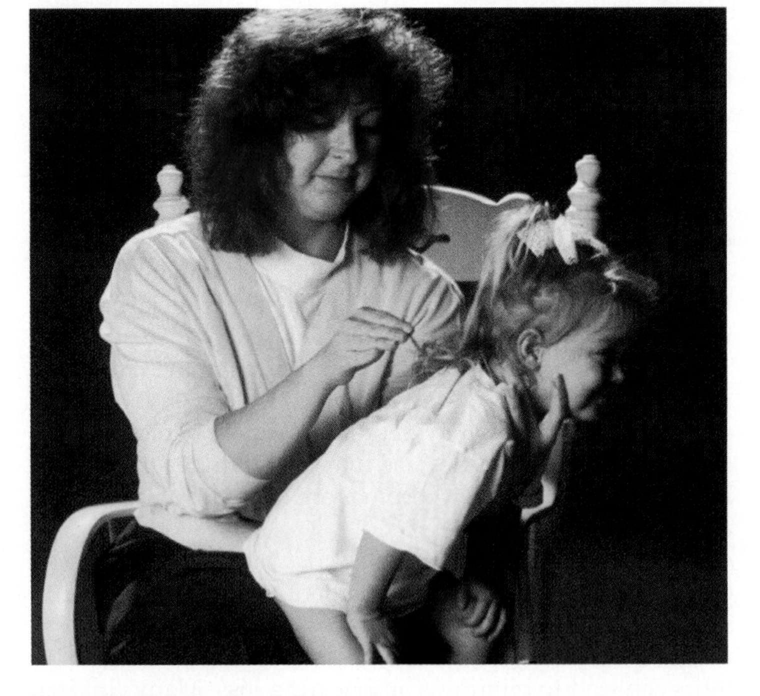

Figure 23.12 Cystic fibrosis.
The mucus in the lungs of a child with cystic fibrosis should be periodically loosened by clapping the back. A new treatment helps break up mucus by digesting long, sticky strands of DNA released by dying cells. Antibiotics control pulmonary infection, and anti-inflammatory drugs are sometimes used to reduce the harmful effects of immune responses.

Cystic Fibrosis

Cystic fibrosis is the most common lethal genetic disease among Caucasians in the United States. About one in 20 Caucasians is a carrier, and about one in 2,500 children born to this group has the disorder. In these children, the mucus in the bronchial tubes and pancreatic ducts is particularly thick and viscous, interfering with the function of the lungs and pancreas. To ease breathing, the thick mucus in the lungs has to be manually loosened periodically (Fig. 23.12), but still the lungs become infected frequently. The clogged pancreatic ducts prevent digestive enzymes from reaching the small intestine, and to improve digestion patients take digestive enzymes mixed with applesauce before every meal.

In the past few years, much progress has been made in our understanding of cystic fibrosis, and new treatments have raised the average life expectancy to 17 to 28 years of age. Research has demonstrated that chloride ions (Cl^-) fail to pass through plasma membrane channel proteins in these patients. Ordinarily, after chloride ions have passed through the membrane, water follows. It is believed that lack of water is the cause of abnormally thick mucus in bronchial tubes and pancreatic ducts. The cystic fibrosis gene, which is located on chromosome 7, has been isolated, and attempts have been made to insert it into nasal epithelium, so far with little success. Genetic testing for the gene in adult carriers and in fetuses is possible; if a disease-causing gene is present, couples have to consider that gene therapy may be possible some day.

Phenylketonuria (PKU)

Phenylketonuria (PKU) occurs once in 5,000 births, so it is not as frequent as the disorders previously discussed. However, it is the most commonly inherited metabolic disorder to affect nervous system development. First cousins who marry are more apt to have a PKU child.

Affected individuals lack an enzyme that is needed for the normal metabolism of the amino acid phenylalanine, and an abnormal breakdown product, a phenylketone, accumulates in the urine. The PKU gene is located on chromosome 12, and a prenatal DNA test can determine the presence of this mutation. Years ago, the urine of newborns was tested at home for phenylketone in order to detect PKU. Presently, newborns are routinely tested in the hospital for elevated levels of phenylalanine in the blood. If necessary, newborns are placed on a diet low in phenylalanine, which must be continued until the brain is fully developed, or else severe mental retardation develops.

There are many autosomal recessive disorders in humans. Among these are Tay-Sachs disease, cystic fibrosis, and phenylketonuria (PKU).

23.3 Beyond Mendel's Laws

Certain traits, such as those just studied, follow the rules of simple Mendelian inheritance. There are, however, others that do not follow these rules.

Polygenic Inheritance

Polygenic inheritance occurs when one trait is governed by two or more sets of alleles, possibly located on many different pairs of chromosomes. Each dominant allele has a quantitative effect on the phenotype, and these effects are additive. The result is a continuous variation of phenotypes, resulting in a distribution of these phenotypes that resembles a bell-shaped curve. The more genes involved, the more continuous the variation and the distribution of the phenotypes. Also, environmental effects cause many intervening phenotypes; in the case of height, differences in nutrition assure a bell-shaped curve (Fig. 23.13).

Skin Color

Just how many pairs of alleles control skin color is not known, but a range in colors can be explained on the basis of two pairs. When a very dark person reproduces with a very light person, the children have medium brown skin; and when two people with medium brown skin reproduce with one another, the children range in skin color from very dark to very light. This can be explained by assuming that skin color is controlled by two pairs of alleles and that each capital letter contributes to the color of the skin:

Phenotype	Genotypes
Very dark	*AABB*
Dark	*AABb* or *AaBB*
Medium brown	*AaBb* or *AAbb* or *aaBB*
Light	*Aabb* or *aaBb*
Very light	*aabb*

Notice again that there is a range in phenotypes and that there are several possible phenotypes in between the two extremes. Therefore, the distribution of these phenotypes is expected to follow a bell-shaped curve—few people have the extreme phenotypes, and most people have the phenotype that lies in the middle between the extremes.

Polygenic Disorders

Many human disorders, such as cleft lip and/or palate, clubfoot, congenital dislocations of the hip, hypertension, diabetes, schizophrenia, and even allergies and cancers, are most likely controlled by polygenes and subject to environmental influences. Therefore, many investigators are in the process of considering the *nature versus nurture* question; that is, what percentage of the trait is controlled by genes and what percentage is controlled by the environment? Thus far, it has not been possible to come to precise, generally accepted percentages for any particular trait.

a.

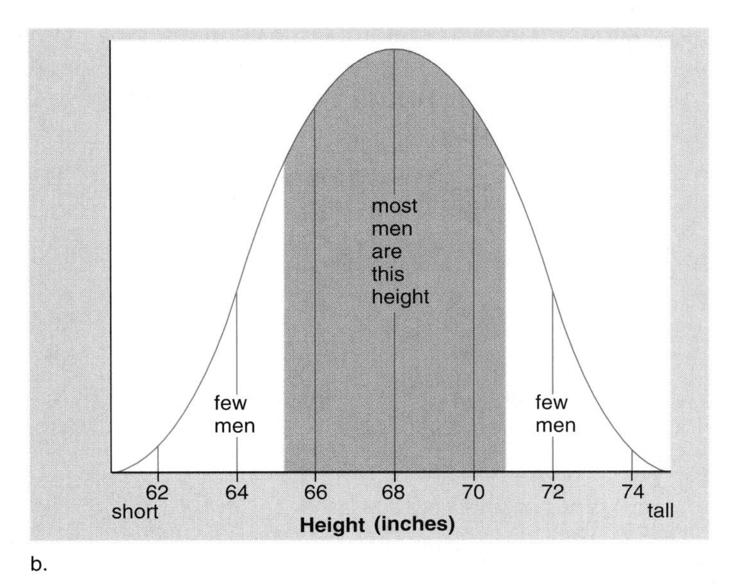

b.

Figure 23.13 Polygenic inheritance.
When you record the heights of a large group of young men **(a)** the values follow a bell-shaped curve **(b)**. Such a continuous distribution is due to control of a trait by several sets of alleles and environmental effects.

In recent years, reports have surfaced that all sorts of behavioral traits such as alcoholism, phobias, and even suicide can be associated with particular genes. No doubt behavioral traits are to a degree controlled by genes, but again, it is impossible at this time to determine to what degree. And very few scientists would support the idea that these traits are predetermined by our genes.

Many human traits most likely controlled by polygenes are subject to environmental influences. The frequency of the phenotypes of such traits follows a bell-shaped curve.

Multiple Alleles

In inheritance by **multiple alleles,** the trait is controlled, for example, by three alleles. However, each person has only two of the three possible alleles.

ABO Blood Types

Three alleles for the same gene control the inheritance of ABO blood types. These alleles determine the presence or absence of antigenic glycoproteins on the red blood cells.

A = A antigen on red blood cells

B = B antigen on red blood cells

O = Neither A or B antigen on red blood cells

Each person has only two of the three possible alleles, and both *A* and *B* are dominant over *O*. Therefore, there are two possible genotypes for type A blood and two possible genotypes for type B blood. On the other hand, alleles *A* and *B* are fully expressed in the presence of the other. This is called **codominance.** Therefore, if a person inherits one of each of these alleles, that person will have type AB blood. Type O blood can only result from the inheritance of two *O* alleles:

Phenotype	Possible Genotype
A	*AA, AO*
B	*BB, BO*
AB	*AB*
O	*OO*

An examination of possible matings between different blood types sometimes produces surprising results; for example,

Genotypes of parents: *AO* × *BO*

Possible genotypes of children: *AB, OO, AO, BO*

Therefore, from this particular mating, every possible phenotype (types AB, O, A, B blood) is possible.

Blood typing can sometimes aid in paternity suits. However, a blood test of a supposed father can only suggest that he might be the father, not that he definitely is the father. For example, it is possible, but not definite, that a man with type A blood (having genotype *AO*) is the father of a child with type O blood. On the other hand, a blood test sometimes can definitely prove that a man is not the father. For example, a man with type AB blood cannot possibly be the father of a child with type O blood. Therefore, blood tests can be used legally only to exclude a man from possible paternity.

Rh Factor The Rh factor is inherited separately from A, B, AB, or O blood types. In each instance, it is possible to be Rh positive (Rh^+) or Rh negative (Rh^-). When you are Rh positive, there is a particular antigen on the red blood cells, and when you are Rh negative, it is absent. It can be assumed that the inheritance of this antigen is controlled by a single allelic pair in which simple dominance prevails: the Rh-positive allele is dominant over the Rh-negative allele. Complications arise when an Rh-negative woman reproduces with an Rh-positive man and the child in the womb is Rh positive. Under certain circumstances, the woman may begin to produce antibodies that will attack the red blood cells of this baby or of a future Rh-positive baby. The Rh problem can be eliminated by giving an Rh-negative woman an Rh immunoglobulin injection from midway through a pregnancy to no later than 72 hours after giving birth to any Rh-positive child.

Degrees of Dominance

The field of human genetics also has examples of codominance and incomplete dominance. We have already mentioned that the multiple alleles controlling blood type are

Practice Problems 5*

1. What is the darkest child that could result from a mating between a light individual and a very light individual?
2. What is the lightest child that could result from a mating between two medium brown individuals?
3. From the following blood types, determine which baby belongs to which parents:

Mrs. Doe	Type A
Mr. Doe	Type A
Mrs. Jones	Type A
Mr. Jones	Type AB
Baby 1	Type O
Baby 2	Type B

4. Prove that a child does not have to have the blood type of either parent by indicating what blood types might be possible when a person with type A blood reproduces with a person with type B blood.

*Answers to Practice Problems appear in Appendix A.

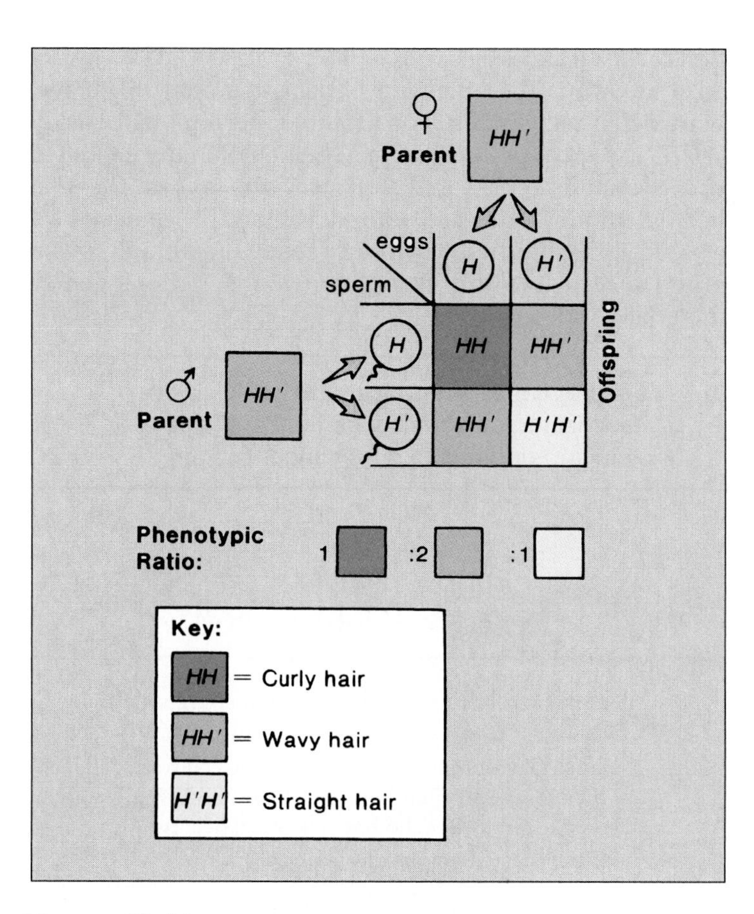

Figure 23.14 Incomplete dominance.
When two persons with the genotype HH' reproduce, a phenotypic ratio of 1:2:1 is expected among the offspring because the alleles H and H' are incompletely dominant.

codominant. An individual with the genotype *AB* has type AB blood. Skin color, recall, is controlled by polygenes, and therefore it is possible to observe a range of skin colors from very dark to very light. An example of **incomplete dominance** is given in Figure 23.14. When two wavy-haired persons reproduce, the expected phenotypic ratio among the offspring is 1:2:1.

Sickle-Cell Disease

Sickle-cell disease is an example of a human disorder that is controlled by incompletely dominant alleles. Individuals with the Hb^AHb^A genotype are normal, those with the Hb^SHb^S genotype have sickle-cell disease, and those with the Hb^AHb^S genotype have the sickle-cell trait. Two individuals with *sickle-cell trait* can produce children with all three phenotypes, as indicated in Figure 23.15.

In persons with sickle-cell disease, the red blood cells aren't biconcave disks like normal red blood cells; they are irregular. In fact, many are sickle shaped. The defect is caused by an abnormal hemoglobin (Hb^S) that accumulates inside the cells. Normal hemoglobin (Hb^A) differs from Hb^S by one amino acid in the protein globin. Although globin is 146 amino acids long, just one change in the sixth position results in sickle-shaped cells.

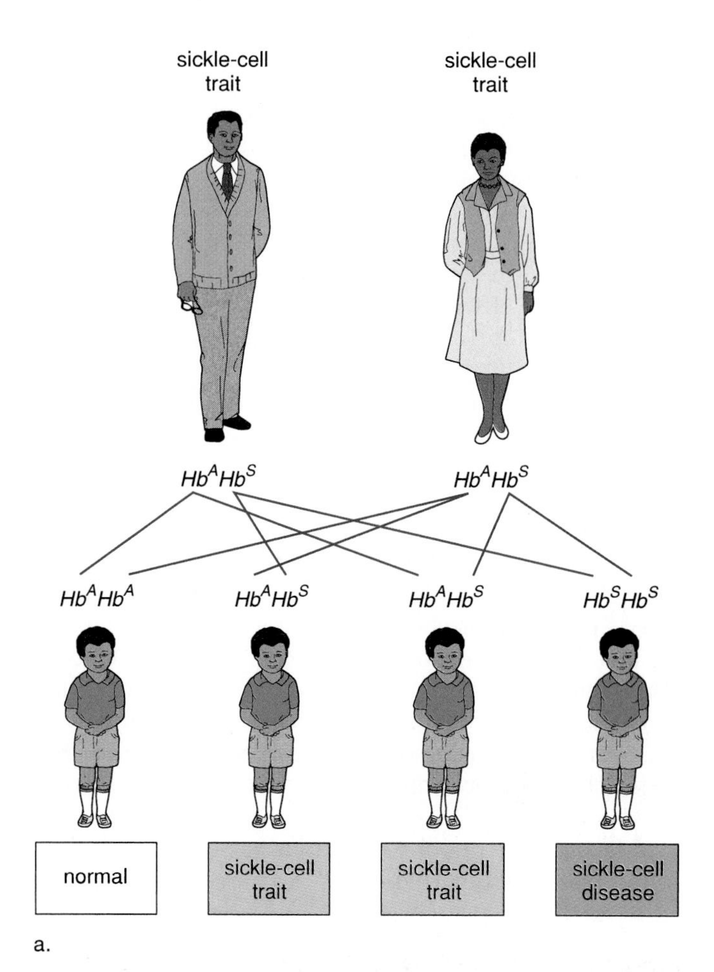

Figure 23.15 Inheritance of sickle-cell disease.
a. In this example, both parents have the sickle-cell trait. Therefore, each child has a 25% chance of having sickle-cell disease or of being perfectly normal, and a 50% chance of having the sickle-cell trait. **b.** Sickled cells. Individuals with sickle-cell disease have sickled red blood cells, as illustrated here.

Because the sickle-shaped cells can't pass along narrow capillary passageways like disk-shaped cells, they clog the vessels and break down. This is why persons with sickle-cell disease suffer from poor circulation, anemia, and poor resistance to infection. Internal hemorrhaging leads to further complications, such as jaundice, episodic pain of the abdomen and joints, and damage to internal organs.

Persons with sickle-cell trait do not usually have any sickle-shaped cells unless they experience dehydration or mild oxygen deprivation. Although a recent study found that army recruits with sickle-cell trait are more likely to die when subjected to extreme exercise, previous studies of athletes do not substantiate these findings. At present, most investigators believe that no restrictions on physical activity are needed for persons with the sickle-cell trait.

Among regions of malaria-infested Africa, infants with sickle-cell disease die, but infants with sickle-cell trait have a better chance of survival than the normal homozygote. The malaria parasite normally reproduces inside red blood cells. But a red blood cell of a sickle-cell trait infant becomes sickle-shaped if it becomes infected with a malaria-causing parasite. As the cell becomes sickle-shaped and it loses potassium, the parasite dies. The protection afforded by the sickle-cell trait keeps the allele prevalent in populations exposed to malaria. As many as 60% of the population in malaria-infected regions of Africa have the allele. In the United States, about 10% of the African-American population carries the allele.

Innovative therapies are being attempted in persons with sickle-cell disease. For example, persons with sickle-cell disease produce normal fetal hemoglobin during development, and drugs that turn on the genes for fetal hemoglobin in adults are being developed. Mice have been genetically engineered to produce sickled red blood cells in order to test new antisickling drugs and various genetic therapies.

There are many exceptions to Mendel's laws. These include polygenic inheritance, multiple alleles, and degrees of dominance.

Practice Problems 6*

1. What is the genotype of a person with straight hair? Could this individual ever have a child with curly hair?
2. A man and a woman both have a sickle-cell trait. What are their chances of having a child who is normal? who has sickle-cell trait? who has sickle-cell disease?

*Answers to the Practice Problems appear in Appendix A.

Bioethical Issue

If the results were absolutely private, would you want to know if you have a genetic disease whose symptoms may not appear for several decades? If the disease were curable, it might be a good idea. Testing before symptoms appear encourages increased vigilance for the onset of breast and colon cancers. If gene therapy is the only way out, you might be first in line if such therapy is developed in the future.

Huntington disease is a particularly frightening genetic disorder, especially if you have seen a loved one suffer the mental and musculoskeletal deterioration that finally results in death. Although there is no cure for Huntington disease, you would be able to warn your children if you test positive, or it might help you decide whether to have a child. If you have this dominant allele, you will develop the condition and your children have a 50% chance of developing the condition. Children usually begin to have symptoms earlier than their parents exhibit symptoms.

Jason Brandt of the Johns Hopkins University School of Medicine has developed a program of pretest and posttest counseling for those who decide they want to be tested for Huntington disease. So far, about 200 persons have been tested, and almost twice that number have dropped out of pretest counseling, signifying no doubt they really didn't want to be tested.

Brandt refuses to perform the test on persons he feels could not psychologically handle the results if they were positive. Surprisingly, he finds that people rarely make major life changes as a result of knowing they have the allele. Among 63 who tested positive, 10 got married to persons who knew they would come down with Huntington disease, and 10 others went on to have more children. One patient made a radical change after she knew she was negative for Huntington disease. She divorced her present husband, remarried, had another child and took up a career as a physical therapist.

Questions

1. Is it a person's right to be tested for genetic diseases or do you think Brandt should forgo testing until he considers the person ready to receive bad news?
2. Is it ethical to discourage people with a dominant allele for Huntington disease from having children? Why or why not?
3. If gene therapy for Huntington disease becomes available should it be made available to adults? children? embryos? Why or why not?

Summarizing the Concepts

23.1 Mendel's Laws

The genes are on the chromosomes; each gene has a minimum of two alternative forms called alleles. Mendel's laws are consistent with the observation that each pair of alleles segregates independently of the other pairs during meiosis when the gametes form.

It is customary to use letters to represent the genotype of individuals. Homozygous dominant is indicated by two capital letters, and homozygous recessive is indicated by two lowercase letters. Heterozygous is indicated by a capital letter and a lowercase letter. In a one-trait cross, each heterozygous individual can form two types of gametes. In a two-trait cross, each individual heterozygous for both traits can form four types of gametes.

Use of the Punnett square allows us to make sure that all possible sperm have fertilized all possible eggs. These results tell us the chances of a child inheriting a particular phenotype. With regard to the monohybrid cross, there is a 25% chance of each child having the recessive phenotype and a 75% chance of each having the dominant phenotype.

Testcrosses are used to determine if an individual with the dominant phenotype is homozygous or heterozygous. If an individual expressing the dominant allele reproduces with an individual expressing the recessive allele and an offspring with the recessive phenotype results, we know that the individual with the dominant phenotype is heterozygous.

23.2 Genetic Disorders

Studies of human genetics have shown that there are many autosomal genetic disorders that can be explained on the basis of simple Mendelian inheritance. When studying human genetic disorders, biologists often construct pedigree charts to show the pattern of inheritance of a characteristic within a family. The particular pattern indicates the manner in which a characteristic is inherited. Sample charts are given for autosomal dominant and autosomal recessive patterns.

Neurofibromatosis and Huntington disease are autosomal dominant disorders that have been well studied. Tay-Sachs disease, cystic fibrosis, and PKU are autosomal recessive disorders that have been studied in detail.

23.3 Beyond Mendel's Laws

There are many exceptions to Mendel's laws. These include polygenic inheritance (skin color), multiple alleles (ABO blood type), and degrees of dominance (curly hair).

Traits controlled by polygenes are subject to environmental effects and show continuous variations whose frequency distribution forms a bell-shaped curve. Several human disorders, such as cleft palate, and human behaviors are most likely controlled by polygenes. Sickle-cell disease is a human disorder that is controlled by incomplete dominant alleles.

Studying the Concepts

1. What is Mendel's law of segregation? 470 What do we call his factors today, and where are they located? 470, 472
2. What is the difference between the genotype and the phenotype of an individual? For which phenotype in a one-trait testcross are there two possible genotypes? 472–73
3. What is Mendel's law of independent assortment? 475 Relate Mendel's laws to one-trait and two-trait testcrosses. 474–77
4. What are the expected results of the following crosses? 476–77
   monohybrid × monohybrid
   monohybrid × recessive
   dihybrid × dihybrid
   dihybrid × recessive in both traits
5. Which of these crosses are called testcrosses? Why? 474
6. What are the chances of the dominant phenotype(s) for each of these crosses? 477 What does the phrase "chance has no memory" mean? 473
7. How might you distinguish an autosomal dominant trait from an autosomal recessive trait when viewing a pedigree chart? 478
8. Describe the symptoms of neurofibromatosis and Huntington disease. For most autosomal dominant disorders, what are the chances of a heterozygote and a normal individual having an affected child? 479
9. Describe the symptoms of Tay-Sachs disease, cystic fibrosis, and PKU. For any one of these, what are the chances of two carriers having an affected child? 480
10. Give an example of these patterns of inheritance: polygenic inheritance, multiple alleles, and degrees of dominance. 481–84

Testing Yourself

Choose the best answer for each question.
For questions 1–4, match the cross with the results in the key.

Key:
a. 3:1
b. 9:3:3:1
c. 1:1
d. 1:1:1:1

1. *TtYy* × *TtYy*
2. *Tt* × *Tt*
3. *Tt* × *tt*
4. *TtYy* × *ttyy*
5. Which of these could be a normal gamete?
   a. *GgRr*
   b. *GRr*
   c. *Gr*
   d. None of these are correct.
6. In humans, pointed eyebrows (*B*) are dominant over smooth eyebrows (*b*). Mary's father has pointed eyebrows, but she and her mother have smooth. What is the genotype of the father?
   a. *BB*
   b. *Bb*
   c. *bb*
   d. *BBbb*
   e. Any one of these is correct.
7. Parents who do not have Tay-Sachs disease (recessive) produce a child who has Tay-Sachs. What are the chances that each child born to this couple will have Tay-Sachs disease?
   a. 100%
   b. 75%
   c. 25%
   d. 0%
   e. All of these are correct.

8. Both Mr. and Mrs. Smith have freckles (dominant) and attached earlobes (recessive). Some of their children do not have freckles. What are the chances that their next child will have freckles and attached earlobes?
   a. 100%
   b. 75%
   c. 25%
   d. 0%
   e. All of these are correct.
9. A woman with very light skin has parents with medium brown skin. If this woman marries a man with light skin, what is the darkest skin color possible for their children?
   a. very dark
   b. dark
   c. medium brown
   d. light
   e. very light
10. A man has type AB blood. If he has a child with type B blood, what blood types could the child's mother have?
   a. Type A
   b. Type B
   c. Type AB
   d. Type O
   e. All of these are correct.
11. Determine if the characteristic possessed by the darkened squares (males) and circles (females) below is autosomal dominant or autosomal recessive.

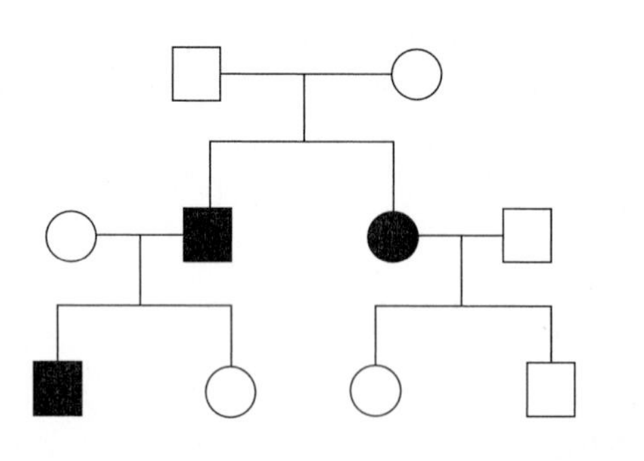

Thinking Scientifically

1. Considering Mendel's laws:
   a. Before Mendel formulated his law of segregation, what two alternative hypotheses might he have formulated about the kinds of gametes for a parent that is *Yy* (*Y* = yellow peas; *y* = green peas)?
   b. How did his results of 3 yellow to 1 green support one of these hypotheses and not the other?
   c. Before Mendel formulated his law of independent assortment, what two alternative hypotheses might he have formulated about the kinds of gametes for a parent in Figure 23.6?
   d. How did his results of 9:3:3:1 support one of these hypotheses and not the other?
2. Some individuals are albinos—they have no melanin in any of their skin cells. Melanin is a molecule produced by a biochemical pathway.
   a. What fault have albinos inherited?
   b. Considering your answer to question 2a, why would you have predicted that albinism is a recessive rather than a dominant disorder?
   c. What possible crosses would produce an offspring that is an albino? Are any of these individuals carriers?
   d. Suppose you want to ensure that children born to an albino woman have normal pigmentation. Which of these—melanin, enzyme to produce melanin, or a normal gene—would you inject into the egg?

Understanding the Terms

allele 470	independent assortment 475
carrier 478	monohybrid 473
codominance 482	multiple allele 482
dihybrid 476	phenotype 472
dominant allele 472	polygenic inheritance 481
genomic imprinting 479	Punnett square 473
genotype 472	recessive allele 472
heterozygous 472	segregation 470
homozygous 472	testcross 474
incomplete dominance 483	trait 470

Match the terms to these definitions:

a. ________ Individual that appears normal but is capable of transmitting an allele for a genetic disorder.

b. ________ Alternative forms of a gene which occur at the same locus on homologous chromosomes.

c. ________ Allele that exerts its phenotypic effect in the heterozygote; it masks the expression of the recessive allele.

d. ________ Cross between an individual with the dominant phenotype and an individual with the recessive phenotype to see if the individual with the dominant phenotype is homozygous or heterozygous.

e. ________ Genes of an individual for a particular trait or traits; for example, *BB* or *Aa*.

Using Technology

Your study of gene inheritance is supported by these available technologies:

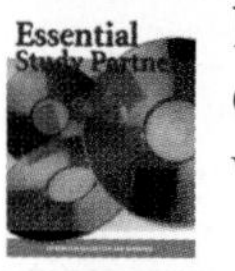

Essential Study Partner CD-ROM

Genetics → Mendelian Genetics

Visit the Mader web site for related ESP activities.

Exploring the Internet

The Mader Home Page provides resources and tools as you study this chapter.

http://www.mhhe.com/biosci/genbio/mader

CHAPTER 24

Patterns of Chromosomal Inheritance

Chapter Concepts

24.1 Inheritance of Chromosomes

- Normally, humans inherit 22 pairs of autosomes and one pair of sex chromosomes for a total of 46 chromosomes. 488
- Abnormalities arise when humans inherit an extra autosome or an abnormal autosome. 489

24.2 Inheritance of Sex Chromosomes

- Normally, males are XY and females are XX. 492
- Abnormalities arise when humans inherit an incorrect number of sex chromosomes. 492

24.3 Sex-Linked Inheritance

- Certain traits, unrelated to the gender of the individual, are controlled by genes located on the sex chromosomes. 496
- Males always express X-linked recessive disorders because they inherit only one X chromosome. 496

24.4 Linked Genes

- Alleles that occur on the same chromosome form a linkage group and tend to be inherited together. 500

This child with Down syndrome has an extra chromosome 21. Our chromosomal inheritance has a marked influence on our physical characteristics because there are many genes per each chromosome.

Susan experienced a tremendous feeling of relief and joy as the genetic counselor explained the significance of the photo of her muscle cells. They were all outlined by stain, which meant that she was not a carrier for Duchenne muscular dystrophy. This was the same disease her brother had slowly and painfully succumbed to before his 18th birthday. She was aware that females can be carriers of the disease, and then their sons have a 50% chance of having the disease. She could hardly wait to tell her husband the good news, because now they could have sons or daughters without fear of them having the disease. Her joy was tempered by the remembrance of her brother's death, and she hoped that sometime in the near future researchers would discover a method of preventing or curing this genetic disease.

24.1 Inheritance of Chromosomes

The chromosomal theory of inheritance says that the genes are on the chromosomes. To view the chromosomes, cells can be treated and photographed just prior to division so that a picture of the chromosomes is obtained. The picture may be entered into a computer and the chromosomes electronically arranged by pairs (Fig. 24.1). The members of a pair not only have the same size and shape, they also have the same banding pattern. The resulting display of pairs of chromosomes is called a **karyotype.** Both males and females normally have 23 pairs of chromosomes, but one of these pairs is of unequal length in males. The larger chromosome of this pair is the **X chromosome** and the smaller is the **Y chromosome.** These are called the **sex chromosomes** because they contain the genes that determine sex. The other

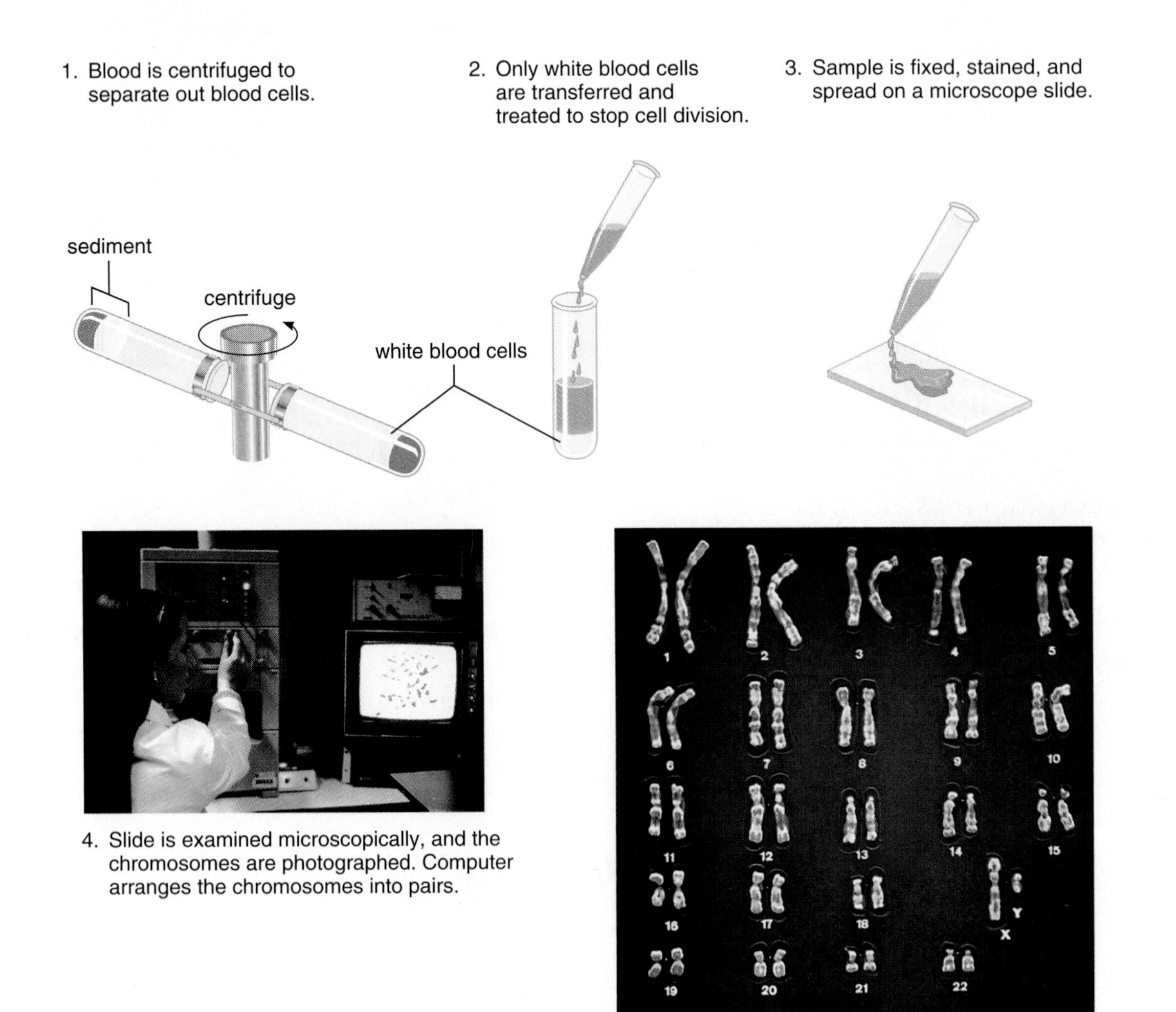

Figure 24.1 Human karyotype preparation.
As illustrated here, the stain used can result in chromosomes with a banded appearance. The bands help researchers identify and analyze the chromosomes.

chromosomes, known as **autosomes,** include all the pairs of chromosomes except the X and Y chromosomes. In a karyotype, autosomes are usually ordered by size and numbered from the largest to smallest.

Nondisjunction

Normally, an individual receives 22 autosomes. Sometimes individuals are born with either too many or too few autosomes, most likely due to nondisjunction during meiosis. **Nondisjunction** occurs during meiosis I when the members of a homologous pair both go into the same daughter cell, or during meiosis II when the sister chromatids fail to separate and both daughter chromosomes go into the same gamete (Fig. 24.2).

Following fertilization, the zygote will have one less chromosome than usual, called a **monosomy,** or one more chromosome than usual, called a **trisomy.**

A study of abortuses (spontaneous abortions) suggests that many autosomal trisomies and nearly all monosomies are fatal. Individuals with trisomy 13 (Patau syndrome) and trisomy 18 (Edward syndrome) have a life span of less than one year. Heart and nervous system defects prevent normal development. A **syndrome** is a group of symptoms that appear together and tend to indicate the presence of a particular disorder. Nondisjunction of sex chromosomes results in XO (Turner syndrome), XXX (triplo-X), XXY (Klinefelter), and XYY (Jacobs). The symbol O means that a sex chromosome is missing.

Nondisjunction causes an abnormal chromosomal number in the gametes. Offspring sometimes inherit an extra chromosome (trisomy) or are missing a chromosome (monosomy).

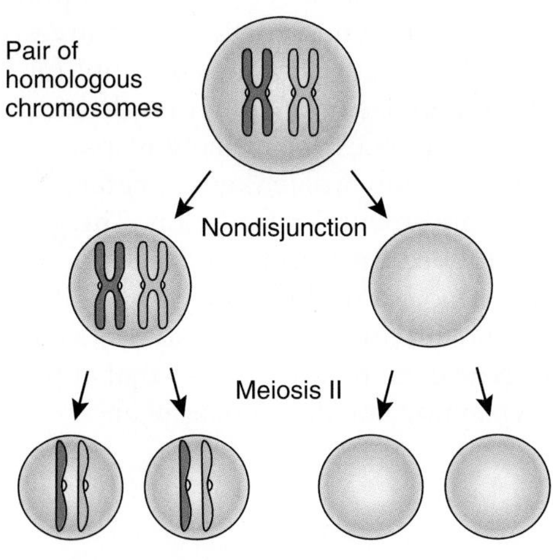

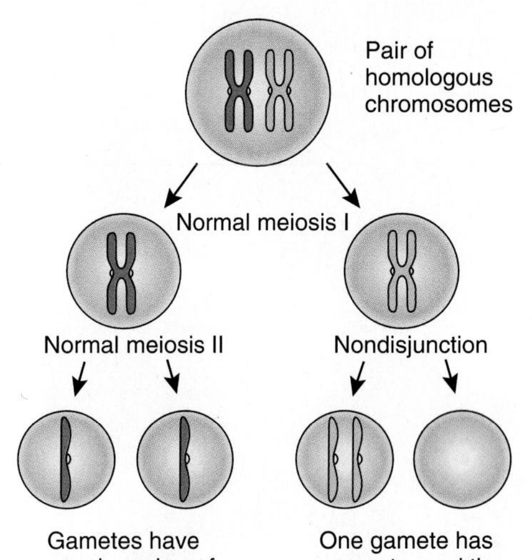

Syndrome	Sex	Chromosomes	Frequency	
			Abortuses	*Births*
Down	M or F	Trisomy 21	1/40	1/800
Patau	M or F	Trisomy 13	1/33	1/15,000
Edward	M or F	Trisomy 18	1/200	1/6,000
Turner	F	XO	1/18	1/6,000
Triplo-X	F	XXX (or XXXX)	0	1/1,500
Klinefelter	M	XXY (or XXXY)	0	1/1,500
Jacobs	M	XYY	?	1/1,000

b.

Figure 24.2 Nondisjunction of chromosomes during meiosis.
a. Nondisjunction occurring during meiosis I if homologous chromosomes fail to separate, and during meiosis II if the daughter chromosomes derived from sister chromatids fail to separate. In either case, abnormal gametes carry an extra chromosome or lack a chromosome.
b. Frequency of syndromes caused by abnormal chromosomal numbers. Abortuses = spontaneous abortions.

Down Syndrome

The most common autosomal trisomy seen among humans is trisomy 21, also called *Down syndrome* (Fig. 24.3). This syndrome is easily recognized by these characteristics: short stature; an eyelid fold; stubby fingers; a wide gap between the first and second toes; a large, fissured tongue; a round head; a palm crease, the so-called simian line; and, unfortunately, mental retardation, which can sometimes be severe.

Persons with Down syndrome usually have three copies of chromosome 21 because the egg had two copies instead of one. (In 23% of the cases studied, however, the sperm had the extra chromosome 21.) The chances of a woman having a Down syndrome child increase rapidly with age, starting at about age 40, and the reasons for this are still being determined.

Although an older woman is more likely to have a Down syndrome child, most babies with Down syndrome are born to women younger than age 40 because this is the age group having the most babies. Karyotyping can detect a Down syndrome child; however, young women are not encouraged to undergo the procedures necessary to get a sample of fetal cells (i.e., amniocentesis or chorionic villi testing) because the risk of complications is greater than the risk of having a Down syndrome child. Fortunately, there is now a test, based on substances in maternal blood, that can help identify fetuses who may need to be karyotyped.

It is known that the genes that cause Down syndrome are located on the bottom third of chromosome 21 (Fig. 24.3*b*), and extensive investigative work has been directed toward discovering the specific genes responsible for the characteristics of the syndrome. Thus far, investigators have discovered several genes that may account for various conditions seen in persons with Down syndrome. For example, they have located genes most likely responsible for the increased tendency toward leukemia, cataracts, accelerated rate of aging, and mental retardation. The gene for mental retardation, dubbed the *Gart* gene, causes an increased level of purines in blood, a finding associated with mental retardation. One day it may be possible to control the expression of the *Gart* gene even before birth so that at least this symptom of Down syndrome does not appear.

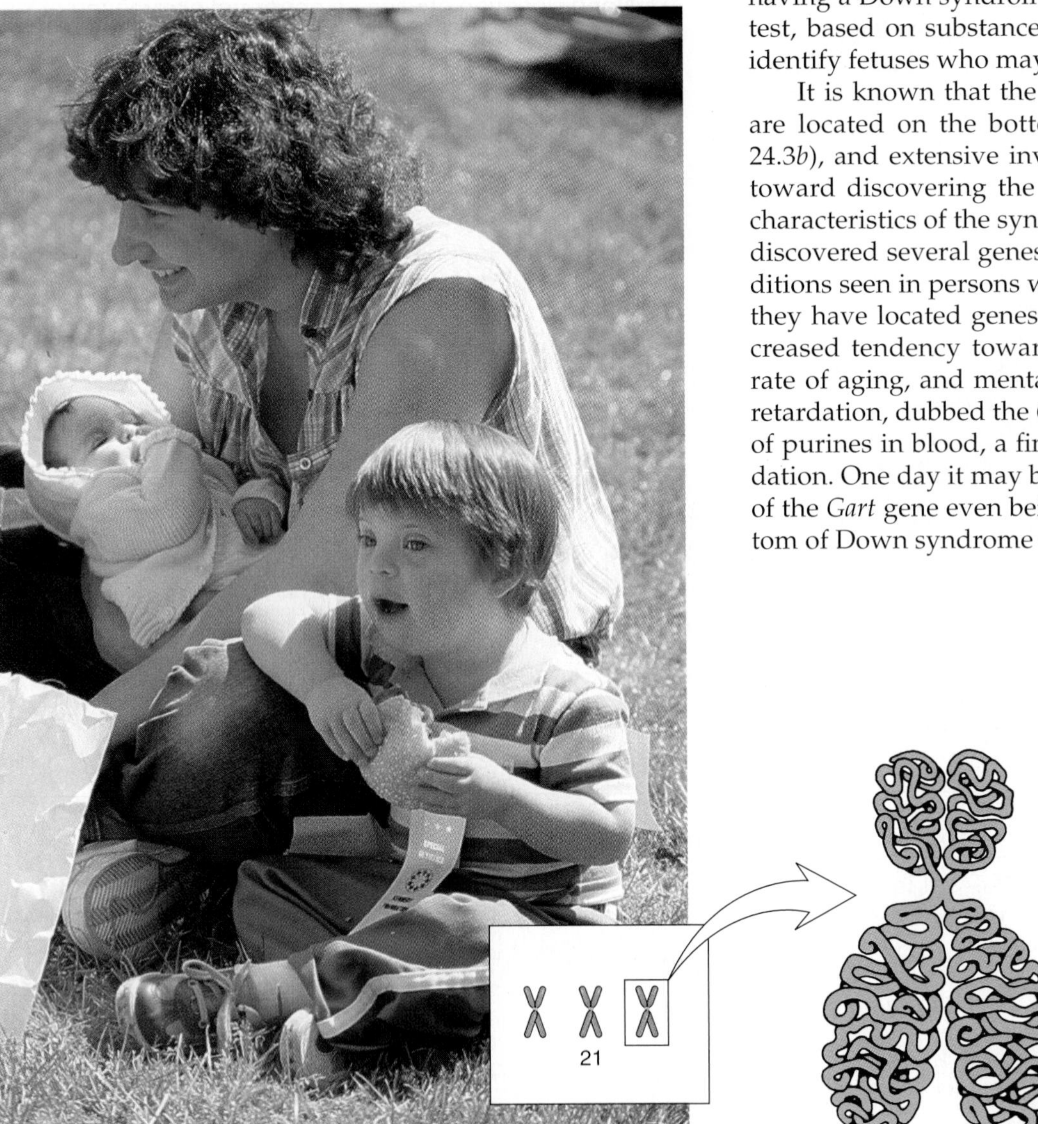

Figure 24.3 Down syndrome.
a. Common characteristics of the syndrome include a wide, rounded face and a fold of the upper eyelids. Mental retardation, along with an enlarged tongue, makes it difficult for a person with Down syndrome to speak distinctly. **b.** Karyotype of an individual with Down syndrome shows an extra chromosome 21. More sophisticated technologies allow investigators to pinpoint the location of specific genes associated with the syndrome. An extra copy of the *Gart* gene, which leads to a high level of purines in blood, may account for the mental retardation seen in persons with Down syndrome.

Chromosomal Mutations

A mutation is a permanent genetic change. A change in chromosomal structure that can be detected microscopically is a **chromosomal mutation.** There are various environmental agents—radiation, certain organic chemicals, or even viruses—that can cause chromosomes to break apart. Ordinarily, when breaks occur in chromosomes, the two broken ends reunite to give the same sequence of genes. Sometimes, however, the broken ends of one or more chromosomes do not rejoin in the same pattern as before, and this results in a chromosomal mutation. Various types of chromosomal mutations occur; an inversion, a translocation, a deletion, and a duplication of chromosomal segments are all illustrated in Figure 24.4.

An *inversion* occurs when a segment of a chromosome is turned around 180 degrees. You might think this is not a problem because the same genes are present, but the reversed sequence of genes can lead to altered gene activity.

A *translocation* is the movement of a chromosomal segment from one chromosome to another, nonhomologous, chromosome. In 5% of cases, a translocation that occurred in a previous generation between chromosomes 21 and 14 is the cause of Down syndrome. In these cases, Down syndrome is not related to the age of the mother, but instead tends to run in the family of either the father or the mother.

A *deletion* occurs when an end of a chromosome breaks off or when two simultaneous breaks lead to the loss of a segment. Even when only one member of a pair of chromosomes is affected, a deletion often causes abnormalities. An example is *cri du chat* (cat's cry) *syndrome* in which a portion of chromosome 5 is deleted. The affected individual has a small head, is mentally retarded, and has facial abnormalities. Abnormal development of the glottis and larynx results in the most characteristic symptom—the infant's cry resembles that of a cat.

A *duplication,* the presence of a chromosomal segment more than once in the same chromosome, can occur in two ways. A broken segment from one chromosome can simply attach to its homologue. The presence of a duplication in the middle of a chromosome is more likely due to unequal crossing-over. This can occur when homologous chromosomes are mispaired slightly; following crossing-over, there is both a gene duplication and a gene deletion:

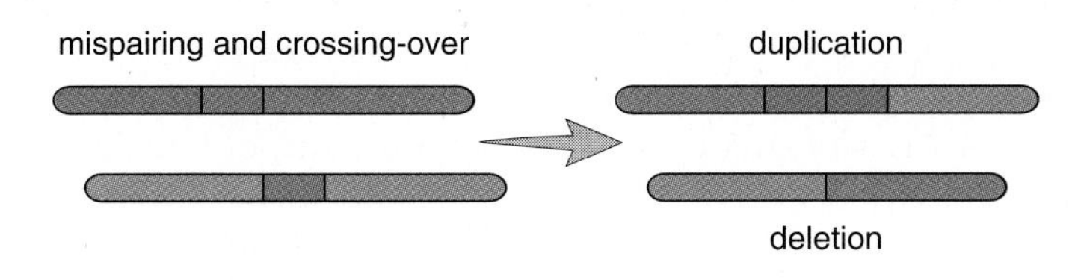

Instances of gene duplication occur in human cells. For example, human cells have multiple copies of the genes that code for the proteins found in hemoglobin. Each of these proteins is slightly different, and some occur only during fetal development.

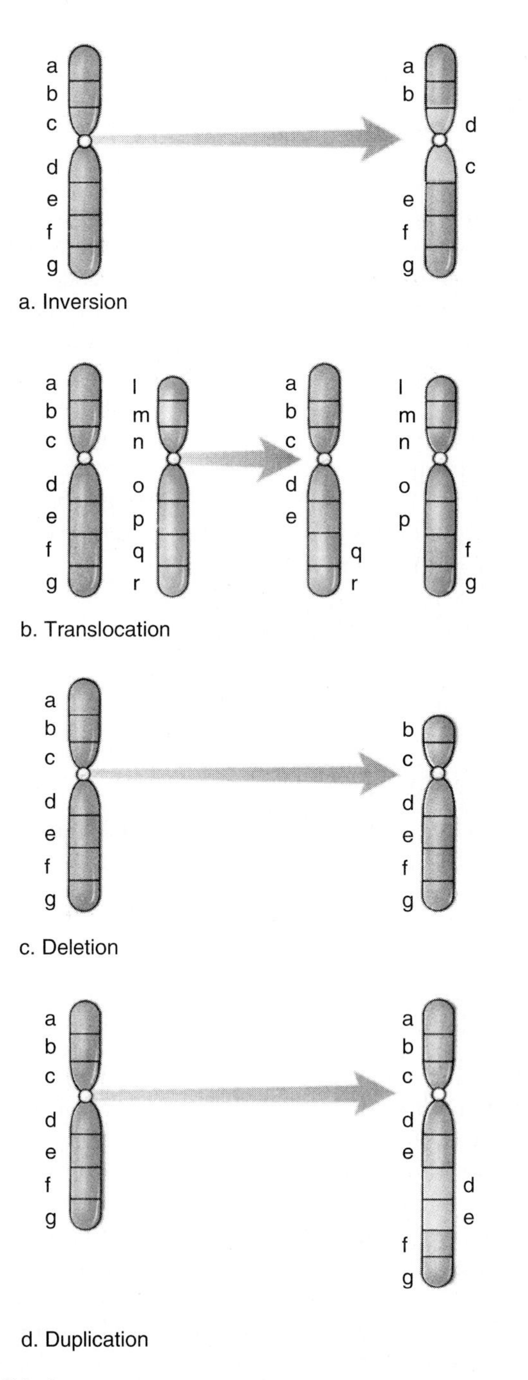

Figure 24.4 Types of chromosomal mutations.
a. Inversion occurs when a chromosomal segment breaks apart and then rejoins in reversed direction. **b.** Translocation is the exchange of chromosomal segments between nonhomologous chromosomes. **c.** Deletion is the loss of a chromosomal segment. **d.** Duplication occurs when the same segment is repeated within the chromosome.

Abnormal chromosomal inheritance can be due to the inheritance of a mutated chromosome. Cri du chat syndrome occurs when a portion of chromosome 5 is deleted.

24.2 Inheritance of Sex Chromosomes

The members of one special pair of chromosomes are called the sex chromosomes because this pair determines the sex of the individual. The sex chromosomes in humans are called X and Y. In humans, but not all animals, females are XX. The egg, therefore, always contains an X. In humans, but not all animals, males are XY, and, therefore, a sperm can contain an X or a Y. The sex of the newborn child is determined by the father: if a Y-bearing sperm fertilizes the egg, then the XY combination results in a male. On the other hand, if an X-bearing sperm fertilizes the egg, the XX combination results in a female. All factors being equal, there is a 50% chance of having a girl or of having a boy. It is possible to illustrate this probability with a *Punnett square:*

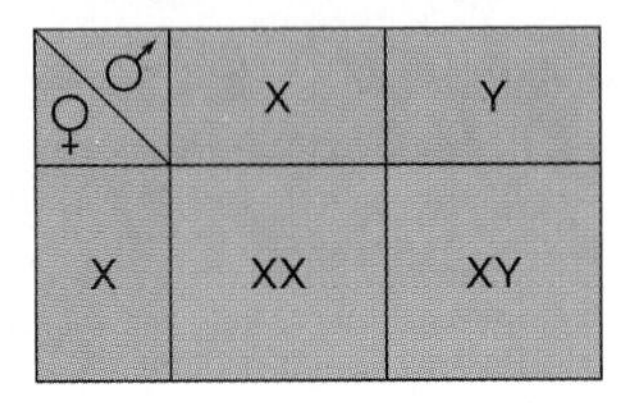

♀ \ ♂	X	Y
X	XX	XY

For reasons that are not known, more males than females are conceived, but from then on, the death rate for males is higher than for females. By age 85 there are twice as many females as males.

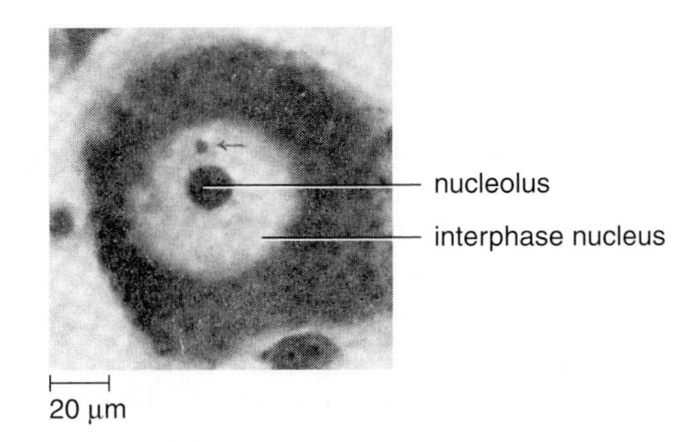

Figure 24.5 Barr bodies.
In the body of females, the nucleus of each cell contains a Barr body (at arrow) which is a condensed, inactive X chromosome. Chance alone dictates which of the two X chromosomes becomes a Barr body.

Barr Bodies

XX females have a darkly staining mass of condensed chromatin called a *Barr body* (after the person who discovered it) in their nuclei (Fig. 24.5). A Barr body is an inactive X chromosome; therefore, female cells function with a single X chromosome just as males do. Chance alone determines which X chromosome is found in a Barr body. If a female is heterozygous, 50% of her cells have Barr bodies containing one allele and the other 50% have Barr bodies with the other allele. This causes the body of heterozygous females to be mosaic, with patches of genetically different cells. The mosaic is exhibited in various ways: some females show columns of both normal and abnormal dental enamel; some have patches of pigmented and nonpigmented cells at the back of the eye; and some have patches of normal and abnormal sweat glands.

X inactivation within Barr bodies can be used to detect female carriers of certain genetic diseases if they are controlled by an allele on the X chromosome. *Lesch-Nyhan syndrome* is a recessive disorder characterized by mental retardation, chewing lips and fingers to the point of mutilation, and the formation of painful urinary stones due to the accumulation of uric acid. These symptoms are due to the lack of an enzyme known by the acronym HGPRT. It is possible to test hair for the presence of this enzyme. If a woman is a carrier, some of the hairs of the head test positive for the enzyme and some test negative. Similarly, a female carrier for Duchenne muscular dystrophy can be detected by testing muscle tissue for the protein dystrophin, which is missing in boys with the disorder. The muscle cells removed by biopsy are exposed to a stain specific for dystrophin, and if the woman is a carrier, only certain cells will be outlined by the stain.

Abnormal Sex Chromosomal Inheritance

Abnormal sex chromosomal inheritance can be due to receiving an abnormal number of sex chromosomes or an abnormal X chromosome. Counting the number of Barr bodies in the nucleus helps determine if there is an abnormal number of X chromosomes. For example, normal males have no Barr body, but XXY males do have a Barr body.

Abnormal Number Syndromes

Abnormal sex chromosomal number inheritance occurs when a person receives too many or too few X or Y chromosomes. Nondisjunction of the sex chromosomes can occur during both oogenesis and spermatogenesis, resulting in abnormal zygotes. After fusion with normal gametes, the syndromes shown in Figure 24.2 occur.

From birth, an XO individual with *Turner syndrome* has only one sex chromosome, an X; the O signifies the absence of a second sex chromosome. Turner females are short, have a broad chest, and webbed neck. The ovaries, oviducts, and uterus are very small and nonfunctional. Turner females do not undergo puberty or menstruate, and there is a lack of breast development (Fig. 24.6*a*). They usually are of normal intelligence and can lead fairly normal lives, but they are always infertile, even if they receive hormone supplements.

A male with *Klinefelter syndrome* has two or more X chromosomes in addition to a Y chromosome. Counting the number of Barr bodies can tell the number of extra X chromosomes. The testes and prostate gland are underdeveloped and there is no facial hair. But there may be some breast development (Fig. 24.6*b*). Affected individuals have large hands and feet and very long arms and legs. They are usually slow to learn but not mentally retarded unless they inherit more than two X chromosomes.

A *triplo-X* individual has more than two X chromosomes and extra Barr bodies in the nucleus. It might be supposed that the XXX female is especially feminine, but this is not the case. Although in some cases there is a tendency toward learning disabilities, most triplo-X females have no apparent physical abnormalities except that they may have menstrual irregularities, including early onset of menopause.

XYY males with *Jacobs syndrome* also can result from nondisjunction during spermatogenesis. Affected males usually are taller than average, suffer from persistent acne, and tend to have speech and reading problems. At one time, it was suggested that these men were likely to be criminally aggressive, but it has since been shown that the incidence of such behavior among them may be no greater than among XY males.

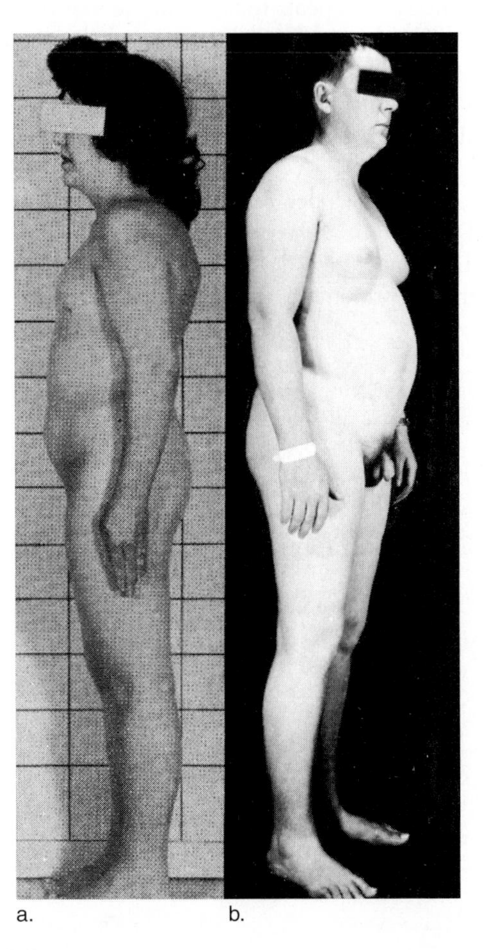

Figure 24.6 Abnormal sex chromosomal inheritance.
a. Female with Turner (XO) syndrome has a short thick neck, short stature, and lack of breast development. **b.** A male with Klinefelter (XXY) syndrome has immature sex organs and some development of the breasts.

Fragile X Syndrome

Males outnumber females by about 25% in institutions for the mentally retarded. In some of these males, the X chromosome is nearly broken, leaving the tip hanging by a flimsy thread. These males are said to have *fragile X syndrome* (Fig. 24.7).

As children, fragile X syndrome individuals may be hyperactive or autistic; their speech is delayed in development and often repetitive in nature. As adults, they have large testes and big, usually protruding, ears. They are short in stature, but the jaw is prominent and the face is long and narrow. Stubby hands, lax joints, and a heart defect may also occur. Fragile X chromosome also occurs in females, but when symptoms do appear in females, they tend to be less severe. Fragile X syndrome is inherited in an unusual pattern, as discussed in the reading on page 494.

Usually females are XX and males are XY. Individuals sometimes inherit an abnormal number of sex chromosomes or an abnormal X chromosome, as in fragile X syndrome.

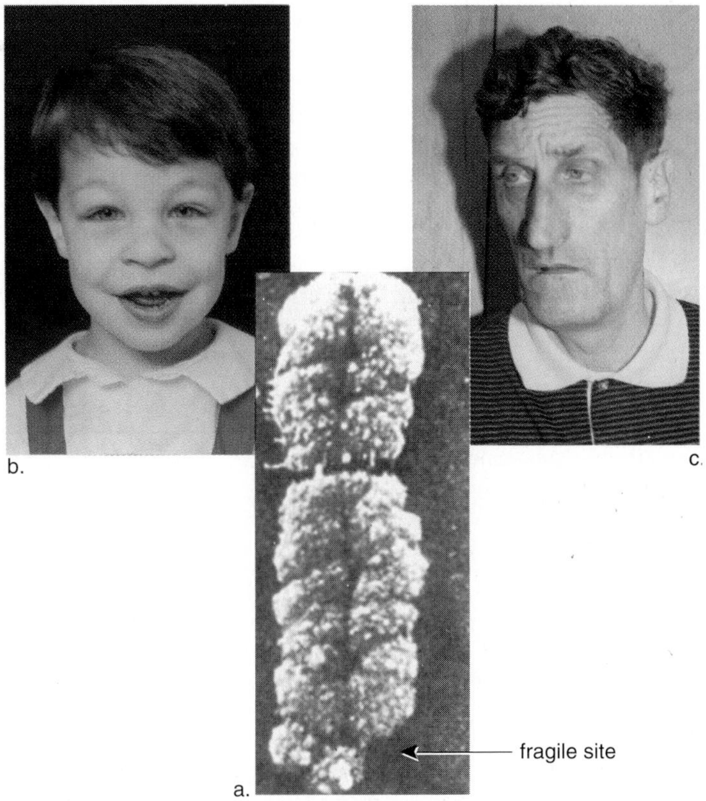

Figure 24.7 Fragile X syndrome.
a. An arrow points out the fragile site of this fragile X chromosome. **b.** A young person with the syndrome appears normal but (**c**) with age, the elongated face has a prominent jaw, and the ears noticeably protrude.

Science Focus

Fragile X Syndrome

Fragile X syndrome is one of the most common genetic causes of mental retardation, second only to Down syndrome. It affects about one in 1,500 males and one in 2,500 females and is seen in all ethnic groups. It is called fragile X syndrome because its diagnosis used to be dependent upon observing an X chromosome whose tip is attached to the rest of the chromosome by only a thin thread.

The inheritance pattern of fragile X syndrome is not like any other pattern we have studied (Fig. 24A). The chance of being affected increases in successive generations almost as if the pattern of inheritance switches from being a recessive one to a dominant one. Then, too, an unaffected grandfather can have grandchildren with the disorder; in other words, he is a carrier for an X-linked disease. This is contrary to what occurs with other mutant alleles on the X chromosome; in those cases, the male always shows the disorder.

In 1991, the DNA sequence at the fragile site was isolated and found to have trinucleotide repeats. The base triplet, CGG, was repeated over and over again. There are about 6–50 copies of this repeat in normal persons but over 230 copies in persons with fragile X syndrome. Carrier males have what is now termed a premutation; they have between 50 and 230 copies of the repeat and no symptoms. Both daughters and sons receive the premutation, but only the daughters pass on the full mutation—that is, over 230 copies of the repeat. Any male, even those with fragile X syndrome and over 230 repeats, passes on at most the premutation number of repeats. It is unknown what causes the difference between males and females.

This type of mutation—called by some a dynamic mutation, because it changes, and by others an expanded trinucleotide repeat, because the number of triplet copies increases—is now known to characterize other conditions (Table 24A). With Huntington disease, the age of onset of the disorder is roughly correlated with the number of repeats, and the disorder is more likely to have been inherited from the father. For autosomal conditions, we would expect the sex of the parent to play no role in inheritance. The present exceptions have led to the genomic imprinting hypothesis—that the sperm and egg carry chromosomes that have been "imprinted" differently. Imprinting is believed to occur during gamete formation, and thereafter, the genes are expressed one way if donated by the father and another way if donated by the mother. Perhaps when we discover why more repeats are passed on by one parent than the other, we will discover the cause of so-called genomic imprinting.

What might cause repeats to occur in the first place? Something must go wrong during DNA replication prior to cell division. The difficulty that causes triplet repeats is not known. But when DNA codes for cellular proteins, the presence of repeats undoubtedly leads to nonfunctioning or malfunctioning proteins.

Scientists have developed a new technique that can identify repeats in DNA, and they expect that this technique will help them find the genes for other human disorders. They expect expanded trinucleotide repeats to be a very common mutation indeed.

Table 24A Human Genetic Disorders Caused by Base Triplet Repeats

Chromosome	Disease	Sex Bias of Parent Donating Severe Form	Repeated Sequence	Normal Number of Copies	Number of Copies Associated with the Disease
Fragile X syndrome	X chromosome	Maternal	CGG	6–50	Premutation = 50–230 Full mutation = 230–2,000
Spinobulbar muscular dystrophy (Kennedy disease)	X chromosome	?	CAG	11–40	40–62
Myotonic dystrophy	Chromosome 19	Maternal	CTG	5–50	Premutation = 50–80 Full mutation = 80–2,000
Huntington disease	Chromosome 4	Paternal	CAG	11–34	Premutation = 34–42 Full mutation = 42–121
Spinocerebellar ataxia type 1	Chromosome 6	Paternal (possibly)	CAG		25–43 43–81

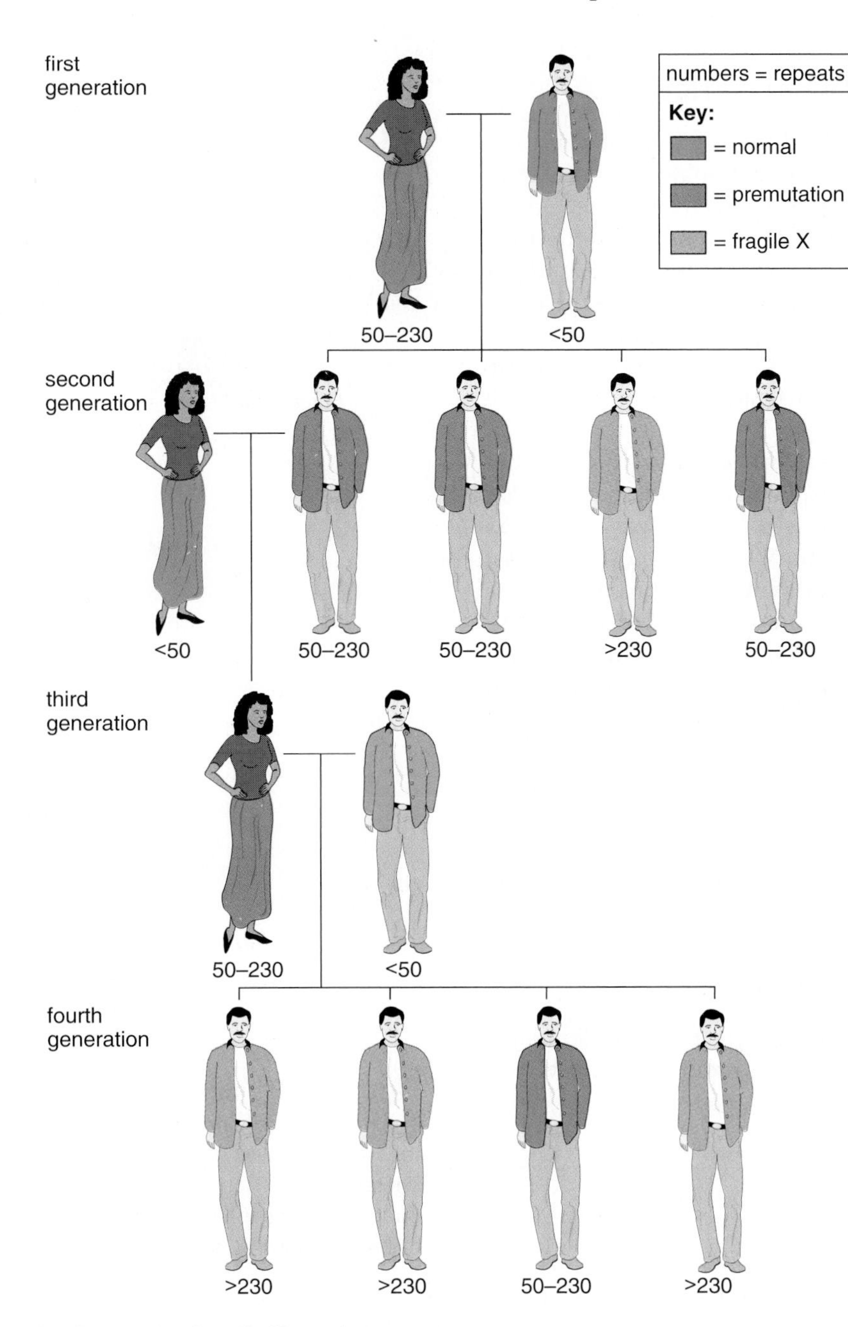

Figure 24A Pattern of inheritance for fragile X syndrome.

Fragile X syndrome is a disorder caused by the presence of base triplet repeats at a particular locus. Affected persons have over 230 repeats, a premutation is 50–230 repeats, and normal persons have fewer than 50 repeats. In each successive generation there are more affected individuals, usually males. The mutation is passed on by women with only a premutation number of repeats; men, even if affected, do not pass on the mutation.

24.3 Sex-Linked Inheritance

Embryos begin life with no evidence of a gender, but by about the third month of development, males can be distinguished from females. Investigators have now discovered a whole series of genes on the Y chromosome that determine the development of male genitals, and at least one on the X chromosome that seems to be necessary for the development of female genitals. It turns out, though, that there are also genes on the sex chromosomes that have nothing to do with sexual development and instead are concerned with other body traits.

Such traits are said to be **sex-linked traits:** an allele that is only on the X chromosome is X-linked, and an allele that is only on the Y chromosome is Y-linked. Very few sex-linked genes have been found on the Y chromosome, most likely because it is much smaller than the X chromosome.

The X chromosomes carry many sex-linked traits, and we will look at a few of these in depth. It would be logical to suppose that a sex-linked trait is passed from father to son or from mother to daughter, but this is not the case. A male always receives a sex-linked condition from his mother, from whom he inherited an X chromosome. The Y chromosome from the father does not carry an allele for the trait. Usually the trait is recessive; therefore, a female must receive two alleles, one from each parent, before she has the condition.

Solving X-Linked Genetics Problems

Recall that when solving autosomal genetics problems, we represent the genotypes of males and females similarly, as shown in the following example for humans.

Key:	Genotypes
W = Widow's peak	WW, Ww, or ww
w = Continuous hairline	

In contrast, when we set up the key for an X-linkage problem, the key looks like this:

Key:

X^B = Normal vision
X^b = Color blindness

The possible genotypes in both males and females are as follows:

X^BX^B = Female who has normal color vision
X^BX^b = Carrier female who has normal color vision
X^bX^b = Female who is color blind
X^BY = Male who has normal vision
X^bY = Male who is color blind

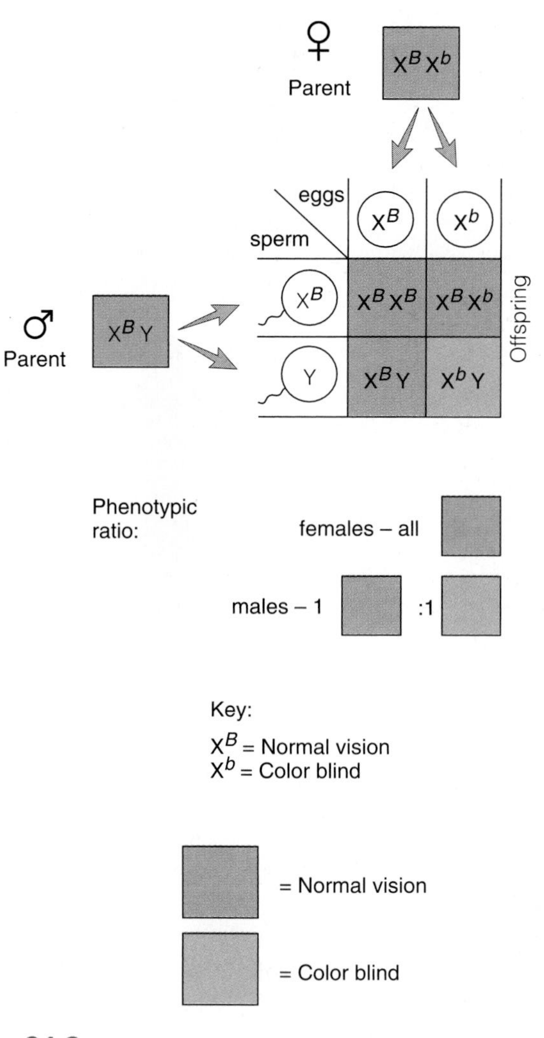

Figure 24.8 Inheritance of a sex-linked trait.
The male parent is normal, but the female parent is a carrier; an allele for color blindness is located on one of her X chromosomes. Therefore, each son stands a 50% chance of being color blind. The daughters will appear to be normal but each one stands a 50% chance of being a carrier.

Note that the second genotype is a *carrier* female, because although a female with this genotype appears normal, she is capable of passing on an allele for color blindness. Color-blind females are rare because they must inherit the allele from both parents; color-blind males are more common since they need only one recessive allele to be color blind. The allele for color blindness has to be inherited from the mother because it is only on the X chromosome; males inherit only the Y chromosome from their father.

Now, let us consider a particular cross. If a heterozygous woman reproduces with a man with normal vision, what are the chances of their having a color-blind daughter? a color-blind son?

Parents: $X^BX^b \times X^BY$

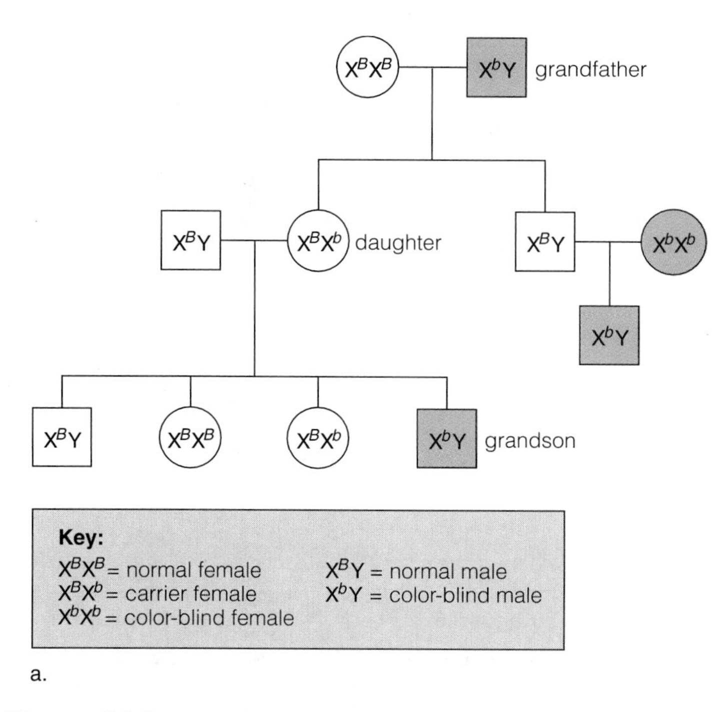

a.

X-linked Recessive Disorders

- More males than females are affected.
- An affected son can have parents who have the normal phenotype.
- For a female to have the characteristic, her father must also have it. Her mother must have it or be a carrier.
- The characteristic often skips a generation from the grandfather to the grandson.
- If a woman has the characteristic, all of her sons will have it.

b.

Among 205 catalogued X-linked recessive disorders are:
- Agammaglobulinemia—lack of immunity to infections
- Color blindness—inability to distinguish certain colors
- Hemophilia—defect in blood-clotting mechanisms
- Muscular dystrophy (some forms)—progressive wasting of muscles
- Spinal ataxia (some forms)—spinal cord degeneration

c.

Figure 24.9 X-linked recessive disorders.
a. Sample pedigree chart. **b.** Ways to recognize X-linked recessive disorders. **c.** Common X-linked recessive disorders.

Inspection indicates that all daughters will have normal color vision because they all will receive an X^B from their father. The sons, however, have a 50% chance of being color blind, depending on whether they receive an X^B or an X^b from their mother. The inheritance of a Y chromosome from their father cannot offset the inheritance of an X^b from their mother.

Figure 24.8 illustrates the use of the Punnett square for solving X-linkage problems. Notice that when a cross involves an X-linked allele, the phenotypic results are given separately for males and females.

Pedigree Charts

Figure 24.9 gives a pedigree chart for an X-linked recessive condition. In pedigree charts, males are designated by squares and females by circles. Shaded circles and squares are affected individuals. A line between a square and a circle represents a sexual union. Children are connected to this line.

Figure 24.9 lists ways to recognize the pattern of X-linked recessive inheritance. Try to explain why each of these is to be expected. For example, why is it be expected that more males than females are affected?

The X chromosome carries alleles that are not on the Y chromosome. Therefore, a recessive allele on the X chromosome is expressed in males.

Muscular Dystrophy

Muscular dystrophy, as the name implies, is characterized by a wasting away of the muscles. The most common form, *Duchenne muscular dystrophy,* is X-linked and occurs in about one out of every 3,600 male births. Symptoms, such as waddling gait, toe walking, frequent falls, and difficulty in rising, may appear as soon as the child starts to walk. Muscle weakness intensifies until the individual is confined to a wheelchair. Death usually occurs by age 20; therefore, affected males are rarely fathers. The recessive allele remains in the population by passage from carrier mother to carrier daughter.

Recently, the gene for muscular dystrophy was isolated, and it was discovered that the absence of a protein now called dystrophin is the cause of the disorder. Much investigative work determined that dystrophin is involved in the release of calcium from the sarcoplasmic reticulum in muscle fibers. The lack of dystrophin causes calcium to leak into the cell, which promotes the action of an enzyme that dissolves muscle fibers. When the body attempts to repair the tissue, fibrous tissue forms, and this cuts off the blood supply so that more and more cells die.

A test is now available to detect carriers for Duchenne muscular dystrophy. Also, various treatments are being attempted. Immature muscle cells can be injected into muscles, and for every 100,000 cells injected, dystrophin production occurs in 30–40% of muscle fibers. The gene for dystrophin has been inserted into the thigh muscle cells of mice, and about 1% of these cells then produced dystrophin.

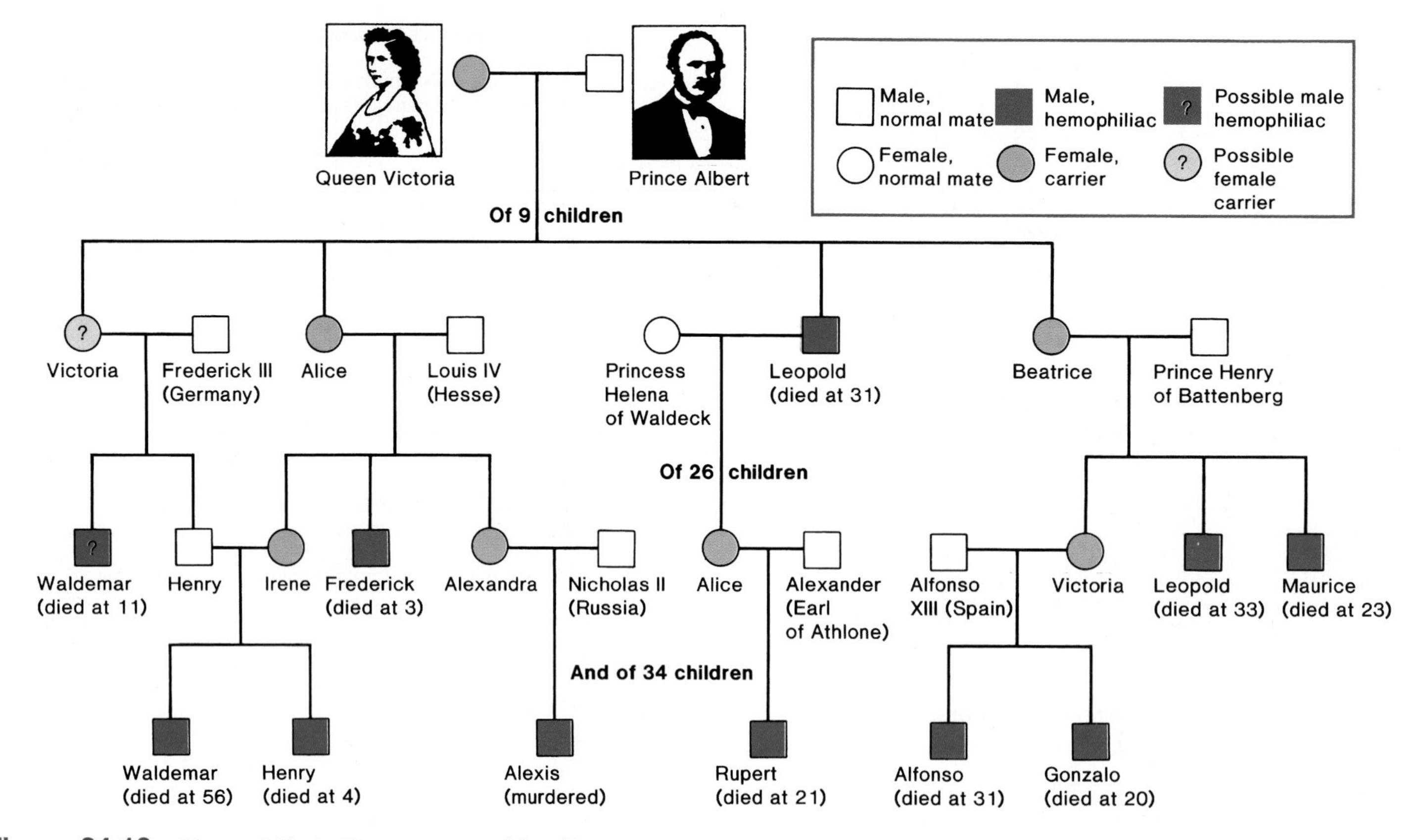

Figure 24.10 Hemophilia in European royal families.
Because Queen Victoria was a carrier, each of her sons had a 50% chance of having the disease, and each of her daughters had a 50% chance of being a carrier. This pedigree shows only the affected individuals. Many others are unaffected, such as the members of the present British royal family.

Hemophilia

About one in 10,000 males is a hemophiliac. The most common type of hemophilia is hemophilia A, due to the absence or minimal presence of a particular clotting factor called factor VIII. *Hemophilia* is called the bleeder's disease because the affected person's blood does not clot. Although hemophiliacs bleed externally after an injury, they also suffer from internal bleeding, particularly around joints. Hemorrhages can be checked with transfusions of fresh blood (or plasma) or concentrates of the clotting protein. Unfortunately, some hemophiliacs have contracted AIDS after receiving blood or using a blood concentrate, but donors are now screened more closely and donated blood is now tested for HIV. Also, factor VIII is now available as a biotechnology product.

At the turn of the century, hemophilia was prevalent among the royal families of Europe, and all of the affected males could trace their ancestry to Queen Victoria of England (Fig. 24.10). Because none of Queen Victoria's forebears or relatives was affected, it seems that the gene she carried arose by mutation either in Victoria or in one of her parents. Her carrier daughters, Alice and Beatrice, introduced the gene into the ruling houses of Russia and Spain, respectively. Alexis, the last heir to the Russian throne before the Russian Revolution, was a hemophiliac. There are no hemophiliacs in the present British royal family because Victoria's eldest son, King Edward VII, did not receive the gene and therefore could not pass it on to any of his descendants.

Color Blindness

In humans, there are three genes involved in distinguishing color because there are at least three different types of cones, the receptors for color vision in the retina of the eyes. Two of these are X-linked; one gene affects the green-sensitive cones, whereas the other affects the red-sensitive cones. About 5% of Caucasian men are color blind due to a mutation involving green perception, and about 2% are color blind due to a mutation involving red perception.

Color blindness is determined by using test plates for the condition (Fig. 24.11).

Certain traits that have nothing to do with the gender of the individual are controlled by genes on the X chromosomes. Males have only one X chromosome, and therefore X-linked recessive alleles are expressed.

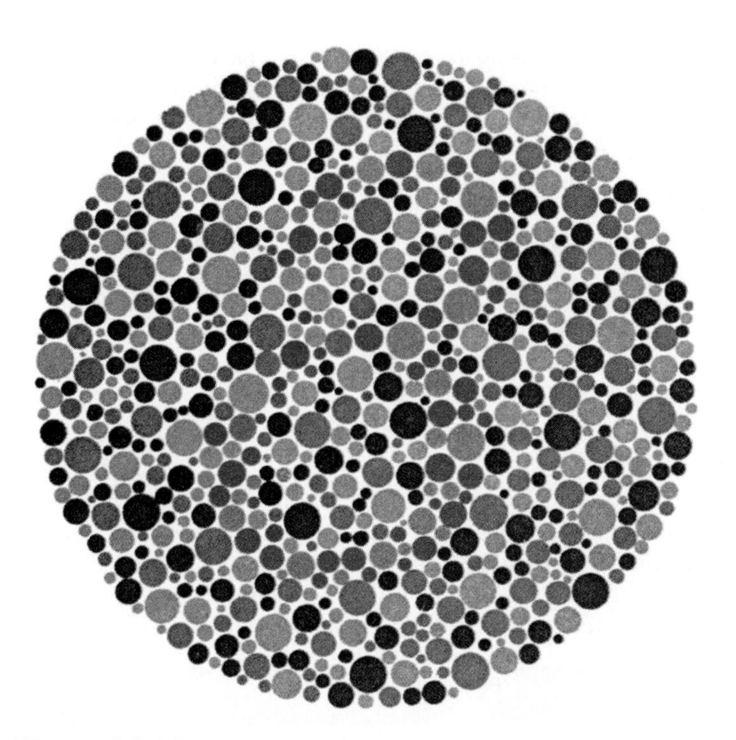

Figure 24.11 Test plate for color blindness.
The most common form of color blindness involves an inability to distinguish reds and greens, determined by an inability to see the number embedded in the dots of a test plate. The above has been reproduced from Ishihara's Test for Colour Blindness published by KANEHARA & CO., LTD., Tokyo, Japan. For accurate testing, the original plate should be used.

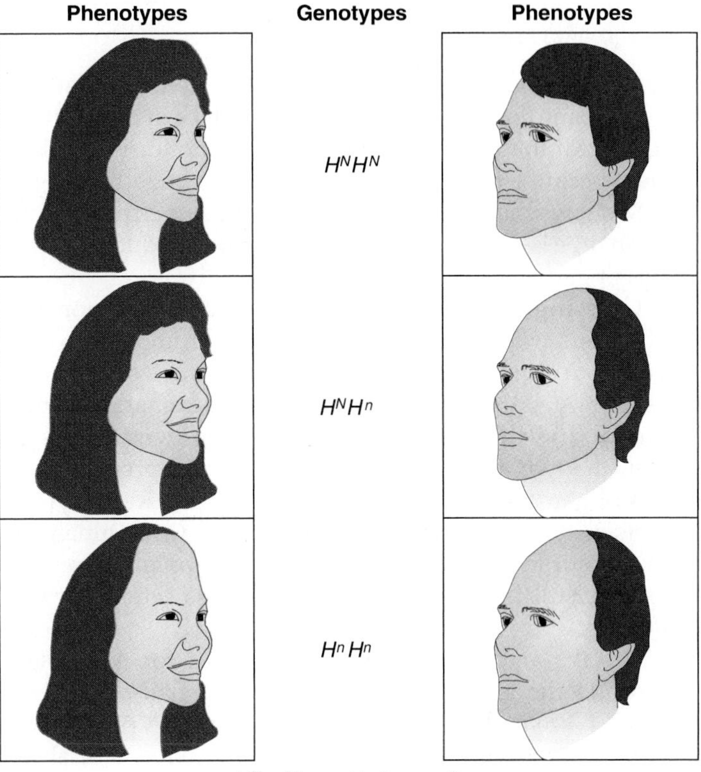

Figure 24.12 Sex-influenced traits.
Pattern baldness is sex-influenced. Due to hormonal influences, the presence of only one allele for baldness causes the condition in the male, whereas the condition does not occur in the female unless she possesses two alleles for baldness.

Practice Problems*

1. Both the mother and the father of a male hemophiliac appear to be normal. From whom did the son inherit the allele for hemophilia? What are the genotypes of the mother, the father, and the son?
2. A woman is color blind. What are the chances that her sons will be color blind? If she is married to a man with normal vision, what are the chances that her daughters will be color blind? will be carriers?
3. Both parents are right handed (R = right handed, r = left handed) and have normal vision. Their son is left handed and color blind. Give the genotypes of all persons involved.
4. Both the husband and the wife have normal vision. The wife gives birth to a color-blind daughter. What can you deduce about the girl's parentage?

*Answers to Practice Problems appear in Appendix A.

Sex-Influenced Traits

Not all traits we associate with the gender of the individual are sex-linked traits. Some are simply **sex-influenced traits;** that is, the phenotype is determined by autosomal genes that are expressed differently in males and females.

It is possible that the sex hormones determine whether these genes are expressed or not. Pattern baldness (Fig. 24.12) is believed to be influenced by the male sex hormone testosterone because males who take the hormone to increase masculinity begin to lose their hair. A more detailed explanation has been suggested by some investigators. It has been reasoned that due to the effect of hormones, males require only one allele for baldness in order for the condition to appear, whereas females require two alleles. In other words, the allele for baldness acts as a dominant in males but as a recessive in females. Another sex-influenced trait of interest is the length of the index finger. In females, an index finger longer than the fourth finger (ring finger) seems to be dominant. In males, an index finger longer than the fourth finger seems to be recessive.

24.4 Linked Genes

The chromosomal theory of inheritance predicts that each chromosome contains a long series of alleles in a definite sequence. All the alleles on one chromosome form a **linkage group** because they tend to be inherited together. Figure 24.13*a* shows that if two alleles are on the same chromosome and linkage is complete, a dihybrid would produce only two types of gametes in equal proportion.

Crossing-over, you recall, occurs between nonsister chromatids when homologous pairs of chromosomes come together prior to separation during meiosis. During crossing-over the nonsister chromatids exchange genetic material and therefore genes. If crossing-over occurs between the two alleles of interest, a dihybrid produces four types of gametes instead of two (Fig. 24.13*b*). **Recombinant** means a new combination of alleles. The recombinant gametes occur in reduced number because crossing-over is infrequent. Still, all possible phenotypes will occur among the offspring.

To take an actual example, the genes for ABO blood types and the gene for a condition called nail-patella syndrome (NPS) are on the same chromosome. A person with NPS has fingernails and toenails that are reduced or absent and a kneecap (patella) that is small. NPS (*N*) is dominant, while the normal condition (*n*) is recessive. In one family, the spouses had these chromosomes, and the results of their mating were predicted as shown in this diagram:

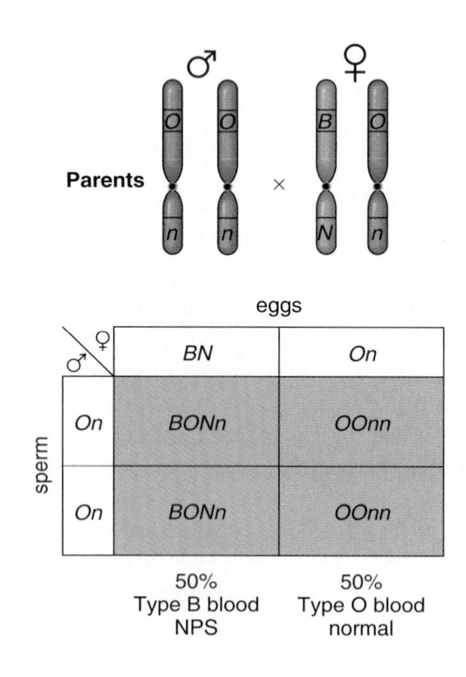

Linkage was not complete; that is, crossing-over occurred, and 10% of the children had recombinant phenotypes: 5% had type B blood and no NPS, and 5% had type O blood and NPS:

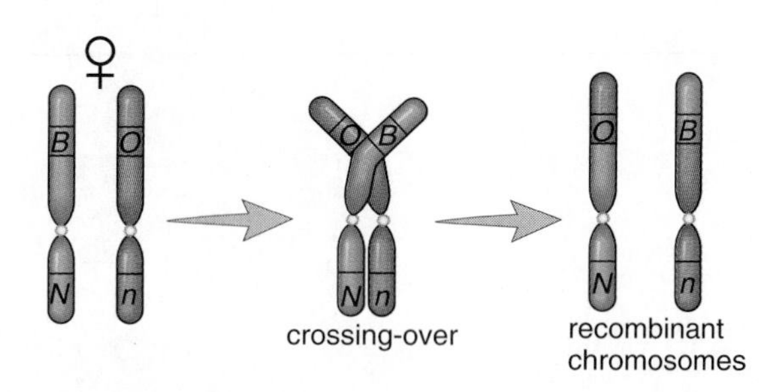

The occurrence of crossing-over helps tell the sequence of genes on a chromosome because crossing-over occurs more often between distant genes than between genes that are close together on a chromosome. For example, consider these homologous chromosomes:

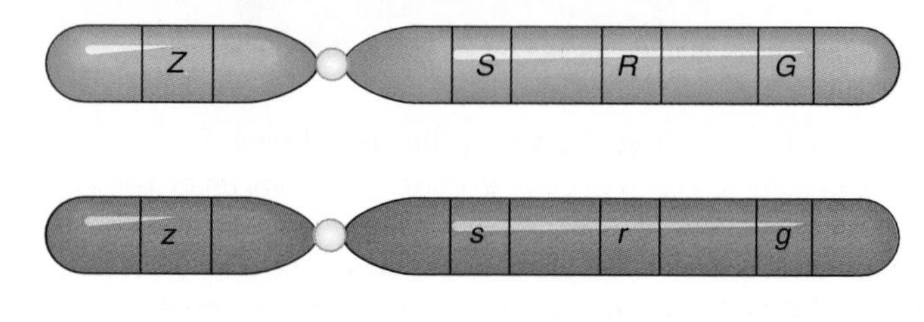

pair of homologous chromosomes

We expect recombinant gametes to include *G* and *z* more often than *R* and *s*. In keeping with this observation, investigators began to use recombination frequencies to map the chromosomes. Each 1% of crossing-over is equivalent to one map unit between genes.

Linkage data have been used to map the chromosomes of the fruit fly *Drosophila,* but the possibility of using linkage data to map human chromosomes is limited because we can only work with matings that have occurred by chance. This, coupled with the fact that humans tend not to have numerous offspring, means that additional methods are used to sequence the genes on human chromosomes. Today, it is customary to also rely on biochemical methods to map the human chromosomes.

The presence of linkage groups changes the expected results of genetic crosses. The frequency of recombinant gametes that occurs due to the process of crossing-over has been used to map the chromosomes.

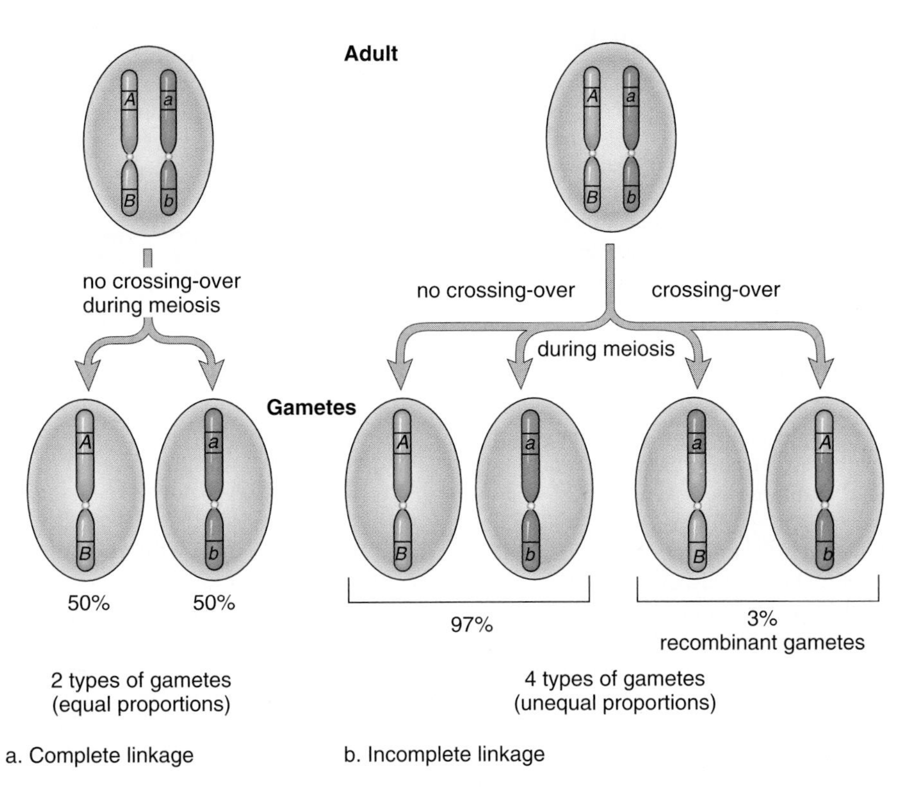

Figure 24.13 Linkage group.
In this individual, alleles *A* and *B* are on one member of a homologous pair, and alleles *a* and *b* are on the other member. **a.** When linkage is complete, this dihybrid produces only two types of gametes in equal proportion. **b.** When linkage is incomplete, this dihybrid produces four types of gametes because crossing-over has occurred. The recombinant gametes occur in reduced proportion because crossing-over occurs infrequently.

Bioethical Issue

Do you approve of choosing a baby's gender even before it is conceived? As you know, the sex of a child is dependent upon whether an X-bearing sperm or a Y-bearing sperm enters the egg. A new technique has been developed that can separate X-bearing sperm from Y-bearing sperm. First, the sperm are dosed with a DNA-staining chemical. Because the X chromosome has slightly more DNA than the Y chromosome, it takes up more dye. When a laser beam shines on the sperm, the X-bearing sperm shine a little more brightly than the Y-bearing sperm. A machine sorts the sperm into two groups on this basis. The results are not perfect. Following artificial insemination, there's about an 85% success rate for a girl and about a 65% success rate for a boy.

Some might argue that while it is acceptable to use vaccines to prevent illnesses or give someone a heart transplant, it goes against nature to choose gender. But what if the mother is a carrier of an X-linked genetic disorder such as hemophilia or Duchenne muscular dystrophy? Or, is it acceptable to bring a child into the world with a genetic disorder that may cause an early death? Would it be better to select sperm for a girl, who at worst would be a carrier like her mother? Previously a pregnant woman with these concerns had to wait for the results of an amniocentesis test, and then abort the pregnancy if it were a boy. Is it better to increase the chances of a girl to begin with?

Some authorities do not find gender selection acceptable for any reason. Even if it doesn't lead to a society with far more members of one sex than another, there could be a problem. Once you separate reproduction from the sex act, they say, it opens the door to children that have been genetically designed in every way.

Questions

1. Do you think it is acceptable to choose the gender of a baby? Even if it requires artificial insemination at a clinic? Why or why not?
2. Do you see any difference between choosing gender and choosing eggs or embryos free of a genetic disease for reproduction purposes ? Explain.
3. As a society, should we accept certain ways of interfering with nature and not accept other ways? Why or why not?

Summarizing the Concepts

24.1 Inheritance of Chromosomes

Humans inherit 22 autosomes from each parent. Nondisjunction during meiosis can result in an abnormal number of autosomes to be inherited. Down syndrome results when an individual inherits three copies of chromosome 21. Also, chromosomal mutations lead to phenotypic abnormalities; for example, in *cri du chat* syndrome, one copy of chromosome 5 has a deletion.

24.2 Inheritance of Sex Chromosomes

The father determines the sex of a child because the mother gives only an X chromosome while the father gives an X or a Y chromosome. Males who inherit a fragile X chromosome are subject to mental retardation. Nondisjunction of the sex chromosomes can also cause abnormal sex chromosomal numbers in offspring. Females who are XO have Turner syndrome, and those who are XXX are triplo-X females. Males with Klinefelter syndrome are XXY. There are also XYY males.

Females have an inactive condensed X chromosome in their nuclei called a Barr body. If heterozygous, their cells differ in which allele is active. Sometimes this allows them to be tested to see if they are carriers for a genetic disease.

24.3 Sex-Linked Inheritance

Because males normally receive only one X chromosome, they are subject to disorders caused by the inheritance of a recessive allele on the X chromosome. For example, in a cross between a normal male and a carrier female, only the male children could have the X-linked disorder color blindness. Other well-known X-linked disorders are hemophilia and Duchenne muscular dystrophy.

24.4 Linked Genes

All the genes on one chromosome form a linkage group, which is broken only when crossing-over occurs. Genes that are linked tend to go together into the same gamete. If crossing-over occurs, a dihybrid cross gives all possible phenotypes among the offspring, but the expected ratio is greatly changed. Crossing-over data is used to map the chromosomes of animals, such as fruit flies, but is not sufficient to map the human chromosomes.

Studying the Concepts

1. What does the normal human karyotype look like? 488
2. Diagram the occurrence of nondisjunction of autosomes during meiosis I and during meiosis II. 489
3. What is the most common autosomal abnormality seen in humans? What causes this abnormality? 490
4. Name and describe four chromosomal mutations. 491
5. Describe fragile X syndrome, Turner syndrome, Klinefelter syndrome, triplo-X individuals, and Jacobs syndrome. 492–95
6. What is a Barr body, and how do Barr bodies make it possible to detect female carriers of genetic diseases? 496
7. Name four ways to recognize an X-linked recessive disorder. Why do males exhibit such disorders more often than females? 496–98
8. Explain the occurrence of sex-influenced traits. How do they differ from sex-linked traits? 499
9. What is a linkage group, and how can the occurrence of linkage groups help to map the human chromosomes? 500
10. Name two methods being used to develop a genetic map of the human genome. What is the physical map of the human genome? 500

Testing Yourself

Choose the best answer for each question.
For questions 1–3, match the conditions in the key with the descriptions below.

Key:
a. Down syndrome
b. Turner syndrome
c. Klinefelter syndrome
d. XYY

1. ________ male with underdeveloped testes and some breast development
2. ________ trisomy 21
3. ________ XO female
4. Down syndrome
   a. is always caused by nondisjunction of chromosome 21.
   b. shows no overt abnormalities.
   c. is more often seen in children of mothers past the age of 40.
   d. Both a and c are correct.
5. A male has a genetic disorder. Which one of these is inconsistent with X-linked recessive inheritance?
   a. Both parents do not have the disorder.
   b. Only males in a pedigree chart have the disorder.
   c. Only females in previous generations have the disorder.
   d. Both a and c are inconsistent.
6. John has hemophilia but his parents do not. Using *H* for normal and *h* for hemophilia, give the genotype of his father, mother, and himself in that order.
   a. *Hh*, *Hh*, *hh*
   b. X^HY, *hh*, X^HY
   c. X^HY, X^HX^h, X^hY
   d. X^hY, X^HX^H, X^hY
   e. X^HY, X^hY, X^hy
7. A woman who is homozygous dominant for widow's peak and is a carrier of color blindness reproduces with a man who is heterozygous for widow's peak and has normal vision. What are the chances a son will be color blind with a widow's peak?
   a. 100%
   b. 50%
   c. 24%
   d. 0%
   e. Both b and d are correct.
8. The ability to curl the tongue (dominant) is linked to a rare form of mental retardation (dominant). The parents are heterozygous with dominant alleles on one chromosome and recessive alleles on the other. What is the expected phenotypic ratio among the offspring if crossing-over does not occur?
   a. 3:1
   b. 1:1
   c. 9:3:3:1
   d. 1:1:1:1

9. Which chromosomal mutation is opposite to a deletion?
   a. inversion
   b. translocation
   c. duplication
   d. Both a and b are correct.
   e. Both b and c are correct.
10. This pedigree chart pertains to color blindness. The genotype of the starred individual is ________ .

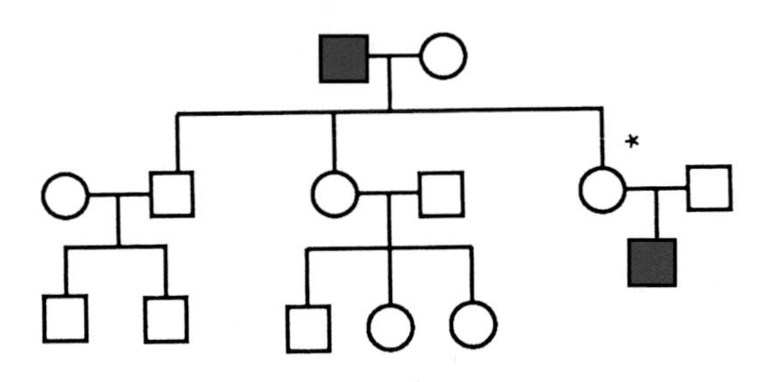

Additional Genetics Problems for Chapter 23 and Chapter 24

Autosomal One-Trait Problems

1. A woman heterozygous for polydactyly (dominant), a condition that produces 6 fingers and 6 toes, is married to a man without this condition. What are the chances that her children will have 6 fingers and 6 toes?
2. A young man's father has just been diagnosed as having Huntington disease (dominant). What are the probable chances that the son will inherit this condition?
3. Black hair is dominant over blond hair. A woman with black hair whose father had blond hair reproduces with a blond-haired man. What are the chances of this couple having a blond-haired child?
4. Your maternal grandmother Smith had Huntington disease. Aunt Jane, your mother's sister, also had the disease. Your mother dies at age 75 with no signs of Huntington disease. What are your chances of getting the disease?
5. Could a person who can curl her tongue (dominant) have parents who cannot curl their tongues? Explain your answer.
6. Parents who do not have Tay-Sachs disease (recessive) produce a child who has Tay-Sachs disease. What is the genotype of each parent? What are the chances each child will have Tay-Sachs disease?
7. One parent has lactose intolerance (recessive), the inability to digest lactose, the sugar found in milk, and the other is heterozygous. What are the chances that their child will have lactose intolerance?
8. A child has cystic fibrosis (recessive). His parents are normal. What is the genotype of all persons mentioned?
9. A woman heterozygous for polydactyly (dominant) reproduces with a homozygous normal man. What are the chances that their children will have 12 fingers and 12 toes?
10. If a woman homozygous for widow's peak (dominant) reproduces with a man homozygous for straight hairline (recessive), what are the chances of their children having a widow's peak? A straight hairline?
11. In humans, the allele for short fingers is dominant over that for long fingers. If a person with short fingers who had one parent with long fingers reproduces with a person having long fingers, what are the chances of each child having short fingers?
12. In fruit flies, gray body (*G*) is dominant over black body (*g*). In a fruit fly experiment, two gray-bodied fruit flies produce mostly gray-bodied offspring, but some offspring have black bodies. If there are 280 offspring, how many do you predict will have gray bodies and how many will have black bodies? How many of the 280 offspring do you predict will be heterozygous? If you wanted to test whether a particular gray-bodied fly was homozygous dominant or heterozygous, what cross would you do?

Autosomal Two-Trait Problems

13. In rabbits, black color (*B*) is dominant over brown(*b*) and short hair (*S*) is dominant over long (*s*). In a cross between a homozygous black, long-haired rabbit and a brown, homozygous short-haired one, what would the F_1 generation look like? the F_2 generation? If one of the F_1 rabbits reproduced with a brown, long-haired rabbit, what phenotypes and in what ratio would you expect?
14. In horses, black coat (*B*) is dominant over brown coat (*b*), and being a trotter (*T*) is dominant over being a pacer (*t*). A black pacer is crossed with a brown trotter. The offspring is a brown pacer. Give the genotypes of all these horses.
15. In fruit flies, gray body (*G*) is dominant over black body (*g*), and long wings (*L*) is dominant over short wings (*l*). The complete genotype of a long-winged, gray-bodied fruit fly is unknown. When this fly is crossed with a short-winged, black-bodied fruit fly, the offspring all have a gray body but about half of them have short wings. What is the genotype of the long-winged, gray-bodied fly.
16. In humans, widow's peak hairline is dominant over straight hairline, and short fingers are dominant over long fingers. If an individual who is heterozygous for both traits reproduces with an individual who is recessive for both traits, what are the chances of their child also being recessive for both traits?

Autosomal Incompletely Dominant Problems

17. What are the chances that a person pure for straight hair who is married to a person pure for curly hair will have children with wavy hair?
18. One parent has sickle-cell disease and the other is perfectly normal. What are the phenotypes of their children?
19. A child has sickle-cell disease but her parents do not. What is the genotype of each parent?
20. Both parents have the sickle-cell trait. What are their chances of having a perfectly normal child?

Autosomal Multiple Alleles Problems

21. The genotype of a woman with type B blood is *BO*. The genotype of her husband is *AO*. What could be the genotypes and phenotypes of the children?
22. A man has type O blood. What is his genotype? Could this man be the father of a child with type A blood? If not, why not? If so what blood types could the child's mother have?
23. Baby Susan has type B blood. Her mother has type O blood. What type blood could her father have?

X-Linked Recessive Problems

24. A boy has severe combined immune deficiency syndrome. What are the genotypes of the parents, who have the normal phenotype?
25. A woman is color blind and her spouse has normal vision. If they produce a son and a daughter, which child will be color blind?
26. If a female who carries an X-linked allele for Lesch-Nyhan syndrome reproduces with a normal man, what are the chances that male children will have the condition? that female children will have the condition?
27. A girl has hemophilia. What is the genotype of her father? What is the genotype of her mother, who has a normal phenotype?
28. In fruit flies X^R = red eye and X^r = white eye.
    a. If a white-eyed male reproduces with a homozygous red-eyed female, what phenotypic ratio is expected for males? for females?
    b. If a white-eyed female reproduces with a red-eyed male, what phenotypic ratio is expected for males? for females?

Mixed Problems

29. What is the genotype of a man who is color blind (X-linked recessive) and has a straight hairline (autosomal recessive)? If this man has children by a woman who is homozygous dominant for normal color vision and widow's peak, what will be the genotype and phenotype of the children?
30. In fruit flies, gray body (*G*) is dominant over black body (*g*). A female fly heterozygous for both gray body and red eyes reproduces with a red-eyed male heterozygous for gray body. What phenotypic ratio is expected for males? for females?

Thinking Scientifically

1. Early in this century, geneticists performed this cross:
   P red-eyed female × white-eyed male
   F_1 red-eyed female red-eyed male
   a. From these results, which characteristic is dominant?

   b. They went on to perform this cross:
   $F_1 \times F_1$ red-eyed female × red-eyed male
   F_2 red-eyed female 1:1 red- to white-eyed male
   Are these results explainable if the allele for red/white eye color is on the Y chromosome but not on the X chromosome? On the X chromosome but not on the Y chromosome? Explain. How do these results support the hypothesis that genes are on the chromosomes?
2. Considering Figure 24.10, you are a geneticist who has been hired to convince the British royal family that Queen Victoria was a carrier for hemophilia, an X-linked gene.
   a. How does Figure 24.10 show that hemophilia is X-linked?
   b. The present members of the royal family are descended from Edward VII, who did not inherit the allele for hemophilia. What were his chances of inheriting the allele?
   c. In hemophiliacs, a clotting factor is defective. Would you expect this factor to be a carbohydrate, lipid, nucleic acid, or protein? Why?
   d. Considering that it is sometimes possible today to test the DNA (taken from bone cells) of persons long dead, what would you expect to find if the DNA of Victoria were examined?

Understanding the Terms

autosome 489
chromosomal mutation 491
crossing-over 500
karyotype 488
linkage group 500
monosomy 489
nondisjunction 489
recombinant 500
sex chromosome 488
sex-influenced trait 499
sex-linked trait 496
syndrome 489
trisomy 489
X chromosome 488
Y chromosome 488

Match the terms to these definitions:

a. ________ Alleles on the same chromosome are linked in the sense that they tend to move together to the same gamete; crossing-over interferes with linkage.
b. ________ Failure of homologous chromosomes or sister chromatids to separate during the formation of gametes.
c. ________ Variation in regard to the normal number of chromosomes inherited or in regard to the normal sequence of alleles on a chromosome; the sequence can be inverted, translocated from a nonhomologous chromosome, deleted, or duplicated.
d. ________ Arrangement of all the chromosomes within a cell by pairs in a fixed order.
e. ________ Phenotype that is controlled by a gene located on a sex chromosome, usually the X chromosome, whose pattern of inheritance differs in males and females.

Using Technology

Your study of patterns of chromosomal inheritance is supported by these available technologies:

Essential Study Partner CD-ROM

Genetics → Chromosomes

Visit the Mader web site for related ESP activities.

Exploring the Internet

The Mader Home Page provides resources and tools as you study this chapter.

http://www.mhhe.com/biosci/genbio/mader

C H A P T E R 25

Molecular Basis of Inheritance

Chapter Concepts

DNA stores genetic information. DNA is a double helix composed of two strands of nucleotides hydrogen-bonded to one another. The genetic information is in the sequence of nucleotides.

The young man in the lawyer's office was there to make a claim. He had come from Europe to the United States to say that he was the rightful heir to millions left behind by an oil magnate who had recently died. The magnate had a son born out of wedlock to a French woman he had met while a soldier in Europe during the war. There were two others before him who also claimed to be the long-separated son of the magnate. The lawyer had told them, just as he was now telling this claimant, that each would have to undergo a blood cell test. DNA fingerprinting would be used to compare DNA taken from his white blood cells with DNA taken from the deceased man's cells. Human DNA can form more patterns than there are people in the world, but even so, the patterns of close relatives are quite similar. Therefore, this DNA test would reveal which of the claimants was the rightful heir.

This chapter discusses the structure of DNA and its function in cells. We will see that DNA fulfills the requirements for the genetic material: it (1) can be replicated prior to cell division and can be transmitted to daughter cells and from generation to generation, and (2) is able to store information that pertains to both the development and metabolic activities of cells and the individual, and (3) is able to undergo rare changes called mutations that are so necessary for evolution to occur.

25.1 DNA Structure and Replication

In the mid-1900s, scientists knew that the chromosomes contained genetic information. But because the chromosomes were composed of both **DNA (deoxyribonucleic acid)** and proteins, they were uncertain which one was the genetic material. They turned to experiments with viruses to resolve this question since they knew that viruses are tiny particles having just two parts: an inner nucleic acid core and an outer protein coat. A virus called the T_2 virus (the T_2 simply means *type* 2) infects bacteria, and if they could determine which part of the virus enters a bacterium and produces more viruses, they would know whether the genes were made up of DNA or protein.

Two batches of viruses were prepared. One batch had ^{32}P-labeled DNA (P = phosphorus), and the other had ^{35}S-labeled protein (S = sulfur). In one experiment, bacteria were exposed to the viruses with labeled DNA, and in the other, bacteria were exposed to the viruses with labeled protein (Fig. 25.1). They found that only labeled DNA enters a bacterium. Therefore, only DNA is needed for the reproduction of viruses, and only DNA is the genetic material.

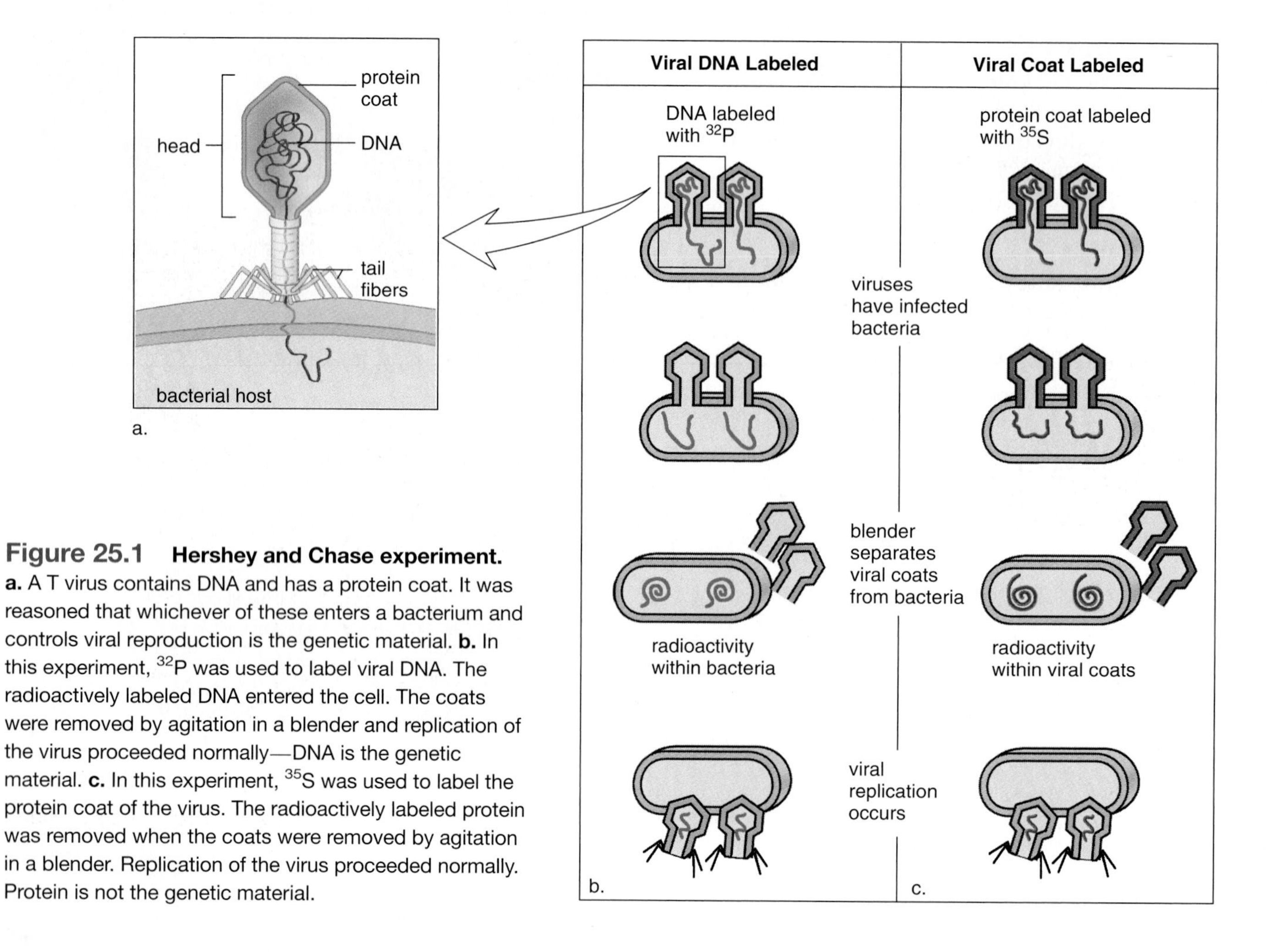

Figure 25.1 Hershey and Chase experiment. **a.** A T virus contains DNA and has a protein coat. It was reasoned that whichever of these enters a bacterium and controls viral reproduction is the genetic material. **b.** In this experiment, ^{32}P was used to label viral DNA. The radioactively labeled DNA entered the cell. The coats were removed by agitation in a blender and replication of the virus proceeded normally—DNA is the genetic material. **c.** In this experiment, ^{35}S was used to label the protein coat of the virus. The radioactively labeled protein was removed when the coats were removed by agitation in a blender. Replication of the virus proceeded normally. Protein is not the genetic material.

Science Focus

Finding the Structure of DNA

In 1951, James Watson, an American biologist, began an internship at the University of Cambridge, England. There he met Francis Crick, a British physicist, who was interested in molecular structures. Together they set out to determine the structure of DNA and to build a model that would explain how DNA, the genetic material, can vary from species to species and even from individual to individual. They also discovered the way DNA replicates (makes a copy of itself) so that daughter cells and offspring can receive a copy.

The bits and pieces of data available to Watson and Crick were like puzzle pieces they had to fit together. This is what they knew from the research of others:

1. DNA is a polymer of nucleotides, each one having a phosphate group, the sugar deoxyribose, and a nitrogen-containing base. There are four types of nucleotides because there are four different bases: adenine (A) and guanine (G) are purines, while cytosine (C) and thymine (T) are pyrimidines.
2. A chemist, Erwin Chargaff, had determined in the late 1940s that regardless of the species under consideration, the number of purines in DNA always equals the number of pyrimidines. Further, the amount of adenine equals the amount of thymine (A = T), and the amount of guanine equals the amount of cytosine (G = C). These findings came to be known as Chargaff's rules.
3. Rosalind Franklin and Maurice Wilkins, working at King's College, London, had just prepared an X-ray diffraction photograph of DNA. It showed that DNA is a double helix of constant diameter and that the bases are regularly stacked on top of one another.

Using these data, Watson and Crick deduced that DNA has a twisted, ladder-type structure; the sugar-phosphate molecules make up the sides of the ladder, and the bases make up the rungs. Further, they determined that if A is normally hydrogen bonded with T, and G is normally hydrogen bonded with C (in keeping with Chargaff's rules), then the rungs always have a constant width (as required by the X-ray photograph).

Watson and Crick built an actual model of DNA out of wire and tin. This double-helix model does indeed allow for differences in DNA structure between species because the base pairs can be in any order. Also, the model suggests that complementary base pairing plays a role in the replication of DNA. As Watson and Crick pointed out in their original paper, "It has not escaped our notice that the specific pairing we have postulated immediately suggests a possible copying mechanism for the genetic material."

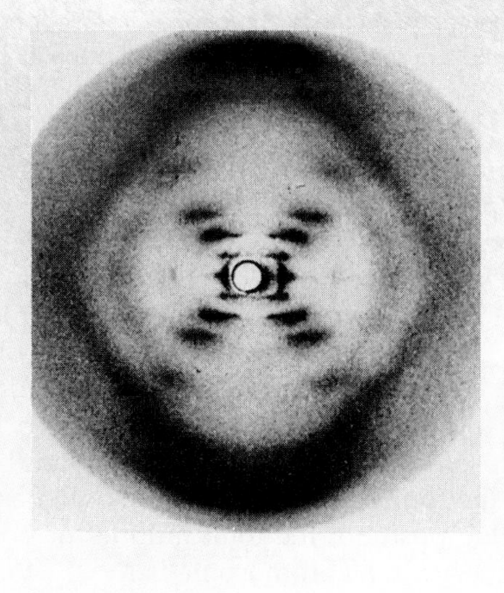

a.

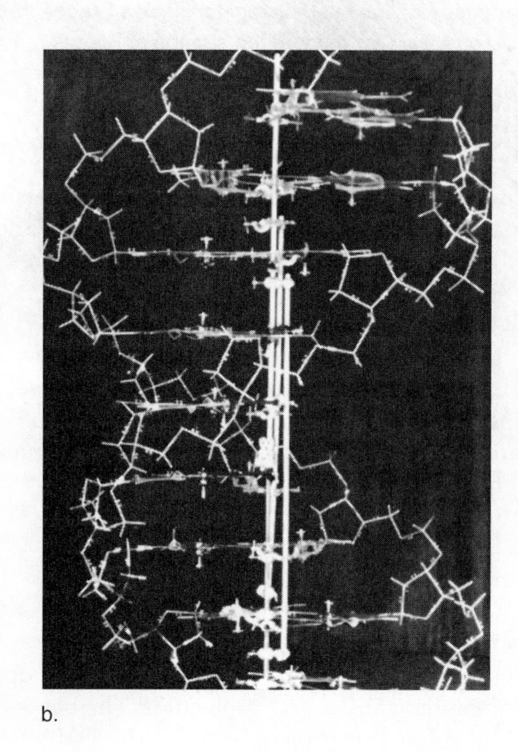

b.

Figure 25A Discovering the structure of DNA.
a. X-ray diffraction photograph of DNA taken by Rosalind Franklin. **b.** A portion of the actual wire and tin model constructed by Watson and Crick.

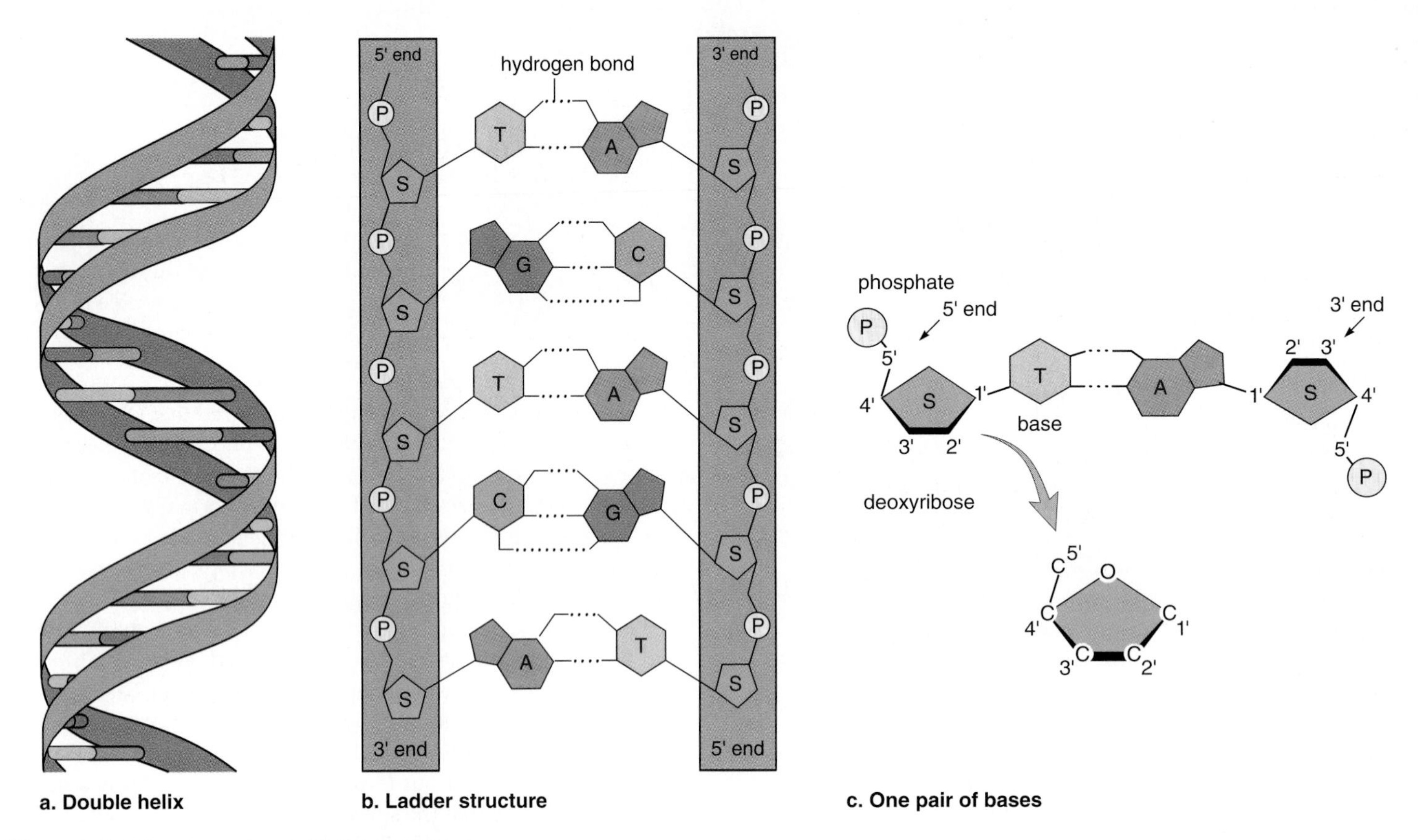

Figure 25.2 **Overview of DNA structure.**
a. DNA double helix. **b.** When the helix is unwound, a ladder configuration shows that the uprights are composed of sugar and phosphate molecules and the rungs are complementary bases. Notice the bases in DNA pair in such a way that the phosphate-sugar groups are oriented in different directions. This means that the strands of DNA end up running antiparallel to one another, with the 3′ end of one strand opposite the 5′ end of the other strand. **c.** When you examine one pair of bases, you see that 3′ and 5′ refer to a numbering system for the carbon atoms.

Structure of DNA

The structure of DNA was determined by James Watson and Francis Crick in the early 1950s. The data they used and how they used the data to deduce DNA's structure are reviewed in the reading on the previous page.

DNA is a polynucleotide; each nucleotide is a complex of three subunits—phosphoric acid (phosphate), a pentose sugar (deoxyribose), and a nitrogen-containing base. There are four possible bases: two are **purines** with a double ring, and two are **pyrimidines** with a single ring. The names of the bases are as follows:

Purines	Pyrimidines
Adenine (A)	Thymine (T)
Guanine (G)	Cytosine (C)

A polynucleotide *strand* has a backbone made up of alternating phosphate and sugar molecules. The bases are attached to the sugar but project to one side. DNA has two such strands, and the two strands twist about one another in the form of a **double helix** (Fig. 25.2*a* and *b*). The strands are held together by hydrogen bonding between the bases: A pairs with T by forming two hydrogen bonds, and G pairs with C by forming three hydrogen bonds, or vice versa. This is called **complementary base pairing.**

When the DNA helix unwinds, it resembles a ladder (Fig. 25.2*b*). The sides of the ladder are the phosphate-sugar backbones, and the rungs of the ladder are the complementary-paired bases. Notice that a purine is always bonded to a pyrimidine (Fig. 25.3). The bases can be in any order, and the variability that can be obtained is overwhelming. For example, a chromosome can have about 140 million base pairs. Since any of the four possible nucleotides can be present at each nucleotide position, the total number of possible nucleotide sequences in a chromosome is $4^{140,000,000}$. No wonder individuals have a unique order which they inherit from their parents.

DNA is a double helix with phosphate-sugar backbones on the outside and paired bases on the inside. Complementary base pairing occurs: adenine (A) pairs with thymine (T), and guanine (G) pairs with cytosine (C).

Replication of DNA

Exact copies of DNA are produced during the replication process. The double-stranded structure of DNA aids replication because each strand can serve as a template for the formation of a complementary strand. A **template** is most often a mold used to produce a shape opposite to itself. In this case, each old (parental) strand is a template for each new (daughter) strand.

Replication has the following steps (Fig. 25.3):

1. *Unwinding.* The two strands that make up DNA unwind and "unzip" (i.e., the weak hydrogen bonds between the paired bases break). A special enzyme called helicase causes the molecule to unwind.
2. *Complementary base pairing.* New complementary nucleotides, always present in the nucleus, fit into place by the process of complementary base pairing.
3. *Joining.* The complementary nucleotides join to form new strands. This step is carried out by an enzyme called DNA **polymerase.**

Because each old strand has produced a new strand through complementary base pairing, there are now two DNA helices identical to each other and to the original molecule. DNA replication is termed *semiconservative* because each new double helix has one old strand and one new strand. In other words, one of the parental strands is conserved, or present, in each new double helix.

DNA replication must occur before a cell can divide. Cancer, which is characterized by rapidly dividing cells, is treated with chemotherapeutic drugs that stop replication and therefore cell division. Some chemotherapeutic drugs are analogs that have a similar, but not identical, structure to the four nucleotides in DNA. When these are mistakenly used by the cancer cells to synthesize DNA, replication stops and the cells die off.

During DNA replication, DNA unwinds and unzips, and new strands that are complementary to the original strands form.

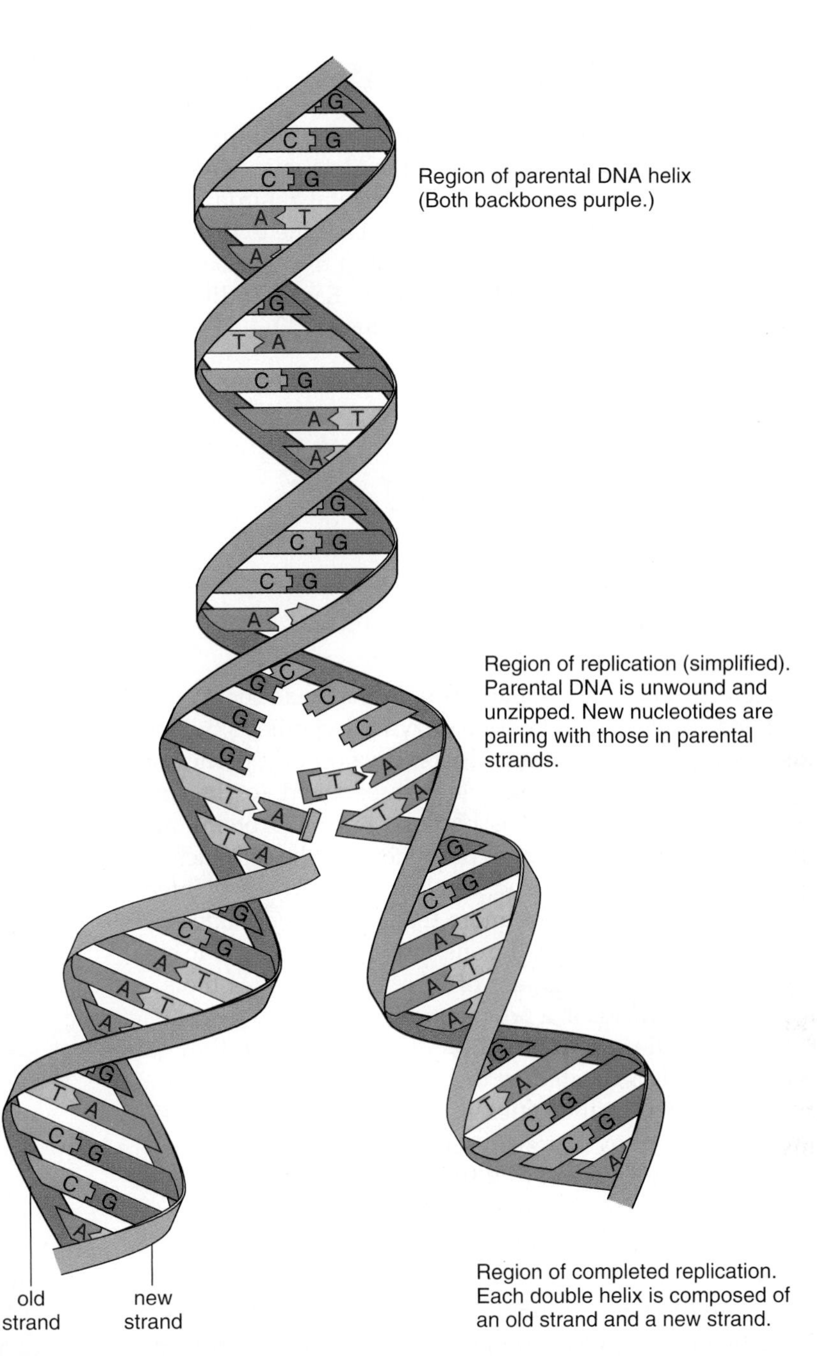

Figure 25.3 DNA replication.
After the DNA molecule unwinds, each old strand serves as a template for the formation of the new strand. Complementary nucleotides available in the cell pair with those of the old strand and then are joined together to form a daughter strand. After replication is complete, there are two daughter strands. Replication is called semiconservative because each new double helix is composed of an old (parental) strand and a new (daughter) strand. Each molecule has the same sequence of base pairs as the parent molecule had before unwinding occurred.

25.2 Gene Expression

The occurrence of inherited metabolic disorders first suggested that genes are responsible for the metabolic workings of a cell. In phenylketonuria (PKU), mental retardation is caused by the inability to convert phenylalanine to tyrosine. In albinism, there is no natural pigment in the skin because tyrosine cannot be converted to melanin. Each condition is caused by the presence of a faulty enzyme:

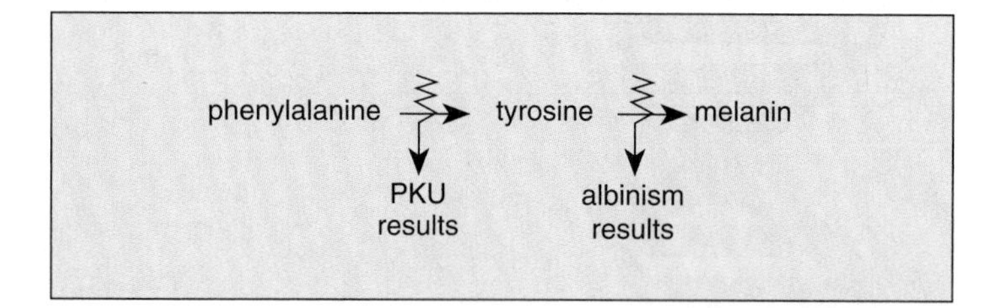

Even in the early 1900s, these conditions were called *inborn errors* of metabolism.

Later laboratory investigations with the bread mold *Neurospora* led to the one gene–one enzyme hypothesis, which stated that each gene controls the production of a particular enzyme. This hypothesis was later broadened to the one gene–one protein hypothesis because not all proteins are enzymes—some are structural components of the cell, such as the muscle proteins, actin and myosin. When it was pointed out that some proteins have more than one polypeptide, the hypothesis was modified to the one gene–one polypeptide hypothesis. A **gene** is a segment of DNA that specifies the sequence of amino acids in a polypeptide of a protein.

DNA specifies the production of proteins, even though in eukaryotes it is located in the nucleus, and proteins are synthesized at the ribosomes in the cytoplasm. **RNA (ribonucleic acid),** however, is not confined to the nucleus; it occurs in both the nucleus and the cytoplasm.

Role of RNA

Like DNA, RNA is a polynucleotide (Fig. 25.4). However, the nucleotides in RNA contain the sugar ribose, not deoxyribose. Also, the bases in RNA are adenine (A), cytosine (C), guanine (G), and uracil (U). In other words, the base uracil replaces thymine found in DNA (Table 25.1). Finally, RNA is single stranded and does not form a double helix in the same manner as DNA.

There are three major classes of RNA, each with specific functions in protein synthesis:

messenger RNA (mRNA): takes a message from DNA in the nucleus to the ribosomes in the cytoplasm.

ribosomal RNA (rRNA): along with proteins, makes up the ribosomes, where polypeptides are synthesized.

transfer RNA (tRNA): transfers amino acids to the ribosomes.

Table 25.1 DNA Structure Compared to RNA Structure

	DNA	RNA
Sugar	Deoxyribose	Ribose
Bases	Adenine, guanine, thymine, cytosine	Adenine, guanine, uracil, cytosine
Strands	Double stranded with base pairing	Single stranded
Helix	Yes	No

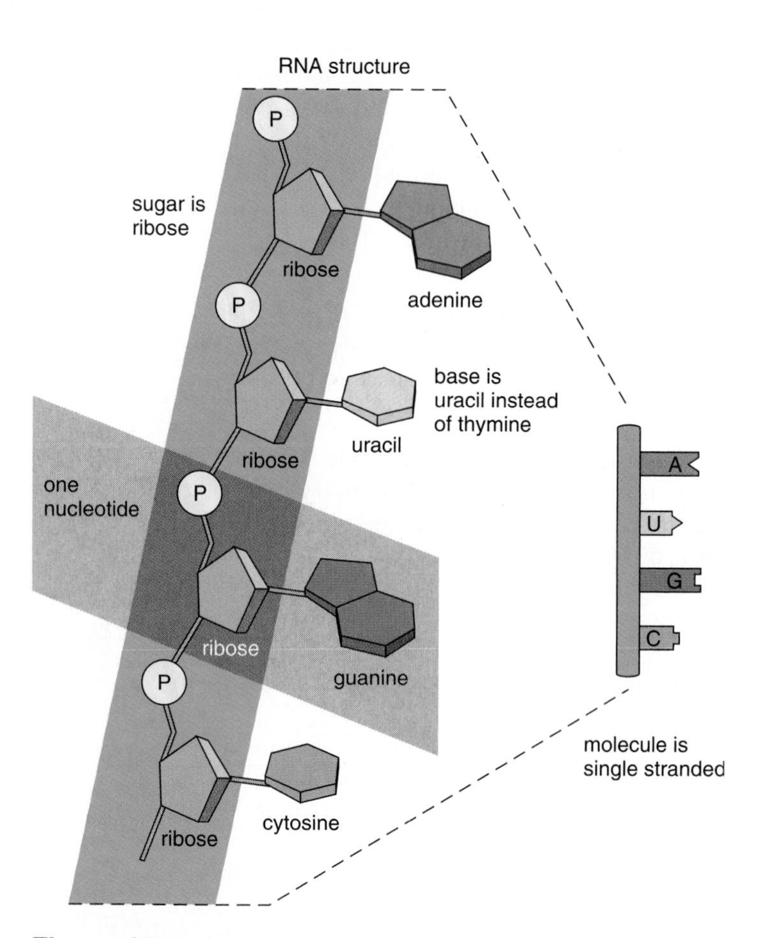

Figure 25.4 Structure of RNA.
RNA is single stranded, the backbone contains the sugar ribose instead of deoxyribose, and the bases are guanine (G), uracil (U), cytosine (C), and adenine (A).

With the help of RNA, a gene (a segment of DNA) specifies the sequence of amino acids in a polypeptide. In this way, genes control the structure and the metabolism of the cell.

From DNA to RNA to Protein

DNA not only serves as a template for its own replication, it is also a template for RNA formation. **Transcription** is making an RNA molecule that is complementary to a portion of DNA. Following transcription, RNA moves into the cytoplasm. There are micrographs showing radioactively labeled RNA moving through a nuclear pore to the cytoplasm, where protein synthesis occurs. *Messenger RNA* (*m*RNA) carries information for the synthesis of a polypeptide. During **translation,** this information is used to sequence the amino acids of a polypeptide (Fig. 25.5).

In ordinary speech, transcription means making a close copy of a document, and translation means putting the document in an entirely different language. In genetics, transcription is making a strand of RNA with the same base sequence as DNA; translation is going from a sequence of nucleotides (bases) to a sequence of amino acids.

The Genetic Code

DNA has a particular sequence of bases, and a polypeptide has a particular sequence of amino acids. This suggests that DNA contains coded information. Can four bases provide enough combinations to code for 20 amino acids? If the code were a doublet (any two bases stand for one amino acid), it would not be possible to code for 20 amino acids, but if the code were a triplet, then the four bases could supply 64 different triplets, far more than needed to code for 20 different amino acids. It should come as no surprise, then, to learn that the code is a **triplet code.**

To crack the code, a cell-free experiment was done: artificial RNA was added to a medium containing bacterial ribosomes and a mixture of amino acids. Comparison of the bases in the RNA with the resulting polypeptide allowed investigators to decipher the code. Each three-letter unit of an mRNA molecule is called a **codon.** All 64 mRNA codons have been determined (Fig. 25.6). Sixty-one triplets correspond to a particular amino acid; the remaining three are stop codons, which signal polypeptide termination. The one codon that stands for the amino acid methionine is also a start codon signaling polypeptide initiation.

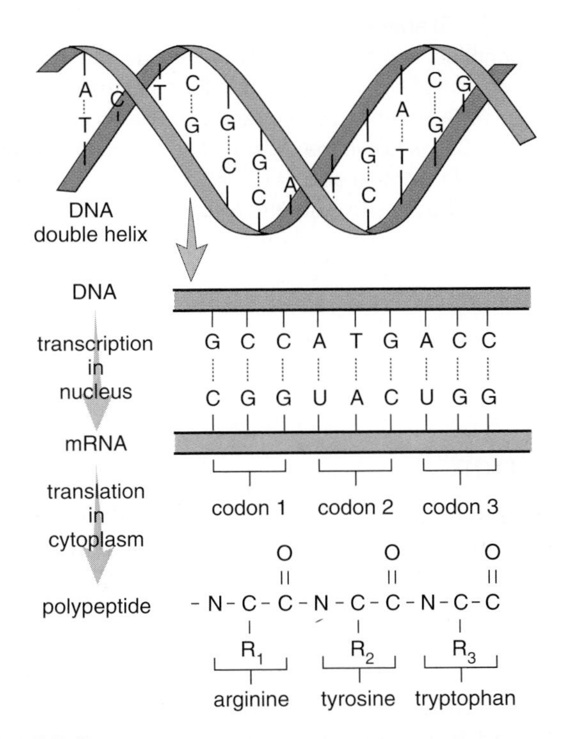

Figure 25.5 Overview of gene expression.
Transcription occurs in the nucleus when DNA acts as a template for mRNA synthesis. Translation occurs in the cytoplasm when the sequence of the mRNA codons determines the sequence of the amino acids in a polypeptide.

First Base	Second Base U	Second Base C	Second Base A	Second Base G	Third Base
U	UUU phenylalanine	UCU serine	UAU tyrosine	UGU cysteine	U
U	UUC phenylalanine	UCC serine	UAC tyrosine	UGC cysteine	C
U	UUA leucine	UCA serine	UAA *stop*	UGA *stop*	A
U	UUG leucine	UCG serine	UAG *stop*	UGG tryptophan	G
C	CUU leucine	CCU proline	CAU histidine	CGU arginine	U
C	CUC leucine	CCC proline	CAC histidine	CGC arginine	C
C	CUA leucine	CCA proline	CAA glutamine	CGA arginine	A
C	CUG leucine	CCG proline	CAG glutamine	CGG arginine	G
A	AUU isoleucine	ACU threonine	AAU asparagine	AGU serine	U
A	AUC isoleucine	ACC threonine	AAC asparagine	AGC serine	C
A	AUA isoleucine	ACA threonine	AAA lysine	AGA arginine	A
A	AUG (*start*) methionine	ACG threonine	AAG lysine	AGG arginine	G
G	GUU valine	GCU alanine	GAU aspartate	GGU glycine	U
G	GUC valine	GCC alanine	GAC aspartate	GGC glycine	C
G	GUA valine	GCA alanine	GAA glutamate	GGA glycine	A
G	GUG valine	GCG alanine	GAG glutamate	GGG glycine	G

Figure 25.6 Messenger RNA codons.
Notice that in this chart, each of the codons (blue squares) is composed of three letters representing the first base, second base, and third base. For example, find the blue square where C for the first base and A for the second base intersect. You will see that U, C, A, or G can be the third base. The three bases CAU and CAC are codons for histidine; the three bases CAA and CAG are codons for glutamine.

Transcription

During transcription, a segment of the DNA helix unwinds and unzips, and complementary RNA nucleotides from an RNA nucleotide pool in the nucleus pair with the DNA nucleotides of one strand. The RNA nucleotides are joined by an enzyme called **RNA polymerase,** and an mRNA molecule results (Fig. 25.7). Therefore, when mRNA forms, it has a sequence of bases complementary to DNA; wherever A, T, G, or C is present in the DNA template, U, A, C, or G is incorporated into the mRNA molecule. In this way, the code is transcribed, or copied. Now mRNA has a sequence of codons, three bases that are complementary to the DNA triplet code.

Following transcription, mRNA has a sequence of bases complementary to one of the DNA strands. Now, mRNA contains codons which are complementary to the DNA triplet code.

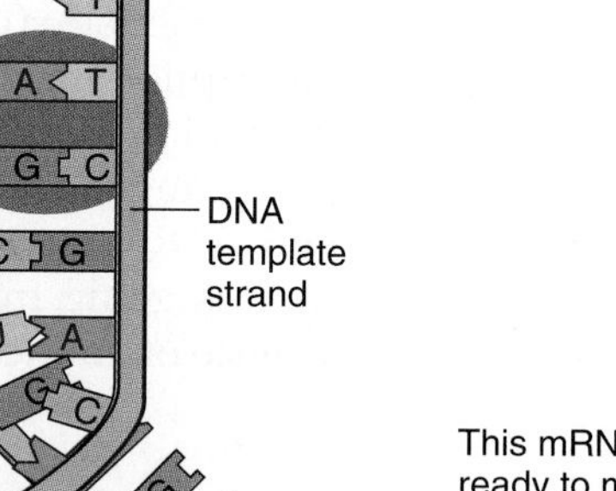

Figure 25.7 Transcription.
During transcription, complementary RNA is made from a DNA template. A portion of DNA unwinds and unzips at the point of attachment of RNA polymerase. A strand of mRNA is produced when complementary bases join in the order dictated by the sequence of bases in DNA. Transcription occurs in the nucleus and the mRNA passes out of the nucleus to enter the cytoplasm.

Processing of mRNA

Most genes in humans are interrupted by segments of DNA that are not part of the gene. These portions are called *introns* because they are intragene segments. The other portions of the gene are called *exons* because they are ultimately *expressed*. They result in a protein product.

When DNA is transcribed, the mRNA contains bases that are complementary to both exons and introns, but before the mRNA exits the nucleus, it is *processed.* During processing, the nucleotides complementary to the introns are spliced out by ribozymes. **Ribozymes** are organic catalysts composed of RNA and not protein. There has been much speculation about the role of introns. It is possible that they allow crossing-over within a gene during meiosis. It is also possible that introns divide a gene into domains that can be joined in different combinations to give novel genes and protein products, facilitating the evolution of new species.

Processing occurs in the nucleus. The newly formed mRNA is called the primary mRNA molecule, and the processed mRNA is called the mature mRNA molecule. The mature mRNA molecule passes from the cell nucleus into the cytoplasm. There it becomes associated with ribosomes.

In humans, the primary mRNA molecule is processed; introns are removed, so that the mature mRNA molecule contains only exons. Mature RNA leaves the nucleus and becomes associated with ribosomes.

Translation

Translation, which takes place in the cytoplasm of eukaryotic cells, is the second step by which gene expression leads to protein synthesis. During translation, the sequence of codons in mRNA specifies the order of amino acids in a polypeptide. This is called translation because the sequence of DNA and then RNA bases is translated into a sequence of amino acids. Translation requires several enzymes and two other types of RNA: transfer RNA and ribosomal RNA.

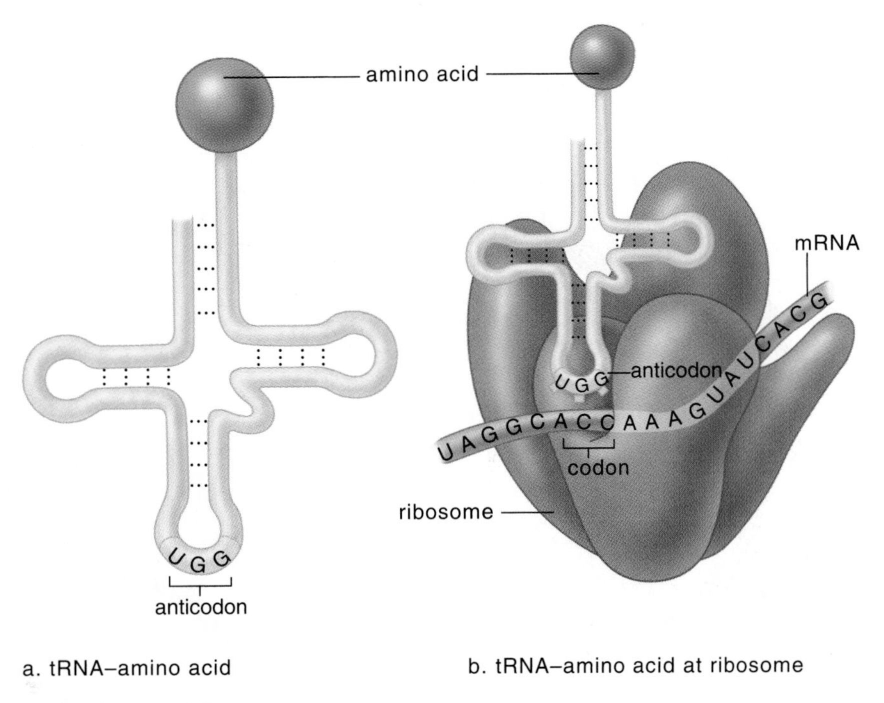

Figure 25.8 Anticodon-codon base pairing.
a. tRNA molecules have an amino acid attached to one end and an anticodon at the other end. **b.** The anticodon of a tRNA molecule is complementary to a codon. The pairing between codon and anticodon ensures that the sequence of amino acids in a polypeptide is that directed originally by DNA. If the codon is ACC, the anticodon is UGG, and the amino acid is threonine.

Transfer RNA

Transfer RNA (tRNA) molecules bring amino acids to the ribosomes. Each is a single-stranded nucleic acid that doubles back on itself to create regions where complementary bases are hydrogen bonded to one another. There is at least one tRNA molecule for each of the twenty amino acids found in proteins. The amino acid binds to one end of the molecule (Fig. 25.8*a*). Attachment requires ATP energy, and the resulting bond is a high-energy bond represented by a wavy line. The entire complex is designated as tRNA–amino acid. One area of active research is to determine how the correct amino acid becomes attached to the correct tRNA molecule. Somehow an enzyme called *tRNA synthetase* recognizes which amino acid should be joined to which tRNA molecule.

At the other end of each tRNA molecule, there is a specific **anticodon,** a group of three bases that is complementary to an mRNA codon (Fig. 25.8*b*). A tRNA molecule comes to the ribosome, where its anticodon pairs with an mRNA codon. Let us consider an example: If the codon is ACC, what is the anticodon, and what amino acid will be attached to the tRNA molecule? Inspection of Figure 25.8 allows us to determine this:

Codon	Anticodon	Amino Acid
ACC	UGG	Threonine

The order of the codons of the mRNA determines the order that tRNA–amino acids come to a ribosome, and therefore the final sequence of amino acids in a polypeptide.

Ribosomal RNA

Ribosomal RNA (rRNA) is called structural RNA because it is found in the **ribosomes,** small structural bodies. Ribosomal RNA is produced in a nucleolus within the nucleus. There it joins with proteins manufactured in the cytoplasm. Ribosomal subunits then migrate to the cytoplasm, where they join just as protein synthesis begins. The small subunit contains one rRNA molecule and many different types of proteins, and the large subunit contains two rRNA molecules and also many different types of proteins. Among these proteins is the enzyme that joins amino acids together by means of a peptide bond.

During translation, the sequence of bases in mRNA determines the order that tRNA amino acids come to a ribosome and therefore the order of amino acids in a particular polypeptide.

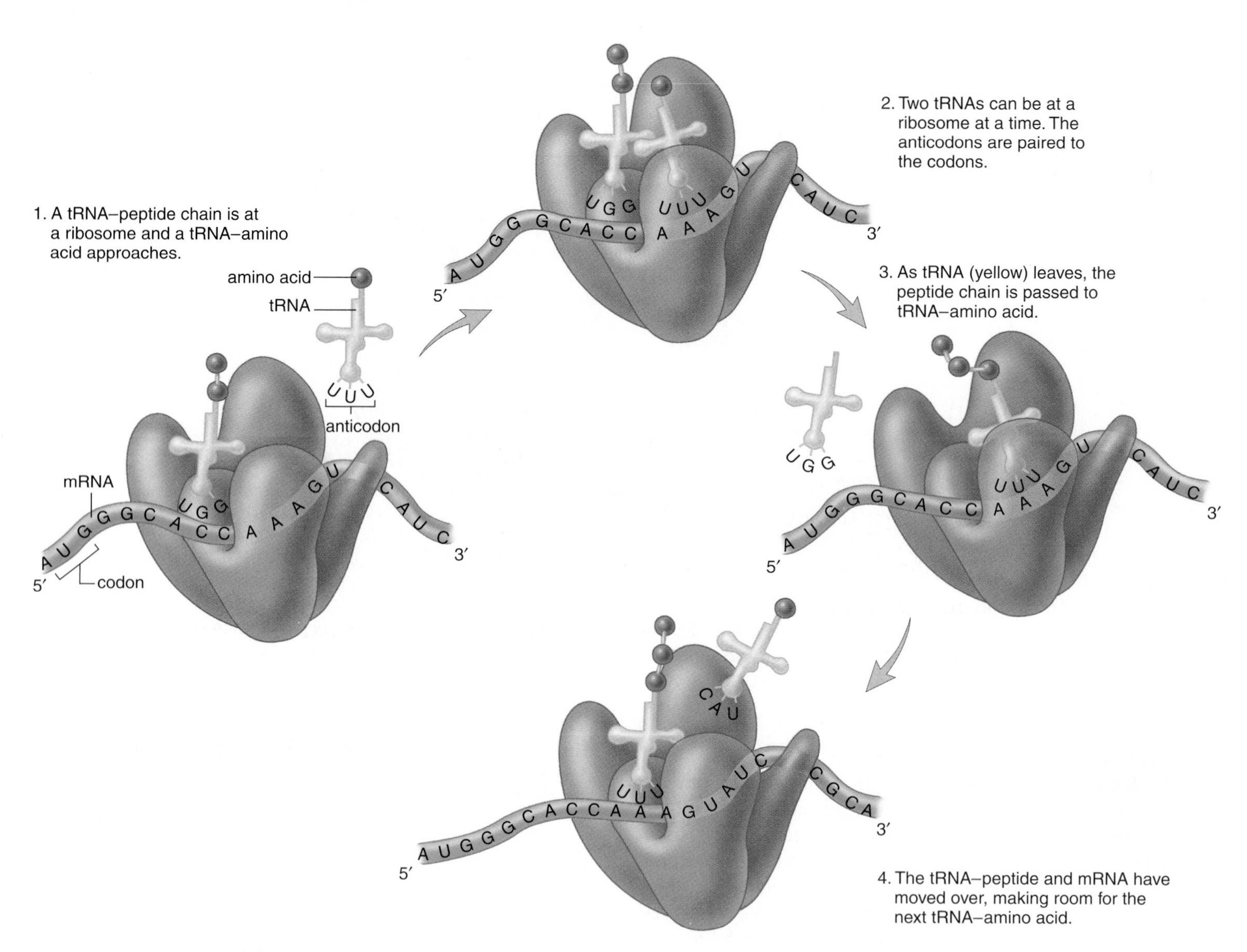

Figure 25.9 Translation.
Transfer RNA (tRNA)–amino acid molecules arrive at the ribosome, and the sequence of messenger RNA (mRNA) codons dictates the order in which amino acids become incorporated into a polypeptide.

Steps of Translation

Polypeptide synthesis requires three steps: initiation, elongation, and termination.

1. During *initiation,* a small ribosomal subunit attaches to the mRNA in the vicinity of the start codon (AUG). The first, or initiator, tRNA pairs with this codon. Then a large ribosomal subunit joins to the small subunit.
2. During *elongation,* the polypeptide lengthens one amino acid at a time (Fig. 25.9). A ribosome is large enough to accommodate two tRNA molecules: the incoming tRNA molecule and the outgoing tRNA molecule. The incoming tRNA–amino acid complex receives the peptide from the outgoing tRNA. The ribosome then moves laterally so that the next mRNA codon is available to receive an incoming tRNA–amino acid complex. In this manner, the peptide grows and the primary structure of a polypeptide comes about. (The secondary and tertiary structures of a polypeptide appear after termination, as the amino acids interact with one another. Some proteins consist of one polypeptide and some have more than one polypeptide chain.)
3. Then *termination* of synthesis occurs at a stop codon on the mRNA. The release factor which binds to this site enzymatically cleaves the polypeptide from the last tRNA. The ribosome dissociates into its two subunits and falls off the mRNA molecule.

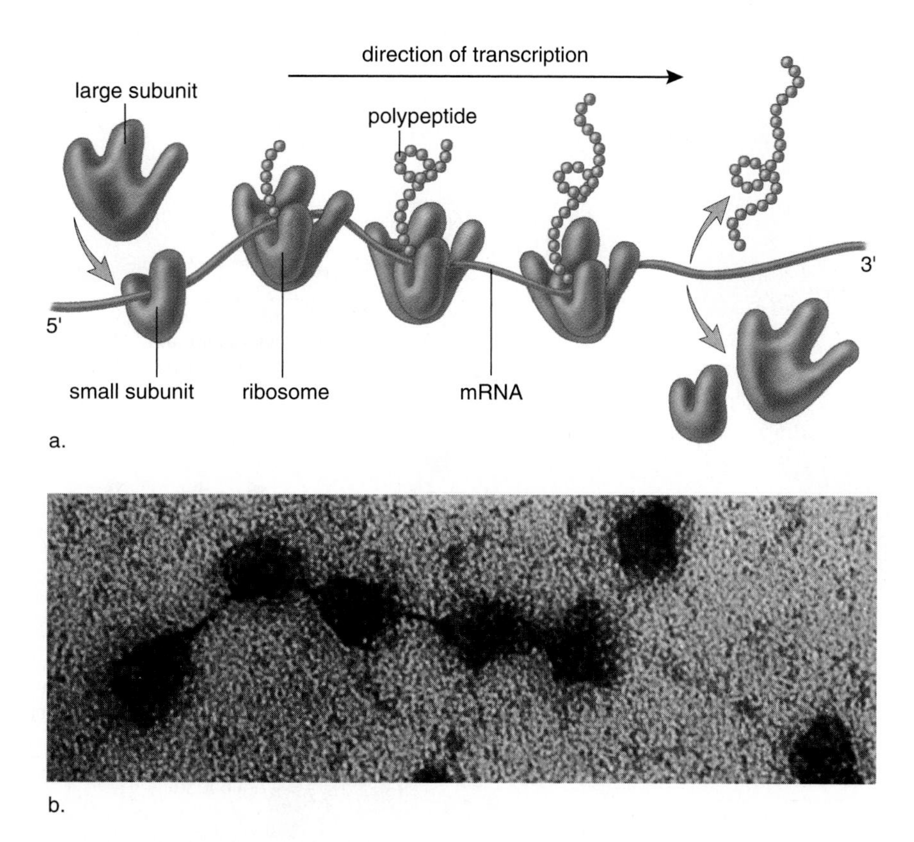

Figure 25.10 Polyribosome structure.
a. Several ribosomes, collectively called a polyribosome, move along a messenger RNA (mRNA) molecule at one time. They function independently of one another; therefore, several polypeptides can be made at the same time. **b.** Electron micrograph of a polyribosome.

As soon as the initial portion of mRNA has been translated by one ribosome, and the ribosome has begun to move down the mRNA, another ribosome attaches to the mRNA. Therefore, several ribosomes, collectively called a **polyribosome,** can move along one mRNA at a time. And several polypeptides of the same type can be synthesized using one mRNA molecule. The life expectancy of an mRNA molecule (how long it exists at a ribosome) can vary; the longer the mRNA stays the more polypeptides form. (Fig. 25.10).

During translation, tRNA molecules, each carrying a particular amino acid, travel to the mRNA. Through complementary base pairing between anticodon and codon, the tRNA molecules and therefore the amino acids in a polypeptide are sequenced in a particular order, the order specified by the DNA triplet code.

Practice Problem*

This is a segment of a DNA molecule. With reference to the transcribed strand, what are (1) the messenger RNA codons, (2) the possible tRNA anticodons, and (3) the sequence of amino acids in the polypeptide?

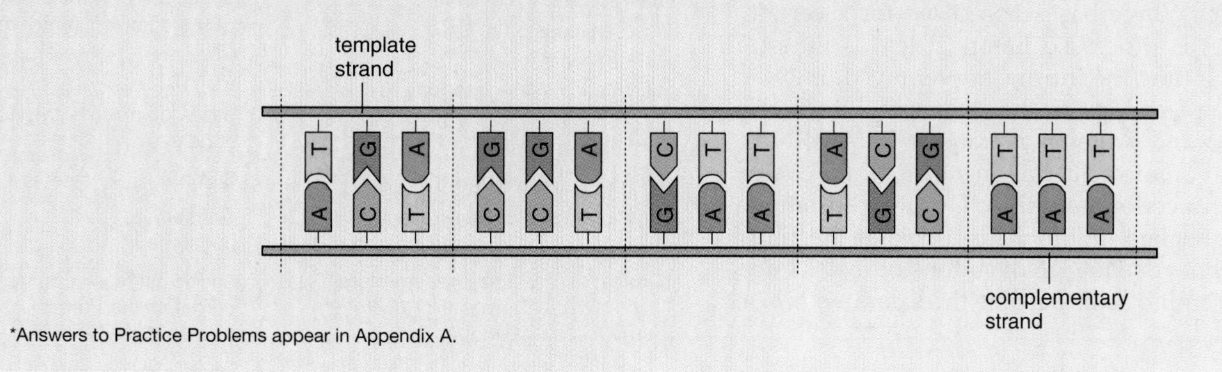

*Answers to Practice Problems appear in Appendix A.

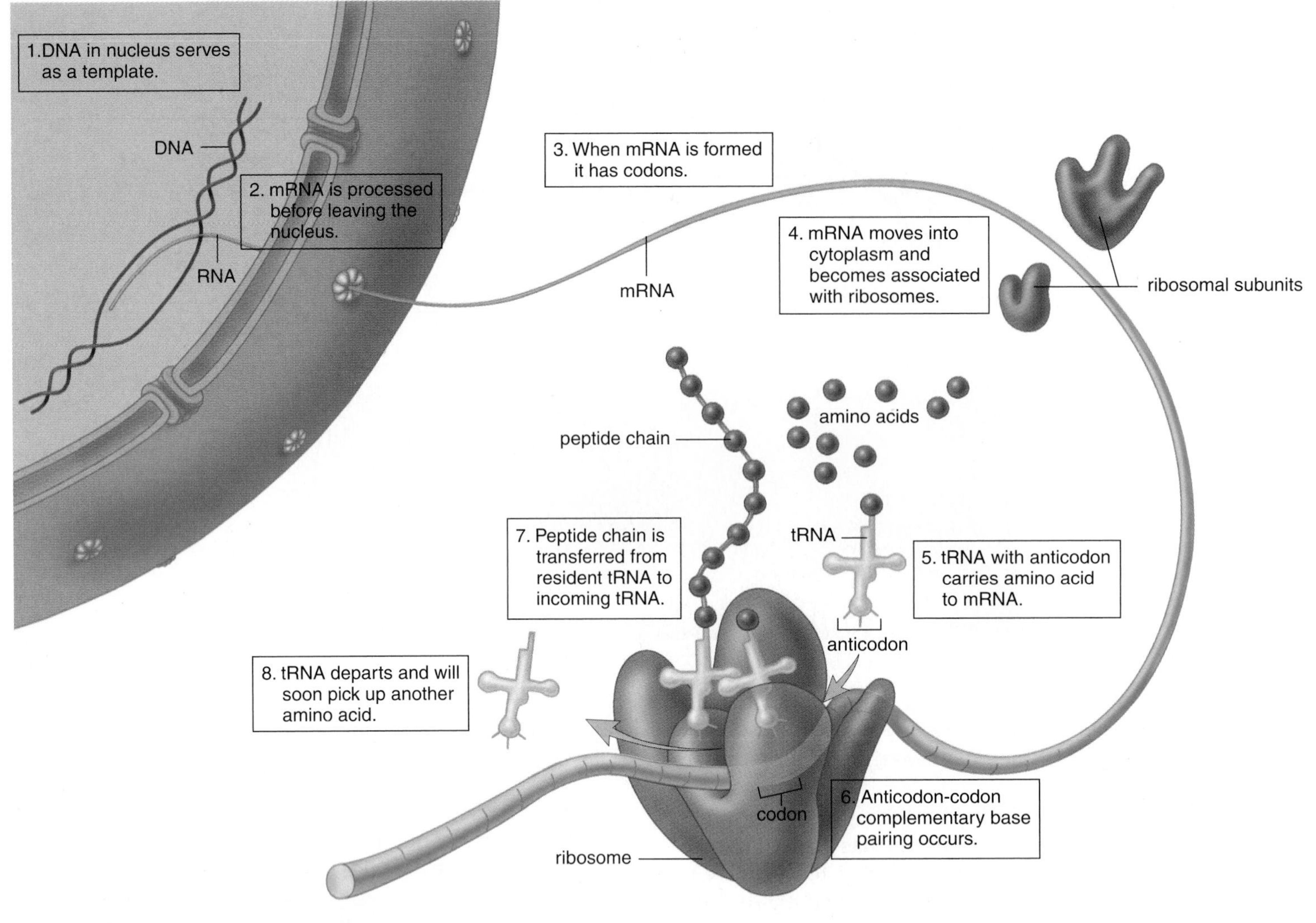

Figure 25.11 Gene expression.
Gene expression leads to the formation of a product, most often a protein. The two steps required for gene expression are transcription, which occurs in the nucleus, and translation, which occurs in the cytoplasm at the ribosomes.

Review of Gene Expression

DNA in the nucleus contains a *triplet code.* Each group of three bases stands for a specific amino acid (Figure 25.11 and Table 25.2). During transcription, a segment of a DNA strand serves as a template for the formation of messenger RNA (mRNA). The bases in mRNA are complementary to those in DNA; every three bases is a *codon* for a certain amino acid. mRNA is processed before it leaves the nucleus, during which time the introns are removed. mRNA carries a sequence of codons to the *ribosomes,* which are composed of rRNA and proteins. A transfer RNA (tRNA) bond to a particular amino acid, has an *anticodon* that pairs complementarily to a codon in mRNA. During translation, tRNAs and their attached amino acids arrive at the ribosomes, where the linear sequence of codons of mRNA determines the order amino acids become incorporated into a protein.

Name of Molecule	Special Significance	Definition
DNA	Genetic information	Sequence of DNA bases
mRNA	Codons	Sequence of three RNA bases complementary to DNA
tRNA	Anticodon	Sequence of three RNA bases complementary to codon
rRNA	Ribosome	Site of protein synthesis
Amino acid	Building block for protein	Transported to ribosome by tRNA
Protein	Enzyme, structural protein, or secretory product	Amino acids joined in a predetermined order

25.3 Control of Gene Expression

All cells receive a copy of all genes; however, cells differ as to which genes are being actively expressed. Muscle cells, for example, have a different set of genes that are turned on in the nucleus and proteins that are active in the cytoplasm than do nerve cells. In eukaryotic cells, a variety of mechanisms regulates gene expression from transcription to protein activity. These mechanisms can be grouped under four primary levels of control, two of which pertain to the nucleus and two of which pertain to the cytoplasm (Fig. 25.12).

1. *Transcriptional control:* In the nucleus, a number of mechanisms serve to control which genes are transcribed and/or the rate at which transcription of the genes occurs. These include the organization of chromatin and the use of transcription factors that initiate transcription, the first step in gene expression.
2. *Posttranscriptional control:* Posttranscriptional control occurs in the nucleus after DNA is transcribed and mRNA is formed. How mRNA is processed before it leaves the nucleus and also the speed with which mature mRNA leaves the nucleus can affect the amount of gene expression.
3. *Translational control:* Translational control occurs in the cytoplasm after mRNA leaves the nucleus and before there is a protein product. The life expectancy of mRNA molecules (how long they exist in the cytoplasm) can vary, as can their ability to bind ribosomes. It is also possible that some mRNAs may need additional changes before they are translated at all.
4. *Posttranslational control:* Posttranslational control, which also occurs in the cytoplasm, occurs after protein synthesis. The polypeptide product may have to undergo additional changes before it is biologically functional. Also, a functional enzyme is subject to feedback control—the binding of an enzyme's product can change its shape so that it is no longer able to carry out its reaction.

Control of gene expression occurs at four levels in eukaryotes. In the nucleus there is transcriptional and posttranscriptional control; in the cytoplasm there is translational and posttranslational control.

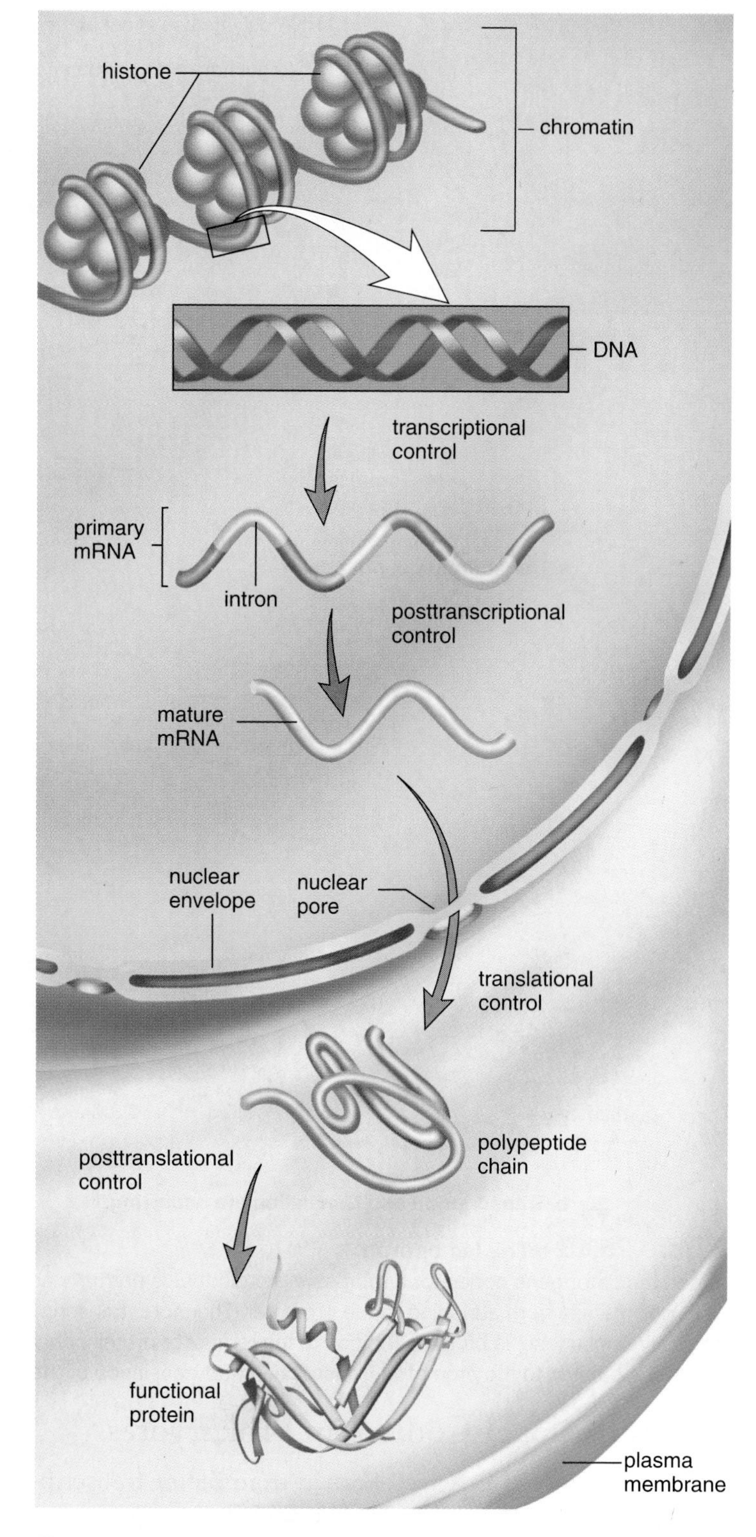

Figure 25.12 Levels at which control of gene expression occurs in eukaryotic cells.
Transcriptional and posttranscriptional control occur in the nucleus. Translational and posttranslational control occur in the cytoplasm.

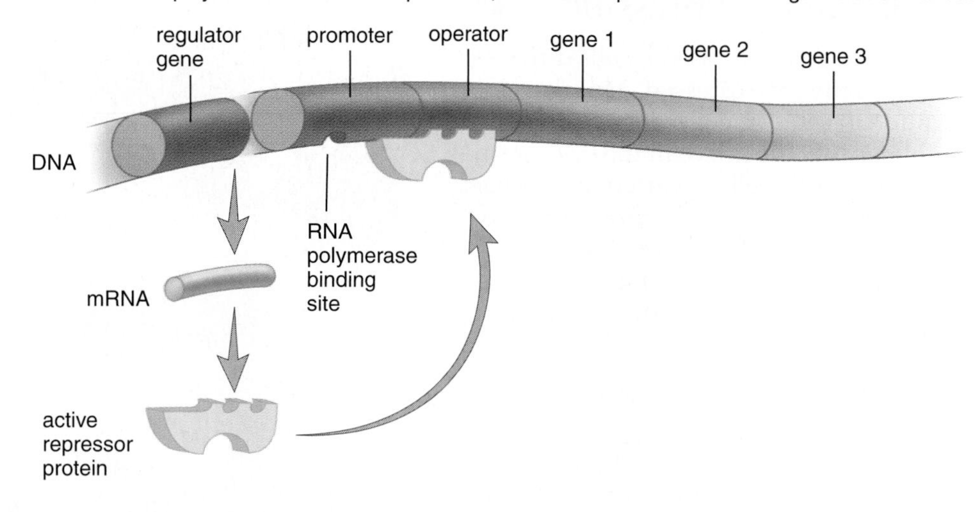

a. Transcription is not occurring.

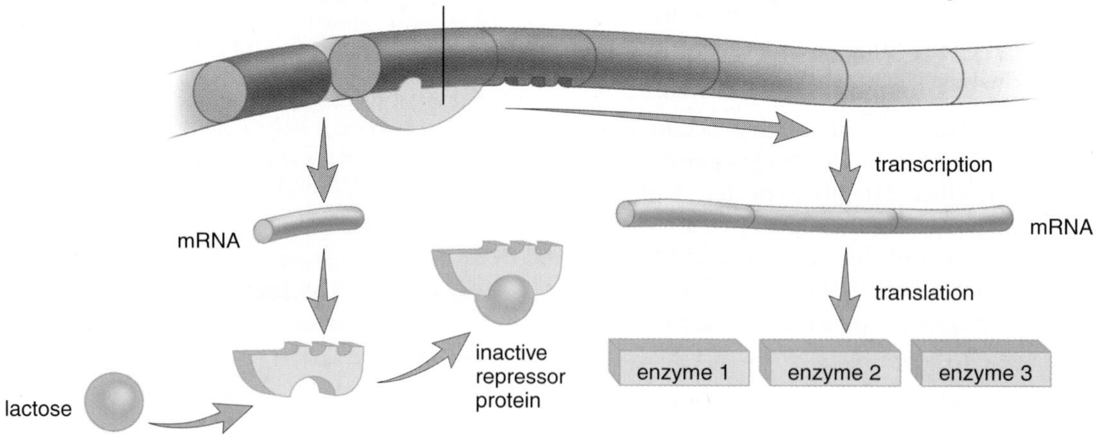

b. Transcription and translation are occurring.

Figure 25.13 **The *lac* operon.**
a. The regulator gene codes for a repressor protein that is normally active. When active, the repressor protein binds to the operator and prevents RNA polymerase from attaching to the promoter. Therefore, transcription of the three structural genes does not occur. **b.** When lactose (or more correctly, allolactose) is present, it binds to the repressor protein, changing its shape so that it can no longer bind to the operator. Now RNA polymerase binds to the promoter; transcription and translation of the three structural genes follow.

Transcriptional Control in Prokaryotes

The operon model is a well-known example of transcriptional control in prokaryotes (Fig. 25.13). An **operon** includes the following elements:

Promoter—a short sequence of DNA where RNA polymerase first attaches when a gene is to be transcribed. When RNA polymerase can bind to the promoter, transcription can occur.

Operator—a short sequence of DNA where the *repressor* protein, coded for by a regulator gene, can bind. When the repressor protein is bound to the operator, RNA polymerase cannot attach to the promoter and transcription cannot occur. Otherwise, transcription does occur.

Structural genes—one to several genes coding for enzymes of a metabolic pathway that are transcribed as a unit. When structural genes are transcribed, a metabolic pathway is active.

A **regulator gene** located outside the operon codes for a repressor protein that can bind to the operator and switch off the operon. Therefore a *regulator gene* regulates the activity of *structural genes*. Each cell contains a full complement of genes but due to the activity of regulator genes only certain genes are active at any one time.

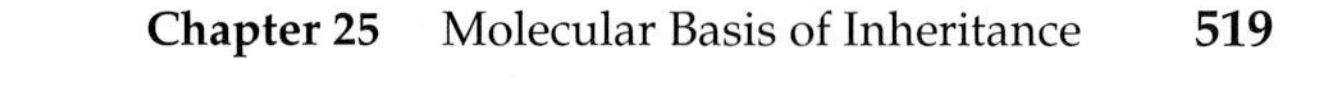

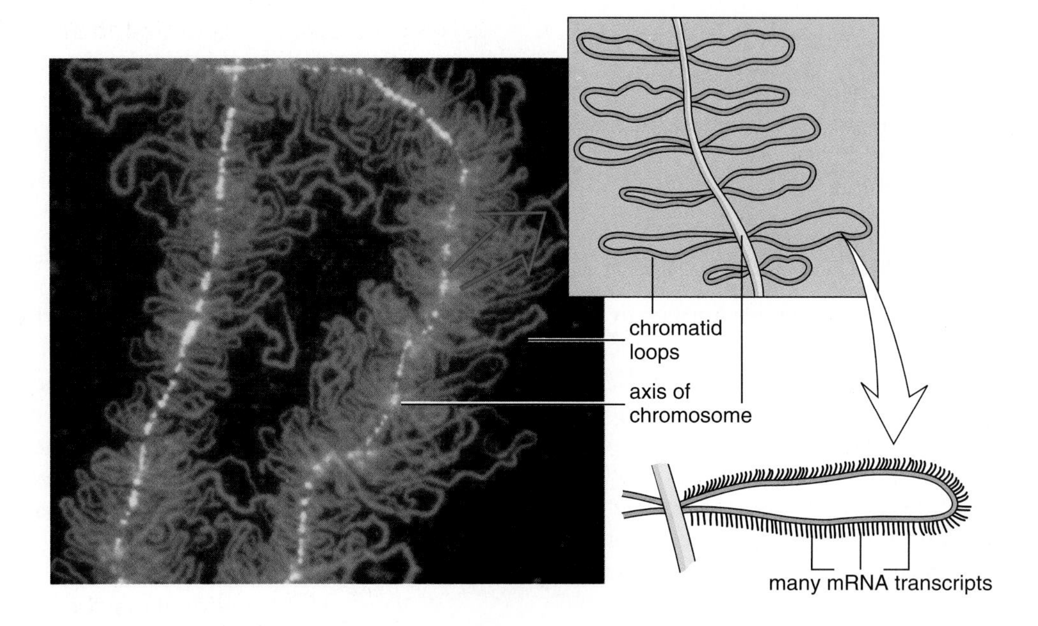

Figure 25.14 Lampbrush chromosomes.
These chromosomes, which are present in maturing amphibian egg cells, give evidence that when mRNA is being synthesized, chromosomes most likely decondense. Each chromatid has many loops extended from the axis of the chromosome (white). Many mRNA transcripts are being made off of these DNA loops (red).

Lac *Operon*

The *lac* operon was the first operon discovered. Ordinarily, the bacterium *Escherichia coli* uses glucose as its energy source; however, if it is denied glucose and is given the milk sugar lactose instead, it immediately begins to make three enzymes needed to metabolize lactose (Fig. 25.13).

Notice that the structural genes in this operon are normally not transcribed because the regulator gene codes for an active repressor protein that automatically attaches to the operator, preventing transcription from occurring. The operon becomes active when the repressor joins with an inducer—lactose—to form an inactive repressor, which is unable to bind to the operator.

The operon model explains one means of transcriptional control in prokaryotes—a way in which genes are turned on or off.

Transcriptional Control in Eukaryotes

Rarely are there operons in eukaryotic cells. Instead, transcriptional control in eukaryotes involves (1) the organization of chromatin and (2) regulator proteins called transcription factors.

Activated Chromatin

For a gene to be transcribed in eukaryotes, the chromosome in that region must first decondense. The chromosomes within the developing egg cells of many vertebrates are called lampbrush chromosomes because they have many loops that appear to be bristles (Fig. 25.14). Here mRNA is being synthesized in great quantity; then protein synthesis can be carried out after fertilization, despite rapid cell division. In the salivary glands and other tissues of larval flies, the chromosomes duplicate and reduplicate many times without dividing mitotically. The homologues, each consisting of about 1,000 sister chromatids, synapse together to form giant chromosomes called polytene chromosomes. It is observed that as a larva develops, first one and then another of the chromosome regions bulge out, forming *chromosome puffs.* The use of radioactive uridine, a label specific for RNA, indicates that DNA is being actively transcribed at these chromosome puffs. It appears that the chromosome is decondensing at the puffs, allowing RNA polymerase to attach to a section of DNA.

Transcription Factors

In eukaryotic cells, *transcription factors* are DNA-binding proteins. Every cell contains many different types of transcription factors, and a specific combination is believed to regulate the activity of any particular gene. After the right combination of transcription factors binds to DNA, an RNA polymerase attaches to DNA and begins the process of transcription.

As cells mature, they become specialized. Specialization is determined by which genes are active, and therefore perhaps by which transcription factors are present in that cell. Signals received from inside and outside the cell could turn on or off genes that code for certain transcription factors. For example, the gene for fetal hemoglobin ordinarily gets turned off as a newborn matures—one possible treatment for sickle-cell disease is to turn this gene on again.

Regulator proteins in eukaryotic cells consist of transcription factors which bind to DNA.

25.4 Gene Mutations

Early geneticists understood that genes undergo mutations, but they didn't know what causes mutations. It is apparent today that a *gene mutation* is a change in the sequence of bases within a gene.

Frameshift Mutations

The term *reading frame* applies to the sequence of codons because they are read from some specific starting point, as in this sentence: THE CAT ATE THE RAT. If the letter C is deleted from this sentence and the reading frame is shifted, we read THE ATA TET HER AT—something that doesn't make sense. *Frameshift mutations* occur most often because one or more nucleotides is either inserted or deleted from DNA. The result of a frameshift mutation can be a completely nonfunctional protein because the sequence of codons is altered.

Point Mutations

Point mutations involve a change in a single nucleotide and therefore a change in a specific codon. When one base is substituted for another, the results can be variable. For example, if UAC is changed to UAU, there is no noticeable effect, because both of these codons code for tyrosine. Therefore, this is called a *silent mutation.* If UAC is changed to UAG, however, the result could very well be a drastic one, because UAG is a stop codon. If this substitution occurs early in the gene, the resulting protein may be too short and may be unable to function. This is called a *nonsense mutation.* Finally, if UAC is changed to CAC, then histidine is incorporated into the protein instead of tyrosine. A change in one amino acid does not necessarily affect the function of a protein, but in this example the polarity of tyrosine and histidine differ. Therefore, this substitution most likely will affect the final shape of the protein and its function. This is called a *missense mutation.* The occurrence of valine instead of glutamate in the β chain of hemoglobin results in sickle-cell disease (Fig. 25.15). The abnormal hemoglobin stacks up inside of cells and their sickle shape makes them clog small vessels. Hemorrhaging leads to pain in internal organs and joints.

Cause and Repair of Mutations

Mutations due to DNA replication errors are rare; a frequency of 10^{-8} to 10^{-5} per cell division is often quoted. DNA polymerase, the enzyme that carries out replication, proofreads the new strand against the old strand and detects any mismatched pairs, which are then replaced with the correct nucleotides. In the end, there is usually only one mistake for every one billion nucleotide pairs replicated.

Mutagens, environmental influences that cause mutations, such as radiation (e.g., radioactive elements, X rays, ultraviolet (UV) radiation) and organic chemicals (e.g., chemicals in cigarette smoke and certain pesticides), are another source of mutations in organisms, including humans. If mutagens bring about a mutation in the gametes, then the offspring of the individual may be affected. On the other hand, if the mutation occurs in the body cells, then cancer may be the result.

A gene mutation is an alteration in the nucleotide sequence of a gene. The usual rate of mutation is low because DNA repair enzymes constantly monitor and repair any irregularities.

Transposons: Jumping Genes

Transposons are specific DNA sequences that have the remarkable ability to move within and between chromosomes. As discussed in the reading on the next page, their movement to a new location sometimes alters neighboring genes, particularly by increasing or decreasing their expression. This can happen if the transposon is a regulator gene. Although "movable elements" in corn were described 40 years ago, their significance was only realized recently. So-called *jumping genes* now have been discovered in bacteria, fruit flies, and humans, and it is likely that all organisms have such elements.

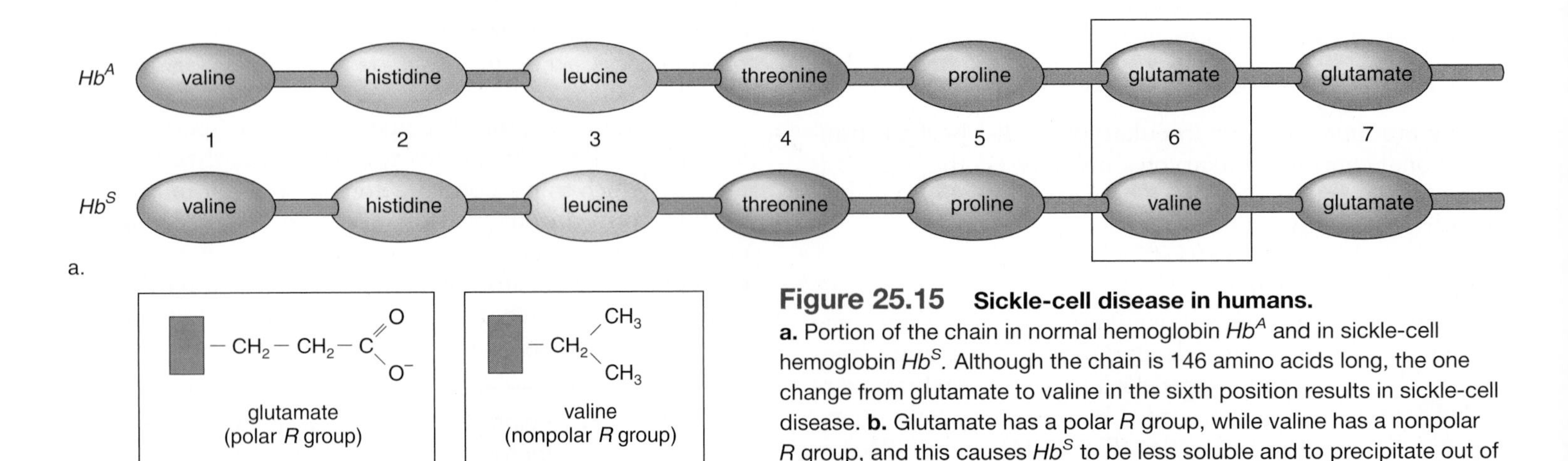

Figure 25.15 Sickle-cell disease in humans.
a. Portion of the chain in normal hemoglobin *Hb*^{A} and in sickle-cell hemoglobin *Hb*^{S}. Although the chain is 146 amino acids long, the one change from glutamate to valine in the sixth position results in sickle-cell disease. **b.** Glutamate has a polar *R* group, while valine has a nonpolar *R* group, and this causes *Hb*^{S} to be less soluble and to precipitate out of solution, distorting the red blood cell into the sickle shape.

Science Focus

Barbara McClintock and the Discovery of Jumping Genes

When Barbara McClintock (Fig. 25B) first began studying inheritance in corn (maize) plants, geneticists believed that each gene had a fixed locus on a chromosome. Thomas Morgan and his colleagues at Columbia University were busy mapping the chromosomes of *Drosophila,* but McClintock preferred to work with corn. In the course of her studies, she came to the conclusion that "controlling elements" could move from one location to another on the chromosome. If a controlling element landed in the middle of a gene, it prevented the expression of that gene. Today, McClintock's controlling elements are called movable genetic elements, transposons, or (in slang), "jumping genes."

Figure 25B **Barbara McClintock at work.**

Based on her experiments with maize, McClintock showed that because transposons are capable of suppressing gene expression, they could account for the pigment pattern of the corn strain popularly known as Indian corn. A colorless corn kernel results when cells are unable to produce a purple pigment due to the presence of a transposon within a particular gene needed to synthesize the pigment. While mutations are usually stable, a transposition is very unstable. When the transposon jumps to another chromosome location, some cells regain the ability to produce the purple pigment, and the result is a corn kernel with a speckled pattern. When McClintock first published her results in the 1950s, the scientific community ignored them. Years later, when molecular genetics was well established, transposons were also discovered in bacteria, yeasts, plants, flies, and humans.

Geneticists now believe that transposons

1. can cause localized mutations, that is, mutations that occur in certain cells and not others.
2. can carry a copy of certain host genes with them when they jump. Therefore, they can be a source of chromosome mutations such as translocations, deletions, and inversions.
3. can leave copies of themselves and certain host genes before jumping. Therefore, they can be a source of a duplication, another type of chromosome mutation.
4. can contain one or more genes that make a bacterium resistant to antibiotics.

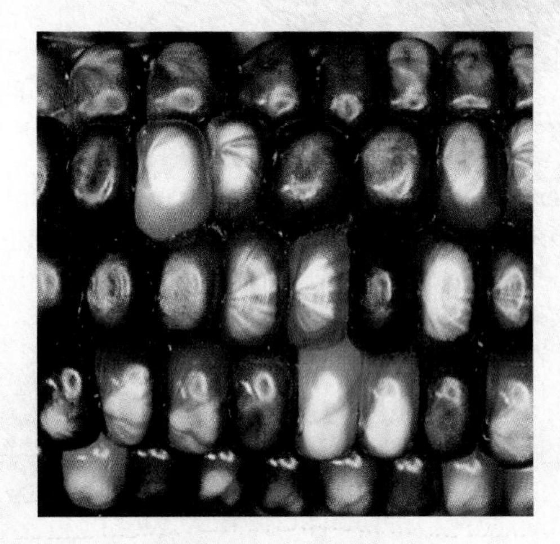

Figure 25C **Corn kernels.** Some kernels are purple, some colorless, and some speckled.

Considering that transposition has a powerful effect on the genotype and phenotype, it most likely has played an important role in evolution. For her discovery of transposons, McClintock was, in 1983, finally awarded the Nobel Prize in Physiology or Medicine. In her Nobel Prize acceptance speech, the eighty-one-year-old scientist proclaimed that "it might seem unfair to reward a person for having so much pleasure over the years, asking the maize plant to solve specific problems and then watching its responses."

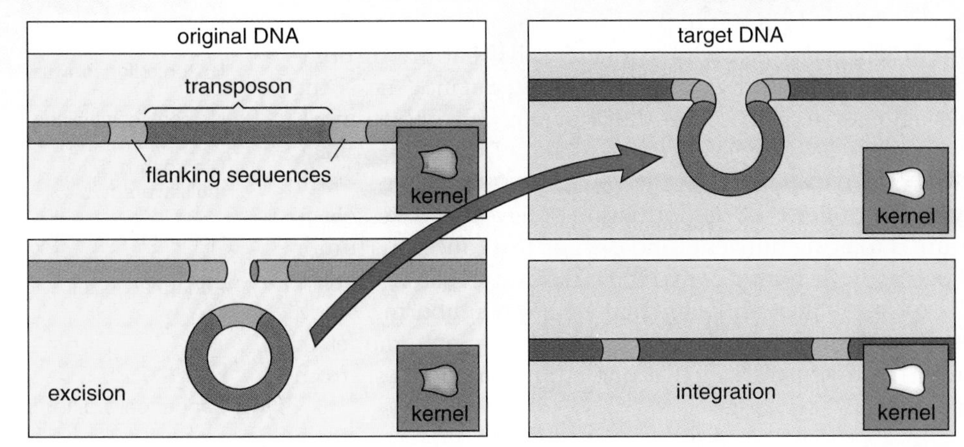

Figure 25D **Transposon.** In its original location, a transposon is interrupting a gene (brown) that is not involved in kernel pigmentation. When the transposon moves to another chromosome position, it blocks the action of a gene (blue) required for synthesis of a purple pigment, and the kernel is now colorless.

25.5 Cancer: A Failure of Genetic Control

Cancer is a genetic disease requiring a series of mutations each propelling cells toward the development of a **tumor,** an abnormal mass of cells. **Carcinogenesis,** the development of cancer, is a gradual, stepwise process and it may be decades before a person notices any sign or symptom of a tumor whose cells have the characteristics listed in Figure 25.16.

Cancer cells lack differentiation. Most cells are specialized; they have a specific form and function that suits them to the role they play in the body. Cancer cells are nonspecialized and do not contribute to the functioning of a body part. A cancer cell does not look like a differentiated epithelial, muscle, nervous, or connective tissue cell; instead, it looks distinctly abnormal. Normal cells can enter the cell cycle about fifty times, and then they die. Cancer cells can enter the cell cycle repeatedly, and in this way they are immortal. In cell tissue culture, they die only because they run out of nutrients or are killed by their own toxic waste products.

Cancer cells have abnormal nuclei. The nuclei of cancer cells are enlarged, and there may be an abnormal number of chromosomes. The chromosomes have mutated; some parts may be duplicated and some may be deleted. In addition, gene amplification (extra copies of specific genes) is seen much more frequently than in normal cells. Ordinarily, cells with damaged DNA undergo **apoptosis,** or programmed cell death. Cancer cells fail to undergo apoptosis involving a series of enzymatic reactions that lead to the death of the cell.

Cancer cells form tumors. Normal cells anchor themselves to a substratum and/or adhere to their neighbors. They exhibit contact inhibition—when they come in contact with a neighbor, they stop dividing. In culture, normal cells form a single layer that covers the bottom of a petri dish. Cancer cells have lost all restraint; they pile on top of one another and grow in multiple layers. They have a reduced need for stimulatory growth factors, such as epidermal growth factor from their neighbors. Conversely, cancer cells no longer respond to inhibitory growth factors such as transforming growth factor beta (TGF-β) from their neighbors. Their growth, termed a *neoplasia,* contains cells that are disorganized, a condition termed *anaplasia.* During carcinogenesis, the most aggressive cell becomes the dominant cell of the tumor.

Cancer cells undergo angiogenesis and metastasis. A *benign tumor* is usually encapsulated, and does not invade adjacent tissue. **Angiogenesis,** the formation of new blood vessels, is required to bring nutrients and oxygen to a cancerous tumor whose growth is not contained within a capsule. Cancer cells release a growth factor that causes neighboring blood vessels to branch into the cancerous tissue. Some modes of cancer treatment are aimed at preventing angiogenesis from occurring.

Cancer in situ is found in its place of origin; there has been no invasion of normal tissue. Malignancy is present when **metastasis** establishes new tumors distant from the primary tumor. To accomplish metastasis, cancer cells must first make their way across a basement membrane and into a blood vessel or lymphatic vessel. Cancer cells produce proteinase enzymes that degrade the basement membrane and allow them to invade underlying tissues. Cancer cells tend to be motile, have a disorganized internal cytoskeleton, and lack intact actin filament bundles. After traveling through the blood or lymph, cancer cells may start tumors elsewhere in the body.

The patient's prognosis (probable outcome) is dependent on the degree to which the cancer has progressed: (1) whether the tumor has invaded surrounding tissues, (2) if so, whether there is any lymph node involvement, and (3) whether there are metastatic tumors in distant parts of the body. With each progressive step of the cancerous condition, the prognosis becomes less favorable.

Cancer cells grow and divide uncontrollably, and then they metastasize, forming new tumors wherever they relocate.

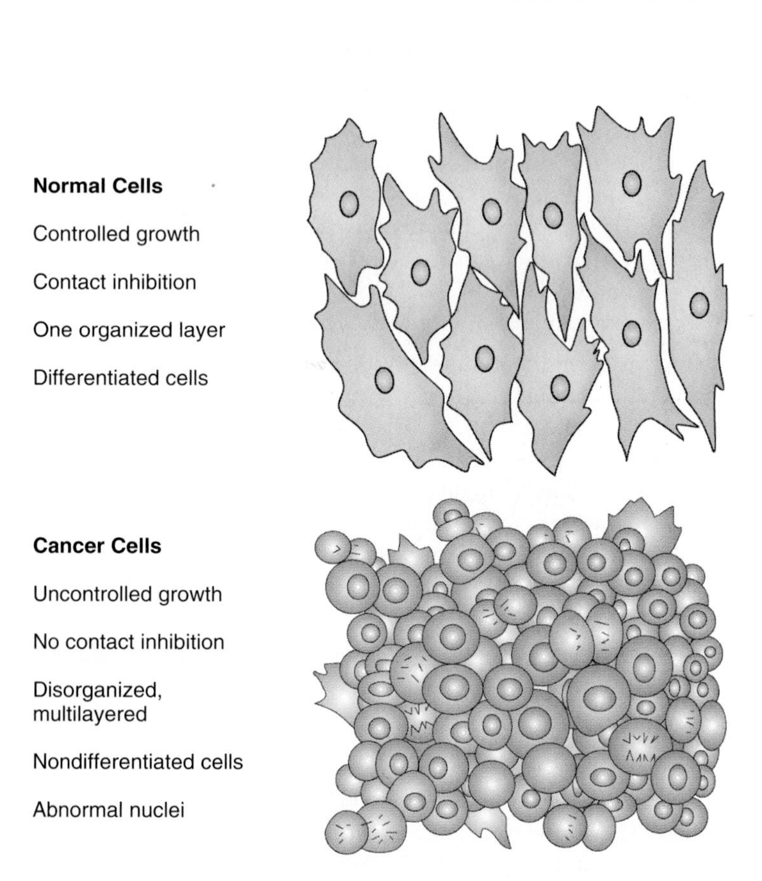

Figure 25.16 Cancer cells.
Cancer cells differ from normal cells in the ways noted.

Health Focus

Prevention of Cancer

There is clear evidence that the risk of certain types of cancer can be reduced by adopting protective behaviors and the right diet.

Protective Behaviors

These behaviors help prevent cancer:

Don't Smoke Cigarette smoking accounts for about 30% of all cancer deaths. Smoking is responsible for 90% of lung cancer cases among men and 79% among women—about 87% altogether. Those who smoke two or more packs of cigarettes a day have lung cancer mortality rates fifteen to twenty-five times greater than nonsmokers. Cigars and smokeless tobacco (chewing tobacco or snuff) increase the risk of cancers of the mouth, larynx, throat, and esophagus.

Don't Sunbathe Almost all cases of basal-cell and squamous-cell skin cancers are considered to be sun related. Further, sun exposure is a major factor in the development of melanoma, and the incidence of this cancer increases for those living near the equator.

Avoid Alcohol Cancers of the mouth, throat, esophagus, larynx, and liver occur more frequently among heavy drinkers, especially when accompanied by tobacco use (cigarettes or chewing tobacco).

Avoid Radiation Excessive exposure to ionizing radiation can increase cancer risk. Even though most medical and dental X rays are adjusted to deliver the lowest dose possible, unnecessary X rays should be avoided. Excessive radon exposure in homes increases the risk of lung cancer, especially in cigarette smokers. It is best to test your home and take the proper remedial actions.

Be Tested for Cancer Do the shower check for breast cancer or testicular cancer. Have other exams done regularly by a physician.

Be Aware of Occupational Hazards Exposure to several different industrial agents (nickel, chromate, asbestos, vinyl chloride, etc.) and/or radiation increases the risk of various cancers. Risk from asbestos is greatly increased when combined with cigarette smoking.

Be Aware of Hormone Therapy Estrogen therapy to control menopausal symptoms increases the risk of endometrial cancer. However, including progesterone in estrogen replacement therapy helps to minimize this risk.

The Right Diet

Statistical studies have suggested that persons who follow certain dietary guidelines are less likely to have cancer. The following dietary guidelines greatly reduce your risk of developing cancer:

Avoid obesity The risk of cancer (especially colon, breast, and uterine cancers) is 55% greater among obese women and 33% greater among obese men, compared to people of normal weight.

Lower total fat intake A high-fat intake has been linked to development of colon, prostate, and possibly breast cancers. Eat plenty of high-fiber foods These include whole-grain cereals, fruits, and vegetables. Studies have indicated that a high-fiber diet protects against colon cancer, a frequent cause of cancer deaths. It is worth noting that foods high in fiber also tend to be low in fat!

Increase consumption of foods that are rich in vitamins A and C Beta-carotene, a precursor of vitamin A, is found in dark green, leafy vegetables, carrots, and various fruits. Vitamin C is present in citrus fruits. These vitamins are called antioxidants because in cells they prevent the formation of free radicals (organic ions that have an unpaired electron) that can possibly damage DNA. Vitamin C also prevents the conversion of nitrates and nitrites into carcinogenic nitrosamines in the digestive tract.

Cut down on consumption of salt-cured, smoked, or nitrite-cured foods Salt-cured or pickled foods may increase the risk of stomach and esophageal cancer. Smoked foods like ham and sausage contain chemical carcinogens similar to those in tobacco smoke. Nitrites are sometimes added to processed meats (e.g., hot dogs and cold cuts) and other foods to protect them from spoilage; as mentioned previously, nitrites are converted to nitrosamines in the digestive tract.

Include vegetables from the cabbage family in the diet The cabbage family includes cabbage, broccoli, brussels sprouts, kohlrabi, and cauliflower. These vegetables may reduce the risk of gastrointestinal and respiratory tract cancers.

Be moderate in the consumption of alcohol People who drink and smoke are at an unusually high risk for cancers of the mouth, larynx, and esophagus.

Visual Focus

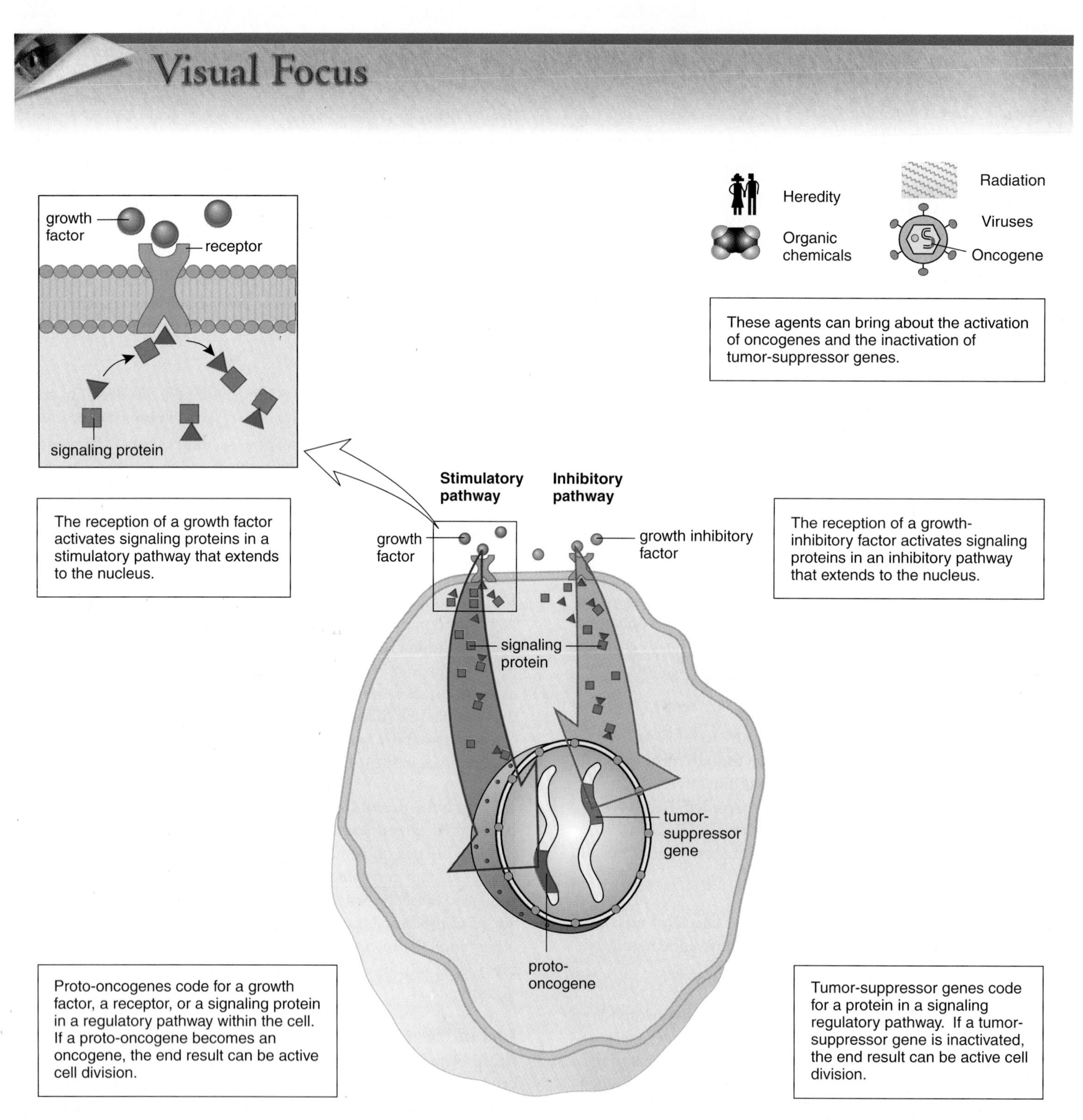

Figure 25.17 Causes of cancer.
Two types of regulatory pathways extend from the plasma membrane to the nucleus. In the stimulatory pathway, plasma membrane receptors receive growth-stimulatory factors. Then, proteins within the cytoplasm and proto-oncogenes within the nucleus stimulate the cell cycle. In the inhibitory pathway, plasma membrane receptors receive growth-inhibitory factors. Then, proteins within the cytoplasm and tumor-suppressor genes within the nucleus inhibit the cell cycle from occurring. Whether cell division occurs or not depends on the balance of stimulatory and inhibitory signals received. Hereditary and environmental factors cause mutations of proto-oncogenes and tumor-suppressor genes. These mutations can cause uncontrolled growth and a tumor.

The Effect of Environmental Factors

Factors that contribute to the origination of cancer are listed in Figure 25.17. It's possible to inherit a gene that tends to cause cancer. The recently isolated *BRCA1* and *BRCA2* genes seem to account for about 20% of premenopausal breast cancer that runs in families and a substantial proportion of such ovarian cancers as well.

Exposure to environmental cancer-causing agents called **carcinogens** can also lead to cancer. Common carcinogens are tobacco smoke, which is well known to cause lung cancer, radiation, such as solar radiation that leads to skin cancer, and certain viruses, such as human papillomavirus which is implicated in cancer of the cervix.

Even our diet, as discussed in the reading on page 523, influences whether we eventually develop cancer or not. Obesity is associated with uterine, postmenopausal breast cancer, as well as cancers of the colon, kidney, and gallbladder.

These factors promote mutations in genes that control the cell cycle. All the cells of any tumor are derived from an ancestral cell whose genes have undergone a series of mutations. These mutations cause the cell to repeatedly enter the cell cycle.

Regulation of the Cell Cycle

The cell cycle is a series of stages that occur in this sequence: G_1 stage—organelles begin to double in number; S stage—replication of DNA occurs and duplication of chromosomes occurs; G_2 stage—synthesis of proteins occurs that prepares the cell for mitosis; M stage—mitosis occurs. The passage of a cell from G_1 to the S stage is tightly regulated, as is the passage of a cell from G_2 to mitosis.

Two classes of genes known as proto-oncogenes and tumor-suppressor genes control the passage of cells through the cell cycle. **Proto-oncogenes** promote the cell cycle and **tumor-suppressor genes** inhibit the cell cycle. Each class of genes is a part of a *regulatory pathway* that involves extracellular *growth factors,* plasma membrane *growth factor receptors,* various *signaling proteins* within the cytoplasm, and a number of *genes* within the nucleus.

In the stimulatory pathway, a growth factor released by a neighboring cell is received by a plasma membrane receptor, and this sets in motion a whole series of enzymatic reactions that ends when proteins that can trigger cell division enter the nucleus (Fig. 25.17). In the inhibitory pathway, a growth-inhibitory factor released by a neighboring cell is received by a plasma membrane receptor, which sets in motion a whole series of enzymatic reactions that ends when proteins that inhibit cell division enter the nucleus. The balance between stimulatory signals and inhibitory signals determines whether proto-oncogenes are active or tumor-suppressor genes are active.

Oncogenes

Proto-oncogenes are so-called because a mutation can cause them to become **oncogenes** (cancer-causing genes). An oncogene may code for a faulty receptor in the stimulatory pathway. A faulty receptor may be able to start the stimulatory process even when no growth factor is present! Or an oncogene may produce an abnormal protein product or else abnormally high levels of a normal product that stimulates the cell cycle to begin or go to completion. In either case, uncontrolled growth threatens.

Researchers have identified perhaps one hundred oncogenes that can cause increased growth and lead to tumors. The oncogenes most frequently involved in human cancers belong to the *ras* gene family. An alteration of only a single nucleotide pair is sufficient to convert a normally functioning ras proto-oncogene to an oncogene. The *ras*K oncogene is found in about 25% of lung cancers, 50% of colon cancers, and 90% of pancreatic cancers. The *ras*N oncogene is associated with leukemias (cancer of blood-forming cells) and lymphomas (cancers of lymphoid tissue), and both *ras* oncogenes are frequently found in thyroid cancers.

Tumor-Suppressor Genes

When a tumor-suppressor gene undergoes a mutation, inhibitory proteins fail to be active and the regulatory balance shifts in favor of cell cycle stimulation. Researchers have identified about a half-dozen tumor-suppressor genes. The *RB* tumor-suppressor gene was discovered when the inherited condition retinoblastoma was being studied. If a child receives only one normal *RB* gene and that gene mutates, eye tumors develop in the retina by the age of three. The *RB* gene has now been found to malfunction in cancers of the breast, prostate, and bladder, among others. Loss of the *RB* gene through chromosome deletion is particularly frequent in a type of lung cancer called small cell lung carcinoma. How the RB protein fits into the inhibitory pathway is known. When a particular growth-inhibitory factor attaches to a receptor, the RB protein is activated. An active RB protein turns off the expression of a proto-oncogene, whose product initiates cell division.

Another major tumor-suppressor gene is called *p53,* a gene that is more frequently mutated in human cancers than any other known gene. It has been found that the p53 protein acts as a transcription factor and as such is involved in turning on the expression of genes whose products are cell cycle inhibitors. p53 can also stimulate apoptosis, programed cell death.

Each cell contains regulatory pathways involving proto-oncogenes and tumor-suppressor genes that code for components active in the pathways.

Apoptosis

Most, if not all, cells contain apoptotic enzymes that can bring about their own self-destruction. These enzymes are known as ICE-like proteases because they structurally resemble interleukin-1 converting enzyme (ICE). Ordinarily, apoptotic enzymes are inactive and cause no harm, but when they are activated, they cleave the structural components of the cell, including its genetic material. Apoptosis is an important aspect of development. For example, when the webbing between the fingers is removed as the hand develops, apoptotic enzymes are involved. Apoptosis in the mature individual is a way to regulate cell population size and constituency. As you may recall, T lymphocytes are continually produced in red bone marrow and mature in the thymus. Any T lymphocytes that bear receptors capable of recognizing the body's own cells undergo apoptosis.

Recently, it has been suggested that the mutation of a proto-oncogene to an oncogene or the mutation of a tumor-suppressor gene ordinarily brings about apoptosis of the cell (Fig. 25.18). This is regarded as a safeguard to prevent the development of tumors in the body. If DNA is damaged in any way, the p53 protein inhibits the cell cycle, and this gives cellular enzymes the opportunity to repair the damage. But if DNA damage should persist, the *p53* gene goes on to bring about apoptosis of the cell. No wonder many types of tumors contain cells that lack an active *p53* gene. Some cancer cells have a *p53* gene but make a large amount of Bcl-2, a protein which binds to and inactivates the p53 protein.

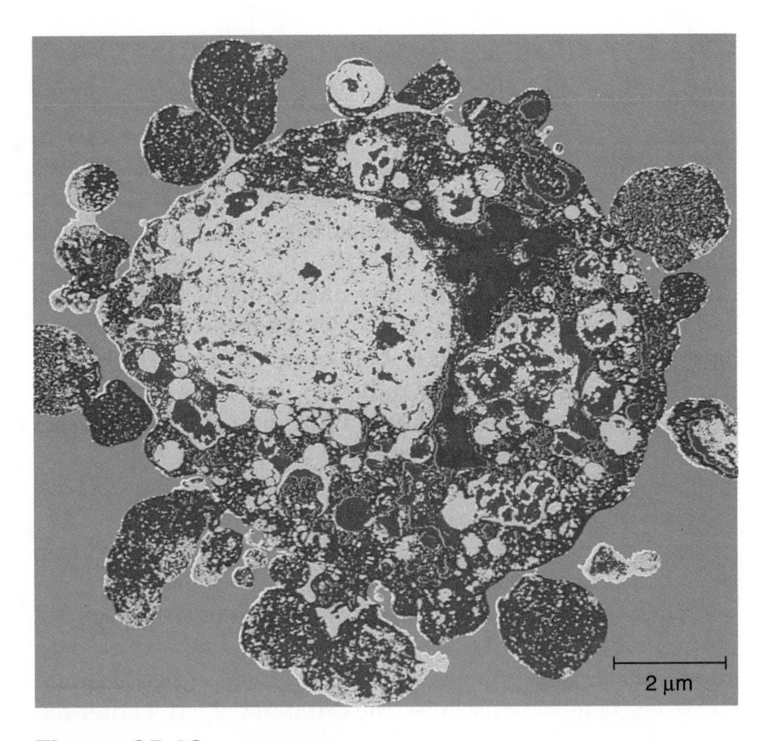

Figure 25.18 Apoptosis.
This cell is undergoing apoptosis, programmed cell death. Outside the nucleus (yellow), the cytoplasm (green/red) is fragmenting into so-called blebs. These cell fragments are digested by phagocytic cells. Failure of a precancer cell to undergo apoptosis is a cause of cancer.

Apoptosis of cells that contain oncogenes and mutated tumor-suppressor genes is a reasonable way for the body to prevent the development of cancer.

Bioethical Issue

Over the past decade, genetic tests have become available for certain cancer genes. If women test positive for defective *BRCA1* and *BRCA2* genes, they have an increased risk for early-onset breast and ovarian cancer. If individuals test positive for the *APC* gene, they are at greater risk for the development of colon cancer. Other genetic tests exist for rare cancers, including retinoblastoma and Wilms' tumor.

Advocates of genetic testing say that it can alert those who test positive for these mutated genes to undergo more frequent mammograms or colonoscopies. Early detection clearly offers the best chance of successful treatment of cancer. Others feel that genetic testing at this point is unnecessary because there is nothing that can presently be done to prevent the disease. Perhaps it is enough for those who have a family history of cancer to schedule more frequent check-ups, beginning at a younger age.

Those opposed to genetic testing worry about the possibility that being predisposed to cancer might threaten one's job or health insurance. They suggest that genetic testing be confined to a research setting, especially since it is not known which particular mutations in these genes predispose one to cancer. They are afraid that a woman with a defective *BRCA1* or *BRCA2* gene might make the unnecessary decision to have a radical mastectomy. The lack of proper counseling concerns many. In a study of 177 patients who underwent *APC* gene testing for susceptibility to colon cancer, less than 20% received counseling before the test. Moreover, physicians misinterpreted the test results in nearly one-third of the cases.

Another concern is that testing negative for a particular genetic mutation may give people the false impression that they are not at risk for cancer. Such a false sense of security can prevent them from having routine cancer screening. Regular testing and avoiding known causes of cancer—such as smoking, a high-fat diet, or too much sunlight—is important for everyone.

Questions

1. Should everyone be aware that testing for cancer is a possibility, or should it be confined to a research setting? Explain.
2. If genetic testing for cancer were offered to you, would you take advantage of it? Why or why not?
3. Do you feel that everyone should do all they can to avoid having cancer, like not smoking, or is it the individual's choice? Explain.

Summarizing the Concepts

25.1 DNA Structure and Replication

DNA, the genetic material, is a double helix containing the nitrogen bases A (adenine) paired with T (thymine) and G (guanine) paired with C (cytosine). During replication, DNA "unzips," and then a complementary strand forms opposite to each original strand.

25.2 Gene Expression

RNA is a single-stranded nucleic acid in which A pairs with U (uracil) while G still pairs with C. DNA specifies the synthesis of proteins because it contains a triplet code: every three bases stand for one amino acid. During transcription, mRNA is made complementary to one of the DNA strands. mRNA, bearing codons, moves to the cytoplasm, where it becomes associated with the ribosomes. During translation, tRNA molecules, attached to their own particular amino acid, travel to a ribosome, and through complementary base pairing between codons of tRNA and codons of mRNA, the tRNAs and therefore the amino acids in a polypeptide are sequenced in a predetermined way.

25.3 Control of Gene Expression

The following levels of control of gene expression are possible in eukaryotes: transcriptional control, posttranscriptional control, translational control, and posttranslational control. The prokaryote operon model explains how one regulator gene controls the transcription of several structural genes—genes that code for proteins. In eukaryotes, the chromosome has to decondense before transcription can begin. Transcription factors attach to DNA and turn on particular genes.

25.4 Gene Mutations

In molecular terms, a gene is a segment of DNA, which codes for a specific polypeptide, and a mutation is a change in the normal sequence of nucleotides of this segment. Frameshift mutations result when a base is added or deleted and the result is a nonfunctioning protein. Point mutations can range in effect, depending on the particular codon change. Gene mutation rates are rather low because DNA polymerase proofreads the new strand during replication and because there are repair enzymes that constantly monitor the DNA.

25.5 Cancer: A Failure of Genetic Control

Cancer is characterized by a lack of control: the cells grow uncontrollably and metastasize. Cancer development is a multistep process involving the mutation of genes. Proto-oncogenes and tumor-suppressor genes are normal genes that code for products involved in cell growth. If they mutate, they can bring on or allow cancer to develop. Usually, cells that bear damaged DNA undergo apoptosis. Apoptosis fails to take place in cancer cells.

Studying the Concepts

1. Describe the experiment that designated DNA rather than protein as the genetic material. 506
2. Describe the structure of DNA including complementary base pairing. Explain how DNA replicates. 507–09
3. How is the structure of RNA different from the structure of DNA? Name the three classes of RNA and give their functions. 510
4. What happens during transcription? during translation? 511–12
5. What are the four levels of control of gene expression in eukaryotes? 517
6. What is the operon model of structural gene control? Describe the *lac* operon. 518–19
7. The substitution of one base for another base in DNA can have what effects on the phenotype? Why would you expect additions and deletions of bases to have a major effect on the phenotype? 520
8. Describe the characteristics of cancer cells that set them apart from normal cells. 522
9. What is a proto-oncogene, and what is a tumor-suppressor gene, and how do they function in normal cells? Explain why cancer develops if they mutate. 524–25
10. What is apoptosis, and how is it involved in the development of cancer? 526

Testing Yourself

Choose the best answer for each question.

1. The double-helix model of DNA resembles a twisted ladder in which the rungs of the ladder are
   a. a purine paired with a pyrimidine.
   b. A paired with G and C paired with T.
   c. A paired with T and G paired with C.
   d. sugar-phosphate paired with sugar-phosphate.
   e. Both a and c are correct.
2. In a DNA molecule, the
   a. backbone is sugar and phosphate molecules.
   b. bases are covalently bonded to the sugars.
   c. sugars are covalently bonded to the phosphates.
   d. bases are hydrogen-bonded to one another.
   e. All of these are correct.
3. If you grew bacteria in heavy nitrogen and then switched them to light nitrogen, how many generations after switching would you have some in which both strands are light?
   a. never, because replication is semiconservative
   b. the first generation
   c. the second generation
   d. the third generation
   e. the fourth generation
4. If the sequence of bases in DNA is TAGCCT, then the sequence of bases in RNA will be
   a. TCCGAT.
   b. ATCGGA.
   c. TAGCCT.
   d. AUCGGU.
   e. Both a and b are correct.
5. RNA processing
   a. takes place in the cytoplasm.
   b. is the same as transcription.
   c. is an event that occurs after RNA is transcribed.
   d. is the rejection of old, worn-out RNA.
   e. Both a and c are correct.
6. During protein synthesis, an anticodon of a transfer RNA (tRNA) pairs with
   a. amino acids in the polypeptide.
   b. DNA nucleotide bases.
   c. ribosomal RNA (rRNA) nucleotide bases.
   d. messenger RNA (mRNA) nucleotide bases.
   e. other tRNA nucleotide bases.
7. If the DNA codons are CAT CAT CAT, and a guanine base is added at the beginning, then which would result?
   a. CAT CAT CAT G
   b. G CAT CAT CAT
   c. GCA TCA TCA T
   d. frameshift mutation
   e. Both c and d are correct.

8. Label this diagram of an operon.

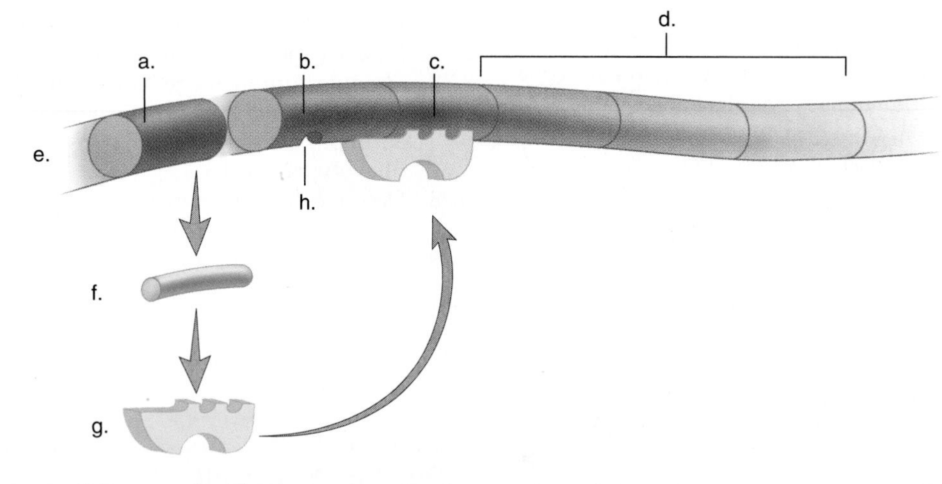

9. RNA processing varies in different cells. This is an example of ________ control of gene expression.
   a. transcriptional
   b. posttranscriptional
   c. translational
   d. posttranslational
   e. enzymatic
10. A cell is cancerous. You might find an abnormality in the
   a. plasma membrane
   b. cytoplasm
   c. genes
   d. All of these are correct.

Thinking Scientifically

1. Considering the structure and function of DNA,
   a. What is your evidence that DNA stores genetic information?
   b. What is your evidence that DNA replicates prior to cell division?
   c. What part of DNA mutates and what evidence in nature shows that DNA mutates?
2. Considering the functions of RNA:
   a. If you supplied an actively metabolizing cell with radioactive uridine, where would you first and then later expect to find the label?
   b. What evidence do you have that mRNA and not rRNA directs the synthesis of a polypeptide?
   c. What physical evidence can you present that ribosomes (rRNA) move along the mRNA?

Understanding the Terms

angiogenesis 522
anticodon 513
apoptosis 522
cancer 522
carcinogen 525
carcinogenesis 522
codon 511
complementary base pairing 508
DNA (deoxyribonucleic acid) 506
DNA polymerase 509
double helix 508
gene 510
messenger RNA (mRNA) 510
metastasis 522
mutagen 520
oncogene 525
operator 518
operon 518
polyribosome 515
promoter 518
proto-oncogene 525
purine 508
pyrimidine 508
regulator gene 518
replication 509
ribosomal RNA (rRNA) 510
ribosome 513
ribozyme 512
RNA polymerase 512
RNA (ribonucleic acid) 510
structural gene 518
template 509
transcription 511
transfer RNA (tRNA) 510
translation 511
triplet code 511
tumor 522
tumor-suppressor gene 525

Match the terms to these definitions:

a. ________ Agent, such as radiation or a chemical, that brings about a mutation.
b. ________ Enzyme that speeds the formation of RNA from a DNA template.
c. ________ Process whereby a DNA strand serves as a template for the formation of mRNA.
d. ________ Process whereby the sequence of codons in mRNA determines (is translated into) the sequence of amino acids in a polypeptide.
e. ________ Three nucleotides on a tRNA molecule attracted to a complementary codon on mRNA.

Using Technology

Your study of molecular basis of inheritance is supported by these available technologies:

Essential Study Partner CD-ROM

Genetics → DNA
→ Protein Synthesis

Visit the Mader web site for related ESP activities.

Exploring the Internet

The Mader Home Page provides resources and tools as you study this chapter.

http://www.mhhe.com/biosci/genbio/mader

C H A P T E R 26

Biotechnology

Chapter Concepts

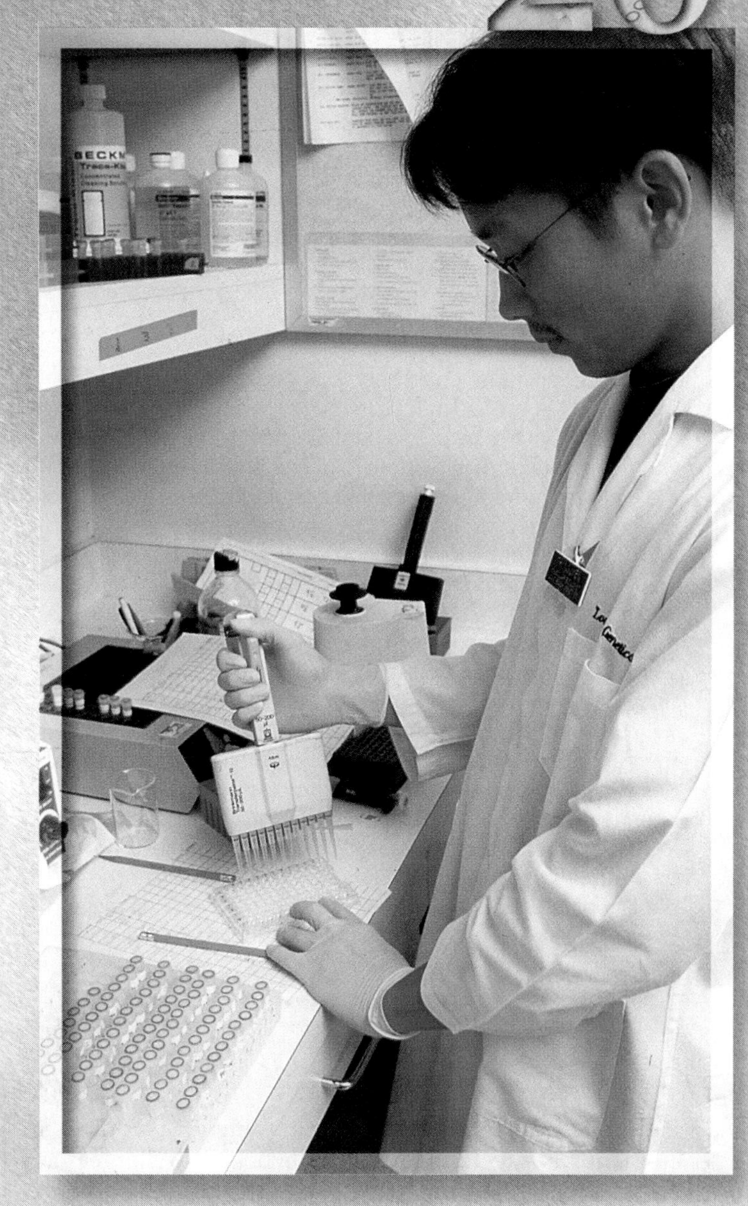

DNA can be extracted from a cell and manipulated using standard laboratory techniques. This discovery has led to the commercial availability of gene products and gene therapy in humans.

The experiment was Fred's last hope. The antiviral drugs had done little to stop the AIDS virus from ravaging his immune system, so Fred's physicians decided to try to build him HIV-proof immune cells. After removing some of Fred's bone marrow, the source of the body's immune cells, they grew these cells in the laboratory and added to them several genes they hoped would make the cells less susceptible to infection by the AIDS virus. They then transplanted the genetically altered cells back into Fred, hoping that they would multiply and reconstitute his weakened immune system with one that HIV could no longer defeat.

Since birth Mary had been plagued with cystic fibrosis, which causes excess mucus in the lungs and digestive tract. Just a few years before, scientists had isolated the gene whose mutant form is responsible for the torturous effects of the disease. A modified virus that normally causes a cold had been genetically engineered to carry the normal gene deep into the cells that line the lungs. A single therapeutic dose would place the modified virus in the lungs of the little girl now racked with cough. Maybe someday Mary would be able to run and play just like all the other children her age. These are just two examples of gene therapy in humans, a field that is burgeoning with new ideas and treatments every day.

Genetic engineering is the use of technology to alter the genome of viruses, bacteria, and other cells for medical or industrial purposes. Biotechnology includes genetic engineering and other techniques that make use of natural biological systems to produce a product or to achieve an end desired by human beings. Genetic engineering has the capability of altering the genotype of unicellular organisms and the genotype of plants and animals, including ourselves.

26.1 Cloning of a Gene

The cloning of a gene produces many identical copies. Recombinant DNA technology is used when a very large quantity of the gene is required. The use of the polymerase chain reaction (PCR) creates a lesser number of copies within a laboratory test tube.

Recombinant DNA Technology

Recombinant DNA (rDNA) contains DNA from two different sources. To make rDNA, a technician often begins by selecting a **vector,** the means by which recombinant DNA is introduced into a host cell. One common type of vector is a plasmid. **Plasmids** are small accessory rings of DNA. The ring is not part of the bacterial chromosome and can be replicated independently. Plasmids were discovered by investigators studying the sex life of the intestinal bacterium *Escherichia coli.*

Two enzymes are needed to introduce foreign DNA into vector DNA (Fig. 26.1). The first enzyme, called a **restriction enzyme,** cleaves plasmid DNA, and the second, called **DNA ligase,** seals foreign DNA into the opening created by the restriction enzyme.

Restriction enzymes occur naturally in bacteria, where they stop viral reproduction by cutting up viral DNA. They are called restriction enzymes because they *restrict* the growth of viruses. Hundreds of different restriction

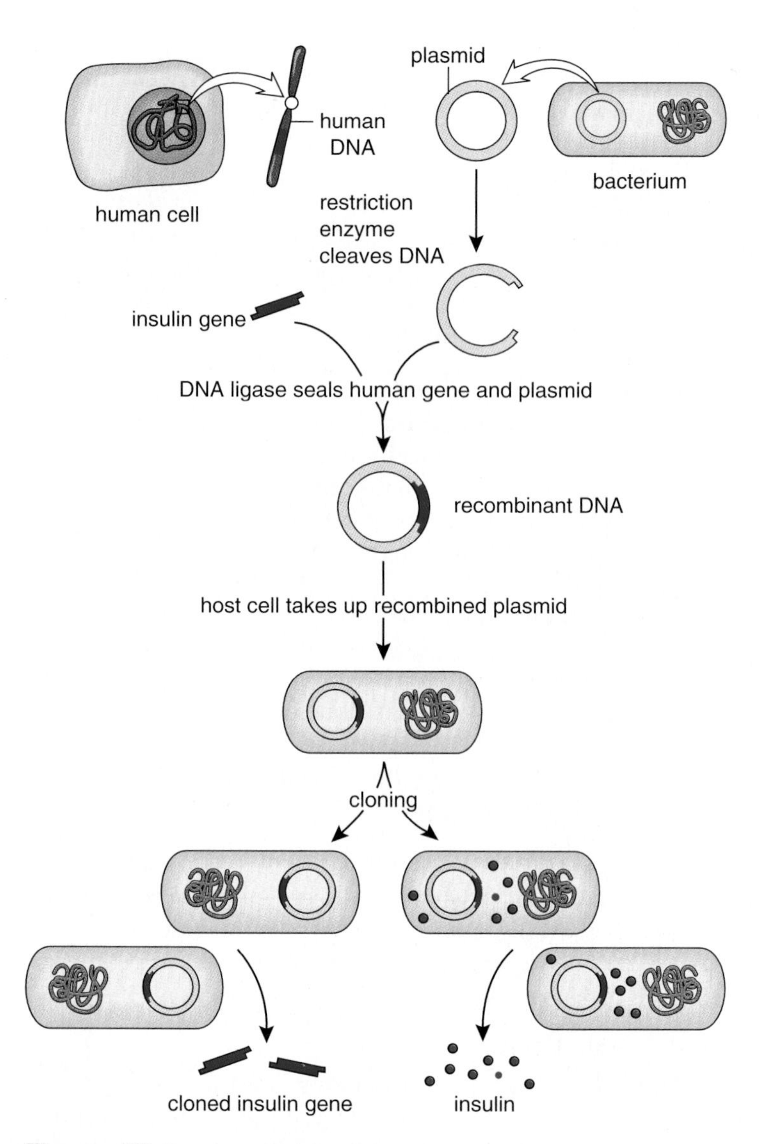

Figure 26.1 Cloning of a gene.
Human DNA and plasmid DNA are cleaved by the same type of restriction enzyme and spliced together by the enzyme DNA ligase. Gene cloning is achieved when a host cell takes up the recombined plasmid and the plasmid reproduces. Multiple copies of the gene are now available to an investigator. If the insulin gene functions normally as expected, the product (insulin) may also be retrieved.

enzymes have been isolated and purified. Each one cuts DNA at a specific cleavage site. For example, the restriction enzyme called *Eco*RI always cuts double-stranded DNA when it has this sequence of bases at the cleavage site:

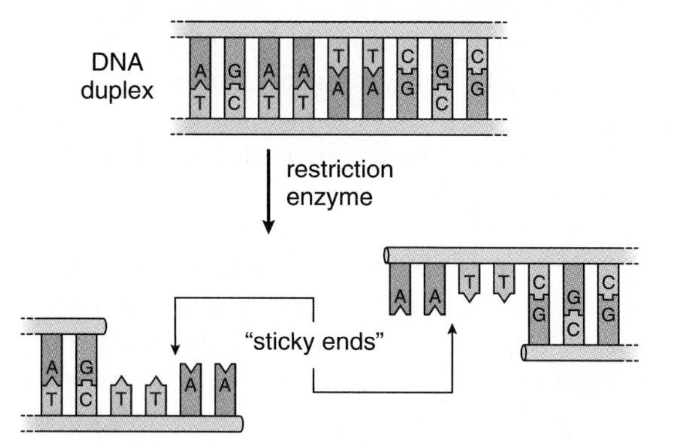

Notice there is now a gap into which a piece of foreign DNA can be placed if it ends in bases complementary to those exposed by the restriction enzyme. To assure this, it is only necessary to cleave the foreign DNA with the same type of restriction enzyme. The single-stranded but complementary ends of the two DNA molecules are called "sticky ends" because they can bind by complementary base pairing. They therefore facilitate the insertion of foreign DNA into vector DNA.

The second enzyme needed for preparation of rDNA, DNA ligase, is a cellular enzyme that seals any breaks in a DNA molecule. Genetic engineers use this enzyme to seal the foreign piece of DNA into the vector. DNA splicing is now complete; an rDNA molecule has been prepared.

Plasmid Vector Compared to Viral Vector

A **clone** can be a large number of molecules (i.e., cloned genes) or cells (i.e., cloned bacteria) or organisms that are identical to an original specimen. Figure 26.2 compares the use of a plasmid and a virus to clone a gene.

Bacterial cells take up recombined plasmids, especially if they are treated with calcium chloride to make them more permeable. Thereafter, as the host cell reproduces, a bacterial clone forms and each new cell contains at least one plasmid. Therefore, each of the bacteria contains the gene of interest which hopefully is expressing itself and producing a product. The investigator can recover either the cloned gene, or the protein product from this bacterial clone (see also Fig. 26.1).

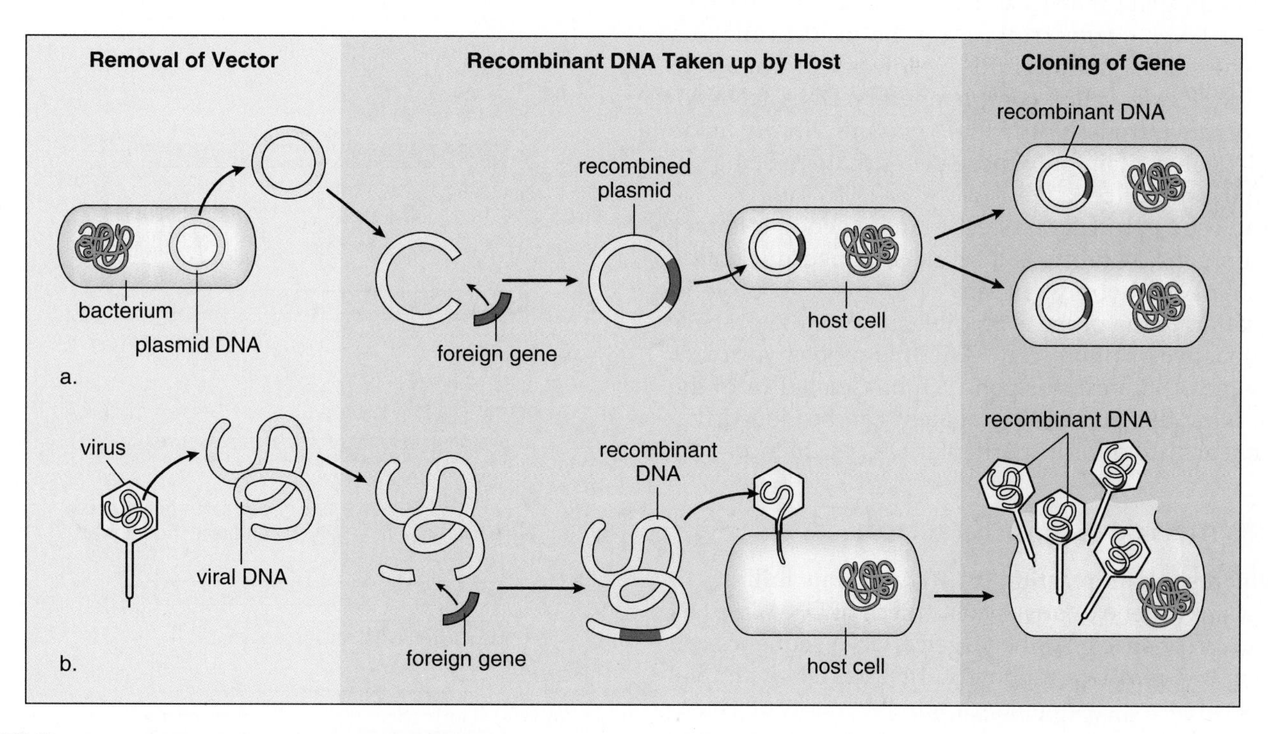

Figure 26.2 Preparation of a genomic library.
Each bacterial or viral clone in a genomic library contains a segment of DNA from a foreign cell. **a.** A plasmid is removed from a bacterium and is used to make recombinant DNA. After the recombined plasmid is taken up by a host cell, replication produces many copies. **b.** Viral DNA is removed from a bacteriophage such as lambda and is used to make recombinant DNA. The virus containing the recombinant DNA infects a host bacterium. Cloning is achieved when the virus reproduces and then leaves the host cell.

Viruses that infect bacteria are called **bacteriophages.** In Figure 26.2*b*, the DNA of a bacteriophage called lambda is being used as a vector. After lambda attaches to a host bacterium, recombined DNA is released from the virus and enters the bacterium. Here, it will direct the reproduction of many more viruses. Each virus in the bacteriophage clone contains a copy of the gene being cloned.

Genomic Library

A **genome** is the full set of genes of an individual. A **genomic library** is a collection of bacterial or bacteriophage clones; each clone contains a particular segment of DNA from the source cell. When you make a genomic library, an organism's DNA is simply sliced up into pieces, and the pieces are put into vectors (i.e., plasmids or viruses) that are taken up by host bacteria. The entire collection of bacterial or bacteriophage clones that result contains all the genes of that organism.

In order for human gene expression to occur in a bacterium, the gene has to be accompanied by the proper regulatory regions. Also, the gene should not contain introns because bacterial cells do not have the necessary enzymes to process primary messenger RNA (mRNA). It's possible to make a human gene that lacks introns, however. The enzyme called reverse transcriptase can be used to make a DNA copy of all the mature mRNA molecules from a cell. This DNA molecule, called **complementary DNA (cDNA),** does not contain introns. Notice that a genomic library made from cDNA will contain only those genes that are being expressed in the source cell.

You can use a particular probe to search a genomic library for a certain gene. A **probe** is a single-stranded nucleotide sequence that will hybridize (pair) with a certain piece of DNA. Location of the probe is possible because the probe is either radioactive or fluorescent. After the probe hybridizes with the gene of interest, the gene can be isolated from the fragment. Now this particular fragment can be cloned further or even analyzed for its particular DNA sequence.

The Polymerase Chain Reaction

The **polymerase chain reaction (PCR)** can create millions of copies of a single gene or any specific piece of DNA in a test tube. PCR is very specific—the targeted DNA sequence can be less than one part in a million of the total DNA sample! This means that a single gene, or smaller piece of DNA, among all the human genes can be amplified (copied) using PCR.

PCR takes its name from DNA polymerase, the enzyme that carries out DNA replication in a cell. It is considered a chain reaction because DNA polymerase will carry out replication over and over again, until there are millions of copies of the desired DNA. PCR does not replace gene cloning, which is still used whenever a large quantity of gene or protein product is needed.

Before carrying out PCR, primers—sequences of about 20 bases that are complementary to the bases on either side of the "target DNA"—must be available. The primers are needed because DNA polymerase does not start the replication process; it only continues or extends the process. After the primers bind by complementary base pairing to the DNA strand, DNA polymerase copies the target DNA (Fig. 26.3).

PCR has been in use for several years, and now almost every laboratory has automated PCR machines to carry out the procedure. Automation became possible after a temperature-insensitive (thermostable) DNA polymerase was extracted from the bacterium *Thermus aquaticus,* which lives in hot springs. This enzyme can withstand the high temperature used to separate double-stranded DNA; therefore, replication need not be interrupted by the need to add more enzyme.

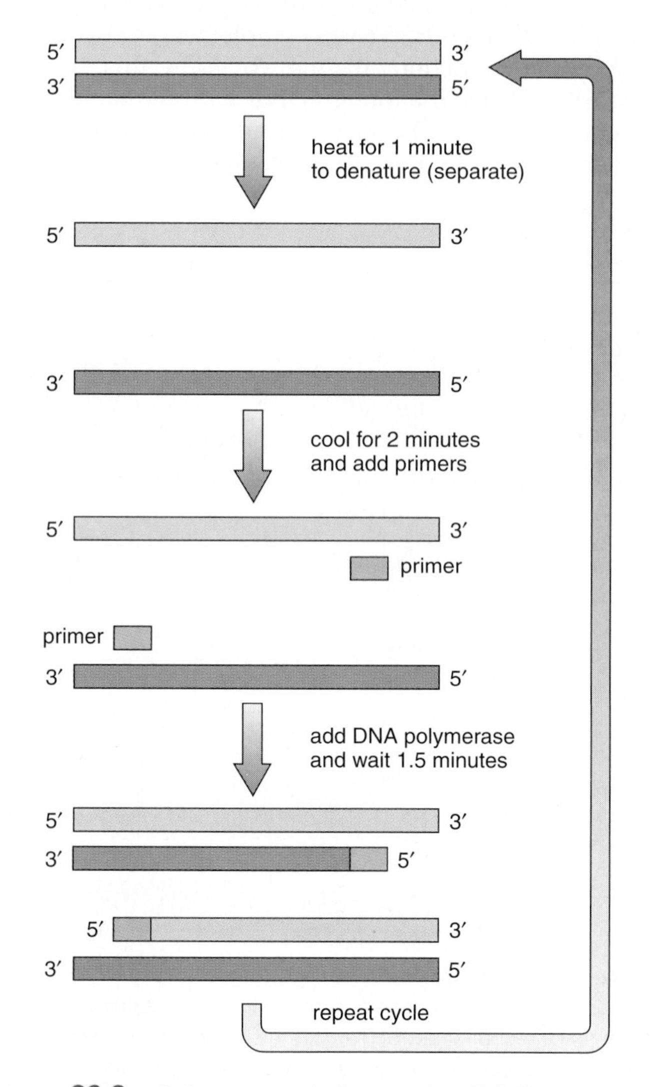

Figure 26.3 Polymerase chain reaction (PCR).
PCR is performed in laboratory test tubes. Primers (red), which are DNA sequences complementary to the 3′ end of the targeted DNA, are necessary for DNA polymerase to make a copy of the DNA strand.

Analyzing DNA

The entire genome of an individual can be subjected to **DNA fingerprinting,** a process described in Figure 26.4. The genome is treated with restriction enzymes, which results in a unique collection of different-sized fragments. Therefore, **restriction fragment length polymorphisms (RFLPs)** exist between individuals. During a process called gel electrophoresis, the fragments can be separated according to their lengths, and the result is a number of bands that are so close together they appear as a smear. However, the use of probes for genetic markers produces a distinctive pattern that can be recorded on X-ray film.

The DNA from a single sperm is enough to identify a suspected rapist. Since DNA is inherited, its fingerprint resembles that of one's parents. DNA fingerprinting successfully identified the remains of a teenager who had been murdered eight years before because the skeletal DNA was similar to that of the parents' DNA. DNA fingerprinting has also been helpful to evolutionists. For example, it was used to determine that the quagga, an extinct zebralike animal, was a zebra rather than a horse. The only remains of the quagga consisted of dried skin.

Following PCR, DNA segments, as opposed to the full genome, can also be cut by restriction enzymes and subjected to gel electrophoresis. A probe is not needed because the restriction fragments will appear as distinctive bands. PCR amplification and analysis can be used to diagnose viral infections, genetic disorders, and cancer. When the amplified DNA matches that of a virus, mutated gene, or oncogene, then we know that a viral infection, genetic disorder, or cancer is present.

Sometimes DNA is sequenced following PCR. Sequencing mitochondrial DNA segments helped determine the evolutionary history of human populations. It has even been possible to sequence DNA taken from a 76,000-year-old mummified human brain and from a 17- to 20-million-year-old plant fossil following PCR amplification. Sequencing can be done quickly with present-day DNA sequencers that make use of computers.

Recombinant DNA technology and the polymerase chain reaction are two ways to clone a gene.

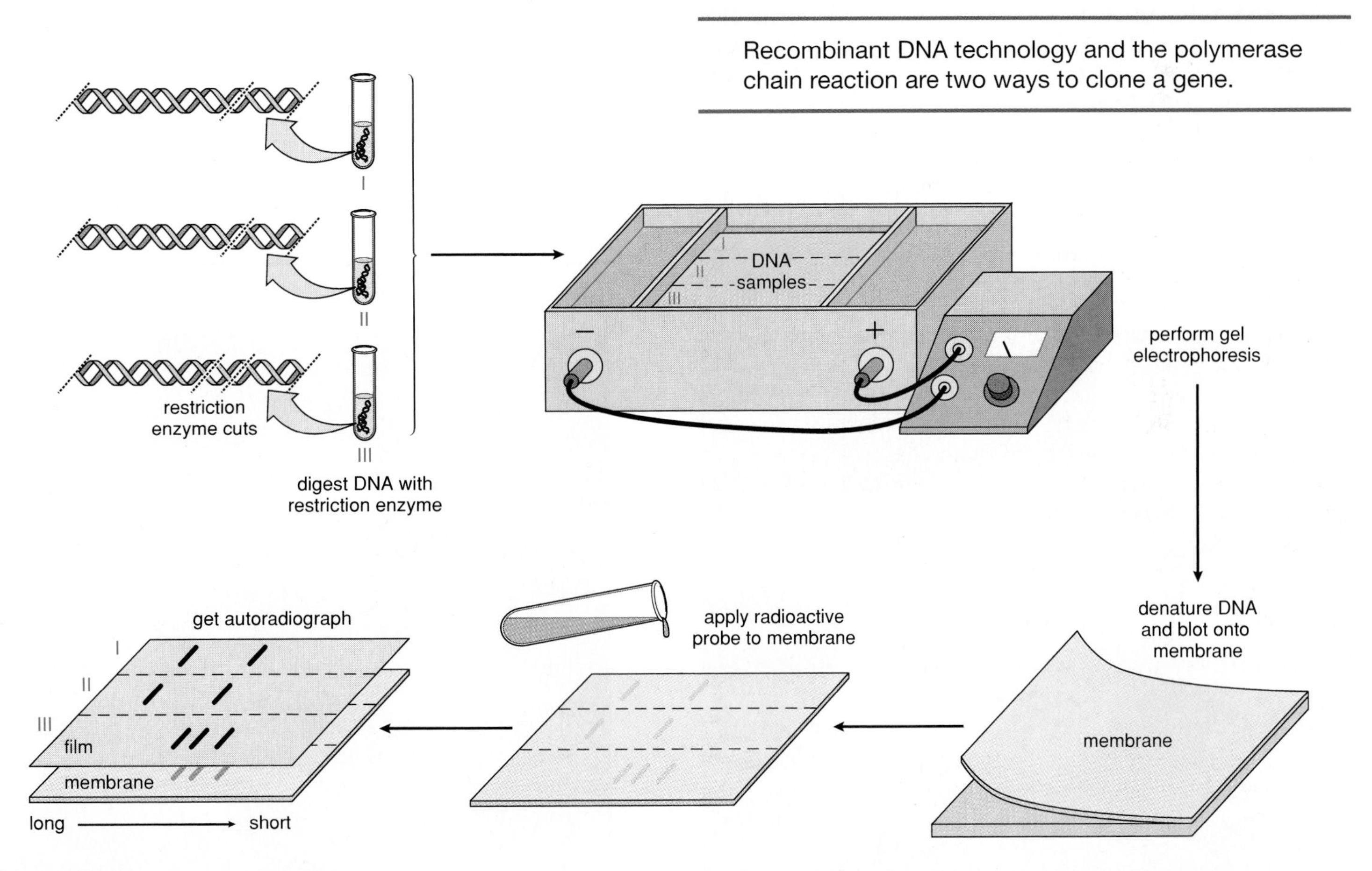

Figure 26.4 DNA fingerprinting.
DNA samples I and II are from the same individual. DNA sample III is from a different individual. Notice, therefore, that the restriction enzyme cuts are different for sample III. Gel electrophoresis separates the DNA fragments according to their length because shorter fragments migrate farther in an electrical field than do longer fragments. The fragments are denatured (separated) and transferred to a membrane where a radioactive probe can be applied. The resulting pattern (the DNA fingerprint) can then be detected by autoradiography. In a theoretical rape case, for example, sample I could be from the suspect's white blood cells, sample II could be from sperm in the victim's vagina, and sample III could be from the victim's white blood cells.

26.2 Biotechnology Products

Today, bacteria, plants, and animals are genetically engineered to produce biotechnology products. Organisms that have had a foreign gene inserted into them are called **transgenic organisms.**

From Bacteria

Recombinant DNA technology is used to produce bacteria that reproduce in large vats called bioreactors. If the foreign gene is replicated and actively expressed, a large amount of protein product can be obtained. Biotechnology products produced by bacteria, such as insulin, human growth hormone, t-PA (tissue plasminogen activator), and hepatitis B vaccine, are now on the market (Fig. 26.5).

Transgenic bacteria have been produced to promote the health of plants. For example, bacteria that normally live on plants and encourage the formation of ice crystals have been changed from frost-plus to frost-minus bacteria. Also, a bacterium that normally colonizes the roots of corn plants has now been endowed with genes (from another bacterium) that code for an insect toxin. The toxin protects the roots from insects.

Bacteria can be selected for their ability to degrade a particular substance, and then this ability can be enhanced by genetic engineering. For instance, naturally occurring bacteria that eat oil can be genetically engineered to do an even better job of cleaning up beaches after oil spills (Fig. 26.6). Industry has found that bacteria can be used as biofilters to prevent airborne chemical pollutants from being vented into the air. They can also remove sulfur from coal before it is burned and help clean up toxic waste dumps. One such strain was given genes that allowed it to clean up levels of toxins that would have killed other strains. Further, these bacteria were given "suicide" genes that caused them to self-destruct when the job had been accomplished.

Organic chemicals are often synthesized by having catalysts act on precursor molecules or by using bacteria to carry out the synthesis. Today, it is possible to go one step further and to manipulate the genes that code for these enzymes. For instance, biochemists discovered a strain of bacteria that is especially good at producing phenylalanine, an organic chemical needed to make aspartame, the dipeptide sweetener better known as NutraSweet. They isolated, altered, and formed a vector for the appropriate genes so that various bacteria could be genetically engineered to produce phenylalanine.

Many major mining companies already use bacteria to obtain various metals. Genetic engineering may enhance the ability of bacteria to extract copper, uranium, and gold from low-grade sources. Some mining companies are testing genetically engineered organisms that have improved bioleaching capabilities.

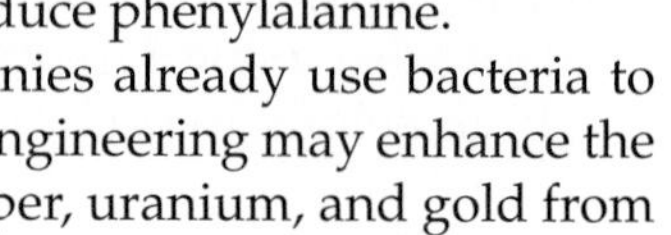

Bacteria are being genetically altered to perform all sorts of tasks, not only in the factory, but also in the environment.

Figure 26.5 Biotechnology products.
Products like clotting factor VIII, which is administered to hemophiliacs, can be made by transgenic bacteria, plants, or animals. After being processed and packaged, it is sold as a commercial product.

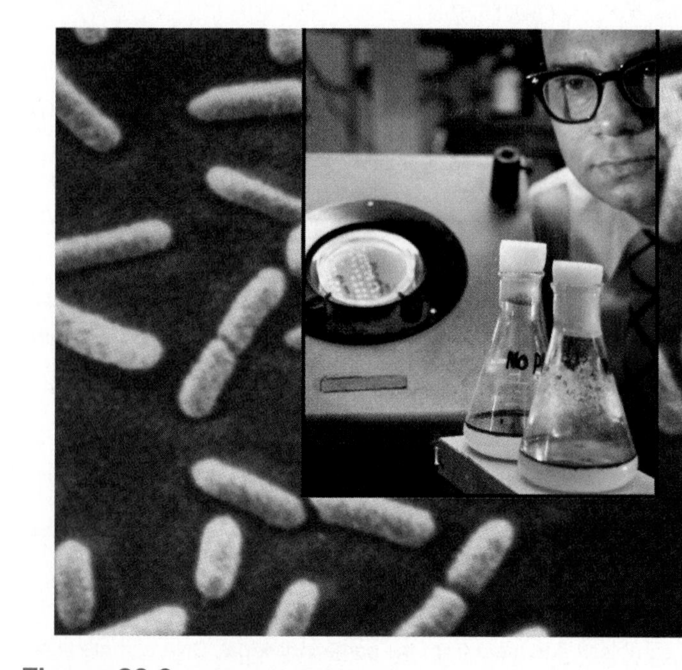

Figure 26.6 Bioremediation.
Bacteria capable of decomposing oil have been engineered and patented by the investigator, Dr. Chakrabarty. In the inset, the flask toward the rear contains oil and no bacteria; the flask toward the front contains the bacteria and is almost clear of oil.

From Plants

Techniques have been developed to introduce foreign genes into immature plant embryos, or into plant cells that have had the cell wall removed and are called protoplasts. It is possible to treat protoplasts with an electric current while they are suspended in a liquid containing foreign DNA. The electric current makes tiny, self-sealing holes in the plasma membrane through which genetic material can enter. Then a protoplast will develop into a complete plant.

Foreign genes transferred to cotton, corn, and potato strains have made these plants resistant to pests because their cells now produce an insect toxin. Similarly, soybeans have been made resistant to a common herbicide. Some corn and cotton plants are both pest and herbicide resistant. These and other genetically engineered crops are now reaching the market place.

Plants are also being engineered to produce human proteins, such as hormones, clotting factors, and antibodies, in their seeds. One type of antibody made by corn can deliver radioisotopes to tumor cells, and another made by soybeans can be used as treatment for genital herpes. A weed called mouse-eared cress has been engineered to produce a biodegradable plastic (polyhydroxybutyrate, or PHB) in cell granules.

Genetically engineered crops are now reaching the market, and medicines made by genetically engineered plants will soon be used to treat cancer and other types of diseases.

Ecology Focus

Biotechnology: Friend or Foe?

In a hungry world you would expect herbicide-resistant crops to be greeted with enthusiasm—genetically engineered wheat and corn offer the possibility of a more bountiful harvest and the feeding of many more people. There are some, however, who see dangers lurking in the use of biotechnology to develop new and different strains of plants and animals. And these doomsayers are not just anybody—they are ecologists.

First, we have to consider that herbicide-resistant crops will allow farmers to use more herbicide than usual in order to kill off weeds. Then, too, suppose this new form of wheat is better able to compete in the wild. Certainly we know of plants that have become pests when transported to a new environment: prickly pear cactus took over many acres of Australia; an ornamental tree, the melaleuca, has invaded and is drying up many of the swamps in Florida. Such plants spread because they are able to overrun the native plants of an area. Perhaps genetically engineered plants will also spread beyond their intended areas and be out of control. Or worse, suppose herbicide-resistant wheat were to hybridize with a weed, making the weed also resistant and able to take over other agricultural fields. As more and different herbicides are used to kill off the weed, the environment would be degraded. And similar concerns pertain to any transgenic organism, whether a bacterium, animal, or plant.

In the past, humans have been quick to believe that a new advance was the answer to a particular problem. The pesticide DDT was going to kill off mosquitoes, making malaria a disease of the past. And this worked until mosquitoes became resistant. Today, we know that DDT accumulates in the tissues of humans, possibly contributing to all manner of health problems, from reduced immunity to reproductive infertility. When antibiotics were first introduced, it was hoped a disease like tuberculosis would be licked forever. Resistant strains of tuberculosis have now evolved to threaten us all.

More and more transgenic varieties have been developed and are being tested in agricultural fields. A few of these have a weedy relative in the wild with which they could hybridize. Agricultural officials point out, however, that genetically engineered plants have been growing in fields since 1994, and although hybridization with weedy relatives may have occurred, a "superweed" has not emerged. Still, say botanists, it could happen in the future. Laboratory studies in Denmark showed that a hybrid of transgenic oilseed rape and field mustard did resist herbicides and was able to produce highly fertile pollen. Ecologists maintain it may be only a matter of time before a "superweed" does appear in the wild.

From Animals

Techniques have been developed to insert genes into the eggs of animals. It is possible to microinject foreign genes into eggs by hand, but another method uses vortex mixing. The eggs are placed in an agitator with DNA and silicon-carbide needles, and the needles make tiny holes through which the DNA can enter. When these eggs are fertilized, the resulting offspring is a transgenic animal. Using this technique, many types of animal eggs have taken up the gene for bovine growth hormone (bGH). The procedure has been used to produce larger fishes, cows, pigs, rabbits, and sheep. Genetically engineered fishes are now being kept in ponds that offer no escape to the wild because there is much concern that they will upset or destroy natural ecosystems.

Gene pharming, the use of transgenic farm animals to produce pharmaceuticals, is being pursued by a number of firms. Genes that code for therapeutic and diagnostic proteins are incorporated into the animal's DNA, and the proteins appear in the animal's milk. There are plans to produce drugs for the treatment of cystic fibrosis, cancer, blood diseases, and other disorders. Antithrombin III, for preventing blood clots during surgery, is currently being produced by a herd of goats, and clinical trials have begun. Figure 26.7*b, c* outlines the procedure for producing transgenic mammals: DNA containing the gene of interest is injected into donor eggs. Following in vitro fertilization, the zygotes are placed in host females where they develop. After female offspring mature, the product is secreted in their milk.

USDA scientists have been able to genetically engineer mice to produce human growth hormone in their urine instead of in milk. They expect to be able to use the same technique on larger animals. Urine is a preferable vehicle for a biotechnology product rather than milk because all animals in a herd urinate—only females produce milk; animals start to urinate at birth—females don't produce milk until maturity; and it's easier to extract proteins from urine than from milk.

Xenotransplantation

Scientists have begun the process of genetically engineering animals to serve as organ donors for humans who need a transplant. **Xenotransplantation** is the use of animal organs instead of human organs in transplant patients. We now have the ability to transplant kidneys, heart, liver, pancreas, lung, and other organs for two reasons. First, solutions have been developed that preserve donor organs for several hours, and second, rejection of transplanted organs can be prevented by immunosuppressive drugs. Unfortunately, however, there are not enough human donors to go around. Fifty thousand Americans needed transplants in 1996, but only 20,000 patients got them. As many as 4,000 died that year while waiting for an organ.

It's no wonder, then, that scientists are suggesting that we should get organs from a source other than another human. You might think that apes, such as the chimpanzee or the baboon might be a scientifically suitable species for this purpose. But apes are slow breeders and probably cannot be counted on to supply all the organs needed. Anyway, many people might object to using apes for this purpose. In contrast, animal husbandry has long included the raising of pigs as a meat source, and pigs are prolific. A female pig can become pregnant at six months and can have two litters a year, each averaging about ten offspring.

Ordinarily, humans would violently reject transplanted pig organs. Genetic engineering, however, can make these organs less antigenic. Scientists have produced a strain of pigs whose organs would most likely, even today, survive for a few months in humans. They could be used to keep a patient alive until a human organ was available. The ultimate goal is to make pig organs as widely accepted by humans as type O blood. A person with type O blood is called a universal donor because the red blood cells carry no AB nor B antigens.

As xenotransplantation draws near, other concerns have been raised. Some experts fear that animals might be infected with viruses, akin to Ebola virus or the virus that causes "mad cow" disease. After infecting a transplant patient, these viruses might spread into the general populace and begin an epidemic. Scientists believe that HIV was spread to humans from monkeys when humans ate monkey meat. Those in favor of using pigs for xenotransplantation point out that pigs have been around humans for centuries without infecting them with any serious diseases.

Cloning of Animals

Imagine that an animal has been genetically altered to produce a biotechnology product or to serve as an organ donor. What would be the best possible way to get identical copies of this animal? If cloning of the animal was possible, you could get many exact copies of this animal. Asexual reproduction through cloning would be the preferred procedure to use. Cloning is a form of asexual reproduction because it requires only the genes of that one animal. For many years it was believed that adult vertebrate animals could not be cloned. Although each cell contains a copy of all the genes, certain genes are turned off in mature specialized cells. Different genes are expressed in muscle cells, which contract, compared to nerve cells which conduct nerve impulses, and to glandular cells, which secrete. Cloning of an adult vertebrate would require that all genes of an adult cell be turned on again if development is to proceed normally. It has long been thought this would be impossible.

But in 1997 scientists at the Raslin Institute in Scotland announced that they achieved this feat and had produced a cloned sheep called Dolly. In 1998, genetically altered calves were cloned in the United States using the same

a.

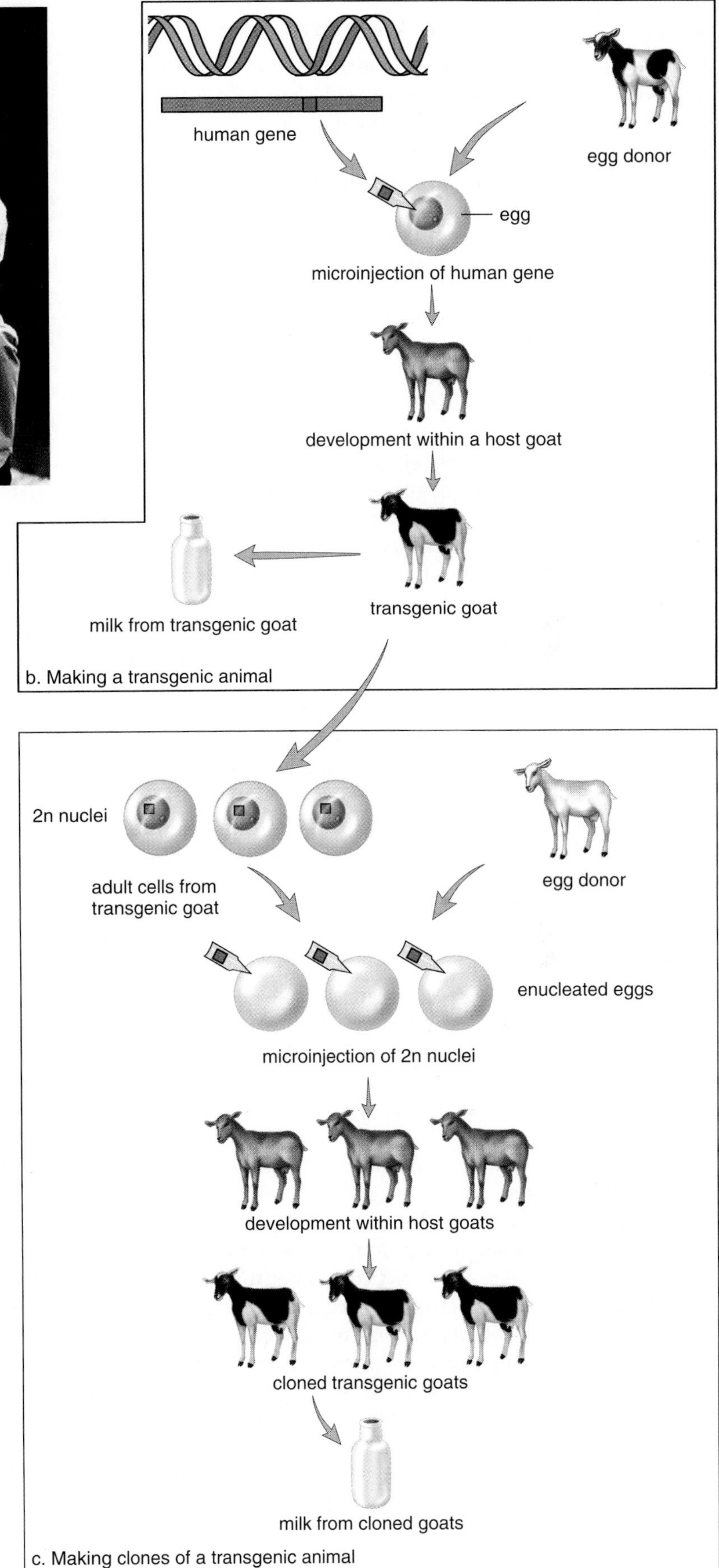

Figure 26.7 Genetically engineered animals. **a.** This goat is genetically engineered to produce antithrombin III, which is secreted in her milk. This researcher and many others are involved in the project. **b.** The procedure to produce a transgenic animal. **c.** The procedure to clone a transgenic animal.

method. An alternate method, used at the University of Hawaii for the cloning of mice, was so successful that clones of clones were produced. Figure 26.7*c* suggests that after enucleated eggs have been injected with 2n nuclei of adult cells, they can be coaxed to begin development. The offspring have the genotype and phenotype of the adult that donated the nuclei; therefore, the adult has been cloned. In the procedure that produced cloned mice, the 2n nuclei were taken from cumulus cells. Cumulus cells are those that cling to an egg after ovulation occurs. A specially prepared chemical bath was used to stimulate the eggs to divide and begin development. Now that scientists have a method to clone mammals, this procedure will undoubtedly be used routinely. Some are even beginning to think about the cloning of humans despite the objections of many and a presidential order that the procedure is not to be developed in the United States.

Transgenic animals, which secrete a biotechnology product in their milk, and pigs, whose organs can be used for xenotransplantation, have been produced. Procedures have been developed to allow the cloning of these animals.

26.3 The Human Genome Project

The Human Genome Project is a massive effort originally funded by the U.S. government and now increasingly by U.S. pharmaceutical companies to map the human chromosomes. Many nonprofit and for profit biochemical laboratories about the world are now involved in the project which has two primary goals.

The first goal is to construct a genetic map of the human genome. The aim is to show the sequence of genes along the length of each type chromosome, such as depicted for the X chromosome in Figure 26.8. If the estimate of 1,000+ human genes is correct, each chromosome on average would contain about 50 alleles.

The map for each chromosome is presently incomplete, and in many instances scientists rely on the placement of RFLPs (see page 533). These sites eventually allow scientists to pinpoint disease-causing genes because a particular RFLP and a defective gene are often inherited together. For example, it is known that persons with Huntington disease have a unique site where a restriction enzyme cuts DNA. The test for Huntington disease relies on this difference from the normal.

The genetic map of a chromosome can be used not only to detect defective genes, but possibly also to tailor treatments to the individual. Only certain hypertension patients benefit from a low-salt diet, and it would be useful to know which patients these are. Myriad Genetics, a genome company, has developed a test for a mutant angiotensinogen gene because they want to see if patients with this mutation are the ones that benefit from a low-salt diet. Several other mutant genes have also been correlated with specific drug treatments (Table 26.1). One day the medicine you take might carry a label that it is effective only in persons with genotype #101!

The second goal is to construct a base sequence map. There are three billion base pairs in the human genome, and it's estimated it would take an encyclopedia of 200 volumes, each with 1,000 pages, to list all of these. Yet, this goal of the Human Genome Project is expected to be reached by the year 2004, if not earlier.

The methodology, thus far, has been to first chop up the genome into small pieces, each just 1,000 to 2,000 base pairs long. PCR instruments copy the pieces many times, and then an automatic DNA sequencer determines the order of the base pairs. You need many DNA copies because of the way the sequencer works. A computer program later strings the sequenced pieces together in the correct order by looking for base sequence overlaps between them. Instrumentation has gradually improved, and recently one scientist, J. Craig Venter, has founded a company which he says will sequence the entire genome in three years.

Venter plans on using what is called a whole-genome shotgun sequencing method. He plans on working with the entire human genome at once. Each overlapping fragment will be about 5,000 bases long, and he will sequence the ends of each fragment using only powerful new sequencing machines. Again, a computer program will string the fragments together by looking for overlapping regions. Why is Venter going off on his own, instead of participating in the worldwide effort by many laboratories and scientists to sequence the human genome? His backers expect to market the whole-genome database to subscribers, and to patent rare but pharmacologically interesting genes. Private enterprise is giving new impetus to the field now called genomics.

Knowing the base sequence of normal genes may make it possible one day to treat certain human ills by administering normal genes and/or their protein products to those who suffer from a genetic disease.

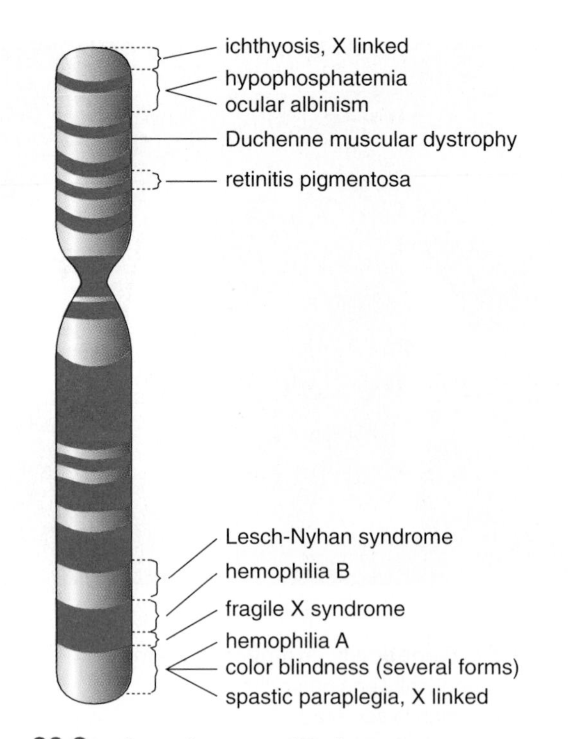

Figure 26.8 Genetic map of X chromosome.
The human X chromosome has been partially mapped, and this is the order of some of the genes now known to be on this chromosome.

Table 26.1 Customizing Drug Treatments

Mutant Gene for	Disease	Treatment
Apolipoprotein E	Alzheimer	Experimental Glaxo Wellcome drug
Cytochrome P-450	Cancer	Amonafide
Chloride gate	Cystic fibrosis	Pulmozyme
Dopamine receptor D4	Schizophrenia	Clozapine
Angiotensinogen	Hypertension	Low-salt diet

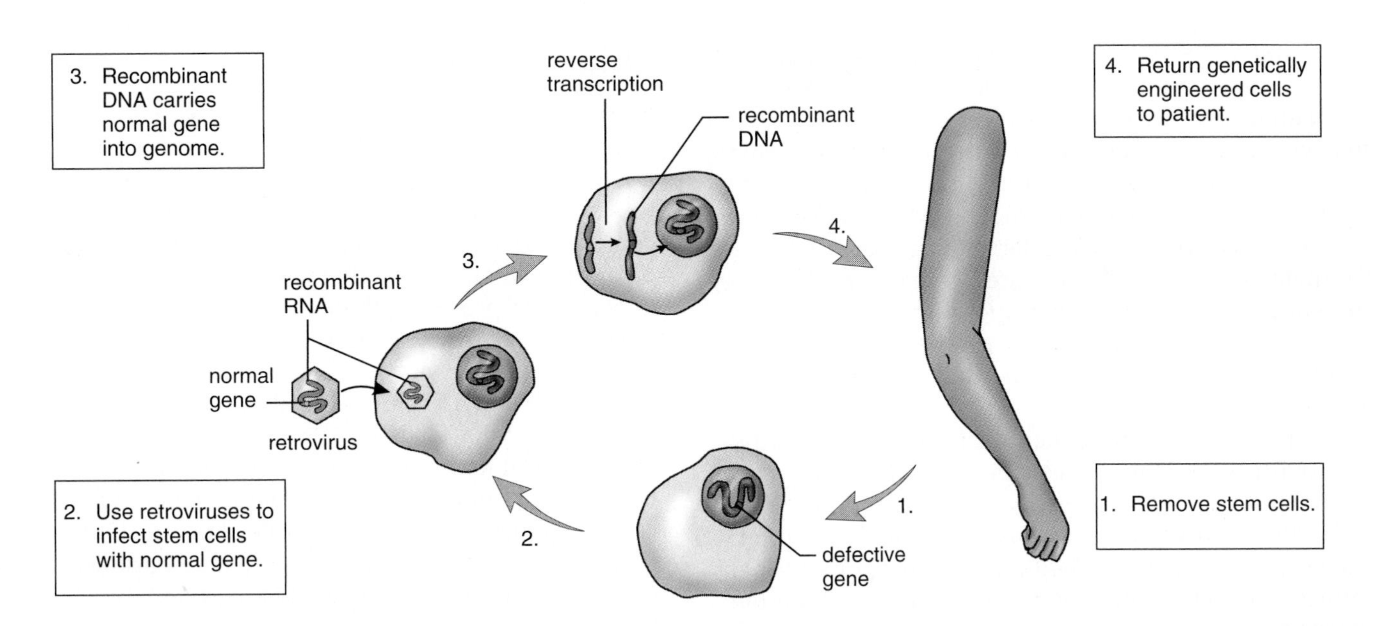

Figure 26.9 Ex vivo gene therapy in humans.
Bone marrow stem cells are withdrawn from the body, a virus is used to insert a normal gene into them, and they are returned to the body.

26.4 Gene Therapy

Gene therapy is the insertion of genetic material into human cells for the treatment of a disorder. It includes procedures that give a patient healthy genes to make up for faulty genes and also includes the use of genes to treat various other human illnesses such as cancer and cardiovascular disease. Currently there are approximately 1,000 patients enrolled in nearly 200 approved gene therapy trials in the United States. There are ex vivo (outside the body) and in vivo (inside the body) methods of gene therapy.

Genetic Disorders

As an example of an ex vivo method of gene therapy, consider Figure 26.9, which describes a methodology for the treatment of children with severe combined immunodeficiency syndrome (SCID). These children lack an enzyme called adenosine deaminase (ADA) that is involved in the maturation of T and B cells, and therefore they are subject to life-threatening infections. Bone marrow stem cells are removed from the blood and infected with a retrovirus (RNA virus) that carries a normal gene for the enzyme. Then the cells are returned to the patient. Bone marrow stem cells are preferred for this procedure because they divide to produce more cells with the same genes. With an ex vivo method of genetic therapy, it is possible to test and make sure gene transfer has occurred before the cells are returned to the patient. Patients who have undergone this procedure do have a significant improvement in their immune function that is associated with a sustained rise in the level of ADA enzyme activity in the blood.

Among the many gene therapy trials, one is for the treatment of familial hypercholesterolemia, a condition that develops when liver cells lack a receptor for removing cholesterol from the blood. The high levels of blood cholesterol make the patient subject to fatal heart attacks at a young age. In a newly developed procedure, a small portion of the liver is surgically excised and infected with a retrovirus containing a normal gene for the receptor. Several patients have experienced a lowering of serum cholesterol levels following this procedure.

Cystic fibrosis patients lack a gene that codes for transmembrane carrier of the chloride ion. Patients often die due to numerous infections of the respiratory tract. An in vivo method of treatment is being tried. Liposomes—microscopic vesicles that spontaneously form when lipoproteins are put into a solution—have been coated with the gene needed to cure cystic fibrosis. Then the solution is sprayed into patients' nostrils. Due to limited gene transfer this methodology has not as yet been successful.

In a recent and surprising move, some researchers are investigating the possibility of directly correcting the base sequence of patients with a genetic disorder. The exact procedure has not yet been published.

Cancer

Chemotherapy in cancer patients often kills off healthy cells as well as cancer cells. In clinical trials, researchers have given genes to cancer patients that either make healthy cells more tolerant of chemotherapy or make tumors more vulnerable to it. In one ex vivo clinical trial, bone marrow stem cells from about 30 women with late-stage ovarian cancer were infected with a virus carrying a gene to make them more tolerant of chemotherapy. Once the bone marrow stems cells were protected, it was possible to increase the level of chemotherapy to kill the cancer cells

In a recent in vivo study, a retrovirus carrying a normal *p53* gene was injected directly into the tumor of lung cancer patients. This gene, which helps regulate the cell cycle and also brings about apoptosis in cells with damaged DNA, is often mutated in tumor cells. No cures were reported but the tumors shrank in three patients and stopped growing in three others. Expression of *p53* is only needed for a short time and an elevated amount in normal cells seems to do them no harm.

Some investigators prefer the use of adenoviruses rather than retroviruses. Adenoviruses can be produced in much larger quantities and they are active even when they are not dividing. (Retroviruses integrate DNA into the host chromosome where it is inactive unless replication occurs.) When adenoviruses infect a cell, they normally produce a protein that binds to *p53* and inactivates it. In a cleverly designed procedure, researchers genetically engineered adenoviruses to lack the gene that produces this protein. Therefore, the virus will kill tumor cells that lack *p53* but not healthy cells that have a *p53* gene. Clinical trials are underway, and it is expected that the virus will spread through the cancer, killing the cancer cells, and stopping when it reaches normal cells.

Other Illnesses

During coronary artery angioplasty, a balloon catheter is sometimes used to open up a closed artery. Unfortunately, the artery has a tendency to close up again. But investigators have come up with a new procedure. The balloon is coated with a plasmid that contains a gene for VEGF (vascular endothelial growth factor). The expression of the gene, which promotes the proliferation of blood vessels to bypass the obstructed area, has been observed in at least one patient.

Perhaps it will be possible also to use in vivo therapy to cure hemophilia, diabetes, Parkinson disease, or AIDS. To treat hemophilia, patients could get regular doses of cells that contain normal clotting-factor genes. Or such cells could be placed in *organoids,* artificial organs that can be implanted in the abdominal cavity. To cure Parkinson disease, dopamine-producing cells could be grafted directly into the brain.

The Human Genome Project produces information useful to gene therapists. Researchers are envisioning all sorts of ways to cure human genetic disorders as well as many other types of illnesses.

Bioethical Issue

Somatic gene therapy attempts to treat or prevent human illnesses. Some day, for example, it may be possible to give children who have cystic fibrosis, Huntington disease, or any other genetic disorder a normal gene to make up for the inheritance of a faulty gene. And gene therapy is even more likely for treatment of diseases like cancer, AIDS, and heart disease. Germ line gene therapy is the term that is now being used to mean the use of gene therapy solely to improve the traits of an individual. It would be the genetic equivalent of procedures like body building, liposuction, or hair transplants. If genetic interference occurred earlier—that is, on the eggs, sperm, or embryo, it's possible that it would indeed affect the germ line—that is, all the future descendants of the individual.

How might germ line gene therapy become routine? Consider this scenario. Presently, a gene for VEGF (vascular endothelial growth factor), a protein produced by cells to grow new blood vessels, is being used to treat atherosclerosis. Improved arterial circulation in the legs of some recent patients did away with the threat of possible amputation. This same gene is now being considered for the treatment of blocked coronary arteries. There may be instances, though, in which healthy people want to grow new blood vessels for enhancement purposes. Runners might want to improve their circulation in order to win races, and parents might think that increased circulation to the brain might increase the intelligence of their children. Bioethicist Eric Juengst of Case Western thinks that as a society, there is nothing we can now do to prevent us from crossing the line between the use of gene therapy for therapeutic purposes and its use for enhancement reasons.

Questions

1. Do you approve of gene therapy to treat or prevent human illnesses? after they occur? or possibly before they occur, if genetic testing shows the likelihood of their development? Why or why not?
2. Do you approve of gene therapy to enhance the characteristics of an individual? after a condition like baldness occurs? or possibly before it occurs if genetic testing shows the likelihood of it occurring? Why or why not?
3. Do you approve of somatic gene therapy of embryos? germ line gene therapy of embryos? Explain.

Summarizing the Concepts

26.1 Cloning of a Gene

Two methods are currently available for making copies of DNA. Recombinant DNA contains DNA from two different sources. A restriction enzyme is used to cleave plasmid DNA and to cleave foreign DNA. The "sticky ends" produced facilitate the insertion of foreign DNA into vector DNA. The foreign gene is sealed into the vector DNA by DNA ligase. Both bacterial plasmids and viruses can be used as vectors to carry foreign genes into bacterial host cells.

A genomic library can be used as a source of genes to be cloned. A radioactive or fluorescent probe is used to identify the location of a single gene among the cloned fragments of an organism's DNA.

Restriction enzymes are used to fragment DNA. If the entire genome is used for fingerprinting, the use of probes is necessary in order to see a pattern following gel electrophoresis. If just a portion of the genome is of interest or available, PCR uses the enzyme DNA polymerase to make multiple copies of target DNA. Analysis of DNA segments following PCR can involve gel electrophoresis to identify the DNA as belonging to a particular organism or can involve determining the base sequence of the DNA segment.

26.2 Biotechnology Products

Transgenic organisms are ones that have had a foreign gene inserted into them. Genetically engineered bacteria, agricultural plants, and farm animals now produce commercial products of interest to humans, such as hormones and vaccines. Bacteria secrete the product. The seeds of plants and the milk of animals contain the product.

Transgenic bacteria have been produced to promote the health of plants, perform bioremediation, extract minerals, and produce chemicals. Transgenic agricultural plants which have been engineered to resist herbicides and pests are commercially available. Transgenic animals have been given various genes, in particular the one for bovine growth hormone (bGH). Pigs have been genetically altered to serve as a source of organs for transplant patients. Cloning of animals is now possible.

26.3 The Human Genome Project

The Human Genome Project has two goals. The first goal is to construct a genetic map which will show the sequence of the genes on each chromosome. Although the known location of genes is expanding, the genetic map still contains many RFLPs (restriction fragment length polymorphisms) as the first step toward finding more genes. The second goal of the project is to sequence the bases. To do this researchers rely on PCR and automatic sequencing instruments. As the sequencers have improved so has the speed with which DNA is sequenced. The hope is that knowing the location and sequence of bases in a gene will promote the possibility of gene therapy.

26.4 Gene Therapy

Gene therapy is used to correct the genotype of humans and to cure various human ills. There are ex vivo and in vivo methods of gene therapy. Gene therapy has apparently helped children with SCID to lead a normal life; the treatment of cystic fibrosis has been less successful. A number of imaginative therapies are being employed in the war against cancer and other human illness such as cardiovascular disease.

Studying the Concepts

1. What is the methodology for producing recombinant DNA to be used in gene cloning? 530
2. Bacteria can be used to clone a gene or produce a product. Explain. 531
3. What is a genomic library, and how do you locate a gene of interest in the library? 532
4. What is the polymerase chain reaction (PCR), and how is it carried out to produce multiple copies of a DNA segment? 532
5. What is DNA fingerprinting, a process that utilizes the entire genome? 533
6. What are some practical applications of DNA segments analysis following PCR? 532–33
7. For what purposes have bacteria, plants, and animals been genetically altered? 534–36
8. What is xenotransplantation, and what drawback is presently preventing pigs from serving as source animals? 536
9. Explain how and why transgenic animals that secrete a product are often cloned. 536–37
10. Explain the two primary goals of the Human Genome Project. What are the possible benefits of the project? 538
11. Explain and give examples of ex vivo and in vivo gene therapies in humans. 539–40

Testing Yourself

Choose the best answer for each question.

1. Which of these is a true statement?
   a. Both plasmids and viruses can serve as vectors.
   b. Plasmids can carry recombinant DNA but viruses cannot.
   c. Vectors carry only the foreign gene into the host cell.
   d. Only gene therapy uses vectors.
   e. Both a and d are correct.
2. Which of these is a benefit to having insulin produced by biotechnology?
   a. It is just as effective.
   b. It can be mass-produced.
   c. It is nonallergenic.
   d. It is less expensive.
   e. All of these are correct.
3. Restriction fragment length polymorphisms (RFLPs)
   a. are achieved by using restriction enzymes.
   b. identify individuals genetically.
   c. are the basis for DNA fingerprints.
   d. can be subjected to gel electrophoresis.
   e. All of these are correct.
4. Which of these would you not expect to be a biotechnology product?
   a. vaccine
   b. modified enzyme
   c. DNA probe
   d. protein hormone
   e. steroid hormone
5. What is the benefit of using a retrovirus as a vector in gene therapy?
   a. It is not able to enter cells.
   b. It incorporates the foreign gene into the host chromosome.
   c. It eliminates a lot of unnecessary steps.
   d. It prevents infection by other viruses.
   e. Both b and c are correct.

6. Gel electrophoresis
   a. cannot be used on nucleotides.
   b. measures the size of plasmids.
   c. tells whether viruses are infectious.
   d. measures the charge and size of proteins and DNA fragments.
   e. All of these are correct.
7. Using this key, put the phrases in the correct order to form a plasmid-carrying recombinant DNA.
   Key:
   (1) use restriction enzymes
   (2) use DNA ligase
   (3) remove plasmid from parent bacterium
   (4) introduce plasmid into new host bacterium
   _____ a. 1,2,3,4
   _____ b. 4,3,2,1
   _____ c. 3,1,2,4
   _____ d. 2,3,1,4
8. Which of these is incorrectly matched?
   a. xenotransplantation—source of organs
   b. protoplast—plant cell engineering
   c. RFLPs—DNA fingerprinting
   d. DNA polymerase—PCR
   e. DNA ligase—mapping human chromosomes
9. The restriction enzyme called *Eco*RI has cut double-stranded DNA in this manner. The piece of foreign DNA to be inserted has what bases from left to the right?

10. Label the following drawings, using these terms: retrovirus, recombinant RNA (twice), defective gene, recombinant DNA, reverse transcription, and human genome.

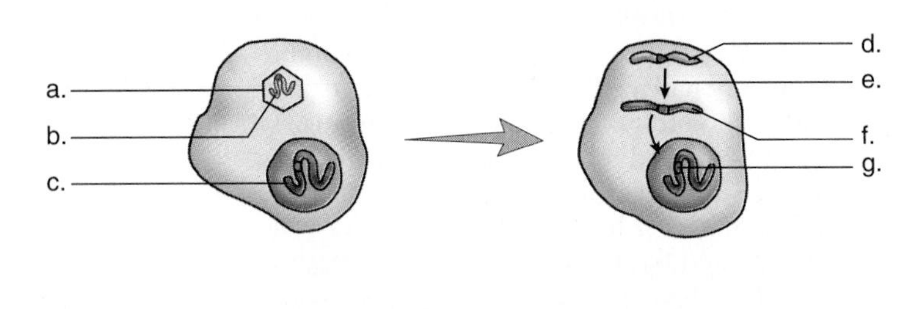

Thinking Scientifically

1. What experiment would you suggest to support the hypothesis that insulin produced by biotechnology will have fewer side effects than insulin taken from the organs of livestock? Why do you expect your results to support the hypothesis?
2. What experiment would you suggest to support the hypothesis that meat from cattle bioengineered to contain extra growth hormone genes is safe for human consumption? Why do you expect your results to support the hypothesis? Your answer should consider the fact that growth hormone is a protein.
3. What experiment would you suggest to support the hypothesis that plants engineered to contain genes for an insect toxin against insects is safe for human consumption? Why might your results not support the hypothesis?
4. What experiment would you suggest to support the hypothesis that bacteria engineered to clean up a pollutant disappear when the pollutant is gone? What would make them disappear?

Understanding the Terms

bacteriophage 532
clone 531
complementary DNA (cDNA) 532
DNA fingerprinting 533
DNA ligase 530
gene therapy 539
genetic engineering 530
genome 532
genomic library 532
plasmid 530
polymerase chain reaction (PCR) 532
probe 532
recombinant DNA (rDNA) 530
restriction enzyme 530
restriction fragment length polymorphism (RFLP) 533
transgenic organism 534
vector 530
xenotransplantation 536

Match the terms to these definitions:

a. _________ Bacterial enzyme that stops viral reproduction by cleaving viral DNA; used to cut DNA at specific points during production of recombinant DNA.
b. _________ Free-living organisms in the environment that have had a foreign gene inserted into them.
c. _________ Known sequences of DNA that are used to find complementary DNA strands; can be used diagnostically to determine the presence of particular genes.
d. _________ Production of identical copies; in genetic engineering, the production of many identical copies of a gene.
e. _________ Self-duplicating ring of accessory DNA in the cytoplasm of bacteria.

Using Technology

Your study of biotechnology is supported by these available technologies.

Essential Study Partner CD-ROM
Genetics → Recombinant DNA

Visit the Mader web site for related ESP activities.

Exploring the Internet

The Mader Home Page provides resources and tools as you study this chapter.

http://www.mhhe.com/biosci/genbio/mader

Life Science Animations 3D Video

24 Polymerase Chain Reaction
25 Gel Electrophoresis
26 Sucrose Gradient Centrifugation
27 Protein Purification/Column Chromatography

Further Readings for Part 5

Alcamo, I. E. 1997. *AIDS: The biological basis.* 2d ed. Dubuque, Iowa: Wm. C. Brown Publishers. This easily understood book focuses on the biology of AIDS.

Alexander, N. J. March/April 1996. Barriers to sexually transmitted diseases. *Scientific American Science & Medicine* 3(2):32. Article discusses the effectiveness of certain contraceptives in protecting women against STDs.

Berns, M. W. April 1998. Laser scissors and tweezers. *Scientific American* 278(62):4. New laser techniques allow manipulation of chromosomes and other structures inside cells.

Blaser, M. J. February 1996. The bacteria behind ulcers. *Scientific American* 274(2):104. Acid-loving pathogens are linked to stomach ulcers and stomach cancer.

Borek, C. November/December 1997. Antioxidants and cancer. *Science & Medicine* 4(6):52. The importance of supplemental antioxidant vitamins depends on factors such as diet and life-style.

Cox, F. D. 1996. *The AIDS booklet.* 4th ed. Dubuque, Iowa: Brown & Benchmark Publishers. This easy to read, informative booklet covers the transmission, prevention, and treatment of AIDS.

Crooks, R., and Baur, K. 1996. *Our sexuality.* 6th ed. Redwood City, Calif.: Benjamin/Cummings Publishing. Introduction to the biological, psychosocial, behavioral, and cultural aspects of sexuality.

Curiel, T. September/October 1997. Gene therapy: AIDS-related malignancies. *Science & Medicine* 4(5):4. The field of AIDS-related gene therapies is advancing.

Duke, R. C., et al. December 1996. Cell suicide in health and disease. *Scientific American* 275(6):80. Failures in the processes of cellular self-destruction may give rise to cancer, AIDS, Alzheimer disease, and some genetic diseases.

Dusenbery, D. B. 1996. *Life at small scale: The behavior of microbes.* New York: Scientific American Library. This easy-to-read, well-illustrated text describes how microbes respond to the physical demands of their environment.

Galili, U. September/October 1998. Anti-Gal antibody prevents xenotransplantation. *Science & Medicine* 5(5):28. Prevention of interaction of the anti-Gal antibody with pig cells is necessary to the progress of xenotransplantation.

Garnick, M. B., and Fair, W. R. December 1998. Combating prostate cancer. *Scientific American* 279(6):74. Article details the recent developments in diagnosis and treatment of prostate cancer.

Glausiusz, J. May 1998. The great gene escape. *Discover* 19(5):90. Genes from genetically engineered plants can escape from crops into the wild, causing resistance in the wild plant.

Goldberg, J. April 1998. A head full of hope. *Discover* 19(4):70. Article discusses a new gene therapy for killing brain cancer cells.

Greider, C. W., and Blackburn, E. H. February 1996. Telomeres, telomerase, and cancer. *Scientific American* 274(2):92. The enzyme telomerase rebuilds the chromosomes of tumor cells; this enzyme is being researched as a target for anticancer treatments.

Haseltine, W. A. March 1997. Discovering genes for new medicines. *Scientific American* 276(3):92. New medical products being developed are a result of recent genetic analyses of the human genome.

Johnson, G. B. 1996. *How scientists think.* Dubuque, Iowa: Wm. C. Brown Publishers. Presents the rationale behind 21 important experiments in genetics and molecular biology that became the foundation for today's research.

Jordan, V. C. October 1998. Designer estrogens. *Scientific American* 279(4):60. Selective estrogen receptor modulators may protect against breast and endometrial cancers, osteoporosis, and heart disease.

Kher, U. January 1998. A man-made chromosome. *Discover* 18(1):40. Researchers announce a promising new gene carrier, a human artificial chromosome, for use in gene therapy.

Leffell, D. J., and Brash, D. E. July 1996. Sunlight and skin cancer. *Scientific American* 275(1):52. Discusses the sequence of changes that may occur in skin cells after exposure to UV rays.

Lyon, J., and Gorner, P. 1995. *Altered fates: Gene therapy and the retooling of human life.* New York: W. W. Norton & Company, Inc. The development of gene therapy, from the scientists to the patients and their families, is discussed.

MacDonald, P. C., and Casey, M. L. March/April 1996. Preterm birth. *Scientific American Science & Medicine* 3(2):42. Article discusses the role of oxytocin, prostaglandins, and infections in the initiation of human labor.

Mader, S. S. 1990. *Human reproductive biology.* 2d ed. Dubuque, Iowa: Wm. C. Brown Publishers. An introductory text covering human reproduction in a clear, easily understood manner.

Markowitz, M. H. June 1998. A new dawn in AIDS treatments. *Discover* 19(6):S-6. A new combination therapy greatly reduces viral replication.

Miller, R. V. January 1998. Bacterial gene swapping in nature. *Scientific American* 278(1):66. The study of the process of DNA exchange between bacteria can help limit the risks of releasing genetically engineered microbes into the environment.

Moses, V., and Moses, S. 1995. *Exploiting biotechnology.* Chur, Switzerland: Harwood Academic Publishers. Provides a general understanding of biotechnology and presents its commercial and industrial applications.

Nicolaou, K. C., et al. June 1996. Taxoids: New weapons against cancer. *Scientific American* 274(6):94. Chemists are synthesizing a family of drugs related to taxol for the treatment of cancer.

Nielson, P. E. September/October 1998. Peptide nucleic acids. *Science & Medicine* 5(5):48. Peptide nucleic acids mimic DNA and can substitute for DNA in gene therapy.

Nusslein-Volhard, C. August 1996. Gradients that organize embryo development. *Scientific American* 275(2):54. Nobel Prize-winning researcher describes how morphogens shape an evolving embryo.

O'Brien, S. J., and Dean, M. September 1997. In search of AIDS-resistance genes. *Scientific American* 277(3):44. Study of genes that deter the AIDS virus may lead to prevention or treatment.

O'Brochta, D. A., and Atkinson, P. W. December 1998. Building a better bug. *Scientific American* 279(6):90. Transgenic insect technology could decrease pesticide use, and prevent certain infectious diseases. Article discusses the production of a transgenic insect.

Packer, C. July/August 1998. Why menopause? *Natural History* 107(6):24. Article addresses possible reasons why menopause occurs so early in life, compared to other aging processes.

Pennisi, E. 13 November 1998. Training viruses to attack cancers. *Science* 282(5392):1244. Certain viruses can replicate in and kill cancer cells, but leave normal tissue intact.

Perera, F. P. May 1996. Uncovering new clues to cancer risk. *Scientific American* 274(5):54. Molecular epidemiology finds biological markers that explain what makes people susceptible to cancer.

Plomerin, R., and DeFries, J. C. May 1998. The genetics of cognitive abilities and disabilities. *Scientific American* 278(5):62. The search is underway for the genes involved in cognitive abilities and disabilities, including dyslexia.

Pool, R. May 1998. Saviors. *Discover* 19(5):52. Genetic engineering may make animal organs compatible for human transplants.

Ricklefs, R. E., and Finch, C. E. 1995. *Aging: A natural history.* New York: Scientific American Library. This text emphasizes the nature of aging and the mechanisms of physiological deterioration.

Ronald, P. C. November 1997. Making rice disease-resistant. *Scientific American* 277(5):100. Genetic engineering is being used to protect rice from disease.

Ross, I. K. 1995. *Aging of cells, humans and societies.* Dubuque, Iowa: Wm. C. Brown Publishers. Presents current concepts on aging.

Russell, P. J. 1996. *Genetics.* 4th ed. New York: HarperCollins College Publishers. This easy-to-read text emphasizes an inquiry-based approach to genetics, exploring many experiments that led to important genetic advances.

Science & Medicine. March/April 1998. Special report: Dynamics of HIV infection. HIV-1 may use a variety of coreceptors to gain entry into cells.

Scientific American editors. June 1997. Special report: Making gene therapy work. 276(6):95. Obstacles must be overcome before gene therapy is ready for widespread use.

Scientific American. July 1998. Defeating AIDS: What will it take? 279(1):81. Nine separate articles address AIDS problems and issues.

Scientific American Special Issue. September 1996. What you need to know about cancer. 275(3). The entire issue is devoted to the causes, prevention, and early detection of cancer, and cancer therapies—conventional and future.

Shcherbak, Y. M. April 1996. Ten years of the Chernobyl Era. *Scientific American* 274(4):44. Article discusses the medical aftermath of the accident.

Stix, G. October 1997. Growing a new field. *Scientific American* 277(4):15. Tissue engineers try to grow organs in the laboratory.

Stolley, P. D., and Lasky, T. 1995. *Investigating disease patterns: The science of epidemiology.* New York: Scientific American Library. The process of epidemiology and its contribution to the understanding of disease is covered in this interesting, easy-to-read book.

Van Noorden, C. J. F., et al. March/April 1998. Metastasis. *American Scientist* 86(2):130. The mechanisms by which cancer cells metastasize are discussed.

Velander, W. H., et al. January 1997. Transgenic livestock as drug factories. *Scientific American* 276(1):70. Farm animals can be bred to produce quantities of medicinal proteins in their milk.

Wallace, D. C. August 1997. Mitochondrial DNA in aging and disease. *Scientific American* 277(2):40. Genes in mitochondria have been linked to certain diseases, and could also be important in age-related disorders.

Weindruch, R. January 1996. Caloric restriction and aging. *Scientific American* 274(1):64. Consuming fewer calories may increase longevity.

Weiss, R. November 1997. Aging—new answers to old questions. *National Geographic* 192(5):2. The mechanics of human aging are studied.

Wills, C. January 1998. A sheep in sheep's clothing? *Discover* 18(1):22. Some pros and cons of cloning are discussed.

Wilmut, I. December 1998. Cloning for medicine. *Scientific American* 279(6):58. Cloning holds many benefits for the advancement of medical science and animal husbandry.

Winkonkal, N. M., and Brash, D. E. September/October 1998. Squamous cell carcinoma. *Science & Medicine* 5(5):18. Mutations of tumor-suppressor gene *p53* are commonly found in squamous cell carcinomas.

Wolffe, A. P. November/December 1995. Genetic effects of DNA packaging. *Scientific American Science & Medicine* 2(6):68. The regulation of DNA coiling in the chromosomes adds to the properties of the genes involved in several genetic diseases.

Zhong, G., and Brunham, G. C. September/October 1998. Chlamydial resistance to host defense. *Science & Medicine* 5(5):38. Chlamydia may produce anti-apoptosis factors.

Zimmer, C. January 1998. Hidden unity. *Discover* 18(1):46. Studies suggest the same basic gene may be involved in the development of certain features in vertebrates and invertebrates.

PART VI

Evolution and Diversity

Taxonomy is the science of identifying and classifying organisms. The scientific name of an organism is a binomial that tells its genus and species. Organisms are grouped in ever larger categories according to their evolutionary history. All sorts of data such as the fossil record, behavior, structure, and increasingly genetic similarities are used to tell how organisms should be classified. In the largest category— kingdom— organisms have just general characteristics in common and are only distantly related. This text recognizes five kingdoms: the unicellular monerans, often unicellular protistans and the multicellular fungi, plants, and animals.

Careers in Botany and Zoology

Paleontologists excavating a dinosaur.

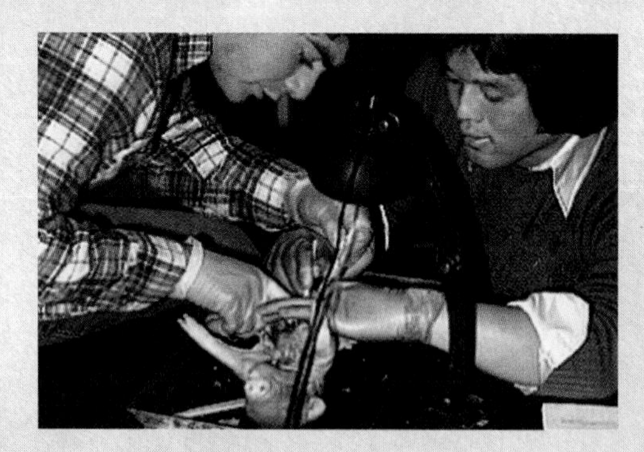

Zoologists dissecting an animal.

Florist arranging flowers.

Paleontologists study the life of the past that is preserved in rock. Both large bones of dinosaurs and small specimens, such as plant parts and pollen grains, are collected to help understand how the earth and its life have changed through time.

Taxonomists classify the enormous diversity of life into logical categories that reflect their line of descent. Taxonomists can also help field researchers decide if a new sample is a new species, and if so, how it is related to known species.

Curators and conservators oversee collections in museums, zoos, aquariums, botanical gardens, and historic sites. Curators also plan and prepare exhibits. Conservators coordinate the activities of workers engaged in the examination, repair, and conservation of museum objects. This may require substantial historical and archaeological research.

Park and zoo staff at both natural regions, such as national parks, and artificial locales, such as zoological parks, help protect species living there and educate visitors about their unique value. Some zoos actively assist the growth of endangered species populations through breeding programs.

Horticulturists and florists grow and care for plants and flowers for retail and wholesale marketing. On horticultural specialty farms, operators oversee the production of ornamental plants, nursery products (such as flowers, bulbs, shrubbery, and sod), and fruits and vegetables grown in greenhouses.

Zoologists Zoologists study the biology of animals and may specialize in a particular group of animals or in a particular discipline, such as the anatomy, ecology, genetics, or physiology of animals. Since there are many different types of animals and many subdisciplines, the choice of specialization is very broad. Zoologists carry out research in their area of expertise, write articles and books, and instruct others about the biology of animals.

C H A P T E R 27

Origin and Evolution of Life

It's hard to imagine a time when the earth was devoid of all living things. The first organisms came into existence in the ocean and only later invaded land as biological evolution occurred.

Chapter Concepts

As lightning bolts crackle through the oxygen-poor atmosphere, some strike warm water on the planet's surface, sparking chemical reactions that transform simple molecules into more complex ones. A few of them have the ability to make copies of themselves. These self-replicating molecules eventually envelop themselves in a protective barrier, forming the first cells. Over millions of years, descendants of those cells generate a stunning array of life forms. Among them is an animal intelligent enough to ponder its fundamental origins. A scenario such as this occurred on earth. Has it elsewhere?

27.1 Origin of Life

The sun and the planets, including earth, probably formed over a 10-billion-year period from aggregates of dust particles and debris. At 4.6 billion years ago, the solar system was in its present form. Intense heat produced by gravitational energy and radioactivity caused the earth to become stratified into several layers. Heavier atoms of iron and nickel became the molten liquid core, and dense silicate minerals became the semiliquid mantle. Upwellings of volcanic lava produced the first crust.

The Primitive Atmosphere

The size of the earth is such that the gravitational field is strong enough to have an atmosphere. If the earth were smaller and lighter, atmospheric gases would escape into outer space. The earth's *primitive atmosphere* was not the same as today's atmosphere. It is now thought that the primitive atmosphere was produced by outgassing from the interior, particularly by volcanic action. In that case, the primitive atmosphere would have consisted mostly of water vapor (H_2O), nitrogen (N_2), and carbon dioxide (CO_2), with only small amounts of hydrogen (H_2) and carbon monoxide (CO). The primitive atmosphere, with little if any free oxygen, was a *reducing atmosphere* as opposed to the *oxidizing atmosphere* of today. This was fortuitous to life because oxygen (O_2) attaches to organic molecules, preventing them from joining to form larger molecules.

At first the earth was so hot that water was present only as a vapor that formed dense, thick clouds. Then as the earth cooled, water vapor condensed to liquid water, and rain began to fall. It rained in such enormous quantity over hundreds of millions of years that the oceans of the world were produced. By chance, the distance of earth from the sun is such that all water does not evaporate from too much heat nor freeze because of too little heat.

Small Organic Molecules

The atmospheric gases, dissolved in rain, were carried down into newly forming oceans. Aleksandr Oparin, a Soviet biochemist, suggested as early as 1938 that organic molecules could have been produced from the gases of the primitive atmosphere in the presence of strong outside *energy sources*. The energy sources on the primitive earth included heat from volcanoes and meteorites, radioactivity from isotopes in the earth's crust, powerful electric discharges in lightning, and solar radiation, especially ultraviolet radiation.

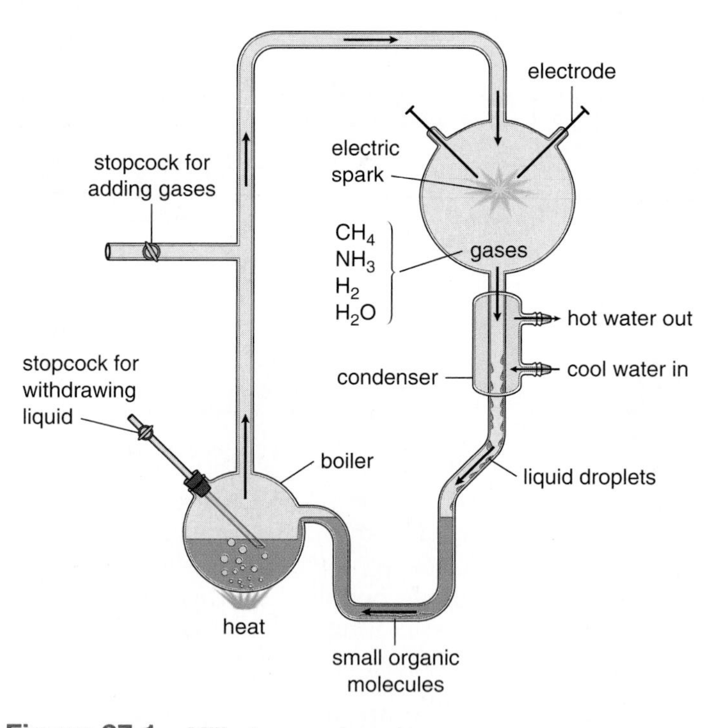

Figure 27.1 Miller's experiment.
In Miller's experiment, gases were admitted to the apparatus, circulated past an energy source (electric spark), and cooled to produce a liquid that could be withdrawn. Upon chemical analysis, the liquid was found to contain various small organic molecules.

In 1953, Stanley Miller provided support for Oparin's ideas through an ingenious experiment (Fig. 27.1). Miller placed a mixture resembling a strongly reducing atmosphere (methane [CH_4], ammonia [NH_3], hydrogen [H_2], and water [H_2O]) in a closed system, heated the mixture, and circulated it past an electric spark (simulating lightning). After a week's run, Miller discovered that a variety of amino acids and organic acids had been produced. Since that time, other investigators have achieved similar results by utilizing other, less-reducing combinations of gases dissolved in water.

These experiments support the hypothesis that the primitive gases could have reacted with one another to produce small organic compounds. Neither oxidation (there was no free oxygen) nor decay (there were no bacteria) would have destroyed these molecules, and they would have accumulated in the oceans for hundreds of millions of years. With the accumulation of these small organic compounds, the oceans became a warm, organic soup containing a variety of organic molecules.

Macromolecules

The newly formed organic molecules likely polymerized to produce still larger molecules and then macromolecules. There are three primary hypotheses concerning this stage in the origin of life. One is the *RNA-first hypothesis,* which suggests that only the macromolecule RNA (ribonucleic acid) was needed at this time to progress toward formation of the first cell or cells. This hypothesis was formulated after the discovery that RNA can sometimes be both a substrate and an enzyme. Such RNA molecules are called *ribozymes.* It would seem, then, that RNA could have carried out the processes of life commonly associated today with DNA (deoxyribonucleic acid, the genetic material) and proteins (enzymes). The first genes could have been composed of RNA—some viruses have only RNA genes. And the first enzymes also could have been RNA molecules, since we now know that ribozymes exist. Those who support this hypothesis say that it was an "RNA world" some 4 billion years ago.

Another hypothesis is termed the *protein-first hypothesis.* Sidney Fox has shown that amino acids polymerize abiotically when exposed to dry heat. He suggests that amino acids collected in shallow puddles along the rocky shore and the heat of the sun caused them to form **proteinoids,** small polypeptides that have some catalytic properties. When proteinoids are returned to water, they form **microspheres,** structures composed only of protein that have many properties of a cell. It is possible that the first polypeptides had enzymatic properties, and some proved to be more capable than others. Those that led to the first cell or cells had a selective advantage. This hypothesis assumes that DNA genes came after protein enzymes arose. After all, it is protein enzymes that are needed for DNA replication.

The third hypothesis is put forth by Graham Cairns-Smith. He believes that clay was especially helpful in causing polymerization of both proteins and nucleic acids at the same time. Clay attracts small organic molecules and contains iron and zinc, which may have served as inorganic catalysts for polypeptide formation. In addition, clay has a tendency to collect energy from radioactive decay and to discharge it when the temperature and/or humidity changes. This could have been a source of energy for polymerization to take place. Cairns-Smith suggests that RNA nucleotides and amino acids became associated in such a way that polypeptides were ordered by, and helped synthesize, RNA. It is clear that this hypothesis suggests that both polypeptides and RNA arose at the same time.

A chemical evolution produced the macromolecules we associate with living things.

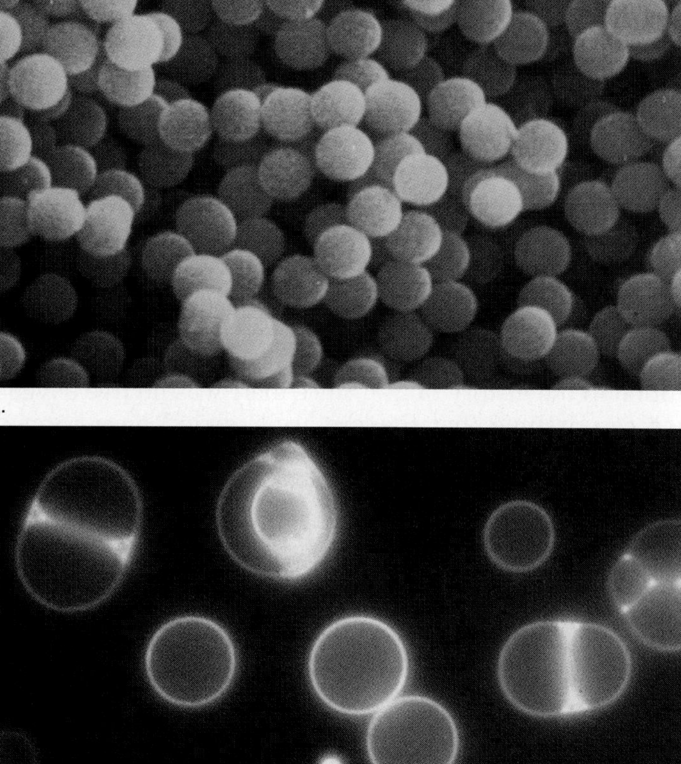

a.

b.

Figure 27.2 Protocell anatomy.
a. Microspheres, which are composed only of protein, have a number of cellular characteristics and could have evolved into the protocell.
b. Liposomes form automatically when phospholipid molecules are put into water. Plasma membrane may have evolved similarly.

The Protocell

Before the first true cell arose, there would have been a **protocell,** a structure that had a lipid-protein membrane and carried on energy metabolism (Fig. 27.2). Fox has shown that if lipids are made available to microspheres, lipids tend to become associated with microspheres, producing a lipid-protein membrane.

Some researchers support the work of Oparin, who was mentioned previously. Oparin showed that under appropriate conditions of temperature, ionic composition, and pH, concentrated mixtures of macromolecules tend to give rise to complex units called *coacervate droplets.* Coacervate droplets have a tendency to absorb and incorporate various substances from the surrounding solution. Eventually, a semipermeable-type boundary may form about the droplet. In a liquid environment, phospholipid molecules automatically form droplets called **liposomes.** Perhaps the first membrane formed in this manner.

The Heterotroph Hypothesis

The protocell would have had to carry on nutrition so that it could grow. Nutrition was no problem because the protocell existed in the ocean, which at that time contained small organic molecules that could have served as food. Therefore, the protocell likely was a **heterotroph,** an organism that takes in preformed food. Notice that this suggests that heterotrophs preceded **autotrophs,** organisms that make their own food.

At first, the protocell may have used preformed ATP, but as this supply dwindled, natural selection favored any cells that could extract energy from carbohydrates in order to transform ADP to ATP. Glycolysis is a common metabolic pathway in living things, and this testifies to its early evolution in the history of life. Since there was no free oxygen, we can assume that the protocell carried on a form of fermentation.

It seems logical that the protocell at first had limited ability to break down organic molecules and that it took millions of years for glycolysis to evolve completely. It is of interest that Fox has shown that a microsphere similar to those from which the protocell may have evolved has some catalytic ability, and that Oparin found that coacervates do incorporate enzymes if they are available in the medium.

The protocell is hypothesized to have had a membrane boundary and to have been a heterotrophic fermenter with some degree of enzymatic ability.

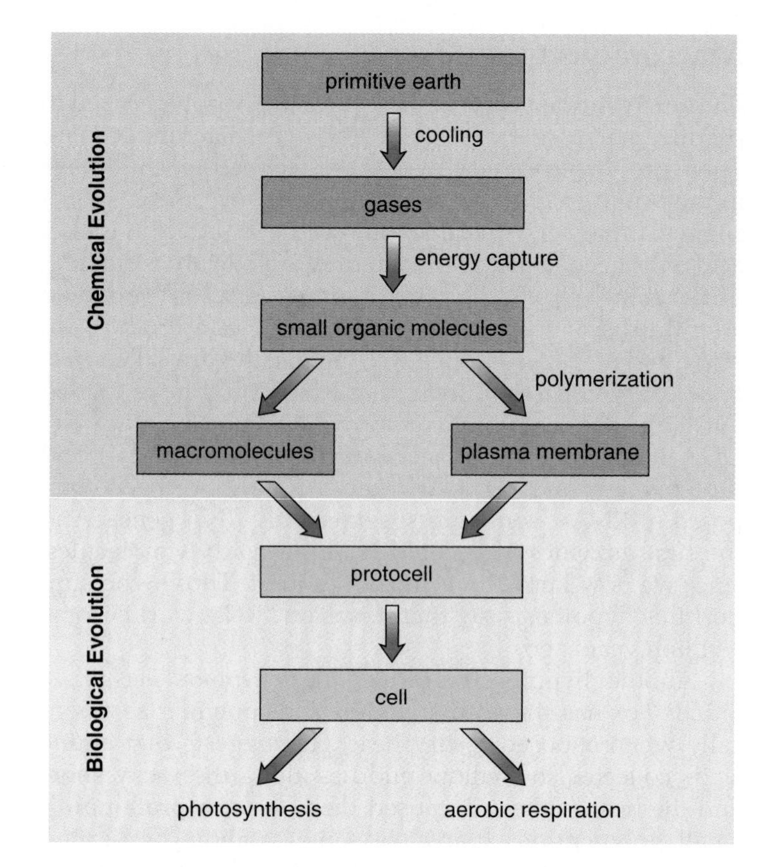

Figure 27.3 A chemical evolution produced the protocell. There was an increase in the complexity of macromolecules, leading to a self-replicating system (DNA → RNA → protein) enclosed by a plasma membrane. The protocell, a heterotrophic fermenter, underwent biological evolution, becoming a true cell, which then diversified.

The True Cell

A true cell is a membrane-bounded structure that can carry on protein synthesis needed to produce the enzymes that allow DNA to replicate (Fig. 27.3). The central dogma of genetics states that DNA directs protein synthesis and that there is a flow of information from DNA → RNA → protein. It is possible that this sequence developed in stages.

According to the RNA-first hypothesis, RNA would have been the first to evolve, and the first true cell would have had RNA genes. These genes would have directed and enzymatically carried out protein synthesis. As mentioned, RNA enzymes called ribozymes have been discovered. Also, today we know there are viruses that have RNA genes. These viruses have a protein enzyme called reverse transcriptase that uses RNA as a template to form DNA. Perhaps with time, reverse transcription occurred within the protocell, and this is how DNA genes arose. Once there were DNA genes, then protein synthesis would have been carried out in the manner dictated by the central dogma of genetics.

According to the protein-first hypothesis, proteins, or at least polypeptides, were the first of the three (i.e., DNA, RNA, and protein) to arise. Only after the protocell developed sophisticated enzymes did it have the ability to synthesize DNA and RNA from small molecules provided by the ocean. Researchers point out that because nucleic acid is a very complicated molecule, the likelihood that RNA arose *de novo* (on its own) is minimal. It seems more likely that enzymes were needed to guide the synthesis of nucleotides and then nucleic acids.

Cairns-Smith proposes that polypeptides and RNA evolved simultaneously. Therefore, the first true cell would have contained RNA genes that could have replicated because of the presence of proteins. This eliminates the baffling chicken-and-egg paradox: which came first, proteins or RNA? But it does mean, however, that two unlikely events would have to happen at the same time.

Once the protocells acquired genes that could replicate, they became cells capable of reproducing, and evolution began.

Once the protocell was capable of reproduction, it became a true cell, and biological evolution began.

27.2 Evidence of Evolution

Evolution is all the changes that have occurred in living things since the beginning of life (Table 27.1). Notice that earth is about 4.6 billion years old and that life evolved in the form of prokaryotes about 3.5 billion years ago. The eukaryotic cell arose about 2.1 billion years ago, but multicellularity didn't begin until 700 million years ago. This means that for 80% of the time since life arose, only unicellular organisms were present. Animal evolution didn't begin in earnest until the Cambrian explosion, the popular term that refers to a burst of evolution at the start of the Cambrian period. Most of the evolutionary events we will be discussing in future chapters occurred in less than 20% of the history of life!

Evolution explains the unity and the diversity of life. All living things share the same fundamental characteristics: they are made of cells; take chemicals and energy from the environment, respond to external stimuli, reproduce, and evolve. Living things are related because they are descended from a common ancestor. Life is diverse because the various types of living things are adapted to different ways of life. Many fields of biology provide evidence that supports the hypothesis of common descent. This is significant because the more varied the evidence supporting a hypothesis, the more certain it becomes.

Fossil Evidence

Fossils are the remains and traces of past life, or any other direct evidence of past life. Traces include trails, footprints, burrows, worm casts, or even preserved droppings. Usually, when an organism dies, the soft parts are either consumed by scavengers or undergo bacterial decomposition. Occasionally, the organism is buried quickly and in such a way that decomposition is never completed or is completed so slowly that the soft parts leave an imprint of their structure. Most fossils, however, consist only of hard parts such as shells, bones, or teeth, because these are usually not consumed or destroyed.

The great majority of fossils are found embedded in, or recently eroded from, sedimentary rock. Sedimentation, a process that has been going on since the earth was formed, can take place on land or in bodies of water. Weathering and erosion of rocks produces an accumulation of particles that vary in size and nature and are called sediment. Sediment becomes a stratum (pl., *strata*), a recognizable layer in a stratigraphic sequence. Any given stratum is older than the one above it and younger than the one immediately below it. Paleontologists are biologists that discover and study the fossil record and from it make decisions about the history of life. Particularly interesting are the fossils that serve as *transitional links* between groups. For example, the famous fossils of *Archaeopteryx* are intermediate between reptiles and birds (Fig. 27.4). The dinosaur-like skeleton of this fossil has reptilian features, including jaws with teeth and a long, jointed tail, but *Archaeopteryx* also had feathers and wings. Other transitional links among fossil vertebrates include the amphibious fish *Eustheopteron,* the reptile-like amphibian *Seymouria,* and the mammal-like reptiles, or therapsids. These fossils allow us to deduce that fishes preceded amphibians, which preceded reptiles, which preceded both birds and mammals in the history of life.

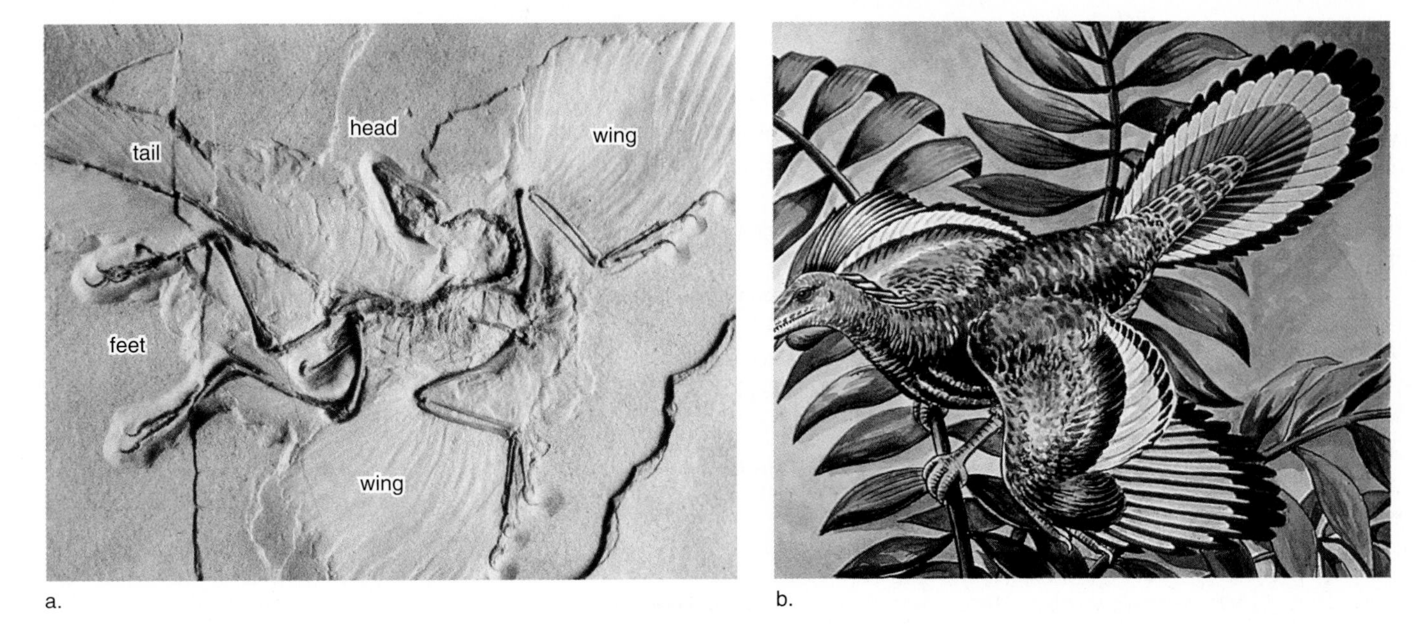

Figure 27.4 Transitional fossils.
a. This is *Archaeopteryx*, a transitional link between reptiles and birds. It had feathers and wing claws. Most likely, it was a poor flier. Perhaps it ran over the ground on strong legs and climbed up into trees with the assistance of these claws. **b.** It also had a feather-covered, reptilian-type tail that shows up well in this artist's representation of the animal.

Table 27.1 The Geological Time Scale: Major Divisions of Geological Time with Some of the Major Evolutionary Events of Each Geological Period

Era	Period	Epoch	Millions of Years Ago	Plant Life	Animal Life
Cenozoic* (from the present to 66.4 million years ago)	Neogene	Holocene	0–0.01	Destruction of tropical rain forests by humans accelerates extinctions	AGE OF HUMAN CIVILIZATION
			Significant Mammalian Extinction		
		Pleistocene	0.01–2	Herbaceous plants spread and diversify	Modern humans appear
		Pliocene	2–6	Herbaceous angiosperms flourish	First hominids appear
		Miocene	6–24	Grasslands spread as forests contract	Apelike mammals and grazing mammals flourish; insects flourish
		Oligocene	24–37	Many modern families of flowering plants evolve	Browsing mammals and monkeylike primates appear
	Paleogene	Eocene	37–58	Subtropical forests thrive with heavy rainfall	All modern orders of mammals are represented
		Paleocene	58–66	Angiosperms diversify	Primitive primates, herbivores, carnivores, and insectivores appear
			Mass Extinction: Dinosaurs and Most Reptiles		
Mesozoic (from 66.4 to 245 million years ago)	Cretaceous		66–144	Flowering plants spread; coniferous trees decline	Placental mammals appear; modern insect groups appear
	Jurassic		144–208	Cycads and other gymnosperms flourish	Dinosaurs flourish; birds appear
			Mass Extinction		
	Triassic		208–245	Cycads and ginkgoes appear; forests of gymnosperms and ferns dominate	First mammals appear; first dinosaurs appear; corals and mollusks dominate seas
			Mass Extinction		
Paleozoic (from 245 to 570 million years ago)	Permian		245–286	Conifers appear	Reptiles diversify; amphibians decline
	Carboniferous		286–360	Age of great coal-forming forests: club mosses, horsetails, and ferns flourish	Amphibians diversify; first reptiles appear; first great radiation of insects
			Mass Extinction		
	Devonian		360–408	First seed ferns appear	Jawed fishes diversify and dominate the seas; first insects and first amphibians appear
	Silurian		408–438	Low-lying vascular plants appear on land	First jawed fishes appear
			Mass Extinction		
	Ordovician		438–505	Marine algae flourish	Invertebrates spread and diversify; jawless fishes, first vertebrates appear
	Cambrian		505–570	Marine algae flourish	Invertebrates with skeletons are dominant
Precambrian time (from 570 to 4,600 million years ago)			700	Multicellular organisms appear	
			2,100	First complex (eukaryotic) cells appear	
			3,100–3,500	First prokaryotic cells in stromatolites appear	
			4,600	Earth forms	

*Many authorities divide the Cenozoic era into the Tertiary period (contains Paleocene, Eocene, Oligocene, Miocene, and Pliocene) and the Quaternary period (contains Pleistocene and Holocene).

Figure 27.5 Dinosaurs.
Triceratops (left) and *Tyrannosaurus rex (right)* were dinosaurs of the Cretaceous period, when flowering plants were increasing in dominance.

Geological Time Scale

As a result of their studying strata, geologists have divided the history of the earth into eras, then periods and epochs (Table 27.1). The fossil record has helped determine the dates given in the table. There are two ways to date fossils. The *relative dating method* determines the relative order of fossils and strata but does not determine the actual date they were formed. The relative dating method is possible because fossil-containing sedimentary rocks occur in layers. Because the top layers are usually younger than the lower layers, the fossils in each succeeding layer are older than the fossils in the layer above.

The *absolute dating method* relies on radioactive dating techniques to assign an actual date to a fossil. All radioactive isotopes have a particular half-life, the length of time it takes for half of the radioactive isotope to change into another stable element. Carbon 14, ^{14}C, is the only radioactive isotope in organic matter. Assuming a fossil contains organic matter, half of the ^{14}C will have changed to nitrogen 14, ^{14}N, in 5,730 years. In order to know how much ^{14}C was in the organism to begin with, it is reasoned that organic matter always begins with the same amount of ^{14}C. (In reality, it is known that the ^{14}C levels in the air—and therefore the amount in organisms—can vary from time to time.) Now we need only compare the ^{14}C radioactivity of the fossil to that of a modern sample of organic matter. The amount of radiation left can be converted to the age of the fossil. After 50,000 years, however, the amount of ^{14}C radioactivity is so low it cannot be used to measure the age of a fossil accurately.

Certain radioactive isotopes can be used to date rocks directly, and from that, the age of a fossil contained in the rock can be inferred. The ratio of potassium 40 (^{40}K) to argon 40 (^{40}Ar) trapped in rock is often used; if the ratio happens to be 1:1, then half of the ^{40}K has decayed and the rock is 1.3 billion years old. The ratio of isotope uranium 238 (^{238}Ur) to lead 207 (^{207}Pb) can be used for rocks older than 100 million years. This isotope has such a long half-life that no perceptible decay will have occurred in a shorter length of time.

Mass Extinctions

Extinction is the death of every member of a species. *Mass extinctions* are times when a large percentage of existing species becomes extinct within a relatively short period of time. There have been at least five mass extinctions throughout history: at the ends of the Ordovician, Devonian, Permian, Triassic, and Cretaceous periods. Mass extinctions are usually followed by at least partial recovery in which the remaining groups of organisms spread out and fill the habitats vacated by those that have become extinct.

Cretaceous clay contains an abnormally high level of iridium, an element that is rare in the earth's crust but more common in meteorites. It was proposed in 1977 that the Cretaceous extinction, which saw the demise of the dinosaurs (Fig. 27.5), was due to an asteroid that exploded, producing meteorites that fell to earth. The result of a large meteorite striking the earth could have produced a cloud of dust that would have mushroomed into the atmosphere, blocking out the sun and causing plants to freeze and die. Recently, a layer of soot has also been identified in the stratum alongside the iridium, and a huge crater that could have been caused by a meteorite was found a few years ago in the Caribbean–Gulf of Mexico region on the Yucatán Peninsula.

In 1984, paleontologists found that the fossil record of marine animals shows that mass extinctions have occurred about every 26 million years and, surprisingly, astronomers can offer an explanation. Our solar system is in the Milky Way, a starry galaxy that is 1,000,000 light-years[1] in diameter and 1,500 or so light-years thick. Our sun moves up and down as it orbits in the Milky Way. Astronomers predict that this vertical movement will cause our solar system to approach certain other members of the Milky Way every 26–33 million years, producing an unstable situation that could lead to the occurrence of a meteorite. This evidence suggests that mass extinctions are very often associated with extraterrestrial events, but these events are not necessarily the only cause of mass extinctions.

[1]One light-year, which is the distance light travels in a year, is about 6 trillion miles.

Biogeographical Evidence

Biogeography is the study of the distribution of plants and animals throughout the world. Such distributions are consistent with the hypothesis that related forms evolve in one locale and then spread out into other regions. For example, there are no rabbits in South America because rabbits originated somewhere else, and they had no means of reaching South America.

Physical factors, such as the location of continents, often determine where a population can spread. **Continental drift** is a hypothesis that states that the continents are not fixed; instead, their position and the position of the oceans have changed over time (Fig. 27.6). Continental drift explains why the coastlines of several continents are mirror images of each other—the outline of the west coast of Africa matches that of the east coast of South America. It also explains the unique distribution patterns of several fossils. Fossils of the same species of seed fern (*Glossopteris*) have been found on all the southern continents. Similarly, the fossil reptile *Cynognathus* is found in Africa and South America, and *Lystrosaurus,* a mammal-like reptile, has now been found in Antarctica, far from Africa and southeast Asia, where it also occurs. With mammalian fossils, the situation is different: Australia, South America, and Africa all have their own distinctive mammals because mammals evolved after the continents separated. The mammalian biological diversity of today's world is the result of isolated evolution on separate continents.

The distribution of organisms on the earth is explainable by assuming that related forms evolved in one locale. They then diversified as they spread out into other accessible areas.

Anatomical Evidence

A common descent hypothesis offers a plausible explanation for anatomical similarities among organisms. Vertebrate forelimbs are used for flight (birds and bats), orientation

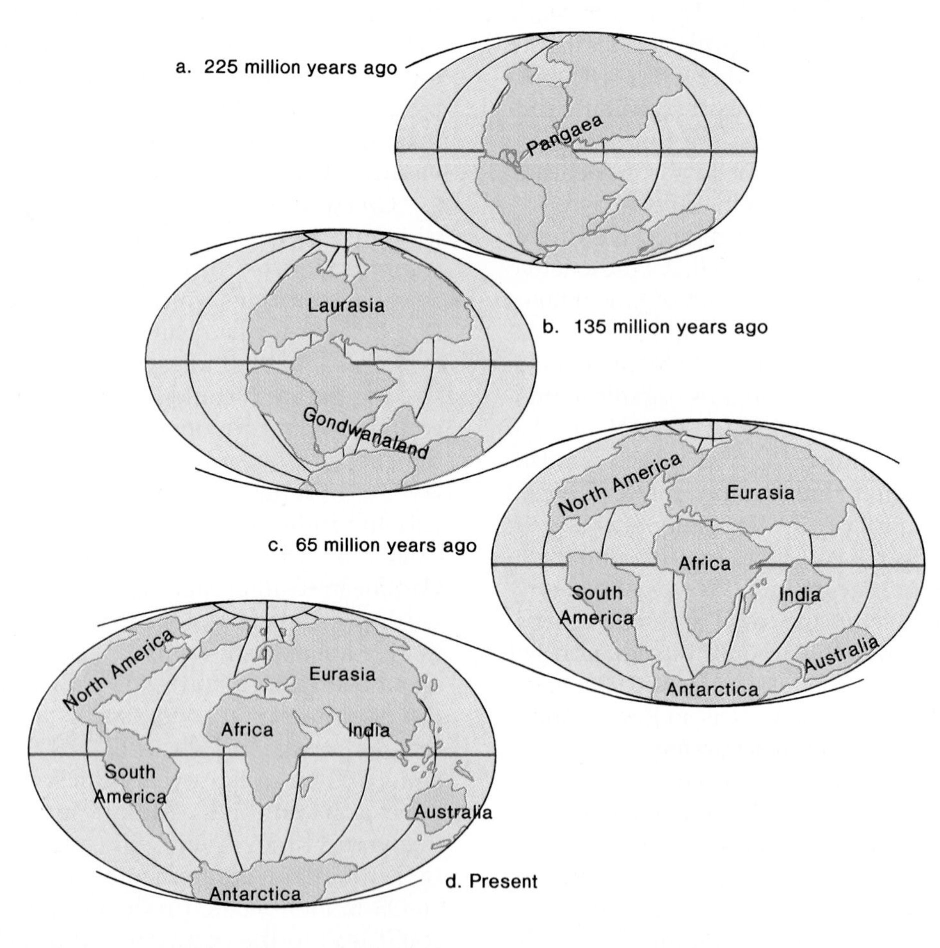

Figure 27.6 History of the placement of today's continents.
a. About 200–250 million years ago, all the continents were joined into a supercontinent called Pangaea. **b.** When the joined continents of Pangaea first began moving apart, there were two large continents called Laurasia and Gondwanaland. **c.** By 65 million years ago, all the continents had begun to separate. This process is continuing today. **d.** North America and Europe are presently drifting apart at a rate of about 2 cm per year.

during swimming (whales and seals), running (horses), climbing (arboreal lizards), or swinging from tree branches (monkeys). Yet all vertebrate forelimbs contain the same sets of bones organized in similar ways, despite their dissimilar functions (Fig. 27.7). The most plausible explanation for this unity is that the basic forelimb plan originated with a common ancestor, and then the plan was modified in the succeeding groups as each continued along its own evolutionary pathway. Structures that are similar because they were inherited from a common ancestor are called **homologous structures.** The wing of a bird and insect are **analogous structures**—they are all adaptations for flying but are structurally unrelated.

Vestigial structures are anatomical features that are fully developed in one group of organisms but are reduced and may have no function in similar groups. Most birds, for example, have well-developed wings used for flight. Some bird species (e.g., ostrich), however, have wings that are greatly reduced, and they do not fly. Similarly, snakes have no use for hind limbs, and yet some have remnants of a pelvic girdle and legs. Humans have a tailbone but no tail. The presence of vestigial structures can be explained by the common descent hypothesis. Vestigial structures are thought to occur because organisms inherit their anatomy from their ancestors; they are traces of an organism's evolutionary history.

The unity of plan shared by vertebrates extends to their embryological development (Fig. 27.8). At some time during development, all vertebrates have a supporting dorsal rod, called a notochord, and exhibit paired pharyngeal pouches. In fishes and some amphibian larvae, these pouches develop into functioning gills. In humans, the first pair of pouches becomes the cavity of the middle ear and the auditory tube. The second pair becomes the tonsils, while the third and fourth pairs become the thymus and parathyroid glands. Why do pharyngeal pouches appear at all during vertebrate development, since they later undergo modification? Fishes are ancestral to other vertebrate groups—evolution doesn't start from scratch; it modifies existing structures.

Organisms share a unity of plan when they are closely related because of common descent. This is substantiated by comparative anatomy and embryological development.

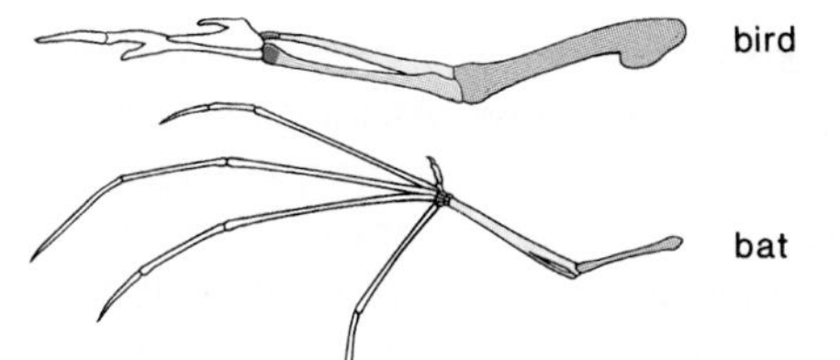

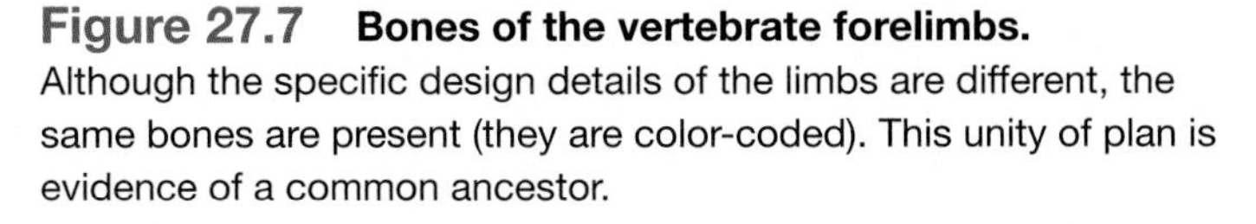

Figure 27.7 Bones of the vertebrate forelimbs.
Although the specific design details of the limbs are different, the same bones are present (they are color-coded). This unity of plan is evidence of a common ancestor.

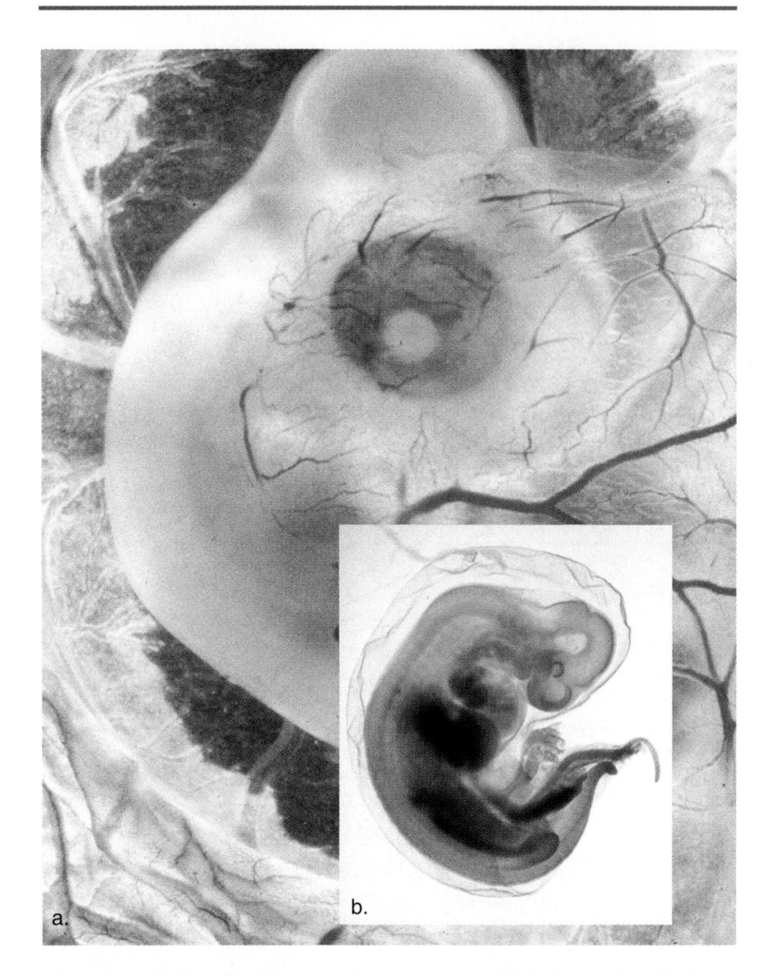

Figure 27.8 Early developmental stages of chick and pig.
a. A chick embryo. **b.** A pig embryo. At these comparable early developmental stages, the two have many features in common, although eventually they are completely different animals. This is evidence that they evolved from a common ancestor.

Biochemical Evidence

Cellular respiration is the common means by which organisms acquire a supply of ATP for cellular metabolism. DNA is the common means by which genetic information is stored. There is no functional reason why biochemical processes need be so similar. But their similarity can be explained by descent from a common ancestor.

From bacteria to humans and from algae to flowering plants, cytochromes are involved in transport of electrons and release of energy to produce ATP molecules. The amino acid sequence of cytochrome *c* is so similar across organisms that it can be used to indicate relatedness. The amino acid sequence of the cytochrome *c* in humans differs from that in a monkey by only one amino acid; from that in a duck by 11, and from that in *Candida,* a yeast, by 51 amino acids.

DNA is the carrier of genetic information in all organisms on earth. Further, organisms utilize the same DNA triplet code for the same 20 amino acids in their proteins. Organisms sometimes even share the same introns and hypervariable regions. When the degree of similarity in DNA base sequences, as well as amino acid sequences of proteins, is examined, the data are as expected, assuming common descent. Humans and chimpanzees have only a 2.5% difference in DNA base sequences, while humans and lemurs have a 42% difference in DNA base sequences.

Classification of organisms is usually based on anatomical and physiological attributes. Biochemical data such as we have been discussing support the same groupings of organisms as do anatomical data, and therefore support the concept of evolutionary relatedness caused by descent from a common ancestor.

All organisms have certain biochemical molecules in common. The degree of similarity between DNA base sequences and amino acid sequences is thought to indicate the degree of relatedness.

27.3 Process of Evolution

The study of macroevolution tells how species or higher levels of classification are related to one another. Many modern evolutionists are interested in the study of population genetics because it allows us to see when and if evolution has occurred. **Microevolution,** which we will now consider, is defined as a change in gene frequencies within a population over time.

Population Genetics

A **population** is all members of a single species occupying a particular area at the same time. A population could be all the green frogs in a frog pond, all the field mice in a barn, or all the English daisies on a hill. The members of a population reproduce with one another to produce the next generation.

Each member of a population is assumed to be free to reproduce with any other member, and when reproduction occurs, the genes of one generation are passed on in the manner described by Mendel's laws. Therefore, in this so-called Mendelian population of sexually reproducing individuals, the various alleles of all the gene loci in all the members make up a **gene pool** for the population. It is customary to describe the gene pool of a population in terms of allele frequencies for the various genes. Using this methodology, two investigators, G. H. Hardy, an English mathematician, and W. Weinberg, a German physician, discovered a law that bears their names.

Hardy-Weinberg law: As long as certain conditions are met, allele frequencies in a sexually reproducing population come to an equilibrium that is maintained generation after generation. The conditions are no mutations, random mating, no gene flow, no genetic drift, and no natural selection.

Practice Problems*

1. In a certain population, 21% are homozygous dominant, 49% are heterozygous, and 30% are homozygous recessive. What percentage of the next generation is predicted to be homozygous recessive, assuming a Hardy-Weinberg equilibrium?
2. Of the members of a population of pea plants, 1% are short. What are the frequencies of the recessive allele *t* and the dominant allele *T*? What are the genotypic frequencies in this population?
3. A student places 600 fruit flies with the genotype *Ll* and 400 with the genotype *ll* in a culture bottle. What will the genotypic frequencies be in the next generation and each generation thereafter, assuming a Hardy-Weinberg equilibrium?

*Answers to Practice Problems appear in Appendix A.

$$p^2 + 2pq + q^2$$

p^2 = % homozygous dominant individuals
p = frequency of dominant allele
q^2 = % homozygous recessive individuals
q = frequency of recessive allele
$2pq$ = % heterozygous individuals

Realize that $p + q = 1$ (there are only 2 alleles)
$p^2 + 2pq + q^2 = 1$ (these are the only genotypes)

Example

An investigator has determined by inspection that 16% of a human population has a straight hairline (recessive trait). Using this information, we can complete all the genotypic and allelic frequencies for the population, provided the conditions for Hardy-Weinberg equilibrium are met.

Given: $q^2 = 16\% = 0.16$ are homozygous recessive individuals

Therefore, $q = \sqrt{0.16} = 0.4$ = frequency of recessive allele
$p = 1.0 - 0.4 = 0.6$ = frequency of dominant allele
$p^2 = (0.6)(0.6) = 0.36 = 36\%$ are homozygous dominant individuals
$2pq = 2(0.6)(0.4) = 0.48 = 48\%$ are heterozygous individuals
or
$= 1.00 - 0.52 = 0.48$

84% have the dominant phenotype

Figure 27.9 **Using the Hardy-Weinberg equation.**

The Hardy-Weinberg law predicts that sexual reproduction alone cannot alter the allele frequencies in a population. For example, suppose it is known that one-fourth of all flies in a *Drosophila* population are homozygous dominant for long wings, one-half are heterozygous, and one-fourth are homozygous recessive for short wings. Therefore, in a population of 100 individuals, we have

25 *LL*, 50 *Ll*, and 25 *ll*

What is the number of the allele *L* and the allele *l* in the population?

Number of L Alleles		Number of l Alleles	
LL (2 *L* × 25)	= 50	*LL* (0 *l*)	= 0
Ll (1 *L* × 50)	= 50	*Ll* (1 *l* × 50)	= 50
ll (0 L)	= 0	*ll* (2 *l* × 25)	= 50
	100 *L*		100 *l*

To determine the frequency of each allele, calculate its percentage of the total number of alleles in the population; in each case, 100/200 = 50% = 0.50. The sperm and the eggs produced by this population also contain these alleles in these frequencies. Assuming random mating (all possible gametes have an equal chance to combine with any other), we can calculate the ratio of genotypes in the next generation using a Punnett square.

		sperm	
		0.5 *L*	0.5 *l*
eggs	0.5 *L*	0.25 *LL*	0.25 *Ll*
	0.5 *l*	0.25 *Ll*	0.25 *ll*

Results:
0.25*LL* + 0.5*Ll* + 0.25*ll*
($\frac{1}{4}$ *LL* + $\frac{1}{2}$ *Ll* + $\frac{1}{4}$ *ll*)

There is an important difference between this Punnett square and one used for a cross between individuals. Here the sperm and the eggs are those produced by the members of a population—not those produced by individuals. As you can see, the results of the Punnett square indicate that the frequency for each allele in the next generation is still 0.50.

G. H. Hardy and W. Weinberg decided to use the binomial expression ($p^2 + 2pq + q^2$) to calculate the genotypic and allele frequencies of a population. Figure 27.9 shows you how this is done. However, it is not necessary to do the mathematics in order to realize that sexual reproduction in and of itself cannot bring about a change in the allele frequencies of a gene pool. Also, the dominant allele does not necessarily increase from one generation to the next. Dominance does not cause an allele to become a common allele.

Sexual reproduction alone cannot affect a Hardy-Weinberg equilibrium, which predicts the same gene pool frequencies generation after generation.

The Hardy-Weinberg law states that an equilibrium of allele frequencies in a gene pool, calculated by using the expression $p^2 + 2pq + q^2$, will remain in effect in each succeeding generation of a sexually reproducing population as long as five conditions are met.

- No mutations. Allelic changes do not occur, or changes in one direction are balanced by changes in the opposite direction.
- No genetic drift. The population is very large, and changes in allele frequencies due to chance alone are insignificant.
- No gene flow. Migration of individuals, and therefore alleles, into or out of the population does not occur.
- Random mating. Individuals pair by chance and not according to their genotypes or phenotypes.
- No selection. No selective force favors one genotype over another.

In real life, these conditions are rarely, if ever, met, and allele frequencies in the gene pool of a population do change from one generation to the next. Therefore, evolution has occurred. *The significance of the Hardy-Weinberg law is that it tells us what factors cause evolution—those that violate the conditions listed.* Microevolution can be detected by noting any deviation from a Hardy-Weinberg equilibrium of allele frequencies in the gene pool of a population.

Figure 27.10 gives an example of microevolution in a population of peppered moths. Peppered moths can be dark-colored or light-colored, and the percentage of each in the population can vary. If tree trunks are light, light-colored moths make up most of the population, and when the tree trunks are dark due to pollution, dark-colored moths make up most of the population. Predatory birds are the selective agent that causes the makeup of the population to vary. When dark-colored moths rest on light trunks in a nonpolluted area, they are seen and eaten by these birds. With the advent of pollution, the trunks of trees darken and it is the light-colored moths that stand out and are eaten. We know that evolution has occurred in Figure 27.10 because the population changes from 10% dark-colored phenotype to 80% dark-colored phenotype over time.

A Hardy-Weinberg equilibrium provides a baseline by which to judge whether evolution has occurred. Any change of allele frequencies in the gene pool of a population signifies that evolution has occurred.

Figure 27.10 Microevolution.
Microevolution has occurred when there is a change in gene pool frequencies—in this case, due to natural selection. On the *far left,* birds cannot see light-colored moths on light tree trunks and, therefore, the light-colored phenotype is more frequent in the population. On the *far right,* birds cannot see dark-colored moths on dark tree trunks, and the dark-colored phenotype is more frequent in the population. The percentage of the dark-colored phenotype has increased in the population because predatory birds can see light-colored moths against tree trunks that are now sooty due to pollution.

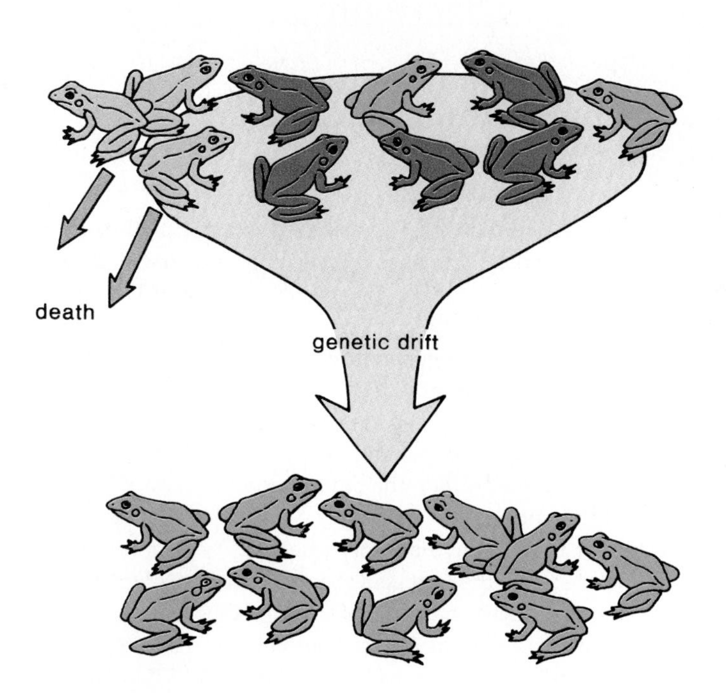

Figure 27.11 Genetic drift.
Genetic drift occurs when by chance only certain members of a population (in this case, green frogs) reproduce and pass on their genes to the next generation. The allele frequencies of the next generation's gene pool may be markedly different from that of the previous generation.

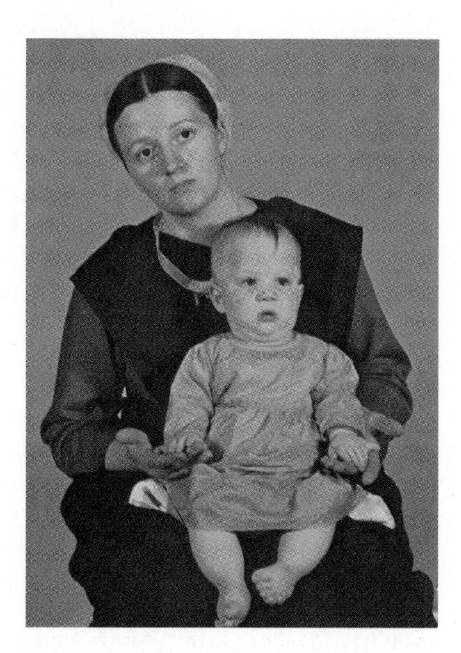

Figure 27.12 Founder effect.
A member of the founding population of Amish in Pennsylvania had a recessive allele for a rare kind of dwarfism. The percentage of the Amish population now carrying this allele is much higher than that of the general population.

Five Agents of Evolutionary Change

The list of conditions for genetic equilibrium stated previously implies that the opposite conditions can cause evolutionary change. These conditions are mutations, genetic drift, gene flow, nonrandom mating, and natural selection.

Mutations

Mutations provide new alleles and therefore underlie all other mechanisms that produce variation, the raw material for evolutionary change. Investigations have demonstrated that high levels of molecular variation are the rule in natural populations. When various *Drosophila* enzymes were extracted and subjected to electrophoresis, it was found that a fly population is polymorphic at no less than 30% of all its gene loci and that an individual fly is likely to be heterozygous at about 12% of its loci.

Mutations alone are not likely to cause evolution—selective agents acting on heritable variation may cause evolution to occur. Even a seemingly harmful mutation can be a source of variation that can help a population adapt to the new surroundings. For example, the water flea *Daphnia* ordinarily thrives at temperatures around 20°C and cannot survive temperatures of 27°C or more. There is, however, a mutation that allows *Daphnia* to live at temperatures between 25°C and 30°C. The adaptive value of this mutation is entirely dependent on environmental conditions.

Genetic Drift

Genetic drift refers to changes in allele frequencies of a gene pool due to chance. In nature, two situations, called *founder effect* and *bottleneck effect,* lead to small populations in which genetic drift drastically affects gene pool frequencies (Figure 27.11).

The **founder effect** occurs when a few individuals found a colony and only a fraction of the total genetic diversity of the original gene pool is represented. Which particular alleles are carried by the founders is dictated by chance alone. The Amish of Lancaster, Pennsylvania, is an isolated religious sect descended from German founders. Today, as many as one in 14 individuals in this group carries a recessive allele that causes an unusual form of dwarfism (it affects only lower arms and legs) and polydactylism (extra fingers) (Fig. 27.12). In the non-Amish population, only one in 1,000 individuals has this allele.

Sometimes a population is subjected to near extinction because of a natural disaster (e.g., earthquake or fire) or because of human interference. The disaster acts as a bottleneck, preventing the majority of genotypes from participating in the production of the next generation. For example, the large genetic similarity found in cheetahs is believed to be due to a **bottleneck effect.** In a study of 47 different enzymes, each of which can come in several different forms, all the cheetahs had exactly the same form. This demonstrates that genetic drift can cause certain alleles to be lost from a population.

Science Focus

Charles Darwin's Theory of Natural Selection

At the age of 22, Charles Darwin signed on as a naturalist with the HMS Beagle, a ship that took a five-year trip around the world in the first half of the nineteenth century. Because the ship sailed in the tropics of the Southern Hemisphere, where life forms are more abundant and varied, Darwin encountered forms of life very different from those in his native England.

Even though it was not his original intent, Darwin began to realize and to gather evidence that life forms change over time and from place to place. He read a book by Charles Lyell, a geologist, who suggested the world is very old and has been undergoing gradual changes for many, many years. Darwin found the remains of a giant ground sloth and an armadillo on the east coast of South America and wondered if these extinct forms were related to the living forms of these animals. When he compared the animals of Africa to those of South America, he noted that the African ostrich and the South American rhea, although similar in appearance, were actually different animals. He reasoned that they had a different line of descent because they were on different continents. When Darwin arrived at the Galápagos Islands, he began to study the diversity of finches (see Fig. 27.17), whose adaptations could best be explained by assuming they had diverged from a common ancestor. With this type of evidence, Darwin concluded that species evolve (change) with time.

When Darwin returned home, he spent the next 20 years gathering data to support the principle of organic evolution. His most significant contribution to this principle was his theory of natural selection, which explains how a species becomes adapted to its environment. Before formulating the theory, he read an essay on human population growth written by Thomas Malthus. Malthus observed that although the reproductive potential of humans is great, there are many environmental factors, such as availability of food and living space, that tend to keep the human population within bounds. Darwin applied these ideas to all populations of organisms. For example, he calculated that a single pair of elephants could have 19 million descendants in 750 years. He realized that other organisms have even greater reproductive potential than this pair of elephants; yet, usually the number of each type of organism remains about the same. Darwin decided there is a constant struggle for existence, and only a few members of a population survive to reproduce. The ones that survive and contribute to the evolutionary future of the species are by and large the better-adapted individuals. This so-called survival of the fittest causes the next generation to be better adapted than the previous generation.

Darwin's theory of natural selection was nonteleological; that is, organisms do not strive to adapt themselves to the environment; rather, the environment acts on them to select those individuals that are best adapted. These are the ones that have been "naturally selected" to pass on their characteristics to the next generation. In order to emphasize the nonteleological nature of Darwin's theory, it is often contrasted with the theory of Jean-Baptiste Lamarck, another nineteenth-century naturalist (Fig. 27A). The Lamarckian explanation for the long neck of the giraffe was based on the assumption that the ancestors of the modern giraffe were trying to reach into the trees to browse on high-growing vegetation. Continual stretching of the neck caused it to become longer, and this acquired characteristic was passed on to the next generation. Lamarck's theory is teleological because, according to him, the outcome is known ahead of time. This type of explanation has not stood the test of time, but Darwin's theory of evolution by natural selection has been fully substantiated by later investigations.

These are the critical elements of Darwin's theory.

- **Variations.** Individual members of a species vary in physical characteristics. Physical variations can be passed from generation to generation. (Darwin was never aware of genes, but we know today that the inheritance of the genotype determines the phenotype.)
- **Struggle for existence.** The members of all species compete with each other for limited resources. Certain members are able to capture these resources better than others.
- **Survival of the fittest.** Humans carry on artificial breeding programs to select which plants and animals will reproduce. Similarly, natural selection by the environment determines which organisms will survive and reproduce. While Darwin emphasized the

- **Adaptation.** Natural selection causes a population of organisms and ultimately a species to become adapted to the environment. The process is slow, but each subsequent generation includes more individuals that are better adapted to the environment.

Can natural selection account for the origin of new species and for the great diversity of life? Yes, if we are aware that life has been evolving for a very long time and that variously adapted populations can arise from a common ancestor.

Darwin was prompted to publish his findings only after he received a letter from another naturalist, Alfred Russel Wallace, who had come to the exact same conclusions about evolution. Although both scientists subsequently presented their ideas at the same meeting of the famed Royal Society in London in 1858, only Darwin later gathered together detailed evidence in support of his ideas. He described his experiments and reasonings at great length in *The Origin of Species by Means of Natural Selection,* a book still studied by many biologists today.

Figure 27A Mechanism of evolution.
This diagram contrasts **(a)** Jean-Baptiste Lamarck's theory as to how evolution occurs, called the theory of acquired characteristics, with that of **(b)** Charles Darwin, called the theory of natural selection. Only Darwin's theory is supported by data.

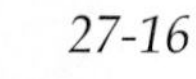

Gene Flow

Gene flow is the movement of alleles between populations by, for example, the migration of breeding individuals from one population to the other. Adult plants are not able to migrate, but their gametes are often either windblown or carried by insects. The wind, in particular, can carry pollen for long distances and can therefore be a factor in gene flow between plant populations.

Gene flow between two populations keeps their gene pools similar. It also prevents close adaptation to a local environment.

Nonrandom Mating

Nonrandom mating occurs when individuals pair up, not by chance, but according to their genotypes or phenotypes. Inbreeding, or mating between relatives to a greater extent than by chance, is an example of nonrandom mating. Inbreeding decreases the proportion of heterozygotes and increases the proportions of both homozygotes at all gene loci. In a human population, inbreeding increases the frequency of recessive abnormalities (see Fig. 27.12).

Natural Selection

Natural selection is the process by which populations become adapted to their environment. The reading on pages 560–61 outlines how Charles Darwin, the father of evolution, explained evolution by natural selection. Here, we restate these steps in the context of modern evolutionary theory. In evolution by natural selection, the **fitness** of an individual is measured by how reproductively successful its offspring are in the next generation.

Evolution by natural selection requires:

1. **Variation.** The members of a population differ from one another.
2. **Inheritance.** Many of these differences are heritable genetic differences.
3. **Differential adaptedness.** Some of these differences affect how well an organism is adapted to its environment.
4. **Differential reproduction.** Individuals that are better adapted to their environment are more likely to reproduce, and their fertile offspring will make up a greater proportion of the next generation.

Random gene mutations are the ultimate source of variation because they provide new alleles. However, in sexually reproducing organisms, recombination of alleles and chromosomes due to crossing-over during meiosis, independent assortment of chromosomes, and fertilization contribute greatly to variation. Recombination may at some time bring a more favorable combination of alleles together. After all, it is the combined phenotype that is subjected to natural selection. In fact, most of the traits on which natural selection acts are polygenic and controlled by more than one pair of alleles. Such traits have a range of phenotypes, the frequency distribution of which usually resembles a bell-shaped curve.

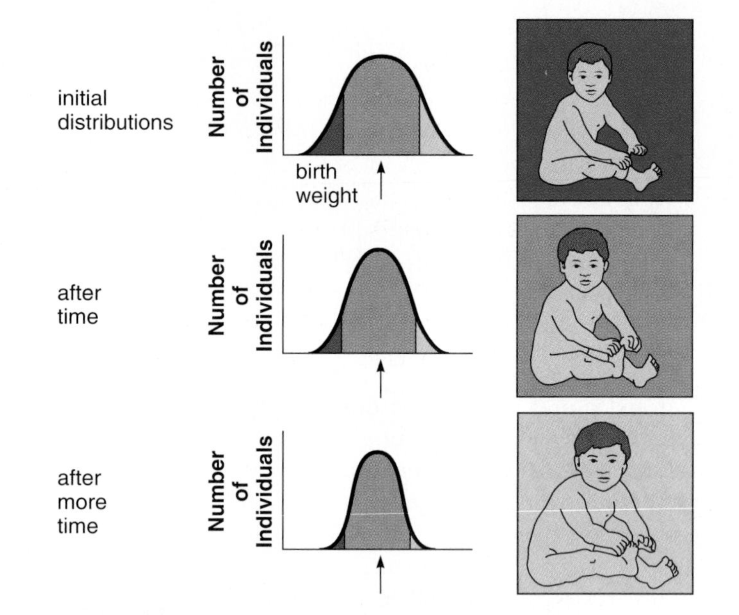

Figure 27.13 Stabilizing selection.
Natural selection favors the intermediate phenotype (see arrows) over the extremes. Today, it is observed that most human babies are of intermediate weight (about 3.2 kg, or 7 lb), and very few babies are either very small or very large.

Three types of natural selection have been described for any particular trait. They are stabilizing selection, directional selection, and disruptive selection.

Stabilizing Selection **Stabilizing selection** occurs when an intermediate phenotype is favored (Fig. 27.13). It can improve adaptation of the population to those aspects of the environment that remain constant. With stabilizing selection, extreme phenotypes are selected against, and individuals near the average are favored. As an example, consider the birth weight of human infants, which ranges from 0.89–4.9 kilograms (2–10.8 lb). The death rate is higher for infants whose birth weights are at these extremes and lowest for babies who have a birth weight between 3.1 kilograms and 3.5 kilograms. Most babies have a birth weight within this range, which gives the best chance of survival. Similar results have been found in other animals, also.

Directional Selection **Directional selection** occurs when an extreme phenotype is favored and the distribution curve shifts in that direction (Fig. 27.14). Such a shift can occur when a population is adapting to a changing environment. For example, the gradual increase in the size of the modern horse, *Equus,* can be correlated with a change in the environment from forestlike conditions to grassland conditions.

Hyracotherium, which was about the size of a dog, was adapted to the forestlike environment of the Eocene, an epoch of the Paleogene period. This animal could have hidden among the trees for protection, and its low-crowned teeth were appropriate for browsing on leaves. In the Miocene and Pliocene epochs, however, grasslands began to replace the forests. Then the ancestors of *Equus* were subject to selective pressure for the development of strength, intelligence, speed, and durable grinding teeth. A larger size provided the strength needed for combat, a larger skull made room for a larger brain, elongated legs ending in hooves gave speed to escape enemies, and the durable grinding teeth enabled the animals to feed efficiently on grasses. Nevertheless, the evolution of the horse should not be viewed as a straight line of descent; there were many side branches that became extinct.

Industrial melanism is another good example of directional selection in which the selective agent is known. Moths rest on the trunks of trees during the day; if they are seen by predatory birds, they are eaten. As long as the tree trunks in the environment are light in color, the light-colored moths live to reproduce. But if the tree trunks turn black from industrial pollution, the dark-colored moths survive and reproduce to a greater extent than the light-colored moths. The dark-colored phenotype then becomes the more frequent one in the population (see Fig. 27.10). However, if pollution is reduced and the trunks of the trees regain their normal color, the light-colored moths again increase in number.

Pesticides and antibiotics are selective agents for insects and bacteria, respectively. The forms that survive exposure to these agents give rise to future generations that are resistant to these toxic substances.

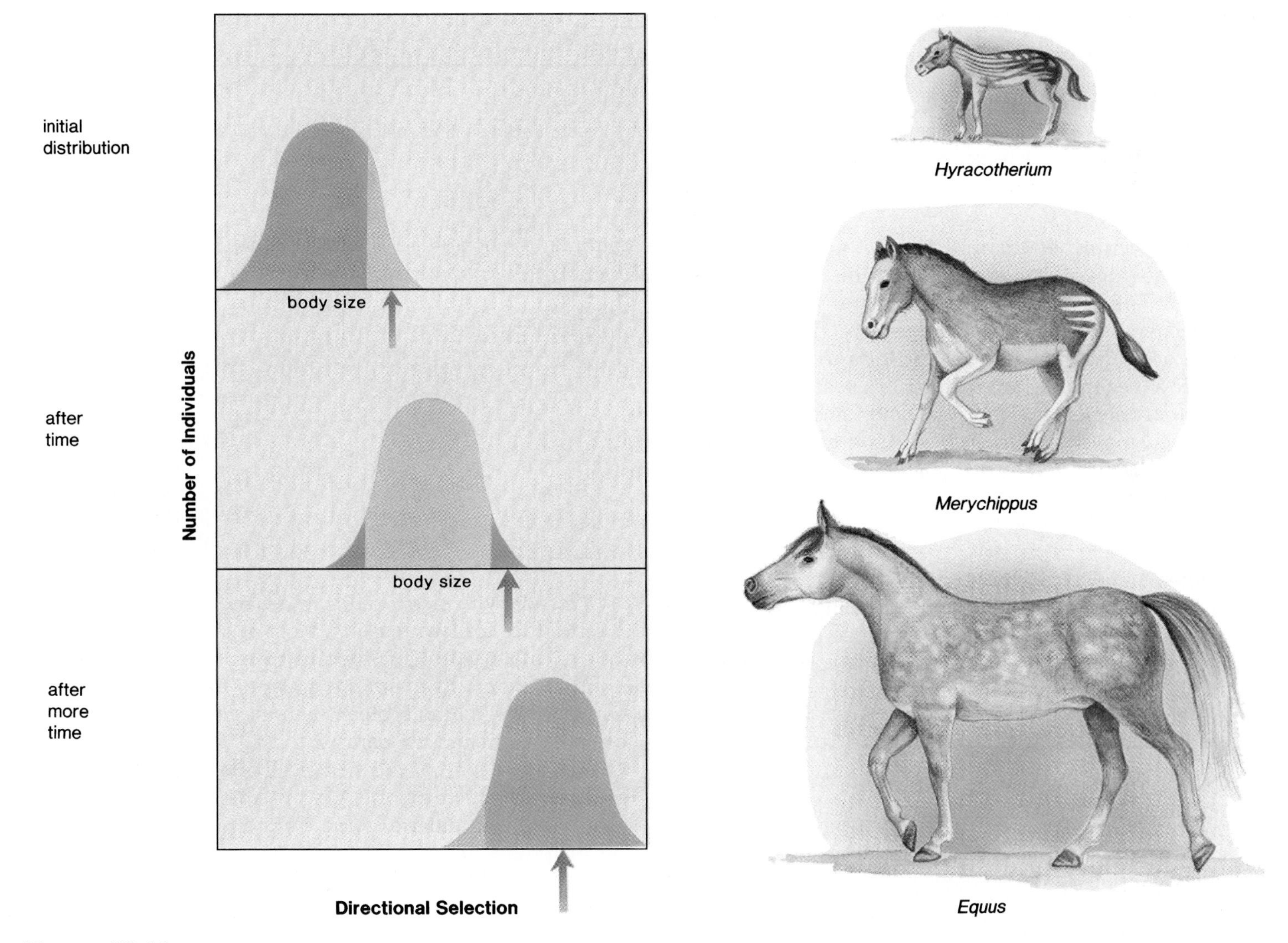

Figure 27.14 Directional selection.
This occurs when natural selection favors one extreme phenotype (see arrows), resulting in a shift in the distribution curve. *Equus,* the modern-day horse, which is adapted to a grassland habitat, evolved from *Hyracotherium,* which was adapted to a forest habitat.

Figure 27.15 Disruptive selection.
Natural selection favors two extreme phenotypes (see arrows). Today, it is observed that British land snails comprise mainly two different phenotypes, each adapted to a particular habitat.

Disruptive Selection In **disruptive selection,** two or more extreme phenotypes are favored over any intermediate phenotype (Fig. 27.15). For example, British land snails (*Cepaea nemoralis*) have a wide habitat range that includes low-vegetation areas (grass fields and hedgerows) and forest areas. In low-vegetation areas, thrushes feed mainly on snails with dark shells that lack light bands, and in forest areas, they feed mainly on snails with light-banded shells. Therefore, the two different habitats have resulted in two different phenotypes in the population.

The agents of evolutionary change are mutations, genetic drift, gene flow, nonrandom mating, and natural selection. These processes cause changes in the gene pool frequencies of a population. Only natural selection results in adaptation to the environment.

Maintenance of Variations

Sickle-cell disease shows how genetic variation is sometimes maintained in a population. Persons with sickle-cell disease have sickle-shaped red blood cells, leading to hemorrhaging and organ destruction. In parts of Africa, there is a high incidence of malaria caused by a parasite that lives in and destroys red blood cells. Sickle-cell disease tends to be more common in such areas. A study of the three genotypes and phenotypes involved explains why.

Genotype	Phenotype	Result
Hb^AHb^A	Normal	Dies due to malarial infection
Hb^AHb^S	Sickle-cell trait	Lives due to protection from both
Hb^SHb^S	Sickle-cell disease	Dies due to sickle-cell disease

Persons with sickle-cell trait are more likely to survive to reproduce for two reasons. Most of the time they do not have circulatory problems because their red blood cells have a normal shape. Even so, the malarial parasite cannot survive in their red blood cells. When the cells are sickled, they lose potassium and the parasite dies.

The frequency of the sickle-cell allele in some parts of Africa is 0.40, while among African-Americans it is only 0.05 due to reduced malaria in the United States. The ability of the heterozygote to survive accounts for the greater frequency of the sickle-cell allele in Africa. The favored heterozygote keeps the two homozygotes present in the population; when the ratio of two or more phenotypes remains the same in each generation, it is called *balanced polymorphism*.

27.4 Speciation

Usually, a species occupies a certain geographical range, within which there are several subpopulations. For our present discussion, **species** is defined as a group of interbreeding subpopulations that share a gene pool and that are isolated reproductively from other species. The subpopulations of the same species exchange genes, but different species do not exchange genes. Reproductive isolation of the gene pools of similar species is accomplished by such mechanisms as those listed in Table 27.2. If **premating isolating mechanisms** are in place, reproduction is never attempted. If **postmating isolating mechanisms** are in place, reproduction may take place, but it does not produce fertile offspring.

Process of Speciation

Whenever reproductive isolation develops, speciation has occurred. Figure 27.16 outlines how it is believed that reproductive isolation usually comes about. In the first frame, a species is represented by two populations which are experiencing gene flow. However, when the populations become separated by a *geographic barrier,* gene flow is no longer possible. A geographic barrier could be a new canal recently built by humans, or an upheaval caused by an earthquake, and so forth. Now different variations arise in the two populations due to independent mutations, drift, and selection so that first postmating, and then, if enough time passes, premating reproductive isolation occurs. Even if the geographic barrier is now removed, the two populations will not be able to reproduce with one another, and therefore, what was one species has become two species. This model of speciation is called **allopatric speciation.**

It is also possible that a single population could suddenly divide into two reproductively isolated groups without the need for geographic isolation. The best evidence for this type of speciation, called **sympatric speciation,** is found among plants, where multiplication of the chromosome number in one plant prevents it from successfully reproducing with others of its kind. But self-reproduction could lead to a number of offspring with the new chromosome number.

Speciation is the origin of species. This usually requires geographic isolation followed by reproductive isolation.

Table 27.2 Reproductive Isolating Mechanisms

Isolating Mechanism	Example
Premating	
Habitat isolation	Species at same locale occupy different habitats
Temporal isolation	Species reproduce at different seasons or different times of day
Behavioral isolation	In animals, courtship behavior differs or they respond to different songs, calls, pheromones, or other signals
Mechanical isolation	Genitalia unsuitable for one another
Postmating	
Gamete isolation	Sperm cannot reach or fertilize egg
Zygote mortality	Fertilization occurs, but zygote does not survive
Hybrid sterility	Hybrid survives but is sterile and cannot reproduce
F_2 fitness	Hybrid is fertile but F_2 hybrid has reduced fitness

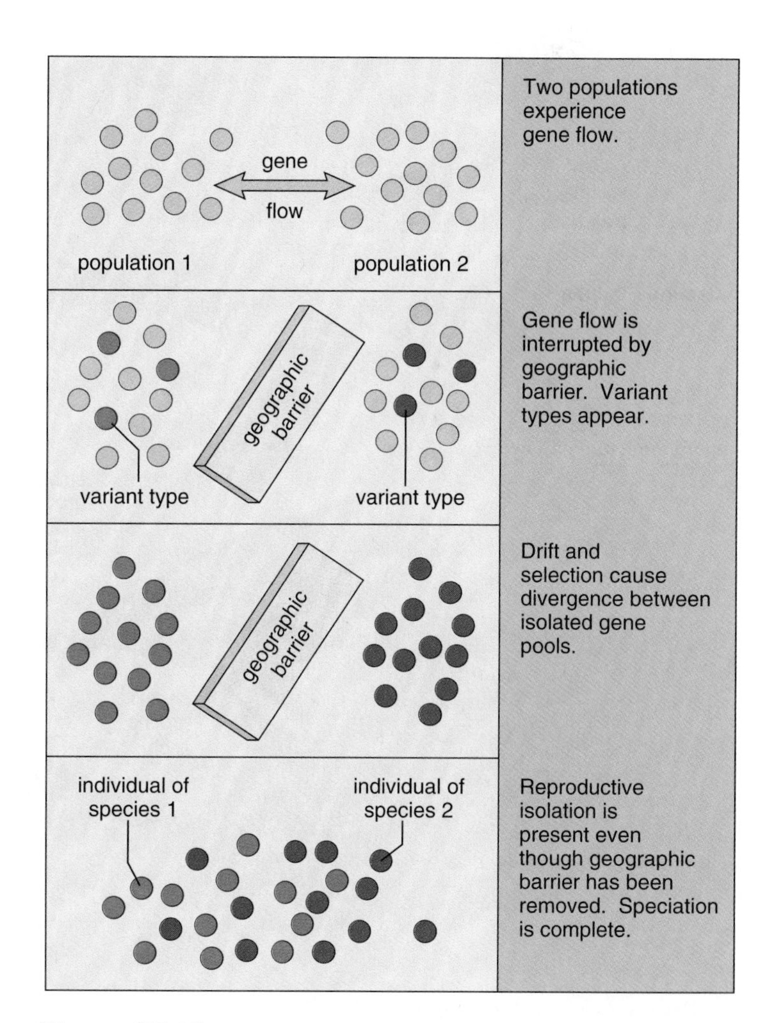

Figure 27.16 Allopatric speciation.
Allopatric speciation occurs after a geographic barrier prevents gene flow between populations that originally belonged to a single species.

Visual Focus

Figure 27.17 The Galápagos finches.
Each of these finches is adapted to gathering and eating a different type of food. Tree finches have beaks largely adapted to eating insects and, at times, plants. The woodpecker-type finch, a tool-user, uses a cactus spine or twig to probe in the bark of a tree for insects. Ground finches have beaks adapted to eating prickly-pear cactus or different size seeds.

Time

new species

divergence

a. Phyletic Gradualism

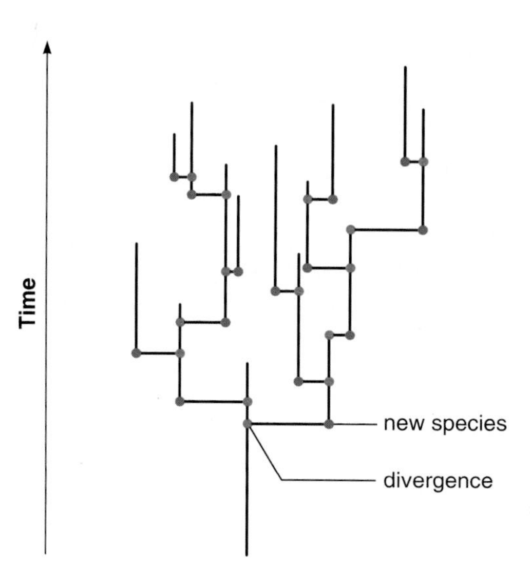

b. Punctuated Equilibrium

Figure 27.18 Modes of evolutionary change.
The differences between phyletic gradualism **(a)** and punctuated equilibrium **(b)** are reflected in these two patterns of time versus structure.

Adaptive Radiation

One of the best examples of speciation is provided by the finches on the Galápagos Islands, which are often called Darwin's finches because Darwin first realized their significance as an example of how evolution works. The Galápagos Islands, located 600 miles west of Ecuador, South America, are volcanic but do have forest regions at higher elevations. The 13 species of finches (Fig. 27.17), placed in three genera, are believed to be descended from mainland finches that migrated to one of the islands. Therefore, Darwin's finches are an example of *adaptive radiation,* or the proliferation of a species by adaptation to different ways of life. We can imagine that after the original population of a single island increased, some individuals dispersed to other islands. The islands are ecologically different enough to have promoted divergent feeding habits. This is apparent because although the birds physically resemble each other in many respects, they have different beaks, each of which is adapted to gathering and eating a different type of food. There are seed-eating ground finches, with beaks appropriate to cracking small-, medium-, or large-sized seeds; cactus-eating ground finches, with beaks appropriate for eating prickly-pear cacti; insect-eating tree finches, also with different-sized beaks; and a warbler-type tree finch, with a beak adapted to insect eating and nectar gathering. Among the tree finches, there is a woodpecker type, which lacks the long tongue of a true woodpecker but makes up for this by using a cactus spine or a twig to ferret out insects.

The evolution of several species of finches on the Galápagos Islands is an example of adaptive radiation.

The Pace of Speciation

The fossil *Archaeopteryx* (see Fig. 27.4) is a transitional form that can be used to link one major group of organisms to another. The expression "missing link," which is sometimes used, is suitable because few such transitional forms have been found. Some of today's organisms are called *living fossils* because they are so similar to their original ancestors. Horseshoe crabs, crocodiles and coelacanth fish are animals that still resemble their earliest ancestors. A time of limited change in a *lineage* (evolutionary line) is called *stasis.*

Currently there are two hypotheses about the pace of speciation and therefore evolution (Fig. 27.18). Traditionally, evolutionists have supported a model called **phyletic gradualism,** which says that change is very slow but steady within a lineage before and after a divergence (splitting of the line of descent). Therefore, it is not surprising that few *transitional links* have been found. Indeed, the fossil record, even if it were complete, may be unable to show when speciation has occurred. A species is defined on the basis of reproductive isolation, and reproductive isolation cannot be detected in the fossil record!

In recent years, a new model of evolution called **punctuated equilibrium** has been proposed. It says that stasis, a period of equilibrium, is punctuated by speciation. With reference to the length of the fossil record (about 3.5 billion years), speciation occurs relatively rapidly, and this explains why few transitional links ever became fossils. Indeed, speciation most likely involves only an isolated population at one locale. Only when a new species evolves and displaces the existing species is the new species likely to show up in the fossil record.

Classification

As mentioned, macroevolution concerns the relationship between groups of organisms above the species level. The study of macroevolution is related to the science of taxonomy.

Taxonomy

Taxonomy, the discipline of describing and classifying organisms, begins when each is assigned a name. In the binomial system, names have two parts. For example, *Homo sapiens* and *Homo erectus* (now extinct) are two different species of humans. The first word, *Homo,* is the genus (pl., genera), a classification category that can contain many species. The second word is the **specific epithet** which may tell something descriptive about the organism. *Sapiens* refers to a large brain, and *erectus* refers to the ability to walk erect. The scientific name is in italics; the genus is capitalized while the specific epithet is not. The species is designated by the full name; the genus name can be used alone to refer to a group of related species.

Taxonomists group species into ever larger categories. Today there are at least seven obligatory categories, often called taxa: **species, genus, family, order, class, phylum** (called divisions in the plant kingdom), and **kingdom.** All species in the same genus have many characteristics in common. We would expect, for example, all species of oak trees in the genus *Quercus* to be very similar. Just as there can be several species within a genus, there can be several genera within a family, and so forth—the higher the category the more inclusive it is. Organisms in the same kingdom have general characteristics in common despite their having a different appearance. In the plant kingdom, rose bushes are easily distinguishable from pine trees although both carry on photosynthesis.

This text recognizes five kingdoms: kingdoms **Monera, Protista, Fungi, Plantae,** and **Animalia.** Figure 27.19 describes and Table 27.3 lists the characteristics of each kingdom. Figure 27.20 depicts an organism from each kingdom. It is customary to show the relationship of organisms in the form of an evolutionary tree; Figure 27.19 gives an evolutionary tree to show how the five kingdoms might be related. In an evolutionary tree, a branching point represents a common ancestor that gave rise to other lineages.

Systematics

Taxonomy is a part of the broader field of systematics, which is the study of the diversity of organisms. One goal of systematics is to determine phylogeny, or the evolutionary history of a group of organisms. In recent years cladistics, which is a relatively new school of systematics, has been faulting the long-accepted traditional school. The **traditional school** classifies organisms according to both common ancestry and the degree of difference among evolved groups. On this basis, birds are not reptiles because birds have feathers and fly. **Cladistics** classifies organisms according to shared, specialized characteristics. On this basis, birds are reptiles because the two groups of animals have many derived characteristics in common. Cladists develop diagrams called cladograms (instead of evolutionary trees) based solely on shared characteristics.

Another area of controversy today concerns the possible need to introduce a higher level of classification called a domain. Ribosomal RNA sequencing data have been interpreted to mean that archaea (types of bacteria living in extreme habitats, possibly resembling those of the primitive earth) and bacteria (all the other types of prokaryotes) should be in separate domains, appropriately called **Archaea** and **Bacteria.** All the eukaryotes (protists, fungi, plants, and animals) would be in the domain **Eukarya.**

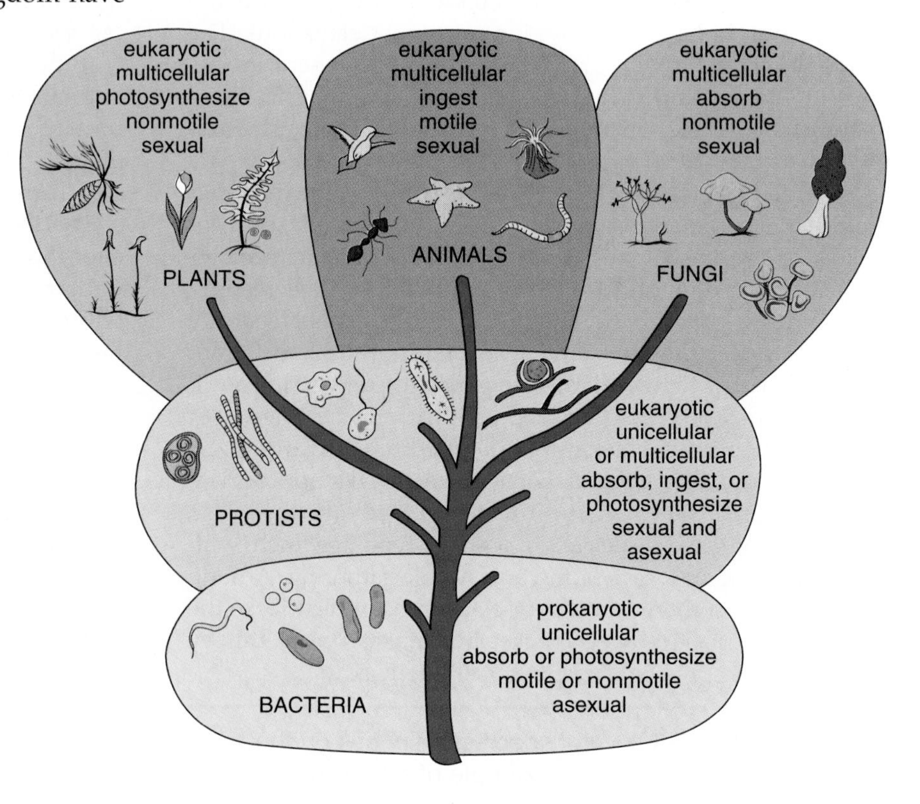

Figure 27.19 The five-kingdom system of classification.
Representatives of each kingdom are depicted in the ovals, and an evolutionary tree roughly indicates the lines of descent.

Figure 27.20 **The five-kingdom system—a pictorial representation.**

Table 27.3 **Classification Criteria for the Five-Kingdom System**

	Monera	Protista	Fungi	Plantae	Animalia
Type of cell	Prokaryotic	Eukaryotic	Eukaryotic	Eukaryotic	Eukaryotic
Complexity	Unicellular	Unicellular	Unicellular or multicellular	Multicellular	Multicellular
Type of nutrition	Autotrophic by various or heterotrophic by various	Photosynthetic or heterotrophic by various	Heterotrophic saprotrophs	Photosynthetic	Heterotrophic by ingestion
Motility	Sometimes by flagella	Sometimes by flagella (or cilia)	Nonmotile	Nonmotile	Motile by contractile fibers
Reproduction	Asexual usual	Asexual/sexual	Sexual usual	Sexual usual	Sexual usual
Internal protection of zygote	No	No	No	Yes	Yes
Nervous system	None	Conduction of stimuli in some forms	None	None	Present

Bioethical Issue

Evolution is a scientific theory. So is the cell theory, which says that all organisms are composed of cells, and so is the atomic theory that says all matter is composed of atoms. Yet, no one argues that schools should teach alternatives to the cell theory or the subatomic theory. Confusion reigns over the use of the expression, "the theory of evolution." But the term theory in science is reserved for those ideas that scientists have found to be all encompassing because they are based on data collected in a number of different fields.

No wonder most scientists in our country are dismayed when state legislatures or school boards rule that teachers must put forward a variety of "theories" on the origin of life, including one that runs contrary to the mass of data that supports the theory of evolution. An institute in California called the Institute for Creation Research advocates that students be taught an "intelligent-design theory" which says that DNA could never have arisen without the involvement of an "intelligent agent," and that gaps in the fossil record mean that species arose fully developed with no antecedents.

Since our country forbids the mingling of church and state—no purely religious ideas can be taught in the schools—the advocates for an "intelligent-design theory" are careful to never mention the Bible nor any ideas like God created the world in seven days. Still, teachers who have a good scientific background do not feel comfortable teaching an "intelligent-design theory" because it does not meet the test of a scientific theory. Science is based on hypotheses that have been tested by observation and/or experimentation. A scientific theory has stood the test of time—no hypotheses have been supported by observation and/or experimentation that run contrary to the theory. On the contrary, the theory of evolution is supported by data collected in such wide-ranging fields as development, anatomy, geology, biochemistry, and so forth.

The polls consistently show that nearly half of all Americans prefer to believe the Old Testament account of how God created the world in seven days. That, of course, is their right, but should schools be required to teach an "intelligent-design theory" that traces its roots back to the Old Testament, and is not supported by observation and experimentation?

Questions

1. Should teachers be required to teach an "intelligent-design theory" of the origin of life in schools? Why or why not?
2. Should schools rightly teach that science is based on data collected by the testing of hypotheses by observation and experimentation? Why or why not?
3. Should schools be required to show that the "intelligent-design theory" does not meet the test of being scientific? Why or why not?

Summarizing the Concepts

27.1 Origin of Life

In the presence of an outside energy source, such as ultraviolet radiation, the primitive atmospheric gases reacted with one another to produce small organic molecules.

Next, macromolecules evolved and interacted. The RNA-first hypothesis is supported by the discovery of ribozymes, RNA enzymes. The protein-first hypothesis is supported by the observation that amino acids polymerize abiotically when exposed to dry heat. The Cairns-Smith hypothesis suggests that macromolecules could have originated in clay. The protocell must have been a heterotrophic fermenter living on the preformed organic molecules in the organic soup. Eventually the DNA → RNA → protein self-replicating system evolved, and a true cell came into being.

27.2 Evidence of Evolution

The fossil record and biogeography, as well as comparative anatomy, development, and biochemistry, all give evidence of evolution. The fossil record gives us the history of life in general and allows us to trace the descent of a particular group. Biogeography shows that the distribution of organisms on earth is explainable by assuming organisms evolved in one locale. Comparing the anatomy and the development of organisms reveals a unity of plan among those that are closely related. All organisms have certain biochemical molecules in common, and similarities indicate the degree of relatedness.

27.3 Process of Evolution

Evolution is described as a process that involves a change in gene frequencies within the gene pool of a sexually reproducing population. The Hardy-Weinberg law states that the gene pool frequencies arrive at an equilibrium that is maintained generation after generation unless disrupted by mutations, genetic drift, gene flow, nonrandom mating, or natural selection. Any change from the initial allele frequencies in the gene pool of a population signifies that evolution has occurred.

27.4 Speciation

Speciation is the origin of species. This usually requires geographic isolation, followed by reproductive isolation.

The evolution of several species of finches on the Galápagos Islands is an example of adaptive radiation because each species has a different way of life, but all species came from one common ancestor.

Currently, there are two hypotheses about the pace of speciation. Traditionalists support phyletic gradualism—slow, steady change leading to speciation. A new model, punctuated equilibrium, says that a long period of stasis is interrupted by speciation.

Classification involves the assignment of species to categories. There are seven obligatory categories of classification: species, genus, family, order, class, phylum, and kingdom. Each higher category is more inclusive; members of the same kingdom share general characteristics, and members of the same species share quite specific characteristics. The five-kingdom system of classification recognizes these kingdoms: Monera (the bacteria), Protista (e.g. algae, protozoa), Fungi, Plantae, Animalia.

Studying the Concepts

1. Contrast the RNA-first hypothesis with the protein-first hypothesis. If polymerization occurred in clay, what macromolecules would have resulted? 549
2. Trace the steps by which a chemical evolution may have produced a protocell. 549–50
3. Why is it likely the protocell was a heterotrophic fermenter? 550
4. How did the protocell become a true cell? 550
5. Show that the fossil record and biogeography, as well as comparative anatomy, development, and biochemistry, all give evidence of common descent. 551–56
6. What is the Hardy-Weinberg law? What is its significance? 556–58
7. Name and discuss the agents of evolutionary change. 559, 562
8. What are the four requirements for evolution by natural selection? 560–62
9. Name and give an example for each type of selection. 560–64
10. Define a species. How do new species originate? 565
11. When is adaptive radiation apt to take place? 567
12. Contrast the basic tenets of phyletic gradualism and punctuated equilibrium with regard to speciation. 567
13. Describe the five-kingdom system of classification, and give examples of each kingdom. 568

Testing Yourself

Choose the best answer for each question.

1. Which of these did Stanley Miller place in his experimental system to show that organic molecules could have arisen from inorganic molecules on the primitive earth?
   a. microspheres
   b. clay and water
   c. purines and pyrimidines
   d. the primitive gases
   e. All of these are correct.
2. Which of these is the chief reason the protocell was probably a fermenter?
   a. It didn't have any enzymes.
   b. It didn't have a nucleus.
   c. The atmosphere didn't have any oxygen.
   d. Fermentation provides the greatest amount of energy.
   e. All of these are correct.
3. Evolution of the DNA → RNA → protein system was a milestone because the protocell
   a. was a heterotrophic fermenter.
   b. could now reproduce.
   c. lived in the ocean.
   d. needed energy to grow.
   e. All of these are correct.
4. Continental drift helps explain the
   a. occurrence of mass extinctions.
   b. distribution of fossils on earth.
   c. geological upheavals like earthquakes.
   d. Only a and b are correct.
   e. All of these are correct.
5. If evolution occurs, we would expect different biogeographical regions with similar environments to
   a. all contain the same mix of plants and animals.
   b. have all land masses connected.
   c. each have its own specific mix of plants and animals.
   d. have plants and animals that have similar adaptations.
   e. Both c and d are correct.
6. The fossil record offers direct evidence for common descent because you can
   a. see that the types of fossils change over time.
   b. sometimes find common ancestors.
   c. trace the ancestry of a particular group.
   d. Only b and c are correct.
   e. All of these are correct.
7. Organisms adapted to the same way of life will
   a. have structures that show they share a unity of plan.
   b. have similarities that need not indicate a unity of plan.
   c. always live in the same biogeographical region.
   d. Both a and c are correct.
   e. Both b and c are correct.
8. Assuming a Hardy-Weinberg equilibrium, 21% of a population is homozygous dominant, 50% is heterozygous, and 29% is homozygous recessive. What percentage of the next generation is predicted to be homozygous recessive?
   a. 21%
   b. 50%
   c. 29%
   d. 25%
   e. 42%
9. A human population has a higher-than-usual percentage of individuals with a genetic disease. The most likely explanation is
   a. gene flow.
   b. stabilizing selection.
   c. directional selection.
   d. genetic drift.
   e. All of these are correct.
10. Which of these is/are necessary to natural selection?
    a. variations
    b. differential reproduction
    c. inheritance of differences
    d. Only b and c are correct.
    e. All of these are correct.
11. Which of these is a premating isolating mechanism?
    a. habitat isolation
    b. temporal isolation
    c. gamete isolation
    d. hybrid sterility
    e. Both a and b are correct.
12. Allopatric but not sympatric speciation requires
    a. reproductive isolation.
    b. geographic isolation.
    c. spontaneous differences in males and females.
    d. prior hybridization.
    e. rapid rate of mutation.
13. The many species of Galápagos finches were each adapted to eating different foods. This is an example of
    a. gene flow.
    b. adaptive radiation.
    c. sympatric speciation.
    d. Only b and c are correct.
    e. All of these are correct.
14. The classification category below the level of family is
    a. class.
    b. order.
    c. species.
    d. phylum.
    e. genus.
15. Which kingdom is mismatched?
    a. Monera—fungi
    b. Protista—multicellular algae
    c. Plantae—flowers and mosses
    d. Animalia—arthropods and humans
    e. Fungi—molds and mushrooms

16. The following diagrams represent a distribution of genotypes (phenotypes) in a population. Superimpose on the diagram in (a) another diagram to show that disruptive selection has occurred; in (b) to show that stabilizing selection has occurred; and in (c) to show that directional selection has occurred.

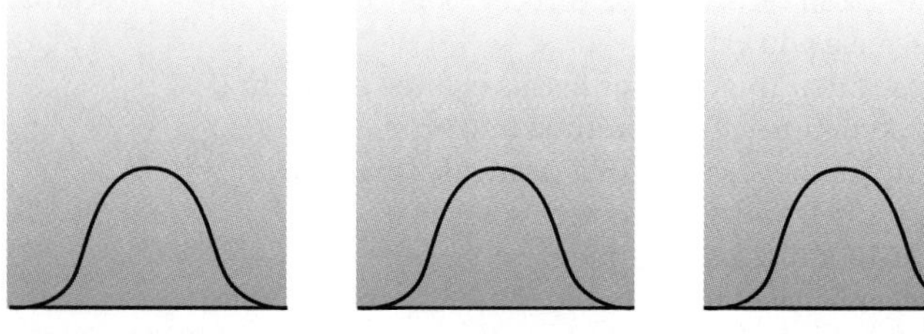

a. Disruptive selection **b.** Stabilizing selection **c.** Directional selection

Thinking Scientifically

1. Considering the origin of life:
   a. Why did Fox show that polymerization of amino acids can occur with enzymes? (page 549)
   b. Why do investigators believe that proteins alone could not form a cell? What property do they lack?
   c. By what mechanism would it be possible for RNA to replicate in the first cell?
   d. Why was it necessary for this RNA to have enzymatic properties?
2. A scientist observes that members of a particular plant species are shorter at the top of a mountain that at the bottom.
   a. Give an explanation based on natural selection.
   b. The scientist gathers seeds from the plants at the top of the mountain and plants them at the base of the mountain. If your explanation is correct, what will be the appearance of the plants?
   c. Using Figure 27.16, explain these experimental results.

Understanding the Terms

allopatric speciation 565
analogous structure 555
Archaea 568
autotroph 550
Bacteria 568
biogeography 554
bottleneck effect 559
cladistics 568
class 568
continental drift 554
directional selection 562
disruptive selection 564
Eukarya 568
evolution 551
family 568
fitness 562
fossil 551
founder effect 559
gene flow 562
gene pool 556
genetic drift 559
genus 568
heterotroph 550
homologous structure 555
kingdom 568
kingdom Animalia 568
kingdom Fungi 568
kingdom Monera 568
kingdom Plantae 568
kingdom Protista 568
liposome 549
microevolution 556
microsphere 549
natural selection 562
nonrandom mating 562
order 568
phyletic gradualism 567
phylum 568
population 556
postmating isolating mechanism 565
premating isolating mechanism 565
proteinoid 549
protocell 549
punctuated equilibrium 567
species 565
specific epithet 568
stabilizing selection 562
sympatric speciation 565
taxonomy 568
traditional school 568
vestigial structure 555

Match the terms to these definitions:

a. __________ Process by which populations become adapted to their environment.

b. __________ Natural selection in which an extreme phenotype is favored, usually in a changing environment.

c. __________ An evolutionary model that proposes there are periods of rapid change dependent on speciation followed by long periods of stasis.

d. __________ Structure that is similar in two or more species because of common ancestry.

e. __________ Movement of genes from one population to another via sexual reproduction between members of the populations.

Using Technology

Your study of the origin and evolution of life is supported by these available technologies.

Essential Study Partner CD-ROM

Evolution and Diversity → History
→ Processes
→ Speciation
→ Classification

Visit the Mader web site for related ESP activities.

Exploring the Internet

The Mader Home Page provides resources and tools as you study this chapter.

http://www.mhhe.com/biosci/genbio/mader

Life Science Animations 3D Video

28 Plate Tectonics
30 Movement to Land
31 Evolution of Gills to Jaws

CHAPTER 28

Microbiology

Chapter Concepts

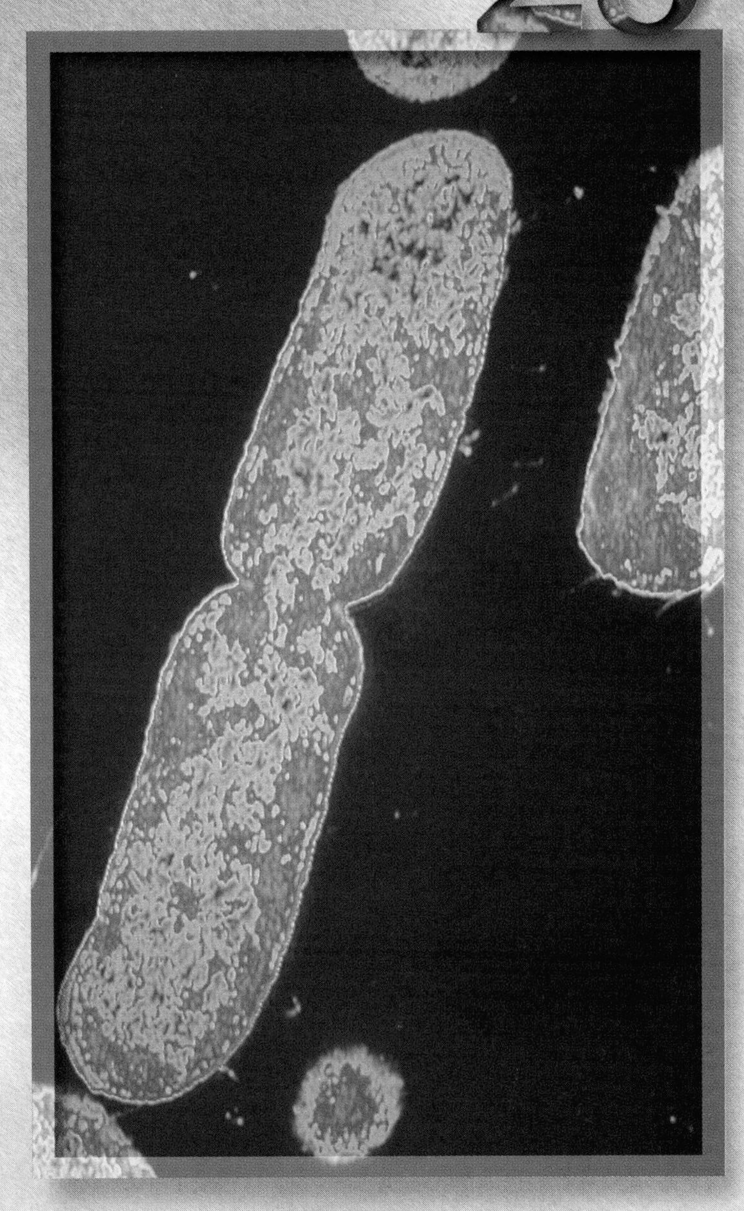

Most bacteria don't cause illness, but Salmonella typhimurium *is responsible for food poisoning. Poultry, eggs, pork, beef, and oysters are all carriers of this bacterium.*

Place a single drop of pond water under a microscope and you can see an amazing menagerie of unicellular organisms, such as protozoans and algae. Fungi and bacteria also abound in nature. Fungi feed on organic matter in the soil, but bacteria are even found in inorganic environments—in water boiling out of the ocean's volcanic vents and within rocks miles below the surface of the earth, for example. Closer to home, microbes can live in the crevices of your kitchen cutting board or within the food you're eating. Set a loaf of bread, free of preservatives, on the counter and you'll observe colonies of bacteria and fungi in no time.

Most microbes are harmless to people, and many are beneficial. Some viruses, bacteria, and fungi, however, make a home inside the tissues or cells of people. To combat these microbial invaders, researchers desperately hunt for drugs, such as antiviral agents, and antibiotics, which kill bacteria and fungi. Some antibiotics are extracted from microbes themselves, and scientists are now hopeful of reviving the decades-old idea of using bacteria-destroying viruses against bacteria.

This chapter will introduce you to the myriad of organisms that inhabit the microscopic world.

28.1 Viruses

Viruses are not included in the classification table found in Appendix B because they are noncellular and should not be classified with cellular organisms. Viruses are generally smaller than 200 nm in diameter and therefore are comparable in size to that of a large protein macromolecule. Many can be purified and crystallized, and the crystals can be stored just as chemicals are stored.

Structure of Viruses

A virus always has at least two parts: an outer capsid composed of protein units, and an inner core of nucleic acid—either DNA or RNA (Fig. 28.1). The viral genome has at most several hundred genes; a human cell contains thousands of genes. A virus may also contain various enzymes for nucleic acid replication. The capsid is often surrounded by an outer membranous envelope, which is actually partially composed of the host's plasma membrane. The classification of viruses is based on (1) type of nucleic acid, including whether it is single stranded or double stranded, (2) viral size and shape, and (3) the presence or absence of an outer envelope.

Parasitic Nature

Viruses are *obligate intracellular parasites.* In order to have a ready supply of animal viruses in the laboratory, they are sometimes injected into live chick embryos. Today, however, it is more customary to infect cells that are maintained in tissue culture. Viruses infect all sorts of cells—from bacterial cells to human cells—but each type is very specific. For example, bacteriophages infect only bacteria, the tobacco mosaic virus infects only plants, and the rabies virus infects only mammals. Some human viruses even specialize in a particular tissue. Human immunodeficiency viruses (HIV) enter specific types of blood cells, the polio virus reproduces in spinal nerve cells, the hepatitis viruses infect only liver cells. What could cause this remarkable parasite-host cell relationship? It is now believed that viruses are derived from the very cell they infect; the nucleic acid of viruses came from their host cell genomes! Therefore, viruses must have evolved after cells came into existence, and new viruses are probably evolving even now.

Viruses, like other organisms, can mutate, and this habit can be quite troublesome because a vaccine that is effective today may not be effective tomorrow. Flu viruses are well known for mutating, and this is why you have to have a flu shot every year—antibodies generated from last year's shot are not expected to be effective this year.

Viruses are nonliving particles that reproduce only inside specific host cells.

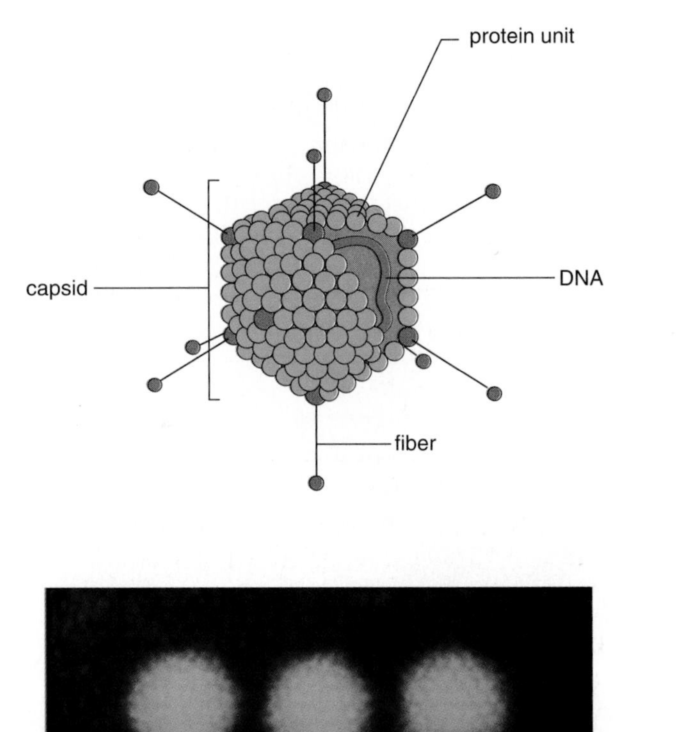

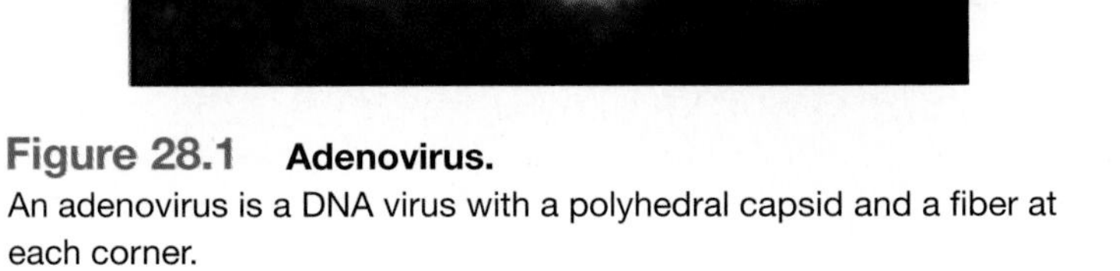

Figure 28.1 Adenovirus.
An adenovirus is a DNA virus with a polyhedral capsid and a fiber at each corner.

Replication of Viruses

Viruses are specific to a particular host cell because portions of the capsid (or the spikes of the envelope) bind in a lock-and-key manner with a receptor on the host cell plasma membrane. After viral nucleic acid enters the cell, it *takes over the metabolic machinery of the host cell* so that more viruses are produced.

Replication of Bacteriophages

Bacteriophages, or simply phages, are viruses that parasitize bacteria; the bacterium in Figure 28.2 could be *Escherichia coli,* which lives in our intestines. In the lytic cycle, the host cell undergoes *lysis,* a breaking open of the cell to release new viruses. In the lysogenic cycle, viral replication does not immediately occur, but replication may take place sometime in the future. The bacteriophage, termed lambda, is capable of carrying out both cycles.

Lytic Cycle The **lytic cycle** may be divided into five stages: attachment, penetration, biosynthesis, maturation, and release. During attachment, portions of the capsid combine with a receptor on the rigid bacterial cell wall in a lock-and-key manner. During penetration, a viral enzyme digests away part of the cell wall, and viral DNA is injected into the bacterial cell. Biosynthesis of viral components begins after the virus brings about inactivation of host genes not necessary to viral replication. The virus takes over the machinery of the cell in order to carry out viral DNA replication and production of multiple copies of the capsid protein subunits. During maturation, viral DNA and capsids are assembled to produce several hundred viral particles. Lysozyme, an enzyme coded for by a viral gene, is produced; this disrupts the cell wall, and the release of new viruses occurs. The bacterial cell dies as a result.

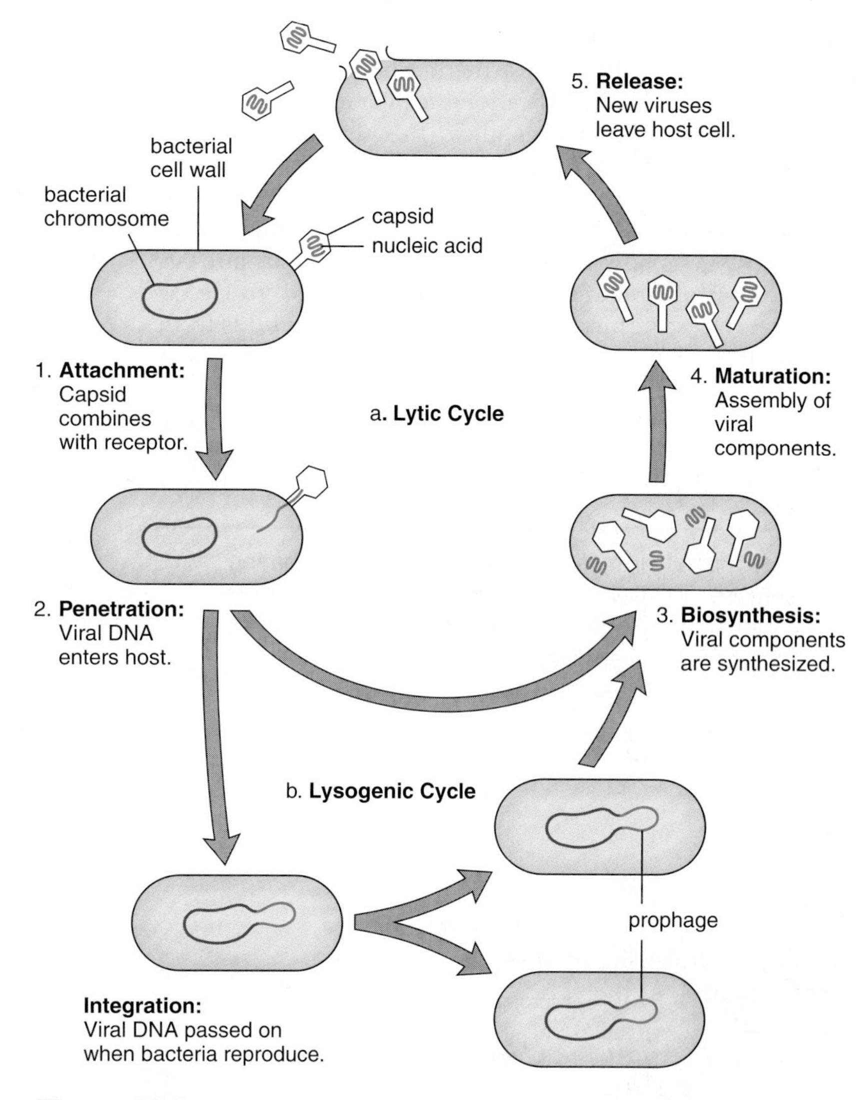

Figure 28.2 Lytic and lysogenic cycles.
a. In the lytic cycle, viral particles escape when the cell is lysed (broken open).
b. In the lysogenic cycle, viral DNA is integrated into host DNA. At some time in the future, the lysogenic cycle can be followed by the lytic cycle.

During the lytic cycle, a bacteriophage takes over the machinery of the cell so that viral replication and release occur.

Lysogenic Cycle In the **lysogenic cycle,** the infected bacterium does not immediately produce viruses but may do so sometime in the future. In the meantime the phage is latent—not actively replicating. Following attachment and penetration, viral DNA becomes integrated into bacterial DNA with no destruction of host DNA. While latent, the viral DNA is called a prophage. The prophage is replicated along with the host DNA, and all subsequent cells, called lysogenic cells, carry a copy of the prophage. Certain environmental factors, such as ultraviolet radiation, can induce the prophage to enter the lytic stage of biosynthesis, followed by maturation and release.

During the lysogenic cycle, the phage becomes a prophage that is integrated into the host genome. At a later time, the phage may reenter the lytic cycle and replicate itself.

Replication of Animal Viruses

Animal viruses replicate in a manner similar to bacteriophages, but there are modifications. If the virus has an envelope, its glycoprotein spikes allow the virus to adhere to plasma membrane receptors. Then the capsid and viral genome penetrate a host cell. Once inside, the virus is uncoated as the capsid is removed. The viral genome, either DNA or RNA, is now free of its coverings and biosynthesis proceeds. Another difference among enveloped viruses is that viral release occurs by budding. As the virus buds from the cell, it acquires an envelope partially consisting of host plasma membrane. Certain envelope components, such as the glycoproteins that allow the virus to enter a host cell, were coded for by viral genes. Budding does not necessarily result in the death of the host cell.

After animal viruses enter the host cell, uncoating releases viral DNA or RNA, and replication occurs. If release is by budding, the viral particle acquires a membranous envelope.

Retroviruses **Retroviruses** are RNA animal viruses that have a DNA stage (Fig. 28.3). A retrovirus contains a special enzyme called reverse transcriptase, which carries out RNA → cDNA transcription. The DNA is called cDNA because it is a DNA copy of the viral genome. Following replication, the resulting double-stranded cDNA is integrated into the host genome. The viral DNA remains in the host genome and is replicated when host DNA is replicated. When and if this DNA is transcribed, new viruses are produced by the steps we have already cited: biosynthesis, maturation, and release—not by destruction of the cell, but by budding.

The enzyme reverse transcriptase allows retroviruses to produce a cDNA copy of their genes which become integrated into the host genome.

Viral Infections

Viruses are best known for causing infectious diseases in plants and animals, including humans. Some animal viruses

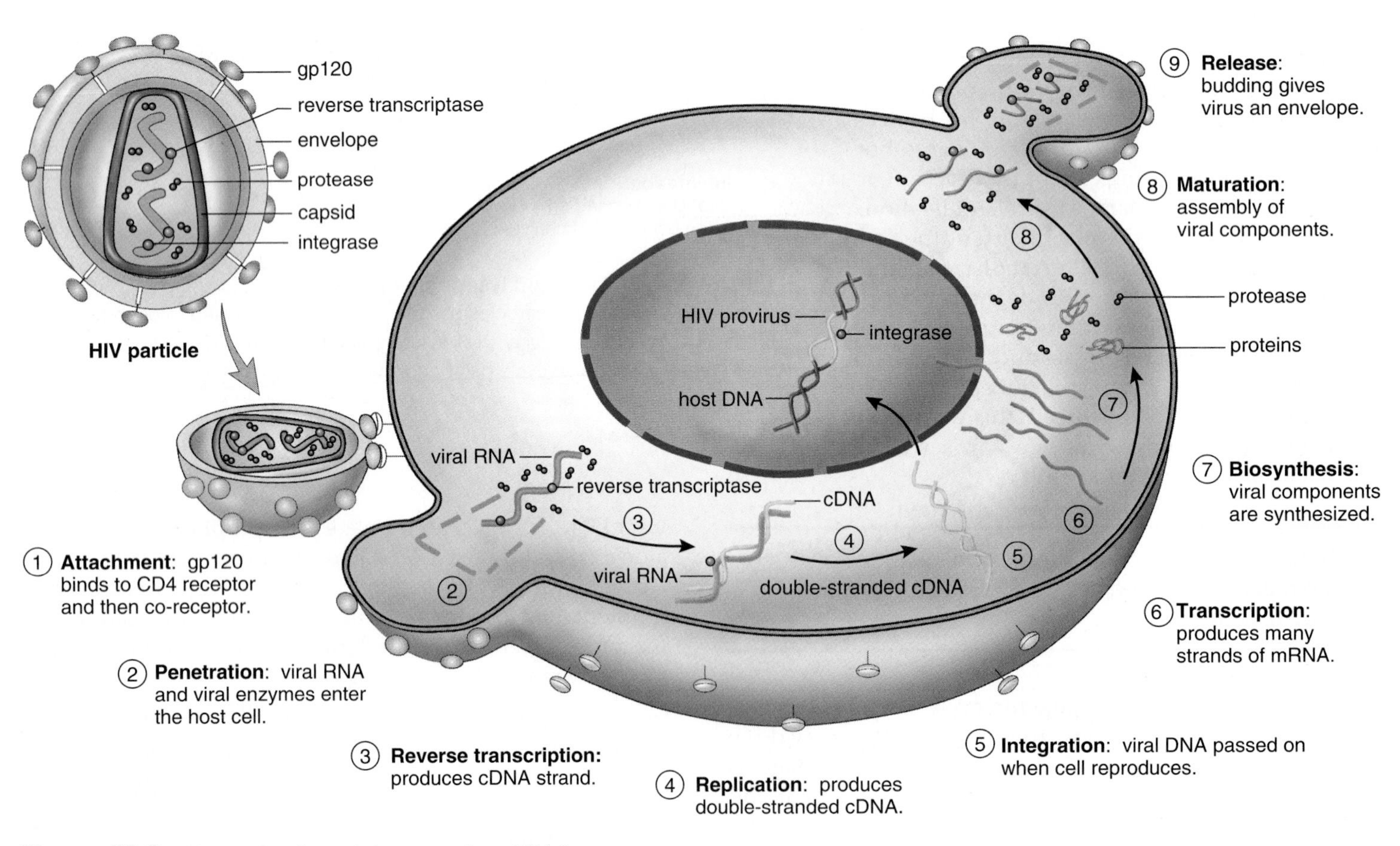

Figure 28.3 Reproduction of the retrovirus HIV-1.
HIV-1 utilizes reverse transcription to produce cDNA (DNA copy of RNA genes). cDNA integrates into the cell's chromosomes before it reproduces and buds from the cell.

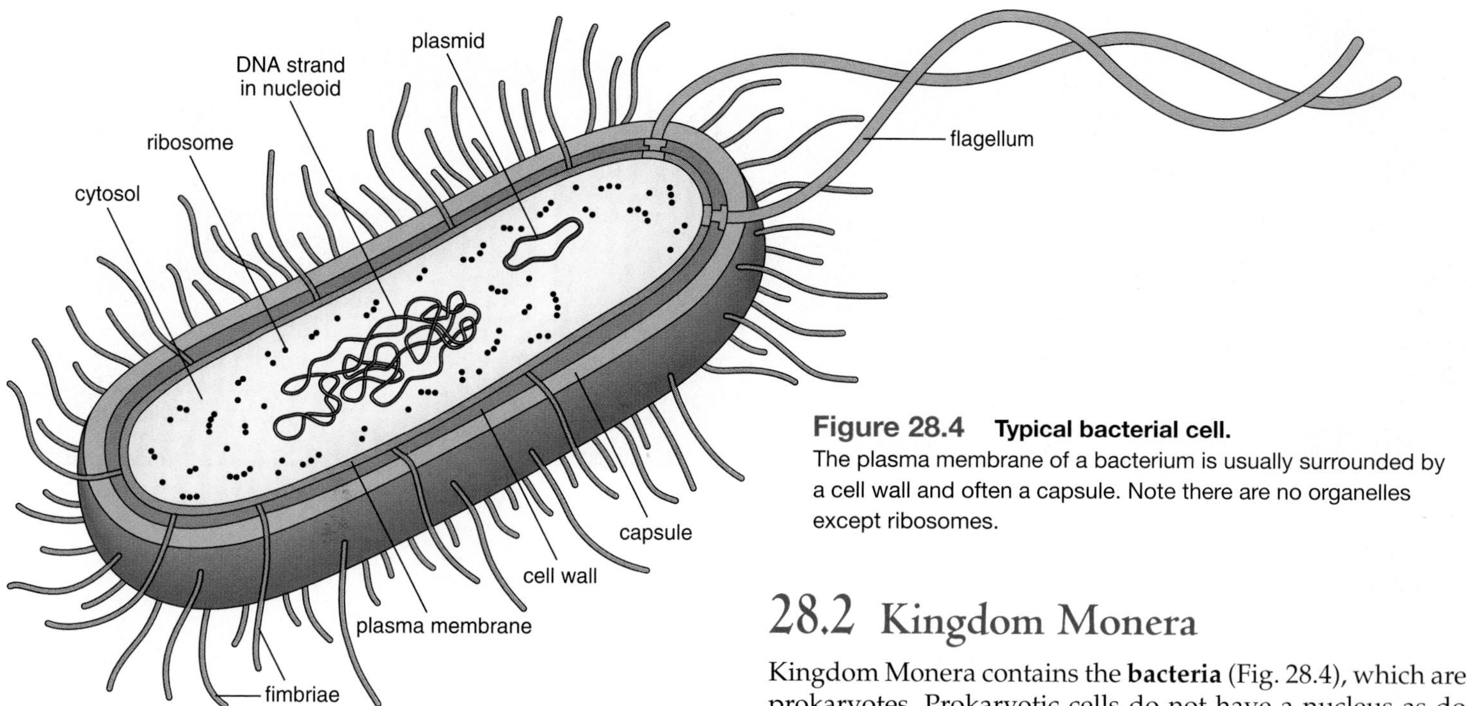

Figure 28.4 Typical bacterial cell.
The plasma membrane of a bacterium is usually surrounded by a cell wall and often a capsule. Note there are no organelles except ribosomes.

are specific to human cells. Of special concern are those such as the papillomavirus, the herpes viruses, the hepatitis viruses, and the adenoviruses, which can cause specific types of cancer. Retroviruses are of interest because human immunodeficiency viruses (HIV), which cause AIDS, are retroviruses. Retroviruses also cause certain forms of cancer.

At least a thousand different viruses cause diseases in plants. About a dozen crop diseases have been attributed not to viruses but to **viroids,** which are naked strands of RNA not covered by a capsid. Like viruses, though, viroids direct the cell to produce more viroids.

Some diseases in humans have been attributed to **prions,** which are protein particles that possibly can convert other proteins in the cell to become prions. It seems that prions may have a misshapen tertiary structure and can cause other similar proteins to convert to this shape also. Prions are believe to cause Creutzfeldt-Jakob disease (CJD) a mental disorder that over a decade can lead to loss of vision and speech before paralysis and death occur. Recently, prions have been linked to a serious outbreak in Great Britain of bovine spongiform encephalopathy (BSE), better known as **mad cow disease.** Cattle feed had been supplemented with the remains of sheep that had died of scrapie, another prion disease. In an effort to prevent an outbreak of BSE in this country, the importation of cattle feed, meat products, and live cattle from Great Britain and any other BSE-affected country has been banned since 1989.

Viruses, viroids, and prions are all known to cause diseases in humans.

28.2 Kingdom Monera

Kingdom Monera contains the **bacteria** (Fig. 28.4), which are prokaryotes. Prokaryotic cells do not have a nucleus as do eukaryotic cells.

Structure of Bacteria

Prokaryotic cells are very small (1–10 μm in length and 0.7–1.5 μm in width), and except for ribosomes, they do not have the cytoplasmic organelles found in eukaryotic cells. They do have a chromosome, but it is contained within a **nucleoid,** which has no nuclear envelope; therefore, bacteria are said to lack a nucleus. Many bacteria have extrachromosomal rings of DNA called *plasmids,* which are often extracted and used as vectors to carry foreign DNA into other bacteria during genetic engineering.

Bacteria have a cell wall containing unique amino sugars cross-linked by peptide chains. The cell wall may be surrounded by a *capsule.* Some bacteria move by means of *flagella,* and some adhere to surfaces by means of short, fine, hairlike appendages called *fimbriae.*

Prokaryotic cells lack a nucleus and most of the other organelles found in eukaryotic cells.

Metabolism of Bacteria

Some bacteria are *obligate anaerobes,* unable to grow in the presence of oxygen. A few serious illnesses, such as botulism and tetanus, are caused by anaerobic bacteria. Some other bacteria, called *facultative anaerobes,* are able to grow in either the presence or the absence of oxygen. Most bacteria, however, are *aerobic* and, like animals and plants, require a constant supply of oxygen to carry out cellular respiration.

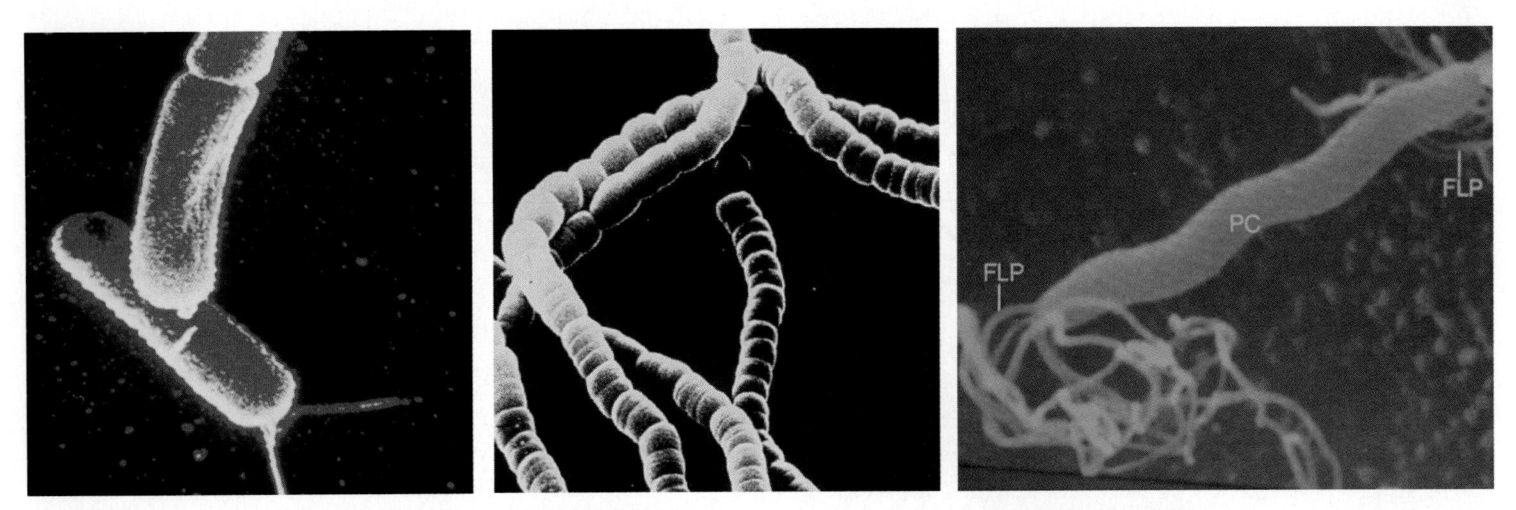

a. Bacilli in pairs

b. Cocci in chains

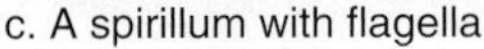

c. A spirillum with flagella

Figure 28.5 Diversity of bacteria.
a. Bacillus (rod-shaped) bacterium. **b.** Coccus (round) bacterium. **c.** Spirillum (spiral-shaped) bacterium.

Every type of nutrition is found among bacteria except heterotrophism by ingestion. Some bacteria are autotrophic by *photosynthesis;* they use light as a source of energy to produce their own food. Cyanobacteria (see Fig. 28.6), which photosynthesize in the same manner as plants, also give off oxygen. Some bacteria are autotrophic by *chemosynthesis;* they oxidize inorganic compounds to obtain the necessary energy to produce their own food. Such bacteria are very important to the cycling of nitrogen in ecosystems.

Most types of bacteria are heterotrophic by absorption. They are **saprotrophs,** organisms that carry on external digestion of organic matter and absorb the resulting nutrients across the plasma membrane. In ecosystems, bacteria are called **decomposers** because they break down organic matter and make inorganic nutrients available to photosynthesizers. The metabolic capabilities of bacteria are utilized for the digestion of sewage and oil and also for the production of such products as alcohol, vitamins, and even antibiotics. By means of genetic engineering, bacteria are now used to produce useful substances, such as human insulin and growth hormone.

Bacteria are often **symbiotic;** they live in association with other organisms. The nitrogen-fixing bacteria in the nodules of legumes are *mutualistic,* as are the bacteria that live within our own intestinal tract. We provide a home for the bacteria, and they provide us with certain vitamins. *Commensalistic* bacteria reside on our skin, where they usually cause no problems. *Parasitic* bacteria are responsible for a wide variety of plant and animal diseases.

The majority of bacteria are heterotrophic by absorption (saprotrophic decomposers) and contribute significantly to recycling matter through ecosystems.

Classification of Bacteria

Bacteria are found in three basic shapes. (Fig. 28.5): rod (bacillus, pl., bacilli); round or spherical (coccus, pl., cocci); and spiral or helical-shaped (spirillum, pl., spirilli). These three basic shapes may occur in particular arrangements. For example, cocci may form clusters (staphylococci, diplococci) or chains (streptococci). Rod-shaped bacteria may appear as very short rods (coccobacilli) or as very long filaments.

Traditionally bacteria are classified as either Gram-positive or Gram-negative. Gram-positive bacteria retain a dye-iodine complex and appear purple under the light microscope, while Gram-negative bacteria do not retain the complex and appear pink. Gram-positive bacteria have a thick layer of peptidoglycan on their cell wall, whereas Gram-negative bacteria have only a thin layer. For the past 75 years bacterial taxonomy has been compiled in *Bergey's Manual of Determinative Bacteriology.* The most recent edition of *Bergey's Manual* divided the bacteria into 10 major groups, which were further subdivided into order, families, genera, and species. The names of the groups—for example, "Nonmotile, Gram-negative, curved bacteria" or "Nonsporeforming Gram-positive rods"—reflect the phenotypic way the manual groups the bacteria.

A new way of classifying bacteria on the basis of rRNA (ribosomal RNA) sequences was introduced in the 1980s. Bacteria that share the same sequence of rRNA bases are put into the same group. Some of the new groups, such as Spirochetes, are essentially identical to early classification systems. However, other groups contain a diverse assortment of bacteria that appear to be phenotypically distant from one another. Gram-negative bacteria and Gram-positive bacteria are placed in the same group as long as they share the same rRNA sequence of bases. The reading on the next page describes how this methodology has led to the suggestion that there are three domains of life, one of which includes the **archaea,** which were formerly considered to be a type of bacteria.

Science Focus

Archaea

Carl Woese at the University of Illinois has championed the hypothesis that sequencing of rRNA (ribosomal RNA) can be used to classify organisms. He chose rRNA because of its involvement in protein synthesis—it's just possible that any changes in rRNA sequence occur in a slow, steady manner as evolution occurs.

On the basis of rRNA sequencing, Woese and his colleagues concluded that there are three domains of life. A domain is a higher classification category than the category kingdom. The first two domains, they maintain, are the Archaea and the Bacteria, both of which contain only unicellular prokaryotes. The third domain, called Eukarya, contains all the eukaryotes (protists, fungi, plants, and animals).

Woese's hypothesis is remarkable because it suggests that the archaea, long considered to be bacteria, should be considered as different from bacteria as bacteria are from eukaryotes. The archaea are found in extreme environments thought to be similar to those of the primitive earth (Figure 28A). The archaea include the methanogens which live in anaerobic environments such as swamps, marshes, and intestinal tracts of animals where they produce methane (CH_4) by a process that gives them energy for ATP formation. Methane produced by methanogenic archaea is believed to contribute significantly to global warming. The halophiles, another type of archaea, are salt lovers that can be isolated from bodies of water like the Great Salt Lake in Utah, the Dead Sea, solar salt ponds, and hypersalty soils. These organisms pump chloride into their cells and synthesize ATP only in the presence of light. The third major type of archaea are the thermoacidophiles, which are both temperature and acid loving. These archaea live in extremely hot, acidic environments such as hot springs, geysers, submarine thermal vents, and around volcanoes. They survive best at temperatures of about 80°C, and some can grow even at 105°C (remember that water boils at 100°C)! Many of the thermoacidophiles use sulfate as a source of energy to make ATP and are found in natural environments with high sulfate concentrations.

Woese and his colleagues also went on to propose a new tree of life. According to their rRNA-based tree, a universal common ancestor gave rise to the Archaea and Bacteria and later, the Archaea gave rise to the Eukarya. This conclusion was based on the finding that archaea have base sequences that are closer to eukaryotic sequences than to bacterial sequences. But in science, as you know, new information, often available because of new technologies, may call into question former beliefs and hypotheses. Recently, new and more powerful sequencing instruments have allowed scientists to sequence the entire genomes of bacteria and eukaryotic genomes, such as those of yeast. And the whole genome data has called into question the tree of life based on rRNA sequencing. The DNA genome data shows, to a surprising degree, a mixture of DNA sequences from both archaea and bacteria in eukaryotes. For example, Russell Doolittle who is with the Canadian Institute for Advanced Research in Halifax, Nova Scotia, has found that 17 of 34 families of eukaryotic proteins look as if they came from bacteria and only eight show a greater similarity to archaea, the supposed ancestor of Eukarya. Even worse, some DNA data show a close relation between archaea and bacteria. Ron Swanson and Robert Feldman of Diversa Corporation in San Diego found that the gene for an enzyme involved in the synthesis of tryptophan , an amino acid, is just about the same in *Aquifex*, domain Archaea, and *Bacillus subtilus*, domain Bacteria. Robert Feldman has said, "I think it's open whether the three domains [of life] will hold up."

Others have gone on to suggest the possibility of widespread gene swapping among the first organisms to evolve. In explanation, Doolittle says, "You are what you eat," meaning that as early unicellular organisms fed on one another they incorporated each other's genes into their genomes. If gene swapping occurred, Woese agrees that it will make it difficult to draw a correct tree of life, but he still has faith that rRNA sequencing data is sufficient to conclude that there are three domains of life: Archaea, Bacteria, and Eukarya.

a.

b.

Figure 28A Habitats of archaea.
a. Halophiles turn the water red in salt collection ponds in California near San Francisco Bay. **b.** The scientific submarine Alvin captures methanogen-containing material around a deep-sea vent.

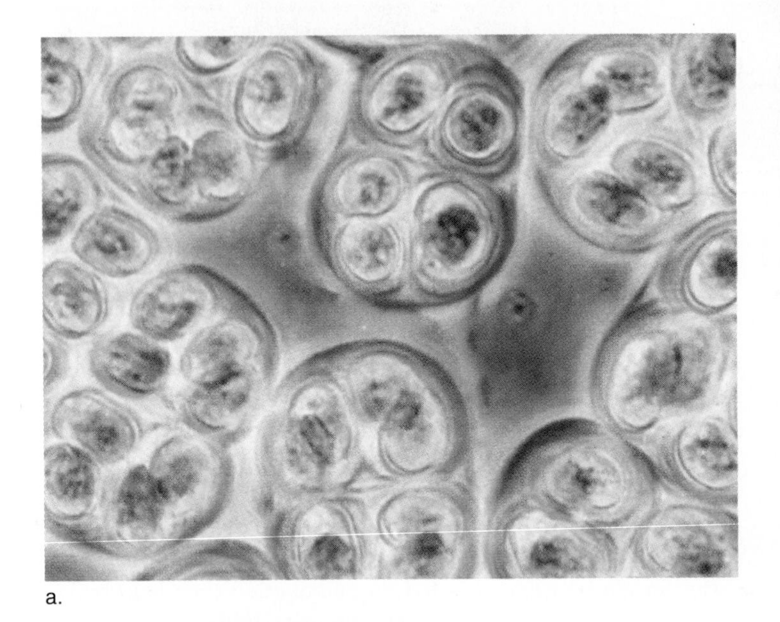

a.

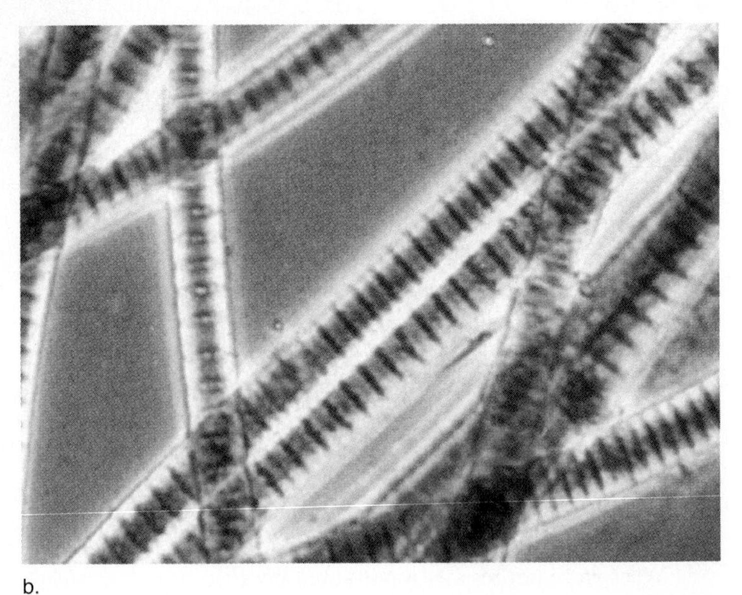

b.

thylakoids
DNA
cell wall
plasma membrane
storage granule

c.

Figure 28.6 Diversity of the cyanobacteria.
a. In *Gloeocapsa,* single cells are grouped in a common gelatinous sheath. **b.** Filaments of cells occur in *Oscillatoria.* **c.** One cell of *Oscillatoria* as it appears through the electron microscope.

Cyanobacteria **Cyanobacteria** are gram-negative rods with a number of unusual traits. They photosynthesize in the same manner as plants and are believed to be responsible for first introducing oxygen into the primitive atmosphere. Formerly, the cyanobacteria were called blue-green algae and were classified with eukaryotic algae, but now we know that they are prokaryotes. They can have other pigments that mask the color of chlorophyll, so that they appear, for example, not only blue-green but also red, yellow, brown, or black.

Cyanobacterial cells are rather large and range in size from 1 µm to 50 µm in length. They can be unicellular, colonial, or filamentous (Fig. 28.6). Cyanobacteria lack any visible means of locomotion, although some glide when in contact with a solid surface and others oscillate (sway back and forth). Some cyanobacteria have a special advantage because they possess heterocysts, which are thick-walled cells without nuclei, where nitrogen fixation occurs. The ability to photosynthesize and also to fix atmospheric nitrogen (N_2) means they can obtain their nutritional requirements from the environment. They serve as food for heterotrophs in aquatic ecosystems.

Cyanobacteria are common in fresh water, in soil, and on moist surfaces, but they are also found in harsh habitats, such as hot springs. They are symbiotic with a number of organisms, such as liverworts, ferns, and even at times invertebrates like corals. In association with fungi, they form *lichens* that can grow on rocks. Lichens help transform rocks into soil. Other forms of life may then follow. The first moneran fossil dated to be 3.5 billion years old is believed to be a cyanobacterium, and there is evidence that cyanobacteria were the first colonizers of land during the course of evolution.

Cyanobacteria are ecologically important in still another way. If care is not taken in the disposal of nutrient-rich industrial, agricultural, and human wastes, phosphates and nitrates drain into lakes and ponds, resulting in a "bloom" of these organisms. The surface of the water becomes turbid, and light cannot penetrate to lower levels. When a portion of the cyanobacteria die off, the decomposing bacteria use up the available oxygen, causing fishes to die from lack of oxygen.

Cyanobacteria are photosynthesizers that sometimes can also fix atmospheric nitrogen. In association with fungi, they form lichens, which are important soil formers.

Reproduction of Bacteria

The singular circular chromosome of bacteria consists only of double-stranded DNA. Bacteria reproduce asexually by means of **binary fission.** First, the chromosome duplicates; then there are two chromosomes attached to the inside of the plasma membrane. The chromosomes are separated by an elongation of the cell, which pushes the chromosomes apart. Then the plasma membrane grows inward and the cell wall forms, dividing the cell into two daughter cells, each of which now has its own chromosome (Fig. 28.7).

Sexual exchange of DNA occurs among bacteria in three ways. *Conjugation* takes place when the so-called male cell passes DNA to the female cell by way of a sex pilus. *Transformation* occurs when a bacterium binds to and then takes up DNA released into the medium by dead bacteria. During *transduction,* bacteriophages carry portions of DNA from one bacterium to another.

When faced with unfavorable environmental conditions, some bacteria form *endospores* (Fig. 28.8). A portion of the cytoplasm and a copy of the chromosome dehydrate and are then encased by three heavy, protective spore coats. The rest of the bacterial cell deteriorates and the endospore is released. When environmental conditions are again suitable for growth, the endospore absorbs water and grows out of the spore coats. In time, it becomes a typical bacterial cell, capable of reproducing once again by binary fission.

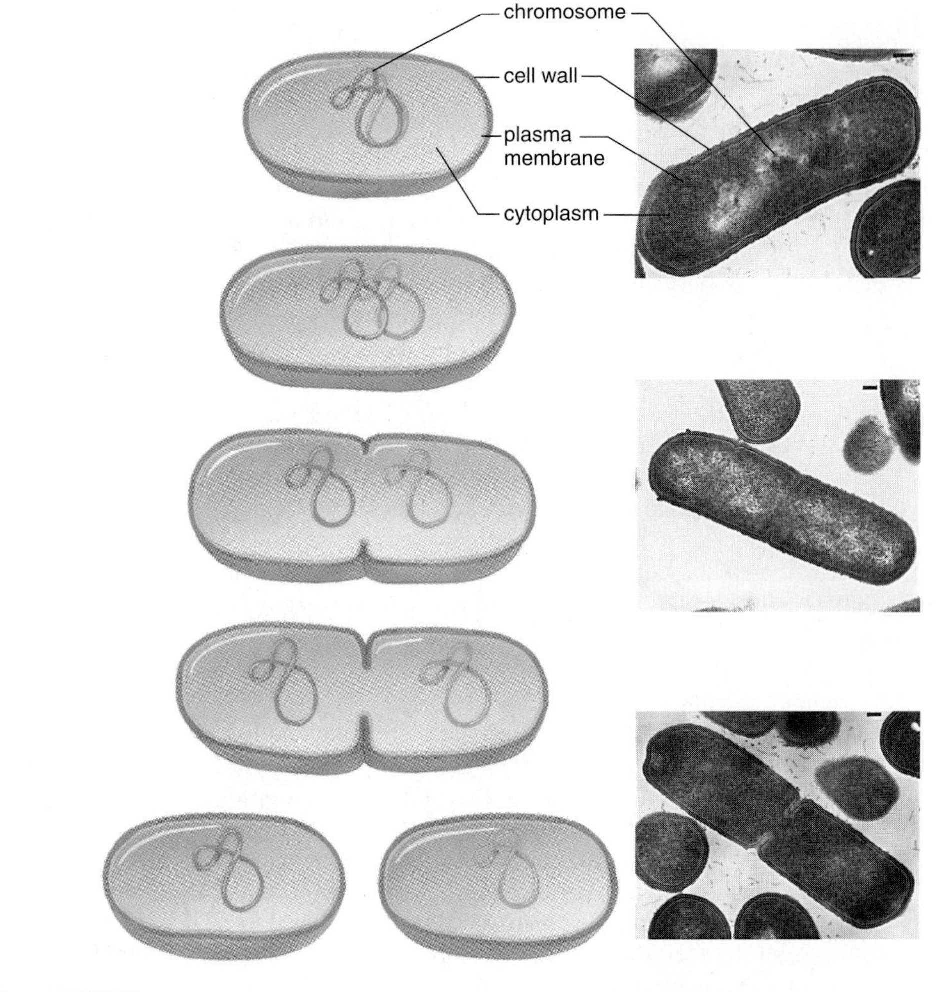

Figure 28.7 Binary fission.
In electron micrographs, it is possible to observe a bacterium dividing to become two bacteria. First DNA replicates, and as the plasma membrane lengthens, the two chromosomes separate. Upon fission, each bacterium has its own chromosome.

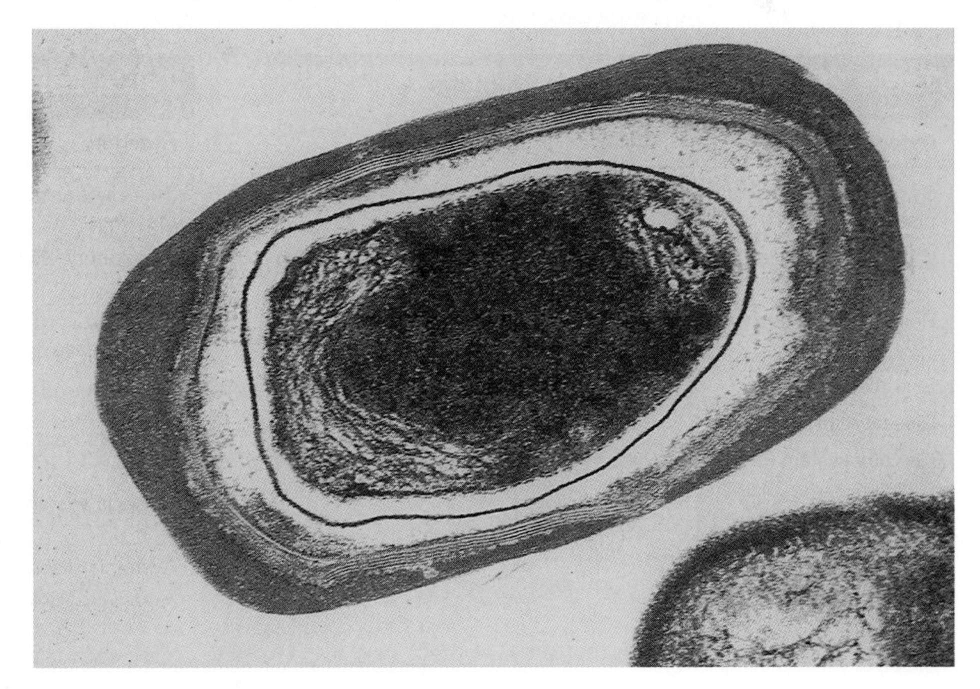

Figure 28.8 The endospore.
An endospore is resistant to extreme environmental conditions. Sterilization, a process that kills all living organisms—even endospores—can be achieved by using an autoclave, a container that maintains steam under pressure. This bacterium, *Bacillus subtilis,* contains an endospore.

Health Focus

Killing Microorganisms

Viruses and bacteria are microbes that cause diseases in humans (Tables 28A and 28B). The development of drugs to kill viruses has lagged far behind the development of those to kill bacteria. Viruses lack most enzymes and, instead, utilize the metabolic machinery of the host cell. Rarely has it been possible to find a drug that successfully interferes with viral reproduction without also interfering with host metabolism. One such drug, however, called vidarabine, was approved in 1978 for treatment of viral encephalitis, an infection of the nervous system. Acyclovir (ACV) seems to be helpful in treating genital herpes, and there are now various drugs (e.g., AZT) for the treatment of AIDS. Since viral drugs are difficult to develop, there is much concern about the possibility of other worldwide epidemics as well as AIDS. The Ebola virus, which begins with flulike symptoms and ends with vomiting and hemorrhaging, is especially feared. Spread by direct contact with a victim's blood or other body fluids, the disease is controllable only via strict hygienic and sanitary controls.

An antibiotic is a drug that selectively kills bacteria. Most antibiotics are produced naturally by soil microorganisms. Penicillin is made by the fungus *Penicillium,* and streptomycin, tetracycline, and erythromycin are all produced by the bacterium *Streptomyces.* Sulfa, a chemotherapeutic agent, is produced in the laboratory. Antibiotics poison bacterial enzymes without harming host enzymes. Penicillin blocks the synthesis of the bacterial cell wall; streptomycin, tetracycline, and erythromycin block protein synthesis; and sulfa prevents the production of a coenzyme. New antibiotics are being developed, but it will be some time before they are ready for general use.

There are problems associated with antibiotic therapy. Some patients are allergic to antibiotics, and the reaction can be fatal. Antibiotics not only kill off disease-causing bacteria, they also reduce the number of beneficial bacteria in the intestinal tract and other locations. These beneficial bacteria hold in check the growth of certain microbes that now begin to flourish. Diarrhea can result, as can a vaginal yeast infection. The use of antibiotics can also prevent natural immunity from occurring, leading to the need for recurring antibiotic therapy. Most important, perhaps, is the growing resistance of certain strains of bacteria to antibiotics. While penicillin used to be 100% effective against hospital strains of *Staphylococcus aureus,* today it is far less effective. Tetracycline and penicillin, long used to cure gonorrhea, now have a failure rate of more than 20% against certain strains of gonococcus. Pulmonary tuberculosis is on the rise, particularly among AIDS patients, the homeless, and the rural poor, and the new strains are resistant to the usual combined antibiotic therapy. A virulent streptococcal infection is now believed to be the cause of the much publicized "flesh-eating" condition more properly called necrotizing fasciitis. About 30% of the people in the U. S. who develop fasciitis usually die.

To keep antibiotics effective, most physicians believe that they should be administered only when absolutely necessary. Some believe that if antibiotic use is not strictly limited, resistant strains of bacteria will completely replace present strains and antibiotic therapy will no longer be effective. They are much opposed to the current practice of adding antibiotics to livestock feed in order to make animals grow fatter. Bacteria that become resistant are easily transferred from animals to humans. Antibiotics have been a boon to humans, but they should be used with care.

Table 28A Viral Diseases in Humans

Category	Disease
Sexually transmitted diseases	AIDS (HIV), genital warts, genital herpes
Childhood diseases	Mumps, measles, chicken pox, German measles
Respiratory diseases	Common cold, influenza, acute respiratory infection
Skin diseases	Warts, fever blisters, shingles
Digestive tract diseases	Gastroenteritis, diarrhea
Nervous system diseases	Poliomyelitis, rabies, encephalitis
Other diseases	Cancer, hepatitis

Table 28B Bacterial Diseases in Humans

Category	Disease
Sexually transmitted diseases	Syphilis, gonorrhea, chlamydia
Respiratory diseases	Strep throat, scarlet fever, tuberculosis, pneumonia, Legionnaires' disease, whooping cough
Skin diseases	Erysipelas, boils, carbuncles, impetigo, infections of surgical or accidental wounds and burns, acne
Digestive tract diseases	Gastroenteritis, food poisoning, dysentery, cholera
Nervous system diseases	Botulism, tetanus, spinal meningitis, leprosy
Systemic diseases	Plague, typhoid fever, diphtheria
Other diseases	Gas gangrene, puerperal fever, toxic shock syndrome, Lyme disease

Classification

Kingdom Protista

Eukaryotic; unicellular organisms and their immediate multicellular descendants; sexual reproduction; flagella and cilia with 9 + 2 microtubules

Algae*

- Phylum Chlorophyta: green algae
- Phylum Phaeophyta: brown algae
- Phylum Chrysophyta: diatoms and allies
- Phylum Dinoflagella: dinoflagellates
- Phylum Euglenophyta: euglenoids
- Phylum Rhodophyta: red algae

Protozoans*

- Phylum Sarcodina: amoebas and allies
- Phylum Ciliophora: ciliates
- Phylum Zoomastigophora: zooflagellates
- Phylum Sporozoa: sporozoa

Slime Molds*

- Phylum Gymnomycota: slime molds

Water Molds*

- Phylum Oomycota: water molds

*Categories which are not used in the classification of organisms, but added here for clarity.

28.3 Kingdom Protista

The protists are eukaryotes; their cells have a nucleus and all of the various organelles. Unicellular organisms are predominant in kingdom Protista, and even the multicellular forms lack the tissue differentiation that is seen in more complex organisms. The protists are grouped according to their mode of nutrition and other characteristics into the categories shown in the classification box. Three different types of life cycles are typical of eukaryotes, and all three are seen in kingdom Protista (Fig. 28.9). In the haplontic cycle, which is typical of protists and fungi, the zygote is the only diploid phase, and it undergoes meiosis to produce haploid spores. In the alternation of generations cycle, which is typical of plants, the sporophyte is a diploid individual that produces spores by meiosis. In the diplontic cycle, which is typical of animals, the diploid adult produces gametes by meiosis, and the only haploid phase consists of the gametes.

Algae

Algae are autotrophic by photosynthesis like plants. However, algae are aquatic, so they do not need to protect the zygote and embryo from drying out. Algae produce the food that maintains communities of organisms in both the oceans and bodies of fresh water. They are commonly named for the type of pigment they contain; therefore, there are green, golden brown, brown, and red algae. All algae contain chlorophyll, but they may also contain other pigments that mask the color of the chlorophyll. Algae are grouped according to their color and biochemical differences, such as the chemistry of the cell wall and the way they store reserve food.

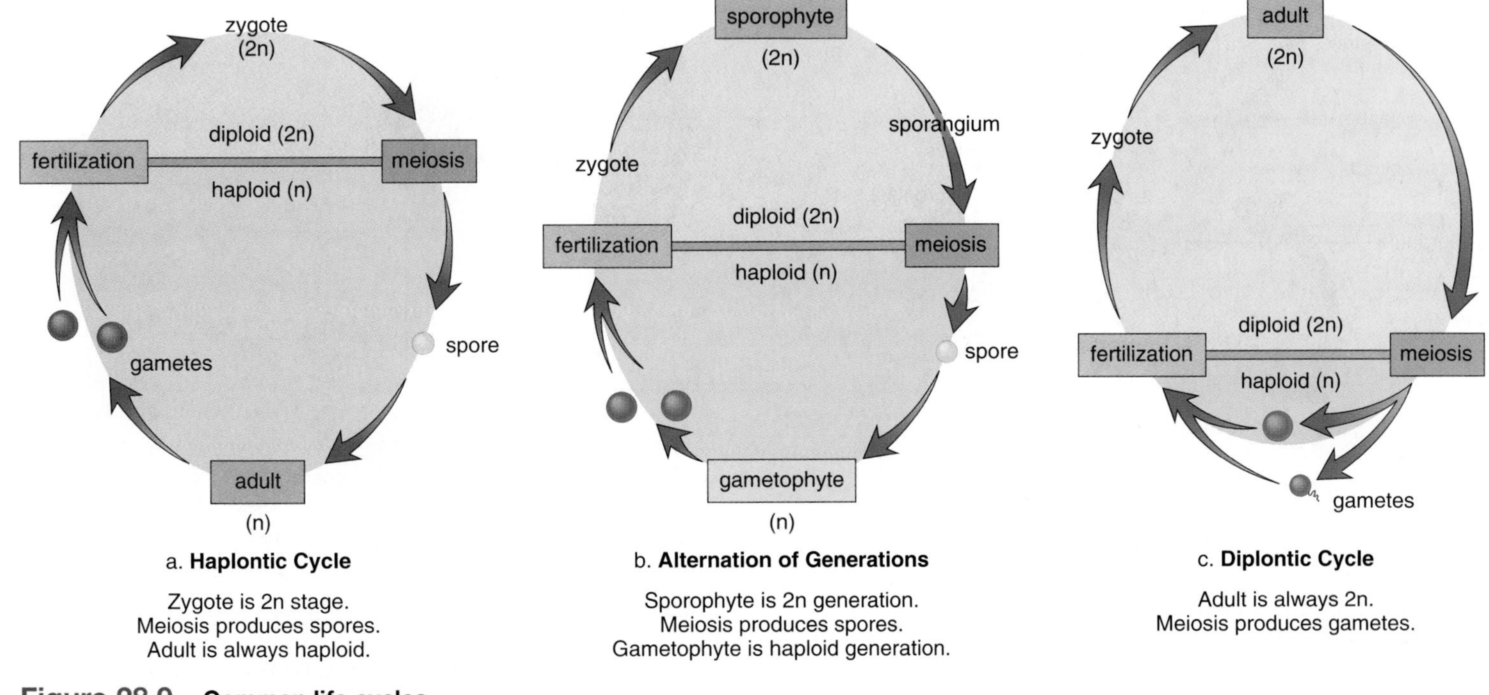

a. **Haplontic Cycle**
Zygote is 2n stage.
Meiosis produces spores.
Adult is always haploid.

b. **Alternation of Generations**
Sporophyte is 2n generation.
Meiosis produces spores.
Gametophyte is haploid generation.

c. **Diplontic Cycle**
Adult is always 2n.
Meiosis produces gametes.

Figure 28.9 Common life cycles.

Green Algae

Green algae (phylum Chlorophyta, 7,000 species) live in the ocean, but they are more likely found in fresh water and can even be found on land, especially if moisture is available. Some, however, have modifications that allow them to live on tree trunks, even in bright sun. Green algae are believed to be closely related to the first plants because both of these groups (1) have a cell wall that contains cellulose, (2) possess chlorophylls *a* and *b,* and (3) store reserve food as starch inside the chloroplast. (Other types of algae store reserve food outside the chloroplast.) Green algae are not always green because some have pigments that give them an orange, red, or rust color.

Unicellular Green Algae *Chlamydomonas* is a unicellular green alga usually less than 25 mm long that has two whiplash flagella (Fig. 28.10). A single, cup-shaped chloroplast contains a pyrenoid, where starch is synthesized. A red-pigmented eyespot (stigma) is sensitive to light and helps the organism detect light, which is necessary for photosynthesis.

When growth conditions are favorable, *Chlamydomonas* reproduces asexually; the adult divides, forming *zoospores* (flagellated spores) that resemble the parent cell. A **spore** is a reproductive cell that develops into a haploid individual when environmental conditions permit. When growth conditions are unfavorable, *Chlamydomonas* reproduces sexually according to the **haplontic life cycle** (see Fig. 28.9*a*). In most species, the gametes are identical and are therefore called *isogametes.* A heavy wall forms around the zygote, and it becomes a resistant zygospore able to survive until conditions are favorable for germination. When a zygospore germinates, it produces four zoospores by meiosis.

Colonial Green Algae *Volvox* is a colony (loose association of cells) in which thousands of flagellated cells are arranged in a single layer surrounding a watery interior. (Each cell of a *Volvox* colony resembles a *Chlamydomonas* cell.) In *Volvox,* the cells cooperate in that the flagella beat in a coordinated fashion. Cells that are specialized for reproduction divide asexually to form a new daughter colony (Fig. 28.11). This daughter colony resides for a time within the parental colony, but then an enzyme that dissolves away a portion of the parental colony allows it to escape. Sexual reproduction among these algae involves heterogametes, that is, a definite sperm and egg.

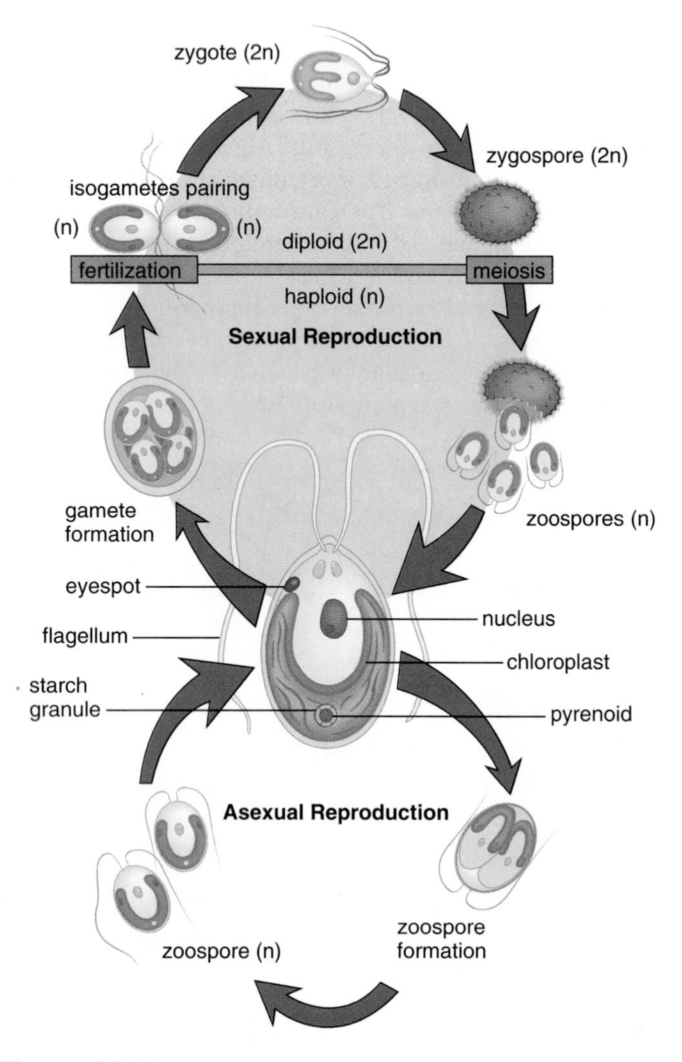

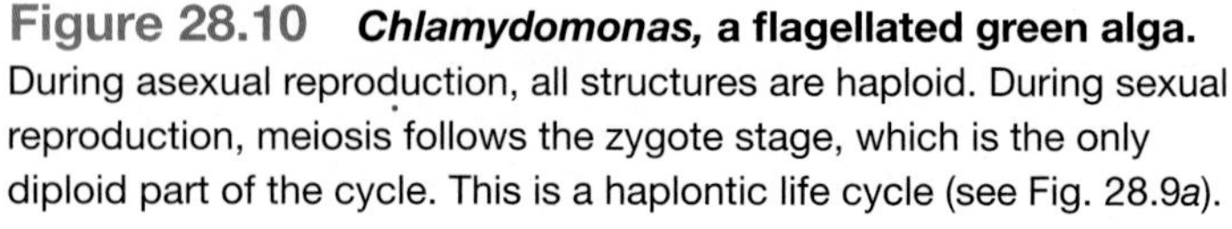

Figure 28.10 ***Chlamydomonas,* a flagellated green alga.** During asexual reproduction, all structures are haploid. During sexual reproduction, meiosis follows the zygote stage, which is the only diploid part of the cycle. This is a haplontic life cycle (see Fig. 28.9*a*).

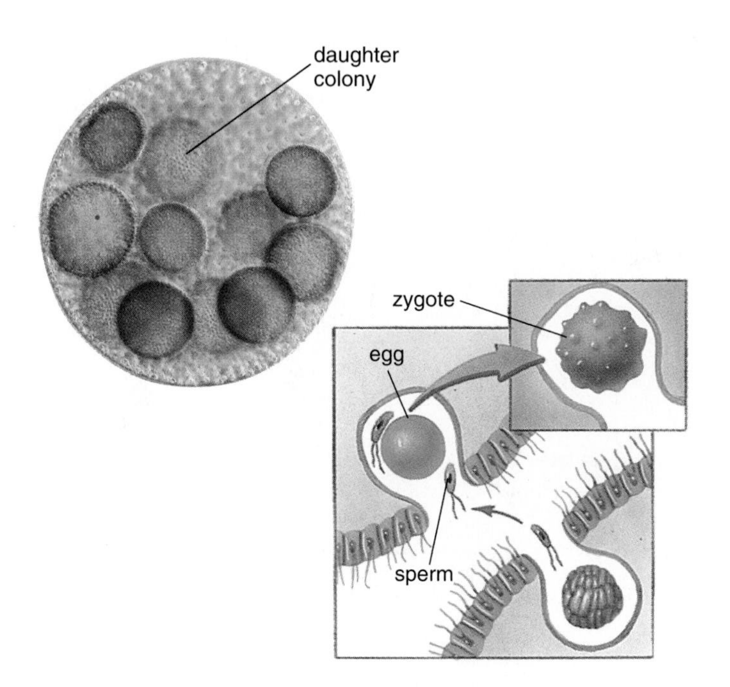

Figure 28.11 ***Volvox,* a colonial green alga.** The adult *Volvox* colony often contains daughter colonies, which are asexually produced by specialized cells. During sexual reproduction, colonies produce a definite sperm and egg.

Filamentous Green Algae **Filaments** are end-to-end chains of cells that form after cell division occurs in only one plane. *Spirogyra,* a filamentous green alga, is found in green masses on the surfaces of ponds and streams. It has chloroplasts that are ribbonlike and are arranged in a spiral within the cell (Fig. 28.12). **Conjugation,** the temporary union of two individuals during which there is an exchange of genetic material, occurs during sexual reproduction. The two filaments line up next to each other, and the cell contents of one filament move into the cells of the other filament, forming diploid zygotes. These zygotes survive the winter, and in the spring they undergo meiosis to produce new haploid filaments.

Multicellular Green Algae Multicellular *Ulva* is commonly called sea lettuce because of its leafy appearance (Fig. 28.13). The thallus is two cells thick and can be a meter long. *Ulva* has an **alternation of generations** life cycle (see Fig. 28.9*b*) like that of plants, except that both generations look exactly alike, the gametes look alike (isogametes), and the spores are flagellated.

Green algae can be unicellular, colonial, filamentous, or multicellular. During sexual reproduction, the zygote usually undergoes meiosis and the adult is always haploid. *Ulva* has an alternation of generations like plants do.

Brown and Golden Brown Algae

Brown and golden brown algae have chlorophylls *a* and *c* in their chloroplasts and a type of carotenoid pigment (fucoxanthin) that gives them their color.

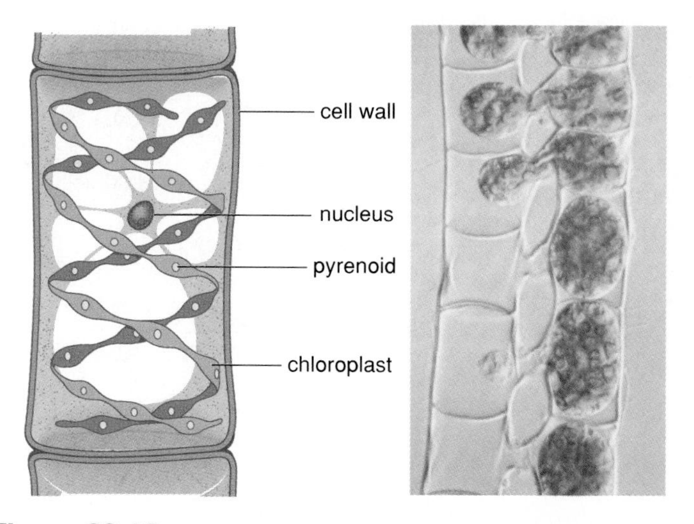

Figure 28.12 ***Spirogyra,* a filamentous green alga.**
During conjugation the cell contents of one filament enter the cells of another filament. Zygote formation follows.

Brown algae (phylum Phaeophyta, 1,500 species) range from small forms with simple filaments to large blade forms between 50 meters and 100 meters in length (Fig. 28.14). Large brown algae, often called *seaweeds,* are observed along the rocky shoreline in the north temperate zone. These plants are firmly anchored by holdfasts, and when the tide is in, their broad flattened blades are buoyed by air bladders. When the tide is out, they do not dry up because their cell walls contain a mucilaginous, water-retaining material. Most brown algae have an alternation of generations life cycle, but some species of rockweed (*Fucus*) are unique in that they have the **diplontic life cycle** (see Fig. 28.9*c*), in which meiosis produces gametes and the adult is always diploid, as in animals.

Brown algae provide food and habitat for marine organisms, and in several parts of the world they have been har-

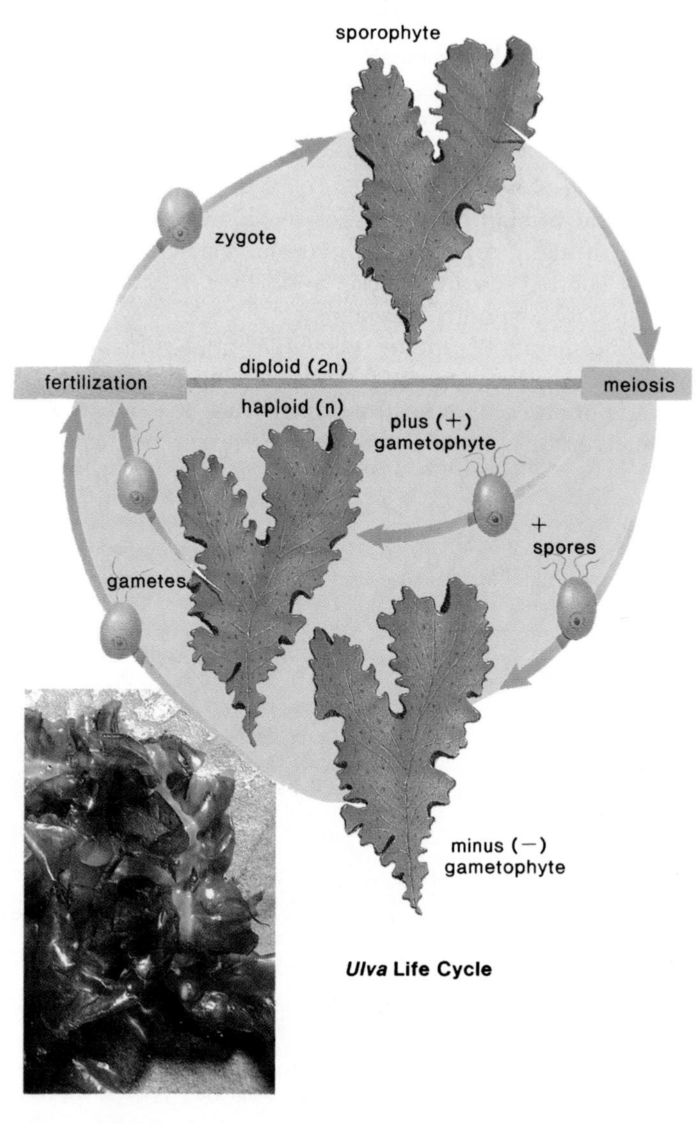

Figure 28.13 ***Ulva,* a multicellular green alga.**
The sporophyte and gametophyte have the same appearance, and the gametophyte produces isogametes. *Ulva* has an alternation of generations life cycle, as do plants (see Fig. 28.9*b*).

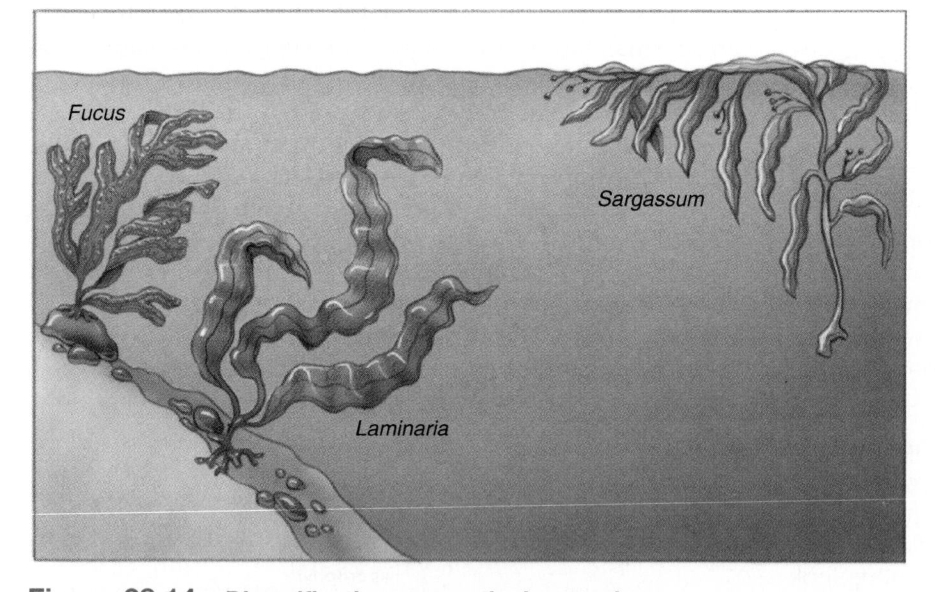

Figure 28.14 Diversification among the brown algae.
Laminaria and *Fucus* are seaweeds known as kelps. They live along rocky coasts of the north temperate zone. *Sargassum*, the other brown alga shown, lives at sea where floating masses form a home for many organisms.

vested for human food and for fertilizer. They are also a source of algin, a pectinlike material that is added to ice cream, sherbet, cream cheese, and other products to give them a stable, smooth consistency.

Diatoms are a type of unicellular golden brown algae (phylum Chrysophyta, 11,000 species). The structure of a diatom is often compared to a box because the cell wall has two halves, or valves, with the larger valve acting as a "lid" for the smaller valve (Fig. 28.15*a*). When diatoms reproduce, each receives only one old valve. The new valve fits inside the old one.

The cell wall of a diatom has an outer layer of silica, a common ingredient of glass. The valves are covered with a great variety of striations and markings, which form beautiful patterns when observed under the microscope. Diatoms are among the most numerous of all unicellular algae in the oceans. As such, they serve as an important source of food for other organisms. In addition, they produce a major portion of earth's oxygen supply. In ancient times, diatoms were also present in astronomical numbers. Their remains, raised above sea level by geological upheavals, are now mined as diatomaceous earth for use as filtering agents, soundproofing materials, and scouring powders.

Dinoflagellates

Many **dinoflagellates** (phylum Dinoflagella, 1,000 species) are bounded by protective cellulose plates (Fig. 28.15*b*). Most have two flagella; one is free, but the other is located in a transverse groove. The beating of the flagella causes the organism to spin like a top. Occasionally, when surface waters are warm and nutrients are high, there are so many of these unicellular organisms in the ocean that they cause a condition called "red tide." Toxins in red tides cause widespread fish kills and can paralyze humans who eat shellfish that have fed on the dinoflagellates.

Dinoflagellates are an important source of food for small animals in the ocean. They also live as symbiotes within the bodies of some invertebrates. For example, because corals usually contain large numbers of dinoflagellates, corals grow much faster than they would otherwise.

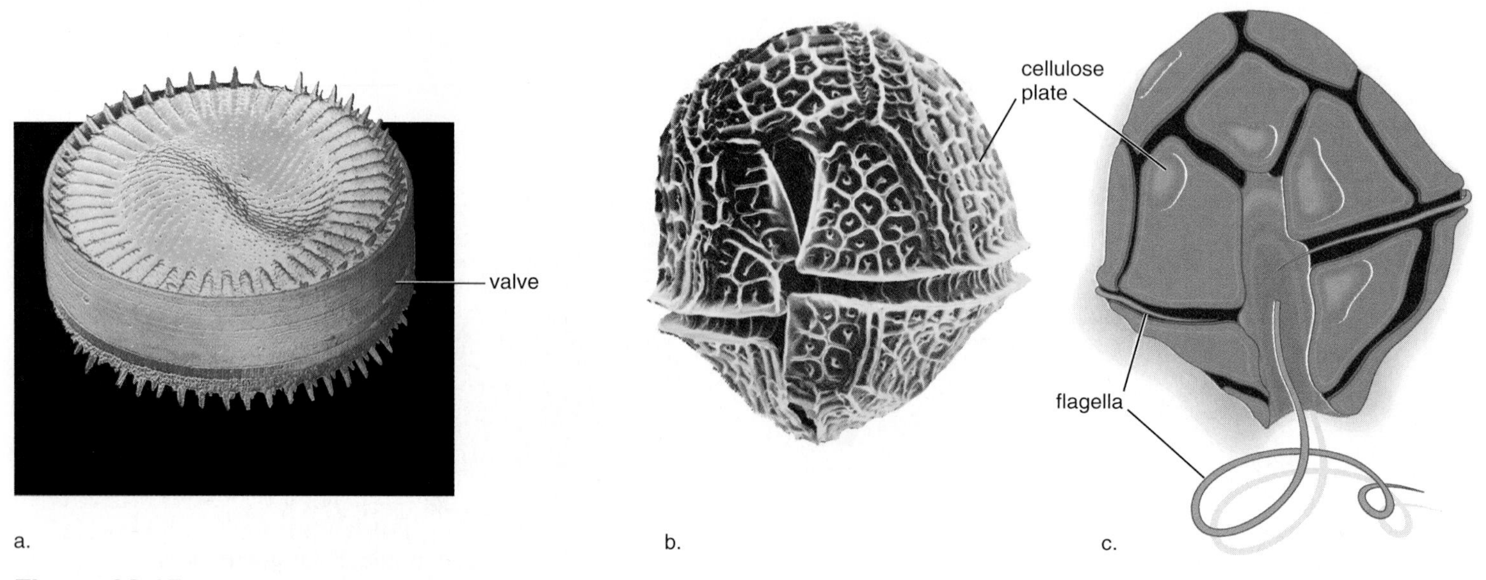

Figure 28.15 Diatoms and dinoflagellates.
a. Diatoms may be variously colored, but their chloroplasts contain a unique golden brown pigment in addition to chlorophylls *a* and *c*. The beautiful pattern results from markings on the silica-embedded wall. **b.** Dinoflagellates have cellulose plates; these belong to *Gonyaulax*, the dinoflagellate that contains a red pigment and is responsible for occasional "red tides."

Euglenoids

Euglenoids (phylum Euglenophyta, 1,000 species) are small (10–500 μm) freshwater unicellular organisms that typify the problem of classifying protists. One-third of all genera have chloroplasts; the rest do not. Those that lack chloroplasts ingest or absorb their food. Euglenoids grown in the absence of light have been known to lose their chloroplasts and become heterotrophic. The chloroplasts are surrounded by three rather than two membranes. The pyrenoid produces an unusual type of carbohydrate polymer (paramylon) not seen in green algae.

Euglenoids have two flagella, one of which typically is much longer than the other and projects out of an anterior vase-shaped invagination (Fig. 28.16). It is called a tinsel flagellum because it has hairs on it. Near the base of this flagellum is an eyespot, which shades a photoreceptor for detecting light. Because euglenoids are bounded by a flexible *pellicle* composed of protein strips lying side by side, they can assume different shapes as the underlying cytoplasm undulates and contracts. As in certain protozoans, there is a contractile vacuole for ridding the body of excess water. Euglenoids reproduce by longitudinal cell division, and sexual reproduction is not known to occur.

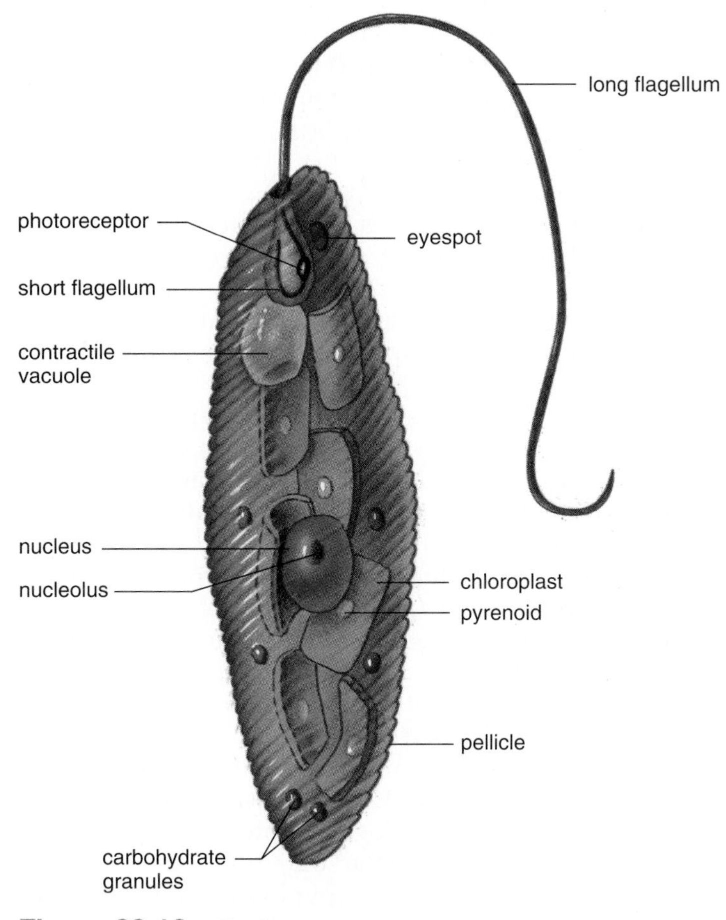

Figure 28.16 ***Euglena.***
A very long flagellum propels the body, which is enveloped by a flexible pellicle. A photoreceptor shaded by an eyespot allows *Euglena* to find light, after which photosynthesis can occur in the numerous chloroplasts. Pyrenoids synthesize a reserve carbohydrate, which is stored in the chloroplasts and also in the cytoplasm.

Euglenoids have both plant- and animal-like characteristics. They have chloroplasts but lack a cell wall and swim by flagella.

Red Algae

Like the brown algae, red algae (phylum Rhodophyta, 4,000 species) are multicellular, but they live chiefly in warmer seawater, growing in both shallow and deep waters. Red algae are usually much smaller and more delicate than the brown algae, although they can be up to a meter long. Some forms of red algae are simple filaments, but more often they are complexly branched, with the branches having a feathery, flat, or expanded ribbonlike appearance (Fig. 28.17). Coralline algae are red algae that have cell walls impregnated with calcium carbonate. In some instances, they contribute as much to the growth of coral reefs as do coral animals.

Like brown algae, red algae are seaweeds of economic importance. The mucilaginous material in the cell walls of certain genera of red algae is a source of agar used commercially to make capsules for vitamins and drugs, as a material for making dental impressions, and as a base for cosmetics. In the laboratory, agar is a culture medium for bacteria. When purified, it becomes the gel for electrophoresis, a procedure that separates proteins and nucleotides. Agar is also used in food preparation—as an antidrying agent for baked goods and to make jellies and desserts set rapidly.

Many red algae have filamentous branches or are multicellular.

Figure 28.17 **Red alga.**
Red algae are smaller and more delicate than brown algae.

Protozoans

Protozoans are typically heterotrophic, motile, unicellular organisms of small size (2–1,000 μm). They are not animals because animals in the classification used by this text are multicellular and undergo embryonic development.

Protozoans usually live in water, but they can also be found in moist soil or inside other organisms. Some protozoans engulf whole food and are termed *holozoic;* others are saprotrophic, and they absorb nutrient molecules across the plasma membrane. Still others are parasitic and are responsible for several significant human infections.

Most protozoans are unicellular, but they should not be considered simple organisms. Each cell alone must carry out all the functions performed by specialized tissues and organs in more complex organisms. They have organelles for purposes we have not seen before. Their food is digested inside food vacuoles, and freshwater protozoans have "contractile" vacuoles for the elimination of water. Although asexual reproduction involving binary fission and mitosis is the rule, many protozoans also reproduce sexually during some part of their life cycle. The protozoans we will study can be placed in four groups according to their type of locomotor organelle:

Name	Locomotion	Example
Amoeboids	Pseudopods	*Amoeba*
Ciliates	Cilia	*Paramecium*
Zooflagellates	Flagella	*Trypanosoma*
Sporozoa	No locomotion	*Plasmodium*

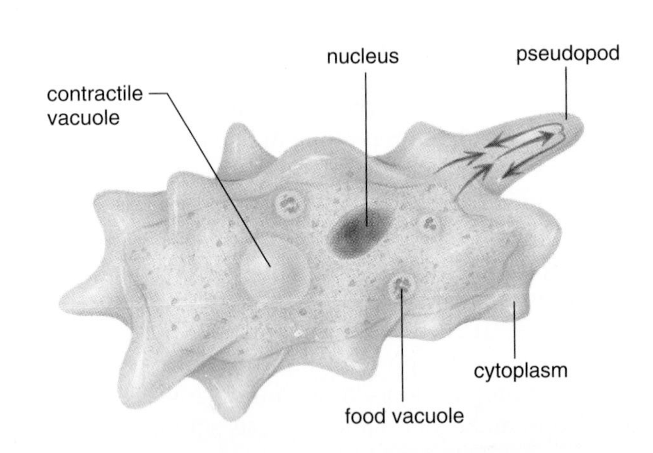

a.

b.

Figure 28.18 Amoeboid protozoans.
a. *Amoeba proteus* is common in freshwater ponds. Bacteria and other microorganisms are digested in food vacuoles, and contractile vacuoles rid the body of excess water. **b.** Pseudopods of a live foraminiferan project through holes in the calcium carbonate shell. These shells were so numerous that they became a large part of the White Cliffs of Dover when a geological upheaval occurred.

Amoeboids

The amoeboids (phylum Sarcodina, 40,000 species) are protists that move and engulf their prey with **pseudopods.** *Amoeba proteus* is a commonly studied freshwater member of this group (Fig. 28.18*a*). When amoeboids feed, they **phagocytize;** the pseudopods surround and engulf the prey, which may be algae, bacteria, or other protozoans. Digestion then occurs within a food vacuole. Some white blood cells in humans are amoeboid, and they phagocytize debris, parasites, and worn-out cells. Freshwater amoeboids, including *Amoeba proteus,* have contractile vacuoles where excess water from the cytoplasm collects before the vacuole appears to "contract," releasing the water through a temporary opening in the plasma membrane.

Entamoeba histolytica is a parasite that can infect the human intestine and cause amoebic dysentery. Complications arise when this parasite invades the intestinal lining and reproduces there. If the parasites enter the body proper, liver and brain impairment can be fatal.

The *foraminifera,* which are largely marine, have an external calcareous shell (made up of calcium carbonate) with foramina, holes through which long, thin pseudopods extend (Fig. 28.18*b*). The pseudopods branch and join to form a net where the prey is digested. Foraminifera live in the sediment of the ocean floor in incredible numbers—there may be as many as 50,000 shells in a single gram of sediment. Deposits for millions of years, followed by a geological upheaval, formed the White Cliffs of Dover along the southern coast of England. Also, the great Egyptian pyramids are built of foraminiferan limestone. Today, oil geologists look for foraminifera in sedimentary rock as an indicator of organic deposits, which are necessary for the formation of oil.

Ciliates

The **ciliates** (phylum Ciliophora, 8,000 species) such as those in the genus *Paramecium* are the most complex of the protozoans (Fig. 28.19). Hundreds of cilia, which beat in a coordinated rhythmic manner, project through tiny holes in a semirigid outer covering, or pellicle. Numerous oval capsules lying in the cytoplasm just beneath the pellicle contain *trichocysts.* Upon mechanical or chemical stimulation, trichocysts discharge long, barbed threads, useful for defense and for capturing prey. When a paramecium feeds, food is swept down a gullet, below which food vacuoles form. Following digestion, the soluble nutrients are absorbed by the cytoplasm, and the indigestible residue is eliminated at the anal pore.

During asexual reproduction, ciliates divide by transverse binary fission. Ciliates have two types of nuclei: a large *macronucleus* and one or more small *micronuclei.* The macronucleus controls the normal metabolism of the cell; during sexual reproduction, two ciliates exchange a micronucleus.

The diversity of ciliates is quite remarkable. Barrel-shaped didinia expand to consume paramecia much larger than themselves. Suctoria have tentacles they use like straws to suck their prey dry. *Stentor* looks like a blue vase decorated with stripes.

Zooflagellates

Protozoans that move by means of flagella are called **zooflagellates** (phylum Zoomastigophora, 1,500 species) to distinguish them from unicellular algae that also have flagella. Many zooflagellates enter into symbiotic relationships. *Trichonympha collaris* lives in the gut of termites; it contains a bacterium that enzymatically converts the cellulose of wood to soluble carbohydrates that are easily digested by the insect. *Giardia lamblia,* whose cysts are transmitted through contaminated water, causes severe diarrhea. *Trichomonas vaginalis,* a sexually transmitted organism, infects the vagina and urethra of women and the prostate, seminal vesicles, and urethra of men. A **trypanosome,** *Trypanosoma brucei,* transmitted by the bite of the tsetse fly, is the cause of African sleeping sickness (Fig. 28.20). The white blood cells in an infected animal accumulate around the blood vessels leading to the brain and cut off circulation. The lethargy characteristic of the disease is caused by an inadequate supply of oxygen to the brain.

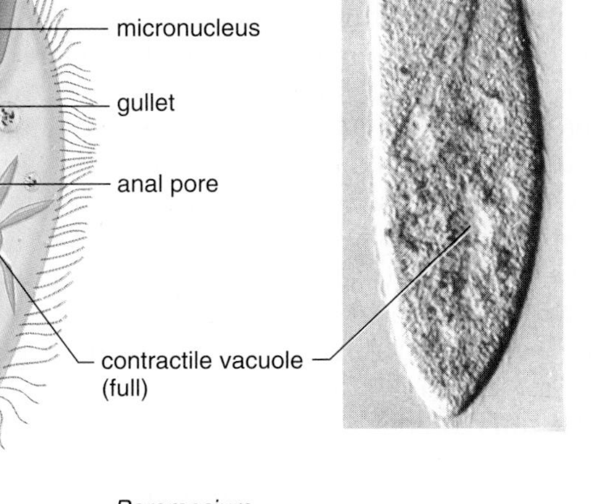

Figure 28.19 Ciliated protozoans.
Structure of *Paramecium*, adjacent to an electron micrograph. Ciliates are the most complex of the protozoans. Note the oral groove, the gullet, and the anal pore.

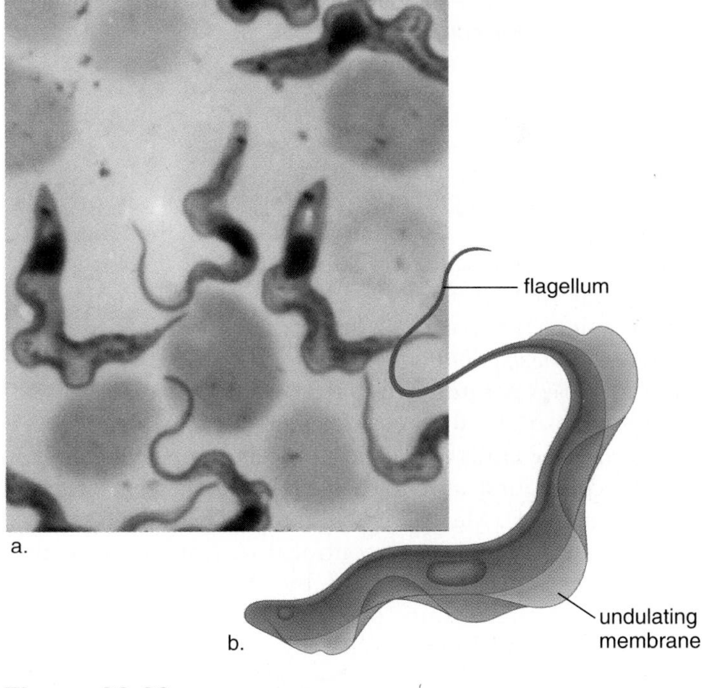

Figure 28.20 Zooflagellates.
a. Photograph of *Trypanosoma brucei*, the cause of African sleeping sickness, among red blood cells. **b.** The drawing shows its general structure.

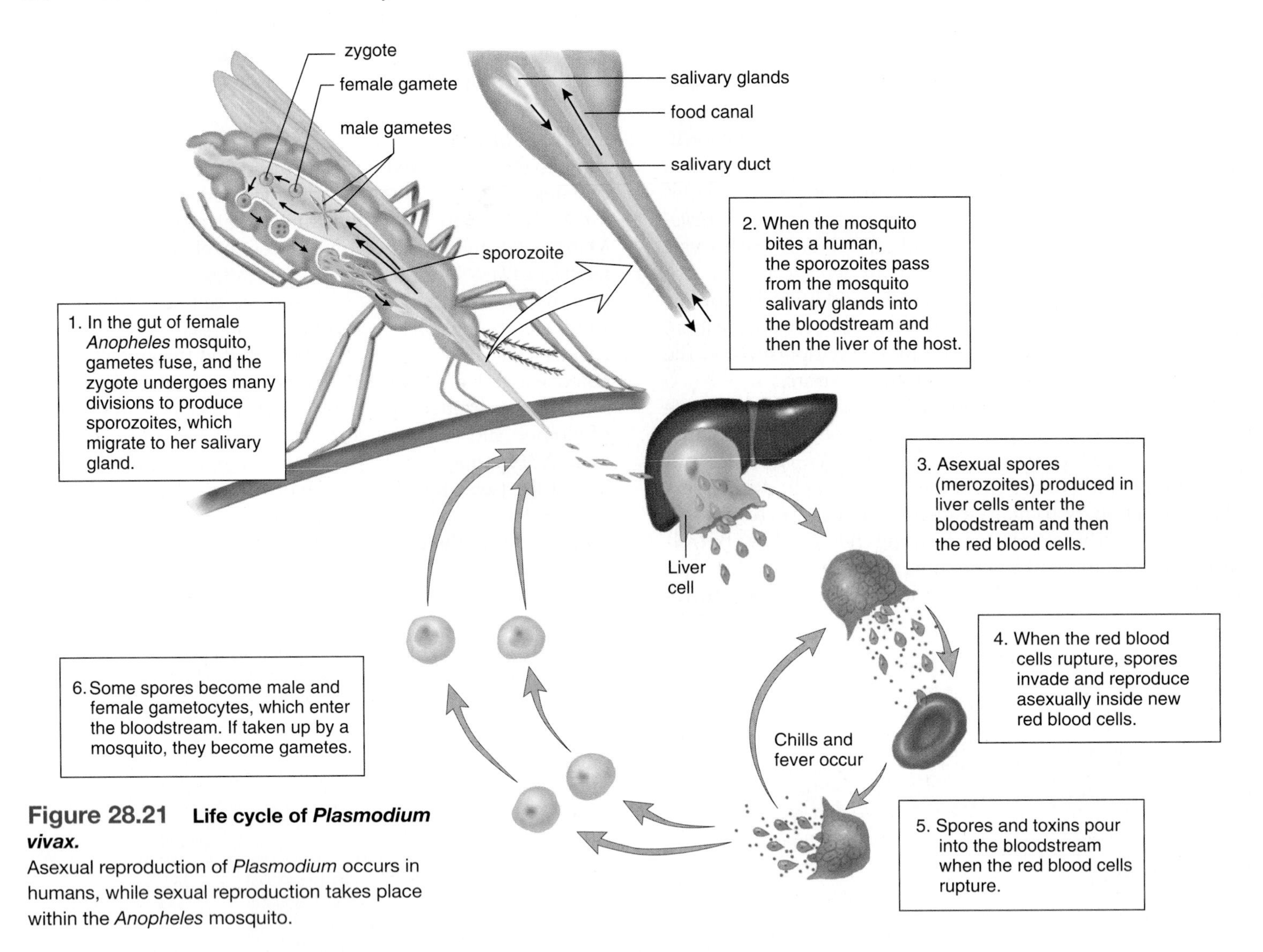

Figure 28.21 Life cycle of *Plasmodium vivax.*
Asexual reproduction of *Plasmodium* occurs in humans, while sexual reproduction takes place within the *Anopheles* mosquito.

Sporozoa

Sporozoa (phylum Sporozoa, 3,600 species) are nonmotile parasites of animals. Their name recognizes that these organisms form spores at some point in their life cycle.

Pneumocystis carinii causes the type of pneumonia seen primarily in AIDS patients. The most widespread human parasite is *Plasmodium vivax,* the cause of one type of malaria. When a human is bitten by an infected female *Anopheles* mosquito, the parasite eventually invades the red blood cells. The chills and fever of malaria appear when the infected cells burst and release toxic substances into the blood (Fig. 28.21). Malaria is still a major killer of humans, despite extensive efforts to control it. A resurgence of the disease was caused primarily by the development of insecticide-resistant strains of mosquitoes and by parasites resistant to current antimalarial drugs.

The protozoans are the animal-like protists—they ingest their food and are motile. Protozoans are classified according to the type of locomotor organelle employed.

Slime Molds and Water Molds

Slime molds (phylum Gymnomycota, 560 species) might look like molds, but their vegetative state is amoeboid, whereas fungi are filamentous. Fungi are saprotrophic, whereas slime molds are heterotrophic by ingestion. When conditions are unfavorable to growth, however, slime molds produce and release spores that are resistant to environmental extremes. Fungi also produce such spores.

Usually *plasmodial slime molds* exist as a **plasmodium,** a diploid multinucleated cytoplasmic mass enveloped by a slime sheath that creeps along, phagocytizing decaying plant material in a forest or agricultural field (Fig. 28.22). Under unfavorable conditions, the plasmodium forms sporangia, structures that produce spores which are dispersed by the wind. The spores germinate to produce gametes that join to form a zygote. This zygote begins the cycle again. *Cellular slime molds,* as you might expect, exist as individual amoeboid cells. Each lives by phagocytizing bacteria and yeast. As the food supply runs out, the cells release a chemical that causes them to aggregate into a pseudoplasmodium that produces spores within sporangia.

Water molds (phylum Oomycota, 580 species) live in the water, where they parasitize fish, forming furry growths on their gills. Others live on land and parasitize insects and plants; a water mold was responsible for the 1840s potato famine in Ireland. Most water molds, like fungi, are saprotrophic and have a filamentous body, but they have the diplontic life cycle (see Fig. 28.9*c*), whereas fungi have the haplontic cycle (see Fig. 28.9*a*).

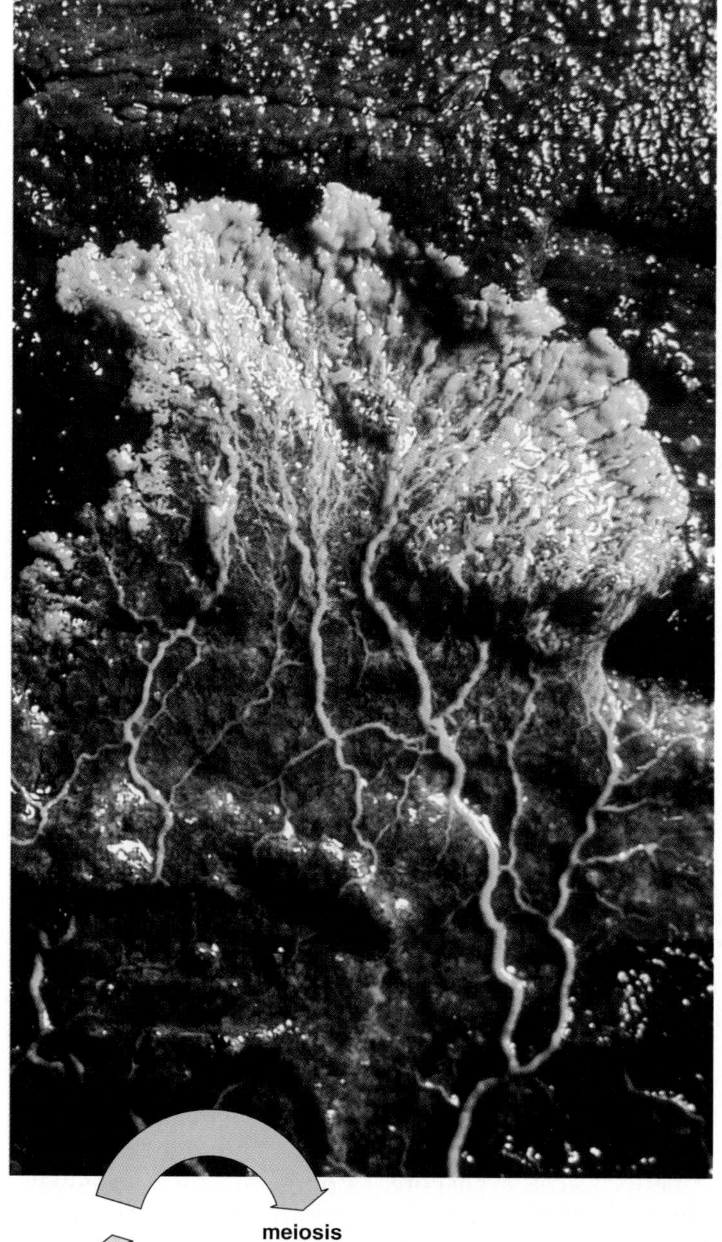

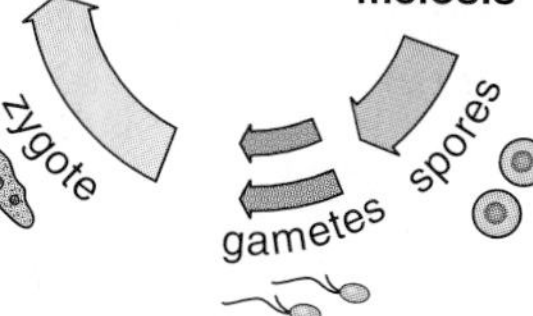

Figure 28.22 Plasmodium and life cycle of a yellow slime mold, *Hemitrichia stipitata*.

28.4 Kingdom Fungi

Fungi are multicellular eukaryotes that are heterotrophic by absorption. They send out digestive enzymes into the immediate environment, and then, when organic matter is broken down, they absorb nutrient molecules. Like bacteria, most fungi are *saprotrophic decomposers* that break down the waste products and dead remains of plants and animals. Some fungi parasitize both plants and animals; in humans, they cause ringworm, athlete's foot, and yeast infections.

Although yeast are unicellular fungi, the body of a fungus is usually a multicellular structure known as a mycelium. A **mycelium** is a network of filaments called hyphae (sing., **hypha**):

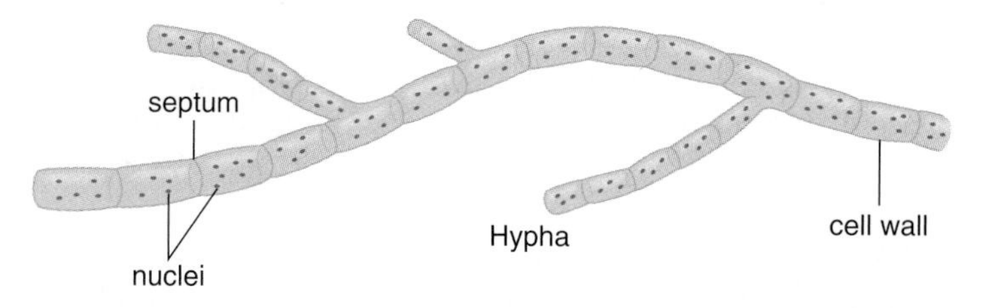

Classification

Kingdom Fungi

Multicellular eukaryote; heterotrophic by absorption; lack flagella; nonmotile spores form during both asexual and sexual reproduction

Division Zygomycota: zygospore fungi

Soil and dung molds, black bread molds (*Rhizopus*).

Division Ascomycota: sac fungi

Many small wood-decaying fungi, yeasts (*Saccaromyces*), molds (*Neurospora*), morels, cup fungi, truffles; plant parasites: powdery mildews, ergots.

Division Basidiomycota: club fungi

Mushrooms, stinkhorns, puffballs, bracket and shelf fungi, coral fungi; plant parasites: rusts, smuts.

Division Deuteromycota: imperfect fungi (i.e., means of sexual reproduction not known)

Athlete's foot, ringworm, candidiasis.

Fungal cells are quite different from plant cells not only by lacking chloroplasts but also by having a cell wall that contains *chitin* and not cellulose. Chitin is a polymer of glucose, but each glucose molecule has an amino group attached to it. (Chitin is also found in the external skeleton of insects and all arthropods.) How can a nonmotile terrestrial organism ensure that the species will be dispersed to new locations? Fungi produce nonflagellate spores during both sexual and asexual reproduction, which are dispersed by the wind.

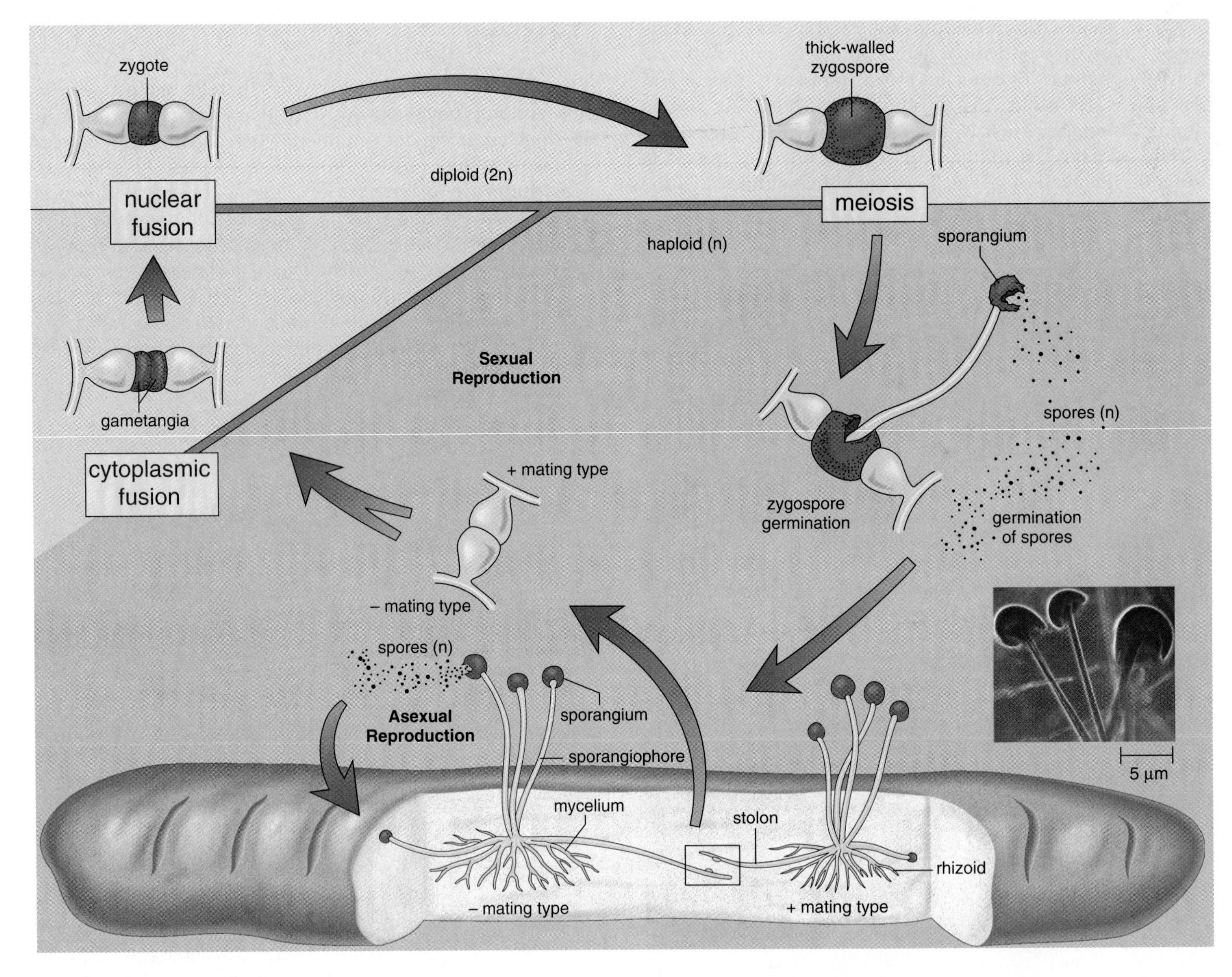

Figure 28.23 Black bread mold, *Rhizopus stolonifer.*
Asexual reproduction is the norm. During sexual reproduction, two compatible mating types make contact: first gametangia fuse, and then nuclei fuse. The zygospore is a resting stage that can survive unfavorable growing conditions. Meiosis occurs when the zygote germinates and spores are released from a sporangium.

Zygospore Fungi

The zygospore fungi (phylum Zygomycota, 600 species) live off plant and animal remains in the soil and also bakery goods in our kitchens. Some, however, are parasites of small soil protists or worms, and even insects such as the housefly.

In *Rhizopus stolonifer,* black bread mold, stolons are horizontal hyphae that exist on the surface of the bread; rhizoids grow into the bread, and *sporangiophores* are stalks that bear sporangia (Fig. 28.23). A **sporangium** is a capsule that produces spores, more properly called sporangiospores. During asexual reproduction all structures involved are haploid; during sexual reproduction there is a diploid zygospore for which the phylum is named. Hyphae of opposite mating types, termed plus (+) and minus (−), grow toward each other until they touch. *Gametangia* form and merge, producing a large cell in which nuclei of the two mating types pair and then fuse. A thick wall develops around the cell, which is now called a **zygospore.** Upon germination, sporangiophores develop, and many spores are produced by meiosis. By now, you will have no trouble in identifying this cycle as a haplontic life cycle (see Fig. 28.9*a*).

Zygospore fungi produce spores within sporangia. During sexual reproduction, a zygospore forms prior to meiosis and production of spores.

a.

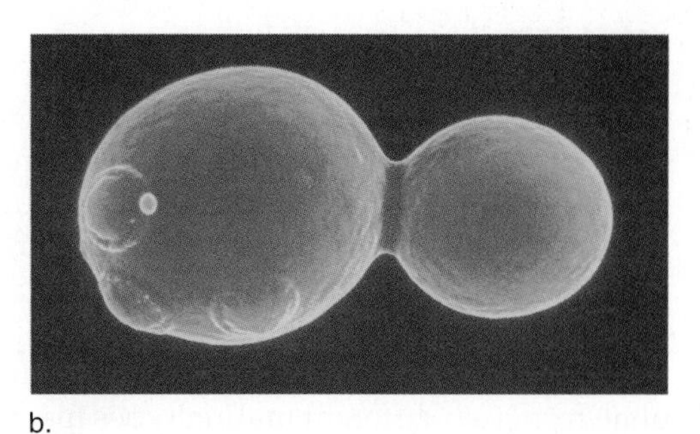

b.

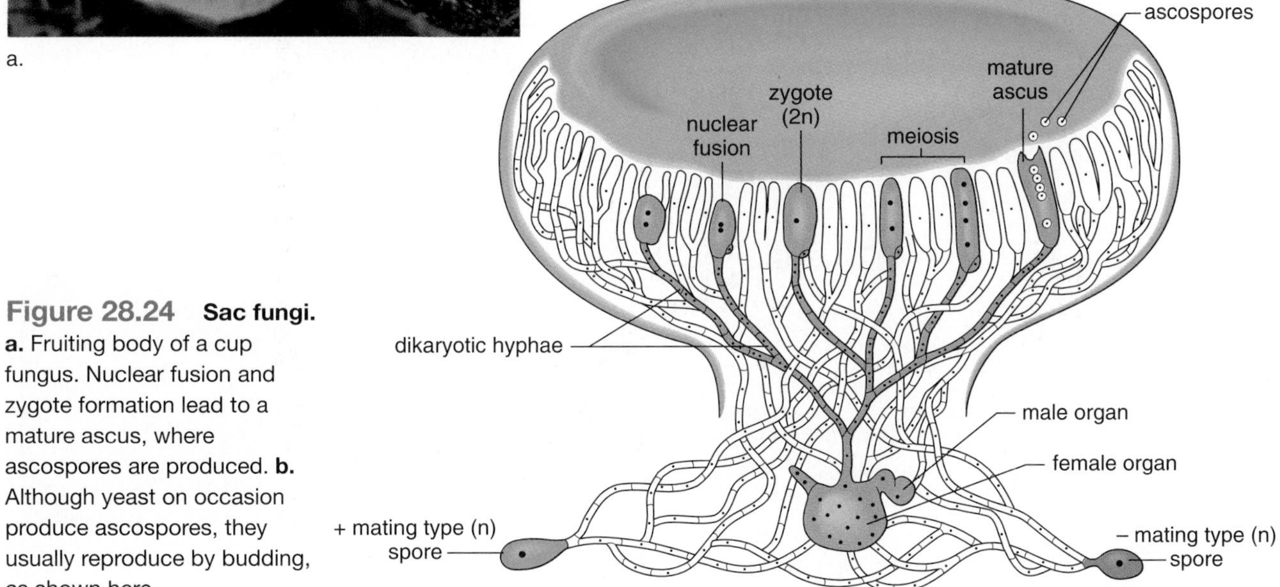

Figure 28.24 Sac fungi. **a.** Fruiting body of a cup fungus. Nuclear fusion and zygote formation lead to a mature ascus, where ascospores are produced. **b.** Although yeast on occasion produce ascospores, they usually reproduce by budding, as shown here.

Sac Fungi

Sac fungi (division Ascomycota, 30,000 species) include red bread molds (e.g., *Neurospora*) and cup fungi (Fig. 28.24*a*). Also, morels and truffles are sac fungi highly prized as gourmet delicacies. A large number of sac fungi are parasitic on plants; powdery mildews grow on leaves, as do leaf curl fungi; chestnut blight and Dutch elm disease destroy the trees named. Ergot is a parasitic sac fungus that infects rye and (less commonly) other grains.

The division name for sac fungi, Ascomycota, refers to the **ascus,** a fingerlike sac that develops after hyphae from two mating strains merge, producing dikaryotic (each cell has two nuclei) hyphae. In an ascus, a zygote forms and undergoes meiosis to produce eight haploid nuclei that become eight ascospores.

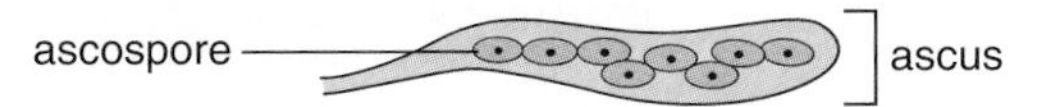

The asci are usually surrounded and protected by sterile hyphae within a **fruiting body**. Asexual reproduction, which is the norm among ascomycetes, involves the production of spores called **conidiospores** (see Fig. 28.26).

Yeasts

Yeasts are unicellular sac fungi that reproduce asexually either by mitosis or by budding (Fig. 28.24*b*). When yeasts ferment, they produce ethanol and carbon dioxide. In the wild, yeasts grow on fruits, and historically the yeasts already present on grapes were used to produce wine. Today selected yeasts are added to relatively sterile grape juice in order to make wine. Also, yeasts are added to prepared grains to make beer. Both the ethanol and the carbon dioxide are retained for beers and sparkling wines; carbon dixoide is released for still wines. In baking, the carbon dioxide given off by yeast is the leavening agent that causes bread to rise.

Yeasts are serviceable to humans in another way. They have become the material of choice in genetic engineering experiments requiring a eukaryote. *Escherichia coli,* the usual experimental material, is a prokaryote and does not function during protein synthesis as a eukaryote would.

When sac fungi reproduce sexually, they produce ascospores within asci, usually within a fruiting body.

Club Fungi

Club fungi (phylum Basidiomycota, 16,000 species) include shelf or bracket fungi on dead trees and mushrooms in lawns and forests. Less well known are puffballs, bird's nest fungi, and stinkhorns. These structures are all fruiting bodies that contain **basidia,** club-shaped structures that give this phylum its name.

Club fungi usually reproduce sexually (Fig. 28.25). Hyphae from two different mating types meet, and cytoplasmic fusion occurs. The resulting dikaryotic mycelium periodically produces fruiting bodies, which are composed of tightly packed hyphae. The fruiting body of a mushroom has a stalk and a cap. The cap of a gilled mushroom contains radiating lamellae lined by basidia where nuclear fusion, meiosis, and spore production occur. A basidium has four projections into which cytoplasm and a haploid nucleus enter. The spores are windblown and germinate to give a haploid mycelium.

Club fungi usually reproduce sexually. The dikaryotic stage is prolonged and periodically produces fruiting bodies where spores are produced in basidia.

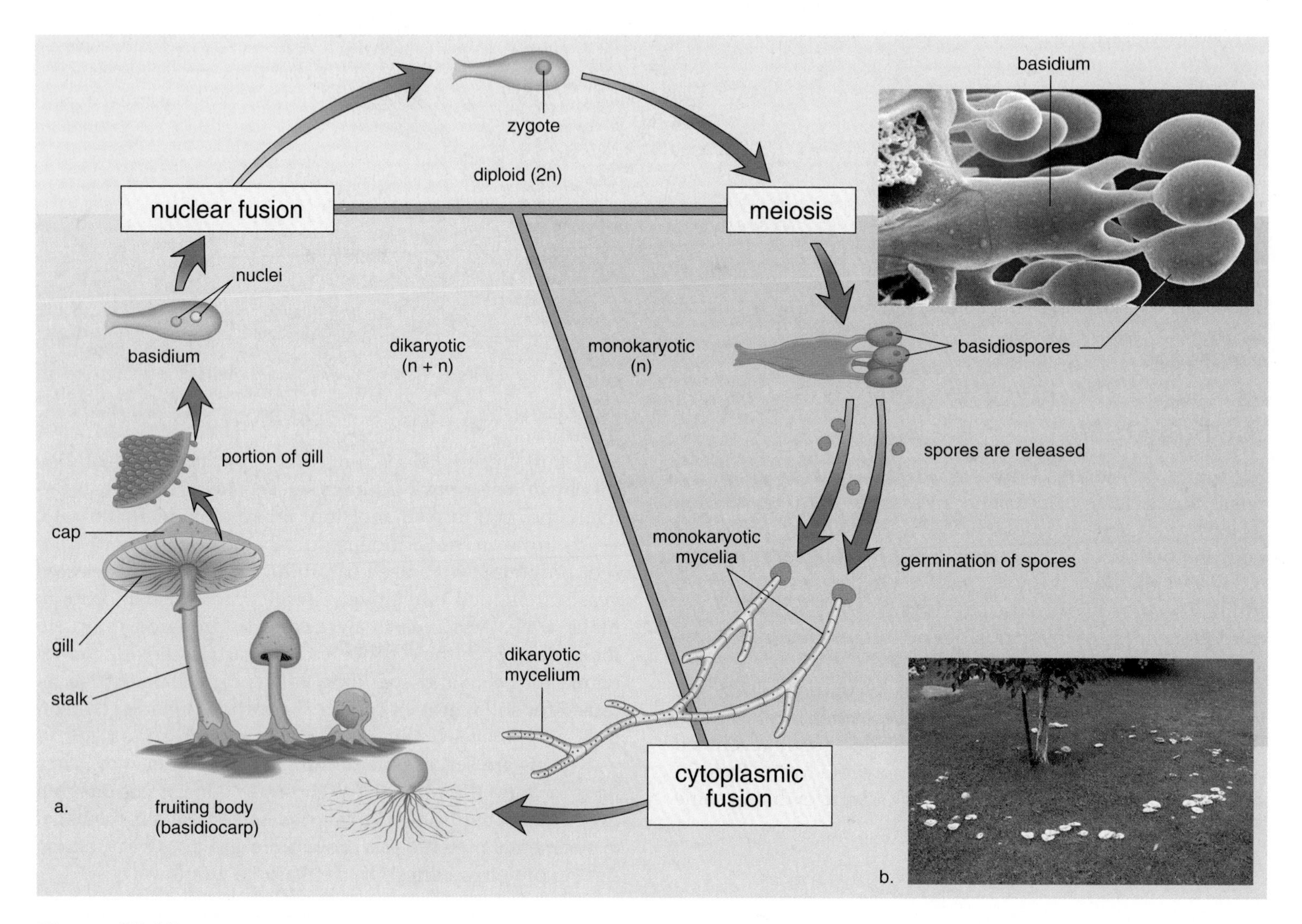

Figure 28.25 Club fungi.
a. Life cycle of a mushroom in which sexual reproduction is the norm. After hyphae from two compatible mating types fuse, the dikaryotic mycelium is long-lasting. Nuclear fusion results in zygotes within basidia on the gills of the fruiting body shown. Meiosis occurs and basidiospores are released. **b.** Fairy ring. Mushrooms develop in a ring on the outer living fringes of a dikaryotic mycelium. The center has used up its nutrients and is no longer living.

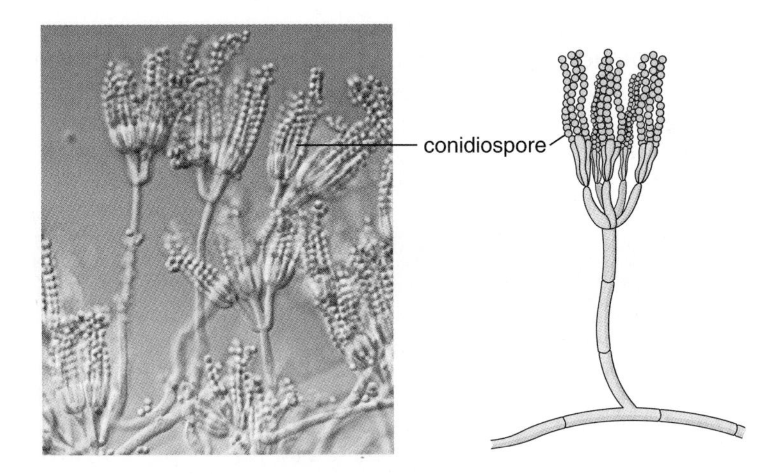

Figure 28.26 Conidiospores.
Sac fungi and imperfect fungi reproduce asexually by producing spores called conidiospores at the ends of certain hyphae. The organism shown here is *Penicillium*, an imperfect fungus.

Imperfect Fungi

The imperfect fungi (phylum Deuteromycota, 25,000 species) always reproduce asexually by forming *conidiospores* (Fig. 28.26). These fungi are "imperfect" in the sense that no sexual stage has yet been observed and may not exist. Without knowing the sexual stage, it is often difficult to classify a fungus as belonging to one of the other phyla.

Several imperfect fungi are serviceable to humans. Some species of the mold *Penicillium* are sources of the antibiotic penicillin, while other species give the characteristic flavor and aroma to cheeses such as Roquefort and Camembert. The bluish streaks in blue cheese are patches of conidiospores. The drug cyclosporine, which is administered to suppress the immune system following an organ transplant operation, is derived from an imperfect fungus found in soil.

Unfortunately, some imperfect fungi cause disease in humans. Certain dust-borne spores can cause infections of the respiratory tract, while athlete's foot and ringworm are spread by direct contact. *Candida albicans* is a yeastlike organism that causes infections of the vagina, especially in women on the birth-control pill. This organism also causes thrush, an inflammation of the mouth and throat.

> The imperfect fungi cannot be classified into one of the other phyla because their mode of sexual reproduction is unknown.

Fungal Relationships

We have already mentioned several instances in which fungi are parasites of plants and animals. Two other associations are of interest.

Lichens are a symbiotic relationship between a fungus and a cyanobacterium or a green alga. The body of a lichen has three layers: the fungus forms a thin, tough upper layer and a loosely packed lower layer that shield the photosynthetic cells in the middle layer (Fig. 28.27*a*). In the past, lichens were assumed to be a relationship of mutual benefit: the fungus received nutrients from the algal cells, and the algal cells were protected from desiccation by the fungus. Actually, lichens may involve a controlled form of parasitism of the algal cells by the fungus and algae may not benefit at all from these associations. This is supported by experiments in which the fungal and algal components are removed and grown separately. It is difficult to cultivate the fungus, which does not usually grow alone.

Three types of lichens are recognized. Compact crustose lichens (Fig. 28.27*b*) are often seen on bare rocks or on tree bark; foliose lichens are leaflike; and fruticose lichens are shrublike. Lichens can live in areas of extreme environmental conditions and are important soil formers. In Arctic ecosystems, a lichen called reindeer moss is a common photosynthetic organism and an important food source for animals.

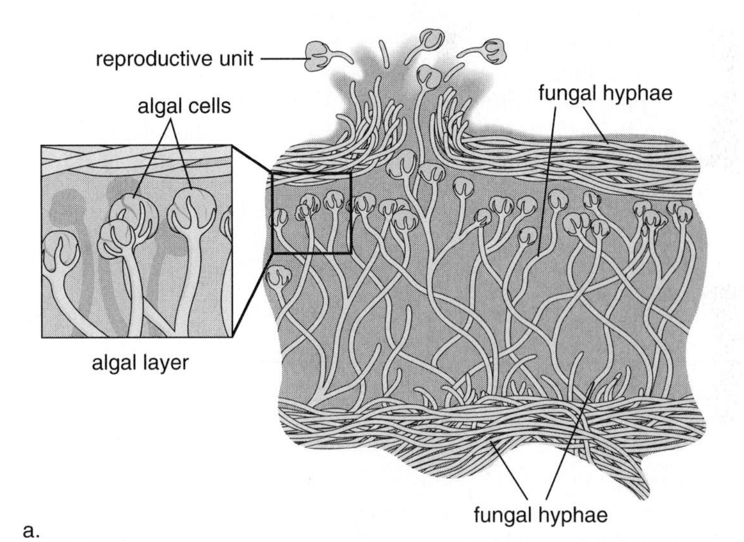

b.

Figure 28.27 Lichen morphology.
a. A section of a lichen shows the placement of the algal cells and the fungal hyphae, which encircle and penetrate the algal cells.
b. Mixture of compact crustose lichens.

Mycorrhizae (fungus roots) are symbiotic relationships of mutual benefit between soil fungi and the roots of most plants. The fungus provides the plants with inorganic nutrients, and the plant provides the fungus with organic nutrients. Plants whose roots are infected with mycorrhizae grow more successfully in poor soils—particularly soils deficient in phosphates—than do plants without mycorrhizae. The fungal partner may enter the cortex of roots but does not enter plant cells.

Bioethical Issue

Carriers of disease are persons who do not appear to be ill but who can nonetheless pass on an infectious disease. Carriers of sexually transmitted diseases can pass on an infection to their partners, and even sometimes to people who have never had sexual relations with them. Hepatitis C and HIV are transmitted by blood to blood contact, as when drug abusers share needles.

The only way society can protect itself is to identify carriers and remove them from areas or activities where transmission of the pathogen is most likely. Sometimes it's difficult to identify activities that might pass on a pathogen like HIV. A few people believe that they have acquired HIV from their dentists, and while this is generally believed to be unlikely, medical personnel are still required to identify themselves when they are carriers of HIV. Magic Johnson is a famous basketball star who is HIV positive. When he made it known that he was a carrier, some of his teammates refused to play basketball with him and he had no choice but to retire. Later, Johnson did play again for a while. There are other sports in which transmission of HIV is more likely. The Centers for Disease Control did a statistical study to try to figure the odds of acquiring HIV from another football player. They figured that the odds were 1 in 85 million.

The odds might be higher for boxing, a bloody sport. When two brothers, one of whom had AIDS, got into a vicious fight, the infected brother repeatedly bashed his head against his brother's. Both men bled profusely and soon after, the previously uninfected brother tested positive for the virus. The possibility of transmission of HIV in the boxing ring has caused several states to require boxers to undergo routine HIV testing. If they are HIV positive, they can't fight.

Questions

1. Should all people who are HIV positive identify themselves? Why or why not? By what method would they identify themselves at school, at work, and other places?
2. What type of contact would you have with an HIV positive individual? Would you play on the same team? Would you have sexual relations with an HIV positive individual if a condom was used? Discuss.
3. Is it ethical to make an HIV positive individual feel stigmatized? Discuss.

Summarizing the Concepts

28.1 Viruses

Viruses are noncellular obligate parasites that have a protein coat called a capsid and a nucleic acid core. Viral DNA must enter a host cell before reproduction is possible. In the lytic cycle, a bacteriophage immediately reproduces, and in the lysogenic cycle, viral DNA integrates into the host genome and may eventually reproduce.

28.2 Kingdom Monera

The kingdom Monera includes prokaryotic unicellular organisms, namely the bacteria. Most bacteria are saprotrophs (heterotrophic by absorption), and along with fungi, fulfill the role of decomposers in ecosystems. The cyanobacteria are photosynthetic in the same manner as plants. Reproduction of bacteria is by binary fission, but sexual exchange occasionally takes place. Some bacteria form endospores, which can survive the harshest of treatment except sterilization.

28.3 Kingdom Protista

Kingdom Protista includes eukaryotic unicellular organisms and some multicellular forms. Algae are aquatic autotrophs by photosynthesis; protozoans are aquatic heterotrophs by ingestion. Slime molds, which are terrestrial, and water molds, which are aquatic, have some characteristics of fungi.

Algae are classified according to their pigments (colors). Green algae are diverse: some are unicellular or colonial flagellates, some are filamentous, and some are multicellular sheets. Diatoms and dinoflagellates are unicellular producers in oceans; brown and red algae are seaweeds; euglenoids are unicellular with both plant and animal characteristics. Every type of life cycle is seen among the algae.

Protozoans are classified according to the type of locomotor organelle. The amoebas phagocytize, the ciliates are very complex, and sporozoa are all animal parasites. Malaria is a significant disease caused by a sporozoan.

Slime molds have an amoeboid stage and then form fruiting bodies, which produce spores that are dispersed by the wind. Water molds have threadlike bodies.

28.4 Kingdom Fungi

Kingdom Fungi includes eukaryotic multicellular organisms that are saprotrophs. Fungi are composed of hyphae, which form a mycelium. Along with heterotrophic bacteria, they are decomposers. The fungi produce windblown spores during both sexual and asexual reproduction. The major groups of fungi are distinguished by type of sexual spore and fruiting body. Zygospore fungi produce spores in sporangia; sac fungi produce ascospores in asci; and club fungi form basidiospores in basidia.

Fungi form two symbiotic associations of interest. Lichens contain both a fungus and an alga; mycorrhizae is a symbiotic relationship of mutual benefit between soil fungi and roots of plants.

Studying the Concepts

1. Describe the structure of viruses and the manner of reproduction of bacteriophages, including the lytic and lysogenic cycles. 574–75
2. Describe the life cycle of a retrovirus and explain why retroviruses are of interest today. 576
3. In general, describe the structure of prokaryotic cells. 577
4. Describe the metabolic diversity of bacteria and how they are classified. 577–80
5. Describe the manner in which bacteria reproduce, exchange genetic material, and form endospores. 581
6. What are the three groups of protists and how is each group distinguishable? 583
7. Describe the different types of green algae and how their means of sexual reproduction differ. Show that the life cycle of *Chlamydomonas* is an example of the haplontic cycle and that the life cycle of *Ulva* is an example of the alternation of generations life cycle. 584–85
8. In what ways are brown and red algae similar? How are they different? 585–87
9. In what ways are diatoms and dinoflagellates similar? How are they different? 586
10. What are the animal-like characteristics of euglenoids? The plantlike characteristics? 587
11. Why are the slime molds and the water molds sometimes called the funguslike protists? 590–91
12. Describe the anatomical features of fungi, and tell how fungi are classified. 591
13. Describe the life cycles of black bread mold and a mushroom. 592, 594
14. Describe the structure of lichens and the importance of both lichens and mycorrhizae. 595–96

Testing Yourself

Choose the best answer for each question.

1. Which of these are found in all viruses?
   a. envelope, nucleic acid, capsid
   b. hyphae and cilia
   c. DNA, RNA, and proteins
   d. proteins and a nucleic acid
   e. proteins, nucleic acids, carbohydrates, and lipids
2. Which step in the lytic cycle follows attachment of the virus and release of DNA into the cell?
   a. production of lysozyme
   b. biosynthesis of viral components
   c. assemblage
   d. integration of viral DNA into host DNA
   e. host DNA replication
3. RNA retroviruses have a special enzyme that
   a. disintegrates host DNA.
   b. polymerizes host DNA.
   c. transcribes viral RNA to cDNA.
   d. translates host DNA.
   e. repairs viral DNA.
4. Facultative anaerobes
   a. require a constant supply of oxygen.
   b. are killed in an oxygenated environment.
   c. do not always need oxygen.
   d. are photosynthetic.
   e. are chemosynthetic.
5. Cyanobacteria, unlike other types of bacteria that photosynthesize, do
   a. not give off oxygen.
   b. give off oxygen.
   c. not have chlorophyll.
   d. not have a cell wall.
   e. form plasmodia.
6. Chemosynthetic bacteria
   a. are autotrophic.
   b. use the rays of the sun to acquire energy.
   c. oxidize inorganic compounds to acquire energy.
   d. Both a and c are correct.
   e. Both a and b are correct.
7. Which is mismatched?
   a. red algae—multicellular, delicate, seaweed
   b. diatoms—silica shell, boxlike, golden brown
   c. euglenoids—flagella, pellicle, eyespot
   d. *Fucus*—adult is diploid, seaweed, chlorophylls *a* and *c*
   e. *Paramecium*—cilia, calcium carbonate shell, gullet
8. Which is a false statement?
   a. Slime molds and water molds are protists.
   b. There are flagellated algae and flagellated protozoans.
   c. Among protozoans, both flagellates and sporozoans are symbiotic.
   d. Among protists, only green algae have a sexual life cycle.
   e. Ciliates exchange genetic material during conjugation.
9. Which is mismatched?
   a. water mold—potato famine
   b. trypanosome—African sleeping sickness
   c. *Plasmodium vivax*—malaria
   d. amoeboid—severe diarrhea
   e. AIDS—*Giardia lamblia*
10. Which is found in slime molds but not fungi?
    a. nonmotile spores
    b. amoeboid adult
    c. zygote formation
    d. photosynthesis
    e. All of these are correct.
11. The taxonomy of fungi is based on
    a. sexual reproductive structures.
    b. shape of the sporocarp.
    c. mode of nutrition.
    d. type of cell wall.
    e. All of these are correct.
12. In the life cycle of black bread mold, the zygospore
    a. undergoes meiosis and produces zoospores.
    b. produces spores as a part of asexual reproduction.
    c. is a thick-walled dormant stage.
    d. is equivalent to asci and basidia.
    e. Both a and c are correct.
13. When sac fungi and club fungi reproduce sexually, they produce
    a. a fruiting body.
    b. spores.
    c. conidiospores.
    d. hyphae.
    e. Both a and b are correct.
14. Lichens
    a. are comprised of bacteria and fungi.
    b. cannot reproduce.
    c. need a nitrogen source to live.
    d. are parasitic on trees.
    e. are able to live in extreme environments.

15. Label this diagram of the *Chlamydomonas* life cycle.

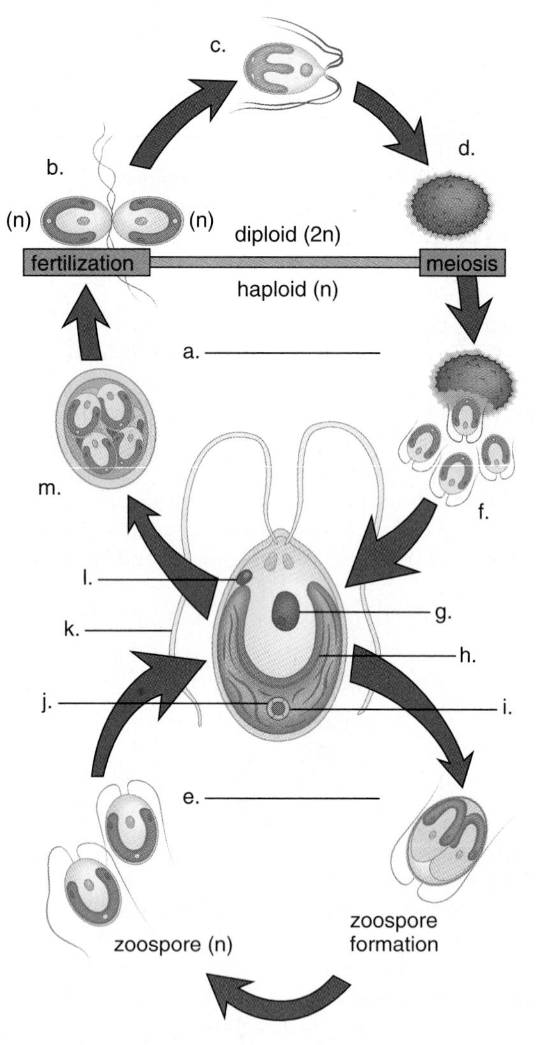

n. Give two reasons why each portion of the cycle is either asexual or sexual.

Thinking Scientifically

1. In reference to the life cycles (Figure 28.9):
   a. The timing of what one element determines whether or not a diploid adult results?
   b. Hypothesize how the haplontic life cycle may have given rise to the alternation of generations life cycle.
   c. Hypothesize how the alternation of generations life cycle may have given rise to the diplontic life cycle.

2. Members of the kingdom Fungi are adapted to living on land.
   a. What characteristic of fungi would make you think that other organisms (such as plants and animals) must have been present on land before fungi?
   b. Both *Chlamydomonas* and fungi follow the haplontic life cycle. What elements in the cycle show that the former is adapted to living in the water and the latter is adapted to living on land?
   c. The mycelia of sac fungi and club fungi are found in the ground, and the fruiting bodies usually appear after a rain. What does this tell you about the adaptation of fungi to living on land?

Understanding the Terms

alga (pl., algae) 583
alternation of generations 585
archaea 578
ascus 593
bacteriophage 575
bacterium 577
basidium 594
binary fission 581
ciliate 589
conidiospore 593
conjugation 585
cyanobacterium 580
decomposer 578
diatom 586
dinoflagellate 586
diplontic life cycle 585
filament 585
fruiting body 593
fungus (pl., fungi) 591
haplontic life cycle 584
hypha 591
lichen 595
lysogenic cycle 575
lytic cycle 575
mad cow disease 577
mycelium 591
mycorrhiza 596
nucleoid 577
phagocytize 588
plasmodium 590
prion 577
protozoan 588
pseudopod 588
retrovirus 576
saprotroph 578
sporangium 592
spore 584
symbiotic 578
trypanosome 589
viroid 577
virus 574
zooflagellate 589
zygospore 592

Match the terms to these definitions:

a. ________ Portion of the bacterial cell that contains its genetic material.
b. ________ Relationship that occurs when two different species live together in a unique way; it may be beneficial, neutral, or detrimental to one and/or the other species.
c. ________ Splitting of a parent cell into two daughter cells; serves as an asexual form of reproduction in bacteria.
d. ________ Organism that secretes digestive enzymes and absorbs the resulting nutrients back across the plasma membrane.
e. ________ Symbiotic relationship between certain fungi and algae, in which the fungi possibly provide inorganic food or water and the algae provide organic food.

Using Technology

Your study of microbiology is supported by these available technologies.

Essential Study Partner CD-ROM
Evolution & Diversity → Viruses
→ Bacteria
→ Protists
→ Fungi

Visit the Mader web site for related ESP activities.

Exploring the Internet
The Mader Home Page provides resources and tools as you study this chapter.

http://www.mhhe.com/biosci/genbio/mader

C H A P T E R 29

Plants

Chapter Concepts

Not all plants produce flowers. This is Equisetum sylvaticum, *a type of horsetail. Spores borne in strobili (conelike structures), not seeds, disperse the species.*

Sally was quite disgruntled as she followed along behind her class. They were on a botanical field trip, and she wasn't much interested in plants in the first place. She was a pre-med student and had only signed up for this course because it offered four hours of general education credit. The professor kept pointing out different types of plants, and Sally felt a certain sense of satisfaction when no one seemed interested. Then he pointed to a plant he called "wild ginger" and explained how early settlers had used the ground-up root for seasoning food. Next he had them break and smell a mint leaf. "This plant is still used by some," he said, "to make mint tea." He then pulled a piece of bark off a willow tree, and while chewing it, he explained that the bark contains salicylic acid, the chemical found in aspirin. In days past, people chewed willow bark when they had a toothache. The professor asked them if they knew any present-day medications that were extracted from plants. One student mentioned that taxol from the yew tree is now used in the treatment of cancer, and the professor added that digitalis, a heart medicine, comes from foxglove, and codeine for pain comes from poppies. Sally noticed that by now everyone seemed interested, and realized that she was, too!

Classification

Kingdom Plantae

Multicellular; primarily terrestrial eukaryotes with well-developed tissues; autotrophic by photosynthesis; alternation of generations life cycle.

Nonvascular plants*

- Division Hepatophyta: liverworts
- Division Bryophyta: mosses
- Division Anthocerotophyta: hornworts

Seedless Vascular Plants*

- Division Psilotophyta: whisk ferns
- Division Lycopodophyta: club mosses
- Division Equisetophyta: horsetails
- Division Pteridophyta: ferns

Gymnosperms*

- Division Pinophyta: conifers
- Division Cycadophyta: cycads
- Division Ginkgophyta: maidenhair tree
- Division Gnetophyta: gnetophytes

Angiosperms*

- Division Magnoliophyta: flowering plants
  - Class Magnoliopsida: dicots
  - Class Liliopsida: monocots

*Category not in the classification of organisms, but added here for clarity. Notice the use of division rather than phylum in the plant kingdom.

29.1 Characteristics of Plants

Plants **(Kingdom Plantae)** are multicellular eukaryotes with well-developed tissues (Fig. 29.1). Most likely plants share a common ancestor with green algae. Both utilize chlorophylls *a* and *b* as well as carotenoid pigments during photosynthesis. The primary food reserve is starch, which can be stored inside chloroplasts. Both have a cell wall that contains cellulose and form a cell plate when cell division occurs. Unlike algae, plants live in a wide variety of terrestrial environments, from lush forest to dry desert or frozen tundra. One advantage of a land existence is the greater availability of light for photosynthesis—water, even if clear, filters light. Also, carbon dioxide diffuses more readily in air than in water.

The land environment requires adaptations to deal with the constant threat of drying out, and this accounts for many of the characteristics of plants. Plants, but not green algae, protect reproductive cells and later the embryo within the body of the plant. Most plants can obtain water from soil by means of roots. To conserve water, leaves and stems are covered by a waxy cuticle that is impervious to water. Many plants have a vascular system that transports water up to, and nutrients down from, the leaves.

Moss, a bryophyte

Fern, a seedless vascular plant

Conifer, a gymnosperm

Flowering plant, an angiosperm

Figure 29.1 Common plants today.

Life Cycle of Plants

Plants have a two-generation life cycle, called **alternation of generations.** (1) The **sporophyte,** the diploid (2n) generation, produces spores by meiosis within structures called sporangia (sing., **sporangium**). **Spores** are reproductive cells that develop directly into gametophytes. (2) The **gametophyte,** the haploid (n) generation, produces gametes (egg and sperm) that unite to form a zygote. The zygote develops into the sporophyte.

In plants, one generation, either the gametophyte or the sporophyte, is *dominant* over the other generation: the dominant generation lasts longer, is larger, and is more conspicuous. In mosses, a type of nonvascular plant, the haploid gametophyte is the dominant generation; therefore more space is allotted to this generation in Figure 29.2*a*. In ferns, the separate sporophyte which has vascular tissue is dominant, so more space is allotted to this generation in Figure 29.2*b*. In nonvascular and seedless vascular plants, flagellated sperm require outside moisture to reach the egg.

Seed plants (Figure 29.2*c*) are well adapted to reproduce on land. They produce **heterospores:** microspores and megaspores. A **microspore** develops into a pollen grain which is carried by wind or insect to the vicinity of the ovule. The **megaspore** develops into an egg-producing gametophyte within an ovule that is still protected by the sporophyte. The ovule becomes a seed, which contains an embryonic sporophyte and stored food enclosed by a protective seed coat.

Plants are adapted to living on land; they all have means of preventing drying out and protecting the embryo during a life cycle called alternation of generations.

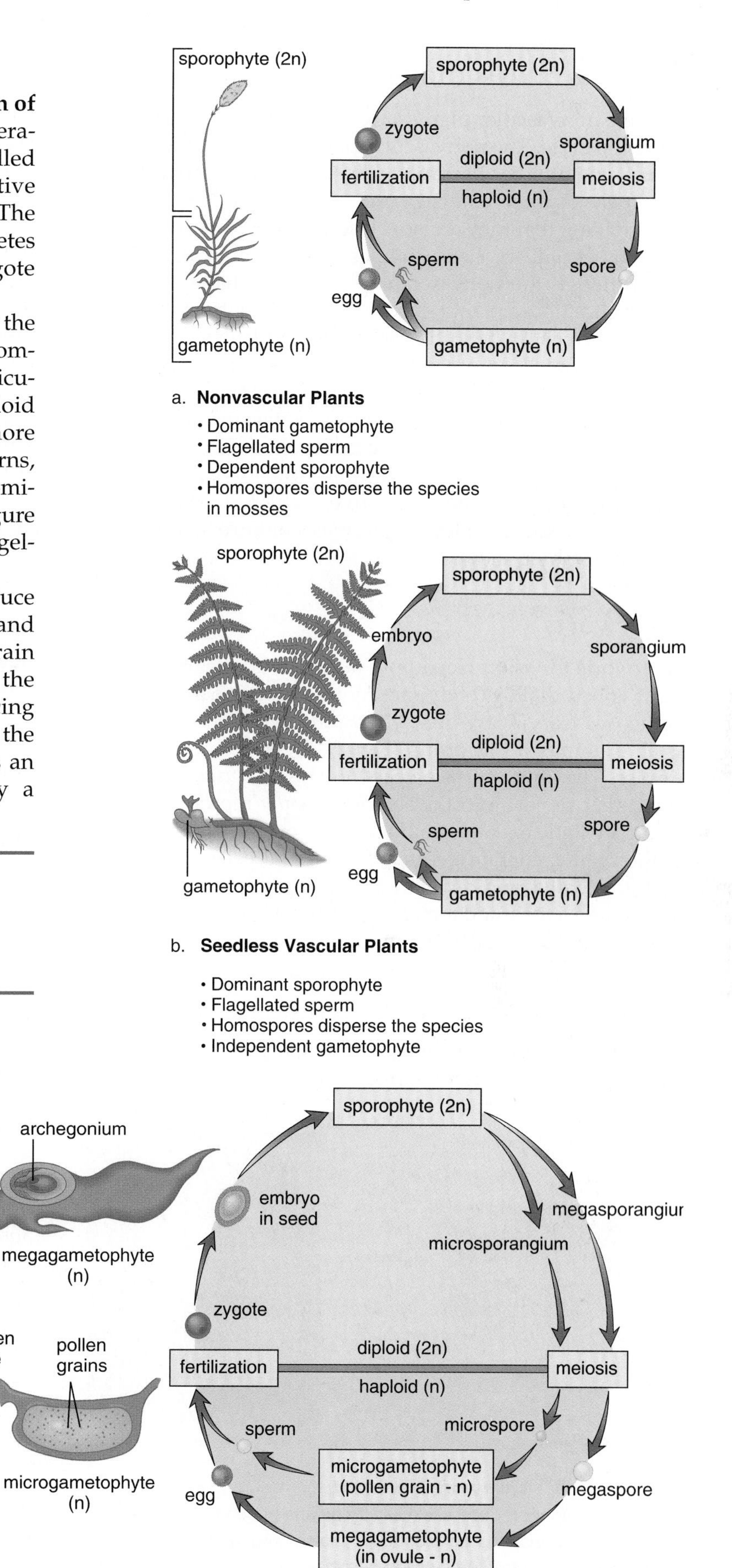

c. **Seed Vascular Plants**

- Dominant sporophyte
- Pollen grains
- Heterosporous
- Seeds disperse the species
- Dependent micro-,mega- gametophytes

Figure 29.2 Plant life cycles.

29.2 Nonvascular Plants

Only the **nonvascular plants** lack vascular tissue during their entire life cycle. The nonvascular plants consist of three divisions—one each for hornworts, liverworts, and mosses. The placement of these plants in separate divisions reflects current thinking that they are not closely related. The liverworts and mosses will serve as examples of the nonvascular plants.

Although the nonvascular plants often have a "leafy" appearance, these plants do not have true roots, stems, and leaves—which by definition must contain true vascular tissue. Therefore, the nonvascular plants are said to have rootlike, stemlike, and leaflike structures.

The gametophyte is the dominant generation in bryophytes—it is the generation we recognize as the plant. Further, flagellated sperm swim in a continuous film of water to the vicinity of the egg. The sporophyte, which develops from the zygote, is attached to, and derives its nourishment from, the gametophyte shoot.

Liverworts

Liverworts (division Hepatophyta, 10,000 species) that have a flat, lobed thallus (body) are more familiar than the more numerous leafy liverworts. *Marchantia,* which is often used as an example of this group (Fig. 29.3), has a smooth upper surface. The lower surface bears numerous *rhizoids* (rootlike hairs) that project into the soil. *Marchantia* reproduces both asexually and sexually. Gemmae cups on the upper surface of the thallus contain gemmae, groups of cells that detach from the thallus and can start a new plant. Sexual reproduction involves male and female umbrella-like structures called gametophores, so-called because they produce the gametes.

Mosses

Mosses (division Bryophyta, 12,000 species) can be found from the Arctic through the tropics to parts of the Antarctic. Although most prefer damp, shaded locations in the temperate zone, some survive in deserts and others inhabit bogs and streams. In forests, they frequently form a mat that covers the ground and rotting logs. Mosses can store large quantities of water in their cells, but if a dry spell continues for long, they become dormant until it rains.

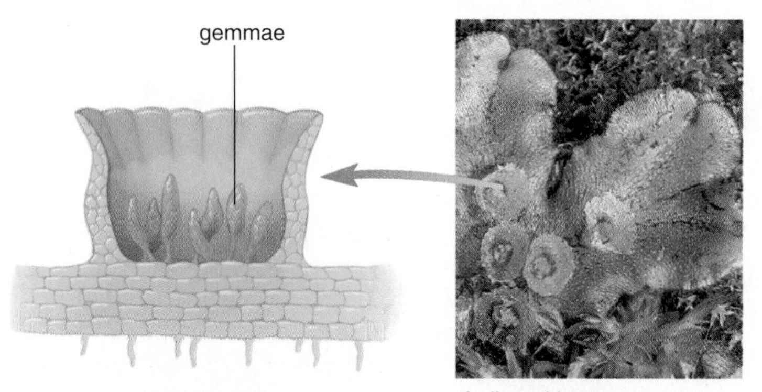

Figure 29.3 Liverwort, *Marchantia*.
Marchantia can reproduce asexually by means of gemmae—minute bodies that give rise to new plants. As shown here, gemmae are located in cuplike structures called gemmae cups.

Most mosses can reproduce asexually by fragmentation. Just about any part of the plant is able to grow and eventually produce leafy shoots. Figure 29.4 describes the life cycle of a typical temperate-zone moss. The gametophyte of mosses has two stages. First, there is the algalike *protonema,* a branching filament of cells. After about three days of favorable growing conditions, upright leafy shoots are seen at intervals along the protonema. Rhizoids anchor the shoots, which bear antheridia and archegonia. An **antheridium** consists of a short stalk, an outer layer of sterile cells, and an inner mass of cells that become the flagellated sperm. An **archegonium,** which looks like a vase with a long neck, has a single egg located inside the base.

The dependent sporophyte consists of a *foot,* which grows down into the gametophyte tissue, a *stalk,* and an upper *capsule,* or *sporangium,* where spores are produced. At first the sporophyte is green and photosynthetic; at maturity it is brown and nonphotosynthetic. Since the gametophyte is the dominant generation, it seems consistent for spores to disperse the species—that is, when the haploid spores germinate, the gametophyte is in a new location.

Adaptations and Uses of Nonvascular Plants

Nonvascular plants are quite small and low lying. This characteristic is linked to the lack of an efficient means to transport water to any height. And because sexual reproduction involves flagellated sperm that must swim, bryophytes are usually found in moist habitats. Nevertheless, mosses are better than flowering plants at living on stone walls and on fences and even in the shady cracks of hot, exposed rocks. For these particular microhabitats, there seems to be a selective advantage to being small and simple. When bryophytes help colonize bare rock, they help to convert the rocks to soil that can be used for the growth of other organisms.

Sphagnum, also called bog or peat moss, has commercial importance. This moss has special nonliving cells that can absorb moisture, which is why peat moss is often used in gardens to improve the water-holding capacity of the soil. In some areas, like bogs, where the ground is wet and acidic, dead mosses, especially sphagnum, do not decay. The accumulated moss, called peat, can be used as fuel.

The nonvascular plants include the inconspicuous liverworts and the mosses, plants that have a dominant gametophyte. Fertilization requires an outside source of moisture; windblown spores disperse the species.

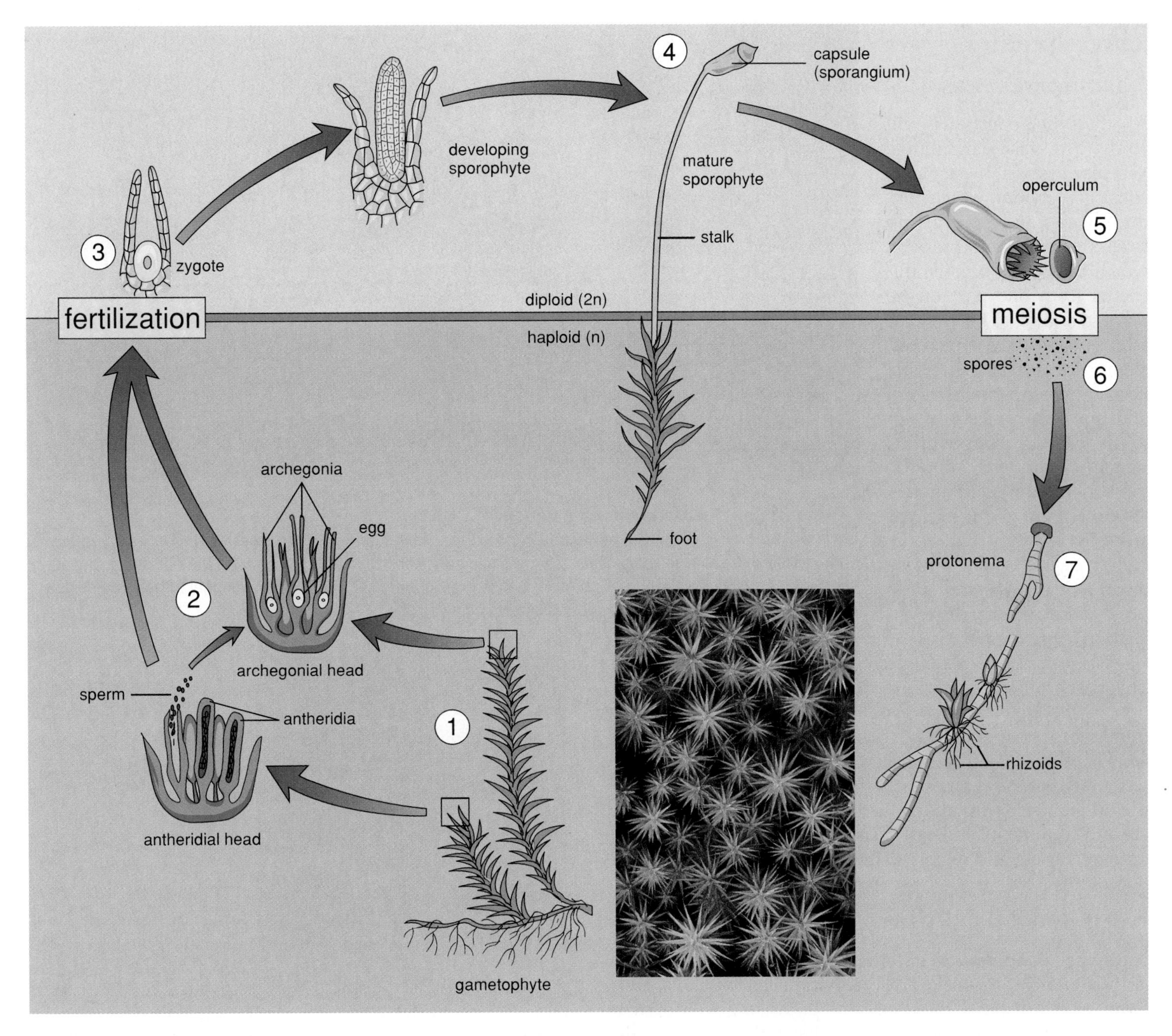

Figure 29.4 Moss life cycle.
The gametophyte is dominant in bryophytes, such as mosses. ① The leafy shoots bear separate antheridia and archegonia. ② Flagellated sperm are produced in **antheridia,** and these swim in external water to an **archegonium** that contains a single egg. ③ When the egg is fertilized, the zygote and developing sporophyte are retained within the archegonium. The drawings show only an enlarged portion of the archegonium. ④ The mature sporophyte growing atop a gametophyte shoot consists of a foot that grows down into the gametophyte tissue, a stalk, and an upper capsule, or **sporangium,** where meiosis occurs and spores are produced. ⑤ When the covering and capsule lid (operculum) fall off, the spores are mature and are ready to escape. The release of spores is controlled by one or two rings of "teeth" that line the margin of the capsule. The teeth close the opening when the weather is wet but project outward to free the spores when the weather is dry. ⑥ Spores are released at times when they are most likely to be dispersed by air currents. ⑦ When a spore lands on an appropriate site, it germinates into a protonema, the first stage of the gametophyte.

29.3 Seedless Vascular Plants

All the other plants we will study are **vascular plants.** Vascular tissue in these plants consists of **xylem,** which conducts water and minerals up from the soil, and **phloem,** which transports organic nutrients from one part of the plant to another. The vascular plants usually have true roots, stems, and leaves. The roots absorb water from the soil and the stem conducts water to the leaves. Xylem, with its strong-walled cells, supports the body of the plant against the pull of gravity. The leaves are fully covered by a waxy cuticle except where it is interrupted by *stomates,* little pores whose size can be regulated to control water loss.

Figure 29.5 The Carboniferous period.
In the swamp forests of the Carboniferous period, there were treelike club mosses *(left),* treelike horsetails *(right),* and fernlike foliage *(left).* When the trees fell, they were covered by water and did not decompose well. Sediment built up and turned to rock, whose pressure caused the organic material to become coal, a fossil fuel which still helps run our industrialized society today.

The sporophyte is the dominant generation in vascular plants. This is advantageous because the sporophyte is the generation with vascular tissue. Another advantage of having a dominant sporophyte relates to its being diploid. If a faulty gene is present, it can be masked by a functional gene. Then, too, the greater the amount of genetic material, the greater the possibility of mutations that will lead to increased variety and complexity. Indeed, vascular plants are complex, extremely varied, and widely distributed.

Ferns and Their Allies

Some vascular plants do not produce seeds. Seedless vascular plants include whisk ferns, club mosses, horsetails, and ferns, which disperse the species by producing windblown spores. When the spores germinate, they will produce a relatively large gametophyte that is independent of the sporophyte for its nutrition. In these plants, flagellated sperm are released by antheridia and swim in a film of external water to the archegonia, where fertilization occurs. Because spores disperse the species and the nonvascular gametophyte is independent of the sporophyte, these plants cannot wholly benefit from the adaptations of the sporophyte to a terrestrial environment.

The seedless vascular plants formed the great swamp forests of the Carboniferous period (Fig. 29.5). A large number of these plants died but did not decompose completely. Instead, they were compressed to form the coal that we still mine and burn. (Oil has a similar origin but most likely formed in marine sedimentary rocks and included animal remains.)

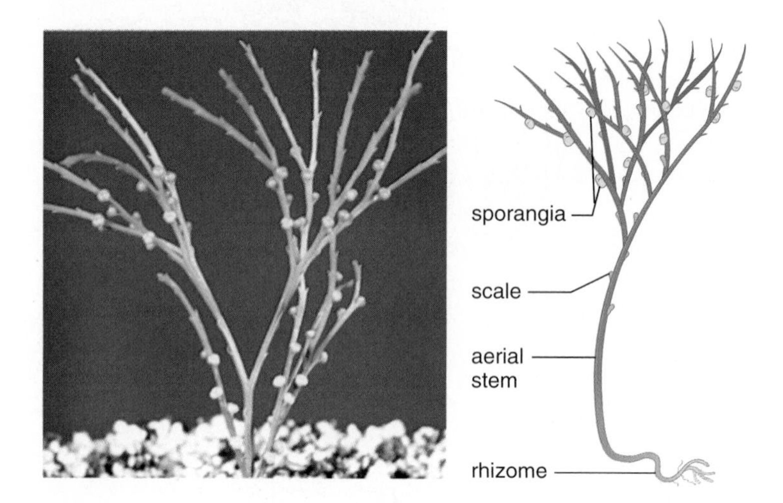

Figure 29.6 Whisk fern, *Psilotum*.
Whisk ferns have no roots or leaves—the branches carry on photosynthesis. The sporangia are yellow.

Whisk Ferns

The **psilotophytes** (division Psilotophyta, several species) are represented by *Psilotum,* a whisk fern (Fig. 29.6). Whisk ferns, named for their resemblance to whisk brooms, are found in Arizona, Texas, Louisiana, and Florida, as well as Hawaii and Puerto Rico. *Psilotum* looks like a rhyniophyte, a vascular plant that is known only from the fossil record. An erect stem forks repeatedly and is attached to a *rhizome,* a fleshy horizontal stem that lies underground. There are no leaves, and the branches carry on photosynthesis. Sporangia

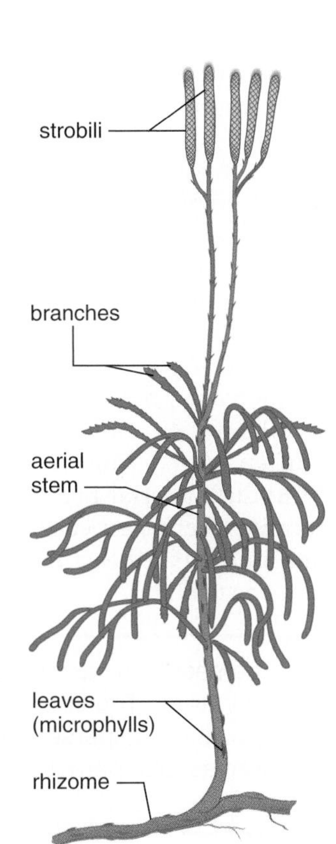

Figure 29.7 Club moss, *Lycopodium*.
Green photosynthetic stems are covered by scalelike leaves, and sporangia are found on leaves arranged as strobili.

Figure 29.8 Horsetail, *Equisetum*.
There are whorls of branches and tiny leaves at the joints of the stem. The sporangia are borne in strobili.

located at the ends of short branches produce spores that disperse the species. The independent gametophyte, which is found underground and is penetrated by a mycorrhizal fungus, produces flagellated sperm.

Club Mosses

The **club mosses** (division Lycopodophyta, 1,000 species) are common in moist woodlands of the temperate zone where they are known as ground pines. Typically, a branching rhizome sends up aerial stems less than 30 cm tall. Tightly packed, scalelike leaves cover stems and branches, giving the plant a mossy look (Fig. 29.7). In club mosses, the sporangia are borne on terminal clusters of leaves, called *strobili* (sing., strobilus), which are club shaped.

The majority of club mosses live in the tropics and subtropics where many of them are epiphytes—plants that live on, but are not parasitic on, trees. The closely related spike mosses (*Selaginella*) are extremely varied and include the resurrection plant, which curls up into a tight ball when dry but unfurls as if by magic when moistened.

Horsetails

Horsetails (division Equisetophyta, 15 species), which thrive in moist habitats around the globe, are represented by *Equisetum*, the only genus in existence today (Fig. 29.8). A rhizome produces aerial stems that stand about 1.3 meters. In some species, the whorls of slender green side branches at the joints (nodes) of the stem make the plant bear a fanciful resemblance to a horse's tail. The leaves are small and scalelike. Many horsetails have strobili at the tips of the stems that bear branches; others send up special buff-colored, naked stems that bear the strobili.

The stems are tough and rigid because of silica deposited in cell walls. Early Americans, in particular, used horsetails for scouring pots and called them "scouring rushes." Today they are still used as ingredients in a few abrasive powders.

Ferns

Ferns (division Pteridophyta, 12,000 species) are a widespread group of plants. They are most abundant in warm, moist, tropical regions, but they are also found in northern regions and in dry, rocky places. They range in size from those that are low growing and resemble mosses to those that are tall trees. The *fronds* (leaves) that grow from a rhizome in particular can vary. The royal fern has fronds that stand 6 feet tall; those of the maidenhair fern are branched with broad leaflets. And those of the hart's tongue fern are straplike and leathery. In nearly all ferns, the leaves first appear in a curled-up form called a fiddlehead, which unrolls as it grows. Figure 29.9 gives examples of fern diversity, and Figure 29.10 shows the life cycle of a typical temperate fern.

The fronds may have evolved in the following way:

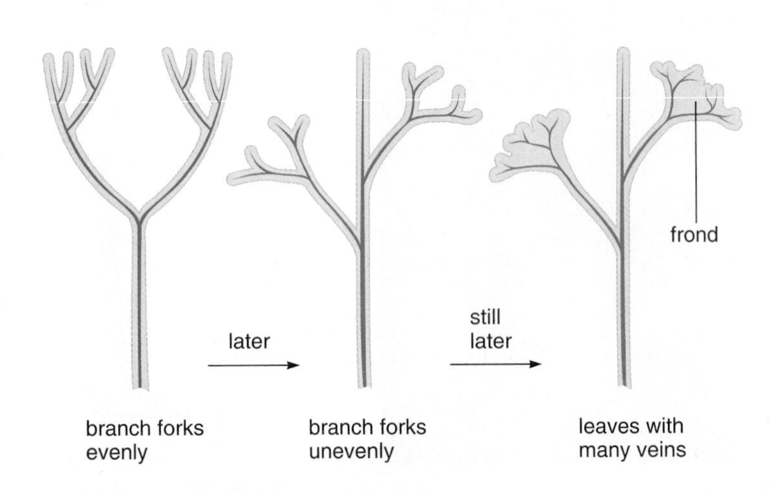

Adaptations and Uses of Ferns Ferns have true roots, stems, and leaves; that is, they contain vascular tissue. The well-developed leaves fan out, capture solar energy, and photosynthesize. The water-dependent gametophyte, which lacks vascular tissue, is separate from the sporophyte. Flagellated sperm require an outside source of moisture in which to swim to the eggs in the archegonia. Once established, some ferns, like the bracken fern *Pteridium aquilinum*, can spread into drier areas by means of vegetative (asexual) reproduction. Ferns also spread by means of the rhizomes growing horizontally in the soil, producing the fiddleheads that grow up as new fronds.

At first it may seem that ferns do not have much economic value, but they are much used by florists in decorative bouquets and as ornamental plants in the home and garden. Wood from tropical tree ferns is often used as a building material because it resists decay, particularly by termites. Ferns, especially the ostrich fern, are used as food and in the Northeast many restaurants feature fiddleheads as a special treat.

The seedless vascular plants include ferns and their allies. The sporophyte is dominant, and the species is dispersed by spores. The independent, nonvascular gametophyte produces flagellated sperm.

Maidenhair fern, *Adiantum pedatum*

Royal fern, *Osmunda regalis*

Hart's tongue fern, *Campyloneurum scolopendrium*

Figure 29.9 Fern diversity.

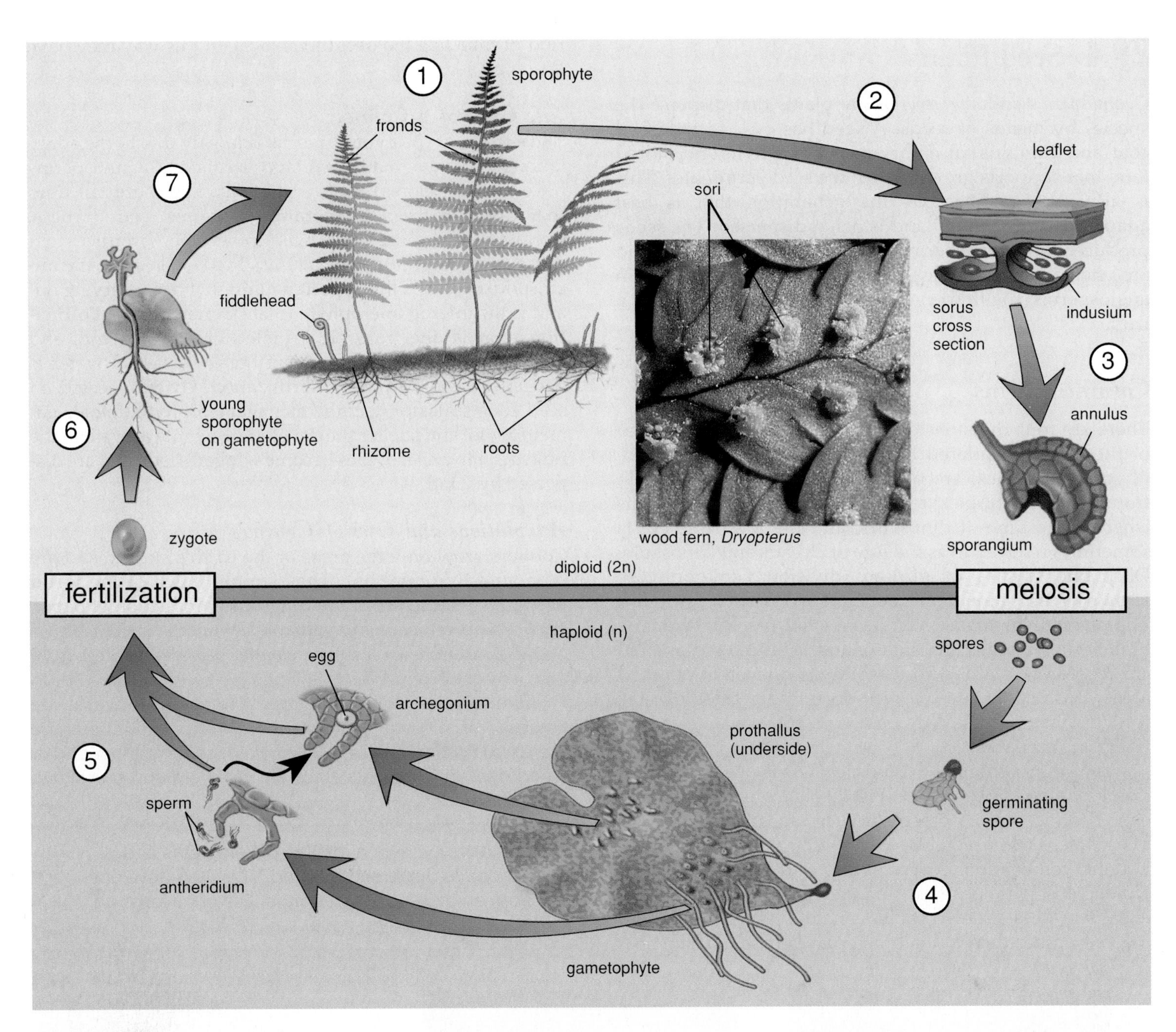

Figure 29.10 Fern life cycle.
① The sporophyte is dominant in ferns. ② In the fern shown here, sori (sing., **sorus**), which contain many sporangia, are located on the underside of the leaflets. Within a sporangium, meiosis occurs and spores are produced. ③ As a band of thickened cells on the rim of a sporangium (the annulus) dries out, it moves backward, pulling the **sporangium** open, and the spores are released. ④ A spore germinates into a prothallus, which bears **antheridia** and **archegonia** on the underside. Typically, the archegonia are at the notch and antheridia are toward the tip, between the rhizoids. ⑤ Fertilization takes place when moisture is present, because the flagellated sperm must swim in a film of water from the antheridia to the egg within the archegonium. ⑥ The resulting zygote begins its development inside an archegonium, but the embryo soon outgrows the available space. As a distinctive first leaf appears above the prothallus, and as the roots develop below it, the sporophyte becomes visible. Often the sporophyte tissues and the gametophyte tissues are distinctly different shades of green. ⑦ The young sporophyte develops a root-bearing rhizome from which the fronds project.

29.4 Seed Plants

Gymnosperms and *angiosperms* are plants that disperse the species by means of seeds. A **seed** has a protective seed coat and contains an embryonic sporophyte and stored food that supports growth when the seed germinates. This is advantageous because the generation that is best adapted to survival on land is being dispersed. The seeds produced by gymnosperms are uncovered (i.e., naked) and are often exposed on the surface of scales within cones. In angiosperms, the flowering plants, seeds are covered by a fruit.

Gymnosperms

There are four divisions of **gymnosperms,** but only three of these are considered here. **Cycads** (division Cycadophyta, 100 species) are palmlike plants found mainly in tropical and subtropical regions (Fig. 29.11*a*). Cycads flourished at the time of dinosaurs, and the Mesozoic era is sometimes referred to as the Age of Cycads and Dinosaurs. Only one species of **ginkgo** (division Ginkgophyta), known as the maidenhair tree (Fig. 29.11*b*), survives today. The maidenhair tree, which has a stout trunk with many branches, is planted in Chinese and Japanese ornamental gardens, and is valued for its ability to do well in polluted areas. Because female trees produce rather smelly seeds, it is customary to use only male trees, propagated vegetatively, in city parks.

The largest group of gymnosperms is the **conifers** (division Pinophyta, 550 species) (Fig. 29.11*c*), which include the cone-bearing pine, cedar, spruce, fir, and redwood trees. These trees have needlelike leaves, which are well adapted not only to hot summers but also to cold winters and high winds. Most gymnosperms, and therefore most conifers, are evergreen trees—they continuously lose leaves throughout the year rather than lose all their leaves within a short period of time like the deciduous trees. In this way they have leaves all year long.

Life Cycle of a Conifer

The success of the gymnosperms is largely due to the adaptations they have made to a land existence. Gymnosperms have well-developed roots and stems. Many are tall trees that can withstand temperature extremes and dryness. Conifers, as exemplified by a pine tree, produce heterospores within cones (Fig. 29.12). In a pollen cone, the microspore develops into a **pollen grain** that resists drying out and is the **microgametophyte** (male gametophyte). **Pollination,** that is, the transfer of pollen grains by wind, and growth of the pollen tube mean that no external water is needed for the purpose of fertilization. **Ovules** located on seed cone scales protect the **megagametophyte** (female gametophyte) and shelter the developing zygote as well. In the pine tree life cycle, ovules become winged seeds that are dispersed by wind.

Adaptations and Uses of Conifers

Conifers grow on large areas of the earth's surface and are economically important. They supply much of the wood used for construction of buildings and production of paper. They also produce many valuable chemicals, such as those extracted from resin, a substance that protects conifers from attack by fungi and insects.

Perhaps the oldest and largest trees in the world are conifers. Bristlecone pines in the Nevada mountains are known to be more than 4,500 years old, and a number of redwood trees in California are 2,000 years old and more than 90 meters tall.

In the life cycle of a conifer, pollen grains and seeds are dispersed by wind. The seed develops from an ovule, a structure that lies uncovered on the scale of a seed cone.

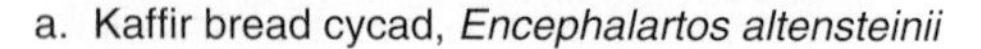

a. Kaffir bread cycad, *Encephalartos altensteinii*

b. Maidenhair tree, *Ginkgo biloba*

c. Eastern hemlock, *Tsuga canadensis*

Figure 29.11 Gymnosperm diversity.
a. Cycads resemble palm trees, but are gymnosperms that produce cones. **b.** Ginkgoes exist only as a single species—the maidenhair tree. **c.** Conifers are the most common gymnosperms.

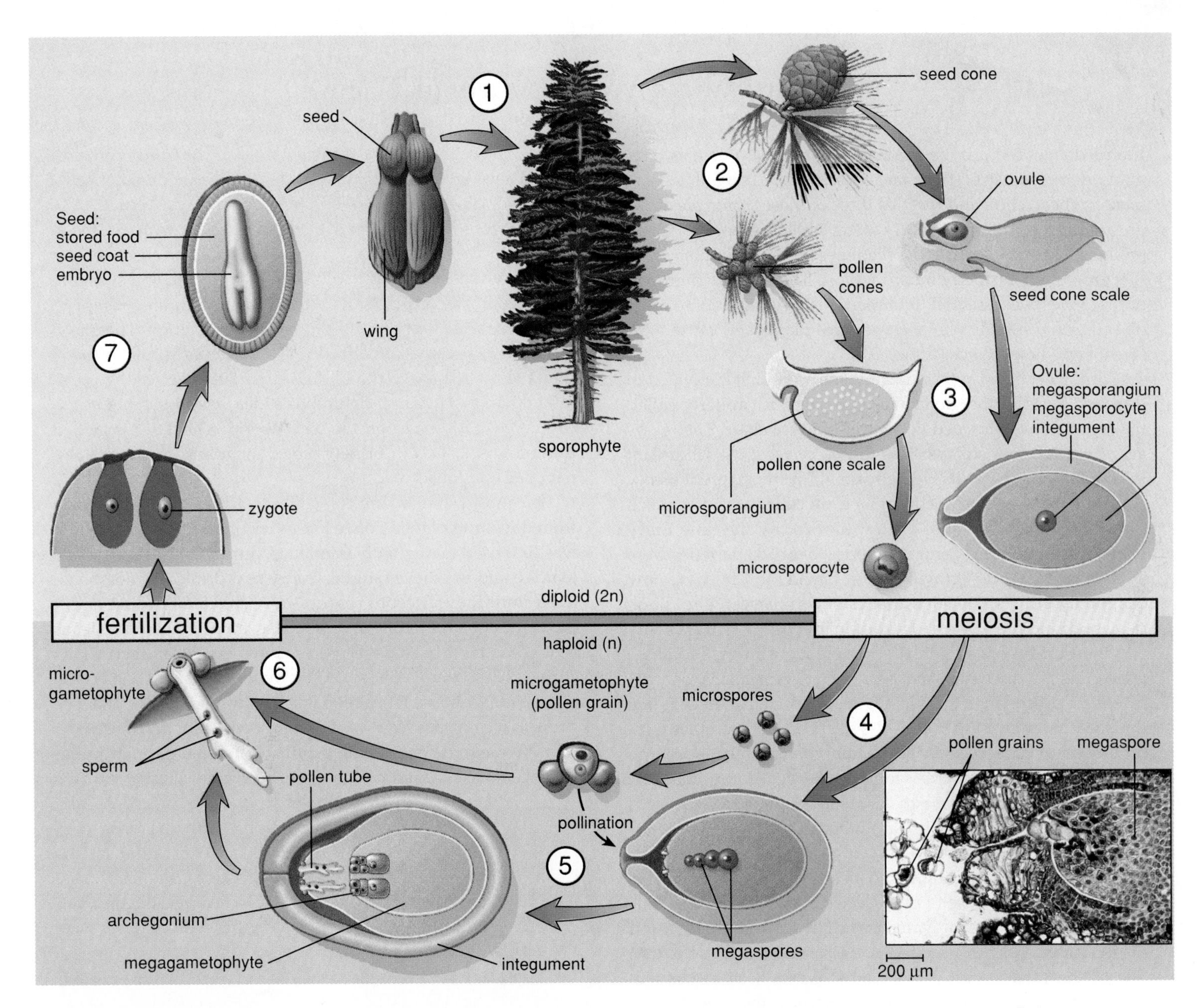

Figure 29.12 Pine life cycle.
① The sporophyte is dominant and its sporangia are borne in cones. ② There are two types of cones: pollen cones and seed cones. Typically, the pollen cones are quite small and develop near the tips of lower branches. ③ Each scale of a pollen cone has two or more microsporangia on the underside. ④ Within these sporangia, each microsporocyte (microspore mother cell) undergoes meiosis and produces four **microspores.** ⑤ Each microspore develops into a **microgametophyte,** which is the pollen grain. The pollen grain has two wings and is carried by the wind to the seed cone during pollination. The seed cones are larger than the pollen cones and are located near the branch tips on the higher branches.③ Each scale of the seed cone has two ovules that lie on the upper surface. Each ovule is surrounded by a thick, layered integument, having an opening at one end. ④ The megasporangium is within the ovule, where a megasporocyte (megaspore mother cell) undergoes meiosis, producing four **megaspores.** ⑤ Only one of these spores develops into a **megagametophyte,** with two to six archegonia, each containing a single large egg lying near the ovule opening. ⑥ Once a pollen grain is enclosed within the seed cone, it develops a pollen tube that digests its way slowly toward a megagametophyte. The pollen tube discharges two nonflagellated sperm. One of these fertilizes an egg in an archegonium and the other degenerates. Fertilization, which takes place one year after pollination, is an entirely separate event from pollination. ⑦ After fertilization, the ovule matures and becomes the seed, composed of the embryo, the reserve food, and the seed coat. Finally, in the fall of the second season, the seed cone, by now woody and hard, opens to release winged seeds. When a seed germinates, the sporophyte embryo develops into a new pine tree, and the cycle is complete.

Ecology Focus

Plants: Could We Do Without Them?

Plants define and are the producers in most ecosystems. Humans derive most of their sustenance from three flowering plants: wheat, corn, and rice. All three of these plants are of the grass family and are collectively, along with other species, called grains. Most of the earth's 5.6 billion people live a simple way of life, growing their food on family plots. The continued growth of these plants is essential to human existence. A virus or other disease could hit any one of these three plants and cause massive loss of life from starvation.

Corn, wheat, and rice originated and were first cultivated in different parts of the globe. Corn, or what is properly called maize, was first cultivated in Central America about 7,000 years ago. Disagreement still exists as to exactly what the wild plant looked like and where it originally grew. The most recent theory is that maize was developed from a plant called teosinte, which grows in the highlands of central Mexico. By the time Europeans were exploring Central America, over 300 varieties were already in existence—growing from Canada to Chile. We now commonly grow six major varieties of corn: sweet, pop, flour, dent, pod, and flint. Rice had its origin in southeastern Asia several thousand years ago, where it grew in swamps. Today we are familiar with white and brown rice, which differ in the extent of processing. Brown rice results when the seeds are thrashed to remove the hulls—the seed coat and complete embryo remain. If the seed coat and embryo are removed leaving only the starchy endosperm, white rice results. Unfortunately, it is the seed coat and embryo that contain the greatest nutritional value. Today rice is grown throughout the tropics and subtropics where water is abundant. It is also grown in some of our Western states by flooding diked fields with irrigation water. Wheat is commonly used in the United States to produce flour and bread. It was first cultivated in the Near East (Iran, Iraq, and neighboring countries) about 8,000 B.C.; hence, it is thought to be one of the earliest cultivated plants. Wheat was brought to North America in 1520 with early settlers; now the United States is one of the world's largest producers of wheat.

Many of us have an "addiction" to sugar. This nifty carbohydrate comes almost exclusively from two plants—sugarcane (grown in South America, Africa, Asia, and the Caribbean) and sugar beets (grown mostly in Europe). Each provides about 50% of the world's sugar.

Numerous other foods are bland or tasteless without spices. In the Middle Ages, wealthy Europeans spared no cost to obtain spices from the Near and Far East. In the fifteenth and sixteenth centuries, major expeditions were launched in an attempt to find better and cheaper routes for spice importation. The queen of Portugal became convinced that Columbus would actually find a shorter route to the Far East by traveling west by ocean rather than east by land. Columbus's idea was sound, but he encountered a little barrier, the New World. This later provided Europe with a wealth of new crops, including corn, potatoes, peppers, and tobacco.

Our most popular drinks—coffee, tea, and cola—also come from flowering plants. Coffee has its origin in Ethiopia, where it was first used (along with animal fat) during long trips for sustenance and to relieve fatigue. Coffee as a drink was not developed until the thirteenth century in Arabia and Turkey, and it did not catch on in Europe until the seventeenth century. Tea is thought to have been developed somewhere in central Asia. Its earlier uses were almost exclusively medicinal, especially among the Chinese, who still drink tea for medical reasons. The drink as we now know it was not developed until the fourth century. By the mid-seventeenth century it became popular in Europe. Cola is a common ingredient in tropical drinks and was used around the turn of the century, along with the drug coca (used to make cocaine), in the "original" Coca-Cola.

Until a few decades ago, cotton and other natural fibers were

Wheat, *Triticum*

Corn, *Zea*

Rice, *Oryza*

Figure 29A Cereal grains.
These cereal grains are the principle source of calories and protein for our civilization.

Tropical Rain Forest: Can We Live Without Them?

our only source of clothing. China is now the largest producer of cotton. Over thirty species of native cotton grow around the world. In the United States, cotton grows as a one-to-two meter annual, but in the tropics, woody shrubs up to 6 meters are not uncommon. The cotton fiber itself comes from filaments that grow on the seed. In sixteenth-century Europe, cotton was a little-understood fiber known only from stories brought back from Asia. Columbus and other explorers were amazed to see the elaborately woven cotton fabrics in the New World. But by 1800 Liverpool was the world's center of cotton trade. Interestingly, when Levi Strauss wanted to make a tough pair of jeans, he needed a stronger fiber than cotton, so he used hemp. No longer used for clothes, hemp (marijuana) is now known primarily as a hallucinogenic drug.

Rubber is another plant that has many uses today. The product had its origin in Brazil from the thick, white sap of the rubber tree. Once collected, the sap is placed in a large vat, where acid is added to coagulate the latex. When the water is pressed out, the product is formed into sheets or crumbled and placed into bales. Much stronger rubber, such as that in tires, was made by adding sulfur and heating in a process called vulcanization; this produces a flexible material less sensitive to temperature changes. Today, though, much rubber is synthetically produced.

Beyond these uses, plants have been used for centuries for a number of important household items, including the house itself. We are most familiar with lumber being used as the major structural portion in buildings. This wood comes mostly from a variety of conifers: pine, fir, and spruce, among others. In the tropics, trees and even herbs provide important components for houses. In rural parts of Central and South America, palm leaves are preferable to tin for roofs, since they last as long as ten years and are quieter during a rainstorm. In the Near East, numerous houses along rivers are made entirely of reeds.

An actively researched area of plant use today is that of medicinal plants. Currently about 50% of all pharmaceutical drugs have their origins from plants. The treatment of cancers appears to rest in the discovery of miracle plants. Indeed, the National Cancer Institute (NCI) and most pharmaceutical companies have spent millions (or, more likely, billions) of dollars to send botanists out to collect and test plant samples from around the world. Tribal medicine men, or shamans, of South America and Africa have already been of great importance in developing numerous drugs.

Over the centuries, malaria has caused far more human deaths than any other disease. After European scientists became aware that malaria can be cured by quinine, which comes from the bark of the cinchona tree, a synthetic form of the drug, chloroquine, was developed. But by the late 1960s, it was found that some of the malaria parasites, which live in red blood cells, had become resistant to the synthetically produced drug. Resistant parasites were first seen in Africa but now are showing up in Asia and the Amazon. Today the only 100% effective drug for malaria treatment must come directly from the cinchona tree, common to northeastern South America.

Numerous plant extracts continue to be misused for their hallucinogenic or other effects on the human body; coca for cocaine and crack, opium poppy for morphine, and yam for steroids.

In addition to all these uses of plants, we should not forget nor neglect the aesthetic value of plants. Flowers brighten any yard; ornamental plants accent landscaping, and trees provide cooling shade during the summer and break the wind of winter days. Plants also produce oxygen, which is so necessary for all plants and animals.

a. Dwarf fan palms, *Chamaerops*

b. Cotton, *Gossypium*

c. Rubber, *Hevea*

Figure 29B Uses of plants.
a. Dwarf fan palms can be used to make baskets. **b.** Cotton becomes clothing worn by all. **c.** A rubber plant provides latex for making tires.

Angiosperms

Angiosperms (division Magnoliophyta) are the flowering plants; their seeds are covered by fruits. With 235,000 species, they are an exceptionally large and successful group of plants. This is six times the number of species of all other plant groups combined. All hardwood trees, including all the deciduous trees of the temperate zone and the broad-leaved evergreen trees of the tropical zone, are angiosperms, although sometimes the flowers are inconspicuous. All herbaceous (nonwoody) plants common to our everyday experience, such as grasses and most garden plants, are flowering plants. Angiosperms provide us with clothing, food, medicines, and commercially valuable products, as discussed in the reading on page 610.

Angiosperms are classified into two groups: dicotyledons and monocotyledons. The dicotyledons (or dicots) are either woody or herbaceous, and they have flower parts, usually in fours and fives, net-veined leaves, vascular bundles arranged in a circle within the stem, and two cotyledons, or seed leaves. Dicot families include many familiar plant groups, such as the buttercup, mustard, maple, cactus, pea, and rose families. The monocotyledons (or monocots) are almost always herbaceous and have flower parts in threes, parallel-veined leaves, scattered vascular bundles in the stem, and one cotyledon, or seed leaf. Monocot families include the lily, palm, orchid, iris, and grass families. The grass family includes wheat, rice, corn (maize), and other agriculturally important plants.

Life Cycle of Angiosperms

Angiosperms produce heterospores within their flowers. The microspores develop into pollen grains within the pollen sacs, and the megaspore develops into an embryo sac within an ovule (Fig. 29.13). Although some flowers disperse their pollen by wind, many flowers attract pollinators such as bees, wasps, flies, butterflies, moths, and even bats, which carry pollen from flower to flower.

The pollinator and the flower have coevolved and, therefore, they are specific to one another. For example, bee-pollinated flowers are usually blue or yellow and have ultraviolet shadings that lead the pollinator to seek nectar at the base of the flower. The mouthparts of bees are fused into a long tube, through which the bee is able to obtain nectar from this location. As the bee collects nectar, pollen is deposited on its body, which it then inadvertently carries to the next flower.

The angiosperm life cycle includes **double fertilization** in which one sperm joins with the egg, and another joins with the polar nuclei, forming a 3n endosperm nucleus. **Endosperm** is stored food for the developing embryo. In dicots, the endosperm is later absorbed by the cotyledons (seed leaves) of the embryo.

As in gymnosperms, an ovule eventually becomes a seed. Unlike gymnosperms, the seeds are covered by a **fruit** derived from the ovary and possibly from adjacent structures as well. Fruits include such structures as milkweed pods; peas and beans, which are legumes; pecans and other nuts; and grains such as rice and wheat. Tomatoes, oranges, and watermelons are examples of fleshy fruits. Fruits aid in the dispersal of seeds. There are fruits that utilize wind, gravity, water, and animals for dispersal. Because animals live in particular habitats and/or have particular migration patterns, they are apt to deliver the fruit-enclosed seeds to a suitable location for germination (when the embryo begins to grow again) and development of the adult sporophyte plant.

Figure 29.13 Flowering plant life cycle.
① The parts of the flower involved in reproduction are the **stamens** and the **pistil.** Reproduction has been divided into development of the megagametophyte, development of the microgametophyte, double fertilization, and the seed. *Development of the megagametophyte.* The **ovary** at the base of the pistil contains one or more ovules. ② Within an **ovule,** a megasporocyte (megaspore mother cell) undergoes meiosis to produce four haploid megaspores. ③ Three of these megaspores disintegrate, leaving one functional **megaspore,** which divides mitotically. ④ The result is the megagametophyte, or **embryo sac,** which typically consists of eight haploid nuclei embedded in a mass of cytoplasm. The cytoplasm differentiates into cells, one of which is an egg and another of which is the endosperm cell with two nuclei (called the polar nuclei). *Development of the microgametophyte.* ① The **anther** at the top of the stamen has pollen sacs, which contain numerous microsporocytes (microspore mother cells). ② Each microsporocyte undergoes meiosis to produce four haploid cells called microspores. When the **microspores** separate, each one becomes a microgametophyte, or pollen grain. ③ At this point, the young microgametophyte contains two nuclei: the generative cell and the tube cell. Pollination occurs when pollen is windblown or carried by insects, birds, or bats to the stigma of the same type of plant. ④ Only then does a pollen grain germinate and produce a long pollen tube. This pollen tube grows within the style until it reaches an ovule in the ovary. Before fertilization occurs, the generative nucleus divides, producing two sperm, which have no flagella. This germinated **pollen grain** with its pollen tube and two sperm is the mature microgametophyte. *Double fertilization.* ⑤ On reaching the ovule, the pollen tube discharges the sperm. One of the two sperm migrates to and fertilizes the egg, forming a zygote; the other unites with the two polar nuclei, producing a 3n (triploid) endosperm nucleus. The endosperm nucleus divides to form **endosperm,** food for the developing plant. This so-called double fertilization is unique to angiosperms. *The seed.* ⑥ The ovule now develops into the **seed,** which contains an embryo and food enclosed by a protective seed coat. The wall of the ovary and sometimes adjacent parts develop into a fruit that surrounds the seeds. Therefore, angiosperms are said to have *covered seeds.*

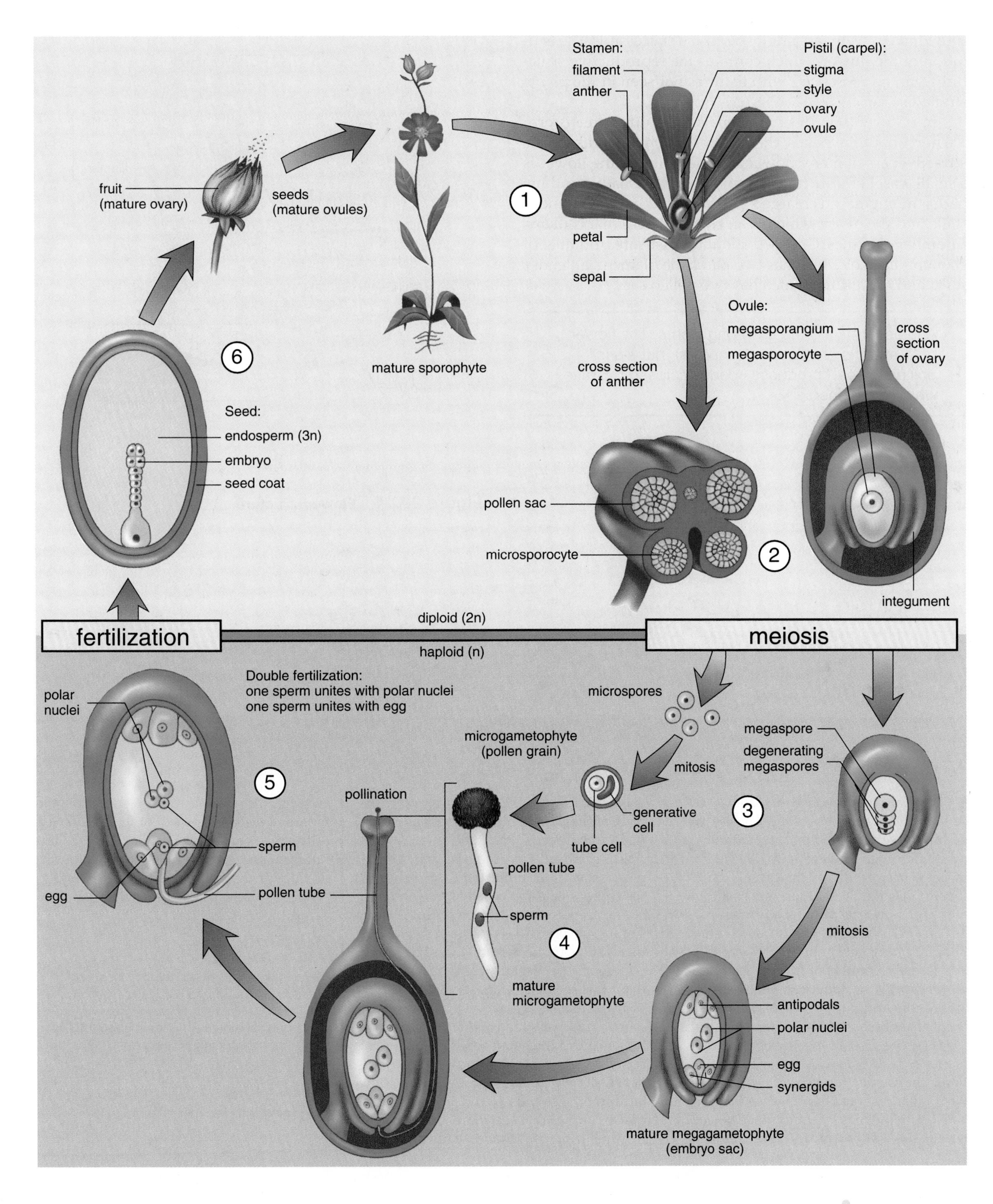
Stamen:
filament
anther
Pistil (carpel):
stigma
style
ovary
ovule
petal
sepal
fruit
(mature ovary)
seeds
(mature ovules)
mature sporophyte
1
Ovule:
megasporangium
megasporocyte
cross
section
of ovary
cross section
of anther
6
Seed:
endosperm (3n)
embryo
seed coat
pollen sac
microsporocyte
2
integument
diploid (2n)
fertilization
meiosis
haploid (n)
Double fertilization:
one sperm unites with polar nuclei
one sperm unites with egg
microspores
polar
nuclei
megaspore
degenerating
megaspores
microgametophyte
(pollen grain)
mitosis
5
pollination
generative
cell
3
tube cell
sperm
pollen tube
egg
pollen tube
sperm
mitosis
4
mature
microgametophyte
antipodals
polar nuclei
egg
synergids
mature megagametophyte
(embryo sac)

The Flower Why are angiosperms so successful? The evolution of the **flower**, which contains the reproductive structures, can be associated with features not seen in the plants studied so far.

In a flower (Fig. 29.14) the sepals, most often green, form a whorl about the petals, the color of which accounts for the attractiveness of many flowers. In the center of the flower is a small, vaselike structure, the **pistil,** which usually has three parts: the stigma, an enlarged sticky knob; the style, a slender stalk; and the **ovary,** which is an enlarged base that contains a number of ovules. Grouped about the pistil are a number of **stamens,** each of which has two parts: the filament, a slender stalk, and the **anther,** which has two pollen sacs.

Angiosperms are the flowering plants. The flower both attracts animals (e.g., insects) that aid in pollination and produces seeds enclosed by fruits, which aid dispersal.

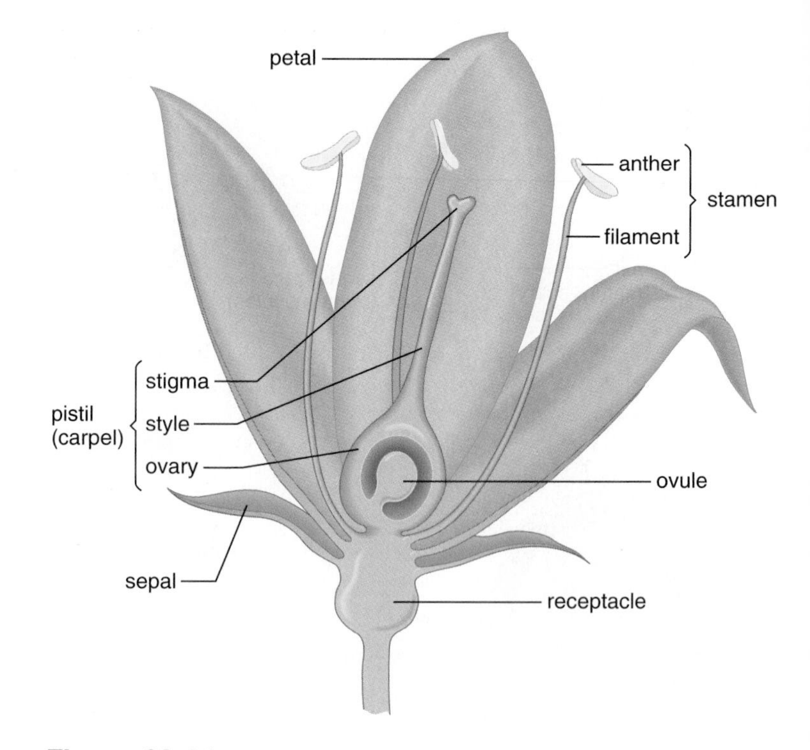

Figure 29.14 Flower structure.

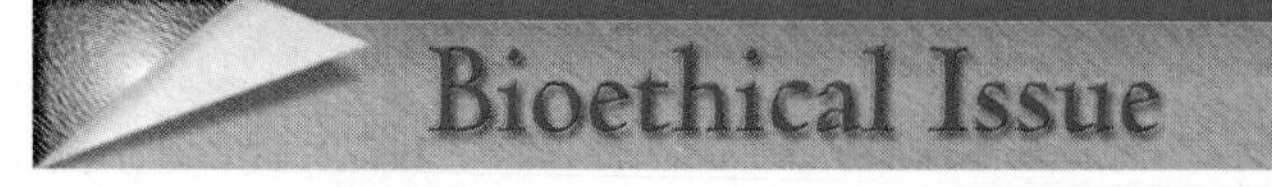
Bioethical Issue

How far should you go to save an endangered species? As a society we seem to have answered that any amount of time, energy, and money is worth the saving of certain species. Think of all the effort to preserve the condor, a magnificent bird that now graces the California skies once more. Would we put so much effort into preserving any animal? any plant?

Largely because of human encroachment, dozens of American plant species are disappearing from their native habitat each year. Botanists estimate that some 3,000 of the 22,000 species of flowering plants in the U.S. may be facing extinction. Enter the Center for Plant Conservation, which has its headquarters at Harvard University's Arnold Arboretum. The center aims to preserve every kind of threatened plant in the U.S. through a network of 18 affiliated botanical gardens and horticultural research facilities in 14 states. Aside from growing the plants in greenhouses and other protected environments, the center stockpiles seeds of most species at the Department of Agriculture's Fort Collins, Colorado, seed storage facility. In that way, despite any disaster that may wipe out a particular greenhouse or garden, the seeds would be available for propagation.

Should we spend money to save the frostweed or the small whorled pogonia? Some say yes because of possible medical benefits. They point out, for example, that antitumor alkaloids found in the Madagascar periwinkle are now used in the treatment of childhood leukemia and Hodgkin's disease. Also, some of the plants resist disease or can survive droughts, characteristics that could be important to agriculture. Through biotechnology techniques, it may be possible to transfer these traits from wild plants to crop species some day.

Questions

1. Are you in favor of putting a financial limit on the preservation of a particular animal species? a particular plant species?
2. Are you in favor of preserving every native plant in America in hopes that any one of them might contain a drug useful to humans?
3. Are you in favor of preserving the habitat of an animal or a plant if, in so doing, it means giving up jobs for humans?

Summarizing the Concepts

29.1 Characteristics of Plants

Plants are multicellular, photosynthetic organisms adapted to a land existence. Among the various adaptations, all plants protect the developing embryo from drying out. Plants have an alternation of generations life cycle, but some have a dominant gametophyte (haploid generation) and others have a dominant sporophyte (diploid generation).

29.2 Nonvascular Plants

The nonvascular plants, which include the liverworts and the mosses, lack true roots, stems, and leaves; that is, these structures do not have vascular tissue. In the moss life cycle, antheridia produce swimming sperm that use external water to reach the eggs in the archegonia. Following fertilization, the dependent moss sporophyte consists of a foot, stalk, and a capsule within which wind-blown spores are produced by meiosis. Each spore germinates to produce a gametophyte.

29.3 Seedless Vascular Plants

Vascular plants have vascular tissue, that is, xylem and phloem. In the life cycle of vascular plants, the sporophyte is dominant. Whisk ferns, club mosses, horsetails, and ferns are the seedless vascular plants that were prominent in swamp forests during the Carboniferous period. In the life cycle of seedless plants, spores disperse the species, and the separate gametophyte produces flagellated sperm. Vegetative (asexual) reproduction is used to a degree to disperse ferns in dry habitats.

29.4 Seed Plants

Seed plants have a life cycle in which there are microgametophytes (male gametophytes) and megagametophytes (female gametophytes). The microgametophyte is the pollen grain, which produces nonflagellated sperm, and the megagametophyte, which is located within an ovule, produces an egg. The pollen grain replaces the flagellated sperm of seedless vascular plants.

Gymnosperms produce seeds that are uncovered. In gymnosperms, the microgametophytes develop in pollen cones. The megagametophyte develops within an ovule located on the scales of seed cones. Following pollination and fertilization, the ovule becomes a winged seed that is dispersed by wind.

Angiosperms produce seeds that are covered by fruits. The petals of flowers attract pollinators, and the ovary develops into a fruit, which aids dispersal of seeds. Angiosperms provide most of the food that sustains terrestrial animals, and they are the source of many products used by humans.

Studying the Concepts

1. What characteristics define plants? 600
2. What are the three types of bryophytes? Mosses are found in varied habitats. Explain. 602
3. Describe the moss life cycle, and point out significant features of this cycle. What is the significance of spores dispersing the species? 603
4. Describe the plants that are included in the seedlesss vascular plants. Include in your discussion their evolutionary significance and their usefulness to humans. 604–06
5. Describe the fern life cycle, and point out significant features of this cycle. Discuss the significance of spores dispersing the species. 607
6. Describe three types of gymnosperms. 608
7. Describe the pine life cycle, and point out significant features of this cycle. Discuss the significance of seeds dispersing the species. 609
8. What are the two classes of angiosperms? Name several types of plants in each class. 612
9. How did the evolution of the flower contribute to the angiosperms' dominance of the terrestrial environment? 612–14
10. Contrast the reproductive adaptations of the nonvascular plants, ferns and their allies, gymnosperms, and angiosperms to a land environment. 602, 604, 608–09, 612–14.

Testing Yourself

Choose the best answer for each question.

1. Which of these are characteristics of plants?
   a. multicellular with specialized tissues and organs
   b. photosynthetic and contain chlorophylls *a* and *b*
   c. protect the developing embryo from desiccation
   d. have an alternation of generations life cycle
   e. All of these are correct.
2. In the moss life cycle, the sporophyte
   a. consists of leafy green shoots.
   b. is microscopic.
   c. is the heart-shaped prothallus.
   d. consists of a foot, a stalk, and a capsule.
   e. is the dominant generation.
3. The rhyniophytes
   a. are a flourishing group of plants today.
   b. had large leaves like today's ferns.
   c. had sporangia at the tips of their branches.
   d. only lived in dry areas.
   e. All of these are correct.
4. You are apt to find ferns in a moist location because they have
   a. a water-dependent sporophyte generation.
   b. flagellated spores.
   c. flagellated sperm.
   d. large, leafy fronds.
   e. All of these are correct.
5. Which of these is mismatched?
   a. pollen grain—microgametophyte
   b. ovule—megagametophyte
   c. flowering plant—mature sporophyte
   d. seed—immature sporophyte
   e. pollen tube—spores
6. In the life cycle of the pine tree, the ovules are found on the
   a. needlelike leaves.
   b. seed cones.
   c. root hairs
   d. pollen cones.
   e. All of these are correct.
7. Which of these is mismatched?
   a. anther—produces microsporangia
   b. pistil—produces pollen
   c. stigma—enlarged sticky knob
   d. ovule—becomes seed
   e. ovary—becomes fruit

8. Which of these plants contributed the most to our present-day supply of coal?
   a. nonvascular plants
   b. seedless vascular plants
   c. rhyniophytes
   d. conifers
   e. angiosperms
9. Which of these is found in seed plants?
   a. complex vascular tissue
   b. megaspores and microspores
   c. pollen grains replace swimming sperm
   d. retention of megagametophyte within the ovule
   e. All of these are correct.
10. Label this diagram of the alternation of generations life cycle.

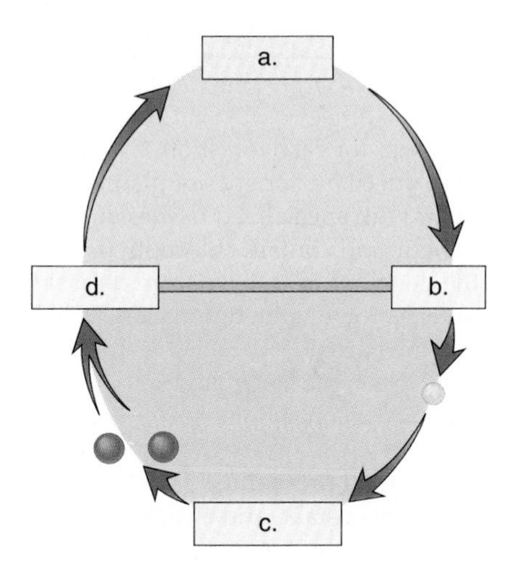

Thinking Scientifically

1. A student has learned that a plant sporophyte has vascular tissue.
   a. Do you expect the student to find vascular tissue in the moss sporophyte?
   b. How could you test that bryophytes are adapted to moist locations?
   c. Why would you expect any mutations in bryophytes to be tested immediately by the environment?
   d. Would diversification be promoted if the rudiments of vascular tissue evolved in bryophytes? Explain.
2. Terrestrial organisms have similar adaptations.
   a. Most terrestrial organisms protect the gametes and the embryo from dying out. Compare the reproductive adaptations of humans to that of trees.
   b. Most terrestrial organisms have a transport system. Compare the transport adaptations of a tree to that of humans.
   c. Many terrestrial organisms have an internal skeleton to oppose the force of gravity. Compare the skeletal adaptations of a tree to that of humans.
   d. In the temperate zone, humans remain active in the winter; deciduous trees do not. Explain.

Understanding the Terms

alternation of generations 601
angiosperm 612
anther 614
antheridium 602
archegonium 602
club moss 605
conifer 608
cycad 608
double fertilization 612
embryo sac 612
endosperm 612
fern 605
flower 614
fruit 612
gametophyte 601
ginkgo 608
gymnosperm 608
heterospore 601
horsetail 605
Kingdom Plantae 600
liverwort 602
megagametophyte 608
megaspore 601
microgametophyte 608
microspore 601
moss 602
nonvascular plant 602
ovary 614
ovule 608
phloem 604
pistil 614
pollen grain 608
pollination 608
psilotophyte 604
seed 608
sorus 607
sporangium 601
spore 601
sporophyte 601
stamen 614
vascular plant 604
xylem 604

Match the terms to these definitions:

a. ________ In seed plants, a large spore that develops into the egg-producing megagametophyte.
b. ________ Seed plant with uncovered (naked) seeds; examples are conifers and cycads, which bear cones.
c. ________ In the life cycle of a plant, the diploid generation that produces spores by meiosis.
d. ________ Mature ovule that contains a sporophyte embryo with stored food enclosed by a protective coat.
e. ________ In seed plants, a structure where the megaspore becomes an egg-producing megagametophyte and which develops into a seed following fertilization.

Using Technology

Your study of plants is supported by these available technologies.

Essential Study Partner CD-ROM
Evolution & Diversity → Plants

Visit the Mader web site for related ESP activities.

Exploring the Internet

The Mader Home Page provides resources and tools as you study this chapter.

http://www.mhhe.com/biosci/genbio/mader

C H A P T E R 30

Animals: Part I

Chapter Concepts

Sea anemones, such as these strawberry sea anemones, Corynachtis californica, *are called the "flowers of the sea," but they are animals, not plants. The projections you see are tentacles, which capture small prey and stuff it down their mouths.*

While Oscar watches a spider close in on a dragonfly trapped by its web, a darting butterfly distracts the young boy. He chases the winged creature down the path to the ocean. At the water's edge, he notices several sea sponges and shells that have washed up and dried out in the hot sun. Oscar then spots his sister who has swum out to a small coral reef to snorkel. She yelps as an unseen jellyfish stings her, causing her brother to laugh. "Oscar! Time to set the table, the lobsters are boiling," shouts out a motherly voice from the porch of the nearby beach house.

Oscar's brief period of play, and his approaching dinner, has brought him into contact with a large crowd of invertebrates, the amazingly diverse group of animals that includes more than a million species of arthropods alone. This chapter will introduce you to some of them, and explain how scientists classify this rich zoo of creatures.

30.1 Evolution and Classification of Animals

Although there are many different types of animals, they have characteristics in common. Animals

1. are heterotrophic and usually acquire food by ingestion followed by digestion.
2. typically have the power of motion or locomotion by means of muscle fibers.
3. are multicellular, and most have specialized cells that form tissues and organs.
4. have a life cycle in which the adult is typically diploid.
5. usually practice sexual reproduction and produce an embryo that undergoes developmental stages.

There are approximately 34 animal phyla, but we will consider only the 9 phyla listed in the classification table. All 9 phyla contain **invertebrates,** which are animals without backbones. The phylum Chordata also contains **vertebrates,** which are animals with backbones.

Animals evolved from unicellular protozoans. The advent of multicellularity, which allows specialization of cells to occur, seems to have been a requirement for the evolution of animals.

Sponges have the *cellular level of organization,* meaning that they have no tissues. One of the main events during the development of the rest of the animal groups is the establishment of *germ layers* from which all other structures are derived. Although a total of three germ layers is seen in most animals, the cnidarians have only two germ layers (ectoderm and endoderm) and the *tissue level of organization.* Animals with three germ layers—ectoderm, mesoderm, and endoderm—have an *organ level of organization.*

Classification

Kingdom Animalia

Multicellular organisms with well-developed tissues; usually motile; heterotrophic by ingestion, generally in a digestive cavity; diplontic life cycle.

Phylum	Some Representatives	Approximate Number of Described Species
Porifera (sponges)	Glass, chalk, bath sponges	5,000
Cnidarians (cnidarians)	Hydrozoans, jellyfishes, sea anemones, corals	9,000
Platyhelminthes (flatworms)	Planarians, flukes, tapeworms	13,000
Nematoda (roundworms)	Pinworms, hook worms	500,000
Mollusca (mollusks)	Snails, clams, squids, octopuses	110,000
Annelida (annelids)	Clam worms, earthworms,leeches	12,000
Arthropoda (arthropods)	Crayfish, insects, millipedes, spiders	One million plus
Echinodermata (echinoderms)	Starfish, sea urchins, sand dollars, sea cucumbers	6,000
Chordata (chordates)		
Cephalochordata (cephalochordates)	Lancelet	23
Urochordata (urochordates)	Tunicates	1,250
Vertebrata (vertebrates)		
Agnatha (jawless fishes)	Lampreys, hagfishes	63
Chondrichthyes (cartilaginous fishes)	Sharks, skates, rays	850
Osteichthyes (bony fishes)	Herring, salmon, cod, eel	20,000
Amphibia (amphibians)	Frogs, toads, salamanders	3,900
Reptilia (reptiles)	Snakes, lizards, turtles	6,000
Aves (birds)	Sparrows, penguins, ostriches	9,000
Mammalia (mammals)	Cats, dogs, horses, rats, humans	4,500
Primates	Prosimians, monkeys, apes	
Anthropoidea	Monkeys, apes, humans	
Hominidae	Apes, humans	
Homo	Humans	

Sponges are asymmetrical. **Asymmetry** means that the animal has no particular symmetry. Other animals are either radially symmetrical or bilaterally symmetrical. **Radial symmetry** means that the animal is organized circularly, and, just as with a wheel, two identical halves are obtained no matter how the animal is sliced longitudinally. **Bilateral symmetry** means that the animal has definite right and left halves; only one longitudinal cut down the center of the animal will produce two equal halves.

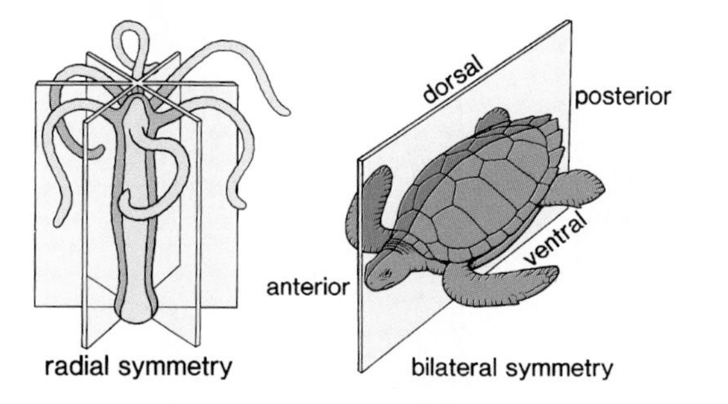

Radially symmetrical animals are sometimes attached to a substrate; that is, they are **sessile.** This type of symmetry is useful to these animals since it allows them to reach out in all directions from one center. Bilaterally symmetrical animals tend to be active and to move forward with an anterior end. During the evolution of animals, bilateral symmetry is accompanied by **cephalization,** localization of a brain and specialized sensory organs at the anterior end of an animal.

Two body plans are observed in the animal kingdom: the *sac plan* and the *tube-within-a-tube plan.* Animals with the sac plan have an incomplete digestive system. It has only one opening, which is used both as an entrance for food and an exit for undigested material. Animals with the tube-within-a-tube plan have a complete digestive system, with a separate entrance for food and exit for undigested material. Having two openings allows specialization of parts to occur along the length of the tube.

A true **coelom** is an internal body cavity completely lined by mesoderm, where internal organs are found. Flatworms are acoelomates—they have mesoderm but no body cavity. The roundworms have a **pseudocoelom,** a body cavity incompletely lined by mesoderm because it develops between the mesoderm and endoderm. There is a layer of mesoderm beneath the body wall but not around the gut. The rest of the phyla in the classification table are true coelomates—they have a coelom that is completely lined with mesoderm. Coelomates are either protostomes or deuterostomes. When the first embryonic opening becomes the mouth, the animal is a **protostome.** When the second opening becomes the mouth, the animal is a **deuterostome.**

Among coelomates, mollusks and echinoderms are nonsegmented, while annelids, arthropods, and chordates are segmented. **Segmentation,** which is the repetition of body parts along the length of the body, leads to specialization of parts because the various segments can become differentiated for specific purposes.

Classification of animals is based on type of level of organization, symmetry, body plan, type of coelom, and presence of segmentation.

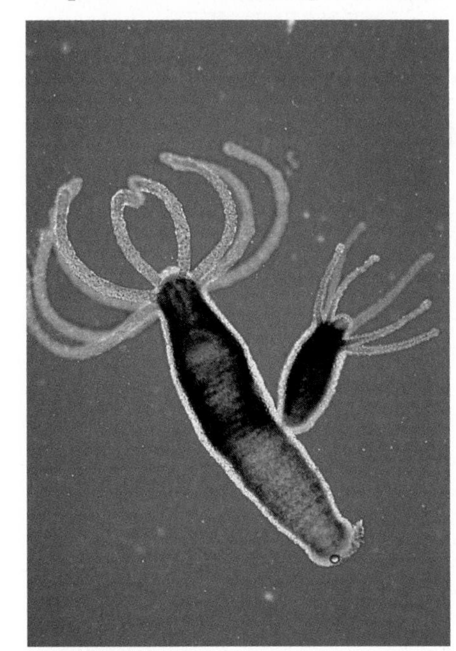

a. Hydra, *Hydra*

b. Green crayfish, *Barbi*

c. Human being, *Homo*

Figure 30.1 Animal diversity.
How do hydras differ from crayfish and humans? They are all multicellular heterotrophic organisms but differ according to their level of organization (hydras have only tissues but crayfishes and humans have organ systems); symmetry (hydras are radially symmetrical but crayfish and humans are bilaterally symmetrical with cephalization); and body plan (hydras have a sac body plan while crayfish and humans have a tube-within-a-tube body plan).

Visual Focus

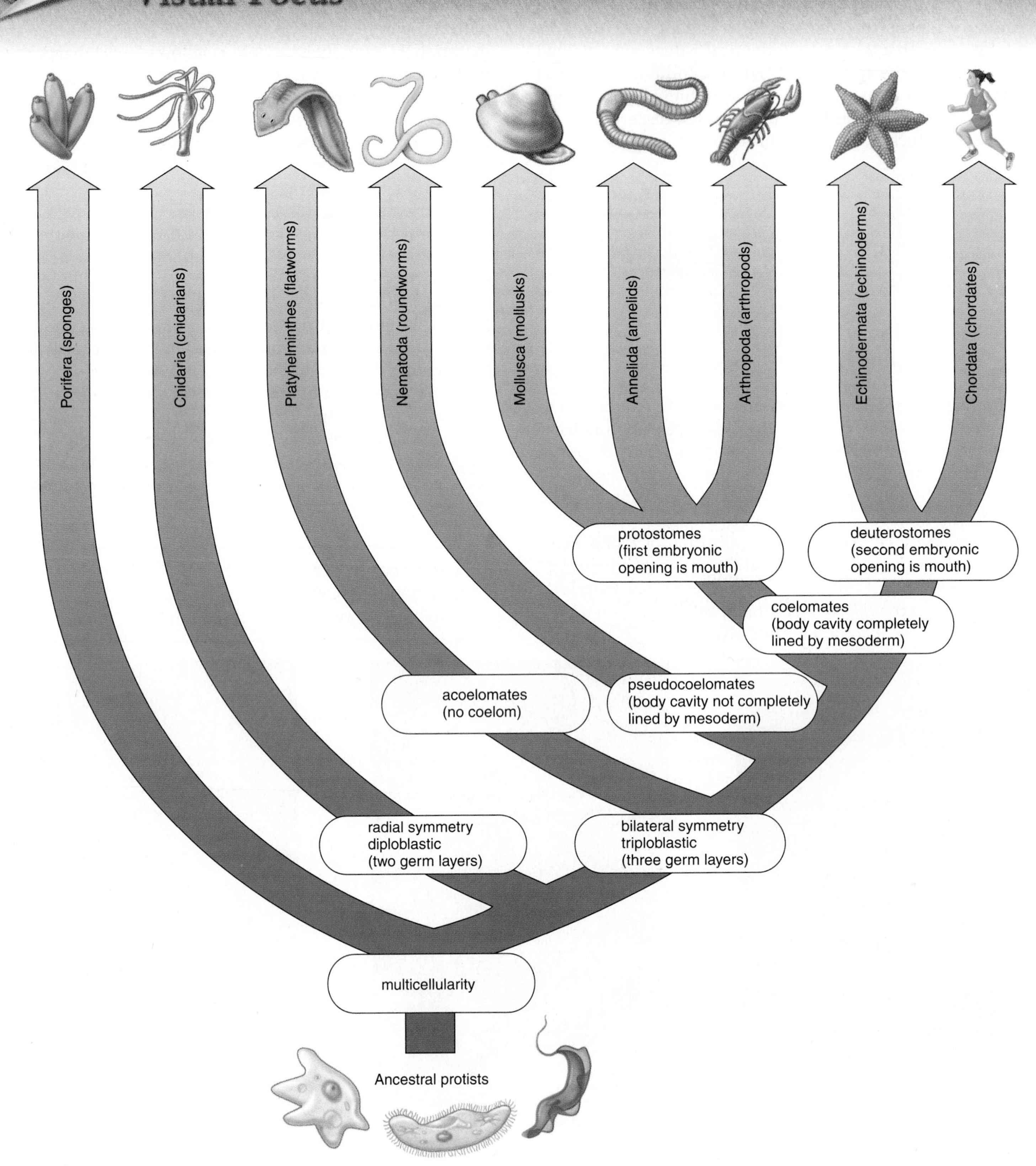

Figure 30.2 **Evolutionary tree.**
All animals are believed to be descended from protists; the Porifera (sponges) with the cellular level of organization may have evolved separately.

30.2 Introducing the Invertebrates

Sponges are asymmetrical—have no definite symmetry—while cnidarians are radially symmetrical. Flatworms and roundworms are bilaterally symmetrical, but differ in that flatworms are acoelomates while roundworms are pseudocoelomates.

Sponges

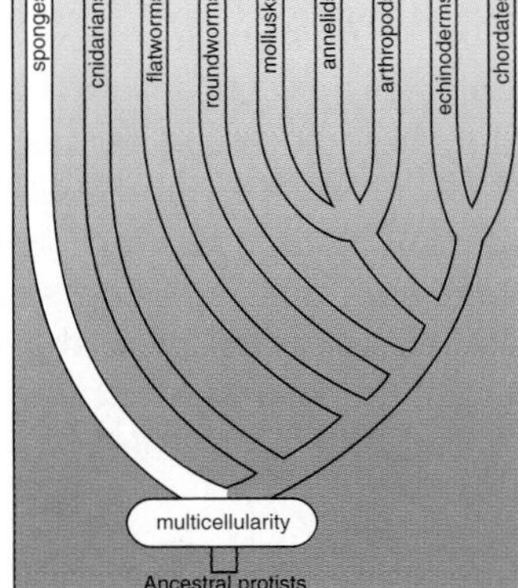

Sponges (phylum Porifera, about 5,000 species) are aquatic, largely marine animals, that vary greatly in size, shape, and color. Their saclike bodies are perforated by many pores; the phylum name, Porifera, means pore bearing. Sponges are multicellular with the *cellular level of organization.* As such, they are believed to be out of the mainstream of animal evolution. Most likely, they evolved separately from protozoan ancestors and represent a dead-end branch of the evolutionary tree.

The outer layer of the wall contains flattened epidermal cells, some of which have contractile fibers; the middle layer is a semifluid matrix with wandering amoeboid cells, and the inner layer is composed of flagellated cells called *collar cells* (or choanocytes) (Fig. 30.3). The beating of the flagella produces water currents that flow through the pores into the central cavity and out through the *osculum,* the upper opening of the body. Even a simple sponge only 10 cm tall is estimated to filter as much as 100 liters of water each day. It takes this much water to supply the needs of the sponge. A sponge is a sessile **filter feeder,** an organism that filters its food from the water by means of a straining device—in this case, the pores of the walls and the microvilli making up the collar of collar cells. Microscopic food particles that pass between the microvilli are engulfed by the collar cells and digested by them in food vacuoles, or are passed to the amoeboid cells for digestion. The amoeboid cells also act as a circulatory device to transport nutrients from cell to cell, and they produce the sex cells (the egg and the sperm) and spicules.

Sponges can reproduce asexually by fragmentation or by *budding.* During budding, a small protuberance appears and gradually increases in size until a complete organism forms. Budding produces colonies of sponges that can become quite large. During sexual reproduction, eggs and sperm are released into the central cavity and the zygote develops into a flagellated larva that may swim to a new location. If the cells of a sponge are mechanically separated, they will reassemble into a complete and functioning organism! Like all less specialized organisms, sponges are also capable of regeneration, or growth of a whole from a small part.

Sponges are classified on the basis of their skeleton. Some sponges have an internal skeleton composed of *spicules,* small needle-shaped structures with one to six rays. Chalk sponges have spicules made of calcium carbonate; glass sponges have spicules that contain silica. Most sponges have fibers of spongin, a modified form of collagen. But some sponges contain only spongin fibers; a bath sponge is the dried spongin skeleton from which all living tissue has been removed. Today, however, commercial "sponges" are usually synthetic.

Sponges have a cellular level of organization and most likely evolved independently from protozoa. They are the only animals in which digestion occurs within cells.

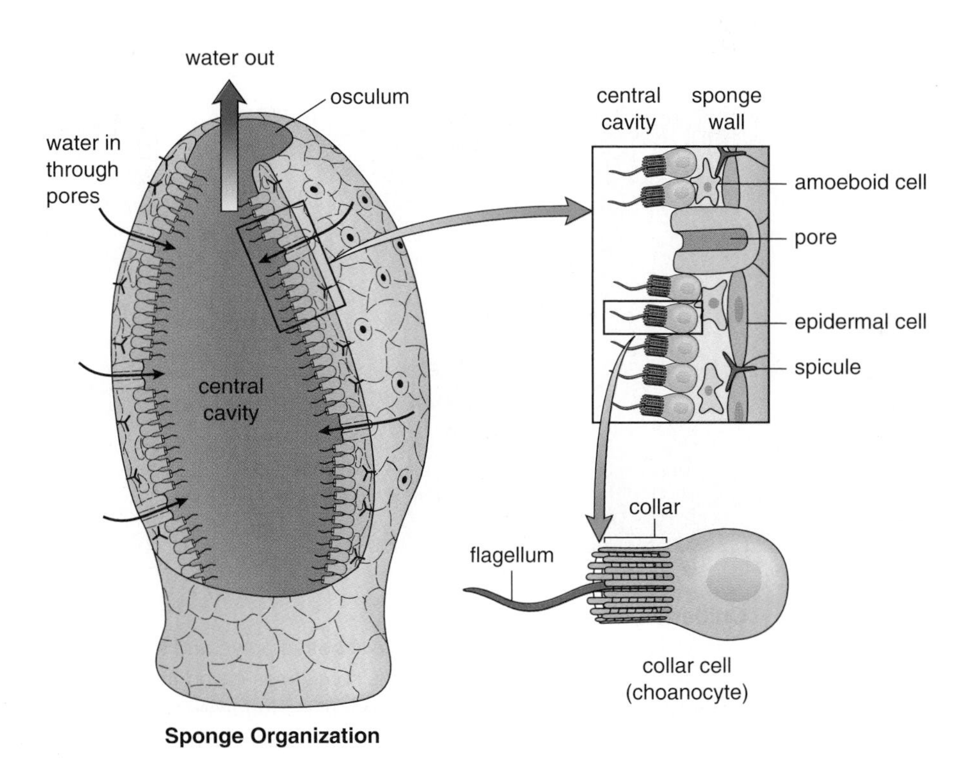

Figure 30.3 Sponge.
In a sponge, the wall contains two layers of cells: the outer epidermal cells and the inner collar cells. The collar cells (enlarged) have flagella that beat, moving the water through pores as indicated by the arrows. Food particles in the water are trapped by the collar cells and digested within their food vacuoles. Amoeboid cells transport nutrients from cell to cell; spicules form an internal skeleton in some sponges.

Cnidarians

Cnidarians (phylum Cnidaria, about 9,000 species) are tubular or bell-shaped animals that reside mainly in shallow coastal waters, except for the oceanic jellyfishes. During development, cnidarians have only two germ layers (ectoderm and endoderm), and as adults they have the *tissue level of organization.* Cnidarians are *radially symmetrical,* meaning that any longitudinal cut produces two identical halves.

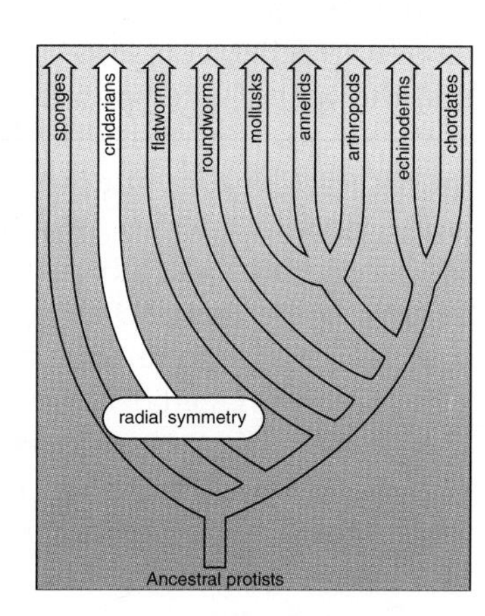

Unique to cnidarians are specialized stinging cells, called cnidocytes, which give the phylum its name. Each cnidocyte has a capsule called a **nematocyst,** which contains a long, spirally coiled hollow thread. When the trigger of the cnidocyte is touched, the nematocyst is discharged. Some threads merely trap a prey or predator; others have spines that penetrate and inject paralyzing toxins.

Two basic body forms are seen among cnidarians. The mouth of a *polyp* is directed upward from the substrate, while the mouth of a jellyfish or *medusa* is directed downward. A medusa has more mesoglea than a polyp, and the tentacles are concentrated on the margin of the bell. At one time, both body forms may have been a part of the life cycle of all cnidarians (Fig. 30.4*a*). When both are present, the sessile polyp stage produces medusae and the motile medusan stage produces egg and sperm; the zygote develops into a ciliated larva that is capable of dispersal. In some cnidarians, one stage is dominant and the other is reduced; in other species one form is absent altogether.

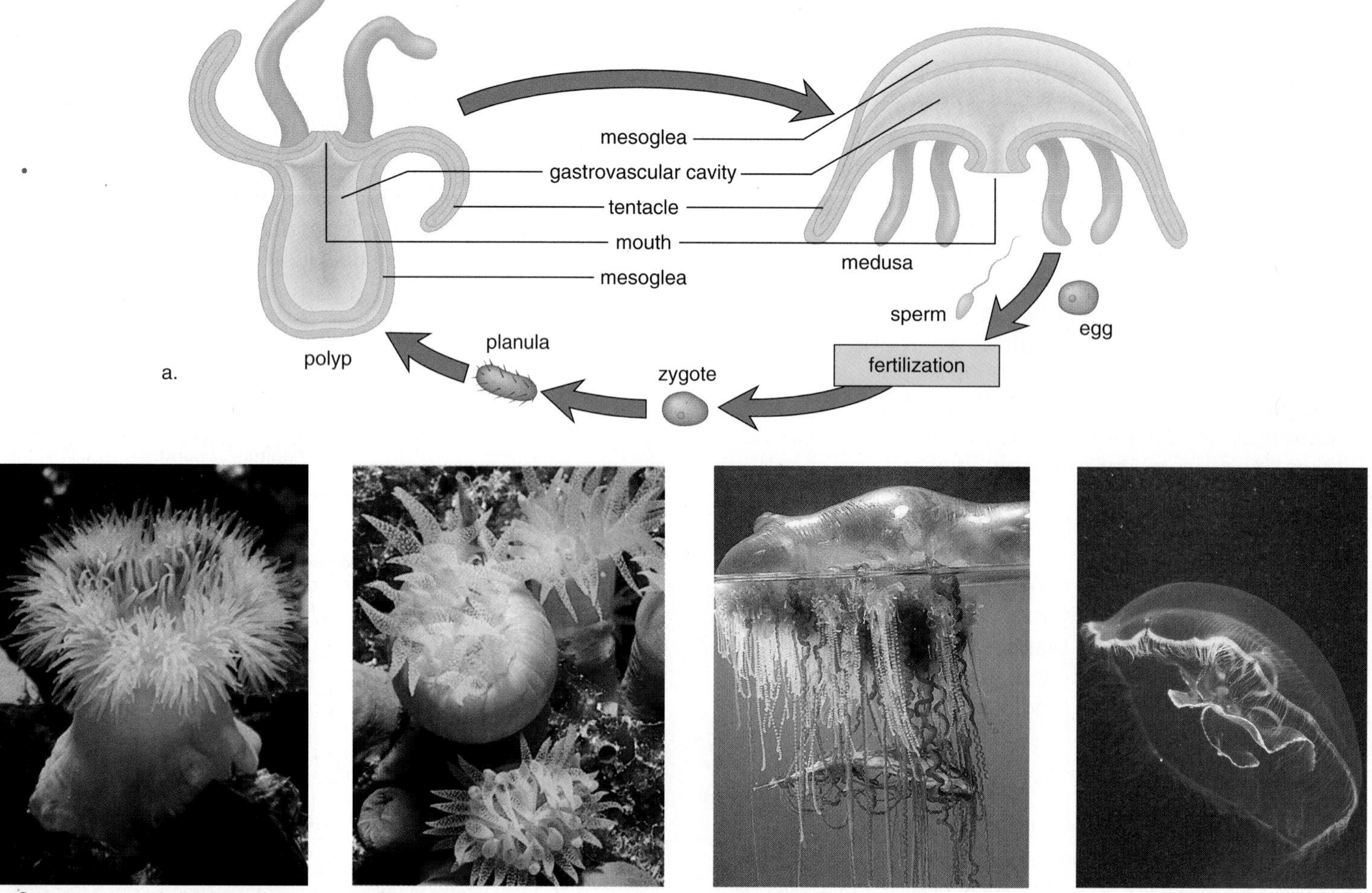

b. Sea anemone, *Apitasia*

c. Coral, *Tubastrea*

d. Portuguese man-of-war, *Physalia*

e. Jellyfish, *Aurelia*

Figure 30.4 Cnidarian diversity.
a. The life cycle of a cnidarian; in some cnidarians there is both a polyp and a medusa stage; in others, one form is dominant, and in still others, one form is absent altogether. **b.** The anemone, which is sometimes called the flower of the sea, is a solitary polyp. **c.** Corals are colonial polyps residing in a calcium carbonate or proteinaceous skeleton. **d.** Portuguese man-of-war is a colony of modified polyps and medusae. **e.** True jellyfish undergo a complete life cycle; the photo shows the medusan stage.

Cnidarians are quite diverse (Fig. 30.4*b–e*). Sea anemones are solitary polyps, often very large and with thick walls. They can be brightly colored, resembling beautiful flowers. **Corals** are similar to sea anemones, but they have calcium carbonate skeletons. Some corals are solitary, but most are colonial, with either flat and rounded or upright and branched colonies. The slow accumulation of coral skeletons forms coral reefs, which are areas of biological abundance in warmer waters. The Portuguese man-of-war is a colony of polyp and medusa types of individuals. One polyp becomes a gas-filled float and the other polyps are specialized for feeding. The medusae are specialized for reproduction. In jellyfishes, the medusa is the primary stage of the life cycle, and the polyp remains quite small and inconspicuous.

Hydra

Hydra is a freshwater cnidarian (Fig. 30.5); it is likely to be found attached to underwater plants or rocks in most lakes and ponds. The body is a small tubular polyp about 7.5 mm in length. Like all cnidarians, a hydra has the *sac body plan;* that is, there is only one opening that serves as both a mouth and anus. The outer tissue layer is a protective epidermis derived from ectoderm. The inner tissue layer, derived from endoderm, is called a gastrodermis. The two tissue layers are separated by a jellylike packing material called *mesoglea.* There are both circular and longitudinal muscle fibers. Nerve cells located below the epidermis near the mesoglea interconnect and form a *nerve net* that communicates with sensory cells throughout the body. The nerve net allows transmission of impulses in several directions at once. Having both muscle fibers and nerve fibers, cnidarians are capable of directional movement; the body can contract or extend, and the tentacles that ring the mouth can reach out and grasp prey.

Digestion begins within the central cavity but is completed within the food vacuoles of gastrodermal cells. Nutrient molecules are passed by diffusion to the other cells of the body. The large central cavity allows gastrodermal cells to exchange gases directly with a watery medium. Because the central cavity carries on digestion and acts as a circulatory system by distributing food and gases, it is called a **gastrovascular cavity.** All cnidarians have a gastrovascular cavity.

Although hydras exist only as polyps, and there are no medusae, still they can reproduce sexually or asexually. When sexual reproduction is going to occur, an ovary or a testis develops in the body wall. Like sponges, cnidarians can regenerate from a small piece. When conditions are favorable, hydras produce small outgrowths, or *buds,* that pinch off and begin to live independently.

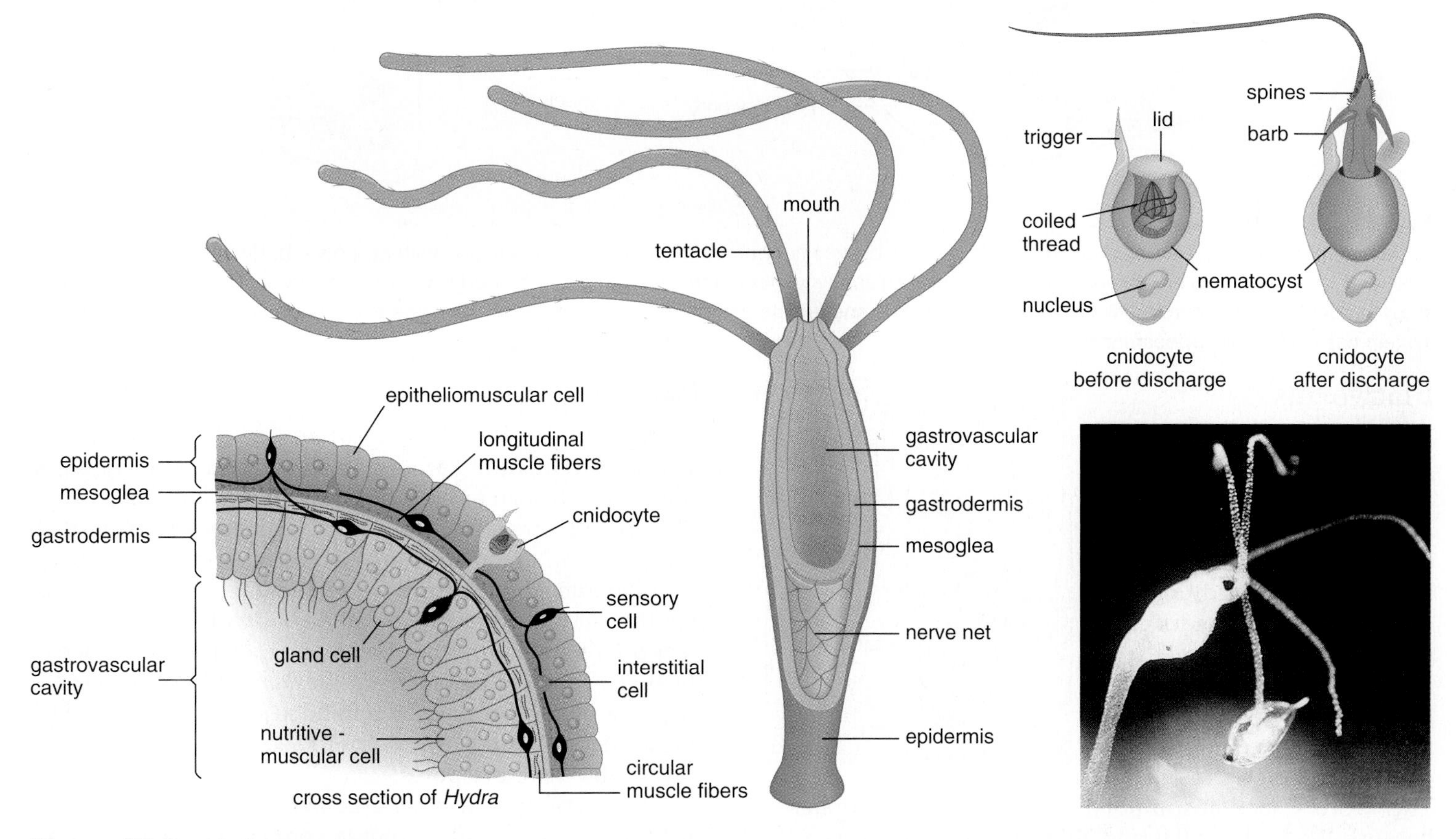

Figure 30.5 Hydrozoan, hydra.
The wall of a hydra, which lines a gastrovascular cavity, has two tissue layers. Special stinging cells (cnidocytes) contain nematocysts that assist in capturing prey. The tentacles deliver the prey to the mouth.

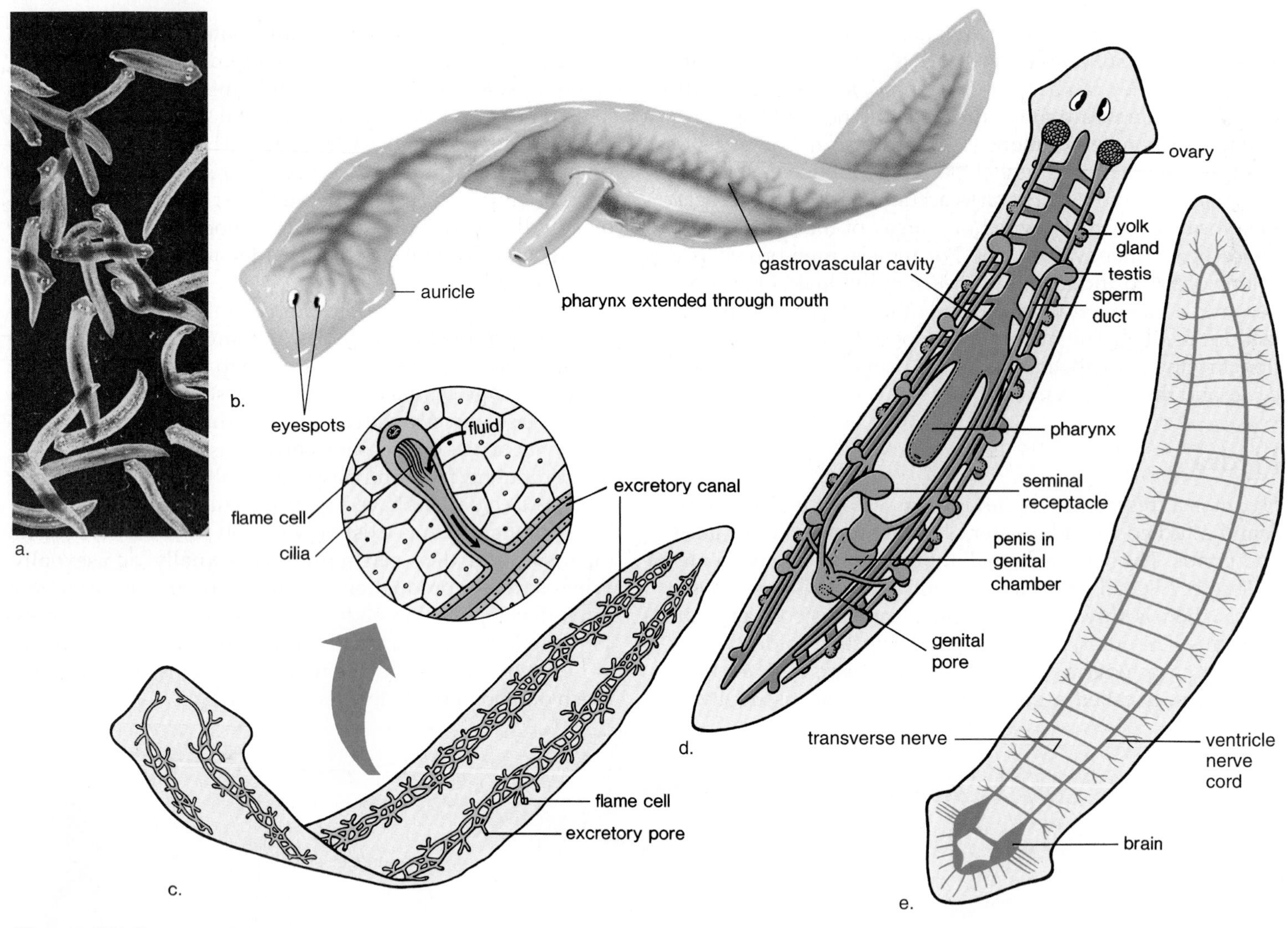

Figure 30.6 Planarian.
a. The micrograph of *Dugesia* shows that this flatworm is bilaterally symmetrical and has a head region with eyespots. **b.** When the pharynx is extended as shown, food is sucked up into a gastrovascular cavity that branches throughout the body. **c.** The excretory system with flame cells is shown in detail. **d.** The reproductive system has both male and female organs, and the digestive system has a single opening. **e.** The nervous system has a ladderlike appearance.

Flatworms

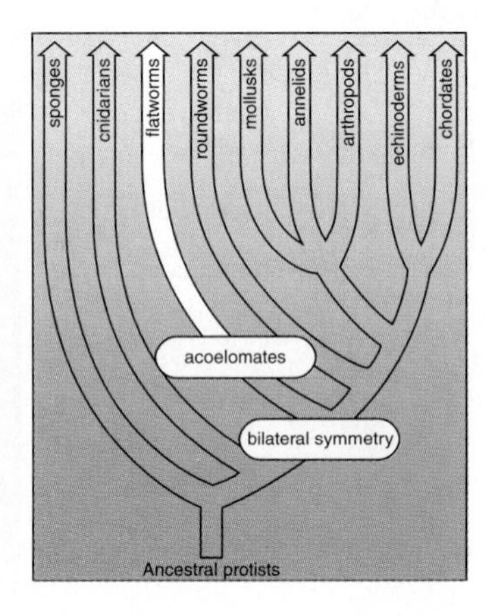

Flatworms (phylum Platyhelminthes, about 13,000 species) also have a *sac body plan.* Flatworms, however, have three germ layers. The presence of mesoderm in addition to ectoderm and endoderm gives bulk to the animal and leads to greater complexity. Free-living flatworms have muscles and excretory, reproductive, and digestive organs. The worms lack respiratory and circulatory organs—because the body is flat and thin, diffusion alone is adequate for the passage of oxygen and other substances from cell to cell.

Planarians

Freshwater *planarians* (Fig. 30.6) are small (several mm to several cm), literally flat worms. Some tend to be colorless; others have brown or black pigmentation. Planarians live in lakes, ponds, streams, and springs, where they feed on small living or dead organisms, such as worms and crustaceans.

Planarians live in fresh water and have an excretory organ that serves primarily to rid the body of excess water. A network of interconnecting canals extends through much of the body. The beating of cilia in the *flame cells* (so named because the beating of the cilia reminded some early investigator of the flickering of a flame) keeps the water moving toward the excretory pores.

Planarians have a *ladder-type nervous organ.* A small anterior brain and two lateral nerve cords are joined by cross-branches. Planarians exhibit *cephalization*—aside from a brain, there are light-sensitive organs (the eyespots) and chemosensitive organs located on the auricles. They have well-developed muscles. There are three kinds of muscle layers—an outer circular layer, an inner longitudinal layer, and a diagonal layer—that allow for quite varied movement. Their ciliated epidermis allows planarians to glide along a film of mucus.

The animal captures food by wrapping itself around the prey, entangling it in slime, and pinning it down. Then a muscular pharynx is extended, and by a sucking motion the food is torn up and swallowed. The pharynx leads into a three-branched gastrovascular cavity in which digestion is both extracellular and intracellular. The digestive tract is incomplete because it has only one opening.

Planarians are **hermaphrodites;** they possess both male and female sex organs. The worms practice cross-fertilization: the penis of one is inserted into the genital pore of the other, and there is a reciprocal transfer of sperm. The fertilized eggs hatch in 2–3 weeks as tiny worms. Like sponges and cnidarians, planarians can regenerate. If a worm is cut crosswise, each piece grows a new head or a new tail, as appropriate.

Parasitic Flatworms

Flukes and **tapeworms** are two classes of parasitic flatworms. The anterior end of these animals carries suckers and sometimes hooks for attachment to the host. The parasite absorbs nutrients from the digestive tract of the host, and in tapeworms the digestive system is essentially absent. The tegument, a specialized body wall resistant to host digestive juices, is covered by the glycocalyx, a mucopolysaccharide coating. The extensive development of the reproductive system, with the production of millions of eggs, may be associated with difficulties in dispersing offspring. Both parasites utilize a *secondary host* to transport an intermediate stage from *primary host* to primary host. The primary host is infected with the sexually mature adult; the secondary host contains the larval stage or stages.

Both flukes and tapeworms cause serious illnesses in humans. The fluke body tends to be oval to elongate. At the anterior end surrounded by sensory papilla there is an oral sucker and at least one other sucker for attachment to the host. Different fluke species infect the digestive tract, the bile duct, blood, and the lungs. Schistosomes are blood flukes that enter the body by active penetration of the skin (Fig. 30.7). *Schistosomiasis,* a serious infection caused by blood flukes, is seen predominantly in the Middle East, Asia, Africa, and South America. About 200 million people are infected worldwide.

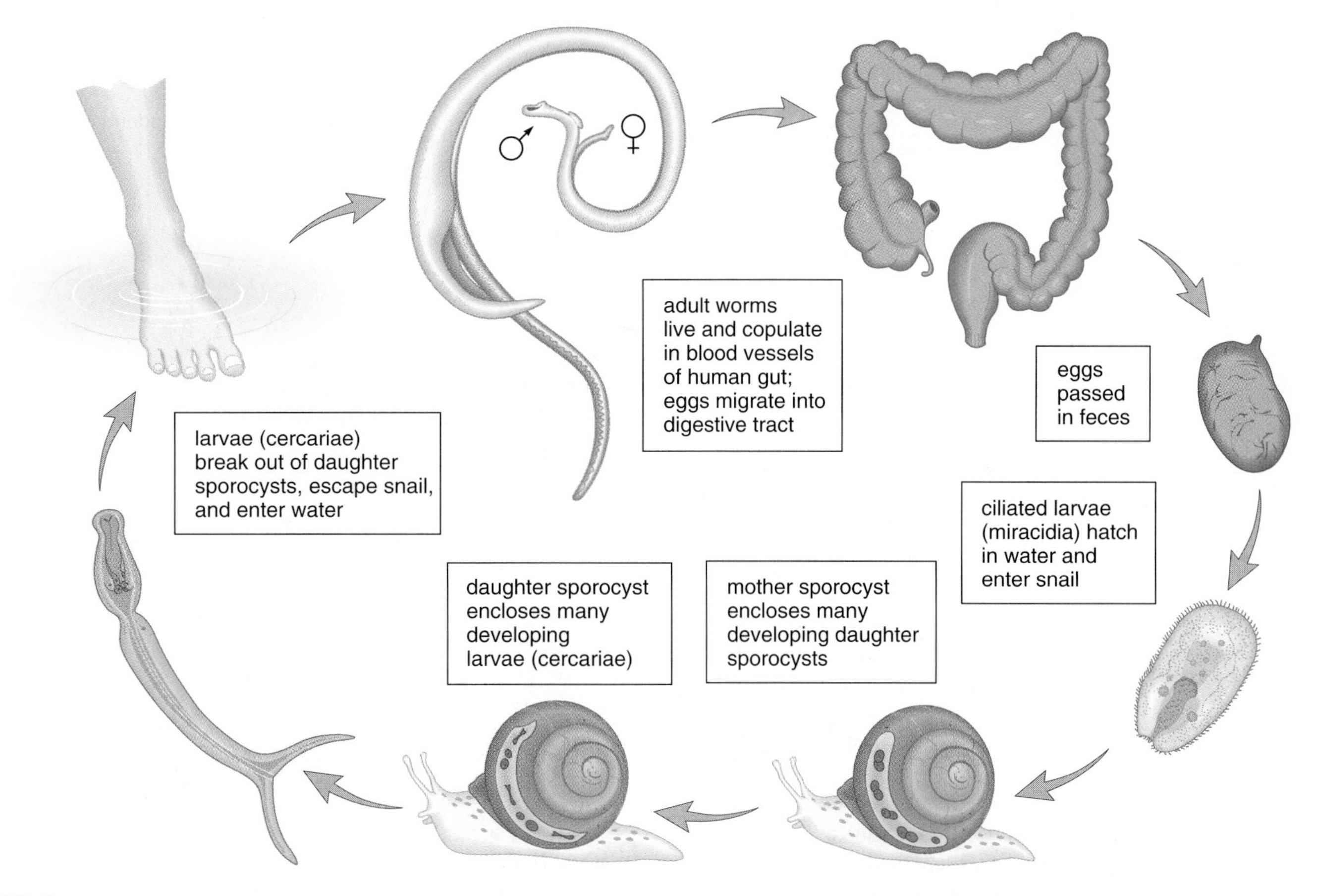

Figure 30.7 Schistosomiasis.
This infection of humans, caused by blood flukes, *Schistosoma,* is an extremely prevalent disease in Egypt—especially since the building of the Aswan High Dam. Standing water in irrigation ditches, combined with unsanitary practices, has created the conditions for widespread infection.

A tapeworm has an anterior region, called a *scolex,* containing hooks and suckers for attachment to the intestinal wall of the host. Behind the scolex, there is a long series of *proglottids,* segments that contain a full set of both male and female sex organs and little else. Mature proglottids, which are nothing but bags of eggs, break off, and as they pass out with the feces, the eggs are released. If feces-contaminated food is fed to pigs or cattle, the larvae escape when the covering of the eggs is digested away. They burrow through the intestinal wall and travel in the bloodstream to finally lodge and encyst in muscle. Here a *cyst* means a small, hard-walled structure that contains a larval worm. When humans eat infected meat, the larvae break out of the cyst, attach themselves to the intestinal wall, and grow to adulthood. Then the cycle begins again.

Also interesting to humans is the fact that fleas, which as larvae have fed on the feces of an infected host, can transmit some types of tapeworms between cats and/or dogs.

Roundworms

Roundworms (phylum Nematoda, 500,000 species) are nonsegmented—they have a smooth outside body wall. These worms, which are generally colorless and less than 5 cm in length, occur almost anywhere—in the sea, in fresh water, and in the soil—in such numbers that thousands of them can be found in a small area.

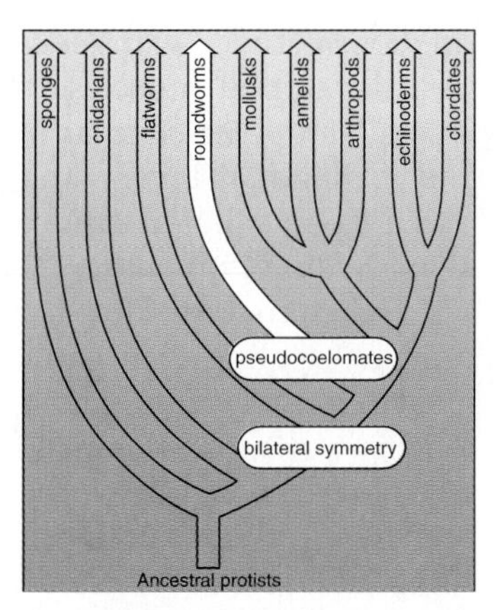

Roundworms possess two anatomical features not seen before: a tube-within-a-tube body plan and a body cavity. With a *tube-within-a-tube body plan,* the digestive tract is complete; there is both a mouth and an anus. The body cavity is a *pseudocoelom,* or a body cavity incompletely lined with mesoderm (Fig. 30.8*b*). The fluid-filled pseudocoelom provides space for the development of organs, substitutes for a circulatory system by allowing easy passage of molecules, and provides a type of skeleton. Worms in general do not have an internal or external skeleton, but they do have a **hydrostatic skeleton,** a fluid-filled interior that supports muscle contraction and enhances flexibility.

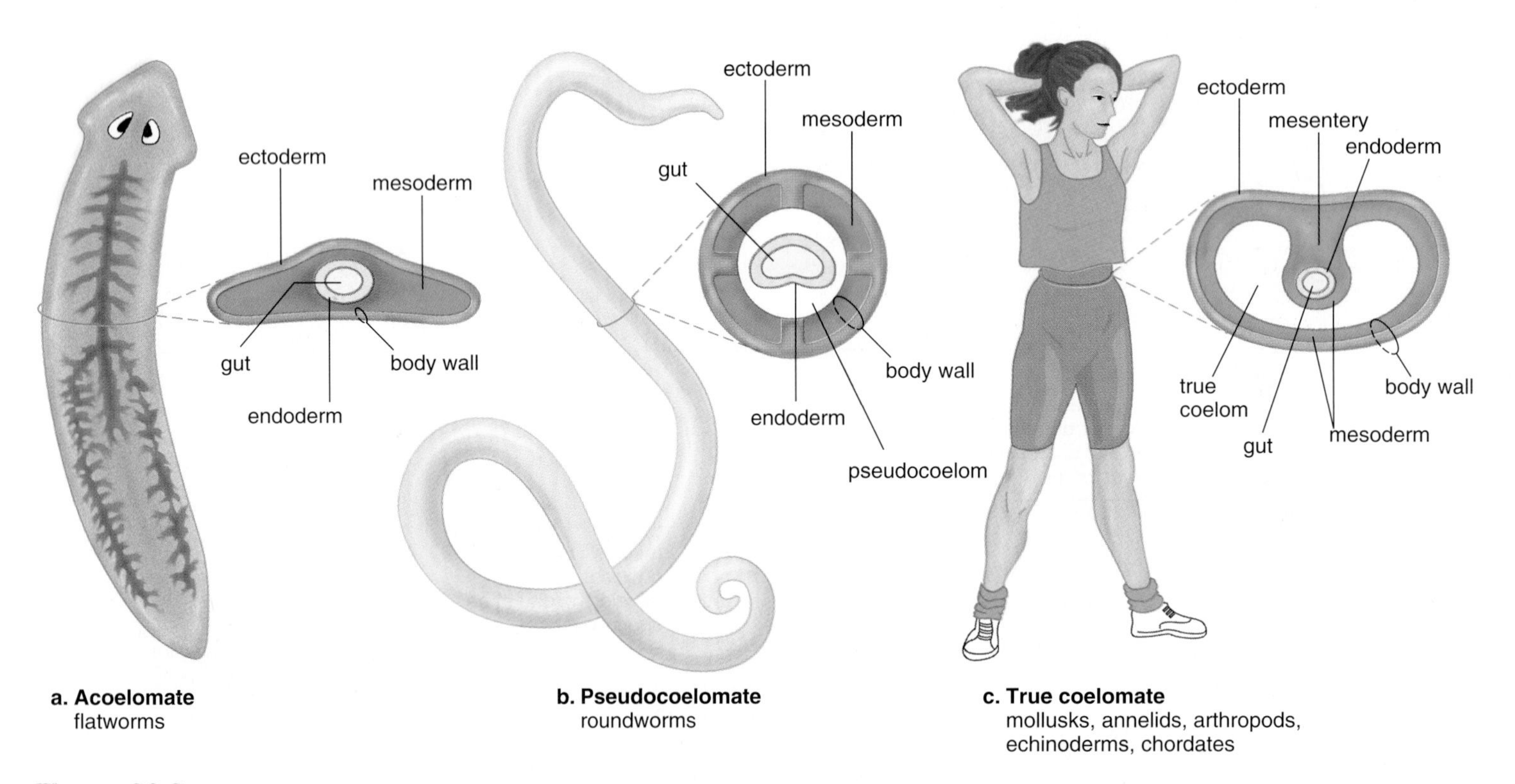

Figure 30.8 Coelom structure and function.
a. Flatworms have no body cavity, and mesodermal tissue fills the interior space. **b.** Roundworms have a pseudocoelom, and the body cavity is incompletely lined by mesodermal tissue. In animals that have no other skeleton, a fluid-filled coelom acts as a hydrostatic skeleton. **c.** Humans are true coelomates, the body cavity is completely lined by mesodermal tissue, and mesentery holds the internal organs in place.

Ascaris

In the roundworm *Ascaris lumbricoides* (Fig. 30.9), females tend to be larger (20–35 cm in length) than males. Both sexes move by means of a characteristic whiplike motion because only longitudinal muscles and no circular muscles lie next to the body wall.

The *Ascaris* life cycle usually begins when larvae within a protective covering are swallowed. The worms then escape from the covering and burrow through the host's intestinal wall. Making their way through the organs of the host, they move from the intestine to the liver, the heart, and then the lungs. While in the lungs for about ten days, they grow in size. The larvae then migrate up the windpipe to the throat, where they are swallowed, allowing them to once again reach the intestine. Once the worms are mature they mate, and the female produces larva-containing eggs, which pass out with the feces.

Other Roundworms

Trichinosis is a fairly serious infection caused by *Trichinella spiralis,* a roundworm that rarely infects humans in the United States. Humans contract the disease when they eat rare pork containing encysted larvae. After maturation, the female adult burrows into the wall of the host's small intestine and produces live offspring, which are carried by the bloodstream to the skeletal muscles, where they encyst (Fig. 30.9). The symptoms of trichinosis include muscular pain, weakness, fever, and anemia.

Elephantiasis is caused by a roundworm called the filarial worm, which utilizes the mosquito as a secondary host. Because the adult worms reside in lymphatic vessels, fluid return is impeded, and the limbs of an infected human can swell to an enormous size, even resembling those of an elephant. When a mosquito bites an infected person, it transports larvae to a new host.

Other roundworm infections are more common in the United States. Children frequently acquire a pinworm infection, and hookworm is seen in the southern states, as well as worldwide. A hookworm infection can be very debilitating because the worms attach to the intestinal wall and feed on blood. Good hygiene, proper disposal of sewage, and cooking meat thoroughly usually protect people from parasitic roundworms.

Roundworms have bilateral symmetry, a tube-within-a-tube body plan, three germ layers, the organ level of organization, and a pseudocoelom. The rest of the animal phyla have these features, except there is a true coelom instead of a pseudocoelom.

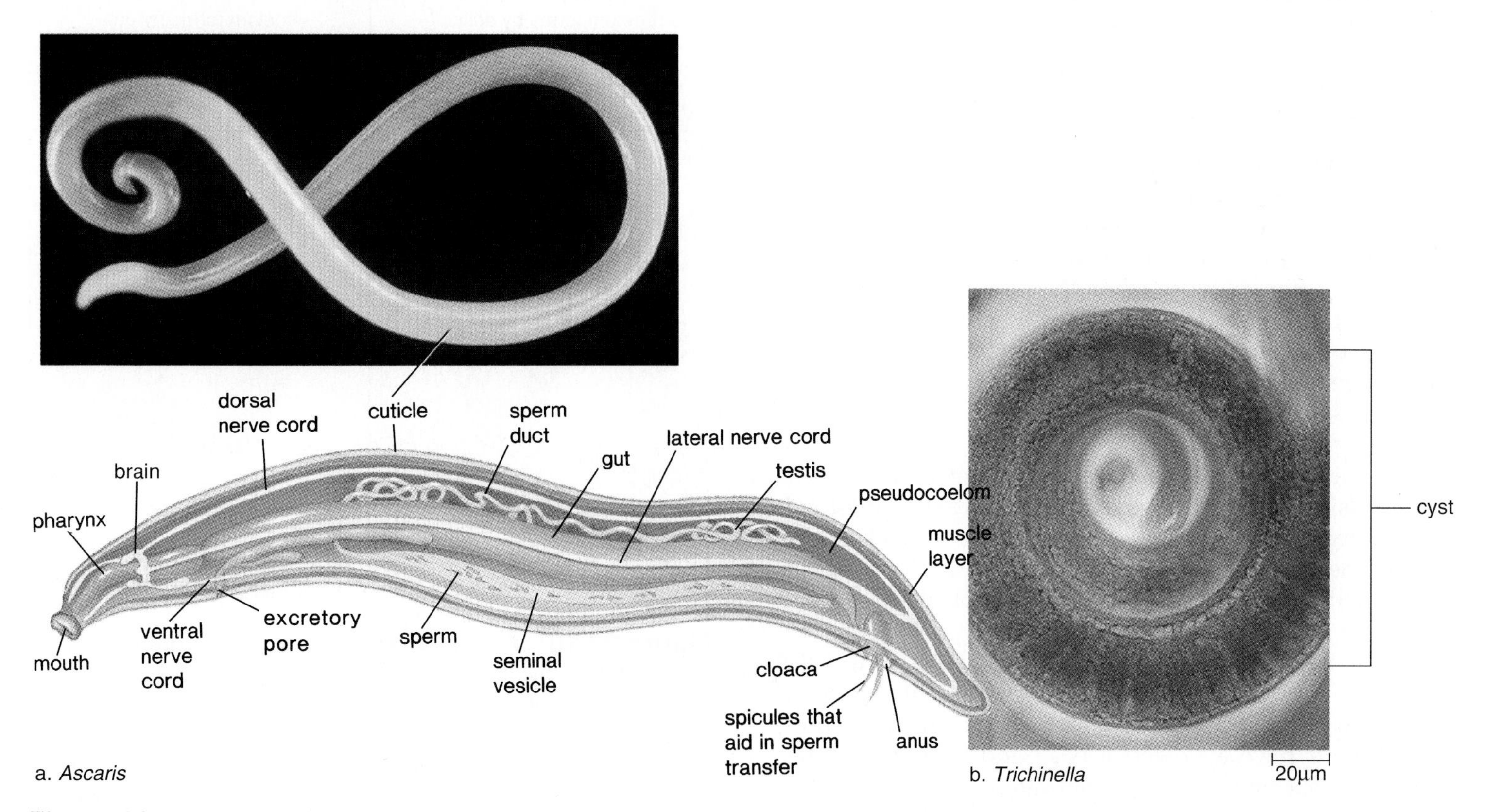

Figure 30.9 Roundworm diversity.
a. Note that roundworms such as *Ascaris lumbricoides* have a pseudocoelom and a complete digestive tract with a mouth and an anus. Therefore, roundworms have a tube-within-a-tube body plan. The sexes are separate; this is a male roundworm. **b.** The larvae of the roundworm *Trichinella spiralis* encyst as larvae in skeletal muscle fibers where they coil in a sheath formed from a muscle fiber. This infection in humans is called trichinosis.

30.3 Mollusks

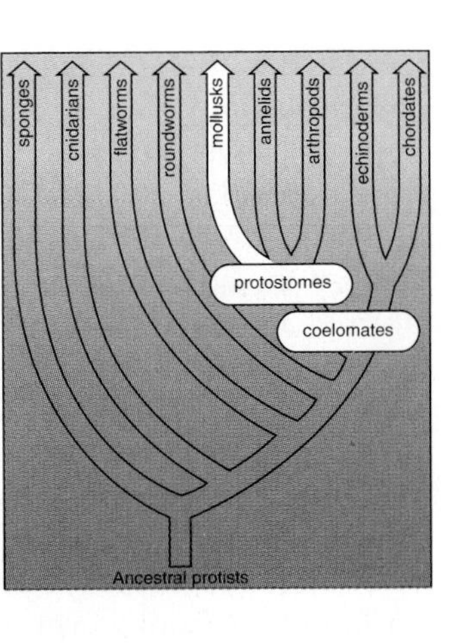

Mollusks, along with annelids and arthropods, are protostomes (Fig. 30.10). In protostomes, the first embryonic opening becomes the mouth. Because the coelom forms by splitting of the mesoderm, protostomes are also schizocoelomates. In certain (but not all) protostomes, there is a trochophore larva (top-shaped with a band of cilia at the midsection).

Characteristics of Mollusks

Mollusks (phylum Mollusca, about 110,000 species) are a very large and diversified group; however, they all have a body composed of at least three distinct parts.

1. *Visceral mass:* the soft-bodied portion that contains internal organs.
2. *Foot:* the strong, muscular portion used for locomotion.
3. *Mantle:* the membranous or sometimes muscular covering that envelops but does not completely enclose the visceral mass. The *mantle cavity* is the space between the two folds of the mantle. The mantle may secrete a shell.

In addition to these three parts, many mollusks have a head region with eyes and other sense organs.

The division of the body into distinct areas may have contributed to diversification of animals in this phyla. There are many different types of mollusks adapted to various ways of life (Fig. 30.11). Molluscan groups can be distinguished by a modification of the foot. In the *gastropods* (meaning stomach-footed), including nudibranchs, conchs, and snails, the foot is ventrally flattened, and the animal moves by muscle contractions that pass along the foot. While nudibranchs, also called sea slugs, lack a shell, conchs and snails have a coiled shell in which the visceral mass spirals. Some types of snails are adapted to life on land. For example, their mantle is richly supplied with blood vessels and functions as a lung when air is moved in and out through respiratory pores.

In *cephalopods* (meaning head-footed), including octopuses and squids, the foot has evolved into tentacles about the head. Aside from the tentacles, which seize prey, cephalopods have a powerful beak and a radula (toothy tongue) to tear prey apart. Cephalization aids these animals in recognizing prey and in escaping enemies. The eyes are superficially similar to those of vertebrates—they have a lens and a retina with photoreceptors. However, its construction is so different from the vertebrate eye that we believe the

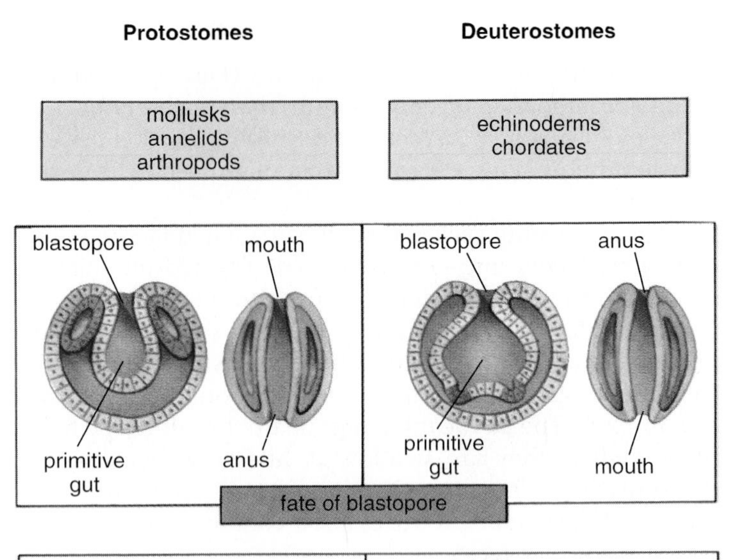

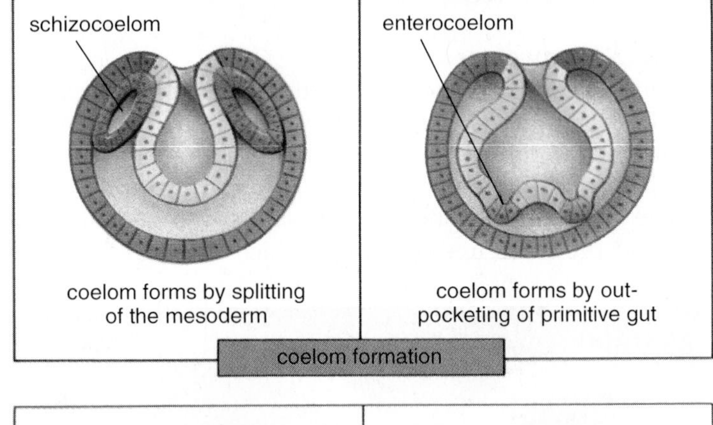

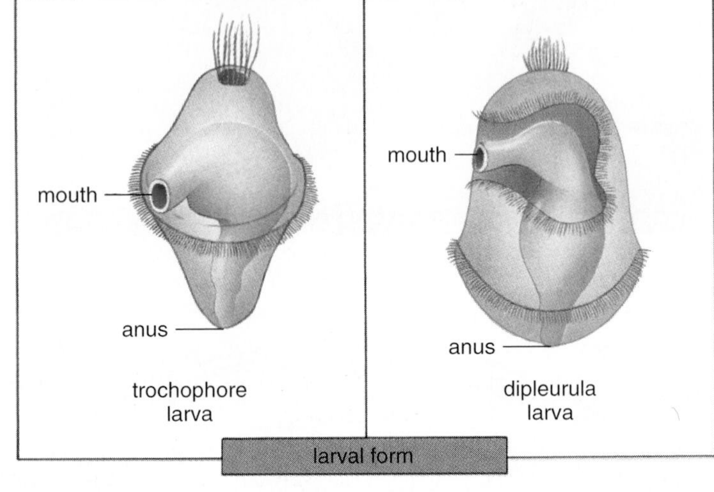

Figure 30.10 Protostomes versus deuterostomes.
In protostomes, the first embryonic opening called the blastopore becomes the mouth, the coelom forms by splitting of the mesoderm (they are schizocoelomates), and the trochophore larva is typical. In deuterostomes, the blastopore becomes the anus, the coelom forms by outpocketing of the primitive gut (they are enterocoelomates), and the dipleurula larva is found among some.

a. Chiton, *Tonicella*

b. Chambered nautilus, *Nautilus*

c. Scallop, *Pecten*

d. Spanish shawl nudibranch, *Flabellina*

Figure 30.11 Molluscan diversity.
a. A chiton has a flattened foot and a shell that consists of eight articulating valves. **b.** A chambered nautilus achieves buoyancy by regulating the amount of air in the chambers of its shell. **c.** A scallop has sensory tentacles extended between the valves. **d.** A nudibranch (sea slug) lacks a shell, gills, and a mantle cavity. Dorsal projections function in gas exchange.

so-called camera-type eye actually evolved twice—once in the mollusks and once in the vertebrates. In cephalopods, the brain is formed from a fusion of ganglia, and nerves leaving the brain supply various parts of the body. An especially large pair of nerves controls the rapid contraction of the mantle, allowing these animals to move quickly by a jet propulsion of water. Rapid movement and the secretion of a brown or black pigment from an ink gland help cephalopods to escape their enemies. Octopuses have no shell, and squid have only a remnant of one concealed beneath the skin.

In **bivalves,** such as clams, oysters, and scallops, the foot is laterally compressed. They are called bivalves because there are two parts to the shell. Notice in Table 30.1 that a clam is adapted to a less active life and a squid is adapted to a more active life.

Table 30.1 Comparison of Clam to Squid

Feature	Clam	Squid
Food Gathering	Filter feeder	Active predator
Skeleton	Heavy shell for protection	No external skeleton
Circulation	Open	Closed
Cephalization	None	Marked
Locomotion	"Hatchet" foot	Jet propulsion
Nervous System	3 separate ganglia	Brain and nerves

Bivalves

In a clam, such as the freshwater clam *Anodonta,* the shell, secreted by the mantle, is composed of protein and calcium carbonate, with an inner layer of *mother-of-pearl.* If a foreign body is placed between the mantle and the shell, pearls form as concentric layers of shell are deposited about the particle.

The adductor muscles hold the valves of the shell together. Within the mantle cavity, the *gills,* an organ for gas exchange in aquatic forms, hang down on either side of the visceral mass, which lies above the foot. The heart of a clam lies just below the hump of the shell within the pericardial cavity, the only remains of the coelom. Therefore, the *coelom is reduced.* The heart pumps blood into a dorsal aorta that leads to the various organs of the body. Within the organs, however, blood flows through spaces, or sinuses, rather than through vessels. This is an *open circulatory system* because the blood is not contained within blood vessels all the time. This type of circulatory system can usually be associated with a relatively inactive animal because it is an inefficient means of transporting blood throughout the body.

The nervous system of a clam (Fig. 30.12) is composed of three *pairs of ganglia* (anterior, foot, and posterior), which are all connected by nerves. Clams lack cephalization. The foot projects anteriorly from the shell, and by expanding the tip of the foot and pulling the body after it, the clam moves forward.

The clam is a *filter feeder.* Food particles and water enter the mantle cavity by way of the *incurrent siphon,* a posterior opening between the two valves. Mucous secretions cause smaller particles to adhere to the gills, and ciliary action sweeps them toward the mouth. This method of feeding does not require rapid movement.

The digestive system of the clam includes a mouth with labial palps, an esophagus, a stomach, and an intestine, which coils about in the visceral mass and then is surrounded by the heart as it extends to the anus. The anus empties at an *excurrent siphon,* which lies just above the incurrent siphon. There is also an accessory organ of digestion called a digestive gland. The two excretory kidneys in the clam (Fig. 30.12), which lie just below the heart, remove waste from the pericardial cavity for excretion into the mantle cavity.

The sexes are usually separate. The gonad (e.g., ovary or testis) is located around the coils of the intestine. While all clams have some type of larval stage, only marine clams have a trochophore larva. The presence of the *trochophore larva* (see Fig. 30.10) among some mollusks indicates a relationship to the annelids, some members of which also have this type of larval stage.

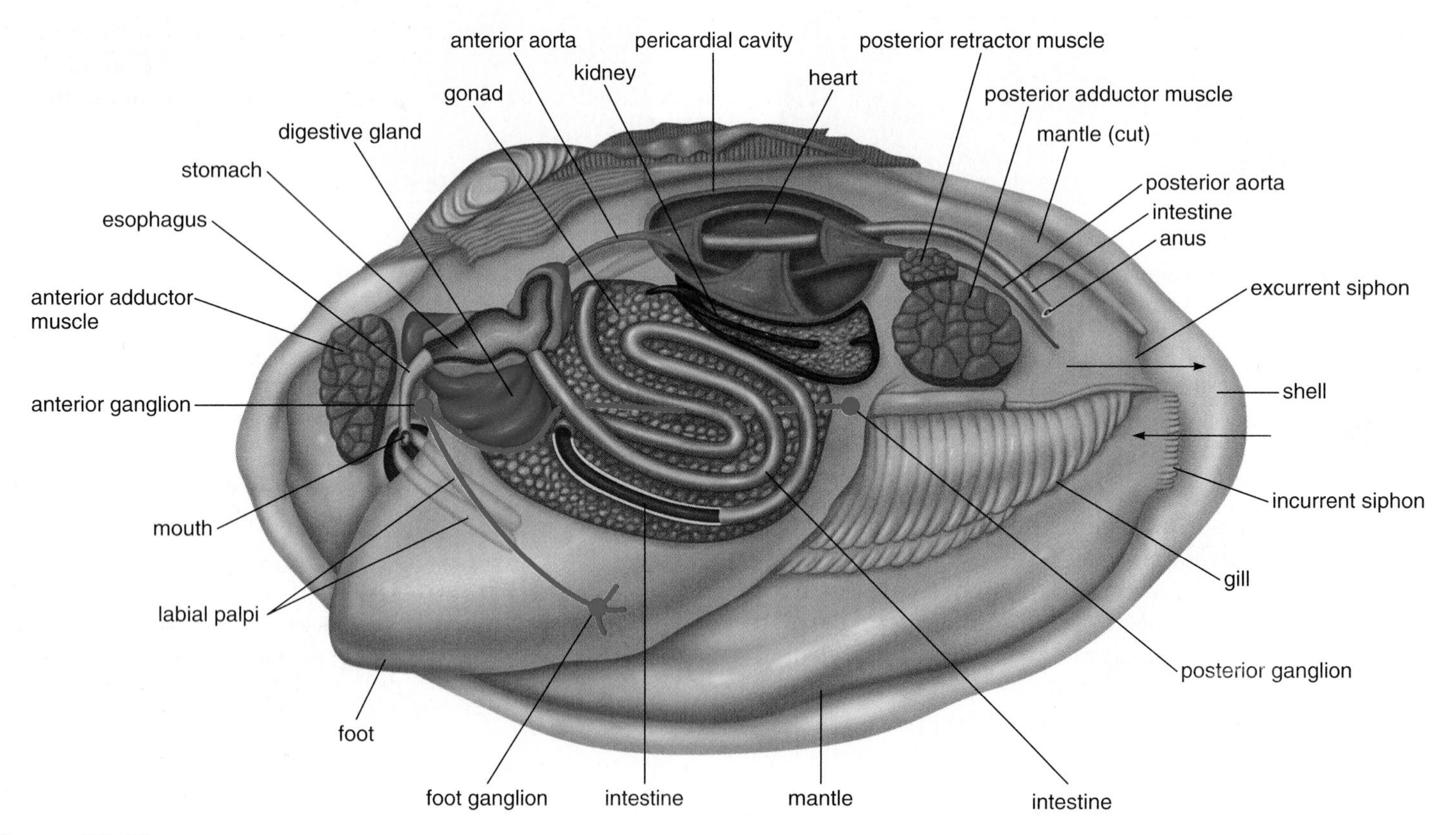

Figure 30.12 Clam.
The shell and the mantle have been removed from one side. Trace the path of food from the incurrent siphon past the gills, to the mouth, the stomach, the intestine, the anus, and the excurrent siphon. Locate the three ganglia: anterior, foot, and posterior. The heart lies in the reduced coelom.

30.4 Annelids

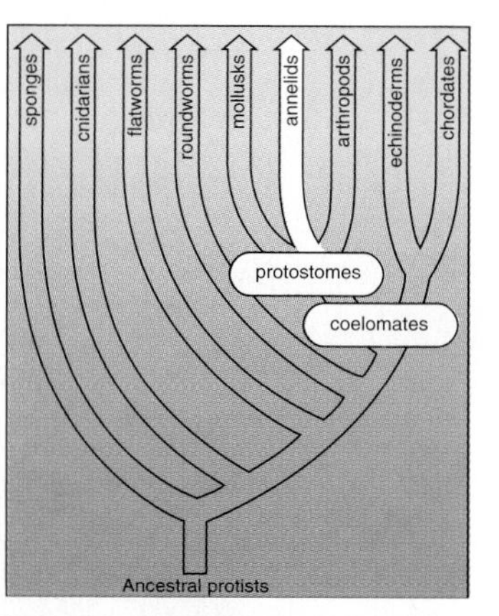

Annelids (phylum Annelida, about 12,000 species) are *segmented,* as is externally evidenced by the rings that encircle the body. Internally, partitions called *septa* (sing., septum) divide the well-developed, fluid-filled coelom, which acts as a hydrostatic skeleton.

Segmentation is a subdivision of the body along its length into repeating units, called segments. Annelids have a hydrostatic skeleton, and partitioning of the coelom permits each body segment independence of movement. Also, each segment of an earthworm has its own set of longitudinal and circular muscles and its own nerve supply, so each segment or group of segments may function independently. When circular muscles contract, the segments become thinner and elongate. When longitudinal muscles contract, the segments become thicker and shorten. By alternating circular muscle contraction and longitudinal muscle contraction, the animal moves forward.

In annelids, the tube-within-a-tube body plan has led to *specialization of the digestive tract.* For example, the digestive system may include a pharynx, esophagus, crop, gizzard, intestine, and accessory glands. They have an extensive *closed circulatory system* with blood vessels that run the length of the body and branch to every segment. The nervous system consists of a *brain connected to a ventral solid nerve cord, with ganglia in each segment.* The excretory system consists of *nephridia* in most segments. A **nephridium** is a tubule that collects waste material and excretes it through an opening in the body wall.

Marine Worms

Most annelids are marine; their class name, Polychaeta, refers to the presence of many setae. *Setae* are bristles that anchor the worm or help it move. In *polychaetes,* the setae are in bundles on *parapodia,* which are paddlelike appendages found on most segments. These are used not only in swimming but also as respiratory organs where the expanded surface area allows for increased exchange of gases. Clam worms (Fig. 30.13) such as *Nereis* are predators. They prey on crustaceans and other small animals, which are captured by a pair of strong, chitinous jaws that extend with a part of the pharynx when the animal is feeding. Associated with its way of life, *Nereis* has a well-defined head region with eyes and other sense organs.

Other polychaetes are sedentary (sessile) tube worms, with tentacles that form a funnel-shaped fan. Water currents, created by the action of cilia, trap food particles that are directed toward the mouth. They are *sessile filter feeders;* a sorting mechanism rejects large particles, and only the smaller ones are accepted for consumption.

Polychaetes have breeding seasons, and only during these times do the worms have functional sex organs. In *Nereis,* many worms concurrently shed a portion of their bodies containing either eggs or sperm, and these float to the surface, where fertilization takes place. The zygote rapidly develops into a *trochophore larva,* just as in marine clams. The existence of this larva in both the annelids and mollusks is evidence that these two groups of animals are related.

Polychaetes are marine worms with bundles of setae attached to parapodia.

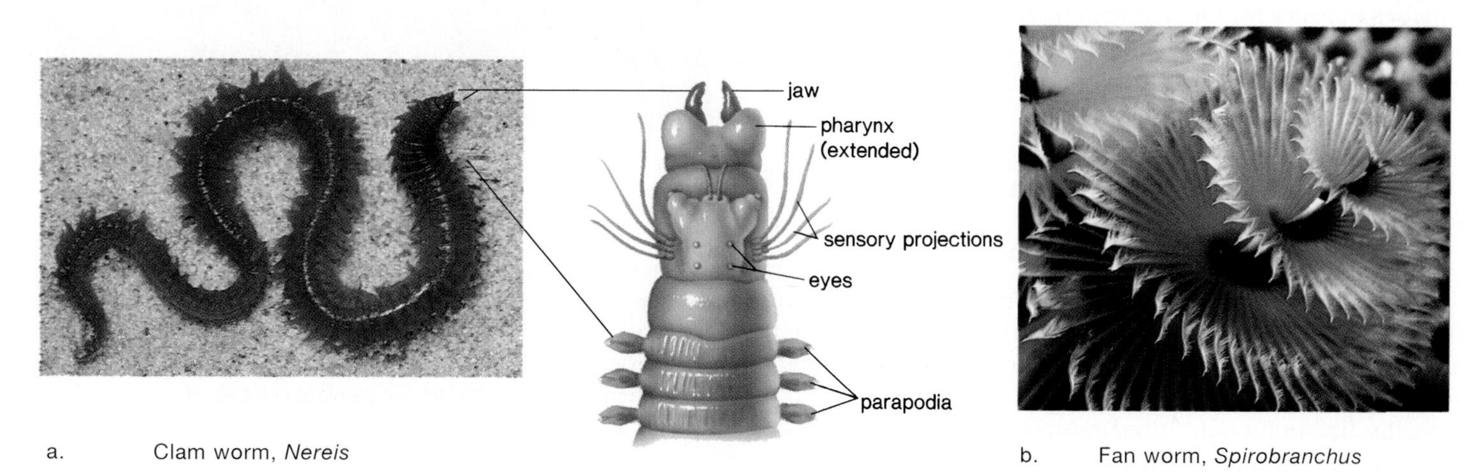

a. Clam worm, *Nereis*

b. Fan worm, *Spirobranchus*

Figure 30.13 Polychaete diversity.
a. Nereis is a predaceous polychaete that has a head region. Note the parapodia, which are used for swimming and as respiratory organs.
b. Fan worms (a type of tube worm) are sessile filter feeders whose cilia-lined tentacles funnel food particles to the mouth.

Earthworms

The *oligochaetes,* which include earthworms, have few setae per segment. Earthworms (e.g., *Lumbricus)* do not have a well-developed head or parapodia (Fig. 30.14). Their setae protrude in pairs directly from the surface of the body. Locomotion, which is accomplished section by section, utilizes muscle contraction and the setae. When longitudinal muscles contract, segments bulge and their setae protrude into the soil; then, when circular muscles contract, the setae are withdrawn, and these segments move forward.

Earthworms reside in soil where there is adequate moisture, which keeps the body wall moist for gas exchange purposes. They are scavengers that do not have an obvious head and feed on leaves or any other organic matter, living or dead, which can conveniently be taken into the mouth along with dirt. Food drawn into the mouth by the action of the muscular *pharynx* is stored in a *crop* and ground up in a thick, muscular *gizzard.* Digestion and absorption occur in a long intestine whose dorsal surface has an expanded region called a *typhlosole* that increases surface for absorption.

Earthworm *segmentation,* which is so obvious externally, is also internally evidenced by septa. The long ventral solid nerve cord leading from the brain has ganglionic swellings and lateral nerves in each segment. The paired *nephridia* in most segments have two openings: one is a ciliated funnel that collects coelomic fluid, and the other is an exit in the body wall. Between the two openings is a convoluted region where waste material is removed from the blood vessels about the tubule. Red blood moves anteriorly in the dorsal blood vessel, which connects to the ventral blood vessel by five pairs of connectives called "hearts." Pulsations of the dorsal blood vessel and the five pairs of hearts are responsible for blood flow. As the ventral vessel takes the blood toward the posterior regions of the worm's body, it gives off branches in every segment. Altogether, segmentation is evidenced by

- body rings
- coelom divided by septa
- setae on most segments
- ganglia and lateral nerves in each segment
- nephridia in most segments
- branch blood vessels in each segment

The worms are *hermaphroditic;* the male organs are the testes, the seminal vesicles, and the sperm ducts, and the female organs are the ovaries, the oviducts, and the seminal receptacles. Two worms lie parallel to each other facing in opposite directions. The fused midbody segment, called a *clitellum,* secretes mucus, protecting the sperm from drying

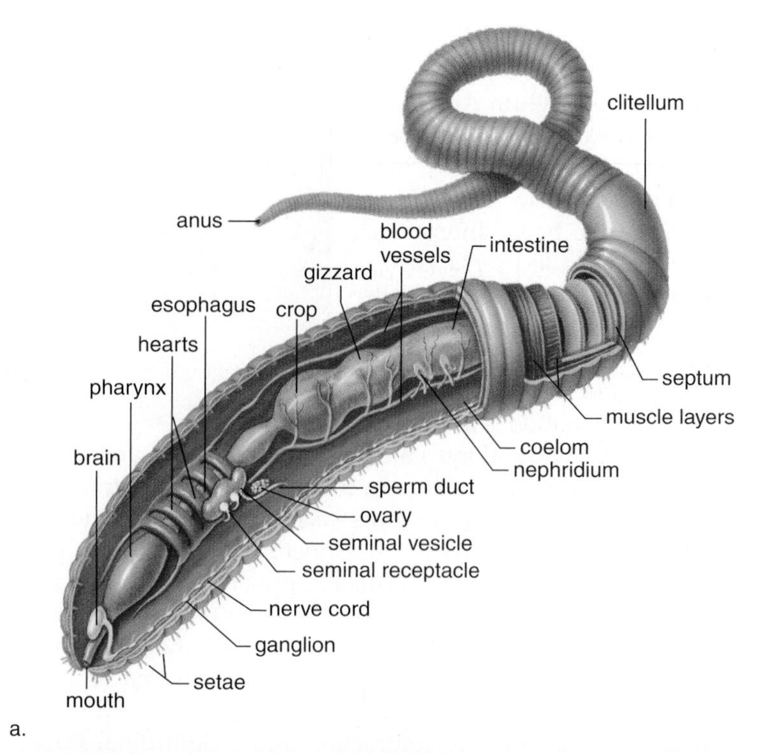

a.

b.

Figure 30.14 Earthworm.
a. Internal anatomy of the anterior part of an earthworm. Notice that each body segment bears four pairs of setae and that internal septa divide the coelom into compartments. **b.** When earthworms, such as *Lumbricus*, mate, they are held in place by a mucus secreted by the clitellum. The worms are hermaphroditic but practice cross-fertilization; sperm pass from the seminal vesicles of each to the seminal receptacles of the other.

Health Focus

Medicinal Bloodsuckers

Parasitic leeches attach themselves to the body of their hosts and suck blood or other body fluids from fishes, turtles, snails, or mammals. The European species of *Hirudo medicinalis* was used in the 1700s and 1800s to "let blood" in feverish patients in order to withdraw any poisons and "excess" blood. This procedure, which is most debilitating for the patient, is no longer practiced, but leeches are still used on occasion to remove blood from bruised skin (Fig. 30A). They are also applied to small body parts like fingers and toes that have just been reattached to the body. The sucking by the leech unclogs small blood vessels, causing blood to flow normally again through the part. The anterior sucker of the medicinal leech has three jaws that saw through the skin, producing a Y-shaped incision through which blood is drawn up by the sucking action of the muscular pharynx. The salivary glands secrete an anticoagulant called hirudin, which keeps the blood flowing while the leech is feeding. Other salivary ingredients dilate the host's blood vessels and act as an anesthetic. A medicinal leech can take up to five times its body weight in blood because the crop has pouches where the blood can be stored as the animal expands in size. When it has taken its fill, the leech drops off and digests its meal. Complete digestion takes a long time, and it's been suggested that a leech needs to feed only once a year.

Not all leeches suck blood or fluids; some are even predaceous and eat a variety of small invertebrates.

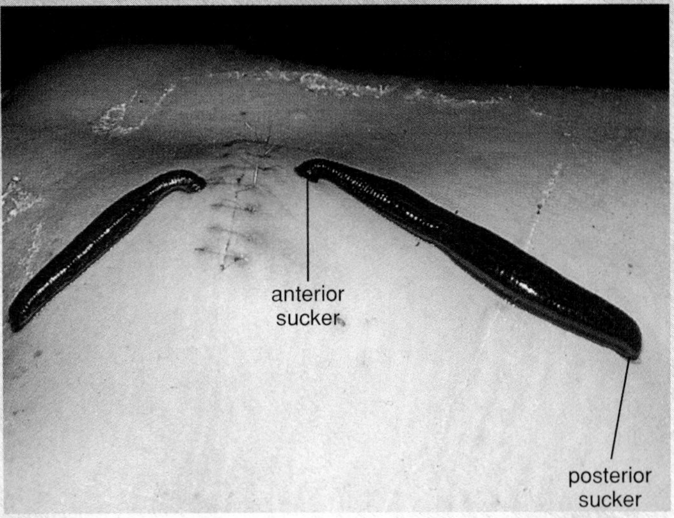

Figure 30A Medicinal leeches, *Hirudo medicinalis*. They are sometimes used medically to remove blood that has accumulated in damaged tissues.

out as they pass between the worms. After the worms separate, the clitellum of each produces a slime tube, which is moved along over the anterior end by muscular contractions. As it passes, eggs and the sperm received earlier are deposited and fertilization occurs. The slime tube then forms a cocoon to protect the worms as they develop. There is no larval stage.

It is interesting to compare the anatomy of marine clam worms to terrestrial earthworms because it highlights the manner in which earthworms are adapted to life on land. Marked cephalization is seen in the nonpredatory earthworms that extract organic remains from the soil they eat. The lack of parapodia helps reduce the possibility of water loss and facilitates burrowing in soil. The clam worm makes use of external water, while the earthworm provides a mucous secretion to aid fertilization. It is the water form that has the swimming, or trochophore larva, and not the land form.

Earthworms, which burrow in the soil, lack obvious cephalization and parapodia. They are hermaphroditic, and there is no larval stage.

Leeches

Leeches are usually found in fresh water, but some are marine or even terrestrial. They have the same body plan as other annelids but they have no setae, and each body ring has several transverse grooves. Most leeches are only 2–6 cm in length, but some, including the medicinal leech, are as long as 20 cm.

Among their modifications are two *suckers*, a small oral one around the mouth and a large posterior one. While some leeches are free-living predators, most are fluid feeders that attach themselves to open wounds. Some bloodsuckers, such as the medicinal leech, are able to cut through tissue. Leeches are able to keep blood flowing and prevent clotting by means of a substance in their saliva known as *hirudin*, a powerful anticoagulant. This has added to their potential usefulness in the field of medicine today, as discussed in the accompanying reading.

Leeches are modifed in a way that lends itself to the parasitic way of life. Some are external parasites known as bloodsuckers.

a. Flat-backed millipede, *Sigmoria*

b. Tarantula, *Aphonopelma*

c. Dungeness crab, *Cancer*

d. Paper wasp, *Polistes*

e. Stone centipede, *Lithobius*

Figure 30.15 Arthropod diversity.
a. A millipede has only one pair of antennae, and the head is followed by a series of segments, each with two pairs of appendages. **b.** The hairy tarantulas of the genus *Aphonopelma* are dark in color and sluggish in movement. Their bite is harmless to people. **c.** A crab is a crustacean with a calcified exoskeleton, one pair of claws and four other pairs of walking legs. **d.** A wasp is an insect with two pairs of wings, both used for flying, and three pairs of walking legs. **e.** A centipede has only one pair of antennae, and the head is followed by a series of segments, each with a single pair of appendages.

30.5 Arthropods

Arthropods (phylum Arthropoda) are extremely diverse. Over one million species have been discovered and described, but some experts suggest that as many as 30 million arthropods could exist—mostly insects.

What characteristics account for the success of arthropods? Arthropod literally means "jointed foot," but actually they have freely movable *jointed appendages*. The exoskeleton of arthropods is composed primarily of **chitin,** a strong, flexible, nitrogenous polysaccharide. The exoskeleton serves many functions such as protection, attachment for muscles, locomotion, and prevention of desiccation. However, because this exoskeleton is hard and nonexpandable, arthropods must undergo **molting,** or shedding of the exoskeleton, as they grow larger. Before molting, the body secretes a new, larger exoskeleton, which is soft and wrinkled, underneath the old one. After enzymes partially dissolve and weaken the old exoskeleton, the animal breaks it open and wriggles out. The new exoskeleton then quickly expands and hardens.

Arthropods are segmented, but some *segments are fused* into regions, such as a head, a thorax, and an abdomen. In trilobites, an early and now extinct arthropod, there was a pair of appendages on each body segment. In modern arthropods, *appendages are specialized* for such functions as walking, swimming, reproducing, eating, and sensory reception. These modifications account for much of the diversity of arthropods. Each of the five major groups of arthropods (Fig. 30.15) contains species that are adapted to terrestrial life.

Arthropods have a *well-developed nervous system.* There is

a brain and a ventral solid nerve cord. The head bears various types of sense organs, including *antennae* (or feelers) and eyes of two types—compound and simple. The *compound eye* is composed of many complete visual units, each of which operates independently. The lens of each visual unit focuses an image on the light-sensitive membranes of a small number of photoreceptors within that unit. Vision is not acute, but it is much better for details of movement than with our eyes.

Crustaceans

Crustaceans are a group of largely marine arthropods that include barnacles, shrimps, lobsters, and crabs. There are also some freshwater crustaceans, including the crayfish, and some terrestrial ones, including the sowbug or "roly-poly bug."

Crustaceans are named for their hard shells; the exoskeleton is calcified to a greater degree in some forms than in others. Although crustacean anatomy is extremely diverse, the head usually bears a pair of compound eyes and *five pairs of appendages.* The first two pairs, called antennae, lie in front of the mouth and have sensory functions. The other three pairs are mouthparts used in feeding.

In crayfish such as *Cambaris,* the thorax bears five pairs of walking legs. The first walking leg is a pinching claw (Fig. 30.16). The *gills* are situated above the walking legs. The head and thorax are fused into a *cephalothorax,* which is covered on the top and sides by a nonsegmented carapace. The abdominal segments are equipped with *swimmerets,* small paddlelike structures. The last two segments bear the uropods and the telson, which make up a fan-shaped tail to propel the crayfish backward.

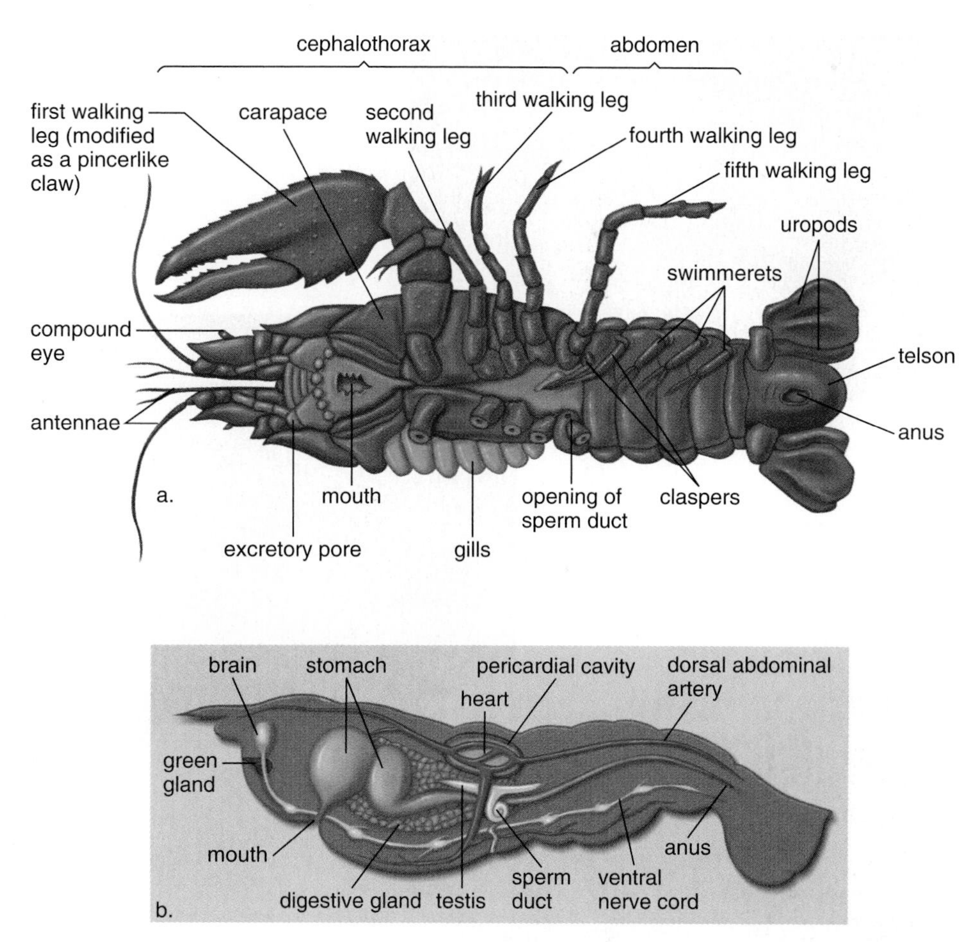

Figure 30.16 Male crayfish.
a. Externally, it is possible to observe the jointed appendages. Note the swimmerets and the walking legs, which include the claws. These appendages, plus a portion of the carapace, have been removed from the right side so that the gills are visible. **b.** Internally, the parts of the digestive system are particularly visible. The circulatory system can also be clearly seen. Note the ventral nerve cord.

Ordinarily, a crayfish lies in wait for prey. It faces out from an enclosed spot with the claws extended and the antennae moving about. The claws seize any small animal, dead or alive, that happens by and carry it to the mouth. When a crayfish moves about, it generally crawls slowly, but it may swim rapidly by using its heavy abdominal muscles and tail.

The digestive system includes a stomach, which is divided into two main regions: an anterior portion called the *gastric mill,* equipped with chitinous teeth to grind coarse food, and a posterior region, which acts as a filter to prevent coarse particles from entering the digestive glands where absorption takes place. *Green glands* lying in the head region, anterior to the esophagus, excrete metabolic wastes through a duct that opens externally at the base of the antennae. The coelom, which is so well developed in the annelids, is *reduced* in the arthropods and is composed chiefly of the space about the reproductive system. A heart within a pericardial cavity pumps blood containing the respiratory pigment hemocyanin into a **hemocoel** consisting of sinuses (open spaces), where the hemolymph flows about the organs. (Whereas hemoglobin is a red iron-containing pigment, hemocyanin is a blue copper-containing pigment.) This is an *open circulatory system* because blood is not contained within blood vessels.

The nervous system is quite similar to that of the earthworm. There is a brain, as well as a *ventral nerve cord* that passes posteriorly. Along the length of the nerve cord, ganglia in segments give off 8 to 19 paired lateral nerves.

The sexes are separate in the crayfish, and the gonads are located just ventral to the pericardial cavity. In the male, a coiled sperm duct opens to the outside at the base of the fifth walking leg. Sperm transfer is accomplished by the first two pairs of swimmerets, which are enlarged and quite strong. In the female, the ovaries open at the bases of the third walking legs. A stiff fold between the bases of the fourth and fifth pairs of walking legs serves as a seminal receptacle. Following fertilization, the eggs are attached to the swimmerets of the female.

a. Walking stick, *Diapheromera*

b. Bee, *Apis*

c. Housefly, *Musca*

d. Dragonfly, *Aeshna*

e. American copper butterfly, *Lycaena*

Figure 30.17 Insect diversity.
a. Walking sticks are herbivorous, with biting and chewing mouthparts. **b.** Bees have four translucent wings and a thorax separated from the abdomen by a narrow waist. **c.** Flies have a single pair of wings and lapping mouthparts. **d.** Dragonflies have two pairs of similar wings. They catch and eat other insects while flying. **e.** Butterflies have forewings larger than hindwings. Their mouthparts form a long tube for siphoning up nectar from flowers.

Insects

There are more different kinds of insects than any other type animal. Examples of insects include crickets and grasshoppers, dragonflies, water striders, beetles, moths and butterflies, flies, bees, and wasps (Fig. 30.17).

Insects have certain features in common. The body is divided into a head, a thorax and an abdomen. The head usually bears a pair of sensory *antennae,* a pair of *compound eyes,* and several *simple eyes.* The mouthparts are adapted to the way of life: a grasshopper has mouthparts that chew, and a butterfly has a long tube for siphoning the nectar of flowers. The thorax bears three pairs of legs and no pairs, one pair, or two pairs of wings, and the abdomen contains most of the internal organs. Wings enhance an insect's ability to survive by providing a way of escaping enemies, finding food, facilitating mating, and dispersing the offspring. The exoskeleton of an insect is lighter and contains less chitin than that of many other arthropods.

In the grasshopper (Fig. 30.18), the third pair of legs is suited to jumping. There are two pairs of wings. The *forewings* are tough and leathery, and when folded back at rest, they protect the broad, thin *hindwings.* On the lateral surface, the first abdominal segment bears a large *tympanum* on each side for the reception of sound waves. The posterior region of the exoskeleton in the female has two pairs of projections that form an ovipositor, which is used to dig a hole in soil where eggs are laid.

The digestive system is suitable for a herbivorous diet. In the mouth, food is broken down mechanically by mouthparts and enzymatically by salivary secretions. Food is temporarily stored in the crop before passing into a gastric mill, where it is finely ground before digestion is completed in the stomach. Nutrients are absorbed into the hemocoel from outpockets called gastric ceca; a *cecum* is a cavity open at one end only. The stomach is followed by an intestine and rectum, which empties by way of an anus. The excretory system consists of **Malpighian tubules,** which extend into the hemocoel and collect nitrogenous wastes that are concentrated and excreted into the digestive tract. The formation of a solid nitrogenous waste, namely uric acid, conserves water.

The respiratory system begins with openings in the exoskeleton called *spiracles.* From here, the air enters small tubules called **tracheae** (Fig. 30.18*a*). The tracheae branch and rebranch until they end intracellularly, where the actual exchange of gases takes place. The movement of air through this complex of tubules is not a passive process; air is pumped by alternate contraction and relaxation of the body wall through a series of several bladderlike structures, which are attached to the tracheae near the spiracles. Air enters the anterior four spiracles and exits by the posterior six spiracles. Breathing by tracheae may account for the small size of insects (most are less than 60 mm in length), since the tracheae are so tiny and fragile that they would be crushed by any significant amount of weight.

The circulatory system contains a slender, tubular heart that lies against the dorsal wall of the abdominal exoskeleton and pumps hemolymph into an aorta that leads to a hemocoel, where it circulates before returning to the heart again. The hemolymph is colorless and lacks a respiratory pigment—the tracheal system exchanges gases.

Reproduction is adapted to life on land. The male has a penis, and sperm passed to the female are stored in a seminal receptacle. Internal fertilization protects both gametes and zygotes from drying out. The female deposits the fertilized eggs in the ground with her ovipositor.

Metamorphosis is a change in form and physiology that occurs as an immature stage, called a larva, becomes an adult. Grasshoppers undergo *gradual metamorphosis,* or a gradual change in form, as the animal matures. The immature grasshopper, called a nymph, is recognizable as a grasshopper, even though it differs somewhat in shape and form from the adult. Other insects, such as butterflies, undergo *complete metamorphosis,* involving drastic changes in form. At first, the animal is a wormlike larva (caterpillar) with chewing mouthparts. It then forms a case, or cocoon, about itself and becomes a *pupa.* During this stage, the body parts are completely reorganized; the *adult* then emerges from the cocoon. This life cycle allows the larvae and adults to typically make use of different food sources. Most eating by insects occurs during the larval stage.

Insects also show remarkable behavior adaptations, exemplified by the social systems of bees, ants, termites, and other colonial insects. Insects are so numerous and so diverse that the study of this one group is a major specialty in biology called entomology.

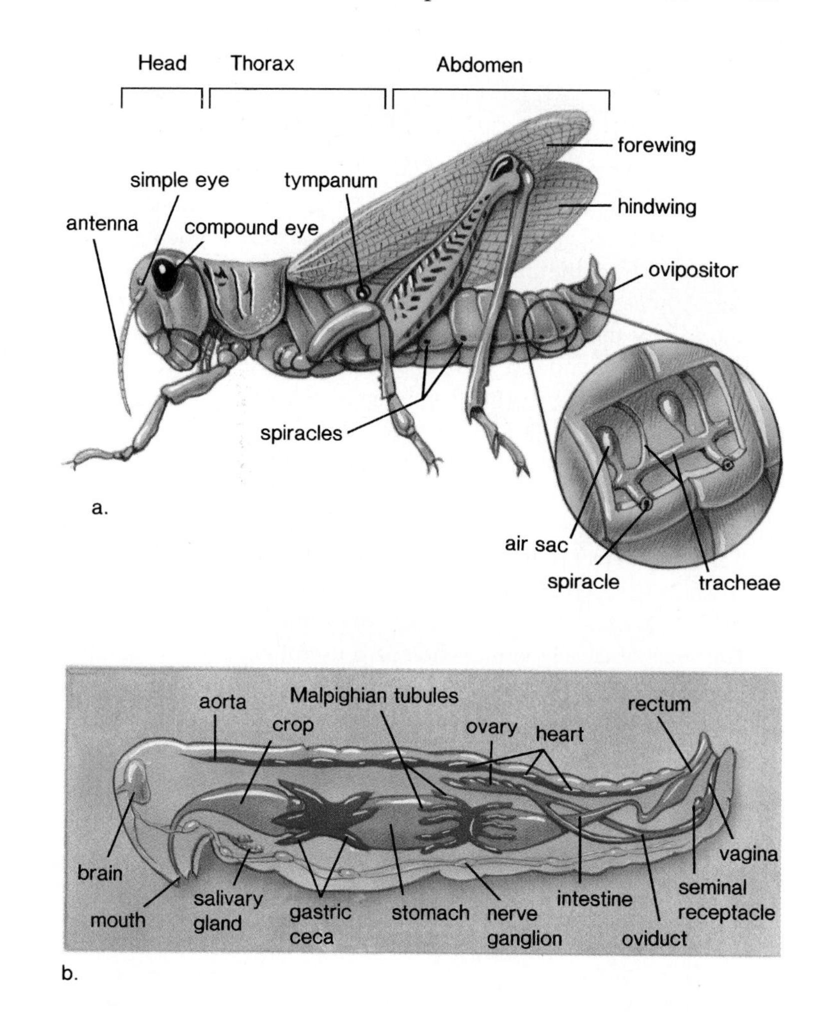

Figure 30.18 **Female grasshopper.**
a. Externally, the tympanum receives sound waves, and the hopping legs and the wings are for locomotion. **b.** Internally, the digestive system is specialized. The Malpighian tubules excrete a solid nitrogenous waste (uric acid). A seminal receptacle receives sperm from the male, which has a penis.

It is of interest to compare the adaptations of a grasshopper to land with the adaptations of a crayfish to an aquatic environment. In crayfish, gills take up oxygen from water, while in the grasshopper, tracheae allow oxygen-laden air to enter the body. Appropriately, the crayfish has an oxygen-carrying pigment, but a grasshopper has no such pigment in its blood. A liquid nitrogenous waste (ammonia) is excreted by a crayfish, while a solid nitrogenous waste (uric acid) is excreted by a grasshopper. Only in grasshoppers (1) is there a tympanum for the reception of sound waves and (2) do males have a penis for passing sperm to females without possible desiccation and females an ovipositor for laying eggs in soil. Crayfish utilize their uropods when they swim; a grasshopper has legs for hopping and wings for flying.

Science Focus

Spider Webs and Spider Classification

By using data from many sources, the evolution, and thus the classification, of an animal group can be clarified. A good example comes from the spiders, members of the class Arachnida, order Araneae. Spiders are distinguished from other similar animals by their ability to weave webs of silk. The silk is produced from glands in their abdomens, and it emerges from modified appendages called spinnerets.

Arachnologists (scientists who study spiders) are largely convinced that spider webs were originally used to line a cavity or a burrow in which an early spider hid. Many spiders that still live in such burrows put a collar of silk around their burrow entrances to detect prey over a wider area than they could otherwise easily search. From this array of threads evolved the basic sheet web, made by a wide variety of primitive spider families. The sheet of closely woven threads is useful not only in signaling the presence of prey, but also in slowing the prey as the insect's legs tangle in the matted silk. However, the appearance of the sheet web in many clearly unrelated families of spiders suggests that it evolved numerous times and cannot be used to answer questions about the true relationships of spider families. Only if some advanced feature of a sheet web is shared by two or more families does it indicate a common ancestry.

The geometric orb web probably evolved from a sheet web. The orb web, which has threads placed in a regular fashion, uses less silk and thus "costs" less. Until recently, the orb web was thought to have arisen at least twice because it is made by two groups of spiders that look very different. The dinopoids (superfamily Dinopoidea) includes spiders with a special spinning apparatus, the cribellum, which produces extremely fine fibers. The araneoids (superfamily Araneoidea), which also make orb webs, lacks the cribellum, and thus are called ecribellates. If the cribellum is a specialization that arose only in the dinopoid lineage, then it seems most logical that the araneoids and dinopoids are not closely related and their orb webs are quite separate developments.

Arachnologists noticed, however, the great similarities of the orbs made by the two families—even extending to the specific movements made by the spiders' legs while weaving them. Perhaps then the two families are closely related despite the lack of a cribellum in the araneoid line. The Finnish biologist Pekka Lehtinen made a sweeping study of all sorts of spiders in 1967 and found numerous examples of spiders that were nearly identical except for the presence or absence of a cribellum. The mass of evidence he accumulated convinced arachnologists that the cribellum could easily have been lost in the araneoid line of descent.

By considering the new data regarding the cribellum, together with observations of orb-web building behavior, most arachnologists are convinced that the orb web originated only once, and that the dinopoids and araneoids do share a common ancestor (Fig. 30B).

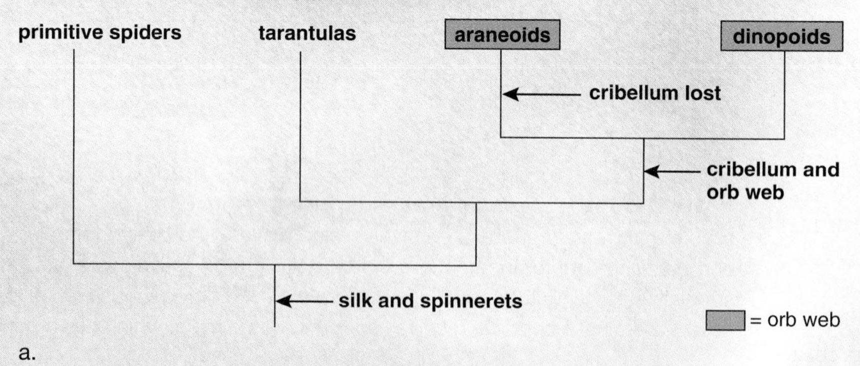

b.

Figure 30B Evolution of orb web.
a. Evolutionary tree of spiders. **b.** Orb web of the garden spider *Araneus diadematus* differs only in detail from **(c)** the orb of a cribellate spider, the New Zealand species *Waitkera waitkerensis.* The leg movements used by both during web construction are very similar, making it likely that there is a close evolutionary relationship between them, despite their anatomical differences.

Arachnids

The **arachnids** (class Arachnida, about 50,000 species) include terrestrial spiders, scorpions, ticks and mites (Fig. 30.19). In this group, the cephalothorax bears six pairs of appendages: the chelicerae and the pedipalps and four pairs of walking legs. The cephalothorax is followed by an abdomen that contains internal organs.

Scorpions are the oldest terrestrial arthropods. Today, they occur in the tropics, subtropics, and temperate regions of North America. They are nocturnal and spend most of the day hidden under a log or a rock. In scorpions the pedipalps are large pincers, and the long abdomen ends with a stinger that contains venom. Ticks and mites are parasites. Ticks suck the blood of vertebrates and are sometimes transmitters of diseases, such as Rocky Mountain spotted fever or Lyme disease. Chiggers, the larvae of certain mites, feed on the skin of vertebrates.

Spiders, the most familiar arachnids, have a narrow waist that separates the cephalothorax from the abdomen. Each chelicera consists of a basal segment and a fang that delivers poison to paralyze or kill prey. Two venom glands are located in the chelicerae or in the head. The pedipalps assist in sensing and holding the prey. Digestive juices from the mouth liquefy the tissues and the resulting broth is sucked into the stomach. Spiders use silk threads for all sorts of purposes, from lining their nests to catching prey. The reading on the previous page tells how biologists have used web-building behavior as a way to discover how spiders are related.

The internal organs of spiders also show how they are adapted to a terrestrial way of life. *Malpighian tubules* work in conjunction with rectal glands to reabsorb ions and water before a relatively dry nitrogenous waste (uric acid) is excreted. Invaginations of the inner body wall form lamellae ("pages") of their so-called *book lungs.* Air flowing into the folded lamellae on one side exchanges gases with blood flowing in the opposite direction on the other side.

The arthropods are a diverse group of animals, of which some members are adapted to an aquatic existence and others are adapted to a terrestrial way of life. Jointed appendages and a water repellent exoskeleton assists locomotion and lessens the threat of drying out on land.

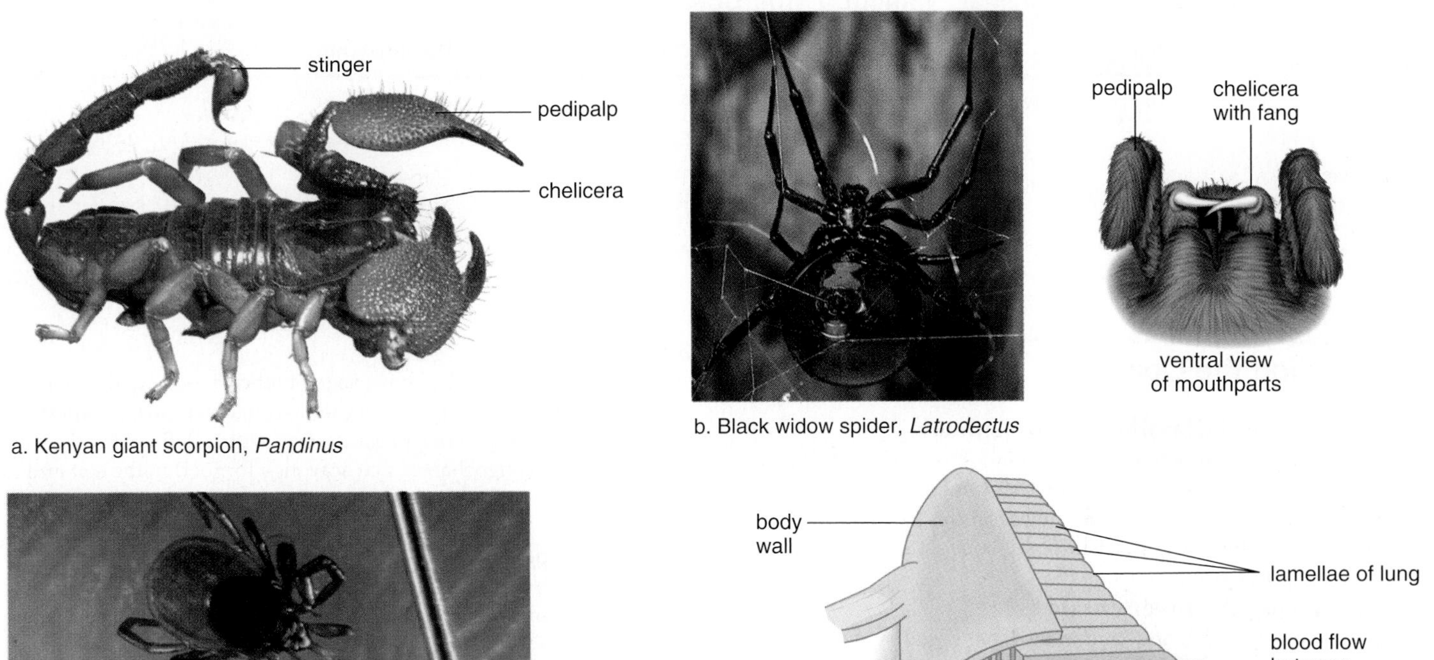

a. Kenyan giant scorpion, *Pandinus*

b. Black widow spider, *Latrodectus*

c. Wood ticks, *Ixodes*

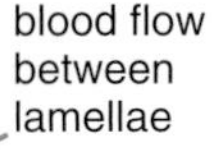

d. Book lung anatomy

Figure 30.19 Arachnid diversity.
a. A scorpion has pincerlike chelicerae and pedipalps, and there is a long abdomen ending with a stinger that contains venom. **b.** Most spiders are harmless, but the venom of the black widow spider is dangerous to humans. **c.** In the western United States, the wood tick carries a debilitating disease called Rocky Mountain spotted fever. **d.** Arachnids breathe by means of book lungs, in which the "pages" are double sheets of thin tissue (lamellae).

Bioethical Issue

The many types of animals in a coral reef form a complex community that is admired by both snorkelers and scuba divers. The various types of fish and shellfish in a coral reef are sources of food for millions of people. Like a tropical forest, coral reefs are most likely sources of medicines yet to be discovered. And a reef serves as a storm barrier that protects the shoreline and provides a safe harbor for ships.

Reefs around the globe are being destroyed. Tons of soil from deforested tracts of land bring nutrients that stimulate the growth of all kinds of algae. This has contributed to a population explosion of the crown-of-thorn starfish that are devouring Australia's 1,200 mile-long Great Barrier Reef. Reefs are also being damaged by pollutants that seep into the sea from factories, farm fields, and sewers. Stress, combined with unusually warm sea water, has caused the corals to expel their colorful, symbiotic algae, which carry on photosynthesis and help sustain the corals. So-called coral bleaching has been noticed in reefs of the Pacific Ocean and the Caribbean. Might worldwide global warming also contribute to coral bleaching and death?

Marine scientist Edgardo Gomez estimates that 90% of coral reefs of the Philippines are dead or deteriorating due to pollution, but especially due to overfishing. The methods are sinister, including the use of dynamite to kill the fish, making it easier to scoop them up, use of cyanide to stun the fish to capture them alive, and using satellite navigation systems to home in on areas where mature fish are spawning to reproduce. If all large herbivores are killed, seaweed overgrows and kills the coral.

Paleobiologist Jeremy Jackson of the Smithsonian Tropical Research Institute near Panama City, Republic of Panama, wonders if he is doing enough to warn the public that reefs around the world are in danger. He estimates that we may lose 60% of all coral reefs by the year 2050.

Questions

1. Do you think it would be possible to make the public care about the loss of coral reefs? Explain.
2. When and under what circumstances do dire predictions help preserve the environment?
3. Considering what is causing the loss of coral reefs, would it be possible to save them? How?

Table 30.2 Comparison of Animals without a True Coelom

	Sponges	Cnidarians	Flatworms	Roundworms
Symmetry	Radial or none	Radial	Bilateral	Bilateral
Type of body plan	—	Sac	Sac	Tube-within-a-tube
Tissue layers	—	Two	Three	Three
Level of organization	Cell	Tissue	Organ	Organ
Body cavity	—	—	Acoelomate	Pseudocoelomate

Summarizing the Concepts

30.1 Evolution and Classification of Animals

Animals are multicellular heterotrophs that ingest their food. The classification of animals is based on symmetry, number of germ layers, type of coelom, and body plan, among other features. An evolutionary tree based on these features depicts a possible evolutionary relationship between the animals.

30.2 Introducing the Invertebrates

Sponges are asymmetrical, and cnidarians have radial symmetry. All other phyla contain bilaterally symmetrical animals (Table 30.2). Flatworms have three germ layers but no coelom. Roundworms have a pseudocoelom and a tube-within-a-tube body plan.

30.3 Mollusks

Mollusks, along with annelids and arthropods, are protostomes because the first embryonic opening becomes the mouth. The body of a mollusk typically contains a visceral mass, a mantle, and a foot. Predaceous squids display marked cephalization, move rapidly by jet propulsion, and have a closed circulatory system. Bivalves (e.g., clams), which have a hatchet foot, are filter feeders. Water enters and exits by siphons, and food trapped on the gills is swept toward the mouth.

30.4 Annelids

Annelids are segmented worms; segmentation is seen both externally and internally. Polychaetes are marine worms that have parapodia. A clam worm is a predaceous marine worm with a defined head region. Earthworms are oligochaetes that scavenge for food in the soil and do not have a well-defined head region.

30.5 Arthropods

Arthropods are the most varied and numerous of animals. Their success is largely attributable to a flexible exoskeleton and specialization of body regions. Like many other arthropods, crustaceans have a head that bears compound eyes, antennae, and mouthparts. The crayfish, a crustacean, also illustrates features such as an open circulatory system, respiration by gills, and a ventral solid nerve cord. Like most other insects, grasshoppers have wings and three pairs of legs attached to the thorax. Grasshoppers also have many other features that illustrate adaptation to a terrestrial life, such as respiration by tracheae.

Spiders are arachnids with chelicerae, pedipalps, and four pairs of walking legs attached to a cephalothorax. Spiders, too, are adapted to life on land, and they spin silk which is used in various ways. Some spiders spin webs, and the type of web can be used to discover the evolutionary relationship among spiders.

Studying the Concepts

1. What does the evolutionary tree (see Fig. 30.2) tell you about the evolution of the animals studied in this chapter? 620
2. List the types of cells found in a sponge, and describe their functions. 621
3. What are the two body forms found in cnidarians? Explain how they function in the life cycle of various types of cnidarians. 622
4. Describe the anatomy of hydra, pointing out those features that typify cnidarians. 623
5. Describe the anatomy of a free-living planarian, pointing out those features that typify nonparasitic flatworms. 624–25
6. Describe the parasitic flatworms, and describe the life cycle of a pork tapeworm. 625–26
7. Describe the anatomy of *Ascaris,* pointing out those features that typify roundworms. 627
8. What are the general characteristics of mollusks? Contrast the anatomy of the clam and the squid, indicating how each is adapted to its way of life. 628–29
9. What are the general characteristics of annelids? Contrast the anatomy of the clam worm and the earthworm, indicating how each is adapted to its way of life. 631–32
10. What are the general characteristics of arthropods? Describe the specific features of crayfish, the grasshopper, and a spider, indicating how each is adapted to its way of life. 634–39

Testing Yourself

Choose the best answer for each question.

1. Label the following diagram of the cnidarian polyp:

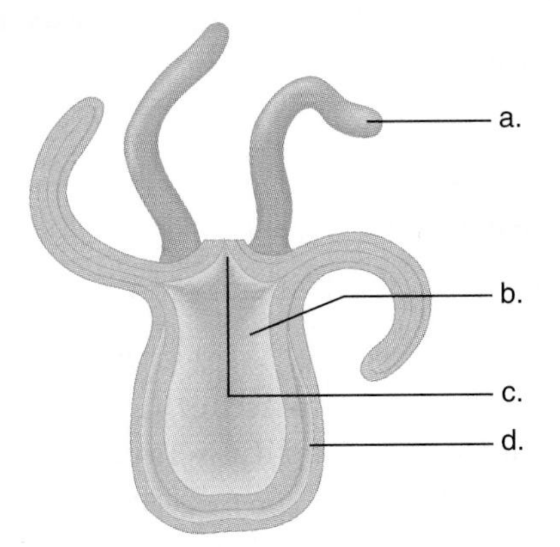

2. Which of these is not a characteristic of animals?
   a. heterotrophic
   b. diplontic life cycle
   c. usually practice sexual reproduction
   d. have contracting fibers
   e. single cells or colonial
3. The evolutionary tree of animals shows that
   a. cnidarians evolved directly from sponges.
   b. flatworms evolved directly from roundworms.
   c. both sponges and cnidarians evolved from protista.
   d. coelomates gave rise to the acoelomates.
   e. All of these are correct.
4. Which of these sponge characteristics is not typical of animals?
   a. They have a type of skeleton.
   b. They practice sexual reproduction.
   c. They have the cellular level of organization.
   d. They are asymmetrical.
   e. Both c and d are correct.
5. Which of these is mismatched?
   a. sponges—spicules
   b. tapeworms—proglottids
   c. cnidarians—nematocysts
   d. crayfish—compound eyes
   e. roundworms—cilia
6. Flukes and tapeworms
   a. show cephalization.
   b. have well-developed reproductive systems.
   c. have well-developed nervous systems.
   d. have a tube-within-a-tube body plan.
   e. are plant parasites.
7. The presence of mesoderm
   a. is necessary to mesoglea formation.
   b. restricts the development of a coelom.
   c. is associated with the organ level of organization.
   d. is associated with the development of muscles.
   e. Both b and c are correct.
8. *Ascaris* is a parasitic
   a. roundworm.
   b. flatworm.
   c. hydra.
   d. sponge.
   e. mollusk.
9. Which of these best shows that annelids and arthropods are closely related? In both,
   a. the first embryonic opening becomes the mouth.
   b. there is a complete digestive tract.
   c. there is a ventral solid nerve cord.
   d. there are segments.
   e. All of these are correct.
10. Which of these is mismatched?
    a. clam—gills
    b. lobster—gills
    c. grasshopper—book lungs
    d. polychaete—parapodia
    e. All of these are matched correctly.
11. Which of these is an incorrect statement?
    a. Planarians are heterotrophs.
    b. Spiders are carnivores in the phylum Arthropoda.
    c. Clams are filter feeders in the phylum Mollusca.
    d. Earthworms are scavengers in the phylum Arthropoda.
    e. Squids are predators in the phylum Mollusca.
12. A radula is a unique organ for feeding found in
    a. mollusks.
    b. roundworms.
    c. annelids.
    d. arthropods.
    e. All of these are correct.
13. Which of these is mismatched?
    a. roundworms—hydrostatic skeleton
    b. crayfish—walking legs
    c. clam—hatchet foot
    d. grasshopper—wings
    e. earthworm—many cilia

14. Which of these is mismatched?
   a. flatworms—acoelomate
   b. mollusk—reduced coelom
   c. insects—hemocoel
   d. crayfish—coelom divided by septa
   e. clam worm—true coelom
15. Label this diagram.

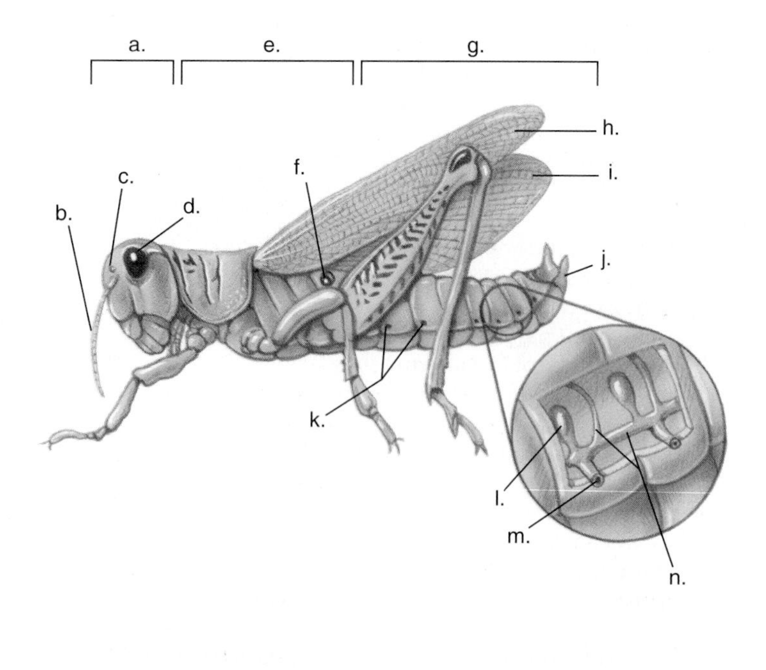

Thinking Scientifically

1. Considering the characteristics of animals (page 618):
   a. Compare corresponding characteristics of plants and animals.
   b. What one animal characteristic can be associated with an animal's need to acquire food?
   c. Why is it reasonable to assume that animals arose from protozoans?
   d. Some animals can regenerate from a small part of the whole. Why would you expect only simple animals, and not those with highly differentiated tissues, to be able to do this?

2. Some animals have a complete (both mouth and anus) digestive tract, and some have an incomplete (only one opening) digestive tract.
   a. In which type of animal would you expect the digestive tract to have specialized parts? Why?
   b. Why is it consistent that a coelomate animal would have a complete digestive tract with specialized parts?
   c. In which type of animal would you expect the animal to be a discontinuous (eat at intervals) feeder?
   d. In more complex animals, blood vessels deliver the product of digestion to the cells. How does a planarian make do without a circulatory system?

Understanding the Terms

annelid 631
arachnid 639
arthropod 634
asymmetry 619
bilateral symmetry 619
bivalve 629
cephalization 619
chitin 634
cnidarian 622
coelom 619
coral 623
crustacean 635
deuterostome 619
filter feeder 621
fluke 625
gastrovascular cavity 623
hemocoel 635
hermaphrodite 625
hydrostatic skeleton 626
insect 636
invertebrate 618
leech 633
Malpighian tubule 637
metamorphosis 637
mollusk 628
molting 634
nematocyst 622
nephridium 631
protostome 619
pseudocoelom 619
radial symmetry 619
segmentation 619
sessile 619
tapeworm 625
trachea 637
vertebrate 618

Match the terms to these definitions:

a. _________ Air tube in insects located between the spiracles and the tracheoles.
b. _________ Body cavity lying between the digestive tract and body wall that is completely lined by mesoderm.
c. _________ Body plan having two corresponding or complementary halves.
d. _________ Periodic shedding of the exoskeleton in arthropods.
e. _________ Segmentally arranged, paired excretory tubules of many invertebrates, as in the earthworm.

Using Technology

Your study of animals is supported by these available technologies:

Essential Study Partner CD-ROM
Evolution & Diversity → Invertebrates

Visit the Mader web site for related ESP activities.

Exploring the Internet

The Mader Home Page provides resources and tools as you study this chapter.

http://www.mhhe.com/biosci/genbio/mader

CHAPTER 31

Animals: Part II

Over half of the vertebrates alive today are fishes. They come in all sizes and shapes—a coral reef is home for these lemon butterfly fish, Chaetodon miliaris.

Chapter Concepts

As Wendy waited in a camouflaged blind for her next target, the wildlife photojournalist thought about her recent assignments. Last month, she had donned scuba gear to photograph the remarkably ugly hagfish, and also worked underwater in a cage to capture pictures of the fierce hammerhead shark. Her cameras had documented alligators chasing down prey and turtles returning to the ocean after burying their eggs on the seashore. Using a hang glider, Wendy had even photographed a bald eagle flying around its nest on the side of a steep mountain. Other jobs had her track kangaroos in Australia and apes in Africa. Today, she was waiting for a lion to attack the herd of antelopes at the watering hole near her blind.

From fish to birds, reptiles to mammals, the vertebrate world offers Wendy a seemingly endless number of photographic subjects. Many of those creatures will make an appearance in this chapter. We'll also detail how, through evolution, one line of descent gave rise to the uniquely intelligent species called *Homo sapiens sapiens,* otherwise known as modern humans.

31.1 Echinoderms

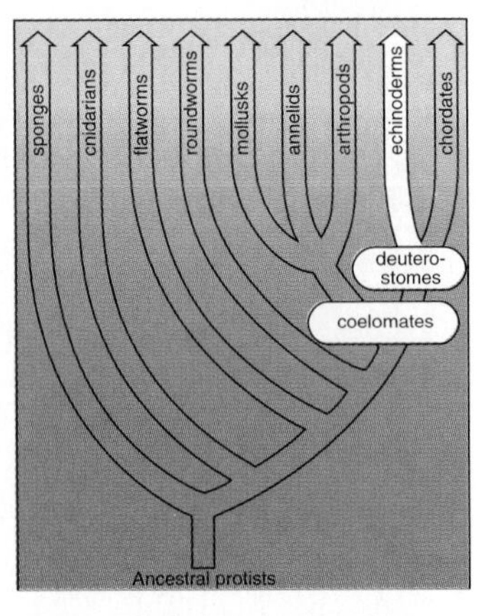

In echinoderms, as well as chordates, the second embryonic opening becomes the mouth. Therefore, echinoderms and chordates are called **deuterostomes.** In these animals, the coelom forms by outpocketing of the primitive gut, and therefore they are also called enterocoelomates. Echinoderms and close relatives of chordates have a dipleurula larva (bands of cilia placed as shown in Fig. 30.10).

Because the chordates include the vertebrates (e.g., human beings), it may seem surprising to learn that chordates are most closely related to the echinoderms (e.g., sea stars) that lack those features we associate with vertebrates. For example, the echinoderms are often radially and not bilaterally symmetrical. However, their larva is a free-swimming filter feeder with bilateral symmetry. Metamorphosis results in the radially symmetrical adult.

Characteristics of Echinoderms

Echinoderms (phylum Echinodermata, about 6,000 species), a diverse group of marine animals, have an *endoskeleton* (internal skeleton) consisting of spine-bearing, calcium-rich plates. The spines, which stick out through their delicate skin, account for their name. The echinoderms include only marine animals—sea stars, sea urchins, sea cucumbers, feather stars, sea lilies, and sand dollars (Fig. 31.1).

a. Purple sea urchin, *Strongylocentrotus*

b. Sea cucumber, *Parastichopus*

c. Feather star, *Comanthus*

Figure 31.1 Echinoderm diversity.
a. Sea urchins have large, colored, external spines for protection. **b.** Sea cucumbers look like a cucumber; they lack arms but have tentaclelike tube feet with suckers around the mouth. **c.** A feather star extends its arms to strain suspended food particles from the sea.

Sea Stars

Sea stars are commonly found along rocky coasts where they feed on clams, oysters, and other bivalve mollusks. The *five-rayed body* has an oral, or mouth, side (the underside) and an aboral, or anus, side (the upper side) (Fig. 31.2). Various structures project through the body wall: (1) spines from the endoskeletal plates offer some protection; (2) pincerlike structures called pedicellariae keep the surface free of small particles; and (3) skin gills, tiny fingerlike extensions of the skin, are used for gas exchange. On the oral surface, each arm has a groove lined by small *tube feet.*

To feed, a sea star positions itself over a bivalve and attaches some of its tube feet to each side of the shell. By working its tube feet in alternation, it pulls the shell open. A very small crack is enough for the sea star to evert its cardiac stomach and push it through the crack, so that it contacts the soft parts of the bivalve. The stomach secretes enzymes, and digestion begins even while the bivalve is attempting to close its shell. Later, partly digested food is taken into the sea star's body, where digestion continues in the pyloric stomach using enzymes from the digestive glands found in each arm. A short intestine opens at the anus on the aboral side.

In each arm, the well-developed coelomic cavity contains not only a pair of digestive glands, but also gonads (either male or female), which open on the aboral surface by very small pores. The nervous system consists of a central nerve ring that gives off radial nerves in each arm. A light-sensitive eyespot is at the tip of each arm. Sea stars are capable of coordinated but slow responses and body movements.

Locomotion depends on the **water vascular system.** Water enters this system through a structure on the aboral side called the *sieve plate,* or madreporite. From there it passes down a stone canal into a ring canal, which surrounds the mouth. The ring canal gives off a radial canal in each arm. From the radial canals, water enters the ampullae. Contraction of the ampulla forces water into the tube foot, expanding it. When the foot touches a surface, the center is withdrawn, giving it suction so that it can adhere to the surface. By alternating the expansion and contraction of the tube feet, a sea star moves slowly along.

Echinoderms don't have a respiratory, excretory, or circulatory system. Fluids within the coelomic cavity and the water vascular system carry out many of these functions. For example, gas exchange occurs across the skin gills and the tube feet. Nitrogenous wastes diffuse through the coelomic fluid and the body wall. Cilia on the peritoneum lining the coelom keep the coelomic fluid moving.

Sea stars reproduce asexually and sexually. If the body is fragmented, each fragment can regenerate a whole animal. Fishermen who try to get rid of sea stars by cutting them up and tossing them overboard are merely propagating more sea stars! Sea stars spawn, releasing either eggs or sperm. The dipleurula larva is bilateral and metamorphoses to become the radially symmetrical adult.

Echinoderms have a well-developed coelom and internal organs despite being radially symmetrical. Spines project from their endoskeleton, and there is a unique water vascular system.

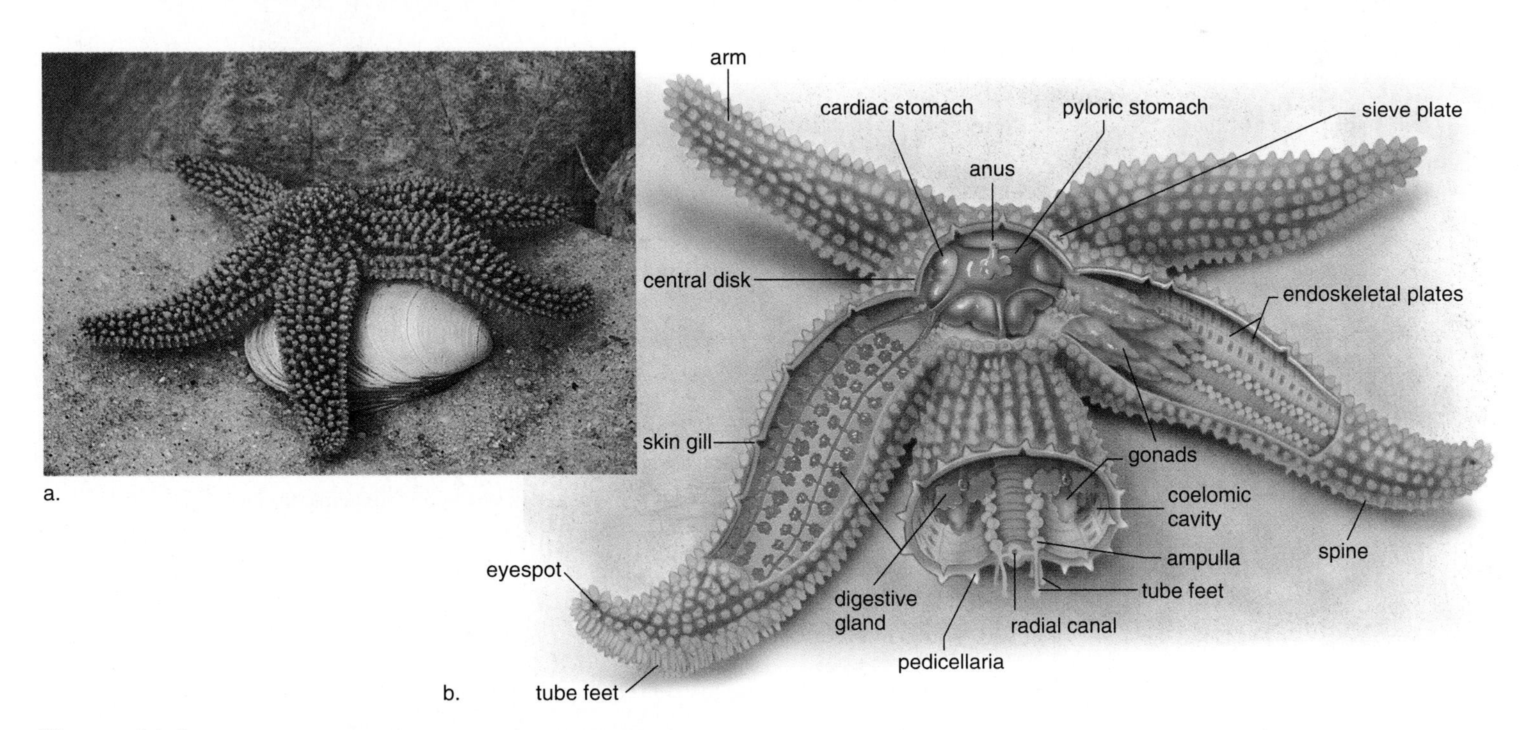

Figure 31.2 Sea star anatomy and behavior.
a. A sea star uses the suction of its tube feet to open a clam, its primary source of food. **b.** Each arm of a sea star contains digestive glands, gonads, and portions of the water vascular system. This system (colored yellow) terminates in tube feet.

31.2 Chordates

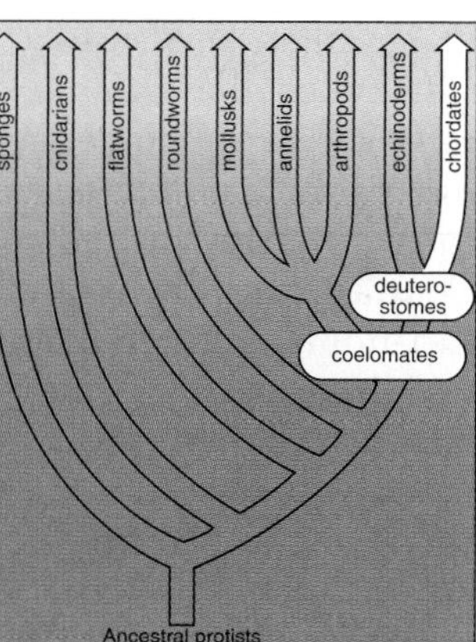

To be considered a **chordate,** (phylum Chordata, about 45,000 species) an animal must have the four basic characteristics listed below at some time during its life history:

1. A *dorsal supporting rod called a notochord.* The **notochord** is located just below the nerve cord. Vertebrates have an embryonic notochord that is replaced by the vertebral column during development.
2. A *dorsal hollow nerve cord.* By hollow, it is meant that the cord contains a canal filled with fluid. In vertebrates, the nerve cord, more often called the spinal cord, is protected by the vertebrae.
3. *Pharyngeal pouches.* These are seen only during embryonic development in most vertebrates. In the invertebrate chordates, the fishes, and some amphibian larvae, the pharyngeal pouches become functioning **gills.** Water passing into the mouth and the pharynx goes through the gill slits, which are supported by gill arches. In terrestrial vertebrates which breathe by lungs, the pouches are modified for various purposes. In humans, the first pair of pouches become the auditory tubes. The second pair become the tonsils, while the third and fourth pairs become the thymus gland and the parathyroids.
4. A *tail* that extends beyond the anus and is therefore called a post-anal tail.

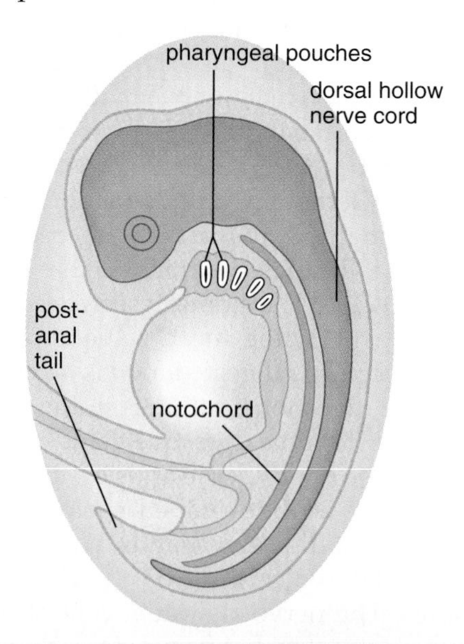

Classification

The Deuterostomes* The second embryonic opening is associated with the mouth; enterocoelomates; dipleurula larva.

Phylum Echinodermata (echinoderms)	Radial symmetry; endoskeleton of spine-bearing plates; water vascular system with tube feet; about 6,000 species.
Phylum Chordata (chordates)	Notochord, dorsal hollow nerve cord, pharyngeal pouches; about 45,000 species.
Subphylum Cephalochordata (chordates)	Marine fishlike animals with three chordate characteristics as adults.
Subphylum Urochordata (urochordates)	Larva free-swimming with three chordate characteristics; adults sessile filter feeders with plentiful gill slits; about 1,250 species.
Subphylum Vertebrata (vertebrates)	Notochord replaced by vertebrae that protect the nerve cord; skull that protects the brain; segmented with jointed appendages; about 43,700 species.
Superclass Agnatha (jawless fishes)	Marine and freshwater fishes; lack jaws and paired appendages; notochord; about 63 species.
Superclass Gnathostomata (jawed fishes and tetrapods)	Hinged jaws; paired appendages.
Class Chondrichthyes (cartilaginous fishes)	Marine cartilaginous fishes; lack operculum and swim bladder; tail fin usually asymmetrical; about 850 species.
Class Osteichthyes (bony fishes)	Marine and freshwater bony fishes; operculum; swim bladder or lungs; tail fin; usually symmetrical; about 20,000 species.
Class Amphibia (amphibians)	Tetrapod with nonamniote egg; nonscaly skin; three-chambered heart; ectothermic; metamorphosis; about 3,900 species.
Class Reptilia (reptiles)	Tetrapod with amniote egg; scaly skin; ectothermic; teeth not differentiated typically; about 6,000 species.
Class Aves (birds)	Tetrapod with feathers; bipedal with wings; double circulation; endothermic; about 9,000 species.
Class Mammalia (mammals)	Tetrapods with hair, mammary glands; double circulation; endothermic; teeth differentiated. Monotremes, marsupials, placental mammals; about 4,500 species.
Order Primates	Large cerebral hemispheres; mostly tree dwellers; opposable thumb and flat nails; about 233 species

*Not in the classification of organisms, but added here for clarity.

Visual Focus

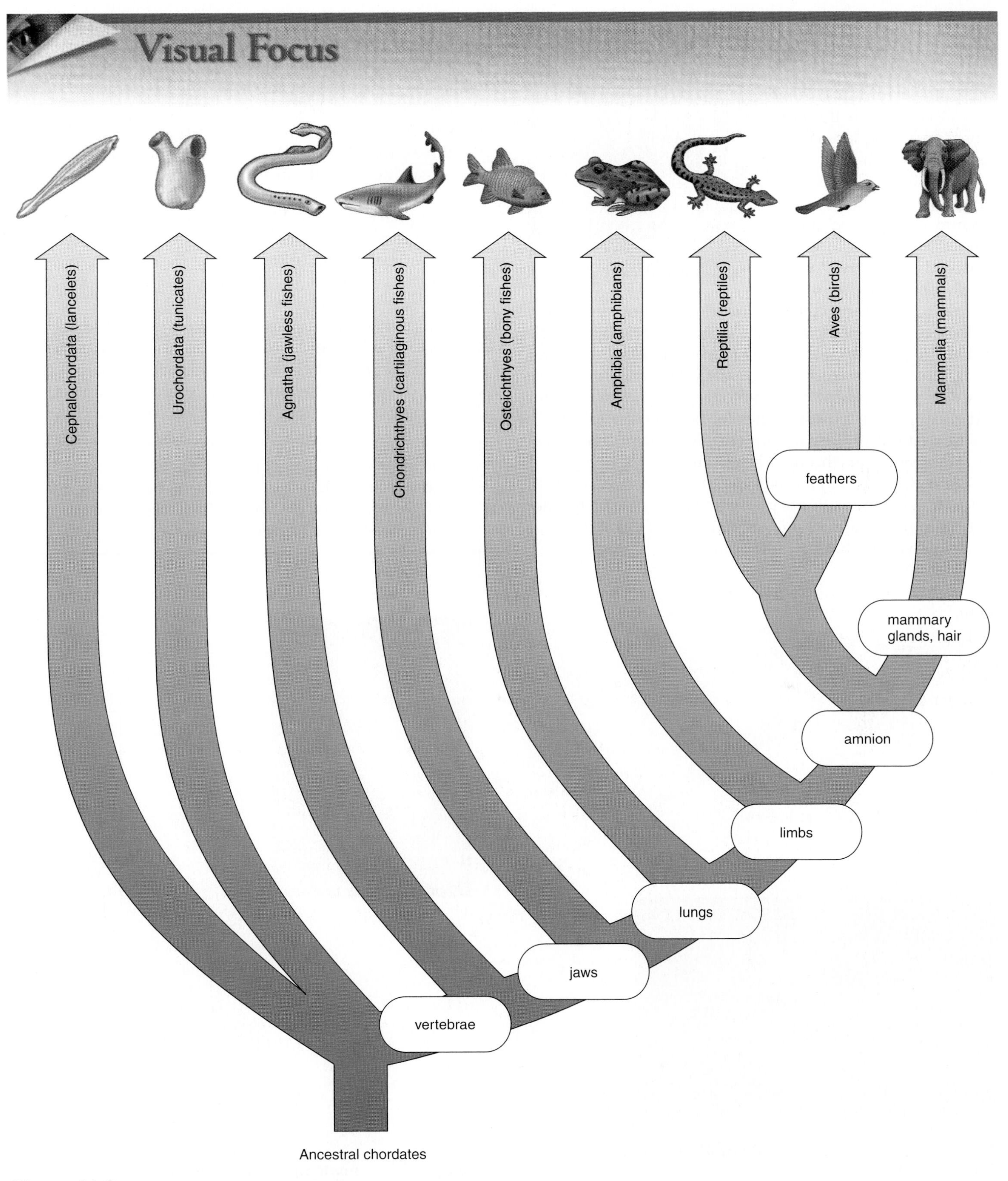

Figure 31.3 Evolutionary tree of chordates.
Each of the highlighted features is an evolved characteristic that is shared by the classes beyond this point.

Invertebrate Chordates

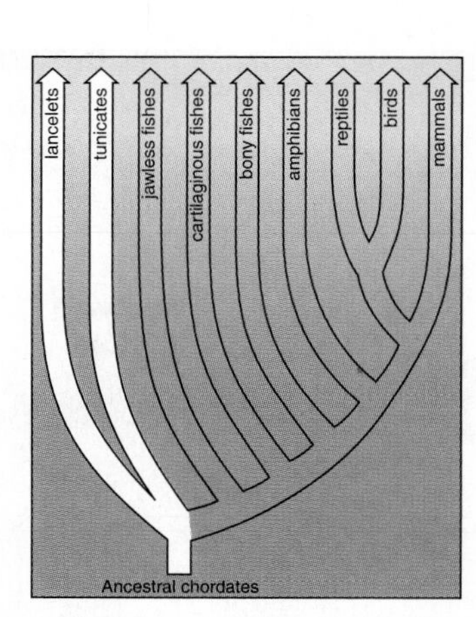

There are a few *invertebrate chordates* in which the notochord is never replaced by the vertebral column.

Lancelets, formerly referred to as amphioxus, (subphylum Cephalochordata, about 23 species), are now classified in the genus *Branchiostoma.* These marine chordates, which are only a few centimeters long, are named for their resemblance to a lancet—a small, two-edged surgical knife (Fig. 31.4). Lancelets are found in the shallow water along most coasts where they usually lie partly buried in sandy or muddy substrates with only their anterior mouth and gill apparatus exposed. They feed on microscopic particles filtered out of the constant stream of water that enters the mouth and exits through the gill slits.

Lancelets retain the four chordate characteristics as an adult. In addition, segmentation is present, as witnessed by the fact that the muscles are segmentally arranged and the dorsal hollow nerve cord has periodic branches.

Tunicates, or sea squirts (subphylum Urochordata, about 1,250 species), live on the ocean floor and take their name from a tunic that makes the adults look like thick-walled, squat sacs. They are also called sea squirts because they squirt out water from one of their siphons when disturbed. The tunicate larva is bilaterally symmetrical and has the three chordate characteristics. Metamorphosis produces the sessile adult with an incurrent and excurrent siphon (Fig. 31.5).

The pharynx is lined by numerous cilia whose beating creates a current of water that moves into the pharynx and out the numerous *gill slits,* the only chordate characteristic that remains in the adult. Microscopic particles adhere to a mucous secretion and are eaten.

Is it possible that the tunicates are directly related to the vertebrates? It has been suggested that a larva with the four chordate characteristics may have become sexually mature without developing the other adult tunicate characteristics. Then it may have evolved into a fishlike vertebrate.

The invertebrate chordates include the tunicates and the lancelets—a lancelet is the best example of a chordate that possesses the three chordate characteristics as an adult.

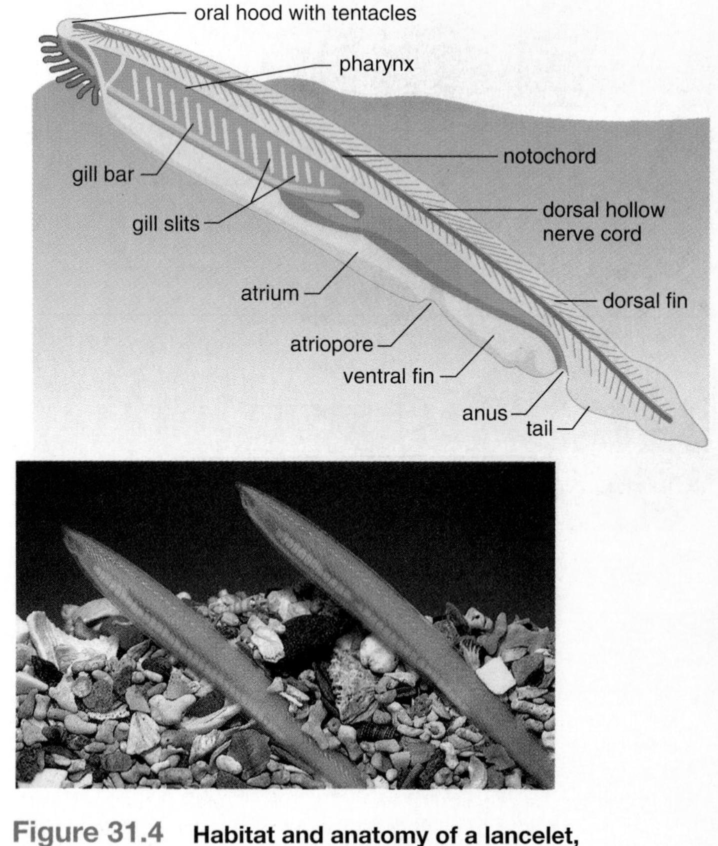

Figure 31.4 Habitat and anatomy of a lancelet, *Branchiostoma.*
Water enters the mouth and exits at the atripore after passing through the gill slits. Lancelets are filter feeders.

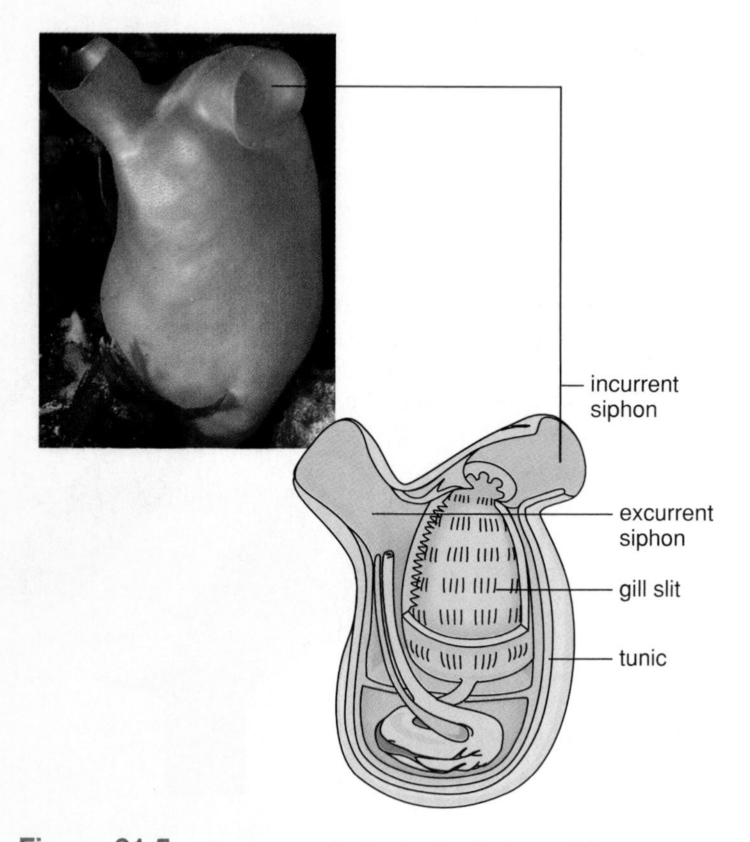

Figure 31.5 Anatomy of a tunicate, *Halocynthia.*
Note that the only chordate characteristic remaining in the adult is gill slits.

31.3 Vertebrates

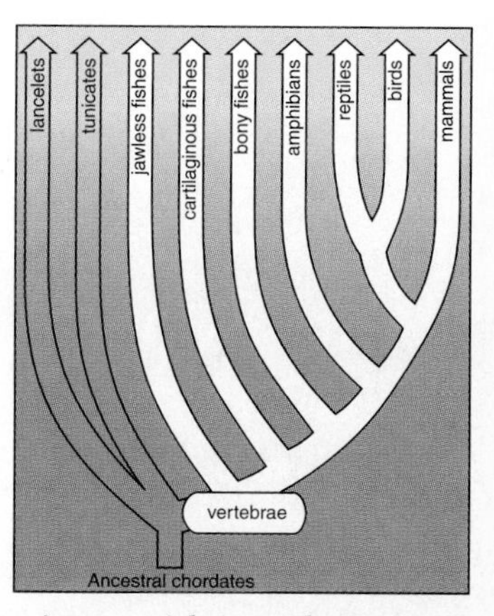

At some time in their life history, **vertebrates** (subphylum Vertebrata, about 43,700 species) have all three chordate characteristics. The embryonic notochord, however, is generally replaced by a *vertebral column* composed of individual vertebrae. The vertebral column, which is a part of the flexible but strong *jointed endoskeleton,* gives evidence that vertebrates are segmented. The skeleton protects internal organs and serves as a place of attachment for muscles. Together the skeleton and muscles form a system that permits rapid and efficient movement. *Two pairs of appendages* are characteristic. The pectoral and pelvic **fins** of fish evolved into the *jointed appendages* that allowed vertebrates to move onto land.

The main axis of the endoskeleton consists of not only the vertebral column, but also a skull that encloses and protects the brain. The high degree of *cephalization* is accompanied by complex sense organs. The eyes develop as outgrowths of the brain. The ears are primarily equilibrium devices in aquatic vertebrates, but they also function as sound-wave receivers in land vertebrates.

The evolution of jaws in vertebrates has allowed some to take up the predatory way of life. Vertebrates have a complete digestive tract and a large coelom. The circulatory system is closed (the blood is contained entirely within blood vessels). Vertebrates have an efficient means of obtaining oxygen from water or air, as appropriate. The kidneys are important excretory and water-regulating organs that conserve or rid the body of water as necessary. The sexes are generally separate, and reproduction is usually sexual. The evolution of the amnion allowed reproduction to take place on land. Reptiles, birds and some mammals lay a shelled egg. In placental mammals the development takes place within the uterus of the female.

Figure 31.6 shows major milestones in the evolutionary history of vertebrates: the evolution of jaws, limbs, and the amnion, an extraembryonic membrane which is first seen in the shelled **amniote egg.**

Vertebrates are distinguished in particular by:

- living endoskeleton
- closed circulatory system
- paired appendages
- efficient respiration and excretion
- high degree of cephalization

In short, vertebrates are adapted to an active lifestyle.

a. Great white shark, *Carcharodon*

b. Spotted salamander, *Ambystoma*

c. Veiled chameleons, *Chamaeleo*

Figure 31.6 Milestones in vertebrate evolution.
a. The evolution of jaws in fishes allows animals to be predators and feed off other animals. **b.** The evolution of limbs in amphibians is adaptive for locomotion on land. **c.** The evolution of a shelled egg in reptiles is adaptive for reproduction on land.

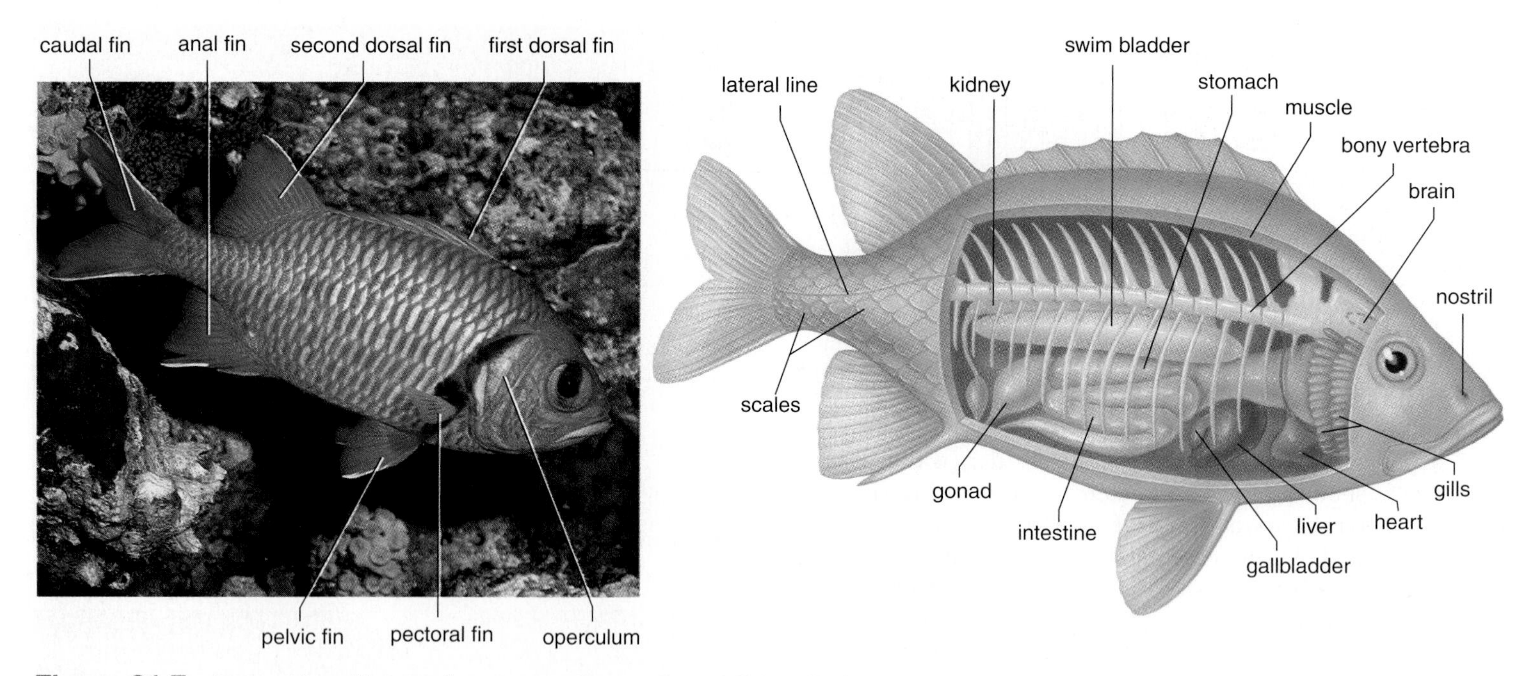

Figure 31.7 **External and internal anatomy of a ray-finned fish called a soldierfish, *Myripristis*.**

Fishes

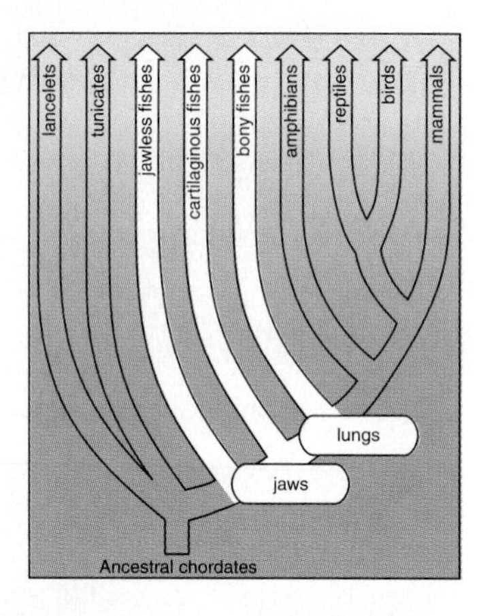

The first vertebrates were fishlike, and today there are three living classes: jawless fishes, cartilaginous fishes, and the bony fishes.

Jawless Fishes

Living representatives of the **jawless fishes** (superclass Agnatha, about 63 species) are cylindrical and up to a meter long. They have smooth, scaleless skin and no jaws or paired fins. There are two groups of living jawless fishes: *hagfishes* and *lampreys*. The hagfishes are scavengers, feeding mainly on dead fishes, while some lampreys are parasitic. When parasitic, the round mouth of the lamprey serves as a sucker. The lamprey attaches itself to another fish and taps into its circulatory system. Water cannot move in through the mouth and out over the gills as is common in all other fishes. Instead, water moves in and out through the gill openings.

Fishes with Jaws

All of the other fishes have jaws, tooth-bearing bones of the head. **Jaws** are believed to have evolved from the first pair of gill arches. The second pair of gill arches became support structures for jaws. The presence of jaws is an adaptation to a predatory way of life.

Cartilaginous Fishes **Cartilaginous fishes** (class Chondrichthyes, about 850 species) are the sharks, the rays, and the skates, which have skeletons of cartilage instead of bone. The small dogfish shark is often dissected in biology laboratories. One of the most dangerous sharks inhabiting both tropical and temperate waters is the hammerhead shark. The largest sharks, the whale sharks, feed on small fishes and marine invertebrates and do not attack humans. Skates and rays are rather flat fishes that live partly buried in the sand and feed on mussels and clams.

Three well-developed senses enable sharks and rays to detect their prey. They have the ability to sense electric currents in water—even those generated by the muscle movements of animals. They have a lateral line system, a series of pressure-sensitive cells that lie within canals along both sides of the body, which can sense pressure caused by a fish or other animal swimming nearby. They also have a keen sense of smell; the part of the brain associated with this sense is twice as large as the other parts. Sharks can detect about one drop of blood in 115 liters (25 gallons) of water.

Bony Fishes **Bony fishes** (class Osteichthyes, about 20,000 species) are by far the most numerous and diverse of all the vertebrates. Most of the fishes we eat, such as perch, trout, salmon, and haddock are a type of bony fish called *ray-finned fishes*. Ray-finned fishes are the most successful and diverse of all the vertebrates. Some, like herrings, are filter feeders; others, like trout, are opportunists, and still others are predaceous carnivores, like piranhas and barracudas.

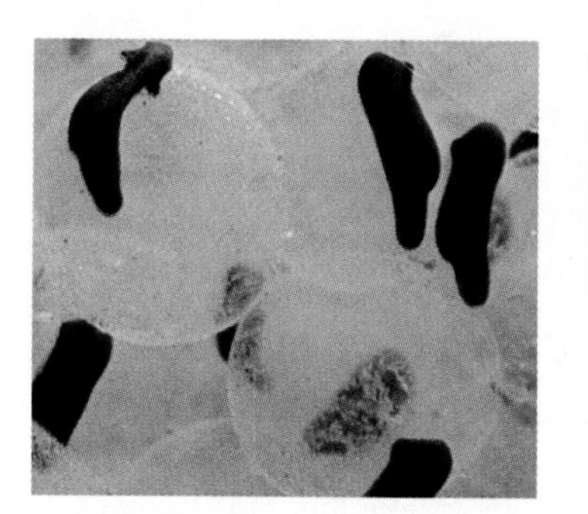
a. Tadpoles hatch

b. Tadpole respires with external gills

c. Front and hind legs are present

d. Frog respires with lungs

Figure 31.8 Frog metamorphosis.
During this time, the animal changes from an aquatic to a terrestrial organism.

Bony fishes have a *swim bladder,* which usually serves as a buoyancy organ (Fig. 31.7). By secreting gases into the bladder or by absorbing gases from it, these fishes can change their density and thus go up or down in the water. Ray-finned refers to the fact that the paired fins, which are paddlelike processes used in balancing and propelling the body, are thin and supported by bony rays.

Fishes are adapted to life in the water. Usually sperm and eggs are shed into the water, where fertilization occurs. The zygote develops into a swimming larva, which can fend for itself until it develops into the adult form. The streamlined shape, fins, and muscle action of most bony fishes are all suited to locomotion in the water. Their skin is covered by bony scales which protect the body but do not prevent water loss. When fishes respire, the *gills* are kept continuously moist by the passage of water through the mouth and out the gill slits. As the water passes over the gills, oxygen is absorbed by blood and carbon dioxide is given off. Fishes have a single-circuit circulatory system. The heart is a simple pump, and the blood flows through the chambers, including a nondivided atrium and ventricle, to the gills. Oxygenated blood leaves the gills and goes to the body proper, eventually returning to the heart for recirculation.

Another type of bony fish called the *lobe-finned fishes* evolved into the amphibians. These fishes not only had fleshy appendages that could be adapted to land locomotion; most also had a **lung,** which was used for respiration. A type of lobe-finned fish called the coelacanth, which exists today, is the only "living fossil" among the fishes. The coelacanth, however, does not have a lung.

Most fishes today are ray-finned fishes. They have the following characteristics:

- bony skeleton and scales
- swim bladder
- two-chambered heart
- paired fins
- jaws
- gills

Amphibians

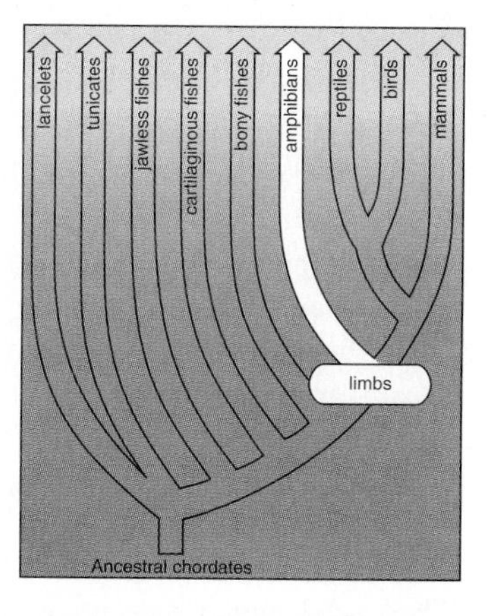

Amphibians (class Amphibia, about 3,900 species), whose class name means living on both land and in the water, are represented today by frogs, toads, newts, and salamanders. Aside from being tetrapods (presence of two sets of paired limbs) and appropriate modifications of the girdles, amphibians have other features not seen in bony fishes. Their eyes have eyelids for keeping eyes moist, ears adapted to picking up sound waves, and a voice-producing larynx. The brain is larger than that of a fish. Adult amphibians usually have small lungs. Air enters the mouth by way of nostrils, and when the floor of the mouth is raised, air is forced into the relatively small lungs. Respiration is supplemented by gas exchange through the smooth, moist, and glandular skin. The amphibian heart has a divided atrium but a single ventricle. The right atrium receives blood that has little oxygen from the body proper, and the left atrium receives blood that has much oxygen from the lungs. These two types of blood are partially mixed in the single ventricle (see page 654). Mixed blood is then sent to all parts of the body; some is sent to the skin, where it is further oxygenated.

Most members of this group lead an amphibious life—that is, the larval stage lives in the water and the adult stage lives on the land. The adult usually returns to the water, however, to reproduce. Figure 31.8 illustrates how the frog tadpole undergoes metamorphosis into the adult before taking up life on the land.

These features in particular distinguish amphibians:

- usually tetrapods
- mostly metamorphosis
- three-chambered heart
- usually lungs in adults
- smooth, moist skin

Science Focus

Real Dinosaurs, Stand Up!

Today's paleontologists are setting the record straight about dinosaurs. Because dinosaurs are classified as reptiles, it is assumed that they must have had the characteristics of today's reptiles. They must have been ectothermic, slow moving, and antisocial, right? Wrong!

First of all, not all dinosaurs were great lumbering beasts. Many dinosaurs were less than 1 meter (3 feet) long and their tracks indicate they moved at a steady pace. These dinosaurs stood on two legs that were positioned directly under the body. Perhaps they were as agile as ostriches, which are famous for their great speed.

Dinosaurs may have been endothermic. Could they have competed successfully with the preevolving mammals otherwise? They must have been able to hunt prey and escape from predators as well as mammals, which are known to be active because of their high rate of metabolism. Some argue that ectothermic animals have little endurance and cannot keep up. They also believe that the bone structure of dinosaurs indicates they were endothermic.

Dinosaurs cared for their young much like birds do today. In Montana, paleontologist Jack Horner has studied fossilized nests complete with eggs, embryos, and nestlings (Fig. 31A). The nests are about 7.5 meters (24.6 feet) apart, the space needed for the length of an adult parent. About 20 eggs are laid in neatly arranged circles and may have been covered with decaying vegetation to keep them warm. Many contain the bones of juveniles as much as a meter long. It would seem then that baby dinosaurs remained in the nest to be fed by their parents. They must have obtained this size within a relatively short period of time, again indicating that dinosaurs were endothermic. Ectothermic animals grow slowly and take a long time to reach this size.

Dinosaurs were also social! An enormous herd of dinosaurs found by Horner and colleagues is estimated to have nearly 30 million bones, representing 10,000 animals in one area measuring about 1.6 square miles. Most likely, the herd kept on the move in order to be assured of an adequate food supply, which consisted of flowering plants that could be stripped one season and grow back the next season. The fossilized herd is covered by volcanic ash, suggesting that the dinosaurs died following a volcanic eruption.

Some dinosaurs, such as the duck-billed dinosaurs and horned dinosaurs, have a skull crest. How might it have functioned? Perhaps it was a resonating chamber, used when dinosaurs communicated with one another. Or, as with modern horned animals that live in large groups, the males could have used the skull crest in combat to establish dominance.

Some zoologists do not agree with the assertions of paleontologists. They believe that dinosaurs did have the characteristics of today's reptiles, and this doesn't make them in any way inferior to endothermic mammals. Ectothermic animals can wait out cold weather conditions, while endothermic animals must always venture forth to find food, no matter what the conditions.

a.

b.

Figure 31A Behavior of dinosaurs.
a. Nest of fossil dinosaur eggs found in Montana, dating from the Cretaceous period. **b.** Bones of a hatchling (about 50 cm [20 inches]) found in the nest. These dinosaurs have been named *Maiosaura*, which means "good mother lizard" in Greek.

Figure 31.9 The tongue as a sense organ.
Snakes wave a forked tongue in the air to collect chemical molecules which are brought back into the mouth and then delivered to an organ in the roof of the mouth. Analyzed chemicals help the snake trail a prey animal, recognize a predator, or find a mate.

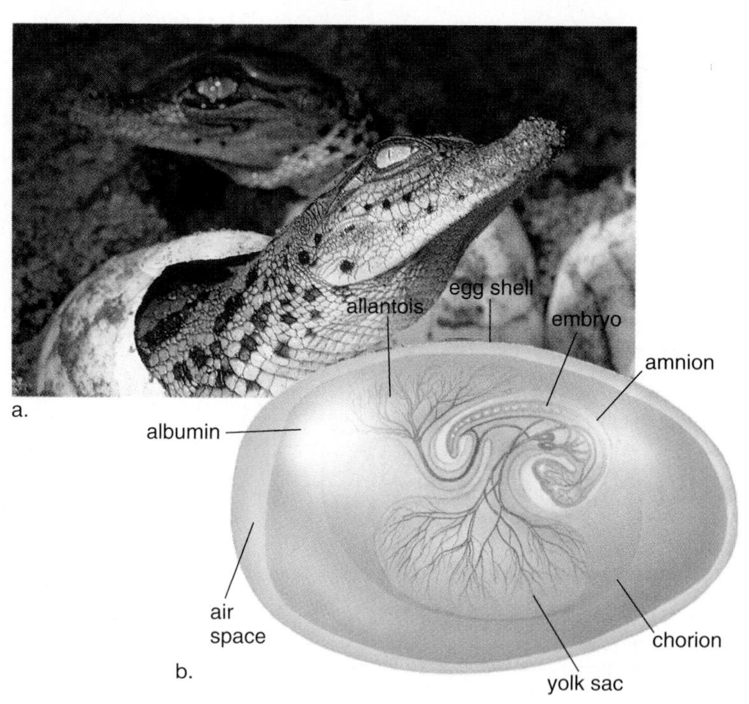

Figure 31.10 The reptilian egg allows reproduction on land.
a. Baby American crocodile, *Crocodylus*, hatching out of its shell. Note that the shell is leathery and flexible, not brittle like birds' eggs. **b.** Inside the egg, the embryo is surrounded by membranes. The chorion aids gas exchange, the yolk sac provides nutrients, the allantois stores waste, and the amnion encloses a fluid that prevents drying out and provides protection.

Reptiles

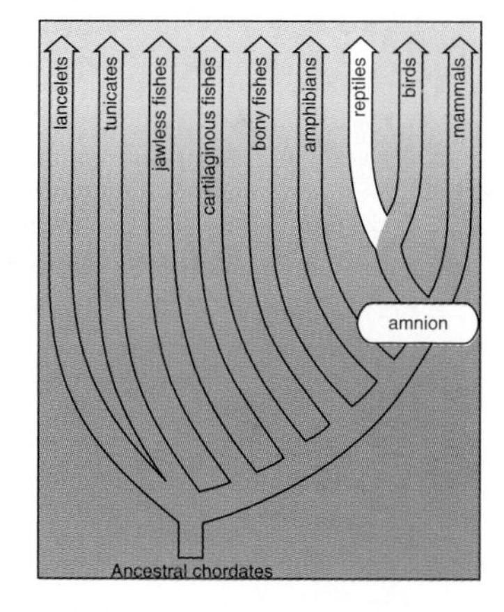

Reptiles (class Reptilia, about 6,000 species) diversified and were abundant some 245 to 66 millions of years ago. Aside from those that lived on to become the living reptiles of today, there were the mammal-like reptiles and the dinosaurs, which became extinct. Some dinosaurs are remembered for their great size. *Brachiosaurus,* a herbivore, was about 23 meters (75 feet) long and about 17 meters (56 feet) tall. *Tyrannosaurus rex,* a carnivore, was 5 meters (16 feet) tall when standing on its hind legs. A bipedal stance was preadaptive for the evolution of wings in birds. Some say birds are actually living dinosaurs; new studies on the behavior of dinosaurs discussed in the accompanying reading give credence to this hypothesis.

The reptiles living today are mainly turtles, alligators, snakes, and lizards. The body is covered with hard, *keratinized scales,* which protect the animal from desiccation and from predators. Another adaptation for a land existence is the manner in which snakes use their tongue as a sense organ (Fig. 31.9). Reptiles have well-developed lungs enclosed by a protective rib cage. When the rib cage expands, the lungs expand and air rushes in. The creation of a partial vacuum establishes a negative pressure, which causes air to rush into the lungs. The atrium of the heart is always separated into right and left chambers, but division of the ventricle varies. An interventricular septum is incomplete in certain species; therefore, there is some exchange of oxygenated and deoxygenated blood between the ventricles.

Perhaps the most outstanding adaptation of the reptiles is that they have a means of reproduction suitable to a land existence. The penis of the male passes sperm directly to the female. *Fertilization* is internal, and the female lays leathery, flexible, shelled eggs. The *amniote egg* made development on land possible and eliminated the need for a swimming-larval stage during development. It provides the developing embryo with atmospheric oxygen, food, and water; it removes nitrogenous wastes; and it protects the embryo from drying out and from mechanical injury. This is accomplished by the presence of *extraembryonic membranes* (Fig. 31.10).

Fishes, amphibians, and reptiles are *ectothermic.* Their body temperature matches the temperature of the external environment. If it is cold externally, they are cold internally; if it is hot externally, they are hot internally. Reptiles try to regulate body temperatures by exposing themselves to the sun if they need warmth or by hiding in the shadows if they need cooling off. This works reasonably well in most areas of the world.

These features in particular distinguish reptiles:

- usually tetrapods
- lungs with expandable rib cage
- shelled egg
- dry, scaly skin

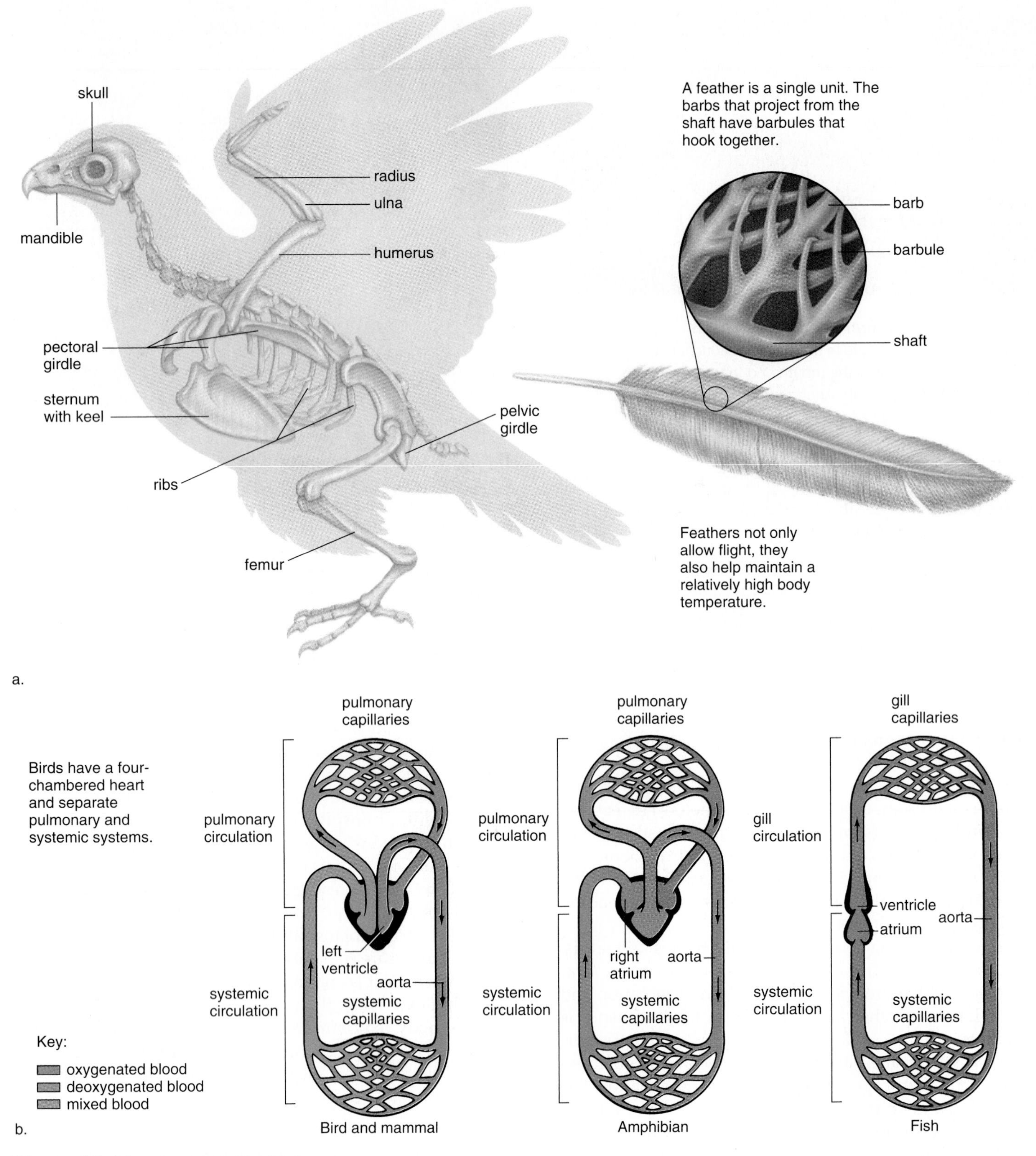

Figure 31.11 Anatomy of a bird.
a. Birds have a large, keeled sternum to which flight muscles attach. Bird bones are strong but weigh very little because they contain air cavities. In feathers, a hollow central shaft gives off barbs and barbules, which interlock in a latticelike array. **b.** In birds and mammals, the heart has four chambers and sends only blood that has little oxygen to the lungs and blood that has much oxygen to the body. In amphibians, the heart has two chambers and there is some mixture of oxygenated and deoxygenated blood. In fishes, a single-looped system utilizes a two-chambered heart.

Birds

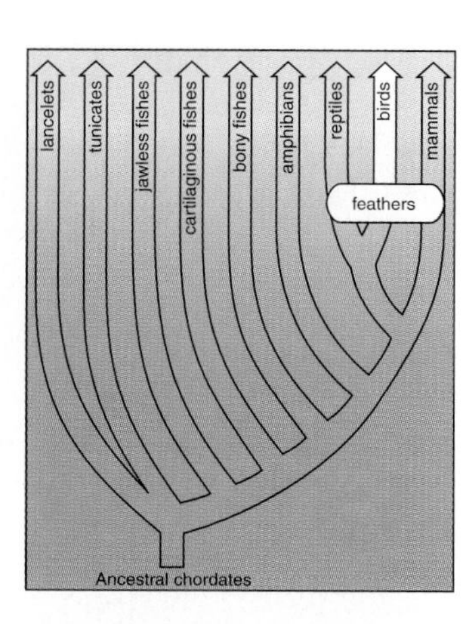

Birds (class Aves, about 9,000 species) are characterized by the presence of **feathers** (Fig. 31.11), which are modified reptilian scales. (Perhaps you have noticed that there are scales on the legs of a chicken.)

Nearly every anatomical feature of a bird can be related to its ability to fly. The anterior pair of appendages (wings) are adapted for flight; the posterior are variously modified, depending on the type of bird. Some are adapted for swimming, some for running, and some for perching on limbs. The breastbone is well developed and has a ridge, the keel, to which the flight muscles are attached (Fig. 31.11). Respiration is efficient since the lobular lungs form *anterior and posterior air sacs.* The presence of these sacs means that the air circulates one way through the lungs and there is a continuous exchange of gases across respiratory tissues (Fig. 31.12). Another benefit of air sacs is that they lighten the body and aid flying.

Birds have a four-chambered heart which completely separates blood with much oxygen from blood that has little oxygen (see page 654). Blood that has a relatively high concentration of oxygen is sent under pressure to the muscles. Birds are *endothermic.* Like mammals, their internal temperature is constant because they generate and maintain metabolic heat. This may be associated with their efficient nervous, respiratory, and circulatory systems. Also, their feathers provide insulation. Birds have no bladder and excrete uric acid in a semidry state.

Birds have well-developed brains; the enlarged portion seems to be the area responsible for instinctive behavior. A ritualized courtship often precedes mating. Birds practice internal fertilization and lay hard-shelled eggs. Many newly hatched birds require parental care before they are able to fly away and seek food for themselves. A remarkable aspect of bird behavior is the seasonal migration of many species over very long distances. Birds navigate by day and night, and whether it's sunny or cloudy, by using the sun and stars and even the earth's magnetic field to guide them.

There are many orders of birds, including birds that are flightless (ostriches), web-footed (penguins), divers (loons), fish eaters (pelicans), waders (flamingos), broad billed (ducks), birds of prey (hawks), vegetarians (fowl, e.g., chickens and turkeys), shorebirds (sandpipers), nocturnal (owls), small (hummingbirds), large (condors), and songbirds, the most familiar of the birds.

These features in particular distinguish birds:

- usually wings for flying
- hard-shelled egg
- four-chambered heart
- feathers
- air sacs
- endothermic

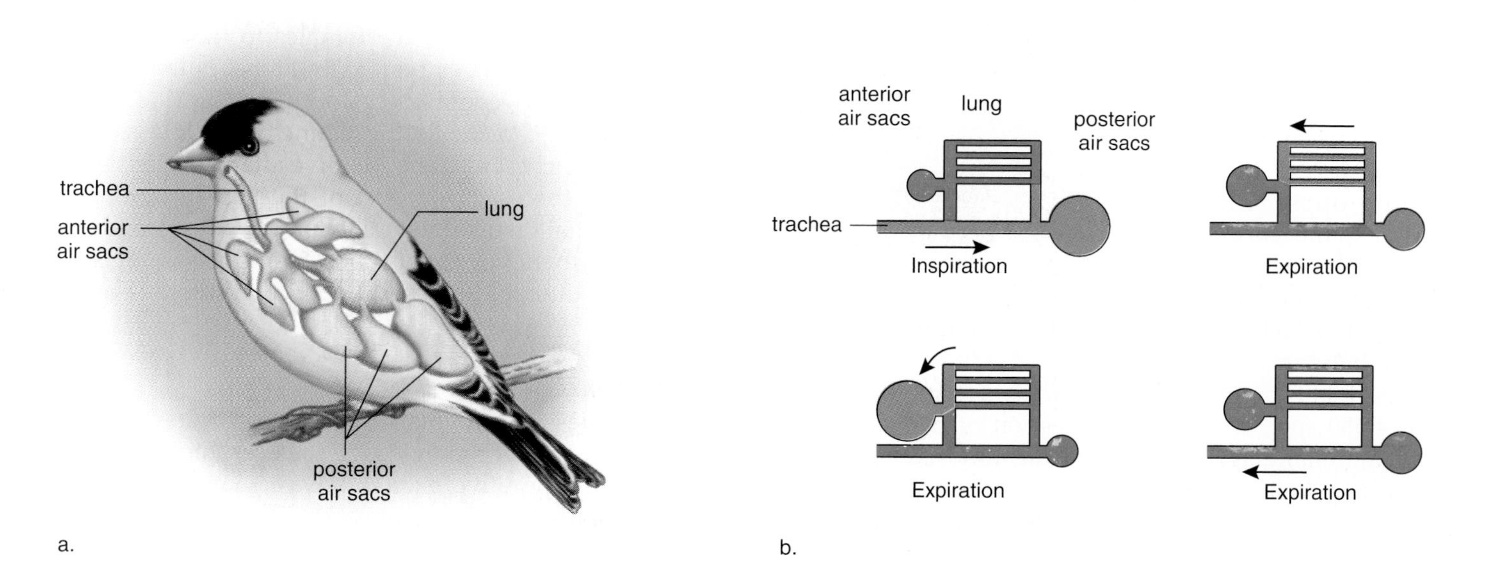

Figure 31.12 **How birds breathe.**
a. Birds have a system of air sacs in their bones. **b.** When a bird inhales, air enters the posterior air sacs, and when a bird exhales, air moves through the lungs to the anterior air sacs before exiting the trachea. This one-way flow of air is efficient and allows more oxygen to be removed from one breath of air.

a. Duckbill platypus, *Ornithorhynchus*

b. Koala, *Phascolarctos*

Figure 31.13 Three types of mammals.
a. The duckbill platypus is a monotreme, which lays shelled eggs. **b.** The koala is a marsupial, whose young are born immature and complete their development within the mother's pouch. **c.** The white-tailed deer is a placental mammal, whose young develop within a uterus.

Mammals

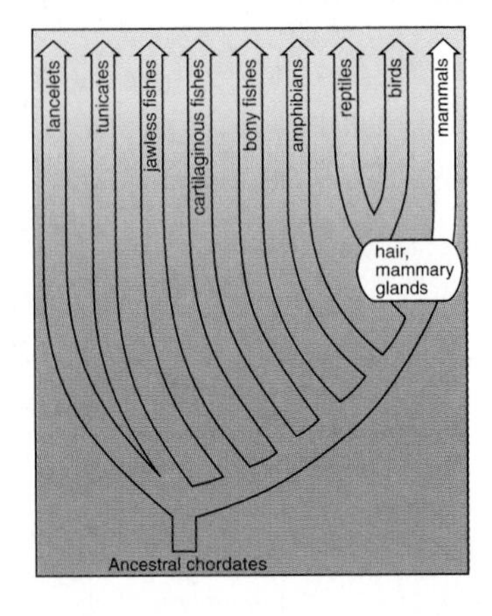

Mammals also evolved from the reptiles. These first mammals were small, about the size of mice. All the time the dinosaurs flourished (165 million years ago), mammals were a minor group that changed little. Some of the earliest mammalian groups, represented today by the monotremes and marsupials, are not abundant today. The placental mammals that evolved later went on to live in many habitats, including air, land, and sea habitats.

The chief characteristics of mammals (class Mammalia, about 4,500 species) are the possession of hair and milk-producing mammary glands. They are almost all endothermic and maintain a constant internal temperature. Many of the adaptations of mammals are related to temperature control. Hair, for example, provides insulation against heat loss and allows mammals to be active, even in cold weather. Like birds, mammals have efficient respiratory and circulatory systems, which assure a ready oxygen supply to muscles whose contraction produces body heat. Like birds, mammals have a double-loop circulation and a *four-chambered heart* (see Fig. 31.11*b*).

Mammary glands enable females to feed (nurse) their young without leaving them to find food. Nursing also creates a bond between mother and offspring that helps ensure parental care while the young are helpless. In most mammals, the young are born alive after a period of development in the uterus, a part of the female reproductive tract. Internal development shelters the young and allows the female to move actively about while the young are maturing. Mammals are classified according to how they reproduce.

Monotremes

Monotremes are mammals that, like birds, have a *cloaca,* a terminal region of the digestive tract serving as a common chamber for digestive, excretory wastes, and sex cells. They also lay hard-shelled amniote eggs. They are represented by the duckbill platypus and the spiny anteater, both of which are found in Australia (Fig. 31.13*a*). The female duckbill platypus lays her eggs in a burrow in the ground. She incubates the eggs and, after hatching, the young lick up milk that seeps from modified sweat glands on the abdomen of

c. White-tailed deer, *Odocoileus*

both males and females. The spiny anteater has a pouch on the belly side formed by swollen mammary glands and longitudinal muscle. The egg moves from the cloaca to this pouch, where hatching takes place, and the young remain for about 53 days. Then they stay in a burrow, where the mother periodically visits and nurses them.

Marsupials

The young of **marsupials** begin their development inside the female's body, but they are born in a very immature condition. Newborns crawl up into a pouch on their mother's abdomen. Inside the pouch, they attach to nipples of mammary glands and continue to develop. Frequently, more are born than can be accommodated by the number of nipples, and it's "first come, first served."

Today, marsupial mammals are found mainly in Australia, but also in Central and South America. They could not compete against placental mammals, and only a few marsupials, such as the American opossum, are found in North America. In Australia, marsupials underwent adaptive radiation for several million years without competition from placental mammals, which arrived there only recently. Among the herbivorous marsupials, koalas are tree-climbing browsers (Fig. 31.13*b*) and kangaroos are grazers. The Tasmanian wolf or tiger, thought to be extinct, was a carnivorous marsupial about the size of a collie dog.

Placental Mammals

The vast majority of living mammals are **placental mammals** (Fig. 31.13*c*). In these mammals, the extraembryonic membranes found in reptiles (see Fig. 31.11) have been modified for internal development within the uterus of the female. The chorion contributes to the fetal portion of the placenta, while a part of the uterine wall contributes to the maternal portion. Here, nutrients, oxygen, and waste are exchanged between fetal and maternal blood.

Mammals are adapted to life on land and have limbs that allow them to move rapidly. In fact, an evaluation of mammalian features leads us to the obvious conclusion that they lead active lives. The brain is well developed; the lungs are expanded not only by the action of the rib cage but also by the contraction of the *diaphragm,* a horizontal muscle that divides the thoracic cavity from the abdominal cavity; and the heart has *four chambers.* The internal temperature is constant, and hair, when abundant, helps to insulate the body.

The mammalian brain is enlarged due to the expansion of the foremost part—the cerebral hemispheres (see Figure 17.12). These have become convoluted and have expanded to such a degree that they hide many other parts of the brain from view. The brain is not fully developed for some time after birth, and there is a long period of dependency on the parents, during which the young learn to take care of themselves.

Mammals have differentiated teeth. Typically, in the front, the incisors and canine teeth have cutting edges for capturing and killing prey. On the sides, the premolars and molars chew food. Classification of placental mammals is based on methods of obtaining food and mode of locomotion. For example, bats (order Chiroptera) have membranous wings supported by digits; horses (order Perissodactyla) have long, hoofed legs; and whales (order Cetacea) have paddlelike forelimbs. The specific shape and size of the teeth may be associated with whether the mammal is an herbivore (eats vegetation), a carnivore (eats meat), or an omnivore (eats both meat and vegetation). For example, mice (order Rodentia) have continuously growing incisors; horses (order Perissodactyla) have large, grinding molars; and dogs (order Carnivora) have long canine teeth.

These features in particular distinguish placental mammals:

- body hair
- differentiated teeth
- infant dependency
- constant internal temperature
- mammary glands
- well-developed brain
- internal development

Visual Focus

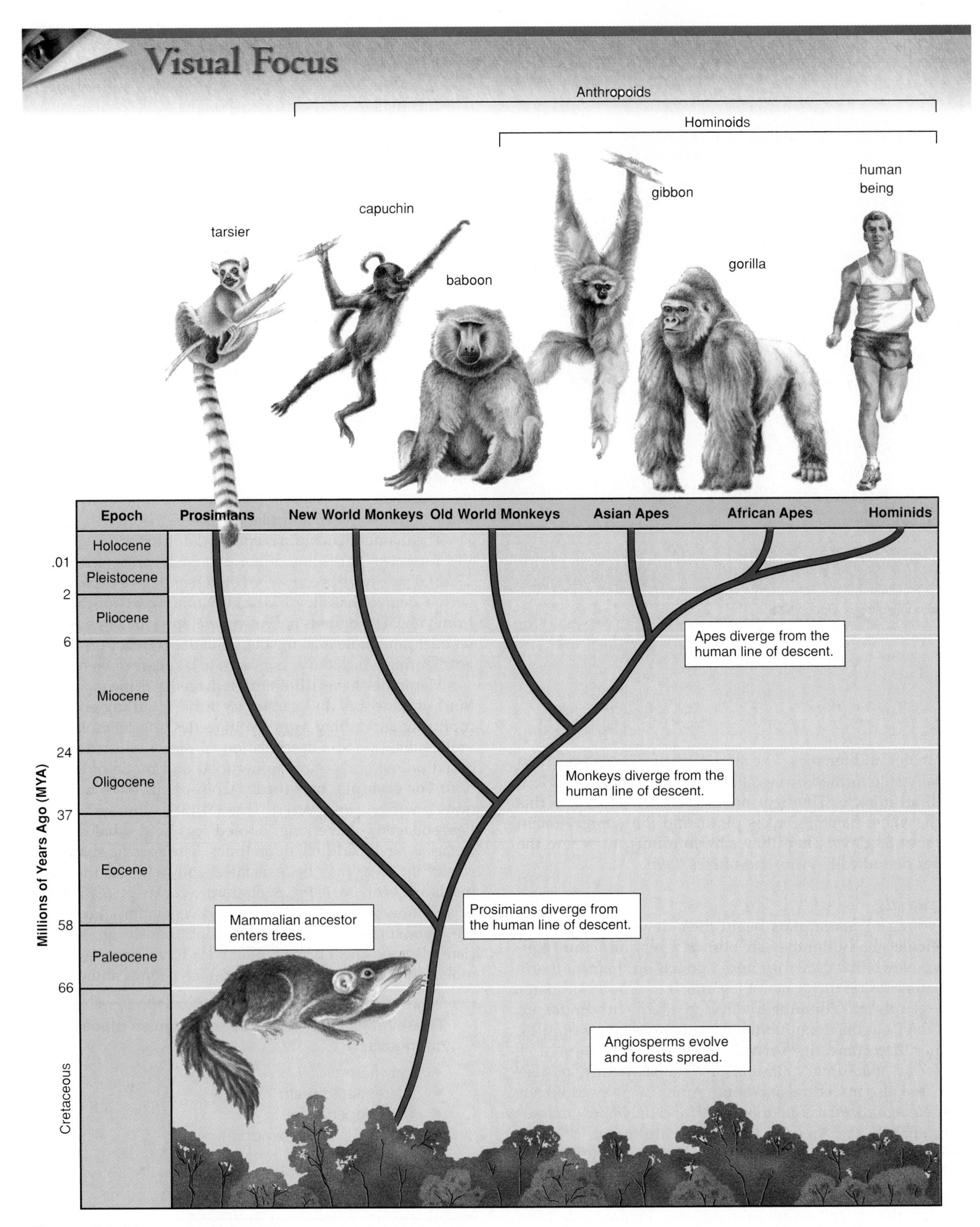

Figure 31.14 Primate evolution.
There are at least four lines of evolution among the primates, which arose from mammalian insectivores: prosimians; monkeys, including New World and Old World monkeys; apes, including Asian apes and African apes (orangutans, gorillas, and chimpanzees); and humans.

31.4 Human Evolution

All of the animals depicted in Figure 31.14 are primates, which are classified as shown in Table 31.1.

Table 31.1 Primate Classification

Phylum Chordata	Invertebrate and vertebrate chordates
Subphylum Vertebrata	Fish, amphibians, reptiles, birds, mammals
Class Mammalia	Monotremes, marsupials, placental mammals
Order Primates	Prosimians and anthropoids
Suborder Anthropoidea	Monkeys, apes, and hominids
Superfamily Hominoidea	Apes and hominids
Family Hominidae	Australopithecines, *Homo habilis*, *Homo erectus*, *Homo sapiens*

Primates

Primates (order Primates, about 180 species) belong to a mammalian order in which the animals are adapted to living in trees. The limbs are mobile, as are the hands, because the thumb (and in nonhuman primates, the big toe as well) is opposable; that is, the thumb can touch each of the other fingers. Therefore, a primate can easily reach out and bring food such as fruit to the mouth. When locomoting, tree limbs can be grasped and released freely because nails have replaced claws.

The sense of smell is of primary importance in animals with a snout. In primates, the snout is shortened considerably, allowing the eyes to move to the front of the head. The stereoscopic vision (or depth perception) that results permits primates to make accurate judgments about the distance and position of adjoining tree limbs.

Gestation is lengthy, allowing time for good forebrain development; the visual portion of the brain is proportionately large, as are those centers responsible for hearing and touch. One birth at a time is the norm in primates; it is difficult to care for several offspring while moving from limb to limb. The juvenile period of dependency is extended, and there is an emphasis on learned behavior and complex social interactions.

The primate order contains two suborders: Prosimii and Anthropoidea. The first primates were the **prosimians,** a term meaning premonkey. The prosimians are represented today by several types of animals. Lemurs have a squirrel-like appearance (Fig. 31.14), and tarsiers are curious mouse-sized creatures with enormous eyes suitable for their nocturnal way of life. The prosimians are believed to have evolved from an insectivore-type mammal.

Monkeys, apes, and humans are all **anthropoids** (suborder Anthropoidea). There are two types of monkeys: New World monkeys, which have long prehensile (grasping) tails and flat noses, and Old World monkeys, which lack such tails and have protruding noses. Two of the well-known New World monkeys are the spider monkey and the capuchin, the "organ grinder's monkey." Some of the better known Old World monkeys are now ground dwellers, such as the baboon and the rhesus monkey.

Humans are more closely related to apes than to monkeys; only apes and humans are *hominoids* (superfamily Hominoidea). There are four types of apes: gibbons, orangutans, gorillas, and chimpanzees. Gibbons, the smallest of the apes, have extremely long arms, which are specialized for swinging between tree limbs. The orangutan is a large ape but nevertheless spends a great deal of time in trees. In contrast, the gorilla, the largest of the apes, spends most of its time on the ground. Chimpanzees, which are at home both in the trees and on the ground, are the most humanlike of the apes in appearance. Molecular data tell us that humans are genetically very similar to chimpanzees and gorillas. Nucleotide sequences in genes, amino acid sequences in proteins, and immunological properties of various molecules all indicate that we are even more closely related to these apes than orangutans are related to them.

Humans can be distinguished from apes, however, by locomotion and posture, dental features, and other characteristics. When moving in a tree, a monkey is quadrupedal (walks on all four limbs) and leaps from branch to branch. In keeping with this method of locomotion, the vertebral column is arched and the shoulder joint has limited mobility. In contrast to monkeys, apes swing from branch to branch. Consistent with this type of locomotion, the vertebral column is straight and the shoulder joint is mobile. Also, the arms are elongated. Humans are bipedal—that is, they walk on two feet. Humans have an S-shaped vertebral column, which provides a way to transmit the upper body's weight to the pelvis. The pelvis is stronger than that of an ape because it is shorter, and the sacrum fits like a keystone between the two pelvic bones. The broad pelvic bones serve as attachments for muscles that maintain stability as first one leg and then the other leaves the ground while walking.

These characteristics especially distinguish primates from other mammals:

- opposable thumb (and in some cases, big toe)
- extended period of parental care
- emphasis on learned behavior
- expanded forebrain
- nails (not claws)
- single birth

Hominids

The **hominid** line of descent begins with the **australopithecines,** which evolved and diversified in eastern Africa. It wasn't until the 1960s that paleontologists, following the lead of the renowned couple Louis and Mary Leakey, began to concentrate their efforts in eastern as opposed to southern Africa. It has proven fruitful for paleontologists. One 1994 find consisting of skull fragments, teeth, arm bones, and a part of a child's lower jaw has been dated at 4.4 MYA (millions of years ago). Called *A. ramidus,* it is believed to represent an early stage in human evolution. In comparison to later australopithecines, the canine teeth are larger, and the skull is more like that of a chimpanzee. Also, this fossil did not walk erect. However, a 1995 discovery, *Australopithecus anamensis,* dated at just about 4 MYA, has jaws like those of an ape but legs like those of humans.

Figure 31.15 ***Australopithecus afarensis.*** Reconstruction of Lucy, who is believed to have walked erect. *A. afarensis* was a dimorphic species, in which males were much larger than the females. Lucy was four feet tall and weighed about 30 kilograms. A male was five feet tall and weighed about 45 kilograms.

More than twenty years ago, a team led by Donald Johanson unearthed nearly 250 fossils of a hominid called *Australopithecus afarensis.* A now-famous female skeleton dated at about 3.4 MYA is known worldwide by its field name, Lucy. (The name derives from the Beatles' song "Lucy in the Sky with Diamonds" which they played in camp.) Although her brain was about the size of a chimp (400 cc), the shapes and relative proportions of her limbs indicate that Lucy walked bipedally (Fig. 31.15). There are even footprints that show how members of her species walked.

The australopithecines were sexually dimorphic. Lucy was about four feet tall and weighed about 30 kilograms. In contrast, the males of the species were five feet tall and weighed up to 45 kilograms. Some have speculated that such size differences may indicate two separate species, but in 1994 new finds, including a more complete skull (dubbed the son of Lucy and dated at just about 3 MYA), confirmed the opinion that the fossils belong to one species. Taking into account their smaller body size, the relative brain size is about one-third of ours and the jaw is heavy. The cheek teeth are enormous, and in males, large canine teeth project forward.

A. afarensis had descendants. After a period of stasis that may have lasted a million years (from 3 to 2 MYA), branching speciation occurred (Fig. 31.16). Therefore, instead of thinking about descent in terms of a straight line, it is far better to envision a bush. Some think there may have been as many as ten species of hominids about 2 MYA in Africa, but we will discuss only four of them, three of which are australopithecines. *A. africanus,* which was first named in southern Africa by Raymond Dart in the 1920s, is a gracile (slender) type. *A. boisei* is a robust form from eastern Africa, and *A. robustus* is a similar form from southern Africa. The robust forms have stronger jaws, larger attachments for larger chewing muscles, and bigger grinding teeth because they most likely fed on tougher foods than the gracile form. The robust forms lived in drier habitats where soft fruits and leaves would be harder to come by. In both forms, the pelvis resembles that of Lucy but the hands were more humanlike; possibly they were capable of making tools. Both forms, which are believed to have eaten meat at least occasionally, may have used the tools to process animal carcasses.

At one time it was believed that the gracile australopithecine form gave rise to the robust forms. However, in 1985, Alan Walker discovered a robust skull, called the black skull, which was old enough (2.5 MYA) to be ancestral to the robust forms we have been discussing. It has been given the name *A. aethiopicus.*

Several species of australopithecines have been identified. After a period of stasis, during which only *A. afarensis* existed, branching speciation occurred.

Homo habilis

The oldest fossils to be classified in the genus *Homo* are known as ***Homo habilis*** (meaning handy man). His remains, dated as early as 2 MYA, are often accompanied by stone tools. Why is this hominid classified within our own genus? *H. habilis* was small—about the size of Lucy—but the brain, at 700 cc, is about 45% larger. In addition, certain portions of the brain thought to be associated with speech areas are enlarged. The cheek teeth are smaller than even those of the gracile australopithecines. Apparently *H. habilis* had a different way of life than the other hominids of this time.

Cut marks on bones that could have been made by stone flakes have been found at many sites throughout eastern Africa, dating from 2 MYA. *H. habilis* could have made and used tools in order to strip meat off these bones and, in keeping with the size of the teeth, could have eaten meat to satisfy protein demands. As a scavenger, *H. habilis* may have depended simply on the kills of other animals; or as a predator, he may have killed small- to medium-sized prey. This new way of life became available to hominids when they had the ability to make and use tools intelligently.

The stone tools made by *H. habilis* are called Oldowan tools because they were first identified as tools by the Leakeys at a place called Olduvai Gorge. Oldowan tools are simple and look rather clumsy, but perhaps this hominid also used stone flakes. The flakes would have been sharp and able to scrape away hide and cut tendons to easily remove meat from a carcass.

H. habilis most likely still ate fruits, berries, seeds, and other plant materials. Perhaps a division of labor arose, with certain members of a group serving as hunters and others as gatherers. Speech would have facilitated their cooperative efforts, and later they most likely shared their food and ate together. In this way, society and culture could have begun. Culture, which encompasses human behavior and products (such as technology and the arts), is dependent upon the capacity to learn and transmit knowledge through the ability to speak and think abstractly.

Prior to the development of culture, adaptation to the environment necessitated a biological change. The acquisition of culture provided an additional way by which adaptation was possible. And the possession of culture by *H. habilis* may have hastened the extinction of the australopithecines.

H. habilis warrants classification as a *Homo* because of brain size, posture, and dentition. Circumstantial evidence suggests the use of tools to prepare meat and also the development of culture.

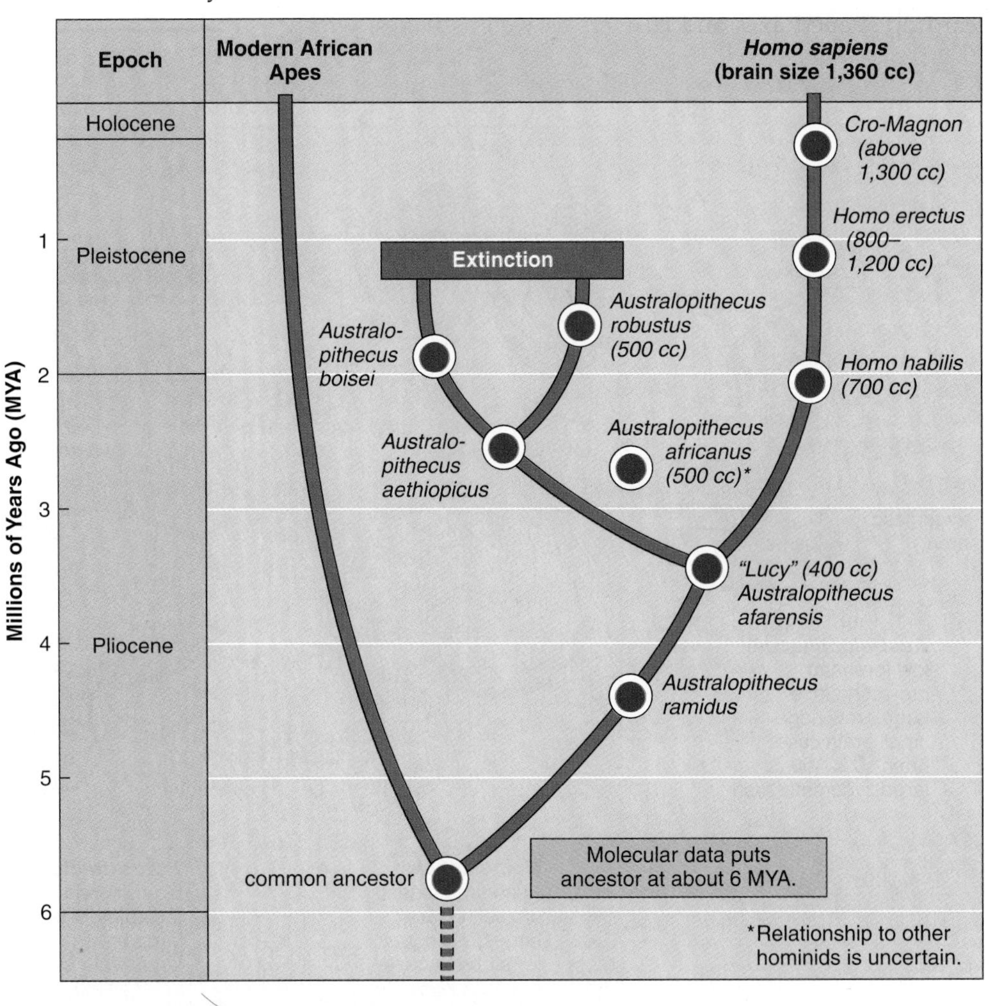

Figure 31.16 Recently constructed hominid evolutionary tree. African apes and hominids split about 6 million years ago (MYA). *A. ramidus* is the oldest known of the hominids; *A. afarensis* lasted about a million years before branching speciation occurred. In particular, there were gracile and robust australopithecine forms and also the first human form, *Homo habilis.*

Homo erectus

Homo erectus is the name assigned to hominid fossils found in Asia, and dated between 1.9 and 0.5 MYA. Compared to *H. habilis, H. erectus* had a larger brain (about 1,000 cc), more pronounced brow ridges, a flatter face, and a nose that projects like ours. These hominids not only stood erect, they most likely had a striding gait like ours.

It is believed by some that *H. erectus* first appeared in Africa and then migrated into Asia and Europe. Such an extensive population movement is a first in the history of humankind and a tribute to the intellectual and physical skills of the species. *H. erectus* was also the first hominid to use fire, and it fashioned more advanced tools called Acheulean tools, named after a site in France (Fig. 31.17). There are heavy teardrop-shaped axes and cleavers as well as flakes, which were probably used for cutting and scraping. Some believe that *H. erectus* was a systematic hunter and brought kills to the same site over and over again. In one location there are over 40,000 bones and 2,647 stones. These sites could have been "home bases" where social interaction occurred and a prolonged childhood allowed time for much learning. Perhaps a language evolved and a culture more like our own developed.

> *H. erectus,* which evolved from *H. habilis,* had a striding gait, made well-fashioned tools (perhaps for hunting), and could control fire. This hominid migrated into Europe and Asia from Africa about 1 MYA.

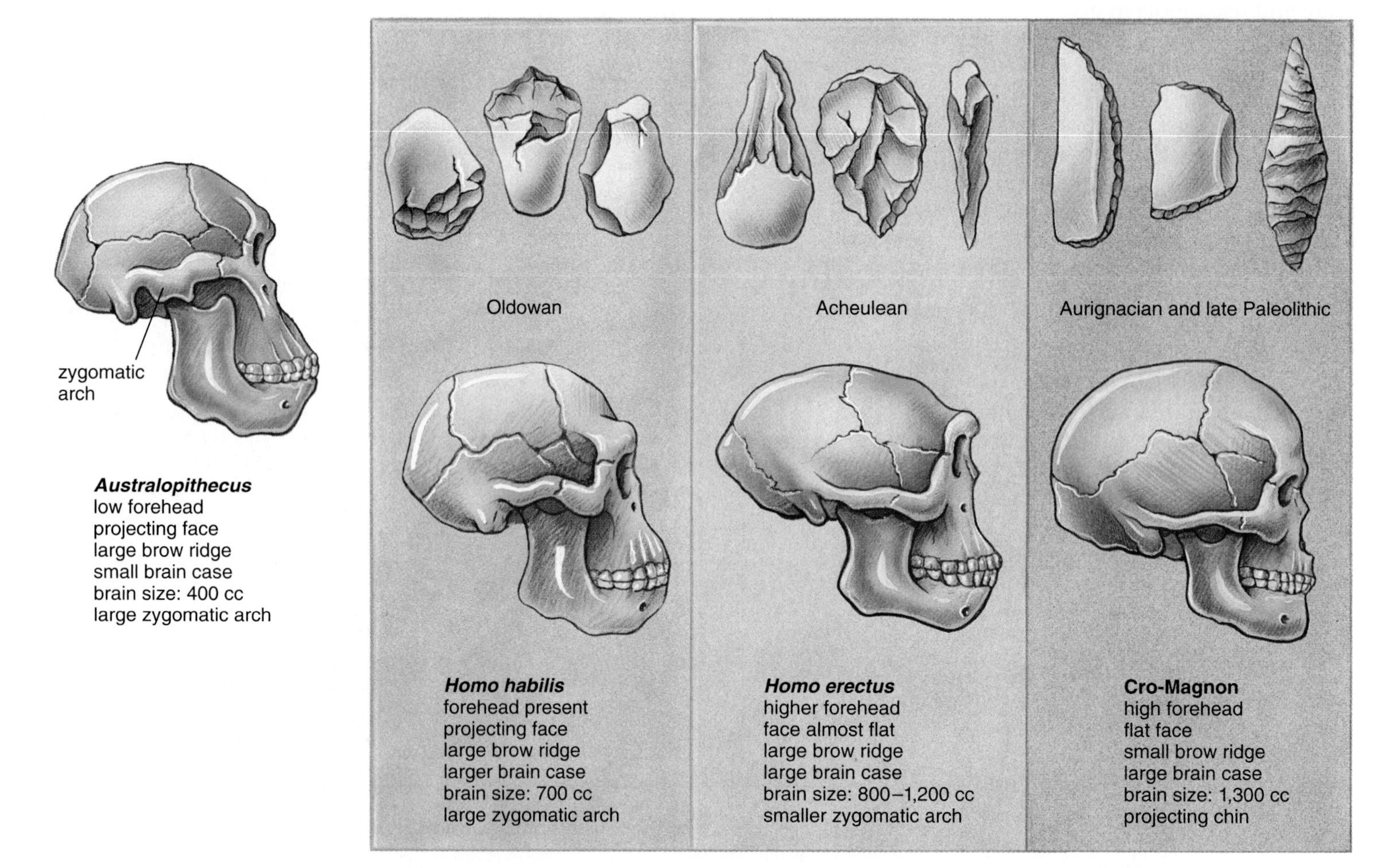

Figure 31.17 **Hominid skull anatomy and tools.**
The listings compare hominid skulls. Oldowan tools are crude. Acheulean tools are better made and more varied, and Aurignacian tools are well designed for specific purposes.

Origin of Modern Humans

It is generally recognized that modern humans originated from *H. erectus* or a closely related species, but where did this occur? About 300,000 years B.P. (before present), so-called "archaic *Homo sapiens*" had evolved in Europe, Asia, and Africa. (Neanderthal is a well-known archaic *H. sapiens.*) The hypothesis that each of these individual populations (with some interbreeding) went on to evolve into modern humans is called the *multiregional continuity hypothesis* (Fig. 31.18*a*). This hypothesis, which proposes no migrations, requires that evolution be essentially similar in several different places. Each region would show a continuity of its own anatomical characteristics from about a million years ago, when *H. erectus* first arrived in Eurasia. Opponents argue that it seems highly unlikely that evolution would have produced essentially the same result in these different places. They suggest, instead, the *out-of-Africa hypothesis,* which proposes that archaic *H. sapiens* became fully modern only in Africa, and thereafter they migrated to Europe and Asia about 100,000 years B.P. Modern humans may have interbred to a degree with archaic populations, but in effect, they supplanted them (Fig. 31.18*b*).

Investigators are currently testing two hypotheses: that modern humans (1) evolved separately in Europe, Africa, and Asia, and (2) evolved in Africa and then migrated.

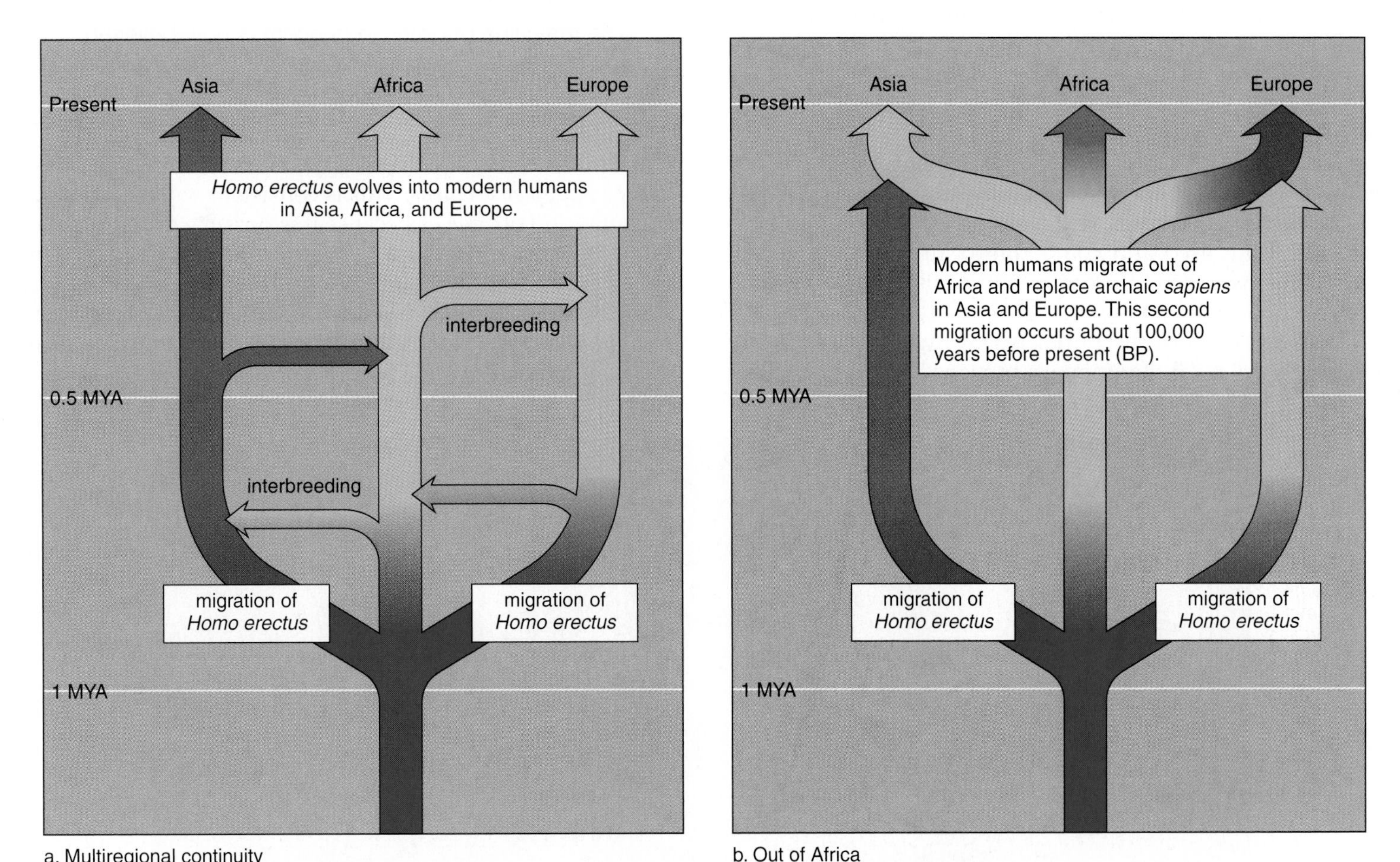

Figure 31.18 Origin of modern humans.
a. The multiregional continuity hypothesis proposes that modern humans evolved separately in at least three different places: Asia, Africa, and Europe. Therefore, continuity of genotypes and phenotypes is expected in these regions. **b.** The out-of-Africa hypothesis proposes that modern humans originated only in Africa; then they migrated and supplanted populations of *Homo* in Asia and Europe about 100,000 years ago.

Homo sapiens

A brain capacity larger than 1,000 cc allows a *Homo* species to be classified as *Homo sapiens.* The **Neanderthals** (*H. sapiens neanderthalensis*) take their name from Germany's Neander Valley, where one of the first Neanderthal skeletons, dated some 200,000 years old, was discovered. The Neanderthals had massive brow ridges, and the nose, the jaws, and the teeth protruded far forward. The forehead was low and sloping, and the lower jaw sloped back without a chin.

At this time, the Neanderthals are thought to be an archaic *H. sapiens* and most likely not in the main line of *Homo* descent. Surprisingly, however, the Neanderthal brain was, on the average, slightly larger than that of modern humans (1,400 cc, compared to 1,360 cc in most modern humans). The Neanderthals were heavily muscled, especially in the shoulders and the neck. The bones of the limbs were shorter and thicker than those of modern humans. It is hypothesized that a larger brain than that of modern humans was required to control the extra musculature. They lived in Eurasia during the last Ice Age, and their sturdy build could have helped conserve heat.

The Neanderthals give evidence of being culturally advanced. They most likely successfully hunted bears, woolly mammoths, rhinoceroses, reindeer, and other contemporary animals. They used and could control fire, and they even buried their dead with flowers and tools, indicating that they may have had a religion.

In keeping with the out-of-Africa hypothesis, it is increasingly believed that modern humans, by custom called **Cro-Magnon** after a fossil location in France, entered Eurasia 100,000 years ago or even earlier. Cro-Magnons (*H. sapiens sapiens*) had a thoroughly modern appearance (Fig. 31.19). They made advanced stone tools called Aurignacian tools (see Fig. 31.17). They were such accomplished hunters that some researchers believe they were responsible for the extinction of many larger mammals, such as the giant sloth, the mammoth, the saber-toothed tiger, and the giant ox during the late Pleistocene epoch.

Cro-Magnons hunted cooperatively and most likely lived in small groups, with the men hunting by day while the women were gatherers and took care of the children. The Cro-Magnon culture included art. They sculpted small figurines out of reindeer bones and antlers. They also painted beautiful drawings of animals on cave walls in Spain and France.

The main line of hominid descent is now believed to include *A. ramidus*, *A. afarensis*, *H. habilis*, *H. erectus*, and Cro-Magnon.

Human beings are diverse, but even so, we are all classified as *H. sapiens sapiens.* This is consistent with the biological definition of species because it is possible for all types of humans to interbreed and to bear fertile offspring. While it may appear that there are various "races," molecular data show that the DNA base sequence varies as much between individuals of the same ethnicity as between individuals of different ethnicity.

Figure 31.19 Cro-Magnon people.
Cro-Magnon people are the first to be designated *Homo sapiens sapiens.* Their toolmaking ability and other cultural attributes, such as their artistic talents, are legendary.

Bioethical Issue

Some people believe that animals should be protected in every way, and should not be used in laboratory research. In our society as a whole, the trend is toward a growing recognition of what is generally referred to as animal rights. In 1985, 63% of Americans polled agreed that "scientists should be allowed to do research that causes pain and injury to animals like dogs and chimpanzees if it produces new information about human health problems." That dropped to 53% in 1995. Psychologists with Ph.D.s earned in the 1990s are half as likely to express strong support for animal research as those who earned their Ph.D.s before 1970.

Those who approve of laboratory research involving animals give examples to show that even today it would be difficult to develop new vaccines and medicines against infectious diseases, develop new surgical techniques for saving human lives, or develop treatments for spinal cord injuries without the use of animals. Even so, most scientists today are in favor of what is now called the "three Rs": replacement of animals by in vitro, or test-tube, methods whenever possible; reduction of the numbers of animals used in experiments; and refinement of experiments to cause less suffering to animals. In the Netherlands, every scientist starting research that involves animals is well trained in the three Rs. After designing an experiment that uses animals, they are asked to find ways to answer the same questions without using animals.

F. Barbara Orlans of the Kennedy Institute of Ethics at Georgetown University says "It is possible to be both pro research and pro reform." She feels that animal activists need to accept that sometimes animal research is beneficial to humans, and all scientists need to consider the ethical dilemmas that arise when animals are used for laboratory research.

Questions

1. Are you opposed to the use of animals in laboratory experiments? under certain circumstances? or completely in favor? Explain.
2. Do you feel that it would be possible for animal activists and scientists to find a compromise they could both accept? Discuss.

Summarizing the Concepts

31.1 Echinoderms

Both echinoderms and chordates are deuterostomes. In deuterostomes, the second embryonic opening becomes the mouth, the coelom forms by outpocketing of the primitive gut (they are enterocoelomates), and the dipleurula larva is found among some. Echinoderms (e.g., sea stars, sea urchins, sea cucumbers, sea lilies) have radial symmetry as adults (not as larvae) and spines from endoskeletal plates. Typical of echinoderms, sea stars have tiny skin gills, a central nerve ring with branches, and a water vascular system for locomotion. Each arm of a sea star contains branches from the nervous, digestive, and reproductive systems.

31.2 Chordates

Chordates (tunicates, lancelets, and vertebrates) have a notochord, a dorsal hollow nerve cord, pharyngeal pouches, and post-anal tail at one time in their life history. Tunicates and lancelets are the invertebrate chordates. Adult tunicates lack chordate characteristics except gill slits, but adult lancelets have the three chordate characteristics.

31.3 Vertebrates

Vertebrates have the three chordate characteristics as embryos but the notochord is replaced by the vertebral column, a part of their endoskeleton. The internal organs are well developed and cephalization is apparent. The vertebrate classes trace their evolutionary history. The first vertebrates, represented by hagfishes and lampreys, lacked jaws and fins. Bony fishes have jaws and two pairs of fins; the bony fishes include those that are ray-finned and a few that are lobe-finned. Some of the lobed-finned fishes have lungs.

Amphibians evolved from the lobe-finned fishes and have two pairs of limbs. They are represented today by frogs and salamanders. Frogs usually return to the water to reproduce; frog tadpoles metamorphose into terrestrial adults.

Reptiles (today's snakes, lizards, turtles, and crocodiles) lay a shelled egg, which contains extraembryonic membranes, including an amnion that allows them to reproduce on land.

Birds are feathered, which helps them maintain a constant body temperature. They are adapted for flight; their bones are hollow with air sacs that allow one-way ventilation; their breastbone is keeled, and they have well-developed sense organs.

Mammals have hair and mammary glands. The former helps them maintain a constant body temperature, and the latter allow them to nurse their young. Monotremes lay eggs; marsupials have a pouch in which the newborn matures, and placental mammals, which are far more varied and numerous, retain offspring inside the uterus until birth.

31.4 Human Evolution

Primates are mammals adapted to living in trees. During the evolution of primates, various groups diverged in a particular sequence from the human line of descent. Prosimians (tarsiers and lemurs) diverged first; then the monkeys, then the apes. Molecular biologists tell us we are most closely related to the African apes, whose ancestry split from ours about 6 MYA.

Human evolution continued in eastern Africa with the evolution of the australopithecines. The most famous australopithecine is Lucy (3.4 MYA) whose brain was small but who walked bipedally. *Homo habilis,* present about 2 MYA, is certain to have made tools. *Homo erectus,* with a brain capacity of 1,000 cc and a striding gate, was the first to migrate out of Africa.

Two contradicting hypotheses have been suggested about the origination of modern humans. The multiregional continuity hypothesis says that modern humans originated separately in Asia, Europe, and Africa. The out-of-Africa hypothesis says that modern humans originated in Africa and, after migrating into Europe and Asia, replaced the archaic *Homo* species found there. Cro-Magnon is a name often given to modern humans.

Studying the Concepts

1. What are the general characteristics of deuterostomes? 644
2. What are the general characteristics of echinoderms? Explain how the water vascular system works in the sea star. 644–45
3. What three characteristics do all chordates have at some time in their life history? Describe the two groups of invertebrate chordates. 646, 648
4. Discuss the distinguishing characteristics of vertebrates and cite the milestones in the evolutionary history of vertebrates. 649
5. Describe the three classes of fishes. Which have jaws? two pairs of fins? lungs? 650–51
6. Explain in what way amphibians are adapted to life on land and ways in which they are not adapted. Describe the metamorphosis of frogs. 651
7. What is the significance of a shelled egg? Explain in what ways reptiles are adapted to life on land. 653
8. What is the significance of wings? In what other ways are birds adapted to flying? 655
9. What is the significance of a constant internal temperature? What are the three groups of mammals, and what are their primary characteristics? 656–57
10. Name several primate characteristics still retained by humans. 659
11. Draw and discuss the evolutionary tree for hominids. 661
12. Contrast the "multiregional continuity" hypothesis and the "out-of-Africa" hypothesis. 663

Testing Yourself

Choose the best answer for each question.

1. The tube feet of echinoderms
   a. are their head.
   b. are a part of the water vascular system.
   c. are found in the coelom.
   d. help pass sperm to females during reproduction.
   e. All of these are correct.
2. Which of these is not a chordate characteristic?
   a. dorsal supporting rod, the notochord
   b. dorsal hollow nerve cord
   c. pharyngeal pouches
   d. post-anal tail
   e. vertebral column
3. Sharks and bony fishes are different in that only
   a. bony fishes have paired fins.
   b. bony fishes have a keen sense of smell.
   c. bony fishes have an operculum.
   d. sharks have a bony skeleton.
   e. sharks are predaceous.
4. Amphibians arose from
   a. tunicates and lancelets.
   b. cartilaginous fishes.
   c. jawless fishes.
   d. ray-finned fishes.
   e. bony fishes with lungs.
5. Which of these is not a feature of amphibians?
   a. dry skin that resists desiccation
   b. metamorphosis from a swimming form to a land form
   c. small lungs and a supplemental way of gas exchange
   d. reproduction in the water
   e. three-chambered heart
6. Which of these is not a difference between reptiles and birds?
   Reptiles—Birds
   a. hard-shelled egg—partial internal development
   b. scales—feathers
   c. tetrapods—wings
   d. cold blooded—warm blooded
   e. no air sacs—air sacs
7. Which of these is a true statement?
   a. In all mammals, offspring develop within the female.
   b. All mammals have hair and mammary glands.
   c. All mammals have one birth at a time.
   d. All mammals are land-dwelling forms.
   e. All of these are true.
8. The most likely line of descent for humans is
   a. Common ancestor with an African ape, (a hominoid), Lucy (a hominid), *Homo habilis* (a human), *Homo erectus*, *Homo sapiens*.
   b. prosimians, New World monkeys, Old World monkeys, Asian apes, African apes, hominids.
   c. chimpanzees, orangutan, *Australopithecus africanus*, marsupials, placental mammals, primates, humans.
   d. Only a and b are correct.
   e. All of these are correct.
9. Lucy is a member of what species?
   a. *Homo erectus*
   b. *Australopithecus afarensis*
   c. Neanderthal
   d. *H. habilis*
   e. *A. robustus*
10. If the multiregional continuity hypothesis is correct, then
    a. hominid fossils in China after 100,000 B.P. are not expected to resemble earlier fossils.
    b. hominid fossils in China after 100,000 B.P. are expected to resemble earlier fossils.
    c. humans evolved in Africa.
    d. humans evolved in Eurasia.
    e. Both b and d are correct.
11. Which of these is incorrectly matched?
    a. *H. erectus*—made tools
    b. Neanderthal—good hunter
    c. *H. habilis*—controlled fire
    d. Cro-Magnon—good artist
    e. *A. afarensis*—walked erect
12. Complete the following diagram.

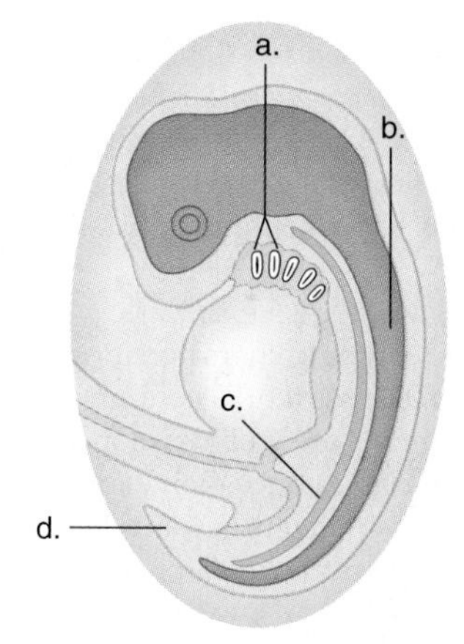

Thinking Scientifically

1. Humans use negative pressure to fill the lungs, and frogs use positive pressure.
   a. What is the difference between negative and positive pressure?
   b. Birds have a flow-through system—the air flows from the lungs into air sacs and then out by way of a separate set of tubes. In what way is this more efficient than human breathing?
   c. Frogs have thin, moist skin, and reptiles have thick skin. Which animal do you predict practices skin breathing? Which animal most likely has better developed lungs?
   d. Humans have a diaphragm, reptiles do not. In what way does a diaphragm assist breathing?
2. Refer to Figure 31.16 to answer the following questions.
   a. What might have caused the evolution of the upright posture of *A. afarensis*?
   b. How does the hominid evolutionary tree illustrate adaptive radiation?
   c. How does the evolution of *Homo erectus* and Cro-Magnon support the concept of punctuated equilibrium (page 567)?
   d. How do you know that all human races should be considered the same species?

Understanding the Terms

amniote egg 649
amphibian 651
anthropoid 659
australopithecines 660
bird 655
bony fish 650
cartilaginous fish 650
chordate 646
Cro-Magnon 664
deuterostome 644
echinoderm 644
feather 655
fin 649
gill 646
hominid 660
Homo erectus 662
Homo habilis 661
jaw 650
jawless fishes 650
lancelet 648
lung 651
mammal 656
marsupial 657
monotreme 656
Neanderthal 664
notochord 646
placental mammal 657
primate 659
prosimian 659
reptile 653
tunicate 648
vertebrate 649
water vascular system 645

Match the terms to these definitions:

a. ________ Series of canals that takes water to the tube feet of an echinoderm, allowing them to expand.

b. ________ Group of primates that includes lemurs and tarsiers and may resemble the first primates to have evolved.

c. ________ Dorsal supporting rod that exists in all chordates sometime in their life history; replaced by the vertebral column in vertebrates.

d. ________ Group of coelomate animals in which first embryonic opening is associated with the anus, and the second embryonic opening is associated with the mouth.

e. ________ Member of a family that contains humans, and their direct ancestors which are known only from the fossil record.

Using Technology

Your study of animals and human evolution is supported by these available technologies:

Essential Study Partner CD-ROM
Evolution & Diversity → Vertebrates → Human Evolution

Visit the Mader web site for related ESP activities.

Exploring the Internet

The Mader Home Page provides resources and tools as you study this chapter.

http://www.mhhe.com/biosci/genbio/mader

Further Readings for Part 6

Agnew, N., and Demas, M. September 1998. Preserving the Laetoli footprints. *Scientific American* 279(3):44. This article recaps the discovery of hominid footprints in East Africa, and explains steps taken to preserve them.

Beckage, N. E. November 1997. The parasitic wasp's secret weapon. *Scientific American* 277(5):82. Some parasitic wasps produce a virus that suppresses a living host's immune system.

Ben-Jacob, E., and Levine, H. October 1998. The artistry of microorganisms. *Scientific American* 279(4):82. Colonies of bacteria form geometric patterns, which reflect survival strategies.

Berger, L. August 1998. The dawn of humans: Redrawing our family tree? *National Geographic* 194(2):90. South African fossils show that *A. africanus* was more apelike than Lucy.

Castro, P., and Huber, M. 1997. *Marine biology.* 2d ed. St. Louis: Mosby-Year Book, Inc. This introductory text is designed to provide a stimulating overview of marine biology.

Chadwick, D. H. March 1998. Planet of the beetles. *National Geographic* 193(3):100. With diverse sizes, forms, and functions, beetles make up one-third of the world's identified insects.

Chiappe, L. M. September 1998. Wings over Spain. *Natural History* 107(7):30. The presence of a first digit on fossil remains of a 115-million-year-old bird shows advanced flying ability.

Diamond, J. September 1998. Evolving backward. *Discover* 19(9):64. Studies of the blind mole rat shows how evolved traits are lost if they are not used.

Erwin, D. E. July 1996. The mother of mass extinction. *Scientific American* 275(1):72. Global sea level decline and volcanic eruptions may have caused mass extinctions.

Foster, K. R., et al. August 1998. The Philadelphia yellow fever epidemic of 1793. *Scientific American* 279(2):88. The history of this epidemic and the possibility of its recurrence are discussed.

Genthe, H. August 1998. The incredible sponge. *Smithsonian* 29(5):50. Sponge biology and the therapeutic uses of sponges in treating cancer are discussed.

Gore, R. January 1996. Neanderthals. *National Geographic* 189(1):2. Archeological finds are providing much information on the Neanderthal culture.

______. September 1997. The dawn of humans. *National Geographic* 192(3):92. A footprint dated about 117,000 years ago was discovered in southern Africa.

Gwynne, D. T. August 1997. Glandular gifts. *Scientific American* 277(2):66. Male insects offer body parts and secretions as a strategy for fertilizing the female's eggs.

Johanson, D. C. March 1996. Face-to-face with Lucy's family. *National Geographic* 189(3):96. New fossils from Ethiopia provide more information about human evolution.

Kellert, S. R. 1996. *The value of life: Biological diversity and human society.* Washington, D.C.: Island Press/Shearwater Books. Explores the importance of biological diversity to the well-being of humanity.

Knols, B. G., and Meijerink, J. September/October 1997. Odors influence mosquito behavior. *Science & Medicine* 4(5):56. Definition of odors may lead to the control of insects that transmit infectious diseases.

Leakey, M., and Walker, A. June 1997. Early hominid fossils from Africa. *Scientific American* 276(6):74. A bone unearthed in 1965 recently proved the existence of a new species of *Australopithecus,* showing ancestral humans existed 4 million years ago.

Levetin, E., and McMahon, K. 1996. *Plants and society.* Dubuque, Iowa: Wm. C. Brown Publishers. Basic botany and the impact of plants on society are topics covered in this introductory text.

Levy, S. B. March 1998. The challenge of antibiotic resistance. *Scientific American* 278(3):46. Misuse and overuse of antibiotics must end in order to preserve their effectiveness.

Lewin, R. 1997. *Patterns in evolution: The new molecular view.* New York: Scientific American Library. This book explores how genetic information provides insights into evolutionary events.

Lim, D. 1998. *Microbiology.* 2d ed. Dubuque, Iowa: WCB/McGraw-Hill. This introductory text shows how microorganisms relate to one another and to other organisms; new material in evolution and biodiversity is included.

Line, L. October/November 1998. Fast decline of slow species. *National Wildlife* 36(6):22. Box turtles are declining in numbers, mainly from habitat loss and over-collecting.

Losick, R., and Kaiser, D. February 1997. Why and how bacteria communicate. *Scientific American* 276(2):68. Bacteria send and receive chemical messages and can organize into structures.

Luoma, J. R. March 1997. The magic of paper. *National Geographic* 191(3):88. The papermaking process is discussed in this article.

Madigan, M. T., and Marrs, B. L. April 1997. Extremophiles. *Scientific American* 276(4):86. Certain microorganisms can withstand extreme environments.

Margulis, L., et al. 1998. *Five kingdoms: An illustrated guide to the phyla of life on earth.* 3d ed. New York: W. H. Freeman & Company. Introduces the kingdoms of organisms.

Martini, F. H. October 1998. Secrets of the slime hag. *Scientific American* 279(4):70. The roles of hagfish in ocean ecosystems.

Monastersky, R. March 1998. The rise of life on earth. *National Geographic* 193(3):54. Article discusses the origins of microbial life, stromatolite reefs, and Stanley Miller's model of the primitive atmosphere.

Moore, R., Clark, W. D., et al. 1998. *Botany.* 2d ed. Dubuque, Iowa: WCB/McGraw-Hill. This introductory botany text stresses the importance of plants and the process of science.

Moore-Landecker, E. 1996. *Fundamentals of the fungi.* 4th ed. Upper Saddle River, N.J.: Prentice-Hall. For intermediate students, this text presents a broad introduction to the field of mycology.

Murawski, D. A. March 1997. Moths come to light. *National Geographic* 191(3):40. Moths display a variety of disguises and survival techniques.

Nester, E. W., Roberts, C. E., et al. 1998. *Microbiology: A human perspective.* 2d ed. Dubuque, Iowa: WCB/McGraw-Hill. This introductory text relates basic microbiology to human health.

Northington, D., and Goodin, J. R. 1996. *The botanical world.* 2d ed. St. Louis: Times-Mirror/Mosby College Publishing. This is an account of plant interactions and basic physiology.

Padian, K., and Chiappe, L. M. February 1998. The origin of birds and their flight. *Scientific American* 278(2):38. Recent fossil discoveries confirm that birds descended from dinosaurs.

Paolella, P. 1998. *Introduction to molecular biology.* Dubuque, Iowa: WCB/McGraw-Hill. This introductory text explores the processes and mechanisms of gene function and control.

Parfit, M. October 1998. Antarctic desert. *National Geographic* 4:120. Microscopic organisms that survive in the Antarctic desert have been discovered.

Schmidt, G. D., and Roberts, L. S. 1996. *Foundations of parasitology.* 5th ed. Dubuque, Iowa: Wm. C. Brown Publishers. For upper-division parasitology classes, this text emphasizes the major parasites of humans and animals.

Seymour, R. S. March 1997. Plants that warm themselves. *Scientific American* 276(3):104. Some plants generate heat to keep blossoms at a constant temperature.

Sharpe, G. W., et al. 1995. *Introduction to forests and renewable resources.* 6th ed. New York: McGraw-Hill, Inc. For forestry students, this text presents policies and practices in forest conservation and management.

Shreeve, J. December 1997. Uncovering Patagonia's lost world. December 1997. *National Geographic* 192(6):120. Recent fossil finds cause scientists to rethink the evolution of dinosaurs.

Stern, K. 1997. *Introductory plant biology.* 7th ed. Dubuque, Iowa: Wm. C. Brown Publishers. This text presents basic botany in a clear, informative manner.

Sumich, J. L. 1996. *An introduction to the biology of marine life.* 6th ed. Dubuque, Iowa: Wm. C. Brown Publishers. This introductory text covers taxonomy, evolution, ecology, behavior, and physiology of selected groups of marine organisms.

Sze, P. 1998. *A biology of the algae.* 3d ed. Dubuque, Iowa: WCB/McGraw-Hill. This concise text introduces algae morphology, evolution, and ecology to the botany major.

Tattersall, I. April 1997. Out of Africa again . . . and again? *Scientific American* 276(4):60. Hominids may have migrated out of Africa several times, with each emigration sending a different species.

Walters, D. R., and Keil, D. J. 1996. *Vascular plant taxonomy.* 4th ed. Dubuque, Iowa: Kendall/Hunt Publishing. Introduces plant families and experimental aspects of taxonomy.

Webster, D. July 1996. Dinosaurs of the Gobi. *National Geographic* 190(1):70. Dinosaur fossils are unearthed in the Gobi desert.

Wenke, R. 1996. *Patterns in prehistory: Humankind's first three million years.* 4th ed. New York: Oxford University Press. Provides a comprehensive review of world prehistory.

Zimmer, C. September 1998. The slime alternative. *Discover* 19(9):86. Occasionally, individual *Dictyostelium* amoebas join together and behave like multicellular organisms.

PART VII

Behavior and Ecology

Ecology began as a descriptive science that has now become an experimental and predictive science. Models have been developed that predict how population sizes within communities change over time due to species interactions. Some ecologists study and also include in their models physical factors that influence energy flow and nutrient cycling within an ecosystem. They know that humans alter the transfer rates of substances within biogeochemical cycles that maintain the biosphere. This accounts for acid rain, global warming, ozone depletion, and other changes that are expected to adversely affect all species.

Careers in Ecology

Behaviorist collaring a coyote.

Ethologists study the behavior of animals in the wild. It is one of the most difficult and rewarding of fields, because it requires the study of animals in their own territory over extended periods of time. Such study requires patience, stamina, and creativity. Ethologists may study behaviors such as mating, territorial protection, feeding, migration, and hibernation.

Ecologists study the relationships between living things and their environment, and among different species. They measure the importance of such interactions to the species being studied. The results of these studies help humans to understand the significance of development on natural areas.

Ecologists measuring tropical trees.

Environmental engineers are multidisciplinary specialists who help preserve the environment and plan the cleanup of polluted areas. These engineers use experience in engineering, chemistry, and biology to determine the extent of the pollution and the costs and benefits of different levels of cleanup effort. They also advise developers regarding the potential impact of new projects on the environment. Many also work to devise better ways to limit air and water pollution by industry.

Wildlife managers determine which species should be available for hunters and fishermen to continue their sports. They identify the behavioral and environmental factors that affect population sizes and develop management programs that will permit a sustained yield. Endangered species management also falls under wildlife management, but in this case, the species are protected and are not to be killed by anyone.

Environmental engineer testing for harmful gases.

Foresters manage, develop, and help protect forest resources. Foresters manage timberland, which involves a variety of duties. Those working in private industry may be responsible for procuring timber from private landowners. Foresters also supervise the planting and growing of new trees. They advise on the type, number, and placement of trees to be planted. Foresters monitor the trees to ensure healthy growth and to determine the best time for harvesting. If foresters detect signs of disease or harmful insects, they decide on the best course of treatment to prevent contamination or infestation of healthy trees.

C H A P T E R 32

Animal Behavior

Chapter Concepts

These western grey kangaroos, Macropus fulginosus, *are trying to push each other to the ground in a battle over females. Aggression between members of a society is ritualized and neither party in the struggle is usually harmed by the conflict.*

At the start of the breeding season, male bowerbirds use small sticks and twigs to build elaborate display areas called bowers. They clear the space around the bowers, removing leaves and debris, and decorate the area with fresh flowers, fruits, moss, mushrooms, pebbles, or shells. Each species has its own preference in decorations. The satin bowerbird of eastern Australia prefers blue objects, a color that harmonizes with the male's glossy blue-black plumage (Fig. 32.1).

After the bower is complete, a male spends most of his time near his bower, calling to females, renewing his decorations and guarding his work against possible raids by other males. After inspecting many bowers and their owners, a female approaches one and the male begins a display. He faces her, fluffs up his feathers, and flaps his wings to the beat of a call. The female enters the bower, and if she crouches, the two mate.

Ethologists (scientists who study behavior) want to know how the male bowerbird is structured to perform this behavior and how the behavior helps him secure a mate. In general, ethologists determine how a behavior is controlled and how a behavior enables an animal to survive and/or reproduce.

Figure 32.1 Mating behavior of satin bowerbirds. A female satin bowerbird has chosen to mate with this male. Most likely, she was attracted by his physique and the blue decorations of his bower.

32.1 Genetic Basis of Behavior

The **behavior** of animals is any action that can be observed and described. All behavior has a genetic base in that the anatomy and physiology of the animal which is inherited is suitable to performing the behavior. In addition, various experiments have been done to show that specific behaviors have a genetic base.

A peach-faced lovebird, *Agapornis roseicollis,* cuts long, regular strips of material with its strong beak and then tucks them in its rump feathers for transport to the nest. A Fisher's lovebird, *Agapornis fischeri,* carries stronger materials, such as sticks, directly in its beak.

A. roseicollis

A. fischeri

If the behavior for obtaining and carrying nesting material is inherited, then hybrids might show intermediate behavior. When the two species of birds were mated, it was observed that the hybrid birds have difficulty carrying nesting materials. They cut strips and try to tuck them in their rump feathers, but they are unsuccessful. After a long period of time (about three years), a hybrid learns to carry the cut strips in its beak but still briefly turns its head toward its rump before flying off. Therefore, these studies support the hypothesis that behavior has a genetic basis.

Several experiments have been done with the garter snake, *Thamnophis elegans,* which has two different types of snake populations in California. Inland populations are aquatic and commonly feed underwater on frogs and fish. Coastal populations are terrestrial and feed mainly on slugs. In the laboratory, inland adult snakes refused to eat slugs while coastal snakes readily did so. To test for possible genetic differences between the two populations, matings were arranged between inland and coastal individuals, and it was found that isolated newborns show an overall intermediate incidence of slug acceptance.

The difference between slug acceptors and slug rejecters appears to be inherited, but what physiological difference have the genes brought about? A clever experiment answered this question. When snakes eat, their tongues carry chemicals to an odor receptor in the roof of the mouth. They

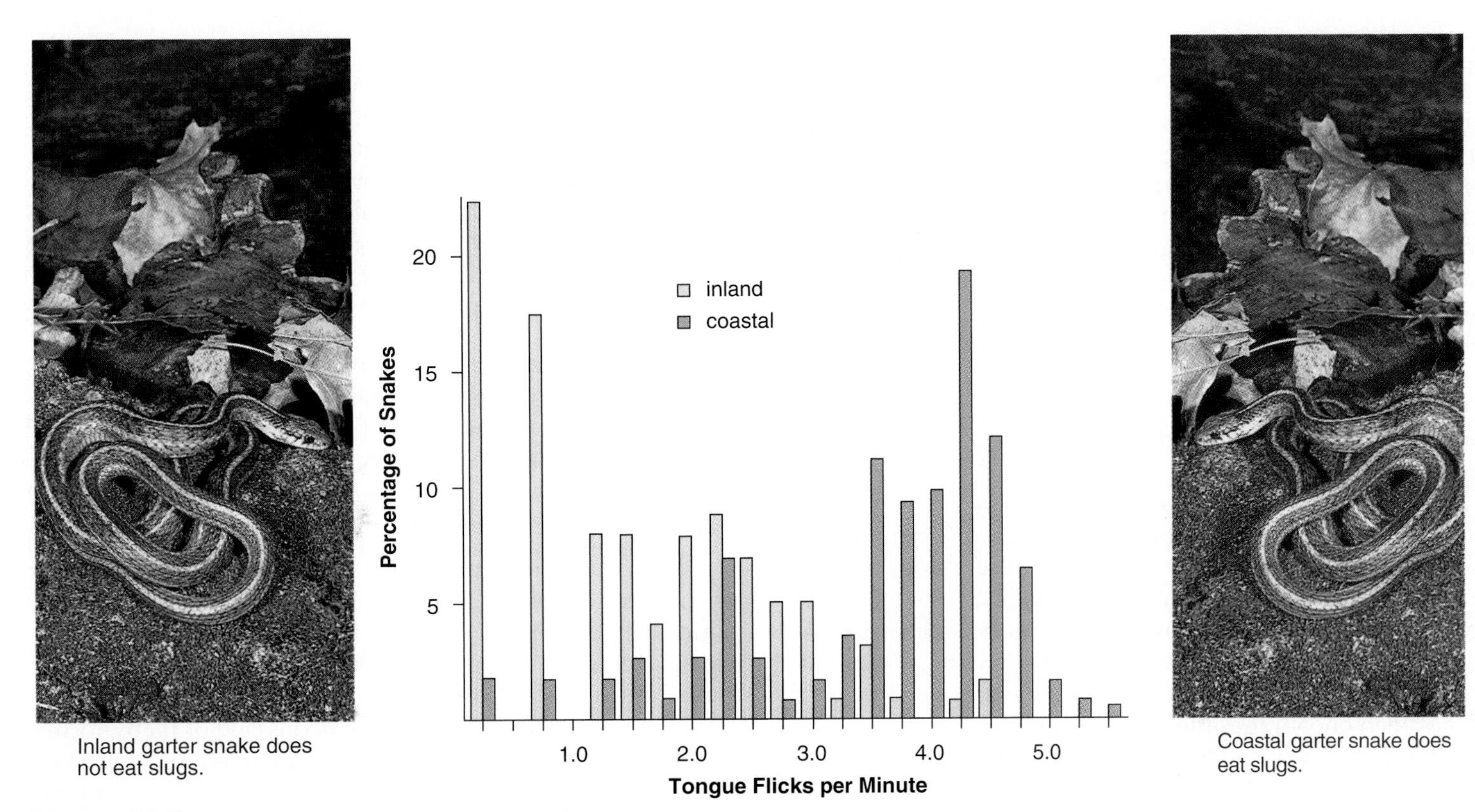

Figure 32.2 Feeding behavior of garter snakes.
The number of tongue flicks by inland and coastal garter snakes as a response to slug extract on cotton swabs. Coastal snakes tongue-flicked more than inland snakes.

use tongue flicks to recognize their prey. Even newborns will flick their tongues at cotton swabs dipped in fluids of their prey. Swabs were dipped in slug extract, and the number of tongue flicks were counted for newborn inland and coastal snakes. Coastal snakes had a higher number of tongue flicks than inland snakes (Fig. 32.2). Apparently, inland snakes do not eat slugs because they are not sensitive to their smell. A genetic difference between the two populations of snakes has resulted in a physiological difference in their nervous systems. Although hybrids showed a great deal of variation in the number of tongue flicks, they were generally intermediate as predicted by the genetic hypothesis.

Both nervous and endocrine systems are responsible for the integration of body systems. Is the endocrine system also involved in behavior? Various studies have been done to show that it is. For example, the egg-laying behavior in the marine snail *Aplysia* involves a set sequence of movements. Following copulation, the animal extrudes long strings of more than a million egg cases. It takes the egg case string in its mouth, covers it with mucus, waves its head back and forth to wind the string into an irregular mass, and attaches the mass to a solid object, like a rock. Several years ago, scientists isolated and analyzed an egg-laying hormone (ELH) that causes the snail to lay eggs even if it has not mated. ELH was found to be a small protein of 36 amino acids that diffuses into the circulatory system and excites the smooth muscle cells of the reproductive duct, causing them to contract and expel the egg string. Using recombinant DNA (deoxyribonucleic acid) techniques, the investigators isolated the ELH gene. The gene's product turned out to be a protein with 271 amino acids. The protein can be cleaved into as many as 11 possible products, and ELH is one of these. ELH alone, or in conjunction with these other products, is thought to control all the components of egg-laying behavior in *Aplysia.*

The results of many types of studies support the hypothesis that behavior has a genetic basis and that genes influence the development of neural and hormonal mechanisms that control behavior.

32.2 Development of Behavior

Given that all behaviors have a genetic basis, we can go on to ask if environmental experiences after hatching or birth also shape the behavior. Some behaviors seem to be stereotyped—they are always performed the same way each time. These were called fixed action patterns (FAP), and it is said that FAPs were elicited by a sign stimulus, a cue that sets the behavior in motion. For example, human babies will smile when a flat but face-sized mask with two dark spots for eyes is brought near them. It's possible that some behaviors are FAPs, but increasingly, investigators are finding that many behaviors, formerly thought to be FAPs, develop after practice.

Laughing gull chicks' begging behavior is always performed the same way in response to the parent's red beak. A chick directs a pecking motion toward the parent's beak, grasps it, and strokes it downward (Fig. 32.3). Sometimes a parent stimulates the begging behavior by swinging its beak gently from side to side. After the chick responds, the parent regurgitates food onto the floor of the nest. If need be, the parent then encourages the chick to eat. This interaction between the chicks and their parents suggests that the begging behavior involves learning. (**Learning** is defined as a durable change in behavior brought about by experience.) To test this hypothesis, diagrammatic pictures of gull heads were painted on small cards and then eggs were collected in the field. The eggs were hatched in a dark incubator to eliminate visual stimuli before the test. On the day of hatching, each chick was allowed to make about a dozen pecks at the model. The chicks were returned to the nest and were each retested. The tests showed that on the average, only one-third of the pecks by a newly hatched chick strike the model. But one day after hatching, more than half of the pecks are accurate, and two days after hatching, the accuracy reaches a level of more than 75%. Investigators concluded that improvement in motor skills, as well as visual experience, strongly affect development of chick begging behavior.

Behavior has a genetic basis, but the development of mechanisms that control behavior is subject to environmental influences, such as practice after birth.

How do chicks recognize a parent? Newly hatched chicks peck equally at any model as long as it has a red beak. Chicks a week old, however, will peck only at models that closely resemble the parent. Perhaps operant conditioning with a reward of food could account for this change in behavior. **Operant conditioning,** which is one of many forms of learning, is often defined as the gradual strengthening of stimulus-response connections. In everyday life, most people know that animals can be taught tricks by giving rewards such as food or affection. The trainer presents the stimulus, say a hoop, and then gives a reward (food) for the proper response (jumping through the hoop). B. F. Skinner is well known for studying this type of learning in the

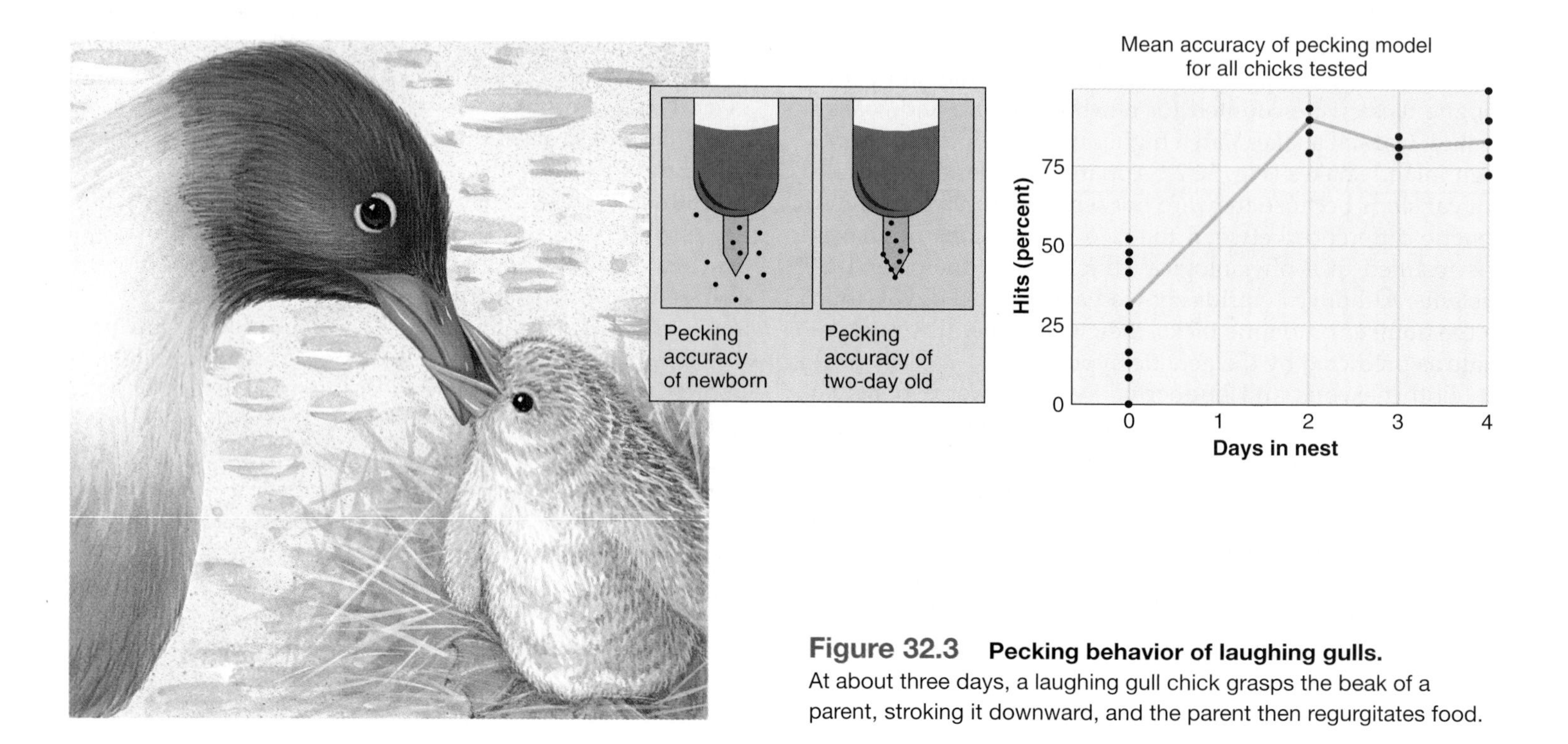

Figure 32.3 **Pecking behavior of laughing gulls.** At about three days, a laughing gull chick grasps the beak of a parent, stroking it downward, and the parent then regurgitates food.

laboratory. In the simplest type of experiment performed by Skinner, a caged rat happens to press a lever and is rewarded with sugar pellets, which it avidly consumes. Thereafter, the rat regularly presses the lever whenever it is hungry. In more sophisticated experiments, Skinner even taught pigeons to play ping-pong by reinforcing desired responses to stimuli.

Imprinting is another form of learning; chicks, ducklings, and goslings will follow the first moving object they see after hatching. This object is ordinarily their mother, but they seemingly can be imprinted on any object—a human or a red ball—if it is the first moving object they see during a sensitive period of two to three days after hatching. The term *sensitive period* means that the behavior only develops during this time. Although the Englishman Douglas Spalding first observed imprinting, the Austrian Konrad Lorenz is well known for investigating it. He found that imprinting not only served the useful purpose of keeping chicks near their mother, it also caused male birds to court a member of the correct species—someone who looks like mother! The goslings who had been imprinted on Lorenz courted human beings later in life. In-depth studies on imprinting have shown that the process is more complicated than originally thought. Eckhard Hess found that mallard ducklings imprinted on humans in the laboratory would switch to a female mallard that had hatched a clutch of ducklings several hours before. He found that vocalization before and after hatching was an important element in the imprinting process. Female mallards cluck during the entire time that imprinting is occurring. Do social interactions influence other forms of learning? Patterns of song learning in birds suggests that they can.

Song Learning in Birds

During the past several decades, an increasing number of investigators have studied song learning in birds. White-crowned sparrows sing a species-specific song, but males of a particular region have their own dialect. Birds were caged in order to test the hypothesis that young white-crowned sparrows learn how to sing from older members of their species (Fig. 32.4). A group of birds that *heard no songs at all* sang a song, but it was not fully developed. Birds that *heard tapes of white-crowns singing* sang in that dialect, as long as the tapes were played during a sensitive period from about age 10 to 50 days. White-crowned sparrows' dialects (or other species' songs) played before or after this sensitive period had no effect on their song. Apparently, their brain is especially primed to respond to acoustical stimuli during the sensitive period. Neurons that are critical for song production have been located, and they fire when the bird's own song is played or when a song of the same dialect is played. Other investigators have shown that birds *given an adult tutor* will sing the song of even a different species—no matter when the tutoring begins! It would appear that social experience has a very strong influence over the development of singing.

Animals have an ability to benefit from experience; learning occurs when a behavior changes with practice.

Figure 32.4 Song learning by white-crowned sparrows.
Three different experimental procedures are depicted and the results noted. These results suggest that there is both a genetic basis and an environmental basis for song learning in white-crowned sparrows.

32.3 Adaptiveness of Behavior

Since genes influence the development of behavior, it is reasonable to assume that behavioral traits (like other traits) are subject to natural selection. Our discussion will focus on reproductive behavior—specifically, the manner in which animals secure a mate. But we will also touch on the other two survival issues—capturing resources and avoiding predators—because these help an animal survive, and without survival, reproduction is impossible. Investigators studying survival value seek to test hypotheses that specify how a given trait might improve reproductive success.

Males can father many offspring because they continually produce sperm in great quantity. We would then expect competition among males to inseminate as many different females as possible. In contrast, females produce few eggs, so the choice of a mate becomes a prevailing consideration. Sexual selection can bring about evolutionary changes in the species. **Sexual selection** is changes in males and females, often due to male competition and female selectivity, leading to reproductive success.

Female Choice

Courtship displays are rituals that serve to prepare the sexes for mating. They help male and female recognize each other so that mating will be successful. They also play a role in a female's choice of a mate.

In a study of satin bowerbirds (see Fig. 32.1), two opposing hypotheses regarding female choice were tested:

Good genes hypothesis: females choose mates on the basis of traits that improve their chances of survival.

Run-away hypothesis: females choose mates on the basis of traits that make them attractive to females. The term "run away" pertains to the possibility that the trait will be exaggerated in the male until its reproductively favorable benefit is checked by the trait's unfavorable survival cost.

Investigators watched bowerbirds at feeding stations and also monitored the bowers. They discovered that although males tend to steal blue feathers and/or actively destroy a neighbor's bower, more aggressive and vigorous males were able to keep their bowers in good condition. These were the males usually chosen as mates by females. These data do not clearly support either hypothesis. It could be that aggressiveness, if inherited, does improve the chances of survival, or it could be that females simply preferred bowers with the most blue feathers.

The raggiana bird of paradise is remarkably dimorphic—the males are larger than females and have beautiful orange flank plumes. In contrast, the females are drab (Fig. 32.5). Courting males, which form a group called a lek, gather and begin to call. If a female joins them, the males raise their orange display plumes, shake their wings and hop from side to side, while continuing to call. They then stop calling and lean upside down with the wings projected forward to show off their beautiful feathers.

Female choice can explain why male birds are so much more showy than females. The remarkable plumes of the male might signify health and vigor to the female, just as a well-constructed bower might. In barn swallows, females also choose those with the longest tails and investigators have shown that males relatively free of parasites have longer tails than otherwise.

Is a difference in reproductive behavior related to a particular food source? Raggiana birds forage far and wide for their food (nutritious, complex fruits), and lekking is one way for males to attract the wide-ranging females. The male raggiana is polygynous (has more than one mate) and does not help raise the offspring. On the other hand, a related

Figure 32.5 Mating behavior in birds of paradise.
In birds of paradise, males have resplendent plumage brought about by sexual selection. The females are widely scattered, foraging for complex fruits; the males form leks that females visit to choose a mate.

Figure 32.6 A male olive baboon displaying full threat. In olive baboons, males are larger than females and have enlarged canines. Competition between males establishes a dominance hierarchy for the distribution of resources.

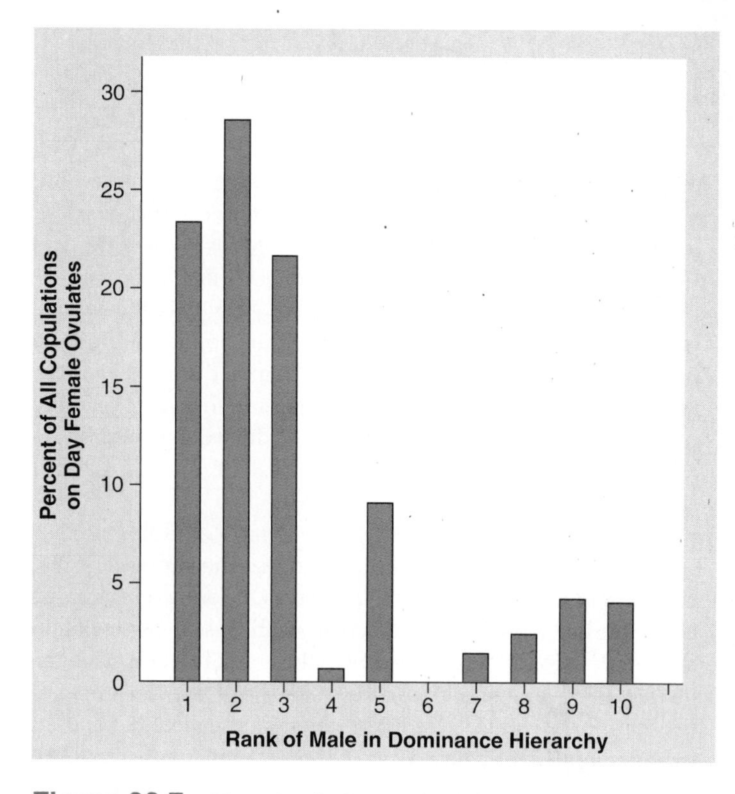

Figure 32.7 Female choice and male dominance among baboons. Although it may appear that females mate indiscriminately, they mate more often with a dominant male when they are most fertile.

species, the trumpet manucode, *Manucodia keraudrenit,* feeds on figs, which are more prevalent but not as nutritious as complex fruits. These birds are monogamous—the pair bonds perhaps for life, and the males are not as colorful as raggiana males. Both sexes are needed to successfully raise the young, and this relaxes pressure on the male to be showy.

Male Competition

Studies have been done to determine if the benefit of mating is worth the cost of competition among males. Only if the positive effects outweigh the negative effects will the animal enjoy reproductive success.

Dominance Hierarchy

Baboons, a type of Old World monkey, live together in a troop. Males and females have separate **dominance hierarchies** in which a higher ranking animal has greater access to resources than a lower ranking animal. Dominance is decided by confrontations, resulting in one animal giving way to the other.

Baboons are dimorphic; the males are larger than the females, and they can threaten other members of the troop with their long, sharp canines (Fig. 32.6). The baboons travel within a home range, foraging for food each day and sleeping in trees at night. The dominant males decide where and when the troop will move and, if the troop is threatened, they cover the troop as it retreats and attack intruders when necessary.

Females undergo a period known as estrus, during which they ovulate and are willing to mate. At this time a female approaches a dominant male and they form a mating pair for several hours or days. The male baboon pays a cost for his dominant position. Being larger means that he needs more food, and being willing and able to fight predators means that he may get hurt, and so forth. Is there a reproductive benefit to his behavior? Yes, in that dominant males do indeed monopolize estrous females when they are most fertile (Figure 32.7).

Nevertheless, there are other avenues to fathering offspring. Some males act as helpers to particular females and her offspring; the next time she is in estrus she may mate preferentially with him instead of a dominant male. Or subordinate males may form a friendship group that can oppose a dominant male, making him give up a receptive female.

Science Focus

Tracking an Animal in the Wild

Miniature radio transmitters that emit a radio signal allow you to track an animal in the wild. The transmitter is encapsulated along with a battery in a protective epoxy resin covering, and the package is either attached to an animal with a collar or a clip, or implanted surgically into the main body cavity of the animal. In either case, it is important to capture the animal carefully to use sedation or anesthesia to calm the animal during tagging or surgical implantation. The animal is permitted to recover fully prior to its release. A radio transmitter device is generally no more than 10% of the animal's body weight, and therefore it should not interfere with normal activities.

To track an animal, a researcher needs a radio receiver equipped with an antenna and earphones. The strongest signal comes from the direction of the animal as the antenna is rotated above the head. The data obtained allow the researcher to obtain a series of fixes (to determine where the animal is) and calculate its rate of movement. Plotting the sequence of fixes on a map of the appropriate scale gives information on the area that is used by the animal in the course of a night or several nights (Fig. 32A).

The distance over which the signal travels varies with the size of the transmitter and the strength of the battery. For small rodents it may be necessary to be within 5–10 meters of the animal to hear the signal; for larger collars like some of those used on elk or caribou, the signal can actually be received by a satellite orbiting above the earth. Later, the signals are sent in a radio beam by antenna to the ground. Satellite transmission allows researchers to work with animals that are far away, and prevents any possibility of interfering with the animal's activities.

A number of important findings have been made utilizing radiotelemetry, and a few examples provide a flavor of the kinds of results obtained with this technology. Elk in Yellowstone National Park in Wyoming have summer home range areas that are 5–10 times larger than in winter. Areas occupied by many mice and voles get larger as the animal ages, from the time the animals are weaned until they are adults, but these increases are greater for males than for females. Winter dens of rattlesnakes (and other snakes) are often some distance from the areas where they spend their summer months. Some crows migrate each spring and fall, whereas others remain as residents in the same locale throughout the year.

Radio transmitters can also be used to obtain data on physiological functions (heart rate, body temperature) in conjunction with observations of behavior while animals continue to engage in normal activities. When physiological measurements are taken, a device is included in the package that is sensitive to temperature or heart rate and can translate that information to the transmitter. Altogether, the use of radio transmitters has added a new dimension to the ability to explore the lives of animals.

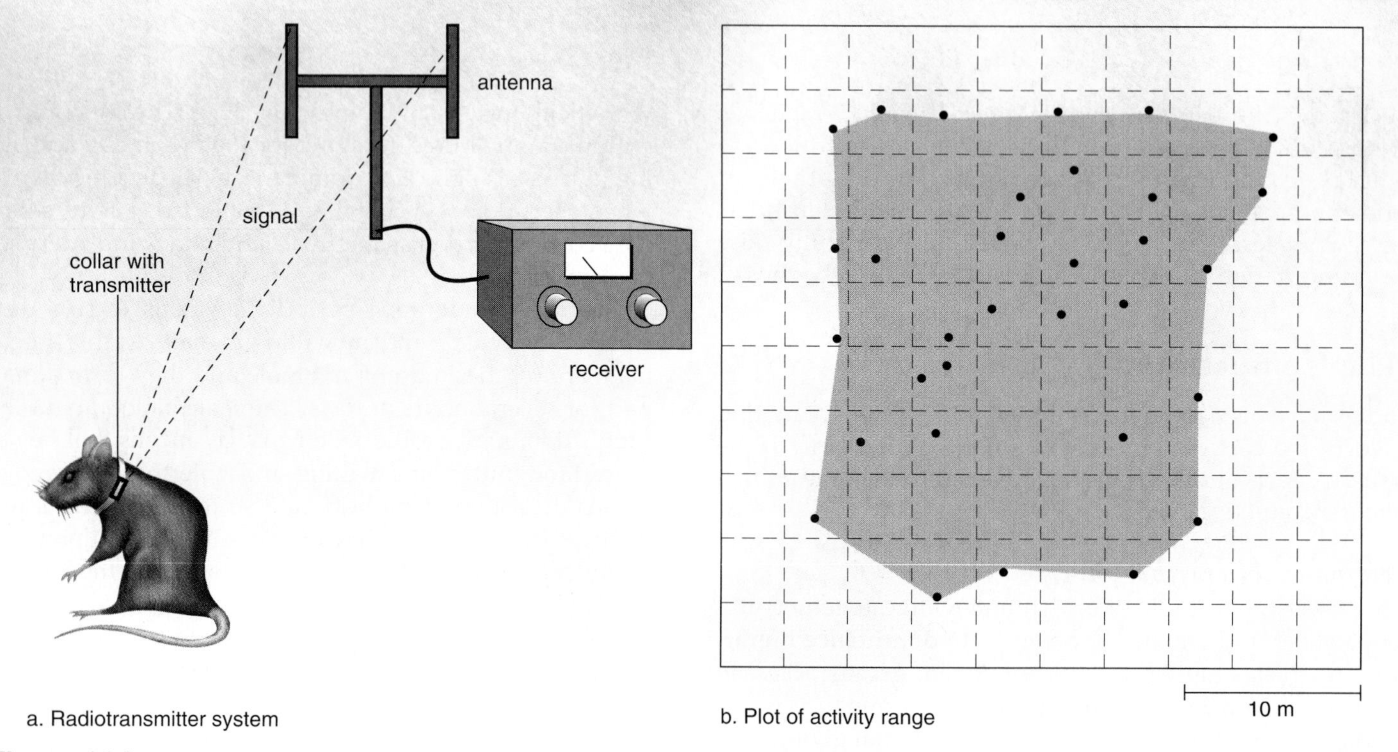

Figure 32A Electronic animal tracking.
a. By using a miniature radio transmitter in a neck collar, an antenna, and a receiver, it is possible to record an animal's location approximately every 15 minutes throughout a 24-hour period. **b.** These data are then used to plot its activity range for that day. (Some data in the plot represent multiple fixes at the same location.)

Territoriality

A territory is an area that is defended against competitors. **Territoriality** includes the type of defensive behavior needed to defend a territory. Vocalization and displays, rather than outright fighting, may be sufficient to defend a territory (Fig. 32.8). Male songbirds use singing to announce their willingness to defend a territory—so other males of the species become reluctant to make use of the same area.

Red deer stags (males) on the Scottish island of Rhum compete to be the harem master of a group of hinds (females) that mate only with them. The reproductive group occupies a territory that the harem master defends against other stags. Harem masters first attempt to repel challengers by roaring. If the challenger remains, the two lock antlers and push against one another. If the challenger now withdraws, the master pursues him for a short distance, roaring the whole time. If the challenger wins, he becomes the harem master.

A harem master can father two dozen offspring at most, because he is at the peak of his fighting ability for only a short time. And there is a cost to be able to father offspring. Stags must be large and powerful in order to fight; therefore, they grow faster and have less body fat. During bad times, they are more likely to die of starvation, and in general, they have shorter lives. The behavior of harem defense by stags will only persist in the population if its cost (reduction in the potential number of offspring because of a shorter life) is less than its benefit (increased number of offspring due to harem access).

Evolution by sexual selection can occur when females have the opportunity to select among potential mates and/or when males compete among themselves for access to reproductive females.

Figure 32.8 Competition between males among red deer.
Male red deer compete for a harem within a particular territory. **a.** Roaring alone may frighten off a challenger. **b.** But outright fighting may be necessary, and the victor is most likely the stronger of the two animals.

32.4 Animal Societies

There is a wide diversity of social behavior among animals. Some animals are largely solitary and join with a member of the opposite sex only for the purpose of reproduction. Others pair, bond, and cooperate in raising offspring. Still others form a **society** in which members of species are organized in a cooperative manner, extending beyond sexual and parental behavior. We have already had occasion to mention the social groups of baboons and red deer. Social behavior in these and other animals requires that they communicate with one another.

Communicative Behavior

Communication is an action by a sender that influences the behavior of a receiver. The communication can be purposeful but does not have to be purposeful. Bats send out a series of sound pulses and listen for the corresponding echoes in order to find their way through dark caves and locate food at night. Some moths have an ability to hear these sound pulses, and they begin evasive tactics when they sense that a bat is near. Are the bats purposefully communicating with the moths? No, bat sounds are simply a cue to the moths that danger is near.

Communication is an action by a sender that affects the behavior of a receiver.

Chemical Communication

Chemical signals have the advantage of working both night and day. The term **pheromone** is used to designate chemical signals in low concentration that are passed between members of the same species. Female moths secrete chemicals from special abdominal glands, which are detected downwind by receptors on male antennae. The antennae are especially sensitive, and this assures that only male moths of the correct species (and not predators) will be able to detect them.

Cheetahs and other cats mark their territories by depositing urine, feces, and anal gland secretions at the boundaries (Fig. 32.9). Klipspringers (small antelope) use secretions from a gland below the eye to mark twigs and grasses of their territory.

Auditory Communication

Auditory (sound) communication has some advantages over other kinds of communication (Fig. 32.10). It is faster than chemical communication, and it also is effective both night and day. Further, auditory communication can be modified not only by loudness but also by pattern, duration, and repetition. In an experiment with rats, a researcher discovered that an intruder can avoid attack by increasing the frequency with which it makes an appeasement sound.

Male crickets have calls, and male birds have songs for a number of different occasions. For example, birds may have one song for distress, another for courting, and still another for marking territories. Sailors have long heard the songs of humpback whales because they are transmitted through the hull of a ship. But only recently has it been shown that the song has six basic themes, each with its own phrases, that can vary in length and be interspersed with sundry cries and chirps. The purpose of the song is probably sexual and serves to advertise the availability of the singer. Language is the ultimate auditory communication, but only humans

Figure 32.9 Use of pheromone to mark a territory. This male cheetah is spraying a pheromone onto a tree in order to mark his territory.

Figure 32.10 A chimpanzee with a researcher. Chimpanzees are unable to speak but can learn to use a visual language consisting of symbols. Some believe chimps only mimic their teachers and never understand the cognitive use of a language. Here the experimenter shows Nim the sign for "drink." Nim copies.

have the biological ability to produce a large number of different sounds and to put them together in many different ways. Nonhuman primates have at most only forty different vocalizations, each having a definite meaning, such as the one meaning "baby on the ground," which is uttered by a baboon when a baby baboon falls out of a tree. Although chimpanzees can be taught to use an artificial language, they never progress beyond the capability level of a two-year-old child. It has also been difficult to prove that chimps understand the concept of grammar or can use their language to reason. It still seems as if humans possess a communication ability unparalleled by other animals.

Visual Communication

Visual signals are most often used by species that are active during the day. Contests between males make use of threat postures and possibly prevent outright fighting that might result in reduced fitness. A male baboon displaying full threat is an awesome sight that establishes his dominance and keeps peace within the baboon troop (see Fig. 32.6). Hippopotamuses have territorial displays that include mouth opening.

The plumage of a male raggiana bird of paradise allows him to put on a spectacular courtship dance to attract females and to give her a basis on which to select a suitable mate (see Fig. 32.5). Defense and courtship displays are exaggerated and are always performed in the same way so that their meaning is clear.

Tactile Communication

Tactile communication occurs when one animal touches another. For example, gull chicks peck at the parent's beak in order to induce the parent to feed them (see Fig. 32.3). A male leopard nuzzles the female's neck to calm her and to stimulate her willingness to mate. In primates, grooming—one animal cleaning the coat and skin of another—helps cement social bonds within a group.

Honeybees use a combination of communication methods, but especially tactile, to impart information about the environment. When a foraging bee returns to the hive, it performs a waggle dance that indicates the distance and the direction of a food source (Fig. 32.11). As the bee moves between the two loops of a figure 8, it buzzes noisily and shakes its entire body in so-called waggles. Outside the hive, the dance is done on a horizontal surface, and the straight run indicates the direction of the food. Inside the hive, the angle of the straight run to that of the direction of gravity is the same as the angle of the food source to the sun. In other words, a 40° angle to the left of vertical means that food is 40° to the left of the sun. Bees can use the sun as a compass to located food because their biological clocks, as discussed in the reading on page 682, allow them to compensate for the movement of the sun in the sky.

Animals use a number of different ways to communicate, and communication facilitates cooperation.

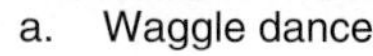

a. Waggle dance

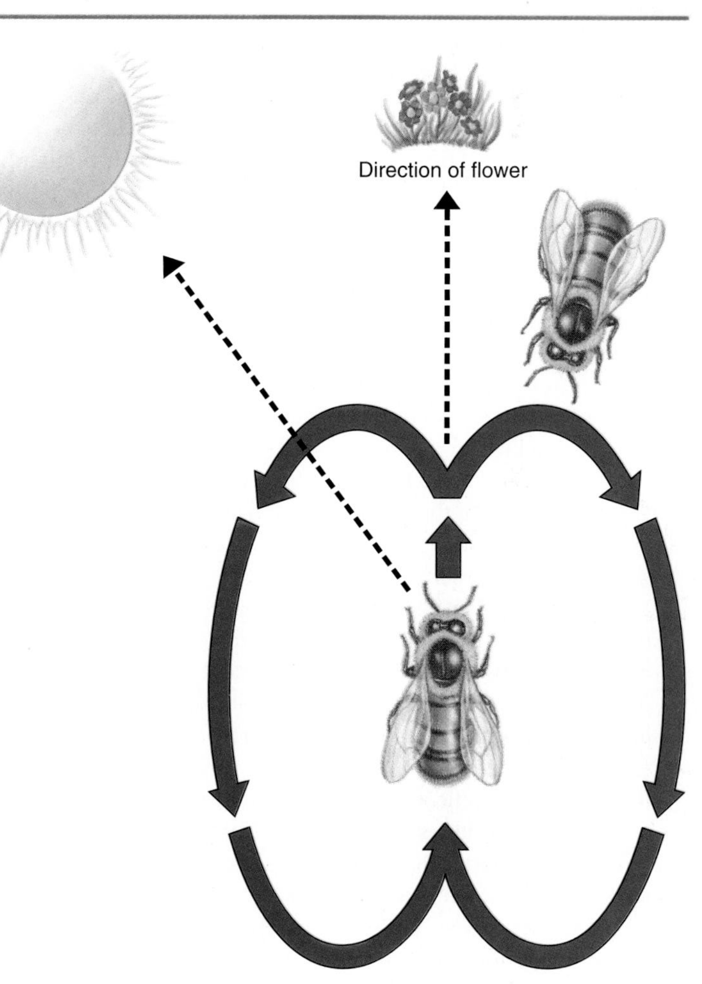

b. Components of dance

Figure 32.11 Communication among bees.
a. Honeybees do a waggle dance to indicate the direction of food. **b.** If the dance is done outside the hive on a horizontal surface, the straight run of the dance will point to the food source. If the dance is done inside the hive on a vertical surface, the angle of the straightaway to that of the direction of gravity is the same as the angle of the food source to the sun.

Science Focus

Biological Rhythms

Certain behaviors in animals recur at regular intervals. Behaviors that occur on a daily basis are said to have a circadian (about a day) rhythm. For example, some animals, like humans, are usually active during the day and sleep at night. Others, such as bats, sleep during the day and hunt at night. There are also behaviors that occur on a yearly basis. For example, in the Northern Hemisphere, birds migrate south in the fall, and the young of many animals are born in the spring. Such behaviors have a circannual rhythm.

There were two hypotheses regarding the control of circadian rhythms: either their timing is controlled externally, or there is an internal timing device, often called a biological clock (Fig. 32B). These alternative hypotheses have been tested in crickets, which regularly call every night to attract females. When laboratory crickets are kept in a room under constant conditions with lights continually on or continually off, they continue to call every night, only calling starts as much as 26 hours later than it did on the day before. But exposure to night and day cycles will right the cycles; therefore, it was possible to conclude that there is a biological clock, but it does not keep perfect time and must be reset by environmental stimuli. Similar results have been obtained with all sorts of animals, from fiddler crabs to humans.

At a minimum, a circadian system must have three components (Fig. 32B). There must be a means to reset a pacemaker according to the current environmental light-dark cycle; a biological clock that keeps time; and the rhythmic behavior itself. By now, there is a general consensus that the mammalian biological clock is a collection of nerve cells in the hypothalamus of the brain known as the suprachiasmatic nucleus (SCN). Electrical and drug stimulation of the SCN upsets circadian rhythms and destruction of the nucleus does away with many rhythmic behaviors. Investigators have also discovered a number of genes whose protein products control the activity of the clock. Less is known about the other components of the circadian system. Whether the receptor that resets the clock is in the eye or not is still to be determined. In one recent study, investigators found that exposing the back of a knee to light reset the clock judged by body temperature and hormone levels. They concluded that any portion of the skin contains photoreceptor proteins that communicate with and reset the biological clock.

A number of investigators are interested in the observation that environmental light suppresses melatonin production by the pineal gland while we are awake. Conversely, as it gets dark, melatonin levels rise and we get sleepy. This knowledge causes some people to take melatonin for the symptoms of insomnia and "jet lag." When we travel by airplane from one part of the world to another it is difficult for our circadian system to adjust, and symptoms like insomnia, fatigue, headache, gastrointestinal distress, and moodiness, occur. A medication for these symptoms would also be helpful to one out of every five people whose work shifts between day and night. Jet lag symptoms are reduced by exposure to daylight in the afternoon after westward flights and in the early morning after eastward flights. Similarly, an adjustment to a nighttime work shift is enhanced by exposure to bright light at night and to darkness during the day. Because light wakes us up and melatonin makes us sleepy, their use at different times of the day may be helpful in resetting the biological clock.

Many people are intrigued by the idea that some psychiatric conditions might be due to a disorder of the circadian system. The best example is seasonal affective disorder (SAD) which affects as much as 5% of the general population and is characterized by depressions during the fall and winter. The hypothesis that short winter days are responsible for SAD has led to successful therapies based on exposure to bright light. Circadian rhythms are clinically significant in other ways. The incidence of heart attacks, sudden cardiac death, and stroke peak in the late morning, and certain cancer drugs are more effective when given in the day or night. This is an exciting time for research concerning the human circadian system especially because such research is expected to have important applications that will help control many human ills, from those that are quite serious to those that are simply a nuisance.

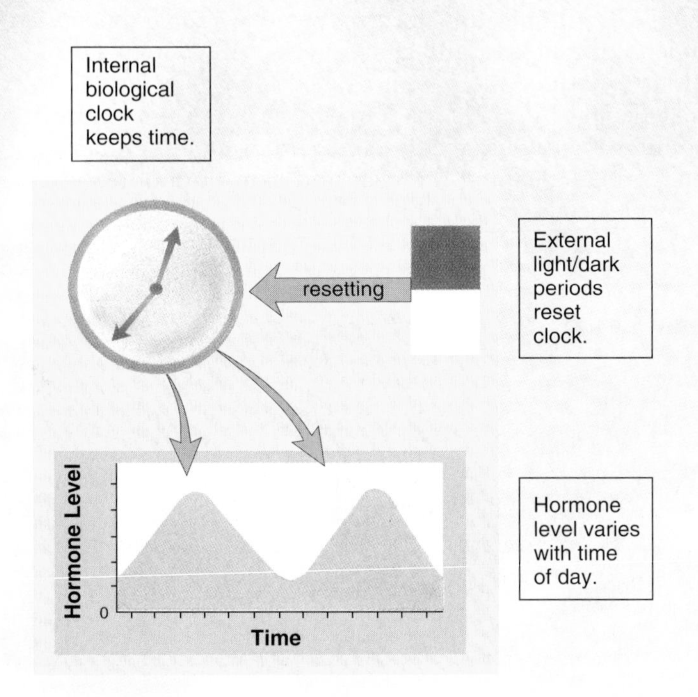

Figure 32B A circadian system.
A circadian system has three components, as shown.

32.5 Sociobiology and Animal Behavior

Sociobiology applies the principles of evolutionary biology to the study of social behavior in animals. Sociobiologists develop hypotheses about social living based on the assumption that a social individual derives more reproductive benefits than costs from living in a society. Then they perform a cost-benefit analysis to see if their hypotheses are correct.

Group living does have benefits under certain circumstances. It can help an animal avoid predators, rear offspring, and find food. A group of impalas is more likely to hear an approaching predator than a solitary one. Many fish moving rapidly in many directions might distract a would-be predator.

Pair bonding of trumpet manucodes helps the birds raise their young. Due to their particular food source, the female cannot rear as many offspring alone as she can with the male's help. Weaver birds form giant colonies that help protect them from predators, but the birds may also share information about food sources. Primate members of the same troop signal to one another when they have found an especially bountiful fruit tree. Lions working together are able to capture large prey, such as zebra and buffalo.

Group living also has its disadvantages. When animals are crowded together into a small area, disputes can arise over access to the best feeding places and sleeping sites. Dominance hierarchies are one way to apportion resources, but this puts subordinates at a disadvantage. Among red deer, the ability of a hind to rear sons is dependent on her dominance. Only large, dominant females can successfully rear sons; small subordinate females tend to rear daughters. From an evolutionary point of view, sons are preferable because, as a harem master, sons will result in a greater number of grandchildren. However, sons, which tend to be larger than daughters, need to be nursed more frequently and for a longer period of time. Subordinate females do not have access to enough food resources to adequately nurse sons and, therefore, they tend to rear daughters and not sons. Still, like the subordinate males in a baboon troop, subordinate females in a red deer harem may be better off in fitness terms if they stay with a group, despite the cost involved.

Living in close quarters means that illness and parasites can pass from one animal to another more rapidly. Baboons and other types of social primates invest much time in grooming one another, and this most likely helps them remain healthy.

Social living has both advantages and disadvantages. Only if the benefits, in terms of individual reproductive success, outweigh the disadvantages will societies evolve.

Altruism Versus Self-Interest

Altruism is behavior that has the potential to decrease the lifetime reproductive success of the altruist while benefiting the reproductive success of another member of the group. In insect societies especially, reproduction is limited to only one pair, the queen and her mate. For example, among army ants the queen is inseminated only during her nuptial flight, and thereafter she spends her time reproducing. The society has three different sizes of sterile female workers. The smallest workers (3 mm), called the nurses, take care of the queen and larvae, feeding them and keeping them clean. The intermediately-sized workers, constituting most of the population, go out on raids to collect food. The soldiers (14 mm), with huge heads and powerful jaws, run along the sides and rear of raiding parties where they can best attack any intruders.

Can the altruistic behavior of sterile workers be explained in terms of reproductive success? A given gene can be passed from one generation to the next in two quite different ways. The first way is direct: a parent can pass the gene directly to an offspring. The second way is indirect: an animal can help a relative reproduce and thereby pass the gene to the next generation via this relative. Direct selection is natural selection that can result in adaptation to the environment when the reproductive success of individuals differs. Indirect selection is natural selection that can result in adaptation to the environment when individuals differ in their effects on the reproductive success of relatives. The **inclusive fitness** of an individual includes personal reproduction and reproduction of relatives.

Among social bees, social wasps, and ants, the queen is diploid but her mate is haploid. If the queen has had only one mate, sister workers are more closely related to each other (sharing on average 75% of their genes) than they are to their potential offspring (with which they would share on average only 50% of their genes). Therefore, a worker can achieve a greater inclusive fitness benefit by aiding her mother (the queen) to produce additional sisters than by directly reproducing herself. Under these circumstances, behavior that appears to be altruistic is more likely to evolve.

Indirect selection can also occur among animals whose offspring receive only a half set of genes from both parents. Consider that your brother or sister shares 50% of your genes, your niece or nephew shares 25%, and so forth. This means that the survival of two nieces (or nephews) is worth the survival of one sibling, assuming they both go on to reproduce.

Among chimpanzees in Africa, a female in estrus frequently copulates with several members of the same group, and the males make no attempt to interfere with each other's matings. How can they be acting in their own self-interest? Genetic relatedness appears to underlie their apparent altruism; members of a group share more than 50% of their genes in common because members never leave the territory in which they are born.

Figure 32.12 Inclusive fitness.
A meerkat is acting as a baby-sitter for its young sisters and brothers while their mother is away. Could this helpful behavior contribute to the baby-sitter's inclusive fitness?

Helpers at the Nest

In some bird species, offspring from one clutch of eggs may stay at the nest helping parents rear the next batch of offspring. In a study of Florida scrub jays, the number of fledglings produced by an adult pair doubled when they had helpers. Mammalian offspring are also observed to help their parents (Fig. 32.12). Among jackals in Africa, pairs alone managed to rear an average of 1.4 pups, whereas pairs with helpers reared 3.6 pups.

Is the reproductive success of the helpers increased by their altruistic behavior? It could be if the chance of their reproducing on their own is limited. A flock of green wood-hoopoes (*Phoeniculus purpurens*), an insect-eating bird of Africa, may have as many as sixteen members but only one breeding pair. The other sexually mature members help feed and protect the fledglings and protect the home territory from invasion by other green wood-hoopoes. Resources are limited, particularly because nest sites are rare. Few acacia trees have suitable cavities to serve as nest sites and those that do are often occupied by other species. Moreover, predation by snakes within the cavities can be intense; even if a pair of birds acquire an appropriate cavity, they would be unable to protect their offspring by themselves. Therefore, the cost of trying to establish a territory is clearly very high.

What are the benefits of staying behind to help? First, a helper is contributing to the survival of its own kin. Therefore, the helper actually gains a fitness benefit (albeit a smaller benefit than it would achieve were it a breeder). Second, a helper is more likely than a nonhelper to inherit a parental territory—including other helpers. Helping, then, involves making a minimal, short-term reproductive sacrifice in order to maximize future reproductive potential. Once again, an apparently altruistic behavior turns out to be an adaptation.

Inclusive fitness is measured by genes an individual contributes to the next generation, either directly by offspring or indirectly by way of relatives. Many of the behaviors once thought to be altruistic turn out, on closer examination, to be examples of kin selection, and are adaptive.

Bioethical Issue

Is it ethical to keep animals in zoos where they are not free to behave as they do in the wild? If we keep animals in zoos are we depriving them of their freedom? Some point out that freedom is never absolute. Even an animal in the wild is restricted in various ways by its abiotic and biotic environment. The so-called five freedoms are to be free of starvation, cold, injury, and fear, as well as free to wander and express one's natural behavior. Perhaps it's worth giving up a bit of the last freedom to achieve the first four? Many modern zoos do keep animals in habitats that nearly match their natural one so that they do have some freedom to roam and behave naturally. Perhaps, too, we should consider the education and enjoyment of the many thousands of human visitors to a zoo compared to the loss to a much smaller number of animals kept in a zoo?

Today, reputable zoos rarely go out and capture animals in the wild—they usually get their animals from other zoos. Most people feel it is never a good idea to take animals from the wild except for very serious reasons. Certainly, zoos should not be involved in the commercial and often illegal trade in wild animals which still goes on today. When animals are captured it should be done by skilled biologists or naturalists who know how to care and transport the animal.

Many zoos today are involved in the conservation of animals. They provide the best home possible while animals are recovering from injury or their numbers are increased until they can be released to the wild. Perhaps we can look upon zoos favorably if they can show that their animals are being kept under good conditions, and that they are also involved in the preservation of animals.

Questions

1. Do you think it is ethical to keep animals in zoos? Under particular circumstances? Explain.
2. Do the animals that are descended from zoo animals have the right to be protected as their parents were? Why or why not?
3. Do the same concerns about zoos also apply to aquariums? Why or why not?

Summarizing the Concepts

Ethologists study how animals are organized to perform a particular behavior and how it benefits them to survive and reproduce.

32.1 Genetic Basis of Behavior

Hybrid studies with lovebirds produce results consistent with the hypothesis that behavior has a genetic basis. Garter snake experiments indicate that the nervous system controls behavior. *Aplysia* DNA studies indicate that the endocrine system also controls behavior.

32.2 Development of Behavior

The environment is influential in the development of behavioral responses, as exemplified by an improvement in laughing gull chick begging behavior and an increased ability of chicks to recognize parents. Modern studies suggest that most behaviors improve with experience. Even behaviors that were formerly thought to be fixed action patterns (FAPs) or otherwise thought to be inflexible sometimes can be modified.

Song learning in birds involves various elements—including the existence of a sensitive period when an animal is primed to learn—and the effect of social interactions.

32.3 Adaptiveness of Behavior

Traits that promote reproductive success are expected to be advantageous overall despite any possible disadvantage. Males who produce many sperm are expected to compete to inseminate females. Females who produce few eggs are expected to be selective about their mates. Experiments with satin bowerbirds and birds of paradise support these bases for sexual selection.

The raggiana bird of paradise males gather in a lek—most likely because females are widely scattered, as is their customary food source. The food source for a related species, the trumpet manucode, is readily available but less nutritious. These birds are monogamous—it takes two parents to rear the young.

A cost-benefit analysis can be applied to competition between males for mates in reference to a dominance hierarchy (e.g., baboons) and territoriality (e.g., red deer).

32.4 Animal Societies

Animals that form social groups communicate with one another. Communications such as chemical, auditory, visual, and tactile signals foster cooperation that benefits both the sender and the receiver.

There are benefits and costs to living in a social group. If animals live in a social group, it is expected that the advantages (e.g., help to avoid predators, raise young, and find food) will outweigh the disadvantages (e.g., tension between members, spread of illness and parasites, and reduced reproductive potential). This expectation can sometimes be tested.

32.5 Sociobiology and Animal Behavior

In most instances, the individuals of a society act to increase their own reproductive success. Sometimes animals perform altruistic acts, as when individuals help their parents rear siblings. There is a benefit to this behavior when one considers inclusive fitness, which involves both direct selection and indirect selection.

In social insects, altruism is extreme but can be explained on the basis that the insects are helping a reproducing sibling survive. A study of wood-hoopoes, an African bird, shows that younger siblings may help older siblings who reared them until the younger get a chance to reproduce themselves.

Studying the Concepts

1. What two aspects of behavior particularly interest ethologists? 672
2. Describe an experiment with lovebirds, and explain how it shows that behavior has a genetic basis. 672
3. An experiment with garter snakes showed that what system is involved in behavior? Explain the experiment and the results. 672–73
4. Studies of *Aplysia* DNA show that the endocrine system is also involved in behavior. Explain. 673
5. Some behaviors require practice before developing completely. How does the experiment with laughing gull chicks support this statement? 674
6. An argument can be made that social contact is an important element in learning. Explain this with reference to imprinting in mallard ducks and song learning in white-crowned sparrows. 675
7. Why would you expect behavior to be subject to natural selection and be adaptive? 676
8. Reproductive behavior sometimes seems tied to how an animal acquires food. Explain with reference to the two bird of paradise species discussed in this chapter. 676
9. Explain how the anatomy and behavior of dominant male baboons is both a benefit and a drawback. 677
10. Give examples of the different types of communication among members of a social group. 680–81
11. What is a cost-benefit analysis and how does it apply to living in a social group? Give examples. 683
12. How can altruism, as defined on page 683, be explained on the basis of self-interest? 683–84

Testing Yourself

Choose the best answer for each question.

1. Which question is least likely to interest a behaviorist?
   a. How do genes control the development of the nervous system?
   b. Why do animals living in the tundra have white coats?
   c. Does aggression have a genetic basis?
   d. Why do some animals feed in groups and others feed singly?
   e. Behaviorists only study specific animals.
2. Female sage grouse are widely scattered throughout the prairie. Which of these would you expect?
   a. A male will maintain a territory large enough to contain at least one female.
   b. Male and female birds will be monogamous, and both will help feed the young.
   c. Males will form a lek where females will choose a mate.
   d. Males will form a dominance hierarchy for the purpose of distributing resources.
   e. All of these are correct.
3. White-crowned sparrows from two different areas sing with a different dialect. If the behavior is primarily genetic, newly hatched birds from each area will
   a. sing with their own dialect.
   b. need tutors in order to sing in their dialect.
   c. sing only when a female is nearby.
   d. learn to sing later.
   e. Both a and c are correct.

4. Orangutans are solitary but territorial. This would mean orangutans defend their territory's boundaries against
   a. other male orangutans.
   b. female orangutans.
   c. all types of animals, whether orangutans or not.
   d. animals that prey on them.
   e. Both a and b are correct.
5. The resplendent plumes of a raggiana bird of paradise are due to the fact that birds with the best display
   a. are dominant over other birds.
   b. have the best territories.
   c. are chosen by females as mates.
   d. are chosen by males and females as companions.
   e. All of these are correct.
6. Subordinate females in a baboon troop do not produce offspring as often as dominant females. It is clear that
   a. the cost of being in the troop is too high.
   b. the dominant males do not mate with subordinate females.
   c. subordinate females must benefit in some way from being in the troop.
   d. Subordinate females should leave the troop.
   e. Both a and b are correct.
7. German blackcaps migrate southeast to Africa, and Austrian blackcaps fly southwest to Africa. The fact that hybrids of these two are intermediate shows that
   a. the trait is controlled by the nervous system.
   b. nesting is controlled by hormones.
   c. the behavior is at least partially genetic.
   d. behavior is according to the sex of animals.
   e. Both a and c are correct.
8. At first laughing gull chicks peck at any model that looks like a red beak; later they will not peck at any model that does not look like a parent. This shows that the behavior
   a. is a fixed action pattern.
   b. undergoes development after birth.
   c. is controlled by the nervous system.
   d. is under hormonal control.
   e. All of these are correct.
9. Which answer is based on anatomy? Males compete because
   a. they have the size and weapons with which to compete.
   b. they produce many sperm for a long time.
   c. the testes produce the hormone testosterone.
   d. only then do females respond to them.
   e. Both a and c are correct.
10. Which answer is in keeping with evolutionary theory? Females are choosy because
   a. they do not have the size and weapons with which to compete.
   b. they invest heavily in the offspring they produce.
   c. ovaries produce the hormones estrogen and progesterone.
   d. they need time to get ready to respond.
   e. All of these are correct.

Thinking Scientifically

1. In an experiment to determine the control of behavior, young guinea pigs are placed in a cage and are fed a type of food they do not ordinarily eat in the wild. When released, the guinea pigs are offered various types of food, including the one type they were fed while caged (page 673).
   a. How might you attempt to show that the nervous system is involved in food gathering and eating?
   b. What will you conclude if all groups of guinea pigs choose only the type of food they were recently fed?
   c. What will you conclude if the guinea pigs choose only the type of food they ordinarily eat in the wild?
2. Rewarding an animal with food helps them learn a behavior. Based on this finding (page 674),
   a. what advice might you have for parents?
   b. what part of the brain must be involved in learning the behavior?
   c. what other experiments might support your hypothesis?

Understanding the Terms

altruism 683
behavior 672
communication 680
dominance hierarchy 677
imprinting 675
inclusive fitness 683
learning 674
operant conditioning 674
pheromone 680
sexual selection 676
society 680
sociobiology 683
territoriality 679

Match the terms to these definitions:

a. ________ Behavior related to the act of marking or defending a particular area against invasion by another species member; area often used for the purpose of feeding, mating, and caring for young.
b. ________ Social interaction that has the potential to decrease the lifetime reproductive success of the member exhibiting the behavior.
c. ________ Signal by a sender that influences the behavior of a receiver.
d. ________ Chemical substance secreted by one organism that influences the behavior of another.
e. ________ Increase in reproduction that results from direct selection and indirect selection.

Using Technology

Your study of animal behavior is supported by these available technologies.

Essential Study Partner CD-ROM
Ecology → Behavior
Visit the Mader web site for related ESP activities.

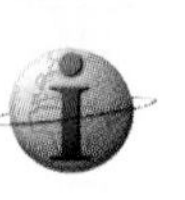

Exploring the Internet
The Mader Home Page provides resources and tools as you study this chapter.

http://www.mhhe.com/biosci/genbio/mader

CHAPTER 33

Population Ecology

Chapter Concepts

These poppies are members of a population whose size is determined by the carrying capacity of the environment.

Imagine a watering hole that can accommodate 100 zebras. If, at first, there are only two zebras, and each pair of zebras produces only four zebras, how many more generations could there be without overtaxing the water hole? You're correct if you say four and incorrect if you say five. The problem is that you can't just consider the newly arrived zebras—you have to add the number of zebras already there.

2 — 4 — 8 — 16 — 32 — 64 — 128

Also, notice that when it's time to stop, there are only 62 zebras (30 + 32). That's one of the unusual things about population growth—at one point it seems as if there is plenty of room, and then all of a sudden, there's not enough room.

Modern ecologists now recognize that knowledge of population growth has almost unlimited application possibilities, including the management of wildlife to prevent extinction and the maintenance of food (organic nutrients) sources for the human population.

33.1 Scope of Ecology

Ecology is the study of the interactions of organisms with each other and with the physical environment. Ecology, like so many biological disciplines, is wide-ranging. At one of its lowest levels, ecologists study how the individual organism is adapted to its environment. For example, they study why fishes in a coral reef live only in warm tropical waters and how the fishes feed (Fig. 33.1). Most organisms do not exist singly; rather, they are part of a population, a functional unit that interacts with the environment. A **population** is defined as all the organisms within an area belonging to the same species. At this level of study, ecologists are interested in factors that affect the growth and regulation of population size.

A **community** consists of all the various populations interacting at a locale. In a coral reef, there are numerous populations of fishes, crustaceans, corals, and so forth. At this level ecologists want to know how interactions like predation and competition affect the organization of a community. An **ecosystem** contains a community of populations and also the *abiotic* (nonliving) environment. Energy flow and chemical cycling are significant aspects of understanding how an ecosystem functions. The **biosphere** is that portion of the entire earth's surface where living things exist.

Modern ecology is not just descriptive, it also develops hypotheses that can be tested. A central goal of modern ecology is to develop models which explain and predict the distribution and abundance of organisms. Ultimately, ecology considers not one particular area, but the distribution and abundance of populations in the biosphere.

Ecology is the study of the interactions of organisms with other organisms and with the physical environment. These interactions determine the distribution and abundance of organisms at a particular locale and over the earth's surface.

Figure 33.1 Ecological levels.
The study of ecology encompasses various levels, from the individual organism to the population, community, and ecosystem.

Density and Distribution of Populations

Population density is the number of individuals per unit area or volume. If we calculated the density of the human population, we would know how many individuals there are per square mile, for example. From this we might get the impression that humans are uniformly distributed, but we know full well that most people live in cities. Even within a city more people live in particular neighborhoods than others. *Population distribution* is the pattern of dispersal of individuals within the area of interest.

There are three patterns of distribution: uniform, random, and clumped (Fig. 33.2). Human beings have the clumped pattern of distribution, which is the most common pattern. Today, ecologists want to discover what causes the spatial distribution of organisms. With reference to human beings, we know that many cities sprung up at the junction of rivers or near inlets that make good harbors for ships. In a study of the distribution of hard clams in a bay on the south shore of Long Island, New York, it was found, as discussed on page 693, that clams are apt to occur where the sediment contains oyster shells. Hopefully, this information can be used to transform areas of low abundance to areas of high abundance of clams.

As with clams, the distribution of organisms can be due to abiotic factors. Physical factors like a particular inorganic nutrient (the oyster shells provide calcium carbonate for the formation of clam shells) can determine where organisms occur. Also important are precipitation and temperature, which can be limiting factors for the distribution of an organism. Limiting factors are those factors that particularly determine whether an organism lives in an area. Trout live only in cool mountain streams where there is a high oxygen content, but carp and catfish are found in rivers near the coast because they can tolerate warm waters, which have a low concentration of oxygen. The timberline is the limit of tree growth in mountainous regions or in high latitudes. Trees cannot grow above the high timberline because of low temperature and the fact that water remains frozen most of the year.

The distribution of organisms can also be due to *biotic* (living) factors. In Australia, the red kangaroo does not live outside inland areas because it is adapted to feeding on the arid grasses that grow there. And there are more humans where the soil is suitable for growing crops than where the soil is rocky and poor in inorganic nutrients.

Ecology as a science includes a study of the distribution of organisms: where and why organisms are located in a particular place at a particular time.

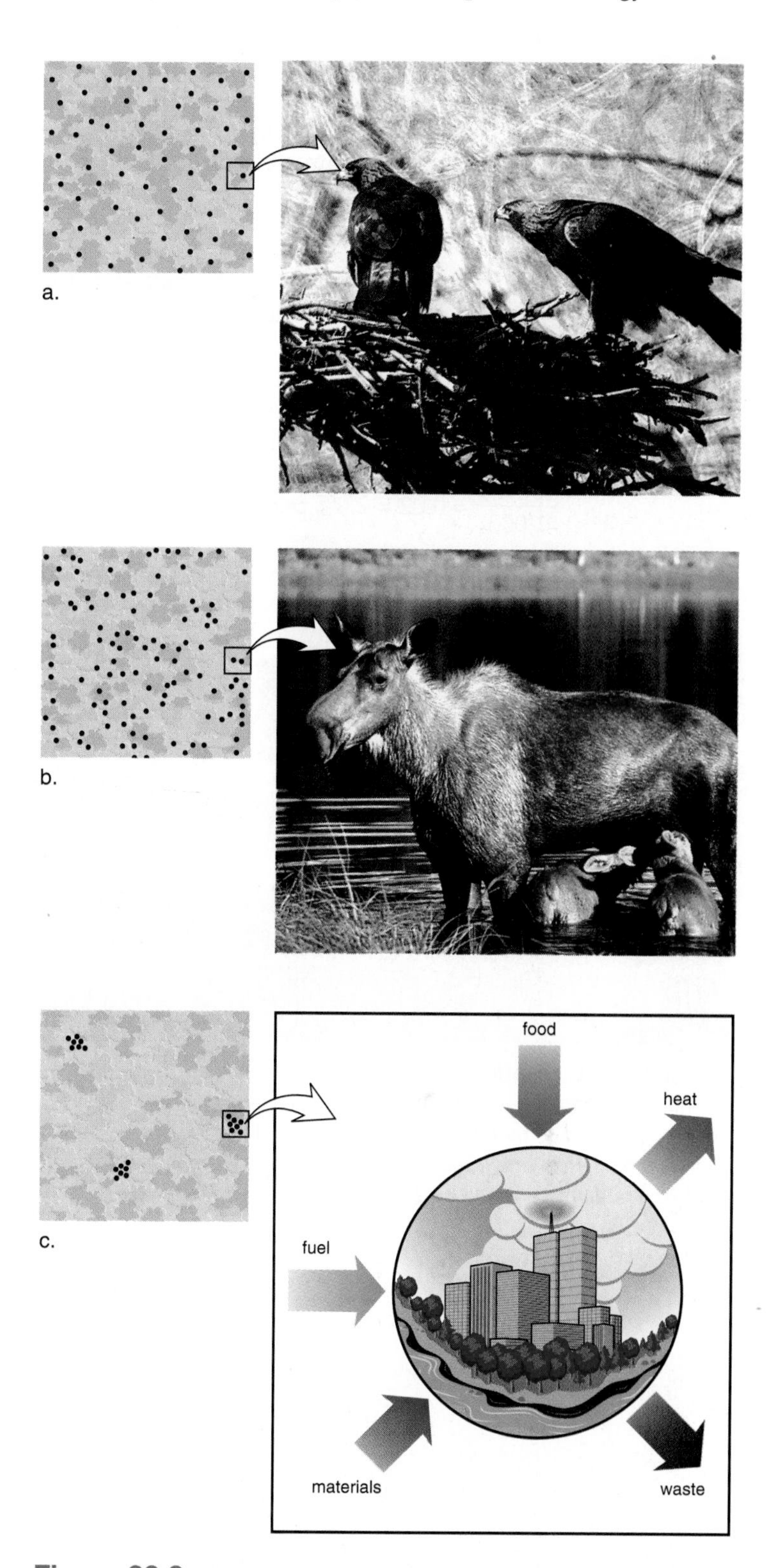

Figure 33.2 Patterns of distribution within a population. Members of a population may be distributed uniformly, randomly, or usually in clumps. **a.** Golden eagle pair distribution is uniform over a suitable habitat area due to the territoriality of the birds. **b.** The distribution of female moose with calves is random over a suitable habitat. **c.** Human beings tend to be clumped in cities where many people take up residence. Cities take resources from and send their waste to surrounding regions.

33.2 Characteristics of Populations

Populations have a particular pattern of growth and survivorship among other possible characteristics.

Patterns of Population Growth

Populations have a certain size and the size can stay the same from year to year, increase, or decrease, according to a *per capita rate of increase.* Suppose, for example, a human population presently has a size of 1,000 individuals, and the birthrate is 30 per year, and the death rate is 10 per year. The per capita rate of increase per year will be:

$$\frac{30 - 10}{1{,}000} = 0.02 = 2.0\% \text{ per year}$$

(Notice that our per capita rate of increase disregarded either immigration or emigration, which for the purpose of our discussion can be assumed to be equivalent.) The highest possible per capita rate of increase for a population is called its **biotic potential** (Fig. 33.3). Whether the biotic potential is high or low depends on such factors as the following:

1. usual number of offspring per reproduction
2. chances of survival until age of reproduction
3. how often each individual reproduces
4. age at which reproduction begins.

Figure 33.3 Biotic potential.
Animal husbandry relies on biotic potential. If a single female pig has her first litter at nine months, and produces two litters a year, each of which contain an average of four females (which in turn reproduced at the same rate), there would be 2,220 pigs by the end of three years.

Suppose we are studying the growth of a population of insects that are capable of infesting and taking over an area. Under these circumstances **exponential growth** is expected. An exponential pattern of population growth results in a J-shaped curve (Fig. 33.4*a*). This pattern of population growth can be likened to compound interest at the bank: as your money increases, the more interest you will get. If the insect population has 2,000 individuals and the per capita rate of increase is 20% per month, then there will be 2,400 insects after one month, 2,880 after two months, and 3,456 after three months, and so forth.

Notice that a J-shaped curve has these phases:

lag phase: during this phase, growth is slow because the population is small.

exponential growth phase: during this phase, growth is accelerating and the population is exhibiting its biotic potential.

Usually, exponential growth cannot continue for long because of environmental resistance. **Environmental resistance** is all those environmental conditions such as a limited supply of food, an accumulation of waste products, increased competition, or predation that prevent populations from achieving their biotic potential. Due to environmental resistance, growth levels off and a pattern of population growth called logistic growth is expected. **Logistic growth** results in an S-shaped growth curve (Fig. 33.4*b*).

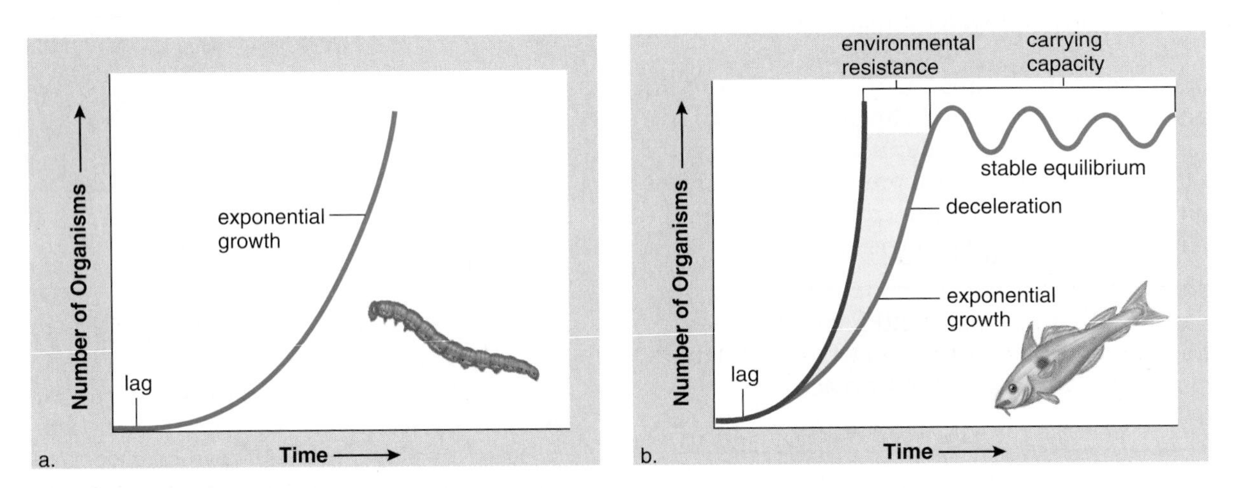

Figure 33.4 Patterns of population growth.
a. Exponential growth results in a J-shaped growth curve because the growth rate is positive. **b.** Logistic growth results in an S-shaped growth curve because environmental resistance causes the population size to level off and be in a steady state.

Notice that an S-shaped curve has these phases:

lag phase: during this phase, growth is slow because the population is small.

exponential growth phase: during this phase, growth is accelerating due to biotic potential.

deceleration phase: during this phase, the rate of population growth slows down.

stable equilibrium phase: during this phase, there is little if any growth because births and deaths are about equal.

The stable equilibrium phase is said to occur at the **carrying capacity** of the environment. The carrying capacity is the number of individuals the environment can normally support.

Our knowledge of logistic growth has practical implications. The model predicts that exponential growth will occur only when population size is much lower than the carrying capacity. So, as a practical matter, if we are using a fish population as a continuous food source, it would be best to maintain the population size in the exponential phase of growth. Biotic potential is having its full effect and the birth rate is the highest it can be during this phase. If we overfish, the population will sink into the lag phase, and it will be years before exponential growth recurs. On the other hand, if we are trying to limit the growth of a pest, it is best to reduce the carrying capacity rather than reduce the population size. Reducing the population size only encourages exponential growth to begin once again. Farmers can reduce the carrying capacity for a pest by alternating rows of different crops rather than growing one type of crop per the entire field.

Exponential growth produces a J-shaped curve because population growth accelerates over time. Logistic growth produces an S-shaped curve because the population size stabilizes when the carrying capacity of the environment has been reached.

Survivorship

Population growth patterns assume that populations are made up of identical individuals. Actually, the individuals are in different stages of their life span. Let us consider how many members of an original group of individuals born at the same time, called a **cohort,** are still alive after certain intervals of time. If we plot the number surviving, a survivorship curve is produced.

For the sake of discussion, three types of idealized survivorship curves are recognized (Fig. 33.5*a*). The type I curve is characteristic of a population like humans in which most individuals survive well past the midpoint, and death does not come until near the end of the life span. On the other hand, the type III curve would be typical for a population of oysters in which most individuals die very young. In the type II curve, survivorship decreases at a constant rate throughout the life span. This has been found typical of a population of song birds.

Sometimes populations do not fit any of these curves exactly. For example, in a cohort of *Poa annua* plants, most individuals survive till six to nine months, and then the chances of survivorship diminish at an increasing rate.

There is much that can be learned about the life history of a species by studying its survivorship curve. Would you predict that most or few members of a population with a type III survivorship curve are contributing offspring to the next generation? Obviously since death comes early for most members, only a few are living long enough to reproduce. What about the other two types of survivorship curves?

Populations have a pattern of survivorship that becomes apparent from studying the survivorship curve of a cohort.

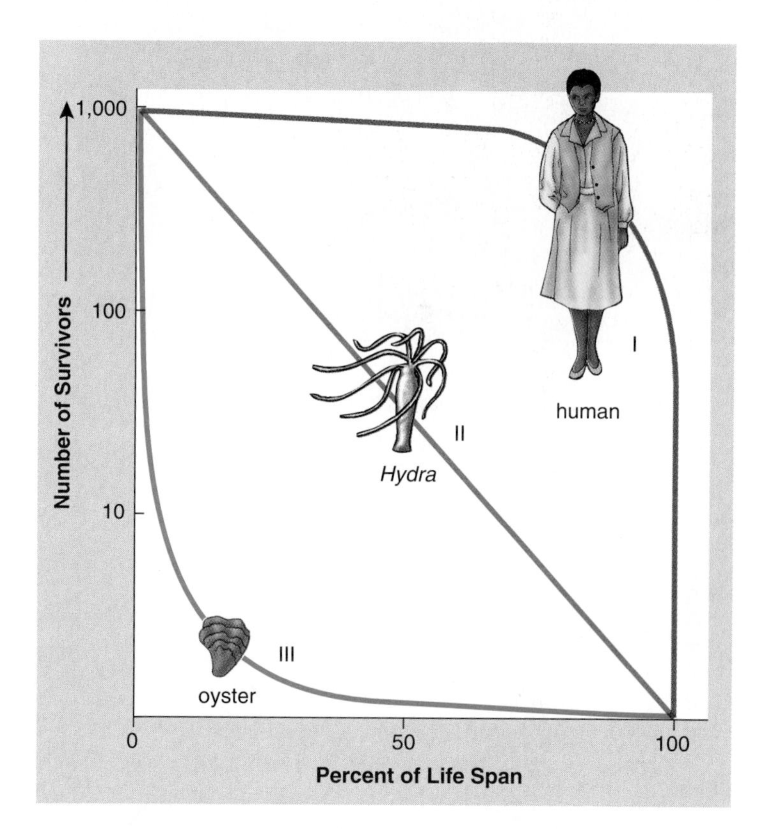

Figure 33.5 Survivorship curves.
Human beings have a type I survivorship curve: the individual usually lives a normal life span and then death is increasingly expected. Hydras have a type II curve in which chances of surviving are the same for any particular age. Oysters have a type III curve: most deaths occur during the free-swimming larva stage, but those that survive to adulthood usually live a normal life span.

33.3 Regulation of Population Size

Ecologists want to determine when and if the two patterns of population growth just discussed occur in nature. They want to know if the factors which regulate population growth are always extrinsic (environmental) or whether there are also intrinsic (based on the anatomy, physiology, or behavior of the organism) factors which regulate population growth.

In one study, 4 male and 21 female reindeer (*Rangifer*) were released on St. Paul Island in the Bering Sea off Alaska in 1911. St. Paul Island had a completely undisturbed environment. The reindeer fed on lichens, which grow slowly and cannot recover quickly from grazing. Normally, reindeer migrate seasonally, giving their food source a chance to regrow, but these reindeer were confined to an island. Also, there was little hunting pressure, and there were no predators. The herd grew exponentially to about 2,000 reindeer; overgrazed the habitat; and then abruptly declined to only 8 animals in 1950. Such a precipitous drop is called a population crash (Fig. 33.6). In this instance it does seem as if environmental factors, such as food source and possibly disease, are regulating population size.

For some time ecologists have recognized that the environment contains both abiotic and biotic components. They suggested that abiotic factors like weather and natural disasters were density-independent. By this they meant that the number of organisms present did not influence the effect of the factor. Fires don't necessarily kill a larger percentage of individuals as the population increases in size (Fig. 33.7). On the other hand, biotic factors like parasitism, competition, and predation were designated as density-dependent. Consider, for example, an area in which there are only 100 holes for crabs to hide in. The greater the number of crabs beyond 100, the better the chance a shorebird will find one and eat it.

In a population study of the great tit, *Parus major,* it was found that the population size fluctuates above and below the carrying capacity. While density-dependent and density-independent factors are involved, the researchers believe that territoriality, an intrinsic factor, also plays a role. Territoriality is apparent when members of a population are spaced out more than would be expected from a random occupation of the area.

> Density-independent and density-dependent factors can often explain the population dynamics of natural populations. Both types of factors are extrinsic to the organism; perhaps intrinsic factors like territoriality also play a role.

a.

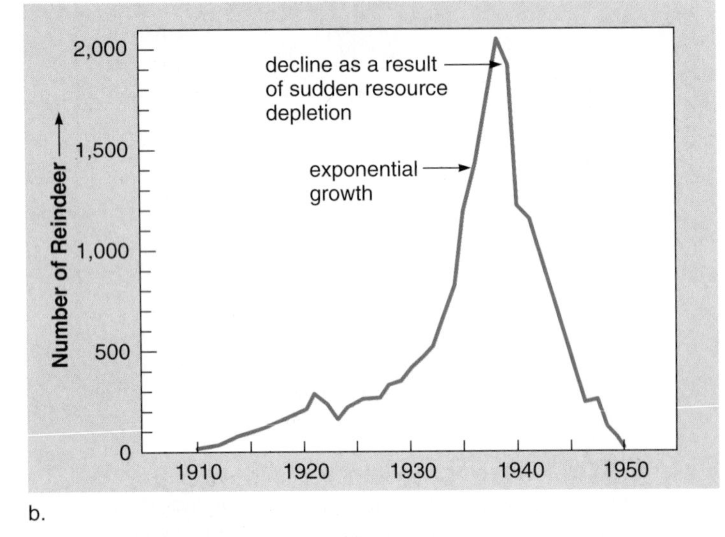

b.

Figure 33.6 Density-dependent effect.
a. A reindeer, *Rangifer.* **b.** On St. Paul Island, Alaska, reindeer grew exponentially for several seasons and then underwent a sharp decline as a result of overgrazing the available range.

Figure 33.7 Density-independent effect.
A fire can start and rage out of control regardless of how many organisms are present.

Science Focus

Distribution of Hard Clams in the Great South Bay

The hard clam is a commercially harvested bivalve mollusk that is found throughout the Great South Bay located on the south shore of Long Island. The clam is eaten raw on the half shell, and is also baked. Hard clams are a necessary ingredient for New England and Manhattan clam chowder. The Great South Bay has often been referred to as a "hard clam factory" because in the 1970s, over half of the hard clams harvested in the United States came from its waters. For the past 20 years, Jeffrey Kassner (Fig. 33A) has been studying the distribution of hard clams in the eastern third of the bay for the Town of Brookhaven.

Knowing the distribution and abundance of hard clams, as well as the responsible environmental factors of this distribution is important to Brookhaven because the hard clam industry has an annual value of $10 million and employs 300 fishermen. Of particular interest to Brookhaven is the possibility of using this information to develop projects, such as utilizing aquaculture technology, that will increase hard clam abundance.

Kassner first took a census of the hard clam population. For several weeks during the summer, a barge-mounted crane with a one square meter clamshell bucket was used to take bottom grabs at 232 stations located throughout the study area. Each bottom sample was placed in a one square meter wire sieve and washed with a high pressure water hose to separate the hard clams from the sediment so that the hard clams could be counted and measured in order to calculate various demographic parameters of the hard clam population. The fieldwork was physically hard, and by the end of the day, the crew of students and biologists were tired, wet, and covered with mud. The results, however, have proven to be well worth the effort.

Working with Dr. Robert Cerrato of the State University of New York, Kassner has drawn a composite census map showing the distribution of and abundance of the clams. He was surprised to find that hard clams are not distributed uniformly throughout the study area, but occur in distinct patches of high and low abundance. A dense assemblage of clams is traditionally referred to as a "clam bed" and six such beds were identified.

Kassner and students found that nearly all of the beds coincided with areas of high shell content sediment associated with "relict" oyster reefs. This observation is of historical note as well as biological interest because up until the early years of this century, the Great South Bay was a major producer of oysters and was the source of the world famous "Blue Point Oyster." Although oysters are no longer found in the Great South Bay because of environmental changes, they left behind a legacy in the sediment that now supports high abundances of hard clams.

The knowledge that hard clam abundance is positively associated with relict oyster reefs might make it possible to transform low-abundance areas into high-abundance areas. Kassner felt he needed a complete sedimentary "portrait" of high- and low-abundance areas. To develop one he borrowed techniques generally associated with other marine science disciplines: from shipwreck hunting, he used a side-scan sonar to map the topography of the bottom; from deep-sea research, he used a ROV (Remotely Operated Vehicle) to photograph the bottom; from commercial fishing, a fathometer to map the bottom; and from pollution studies, a sediment profile camera to photograph the sediment-water interface where hard clams live and feed.

Kassner, with the help of students, is now in the process of putting all this different information together on a single map. Because it will take time and perhaps more studies to develop the portrait, he is concurrently exploring the feasibility of having shells placed on low-abundance areas in order to create new relict oyster reefs. If this strategy works, it would be a tremendous boon to the shellfish industry because nearly three-quarters of the study area is low abundance.

Kassner feels his work is not only scientifically interesting, it is also personally rewarding. Over the years, he has become friends with many fishermen and knows that if he is successful, he will be helping them to continue an occupation that in some cases goes back generations.

Figure 33A
Jeffrey Kassner, with the help of students, is preparing a sedimentary "portrait" of high and low clam abundance areas in Great South Bay, Long Island, New York. This information will be used to increase the yield of clams for local fishermen.

33.4 Life History Patterns

We have already had an opportunity to point out that populations vary on such particulars as their rate of growth and life span. Such particulars are a part of a species' life history pattern, which is based on genetically determined variations that are subject to natural selection.

The logistic growth model has been used to suggest that one possible pattern is an *opportunistic pattern* and the other is an *equilibrium pattern* (Fig. 33.8). These are also called *r*-selected and *K*-selected because in mathematical formulas for population growth, *r* represents the per capita rate of increase and *K* represents the carrying capacity. Members of opportunistic populations are small in size, mature early, and have a short life span. They tend to produce many relatively small offspring and to forego parental care in favor of a greater number of offspring. The more offspring, the more likely some of them will survive a population crash. Because of their short life span and ability to disperse to new locales, density-dependent mechanisms such as predation and competition are unlikely to play a major role in regulating population size and growth rates. Classic examples of such opportunistic species are many insects and weeds.

In contrast, we know there are populations whose size remains pretty much at the carrying capacity. Resources such as food and shelter are relatively scarce for these individuals, and those who are best able to compete will have the largest number of offspring. These organisms allocate energy to their own growth and survival and to the growth and survival of their offspring. Therefore they are fairly large, are slow to mature, and have a fairly long life span. They are specialists rather than colonizers and tend to become extinct when their normal way of life is destroyed. The best possible examples of equilibrium species are found among birds and mammals. The Florida panther is the largest of the animals in the Florida Everglades, requires a very large range, and produces few offspring, which must be cared for. Currently, the Florida panther is unable to compensate for a reduction in its range, and is therefore on the verge of extinction.

Nature is actually more complex than these two possible life history patterns and most populations lie somewhere in between these two extremes. For example a cod is a rather large fish weighing 10–25 pounds and measuring up to 3 feet in length—but the cod releases gametes in vast numbers, the zygotes form in the sea, and the parents make no further investment in developing offspring. Of the 6 to 7 million eggs released by a single female cod, only a few will become adult fish.

Differences in the environment result in different selection pressures and a range of life history characteristics.

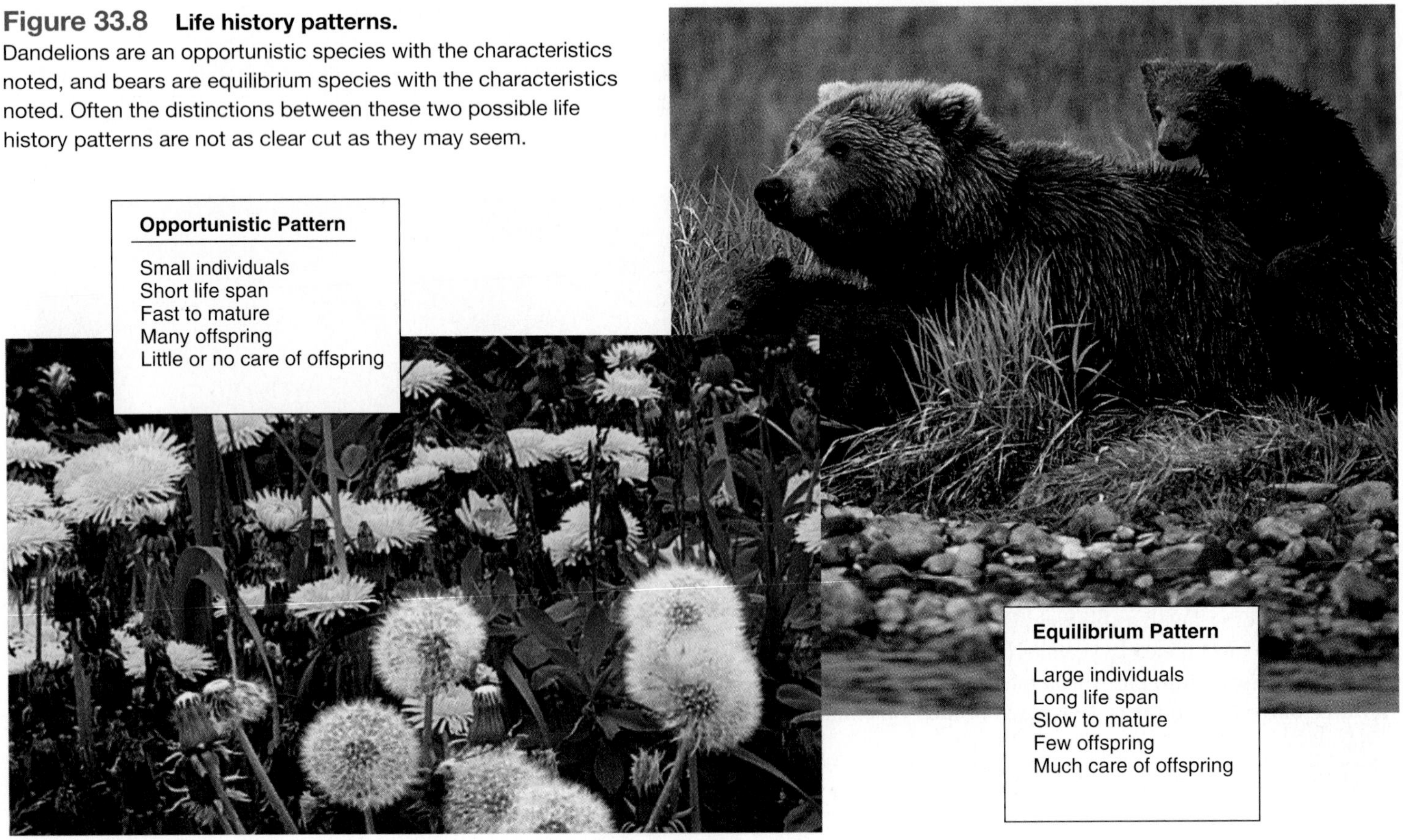

Figure 33.8 Life history patterns.
Dandelions are an opportunistic species with the characteristics noted, and bears are equilibrium species with the characteristics noted. Often the distinctions between these two possible life history patterns are not as clear cut as they may seem.

Opportunistic Pattern

Small individuals
Short life span
Fast to mature
Many offspring
Little or no care of offspring

Equilibrium Pattern

Large individuals
Long life span
Slow to mature
Few offspring
Much care of offspring

33.5 Human Population Growth

The human population has an exponential pattern of growth and a J-shaped growth curve (Fig. 33.9*c*). It is apparent from the position of 1999 on the growth curve in Figure 33.9*a* that growth is still quite rapid. The equivalent of a medium-sized city (200,000) is added to the world's population every day, and 88 million (the equivalent of the combined populations of the United Kingdom, Norway, Ireland, Iceland, Finland, and Denmark) are added every year.

The present situation can be appreciated by considering the doubling time. The **doubling time**—the length of time it takes for the population size to double—is now estimated to be 47 years. Such an increase in population size will put extreme demands on our ability to produce and distribute resources. In 47 years, the world will need double the amount of food, jobs, water, energy, and so on just to maintain the present standard of living.

Many people are gravely concerned that the amount of time needed to add each additional billion persons to the world population has taken less and less time. The first billion didn't occur until 1800; the second billion arrived in 1930; the third billion in 1960, and today there are nearly 6 billion. Only if the per capita rate of increase declines can there be zero population growth, when the birthrate equals the death rate and population size remains steady. The world's population may level off at 8, 10.5, or 14.2 billion, depending on the speed with which the per capita rate of increase declines.

More-Developed Versus Less-Developed Countries

The countries of the world can be divided into two groups. The **more-developed countries (MDCs),** typified by countries in North America and Europe, are those in which population growth is low and the people enjoy a good standard of living (Fig. 33.9*a*). The **less-developed countries (LDCs),** such as countries in Latin America, Africa, and Asia, are those in which population growth is expanding rapidly and the majority of people live in poverty (Fig. 33.9*b*). (Sometimes the term *third-world countries* is used to mean the less-developed countries. This term was introduced by those who thought of the United States and Europe as the first world and the former USSR as the second world.)

The more-developed countries (MDCs) doubled their populations between 1850 and 1950. This was largely due to a decline in the death rate, the development of modern medicine, and improved socioeconomic conditions. The decline in the death rate was followed shortly thereafter by a decline in the birthrate, so that populations in the MDCs experienced only modest growth between 1950 and 1975. This sequence of events (i.e., decreased death rate followed by decreased birthrate) is termed a **demographic transition.**

Yearly growth of the MDCs as a whole has now stabilized at about 0.1%. The populations of a few of the MDCs—Italy, Denmark, Hungary, Sweden—are not growing or are actually decreasing in size. In contrast, there is no leveling

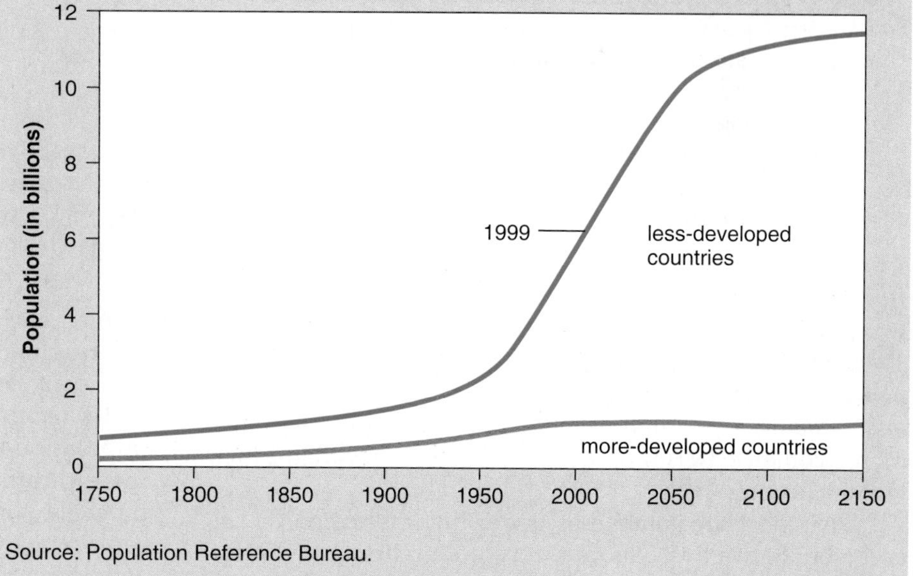

Figure 33.9 World population growth.
People in the **(a)** more-developed countries have a high standard of living and will contribute least, while people in the **(b)** less-developed countries have a low standard of living and will contribute most to the world population growth. **c.** World population growth to 1998 with estimates to 2150.

off and no end in sight to U.S. population growth, as discussed in the reading on page 698. Although yearly growth of the United States is only 0.6%, many people immigrate to the United States each year. In addition, there was an unusually large number of babies born between 1947 and 1964 (called a baby boom). Therefore, a large number of women are still of reproductive age.

Although the death rate began to decline steeply in the LDCs following World War II with the importation of modern medicine from the MDCs, the birthrate remained high. The yearly growth of the LDCs peaked at 2.5% between 1960 and 1965. Since that time, a demographic transition has begun: the decline in the death rate has slowed and the birthrate has fallen. A yearly growth of 1.8% is expected by the end of the century. Still, because of exponential growth, the population of the LDCs may explode from 4.4 billion today to 10.2 billion in 2100. Most of this growth will occur in Africa, Asia, and Latin America. Ways to greatly reduce the expected increase have been suggested:

1. Establish and/or strengthen family planning programs. A decline in growth is seen in countries with good family planning programs supported by community leaders. Currently, 25% of women in the sub-Saharan Africa say they would like to delay or stop childbearing, yet they are not practicing birth control; likewise, 15% of women in Asia and Latin American have an unmet need of birth control.
2. Use social progress to reduce the desire for large families. Many couples in the LDCs presently desire as many as four to six children. But providing education, raising the status of women, and reducing child mortality are desirable social improvements that could cause them to think differently.
3. Delay the onset of childbearing. A delay in the onset of childbearing and wider spacing of births could cause a temporary decline in the birthrate and reduce the present reproductive rate.

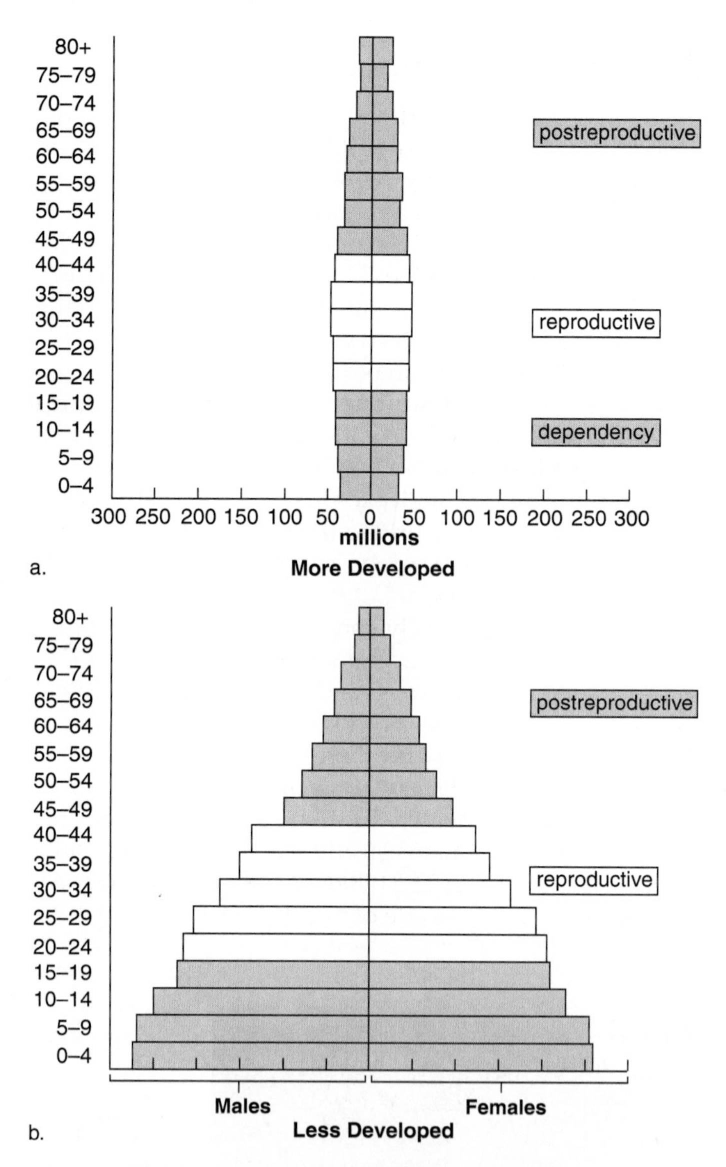

Figure 33.10 Age-structure diagrams (1998). The diagrams illustrate that **(a)** the MDCs are approaching stabilization, whereas **(b)** the LCDs will expand rapidly due to their age distributions. Source: United Nations Population Division, 1998.

Age Distributions

The **age-structure diagrams** of MDCs and LDCs in Figure 33.10 divide the population into three age groups: dependency, reproductive, and postreproductive. The LDCs are experiencing a population momentum because they have more women entering the reproductive years than older women leaving them.

Laypeople are sometimes under the impression that if each couple has two children, **zero population growth** (no increase in population size) will take place immediately. However, **replacement reproduction,** as it is called, will still cause most countries today to continue growing due to the age structure of the population. If there are more young women entering the reproductive years than there are older women leaving them, then replacement reproduction will still result in growth of the population.

Many MDCs have a stable age structure, but most LDCs have a youthful profile—a large proportion of the population is younger than the age of 15. This means that their populations will still expand greatly, even after replacement reproduction is attained. The more quickly replacement reproduction is achieved, however, the sooner zero population growth will result.

Currently, the less-developed countries are expanding dramatically because of exponential growth.

A Sustainable World

While we are sometimes quick to realize that the growing populations of the LDCs are putting a strain on the environment, we should realize that the excessive resource consumption of the MDCs also stresses the environment. Environmental impact is measured not only in terms of population size, but in terms of resource consumption and the pollution caused by each person in the population. An average American family, in terms of per capita resource consumption and waste production, is the equivalent of 30 people in India (Fig. 33.11).

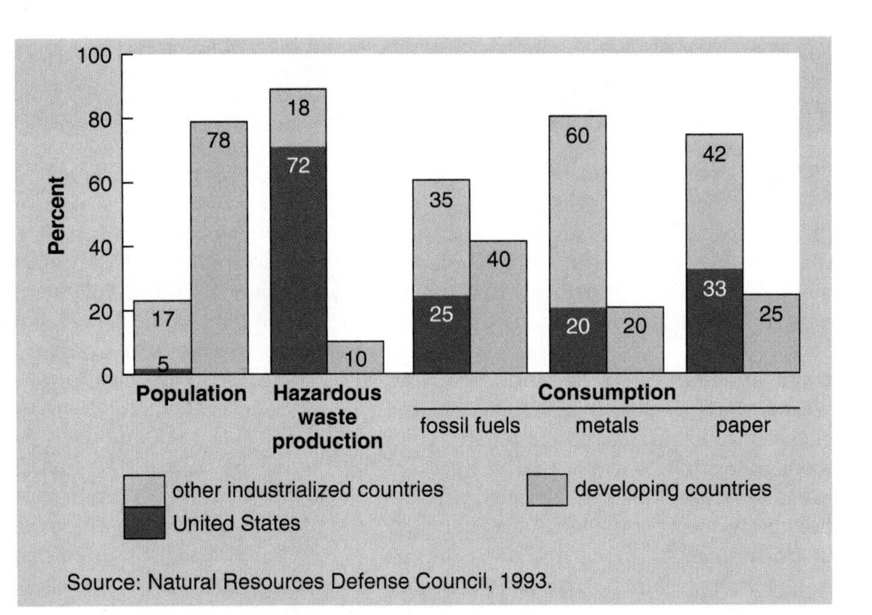

Figure 33.11 Resource consumption for MDCs and LDCs. The populations of MDCs are smaller than LDCs. Yet, the MDCs produce most of the hazardous wastes because their consumption of fossil fuels, metals, and paper, for example, is much greater than the LDCs.

Before the industrial revolution, people felt better connected to the plants and animals on which they depended, and they were better able to live in a sustainable way. After the industrial revolution, we especially began to think of ourselves as separate from nature and endowed with the right to exploit nature as much as possible. But our industrial society lives on *borrowed carrying capacity*—our cities not only borrow resources from the country, our entire population borrows from the past and future. The forests of the Carboniferous period have become the fossil fuels that sustain our way of life today, and the environmental degradation we cause is going to be paid for by our children.

Ecologists have two favorite sayings: (1) Everything is connected to everything else, and (2) there is no free lunch. We have seen that if you affect one part of the carbon cycle, you affect the entire balance of carbon in the entire world. Ecological effects know no boundaries. Coal that is burned in the Midwest releases acids into the atmosphere that affect lakes in the Northeast. And plants and animals aren't the only organisms affected. Humans are dependent on natural cycles just as much as any organism in the biosphere. What we do to natural ecosystems will eventually be felt by us also. The second saying means that we have to pay for what we do. If we build a home on a floodplain, we can expect that it will be flooded once in a while. When we burn fossil fuels, we can expect acid rain and global warming as a consequence. Many times it is difficult to predict the particular consequences, but we can be assured that eventually they will become apparent.

Overpopulation and overconsumption account for increased pollution, and also for the present mass extinction of wildlife. We are expected to lose one-third to two-thirds of the earth's species, any one of which could possibly have made a significant contribution to agriculture or medicine. It should never be said, "What use is this organism?" Aside from its contribution to the ecosystem in which it lives, one never knows how a particular organism might someday be useful to humans. Adult sea urchin skeletons are now used as molds for the production of small artificial blood vessels, and armadillos are used in leprosy research.

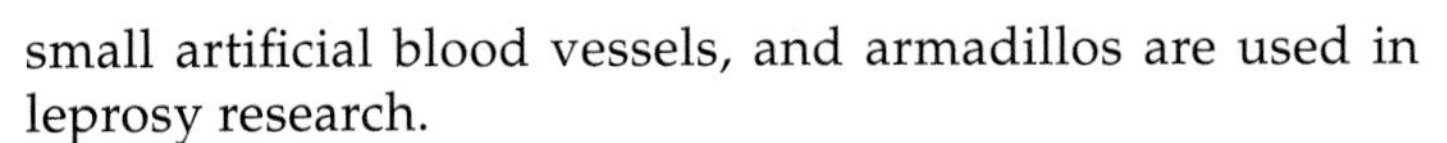

It is clearly time for a new philosophy. In a **sustainable world,** development will meet economic needs of all peoples while protecting the environment for future generations. Various organizations have singled out communities to serve as models of how to balance ecological and economic goals. For example, in Clinch Valley of southwest Virginia, the Nature Conservancy is helping to revive the traditional method of logging with draft horses. This technique, which allows the selective cutting of trees, preserves the forest and prevents soil erosion, which is so damaging to the environment. The United Nations has an established bioreserve system, a global network of sites that combine preservation with research on sustainable management for human welfare. More than 100 countries are now participants in the program. However, sustainability is more than likely incompatible with the kinds of consumption/waste patterns currently practiced in developed countries.

All peoples can benefit from a sustainable world where economic development and environmental preservation are considered complementary, rather than opposing, processes.

Bioethical Issue

The answer to how to curb the expected increase in the world's population lies in discovering how to curb the rapid population growth of the less-developed countries. In these countries, population experts have discovered what they call the "the virtuous cycle." Family planning leads to healthier women, and healthier women have healthier children, and the cycle continues. Women no longer have to have many babies for a few to survive. More education is also helpful because better educated people are more interested in postponing childbearing and promoting women's rights. Women who have equal rights with men tend to have fewer children.

"There isn't any place where women have had the choice that they haven't chosen to have fewer children," says Beverly Winikoff at the Population Council in New York City. "Governments don't need to resort to force." Bangladesh is a case in point. Bangladesh is one of the densest and poorest countries in the world. In 1990 the birthrate was 4.9 children per woman and now it is 3.3. This achievement was due in part to the Dhaka-based Grameen Bank, which loans small amounts of money mostly to destitute women to start a business. The bank discovered that when women start making decisions about their lives, they also start making decisions about the size of their families. Family planning within Grameen families is twice as common as the national average; in fact, those women who get a loan promise to keep their families small! Also helpful has been the network of village clinics that counsel women who want to use contraceptives. The expression "contraceptives are the best contraceptives" refers to the fact that you don't have to wait for social changes to get people to use contraceptives—the two feed back on each other.

Unfortunately, some of the less-developed countries faced with economic crisis have cut back on their family planning programs, and the more-developed countries have not taken up the slack. Indeed, some foreign donors have also cut back on aid—the U.S. by one-third.

Questions

1. Do you think less-developed countries should simply make contraception available, or should more persuasive methods be employed? Explain.
2. Do you think that more-developed countries should be concerned about population growth in the less-developed countries? Why or why not?
3. Are you in favor of foreign aid to help countries develop family planning programs? Why or why not?

Summarizing the Concepts

33.1 Scope of Ecology

Ecology is the study of the interactions of organisms with other organisms and with the physical environment. Ecology encompasses several levels of study: organism, population, community, ecosystem, and finally the biosphere. Population density is simply the number of individuals per unit area or volume. Distribution of these individuals can be uniform, random, or clumped. Most members of a population are clumped as are the members of a human population. Limiting factors such as water, temperature, and availability of organic nutrients often determine a population's distribution.

33.2 Characteristics of Populations

Future population size is dependent upon the per capita rate of increase. The per capita rate of increase is calculated by subtracting the number of deaths from the number of births and dividing by the number of individuals in the population. (Immigration and emigration are usually considered to be equal.) Every population has a biotic potential, the greatest possible per capita rate of increase under ideal circumstances.

Two possible patterns of population growth are considered. Exponential growth results in a J-shaped curve because as the population increases in size so does the expected increase in new members. Most environments restrict growth, and exponential growth cannot continue indefinitely. Under these circumstances logistic growth occurs and an S-shaped growth curve results. When the population reaches carrying capacity, the population stops growing because environmental resistance opposes biotic potential.

Populations tend to have one of three types of survivorship curves, depending on whether most individuals live out the normal life span, die at a constant rate regardless of age, or die early.

33.3 Regulation of Population Size

Population growth is limited by density-independent (e.g., weather) and density-dependent factors (predation, competition, and resource availability). Do some populations have an intrinsic means of regulating population growth as opposed to density-independent and density-dependent factors, which are extrinsic means? Territoriality is given as an example of a possible intrinsic means of regulation.

33.4 Life History Patterns

The logistic growth model has been used to suggest that life history patterns depend on natural selection and vary from those species that are opportunists to those that are in equilibrium with the carrying capacity of the environment. Opportunistic species produce many young within a short period of time and rely on rapid dispersal to new, unoccupied environments. Population size is regulated by density-independent factors. Equilibrium species produce a limited number of young, which they nurture for a long time, and population size is regulated by density-dependent factors.

33.5 Human Population Growth

The human population is expanding exponentially, and it is unknown when the population size will level off. Most of the expected increase will occur in certain LDCs (less-developed countries) of Africa, Asia, and Latin America. Support for family planning, human development, and delayed childbearing could help prevent an expected increase.

Studying the Concepts

1. What are the various levels of ecological study? 688
2. What are three types of distribution patterns for a population? Explain why the human population has a clumped pattern. 689
3. How do you calculate the per capita rate of increase for a population? What is biotic potential? 690
4. What type growth curve indicates that exponential growth is occurring? What are the environmental conditions for exponential growth? 690
5. What type growth curve indicates that biotic potential is being opposed by environmental resistance? What environmental conditions are involved in environmental resistance? 690
6. What is the carrying capacity of an area? 691
7. Describe the three general types of survivorship curves. 691
8. Give examples of extrinsic density-independent and density-dependent factors that regulate population size. 692
9. Give support to the belief that intrinsic factors might regulate population size in some populations. 692
10. Name and give five contrasting characteristics for the two extreme life history patterns. 694
11. Why would you expect the life histories of natural populations to have a mixture of characteristics from these two patterns? 694
12. What type of growth curve presently describes the population growth of the human population? 695
13. Distinguish between MDCs and LDCs. Include a reference to age-structure diagrams. 695–96
14. Explain why the population of LDCs is expected to increase tremendously. What steps could be taken to prevent this from occurring? 696

Testing Yourself

Choose the best answer for each question.

1. Which of these levels of ecological study involves both abiotic and biotic components?
   a. organisms
   b. populations
   c. communities
   d. ecosystem
   e. All of these are correct.
2. When phosphorus is made available to an aquatic community, the algal populations suddenly bloom. This indicates that phosphorus is
   a. a density-dependent regulating factor.
   b. gaseous.
   c. a reproductive factor.
   d. a limiting factor.
   e. All of these are correct.
3. A J-shaped growth curve should be associated with
   a. exponential growth.
   b. biotic potential.
   c. no environmental resistance.
   d. high per capita rate of increase.
   e. All of these are correct.
4. An S-shaped growth curve
   a. occurs when there is no environmental resistance.
   b. includes an exponential growth phase.
   c. occurs if survivorship is short-lived.
   d. occurs in natural populations but not laboratory ones.
   e. All of these are correct.
5. If a population has a type I survivorship curve (most live the entire life span), which of these would you also expect?
   a. a single reproductive event per adult
   b. overlapping generations
   c. sporadic reproductive events
   d. reproduction occurring near the end of the life span
   e. None of these are correct.
6. A pyramid-shaped age distribution means that the
   a. prereproductive group is the largest group.
   b. population will grow for some time in the future.
   c. more young women are entering the reproductive years than older women leaving theirs.
   d. country is more likely an LDC rather than an MDC.
   e. All of these are correct.
7. Which of these is a population-independent regulating factor?
   a. competition
   b. predation
   c. size of population
   d. weather
   e. resource availability
8. Fluctuations in population growth can correlate to changes in
   a. predation.
   b. weather.
   c. resource availability.
   d. population regulating factors.
   e. All of these are correct.
9. An equilibrium life history pattern includes all but
   a. large individuals.
   b. long life span.
   c. individuals slow to mature.
   d. few offspring.
   e. little or no care of offspring.
10. The human population
    a. is undergoing exponential growth.
    b. is not subject to environmental resistance.
    c. fluctuates from year to year.
    d. only grows if emigration occurs.
    e. All of these are correct.
11. Label this S-shaped growth curve.

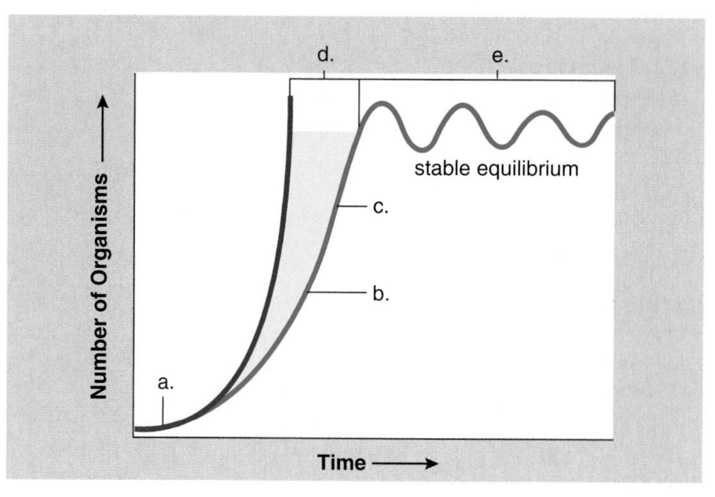

Thinking Scientifically

1. Considering the population of zebras described in the first paragraph:
   a. Plot the growth curve for this zebra population using number of zebras versus time.
   b. How would you describe the shape of your curve?
   c. If the zebra population happens to outstrip the carrying capacity of the environment, what would happen to the curve?
2. Consider this definition of overpopulation: "...where there are more people than can live on the earth in comfort, happiness, and health, and still leave the world a fit place for future generations." [1]
   a. Do comfort and happiness mean the typical standard of living seen in developed countries or in less-developed countries? Should everyone in the world have the same standard of living? Why or why not?
   b. What standard of health is acceptable for the developed countries? Whose responsibility is it to achieve this end?
   c. Should citizens and private industry work to find ways to make the world an ecologically fit place for future generations? Why?
   d. When discussing overpopulation, should we think in terms of the world, the country, or the area?

[1]From George Morris, 1973. *Overpopulation: Everyone's Baby.* London: Priory Press Limited, p. 24.

Understanding the Terms

age-structure diagram 696
biosphere 688
biotic potential 690
carrying capacity 691
cohort 691
community 688
demographic transition 695
doubling time 695
ecology 688
ecosystem 688
environmental resistance 690
exponential growth 690
less-developed country (LDC) 695
logistic growth 690
more-developed country (MDC) 695
population 688
replacement reproduction 696
sustainable world 697
zero population growth 696

Match the terms to these definitions:

a. __________ Group of organisms of the same species occupying a certain area.
b. __________ Maximum population growth rate under ideal conditions.
c. __________ Growth, particularly of a population, in which the increase occurs in the same manner as compound interest.
d. __________ Due to industrialization, a decline in the birthrate following a reduction in the death rate so that the population growth rate is lowered.
e. __________ Largest number of organisms of a particular species that can be maintained indefinitely by a given environment.

Using Technology

Your study of population ecology is supported by these available technologies:

Essential Study Partner CD-ROM
Ecology → Populations
→ Human Impact

Visit the Mader web site for related ESP activities.

Exploring the Internet

The Mader Home Page provides resources and tools as you study this chapter.

http://www.mhhe.com/biosci/genbio/mader

C H A P T E R 34

Community Ecology

Chapter Concepts

Monarch butterflies, Danus plexipus, *are brightly colored. This serves as a warning to predators that they are to be avoided. Monarch butterflies are poisonous, and after a bird has experienced one, it leaves the others alone.*

Step into a forest and look around. You will see animals interacting with both plants and other animals. Insects may be feeding on the leaves of trees and grasses or visiting flowers. If you are lucky, you may see a hawk dart from a tree on the edge of the forest and grab a rodent running through an adjoining meadow. All the members of a particular species living in a location like a forest make up a population, and all the populations interacting there form a **community.**

This chapter examines the various types of community interactions and their importance to the structure of a community. Such interactions illustrate some of the most important selection pressures impinging on individuals. They also help us understand how biodiversity can be preserved.

34.1 Community Composition and Diversity

It is sometimes difficult to decide where one community ends and another begins. A fallen log, for example, can be considered a community. The fungi of decay break down the log and provide food (organic nutrients) for the various invertebrates living in the log that may feed on one another. Yet it is quite possible that bugs and worms living in the log could be eaten by a passing bird that flies about the entire forest. This interaction means that the log is a part of the forest community. And where does the forest end? Doesn't it gradually fade into the surrounding area? So the demarcation of any community turns out to be somewhat arbitrary.

Two characteristics of communities—their composition and diversity—allow us to compare communities. The *composition* of a community is simply a listing of the various species in the community. The diversity includes both species richness (the number of species) and evenness (the relative abundance of individuals of different species).

Just glancing at Figure 34.1 makes it easy for us to see that a coniferous forest has a different composition from a tropical rain forest. Pictorially we can see that narrow-leaved evergreen trees are present in a coniferous forest, and broad-leaved evergreen trees are numerous in a tropical rain forest. Mammals also differ between the two communities, as those listed demonstrate.

Diversity of a community goes beyond composition because it includes the number of species and also the abundance of each species. To take an extreme example: a deciduous forest in West Virginia has, among other species, 76 yellow poplar trees but only one American elm. If we walked through this forest, we might miss seeing the American elm. If, instead, the forest had 36 poplar trees and 41 American elms, the forest would seem more diverse to us and indeed would be more diverse. The greater the diversity, the greater the number and the more even the distribution of species. Ecologists can determine diversity by counting the number and abundance of each type of species in the community.

Figure 34.1 Community structure.
Communities differ in their composition, as witnessed by their predominant plants and animals. Diversity of communities is described by the richness of species and their relative abundance. **a.** A coniferous forest. Some mammals found here are listed to the *left*. **b.** A tropical rain forest. Some mammals found here are listed to the *right*.

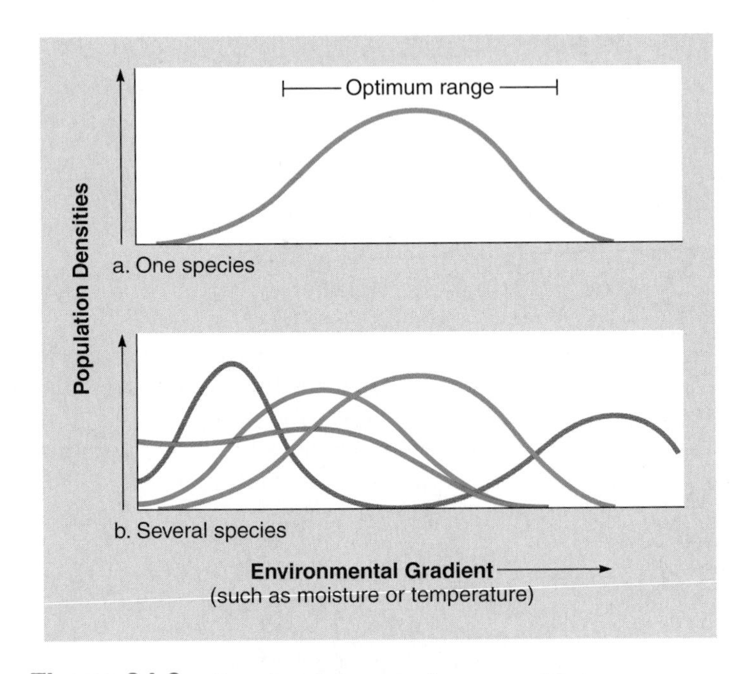

Figure 34.2 Species richness of communities.
According to the individualistic hypothesis **(a)** each species is distributed along environmental gradients according to its own tolerance for abiotic factors, and **(b)** a community is believed to be an assemblage of species that happen to occupy the same area because of similar tolerances.

Why do species assemble together in the same place at the same time? For many years, most ecologists supported the *interactive model* of community structure. According to this model, a community is the highest level of organization arising from cell to tissue, to organism, to populations, and finally, to a community. Just as the parts of an organism are dependent on one another, so species are dependent on biotic interactions such as their food source. Further, like an organism, a community remains stable because of homeostatic mechanisms. This theory predicts that the same species will recur in communities whose boundaries are distinct from one another.

The *individualistic model* of community structure instead hypothesizes that species assemble according to their tolerance for abiotic factors. The number of species in terrestrial communities increases as we move from the northern latitudes to the equator most likely because the weather at the equator is warmer and there is more precipitation. A species' range is based on its tolerance for such abiotic factors as temperature, light, water availability, salinity, and so forth (Fig. 34.2*a*). It's possible that species may assemble because their tolerance ranges simply overlap (Fig. 34.2*b*). The individualistic model predicts that species will have independent distributions, and that the boundaries between communities will not be distinct from one another.

Community composition is probably dependent on both biotic interactions (e.g., organic food source) and abiotic gradients (e.g., climate, inorganic nutrients).

Habitat and Ecological Niche

Each species occupies a particular position in the community, both in a spatial sense (where it lives) and in a functional sense (what role it plays). Its **habitat** is where an organism lives and reproduces in its environment. The habitat of an organism might be the forest floor, a swift stream, or the ocean's edge. The **ecological niche** of an organism is the role it plays in its community, including its habitat and its interactions with other organisms. The niche includes the resources an organism uses to meet its energy, nutrient, and survival demands. For a backswimmer, home is a pond or lake where it eats other insects. The pond must contain vegetation where the backswimmer can hide from its predators such as fish and birds. On the other hand, the water must be clear enough for the backswimmer to see its prey and warm enough for it to be in active pursuit. Since it's difficult to study the total niche of an organism, some observations focus on a certain aspect of an organism's niche, as with the bird featured in Figure 34.3.

An organism's niche is affected by extenuating circumstances; therefore, ecologists distinguish between the fundamental and realized niches. An organism's *fundamental niche* comprises all conditions under which an organism can potentially survive and reproduce; the *realized niche* is the set of conditions under which it actually exists in nature.

Habitat is where an organism lives, and ecological niche is the role an organism plays in its community, including its habitat and its interactions with other organisms.

Figure 34.3 Aspects of niche.
A description of an organism's niche includes its resource requirements and its activities within the community.

34.2 Competition

Competition occurs when members of different species try to utilize a resource (like light, space, or nutrients) that is in limited supply. According to the **competitive exclusion principle,** no two species can occupy the same niche at the same time, if resources are limiting. While it may seem as if several species living in the same area are occupying the same niche, it is usually possible to find slight differences. For example, the various species of monkeys in Figure 34.4 have no difficulty living in close proximity because they have different, although sometimes overlapping, habitats and food requirements.

So, for example, notice that even though the red colobus

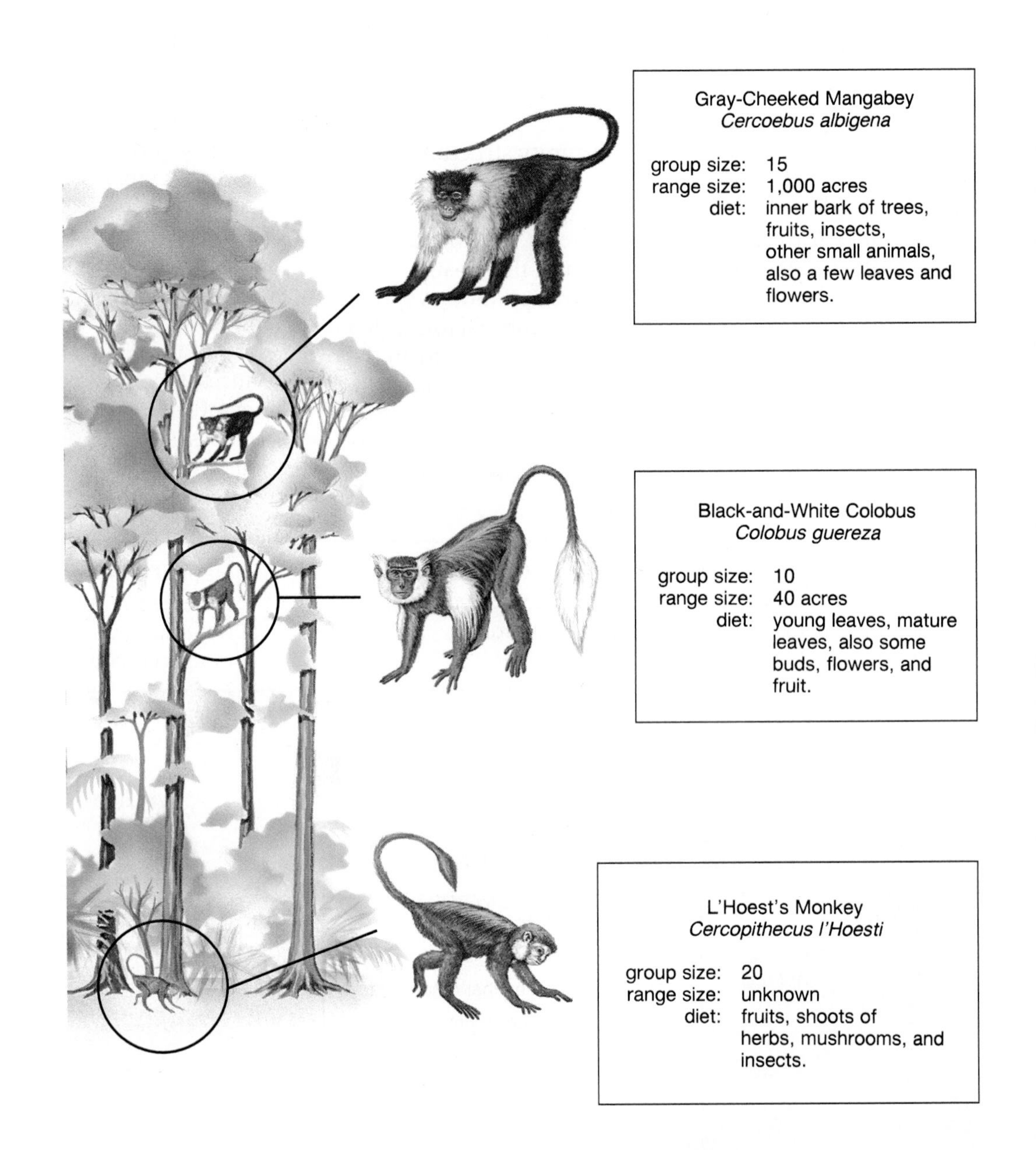

Figure 34.4 Diversity of monkey species in a tropical rain forest.
All of these monkey species can coexist in a tropical rain forest because they have different niches. Each prefers to live at a different height above ground, and each feeds on slightly different foods.

monkey and the gray-cheeked mangabey monkey both prefer the upper canopy of a tropical rain forest, the first primarily feeds on young leaves and flower buds while the second primarily feeds on the inner bark of trees. The ranges of these monkeys also differ. The red colobus monkey normally occupies only 90 acres, while the gray-cheeked mangabey occupies as many as 1,000 acres. The blue monkey and the black and white colobus monkey both prefer the mid-canopy area, but the first primarily feeds on fruit and small insects while the second primarily feeds on leaves. Similarly, the redtail monkey and L'Hoest's monkey live in the lower canopy, but the first primarily feeds on small insects and fruit while the second feeds primarily on fruits and shoots of herbs.

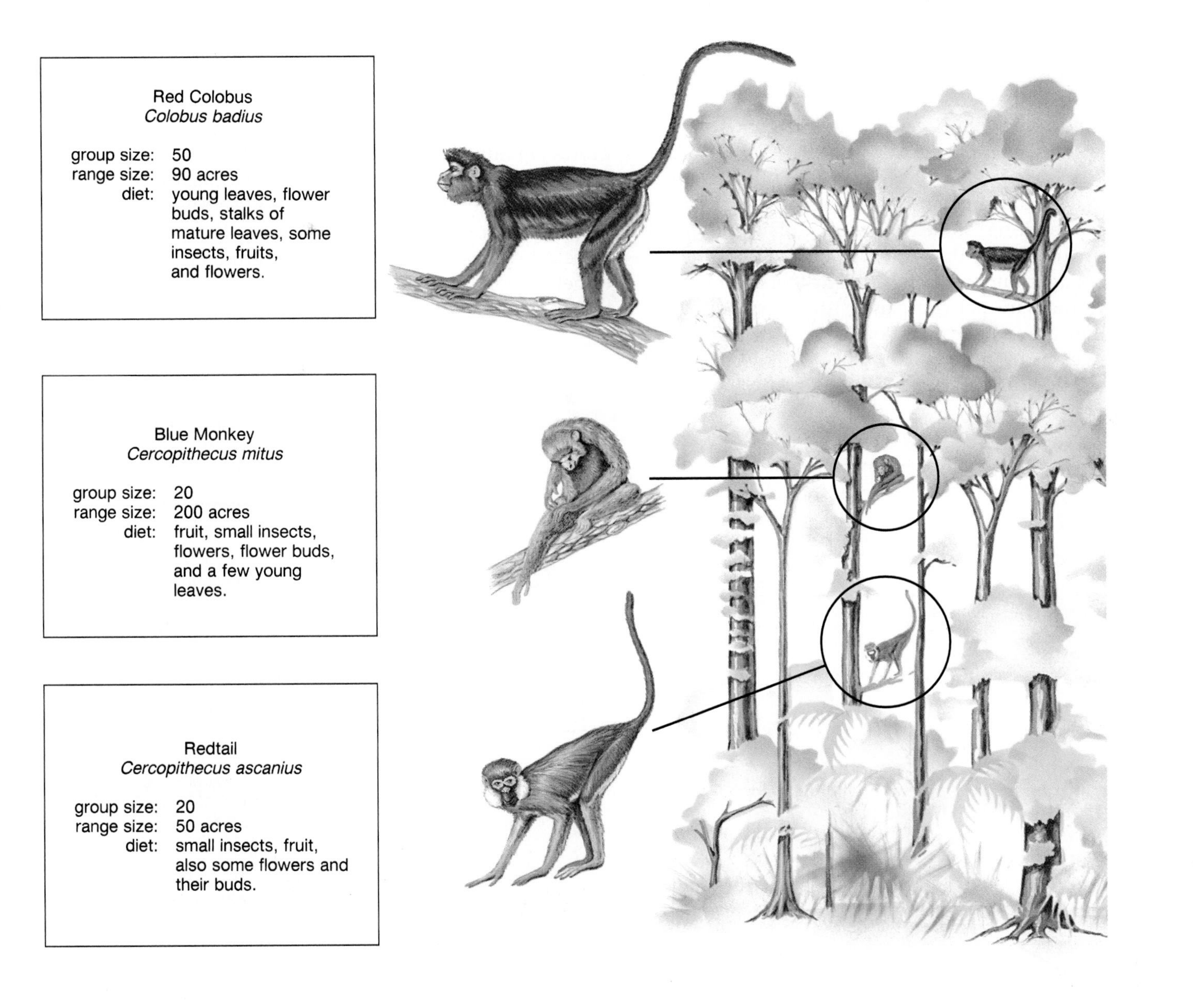

The competitive exclusion principle was developed following experimentation with paramecia. When two species of paramecia were grown separately, each survived; but when they were grown in one test tube, resources were limited and only one species survived (Fig. 34.5). What does it take to have different ecological niches so that extinction of one species is avoided? In another laboratory experiment, two species of paramecia did continue to occupy the same tube when one species fed on bacteria at the bottom of the tube and the other fed on bacteria suspended in solution. Under these circumstances, **resource partitioning** decreased competition between the two species. What could have been one niche became two niches because of a divergence of behavior in this case. The monkeys illustrated in Figure 34.4 are an example of resource partitioning. As another example, consider that swallows, swifts, and martins all eat flying insects and parachuting spiders. These birds even frequently fly in mixed flocks. But each type of bird has different nesting sites and migrates at a slightly different time of year. Therefore they are not competing for the same food source when they are feeding their young.

In all these cases of niche partitioning, we have merely supposed that what we observe today is due to competition in the past. Some ecologists are fond of saying that in doing so we have invoked the "ghosts of competition past." But there are instances in which competition has actually been observed. On the Scottish coast, a small barnacle (*Chthamalus stellatus*) lives on the high part of the intertidal zone, and a large barnacle (*Balanus balanoides*) lives on the lower part (Fig. 34.6). Free-swimming larvae of both species attach themselves to rocks at any point in the intertidal zone, where they develop into the sessile adult forms. In the lower zone, the large *Balanus* barnacles seem to either force the smaller *Chthamalus* individuals off the rocks or grow over them. If the larger barnacle is removed, the smaller barnacle grows equally well on all parts of the rock. The entire intertidal zone is the fundamental niche for *Chthamalus*, but competition is restricting the range of *Chthamalus* on the rocks. *Chthamalus* is more resistant to drying out than is *Balanus;* therefore, it has an advantage that permits it to grow in the upper intertidal zone. The upper intertidal zone becomes the realized niche for *Chthamalus*.

Competition may result in resource partitioning. When similar species seem to be occupying the same ecological niche, it is usually possible to find differences that indicate resource partitioning has occurred.

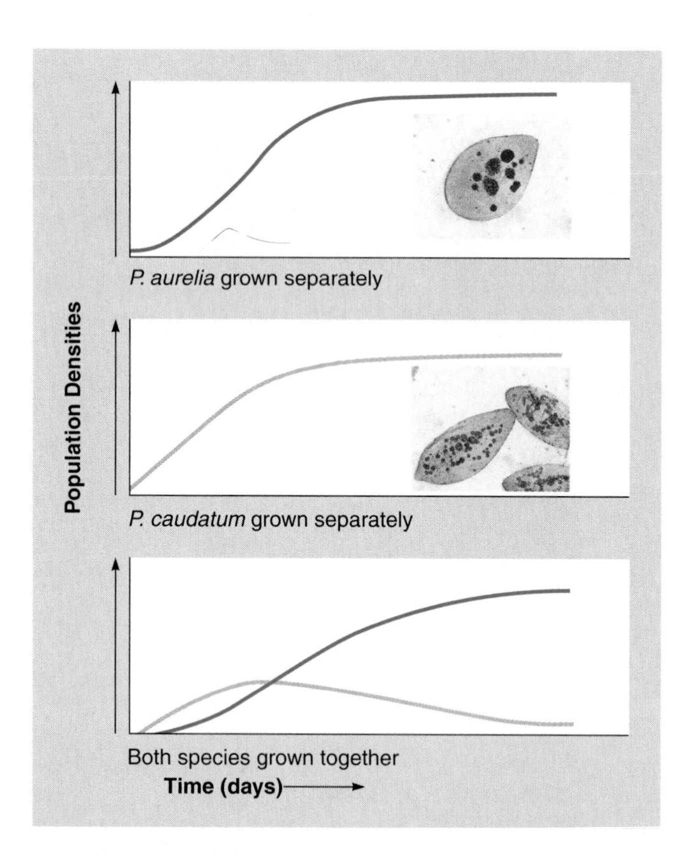

Figure 34.5 Competition between two laboratory populations of *Paramecium*.
When grown alone in pure culture, *Paramecium caudatum* and *Paramecium aurelia* exhibit sigmoidal growth. When the two species are grown together in mixed culture, *P. aurelia* is the better competitor, and *P. caudatum* dies out. Source: Data from G. F. Gause, *The Struggle for Existence*, 1934, Williams & Wilkins Company, Baltimore, MD.

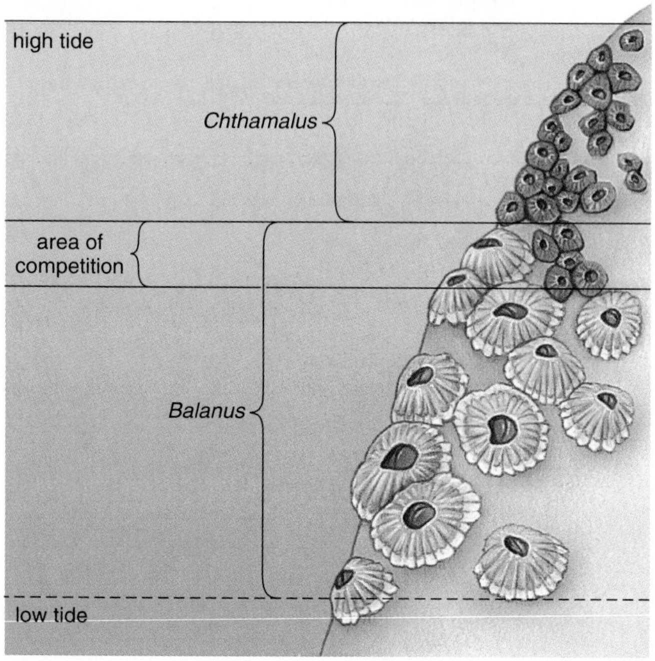

Figure 34.6 Competition between two species of barnacles.
Competition prevents two species of barnacles from occupying as much of the intertidal zone as possible. Both exist in the area of competition between *Chthamalus* and *Balanus*. Above this area only *Chthamalus* survives, and below it only *Balanus* survives.

34.3 Predation

Predation occurs when one living organism, called the **predator,** feeds on another, called the **prey.** In the broadest sense, predaceous consumers include not only animals like lions that kill zebras, but also filter-feeding blue whales that strain krill from ocean waters, parasitic ticks that suck blood from victims, and even herbivorous deer that browse on trees and bushes.

Predator-Prey Population Dynamics

Do predators reduce the population density of prey? On the face of it, you would probably hypothesize that they do, and your hypothesis would certainly be supported by a laboratory study in which the protozoan *Paramecium caudatum* (prey) and *Didinium nasutum* (predator) were grown together in a culture medium. *Didinium* ate all the *Paramecium* and then died of starvation. In nature, we can find a similar example. When a gardener brought prickly-pear cactus to Australia from South America, the cactus spread out of control until millions of acres were covered with nothing but cacti. The cacti were brought under control when a moth from South America, whose caterpillar feeds only on the cactus, was introduced. Now both cactus and moth are found at greatly reduced densities in Australia.

There are mathematical formulas that predict a cycling of predator and prey populations instead of a steady state. Cycling could occur if (1) the predator population overkills the prey and then the predator population also declines in number and (2) if the prey population overshoots the carrying capacity and suffers a crash and then the predator population follows suit because of a lack of food. In either case, the result would be a series of peaks and valleys with the predator population lagging slightly behind the prey.

A famous case of predator/prey cycles occurs between the snowshoe hare and the Canadian lynx, a type of small cat (see Fig. 34.7). The snowshoe hare is a common herbivore in the coniferous forests of North America, where it feeds on terminal twigs of various shrubs and small trees. The Canadian lynx feeds on snowshoe hares but also on ruffed grouse and spruce grouse, two types of birds. Investigators at first assumed that the first explanation given above was sufficient to explain the cycling between the hare and lynx populations. In other words, the lynx had brought about the decline of the hare population. But others noted that the decline in snowshoe hare abundance was accompanied by low growth and reproductive rates that could be signs of a food shortage. It appears that both explanations apply to the data in Figure 34.7. In other words, both a predator-hare cycle and a hare-food cycle have combined to produce an overall effect, which is observed in Figure 34.7. It's interesting to note that the population densities of the grouse populations also cycle, perhaps because the lynx switches to this food source when the hare population declines. Predators and prey do not normally exist as simple, two-species systems, and therefore abundance patterns should be viewed with the complete community in mind.

Interactions between predator/prey and between prey and its own food source can interact to produce complex cycles.

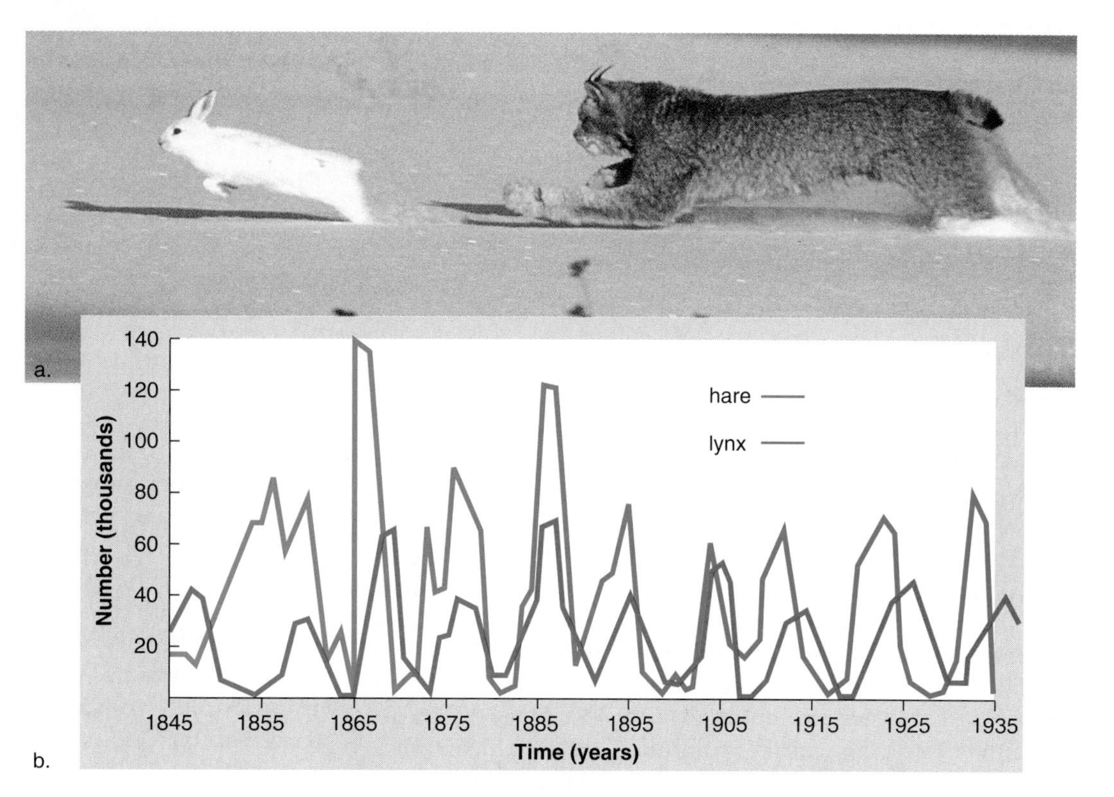

Figure 34.7 Predator-prey interaction between a lynx and a snowshoe hare. **a.** A Canadian lynx (*Lynx canadensis*) is a solitary predator. A long, strong forelimb with sharp claws grabs its main prey, the snowshoe hare (*Lepus americanus*). **b.** The number of pelts received yearly by the Hudson Bay Company for almost 100 years shows a pattern of ten-year cycles in population densities. The snowshoe hare population reaches a peak abundance before that of the lynx by a year or more.

Figure 34.8 Antipredator defenses.
a. Concealment. Flounders can take on the same coloration as their background. **b.** Fright. The South American lantern fly has a large false head that resembles that of an alligator. This may frighten a predator into thinking it is facing a dangerous animal **c.** Warning coloration. The skin secretions of dart-poison frogs are so poisonous that they were used by natives to make their arrows instant lethal weapons. The coloration of these frogs warns others, like birds, to beware.

Prey Defenses

While predators have evolved strategies to secure the maximum amount of food with minimal expenditure of energy, prey organisms have evolved strategies to escape predation. **Coevolution** is present when two species adapt in response to selective pressure imposed by the other.

In plants, the sharp spines of the cactus, the pointed leaves of holly, and the tough, leathery leaves of the oak tree all discourage predation by insects. Plants even produce poisonous chemicals—some are hormone analogues that interfere with the development of insect larvae. Animals have varied antipredator defenses. Some of the more effective defenses are concealment, fright, and warning coloration (Fig. 34.8).

The ecology reading on page 709 tells what happens when exotic (foreign) species are introduced into a community. Prey species are devastated because they have no adequate defense mechanism.

Figure 34.9 Mimicry.
Flies of the family Syrphidae are called flower flies because they are likely to be found on flowers, where they drink nectar and eat pollen. Some species mimic a wasp, which is protected from predation by its sting.

Mimicry

Mimicry occurs when one species resembles another that possesses an antipredator defense. Mimicry can help a predator capture food or it can help a prey avoid capture. For example, snapping turtles have tongues and angler fishes have lures that resemble worms for the purpose of bringing fish within reach. To avoid capture, there are inchworms that resemble twigs and caterpillars that can transform themselves into shapes that resemble snakes.

Batesian mimicry (named for Henry Bates, who discovered it) occurs when a prey mimics another species that has a successful antipredator defense. Many examples of Batesian mimicry involve warning coloration. Among flies of the family Syrphidae, which feed on the nectar and pollen of flowers, one species resembles the wasp *Vespula arenaria* so closely that it is difficult to tell them apart (Fig. 34.9). Once a predator experiences the defense of the wasp, it remembers the coloration and avoids all animals that look similar. There are also examples of species that have the same defense and resemble each other. For example, many coral snake species have brilliant red, black, and yellow body rings. And the stinging insects—bees, wasps, and hornets—all have the familiar black and yellow color bands. Mimics that share the same protective defense are called Müllerian mimics after Fritz Müller, who discovered this form of mimicry.

Just as with other prey defenses, behavior plays a role in mimicry. Mimicry works better if the mimic acts like the model. For example, beetles that resemble a wasp actively fly from place to place and spend most of their time in the same habitat as the wasp model. Their behavior makes them resemble a wasp to an even greater degree.

Prey escape predation by utilizing camouflage, fright, flocking together, warning coloration, and mimicry.

Ecology Focus

Exotic Species Wreak Havoc

While some foreign species live peaceably amid their new neighbors, many threaten entire economies and ecosystems. The brown tree snake slipped onto Guam from southwestern Pacific islands in the late 1940s. Since then, it has wiped out 9 of 11 native bird species, leaving the forests eerily quiet. . . . Indeed, some conservationists now rank invasive species among the top menaces to endangered species. "We're losing more habitat here to pests than to bulldozers," says Alan Holt of the Nature Conservancy of Hawaii. . . . As a remote archipelago, Hawaii was particularly vulnerable to ecological disruption when Polynesian voyagers arrived 1,600 years ago. Having evolved in isolation for millions of years, many native species had discarded evolutionary adaptations that deter predators. The islands abounded with snails with no shells, plants with no thorns, and birds that nested on the ground. Polynesian hunters promptly wiped out several species of large, flightless birds, but a stowaway in their canoes also did serious damage. The Polynesian rat flourished, decimating dozens of species of ground-nesting birds. The first Europeans and their many plant and animal companions unleashed an even larger wave of extinctions. In 1778, for instance, Capt. James Cook brought ashore goats, which soon went feral, devastating native plants. . . .

Biologists Art Medeiros and Lloyd Loope of Haleakala National Park on Maui often feel they are fighting an endless ground war. The park is home to a legion of endangered plants and animals found nowhere else in the world. Six years ago, officials completed a 50-mile, $2.4 million fence to keep out feral goats and pigs. But just as the forest understory was beginning to recover, rabbits released into the park by a bored pet owner launched their own assault. Staffers got to work with rifle and snare, but soon after they'd bagged the last of the rabbits, axis deer—miniature elk from India—began hopping the fence.

Those are just the warmblooded invaders. Medeiros and Loope also are developing chemical weapons for their, thus far, losing battle against the Argentine ant. This tiny terminator threatens to wipe out the park's native insects and the rare native plants that depend on the insects for pollination. "It's an eraser," says Loope. "Shake down a flowering bush inside ant territory and you'll get five species [of native insects]; outside their range, you'll get 10 times that."

Another potential "eraser" at the park gates is the Jackson's chameleon, a colorful Kenyan native that dines on insects and snails. Then there's the dreaded miconia. Since it was introduced to Tahiti as an ornamental in 1937, the tree that locals call the "green cancer" has overrun more than half the island. Its dense foliage shades out other plants—from competing trees to the mosses that anchor soils and hold rainwater. As a result, many plants have been pushed to the brink of extinction, and the mountainsides are eroding, silting over coral reefs that help sustain fisheries. . .

. . . Siccing pests on pests poses its own problems. Some recruits have run amok, doing as much ecological damage as the pests they were meant to control. In the 1880s, for instance, Hawaiian sugar-cane growers brought in mongooses to prune mice and rat populations (Fig. 34A). Prune they did, but they also preyed heavily on native birds. Happily, biocontrol efforts have been more successful in recent years. Nearly 90% of the agents released in the past two decades have been known to attack only the target pest, according to a study conducted jointly by researchers at the Hawaii Department of Agriculture and the University of Hawaii.

By all accounts, preventing invaders from gaining a foothold in the first place is an even better strategy. A recent report by Chris Bright of the Washington D.C.-based Worldwatch Institute recommends, among other measures, that importers be made liable for damages caused by the exotics they introduce and that emergency-response teams be established to jump on new infestations. . . .

Figure 34A Mongoose catching a bird.
Mongooses were introduced to Hawaii in the last century to control rats, but they also prey on native birds.

Table 34.1 Symbiosis

	Species 1	Species 2
Parasitism*	Benefited	Harmed
Commensalism	Benefited	No effect
Mutualism	Benefited	Benefited

*Can be considered a type of predation.

34.4 Symbiosis

Symbiosis refers to interactions in which there is a close relationship between members of two populations.

Parasitism

Parasitism is a symbiotic relationship in which the *parasite* derives nourishment from another called the *host.* Therefore, the parasite benefits and the host is harmed (Table 34.1). Parasites occur in all kingdoms of life. Bacteria (e.g., strep infection), protists (e.g., malaria), fungi (e.g., rusts and smuts), plants (e.g., mistletoe), and animals (e.g., leeches) all contain parasitic members. The effects of parasites on the health of the host can range from a slight weakening effect to actually killing them over time.

In addition to providing nourishment, host organisms also provide their parasites with a place to live and reproduce, as well as a mechanism for dispersing offspring to new hosts. Many parasites have both a primary and secondary host. The secondary host may be a vector that transmits the parasite to the next primary host. As an example, we will consider the deer tick called *Ixodes dammini* in the eastern and *I. ricinus* in the western United States. Deer ticks are arthropods that go through a number of stages (egg, larva, nymph, adult). They are so named because adults feed and mate on white-tailed deer in the fall. The female lays her eggs on the ground, and when the eggs hatch in the spring they become larvae that feed primarily on white-footed mice. If a mouse is infected with the bacterium *Borrelia burgdorfei,* the larvae become infected also. The fed larvae overwinter and molt the next spring to become nymphs that can, by chance, take a blood meal from a human. It is at this time that the tick may pass the bacterium on to a human who subsequently comes down with Lyme disease characterized by arthritic-like symptoms. The fed nymphs develop into adults and the cycle begins again (Fig. 34.10).

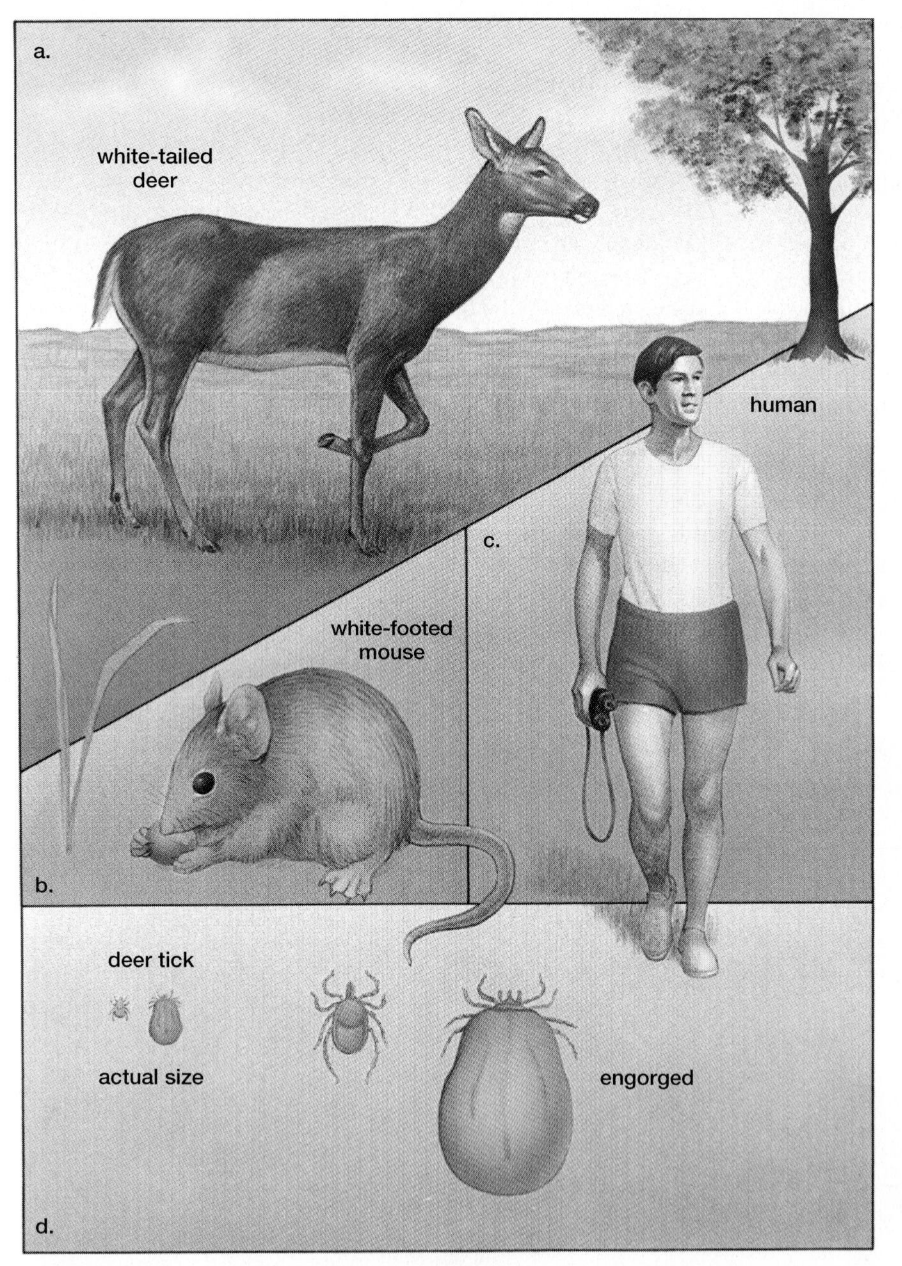

Figure 34.10 The life cycle of a deer tick.
a. Adult ticks feed and mate on white-tailed deer, which accounts for the common name of the ticks. Female adults lay their eggs in soil and then die. **b.** In the spring and summer, larvae feed mainly on white-footed mice. They then overwinter. **c.** During the next summer, nymphs feed on white-footed mice or other animals, including humans. If a nymph infected with the bacterium *Borrelia burgdorfei* feeds on a human, the human gets Lyme disease. **d.** Deer tick before feeding and after feeding. Actual size is shown along with enlarged size.

Commensalism

Commensalism is a symbiotic relationship between two species in which one species is benefited and the other is neither benefited nor harmed. Often one species provides a home and/or trans-

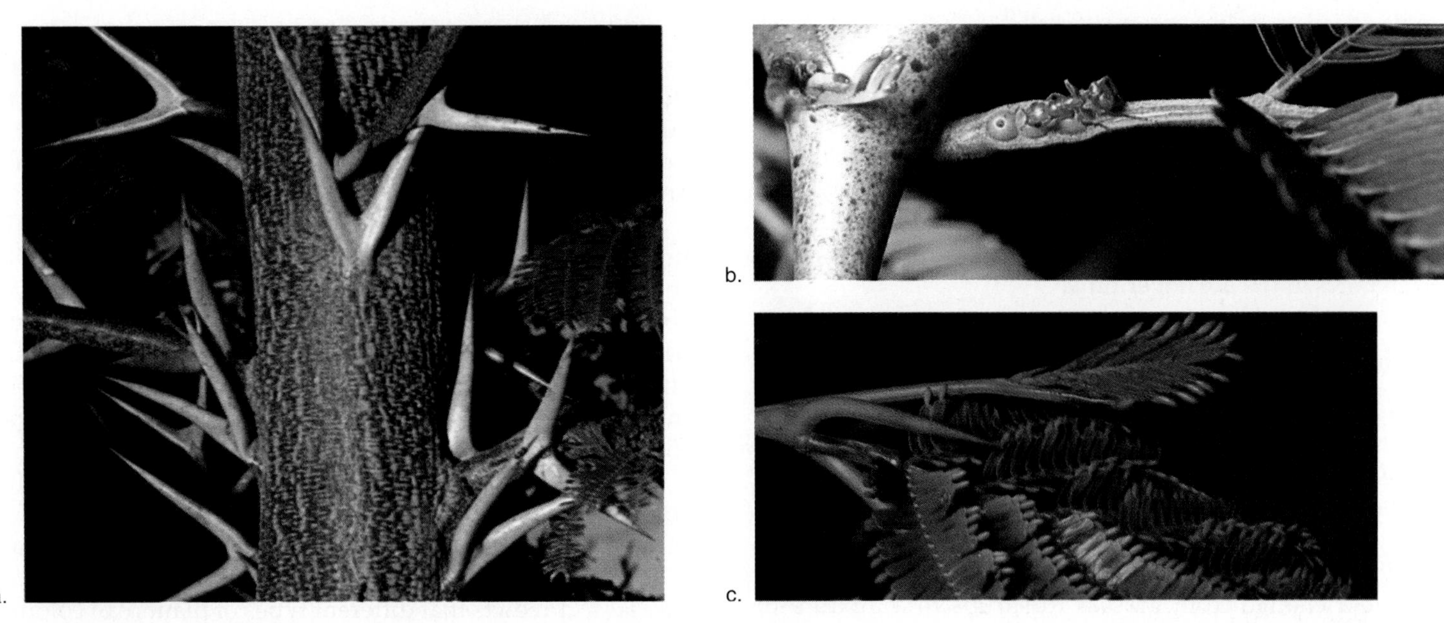

Figure 34.11 Mutualism between the bullhorn acacia tree and ants.
The bullhorn acacia tree (*Acacia*) is adapted to provide nourishment for ants *(Pseudomyrmex ferruginea).* **a.** The thorns are hollow and the ants live inside. **b.** The base of leaves has nectaries (openings) where ants can feed. **c.** Leaves of the bullhorn acacia have bodies at the tips that ants harvest for larval food.

portation for the other species such as when barnacles attach themselves to the backs of whales. Remoras are fishes that attach themselves to the bellies of sharks by means of a modified dorsal fin acting as a suction cup. The remoras obtain a free ride and also feed on the remains of the shark's meals. Clownfishes live within the waving mass of tentacles of sea anemones. Because most fishes avoid the poisonous tentacles of the anemones, clownfishes are protected from predators.

If clownfishes attract other fishes on which the anemone can feed, this relationship borders on mutualism. Other examples of commensalism may also be mutualistic. For example, cattle egrets benefit from grazing near cattle because the cattle flush insects and other animals from the vegetation as they graze. To some it seems like wasted effort to try to classify symbiotic relationships into the three categories of parasitism, commensalism, and mutualism. The amount of harm or good two species seem to do one another is dependent on what the investigator chooses to measure.

Mutualism

Mutualism is a symbiotic relationship in which both members of the association benefit. Mutualistic relations need not be equally beneficial to both species. We can imagine that the relationship between plants and their animal pollinators began when herbivores, like insects, feasted on pollen. The provision of nectar may have spared the pollen and at the same time allowed the animal to become an instrument of pollination. Lichens can grow on rocks because their fungal member conserves water and leaches minerals that are provided to the algal partner, which photosynthesizes and provides organic food for both populations. When each is grown separately in the laboratory, the algae seem to do fine but the fungus does poorly. For that reason, it's been suggested that the fungus is parasitic at least to a degree on the algae.

Ants form mutualistic relationships with both plants and insects. In tropical America, the bullhorn acacia tree is adapted to provide a home for ants of the species *Pseudomyrmex ferruginea* (Fig. 34.11). Unlike other acacias, this species has swollen thorns with a hollow interior, where ant larvae can grow and develop. In addition to housing the ants, the acacias provide them with food. The ants feed from nectaries at the base of the leaves and eat fat and protein-containing nodules called Beltian bodies, which are found at the tops of some of the leaves. The ants constantly protect the plant from caterpillars of moths and butterflies and other plants that might shade it because, unlike other ants, they are active twenty-four hours a day. Indeed, when the ants on experimental trees were poisoned, the trees died.

Cleaning symbiosis is a symbiotic relationship in which the individual being cleaned is often a vertebrate. Crustaceans, fish, and birds act as cleaners and are associated with a variety of vertebrate clients. Large fish in coral reefs line up at cleaning stations and wait their turn to be cleaned by small fish that even enter the mouths of the large fish. It's been suggested that cleaners may be exploiting the relationship by feeding on host tissues as well as on ectoparasites. On the other hand, cleaning could ultimately lead to net gains in client fitness.

Symbiotic relationships do occur between species, but it may be too simplistic to divide them into parasitic, commensalistic, and mutualistic relationships.

34.5 Community Stability and Diversity

Communities are subject to disturbances that can range in severity from a storm blowing down a patch of trees to a beaver damming a pond to a volcanic eruption. We know from observation that following these disturbances, we'll see changes in the area over time.

Ecological Succession

Ecological succession is a change in community composition over time. On land, *primary succession* occurs in areas where there is no soil formation such as following a volcanic eruption or a glacial retreat. *Secondary succession* begins in areas where soil is present, as when a cultivated field like the cornfield returns to a natural state. Notice that we roughly observe a change from grasses to shrubs to a mixture of shrubs and trees.

The first species to begin secondary succession are called *pioneer species,* that is, plants that are invaders of disturbed areas, and then the succession progresses through a series of stages that are also described in Figure 34.12. Again we observe a series that begins with grasses and proceeds from shrub stages to a mixture of shrubs and trees, until finally there are only trees.

Succession also occurs in aquatic communities, as when lakes and ponds undergo a series of stages by which they disappear and become filled in.

Models of Succession

The *climax-pattern model* of succession says that particular areas will always lead to the same type of community, which is called a **climax community.** This model is based on the observation that climate, in particular, determines whether a desert, a type of grassland, or a particular type of forest results. Therefore, for example, there is a coniferous forest in northern latitudes, a deciduous forest in temperate zones, and a tropical rain forest in the tropics.

Does each stage in succession facilitate or inhibit the next stage? To support a *facilitation model* it can be observed that shrubs can't grow on dunes until dune grass has caused soil to develop. Similarly, in the example given in Figure 34.12, shrubs can't arrive until grasses have made the soil suitable to them. So, it's possible that each successive community prepares the way for the next, so that grass-shrub-forest development occurs in a sequential way.

On the other hand, the *inhibition model* says that colonists hold on to their space and inhibit the growth of other plants until the colonists die or are damaged. Still another possible model is called a tolerance model. The *tolerance model* predicts that different types of plants can colonize an area at the same time. Sheer chance determines which seeds arrive first, and successional stages may simply reflect the length of time it takes species to mature. This alone could account for the herb-shrub-forest development one often sees (Fig. 34.12). The length of time it takes for trees to develop might simply give the impression that there is a recognizable series of plant communities from the simple to the complex. In reality, the models we have mentioned are not mutually exclusive and succession is probably a complex process.

Ecological succession which occurs after a disturbance probably involves complex processes, and the end result cannot always be foretold.

Figure 34.12 Secondary succession in a forest.
In secondary succession in a large conifer plantation in central New York State, certain species are common to particular stages. However, the process of regrowth shows approximately the same stages as secondary succession from a cornfield.
(After R. L. Smith, 1960.)

Intermediate Disturbance Hypothesis

Increasingly, it has become apparent that the most complex communities most likely consist of habitat patches that are at various stages of succession. Each successional stage has its own mix of plants and animals, and if a sample of all stages is present, community diversity would be greatest.

The *intermediate disturbance hypothesis* states that a moderate amount of disturbance is required for a high degree of community diversity. Fires, for example, promote understory plant diversity by preventing longer-lived shrub species from outcompeting annuals. Naturally, if widespread disturbances occur frequently, diversity is expected to be limited; on the other hand, if disturbances such as fires are very infrequent, diversity will also be limited (Fig. 34.13).

Island Biogeography

According to the equilibrium hypothesis of island biogeography, every island has a balance between rate of immigration of new species and the rate of extinction of species that are already on the island (Fig. 34.14). What could affect these two rates? The size of the island is certainly one important variable: a large island would get more colonists and suffer the least amount of extinction, while a small island would get fewer colonists and suffer the most amount of extinction.

This theory applies to biodiversity! In Panama, Barro Colorado Island (BCI) was created in the 1910s when a river was dammed to form a lake. As predicted by the equilibrium hypothesis of island biogeography, BCI lost species because it was an island now cut off from the mainland. Among those species that became extinct were the top predators (e.g., the jaguar, puma, and ocelot). Therefore, medium-size terrestrial mammals (e.g., coatimundi) increased in number and preyed on birds, their eggs, and nestlings among other animals. Not surprising, there are now fewer bird species on BCI than previously.

Conservationists point out that humans create "islands" when natural areas are surrounded by farms, towns, and cities. Unless these areas are large and connected in some way to other natural areas that provide colonists, more extinctions will necessarily occur.

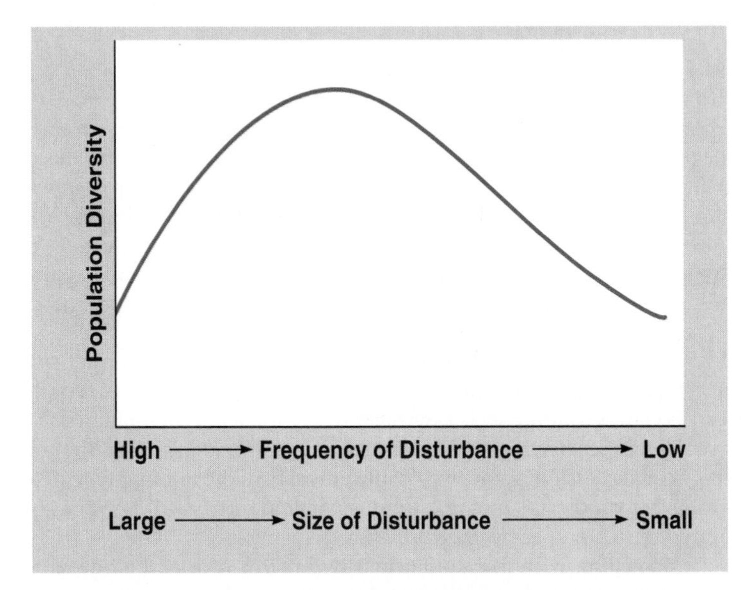

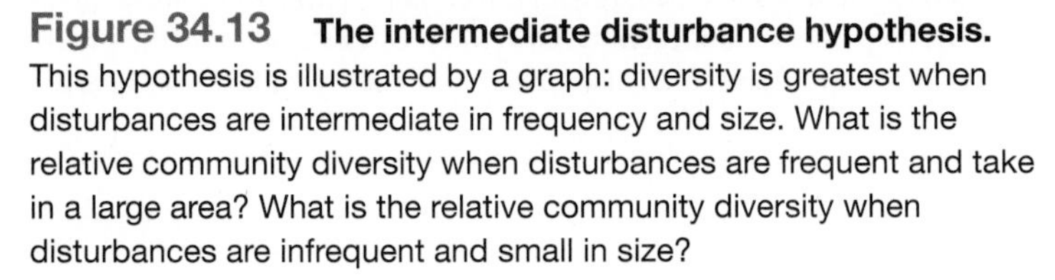

Figure 34.13 The intermediate disturbance hypothesis. This hypothesis is illustrated by a graph: diversity is greatest when disturbances are intermediate in frequency and size. What is the relative community diversity when disturbances are frequent and take in a large area? What is the relative community diversity when disturbances are infrequent and small in size?

Predation and Biodiversity

In certain communities, predation by a particular species reduces competition and increases diversity. If the starfish *Pisaster* is removed from areas along the rocky intertidal zone on the west coast of North America, the mussel *Mytilus* increases in number and excludes other invertebrates and algae from attachment sites on the rocks. The species richness declines drastically. Predators that regulate competition and maintain the diversity of a community are now called **keystone predators.** The elephant is a possible keystone predator in Africa. Elephants feed on shrubs and small trees, causing woodland habitats to become open grassland. This is not beneficial to the elephant, which needs woody species in its diet, but it is beneficial to other ungulates that graze on grasses.

Data from various sources suggest that biodiversity is increased when an area is large with many habitats and when normal ecological interactions, such as predation, are maintained.

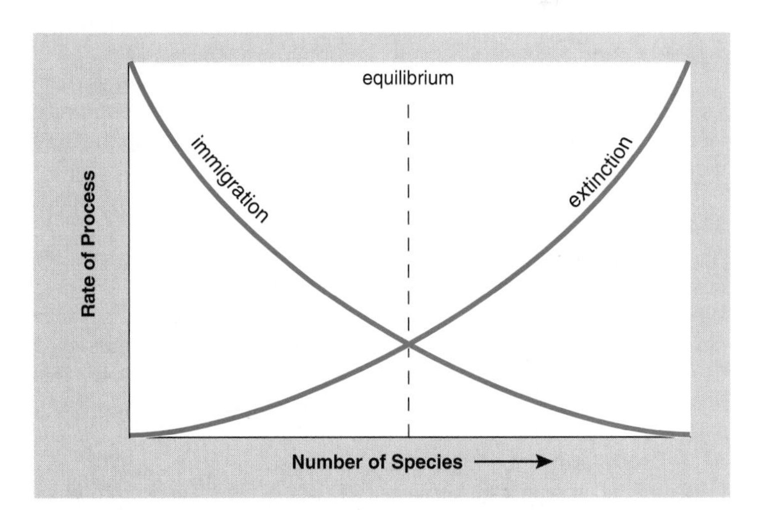

Figure 34.14 Theory of island biogeography. The biodiversity of any island whether surrounded by water or by human society is a balance between immigration and extinction. Large islands will be more diverse than small islands. Similarly, large terrestrial areas that are in contact with other natural areas will be more diverse than those that have limited access.

Bioethical Issue

Today, we are very much concerned about emerging diseases caused by parasites. Emerging diseases are ones like AIDS and Ebola, which emerge from their natural host to cause illness in humans. In 1993, the hantavirus strain emerged from the common deer mouse and killed about 60 young people in the Southwest. In the case of hantavirus, we know that climate was involved. An unusually mild winter and wet spring caused piñon trees to bloom well and provide pine nuts to the mice. The increasing deer mouse population came into contact with humans, and the hantavirus leaped easily from mice to humans. The prediction is that global warming will upset normal weather cycles and result in outbreaks of hantavirus as well as malaria, dengue and yellow fevers, filariasis, encephalitis, schistosomiasis, and cholera. Clearly any connection between global warming and emerging diseases offers another reason why greenhouse gases should be curtailed.

In December of 1997, 159 countries met in Kyoto, Japan, to work out a protocol that would reduce greenhouse gases worldwide. Greenhouse gases are those like carbon dioxide and methane, which allow the sun's rays to pass through but then trap the heat from escaping. It is believed that the emission of greenhouse gases, especially from power plants, will cause earth's temperature to rise 1.5°–4.5° by 2060. The U.S. Senate does not want to ratify the agreement because it does not include a binding emissions commitment from the less-developed countries. While the U.S. presently emits a large proportion of the greenhouse gases, China is expected to pass the United States in about 2020 to become the biggest source of greenhouse emissions.

Negotiations with the developing countries is still going on and some creative ideas have been put forward. Why not have a trading program that allows companies to buy and sell emission credits across international boundaries? Accompanying that would be a market in greenhouse reduction techniques. If it became monetarily worth their while, companies in developing countries would have an incentive to reduce greenhouse emissions.

Questions

1. Should the less-developed countries as well as the more-developed countries be expected to reduce their greenhouse emissions? Why or why not?
2. Do you approve of giving companies monetary incentives to reduce greenhouse emissions? Why or why not?
3. If you were a CEO, would you be willing to reduce greenhouse emissions simply because they cause a deterioration of the environment and probably cause human illness? Why or why not?

Summarizing the Concepts

34.1 Community Composition and Diversity

A community is an assemblage of populations interacting with one another within the same environment. Communities differ in their composition (species found there) and their diversity (species richness and relative abundance). Abiotic factors such as latitude and environmental gradients seem to largely control community composition.

An organism's habitat is where it lives in the community. An ecological niche is defined by the role an organism plays in its community, including its habitat and how it interacts with other species in the community.

34.2 Competition

According to the competitive exclusion principle, no two species can occupy the same niche at the same time when resources are limiting. When resources are partitioned between two or more species, resource partitioning has occurred. Often we have to assume that we are seeing the results of competition. Barnacles competing on the Scottish coast may be an example of present ongoing competition.

34.3 Predation

Predator-prey interactions between two species are influenced by environmental factors. Sometimes predation can cause prey populations to decline and remain at relatively low densities, or a cycling of population densities may occur.

Prey defenses take many forms: camouflage, use of fright, and warning coloration are three possible mechanisms. Batesian mimicry occurs when one species has the warning coloration but lacks the defense. Müllerian mimicry occurs when two species with the same warning coloration have the same defenses.

34.4 Symbiosis

In a parasitic relationship, the parasite benefits and the host is harmed. In a commensalistic relationship, neither party is harmed. And in a mutualistic relationship, both partners benefit. Parasites often utilize more than one host as is the case with deer ticks. In a commensalistic relationship, one species often provides a home and/or transportation for another species. Mutualistic relationships are quite varied. Flowers and their pollinators, algae and fungi in a lichen, ants who have a plant partner, and cleaning symbiosis are all examples of mutualistic relationships.

34.5 Community Stability and Diversity

A change in community composition over time is called ecological succession. A climax community is associated with particular geographic areas.

Heterogeneity caused by the presence of patches in different stages of succession results in the most diversity. The intermediate disturbance hypothesis states that a moderate amount of disturbance is required for a high degree of community diversity.

In keeping with the equilibrium hypothesis of island biogeography, a natural area must be large and in contact with other natural areas in order to preserve biodiversity. Only then can interactions like predation and competition which also preserve diversity continue as before.

Studying the Concepts

1. Compare and contrast community composition with community diversity. What are the two aspects of community diversity? 702
2. Describe the habitat and ecological niche of some particular organism. 703
3. What is the competitive exclusion principle? How does the principle relate to resource partitioning? 704–06
4. What is the special significance of the barnacle experiment off the coast of Scotland? 706
5. Why might you expect a cycling between predator and prey populations? 707
6. Give examples of prey defenses. What is mimicry, and why does it work as a prey defense? 708
7. Give examples of parasitism, commensalism, and mutualism in order to show the differences between these interactions. 710–11
8. What is ecological succession? Discuss three models to explain the process of succession. 712
9. What is the intermediate disturbance hypothesis, and how does it relate to community diversity? 713
10. How can predation increase biodiversity? 713

Testing Yourself

1. Place a dot on the following graph to indicate "intermediate disburbance." What does the graph tell you about the relationship between community diversity and disturbance?

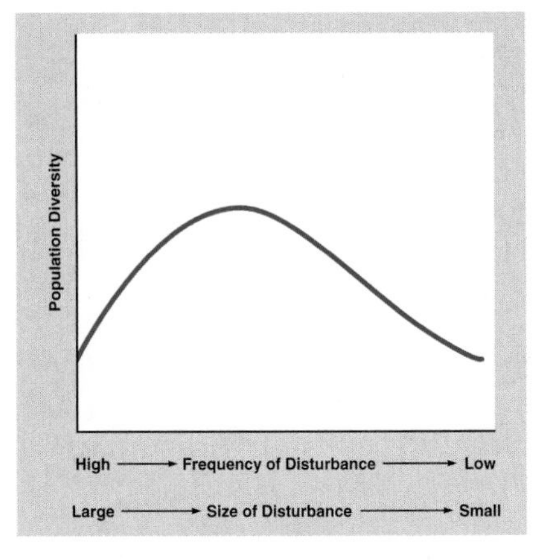

Choose the best answer for each question.

2. Six species of monkeys are found in a tropical forest. Most likely, they
   a. occupy the same ecological niche.
   b. eat different foods and occupy different ranges.
   c. spend much time fighting each other.
   d. are from different stages of succession.
   e. All of these are correct.
3. Leaf cutter ants keep fungal gardens. The ants provide food for the fungus but also feed on the fungus. This is an example of
   a. competition.
   b. predation.
   c. commensalism.
   d. parasitism.
   e. mutualism.
4. Clownfishes live among sea anemone tentacles, where they are protected. If the clownfish provides no service to the anemone, this is an example of
   a. competition.
   b. predation.
   c. parasitism.
   d. commensalism.
   e. mutualism.
5. Two species of barnacles vie for space in the intertidal zone. The one that remains is
   a. the better competitor.
   b. better adapted to the area.
   c. the better predator on the other.
   d. the better parasite on the other.
   e. Both a and b are correct.
6. A bullhorn acacia provides a home and nutrients for ants. Which statement is correct?
   a. The plant is under the control of pheromones produced by the ants.
   b. The ants protect the plant.
   c. They compete with one another.
   d. They have coevolved to occupy different ecological niches.
   e. All of these are correct.
7. The ecological niche of an organism
   a. is the same as its habitat.
   b. includes how it competes and acquires food.
   c. is specific to the organism.
   d. is usually occupied by another species.
   e. Both b and c are correct.
8. The frilled lizard of Australia suddenly opened its mouth wide and unfurled folds of skin around its neck. Most likely this was a way to
   a. conceal itself.
   b. warn that it was noxious to eat.
   c. scare a predator.
   d. scare its prey.
   e. All of these are correct.
9. When one species mimics another species, the mimic sometimes
   a. lacks the defense of the model.
   b. possesses the defense of the model.
   c. is brightly colored.
   d. competes with another mimic.
   e. All of these are correct.
10. Which of these models of succession are mismatched?
    a. climax model—last stage is the one typical of that community.
    b. facilitation model—each stage helps the next stage to take place.
    c. inhibition model—each stage hinders the next stage from occurring.
    d. tolerance model—chance determines which plants are present, and life cycles determine how long each stage is present.
    e. None of these are mismatched.
11. Which of these most likely would help account for diversity in a coral reef?
    a. warm temperatures
    b. presence of predation
    c. intermediate disturbance
    d. prey defenses
    e. All of these are correct.

Thinking Scientifically

1. There are two types of competition: scramble competition, in which all organisms have equal access to resources, and contest competition, in which a contest decides which organism will have access to resources.
   a. When a large population of blowfly larvae are provided with a limited amount of food in the laboratory, most larvae die from lack of food. Which type of competition most likely took place? Why?
   b. In a baboon troop, certain males receive food and mate first. These males receive this treatment because they successfully fought other males. Which type of competition occurred? Why?
   c. In nature, blowflies are not restricted to a particular territory, although baboons are. How does this information help to explain why scramble competition is seen in the one species and contest competition is seen in the other?
2. Sometimes it is difficult to determine what type of symbiotic relationship organisms have.
   a. In lichens, fungal hyphae characteristically penetrate algal cells via specialized organs called haustoria. What type of symbiotic relationship is suggested by this information? Why?
   b. In lichens, algae are surrounded and mechanically protected by the meshwork of fungal hyphae, which absorb water and minerals from the substrate. If the algae benefit from this arrangement, what type of relationship do the two organisms have? Why?
   c. Both the fungi and the algae found in a lichen can exist separately. Is this counter to a parasitic relationship, but consistent with a mutualistic one?

Understanding the Terms

Match the terms to these definitions:

a. __________ Theory that no two species can occupy the same niche.
b. __________ Directional pattern of change in which one community replaces another until a community typical of the area results.
c. __________ Relationship that occurs when two different species live together in a unique way; it may be beneficial, neutral, or detrimental to one and/or the other species.
d. __________ Role an organism plays in its community, including its habitat and its interactions with other organisms.
e. __________ Symbiotic relationship in which one species is benefited, and the other is neither harmed nor benefited.

Using Technology

Your study of community ecology is supported by these available technologies:

Essential Study Partner CD-ROM

Ecology → Communities

Visit the Mader web site for related ESP activities.

Exploring the Internet

The Mader Home Page provides resources and tools as you study this chapter.

http://www.mhhe.com/biosci/genbio/mader

C H A P T E R 35

Biosphere

Chapter Concepts

Coral reefs located in shallow tropical waters are among the most productive of the biological communities. Some fish feed on plankton brought by the tides, but photosynthesizing algae also produce food (organic nutrients) for themselves and other members of the reef.

As the space shuttle orbits the earth, astronauts inside train a variety of cameras and other survey instruments on the planet below them. The first target is the frigid land near the North Pole. Having little rainfall and devoid of light most of the year, this Arctic tundra only teems with life in summer, when animals such as caribou migrate to the region. The shuttle scientists next concentrate their attention on a tropical rain forest in South America, home to a tremendous diversity of animal and plant species—many still undocumented. The astronauts then direct their devices toward the Sahara, the massive desert that spans Northern Africa. As dry as the Arctic tundra, but considerably warmer, the desert is nevertheless home to an impressive variety of hardy plants and animals. Finally, the shuttle passes above the waters of the south Pacific, where the scientists try to determine if the weather phenomenon called El Niño will strike again. In its brief time in orbit, the shuttle has studied strikingly different parts of the earth's **biosphere,** the thin layer of water, land, and air inhabited by living organisms.

35.1 Climate and the Biosphere

The distribution of biomes in the biosphere is dependent upon (1) variations in reception of solar radiation due to a spherical earth, (2) the tilt of the earth's axis as it rotates about the sun, (3) distribution of land masses and oceans, and (4) topography (landscape) features. Due to these factors, **climate** is particularly dictated by temperature and rainfall differences throughout the biosphere.

Air Circulation

Because the earth is a sphere, the sun's rays are more direct at the equator and more spread out at polar regions. Therefore, the tropics are warmer than temperate regions (Fig. 35.1*a*). The tilt of the earth as it orbits around the sun causes one pole or the other to be closer to the sun (except at the spring and fall equinoxes), and this accounts for the seasonal changes in climate in all parts of the earth except the equator (Fig. 35.1*b*). When the Northern Hemisphere is having winter, the Southern Hemisphere is having summer and vice versa.

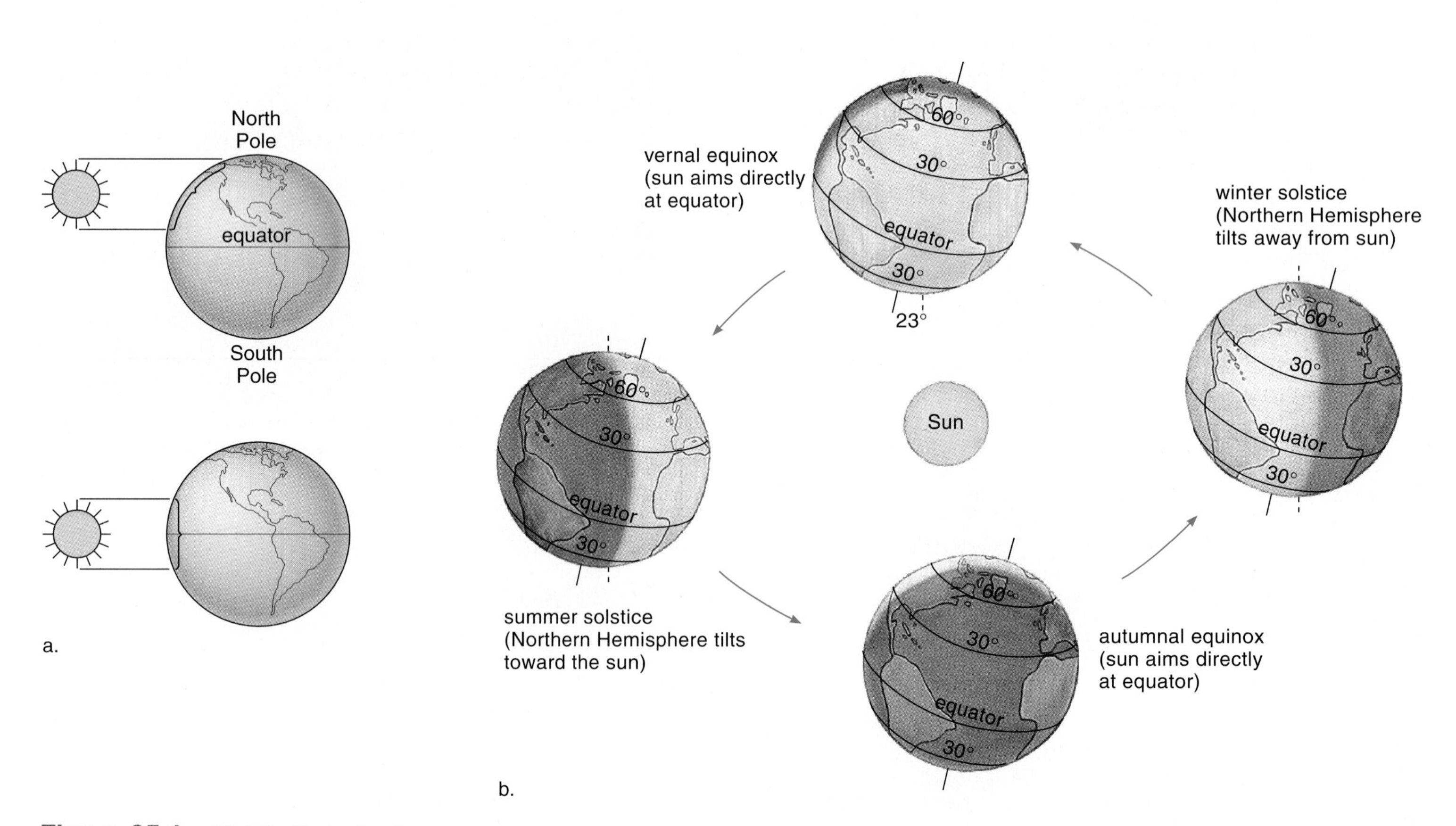

Figure 35.1 Distribution of solar energy.
a. Since the earth is a sphere, beams of solar energy striking the earth near one of the poles is spread over a wider area than similar beams striking the earth at the equator. **b.** The seasons of the Northern and Southern Hemispheres are due to the tilt of the earth on its axis as it rotates about the sun.

In the atmosphere, heat always passes from warm areas to colder areas. If the earth were standing still, and were a solid, uniform ball, all air movements—which we call winds—would be in two directions. Warm equatorial air would rise and move directly to the poles, creating a zone of lower pressure that would be filled by cold polar air moving equatorward.

Because the earth rotates, and because its surface consists of continents and oceans, the flows of warm and cold air are modified into three large circulation cells in each hemisphere (Fig. 35.2). At the equator, the sun heats the air and evaporates water. The warm moist air rises, cools, and loses most of its moisture as rain. The greatest amounts of rainfall on earth are near the equator. The rising air flows toward the poles, but at about 30° north and south latitude it sinks toward the earth's surface and reheats. As the air descends and warms, it becomes very dry, creating zones of low rainfall. The great deserts of Africa, Australia, and the Americas occur at these latitudes. At the earth's surface, the air flows both poleward and equatorward. At about 60° north and south latitude, the air rises and cools, producing another zone of high rainfall. This moisture supports the great forests of the Temperate Zone. Part of this rising air flows equatorward, and part continues poleward, descending near the poles, which are zones of low precipitation.

The earth is not standing still; it is rotating on its axis daily. The spinning of the earth affects the winds, so that the major global circulation systems flow toward the east or west rather than directly north or south (Fig. 35.2). Between about 30° north latitude and 30° south latitude, the winds blow from the southeast to the west in the Southern Hemisphere and from the northeast to the west in the Northern Hemisphere (the east coasts of continents at these latitudes are wet). These are called trade winds because sailors depended upon them to fill the sails of their trading ships. Between 30° and 60° north and south latitude, strong winds, called the prevailing westerlies, blow from west to east. The west coasts of the continents at these latitudes are wet, as is the Pacific Northwest where a massive evergreen forest is located. Weaker winds, called the polar easterlies, blow from east to west at still higher latitudes of their respective hemispheres.

Topography means the physical features or "the lay" of the land. One topographical feature that affects climate is the presence of mountains. As air blows up and over a mountain range, it rises and cools. This side of the mountain, called the windward side, receives more rainfall than the other side, called the leeward side. On the leeward side, the air descends, picks up moisture and produces clear weather (Fig. 35.3). The difference between the windward side and the leeward side can be quite dramatic. In the Hawaiian Islands, for example, the windward side of the mountains receives more than 750 cm of rain a year, while the leeward side, which is in a **rain shadow,** gets on the average only 50 cm of rain and is generally sunny. In the United States, the western side of the Sierra Nevada Mountains is lush, while the eastern side is a semidesert.

The distribution of solar energy and the rotation and path of the earth about the sun affect how the winds blow and the amount of rainfall that regions of the biosphere receive. Topography also affects rainfall.

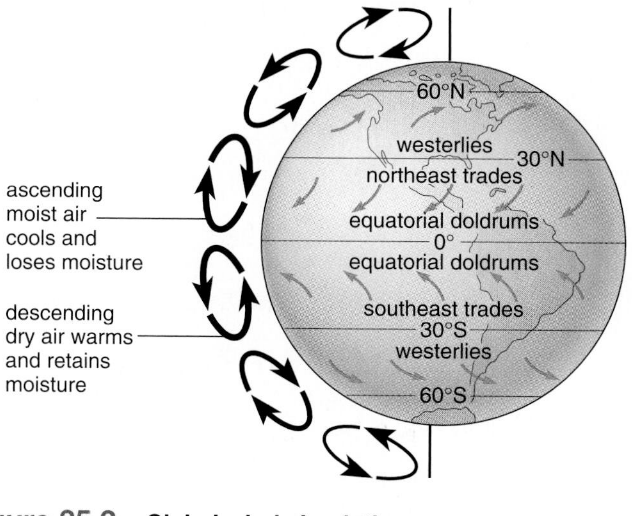

---

ascending moist air cools and loses moisture
descending dry air warms and retains moisture
60°N
westerlies
30°N
northeast trades
equatorial doldrums
0°
equatorial doldrums
southeast trades
30°S
westerlies
60°S

Figure 35.2 Global wind circulation.
At the equator, warm air rises and loses its moisture. At 30°, dry air descends; therefore, deserts occur at 30° latitude around the world. Because the earth is rotating on its axis, the trade winds move from the northeast to west in the Northern Hemisphere, and from the southeast to the west in the Southern Hemisphere. The westerlies move toward the east.

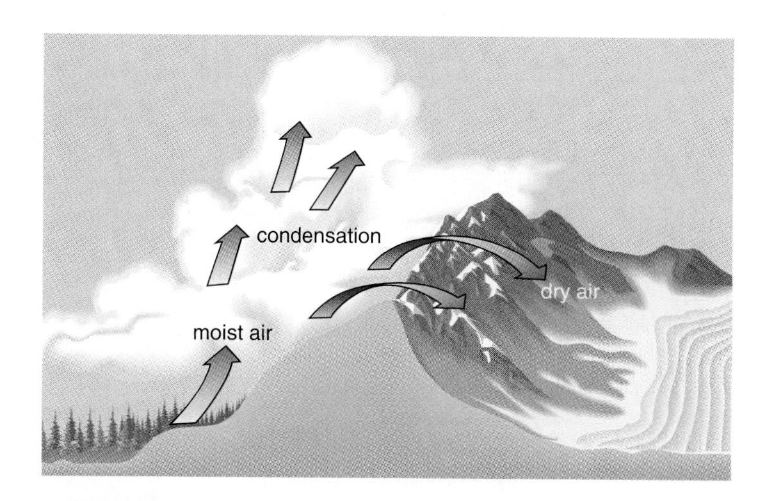

Figure 35.3 Formation of a rain shadow.
When winds from the sea cross a coastal mountain range, they rise and release their moisture as they cool this side of a mountain, which is called the windward side. The leeward side of mountains receives relatively little rain and is therefore said to lie in a "rain shadow."

Figure 35.4 Pattern of biome distribution. **a.** Pattern of world biomes in relation to temperature and moisture. The dashed line encloses a wide range of environments in which either grasses or woody plants can dominate the area, depending on the soil type. **b.** The same type of biome can occur in different regions of the world, as shown on this global map.

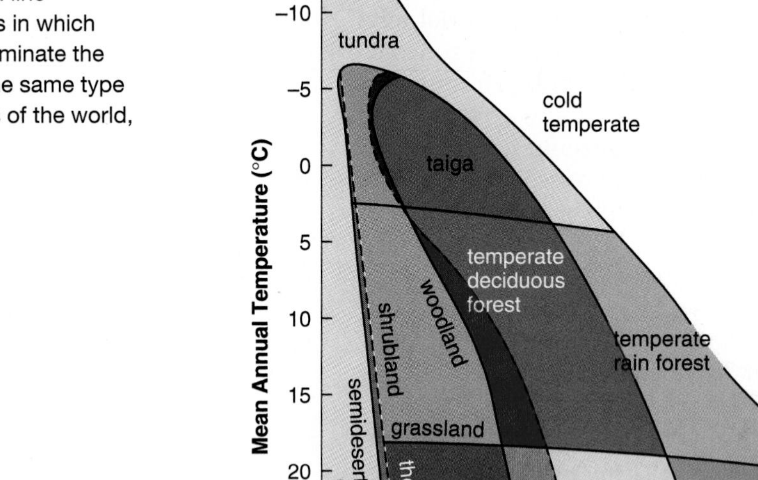

a.

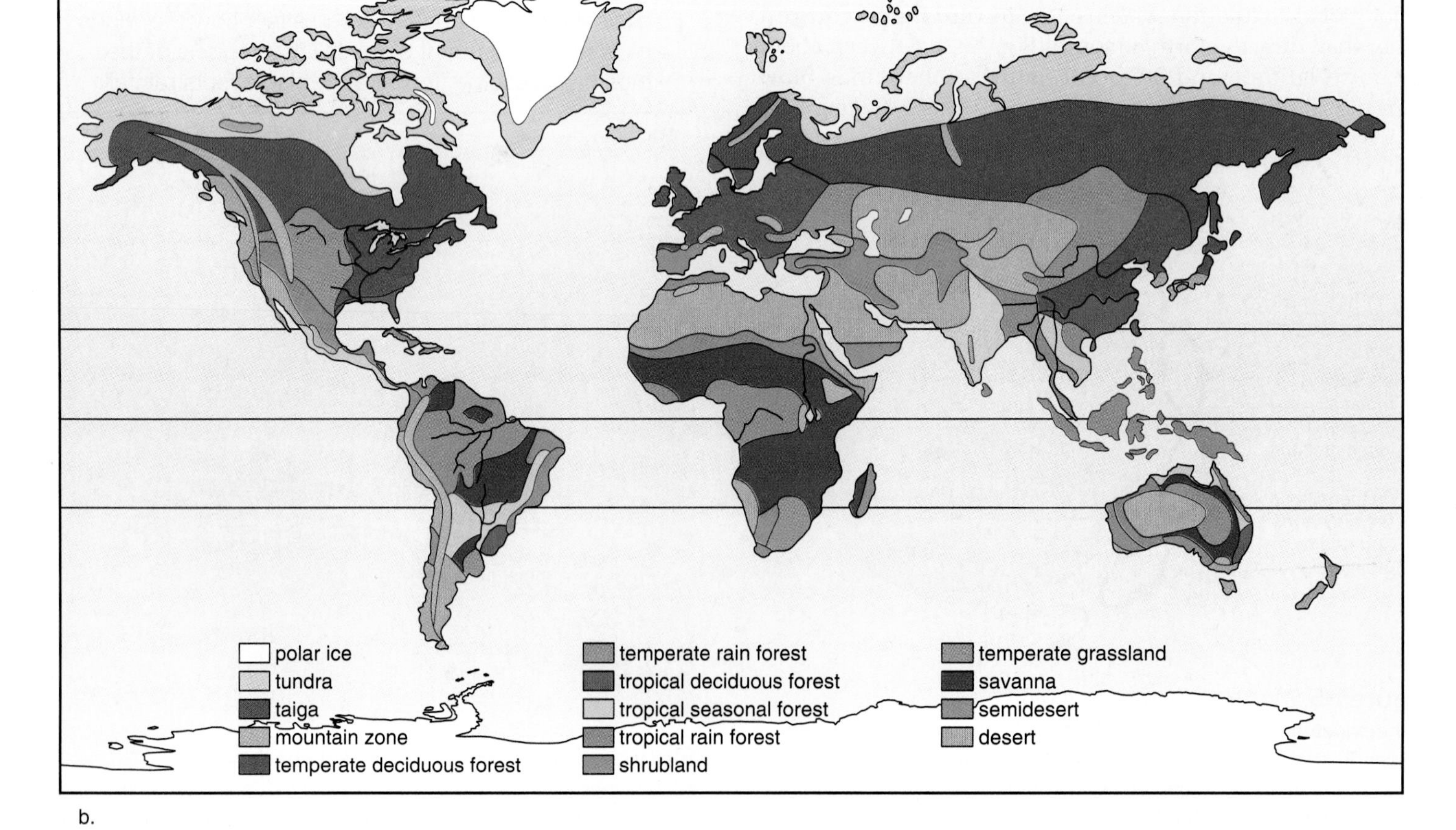

b.

35.2 Biomes of the World

A **biome** is the largest biogeographical unit of the biosphere. Although the term arose with reference only to terrestrial communities, we will be using it for both terrestrial and aquatic communities. A biome has a particular mix of plants and animals that are adapted to living under certain environmental conditions, of which climate has an overriding influence. For example, when terrestrial biomes are plotted according to their mean annual temperature and mean annual rainfall, a particular pattern results (Fig. 35.4*a*). The distribution of biomes is shown in Figure 35.4*b*. Even though Figure 35.4 shows definite demarcations, the biomes gradually change from one type to the other. Also, although we will be discussing each type of biome separately, we should remember that each biome has inputs from and outputs to all the other terrestrial and aquatic biomes of the biosphere.

The pattern of life on earth is determined principally by climate, which is influenced also by topographical features. The effect of a temperature gradient can be seen not only when we consider latitude but also when we consider altitude. If you travel from the equator to the North Pole, it is possible to observe first a tropical rain forest, followed by a temperate deciduous forest, a coniferous forest, and tundra, in that order, and this sequence is also seen when ascending a mountain (Fig. 35.5). The coniferous forest of a mountain is called a **montane coniferous forest,** and the tundra near the peak of a mountain is called an **alpine tundra.** When going from the equator to the South Pole, you would not reach a region corresponding to a coniferous forest and tundra of the Northern Hemisphere. Why not? Look at the distribution of the land masses—they are shifted toward the north.

The distribution of biomes is determined by physical factors such as climate (principally temperature and rainfall), which varies according to latitude and altitude.

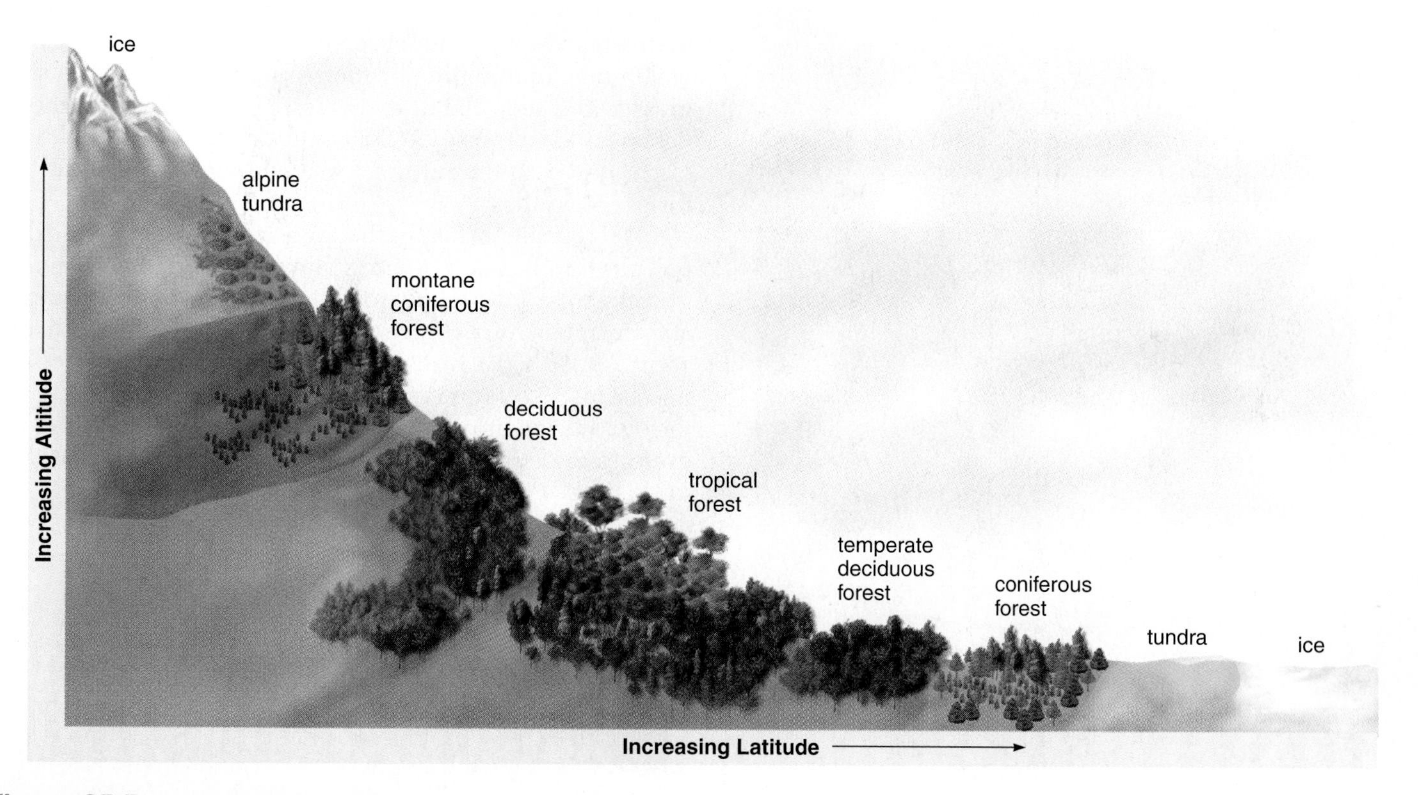

Figure 35.5 Climate and biomes.
Biomes change with altitude just as they do with latitude because vegetation is partly determined by temperature. Rainfall also plays a significant role, which is one reason why grasslands, instead of tropical or deciduous forests, are sometimes found at the base of mountains.

35.3 Terrestrial Biomes

There are many major terrestrial biomes, and we will consider these: tundra, coniferous forests (taiga, temperate rain forest), temperate deciduous forest, tropical rain forest, savanna, temperate grasslands, shrubland, and desert.

Tundra

The **Arctic tundra** biome, which encircles the earth just south of ice-covered polar seas in the Northern Hemisphere, covers about 20% of the earth's land surface (Fig. 35.6). (A similar community, called the alpine tundra, occurs above the timberline on mountain ranges.) The Arctic tundra is cold and dark much of the year. Because rainfall amounts to only about 20 cm a year, the tundra could possibly be considered a desert, but melting snow creates a landscape of pools and mires in the summer, especially because so little evaporates. Only the topmost layer of earth thaws; the **permafrost** beneath this layer is always frozen, and, therefore, drainage is minimal.

Trees are not found in the tundra because the growing season is too short, their roots cannot penetrate the permafrost, and they cannot become anchored in the boggy soil of summer. In the summer, the ground is covered with short grasses and sedges, but there also are numerous patches of lichens and mosses. Dwarf woody shrubs, such as dwarf birch, flower and seed quickly while there is plentiful sun for photosynthesis.

A few animals live in the tundra year-round. For example, the mouselike lemming stays beneath the snow; the ptarmigan, a grouse, burrows in the snow during storms; and the musk ox conserves heat because of its thick coat and short, squat body. In the summer, the tundra is alive with numerous insects and birds, particularly shorebirds and waterfowl that migrate inland. Caribou and reindeer also migrate to and from the tundra, as do the wolves that prey upon them. Polar bears are common near the coast.

Figure 35.6 The tundra.
a. In this biome, which is nearest the polar regions, the vegetation consists principally of lichens, mosses, grasses, and low-growing shrubs. **b.** Pools of water that do not evaporate nor drain into the permanently frozen ground attract many birds that feed on the plentiful insects in the summer. **c.** Caribou, more plentiful in the summer than the winter, feed on lichens, grasses, and shrubs.

Coniferous Forests

Coniferous forests are found in three locations: in the **taiga,** which extends around the world in the northern part of North America and Eurasia; near mountain tops (where it is called a montane coniferous forest), and also along the Pacific coast of North America, as far south as northern California.

The taiga (Fig. 35.7*a, b*) typifies the coniferous forest with its cone-bearing trees, such as spruce, fir, and pine. These trees are well adapted to the cold because both the leaves and bark have thick coverings. Also, the needlelike leaves can withstand the weight of heavy snow. There is a limited understory of plants, but the floor is covered by low-lying mosses and lichens beneath the layer of needles. Birds harvest the seeds of the conifers, and bears, deer, moose, beaver, and muskrat live around the cool lakes and along the streams. Wolves prey on these larger mammals. A montane coniferous forest also harbors the wolverine and mountain lion.

The coniferous forest that runs along the west coast of Canada and the United States is sometimes called a **temperate rain forest.** The prevailing winds moving in off the Pacific Ocean lose their moisture when they meet the coastal mountain range. The plentiful rainfall along with a rich soil have produced some of the tallest conifer trees ever in existence, including the coastal redwoods. This forest is also called an old-growth forest because some trees are as old as 800 years. It truly is an evergreen forest because all trees are covered with mosses, ferns, and other plants that grow on their trunks. Whether the limited portion remaining should be preserved from logging has been quite a controversy. Unfortunately, the controversy has centered around the northern spotted owl, which is endemic to this area. The actual concern is conservation of this particular ecosystem.

Figure 35.7 The taiga.
The taiga, which means swampland, spans northern Europe, Asia, and North America. The appellation "spruce-moose" refers to the dominant presence of spruce trees and moose, which frequent the ponds.

Temperate Deciduous Forests

Temperate deciduous forests are found south of the taiga in eastern North America (Fig. 35.8), eastern Asia, and much of Europe. The climate in these areas is moderate, with relatively high rainfall (75–150 cm per year). The seasons are well defined, and the growing season ranges between 140 and 300 days. The trees, such as oak, beech, and maple, have broad leaves and are termed deciduous trees; they lose their leaves in the fall and grow them in the spring.

The tallest trees form a canopy, an upper layer of leaves that are the first to receive sunlight. Even so, enough sunlight penetrates to provide energy for another layer of trees called understory trees. Beneath these trees are shrubs that may flower in the spring before the trees have put forth their leaves. Still another layer of plant growth—mosses, lichens, and ferns—resides beneath the shrub layer. This stratification provides a variety of habitats for insects and birds. Ground life is also plentiful. Squirrels, cottontail rabbits, shrews, skunks, woodchucks, and chipmunks are small herbivores. These and ground birds such as turkeys, pheasants, and grouse are preyed on by red foxes. White-tail deer and black bears have increased in number of late. In contrast to the taiga, amphibians and reptiles occur in this biome because the winters are not as cold. Frogs and turtles prefer an aquatic existence, as do the beaver and muskrat, which are mammals.

Autumn fruits, nuts, and berries provide a supply of food for the winter, and the leaves, after turning brilliant colors and falling to the ground, contribute to the rich layer of humus. The minerals within the rich soil are washed far into the ground by the spring rains, but the deep tree roots capture these and bring them back up into the forest system again.

Figure 35.8 Temperate deciduous forest.
A temperate deciduous forest is home to many and varied plants and animals. Millipedes can be found among leaf litter; chipmunks feed on acorns; and bobcats prey on these and other small mammals.

Science Focus

Wildlife Conservation and DNA

After DNA analysis, scientists were amazed to find that some 60% of loggerheads drowning in the nets and hooks of fisheries in the Mediterranean Sea were from U.S. Southeast beaches. Since the unlucky creatures were a good representative sample of the turtles in the area, that meant that more than half the young turtles living in the Mediterranean Sea had hatched from nests on beaches in Florida, Georgia, and South Carolina. Some 20,000 to 50,000 loggerheads die each year due to the Mediterranean fisheries, which may partly explain the decline of loggerheads nesting on U.S. Southeast beaches observed for the last 25 years. . . .

At the Institute of Arctic Biology at the University of Alaska, Fairbanks, graduate student Sandra Talbot recently finished sequencing DNA by hand from Alaskan brown bears. Wildlife geneticist Gerald Shields, who heads the program, and Talbot have determined that there are two types of brown bears in Alaska. One type resides only on southeastern Alaska's Admiralty, Baranof, and Chichagof Islands, known as the ABC Islands. . . .

The other brown bear in Alaska is found throughout the rest of the state, as well as in Siberia and western Asia. A third distinct type of brown bear, known as the Montana grizzly, resides in other parts of North America. The three distinct types comprise all of the known brown bears in the New World.

The ABC bears' uniqueness may be bad news for the timber industry, which has expressed interest in logging parts of the ABC Islands. Says Shields, "Studies show that when roads are built and the habitat is fragmented, the population of brown bears declines. Our genetic observations suggest they are truly unique, and we should consider their heritage. They could never be replaced by transplants. . . ."

In what will become a classic example of how DNA analysis might be used to protect endangered species from future ruin, scientists from the United States and New Zealand recently carried out discreet experiments in a Japanese hotel room on whale sushi bought in local markets. A staple of the Japanese diet, sushi is a rice and meat concoction wrapped in seaweed. Armed with a miniature DNA sampling machine, the scientists found that of the 16 pieces of whale sushi they examined, many were from whales that are endangered or protected under an international moratorium on whaling. "Their findings demonstrated the true power of DNA studies," says David Woodruff, a conservation biologist at the University of California, San Diego.

One sample was from an endangered humpback, four were fin whale, one was from a northern minke, and another from a beaked whale. Stephen Palumbi, of the University of Hawaii, says the technique could be used for monitoring and verifying catches. Until then, he says, "no species of whale can be considered safe."

Meanwhile, Ken Goddard, director of the unique U.S. Fish and Wildlife Service Forensics Laboratory in Ashland, Oregon, is already on the watch for wildlife crimes in the United States and 122 other countries that send samples to him for analysis. "DNA is one of the most powerful tools we've got," says Goddard, a former California police crime-lab director.

The lab has blood samples, for example, for all of the wolves being released into Yellowstone Park— "for the obvious reason that we can match those samples to a crime scene," says Goddard. The lab has many cases currently pending in court that he cannot discuss. But he likes to tell the story of the lab's first DNA-matching case. Shortly after the lab opened in 1989, California wildlife authorities contacted Goddard. They had seized the carcass of a trophy-size deer from a hunter. They believed the deer had been shot illegally on a 3,000-acre preserve owned by actor Clint Eastwood. The agents found a gut pile on the property but had no way to match it to the carcass. The hunter had two witnesses to deny the deer had been shot on the preserve.

Goddard's lab analysis made a perfect match between tissue from the gut pile and tissue from the carcass. Says Goddard: "We now have a cardboard cutout of Clint Eastwood at the lab saying 'Go ahead: Make my DNA.'"

Figure 35A Brown bear diversity. These two brown bears appear to be similar, but DNA studies recently revealed that one type known as an ABC bear resides only on southeastern Alaska's Admiralty, Baranof, and Chichagof Islands.

Tropical Forests

In the **tropical rain forests** of South America, Africa, and the Indo-Malayan region near the equator, the weather is always warm (between 20° and 25°C), and rainfall is plentiful (with a minimum of 190 cm per year). This may be the richest biome, both in terms of number of different kinds of species and their abundance.

A tropical rain forest has a complex structure, with many levels of life (Fig. 35.9). Some of the broadleaf evergreen trees grow from 15 to 50 meters or more. These tall trees often have trunks buttressed at ground level to prevent their toppling over. Lianas, or woody vines, that encircle the tree as it grows, also help to strengthen the trunk. The diversity of species is enormous—a 10-km^2 area of tropical rain forest may contain 750 species of trees and 1,500 species of flowering plants.

Although there is animal life on the ground (e.g., pacas, agoutis, peccaries, and armadillos), most animals live in the trees (Fig. 35.10). Insect life is so abundant that the majority of species have not been identified yet. Termites play a vital role in the decomposition of woody plant material, and ants are found everywhere, particularly in the trees. The various birds, such as hummingbirds, parakeets, parrots, and toucans, are often beautifully colored. Amphibians and reptiles are well represented by many types of frogs, snakes, and lizards. Lemurs, sloths, and monkeys are well-known primates that feed on the fruits of the trees. The largest carnivores are the big cats—the jaguars in South America and the leopards in Africa and Asia.

Many animals spend their entire life in the canopy, as do some plants. **Epiphytes** are plants that grow on other plants but usually have roots of their own that absorb moisture and minerals leached from the canopy; others catch rain and debris in hollows produced by overlapping leaf bases. The most common epiphytes are related to pineapples, orchids, and ferns.

Figure 35.9 Tropical rain forest.
Levels of life in a tropical rain forest. Even the canopy (solid layer of leaves) has levels, and some organisms spend their entire life in one particular level. Long lianas (hanging vines) climb into the canopy, where they produce leaves. Epiphytes are air plants that grow on the trees but do not parasitize them.

While we usually think of tropical forests as being nonseasonal rain forests, there are tropical forests with wet and dry seasons in India, Southeast Asia, West Africa, South and Central America, the West Indies, and northern Australia. Here, there are deciduous trees, with many layers of growth beneath the trees. In addition to the animals just mentioned, certain of these forests also contain elephants, tigers, and hippopotamuses.

Whereas the soil of a temperate deciduous forest biome is rich enough for agricultural purposes, the soil of a tropical rain forest biome is not. Nutrients are cycled directly from the litter to the plants again. Productivity is high because of high temperatures, a yearlong growing season, and the rapid recycling of nutrients from the litter. (In humid tropical forests, iron and aluminum oxides occur at the surface, causing a reddish residue known as laterite. When the trees are cleared, laterite bakes in the hot sun to a bricklike consistency that will not support crops.) Swidden agriculture, often called slash-and-burn agriculture, is a type of agriculture that has been successful, but also destructive, in the tropics. Trees are felled and burned, and the ashes provide enough nutrients for several harvests. Thereafter, the forest must be allowed to regrow, and a new section must be cut and burned.

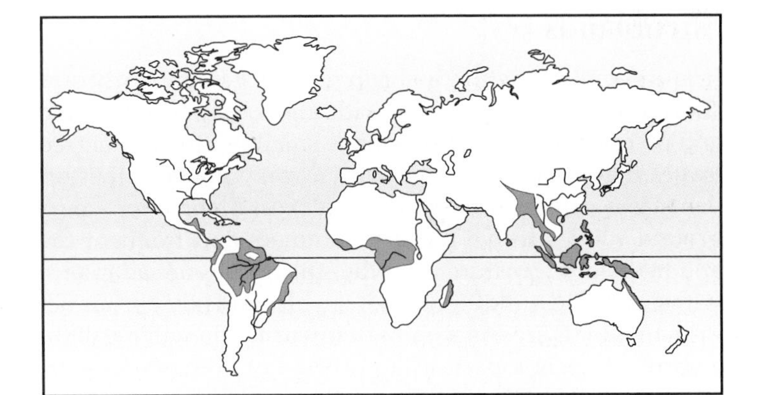

Figure 35.10 **Animals of the tropical rain forest.**

Shrublands

It is difficult to define a shrub, but in general shrubs are shorter than trees (4.5–6 m) with a woody persistent stem, and no central trunk. Shrubs have small but thick evergreen leaves that often are coated with a waxy material that prevents loss of moisture from the leaves. Their thick underground roots survive the dry summers and frequent fires and take deep moisture from the soil. Shrubs are adapted to withstand arid conditions and can also quickly sprout new growth after a fire. As a point of interest, you will recall that a shrub stage is a part of the process of both primary and secondary succession.

Shrublands tend to occur along coasts that have dry summers and receive most of their rainfall in the winter. A shrubland is found along the cape of South Africa, the western coast of North America, and the southwest and southern shores of Australia, around the Mediterranean Sea, and in central Chile. The dense shrubland that occurs in California is known as **chaparral** (Fig. 35.11). This type of shrubland, called the Mediterranean type, lacks an understory and ground litter, and is highly flammable. The seeds of many species require the heat and scarring action of fire to induce germination. Other shrubs sprout from the roots after a fire.

There is also a northern shrub area that lies west of the Rocky Mountains. This area is sometimes classified as a cold desert, but the region is dominated by sagebrush and other hardy plants. Some of the birds found here are dependent upon sagebrush for their existence.

Figure 35.11 Shrubland.
Shrublands, such as chaparral in California, are subject to raging fires, but the shrubs are adapted to quickly regrow.

Grasslands

Grasslands occur where rainfall is greater than 25 cm but is generally insufficient to support trees. Natural grasslands once covered more than 40% of the earth's land surface, but many areas that once were grasslands are now used for the cultivation of crops, such as wheat and corn. In temperate areas, where rainfall is between 10 and 30 inches a year, grasslands occur. Here, it is too dry for forests and too wet for deserts to form.

The grasses are well adapted to a changing environment and can tolerate a high degree of grazing, flooding, drought, and sometimes fire. Where rainfall is high, large tall grasses that reach more than 2 meters in height (e.g., pampas grass) can flourish. In drier areas, shorter grasses between 5 and 10 cm are dominant. Low-growing bunch grasses (e.g., grama grass) grow in the United States near deserts. Grasses also generally grow in different seasons; some grassland animals migrate, and ground squirrels hibernate, when there is little grass for them to eat.

The temperate grasslands include the Russian steppes, the South American pampas, and the North American prairies (Fig. 35.12). When traveling across the United States from east to west, the line between the temperate deciduous forest and a tall-grass prairie is roughly along the border between Illinois and Indiana. The tall-grass prairie requires more rainfall than does the short-grass prairie that occurs near deserts. Large herds of bison—estimated at hundreds of thousands—once roamed the prairies, as did herds of pronghorn antelope. Now, small mammals, such as mice, prairie dogs, and rabbits, typically live belowground, but usually feed aboveground. Hawks, snakes, badgers, coyotes, and foxes feed on these mammals. Virtually all of these grasslands, however, have been converted to agricultural lands.

Savannas, which are grasslands that contain some trees, occur in regions where a relatively cool dry season is followed by a hot, rainy one (Fig. 35.13). One tree that can survive the severe dry season is the flat-topped acacia, which sheds its leaves during a drought. The African savanna supports the greatest variety and number of large herbivores of all the biomes. Elephants and giraffes are browsers that feed on tree vegetation. Antelopes, zebras, wildebeests, water buffalo, and rhinoceroses are grazers that feed on grasses. Any plant litter that is not consumed by grazers is attacked by a variety of small organisms, among them termites. Termites build towering nests in which they tend fungal gardens, their source of food. The herbivores support a large population of carnivores. Lions and hyenas hunt in packs, cheetahs hunt singly by day, and leopards hunt singly by night.

Figure 35.12 The prairie.
Tall-grass prairies are seas of grasses dotted by pines and junipers. Bison, once abundant, are now being reintroduced into certain areas.

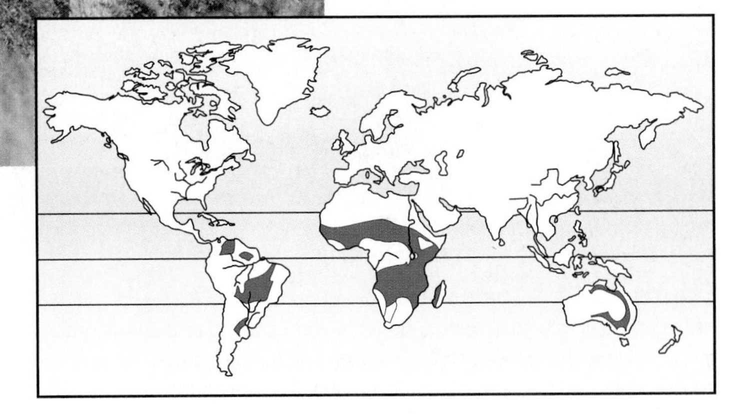

Figure 35.13 The savanna.
The African savanna varies from grassland to widely spaced shrubs and trees because the soil is low in moisture and nutrients. This biome supports a large and varied assemblage of grazers (e.g., zebras and wildebeests) and browsers (e.g., giraffes). Cheetahs and lions prey on these.

Deserts

As discussed previously, **deserts** are usually found at latitudes of about 30°, in both Northern and Southern Hemispheres. The winds which descend in these regions lack moisture. Therefore, the annual rainfall is less than 25 cm. Days are hot because a lack of cloud cover allows the sun's rays to penetrate easily, but the nights are cold because heat escapes easily into the *atmosphere.*

The Sahara, which stretches all the way from the Atlantic coast of Africa to the Arabian Peninsula, and a few other deserts have little or no vegetation. But most have a variety of plants (Fig. 35.14). The best-known desert perennials in North America are the succulent, spiny-leafed cacti, which have stems that store water and carry on photosynthesis. Also common are nonsucculent shrubs, such as the many-branched sagebrush with silvery gray leaves and the spiny-branched ocotillo that produces leaves during wet periods and sheds them during dry periods.

Some animals are adapted to the desert environment. Reptiles and insects have waterproof outer coverings that conserve water. A desert has numerous insects, which pass through the stages of development from pupa to the next pupa again when there is rain. Reptiles, especially lizards and snakes, are perhaps the most characteristic group of vertebrates found in deserts, but running birds (e.g., the roadrunner) and rodents (e.g., the kangaroo rat) are also well known (Fig. 35.14). Larger mammals, like the coyote, prey on the rodents, as do the hawks.

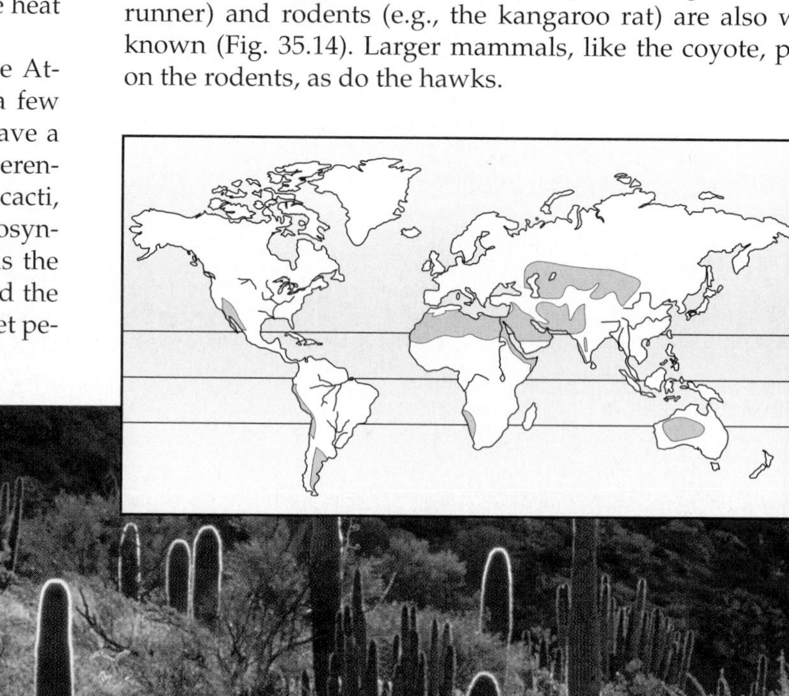

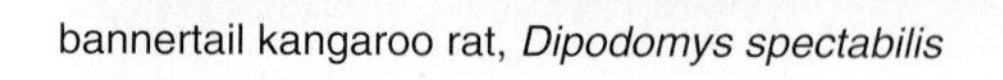

bannertail kangaroo rat, *Dipodomys spectabilis*

greater roadrunner, *Geococcyx californianus*

Figure 35.14 The desert.
Plants and animals that live in a desert are adapted to arid conditions. The plants are either succulents that retain moisture or shrubs with woody stems and small leaves that lose little moisture. The kangaroo rat feeds on seeds and other vegetation; the roadrunner preys on insects, lizards, and snakes.

35.4 Aquatic Biomes

Aquatic biomes are classified as two types: freshwater (inland) or saltwater (usually marine). Brackish water, however, is a mixture of fresh and salt water. Figure 35.15 shows how these communities are joined physically. In the water cycle, the sun's rays cause seawater to evaporate and the salts are left behind. The vaporized fresh water rises into the atmosphere, cools, and falls as rain either over the ocean or over the land. A lesser amount of water also evaporates from and returns to the land. Since land lies above sea level, gravity eventually returns all fresh water to the sea, but in the meantime, it is contained within standing waters (lakes and ponds), flowing waters (streams and rivers), and groundwater.

When rain falls, some of the water sinks or percolates into the ground and saturates the earth to a certain level. The top of the saturation zone is called the groundwater table, or simply the water table. Wherever the earth contains basins or channels, water will appear to the level of the water table. The water within basins is called lakes and ponds, and the water within channels is called streams or rivers. Sometimes groundwater is also located in underground rivers called aquifers.

Humans have the habit of channeling aboveground rivers and filling in wetlands (lands that are wet for at least part of the year). These activities degrade ecosystems and eventually cause seasonal flooding. Wetlands provide food and habitats for fish, waterfowl, and other wildlife. They also purify waters by filtering them and by diluting and breaking down toxic wastes and excess nutrients. Wetlands directly absorb storm waters and also absorb overflows from lakes and rivers. In this way they protect farms, cities, and towns from the devastating effects of floods. There are now federal and local laws for the protection of wetlands, but they are not always enforced.

Aquatic biomes can be classified as freshwater or saltwater. The two sets of communities interact and are joined by the water cycle.

stonefly larva, *Plecoptera* sp.

red banded trout, *Salmo gairdneri*

carp, *Cyprinus carpio*

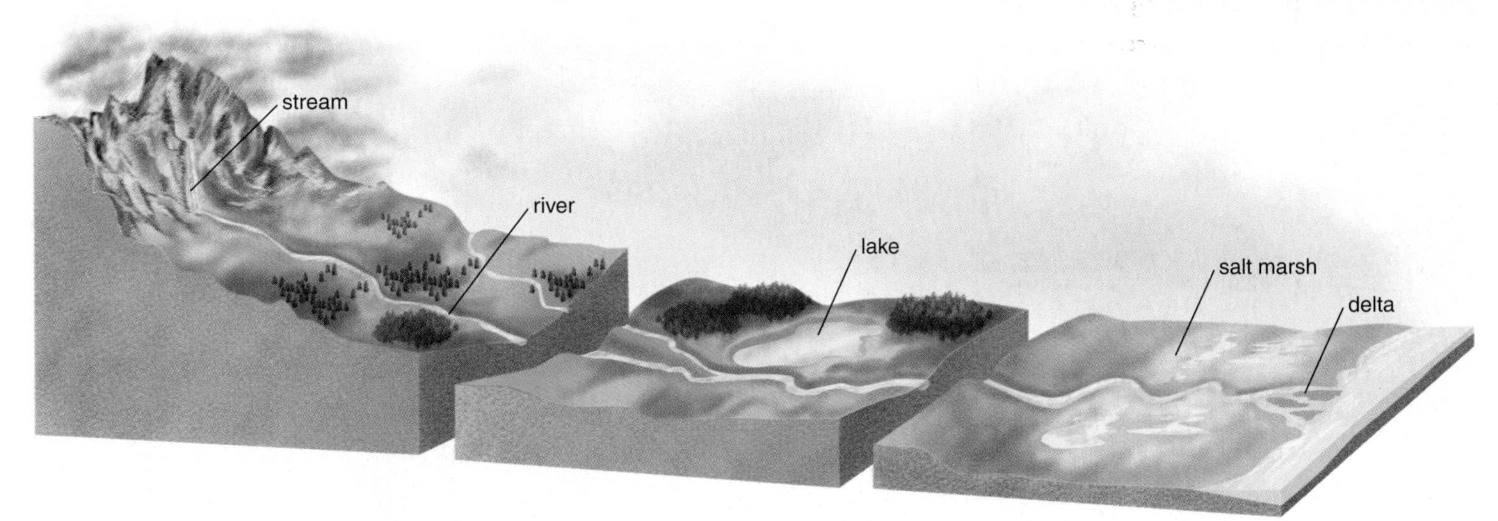

Figure 35.15 Streams and rivers.
Mountain streams have cold, clear water that flows over waterfalls and rapids. The feet of this long-legged stonefly insect larva are clawed, helping it to hold on to stones. Trout are found in occasional pools of the highly oxygenated water. As the streams merge, a river forms that gets increasingly wider and deeper until it meanders across broad, flat valleys. Carp are adapted to water that contains little oxygen and has much sediment. At its mouth, a river may divide into many channels where wetlands and estuaries are located.

Lakes

Lakes are bodies of fresh water often classified by their nutrient status. Oligotrophic (nutrient-poor) lakes are characterized by low organic matter and low productivity. Eutrophic (nutrient-rich) lakes are characterized by high organic matter and high productivity. Such lakes are usually situated in naturally nutrient-rich regions or are enriched by agricultural or urban and suburban runoffs. Oligotrophic lakes can become eutrophic through large inputs of nutrients (Fig. 35.16). This process is called **eutrophication.**

In the temperate zone, deep lakes are stratified in the summer and winter. In summer, lakes in the temperate zone have three layers of water that differ in temperature (Fig. 35.17). The surface layer, the epilimnion, is warm from solar radiation; the middle thermocline experiences an abrupt drop in temperature; and the hypolimnion is cold. These differences in temperature prevent mixing. The warmer, less dense water of the epilimnion "floats" on top of the colder, more dense water of the hypolimnion.

As the season progresses, the epilimnion becomes nutrient-poor, while the hypolimnion begins to be depleted of oxygen. The phytoplankton found in the sunlit epilimnion use up nutrients as they photosynthesize. Photosynthesis releases oxygen, giving this layer a ready supply. Detritus naturally falls by gravity to the bottom of the lake, and here oxygen is used up as decomposition occurs. Decomposition releases nutrients, however.

In the fall, as the epilimnion cools, and in the spring, as it warms, an overturn occurs. In the fall, the upper epilimnion waters become cooler than the hypolimnion waters. This causes the surface water to sink and the deep water to rise. The **fall overturn** continues until the temperature is uniform throughout the lake. At this point, wind aids in the circulation of water so that mixing occurs. Eventually, oxygen and nutrients become evenly distributed.

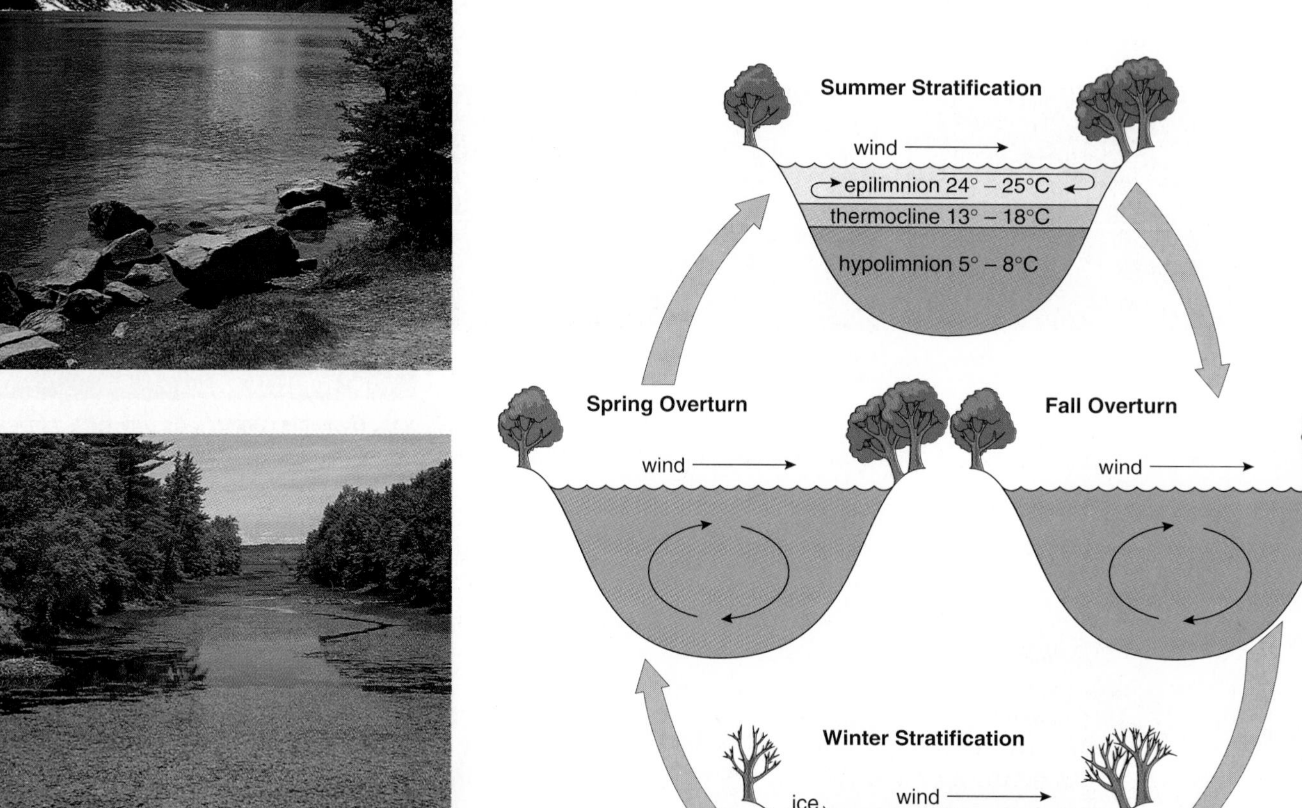

Figure 35.16 Types of lakes.
Lakes can be classified according to whether they are **(a)** oligotrophic (nutrient-poor) or **(b)** eutrophic (nutrient-rich). Eutrophic lakes tend to have large populations of algae and rooted plants, resulting in a large population of decomposers that use up much of the oxygen, leaving little oxygen for fishes.

Figure 35.17 Lake stratification.
Temperature profiles of a large oligotrophic lake in a temperate region vary with the season. During spring and fall overturn, the deep waters receive oxygen from surface waters, and surface waters receive inorganic nutrients from deep waters.

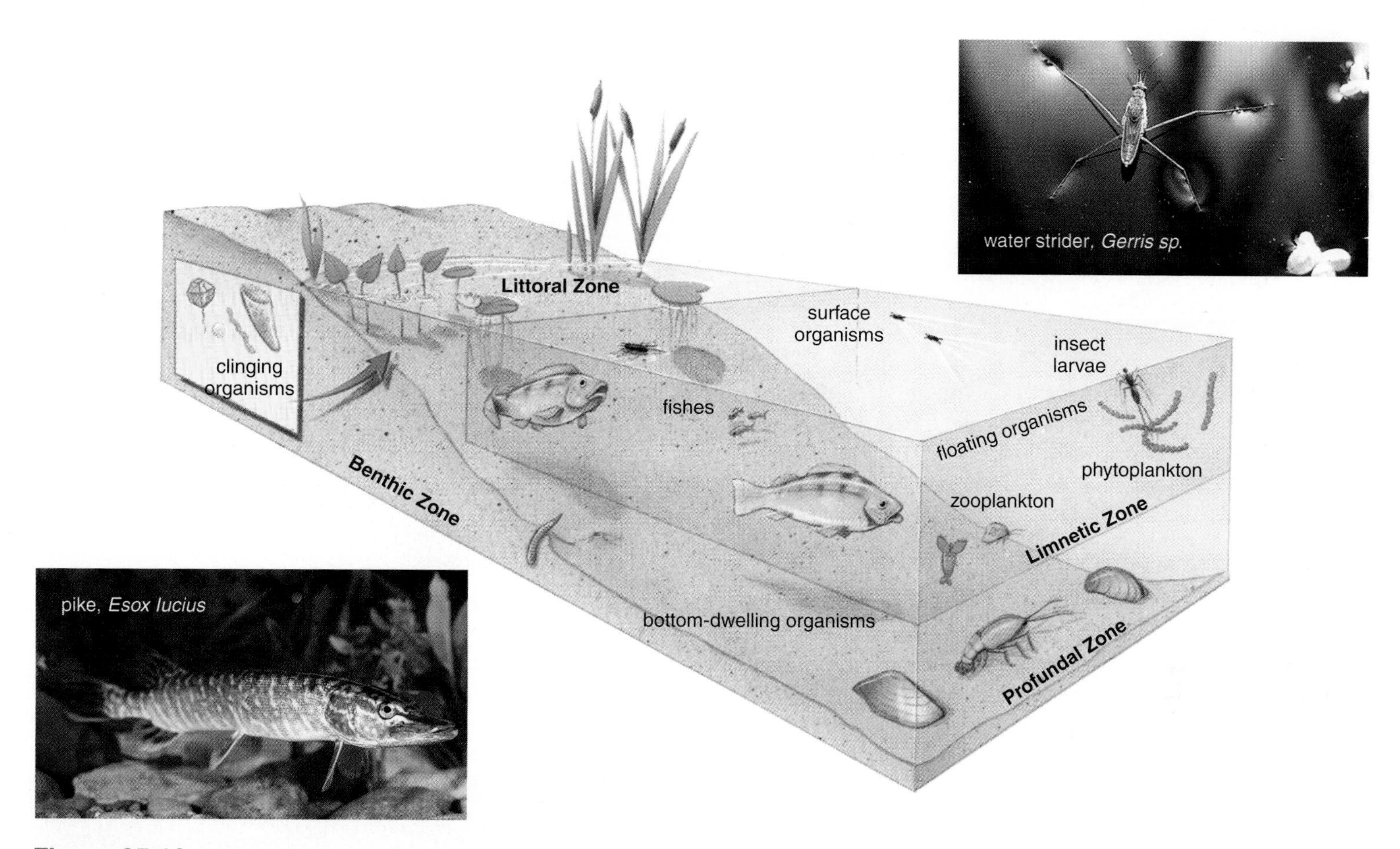

Figure 35.18 **Zones of a lake.**
Rooted plants and clinging organisms live in the littoral zone. Phytoplankton, zooplankton, and fishes are in the sunlit limnetic zone. Water striders stand on the surface film of water with water-repellent feet. Crayfishes and mollusks are in the profundal zone and also the littoral zone. Pike are top carnivores prized by fishermen.

As winter approaches, the water cools. Ice formation begins at the top, and the ice remains there because ice is less dense than cool water. Ice has an insulating effect, preventing further cooling of the water below. This permits aquatic organisms to live through the winter in the water beneath the surface of the ice.

In the spring, as the ice melts, the cooler water on top sinks below the warmer water on the bottom. The **spring overturn** continues until the temperature is uniform through the lake. At this point, wind aids in the circulation of water as before. When the surface waters absorb solar radiation, thermal stratification occurs once more.

This vertical stratification and seasonal change of temperatures in a lake basin influence the seasonal distribution of fish and other aquatic life in the lake basin. For example, coldwater fish move to the deeper water in summer and inhabit the upper water in winter. In the fall and spring just after mixing occurs, phytoplankton growth at the surface is most abundant.

Life Zones

In both fresh and salt water, free-drifting microscopic organisms, called **plankton,** are important components of the community. **Phytoplankton** are photosynthesizing algae that become noticeable when a green scum or red tide appears on the water. **Zooplankton** are animals that feed on the phytoplankton. Lakes and ponds can be divided into several life zones. The *littoral zone* is closest to the shore, the *limnetic zone* forms the sunlit body of the lake, and the *profundal zone* is below the level of light penetration (Fig. 35.18). The *benthic zone* includes the sediment at the soil-water interface. Aquatic plants are rooted in the shallow littoral zone of a lake, and various microscopic organisms cling to these plants and to rocks. Some organisms such as the water strider live at the water-air interface and can literally walk on water. In the limnetic zone, small fishes, such as minnows and killifish feed on plankton and also serve as food for large fishes. In the profundal zone, there are zooplankton and fishes such as whitefish that feed on debris that falls from above. Pike species are "lurking predators." They wait among vegetation around the margins of lakes and surge out to catch passing prey.

A few insect larvae are in the limnetic zone, but they are far more prominent in both the littoral and profundal zones. Midge larvae and ghost worms are common members of the benthos. The benthos are animals that live on the bottom in the benthic zone. In a lake, the benthos include crayfish, snails, clams, and various types of worms and insect larvae.

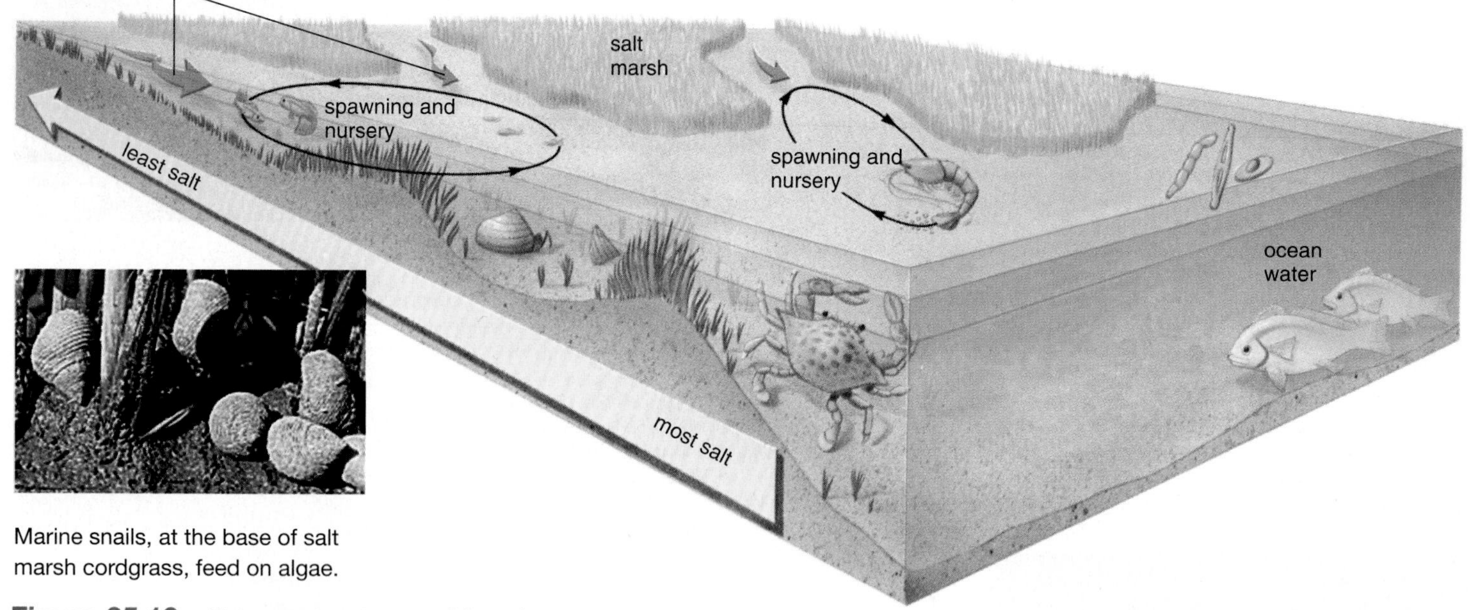

Marine snails, at the base of salt marsh cordgrass, feed on algae.

Figure 35.19 Estuary structure and function.
Since an estuary is located where a river flows into the ocean, it receives nutrients from land. Estuaries serve as a nursery for the spawning and rearing of the young for many species of fishes, shrimp and other crustaceans, and mollusks.

Coastal Communities

Near the mouth of a river, a *salt marsh* in the temperate zone and a *mangrove swamp* in the subtropical and tropical zones are likely to develop. Also, the silt carried by a river may form mudflats. It is proper to think of seacoasts and mudflats, salt marshes, and mangrove swamps as belonging to one ecological system.

Estuaries

An **estuary** is a partially enclosed body of water where fresh water and seawater meet and mix (Fig. 35.19). A river brings fresh water into the estuary, and the sea, because of the tides, brings salt water. Coastal bays, tidal marshes, fjords (an inlet of water between high cliffs), some deltas (triangular-shaped areas of land at the mouths of rivers), and lagoons (a body of water separated from the sea by a narrow strip of land) are all examples of estuaries.

Organisms living in an estuary must be able to withstand constant mixing of waters and rapid changes in salinity. Not many organisms are suited to this environment, but for those that are suited, there is an abundance of nutrients. An estuary acts as a nutrient trap because the sea prevents the rapid escape of nutrients brought by a river.

Although only a few small fish permanently reside in an estuary, many develop there, so that there is always an abundance of larval and immature fish. It has been estimated that well over half of all marine fishes develop in the protective environment of an estuary, which explains why

a.

b.

Figure 35.20 Types of estuaries.
Many types of regions qualify as estuaries, such as the salt marsh depicted in Figure 35.19 and **(a)** mudflats, which are frequented by migrant birds, and **(b)** mangrove swamps skirting the coastlines of many tropical and subtropical lands. The tangled roots of mangrove trees trap sediments and nutrients that sustain many immature forms of sea life.

Figure 35.21 Seacoasts.
a. The littoral zone of a rocky coast, where the tide comes in and out, has different types of shelled and algal organisms at its upper, middle, and lower portions. **b.** Some organisms of a rocky coast live in tidal pools. **c.** A sandy shore looks devoid of life except for the birds that feed there. However, a number of invertebrate species burrow in the sand and sediment beneath the sand.

estuaries are called the *nurseries of the sea.* Estuaries are also feeding grounds for many birds, fish, and shellfish because they offer a ready supply of food.

Salt marshes dominated by salt marsh cordgrass are often associated with estuaries. So are mudflats and mangrove swamps, where sediment and nutrients from the land collect (Fig. 35.20).

Seashores

Both rocky and sandy shores are constantly bombarded by the sea as the tides roll in and out (Fig. 35.21). The **littoral zone** lies between the high and low water marks. The littoral zone of a rocky beach is divided into subzones. In the upper portion of the littoral zone, barnacles are glued so tightly to the stone by their own secretions that their calcareous outer plates remain in place even after the enclosed shrimplike animal dies. In the midportion of the littoral zone, brown algae known as rockweed may overlie the barnacles. In the lower portions of the littoral zone, oysters and mussels attach themselves to the rocks by filaments called *byssal threads.* Also present are snails called limpets and periwinkles. But periwinkles have a coiled shell and secure themselves by hiding in crevices or under seaweeds, while limpets press their single flattened cone tightly to a rock. Below the littoral zone, macroscopic seaweeds, which are the main photosynthesizers, anchor themselves to the rocks by holdfasts.

Organisms cannot attach themselves to shifting, unstable sands on a *sandy beach;* therefore, nearly all the permanent residents dwell underground. They either burrow during the day and surface to feed at night, or they remain permanently within their burrows and tubes. Ghost crabs and sandhoppers (amphipods) burrow themselves above the high tide mark and feed at night when the tide is out. Sandworms and sand (ghost) shrimp remain within their burrows in the littoral zone and feed on detritus whenever possible. Still lower in the beach, clams, cockles, and sand dollars are found.

Oceans

Climate is driven by the sun, but the oceans play a major role in redistributing heat in the biosphere. Water tends to be warm at the equator and much cooler at the poles because of the distribution of the sun's rays, as we have discussed before (see Fig. 35.1*a*). Air takes on the temperature of the water below, and warm air moves from the equator to the poles. In other words, the oceans make the winds blow. (The land masses also play a role, but the oceans hold heat longer and remain cool longer during periods of changing temperature than do solid continents.)

When the wind blows strongly and steadily across a great expanse of ocean for a long time, friction from the moving air begins to drag the water along with it. Once the water has been set in motion, its momentum, aided by the wind, keeps it moving in a steady flow we call a current. Because the ocean currents eventually strike land, they move in a circular path—clockwise in the Northern Hemisphere and counterclockwise in the Southern Hemisphere (Fig. 35.22). As the currents flow, they take warm water from the equator to the poles. One such current, called the Gulf Stream, brings tropical Caribbean water to the east coast of North America and the higher latitudes of western Europe. Without the Gulf Stream, Great Britain, which has a relatively warm temperature, would be as cold as Greenland. In the Southern Hemisphere, another major ocean current warms the eastern coast of South America.

Also, in the Southern Hemisphere a current called the Humboldt Current flows toward the equator. The Humboldt Current carries phosphorus-rich cold water northward along the west coast of South America. During a process called **upwelling,** cold offshore winds cause cold nutrient-rich waters to rise and take the place of warm nutrient-poor waters. In South America, the enriched waters cause an abundance of marine life that supports the fisheries of Peru and northern Chile. Birds feeding on these organisms deposit their droppings on land, where it is mined as guano, a commercial source of phosphorus. When the Humboldt Current is not as cool as usual, upwelling does not occur, stagnation results, the fisheries decline, and climate patterns change globally. This phenomenon, which is discussed in the accompanying reading, is called **El Niño–Southern Oscillation.**

Major ocean currents move heat from the equator to cooler parts of the biosphere. The Gulf Stream warms the east coast of North America and parts of Europe.

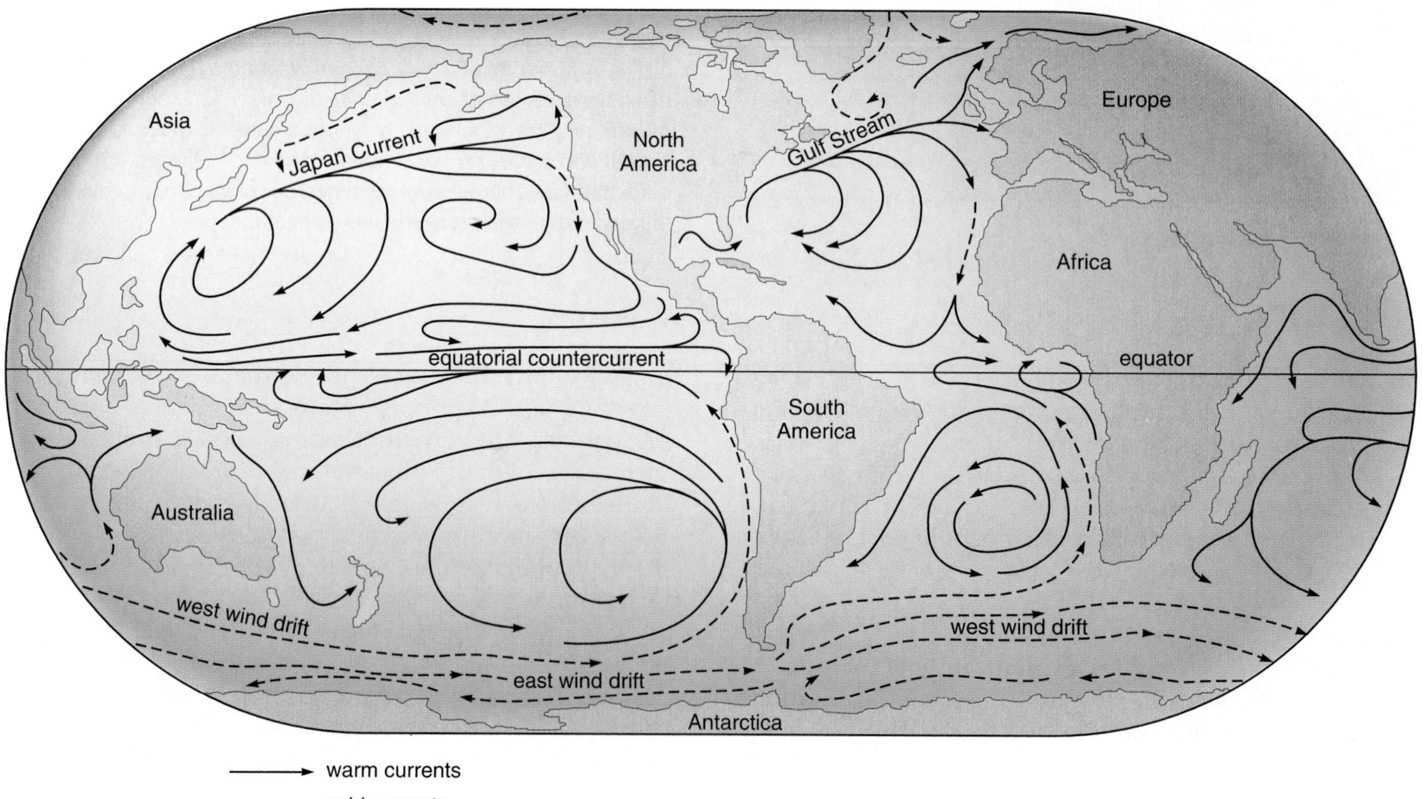

Figure 35.22 Ocean currents.
The arrows on this map indicate the locations and directions of the major ocean currents set in motion by the global wind circulation. By carrying warm water to cool latitudes (e.g., Gulf Stream) and cool water to warm latitudes (e.g., Humboldt Current), these currents have a major effect on the world's climates.

Ecology Focus

El Niño—Southern Oscillation

Climate largely determines the distribution of life on earth. Short-term variations in climate, which we call weather, also have a pronounced effect on living things. There is no better example than an El Niño. Originally, El Niño referred to a warming of the seas off the coast of Peru at Christmas time—hence its name El Niño, "the boy child" for the Christ child Jesus.

Now scientists prefer the term El Niño-Southern Oscillation (ENSO) for a severe weather change brought on by an interaction between the atmosphere and ocean currents. Ordinarily the southeast trade winds move along the coast of South America and turn west because of the earth's daily rotation on its axis. As the winds drag warm ocean waters from east to west, there is an upwelling of nutrient-rich cold water from the ocean's depths that results in a bountiful Peruvian harvest of anchovies. When the warm ocean waters reach their western destination, the monsoons bring rain to India and Indonesia. Scientists have noted that these events correlate with a difference in the barometric pressure over the Indian Ocean and the southeastern Pacific—the barometric pressure is low over the Indian Ocean and the barometric pressure is high over the southeastern Pacific. But when a "southern oscillation" occurs and the barometric pressures switch, an El Niño begins.

During an El Niño, both the northeast and the southeast trade winds slacken. Upwelling no longer occurs and the anchovy fishery off the coast of Peru plummets. During a severe El Niño, waters from the east never reach the west and the winds lose their moisture in the middle of the Pacific instead of over the Indian Ocean. The monsoons fail and there is drought in India, Indonesia, Africa, and Australia. Harvests decline, cattle must be slaughtered, and famine is likely in highly populated India and Africa, where funds to import replacement supplies of food are limited.

There can even be a backward movement of winds and ocean currents so that the waters can warm to more than 14° above normal along the west coast of the Americas. Now a severe El Niño has occurred, and the weather changes are dramatic in the Americas also. Southern California is hit by storms and even hurricanes, and the deserts of Peru and Chile receive so much rain that flooding occurs. A jet stream (strong wind currents) can carry moisture into Texas, Louisiana, and Florida, with flooding a near certainty. Or the winds can turn northward and deposit snow in the mountains along the west coast so that flooding occurs here in the spring. Some parts of the United States, however, benefit from an El Niño. The Northeast is warmer than usual, few if any hurricanes hit the east coast, and there is a lull in tornadoes throughout the Midwest. Altogether, a severe El Niño affects the weather over three-quarters of the globe.

Eventually an El Niño dies out and normal conditions return. The normal cold-water state off the coast of Peru is known as La Niña (the girl). Figure 35B contrasts the weather conditions of a La Niña with those of an El Niño. Since 1991, the sea surface has been almost continuously warm and there have been two recording-breaking El Niños. What could be causing more of the El Niño state rather than the La Niña state? Some scientists are seeking data to relate this environmental change to global warming. Global warming is a rise in environmental temperature due to greenhouse gases, like carbon dioxide, in the atmosphere. Like the glass of a greenhouse, the gases allow the sun's rays to pass through, but trap the heat. Greenhouse gases are pollutants that human beings have been pumping into the atmosphere since the industrial revolution.

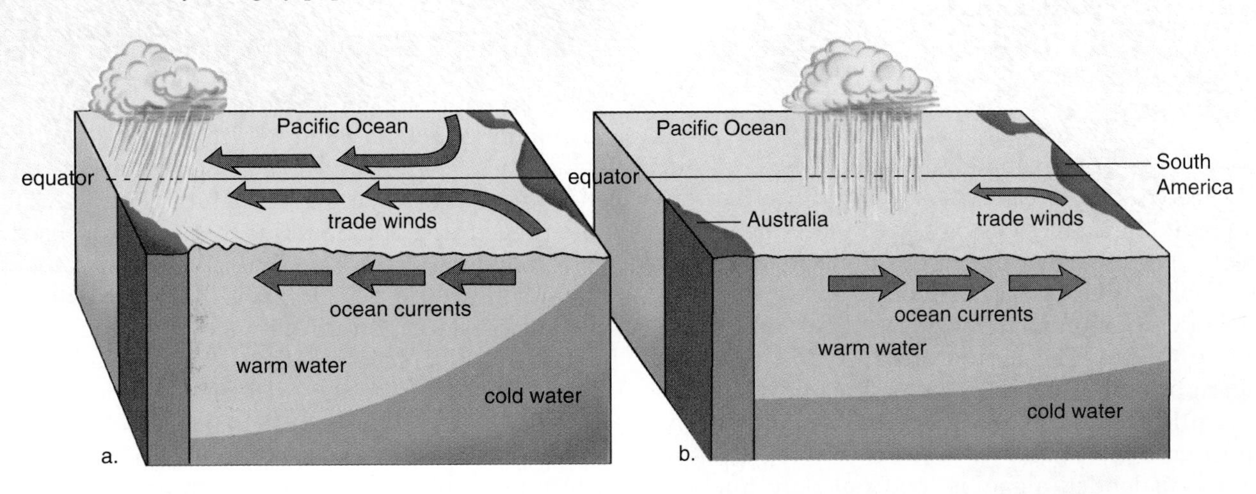

Figure 35B **a. La Niña**

- Upwelling off the west coast of South America brings cold waters to the surface.
- High barometric pressure over southeastern Pacific.
- Monsoons associated with Indian Ocean occur.
- Hurricanes off the east coast of the U.S.

b. El Niño

- Great ocean warming off the west coast of the Americas.
- Low barometric pressure over southeastern Pacific.
- Monsoons associated with Indian Ocean fail.
- Hurricanes off the west coast of the U.S.

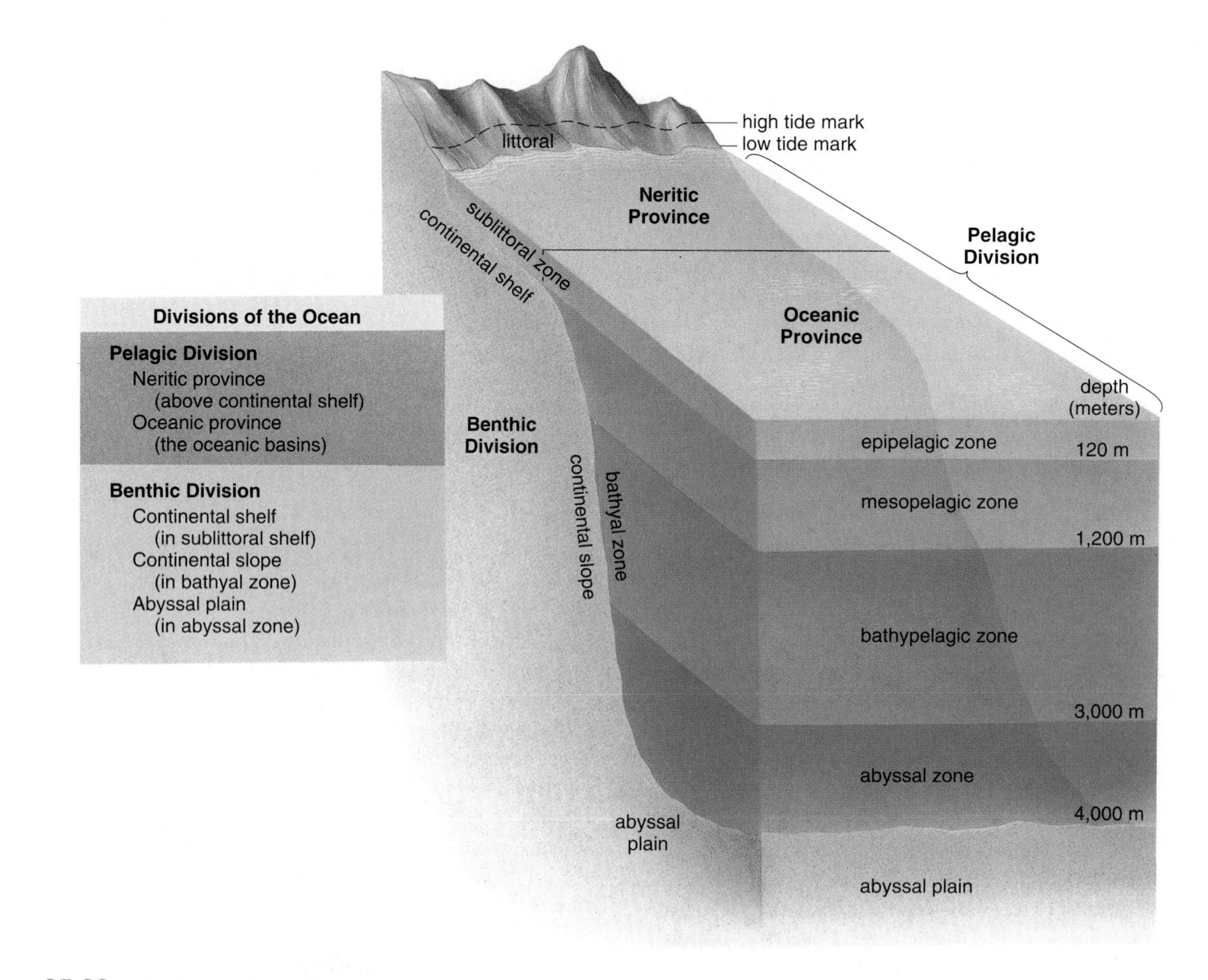

Figure 35.23 Marine environment.
Organisms reside in the pelagic division (blue), where waters are divided as indicated. Organisms also reside in the benthic division (brown), with surfaces divided and in the zones indicated.

Pelagic Division

The oceans cover approximately three-quarters of our planet. The geographic areas and zones of an ocean are shown in Figure 35.23. It is customary to place the organisms of the oceans into either the pelagic division (open waters) or the benthic division (ocean floor).

The **pelagic division** includes the neritic province and the oceanic province. The neritic province has a greater concentration of organisms than the oceanic province because the neritic province is sunlit and has a supply of inorganic nutrients for photosynthesizers. Phytoplankton, consisting of suspended algae, is food not only for zooplankton but also for small fishes. These small fishes in turn are food for commercially valuable fishes—herring, cod, and flounder.

The oceanic province lacks the inorganic nutrients of the neritic province, and therefore does not have as high a concentration of phytoplankton, even though the epipelagic zone is sunlit. Still, the photosynthesizers are food for a large assembly of zooplankton, which are food for herrings and bluefishes. These, in turn, are eaten by larger mackerels, tunas, and sharks. Flying fishes, which glide above the surface, are preyed upon by dolphins (not to be confused with porpoises, which also are present). Whales are other mammals found in the epipelagic zone. Baleen whales strain krill (small crustacea) from the water, and the toothed sperm whales feed primarily on the common squid.

Animals in the mesopelagic zone are carnivores, are adapted to the absence of light, and tend to be translucent, red colored, or even luminescent. There are luminescent shrimps, squids, and fishes, such as lantern and hatchet fishes.

The bathypelagic zone is in complete darkness except for an occasional flash of bioluminescent light. Carnivores and scavengers are found in this zone. Strange-looking fishes with distensible mouths and abdomens and small, tubular eyes feed on infrequent prey.

Benthic Division

The **benthic division** includes organisms that live on or in the soil of the continental shelf (sublittoral zone), the continental slope (bathyal zone), and the abyssal plain (abyssal zone) (Fig. 35.24). These are the organisms of the sublittoral, bathyal, and abyssal zones.

Seaweed grows in the sublittoral zone, and it can be found in batches on outcroppings as the water gets deeper. There is more diversity of life in the sublittoral and bathyal zones than in the abyssal zone. In these first two zones, clams, worms, and sea urchins are preyed upon by starfishes, lobsters, crabs, and brittle stars. Photosynthesizing algae occur in the sunlit sublittoral zone, but benthic organisms in the bathyal zone are dependent on the slow rain of detritus from the waters above.

The abyssal zone is inhabited by animals that live at the soil-water interface of the abyssal plain (Fig. 35.24). It once was thought that few animals exist in this zone because of the intense pressure and the extreme cold. Yet many invertebrates live here by feeding on debris floating down from the mesopelagic zone. Sea lilies rise above the seafloor; sea cucumbers and sea urchins crawl around on the sea bottom; and tube worms burrow in the mud.

The flat abyssal plain is interrupted by enormous underwater mountain chains called oceanic ridges. Along the axes of the ridges, crustal plates spread apart and molten magma rises to fill the gap. At **hydrothermal vents,** seawater percolates through cracks and is heated to about 350°C, causing sulfate to react with water and form hydrogen sulfide (H_2S). Chemosynthetic bacteria that obtain energy from oxidizing hydrogen sulfide exist freely or mutualistically within the tissues of organisms. They are the start of food chains for a community that includes huge tube worms and clams. It was a surprise to find communities of organisms living so deep in the ocean, where light never penetrates. Unlike photosynthesis, chemosynthesis does not require light energy.

The neritic province and the epipelagic zone of the oceanic province receive sunlight and contain the organisms with which we are most familiar. The organisms of the benthic division are dependent upon debris that floats down from above.

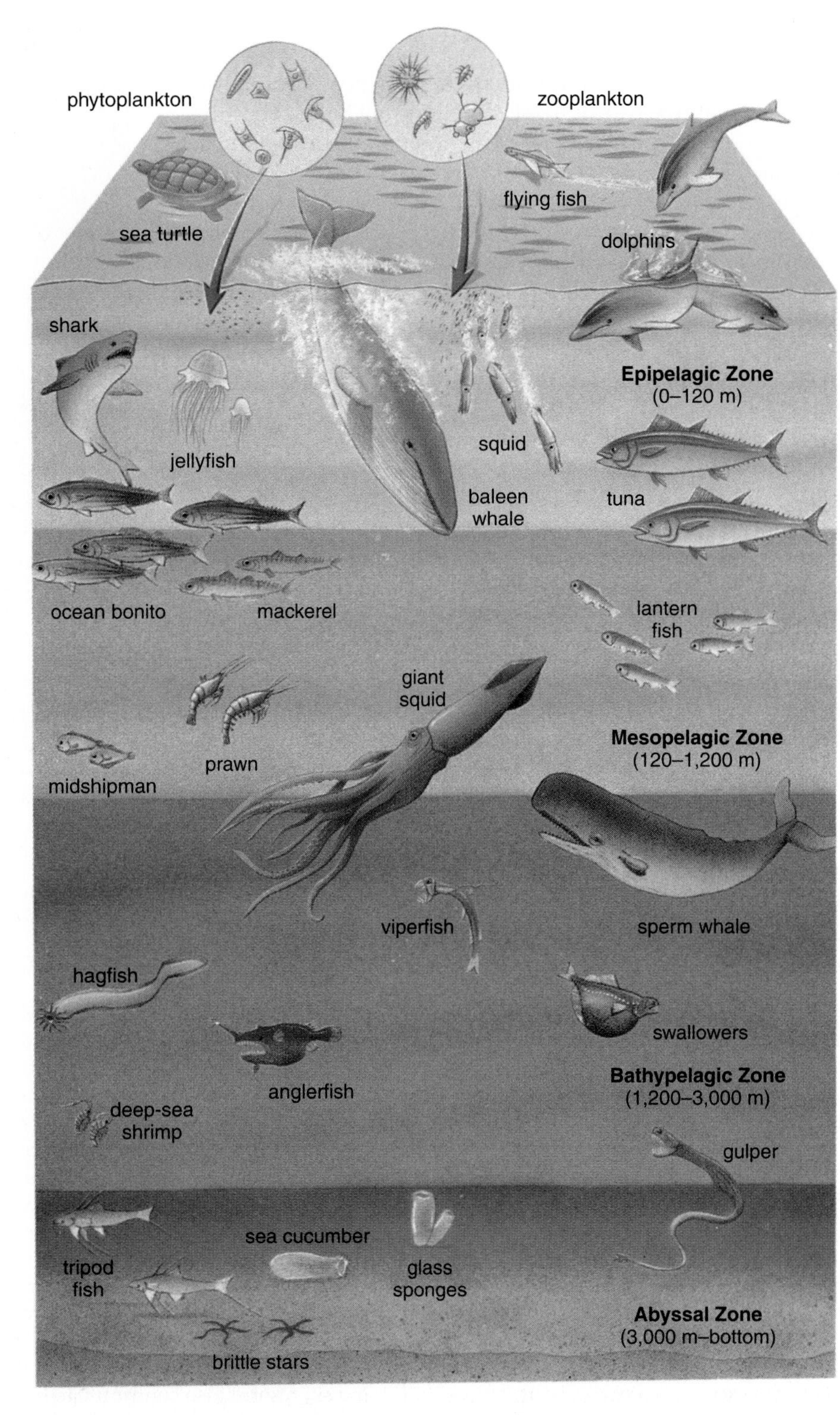

Figure 35.24 **Pelagic division.**
Organisms of the epipelagic, mesopelagic, and bathypelagic zones are shown. The abyssal zone is a part of the benthic division.

Coral Reefs

Coral reefs (Fig. 35.25) are areas of biological abundance found in shallow, warm, tropical waters just below the surface of the water. Their chief constituents are stony corals, animals that have a calcium carbonate (limestone) exoskeleton, and calcareous red and green algae. Corals do not usually occur individually; rather, they form colonies derived from an individual coral that has reproduced by means of budding. Corals provide a home for a microscopic alga called zooxanthellae. The corals, which feed at night, and the algae, which photosynthesize during the day, are mutualistic and share materials and nutrients. The close relationship of corals and zooxanthellae may be the reason why coral reefs form only in shallow sunlit water—the algae utilize sunlight for photosynthesis.

A reef is densely populated with life, perhaps because reefs are areas of intermediate disturbance (p. 713). The large number of crevices and caves provide shelter for filter feeders (sponges, sea squirts, and fan worms) and for scavengers (crabs and sea urchins). The barracuda, moray eel, and sharks are top predators in coral reefs. There are many types of small beautifully colored fishes. Parrot fishes feed directly on corals, and others feed on plankton or detritus. Small fishes become food for larger fishes like snappers that are caught for human consumption.

Figure 35.25 Coral reefs.
A coral reef is a rich community of marine organisms.

Coral reefs, which are areas of biological abundance, occur in tropical seas.

Bioethical Issue

Pollution of fresh water comes from two sources. Point sources, like paper mills or other industries that discharge pollutants directly into nearby rivers and bodies of water, are easily identifiable. Nonpoint sources, which are simply due to runoff from the land, are not easily identifiable. Although the public generally believes that big corporations are still causing most of the freshwater pollution, this is not the case. Instead, nonpoint sources are causing most of the problem. Agricultural fertilizers are the chief cause of nitrate contamination of drinking-water wells. Excessive nitrates in a baby's bloodstream can lead to a slow suffocation known as blue-baby syndrome. Agricultural herbicides are suspected carcinogens in the tap water of scattered communities coast to coast. A 1993 runoff from a dairy farm carried a parasite called cryptosporidium to the drinking water of Milwaukee, Wisconsin, sickening thousands of people, and killing a few. What can be done? Some farmers are already using irrigation methods that deliver water directly to plant roots, no-till agriculture that reduces the loss of top soil and cuts back on herbicide use, and integrated pest management that relies heavily on good bugs to kill bad bugs. Perhaps more should do so. Encouraged—in some cases compelled—by state and federal agents, dairy farmers have built sheds, concrete containments, and underground liquid storage tanks to hold the wastes from rainy days. Then, the manure can be trucked to fields and spread as fertilizer.

Home owners, like business golf clubs and ski resorts, also contribute to the problem. The manicuring of lawns, the use of motor vehicles, and the construction and use of roads and buildings all add contaminants to streams, lakes, and aquifers. Citizens around Grand Traverse Bay on the eastern shore of Lake Michigan have gotten the message, especially because they want to keep on enjoying water-dependent activities like boating, bathing, and fishing. James Haverman, a concerned member of the Traverse Bay Watershed Initiative says, "If we can't change the way people live their everyday lives, we are not going to be able to make a difference." Builders in Traverse County are already required to control soil erosion with filter fences, steer rainwater away from exposed soil, build sediment basins, and plant protective buffers. Should similar restrictions hold for home builders? Presently they must have a 25-foot setback from wetlands and a 50-foot setback from lakes and creeks. They are also encouraged to pump out their septic systems every two years.

Questions

1. Do you approve of legislation that requires farmers and homeowners to protect freshwater supplies? Why or why not?
2. Should landscapers and gardeners be required to follow certain restrictions? Why or why not?
3. Do you approve of legislation that restricts the number of homes in a particular area? Why or why not?

Summarizing the Concepts

35.1 Climate and the Biosphere

Because the earth is a sphere, the sun's rays at the poles are spread out over a larger area than the vertical rays at the equator. The temperature at the surface of the earth therefore decreases from the equator to each pole. The earth is tilted on its axis, and the seasons change as the earth rotates annually about the sun.

Warm air rises near the equator and loses its moisture and then descends at about 30° north and south latitude, and so forth to the poles. When the air descends, it is dry, and therefore the great deserts of the world are formed at 30° latitudes. Because the earth rotates on its axis daily, the winds blow in opposite directions above and below the equator. Topography also plays a role in the distribution of moisture. Air rising over coastal ranges loses its moisture on the windward side, making the leeward side arid.

35.2 Biomes of the World

The term biome is used to refer to terrestrial and aquatic communities. Biomes are distributed according to climate; that is, temperature and rainfall influence the pattern of biomes about the world. The significance of temperature is also seen by observing that the same sequence of biomes is seen when traveling to northern latitudes as traveling up a mountain.

35.3 Terrestrial Biomes

The tundra is the northernmost biome and consists largely of short grasses and sedges and dwarf woody plants. Because of the cold winters and short summers, most of the water in the soil is frozen the year round. This is called the permafrost.

The taiga, a coniferous forest, has less rainfall than other types of forests. The temperate deciduous forest has trees that gain and lose their leaves because of the alternating seasons of summer and winter. Tropical rain forests are the most complex and productive of all biomes.

Shrublands usually occur along coasts that have dry summers and receive most of their rainfall in the winter. Among grasslands, the savanna, a tropical grassland, supports the greatest number of different types of large herbivores. The prairie, found in the United States, has a limited variety of vegetation and animal life.

Deserts are characterized by a lack of water—they are usually found in places with less than 25 cm a year. Some plants, such as cacti, are succulents, and others are shrubs with thick leaves they often lose during dry periods.

35.4 Aquatic Biomes

Rain falls in mountains, and streams develop that join to form a river which runs into the sea. Streams, rivers, lakes, and wetlands are different communities.

In deep lakes of the temperate zone, the temperature and the concentration of nutrients and gases in the water vary with depth. The entire body of water is cycled twice a year, distributing nutrients from the bottom layers. Lakes and ponds have three life zones. Rooted plants and clinging organisms live in the littoral zone, plankton and fishes are in the sunlit limnetic zone, and bottom-dwelling organisms like crayfishes and mollusks are in the profundal zone.

Estuaries (e.g., salt mashes, mudflats, mangrove forests) are near the mouth of a river. Estuaries are the nurseries of the sea. Marine communities are divided into coastal communities and the oceans. The coastal communities, especially estuaries, are more productive than the oceans.

An ocean is divided into the pelagic division and the benthic division. The oceanic province of the pelagic division (open waters) has three zones. The epipelagic zone receives adequate sunlight and supports the most life. The mesopelagic zone contains organisms adapted to minimum and no light, respectively. The benthic division (ocean floor) includes organisms living on the continental shelf in the sublittoral zone, the continental slope in the bathyal zone, and the abyssal plain in the abyssal zone. Coral reefs are extremely productive communities found in shallow tropical waters.

Studying the Concepts

1. Tell how a spherical earth and the path of the earth about the sun affect climate. 718–19
2. Describe the air circulation about the earth; tell why deserts are apt to occur at 30° north and south of the equator. Why does the Pacific coast of California get plentiful rainfall? 719
3. Name the terrestrial biomes you would expect to find when going from the base to the top of a mountain. 721
4. Describe the location, the climate, and the populations of the Arctic tundra, coniferous forests (both taiga and temperate rain forest), temperate deciduous forests, tropical forests, shrubland, grasslands (both prairie and savanna), and deserts. 722–24, 726–30
5. Describe the overturn of a temperate lake, and the life zones of a lake, and the organisms you would expect to find in each zone. 732–33
6. Describe the coastal communities, and discuss the importance of estuaries to the productivity of the ocean. 734–35
7. Describe the ocean currents and how the Gulf Stream accounts for Great Britain having a mild temperature. 736
8. Describe the zones of the open ocean and the organisms you would expect to find in each zone. 738–39
9. Describe a coral reef, including the varied organisms found there. 740

Testing Yourself

Choose the best answer for each question.

1. The seasons are best explained by
   a. the distribution of temperature and rainfall in biomes.
   b. the tilt of the earth on its axis as it rotates about the sun.
   c. the daily rotation of the earth on its axis.
   d. the fact that the equator is warm and the poles are cold.
2. The mild climate of Great Britain is best explained by
   a. the winds called the westerlies.
   b. the spinning of the earth on its axis.
   c. Great Britain being a mountainous country.
   d. the flow of ocean currents.
3. The location of deserts at 30° is best explained by which statement?
   a. Warm air rises and loses its moisture; cool air descends and becomes warmer and drier.
   b. Ocean currents give up heat as they flow from the equator to the poles.
   c. Cool ocean breezes cool the coast during the day.
   d. All of these are correct.

4. Which of these is mismatched?
   a. tundra—permafrost
   b. savanna—acacia trees
   c. prairie—epiphytes
   d. coniferous forest—evergreen trees
5. All of these phrases describe the tundra except
   a. low-lying vegetation.
   b. northernmost biome.
   c. short growing season.
   d. many different types of species.
6. The forest with a multilevel understory is the
   a. tropical rain forest.
   b. coniferous forest.
   c. tundra.
   d. temperate deciduous forest.
7. All of these phrases describe a tropical rain forest except
   a. nutrient-rich soil.
   b. many arboreal plants and animals.
   c. canopy composed of many layers.
   d. broad-leaved evergreen trees.
8. Phytoplankton are more likely to be found in which life zone of a lake?
   a. limnetic zone
   b. profundal zone
   c. benthic zone
   d. All of these are correct.
9. An estuary acts as a nutrient trap because of the
   a. action of rivers and tides.
   b. depth at which photosynthesis can occur.
   c. amount of rainfall received.
   d. height of the water table.
10. Which area of an ocean has the greatest concentration of nutrients?
    a. epipelagic zone and benthic zone
    b. epipelagic zone only
    c. benthic zone only
    d. neritic province

Thinking Scientifically

1. Concerning biome structure, use the data given in Figure 35.4*a*
   a. to hypothesize what two environmental conditions are most influential in determining the type of biome.
   b. If a deciduous forest is destroyed by fire, what less complex biome might take its place? If a grassland biome is overgrazed, what less complex biome might takes its place?
   c. Hypothesize that the greater amount of life in tropical rain forest is related to a plentiful supply of sunlight. Explain why this might be so.
   d. Hypothesize that the great diversity of a tropical rain forest is explained by the competitive exclusion principle. Explain why this might be so.
   e. Hypothesize that the great diversity of a tropical rain forest is related to a plentiful supply of varied foods. Explain why this might be so.
2. Concerning the mix of plants and animals in a biome,
   a. give two reasons based on the principle of adaptation to the environment why you would expect to find a squid in the ocean and not in tropical rain forest.
   b. give two reasons why you would expect to find a monkey in a tropical rain forest and not in a grassland.
   c. give two reasons why you would expect to find a zebra in a grassland and not in a tropical rain forest.
   d. give two reasons why you would expect to find a polar bear in the Arctic tundra along the coast and not in a desert.
   e. What general conclusions can you draw from these examples?

Understanding the Terms

alpine tundra 721
Arctic tundra 722
benthic division 739
biome 721
biosphere 718
chaparral 728
climate 718
coral reef 740
desert 730
El Niño–Southern Oscillation 736
epiphyte 726
estuary 734
eutrophication 732
fall overturn 732
hydrothermal vent 739
lake 732
montane coniferous forest 721
pelagic division 738
permafrost 722
phytoplankton 733
plankton 733
rain shadow 719
savanna 728
shrubland 728
spring overturn 733
taiga 723
temperate deciduous forest 724
temperate rain forest 723
tropical rain forest 726
upwelling 736
zooplankton 733

Match the terms to these definitions:

a. ________ End of a river where fresh water and salt water mix as they meet.
b. ________ Major terrestrial community characterized by certain climatic conditions and dominated by particular types of plants.
c. ________ Ocean floor, which supports a unique set of organisms in contrast to the pelagic division.
d. ________ Terrestrial biome that is a grassland in Africa, characterized by few trees and a severe dry season.
e. ________ Thin layer around the earth that supports life.

Using Technology

Your study of the biosphere is supported by these available technologies:

Essential Study Partner CD-ROM

Ecology → Biosphere

Visit the Mader web site for related ESP activities.

Exploring the Internet

The Mader Home Page provides resources and tools as you study this chapter.

http://www.mhhe.com/biosci/genbio/mader

Life Science Animations 3D Video

42 Nutrient Cycling

C H A P T E R 36

Ecosystems and Human Interferences

Humans usually live in developed areas with a limited variety of species, and these areas are sources of pollution that is harmful to all forms of life.

Chapter Concepts

A low rumble shook the ground and Dan smiled with quiet satisfaction. He had led the successful effort to persuade the state to dynamite the eroding dam on the river. Soon, for the first time in three decades, salmon would again swim upstream to spawn, one small sign that the region was returning to its natural state. Although he worked for the state as a conservation biologist, Dan thought his parents, both environmental activists, would be proud. As a child, they had dragged him around the globe to protest the destruction of rain forests in Brazil, campaign against acid rain in Europe, and expose the dumping of toxic waste into New England rivers. Such journeys had instilled in Dan an appreciation for the complex interactions between animals and their surroundings and a concern about how quickly people's actions could disrupt this natural order. This chapter will describe some of what Dan has learned about ecosystems and the upheaval in them that people can cause.

36.1 The Nature of Ecosystems

An **ecosystem** is a community along with its physical and chemical environment. The populations are the biotic component, and the physical and chemical environment make up the abiotic component of an ecosystem. In ecosystems, populations are classified according to how they get their food (organic nutrients) (Fig. 36.1). **Autotrophs** make their own food and heterotrophs feed on other organisms. **Heterotrophs** include decomposers which feed on organic material in the soil.

Autotrophs

Because autotrophs produce their own food (organic nutrients) for themselves and other members of the community, they are called **producers.** Chemoautotrophs are bacteria that obtain energy by oxidizing inorganic compounds, such as ammonium (NH_4^+) and hydrogen sulfide (H_2S), and they use this energy to synthesize carbohydrates. The chemoautotrophs that function in the nitrogen cycle will be discussed on page 752. Photoautotrophs are photosynthesizers that produce most of the food for the biosphere. Algae and sea grasses carry on photosynthesis in aquatic habitats. Algae make up the phytoplankton, which are photosynthesizing organisms suspended in water. Green plants are the dominant photosynthesizers on land.

Heterotrophs

Heterotrophs need a preformed source of food. They are the **consumers**—they consume food. **Herbivores** are animals that graze directly on plants or algae. In terrestrial habitats, insects are small herbivores, while in aquatic habitats, zooplankton, such as many types of protozoa, play that role. **Carnivores** feed on other animals; birds that feed on insects are carnivores, and so are hawks that feed on birds. This example illustrates that there are *primary consumers* (e.g., insects), *secondary consumers* (e.g., birds), and *tertiary consumers* (e.g., hawks). **Omnivores** are animals that feed both on plants and animals. As you most likely know, humans are omnivores.

The bacteria and fungi of decay are **decomposers** that break down dead organic matter, including animal wastes. Some animals feed on **detritus,** which is decomposing particles of organic matter. Earthworms and some beetles, termites, and maggots consume terrestrial detritus. Decomposers perform a very valuable service because they release inorganic substances that are taken up by plants once more.

The biotic components of an ecosystem can be classified according to the way they get their food.

a.

b.

Figure 36.1 Biotic components of an ecosystem.
In the savanna biome, **(a)** the grasses, which are producers, are eaten by many types of herbivores, such as these zebras. **(b)** Lions are carnivores that feed on zebras.

Energy Flow and Chemical Cycling

When we diagram all the biotic components of an ecosystem, as in Figure 36.2 it is possible to illustrate that every ecosystem is characterized by two fundamental phenomena: energy flow and chemical cycling. *Energy flow* begins when producers absorb solar energy, and chemical cycling begins when producers take in inorganic nutrients from the physical environment. Thereafter, producers make food for themselves and indirectly for the other populations of the ecosystem. Energy flow occurs because all the energy content of organic nutrients is eventually converted to heat, which dissipates in the environment. Therefore most ecosystems cannot exist without a continual supply of solar energy. *Chemicals cycle* when inorganic nutrients are returned to the producers from the atmosphere or soil, as appropriate.

Only a portion of the food made by autotrophs is passed on to heterotrophs because plants use organic molecules to fuel their own cellular respiration. Only about 55% of the food made by producers is available to heterotrophs. Similarly, only a small percentage of food taken in by heterotrophs is available to higher level consumers. Figure 36.3 shows why. A certain amount of the food eaten by a herbivore is never digested and is eliminated as feces. Metabolic wastes are excreted as urine. Of the assimilated energy, a large portion is utilized during cellular respiration and thereafter becomes heat. Only the remaining food which is converted into increased body weight (or additional offspring) becomes available to carnivores.

The elimination of feces and urine by a heterotroph, and indeed the death of all organisms, does not mean that substances are lost to an ecosystem. They represent the food made available to decomposers. Since decomposers can be food for other heterotrophs of an ecosystem, the situation can get a bit complicated. Still, we can conceive that all the solar energy that enters an ecosystem eventually becomes heat. And this is consistent with the observation that ecosystems are dependent on a continual supply of solar energy.

The laws of thermodynamics support the concept that energy flows through an ecosystem. The first law states that energy cannot be created (nor destroyed). This explains why ecosystems are dependent on a continual outside source of energy, usually solar energy, which is used by photosynthesizers to produce food. The second law states that with every transformation some energy is degraded into a less available form such as heat. Because plants carry on cellular respiration, for example, only about 55% of the original energy absorbed by plants is available to an ecosystem.

Energy flows through an ecosystem, while chemicals cycle within and between ecosystems.

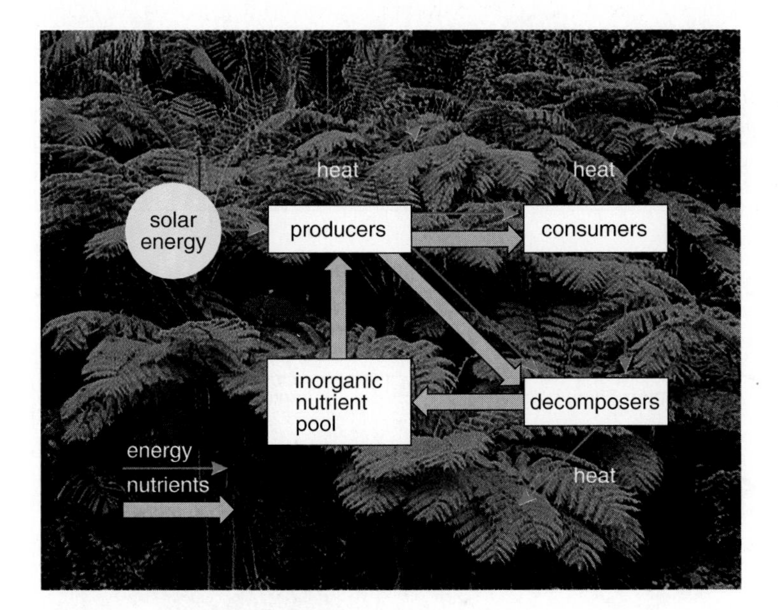

Figure 36.2 Nature of an ecosystem.
Chemicals cycle but energy flows through an ecosystem. All the energy derived from the sun eventually dissipates as heat as energy transformations repeatedly occur.

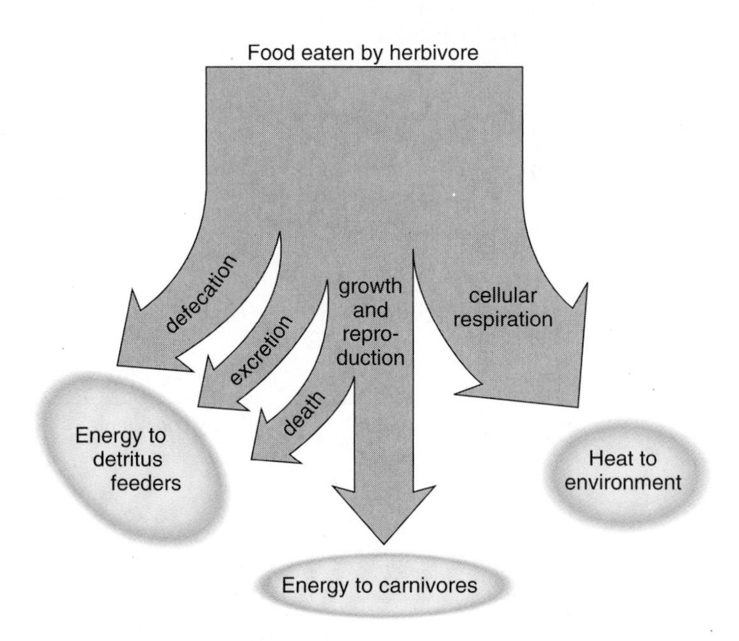

Figure 36.3 Energy balances.
Only about 10% of the food energy taken in by a herbivore is passed on to carnivores. A large portion goes to detritus feeders in the ways indicated, and another large portion is used for cellular respiration.

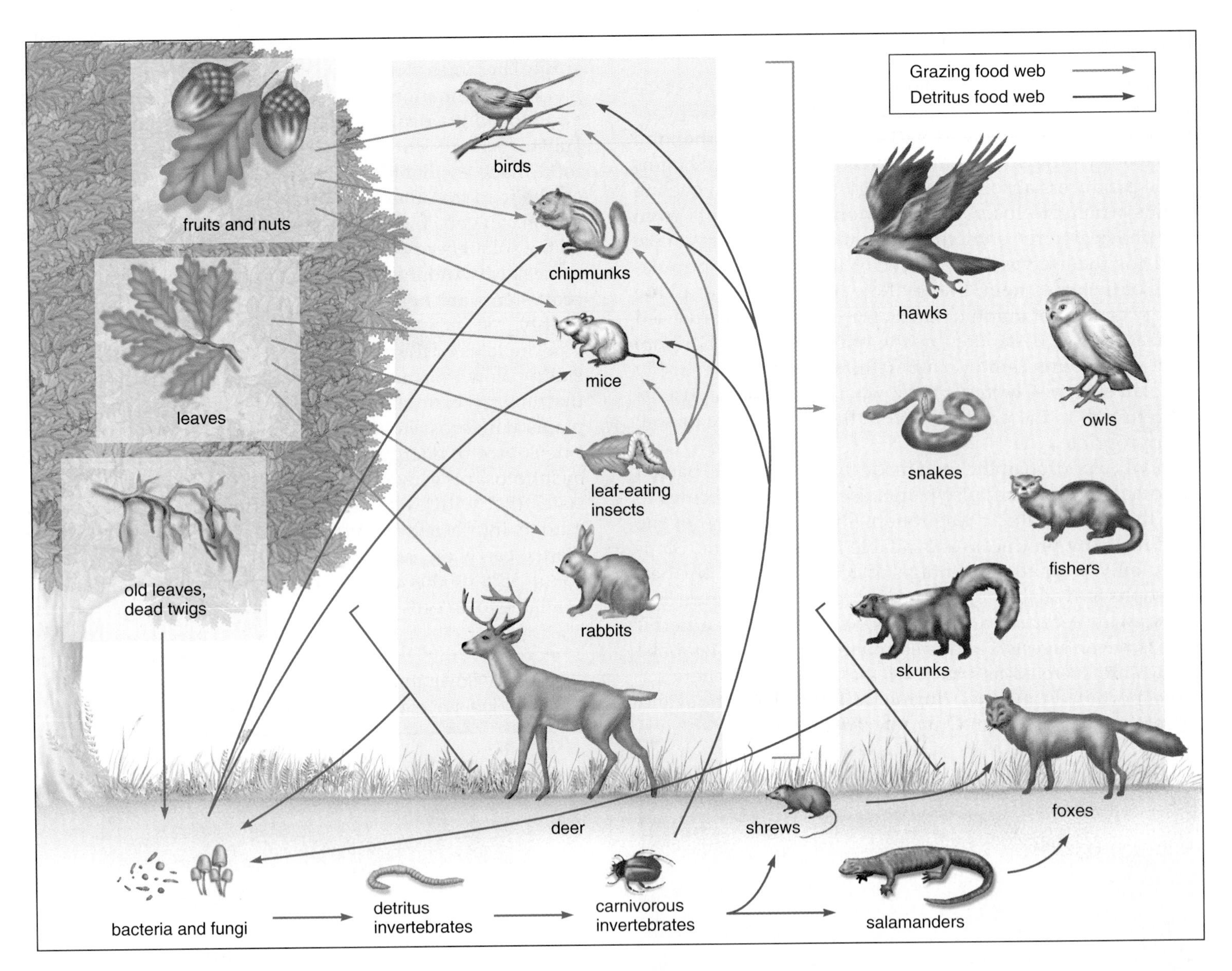

Figure 36.4 Forest food webs.
Two linked food webs are shown for a forest ecosystem: a grazing food web and a detrital food web.

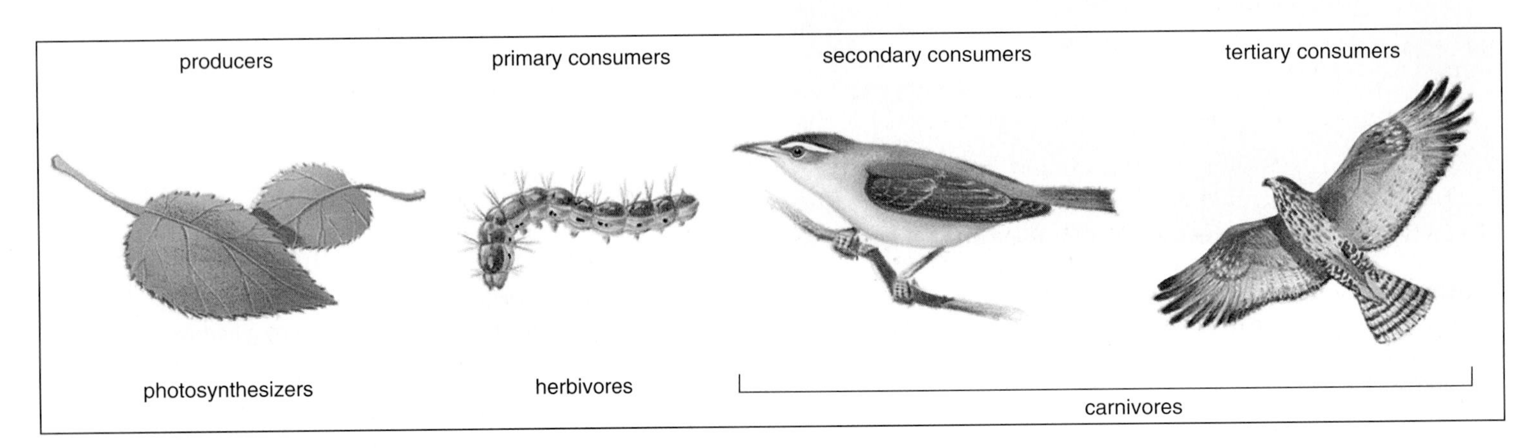

Figure 36.5 Food chain.
Trace this grazing food chain in the grazing food web depicted in Figure 36.4.

Food Webs and Trophic Levels

The principles we have been discussing can now be applied to an actual example—a forest in New Hampshire. In this forest, the producers include sugar maple, beech, and yellow birch trees. The complicated feeding relationships that exist in natural ecosystems are called food webs. A **food web** shows how organisms acquire their food. For example, Figure 36.4 shows that insects in the form of caterpillars feed on leaves, while mice, rabbits, and deer feed on leaf tissue at or near the ground. Birds, chipmunks, and mice feed on fruits and nuts, but they are in fact omnivores because they also feed on caterpillars. These herbivores and omnivores all provide nutrients for a number of different carnivores. This portion of the diagram is called a **grazing food web** because it begins with aboveground plant material.

The lower half of Figure 36.4 is devoted to the **detrital food web.** Detritus, along with the bacteria and fungi of decay, can be food for larger decomposers. Because some of these, like shrews and salamanders, become food for aboveground carnivores, the detrital and the grazing food webs are connected.

We naturally tend to think that aboveground vegetation like trees are the largest storage form of organic matter and energy, but this is not necessarily the case. In this particular forest, the organic matter lying on the forest floor and mixed into the soil contains much more energy than does the leaf matter of living trees. The soil contains over twice as much energy as the forest floor. Therefore, more energy in a forest may be funneling through the detrital food web than through the grazing food web.

Trophic Levels

You can see that Figure 36.4 would allow us to link organisms one to another in a straight line manner, according to who eats whom. Such diagrams are called **food chains** (Fig. 36.5). For example, in the grazing food web we can find this **grazing food chain:**

leaves → caterpillars → tree birds → hawks

And in the detrital food web we could find this **detrital food chain:**

dead organic matter → soil microbes → earthworms → etc.

A **trophic level** is all the organisms that feed at a particular link in a food chain. In the grazing food web, going from left to right, the trees are primary producers (first trophic level), the first series of animals are primary consumers (second trophic level), and the next group of animals are secondary consumers (third trophic level) and so forth.

Ecological Pyramids

Ecologists portray the energy relationships between trophic levels in the form of **ecological pyramids,** diagrams whose building blocks designate the various trophic levels (Fig. 36.6). (We need to keep in mind that sometimes organisms don't fit into one trophic level. For example, chipmunks feed on fruits and nuts, but they also feed on leaf-eating insects.)

A *pyramid of numbers* simply tells how many organisms there are at each trophic level. It's easy to see that a pyramid of numbers could be completely misleading. For example, in Figure 36.4 you would expect each tree to contain numerous caterpillars; therefore there would be more herbivores than autotrophs! The problem, of course, has to do with size. Autotrophs can be tiny, like microscopic algae, or they can be big like beech trees; similarly, herbivores can be small like caterpillars, or they can be large like elephants.

Pyramids of biomass eliminate size as a factor since biomass is the number of organisms multiplied by their weight. You would certainly expect the biomass of producers to be greater than the biomass of the herbivores, and that of the herbivores to be greater than the carnivores. In some aquatic ecosystems such as lakes and open seas, where algae are the only producers, the herbivores may have a greater biomass than the producers when you take their measurements. Why? The reason is that over time, the algae reproduce rapidly, but they are also consumed

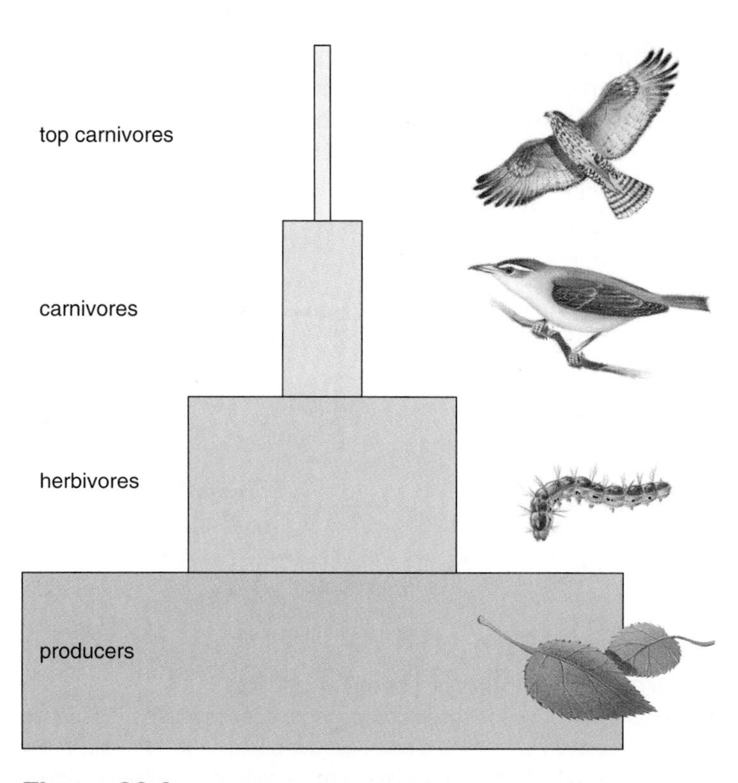

Figure 36.6 Ecological pyramid.
An ecological pyramid shows the relationship between either the number of organisms, the biomass, or the amount of energy theoretically available at each trophic level.

at a high rate. Pyramids, like this one, that have more herbivores than producers are called inverted pyramids:

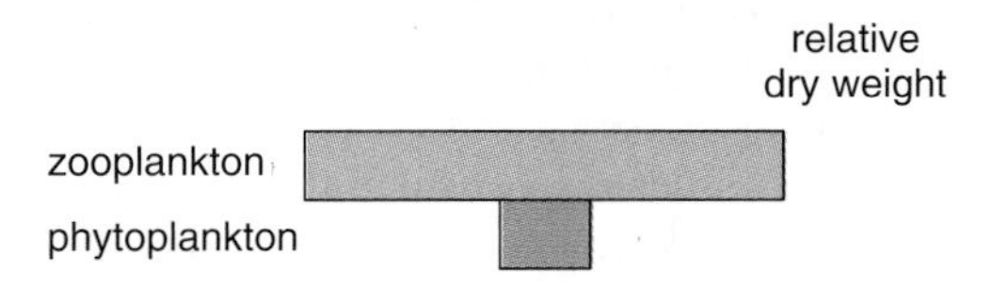

There are ecological *pyramids of energy* also, and they generally have the appearance of Figure 36.6. Ecologists are now beginning to rethink the usefulness of utilizing pyramids to describe energy relationships. One problem is what to do with the decomposers, which are rarely included in pyramids, and yet a large portion of energy becomes detritus in many ecosystems.

There is a rule of 10% with regard to biomass (or energy) pyramids. It says that, in general, the amount of biomass (or energy) from one level to the next is reduced by a magnitude of 10. Thus, if an average of 1,000 kg of plant material is consumed by herbivores, about 100 kg is converted to herbivore tissue, 10 kg to first-level carnivores, and 1 kg to second-level carnivores. The rule of 10% suggests that few carnivores can be supported in a food web. This is consistent with the observation that each food chain has from three to four links, rarely five.

36.2 Global Biogeochemical Cycles

All organisms require a variety of organic and inorganic nutrients. Carbon dioxide and water are necessary for photosynthesis. Nitrogen is a component of all the structural and functional proteins and nucleic acids that sustain living tissues. Phosphorus is essential for ATP and nucleotide production. In contrast to energy, inorganic nutrients are used over and over again by autotrophs.

Since the pathways by which chemicals circulate through ecosystems involve both living (biosphere) and nonliving (geological) components, they are known as **biogeochemical cycles.** For each element, chemical cycling may involve (1) a reservoir—a source normally unavailable to producers, such as fossilized remains, rocks, and deep-sea sediments; (2) an exchange pool—a source from which organisms do generally take chemicals, such as the atmosphere or soil; and (3) the biotic community—through which chemicals move along food chains, perhaps never entering a pool (Fig. 36.7).

There are two general categories of biogeochemical cycles. In a *gaseous cycle,* exemplified by the carbon and nitrogen cycles, the element returns to and is withdrawn from the atmosphere as a gas. In the *sedimentary cycle,* exemplified by the phosphorus cycle, the element is absorbed from the sediment by plant roots, passed to heterotrophs, and is eventually returned to the soil by decomposers, usually in the same general area.

The diagrams on the next few pages make it clear that nutrients can flow between terrestrial and aquatic ecosystems. In the nitrogen and phosphorus cycles, these nutrients run off from a terrestrial to an aquatic ecosystem and in that way enrich aquatic ecosystems. Decaying organic material in aquatic ecosystems can be a source of nutrients for intertidal inhabitants like fiddler crabs. Sea birds feed on fish but deposit guano (droppings) on land, and in that way phosphorus from the water is deposited on land. It would seem that anything put into the environment in one ecosystem could find its way to another ecosystem. Scientists find the soot from urban areas and pesticides from agricultural fields in the snow and animals of the Arctic.

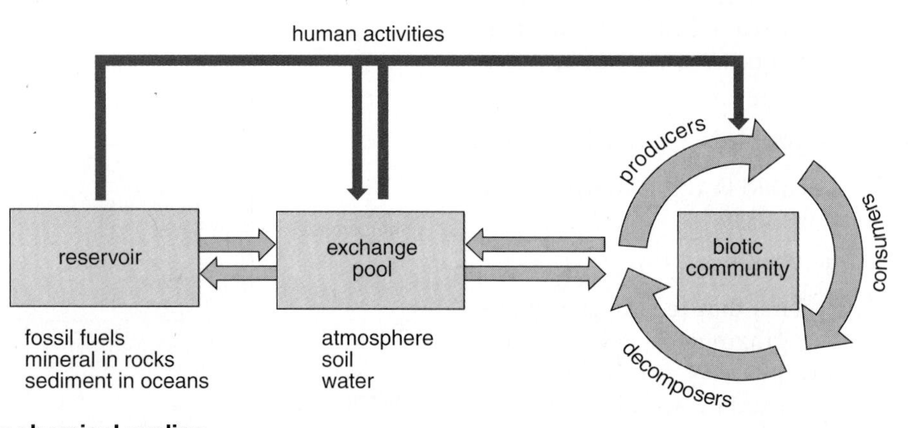

Figure 36.7 Model for chemical cycling.
Nutrients cycle between these components of ecosystems: Reservoirs such as fossil fuels, minerals in rocks, and sediments in oceans are normally relatively unavailable sources, but pools such as those in the atmosphere, soil, and water are available sources of chemicals for the biotic community. Human activities remove chemicals from reservoirs and pools and make them available to the biotic community, and the result can be pollution.

The Water Cycle

The **water (hydrologic) cycle** is described in Figure 36.8. Fresh water is distilled from salt water. The sun's rays cause fresh water to evaporate from seawater, and the salts are left behind. Vaporized fresh water rises into the atmosphere, cools, and falls as rain over the oceans and the land.

Water evaporates from land and from plants (evaporation from plants is called transpiration). It also evaporates from bodies of fresh water, but since land lies above sea level, gravity eventually returns all fresh water to the sea. In the meantime, water is contained within standing waters (lakes and ponds), flowing water (streams and rivers), and groundwater.

When rain falls, some of the water sinks or percolates into the ground and saturates the earth to a certain level. The top of the saturation zone is called the groundwater table, or simply, the water table. Sometimes groundwater is also located in **aquifers,** rock layers that contain water and will release it in appreciable quantities to wells or springs. Aquifers are recharged when rainfall and melted snow percolate into the soil. In some parts of the country, especially arid areas and southern Florida, withdrawals from aquifers exceed any possibility of recharge. This is called "groundwater mining." In these locations the groundwater is dropping, and residents may run out of groundwater, at least for irrigation purposes, within a few short years. Fresh water, which makes up only about 3% of the world's supply of water, is called a renewable resource because a new supply is always being produced. But it is possible to run out of fresh water when the available supply is not adequate and/or is polluted so that it is not usable.

In the water cycle, fresh water evaporates from the bodies of water. Water that falls on land enters the ground, surface waters, or aquifers. Water ultimately returns to the ocean—even the quantity that remains in aquifers for some time.

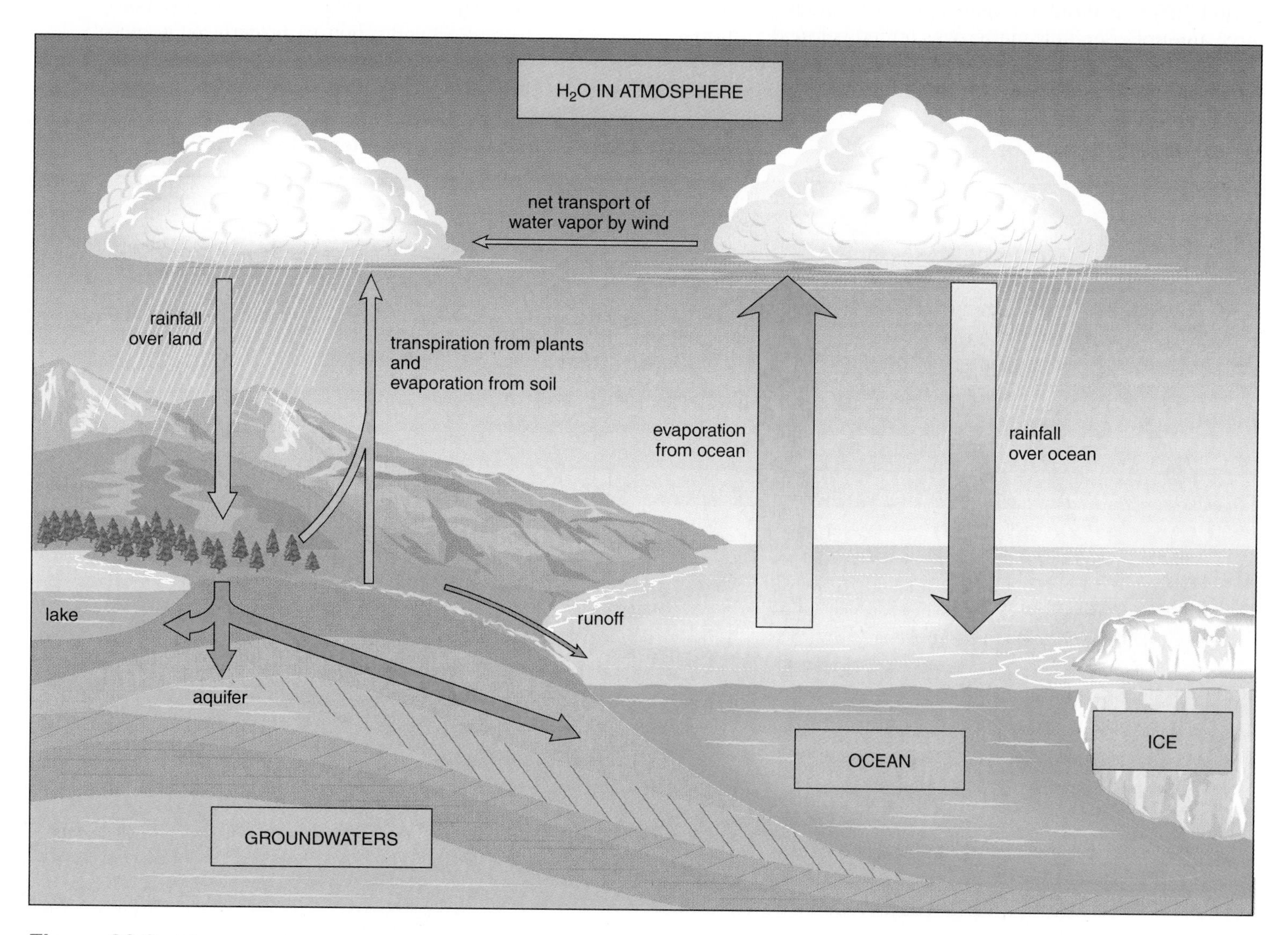

Figure 36.8 **The water (hydrologic) cycle.**

The Carbon Cycle

In the **carbon cycle**, both terrestrial and aquatic organisms exchange carbon dioxide with the atmosphere (Fig. 36.9). On land, plants take up carbon dioxide from the air, and through photosynthesis they incorporate carbon into food that is used for other living things. When organisms (e.g., plants, animals, and decomposers) respire, a portion of this carbon is returned to the atmosphere as carbon dioxide. In aquatic ecosystems, the exchange of carbon dioxide with the atmosphere is indirect. Carbon dioxide from the air combines with water to produce bicarbonate ion (HCO_3^-), a source of carbon for algae, which also produce food through photosynthesis. And when aquatic organisms respire, the carbon dioxide they give off becomes bicarbonate ion.

Living and dead organisms are reservoirs for carbon. If decomposition of dead remains fails to occur, they are subject to physical processes that transform them into coal, oil, and natural gas. We call these reservoirs for carbon the **fossil fuels.** Most of the fossil fuels were formed during the Carboniferous period, 286 to 360 million years ago, when an exceptionally large amount of organic matter was buried before decomposing. Another reservoir for carbon is calcium carbonate shells, which accumulate in ocean bottom sediments.

Carbon Dioxide and Global Warming

A **transfer rate** is defined as the amount of a nutrient that moves from one component of the environment to another within a specified period of time. The width of the arrows in Figure 36.9 indicates the transfer rate of carbon dioxide. The transfer rates due to photosynthesis and respiration, which includes decay, are just about even. However, there is now more carbon dioxide being deposited in the atmosphere than being removed. In 1850, atmospheric carbon dioxide was about 280 parts per million (ppm) and today it is about 350 ppm. This increase is largely due to the burning of fossil fuels and the destruction of forests to make way for farmland and pasture.

The emission of other gases due to human activities is also taking place. Altogether the following gases are expected to contribute significantly to global warming:

Gas	From
Carbon dioxide (CO_2)	Fossil fuel and wood burning
Nitrous oxide (N_2O)	Fertilizer use and animal wastes
Methane (CH_4)	Biogas (bacterial decomposition, particularly in the guts of animals, in sediments, and in flooded rice paddies)

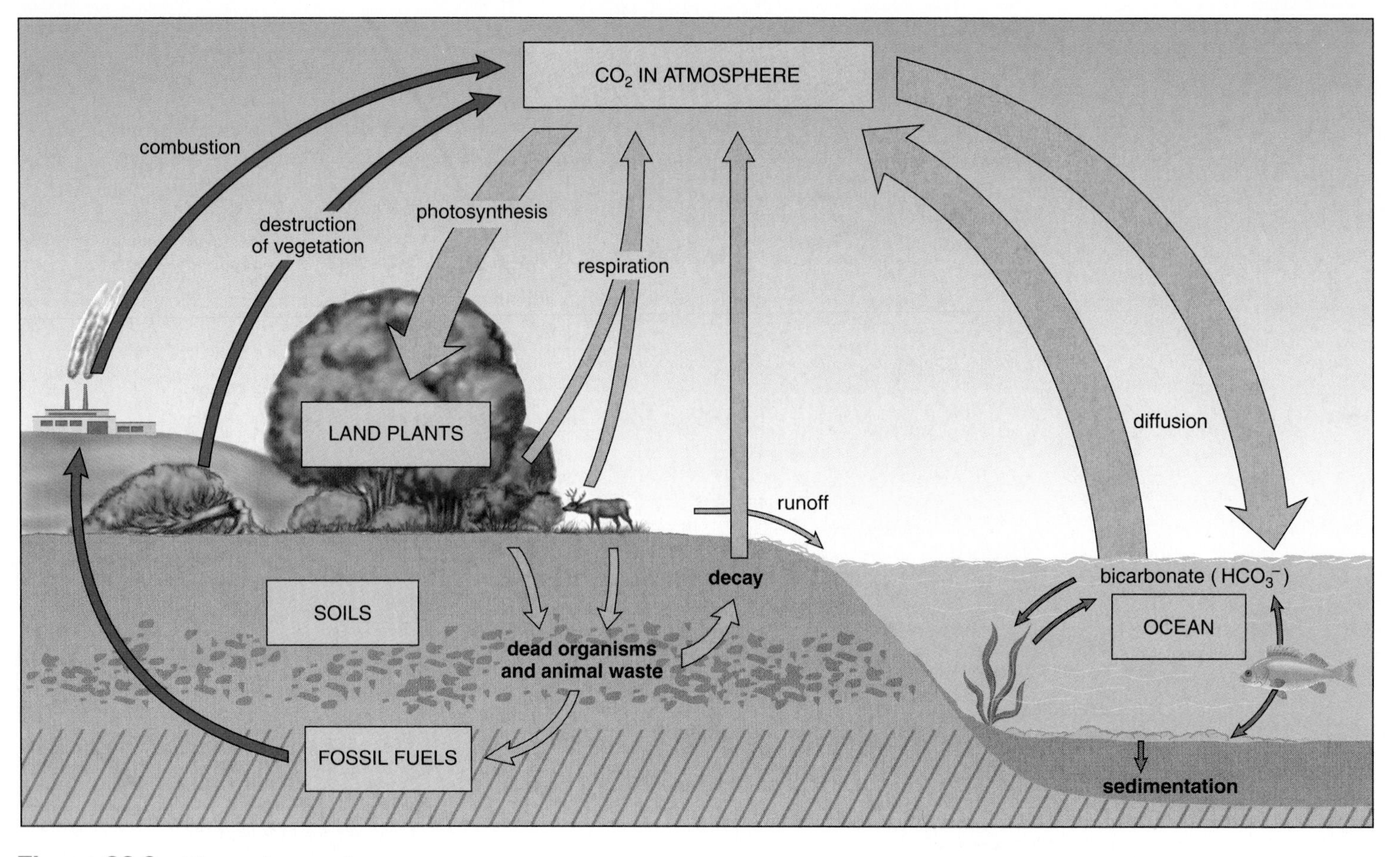

Figure 36.9 **The carbon cycle.**

These gases are called greenhouse gases because just like the panes of a greenhouse they allow solar radiation to pass through but hinder the escape of infrared rays (heat) back into space. Figure 36.10 shows the earth's radiation balances. One thing to be learned from this diagram is that water vapor is a greenhouse gas: clouds also reradiate heat back to earth. If the earth's temperature rises due to the **greenhouse effect,** more water will evaporate, forming more clouds, setting up a positive feedback effect that could increase global warming.

Today, data collected around the world show a steady rise in the concentration of the various greenhouse gases. Methane, another significant greenhouse gas, given off by oil and gas wells, rice paddies, and organisms, is increasing by about 1% a year. Such data are used to generate computer models that predict the earth may warm to temperatures never before experienced by living things. The global climate has already warmed about 0.6°C since the industrial revolution. Computer models are unable to consider all possible variables, but the earth's temperature may rise 1.5–4.5°C by 2100 if greenhouse emissions continue at the current rates.

Global warming will bring about other effects, which computer models attempt to forecast. It is predicted that as the oceans warm, temperatures in the polar regions will rise to a greater degree than other regions. If so, glaciers would melt, and sea levels will rise, not only due to this melting but also because water expands as it warms. Water evaporation will increase, and most likely there will be increased rainfall along the coasts and dryer conditions inland. The occurrence of droughts will reduce agricultural yields and also cause trees to die off. Expansion of forests into Arctic areas might not offset the loss of forests in the temperate zones. Coastal agricultural lands such as the deltas of Bangladesh, India, and China would be inundated, and billions of dollars will have to be spent to keep coastal cities, like New York, Boston, Miami, and Galveston in the United States from disappearing into the sea.

The atmosphere is an exchange pool for carbon dioxide. Fossil fuel combustion in particular has increased the amount of carbon dioxide in the atmosphere. Global warming is predicted because carbon dioxide and other gases impede the escape of infrared radiation from the surface of the earth.

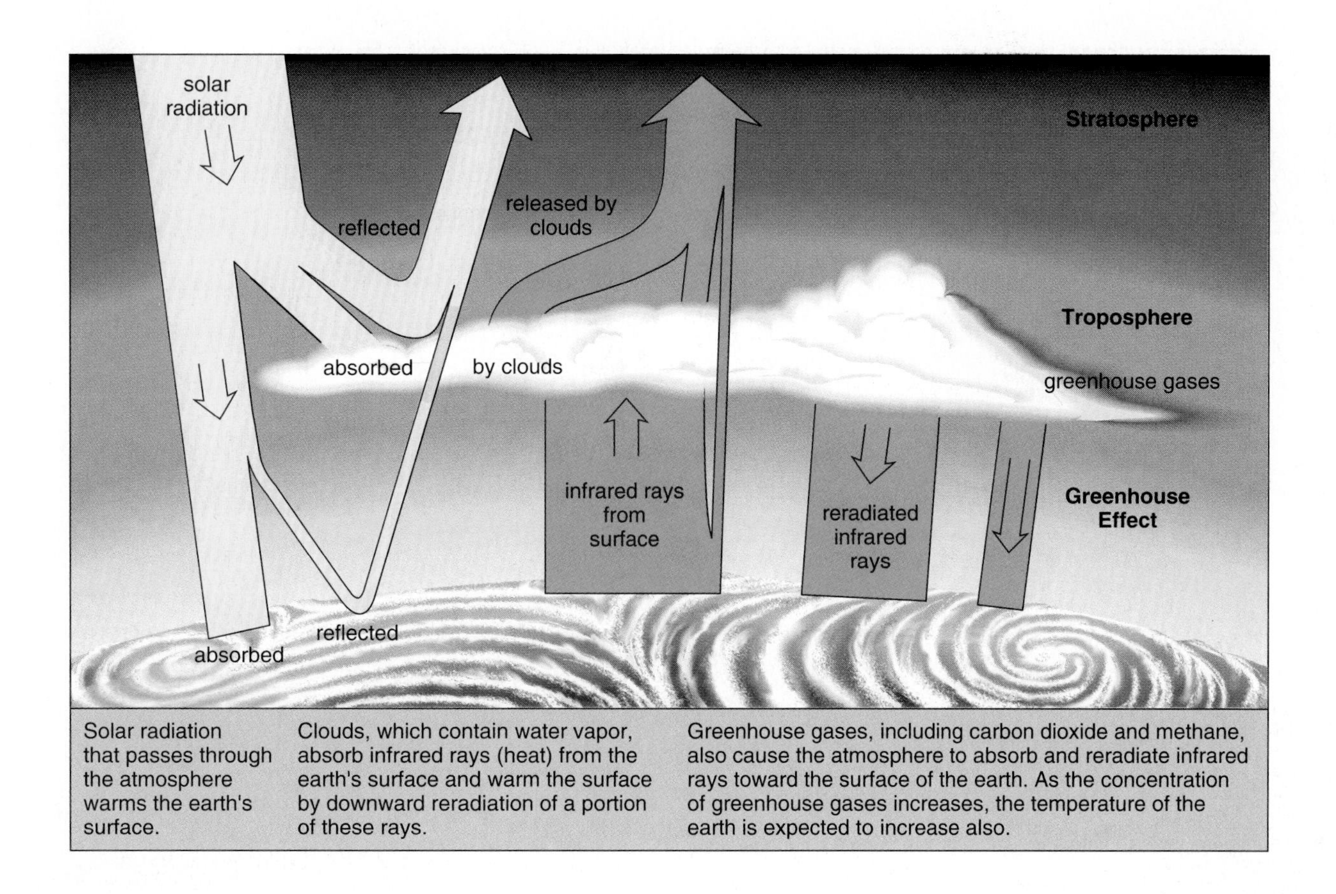

Figure 36.10 Earth's radiation balances.
The contribution of greenhouse gases (*far right*) to the earth's surface is called global warming.

The Nitrogen Cycle

Nitrogen is an abundant element in the atmosphere. Nitrogen (N_2) makes up about 78% of the atmosphere by volume, yet nitrogen deficiency sometimes limits plant growth. Plants cannot incorporate nitrogen gas into organic compounds and therefore depend on various types of bacteria to make nitrogen available to them in the **nitrogen cycle** (Fig. 36.11).

Nitrogen fixation occurs when nitrogen (N_2) is converted to a form that plants can use. Some nitrogen-fixing bacteria live in nodules on the roots of legumes. They make nitrogen-containing organic compounds available to a host plant. Cyanobacteria in aquatic ecosystems and free-living bacteria in soil are able to fix nitrogen gas as ammonium (NH_4^+). Plants can use NH_4^+ and nitrate (NO_3^-) from the soil. After NO_3^- is taken up, it is enzymatically reduced to NH_4^+, which is used to produce amino acids and nucleic acids.

Nitrification is the production of nitrates. Nitrogen gas (N_2) is converted to nitrate (NO_3^-) in the atmosphere when cosmic radiation, meteor trails, and lightning provide the high energy needed for nitrogen to react with oxygen. Ammonium (NH_4^+) in the soil is converted to nitrate by chemoautotrophic soil bacteria in a two-step process. First, nitrite-producing bacteria convert ammonium to nitrite (NO_2^-), and then nitrate-producing bacteria convert nitrite to nitrate. Notice the subcycle in the nitrogen cycle that involves dead organisms and animal wastes, ammonium, nitrites, nitrates, and plants. This subcycle does not necessarily depend on nitrogen gas at all (Fig. 36.11).

Denitrification is the conversion of nitrate to nitrous oxide and nitrogen gas. There are denitrifying bacteria in both aquatic and terrestrial ecosystems. Denitrification balances nitrogen fixation, but not completely.

Nitrogen and Air Pollution

Human activities significantly alter transfer rates in the nitrogen cycle. Because we produce fertilizers, thereby converting N_2 to NO_3^-, and burn fossil fuels, the atmosphere contains three times the nitrogen oxides (NO_x) than it would otherwise. Fossil fuel combustion also pumps much sulfur dioxide (SO_2) into the atmosphere. Both nitrogen oxides and sulfur dioxide are converted to acids when they combine

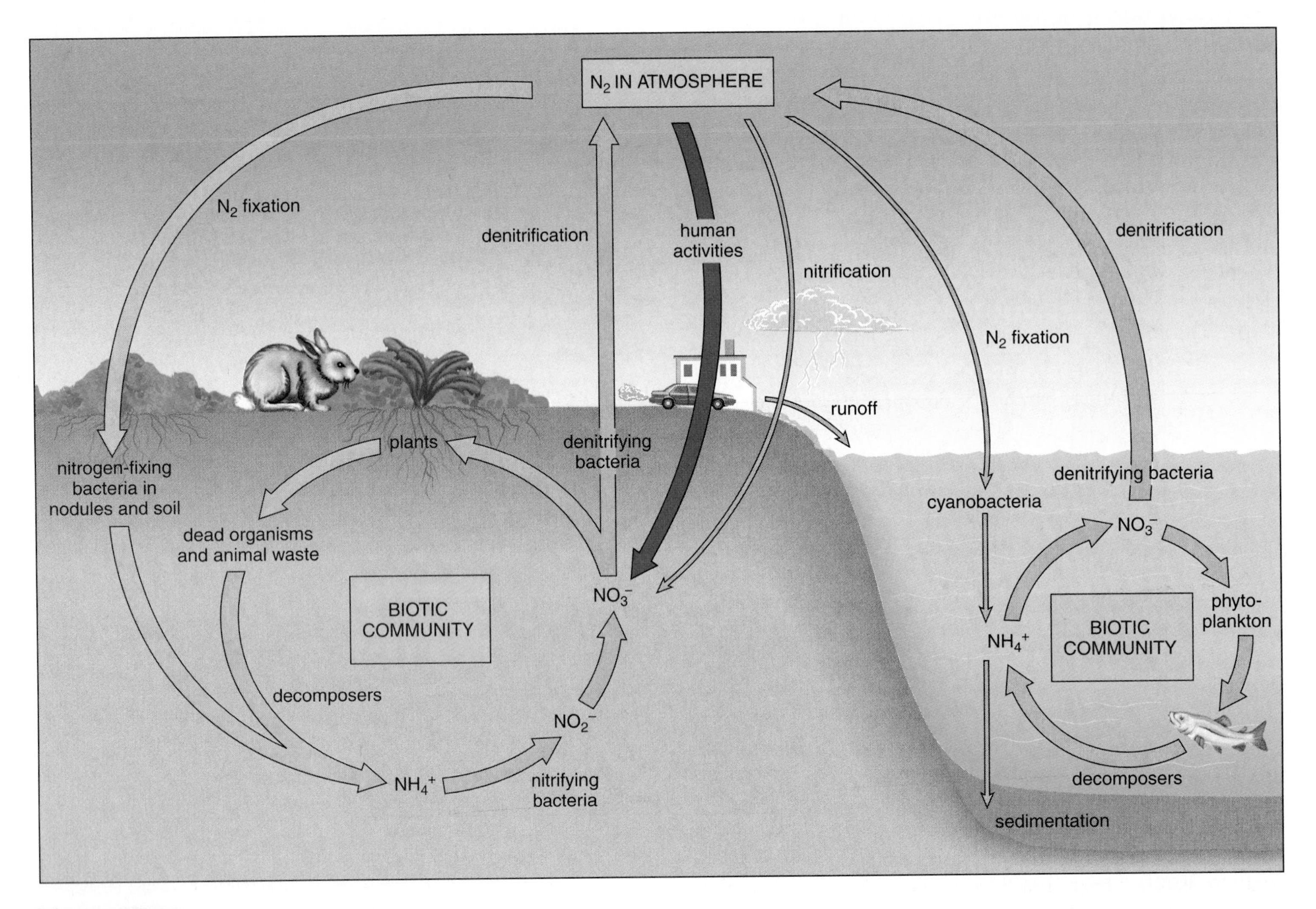

Figure 36.11 The nitrogen cycle.

with water vapor in the atmosphere. These acids return to earth as either wet deposition (acid rain or snow) or dry deposition (sulfate and nitrate salts).

Increased deposition of acids has drastically affected forests and lakes in northern Europe, Canada, and northeastern United States because their soils are naturally acidic and their surface waters are only mildly alkaline (basic) to begin with. The forests in these areas are dying (Fig. 36.12), and their waters cannot support normal fish populations. **Acid deposition** reduces agricultural yields and corrodes marble, metal, and stonework, an effect that is noticeable in cities.

Nitrogen oxides (NO_x) and hydrocarbons (HC) react with one another in the presence of sunlight to produce **photochemical smog,** which contains ozone (O_3) and **PAN (peroxyacetylnitrate).** Hydrocarbons come from fossil fuel combustion, but additional amounts come from various other sources as well, including paint solvents and pesticides. Breathing ozone affects the respiratory and nervous systems, resulting in respiratory distress, headache, and exhaustion. These symptoms are particularly apt to appear in young people. Ozone is especially damaging to plants, resulting in leaf mottling and reduced growth.

Normally, warm air near the ground is able to escape into the atmosphere. Sometimes, however, air pollutants, such as those in smog and soot, trap warm air near the earth. During a **thermal inversion** there is cold air at ground level beneath a layer of warm stagnant air above. Some areas surrounded by hills are particularly susceptible to the effects of a temperature inversion because the air tends to stagnate, and there is little turbulent mixing (Fig. 36.13).

Fertilizer use also results in the release of nitrous oxide (N_2O), a greenhouse gas and a contributor to ozone shield depletion in the stratosphere, a topic to be discussed later.

Atmospheric N_2, a reservoir and exchange pool for nitrogen, must be fixed by bacteria in order to make nitrogen available to plants. Environmental problems are associated with the release of nitrous oxide (N_2O) and nitrogen oxides (NO_x) due to the action of bacteria on fertilizers and fossil fuel combustion, respectively.

Figure 36.12 Acid deposition.
a. Many forests in higher elevations of northeastern North America and Europe are dying due to acid deposition. **b.** Air pollution due to emissions from factories and fossil fuel burning is the major cause of acid deposition, which contains nitric acid (H_2NO_3) and sulfuric acid (H_2SO_4).

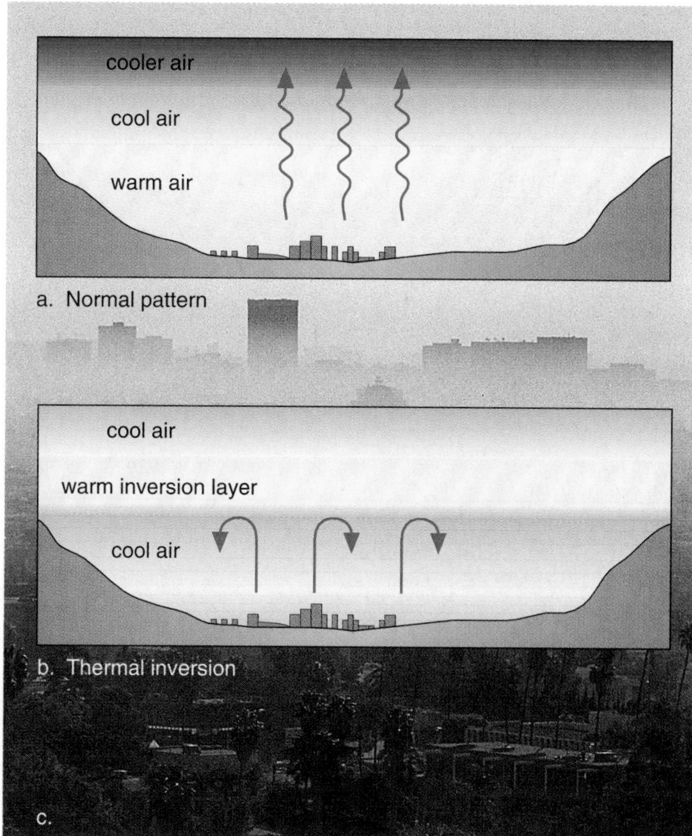

Figure 36.13 Thermal inversion.
a. Normally, pollutants escape into the atmosphere when warm air rises. **b.** During a thermal inversion, a layer of warm air (warm inversion layer) overlies and traps pollutants in cool air below. **c.** Los Angeles is particularly susceptible to thermal inversions, and this accounts for why this city is the "air pollution capital" of the United States.

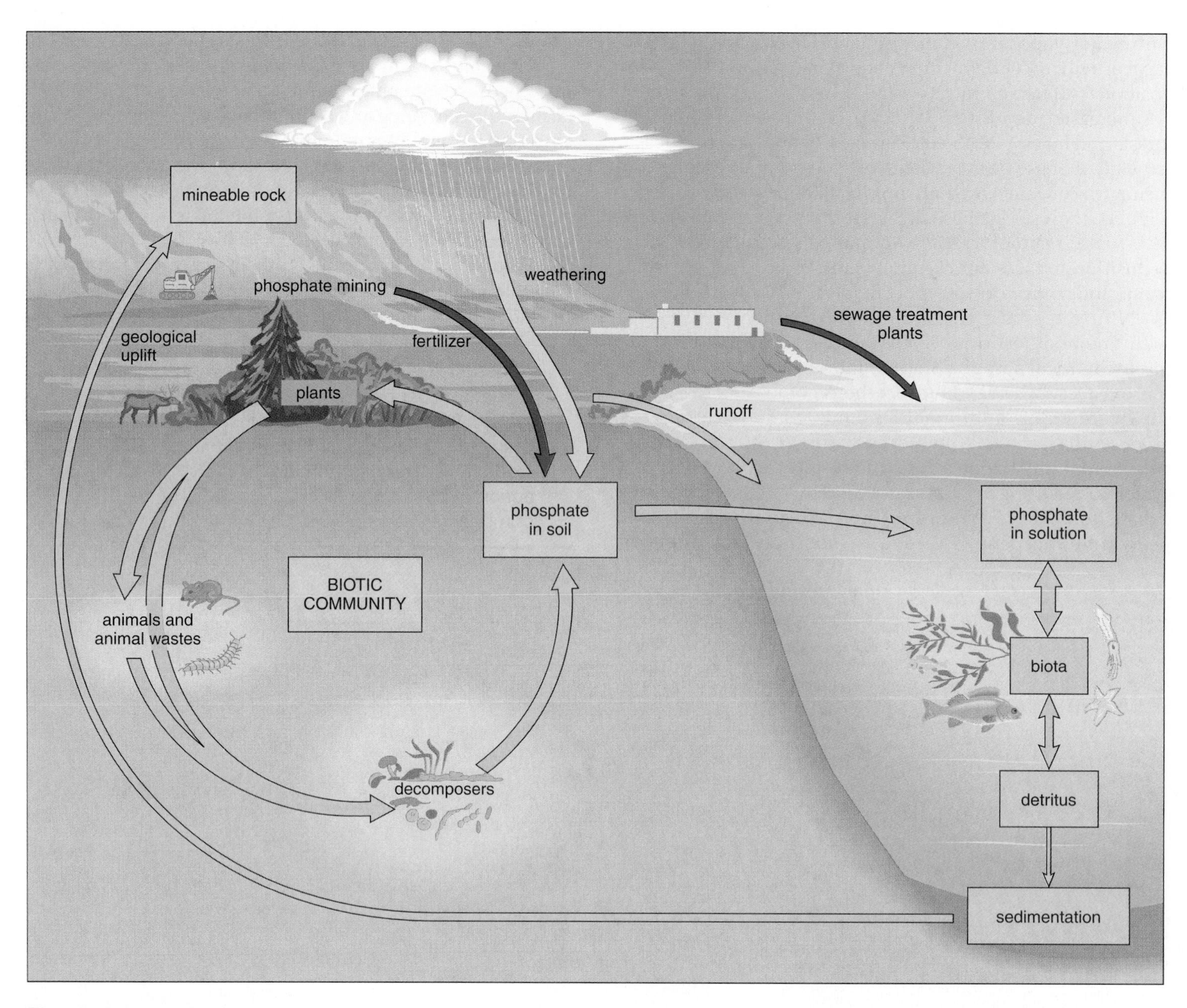

Figure 36.14 The phosphorus cycle.

The Phosphorus Cycle

On land, the weathering of rocks makes phosphate ions (PO_4^{3-} and HPO_4^{2-}) available to plants, which take up phosphate from the soil (Fig. 36.14). Some of this phosphate runs off into aquatic ecosystems where algae take phosphate up from the water before it becomes trapped in sediments. Phosphate in sediments only becomes available when a geological upheaval exposes sedimentary rocks to weathering once more. Phosphorus does not enter the atmosphere; therefore, the **phosphorus cycle** is called a sedimentary cycle.

The phosphate taken up by producers is incorporated into a variety of molecules, including phospholipids and ATP or the nucleotides that become a part of DNA and RNA. Animals eat producers and incorporate some of the phosphate into teeth, bones, and shells that do not decompose for very long periods. Death and decay of all organisms, and also decomposition of animal wastes, do, however, make phosphate ions available to producers once again. Because available phosphate is generally taken up very quickly, it is often a limiting inorganic nutrient in most ecosystems. A limiting nutrient is one that regulates the growth of organisms because it is in shorter supply than other nutrients in the environment.

Phosphorus and Water Pollution

Human beings boost the supply of phosphate by mining phosphate ores for fertilizer and detergent production.

Runoff of phosphate and nitrogen due to fertilizer use, animal wastes from livestock feedlots, as well as discharge from sewage treatment plants results in *eutrophication* (overenrichment). Eutrophication can lead to an algal bloom, apparent when green scum floats on the water. When the algae die off, decomposers use up all available oxygen during cellular respiration. The result is a massive fish kill.

Figure 36.15 lists the various sources of water pollution. Point sources are sources of pollution that are specific, and nonpoint sources are those caused by runoff from the land. Industrial wastes can include heavy metals and organochlorides, such as those in some pesticides. These materials are not degraded readily under natural conditions nor in conventional sewage treatment plants. They enter bodies of water and are subject to **biological magnification** because they remain in the body and are not excreted. Therefore, they become more concentrated as they pass along a food chain. Biological magnification occurs more readily in aquatic food chains because aquatic food chains have more links than terrestrial food chains. Humans are the final consumers in food chains, and in some areas, human milk contains detectable amounts of DDT and PCBs, which are organochlorides.

Coastal regions are the immediate receptors for local pollutants, and are the final receptors for pollutants carried by rivers that empty at a coast. Waste dumping occurs at sea, but ocean currents sometimes transport both trash and pollutants back to shore. Offshore mining and shipping add pollutants to the oceans. Some 5 million metric tons of oil a year—or more than one gram per 100 square meters of the oceans' surfaces—end up in the oceans. Large oil spills kill plankton, fish fry, and shellfishes, as well as birds and marine mammals. The largest tanker spill in U.S. territorial waters occurred on March 24, 1989, when the tanker *Exxon Valdez* struck a reef in Alaska's Prince William Sound and leaked 44 million liters of crude oil.

In the last 50 years, we have polluted the seas and exploited their resources to the point that many species are at the brink of extinction. Fisheries once rich and diverse, such as George's Bank off the coast of New England, are in severe decline. Haddock was once the most abundant species in this fishery, but now it accounts for less than 2% of the total catch. Cod and bluefin tuna have suffered a 90% reduction in population size. In warm, tropical regions, many areas of coral reefs are now overgrown with algae because the fish that normally keep the algae under control have been killed off.

Sedimentary rock is a reservoir for phosphorus; for the most part producers are dependent on decomposers to make phosphate available to them. Fertilizer production and other human activities add phosphate to aquatic ecosystems, contributing to water pollution.

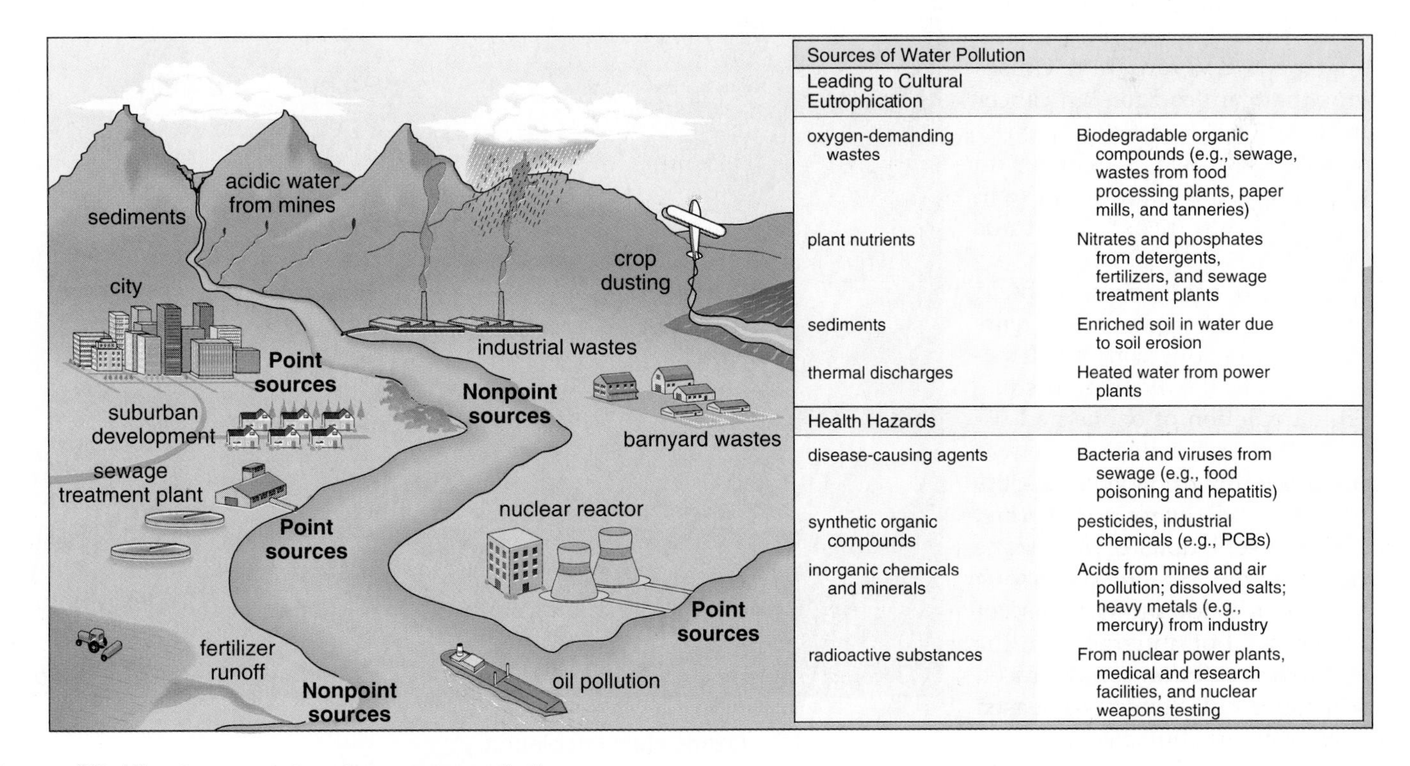

Sources of Water Pollution	
Leading to Cultural Eutrophication	
oxygen-demanding wastes	Biodegradable organic compounds (e.g., sewage, wastes from food processing plants, paper mills, and tanneries)
plant nutrients	Nitrates and phosphates from detergents, fertilizers, and sewage treatment plants
sediments	Enriched soil in water due to soil erosion
thermal discharges	Heated water from power plants
Health Hazards	
disease-causing agents	Bacteria and viruses from sewage (e.g., food poisoning and hepatitis)
synthetic organic compounds	pesticides, industrial chemicals (e.g., PCBs)
inorganic chemicals and minerals	Acids from mines and air pollution; dissolved salts; heavy metals (e.g., mercury) from industry
radioactive substances	From nuclear power plants, medical and research facilities, and nuclear weapons testing

Figure 36.15 Sources of surface water pollution.
Many bodies of water are dying due to the introduction of pollutants from point sources, which are easily identifiable, and nonpoint sources, which cannot be specifically identified.

36.3 Human Impact on Biodiversity

All of the human activities we have discussed thus far have a negative impact on biodiversity. Global warming may mean that coastal ecosystems, such as marshes, swamps, and bayous, will have to move inland to higher ground as the sea level rises, but many of these are blocked in by artificial structures and may be unable to move inland. Acid deposition is associated with dead or dying lakes and forests, particularly in North America and Europe. We have polluted the seas and exploited their resources to the point that many species are on the brink of extinction. Still there are other human activities that will adversely affect the number of species on earth.

Stratospheric Ozone Depletion

The earth's atmosphere is divided into layers. The troposphere envelops us as we go about our day-to-day lives. Ozone in the troposphere is a pollutant, but in the stratosphere, some 50 km above the earth, ozone (O_3) forms a layer, called the **ozone shield,** that absorbs most of the wavelengths of harmful ultraviolet (UV) radiation so that they do not strike the earth. Life on earth is threatened if the ozone shield is reduced. UV radiation impairs crop and tree growth and also kills off plankton (microscopic plant and animal life) that sustain oceanic life. Without an adequate ozone shield, living things, our food sources, and health are threatened. UV radiation causes mutations that can lead to skin cancer and can make the lens of the eyes develop cataracts. It also is believed to adversely affect the immune system and our ability to resist infectious diseases.

Depletion of the ozone shield within the stratosphere in recent years is, therefore, of serious concern. It became apparent in the 1980s that some worldwide depletion of ozone had occurred, and that by the 1990s there was a severe depletion of some 40–50% above the Antarctic every spring (Fig. 36.16). Severe depletions of the ozone layer are commonly called **"ozone holes."** Nitrous oxide is one cause of ozone depletion, but, in large part, the cause of ozone depletion can be traced to chlorine atoms (Cl) that are released in the troposphere but rise into the stratosphere. Chlorine atoms combine with ozone and strip away the oxygen atoms one by one. One atom of chlorine can destroy up to 100,000 molecules of ozone before settling to the earth's surface as chloride many years later. These chlorine atoms come from the breakdown of **chlorofluorocarbons (CFCs),** chemicals much in use by humans from 1955 to 1990. The best-known CFC is Freon, a heat transfer agent still found in refrigerators and air conditioners today. CFCs were used as cleaning agents and during the production of styrofoam found in coffee cups, egg cartons, insulation, and paddings. Their use as a propellent in spray cans has been outlawed in the United States and several other countries but not in western Europe. Although most countries of the world have agreed to stop using CFCs by the year 2000, CFCs already in the atmosphere will be there for over a hundred years before they stop their destructive activity.

Ozone depletion is one of our air pollution problems. The others are global warming, acid deposition and photochemical smog.

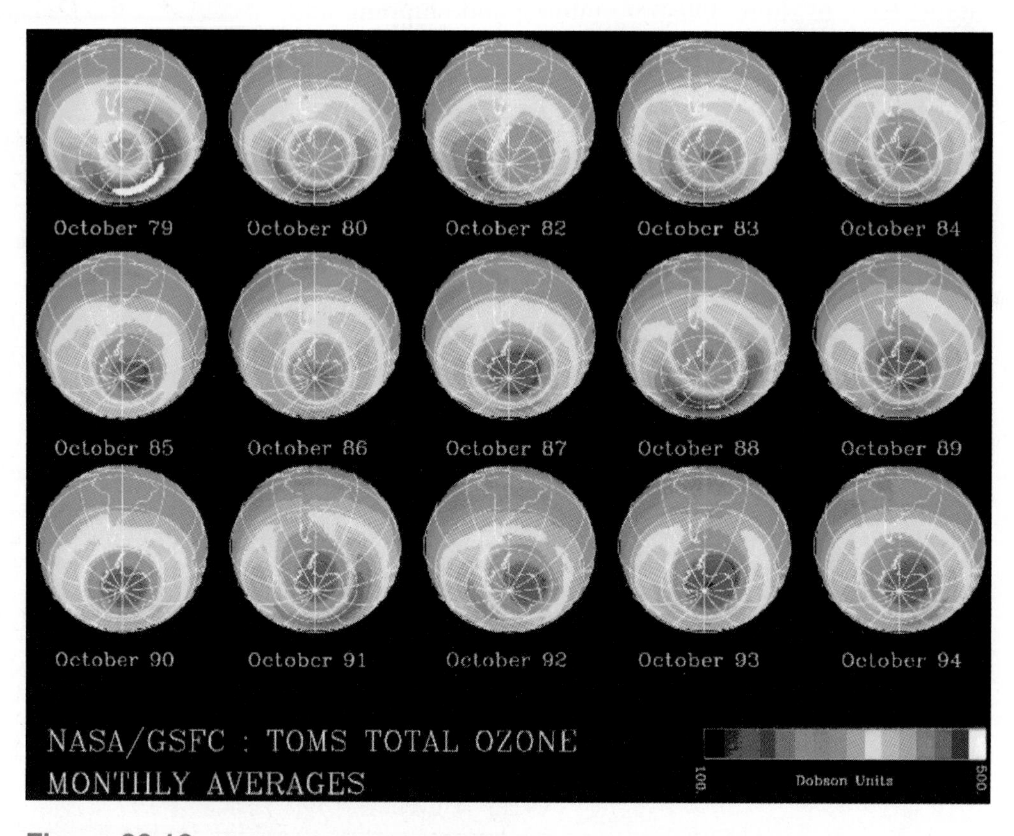

Figure 36.16 Ozone shield depletion.
These satellite observations show that the amount of ozone over the South Pole between October 1979 and October 1994 fell by more than 50%. Green represents an average amount of ozone, blue less, and purple still less.

Tropical Rain Forest Destruction

Tropical rain forests are much more biologically diverse than temperate forests (see Fig. 36.4). For example, temperate forests across the entire United States contain about 400 tree species. In the rain forest, a typical ten-hectare area holds as many as 750 types of trees. Tropical rain forests are also noted for their animal diversity. On the eastern slopes of the Andes, there are 80 or more species of frogs and toads, and in Ecuador, there are more than 1,200 species of birds—roughly twice as many as those inhabiting all of the United States and Canada. Therefore, a very serious side effect of deforestation in tropical countries is a loss of biological diversity.

A National Academy of Sciences study estimated that a million species of plants and animals are in danger of disappearing within 20 years as a result of **deforestation** in tropical countries. Many of these life forms have never been studied, and yet they may be useful sources of food or medicines. Figure 36.17 lists other deleterious effects of deforestation.

Logging of tropical forests occurs because industrialized nations prefer furniture made from costly tropical woods and because people want to farm the land. In Brazil, the government allows citizens to own any land they clear in the Amazon forest (along the Amazon River). When they arrive, the people practice slash-and-burn agriculture, in which trees are cut down and burned to provide inorganic nutrients and space to raise crops. Unfortunately, the fertility of the land is sufficient to sustain agriculture for only a few years. Once the cleared land is incapable of sustaining crops, the farmer moves on to another part of the rain forest to slash and burn again. In the meantime, cattle ranchers move in. Cattle ranchers are the greatest beneficiaries of deforestation, and increased ranching is therefore another reason for tropical rain forest destruction. A newly begun pig-iron industry in Brazil also indirectly results in further exploitation of the rain forest. The pig iron must be processed before it is exported, and smelting the pig iron requires the use of charcoal (burnt wood).

There is much concern worldwide about the loss of biological diversity due to the destruction of tropical rain forests.

Conservation Biology

Conservation biology is a relatively new scientific discipline that brings together people and knowledge from many different fields to attempt to solve the biodiversity crisis. Conservation biology wants to understand the effects of human activities on species, communities, and ecosystems, and develop practical approaches to preventing the extinctions of species and the destruction of ecosystems. In the past, ecologists have preferred to study the workings of ecosystems not tainted by human activities, and wildlife managers have been concerned with managing a small number of species for the marketplace and for recreation. Therefore, neither endeavor has addressed the possibility of preserving entire biological communities, although humans are active in the area. Conservation biologists want to draw from scientific research and experience in the field to develop a management program that will preserve an ecosystem.

Many conservation biologists believe that each species has a value all its own, regardless of its direct material value to humans. Other conservation biologists are willing to test the hypothesis of sustainability; that it is possible to manage ecosystems so that biodiversity is preserved while still meeting the economic needs of humans.

Much scientific research is being directed on how to preserve ecosystems and, therefore, biodiversity. The reading on the next page describes work being done at the University of Rhode Island.

Conservation biology is the scientific study of biodiversity, leading to the preservation of species and the management of ecosystems for sustainable human welfare.

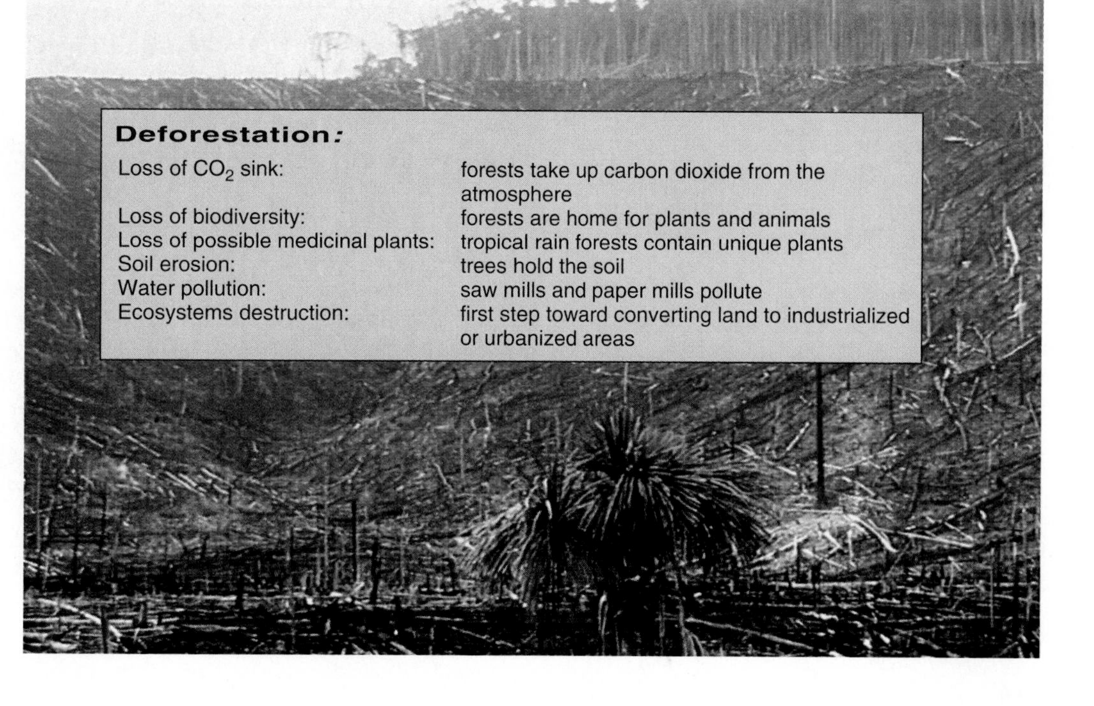

Figure 36.17 Tropical rain forest. Forest destruction leads to the detrimental effects listed.

Science Focus

Marine Enclosures for Whole-Ecosystem Studies

The MERL (Marine Ecosystem Research Laboratory) enclosures shown in Figure 36A provide marine researchers with a unique means to experiment with an entire marine ecosystem. The tanks measure 1.8 meters in diameter and 5 meters deep and are located outdoors, exposed to natural sunlight. To initiate a typical experiment, a benthic (bottom of the ocean) community is collected, usually from a silt-clay area in Narragansett Bay, off the coast of Rhode Island. A 37-cm-thick bed of sediment weighing roughly a ton is placed into each tank. Thirteen cubic meters of unfiltered seawater are transferred from the adjacent bay with nondisruptive displacement pumps. Mechanical mixing provides water movement in the enclosures to simulate wind and wave movement in the field. In the summer, cooling is provided, and in the winter, heat is provided to keep the enclosure temperatures similar to those in the bay.

When set up in this manner, unmanipulated enclosures maintain healthy ecosystems for many months and with properties that are similar to those actually found in the bay. Their large size and proximity to laboratory facilities allows repetitive sampling of all biological populations at relatively short time intervals. Because there are 14 enclosures, replication of experiments with controls is possible.

Experiments are done to determine the effect of a contaminant and to study the fate of chemicals within an entire coastal ecosystem. One set of MERL experiments addressed the problem of chronic additions of oil hydrocarbons to coastal waters. Water runoff from land, especially in urban areas, carries a continuous trickle of oil to coastal environments. The total amounts of petroleum hydrocarbons introduced into coastal waters through urban runoff and river runoff are believed to be greater than introductions through oil spills. Daily additions of fuel oil, of the type regularly used in home furnaces, were made into three replicate enclosures. Two experiments were conducted. The first 5.5-month experiment added oil to achieve about 0.2 ppm total hydrocarbons in the water, while the second 4-month experiment achieved about 0.1 ppm total hydrocarbons. Thereafter, recovery from the oil additions was studied for one year.

These levels of fuel hydrocarbons are below those that cause most tested marine species to die. One objective of the experiments was to develop an index system that could be used to indicate the health of an ecosystem. It was hypothesized that community diversity and primary productivity would be lower in the treated enclosures compared to the controls. The oil additions had a clear effect on the populations in the enclosures. Zooplankton and benthic macroorganisms were greatly reduced in abundance. Benthic populations remained depressed for at least a year after the treatments were stopped.

Even though the additions of oil clearly had a major impact on the communities in the enclosures, neither of the original hypotheses turned out to be correct. Although the population levels were quite different in treated and control tanks, measures of the diversity of benthic organisms in treated and control enclosures were indistinguishable. With oil treatments, there were increases in phytoplankton abundance and primary productivity instead of the expected decrease. In hindsight, the reason for the increase in phytoplankton abundance was clear. The oil was more toxic to the organisms that may graze the phytoplankton than to the phytoplankton itself. With the population of grazers reduced, the abundance and production of phytoplankton increased.

This is a good example of the interactions within an ecosystem, and the inherent difficulty of predicting how any component of an ecosystem will respond to stress.

Figure 36A Marine Ecosystem Research Laboratory (MERL). MERL enclosures at the University of Rhode Island's Graduate School of Oceanography. Students are shown taking samples and measurements.

Bioethical Issue

Peter Jutro, a scientist working for the U.S. Environmental Protection Agency, wants to research native traditions for clues on how to preserve ecosystems. He has run into opposition from the indigenous groups because they mistrust conservationists. Take, as an example, the fact that the Kuna people in Panama refused to renew the lease for a Smithsonian Institution studying reef ecology for the past 21 years. Much of the trouble seems to have come from the failure of scientists to explain their program to local communities. In a meeting of the Kuna congress, the scientists were accused by the Kuna of "stealing their knowledge, stealing their reefs, stealing their sand." Local people find it hard to see a difference between a scientific study and commercial ventures which exploit their areas for minerals, timber, and other resources.

Laura Snook, a forester from Duke University, researches the growing habits of mahogany trees in Mexico. She came to the conclusion that because the trees grow slowly, logging shouldn't be done too fast. Most local foresters resented her findings, but she was able to establish a good relationship with two women foresters who were open to her ideas. Now local people are changing the way they replant mahogany trees so that the resource will be there for some time to come. The point is that conservationists working with people from different cultural backgrounds probably need to communicate their goals more clearly and involve local people in project planning. Perhaps they should learn to accept slower time scales and styles of decision-making different from our own.

Questions

1. Should U.S. scientists be studying ecosystems in such far-flung places as Panama, Alaska, Mexico, and the Amazon? Why or why not?
2. Should we insert political correctness into negotiations with local groups in order to bring about conservation? Why or why not?
3. How far should a scientist go to establish communication with local groups in order to preserve the environment in other countries? Explain.

Summarizing the Concepts

36.1 The Nature of Ecosystems

Ecosystems contain biotic (living) components and abiotic (physical) components. The biotic components of ecosystems are either producers or consumers. Producers are autotrophs that produce their own food. Consumers are heterotrophs that take in preformed food. Consumers may be herbivores, carnivores, omnivores, or decomposers.

Energy flows through an ecosystem. Producers transform solar energy into food for themselves and all consumers. As herbivores feed on plants (or algae), and carnivores feed on herbivores, some energy is converted to heat. Feces, urine, and dead bodies become food for decomposers. Eventually, all the solar energy that enters an ecosystem is converted to heat, and thus ecosystems require a continual supply of solar energy.

Chemicals are not lost from the biosphere as is energy. They recycle within and between ecosystems. Decomposers return some proportion of inorganic nutrients to autotrophs, and other portions are imported or exported between ecosystems in global cycles.

Ecosystems contain food webs, and a diagram of a food web shows how the various organisms are connected by eating relationships. Grazing food chains begin with vegetation that is fed on by a herbivore, which becomes food for a carnivore, and so forth. In detrital food chains, a decomposer acts on organic material in the soil, and when it is fed on by a carnivore, the two food webs are joined. A trophic level is all the organisms that feed at a particular link in food chains. Ecological pyramids show trophic levels stacked one on the other like building blocks. Generally they show that biomass and energy content decrease from one trophic level to the next. Most pyramids pertain to grazing food webs and largely ignore the detrital food web portion of an ecosystem.

36.2 Global Biogeochemical Cycles

Biogeochemical cycles contain reservoirs, components of ecosystems like fossil fuels, sediments, and rocks that contain elements available on a limited basis to living things. Pools are components of ecosystems like the atmosphere, soil, and water—which are ready sources of nutrients for living things. Nutrients cycle among the members of the biotic component of an ecosystem.

In the water cycle, evaporation over the ocean is not compensated for by rainfall. Evaporation from terrestrial ecosystems includes transpiration from plants. Rainfall over land results in bodies of fresh water plus groundwater, including aquifers. Eventually all water returns to the oceans.

In the carbon cycle, organisms add as much carbon dioxide to the atmosphere as they remove. Shells in ocean sediments, organic compounds in living and dead organisms, and fossil fuels are reservoirs for carbon. Human activities such as the burning of fossil fuels and trees are adding carbon dioxide to the atmosphere. Like the panes of a greenhouse, carbon dioxide and other gases allow the sun's rays to pass through but impede the release of infrared wavelengths. It is predicted that a buildup of these "greenhouse gases" will lead to a global warming. The effects of global warming could be a rise in sea level and a change in climate patterns with disastrous effects.

In the nitrogen cycle, the biotic community, which includes several types of bacteria, keeps nitrogen recycling back to the producers. A few organisms (cyanobacteria in aquatic habitats and bacteria in soil and root nodules) can fix atmospheric nitrogen. Other bacteria return nitrogen to the atmosphere. Human activities convert atmospheric nitrogen to fertilizer which is broken down by soil bacteria and they burn fossil fuels. In this way a large quantity of nitrogen oxide (NO_x) and sulfur dioxide (SO_2) is added to atmosphere where it reacts with water vapor to form acids that contribute to acid deposition. Acid deposition is killing lakes and forests and also corrodes marble, metal, and stonework. Nitrogen oxides and hydrocarbons (HC) react to form smog, which contains ozone and PAN (peroxyacetylnitrate). These oxidants are harmful to animal and plant life.

In the phosphorus cycle, the biotic community recycles phosphorus back to the producers, and only limited quantities are made available by the weathering of rocks. Phosphates are mined for fertilizer production; when phosphates and nitrates enter lakes and ponds, over-enrichment occurs. Many kinds of wastes enter rivers which flow to the oceans now degraded from added pollutants.

36.3 Human Impact on Biodiversity

Global warming, acid deposition, and water pollution all act to reduce biodiversity. Ozone shield destruction, which is particularly associated with CFCs, is expected to result in decreased productivity of the oceans. The tropical rain forests are being cut to provide wood for export. Slash-and-burn agriculture also reduces tropical rain forests. The loss of biological diversity due to the destruction of tropical rain forests will be immense. Many of these threatened organisms could possibly be of benefit to humans if we had time to study and domesticate them. Conservation biology is a new discipline that pulls together information from a number of biological fields to determine how best to manage ecosystems for the benefit of all species, including humans.

Studying the Concepts

1. Distinguish between autotrophs and heterotrophs, and describe four different types of heterotrophs found in natural ecosystems. Explain the terms producer and consumer. 744
2. Tell why energy must flow but chemicals can cycle in an ecosystem. 745
3. Describe two types of food webs and two types of food chains typically found in terrestrial ecosystems. Which of these typically moves more energy through an ecosystem? 746–47
4. What is a trophic level? an ecological pyramid? 747–48
5. Give examples of reservoirs and pools in biogeochemical cycles. Which is less accessible to biotic communities? 748
6. Draw a diagram to illustrate the water cycle and the carbon cycle. 749–50
7. How and why is the global climate expected to change, and what are the predicted consequences of this change? 750–51
8. Draw a diagram of the nitrogen cycle. What types of bacteria are involved in this cycle? 752
9. What causes acid deposition, and what are its effects? 752–53
10. How does photochemical smog develop, and what is a thermal inversion? 753
11. Draw a diagram of the phosphorus cycle. 754
12. What are several ways in which fresh water and marine waters can be polluted? What is biological magnification? 754–55
13. Of what benefit is the ozone shield? What pollutant in particular should be associated with ozone shield depletion, and what are the consequences of this depletion? 756
14. What are the primary ecological concerns associated with the destruction of rain forests? 757
15. Explain the primary causes of the biodiversity crisis and the goals of conservation biology. 756–57

Testing Yourself

Choose the best answer for each question.

1. Of the total amount of energy that passes from one trophic level to another, about 10% is
   a. respired and becomes heat.
   b. passed out as feces or urine.
   c. stored as body tissue.
   d. All of these are correct.
2. Compare this food chain:
   algae → water fleas → fish → green herons
   to this food chain:
   trees → tent caterpillars → red-eyed vireos → hawks.
   Both water fleas and tent caterpillars are
   a. carnivores.
   b. primary consumers.
   c. detritus feeders.
   d. Both a and b are correct.
3. Which of the following contribute(s) to the carbon cycle?
   a. respiration
   b. photosynthesis
   c. fossil fuel combustion
   d. All of these are correct.
4. How do plants contribute to the carbon cycle?
   a. When they respire, they release CO_2 into the atmosphere.
   b. When they photosynthesize, they consume CO_2 from the atmosphere.
   c. They do not contribute to the carbon cycle.
   d. Both a and b are correct.
5. How do nitrogen-fixing bacteria contribute to the nitrogen cycle?
   a. They return nitrogen (N_2) to the atmosphere.
   b. They change ammonium to nitrate.
   c. They change N_2 to ammonium.
   d. They withdraw nitrate from the soil.
6. In what way are decomposers like producers?
   a. Either may be the first member of a grazing food chain.
   b. Both produce oxygen for other forms of life.
   c. Both require a source of nutrient molecules and energy.
   d. Both supply organic food for the biosphere.
7. Which statement is true concerning this food chain: grass → rabbits → snakes → hawks?
   a. Each predator population has a greater biomass than its prey population.
   b. Each prey population has a greater biomass than its predator population.
   c. Each population is omnivorous.
   d. Both a and c are correct.

For questions 8–11, match the terms with those in the key:

Key:
a. sulfur dioxide
b. ozone
c. carbon dioxide
d. chlorofluorocarbons (CFCs)

8. acid deposition
9. ozone shield destruction
10. greenhouse effect
11. photochemical smog
12. Which of these is mismatched?
    a. fossil fuel burning—carbon dioxide given off
    b. nuclear power—radioactive wastes
    c. solar energy—greenhouse effect
    d. biomass burning—carbon dioxide given off
13. Acid deposition causes
    a. lakes and forests to die.
    b. acid indigestion in humans.
    c. the greenhouse effect to lessen.
    d. All of these are correct.
14. Water is a renewable resource, and
    a. there will always be a plentiful supply.
    b. the oceans can never become polluted.
    c. it is still subject to pollution.
    d. primary sewage treatment plants assure clean drinking water.

15. Label this diagram.

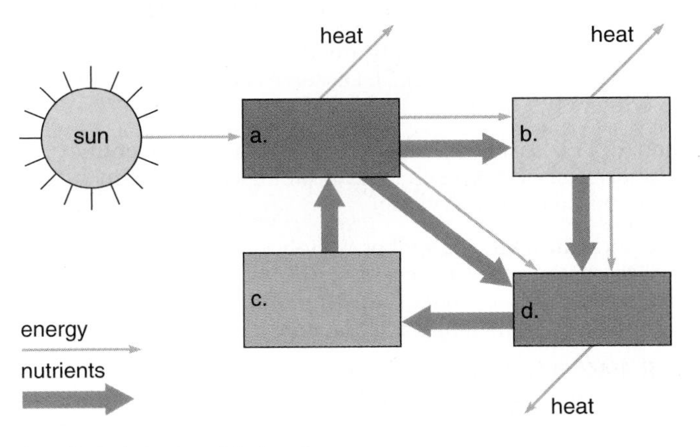

16. Label the trophic levels.

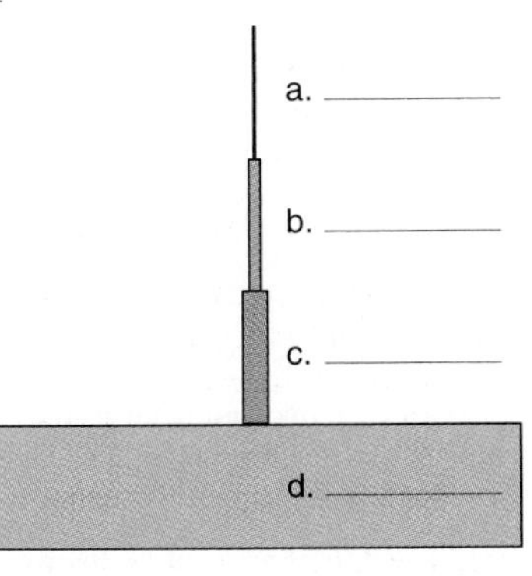

Thinking Scientifically

1. Considering an ecological pyramid:
   a. Why would you expect mice (herbivores) to be more common than weasels, foxes, or hawks (carnivores) in the environment?
   b. Why you would expect food chains to be short—4 or 5 links at most?
   c. The population size of a top predator is not held in check by another predator population. Why does a top predator population not increase constantly in size?
   d. What would you expect to happen to an ecosystem if one of the secondary consumer populations suffered a collapse?
2. You are an ecologist who has been hired by a less-developed country to help them increase their agricultural yield per acre. Why might you recommend that they
   a. retain labor intensive methods instead of adopting mechanized means of growing food?
   b. limit their consumption of meat and use grains and legumes as a source of protein?
   c. not keep cattle in feedlots and feed them grain?
   d. plant as many different varieties of crops as possible?
   e. grow crops that require as little irrigation as possible?
   f. begin a program of population control?

Understanding the Terms

acid deposition 753
aquifer 749
autotroph 744
biogeochemical cycle 748
biological magnification 755
carbon cycle 750
carnivore 744
chlorofluoro-carbons 756
conservation biology 757
consumer 744
decomposer 744
deforestation 757
denitrification 752
detrital food chain 747
detrital food web 747
detritus 744
ecological pyramid 747
ecosystem 744
food chain 747
food web 747
fossil fuel 750
global warming 751
grazing food chain 747
grazing food web 747
greenhouse effect 751
herbivore 744
heterotroph 744
nitrification 752
nitrogen cycle 752
nitrogen fixation 752
omnivore 744
ozone hole 756
ozone shield 756
PAN (peroxyacetylnitrate) 753
phosphorus cycle 754
photochemical smog 753
producer 744
thermal inversion 753
transfer rate 750
trophic level 747
water (hydrologic) cycle 749

Match the terms to these definitions:

a. ________ Partially decomposed remains of plants and animals found in soil and on the beds of bodies of water.
b. ________ Formed from oxygen in the upper atmosphere, it protects the earth from ultraviolet radiation.
c. ________ Remains of once living organisms that are burned to release energy, such as coal, oil, and natural gas.
d. ________ Process by which atmospheric nitrogen gas is changed to forms that plants can use.
e. ________ Complex pattern of interlocking and crisscrossing food chains.

Using Technology

Your study of ecosystems and human interferences is supported by these available technologies.

Essential Study Partner CD-ROM
Ecology → Ecosystems
Visit the Mader web site for related ESP activities.

Exploring the Internet
The Mader Home Page provides resources and tools as you study this chapter.
http://www.mhhe.com/biosci/genbio/mader

Further Readings for Part 7

Bavendam, F. July 1998. Lure of the frogfish. *National Geographic* 194(1):40. Use of camouflaging coloration to obtain prey.

Begon, M., et al. 1996. *Ecology: Individuals, populations, and communities.* London: Blackwell Science Ltd. The distribution and abundance of organisms and their physcial and chemical interactions within ecosystems is discussed.

Bennet-Clark, H. C. May 1998. How cicadas make their noise. *Scientific American* 278(5):58. This 2.3-inch-long insect can produce mating calls at 100 decibels.

Boyd, C. E., and Clay, J. W. June 1998. Shrimp aquaculture and the environment. *Scientific American* 278(6):58. Building shrimp ponds for shrimp farming can result in destructive flooding.

Cunningham, W. P., and Saigo, B. W. 1997. *Environmental science: A global concern.* Dubuque, Iowa: Wm. C. Brown Publishers. Provides scientific principles plus insights into the social, political, and economic systems impacting the environment.

Dobson, A. P. 1996. *Conservation and biodiversity.* New York: Scientific American Library. Discusses the value of biodiversity; describes endangered species management.

Dugatkin, L. A., and Godin, J. J. April 1998. How females choose their mates. *Scientific American* 278(4):56. Female choice is studied in relation to a number of fish and bird species.

Fox, G. 1997. *Conservation ecology.* 2d ed. Dubuque, Iowa: Wm. C. Brown Publishers. Discusses the nature of the biosphere, the threats to its integrity, and ecologically sound responses.

Gorman, J. January 1998. They saw it coming. *Discover* 18(1):82. Article discusses how El Niño was forecast.

Hedin, L. O., and Likens, G. E. December 1996. Atmospheric dust and acid rain. *Scientific American* 274(6):88. Despite pollution reduction, acid rain continues to be a problem.

Kovacs, K. March 1997. Bearded seals. *National Geographic* 191(3):124. This article describes a behavioral study of bearded seals in their natural environment.

Long, M. E. April 1998. The vanishing prairie dog. *National Geographic* 193(4):116. Prairie dogs and their ecosystems are disappearing from the American West.

McClintock, J. B., and Baker, B. J. May/June 1998. Chemical ecology in Antarctic seas. *American Scientist* 86(3):254. Sessile benthic dwellers of polar seas use chemical defenses to ward off predators.

Miller, G. T. 1996. *Living in the environment.* 9th ed. Belmont, Calif.: Wadsworth Publishers. This introductory environmental science text discusses how the environment is being abused, and what can be done to protect it.

Mitchell, J. G. February 1996. Our polluted runoff. *National Geographic* 189(2):106. 80% of U.S. water pollution is due to land runoff not resulting from industrial sources.

Morgan, M., et al. 1997. *Environmental science: Managing biological and physical resources.* Dubuque, Iowa: Wm. C. Brown Publishers. Written for the undergraduate, this book explains how various environmental issues are linked.

National Geographic. October 1998. Millennium supplement: Population. Articles survey the needs of the worldwide population, and address issues such as birthrate, global food production, and migration.

Natural History Magazine. July/August 1998. 107(6):34–51. Articles address the preservation of Amazon rain forest diversity.

Nemecek, S. August 1997. Frankly, my dear, I don't want a dam. *Scientific American* 277(2):20. Discusses how dams affect biodiversity.

Newman, E. 1997. *Applied ecology.* Oxford: Blackwell Scientific Publications. Presents the role of biological science in environmental preservation.

Nicol, S., and Allison, I. September/October 1997. The frozen skin of the southern ocean. *American Scientist* 85(5):426. Sea-ice and the organisms that occupy it interact with the ocean-atmosphere system in ways that may influence climate.

Odum, E. 1997. *A bridge between science and society.* 3d ed. Sunderland, Mass.: Sinauer Associates. Introduces the principles of modern ecology as they relate to threats to the biosphere.

Ostfeld, R. S. July/August 1997. The ecology of Lyme-disease risk. *American Scientist* 85(4):338. Article discusses the history of Lyme disease, its symptoms and diagnosis, and the life cycle of the deer tick.

Pitelka, L. F., et al. September/October 1997. Plant migration and climate change. *American Scientist* 85(5):464. There may be a relationship between plant migration and climate change, as evidenced by the fossil record and computer models.

Rice, R. E., et al. April 1997. Can sustainable management save tropical forests? *Scientific American* 276(4):44. The strategy of replacing harvested trees in rain forests often fails.

Robinson, G. E. September/October 1998. From society to genes with the honey bee. *American Scientist* 86(5):456. The life stages of a honey bee are regulated by hormones, neurobiology, genes, and environment.

Rutowski, R. L. July 1998. Mating strategies in butterflies. *Scientific American* 279(1):64. Visual attributes (colorful wing patterns) and chemical signals (pheromones) play important roles in butterfly mating.

Schmidt, M. J. January 1996. Working elephants. *Scientific American* 274(1):82. In Asia, teams of elephants serve as an alternative to destructive logging equipment.

Schoech, S. J. January/February 1998. Physiology of helping in Florida scrub jay. *American Scientist* 86(1):70. Birds, which help rear the offspring of others, apparently experience delayed reproduction due to hormonal effects.

Scientific American Quarterly. Fall 1998. The oceans. *Scientific American* 9(3). This issue's articles discuss the origins of earth's water, polar ice cap melting, weather, pollution and legal issues, aquaculture, mineral mining, and marine diversity.

Simmons, L. M. August 1998. Indonesia's plague of fire. *National Geographic* 194(2):100. Slash-and-burn agricultural techniques result in air pollution and respiratory disease, as well as deforestation.

Steiner, R. September/October 1998. Resurrection in the wind. *International Wildlife* 28(5):12. The short-tailed albatross is recovering from near-extinction.

Suplee, C. May 1998. Unlocking the climate puzzle. *National Geographic* 193(5):38. Our use of fossil fuels may be altering the earth's natural warming and cooling cycles.

Appendix A

Answer Key

This appendix contains the answers to the Testing Yourself, Thinking Scientifically, and Understanding the Terms questions, which appear at the end of each chapter, and the Practice Problems and Additional Genetics Problems, which appear in Chapters 23–25.

Chapter 1

Testing Yourself

1. c; **2.** b; **3.** b; **4.** d; **5.** a; **6.** c; **7.** e; **8.** d; **9.** a; **10. a.** Dye is spilled on culture plate. Investigator notices that bacteria live despite exposure to sunlight; **b.** Dye protects bacteria against death by UV light; **c.** Expose all culture plates to UV light. One set of plates contains bacteria and dye; the other set contains only bacteria. The bacteria on both sets die; **d.** Dye does not protect bacteria against death by UV light.

Thinking Scientifically

1. a. Sweetener S is being varied in order to determine its effects on the body. **b.** The control group is subjected to all conditions (i.e., living in a cage kept at a certain temperature and eating the same food) except it is not receiving sweetener S. **c.** Chance of bladder cancer development is dependent upon quantity of sweetener S consumed. **d.** Yes, the conditions mentioned in answer b are constant. **e.** A control group increases confidence in results. For example, if there is no bladder cancer in control mice, it is more likely that development of bladder cancer is caused by sweetener S.

2. a. There is no observation or experiment that can be performed to prove the hypothesis false. **b.** You can perform a controlled experiment. The experimental group is fed biotin-free food. The control group is fed biotin-free food. The control group is expected to remain healthy, while the experimental group is expected to get sick. **c.** Religious beliefs are not subjected to the process described in Figure 1.3: scientific beliefs arise from this process.

Understanding the Terms

a. theory; **b.** hypothesis; **c.** energy; **d.** adaptation; **e.** homeostasis

Chapter 2

Testing Yourself

1. b; **2.** b; **3.** c; **4.** c; **5.** a; **6.** b; **7.** c; **8.** b; **9.** c; **10.** e; **11.** a; **12. a.** monomers; **b.** condensation synthesis; **c.** polymer; **d.** hydrolysis

Thinking Scientifically

1. a. Water does not contain carbon. Water does contain covalent bonds. **b.** Water absorbs heat and gives off heat—this helps to keep the body warm; water is cohesive—this helps in fluid transport; and water is a solvent in bodies—this helps in chemical reaction.

2. a. Starch is not as branched as glycogen. **b.** The length of the chain and the placement of unsaturated bonds in the two acids may be different. **c.** Actin and myosin differ in the sequence of their amino acids.

Understanding the Terms

a. protein; **b.** enzyme; **c.** cellulose; **d.** neutron; **e.** acid

Chapter 3

Testing Yourself

1. c; **2.** a; **3.** c; **4.** c; **5.** a; **6.** d; **7.** b; **8.** e; **9.** c; **10.** a; **11. a.** nucleus—DNA specifies protein synthesis; **b.** nucleolus—RNA helps form ribosomes; **c.** rough ER produces; **d.** smooth ER transports; **e.** Golgi apparatus packages and secretes.

Thinking Scientifically

1. a. Carbon (C) is found in carbohydrates and fats, while sulfur is unique to the amino acids cysteine and methionine, and therefore proteins. **b.** Radiation will appear first at the ribosomes (polysomes) and subsequently at the region of the nuclear pore and then at the nucleolus. **c.** Radiation will appear first at the rough ER and subsequently at the Golgi apparatus and then at the plasma membrane.

2. a. There are no centrioles in plant cells. **b.** Centrioles are present in the microtubule organizing center. **c.** Centrioles give rise to basal bodies that organize the microtubules in cilia and flagella. Cilia and flagella are associated with animal cells.

Understanding the Terms

a. nucleoid region; **b.** nucleolus; **c.** cytoskeleton; **d.** Golgi apparatus; **e.** endoplasmic reticulum

Chapter 4

Testing Yourself

1. a. glycolipid; **b.** glycoprotein; **c.** carbohydrate chain; **d.** hydrophilic head; **e.** hydrophobic tails; **f.** phospholipid bilayer; **g.** filaments of the cytoskeleton; **h.** peripheral protein; **i.** cholesterol; **j.** integral protein; **2.** b; **3.** b; **4.** c; **5.** c; **6.** c; **7.** e; **8.** e; **9.** c; **10.** b; **11. a.** hypertonic—net movement of water toward outside, cell shrivels. **b.** hypotonic—net movement of water toward inside; vacuoles fill with water and turgor pressure develops, causing chloroplasts to push against cell wall.

Thinking Scientifically

1. a. Alcohol is lipid soluble, and water is small enough to pass through. **b.** Na^+ is pumped across by the sodium-potassium pump. Cl^- diffuses out by way of a channel protein. **c.** Amino acids enter by facilitated diffusion. Proteins enter by endocytosis. **d.** The proteins would be digested after the endocytic vesicle fuses with a lysosome containing hydrolytic enzymes.

2. a. The number of exocytic and endocytic cells must be about the same, assuming equal sizes of vacuoles. **b.** The exocytic vesicles must be adding membrane at the forward end (right), and the endocytic vesicles must be forming at the hind end (left). **c.** The virus must somehow get out of the vesicle in order to enter the cytoplasm.

Understanding the Terms

a. diffusion; **b.** turgor pressure; **c.** isotonic; **d.** facilitated transport; **e.** exocytosis

Chapter 5

Testing Yourself

1. e; **2.** e; **3.** d; **4.** a; **5.** c; **6.** b; **7.** c; **8.** c; **9.** c; **10.** c; **11.** c; **12. a.** chromatid; **b.** centrosome (or centriole); **c.** spindle fiber or aster; **d.** nuclear envelope; **13.** The right cell represents metaphase I because homologous pairs are at the metaphase plate.

Thinking Scientifically

1. a. You can tell that chromatin is metabolically active because it is the extended form of the genetic material seen in metabolically active cells. **b.** Yes, they might differ. It is not necessary to inherit exactly the same form of a gene from each parent. **c.** Various disorders result because there are too many proteins (enzymes) of the same kind. **d.** If a particular gene is not needed for the maturation of the sperm and the egg, a defective form cannot have an effect.

2. a. Colchicine disrupts the spindle apparatus. Specifically, it prevents microtubule assembly. **b.** Asexual reproduction, requiring only mitosis, produces cells (offspring) that have the same kinds of chromosomes as the other cell (parent).**c.** Meiosis is a part of sexual reproduction. Because of meiosis, the daughter cells can have any combination of the haploid number of chromosomes, and the zygote has a different combination of the haploid number of chromosomes than either parent. **d.** Yes, the production of variation allows new types of organisms to evolve.

Understanding the Terms

a. spermatogenesis; **b.** polar body; **c.** secondary oocyte; **d.** synapsis; **e.** oogenesis

Chapter 6

Testing Yourself

1. a; **2.** e; **3.** e; **4.** d; **5.** e; **6.** d; **7.** d; **8.** a; **9.** b; **10.** e; **11. a.** active site; **b.** substrates; **c.** product; **d.** enzyme; **e.** enzyme-substrate complex; **f.** enzyme. The shape of an enzyme is important to its activity because it allows an enzyme-substrate complex to form.

Thinking Scientifically

1. a. Correct pH and a warm temperature are the optimal conditions. **b.** The yield could be increased if more pepsin or more egg white, whichever is in short supply, was added. **c.** With irreversible inhibition, the reaction stops; with reversible inhibition, the reaction continues at a reduced rate.

2. a. During aerobic cellular respiration, glucose is oxidized to carbon dioxide and water with the concomitant buildup of ATP. **b.** ATP is necessary to muscle contraction and without ATP the heart action will stop. **c.** Without the pumping action of the heart, blood flow will cease, and cells will not receive nutrients and oxygen. **d.** Without a supply of ATP, brain cells will cease to function and nerve conduction will stop.

Understanding the Terms

a. metabolism; **b.** cofactor; **c.** kinetic energy; **d.** vitamin; **e.** denatured

Chapter 7

Testing Yourself

1. b; **2.** a; **3.** c; **4.** c; **5.** a; **6.** b; **7.** c; **8.** b; **9.** d; **10.** c; **11 a.** cristae; **b.** matrix; **c.** outer membrane; **d.** intermembrane space; **e.** inner membrane

Thinking Scientifically

1. a. Oxygen (O_2) is the final acceptor for hydrogen (H) atoms at the end of the respiratory chain. The chain can continue to produce ATP only if oxygen is present. **b.** Carbon dioxide (CO_2) is produced as molecules are broken down by the transition reaction and the Krebs cycle. **c.** Muscles store glycogen and glycogen supplies glucose for ATP buildup.

2. a. From the intermembrane space to the matrix because H^+ is pumped to the intermembrane space. **b.** Without an adequate H^+ gradient, ATP production will be reduced and may stop. **c.** Glycogen and lipid supplies would be called upon for ATP production. **d.** No, the person could die due to the lack of ATP which is necessary to the life of cells.

Understanding the Terms

a. Krebs cycle; **b.** anaerobic; **c.** fermentation; **d.** pyruvate; **e.** electron transport system

Chapter 8

Testing Yourself

1. e; **2.** e; **3.** a; **4.** e; **5.** e; **6.** e; **7.** d; **8.** e; 9. **a.** outer membrane; **b.** inner membrane; **c.** stroma; **d.** thylakoid; **e.** thylakoid space; **f.** thylakoid; **g.** stroma; **10. a.** water; **b.** oxygen; **c.** carbon dioxide; **d.** carbohydrate; **e.** ADP +Ⓟ, ATP; **f.** $NADP^+$, NADPH

Thinking Scientifically

1. a. A plant takes (CO_2) and (H_2O) from the environment and gives (O_2) to the environment. **b.** Oxygen comes from the thylakoids and the breakdown of water. Carbon dioxide goes into the stroma and becomes carbohydrate. **c.** Oxygen is used for cellular respiration, and carbon dioxide comes from cellular respiration. **d.** Some glucose molecules are used in cellular respiration; some become starch; and some become cellulose and other molecules needed by plants.

2. a. The energy required is 686 Kcal/mole. This energy comes from the sun. **b.** This energy is used to make ATP molecules for cellular metabolism. **c.** The energy captured by photosynthesis produces glucose; the energy released by cellular respiration comes from glucose breakdown. **d.** Glucose is an energy source within organisms.

Understanding the Terms

a. light-dependent reaction; **b.** thylakoid; **c.** C_3 plant; **d.** chlorophyll; **e.** photosystem

Chapter 9

Testing Yourself

1. c; **2.** b; **3.** c; **4.** c; **5.** b; **6.** b; **7.** b; **8.** c; **9.** c; **10.** b; **11. a.** epidermis; **b.** cortex; **c.** endodermis; **d.** phloem; **e.** xylem; **12. a.** cork; **b.** phloem; **c.** vascular cambium; **d.** bark; **e.** xylem (wood); **f.** pith; **g.** annual ring; **13. a.** upper epidermis; **b.** palisade mesophyll; **c.** leaf vein; **d.** spongy mesophyll; **e.** lower epidermis

Thinking Scientifically

1. a. Plants have meristem tissue. **b.** A woody plant grows ever taller and humans stop growing when they reach their adult height. Plants grow new branches and leaves but humans do not grow new parts. **c.** Wood supports the plants as it grows taller. **d.** Deciduous plants grow new leaves each season but they are always about the same size.

2. a. Leaf epidermis prevents drying out; it is covered by a waxy cuticle. Leaf epidermis allows gas exchange; it contains stomates. **b.** The spongy layer carries on gas exchange; there are air spaces next to these cells. The spongy layer carries on photosynthesis; the cells contain chloroplasts. **c.** Leaf veins transport water and minerals and organic substances; they contain xylem and phloem. **d.** C_3 leaves have a palisade layer and a spongy layer; C_4 leaves have mesophyll cells in a ring around bundle sheath cells. In C_4 leaves, but not in C_3 leaves, mesophyll cells pass CO_2 to bundle sheath cells.

Understanding the Terms

a. mesophyll; **b.** cork; **c.** root hair; **d.** phloem; **e.** palisade mesophyll

Chapter 10

Testing Yourself

1. b; **2.** d; **3.** c; **4.** d; **5.** b; **6.** d; **7.** b; **8.** d; **9.** c; **10.** a; **11.** (Right) After K^+ (dots) enters guard cells, water follows by osmosis, causing guard cells to become turgid, opening the stomate; (Left) After K^+ (dots) exit guard cells, water follows by osmosis, causing guard cells to become flaccid and close. **12. a.** sporophyte; **b.** meiosis; **c.** microspore; **d.** megaspore; **e.** microgametophyte (pollen grain); **f.** megagametophyte (embryo sac); **g.** egg and sperm; **h.** fertilization; **i.** zygote; **j.** seed

Thinking Scientifically

1. a. Atmospheric pressure is the pressure of the air. This shows that atmospheric pressure cannot raise water to the height of a tall tree. **b.** Transpiration occurs and pulls on water. **c.** This suggests that transpiration could raise water to the top of trees.

2. a. Corn cells are easier to acquire than the eggs and sperm of plants. **b.** They need only extract meristem tissue from corn plants. Protoplasts which are plant cells that lack a cell wall can then be prepared. **c.** Plant cells will develop into seedlings in tissue culture. **d.** Most likely the trait will be retained if the plants are allowed to self pollinate.

Understanding the Terms

a. tropism; **b.** gametophyte; **c.** pressure-flow theory; **d.** microspore; **e.** phytochrome

Chapter 11

Testing Yourself

1. b; **2.** b; **3.** a; **4.** e; **5.** e; **6.** e; **7.** b; **8.** d; **9.** e; **10.** c; **11. a.** columnar epithelium, lining of intestine (digestive tract), protection and absorption; **b.** cardiac muscle, wall of heart, pumps blood; **c.** compact bone, skeleton, support and protection.

Thinking Scientifically

1. a. Epithelial cells have flat surfaces; therefore, they can be placed easily next to one another. This makes them suitable for covering a surface. **b.** Casparian strip. Both tight junctions and the Casparian strip prevent materials from moving between the cells. **c.** The long, tubular cells contain actin filaments and myosin filaments, and these account for their ability to contract. Because the cells run the length of a muscle, when they contract, the muscle contracts. The object would move left. **d.** Nerve cells conduct nerve impulses sometimes over long distances. The long, skinny process, or fiber, of a nerve cell makes this possible.

2. a. Humans have many more neurons than other animals. **b.** The human brain weighs about three pounds, much larger than that of other vertebrates. **c.** The human brain is more complex. In particular the human has a more highly convoluted cerebral cortex (outer layer) than any other animal. **d.** The human nervous system is highly organized; all nerves communicate directly with the brain or with the spinal cord. The spinal cord is in communication with the brain.

Understanding the Terms

a. ligament; **b.** epidermis; **c.** striated; **d.** homeostasis; **e.** spongy bone

Chapter 12

Testing Yourself

1. d; **2.** b; **3.** a; **4.** d; **5.** d; **6.** d; **7.** e; **8.** c; **9.** d; **10. a.** salivary glands; **b.** esophagus; **c.** stomach; **d.** liver; **e.** gallbladder; **f.** pancreas; **g.** small intestine; **h.** large intestine; **i.** sugar and amino acids; **j.** lipids; **k.** water

Thinking Scientifically

1. a. The epithelial portion of mucous membrane produces the enzymes. **b.** The digestive system provides nutrients needed by cells. **c.** The liver regulates the output of glucose to keep the amount in blood fairly constant.

2. a. No. Experiment is as follows: tube 1—pepsin, water, and egg white so pH is neutral; tube 2—pepsin, water, egg white, $NaHCO_3$ so pH is basic; tube 3—same as tube 4 in Figure 12.12, which is expected to show the best digestion. **b.** No. Experiment is as follows: tube 1—pepsin, HCl, water (control); tube 2—pepsin, HCl, water, starch; tube 3—same as tube 4 in Figure 12.12. Plate one in the cold, one at room temperature, and one in the incubator at body temperature. The last tube is expected to show the best digestion. **c.** No. Experiment is as follows: 3 tubes, all having the contents of test tube 4 in Figure 12.12. Place one in the cold, one at room temperature, and one in the incubator at body temperature. The last tube is expected to show the best digestion.

Understanding the Terms

a. duodenum; **b.** sphincter; **c.** defecation; **d.** lipase; **e.** gallbladder

Chapter 13

Testing Yourself

1. e; **2.** b; **3.** c; **4.** e; **5.** a; **6.** c; **7.** b; **8.** b; **9.** e; **10.** e; **11. a.** blood pressure; **b.** osmotic pressure; **c.** blood pressure; **d.** osmotic pressure; **12. a.** aorta; **b.** left pulmonary arteries; **c.** pulmonary trunk; **d.** left pulmonary veins; **e.** left atrium; **f.** semilunar valves; **g.** atrioventricular (mitral) valve; **h.** left ventricle; **i.** septum; **j.** inferior vena cava; **k.** right ventricle; **l.** chordae tendineae; **m.** atrioventricular (tricuspid) valve; **n.** right atrium; **o.** right pulmonary veins; **p.** right pulmonary arteries; **q.** superior vena cava. See also Figure 13.4, page 243.

Thinking Scientifically

1. a. lungs; **b.** brain; **c.** liver; **d.** Coronary arteries are the first blood vessels off the aorta; most likely, any clots in the coronary arteries or the capillaries formed right there.

2. Smoking **a.** increases the heartbeat; **b.** decreases the bore of arteries; **c.** closes capillary beds in fingers and toes; **d.** increases resistance of blood flow.

Understanding the Terms

a. diastole; **b.** vena cava; **c.** fibrinogen; **d.** hemoglobin; **e.** systemic circuit

Chapter 14

Testing Yourself

1. a. antigen-binding sites; **b.** light chain; **c.** heavy chain; **d.** constant region; **e.** variable region; **f.** spherical; **2.** e; **3.** e; **4.** a; **5.** b; **6.** c; **7.** b; **8.** a; **9.** a; **10.** b; **11.** d; **12.** b

Thinking Scientifically

1. a. B cells produce antibodies, when stimulated by helper T cells. T cell maturation occurs in the thymus. **b.** No, all B cells do not bind because each B cell is specific for only one type of antigen. **c.** T cells do not recognize an antigen unless it is presented by an APC. **d.** They communicate by surface-to-surface interaction and by chemical signals (i.e., cytokine).

2. a. An antigen is a foreign substance that induces an immune response by B or T cells. **b.** First the process described in Figure 14.11 has to occur two times—once for the two types of T cell. Tag each prepared monoclonal antibody with a different dye. Apply the two types of antibodies to a blood sample, and view the sample with a microscope, looking for the different types of dyes. **c.** Prepare a monoclonal antibody against HSV-2. Apply it to a sample of HSV-2 and HSV-1. The antibody should combine only with HSV-2. **d.** If the desired substance combines with the antibody, and it will be separated out. (There is a chemical process that later releases the described substance from the tube.)

Understanding the Terms

a. vaccine; **b.** lymph; **c.** antigen; **d.** apoptosis; **e.** T lymphocyte

Chapter 15

Testing Yourself

1. d; **2.** d; **3.** b; **4.** b; **5.** e; **6.** e; **7.** b; **8.** b; **9.** d; **10.** a; **11. a.** nasal cavity; **b.** nose; **c.** pharynx; **d.** epiglottis; **e.** glottis; **f.** larynx; **g.** trachea; **h.** bronchus; **i.** bronchiole. See also Figure 15.1, p. 284.

Thinking Scientifically

1. a. When CO_2 is added in the tissues, the reaction is driven to the right. **b.** When CO_2 diffuses out of the lungs, the reaction is driven to the left. **c.** When H^+ is added, more H_2CO_3 forms and since it dissociates little, the H^+ is taken up. **d.** When OH^- is added, it combines with the H^+, forming water. Since water dissociates little, the OH^- is taken up.

2. a. CO_2 stimulates breathing; O_2 does not. **b.** The buildup of CO_2 stimulates breathing. **c.** CO_2 in blood raises the pH of blood. **d.** Sense receptors usually are stimulated by the presence of something.

Understanding the Terms

a. pharynx; **b.** diaphragm; **c.** bicarbonate ion; **d.** expiration; **e.** alveolus

Chapter 16

Testing Yourself

1. c; **2.** c; **3.** e; **4.** a; **5.** c; **6.** b; **7.** d; **8.** a; **9.** b; **10. a.** glomerulus; **b.** efferent arteriole; **c.** afferent arteriole; **d.** proximal convoluted tubule; **e.** loop of the nephron; **f.** descending limb; **g.** ascending limb; **h.** peritubular capillary network; **i.** distal convoluted tubule; **j.** renal vein; **k.** renal artery; **l.** collecting duct

Thinking Scientifically

1. a. Urea is a single molecule, urine is a mixture of molecules and ions. **b.** This increases blood pressure in the glomerulus. **c.** The force is osmotic pressure. **d.** The rate increases because osmotic pressure would decrease in the glomerulus.

2. a. To say that urine is 95% water only indicates how much water per solutes there is. **b.** Carriers can only work so fast. The fluid is moving in the proximal convoluted tubule, and in the meantime, glucose has gone by. **c.** Refer to this equation:

$$CO_2 + H_2O \rightleftharpoons H_2CO_3 \rightleftharpoons H^+ + HCO_3^-$$

When lungs excrete CO_2, the equation is driven to the left and blood becomes more basic. When kidneys excrete the HCO_3^-, the equation is driven to the right and blood becomes more acidic. **d.** The pH affects enzymes, causing a change in shape so that they do not function as well.

Understanding the Terms

a. loop of the nephron; **b.** tubular reabsorption; **c.** aldosterone; **d.** renal cortex; **e.** peritubular capillary network

Chapter 17

Testing Yourself

1. b; **2.** c; **3.** a; **4.** a; **5.** c; **6.** d; **7.** b; **8.** c; **9.** b; **10.** d; **11.** d; **12.** c; **13.** e; **14. a.** sensory neuron (or fiber); **b.** interneuron; **c.** motor neuron (or fiber); **d.** sensory receptor; **e.** cell body; **f.** dendrites; **g.** axon; **h.** nucleus of Schwann cell (neurolemmocyte); **i.** node of Ranvier (neurofibril node); **j.** effector

Thinking Scientifically

1. a. The nerve impulse travels along a membrane and is dependent on the Na^+ and K^+ across the membrane. **b.** A reading lower than -65mV is expected. The resting potential is -65mV, and inhibitory neurotransmitters increase the polarity. **c.** Synaptic vesicles occur only at one end of an axon. **d.** The degree of contraction depends on the number of neurons that are stimulating muscle fibers.

2. a. Interneurons can take nerve impulses across the spinal cord from one side to the other. **b.** Neither leg would respond because nerve impulses would never reach the cord. **c.** The right leg would still be able to respond. **d.** Neither leg would respond because interneurons would be destroyed.

Understanding the Terms

a. nerve impulse; **b.** cerebral hemisphere; **c.** integration; **d.** motor neuron; **e.** reticular formation

Chapter 18

Testing Yourself

1. d; **2.** e; **3.** e; **4.** c; **5.** d; **6.** d; **7.** c; **8.** b; **9.** e; **10.** d; **11. a.** retina—contains receptors; **b.** choroid—absorbs stray light; **c.** sclera—protects and supports eyeball; **d.** optic nerve—transmits impulses to brain; **e.** fovea centralis—makes acute vision possible; **f.** ciliary body—holds lens in place, accommodation; **g.** lens—refracts and focuses light rays; **h.** iris—regulates light entrance; **i.** pupil—admits light; **j.** cornea—refracts light rays

Thinking Scientifically

1. a. One possible categorization: focusing—lens, cornea, humors, ciliary body; vision—retina (rods, cones, fovea), optic nerve; other—iris, pupil, choroid, sclera (except for cornea). Justification: Some parts of the eye are concerned with focusing the light, some with bringing about vision, and some have neither of these functions. Glasses usually correct focusing. **b.** Pigments are all colored molecules that usually are capable of absorbing energy. **c.** There must be a neural pathway between the eyes and the pineal gland.

2. a. The evolution of the ear in a sequence of animals is needed to support the hypothesis. You would expect to find stages by which changes led to the mammalian ear. **b.** Most likely, the inner ear evolved from the lateral line. Most likely, the outer ear and the middle ear evolved otherwise. The human inner ear has mechanoreceptors sensitive to fluid pressure waves. The outer ear receives sound waves in the air, and the middle ear transmits and amplifies these.

Understanding the Terms

a. cochlea; **b.** chemoreceptor; **c.** rhodopsin; **d.** sensation; **e.** spiral organ

Chapter 19

Testing Yourself

1. b; **2.** f; **3.** c; **4.** e; **5.** b; **6.** c; **7.** b; **8.** d; **9.** d; **10.** a; **11.** b; **12. a.** T tubule; **b.** sarcoplasmic reticulum; **c.** myofibril; **d.** Z line; **e.** sarcomere; **f.** sarcolemma of muscle fiber

Thinking Scientifically

1. a. Bone is living tissue: it grows and heals; it is supplied with blood and nerves; and it contains cells. **b.** Bone strength has to equal muscle strength or movement of muscles can cause bones to crack or to break. **c.** These are for attachment of muscles. **d.** The wide pelvis is associated with childbirth.

2. a. Yes, all myofibrils contract because a muscle fiber does not have degrees of contraction. **b.** Yet, it gets closer to the center. **c.** Myoglobin has the higher affinity or else it could never receive O_2 from hemoglobin. **d.** Mitochondria in muscle fibers use the oxygen to receive electrons at the electron transport system.

Understanding the Terms

a. tetanus; **b.** ligament; **c.** sarcolemma; **d.** creatine phosphate; **e.** red bone marrow

Chapter 20

Testing Yourself

1. a; **2.** a; **3.** f; **4.** b; **5.** c; **6.** a; **7.** e; **8.** d; **9.** c; **10.** d; **11.** b; **12.** e; **13.** d; **14. a.** inhibits; **b.** inhibits; **c.** releasing hormone; **d.** stimulating hormone; **e.** target gland hormone

Thinking Scientifically

1. a. You might want to change the ending of the definition to "target cell, organ or organism." Figure 20.1 shows the liver as a target organ for insulin and a moth as a target organism for a pheromone. **b.** A target cell must have receptor proteins for the environmental signal. **c.** Behavior is a good term because it encompasses a wide variety of responses. For example, a target cell changes its metabolism and might begin to produce a product; the liver begins to store glucose as glycogen in response to insulin; and the male moth begins to fly toward the female. **d.** Low blood sugar is the stimulus for the secretion of insulin, and negative feedback shuts down its production. When a nerve impulse is crossing a synapse, the neurotransmitter then either is broken down or is taken up by the presynaptic membrane.

2. a. You would expect to find sugar in the urine because the body would contain no insulin; blood sugar would rise and spill over into the urine. **b.** No, your findings only prove that blood sugar rises when the pancreas is missing. **c.** You have to get a supply of pure insulin, inject it in an animal, and show that the blood sugar lowers. **d.** Yes, you now know that the presence of both the pancreas and insulin lowers blood sugar. The logical conclusion is that the pancreas is the source of insulin.

Understanding the Terms

a. thyroid gland; **b.** diabetes mellitus; **c.** adrenocorticotropic hormone (ACTH); **d.** nonsteroid hormone; **e.** oxytocin

Chapter 21

Testing Yourself

1. a. seminal vesicle; **b.** ejaculatory duct; **c.** prostate gland; **d.** bulbourethral gland; **e.** anus; **f.** vas deferens; **g.** epididymis; **h.** testis; **i.** scrotum; **j.** foreskin; **k.** glans penis; **l.** penis; **m.** urethra; **n.** vas deferens; **o.** urinary bladder; **2.** c; **3.** d; **4.** c; **5.** c; **6.** d; **7.** e; **8.** c; **9.** c; **10.** c; **11.** b; **12.** e

Thinking Scientifically

1. a. Due to negative feedback, the intake of anabolic steroids causes the anterior pituitary to stop producing gonadotropic hormones, leading to atrophy of the interstitial cells of the testes. **b.** To test the hypothesis, administer anabolic steroids to mice and collect data on resulting blood levels of gonadotropic hormone. Remove the testes and using a microscope look for atrophy of tissues. **c.** Anabolic steroids raise the level of LDL in the blood. This could lead to increased risk of heart disease. **d.** To test the hypothesis, administer anabolic steroids to mice and collect data on resulting blood levels of LDL. Remove coronary blood vessels and look for the presence of plaque.

2. a. Due to negative feedback, the administration of estrogen and progesterone causes the anterior pituitary to stop producing FSH and LH, and no follicles or oocytes mature. Without egg production, there can be no pregnancy. **b.** Administer this birth-control pill to mice. Collect data on blood levels of FSH and LH. Remove the ovaries, and examine them for the presence of mature follicles. **c.** Postmenopausal women who take birth-control pills should have minor or no levels of FSH and LH in the blood. **d.** Administer birth-control pills to postmenopausal women and collect data on blood levels of FSH and LH.

Understanding the Terms

a. vagina; **b.** endometrium; **c.** prostate gland; **d.** follicle-stimulating hormone; **e.** vulva

Chapter 22

Testing Yourself

1. b; **2.** b; **3.** b; **4.** a; **5.** e; **6.** b; **7.** e; **8.** a; **9.** e; **10.** a; **11. a.** chorion (contributes to forming placenta where wastes are exchanged for nutrients with the mother); **b.** amnion (protects and prevents desiccation); **c.** embryo; **d.** allantois (blood vessels become umbilical blood vessels); **e.** yolk sac (first site of blood cell formation); **f.** fetal portion of placenta; **g.** maternal portion of placenta; **h.** umbilical cord (connects developing embryo to the placenta). See also Figure 22.9, page 451.

Thinking Scientifically

1. a. When a cell inherits a certain cytoplasmic composition, only certain genes are activated. **b.** These activated genes begin to direct the synthesis of particular proteins. **c.** Some of these proteins may be secreted to act as specific signals for other cells.

2. a. Tissue A gives off certain signals that influence the morphogenesis of tissue. **b.** Because of this, tissue B gives off certain signals that influence the morphogenesis of tissue. **c.** Tissue C then gives off signals, and so forth.

Understanding the Terms

a. induction; **b.** germ layer; **c.** morphogenesis; **d.** placenta; **e.** cleavage

Chapter 23

Testing Yourself

1. b; **2.** a; **3.** c; **4.** d; **5.** c; **6.** b; **7.** c; **8.** b; **9.** d; **10.** e; **11.** autosomal recessive

Thinking Scientifically

1. a. Alternative hypotheses: Factors do not segregate, therefore, all parental gametes would be the same—that is, *Yy*; factors do segregate; therefore, two parental gametes are possible—that is, *Y* and *y*. **b.** If the gametes were always *Yy*, then the phenotype green would not have appeared. Since green does appear, then the second hypothesis is supported. **c.** Alternative hypotheses for Figure 23.6: (a) factors do not assort independently; therefore, the gametes always will be, for example, either *WS* or *ws*; (b) factors do assort independently; therefore, 4 gametes are possible—*WS, Ws, wS, ws*. **d.** If the factors do not assort, then there would be fewer phenotypes among the offspring. Since there are all four possible phenotypes among the offspring, the factors had to assort independently of one another.

2. a. The fault is an inability to produce melanin because of an enzyme defect. **b.** An inability to produce a normal enzyme is most likely recessive. **c.** The possible crosses are *aa* × *aa*; *Aa* × *aa*; *Aa* × *Aa*. Yes, *Aa* individuals are carriers. **d.** A functioning gene would ensure that all skin cells are capable of producing melanin.

Understanding the Terms

a. carrier; **b.** allele; **c.** dominant allele; **d.** testcross; **e.** genotype

Practice Problems 1

1. a. *W*; **b.** *WS, Ws*; **c.** *T, t*; **d.** *Tg, tg*; **e.** *AB, Ab, aB, ab*
2. a. gamete; **b.** genotype; **c.** gamete; **d.** genotype

Practice Problems 2

1. 75% or 3:1; **2.** Both heterozygous; **3.** *DD* × *dd*; *Dd*

Practice Problems 3

1. Dihybrid; **2.** 1/16; **3.** *DdFf* × *ddff*; *ddff*

Practice Problems 4

1. 50%; **2.** Homozygous dominant or heterozygous

Practice Problems 5

1. light; **2.** very light; **3.** baby 1 = Doe; baby 2 = Jones; **4.** types AB, O, A, B

Practice Problems 6

1. $H'H'$; no; **2.** normal–25%; sickle-cell trait–50%; sickle-cell disease–25%

Chapter 24

Testing Yourself

1. c; **2.** a; **3.** b; **4.** c; **5.** c; **6.** c; **7.** b; **8.** a; **9.** c; **10.** X^AX^a

Thinking Scientifically

1. a. Red eye is dominant. **b.** No, because females do not have a Y chromosome, and yet they have red eye color. Yes, because males have only one X chromosome, and this explains why only males have white eyes in the F_2 generation. The results are explainable only on the basis that the red/white allele is on the X chromosome.

2 a. The best evidence is passage of the trait from Leopold, a son of Victoria's, through Alice, his carrier daughter, to his grandson Rupert. Also, only males have the trait and they all have mothers who can trace their ancestry to Victoria. **b.** 50% **c.** It is a protein because genes specify protein synthesis. **d.** Abnormal DNA for a blood clotting factor.

Understanding the Terms

a. linkage group; **b.** nondisjunction; **c.** chromosomal mutation; **d.** karyotype; **e.** sex-linked trait

Practice Problems

1. His mother; X^HX^h, X^HY, X^hY
2. 100%; none; 100%
3. $RrX^BX^b \times RrX^BY$; rrX^bY
4. The husband is not the father.

Additional Genetics Problems for Chapter 23 and Chapter 24

1. 50%; **2.** 50% (most likely father is heterozygous); **3.** 50%; **4.** None; **5.** No. Only homozygous recessives cannot curl the tongue, and persons with this genotype cannot pass on the ability to curl the tongue. **6.** *Tt*, 25%; **7.** 50%; **8.** Child *aa*, parents *Aa*; **9.** 50%; **10.** 100% chance for widow's peak and 0% chance for continuous hairline; **11.** 50%; **12.** 210 gray bodies and 70 black bodies; 140 = heterozygous; cross fly with recessive (black body); **13.** F_1 = all black with short hair; F_2 = 9 black, short : 3 black, long : 3 brown, short : 1 brown, long; offspring would be 1 brown long : 1 brown short : 1 black long : 1 black short; **14.** *Bbtt* × *bbTt* and *bbtt*; **15.** *GGll*; **16.** 25%; **17.** 100%; **18.** sickle-cell trait; **19.** *Ss*; **20.** 25%; **21.** *BO* (type B blood), *AO* (type A blood), *AB* (type AB) blood, *OO* (type O blood); **22.** *OO*, yes, *A* or *AB*; **23.** Father could be *BB* or *BO* or *AB*; **24.** father: X^sY; mother: X^SX^s; **25.** son; **26.** 50%, none; **27.** X^hY, X^HX^h; **28. a.** Males: all red eyes; females: all red eyes; **b.** Males: 1 red eye: 1 white eye; females: all red eyes; **c.** autosomal dominant, *Aa*; **29.** X^bYww, boys = X^BYWw, girls = X^BX^bWw; **30.** Males: 3 gray body with red eyes: 1 black body with red eyes: 3 gray body with white eyes: 1 black body with white eyes; females: 3 gray body with red eyes: 1 black body with red eyes.

Chapter 25

Testing Yourself

1. e; **2.** e; **3.** c; **4.** d; **5.** c; **6.** d; **7.** e; **8. a.** regulator gene; **b.** promoter; **c.** operator; **d.** structural genes; **e.** DNA; **f.** mRNA; **g.** active repressor protein; **h.** RNA polymerase binding site; **9.** b; **10.** d

Thinking Scientifically

1. a. The evidence is that the sequence of amino acids in a protein parallels the code in DNA. **b.** The evidence is the existence of duplicated chromosomes prior to cell division. **c.** The sequence of bases in DNA can change, and this allows mutation.

2. a. It would show up first in the nucleus and then in the cytoplasm. **b.** The evidence is that the sequence of amino acids in a polypeptide parallels the codons in mRNA, not rRNA. **c.** Figure 25.10 shows ribosomes moving along the mRNA.

Understanding the Terms

a. mutagen; **b.** RNA polymerase; **c.** transcription; **d.** translation; **e.** anticodon

Practice Problems

1. ACU′ CCU′ GAA′ UGC′ AAA
2. UGA′ GGA′ CUU′ ACG′ UUU
3. threonine–proline–glutamate–cysteine–lysine

Chapter 26

Testing Yourself

1. a; **2.** e; **3.** e; **4.** e; **5.** b; **6.** d; **7.** c; **8.** e; **9.** AATTTTAA; **10. a.** retrovirus; **b.** recombinant RNA; **c.** human genome; **d.** recombinant RNA; **e.** reverse transcription; **f.** recombinant DNA; **g.** defective gene. See also Figure 26.9, page 539.

Thinking Scientifically

1. Inject a large number of diabetics with both types of insulin (at different times), and observe any effects. The biotechnology insulin is expected to show fewer side effects because it is human insulin, not cattle or pig insulin—the sequence of amino acids is expected to be closer to that of the individual receiving the insulin. Also, it might be pure—it does not contain any substances other than insulin.

2. Feed the meat to two groups of human volunteers, and observe any effects. Since growth hormone is a protein, any present in the meat is denatured upon cooking or digested upon eating.

3. Same experiment as described in answer 2. First, feed the plants to animals, and if no effects are observed, then feed the plants to humans. The toxin might be harmful to humans.

4. Keep testing for the presence of the pollutant and the bacteria. See if the bacteria disappears. They run out of food, that is, the pollutant.

Understanding Terms

a. restriction enzyme; **b.** transgenic organism; **c.** probe; **d.** clone; **e.** plasmid

Chapter 27

Testing Yourself

1. d; **2.** c; **3.** b; **4.** e; **5.** e; **6.** e; **7.** b; **8.** c; **9.** d; **10.** e; **11.** e; **12.** b; **13.** b; **14.** e; **15.** a; **16.**

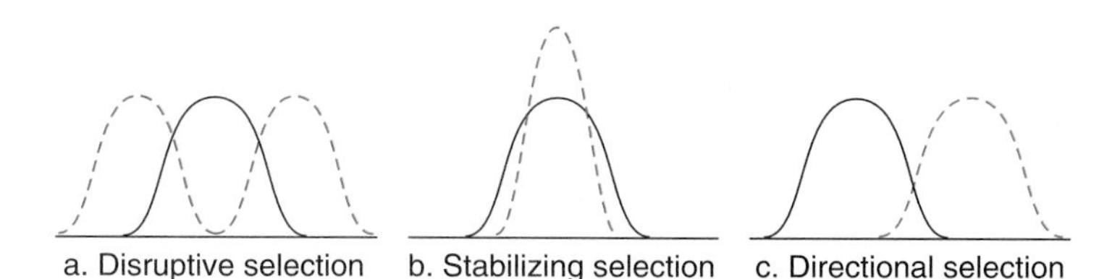

Thinking Scientifically

1. a. There would have been no enzymes present before the first protein formed. **b.** Proteins are unable to store genetic information or to replicate. **c.** By the same mechanism as DNA—complementary base pairing. **d.** Enzymes are needed for replication to occur and for other metabolic processes.

2. a. The plant species contains variations that are inheritable. On the mountain top, plants that were shorter tended to survive and to reproduce, until only shorter plants were observed there. **b.** You expect the plants still to be short because only these genes now are present in the gene pool. **c.** Directional selection (toward shorter plants) was followed by stabilizing selection (most plants, then, tend to have genes for shortness).

Understanding Terms

a. natural selection; **b.** directional selection; **c.** punctuated equilibrium; **d.** homologous structure; **e.** gene flow

Practice Problems

1. 30%; **2.** $q = 0.1$, $p = 0.9$; homozygous recessive = 1%; homozygous dominant = 81%; heterozygous = 18%; **3.** homozygous recessive = 49%; homozygous dominant = 9%; heterozygous = 42%

Chapter 28

Testing Yourself

1. d; **2.** b; **3.** c; **4.** c; **5.** b; **6.** d; **7.** e; **8.** d; **9.** e; **10.** b; **11.** a; **12.** c; **13.** e; **14.** e; **15. a.** sexual reproduction; **b.** isogametes pairing; **c.** zygote (2n); **d.** zygospore (2n); **e.** asexual reproduction; **f.** zoospores (n); **g.** nucleus; **h.** chloroplast; **i.** pyrenoid; **j.** starch granule; **k.** flagellum; **l.** eyespot; **m.** gamete formation. See also Figure 28.10, page 584. **n.** Asexual: one parent, no gametes produced, no genetic recombination. Sexual: two parents, gametes produced, genetic recombination.

Thinking Scientifically

1. a. If the zygote undergoes meiosis, there is no diploid adult. **b.** If meiosis is delayed until there is a diploid adult, the alternation of generations life cycle occurs. **c.** If haploid spores join (then they act like gametes), the haploid generation is eliminated and the diplontic life cycle occurs.

2. Fungi live on dead organic matter. **a.** Without other living things, there is no dead organic matter. **b.** In the *Chlamydomonas* cycle, the adult, the spores, and the gametes are flagellated. In the fungal cycle, the adult, the spores, and the gametes are not flagellated. **c.** The body of a fungus may not have the ability to withstand dryness.

Understanding Terms

a. nucleoid; **b.** symbiotic; **c.** binary fusion; **d.** saprotroph; **e.** lichen

Chapter 29

Testing Yourself

1. e; **2.** d; **3.** c; **4.** c; **5.** e; **6.** b; **7.** b; **8.** b; **9.** e; **10. a.** sporophyte (2n); **b.** meiosis; **c.** gametophyte (n); **d.** fertilization. See also Figure 29.2*a*, page 601.

Thinking Scientifically

1. a. No, vascular tissue would not be found because bryophytes lack vascular tissue. **b.** Try to grow them in a dry location. Observe results. **c.** The gametophyte (n) is dominant, and it bears the burden of adaptation to the environment. **d.** If vascular tissue evolved, it is possible the bryophytes might spread into more habitats on land, but only if they also lost their dependence on water for reproduction.

2. a. In humans, the male passes sperm directly to the female, who retains the egg within her body. In trees, pollen, which can resist drying out, carries the sperm to the vicinity of the egg. Then the pollen germinates to give a pollen tube, through which the sperm passes. **b.** Humans have a blood vascular system; trees have xylem. **c.** Humans have an internal skeleton of bone; trees are supported by xylem. **d.** Humans can maintain a warm internal temperature; deciduous trees lose their leaves and become inactive.

Understanding Terms

a. megaspore; **b.** gymnosperm; **c.** sporophyte; **d.** seed; **e.** ovule

Chapter 30

Testing Yourself

1. a. tentacle; **b.** gastrovascular cavity; **c.** mouth; **d.** mesoglea; **2.** e; **3.** c; **4.** e; **5.** e; **6.** b; **7.** e; **8.** a; **9.** e; **10.** c; **11.** d; **12.** a; **13.** e; **14.** d; **15. a.** head; **b.** antenna; **c.** simple eye; **d.** compound eye; **e.** thorax; **f.** tympanum; **g.** abdomen; **h.** forewing; **i.** hindwing; **j.** ovipositor; **k.** spiracles; **l.** air sac; **m.** spiracle; **n.** tracheae. See also Figure 30.18, page 637.

Thinking Scientifically

1. a. Animals are heterotrophic, locomote by means of contracting fibers, are multicellular with specialized tissues, have a diplontic life cycle, have sex organs, and do not always protect the zygote and the embryo. **b.** Animals locomote by means of contracting fibers. **c.** Protozoans locomote and are heterotrophic. **d.** Complex animals have tissues and organs that are highly differentiated.

2. a. An animal with a complete digestive tract because the food can be processed as it moves in one direction from mouth to anus. **b.** Coelomate animals have a cavity for the location of specialized organs. **c.** An animal with a complete digestive tract because one of the organs can be specialized for storage of food. **d.** In a planarian, the digestive system is branched and this allows nutrients to be delivered to all parts of the body.

Understanding the Terms

a. trachea; **b.** coelom; **c.** bilateral symmetry; **d.** molt; **e.** nephridia

Chapter 31

Testing Yourself

1. b; **2.** e; **3.** c; **4.** e; **5.** a; **6.** a; **7.** b; **8.** a; **9.** b; **10.** e; **11.** c; **12. a.** pharyngeal pouches; **b.** dorsal hollow nerve cord; **c.** notochord; **d.** post-anal tail.

Thinking Scientifically

1. a. With negative pressure, air is drawn in, and with positive pressure, air is pushed in. Frogs force air into the lungs by gulping it. **b.** There is no mixing of used air with new air coming in; therefore, more O_2 enters the blood. **c.** Frogs practice skin breathing. Reptiles have better developed lungs. **d.** Movement of the diaphragm assists in creating the negative pressure that draws air into the lungs.

2. a. Adaptation of the environment. **b.** Branching occurs where the hominids are adapted variously. For example, *A. africanus* and *A. robustus* had different diets, and Neanderthals in Europe and *H. sapiens* in Africa were built differently. **c.** *H. erectus* is markedly different from the previous fossils; many advances are evident. Similarly, Cro-Magnon is much more advanced than previous fossils. **d.** All races can interbreed.

Understanding the Terms

a. watervascular system; **b.** prosimian; **c.** notochord; **d.** deuterostomes; **e.** hominid

Chapter 32

Testing Yourself

1. b; **2.** c; **3.** a; **4.** e; **5.** c; **6.** c; **7.** c; **8.** b; **9.** e; **10.** b

Thinking Scientifically

1. a. Have at least two group of guinea pigs. The first group is allowed to sense their food and the second group is prevented from seeing or smelling the food. If only the first group eats willingly, the nervous system must be involved in acquiring food. **b.** Food selection is controlled by immediate environmental influences. **c.** Food selection is controlled by inheritance of genes.

2. a. If you make learning pleasurable, and reward (not punish) children, they are more likely to behave as you wish. **b.** Assuming that the food is giving the animal pleasure, the limbic system is most likely involved. An emotional experience is remembered best. **c.** Find a way to stimulate the limbic system directly (e.g., implanted electrodes in brain) and see if the animal has an improved ability to learn the behavior.

Understanding the Terms

a. territoriality; **b.** altruism; **c.** communication; **d.** pheromone; **e.** inclusive fitness

Chapter 33

Testing Yourself

1. d; **2.** d; **3.** e; **4.** b; **5.** b; **6.** e; **7.** d; **8.** e; **9.** e; **10.** a; **11. a.** lag; **b.** exponential growth; **c.** deceleration; **d.** environmental resistance; **e.** carrying capacity

Thinking Scientifically

1. a. and **b.** It is a J-shaped curve. **c.** The curve would fall dramatically.

2. a. Formerly, it was assumed that each country was responsible for its own standard of living. This attitude is changing because planners now think in global terms. **b.** Increasingly, people in the developed countries believe that health care for all people should be the same. **c.** Most people now believe that private citizens and industry should find ways to ensure an ecologically fit world for future generations. If we do not, the standard of living will decrease dramatically. **d.** Increasingly, people are beginning to think in global terms.

Understanding the Terms

a. population; **b.** biotic potential; **c.** exponential growth; **d.** demographic transition; **e.** carrying capacity

Chapter 34

Testing Yourself

1. Place a dot on the highest point of the graph. An intermediate level of disturbance results in the greatest amount of diversity. **2.** b; **3.** e; **4.** d; **5.** e; **6.** b; **7.** e; **8.** c; **9.** e; **10.** e; **11.** e

Thinking Scientifically

1. a. This is scramble competition. The larvae all compete for the same resource, and each receives minimal food. Few, then, survive. **b.** This is contest competition. With this type of competition, the unsuccessful may not survive and reproduce but the successful will survive and reproduce. **c.** In nature, female blowflies lay many eggs in various locations. The larvae are likely to survive in some of these locations. Scramble competition is not a disadvantage under these circumstances, contest competition ensures that some members of the population survive.

2. a. This suggests that the fungi are parasitic on the algae in a lichen because organic molecules can pass from algae to fungi via these organs. **b.** This suggests that the fungi and the algae are mutualistic in a lichen because each member of the relationship benefits. **c.** Yes, both parasitism and mutualism involve some sort of dependency.

Understanding the Terms

a. competitive exclusion principle; **b.** ecological succession; **c.** symbiosis; **d.** niche; **e.** commensalism

Chapter 35

Testing Yourself

1. b; **2.** d; **3.** a; **4.** c; **5.** d; **6.** d; **7.** a; **8.** a; **9.** a; **10.** d

Thinking Scientifically

1. a. The conditions are temperature and amount of rainfall. **b.** A grassland might replace a deciduous forest and a desert might replace a grassland. **c.** The more sunlight; the more photosynthesis; the more life supported. **d.** Similar species cannot play the same role (occupy the same niche) in a community; therefore, diversification occurs. **e.** Varied and plentiful food sources provide different ways of getting food.

2. a. A squid has adaptations for a water environment, not a land environment. It is streamlined and moves by jet propulsion—both of which are adaptations for locomotion in water. A squid breathes by means of gills. **b.** A monkey has adaptations for a forest environment, not a grassland environment. It swings from limb to limb of a tree and mainly eats fruits.

Understanding the Terms

a. estuary; **b.** biome; **c.** benthic division; **d.** savanna; **e.** biosphere

Chapter 36

Testing Yourself

1. c; **2.** b; **3.** d; **4.** d; **5.** c; **6.** c; **7.** b; **8.** a; **9.** d; **10.** c; **11.** b; **12.** c; **13.** a; **14.** c; **15. a.** producers; **b.** consumers; **c.** inorganic nutrient pool; **d.** decomposers; **16. a.** top carnivores; **b.** carnivores; **c.** herbivores; **d.** producers

Thinking Scientifically

1. a. Primary consumers have more energy available to them (in the form of food) than higher level consumers. **b.** The pyramid indicates that there is less energy available at each trophic level. Eventually, there is not enough energy to support another population. **c.** The size of a top predator population is controlled by the amount of food energy available to it. **d.** The other secondary consumer populations would increase in size due to less competition, and the ecosystem would remain about the same.

2. a. Consumption of fuel is costly and combustion of fuel leads to air pollution. **b.** As a general rule only 10% of the energy within one tropic level is passed on to the next level; therefore it wastes energy to feed a population on cattle. **c.** Allowing cattle to graze is cheaper, the meat is better for you because it contains less fat, and you have reduced the probability of sewage ending up in the water supply. **d.** Growing different varieties of crops is a safety measure—if one variety is wiped out by a parasite, for example, you would still have the other types of crops. **e.** Irrigation uses much water and results in water pollution when pesticides and fertilizers get washed into water supplies. f. The fewer the people, the less food required to feed them.

Understanding the Terms

a. detritus; b. ozone shield; c. fossil fuel; d. nitrogen fixation; e. food web

Appendix B

Classification of Organisms

The classification system given here is a simplified one, containing all the major kingdoms, as well as the major divisions (called phyla in the kingdom Protista and the kingdom Animalia). The text does not discuss all the divisions and phyla listed here.

Kingdom Monera

Prokaryotic, unicellular bacteria. Nutrition principally by absorption, but some are photosynthetic, chemosynthetic, or parasitic.

Phylum Chemoautotrophs: chemosynthetic bacteria

Phylum Photoautotrophs: photosynthetic bacteria that do not evolve oxygen

Phylum Enterobacteria: facultatively anaerobic rod-shaped bacteria that live in intestinal tracts

Phylum Myxobacteria: aerobic, gliding rod-shaped bacteria, found in soil or are parasitic

Phylum Pseudomonads: aerobic curved or rod-shaped bacteria, usually found in soil

Phylum Rickettsias: intracellular rod-shaped parasites

Phylum Spirochaetes: facultatively anaerobic helical-shaped bacteria

Phylum Cyanobacteria: photosynthetic bacteria that do evolve oxygen

Phylum Actinobacteria: filamentous soil-dwelling bacteria

Kingdom Protista

Eukaryotic, unicellular organisms and their immediate multicellular descendants; sexual reproduction; flagella and cilia with 9 + 2 arrangement of microtubules

Phylum Chlorophyta: green algae 7,000 species

Phylum Phaeophyta: brown algae 1,500 species

Phylum Chrysophyta: diatoms 11,000 species

Phylum Dinoflagellata: dinoflagellates 1,000 species

Phylum Euglenophyta: euglenoid flagellates 1,000 species

Phylum Rhodophyta: red algae 4,000 species

Phylum Sarcodina: amoeboid protozoa 40,000 species

Phylum Ciliophora: ciliated protozoa 8,000 species

Phylum Zoomastigophora: flagellated protozoa thousands of species

Phylum Sporozoa: parasitic protozoa 3,600 species

Phylum Myxomycota: slime molds 560 species

Phylum Oomycota: water molds 580 species

Kingdom Fungi

Multicellular eukaryotes: heterotrophic by absorption; lack flagella; nonmotile spores form during both asexual and sexual reproduction

Division Zygomycota: zygospore fungi 600 species

Division Ascomycota: sac fungi 30,000 species

Division Basidiomycota: club fungi 16,000 species

Division Deuteromycota: imperfect fungi 25,000 species

Kingdom Plantae

Multicellular, primarily terrestrial eukaryotes with well-developed tissues; autotrophic by photosynthesis; alternation of generations life cycle

Division Hepatophyta: liverworts 10,000 species

Division Bryophyta: mosses 12,000 species

Division Anthocerotophyta: hornworts 100 species

Division Psilotophyta: whisk ferns several species

Division Lycopodophyta: club mosses 1,000 species

Division Equisetophyta: horsetails 15 species

Division Pteridophyta: ferns 12,000 species

Division Coniferophyta: conifers 550 species

Division Cycadophyta: cycads 100 species

Division Ginkgophyta: maidenhair tree one species

Division Gnetophyta: gnetophytes 70 species

Division Magnoliophyta: flowering plants

- Class Magnoliopsida: dicots 170,000 species
- Class Liliopsida: monocots 65,000 species

Kingdom Animalia

Multicellular organisms with well-developed tissues; usually motile; heterotrophic by ingestion, generally in a digestive cavity; diplontic life-style

Phylum Porifera: sponges 5,000 species

Phylum Cnidaria: radially symmetrical marine animals 9,000 species

- Class Hydrozoa: hydras, Portuguese man-of-war
- Class Scyphozoa: jellyfish
- Class Anthozoa: sea anemones and corals

Phylum Platyhelminthes: flatworms 13,000 species

- Class Turbellaria: free-living flatworms
- Class Trematoda: parasitic flukes
- Class Cestoda: parasitic tapeworms

Phylum Nematoda: roundworms 12,000 species

Phylum Rotifera: rotifers 2,000 species

Phylum Mollusca: softbodied, unsegmented animals 110,000 species

- Class Polyplacophora: chitons
- Class Monoplacophora: Neopilina
- Class Gastropoda: snails and slugs
- Class Cephalopoda: squid and octopuses
- Class Bivalvia: clams and mussels

Phylum Annelida: segmented worms 12,000 species

- Class Polychaeta: sandworms
- Class Oligochaeta: earthworms
- Class Hirudinea: leeches

Phylum Arthropoda: animals with chitinous exoskeleton and jointed appendages over 6 million species

Class Crustacea: lobsters, crabs, barnacles

Class Arachnida: spiders, scorpions, ticks

Class Chilopoda: centipedes

Class Diplopoda: millipedes

Class Insecta: grasshoppers, termites, beetles

Phylum Echinodermata: spiny, radially symmetrical marine animals 6,550 species

Class Crinoidea: sea lilies and feather stars

Class Asteroidea: starfishes

Class Ophiuroidea: brittle stars

Class Echinoidea: sea urchins and sand dollars

Class Holothuroidea: sea cucumbers

Phylum Hemichordata: acorn worms 90 species

Phylum Chordata: dorsal supporting rod (notochord) at some stage; dorsal hollow nerve cord; pharyngeal pouches or slits 45,000 species

Subphylum Urochordata: tunicates 1,250 species

Subphylum Cephalochordata: lancelets 23 species

Subphylum Vertebrata: vertebrates over 44,000 species

Class Agnatha: jawless fishes (lampreys, hagfishes) 63 species

Class Chondrichthyes: cartilaginous fishes (sharks, rays) 850 species

Class Osteichthyes: bony fishes 20,000 species

Subclass Sarcopterygii: lobe-finned fishes

Subclass Actinopterygii: ray-finned fishes

Class Amphibia: frogs, toads, salamanders 3,900 species

Class Reptilia: snakes, lizards, turtles 6,000 species

Class Aves: birds 9,000 species

Class Mammalia: mammals 4,500 species

Subclass Prototheria: egg-laying mammals

Order Monotremata: duck-billed platypuses, spiny anteaters

Subclass Metatheria: marsupial mammals

Order Marsupialia: opossums, kangaroos

Subclass Eutheria: placental mammals

Order Insectivora: shrews, moles

Order Chiroptera: bats

Order Edentata: anteaters, armadillos

Order Rodentia: rats, mice, squirrels

Order Lagomorpha: rabbits, hares

Order Cetacea: whales, dolphins, porpoises

Order Carnivora: dogs, bears, weasels, cats, skunks

Order Proboscidea: elephants

Order Sirenia: manatees

Order Perissodactyla: horses, hippopotamuses, zebras

Order Artiodactyla: pigs, deer, cattle

Order Primates: lemurs, monkeys, apes, humans

Suborder Prosimii: lemurs, tree shrews, tarsiers, lorises, pottos

Suborder Anthropoidea: monkeys, apes, humans

Superfamily Ceboidea: New World monkeys

Superfamily Cercopithecoidea: Old World monkeys

Superfamily Hominoidea: apes and humans

Family Hylobatidae: gibbons

Family Pongidae: chimpanzees, gorillas, orangutans

Family Hominidae: *Australopithecus*, Homo habilis*, Homo erectus*, Homo sapiens sapiens**

*extinct

Glossary

A

acetyl-CoA (uh-SEET-ul) Molecule made up of a two-carbon acetyl group attached to coenzyme A. The acetyl group enters the Krebs cycle for further oxidation. 122

acetylcholine (ACh) (uh-seet-ul-KOH-leen) Neurotransmitter active in both the peripheral and central nervous systems. 327

acetylcholinesterase (AChE) (uh-SEET-ul-koh-luh-nes-tuh-rays, -rayz) Enzyme that breaks down acetylcholine bound to postsynaptic receptors within a synapse. 327

acid Solution in which pH is less than 7; a substance that contributes or liberates hydrogen ions (protons) in a solution. 28

acrosome (AK-ruh-sohm) Covering on the tip of a sperm cell's nucleus that is believed to contain enzymes necessary for fertilization. 423

actin (AK-tin) One of two major proteins of muscle; makes up thin filaments in myofibrils of muscle fibers. See myosin. 385

action potential Polarity changes due to the movement of ions across the plasma membrane of an active neuron; nerve impulse. 324

active site Region on the surface of an enzyme where the substrate binds and where the reaction occurs. 109

active transport Transfer of a substance into or out of a cell from a region of lower concentration to a region of higher concentration by a process that requires a carrier and an expenditure of energy. 77

adaptation Fitness of an organism for its environment, including the process by which it becomes fit and is able to survive and to reproduce. Also, decrease in the excitability of a receptor in response to continuous constant-intensity stimulation. 9

adenine (A) (AD-un-een) One of four nitrogen bases in nucleotides composing the structure of DNA and RNA. 508

ADP (adenosine diphosphate) (ah-DEN-ah-seen dy-FAHS-fayt) Nucleotide with two phosphate groups that can accept another phosphate group and become ATP. 41, 107

adrenal cortex (uh-DREE-nul kor-teks) Outer portion of the adrenal gland; secretes hormones such as mineralocorticoid aldosterone and glucocorticoid cortisol. 405

adrenal gland (uh-DREE-nul) Gland that lies atop a kidney; the adrenal medulla produces the hormones epinephrine and norepinephrine, and the adrenal cortex produces the glucocorticoid and mineralocorticoid hormones. 405

adrenal medulla (uh-DREE-nul muh-dul-uh) Inner portion of the adrenal gland; secretes the hormones epinephrine and norepinephrine. 405

adrenocorticotropic hormone (ACTH) (uh-DREE-noh-kawrt-ih-koh-troh-pik) Hormone secreted by the anterior lobe of the pituitary gland that stimulates activity in the adrenal cortex. 400

aerobic (air-OH-bik, uh-roh-) Growing or metabolizing only in the presence of oxygen, as in aerobic respiration. 118

aerobic cellular respiration Metabolic reactions that provide energy to cells by the step-by-step oxidation of substrates with the concomitant buildup of ATP molecules. 118

agglutination (uh-gloot-un-ay-shun) Clumping of cells, particularly in reference to red blood cells involved in an antigen-antibody reaction. 278

aldosterone (al-DAHS-tuh-rohn) Hormone secreted by the adrenal cortex that regulates the sodium and potassium balance of the blood. 313, 406

alga (AL-guh) Aquatic organism that carries on photosynthesis. 583

allantois (uh-LAN-toh-is) Extraembryonic membrane that accumulates nitrogenous wastes in reptiles and birds and contributes to the formation of umbilical blood vessels in mammals, including humans. 451

allele (uh-LEEL) Alternative form of a gene located at a particular chromosome site (locus). 470

allergen (AL-ur-jun) Foreign substance capable of stimulating an allergic response. 276

allergy Immune response to substances that usually are not recognized as foreign. 276

allopatric speciation (al-uh-PAT-rik) Origin of new species in populations that are separated geographically. 565

alpine tundra Tundra near the peak of a mountain. 721

alternation of generations Life cycle, typical of plants, in which a diploid sporophyte alternates with a haploid gametophyte. 180, 585, 601

altruism Social interaction that has the potential to decrease the lifetime reproductive success of the member exhibiting the behavior. 683

alveolus (pl., alveoli) (al-VEE-uh-lus) Air sac of a lung. 287

amino acid Monomer of a protein; takes its name from the fact that it contains an amino group (—NH_2) and an acid group (—COOH). 37

amnion (AM-nee-ahn) Extraembryonic membrane of reptiles, birds, and mammals that forms an enclosing, fluid-filled sac. 451

amniote egg (AM-nee-oht) Egg that has an amnion, as seen during the development of reptiles, birds, and mammals. 649

amylase (AM-ih-lays) Starch-digesting enzyme secreted by the salivary glands (salivary amylase) and the pancreas (pancreatic amylase). 222, 224

anabolic steroid (a-nuh-BAHL-ik) Synthetic steroid that mimics the effect of testosterone. 411

anaerobic (AN-uh-roh-bik) Growing or metabolizing in the absence of oxygen. 120

analogous structure Structures that have a similar function in separate lineages and differ in structure and origin. 555

anaphase Mitosis phase during which daughter chromosomes move toward poles of spindle. 89

androgen (AN-druh-jun) Male sex hormone (e.g., testosterone). 411

angiosperm (AN-jee-uh-spurm) Flowering plant that produces seeds within an ovary that develops into a fruit; therefore, the seeds are covered. 152, 612

annual ring Layer of wood (secondary xylem) usually produced during one growing season. 161

anterior pituitary (pih-TOO-ih-tair-ee) Portion of the pituitary gland that produces six types of hormones and is controlled by hypothalamic-releasing and release-inhibiting hormones. 400

anther (AN-thur) Part of stamen where pollen grains are produced in pollen sacs. 80, 614

antheridium (an-thuh-RID-ee-um) Structure in seedless plants that produces flagellated sperm. 602

anthropoid (AN-thruh-poyd) Group of primates that includes monkeys, apes, and humans. 659

antibody (AN-tih-bahd-ee) Protein produced in response to the presence of an antigen; each antibody combines with a specific antigen. 268

antibody-mediated immunity Specific mechanism of defense in which plasma cells derived from B cells produce antibodies that combine with antigens. 269

anticodon (an-tih-KOH-dahn) "Triplet" of bases in tRNA that pairs with a complementary triplet (codon) in mRNA. 513

antidiuretic hormone (ADH) (an-tih-dy-uh-RET-ik) Hormone secreted by the posterior pituitary that promotes the reabsorption of water from the collecting ducts, which receive urine produced by nephrons within the kidneys. 312, 400

antigen (AN-tih-jun) Foreign substance, usually a protein or a polysaccharide, that stimulates the immune system to react, such as to produce antibodies. 268

antigen-presenting cell (APC) Cell that displays the antigen to the cells of the immune system so they can defend the body against that particular antigen. 272

anus Outlet of the digestive tube. 220

aorta (ay-OR-tuh) Major systemic artery that receives blood from the left ventricle. 243, 246

aortic bodies Sensory receptors in the aortic arch sensitive to oxygen content, carbon dioxide content, and blood pH. 290

apoptosis (ap-ohp-TOH-sis, -ahp-) Programmed cell death involving a cascade of specific cellular events leading to death and destruction of the cell. 269, 522

appendicular skeleton (ap-un-DIK-yuh-lur) Portion of the skeleton forming the pectoral girdle and upper extremities, and the pelvic girdle and lower extremities. 378

appendix (uh-PEN-diks) Small, tubular appendage that extends outward from the cecum of the large intestine. 220

Archaea (AR-kee-uh) Cannot confirm pronunciation. One of three domains of life; consists of organisms (formerly considered bacteria) found living in extreme habitats. 568, 578

archegonium (ar-kih-GOH-nee-um) Structure in plants that produces an egg; occurs in seedless plants and in gymnosperms. 602

arterial duct (ar-TEER-ee-ul) Fetal connection between the pulmonary artery and the aorta; ductus arteriosus. 458

arteriole (ar-TEER-ee-ohl) Vessel that takes blood from an artery to capillaries. 240

artery Vessel that takes blood away from the heart to arterioles; characteristically possessing thick elastic and muscular walls. 240

association area Region of the cerebral cortex related to memory, reasoning, judgment, and emotional feelings. 337

aster Short microtubule that extends outward from a spindle pole in animal cells during cell division. 88

atherosclerosis (ath-uh-roh-skluh-ROH-sis) Condition in which fatty substances accumulate abnormally beneath the inner linings of the arteries. 257

atom Smallest unit of matter that cannot be divided by chemical means. 20

atomic number Number of protons within the nucleus of an atom. 20

atomic weight Weight of an atom equal to the number of protons plus the number of neutrons within the nucleus of an atom. 21

ATP (adenosine triphosphate) (uh-DEN-uh-seen try-FAHS-fayt) Nucleotide with three phosphate groups. The breakdown of ATP into ADP + Ⓟ makes energy available for energy-requiring processes in cells. 41, 107

atrial natriuretic hormone (ANH) (AY-tree-ul nay-tree-yoo-RET-ik) Substance secreted by the atria of the heart that accelerates sodium excretion so that blood volume and blood pressure decreases. 313, 406

atrioventricular bundle (ay-tree-oh-ven-TRIK-yuh-lur) Group of specialized fibers that conduct impulses from the atrioventricular node to the ventricular muscle of the heart; A-V bundle. 244

atrioventricular valve Valve located between the atrium and the ventricle. 242

atrium (AY-tree-um) Chamber; particularly an upper receiving chamber of the heart lying above the ventricles; either the left atrium or the right atrium. 242

autonomic system (awt-uh-NAHM-ik) Branch of the peripheral nervous system that has control over the internal organs; consists of the sympathetic and parasympathetic systems. 331

autosome (AW-tuh-sohm) Chromosome other than a sex chromosome. 489

autotroph (AW-tuh-trahf) Organism that is capable of making its food (organic molecules) from inorganic molecules. 136, 550, 744

auxin (AHK-sin) Plant hormone regulating growth, particularly cell elongation; most often indoleacetic acid (IAA). 176

AV (atrioventricular) node Small region of neuromuscular tissue that transmits impulses received from the SA node to the ventricular walls. 244

axial skeleton (AK-see-ul) Portion of the skeleton that supports and protects the organs of the head, the neck, and the trunk. 374

axillary bud (AK-suh-lair-ee) Bud located in the axil of a leaf. 158

axon (AK-sahn) Fiber of a neuron that conducts nerve impulses away from the cell body. 322

B

B lymphocyte (LIM-fuh-syt) Lymphocyte that matures in the bone marrow and, when stimulated by the presence of a specific antigen, gives rise to antibody-producing plasma cells. 268

bacteria One of three domains of life; prokaryotic cells other than archaea with unique genetic, biochemical, and physiological characteristics. 62, 568, 577

bacteriophage (bak-TEER-ee-uh-fayj) Virus that infects a bacterial cell. 532, 575

bark External part of a tree, containing cork, cork cambium, and phloem. 161

basal body Structure in the cell cytoplasm, located at the base of a cilium or flagellum. 60

base Solution in which pH is greater than 7; a substance that contributes or liberates hydroxide ions (OH^-) in a solution; alkaline; opposite of acid. Also, a term commonly applied to one of the components of a nucleotide. 28

behavior Observable, coordinated responses to environmental stimuli. 8, 672

bile (byl) Secretion of the liver that is temporarily stored in the gallbladder before being released into the small intestine, where it emulsifies fat. 219

binary fission Bacterial reproduction into two daughter cells without the utilization of a mitotic spindle. 581

biodiversity Total number of species, the variability of their genes, and the ecosystems in which they live. 12

biogeochemical cycle (by-oh-jee-oh-KEM-ih-kul) Circulating pathway of an element through the biotic and abiotic components of ecosystems within the biosphere. 748

biogeography Study of the geographical distribution of organisms. 554

biological magnification Process by which nonexcreted substances like DDT become more concentrated in organisms in the higher trophic levels of the food chain. 755

biome (BY-ohm) Major terrestrial community characterized by certain climatic conditions and dominated by particular types of plants. 721

biosphere (BY-uh-sfeer) That portion of the surface of the earth (air, water, and land) where living things exist. 10, 688, 718

biotic potential (by-AHT-ik) Maximum population growth rate under ideal conditions. 690

blastocyst (BLAS-tuh-sist) Early stage of human embryonic development that consists of a hollow fluid-filled ball of cells. 452

blastula (BLAS-chuh-luh) Hollow ball of cells occurring during animal development prior to gastrula formation. 445

blood pressure Force of blood pushing against the inside wall of an artery. 248

bottleneck effect Type of genetic drift, in which a majority of genotypes is prevented from participating in the production of the next generation as a result of a natural disaster or human interference. 559

brain Enlarged superior portion of the central nervous system located in the cranial cavity of the skull. 334

brain stem Portion of the brain consisting of the medulla oblongata, pons, and midbrain. 335

bronchiole (BRAHNG-kee-ohl) Smaller air passages in the lungs that eventually terminate in alveoli. 287

bronchus (pl., bronchi) (BRAHNG-kus) One of two major divisions of the trachea leading to the lungs. 287

budding Asexual reproduction in which a new organism forms from the body wall of a parent and then detaches itself. 587

buffer Substance or compound that prevents large changes in the pH of a solution. 29

bulbourethral gland (bul-boh-yoo-REE-thrul) Either of two small structures located below the prostate gland in males; adds secretions to semen. 421

C

C_3 plant Plant that directly uses the Calvin cycle; the first detected molecule during photosynthesis is PGA, a three-carbon molecule. 146

C_4 plant Plant that fixes carbon dioxide to produce a C_4 molecule that releases carbon dioxide to the Calvin cycle. 146

calorie Amount of heat energy required to raise the temperature of water 1°C. 27

Calvin cycle Primary pathway of the light-independent reactions of photosynthesis; converts carbon dioxide to carbohydrate. 138

cambium (KAM-bee-um) Lateral layer of meristematic cells that divide to provide girth to vascular plants. 154

cancer Malignant tumor whose nondifferentiated cells exhibit loss of contact inhibition, uncontrolled growth, and the ability to invade tissues and metastasize. 522

capillary (KAP-uh-lair-ee) Microscopic vessel connecting arterioles to venules and through the thin walls of which substances either exit or enter blood. 240

carbaminohemoglobin (kar-buh-MEE-noh-hee-muh-gloh-bun) Hemoglobin carrying carbon dioxide. 292

carbohydrate Organic compound characterized by the presence of CH_2O groups; includes monosaccharides, disaccharides, and polysaccharides. 32

carbon dioxide fixation Photosynthetic reaction in which carbon dioxide is attached to an organic compound. 144

carbonic anhydrase (kar-BAHN-ik an-HY-drays, -drayz) Enzyme in red blood cells that speeds the formation of carbonic acid from water and carbon dioxide. 292

carcinogen (kar-SIN-uh-jun) Environmental agent that causes mutations leading to the development of cancer. 525

cardiac cycle (KAR-dee-ak) Series of myocardial contractions that constitute a complete heartbeat. 244

carnivore (KAR-nuh-vor) Secondary or higher consumer in a food chain, that therefore eats other animals. 744

carotenoid (kuh-RAHT-n-oyd) Yellow or orange pigment that serves as an accessory to chlorophyll in photosynthesis. 137

carotid bodies (kuh-RAHT-id) Structures located at the branching of the carotid arteries and that contain chemoreceptors sensitive to the hydrogen ion concentration but also the level of carbon dioxide and oxygen in blood. 290

carrier Individual that appears normal but is capable of transmitting an allele for a genetic disorder. 478

carrier protein Protein molecule that combines with a substance and transports it through the plasma membrane. 70, 76

carrying capacity Largest number of organisms of a particular species that can be maintained indefinitely in an ecosystem. 691

cecum (SEE-kum) Small pouch that lies below the entrance of the small intestine, and is the blind end of the large intestine. 220

cell Structural and functional unit of an organism; the smallest structure capable of performing all the functions necessary for life. 7, 46

cell body Portion of a neuron that contains a nucleus and from which the fibers extend. 322

cell cycle Repeating sequence of events in eukaryotic cells consisting of the phases of interphase, when growth and DNA synthesis occurs, and the stages of mitosis, when cell division occurs. 86

cell-mediated immunity Specific mechanism of defense in which T cells destroy antigen-bearing cells. 272

cell plate Structure that precedes the formation of the cell wall as a part of cytokinesis in plant cells. 90

cell theory One of the major theories of biology which states that all organisms are made up of cells; cells are capable of self-reproduction and cells come only from pre-existing cells. 46

cell wall Protective barrier outside the plasma membrane of plant and certain other cells. 49, 62

cellular differentiation Process and the developmental stages by which a cell becomes specialized for a particular function. 448

cellulose (SEL-yuh-lohs, -lohz) Polysaccharide composed of glucose molecules; the chief constituent of a plant's cell wall. 33

centriole (SEN-tree-ohl) Short, cylindrical organelle in animal cells that contains microtubules in a 9 + 0 pattern; present in a centrosome and associated with the formation of basal bodies. 60

centromere (SEN-truh-meer) Constricted region of a chromosome where sister chromatids are attached to one another and where the chromosome attaches to a spindle fiber. 84

centrosome (SEN-truh-sohm) Major microtubule organizing center of cells, consisting of granular material. In animal cells, it contains two centrioles. 58

cephalization (sef-uh-lih-ZAY-shun) Development of a well-recognized anterior head with concentrated nerve masses and sensory receptors. 619

cerebellum (ser-uh-BEL-um) Part of the brain located posterior to the medulla oblongata and pons that coordinates skeletal muscles to produce smooth, graceful motions. 335

cerebral cortex (suh-REE-brul, SER-uh-brul KOR-teks) Outer layer of cerebral hemispheres; receives sensory information and controls motor activities. 336

cerebral hemisphere One of the large, paired structures that together constitute the cerebrum of the brain. 335

cerebrospinal fluid (sair-uh-broh-SPY-nul, suh-REE-broh-) Fluid found in the ventricles of the brain, in the central canal of the spinal cord, and in association with the meninges. 332

cerebrum (SAIR-uh-brum, suh-REE-brum) Main part of the brain consisting of two large masses, or cerebral hemispheres; the largest part of the brain in mammals. 335

cervix (SUR-viks) Narrow end of the uterus, which leads into the vagina. 425

channel protein Forms a channel to allow a particular molecule or ion to cross the plasma membrane. 70

chemiosmosis (kem-ee-ahz-MOH-sis) Production of ATP due to the buildup of a hydrogen ion gradient across a membrane by an electron transport system. 125

chemoreceptor (kee-moh-rih-SEP-tur) Sensory receptor that is sensitive to chemical stimulation—for example, sensory receptors for taste and smell. 348

chlorophyll (KLOR-uh-fil) Green pigment that captures solar energy during photosynthesis. 137

chloroplast (KLOR-uh-plast) Membranous organelle that contains chlorophyll and is the site of photosynthesis. 56, 138

cholesterol (kuh-LES-tuh-rawl) Steroid that occurs in animal plasma membranes; is known to contribute to the development of plaque on blood vessel walls. 68

chordae tendineae (KOR-dee TEN-din-ee-ee) Tough bands of connective tissue that attach the papillary muscles to the atrioventricular valves within the heart. 242

chordate (KOR-dayt) Member of the phylum Chordata, which includes lancelets, tunicates, fishes, amphibians, reptiles, birds, and mammals; characterized by a notochord, dorsal hollow nerve cord, and pharyngeal pouches or gill slits. 646

chorion (KOR-ee-ahn) Extraembryonic membrane functioning for respiratory exchange in reptiles and birds; contributes to placenta formation in mammals. 451

chorionic villi (kor-ee-AHN-ik VIL-eye) Treelike extensions of the chorion of the embryo, projecting into the maternal tissues at the placenta. 455

chromatin (KROH-muh-tin) Threadlike network in the nucleus that is made up of DNA and proteins. 52, 84

chromosomal mutation (kroh-muh-SOH-mul) Variation in regard to the normal number of chromosomes inherited or in regard to the normal sequence of alleles on a chromosome; the sequence can be inverted, translocated from a nonhomologous chromosome, deleted, or duplicated. 491

chromosome (KROH-muh-som) Rodlike structure in the nucleus seen during cell division; contains the hereditary units, or genes. 52, 84

chyme (kym) Thick, semi-liquid food material that passes from the stomach to the small intestine. 218

cilium (pl., cilia) (SIL-ee-um) Short, hairlike extension from a cell, occurring in large numbers and used for cell mobility. 60, 194

circadian rhythm (sur-KAY-dee-un) Regular physiological or behavioral event that occurs on an approximately 24-hour cycle. 411

cladistics (kluh-DIS-tiks) School of systematics that determines the degree of relatedness by analyzing primitive and derived characters and constructing cladograms. 568

class One of the seven categories, or taxa, used by taxonomists to group species. 568

cleavage (KLEE-vij) Cell division without cytoplasmic addition or enlargement; occurs during first stage of animal development. 445

cleavage furrow Indentation that begins the process of cleavage, by which animal cells undergo cytokinesis. 91

climax community In ecology, community that results when succession has come to an end. 712

clonal selection theory States that the antigen selects which lymphocyte will undergo clonal expansion and produce more lymphocytes bearing the same type of receptor. 269

clone Production of identical copies; in genetic engineering, the production of many identical copies of a gene. 531

clotting Process of blood coagulation, usually when injury occurs. 254

cochlea (KOHK-lee-uh, koh-klee-uh) Portion of the inner ear that resembles a snail's shell and contains the spiral organ, the sense organ for hearing. 362

codominance Pattern of inheritance in which both alleles of a gene are equally expressed. 482

codon (KOH-dahn) "Triplet" of bases in mRNA that directs the placement of a particular amino acid into a polypeptide. 511

coelom (SEE-lum) Embryonic body cavity lying between the digestive tract and body wall that is completely lined by mesoderm; in humans, the embryonic coelom becomes the thoracic and abdominal cavities. 201, 417, 619

coenzyme (koh-EN-zym) Nonprotein organic molecule that aids the action of the enzyme to which it is loosely bound. 111

coevolution Interaction of two species such that each influences the evolution of the other species. 708

cofactor Nonprotein adjunct required by an enzyme in order to function; many cofactors are metal ions, while others are coenzymes. 111

cohesion-tension theory Explanation for upward transportation of water in xylem based upon transpiration-created tension and the cohesive properties of water molecules. 170

collecting duct Tube that receives urine from the distal convoluted tubules of several nephrons within a kidney. 309

colon (KOH-lun) Portion of the large intestine that extends from the cecum to the rectum. 220

colostrum (kuh-LAHS-trum) Thin, milky fluid rich in proteins, including antibodies, that is secreted by the mammary glands a few days prior to or after delivery before true milk is secreted. 430, 461

commensalism (kuh-MEN-suh-liz-um) Symbiotic relationship in which one species is benefited and the other is neither harmed nor benefited. 710

community Group of many different populations that interact with one another. 7, 10, 688, 702

companion cell Cell associated with sieve-tube elements in phloem of vascular plants. 155

competitive exclusion principle Theory that no two species can occupy the same niche. 704

complement system Group of plasma proteins that form a nonspecific defense mechanism, often by puncturing microbes; it complements the antigen-antibody reaction. 268

complementary base pairing Pairing of bases between nucleic acid strands;

adenine pairs with either thymine (DNA) or uracil (RNA), and cytosine pairs with guanine. 508

complementary DNA (cDNA) DNA that has been synthesized from mRNA by the action of reverse transcriptase. 532

condensation synthesis Chemical change resulting in the covalent bonding of two monomers with the accompanying loss of a water molecule. 32

cone cell Bright-light sensory receptor in the retina of the eye that detects color and provides visual acuity. 356

conifer (KAHN-uh-fur) Cone-bearing gymnosperm plants that include pine, cedar, and spruce trees. 608

conjugation (kahn-juh-GAY-shun) Sexual union between organisms in which the genetic material of one cell enters another. 585

connective tissue Type of tissue characterized by cells separated by a matrix that often contains fibers. 196

conservation biology Scientific discipline that seeks to understand the effects of human activities on species, communities, and ecosystems and to develop practical approaches to preventing the extinction of species and the destruction of ecosystems. 757

consumer Organism that feeds on another organism in a food chain; primary consumers eat plants, and secondary (or higher) consumers eat animals. 744

continental drift Movement of continents with respect to one another over the earth's surface. 554

contraceptive (kahn-truh-SEP-tiv) Medication or device used to reduce the chance of pregnancy. 431

control group In experimentation, a sample that goes through all the steps of an experiment except the one being tested; a standard against which results of an experiment are checked. 3

coral reef Structure found in tropical waters that is formed by the buildup of coral skeletons and where many and various types of organisms reside. 623, 740

cork Outer covering of bark of trees; made up of dead cells that may be sloughed off. 154

coronary artery (KOR-uh-nair-ee) Artery that supplies blood to the wall of the heart. 247

corpus luteum (KOR-pus LOOT-ee-um) Yellow body that forms in the ovary from a follicle that has discharged its egg; it secretes progesterone. 427

cortisol (KOR-tuh-sawl) Glucocorticoid secreted by the adrenal cortex that results in increased blood glucose. 406

cotyledon (kaht-ul-EED-n) Seed leaf for embryonic plant, providing nutrient molecules for the developing plant before its mature leaves begin photosynthesis. 153, 182

covalent bond (coh-VAY-lent) Chemical bond between atoms that results from the sharing of a pair of electrons. 24

cranial nerve (KRAY-nee-ul) Nerve that arises from the brain. 328

creatine phosphate (KREE-uh-teen FAHS-fayt) Compound unique to muscles that contains a high-energy phosphate bond. 389

creatinine (kree-AH-tuhn-een) Nitrogenous waste, the end product of creatine phosphate metabolism. 305

cristae (sing., crista) (KRIS-tee) Short, fingerlike projections formed by the folding of the inner membrane of mitochondria. 57

crossing-over Exchange of corresponding segments of genetic material between nonsister chromatids of homologous chromosomes during synapsis of meiosis I. 93, 500

cyanobacterium (SY-ah-noh-bak-TEER-ee-um) Photosynthetic prokaryote that contains chlorophyll and releases O_2; formerly called a blue-green alga. 580

cyclic AMP (SY-klik, SIH-klik) ATP-related compound that acts as the second messenger in peptide hormone transduction; it initiates activity of the metabolic machinery. 397

cyclin (SY-klin) Protein that cycles in quantity as the cell cycle progresses; combines with and activates the kinases that function to promote the events of the cycle. 86

cytochrome (SY-tuh-krohm) Protein within the inner mitochondrial membrane that is an electron carrier in aerobic respiration (electron transport system). 124

cytokine (SY-tuh-kyn) Type of protein secreted by a T lymphocyte that attacks viruses, virally infected cells, and cancer cells. 275

cytokinesis (sy-toh-kih-NEE-sis) Division of the cytoplasm of a cell during telophase of mitosis or meiosis I and II. 84

cytoplasm (SY-tuh-plaz-um) Semifluid medium between the nucleus and the plasma membrane and that contains the organelles. 49

cytosine (C) (SY-tuh-seen) One of four nitrogen bases in nucleotides composing the structure of DNA and RNA. 508

cytoskeleton Fibrous protein elements found throughout the cytoplasm that help maintain the shape of the cell, anchor the organelles, and allow the cell and its organelles to move. 58

cytotoxic (sy-tuh-TAHK-sik) **T cell** T lymphocyte that attacks and kills antigen-bearing cells. 272

D

data Facts that are derived from observations and experiments pertinent to the matter under study. 5

deamination (dee-am-uh-NAY-shun) Removal of an amino group (—NH_2) from an amino acid or other organic compound. 127

decomposer Organism, usually a bacterium or fungus, that breaks down organic matter into inorganic nutrients that can be recycled in the environment. 578, 744

demographic transition Decline in the birthrate following a reduction in the death rate so that the population growth rate is lowered. 695

denaturation (dee-nay-chuh-RAY-shun) Loss of normal shape by an enzyme so that it no longer functions; caused by a less than optimal pH and temperature. 38, 110

dendrite (DEN-dryt) Fiber of a neuron, typically branched, that conducts signals toward the cell body. 322

denitrification (dee-ny-truh-fuh-KAY-shun) Conversion of nitrate or nitrite to nitrogen gas by bacteria in soil. 752

dense fibrous connective tissue Type of connective tissue containing many collagen fibers packed together, and found in tendons and ligaments, for example. 196

detrital food chain (dih-TRYT-ul) Straight-line linking of organisms in the detrital food web according to who eats whom. 747

detrital food web Flow of nutrients and energy requiring decomposition of dead organic matter as a source food for other organisms. 747

detritus (dih-TRYT-us) Partially decomposed remains of plants and animals found in soil and on the beds of bodies of water. 744

deuterostome (DOO-tuh-roh-stohm) Group of coelomate animals in which the second embryonic opening is associated with the mouth; the first embryonic opening, the blastopore, is associated with the anus. 644

diabetes mellitus (dy-uh-BEE-teez MEL-ih-tus, muh-ly-tus) Condition characterized by a high blood glucose level and the appearance of glucose in the urine, due to a deficiency of insulin production or glucose uptake by cells. 409

diastole (dy-AS-tuh-lee) Relaxation of a heart chamber. 244
diastolic pressure (dy-uh-STAHL-ik) Arterial blood pressure during the diastolic phase of the cardiac cycle. 248
dicot (DY-kaht) Abbreviation of dicotyledon. Flowering plant group; members have two embryonic leaves (cotyledons), net-veined leaves, cylindrical arrangement of vascular bundles, flower parts in fours or fives, and other characteristics. 153
diencephalon (dy-en-SEF-uh-lahn) Portion of the brain in the region of the third ventricle that includes the thalamus and hypothalamus. 335
differentially permeable (dif-uh-RENT-shuh-lee pur-mee-uh-bul) Having degrees of permeability; the cell is impermeable to some substances and allows others to pass through at varying rates. 72
diffusion (dih-FYOO-zhun) Movement of molecules from a region of higher concentration to a region of lower concentration. 73
dihybrid (dy-HY-brud) Individual that is heterozygous for two traits; shows the phenotype governed by the dominant alleles but carries the recessive alleles. 476
diploid (DIP-loyd) 2n number of chromosomes; twice the number of chromosomes found in gametes. 84
diplontic life cycle (dip-LAHN-tik) Life cycle typical of animals in which the adult is always diploid and meiosis produces the gametes. 585
directional selection Natural selection in which an extreme phenotype is favored, usually in a changing environment. 562
disaccharide (dy-SAK-uh-ryd) Sugar that contains two units of a monosaccharide; e.g., maltose. 32
disruptive selection Natural selection in which extreme phenotypes are favored over the average phenotype, leading to more than one distinct form. 564
distal convoluted tubule (DIS-tul TOO-byool) Highly coiled region of a nephron , that is distant from the glomerular capsule, where tubular secretion takes place. 309
DNA (deoxyribonucleic acid) Nucleic acid found in cells; the genetic material that specifies protein synthesis in cells. 40, 506
DNA fingerprinting Using DNA fragment lengths, resulting from restriction enzyme cleavage, to identify particular individuals. 533
DNA ligase (LY-gays) Enzyme that links DNA fragments; used during production of recombinant DNA to join foreign DNA to vector DNA. 530
DNA polymerase (PAHL-uh-muh-rays) During replication, an enzyme that joins the nucleotides complementary to a DNA template. 509
dominance hierarchy Social ranking within a group in which a higher-ranking individual acquires more resources than a lower-ranking individual. 677
dominant allele Hereditary factor that expresses itself in the phenotype when the genotype is heterozygous. 472
dorsal-root ganglion (GANG-glee-un) Mass of sensory neuron cell bodies located in the dorsal root of a spinal nerve. 328
double fertilization In flowering plants, one sperm joins with polar nuclei within the embryo sac to produce a 3n endosperm nucleus, and another sperm joins with an egg to produce a zygote. 181, 612
double helix (HEE-liks) Double spiral; describes the three-dimensional shape of DNA. 508
doubling time Number of years it takes for a population to double in size. 695
duodenum (doo-uh-DEE-num) First portion of the small intestine into which secretions from the liver and pancreas enter. 219
dynamic equilibrium Maintenance of balance when the head and body are suddenly moved or rotated. 365

E

ecological niche (ee-kuh-LAH-jih-kul, ek-uh-lah-jih-kul nich) Role an organism plays in community, including its habitat and its interactions with other organisms. 703
ecological pyramid Pictorial graph representing biomass, organism number, or energy content of each trophic level in a food web—from the producers to the final consumer populations. 747
ecological succession Sequential change in the relative dominance of species within a community; primary succession begins on bare rock, and secondary succession begins where soil already exists. 712
ecosystem (EK-uh-sis-tum, ee-kuh-) Region in which populations interact with each other and with the physical environment. 7, 10, 688, 744
ectoderm (EK-tuh-durm) Outer germ layer of the embryonic gastrula; it gives rise to the nervous system and skin. 445, 453
edema (ih-DEE-muh) Swelling due to tissue fluid accumulation in the intercellular spaces. 264
egg Nonflagellate female gamete; also referred to as ovum. 181, 424
electrocardiogram (ECG or EKG) (ih-lek-troh-KAR-dee-uh-gram) Recording of the electrical activity associated with the heartbeat. 245
electron Subatomic particle that has almost no weight and carries a negative charge; orbits in a shell about the nucleus of an atom. 20
electron transport system Chain of electron carriers in the cristae of mitochondria and thylakoid membrane of chloroplasts. The electrons release energy that is used to establish a hydrogen ion gradient. This gradient is associated with the production of ATP molecules. 119, 124, 140
element Substance that cannot be broken down into substances with different properties; composed of only one type atom. 20
embolus (EM-buh-lus) Moving blood clot that is carried through the bloodstream. 257
embryo (EM-bree-oh) Stage of a multicellular organism that develops from a zygote and before it becomes free living; in seed plants the embryo is part of the seed. 445
embryo sac Megagametophyte of flowering plants that produces an egg cell. 180
embryonic period From approximately the second to the eighth week of human development, during which the major organ systems are organized. 451
endergonic reaction (en-dur-GAHN-ik) Chemical reaction that requires an input of energy; opposite of exergonic reaction. 106
endocrine gland (EN-duh-krin) Ductless organ that secretes (a) hormone(s) into the bloodstream. 399
endocytosis (en-duh-sy-TOH-sis) Process in which a vesicle is formed at the plasma membrane to bring a substance into the cell. 78
endoderm (EN-duh-durm) Inner germ layer that lines the archenteron/gut of the gastrula; it becomes the lining of the digestive and respiratory tract and associated organs. 445, 453
endometrium (en-doh-MEE-tree-um) Lining of the uterus, which becomes thickened and vascular during the uterine cycle. 425
endoplasmic reticulum (ER) (en-duh-PLAZ-mik reh-TIK-yuh-lum) Membranous system of tubules, vesicles, and sacs in cells, sometimes having attached ribosomes. Rough ER has ribosomes; smooth ER does not. 53

endosperm (en-DOH-spurm) In angiosperms, the 3n tissue that nourishes the embryo and seedling and is formed as a result of a sperm joining with two polar nuclei. 181, 612

energy Capacity to do work and bring about change; occurs in a variety of forms. 8, 104

energy of activation Energy that must be added to cause molecules to react with one another. 108

entropy (en-TRUH-pee) Measure of disorder or randomness. 104

environmental resistance Sum total of factors in the environment that limit the numerical increase of a population in a particular region. 690

enzyme (EN-zym) Organic catalyst, usually a protein that speeds up a reaction in cells due to its particular shape. 37, 108

enzyme inhibition Means by which cells regulate enzyme activity; there is competitive and noncompetitive inhibition. 110

epididymis (ep-ih-DID-uh-mus) Coiled tubule next to the testes where sperm mature and may be stored for a short time. 420

epiglottis (ep-uh-GLAHT-us) Structure that covers the glottis and closes off the air tract during the process of swallowing. 216, 286

epinephrine (ep-uh-NEF-rin) Hormone secreted by the adrenal medulla in times of stress; also called adrenaline. 405

epithelial tissue (ep-uh-THEE-lee-ul) Type of tissue that covers the external surface of the body and lines its cavities. 194

erythrocyte (ih-RITH-ruh-syt) See red blood cell. 198, 250

erythropoietin (ih-rith-roh-poy-EE-tin) Hormone, produced by the kidneys, that speeds the maturation of red blood. 250, 305

esophagus (ih-SAHF-uh-gus) Tube that transports food from the mouth to the stomach. 217

essential amino acids Amino acids required in the human diet because the body cannot make them. 228

estrogen (ES-truh-jun) Female sex hormone, which, along with progesterone, maintains the primary sex organs and stimulates development of the female secondary sex characteristics. 411, 429

Eukarya (yoo-KAR-ee-uh) One of the three domains of life, consisting of organisms in the kingdoms Protista, Fungi, Plantae, and Animalia that have unique genetic, biochemical, and physiological characteristics. 568

eukaryotic cell (yoo-kair-ee-AHT-ik) Cell that possesses a nucleus and the other membranous organelles characteristic of complex cells. 49

eutrophication (yoo-trahf-uh-KAY-shun, yoo-troh-fuh-kay-shun) In aquatic ecosystems, rapid nutrient cycling, high productivity by phytoplankton, and relatively few species. 732

evolution Changes that occur in the members of a species with the passage of time, often resulting in increased adaptation of organisms to the environment. 9, 551

excretion Removal of metabolic wastes from the body. 305

exergonic reaction (ek-sur-GAHN-ik) Chemical reaction that releases energy; opposite of endergonic reaction. 106

exocytosis (ek-soh-sy-TOH-sis) Process in which an intracellular vesicle fuses with the plasma membrane so that the vesicle's contents are released outside the cell. 78

experiment Artificial situation devised to test a hypothesis. 3

expiration (ek-spuh-RAY-shun) Act of expelling air from the lungs; exhalation. 284

expiratory reserve volume (ik-SPY-ruh-tor-ee) Volume of air that can be forcibly exhaled after normal exhalation. 288

exponential growth Growth, particularly of a population, in which the increase occurs in the same manner as compound interest. 690

external respiration Exchange of oxygen and carbon dioxide between alveoli and blood. 292

extraembryonic membrane (ek-struh-em-bree-AHN-ik) Membrane that is not a part of the embryo but is necessary to the continued existence and health of the embryo. 451

F

facilitated transport Passive transfer of a substance into or out of a cell along a concentration gradient by a process that requires a carrier. 76

FAD (flavin adenine dinucleotide) (FLAY-vun ad-un-een dy-NOO-klee-uh-tyd) Coenzyme that functions as a carrier of electrons, especially in aerobic cellular respiration. 123

falsify To show a hypothesis to be untrue. 5

family One of the seven categories, or taxa, used by taxonomists to group species. 568

fat Organic molecule that contains glycerol and fatty acids and is found in adipose tissue of vertebrates. 34

fatty acid Molecule that contains a hydrocarbon chain and ends with an acid group. 34

feedback inhibition Mechanism by which an enzyme is rendered inactive by combining with the product of an enzymatic reaction or pathway. 111

fermentation Anaerobic breakdown of carbohydrates that results in organic end products such as alcohol and lactic acid. 129

fertilization Union of a sperm nucleus and an egg nucleus, which creates the zygote with the diploid number of chromosomes. 92, 444

fiber Plant material that is nondigestible. Insoluble fiber has a laxative effect; soluble fiber prevents bile acids and cholesterol from being absorbed. 227

fibrin (FY-brun) Insoluble protein threads formed from fibrinogen during blood clotting. 254

fibrinogen (fy-BRIN-uh-jun) Plasma protein that is converted into fibrin threads during blood clotting. 254

fight or flight Response in which the sympathetic division accelerates heartbeat and dilates bronchi to provide muscles with a ready supply of glucose and oxygen. 331

fitness Ability of an organism to survive and reproduce in its environment. 562

flagellum (pl., flagella) (fluh-JEL-um) Slender, long extension that propels a cell through a fluid medium. 60, 62

flower Reproductive organ of plants that contains the structures for the production of pollen grains and covered seeds. 180, 614

fluid-mosaic model of membrane structure Proteins form a pattern within a bilayer of lipid molecules having a fluid consistency. 68

focus Manner by which light rays are bent by the cornea and lens, creating an image on the retina. 355

follicle (FAHL-ih-kul) Structure in the ovary that produces the egg and, in particular, the female sex hormones, estrogen and progesterone. 427

follicle-stimulating hormone (FSH) Hormone secreted by the anterior pituitary gland that stimulates the development of an ovarian follicle in a female or the production of sperm in a male. 423

food chain Portion of a food web. Sequence that describes the energy flow from one population to the next in an ecosystem. Grazing food chains begin with a producer, and detritus food chains begin with partially decomposed organic matter. 747

food web Complex pattern of interlocking and crisscrossing food chains. 747

formed element Constituent of blood that is either cellular (red blood cells and white blood cells) or at least cellular in origin (platelets). 249

fossil Any past evidence of an organism that has been preserved in the earth's crust. 551

fossil fuel Remains of once living organisms that are burned to release energy, such as coal, oil, and natural gas. 750

founder effect Type of genetic drift, in which only a fraction of the total genetic diversity of the original gene pool is represented as a result of a few individuals founding a colony. 559

fovea centralis (FOH-vee-uh sen-TRA-lis, sen-TRAY-lis) Region of the retina consisting of densely packed cones that is responsible for the greatest visual acuity. 355

free energy System's useful energy that is capable of performing work. 106

fruit In flowering plants (angiosperms), the structure that forms from an ovary and associated tissues and encloses seeds. 183, 612

fruiting body Spore-bearing structure found in certain types of fungi, such as mushrooms. 593

functional group Cluster of atoms that always behaves in a certain way. 31

fungus (*pl.*, fungi) (FUNG-gus) Multicellular eukaryote, usually composed of filaments called hyphae, and usually saprotrophic; e.g., mushroom and mold. 591

G

gallbladder Saclike organ, associated with the liver, that stores and concentrates bile. 223

gamete (GA-meet, guh-MEET) Haploid reproductive cell. Two gametes, most often an egg or a sperm, join in fertilization to form a zygote. 92

gametophyte (guh-MEET-uh-fyte) In the life cycle of a plant, the haploid generation that produces gametes. 180, 601

ganglion (GANG-glee-un) Collection of neuron cell bodies within the peripheral nervous system. 328

gastrovascular cavity (gas-troh-VAS-kyuh-lur) Blind digestive cavity that also serves a circulatory (transport) function in animals that lack a circulatory system. 623

gastrula (GAS-truh-luh) Stage of animal development during which the germ layers form. 445

gene (jeen) Unit of heredity that codes for a polypeptide and is passed on to offspring. 9, 510

gene flow Movement of genes from one population to another via sexual reproduction between members of the populations. 562

gene pool Total of all the genes of all the individuals in a population. 556

gene therapy Use of genetically engineered cells or other biotechnology techniques to treat human genetic and other disorders. 539

genetic drift Evolution by chance processes alone. 559

genetic engineering Use of technology to alter the genome of a living cell for medical or industrial use; bioengineered. 530

genome (JEE-nohm) Full set of genes in an individual whether a haploid or diploid organism or a virus. 532

genomic library Collection of bacterial or bacteriophage clones, each containing a particular segment of DNA from the source cell. 532

genotype (JEE-nuh-typ) Genes of any individual for (a) particular trait(s). 472

genus (jee-nus) One of the seven categories, or taxa, used by taxonomists to group species. 568

germ layer Developmental layer of the body—that is, ectoderm, mesoderm, or endoderm. 445

gill Respiratory organ in most aquatic animals; in fish, an outward extension of the pharynx. 646

gland Epithelial cell or group of epithelial cells that are specialized to secrete a substance. 194

global warming Predicted increase in the earth's temperature, due to the greenhouse effect, which will lead to the melting of polar ice and a rise in sea levels. 751

glomerular capsule (gluh-MAIR-yuh-lur) Double-walled cup that surrounds the glomerulus at the beginning of the nephron. 309

glomerular filtrate Filtered portion of blood contained within the glomerular capsule. 311

glomerular filtration Movement of small molecules from the glomerulus into the glomerular capsule due to the action of blood pressure. 311

glomerulus (gluh-MAIR-uh-lus, gloh-mair-yuh-lus) Cluster; for example, the cluster of capillaries surrounded by the glomerular capsule in a nephron, where glomerular filtration takes place. 308

glottis (GLAHT-us) Opening for airflow into the larynx. 216, 286

glucagon (GLOO-kuh-gahn) Hormone secreted by the pancreas which causes the liver to break down glycogen and raises the blood glucose level. 408

glucocorticoid (GLOO-koh-KOR-tih-koyd) Any one of a group of hormones secreted by the adrenal cortex that influences carbohydrate, fat, and protein metabolism. 405

glucose (GLOO-kohs) Six-carbon sugar that organisms degrade as a source of energy during cellular respiration. 32

glycogen (GLY-koh-jun) Storage polysaccharide, found in animals, that is composed of glucose molecules joined in a linear fashion but having numerous branches. 32

glycolipid (gly-koh-LIP-ud) Component of cell membranes composed of a hydrophilic head bonded to a hydrophobic lipid tail. 68

glycolysis (gly-KAHL-uh-sis) Metabolic pathway found in the cytoplasm that participates in aerobic cellular respiration and fermentation; it converts glucose to two molecules of pyruvate. 119, 120

glycoprotein (gly-koh-PROH-teen) Compound composed of a carbohydrate combined with a protein. 69

Golgi apparatus (GAHL-jee) Organelle, consisting of concentrically folded saccules, which functions in the packaging, storage, and distribution of cellular products. 54

gonad (GOH-nad) Organ that produces sex cells; the ovary, which produces eggs, and the testis, which produces sperm. 420, 424

gonadotropic hormone (goh-nad-uh-TRAHP-ic, -troh-pic) Substance secreted by anterior pituitary that regulates the activity of the ovaries and testes; principally, follicle-stimulating hormone (FSH) and luteinizing hormone (LH). 400

granum (GRAY-num) Stack of thylakoids within chloroplasts. 57

gray matter Nonmyelinated nerve fibers in the central nervous system. 332

greenhouse effect Reradiation of solar heat toward the earth, caused by gases in the atmosphere. 751

growth factor Chemical messenger that stimulates mitosis and differentiation of target cells that have receptors for it; important in such processes as fetal development, tissue maintenance and repair, and hemopoiesis; sometimes a contributing factor in cancer. 412

growth hormone (GH) Substance secreted by the anterior pituitary; it promotes cell division, protein synthesis, and bone growth. 400

guanine (G) (GWAH-neen) One of four nitrogen bases in nucleotides composing the structure of DNA and RNA. 508

guard cell Bean-shaped epidermal cell; one found on each side of a leaf stomate; their activity controls stomate size. 154, 172

gymnosperm (JIM-nuh-spurm) Seed plant with uncovered (naked) seeds; examples are conifers and cycads, which bear cones. 608

H

habitat Place where an organism lives and is able to survive and reproduce. 703

haploid (HAP-loyd) n number of chromosomes; half the diploid number; the number characteristic of gametes which contain only one set of chromosomes. 84

haplontic life cycle (hap-LAHN-tik) Life cycle typical of protists in which the adult is always haploid because meiosis occurs after zygote formation. 584

hard palate (PAL-it) Bony, anterior portion of the roof of the mouth. 214

helper T cell T lymphocyte that releases cytokines and stimulates certain other immune cells to perform their respective functions. 272

hemocoel (HEE-muh-seel) Residual coelom found in arthropods that is filled with hemolymph. 635

hemoglobin (HEE-muh-gloh-bun) Red iron-containing pigment in blood that combines with and transports oxygen. 250, 292

hepatic portal system (hih-PAT-ik) Portal system that begins at the villi of the small intestine and ends at the liver. 247

herbaceous stem (UR-bay-shus) Nonwoody stem. 158

herbivore (UR-buh-vor) Primary consumer in a food chain; a plant eater. 744

hermaphrodite (hur-MAF-ruh-dyt) Animal having both male and female sex organs. 625

heterospore (HET-ur-uh-spor) Spore that is dissimilar from another produced by the same plant; microspores and megaspores are produced by a seed plant. 601

heterotroph (HET-ur-uh-trohf, -trahf) Organism that takes in preformed foods. 136, 550, 744

heterozygous (het-uh-roh-ZY-gus) Having two different alleles (as *Aa*) for a given trait. 472

hexose (HEK-sohs) Six-carbon sugar. 32

hippocampus (hip-uh-KAM-pus) Part of the cerebral cortex where memories form. 338

histamine (HIS-tuh-meen, -mun) Substance, produced by basophils and mast cells in connective tissue, that causes capillaries to dilate. 266

HLA (human leukocyte associated) antigen Protein in a plasma membrane that identifies the cell as belonging to a particular individual and acts as a self-antigen. 272

homeobox (HOH-mee-uh-bahks) nucleotide sequence located in all homeotic genes and serving to identify portions of the genome, in many different types of organisms, that are active in pattern formation. 180, 450

homeostasis (hoh-mee-oh-STAY-sis) Maintenance of the internal environment, such as temperature, blood pressure, and other body conditions, within narrow limits. 8, 203, 208

homologous chromosome (hoh-MAHL-uh-gus, huh-mahl-uh-gus) Similarly constructed chromosomes with the same shape and that contain genes for the same traits; also called homologues. 92

homologous structure Structure that is similar in two or more species because of common ancestry. 555

homozygous (hoh-moh-zy-gus) Having identical alleles (as *AA* or *aa*) for a given trait; pure breeding. 472

hormone (HOR-mohn) Chemical messenger produced in low concentrations that has physiological and/or developmental effects, usually in another part of the organism. 176, 220, 396

human chorionic gonadotropin hormone (HCG) (kor-ee-AHN-ik, goh-nad-uh-TRAHP-in, -TROH-pin) Gonadotropin hormone produced by the chorion that functions to maintain the uterine lining. 429, 452

hydrogen bond Weak attraction between a hydrogen atom carrying a partial positive charge and an atom of another molecule carrying a partial negative charge. 26

hydrolysis (hy-DRAHL-ih-sis) Splitting of a covalent bond by the addition of water. 32

hydrophilic (hy-druh-FIL-ik) Type of molecule that interacts with water by dissolving in water and/or forming hydrogen bonds with water molecules. 26

hydrophobic (hy-druh-FOH-bik) Type of molecule that does not interact with water because it is nonpolar. 26

hydrostatic skeleton (hy-druh-STAT-ik) Internal body fluid that offers resistance to muscles and allows movement to occur. 626

hypertonic solution Solution that has a higher concentration of solute and a lower concentration of water than the cell. 75

hypothalamus (hy-poh-THAL-uh-mus) Part of the brain located below the thalamus that helps regulate the internal environment of the body and produces releasing factors that control the anterior pituitary. 335, 400

hypothesis (hy-PAHTH-ih-sis) Statement that is capable of explaining present data and is used to predict the outcome of future experimentation. 2

hypotonic solution Solution that has a lower concentration of solute and a higher concentration of water than the cell. 74

I

immunity Ability of the body to protect itself from foreign substances and cells, including infectious microbes. 266

immunoglobulin (Ig) (im-yuh-noh-GLAHB-yuh-lin, -yoo-lin) Globular plasma protein that functions as an antibody. 270

implantation Attachment and penetration of the embryo into the lining of the uterus (endometrium). 432, 451

inclusive fitness Fitness that results from direct selection and indirect selection. 683

independent assortment Alleles of unlinked genes assort independently of each other during meiosis so that the gametes contain all possible combinations of alleles. 93, 475

induction Ability of a chemical or a tissue to influence the development of another tissue. 449

inferior vena cava (VEE-nuh KAY-vuh) One of the largest veins in the systemic circuit, it collects blood from the lower body regions. 246

insertion End of a muscle that is attached to a movable bone. 381

inspiration (in-spuh-RAY-shun) Act of taking air into the lungs; inhalation. 284

insulin (IN-suh-lin) Hormone secreted by the pancreas that lowers the blood glucose level by promoting the uptake of glucose by cells and the conversion of glucose to glycogen by the liver and skeletal muscles. 408

integration Summing up of excitatory and inhibitory signals by a neuron. 327, 349

interferon (in-tur-FEER-ahn) Protein formed by a cell infected with a virus that can increase the resistance of other cells to the virus. 268

interkinesis (in-tur-kuh-NEE-sis) Period of time between meiosis I and meiosis II during which no DNA replication takes place. 94

internal respiration Exchange of oxygen and carbon dioxide between blood and tissue fluid. 292

interneuron Neuron found within the central nervous system that takes nerve impulses from one portion of the system to another. 322

interphase Interval between successive cell divisions; during this time, the chromosomes are extended and DNA replication and growth are occurring. 86

invertebrate (in-VUR-tuh-brit, -brayt) Referring to an animal without a serial arrangement of vertebrae, or a backbone. 618

ion (eye-un, -ahn) Atom or group of atoms carrying a positive or negative charge. 23

ionic bond (eye-AHN-ik) Bond created by an attraction between oppositely charged ions. 23

isotonic solution (eye-suh-tahn-ik) One that contains the same concentration of solute and water as the cell. 74

isotope (EYE-suh-tohp) One of two or more atoms with the same atomic number that differ in the number of neutrons and therefore in weight. 21

J

joint Union of two or more bones; an articulation. 370

juxtaglomerular apparatus (juk-stuh-gluh-MER-yuh-lur) Structure located in the walls of arterioles near the glomerulus that regulates renal blood flow. 313

K

karyotype (KAR-ee-uh-typ) Arrangement of all the chromosomes within a cell by pairs in a fixed order. 488

kinase (KY-nays) Any one of several enzymes that phosphorylate their substrates. 86

kingdom One of the seven categories, or taxa, used by taxonomists to group species. 568

kinin (KY-nen) Chemical mediator, released by damaged tissue cells and mast cells, which causes the capillaries to dilate and become more permeable. 266

Krebs cycle (krebz) Cyclical metabolic pathway found in the matrix of mitochondria that participates in aerobic cellular respiration; breaks down acetyl groups to carbon dioxide and hydrogen. Also called the citric acid cycle because the reactions begin and end with citrate. 119, 123

L

lacteal (LAK-tee-ul) Lymphatic vessel in a villus of the intestinal wall. 219

large intestine Last major portion of the digestive tract, extending from the small intestine to the anus and consisting of the cecum, the colon, the rectum, and the anal canal. 220

larynx (LAR-ingks) Cartilaginous organ located between pharynx and trachea that contains the vocal cords; voice box. 286

law Theory that is generally accepted by an overwhelming number of scientists. 5

learning Relatively permanent change in an animal's behavior that results from practice and experience. 674

lens Clear membranelike structure found in the eye behind the iris; brings objects into focus. 355

less-developed country (LDC) Country in which population growth is expanding rapidly and the majority of people live in poverty. 695

leukocyte (LOO-kuh-syt) See white blood cell.

lichen (LY-kun) Fungi and algae coexisting in a symbiotic relationship. 595

ligament (LIG-uh-munt) Dense fibrous connective tissue that joins bone to bone at a joint. 196, 370

limbic system (LIM-bik) Portion of the brain concerned with memory and emotions. 337

linkage group Alleles on the same chromosome are linked in the sense that they tend to move together to the same gamete; crossing-over interferes with linkage. 500

lipase (LY-pays, LY-payz) Fat-digesting enzyme secreted by the pancreas. 222, 224

lipid (LIP-id, LY-pid) Organic compound that is insoluble in water; notably fats, oils, and steroids. 34

logistic growth Population increase that results in an S-shaped curve; growth is slow at first, steepens, and then levels off due to environmental resistance. 690

loop of the nephron (NEF-rahn) Portion of the nephron lying between the proximal convoluted tubule and the distal convoluted tubule that functions in water reabsorption. 309

loose fibrous connective tissue Tissue composed mainly of fibroblasts that are widely separated by a matrix containing collagen and elastic fibers and found beneath epithelium. 196

lumen (LOO-mun) Cavity inside any tubular structure, such as the lumen of the digestive tract. 217

lungs Paired, cone-shaped organs within the thoracic cavity, functioning in internal respiration and containing moist surfaces for gas exchange. 287, 651

luteinizing hormone (LH) (LOO-tee-uh-ny-zing, loo-tee-ny-zing) Hormone produced by the anterior pituitary gland that stimulates the development of the corpus luteum in females and the production of testosterone in males. 423

lymph (limf) Fluid, derived from tissue fluid, that is carried in lymphatic vessels. 255, 264

lymph nodes Mass of lymphoid tissue located along the course of a lymphatic vessel. 265

lymphocyte (LIM-fuh-syt) Specialized white blood cell; occurs in two forms—T lymphocyte and B lymphocyte. 252

lysogenic cycle (LY-suh-jen-ik) Bacteriophage life cycle in which the viral DNA is incorporated into host cell DNA without viral reproduction and lysis. 575

lysosome (LY-suh-sohm) Membrane-bounded organelle containing digestive enzymes. 55

lytic cycle (lit-ik) Bacteriophage life cycle in which viral reproduction takes place and host cell lysis does occur. 575

M

macrophage (MAK-ruh-fayj) Large phagocytic cell derived from a monocyte that ingests microbes and debris. 252, 266

Malpighian tubule (mal-PIG-ee-un) Blind, threadlike excretory tubule attached to the gut of an insect. 637

maltase (MAHL-tays, -tayz) Enzyme produced in small intestine that breaks down maltose to two glucose molecules. 224

mast cell Cell to which antibodies, formed in response to allergens, attach, bursting the cell and releasing allergy mediators, which cause symptoms. 266

matter Anything that takes up space and has weight. 20

mechanoreceptor (mek-uh-noh-rih-SEP-tur) Sensory receptor that is sensitive to mechanical stimulation, such as that from pressure, sound waves, and gravity. 348

medulla oblongata (muh-DUL-uh ahb-lawng-gah-tuh) Part of the brain stem controlling heartbeat, blood pressure, breathing, and other vital functions. It also serves to connect the spinal cord to the brain. 335

megagametophyte (meg-uh-guh-MEE-tuh-fyt) In the life cycle of seed plants, the female gametophyte that produces an egg; in flowering plants, also the embryo sac. 608

megakaryocyte (meg-uh-KAR-ee-oh-syt, -uh-syt) Large bone marrow cell that gives rise to blood platelets. 254

megaspore (MEG-uh-spor) In seed plants, a large spore that develops into the egg-producing megagametophyte. 180, 601

meiosis (my-OH-sis) Type of cell division that in animals occurs during the production of gametes and results in four daughter cells with the haploid number of chromosomes. 92

melanocyte (MEL-uh-noh-syt) Specialized cell in the epidermis that produces melanin, the pigment responsible for skin color. 205

memory Capacity of the brain to store and retrieve information about past sensations and perceptions; essential to learning. 338

meninges (sing., meninx) (muh-NIN-jeez) Protective membranous coverings about the central nervous system. 201, 332

menopause (MEN-uh-pawz) Termination of the ovarian and uterine cycles in older women. 430

menstruation (men-stroo-AY-shun) Loss of blood and tissue from the uterus at the end of a uterine cycle. 429

meristem (mer-ih-stem) Undifferentiated, embryonic tissue in the active growth regions of plants. 154

mesoderm (MEZ-uh-durm, mes-) Middle germ layer of embryonic gastrula; gives rise to the muscles, the connective tissue, and the circulatory system. 445

messenger RNA (mRNA) Ribonucleic acid whose sequence of codons specifies the sequence of amino acids during protein synthesis. 510

metabolic pathway Series of linked reactions, beginning with a particular reactant and terminating with an end product. 108

metabolic pool Metabolites that result from catabolism and subsequently can be used for anabolism. 127

metabolism All of the chemical changes that occur within a cell. 9, 106

metamorphosis (met-uh-MOR-fuh-sis) Change in shape and form that some animals, such as amphibians and insects, undergo during development. 637

metaphase Mitosis phase during which chromosomes are aligned at the metaphase plate (equator) of the mitotic spindle. 89

metastasis (muh-TAS-tuh-sis) Spread of cancer from the place of origin throughout the body; caused by the ability of cancer cells to migrate and invade tissues. 522

microevolution Change in gene frequencies within a population over time. 556

microgametophyte (-guh-MEE-tuh-fyt) In seed plants, the sperm-producing male gametophyte. 608

microspore In seed plants, a small spore that develops into the sperm-producing microgametophyte; pollen grain. 180, 601

microtubule (-TOO-byool) Organelle composed of 13 rows of globular proteins; found in multiple units within other organelles, such as the centriole, cilia, flagella, as well as spindle fibers. 58

midbrain Most superior part of the brain stem; contains traits and reflex centers. 335

mimicry (MIM-ih-kree) Superficial resemblance of one organism to another organism of a different species; often used to avoid predation. 708

mineralocorticoid (min-ur-uh-loh-KOR-tih-koyd) Hormones secreted by the adrenal cortex that regulate salt and water balance, leading to increases in blood volume and blood pressure. 405

mitochondrion (my-tuh-KAHN-dree-un) Membranous organelle in which aerobic cellular respiration produces the energy carrier ATP. 56, 122

mitosis (my-TOH-sis) Type of cell division in which daughter cells receive the exact chromosome and genetic makeup of the parent cell; occurs during growth and repair. 84

molecule Like or different atoms joined by a bond; the unit of a compound. 23

molting Periodic shedding of the exoskeleton in arthropods. 634

monoclonal antibody Antibody of one type produced by a single plasma cell. 276

monocot (MAHN-uh-kaht) Abbreviation of monocotyledon. Flowering plant group; members have one embryonic leaf (cotyledon), parallel-veined leaves, scattered vascular bundles, and other characteristics. 153

monocyte (MAHN-uh-syt) Type of a granular leukocyte that functions as a phagocyte. 252

monohybrid Individual that is heterozygous for one trait; shows the phenotype of the dominant allele but carries the recessive allele. 473

monosaccharide (mahn-uh-SAK-uh-ryd) Simple sugar; a carbohydrate that cannot be decomposed by hydrolysis. 32

monosomy (MAHN-uh-soh-mee) One less chromosome than usual. 489

more-developed country (MDC) Country in which population growth is low and the people enjoy a good standard of living. 695

morphogenesis (mor-foh-JEN-uh-sus) Movement of early embryonic cells to establish body outline and form. 448

morula (MOR-yuh-luh) Spherical mass of cells resulting from cleavage during animal development prior to the blastula stage. 445

motor molecule Protein that moves along either actin filaments or microtubules by attaching, detaching, and reattaching. 58

motor neuron Neuron that takes nerve impulses from the central nervous system to the effectors. 322

multiple allele Pattern of inheritance in which there are more than two alleles for a particular trait, although each individual has only two of these alleles. 482

mutagen (MYOO-tuh-jun) Agent, such as radiation or a chemical, that brings about a mutation in DNA. 520

mutualism Symbiotic relationship in which both species benefit. 711

mycelium (my-SEE-lee-um) Mass of hyphae that makes up the body of a fungus. 591

mycorrhiza (my-kuh-RY-zuh) Mutually beneficial symbiotic relationship between a fungus and the roots of vascular plants. 596

myelin sheath (MY-uh-lin) Schwann plasma membranes that cover long neuron fibers, giving them a white, glistening appearance. 323

myocardium (my-oh-kar-dee-um) Cardiac muscle in the wall of the heart. 242

myofibril (my-uh-FY-brul) Contractile portion of muscle fibers. 385

myoglobin (MY-uh-gloh-bin) Pigmented compound in muscle tissue that stores oxygen. 389

myogram Recording of a muscular contraction. 388

myosin (MY-uh-sin) One of two major proteins of muscle; makes up thick filaments in myofibrils and is capable of breaking down ATP. See actin. 385

N

NAD^+ (nicotinamide adenine dinucleotide) (nik-uh-TEE-nuh-myd ad-un-een dy-noo-klee-uh-tyd) Coenzyme that functions as a carrier of electrons, especially in aerobic cellular respiration. 112, 120

$NADP^+$ (nicotinamide adenine dinucleotide phosphate) (FAHS-fayt) Coenzyme that becomes $NADPH + H^+$ during the light-dependent reactions of photosynthesis and reduces participants in the Calvin cycle during the light-independent reactions. 112

nasopharynx (nay-zoh-FAR-ingks) Region of the pharynx associated with the nasal cavity. 216

natural killer (NK) cell Lymphocyte that causes an infected or cancerous cell to burst. 268

natural selection Process by which populations become adapted to their environment. 562

negative feedback Self-regulatory mechanism that is activated by an imbalance and results in a fluctuation above and below a mean. 209

nephridium (pl., nephridia) (nuh-FRID-ee-um) Excretory tubule of many invertebrates (e.g., earthworm) with an opening in the body wall. 631

nephron (NEF-rahn) Anatomical and functional unit of the kidney; kidney tubule. 307

nerve Bundle of nerve fibers outside the central nervous system. 200, 328

nerve impulse Action potential (electrochemical change) traveling along a neuron. 324

neural tube During development, tube formed by closure of the neural groove. Neural tube develops into the spinal cord and brain. 447

neuroglial cell (noo-RAHG-lee-ul, noo-rohg-lee-ul) One of several types of cells found in nervous tissue that supports, protects, and nourishes neurons. 200, 322

neuromuscular junction Point of communication between a nerve cell and a muscle fiber. 386

neuron (NOOR-ahn, NYOOR-) Nerve cell that characteristically has three parts: dendrites, cell body, and axon. 200, 322

neurotransmitter Chemical stored at the ends of axons that is responsible for transmission across a synapse. 327

neurula (noor-uh-luh, nyoo-) Name for the embryo when the neural tube has formed. 447

neutron (NOO-trahn) Subatomic particle that has a weight of one atomic mass unit, carries no charge, and is found in the nucleus of an atom. 20

neutrophil (NOO-truh-fil) Granular leukocyte that is the most abundant of the white blood cells; first to respond to infection. 252

nitrification (ny-truh-fuh-KAY-shun) Process by which nitrogen in ammonia and organic compounds is oxidized to nitrites and nitrates by soil bacteria. 752

nitrogen fixation Process whereby nitrogen gas is reduced and nitrogen is added to organic compounds. 752

node In plants, the place where one or more leaves attach to a stem. 152

node of Ranvier (RAHN-vee-ay) Gap in the myelin sheath around a nerve fiber. 323

nondisjunction Failure of homologous chromosomes or sister chromatids to separate during the formation of gametes. 489

nonsteroid hormone Type of hormone that is not a steroid, such as a peptide, that is received by a plasma membrane receptor and brings about a change in metabolic reactions within a cell. 397

nonvascular plant Bryophytes such as mosses and liverworts that have no vascular tissue and either occur in moist locations or have special adaptations for living in dry locations. 602

norepinephrine (NE) (NOR-ep-uh-NEF-rin) Neurotransmitter active in the peripheral and central nervous systems; also a hormone secreted by the adrenal medulla in times of stress. 327, 405

notochord (noh-tuh-kord) Dorsal supporting rod that exists in all chordates sometime in their life history; replaced by the vertebral column in vertebrates. 447, 646

nuclear envelope Double membrane that surrounds the nucleus and is continuous with the endoplasmic reticulum. 52

nuclear pore Opening in the nuclear envelope which permits the passage of proteins into the nucleus and ribosomal subunits out of the nucleus. 52

nuclei (NOO-klee-eye, NYOO-) In the reticular formation, masses of cell bodies in the CNS. 335

nucleoid region (NOO-klee-oyd) Region of a bacterium where the bacterial chromosome is found; it is not bounded by a nuclear envelope. 62, 577

nucleolus (noo-KLEE-uh-lus) Organelle found inside the nucleus where rRNA is produced for ribosome formation. 52

nucleoplasm (NOO-klee-uh-plaz-um) Semifluid medium of the nucleus, containing chromatin. 52

nucleotide Monomer of a nucleic acid that forms when a nitrogen base, a pentose sugar, and a phosphate join. 40

O

oil Substance, usually of plant origin and liquid at room temperature, formed when a glycerol molecule reacts with three fatty acid molecules. 34

olfactory cell (ahl-FAK-tuh-ree, -tree, ohl-) Neuron modified as a sensory receptor for the sense of smell. 353

omnivore (AHM-nuh-vor) Organism in a food chain that feeds on both plants and animals. 744

oncogene (AHNG-koh-jeen) Gene that contributes to the transformation of a normal cell into a cancerous cell. 525

oogenesis (oh-uh-JEN-uh-sis) Production of an egg in females by the process of meiosis and maturation. 98, 424

operon (AHP-uh-rahn) Group of structural and regulating genes that function as a single unit. 518

opportunistic infection Infection that has an opportunity to occur because the immune system has been weakened. 436

order One of the seven categories, or taxa, used by taxonomists to group species. 568

organ Combination of two or more different tissues performing a common function. 7

organelle (or-guh-NEL) Specialized structure within cells (e.g., nucleus, mitochondria, and endoplasmic reticulum). 49

organic molecule Molecule that always contains carbon (C) and hydrogen (H); organic molecules are associated with living things. 31

origin End of a muscle that is attached to a relatively immovable bone. 381

osmosis (ahz-MOH-sis, ahs-) Movement of water from an area of higher concentration of water to an area of lower concentration of water across a differentially permeable membrane. 74

osmotic pressure (ahz-MAHT-ik) Pressure generated by and due to the flow of water across a semipermeable membrane. 74

ossicle (AHS-ih-kul) One of the small bones of the middle ear—malleus, incus, stapes. 362

osteocyte (AHS-tee-uh-syt) Mature bone cell. 370

osteon (AHS-tee-ahn) Cylindrical-shaped unit containing bone cells that surround an osteonic canal; Haversian system. 370

otolith (OH-tuh-lith) Calcium carbonate granule associated with ciliated cells in the utricle and the saccule. 365

ovarian cycle (oh-VAIR-ee-un) Monthly changes occurring in the ovary that determine the level of sex hormones in the blood. 427

ovary In animals, the female gonad, the organ that produces eggs, estrogen, and progesterone; in flowering plants, the base of the pistil that protects ovules and along with associated tissues becomes a fruit. 180, 411, 424, 614

oviduct (OH-vuh-dukt) Tube that transports eggs to the uterus; also called uterine tube. 424

ovulation (ahv-yuh-LAY-shun, ohv-) Discharge of a mature egg from the follicle within the ovary. 424

ovule (AHV-yool, OHV-yool) In seed plants, a structure where the megaspore becomes an egg-producing megagametophyte and which develops into a seed following fertilization. 180, 608

oxidation (ahk-sih-DAY-shun) Chemical reaction that results in removal of one or more electrons from an atom, ion, or compound; oxidation of one substance occurs simultaneously with reduction of another. 25, 112

oxidative phosphorylation (ahk-sih-DAY-tiv fahs-fuh-ruh-LAY-shun) Process by which ATP production is tied to an electron transport system that uses oxygen as the final acceptor; occurs in mitochondria. 124

oxygen debt Oxygen that is needed to metabolize lactate, a compound that accumulates during vigorous exercise. 129, 389

oxyhemoglobin (ahk-see-HEE-muh-gloh-bin) Compound formed when oxygen combines with hemoglobin. 292

oxytocin (ahk-sih-TOH-sin) Hormone released by the posterior pituitary that causes contraction of uterus and milk letdown. 400

ozone shield Formed from oxygen in the upper atmosphere, it protects the earth from ultraviolet radiation. 756

P

pacemaker See SA (sinoatrial) node. 244

pain receptor Sensory receptor that is sensitive to chemicals released by damaged tissues or excess stimuli of heat or pressure. 348

palisade mesophyll (pal-ih-SAYD) In a plant leaf, the layer of mesophyll containing elongated cells with many chloroplasts. 163

PAN (peroxyacetylnitrate) (puh-rahk-SEE-uh-see-tul) Type of chemical found in photochemical smog. 753

pancreas (PANG-kree-us, pan-) Elongate, flattened organ in the abdominal cavity that secretes enzymes into the small intestine (exocrine function) and hormones into the blood (endocrine function). 222, 408

pancreatic islets (of Langerhans) Distinctive group of cells within the pancreas that secretes insulin and glucagon. 408

parasympathetic division (par-uh-sim-puh-THET-ik) That part of the autonomic system that usually promotes activities associated with a restful state. 331

parathyroid gland (par-uh-THY-royd) One of four glands embedded in the posterior surface of the thyroid gland; produces parathyroid hormone. 404

parathyroid hormone (PTH) Hormone secreted by the four parathyroid glands that increases the blood calcium level and decreases the blood phosphate level. 404

parenchyma (puh-RENG-kuh-muh) **cell** Thin-walled, minimally differentiated cell that photosynthesizes or stores the products of photosynthesis. 154

pathogen (PATH-uh-jun) Disease-causing agent. 194, 250, 266

penis External organ in males through which the urethra passes and that serves as the organ of sexual intercourse. 421

pentose (PEN-tohs, -tohz) Five-carbon sugar; deoxyribose is the pentose sugar found in DNA; ribose is a pentose sugar found in RNA. 32

pepsin (PEP-sin) Protein-digesting enzyme secreted by gastric glands. 218, 224

peptidase (PEP-tih-days, -dayz) Intestinal enzyme that breaks down short chains of amino acids to individual amino acids that are absorbed across the intestinal wall. 224

peptide bond Covalent bond that joins two amino acids. 38

perception Mental awareness of sensory stimulation. 348

peripheral nervous system (PNS) (puh-RIF-ur-ul) Nerves and ganglia that lie outside the central nervous system. 322

peristalsis (paier-ih-STAWL-sis) Rhythmic contraction that serves to move the contents along in tubular organs, such as the digestive tract. 217

peritubular capillary network (paier-ih-TOO-byuh-lur) Capillary network that surrounds a nephron and functions in reabsorption during urine formation. 308

peroxisome Enzyme-filled vesicle in which fatty acids and amino acids become hydrogen peroxide which is broken down to harmless products. 55

PGAL (glyceraldehyde-3-phosphate) (glis-uh-RAL-duh-hyd) Significant metabolite in both the glycolytic pathway and the Calvin cycle; PGAL is oxidized during glycolysis and reduced during the Calvin cycle. 143

pH scale Measure of the hydrogen ion concentration $[H^+]$; any pH below 7 is acidic and any pH above 7 is basic. 29

phagocytosis (fag-uh-sy-TOH-sis) Taking in of bacteria and/or debris by engulfing; cell eating (verb, phagocytize). 78, 588

pharynx (FAR-ingks) Portion of the digestive tract between the mouth and the esophagus which serves as a passageway for food and also air on its way to the trachea. 216, 285

phenomenon Observable event. 2

phenotype (FEE-nuh-typ) Outward appearance of an organism caused by the genotype and environmental influences. 472

pheromone (FER-oh-mohn) Chemical substance secreted by one organism that influences the behavior of another. 396, 680

phloem (FLOH-em) Vascular tissue that conducts organic solutes in plants; contains sieve-tube elements and companion cells. 154, 174, 604

phospholipid (fahs-foh-LIP-id) Molecule having the same structure as a neutral fat except one bonded fatty acid is replaced by a group that contains phosphate; an important component of plasma membranes. 35

photoperiodism (foh-toh-PIR-ee-ud-iz-um) Response to light and dark; particularly in reference to flowering in plants. 178

photoreceptor Light-sensitive sensory receptor. 348

photosynthesis (foh-toh-SIN-thuh-sis) Process by which plants make their own food using the energy of the sun. 136

photosystem Cluster of light-absorbing pigment molecules within thylakoid membranes. 140

phototropism (foh-TAH-truh-piz-um) Directional growth of plants in response to light; stems demonstrate positive phototropism. 177

phyletic gradualism (fy-LET-ik) Evolutionary model that proposes evolutionary change resulting in a new species can occur gradually in an unbranched lineage. 567

phylum (FY-lum) One of the seven categories, or taxa, used by taxonomists to group species. 568

phytochrome (fy-tuh-krohm) Plant pigment that induces a photoperiodic response in plants. 179

phytoplankton (fy-toh-PLANGK-tun) Part of phytoplankton containing organisms that (1) photosynthesize and produce much of the oxygen in the atmosphere and (2) serve as food producers in aquatic ecosystems. 733

pineal gland (PIN-ee-ul, PY-nee-ul) Gland—either at the skin surface on the dorsal side of the head (fish, amphibians) or in the third ventricle of the brain, (mammals)—that produces melatonin. 411

pinocytosis (pin-uh-sy-TOH-sis) Taking in of fluid along with dissolved solutes by engulfing; cell drinking. 78

pistil (PIS-tul) Structure in a flower that consists of a stigma, a style, and an ovule-containing ovary; the ovule becomes a seed and the ovary becomes fruit. 180, 614

pith Parenchyma tissue in the center of some stems and roots. 157

pituitary gland Small gland that lies just inferior to the hypothalamus; the anterior pituitary produces several hormones, some of which control other endocrine glands; the posterior pituitary stores and secretes oxytocin and antidiuretic hormone. 400

placenta (pluh-SEN-tuh) Structure that forms from the chorion and the uterine wall and allows the embryo, and then the fetus, to acquire nutrients and rid itself of wastes. 429, 455

placental mammal Member of a mammalian subclass characterized by a placenta, an organ of exchange of nutrients, gases, and wastes between maternal and fetal blood. 657

plankton (PLANGK-tun) Freshwater and marine organisms that float on or near the surface of the water. 733

plaque (plak) Accumulation of soft masses of fatty material, particularly cholesterol, beneath the inner linings of the arteries. 229

plasma (PLAZ-muh) Liquid portion of blood. 198, 249

plasma cell Cell derived from a B-cell lymphocyte that is specialized to mass-produce antibodies. 269

plasma membrane Membrane surrounding the cytoplasm that consists of a phospholipid bilayer with embedded proteins; functions to regulate the entrance and exit of molecules from cell. 49, 62

plasmid (PLAZ-mid) Self-duplicating ring of accessory DNA in the cytoplasm of bacteria. 62, 530

plasmodesma (pl., plasmodesmata) (plaz-muh-DEZ-muh) Cytoplasmic strand that extends through a pore in the cell wall and connects the cytoplasms of adjacent cells. 174

plasmolysis (plaz-MAHL-ih-sis) Contraction of the cell contents due to the loss of water. 75

platelet (PLAYT-lit) Cell fragment that is necessary to blood clotting; also called a thrombocyte. 198, 254

polar body Nonfunctioning daughter cell, formed during oogenesis, that has little cytoplasm. 98

pollen grain In seed plants, the microspore that develops into the sperm-producing microgametophyte. 180, 608

pollination In seed plants, the delivery of pollen to the vicinity of the egg-producing megagametophyte. 180, 608

polygenic inheritance (pahl-ee-JEN-ik) Pattern of inheritance in which many allelic pairs control a trait; each dominant allele has a quantitative effect on the phenotype. 481

polymerase chain reaction (PCR) (PAHL-uh-muh-rays, -rayz) Technique that uses the enzyme DNA polymerase to produce millions of copies of a particular piece of DNA. 532

polypeptide Polymer of many amino acids linked by peptide bonds. 38

polyribosome (pahl-ih-RY-buh-sohm) Cluster of ribosomes attached to the same mRNA molecule; each ribosome is producing a copy of the same polypeptide. 53, 515

polysaccharide (pahl-ee-SAK-uh-ryd) Carbohydrate composed of many bonded glucose units—for example, glycogen. 32

pons (pahnz) Portion of the brain stem above the medulla oblongata and below the midbrain; assists the medulla oblongata in regulating the breathing rate. 335

population All the members of the same species that inhabit a particular area. 7, 10, 556, 688

positive feedback Mechanism of homeostatic response in which the output intensifies and increases the likelihood of response, instead of countering it and canceling it. 209

posterior pituitary Portion of the pituitary gland that stores and secretes oxytocin and antidiuretic hormone which are produced by the hypothalamus. 400

potential energy Stored energy as a result of location or spatial arrangement. 104

prefrontal area Association area in the frontal lobe that receives information from other association areas and uses it to reason and plan actions. 337

pressure-flow theory Explanation for phloem transport; osmotic pressure following active transport of sugar into phloem brings about a flow of sap from a source to a sink. 174

primary motor area Area in the frontal lobe where voluntary commands begin; each section controls a part of the body. 336

primary somatosensory area (soh-mat-uh-SENS-ree, -suh-ree) Area dorsal to the central sulcus where sensory information arrives from skin and skeletal muscles. 336

primate (PRY-mayt) Animal that belongs to the order Primates, the order of mammals that includes prosimians, monkeys, apes, and humans. 659

primitive streak Ectodermal ridge in the midline of the embryonic disk from which arises the mesoderm by inward and then lateral migration of cells. 447

principle Theory that is generally accepted by an overwhelming number of scientists. Also called law. 5

prion (PREE-ahn) (contraction for a proteinlike infectious agent) Subviral pathogen consisting of only a glycoprotein molecule; believed to be linked to several diseases of the central nervous system in humans. 577

probe Known sequences of DNA that are used to find complementary DNA strands; can be used diagnostically to determine the presence of particular genes. 532

producer Organism at the start of a food chain that makes its own food (e.g., green plants on land and algae in water). 744

product Substance that forms as a result of a reaction. 106

progesterone (proh-JES-tuh-rohn) Female sex hormone secreted by the corpus luteum of the ovary and by the placenta. 411, 429

prokaryotic cell (proh-kar-ee-AHT-ik) Cell lacking a nucleus and the membranous organelles found in complex cells; bacteria, including cyanobacteria. 62

prolactin (PRL) (proh-LAK-tin) Hormone secreted by the anterior pituitary that stimulates the production of milk from the mammary glands. 400

prophase (PROH-fayz) Mitosis phase during which chromatin condenses so that chromosomes appear. 88

proprioceptor (proh-pree-oh-SEP-tur) Sensory receptor that assists the brain in knowing the position of the limbs. 350

prostaglandin (PG) (prahs-tuh-GLAN-din) Hormone that has various and powerful local effects. 412

prostate gland (PRAHS-tayt) Gland located around the male urethra below the urinary bladder; adds secretions to semen. 420

protein Organic compound that is composed of either one or several polypeptides. 37

proteinoid (PROHT-en-oyd) Abiotically polymerized amino acids that are joined in a preferred manner; possible early step in cell evolution. 549

prothrombin (proh-THRAHM-bin) Plasma protein that is converted to thrombin during the steps of blood clotting. 254

prothrombin activator Enzyme that catalyzes the transformation of the precursor prothrombin to the active enzyme thrombin. 254

proto-oncogene (PROH-toh-AHNG-koh-jeen) Normal gene that can become an oncogene through mutation. 525

protocell Structure that preceded the true cell in the history of life. 549

proton Subatomic particle found in the nucleus of an atom that has a weight of one atomic mass unit and carries a positive charge; a hydrogen ion. 20

protostome (PROH-tuh-stohm) Group of coelomate animals in which the first embryonic opening (the blastopore) is associated with the mouth. 619

protozoan (proh-tuh-ZOH-un) Animal-like protist that is classified according to means of locomotion: pseudopods, flagella, cilia, or none. 588

proximal convoluted tubule Highly coiled region of a nephron near the glomerular capsule, where tubular reabsorption takes place. 309

pseudocoelom (soo-duh-SEE-lum) Body cavity lying between the digestive tract and body wall that is incompletely lined by mesoderm. 619

pulmonary artery (POOL-muh-naier-ee, puul-) Blood vessel that takes blood away from the heart to the lungs. 243, 246

pulmonary circuit That part of the circulatory system that takes deoxygenated blood to and oxygenated blood away from the gas-exchanging surfaces in the lungs. 246

pulmonary vein Blood vessel that takes blood to the heart from the lungs. 243, 246

pulse Vibration felt in arterial walls due to expansion of the aorta following ventricle contraction. 244

punctuated equilibrium Evolutionary model that proposes there are periods of rapid change dependent on speciation followed by long periods of stasis. 567

Punnett square (PUN-ut) Gridlike device used to calculate the expected results of simple genetic crosses. 473

purine (pyoor-een) Nitrogen base found in DNA and RNA that has two interlocking rings, as in adenine and guanine. 508

Purkinje fibers (pur-KIN-jee) Specialized muscle fibers that conduct the cardiac impulse from AV bundle into the ventricular walls. 244

pyrimidine (py-RIM-ih-deen, pih-) Nitrogen base found in DNA and RNA that has just one ring, as in cytosine, uracil, and thymine. 508

pyruvate (py-ROO-vayt) End product of glycolysis; pyruvic acid. 119

R

reactant (re-AK-tunt) Substance that participates in a reaction. 106

receptor-mediated endocytosis Selective uptake of molecules into a cell by binding to a specific receptor. 79

recessive allele (uh-LEEL) Hereditary factor that expresses itself in the phenotype only when the genotype is homozygous. 472

recombinant (ree-KAHM-buh-nunt) New combination of alleles as a result of crossing-over. 500

recombinant DNA (rDNA) DNA that contains genes from more than one source. 530

red blood cell (erythrocyte) Formed element that contains hemoglobin and carries oxygen from the lungs to the tissues. 198, 250

red bone marrow Blood cell-forming tissue located in the spaces within spongy bone. 266, 370

redox reaction (REE-doks) Oxidation-reduction reaction; one molecule loses electrons (oxidation) while another molecule simultaneously gains electrons (reduction). 25

reduced hemoglobin (HEE-muh-gloh-bun) Hemoglobin that is carrying hydrogen ions. 292

reduction Chemical reaction that results in addition of one or more electrons to an atom, ion, or compound. Reduction of one substance occurs simultaneously with oxidation of another. 25, 112

referred pain Pain perceived as having come from a site other than that of its actual origin. 351

reflex action Automatic, involuntary response of an organism to a stimulus. 216, 329

refractory period (rih-FRAK-tuh-ree) Time following an action potential when a neuron is unable to conduct another nerve impulse. 324

regulator gene Gene that codes for a protein involved in regulating the activity of structural genes. 518

renin (ren-in) Enzyme released by kidneys that leads to the secretion of aldosterone and a rise in blood pressure. 313, 406

replacement reproduction Population in which each person is replaced by only one child. 696

replication Making an exact copy, as in the duplication of DNA. 509

reproduce To make a copy similar to oneself, as when one-celled organisms divide or humans have children. 9

resource partitioning Apportioning the supply of a resource such as food and living space between species as a means to increase the number of niches following competition between species. 706

respiratory center Group of nerve cells in the medulla oblongata that send out nerve impulses on a rhythmic basis, resulting in inspiration. 290

resting potential Polarity across the plasma membrane of a resting neuron due to an unequal distribution of ions. 324

restriction enzyme Bacterial enzyme that stops viral reproduction by cleaving viral DNA; used to cut DNA at specific points during production of recombinant DNA. 530

restriction fragment length polymorphism (RFLP) (pahl-ee-mor-fiz-um) Differences in DNA sequence between individuals that result in different patterns of restriction fragment lengths (DNA segments resulting from treatment with restriction enzymes). 533

reticular formation Complex network of nerve fibers within the brain stem that arouses the cerebrum. 335

retina (RET-n-uh, RET-nuh) Innermost layer of the eyeball, which contains the rod cells and the cone cells. 355

retinal (RET-n-al, -awl) Light-absorbing molecule, which is a derivative of vitamin A and a component of rhodopsin. 356

retrovirus (REH-tro-vy-rus) Virus that contains only RNA and carries out RNA → cDNA transcription, called reverse transcription. 576

rhodopsin (roh- DAHP-sun) Visual pigment found in the rods whose activation by light energy leads to vision. 356

ribosomal RNA (rRNA) (ry-buh-SOH-mul) RNA occurring in ribosomes, which are the structures involved in protein synthesis. 510

ribosome (RY-buh-sohm) Minute particle that is attached to endoplasmic reticulum or occurs loose in the cytoplasm and is the site of protein synthesis. 53, 62, 513

ribozyme (RY-buh-zym) Enzyme that carries out mRNA processing. 512

ribulose bisphosphate (ry-byoo-lohs bis-FAHS-fayt) See RuBP.

RNA (ribonucleic acid) (ry-boh-noo-KLEE-ik) Nucleic acid found in cells that assists DNA in controlling protein synthesis. 40, 510

RNA polymerase (PAHL-uh-muh-rays) Enzyme that speeds the formation of RNA from a DNA template. 512

rod cell Photoreceptor in vertebrate eyes that responds to dim light. 356

root hair Extension of a root epidermal cell that increases the surface area for the absorption of water and minerals. 154

RuBP (ribulose bisphosphate) Five-carbon compound that combines with and fixes carbon dioxide during the Calvin cycle and is later regenerated by the same cycle. 143

S

SA (sinoatrial) node (sy-noh-AY-tree-ul) Small region of neuromuscular tissue that initiates the heartbeat; also called the pacemaker. 244

saccule (SAK-yool) Saclike cavity in the vestibule of the inner ear; contains sensory receptors for static equilibrium. 365

salivary gland Gland associated with the oral cavity that secretes saliva. 214

saprotroph (SAP-ruh-trohf, -trahf) Heterotroph such as a bacterium or a fungus that externally digests dead organic matter before absorbing the products. 578

sarcolemma (sar-kuh-LEM-uh) Membrane that surrounds striated muscle cells. 385

sarcomere (SAR-kuh-mir) Structural and functional unit of a myofibril; contains actin and myosin filaments. 385

sarcoplasmic reticulum (sar-kuh-PLAZ-mik rih-tik-yuh-lum) Smooth endoplasmic reticulum of skeletal muscle cells; surrounds the myofibrils and stores calcium ions needed for myosin to bind to actin and therefore muscle contraction. 385

saturated fatty acid Fatty acid molecule that lacks double bonds between the atoms of its carbon chain. 34

Schwann cell (shwahn) Cell that surrounds a fiber of a peripheral nerve and forms the neurilemmal sheath and myelin. 323

scientific method Process by which scientists formulate a hypothesis, gather data by observation and experimentation, and come to a conclusion. 6

scientific theory Concept that joins together well-supported and related hypotheses; a conceptual scheme supported by a broad range of observations, experiments, and data. 5

scrotum (SKROH-tum) Pouch of skin that encloses the testes. 420

secretion Releasing of a substance by exocytosis from a cell that may be gland or part of gland. 54

seed Mature ovule that contains a sporophyte embryo with stored food enclosed by a protective coat. 181, 608

seed coat Protective covering for the seed, formed by the hardening of the ovule wall. 181

segmentation (seg-mun-TAY-shun) Repetition of body parts as segments along the length of the body; seen in annelids, arthropods, and chordates. 619

segregation (seg-rih-GAY-shun) Separation of alleles from each other during meiosis so that the gametes contain one from each pair. Each resulting gamete has an equal chance of receiving either allele. 470

semen (SEE-mun) Thick, whitish fluid consisting of sperm and secretions from several glands of the male reproductive tract. 420

semicircular canal (sem-ih-SUR-kyuh-lur) One of three tubular structures within the inner ear that contain sensory receptors responsible for the sense of dynamic equilibrium. 365

semilunar valve (sem-ee-LOO-nur) Valve resembling a half moon located between the ventricles and their attached vessels. 242

seminal vesicle (SEM-uh-nul) Convoluted, saclike structure attached to the vas deferens near the base of the urinary bladder in males; adds secretions to semen. 420

seminiferous tubule (sem-uh-NIF-ur-us) Highly coiled duct within the male testes that produces and transports sperm. 422

sensation Conscious awareness of a stimulus due to nerve impulses sent to the brain from a sensory receptor by way of sensory neurons. 348

sensory adaptation Phenomenon of a sensation becoming less noticeable once it has been recognized by constant repeated stimulation. 349

sensory neuron Neuron that takes nerve impulses to the central nervous system and typically has a long dendrite and a short axon; afferent neuron. 322

sensory receptor Structure specialized to receive information from the environment and to generate nerve impulses. 348

septum (sep-tum) Partition or wall that divides two areas; the septum in the heart separates the right half from the left half. 242

serum (seer-um) Light yellow liquid left after clotting of blood. 254

sex chromosome Chromosome responsible for the development of characteristics associated with gender; an X or Y chromosome. 488

sex-influenced trait Autosomal trait that is expressed differently in the two sexes. 499

sex-linked trait Phenotype that is controlled by a gene located on a sex chromosome, usually the X chromosome, whose pattern of inheritance differs in males and females. 496

sexual selection Changes in males and females due to male competition and female selectivity. 676

sieve plate (siv) Perforated end wall of a sieve-tube element. 155

sieve-tube element Member that joins with others in the phloem tissue of plants as a means of transport for nutrient sap. 155

sinus (sy-nus) Cavity or hollow space in an organ such as the skull. 374

sister chromatid (KROH-muh-tid) One of two genetically identical chromosomal units that are the result of DNA replication and are attached to each other at the centromere. 84

sliding filament theory Movement of actin in relation to myosin; accounts for muscle contraction. 385

small intestine Long, tubelike chamber of the digestive tract between the stomach and large intestine. 219

soap Salt formed from a fatty acid and an inorganic base. 34

society Group in which members of species are organized in a cooperative manner, extending beyond sexual and parental behavior. 680

sociobiology (soh-see-oh-by-AH-luh-jee, soh-shee-) Application of evolutionary biology principles to the study of social behavior in animals. 683

sodium-potassium pump Transport protein in the plasma membrane that moves sodium ions out of and potassium ions into animal cells; important in nerve and muscle cells. 77, 324

soft palate (PAL-it) Entirely muscular posterior portion of the roof of the mouth. 214

solute (SAHL-yoot) Substance dissolved in a solvent to form a solution. 73

solvent (SAHL-vunt) Fluid, such as water, that dissolves solutes. 73

somatic cell (soh-MAT-ik) In animals, a body cell, excluding those that undergo meiosis and become a sperm or egg. 84

somatic system That portion of the peripheral nervous system containing motor neurons that control skeletal muscles. 329

species Group of similarly constructed organisms capable of interbreeding and producing fertile offspring; organisms that share a common gene pool. 9, 565, 568

sperm Male sex cell with three distinct parts at maturity: head, middle piece, and tail. 423

spermatogenesis (spur-mat-uh-JEN-ih-sis) Production of sperm in males by the process of meiosis and maturation. 98, 423

sphincter (SFINGK-tur) Muscle that surrounds a tube and closes or opens the tube by contracting and relaxing. 217

spinal cord Part of the central nervous system; the nerve cord that is continuous with the base of the brain and housed within the vertebral column. 332

spinal nerve Nerve that arises from the spinal cord. 328

spindle Structure consisting of fibers, poles, and asters (if animal cell) that brings about the movement of chromosomes during cell division. 88

spiral organ Portion of inner ear that permits hearing and consists of hair cells located on the basilar membrane within the cochlea; also called organ of Corti. 363

spleen Large, glandular organ located in the upper left region of the abdomen that stores and purifies blood. 265

sporangium (spuh-RAN-jee-um) Structure within which spores are produced. 592, 601

spore Haploid reproductive cell, sometimes resistant to unfavorable environmental conditions, that is capable of producing a new individual which is also haploid. 180, 584, 601

sporophyte (SPOR-uh-fyt) In the life cycle of a plant, the diploid generation that produces spores by meiosis. 180, 601

stabilizing selection Outcome of natural selection in which extreme phenotypes are eliminated and the average phenotype is conserved. 562

stamen (STAY-mun) Structure in a flower that consists of a filament and an anther with pollen sacs. 180, 614

starch Storage polysaccharide found in plants that is composed of glucose molecules joined in a linear fashion. 32

static equilibrium Maintenance of balance when the head and body are motionless. 365

stem cell Any undifferentiated cell that can divide and differentiate into more functionally specific cell types such as blood cells and germ cells. 250

steroid (STEER-oyd) Type of lipid molecule having four interlocking rings; examples are cholesterol, progesterone, and testosterone. 35

steroid hormone Chemical messenger that is lipid soluble and therefore passes through the plasma and nuclear envelope to bind with a receptor inside the nucleus; the complex turns on specific genes leading to the production of particular proteins. 397

stimulus Change in the internal or external environment that a sensory receptor can detect leading to nerve impulses in sensory neurons. 348

stomach Muscular sac that mixes food with gastric juices to form chyme, which enters the small intestine. 218

stomate (STOH-mayt) Microscopic opening bordered by guard cells in the leaves of plants through which gas exchange takes place. 154, 170

stream Channel of running water; also called river. 731

striated (STRY-ayt-ud) Having bands; cardiac and skeletal muscle are striated with light and dark bands. 199

structural gene Gene that directs the synthesis of an enzyme or a structural protein in the cell. 518

substrate Reactant in a reaction controlled by an enzyme. 108

substrate-level phosphorylation (fahs-fur-uh-LAY-shun) Process in which ATP is formed by transferring a phosphate from a metabolic substrate to ADP. 120

superior vena cava (VEE-nuh KAY-vuh) Vein that returns blood from the lower limbs and the greater part of the pelvic and abdominal organs to the right atrium. 246

sustainable world Global way of life that can continue indefinitely, because the economic needs of all peoples are met while still protecting the environment. 697

sustentacular (Sertoli) **cell** (sus-tun-TAK-yuh-lur) Elongated cell in the wall of the seminiferous tubules to which spermatids are attached during spermatogenesis. 423

symbiotic (sim-bee-AH-tik) Relationship that occurs when two different species live together in a unique way; it may be mutually beneficial, neutral, or detrimental to one and/or the other species. 578, 710

sympathetic division That part of the autonomic system that usually promotes activities associated with emergency (fight or flight) situations. 331

sympatric speciation (sim-PAT-rik spee-shee-AY-shun, -see-) Origin of new species in populations that overlap geographically. 565

synapse (SIN-aps, si-NAPS) Region between two nerve cells where the nerve impulse is transmitted from one to the other, usually from axon to dendrite. 327

synapsis (sih-NAP-sis) Pairing of homologous chromosomes during prophase I of meiosis. 92

synaptic cleft (sih-NAP-tik) Small gap between presynaptic and postsynaptic membranes of a synapse. 327

syndrome Group of symptoms that appear together and tend to indicate the presence of a particular disorder. 489

synovial joint (sih-NOH-vee-ul) Freely movable joint. 380

systemic circuit That part of the circulatory system that serves body parts other than the gas-exchanging surfaces in the lungs. 246

systole (SIS-tuh-lee) Contraction of a heart chamber. 244

systolic pressure (sis-TAHL-ik) Arterial blood pressure during the systolic phase of the cardiac cycle. 248

T

T lymphocyte (LIM-fuh-syt) Lymphocyte that matures in the thymus and exists in four varieties, one of which kills antigen-bearing cells outright. 268

T (transverse) tubule Membranous channel which extends inward from a muscle fiber membrane and passes through the fiber. 385

taste bud Sense organ containing the receptors associated with the sense of taste. 352

taxonomy (tak-SAHN-uh-mee) Science of naming and classifying organisms into various categories. 13, 568

telophase (TEL-uh-fayz) Mitosis phase during which the diploid number of daughter chromosomes are located at each pole. 89

template (TEM-plit) Pattern that serves as a mold for the production of an oppositely shaped structure; one strand of DNA is a template for a complementary strand. 509

tendon (TEN-dun) Fibrous connective tissue that joins muscle to bone. 196, 370

territoriality Behavior related to the act of marking or defending a particular area against invasion by another species member; area often used for the purpose of feeding, mating, and caring for young. 679

testcross Crossing a heterozygous individual with one who has the recessive phenotype in order to determine the genotype of the heterozygote. 474

testes (sing., testis) (TES-teez) Male gonads, the organ that produces sperm and testosterone. 411, 420, 422

testosterone (tes-TAHS-tuh-rohn) In mammals, major male sex hormone produced by interstitial cells in the testes; it stimulates development of primary sex organs and maintains secondary sexual characteristics in males. 411, 423

tetanus (TET-n-us) Sustained muscle contraction without relaxation. 388

thalamus (THAL-uh-mus) Part of the brain located in the lateral walls of the third ventricle that serves as the integrating center for sensory input; it plays a role in arousing the cerebral cortex. 335

thermoreceptor Sensory receptor that is sensitive to changes in temperature. 348

threshold Level of potential at which an action potential or nerve impulse is produced. 324

thrombin (THRAHM-bin) Enzyme that converts fibrinogen to fibrin threads during blood clotting. 254

thylakoid (THY-luh-koyd) Flattened sac within a granum whose membrane contains the photosynthetic pigments (e.g., chlorophyll); where the light-dependent reactions occur. 57, 62, 138

thymine (T) (THY-meen) One of four nitrogen bases in nucleotides composing the structure of DNA. 508

thymus gland Organ that lies in the neck and chest area and is absolutely necessary to the development of immunity. 266, 411

thyroid gland Organ that lies in the neck and produces several important hormones, including thyroxin and calcitonin. 403

thyroid-stimulating hormone (TSH) Substance produced by the anterior pituitary that causes the thyroid to secrete thyroxin. 400

thyroxine (thy-RAHK-sin) Hormone secreted from the thyroid gland that promotes growth and development; in general, it increases the metabolic rate in cells. 375

tissue Group of similar cells combined to perform a common function. 7, 194

tissue fluid Solution that bathes and services every cell in the body; also called interstitial fluid. 255

tone Continuous, partial contraction of muscle. 388

tonicity (toh-NIS-ih-tee) Degree to which the concentration of solute versus solvent causes fluids to move into or out of cells. 74

tonsils Partially encapsulated lymph nodules located in the pharynx. 265, 295

trachea (TRAY-kee-uh) Air tube in insects that transports oxygen in air to the tissues; also the windpipe in terrestrial vertebrates that takes air to the lungs. 286, 637

tracheid (TRAY-kee-id, -keed) In flowering plants, type of cell in xylem that has tapered ends and pits through which water and minerals flow. 154

trait Specific term for a distinguishing phenotypic feature studied in heredity. 470

transcription Process resulting in the production of a strand of mRNA that is complementary to a segment of DNA. 511

transfer RNA (tRNA) Molecule of RNA that carries an amino acid to a ribosome engaged in the process of protein synthesis. 510

transgenic organism Free-living organisms in the environment that have had a foreign gene inserted into them. 534

transition reaction Reaction within aerobic cellular respiration during which electrons and carbon dioxide are removed from pyruvate; results in acetyl groups that enter the Krebs cycle. 119, 122

translation Process by which the sequence of codons in mRNA dictates the sequence of amino acids in a polypeptide. 511

transpiration Evaporation of water from a leaf; pulls water from the roots through a stem to leaves. 170

triglyceride (trih-GLIS-uh-ryd) Neutral fat composed of glycerol and three fatty acids. 34

triplet code Genetic code (mRNA, tRNA) in which sets of three bases call for specific amino acids in the formation of polypeptides. 511

trisomy (TRY-soh-mee) One more chromosome than usual. 489

trophic level Feeding level of one or more populations in a food web. 747

trophoblast (TROH-fuh-blast) Outer membrane surrounding the embryo in mammals; when thickened by a layer of mesoderm, it becomes the chorion, an extraembryonic membrane. 452

tropism (TROH-piz-um) In plants, a growth response toward or away from a directional stimulus. 176

tropomyosin (trahp-uh-MY-uh-sin, trohp-) Protein that blocks muscle contraction until calcium ions are present. 387

troponin (TROH-puh-nin) Protein that functions with tropomyosin to block muscle contraction until calcium ions are present. 387

trypsin (TRIP-sin) Protein-digesting enzyme secreted by the pancreas. 222

tubular reabsorption Movement of nutrient molecules, as opposed to waste molecules, from the contents of the nephron into blood at the proximal convoluted tubule. 311

tubular secretion Movement of certain molecules from blood into the distal convoluted tubule of a nephron so that they are added to urine. 311

tumor (TOO-mur) Cells derived from a single mutated cell that has repeatedly undergone cell division; benign tumors remain at the site of origin and malignant tumors metastasize. 522

tumor-suppressor gene Gene that suppresses the development of a tumor; the mutated form contributes to the development of cancer. 525

turgor pressure (TUR-gur, -gor) Internal pressure that adds to the strength of a cell and builds up when water moves by osmosis into a cell. 75

U

umbilical arteries and **vein** (um-BIL-ih-kul) Fetal blood vessels that travel to and from the placenta. 458

umbilical cord Acts as a tube connecting the fetus to the placenta, through which blood vessels pass. 454

unsaturated fatty acid Fatty acid molecule that has one or more double bonds between the atoms of its carbon chain. 34

uracil (U) (YOOR-uh-sil) One of four nitrogen bases in nucleotides composing the structure of RNA. 480

urea (yoo-REE-uh) Primary nitrogenous waste of humans derived from amino acid breakdown. 305

ureter (YOOR-uh-tur) One of two tubes that take urine from the kidneys to the urinary bladder. 304

urethra (yoo-REE-thruh) Tube that takes urine from the bladder to outside. 304, 420, 425

uric acid (YOOR-ik) Waste product of nucleotide metabolism. 305

urinary bladder Organ where urine is stored before being discharged by way of the urethra. 304

uterine cycle (YOO-tur-in, -tuh-ryn) Monthly occurring changes in the characteristics of the uterine lining (endometrium). 429

uterus (YOO-tur-us) Organ located in the female pelvis where the fetus develops; the womb. 424

utricle (YOO-trih-kul) Saclike cavity in the vestibule of the inner ear that contains sensory receptors for static equilibrium. 365

V

vaccine Antigens prepared in such a way that they can promote active immunity without causing disease. 274

vacuole (VAK-yoo-ohl) Membranous cavity, usually filled with fluid. 55

vagina Organ that leads from the uterus to the vestibule and serves as the birth canal and organ of sexual intercourse in females. 425

valve Membranous extension of a vessel or the heart wall that opens and closes, ensuring one-way flow. 241

vas deferens (vas DEF-ur-unz, -uh-renz) Tube that leads from the epididymis to the urethra in males. 420

vascular bundle In plants, primary phloem and primary xylem enclosed by a bundle sheath. 153

vascular cambium (KAM-bee-um) Meristem tissue that produces secondary phloem and secondary xylem, which add to the girth of a plant. 158

vascular plant Plant that has vascular tissue (xylem and phloem); includes seedless vascular plants (e.g., ferns) and seed plants (gymnosperms and angiosperms). 604

vascular tissue Transport tissue in plants consisting of xylem and phloem. 154

vector (VEK-tur) In genetic engineering, a means to transfer foreign genetic material into a cell—for example, a plasmid. 530

vein Vessel that takes blood to the heart from venules; characteristically having nonelastic walls; in a leaf, veins are vascular bundles. 153, 240

vena cava (VEE-nuh KAY-vuh) Large systemic vein that returns blood to the right atrium of the heart; either the superior or inferior vena cava. 243

venous duct (VEE-nus) Fetal connection between the umbilical vein and the inferior vena cava; ductus venosus. 458

ventricle (VEN-trih-kul) Cavity in an organ, such as a lower chamber of the heart; or the ventricles of the brain, which are interconnecting cavities that produce and serve as a reservoir for cerebrospinal fluid. 242, 332, 334

venule (VEN-yool, VEEN-) Vessel that takes blood from capillaries to a vein. 241

vertebral column (vur-tuh-brul) Backbone of vertebrates through which the spinal cord passes. 376

vertebrate (VUR-tuh-brit, -brayt) Referring to an animal with a backbone composed of vertebrae. 447, 618, 649

vesicle (VES-ih-kul) Small, membranous sac that stores substances within a cell. 53

vessel element Cell which joins with others to form a major conducting tube found in xylem. 154

vestigial structure (veh-STIJ-ee-ul, STIJ-ul) Underdeveloped structure that was functional in some ancestor but is no longer functional in a particular organism. 555

villus (pl., villi) (VIL-us) Fingerlike projection from the wall of the small intestine that functions in absorption. 219

viroid (VY-royd) Infectious agent consisting of a small strand of RNA that is apparently replicated by host cell enzymes. 577

virus Noncellular obligate parasite of living cells consisting of an outer capsid and an inner core of nucleic acid. 574

vital capacity Maximum amount of air moved in or out of the human body with each breathing cycle. 288

vitamin Essential requirement in the diet, needed in small amounts; often a part of a coenzyme. 111, 230

vocal cord Fold of tissue within the larynx; creates vocal sounds when it vibrates. 286

vulva External genitals of the female that surround the opening of the vagina. 425

W

white blood cell (leukocyte) Formed element of which there are several types, each having a specific function in protecting the body from invasion by foreign substances and organisms. 252

white matter Myelinated nerve fibers in the central nervous system. 333

wood Secondary xylem which builds up year after year in woody plants and becomes the annual rings. 160

X

X chromosome Female sex chromosome that carries genes involved in sex determination; see Y chromosome. 488

xenotransplantation (zen-uh-trans-plan-TAY-shun) Use of animal organs, instead of human organs, in human transplant patients. 536

xylem (ZY-lum) Vascular tissue that transports water and minerals upward through the plant body. 154, 170, 604

Y

Y chromosome Male sex chromosome that carries genes involved in sex determination; see X chromosome. 488

yolk sac Extraembryonic membrane that encloses yolk in reptiles and birds; in placental mammals it is the first site of blood cell formation. 451

Z

zero population growth No increase in population size. 696

zooplankton (zoh-uh-PLANGK-tun) Part of plankton containing protozoa and other types of microscopic animals. 733

zygote (ZY-goht) Diploid cell formed by the union of two gametes; the product of fertilization. 98, 181, 424

Credits

Line Art and Readings

Chapter 2

Ecology Box, p. 30 Sources: Data from G. Tyler Miller, *Living in the Environment*, 1993, Wadsworth Publishing Company, Belmont, CA; and Lester R. Brown, *State of the World*, 1992, W. W. Norton and Company, Inc., New York, NY.

Chapter 7

Health Focus, p. 128 Source: Data from S.K. Powers and E.T. Howley, *Exercise Physiology*, 2d edition, 1994, Times Mirror Higher Education Group, Inc., Dubuque, Iowa.

Chapter 10

10.15 From Kingsley R. Stern, *Introductory Plant Biology*, 6th edition. Copyright © 1994 Times Mirror Higher Education Group, Inc., Dubuque, Iowa. All Rights Reserved. Reprinted by permission.
10.16 From Kingsley R. Stern, *Introductory Plant Biology*, 6th edition. Copyright © 1994 Times Mirror Higher Education Group, Inc., Dubuque, Iowa. All Rights Reserved. Reprinted by permission.

Chapter 12

12.15 Source: Data from T. T. Shintani, *Eat More, Weigh Less™ Diet*, 1983.
Health Focus, p. 236 Reprinted by permission from page 374 of *Understanding Nutrition*, Fifth Edition, by W.N. Whitney, et al.; Copyright © 1990 by West Publishing Company. All Rights Reserved.
12.17 Reprinted by permission from Nutrition: *Concepts and Controversies*, 6th edition, by Frances Sienkiewicz Sizer and Elanor Noss Whitney. Copyright © 1994 by West Publishing Company. All Rights Reserved.

Chapter 13

13.2 From Kent M. Van De Graaff and Stuart Ira Fox, *Concepts of Human Anatomy & Physiology*, 3rd edition. Copyright © 1992 Times Mirror Higher Education Group, Inc., Dubuque, Iowa. All Rights Reserved. Reprinted by permission.

Chapter 14

14.9b From David Shier, et al., *Hole's Human Anatomy & Physiology*, 7th edition. Copyright © 1996 Times Mirror Higher Education Group, Inc., Dubuque, Iowa. All Rights Reserved. Reprinted by permission.

Chapter 15

Health Box, p. 299 From "The Most Often Asked Questions About Smoking Tobacco, and Health and . . . The Answers," revised July 1993. © American Cancer Society, Inc., Atlanta, GA. Reprinted by permission.

Chapter 19

Figure 19.12 From Kent Van De Graaff and Stuart Ira Fox, *Concepts of Human Anatomy and Physiology*, 4th edition. Copyright © 1995 The McGraw-Hill Companies, Inc. All Rights Reserved. Reprinted by permission.

Chapter 21

21.11 From Kent M. Van De Graaff and Stuart Ira Fox, *Concepts of Human Anatomy & Physiology*, 4th edition. Copyright © 1995 Times Mirror Higher Education Group, Inc., Dubuque, Iowa. All Rights Reserved. Reprinted by permission.
21.15, 21.16, 21.17, 21.18 Source: Data from Division of STD Prevention. Sexually Transmitted Disease Surveillance, 1996. U.S. Department of Health and Human Services, Public Health Service. Atlanta: Centers for Disease Control and Prevention, September 1997.

Chapter 29

Ecology Focus, pp. 610-611 Charles N. Horn.

Chapter 32

32.2 (graph) Source: Data from S. J. Arnold, "The Microevolution of Feeding Behavior" in *Foraging Behavior: Ecological, Ethological, and Psychological Approaches*, edited by A. Kamil and T. Sargent, 1980, Garland Publishing Company, New York, NY.

Chapter 34

Ecology Focus, p. 709 From Betsy Carpenter, "Biological Nightmares," in *U. S. News & World Report*. Copyright, November 20, 1995, U.S. News & World Report.

Chapter 35

Ecology Focus, p. 725 Copyright 1995 by the National Wildlife Federation. Reprinted with permission from *National Wildlife*, magazine's October/November, 1995 issue.

Chapter 36

Ecology Focus, p. 758 Kenneth R. Hinga.

Photographs

History of Biology

(McClintock): AP/Wide World Photos; (Franklin): Cold Springs Harbor Laboratory; (rest): © Corbis-Bettmann.

Table of Contents

Part 1: © CNRI/SPL/Photo Researchers, Inc.; Part 2: © Ed Reschke/Peter Arnold, Inc.; Part 3: © CNRI/SPL/Photo Researchers; Part 4: © SIU/Visuals Unlimited; Part 5: © P. Motta/SPL/Photo Researchers, Inc.; Part 6: © Norbert Wu/Peter Arnold, Inc.; Part 7: © Richard R. Hansen/Photo Researchers, Inc.

Chapter One

Opening Photo: © Roger Tully/Tony Stone Images; 1.1(protozoa): © A.M. Siegelman/Visuals Unlimited; 1.1(hibiscus): © Rosemary Calvert/Tony Stone Images; 1.1(crab): © Tui DeRoy/Bruce Coleman, Inc.; 1.1(leopard): © James Martin/Tony Stone Images; 1.5: © Joe McDonald/Visuals Unlimited; 1.6: © Dennis Schmidt/Valan Photos; 1.7(left): © Hermann Eisenbeiss/Photo Researchers, Inc.; 1.7(middle): © John D. Cunningham/Visuals Unlimited; 1.7(right): © Hermann Eisenbeiss/Photo Researchers, Inc.; 1.9(background): © Barbara von Hoffman/Tom Stack & Associates; 1.9(toucan): © Ed Reschke/Peter Arnold, Inc.; 1.9(morpho): © Kjell Sandved/Butterfly Alphabet; 1.9(jaguar): © BIOS (Seitre)/Peter Arnold, Inc.; 1.9(orchid): © Max & Bea Hunn/Visuals Unlimited; 1.9(frog): © Kevin Schafer & Martha Hill/Tom Stack & Associates.

Part Opening One

2: © CNRI/SPL/Photo Researchers, Inc.; 3: © Francis Leroy/Biocosmos/Photo Researchers, Inc.; 4: © Hank Morgan/Photo Researchers, Inc.; 5: © David M. Phillips/Visuals Unlimited; 6: © Newcomb & Wergin/BPS/Tony Stone Images; 7: © Pete Saloutos/The Stock Market.

Careers in Biology

Page 18 (top): © Edgar Bernstein/Peter Arnold, Inc.; p.18 (middle): © Jerry Mason/SPL/Photo Researchers, Inc.; p. 18(bottom): © Lawrence Migdale/Photo Researchers, Inc.

Chapter Two

Opening Photo: © CNRI/SPL/Photo Researchers, Inc.; 2.4: © Charles M. Falco/Photo Researchers, Inc.; 2.9a: © Martin Dohrn/SPL/Photo Researchers, Inc.; 2.9b: © Comstock, Inc.; 2.9c: © Marty Cooper/Peter Arnold, Inc.; 2A(left): © Ray Pfortner/Peter Arnold, Inc.; 2A(middle): © John Millar/Tony Stone Images; 2A(right): © Frederica Georgia/Photo Researchers, Inc.; 2.14, 2.15, 2.16: © Dwight Kuhn; 2.19: © Jeremy Burgess/SPL/Photo Researchers, Inc.; 2.20: © Don W. Fawcett/Photo Researchers, Inc.; 2.21: © Richard C. Johnson/Visuals Unlimited.

Chapter Three

Opening Photo: © Francis Leroy/Biocosmos/Photo Researchers, Inc.; 3A(left): © David M. Phillips/Visuals Unlimited; 3A(middle): © Robert Caughey/Visuals Unlimited; 3A(right): © Warren Rosenberg/BPS/Tony Stone Images; 3.2b: © Alfred Paisieka/SPL/Photo Researchers, Inc.; 3.3b: © Newcomb/Wergin/BPS/Tony Stone Images; 3.4(top): Courtesy E.G. Pollock; 3.4(bottom): Courtesy Ron Milligan/Scripps Research Institute; 3.5b: © Barry F. King/Biological Photo Service; 3.6(top): © R. Rodewald/Biological Photo

Service; 3.6(bottom): © K.G. Murti/Visuals Unlimited; 3.7: © E.H. Newcomb & S.E. Frederick/Biological Photo Service; 3.8a: Courtesy Herbert W. Israel, Cornell University; 3.9a: Courtesy Keith Porter; 3.10b: © M. Schliwa/Visuals Unlimited; 3.10c,d: © K.G. Murti/Visuals Unlimited; 3.11(top): Courtesy Kent McDonald, University of Colorado-Boulder; 3.11(bottom): From Manley McGill, D.P. Highfield, T.M. Monahan, and B.R. Brinkley, Journal of Ultrastructure Research 57, 43-53 pg. 48, fig. 6, (1976) Academic Press; 3.12(sperm): © David M. Phillips/Photo Researchers, Inc.; 3.12(flagellum): © W.L. Dentler/Biological Photo Service; 3.12(b.body): © W.L. Dentler/Biological Photo Service; 3.13a: © David M. Phillips/Visuals Unlimited; 3.13b: © Biophoto Associates/Photo Researchers, Inc.

Chapter Four

Opening Photo: © Hank Morgan/Photo Researchers, Inc.; 4.A: Courtesy Robert P. Lanza, MD; 4.13(both): Courtesy Mark Bretscher.

Chapter Five

Opening Photo: © David M. Phillips/Visuals Unlimited; 5B: Courtesy Dr. Stephen Wolfe; 5.4a: © Ed Reschke; 5.5(late prophase): © Michael Abbey/Photo Researchers, Inc.; 5.5(rest): © Ed Reschke; 5.6(all): © Ed Reschke; 5.7: © B.A. Palevitz & E.H. Newcomb/BPS/Tom Stack & Associates; 5.8(both): © R.G. Kessel and C.Y. Shih, "Scanning Electron Microscopy in Biology: A Student's Atlas on Biological Organization," 1974 Springer-Verlag, New York.

Chapter Six

Opening Photo: © Newcomb & Wergin/BPS/Tony Stone Images; 6.1: © Gregg Adams/Tony Stone Images; 6.8(left): © John Kieffer/Peter Arnold, Inc.; 6.8(right): © Sunset Moulu (20)/Peter Arnold, Inc.

Chapter Seven

Opening Photo: © Pete Saloutos/The Stock Market; 7.4: Courtesy Keith Porter; p.128: © Tim Davis/Photo Researchers, Inc.

Part Opening Two

8: © Breck Kent/Animals Animals/Earth Scenes; 9: © Dwight R. Kuhn; 10: © Ed Reschke/Peter Arnold, Inc.

Careers in Biology

Page 134(top): © Paul Von Baich/The Image Works; p.134(middle) : © Kathy Sloane/Photo Researchers, Inc.; p.134(bottom): © Inga Spence/Tom Stack & Associates.

Chapter Eight

Opening Photo: © Breck Kent/Animals Animals/Earth Scenes; 8.3: © Ed Degginger/Color-Pic, Inc.; 8.4: Courtesy Herbert W. Israel, Cornell University; 8.9: © The McGraw Hill Companies, Inc./Bob Coyle, photographer; 8.11: © Peter Arnold/Peter Arnold, Inc.

Chapter Nine

Opening Photo: © Dwight R. Kuhn; 9.4: © Ed Reschke/Peter Arnold, Inc.; 9.5a: © J. Robert Waaland/Biological Photo Service; 9.5b: © Biophoto Associates/Photo Researchers, Inc.; 9.6a: © J. Robert Waaland/Biological Photo Service; 9.6b: © George Wilder/Visuals Unlimited; 9.7b: © CABISCO/Phototake; 9.8: © Dwight Kuhn; 9.9: Courtesy G.S. Ellmore; 9.10a: © J. Robert Waaland/Biological Photo Service; 9.11a, 9.12a: © CABISCO/Phototake; 9.12b,c: © Runk/Schoenberger/Grant Heilman Photography; 9.13b: © Ardea London; 9A: © James Schnepf Photography, Inc.; 9.15: © Jeremy Burgess/SPL/Photo Researchers, Inc.; 9.16a: © John D. Cunningham/Visuals Unlimited; 9.16b: © William E. Ferguson; 9.16c,d: © The McGraw-Hill Companies, Inc./Carlyn Iverson, photographer; 9.17a: © Patti Murray/Animals Animals/Earth Scenes; 9.17b: © Michael Gadomski/Photo Researchers, Inc.; 9.17c: © CABISCO/Phototake.

Chapter Ten

Opening Photo: © Ed Reschke/Peter Arnold, Inc.; 10.1a,b: Courtesy W.A. Cote, Jr., N.C. Brown Center for Ultrastructure Studies, SUNY-ESF; 10.3(both): © Jeremy Burgess/SPL/Photo Researchers, Inc.; 10.4: © BioPhot; 10.6: © Runk/Schoenberger/Grant Heilman Photography; 10.9: Courtesy of Scott R. Bauer, *Agricultural Research Magazine*, USDA; 10.14a(top): © Kingsley R. Stern; 10.14a(bottom): © Joe Munroe/Photo Researchers, Inc.; 10.14b(left): © Ralph A. Reinhold/Animals Animals/Earth Scenes; 10.14b(right): © W. Ormerod/Visuals Unlimited; 10.14c(left): © C.S. Lobban/Biological Photo Service; 10.14c(right): © John N.A. Lott/Biological Photo Service; 10.17a-d: © Runk/Schoenberger/Grant Heilman Photography.

Part Opening Three

11: © CNRI/SPL/Photo Researchers, Inc.; 12: © R.G. Kessel and R.H. Kardon, "Tissues and Organs: A Text-Atlas of Scanning Electron Microscopy". W.H. Freeman, 1979, All Rights Reserved; 13: © P. Motta & S. Correr/Photo Researchers, Inc.; 14: © Nina Lampen/Photo Researchers, Inc.; 15: © CNRI/SPL/ Photo Researchers, Inc.; 16: © CNRI/ SPL/Photo Researchers.

Careers in Biology

Page 192(top): © The McGraw-Hill Companies, Inc./Carlyn Iverson, photographer; p.192(middle): PhotoDisc/Health & Medicine; p.192(bottom): © The McGraw-Hill Companies, Inc./Carlyn Iverson, photographer.

Chapter Eleven

Opening Photo: © CNRI/SPL/Photo Researchers, Inc.; 11.1a-d,11.3a-d,11.5a-c: © Ed Reschke; 11Ba: © Ken Greer/Visuals Unlimited; 11Bb: © Dr. P. Marazzi/SPL/Photo Researchers, Inc.; 11Bc: © James Stevenson/SPL/Photo Researchers, Inc.

Chapter Twelve

Opening Photo: © R.G. Kessel and R.H. Kardon, "Tissues and Organs: A Text-Atlas of Scanning Electron Microscopy". W.H. Freeman, 1979, All Rights Reserved; 12.4b: © Biophoto Associates/Photo Researchers, Inc.; 12.5b: © Ed Reschke/Peter Arnold, Inc.; 12.5c: © St. Bartholomew's Hospital/SPL/Photo Researchers, Inc.; 12.6(right): © Manfred Kage/Peter Arnold, Inc.; 12.6(left): Photo by Susumu Ito, from Charles Flickinger, *Medical Cellular Biology*, W.B. Saunders, 1979; 12.14: © The McGraw-Hill Companies, Inc./Bob Coyle, photographer; 12.16a: © Biophoto Associates/Photo Researchers, Inc.; 12.16b: © Ken Greer/Visuals Unlimited; 12.16c: © Biophoto Associates/Photo Researchers, Inc.; 12.18: © Carl Centineo/Medichrome/The Stock Shop, Inc.; 12.19: © Donna Day/Tony Stone Images; 12.20: © Tony Freeman/Photo Edit.

Chapter Thirteen

Opening Photo: © P. Motta & S. Correr/Photo Researchers, Inc.; 13.1d,13.6b,13.6c: © Ed Reschke; 13.11a: © Lennart Nilsson, "Behold Man" Little Brown and Company, Boston; 13.13: © Boehringer Ingelheim International GmbH, photo by Lennart Nilsson; 13.14: © Manfred Kage/Peter Arnold, Inc.; 13B: © Biophoto Associates/Photo Researchers, Inc.

Chapter Fourteen

Opening Photo: © Nina Lampen/Photo Researchers, Inc.; 14.2(thymus, spleen): © Ed Reschke/Peter Arnold, Inc.; 14.2(lymph): © Fred E. Hossler/Visuals Unlimited; 14.2(bone marrow): © R. Calentine/Visuals Unlimited; 14.6b: Courtesy Dr. Arthur J. Olson, Scripps Institute; 14A: © AP/Wide World Photo; 14.7a: © Boehringer Ingelheim International GmbH, photo by Lennart Nilsson; 14.9: © Matt Meadows/Peter Arnold, Inc.; 14.10: © Estate of Ed Lettau/Peter Arnold, Inc.; 14B: © Martha Cooper/Peter Arnold, Inc.; 14.12(both): Courtesy of Stuart I. Fox.

Chapter Fifteen

Opening Photo: © CNRI/SPL/Photo Researchers, Inc.; 15.5: © SIU/Visuals Unlimited; 15Ac: © Bill Aron/Photo Edit; 15.12a,b: © Martin Rotker/Martin Rotker Photography.

Chapter Sixteen

Opening Photo: © CNRI/SPL/Photo Researchers; 16.4(top left): © David M. Phillips/Visuals Unlimited; 16.4(top right & bottom): © 1966 Academic Press, from A.B. Maunsbach, J. Ultrastruct. Res. 15:242-282; 16.5a: © Gennaro/Photo Researchers, Inc.

Part Opening Four

17: © Dan McCoy/Rainbow; 18: © Jon Riley/The Stock Shop; 19: © SIU/Visuals Unlimited; 20: © Manfred Kage/Peter Arnold.

Careers in Biology

Page 320(top): © Michael Newman/Photo Edit; p.320(middle): © Michael Newman/ PhotoEdit; p.320(bottom): © Bonnie Kamin/PhotoEdit.

Chapter Seventeen

Opening Photo: © Dan McCoy/Rainbow; 17.1: © Gerhard Gscheidle/Peter Arnold, Inc.; 17.3b: © M.B. Bunge/Biological Photo Service; 17.4: © Linda Bartlett; 17.5: Courtesy Dr. E.R. Lewis, University of California Berkeley; 17.11c: © Manfred Kage/Peter Arnold, Inc.; 17.12b: © Colin Chumbley/Science Source/Photo Researchers, Inc.; 17A: © Marcus Raichle/Peter Arnold, Inc.; 17.19: © Lawrence Migdale/Photo Researchers, Inc.

Chapter Eighteen

Opening Photo: © Jon Riley/The Stock Shop; 18.4: © Omikron/SPL/Photo Researchers, Inc.; 18.8: © Lennart Nilsson, "The Incredible Machine"; 18.9: © Biophoto Associates/Photo Researchers, Inc.; 18A: Robert S. Preston and Joseph E. Hawkins, Kresge Hearing Research Institute, University of Michigan; 18.13: © P. Motta/SPL/Photo Researchers, Inc.

Chapter Nineteen

Opening Photo: © SIU/Visuals Unlimited; 19.1(top two): © Ed Reschke; 19.1(bottom): © Biophoto Associates/Photo Researchers, Inc.; 19.5b: © The McGraw-Hill Companies, Inc./Joe DeGrandis, photographer; 19.10(fibrous): © Ed Reschke/Peter Arnold, Inc.; 19.10(rest): © Ed Reschke; 19.13: © Ed Reschke/Peter Arnold, Inc.; 19.14: © Ed Reschke; 19.15: © Victor B. Eichler; 19.18b: © G.W. Willis/Biological Photo Service.

Chapter Twenty

Opening Photo: © Manfred Kage/Peter Arnold; 20.5(left): © Bob Daemmrich/Stock Boston; 20.5(right): © Ewing Galloway, Inc.; 20.6: Courtesy Department of Illustrations, Washington University School of Medicine. From Clinical Pathological Conference, "Acromegaly, Diabetes, Hypermetabolism, Proteinura and Heart Failure," American Journal of Medicine, 20 (1956) 133. Reprinted with permission from Excerpta Medica Inc.; 20.7: © Biophoto Associates/Photo Researchers, Inc.; 20.8: © John Paul Kay/Peter Arnold, Inc.; 20.12a: © Custom Medical Stock Photos; 20.12b: © NMSB/Custom Medical Stock Photos; 20.13a,b: "Atlas of Pediatric Physical Diagnosis," Second Edition by Zitelli & Davis, 1992. Mosby-Wolfe Europe Limited, London, UK.

Part Opening Five

21: © P. Motta/SPL/Photo Researchers, Inc.; 22: © Camera M.D. Studios; 23: © Rosanne Olson/Tony Stone Images; 24: © Lawrence Migdale/Photo Researchers, Inc.; 25: © Ken Edward/Biografx/Photo Researchers, Inc.; 26: © Michael Newman/Photo Edit.

Careers in Biology

Page 418(top): © David Parker/SPL/Photo Researchers, Inc.; p.418(middle): © Richard Choy/Peter Arnold, Inc.; p.418(bottom): © Will & Deni McIntyre/Photo Researchers, Inc.

Chapter Twenty One

Opening Photo: © P. Motta/SPL/Photo Researchers, Inc.; 21.3b: © Biophoto Associates/Photo Researchers, Inc.; 21.7: © Ed Reschke/Peter Arnold, Inc.; 21.10: © Dr. Landrum B. Shettles; 21.12a: © The McGraw-Hill Companies, Inc./Bob Coyle, photographer; 21.12b: © The McGraw-Hill Companies, Inc./Vincent Ho, photographer; 21.12c,d: © The McGraw-Hill Companies, Inc./Bob Coyle, photographer; 21.12e: © Hank Morgan/Photo Researchers, Inc.; 21.12f: © The McGraw-Hill Companies, Inc./Bob Coyle, photographer; 21.13: © M. Long/Visuals Unlimited; 21.15: © Charles Lightdale/Photo Researchers, Inc.; 21.16: © CDC/Peter Arnold, Inc.; 21.17: © CNRI/SPL/Photo Researchers, Inc.; 21.18: © G.W. Willis/BPS/Tony Stone Images.

Chapter Twenty Two

Opening Photo: © Camera M.D. Studios; 22.8a,b: Courtesy of E.B. Lewis; 22.12a: Lennart Nilsson, "A Child is Born," Dell Publishing Company; 22.14: © John Watney/Photo Researchers, Inc.; 22.19: © John Cunningham/Visuals Unlimited; 22.20: © Richard Hutchings/Photo Edit.

Chapter Twenty Three

Opening Photo: © Rosanne Olson/Tony Stone Images; 23.1: © Corbis-Bettmann; 23.3a: © SuperStock, Inc.; 23.3b: © Michael Grecco/Stock Boston; 23.11(both): © Steve Uzzell; 23.12: Courtesy the Cystic Fibrosis Foundation; 23.13a: © Kevin Fleming/Corbis; 23.15b: © Bill Longcore/Photo Researchers, Inc.

Chapter Twenty Four

Opening Photo: © Lawrence Migdale/Photo Researchers, Inc.; 24.1(4): © CNRI/SPL/Photo Researchers, Inc.; 24.1(5): © CNRI/SPL/Photo Researchers, Inc.; 24.3a: © Jill Cannefax/EKM-Nepenthe; 24.5 : Courtesy NATURE 163:676 (1949); 24.6a,b: From R. Kampheier, "Physical Examination of Health Diseases," © 1958 F.A. Davis Company; 24.7a: © David M. Phillips/Visuals Unlimited; 24.7b,c: From R. Simensen and R. Curtis Rogers, "Fragile X Syndrome," AMERICAN FAMILY PHYSICIAN 39(5):186, May 1989. © American Academy of Family Physicians.

Chapter Twenty Five

Opening Photo: © Ken Edward/Biografx/Photo Researchers, Inc.; 25Aa: Courtesy of Biophysics Department, King's College, London; 25.Ab: From "The Double Helix" by James D. Watson, Antheneum Press, New York, 1968. Courtesy of Cold Spring Harbor Laboratory; 25.10b: Courtesy of Alexander Rich; 25.14: From M.B. Roth and J.G. Gall, "Cell Biology" 105:1047-1054, 1987. © Rockefeller University Press; 25B: Cold Spring Harbor Laboratory; 25C: © John N.A. Lott/Biological Photo Service; 25.18: © Dr. Gopal Murti/SPL/Photo Researchers, Inc.

Chapter Twenty Six

Opening Photo: © Michael Newman/Photo Edit; 26.5: © Will & Deni McIntyre/Photo Researchers, Inc.; 26.6(both): Courtesy of General Electric Research and Development Center; 26.7a: Courtesy of Genzyme Corporation and Tufts University School of Medicine.

Part Opening Six

27: © Jan-Peter Lahall/Peter Arnold, Inc.; 28: © Morendun Animal Health LTD/SPL/Photo Researchers, Inc.; 29: © Ken Brate/Photo Researchers, Inc.; 30: © Norbert Wu/Peter Arnold, Inc.; 31: © Mike Severns/Tony Stone Images.

Careers in Biology

Page 546(top): © The McGraw-Hill Companies, Inc./Carlyn Iverson, photographer; p.546(middle): © William E. Ferguson; p.546(bottom): © Paul Murphy/Unicorn Stock Photos.

Chapter Twenty Seven

Opening Photo: © Jan-Peter Lahall/Peter Arnold, Inc.; 27.2a: © Science VU/Visuals Unlimited; 27.2b: Courtesy of Dr. David Deamer; 27.4a: © John Cunningham/Visuals Unlimited; 27.4b: Transparency #213, Courtesy Department of Library Services, American Museum of Natural History; 27.5: The Field Museum of Natural History, Neg. #CK9T, Chicago; 27.8a,b: © Carolina Biological Supply/Phototake; 27.10(left): © Breck Kent/Animals Animals/Earth Scenes; 27.10(right): © Michael Tweedie/Photo Researchers, Inc.; 27.12: Courtesy of Dr. Mckusick; 27.15: © Bob Evans/Peter Arnold, Inc.; 27.20(wolf): © Art Wolfe/Tony Stone Images; 27.20(flower): © Ed Reschke/Peter Arnold, Inc.; 27.20(paramecium): © M. Abbey/Visual Unlimited; 27.20(mushroom): © Rod Planck/Tom Stack & Associates; 27.20(bacteria): © David M. Phillips/Visuals Unlimited.

Chapter Twenty Eight

Opening Photo: © Morendun Animal Health LTD/SPL/Photo Researchers, Inc.; 28.1: © Robert Caughey/Visuals Unlimited; 28.5a,b: © David M. Phillips/Visuals Unlimited; 28.5c: © R.G. Kessel - C.Y. Shih/Visuals Unlimited; 28Aa: Courtesy NASA Media Resource Center; 28Ab: © Woods Hole Oceanographic Institution; 28.6a: © R. Knauft/Biology Media/Photo Researchers, Inc.; 28.6b: © Eric Grave/Photo Researchers, Inc.; 28.7(all): © S.C. Holt/Biological Photo Service; 28.8: © Dr. Tony Brain/SPL/Photo Researchers, Inc.; 28.11: © R. Knauft/Photo Researchers, Inc.; 28.12: © M.I. Walker/Science Source/Photo Researchers, Inc.; 28.13: © William E. Ferguson; 28.15a: © Dr. Ann Smith/SPL/Photo Researchers, Inc.; 28.15b: © Biophoto Associates/Photo Researchers, Inc.; 28.17: © Walter H. Hodge/Peter Arnold, Inc.; 28.18b: © Manfred Kage/Peter Arnold, Inc.; 28.19: © Carolina Biological Supply/Phototake; 28.20a: © Ed Reschke/Peter Arnold, Inc.; 28.22: © Ray Simons/Photo Researchers, Inc.; 28.23: © David M. Phillips/Visuals Unlimited; 28.24: © Walter H. Hodge/Peter Arnold, Inc.; 28.24b: © J. Forsdyke/Gene Cox/SPL/Photo Researchers, Inc.; 28.25a: © Biophoto Associates; 28.25b: © Glenn Oliver/Visuals Unlimited; 28.26: Courtesy Dr. G.L. Barron, University of Guelph; 28.27b: © Stephen Krasemann/Peter Arnold, Inc.

Chapter Twenty Nine

Opening Photo: © Ken Brate/Photo Researchers, Inc.; 29.1(top to bottom): © John Shaw/Tom Stack and Associates, © William E. Ferguson, © Kent Dannen/Photo Researchers, Inc., © Leonard Lee Rue III/Bruce Coleman, Inc.; 29.3: © Ed Reschke/Peter Arnold, Inc.; 29.4: © John Gerlach/Visuals Unlimited; 29.5: The Field Museum of Natural History, Neg. #Geo 75400C, Chicago; 29.6: © Carolina Biological Supply/Phototake; 29.7: © Steve Solum/Bruce Coleman, Inc.; 29.8: © Robert P. Carr/Bruce Coleman, Inc.; 29.9(left): © John Gerlach/Visuals Unlimited; 29.9(middle): © W.H. Hodge/Peter Arnold, Inc.; 29.9(right): © Forest W. Buchanan/Visuals Unlimited; 29.10b: © Matt Meadows/Peter Arnold, Inc.; 29.11a: © D. Cavanaro/Visuals Unlimited; 29.11b: © Kingsley R. Stern; 29.11c: © Karlene Schwartz; 29.12: © Carolina Biological Supply/Phototake; 29.A(wheat plants): © C.P. Hickman/Visuals Unlimited; 29.A(wheat grains): © Philip Hayson/Photo Researchers, Inc.; 29.A(corn plants): © Adam Hart-Davis/SPL/Photo

Researchers, Inc.; 29.A(corn grains): © Mark S. Skalny/Visuals Unlimited; 29.A(rice plants): © Scott Camazine; 29.A(rice grains): © John Tiszler/Peter Arnold, Inc.; 29.B(dwarf palms and baskets): © Heather Angel; 29.B(cotton plants): © Dale Jackson/Visuals Unlimited; 29.B(man in cotton shirt): © Bob Daemmrich/The Image Works, Inc.; 29.B(rubber tree): © Steven King/Peter Arnold, Inc.; 29.B(tires): © Will & Deni McIntyre/Photo Researchers, Inc.

Chapter Thirty

Opening Photo: © Norbert Wu/Peter Arnold, Inc.; 30.1a: © Biophoto Associates/Photo Researchers, Inc.; 30.1b: © David M. Dennis/Tom Stack & Associates; 30.4b: © Caroline Biological Supply/Phototake; 30.4c: © Ron Taylor/Bruce Coleman, Inc.; 30.4d: © Runk/Schoenberger/Grant Heilman Photography; 30.4e: © Lloyd Rye/Photo Researchers, Inc.; 30.5: © CABISCO/Visuals Unlimited; 30.6a: © Carolina Biological Supply/Phototake; 30.9a: © Arthur Siegelman/Visuals Unlimited; 30.9b: © James Solliday/Biological Photo Service; 30.11a: © Fred Bavendam/Peter Arnold, Inc.; 30.11b: © Douglas Faulkner/Photo Researchers, Inc.; 30.11c: © Biophoto Associates; 30.11d: © Marty Snyderman/Visuals Unlimited; 30.13a: © Michael DiSpezio; 30.13b: © W.H. Hughes/Visuals Unlimited; 30.14b: © Roger K. Burnard/Biological Photo Service; 30A: © St. Bartholomew's Hospital/SPL/Photo Researchers, Inc.; 30.15a: © John MacGregor/Peter Arnold, Inc.; 30.15b: © G.C. Kelley/Photo Researchers, Inc.; 30.15c: © Robert Evans/Peter Arnold, Inc.; 30.15d: © Herbert Parsons/Arees Photography; 30.15e: © Dwight Kuhn; 30.17a: © Charles E. Schmidt/Unicorn Stock Photos; 30.17b: © John Shaw/Tom Stack & Associates; 30.17c: © L. West/Bruce Coleman, Inc.; 30.17d: © William E. Ferguson; 30.17e: © Bill Beatty/Visuals Unlimited; 30Bb: © Max Meier/Tierfotografie, Bauen, Switzerland; 30Bc: © Brent Opell; 30.19a: © Tom McHugh/Photo Researchers, Inc.; 30.19b: © Scott Camazine/Photo Researchers, Inc.; 30.19c: © E.R. Degginger/Bruce Coleman, Inc.

Chapter Thirty One

Opening Photo: © Mike Severns/Tony Stone Images; 31.1a: © Randy Morse/Animals Animals/Earth Scenes; 31.1b: © E.R. Denninger/Animals Animals/Earth Scenes; 31.1c: © Mike Bacon/Tom Stack & Associates; 31.2a: © Michael DiSpezio; 31.4: © Heather Angel; 31.5: © Rick Harbo; 31.6a: © James Watt/Animals Animals/Earth Scenes; 31.6b: © Dwight Kuhn; 31.6c: © Zig Leszczynski/Animals Animals/Earth Scenes; 31.7: © Ron & Valerie Taylor/Bruce Coleman, Inc.; 31.8a-d: © Jane Burton/Bruce Coleman, Inc.; 31.A(both): Courtesy Museum of the Rockies; 31.9: © Zig Leszczynski/Animals Animals/Earth Scenes; 31.10a: © Bruce Davidson/Animals Animals/Earth Scenes; 31.13a: © Tom McHugh/Photo Researchers, Inc.; 31.13b: © Tony Stone Images; 31.13c: © Leonard Lee Rue Enterprises; 31.15: © Dan Dreyfus and Associates; 31.19: Transp. #608 Courtesy Dept. of Library Services, American Museum of Natural History.

Part Opening Seven

32: © David C. Fritts/Animals Animals/Earth Scenes; 33: © Jan-Peter Lahall/Peter Arnold, Inc.; 34: © Richard R. Hansen/Photo Researchers, Inc.; 35: © David Higgs/Tony Stone Images; 36: © Jim Wark/Peter Arnold, Inc.

Careers in Biology

Page 670(top): © Jeff Henry/Peter Arnold, Inc.; p.670(middle): © Foto Luiz Claudio Marijo/Peter Arnold, Inc.; p.670(bottom): © Adam Hart-Davis/SPL/Photo Researchers, Inc.

Chapter Thirty Two

Opening Photo: © David C. Fritts/Animals Animals/Earth Scenes; 32.1: © Frank Lane/Bruce Coleman, Inc.; 32.2(both): © R. Andrew Odum/Peter Arnold, Inc.; 32.6: © Frans Lanting/Minden Pictures; 32.8a: © Y. Arthus-Bertrand/Peter Arnold, Inc.; 32.8b: © FPG International; 32.9: © Jonathan Scott/Planet Earth Pictures; 32.10: © Susan Kuklin/Photo Researchers, Inc.; 32.11a: © OSF/Animals Animals/Earth Sciences; 32.12: © J & B Photo/Animals Animals/Earth Scenes.

Chapter Thirty Three

Opening Photo: © Jan-Peter Lahall/Peter Arnold, Inc.; 33.1: © Mike Bacon/Tom Stack & Associates; 33.2(top): © C. Palck/Animals Animals/Earth Scenes; 33.2(bottom): © S.J. Krasemann/Peter Arnold, Inc.; 33.3: © The McGraw-Hill Companies, Inc./Bob Coyle, photographer; 33.6a: © Paul Janosi/Valan Photos; 33.7: © Kent & Donna Dannen/Photo Researchers, Inc.; 33.A: Courtesy Jeffrey Kassner; 33.8(top): © Michio Hoshino/Minden Pictures; 33.8(bottom): © Ted Levin/Animals Animals/Earth Scenes; 33.9a: © Jeff Greenberg/Peter Arnold, Inc.; 33.9b: © Ben Osborne/OSF/Animals Animals/Earth Scenes.

Chapter Thirty Four

Opening Photo: © Richard R. Hansen/Photo Researchers, Inc.; 34.1a: © Charlie Ott/Photo Researchers, Inc.; 34.1b: © Michael Graybill and Jan Hodder/Biological Photo Service; 34.3(left): © David Madison/Bruce Coleman, Inc.; 34.3(right): © Martin Rogers/Tony Stone Images; 34.5(both): © Ken Wagner/Phototake; 34.7a: © Alan Carey/Photo Researchers, Inc.; 34.8a: © Runk/Schoenberger/Grant Heilman Photography; 34.8b: © National Audubon Society/A. Cosmos Blank/Photo Researchers, Inc.; 34.8c: © Z. Leszczynski/Animals Animals/Earth Scenes; 34.9: © Hans Pfletschinger/Peter Arnold, Inc.; 34A: © Chris Johns/National Geographic Society Image Collection; 34.11a-c: Courtesy of Daniel Janzen.

Chapter Thirty Five

Opening Photo: © David Higgs/Tony Stone Images; 35.6a: © John Shaw/Tom Stack & Associates; 35.6b: © John Eastcott/Animals Animals/Earth Scenes; 35.6c: © John Shaw/Bruce Coleman, Inc.; 35.7(biome): © Norman Owen Tomalin/Bruce Coleman, Inc.; 35.7(moose): © Bill Silliker, Jr./Animals Animals/Earth Scenes; 35.8(chipmunk): © Zig Lesczynski/Animals Animals/Earth Scenes; 35.8(marigolds): © Virginia Neefus/Animals Animals/Earth Scenes; 35.8(bobcat): © Tom McHugh/Photo Researchers, Inc.; 35.8(millipede): © OSF/Animals Animals/Earth Scenes; 35.8(forest): © E.R. Degginger/Animals Animals/Earth Scenes; 35.A: © Michio Hoshino/Minden Pictures; 35.10(lemur): © Erwin & Peggy Bauer/Bruce Coleman, Inc.; 35.10(lizard): © Kjell Sandved/Butterfly Alphabet; 35.10(butterfly): © Kjell Sandved/Butterfly Alphabet; 35.10(frog): © James Castner; 35.10(macaw): © Kjell Sandved/Butterfly Alphabet; 35.10(katydid): © James Castner; 35.10(ocelot): © Martin Wendler/Peter Arnold, Inc.; 35.11(background): © Bruce Iverson; 35.11(inset): © Kathy Merrifield/Photo Researchers, Inc.; 35.12(bison): © Steven Fuller/Animals Animals/Earth Scenes; 35.12(prairie): © Jim Steinberg/Photo Researchers, Inc.; 35.13(zebra, cheetah, wildebeest): © Darla G. Cox; 35.13(giraffe): © George W. Cox; 35.14(kangaroo rat): © Bob Calhoun/Bruce Coleman, Inc.; 35.14(roadrunner): © Jack Wilburn/Animals Animals/Earth Scenes; 35.14(desert): © John Shaw/Bruce Coleman, Inc.; 35.15(stonefly): © Kim Taylor/Bruce Coleman, Inc.; 35.15(trout): © William H. Mullins/Photo Researchers, Inc.; 35.15(carp): © Robert Maier/Animals, Animals/Earth Scenes; 35.16a: © Roger Evans/Photo Researchers, Inc.; 35.16b: © Michael Gadomski/Animals, Animals/Earth Scenes; 35.18(pike): © Robert Maier/Animals, Animals/Earth Scenes; 35.18(water strider): © G.I. Bernard/Animals, Animals/Earth Scenes; 35.19: © Fred Whitehead/Animals Animals/Earth Scenes; 35.20a: © John Eastcott/Yva Momatiuk/Animals, Animals/Earth Scenes; 35.20b: © James Castner; 35.21a: © Anne Wertheim/Animals, Animals/Earth Scenes; 35.21b: © Anne Wertheim/Animals, Animals/Earth Scenes; 35.21c: © Jeff Greenburg/Photo Researchers, Inc.; 35.25: © Mike Bacon/Tom Stack & Associates.

Chapter Thirty Six

Opening Photo: © Jim Wark/Peter Arnold, Inc.; 36.1(left): © Spencer Swanger/Tom Stack & Associates; 36.1(right): © Norbert Rosing/OSF/Animals Animals/Earth Scenes; 36.2: /cW/ Barbara J. Miller/Biological Photo Service; 36.12a: © John Shaw/Tom Stack & Associates; 36.12b: © Thomas Kitchin/Tom Stack & Associates; 36.13: © Bill Aron/Photo Edit; 36.16: Courtesy Arlin J. Krueger/Goddard Space Flight Center/NASA; 36.17: © G. Prance/Visuals Unlimited; 36.A: Courtesy of C. Oviatt.

Index

Note: Page numbers in *italics* indicate material presented in figures and tables. Text and page numbers in magenta indicate topics of human interest.

A

C

D

E

H

T

U

V